31192022665424

MANUAL
OF
REMOTE SENSING

Second Edition
Volume I

MANUAL OF REMOTE SENSING

Second Edition

In two volumes

Volume I
Theory, Instruments and Techniques

Editor-in-Chief

Robert N. Colwell

Volume I Editor : David S. Simonett
Associate Editor : Fawwaz T. Ulaby

Volume II Editor : John E. Estes
Associate Editor : Gene A. Thorley

AMERICAN SOCIETY
OF
PHOTOGRAMMETRY

Library of Congress Cataloging in Publication Data
Main entry under title:

Manual of remote sensing.

  Includes bibliographies and index.
  1. Remote sensing.   I. Colwell, Robert N.
II. American Society of Photogrammetry.
G70.4.M36   1983   621.36′78   83-6055
ISBN 0-937294-41-1 (v. 1)
ISBN 0-937294-42-X (v. 2)

PUBLISHED BY

**AMERICAN SOCIETY OF
PHOTOGRAMMETRY**
210 Little Falls Street
Falls Church, Virginia 22046

Printed in the United States of America by

The Sheridan Press

Two color renditions, scale 1/250,000, made from the same scene of Landsat Thematic Mapper imagery acquired on July 20, 1982. This subscene is centered over the Detroit/Windsor area on the boundary between the United States and Canada. *Top:* False color composite image formed by using TM bands 2 (green), 3 (red) and 4 (near infrared) printing them in blue, green and red, respectively. *Bottom:* Natural color composite image formed by printing bands 1 (blue), 2 (green) and (3) red, in blue, green, and red, respectively. Nominally, both color composites have the same spatial resolution, viz. approximately 30 meters. Various kinds of features differ, however, in their multiband spectral signatures. Therefore, the color contrasts among such features usually differ on these two image types. Consequently, the top figure is superior for detecting row patterns at (A), riparian vegetation along stream channels at (C); water channels at (E); vegetated islands at (G); differences in soil and vegetative conditions at (I), and the distinction between ponds, such as (J) (near the Detroit Airport) and cloud shadows, such as (K). The natural color photo is better, however, for detecting sediment plumes at (B); primary and secondary roads at (D); runways and walls at (F) and areas of bare soil at (H). (Imagery courtesy of NASA and Earth Satellite Corporation).

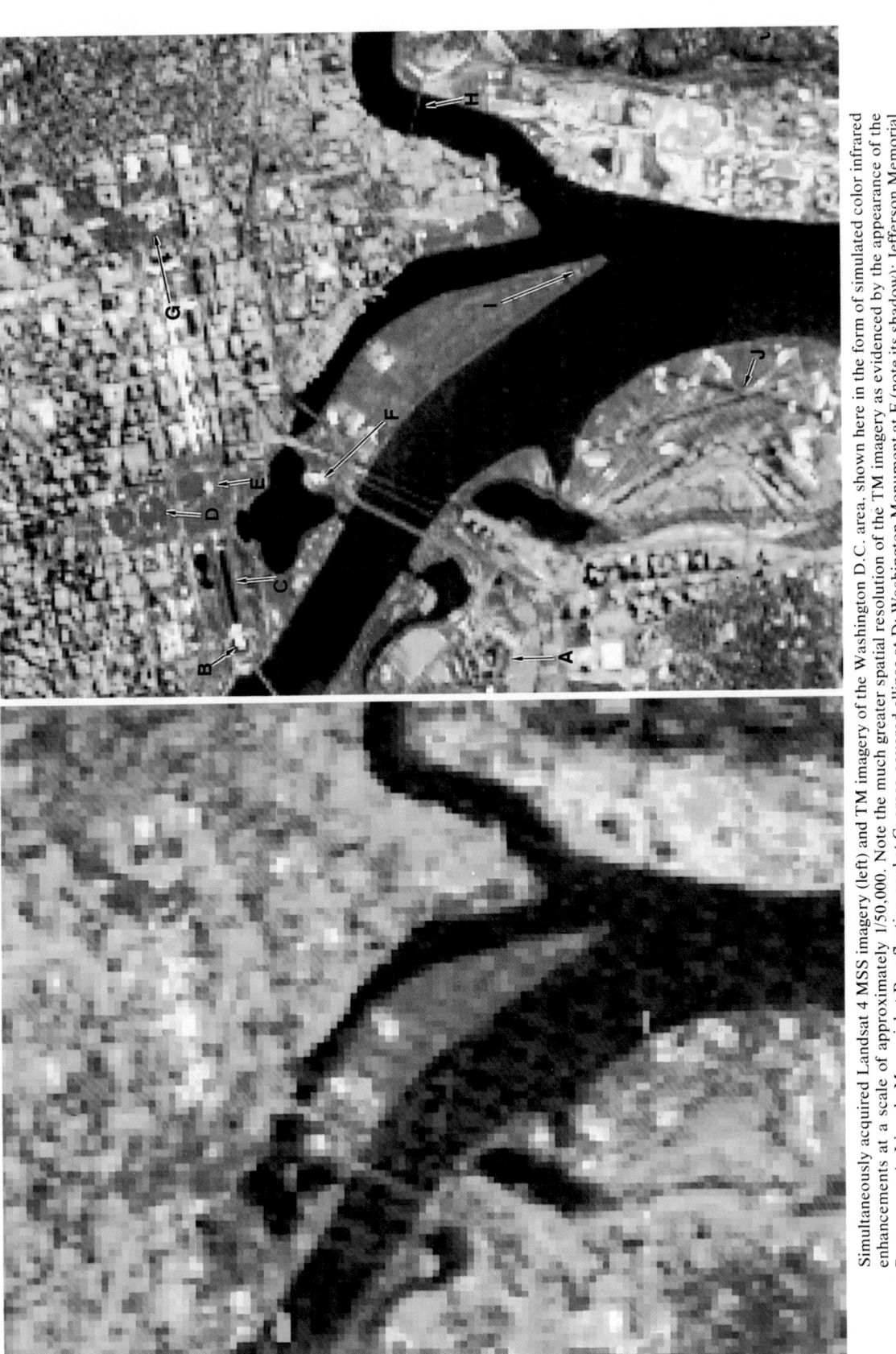

Simultaneously acquired Landsat 4 MSS imagery (left) and TM imagery of the Washington D.C. area, shown here in the form of simulated color infrared enhancements at a scale of approximately 1/50,000. Note the much greater spatial resolution of the TM imagery as evidenced by the appearance of the Pentagon at A; Lincoln Memorial at B; reflecting pool at C; grass-covered ellipse at D; Washington Monument at E (note its shadow); Jefferson Memorial at F; the Capitol at G; Anacostia Bridge at H; Hains Point at I; and National Airport at J. (Courtesy of NASA and the EROS Data Center.)

For some resource inventory purposes, one of Landsat's most important capabilities is that of obtaining multidate data of the same geographic area. One example is in the inventory and monitoring of agricultural crops in a vast area throughout each growing season. Illustrated in the above three Landsat images is a representative area of the state of California (total area of 100 million acres) in which the objective is to determine the location and acreage of all irrigated lands, each year, the better to manage critical supplies of water. (A), above, was acquired in midsummer, by which time all crops except those being irrigated had matured. Hence on this false color infrared enhancement the red areas are under irrigation and all other areas are not. (B) was acquired in spring so that additional irrigated areas, (mainly early-maturing small grains) that would be missed in (A) are included. Finally (C) was acquired in fall to identify additional irrigated areas that are cropped more than once during the growing season. Some of these areas might have been in a fallow state (between crops) on the earlier coverages. This 3-date direct visual interpretation, when combined with a stratified sample of "ground truth" data, provides estimates of irrigated acreage, year after year, to the required accuracy standard, viz. accurate to within three percent at the 99% probability level. A digital analysis system by means of which a computer can provide estimates of irrigated land using multidate Landsat data has also been developed. This system uses the three dates of Landsat described above. The digital data are registered to a 7½ minute USGS quadrangle base. Through use of a simple and inexpensive computer program, the MSS band 7-to-MSS band 5 ratio is used as a vegetation indicator. As a result, irrigated land is classified by interactively selecting threshold values above which land is designated as irrigated and below which it is designated as not irrigated. Figure (D) shows for the entire Sacramento Valley (including the area shown in A, B and C), all land that was irrigated. The sequence of irrigation is color coded as follows:

Black	—not irrigated	Purple —irrigated fall only	Red —irrigated summer and fall
Blue	—irrigated spring only	Green —irrigated spring & summer	
Pink	—irrigated summer only	Tan —irrigated spring & fall	White —irrigated spring summer & fall

The results of this classification are also combined with a stratified sample of ground data to provide estimates of irrigated land. Map-like products are also derivable from this classification. With the current interest and development of geo-based information systems, the value of Landsat digital data becomes increasingly significant. (Photos courtesy of NASA and Remote Sensing Research Program, University of California, Berkeley).

To an increasing extent, remote sensing from within the earth's atmosphere is being accomplished through the use of either the U-2 aircraft, shown here, or a greatly improved version of it, the ER-2, specifically designed for the remote sensing of Earth Resources. For any given U-2 or ER-2 mission, a choice can be made from among a great many types of cameras and other remote sensing devices, including those shown here, which have been letter-coded as follows:

A TIRS (Thermal Infrared Scanner)
B IRR (Infrared Radiometer)
C & D HCMR (Heat Capacity Mapping Radiometer)
E Filter Sampler
F OCS (Ocean Color Scanner)
G ASISGS (Ames Stratospheric In-Situ Gas Sampler)
H Aether Drift Radiometer
I SCS (Stratospheric Cryogenic Sampler)
J HR-73B Camera
K RC-10/HR-732 Camera System
L HR-732 Tri-Vertical Cameras

M KA-80A Optical Bar Panoramic Camera
N Dual RC-10 Camera System
O RC-10/Vinten Multispectral Camera System
P FLO (Infrared Spectrometer)
Q REFLEX (Resonance Fluorescence Experiment)
R SEMIS (Solar Energy Measurements in Space)
S CO_2 Collector
T H_2O Vapor Radiometer
U HSI (High Speed Interferometer)
V APS (Aerosol Particulate Sampler)

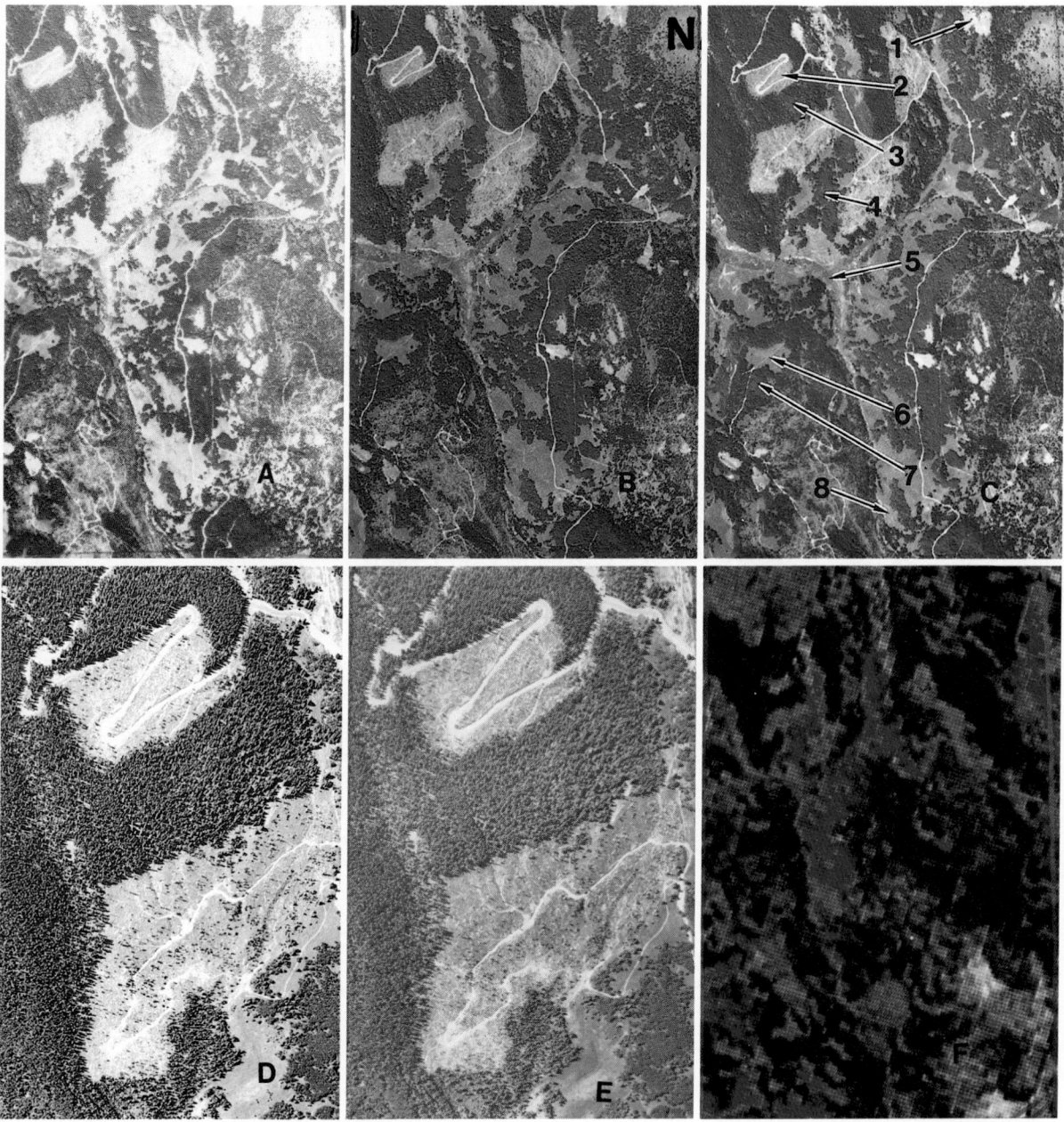

Some kinds of multiband, multistage imagery are proving to be more useful than others for the making of resource inventories, as evidenced by the above image examples from a test site on the San Juan National Forest, Colorado. Images A, B, and C were all acquired simultaneously from an altitude of 60,000 feet. A is panchromatic minus-blue photography from an AMPS Hasselblad camera; B is Color Infrared photography from a second Hasselblad camera; C is Color Infrared photography from a Wild RC-10 camera. All three are reproduced here at a scale of approximately 1/60,000. In D and E are matching color and black-and-white images, both made from the same frame of aeroneg imagery at a scale of 1/24,000 taken with a Zeiss precision mapping camera. Image F is a Geopic-enhanced Landsat MSS color composite image, scale 1/100,000. As reported on page 1967 of this manual, nearly 40,000 individual photo-interpreter responses were obtained in determining the interpretability of these image types with respect to the resource categories that are labelled in the upper right photo above, viz. 1 = rock, 2 = cut-over (spruce-fir), 3 = spruce-fir, 4 = aspen, 5 = willow, 6 = mesic meadow, 7 = spruce-fir/aspen, and 8 = xeric meadow. (Photos courtesy of NASA and the Nationwide Forestry Applications Program of the U.S. Forest Service.)

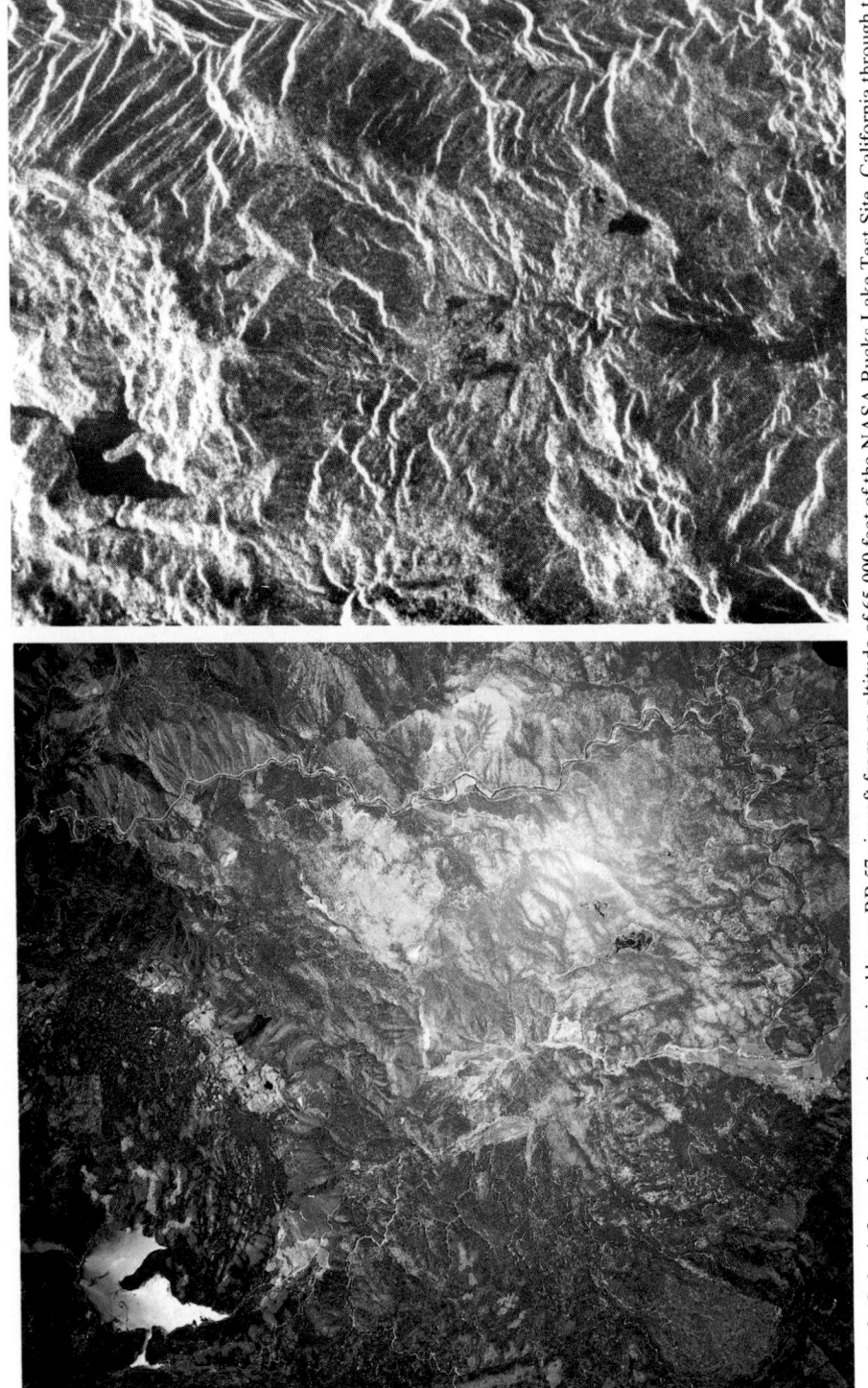

Left: Color infrared photography acquired by an RB 57 aircraft from an altitude of 65,000 feet of the NASA Bucks Lake Test Site, California through the use of a Zeiss 6-inch-focal-length metric mapping camera. *Right:* Seasat SAR (synthetic aperture radar) imagery of the same area acquired from an altitude of approximately 850 km (570 mi.) Both images are reproduced here at a scale of about 1/200,000. Note that the SAR imagery not only accentuates topographic features but also exhibits high spatial resolution for many features including lakes and moist areas. If this figure is rotated 180 degrees, radar shadows are made to fall away from the observer and the terrain appears to be reversed in that high features appear low and low features appear high.

Top: Radar imagery of the Enos Lake Quadrangle and environs of Idaho's Payette National Forest, acquired from an altitude of 60,000 feet with the AN/APQ 97 side-looking airborne radar (SLAR) system. Note the accentuation of terrain features in this highly glaciated topography. *Bottom:* A low-altitude aerial oblique stereogram that was taken (as indicated by matching arrows on the top and bottom figures) while looking through the notch between North Loon Mountain and South Loon Mountain. Note that while some features are better seen on this low-altitude stereogram, others are better seen on the very high altitude radar imagery. (Courtesy of NASA and the Nationwide Forestry Applications Program of the U.S. Forest Service).

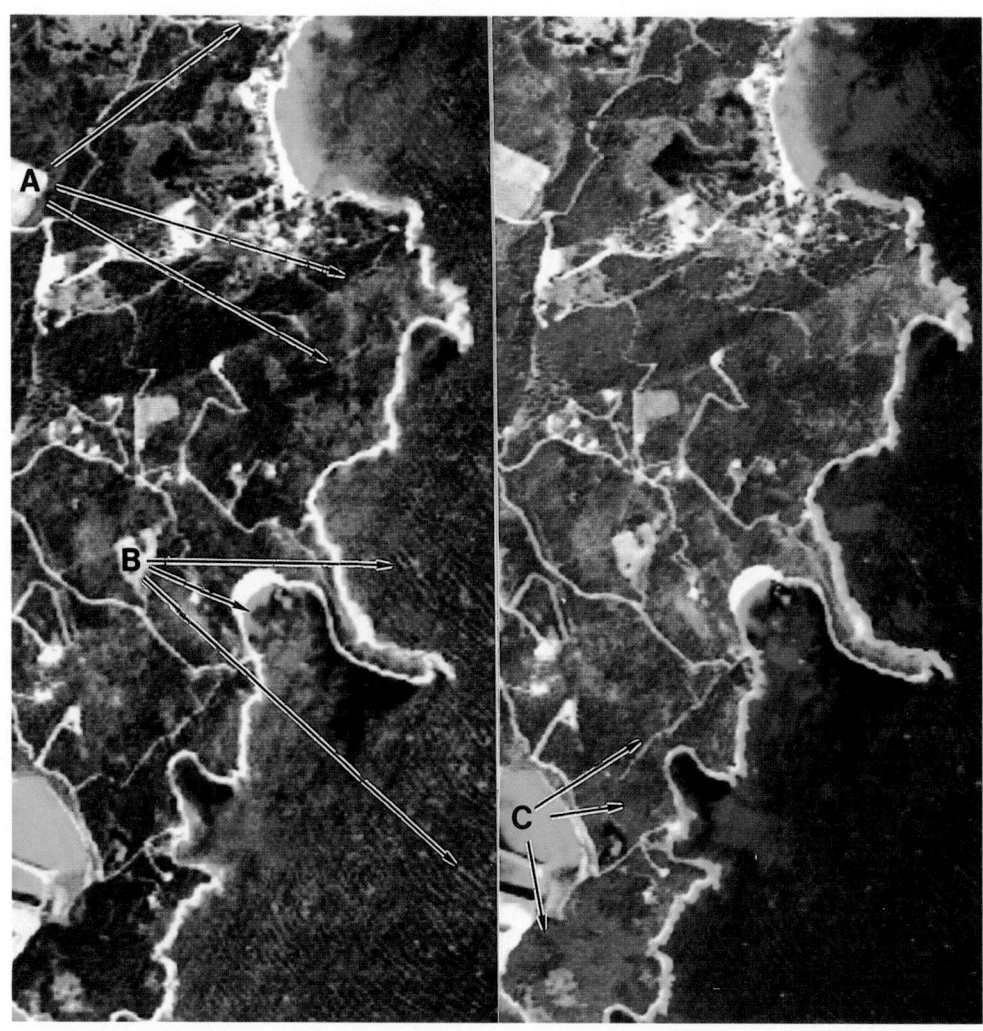

The potential value of applying multi-enhancement techniques to space-acquired imagery is well illustrated by these two simulated SPOT images. The area shown is on the island of Corsica in the Mediterranean Sea. These two enhancements were prepared by applying different degrees of color stretching to Bands 1, 2 and 3 of Daedalus scanner data. The imagery was acquired at 10 meter resolution and preprocessed by the French Space Center, CNES. It was then resampled to obtained the 20 meter resolution used in producing the two color renditions shown above. The *left* image is superior for showing roads at A as well as the details of various water surface and subsurface features, as at B. The *right* image is superior, however, for showing significant vegetation boundaries at C. A comparison of this "subsample" figure with Color Figure 31-261 of Volume II demonstrates the feasibility of extrapolating these detailed interpretations to imagery that is of much smaller scale but covering a much larger geographic area. The original scale of the above figure was 1/62,500. (Imagery courtesy of CNES). See also the excellent simulated SPOT imagery on Frontispiece page viii of Volume II.

ALL OF THE REMAINING 4-COLOR FIGURES FOR THIS VOLUME APPEAR IN A SINGLE 32-PAGE COLOR SIGNATURE. IN THE TEXT EACH SUCH FIGURE IS REFERRED TO AS A "COLOR FIGURE" AND IS DESIGNATED BY BOTH THE NUMBER OF THE CHAPTER TO WHICH IT PERTAINS AND THE FIGURE NUMBER WITHIN THE CHAPTER. THE COLOR PLATES BEGIN FOLLOWING PAGE 590 AND APPEAR IN SEQUENCE BY CHAPTER AND FIG. NO.

FOREWORD

With the publication of the *Manual of Remote Sensing, Second Edition,* the American Society of Photogrammetry completes the ninth in a series of book projects aimed at advancing the science of photogrammetry and remote sensing. Since 1944, the Society has produced this series at an average rate of one major book every five years, each new book bringing forth a text that reflects the most modern aspects of the art. Considering that these landmark publications are produced entirely with volunteer authors and editors, the American Society of Photogrammetry can justifiably take pride in this remarkable series of achievements by its members. The list of major publications includes:

Manual of Photogrammetry, First Edition, 1944
Manual of Photogrammetry, Second Edition, 1952
Manual of Photographic Interpretation, 1960
Manual of Photogrammetry, Third Edition, 1966
Manual of Color Aerial Photography, 1968
Manual of Remote Sensing, First Edition, 1975
Handbook of Non-Topographic Photogrammetry, 1979
Manual of Photogrammetry, Fourth Edition, 1980
Manual of Remote Sensing, Second Edition, 1983

The publication of the second Edition of the Manual of Remote Sensing is especially gratifying because it represents a triumph over unfortunate circumstances that beset the project in its early stages. The formidable task of overseeing the preparation of the Second Edition, as Editor-in-Chief, was assigned originally to Professor Leonard Bowden of the University of California at Riverside. The talented Professor Bowden promptly proceeded to the appointment of editors and associate editors for the two volumes of the manual. The organization of the manual, chapter-by-chapter, was developed, a production time table was worked out and an author-editor was selected for each chapter. Then, in May 1979, came a sad and untimely event: Professor Bowden's sudden death of a heart attack. Although he died before the task had gained full momentum, Professor Bowden laid the groundwork for the project and his successor was able to build on that foundation. For the work that Professor Bowden accomplished in initiating the project, the American Society of Photogrammetry acknowledges a profound debt of gratitude and recognizes that the publication of this manual is, in considerable measure, a tribute to his initiative.

Following Professor Bowden's death, careful consideration was given to the selection of a successor with the competence and willingness to assume the formidable task of carrying on a major project initiated by someone else. Fortunately, a man with ideal qualifications and outlook was active in the society, and more fortunately still, he accepted the position when it was offered to him. Professor Robert N. Colwell of the University of California, Berkeley, brought to the position of Editor-in-Chief a distinguished technical background in photogrammetry and remote sensing plus a brilliant record as a writer and editor. As Editor-in-Chief of the society's *Manual of Photographic Interpretation,* published in 1960, he had acquired valuable experience in successfully managing the publication of that popular book. Above all, he had the talent and ability to adjust his outlook to prevailing conditions; this was a paramount consideration in maintaining the continuity and momentum that already existed within the team of some 300 people who were involved in preparing the Manual. Timely grants provided by the U.S. Geological Survey helped to defray certain editorial costs and also some of the costs entailed in preparing the chapter on Geological Applications. The

American Society of Photogrammetry hereby expresses its sincere thanks to the U.S.G.S. for this valuable assistance.

The extent of the trials and tribulations that Bob Colwell endured and conquered during the preparation of this Manual are completely known only to himself. Perhaps Executive Director Bill French and Publications Consultant Morris Thompson, who managed the business and production aspects of the project, have a substantial awareness of the problems that had to be solved. But that is all now history and the *Second Edition of the Manual of Remote Sensing* is now a reality. For this we, the presidents of the American Society of Photogrammetry during development of the Manual of Remote Sensing, Second Edition, congratulate and thank the Editor-in-Chief, the Volume Editors, the Associate Editors, the Author-Editors and all the contributors to this outstanding publication.

Francis H. Moffitt, 1979
Rex R. McHail, 1980
George J. M. Zarzycki, 1981
Allan C. Bock, 1982
William G. Hemple, 1983

PREFACE TO THE SECOND EDITION

As defined in Webster's dictionary, a "Manual" is "a book containing, in concise form, the principles, rules and directions needed for the mastery of an art, science or skill". That definition describes exactly what this 2-volume publication purports to be with respect to the art, science and skill known as "remote sensing". The term "concise" as used in the above definition connotes "the cutting out of all superficialities and the avoidance of elaboration". Nevertheless, Webster asserts that a Manual should be "only as summary as is compatible with an adequate statement of the available information". The authors of this Manual have attempted to be both concise and summary. The fact that, in so doing, they have produced a document of nearly 2700 pages indicates that the multifaceted field of remote sensing has truly come of age.

This maturing process has been due in no small measure to a perception by successive governing boards of the American Society of Photogrammetry over the years that progressively greater recognition should be given to various aspects of a fast-growing field that is now encompassed by the term "remote sensing—the progenitor of which was" photographic interpretation". Briefly stated: (1) in 1952 the Society added greatly to the prestige of this field by devoting an entire chapter in its Manual of Photogrammetry—Second Edition to the subject of "photographic interpretation; (2) in 1960 the Society gave this field another quantum jump in prestige by devoting an entire Manual to the subject of photographic interpretation; (3) by 1975 the Society acknowledged, through publication of its "Manual of Remote Sensing", that the advent of space age platforms, multispectral scanners, and computer-assisted analysis techniques was bringing about a far more encompassing science than had been dealt with in its comprehensive Manual of Photographic Interpretation; and (4) in 1978 the Society formally adopted the recommendation of several of its most visionary remote sensing scientists, including the late Professor Bowden, that work should begin immediately on the preparation of this, the definitive Second Edition of the Manual of Remote Sensing—an effort that was to require four more years to complete after Professor Bowden's untimely death. His initiative and the organizational ability that he demonstrated in getting this second edition underway are hereby gratefully acknowledged. In keeping with the previously mentioned tradition of high level support, two officials at Society headquarters who have greatly facilitated the preparation of this Manual are Morris M. Thompson and William D. French who have served, respectively, as Production Editor and Publications Director.

Throughout much of the time that was spent in producing this Manual the Editor-in-Chief had almost daily dealings with the highly competent Editors of Volumes I and II, Professors David Simonett and John Estes, respectively. In so doing, he gained great respect for their editorial abilities, their maturity of judgment and the great team spirit that they demonstrated in their dealings both with him and with the author-editors of the 36 chapters. Their jobs, in turn, were greatly facilitated by yet another pair of very competent and dedicated individuals, Professor Fawwaz Ulaby and Dr. Gene Thorley, who served as Associate Editors for Volumes I and II, respectively and whose valuable roles in producing this Manual are likewise gratefully acknowledged. Furthermore, our collective tasks were facilitated and the Manual's quality was significantly improved through the editorial reviews, chapter-by-chapter, that were made by members of the Society's Remote Sensing Applications Division. The uniformly high quality that will be found in the figures, tables and page layout throughout this Manual is due very substantially to the dedication and effectiveness of Jane Schott, the printer's representative and Accounts Director for The Sheridan Press.

But in the hierarchy that is required to produce a Manual such as this, the primary people in the trenches are the author-editors and contributing authors for the individual chapters. In a very major sense the high quality that will be found throughout this Manual is a tribute to the professional expertise and dedication of these individuals. We all are greatly indebted to them.

The potential usefulness of an index tends to increase with the size and comprehensiveness of the book to which it applies. It is especially gratifying to know, therefore, that this voluminous Manual contains an index of unusually high quality and thoroughness, thanks to the efforts of one dedicated individual, G. Carper Tewinkel,—a former president of the Society and an indefatigable indexer of its many publications.

There were many other friendly faces that I encountered in the trenches on a daily basis during the chapter-by-chapter editorial skirmishes. I refer to several coworkers of mine in the University's Space Sciences Laboratory, including James Hardin, Kevin Dummer and Betsy Ross. Their professional response to my many requests also is gratefully acknowledged.

When compared with the First Edition, this Second Edition will be found to have major parts that are entirely new, including the comprehensive chapter dealing with Geological Applications—a field in which very substantial advances have been made in recent years. Special thanks are hereby extended to Dr. Richie Williams and his team of more than fifty contributors for preparing the Geology Chapter under difficult circumstances. Most other parts of the Manual are very substantially new. There are some parts however, including those dealing with remote-sensing-related theory, instruments and techniques, which draw extensively on the text and illustrations that appeared in the First Edition of the Manual or even in its predecessor by 15 years—the Manual of Photographic Interpretation. In those limited cases in which the earlier material seemed to be the best, there obviously was little point in replacing it with something that was only second best, however novel it might be.

An overview of some of the most important recent advances in remote sensing technology is provided by the illustrations that appear on the dust jackets and in a preface of color photos for each of the two volumes. It has been possible to greatly increase the number of color photos throughout this Manual through cost savings achieved by placing them in ''signatures''. Within each chapter, the color figures have been numbered serially with all other figures, even though separated from them. Throughout the text, use of the term ''Color Figure'' is made where necessary to direct the reader to the nearest color signature.

Now that this tome finally has been completed, a glance at its table of contents will show even the uninitiated person that remote sensing, in its non-military aspects, is primarily useful in the discovery, evaluation, and development of the earth's natural resources and in the intelligent planning for human occupancy of the earth's surface. In the book of Genesis we learn that it was God's plan to give man ''dominion over all the Earth . . . and to subdue it''. This Manual presents an effort to describe one of the most valuable means to that end that man has yet devised. But quite probably more than a mere search of the Scriptures would be needed to convince the average contributor to this Manual, frustrated as he/she has been by tight deadlines and perplexed by rigid requirements imposed by an uncompromising editor, that preparation of The Second Edition of the Manual of Remote Sensing was in response to a Divine calling. Nevertheless, it is hoped that this, the published product, will convince both the contributors and the readers that the effort was eminently worthwhile.

Among the primary aims, slightly paraphrased, of the American Society of Photogrammetry are these: to advance knowledge concerning and stimulate interest in the science and art of photogrammetry and remote sensing; to provide means for the dissemination of new knowledge and information and thus to encourage the free exchange of ideas and intercourse among those contributing to the advancement of the art; to stimulate student interest in the fields of photogrammetry and remote sensing; and to foster a spirit of understanding and cooperation among the users of aerospace photography in the United States and throughout the world. It is hoped that those using this Manual will share the conviction of those who produced it, that it represents a furthering of all of these aims and that it is an excellent example of cooperative accomplishment that could only have been achieved under the sponsorship of the American Society of Photogrammetry.

Robert N. Colwell
Editor-in-Chief

Berkeley, California
January, 1983

PREFACE TO VOLUME I

Volume I of the Second Edition of the Manual of Remote Sensing differs from the first edition in the following respects:

1) To make the volume more readable and accessible to students we decided to keep the length of each chapter to between 20 and 60 final printed pages (the mean length is 45 pages).
2) We start—in some cases finish—most chapters with a list of nomenclature.
3) MKS units are employed throughout insofar as feasible.
4) We eliminated most of the possible overlaps between chapters except in a few instances. In these we deliberately retained overlap (e.g. between Chapters 17 and 21), because of differences in viewpoint and treatment of the same material.
5) The number of chapters has approximately been doubled to 25 and each new chapter has been halved in length and has a narrower focus.
6) The material in three chapters has no counterpart in the First Edition: Chapter 16 on Orbital Mechanics; Chapter 19 on Remote Sensing Software Systems; and Chapter 20 on Digital Hardware.
7) Eight chapters are so completely revised as to be essentially new: Chapter 1 on the Development and Principles of Remote Sensing; Chapter 12 on Landsat satellites; Chapter 13 on Microwave and Infrared Satellite Remote Sensors; Chapter 14 on Meteorological Satellites; Chapter 17 on Data Processing and Reprocessing; Chapter 18 on Pattern Recognition and Classification; Chapter 21 on Image Geometry and Rectification; and Chapter 22 on Geographic Information Systems and Remote Sensing.
8) All the material in the remaining chapters (2, 3, 4, 5, 6, 7, 8, 9, 10, 11, 15, 23, 24 and 25) has undergone substantial revision and updating, and much new material has been added. However, these chapters also incorporate considerable material from the first edition, where that material has not become outdated.

We wish to thank the author-editors and contributors to each chapter. It was a privilege to work with such an outstanding group of professionals, and to learn from them in our editing chores. We also thank Lee Blackledge and Robert E. Davis for their able assistance in editing.

David S. Simonett   Fawwaz T. Ulaby
Editor, Volume I   Associate Editor, Volume I

CONTENTS

VOLUME I (THEORY, INSTRUMENTS AND TECHNIQUES)

Chapter 1. The Development and Principles of Remote Sensing

Chapter 2. The Nature of Electromagnetic Radiation

Chapter 5. Interaction Mechanisms Within the Atmosphere

Chapter 9. *Radar Fundamentals and Scatterometers*

Chapter 10. *Imaging Radar Systems*

Chapter 11. *Passive Microwave Radiometry*

Chapter 12. Landsat Satellites

Chapter 13. Microwave and Infrared Satellite Remote Sensors

Chapter 14. Meteorological Satellites

Chapter 15. Communication and Data Transmission Systems

Chapter 16. Orbital Mechanics for Remote Sensing

Chapter 17. Data Processing and Reprocessing

Chapter 18. *Pattern Recognition and Classification*

Chapter 19. *Remote Sensing Software Systems*

Chapter 20. Digital Hardware

Chapter 21. Image Geometry and Rectification

Chapter 22. Geographic Information Systems and Remote Sensing

Chapter 23. Ground Investigations in Support of Remote Sensing

Chapter 24. Fundamentals of Image Analysis: Analysis of Visible and Thermal Infrared Data

Chapter 25. *Image Analysis-Microwave Region*

VOLUME II (INTERPRETATION AND APPLICATIONS)

Chapter 26. Archaeology, Anthropology, and Cultural Resources Management

Chapter 27. Remote Sensing of Weather and Climate

Chapter 28. The Marine Environment

Chapter 31. Geological Applications

Chapter 32. Engineering Applications

Chapter 33. Remote Sensing Applications In Agriculture

Chapter 34. Forest Resources Assessments

Chapter 35. Rangeland Applications

Chapter 36. *Terrestrial Moons and Planets*

Combined Index to Volumes I and II *2417*

CHAPTER 1

The Development and Principles of Remote Sensing

Author/Editor: DAVID S. SIMONETT

Contributing Authors: ROBERT G. REEVES, JOHN E. ESTES, SUSAN E. BERTKE, CHARLENE T. SAILER

GENERAL CONTENTS: Development of remote sensing prior to 1960; since 1960; systems of the near future; other national systems under consideration; geographic information systems; making remote sensing work; technology transfer; political and legal implications of remote sensing; a systems view of remote sensing; the multi concept; references.

INTRODUCTION

The earth is a finite planet with limited resources. As population continues to grow, placing ever increasing demands on this resource base, wise and prudent management of these resources is becoming increasingly important. Such management is best achieved if accurate resource inventories are made available to the resource manager at suitably frequent intervals. Although the identification, measurement, and inventory of the earth's resources is a considerable task, the technology of remote sensing does offer mankind the potential to produce a broadly consistent data base at a spatial, spectral, and temporal resolution, that is useful for resource managers.

In a recent article, Estes, Jensen, and Simonett (1980) argue that remote sensing is a reality whose time has come. It is a powerful tool that cannot be ignored because of its information potential and the logic implicit in the reasoning processes employed to analyze remote sensing data. When allied with cartography through the use of information systems, remote sensing techniques can rise above the level of a mere technology. This coupling can change our perceptions, our methods of data analysis, our models, and our paradigms. At its most fundamental level, remote sensing provides a means by which data can be produced and analyzed for an area and then incorporated in decision-making or problem-solving procedures.

Remote sensing is the acquisition of information about an object without physical contact. The term "remote sensing" was coined in the early 1960's by geographers in the Office of Naval Research to apply to the information derived from photographic and nonphotographic instruments. In the simplest case the human eye can be considered a remote sensor because it visually senses information from the surrounding world. However, the use of the term remote sensing usually refers to the gathering and processing of information about the earth's environment, particularly its natural and cultural resources, through the use of photographs and related data acquired from an aircraft or satellite.

This two-volume second edition of the *Manual of Remote Sensing* has been written and published to provide a definitive and authoritative up-to-date text on the theory, systems hardware and software, analysis techniques and methodologies, and applications of remote sensing of the environment. The objective of this chapter is to provide an overview and historical perspective of remote sensing, and to introduce the more important basic concepts, sensors, and applications which are discussed in detail elsewhere in the manual. We first examine the development of remote sensing, and follow this with details on remote sensing systems now being planned for operational use in the near future. The growth of remote sensing and its applications in both developed and developing countries is then examined with attention given to problems of technology transfer, the tradeoffs involved in the decision to implement the technology in a given country, and the political, legal, and socio-economic implications that may result from this decision. Next follows a brief discussion of resolution as a critical multifaceted constraint on the use of remotely sensed data. Questions concerning important aspects of scale and data aggregation as they impact the information extraction of remotely sensed data are then addressed. Next, the "multi" concept is presented. First enunciated by R. N. Colwell, in 1975, this concept encompasses the variety of remote sensing platforms and sensors that can be used in various combinations, to acquire remote sensing data and also takes into consideration the processing, analysis and utilization methodologies available to practitioners in the field of remote sensing. Finally, we summarize the state-of-the-art of remote sensing within a broadly philosophical context.

TABLE 1-1

Comparison of the Two Major Time Periods in Remote Sensing Development. (Colwell, 1979)

Prior to the Space Age (1860–1960)		Since 1960	
A.	Only one kind and date of photography	A.	Many kinds and dates of remote sensing data
B.	Heavy reliance on the human analysis of un-enhanced images	B.	Heavy reliance on the machine analysis and enhancement of images
C.	Extensive use of photo interpretation keys	C.	Minimal use of photo interpretation keys
D.	Relatively good military/civil relations with respect to remote sensing	D.	Relatively poor military/civil relations with respect to remote sensing
E.	Few problems with uninformed opportunists	E.	Many problems with uninformed opportunists
F.	Minimal applicability of the "multi" concept	F.	Extensive applicability of the "multi" concept
G.	Equipment simple and inexpensive; readily operated and maintained by resource-oriented workers	G.	Equipment complex and expensive; not readily operated and maintained by resource-oriented workers
H.	Little concern about the renewability of resources, environmental protection, global resource information systems, and associated problems related to "signature extension", "complexity of an area's structure", and/or the threat imposed by "economic weaponry"	H.	Much concern about the renewability of resources, environmental protection, global resource information systems, and associated problems related to "signature extension", "complexity of an area's structure", and/or the threat imposed by "economic weaponry"
I.	Heavy resistance to "technology acceptance" by potential users of remote sensing-derived information	I.	Continuing heavy resistance to "technology acceptance" by potential users of remote sensing-derived information

THE DEVELOPMENT OF REMOTE SENSING

The development of remote sensing can be divided into two general areas (Colwell, 1979). Prior to about 1960, aerial photography was the sole system used in remote sensing. With the advent of the space program in the early 1960's and the first photographs from the Mercury program, the pace of technological development for remote sensing accelerated. Table 1-1 shows a comparison of the two time periods.

REMOTE SENSING PRIOR TO 1960

The development of cameras stemmed from experiments over 2,300 years ago by Aristotle with a "camera obscura". This led to attempts to permanently record the images so formed. Advances in chemistry provided the knowledge of photosensitive chemicals, which is essential to photography. Continued experimentation with the camera obscura from the 13th through the 19th centuries by such noted scientists and scholars as Leonardo da Vinci, Levi ben Gerson, Roger Bacon, Daniel Barbara, Johan Zahr, Athanins Kircher, Johannes Kepler, Robert Boyle, Robert Hooke, William Wollaston, and George Airy resulted in many improvements, including the use of mirrors (Zahr) and lenses (Barbara). This experimentation set the stage for development of the photographic process by Louis Jacques Mande Daguerre and Joseph Nicephoce Niepce, which they jointly announced in August 1839, as the "daguerrotype" The excess unexposed silver halides could be removed without damaging the exposed light-sensitive material. However, it was not until 1839

that Daguerre and Niepce learned of Herschel's earlier (1819) discovery that thiosulfate chemicals could be used to fix the enclosed halide and that fixing was a key to the process. Shortly after the initial announcement of the daguerrotype process in 1839, additional work was done on the process, optical glasses were developed for lenses, and lens design was incorporated into the design of camera systems. These developments so improved the level of knowledge that, by the end of the century, the basis for modern photography as now known was established. For an excellent summary of the development of photography, and an extensive bibliography see Sturge (1977).

Not content with photographing subjects from the ground, photographers soon took to balloons. The first known balloon photograph was taken in 1859 by Gaspard Felix Tournachon (later known as "Nadar") of the French village of Petit Becetre near Paris. Shortly afterwards, in 1860, Samuel A. King and James W. Black took photographs of Boston, Massachusetts, from a captive balloon 1,200 feet above the terrain.

As with many other scientific and technical endeavors, wartime needs spurred the development of photography, especially aerial photography. During the U.S. Civil War photographs of Confederate positions defending Richmond, Virginia were reportedly taken by photographers in balloons for the Union forces. Experiments on the acquisition of aerial photographs continued through the end of the century, mostly in the U.S., England, France, Germany, and Russia. The early airplanes greatly stimulated aerial photography because they provided more reliable and much more controllable platforms for still and motion picture cameras than was possible using balloons.

Again it was war—World War I—that gave a quantum jump in the use of aerial photography. Although research in the military applications of aerial photography was at a low ebb between World Wars I and II, civilian use for geology, forestry, agriculture, and cartography, among other purposes, increased dramatically and led to the development of much improved cameras, films, and interpretation equipment. The reliability and performance of aircraft also increased dramatically during this period. By the outbreak of World War II the principal combatants, particularly the Germans, had excellent photoreconnaissance and photointerpretation capabilities. The British likewise made excellent use of photoreconnaissance and, from the detection of boats that were being concentrated across the English Channel on the mainland of Europe, were able to forestall the German invasion of England, planned for the summer of 1940. American photoreconnaissance was slow to develop, but by the end of 1942 the U.S. Navy was extensively and successfully using aerial photographs in the Pacific. The Navy led in the use of aerial photography, largely through the establishment of the Naval Photographic Interpretation School. Largely because of the initiative of one Navy lieutenant (now Rear Admiral) Robert S. Quackenbush, that school opened its doors on January 5, 1942, less than a month after the attack on Pearl Harbor. It is no accident that many of the leaders in photographic interpretation in the United States graduated from that school. The U.S. Army also came to rely heavily on aerial photographs and photointerpretation.

Exemplary photoreconnaissance and skillful interpretation contributed markedly to successful campaigns during World War II, first by the Germans and subsequently by the Allies. Many of the civilian photographic pilots and interpreters in England, France, Germany, Italy, and the United States started as military reconnaissance pilots and military interpreters during World War II, and continued on in a civilian capacity following the war's end.

World War II not only greatly advanced the art and science of aerial photography and photointerpretation, but also spurred the development and use of other aerial reconnaissance devices, including thermal infrared and active microwave (radar) systems. The development of both infrared and radar systems depended on the results of basic research that had been performed many years before they came into being—for infrared, on research that commenced over a century and a quarter before the development of imaging systems, and for radar, on research that started over a half-century before the development of operational airborne systems.

During World War II, the United States, Great Britain, and Germany developed infrared sensing devices, but it was not until well after the war that sensitive detector elements with rapid response times were developed. It was these detectors that allowed the development of modern airborne optical-mechanical scanners, radiometers, and spectrometers.

The first successful airborne imaging radar, the "plan position indicator" (PPI), was developed in Great Britain as an aid to nighttime bombing. Although the initial research and development of radar took place in Germany, the United States, and Great Britain, it was the British scientists and military officers who earliest saw its military possibilities, and Great Britain had developed operational aircraft- and ship-detection radar systems prior to the outbreak of World War II. With the development of the multicavity magnetron (tube) in England in 1939, high-power microwave radars were possible.

The resolution of real aperture or "brute force" radar systems is a function of the ratio of the aperture (antenna) size to wavelength; the larger the aperture/wavelength ratio, the higher the resolution. Thus, decreasing the wavelength provides better resolution for the same size aperture, or permits a smaller aperture to achieve the same resolution. The development of high-power microwave transmitters enabled the propagation of much shorter wavelengths, permitting the use of shorter antennas to achieve adequate resolution.

The smaller antennas allowed the development of airborne radar, first for aircraft detection by British night-fighters then, shortly thereafter, of imaging systems as an aid to nighttime or inclement-weather bombing. It was the latter that evolved into the "mapping" radar systems in use today. The development of synthetic aperture radar, in which an artificially long antenna can be synthetically created, permits fine resolution with long wavelengths and low power.

During the 1950's, remote sensing systems continued to evolve from the systems developed for the war effort. Color-infrared photography (CIR), originally developed as a military reconnaissance tool, was found to be of great use within the plant sciences. Colwell (1956) performed some of the early experiments on the use of special-purpose aerial photography for the classification and recognition of vegetation types and the detection of diseased and damaged (stressed) vegetation. For thermal infrared (IR), emphasis was placed on developing "thermal" or "radiation" models, using as a basis the spectral reflectance work by Krinov (1947). The key technology of the IR detectors was still years away. Krinov's work was also used for the continued development of multispectral imagery, a remote sensing technology which received attention near the end of World War II.

Various aspects of the progress in radar technology occurred in parallel because of the two types of radar that were evolving—side-looking air-borne radar (SLAR) and synthetic aperature radar (SAR). The SLAR development centered around acquisition of the highest possible resolution for two kinds of systems. One system, initiated at Westinghouse under sponsorship of the USAF Strategic Air Command, was designed for the high-altitude aircraft of the era, operating at relatively long stand-off ranges. The other system, resulting from Texas Instrument/USAF research and development efforts, was designed to operate at short ranges and from low-altitude aircraft. Crucial to the SAR development, under military security restrictions, was the ability to finely resolve Doppler frequencies via a frequency analysis of the returning or reflected signal (Sherwin et al., 1962). By the late 1950's, a successful optical processor using a combination of spherical, cylindrical and conical lenses to perform the required processing of the return signal had been demonstrated for the SAR applications (Cutrona et al., 1966).

REMOTE SENSING SINCE 1960

The placing of remote sensors in space by NASA, for the purpose of making earth observations, began in the early 1960s, and came about as an off-shoot of the decision to land men on the moon. In due time remote sensors were placed on lunar-orbiting satellites so that data could be collected to ensure that the manned lunar spacecraft could be safely landed. These instruments were tested over terrestrial sites and interpretations correlated with the physical conditions of those test sites. Although the first test sites chosen were lunar analog sites, the value of the information extracted for general geologic studies soon became apparent. As a result, the program was broadened to include test sites for agriculture, forestry, geography, geology and mineral resources, hydrology and water resources, oceanography and marine resources, and urban and regional studies. The recognition of the value of remote sensing as a means of collecting data for the study of the earth led to the establishment of the Earth Resources Survey Program in NASA.

Supporting the instrument-development program and test-site studies was the remote sensing aircraft-program of NASA. This program contributed substantially to the collection not only of intermediate and high altitude photography but also of thermal infrared and radar imagery covering very large areas of the United States. The data, particularly those covering the test sites established by the NASA Earth Resources Survey Program, provided training and experience for those working on instrument design, data storage and retrieval, image processing and interpretation, and other aspects of the space program.

The material which follows provides an overview of some of the satellite remote sensor systems and space platforms launched by the United States and having a major impact on earth resources remote sensing.

Systematic earth orbital observations by United States satellites began in 1960 with the launch of the Television Infrared Observation Satellite (TIROS-I). TIROS-I was the first meteorological satellite and carried a low resolution imaging system. Since the launch of TIROS-I, NASA has launched more than 40 meteorological and environmental satellites with steadily improving sensor data-collection capabilities. (See Chapter 14.) TIROS satellites were originally supported by the Environmental Sciences Services Administration (ESSA) which was absorbed into the National Oceanic and Atmospheric Administration (NOAA) in October, 1970. A second series of the TIROS satellites called the Improved TIROS Operational System (ITOS) has been developed and launched by NASA. Since ITOS satellites proved successful, these were also designated as members of the NOAA satellite series.

In December, 1981, the two most recent of the satellites in this series were put in operation. Both of these satellites, called TIROS N and NOAA-C (also called NOAA 7) operate at an altitude of 830 km and have identical payloads but in different orbits to provide better coverage of global meteorological conditions than previous NOAA polar orbiting satellites were able to provide. Among the sensor systems on these satellites are:

1) The Advanced Very High Resolution Radiometer (AVRR). Data acquired by this device provide a one kilometer resolution for six spectral channels from the visible to near infrared. These data are used for daytime and nighttime mapping of sea surface temperature, snow cover extent, ice sheets, vegetation, and cloud cover.
2) The TIROS Operational Vertical Sounder. This is a package of three sensors, a High Resolution Infrared Sounder, a Stratosphere Sounding Unit, and a Microwave Sounding Unit. Data from this sensor package are designed to measure radiance as required to calculate temperature and humidity profiles in the atmosphere up to the stratosphere, stratospheric temperature profiles, and water vapor soundings under cloudy and cloud-free conditions.
3) The Data Collection and Platform Location system (DCS). This system receives low duty-cycle transmissions of meteorological observations from free-floating balloons, ocean buoys, and other satellites and fixed ground based sensor platforms distributed world-wide.
4) The Space Environment Monitor. This device consists of four detector systems and a data processing unit. The Space Environment Monitor system is used for measuring the proton and electron flux near the earth.

The first space photograph of the earth was transmitted by Explorer-6 in August of 1959. In

1960 the first orbital photography also became available. A series of several hundred 70-mm color photographs were taken by an automatic camera in the unmanned MA-4 (Mercury-Atlas) spacecraft. Although taken primarily to monitor the spacecraft's altitude, the pictures of North Africa demonstrated the value of orbital (as opposed to hyperaltitude) photography for viewing the earth's surface.

J. A. McDivitt and E. H. White II conducted the first formal geologic photography experiment on the GT-4 (Gemini-Titan) mission on June 3–7, 1965 (Lowman et al., 1967). The four-day mission produced a spectacular series of 39 overlapping, nearly vertical photographs of the southwestern United States and northern Mexico. The mission demonstrated that useful data could be acquired for earth-resource analysis. In addition the GT-4 astronauts acquired more than 60 vertical and oblique photographs of other areas of North America, Africa, and Asia. Because of the intense interest generated by the GT-4 and subsequent GT-5 pictures among earth scientists, the scope of the experiment was enlarged for GT-7 to include photography of the earth's surface for geographic and oceanographic study. Subsequently, crews on all remaining Gemini missions were given a standing request to photograph glitter patterns on the ocean, and a variety of terrain features. By the end of the Gemini program, 1,100 usable color photographs had been obtained, in addition to those of various spacecraft and extravehicular activities. Many of these photographs have been published in two excellent books by NASA (1967a, 1967b).

The early space photographs led the U.S. Geological Survey to formulate and publish a set of performance specifications and a general plan for repetitive surveys of the earth for resource and environmental investigations (Fischer and Robinove, 1968). The efforts by the Survey to establish an Earth Resources Observation Satellite Program formed the basis for the current USGS EROS program and contributed to the development and flight of the NASA ERTS-1 satellite, now known as Landsat 1.

The Apollo program was another step forward in acquiring orbital terrain photography. The unmanned Apollo-6 mission produced a remarkable series of overlapping vertical photographs across North America, the Atlantic Ocean, and West Africa. Hand-held 70-mm cameras were used by the Apollo-7 and Apollo-9 crews for terrain photography as in the Gemini flights. Several hundred high-quality color photographs were obtained during these missions.

During the Apollo-9 mission, the first controlled multispectral experiment from space, from SO65 Multispectral Terrain Photography Experiment, was conducted. The SO65 experiment demonstrated the feasibility and the advantages of employing multispectral imaging techniques. The astronauts used a series of four electrically-driven coaxially-mounted 70-mm cameras, to photograph terrain with four different film/filter combinations: panchromatic film with red and green filters, black-and-white infrared film, and color infrared film. The spectral bands employed were close to those planned for the NASA Earth Resources Technology Satellite (ERTS-1). Over a period of four days, the crew obtained 350 pictures of good to outstanding quality. Many of these are contained in a book that summarizes results of the SO65 Experiment, as obtained by a large team of remote sensing scientists (NASA, 1969).

Landsat

The first satellite designed specifically to collect data of the earth's surface and resources was the Earth Resources Technology Satellite (ERTS-1). The first three of this series of modified Nimbus satellites were launched into near-polar circular orbits at an altitude which varied between 897 and 918 km. ERTS-1 (later renamed Landsat 1) far exceeded its design life of one year. Launched on July 23, 1972, it carried a four-channel Multispectral Scanner (MSS), a three-camera return beam vidicon (RBV), a data collection system, and two video tape recorders. The MSS operated in the following spectral intervals: band 4 $(0.5-0.6\ \mu)$, band 5 $(0.6-0.7\ \mu)$, band 6 $(0.7-0.8\ \mu)$, and band 7 $(0.8-1.1\ \mu)$. Figure 1-1 is a Landsat image of Santa Barbara County, California.

The three independent cameras of the RBV covered three spectral bands, viz. blue-green $(0.47-0.575\ \mu)$, yellow-red $(0.58-0.68\ \mu)$, and near-infrared $(0.69-0.83\ \mu)$. Both of these systems viewed a ground scene of approximately 185 km by 185 km in area, with a ground resolution of about 80 meters. Because of a malfunction, the RBV was deactivated after only 130 satellite orbits.

With the launch of Landsat 2 on January 22, 1975, the name of the series was changed from ERTS to Landsat. This satellite carried a payload similar to that of Landsat 1. Landsat 3 launched on March 5, 1978, added a fifth band in the thermal infrared $(10.4-12.6\ \mu)$ to the MSS or Landsat 1 and 2. The RBV camera system was also significantly changed: it employed two identical cameras covering the spectral band from 0.53 to 0.75 μ, so aligned as to simultaneously view adjacent 84 km-square ground segments, giving a linear swath of about 185 km. Its resolution of 40 meters, was better by a factor of 2 than that of the previous MSS and RBV systems.

The RBV and MSS data were transmitted directly to a ground receiving station when the satellite was within range. When the satellite was not within range of a ground receiving-station, the sensors were turned on according to a preset program and the data were temporarily stored on an on-board magnetic tape. A short time later, when the ground receiving station was within range,

Fig. 1-1. Band 5 Landsat 2 image of the Santa Barbara Channel, California.

playback of this magnetic tape record permitted the previously-acquired data to be transmitted to earth.

Landsat 2 ceased operation in February 1982. Landsat 3 is beginning to show signs of failure but at the time of this writing, is still supplying RBV data. To prolong the operation of the Landsat 3 RBV and MSS, the system is now activated only on special request.

Often-cited limitations of the Landsat series are its modest spatial resolution, spectral channels not precisely aligned to known absorption bands, and the inordinate time delay before the data become available to the user. Landsat D has been designed to alleviate these problems as well as some others. Landsat D continues to employ the same four bands of the Landsat 3 MSS but adds an advanced MSS called a Thematic Mapper (TM). The TM is intended for use in providing unique tone signatures for various kinds of resource-related features, thereby facilitating the production of thematic maps pertaining to those categories. The TM has seven spectral bands covering four regions of the electromagnetic spectrum: bands 1 through 3, the visible range (0.45 – 0.69 μ), band 4, the near-infrared (1.55 – 2.35 μ), and band 7, the thermal infrared (10.40 – 12.50 μ). The resolution of the first six bands is approximately 30 meters, while that of band 7 is 120 meters. (For a more detailed discussion, please refer to chapter 12).

One fundamental difference between the 4-band MSS and the TM is that the MSS scans in only one direction, while the TM scans and obtains data in two directions. Furthermore, the TM detector arrays are located in the instrument's primary focal plane, thus allowing the incoming radiation to be reflected directly onto the detector while, in the MSS, the incoming light is transmitted through fiber optics before being reflected onto the detector arrays, with some loss and degradation of the signal.

Major improvements in data acquisition will be achieved through the use of a series of communications satellites (DOMSAT). The Tracking and Data Relay Systems (TDRSS) satellites will eliminate the need for on-board tape recorders as used by the previous Landsats. When the Landsat vehicle is within range of a ground receiving-station, its data will be transmitted directly. MSS data will be transmitted at 15 megabits/second on S-band, and the TM data will be transmitted at 85 megabits/second on X-band. After preliminary processing, the data will be relayed by DOMSAT to the Goddard Space Flight Center (GSFC). When the Landsat vehicle is not within range of a ground receiving-station, the TDRSS will be used to transmit the MSS and TM data to the ground receiving-station in White Sands, New Mexico. The site at White Sands was chosen for its relatively cloud-free location because the TDRSS uses a frequency which is affected by rain droplets and cloud particles. From the White Sands receiving-station, DOMSAT will be used to transmit the data to GSFC, thereby reducing previous delays in data transfer. From that location the data will then be shipped to a data-distribution center.

Until the TDRSS are operational in late 1983, the MSS data will be transmitted on S-band directly to GSFC and to existing foreign stations. Because there are no tape recorders on Landsat D, ground tape recorders will be provided by NASA to select foreign stations. The TM will initially be operated as an experimental sensor. The data will be transmitted on X-band directly to GSFC until the TDRSS is operational. These data will not be transmitted to any foreign stations because such stations are not yet configured to receive and process the X-band 85 megabit/second data.

Approximately one year after the launch of Landsat D, an identical satellite, Landsat D′, will be ready for launch. In an effort to assure data continuity, this spacecraft will not be launched until Landsat D shows signs of failure. The MSS has a specification lifetime of three years, and the TM has a specification lifetime of two years, so this plan should provide data through 1986 to 1988.

Much more complete information with respect to satellites in the Landsat series will be found in Chapter 12 of this Manual.

Skylab

Skylab, the first American space station, was launched on May 14, 1973 into a near circular orbit carrying with it the Earth Resources Experiment Package (EREP). Initially unmanned, Skylab was occupied sequentially by three crews of three astronauts each, from the period May 25, 1973 to November 16, 1973. At the end of the Skylab 4 mission the spacecraft was deactivated, and in 1979 Skylab reentered the earth's atmosphere and disintegrated over Australia. The EREP sensor system included two photographic and four electronic sensor packages. The multispectral photographic camera system (S190A) used six identical cameras with different film/filter combinations in order to view the same ground area simultaneously over the 0.4 to 0.9 μm range. A second photographic system, the Earth Terrain Camera (S190B), was also carried. This was a single 127-mm focal length camera covering the 0.4 to 0.88 μ range. Electronic sensor systems included an infrared spectrometer, (S191). The S191 system was a filter-wheel spectrometer, which scanned the radiation entering its aperture, and a tracking telescope that enabled the crew to track the test site. The S191 system did not produce images; instead it provided separate digital records of incoming radiation that had been split into short (0.4−2.5 μ) and long (6.6 to 16.0 μ) wavelength bands. The multispectral scanner element of the EREP package (S192) was a thirteen-channel optical-mechanical scanner operating in twelve bands from the visible to the middle infrared (0.41 to 2.35 μ) and in one thermal infrared band (10.2 to 12.5 μ). Still another EREP sensor was the Radscat system (S193), which operated at the Ke-band (2.2 cm) as a passive microwave radiometer, active scatterometer, and radar altimeter. This system was the first implementation of a combined active and passive microwave sensor and was used to record data from which the wind speed of hurricanes and tropical storms could be derived. This concept was later incorporated in the Seasat microwave sensor design. The fourth sensor system was a 21 cm (L-band) radiometer employed to measure the radiant energy from a 111-km wide swath centered around the nadir.

A substantial volume of data (35,000 photographs and 238,600 feet of magnetic tape) was obtained from the three Skylab manned missions, even though only 25 minutes per day were devoted to remote sensing. The time spent in the collection of earth resources data was constrained by the amount of film and magnetic tape that could be carried by the astronauts, the data-collection requirements of other experiments, the limitations imposed by cloud cover, and the activity schedules of the astronauts. Nevertheless, the imagery and related data acquired did prove to be of excellent quality and were in some cases, unique. For an excellent and well illustrated book that summarizes the major remote sensing-related findings of the various Skylab experiments see NASA (1977).

HCMM

Launched April 26, 1978, the HCMM (Heat Capacity Mapping Mission) was a test of thermal inertia to help discriminate between different surface materials, and to identify different states, such as degree of soil wetness. Thermal inertia is the property of a substance to resist temperature changes when incident energy varies, for example, over a daily cycle. Since the most common thermal inertia effect exhibited by features on the earth's surface is related to diurnal temperature variation, which has a daily maximum and minimum, the mission was designed to acquire repetitive thermal data twice a day at times close to the expected daily surface temperature maximum and minimum. Thus the satellite was placed in a nearly sun-synchronous circular orbit to match as closely as possible the respective maximum (1:30 P.M.), and minimum (4:30 A.M.) of the diurnal cycle. The HCMM was the first of a series of NASA Application Explorer Missions (AEM) which, compared with satellites of the Landsat series, were smaller, less expensive, and with less precise orbit accuracies and attitude stabilization.

The only sensor carried on board the HCMM was the Heat Capacity Mapping Radiometer (HCMR), which sensed in two channels. The HCMR was developed from a modified spare of the Nimbus-5 Surface Composition Mapping Radiometer. The first channel was matched to band 7 of the Landsat MSS, and thus responded to near-infrared energy, (0.8−1.1 μ). This channel measured the near-infrared solar reflection from the earth's surface. The second channel recorded thermal infrared radiation (10.5 to 12.5 μ) which is a measure of the thermal region emission from the surface. The HCMM transmitted analog data in real-time to six NASA receiving stations when the satellite was within range.

With a spatial resolution of 600 m at the nadir, the HCMM was oriented to broad applications, particularly in the areas of geology, hydrology, and agriculture. (See Figure 1-2) Although the HCMM was originally planned to have a duration of one year, this satellite operated for 2½ years, during which time it provided nearly 6000 data passes and 26,500 frames of imagery. Flight operations for the HCMM were terminated on September 30, 1980.

The HCMM data have been used to distinguish rock types even through vegetation cover, discriminate geologic units of similar albedo from determinations of their thermal inertia differences, identify shallow water tables, estimate moisture availability, map industrial thermal pollution, and map the areal extent of snow.

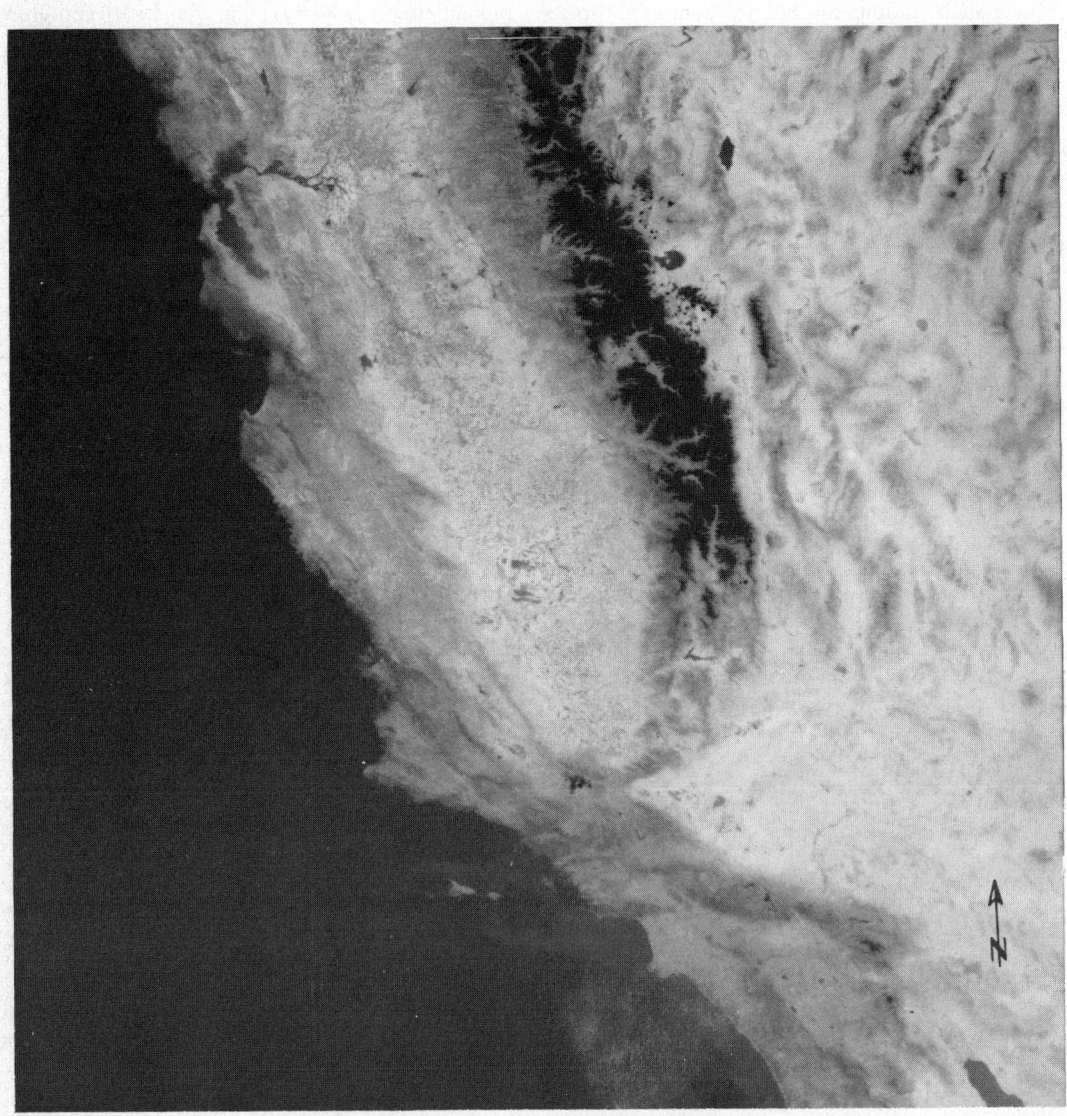

Fig. 1-2. HCMM day-time infrared image of Santa Barbara, California.

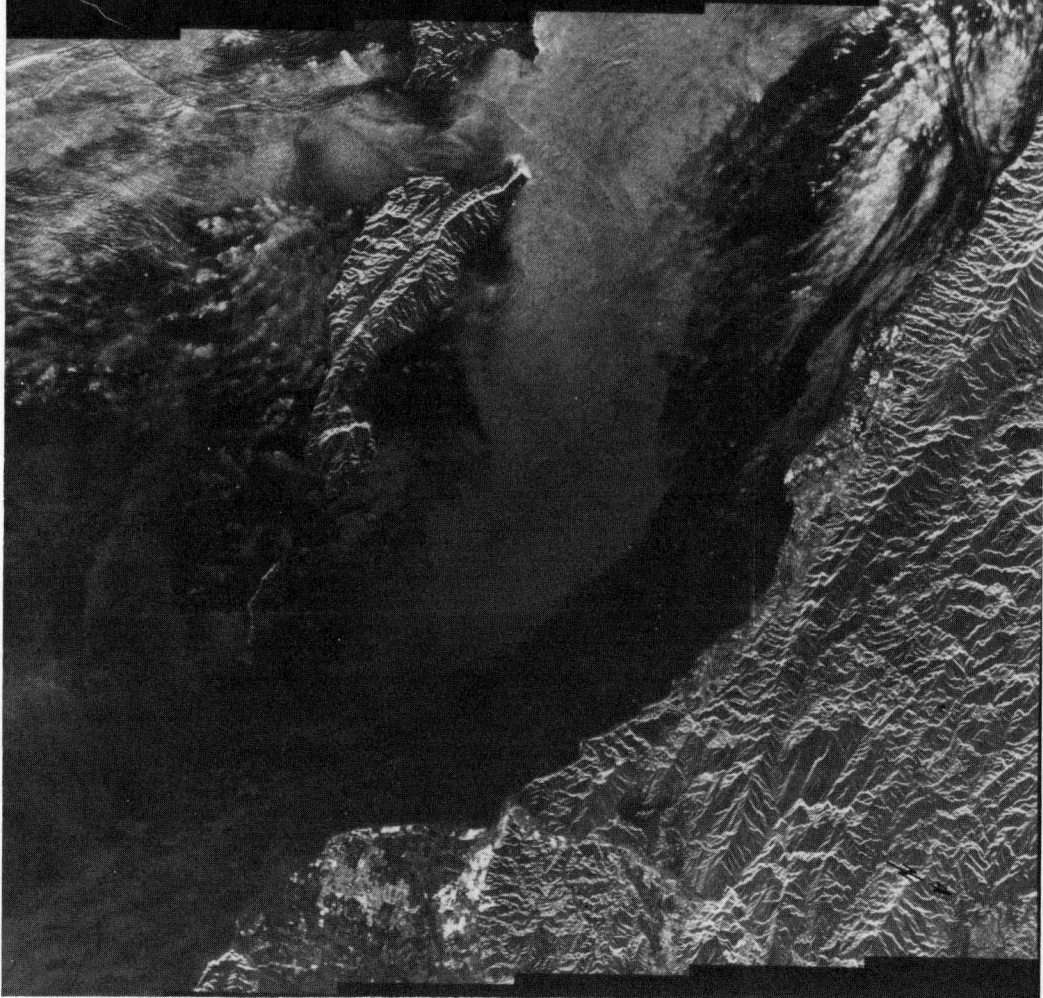

Fig. 1-3. Digitally processed SAR image from Seasat-1 of the Santa Barbara/Ventura coast, California.

Seasat

Launched June 26, 1978, Seasat-1 was the first of a proposed series of oceanographic research satellites. Placed in a near-polar orbit at 800 km, it provided alternating day and night coverage of 95 percent of the earth's oceans every 36 hours. There were five sensors aboard Seasat-1:

1) a compressed-pulse radar altimeter to provide precision altimetry for marine geoid and sea-surface topography studies;
2) a microwave scatterometer to measure global wind speeds and directions;
3) a two-channel scanning radiometer operating in the visible (0.52 to 0.73 μ) and infrared (10.5 to 12.5 μ) portions of the spectrum to monitor ocean color and temperature;
4) a scanning multifrequency microwave-radiometer which had five bands imaging between 6.6 and 0.8 cm; and

5) an L-band (25 cm) synthetic-aperture imaging radar (SAR) to provide a determination of wave patterns and sea ice in selected areas.

The SAR on Seasat-1 provided the first synoptic high-resolution radar images of the earth's surface (See Figure 1-3). Due to the high rate of data acquisition (110 megabits per second), SAR data were not recorded on board the satellite, but were transmitted to the earth when within range of a receiving station, and recorded on the ground. Since there were only five stations able to receive SAR data, (Goldstone, California; Fairbanks, Alaska; Merritt Island, Florida; Shoe Cove, Newfoundland, Canada; and Oakhanger, United Kingdom), coverage of the oceans was limited to the North Atlantic, Gulf of Mexico, Caribbean, Mediterranean, North Sea, Norwegian Sea, eastern North Pacific, Gulf of California, and Beaufort Sea. During its 98 days of operation, the Seasat

SAR acquired images covering approximately 100 million square miles. Although the rationale for placing a SAR on board Seasat was its potential for monitoring the global surface wave-field and polar sea-ice conditions, the images of the oceans revealed a very wide spectrum of oceanic and atmospheric phenomena, including internal waves, current boundaries, eddies, fronts, bathymetric features, rainfalls, and storms. For more information, please refer to Chapter 13. With the exception of the SAR, all of the sensors were designed for continuous operation. The SAR was to operate only when it was over selected high-data-rate ground stations. The incidence angles for the SAR ranged from 70 degrees in high-relief terrain, to 20 degrees in low-relief terrain so that shadowing was minimized. However, this created some difficulty in discriminating landforms in high relief areas owing to foreshortening and layover.

Unfortunately on October 10, 1978, a massive short circuit in the electrical system aboard the Seasat-1 platform terminated any subsequent acquisition of data.

Space Shuttle

The first launch of a Space Shuttle vehicle was in April, 1981. That event marked the beginning of a new era of manned space flight. Unlike other launch vehicles Space Shuttle may be used repeatedly. This new space-transportation system can carry a maximum crew of seven comprised of the commander, pilot, and mission specialist, who are NASA astronauts, and four payload specialists who conduct the experiments and who may or may not be NASA astronauts. The Shuttle will be used as an orbital laboratory to conduct highly specialized experiments under the weightless and vacuum conditions of space, and to retrieve and strategically place earth-orbiting satellites.

The Space Shuttle flight system consists of the Orbiter, an external tank containing the ascent propellant used by the main engines of the Orbiter, and two solid rocket blasters, each with a sea-level thrust of 11.8 million newtons. The external tank is the only part of the system which is not reusable.

Seven experiments have been selected to demonstrate the potential of the Shuttle for research in earth resources. These experiments will investigate continental geology, atmospheric chemistry, meteorology, marine biology, and plant physiology. Collectively they are referred to as OSTA 1 because the Office of Space and Terrestrial Applications will be managing them (Taranik and Settle, 1981). The data from these experiments should be available to the public within six to twelve months after completion of the first mission.

The sensor systems of the Shuttle include the Shuttle Imaging Radar (SIR-A), the Shuttle Multispectral Infrared Radiometer (SMIRR), the Measurement of Air Pollution from Satellites (MAPS), the Night/Day Optical Survey of Lightning (NOSL), the Ocean Color Experiment (OCE), the Feature Identification and Location Experiment (FILE), and the Helianthus annuus Flight Experiment (Heflex) Bioengineering Test (HBT). The SIR-A experiment will be used to evaluate the potential of orbital radar for geological mapping, and has been developed from the Seasat L-band radar system. The system has a 47 degree depression angle, a 56 km swath width, and a 40 meter resolution (see Figure 1-4). The SMIRR experiment will be used to evaluate the potential for rock type discrimination and mapping employing narrow band absorptions of rock and clay minerals in the infrared region. The system includes a telescope, a rotating filter wheel, and a detector to record surface reflectivity in 10 bands between 0.5 and 2.5 micrometers, with a 100 meter instantaneous field of view (IFOV). The MAPS project is designed to measure the CO_2 concentration in the middle and upper troposphere from the low to middle latitudes so that air mass transport may be monitored. The NOSL experiment will be an effort to obtain films of lightning storms to attempt the calculation of the path and intensity of thunderstorms. The OCE has been developed to collect ocean color data in the deep sea at different solar azimuths and inclinations. It is an outgrowth of the NIMBUS experiments. This system consists of a rotating-mirror imaging scanner with a 3 km IFOV and a 550 km swath. The OCE has eight spectral bands between 0.49 and 0.79 micrometers, which are used to determine chlorophyll concentrations as a measure of ocean bioproductivity. The FILE experiment is designed to classify, onboard the spacecraft, such resources as vegetation, rocks and soil, water, clouds, snow and ice. Classes are determined from scene brightness ratios based on a visible red image, and a near-infrared image. The HBT experiment will study plant-growth behavior in weightless conditions in an effort to determine the optimal soil moisture conditions for the germination and growth of plants in space.

The 16th mission of the Shuttle, scheduled for launch in mid-1984, is to be dedicated to the Space Transportation System-16 (STS-16), and will include the Large Format Reference System (ARS), (being built as of March 1982) and an imaging radar, SIR-B, (also under development), as well as the FILE, SMIRR, and MAPS experiments.

A Large Format Camera (LFC) is also to be carried. It has a 30.5 cm focal length with a 23 by 46 cm format. It is equipped with forward motion compensation and automatic exposure sensors, permitting the use of high resolution fine grain film. A ground resolution of 10 to 15 meters can be obtained at nominal Shuttle altitudes.

The ARS will consist of two stellar cameras of 152 cm focal length. These cameras will be directed horizontally when the LFC is looking vertically and will be aimed 45 degrees fore and aft of the normal to the satellite flight path.

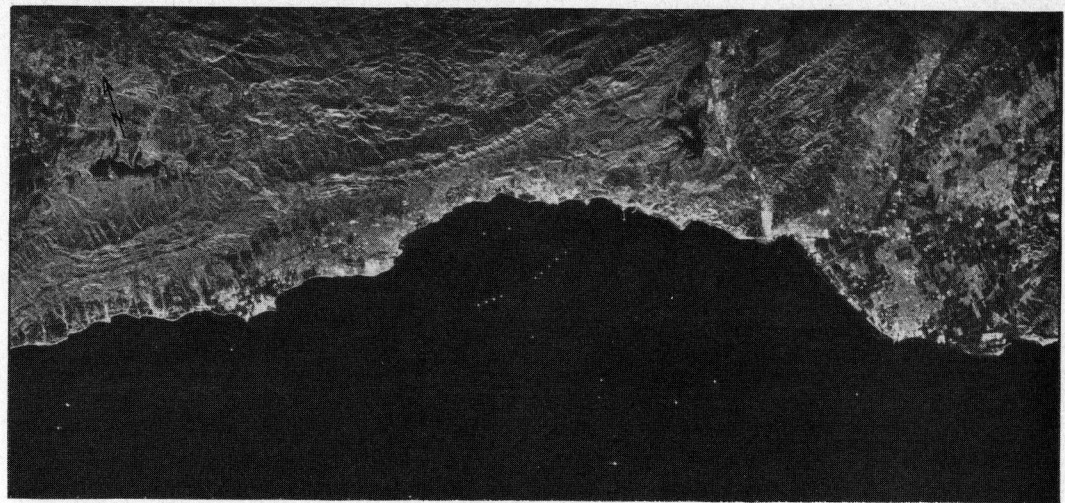

Fig. 1-4. Digitally processed Shuttle Imaging Radar (SIR-A) image of the Santa Barbara/Ventura Coast.

SIR-B has been designed to investigate the effect of various radar parameters, such as look angle, resolution, and stereoscopy. This sensor will produce data at the rate of 150 megabits/second, which will be transmitted to the TDRSS. The ground resolution of this radar is anticipated to be 15 meters.

SYSTEMS OF THE NEAR FUTURE

SPOT

The European Space Agency (ESA) has developed an expendable launch vehicle called ARIANE, which is capable of delivering a payload of 1700 kg into a geostationary orbit.[1] This vehicle is scheduled for operational use in 1982. Its first remote sensing payload, scheduled to be launched in April 1984, is the French Systeme Probatoire d 'Observation de la Terre (SPOT 1). It will be an earth observation satellite which will have finer ground resolution than any of the Landsat series to date (1,2,3,D).

This satellite will be placed in a sun-synchronous orbit at an altitude which will vary between 815 and 829.6 km. The satellite payload consists of two identical High Resolution Visible (HRV) imaging instruments, two magnetic tape recorders, and a telemetry transmitter.

Each HRV unit will operate in either of two modes; 1) a three-band multispectral color mode with proposed spectral bands of (0.50 through 0.59 micrometers), (0.61 through 0.68 micrometers), and (0.79 through 0.89 micrometers), and a ground

resolution of 20 meters. 2) a black-and-white panchromatic mode with a ground resolution of 10 meters (see Color Figure 1-5).

The detector arrays of the panchromatic and multispectral modes consist of 3000, and 6000 detectors respectively, which are arranged in a one dimensional linear array located at the focal plane of the optical systems. The process, called 'push broom' scanning, obtains successive lines of ground coverage as the satellite moves over the earth. This system has the advantages of no moving parts, higher geometric fidelity, and a longer life expectancy than the mechanical line scanners of the Landsat series. At the present level of technology, this system has two distinct disadvantages: 1) Each detector in a linear array must be individually calibrated in order to produce a uniform response over an imaged scene. 2) The detectors presently available are not suitable for use with wavelengths longer than about 1.05 micrometers.

The viewing axis of the satellite is movable for off-nadir as well as nadir viewing, and covers a range of plus or minus 27 degrees fore and aft with respect to the vertical. The width of the observed swath varies between 60 km for nadir viewing to 80 km at the most extreme off-nadir viewing.

The operation sequences of the HRV will be loaded daily by the Toulouse ground station when the satellite is within range. The operation sequences of the two HRV are independent. The data and images will be processed at the Center de Rectification des Images Spatials.

MOS 1

The National Space Agency of Japan (NASDA) has been using MSS and RBV data from the Landsat satellites since January, 1972. Since these data have been widely used by government agen-

---

[1] ESA consists of the following states: Switzerland (only non-UN member), Spain, Denmark, the Netherlands, Belgium, Federal Republic of Germany, France, Italy, Sweden and the United Kingdom. Canada has observer status with the ESA council.

cies, universities, and private industry, NASDA has proposed the development of a series of earth-observation satellites.

The first of this series, Marine Observation Satellite 1 (MOS 1), is scheduled to be launched in fiscal year 1984. The spacecraft will be launched to an orbital altitude of 909 km at an inclination of 99.1 degrees. MOS 1 will carry a payload of three instruments; a multispectral electronic self-scanning radiometer (MESSR), a visible and thermal infrared radiometer (VTIR), and a two-frequency microwave scanning radiometer (MSR). The MESSR will have four spectral bands between 0.51 and 1.10 micrometers. It will have a 50 meter IFOV for a 100 km swath, and will be used to measure sea surface color. The VTIR will have a 500 km swath, one band operating in the visible with a 0.9 km IFOV, and three bands operating in the infrared between 6.0 and 12.5 micrometers with an IFOV of 2.6 km. This sensor will be used to measure sea surface temperature. The MSR will be used to measure atmospheric water content.

If this satellite is successful, a second and third spacecraft may be launched. Based on preliminary studies by the Science and Technology Agency, and the Ministry of International Trade and Industry, NASDA has directed the design of an Earth Resources Satellite (ERS 1), which would carry a liner-array stereo camera with a 30 km IFOV operating in four spectral bands between 0.51 and 1.10 micrometers, and an L-band synthetic aperture radar providing a 25 m ground resolution and a 75 km swath. The spacecraft has a probable launch date of 1987. Note that this satellite has the same designation as the European Space Agency resources satellite (ERS 1) scheduled for launch in 1986–1987.

SPACELAB

Scheduled for launch on the 9th mission of the Space Transportation System (STS-9) in September 1983, is a payload called Spacelab. Spacelab is a joint venture between the European Space Agency (ESA) and NASA. It consists of a pressurized compartment for the housing of personnel and equipment, and a space-exposed platform to accommodate the sensing instruments. The compartment and sensor platform will be transported inside the payload compartment of the space shuttle orbiter. While in space, the payload compartment doors will be opened to allow viewing of the earth, sun, and deep space. The mission is scheduled to last seven days.

The payload will consist of manned experiments, and platforms for other instruments. There are forty scheduled experiments for the first Spacelab mission. Of particular interest are the Metric Camera Facility (MC), and the Microwave Remote Sensing Experiment (MRSE).

The topics of research with MC include topographic mapping, orthophoto mapping, resolution experiments, thematic mapping, and interpretation. The MC will be a standard Zeiss 30/23 aerial camera with a 30 cm focal length, and a 23 by 23 cm format. It will be operated through a window in the manned module. Based on the anticipated altitude of Spacelab of 250 km, MC images will cover an approximate ground area of 190 by 190 km, with a ground resolution of approximately 20 meters. This camera can use black-and-white, color, or color infrared films. In order to obtain an eighty percent longitudinal overlap of consecutive photographs at a spacelab velocity of 7.7 km/sec., a time interval of about five seconds between two successive exposures is required. Strips 1800 to 2300 km on the ground can then be covered in each sequence.

The first camera is part of Phase A of the ATLAS camera development program conducted by the West German research agency Deutsche Forschungs und Versuchsanstalt fur Luft und Raumfahrt (DFVLR). Phase B will add motion compensation to the MC, using a similar camera but with a focal length of 60 cm. Phase C proposes to mount the camera system on free-flying satellites which will operate independently.

The MRSE will operate in different modes; as a two-frequency scatterometer to measure backscatter from the ocean surface at two adjacent frequencies, and as a thermal radiometer to measure surface temperature.

Spacelab 2, scheduled for launch in October, 1983, will have a geocentric orbit at approximately 400 km. It will consist of three platforms and a unique structure called the igloo, on which several instruments will be exposed to the space environment. Included in the payload is the ESA instrument-pointing system.

Spacelab 3, scheduled for launch in April, 1984, has been designed for experiments requiring a low-gravity and motion-stable environment. These experiments will emphasize materials processing.

Other ESA Satellites

The European Space Agency (ESA) has designed a space vehicle for ocean observations, the ESA Resources Satellite-1 (ERS-1). This satellite will use the SPOT vehicle launched by ARIANE, and is anticipated to be launched in the period 1986–1987. The sensor payload is expected to include five sensors; 1) a synthetic aperture radar (SAR) with 30 meter resolution and a 100 km swath, 2) an Ocean Color Monitor (OCM) with 10 spectral bands between 0.4 and 11.5 micrometers, 3) an imaging microwave radiometer (IMR) operating in six frequencies, 4) a two-frequency scatterometer for wind direction and velocity, and 5) a radar altimeter for sea-state determination.

An advanced ESA Resource Satellite (AERS) has been proposed for launch around 1989, geared principally for land observations. It would use the SPOT vehicle and its payload would include the

SAR from ERS-1, an Optical Imaging Instrument (OII) which would have six spectral bands from 0.52 to 2.35 micrometers with an IFOV of 30 meters, and one panchromatic band with a 15 meter IFOV, and a 175 km swath.

Remote Sensing Satellites

Canadian Radarsat, scheduled for launch in 1990, will consist of a 3-axis stabilized platform of sufficient power and weight capacity to carry a C-band or L-band Synthetic Aperture Radar (SAR) (Raney, 1982). This satellite has been designed to provide information on frozen and open ocean and to provide remotely sensed data for application in forestry, geology, hydrology, and agriculture. The frequency selection and data processing are under consideration; presently the choices of frequencies are either L-band (23.5 cm wavelength) or C-band (5.7 cm wavelength). Research is currently under way to develop a processor that can process raw data received in real time to high quality imagery in 4 times real time or faster (Raney, 1982).

OTHER NATIONAL SYSTEMS UNDER CONSIDERATION

In the preceding section, only some of the many active and planned satellites have been discussed. The Peoples Republic of China has launched several satellites since the launch of Chinasat 1 in July 1975. No images from Chinasat 1 through Chinasat 7 (launched December 1976) have been made available to the international community. Additional Chinasats are scheduled in the near future; for example, Chinasat 10 plans to carry a two channel meteorological radiometer with visible and infrared bands. In November, 1981, China announced that it is proceeding with the development of an 11-band multispectral scanner, linear array sensor, and synthetic aperture radar (Doyle, 1982).

The Indian Space Research Organization (ISRO) developed an Earth Observation satellite, Bhaskara, which was placed in orbit by a U.S.S.R. vehicle launched from a cosmodrome in the U.S.S.R. This satellite was designed to conduct earth resources observations in forestry, hydrology, and geology using a two-band television camera system. This satellite also conducts ocean surface studies using a two-frequency microwave-radiometer system. An identical satellite, Bhaskara-2 was launched by the U.S.S.R. in November 1981. Since that time, ISRO has announced its intention of developing second generation Indian Resources Satellites (IRS), and is negotiating with both NASA and the USSR for launch service.

The Netherlands Agency for Aerospace Programmes (NIRV) in cooperation with Indoneasia, has planned the development of a Tropical Earth Resources Satellite (TERS) which would carry a Dutch-built multispectral linear-array sensor into a low inclination orbit. The sensor design will be adapted to the weather conditions, and vegetation types in the tropical areas.

GEOGRAPHIC INFORMATION SYSTEMS

The data accumulated from the previously described remote sensing systems can be put to their best use if they are incorporated in a system capable of efficient data storage and expedient data processing and retrieval.

Geographic or Geo-based Information Systems (GIS) have evolved as a means of assembling and analyzing diverse data pertaining to specific geographic areas, using the spatial locations of the data as the basis for the information system (Shelton and Estes, 1979). In the broadest context, Geographic Information Systems identify a user's data needs, and channel the data to the intelligence level at which decision-making takes place. Remote sensing provides a source of data for such systems, and has the potential to improve the quality and quantity of data available. In some ways remote sensing can contribute data never before available to such systems (Estes, 1981).

In discussing the role of information systems in developed and developing countries, Simonett (1981) states:

"At the highest levels of government, decisions on capital investment and resource development are made in an atmosphere of greater or lesser uncertainty, and with value judgments strongly conditioned by political circumstances. The uncertainty is almost always greater in the developing than the developed world, because of the weaker information base. As a result, with almost all development projects in the developing world, unexpected, and usually undesirable outcomes are found economically, socially, and environmentally. Since the major bases for economic development in most developing nations lie in their natural resources, improved information on these resources, derived from remote sensing, will reduce some of the uncertainty of decision making."

The true effectiveness of remote sensing inputs to information systems, then, can be measured by the appropriateness, timeliness, and accuracy of data provided to the system. These in turn are functions of the effectiveness of the GIS as a whole. For a more detailed discussion of GIS, refer to Estes, (1981), Shelton and Estes, (1979), Tom and Miller, (1979), Richason, (1981), and also to Chapter 22. An example of the use of remote sensing as input to an information system for analytical modelling in the developing world is the Comprehensive Resource Inventory and Evaluation System (CRIES). CRIES originated in a 1975 proposal by the U.S. Department of Agriculture to research a means for estimating the technological and economic potentials for agricultural production in developing countries (Shelton and Tilman, 1979). This system is composed of a resource information system which supports an analytical model. Remote sensing data can be used as input to this information base.

In the CRIES system, the analytical needs of the users are incorporated into the system design. The GIS serves two functions in this system: it supports the analytical model, and it functions independently for other analyses of land use and natural resource data. The system has been successfully implemented in the Dominican Republic, and the program is beginning in Costa Rica and Nicarague (Shelton and Tilman, 1979). As soon as the necessary data are assembled, and training of nationals in each country in its use is completed, it will be transferred to each country.

To be effective, geographic information systems are dependent upon the accuracy and timeliness of their input data. Remote sensing offers the potential to meet these criteria. Yet, the extent to which remote sensing, itself, can be most useful in the management decision process usually will depend upon the degree of integration of collateral or ancillary data. Integration of the parallel technologies of Geographic Information Systems and remote sensing will be important for the fullest maturation of both areas.

MAKING REMOTE SENSING WORK

The evolution of remote sensing as a discipline has depended not only on the developments discussed above but also on the degree to which remotely sensed data are assimilated into the information base and decision-making processes of public and private organizations. This section will deal with the assimilation of remote sensing into the resource management decision processes of developed and developing nations and the factors which affect the information needs of a country.

The problems in technology transfer to these nations are basically similar to those of any large organization and serve to illustrate the problems and prospects influencing the acceptance of new ways of doing business. In developing countries, which may stand to benefit economically and socially from remote sensing technology, the technology has been approached with trepidation in many cases. Many technically advanced countries are familiar with the principles underlying remote sensing so that the user community realizes its potential and limitations. However, in much of the developing world, countries are not as familiar with the technology and therefore are less able to integrate it into their decision-making processes. The assimilation process has proceeded in several, but not all of the developed countries.

TECHNOLOGY TRANSFER

The path toward economic development for many third world nations is greatly influenced by a number of fundamentally interconnected factors such as geographic location, political and economic system, social structure, cultural heritage, and human and natural resources available. For many countries, the primary basis for economic development and growth as outlined in a 1970 United Nations report, lies in "the process of economic development, (which) consists largely of organizing the development and productive exploitation of natural resources in the interest of the whole community."

Technology transfer may be defined as the process by which scientific knowledge and the skill to apply that knowledge are transferred. Technology transfer was once thought to be a straightforward process in which technological advances used to solve or assist in the solution of one's problems would be recognized and quickly incorporated by others with similar needs (Wagner and Lowe, 1978). Unfortunately, this is not always a valid assumption, since the use of a given technology assumes the existence of certain institutional and societal frameworks. For this and other related reasons, the transfer of technology is often a complex process that involves significant differences in perceived needs, institutional structures, and available resources. Because remote sensing has such a wide range of potential applications, and highly sophisticated components, the technology should be transferred to other societies and nations with care. There is evidence that traditional technology transfer can not only be ineffectual but may be as destructive to the recipient state as it is beneficial (Wagner and Lowe, 1978).

The role of technology in the development of a country is often both under- and overestimated (Arnold, 1978). Those who overestimate technology are generally those who practice or follow science and technology routinely. In this case there is a large base of supporting educated people, and a supporting data infrastructure that readily permits the application of science and technology. However, in most developing countries, this is not the case. Problems related to the acquisition of basic human needs are not only more urgent but generally less controllable. The solutions which need to be implemented are, in general, dependent upon local conditions and must be compatible with the cultural environment of the country. This environment may be radically different from the source of the technology.

Remote sensing is a powerful tool by means of which developing countries can obtain the information needed to knowledgably manage their resources. In many instances developing countries may have the advantage in accepting remote sensing approaches that they will not compete to the same degree as in developed countries with established systems of data acquisition, analysis, and management. For this reason, many may be in a position to design new bureaucracies around remote sensing and reap its benefits without impinging on the prerogatives of entrenched interests.

Nevertheless, many developing countries do have apprehensions about acquiring the new technology because of the perceived expense and its mismatch to an often-preferred labor-intensive solution.

Care must be taken in extrapolating the results of U.S. applications to potential net benefits in the developing world. The practices employed in the United States in such fields as agriculture, forestry and mineral exploration are different from those employed in many other parts of the world. Information systems that are operational in the U.S. do not always have counterparts in developing countries.

The need for a balanced approach to the transfer of technology to developing countries was recognized by the Agency for International Development (AID). In 1970, AID realized that remote sensing was a technology that could have a substantial impact on the development of a country (Conitz, 1978). The office of Science and Technology was established in that year to explore new ways to achieve a proper balance in the transfer of technology. In 1976, AID and NASA jointly set up a world-wide demonstration of the benefits of space technologies in 27 countries. This demonstration committed AID to a program for the transfer of remote sensing technology to developing countries. This commitment had previously been made by then Secretary of State, Henry Kissinger, at the United Nations Conference on Trade and Development in Nairobi when he said: ''Satellite technology offers enormous promise as an instrument for development. Remote sensing satellites can be applied to survey crops, forecast crops, and monitor land use. We are prepared to cooperate with developing countries in establishing centers, training personnel and, where possible, adapting our civilian satellite program to their needs'' (Conitz, 1978).

The major elements of the AID program are training, technical assistance, grants to support local initiatives and institutional development, and research on applications areas which are of particular importance to developing countries.

Figure 1-5 is a map showing Landsat receiving stations now operating or under construction (black circles), and those which are planned for the near future (white circles) (Paul and Mascarenhas, 1981). The circles represent the Landsat reception range from the receiving station (approximately 2700 km). Those applications which were primarily responsible for the adoption of remote sensing technology in that country on an operational basis are represented by symbols inside the circle. As seen in Figure 1-5, remote sensing

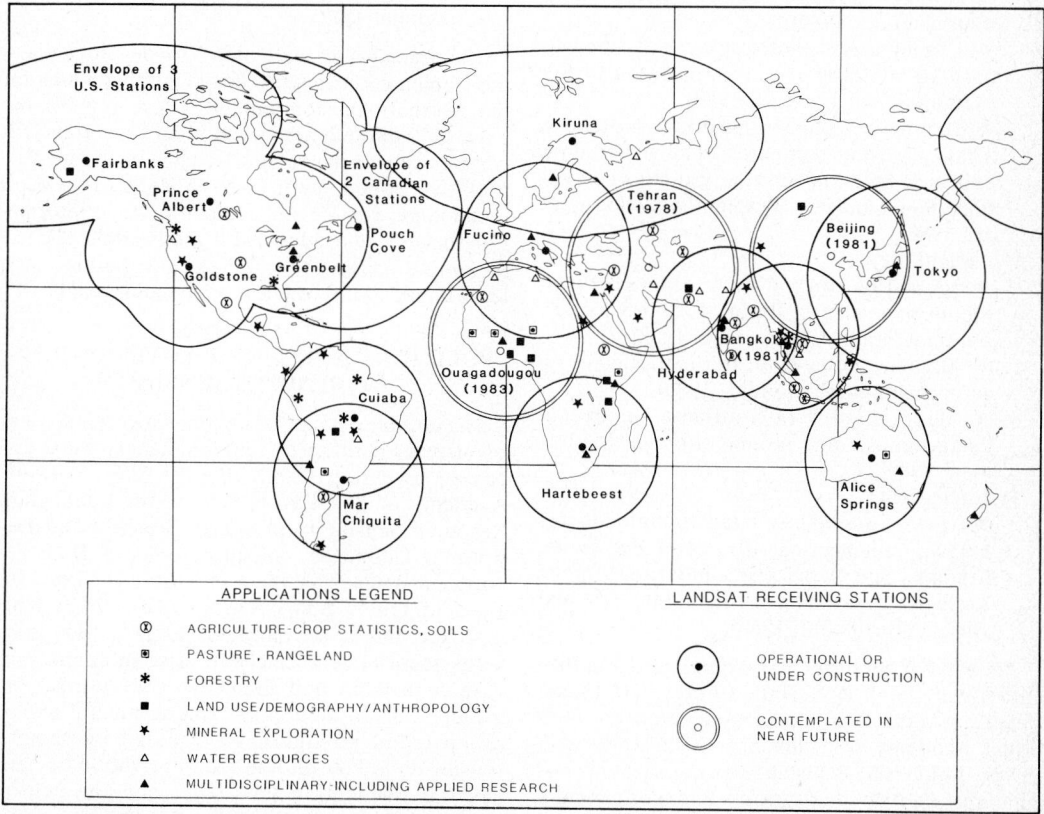

Fig. 1-5. Illustration of Landsat receiving stations that are operational, under construction, or planned. Circle size indicates the range within which the station can receive Landsat data. (adapted from Paul and Mascarenhas, 1981, and Science, Vol. 214, October 9, 1981, pp. 139–145).

imagery has proven useful in many applications in the developing world. Examples of applications include:

1) Forestry
   - Identification of degraded pastureland after deforestation, and the identification of pine and eucalyptus plantations in Brazil (Filho et al., 1980).
   - Identification of mangrove forests and forest reserve and fishpond areas in the Philippines (Roque et al., 1980).
2) Geology
   - Geologic and mineral mapping in the Philippines using Landsat imagery (Guerrero, et al., 1980).
   - Identification of major geologic rock bodies of the Ladakh Himalaya using Landsat data (Francica, et al., 1980).
   - Geologic mapping through the use of aerial photography in Tenchong, Yunnan, China (Junchaeo, et al., 1980).
   - Use of Landsat data in the identification of geologic anomalies indicating the location of tin in Brazil (Keighley et al., 1980).
3) Land Use Planning
   - Identifying environmental factors which have potential effects on land cover, and evaluation of land cover change in the Philippines through use of Landsat data (Santos et al., 1980).
   - Use of aerial photography and Landsat data to develop land cover inventories (Schultink and Karteris, 1980).
   - Landcover classification in Northern Thailand using computer analysis of data collected by aerial survey, ground observation, and Landsat, (Prapinmongkolkarn, et al., 1980).
4) Agriculture
   - Identification of host plants of the Mediterranean fruit fly in Mexico using aerial photography (Diez, et al, 1980).
   - Mapping coffee lands in Costa Rica using aerial photography (Sader, 1980).
   - Mapping soils in Southern Venezuela using Landsat and ground surveys (Siefferman, 1980).
5) Engineering
   - Use of remote sensing techniques for highway engineering surveys in Botswana, Ethiopia, and Nepal, (Beaumont, 1978), and for highway planning in the Middle East and Africa, (Beaumont, 1980).

Almost 120 countries have purchased Landsat data from the EROS Data Center. Of these, roughly two-thirds are developing countries, (National Academy of Sciences, 1977). In Brazil, Egypt, Indonesia, Pakistan, Peru, Tanzania, and Thailand central institutions and government agencies have taken root to conduct and/or control remote sensing operations. In some of these agencies there is an advisory committee to assure that user interests are represented by the agency. Although it is difficult to establish a formula for a national remote sensing effort, as needs and political structures differ from country to country, there are certain functional principles which must be considered if remote sensing is to be effectively and efficiently employed. According to a study by the National Academy of Sciences (1977) the following are the principles of greatest importance:

1) A remote sensing center should have trained personnel with a concentration of scientific skills.
2) Data analysis should be conducted through an interaction among personnel with diverse disciplinary training.
3) There should be strong ties between the remote sensing center and the user agencies concerning the level of data flow, and the priorities of data collection and processing.
4) The remote sensing center should have ground truth studies for confirmation of remote sensing studies.
5) The center should have the capability of handling remote sensing data from a variety of platforms.
6) The center should have the facilities for efficient data storage and retrieval, and
7) the center should have the capacity for making an overview analysis on a national or regional scale.

The transfer of technology to developed and developing countries has posed complications for the international community. These involve the extremely difficult area of national sovereignty. It can be argued that remote sensing is the proverbial two-edged sword. As a data-generation technology, remote sensing can provide information that can be used for both good and ill. The use of remote sensing thus raises both political and legal implications which must be addressed.

POLITICAL AND LEGAL IMPLICATIONS OF REMOTE SENSING

The complex issues of national sovereignty and security in relation to remote sensing were discussed by Estep (1968), by the U.S. National Academy of Sciences (1977), in the publication Resource Sensing from Space: Prospects for Developing Countries, and in a series of U.N. reports on debates in the Committee on Peaceful Uses of Outer Space (See Umali, 1979; Kaltenecker and Lafferranderie, 1977). These concerns seem most strongly to arise in connection with petroleum and mineral exploitation. For example, given a situation where mineral exploration could be significantly aided by remote sensing data, how should the information be handled? If one nation has developed the sensor platform and payload, put the satellite into orbit, and recorded and analyzed the data received from the satellite, how should that nation be able to use the data? What should be the rights of equity between the parties? Can one nation be required to seek permission to remotely sense another from space? What constitutes invasion of privacy? What limits on the spatial resolution of sensing devices should there be, if any? In his foreword to Resource Sensing from Space, Harland Cleveland addresses these difficult topics in the following words:

"Because we are dealing with information, we are dealing with an element of power—in this case (if combined with capital and know-how), the power to develop resources, the power to meet people's requirements, the power to be self-reliant in an interdependent world. As of 1976, the United States of America has a quasi-monopoly of this new power, a situation not destined to last. What the American people through their government do about this made-in-America technology, how and when and with what conditions we share its capabilities with others, will very largely determine whether remote sensing becomes a source of international conflict or a major assist to meeting human needs through international development.

Our present monopoly of decision making about remote sensing from space seems bound to be eroded from two sides—by other industrialized nations' space programs, and by participatory demands from the nations whose territory we are sensing from space. The European Community, Japan, the Soviet Union, and India, among others, are considering experiments with earth-sensing satellites, and a number of "sensed" countries are restive. Judging from bilateral discussions and by the debates in the UN Outer Space Committee and its Legal Subcommittee, these nations are variously worried that (a) the procurement of resource information by remote sensors is itself an invasion of their national sovereignty, (b) the dissemination of such information to other countries and multinational companies may weaken their position in economic bargaining, and (c) remote sensing images may tell their neighbors something about terrain, security arrangements, or economic resources that they are not supposed to know.

These doubts and fears have not yet made remote sensing a major issue in "planetary bargaining." But one way or another, it seems inevitable and probably desirable that "consumers" along with other "producers" of remote sensing data will press to participate in decisions about the disposition of satellite imagery, the configuration of new sensors, and the management of reception and processing facilities on the ground. Our report recommends the formal recognition soon of the inherently international character of remote sensing technology. *The United States Governments should declare soon that remote sensing systems con-*

stitute in effect an international public utility, destined for international governance.

. . . . Our report discusses the varying modes of international organization that might prove to be appropriate to the case of remote sensing technology. For the system's ground segment, the apparent options are bilateral agreements with national ground stations, the establishment of regional stations, coordination through a specialized international agency (inside or outside the United Nations), or a unitary system owned and managed by an international body (inside or outside the United Nations). The space segment could continue to be owned and operated by the United States; it could be taken over by a consortium of those willing to invest in it (on the INTELSAT model); it could take the form of a multination arrangement in which several countries own remote sensing satellites but manage them as a single system (the World Weather Watch model); or it could be a fully international operating utility.

It may be premature to make a firm selection now among these options for several reasons:

• Other nations will remain reluctant to harness themselves to remote sensing technology as long as the United States has not given formal assurance of the system's continuity.

• Thus far the United States alone has financed the R&D and operation of the space segment of the system. But once it is clearly a permanent part of the international picture, some financial contributions could probably be secured in return for a real voice in decisions about management and about follow-on technologies. As new technologies go, remote sensing is remarkably inexpensive so far: total investment in the three Landsats (one yet to be launched) and the related U.S. ground systems will run only to $250 million, which would barely buy two modern destroyers or one nuclear-powered attack submarine. NASA's proposed budget for research and development in the area of resource sensing and environmental monitoring is $67.3 million for fiscal year 1977.

• Remote sensing technology is still in its infancy, and some important decisions yet to be taken need to be carefully considered to make sure they fit into a phased approach to internationalizing the system. One example will make the point: NASA is planning a communications relay satellite to pick up data from a variety of U.S. data-gathering satellites (meteorological, atmospheric, resource sensing, space research), and for purposes of economy and efficiency, to relay them to a single reception point in the United States. The data could then be retransmitted to users elsewhere in the United States and around the world. The costs and the technical problems that might be involved in speedy retransmission of resource sensing data to other countries are still unclear, and the likely consequence of this relay system for direct reception in other countries and regions (not to speak of their reaction to so centralized and American a system) is, at this point, unpredictable."

In the United Nations the concerns of developing nations on exploitation have spilled over into the remote sensing debates. The following issues are the focus of such debates, and as yet there are no internationally accepted principles governing these issues:

1) the need for countries operating satellites to obtain prior approval of other nations before conducting surveys,
2) the right of the surveyed state or states to obtain access to all information collected about their country from remote sensing by another country and,
3) the obligation to secure prior consent of the sensed country before transferring to a third country any of the information collected.

The United States, as well as a number of other countries, has maintained that the policy of open data dissemination to all interested parties is more likely to enhance rather than diminish the ability of a nation to control its natural resources (Stowe, 1978). In point of fact, had the presently proposed restrictions on data dissemination been in force in 1972, Landsat data would have only been available to the United States and its citizens. However, the U.S. delegation to the United Nations did agree that the rights of the sensed state should be maintained, and that states with remote sensing satellite systems should provide information about any program before that program actually commences. The U.S. further recommended that, in cases of disagreements, every effort should be made to reach a mutually acceptable settlement through consultation between the sensed and sensing states (Umali, 1979). At present, the United Kingdom, Japan, and a number of West European nations agree with the United States, while the East European states, the U.S.S.R., and Indonesia have been pressing for more restrictive measures.

The Soviet Union has proposed that the resolving power of remote sensing instruments be restricted to 50 meters (The French SPOT system will have 10 meter resolution). This suggestion has been supported by India and several other nations. The Indian delegate to the United Nations Committee on the Peaceful Uses of Outer Space stated that he found it "inconceivable that there is no resolution limit below which any country here would have reservations about the dissemination of data. You can have resolutions of the order of a meter. In theory, it might be said that you could have them on the order of centimeters. I am sure it would be impossible for any country to say that everything right down to the finest possible resolution should be freely disseminated. If we do not

come up with (the principle of) sovereignty, we at least come up against personal privacy" (Umali, 1979). The Indian delegate, however, did not want to commit himself to any proposal which included a specific resolution limit as did the Soviet Union. This issue of sovereignty and national security, then, remains largely unresolved.

In 1967, the Legal Subcommittee of the United Nations Committee on the Peaceful Uses of Outer Space reviewed the legal implications of remote sensing from space. The result of this investigation was the Treaty on Principles Governing the Activities of States in the Exploration and Use of Outer Space, Including the Moon and Other Celestial Bodies. Article I of that treaty states:

"Outer space, including the moon and other celestial bodies, shall be free for exploitation and use by all states . . . and there shall be free access to all areas of celestial bodies." The Article goes on to say that: "There shall be freedom in outer space . . . (and that) states shall facilitate and encourage international cooperation in such investigations"
(Kaltenecker and Lafferranderie, 1977)

This rule proclaims freedom of the use of outer space which is stated in the Charter of the United Nations with respect to the competence of a state to govern its own internal affairs, as well as the soverign equality of all states. Article II states:

"Outer space, including the moon and other celestial bodies, is not subject to national appropriation by claim or sovereignty, by means of use or occupation, or by any other means."

Since 1968, the primary role of the United Nations in this regard has been an attempt by the Committee on the Peaceful Uses of Outer Space to define the political, legal, economic, and organizational implications that are involved in the internationalization of remote sensing (Matte and DeSaussure, 1976).

Currently, United States policy seems to favor the transfer of the U.S. Environmental Satellite program from the public to the private sector. In 1979, President Carter issued Presidential Directive 54 which gave NOAA stewardship over the Landsat system, starting in 1983 when the system is anticipated to be operational. Implicit in this directive was a commitment by the Federal government to provide data continuity until the private sector was in a position to take control. This continuity would be provided by Landsat D and D'. These satellites were to be followed by two additional satellites, D'' and D'''. This combination of spacecraft would provide continuous Landsat coverage into the 1990's. However, in 1981, President Reagan cut Landsat D'' and D''' completely from the budget. Thus, by definition, the Landsat program is in the hands of the private sector when Landsat D' ceases to function (anticipated to be around 1987). If, indeed, this transfer of control to the private sector does occur then perhaps that organization should have some element of a public trust or utility.

Joint hearings by the U.S. Senate and Congressional Subcommittee on the possible transfer by the U.S. Government of civil land remote sensing to the U.S. private sector for commercial operation were held July 22 and 23, 1981. Among the many issues examined in the hearings were the following:

Single company, competitive selection, or consortia operation;

Licensing, joint ventures, and internationalization;

Marketing; improved timeliness of data delivery; pricing options, including price subsidies for developing countries and less sophisticated users; relationship with foreign systems, such as SPOT; combination of weather and land-resource satellites; guarantees of government business; non-discriminatory access to data; proprietary rights to private satellite data; regulatory mechanisms and procedures; and division between operations and research activities, giving due consideration to government and the roles of the private sector and the government in each.

A SYSTEMS VIEW OF REMOTE SENSING

The earlier account of the development of remote sensing correctly suggests that the potential of remote sensing has only begun to be tapped. It provides new and exciting "windows" to reality and serves to stimulate thought in a broader context on the formulation of hypotheses and methodologies. In order to maximize the use of this extraordinary new expansion of our capability to monitor earth resources, it is important that remote sensing applications be viewed within a holistic system framework and within an environmental framework. A total analysis should not be restricted to the engineering aspects of design, but should be an integration of environmental engineering, scientific, economic, human, and social considerations.

Earlier in this chapter an overview also was presented describing the remote sensing systems that have collected data in the past, those presently in operation, and the anticipated design of future sensors. With the increasing number of sensors available for use, the decision to use any one sensor should result from a total systems analysis in which the trade-offs between sensor systems and the data they deliver are examined.

The extent to which the data/information derived from various sensors supplement and complement one another will help to determine how and when trade-offs occur. Sensors will have primary and secondary roles that are not static and that may be determined in part by inherent information potential and the cost of acquiring the desired data. The label of primary or secondary may

change from environment to environment, from application to application.

There are trade-offs between the high spatial and geometric fidelity available from vertical photographs obtained frequently from either aircraft or spacecraft and the lower spatial resolution and geometric fidelity presently obtainable from line scanning devices such as the Landsat MSS. A case in point would be the use of high-resolution aerial photography to generate maps for regional planning. It may take several months to obtain coverage over a large area (e.g. statewide). It may then take several years to analyze the photographs, employing standard photographic interpretation techniques, and to produce a final product going through all of the steps in the cartographic production process. When the product has been published, the utility of the information portrayed can be questionable, particularly in areas experiencing rapid changes such as urban areas. For areas at the urban fringe, such maps can be quite out-of-date. Yet for areas that are unchanging or changing very slowly, such as forest or wilderness areas, the maps may be excessively detailed. So both time and scale (level of detail) may be mismatched for different uses within the same presentation of data/information on which planning decisions are to be based. However, it might have been more useful to use the timely Landsat MSS data with their coarse resolution to gain a regional overview of an area and then to selectively target areas where high spatial resolution is needed.

Again, there are trade-offs to be considered when the need for total area coverage becomes an issue. If the presence of clouds within a scene becomes a problem when using Landsat MSS data, the use of radar can guarantee complete area coverage in most instances. However, presently radar remains a costly form of imagery to acquire and process, and we have yet to adequately document its potential to accurately provide a variety of specific categories of information of value to users. For these reasons radar has yet to gain widespread use outside the humid tropics. Nevertheless, in other areas where cloud cover likewise is of major concern, the use of radar should be carefully considered.

Perhaps of most importance is the trade-off between the quality of information and the cost of obtaining that information. For any system selected, there is a residual level of error within the data beyond which it could not be economically justified to attempt improvement. A question that must constantly be posed is: which is more cost-effective; to increase the number of aircraft and/or ground samples, or to rely on the timeliness of satellites with coarse resolution? If the number of aircraft and/or ground samples is doubled the associated costs of obtaining those samples may triple, and there is no guarantee that the accuracy will improve in proportion. What is needed within a total systems context is to define a level of accuracy for each system that will be acceptable to private enterprise and to government-funded research institutions. This has not happened, although the movement of remote sensing into the private sector will probably force problems to be analyzed within a systems context in order to determine the most cost-effective approaches. This cost-effective end-to-end systems approach has been used by the St. Regis Paper Company (Goodrick, 1981). In their analysis, remote sensing provided only a small but nevertheless crucial part of their forest-inventorying procedure.

The question of trade-offs, therefore, is a key to effective and efficient use of remote sensing. But, because man's knowledge of the world is imperfect, we will continue to make errors in judgement concerning the application of this tool. As long as there are voids in our knowledge about the environment, those voids will ensure that there will be a residual level of uncertainty which will inhere in any systems analysis.

RESOLUTION

The decision as to the most appropriate data will depend on the study objectives, scale of the subject, temporal and financial constraints, and the nature of the anticipated results. In many instances the use of multiple data types (multi-date images, multispectral images and specially enhanced images) will aid the analysis. However, there are underlying concepts that must be kept in mind when remote sensing is considered. A number of these concepts reduce to the basic constructs of resolution and scale, each complexly and somewhat interdependently linked to the other. In order to present these ideas here in an orderly way, resolution will be discussed before scale. However, it should be clear that one cannot address topics associated with either resolution or scale in isolation from the other, as they are interdependent characteristics.

The amount of data/information contained within an image depends in large part on the image resolution. Definitions of spatial, spectral, and radiometric resolution are not unique, or single-valued, since both technical and broader application-oriented usages exist for each. When the technical and applications-oriented contexts are considered, the resulting definitions are necessarily broad and illustrate the complexities which arise in attempting to provide a concise discussion of a multi-dimensional concept such as "resolution."

An analogy which illustrates the complex nature of resolution may be drawn from the problems which arise when the attempt is made to map information from the round earth on a flat surface. It is only on a globe that true shape, area, and direction can accurately be represented for features on the surface of the earth. When this information is transferred to a "two-dimensional"

map, particularly for small scales, one can only get a somewhat generalized, fairly accurate representation for two of the three components. However, it is frequently the case that shape, area, and direction are each sub-optimized in the final map representation in the process of cartographic generalization. In a similar manner for a given remote sensing system, an improvement in resolution in either the spatial, spectral or radiometric domains will result in some degradation in resolution in one or both of the other domains.

SPATIAL RESOLUTION—TECHNICAL ASPECTS

Spatial resolution is a complex concept, though it is often misconceived as elementary. Part of the misconception surrounding spatial resolution arises from the variety of ways that can be used to describe the resolution characteristics of a given system or image. In the simplest case, spatial resolution may be defined as the minimum distance between two objects that a sensor can record distinctly. Another way of stating this is in terms of the overall fineness of detail characterizing an image. However, it is the format of the sensor system that determines how spatial resolution is measured. The measurements from one system may not be readily convertible to those obtained for another and a single measure of spatial resolution can not be satisfactorily applied to all sensor systems.

It has been suggested by Forshaw et al. (1980) that definitions of spatial resolution be placed into one of four categories: 1) the geometric properties of the imaging system, 2) the ability to distinguish between point targets, 3) the ability to measure the periodicity of repetitive targets, and 4) the ability to measure the spectral properties of small finite objects. Townshend (1980) provides an excellent review of spatial resolution, including the measures listed above, and much of the following material comes from his paper. Additional discussion of spatial resolution may be found in Chapter 24.

Table 1-2 summarizes the different estimates of resolving power for the Landsat MSS that have been calculated over the past few years. Spatial resolution, in terms of the *geometric properties of the imaging system,* is usually described as the instantaneous field of view (IFOV). IFOV is a function of satellite orbital altitude, detector size, and the focal length of the optical system. For the visible and near infrared bands of Landsat, the IFOV is most commonly quoted as being 79 meters. However, if the effects of cladding (using walls and adhesives) around the fiber optics are taken into consideration, the IFOV is reduced to 73.4 m (Colvocoresses, 1979) or to 76.2 m according to calculations by Slater (1979). Changes in the orbital altitude will also cause the IFOV to change. Since Landsat altitude has varied from 880 to 940 km, the IFOV has varied from 76 to 81 m (Gordon, 1980; also see Table 1-2).

TABLE 1-2

Estimates of the Resolving Power of the Landsat MSS. (Townshend, 1980)

Resolution Measure	Source	Resolution (meters)
1. IFOV–geometric	NASA 1972	79
2. IFOV–geometric	Slater, 1979	76.2
3. IFOV–geometric	Colvocoresses, 1979	73.4
4. IFOV–geometric (min. altitude)	Gordon 1980	76
5. IFOV–geometric (max altitude)	Gordon 1980	81
6. Pixel size	General Electric (undated)	79×56
7. Pixel size–resampled (Landsat 3 CCT's)	Holkenbrink, 1978	57×57
8. IFOV–point spread	Landgrebe, et al., 1977	90
9. EIFOV–half cycle	Welch, 1977	66
10. EIFOV–full cycle	Welch, 1977	135
11. ERE	Colvocoresses, 1979b	87
12. Modified ERE–estimate for Channel 4	Norwood, 1974	125
13. Minimum classifiable area	Shay et al., 1975	500×350
	General Electric, 1975	320×220

An IFOV value is not in all cases a true indication of the size of the smallest object that can be detected. An object of sufficient contrast with respect to its background, either brighter or darker, can change the overall radiance of a given pixel so that the object becomes detectable. Because of this features smaller than 79 m, such as transportation routes and low order drainage patterns, may be detectable on Landsat images. In the Shadow Mountain Eye Project small mirrors, 24 inches in diameter, were arranged in the pattern of an eye and oriented to Landsat's view angle, one mirror per pixel. The ninety mirrors formed a pattern 1.4 miles across and were detected because the reflection was sufficient to saturate the associated pixels (Colvocoresses, 1980; see Figure 1-6). Conversely, on Landsat imagery, objects of medium to low contrast may only be detectable if they occupy an area of 250 m or more. Color Figure 1-7 shows imagery exhibiting spatial resolutions of 20 m, 30 m, 40 m, and 80 m for the same area in Maryville, Tennessee. Using the airport runway in the lower left corner of each of the images as a reference, one can see how changes in spatial resolution (IFOV in this case) can affect interpretability.

For measurements based on the *distinguishability between two point targets*, the Rayleigh criterion is used (Perrin, 1966; Slater, 1975). A point source, even if viewed with a perfect or aberration-free lens, would not appear as a point source but as a central disk surrounded by faint dark and light rings. These features known as Airy rings, are the result of diffraction. The Rayleigh criterion for distinguishing between two equal-intensity point-source targets with a perfect lens says that the two will just be resolved if the central peak of the image of one source lies on the dark ring of the second. Once the angular separation is calculated and the height of the sensor above the ground is known, a measure of ground resolution may be derived. Since aberration-free lenses do not exist, actual lens resolutions must be less than the Rayleigh criterion.

Otterman (1969) derived a measure for extended circular targets ("blobs") because most remote sensing targets are neither points nor circular. Because of these reasons, means for making resolution estimates for square and rectangular objects were also required. Otterman showed that the diffraction-limited resolution for such sources is nearly six times coarser than for point sources.

Measures of resolution using the *periodicity of repetitive targets*, or "resolution targets" arose primarily from work using photographic images. On one commonly used form of resolution target, a pattern of parallel black lines on a white background provides a 100-to-1 contrast in the amount of light reflected. For each pattern, three black lines are used. The width of the space between adjacent lines is equal to the width of the lines in the pattern, and the length of each line is five times the width. Successive patterns on a target are composed of black lines and spacings of progressively narrower width. As the spacing between the parallel lines and the white background decreases, the contrast between them will appear to be less until a point is reached when the contrast is so small that the parallel lines are indistinguishable from each other. Values of resolution derived in this way are consequently expressed by spatial frequency measures such as line pairs/mm which is sometimes abbreviated to lines/mm.

Resolution may be expressed in cycles/mm since the linear targets used often have a sinusoidally-varying tone. The spatial frequencies of a linear target can be further characterized by the modulation transfer function (MTF). Modulation, a measure of contrast, is unity or near unity at low spatial frequencies and declines with higher spatial frequencies. The spatial frequency at which the modulation drops to zero is referred to

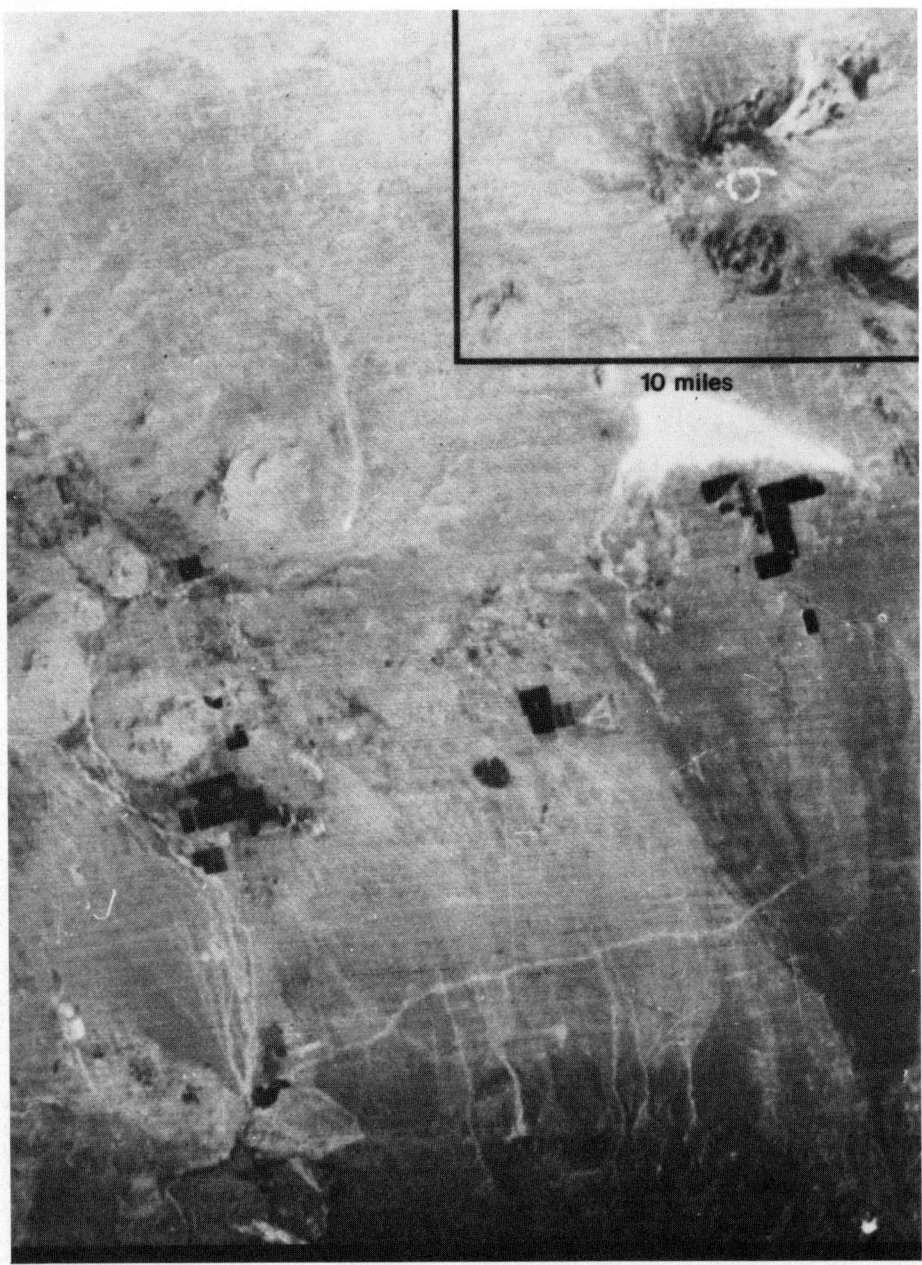

10 miles

Fig. 1-6. Landsat image, June 1980, of a portion of the California Desert. Within the delineated area are the Shadow Mountains with ninety mirrors arranged in the pattern of an eye and oriented to Landsat's view angle. (Colvocoresses, 1980)

as the cut-off frequency. The MTF of an image-forming system can be calculated from its design or measured in a laboratory. There are a number of MTF's which must be considered when examining the spatial resolution characteristics of a given remote sensor system, (Slater, 1980) the two principal ones being related to the size of the detector or to the image displacement caused by linear image motion. The reader is referred to Chapter 6 for an in-depth discussion of the MTF.

Through the use of MTF curves (see Chapter 6) different measures of resolution can be derived at various frequencies. For example, a spatial frequency can be calculated where the modulation falls to a set proportion of its maximum value. The effective instantaneous field of view (EIFOV) is half the value of the spatial frequency for which the modulation of an object with a sinusoidal distribution of radiance has dropped to 50 percent of its original value as a result of the MTF of the system (Townshend, 1980; NASA, 1973). In Table 1-2, the EIFOV for the Landsat MSS as calculated by Welch (1977) is 66 m. If the full cycle definition is used a value of 132 m is obtained.

A resolution employing *spectral properties of the target* is the effective resolution element (ERE). This measure of spatial resolution is of interest because of the increasing importance of automated classification procedures, which are highly dependent upon the fidelity of the spectral measurements recorded by the sensor system (Swain and Davis, 1978). Colvocoresses (1979) defines an ERE as the size of an area for which a single radiance value can be assigned with reasonable assurance that the response is within 5 percent of the value representing the actual relative radiance. Using this criterion, the calculated spatial resolution for the Landsat MSS is 8 m (see Table 1-2) (Colvocoresses, 1979). A refinement of this idea (Strome, 1979), defines a modified ERE as the minimum area for which the spectral properties of the center can be assigned with at least 95 percent confidence that the values differ from the actual parameter values by no more than 5 percent of the full scale of the measuring instrument. Work done earlier by Norwood (1974) using Landsat MSS data for agricultural scenes reflects a somewhat modified ERE concept (see Table 1-2). Norwood's results show that a 5 percent error in radiance values for Landsat bands 4 and 6 can be expected when the field size is approximately 125 m and 200 m, respectively.

From the preceding discussion of technically derived measures of spatial resolution it is clear that no single definition suffices. Table 1-2 shows that very different estimates of resolution can be obtained for the same sensor, in this case the Landsat MSS. Which of the entries in this table represents the "actual" or "true" resolution? The answer depends on which image properties are of interest to the user as determined not only by the application but also by the method of analysis.

BROADER ASPECTS OF SPATIAL RESOLUTION

The spatial resolution of a system must be appropriate if one is to discern and analyze the phenomena of interest. The phenomena of interest may be natural or cultural features, and these may coexist within the environment at macro-, meso-, or micro-scales. Each drop in scale requires progressively finer resolution if identification and analysis are to be successfully accomplished. Thus resolution must be "germane to the task" (Simonett, 1976). Simonett's idea of resolution germane to the task is not limited to spatial resolution, but can be applied to any of the resolution types.

The spatial resolution that is germane to a given task will be determined, in part, by the specific needs of the resource application in question: 1) the detection of objects, 2) the identification, or 3) their analysis. *Detection* determines whether something is present or absent. For example, within an image of large open rangeland areas objects are detected that are not consistent with rangeland vegetation. *Identification* results when enough information is available from the remote sensing data or from other sources to identify the objects (e.g. there are clumps of trees in the rangeland). At the *analysis* level of resolution information about the object is obtained beyond its initial identification (e.g. the clumps of trees in the rangeland are quaking aspen lining streams). To move from detection to identification, the spatial resolution must improve by about 3 times. To pass from identification to analysis, a further improvement in spatial resolution of 10 or more times may be needed.

Since image information-content is resolution-dependent, a trade-off exists between the levels of resolution for most remote sensing applications. High spatial resolutions provide small area observations, but regional patterns may be difficult to characterize because generalization is required of the high frequency information inherent in such images. Low or coarse spatial resolution allows regional patterns to be readily observed, interpreted and analyzed. However, detail may be averaged within a pixel, implying a loss of information: Every system is sub-optimal for some or all of the objects in a scene.

SPECTRAL AND RADIOMETRIC RESOLUTION

The spectral resolution of a remote sensing instrument is determined by the band-widths of the channels used. High spectral resolution, thus, is achieved by narrow band widths which, collectively, are likely to provide a more accurate spectral signature for discrete objects than are broad bandwidths. However, narrow-band instruments tend to acquire data with a low signal-to-noise ratio, lowering the system's radiometric resolution. This problem may be alleviated if relatively long look (or dwell) times are used during imaging. In contrast, broad-band sensors usually have good spatial and radiometric resolution.

In the broader usages of spectral resolution, there are also trade-offs between application and spectral and radiometric resolution (Figure 1-8): Crop growth and development are probably better monitored using very narrow bands in the visible and reflective infrared regions (Chapter 33). Monitoring sea state is probably best accomplished with multichannel active and passive microwave systems (Chapter 28). Even within a discipline such as geology, (Chapter 31), specific applications are best studied using specific spectral ranges: Volcanism may be explored with broad thermal infrared bands of high radiometric resolution; surface geology may be mapped with Landsat images, aerial photographs or imagery acquired by means of active microwave systems, depending on the situation.

Radiometric resolution is determined by the number of discrete levels into which a signal may be divided. Considering the effects of varying illumination, the radiometric dynamic range of a sensor is determined by the maximum radiance value that the sensor system can experience for a given band. For example, the initial analog voltage signal of the Landsat MSS detectors is converted to digital count outputs ranging from 0 to 63 for a total of 64 quantizing levels. However, the maximum number of quantizing levels possible from a sensor system depends on the signal-to-noise ratio and the confidence level that can be assigned when discriminating between levels (Slater, 1980).

With a given spatial resolution, increasing the number of quantizing levels or improving the radiometric resolution will improve discrimination between scene objects. Tucker (1979) found that

the number of quantizing levels had a decided effect upon the ability to resolve spectral radiances that were related to plant-canopy status. Specifically, his analysis showed a per-channel improvement of 2 to 3 percent for 256 levels (Thematic Mapper bands 3 and 4) versus 64 levels (Landsat MSS). With such comparisons it is essential to assess the cost/benefit ratio of improved, but more costly procedures; a 2 to 3 percent gain is modest for the extra processing costs involved!

Interdependencies between spatial, spectral, and radiometric resolutions for each remote sensing instrument affect the various compromises and trade-offs. These trade-offs will be governed by the particular application or group of applications. To study housing quality within an urban area, high spatial resolution is needed. The information could be obtained through the use of panchromatic film (low spectral resolution). Forest inventories require fine spatial resolution to discriminate between species, but coarser resolutions, such as those available with some space sensors, may be adequate for discriminating between plant communities.

TEMPORAL RESOLUTION

When choices become difficult as to where trade-offs between three resolution types occur, the fourth type of resolution—temporal resolution—may achieve the needed discrimination. There are very few objects and/or phenomena in nature that do not change with respect to one another or to themselves throughout the course of time. For many of the physical and cultural features on the landscape there are optimal time periods during which these features may best be observed. These optimal periods might be seasonal, or only a few days or weeks. With some applications the time interval at which remotely sensed data are acquired becomes an important factor. To monitor crop growth, for example, images should be obtained at a predetermined time interval, perhaps every 10 days. However, to monitor urban growth patterns imagery acquired at time intervals of a year or more may be appropriate. Thus, in sensing a substantial number of dynamic events, such as crop growth, rangeland development, hydrologic processes, earthquake damage, urban change, and marine processes, *time* often may be used as a key discriminant. Figure 1-9 shows the relationship of time and space to monitor hydrologic events.

Since one of the major premises of remote sensing is to monitor change through time, temporal resolution is an important consideration when determining the resolution characteristics of a sensor-system. Perhaps of greater importance are the trade-offs between the time dimension and the other three dimensions of resolution in the quest for the most economical remote sensing procedures. For an agricultural application, the

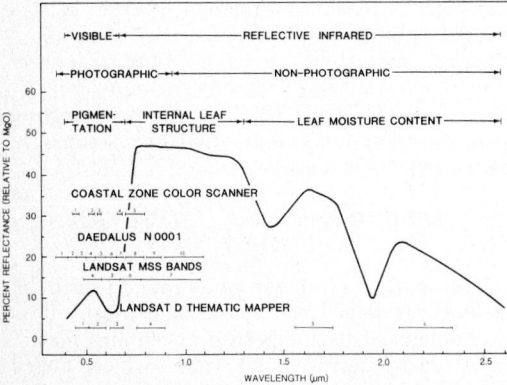

Fig. 1-8. Generalized laboratory reflectance curve of green vegetation, superimposed on a diagram showing the spectral coverage of satellite and aircraft sensing systems. This diagram covers the spectral range of 0.4–2.6 μm in the visible and reflective infrared regions. (Modified from Lintz and Simonett, 1976)

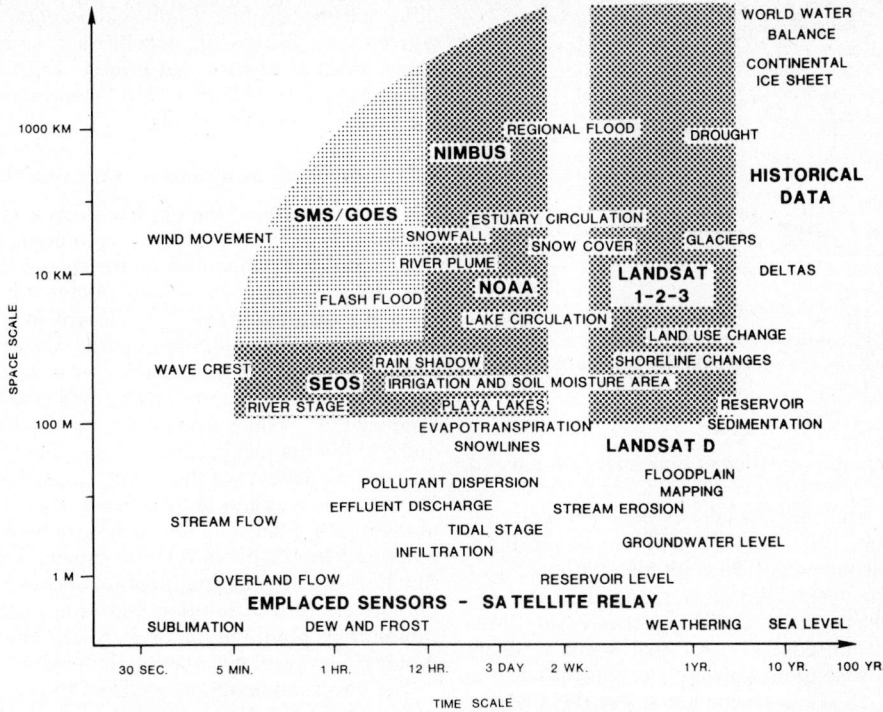

Fig. 1-9. Time and space scales for representative hydrologic parameters with relevant observational capabilities of satellite systems represented by shaded areas. (modified from Salomonson and Rango, 1980)

use of time as a discriminant (together with local *a priori* data, such as can be obtained from an accurate crop calendar) may allow crops over large areas to be identified with sensors possessing spatial and spectral resolutions that are too coarse to identify the same crops on the basis of spectral and morphological characteristics alone. Use of the temporal resolution of a sensor system for a given application may have the effect of reducing the cost of data acquisition and processing.

SCALE

Scale may be defined as the mathematical relationship between the size of objects as represented on maps, or on air photos or other remotely sensed data, and the actual size of the objects, themselves. Scale should not be confused with spatial resolution, as applied to a remote sensing system. As previously stated, the spatial resolution of the Landsat MSS is approximately 79 m, but Landsat MSS data may be presented at different scales. The standard Landsat MSS product, viz. a 9 × 9-inch print of an entire scene, is at a scale of approximately 1:1,000,000. However, because the data comprising the scene are digital, they can be presented at virtually any scale, such as 1:24,000—a scale that is compatible with the 7.5-minute topographic quadrangles produced by the U.S.G.S. Selection of the appropriate scale is influenced by the resolution of the data, and both scale and resolution are application dependent.

INFORMATION AS A SCALE-DEPENDENT PHENOMENON

The scales at which data are collected and analyzed directly influence the level and kinds of information that may be obtained. At this point it is appropriate to examine the relationships between data and information. Figure 1-10 shows the associations between remotely sensed data, ancillary data, and algorithm and theory development. The intersections of A and C, and B and C and, if it exists, A, B, and C, are called scientific information; that is, a composite of quantative observations and descriptions. Remotely sensed data alone cannot become scientific information as the information is fundamentally related to algorithm/theory. It is only through the connections previously identified that remotely sensed data may become information; otherwise the data gathered from the sensing instruments can be nothing more than possible building stones for algorithms and theory (Morgenstern, 1950). Information derived from data collected at a particular scale is dependent on that scale. McCarthy et al. (1956) strongly emphasize the scale-dependent nature of information as follows: " . . . conclusions derived from studies made at one scale should not be expected to apply to problems whose data are expressed at other scales. Every change in scale will bring about a statement of a new problem, and there is no basis for assuming

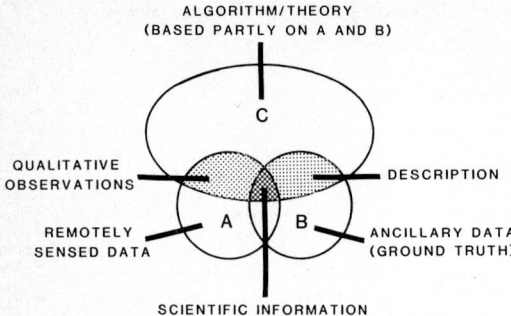

Fig. 1-10. Relationship between remotely sensed data and information. (modified from Morgenstern, 1950)

that associations existing at one scale will exist at another."

What causes the dependency between scale and information? Within the realm of remote sensing it is the environment which is ultimately the subject of most analyses. Within a scene the size and surface characteristics of objects may vary considerably; consequently, for that scene a single scale and resolution will not be appropriate for observing all scene components. For the analysis of large continuous geologic features in a region, the scale of a full-frame Landsat scene may be appropriate. However, difficulties will occur when one tries to identify individual bedding planes, discontinuous intrusive elements, and other features for which the combined scale and resolution are too coarse.

The appearance of objects often changes with changing resolution and scale. Use of large-scale aerial photographs for a forested area can show individual crown shapes from which species, crown size, and closure may be determined. These data can then be used to extract measures of stand density and timber volumes. Individual trees cannot be discriminated on Landsat imagery. Forest stands appear as broad classes of vegetation. Nevertheless, general physical parameters of the forest stand, such as tree height and stand density, may be determined from the Landsat MSS data when these data are augmented by an acceptable sample design including ground or large-scale photographic observations (Strahler et al., 1981).

The importance of being able to recognize the same object at different scales and resolutions was noted by the American Society of Photogrammetry in 1965. "A necessary task of the photointerpreter will be to gain experience in the recognition of common features at widely different scales and resolutions in order to make use of space photography." The ability to discern objects will change from one scale to another, and the scale appropriate for observing an object at a given location within a scene may be inappropriate for another area. Thus the within- and between-scene variability, a result of environmental complexity,

influences the quality of information obtainable at a given scale. This scale dependency, common to many areas of science, led Everett and Simonett (1976) to suggest that there is a "scale germane to the task" in remote sensing.

Environmental Modulation Transfer Function

The variability of the earth's surface is one of the most pervasive problems encountered in remote sensing. Commenting on the spatial nature of the environment Townshend (1980) noted that there is a clear need for " . . . improved quantitative information about the spatial variability of terrain attributes if their significance in affecting the usefulness of remote sensing data is to be fully appreciated." Not surprisingly, the environment and its fluctuating parameters remain the least known components of the remote sensing system. From the discussion above it is evident that the amount and quality of information which can be extracted from a given remote sensing image is a function of complex interactions between sensor parameters (e.g. resolution and scale) and scene dependent phenomena (e.g. socio-economic, biologic, and geologic spatial complexity).

The information obtainable from the analysis of remotely sensed data is dependent upon the variable characteristics of the given environment. Just as a sensor modulates the energy it receives, the environment modulates the information content of the scene. The concept of an environmental modulation transfer function (EMTF) was first proposed by Everett and Simonett in 1976. They attempted to show how the complexity of the environment affects the information potentially extractable from remotely sensed data.

Fundamentally, Everett and Simonett contend that the environment is not passive to the electromagnetic energy that impinges upon it. Its variability in time, space, and number of functional categories affects information-extraction potential. This concept may be formalized by the equation

$$S = f(I,E)$$

where

S = system information (modulation) transfer function,
I = instrument information (modulation) transfer function,
E = environment information (modulation) transfer function.

The instrumental and environmental parts of this equation may be expanded with the result that

$$I = f(\lambda_R, R_R, S_R, T_R, N)$$

where

λ_R = spectral resolution or band width of a single channel,
R_R = radiometric resolution of either reflected or emitted energy

S_R = spatial resolution of the system,

T_R = frequency of observations potentially available with the system

N = number of channels of information.

The contributions to the environmental modulation transfer function may be expressed as follows:

$$E = f(C_R, S_R, T_R, A_R, \ldots \ldots X_R)$$

where

C_R = number of distinct land-use or mapping categories present,

S_R = size and spatial arrangement of the homogeneous categories of an area

T_R = a measure of the time-variable nature of the environment,

A_R = a measure of the atmospheric constraints on sensing, and

X_R = other aspects of the environment.

The equations shown above were developed to begin to provide a formalized means of illustrating the interaction between sensor-system resolution and the complexities of the environment which, together, modulate the data and ultimately the information. The EMTF emphasizes the previously stated concept that a single resolution used for multiple environments cannot and indeed will not produce uniform classes of information for all environments, or scene components.

The complexity and modulation of information from a given environment become even more evident when the consequences arising from the aggregation of data are taken into account. (Issues relating to aggregation will be more completely addressed later in this chapter.) The effects of environmental complexity are notably evident in Landsat data with their relatively coarse spatial resolution. Within a single resolution cell (pixel), two, three, or more land use or land cover categories may occur—each with its peculiarities of spatial size and location, class, spectral response and temporal nature. Therefore, depending on the environment (e.g. large agricultural fields versus urban patterns) many of the pixels may represent mixtures of various functional information categories.

It is the presence of internal heterogeneities within individual functional categories that Wiersma and Landgrebe (1978) have termed "scene noise." As an example, consider a residential area where imagery having high spatial resolution may reveal many separate land-cover components such as, roofs, road surfaces, grasses, trees, and others. When imaged at a coarse spatial resolution, such as that available from Landsat, these land-cover components become mixed within a pixel.

This mixed pixel problem becomes compounded in small complex areas, such as the residential example above, where multidate registration is required. Misregistration by a single pixel could be enough to ensure that all land-cover components would be improperly mixed in time confounding accurate classification employing machine assisted analysis techniques.

CONSISTENCY

The EMTF attempts to illustrate in a more precise fashion how the dynamic nature of the environment affects the nature of the data recorded and the potential information. The potential for change within the domains of space, class, and time must be kept in mind to ensure that the types of remotely sensed data and the respective ground truth remain mutually consistent. In other words, it is important to retain a certain level of internal consistency when interfacing data sets of various scales.

The level of consistency achievable for analysis using remotely sensed data will be dependent upon two factors: the system resolution and the complexity of the environment being studied (Everett and Simonett, 1976). This idea may be expressed by

$$C = f\left[SR(s, \lambda, r, t), \text{Env.Compl.}(S, \lambda, Q, T)A \right]$$

where

C = level of consistency

SR = system resolution with respect to spatial (s), spectral (λ), radiometric (r) and temporal (t) components,

Env. Compl. = environmental complexity with respect to the frequency distribution of the spatial (S), categoric (λ), quantitative (Q), and temporal (T) domains of the environment for a given area (A).

Consistency is directly related to the system resolution in the sense that improved resolution will generally yield a higher consistency in the results. On the other hand, consistency is inversely related to the degree of environmental complexity in space and time, and in size of the study area. Suggested by these relationships is the idea that a high level of consistency may be obtained for an area of continental size only if very general categories are specified. For example, Level I of the Anderson et al. (1972) land-use classification is the level at which high consistency may be obtained with Landsat imagery over large areas. Level II is distinctly less consistent, and Level III is not at all consistent from locale to locale.

SCALE AND AGGREGATION

When one deals with information obtained at different scales, it is important to consider the degree of aggregation represented by each scale. Broadly stated, small-scale information requires an aggregation of data while large-scale information requires sub-division. Data aggregation occurs in both the temporal and categoric domains and in the spatial domain. The same consequences occur regardless of the domain in which the aggregation occurs. Two examples will be given to illustrate what happens when data are aggregated 1) to the same level but according to different criteria, and 2) to different levels for the same phenomenon. The first comes from the area of economic statistics and the second from geographical analysis; both examples have analogs in remote sensing.

The classic critique on the accuracy of economic statistics by Morgenstern (1950) showed that figures reported for various national economic indices differed substantially between government agencies. These differences were a result of the way in which the data were aggregated. Aggregation was based on criteria developed separately by each agency.

Agricultural statistics for acreage and quantity of production in 1945 were reported by the Bureau of the Census and the Department of Agriculture for major crops. The production figures showed seven crops for which estimates by the Department of Agriculture differed from the Census by more than 5 percent, to as much as 13 percent. For acreage there were ten crops where the discrepancy was more than 5 percent, ranging up to 15 percent for potatoes.

In most cases the Department of Agriculture estimates were higher than those of the Census, since the Census methods tended to underestimate for several reasons (shifting of farmers, omission of hard-to-reach farms, withholding of data). The revisions of the Department of Agriculture estimates in most cases brought them closer to the Census statistics, although they never were made identical because of underestimation by the Census.

In a geographical context of scale and aggregation, Harvey (1968) states that "geographers have long recognized that different processes are relevant for understanding areal differentiation at different scales—local, regional, (and) national." An illustration of Harvey's statement is seen in a study of early 19th century French settlement patterns (Baker, 1968). From analysis done at 1:25,000, the settlements occuring within a dispersed pattern were randomly distributed. Those within a grouped pattern type were uniform. However, at 1:100,000 the location of the major settlements tended to be uniformly distributed. Thus in the above example the information about the distribution of settlement patterns varied according to the degree to which the data were aggregated as reflected by the scale used in the investigation.

It follows from the preceding that data-aggregation levels must be appropriate for the phenomenon being studied. It is unnecessary and costly to use data that are more specific than the level of analysis requires. Also, there may be relationships or classes contained within the data that might require that the data be "smoothed" or aggregated in order to detect trends that could be lost in highly divided data. For example, a scale at which one can determine individual tree crowns may be useless for the study of vegetation associations because the pattern of the associations may be lost in the detail.

On the other hand, if the available data are too general for the problem of interest, new data must be acquired since aggregated data cannot always be disaggregated to achieve greater detail. From this standpoint, the more general the level of aggregation of data the more limited its potential usefulness to a variety of users. The data obtained at larger scales may be aggregated in an appropriate hierarchy. As noted by Stone (1972) "It is a delusion to assume that large-scale study in the field can add up to small-scale conclusions later in the office."

ARTIFACTING

Another problem that may result from the generalizations and aggregation of data is artifacting. Broad area coverage requires small scales, which means that data must be aggregated. When the aggregation of data occurs to a high level, known relationships occuring within the data may be suppressed, or spurious relationships developed.

Within the context of this discussion, an artifact thus refers to structure that is normally not present in the data. Artifacting may be produced by aggregation or disaggregation of data in the domains of space, time, class or quantity, and by undersampling. The possibility of artifacts occurring within remotely sensed data is not in question; they undoubtedly do. Rather there are a series of concerns as follows: 1) are the artifacts of sufficient magnitude and extent to cause severe problems in terms of loss of data and information content? 2) could unknown artifacts be present without our knowledge?, and 3) how widespread is artifacting? There are three general classes of artifacts that may occur; 1) those arising from excessively fine subdivisions of classes, 2) those occurring from excessive generalization of classes, and 3) those that develop from external factors, including undersampling. A discussion of each follows.

Excessively fine subdivisions

This type of problem occurs in level-slicing in which noise may be contoured. Related to this quantization is noise which arises from dividing a continuous distribution into steps. These examples are taken from the radiometric domain, but artifacting of this nature may take place in any of the other resolution domains. In a spatial context, for example, studies by Clark (1980), Markham et al. (1981), Sadowski and Sarno (1976), and Wharton (1981), have shown that overall classification accuracy for urban and forestry applications generally *decreases* as spatial resolution improves. This is the opposite of what one would expect based on photointerpretation. This paradoxical situation may be explained by the "scene noise" concept of Wiersma and Landgrebe (1978), previously discussed in the section concerning the EMTF. In this situation, the high spatial resolutions are mismatched with the single pixel classification algorithms. Because of the noise within the scene, relative to the capabilities of the classification algorithm, labelling accuracy decreases and an unacceptable level of artifacting has a strong possibility of occurring.

Excessive Generalization

When too many data are aggregated into a single resolution cell, relationships may begin to appear that are artifacts. Results based upon these artificially created relationships can misrepresent the actual conditions in the environment. On imagery of coarse spatial resolution, chance alignments of geologic features might erroneously indicate the presence of major lineaments. Or, without proper consideration of the effects of particular sun-angle illuminations within a scene, false boundaries may appear causing incorrect conclusions about the spatial arrangement of an area.

Artifacts Arising from External Factors

Artifacts of this nature often occur because of the indiscriminant application of man- or machine-based decision rules. As previously noted, overall classification accuracies of pixel classifiers can decrease rather than improve, depending upon the application, as spatial resolution improves. A serious issue in this respect is the possibility that any unsupervised clustering algorithm, through imposition of its structure and logic on data at any level of resolution, may itself produce a whole range of artifacts. An algorithm that one of us (D.S.S.) uses in class may be appropriate: "the data are at the mercy of the algorithm, and to some degree the algorithm is at the mercy of the data." Gould (1980) addresses this troubling issue in his critical paper entitled, "Let the Data Speak for Themselves."

A study by Campbell (1981) showed that there are reasons also to suspect inherent errors in the usual applications of supervised classification algorithms to Landsat digital data. He observed that the use of contiguous pixels for training sites to represent a land-cover category tends to underestimate variation, leading one to believe that an area is more homogeneous and more distinct from other categories than it really is. The resulting classification errors were shown to be related to the interaction between numerous scene-related elements that operate simultaneously and with varying influence throughout the growing season. Another implication of Campbell's study is that classification errors may cluster in space, causing spatially-aggregated errors to be falsely interpreted as genuine land-cover parcels (artifacts),

whereas randomly distributed errors are more likely to be recognized as errors.

Finally, undersampling introduces artifacting in a number of ways: 1) with regular patterns sampled at less than the Nyquist frequency, Moiré effects can occur, 2) bias can occur for categories that represent only a small fraction of an area, and 3) an improper or spatially unsuitable sample design can lead to spatial clumping in sampling and thereby to bias and artifacts.

EXPERIMENTAL DESIGN AND THE MULTIPLE-SCALE APPROACH

With the many considerations of scale and resolution a guide is needed to help ensure that the basic framework of the experimental design formulated for a study will be appropriate.

1. *Determine the scale of the study.* The study subject must be defined. To do this the apparent forms of the phenomenon may be identified with respect to its occurrence within the domains of time, space, category, and quantity of data. Does the phenomenon on interest change daily, weekly, monthly, annually or over periods of years? What is the geographical extent of the phenomenon? What are the classes or subdivisions necessary to describe the study subject? What quantity of data will be needed to describe the phenomenon?

2. *Aims of the study.* What aspects of the problem are to be explored? The purpose of the study must be clearly defined and understood to serve as a basis on which the analysis may take place.

3. *Nature of anticipated results.* If the nature of the study is thoroughly understood then there should be some idea about the outcome of the analysis.

4. *Nature of available data bases.* Will the data be obtained from primary or secondary sources? Data from a primary source have been directly collected in accordance with the scale of the phenomenon, aims of the study, and nature of the anticipated results. Secondary sources are bases of existing data. Use of existing data requires a familiarity with data-collection methodology for meaningful analysis because some of the available data may be aggregated rather than in the form of point data. Also, the variability in size and shape of areas within and between regions of interest must be taken into consideration.

The crucial aspect of this basic framework for experimental design is the multiple-scale approach described by Stone (1972). This approach entails the division of data on a given topic or area into significantly different groups by the scales of information needed to describe, analyze, and present various distributions of the data. According to Stone a determination of the number and limits of the scale classes is the key step of the multiple-scale approach. The method by which the scale classes are selected will depend on objective and complete field observations, followed by careful analytical testing of various scales in comparison with data available from all other sources, and selection of the smallest scales wherein faithful generalizations may be made toward the initial objective of the study. Experience will play a major role in the amount of time and expense necessary for such an approach. However, the thrust of this approach is to develop a methodical procedure that guarantees consideration of all scales.

THE MULTI CONCEPT

From the preceding sections one can begin to recognize the various facets of remote sensing. The wide range of possible data combinations may provide analysts with more insight, and this has become a commonly accepted article of faith in the remote sensing community. However, there is reason for concern that multiple opportunities may contain not only ordinary artifacting, but also the artifacting of particular circumstances, i.e. that each case becomes idiosyncratic or special, and generality is lost. Such a process is by nature completely empirical, and is almost anti-science. In Chapter 1 of the first edition of the *Manual of Remote Sensing,* Robert N. Colwell stresses the possibilities of the "multi" concept in remote sensing which he identified as: multidate, multispectral, multistation, multistage, multipolarization, multidirectional, multienhancement, multidisciplinary, and multithematic (Table 1-3). With the passage of time both the efficacy and weaknesses of these multiple techniques have begun to be demonstrated, though it is still too early to be definitive about the balance between them for all the "multi" concepts involved. Certainly the capability to carry out "multi" remote sensing has increased more rapidly than the depth of our knowledge about the limits of these concepts. The various techniques may be briefly examined.

MULTISTAGE

Remote sensing is multistage in the sense that "progressively more information is obtained for progressively smaller subsamples of the areas being studied" (Colwell, 1975). The spaceborne instruments now in widespread use obtain small-scale, relatively coarse resolution imagery and photography. These data may be invaluable for their synoptic coverage of large areas. The identification and subsequent study of small-scale features and their relationships with the surrounding environment can be carried out with one or a composite of a few contiguous images. One of the principal uses of the small-scale (1:500,000) imagery is to locate potential areas of interest to be examined in greater detail by progressively larger scale remotely sensed data and/or other means. Small-scale data may often be meaningfully interpreted by the use of information obtained from progressively larger scale remotely sensed data or from ground or shipboard surveys of the area. This sampling scheme, using smaller scale to progressively larger scale data, formed the cornerstone of the extensive test site programs carried out in the 1960s (Lee, 1975).

MULTIDATE

The use of imagery for the same area obtained at various instants in time can be likened to time-lapse photography. Events which can be monitored by remote sensors (with the exception of natural disasters) change over longer periods of time, ranging from days to weeks to seasonal or annual time scales. As these phenomena change so do their spectral responses. By analyzing an area through time one may find it possible to develop analysis procedures based on these variations through time, and to differentiate classes of information accordingly. The principal advantage

TABLE 1-3

Summarization of the "multi" Concept. (Colwell, 1978)

1. More information usually is obtainable from *multistation* photography than that obtained from only one station.*
2. More information usually is obtainable from *multiband* photography than from that taken in only one wave-length band.
3. More information is obtainable from *multidate* photography than from that taken on only one date.
4. More information usually is obtainable from *multipolarization* photography than that taken on only one polarization.
5. More information usually is obtainable from *multistage* photography than from that taken from only one stage or flight altitude.
6. More information usually is obtainable through the *multienhancement* of this photography than from only one enhancement.
7. More information usually is obtainable through the *multidisciplinary analysis* of this photography than if it is analyzed by experts from only one discipline.
8. The wealth of information usually derivable through intelligent use of these various means usually is better conveyed to the potential user of it through *multithematic maps,* i.e., through a series of maps, each dedicated to the portraying of one particular theme, rather than through only one map.

* The term "multistation photography" (not to be confused with "multistage photography") pertains primarily to successive overlapping photographs, taken along any given flight line as flown by a photographic aircraft or spacecraft. When two such photographs are studied stereoscopically, the photo interpreter is better able to perceive features than if a photo from only one of the two stations was available.

from the use of multidate analysis is the increased amount of information for the study area.

The seasonal changes for an area in the northern midwest were documented on a month by month basis using Landsat images obtained from 1973 to 1976 (Lucas and Taranik, 1977). This sequence of images documented the seasonal changes that may take place on an annual cycle as different landscape components slip into and out of prominence on the images. It is thus possible to use time as a discriminant function, and take advantage of the way in which time modulates the different parts of the natural spatial system to selectively enhance them.

Within an agricultural context, multidate imagery can help to record the phenology or developmental stages of crops. For identification, crops within an area can be partitioned according to their respective phenologies, resulting in an increase in the probability of correct identification.

The use of multidate imagery has long been recognized as a valuable tool in photointerpretation. Yet, the true power of incorporating multidate imagery into an analysis scheme lies in the capabilities of computer-assisted image-analysis techniques. Presently, the hardware and software of some systems allow several images to be viewed sequentially or simultaneously on a display screen. As the technology of computer systems advances, the manipulation of multidate images will become an increasingly touted tool for analysis. It is appropriate to be skeptical and cautious as well as creative in pushing the multidate idea to its limits.

MULTIBAND AND MULTISPECTRAL

The electromagnetic energy that emanates from the earth's surface can be recorded by instruments sensitive to various parts of the spectrum. Although the radiation measured by a sensor is limited by the spectral sensitivities of the sensor, the object being sensed reflects and emits energy at many wavelengths. The characteristics of the energy emanating from a given feature depends largely on its atomic, molecular and macromolecular composition. From the earliest days of remote sensing, the field has borrowed a simple analogy from spectral analysis in physics and chemistry to suggest that each feature tends to exhibit a unique "spectral signature." We have learned, with great pain to ourselves and great cost to the Federal government, that the multispectral analogy must be used with great caution: it is neither totally true, nor totally false, and its relative degree of indeterminancy varies from place to place and circumstance to circumstance. Despite this caveat, it is still correct to state that having many channels of data has proven more valuable, in researching problems, than employing only a single channel: panchromatic black-and-white photography is less useful than color photography; we expect the thematic mapper to

be more useful than the Landsat multispectral scanner.

The selection of the proper spectral channels for analysis is of great importance. The further apart in the spectrum channels are located, the higher the degree of independence of the potential information; the closer they are, the greater the redundancy. There is some merit, therefore, to obtaining multispectral data from bands that are far apart.

MULTIPOLARIZATION

When treated collectively, the various waves which comprise a beam of energy from the sun can be considered to be vibrating in all planes that are parallel to that beam as they travel through the atmosphere and impinge upon the earth's surface. However, the energy that is reflected back into the atmosphere from one kind of surface feature, such as a body of water, may be strongly polarized (i.e., vibrating primarily in one plane) while that reflected back from some other kind of feature, such as vegetation or fractured rock, may be polarized only slightly, if at all.

The simplest case of multipolarization photography is that obtained with two cameras, each equipped with a polarizing filter that preferentially transmits light waves that are vibrating in one particular plane. Cross-polarization photography is obtained when the polarizing filter of one of these cameras is oriented in one plane (ideally in a plane that includes the sun, camera and feature of interest on the ground) while the polarizing filter of the other camera is oriented perpendicular to that plane. In that event, strongly polarizing features appear much brighter as imaged with the first camera than the second, but not so for non-polarizing features. Since the polarizing capability of a feature is a clue to its identity, more information about earth resources can be obtained from imagery characterized by more than one polarization (Colwell, 1975).

With radar systems the polarization of the transmitted energy is an important consideration. (See Chapters 9, 10, and 25 for detailed explanations of radar fundamentals and systems.) It is less costly to have two receiving antennas than to have two transmitters. One possible system configuration would be to have a single transmitter that transmits in one plane of polarization and two receivers that would receive the parallel and orthogonal components. (Provision needs to be made so that the antennae and receivers are sufficiently separated to avoid cross-talk.) The above-described configuration provides a useful and cost effective system since, in some circumstances, the quality of information available from the HH and HV polarizations may be as much as from a polarization combination of HH and VV.

Within a radar context, information is especially wavelength- and polarization-dependent. The first consideration would be to select the wavelength that is most suited to the phenomenon being investigated.

MULTIDIRECTIONAL

Although most modern-day remote sensing is done with the sensor pointed vertically downward, there are instances in which more information may be obtained if images of the same area from one or more oblique orientations of the sensor are available. The realization of the need for sensors with multidirectional or multi ''look'' capabilities has been recognized and satellites with variable look angles are now being designed.

Multidirectional imagery for radar analyses is of particular usefulness in that it allows the variety of features within a scene to be differentially enhanced; it also eliminates the bias due to enhancement of features aligned parallel to the flight path. Differential and biased enhancements of structured features in geology have been observed with low sun-angle illumination in Landsat images.

MULTIENHANCEMENT

There frequently is a need to facilitate an analyst's task of gleaning the appropriate information contained either in a set of multiband images of an area taken at one date or from a set of multidate imagery of that area using one spectral channel. There are two ways in which imagery may be enhanced—optically and digitally. Within the digital domain, images may be enhanced by a number of techniques in order to aid visual analysis. Some of the different enhancements commonly used include contrast stretching, edge enhancements, and principal component images. These various types of enhancement techniques, especially when done interactively on a computer, may help to identify areas of features where further attention should be directed. Color Figures 1-11 and 1-12a and 1-12b have been enhanced respectively 1) to show variations in water depth, 2) to give a full color range in a standard Landsat color composite, and 3) to show the value in geologic reconnaissance of principal component (eigen) images. Further explanations of different enhancement may be found in Chapters 17 and 24.

REFERENCES

American Society of Photogrammetry, 1965, Photo interpretation in the space sciences; Photogrammetric Engineering, vol. 31, pp. 1060–1075.

Anderson, J. R., E. E. Hardy, and J. T. Roach, 1972, A land use classification system for use with remote-sensor data; U.S. Geological Survey Circular 671, pp. 1–16.

Arnold, A., 1978, The role of science and technology in development; Proceedings of the Twelfth International Symposium on Remote Sensing of Environment, vol. 1, pp. 175–178.

Baker, A. R., 1968, Establissements ruraux sur la marge sudouest du bassin parisien dans les premieres annees du xixe. Siede; Norois, vol. 60, pp. 481–492.

Beaumont, T. E., 1978, Remote Sensing for Transport Planning and Highway Engineering in Developing Countries; Transport and Road Research Lab, Crowthorne, England, 22pp.

――――, 1980, Remote sensing for route location and the mapping of highway construction materials in developing countries; Proceedings of the Fourteenth International Symposium on Remote Sensing of Environment, San Jose, Costa Rica, vol. III, pp. 1429–1441.

Campbell, J. B., 1981, Spatial correlation effects upon accuracy of supervised classification of land cover; Photogrammetric Engineering and Remote Sensing, vol. 47, no. 3, pp. 355–362.

Chevrel, M., M. Courtois, and G. Weill, 1981, The SPOT satellite remote sensing mission; Photogrammetric Engineering and remote sensing, Vol. 47, no. 8, pp. 1163–1171.

Clark, J., 1980, The Effect of Resolution in Simulated Satellite Imagery on Spectral Characteristics and Computer-Assisted Land Use Classification; Jet Propulsion Laboratory Publication 715-22, Pasadena, CA, 125pp.

Colvocoresses, A. P., 1979, Effective Resolution Element (ERE) of Remote Sensors; Memorandum for the Record, Feb. 8, United States Department of Interior, Geologic Survey, Reston, Virginia.

――――, 1980, Los Angeles Solar Mirror Project; Memorandum for the Record (EC-77-Landsat), United States Department of Interior, Geologic Survey, Reston, Virginia.

Colwell, R. N., 1956, Determining the prevalence of certain cereal crop diseases by means of aerial photography; Hilgardia, vol. 26, no. 5, pp. 223–286.

――――, 1975, The ''Multi'' concept as applied to the acquisition and analysis of remote sensing data; Manual of Remote Sensing, R. G. Reeves (ed.), American Society of Photogrammetry, Falls Church, Virginia, pp. 5–11.

――――, 1978, Keynote address: History and Future of Remote Sensing, Technology and Education; Conference of Remote Sensing Educators; A Workshop held at Stanford University by NASA Ames Research Center, Moffett Field, California, June 26–30, pp. 3–92.

――――, 1979, Remote sensing of natural resources—retrospect and prospect; Proceedings of Remote Sensing for Natural Resources, an International View of Problems, Promises and Accomplishments, University of Idaho, Moscow, Idaho, pp. 48–68.

Conitz, M., 1978, A development assistance program in remote sensing; Photogrammetric Engineering and Remote Sensing, vol. 44, no. 2, February 1978, pp. 177–182.

Cutrona, L. J., E. N. Leith, L. J. Procello, and W. E. Vivian, 1966, Paper on the application of coherent optical processing techniques to synthetic aperture radar; Proc. of the IEEE, vol. 54, no. 8, pp. 1026–1032.

Diez, J. A., W. G. Hart, S. J. Ingle, M. R. Davis, and S. Rivera, 1980, The use of remote sensing in detection of host plants of Mediterranean fruit flies in Mexico; Proceedings of the Fourteenth International Symposium on Remote Sensing of Environments; San Jose, Costa Rica, vol. II, pp. 675–683.

Doyle, F., 1982, Status of Satellite Remote Sensing Programs, unpublished report, 21p.

Dubois, P. and W. D. Bruce, 1978, Remote sensing

technology transfer, the Canadian-Peruvian approach; Proceedings of the Twelfth International Symposium on Remote Sensing of Environment, vol. 1, pp. 203–211.

Estep, S. D., 1968, Legal and social policy ramifications of remote sensing techniques; Proc. of the Fifth Symposium on Remote Sensing of Environment, pp. 197–217.

Estes, J. E., 1981, Remote sensing and geographic information systems: coming of age in the eighties; Proceedings of the Seventh Annual William T. Pecora Memorial Symposium: Remote Sensing and Geographic Information Systems, pp. 23–40.

————, 1982, United States remote sensing policy: A perspective; Proceedings of the International Geoscience and Remote Sensing Symposium, Munich, Germany.

Estes, J. E., J. R. Jensen and D. S. Simonett, 1980, Impacts of remote sensing on U.S. geography; Remote Sensing of Environment, vol. 10, 72 pp.

Everett J. and D. S. Simonett, 1976, Principles, Concepts and Philosophies in Remote Sensing; in Remote Sensing of Environment, J. Lintz and D. S. Simonett, eds., Addison-Wesley Pub. Co. Reading, Massachusetts. pp. 85–127.

Filho, P., Armando Pacheco dos Santos, Evlyn Marics Leao de Morales Novo, Yosio Edemer Shimabukuro, and Valdete Duarte, 1980, The use of Landsat data for evaluation and characterization of deforested pastureland and reforested areas in Brazil; Proceedings of the Fourteenth International Symposium on Remote Sensing of Environment, San Jose, Costa Rica, vol. III, pp. 1723–1729.

Fischer, W. A. and C. J. Robinove, 1968, A rationale for a general purpose earth resources observation satellite; Proc. of the Remote Sensing Symp., Univ. of Washington, Feb. 1968.

Forshaw, M. R., A. Haskell, P. F. Miller, D. J. Stanley, and J. R. Townshend, 1980, A Review Paper: Spatial Resolution of Remotely Sensed Imagery; Paper submitted to U.S. Committee for Peaceful Uses of Outer Space, 53 pp.

Francica, J. R., R. W. Bernie and G. D. Johnson, 1980, Geologic mapping of the Lodakh Himalaya by computer processing of Landsat data; Proceedings of the Fourteenth International Symposium on Remote Sensing of Environment, San Jose, Costa Rica, vol. II, pp. 773–782.

General Electric, (undated), Landsat 3 Reference Manual; Valley Forge, Pennsylvania.

————, 1975, Definition of the Total Earth Resources System for the Shuttle ERA: TOSS-TERSSE Optional System Study 10, Valley Forge, Philadelphia, PA.

Goodrick, F. E., 1981, The use of Landsat data in an operational forest resource information system (FRIS); Proceedings, Machine Processing of Remotely Sensed Data, Laboratory for Applications of Remote Sensing, Purdue University, W. Lafayette, Indiana, pp. 543–544.

Gordon, F., 1980, The Space-Time Relationships of Data-Points (Pixels) of the Thematic Mapper and the Multispectral Scanner or the "Myth of Simultaneity"; NASA Technical Paper 1715, Goddard Space Flight Center, Greenbelt, Maryland.

Gould, P., 1981, Letting the data speak for themselves. Annals, Association of American Geographers, vol. 71, #2, pp. 166–176.

Guerrero, D. R., E. Bate, R. Punongbayan and O. Mendoza, 1980, Geologic and mineral potential mapping in the Philippines using Landsat imageries; Proceedings of the Fourteenth International Symposium on Remote Sensing of Environment, San Jose, Costa Rica, vol. II, pp. 941.

Junchao, Wei, Zhou Zhengjie and Liu Zigui, 1980, Geologic application of remote sensing in Tengchong, Yunnan, China; Proceedings of the Fourteenth International Symposium on Remote Sensing of Environment, San Jose, Costa Rica, vol. III, pp. 1891–1900.

Harvey, D. W., 1968, Processes, patterns and scale problems in geographical research; Transactions of the Institute of British Geographers, vol. 45, pp. 71–78.

Holkenbrink, P. E., 1978, Manual on Characteristics of Landsat Computer-Compatible Tapes Produced by the EROS Data Center Digital Image Processing System; U.S. Department of Interior, Geologic Survey, Washington, D.C.

Kaltenecker, H. and G. Lafferranderie, 1977, Thoughts on the legal aspects of remote sensing of the Earth by satellites; Environmental Remote Sensing 2: Practices and Problems; papers presented at the Second Bristol Symposium on Remote Sensing, University of Bristol, Barrett and Curtis, eds., Crane Russak, New York, pp. 72–81.

Keighley, J. R., W. W. Lynn, and K. R. Nelson, 1980, Use of Landsat images in the exploration Brazil; Proceedings of the Fourteenth International Symposium on Remote Sensing of Environment, San Jose, Costa Rica, vol. I, pp. 341–343.

Krinov, E. L., 1947, Spectral reflectance of natural formations; Akad. Nauk, USSR, Laboratorica Aerometodov, Moscow (Translation from NEC of Canada, TT439, G. Belkov).

Krinov, E. L. 1947, Speltra'naia Otrazhatel'naia Sposobnost' Prirodnykh Obrazovanii (Spectral reflectance properties of natural formations), Laboratoriia Aerometodov, Akad. Nauk USSR, Moscow.

Landgrebe, P. A., L. Biehl, and W. Simmons, 1977, An empirical study of scanner system parameters; Institute of Electrical and Electronic Engineers Transaction on Geoscience Electronics, GE-15, p. 120–130.

Lee, K., 1975, Ground investigations in support of remote sensing; Manual of Remote Sensing, R. G. Reeves (ed.), American Society of Photogrammetry, Falls Church, Virginia, p. 805–856.

Lintz, J. Jr., and D. S. Simonett (eds.), 1976, Remote Sensing of Environment; Addison-Wesley Publishing Company, Reading, Massachusetts.

Lowman, P. D., Jr., J. A. McDivitt, and E. H. White II, 1967, Terrain Photography on the Gemini IV Mission, Preliminary Report; NASA Technical Note D-3982.

Lucas, J. R. and J. V. Taranik, 1977, Late Wisconsin deglaciation of the Northern Midwest interpreted from a springtime Landsat color mosaic; Proceedings of the Eleventh International Symposium on Remote Sensing of Environment; Ann Arbor, Michigan, pp. 991–992.

Markham, B., J. Townshend, and M. Labovitz, 1981, Land Cover Classification Accuracy as a Function of Sensory Spatial Resolution; NASA Technical Memorandum 82071, Goddard Space Flight Center, Greenbelt, Maryland, pp. 8.14–8.16.

Matte and DeSaussure (eds.), 1976, Legal Implications of Remote Sensing from Outer Space, A. W. Sitjthoff-Leyden Vitgeversmaatschappy, N.V.

McCarthy, H. H., J. C. Hook, and D. S. Know, 1956, The Measurement of Association in Industrial Geography; Report 1, Department of Geography, University of Iowa, Iowa City.

Morain, S. A., 1978, Transferring remote sensing technology: Activities at TAC; Proceedings of the Twelfth International Symposium on Remote Sensing of Environment, pp. 1199–1205.

Morgenstern, O., 1950, On the Accuracy of Economic Observations; Princeton University Press, Princeton, New Jersey, 101 p.

National Academy of Sciences, 1977, Remote Sensing from Space: Prospects for Developing Countries; Report of the Ad Hoc Committee on Remote Sensing for Development Board on Science and Technology for International Development Commission on International Relations National Research Council.

NASA, 1967a, Earth Photographs from Gemini 3,4,5: NASA SP-129.

NASA, 1967b, Terrain Photography on the Gemini IV Mission: preliminary report: NASA TND-3982.

NASA, 1972, Earth Resources Technology Satellite. Data Users Handbook; Goddard Space Flight Center, Greenbelt, Maryland.

NASA, 1973, Advanced Scanners and Imaging Systems for Earth Observations; NASA SP-335, U.S. Government Printing Office, Washington, D.C.

NASA, 1977, Skylab Explores the Earth, NASA SP-380, U.S. Government Printing Office, Washington, D.C., 517p.

NASA, 1978, SKYLAB EREP investigation summary prepared by NASA Lyndon R. Johnson Space Center, NASA SP-399.

NASA, Heat Capacity Mapping Mission (HCMM) Data Users Handbook for AEM-A, 2nd Revision, 10/80.

NASA, Goddard Space Flight Center, HCMM Data Users Bulletin No. 9, March 1, 1982.

Nicholas and DeSaussure (eds.), 1976, Legal Implications of Remote Sensing from Outer Space, A. W. Sijtthoff-Leyden.

Norwood, V. T., 1974, Balance between resolution and signal-to-noise ratio in scanner design for earth resources systems; Proceedings Society of Photoinstrumentation Engineers, vol. 51, pp. 37–42.

Otterman, J., 1969, Diffraction-limited resolution for geoscience imagery; Applied Optics, no. 9, pp. 1887–1889.

Paul, K. and C. Mascarenhas, 1981, Remote sensing in development; Science, vol. 214, no. 4517, pp. 139–145.

Perrin, F. H., 1966, The Structure of the Developed Image; The Theory of the Photographic Process, T. H. James (ed.), MacMillan, New York, pp. 499–551.

Prapinmong kolkarn, P., C. Thisayakorn, N. Kattiyakulwanich and others, 1980, Remote sensing applications to land cover classification in Northern Thailand; Proceedings of the Fourteenth International Symposium on Remote Sensing of Environment, vol. III, pp. 1273–1284.

Raney, R. K., 1982, Radarsat-Canada's national radar satellite program; Institute of Electrical and Electronics Engineers Geoscience and Remote Sensing Society Newsletter, vol. XXI, no. 1, pp. 5–9.

Richason, F., Jr. (ed.), 1981, Remote Sensing: An Input to Geographic Information Systems in the 1980's; Pecora VII Symposium, Sioux Falls, South Dakota, 619 pp.

Roque, C., R. R. Bina, R. S. Jara and N. Lorenzo, 1980,

Application of Landsat data and selective aerial reconnaissance surveys to mangrove forest resource management in the Philippines; Proceedings of the Fourteenth International Symposium on Remote Sensing of Environment, San Jose, Costa Rica, vol. II, pp. 1225–1237.

Sader, A., 1980, Renewable resource analysis and monitoring in Costa Rica; Proceedings of the Fourteenth International Symposium on Remote Sensing of Environment, April 1980 San Jose, Costa Rica, vol. 2, pp. 1251–1260.

Sadowski, F. and J. Sarno, 1976, Forest Classification Accuracy as Influenced by Multispectral Scanner Spatial Resolution; Report No. 109600-71-F, Environmental Research Institute of Michigan, Ann Arbor, Michigan, 130 p.

Salomonson, V. V. and A. Rango, 1980, Water resources; Remote Sensing in Geology, B. S. Siegal and A. R. Gillespi (eds.), John Wiley & Sons, New York, pp. 607–633.

Santos, V. S., S. Silonga, A. F. Nacv and E. Salvador, The use of multi-temporal Landsat data to monitor land use and land cover change in the Philippines; Proceedings of the Fourteenth International Symposium on Remote Sensing of Environment, vol. II, pp. 1081–1088.

Schultink, G. and A. Karteris, 1980, Resource inventory procedures in the Dominican Republic using linear random sample data acquired by Legat aircraft survey techniques; Proceedings of the Fourteenth International Symposium on Remote Sensing of Environment, vol. III, pp. 1455–1467.

Shay, R., A. Potter, M. Bauer, R. Berstein, R. Haralick, A. Koso, and V. Salomonson, 1975, Landsat-D Thematic Mapper Technical Working Group Final Report (JSC-09797), J. Harnage and D. Landgrebe (eds.), NASA, Houston, Texas.

Shelton, R. L. and J. E. Estes, 1979, Integration of remote sensing and geographic information systems; Proceedings of the Thirteenth International Symposium on Remote Sensing of Environment, vol. I, pp. 675–692.

Shelton, R. L. and E. Tilman, 1979, Remote sensing, geographic information systems, and national land planning: Some Central American and Caribbean experiences; Proceedings of the Thirteenth International Symposium on Remote Sensing, vol. I, pp. 567–585.

Sherwin, C. W., J. P. Ruina, and R. D. Rawclisse, 1962, Some early developments in synthetic aperture radar systems; IRE Transactions on Military Electronics, MIL-6, No. 2, pp. 111–115.

Schultink, G., and M. A. Karteris, 1980, Resource inventory procedures in the Dominican Republic using linear random sample data acquired by light aircraft survey techniques; Proceedings of the Fourteenth International Symposium on Remote Sensing of the Environment, vol. 2, pp. 1455–1468.

Siefferman, G., 1980, Contribution to soil mapping of Southern Venezuela using computer analysis of Landsat II data; Proceedings of the Fourteenth International Symposium on Remote Sensing of Environment, San Jose, Costa Rica, pp. 1221–1222.

Simonett, D. S., 1981, Remote Sensing and the Developing World: Examples of Major Benefits; The Environment: Chinese and American Views, Laurence J. C. Ma and Allen G. Noble (eds.), Methuen and Co. Ltd., New York and London.

Slater, P. N., 1979, A re-examination of the Landsat mote sensing; Manual of Remote Sensing, R. A.

Reeves (ed.); American Society of Photogrammetry, Falls Church, Virginia, pp. 235–321.

Slater, P. N., 1979, A Re-examination of the Landsat MSS: Photogrammetric Engineering and Remote Sensing, vol. 45, no. 11, pp. 1479–1485.

Slater, P. N., 1980, Remote Sensing: Optics and Optical Systems; Addison-Wesley Publishing Company, Reading, Massachusetts.

Stone, K., 1972, A geographer's strength: The multiple-scale approach; The Journal of Geography, vol. 61, no. 6, p. 354–362.

Stowe, R. F., 1978, Legal implications of remote sensing; Photogrammetric Engineering and Remote Sensing, vol. 44, no. 2, pp. 183–188.

Strahler, A. H., J. Franklin, C. E. Woodcock, and T. L. Logan, 1981, FOCIS: A Forest Classification and Inventory System Using Landsat and Digital Terrain Data; Report NFAP-255, U.S. Department of Agriculture, Forest Service, Houston, Texas, 60 p.

Strome, W. M., 1979, Communication to A. P. Colvocoresses; reference 1372-113-2-2, February 19, 1979, Canada Centre for Remote Sensing.

Sturge, J. M. (ed.), 1977, Neblette's Handbook of Photography and Reprography; 7th Edition, New York, Van Nostrand Reinhold, 641 pp.

Swain, P. H., and S. M. Davis (eds.), 1978, Remote Sensing: The Quantitative Approach; McGraw-Hill, New York.

Swain, P. H., and M. Davis, 1978, Remote sensing decoded: Meeting the challenges of multidisciplinary and international technology transfer; Proceedings of the Fourteenth International Symposium on Remote Sensing of Environment, vol. I, pp. 205–218.

Taranik, J. V. and M. Settle, 1981, Space Shuttle: A new era in terrestrial remote sensing; Science, vol. 214, pp. 619–626.

Tom, C. H., and L. D. Miller, 1979, Forest site productivity mapping in the coniferous forests of Colorado with Landsat imagery and landscape variables; Proceedings of the Thirteenth International Symposium on Remote Sensing of Environment, vol. 1, pp. 675–692.

Townshend, I. R., 1980, The Spatial Resolving Power of Earth Resources Satellites: A Review; NASA Technical Memorandum 82020, Goddard Space Flight Center, Greenbelt, Maryland, 36 p.

Tucker, C. J., 1979, Radiometric Resolution for Monitoring Vegetation—How Many Bits are Needed?; NASA Technical Memorandum 80293, Goddard Space Flight Center, Greenbelt, Maryland, 25 p.

Wagner, W. and S. Lowe, 1978, A small grant program for remote sensing technology transfer; Proceedings of the Twelfth International Symposium on Remote Sensing of Environment, vol. 1, pp. 895–906.

Welch, R., 1977, Progress in the specification and analysis of image quality; Photogrammetric Engineering and Remote Sensing, vol. 43, no. 6, pp. 709–719.

Wharton, S., 1981, Classification of High Resolution Remotely Sensed Data via Component Frequency Analysis; NASA Technical Memorandum 82071, Earth Survey Applications Division Report, Goddard Space Flight Center, Greenbelt, Maryland.

Wiersma, W., and D. Landgrebe, 1978, The Analytical Design of Spectral Measurements for Multispectral Remote Sensor Systems; LARS Technical Report 122678, West Lafayette, Indiana.

Umali, R., 1979, East-West perspective office of public affairs; East-West Center, Honolulu, Hawaii, vol. 1, no. 5, pp. 12–21.

United Nations Department of Economic and Social Affairs, 1970, Natural Resources of Developing Countries: Investigation, Development and Rational Utilization, Department of the Advisory Committee on the Application of Science and Technology Development (E/4608/REV. 1 ST/ECA/122), NY: United Nations.

U.S. Government, 1981, Civil Land Remote Sensing Systems; Joint Hearings for the Committee on Science and Technology and the Senate Committee on Commerce, Science, and Transportation, July 22, 23, no. 40 (CST), ser. no. 97-71 (CCST), U.S. Gov. Printing Office, 366 p.

CHAPTER 2

The Nature of Electromagnetic Radiation

Author: GWYNN H. SUITS

GENERAL CONTENTS: Fundamental properties of electromagnetic radiation; nomenclature relating to electromagnetic radiation; measurement of electromagnetic radiation; production of electromagnetic radiation; references.

NOMENCLATURE

To conserve and eliminate repetition in text and references, the following symbols, units and names have been used in this chapter.

Symbol	SI Units	Name
A	m^2	Area
a	m^2	Area
$Å$	10^{-8} cm	Angstrom unit
A_e	m^2	Eye pupil area
c	3×10^8 ms^{-1}	Speed of light
C	2.898×10^{-3} mK	Wein displacement constant
c_1	3.74×10^{-16} Wm^2	First Planck's law constant
c_2	1.44×10^{-2} mK	Second Planck's law constant
E	Wm^{-2}	Irradiance
E_e	Wm^{-2}	Irradiance
E_λ	Wm^{-2} μm^{-1}	Spectral irradiance
E_ν	Wm^{-2} Hz^{-1}	Spectral irradiance
E_v	lm m^{-2}	Illuminance
f	Hz	Frequency
$f(t)$	—	Wave displacement
$g(t)$	—	Wave displacement
h	6.625×10^{-34} Js	Planck's constant
I	Wsr^{-1}	Intensity
I_e	Wsr^{-1}	Intensity
I_λ	Wsr^{-1} μm^{-1}	Spectral intensity
I_ν	Wsr^{-1} Hz^{-1}	Spectral intensity
I_v	lm sr^{-1}	Luminous intensity
L	m	Arc length
L	Wm^{-2} sr^{-1}	Radiance
L_e	Wm^{-2} sr^{-1}	Radiance
L_λ	Wm^{-2} sr^{-1} μm^{-1}	Spectral radiance
L_ν	Wm^{-2} sr^{-1} Hz^{-1}	Spectral radiance
L_v	lm m^{-2} sr^{-1}	Luminance
M	Wm^{-2}	Exitance
M_e	Wm^{-2}	Exitance
M_λ	Wm^{-2} μm^{-1}	Spectral exitance
M_ν	Wm^{-2} Hz^{-1}	Spectral exitance
M_v	lm m^{-2}	Luminous exitance
Q	J	Radiant energy
Q_e	J	Radiant energy
Q_λ	J μm^{-1}	Spectral radiant energy
Q_ν	J Hz^{-1}	Spectral radiant energy
Q_v	lms	Luminous energy
R	m	Range
R	Signal units W^{-1}	Responsivity
$R(\lambda)$	Signal units W^{-1}	Spectral responsivity
S	Signal units	Signal from detector
T	K	Temperature
t	s	Time
u	ms^{-1}	Velocity
v	ms^{-1}	Velocity
$V(\lambda)$	—	Luminous visual efficiency
$W_1(t)$	W	Power delivered
$W_2(t)$	W	Power delivered
W_1	W	Time average power
W_2	W	Time average power
$W_{combined}$	W	Combined power delivered
Z	—	Proportionality constant
α	—	Absorptance
$\alpha(\lambda)$	—	Spectral absorptance
β	—	Relative velocity, v/c
Δ	—	Increment
λ	μm	Wavelength
ν	Hz	Frequency
ν'	Hz	Doppler shifted frequency
ν''	Hz	Doppler shifted frequency
Ω	sr	Solid angle
ω	sr	Solid angle
Φ	W	Radiant flux
Φ_e	W	Radiant flux
Φ_λ	W μm^{-1}	Spectral flux
Φ_ν	W Hz^{-1}	Spectral flux
Φ_v	lm	Luminous flux
ρ	—	Reflectance
$\rho(\lambda)$	—	Spectral reflectance
σ	5.669×10^{-8} Wm^{-2}K^{-4}	Stefan-Boltzmann constant
τ	—	Transmittance
$\tau(\lambda)$	—	Spectral transmittance
Θ	—	Angle

FUNDAMENTAL PROPERTIES OF ELECTROMAGNETIC RADIATION

Energy is propagated by electromagnetic radiation (EMR) with a velocity of 3×10^8 m s^{-1} from the source, directly through free space; or indirectly by reflection and reradiation to the remote sensor. EMR is also one of the most useful force fields for remote sensing, forming a high-speed communications link between the sensor and remotely located substances. Changes in the amount and properties of the EMR become, upon detection, a valuable source of data for interpreting important properties of the media with which it interacts. An understanding of the fundamental properties of this EMR communications link forms the foundation for its use in more sophisticated ways than normal human vision permits.[1]

[1] More detailed discussions of the properties of electromagnetic radiation presented here may be found in Jenkins and White, 1950, *Fundamentals of Optics.*

WAVE PROPERTIES

Many remote sensing techniques utilize wave motion for the detection and identification of distant objects. Seismographs utilize seismic waves or stress waves in the crust of the Earth; microphones and human hearing utilize sound waves in air; sonar sensors utilize sound waves in water; and electromagnetic (EM) sensors utilize EM waves. Some properties of wave motion are independent of the type of wave considered, while other properties are specific to the type. Although this discussion will cover only the EM wave, a number of similarities may be recognized between the properties of these waves and of other waves—sonic, seismic, etc.

Maxwell's Formulation

EMR is a dynamic form of energy made manifest only by its interaction with matter. After many years of experimentation by numerous investigators, James Clerk Maxwell, a British physicist, produced a mathematical formulation of electric and magnetic phenomena that apparently explained not only the experimental results of previous investigators, and thus unified the extant theories, but in addition provided a logical basis for predicting the possible existence of the dynamic form of electric and magnetic phenomena; i.e., EMR in the form of wave motion. Maxwell considered EMR on a macroscopic scale, the interaction with matter depending upon electric and magnetic properties of matter. His formulation imposed no limitations on the possible frequencies, wavelengths, or amplitudes with which such radiation could occur.

Maxwell's formulation and our modern theory of EMR are both based upon the concept of a force field. For example, when a compass needle is held near a magnetized piece of iron, the compass needle becomes oriented in a particular direction, depending upon the location of the needle relative to the magnetized iron. Based upon the mechanical activity of the needle, one infers that a force is being applied to the two ends of the needle and that the force is caused by the magnetized iron. Yet nothing material connects the iron with the compass needle. When two compass needles are tied together with like poles in the same place, the effect of the iron on the pair is doubled. One can take the viewpoint that the iron produces something real but nonmaterial in the surrounding space which affects everything that is placed there and that, even if the compass needle is removed, the effect of the iron will still exist at that point, and is ready to cause a force on each compass needle, with magnitude and direction depending upon the location of that point relative to the piece of iron. This nonmaterial cause of force is a force field; a similar cause of force on electric charges is observed to surround electrified materials.

Maxwell's concept of EMR was that a mathematically smooth wave motion existed in the magnetic and electric force fields; a concept he formulated as a set of differential equations expressing the interrelationship of electric and magnetic fields. Thus, in any region where there is a time rate of change of electric field, a magnetic field appears automatically in that same region as a conjugal partner. Similarly in a region where there is a time rate of change in magnetic field, an electric field appears.

Therefore, the dynamic fields always occur together as inseparable partners, so that neither purely electric nor purely magnetic radiated waves will occur separately from the other. When EM waves are intercepted by matter, the result will depend upon both the magnetic and electric properties of the matter.

Radiant flux density[2] of radiated waves is proportional to the squares of wave amplitudes. These waves propagate through empty space at a fixed velocity, c approximating 3×10^8 meters per second. When the waves propagate through material, the velocity of propagation depends upon the material properties and the frequency of the wave. In all cases, the relation between the velocity of propagation (v) the wavelength λ, and the wave frequency ν, is[3]

$$v = \lambda\nu. \qquad (2\text{-}1)$$

As the frequency does not change when the waves penetrate matter, the wavelength must therefore change as the velocity of propagation changes. For instance, visible light propagates through glass with a velocity, v approximately $c/1.5$, so that the wavelength of the wave in the glass is shorter than in free space by a factor of 1.5. This factor is called the index of refraction and is usually designated by the Greek letter η.

The Superposition Principle and Wave Analysis

The *principle of superposition of wave motion* states that the wave motion produced in a region where two separate waves cross each other has an amplitude which is the sum of the amplitudes of the two separate waves. For waves in force fields, the amplitude is a sum of forces (or a vector sum).

It is not self evident that all wave motion should obey the superposition principle; indeed, some wave motions, such as shock waves in air caused

[2] Flux—the time-rate with which radiation passes a spatial position.
[3] An early visualized description of wavelength is the distance through which an EM wave will travel in the time period of one cycle in frequency f^{-1} with a basic wavelength unit of one meter.
[4] Alternative form for Equation 2-1, $v = f\lambda$. Texts on electromagnetic theory in the microwave region commonly use script "f" as the symbol for frequency instead of ν, which is used herein. Script "f" is also used for focal length and f number (f/no.). No attempt will be made herein to change these, since they have been in the literature for many years. The context of the material is sufficient to indicate the distinction.

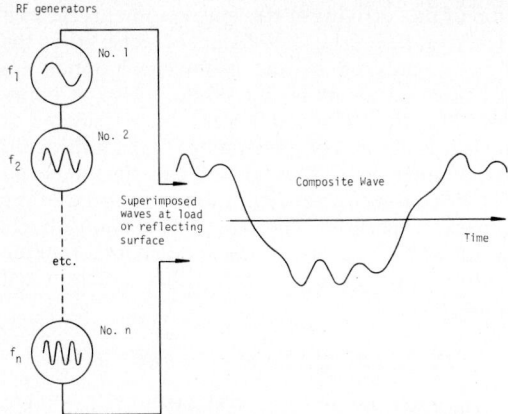

Fig. 2-1. Principles of electromagnetic wave superposition.

by supersonic vehicles, do not. However, experimental observation clearly indicates that EM waves in free space do move in accordance with this principle. Moreover, EM waves propagate through most common materials in conformance with the superposition principle, provided that the amplitudes of the force fields are not so great as to alter the material properties of the matter. We can expect that the superposition principle will apply to all EM waves to be encountered in remote sensing. This principle is illustrated by Figure 2-1.

This principle forms the foundation for the analysis of very complicated wave motions by the superposition of many simple and more easily understood waves; those most frequently used for analysis are sinusoidal or simple harmonic waves. The French mathematician, Jean Baptiste Fourier, showed mathematically that any complicated wave form could be constructed by the

superposition of an infinite number of these sinusoidal waves as components if the sinusoidal waves had the proper amplitudes, frequencies, and phases (or starting times).

To illustrate the analytical procedure which the superposition principle permits, assume that some particular wave impinges upon a material. Let the amplitude of the wave at the surface of the material change in time in some complicated fashion. Conceptually, a graph could be plotted of the manner in which this amplitude changes in time. Through use of Fourier's analysis, a person could determine from this graph the required amplitudes, frequencies, and phases of sinusoids necessary to insure that the graph of the sum of these sinusoids would exactly match the graph of the wave amplitude. Furthermore, if one were to make a set of sinusoidal wave generators, each with the specified frequency, amplitude, and phase, and cause them to launch their waves together towards the material, the resultant wave from these generators would, in every way, be indistinguishable from the original wave by virtue of the superposition principle. Therefore, the view that complicated wave forms are actually composed of superposed sinusoids is quite justifiable (Fig. 2-1), regardless of the manner in which the wave form was generated. Those sinusoids which are presumed to compose a complicated wave are called the *spectral components* of the wave. Figure 2-2 illustrates the composition of a complex wave form with only two spectral components.

It is possible to predict the kind of wave which will result when an incident wave reflects or penetrates matter if it is known which sinusoidal component waves compose the incident wave and how each component will reflect or penetrate separately. The reflected or penetrating wave will be the superposition of the reflecting or penetrating component waves. The material properties which specify the response of the material to sinusoidal component waves at every frequency or wavelength are called *spectral properties*.

The purpose of spectrographic apparatus is to measure the amount of radiant flux carried by each of the sinusoidal components of any wave forms. In remote sensing applications, the actual mathematical analysis of wave forms is rarely performed by numerical calculations from graphs; instead, direct measurements are made with spectrographic devices.

Diffraction

If EM waves impinge upon a finite material obstacle through which the wave cannot penetrate, part of the wave passes by the side of the obstacle. However, as the passing part of the wave enters the region beyond the obstacle, some of the flux near the side of the obstacle changes direction and propagates into the *shadow* region behind the obstacle. This change in direction of some of the flux due to propagation of a wave past an opaque edge is called *diffraction*.

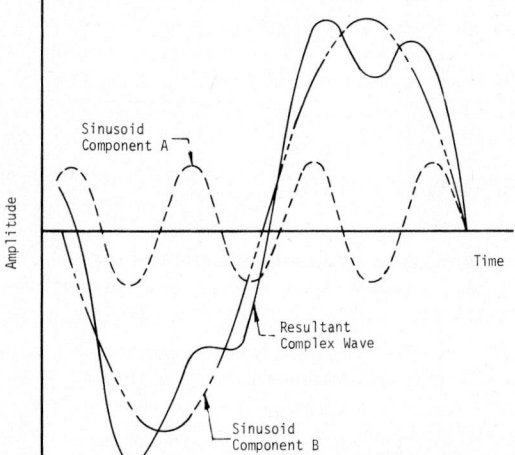

Fig. 2-2. Composition of a complex wave with only two spectral components.

Diffraction also occurs with water surface waves, as well as with sound waves in water and in air, and it is not difficult to understand why such waves will diffract. Imagine a water surface wave approaching the end of a breakwater. As the wave passes the end, the wave which passes is cut off from the wave which strikes the breakwater. The crest and trough of the wave are supported by the end of the breakwater as the wave passes, but, if the wave were to continue past the end without diffraction, one would observe that the sliced-off side of the wave would consist of a vertical wall of water—upward from sea level for the crest, and downward from sea level for the trough. As every one knows, water simply will not stand up like that without some containing force or wall. What happens instead is that, as the crest of the wave passes the end, water spills from the crest to the side behind the breakwater in exact time with the wave frequency. In this way, a new wave source is created at the end by the sideward spilling action which sends flux to the side of the parent wave direction, its energy being derived from the parent wave. It is easy to see the analogous argument for sound-wave diffraction. It must be simply asserted, and believed, that EM waves cannot maintain a sheared-off edge without a suitable EM barrier, and therefore must also diffract around the edge of obstacles.

The direction and amount of EM flux received by a sensor is of major importance in remote sensing, since this flux forms the communications link between objects of interest and the remotely located sensor. The quantity, quality, and direc-tion of flux arriving at the sensor contains the evidence of the presence of distant objects. When the flux is analyzed by the geometrical optics approximation (Jenkins and White, 1950), it is implied that the diffracted flux is to be considered as insignificant in comparison to the undiffracted flux. However, diffraction of EM flux must always occur wherever a wave is cut or sliced-off by a sensor aperture. Therefore, diffraction effects will always form one of the fundamental limiting factors in the measurement of flux direction and quality.

ELECTROMAGNETIC SPECTRUM

The EMR found in the EM spectrum is a limitless source of energy capable of conveying and propagating information. Remote sensing deals largely with the manner by which this energy can be utilized by mankind.

Narrow and broad bands of this EMR are selected by remote sensors for the observation and study of the Earth and its atmosphere. These same sensors are found on satellites launched to observe the surface and atmosphere of various planets in the solar system. Figure 2-3 shows the extent of the EM spectrum, the various named bands (ultraviolet, visible, infrared, radar, etc.), the transmittances of the Earth's atmosphere to this EMR, and the effects caused by its interaction or presence.

The visible, infrared (IR), and microwave regions of the EM spectrum are emphasized in this manual; however, the X-ray and ultraviolet re-

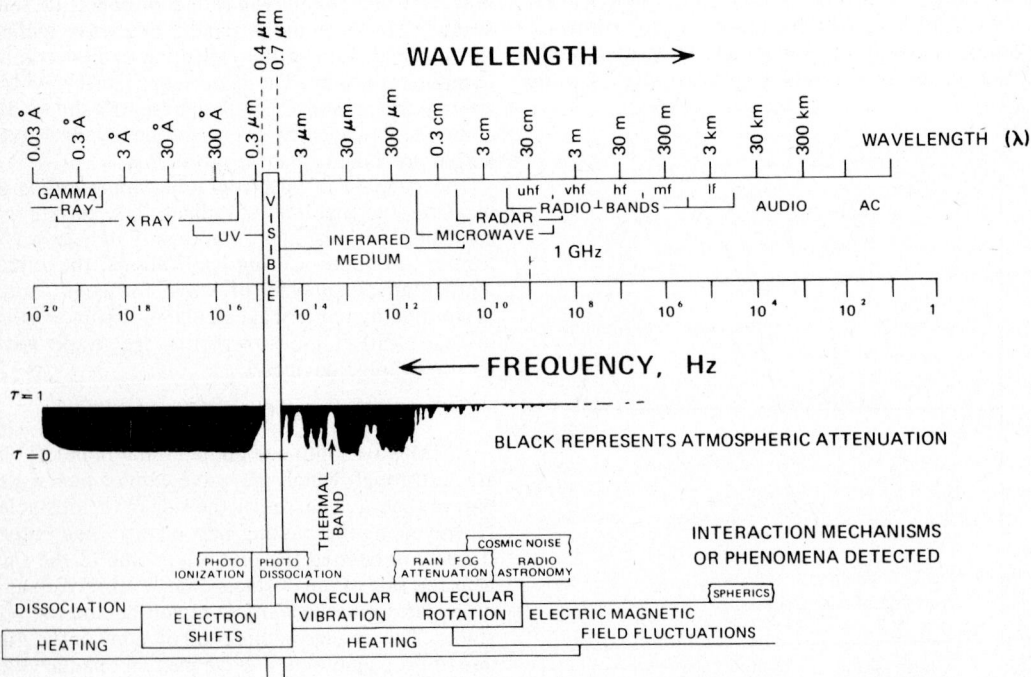

Fig. 2-3. The electromagnetic spectrum.

TABLE 2-1
Remote sensor types and applications

Remote Sensors	Part of EM Spectrum Utilized	Comments
Scintillation counters		
Gamma-ray spectrometer	$<0.03-100$ Å	Measurement of emitted natural radiation by
Geiger counters		gamma ray detectors: NaI, film, etc.
Scanners with photomultipliers		
Image orthicons and cameras with	100 Å-0.4 μm	Records incident natural radiation.
filtered infrared film >2900 Å		Imaged ultraviolet spectroscopy available
Cameras		
Using conventional B&W and color	$4,000-7,000$ Å	B&W film for high spatial detail.
film		
	$(0.4-0.7$ μm$)$	Improved spatial detail through contrast
Using infrared film (B&W and IR	$6,000-9,000$ Å	Greater reflectance gradients useful for vege-
color)	$(0.6-0.9$ μm$)$	tation surveys
Multispectral units	$3,000-9,000$ Å	Individual narrow band scenes available with
	$(0.3-0.9$ μm$)$	multi-camera systems
Lidar		
Laser radar	$4,000-11,000$ Å	Monochromatic active system for measuring
	$(0.3-1.1$ μm$)$	backscattered EMR from atmosphere, par-
		ticularly particulates
Radiometers	Usually IR and micro-wave bands	Generally measures total radiation in a wide band in the infrared or microwave regions. Imagery obtained by scanning techniques.
Photometers	$4,000-7,000$ Å $0.4-0.7$ μm	Measures luminous flux in various bands of the optical region for distribution, color, etc.
Spectrometers	In any spectral region	Narrow-band data available sequentially— EMR amplitude vs frequency.
Solid-state detectors		
Single detectors	1 μm-1 mm	Single detecting element used in scanners,
Linear arrays		radiometers. One- and two-dimensional ar-
Matrices		rays for sequential data gathering.
Radars	1 mm-0.8 m	Narrow band active systems. Both analog data and imagery available
Radiometers (microwave)	1 mm-0.8 m	Passive systems. Both analog and imagery available.

gions have been currently utilized in significant earth resources surveys and thus receive attention.

The EMR collected by the remote sensor is characterized by the spectrum of the generating source, the atmosphere through which it propagates, and the reflecting surface from which it is largely received. Changes in the source EMR due to the reflecting surfaces and the atmosphere are sources of data which, after processing, can provide information about each. Thus, knowledge of the properties of this EMR as a function of wavelength and other pertinent variables is vital to the analysis of remote-sensor data.

Table 2-1 is a preview of but a few of the many remote sensors which utilize the EM spectrum. Included are comments on what is measured and data presentation. EMR particulars and details of remote sensors for various spectral regions are found in subsequent chapters.

QUANTUM PROPERTIES OF ELECTROMAGNETIC RADIATION

Maxwell's formulation of EMR as a mathematically smooth wave motion of EM fields fails to account for certain significant phenomena when EMR interacts with matter. These phenomena become more evident and consequential as the wavelength of the radiation becomes very small. Experiments since the time of Maxwell have shown that the generation of EM waves always occurs in short wave trains or bursts of radiation, (Fig. 2-4), in which each wave train or burst carries a radiant energy, Q, in proportion to the frequency, ν, of the wave, so that

$$Q = h\nu \qquad (2-2)$$

where:
    h is the Universal, or Planck's constant, with a value of 6.625×10^{-34} joule second.

Moreover, the radiant energy carried by these wave trains is not delivered to a receiver as if it were spread evenly over the wave (as Maxwell had visualized), but is delivered all at some one place or other along the wave on a probabilistic basis. The probability that a wave train will make full delivery of its radiant energy at some place along the wave is proportional to the flux density of the wave at that place. If very large numbers of

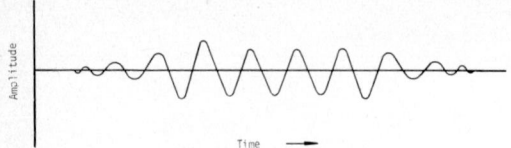

Fig. 2-4. Wave train.

wave trains are incident on a receiving plate, then the time-average rate of energy delivered will be exactly that which Maxwell's formulation would have predicted for that flux density. However, from instant to instant, the rate of energy flow to the plate will be continuously fluctuating because of the inherently quantized and statistical nature of EMR. The nature of quantization of radiation is illustrated by Figure 2-5.

The term *photon* is used to emphasize the quantized and statistical properties of radiation, while the term *wave* is used to emphasize the overall average effects of radiation. When a measurement of radiation lacks precision primarily because of the inherent statistical nature of energy delivery by radiation, the measurement is said to be *photon-noise limited.*

Because of this quantized and statistical energy delivery property of EMR, there is a built-in limit to the precision of the measurement of radiant-energy flow rates. The precision becomes less as the number of incident wave trains arriving in the measurement time interval also become less.

Remote sensing techniques utilize quantitative radiant flux measurements to infer "how much of what is where." The inherent limit to the precision of radiant-flux measurement frequently forces a compromise between the rapidity of making radiant-flux measurements and the precision of these measurements.

POLARIZATION OF ELECTROMAGNETIC RADIATION

As previously discussed *EMR consists of an electric force field and a magnetic force field merged in a close relationship.* The directions of these force fields are important factors in determining the nature of the response of the matter which intercepts the radiation. Both fields are transverse to the direction of propagation, and are at right angles to each other in free space. The direction and magnitude of only one of the fields in a given medium are all that is needed to specify the direction and magnitude of the other (using Maxwell's relations) so that only one field need be considered for analytical purposes. The direction of the electric field is conventionally used to define the direction of wave polarization in electronics. In this discussion, the direction of the electrical vector will be used.

An EMR which produces an electric field in a fixed direction as the wave passes is said to be *plane polarized* in that direction. Two EM waves having the same frequency and direction of propagation, but with a different direction of polarization, superpose or add together to make a resultant EM wave with an electric field equal to the vector sum of the two component electric fields. The direction of polarization of the resultant wave is the direction of the resultant electric field.

The radiation from some sources (mostly in the optical region) does not have any clearly defined polarization, the electric field of the wave assuming different directions at random as the wave is received. This kind of wave is said to be *randomly polarized.* It is sometimes said that such a wave is *unpolarized,* but it is an incorrect implication that the electric field has no direction at all. On the contrary, the electric field has some direction at

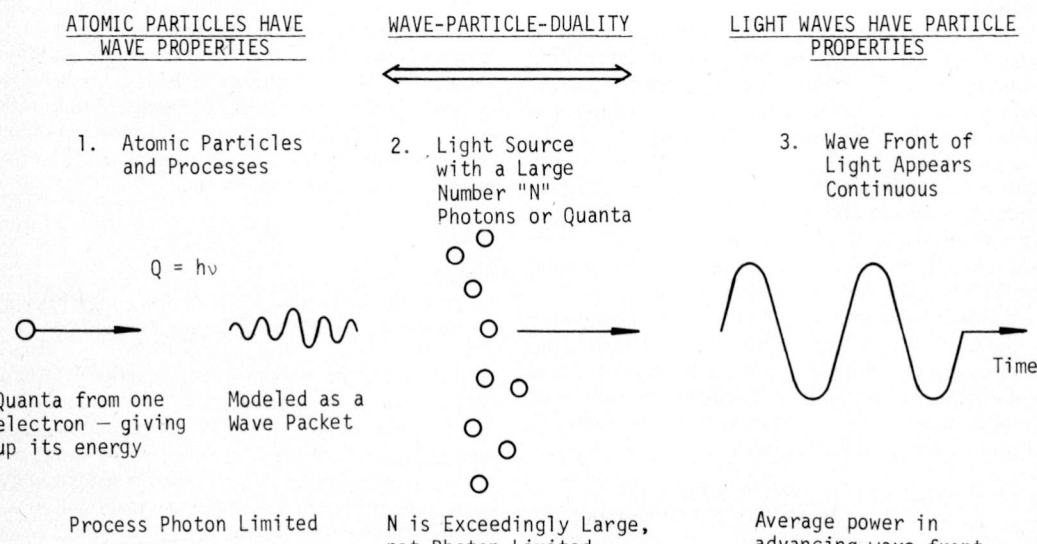

Fig. 2-5. Quantization of radiation.

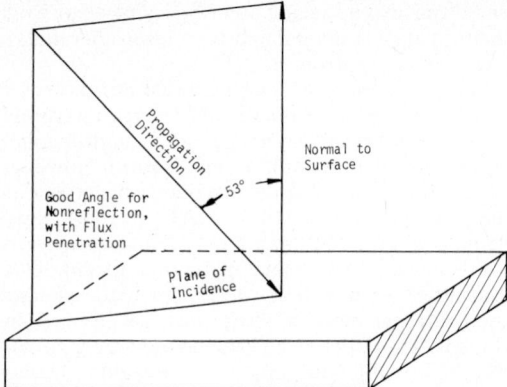

Fig. 2-6. Plane of incidence.

every moment, except when the field has zero magnitude. In randomly polarized radiation, there is just no system to the polarization direction.

There are a number of kinds of polarization, including plane, which are perfectly systematic and are important in remote sensing. Plane polarized waves incident upon a plane surface of material reflect from the surface in different amounts depending upon the polarization of the wave relative to the surface. Plane polarization, the polarization direction of which is parallel to the plane surface upon which the wave is incident, is commonly called "horizontal" polarization. This term is frequently used in literature on microwave radar, in which the plane of interest is the surface of the ground or the horizontal surface. The plane polarization which is at right angles to the horizontally polarized direction is then called *vertical* polarization. It is not vertical to the horizontal plane except when the wave is incident at the grazing angle to the plane.

In the literature on optics, however, a different reference plane is most frequently used. The plane formed by the normal to the surface and the direction of propagation of the incident wave is called the *plane of incidence* (Fig. 2-6), which is perpendicular to the plane receiving the wave. *Parallel polarization* is the polarization direction parallel to the plane of incidence, and *perpendicular* polarization is the polarization direction perpendicular to the plane of incidence.

Through use of these two plane polarizations and of the modes by which such polarized waves reflect from or penetrate into various surface materials, predictions can be made as to the action of any other kind of polarized wave under specified circumstances. Every other type of polarized wave can be considered to be a superposition of two component plane polarized waves, one horizontally polarized and one vertically polarized, propagating in the same direction.

There is no way to tell whether a wave is one large original wave or is a superposition of two smaller ones but the result is the same. Therefore, the original wave will reflect from a surface in such a way that the reflected wave will be the resultant of the two reflected components. For example, at an incident angle of about 53° from zenith (as shown in Fig. 2-5), the vertically polarized wave in the visible part of the spectrum will not reflect from a smooth water surface, but all flux will penetrate. Horizontally polarized waves do reflect partially to the extent of about 7.8% of the flux specularly reflecting, with about 92.2% penetrating the surface. Now, if randomly polarized solar radiation is incident upon a smooth wave surface at that angle, it can be predicted that about 3.9% of the flux will reflect, and will be purely horizontally polarized. This randomly polarized solar radiation can be considered to be the superposition of two waves, one horizontally polarized and carrying half of the flux, and one vertically polarized, and carrying the balance of the flux. Then 7.8% of half of the flux (with horizontal polarization) reflects, yielding only 3.9% of the total in the reflection.

COHERENT AND INCOHERENT RADIATION

Coherent Radiation

Two waves are said to be *coherent* with each other if there is a regular, or systematic, relationship between their amplitudes.

Suppose two waves are incident upon a detector so that the power delivered to the detector by these waves can be measured. Let the time variation of one amplitude be expressed by the formula $f(t)$ and the other by $g(t)$. Then, the power, W[5], delivered by the first is

$$W_1(t) = Z f^2(t) \qquad (2\text{-}3)$$

and by the second,

$$W_2(t) = Z g^2(t), \qquad (2\text{-}4)$$

where:

Z = proportionality constant between delivered power and square of the amplitudes.

For any reading of average power to be made, it is necessary to take the time-average value of the instaneous powers delivered at each moment of time. For wave number one, we have

$$W_1 = Z \overline{f^2(t)} \qquad (2\text{-}5)$$

and for wave number two,

$$W_2 = Z \overline{g^2(t)}, \qquad (2\text{-}6)$$

where:

the bar denotes the average value over a measurement time interval, τ.

[5] Unfortunately various symbologies for power are used in the texts relating to different parts or applications of the EM spectrum. As an attempt for consistency in this manual, the symbol W (for watt. Φ also is used in the optical region, and P in the radio portion of the spectrum) will be used throughout to indicate the electromagnetic relationship whenever plausible along with the unit designation W.

If both waves are now considered as incident together upon the detector, the resultant wave will have an amplitude of $f(t) + g(t)$ on the detector. The power delivered is

$$W(t) = Z[f(t) + g(t)]^2 \qquad (2\text{-}7)$$
$$= Z[f^2(t) + 2f(t)g(t) + g^2(t)],$$

and an average power, W, after time, τ, of

$$W_{\text{combined}} = Z\overline{[f^2(t) + 2f(t)g(t) + g^2(t)]}. \qquad (2\text{-}8)$$

The average of a sum is the same as the sum of the averages, so that

$$W_{\text{combined}} = Z\overline{f^2(t)} + Z\overline{g^2(t)} + 2Z\overline{f(t)g(t)}. \qquad (2\text{-}9)$$

Therefore,

$$W_{\text{combined}} = W_1 + W_2 + 2Z\overline{f(t)g(t)}. \qquad (2\text{-}10)$$

Incoherent Radiation

If the waves are *incoherent*, their amplitudes are related in random fashion, so that the amplitude $f(t)$ could be positive, while $g(t)$ could be negative, or both could be positive or negative together, so that the third term in Eq. 2-10 would average to zero or be insignificantly small. The power of the combined wave is merely the sum of the powers of each separate wave. If the two waves are coherent, then a regular, or systematic relation exists between the value of the amplitudes $f(t)$ and $g(t)$, so that the combined power may be more or less than the sum of the powers of the waves incident separately. The third term need not average to zero since, when $f(t)$ has positive values, $g(t)$ could also have regularly positive values, or vice versa, because of the coherent relationship between the two waves.

For incoherent radiation, a receiving detector will always indicate an average power equal to the sum of the average powers of each wave separately, for all detector positions. In the case of coherent radiation, a receiving detector will indicate more power at some locations, and less power at other locations.

Monochromatic Radiation

Microwave radar and lasers normally utilize *monochromatic* waves for illumination of remote objects. The reflected waves from two distant objects close together are highly coherent, so that an image made by this type of sensor may sometimes indicate no power received from these two objects, and sometimes four times the average power of one object from these two objects, depending upon the position of the sensor relative to these objects. For this reason (the coherence of the waves), microwave radar images and photographs of objects using laser illumination appear to have a grainy or speckled appearance as an inherent feature. This speckled or grainy appear-

ance can also be seen if an ordinary piece of white matte paper is illuminated by a laser operating in the visible spectrum.

This inherent grainy or speckled appearance is usually disturbing to the human image-interpreter since he is trained on images made with incoherent waves that do not have this feature. However, the characteristics of the speckled appearance do have a significance. The received average power from a given direction not only depends upon the reflectance of many small parts of a distant object but also, for coherent waves, upon the geometrical arrangement of these parts relative to the sensor. Those details of arrangement are generally not resolvable but might be revealed through analysis of the speckled character of the image.

DOPPLER EFFECT

If a source of EMR having a fixed frequency, ν, either approaches or recedes from an observer, the observer will receive radiation from the source at a different frequency, ν', where ν' is greater than ν for approaching sources and is less than ν for receding sources. This alteration of EMR frequency caused by relative motion between observer and source is called the *Doppler Effect*. The relationship between the frequency of the source radiation, ν, and the received radiation, ν', is

$$\nu' = \nu \frac{(1 - \beta^2)^{1/2}}{1 - \beta \cos \theta}, \qquad (2\text{-}11)$$

where:
  β = ratio of source velocity to propagation velocity of EMR,
  θ = angle between direction of motion of the source and a line connecting the source and observer.

An approximate relationship between emitted and received frequencies may be used when the relative velocity between source and observer is much less than the velocity of light as is the case for velocities of aircraft and satellites. For such relatively low velocities, the change in frequency, $\Delta\nu$, from that which is emitted is given by

$$\Delta\nu = \frac{\nu u}{c}\cos \theta, \qquad (2\text{-}12)$$

where:
  u = relative velocity between source and receiver,
  c = velocity of EMR.

For approaching sources, $\Delta\nu$ is positive, since θ is an acute angle between 0 and $\pm\pi/2$ radians. For receding sources, $\Delta\nu$ is negative, because θ is an obtuse angle; the cosine of which is a negative number.

The Doppler effect is most frequently used for remote sensing in connection with microwave radar mounted in a moving aircraft. The radar emits frequency, ν, but because of the relative motion of the aircraft to the ground, the microwave frequency observed at the ground is ν'. The radiation reflects from some position on the ground and in turn acts as a source of fixed frequency equal to ν'.

The frequency of the reflected radiation observed at the aircraft is shifted by the Doppler effect once more to ν''. The total change in frequency from that emitted to that received back at the aircraft is twice as much as the single Doppler shift,

$$\Delta \nu = \frac{2u}{\lambda} \cos \theta. \qquad (2\text{-}13)$$

Although the change in frequency will be quite small for aircraft and spacecraft velocities, the use of highly monochromatic radiation and sharply tuned superheterodyne reception makes the change in frequency quite easily measurable and useful for remote sensing purposes. The magnitude of the Doppler effect may be illustrated by a microwave radar mounted in a fast aircraft moving at 300 meters per second. If the radar in the aircraft transmits with a frequency of 3×10^{10} Hz, or a wavelength of about 1 cm, the frequency of the radiation received back at the aircraft from a ground position directly in front of the aircraft would be greater by 6×10^4 Hz. The relative change is only two parts per million in frequency, yet it is easily measurable using heterodyne techniques.

NOMENCLATURE RELATING TO ELECTROMAGNETIC RADIATION

The information to be used for remote sensing is derived from quantitative measurements of the properties of EMR arriving at a sensor, the principle function of which is to make these measurements.

It is helpful to use a common nomenclature and symbolism for expressing these quantitative measurements unambiguously. The nomenclature and symbolism that comes closest to being internationally accepted is that adopted by the International Commission on Illumination (U.S. Stds. Institute, 1967), together with the metric system of units based upon the meter, kilogram, second, and temperature in degrees Kelvin (Mechtly, 1969). Like any living language, nomenclature will surely but gradually incorporate alterations in time, especially as new concepts concerning radiation and its uses become popular; however, the present diversity of nomenclature, symbolism, and units is little better for communication between people than the mixed use of the multitudes of national languages. The detrimental effect of diverse nomenclatures is particularly serious in a cross-disciplinary subject such as remote sensing, in which many persons tend to utilize their own separate technical jargon and symbols, many of which are largely unknown to the others who sincerely wish to know what is being discussed. *The following nomenclature is presented here, not as a rigid system which must never be breached, but as the beginnings of a common language for people of diverse backgrounds who are brought together by their common interest in remote sensing.* Table 2-2 illustrates the nomenclature discussed herein.

RADIATION QUANTITIES

Radiant Energy, Q

Radiant energy, the energy carried by EMR is a measure of the capacity of the radiation to do physical work by moving something by a force, to heat an object, or to cause a change of state of matter. Although radiation is not material, one sometimes speaks of *quantities* of radiation as if it were something tangible to hold and carry about like a bucket of water. Radiant energy causes a sensor-detecting element to change physically in some appropriate manner, that physical change being taken as the evidence for the cause. There is generally no confusion in using the symbol Q to designate energy in all forms rather than just the radiant form. The unit of energy is the joule (unit symbol J).

Radiant Flux, Φ

The term *flux* is defined in Webster's dictionary as "a continuous moving on or passing by." As the term is applied to radiation, it is defined to be the time rate with which radiant energy passes a spatial position. The relationship to the technical term, *power,* which refers to the time rate of doing work or of expending energy, is quite close. Flux is restricted to the flow rate, which conceptually corresponds to water-flow rate past a position along a pipe. The units of radiant flux are either joule/second (unit symbols $J\ s^{-1}$) or most commonly the watt (unit symbol W). The units are the same as those for power (again, both the symbols W and P are commonly found to designate power).

Use of the term *flux* in remote sensing can apply to more than the time rate of flow of energy. For example, one may wish to consider the time rate of flow of the number of quantized energy deliveries (*photons*) to a detector, expressed by the term *photon flux,* or the rate of flow of some visible part of the radiant energy by the term, *luminous flux,* which will be discussed later. When the context makes clear what is flowing, it is not necessary to repeat the modifying terms: radiant, photon, or luminous.

When the radiant flux, Φ, is constant, the total radiant energy, Q, which passes in time, t, is simply

$$Q = \Phi t. \qquad (2\text{-}14)$$

When Φ changes from time to time the total energy is then

$$Q = \int_{t_1}^{t_2} \Phi(t)dt. \qquad (2\text{-}15)$$

TABLE 2-2
Standard Units, Symbols, and Defining Equations for Fundamental Photometric and Radiometric Quantities

Quantity	Symbol	Defining Equation	Commonly Used Units	SI Unit	Symbol
Radiant energy	Q, (Q_e)		erg		
			joule	X	J
			kilowatt-hour		kWh
Radiant energy density	W, (W_e)	$W = dQ/dV$	joule per cubic meter	X	J m^{-3}
			erg per cubic centimeter		erg cm^{-3}
Radiant flux	Φ, (Φ_e)	$\Phi = dQ/dt$	erg per second		erg s^{-1}
Radiant power			watt = joule per second	X	W
Incident flux	(Φ_i)				
Absorbed flux	(Φ_a)				
Reflected flux	(Φ_r)				
Transmitted flux	(Φ_t)				
Radiant flux density at surface					
Irradiance	E, (E_e)	$E = d\Phi/dA$	watt per square meter, etc.	X	W m^{-2}
Radiant exitance (Radiant emittance)	M, (M_e)	$M = d\Phi/dA$	watt per square centimeter		W cm^{-2}
Radiant intensity	I, (I_e)	$I = d\Phi/d\omega$ (ω = solid angle through which flux from point source is radiated)	watt per steradian	X	W sr^{-1}
Radiance	L, (L_e)	$L = d\Phi/d\omega$ $(dA \cos\theta)$ $= dI/(dA \cos\theta)$ (Θ = angle between line of sight and normal to surface considered)	watt per steradian and square centimeter		W sr^{-1} cm^{-2}
			watt per steradian and square meter	X	W sr^{-1} m^{-2}
Emissivity	ϵ	$\epsilon = M/M_{blackbody}$ (M and $M_{blackbody}$ are respectively the radiant exitance of the measured specimen and that of a blackbody at the same temperature as the specimen)	one (numeric)		—

Notes: The symbols for photometric quantities are the same as those for the corresponding radiometric quantities (see above). When it is necessary to differentiate them, the subscripts v and e respectively should be used; e.g., Q_v and Q_e.

Quantities may be restricted to a narrow wavelength band by adding the word spectral and indicating the wavelength. The corresponding symbols are changed by adding a subscript λ, [e.g., Q_λ] for a spectral concentration, or a λ in parentheses; [e.g., $K(\lambda)$] for a function of wavelength].

The units for wavelength in the EM spectrum of interest vary considerably. Early texts used the micron (μ), or 10^{-6} meters; recent texts use the micrometer (μm), also 10^{-6} meters, (that is, the visible spectrum ranges from 0.4 to 0.7 μ or 0.4 to 0.7 μm). Some authors prefer nanometers (nm) or 10^{-9} meters; thus, for the visible spectrum, 400 nm to 700 nm is appropriate. For the infrared spectrum, the use of micrometers largely avoids the need of decimal values, ranging from 0.7 μm to 100 μm. Centimeters (cm) or meters (m) are preferred in the microwave spectrum. The Angstrom unit (Å), or 10^{-10} meters, is used for the X-ray spectrum.

Units symbology will be expressed in this Manual, if possible, in a format to eliminate excessive typesetting. Denominators will be identified by a minus sign with superscript designating the power of denominator quantities; square roots will be identified by superscript of ½. As an example, take the relationship L = 2W sr^{-1} cm^{-2}. This means the radiance of 2 watts per steradian per square centimeter. This *can* be written in several ways:

$$L = 2W/sr/cm^2, \text{ or} \qquad L = 2W/(sr\ cm^2), \text{ or} \qquad L = 2W\ sr^{-1}\ cm^{-2}, \text{ etc.}$$

However, the last form, L = 2W sr^{-1} cm^{-2}, is more direct and flexible.
X indicates that the adjacent unit is an SI unit.

As was stated earlier, the energy which is delivered by radiant flux causes a detector to operate. A certain minimum quantity of energy is needed for a detector to provide the evidence that radiant flux was intercepted. In remote sensing applications, there is concern with both economical and timely methods of knowing "how much of what is where." Therefore, the magnitude of the radiant flux at a point is always of primary interest because of the limitation on the amount of time allowed for each measurement in order to receive the needed minimum energy before moving on to the next point. Thus,

$$\Phi_{min} = Q_{min}/t_{allowed}. \qquad (2\text{-}16)$$

Radiant Flux Density, *E* and *M*

When radiant flux is intercepted by a plane surface, there is a frequent need to know the amount of radiant flux intercepted per unit area of plane surface. The radiant flux intercepted, divided by the area of the plane, is the *average radiant flux density* at the plane. Each small segment of the plane may be visualized as intercepting a small part of the flux; the flux density on each such small segment is the amount of flux intercepted by the segment, divided by the area of the segment. When the segments are taken as infinitesimally small, this ratio can be taken to be the radiant flux density at each point on the surface where that segment is located.

The radiant flux density for flux *incident* upon a plane surface is called *irradiance*, and is represented by the symbol, E (unit symbols W m^{-2}). It is important to note that the direction of flux is not specified as long as it arrives at the point on the surface from any or all directions within a hemisphere over the surface (Fig. 2-7). The irradiance at a surface may be different from point to point. If the irradiance is constant from point to point, then the intercepted flux is simply

$$\Phi = EA, \qquad (2\text{-}17)$$

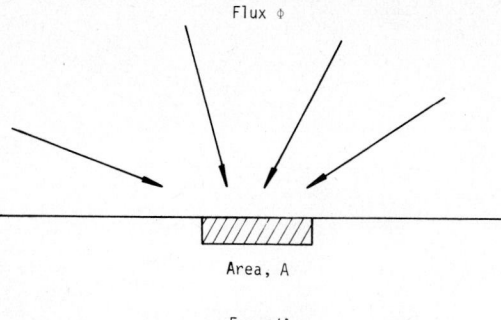

Fig. 2-7. The concept of radiant flux density (irradiance).

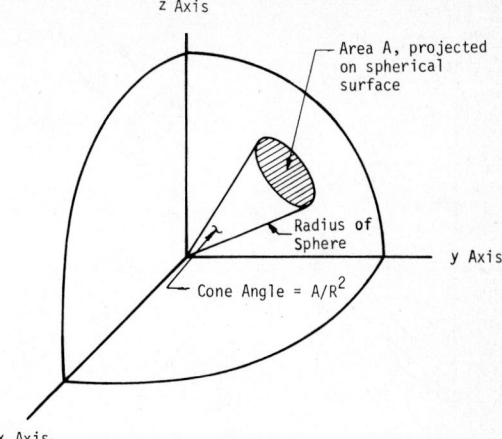

Fig. 2-8. The concept of the solid angle in angular measurement.

where:

A = area of the surface.

While irradiance refers to flux density of radiant flux arriving at a surface, the flux density of radiant flux, *leaving* a surface is called *exitance*[6] and is represented by the symbol, M (units symbols W m^{-2}). The directions taken by the flux are not specified so long as it leaves the point on the surface in any or all directions within a hemisphere over the surface.

The symbol E (which has been adopted for irradiance), is frequently used to represent other concepts. In photographic literature E is used to represent exposure; this is an unfortunate situation which can lead to much confusion. There is a close relationship between irradiance and exposure, but they are different concepts. Exposure is a measure of the radiant energy intercepted by a surface per unit area of that surface.

To correspond with CIE Standards, the reader may wish to substitute H_e for radiant exposure in J m^{-2}; and H_V, for light exposure in lux-second (lx s) where a lux is a lumen m^{-2} (lm m^{-2}).

The symbol, E, also frequently appears in microwave literature to represent the amplitude of the electric field of an EM wave. Here also, considerable confusion can result since the irradiance due to an EM wave will be proportional to the square of the wave amplitude. This author uses (and the text herein may use) the corresponding letter E in script form for the electric field, to remove the possible confusion.

Angular Measure

Before proceeding, it will be worthwhile to briefly review angular measure. Two systems of angular measure are traditional, degrees and radians. The *degree* is the angle subtended by an arc of a circle having 1/360th of the circumference. The *radian* is the angle subtended by an arc of a circle having a length equal to the radius of the circle. Since the circumference of a circle has length equal to 2π times the radius, there are 2π radians in a full circle. The angle subtended by an arc of length, L, in a circle of radius, r, is simply L/r. Both angular measures are dimensionless, but since there are two distinct systems, a *unit* measure is used to distinguish them: for the radian, the unit symbol "*rad*" and for degrees, the unit symbol "°," or the word *degrees*.

The concept of the solid angle of a cone is fundamental to two quantitative measures of radiant flux. It is most common to use the extension of radian measure for this purpose. The cone angle subtended by a part of a spherical surface of area, A, is equal to the area, A, divided by the square of the radius of the sphere. The unit of cone angle or solid angle is the *steradian;* unit symbol, "sr." Solid angle is a dimensionless quantity, but other possible measures could be used, and the use of a unit symbol aids in unit analysis for computations. Since the surface area of a sphere is $4\pi R^2$, there are 4π steradians of solid angle in a sphere. The solid angle concept is illustrated by Figure 2-8.

In remote sensing, there is frequent need to know the solid angle subtended by a flat plate as viewed from a long distance, R. The distance, R, is the radius of the reference sphere, but the area of the flat plate is not exactly the same as the area of the sphere cut by the rays passing the edge of the plate from the center of the sphere. When the distances are large compared to the size of the plate, the area of the plate is sufficiently close to the area of that part of the sphere, so that no significant error is made if the area of the plate is used to compute the solid angle. To illustrate this point, the solid angle of a circular plate with 1-meter radius, as viewed normally from a distance of 10 meters, is $0.993\pi/100$ sr. The area of the plate divided by the square of the range is $1.00\pi/100$. The error is about 0.7%, which is usually very good accuracy for many remote-sensing purposes.

[6] "Exitance" was termed "emittance" in previous works on the subjects.

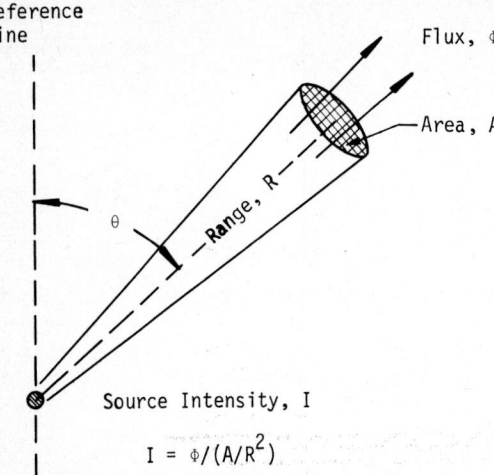

Fig. 2-9. The concept of radiant intensity.

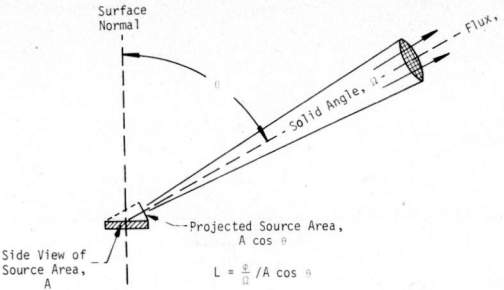

Fig. 2-10. The concept of radiance.

Radiant Intensity, *I*

The *radiant intensity* of a point source in a given direction is the radiant flux per unit solid angle leaving the source in that direction. The symbol for radiant intensity is *I*, (units symbols W sr⁻¹). If a source is isotropic (i.e., it radiates equally well in all directions), then the radiant intensity of an isotropic source emitting flux Φ has a radiant intensity, $I = \Phi/4\pi$ watts per steradian.

The term, radiant intensity, applies to a point source, as shown in Figure 2-9. However, when one states that a source of radiant flux is designated as a point source, it is implied only that the physical extension of the source is not consequential to the discussion. For example, in estimating the irradiance due to a flat circular-plate source of 1-meter radius at a point 10 meters away along the normal to the plate, that plate could be considered as a point source. On the other hand, if the irradiance were to be estimated in the film plane of a camera taking a picture of the plate, the plate could not be considered to be a point source because the spatial extension of the plate would be consequential in the calculation.

The term *intensity* appears in literature on radiation in different ways. In some elementary physics texts it is used both as it is defined here and also to express the concept of irradiance. In some photographic literature the term intensity, and the symbol, *I*, are used for the concept and symbol of irradiance.

Radiance, *L*

Radiance is defined as the radiant flux per unit solid angle leaving an extended source in a given direction per unit projected source area in that direction. See Figure 2-10. The symbol for radiance is *L* (unit symbols W m⁻² sr⁻¹). The concept of radiance is intended to correspond to the concept of brightness. Much of remote sensing is concerned with radiant flux leaving extended

areas in a direction toward the sensor. Consequently, the concept of radiance is very frequently used. Unfortunately, this concept is also the one concept which the novice finds difficult to master.

The projected area in a direction which makes a polar angle, θ, with the normal to an elementary plane segment of area, A, is $A \cos \theta$. If the plane segment is small and can be considered to be a point source, then the radiant intensity of the plane segment in the direction, θ is

$$I = LA \cos \theta. \qquad (2\text{-}18)$$

The plane source for which the radiance, *L*, does not change value as a function of angle of view is called a *Lambertian* source (with reference to Lambert's Laws of Illumination). A piece of white matte paper, when illuminated by diffuse skylight, is a good approximation to a Lambertian source, since the perceived visual brightness of such a piece of paper does not change with angle of view.

It is worthwhile to show radiometrically what is happening when Lambertian sources are viewed at different angles. Consider a Lambertian panel of area, A, viewed at a constant range, R, with a polar angle, θ, from the normal to the panel. Let the pupil of the eye have area A_p. When the panel is viewed at a polar angle, θ, the image on the retina will have only an area $a \cos \theta$, since only a projected area is seen. The radiant flux intercepted by the pupil is

$$\Phi_{\text{intercepted}} = LA \cos \theta \, (A_p/R^2). \qquad (2\text{-}19)$$

Here, *L* is the radiance of the panel in the direction, θ, but *L* is constant with angle because the panel is Lambertian; $A \cos \theta$ is the projected area of source in the direction of view. Thus, $LA \cos \theta$ is the radiant intensity of the panel in the direction of view. The factor (A_p/R^2) is the solid angle subtended by the pupil of the eye from the center of the panel. The radiant flux within that solid angle is intercepted by the eye and is focused evenly over the image of the retina. The irradiance, E_{image}, in the image on the retina is therefore,

$$E_{\text{image}} = \frac{\Phi_{\text{intercepted}}}{a \cos \theta}, \qquad (2\text{-}20)$$

which shows that, with substitution for $\Phi_{\text{intercepted}}$,

$$E_{image} = \frac{LA \cos \theta \, A_p}{R^2 a \cos \theta}$$

$$= LAA_p/(R^2 a). \qquad (2\text{-}21)$$

Therefore, the irradiance in the retinal image does not change with viewing angle for a Lambertian panel. Constant irradiance in the retinal image leads to the perception of constant brightness.

SPECTRAL QUANTITIES

Any complicated wave form can be considered as being composed of sinusoidal component waves or spectral components, each of which transports some share of the radiant flux of the complicated wave form. In discussing the quantities Q, Φ, M, I, and L, it was not stated whether all or only some of these components were being considered. In remote sensing it is usual to consider restricted spectral bands of such spectral components rather than all components together. Sometimes the wavelength or frequency interval under consideration is clear from context. Multispectral sensing, wherein a large number of different spectral bands are considered at one time, is growing in popularity. In such cases, it is important to designate the spectral band of components to which each radiometric symbol applies. If it is desired to denote radiant flux for all components with wavelengths between 0.4 μm and 0.7 μm, as compared to the radiant flux for all components between 0.7 μm and 1.0 μm, then the symbols Φ(0.4 to 0.7 μm) and Φ(0.7 to 1.0 μm) may be used. From this, it is quite clear that the radiant flux for all components in the wavelength interval of 0.4 to 1.0 μm is simply

$$\Phi(0.4 \text{ to } 1.0 \ \mu m) = \Phi \ (0.4 \text{ to } 0.7 \ \mu m) \quad (2\text{-}22)$$
$$+ \ \Phi \ (0.7 \text{ to } 1.0 \ \mu m).$$

An alternative is to designate the bands by band number—such as band 1, band 2, etc., and then the radiant flux for all the spectral components of band 1 is symbolized as $\Phi(1)$.

The use of modifying subscripts to designate bands is not a wise practice for two reasons. The first is that such subscripts could become so complicated as to be unreadable, untypeable, and unprintable. Second, as will be pointed out later, the recommended nomenclature requires the use of certain subscripts to extend the meaning of the nomenclature in a systematic fashion. It is wise to avoid subscript entanglements.

Spectral Radiant Flux, Φ_λ or Φ_ν

One of the important properties of a complicated wave form of EMR is the share of the radiant flux contributed by each sinusoidal or spectral component. The manner in which this flux is distributed among the components of different wavelengths or frequencies is called the *spectral distribution*. The spectral distribution of radiant flux is found by measuring the radiant flux in *a narrow wavelength or frequency* interval

about each wavelength or frequency value through use of *spectrographic* equipment. For example, at wavelength, λ, with wavelength interval, $\Delta\lambda$, $\Phi[\lambda - (\Delta\lambda/2)$ to $\lambda + (\Delta\lambda/2)]$ is measured. It should be apparent that, as the interval becomes smaller, the number of spectral components which are measured also become fewer. The value of $\Phi[\lambda - (\Delta\lambda/2)$ to $\lambda + (\Delta\lambda/2)]$ will become smaller and dependent upon the interval size $\Delta\lambda$. However, by dividing this narrow-band radiant flux by the wavelength interval, the ratio will approach a finite limit as the wavelength interval is made smaller and smaller. This ratio is called the *spectral radiant flux*, and is designated by the symbol Φ_λ for wavelength intervals, or Φ_ν for frequency intervals. (Unit symbols for Φ_λ are W/wavelength unit and, for Φ_ν, W/frequency unit.) The spectral radiant flux values for all wavelengths specify the spectral distribution of the radiant flux.

The spectral radiant flux values at all wavelengths provide all that is needed to calculate the radiant flux in any spectral band. The radiant flux in each narrow band is merely the value of the spectral radiant flux at the band center times the narrow band wavelength interval. The sum of the appropriate adjacent narrow band contribution yields the radiant flux of the wide band.

Using the symbols of calculus,

$$\Phi(\lambda_1 \text{ to } \lambda_2) = \int_{\lambda_1}^{\lambda_2} \Phi_\lambda(\lambda) d\lambda. \qquad (2\text{-}23)$$

Other Spectral Quantities

Just as it may be desirable to designate radiant flux with reference to a restricted band of spectral components, so this may be done for the other radiometric quantities Q, E, M, I, and L. Additionally, the above applies also for the spectral distribution of each of these quantities. When the spectral distribution is to be specified, the modifying word, *spectral*, is used with appropriate quantity. Thus,

Q_λ = spectral radiant energy,
Φ_λ = spectral radiant flux,
E_λ = spectral irradiance,
M_λ = spectral exitance,
I_λ = spectral radiant intensity,
L_λ = spectral radiance.

In each case the subscript, λ, is used to indicate wavelength interval and the subscript, ν, is used to indicate frequency interval. The units of the spectral quantities incorporate the appropriate wavelength or frequency units in addition to the units previously given.

Spectral quantities are most often expressed as functions of wavelength per unit of wavelength interval [indicated by, for example, $\Phi_\lambda(\lambda)$]. For some special reason it may be desired to express the same quantity as a function of frequency per unit frequency interval. The relation between

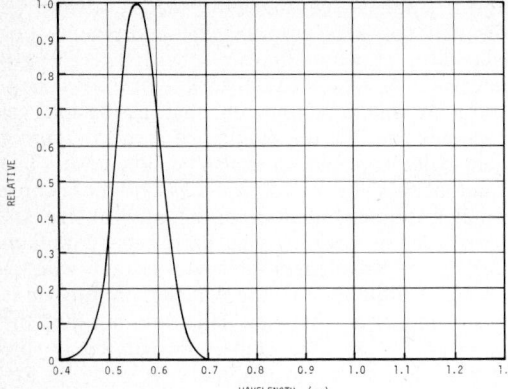

Fig. 2-11. Spectral luminous efficiency, $V(\lambda)$.

$\Phi_\lambda(\lambda)$ and $\Phi_\nu(\nu)$ is easily found from the fundamental relationship between frequency and wavelength, and between frequency interval and wavelength interval, thus,

$$\lambda\nu = c, \text{ and } |\Delta\lambda| = (c/\nu^2) |\Delta\lambda| \qquad (2\text{-}24)$$

and,

$$\Phi_\lambda(\lambda) = \Phi_\nu(\nu)\nu^2 /c \qquad (2\text{-}25)$$

or conversely,

$$\Phi_\nu(\nu) = \Phi_\lambda(\lambda) \lambda^2 /c, \qquad (2\text{-}26)$$

where:

$$c = \text{velocity of light.}$$

In using these relations, care must be given to assure that the units of frequency, wavelength, and velocity of light are compatible.

For many years the units of frequency were given in cycles per second so that the dimensions are fundamentally reciprocal seconds. Recently, a widely accepted change was made to denote the frequency unit by the name *Hertz,* in recognition of the contributions of Heinrich R. Hertz, a German physicist, with a unit symbol of Hz, 1 Hz = 1 cycle per second. Thus, when it is desired to make a dimensional check or a conversion of units, the Hertz should be equivalent to a reciprocal second.

LUMINOUS QUANTITIES

Spectral Luminous Efficiency, $V(\lambda)$

Human vision constitutes the most commonly used remote-sensing system, the human eye being the sensor. The eye responds to EMR spectral components in the 0.4 to 0.7 μm range by photochemical changes in the retina. Some radiant flux with wavelengths longer than 0.7 μm penetrates the eye lens and will be brought to approximate focus on the retina, but the photochemical change will not occur. Nevertheless, large values of irradiance in the spectral range from 0.7 to 0.9 μm can cause permanent thermal damage to the retina even though vision is not stimulated. Radiant flux with wavelengths shorter than 0.4 μm (but with

longer wavelengths than X-rays) do not penetrate to the retina, most of the flux being absorbed in the cornea. Excessive exposure of the eye to this radiation can cause a very serious and painful burn of the cornea.

Within the spectral range 0.4 to 0.7 μm, the photochemical response of the eye is not uniformly efficient in converting radiant flux to a visual response. The relative effectiveness of the daylight-adapted eye in converting radiant flux of different wavelengths to visual response is called the *spectral luminous efficiency* and uses the symbol $V(\lambda)$. Spectral luminous efficiency is a dimensionless quantity having a maximum of unity at about 0.55 μm, and decreasing to small fractions at 0.4 μm and at 0.7 μm. A graph of $V(\lambda)$ illustrating this range of values is shown in Figure 2-11. The visual response capability that can be expected from spectral flux, Φ_λ, can be computed by using this efficiency factor in the relation

$$\Phi_v = (683 \; lm \; W^{-1}) \int_0^\infty \Phi_{e\lambda}(\lambda) \; V(\lambda) \; d\lambda, \qquad (2\text{-}27)$$

where:

$lm \; W^{-1}$ = lumens/watt, and

subscripts v and e are inserted to identify visual or luminous flux, Φ_v, from radiant flux, Φ_e.

Luminous Flux, Φ_v

Luminous flux is a measure of the capability of the radiant flux to produce a visual response. The factor, 683 lumens/watt, converts the radiant unit of flux, the watt, to the luminous unit of flux, the lumen (unit symbol, lm). The lumen is a visual efficiency-weighted flow of radiant flux. While radiant flux is the flow of the capability of radiation to do work or heat an object, luminous flux, as stated, is the flow of the capability to produce a visual response in humans.

USE OF SUBSCRIPTS, SUPERSCRIPTS, AND FUNCTIONAL NOTATIONS

Based on the radiant quantities, Q, Φ, E, M, I, and L, the corresponding luminous quantities may be defined, using the fundamental relationship between Φ_v and Φ_e above, and the same set of symbols. The distinction appears in the use of the subscript "v" for *visual* or luminous quantities, and the subscript "e" for *energy* or radiant quantities. *These subscripts are not needed when context makes clear which quantities are used.* The names for the luminous quantities correspond in most cases to those for radiant quantities:

Q_e = radiant energy $\rightarrow Q_v$ = luminous energy
Φ_e = radiant flux $\rightarrow \Phi_v$ = luminous flux
E_e = irradiance $\rightarrow E_v$ = illuminance
M_e = radiant exitance $\rightarrow M_v$ = luminous exitance
I_e = radiant intensity $\rightarrow I_v$ = luminous intensity
L_e = radiance $\rightarrow L_v$ = luminance

In this context, the following generalized comments may be in order. A numerical subscript generally denotes a specific parameter, subject, or event; e.g., ω_1 = a specific frequency; τ_1 a specific time; s_2, a specific area out of 1, 2, . . . n areas, etc. A variable, such as ω, λ, or θ as a subscript signifies that the parameter or phenomenon takes on a range of values; e.g., E_λ indicates that the emittance (irradiance) value depends on the wavelength λ; whereas a specific value at a given wavelength would be given by $E_{\lambda 1}$. The emittance at a particular wavelength would be a function of the angle of incidence θ, and would therefore be expressed as $E_{\lambda 1}(\theta)$. Superscripts carry the same information and are used primarily for convenience and to reduce mathematical symbology clutter.

RELATIONSHIP BETWEEN LUMINOUS AND RADIANT QUANTITIES

Spectral luminous quantities are defined following a similar pattern, and are related to the spectral radiant quantities through the fundamental relationship of

$$\Phi_{v\lambda} = (683 \ lm \ W^{-1})\Phi_{e\lambda} \ V(\lambda). \qquad (2\text{-}28)$$

The units of the luminous quantities may be formed by interchanging lumens for watts in the corresponding radiometric quantity. Since there is no corresponding luminous unit for the joule, the luminous energy unit is the lumen second.

The great importance of luminous quantities to remote sensing is due to the fact that these quantities appear in descriptive literature relating to sensors other than the human eye. Specifications of photographic film, certain photoemissive detectors, and TV camera tubes employ these quantities in spite of the fact that these sensors rarely have the same spectral response to radiation as does the human eye. One has a right to be confused as to the meaning of such specifications when the spectral responsivity of the sensor is not the same as the human eye. However, the application of luminous quantities in this way is backed

by about one hundred years of tradition in the case of photography, so there is little hope of extrication.

When luminous quantities are used to describe the response of a sensor which is not the human eye, the luminous quantity is used only as a means of specifying the magnitude of a corresponding radiant quantity from a standard lamp. To illustrate, assume that a standard point source lamp has a spectral radiant intensity of $I_{e\lambda}(\lambda)$. The spectral irradiance, $E_{e\lambda}(\lambda)$, on a panel at range, R, with its normal directed at the lamp, is $I_{e\lambda}/R^2$. The illuminance on the panel, E_v, is then

$$E_v = (683 \ lm \ W^{-1}) \int_0^\infty E_{e\lambda}(\lambda) \ V(\lambda) \ d\lambda$$

$$(2\text{-}29)$$

or

$$E_v = \frac{(683 \ lm \ W^{-1})}{R^2} \int_0^\infty I_{e\lambda}(\lambda) \ V(\lambda) \ d\lambda.$$

$$(3\text{-}30)$$

The irradiance from the same source at the same range in some spectral band of interest between wavelengths λ_1 and λ_2 is

$$E_e(\lambda_1 \text{ to } \lambda_2) = \int_{\lambda_1}^{\lambda_2} E_{e\lambda}(\lambda) \ d\lambda \qquad (2\text{-}31)$$

or

$$E_e(\lambda_1 \text{ to } \lambda_2) = \frac{1}{R^2} \int_{\lambda_1}^{\lambda_2} I_{e\lambda}(\lambda) \ d\lambda \qquad (2\text{-}32)$$

Therefore the ratio of the luminous to the radiant quantity is

$$\frac{E_v}{E_e(\lambda_1 \text{ to } \lambda_2)} = \frac{(683 \ lm \ W^{-1}) \int_0^\infty I_{e\lambda}(\lambda) \ V(\lambda) \ d\lambda}{\int_{\lambda_1}^{\lambda_2} I_{e\lambda}(\lambda) \ d\lambda}.$$

As long as the same standard source lamp and wavelengths λ_1 and λ_2 are used, this ratio has a fixed value. The illuminance E_v, is proportional to $E_e(\lambda_1 \text{ to } \lambda_2)$ and can be used as a substitute measure for it. The standard lamp for photographic film tests has the spectral radiant intensity of a 6,000°K blackbody in the visible and near infrared spectral ranges. Such a source has a spectral distribution resembling daylight, as illustrated by Figure 2-12, which shows the spectral distribution for blackbodies at 6,000°, 4,000°, 2,000°, and 1,000°K. However, standard tungsten lamps operating at temperatures between 3,200°K and 2,850°K are sometimes used in tests of photoemissive tubes. Unless the spectral radiant intensity of the standard lamp and the wavelength interval of sensor operation are specified, the relation between the luminous quantity and the radiometric quantity, which it is supposed to measure, is unknown.

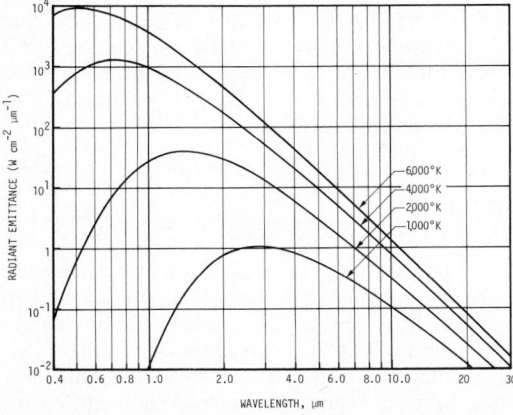

Fig. 2-12. Spectral exitance for blackbodies at 6000, 4000, 2000, and 1000K.

HEMISPHERICAL REFLECTANCE, TRANSMITTANCE, AND ABSORPTANCE

Although a later chapter will deal with the interaction of radiation with matter in more detail, there are three important ratios of radiometric quantities which should be mentioned here. These are the hemispherical reflectance, transmittance, and absorptance for the entire spectrum of radiation.

The *hemispherical reflectance* (symbol ρ), is defined by the dimensionless ratio of the reflected exitance of a plane of material to the irradiance on that plane:

$$\rho = M_{\text{reflected}}/E. \qquad (2\text{-}34)$$

The *hemispherical transmittance* (symbol τ), is defined by the dimensionless ratio of the transmitted exitance, leaving the opposite side of a plane, to the irradiance:

$$\tau = M_{\text{transmitted}}/E. \qquad (2\text{-}35)$$

The *hemispherical absorptance* (symbol α) is defined by the dimensionless relation:

$$\alpha = 1 - \tau - \rho. \qquad (2\text{-}36)$$

These definitions imply that radiant energy must be conserved. Incident flux is either returned back by reflection, transmitted through, or is transformed into some other form of energy inside a panel. The corresponding spectral hemispherical quantities may be defined with use of the spectral exitance and spectral irradiance, so that spectral hemispherical reflectance; $\rho(\lambda)$, is given by

$$\rho(\lambda) = M_{\lambda\text{reflected}}/E_{\lambda}. \qquad (2\text{-}37)$$

Spectral hemispherical transmittance, $\tau(\lambda)$, is given by

$$\tau(\lambda) = M_{\lambda\text{transmitted}}/E_{\lambda}. \qquad (2\text{-}38)$$

and the spectral hemispherical absorptance, $\alpha(\lambda)$, is given by

$$\alpha(\lambda) = 1 - \rho(\lambda) - \tau(\lambda). \qquad (2\text{-}39)$$

In this latter case, one must assume that the internal conversion by fluorescence of radiation of short wavelength to long wavelength does not occur.

The spectral reflectance, transmittance, and absorptance are also dimensionless quantities. The implied wavelength dependence is indicated by the usual function symbol (λ), rather than by subscript, since the use of the subscript λ has been reserved to indicate a differential quantity with a change in units. The radiometric concepts are summarized in Table 2-3.

FACTORS OF TEN NOMENCLATURE

The General Conference on Weights and Measures (Mechtly, 1969) has agreed upon the set of prefixes to units to designate factors of ten, shown in Table 2-4.

For many years, the unit of wavelength equal to 10^{-6} m was the micron. To be consistent with the General Conference, the micron should be changed to micrometer (symbol μm). In addition, the unit of wavelength equal to 10^{-9} m changed from *millimicron* to *nanometer* (symbol nm) which should be used for consistency.

MEASUREMENT OF ELECTROMAGNETIC RADIATION

The means by which the existence of remotely located items or conditions of interest may be inferred is by the measurement of the quantity, quality, and direction of EMR at the position of the observer. Any such measurement is a radiometric measurement by definition. The human visual system is a radiometric system which relies upon the subconscious inferences connecting certain properties of radiation, primarily the direction of flow (shape recognition), and the claim as to the category of item or conditions of interest being viewed. Humans are wholly unconscious of an intermediary radiometric measurement at the retina of the eye. To make it possible to utilize measurements of quantity and quality of radiation which are beyond the capacity of the human visual system, the inferential connection between radiometric measurement by a nonvisual sensor and the category of items (or conditions of interest) must be a conscious and intellectual process for which an extensive and precise system of nomenclature and units is essential.

RESPONSIVITY, R

The output of a radiation detector is an observable physical change, response, or signal indicative of the character of the incident flux on the sensor. The ideal radiometer for incoherent radiation would be one that produces a signal in proportion to the radiant flux arriving at the sensor from a well defined direction within a known solid angle or cone angle including that direction, and which contains spectral components only in some spectral band between two wavelengths, λ_1 and λ_2. The proportionality constant, R, relating signal out of the radiometer to incident flux having these qualities, is called the *radiometer responsivity*, also symbolized as R. Thus

$$S = R\Phi(\lambda_1 \text{ to } \lambda_2), \qquad (2\text{-}40)$$

where:
   S = value of the signal.

Responsivity can be determined by a calibration experiment wherein a source supplies a known incident flux, $\Phi(\lambda_1 \text{ to } \lambda_2)$, and the radiation detector responds with the corresponding signal for that flux. The value of R is determined so that one may infer from the signal the value of incident flux for unknown sources.

Actually, radiometers are not equally respon-

TABLE 2-3
Summary of Radiometric Concepts

Name	Symbol	Units	Concept
Radiant energy	Q_e	Joules, J	Capacity of radiation within a specified spectral band to do work.
Radiant flux	Φ_e	watts, W	Time rate of flow of energy on to, off of, or through a surface.
Radiant flux density at surface Irradiance	E_e	watts per square meter, W m^{-2}	Radiant flux incident upon a surface per unit area of that surface.
Radiant exitance	M_e	watts per square meter, W m^{-2}	Radiant flux leaving a surface per unit area of that surface.
Radiant intensity	I_e	watts per steradian, W sr^{-1}	Radiant flux leaving a small source per unit solid angle in a specified direction.
Radiance	L_e	watts per steradian per square meter, W sr^{-1} m^{-2}	Radiant intensity per unit of projected source area in a specified direction.
Hemispherical reflectance	ρ_e	dimensionless	Φ_e reflected/Φ_e incident for any surface.
Hemispherical transmittance	τ_e	dimensionless	Φ_e trans./Φ_e incident for any surface.
Hemispherical absorptance	α_e	dimensionless	Φ_e absorbed/Φ_e incident for any surface.

Note: The subscript, e, is not used when context indicates that the symbols represent radiometric quantities. The word "Spectral" precedes the name of every radiometric quantity to obtain the names of the spectral quantities. Subscripts λ or ν are required on spectral quantity symbols to denote the differential properties and to emphasize the dimensional differences, except in the case of spectral reflectance, transmittance, and absorptance, which remain dimensionless.

sive to radiant flux at all wavelengths within their operating range. Their responsivity to radiant flux depends upon the wavelength of the incident flux. The calibration of a radiometer can be done for flux at every wavelength by laboratory test. The set of responsivity values for each wavelength is called the *spectral responsivity*, and is symbolized as $R(\lambda)$. A small amount of radiant flux in a narrow spectral band, $\Delta\lambda$, centered at wavelength, λ_1, when applied to a radiometer produces a small signal, ΔS, so that

$$\Delta S_1 = R(\lambda_1)\, \Phi_\lambda(\lambda_1)\, \Delta\lambda. \qquad (2\text{-}41)$$

By changing the center wavelength to λ_2, a small signal, ΔS_2, is obtained:

$$\Delta S_2 = R(\lambda_2)\, \Phi_\lambda(\lambda_2)\, \Delta\lambda. \qquad (2\text{-}42)$$

An important assumption is made concerning the simultaneous response to flux in two different spectral regions by actual radiometers. That is, the signal resulting from simultaneous exposure to flux in different spectral bands is the sum of the signals which would result from exposure to each separately.

Therefore, it is assumed that when the flux from the two spectral bands centered at λ_1 and λ_2 are both incident upon the radiometer, the resulting signal, $\Delta S_{\text{combined}}$, is

$$\Delta S_{\text{combined}} = \Delta S_1 + \Delta S_2, \text{ or} \qquad (2\text{-}43)$$

$$\Delta S_{\text{combined}} = \big[R(\lambda_1)\, \Phi_\lambda(\lambda_1) \\ + R(\lambda_2)\, \Phi_\lambda(\lambda_2) \big]\, \Delta\lambda. \qquad (2\text{-}44)$$

The signal for incident broad-band spectral flux is then, in general,

$$S = \int_0^\infty R(\lambda)\, \Phi_\lambda(\lambda)\, d\lambda. \qquad (2\text{-}45)$$

It is important to note that the signal, S, is not uniquely indicative of the value of the incident flux, $\Phi(\lambda_1 \text{ to } \lambda_2)$, which is given by

$$\Phi = \int_{\lambda_1}^{\lambda_2} \Phi_\lambda(\lambda)\, d\lambda \qquad (2\text{-}46)$$

TABLE 2-4
Factors-of-Ten Nomenclature

Factors of Ten	Prefix	Symbol
10^{12}	tera	T
10^9	giga	G
10^6	mega	M
10^3	kilo	k
10	deka	da
10^{-1}	deci	d
10^{-2}	centi	c
10^{-3}	milli	m
10^{-6}	micro	μ
10^{-9}	nano	n
10^{-12}	pico	p
10^{-15}	femto	f
10^{-18}	atto	a

unless $R(\lambda)$ does not change with wavelength in the range λ_1 to λ_2 and is otherwise zero. Yet, it is the value of signal, S, which the remote sensor applies rather than the value of Φ.

NORMALIZATION

Therefore, real radiometers, the spectral responsivities of which are not constant over the spectral range in which they respond, have an inherent inaccuracy. It is important to devise a measurement procedure which minimizes this inaccuracy. The plan is to find a means of calculating an average or effective spectral band of operation. Such a calculation is said to *normalize* the spectral responsivity.

The most frequently used normalization is called *normalization to the peak*. The spectral responsivity of a real radiometer generally has a maximum (or peak) value for flux at one wavelength with reduced values for flux at other wavelengths. Let the maximum responsivity be symbolized as R_{peak}. Now one might infer the value of incident flux from the signal value by the approximate relation

$$S \cong R_{peak}\, \Phi(\lambda_1 \text{ to } \lambda_2), \qquad (2\text{-}47)$$

where:

λ_1 and λ_2 are the effective spectral limits of response of the radiometer to be determined by normalization.

The effective spectral bandwidth, $(\lambda_2 - \lambda_1)$, of a real radiometer is found by the relation,

$$R_{peak}(\lambda_2 - \lambda_1) = \int_0^\infty R(\lambda)\, d\lambda. \qquad (2\text{-}48)$$

The actual limits of response of a real radiometer are spectrally farther apart than the effective bandwidth. Figure 2-13A shows a hypothetical graph of spectral flux incident upon a radiometer; Figure 2-13B shows the spectral responsivity, $R(\lambda)$, of the same radiometer by the broken line and the effective band width by the solid line, using the relationship expressed by Eq. 2-48. The signal produced by the radiometer, according to the relationship of Eq. 2-45, is equal to the area under the broken-line curve in Figure 2-13C. The area under the solid curve shown as cross-hatched, is the quantity $R_{peak}\, \Phi(\lambda_1 \text{ to } \lambda_2)$. It can be seen that the signal is very nearly equal to the value of $R_{peak}\, \Phi(\lambda_1 \text{ to } \lambda_2)$, so that the quantity $\Phi(\lambda_1 \text{ to } \lambda_2)$ may be deduced from the value of S/R_{peak} with good accuracy.

To illustrate how the same radiometer would operate against some quite different spectral flux, consider the spectral flux shown in Figure 2-13A′ in conjunction with the same radiometer response and effective bandwidth shown in Figure 2-13B′. The area under the broken curve of Figure 2-13C′ is the signal, S, and the cross-hatched area is $R_{peak}\Phi(\lambda_1 \text{ to } \lambda_2)$. Again, these two are nearly equal, in spite of the difference in shape of the two

areas, so that $\Phi(\lambda_1 \text{ to } \lambda_2) = S/R_{peak}$ with good accuracy.

PRODUCTION OF ELECTROMAGNETIC RADIATION

RADIATING STRUCTURES

Fundamentally, EMR is generated by changing the size or direction of an electric or magnetic field with time; this is accomplished by changes in the source of the field. For instance, one can twirl a simple bar magnet about an axis and generate EMR which has the frequency of rotation. However, the radiant intensity of such a weak source would be practically unmeasureable. In similar fashion, two separate electrically charged spheres, one with positive charge and one with negative charge, can be rapidly moved cyclically, and, by the interchanging of their positions, causing a cyclic reversal of electric field direction near the spheres. Again, such a crude mechanical radiator would not be very effective but, nevertheless, would radiate.

Radiation from Wire Antennas and Metallic Structures

A much more effective means is commonly used to obtain radiation. Two straight metallic wire segments are placed in a line with the two adjacent ends connected to the two terminals of an alternating current generator. The generator removes electric charge from one wire and pushes it onto the other, then reverses and charges the wires oppositely in a cyclical manner. Instead of moving the materials holding the electric charge, the electric charge only is moved, which requires considerably less force. In addition, the center of excess charge moves onto and down the wires at nearly the speed of light. This arrangement also creates a rapidly changing magnetic field to produce radiation. The motion of the center of charge down the wire constitutes an electric current that creates a magnetic field around the wire; the magnetic field changes direction with the change in direction of the current. This simple metallic structure is called a dipole antenna.

When the overall length of the structure is one-half the wavelength or an integer multiple of one-half wavelengths of the radiation being radiated, the structure is electrically resonant and radiates energy supplied by the generator copiously. When the structure is much shorter than one-half wavelength of the wave to be radiated, the conversion efficiency is reduced; the generator must supply much more current to the structure to obtain the same quantity of radiation. The result is the expenditure of considerable excess energy in heating the structure, while converting generator power to radiant flux.

The polarization of the radiated wave will be governed by the orientation of the dipole, the electric field having the same direction as the separated electric charges on the two wire elements.

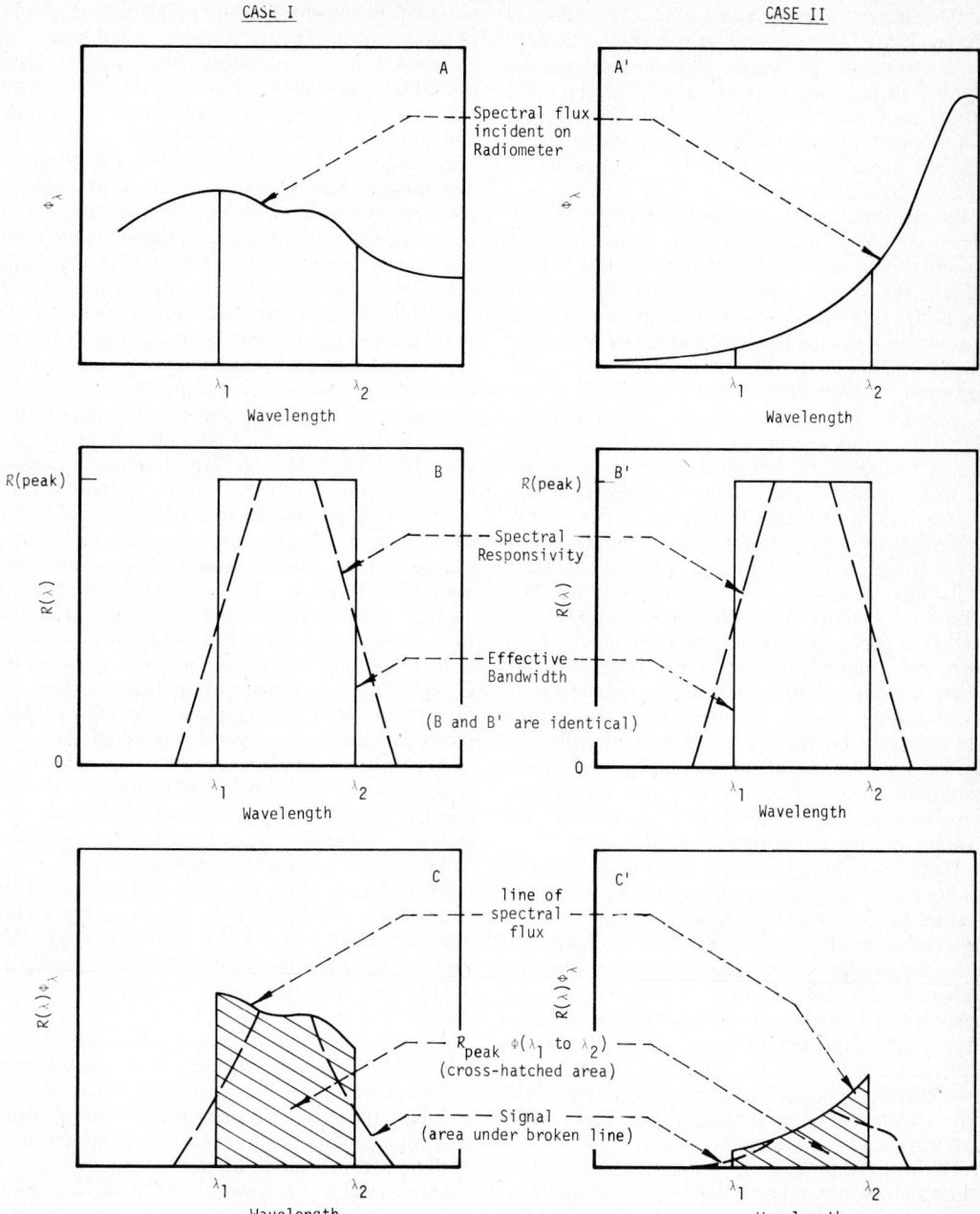

Fig. 2-13. Effect of normalization in maintaining radiometric accuracy.

Thus, a dipole radiator oriented horizontally to the terrain will radiate horizontally polarized EM waves.

Microwave radiators can be expected to be small structures because the wavelengths are short. For wavelengths of the order of a millimeter, simple dipoles become unfeasible to manufacture with good mechanical tolerance and fail to perform well for other practical reasons. Consequently, much more elaborate metallic struc-

tures are used, which may be larger than one half wavelength.

Radiation from Atomic Structures

All of the atoms of which matter is composed are, themselves, composed of moving charged particles—the negatively charged orbiting electrons and the more massive positively charged nuclei. The orbiting electrons can move about the nucleus in certain particular ways so that there is no time rate of change of magnetic or electric field around the atom. For such particular motions, the atoms will not radiate. However, other possible orbital motions do result in a rapidly changing electric and/or magnetic field around the atom, so that an atom must radiate EM waves while executing such motions.

Every orbital motion of electrons about the nucleus contains a certain amount of total energy, potential plus kinetic. In order to change orbital motion, the total energy must change. Since there are no friction bearings or brakes in an atom, the only way to change the total energy of motion of atoms is by collision with other particles, or by the radiation or absorption of EM force fields.

An important discovery concerning the radiation from atoms was the fact that the frequency of atomic radiation did not depend upon the details of the orbital motion, but only upon the total energy loss in changing from a high-energy nonradiating orbit to a lower energy nonradiating orbit through the sequence of intermediate radiating orbits. Thus, the frequencies of radiation with which atoms can radiate are uniquely connected to the set of nonradiating orbits which, in turn, are uniquely related to the type of atom. When a cloud of atoms is made to radiate by intense heating, the set of frequencies of the emitted radiations is all that is needed to identify the atomic types in the cloud. Typically, the atomic emissions which are due to electron orbit changes have wavelengths in the short wave infrared, the visible, and the ultraviolet spectral ranges.

There is nothing in the radiation process of atoms which cannot operate in reverse. If the correct frequency of EM force field is made incident upon an atom, the electrons may be accelerated to a higher energy orbit. The atom may then reradiate later or lose energy by collision with other particles. If a spectrogram is made of the radiation from a radiating cloud of atoms, the spectrograph will record a line on the spectrogram for every characteristic emitted frequency. Because of the appearance of such spectrograms, the atomic gases are said to radiate *line spectra*.

The reverse process of absorption also occurs when radiation is passed through an atomic gas. For example, sunlight leaving the surface of the Sun contains all spectral components. Surrounding the Sun is an atmosphere of atomic gas which absorbs certain spectral components having the characteristic frequencies for the atoms in that atmosphere. Thus, we receive on Earth the sunlight with missing spectral components, and a spectrogram will show *dark* or missing spectral lines for those wavelengths which are characteristic of the atoms in the solar atmosphere. These missing spectral components in sunlight are called *Fraunhofer Lines*—named after the man who discovered them.

Radiation from Molecules

Molecules are two or more atoms joined together by orbiting electrons or by other types of bonding. Because the orbits of the outer electrons of atoms are markedly changed when atoms are combined into molecules, the characteristic radiating frequencies are also markedly changed. For example, the spectrum of atomic sodium gas and atomic chlorine gas does not resemble the spectrum of sodium chloride gas, which has its own unique characteristic frequencies. In addition to these new characteristic frequencies arising from electron motion changes, the molecule offers additional possibilities for radiating because of the relative motion between the atoms of the molecule. Two atoms in chemical bond can vibrate along a line joining them; and just as in electron orbital motion, this change in vibrational motion can also produce radiation. Likewise, a change in rotational motion of atom about atom offers a third means of radiating. The number of possible characteristic frequencies increases rapidly as the number of atoms in the molecule increases. Even simple molecules, such as carbon dioxide and water, have an extremely great number of possible different characteristic frequencies. Generally, these characteristic frequencies are very close together in spectral groups so that a spectrometer of the ordinary sort will not be able to resolve the spectral lines in a group. Groups of characteristic frequencies occupy certain bands of the spectrum; for this reason, molecules are said to produce band spectra.

Absorption in molecules (the reverse action of radiation) occurs with equal facility. Molecular spectroscopy for the identification of molecular gases almost invariably utilizes absorption spectra rather than emission spectra for a very practical reason. Many molecules, particularly large organic molecules, break apart upon heating, so that emission spectra would be the result of the broken parts rather than the whole molecule. However, absorption spectra can be obtained with cool gases.

The polarization of radiation from gaseous molecules and atoms corresponds to the direction of the changing motion causing the emission. But, in a gas, there is no regular orientation of atoms and molecules; for this reason, polarization of radiation from a cloud of gas is random.

In molecules, electron orbital motion changes produce the shortest wavelength radiation; the vibrational motion changes produce typically short- and intermediate-range infrared radiation; the rotational motion changes produce long wavelength infrared radiation extending into the short wavelength microwave range.

Radiation from Liquids and Solids

Liquids and solids are composed of closely packed atoms or molecules. In solids, the atoms or molecules tend to pack together in an ordered array called a crystal lattice. Liquids are not orderly arrangements and permit atoms to move from place to place with more freedom than in a solid. Because of the close proximity of atoms in liquids and solids, the force fields of the neighboring atoms distort the electron orbits of each other so that a large number of different characteristic frequencies may be generated. Just as in molecules, there are a vast number of ways in which the atoms of solids and liquids may vibrate relative to each other. Consequently, the spectral properties of liquids and solids differ from those of the gaseous state because of the influence of the neighboring members of the ensemble. The spectral lines of a solid and liquid are so closely spaced together in wavelength that they generally overlap and cannot be resolved at all. In some crystalline solids the ordered array may result in different radiative properties, depending upon the polarization of radiation incident upon them.

The unresolvable spectra of solids and liquids are called *continuous spectra;* they have fewer resolvable features by which identification can be accomplished. Although absorption or emission band edges may be well defined, within such bands the spectral detail is slight as compared to the spectra of gases.

Radiation from Plasmas

Matter can be heated to such high temperatures that the electrons which normally move in captured nonradiating orbits are broken free. Every encounter of one of the free electrons with a positively charged nucleus is similar to the encounter of a comet with the Sun, rapidly changing electric and magnetic fields being caused by each encounter, so that radiation at all wavelengths can be produced. The hot surface of the Sun is largely a plasma in which radiation of all wavelengths is produced. The spectra of a plasma is a continuous spectrum.

THERMAL EMISSION

Heat energy is the kinetic energy of random motion of the particles of matter, and the concentration of heat energy in a substance is measured by its temperature. The random motion results in particle collisions causing changes in the electron orbital motions or in the vibrational and rotational motions of the molecular or atomic particles. Higher energy states of motion caused by collisions may spontaneously change to lower energy motions with the emission of EMR. In this way, heat energy may be changed into radiant energy.

Blackbodies

An ideal thermal emitter, called a *blackbody,* is one which transforms heat energy into radiant energy with the maximum rate permitted by thermodynamic laws; this is a useful concept since it establishes the maximum conversion rate possible when emission is due to heat energy conversion. Planck derived a formula for the spectral exitance which a blackbody should have, based upon theoretical thermodynamic reasoning. Any real material at the same temperature cannot emit thermally at a rate in excess of that of a blackbody. The reverse process of absorption must

likewise be maximum so that a blackbody must absorb and convert all incident radiant energy into heat energy regardless of the spectral band of the radiant energy. Planck's formula for blackbody spectral exitance, M_λ, is given as,

$$M_\lambda = c_1 \lambda^{-5} \left[\exp(c_2/\lambda T) - 1 \right]^{-1}, \quad (2\text{-}49)$$

where:
  $c_1 = 3.74 \times 10^{-16}$ W m^2
  $c_2 = 1.44 \times 10^{-2}$ m °K
  λ = wavelength in meters
  T = absolute temperature in degrees Kelvin.

The spectral exitance of a blackbody at a given temperature is not the same at all wavelengths. For very long and very short wavelengths, the spectral exitance is low, as shown in Figure 2-13. For some wavelength in between, the spectral exitance reaches a maximum value depending upon the temperature of the blackbody. That wavelength, λ_m, for which the spectral exitance has its maximum value, is given by the Wien Displacement Law

$$\lambda_m = C/T, \quad (2\text{-}50)$$

where:
  $C = 2.898 \times 10^{-3}$ m °K.

The exitance of a blackbody for the spectral band of all wavelengths is given by the Stefan-Boltzmann Law,

$$M_{(\text{blackbody at } T)} = \sigma T^4, \quad (2\text{-}51)$$

where:
  $\sigma = 5.669 \times 10^{-8}$ W M^{-2} ° K^{-4}.

Spectral Emissivity, $\epsilon(\lambda)$

Any body of material at a specific temperature emits radiation in accordance with its own characteristics. It is convenient and customary to express the capability to emit radiation due to thermal-energy conversion as the ratio of the spectral exitance of a material to the spectral exitance of a blackbody at the same temperature. This ratio is called the *spectral emissivity* of the material, and is given by the symbol, $\epsilon(\lambda)$. Thus

$$\epsilon(\lambda) = \frac{M_{\lambda(\text{material, °K})}}{M_{\lambda(\text{blackbody, °K})}}. \quad (2\text{-}52)$$

The spectral emissivity is nearly independent of temperature for most common materials and for temperatures of a terrestrial environment. Significant changes in spectral emissivity can be expected whenever the material undergoes a change of state such as melting, vaporization, oxidation, or any other change which alters the fundamental arrangement of the atomic and molecular components.

For those conditions where the spectral emissivity does not change significantly with temperature, the spectral exitance of any material at any temperature can be calculated if the spectral emissivity is known for the materials. These calculations are simply done by use of the defining relations above and the blackbody spectral exitance

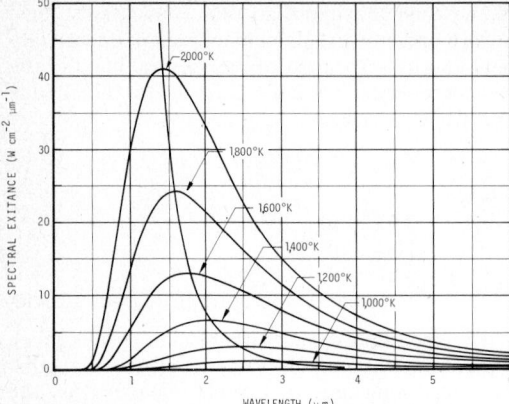

Fig. 2-14. Spectral exitance for blackbodies at 1,000, 1,200, 1,400, 1,600, and 2,000K.

formula, or a suitable blackbody spectral exitance table of values. For example, the spectral exitance of blackbodies at 1,000°, 1,200°, 1,400°, 1,600°, 1,800°, and 2,000°K is illustrated by Figure 2-14.

Kirchhoff's law states that, under conditions of thermal equilibrium, the spectral emissivity of a material must be equal to the spectral absorptance of that material. Experience has shown that Kirchhoff's law holds with good accuracy, even in conditions where thermal equilibrium is not present, provided that temperature differences are not extreme. One can expect Kirchhoff's Law to apply to most terrestrial conditions. Therefore, it is common practice to determine the spectral emissivity of a substance by measuring the spectral absorptance of that substance.

The thermal emission spectrum contains the same spectral details as does the spectral absorptance. Thus, the thermal emission spectra of solids and liquids have fewer spectral details by which they can be identified. Many common materials are spectrally similar in the 8 to 14-μm infrared range, with a spectral emissivity between 0.85 and 0.95. In the spectral range from 3 to 6 μm, the spectral emissivity of those materials tends to differ considerably. When a temperature measurement is made of a terrestrial material by remote means, it is wise to use the 8 to 14-μm band, where it may be assumed that the emissivity of the various terrestrial materials are similar; and an estimated emissivity of 0.90 is probably not far from the truth.

Blackbody Standards

It is frequently necessary to utilize a source of radiation, the spectral radiance of which is known, in order to calibrate a remote sensing apparatus. For microwave systems, these sources are signal generators with known frequency and power output. For the infrared and visible spectral ranges, blackbody cavities at a known temperature are used in conjunction with Planck's formula. Any closed cavity with opaque walls coated

with an absorbing material produces blackbody radiation within the cavity, provided only that the walls are all at the same temperature. A small hole in one wall will allow a small part of the radiation to escape, the spectral radiant exitance of that hole being given by Planck's formula if the hole is very small. Even a novice can construct reliable blackbody standards out of old tin cans with constant temperature water jackets if the blackbody temperature is near the ambient temperature.

If considerable flux is to be generated in the visible spectral range, the blackbody temperature should be above 1,000°K. A blackbody cavity operating at these temperatures poses numerous practical problems with which the novice should not tangle. Commercial blackbody sources are available which may be checked against the accurate high-temperature blackbody source of the National Bureau of Standards.

A true blackbody surface is necessarily a Lambertian radiator, so that the relation between spectral radiance and spectral exitance is simply

$$L_{\lambda \text{ blackbody}} = M_{\lambda \text{ blackbody}}/\pi. \qquad (2\text{-}53)$$

Practical blackbody standards are not necessarily Lambertian radiators unless the cavity hole perimeter lies in a plane and the depth of the hole in the cavity wall is small as compared with the width of the hole.

Apparent, Brightness, and Color Temperatures

Remotely sensed temperature readings form one of the important remote-sensing tasks. The radiant flux arriving at a sensor from a remote object is measured.

The *apparent temperature* of a surface is the temperature of a blackbody surface which, when placed in front of the received aperture, would produce the same received flux within the spectral band of the receiver. Thus, the apparent temperature of a surface is determined not only by the actual temperature of the surface material but also by the emissivity of the material, the atmospheric effects between surface and sensor, and the spectral band used by the receiver.

The *brightness temperature* of a surface is the same as the apparent temperature. The former term is frequently used in literature relating to microwaves, while apparent temperature is more often used in literature relating to infrared applications.

Color temperature may have a number of meanings depending upon context. If the ratio of emitted spectral radiance of a surface is measured at two different wavelengths, the color temperature is the temperature of a blackbody surface which will produce that same spectral radiance ratio at these two wavelengths. For visual contexts the color temperature of a source is the temperature of a blackbody which will produce the same visual color as that source.

It is important to remember that most materials in a natural environment do not produce blackbody radiation because of the variations of the spectral emissivity with wavelength and, in addition, most objects do not have one temperature only; the temperature varies over the surface and below the surface. Radiant flux may originate within the material if it is partially transparent to the emitted radiation. When such temperature and emissivity variations are not large, the emitted flux is spectrally similar to blackbody radiation in certain restricted spectral bands. Under such conditions the three concepts of temperature-apparent, brightness and color are useful.

FLUORESCENT AND LASER EMISSION

Fluorescence

Many materials are capable of converting absorbed radiant energy at one wavelength to emitted radiation at a longer wavelength without first converting the absorbed energy into thermal energy; a process called *fluorescence*.

Absorption of radiant energy causes electrons orbiting in low-energy orbits of an atom or molecule to change to higher energy orbits. Suppose that, in a molecule, the absorption of radiation causes an electron to change from nonradiating orbit number 1 to orbit number 3. The radiant flux supplies energy Q(1 to 3). Since radiant flux is delivered in quantized units of energy (photons), the frequency of radiant energy absorbed must have been ν(1 to 3), as given by

$$Q(1 \text{ to } 3) = h\nu(1 \text{ to } 3), \qquad (2\text{-}54)$$

where:
  h = Planck's constant.

The extra energy can leave the molecule in three ways. The first, and most common, is to release the extra energy in collision with a neighboring molecule, so that Q(1 to 3) becomes thermal energy of random motion. The second is that the electron immediately radiates the energy by making the reverse change. Except for a slight delay, this amounts to no net absorption. The third is that the reverse change occurs in steps—first a change from orbit number three to an intermediate orbit number two, releasing energy Q(3 to 2), and then from orbit two to orbit one, releasing energy Q(2 to 1). If either of these intermediate changes releases energy by radiation rather than by collision then a lower frequency radiation will be emitted so that

$$Q(3 \text{ to } 2) = h\nu(3 \text{ to } 2), \qquad (2\text{-}55)$$

or

$$Q(2 \text{ to } 1) = h\nu(2 \text{ to } 1). \qquad (2\text{-}56)$$

Since all energy must be accounted for,

$$Q(3 \text{ to } 1) = Q(3 \text{ to } 2) + Q(2 \text{ to } 1) \qquad (2\text{-}57)$$

so that $h\nu$(3 to 2) or $h\nu$(2 to 1) must be smaller in value than $h\nu$(3 to 1).

In order for an observer to observe fluorescence, the emitted radiation must find its way from the fluorescing molecule through neighboring molecules to the observer. If neighboring molecules are of the same kind, they are capable of absorbing whatever the fluorescing molecule emits. With the reabsorption of fluorescent radiation, there is increased probability that the energy will eventually be lost by orbital changes due to collisions before it reaches the outside boundary of the material. If the fluorescing molecule is a trace impurity dissolved in a solid or liquid solution of another material which is transparent to the fluorescent radiation, then fluorescence from deep within the material can be observed. The fluorescence is not characteristic of the solvent but of the trace materials as influenced by the solvent. Many minerals are fluorescent in the visible spectral range when exposed to ultraviolet light because of the trace elements dissolved in the mineral.

Chlorophyll in green plants absorbs visible radiation in the 0.4- to 0.68-μm range and converts most of that energy into chemical energy. However, some of that absorbed energy is converted to infrared fluorescence in the 0.7- to 0.85-μm range. The fluorescent radiation is quite small compared to the reflected infrared radiation from green leaves, so that the cause of the bright white appearance of foliage in infrared black and white photography is not due to fluorescence.

The fluorescence of materials of the environment is found in the ultraviolet, visible, and near infrared spectral ranges characteristic of radiation caused by electron orbital motion changes. Fluorescence in the longer wavelength infrared and in the microwave spectral bands would arise from changes in molecular vibration and rotation motions. However, these motions are subjected to collision interference in considerable degree, so that fluorescence is rarely achieved in these spectral ranges except under artificial circumstances. Radiant energy is absorbed and transformed into thermal kinetic energy before reradiation occurs on these spectral ranges.

Lasers

Specialized fluorescent materials can be used as the basis for *lasers,* a term derived from the acronym for "Light Amplification by Stimulated Emitted Radiation." The two major requirements are that loss of absorbed energy by an atom or molecule by collision should be much less likely than normal, and that the step-wise orbital changes from high-energy orbit to low-energy orbit should pause at some intermediate orbit longer than normal, unless radiation having the characteristic frequency for the next step-wise change is present.

When these conditions are met, and the fluorescent material is placed in a cavity which favors the maintenance of a high level of the needed characteristic radiation, then fluorescence occurs readily and the contributions of each atom or molecule are in time with, in the direction of, and with the same polarization as the cavity radiation. Coherent optical radiation can be generated in this way. Thus, a laser can produce coherent, very nearly monochromatic, polarized optical radiation with a precise direction of propagation limited only by the diffraction of the wave as it exits through the laser cavity aperture. A laser permits the application of radar microwave techniques to the infrared, visible, and ultraviolet spectral ranges.

REFERENCES

Jenkins, F., and H. White, 1950, Fundamentals of Optics, Second Edition, McGraw-Hill Book Co.

Mechtly, E. A., 1969, The International Systems of Units, NASA SP-7012, Office of Technology Utilization, NASA, Washington, D.C.

U.S. Standards Institute, 1967 (August 16), Standard Nomenclature and Definition for Illuminating Engineering, RP-16 (USAS 27.1-1967).

CHAPTER 3

Matter-Energy Interaction in the Optical Region

Author: JAMES A. SMITH

GENERAL CONTENTS: The energy flow process; mathematical framework; radiative transfer and energy budget equations; direct solution and difficulties in specifying the terrain parameters; modeling approximations; modeling summary; indirect sensing of terrain characteristics; overview of measured terrain feature response parameters; vegetation; non-vegetation; other sources of data; future research directions; references.

NOMENCLATURE

To conserve and eliminate repetition in text and references, the following symbols, units and names have been used in this chapter.

Symbol	SI Units	Name
A	m^2	Geometric or cross-sectional area
A	W m^{-2}	Shortwave absorption component
A	—	Shortwave albedo
A_1	m^2	Source area
A_2	m^2	Receiver area
a	—	Crown-to-length ratio
a	m^{-1}	Attenuation coefficient
a	m^2	Surface area
a_d	—	Diffuse albedo
a_{net}	—	Total net albedo
a_s	—	Hemispherical albedo
$B(X)$	W m^{-2}	Longwave exitance component
b	—	Bole-to-crown ratio
b	m^{-1}	Attenuation coefficient
C	J Kg^{-1} K^{-1}	Heat capacity
C_d	—	Drag coefficient
C_p	J Kg^{-1} K^{-1}	Heat capacity at constant pressure
c	m^{-1}	Attenuation coefficient
c'	m^{-1}	Attenuation coefficient
D	m	Depth of medium
d	m	Diameter or thickness
d	W m^{-2} sr^{-1}	Diffuse sky radiance
$d_{mn}(i,j)$	—	Geometric form-factor matrix
E	m^{-1}	Attenuation factor
E	W m^{-2}	Evaporation heat exchange component
E	—	Exponential function
E	W m^{-2}	Irradiance
E	W m^{-2}	Power per unit area
E_b	W m^{-2}	Longwave flux from tree bole
E_c	W m^{-2}	Longwave flux from tree crown
e	bar	Vapor pressure of air
F	W m^{-2}	Incident beam irradiance; Flux
F_0	W m^{-2}	Solar irradiance
$F_{1,2}$	—	View factor
f	sr^{-1}	Bidirectional reflectance distribution function (BRDF)
$f_r(\theta_i,\phi_i;$ $\theta_r,\phi_r)$	sr^{-1}	BRDF
$f(\theta_0,\phi_0)$	—	Hemispherical-directional reflectance
f	—	Volume fraction
G	W m^{-2}	Soil conduction term
G_b	W m^{-2}	Longwave flux from bole
G_c	W m^{-2}	Longwave flux from crown
G_s	W m^{-2}	Longwave flux received by snowpack
G_z	W	Total radiant power received
g	—	Asymmetry factor
g	m^{-1}	Mean foliage projection
g	W m^{-2}sr^{-1}	Background radiance
H	m^{-1}	Horizontal leaf projection factor
H	W m^{-2}	Sensible heat flux
h	m	Height

Symbol	SI Units	Name
I	W m^{-2} sr^{-1}	Radiance
I	W sr^{-1}	Intensity
I_+	W m^{-2} sr^{-1}	Upwelling radiance
I_-	W m^{-2} sr^{-1}	Downwelling radiance
J	W m^{-2} sr^{-1}	Source function
J	—	System Jacobian
K	K$_g$m^{-3} s^{-1}	Darcy air permeability
k	m^{-1}	Absorption coefficient
K	W m^{-1} K^{-1}	Thermal conductivity
L	—	Cumulative leaf area index
L	W m^{-2}	Latent heat flux
L	W m^{-2} sr^{-1}	Radiance
LE	W m^{-2}	Evapotranspiration
$L_s(\omega_r)$	W m^{-2} sr^{-1}	Radiance of scene
$L_w(\omega_s)$	W m^{-2} sr^{-1}	Radiance of standard
l	J Kg^{-1}	Latent heat of evaporation
l	m	Length
N	m^{-3}	Number of particles per unit volume
N	—	Number of spheres
n,m	—	Index of refraction
P	—	Phase function
P	—	Probability density function
P	J m^{-2} s$^{-1/2}$	Thermal inertia
	K^{-1}	
P_0	—	Foliage gap probability function
Q_0	W m^{-2}	Solar flux
Q_1	W m^{-2}	Sky flux
Q_{ext}	—	Extinction factor
q	W m^{-2}	Mean radiance penetrating canopy
q_a	—	Mixing ratio of air at 1.5 m
q_g	—	Mixing ratio of air at surface
q_{sky}	W m^{-2}	Longwave sky energy component
q_{solar}	W m^{-2}	Solar source energy component
R	—	Diffuse irradiance ratio
R	W m^{-2}	Net thermal radiation flux
R	m	Radius
R	—	Reflectance factor
R	—	Snow reflectance
R_g	—	Background albedo or reflectance
R_i	—	Layer reflection operator
R_{ji}	—	Probability transition matrix
R_0	—	Surface Fresnel reflection coefficient
R_s	—	Surface Fresnel reflection coefficient
R_s	W m^{-2}	Downwelling thermal flux from the sky
R_n	W m^{-2}	Net longwave flux
R_v	—	Volume reflection coefficient
R_w	—	Reflectance factor of standard
R_∞	—	Infinite reflection coefficient
$R(\omega_i,\omega_r)$	—	Reflectance factor
$R\uparrow$	W m^{-2}	Upwelling thermal flux
$R\downarrow$	W m^{-2}	Downwelling thermal flux
r	m	Grain size
r	—	Linear correlation
$\mathcal{R}_i$	—	Composite reflection operator
S	m^{-1}	Area-to-volume ratio for ice-air interface
S	W m^{-2}	Net shortwave flux
S	W m^{-2}	Longwave flux emitted
S_{diff}	W m^{-2}	Diffuse solar irradiance
s	—	Distributed scattering coefficient
s	W m^{-2} sr^{-1}	Pixel radiance
s	—	Variance of slope

Symbol	Units	Description
s	m	Row spacing
s	m^{-1}	Scattering coefficient
s_i	W m^{-2}	Initial incident flux
s_t	—	Total transverse cross-section
s_b	—	Total backward cross-section
T	K	Temperature
T_a	K	Air temperature
T_b	K	Bole temperature
T_c	K	Crown temperature
T_f	K	Foliage temperature
T_g	K	Ground temperature
T_i	—	Layer transmission operator
T_{sku}	K	Sky temperature
t	W m^{-2} sr^{-1}	Sunlit canopy radiance
t	s	Time
u	—	Parameter of Suits' model (phase function)
u	m s^{-1}	Wind velocity
V	m^{-1}	Vertical leaf projection factor
V	m^3	Volume of a sphere
v	—	Parameter of Suits' model (phase function)
W	m s^{-1}	Wind speed
W	—	Ratio of molecular weight of water to air
W_n	Rad$^{1/2}$ m^{-1}	Fourier constant
w	—	Parameter of Suits' model (phase function)
w	m	Row width
X	m	Cartesian coordinate
X	K	Canopy temperature profile
Y	m	Cartesian coordinate
Z	m	Cartesian coordinate
Z	rad	Solar zenith angle
z	m	Cartesian coordinate
z	W m^{-2} sr^{-1}	Shadowed pixel radiance
α_f	—	Absorptivity of foliage
α_g	—	Absorptivity of ground
α	rad	Angle
β	m^{-1}	Volumetric extinction coefficient
γ	rad	Angle
γ	m	Distance
γ	m	Radius of absorbing sphere
γ	—	Single leaf phase function
γ_0	—	Ice-air interface reflectivity
γ_b	m	Radius of bole
γ_c	m	Radius of crown
γ_s	—	Diffuse reflectance
Δ	—	Finite difference operator
δ	—	Dirac delta function
δ_n	rad	Fourier constant
ϵ	—	Emissivity
η	kg m^{-3}	Density
η	W m^{-2} sr^{-1}	Source term
θ	rad	Angle
κ	m^2 sec^{-1}	Thermal diffusivity
λ	μm	Wavelength
μ	—	Cosine of the zenith angle
ρ	—	Leaf hemispherical reflectance
ρ	Kg m^{-3}	Snow density
ρ_a	Kg m^{-3}	Air density
ρ_c	Kg m^{-3}	Canopy density
ρ_i	Kg m^{-3}	Density of pure ice
$\rho(\psi)$	—	Fresnel reflection coefficient at angle ψ
σ	W m^{-2} K^{-4}	Stefans-Boltzmann constant
σ^2	—	Variance
σ_f	—	Foliage cover fraction
τ,τ_0	—	Optical depth
τ	—	Hemispherical transmittance
ϕ	rad	Angle
ψ	rad	Angle
ω	rad s^{-1}	Angular frequency
$\tilde{\omega}$	—	Single scattering albedo

INTRODUCTION
OVERVIEW

This chapter is concerned with describing our current understanding of and mathematical treatment for the interaction mechanisms of electromagnetic radiation with terrain elements at optical wavelengths. This wavelength regime spans the reflective and thermal emissive portion of the electromagnetic spectrum. In the context of this chapter, terrain elements may be taken to mean either individual subelements, such as individual leaves, or aggregates of such subelements to comprise more complex media, for example, vegetation canopies. In all cases, however, sufficient numbers of scatterers or emitting elements comprise the media under study to generally warrant the application of macroscopic approaches such as radiative transfer theory and energy budget relationships. In contrast to the microwave regime discussed in the next chapter, this means that the optical properties of the terrain elements are such that phase coherence can usually be ignored.

Extensive mathematical analysis has led to the development of equations describing energy interactions with bulk matter. In general, analytical solutions are not possible and numerical approaches must be employed. The main difficulty in describing these interaction mechanisms with both renewable and nonrenewable resources, however, is the problem of linking the biological and physical properties of the target elements to such bulk average quantities as the "volume attenuation coefficient" or the "phase function." It is precisely with these problems that significant progress has been made since publication of the previous edition of this manual. In the reflective region there are now several process-oriented models that are available to describe the bidirectional reflectance distribution-function for different resource classes. The thermal regime has seen the advent of more complex models which are now beginning to include vegetation layers superimposed upon soil substrates.

An important factor in increasing our knowledge of interaction mechanisms since the publication of the first edition of this manual has been the increase in available experimental data, particularly from field studies. The Large Area Crop Inventory Experiment (LACIE) and the present Agriculture and Resources Inventory Surveys Through Aerospace Remote Sensing (Agristars) have been a significant stimulus, for example, for agricultural targets. The availability of high-quality experimental data with sufficient supporting target parameter determinations for modeling purposes is, however, still limited. Correspondingly, intercomparisons of the various modeling approximations have also been limited. Further theoretical work is also required for heterogeneous target categories and terrain elements with irregular geometries.

The primary objective of this chapter is to orient the reader to the various descriptions that one could use to predict the optical reflective or thermal radiance properties of terrain materials as a function of intrinsic target or scene characteristics. These predictive relationships are useful for sensor and algorithm design studies and as a guide for field studies. A second objective is to discuss the inverse problem of inferring target character-istics from measured responses. This particular topic will, however, be addressed much more fully in practical contexts in later chapters by other authors.

The chapter is organized in the following fash-ion. First, a narrative discussion of the scene radi-ation dynamic processes and modeling perspec-tive in the optical regime is given. This is followed by a detailed mathematical framework for these energy processes and their relationships to remote sensing. Examples of modeling approximations for various resource classes are included. Next, some applications of these relationships to ter-rain-feature response are presented. Some exam-ples of measured terrain-feature response in the different energy regimes then follow. Primarily, however, other sources of information are indi-cated and measurement concerns in the interpre-tation of literature data highlighted. The chapter concludes with a brief discussion of future re-search directions.

THE ENERGY FLOW PROCESS

This chapter is primarily concerned with the radiative interactions of scene elements that are of a size ranging from a few meters to tens of meters. These scene elements, then, may comprise rela-tively simple, so-called "homogeneous targets," or conversely include several different element types. In either case the scene element under con-sideration is embedded in its environment, sur-rounded by other interacting elements, and sub-jected to both time and space-varying energy sources and sinks.

In the optical-reflective regime the classical de-scription of the energy-flow process would trace the flow of incident flux from the sun (and dif-fuse sky sources) through the atmosphere to the scene element (target/background composite) thence from the scene element back through the atmosphere and into a sensor. With the advent of scanning laser systems this passive interpretation of the optical reflective regime must be broadened to include the active mode. In the reflective re-gime it is generally implicitly assumed that the target temporal variations are generally slowly varying with respect to measurement integration intervals. Measured radiance variations are thus attributed to fluctuating irradiance conditions, atmospheric state or to true "intrinsic" variations in scene-element composition or condition. It should be noted that not all intrinsic target varia-tions may be meaningful; for example, a gust of wind during measurement acquisition could change the geometric structure of a canopy target and, hence, its radiance. Radiance variations include changes in intensity, spectral composition, angular pattern, and polarization. In the reflective regime we will, in fact, adopt this paradigm and

focus our attention on describing the radiative in-teractions occurring at the surface; that is, other chapters will address atmospheric and sensor considerations.

Neglecting polarization, the reflective spectral radiance for a scene element viewed from the di-rection μ_r, ϕ_r, (where μ_r is the cosine of the zenith view angle and ϕ_r the azimuth when the corre-sponding sun angles are μ_0, ϕ_0), may, therefore, be conveniently summarized by the following ex-pression:

$$L(\mu_0,\phi_0;\mu_r,\phi_r) = \frac{1}{\pi}\int\int f_r(\mu,\phi;\mu_r,\phi_r;\boldsymbol{p})$$
$$E_{TOTAL}(\mu,\phi)\,d\mu\,d\phi$$
$$= \frac{1}{\pi}\delta(\mu - \mu_0)\,\delta(\phi - \phi_0)$$
$$f_r(\mu_0,\phi_0;\mu_r,\phi_r;\boldsymbol{p})\,E_{SUN} \qquad (3\text{-}1)$$
$$+ \frac{1}{\pi}\int\int f_r(\mu,\phi;\mu_r,\phi_r;\boldsymbol{p})$$
$$E_{DIFFUSE}\,d\mu d\,\phi$$

where

f_r = bidirectional reflectance distribu-tion function,

$\boldsymbol{p}$ = a set of parameter biophysical de-scriptors for the target (parameter vector),

L = (spectral) radiance,

E_{TOTAL} = total irradiance at the surface,

E_{SUN} = solar irradiance at the surface,

$E_{DIFFUSE}$ = sky irradiance incident on the ter-rain element,

δ = represents the Dirac delta func-tion.

In the above expressions the wavelength sub-script has been suppressed. It can be seen that the fundamental property governing the reflective be-havior of a scene element is its bidirectional re-flectance distribution-function, BRDF, which is assumed to be an intrinsic property (Nicodemus, 1970; Kasten and Raschke, 1974; and Chapter 2 of this Manual). Later in this chapter various models that have been developed to relate f_r to underlying scene biophysical attributes, $\boldsymbol{p}$ will be described. Typically, these models will consider multiple scattering and absorption of shortwave radiation within the media under consideration.

In the optical thermal regime the classical de-scription of the energy-flow response becomes more complex depending upon the scope of the discussion and level of detail considered. These factors, in turn, depend upon the complexity of the scene element considered. For example, the discussion appropriate to the energy-budget for-mulation for a "simple" soil or snow surface is considerably different than that for dense vegeta-tion. In the thermal regime the scattering and, now also, the emission of long-wave radiative flux within the terrain element are treated. However, in order to determine the effective equilibrium temperature for the emission calculations, several energy source and sink terms must be estimated. These could include long-wave flux transfers from surrounding terrain elements and the sky, convective transfers, conduction, sensible heat, evapotranspiration, condensation, and other similar processes. Many of these processes are very tightly coupled to the environment through several time-dependent control variables. The instantaneous measurement of thermal radiance, $L(\mu_r, \phi_r)$, is thus very dependent upon factors other than the intrinsic properties of the terrain element under consideration. The combined multispectral measurement of terrain elements in both optical reflective and optical thermal bands has, nevertheless, proved valuable as indicated in later chapters of this manual.

Rather than report the thermal radiance from terrain elements, it is also common to convert the measurement to an effective or equivalent blackbody temperature, T, for the target, i.e.,

$$T(\mu_r, \phi_r) = \left[\pi \, L(\mu_r, \phi_r)\right]^{1/4} \sigma^{-1} \qquad (3\text{-}2)$$

where

σ = the Stefan-Boltzmann constant.

The time-dependent nature of thermal radiance and its dependence upon both intrinsic target properties, p, and meteorological driving variables, u, (including antecedent conditions) is usually indicated through an energy-budget conservation equation (or, possibly, a purely statistical relationship) of the form:

$$F(T, \mu; p) = \frac{dT}{dt} \qquad (3\text{-}3)$$

where t represents time, and F is the sum of the energy terms discussed above, or by a heat-flow equation for layered media in the direction z, which can be characterized by a thermal diffusivity, κ, subject to boundary conditions given by equations of the form in Eq. 3-3 above.

$$\frac{d^2T}{dz^2} = \frac{1}{\kappa}\frac{dT}{dt} . \qquad (3\text{-}4)$$

Examples of different modeling approximations that have been developed to describe the energy flow process in the thermal regime for various resource terrain elements, Eqs. 3-3 and 3-4, will be given in a later section.

RELATING REMOTE SENSING OBSERVABLES TO SCENE PHENOMENA

The fundamental measurable remote-sensing quantity for a scene element is its spectral radiance. A prime objective of applied remote-sensing programs is to infer scene status, either identity or condition, from these observable radiance measurements. The most frequently used data analysis or information-extraction procedure is the mapping of scene elements, based on their radiance measurements, into information classes using the techniques of pattern recognition (Swain and Davis, 1978; and Chapter 18 of this manual.) In addition the direct mapping of such agronomic variables as leaf area index (LAI) or biomass is being attempted using correlations between these desired variables and various functions of spectral radiance. Because of variability induced into the radiance measurements from such factors as differing view and illumination geometries, statistical regression techniques are widely employed. Also, from experience, investigators have determined that transformation of the radiances using, for example, spectral channel ratios, often yield better predictive relationships (Pearson, et al., 1976). In this chapter other examples of information-extraction methods based on an understanding (models) of the physical radiative interactions with terrain elements will be explored. This approach is based on various indirect sensing techniques for the underlying scene parameters which are akin to the standard techniques employed by atmospheric scientists in the deduction of aerosol or temperature profiles in the atmosphere from the integrated radiances of the media (Twomey, 1977; Fymat and Zvev, 1978). The models described in this chapter are fundamental to such techniques and are also useful in deriving various function transformations of the spectral radiances which are better related to intrinsic scene information (Bunnik, 1978; Park and Deering, 1982).

With the advent of satellite observational systems capable of obtaining multiple measurements of terrain elements both seasonally and diurnally, additional temporal methods for relating observed radiances to underlying scene characteristics have been made possible (Chevrel, et al., 1980). New satellite systems are also being proposed which would capitalize on the angular patterns of radiance and, possibly, on scene polarization behavior (Schnetzler and Thompson, 1979). These emerging methods are based on the development of logical inference steps between measured response and scene state. Considerable ancillary information is also required. In the reflective regime, for example, statistical characterization of seasonal crop profiles is being attempted which can be used to estimate date of emergence or other phenology states (Badhwar, 1980). The next logical step in these methods is to link the statistical curve fit parameters of these crop profile patterns to fundamental canopy characteristics using submodels such as those described in this chapter. In the thermal regime, multiple diurnal thermal radiance determinations of terrain elements are being viewed as a sampling of the response of scene elements to their surrounding environment. The use of thermal inertia in geological applications is one example (Watson, 1975; Kahle, 1977). Again, considerable understanding of the physical interactions of the medium with its environment, and additional ancillary information, are required.

Finally, many advanced potential applications of remote sensing have as their objective the inference of spatially or temporally distributed processes of scene elements rather than their status. Such processes could include the cumulative amount of photosynthetic activity or degree of evapotranspiration occurring within a crop (Wiegand, et al., 1979). These processes are closely tied to such crop characteristics as leaf-area index, a typical radiative transfer-model input, or represent components in a process-oriented energy-balance model for terrain elements. These models then may serve as one mechanism for relating remotely sensed radiance to the scene phenomena of interest.

THE DYNAMIC NATURE OF REMOTE SENSING MEASUREMENTS AND SOURCES OF VARIATION

A useful perspective for sorting the various concepts discussed in previous sections into a logical framework is to consider a classical state space description of the remote sensing process (Luenberger, 1979). This state-space description may be used to separate extrinsic effects, which influence scene radiance but which are generally not of interest, from intrinsic scene properties. A state-space formulation is applicable to both the optical reflective and optical thermal regimes. In addition to its being a useful way to organize concepts, a considerable body of literature relevant to parameter estimation within this context is thus made available (Gelb, 1974).

An abstract state-space description of the remote sensing process will first be formulated; specific example interpretations will then be given. As a first step the system under investigation is first identified or isolated. In our case the system is taken to be the collection of terrain elements being measured or sensed. The system is characterized by a state vector and an underlying set of governing equations that describe how the system (state vector) evolves in time, space, or, perhaps, some other generalized coordinate. The system is subject to some external controlling factors that affect the state of the system. The system is also characterized by a set of underlying properties or parameters that influence its state. That is to say, in general, the underlying governing equations describing the evolution of the system are a function of both extrinsic control factors and intrinsic properties. Finally, the system is observed, in our case, by various remote sensing devices. However, it is possible that not all, in fact, not any, state vector components of the system are directly observable but rather that only some function of the underlying state vector can be observed. Further, it is possible that the system is observed or sampled only along some portions of the system trajectory in state space. One can characterize the observations that are made by another set of descriptive observation equations which, in general, depend upon the state of the system, the extrinsic factors and, possibly, the intrinsic system properties. Classical state-space theory is concerned with the estimation of the system state or system parameters, in the presence of measurement errors and with uncertainty in the underlying governing equations.

Mathematically, the state-space interpretation of the remote sensing problem may be summarized as follows:

$$\dot{X} = F(X,u;p) + e$$
$$Z = G(X,u;p) + v \qquad (3\text{-}5)$$

where

X is the state vector for the system; $\dot{X}$ is the variation of the state vector along some generalized coordinate, e.g. time, space, or angle; p is a list of underlying parameters that affect the state of the system; u is the extrinsic and variable control vector that generally drives the state of the system; Z is the set of observations that are made on the system; and F and G are a set of state- and observation-equations for the system, respectively. In the above formulation, measurement uncertainty, v, and imperfect state equations, e, have been included.

The governing system equations, F, and the observation equations, G, may be nonlinear as is, unfortunately, generally true for remote-sensing applications. As in all nonlinear situations this generally complicates the remote-sensing estimation or mapping problem. Some specific examples of this state-space interpretation will now be given.

First, consider the case where the state vector consists of the radiative flux, i.e., radiances, within different layers of a vegetation canopy, i.e., the system. The canopy radiative balance is subject to external controlling factors, i.e., irradiance, and depends on underlying canopy properties such as canopy architecture and leaf and soil optical properties. Observations of upward radiance from only the top layer are made, perhaps over time or along a scan direction. The system equations, F, then correspond to the various canopy reflectance models given later and G would correspond to a selection function for the top nadir exiting radiance. A simplified case would replace the state vector above by a constant vector of canopy parameters, e.g., biomass levels, and the observations by radiance measurements of the canopy. The observation equations would then relate the measured radiance to the biomass levels and irradiance. Classical state-space estimation procedures would then reduce to classical linear or nonlinear regression techniques. Another example in the reflective regime would be to take the state vector as the angular variation in scene radiance and the observations as some subset of these measurements.

In the thermal regime a state-space example

would be to take the state vector to be the temperature distribution within the media; the control vector to consist of such meteorological factors as the wind speed, relative humidity, and irradiance; and the parameter vector to correspond to such factors as element area and flux transfer coefficients of the media for various energy interactions. The observational equations could correspond to the radiance or effective temperature of the top layer, perhaps, sampled through time, and the state equations might correspond to a statistical description of the temperature behavior over time.

This state-space description of the remote sensing process also is a good illustration of the basic difficulties encountered in the field. The observations, Z, are hopefully a function of the desired scene properties, p, but usually in an indirect sense. Z is more generally directly a function of the state vector, X (which is a function of p), and the extrinsic controlling factors, u. Especially, since F and G are nonlinear functions one cannot assess *a priori* whether or not the system is "observable" or whether the estimation of p is feasible. Many of the preprocessing techniques utilized in machine processing of remotely sensed data are designed to minimize the effects of u, for example due to varying sun- and view-angle, or to develop new observation variables, i.e., feature vector transformations, which are more amenable to estimation procedures.

MATHEMATICAL FRAMEWORK

APPROACHES

It is apparent from the previous discussion that the analysis of interaction mechanisms at optical wavelengths must take into consideration the spatial and temporal scales of the terrain elements under study. These scales define the nature of the targets to be modeled, the complexity of both intrinsic and extrinsic factors to be considered, and indicate appropriate mathematical procedures to be used.

With some exceptions, virtually all the techniques described in this chapter are concerned with the radiance patterns from representative terrain elements at a particular point in time. The utilization of such predictions with respect to diurnal or seasonal trajectories is referenced, but full discussion is left to the application chapters. It is instructive, however, to consider spatial modeling scales involved in the analysis of terrain radiation patterns. In analogy with the atmospheric radiative problem one might organize these scales into microscopic, microscale, mesoscale, and macroscale levels. Roughly speaking, each of these levels would vary by factors of ten. At the optical wavelengths appropriate to this chapter, typical sizing dimensions for each of these scales might be as follows:

The microscopic scale would correspond to elements of less than 0.1 meter. This level would include elements as large as individual leaves and as small as mineral grains or snow crystals, or even smaller. For large terrain elements, empirical scattering coefficients are assumed or geometrical optics are used to analyze their scattering patterns. For elements of the same general dimension as the optical wavelengths under consideration, Rayleigh or Mie analysis procedures are employed. These microscopic terrain elements often serve as building blocks for higher canopy level models. As such, therefore, it is advantageous if their response patterns can be related to biophysical attributes of interest.

A sizing dimension for the microscale might be on the order of ten meters. Such elements would correspond to uniform stretches of vegetation, snow, or soil areas. Aircraft multispectral scanning systems have instantaneous fields of view of this order of magnitude. Plane-parallel abstractions of the terrain media are typically used.

The mesoscale would then correspond to terrain elements on the order of 100 meters. At this scale, mixtures of vegetation, soil, and other components would frequently occur. Community associations would appear in natural vegetation cases. Multidimensional modeling would generally be required; however, the sophistication of such approaches would depend upon the level of detail required.

Finally, the macroscale might then correspond to terrain elements on the order of 1000 meters. At these large scales, large vegetation associations would prevail. Topographic undulations would be evident. From a remote sensing perspective these regions would be imaged by meteorological satellites; fairly simple, quasi-empirical characterizations would be employed to estimate general surface properties.

The boundaries between each of these scales are not distinct but depend upon geographic region, cultural practices, natural variability, and other factors. For example, at the microscale level, multidimensional models would probably be required for terrain elements corresponding to row crops with incomplete plant cover, resulting in a mixture of soil and vegetation. It is also probably true that as one moves to larger spatial scales, the resolution of the processes captured by the modeling approaches decreases primarily because of the difficulty of obtaining necessary parameters to drive the analyses at these scales.

In all cases, theories that describe the observed phenomena of light scattering, absorption, polarization, or thermal-flux transfers in terms of the bulk properties of the media would be appropriate. In contrast to the microwave discussions in the next chapter, recourse to detailed classical electromagnetic theory is generally not appropriate here. Various media can be treated either as continuums or as collections of discrete scatterers. Coherent effects are usually ignored. In essence, the appropriate methods are those of the classical radiative transfer theory, geometric optics, and energy-budget analysis with associated simplifying approximations. The basic starting point is essentially the principle of energy conservation.

RADIATIVE TRANSFER AND ENERGY BUDGET EQUATIONS

The radiative transfer theory is a convenient framework within which to discuss the modeling of the scattering and absorption of radiation by terrain elements at optical wavelengths. Further, it provides a bridge between the approximations employed in remote sensing and classical theory. The development of the appropriate energy conservation equations considering elementary volumes and arbitrary bounded media for both the time-dependent and time-independent or steady-state situation are described in several texts, e.g., Chandrasekhar (1960) or Preisendorfer (1965). For terrestrial surface applications additional energy budget terms may need to be appended to include other energy source terms such as convection and sensible heat. In practice, however, various energy budget formulations emphasizing those particular components important to the specific terrain elements under study are used. Given the current state of the art it is most fruitful to consider the time-independent equation for the scalar intensity or radiance of a plane parallel medium. The equation could be expanded to include polarization effects or to account for arbitrarily bounded media. In fact, the central problem for the modeling of terrain elements is not in the writing down of the radiative transfer equations, but in developing appropriate representations for the media phase function in terms of biogeophysical attributes and in estimating the required parameters to drive the models.

The time-independent equation for the plane parallel media may be written in the usual notation (as a casual reader soon surmises there are actually many forms of the radiative transfer equation, normalizations and conventions in the literature, e.g., van de Hulst 1980).

$$\mu \frac{dI}{d\tau}(\tau;\mu,\phi) = -I(\tau;\mu,\phi) + J(\tau;\mu,\phi) \qquad (3\text{-}6)$$

where μ and ϕ have been previously defined, $d\tau$ is the differential optical depth proportional to the total extinction coefficient, i.e., to absorption and scattering; the number density of individual scatters and depth, I, is the intensity or radiance; and J is a source term comprised of the scattering of surrounding flux and attenuation of an external source, πF_0 i.e.,

$$J(\tau;\mu,\phi) = \frac{\tilde{\omega}}{4\pi} \int_0^{2\pi} \int_{-1}^{1}$$

$$P(\mu',\phi';\mu,\phi)\, I(\tau;\mu',\phi')\, d\mu'\, d\phi' \qquad (3\text{-}7)$$

$$+ \frac{\tilde{\omega}}{4\pi} P(\mu_0,\phi_0;\mu,\phi)\, \pi F_0\, E(\tau,\tau_0,\mu_0)$$

where $\tilde{\omega}$ is the single-scattering albedo and is equal to the ratio of the scattering cross-section to the total extinction cross-section, τ_0 is the total optical depth of the medium, P is the phase function, which describes how flux is scattered from the direction μ', ϕ' to the direction μ, ϕ, and $E(\tau,\tau_0,\mu)$ is the attenuation factor whose specific form actually depends upon the media geometry and other properties of the medium but is generally taken to be of the exponential form:

$$E(\tau,\tau_0,\mu) = e^{-(\tau-\tau_0)/\mu} \qquad (3\text{-}8)$$

In thermal applications other energy terms become dominant and the detailed radiative scattering treatment of longwave flux within the medium is usually not emphasized. The simplest expressions are for steady state conditions and for systems that possess no heat-storage capability. Generally, more complex expressions are required. For example, the equation below, taken from the first edition of the *Manual of Remote Sensing,* could be applicable for a vegetation canopy:

$$S + R_n + G + LE$$

$$+ \int_0^z C_a\, \nabla(\rho_a\, u\, T)\, dz + \int_0^z \frac{L_v E_w}{R} \frac{\nabla(ue)}{T}\, dz$$

$$+ \int_0^z C\, \rho_c\, \frac{\partial T}{\partial t}\, dz \qquad (3\text{-}9)$$

$$+ \int_0^z C_a\, \rho_a\, \frac{\partial T}{\partial t}\, dz + \int_0^z \frac{L_v^w}{R\, T} \frac{\partial e}{\partial t}\, dz = 0$$

where

S = net shortwave flux
R_n = net longwave flux
G = soil conduction term
LE = evapotranspiration
z = vertical height of the canopy
∇ = horizontal gradient
C_a = specific heat of air
u = wind velocity
T = canopy temperature
ρ_a = density of air
L_v = latent heat of vaporization
E_w = water vapor radiation
R = universal gas constant
e = vapor pressure of air
C = heat capacity of the canopy
ρ_c = canopy density
w = ratio of the molecular weight of water to air.
t = time

The complete analysis of this equation is given in the earlier manual and is not further elaborated here.

Finally, for soil layers, the thermal diffusion equation is widely employed subject to appropriate boundary conditions, as will be discussed later.

DIRECT SOLUTION AND DIFFICULTIES IN SPECIFYING THE TERRAIN PARAMETERS

It is deceptively easy to write down a formal solution to the radiative transfer equation. Assuming the medium is such that it is meaningful to utilize an integrating function, i.e., that there are no discontinuities, then the formal solution is given by:

$$I(\tau;\mu,\phi) = I(\tau_0;\mu,\phi)\, e^{-(\tau_0-\tau)/\mu}$$

$$+ \int J(\tau';\mu,\phi)\, e^{-(\tau_0-\tau')}\frac{d\,\tau'}{\mu} . \qquad (3\text{-}10)$$

The first term is the radiance from the bottom boundary, attenuated through the medium, but the second term is more complex. It is an integral which depends on the total radiance field at all levels within the medium. In essence, the general radiative transfer problem reduces to that of solving an infinite set of coupled integral differential equations. For arbitrary variations of the single scattering albedo with depth and complicated phase functions, an analytic solution cannot be developed. Numerical techniques are widely utilized.

In the analysis of terrain elements the phase functions and other parameters must be developed based on various abstractions for the individual scattering elements and the way in which they are combined to form the overall medium. In the case of snow and soil, techniques very similar to those of the atmospheric radiative transfer problem have been invoked; i.e., treatment of spherical grains or snow crystals and the use of Mie scattering or derivatives thereof. In the case of vegetation canopies, the orientations of individual scattering elements, generally taken to be leaves and twigs, and the various morphological forms, must be incorporated. For all the media, vertical stratification and inhomogeneities are evident, and all these considerations generally lead to anisotropic phase functions. For many of the terrain elements a limiting condition on the development of appropriate mathematical models and solutions has been the difficulty of obtaining the required terrain parameters and characterizing their spatial and temporal variability. Such terrain-element characterization is required in order to validate the hypotheses of the various models. Once the limits of such models are determined, a major application of these models would, in fact, be to serve as a tool for inferring the spatial and temporal distributions of the underlying biogeophysical attributes.

For steady-state energy-budget equations the problem reduces to that of solving a nonlinear system of algebraic equations. Such methods are obviously readily available and, again, the limiting factors are not so much the mathematical techniques but their sensitivity to each of the individual terms. The solution of more complex ener-gy-budget equations is obviously much more difficult.

Finally, well-developed and well-understood numerical solution approaches to the thermal conduction equation, using either finite difference or possibly finite element techniques, are widely available. Analytic solutions are possible only for simplified boundary conditions.

In summary, then, classical mathematics and modern computer technology offer appropriate tools for the solution of the underlying equations governing the interaction mechanisms of terrain elements. The application of these tools to specific terrain elements requires the development of useful abstractions for such targets so that the model parameters may be related to target characteristics. Several examples for specific targets have now appeared in the literature and will be discussed in the following sections.

MODELING APPROXIMATIONS

A wide variety of models have been developed to explain the radiance patterns for vegetation, snow, soil, and some combinations of the above. Primarily because of time constraints, the modeling of water bodies or concentrations of substances within water will not be addressed. The theory for these kinds of problems and its application is well established. The pioneering work of Duntley and the excellent treatise by Preisendorfer (1965) should be noted.

In the following discussions the reader will be exposed to the details of several calculational procedures and the different abstractions that have been made for various media. Perhaps a few overall comments should initially be made. First, it may be remarked that while some modeling efforts from a remote-sensing perspective were initiated in the 1960s or before, most of the efforts reported really began to appear in the reviewed literature in the decade of the 70s with further intensive efforts even now taking place. The launching of various satellite observational platforms and the wider availability of aircraft multispectral data may have played a role in fostering such recent developments. In the reflective regime there has been a remarkable similarity in the techniques employed in the modeling of vegetation, snow, and soil. For example, it will be seen that for all of these terrain elements the first modeling attempts were to treat each of the media as a diffusing medium, e.g., employing Kulbelka-Munk or Duntley equation theory. In these early efforts the bulk absorption and scattering parameters were treated as semi-empirical parameters to be estimated. The next logical sequence then was to express these parameters in terms of biophysical attributes that were both of interest and that could be estimated directly from field observations. The estimation of the scattering properties of individual components within the media took very similar turns, e.g., quasi-Mie scattering theory and geometric optic approximations. Classical solution approaches to the radiative transfer equation, including the Monte Carlo and method of discrete ordinates, have been applied in all three media. The thermal regime does not show the same degree of similarity across terrain elements primarily because of the major difference in the physics required for modeling soils versus vegetation. However, the modeling of soils with a "simple" vegetation cover represents a transition case.

The applicability of radiative transfer theory to a wide variety of terrain media has been demonstrated. It is evident that the plane-parallel case is well studied but it has not really been demonstrated when this approximation is applicable in a pragmatic context. Multi-dimensional targets are being studied primarily within the context of vegetation terrain elements.

Reflective Regime

For the purpose of the discussions below, terrain elements have been organized into three categories: vegetation, soil and rocks, and snow. While all of these terrain elements offer a wide spectrum of complexity, depending upon the condition and type of target modeled, the biological arena is notorious for its diversity. This diversity is further compounded by cultural effects. Radiatively, vegetation is also a difficult medium as compared to some of the more solid terrain elements in that it is permeated by gaps and possesses semi-diffusing elements with asymmetrical scattering properties. Consequently, there are many cases where vegetation may not be taken as an optically thick medium with resulting simplified approximations employed.

In each of the major sections below a brief overview will be given of the modeling activities for the terrain-element category that is being considered. The overview will then be followed by selected examples.

Vegetation

Vegetation exhibits a significant hierarchical pattern in terms of its component parts and the manner in which they are organized. Similarly, a series of hierarchical models have been developed to address these several cases. Specifically, one may be concerned with individual leaves or other plant parts, with fairly homogeneous canopies, and with structured agricultural or natural vegetation canopies having various percent covers, e.g., row crops and forest canopies.

Individual Leaf Modeling

Willstätter and Stoll presented the earliest description of a theory to explain leaf reflectance in 1918. Their model generally treated the critical reflection of visible light between inter-cellular air spaces and cell walls. Although specific details of this early model have since been called into question and subsequently modified, nevertheless it formed a convenient thinking tool to organize much of the early experimental work. The year 1971 was a key year for experimental literature, with particular reference being made to Breece and Holmes (1971) and Woolley (1971). Partly under the impetus of this experimental work and the earlier work of Gates et al. (1965) a reexamination of many of the fundamental concepts of the Willstätter-Stoll theory was undertaken. In 1973, two papers appeared in *Applied Optics* and one in the *Journal of Agronomy*. The first paper was by Allen, Gausman and Richardson (1973) and developed an optical ray-tracing model fundamentally based on the Willstätter-Stoll theory using solid wall and cellular air interaction. The predicted transmittances were too high and the reflectances were too low but the authors suggested avenues for improvement. The paper generated predictions of the indicatrices for leaves and incorporated novel geometrical procedures for the ray tracing. The paper by Kumar and Silva, which also appeared in *Applied Optics* (1973), extended the Willstätter-Stoll theory by including other components such as chloroplasts in their optical ray tracing. The authors employed Fresnel's equations at all the appropriate media interfaces, utilized geometric optics, but neglected Rayleigh and Mie scattering. Negligible absorption of light in the 0.7 to 1.3 μm region simulated was also assumed. The authors verified the results of Allen et al., that the considering of only cell-wall–air and air–cell-wall interfaces leads to underestimation of reflection and over-estimation of transmission but that including the other plant media constituents improves the predictions. The authors conclude that, while there is some contribution due to Rayleigh and Mie scattering, the major reflection contributions are caused by leaf constituents as predicted by geometric optics. The Willstätter and Stoll theory was also further examined in a third paper appearing in 1973 by Sinclair et al.

All of the above papers were instrumental in providing insight into the general mechanisms of radiative interactions within leaves relative to their physiological structure. However, geometric ray-tracing methods through the leaves' structure is a tedious and time-consuming operation, particularly if angular scattering properties are desired. The most recent paper on leaf modeling is that by Tucker and Garratt (1977), which treats the leaf optical system as a stochastic process and employs Markov transition probabilities together with a compartment flow model. The model provides fair agreement, except possibly in the 0.5–0.7 μm region. It is a very efficient model from a calculation perspective.

Vegetation Canopies—One-Dimensional (Plane Parallel Abstraction)

The modeling of plant canopies from a remote-sensing perspective was initiated by Allen, Gayle and Richardson (1970), who applied the Kubelka-Munk theory and, later, Duntley's five-parameter, diffuse-scattering theory to agricultural crops. These diffuse reflectance theories parameterize the optical properties of the canopy in terms either of the absorption and scattering parameters of the Kubelka-Munk theory or the five Duntley parameters corresponding to upward and downward diffuse flux and specular flux. These parameters were treated as empirical coefficients to be estimated from measurements of overall canopy radiation transfers. The Duntley approach was further expanded by Suits (1972) who developed an abstraction of the canopy in terms of horizontal and vertical leaf facets with individual reflectances and transmittances. Given this abstraction for the canopy media, Suits was able to derive explicit expressions for the five Duntley optical parameters in terms of measurable canopy attributes. Conversely, the way was paved for further application studies that could relate overall canopy reflectance to changes in canopy architecture arising from stress and other factors. Also in 1972, the method of discrete ordinates was used by Weinman and Guetter to describe overall radiance patterns from a combined atmosphere and/or canopy. However, in this model, empirical coefficients for the corn canopy were again employed. The Monte-Carlo model was applied by Smith and Oliver (1972) and the canopy was abstracted in terms of leaf facets possessing a complete leaf-slope distribution. A Monte Carlo model was also employed by Szwarcbaum and Shaviv (1976). In 1975 Ross and Nilson solved the canopy problem using first-order scattering theory and again incorporated a complete leaf-slope distribution in their analyses. A completely different approach to the canopy radiation problem, based on what Preisendorfer termed the interaction principle, was the use of the adding method by Cooper et al. (1982), which provides an efficient method of calculating radiance patterns based on the canopy abstraction of Smith and Oliver. Park and Deering (1982) report a recent extension of the Duntley approach with asymptotic relationships developed for the optically thick case.

Extensive applications of one-dimensional modeling, particularly applying the Suits models, and descriptions of other models developed outside the U.S. appropriate to the one-dimensional canopy case, are given by Bunnik (1978). Chance and Lemaster (e.g., 1978) also give numerous evaluations of the Suits model.

Both the Suits model and the Smith-and-Oliver model have been applied to forested canopies. Ross is also developing a Monte Carlo model appropriate for forest canopies (personal communication).

In summary, numerous calculational procedures have been applied to the one-dimensional canopy case. There is, however, a major dichotomy in the abstraction of the canopy media in terms of either horizontal or vertical leaf projections or utilization of the complete-leaf-slope distribution. Most of the more recent one-dimensional canopy models can account for a multicomponent and multi-layered structure.

Canopies—(Multi-Dimensional Problem)

It is evident from a consideration of agricultural row crops with incomplete plant covering that the one-dimensional canopy abstraction would not be applicable. This problem is currently being addressed and probably the most dominant method of attack is that employing a strictly geometric optics analysis of the gross morphological features. Usually these approaches provide fairly careful analyses of the critical angles pertaining to soil and vegetation surfaces that are in shadow or in sunlight. Generally, measured input optical properties for each of these surface types are assumed. Richardson et al. (1975), Egbert (1977) and Jackson et al. (1979) are examples. Egbert's model is particularly intriguing in that it considers a general ground plane covered with either spherical or cylindrical perturbations. These perturbations or volume elements possess both diffuse and specular reflection properties. Egbert then sums the diffuse and specular contributions of both the perturbations and plane, i.e., five basic components, and the diffuse response from shadows. Canopy reflectance depends primarily on the density of the perturbations and their size. A similar approach has been taken by Strahler and Li (1981) in the modeling of forest canopies using cones. Otterman (1981) similarly addresses natural rangeland canopies. The second major approach to the multidimensional problem is to combine the geometric optics analyses of the gross morphological features of canopies, e.g., row structure, with multiple scattering considerations for interaction within the canopy clumps, as exemplified by the work of Verhoef and Bunnik (1976), Welles and Norman (1979), and Suits (1981). The model by Verhoef and Bunnik is an extension of the Suits model and considers the canopy components as being packed in rows with regular rectangular cross-sections of fixed dimensions. Detailed geometric analyses of the canopy-phase function relative to direct solar flux and canopy row-structure is undertaken and shading is allowed. However, it is assumed that diffuse flux can be treated as normal in the Suits model. Both the direct and diffuse flux contributions and appropriate view probabilities are developed. These are consistent with the row structure and incorporate the soil contribution. The soil contribution is, however, difficult to include in a self-consistent fashion. The authors utilize horizontal and vertical projections as in the Suits model but indicate the inclusion of the leaf-angle distribution and the use of non-Lambertian scatterers as a possible extension. Basically Suits' extension of his model to row crops follows a very similar pattern except that the sharply defined rectangular boundary of the row crop is relaxed and a modulation transfer function is allowed. The model by Welles and Norman treats the canopy as consisting of a finite number of regularly spaced ellipsoids. The ellipsoids are strategically positioned to represent canopy architecture. Each point in the finite array of ellipsoids is transformed to an equivalent-plane parallel canopy by choosing an appropriate equivalent optical depth that has the same diffuse penetration probability considering both upwelling and downwelling radiative flux. Interactions between ellipsoids are accounted for by geometric optic ray tracing.

Canopies—Empirical Techniques

Particularly in natural vegetation communities where topographic relief is significant, several approaches based upon photometric analyses have been suggested. Horn and Bachman (1978) give a practical application of this technique for registering Landsat images to surface terrain and an overall review is given by Hugli and Frei (1982). Generally these approaches incorporate various assumed specular and diffuse surface models with empirical parameters to be estimated from the data. This approach is probably the most practical, particularly for regional problems.

The models of Tucker and Garrett, Suits, Ross and Nilson, and the Adding model of Cooper, et al., will be briefly discussed as examples of calculational procedures and media abstractions. The model of Strahler and Li will be reviewed in a later section of this chapter as an example of model invertibility.

Tucker and Garratt

The Tucker and Garratt model simulates the absorbed, reflected, and transmitted radiation, given normal incidence, in the spectral interval 0.40 to 2.50 micrometers for dicotyledonous (broad) leaves. The model, computes a Markov transition matrix at each 0.01-micrometer increment based on leaf thickness, structure, pigment composition, and water content. The leaf is abstracted as a ten-compartment flow model as indicated in Figure 3-1. The model basically incorporates scattering and absorption processes as a function of wavelength. Scattering is assumed to be proportional to the cellular density in the palisade parenchyma and spongy mesophyll tissues, based on cell-wall airspace reflective index differences. Absorption also occurs in the palisade parenchyma and spongy mesophyll tissues; it arises from the four fundamental constituents: leaf water, chlorophyll a, chlorophyll b, and carotenoid pigments (assumed to be essentially lutein). In addition, Figure 3-1 also indicates that initial reflection of the solar input from the cuticle occurs (compartment 2). Diffuse reflected energy is accumulated in compartment 6 and diffuse transmitted radiation is accumulated in compartment 10. The sum of compartments 4 and 8 represents total absorbed flux. The flow rates between any two compartments, i, j, is summarized in a 10-by-10 probability transition matrix, R_{ji}. However, only 21 transitions are permissible, resulting in a matrix of the following form:

$$P = \begin{bmatrix} 0 & R_{21} & R_{31} & 0 & 0 & 0 & 0 & 0 & 0 & 0 \\ 0 & R_{22} & 0 & 0 & 0 & 0 & 0 & 0 & 0 & 0 \\ 0 & 0 & 0 & R_{43} & R_{53} & 0 & R_{73} & 0 & 0 & 0 \\ 0 & 0 & 0 & R_{44} & 0 & 0 & 0 & 0 & 0 & 0 \\ 0 & 0 & 0 & R_{45} & R_{55} & R_{65} & R_{75} & 0 & 0 & 0 \\ 0 & 0 & 0 & 0 & 0 & R_{66} & 0 & 0 & 0 & 0 \\ 0 & 0 & 0 & 0 & 0 & 0 & 0 & R_{87} & R_{97} & R_{10.7} \\ 0 & 0 & 0 & 0 & 0 & 0 & 0 & R_{88} & 0 & 0 \\ 0 & 0 & R_{39} & 0 & 0 & 0 & 0 & R_{89} & R_{99} & R_{10.9} \\ 0 & 0 & 0 & 0 & 0 & 0 & 0 & 0 & 0 & R_{10.10} \end{bmatrix}$$

$$(3\text{-}11)$$

The authors assume that 1 percent of the radiation is initially reflected by the cuticle, i.e., $R_{21} = 0.01$. The attenuation of radiation due to absorption by plant pigments and leaf water is assumed to follow a Lambert-Beer relationship. The specific transition probabilities used by the authors are given in Table 3-1. As the authors caution, many of these values are based on their specific assumptions and should be appropriately modified depending upon the leaf being modeled.

The model determines the steady-state distribution, p, of radiation at each wavelength, given an initial state, p_0. The model presumes that all radiation entering the system comes from the solar input. The model then consists of the following matrix operation:

$$\lim_{n \to \infty} P^n p_0 = p \qquad (3\text{-}12)$$

where n represents the number of discrete time depths. In practice, the steady-state convergence occurs for n between 25 and 40.

Suits

The Suits model abstracts a plant canopy as consisting of a series of horizontal layers within which leaf facets are distributed randomly. The canopy overlays a reflecting background with reflectance, R_g. The leaf facets possess a hemispherical reflectance, ρ, and transmittance, τ. A key element in the Suits model is the assumption that, radiatively, the leaf-facet orientations may be approximated by horizontal, H, and vertical, V, projections. The use of this assumption leads to tractable expressions in the derivation of media-scattering and -absorption coefficients. Further, closed-form solutions for canopy reflectance can be developed. A modification of the Suits approach, utilizing the complete leaf-slope distribution, has been developed by Youkhana and Smith (1982) and by Bunnik (1982).

The Suits model iteratively solves the radiative transfer equation by first factoring the radiance field into upwelling, $I(+d,z)$, downwelling,

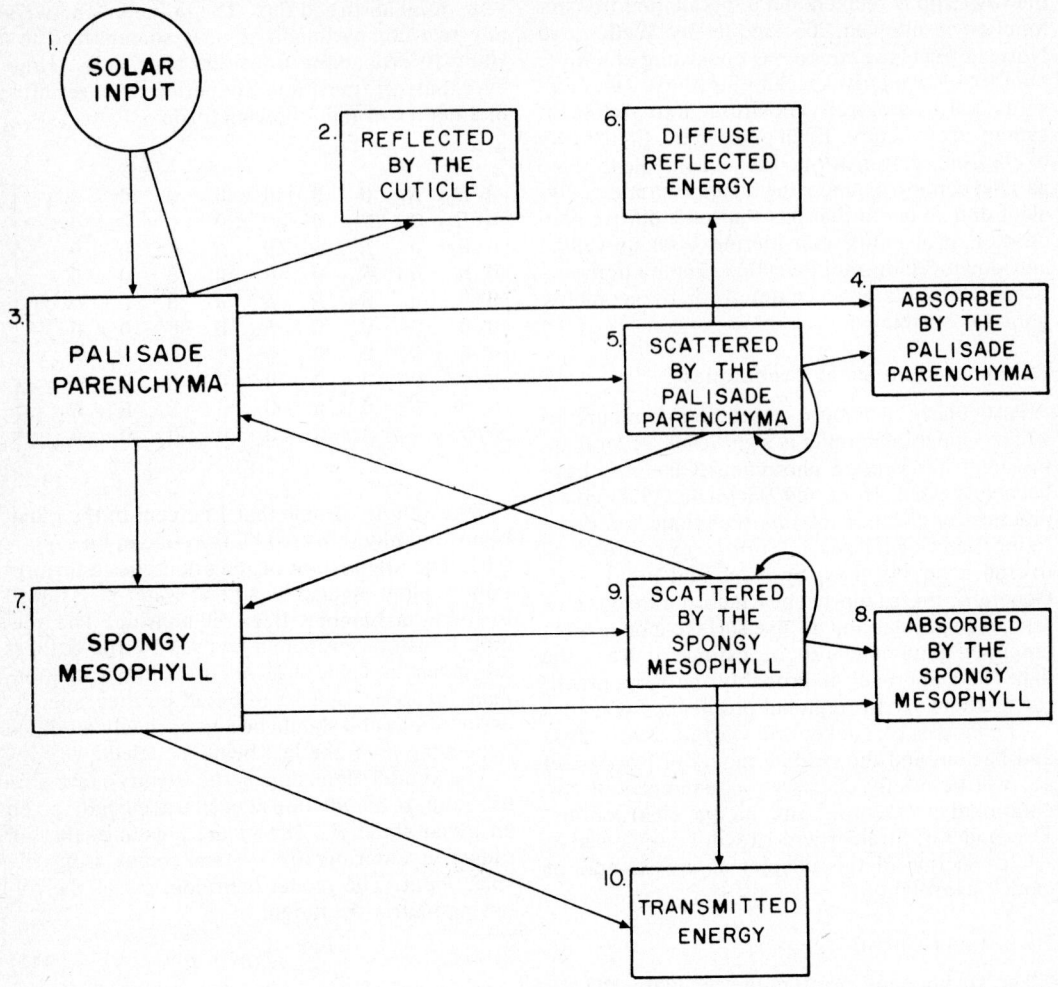

Fig. 3-1. Compartment model of the optical system for a broad leaf, i.e. the leaf of a dicotyledonous plant. The arrows between compartments represent the flow(s) and their direction. The processes or leaf cellular aggregates are indicated within each compartment (from Tucker and Garratt, 1977).

$I(-d,z)$, and specular, $I(s,z)$, terms to calculate an estimate of the source term. This estimate of the source term is then used to obtain an improved estimate for the composite radiance field. The initial factorization leads to the Duntley equations. The coefficients in these equations, as well as the required phase functions for the media, are expressed in terms of the horizontal and vertical foliage projections and their optical properties.

The upwelling radiance at the surface is equal to the attenuated soil-background contribution plus the multiple scattering contribution of the canopy, i.e.:

$$I(o;\mu;\phi) = L(\mu,\phi) = \frac{1}{\pi} R_g\, I(-d,z_0)\, e^{-kz_0}$$

$$+ \int_0^{z_0} J(z';\mu,\phi)\, e^{-k(z_0-z')}\, dz' \quad (3\text{-}13)$$

where the optical depth has been expressed in

terms of the vertical canopy depth coordinate, z, and an effective absorption parameter, k, defined later.

After factorization of the radiance field into the three terms discussed above, the source term can be expressed as:

$$J(z';\mu,\phi) = \frac{1}{4\pi} \int_0^{2\pi} \int_{-1}^1 \{I(+d,z')\, P_+(\mu',\phi';\mu,\phi)$$
$$+ I(-d,z')\, P_-(\mu',\phi';\mu,\phi)$$
$$+ I(s,z')\, P_s(\mu',\phi';\mu,\phi)\}\, d\mu'\, d\phi'.$$
$$(3\text{-}14)$$

The phase functions are given by the u, v, w parameters of the Suits model:

$$P_+ = u = H\tau + V\frac{\rho+\tau}{\pi}\, tan\,\theta$$

$$P_- = v = H\rho + V\frac{\rho+\tau}{\pi}\, tan\,\theta \quad (3\text{-}15)$$

TABLE 3-1

Radiation Transition Probabilities of Various Leaf Pigments is Derived by Tucker and Garratt

$$R_{43} = \sum_{I=1}^{4} 1.0 - \exp[-\alpha(I)X_{PP}],$$

$$R_{45} = R_{43}/2.0,$$

$$R_{87} = \sum_{I=1}^{4} 1.0 - \exp[-\alpha(I)X_{SM}],$$

$$R_{89} = R_{87}/2,$$

$$R_{53} = R_{73}(1.0 - R_{43})/2.0,$$

$$R_{97} = 1.0 - R_{87},$$

$$R_{65} = R_{53} * \frac{PP}{PP + SM} /2.0,$$

$$R_{75} = R_{53} * \frac{SM}{PP + SM} /2:0,$$

$$R_{55} = 1.0 - R_{45} - R_{65} - R_{75},$$

$$R_{109} = 0.12,$$

$$R_{39} = 0.08,$$

$$R_{21} = 0.01,$$

$$R_{31} = 1.0 - R_{21},$$

$$R_{99} = 1.0 - R_{39} - R_{109} - R_{89},$$

$$R_{22} = R_{44} = R_{66} = R_{88} = R_{1010} = 1.0,$$

$R_{ij} \neq$ one of the above, then transition probability = 0.0.

Note: R_{ij} = flow function from compartment j to compartment i, $\alpha(I)$ = extinction coefficient for material I, X = thickness of material, PP = thickness of palisade parenchyma, SM = thickness of spongy mesophyll, and e = Napier's number. These probabilities are recalculated at every 0.01-μm interval between 0.40 μm and 2.50 μm.

$$P_s = w = H\rho + \frac{V}{2\pi}\left[\rho(\sin\psi + (\pi - \psi)\cos\psi)\right.$$

$$\left. + \tau(\sin\psi - \psi\cos\psi)\right]\tan\theta\tan\theta_0$$

where ψ is the relative azimuth between sensor and sun directions and θ_0 is zenith sun angle.

The initial radiances are obtained by solving the Duntley equations, setting $E = \pi L$:

$$\frac{dE(+d,z)}{dz} = -aE(+d,z) + bE(-d,z) + cE(s,z)$$

$$\frac{dE(-d,z)}{dz} = aE(-d,z) - bE(+d,z) - c'E(s,z)$$

$$\frac{dE(s,z)}{dz} = kE(s,z). \tag{3-16}$$

The respective coefficients were shown by Suits to be given by the following expressions:

$$a = H(1 - \tau) + V\left(1 - \frac{\rho + \tau}{2}\right)$$

$$b = H\rho + V\left(\frac{\rho + \tau}{2}\right)$$

$$c = H\rho + \frac{2}{\pi} V\left(\frac{\rho + \tau}{2}\right)\tan\theta_0$$

$$c' = H\tau + \frac{2}{\pi} V\left(\frac{\rho + \tau}{2}\right)\tan\theta_0$$

$$k = H + \frac{2}{\pi} V \tan\theta_0. \tag{3-17}$$

After iteration, canopy bidirectional reflectance is obtained by appropriate normalization. For example, the bidirectional reflectance function with respect to the solar irradiance is given by:

$$f_r(\mu_0,\phi_0;\mu,\phi) = \frac{\pi L}{E(s,0)}. \tag{3-18}$$

In the original papers, Suits generalizes the expressions for multilayer, multicomponent canopies. Bunnik (1978) has performed an exhaustive analysis of the Suits model and its properties. The following closed-form solution is given for the one-layer case:

$$L(\mu_0,\mu,\psi) = \frac{A}{\pi}(u + hv)\frac{1 - e^{-(k'+m)z}}{k' + m}$$

$$+ \frac{B}{\pi}(u + h^{-1}v)\frac{(1 - e^{-(k'-m)z})}{k' - m}$$

$$+ \frac{1}{\pi}(uS + vD + w)E(s,0)$$

$$\frac{1 - e^{-(k'+k)z}}{k' + k}$$

$$+ \frac{R_g}{\pi}\left\{hAe^{-(k'+m)z} + h^{-1}Be^{-(k'-m)z}\right.$$

$$\left. + (D + 1)E(s,0)e^{-(k'+k)z}\right\} \tag{3-19}$$

where

$$A = \frac{-D(1 - h^{-1}R_g)e^m - h^{-1}\left\{R_g(D + 1) - S\right\}e^{-k}}{N} E(s,0)$$

$$B = \frac{D(1 - hR_g)e^{-m} + h\left\{R_g(D + 1) - S\right\}e^{-k}}{N} E(s,0)$$

$$N = (h - R_g)e^m - (h^{-1} - R_g)e^{-m}$$

$$D = \left\{cb + c'(k + a)\right\}/(m^2 - k^2)$$

$$S = \{c'b - c(k - a)\}/(m^2 - k^2)$$
$$h = (a + m)/b$$
$$m = (a^2 - b^2)^{1/2}$$

and

$$k' = H + \frac{2}{\pi} V \tan \theta. \tag{3-20}$$

Ross and Nilson

The Ross and Nilson model abstracts a plant canopy as consisting of a series of randomly placed leaf facets with a given vertical distribution of leaf area per unit volume, $u(z)$, and a statistical distribution of leaf orientations, $a = (\theta_l, \phi_l)$, given by $f(a)$. The canopy overlays a reflecting ground surface with spectral response, R_g, and the optical properties of the leaves are described by a single-leaf phase function, γ. The authors point out that in contrast to the usual assumption in the atmospheric case, the leaf phase function depends specifically on the angles of incidence, exitance, and leaf orientation rather than on a total scattering angle. The canopy is illuminated by direct solar irradiance, F_0, from direction $r_0 = (\theta_0, \phi_0)$, and by diffuse sky radiance, d, from directions $r = (\theta, \phi)$. The model first solves the radiative transfer equation for intercepted flux. Additional correction terms are then appended to account for direct and diffuse flux, penetrating into the canopy through gaps, by utilizing a foliage-gap probability function that varies with height and direction, $P_0(z,r)$.

The pertinent equation of radiative transfer is:

$$\frac{1}{u(z)} \frac{dI(z,r)}{dz} = -g(r) I(z,r) + \frac{1}{\pi} \eta(z,r) \tag{3-21}$$

where $g(r)$ is the mean projection of total canopy-foliage area in the direction r and is calculated by averaging the projection of all leaf orientations in the direction r, weighted by their relative probability of occurrence, i.e.:

$$g(r) = \frac{1}{2\pi} \iint f(a) |a \cdot r| \sin \theta_l \, d\theta_l \, d\phi_l \tag{3-22}$$

where η, the source term, is separated into a multiple-scattering contribution and a first-order contribution:

$$\eta(z,r) = \iint P(r',r) I(z,r') \sin \theta' \, d\theta' \, d\phi' + \iint P(r',r) q(z,r') \sin \theta' \, d\theta' \, d\phi' \tag{3-23}$$

where q is the mean radiance penetrating the canopy to depth z without interception. $P(r',r)$ is the canopy phase function and is given by:

$$P(r',r) =$$
$$\frac{1}{2\pi} \iint \gamma(r',r,a) f(a) |a \cdot r| \, |a \cdot r'| \sin \theta_l \, d\theta_l \, \phi_l. \tag{3-24}$$

If I_+ represents upwelling flux and I_- downwelling flux, then the boundary conditions are taken to be:

$$I_-(z_0,r) = 0$$
$$I_+(0,r) =$$
$$R_g \iint [I_- (0,r') + q(0,r')] \cos \theta' \sin \theta' \, d\theta' \, d\phi'. \tag{3-25}$$

An iterative solution method is applied. First-order scattering yields the following expression for upwelling flux:

$$I_+ = \frac{F_0 \cos \theta_0}{\pi} \{l(r_0,r) \, e^{-K_0 L}$$
$$+ [R_g - l(r_0,r)] \, e^{-(K_0-K)L_0-KL} \tag{3-26}$$

where L is the cumulative foliage area index at height z, L_0 the total leaf area index and:

$$K = g(r)/\cos \theta$$
$$K_0 = g(r_0)/\cos \theta_0$$
$$g_0 = g(r_0)$$
$$l(r_0,r) = P(r_0,r)/(g \cos \theta_0 - g_0 \cos \theta). \tag{3-27}$$

Successive orders of scattering are applied as necessary, particularly in the infrared region. To correct for unintercepted direct and diffuse irradiance the following term is added to the downwelling flux:

$$d(r) \, P_0(z,r) + F_0 \, P_0(z,r_0)(r - r_0). \tag{3-28}$$

In practice, the Poisson distribution is often taken as the gap probability function, i.e.:

$$P_0(z,r) = e^{-K(r)L(z)}. \tag{3-29}$$

Finally, canopy reflectance is obtained by appropriate normalization.

Adding Method—Cooper, Smith, Pitts

The adding method assumes basically the same canopy abstraction as described in the Ross-Nilson model. This is also the same abstraction used in an earlier Monte-Carlo canopy reflectance model (Smith and Oliver, 1972). The model explicitly includes reflectances and transmittances for both sides of a leaf scattering-element. The adding method used by the authors is a modification of that described by Van de Hulst (1980) and is an example of the interaction principle alluded to by Preisendorfer (1965). The model is an efficient calculational procedure for all scattering orders and explicitly factors the pertinent operators

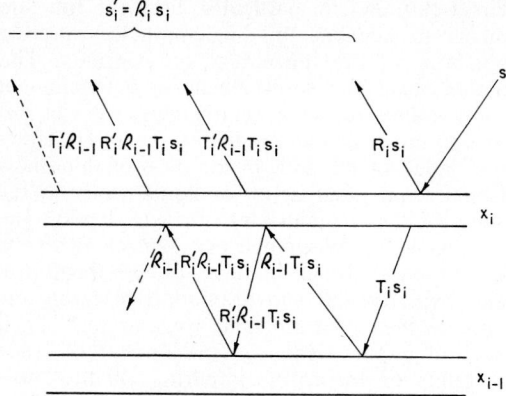

$s_i' = R_i s_i$

Fig. 3-2. Pictorial representation of the scattering of flux back and forth between two arbitrary layers, x_i and x_{i-1}, of a leaf of a dicotyledonous plant.

into a mathematical product of the geometrical terms and optical properties of the canopy. This factorization permits the precalculation of terms dependent upon canopy architecture, which can then be convolved with appropriate optical properties as desired. Further, the authors suggest that estimation techniques for the determination of the geometrical properties from measured canopy reflectance may thus be developed.

The model consists of two parts. First a recursive procedure to calculate the response from a composite of several canopy or understory layers is established. Secondly, the required matrix scattering-operators are related to the geometrical and optical properties of the leaf. Figure 3-2 is a pictorial representation of the scattering of flux back and forth between two arbitrary layers X_i and X_{i-1} with the following properties. First all radiation incident upon layer X_{i-1} is either absorbed or backscattered upwards by means of the reflection operator $\mathcal{R}_{i-1}$. Secondly, the flux incident on either side of layer X_i may be reflected, transmitted, or absorbed. The reflection, R, and transmission, T, operators for layer X_i are assumed to be different depending upon whether the flux is incident from above or below the layer. The unprimed operators are used for flux incident from above while the primed operators are applicable for flux incident from below. The quantity s_i represents the initial incident flux, and s_i' represents the total scattered flux from the composite of both layers. As usual, the wavelength subscript has been suppressed. The dimensions of the operators are defined upon implementation and depend upon the discretizing intervals used for the angular variables.

The authors show that

$$s_i' = \mathcal{R}_i s_i \qquad (3\text{-}30)$$

where

$$\mathcal{R}_i = R_i + T_i' \mathcal{R}_{i-1}(I - R_i' \mathcal{R}_{i-1})^{-1} T_i \qquad (3\text{-}31)$$
$$I = \text{Identity Matrix}$$

which indicates that layer X_{i-1} has been added to layer X_i forming a new composite layer incorporating the properties of the original two layers. The definition is recursive in that if an additional X_{i+1} layer is required to represent the medium, then an appropriate $\mathcal{R}_{i+1}$ can similarly be found.

The authors derive the following expressions for the R, T, R', and T' matrices:

$$R = \text{Diag}\big[(1 - P(\theta_i))^{1/2}\big]\big[\rho_u D_{11} - \tau_u D_{12} + \tau_l D_{21} + \rho_l D_{22}\big]\,\text{Diag}\big[(1 - P(\theta_j))^{1/2}\big]$$

$$T = \text{Diag}\big[(1 - P(\theta_i))^{1/2}\big]\big[\tau_u D_{11} + \rho_u D_{12} + \rho_l D_{21} + \tau_l D_{22}\big]\,\text{Diag}\big[(1 - P(\theta_j))^{1/2}\big] + \text{Diag}\big[P(\theta_i)\big]$$

$$R' = \text{Diag}\big[(1 - P(\theta_i))^{1/2}\big]\big[\rho_l D_{11} + \tau_l D_{12} + \tau_u D_{21} + \rho_u D_{22}\big]\,\text{Diag}\big[(1 - P(\theta_j))^{1/2}\big]$$

$$T' = \text{Diag}\big[(1 - P(\theta_i))^{1/2}\big]\big[\tau_l D_{11} + \rho_l D_{12} + \rho_u D_{21} + \tau_u D_{22}\big]\,\text{Diag}\big[(1 - \rho(\theta_j))^{1/2}\big] + \text{Diag}\big[\rho(\theta_i)\big]$$

where u,l refer to the upper/lower leaf surface and $P(\theta_i)$ is the same as in the Ross-Nilson model, i.e.:

$$P(\theta_i) = e^{-Lg(\theta_i)/\cos\,\theta_i}. \qquad (3\text{-}33)$$

Other expressions may be more appropriate, e.g., DeWit (1965) or Nilson (1971). The product of the asymmetrical optical properties of the leaves on upper and lower sides and the D matrices essentially correspond to the phase function for the media. Because of the probability that the reflectances and transmittances of leaves are not equal, except at some wavelengths, and that the optical properties of the leaves may be different depending upon the side of the leaf, some care must be taken in defining the phase function. In their implementation the authors assume an azimuthally symmetric distribution of leaf orientations and a leaf slope distribution $f(\theta_l)$. With these assumptions there is a critical azimuth angle, ϕ_c, relative to the incident and exitant flux, which defines on which side of the leaf incident flux, a, will be intercepted. If Z1 and Z2 are defined as:

$$Z_1(\theta_i, \theta_l) = \frac{1}{\pi}\int_0^{\phi_c} |a \cdot r|\, d\phi$$

$$Z_2(\theta_i, \theta_l) = \frac{1}{\pi}\int_{\phi_c}^{\pi} |a \cdot r|\, d\phi \qquad (3\text{-}34)$$

then the i, jth element of the D matrices is defined according to the following expression:

$$d_{mn}(i,j) = \int_0^{\pi/2} f(\theta_l)\int_{\Delta\theta_j} Z_n(\theta_j, \theta_l)\,\sin\,\theta_j$$

$$\int_{\Delta\theta_i} Z_m(\theta_i,\theta_l)\, \sin\theta_i\, d\theta_j\, d\theta_i$$

$$m,\, n = 1,2.$$

where incident and exitant directions have been discretized in finite directional bands. The reflection operator for the initial layer is taken to be the background reflectance, R_g.

The radiance from any arbitrary number of layers may then be computed according to the previous recursive definition. Upon appropriate normalization, canopy reflectance is obtained. Finally, as a parenthetical note, the authors utilize this particular representation of canopy reflectance to specify the conditions under which reciprocity is satisfied (Helmholtz, 1909).

Snow Reflectance Models

The development of theoretical snow-reflectance models closely parallels that for vegetation canopies. The prominent early models include the model developed by Middleton and Mungall (1952) to explain the photometric properties of snow and the model of Dunkle and Bevans (1956) to explain diffuse reflectance and transmittance. The paper by Giddings and LaChapelle (1961) describes an extension of the treatment for diffuse radiation in the snow medium by permitting a "certain amount of nondiffuse radiation." In contrast to Dunkle and Bevans, the authors employ the diffusion approximation to describe multiple scattering within the layer and subsequently a random-walk model to estimate key parameters in the diffusion approximation. The decade of the 1970s saw the development of more sophisticated models, partially spurred by the widespread availability of satellite observations, e.g., Landsat and the NOAA TIROS-N series, and by the recognition of snow as an important resource to be assessed and managed. Early snow models included adjustable parameters to characterize the electromagnetic properties of the snow medium. Later efforts, e.g., Bergan (1975), sought to relate these parameters to measurable snow characteristics such as grain size, density, and refractive index.

One of the earliest efforts to apply modern radiative transfer theory to snow was that of Bohren and Barkstrom (1974). In their paper the authors treat the snow medium as consisting of a random distribution of spherical scatterers (transparent ice spheres) separated by sufficiently large distances so that phase coherence can be neglected. The authors then used geometric optics to analyze the scattering off these spheres in order to calculate absorption, scattering, and phase function for the medium by suitable averaging of the individual sphere properties. The authors also related these coefficients to snowpack density and grain size.

More recently Wiscombe and Warren (1980) employed a delta-Eddington approximation for the radiative transfer analysis of both diffuse and direct flux and, in particular, used the full Mie theory to calculate the single-particle scattering behavior of the individual snow grains. The authors calculate snow albedo as a function of snow grain-size, solar zenith-angle, ratio of diffuse to direct incident radiation, complex refractive index of ice and, in the case of thin snow layers, equivalent depth of liquid water in the snowpack and underlying surface albedo. The first paper by Wiscombe and Warren is an excellent and thorough summary of both theoretical and experimental snow-radiation research and outlines their own modeling work as well. Their second paper discusses the impact of trace amounts of absorptive impurities in the snow medium on overall snow albedo.

Choudhury and Chang (1981) also describe a snow-reflectance model that includes angular considerations. The volume scattering contribution of their model is similar to that of Wiscombe and Warren (Dozier, 1981). However, they include a specular surface reflection term that is derived from isotropic Gaussian facets using rough-surface scattering theory (Barrick, 1970; see also Beckman and Spizzichino, 1963).

It is apparent from the foregoing discussion that no single model has been developed that is valid for all of the considerations of interest (Wiscombe and Warren, 1980). It is also apparent that the modeling of heterogeneous snow surfaces with more complex geometries (other than plane-parallel) has not been significantly addressed. This may reflect the lack of sufficient high-quality field data to support such model development.

Below we briefly present in more detail the model of Dunkle and Bevans, Bergan's analysis, and the models of Wiscombe and Warren and Choudhury and Chang as examples of the types of calculational procedures available.

Dunkle and Bevans

The Dunkle and Bevans model (1956) makes the following assumptions:

1) The medium is homogeneous and of constant optical density.
2) The medium is perfectly diffuse.
3) The distributed scattering coefficient, s, is constant, i.e., isotropic scattering is assumed.
4) The absorption coefficient, k, is also constant and independent of depth.
5) The medium is infinite in width and breadth, and of finite depth, D.

Given these assumptions, a set of equations exactly analogous to the Allen and Richardson Kubelka-Munk approximation employed for vegetation canopies may be written to describe the scattering and absorption with depth, z, of the downwelling radiance, I, and upwelling radiance, J:

$$\frac{dI}{dz} = -(k+s)I + sJ$$

$$\frac{dJ}{dz} = -sI + (k+s)J$$

$$(3\text{-}36)$$

where z is measured positive downward from the surface, where $z = 0$. The Dunkle and Bevans model assumes that the snow layer overlays a reflecting boundary at $z = D$ whose diffuse reflectance is given by R_g. In contrast to the Allen and Richardson canopy model, and Dunkle and Bevans model also allows for an upper surface boundary with diffuse reflectance r_s. The boundary conditions are thus taken to be:

$$J(z = D) = R_g I(z = D)$$

$$I(z = 0) = (1 - r_s) I_0 + r_s J(z = 0)$$

$$(3\text{-}37)$$

where I_0 is the incident radiance.

The authors present two solutions for $J(z = 0)$, the volume or multiple scattering contribution to reflectance, corresponding to the general case, i.e., a thin layer of snow, and an optically thick, i.e., infinite, snow layer. In practice, the difference between these two cases is whether or not the bottom boundary surface makes a contribution. For snow the optically thick case is generally reached for depths on the order of 30 cm.

The general solution is given by:

$$J(z) = \left\{ (1 - r_s) I_0 \left[(1 - \alpha_1 R_g) e^{\beta(D-z)} \right. \right.$$
$$\left. \left. - (1 - \alpha_2 r_s) e^{-\beta(D-z)} \right] \right\}$$
$$\left[D_1 e^{\beta D} - D_2 e^{-\beta D} \right]^{-1}$$

$$(3\text{-}38)$$

where

$$\alpha_1 = (k + s - \beta)/s$$
$$\alpha_2 = (k + s + \beta)/s$$
$$\beta = \sqrt{k(k + 2s)}$$
$$D_1 = (\alpha_2 - R_g)(1 - r_s \alpha_1)$$
$$D_2 = (\alpha_1 - R_g)(1 - r_s \alpha_2)$$

$$(3\text{-}39)$$

The optically thick solution is given by:

$$J(z) = \left[\alpha_1 (1 - r_s) I_0 e^{-\beta z} \right] \left[1 - r_s \alpha_1 \right]^{-1}.$$

$$(3\text{-}40)$$

The reflectance of the snow layer, or albedo, is given by the ratio of the reflected radiance to incident flux. The model also differs from the Allen and Richardson model in that the authors explicitly include both the surface reflectance and volume reflectance contributions, i.e., the snow reflectance is given by:

$$R = \left[r_s I_0 + (1 - r_s) J(z = 0) \right]/I_0$$

or if $I_0 = 1$,

$$R = r_s + (1 - r_s) R_v.$$

$$(3\text{-}41)$$

The thick layer reflectance is given by the rather simple expression:

$$R = r_s + \alpha_1 (1 - r_s)^2 (1 - r_s \alpha_1)^{-1}.$$

$$(3\text{-}42)$$

Note that, in the Dunkle and Bevans model, the surface-reflectance term and the underlying bottom reflectance are assumed to be known. In contrast, Middleton and Mungall (1952) calculated the surface-reflectance term but ignored the volume-reflectance contribution.

The authors also address the apparent confusion in the measurement arena and the resultant difficulty of comparing their predictions to existing measurements.

Bergan

The Dunkle and Bevans model made a major contribution to the snow-reflectance problem in that it was the first attempt to calculate snow reflectance using the radiative transfer method. As with the Allen and Richardson model, however, no attempt was made to relate the absorption and scattering parameters to more traditional descriptors of snow, e.g., density or grain size. Several later papers by different authors sought to remedy this deficiency. The work of Bergan (1975) is illustrative of these efforts although Bohren and Barkstrom (1974) have questioned Bergan's estimation of the scattering parameter.

Bergan argues that the absorption parameter k for the snow medium may be taken to be:

$$k = k'(\rho/\rho_i)$$

$$(3\text{-}43)$$

where

k' = the absorption coefficient for solid ice
ρ = snow density
ρ_i = density of ice

and that the scattering parameter may be taken to be:

$$s = r_0 S/3$$

$$(3\text{-}44)$$

where

r_0 = reflectivity of a plane ice–air interface
S = area of the ice–air interface per unit volume

Empirical relations are then taken to relate S to the Darcy air permeability constant, K, and K to grain size; that is:

$$K = \left(\frac{1}{5} S^2 \mu \right) (1 - \rho/\rho_i)^3$$

$$(3\text{-}45)$$

and

$$K = 3.03\, r^{1.63} (1 - \rho/\rho_i)\, 10^{-3}$$

$$(3\text{-}46)$$

where

r = grain size
μ = reference viscosity.

The final result, therefore, is that the snow albedo may be inferentially related to density and grain size.

Wiscombe and Warren

The model of Wiscombe and Warren first abstracts the snow medium to consist of spherical scattering snow-grains characterized by a complex index of refraction, m, and a grain size r. The Mie scattering formulas of van de Hulst (1957) are then used to calculate the extinction coefficient, Q_{ext}, the single scattering albedo, $\bar{\omega}$, and the asymmetry factor, g, i.e., mean value of the cosine of the scattering angle. Given these values, the snowpack water-equivalence, W, and the density of pure ice, ρ_i, the medium optical depth, τ_0, is calculated as:

$$\tau_0 = \frac{3\ W\ Q_{ext}}{4\ r\ \rho_i}. \tag{3-47}$$

Finally, the delta-Eddington approximation is employed to calculate the strong forward scattering occurring within snow. The approximation begins by transforming the optical depth, single scattering albedo, and asymmetry factors as follows:

$$\tau_0^* = (1 - \bar{\omega}\ g)\ \tau_0$$

$$\bar{\omega}^* = \frac{(1 - g^2)\ \bar{\omega}}{1 - g^2\ \bar{\omega}} \tag{3-48}$$

$$g^* = \frac{g}{1 + g}.$$

The ordinary Eddington approximation is then applied to a snow layer described by the above transformed variables. The directional hemispherical albedo for the case of only direct-beam incidence at a zenith angle of $\theta_0 = \cos^{-1} \mu_0$ is given by:

$$a_s(\mu_0) = \frac{2}{Q} \left[P(1 - \gamma + \bar{\omega}^*b^*) \right.$$

$$+ \bar{\omega}^*(1 + b^*)\ \frac{\gamma\zeta\mu_0 - P}{1 - \zeta^2\ \mu_0^2} \left.\right]_e - \tau_0^*/\mu_0$$

$$- \bar{\omega}^*b^*(Q^+ - Q^-) + \bar{\omega}^*(1 + b^*)$$

$$\left(\frac{Q^+}{1 + \zeta\ \mu_0} - \frac{Q^-}{1 - \zeta\ \mu_0} \right)$$

where

$$a^* = 1 - \bar{\omega}^*g^*$$
$$b^* = g^*/a^*$$
$$\zeta = \left[3a^*(1 - \bar{\omega}^*) \right]^{1/2}$$
$$P = \frac{2\ \zeta}{3\ a^*} \tag{3-50}$$
$$\gamma = \frac{1 - R_g}{1 + R_g}$$
$$Q^{\pm} = (\gamma \pm P)\ e^{\pm\zeta\tau_0^*}$$
$$Q = (1 + P)\ Q^+ - (1 - P)\ Q^-$$
$$R_g = \text{Background Albedo.}$$

For the special case of a very thick layer of snow, i.e., generally over 30 cm, the albedo is given by the simplified expression:

$$a_s(\mu_0) = \frac{\bar{\omega}^*}{1 + P}\ \frac{1 - b^*\ \zeta\ \mu_0}{1 + \zeta\ \mu_0}. \tag{3-51}$$

However, if the diffuse sky field is also to be considered, then the direct beam albedo, a_s, must be integrated over the hemisphere yielding the diffuse albedo, a_d. For the sake of brevity the reader is referred to the authors' original paper for the complete expression. However, the total net albedo is then given by:

$$a_{net} = R\ a_d + (1 - R)\ a_s(\mu_0) \tag{3-52}$$

where

R = the diffuse irradiance fraction.

While the expressions may appear somewhat formidable, the reader is reminded that the model is actually quite practical since it requires only a limited number of inputs. The authors indicate that their model compares favorably with experimental data in the near infrared but generally calculates apparently high albedo values for old and melting snow. However, their second paper (1980) examines these questions further.

Choudhury and Chang

Through use of the same nomenclature as employed in the Wiscombe and Warren model above, the spectral directional hemispherical reflectance according to Choudhury and Chang may be written as the sum of a surface scattering term $f(\theta_0,\phi_0)$ and a volume or multiple scattering term: a_d

$$a_s(\mu_0) = f(\theta_0,\phi_0)\ (1 - R)$$
$$+ \left[1 - (1 - R)\ f(\theta_0,\phi_0) \right] a_d \tag{3-53}$$

where $f(\theta_0,\phi_0)$ is the directional hemispherical surface reflectance, i.e.:

$$f(\theta_0,\phi_0) = \int_0^{2\pi} \int_0^1 f_r(\theta,\phi;\theta_0,\phi_0)\ \mu\ d\mu\ d\phi. \tag{3-54}$$

The (surface) bidirectional reflectance distribution function, f_r, in turn is taken to be:

$$f_r(\theta,\phi;\theta',\phi') = \frac{\sec^4 \alpha\ \exp\left[-\dfrac{\tan^2 \alpha}{2s^2} \right] S(\theta,\theta')\ \rho(\psi)}{8\pi s^2 \cos \theta \cos \theta'} \tag{3-55}$$

where

s = the variance of surface slope

$$\tan \alpha = \frac{\left[\sin^2 \theta + \sin^2 \theta' - 2 \sin \theta \sin \theta' \cos \phi' \right]^{1/2}}{(\cos \theta + \cos \theta')}$$

$$\cos \psi = \frac{1}{\sqrt{2}}(1 + \cos \theta \cos \theta' - \sin \theta \sin \theta' \cos \theta')^{1/2}$$

$$(3\text{-}56)$$

$\rho(\psi)$ = the Fresnel reflectivity for angle ψ, i.e.:

$$= \frac{1}{2}\left[\left(\frac{\sqrt{\epsilon - \sin^2 \psi} - \cos \psi}{\sqrt{\epsilon - \sin^2 \psi} + \cos \psi} \right)^2 \right.$$

$$\left. + \left(\frac{\epsilon \cos \psi - \sqrt{\epsilon - \sin^2 \psi}}{\epsilon \cos \psi + \sqrt{\epsilon - \sin^2 \psi}} \right)^2 \right] \quad (3\text{-}57)$$

$$S(\theta,\theta') = \frac{1}{C_0 + C_1 + 1}$$

$$2C_0 = \left(\frac{2s^2}{\pi} \right)^{1/2} \tan \theta \; e^{-\cot^2 \theta / 2s^2} - erfc\left(\frac{\cot \theta}{s\sqrt{2}} \right)$$

$$(3\text{-}58)$$

$$2C_1 = \left(\frac{2s^2}{\pi} \right)^{1/2} \tan \theta' \; e^{-\cot^2 \theta' / 2s^2} - erfc\left(\frac{\cot \theta'}{s\sqrt{2}} \right)$$

$erfc$ = the complementary error function, and
ϵ = the dielectric constant for the ice grains.

The volume-scattering diffuse contribution, a_d, is calculated in a manner very similar to that of Wiscombe and Warren and, in fact, yields basically the same result. The authors use Mie theory to calculate the single-scattering albedo and phase function and then factor the phase function into a strongly forward-scattering term and a term that corresponds to a category that can be regarded as being "everything else." Details of the calculation and the specific approximations used to estimate the forward-scattering fraction and single-scattering albedo are given in the original reference (Choudhury and Chang, 1981).

Soil/Mineral Reflectance Approaches

The radiative transfer modeling of rock spectral-reflectance properties poses several additional problems over those of both vegetation and snow discussed earlier. First, it is evident that particle sizes may be both fine- and coarse-grained with respect to the optical wavelengths considered. In particular, the problem is exacerbated at thermal wavelengths, where reflectance properties are calculated in order to infer the spectral emittance properties of materials. The particles are also irregularly shaped and in physical contact. In addition to these textural considerations, the variable chemical composition, i.e., mineral content, of the medium plays a role.

A review of the literature indicates that theoretical model development was initiated in the late 1960s, perhaps under the impetus of the lunar and planetary exploration programs. Development has been sporadic; first, because of the theoretical dificulties encountered and secondly, perhaps, because of the competing thermal approaches concerned, for example, with the estimation of such bulk thermal properties as thermal inertia. With the advent of the Thematic Mapper and the extension of multilinear-array sensors to include both thermal wavelengths and narrow-spectral-band wavelengths in the visible and near-infrared regions, it is expected that new modeling interest will be stimulated.

An analysis of the approaches used to calculate rock or mineral reflectance shows a striking resemblance to the approaches discussed earlier for snow and vegetation. The Kubelka-Munk theory was utilized by Vincent and Hunt (1968) to calculate volume reflectance from a mineral or rock powder. They also included a specular component using Fresnel arguments and showed that the total spectral reflectance would vary with particle size. Emslie and Aronson (1973) extended the Kubelka-Munk theory to include both coarse- and fine-grained particles. For the individual particles composing the minerals, both geometric optics and wave theory were used appropriately, depending upon the particle-size regime being considered. Egan and Hilgeman (1978) described the use of the Monte Carlo technique to calculate individual particle- or asperity-scattering from irregularly shaped grains. Finally, Conel (1969) used general radiative transfer theory to calculate the response from a "plane-parallel cloud" of particles. Mie theory was used to calculate individual particle behavior assuming spherical scatterers within the cloud.

The present state of the theory for the spectral reflectance from rock surfaces indicates that no one model is capable of providing accurate predictions over the complete range of particle sizes encountered. Multi-mineral media and bifringence effects have not been addressed, and weathering and vegetative surface layers have not been included in the analyses. The models of Conel and of Emslie and Aronson are briefly discussed.

Conel

The Conel model was used to infer infrared spectral emissivities of silicates and treats a particulate medium as a plane-parallel layer of independent, non-phase coherent scatterers. Anisotropic and nonconservative multiple scattering arises from the strongly foward Mie diffraction behavior for particle sizes comparable to the thermal wavelengths simulated. Inputs to the model consist primarily of particle size and the complex index of refraction as a function of wavelength. The model was useful for gaining insight into the variation of spectral emissivity with particle size.

The author begins with the equation of radiative transfer:

$$\mu \frac{dI}{d\tau} = I - \frac{\bar{\omega}}{4\pi} \iint P(\mu,\phi;\mu',\phi') \, d\mu' \, d\phi'.$$

$$(3\text{-}59)$$

The phase function is expressed as a truncated Legendre polynomial series:

$$P(\Theta) = (1 - g \cos \Theta) \qquad (3\text{-}60)$$

where:

g = the anisotropy factor and
Θ = the scattering angle between incident and existance directions.

The optical depth, τ, is related to the number of particles per unit volume, N, the geometrical cross-section per particle, A, the efficiency factor for extinction, Q_{ext}, and the incremental depth, dz, measured normal to the surface:

$$d\tau = - NA \, Q_{ext} \, dz. \qquad (3\text{-}61)$$

The author points out that an exact solution to the above equation for nonconservative and linearly anistropic scattering is available from Chandrasekhar (1960). However, he states that simpler approximate solutions using the method of discrete ordinates and expanding the integral in two-point Gaussian sum yields reasonable insight. The transformed radiative transfer equations for upward-going flux, I_+, at a zenith angle of approximately 55 degrees (i.e., $\mu = 1/\sqrt{3}$) and the downward-going flux at the corresponding angle, I_-, are:

$$\frac{1}{\sqrt{3}} \frac{dI_+}{d\tau} = I_+ - \tfrac{1}{2}\bar{\omega}(1 + g/3) \, I_+$$

$$- \tfrac{1}{2}\bar{\omega}(1 - g/3) \, I_-$$

$$-\frac{1}{\sqrt{3}} \frac{dI_-}{d\tau} = I_- - \tfrac{1}{2}\bar{\omega}(1 - g/3) \, I_+$$

$$- \tfrac{1}{2}\bar{\omega}(1 + g/3) \, I_-. \qquad (3\text{-}62)$$

The boundary conditions are:

$$I_- = I_-(o)$$

$$(3\text{-}63)$$

$$I_+ = I_+(\tau).$$

Thus, ignoring surface specular reflection, the following solution for reflectance is obtained:

$$R = \frac{I_+(o)}{I_-(o)} = \frac{(u^2 - 1) \, (e^{\tau*} - e^{-\tau*})}{(u + 1)^2 \, e^{\tau*} - (u - 1)^2 \, e^{-\tau*}}$$

$$(3\text{-}64)$$

where:

$$u^2 = (1 - \bar{\omega} \, g/3) \, (1 - \bar{\omega})^{-1}$$
$$\tau* = \sqrt{3} \, (1 - \bar{\omega}) \, u \, \tau \qquad (3\text{-}65)$$
$$\tau_0 = \text{total optical thickness.}$$

For an infinitely thick medium, the reflectance simplifies to:

$$R_\infty = (u - 1)/(u + 1). \qquad (3\text{-}66)$$

Consequently, the emissivity is given by:

$$\epsilon = 1 - R_\infty = 2/(u + 1). \qquad (3\text{-}67)$$

Solutions for other directions can similarly be found.

Through use of the assumption of spherical particles of fixed dimension and their complex index of refraction, Mie scattering theory was used to calculate $\bar{\omega}$ and g. The author discusses particular problems induced by large values of the imaginary part of the refractive index and his criteria for truncation of the appropriate series. The geometric optics approximation for large particles was not found to be useful.

Emslie and Aronson

After some manipulation, the authors express the radiative flux balance within the particulate medium by Kubelka-Munk type equations in which the scattering, S', and absorption, K', coefficients have been expanded to include the transverse as well as the backward directions. The central part of their model is then designed to calculate appropriate values for K' and S'. Depending upon the relative size of the individual scatterers within the medium with respect to the wavelengths considered, coherent or incoherent scattering may occur. Consequently, two different methods of calculation, termed a coarse-particle theory and a fine-particle theory, are employed to estimate these coefficients. Essentially, geometric optic analysis is applied to spherical particles with complex indices of refraction in the coarse-grained theory. Wave optics or Rayleigh scattering is utilized in the fine-grained theory.

The expanded absorption and scattering coefficients for the medium are given by:

$$S' = (4K \, S_b + 2 \, S_b S_t + S_t^2)/(4K + 2S_t)$$
$$K' = (4K^2 + 6KS_t)/(4K + 2S_t). \qquad (3\text{-}68)$$

where K is the normal absorption parameter for the medium and S_b and S_t are the backscattering and transverse scattering coefficients, respectively. These coefficients, in turn, are proportional to the product of N, (the number of particles per unit volume) and the appropriate single-particle cross-sections for absorption or scattering as calculated by the appropriate theory. N is a function of the volume fraction, f, and the diameter, d, of an equivalent sphere:

$$N = 6f/\pi \, d^3. \qquad (3\text{-}69)$$

The reflectance of a coarse-grained particulate medium is given by:

$$R_v = 1 + \frac{K'}{S'} - \sqrt{\left(\frac{K'}{S'}\right)^2 + \frac{2K'}{S'}}$$

$$(3\text{-}70)$$

where the following values are substituted in the expressions for S' and K'.

$$S_t = N \sigma_t$$
$$S_b = N \sigma_b \qquad (3\text{-}71)$$
$$K = N \sigma_{abs}.$$

The total transverse and backward scattering cross sections, σ_t and σ_b, are the sum of the individual cross-sections arising from reflection, refraction, and internal reflection processes. Given a sphere of radius a with complex index of refraction, $m = n - i\,k$, the cross-sections for reflection are given, for example, by:

$$\sigma_t = \pi a^2 \int_0^{\pi/2} R_0 \sin^2 2\theta \, \sin 2\theta \, d\theta$$

$$(3\text{-}72)$$

$$\sigma_b = \pi a^2 \int_0^{\pi/4} R_0 \cos^2 2\theta \, \sin 2\theta \, d\theta$$

where R_0 is the surface Fresnel reflection coefficient (actually, the authors derive expanded expressions to include the effects of edges and asperities). The single-particle absorption coefficient is given by:

$$\sigma_{abs} = \pi a^2 \int_0^{\pi/2} \left(1 - R_0 - \frac{T_0^2 \, T}{1 - R_0 \, T} \right) \sin 2\theta \, d\theta$$

where T_0 is the surface transmittance and T is the transmittance factor given by an exponential attenuation proportional to the complex part of the index of refraction, k.

The reflectance of the medium in the fine-particle case is given by:

$$R = R_s + (1 - R_s)^2 \, R_v/(1 - R_s R_v)$$

$$(3\text{-}74)$$

where R_s is the Fresnel surface reflectance for the medium and is of the same form as before, only now the following expressions for S_t, S_b, and K must be used:

$$S_t = 4/6 \, N \, \sigma$$
$$S_b = 1/6 \, N \, \sigma \qquad (3\text{-}75)$$
$$K = 4\pi \, \bar{k}/\lambda$$

where σ is the Rayleigh scattering coefficient, which is a function of the individual particle index of refraction and the average Lorentz-Lorenz index of refraction for the medium. $\bar{k}$ is the complex part of the average Lorentz-Lorenz coefficient.

In both the fine- and coarse-particle theories, the effect of mixtures of different mineral particles or of different particle sizes may be treated by appropriate summation over particle-volume fractions, particle sizes, and complex indices of refraction.

Thermal Regime

Beginning with the mid-1970s there was considerable activity in the modeling of thermal radiance from soil and vegetation-terrain elements. Modeling of the thermal regime for soil and rock targets received impetus from geologic applications and the desire to infer soil moisture. Similarly, the modeling of vegetation surfaces has been partially geared towards the inference of water status, particularly as related to crop yields. Numerous references are given in the applications chapters of this manual. Example references include Heilman, Kanemasu, Rosenberg and Blad (1976); and to Idso, Schmugge, Jackson, and Reginato (1975). The mapping of thermal inertia properties in semi-arid regions received significant impetus from the Heat Capacity Mapping Mission (HCMM). Aircraft thermal data are also readily available. Indeed, refinements and extensions to the basic techniques are being addressed in the case of soils (Pratt, 1980; Rosema, 1975).

A useful way to organize modeling in the thermal regime is by terrain type: soil, soil/vegetation, vegetation, and snow/vegetation.

Considerable advancement has been made in the physical process-oriented modeling of soils and rocks from a remote-sensing perspective. The basic approach is to abstract the soil as a multi-layered stratum and apply the thermal conduction equation to describe heat transfer through the layers. At the surface, various boundary conditions are applied including short- and long-wave fluxes, sensible heat, and evaporation. There are two basic approaches. One is the analytic method (Watson, 1975) in which further assumptions for the top boundary condition are made so that analytic solutions may be derived. The second approach (Kahle, 1977) allows more general forms for the top boundary condition and consequently employs finite-difference techniques for solution of the heat-conduction equation. Both of these approaches are really geared towards geologic applications in semi-arid environments, i.e., sparse vegetation cover.

The thermal-conduction approach has also recently been extended to incorporate simple vegetative cover by appropriately modifying the top boundary-surface condition. Two examples are Deardorff (1978) and Balrick et al. (1981). The emphasis in these cases remains with the heat-transfer mechanisms in the soil layers.

Modeling of the thermal behavior of vegetation canopies requires different approaches. Perhaps the most dominant techniques include the derivation of statistical empirical relationships between the time-dependent behavior of canopies and such driving mechanisms as solar illumination. The second broad approach is to apply energy-budget analyses either to single leaves (Gates, 1968) or canopies as a whole, although few such models incorporate canopy geometry (Goudriaan, 1977; Norman, 1979; Kimes, Smith and Link, 1981; and Smith, et al., 1981). These models generally treat the time-independent or steady-state behavior of a canopy. Recently an attempt has been made to use the statistical approach for diurnal variations in canopy temperature conditioned by physical treatment of flux transfers using a Kalman filtering approach (Smith and Randolph, 1982). The influence of canopy structure on the interpretation of remotely sensed canopy temperatures has been discussed by Kimes, Idso, Pinter, Jackson and Reginato (1980).

Finally, the theory of radiation heat transfer between forest canopies and snowpacks has been studied by Bohren (1971) and Bohren and Thorud (1971) from a radiative-transfer perspective, including the multidimensional problem of the radiative regime around a tree bole.

A comparison of several different modeling approaches has been performed by Dodd and Conrow (1981). In general, significant predictive differences arose primarily due to variations in the surface-boundary conditions employed. As an illustration of the calculational procedures available for physical process-oriented thermal modeling, an example from each of the four broad terrain-element categories discussed above will be given.

The Analytic Method— Watson (1975; 1979)

Watson and co-workers performed some of the earliest investigations of the use of thermal models to discriminate geologic materials based on such thermal property differences as thermal inertia. In their earliest work the investigators treated the one-dimensional heat-conduction equation, subject to a surface boundary condition that ignored sensible and latent heat fluxes. The remaining surface energy-budget terms were then expressed as linear functions of ground temperature. In such a case a complete analytic solution based on Fourier series techniques was possible. In subsequent work the authors included sensible heat components and, again, linearized the resulting boundary equations in order to apply the Fourier series technique. The investigators were well aware of such numerical solution techniques as the finite-difference method discussed in the following sections. Nevertheless, they have become strongly associated with the analytic technique because of the depth of their analyses using this method. While the necessity of linearizing the surface energy-budget relationship often leads to errors in magnitude, significant insight may be derived from the resulting analytic solutions. The following discussion on the analytic method is drawn heavily from Watson's 1979 paper.

The author first summarizes the analytic mathematical-solution techniques into four simple cases. If the thermal conduction equation is given by:

$$\frac{\partial^2 T}{\partial z^2} = \frac{1}{\kappa} \frac{\partial T}{\partial t} \qquad (3\text{-}76)$$

subject to the boundary condition:

$$-K \frac{\partial T}{\partial t} = (1 - A)\, q_{solar} + \epsilon\, q_{sky} - \epsilon\sigma T^4$$

$$(3\text{-}77)$$

where
  A, ϵ are the ground shortwave albedo and emissivity,
  K, κ the thermal conductivity and diffussivity,
  q_{solar}, q_{sky} the solar source flux and long wave sky component,

then three of the cases correspond to the thermal conduction boundary term equal to a constant, a time-variable flux and a simple periodic time-variable flux. The fourth case corresponds to the conduction being expressed as a linear function of temperature. All of these cases and their resulting solutions are given below. In all cases the initial temperature distribution is taken to be zero.

Case 1—constant flux

Boundary condition:

$$-K \frac{\partial T}{\partial z} = F_0 \qquad (3\text{-}78)$$

Solution:

$$T(0,t) = \frac{2F_0}{K} \sqrt{\frac{\kappa t}{\pi}}. \qquad (3\text{-}79)$$

Case 2—time-variable flux

Boundary condition:

$$-K \frac{\partial T}{\partial z} = F(t) \qquad (3\text{-}80)$$

Solution:

$$T(0,t) = \frac{\sqrt{\kappa}}{K\sqrt{\pi}} \int_0^t \frac{F(t - \tau)}{\sqrt{\tau}}\, d\tau. \qquad (3\text{-}81)$$

Case 3—time-periodic flux

Boundary condition:

$$-K \frac{\partial T}{\partial z} = F_0 \cos(\omega t + \phi) \qquad (3\text{-}82)$$

Solution:

$$T(0,t) = \frac{F_0\sqrt{\kappa}}{K\sqrt{\omega}} \cos\left(\omega t + \phi - \frac{\pi}{4}\right). \qquad (3\text{-}83)$$

Case 4—linear temperature dependence

Boundary condition:

$$-K \frac{\partial T}{\partial z} = T - h + hf(t) \qquad (3\text{-}84)$$

where h is a constant
  Solution:

$$T(0,t) = \sum_{n=0} \frac{A_n \, h \, \sin(n\omega t + \phi_n - \delta_n)}{\sqrt{(h + w_n)^2 + w_n{}^2}} \tag{3-85}$$

where

$$w_n = \sqrt{\frac{n\omega}{2\kappa}}$$

$$j_n = \tan^{-1}\left[\frac{w_n}{(h + w_n)}\right] \tag{3-86}$$

$$f(t) = \sum_{n=0} A_n \sin(n\omega t + \phi_n)$$

(Fourier Series Expansion).

For the case where sensible and latent heat-source terms are omitted from the surface energy-budget condition, the main terms consist of absorbed shortwave radiation, downwelling longwave radiation, and upwelling surface longwave flux. Nominally even this simplified boundary condition varies as the fourth power of ground temperature but this term may be linearized about a reference temperature, T_0, to yield the resulting expression:

$$-K\frac{\partial T}{\partial z} = (1 - A)q_{solar}(t) + \epsilon q_{sky}(t)$$

$$+ 3\epsilon\sigma T_0{}^4 - 4\epsilon\sigma T_0{}^3 T(z,t) \tag{3-87}$$

where h, termed the radiation constant, is given by:

$$h = \frac{4\epsilon\sigma T_0{}^3}{K} \tag{3-88}$$

A simplified solution yielding considerable insight can be derived by assuming that the solar flux varies as $Q_0 \cos \omega t$ during the daytime and that the sky flux is a constant, Q_1. The solution for Case 4 may then be expressed as:

$$T(0,t) = \frac{(1 - A)Q_0'}{\pi} + \epsilon Q_1' + \tfrac{3}{4} T_0$$

$$+ \tfrac{1}{2}\frac{(1 - A)Q_0' \, h \, \cos(\omega t - \delta_1)}{\sqrt{(h + w_1)^2 + w_1{}^2}}$$

$$- \frac{2}{\pi}\sum_{n=2,4,6..} \frac{(1 - A)Q_0'(-1)^{n/2} \cos(n\omega - \delta_n)}{(n^2 - 1)\sqrt{(h + w_n)^2 + w_n{}^2}}$$

$$\tag{3-89}$$

where

$$Q_0' = \frac{Q_0}{4\epsilon\sigma T_0{}^3} \tag{3-90}$$

$$Q_1' = \frac{Q_1}{4\epsilon\sigma T_0{}^3}.$$

Thus, the analytic solution is expressed as the sum of three terms. These include a constant term, which is a function of the albedo, and emissivity and oscillatory terms which are proportional to the product of the co-albedo, $(1 - A)$, and a function of h/w_n which is given by:

$$\frac{h}{w_n} = \frac{4\epsilon\sigma \, T_0{}^3\sqrt{2}}{P\sqrt{n\omega}} \tag{3-91}$$

where P = Thermal Inertia.

The ratio of the amplitude of the periodic variation divided by the co-albedo, is a function of the ratio of the thermal inertia to emissivity. Watson suggests, therefore, that it should be possible to map variations in the thermal inertia by measuring the albedo and diurnal temperature amplitude. The Heat Capacity Mapping Mission satellite was designed to capitalize on this argument. Later work by Watson and other investigators led to more formal regression analyses relating thermal inertia to temperature differences based on numerical solutions, particularly when more complex surface-boundary terms must be included.

Finite Difference Technique—Kahle (1977)

The Kahle model for the thermal behavior of geological surfaces utilizes the one-dimensional heat flow equation but modifies the top boundary conditions from Watson's earlier model, discussed above, by including sensible heat flux and latent heat flux components. The boundary condition is thus given by:

$$S + R + H + L + G = 0 \tag{3-92}$$

where

S = net solar radiation,
R = net thermal radiation,
H = sensible heat flux between the atmosphere and the ground,
L = latent heat flux,

and

G = heat flux into the soil.

Simple formulations for the various terms, based on various theoretical or empirical expressions, were utilized by the author. Nevertheless, analytic solutions to the heat conduction equation are no longer possible and an explicit finite difference form of the equation is utilized. Specifically,

$$T(t + 1,z) = T(t,z) + \kappa\frac{\Delta t}{\Delta z^2} \tag{3-93}$$

$$\cdot \left[T(t,z + 1) - 2T(t,z) + T(t,z - 1)\right].$$

The above formulation will calculate the temperature profile throughout the soil layers at time $t + 1$, subject to the boundary condition at the surface in time-step t. The method begins by assuming an initial temperature profile through the surface at time-step $t = 0$ and solving the finite difference equations for time-step $t = 1$. Then the boundary-condition equation, which can be developed as a function of the surface ground temperature, is solved to yield $T(t + 1,0)$. The process is then iterated, alternating between solutions of the finite-difference equation for the next time step, based on the algebraic solution of the top-surface temperature at the previous time step.

The specific forms used by Kahle for the various energy-budget terms are given as follows. The solar radiation term, S, is composed of both a diffuse flux and a direct flux. An allowance for non-horizontal surfaces is also made. The solar radiation term is given by the following expression:

$$S_T = S_{diff} + (S - S_{diff}) \frac{\cos Z'}{\cos Z} \quad (3\text{-}94)$$

where S_{diff} represents the diffuse radiation, Z' is the angle between the normal to a sloping surface and the sun, and Z is the solar zenith angle. The diffuse radiation was taken equal to the following expression:

$$S_{diff} = 0.05\, S + 0.10(1 - \cos Z)\, S \quad (3\text{-}95)$$

where S, under cloud-free sky conditions, is equal to the solar radiation incident upon a horizontal ground surface.

The net thermal radiation, R, is composed of the difference between the emitted ground radiation, $\epsilon \sigma\, T_g^4$ and the sky radiation, $\sigma\, T_{sky}^4$ where

$$T_{sky} = 263° + 10° \cos t \quad (3\text{-}96)$$

and t is the time measured from 1400 hours local time.

The sensible-heat term, representing the flux transfer between the surface and the atmosphere due to molecular conduction and turbulent mixing, was taken equal to the following expression:

$$H = \rho_a\, C_p\, C_D\, W(T_a - T_g) \quad (3\text{-}97)$$

where

ρ_a = the density of air at the surface,
C_p = specific heat of dry air at constant pressure,
C_D = drag coefficient,
W = wind-speed factor
T_a = air temperature.
T_g = ground temperature

Evapotranspiration terms are not included, since the author does not deal with vegetation cover. However, a latency term is included, applicable when the surface is wet. The author states that evaporation can represent the largest upward term in the heat-balance equation. The following expression was employed where 1 is the latent heat of evaporation, q_g is the mixing ratio of the air at the ground surface, and q_a is the mixing ratio near the ground, (viz at 1.5 m).

$$L = \rho_a\, C_d\, W_l(q_g - q_a). \quad (3\text{-}98)$$

Finally, the conduction determined at the ground is also expanded in a finite difference formulation as follows:

$$G = K(T_1 - T_g)/\Delta z$$
$$T_1 = T(t + 1, z = 1). \quad (3\text{-}99)$$

In this particular analysis Kahle used a 1-cm vertical step-size and a time step-size fixed by the following stability criterion:

$$\kappa\, \frac{\Delta t}{\Delta z^2} < 0.5. \quad (3\text{-}100)$$

In addition to predicting the surface ground temperature, this investigator generated a lookup table, giving the diurnal surface range, ΔT, as a function of thermal inertia, which could then be inverted to give thermal inertia as a function of ΔT. Evaluations of the model were made at the Pisgah Crater–Lavic Lake area of the Mojave desert in California where thermal overflight data and ground meteorological measurements were available.

Inclusion of a Simple Vegetation Layer—Balick, et al. (1981)

The Balick model employs a finite difference approximation to the thermal conduction equation, as in the Kahle model discussed above, but significantly modifies the surface boundary condition. In particular, a simple vegetation layer is present which can be described by an emissivity, ϵ_f, absorptivity, α_f, foliage height, Z_f, foliage cover fraction, σ_f (which can be related to leaf-area index) and a relative state parameter, χ, related to moisture stress. An energy-budget analysis for the foliage is performed. Similarly, a surface ground energy-budget term, which includes contributions from the foliage layer, is also computed to yield an average or effective boundary-condition temperature for the iteration process used to solve the conduction equation. The basic energy-budget condition at the surface is taken to be:

$$S + R_s \downarrow + R \uparrow + H + E + G = 0 \quad (3\text{-}101)$$

where

S = solar radiation absorbed at the surface,
$R_s \downarrow$ = downwelling thermal energy from the sky,
$R \uparrow$ = thermal flux emitted by the surface,
H = sensible heat exchange,
E = evaporative heat exchange, and
G = conductive transfer between the surface and terrain material.

The ground energy-budget expression when foliage is present is given by:

$$F_g = (1 - \sigma_f)\, \alpha_g\, S + R_g \downarrow - R_g \uparrow - H_g$$
$$+ E_g - G = 0 \quad (3\text{-}102)$$

where α_g is the absorptivity for the ground and the downwelling and upwelling longwave terms include a coupling between the foliage and ground emissions. The downwelling longwave-flux term is given by:

$$R_g \downarrow = (1 - \sigma_f)\, R_s \downarrow + \sigma_f\, \epsilon_g\, \sigma\, T_g^4$$
$$+ (1 - \epsilon_g)\, \epsilon_f\, \sigma\, T_f^4/\epsilon_1 \quad (3\text{-}103)$$

and the upwelling expression by:

$$R_g \uparrow = (1 - \sigma_f) \epsilon_g \sigma T_g{}^4 + (1 - \epsilon_g) R_s \downarrow$$
$$+ \sigma_f \epsilon_g \sigma T_g{}^4 + (1 - \epsilon_g) \epsilon_f T_f \qquad (3\text{-}104)$$

where

$$\epsilon_1 = \epsilon_f + \epsilon_g - \epsilon_f \epsilon_g. \qquad (3\text{-}105)$$

These expressions as well as other component formulations were taken from Deardorff (1978).

The nonlinear algebraic equation is solved for the ground temperature, T_g, using a root-finding algorithm. This ground temperature is then used to continue the iterative thermal-conduction equation solution for all ground layers. With foliage present, the equilibrium foliage-temperature must be determined in order to predict the composite effective thermal temperature for the scene. The foliage energy-budget equation is given by:

$$F_f = \sigma_f(\alpha_f S + \epsilon_f R_s + R_n) - H_f - E_f = 0$$
$$(3\text{-}106)$$

where the net longwave radiation term again considers the ground-foliage coupling. It is given by:

$$R_n = R_1 - R_2$$
$$R_1 = (\epsilon_f \epsilon_g / \epsilon_1) \sigma T_g{}^4 \qquad (3\text{-}107)$$
$$R_2 = \left[(\epsilon_1 + \epsilon_g)/\epsilon_1 \right] \epsilon_f \sigma T_f{}^4.$$

The remaining energy-budget terms for the foliage layer utilize expressions incorporating canopy resistance to water-vapor diffusion and stomatal resistance. The vegetation layer is termed simple in the sense that neither stratification of vegetation layers nor consideration of detailed canopy geometry is treated.

The foliage temperature determined from the solution of the above expression, T_f, and the ground temperature, T_g, are then used to develop a weighted average of each material to yield the overall radiant exitance that will be observed from a remote sensing device, i.e.:

$$T = \left[\sigma_f \epsilon_f T_f{}^4 + (1 - \sigma_f) \epsilon_g T_g{}^4 \right]^{1/4}.$$
$$(3\text{-}108)$$

The author performs extensive sensitivity analyses for the model and compares it to measurements over forest and grass canopies. It is expected that the simple treatment of vegetation will limit the model's applicability to irrigated crops in arid and semi-arid environments. However, it is likely to be applicable to pastures and rangelands also.

Vegetation Canopy Example—Smith, et al. (1981)

When detailed canopy structure must be considered in the longwave-flux transfers, a steady-state approach is utilized. In this model the authors incorporate the orientations and distributions of leaves in a multilayered canopy. The model assumes a simplified plane-parallel abstraction for the canopy and incorporates simplified expressions for the sensible heat and evapotranspiration terms. The model provides a convenient organization of the energy-flow processes in a self-consistent manner that relates measured canopy temperatures to intrinsic canopy characteristics.

Most physical-process oriented thermal-canopy models begin essentially with the nonlinear algebraic equation representing the steady-state energy-balance relationship. The expression used in this model is given by the vector-matrix equation below, which explicitly factors geometrical properties of the canopy designated by the matrix, S, from the remaining thermal properties.

$$F = \tfrac{1}{2} \alpha \sigma B^{\mathrm{T}}(X) S - \sigma B(X) + A + H(X)$$
$$+ LE(X) = 0 \qquad (3\text{-}109)$$

where:

B = vector of longwave emission terms,
H = vector of sensible heat,
LE = vector of evapotranspiration terms,
α = vector of layer absorptivities, and
X = temperature-profile vector throughout the canopy layers.
A = shortwave flux

Standard expressions are used for the energy-budget components except that a Monte-Carlo model is used to include multiple scattering of the short-wave flux to account for the shortwave absorption-coefficient. The matrix S incorporates view-factor considerations for accounting for the longwave-flux transfers among foliage elements and to the bottom soil-layer and top atmospheric-layer. S_{ij} represents the fraction of longwave flux emitted from a source layer, j, which is intercepted by a sink layer, i. S is obtained by integrating (over all emitting directions) the total flux escaping from the layer plus that which is intercepted by foliage elements possessing a distribution of foliage orientations. The specific expressions are fairly complex and are given in the original paper.

The steady-state equations are solved for the profile temperature-distribution throughout the canopy using an iterative Newton-Raphson technique, i.e.,

$$\delta X = (X - X_0) = (J^{\mathrm{T}}J)^{-1} J^{\mathrm{T}} \left[-F(X_0) \right]$$
$$(3\text{-}110)$$

where J is the system Jacobian. The authors were explicitly able to evaluate the Jacobian in terms of the selected energy budget representations. The distribution of profile temperatures is then convolved with the view-factor matrix for the effective area of radiating surfaces observed from any arbitrary view direction above the canopy.

Heterogeneous Extension—Bohren and Thorud (1971)

Bohren and co-workers performed several theoretical analyses of radiant heat transfer between forest canopies and snowpack. Two separate analyses are of interest. These consist of (1) the formal solution to the radiative transfer equations for longwave-flux transfers within a homogeneous plane-parallel medium composed of individual spheres, and (2) a sample treatment of the spatial variation of the longwave radiation flux from an isolated tree to the surrounding snow-pack. The second analysis represents an alternative method for calculating the S matrices discussed in the previous vegetation example for more complex geometries.

In the first case, the time-independent equation of radiative transfer applied to a homogeneous layer is given by:

$$\mu \frac{dI(\mu,z)}{dz} + \beta\,I(\mu,z) = \frac{S}{4\pi}$$

$$(3\text{-}111)$$

where S is the amount of longwave flux that is isotropically emitted by a unit volume per unit time.

The medium is considered to be composed of perfectly absorbing spheres of radius r, at a density of N/V spheres per unit volume of the medium. The volumetric extinction coefficient, β, which for longwave flux is taken to be solely an absorption coefficient, is calculated by analyzing the attenuation of a beam of radiation, F, incident on the collection of spheres. The extinction coefficient is proportional to the loss of flux that occurs in traversing the medium layer of thickness, d, i.e.:

$$\Delta F = -\beta F\,d. \qquad (3\text{-}112)$$

The cross-sectional area of spheres per unit area of slab of thickness d is the effective cross-section for removing radiation from the incident beam, i.e.:

$$\Delta F = -F\,N\pi r^2 \qquad (3\text{-}113)$$

where the total surface area of the spheres is given by $N\pi r^2$. The extinction coefficient is thus proportional to the ratio of the surface area of the spheres, i.e., the ratio of the foliage surface to the volume of the canopy:

$$\beta = \frac{a}{4V}$$

$$(3\text{-}114)$$

or

$$\beta d = \tfrac{1}{4}\,\frac{a}{A}\,.$$

That is, β is proportional to the foliage area index, a/A (the ratio of the foliage area of the canopy, a, to its cross-sectional area, A). The longwave radiation emitted per unit area of canopy surface is given by σT^4 where T represents the equilibrium temperature of the canopy. The isotropic longwave-source term itself is given by $\beta\sigma T^4/\pi$. The radiative transfer equation for downwelling flux is thus given by:

$$\mu \frac{dI^-}{dz} + \beta I^- = \frac{1}{\pi}\,\beta\sigma T^4 \qquad (3\text{-}115)$$

subject to appropriate boundary conditions. Analysis proceeds as in earlier sections.

Longwave flux transfers from a multidimensional object, such as an isolated tree, to surrounding media are analyzed by the authors using the concept of configuration or geometrical view factors. These roughly correspond to the S matrices in the vegetation model discussed above. Consider a unit surface with area A_1 at location x_1, y_1 radiating energy according to Lambert's law. The total radiant energy, $G_2\,(x_2, y_2)$ received from A_1 per unit time at a point with coordinates (x_2, y_2) and a surface area A_2 is given by:

$$G_2 = E_1 \iint_{A_1} \frac{\cos\theta_1\,\cos\theta_2\,dA_1}{\pi r^2}$$

$$(3\text{-}116)$$

where r is the distance between the center points (x_1, y_1) and (x_2, y_2), θ_1 is the angle between the outward normal to A_1 and the vector between the areas, and θ_2 is the corresponding angle between the outward normal to A_2 and the vector. E_1 is the uniform flux emitted by A_1. The total flux intercepted by A_2 is given by:

$$\iint_{A_2} G_2\,dA_2 = E_1 A_1 F_{1,2} \qquad (3\text{-}117)$$

where the view factor introduced by Reifsnyder (1967) is defined by:

$$F_{1,2} = \frac{1}{A_1} \iint_{A_1} \iint_{A_2} \frac{\cos\theta_1\,\cos\theta_2}{\pi r^2}\,dA\,{}_1 dA_2$$

$$(3\text{-}118)$$

and is equal to the fraction of the total radiation emitted by A_1 which is received by A_2. Extensions to an arbitrary number of radiating and receiving surfaces is accomplished by appropriate summation. The authors abstract an isolated tree by a simplified model that represents the tree as a bole consisting of a line source of length l which emits energy per unit length per unit time $2\pi r_b E_b$ where r_b is the radius of the bole and E_b is the blackbody flux corresponding to a bole temperature T_b and a crown disk of radius r_c emitting blackbody radiation at temperature T_c. The amount of energy received per unit time per unit area by the surrounding snowpack, at a distance r from the bole, is equal to the sum of the contributions from the bole and the crown, i.e.:

$$G_s(r) = G_b(r) + G_c(r). \qquad (3\text{-}119)$$

The authors show that this is given by the expression:

$$G_b(r) = \frac{E_b}{\pi}\,\frac{b}{y}\,(1 + a^2 y^2)^{-1}$$

$$(3\text{-}120)$$

$$G_c(r)$$

$$= \frac{E_c}{2}\left[1 - \frac{1 + (ay)^{-2} - y^{-2}}{\left[(1 + (ay)^{-2} + y^{-2})^2 - 4y^{-2}\right]^{1/2}} \right]$$

Where

$$y = r/r_c$$

$$a = r_c/l \qquad (3\text{-}121)$$
$$b = r_b/r_c.$$

MODELING SUMMARY

Examples of reflective and thermal modeling approximations for describing electromagnetic interactions with terrain elements at optical wavelengths have been presented. Other modeling approximations will be developed and the diversity of terrain types analyzed will be broadened. Models that are useful from a remote-sensing perspective, however, will possess the following two attributes: First, appropriate electromagnetic radiation or energy-budget governing equations for the media will be formulated and solution approaches outlined. Secondly, appropriate descriptions of the media, which relate necessary electromagnetic parameters to biophysical, geometric, or cultural properties, will be developed. Depending upon the scale of the modeling task and the available knowledge to characterize the media, either detailed submodels will be formulated or statistical approximations will be used.

Several examples of one-dimensional reflective and thermal models have been given but process-oriented time-dependent thermal models for vegetated surfaces have only recently been developed. The heterogeneous modeling problem arises when significant spatial variation occurs in the horizontal directions such that plane-parallel approximations to the scattering and emissive terrain elements are no longer valid. From an applications perspective the term "significant" is taken to mean "with respect to the resolution of the sensor," although every sensor is ultimately faced with this problem because of the difficulties posed by edges and boundaries. Horizontal spatial variation arises because of incomplete canopy cover, heterogeneity in composition and structure, and isolated targets. The spatial variability may be regular, as with row crops, or more random (or apparently random), as may occur in natural vegetation communities and associations. In essence the three-dimensional structure of terrain elements becomes important and leads to the casting of distinct shadows resulting from the macro-structure or morphology of the elements. For vegetation targets a merging of radiative transfer theory and geometric optics is evident. The three-dimensional thermal-modeling problem for natural terrain elements has not been attempted for remote-sensing applications.

As is often the case, the transition from one domain, that of one-dimensional approximations, to another, consideration of three-dimensional interactions, is not distinct. In fact, it is presently poorly understood and very little experience has been obtained to determine when one approximation will break down or when a more comprehensive model is required.

Because of the complexity and diversity of terrain targets of interest in remote sensing, a corresponding diversity of modeling abstractions have been and will continue to be developed. The models serve as tools for understanding and predicting optical response patterns. Models also serve to interpret experimental measurements. Of equal importance with the development of these models is an understanding of their limitations and their range of applicability.

INDIRECT SENSING OF TERRAIN CHARACTERISTICS

The modeling problem may be viewed as that of predicting optical radiance given a set of target biophysical descriptors. Naturally, the question arises as to whether or not the problem may be inverted; that is, given measured optical irradiance, can target descriptors be inferred? In a deterministic sense the answer is no. However, reasonable estimates for the parameters may be made, under certain conditions, such as those involving the use of prior estimates (Gelb, 1974). The problem is compounded by the presence of nonlinear equations, multiple scattering, and the potential presence of unknown populations other than the ones desired. This general problem has been addressed by the meteorologists and atmospheric scientists in passive remote sensing where temperature profiles or ozone concentrations are estimated from measured radiance (Barkstrom, 1978). Such applications have been very successful and considerable experience with the mathematical subtleties has been developed. In the terrestrial regime the application of these techniques is only now underway.

There are several reasons for this circumstance. First, the necessary models describing interactions with terrain materials, e.g., those discussed earlier, have only recently been developed. Further, the phase functions that relate terrain-element biogeophysical attibutes to radiation properties are complex. Secondly in terrestrial remote sensing the problem of inferring the horizontal variation of attributes in addition to vertical profiles is of interest and presents a more difficult problem.

However, it is particularly with the inference of spatially or temporally varying terrain characteristics that remote sensing has a great deal to offer. The use of models to infer such characteristics as thermal inertia, described in Chapter 13 of this manual, and applications dealing with soil moisture or evapotranspiration, described in Chapter 29, are examples. These applications-oriented models often correspond to different components in the specification of the boundary conditions for the radiation models discussed earlier.

While the terrestrial application of these indirect sensing techniques has just begun, certain observations can be made. First, it appears that for the radiative transfer models, it is primarily the optically thick case that lends itself to analysis. For example, if only spectral measurements as a function of wavelength are available, then in the general case both upwelling and downwelling flux must be estimated. In essence, the number of measurements cannot keep up with the number of unknowns. With the use of geometric optics models two additional strategies are evident. The first is to use additional measurement attributes such as the angular variation in response. The second strategy uses covariance statistics of estimated mixtures across pixels.

Five specific examples that have appeared in the literature and illustrate the above comments are discussed below. The first example is the use of a model to predict target separability as a function of sun/sensor geometry. The next three deal with the sensing of vegetation-canopy geometrical attributes including biomass or leaf area for a "homogeneous" canopy, estimation of row height and spacing for an agricultural crop, and estimaton of tree height and spacing for a forest canopy. The fifth example deals with estimation of grain size or snow water-equivalent. All of these examples utilize, more or less, explicit model inversions, i.e., the deterministic approach, conditioned on over-determined statistics, i.e., measurements. Alternative approaches, which linearize the radiative equation by expanding the radiance in appropriate partial derivatives of the desired biophysical attributes, or which use Kalman filtering techniques, are under investigation but have not appeared in the literature. Particularly for large-scale problems, e.g., those of a global ecosystem variety, all of the techniques alluded to in this discussion may well be viable additions to the repertoire of remote-sensing assessment procedures.

FEATURE SELECTION—SMITH AND OLIVER (1974)

In this study, a Monte Carlo model, SRVC, was used to calculate both the mean vector and the covariance matrix as a function of wavelength for the bidirectional reflectance-distribution function for a dense (leaf area index of 6.5) and a less dense (leaf area index of 2.0) theoretical short-grass prairie canopy. The wavelengths simulated were 0.40, 0.45, 0.50, 0.55, 0.60, 0.65 and 0.70 micrometers (channels 1 through 7). The two targets differed only in their relative biomass levels and the model was used to predict which sensor look-angles and which combination of two wavelengths would best discriminate the targets at a high sun zenith-angle, 44.5 degrees, and a low sun zenith-angle, 22.3 degrees. The capability of a Monte Carlo model to simulate bivariate distributions allows statistical distance measures between the target spectral distributions to be calculated. The authors used the simple divergence measure, calculated for all possible wavelength combinations

at each sensor view angle. The results are plotted in Figure 3-3. From the figure it is evident that biomass discrimination is best achieved at off-angle viewing with maximum divergence measures near 55 degrees. This result has also been predicted by Bunnik (1978).

BIOMASS ESTIMATION—PARK AND DEERING (1982)

Park and Deering derived the generalized solutions to the Kubelka-Munk equations for diffuse reflectance by permitting asymmetrical scattering and absorbing properties for the medium relative to upward and downward flux transfers. The general solution is complex. However, the authors derive several useful asymptotic relationships that have practical application in the biomass estimation problem. For example, for the case where the forward-scattering diffuse flux coefficient is greater than the backward scattering coefficient the following result is obtained:

$$R = R_\infty + (R_\infty - R_g) \, e^{-2KB} \qquad (3-122)$$

where
R_g = background reflectance,
R_∞ = so-called "infinite canopy reflectance" (Allen and Richardson, 1968),
B = biomass level in kg per hectare, and
K = a bulk attenuation property of the canopy medium.

The applicability of this simplified relationship was evaluated for field observations obtained for a series of alfalfa plots ranging in biomass from 0 to 3,850 kg/ha under a variety of illumination conditions and solar zenith angles. Nonlinear regression techniques were used to estimate R_∞, R_g, and K. The derived equations are plotted in Figure 3-4 and compared with the experimental data. The results indicate that the use of such simplified relationships may sometimes be sufficient. In practice, the background soil and infinite canopy reflectance parameters may be estimated independently from calibration areas or possibly by curve extrapolation from measured data, simplifying the estimation procedure.

ROW CROP CHARACTERISTICS— KIMES (1982)

Jackson, et al. (1979) developed a geometric optics approximation for the composite reflectance of a row crop. This model considered the rows as extended rectangular solids, possessing no gaps, and calculated the proportions of the projected surface area of sunlit and shaded soil and vegetation components in the direction of view. These projected proportions were then weighted by the relative optical properties of each component and summed to yield composite reflectance. In the reflective regime, the authors

CHANGES IN FEATURE SELECTION

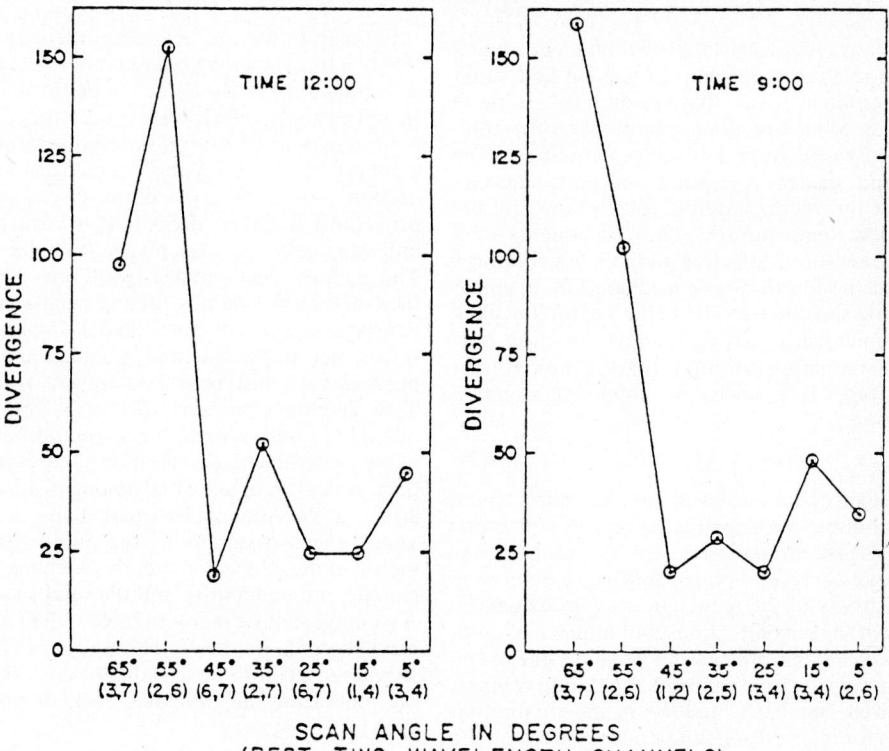

Fig. 3-3. Variation in maximum divergence with scan angle and best two wavelengths out of seven. Results are given for 12:00 (solar zenith angle 22.3 degrees) and 9:00 (solar zenith angle 44.5 degrees) computer simulations (after Smith and Oliver, 1974).

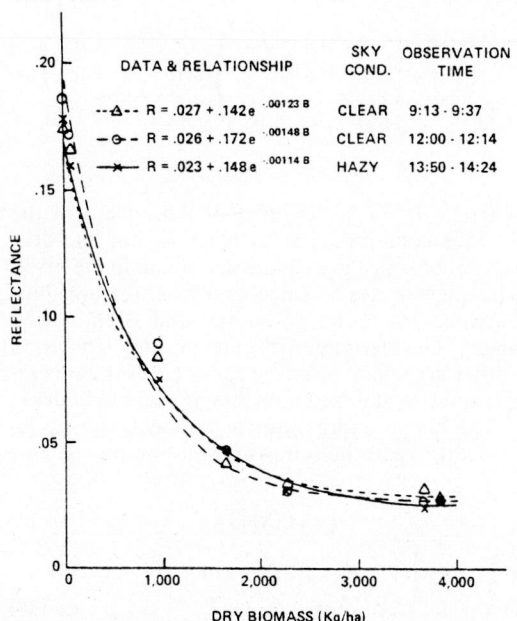

Fig. 3-4. Spectral reflectance vs. dry biomass of alfalfa canopies at 0.68 μm (after Park and Deering, 1982).

gave several examples of inverting the measured response to yield canopy characteristics. Kimes extended this geometric optics analysis to the thermal regime and applied more formal statistical inversion techniques. In Kimes' work the projected proportions of sunlit and shaded soil and vegetation are weighted by the respective component temperatures in degrees Kelvin raised to the fourth power. The fourth root of the sum of these four terms represents the overall composite effective radiant temperature as measured by a thermal scanner in a particular view direction. Kimes considers two cases. First, when knowledge of the row structure is available, a set of linear equations relating canopy and soil temperatures to the *a priori* row structure can be developed. Then, given measurements of composite canopy temperature at a series of view angles, these equations can be inverted to yield soil and vegetation temperatures separately. The second case considered is when the row characteristics must also be inferred from the measurements. A set of non-linear algebraic equations must then be solved to extract both component temperatures and crop geometry.

If the crop geometry is known *a priori* then the

following system of linear equations is appropriate:

$$A t = S + \epsilon \qquad (3\text{-}123)$$

where matrix A consists of probabilities p_{ij}, which give the relative proportion of shaded and sunlit canopy components, j, that are directly visible in the sensor view direction, i. In practice the components j range from $1-4$ corresponding to the sunlit and shaded vegetation and soil. The elements of the vector t are the fourth power of the component temperatures. The components of S are the measured effective radiant temperatures raised to the fourth power measured by a sensor with view direction i. The error vector for each measurement in a particular view direction is ϵ. The least-squares solution for the component temperatures is given by the following standard expression:

$$\hat{t} = (A^T A)^{-1} A^T S. \qquad (3\text{-}124)$$

This expression, as well as the more trivial exact solution, was applied to cotton row-crops with a 48-percent ground cover. A variety of experimental analyses were performed to determine the sensitivity of the technique to various combinations of viewing directions and number of component equations utilized. In general, the mean squared deviations of inferred vegetation temperatures were about 1°C and the mean-squared deviations for inferred soil temperatures were on the order of 2°C.

When *a priori* information about the row, width, height, and spacing of the crops is not available, the following set of nonlinear algebraic equations may be developed:

$$\begin{pmatrix} p_{11}(h,w,s)\, t_1^4 + p_{12}(h,w,s)\, t_2^4 + \ldots + p_{1m}(h,w,s)\, t_m^4 - S_1 \\ p_{21}(h,w,s)\, t_1^4 + p_{22}(h,w,s)\, t_2^4 + \ldots + p_{2m}(h,w,s)\, t_m^4 - S_2 \\ \vdots \\ p_{n1}(h,w,s)\, t_1^4 + \qquad\qquad \ldots \qquad\qquad + p_{nm}(h,w,s)\, t_m^4 - S_n \end{pmatrix} = 0 \qquad (3\text{-}125)$$

where n is the number of view directions utilized and m is the number of scene components analyzed.

The elements of the A matrix, p_{ij}, have now been explicitly identified as functions of the row spacing, s, width, w, and height, h, for canopy component j visible in direction i. Exact 3-by-3 systems, corresponding to various combinations of view angles such as $0-20-40$, $0-20-60$ and $0-40-60$, and 3 components consisting of sunlit vegetation, soil, and the row-height-to-width parameter were analyzed. The root-mean-square deviations of the inferred component temperatures were again on the order of 1 to 2 degrees centigrade. The row-height-to-width parameter was inferred to within ±10% of its mean value, 5 cm.

FOREST CANOPY CHARACTERISTICS—STRAHLER AND LI (1981)

Estimators for the height and density parameters of a forest canopy based on overall composite reflectance may also be developed using geometric optics analyses. However, because of the more random nature of natural vegetation communities as opposed to the regular geometrical structures encountered in the row crops discussed above, probability functions describing the distribution of individual trees across pixels must be invoked. The authors developed a coniferous canopy reflectance model by abstracting such a canopy to consist of a series of cones possessing Lambertian reflectance properties and a background surface possessing a different Lambertian reflectance. Tree heights were assumed to be log-normally distributed with a fixed mean and variance and a known coefficient of variation. Individual trees, (i.e., cones) were allocated among pixels based on either a Poisson or Neyman Type A spacing. Under these assumptions the authors used geometric optics to calculate the amount of sunlit canopy and understory and the total shaded area. Assuming sunlit canopy to have radiance $t(\lambda)$, the background to have radiance $g(\lambda)$, and a shadowed pixel to have radiance $z(\lambda)$, the following model for the radiance, $s(\lambda)$, of pixel i was derived:

$$s_i(\lambda) = g(\lambda) - \eta_i R_i^2 \left[\left(\frac{\pi}{2} + \gamma \right)(g(\lambda) - t(\lambda)) \right.$$
$$\left. + \cot \gamma\, (g(\lambda) - z(\lambda)) \right] \qquad (3\text{-}126)$$

where ηR^2 is the product of the square of the average cone radius at its base, R, and the density, η, of cones per square aerial unit in the pixel. The angle γ can be calculated from relationships between the cone geometry and illumination angle. The derivation arguments are similar to earlier arguments used by Egbert (1976) in his reflectance model based on spheres and cylinders.

The above expression for radiance can be inverted to yield the following solution for $\eta_i R_i^2$:

$$\eta_i R_i^2 = \frac{1 - s_i(\lambda)/g(\lambda)}{\left(\dfrac{\pi}{2} + \gamma \right)(1 - t(\lambda)/g(\lambda)) + \cot \gamma\, (1 - z(\lambda)/g(\lambda))} .$$
$$\qquad (3\text{-}127)$$

The factor ηR^2 is a dimensionless parameter

that reflects both the size and spacing of the cones. R can be directly related to h, the height of the cone, and η to the spacing between the trees. Thus, if η and R could be estimated separately, then the distribution of tree height and spacing could be determined. This separation is accomplished by analyzing collections of pixel responses for an individual timber stand. Given a timber stand consisting of k pixels, the sample mean and variance for ηR^2 can be estimated:

$$\overline{\eta R^2} = \frac{1}{k} \sum_i \eta_i R_i^2$$

$$\text{Var}(\eta R^2) = \frac{1}{k} \sum_i \left(\eta_i R_i^2 - \overline{\eta R^2} \right)^2 \qquad (3\text{-}128)$$

Using the statistical properties of the lognormal height distribution and the Poisson spatial distribution, these estimators may be used to derive the following expression for the maximum-likelihood estimator of the sample variance for $\overline{R^2}$:

$$\hat{R}^2 = \frac{\text{Var}(\eta R^2) - W(\overline{\eta R^2})}{\overline{\eta R^2}(1 + W)} \qquad (3\text{-}129)$$

where

$$W = e^{4\sigma_h^2} - 1 \qquad (3\text{-}130)$$

and σ_n^2 is the assumed variance of tree height. Consequently, the estimator for the density is given by:

$$\hat{\eta} = \overline{\eta R}/\hat{R}^2 \qquad (3\text{-}131)$$

Several assumptions are used in the above derivations; for example, the assumption is made that height and density vary independently. The authors discuss several modifications to the theory to account for such effects.

The basic approach outlined above was evaluated by the authors for stands possessing between 6 and 12 percent tree cover and yielded a standard error in the height and spacing for the stands of roughly 10 percent.

SNOW PROPERTIES—DOZIER (1981)

Dozier, et al utilized the snow model of Warren and Wiscombe, discussed earlier, to develop relationships between albedo, as inferred from the NOAA-6 AVHRR system, snow grain size, and snow water equivalence. Preliminary evaluation of these relationships was then performed for AVHRR data collected over the Lake Winnipeg region in Canada.

The Warren and Wiscombe model predicts the spectral directional-hemispherical reflectance from a snow layer, a_s. The reflectance may be related to the measurements obtained by the AVHRR system by convolution with the spectral response functions of the satellite radiometer and the incident irradiance. That is, the AVHRR inferred albedo is given by:

$$\text{Albedo} = \frac{\int_0^\infty a_s(\theta_0;\lambda)\, E_s(\lambda)\, \Phi(\lambda)\, d\lambda}{\int_0^\infty E_s(\lambda)\, \Phi(\lambda)\, d\lambda}. \qquad (3\text{-}132)$$

where $\Phi(\lambda)$ is the relative spectral response function for the visible band (channel 1), 0.56 to 0.72, and the near-infrared band (channel 2), 0.71 to 0.98, of the radiometer

and

$E_s(\lambda)$ is the beam irradiance as calculated by a radiative transfer model for solar angle θ_0.

The authors consider two cases—an optically thick snow layer and a finite layer. If the snow layer can be assumed to be semi-infinite, then Eq. 3-51 of this chapter is applicable and a unique relationship between AVHRR-inferred albedo and snow grain-size can be developed. The results of these calculations for solar zenith angles of 15, 30, 45, 60, and 75 degrees are given in Figure 3-5 for both the visible and infrared channels. The visible band is relatively insensitive to grain size; however, the near-IR channel could yield useful diagnostic information.

For shallow snow layers, the more complex equation (3-49) of this chapter must be used, although this makes it more difficult to separate the effects of grain size, water equivalence, and density. Figure 3-6 shows the results of the calculations for AVHRR-inferred albedo versus snow water equivalence for two zenith sun-angles, 30 and 60 degrees, and for grain radii of 50, 200, and 1000 micrometers in each spectral channel. The visible channel is somewhat more sensitive to snow water equivalence for larger grain sizes.

OVERVIEW OF MEASURED TERRAIN FEATURE RESPONSE PARAMETERS

In this section, terrain-feature response parameters are taken to mean the bidirectional reflectance distribution function, or its derivatives appropriate for the reflective regime and either the spectral emissivity or the thermal diffusivity. The objective of this section is to give an overview of typical values for these parameters for different types of terrain elements, and their variability as a function of various factors. In fact, while it might be expected that these parameters would be fundamental constants, they are never measured directly and must, therefore, be inferred in one way or another. Their indirect measurement is susceptible to several complicating factors as discussed below. Further, as should be evident from the earlier discussion of modeling of terrain elements, it would be very misleading to quote definitive

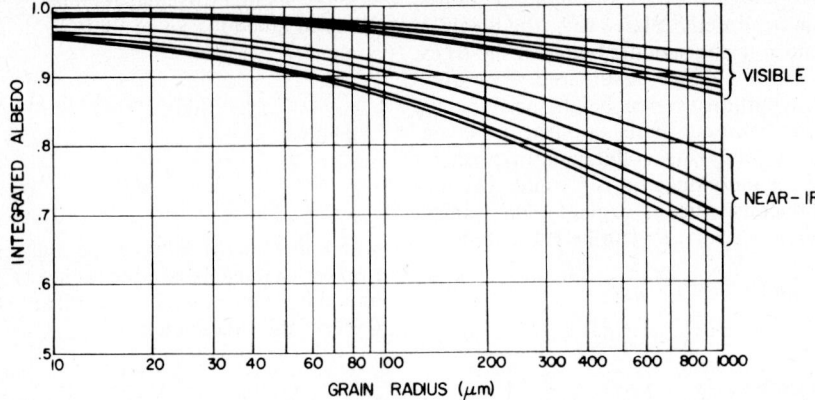

Fig. 3-5. Integrated semi-infinite albedo versus grain radius for the visible and near-infrared channels of the NOAA-6 AVHRR, for solar zenith angles of 15, 30, 45, 60, and 75°. In each group of lines the upper line represents the integrated albedo at the largest zenith angle and successively lower lines represent successively smaller zenith angles (after Dozier, et al., 1981).

values for terrain elements without specifying the precise target and its characteristics. All terrestrial targets are subject to extrinsic environmental factors. Thus, quoting the spectral reflectance value even for a particular soil type is misleading in that it fails to capture the dynamic spread of this parameter based on the surface texture of the samples, the degree of moisture content, the mineral composition, and so forth. Although insufficient measurements have been made to replace such single values with statistical distributions conditioned on intrinsic properties, it is likely that these distributions would encompass a spread that overlaps several other soil types. In the following sections, therefore, any values quoted should not be strictly relied upon. While the examples given may be useful for preliminary design studies, one should either refer to the original sources for complete descriptions of the measurement and target characteristics or, indeed, make one's own measurements as required.

Generally, there are more reflectance data available than data on either emissivities or diffusivities. Since publication of the previous edition of the manual, extensive measurement programs have been undertaken, particularly with regard to vegetation targets, and within this category special attention has been focused on crops. Laboratory, field, and aircraft observations have been used to estimate the terrain response parameters. Generally, when the base of operations is moved from the laboratory to the aircraft platform, measurement difficulty increases. On the other hand, the representativeness of the terrain elements measured is generally better in the reverse order. For example, while the laboratory measurement of reflectance for an individual leaf may be performed very precisely, the condition of the leaf when removed from its natural environment may be so changed as to make this precise measurement essentially meaningless. Similar

comments could also be made, for example, for soil samples taken into the laboratory. This leads us naturally into the discussion in the next section.

CAUTIONARY NOTE IN THE INTERPRETATION OF DATA FROM THE LITERATURE

Even a casual perusal of the literature would quickly indicate to the reader that the terms reflectance and emissivity are used very freely and often contradictorily. In fact, however, their accepted definitions are stated fairly clearly, e.g., Suits (1983) and Nicodemus (1970). The reason for this apparent contradiction arises from four considerations. First, both emissivity and reflectance vary spectrally, that is, with wavelength, and are directionally dependent. Reflectance varies also with the direction of incident energy. Rarely are these parameters measured at single angular values but rather over ranges of these variables. Additional adjectives are usually added to the fundamental terms depending upon the limits and the manner in which they are estimated; for example, the term hemispherical conical reflectance is used to describe the reflectance of a terrain element determined by integration of the incident flux over the complete hemisphere and by integration of the exitance over a discrete solid angle. The second consideration that leads to confusion comprises the ingenious but consequently highly diverse techniques used to estimate these parameters. These techniques may be based upon various assumptions concerning the properties of the surface or ambient conditions, and upon various normalization or calibration strategies that are also applied. The third consideration leading to confusion and difficulty in interpreting "literature" data is the determination of just exactly what target has been measured. This is a seri-

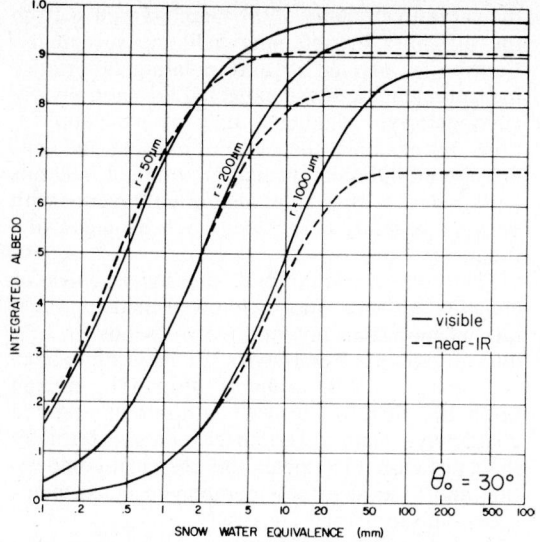

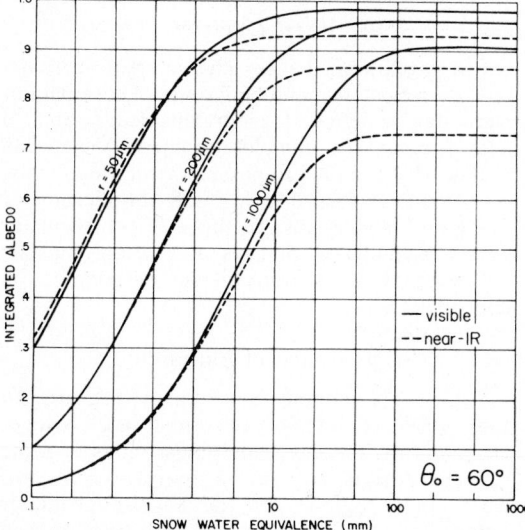

Fig. 3-6. Integrated finite depth, direct beam albedo versus snow water-equivalence (mm) for the visible (solid lines) and near-infrared (dashed lines) channels of the NOAA-6 AVHRR, for grain radii of 50 μm, 200 μm, and 1000 μm, for solar zenith angle 30° (upper graph) and 60° (lower graph). For 1000-m snow particularly, the effect of finite depth can be detected at a much larger snow water-equivalence in the visible than in the near-infrared channel (after Dozier, et al., 1981).

ous limitation in that the first two considerations can usually be determined by a careful reading of the paper. However, few authors sufficiently document the target and background conditions or the other ancillary parameters needed to understand the sources of variation. Finally, the fourth factor is that reflectance and emissivity of terrain elements must really be considered stochastic variables in the natural world. Consequently, one is sampling from a distribution whose properties, in fact, are really poorly understood in relation to target variability.

Reflectance Considerations

The bidirectional-reflectance distribution function was defined earlier. More commonly, it is the integration of this quantity over finite solid angles of incidence and exitance, yielding what are termed reflectance factors, that is estimated. These factors are dimensionless quantities equivalent to π times the bidirectional-reflectance distribution function averaged over the solid angles of incidence ω_i and exitance ω_r. That is, the reflectance factor is defined as

$$R(\omega_i,\omega_r) \tag{3-133}$$
$$= \frac{\pi \int_{\omega_r} \int_{\omega_i} f_r(\theta_i,\phi_i;\theta_r,\phi_r)\cos\theta_i\,d\omega_i\cos\theta_r\,d\omega_r}{\int_{\omega_r}\int_{\omega_i}\omega_i\cos\theta_i\,d\omega_i\cos\theta_r\,d\omega_r}.$$

The precise term for the above definition would be *biconical reflectance factor*. Field measurements generally perform the integration over all incident angles; that is, they include both the anisotropic sky distribution and a direct solar source. A commonly used technique is to ratio the radiance of a target, $L_s(\omega_r)$, to that of a standard Lambertian reference panel, such as barium sulfate, $L_{ir}(\omega_r)$. In this case, the hemispherical conical reflectance factor is given by the following expression:

$$\hat{R}(\omega_r) = \frac{L_s(\omega_r)}{L_{ir}(\omega_r)}$$

$$= \frac{\pi^{-1}\sum_i R(\omega_i,\omega_r)\,L(\omega_i)\,P_i}{\pi^{-1}\sum_i R_{ir}(\omega_i,\omega_r)\,L(\omega_i)\,P_i} \tag{3-134}$$

where $P_i = \int \cos\theta_i\,d\omega_i$ and R_{ir} = the reflectance factor of the standard.

Several other terms are freely used in the literature with different normalizations, but for the sake of limiting confusion they will not be discussed here.

From the above expressions it can be seen that commonly employed techniques for estimating reflectance factors should be concerned with the angular irradiance field, particularly when confounded with non-Lambertian targets, and with normalization properties of the standard calibration panel. The time-varying nature of the irradiance field must also be considered.

The review article by Robinson and Biehl (1979) is an excellent summary of many of the confounding factors that affect the measurement of reflectance factors. The temporal fluctuation of solar radiance and its effect on reflectance factors have been discussed by Duggin (1980; 1981) and by Richardson (1981). The effect of clouds or of partly cloudy conditions has been shown to be dramatic, e.g., Gordon and Church (1966), Kimes et al. (1980). This effect, in particular, has not been fully recognized by most investigators. Fortunately, the influence of sky radiance for clear conditions has been shown to be small (Kirchner, et al., 1982). The variation in solar spectral irradiance as a function of atmospheric turbidity and cloud cover for direct normal, diffuse, and global irradiance has been summarized by Bird, et al. (1982). The authors state that it is not uncommon to see rapid fluctuations in turbidity shortly before and after local solar noon. Discussion of the variability of vegetation reflectance measurements as a function of solar zenith-angle is given by Kimes, et al. (1980).

Even assuming perfect measurements and an understanding of all of the factors discussed above, there is still the problem of just exactly what terrain element is being measured. Many of the commonly used radiometric sensors are non-imaging and, therefore, it is difficult to know precisely where the instrument is pointing and how much of the terrain element has been included within the field of view. The literature is also replete with rather simplistic descriptions of the targets measured. Examples include "a tree" or a "soil surface."

It should be noted, however, that the more recent literature indicates a growing awareness of the considerations discussed above. Perhaps, future editions of the *Manual of Remote Sensing* will not need to caution the reader in the interpretation of literature data.

Emissivity Considerations

Emissivity may be a function of both view angle and wavelength. In addition, it is a function of the temperature of the emitting terrain element. This is usually taken to be the surface; however, it is also possible to have temperature gradients within the surface that could give rise to a volumetric emissivity. Most emissivity measurements are obtained in the laboratory on small samples of polished minerals or powders. Few vegetative measurements are available. A commonly used technique is to utilize Kirchhoff's law for opaque surfaces and to infer emissivity from measured reflectivity. Field and aircraft measurements have also been obtained, but results are contradictory and highly dependent upon the nature of the targets measured.

Emissivity values quoted in the literature are often integrals of the spectral emissivity over some wavelength bandpasses, e.g., typically the 3-to-5 or 8-to-14 micrometer range. The wavelength intervals are not always clearly identified. While emissivity measurements may be quoted under specific angles of incidence, apparently it is generally assumed that directional dependencies are minimal.

In the case of laboratory measurements, variability among emissivity measurements arises from the preparation of samples; whereas in the case of field measurements, variability arises from the natural variability of the target as well as from the difficulty in obtaining field measurements. Caution is advised in extrapolating laboratory measurements to those expected for naturally occurring terrain elements. In geological applications, for example, the surfaces of rocks and soils are altered into complicated mixtures of minerals and oxides. Surfaces are also often covered with various vegetative surfaces such as algae and fungi.

The interpretation of bulk emissivity values for complex surfaces, such as plowed fields or vegetation canopies, is not intuitively obvious. In fact, interpretations of emissivity for naturally occurring materials will probably ultimately depend upon the development of suitable models to characterize natural terrain elements in terms of their physical, biological, and chemical composition and to explain how component constituents contribute to overall emissivity.

VEGETATION

For vegetation, the key characteristic parameters are reflectance and emissivity. These parameters may be defined for individual leaves and, to a lesser extent, for complete canopies. Vegetation is generally assumed to possess a very high emissivity: greater than 0.95 in the 8–14-micrometer range and slightly less in the 3–5.5-micrometer range. Reflectance changes of greater than 200 percent may arise from angular or polarization dependences.

Discussion of Emissivity

Table 3-2 is a compilation summary prepared by Link (1979 from a variety of sources, e.g., Idso, et al., 1969) for the integrated emissivities of some representative vegetative materials in two wavelength regimes. In the 8–14-micrometer range, the emissivities are generally greater than 0.95 for the samples presented except in those cases where dehydration has occurred and the emissivity falls to 0.90. Generally the emissivities are less in the 3.0–5.5-micrometer region but all are greater than 0.85. The next two tables are extracted from the article by Wong and Blevin (1967), who performed some fairly extensive measurements on the infrared reflectances of plant leaves, particularly with regard to the top and bottom surfaces of the leaves and their angular scattering properties at long wavelengths. The authors indicate that they measured the total reflectances, including both specular and diffuse components. Consequently, the emissivity values may be taken as one minus the quoted reflectances. This application of Kirchhoff's law is strictly valid, of course, only if the leaves are opaque; however, this assumption is also excellent for wavelengths greater than 3 micrometers (although not as good between 2 and 3 micrometers).

TABLE 3-2

Integrated Emissivities of Some Representative Vegetation Materials in Two Wavelength Regions.

Feature	Emissivity by Wavelength band, μm	
	3.0–5.5	8.0–14.0
Green mountain laurel	0.90	0.92
Young willow leaf (dry, top)	0.94	0.96
Holly leaf (dry, top)	0.90	0.90
Holly leaf (dry, bottom)	0.86	0.94
Pressed dormant maple leaf (dry, top)	0.87	0.92
Grean leaf winter color—oak leaf (dry, top)	0.90	0.92
Green coniferous twigs (jack pine)	0.96	0.97
Grass—meadow fescue (dry)	0.82	0.88
Bark—northern red oak	0.90	0.96
Bark—northern American jack pine	0.88	0.97
Bark—Colorado spruce	0.87	0.94
Corn		0.94
Indian-fis cactus		0.96
Prickly pear cactus		0.96
Cotton (upland)		0.96
Tobacco		0.97
Blind-pear cactus		0.98
Fremont cottonwood		0.98
Philodendron		0.99
Sugarcane		0.99

* From Link (1979).

The authors summarize results for the spectral reflectance (emissivity) for leaves from plant species for continuous wavelengths from 2.0 to 15 micrometers. Measurements for 32 species at the specific wavelengths of 8, 10, and 12 micrometers are given in Table 3-3. Generally the inferred emissivities of all species are very similar and greater than 0.9. The lowest of the inferred emissivity values, 0.92, corresponds to leaves that have hairs or other morphological features present. At these infrared wavelengths no significant differences exist between the upper and lower surface properties.

Finally, Table 3-4 summarizes the directional properties of the infrared reflectances (and hence, emissivities) at 2 and 10 micrometers. It is evident that the leaves are neither perfectly specular nor diffuse reflectors in the infrared region. Given this note it should also be remarked that all of the measurements reported in the paper were made at 10 degrees from the normal; thus, inferred emissivities really correspond to the directional emissivity at an incidence angle of 10 degrees.

Reflectance Properties—Leaves

Some of the earliest work on the spectral properties of leaves was that performed by Gates and co-workers (1965) and for agricultural crops the exhaustive work of Gausman and co-workers (1970, 1971, 1972, 1978). Measurements of tree leaves may be found in Olsen (1969). Angular scattering and polarization properties of leaves are discussed in Coulson (1965, 1966), the various references by Egan (e.g., 1970), Woolley (1971) and the classic paper by Breece and Holmes (1971).

The general optical properties of leaves and underlying mechanisms are fairly well understood. The low reflectance and transmittance of leaves in the 0.35 to 0.7 region is generally attributed to chlorophyll absorption and other pigments at the blue and red wavelengths, with a slight rise at 0.55 μm in the green region. In the near infrared, 0.75–1.35, because of the leaf's internal mesophyll structure, multiple scattering occurs and reflectances and transmittances tend to be in the 40–50-percent range. Strong water-absorption bands occur in the middle infrared region up to 2.0 μm with specific water absorption centered at roughly 1.4, 1.9, and 2.7 μm. There are considerable differences in leaf reflectance according to leaf type, particularly in the near-infrared region. As leaves mature, their visible reflectance tends to decrease while their near-infrared reflectance increases. On the other hand, senescence produces generally higher values in the visible region and lower values in the infrared. The optical behavior of leaves is also affected by various kinds of stresses, as well as by nutrients, salinity, and geobotanical considerations. Figure 3-7 may be taken as a generalized example of the interplays between reflectance, transmittance, and absorption for healthy green vegetation. The figure is taken from Gausman, et al. (1971) who gives detailed examples of the leaf mesophylls and optical properties of over 20 agricultural species. Tables 3-5a and 3-5b, extracted from this report, summarize reflectances and transmittances, respectively, for bean, corn, lettuce, sorghum, soybean, sunflower and wheat leaves and also include measurements from Smith and Oliver (1972) for Blue Grama or Bouteoula gracilis, a native prairie grass.

Generally, at strongly absorbing wavelengths leaf reflectance tends to become specular. At wavelengths greater than 0.75 μm, both reflectance and transmittance exhibit Lambertian behavior although, as noted earlier, at the long infrared wavelengths the non-Lambertian behavior of leaves is evident. Several authors have found that the specular reflectance of leaves increases with angles of incidence.

The top and bottom surfaces of leaves may exhibit differences in reflectance and transmittance. Although Gausman, in his measurements of crop leaves, found that these differences were generally small, Table 3-6, which lists the integrated shortwave properties of several tree leaves, demonstrates that such differences may occur (from Birkebak and Birkebak, 1964).

Table 3-7, taken from Egan (1970), summarizes polarization measurements for a variety of vegetation samples. The table gives the maximum polarizations observed at their maximum measured phase angle of 96 degrees for incident

TABLE 3-3

Total Spectral Reflectances of Leaves at 8, 10, and 12 μm*
Angle of Incidence 10°.–Indicates that the Measurement Was Not Made

Plant Species	Leaf Surface	Reflectance at		
		8 μm	10 μm	12 μm
ANGIOSPERMAE DICOTYLEDONEAE				
Acacia ancura	Upper	0.05	0.05	0.04
Banksia serrata	Upper	0.03	0.05	0.04
Brassica olcracca				
Waxy leaf	Upper	0.01	0.01	0.02
Waxy leaf	Lower	0.01	0.02	0.02
Green leaf	Upper	0.02	0.02	0.02
Camellia japonica cv. Don Pedro	Upper	0.02	0.04	0.05
Ceraslium tomentosum	Upper	0.04	0.06	0.04
Chenopodium album	Upper	—	0.03	—
Eucalyptus blaxlandii	Upper	0.03	0.04	0.04
Eucalyptus cinerea (glaucous)	Upper	—	0.03	—
Eucalyptus nitens (green)	Upper	—	0.03	—
Eucalyptus niphophila (glaucous)				
5500 ft, Mt. Kosciusko	Upper	—	0.06	—
2000 ft, Goulburn	Upper	—	0.04	—
Eucalyptus saligna (glaucous)	Upper	—	0.05	—
Homalocladium platycladum	Upper	0.03	0.03	0.04
Kalanchoe sp.	Upper	0.02	0.02	0.02
Lychnis coronaria	Upper	0.02	0.07	0.07
Magnolia grandiflora	Lower	0.03	0.04	0.04
	Upper	0.04	0.06	0.05
Mentha rotundifolia	Upper	0.02	0.03	0.02
Nicotiana tabacum cv. Hicks	Upper	0.02	0.03	0.01
Nymphaca sp.	Upper	0.02	0.03	0.01
Populus alba	Upper	—	0.05	—
	Lower	—	0.05	—
Protea grandiflora	Upper	0.03	0.01	0.01
Ruscus aculeatus	Upper	0.03	0.04	0.03
MONOCOTYLEDONEAE				
Avena saliva cv. Algerian	Upper	0.02	0.03	0.01
Eichhornia crassipes	Upper	0.02	0.02	0.01
Spinifex hirsutus	Upper	0.02	0.02	0.02
	Lower	0.02	0.06	0.05
Trachycarpus fortunci	Upper	0.05	0.05	0.05
Xanthorrhoca resinosa	Upper	0.02	0.03	0.03
GYMNOSPERMAE				
Encephalarlos altensteinii	Upper	0.04	0.05	0.04
FILICINAE				
Angiopteris sp.	Upper	0.01	0.02	0.02
Marsilca drummondii	Upper	0.02	0.02	0.02
Phyllophyte				
Parmelia sp.	Upper	0.03	0.04	0.03
Fruits				
Citrus sinensis	Outer skin	0.03	0.03	0.04
Pyrus malus	Outer skin	0.04	0.04	0.04

* From Wong and Blevin, 1967.

energy at 30 degrees from the normal, as well as the bidirectional reflectance factors for measurements made within the principal plane defined by the source and sensor directions and for the target both parallel and perpendicular to this plane. Egan found similar results for forest foliage (1970). In general, several authors have indicated that polarization decreases at longer wavelengths and is tied to moisture content.

Reflectance Properties—Vegetation Canopies

Vegetation-canopy bidirectional reflectance is a consequence of the optical properties of the leaves, understory or soil, and the geometrical arrangement of the constituent elements; that is, canopy structure or architecture. Earlier, various models for relating these factors to canopy reflectance were discussed. It should be evident that spectral and angular variations of overall canopy reflec-

TABLE 3-4

Angular Distribution of Infrared Radiation Reflected from Leaf Surfaces*
Angle of incidence 10°

Plant Species	Surface	Radiant Flux in each Angular Zone (expressed as a fraction of the total for four zones)							
		Wavelength = 2 μm				Wavelength = 10 μm			
		10–30°	30–45°	45–60°	>60°	10–30°	30–45°	45–60°	>60°
—	Specular reflector	1.00	0.00	0.00	0.00	1.00	0.00	0.00	0.00
—	Perfect diffuser	0.22	0.25	0.25	0.27	0.22	0.25	0.25	0.27
Citrus limon									
Juvenile leaf	Upper	0.3	0.3	0.2	0.2	0.9	0.1	0.1	0.0
Mature leaf	Upper	0.4	0.3	0.2	0.2	0.9	0.0	0.1	0.0
Gazania hybrid	Lower	0.2	0.2	0.2	0.3	0.2	0.3	0.3	0.2
Monstera deliciosa									
Juvenile leaf	Upper	0.3	0.3	0.2	0.2	0.8	0.1	0.1	0.0
Mature leaf	Upper	0.3	0.3	0.2	0.2	0.5	0.2	0.1	0.2
Phaseolus vulgaris cv. Brown Beauty	Upper	0.3	0.4	0.1	0.2	0.3	0.6	0.1	0.0

* From Wong and Blevin, 1967.

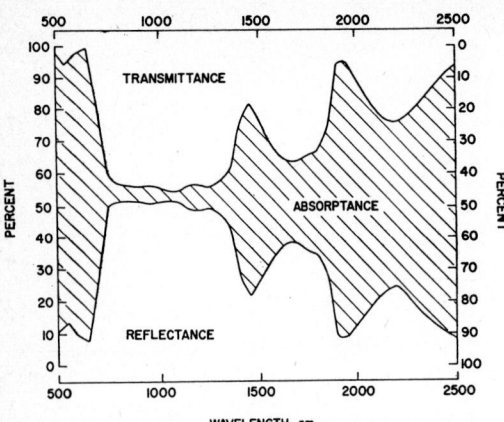

Fig. 3-7. Generalized example of the amount of transmittance, reflectance, and absorptance from healthy green foliage (Gausman et al., 1971).

tance are a complex interplay of all of these factors and that virtually any behavior can result. For example, canopy bidirectional reflectance can show an increasing or a decreasing trend with increasing solar zenith angles or view angles. Indeed, reflectance may also exhibit local maxima and minima. These behavioral trends depend upon whether or not vegetation overlays a dark or light soil relative to the optical properties of the leaves, the sparseness of the vegetation, and so forth. In agricultural crops there are also cultural impacts, e.g., planting in rows, and planting dates and densities, which may be manifested in the overall patterns; conversely, natural vegetation communities exhibit different heterogeneous mixtures and associations.

Nevertheless, particularly for agricultural crops, a considerable body of literature, experimentation, and understanding exists, although it is often difficult to unravel the various confounding interactions. Fewer field data are available for natural vegetation communities, partly because of the difficulty in obtaining such measurements.

The dependence of vegetation-canopy bidirectional reflectance on cultural, economic, and intrinsic biophysical properties has been extensively studied from a pragmatic viewpoint. Indeed, later chapters in this manual will review the status and use of such relationships. An early example of correlations derived for natural shortgrass prairie canopies is given in Table 3-8, which relates the coefficient of determinations for total wet biomass, dry biomass, total dry biomass, leaf water content, dry green biomass, dry brown biomass, and chlorophyll content for a series of six narrow-band wavelengths. The table is taken from the paper by Tucker (1977). The same types of relationships summarized from the LACIE Field Database (Bauer, et al., 1979) are shown in Table 3-9, which gives linear correlations for the

thematic mapper and Landsat bands for percent soil cover, leaf area index, fresh biomass, dry biomass, and plant water content. The work of both Tucker and Bauer, in fact, has been used to help define spectral bandpasses for proposed satellite systems. Another example of the variability resulting from cultural practices and other factors is given in Table 3-10 taken from Bauer, Daughtry, and Vanderbilt (1981).

The variation of bidirectional reflectance with solar zenith angle has been well studied. Table 3-11, taken from Kriebel (1978), summarizes average percent changes in the reflected radiance per degree change in solar zenith angle. Optical depth dependencies are also included. Asymmetrical reflectance behavior with respect to solar noon has also been well documented by Kimes, Smith and Ranson (1980) and Duggin (1977). Diurnal variations exceeding 150 percent are not uncommon particularly in canopies with incomplete cover. Even for nonagricultural crops strong azimuthal effects, e.g., variations by a factor of two or more, are possible (Egbert and Ulaby, 1972). The strong anisotropic variations also are apparent from satellite observations (Salomonson and Marlatt, 1971). Generally, various wavelength-ratio combinations of bidirectional reflectance show decreased sensitivity to sun angle effects (Colwell, 1974).

Limited data are available for variations of bidirectional canopy reflectance with off-angle viewing. Both increasing and decreasing trends were found, e.g., Smith and Oliver (1972), and Kriebel (1978).

As mentioned earlier, measurements of agricultural crops are widely available. Field measurements of other canopies, however, are less so. Some example data taken from Ranson, Kirchner and Smith (1978) for a lodgepole pine and mountain-meadow canopy are given in Table 3-12. Complete descriptions of the target characteristics as well as numerous measurements for several naturally occurring vegetation canopies are given in the original reference.

NON-VEGETATION

In this section, non-vegetation will primarily include soils and rock materials. The reflective properties of snow will also be addressed.

Thermal Properties

For terrestrial remote sensing the thermal wavelength region, 8 to 14 μm, appears to be more advantageous for non-vegetative material than the 3- to 5-μm region. Generally the emission strengths are weaker at the lower wavelengths and, furthermore, few minerals possess diagnostic spectral features in the 3- to 5-μm region. In the future, the multispectral analysis of the thermal regime for geologic applications will likely prove

TABLE 3-5a

Average Percent Reflectances of Top Leaf Surfaces of 10 Leaves for Each of 8 Crops for 41 Wavelengths (nm) Over the 500- to 2500-nm Wavelength Interval

Crop	500	550	600	650	700	750	800	850	900	950	1000	1050	1100	1150
Beans	15.2	18.5	12.0	10.7	37.3	55.7	56.9	56.9	56.5	55.8	56.2	56.6	56.0	53.6
Corn	12.7	16.2	12.0	9.3	24.8	45.4	46.3	46.4	46.2	45.5	45.7	46.0	45.5	43.3
Lettuce	27.6	30.3	26.8	23.6	33.7	37.6	37.6	37.5	36.7	34.6	35.3	36.3	35.0	30.3
Sorghum	15.0	17.2	13.3	11.3	28.2	45.8	47.3	47.4	47.3	46.9	47.0	47.0	46.8	45.5
Soybeans	10.9	13.1	8.7	7.9	28.8	45.6	46.6	46.5	46.3	45.9	46.0	46.2	45.8	44.5
Sunflowers	9.6	11.0	8.4	8.5	27.5	45.4	47.3	47.3	47.1	46.5	46.9	47.2	46.6	44.1
Wheat	10.3	13.4	9.6	7.7	27.3	50.2	51.5	51.7	51.4	51.0	51.2	51.5	51.0	48.9
Blue Grama	5.9	9.8	7.4	5.3	11.6	52.0	61.0	67.9	69.3	73.6	77.5	78.9	—	—

Crop	1200	1250	1300	1350	1400	1450	1500	1550	1600	1650	1700	1750	1800	1850
Beans	53.5	53.6	50.8	44.9	25.6	18.5	24.6	33.1	38.4	40.9	40.6	37.5	35.2	24.2
Corn	43.2	43.5	41.8	38.3	23.4	16.8	21.0	27.1	31.0	32.9	32.6	30.1	28.8	23.1
Lettuce	29.6	29.8	26.4	21.4	11.8	9.1	10.4	13.0	15.4	16.8	16.8	15.0	13.8	10.6
Sorghum	45.3	45.4	44.3	41.7	30.9	24.7	28.2	33.2	36.1	37.4	36.9	35.3	34.2	28.2
Soybeans	44.5	44.4	43.1	40.1	27.7	21.8	26.1	31.9	35.2	36.6	36.3	34.5	33.3	25.5
Sunflowers	44.0	44.2	41.7	36.4	20.4	14.3	18.4	24.9	29.3	31.3	30.5	28.1	26.6	18.9
Wheat	48.8	49.2	47.2	43.5	27.7	21.7	26.5	32.7	36.4	38.2	37.4	35.2	34.3	27.3

Crop	1900	1950	2000	2050	2100	2150	2200	2250	2300	2350	2400	2450	2500
Beans	8.0	6.0	9.4	14.1	18.9	22.6	24.0	21.5	17.2	12.8	9.5	7.2	5.9
Corn	7.9	7.2	9.7	12.6	15.8	18.3	19.8	17.6	14.4	11.6	9.3	7.5	6.7
Lettuce	6.2	5.6	6.4	7.4	8.4	9.2	9.4	8.8	7.7	6.6	5.8	5.2	4.9
Sorghum	14.1	12.0	15.6	19.1	22.1	24.5	25.8	23.7	20.4	17.4	14.7	12.4	11.3
Soybeans	10.2	8.1	12.1	16.6	20.6	23.5	24.8	22.7	19.1	15.4	12.1	9.5	8.2
Sunflowers	8.0	6.5	8.1	10.4	13.2	15.4	16.2	14.4	11.6	9.3	7.6	6.5	6.0
Wheat	9.7	9.0	12.8	16.6	20.2	22.6	24.4	21.7	18.2	15.0	12.2	9.7	8.5

TABLE 3-5b

Average Percent Transmittances of Top Leaf Surfaces of 10 Leaves for Each of 8 Crops for 41 Wavelengths (nm) Over the 500- 2500-nm Wavelength Interval

Crop	500	550	600	650	700	750	800	850	900	950	1000	1050	1100	1150
Beans	6.9	10.9	5.5	3.6	26.6	40.9	42.0	42.2	42.0	41.5	42.2	42.4	41.9	39.9
Corn	3.7	9.8	3.7	0.7	22.6	48.9	50.5	50.9	51.1	50.7	51.2	51.7	51.6	49.7
Lettuce	38.4	44.3	39.5	34.0	49.5	55.3	55.6	55.5	54.8	52.6	53.7	54.9	53.7	48.2
Sorghum	5.0	9.0	4.2	2.1	24.4	46.7	49.1	49.6	49.8	49.9	50.3	50.8	50.7	49.8
Soybeans	10.0	15.6	8.7	5.4	32.5	50.0	51.4	51.8	51.9	51.8	52.2	52.6	52.4	51.4
Sunflowers	6.3	9.1	5.7	5.1	27.8	46.4	48.4	48.8	48.8	48.4	49.1	49.7	49.2	46.8
Wheat	1.9	5.8	2.1	0.7	20.3	41.8	43.4	43.9	44.1	43.9	44.6	45.2	45.1	43.4
Blue Grama	3.4	9.5	6.4	3.5	8.3	26.0	26.5	26.0	24.7	23.0	22.0	21.0	—	—

Crop	1200	1250	1300	1350	1400	1450	1500	1550	1600	1650	1700	1750	1800	1850
Beans	40.0	40.2	38.1	33.5	17.3	11.8	17.3	24.9	29.6	32.2	32.2	29.5	27.9	18.5
Corn	49.8	50.5	49.0	45.9	28.8	20.5	26.8	35.1	40.2	43.0	43.1	40.6	39.6	32.0
Lettuce	47.4	48.0	43.7	35.9	14.6	6.2	11.1	19.9	26.6	30.5	31.0	27.0	24.4	15.2
Sorghum	50.0	50.4	49.6	47.3	35.1	28.2	33.2	39.9	44.0	46.2	46.3	44.8	44.1	36.6
Soybeans	51.6	51.9	50.8	48.0	34.9	28.7	34.3	41.3	45.3	47.4	47.5	45.8	44.8	35.5
Sunflowers	46.8	47.3	45.1	40.0	22.2	15.0	21.0	29.1	34.4	37.0	36.6	34.0	32.7	22.5
Wheat	43.6	44.2	42.8	39.7	24.3	18.5	23.9	30.7	34.7	36.8	36.3	34.3	33.7	26.7

Crop	1900	1950	2000	2050	2100	2150	2200	2250	2300	2350	2400	2450	2500
Beans	3.7	1.9	5.4	10.3	15.3	18.6	19.7	18.4	15.2	11.3	7.8	4.9	3.5
Corn	6.5	5.0	11.8	11.6	24.6	28.5	30.3	28.3	24.4	19.9	14.8	9.7	7.0
Lettuce	2.1	0.5	1.7	4.5	8.8	12.2	13.5	12.2	9.0	5.6	2.9	1.4	0.8
Sorghum	15.4	12.2	19.9	26.7	31.9	35.3	36.9	35.4	32.1	28.2	23.6	18.4	15.6
Soybeans	14.6	11.7	19.3	26.7	32.7	36.3	37.7	36.3	33.0	28.6	23.5	18.5	15.8
Sunflowers	6.0	2.3	6.5	11.9	17.1	20.5	21.6	18.7	16.1	12.1	8.2	5.0	3.3
Wheat	6.0	5.2	10.7	15.9	20.4	23.3	24.7	22.8	19.6	16.2	12.3	8.6	6.5

* From Gausman (1971) and Smith and Oliver (1972).

TABLE 3-6

Total Solar Radiation Properties of Tree Leaves*

Common and technical name	Reflectance		Transmittance		Absorptance	
	Upper	Lower	Upper	Lower	Upper	Lower
Bollcana poplar (*Populus alba* L. var. *pyramidalis*, Bunge) .	0.285	0.44	0.18	0.18	0.535	0.38
Quaking aspen (*Populus tremuloidea* Michx.)....................	0.32	0.355	0.195	0.195	0.485	0.45
Eastern cottonwood (*Populus deltoides* Bartr.).................	0.280	0.295	0.275	0.26	0.445	0.445
Peachleaf willow (*Salix amygdoloides* anderss.)................	0.27	0.31	0.18	0.18	0.55	0.51
Weeping willow (*Salix babylonica* L.).............................	0.285	0.315	0.22	0.20	0.515	0.485
Butternut (*Juglans cinerea* L.)....................................	0.28	0.34	0.24	0.245	0.48	0.415
Common white birch (*Betula papyrifera* Marsh.)	0.32	0.34	0.24	0.235	0.44	0.425
European white birch (*Betula alba* S.)	0.30	0.33	0.22	0.24	0.48	0.43
Common alder (*Alnus rugosa* var. *americana* (Regel) Fern.)	0.23	0.27	0.21	0.20	0.56	0.53
Northern red oak (*Quercus rubra* L.).............................	0.27	0.29	0.24	0.255	0.49	0.455
Northern pine oak (*Quercus ellipsoidalis* E. J. Hill)...........	0.235	0.275	0.17	0.205	0.595	0.52
White oak (*Quercus alba* L.)......................................	0.28	0.325	0.235	0.20	0.485	0.475
Bur oak (*Quercus macrocarpa* Michx., var. *olivaeformis* (Michx.) A. Gray)..	0.24	0.27	0.23	0.22	0.53	0.51
American elm (*Ulmus americana* L.)..............................	0.235	0.285	0.18	0.255	0.585	0.46
Slippery elm (*Ulmus rubra* Muhl.).................................	0.24	0.31	0.275	0.22	0.485	0.47
Black cherry (*Prunus serotina* Ehrh.)	0.30	0.35	0.21	0.185	0.49	0.465
Silver maple (*Acer saccharinum* L.)	0.24	0.33	0.19	0.21	0.57	0.46
Norway maple (*Acer platanoides* L.).............................	0.25	0.29	0.24	0.25	0.51	0.46
Boxelder (*Acer negundo* L.).......................................	0.31	0.32	0.22	0.19	0.50	0.49
Ohio buckeye (*Aesculus glabra* Willd.)...........................	0.27	0.27	0.19	0.19	0.54	0.54
Common locust (*Robinia pseudoacacia* L.).......................	0.325	0.38	0.255	0.265	0.42	0.355
Green ash (*Fraxinus pennsylvanica* Marsh. var. *subintegcrrima* (Vahl) Fern.).......................................	0.29	0.33	0.21	0.15	0.50	0.52

* From Birkebak and Birkebak (1964).

Each value is that fraction of the radiation incident upon a leaf surface either reflected, transmitted or absorbed when total radiation on that surface is unity.

useful. For these applications the spectral emissivity would be of interest. There has been considerable use of the thermal regime in geologic applications to infer bulk thermal properties of the medium as discussed earlier in the modeling sections. For these applications thermal conductivity and diffusivity are important.

Table 3-13, taken from Link (1979), is a compendium of emissivity values for soil and rock types in the 3- to 5.5- and 8- to 14-μm region. Table 3-14 is taken from Buettner and Kern (1965) and shows emissivity values in the 8- to 12-μm for polished minerals illuminated under 30 degrees of incidence. Other examples are given in Taylor (1979).

Table 3-15 gives the conductivity and diffusivity for sandy and clayey soils for various moisture contents and conditions. The thermal properties of other geologic materials are given in Table 3-16.

Reflective Characteristics

The reflectance of snow depends not only on the intrinsic properties of snow itself relative to such factors as grain size and liquid water content but also on surface contaminants. Generally, snow reflectance decreases, at all wavelengths, with impurities and with increasing grain size. Consequently, as snow ages, reflectance de-

creases. Melting conditions also tend to decrease snow reflectance. Snow has been found to be an anisotropic reflector (Dirmhirn and Eaton, 1975) with higher reflectance values at larger solar zenith angles.

Generally, soil reflectance curves exhibit a gentle increase with increasing wavelength. Soil reflectance depends upon the chemical and physical properties of the components, moisture content, organic matter content, iron oxide content, texture and surface roughness. Table 3-17, taken from the exhaustive report by Stoner and Baumgardner (1980), summarizes some of the linear correlations between these factors and reflectance in the visible and near IR. Further soil reflectance examples are given in the report by Stoner, Baumgardner, Biehl and Robinson (1980). Generally, wetter soils appear darker. Also, as organic-matter content increases soil reflectance decreases, at least in the visible region. Usually, soil reflectance also decreases with increasing iron-oxide content. There is an interaction between soil texture or particle size and surface roughness. Generally, soil reflectance increases for smaller particle size but this dependence is not always so straightforward in the field, where surface roughness and shadowing effects may predominate. The classic work by Condit (1970, 1972) identifies three general shapes of reflectance

TABLE 3-7

Bidirectional Reflectance Factors and Maximum Polarization Observed at Large Phase Angles*
(In All Cases, First Value is for Target in Parallel Direction; Second Value is for Target in Perpendicular Case)

	Alfalfa Leaves		Potato Leaves	
Wavelength	Percent Polarization	Reflectance Factors	Percent Polarization	Reflectance Factors
0.350	21.2	5.9	54.8	3.9
	21.0	5.4	43.8	3.8
0.433	19.6	5.9	59.2	3.9
	2.8	5.4	49.6	3.7
0.533	5.9	12.7	33.5	8.9
	6.2	12.1	22.0	8.3
0.566	8.4	12.4	35.1	8.5
	9.9	11.8	20.0	8.0
0.633	13.3	7.7	82.2	5.6
	13.3	7.2	64.8	5.0
0.680	1.8	52.2	4.29	5.8
	1.9	47.6	3.31	5.5
1.0	1.6	51.0	3.69	5.3
	1.8	47.5	2.91	4.7

	Corn Leaves		Corn Tassels	
Wavelength	Percent Polarization	Reflectance Factors	Percent Polarization	Reflectance Factors
0.350	76.9	5.5	12.0	14.9
	50.3	3.9	22.5	8.1
0.433	75.2	6.9	15.4	13.7
	37.8	4.7	21.9	9.5
0.533	35.7	12.2	4.4	25.4
	23.5	10.1	7.0	17.7
0.566	14.2	10.8	5.7	26.8
	18.8	9.0	10.1	18.1
0.633	49.2	7.6	9.2	24.7
	30.9	6.3	12.0	16.2
0.80	20.6	37.5	2.4	55.7
	9.3	39.6	8.2	45.1
1.0	17.6	39.4	5.0	58.5
	10.9	40.2	7.0	47.2

	Rye Stalks		Wheat Stalks	
Wavelength	Percent Polarization	Reflectance Factors	Percent Polarization	Reflectance Factors
0.350	20.4	18.0	10.8	36.1
	8.5	6.5	8.3	6.3
0.433	30.9	27.1	47.1	22.7
	11.5	7.4	6.2	4.1
0.533	8.7	23.7	34.1	44.5
	2.5	16.3	0.9	14.7
0.566	7.9	38.3	16.9	51.5
	2.6	19.6	1.6	16.7
0.633	11.5	53.8	13.0	68.2
	2.8	25.3	2.3	26.0
0.80	8.1	63.2	2.1	86.3
	0.0	34.7	1.4	35.6
1.0	1.4	71.7	4.9	95.0
	0.6	39.3	3.0	38.7

TABLE 3-7

Continued

Wavelength	Rye Heads		Wheat Heads	
	Percent Polarization	Reflectance Factors	Percent Polarization	Reflectance Factors
0.350	19.5	9.3	20.5	10.7
	23.8	7.2	17.8	8.3
0.433	12.6	11.1	7.1	13.0
	32.4	7.7	12.3	9.1
0.533	5.5	18.5	9.1	20.5
	15.6	14.2	10.8	16.6
0.566	7.1	20.9	9.8	23.8
	15.8	15.9	15.0	18.8
0.633	4.2	25.3	5.7	30.8
	14.4	19.2	11.7	24.3
0.80	2.5	35.3	4.6	39.6
	12.9	27.7	8.9	32.1
0.1	3.6	37.1	3.1	42.2
	9.8	29.4	8.7	33.9

* From Egan (1970).

TABLE 3-8

Coefficients of Determination for the Regressions between Reflectance at Six Wavelengths and Total Wet Biomass, Total Dry Biomass, Leaf Water Content, Dry Green Biomass, Dry Brown Biomass, and Chlorophyll Content*

λ (μm)	r^2 (p vs TWB)	r^2 (p vs TDB)	r^2 (p vs HOH)	r^2 (p vs DGB)	r^2 (p vs DBB)	r^2 (p vs CHL)
.385	.78	.80	.65	.74	.56	.65
.465	.60	.54	.72	.65	.28	.48
.550	.13	.08	.20	.11	.03	.07
.675	.49	.31	.74	.47	.10	.40
.720	.00	.08	.02	.04	.01	.00
.765	.71	.57	.78	.67	.31	.60

* From Tucker, 1977.

TABLE 3-9

The Linear Correlations (r) of Reflectances in the Proposed Thematic Mapper and Landsat MSS Wavelength Bands with Percent Soil Cover, Leaf Area Index, Fresh and Dry Biomass, and Plant Water Content*

Wavelength band, μm	Percent soil cover	Leaf area index	Fresh biomass	Dry biomass	Plant water content
		Thematic mapper			
0.45 to 0.52	−0.82	−0.79	−0.75	−0.69	−0.76
0.52 to 0.60	−.82	−.78	−.81	−.77	−.82
0.63 to 0.69	−.91	−.86	−.80	−.73	−.81
0.76 to 0.90	.93	.92	.76	.67	.79
1.55 to 1.75	−.85	−.80	−.83	−.79	−.84
2.08 to 2.35	−.91	−.85	−.86	−.81	−.86
		Landsat MSS			
0.5 to 0.6	−0.82	−0.79	−0.81	−0.76	−0.81
0.6 to 0.7	−.90	−.85	−.81	−.74	−.82
0.7 to 0.8	.84	.84	.57	.46	.60
0.8 to 1.1	.91	.90	.77	.68	.79

* From Bauer et al., 1979.

TABLE 3-10

Percent of Variation in Red (0.6–0.7 μm) and Near-Infrared Reflectance (0.8–1.1 μm) of Corn, Soybean, and Spring-Wheat Canopies Associated with Soil Type and Cultural Practices on Several Dates (Development Stages)*

	Spectral Variable							
Agronomic Factor	Red Reflectance				Infrared Reflectance			
	Corn							
	6/11	7/15	8/22	9/26	6/11	7/15	8/22	9/26
Soil type	56	25	3	2	51	21	5	—
Plant population	—	8	22	—	1	33	36	4
Planting date	12	38	7	53	24	8	14	76
	Soybeans							
	6/18	7/17	8/22	9/26	6/18	7/17	8/22	9/26
Soil type	13	15	—	—	15	11	—	—
Planting date	13	63	52	87	44	69	10	85
Row width	8	—	9	—	2	5	29	1
Cultivar	1	—	6	—	2	—	16	10
	Spring Wheat							
	6/1	6/23	7/7	7/20	6/1	6/23	7/7	7/20
Soil moisture	2	16	73	52	9	28	69	36
Cultivar	—	1	—	2	—	10	4	4
N fertilizer	—	9	3	1	—	1	3	4
Planting date	36	5	7	27	85	35	4	6

* From Bauer, Daughtry and Vanderbilt, 1981.

curves in the 0.3–1.0-μm range and gives relationships to predict the soil reflectance curves from measurements at five characteristic wavelengths. The angular scattering properties of soils have been perhaps best characterized by the work of Coulson (1966) and Coulson and Reynolds (1971). Generally, soils exhibit a more Lambertian-type behavior than vegetation canopies. However, large increases in reflectance are also evident at steep viewing and illumination angles. The polarizing properties of soils, particularly with regard to soil-moisture conditions, have been pointed out by Curran (1978, 1979).

From a remote-sensing perspective, probably the most exhaustive work relating to the measurement of reflectance properties for minerals and mineral complexes in the 0.4- to 1.1-μm region is the work by Hunt and co-workers. The two references of Hunt and Salisbury (1976) relating to the visible and near-infrared spectra for both

TABLE 3-11

Percent Change of the Reflected Radiance due to a Change of the Distribution of the Irradiation Either by One Degree of the Solar Zenith Angle or by 10% Change of the Optical Depth of the Atmosphere, Averaged over all Directions of Reflection and over all Distributions of the Irradiation*

	Average change of the reflected radiance	
Surface type	Per degree change of the solar zenith angle	Per 10% change of the optical depth
Savannah	±1.0%	±1.6%
Bog	±0.9%	±0.7%
Pasture land	±1.7%	±1.0%
Coniferous forest	±2.3%	±1.5%
Average over the four surfaces	±1.5%	±1.2%

* From Kriebel, 1978.

TABLE 3-12

Mean Reflectances for a Mountain-Meadow Lodgepole Pine for Various Seasonal Measurement Periods. Standard Deviations for Each Spectral Band are Enclosed in Parentheses*

	Spectral Band (μm)				
	.48	.55	.68	.80	.96
Meadow					
May (Snow)	.724(.023)	.746(.007)	.757(.024)	.699(.009)	.566(.012)
July	.074(.005)	.104(.015)	.125(.008)	.328(.033)	.377(.042)
August	.061(.003)	.127(.007)	.129(.001)	.466(.044)	.529(.031)
September	.049(.007)	.076(.008)	.105(.009)	.266(.044)	.329(.040)
Lodgepole					
May	.043(.037)	.064(.023)	.019(.004)	.326(.053)	.304(.067)
July–Oct	.070(.018)	.103(.034)	.144(.048)	.292(.019)	.365(.040)

* From Ranson et al., 1978.

sedimentary rocks and metamorphic rocks can be used as an entree into their other reports. Reflectance for rocks and minerals depends on both surface external effects, such as surface roughness, and internal effects related to their chemistry and internal geometry, giving rise to various electronic processes. These processes include charge-transfer modes, crystal field transitions, and transitions into the conduction band. Charge-transfer and conduction-band transitions are intense, whereas the crystal field transitions are weak. Particularly with sedimentary rocks, spectral features may be associated with the constituents appearing as cements or impurities and in most cases reflectance is an indirect indication of rock composition.

Other summary analyses of soil properties may be found in Lyon and Green (1975) and Hovis, et al. (1966).

OTHER SOURCES OF DATA

The first edition of the *Manual of Remote Sensing* gives numerous examples of representative reflectance, emittance, and diffusivity values for terrain elements reported in the literature prior to 1970. For the most part these data have not been repeated here. Other early summary sources of representative values can be found in the Target Signatures Library of the Willow Run Laboratories, now ERIM, (Earing and Smith, 1966) and in the reports by Kondratiev, et al. (1964) and Steiner and Cuterman (1966).

A second major source of data is the *Infrared Handbook* (Wolfe and Zissis, 1978). A recent literature review and source of reported laboratory, field, and aircraft/satellite observations for bidirectional reflectance is the report by Smith and Ranson (1979).

Within the U.S., probably the most significant data sources for field measurements are the Large Area Crop Inventory Experiment (LACIE) and the follow-on AgRISTARS effort. The Laboratory for Applications of Remote Sensing (LARS) at Purdue University is the repository of this infor-

mation and will distribute the data base in computer-compatible form (Bauer, et al., 1978). Purdue University is also the source of numerous summary analyses of this data base. Ungar (1977) has prepared an atlas of narrow-band crop spectra in the 0.4- to 1.1-micrometer range. A good source of data for non-vegetative materials, especially in the thermal regime, is that collected by the U.S. Geological Survey.

Significant measurement activities have been undertaken in other countries as well. For example, the Canada Centre for Remote Sensing has initiated a field spectral measurements program (Brown, et al., 1980). The program undertaken by the NIWARS is particularly noteworthy (Bunnik, 1978).

FUTURE RESEARCH DIRECTIONS

This chapter has summarized several current modeling approximations to describe the interac-

TABLE 3-13

Emissivity Values for General Soil and Rock Types*

Material	Emissivity by Wavelength Band μm	
	3–5.5	8–14
Granite, rough		0.89
Dunite, rough		0.89
Basalt, rough		0.94
Sand, large grains		0.91
Sand, large grains, wetted		0.93
Sand (Monterey), small grains		0.92
Hainamanu silt loam—Hawaii	0.84	0.94
Barnes fine silt loam—South Dakota	0.78	0.93
Gooah fine silt loam—Oregon	0.80	0.98
Vereiniging—Africa	0.82	0.94
Maury silt loam—Tennessee	0.74	0.95
Dublin clay loam—California	0.88	0.97
Pullman loam—New Mexico	0.78	0.93
Grady silt loam—Georgia	0.85	0.94
Colt's neck loam—New Jersey	0.90	0.94
Mesita negra—lower test site	0.75	0.92

* From Link, 1979.

TABLE 3-14

Emissivity of Polished Minerals Reflectivity Under 30° Angle of Incidence Has Been Measured Spectrally. Data Give Average Emissivity for Area 8 to 12μ. Last column: Wavelengths of Minimum Emission*

Mineral	Temperature for Integration				λ for ϵ_λ Minimum
	253°K	273°K	293°K	313°K	
Quartz (agate) SiO_2	0.694	0.682	0.672	0.664	(8.55), 9.05
Granite SiO_2-$KAlSi_3O_8$-$NaAlSi_3O_8$	0.787	0.783	0.780	0.777	(8.55), 8.9
Feldspar $KAlSi_3O_8$	0.826	0.822	0.819	0.817	8.6, (9.5, 9.9)
Lavrotite $CaMgSi_2O_6$ + Vanadium	0.813	0.812	0.812	0.811	(8.65), 9.65
Obsidian SiO_2-$NaAlSi_3O_4$-$CaAl_2Si_2O_8$	0.832	0.830	0.828	0.826	9.2
Basalt $CaMg(SiO_3)_2$	0.904	0.905	0.906	0.907	10 (very broad)
Dunite Mg_2SiO_4	0.851	0.857	0.861	0.865	11.1
Marble $CaMg(CO_3)_2$	0.941	0.942	0.942	0.943	6.65, (11.4)

* From Buettner and Kern, 1965.

TABLE 3-15

Physical and Thermal Properties of Soils

Soil Physical Characteristics		Soil Thermal Characteristics			
		Conductivity cal/cm-hour-°C		Diffusivity cm²/hour	
Dry Density kg/m³ $\times$ 10⁻³	Moisture Content by Weight Percent	Thawed	Frozen	Thawed	Frozen
		Sandy Soils			
1.08	2.0	2.6	2.8	13.0	14.7
1.05	4.0	3.8	4.2	17.3	21.0
1.00	8.0	5.0	6.2	19.2	28.2
1.18	2.0	3.4	3.8	15.5	18.1
1.15	4.0	4.5	5.3	18.0	24.1
1.10	8.0	6.1	7.6	21.0	31.7
1.27	2.0	4.2	4.8	16.8	20.9
1.25	4.0	5.4	6.4	20.0	26.7
1.20	8.0	7.1	9.0	22.2	34.6
1.10	15.0	7.7	10.5	19.2	36.2
1.37	2.0	5.2	5.9	19.2	23.6
1.35	4.0	6.5	7.6	21.7	25.3
1.30	8.0	8.4	10.7	24.0	38.2
1.20	15.0	8.9	12.3	21.2	41.0
1.20	20.0	9.4	13.3	20.0	41.6
1.47	2.0	6.3	7.1	21.7	26.3
1.45	4.0	7.7	9.0	23.3	32.1
1.40	8.0	9.6	12.3	24.6	39.7
1.30	15.0	10.3	14.5	22.9	43.9
1.25	20.0	10.8	15.5	21.0	44.3
1.57	2.0	7.2	8.4	23.2	29.0
1.55	4.0	8.9	10.7	25.4	35.7
1.50	8.0	10.9	14.1	25.3	44.1
1.40	15.0	11.7	16.6	23.9	47.4
1.35	20.0	12.3	17.7	23.2	47.8
1.30	25.0	12.8	19.0	22.1	48.7

TABLE 3-15

Continued

Soil Physical Characteristics		Soil Thermal Characteristics			
		Conductivity cal/cm-hour-°C		Diffusivity cm²/hour	
Dry Density kg/m³ × 10⁻³	Moisture Content by Weight Percent	Thawed	Frozen	Thawed	Frozen
Sandy Soils (Continued)					
1.60	8.0	12.4	16.2	27.6	46.3
1.50	15.0	13.4	19.2	25.8	51.9
1.40	20.0	14.0	20.5	25.0	51.3
1.35	25.0	14.6	22.0	23.9	52.4
1.60	15.0	15.3	22.1	28.3	56.7
1.50	20.0	16.0	23.7	27.1	56.4
1.45	25.0	16.6	25.2	25.9	58.6
1.65	15.0	17.3	25.4	30.4	62.0
1.60	20.0	18.0	27.2	29.0	63.3
1.50	25.0	18.6	28.5	27.8	62.0
1.75	15.0	19.2	28.9	32.5	67.2
1.70	20.0	20.0	30.7	30.8	66.7
1.65	25.0	20.5	31.5	28.9	65.6
1.85	15.0	21.5	32.5	34.1	51.6
1.75	20.0	22.0	33.9	32.4	70.6
1.70	25.0	22.3	34.4	30.1	68.8
Clayey Soils					
1.00	8.0	3.4	4.0	12.1	16.7
1.10	8.0	4.2	5.0	13.1	19.2
1.20	8.0	5.0	6.0	14.3	20.7
1.10	18.0	5.9	7.5	12.8	22.7
1.30	8.0	6.2	7.3	16.3	23.5
1.20	18.8	7.3	9.3	14.9	25.8
1.10	27.0	8.1	10.9	15.0	28.7
1.40	8.0	7.3	8.8	17.4	25.9
1.30	18.0	8.5	10.8	16.3	28.4
1.20	27.0	9.3	12.8	16.3	32.0
1.10	40.0	10.1	14.3	14.9	32.5
1.50	8.0	8.6	10.3	18.7	28.6
1.35	18.0	9.8	12.8	17.8	32.0
1.25	27.0	10.6	14.8	17.1	35.2
1.15	40.0	11.4	16.2	15.8	35.2
1.60	8.0	9.7	11.9	19.4	30.5
1.45	18.0	11.2	14.5	19.3	34.5
1.35	27.0	12.0	16.8	18.2	37.3
1.20	40.0	12.9	18.3	17.2	37.3
1.50	18.0	12.5	16.5	20.5	36.7
1.40	27.0	13.4	18.9	19.4	40.2
1.30	40.0	14.3	20.3	17.9	39.0
1.60	18.0	14.2	18.8	22.2	39.2
1.50	27.0	15.0	21.3	20.5	42.6
1.35	40.0	15.8	22.5	18.8	40.9
1.70	18.0	15.9	21.4	23.7	42.8
1.60	27.0	16.6	23.6	21.6	42.9
1.45	40.0	17.2	24.4	19.5	42.1
1.60	18.0	17.8	24.0	25.8	48.0
1.65	27.0	18.3	26.0	22.6	47.3
1.50	40.0	18.5	26.3	20.1	43.8

TABLE 3-16

Thermal Properties of Common Geologic Materials

Material	Thermal Conductivity cal/cm-hour-°C	Thermal Diffusivity cm²/hour
Basalt	18.0	32.4
Clay soil (moist)	10.8	18.0
Dolomite	43.2	93.6
Gabbro	21.6	43.2
Granite (granite rocks)	27.0 / 23.4	57.6
Gravel	10.8	28.8
Limestone	17.3	39.6
Marble	19.8	36.0
Obsidian	10.8	25.2
Peridotite	39.6	61.2
Pumice, loose (dry)	2.16	14.4
Quartzite	43.2	93.6
Rhyolite	19.8	50.4
Sandy gravel	21.6	50.4
Sandy soil	5.04	10.8
Sandstone, quartz	43.2 / 22.3	46.8
Serpentine	22.7 / 25.9	46.8
Shale	15.1 / 10.8	28.8
Slate	18.0	39.6
Syenite	27.7 / 15.8	32.4
Tuff, welded	10.1	28.8

tion of electromagnetic radiation with terrain elements at optical wavelengths. Certain trends are evident and it is perhaps reasonable to try to extrapolate these trends into the next ten years.

The most obvious future research direction, which will undoubtedly be aggressively addressed, is the continued development of modeling approaches to handle the three-dimensional terrain problem at various spatial and temporal scales. In the reflective regime we are currently beginning to see the coupling of geometric optics and radiative transfer approaches when applied to these problems. These efforts will be intensified and applied to natural communities as well as to agricultural targets. In the thermal regime the problem appears more complex because of the highly variable environmental influences with which the targets are coupled. Further, the required supporting measurements are also more numerous. Thus, it is not quite so obvious what avenues are likely to be fruitful. Because of the large random component to the radiation problem, it is likely that tractable model formulations will include a statistical component. Geometric complexities in the targets modeled will most likely be handled by the application of finite-element techniques for those media, e.g., soils, where appropriate transfer coefficients can be formulated.

It is evident from the discussion in this chapter that a variety of calculational procedures are available to describe the radiation interactions for one-dimensional or multidimensional homogeneous targets. While some additional experimentation and sensitivity analyses with these models will be performed, the main thrust in the future will probably be to investigate linkages between the electromagnetic parameters in the models and the more traditional biophysical and/or cultural descriptors. These linkages are likely to take the form of submodels, relating attributes or processes to the remote-sensing modeling parameters. These new, more comprehensive, models will offer the potential for the development of overall system simulations for particular sensor/ target application trade-offs. The development of complete system-simulation models will be an iterative process and will probably be done with increasing sophistication and fidelity as our knowledge and capabilities increase. The material presented in these volumes concerning atmospheric, sensor, and other interactions will form the basis for such analyses. Such over-all simulations, which capture the phenomena of interest, should become more important as an aid in sensor design trade-offs.

In this chapter we have indicated some approaches for the extraction of underlying scene parameters. In the near future the main technique for developing the association between remote-sensing observables and scene attributes of interest will continue to be those described in the application chapters of this manual. However, if more experience, fidelity, and confidence are developed in the types of models described in this chapter, a greater number of formal, inverse-scattering-type approaches will probably be pursued. As in the atmospheric radiative transfer problem that applies to meteorological studies, it is not clear at this stage within what bounds these inverse-scattering attacks will be successful.

Applications of parameter-estimation techniques will probably, in fact, be appropriate for the experimental programs designed to support model validation and verification studies. It may very well be that the only feasible methods for estimating the spatially and temporally distributed parameters required for model studies will be the use of the models themselves to interpret the remote-sensing measurements for calibration sites. Also, as the need for and applicability of the models described in this chapter become evident, new direct experimental measurement techniques will also be tested.

Some interesting theoretical questions remain relative to the relationship of electromagnetic modeling at optical wavelengths to the modeling approaches in the microwave regimes discussed in

TABLE 3-17

Simple Correlation Coefficients Between Five Soil Parameters and Reflectance in Individual Bands by Climatic Zone*
r Values Are Given for Most Highly Correlated Bands (Parentheses)

Climatic Subgroup No. of Soils in Class	Soil Parameter**				BANDS
	Moisture Percentage by Weight	Particle Size Distribution	Cation Exchange Capacity	Iron Oxide Content	
humid frigid 38	−.43(1)	fine sand .37(10)	−.45(1)	.56(2)	1 0.52−0.62
humid mesic 75	−.29(10)	fine silt .58(9)	−.73(1)	.30(2)	2 0.62−0.72
humid thermic 60	−.65(10)	clay −.53(10)	−.73(9)	−.18(10)	3 0.72−0.82
subhumid frigid 42	−.75(10)	clay −.67(9)	−.86(9)	−.20(10)	4 0.82−0.92
subhumid mesic 46	−.64(10)	clay −.63(10)	−.71(10)	−.52(2)	5 0.92−1.02
subhumid thermic 36	−.82(10)	sand .76(2)	−.63(10)	−.44(6)	6 1.02−1.12
semiarid frigid 18	−.48(10)	clay −.67(10)	−.60(10)	.26(2)	7 1.12−1.22
semiarid mesic 46	−.34(10)	very fine sand .43(2)	−.44(1)	−.42(9)	8 1.22−1.32
semiarid thermic 20	−.55(10)	medium sand .66(10)	−.40(3)	−.67(9)	9 1.55−1.75
arid mesic 32	−.79(3)	clay −.62(3)	−.73(4)	.39(4)	10 2.08−2.32
arid thermic 24	−.75(10)	fine sand .90(4)	−.47(10)	−.73(9)	

* From Stoner and Baumgardner, 1980.
** Numbers in parentheses indicate wavelength bands, as defined at right edge of table.

the next chapter. For example, when does one approach, applicable to one regime, make the transition into an approach that is applicable to another regime? What common model-parameters and what new parameters must be measured or extracted? Future sensor systems will continue to utilize the multi-wavelength concept and will, therefore, encourage efforts to generate combined models of the electromagnetic terrain phenomena in the various regimes.

REFERENCES

Allen, W. A., H. W. Gausman, and A. J. Richardson, 1973, Willstätter-Stoll theory of leaf reflectance evaluated by ray tracing. Appl. Opt. 12(10): 2448–2453.

Allen, W. A., and A. J. Richardson, 1968, Interaction of light with a plant canopy. Proc. 5th Symp. on Remote Sensing of Environment, Univ. of Michigan, Ann Arbor, 219–232.

Allen, W. A., T. V. Gayle, and A. J. Richardson, 1970, Plant canopy irradiance specified by the Duntley equations. J. Opt. Soc. Am. 60(3):372–376.

Badhwar, G. D., 1980, Crop emergence data determination from spectral data. Photog. Eng. and Rem. Sens. 46(3):369–377.

Balick, L. K., R. K. Scoggins, and L. E. Link, 1981, Inclusion of a simple vegetation layer in terrain temp. models for thermal IR signature prediction. IEEE Geo. and Rem. Sens. GE-19(3):143–152.

Barkstrom, B. R., 1978, Passive remote sensing in the presence of multiple scattering: a numerical inversion method. From Developments in Atmospheric Sciences 9 (Remote Sensing of the Atmosphere, Elsevier Sci. Pub. Co., N.Y.)

Barrick, E. E., 1970, Rough surfaces, in Radar Cross-Section Handbook, edited by G. T. Ruck, Plenum, New York.

Bauer, M. E., M. M. Hixson, L. L. Biehl, C. S. T. Daughtry, B. F. Robinson, and E. R. Stoner, 1978, Vol. I Agricultural Scene Understanding. Final Report. Principal Investigator D. A. Landgrebe. LARS Contract Report 112578. Laboratory for Applications of Remote Sensing, Purdue University, West Lafayette, Indiana. 106 p.

Bauer, M. E., M. C. McEwen, W. A. Malila and J. C. Harlan, 1979, Design, implementation and results of LACIE field research. Proc. LACIE Symp., NASA JSC, Houston, Texas.

Bauer, M. E., C. S. T. Daughtry, and V. Vanderbilt, 1981, Spectral-Agronomic Relationships of Corn, Soybean and Wheat Canopies. Report No. SR-P1-04187. Laboratory for Applications of Remote Sensing, Purdue University, West Lafayette, Indiana.

Beckman, P. and A. M. Spizzichino, 1963, The Scattering of Electromagnetic Waves from Rough Surfaces. Pergamon, London, and Macmillan, New York.

Bergan, J. D., 1975, A possible relation of albedo to the density and grain size of natural snow cover. Water Resources Research. 11(5):745–746.

Bird, R. E., R. L. Hulstrom, A. W. Kliman, and H. G. Eldering, 1982, Solar spectral measurements in the terrestrial environment. Appl. Opt. 21(8): 1430–1436.

Birkebak, R. and R. Birkabak, 1964, Solar radiation characteristics of tree leaves. Ecology 34(3):646–649.

Bohren, C. F., 1971, Theory of radiation heat transfer between forest canopy and snowpack. Univ. Ariz. Agr. Expt. Sta. Journal Ser. No. 28.

Bohren, C. F. and D. B. Thorud, 1971, Two theoretical models of radiation heat transfer between forest trees and snowpacks. Univ. Ariz. Agr. pt. Sta. Journal Ser. No. 1866.

Bohren, C. F. and B. R. Barkstrom, 1974, Theory of the optical properties of snow. J. Geophys. Res., 79, 4527–4535.

Breece, H. T. and R. A. Holmes, 1971, Bidirectional scattering characteristics of healthy green soybean and corn leaves in vivo. Appl. Optics 10(1):119–127.

Brown, R. J. and F. J. Ahern, 1980, The field spectral measurements program of the Canada Centre for Remote Sensing. Canadian Journal of Remote Sensing. 6(1):26–37.

Buetter, K. J. K. and C. D. Kern, 1965, The determination of infrared emissivities of terrestrial surfaces. J. of Geophysical Research. 70(6):1329–1337.

Bunnik, N. J. J., 1978, The multispectral reflectance of shortwave radiation by agricultural crops in relation with their morphological and optical properties. Wageningen. Mededelingen Landbouwhogeschool. Nederland 78-1. 175 P.

Bunnik, N. J. J., 1982, Influence of crop geometry on multispectral reflectance determined by the use of canopy reflectance models. Signatures spectrales d'objets en teledetection, Avignon, 8–11. Sept. 1981.

Chance, J. E. and E. W. LeMaster, 1978, Plant canopy light absorption model with application to wheat. Appl. Optics 17(16):2629–2636.

Chandrasekhar, S., 1960, Radiative Transfer. New York: Dover Publications, Inc. 393 p.

Chevrel, M., M. Courtois, and G. Weill, 1980, The SPOT satellite remote sensing mission. Proc. 1980 ACSM-ASP Convention, St. Louis, MO.

Choudhury, B. J. and A. T. C. Chang, 1981, On the angular variation of solar reflectance of snow. J. Geo. Res. 86(C1):465–472.

Colwell, J. E., 1974, Vegetation canopy reflectance. Remote Sens. Environ. 3:175–183.

Condit, H. R., 1970, The spectral reflectance of American soils. Photogram. Eng. 36:955–966.

Condit, H. R., 1972, Application of characteristic vector analysis to the spectral energy distribution of daylight and the spectral reflectance of American soils. Appl. Optics. 11:74–86.

Conel, J. E., 1969, Infrared emissivities of silicates: experimental results and a cloudy atmosphere model of spectral emission from condensed particulate mediums. J. Geo. Res. 74:1614–1634.

Cooper, K., J. A. Smith and D. Pitts, 1982, Reflectance of a vegetation canopy using the Adding method. Appl. Optics. 21(21).

Coulson, K. L., G. M. Bouricius, and E. L. Gray, 1965, Optical reflection properties of natural surfaces. J. Geophy. Res. 70(18):4601–4611.

Coulson, K. L., 1966, Effects of reflection properties of natural surfaces in aerial reconnaissance. Appl. Optics 5(6):905–917.

Coulson, K. L. and D. W. Reynolds, 1971, The spectral reflectance of natural surfaces. J. Appl. Meteorology 10:1285–1295.

Curran, P. J., 1978, A photographic method for the recording of polarized visible light for soil surface moisture indications. Rem. Sens. Envir. 7:305–322.

Curran, P. J., 1979, The use of polarized panchromatic and false-color infrared film in the monitoring of soil surface moisture. Rem. Sens. Envir. 8:249–266.

Deardorff, J. W., 1978, Efficient prediction of ground surface temperature and moisture with inclusion of a layer of vegetation. J. Geophys. Res., pp. 1889–1902.

DeWit, C. T., 1965, Photosynthesis of leaf canopies, Wageningen: Center for Agricultural Publications and Documentation. Pp. 1–26.

Dirmhirn, I. and F. D. Eaton, 1975, Some characteristics of the albedo of snow. J. Appl. Meteorology 14:375–379.

Dodd, J. K. and E. H. Conrow, 1981, An evaluation of four thermal models used in thermal inertia analysis. Int'l. Geoscience and Remote Sensing Symposium (IGARSS '81) Washington, D.C. Pp. 1172–1181.

Dozier, J., S. Schneider, and D. M. Ginnis, Jr., 1981, Effect of grain size and snowpack water equivalence on visible and near-infrared satellite observations of snow. Water Resources Research 17(4):1213–1221.

Duggin, M. J., 1977, Likely effects of solar elevation on the quantification of changes in vegetation with maturity using sequential LANDSAT imagery. Appl. Optics 16:521–523.

Duggin, M. J., 1980, The field measurement of reflectance factors. Photog. Eng. Rem. Sens. 46(5):643–647.

Duggin, M. J., 1981, Simultaneous measurement of irradiance and reflected radiance in field determination of spectral reflectance. Appl. Opt. 20(2):3816–3818.

Dunkle, R. V., and J. T. Bevans, 1956, An approximate analysis of the solar reflectance and transmittance of a snow cover. J. Meteor., 13:212–216.

Earing, D. G. and J. A. Smith, 1966, Data compilation of target and background characteristics: target signature analysis center: data compilation. The University of Michigan, prepared for Air Force Avionics Laboratory, Wright-Patterson Air Force Base, Ohio, AD 489 968.

Egan, W. G., 1970, Optical stokes parameters for farm crop identification. Rem. Sens. of Envir. 1:165–180.

Egan, W. G., 1970, Nonimaging optical differentiation of forest foliage. Forest Science 16(1):79–94.

Egan, W. G. and T. Hilgeman, 1978, Spectral reflectance of particulate materials: a Monte Carlo model including asperity scattering. Appl. Optics. 17(2):245–252.

Egbert, D. D., 1976, Determination of the optical bidirectional reflectance from shadowing parameters, Ph.D. Dissertation, University of Kansas.

Egbert, D. D., 1977, A practical method for correcting bidirectional reflectance variations. Machine Processing of Remote Sensing Data Symposium. Purdue Univ. West Lafayette, Ind.

Egbert, D. D. and F. T. Ulaby, 1972, Effect of angles of reflectivity. Photog. Eng. 38(6):556–564.

Elachi, C., 1983, Microwave and infrared satellite sensors in Chapter 13, ASP Manual of Remote Sensing—Second Edition.

Emslie, A. G. and J. R. Aronson, 1973, Spectral reflectance and emittance of particulate materials. 1:Theory. Appl. Optics. 12(11)2563–2566.

Fymat, A. L. and E. Zuev, 1978, Remote Sensing of the Atmosphere: Inversion Methods and Applications. Elsevier Scientific Pub. Co. New York.

Gates, D. M., H. J. Keegan, J. C. Schleter, and V. R. Weidner, 1965, Spectral properties of plants. Appl. Optics 4(1):11–20.

Gates, D. M., 1968, Energy Exchange in The Biosphere Harper and Row, Inc. New York, 131 p.

Gausman, H. W., W. A. Allen, R. Cardenas, and A. J. Richardson, 1970, Relation of light reflectance to histological and physical evaluations of cotton leaf maturity. Appl. Optics. 9:545–552.

Gausman, H. W., W. A. Allen, C. L. Wiegand, D. E. Escobar, and R. R. Rodriguez, 1971, Leaf light reflectance, transmittance, absorptance, and optical and geometrical parameters for eleven plant genera with different leaf mesophyll arrangements. Proc. 7th Symp. on Remote Sens. Environ., Univ. Mich., Ann Arbor. III:1599–1626.

Gausman, H. W., W. A. Allen, R. Cardenas, and A. J. Richardson, 1972, Age effects of leaves within four growth states of cotton and corn plants on reflectance, leaf thickness, water and chlorophyll, and wavelength selection for crop discrimination. Agron. J.

Gausman, H. W., 1974, Leaf reflectance of near-infrared. Photogram. Eng. and Rem. Sens. 40:183–191.

Gausman, H. W., D. E. Escobar, J. H. Everitt, A. J. Richardson, and R. R. Rodriquez, 1978, The Leaf Mesophylls of Twenty Crops, their Light Spectra, and Optical and Geometrical Parameters. SWC Research Report 423 Rio Grande Soil and Water Research Center, Weslaco, Texas. 88 p.

Gausman, H. W., D. E. Escobar, and R. R. Rodriguez, 1978, Effects of stress and pubescence on plant leaf and canopy reflectance. Int'l. Archives of Photogram. XXII:719–749. (Vol. I of Proc. Int'l. Symp., Preiburg, Germany: G. Hildebrandt and H. J. Boehnel, eds.).

Gelb, A., 1974, Applied Optimal Estimation. The M.I.T. Press, Mass. Institute of Technology. Cambridge, Mass.

Giddings, J. C. and E. R. LaChapelle, 1961, Diffusion theory applied to radiant energy distribution and albedo of snow. J. Geophys. Res., 66:181–189.

Gordon, J. I. and P. V. Church, 1966, Overcast sky luminances and directional luminous reflectances of objects and backgrounds under overcast skies. Appl. Opt. 5:919–923.

Goudriaan, J., 1977, Crop micrometeorology: a simulation study. Wigeningen, the Netherlands: Centre for Agricultural Publishing and Documentation. 249 p.

Haralick, R. M. and A. K. Fung, 1983, Pattern Recognition and Classification. In Chapter 18, ASP Manual of Remote Sensing—Second Edition.

Heilman, J. L., E. T. Kanemasu, N. J. Rosenberg, and B. D. Blad, 1976, Thermal scanner measurement of canopy temperatures to estimate evapotranspiration. Rem. Sens. Envir. 5:132–145.

von Helmholtz, H., 1909, Physiological Optics. Trans. by T. P. C. Southhall. Opt. Soc. America.

Horn, B. K. P. and B. Bachman, 1978, Using synthetic images to register real images with surface models. Comm. AC M 21:914–924.

Hovis, W. A., Jr. and W. R. Callahan, 1966, Infrared reflectance spectra of igneous rocks, tuffs, and red sandstone from 0.5 to 22 incrons, J. Opt. Soc. Am., 56:630–643.

Hugli, H. and W. Frie, 1982, Understanding anisotropic reflectances in mountainous terrains. Submitted to Photog. Eng. and Rem. Sens.

Hunt G. R. and J. W. Salisbury, 1976, Visible and near infrared specta of minerals and rocks. XI Sedimentary Rocks. Mod. Geol. 5(4):211–217.

Hunt, G. R. and J. W. Salisbury, 1976, Visible and near infrared spectra of minerals and rocks. XII. Metaporphic Rocks. Mod. Geol. 5(4):219–228.

Idso, S. B., R. D. Jackson, W. L. Ehrler, and S. T. Mitchell, 1969, A method for determination of infrared emittance of leaves. Ecology 50(5):899–902.

Idso, S. B., T. J. Schmugge, R. Jackson, and R. J. Reginato, 1975, The utility of surface temperature measurements for the remote sensing of soil water status. J. Geophys. Res. 80:3044–3049.

Jackson, R. D., R. J. Reginato, P. J. Pinter, Jr. and S. B. Idso, 1979, Plant canopy information extraction from composite scene reflectance of row crops. Appl. Opt. 18(22):3775–3782.

Kahle, A. B., 1977, A simple thermal model of the earth's surface for geological mapping by remote sensing. J. Geophys. Res. 82(11):1673–1680.

Kasten, F. and Raschke, E., 1974, Reflection and transmission terminology by analogy with scattering. Appl. Optics. 13:460.

Kimes, D. S., S. B. Idso, P. J. Pinter, Jr., R. D. Jackson, and R. J. Reginato, 1980, Complexities of nadir-looking radiometric temperature measurements of plant canopies. Appl. Optics 19:2162–2168.

Kimes, D. S., J. A. Smith and K. J. Ranson, 1980, Vegetation reflectance measurements as a function of solar zenith angle. Photog. Eng. and Rem. Sens. Vol. 46:1563–1573.

Kimes, D. S., J. A. Smith, and L. E. Link, 1981, Thermal IR exitance model of a plant canopy. Appl. Optics. 20(4):623–632.

Kimes, D. S., 1982, Remote sensing of row crop structure and component temperatures using directional radiometric temperatures and inversion techniques. Submitted to Rem. Sens. of Envir.

Kirchner, J. A., J. A. Smith, and S. Youkhana, 1982, Influence of sky radiance distribution on the ratio technique for estimating bidirectional reflectance. Photog. Eng. Rem. Sens. 48(6):955–959.

Kriebel, K. T., 1978, Measured spectral bidirectional reflection properties of four vegetated surfaces. Appl. Optics 17(2):253–259.

Kondratiev, K. Y. Z., F. Mironova, and A. N. Otto, 1964, Spectral albedo of natural surfaces. Pure and Appl. Geophysics 59:207–216.

Kumar, R. and L. Silva, 1973, Light ray tracing through a leaf cross section. Appl. Opt. 12(12):2950–2954.

Link, L. E., 1979, Thermal modeling of battlefield scene components, Miscellaneous Paper EL-79-5, U.S. Army Engineer Waterways Experiment Station, CE, Vicksburg, Miss.

Luenberger, D. G., 1979, Introduction to Dynamic Systems. John Wiley & Sons. New York. 446 p.

Lyon, R. J. P. and A. A. Green, 1975, Reflectance and emittance of terrain in the mid-infrared (6-25 micrometer) region in Infrared and Raman Spectroscopy of Lunar and Terrestrial Minerals. Academic Press, Inc., New York.

Middleton, W. E. K. and A. G. Mungall, 1952, The luminous directional reflectance of snow. J. Opt. Soc. Am. 42(8):572–579.

Nicodemus, F. E., 1970, Reflectance nomenclature and directional reflectance and emissivity. Appl. Optics 9(6):1474–1475.

Nilson, T., 1971, A theoretical analysis of the frequency of gaps in plant stands. Agricultual Meteorology, Vol. 8, pp. 25–38.

Norman, J., 1979, Modeling the complete crop canopy. In: Modification of the Aerial Environment of Plants (B. J. Barfield and J. F. Gerber, Eds.), ASAE Monograph Number 2, Amer. Soc. of Ag. Engs., St. Joseph, MI, 539 p.

Olsen, C. E., Jr., 1969, Spectral reflectance measurements compared with panchromatic and infrared aerial photographs: IST Rept. 4864-8-T, Univ. of Mich., Ann Arbor, Mich.

Otterman, J., 1981, Reflection from soil with sparse vegetation. Adv. Space Res. Vol. 1, pp. 115–119.

Park, J. K. and D. W. Deering, 1982, Simple radiative transfer model for relationships between canopy biomass and reflectance. Appl. Optics. 21(2):303–309.

Pearson, R. L., L. D. Miller and C. J. Tucker, 1976, Hand-held spectral radiometer to estimate gramineas biomass. Appl. Opt. 15(2):416–418.

Pratt, D. A., 1980, Two-dimensional model variability in thermal inertia surveys. Rem. Sens. of Envir. 9:325–338.

Preisendorfer, R. W., 1965, Radiative Transfer on Discrete Spaces. Pergamon Press. 462 p.

Ranson, K. J., J. A. Kirchner, and J. A. Smith, 1978, Scene Radiation Dynamics, Vol. II. Data Library. Final report under contract DACW 39-77-C-0073. Colorado State Univ. Ft. Collins, CO.

Richardson, A. J., E. C. Wiegand, H. Gausman, J. Cuellar, and A. Gerbermann, 1975, Plant, soil, and shadow reflectance components of raw crops. Photog. Eng. and Rem. Sens. 41(11):1401–1407.

Richardson, A. J., 1981, Measurement of reflectance factors under daily and intermittent irradiance variations. Appl. Optics. 20(19):3336–3340.

Reifsnyder, W. E. and H. W. Lull, 1965, Radiant energy in relation to forests. U.S. Dept. Agr., Tech. Bull., 1344: 111 pp.

Robinson, B. F. and L. L. Biehl, 1979, Calibration procedures for the measurement of reflectance factor in remote sensing field research. S.P.I.E. 196:16–26.

Rosema, A., 1975, A mathematical model for the simulation of the thermal behaviour of bare soils, based on heat and moisture transfer, Niwars publ. no. 11, NIWARS, Delft, The Neth.

Ross, J. and T. Nilson, 1975, Radiative exchange in plant canopies, in Heat and Mass Transfer in the Biosphere.

Salomonson, V. V. and W. E. Marlatt, 1971, Airborne measurements of reflected solar radiation. Remote Sens. Environ. 2:1–8.

Salomonson, V. V., 1983, Water Resources Applications in Chapter 29, ASP Manual of Remote Sensing—Second Edition.

Schnetzler, C. C. and L. T. Thompson, 1979, Multispectral resource sampler: an experimental sensor for the mid-1980's. Proc. SPIE Tech. Symp., Huntsville, AL May 22–24, Vol. 183. 8 p.

Sinclair, T. R., M. M. Schreiber, and R. M. Hoffer, 1973, Diffuse reflectance hypothesis for the pathway of solar radiation through leaves. Agronomy Journal, Vol. 65:276–283.

Smith, J. A. and R. E. Oliver, 1972, Plant canopy models for simulating composite scene spectroradiance in the 014 to 1.05 micrometer region. Proc. Eighth Int. Symp. Remote Sensing Envir. University of Michigan, Center for Remote Sensing—Information and Analysis. Ann Arbor, Michigan, pp. 1333–1353.

Smith, J. A. and K. J. Ranson, 1979, MRS Literature Survey of Bidirectional Reflectance. Final Report. ORI, Inc., Silver Spring, Maryland. 200 pp.

Smith, J. A. and R. E. Oliver, 1974, Effects of changing canopy directional reflectance on feature selection. Appl. Opt. 13:1599–1604.

Smith, J. A., K. J. Ranson, D. Nguyen, L. K. Balick, L. E. Link, L. Fritschen, and B. Hutchison, 1981, Thermal vegetation canopy model studies. Rem. Sens. Envir. 11:311–326.

Smith, J. A. and H. Randolph, 1982, A thermal vegetation canopy model using the Kalman filter. (In preparation).

Steiner, D. and T. Cuterman, 1966, Russian Data on Spectral Reflectance of Vegetation, Soil and Rock Types. University of Zurich, Switzerland. Final Technical Report.

Strahler, A. H. and X. Li, 1981, An invertible coniferous forest canopy reflectance model. Proc. Fifteenth Int. Symp. on Remote Sensing of Environ., Ann Arbor, MI, in press.

Stoner, E. R., M. F. Baumgardner, L. L. Biehl and B. F. Robinson, 1980, Atlas of Soil Reflectance Properties. Research Bull. 962. Agricultural Experiment Station, Purdue Univ., W. Lafayette, Ind.

Stoner, E. R. and M. F. Baumgardner, 1980, Physicochemical, Site, and Bidirectional Reflectance Factor Characteristics of Uniformly Moist Soils. LARS Technical Report 111679. Laboratory for Applications of Remote Sensing. Purdue Univ. West Lafayette, Ind.

Suits, G. H., 1972, The cause of azimuthal variations in directional reflectance of vegetative canopies. Remote Sens. Environ. 22:175–182.

Suits, G. H., 1981, The extension of a uniform canopy reflectance model to include row effects. Environmental Research Institute of Michigan. Ann Arbor, Mich.

Suits, G. H., 1983, Electromagnetic Radiation. In Chapter 2, ASP Manual of Remote Sensing—Second Edition.

Swain, P. H. and S. M. Davis, 1978, Remote Sensing: The Quantitative Approach. McGraw-Hill, Inc. New York.

Szwarcbaum and G. Shaviv, 1976, A Monte-Carlo model for the radiation field in plant canopies. Agric. Meteorol., 17:333–352.

Taylor, S. E., 1979, Measured emissivity of soils in the Southeastern United States. Rem. Sens. Environ. 8:359–364.

Tucker, C. L., 1977, Spectral estimation of grass canopy variables. Rem. Sens. of Environ. 6:11–26.

Tucker, C. J. and M. W. Garratt, 1977, Leaf optical system modeled as a stochastic process. Appl. Opt. 16(3):635–642.

Twomey, S., 1977, Introduction to the mathematics of inversion in Remote Sensing and Indirect Measurements. Elsevier. 243 p.

Ungar, S. G., et al., 1977, Atlas of selected crop spectra, Imperial Valley, California: NASA Institute for Space Studies, Goddard Space Flight Center.

van de Hulst, H. C., 1957, Light Scattering by Small Particles. Wiley, 470 pp.

van de Hulst, H. C., 1980, Multiple Light Scattering, Vol. I. Academic Press, New York.

Verhoef, W., and N. J. J. Bunnick, 1976, The Spectral Directional Reflectance of Row Crops. Part 1: Consequences of Non-Lambertian Behavior for Automatic Classification. Part 2: Measurements on Wheat and Simulations by Means of a Reflectance Model for Row Crops. Tech. Rept. No. NIWARS-PUBL-35. Netherlands Interdepartmental Working Group on the Application of Remote Sensing, Delft.

Vincent, R. K., and G. R. Hunt, 1968, Infrared reflectance from mat surfaces. Appl. Opt. 7(1):53–59.

Watson, K., 1975, Geologic applications of thermal infrared images, Proc. IEEE 63:128–137.

Watson, K., 1979, Thermal phenomena and energy exchange in the environment. Strasbourg, France, Summer School of Space Physics on Mathematical and Physical Principles of Remote Sensing. Aug. 1978.

Weinman, J. A., P. J. Guetter, 1972, Penetration of solar irradiances through the atmosphere and plant canopies. J. Of Applied Meteorology, Vol. 11:136–140.

Welles, J. M. and J. M. Norman, 1979, General radiative transfer model for random and non-random canopies. Fourteenth Conf. on Agriculture & Forest Meteorology and Fourth Conf. On Biometeorology. 205–206.

Wiegand, C. L., A. J. Richardson, and E. T. Kanemasu, 1979, Leaf area index estimates for wheat from Landsat and their implications for evapotranspiration and crop modeling. Agron. J. 71:336–342.

Wiscombe, W. J. and S. G. Warren, 1980, A model for the spectral albedo of snow. 1: Pure Snow. J. of the Atmospheric Sciences. Vol. 37:2712–2733.

Willstätter, R., and A. Stoll, 1918, Untersuchungen uber die Assimilation der Kohlensaure. Springer, Berlin.

Wolfe, W. L. and G. J. Zissis, 1978, The Infrared Handbook. Infrared Information and Analysis Center, ERIM, University of Michigan.

Wong, C. L. and W. R. Blevin, 1967, Infrared reflectances on plant leaves. Aust. J. Biol Sci., 20:501–508.

Woolley, J. T., 1971, Reflectance and transmittance of light by leaves. Plant Physiol. 47:656–662.

Youkhana, S., J. A. Smith, 1982, The Suits prime canopy reflectance model (In preparation).

CHAPTER 4

Matter-Energy Interaction in the Microwave Region

Authors: ADRIAN K. FUNG and FAWWAZ T. ULABY

GENERAL CONTENTS: The radar equation; basic characteristics of radar return; dielectric properties; scattering models; imagery gray tones; radar backscattering from water surfaces; radar backscattering from land surfaces and sea ice; passive microwave radiation; microwave emission from land surfaces and sea ice; microwave radiometric characterstics of the ocean.

NOMENCLATURE

To conserve and eliminate repetition in text and references, the following symbols, units and names have been used in this chapter.

Symbol	SI Units	Name
A_r	m²	effective antenna aperture
e_v, e_h	—	emissivity for vertical, horizontal polarization
G_t	—	antenna gain at polarization t
k	—	wave number
n	—	refractive index
P_r	W	received power at polarization r
P_t	W	transmitted power at polarization t
R	m	range between antenna and target
ϵ	—	complex relative dielectric constant
ϵ'	—	real part of ϵ
ϵ''	—	imaginary part of ϵ
θ	rad	polar angle
λ	m	wavelength
$\rho(u,v)$	—	surface correlation coefficient
σ	m	rms surface height
σ_{rt}	m²	radar cross-section
σ_{rt}°	m² m⁻²	scattering coefficient
ϕ	rad	azimuth angle
ω	rad s⁻¹	radian frequency

INTRODUCTION

This chapter deals with energy/matter interaction mechanisms in the microwave region as they relate to the active and passive remote sensing of land- and sea-surfaces or media. Our objective is to provide the reader with a qualitative understanding of the scattering and emission characteristics of a variety of targets of interest and to indicate the possible scattering or emission mechanisms responsible for them. Where appropriate, special scattering or emission characteristics relating to the remote sensing of special target parameters (e.g., wind-speed over ocean) are pointed out. The chapter begins by showing the scattering or reflection characteristics of special geometric objects in order to provide a basic reference point for the reader and includes a summary of the dielectric properties and scattering characteristics of random surfaces and inhomogeneous media. These results provide the background information necessary for radar im-

agery interpretation and radar scatterometry. The rest of the chapter summarizes the influences of target parameters on emission and the emission characteristics of various surfaces or media of interest.

THE RADAR EQUATION

The performance of a radar is conveniently described by the radar equation, which relates the received power to the target parameters and to the parameters of the radar. With the path losses neglected, the received power may be written as one form of the radar equation,

$$P_r = \frac{P_t G_t}{4\pi R^2} \sigma_{rt} \frac{A_r}{4\pi R^2} \qquad (4\text{-}1)$$

where: P_r = received power at polarization r,
P_t = transmitted power at polarization t,
G_t = gain of the transmitting antenna in the direction of the target at polarization t,
R = distance between radar and target,
σ_{rt} = radar cross-section, the area intercepting that amount of incident power of polarization t which, when scattered isotropically, produces an echo at polarization r equal to that observed from the target,
A_r = effective receiving area of the receiving antenna aperture at polarization r.

The above equation assumes that the transmitter and the receiver are in the same location.

The transmitted electromagnetic radiation (EMR) received at the target is given by $P_t G_t / 4\pi R^2$. This expression, multiplied by σ_{rt}, indicates the total EMR scattered by the target, which can be denoted by P_s. Thus the power density received at the receiver is $P_s/4\pi R^2$, and the received power should be the density of EMR multiplied by the effective size of the receiving antenna aperture, A_r.

Most geoscience applications involve extended

targets that usually are much larger than a resolution cell of the radar. In this case it is more convenient to define an average differential cross-section or radar cross-section per unit area, and to consider the average return power. The total average received power is then given by integrating the return from each differential area over the entire irradiated area. The average differential cross-section is also known as the scattering coefficient, and is commonly denoted by σ°_{rt}. Thus, the average received power, P_{ar}, when written in terms of the scattering coefficient, is

$$P_{ar} = \int_{A_0} \frac{P_t G_t}{4\pi R^2} \, \sigma^{\circ}_{rt} \, \frac{A_r}{4\pi R^2} \, dS \qquad (4\text{-}2)$$

where: S = surface; the surface integral is taken over the irradiated area A_0,
σ°_{rt} = scattering coefficient.

The effective size of the antenna aperture is related to the antenna gain by

$$A_r = G_r \lambda^2/4\pi. \qquad (4\text{-}3)$$

Hence, P_{ar} may be written more compactly as

$$P_{ar} = \frac{\lambda^2}{(4\pi)^3} \int_{A_0} \frac{P_t G_t G_r}{R^4} \, \sigma^{\circ}_{rt} \, dS. \qquad (4\text{-}4)$$

When a radar is used to observe a volume distribution of materials rather than a surface distribution, the scattering cross-section per unit volume should be used in place of σ° in Eq. 4-4. Of course, the integral should also be changed to a volume integral over the illuminated volume. Examples of volume scattering are found in radar observations of clouds, rain, and any non-homogeneous layer, such as snow-covered or vegetated terrain.

The equation for the received power indicates that P_t, G, and λ are parameters of the radar system, and R is determined by the location of the radar with respect to the target. The design and operation of radars is such that these quantities normally either remain constant or are known during use of the radar. The factor that governs the average return-power strength as a function of the way in which the incident EMR interacts with the surface is, therefore, σ°_{rt}.

BASIC CHARACTERISTICS OF RADAR RETURN

The scattering coefficient, σ°, in general, is a function of polarization, look angle, wavelength, and interaction properties of the target: geometric, dielectric, and conductive. When transmitted power, antenna gain, wavelength, and polarization are fixed by system design, the average return power strength varies only with σ°.

The gray tone over a homogeneous area on a radar image is proportional to the average return power strength and, hence, to σ°. If the imaging radar system is amplitude-calibrated and enough levels of calibration signals are placed on the same

film that records the radar signal, a microdensitometer reading of the film may be related to the average received power and σ°. The operation of imaging-radar systems is covered in Chapter 10.

Radar returns from a target depend upon both the strength of the transmitted energy and the reflecting or scattering capability of the target. Most radar targets scatter with different strengths along different directions, and their scattering characteristics may be conveniently described in terms of their scattering, or reradiation, patterns (plots of signal strength versus angle). The backscattering pattern is of particular interest to radar; it shows the return-signal strength in the opposite direction of the incident radiation.

For an extended target (a cornfield for instance) the instantaneous return fluctuates (increases and fades) over a wide range as the radar beam moves over it. This is due to changes in the relative phase of signals from different parts of elements forming the target. Despite the rapid signal fluctuation, the radar is supposed to be looking at the same target under similar conditions. Hence, it is meaningful only to speak of the average return rather than the instantaneous return for extended targets. Such an averaged backscattered power is related directly to the scattering coefficient (see Eq. 4-2). The purpose of the averaging is to smooth out the fluctuation of the return signal. If the amount of averaging is not enough to produce a smooth signal, "speckles" will show up on the images.

Many complicated radar targets can be approximated by a collection of facets, spheres, or cylinders. Corner reflectors may also be needed in modeling man-made ground targets, such as a side wall located on a flat piece of ground. To obtain an intuitive understanding of the reradiation patterns for targets of complicated shapes, it is helpful to know the scattering patterns of these simple geometries. In the following discussion, no change is assumed in permeability from one medium to the other, and the medium surrounding the target is assumed to be air.

FACETS

For a perfectly conducting rectangular facet with total dimensions of L and b parallel to the x- and y-axes, respectively, an incident-plane wave polarized either vertically or horizontally in the x-z plane (Figure 4-1), gives rise to a backscattering cross-section of the form

$$\sigma(\theta) = \frac{4\pi b^2 L^2}{\lambda^2} \left[\frac{\sin(kL\,\sin\theta)}{kL\,\sin\theta} \right]^2 \cos^2\theta, \quad (4\text{-}5)$$

where: $\sigma(\theta)$ = backscattering coefficient of incident wave,
λ = wavelength of the incident radiation,
k = wave number $(2\pi/\lambda)$,
and b and L are several wavelengths in length.

This formula is good for θ in the range of approximately $0-45°$. Beyond this range, according

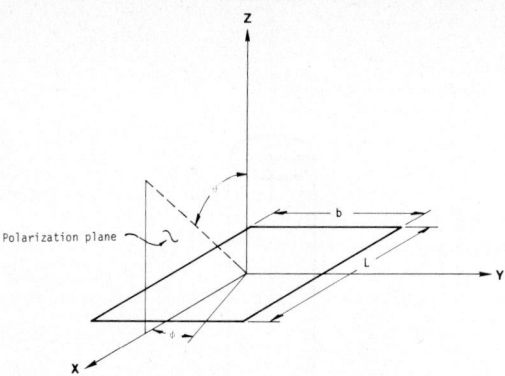

Fig. 4-1. Plane of polarization of incident EM wave.

to Ross (1966), polarization effects may become significant (Figure 4-2). A qualitative illustration of $\sigma(\theta)$ as a function of θ for different values of L in wavelength scale is shown in Figure 4-3. If the incident wave had come from the direction defined by θ and ϕ (Figure 4-1), the backscattering cross-section would take the form (Kerr, 1951) of

$$\sigma(\theta,\phi) = \frac{4\pi b^2 L^2}{\lambda^2} \left[\frac{\sin(kL \sin\theta \cos\phi)}{kL \sin\theta \cos\phi} \right.$$
$$\left. \frac{\sin(kb \sin\theta \sin\phi)}{kb \sin\theta \sin\phi} \right]^2 \cos^2\theta. \quad (4-6)$$

From Eq. 4-6, it is seen that the variations of σ versus θ in planes of constant ϕ are similar, except that the widths of the main lobes are now governed by both b and L. The maximum cross-section occurs only at normal incidence and is given by

$$\sigma_{\max} = 4\pi b^2 L^2/\lambda^2. \quad (4-7)$$

A collection of facets may be used to model any curved surface, such as the surface of the sea (Katzin, 1957).

SPHERES

Due to geometrical symmetry, the scattering pattern of a sphere must be isotropic (i.e., inde-

pendent of the look angle), and must be the same for both horizontal and vertical polarizations. If the radius, r, of the sphere is less than about one-tenth of the wavelength, the Rayleigh scattering expression applies, and the radar cross-section (Kerr, 1951) is given by

$$\sigma = 64\pi^5 \left(\frac{\epsilon - 1}{\epsilon + 2} \right)^2 \frac{r^6}{\lambda^4}, \quad (4-8)$$

$$\epsilon = \epsilon'_r - \frac{jg}{\omega\epsilon_0} \quad (4-9)$$

where: r = radius of sphere,
ϵ'_r = relative dielectric constant of the sphere,
g = conductivity of the sphere,
ω = radian frequency of the radiation,
ϵ_0 = dielectric constant of vacuum.

For a sphere with a radius larger than about 1.6 times the wavelength, its radar cross-section is equal to its geometric cross-section, i.e.,

$$\sigma = |R|^2 \pi r^2, \quad (4-10)$$

where

$$R = \frac{(\epsilon^{1/2} - 1)}{(\epsilon^{1/2} + 1)}, \quad (4-11)$$

and ϵ is the complex dielectric constant.

For the intermediate case where $0.1\lambda < r < 1.6\lambda$, the radar cross-section for a perfectly conducting sphere is illustrated in Figure 4-4.

The case of small spheres is of special importance in applications since rain and clouds may be modeled by a collection of spheres (Kerr, 1951).

CYLINDERS

When the radius r of a perfectly conducting cylinder of length $L(L \gg r)$ (Figure 4-5) is equal to or larger than a wavelength, the backscattering cross-section for both the transverse magnetic mode (TM) case (when the electric-field vector of the incident radiation lies parallel to the axis of the cylinder) and the transverse electric mode

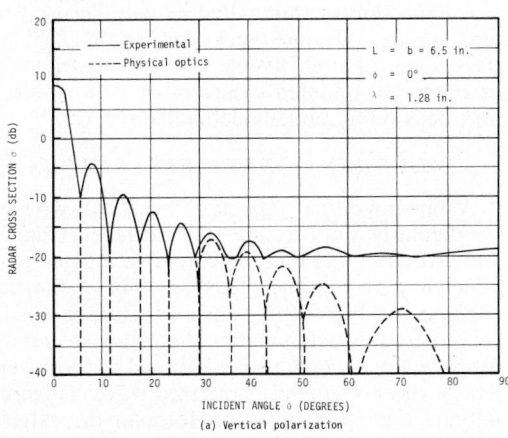

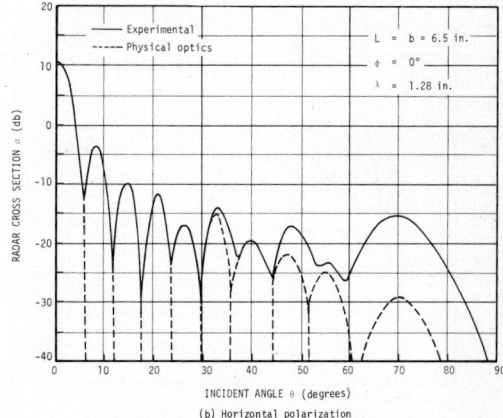

Fig. 4-2. Polarization effects of a plane facet: (a) left side, vertical polarization, (b) right side, horizontal polarization (after Ross, 1966).

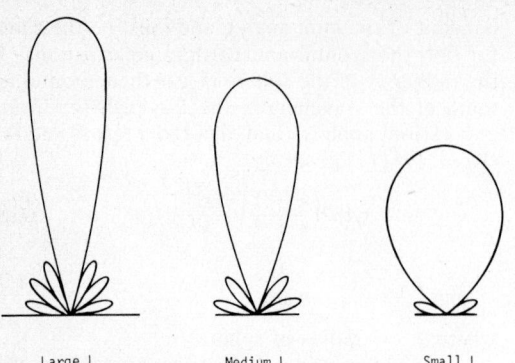

Fig. 4-3. Backscattering cross-sections as functions of the angle of incidence and ground facet size.

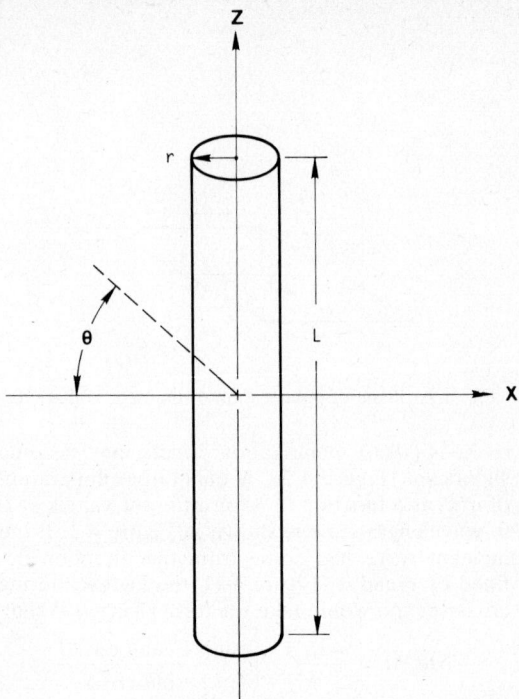

Fig. 4-5. Geometry of a cylinder of perfect conductivity.

(TE) case (when the electric-field vector lies perpendicular to the cylinder axis) is given approximately (Kerr, 1951) by

$$\sigma(\theta) \approx \frac{2\pi r L^2 \cos\theta}{\lambda} \left[\frac{\sin(kL \sin\theta)}{kL \sin\theta} \right]^2 . \quad (4\text{-}12)$$

As expected from symmetry considerations (Figure 4-5), $\sigma(\theta)$ does not vary in the plane perpendicular to the axis of the cylinder. In the plane parallel to the cylinder axis, the angular variation of $\sigma(\theta)$ is quite similar to that of a plane facet. Figure 4-6 presents a qualitative picture of this variation.

If the radius of the cylinder is about one-tenth of the wavelength or smaller, the backscattering cross-section for transverse magnetic (TM) propagation is given approximately by

$$\sigma_m(\theta) \approx \pi L^2 \frac{\left[\dfrac{\sin(kL \sin\theta)}{kL \sin\theta} \right]^2}{\left[\log_e \dfrac{2\pi r \cos\theta}{\lambda} \right]^2} \quad (4\text{-}13)$$

where: $\sigma_m(\theta)$ = backscattering cross-section, TM case.

For transverse electric (TE) propagation, the general result is more involved, except when the

scattering is in the plane perpendicular to the cylinder axis. Then, the backscattering cross-section is approximately given by

$$\sigma_e(\theta) \approx (9\pi/4)\, L^2\, (2\pi r/\lambda)^4, \quad (4\text{-}14)$$

where: $\sigma_e(\theta)$ = backscattering cross-section for TE.

When the radius of the cylinder is neither much greater nor much less than the wavelength, the angular variation of σ versus θ is illustrated by Figure 4-6 for TE polarization and for different cylinder lengths. Figures 4-7(a) and (b) (Ufintsev, 1962), illustrate backscattering for both the TM and TE polarizations for a specific case.

A collection of thin cylinders may be used to model grass and some types of vegetation such as flags (Cosgriff et al., 1960). A volume distribution of randomly oriented cylinders of various sizes may be used to model a defoliated forest.

DIHEDRAL CORNER REFLECTORS

A dihedral corner reflector consists of two perpendicularly intersecting, flat surfaces (Figure 4-8). Any ray entering the corner in a plane perpendicular to the line of intersection of the flat surfaces will be reflected and will return in the direction from which it came. The response pattern for reflection in such planes is broad for both vertically and horizontally polarized waves (Figures 4-9 and 4-10, respectively). However the pattern may change significantly for other polarizations (Figure 4-11). The angular behavior of the cross-

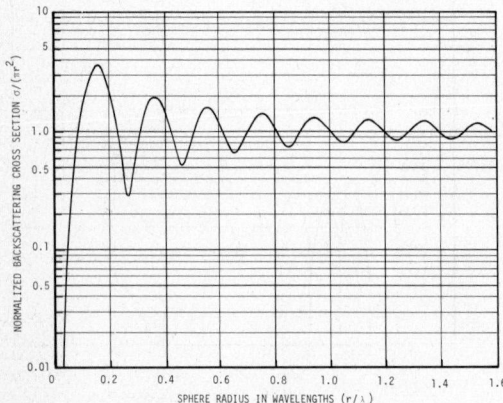

Fig. 4-4. Radar cross-sections for perfectly conducting spheres as functions of radius and wavelength.

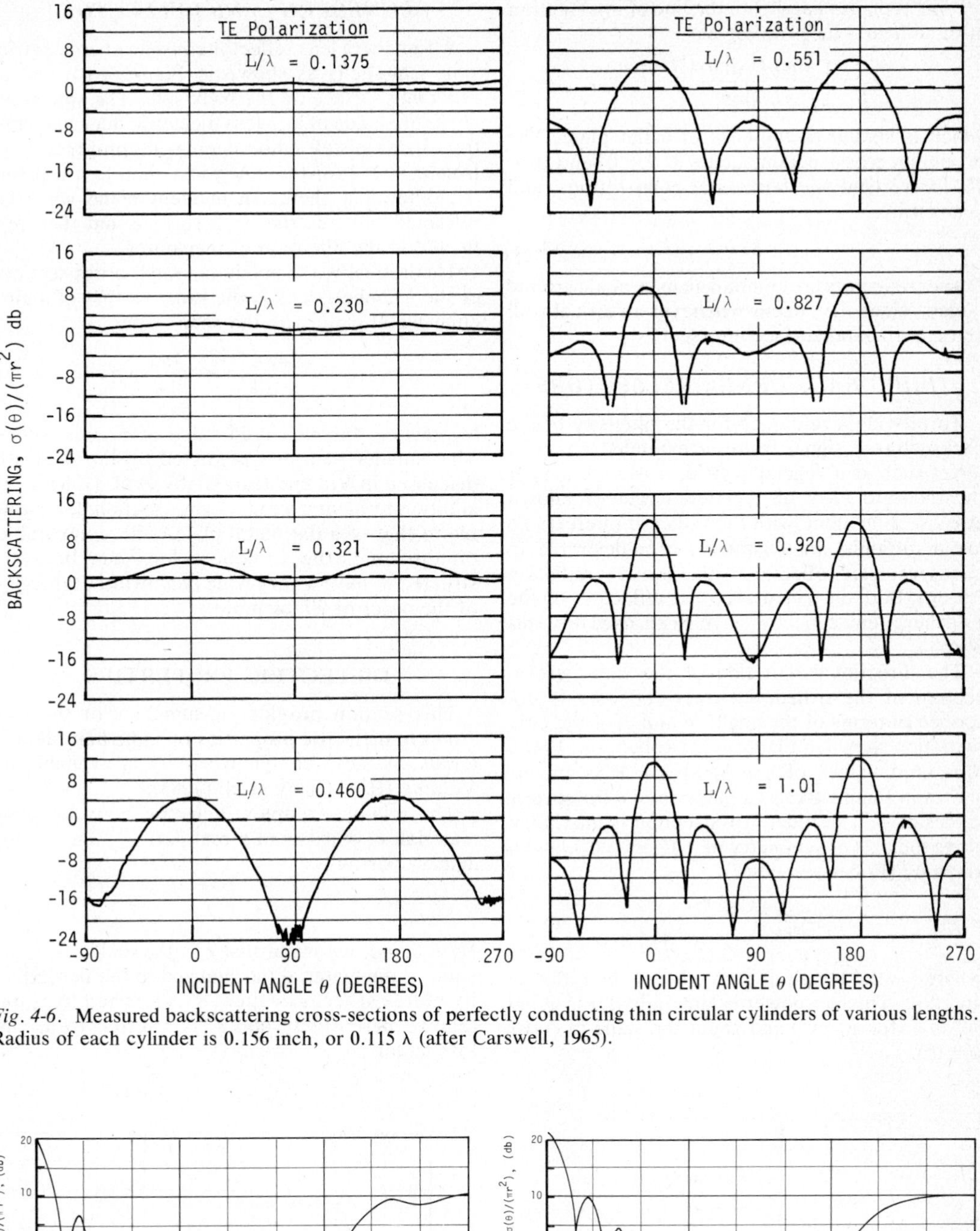

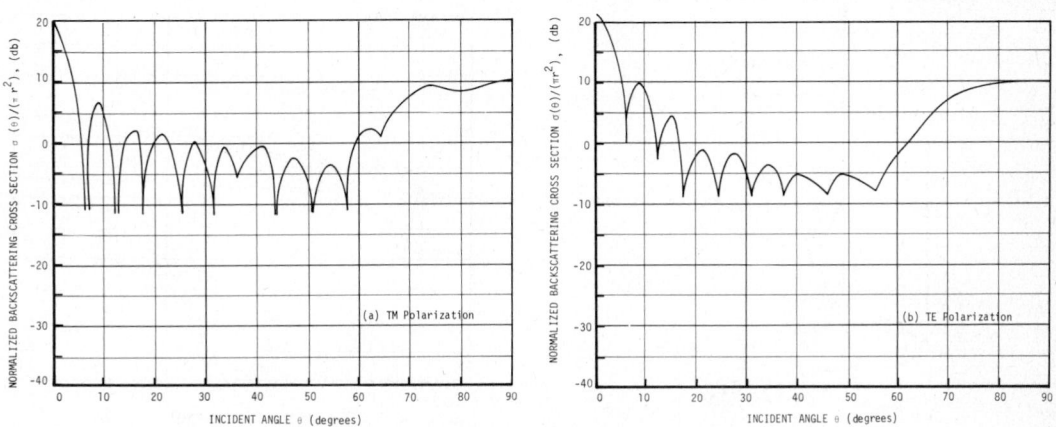

Fig. 4-6. Measured backscattering cross-sections of perfectly conducting thin circular cylinders of various lengths. Radius of each cylinder is 0.156 inch, or 0.115 λ (after Carswell, 1965).

Fig. 4-7(a) - (b). Measured backscattering cross-sections of perfectly conducting circular cylinders of average radius for TM and TE polarization (after Ufintsev, 1962).

section in planes parallel to the line of intersection of the dihedral faces varies as

$$\left[\frac{\sin(kL \sin\theta)}{kL \sin\theta} \right]^2 \qquad (4\text{-}15)$$

which is similar to the case of a flat plate. The maximum cross-section occurs at $\theta = 0°$ and $\phi = 45°$ for vertical and horizontal polarizations, and is given by

$$\sigma_{max} = 8\pi b^2 L^2/\lambda^2. \qquad (4\text{-}16)$$

Corner reflectors are common in man-made ground targets, since they occur whenever a vertical wall and a horizontal surface intersect.

TRIHEDRAL CORNER REFLECTORS

To provide a reference for the intensity of the backscattered signal from an extended complex target such as a vegetated field, it is common for the radar to view an isolated target of known cross-section along with the target of interest. To avoid difficulty in alignment, it is desirable to choose isolated reflectors with a broad reradiation pattern. Both the trihedral corner reflector and the Luneberg lens reflector have been used for this purpose.

The directional variation of the radar cross-section of the trihedral corner reflector is described in terms of the angles θ and ϕ of the configuration shown in Figure 4-12 (Robertson, 1947). Measured values of $\sigma(\theta,\phi)$ at $\lambda = 1.25$ cm are shown in Figure 4-13 as a function of θ for several values of ϕ. The radar cross-section is maximum along the axis of symmetry ($\theta = 0°$, $\phi = 0°$), and is given by (Robertson, 1947):

$$\sigma_{max} = \frac{\pi}{3} \frac{l^4}{\lambda^2} \qquad (4\text{-}17)$$

where l is the length of each edge of the reflector aperture. The half-power beamwidth of σ is about $30°$ in θ (for $\phi = 0°$) and about the same in ϕ (for $\theta = 0°$).

LUNEBERG LENS REFLECTOR

A Luneberg lens reflector consists of concentric spherical shells as shown in Figure 4-14, and a reflecting surface on the back side. The spherical shells are made of low-loss dielectric material with their index of refraction decreasing progressively from $n = \sqrt{2}$ for the innermost shell to $n \cong 1$ for the outermost shell. An incident plane wave is refracted towards the back surface and then reflected in the direction of the source.

Theoretically, the maximum radar cross-section of the Luneberg lens is the same as that of a circular metal plate with the same radius,

$$\sigma_{max} = \frac{4\pi A^2}{\lambda^2} = \frac{4\pi^3}{\lambda^2} r_0^4. \qquad (4\text{-}18)$$

In practice, the measured value of σ_{max} is somewhat smaller than that predicted by Eq. 4-18, as discussed in Volume II of Ulaby et al. (1982a). In addition to having a radar cross-section comparable to that of a flat metal plate (with comparable dimensions), the Luneberg lens can be constructed to have a very wide beamwidth, typically of the order of $90°$ or more.

DIELECTRIC PROPERTIES

This section provides a summary of the microwave dielectric properties of some natural materials. A more complete survey is available in Volume III of Ulaby et al. (1983).

The relative complex dielectric constant of a material, ϵ, consists of a real part, ϵ', and an imaginary part, ϵ'',

$$\epsilon = \epsilon' - j \epsilon''. \qquad (4\text{-}19)$$

By relative, we mean that ϵ is the dielectric constant of the material, normalized to the permittivity of free space, ϵ_0. Often, ϵ' is referred to as the relative permittivity and ϵ'' is referred to as the loss factor.

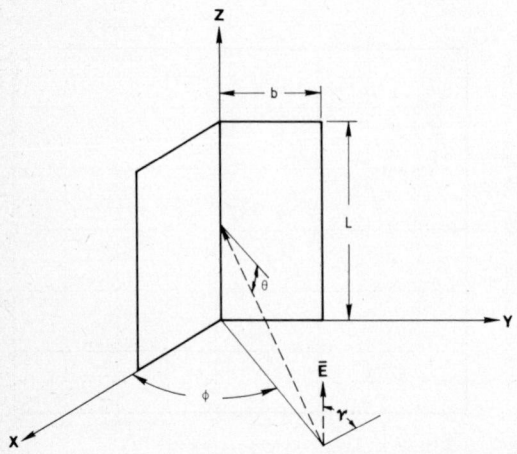

Fig. 4-8. Scattering geometry of dihedral corner reflector.

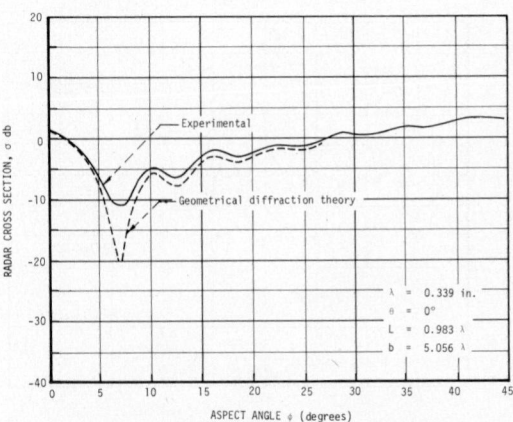

Fig. 4-9. Radar cross-section of a dihedral corner reflector, for vertical polarization, $\gamma = 90°$ (Skolnik, 1970).

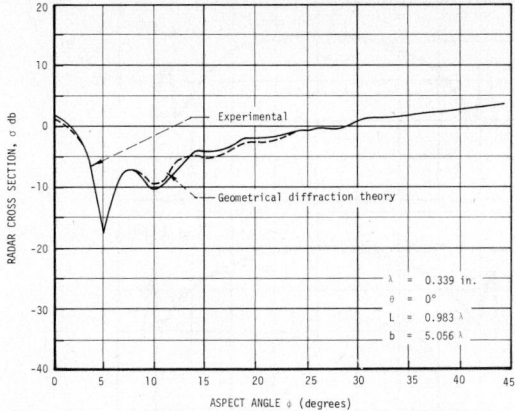

Fig. 4-10. Radar cross-section of a dihedral corner reflector, for horizontal polarization, $\gamma = 0°$ (Skolnik, 1970).

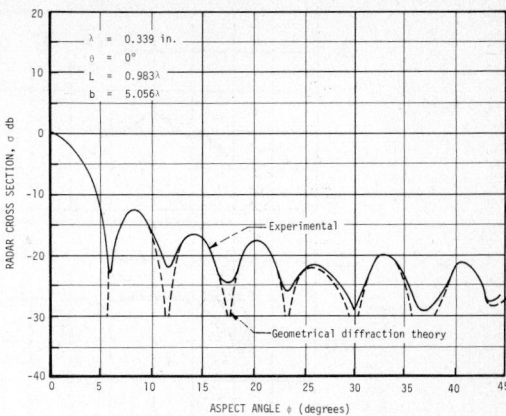

Fig. 4-11. Radar cross-section of a dihedral corner reflector, for parallel polarization, $\gamma = 45°$ (Skolnik, 1970).

LIQUID WATER

As a polar molecule, water exhibits a relaxation phenomenon in its electrical properties. The complex dielectric constant of pure water (water containing no salt) was first formulated by Debye (1929), and later extended by Lane and Saxton (1952) to include the effects of ionic conductivity for salt solutions. Their expressions can be put in the form:

$$\epsilon_w' = \epsilon_{w\infty} + \frac{\epsilon_{wo} - \epsilon_{w\infty}}{1 + (2\pi f \tau_w)^2} \qquad (4\text{-}20a)$$

$$\epsilon_w'' = \frac{2\pi f \tau_w (\epsilon_{wo} - \epsilon_{w\infty})}{1 + (2\pi f \tau_w)^2} + \frac{\sigma_i}{2\pi f \epsilon_0}, \qquad (4\text{-}20b)$$

where

ϵ_{wo} = static relative permittivity of water,
$\epsilon_{w\infty}$ = high-frequency (or optical) limit of ϵ_w,

τ_w = relaxation time of water, s,
f = frequency, Hz,
σ_i = ionic conductivity of the salt solution, Siemens per meter,
ϵ_0 = permittivity of free space.

In general, ϵ_{wo}, τ_w, and σ_i are functions of the temperature and salinity of the water solution; detailed expressions are available in Volume III of Ulaby et al. (1983). The quantity $\epsilon_{w\infty}$ has the constant value of 4.9.

The frequency dependence of ϵ_w' and ϵ_w'' is shown in Figures 4-15 and 4-16, respectively. The curves shown illustrate the behavior for pure water ($\sigma_i = 0$), fresh water ($\sigma_i = 0.01$ Siemens m^{-1}), and sea water with a salinity of 36 $^0/_{00}$.

For pure water, the loss factor ϵ_w'' is maximum at $f = f_0$ where $f_0 = 1/(2\pi\tau_w)$ is known as the relaxation frequency. Figure 4-16 shows that f_0 increases with increasing temperature.

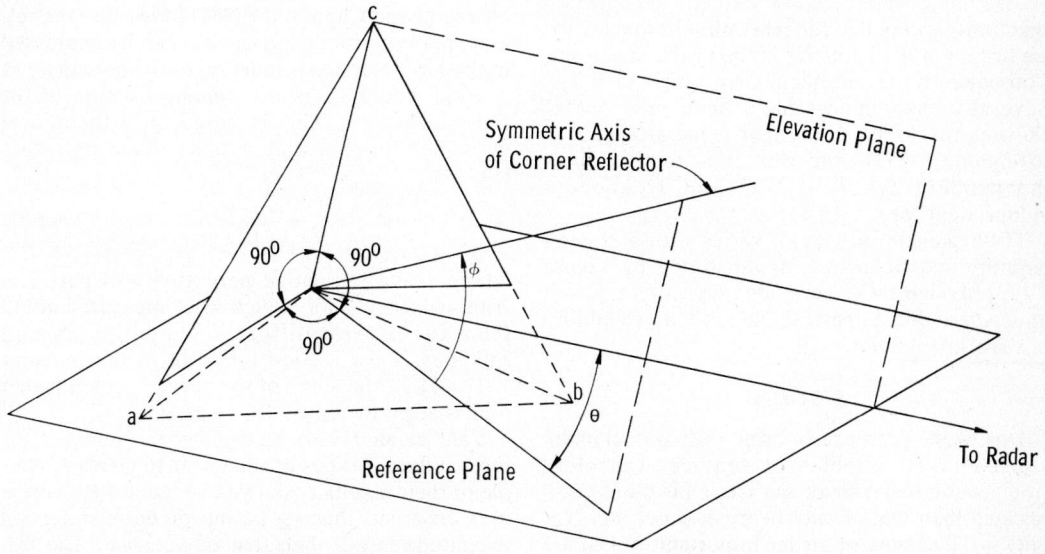

Fig. 4-12. Coordinate system showing the angles θ and ϕ, which define the radar direction relative to the symmetric axis of the trihedral corner reflector.

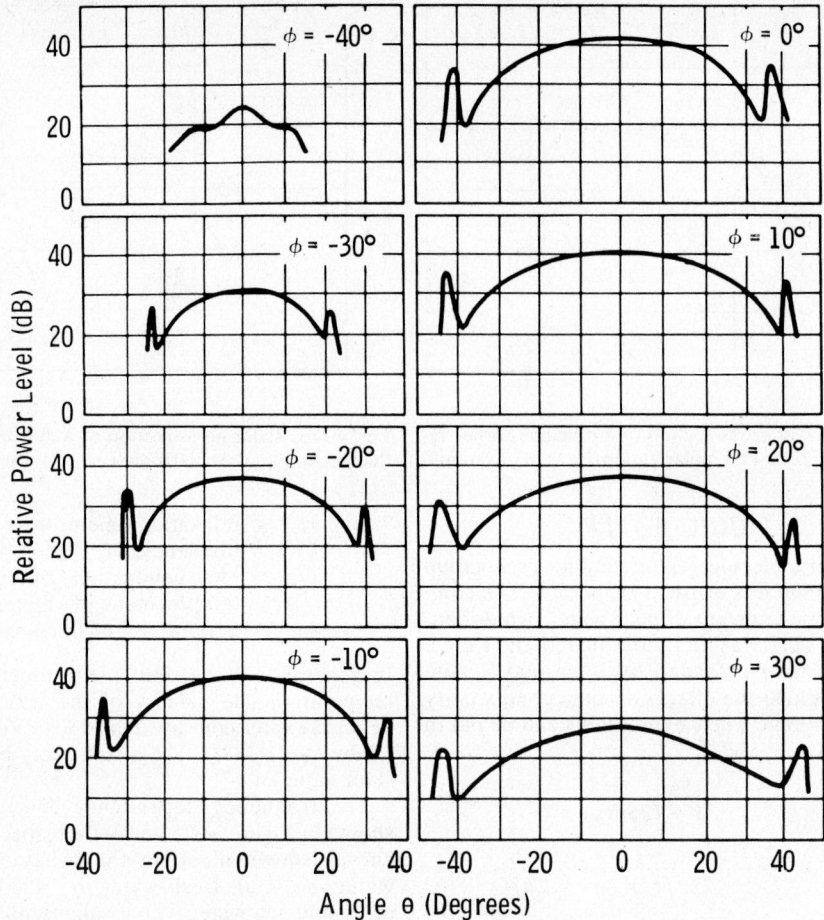

Fig. 4-13. Angular pattern of the radar cross-section (relative scale) of a trihedral corner reflector with $l = 24$ inches. Wavelength is 1.25 cm, and the angles θ and ϕ are defined in Fig. 4-12 (from Robertson, 1947).

FRESH-WATER ICE

Like liquid water, ice also exhibits a dispersion spectrum, except that the relaxation frequency for ice occurs in the kilohertz range of the spectrum. Consequently, in the microwave range $\epsilon'_i = \epsilon_{i\infty}$. Several measurements have been made of ϵ'_i showing that it has a constant value around 3.15 (Cumming, 1952), and that it is temperature-independent for $T \leqslant 0°C$ and frequency-independent for $f \geqslant 10$ MHz.

The loss factor of ice, ϵ''_i, varies with both temperature and frequency, as illustrated by Figure 4-17, although these variations are not very well understood due, in part, to the limited availability of experimental data.

SEA ICE

Due to the presence of brine (salt water) inclusions and air bubbles in sea ice, the electromagnetic behavior of sea ice is far more complicated than that of pure or fresh-water ice. Not only is the salinity of the ice important, but so are the shape and orientation (relative to the electric field of the incident wave) of the inclusions.

Based on dielectric measurements (Figure 4-18) made for artificially grown sea ice samples, Hoekstra and Cappillino (1971) found that the relative permittivity of sea ice, ϵ'_{si}, can be expressed in the form of a linear function of the quantity $1/(1 - 3V_b)$ where V_b is the volume fraction of the brine inclusions. This is based on a theoretical derivation for spherical particles (inclusions) that leads to:

$$\epsilon'_{si} = \frac{\epsilon'_i}{1 - 3V_b} \tag{4-21}$$

where ϵ'_i is the relative permittivity of pure ice. Although their data, which were measured at 9.8 GHz for several different temperatures and salinities, show a good linear fit to the function $1/(1 - 3V_b)$, the slope of the curve is much higher than that predicted by Eq. 4-21.

Vant et al. (1974) reported measurements for three different types of sea ice at 10 GHz. A sample of their results is shown in Figure 4-19, where it is observed that ϵ'_{si} is one or more orders of magnitude larger than that of pure ice. The frequency dependence of ϵ''_{si} is shown in Figure 4-20 for artificially grown samples of sea ice.

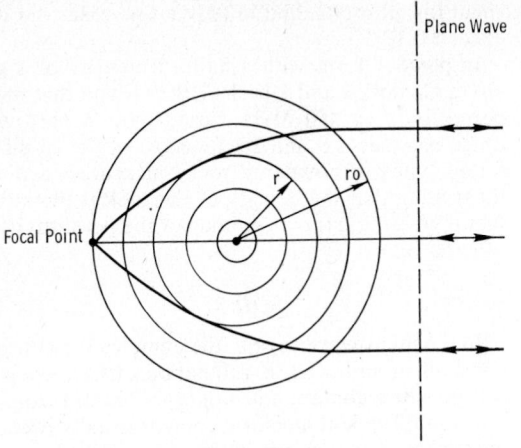

Plane Wave

Fig. 4-14. Luneberg lens reflector.

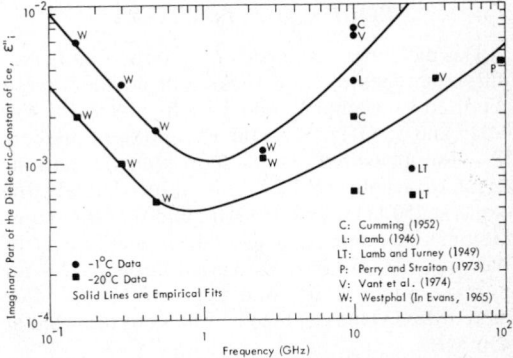

Fig. 4-16. Imaginary part, dielectric constant of water as function of frequency for various temperatures (from Paris, 1969).

SNOW

The microwave dielectric constant of snow is strongly dependent on the presence of water in liquid form in the snow volume. When liquid water is not present, the snow usually is referred to as "dry snow"; otherwise, it is referred to as wet snow.

Dry snow consists of ice crystals and air, and therefore its dielectric constant, ϵ_{ds}, is governed by the dielectric properties of ice and the snow density ρ_s. Since the real part of the relative dielectric constant of ice, ϵ'_i, is independent of temperature and frequency in the microwave region, so is the real part of the dielectric content of dry snow ϵ'_{ds}. Based on measurements of ϵ'_{ds} as a function of snow density ρ_s by Cumming (1952), Glen and Paren (1975) developed the simple expression:

$$\epsilon'_{ds} = (1 + 0.47V_i)^3 \qquad (4\text{-}22a)$$

$$= (1 + 0.51\rho_s)^3 \qquad (4\text{-}22b)$$

where V_i, the ice-volume fraction, is related to the snow and ice densities, ρ_s and ρ_i, respectively, through

$$V_i = \rho_s/\rho_i. \qquad (4\text{-}23)$$

In Eq. 4-22b, $\rho_i = 0.916$ g cm^{-3} was used.

The loss factor of dry snow, ϵ''_{ds}, depends on the loss factor of ice, ϵ''_i, whose behavior as a function of temperature and frequency is not very well understood (see previous section on the dielectric properties of ice). Few measurements of the dielectric loss due to snow have been reported in the literature. The loss tangent of dry snow, $\tan \delta_{ds} = \epsilon''_{ds}/\epsilon'_{ds}$, is shown in Figure 4-21 as a function of temperature for different values of the snow density ρ_s. These curves are based on measurements at 9.375 GHz (Cumming, 1952). It is worth noting that the magnitude of $\tan \delta_{ds}$ is quite small, thereby allowing significant penetration into the snow medium at microwave frequencies.

If conditions are such that the snow volume contains water in liquid form, the magnitudes of ϵ'_{ws} and ϵ''_{ws} (of wet snow) increase rapidly with the liquid-water content (snow wetness), W_v. Mea-

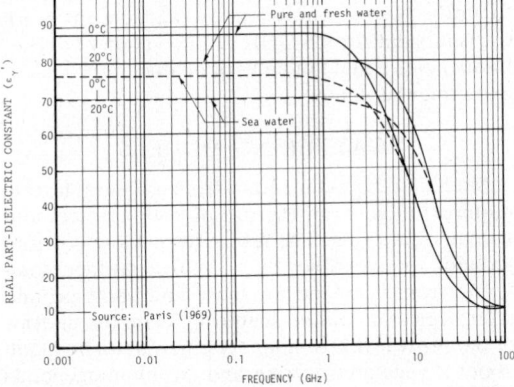

Fig. 4-15. Real part, dielectric constant of water as function of frequency for various temperatures (from Paris, 1969).

Fig. 4-17. Imaginary part of the relative dielectric constant of pure- and fresh-water ice (from Ulaby et al., 1983).

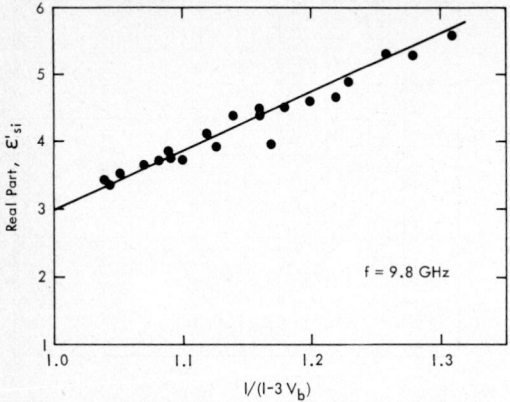

Fig. 4-18. The dielectric constant of sea ice at different temperatures and salinities versus a function of the brine volume V_b (from Hoekstra and Cappillino, 1971).

surements conducted by Linlor (1980) over the 4-12-GHz range lead to the expressions:

$$\epsilon'_{ws} = 1 + 2\,\rho_s + b\,W_v{}^{3/2} \qquad (4\text{-}24a)$$

$$\epsilon''_{ws} = \frac{1.0994}{f}\,\alpha\,\sqrt{\epsilon'_{ws}} \qquad (4\text{-}24b)$$

where ρ_s(g cm^{-3}) is the snow density, f(GHz) is the frequency, W_v (percent by volume) is the liquid water constant, α(dB cm^{-1}) is the attenuation coefficient (Figure 4-22), and b is a frequency-dependent constant given by,

$$b = 5.87 \times 10^{-2} - 3.10 \times 10^{-4}(f - 4)^2. \quad (4\text{-}25)$$

The above expressions were derived by Linlor (1980) on the basis of measurements made by mixing a known amount of water with initially dry snow. It is suspected that the mixing process resulted in spherical ice particles having a surface layer of water, which is different from the geometry that exists for natural snow (Colbeck, 1979). This difference may be the factor responsible for the fact that Linlor's expressions produce lower values for ϵ'_{ws} and ϵ''_{ws} than those reported by Sweeney and Colbeck (1974) and by Tobarias et al. (1978), as discussed in Ulaby et al. (1983).

ROCKS AND POWDERS

The dielectric properties of a large number of different types of solid rocks and powders were reported by Campbell and Ulrichs (1969) for 450 MHz and 35 GHz. With the exception of meteorites, the measured results show that the permittivity of a solid rock, ϵ'_{sr}, is approximately the same at 450 MHz and 35 GHz, and that the major factor influencing its magnitude is the density of the rock. At 35 GHz, ϵ'_{sr} varied between 2.4 for pumice and 9.2 for one type of basalt. For meteorites, values as high as 150 were reported at 450 MHz.

No relationship between ϵ'_{sr} and the loss tangent δ_{sr} was evident from the data; however, at both 450 MHz and 35 GHz, the magnitude of δ_{sr} was small for all rocks, and in only a few cases did it exceed 0.1.

In powder form with a uniform density of 1 g cm^{-3}, Campbell and Ulrichs (1969) found that the permittivity at 450 MHz varied over a narrow range between 1.9 and 2.1 for most of the 25 different types of powdered rocks measured, again illustrating that the density of the rock is the key factor governing the magnitude of the permittivity of solid rocks.

SOILS

In the microwave region, the complex dielectric constant of soil, ϵ_{soil}, is a function of frequency, soil moisture content and soil type (textural composition). The soil moisture content usually is expressed on a gravimetric basis, m_g (percent by dry weight), or on a volumetric basis, m_v (g cm^{-3}), although the latter is preferred from the standpoint of comparing the dielectric properties of different soil samples because it takes soil density into account while m_g does not. Also, dielectric mixing models usually are based on volume fractions of the various constituents in the mixture, rather than their relative weights.

Figures 4-23 and 4-24 show the variation of ϵ'_{soil} and ϵ''_{soil} as a function of m_v at 1.4 and 5 GHz for different types of soil. It is noted that for a given value of m_v, ϵ'_{soil} generally decreases with increasing clay content while an opposite trend is observed for ϵ''_{soil}.

VEGETATION MATERIAL

Although several experimental investigations have been conducted to determine the microwave dielectric properties of timber (Tinga et al., 1973; Tiuri et al., 1980; Meyer and Schilz, 1981), and of some food materials such as potatoes and starch (de Loor, 1968), measurements of the dielectric properties of plant-leaves and -needles are limited to a single study conducted by Carlson (1967) at 8.5 GHz. Curves based on his data are shown in Figure 4-25. A dielectric model developed by Fung and Ulaby (1978) using de Loor's mixing formula (1968) shows good agreement with Carlson's data at 8.5 GHz, but its validity at other frequencies could not be verified due to the lack of experimental data.

SCATTERING MODELS

Modeling the radar return from a natural terrain requires, in general, the use of both surface- and volume-scattering. The return from snow-covered ground, for example, may involve surface scattering from both the air-snow and snow-ground interfaces and volume scattering from the interior of the snow layer. Similar statements can be made about a vegetated terrain and an inhomogeneous ground surface. Hence, in what follows, the most commonly used surface- and volume-scattering models are described.

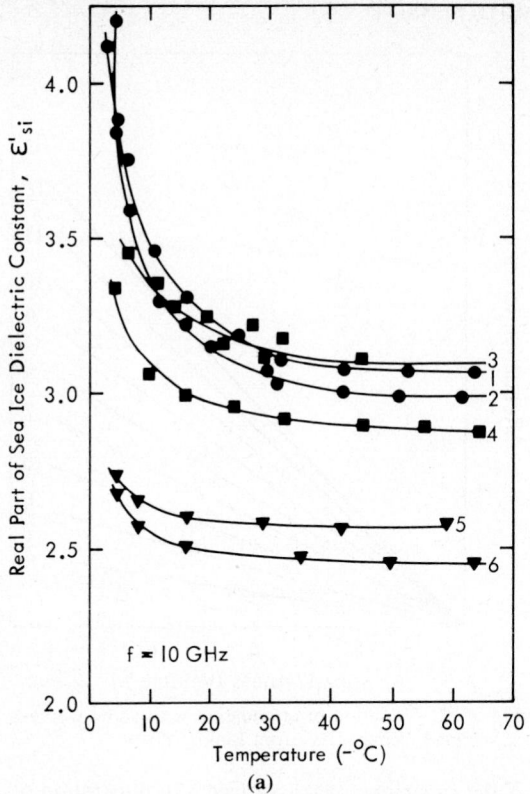

(a)

Variation of relative dielectric constant ϵ'_{si} with temperature plotted for ● frazil, ■ columnar, ▼ multiyear ice types for various salinities and densities. 1. S_i (salinity) = 4.4 $^0/_{00}$, ρ (density) = 0.836; 2. S_i = 3.2 $^0/_{00}$, ρ = 0.836; 3. S_i = 3.2 $^0/_{00}$, ρ = 0.878; 4. S_i = 4.6 $^0/_{00}$, ρ = 0.896; 5. S_i = 0.61 $^0/_{00}$, ρ = 0.771; 6. S_i = 0.70 $^0/_{00}$, π = 0.770.

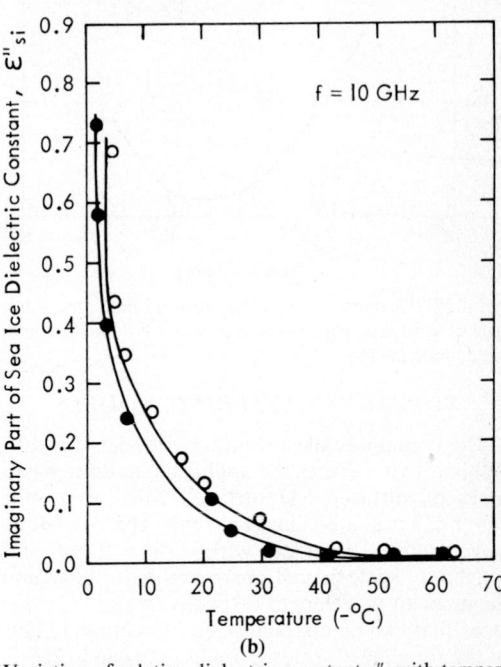
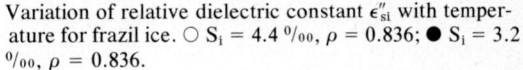

(b)

Variation of relative dielectric constant ϵ''_{si} with temperature for frazil ice. ○ S_i = 4.4 $^0/_{00}$, ρ = 0.836; ● S_i = 3.2 $^0/_{00}$, ρ = 0.836.

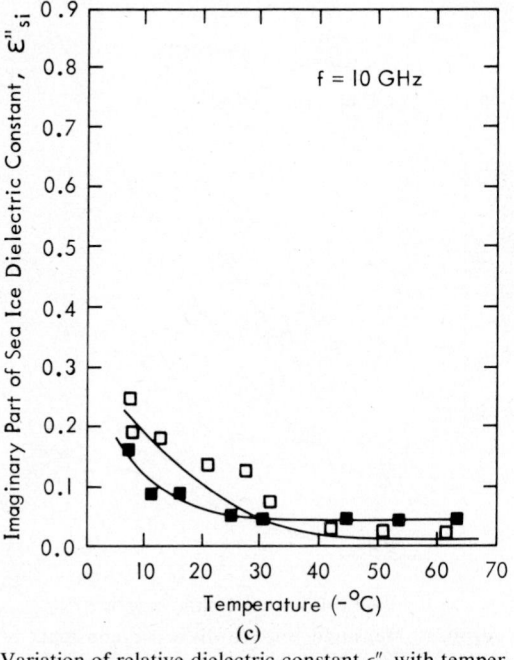

(c)

Variation of relative dielectric constant ϵ''_{si} with temperature for columnar ice. □ S_i = 3.2 $^0/_{00}$, ρ = 0.878; ■ S_i = 4.6 $^0/_{00}$, ρ = 0.896.

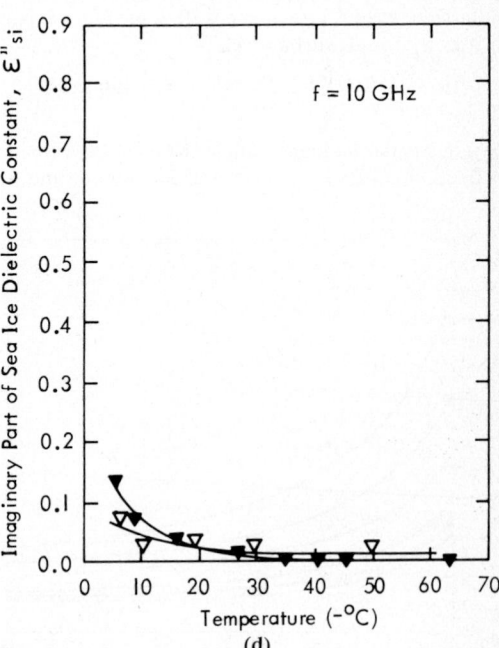

(d)

Variation of relative dielectric constant ϵ''_{si} with temperature for multiyear ice. ▽ S_i = 0.61 $^0/_{00}$, ρ = 0.771; ▼ S_i = 0.70 $^0/_{00}$, ρ = 0.770.

Fig. 4-19. Variation of the real (a) and imaginary (b-d) parts of the relative dielectric constant of sea ice, ϵ'_{si} and ϵ''_{si} with temperature at 10 GHz (from Vant et al., 1974).

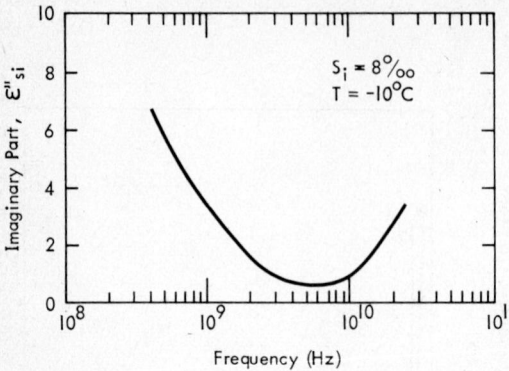

Fig. 4-20. Imaginary part of the relative dielectric constant of sea ice as a function of frequency (Hoekstra and Cappillino, 1971).

SURFACE SCATTERING MODELS

The commonly used scattering models for randomly rough surfaces are applicable to three basic types of surfaces: (1) surfaces with horizontal roughness scales larger than the incident wavelength, (2) surfaces with surface slopes and roughness scales small compared with unity and the incident wavelength respectively, and (3) surfaces that can be characterized by a combination of the roughness scales defined in (1) and (2).

The scattering model for the first type of surface can be derived from the physical-optics formulation (Fung, 1967; Beckmann and Spizzichino, 1963). When the product of the surface standard deviation and the electromagnetic wave number is large enough to permit asymptotic evaluation, the polarized backscattering coefficient for an isotropically rough surface is

$$\sigma^\circ(\theta) = (|R(O)|^2/2M^2\cos^4\theta) \exp\left[-\tan^2\theta/2M^2\right]$$
$$(4\text{-}26)$$

where θ is the incidence angle; $R(O)$ is the Fresnel reflection coefficient at normal incidence; and M

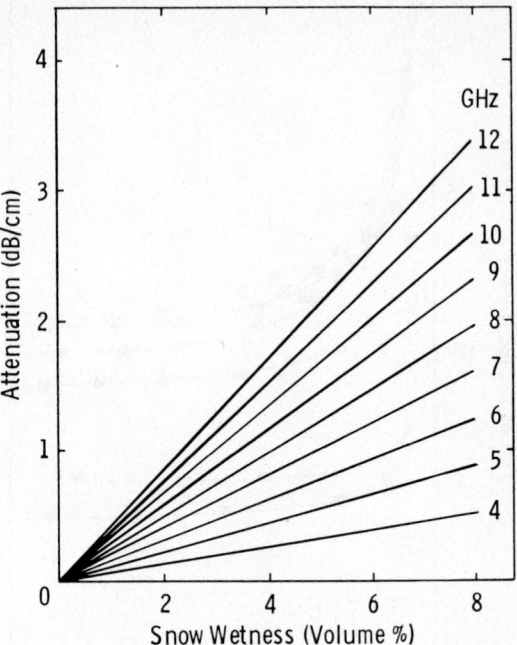

Fig. 4-22. Variation of attenuation with snow wetness at selected frequencies (from Linlor, 1980).

is the rms slope of the surface. An illustration of Eq. 4-26 is shown in Figure 4-26. When the product of the surface standard deviation and the electromagnetic wave number, $k\sigma$, is not large

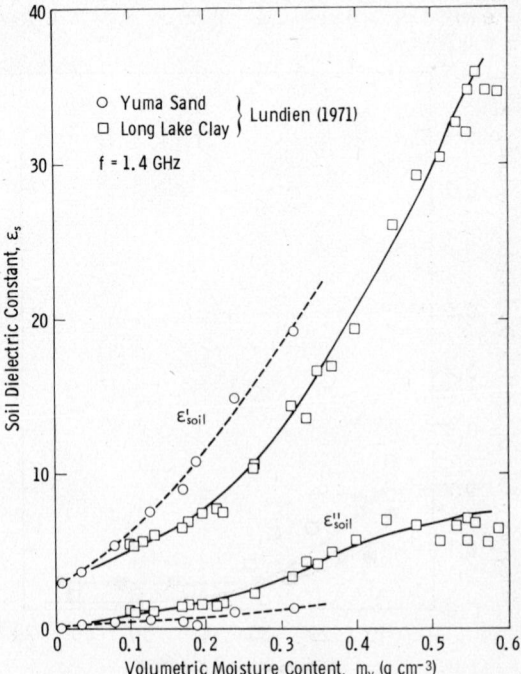

Fig. 4-23. Measured relative dielectric constants of sandy and high-clay soils as a function of volumetric moisture content at 1.4 GHz. The curves shown were calculated by Wang (1980) on the basis of a theoretical model.

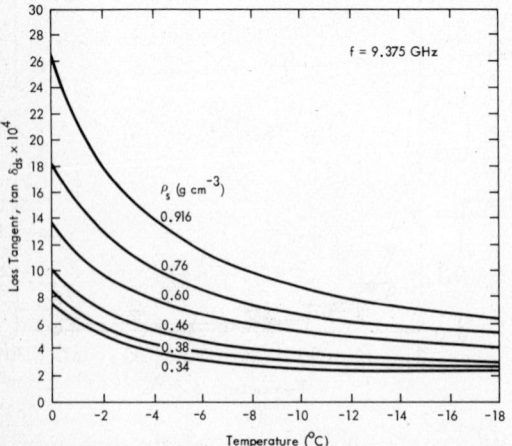

Fig. 4-21. The loss tangent of snow as a function of temperature at a frequency of 9.375 GHz (from Cumming, 1952).

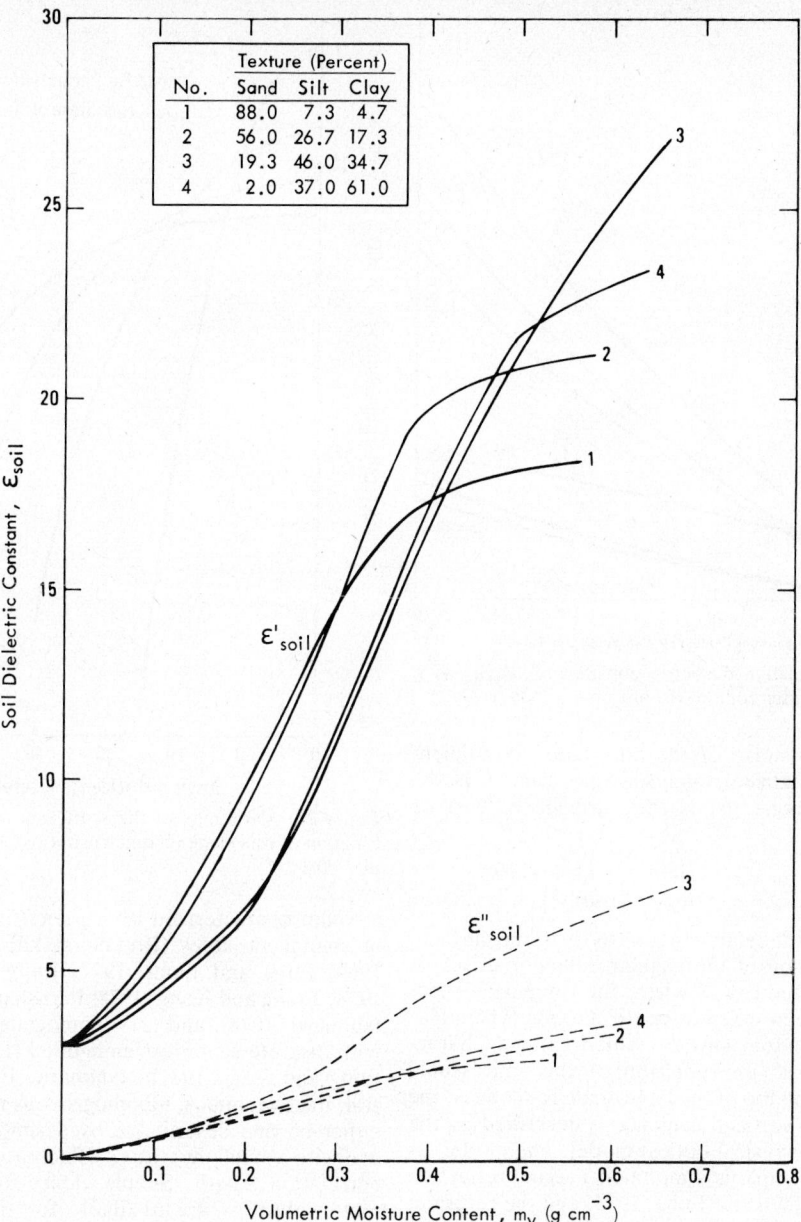

No.	Texture (Percent)		
	Sand	Silt	Clay
1	88.0	7.3	4.7
2	56.0	26.7	17.3
3	19.3	46.0	34.7
4	2.0	37.0	61.0

Fig. 4-24. The relative dielectric constants versus volumetric water content for four soils measured at 5 GHz (from Wang and Schmugge, 1980).

enough, an approximate series solution can be obtained for polarized backscattering (Fung and Eom, 1981a; Beckmann and Spizzichino, 1963),

$$\sigma_{pp}^{\circ}(\theta) = \left[(2k \left| R_{pp}(\theta) \right| \cos\theta)^2/4\pi \right] \exp\left[-(2k\sigma\cos\theta)^2 \right]$$

$$\sum_{n=1}^{\infty} \frac{(2k\sigma\cos\theta)^{2n}}{n!} \int_{-\infty}^{\infty} \int_{-\infty}^{\infty} \rho^n(u,v) \, e^{j2ku\sin\theta} \, dudv, \quad (4\text{-}27)$$

where pp stands for either vertical or horizontal polarization; $R_{pp}(\theta)$ is the Fresnel reflection coefficient; and $\rho(u,v)$ is the surface correlation coefficient. An illustration of Eq. 4-27 assuming an isotropic correlation coefficient, $\rho(u,v) = [1 +$

$(u^2 + v^2)/l^2]^{-1.5}$ and an rms slope of 0.1 is shown in Figure 4-27. It should be noted that the above models are usually not suitable for large incidence angles or steep slopes (i.e., $\theta > 30°$).

For slightly rough surfaces the method of small perturbation (Valenzuela, 1967; Fung, 1968) is usually used to derive a first-order scattering coefficient. For polarized backscattering the scattering coefficient is

$$\sigma_{pp}^{\circ}(\theta) = 8k^4\sigma^2 \cos^4\theta \left| \alpha_{pp} \right|^2 W(2k\sin\theta,0) \quad (4\text{-}28)$$

where $\alpha_{hh} = R_\perp$, the Fresnel reflection coefficient for horizontal polarization; $W(2k\sin\theta,0)$ is the

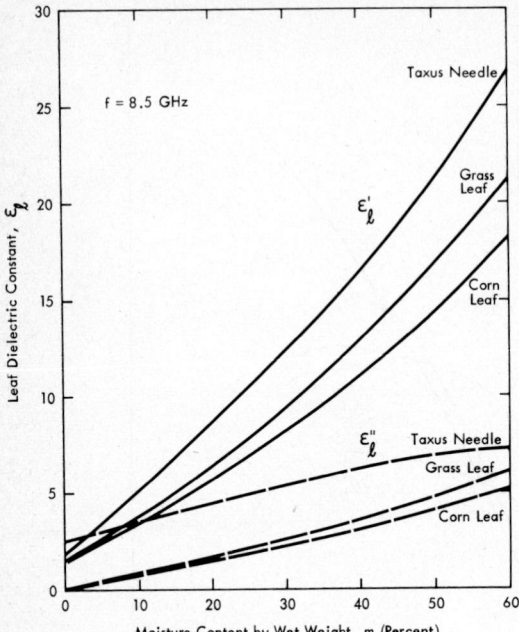

Fig. 4-25. Relative dielectric constant of leaves as a function of water content (from Carlson, 1967).

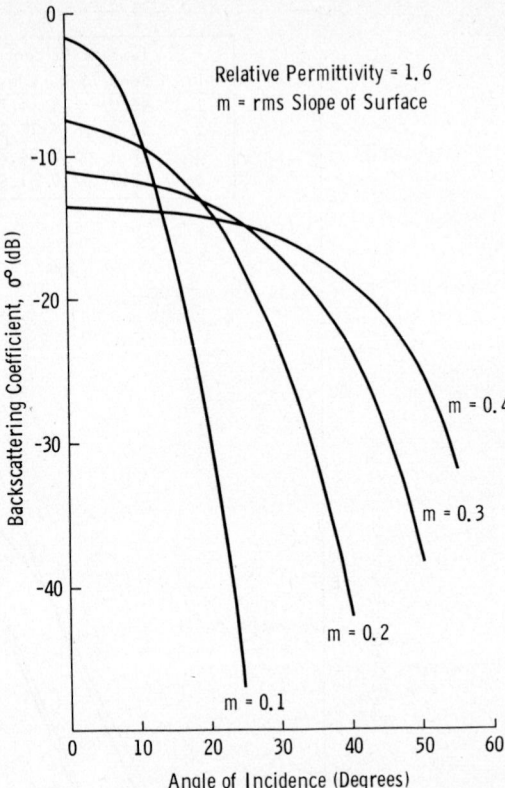

Fig. 4-26. Variations of the scattering coefficient as a function of rms slope (geometric optics), (from Ulaby et al., 1982).

Fourier transform of the correlation coefficient or the normalized roughness spectrum; θ is the incidence angle; σ^2 is the variance of surface heights and

$$\alpha_{vv} = (\epsilon_r - 1) \left[\sin^2\theta - \epsilon_r(1 + \sin^2\theta) \right]$$
$$\cdot \left[\epsilon_r\cos\theta + (\epsilon_r - \sin^2\theta)^{1/2} \right]^{-2},$$

where ϵ_r is the relative permittivity of the surface. An illustration of the angular behavior of σ_{pp}° is shown in Figure 4-28 where the roughness spectrum has been taken to be $(l^2/2) \exp[-(kl\sin\theta)^2]$.

For a random surface which can be characterized as a superposition of the small-scale roughness on top of the large-scale roughness, the return near vertical incidence is described by the Kirchhoff (physical optics) model. The model, as used here, employs a modified Fresnel reflection coefficient (Wu and Fung, 1973) and the return at large angles of incidence ($\theta \geqslant 30°$) is taken to be that from the small perturbation model averaged over the slope distribution of the large-scale roughness (Wu and Fung, 1973). Since the angular-response curve for vertical polarization computed from the small perturbation method is fairly flat, and hence not very sensitive to averaging, a simple approximate description for the vertically polarized return is obtained by using the Kirchhoff scattering model for the near-vertical angular region and the small-perturbation model for incidence angles larger than about 30°.

VOLUME SCATTERING MODELS

The volume-scattering properties of two basic types of inhomogeneous media have been studied extensively: (1) a continuous inhomogeneous medium characterized by a permittivity function of spatial variables (Brekhovskikh, 1960; Fung, 1969; Fung and Fung, 1977; Fung and Ulaby, 1978; Tsang and Kong, 1978; Parashar et al., 1978; Ishimaru, 1978), and (2) a homogeneous medium with discrete scatterers embedded (Leader, 1975; Fung and Eom, 1981b; Ishimaru, 1978). In general, the continuous, inhomogeneous medium must either be one describable by a simple profile or one with a small permittivity fluctuation or a small correlation length. Simple closed-form approximate solutions are available for the following cases:

(i) Reflection from a half-space with a slowly varying linear permittivity profile (Brekhovskikh, 1960).

Let the boundary between the homogeneous and the inhomogeneous media be at $z = 0$ and the value of the refractive index be

$$n = 1, z \leqslant 0$$
$$n = (1 + \alpha z)^{1/2}, z \geqslant 0.$$

Then for an incident-plane wave making an angle θ with the z-axis in the region $z \leqslant 0$, the first-order solution of the reflection coefficient at $z = 0$ is

$$R = \lambda\alpha\left[j \, 16\pi \cos^3\theta \right]^{-1} \qquad (4\text{-}29)$$

where λ is the wavelength of the electromagnetic wave in region $z \leqslant 0$ and α is the rate of change

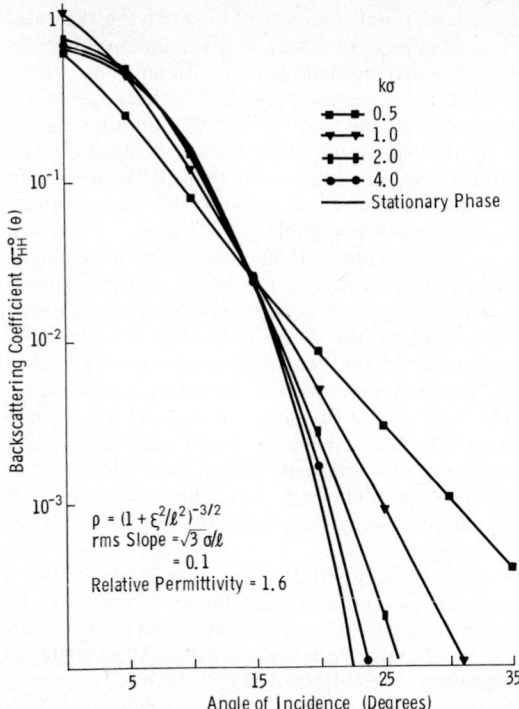

Fig. 4-27. Variations of the scattering coefficient (Kirchhoff method) as a function of surface roughness (from Ulaby et al., 1982).

of permittivity per unit length. It should be noted that the above formula accounts for the profile effect only. Any discontinuity in the refractive index at $z = 0$ must be handled separately. Also, the formula is restricted by the condition that

$$\lambda\alpha(2\pi\cos^3\theta)^{-1} << 1. \qquad (4\text{-}30)$$

For reflection from media with more complex profiles see Brekhovskikh (1960).

(ii) Scattering from an inhomogeneous half-space ($z \leqslant 0$) characterized by the random permittivity function, $\epsilon(x,y,z) = \epsilon_a + \epsilon_1(x,y)$.

The average value of $\epsilon(x,y,z)$ is ϵ_a, which is taken to be a constant, and it is assumed that $|\epsilon_1| << |\epsilon_a|$. The horizontally polarized backscatter scattering coefficient computed from the first-order scattered-field expression (Fung, 1969) due to a horizontally polarized plane wave incident at an angle, θ, in air ($z > 0$) is

$$\sigma_{hh}^{\circ}(\theta) = 2\pi^2 \left| \frac{k\, T_\perp \cos\theta\, \cos(\theta - \phi)}{\cos\phi\, (\epsilon_a\cos\theta + \sqrt{\epsilon_a}\, \cos\phi)} \right|^2$$
$$W(2k\sin\theta,0), \qquad (4\text{-}31)$$

where ϕ is related to θ by Snell's law; k is the wave number in air; $T_\perp$ is the Fresnel transmission coefficient for horizontal polarization; and $W(2k\sin\theta,0)$ is the spectrum of the random permittivity function, $\epsilon_1(x,y)$, related to the correlation function of $\epsilon_1(x,y)$ by the Fourier transform.

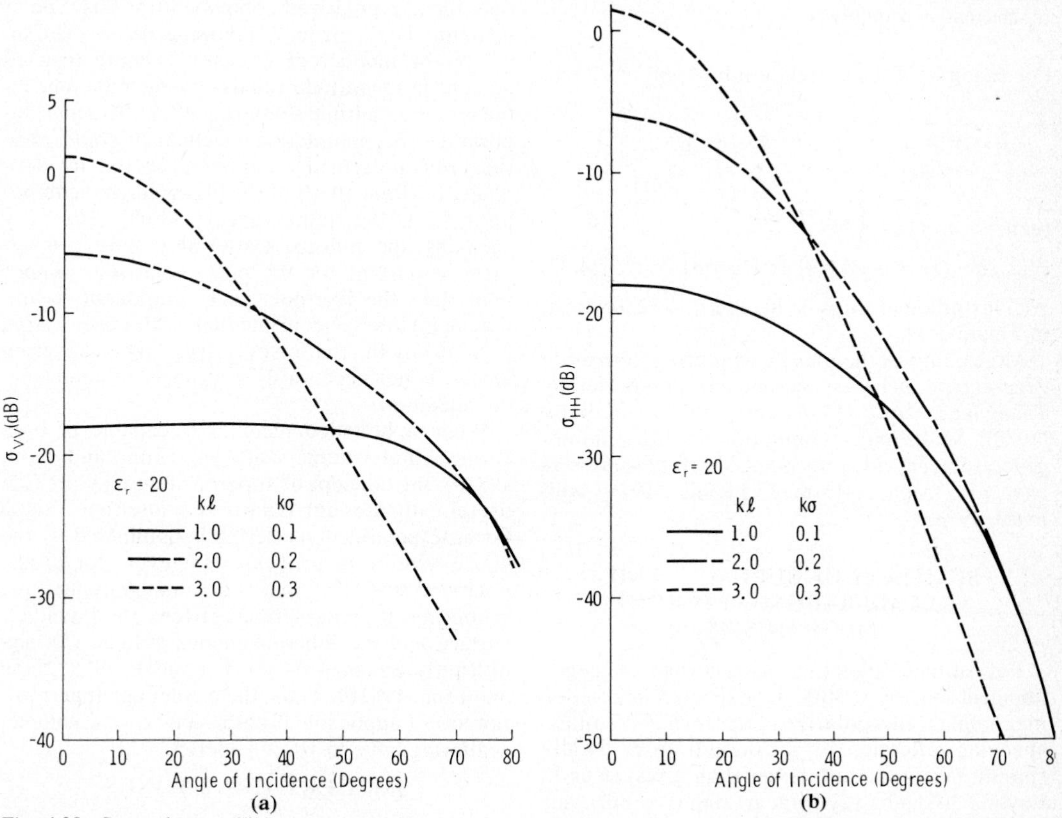

Fig. 4-28. Scattering coefficient (perturbation model) as a function of (a) vertical polarization and (b) horizontal polarization (from Ulaby et al., 1982).

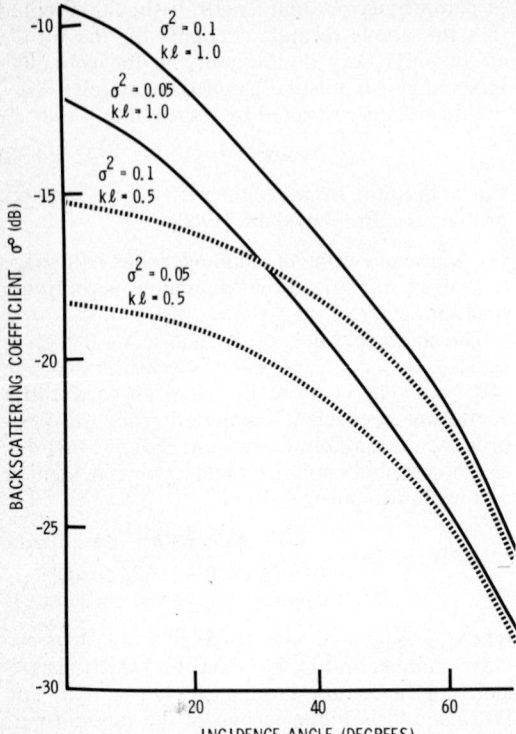

Fig. 4-29. Dependence of the angular backscattering coefficient on the variance σ^2 and correlation length l of the fluctuating permittivity.

For example, if the correlation function of $\epsilon_1(x,y)$ is

$$R(x - x', y - y') = \sigma^2 \exp \left\{ -\left[(x - x')^2 \right. \right.$$
$$\left. \left. + (y - y')^2 \right]^{1/2} \right\}/l$$

then,

$$W(2k\sin\theta,O) = \sigma^2 l^2 \left[1 + 4k^2l^2\sin^2\theta \right]^{-3/2}. \quad (4\text{-}32)$$

An illustration of Eq. 4-31 using Eq. 4-32 is shown in Figure 4-29.

More complex theories for scattering from more general random half-spaces have been reported by Tsang and Kong (1978) and Fung and Ulaby (1978). Scattering computations from a homogeneous layer with embedded, discrete scatterers have also been performed (Leader, 1975; Ishimaru, 1978).

SUMMARY OF SURFACE- AND VOLUME-BACKSCATTERING MECHANISMS

For surface scattering near normal incidence (incident angle $\theta < 30°$), the expected interaction mechanism for polarized scattering is quasi-specular reflection or diffraction from locally smooth facets that are larger than a wavelength. Beyond 30° and away from grazing ($\theta < 80°$), any significant return is likely to be dominated by Bragg scattering, (i.e., only the roughness scale

with horizontal dimension, l, satisfying the relation, $2l \sin\theta = \lambda$, where λ is the electromagnetic wavelength) contributes significantly to backscattering. For cross-polarized backscattering the most likely mechanism in the angular range $0° \leq \theta \leq 80°$ is multiple surface scattering. The contribution is usually more than 10 dB lower than the like-polarized return from the same surface and becomes negligible for surfaces with small permittivity values. Other important mechanisms for cross-polarization are an anisotropic permittivity or a geometrically anisotropic roughness. The level of the cross-polarized returns from either one of the two mechanisms depends upon the azimuth angle (or the direction of incidence). The importance of these mechanisms in applications has yet to be established. At near-grazing incidence, no single dominant scattering mechanism can be identified. It is believed that possible mechanisms are highly dependent upon frequency and target.

In a weakly inhomogeneous medium (or a medium with sparsely distributed scatterers), the coherent propagating intensity is much larger than the incoherent scattered intensity. At all angles of incidence, the polarized radar return is due mainly to the single scattering process; while the cross-polarized return (in the plane of incidence) is due to the second bounce. Usually, the level of the cross-polarized component is 15 dB or more lower than the like-polarized component for this type of medium. For a strongly inhomogeneous medium where the incoherent intensity is more than 10 percent in magnitude relative to the coherent intensity, the multiple scattering effect becomes important. This interaction mechanism could raise the cross-polarized component in the incident plane to within 10 dB of the like-polarized component from the same target. Thus, relatively speaking, the multiple scattering process is much more important for the cross-polarized component than the like-polarized component. (Note that an inhomogeneous medium with either a large permittivity fluctuation or a large single-scattering albedo is not necessarily a strongly inhomogeneous medium.)

When a bounded medium is capable of both surface- and volume-scattering (Fung and Eom, 1981b), the concept of superposition may provide useful estimates for polarized scattering. This is because polarized scattering is dominated by the single scattering process. However, for cross-polarized scattering that is dominated by multiple scattering, the interaction between the boundary surface and the inhomogeneous volume can significantly enhance its level (Leader, 1975; Fung and Eom, 1981b). Thus, the concept of superposition is not applicable if both surface- and volume-scattering have to be considered.

IMAGERY GRAY TONES

System parameters (such as polarization, wavelength, and look angle), and target parameters (such as complex dielectric constant, surface roughness, and subsurface roughness or volume scatterers) can significantly influence the scattering behavior of a surface or volume. Some of these parameters are discussed below, from the standpoint of their influence on the strength of the backscattered signal or the change in the gray tone of a radar image.

INFLUENCE OF POLARIZATION

Imaging radars usually transmit a horizontal electric-field vector, and receive either horizontal or vertical return signals, or both. The HH (transmit and receive horizontal electric-field vector) return is also known as the like-polarized return; the HV (transmit horizontal electric-field and receive vertical electric-field vector) return is known as the cross-polarized return. Radar images produced by the like-polarized return signal may be different from those produced by the cross-polarized return, because of the differences in the physical processes responsible for the two types of returns. Such differences have been shown to provide potentials for differentiation of geologic (Moore and Dellwig, 1966) and geographic features (Waite and MacDonald, 1970).

The basic physical processes responsible for the like-polarized return are (1) quasi-specular surface reflection, and (2) surface or volume scattering. The quasi-specular reflection normally accounts for the high return near vertical incidence. Since most imaging radars transmit at moderate to large incidence angles, the scattering process plays a dominant role in polarized returns, both surface- and volume-scattering being commonly encountered in applications. Returns due to surface scattering are normally stronger near vertical incidence and decrease with increasing incidence angle, with a slower rate of decrease for rougher surfaces. Returns due to volume scattering from an inhomogeneous medium with small average dielectric constant tend to be uniform for all incidence angles away from grazing. The angular behavior of the like-polarized return is illustrated by the solid lines in Figure 4-30; the behavior of the cross-polarized return is shown by the dashed lines in that figure. Note that in Figure 4-30 no significance has been attached to the relative levels between the like- and cross-polarized returns. In practice, the cross-polarized return is usually weaker than the like-polarized return and the receiver channel for the cross-polarized return is usually set higher to compensate for the weaker return signal. Hence, in comparing the like-polarized radar image with that of the cross-polarized image, one should examine the contrast in gray tones between the same targets rather than the absolute gray-tone levels between images.

Depolarization Mechanisms

Four mechanisms known to cause depolarization of EMR, (that is, to produce cross-polarized

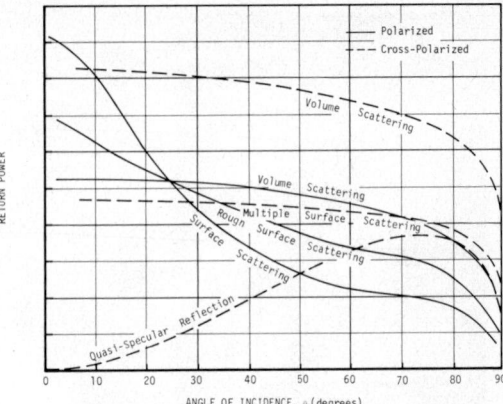

Fig. 4-30. Return power as function of angle of incidence, for polarized and cross-polarized conditions.

radar returns), are: (1) quasi-specular reflection as a result of the difference between the Fresnel reflection coefficients (Fung, 1967) for a homogeneous, two-dimensional, smoothly undulating surface; (2) multiple scattering as a result of target surface roughness (Fung and Eom, 1979); (3) multiple volume scattering due to inhomogeneities, especially those embedded within a skin-depth of the target surface (Leader, 1975); and (4) anisotropic properties (physical or geometric) of the targets (Tan and Fung, 1979).

Of these four mechanisms, the first three are commonly encountered in remote sensing applications. The first mechanism, applicable only to smoothly undulating surfaces, predicts essentially no cross-polarized returns near the vertical, and increasingly larger returns at larger incidence angles, except near grazing incidence. However, the level of the returns remains low as compared with the levels predicted by the second and the third mechanisms, both of which are applicable to rough terrains with or without vegetation cover (regardless of type). These mechanisms predict fairly uniform returns over all incidence angles, except near grazing incidence, where the return decreases to zero. In general, the third mechanism produces higher-level returns than the second. The fourth mechanism can result from either permittivity or geometry. In either case, the presence of the fourth mechanism follows directly from measured target properties.

A significant point about the cross-polarized scattering is that the level difference (or contrast in imagery) between two types of terrain remains fairly constant as the incidence angle increases, until close to the grazing angle, at which point the level-difference begins to vanish. The level of the cross-polarized return may be 8- to 25-dB lower than that of the polarized return. For this reason, the power gain of the cross-polarized receiver is set higher than that of the polarized receiver, and care must be exercised when comparing gray tones on the polarized and cross-polarized imagery.

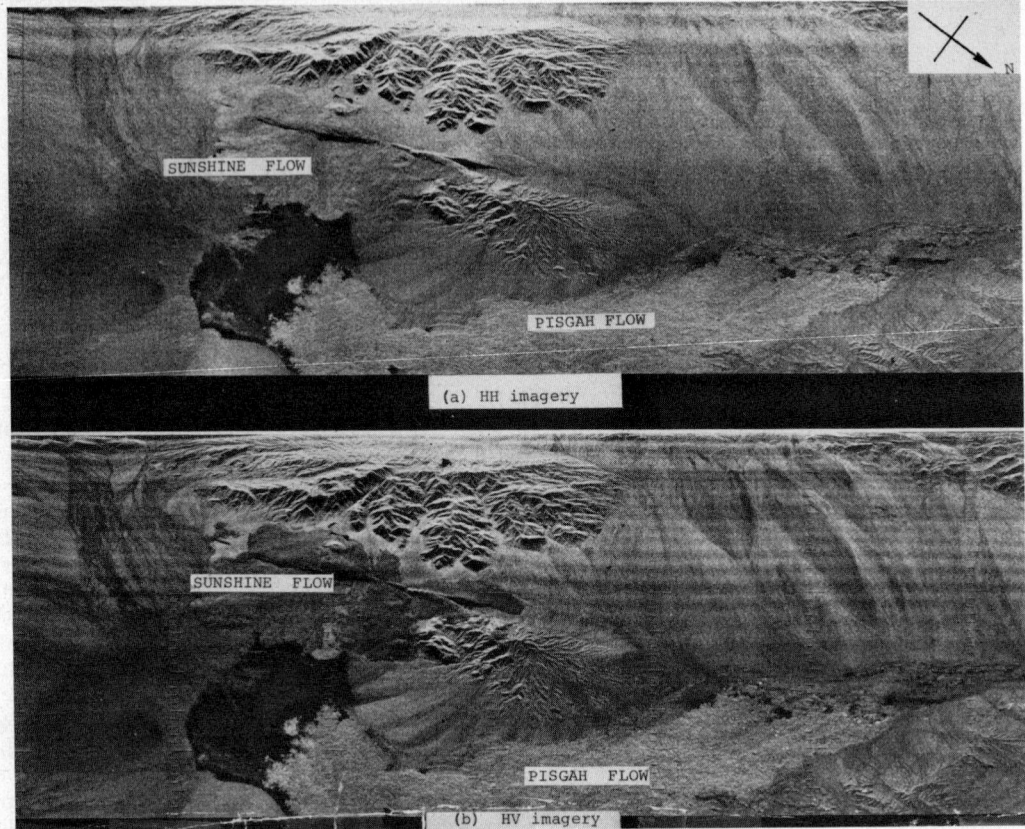

Fig. 4-31. Polarized and cross-polarized radar imagery of the Pisgah Crater Area, Barstow, California.

Comparison of Like- and Cross-Polarization

Figure 4-31 shows an example used by Moore and Dellwig (1966) to illustrate the application of like- and cross-polarized radar imagery to geologic mapping. This example also indicates that volume scattering, as a result of inhomogeneities due to multiple scattering contributes more to cross-polarized return than does surface roughness.

The test site is the Pisgah Crater area in California's Mojave Desert. The test area is dominated by Tertiary and Quaternary lava flows and Quaternary alluvium. Lavic Lake, a playa (seasonal lake), is at the south end of the Pisgah Flow lava field. The most obvious contrast between the polarized and the cross-polarized images is along contacts between the Sunshine Flow lava field and alluvium, and within alluvial areas between adjacent distributary channels on the west edge of Lavic Lake. The Sunshine Flow has a homogeneous rough surface in comparison with the alluvium. The multiple scattering it can produce as a result of surface roughness is insignificant in relation to the volume scattering from the inhomogeneous alluvium. For this reason, the Sunshine Flow appears dark on the cross-polarized imagery. On the like-polarized imagery, however, the Sunshine Flow shows up much brighter, and

cannot readily be distinguished from the alluvium, as direct surface scattering (which contributes insignificantly to the cross-polarized return) contributes significantly to the polarized return. It may appear strange that the Pisgah Flow can be so much brighter than the Sunshine Flow on the cross-polarized imagery. However, samples (Figure 4-32) indicate that the Pisgah Flow is really an inhomogeneous body with many air pockets enabling multiple volume scattering, while the Sunshine Flow is comparatively homogeneous and capable mainly of surface scattering.

An analogous problem is the case of the cross-polarized return from old snow as compared with glacier ice (Waite and MacDonald, 1970). The old snow is found to act like a volume scatterer analogous to the Pisgah Flow, while the glacier ice is analogous to the Sunshine Flow in its capability to produce cross-polarization.

INFLUENCE OF MULTIPLE VOLUME SCATTERING

An example that indicates that multiple scattering can cause a stronger cross-polarized return than surface scattering by an undulating surface has been shown by Morain and Simonett (1967) (see Figure 4-33). Vegetation mapping was examined at Horsefly Mountain area in Oregon. Figure

Fig. 4-32. Samples of surface lava composition from Pisgah Crater.

4-33 shows the radar imagery used for the study, and Figure 4-34 indicates the various types of vegetation found in the area. The dark area, which shows up on the cross-polarized imagery but not on the polarized imagery, was originally covered with sagebrush. On the radar imagery of Figure 4-33, more than 80 percent of this area ranged from heavily grazed to essentially bare ground. The regions surrounding this area are covered mostly with pines, and to a lesser extent, with junipers, as shown in Figure 4-34. Since depolarization due to bare ground is insignificant as compared with that due to vegetation (as a result of multiple volume-scattering), the tonal difference shows up as expected. This interpretation is

further confirmed by a vegetation-free area that resulted from recent burns (Figure 4-33), which shows up just as dark on the cross-polarized imagery.

INFLUENCE OF LOOK ANGLE

Figures 4-35 and 4-36 illustrate the influence of angle on the return signal. In Figure 4-35 the radar beam at point P has a low incidence (high depression) angle; in Figure 4-36, the reverse is true (high incidence angle, low depression angle). The swath width from the imagery is about 21 km, and the incidence angles across it range from approximately 15° to about 75°.

The radial pattern originating from area P (HH imagery, Figure 4-35) coincides with natural levees and consists of narrow bands of relatively short vegetation. Grass, short shrubs, and trees less than ten feet in height predominate. In marked contrast, the light-toned regions that lie between elements of the radial pattern are wooded swamps. The vegetation on top of the swamps has lost most of its leaves, and is much taller than the vegetation on top of the levees. On the HH imagery, the tonal contrast between the swamp regions (wet and smooth compared with the natural levees) and the natural levees is strong at small incidence angles, but decreases gradually as the incidence angle increases. This means that surface scattering from the swamp is stronger than volume scattering from the levees at small angles of incidence. It is interesting to see the tonal reversal appearing on the HV image (Figure 4-35) versus the HH image (Figure 4-35) in area P. This is because leafy vegetations are capable of multiple volume scattering, which can generate much more cross-polarized return than the surface scattering from the swamp. Thus, on the HV image the radial pattern is brighter than the regions in between while on the HH image the reverse is true. At large angles of incidence (Figure 4-36), a tonal reversal would occur on the HH image if returns were still dominated by the swamp and the levees. However, as both areas are actually covered with vegetation, the tonal contrast should simply de-

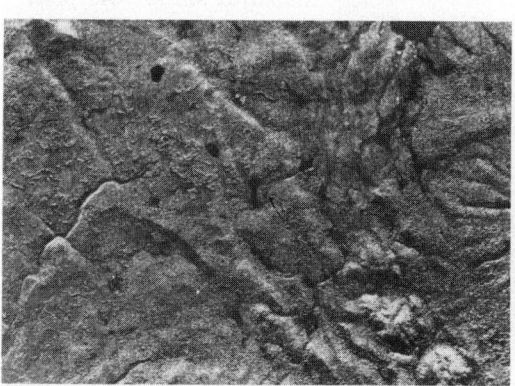

HH Imagery   HV Imagery

Fig. 4-33. K-band radar vegetative imagery of Horsefly Mountain, Oregon.

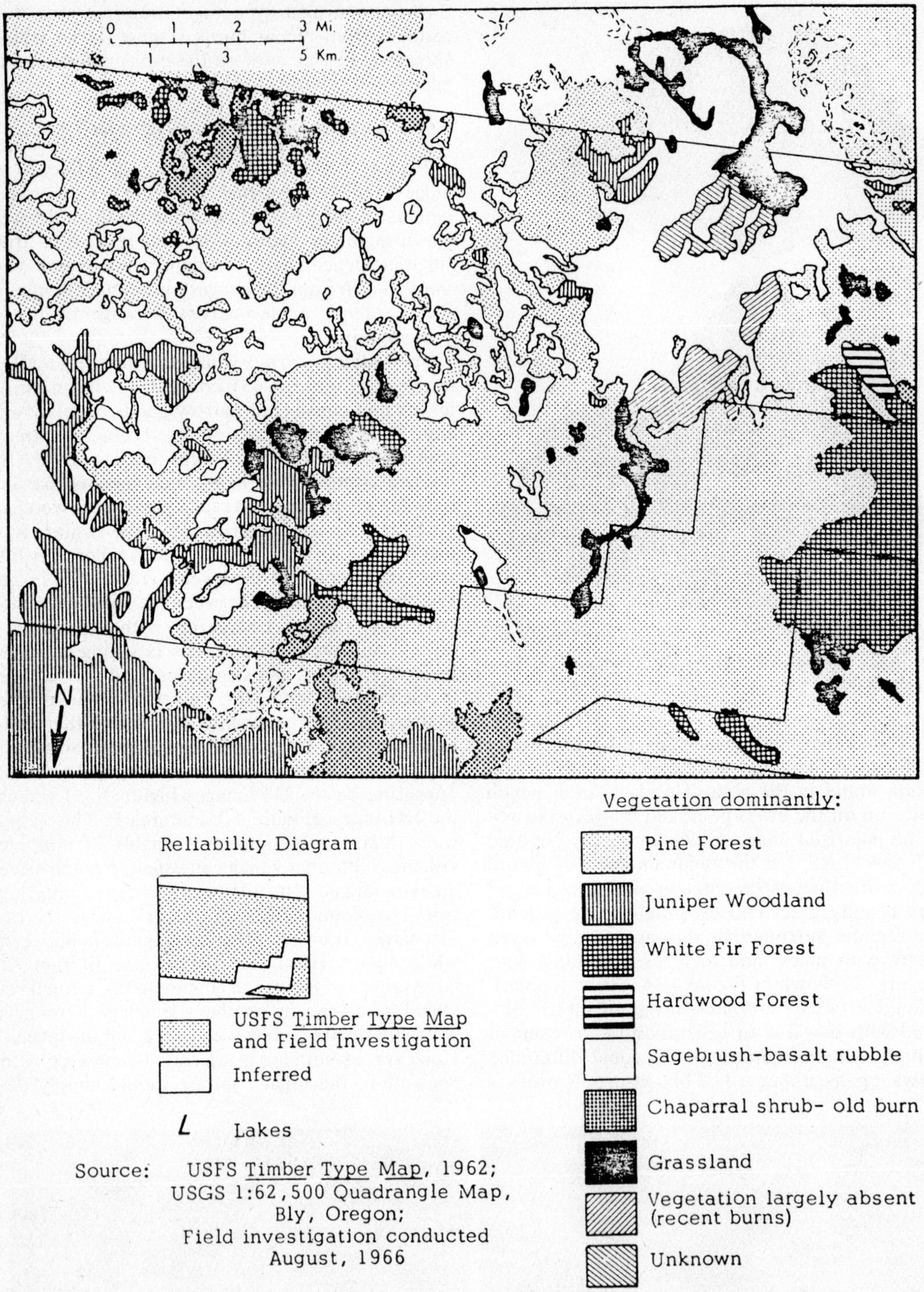

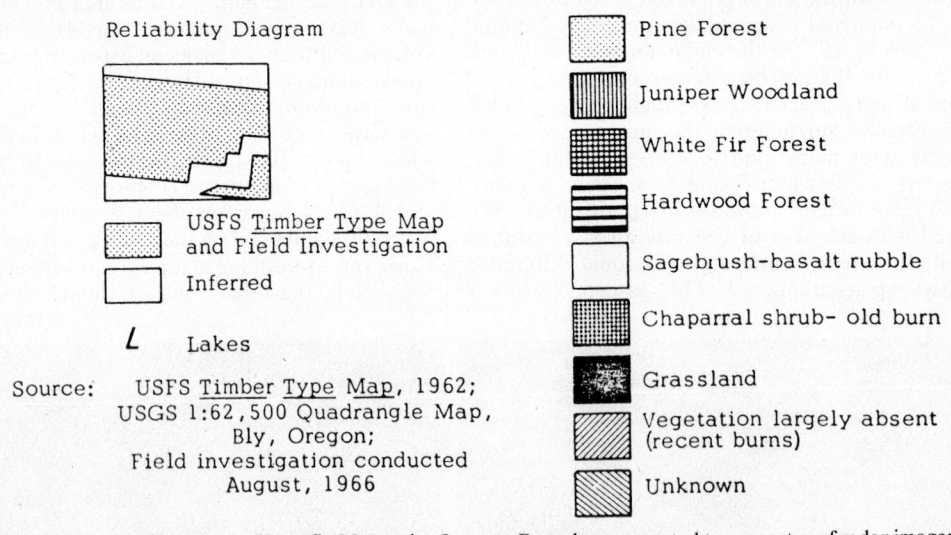

Reliability Diagram

USFS Timber Type Map and Field Investigation

Inferred

L   Lakes

Source: USFS Timber Type Map, 1962;
USGS 1:62,500 Quadrangle Map,
Bly, Oregon;
Field investigation conducted
August, 1966

Vegetation dominantly:

Pine Forest

Juniper Woodland

White Fir Forest

Hardwood Forest

Sagebrush-basalt rubble

Chaparral shrub- old burn

Grassland

Vegetation largely absent
(recent burns)

Unknown

Fig. 4-34. Major vegetative types at Horsefly Mountain, Oregon. Boundary corrected to geometry of radar imagery.

crease to some insignificant value. On the HV imagery, the tonal contrast between the swamp area and the levees remains fairly constant as the incidence angle increases. The tonal contrast at large incidence angles (Figure 4-36) is still clearly discernible, although smaller, as would be expected of volume-scattering processes.

INFLUENCE OF RESOLUTION CELL SIZE

The term "resolution cell" is often used in imaging radar to mean the region on the ground contributing to the return signal that generates a point on the image at a particular instant. The width of the resolution cell is referred to simply as the

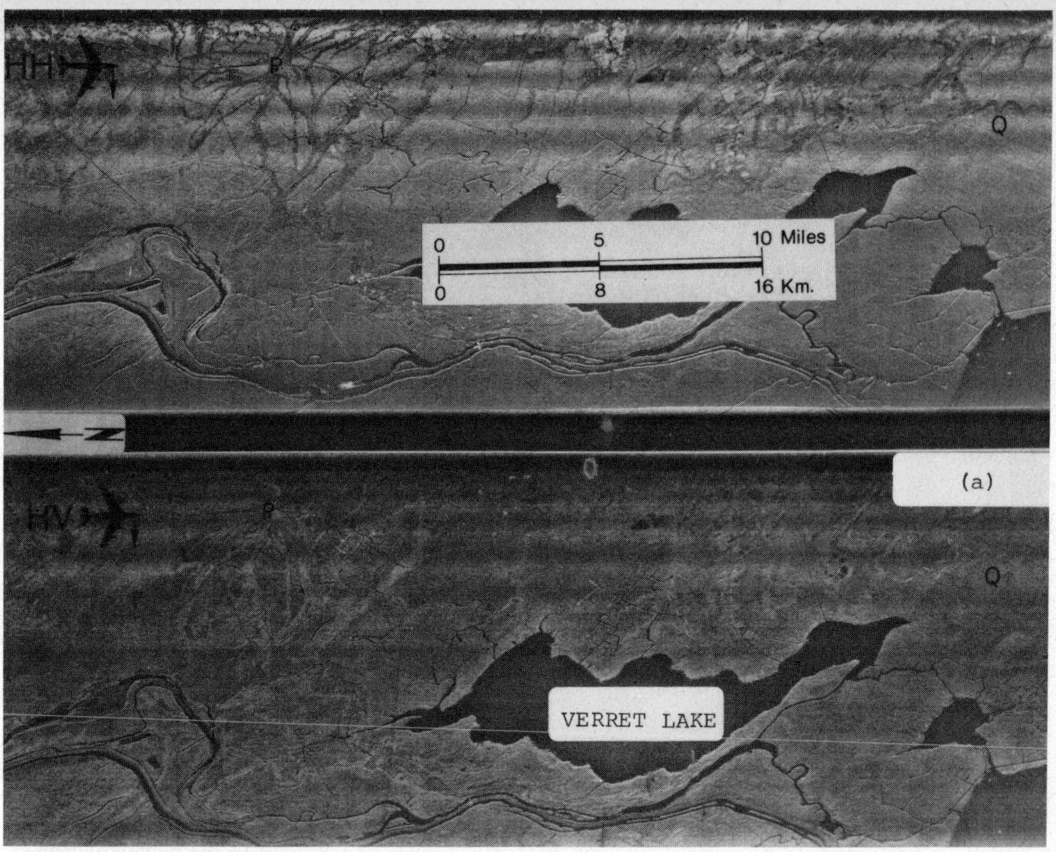

Fig. 4-35. K-band radar imagery, southwest area of Baton Rouge, Louisiana: looking at Region P at small incident angles. Southerly flight direction; look direction, west.

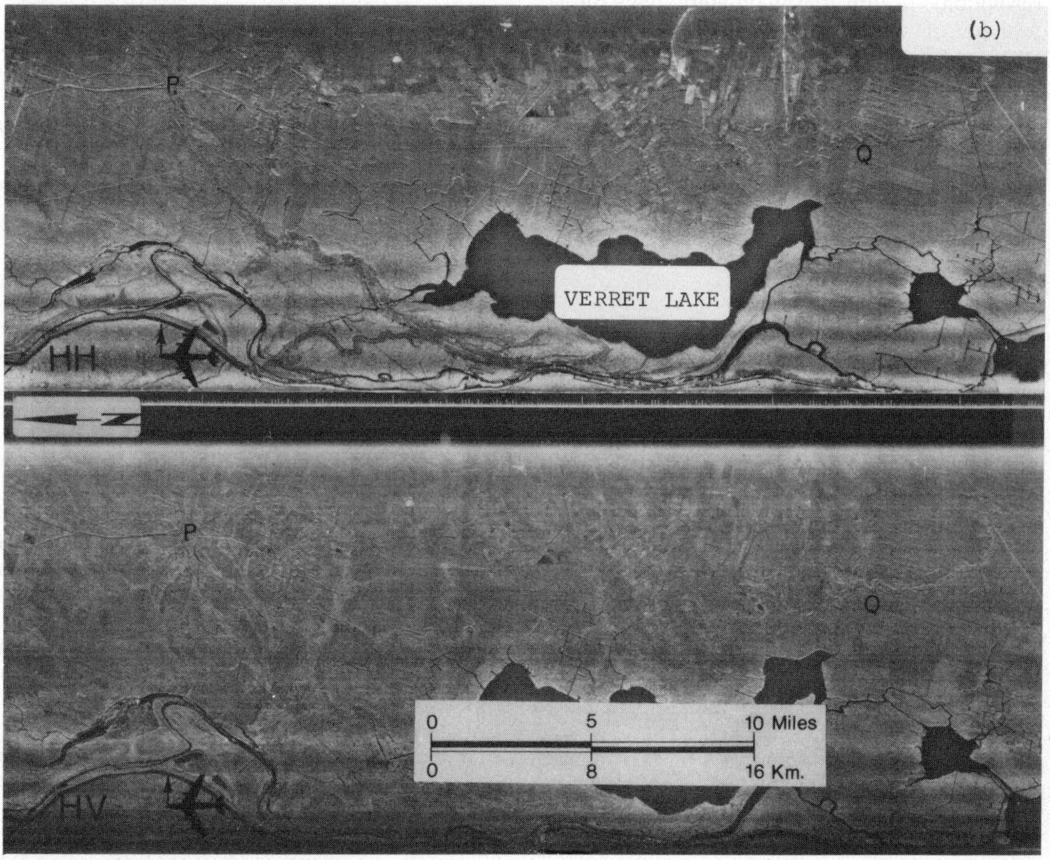

Fig. 4-36. K-band radar imagery, southwest area of Baton Rouge, Louisiana: looking at Region P at large incident angles. Northerly flight direction; look direction, east.

"resolution." The two sides of the resolution cell are normally determined by the half-power angular width of the antenna pattern, the time width of the transmitted pulse, or the bandwidth of the Doppler filter (see Chapters 9 and 10).

Since all the scatterers in a resolution cell contribute together to produce a point on the radar image at a certain gray tone level, it follows that the individual scatterers within the resolution cell are not necessarily identifiable on the image. Thus, a tree comparable in size to the resolution cell might show up on the radar image, but its individual leaves and branches would not. For an individual object of area A_0 and scattering coefficient σ_0° to appear on an image (against a background with scattering coefficient σ_b°) its radar cross-section, $\sigma_0^\circ A_0$, must be much larger than the radar cross-section of a background resolution cell, $\sigma_b^\circ A_c$, where A_c is the resolution area. A strongly reflecting object, such as the roof of a house with the proper orientation relative to the radar location, may appear on an image even if its physical cross-section is smaller than a resolution-cell size.

It seems obvious that as the resolution gets finer, more details of the target will appear on the image. However, a finer resolution image usually contains fewer independent samples per resolution cell (see Chapter 9), which results in greater fluctuation of the return signal (poorer radiometric resolution). In turn, the fluctuation gives the appearance of "speckle" on the image, which makes interpretation more difficult. Trade-offs between spatial and radiometric resolution are discussed in Chapter 10.

INFLUENCE OF COMPLEX DIELECTRIC CONSTANT

A significant change in the dielectric properties of natural materials usually results from a change in moisture content. An increase in the dielectric value increases the reflectivity, and therefore the level of the return signal. Thus, radar returns from vegetation are stronger when there is more moisture in the vegetation. Similarly, a wet ground-surface is capable of stronger scattering than a drier one of the same geometry. In the study by MacDonald and Waite (1971), it was shown that the near-range part of the HH imagery (Figure 4-37, areas A and B) had strong tonal contrasts due to strong returns from the wet soil surface and relatively weaker returns from the vegetation cover. Since volume scattering from vegetation is much more isotropic than surface scattering from the soil surface, these tonal contrasts gradually decrease towards the far range (or larger angles of incidence) of the imagery. In addition, they cited other evidence that the wet soil surface was responsible for the tonal contrasts on HH imagery by observing that similar tonal contrasts were ab-

sent on the HV imagery, since the wet soil surface was not capable of strong depolarization relative to the vegetation. Note that the horizontal dark and brightlines across the image are antenna-pattern effects. These lines are also present on the radar images shown in Figures 4-35 and 4-36.

EFFECTS OF WAVELENGTH, SURFACE ROUGHNESS, AND PENETRATION

The wavelength of the incident radiation can affect the scattered signal in two ways: (1) by defining an effective surface roughness, and (2) by determining the depth of penetration, and therefore the possible volume effect for a given terrain.

That the surface roughness should really be measured in terms of the incident wavelength follows directly from the fact that the terrain is sensed by the incident radiation at a given wavelength. Hence, discussion of the roughness of a terrain always implies effective roughness. For a given terrain, the depth of penetration of the incident radiation depends on its wavelength. The term "skin depth" is commonly used to define the depth below the surface at which the amplitude of the incident wave will have decreased to about 37 percent of its value at the surface. This quantity, denoted by δ, is related to the wavelength in the medium in this manner.

$$\delta = \left(\frac{\lambda}{\pi g \eta} \right)^{1/2} , \qquad (4\text{-}33)$$

where: δ = skin depth
g = conductivity of terrain
$\eta = (\mu/\epsilon)^{1/2}$,
in which ϵ = permittivity of the terrain
μ = permeability of the terrain.

The above expression shows that, for a given terrain, the longer the wavelength, the deeper the penetration.

Examples to illustrate the effects of surface roughness are numerous, and can also be seen from the radar images discussed previously. For instance, Verret Lake on Figures 4-35 and 4-36 is a smoother surface than its surroundings, and therefore shows up dark on the imagery when the angle of incidence (of the radar EMR) is large.

The penetration effect due to a change in wavelength has been discussed by Dellwig (1969) and by Waite and MacDonald (1970 and 1971). A striking example is also available in the Baton Rouge, Louisiana, area where both photographic and K-band radar types of imagery are available over the wooded swamp area. Figure 4-38 shows that, at optical wavelengths (very small compared to K-band EMR), the leafless vegetation on top of the swamp can be seen, while at K-band and small incidence angles, only the surface of the swamp (not the vegetation) shows up on the HH radar imagery.

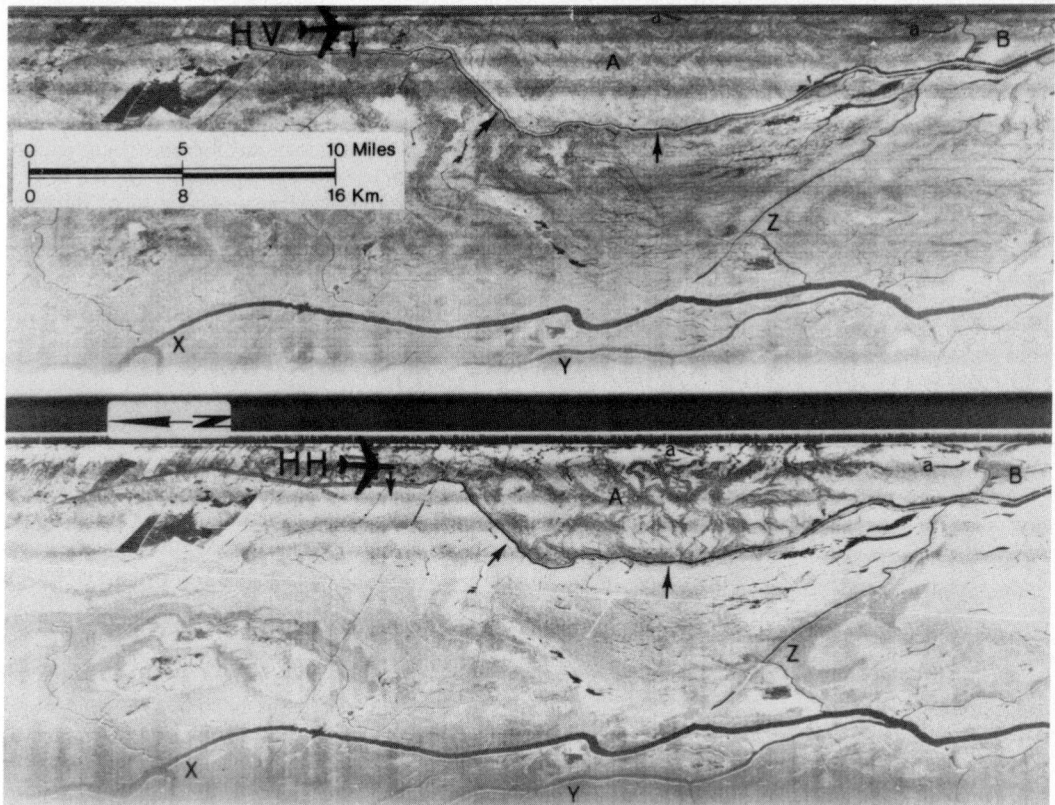

Fig. 4-37. K-band radar imagery, southwest area of Baton Rouge, Louisiana (from McDonald and Waite, 1971).

SUMMARY

The appearance of boundaries on radar images as a result of polarization, frequency, surface roughness, penetration, etc., offers the potential for generating natural-resource maps. The next step is to categorize the various entities detectable on the imagery. Usually categorization is performed on radar imagery on the basis of prior knowledge, ecologic inference, published maps and other documents, or by extracting information from other remote-sensor images in the optical or near-infrared parts of the spectrum. An understanding of the interaction mechanisms provides a foundation for interpreting both the delineation of boundaries and the categorization of entities.

RADAR BACKSCATTERING FROM WATER SURFACES

Radar return from surfaces, as discussed in the previous section, is expressed in terms of the average radar-scattering cross-section per unit area, or the scattering coefficient, $\sigma°$. The scattering coefficient is normally presented in terms of decibels as a function of the incidence angle for each associated wavelength and polarization condition. Curves of this nature consist of different angular regions in which the scattering process, or interaction mechanism, varies (Figure 4-39). For example, the region near the vertical is the "quasi-specular" region, where $\sigma°$ can be quite large. The scattering in this region is dominated by "specular" (mirror-like) reflection from facet-like surfaces oriented normally to the direction of the incident radiation. For the mid-range of angles ($30°-60°$), the typical curve is characterized by a "plateau," or "diffuse" region, which is slowly decreasing with angle. In this region, scattering is dominated by small-scale roughnesses, particularly those satisfying Bragg scattering (Chan and Fung, 1977). Between 60° and approximately 85°, the small roughness is still dominant but the effect of tilting and shadowing on the small-scale roughness by the large gravity waves becomes increasingly important. Beyond 85° other scatterers such as cusped wave crests (Wetzel, 1977), breaking waves (Kalmykov et al., 1976) and spray may become important depending upon the observation wavelength.

Radar backscattering data of quality are most abundant for water surfaces. For remote sensing

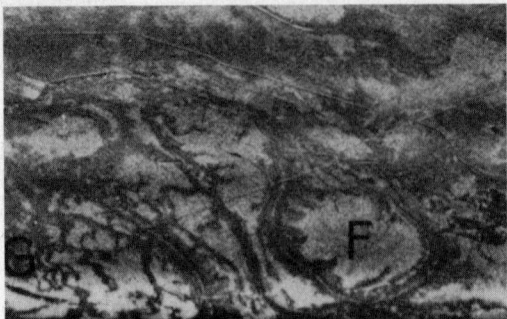

Radar Imagery (HH)

Oblique Photography

Fig. 4-38. K-band radar imagery, southwest area of Baton Rouge, Louisiana: compared with photographic imagery of wooded swamp area.

purposes, the most significant data are the measurements taken by the four-frequency airborne systems of the Naval Research Laboratory (Guinard and Daley, 1970) and the airborne systems of NASA (Krishen et al., 1971; Jones et al., 1977).

One of the first experiments to yield information concerning the angular variation of radar backscatter from water as a function of transmitter frequency [as reported by Grant and Yaplee (1957)] indicated significant dependence on wind velocity.

The first two regions previously mentioned are evident in the curves shown in Figures 4-40, 4-41, and 4-42. In general, their measurements, recorded for wavelengths, λ, of 0.86, 1.25, and 3.2 cm, respectively, established that the average radar scattering coefficient, $\sigma°$, increased with frequency and wind velocity. This is because the sea waves in the high-frequency portion of the sea spectrum, which grows with the wind, are dominating the radar returns.

Moore and Pierson (1971), reporting on 2.25-cm wavelength measurements made by NASA, found a similar wind-speed dependence (see Figure 4-43). However, corresponding measurements at 75 cm fail to exhibit correlation with wind ve-

locities in the range between 12.5 and 49 knots. Data at each of the three wind speeds within this range are about equal for incidence angles from 2.5° to 35°. This is because the portion of the sea spectrum contributing to radar backscatter at the 75-cm wavelength does not increase significantly with an increase in wind speed.

The question of the dependence of $\sigma°$ on wind velocity was studied in some detail by Daley (1973), Guinard and Daley (1970), and more recently by Jones et al. (1977). Their data, as shown in Figure 4-44, indicate that $\sigma°$ increases continuously over the observed wind-speed range (Moore et al., 1978). The wind-speed dependence is strongest for upwind observations, somewhat weaker for the downwind, and weakest for cross-wind observations (Chan and Fung, 1977).

Several attempts have been made to develop a mathematical model that would embody simultaneously the physical characteristics of a water surface and the radar backscattering behavior. Katzin (1957) visualized the sea surface as being composed of randomly oriented facets of various sizes. This model gives a realistic physical picture of the sea surface and, like the Kirchhoff surface-scatter model (Beckmann and Spizzichino, 1963; Fung, 1967), can provide adequate description of radar return at near-vertical incidence. Wind-dependence measurements, as shown in Figures 4-41 through 4-43, indicate that strong wind dependence occurs for incidence angles larger than 30°. This is in agreement with the fact that small waves are more sensitive to the local wind and that $\sigma°$ is dominated by the small-scale roughness at large incidence angles excluding the near-grazing region. Hence, later theoretical modeling studies by Wright (1968), and Chan and Fung (1977) are concentrated at small roughness effects and large incidence angles. The dominant effect of the large gravity waves is to tilt the small capillary waves, which results mainly in a change of the local incidence angle. This is accounted for by averaging the return due to the small-scale waves over the slope distribution of the large-scale waves (Wright, 1968; Chan and Fung, 1977). This two-scale roughness approach coupled with an adequate two-dimensional sea spectrum explains the dominant effects of polarization, incidence angle (over 30°−80°), azimuth angle, and wind speed (Moore and Fung, 1979). Comparisons between the theoretical scattering coefficient at 50° incidence angle versus wind speed at different polarizations and wind directions are shown in Figure 4-45. The effect of the incidence angle on wind-speed dependence is shown in Figure 4-46 and the effect of the azimuth angle is shown in Figure 4-47. In all cases shown, the theoretical predictions of the scattering coefficient by Chan and Fung (1977) provide reasonable magnitude- and trend-estimates of the measured data.

The extent of the depolarization of the incident

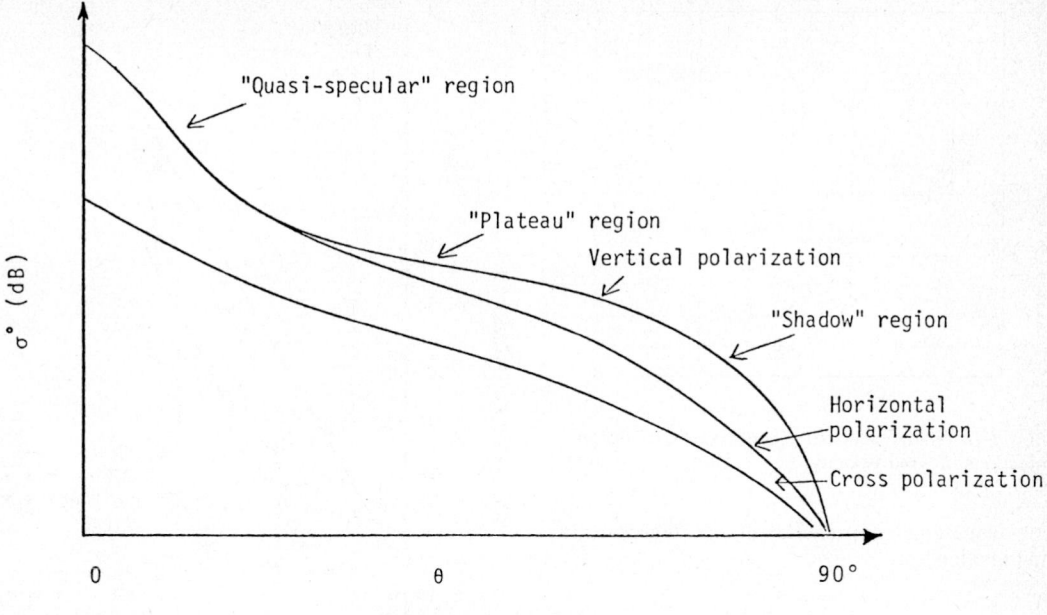

Fig. 4-39. General characteristics of radar-return variation with incidence angle.

signal by an irregular water surface is not well documented. For the sea surface the most likely causes are multiple scattering (Fung, 1968; Valenzuela, 1967), and geometrical anisotropy (Tan and Fung, 1979). Both of these mechanisms are known to produce depolarized signals that are 10- to 30-dB lower than the corresponding like-returns. At large incidence angles ($\theta > 30°$) the depolarized return curve usually varies as the cosine of the incidence angle when multiple scattering is the scatter mechanism.

The least-understood returns from water surfaces are those near grazing incidence. Studies conducted over the sea surface have indicated that different scattering mechanisms are likely to exist in the angular region between 0° and 5° from grazing (Long, 1974; Kalmykov and Pustovoytenko, 1976; Wetzel, 1977). In particular, scattering mechanisms around 1° or less from grazing may be different from those around 3° to 5° (Wetzel, 1977). It is quite possible that, at near-grazing angles, the return is due to isolated scatterers that pop into the radar beam rather than to scattering from a continuous surface (Wetzel, 1977). Intuitively, backscattering from a continuous differentiable surface must vanish as the

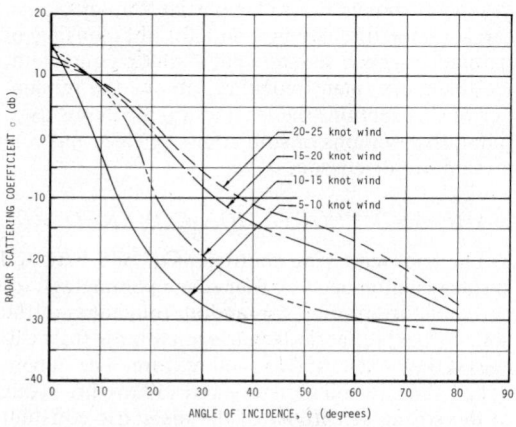

Fig. 4-40. Radar scattering coefficients of ocean, as functions of wind velocity, $\lambda = 0.86$ cm.

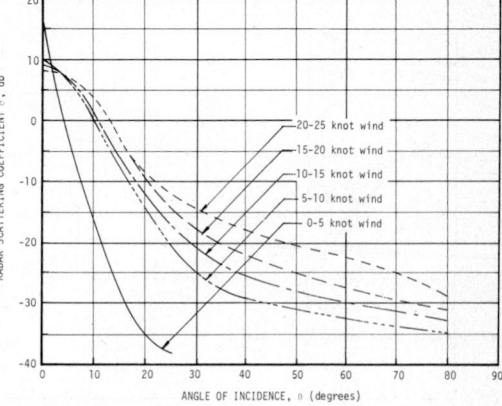

Fig. 4-41. Radar scattering coefficients of ocean, as functions of wind velocity, $\lambda = 1.25$ cm.

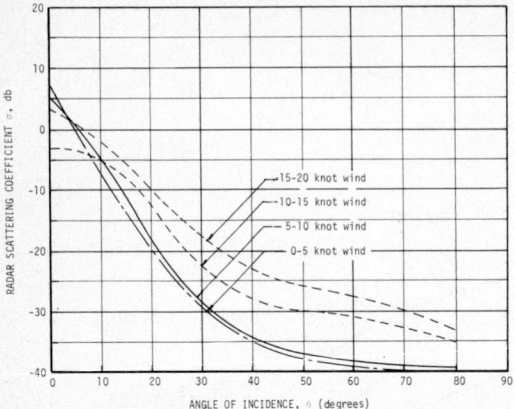

Fig. 4-42. Radar scattering coefficients of ocean, as functions of wind velocity, λ = 3.2 cm.

incidence angle approaches ninety degrees (grazing incidence).

RADAR BACKSCATTERING FROM LAND SURFACES AND SEA ICE

Concentrated efforts to interpret backscattering from water surfaces in recent years have significantly improved our understanding of the interaction mechanisms involved. Even the open questions concerning frequency, wind direction, and polarization dependencies are much better understood now than in the early 1970s. This progress has been facilitated by the homogeneity offered by a water surface. The wave structures (except for confused sea conditions) are products of reasonably well understood physical characteristics and processes, such as fetch, wind, etc.; that is, the surface can be described statistically with repeatable characteristics for similar driving functions, such as wind velocity. However, land is an entirely different matter, since it encompasses an infinite variety of surfaces and compositions. In

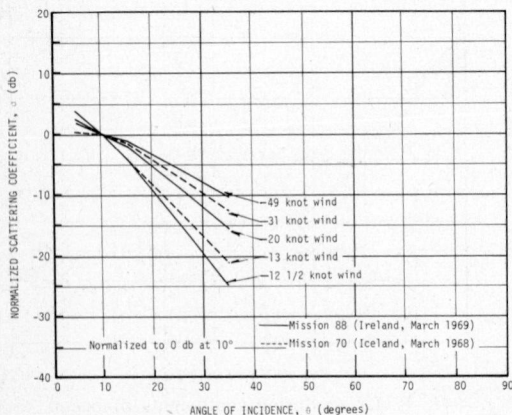

Fig. 4-43. Differential scattering coefficients of ocean, normalized to 0 dB at 10°, as functions of wind velocity, λ = 2.25 cm.

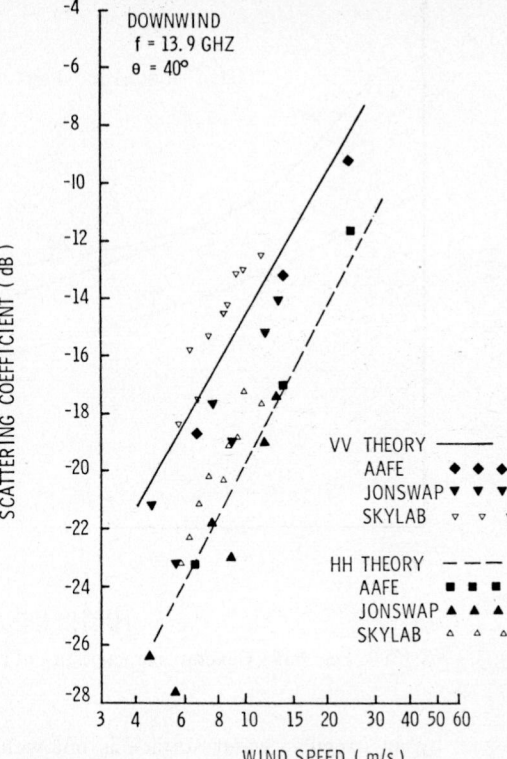

Fig. 4-44. Theoretical and combined experimental values of $\sigma°$ plotted versus wind speed, $\theta = 40°$ (Moore et al., 1978).

general, some degree of penetration is expected at microwave frequencies when targets such as vegetation, soil, snow, and ice are being observed by radar. Thus, for land observations both surface- and volume-scattering are usually present, although it is practical to ignore one or the other whenever possible. On the other hand, this penetration capability permits radar sensors to operate virtually independently of weather and time of day. This makes the radar particularly attractive for soil-moisture studies and for the sensing of parameters such as snowpack water equivalent, ice thickness, plant moisture, etc. In the remainder of this section, the interaction mechanisms in radar observations of soil, snow, sea ice, and vegetation are discussed.

BARE AND VEGETATION-COVERED SOIL

The backscattering coefficient $\sigma°$ of a bare soil surface is influenced by four scene parameters: (a) soil-moisture profile, (b) random roughness of the soil surface, (c) periodic surface patterns (row tillage) if they exist, and (d) soil texture. The dependence of $\sigma°$ on soil moisture and soil texture is due to the strong sensitivity of the dielectric constant to the water content of the soil medium, as discussed earlier (see Figures 4-23 and 4-24).

Electromagnetically, surface roughness refers to the spectrum of surface-height variations, mea-

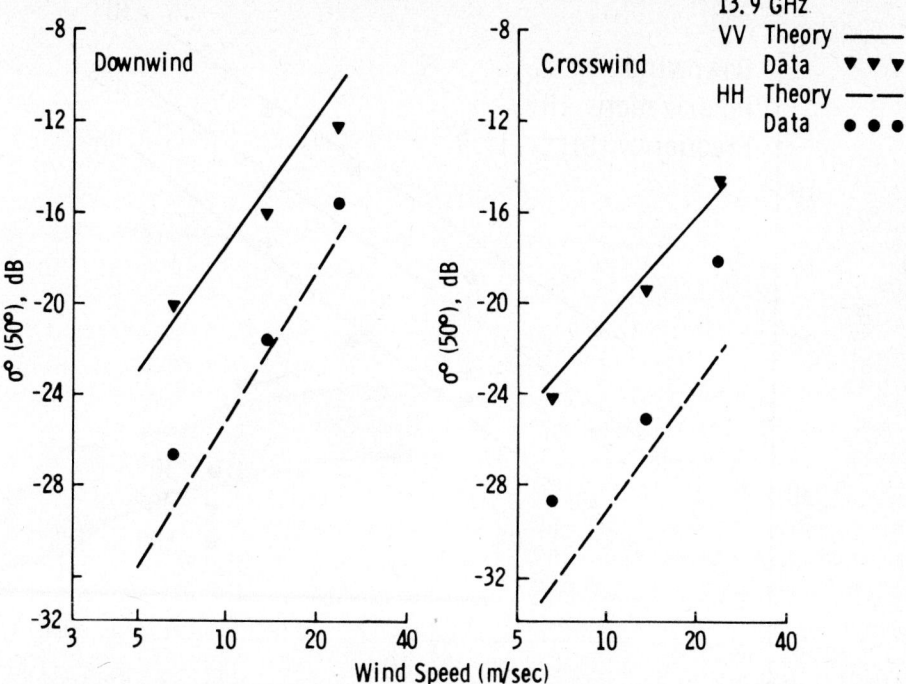

Fig. 4-45. Theoretical and AAFE values of $\sigma°$ plotted versus the wind speed, $\theta = 50°$ (Chan and Fung, 1977).

sured in wavelength units. Thus, a surface that appears relatively rough at one microwave frequency may appear relatively smooth at another frequency. This is illustrated to some extent in Figures 4-48a and b, which show measured angular responses of $\sigma°$ at 1.1 GHz and 7.24 GHz, respectively. The five curves shown in each figure correspond to five different bare soil surfaces with rms heights between 1.1 cm for the smoothest surface and 4.1 cm for the roughest. All five fields had the same soil texture and approximately the same surface soil-moisture content. As expected, the $\sigma°$ versus θ curve drops off more slowly for the rougher fields examined in the above experiment. Another significant observation that should be noted is the presence of an angular region where the various curves shown in Figure 4-48 intersect, implying that radar observations of bare soil surfaces would be approximately independent of surface roughness if made at angles within this region. An optimization analysis performed by Ulaby et al. (1978) has led to the conclusion that the optimum set of radar parameters for monitoring soil moisture content consists of a microwave frequency in the 4- to 5-GHz region, an incidence-angle range between 7° and 17° (from nadir), and HH polarization. Figure 4-49 shows $\sigma°$ plotted against volumetric moisture content for $\theta = 10°$ and $f = 4.25$ GHz. Although the data points shown in the figure include all measurements made for all five different surface-roughness conditions, a high linear correlation (0.86) is obtained between $\sigma°$ (dB) and the moisture content of the top 1-cm surface layer.

If the soil surface is characterized by a periodic pattern with a random roughness component superimposed, the backscattering coefficient of such a surface is given by a correlation of the angular response of the random roughness (small scale), denoted σ^{ss}, with the spatial function describing the periodic pattern in the direction of observation. For the sinusoidal surface sketched in Figure 4-50, the random roughness is that of a smoothly undulating surface characterized by $k\sigma = 0.2$ (where $k = 2\pi/\lambda$ and σ is the standard deviation of surface height) and $kl_e = 9$ (where l_e is the surface correlation length), which leads to the steep σ^{ss} curve shown in the figure. The combined effects of σ^{ss} and the indicated periodic pattern result in the five curves shown in the figure for $\sigma°(\theta)$ corresponding to different azimuth observation angles; $\Phi = 0°$ corresponds to the radar look-direction being parallel to the row direction, while $\Phi = 90°$ corresponds to the radar look-direction being orthogonal to the row direction. For $\Phi = 90°$, $\sigma°$ is maximum at $\theta = \gamma$ where $\tan \gamma$ is the slope of the sinusoidal pattern at the inflection point.

The dependence of $\sigma°$ on row direction, as discussed above, is based on theoretical models that are in good agreement with experimental observations (Ulaby et al., 1981a). The example shown in Figure 4-50 is for a relatively smooth sinusoidal surface (small $k\sigma$). If the microwave frequency is made higher and/or the standard deviation of surface height σ is made larger, the backscattering coefficient $\sigma°(\theta)$ becomes less sensitive to the azimuth angle Φ. That is, for a very rough surface,

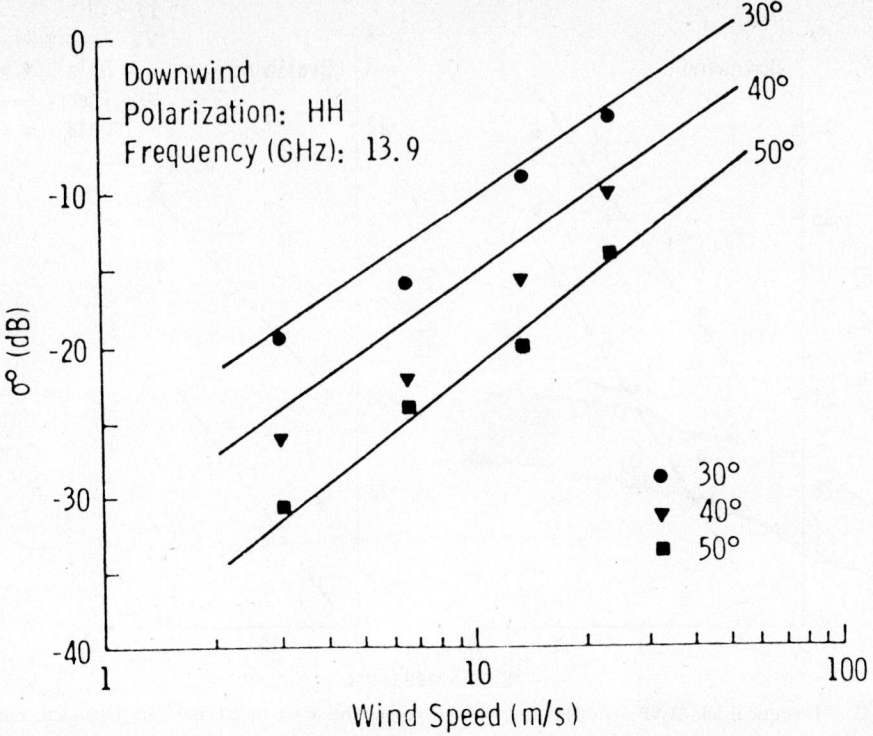

Fig. 4-46. An illustration of the wind-dependence of the backscattering of coefficient at various incidence angles (from Moore and Fung, 1979).

the dependence on row direction essentially disappears (Ulaby and Bare, 1979; Ulaby et al., 1981a).

At angles close to nadir and frequencies below about 8 GHz, the presence of vegetation cover (crops and grasses) exercises a minor influence on σ° of the soil surface. This is illustrated in Figure 4-51 in which σ°(dB) is plotted against M_{FC} (percent of field capacity), a soil moisture quality that incorporates differences due to soil textural composition (Schmugge, 1980).

SNOW

Dry snow is a mixture of air and ice particles. The real part of its relative dielectric constant is around 1.6 and its density is between 0.1 and 0.50 gm/cm³ in its natural state. Clearly, the discontinuity at the air-snow boundary is small and radar backscattering should be dominated by volume scattering. Multiple scattering effects are expected to be strong and this is evidenced by the fact that the cross-polarized return is only around 10 dB below the like-polarized return (Figure 4-52a). In Figure 4-52a, the dots and triangles correspond to two different roughnesses at the air-snow interface (Stiles and Ulaby, 1980b). The fact that the σ° curves of the two surface roughnesses are close to one another indicates that the boundary effects are small.

Wet snow is a mixture of water and dry snow. Its relative dielectric constant is around 2 or more.

As a result the air-snow boundary is expected to have more influence on the radar response. In Figure 4-52b scattering coefficients of two wet snow layers similar in geometric properties to those in Figure 4-52a are shown. It is seen that the two scattering curves are different in their angular trends. The snow layer with a smoother interface (shown by dots) has a scattering coefficient that drops off faster at near-vertical incidence (due to quasi-specular surface scattering) and slower at larger incident angles (due to the volume scattering's interaction with a smoother boundary). This behavior is consistent with the combined surface- and volume-scattering theory by Fung and Eom (1981b).

The angular behavior of σ° is a function of frequency, as illustrated in Figure 4-53 for dry snow. At 1.6 GHz, the radar return is primarily due to the soil surface underneath the 58-cm dry snow layer due to the low attenuation effects of snow at that frequency; as the frequency is increased towards 35.6 GHz, the attenuation by the snow layer increases and its volume-backscattering component increases; at 35.6 GHz, σ° exhibits a slowly varying angular response that is characteristic of volume scattering.

The variation of σ° with snow water equivalent W is illustrated in Figure 4-54 for 9 GHz, where W is the height in cm of melted snow that is contained in a vertical column of the snowpack. If the snowpack has a uniform density ρ_s, $W = \rho_s d$,

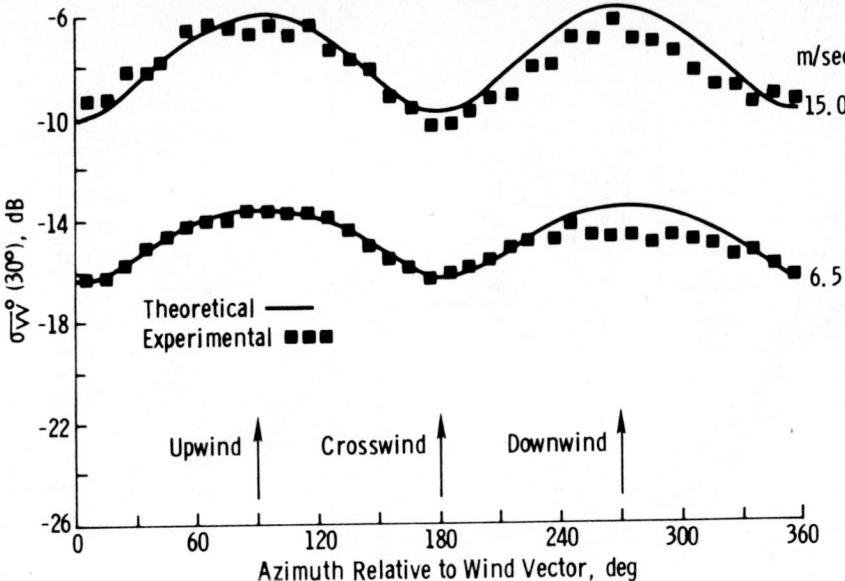

Fig. 4-47. Theoretical and observed values of $\sigma°$ plotted versus the azimuth angle (Chan and Fung, 1977).

where d is the depth. Another snow parameter that exercises a strong influence on $\sigma°$ is the snow liquid water content m_v. Figure 4-55 shows $\sigma°$ as a function of m_v of the surface 2-cm layer at 17 GHz, along with empirically derived expressions.

SEA ICE

Sea ice is a layered structure and is categorized into two major types depending upon its thickness. When it is less than 2 meters thick, it is called first-year ice and when it is over 2 meters thick it is called multi-year ice. First-year ice has a very lossy top layer made up of frazil ice. At a frequency of 10 GHz its attenuation coefficient is from 300 to 500 dB/m and its skin depth is between 2 and 3 cm. The air-ice interface is slightly rough. Its second layer contains many vertically oriented brine inclusions of 3- to 5-mm length with an average radius of 0.025 mm (Vant et al., 1978). The real part of its relative dielectric constant is expected to be around 3.5. Small air bubbles of a radius around 0.5 mm are distributed throughout the ice layers. Thus, around X-band the scattering mechanism for the first-year ice is expected to be a combined surface- and volume-scattering from the top layer mainly. Multiple-scattering effects are important only for the cross-polarized component. In Figure 4-56 backscattering coefficients for the first-year sea ice are shown. Within the first 20° off the vertical a fairly steep drop-off of the scattering curves indicates quasi-specular surface scattering. Beyond 20° the slower drop-off shows volume scattering. The cross-polarized return is more than 12 dB lower than the like returns. Thus, the scattering albedo is not very large, which is consistent with the small air bubbles.

The major differences between multi-year ice and first-year ice are (1) the attenuation for multi-year ice is much less (less than 10 dB/m at 10 GHz), (2) its air bubbles are about 3 times larger, and (3) its air-ice interface is much rougher. These differences in properties are reflected in the scattering coefficients shown in Figure 4-56. It is seen that the absolute level of the multi-year scattering coefficients is higher and that the angular trends are less pronounced as a result of stronger volume- and surface-scattering. The larger air bubbles also imply a larger albedo, which is responsible for smaller spacing between the levels of the like- and cross-polarized returns.

VEGETATION

A vegetation medium may be viewed as a mixture of leaves, fruit (when present), branches, and stalks embedded in air. In most cases, stalks are oriented vertically to the horizon. Thus, radar backscattering away from grazing incidence is not sensitive to stalks but is sensitive to leaves. Radar returns from deciduous trees in the spring season (Figure 4-57) are substantially higher than those in autumn (Bush et al., 1976). This suggests that when leaves are in their mature stage, radar returns from leaves will have dominance over the contributions from branches. It is also possible that returns from fruit may be the dominant contributors or may at least be as significant as the contributions from leaves. According to Figure 4-58, $\sigma°$ of wheat at 17 GHz is characterized by two temporal peaks. The first one is at approximately Day 130 and appears to be driven by the green-leaf area index (LAI) of the wheat canopy, while the second peak occurs shortly before har-

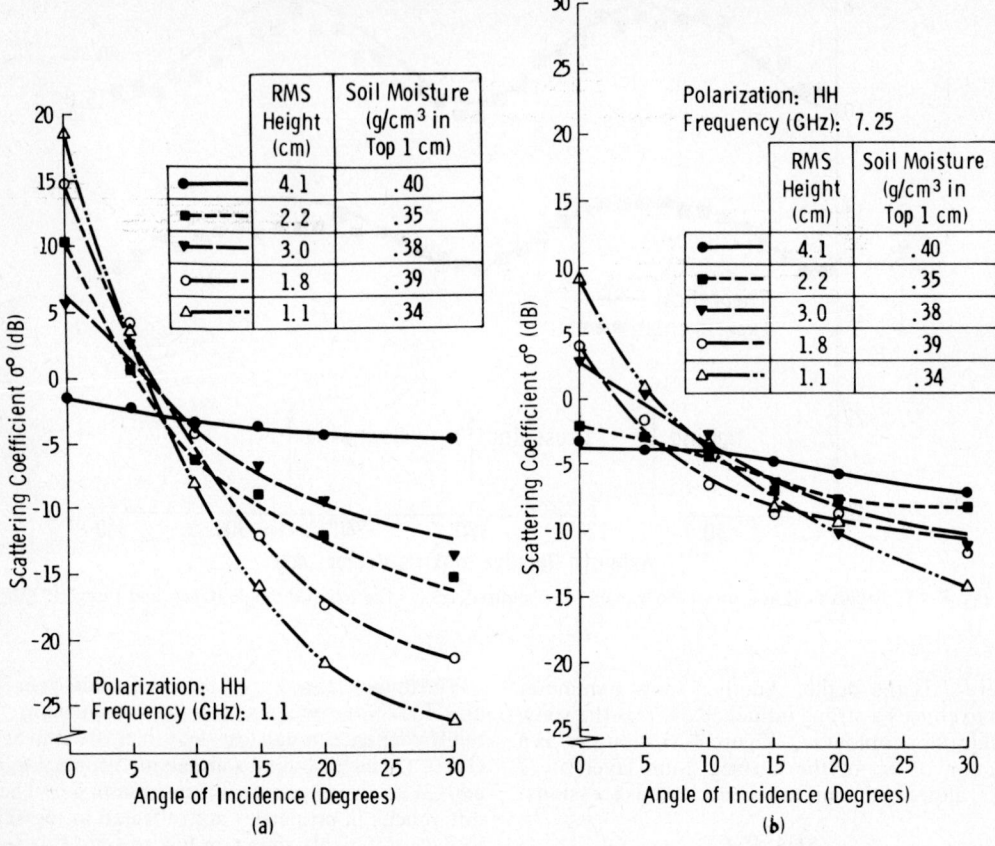

Fig. 4-48. Angular response of $\sigma°$ for five bare-soil surfaces with different rms heights at (a) 1.1 GHz and (b) 7.25 GHz (from Ulaby et al., 1978).

vest and appears to be responding to the dry matter of the canopy, which is contained primarily in the wheat heads.

When leaves are viewed as a random collection of scatterers, volume scattering is clearly the mechanism for generating radar returns. Thus, it is expected that large cross-polarized returns from a vegetation medium are possible (see Figure 4-59). For defoliated plants, branches may be viewed as a collection of finite cylinders with some spatial distribution. The scattering mechanism here is again volume scattering.

PASSIVE MICROWAVE RADIATION

Microwave radiometry is governed by the same theory of radiative transfer that applies to thermal infrared radiometry. Because of the difference in wavelength between the microwave and thermal infrared bands, however, the corresponding emission, scattering, and absorption characteristics of terrestrial and atmospheric media are quite different. The microwave absorption spectrum of the atmosphere includes absorption bands around 22 and 183 GHz due to water vapor and around 60 and 118 GHz due to oxygen (see Chapter 5). Radiometric observations made by satellite-borne

sensors at frequencies in and around these absorption bands are used to estimate the water vapor and temperature profiles of the atmosphere and the water content of clouds and rain, when present. To date, these meteorological applications have been limited primarily to areas where the background emission is spatially uniform, such as the ocean. Several of these systems are described in some detail in Chapter 14.

For terrestrial observations, the frequency ranges of interest are the 1–20 GHz band and the 37-GHz atmospheric transmission window. At higher frequencies, absorption and emission by clouds and rain pose a serious problem to the interpretation of radiometric measurements.

The operation of microwave radiometers and the relationship between the radiometer output voltage and the radiation incident upon its antenna are covered in Chapter 11. This section deals primarily with the identification of the interaction mechanisms responsible for the behavior of microwave emission from solid and liquid surfaces and volumes in the microwave region. Although atmospheric interactions are discussed in Chapter 5, the influence of the atmosphere on microwave measurements is of necessity considered in the sections that follow.

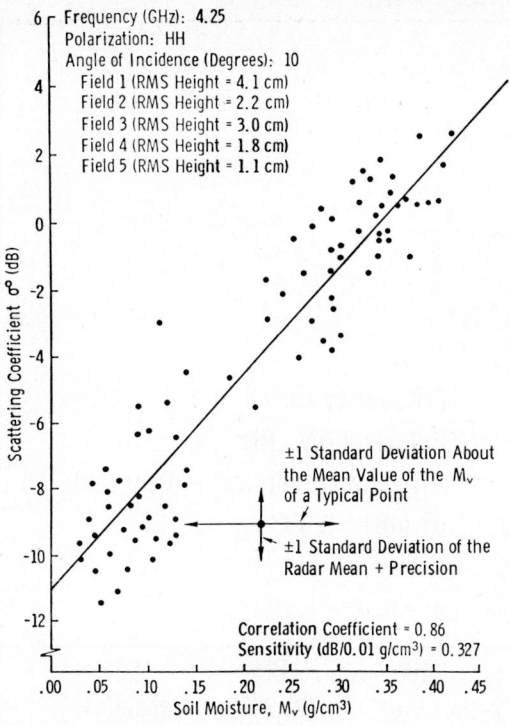

Fig. 4-49. Radar response to 0-1-cm volumetric soil moisture (from Ulaby et al., 1978).

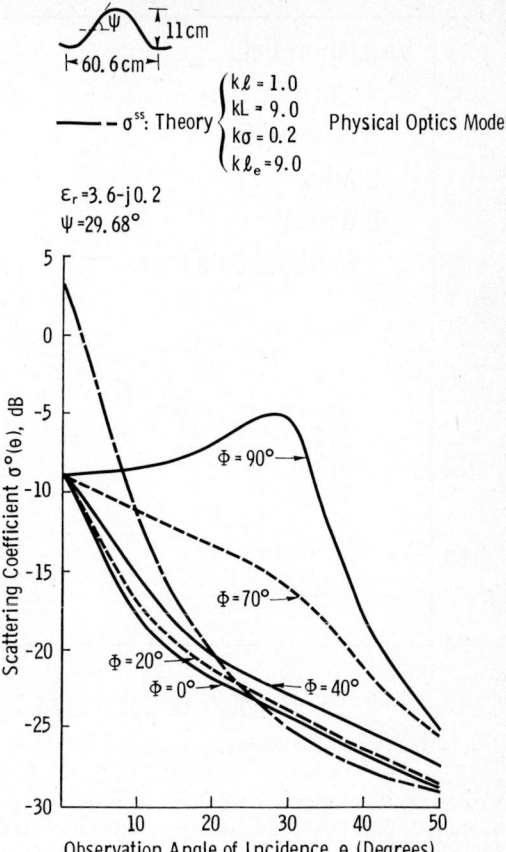

Fig. 4-50. Azimuthal dependence of the backscattering coefficient from a randomly perturbed one-dimensional periodic surface. (from Ulaby et al., 1981a).

FUNDAMENTALS OF MICROWAVE RADIOMETRY

As stated in Chapter 11, in the microwave region the brightness of a blackbody radiator may be approximated by the Rayleigh-Jeans formula,

$$B_{bb} = \frac{2kT}{\lambda^2} \qquad (4\text{-}34)$$

where k is Boltzmann's constant, T is the physical temperature of the blackbody and λ is the wavelength. It was also shown in Chapter 11 that the power available at the output of an ideal bandpass filter, connected to a lossless, linearly-polarized antenna with an incident brightness distribution $B(\theta,\phi)$ is given by Eq. 11-24,

$$P = \frac{1}{2} \int_f^{f+\Delta f} \frac{\lambda^2}{4\pi} \int_{4\pi} G(\theta,\phi)B(\theta,\phi)d\,\Omega\,df \qquad (4\text{-}35)$$

where $G(\theta,\phi)$ is the antenna gain pattern. For a blackbody, use of Eq. 4-34 is shown in Chapter 11 to lead to:

$$P_{bb} = kT\Delta f. \qquad (4\text{-}36)$$

Since a blackbody radiates more energy at a given temperature T than any other material at that same temperature, the brightness $B(\theta,\phi)$ of a real material may be considered as the brightness of an equivalent blackbody at a cooler temperature, $T_B(\theta,\phi)$,

$$B(\theta,\phi) = \frac{2k}{\lambda^2}\,T_B(\theta,\phi), i = h \text{ or } v \qquad (4\text{-}37)$$

where $T_B(\theta,\phi)$, called the brightness temperature of the material, represents emission in the direction (θ,ϕ) (with respect to a specified coordinate system). In addition to being direction-dependent, the brightness (and therefore, the brightness temperature also) may be polarization-dependent. For radiometric observations with a linearly polarized antenna with polarization i (h for horizontal or v for vertical), the polarized emissivity of the material is defined as

$$e_i(\theta,\phi) = \frac{B_i(\theta,\phi;T)}{B_{bb}(T)}$$
$$= \frac{T_{Bi}(\theta,\phi)}{T}; i = h \text{ or } v. \qquad (4\text{-}38)$$

For a lossless narrow-beam antenna observing a material with a brightness temperature $T_{Bi}(\theta,\phi)$ along the direction (θ,ϕ), the power received over a bandwidth Δf is given by an expression similar to Eq. 4-36, namely

$$P_i(\theta,\phi) = kT_{Bi}(\theta,\phi)\Delta f. \qquad (4\text{-}39)$$

EMISSIVITY AND REFLECTIVITY

The external EM radiation is generally considered to be propagating in a plane wavefront to an

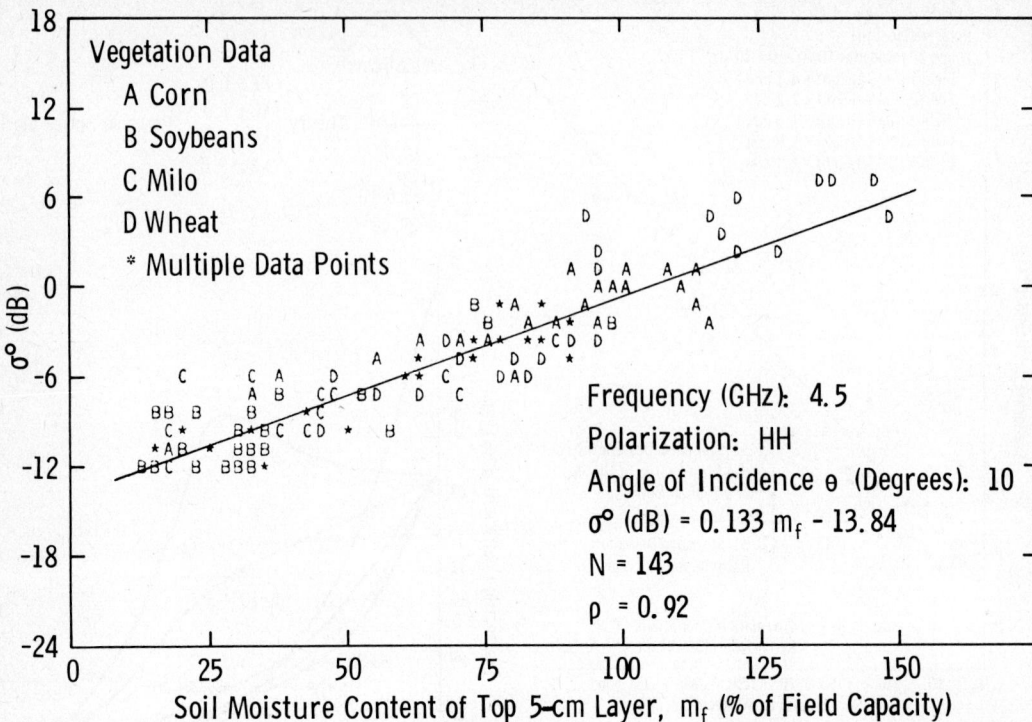

Fig. 4-51.   Radar backscatter response to soil moisture for vegetation-covered fields (from Ulaby et al., 1982).

extensive flat layer where part is reflected or scattered, part absorbed, and the rest transmitted into the subsurface below the layer. Based upon the laws of conservation of energy, the condition of equilibrium on the incident power ρ_i becomes

$$\rho_i^i = \rho_i^r + \rho_i^a + \rho_i^t; i = h \text{ or } v \qquad (4\text{-}40)$$

where the superscripts r, a, and t represent reflected, absorbed, and transmitted power, respectively, and the subscript i denotes the polarization configuration. By normalizing with ρ_i^i, it follows that

$$1 = \rho_i + \alpha_i + \tau_i; i = h \text{ or } v \qquad (4\text{-}41)$$

where

$$\rho_i = \text{reflectivity}$$
$$\alpha_i = \text{absorptivity}$$
$$\tau_i = \text{transmissivity}.$$

For a sufficiently lossy layer (or a half-space) the power transmitted through the layer is negligible. In this case,

$$1 = \rho_i + \alpha_i. \qquad (4\text{-}42)$$

Under thermal-equilibrium conditions, power absorbed must be equal to power emitted so that a given temperature or temperature distribution can be maintained. This means that the emissivity, e, must be equal to absorptivity and therefore,

$$e_i = 1 - \rho_i \text{ (half-space)}. \qquad (4\text{-}44)$$

RADIOMETRIC TEMPERATURE MODELS

A simple two-dimensional model is useful in making predictions of radiometric temperatures.

To keep the model simple, radiation is considered only from direct ground emission and surface reflection from a few external sources. Such a basic model provides valuable families of plots of radiometric (or brightness) temperatures, T_B, as functions of the angle of incidence, with variations of polarization and frequency. A model of this nature supplies preliminary data for a number of applications to land and ocean surfaces, provided that surface functions are also included to account for roughness, periodic structure, and changes in electrical properties. For the configuration shown in Figure 4-60, the radiometric temperature, T_B, at the antenna, can be specified by

$$T_B(\theta,\phi;i) = \frac{1}{L_a(\theta)} \left[T_{BS}(\theta,\phi;i) + T_{SC}(\theta,\phi;i) \right] + T_U(\theta) \qquad (4\text{-}45)$$

where

$T_B(\theta,\phi;i)$ = brightness temperature of the scene, representing energy incident upon the antenna from direction (θ,ϕ) with polarization i.

$L_a(\theta)$ = atmospheric loss factor accounting for extinction between the ground surface and the radiometer antenna, given by Eq. 11-50.

$T_{BS}(\theta,\phi;i)$ = brightness temperature of the ground surface.

$T_{SC}(\theta,\phi;i)$ = scattered brightness temper-

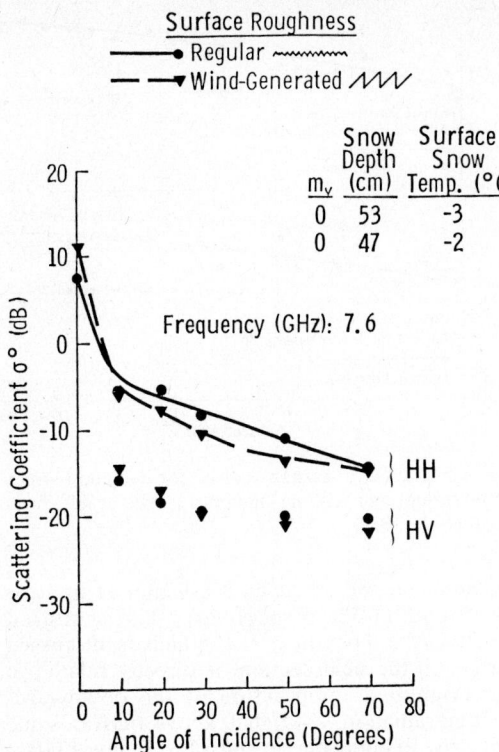

Fig. 4-52a. Effect of surface roughness of $\sigma°$ of dry snow at 7.6 GHz (from Stiles and Ulaby, 1980b).

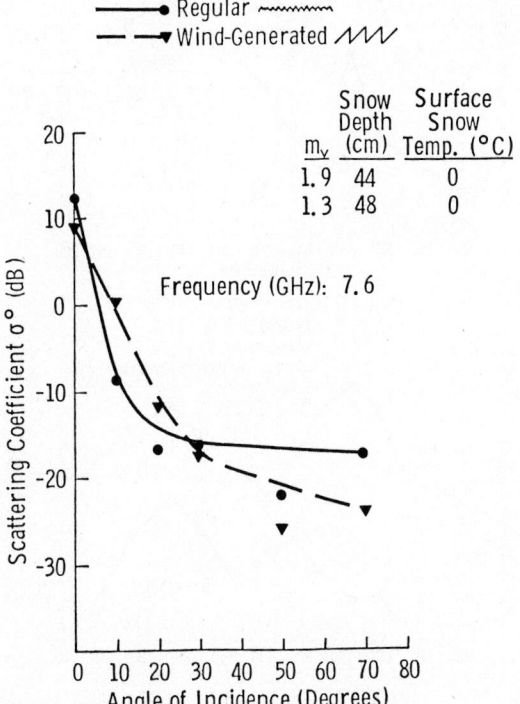

Fig. 4-52b. Effect of surface roughness on $\sigma°$ of wet snow at 7.6 GHz (from Stiles and Ulaby, 1980b).

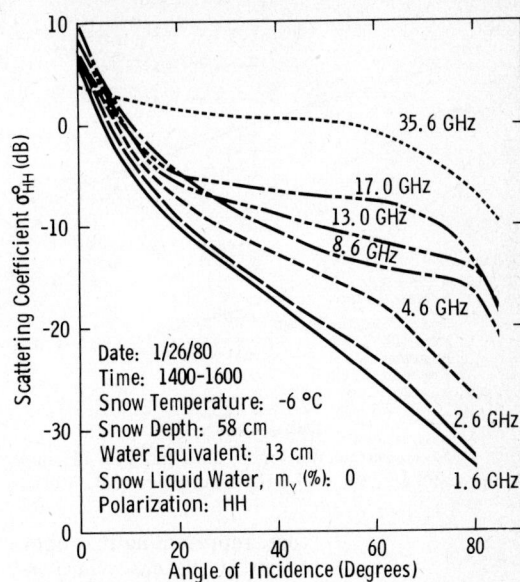

Fig. 4-53. Angular response of $\sigma°$ for dry snow at several microwave frequencies (from Stiles et al., 1981).

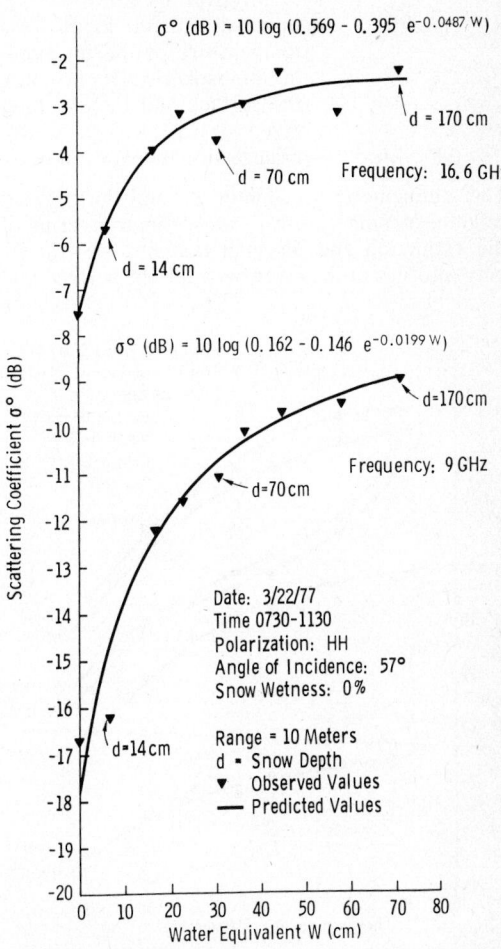

Fig. 4-54. Scattering coefficient response to snow water equivalent (from Ulaby and Stiles, 1980).

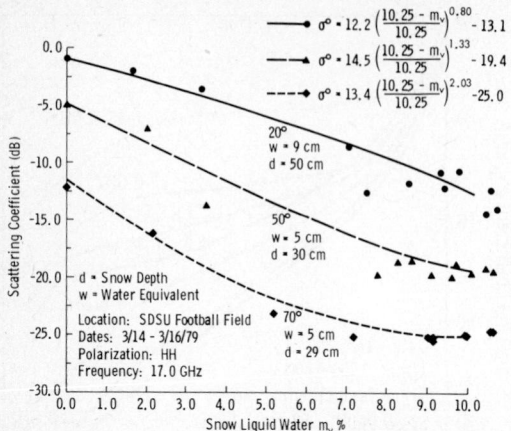

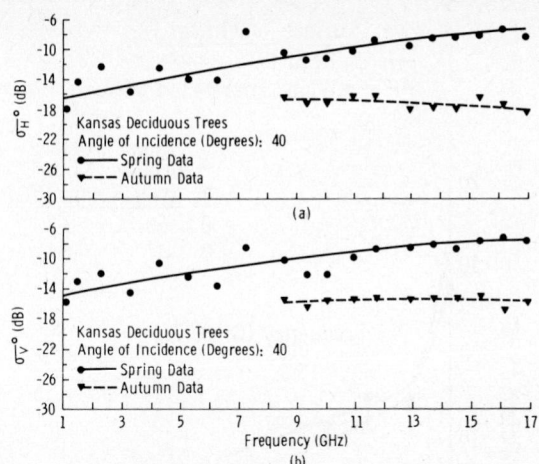

Fig. 4-55. Empirical fits for $\sigma°$ as a function of snow liquid-water content at 17 GHz (from Stiles et al., 1981).

Fig. 4-57. Spectral response of $\sigma°$ for deciduous trees with (spring) and without (autumn) leaves at 40° (from Bush et al., 1976).

ature, representing that component of the downward incident energy upon the ground surface that is scattered by the surface into the direction (θ,ϕ), given by Eq. 4-49.

$T_U(\theta)$ = upward-emitted brightness temperature of the intervening atmospheric layer between the surface and the antenna, given by Eq. 11-53.

i = polarization index, $i = h$ or v.

The atmospheric loss factor L_a and the upward brightness temperature T_U are defined in terms of the extinction and physical temperature (for T_U only) profiles of the layer between the surface and

the antenna, and are given in Chapter 11 by Eqs. 11-50 and 11-53, respectively. The scattered brightness temperature T_{SC}, which is discussed further in the next section, is directly related to the brightness temperature of the downward-emitted radiation, $T_D(\theta)$ that, above 1 GHz, is due primarily to atmospheric emission. Below 1 GHz, incident galactic radiation becomes significant.

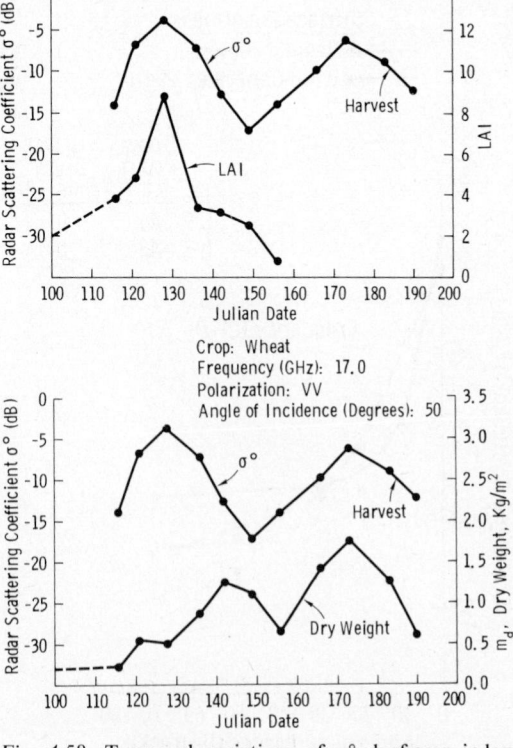

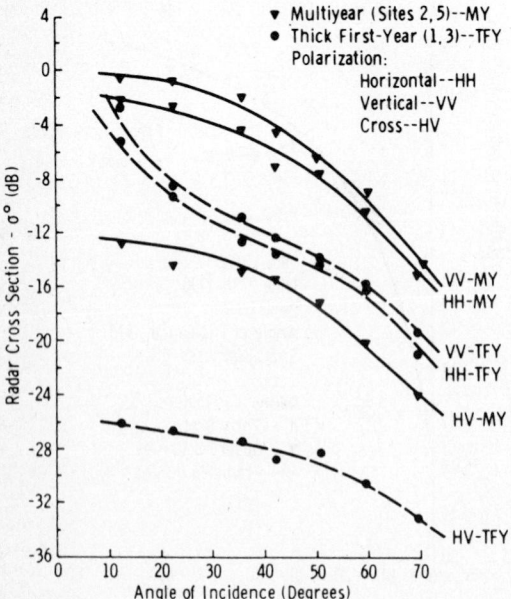

Fig. 4-56. Scattering coefficient of thick first-year and multi-year ice at 13 GHz (from Onstott et al., 1979).

Fig. 4-58. Temporal variations of $\sigma°$, leaf-area index (LAI) and canopy dry mass m_d of wheat (from Ulaby, 1980).

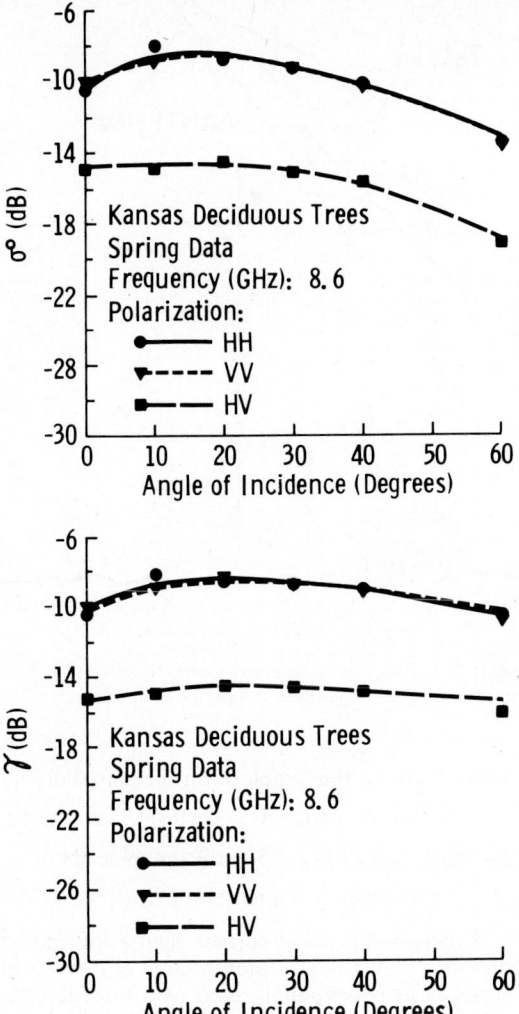

Fig. 4-59. Angular variation of (a) $\sigma°$ and (b) $\gamma = \sigma°/\cos \theta$, for HH, VV, and HV polarizations for deciduous trees at 8.6 GHz. Note that $\gamma(\theta)$ is approximately constant over the 0 to 60° range, which is characteristic of volume scattering (from Bush et al., 1976).

For a lossless atmosphere, $L_a = 1$, $T_U = T_D = T_{SC} = 0$. Hence, Eq. 4-45 reduces to:

$$T_B(\theta,\phi;i) = T_{BS}(\theta,\phi;i) \text{ (lossless atmosphere).}$$
$$(4\text{-}46)$$

Additionally, if the ground medium has a constant physical temperature T_g, use of Eq. 4-38 leads to:

$$T_B(\theta,\phi;i) = e_i(\theta,\phi) \, T_g. \qquad (4\text{-}47)$$

In practice, the physical temperature of the medium under observation usually is not a constant; a soil medium, for example, has a temperature profile with depth. However, in the majority of cases, the observed radiation is emitted primarily by the top surface-layer of the medium, which may vary from about 1 cm to a few tens of

centimeters. Hence, T_g is an effective temperature of the surface layer responsible for the majority of the emitted radiation. Exact solutions for a medium with a nonuniform temperature profile are treated in Ulaby et al. (1981b).

Although Eq. 4-46 applies only in the ideal case of a lossless atmosphere, it is often used in the interpretation of radiometric data as a first-order approximation. The approximation is valid, provided the observations are made (a) under clear-sky conditions and (b) at frequencies in the 1-10-GHz region. Under these conditions, typical values for the atmospheric parameters of interest are: $L_a^{-1} \cong 0.95-0.98$, $T_U \cong 0-10$ K, and $T_{SC} \cong 1-5$ K, all for $\theta < 30°$. For much larger values of θ, atmospheric effects become more significant. The limits on T_U correspond to observations from a ground platform ($T_U = 0$ K) and observations from a satellite platform ($T_U \cong 10$ K).

If the observations are made under cloud-cover or rain conditions and/or at frequencies much higher than 10 GHz, atmospheric corrections are needed to retrieve the ground brightness temperature T_{BS} from the brightness temperature T_B at the antenna (which itself is related to the antenna temperature T_A measured by the radiometer, as discussed in Chapter 11).

For a semi-infinite medium with an effective physical temperature T_g, the brightness temperature is given by Eq. 4-47. The emissivity $e_i(\theta,\phi)$ is governed by the scattering properties of the surface and is given by an expression that was derived by Peake (1959) in the form of an integral of the bistatic scattering coefficient $\sigma°_{ij}$ of the ground surface. The scattering properties of the surface also relate the scattered brightness temperature $T_{SC}(\theta,\phi;i)$ to the downward-emitted atmospheric temperature $T_D(\theta)$. Peake's expressions are given by:

$$e_i(o) = 1 - \frac{1}{4\pi\cos\theta} \int_o^{2\pi} \int_o^{\pi/2} \qquad (4\text{-}48)$$
$$\left[\sigma°_{ii}(o,s) + \sigma°_{ij}(o,s) \right] d\Omega_s$$

and

$$T_{SC}(o) = \frac{1}{4\pi\cos\theta} \int_o^{2\pi} \int_o^{\pi/2} T_D(s) \qquad (4\text{-}49)$$
$$\left[\sigma°_{ii}(o,s) + \sigma°_{ij}(o,s) \right] d\Omega_s$$

where i and j are polarization subscripts (i, j = horizontal or vertical); o stands for (θ,ϕ) and s stands for (θ_s, ϕ_s), which denote the incident and scattered directions, respectively, in a standard spherical coordinate system. It is of interest to consider a few special cases of Eq. 4-48. In particular, we shall first derive expressions for a perfectly smooth surface and a perfectly rough surface. Other more general surface-emission models have been reported by Stogryn (1967), Wu and Fung (1972) and Peake et al. (1966). Emission from an inhomogeneous medium with an irregular surface boundary has been reported by Fung and

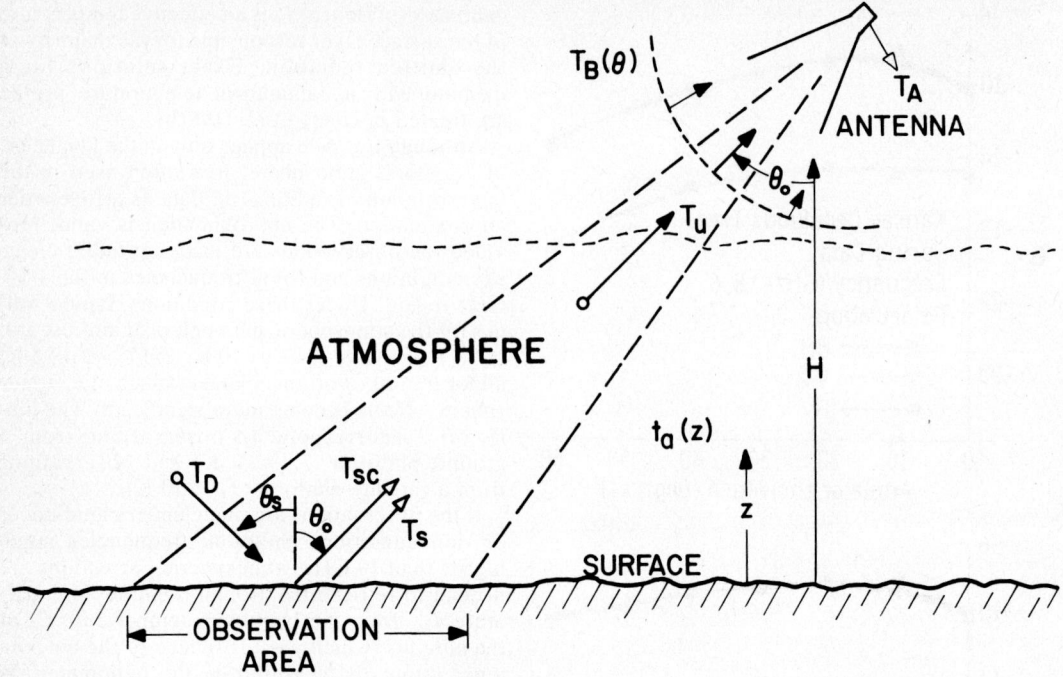

Fig. 4-60. Illustrating the principal contributors to the apparent brightness temperature T_B (θ_o).

Chen (1981) using a bistatic surface scattering coefficient which satisfies energy conservation requirements (Fung and Eom, 1979).

EMISSIVITY MODEL OF A SPECULAR SURFACE

A surface can be defined as smooth if its height variations are much smaller than the radiation wavelength, usually by an order of magnitude or more. Under this condition, the differential scattering coefficient will have only a specular component that can be described mathematically by a Dirac delta function δ as follows (Peake, 1959):

$$\sigma^{\circ}_{ii}(\theta,\phi;\theta_s,\phi_s) = \frac{4\pi\rho_i}{\sin\theta}\,\delta(\theta_s - \theta)\delta(\phi_s - \phi)\cos\theta$$
(4-50)

where

ρ_i = i-polarized power reflection coefficient, $|R_i(\theta)|^2$
θ and ϕ = the specular direction
δ = Dirac delta function
R_i = Fresnel reflection coefficient for the i-polarization.

Furthermore, for $i \neq j$, $\sigma^{\circ}_{ij} = 0$; that is, in the absence of roughness, no depolarization by the surface takes place.

Inserting Eq. 4-50 into 4-48, and using the definition of $d\Omega_s = \sin\theta_s\,d\theta_s\,d\phi_s$, we obtain:

$$e_i(\theta,\phi) = 1 - \frac{1}{4\pi}\int\!\!\int \frac{4\pi\rho_i}{\sin\theta}\,\delta(\theta_s - \theta)\delta(\phi_s - \phi)$$
$$\sin\theta_s\,d\theta_s\,d\phi_s$$
(4-51)

which leads to the simple form for emissivity as

$$e_i(\theta,\phi) = 1 - |R_i(\theta)|^2.$$
(4-52)

Similarly, use of Eq. 4-50 in Eq. 4-49 leads to

$$T_{sc}(\theta,\phi) = T_D(\theta)\,(1 - |R_i(\theta)|^2).$$
(4-53)

Although the plane-surface scattering model provides only a poor representation of real-world surfaces, it is commonly used as a first-step approximation because of its simplicity and ease of calculation. The sensitivity of e_i to the dielectric properties of the emitting medium is illustrated in Figure 4-61, where the emissivity was calculated for two soil conditions, sea water and sea ice, and for both horizontal and vertical polarizations in each case.

Returning to Eq. 4-45, the brightness temperature of a specular surface as observed by an antenna above an intervening atmospheric layer of loss L_a, is given by

$$T_B(\theta,\phi;i) = \frac{1}{L_a(\theta)}\{e_i(\theta,\phi)\,T_g$$
$$+ [1 - e_i(\theta,\phi)]\,T_D(\theta)\} + T_U(\theta)$$
(4-54)

where $e_i(\theta,\phi)$ is given by Eq. 4-52.

EMISSIVITY MODEL OF A PERFECTLY ROUGH SURFACE

According to Lambert's Law, a perfectly rough surface scatters radiation incident upon it in such a way that the angular variation of the bistatic

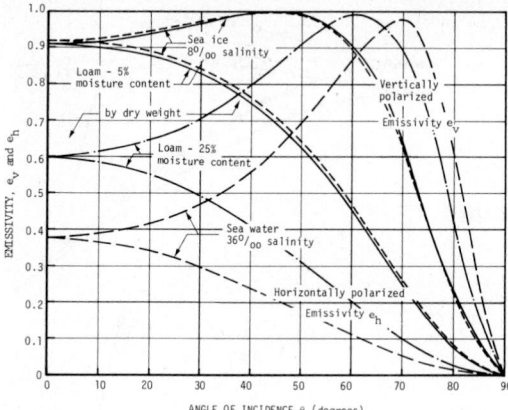

Fig. 4-61. Calculated emissivities as a function of incidence angle at 10 GHz. Calculation based on plane-surface model.

scattering cross-section per unit area is solely dependent on the product of the cosine of the incident and scattering angles:

$$\sigma^o_{ii}(o,s) + \sigma^o_{ij}(o,s) = \gamma_o \cos\theta \, \cos\theta_s, \quad (4\text{-}55)$$

where γ_o is a constant related to the dielectric properties of the scattering surface.

Under this condition, the emissivity of a Lambertian surface becomes:

$$e_i(\theta,\phi) = 1 = \frac{1}{4\pi} \int_{\phi_s=0}^{2\pi} \int_{\theta_s=0}^{\pi/2}$$

$$\gamma_o \cos\theta_s \, \sin\theta_s \, d\theta_s \, d\phi_s, \; = \; 1 - \frac{\gamma_o}{4}.$$

(4-56)

which is a constant in all directions. Experimental measurements of large blocks of pumice at 10 and 35 GHz have been reported by Peake et al. (1966) to exhibit a radar return (σ^o_{ii}) similar to that of a Lambertian surface, and a brightness temperature almost independent of the incidence angle θ.

EFFECTS OF SURFACE ROUGHNESS AND INHOMOGENEITIES ON EMISSION

Comparison of a rough-surface boundary with a plane boundary shows that one significant difference is that the former is capable of multiple scattering. The shadowing of part of the backward-scattered energy means that there is an additional opportunity for the scattered wave to penetrate into the other medium. Thus, a rougher surface should have a larger emission than a plane surface would.

When a lossy medium without inhomogeneities is compared with the same medium containing, say, air bubbles, more emission is expected from the homogeneous medium because the presence of inhomogeneities (air bubbles) causes volume scattering. Thus, a medium with a larger albedo value is expected to emit less than a corresponding medium with a smaller albedo value will.

The emission properties resulting from a combination of an irregular boundary and an inhomogeneous medium have been studied by Fung and Chen (1981). The geometry of the problem is shown in Figure 4-62. When the rms slope of the bottom-boundary roughness is increased, brightness temperature increases (Figure 4-63). The increase for rms slopes in the range 0.15 to 0.45 is more significant at small nadir angles than at large nadir angles. When the rms slope of the top boundary increases (Figure 4-64), the brightness temperature increases for horizontal polarization, especially at large nadir angles. However, for vertical polarization the increase is restricted to small nadir angles. At large nadir angles the peak value actually decreases with an increase in the rms slope. The overall effect of a rough top-boundary is to decrease the spacing between the vertical- and horizontal-brightness temperature curves. When both layer interfaces are rough, the brightness temperatures tend to follow the bottom-rough case at small nadir angles and the top-rough case at large nadir angles (Figure 4-65).

The degree of layer inhomogeneity may be characterized by the albedo of the layer. In general, an increase in albedo causes the brightness temperature curve to decrease at all nadir angles. An increase in layer permittivity has the same effect, since it causes an increase in reflectivity at the air-layer boundary (assuming that the discontinuity in the dielectric constant at the bottom interface remains unchanged).

MICROWAVE EMISSION FROM LAND SURFACES

BARE AND VEGETATION-COVERED SOIL

For a relatively smooth bare-soil surface, the emissivity is strongly influenced by the moisture content of the soil, as illustrated in Figures 4-66

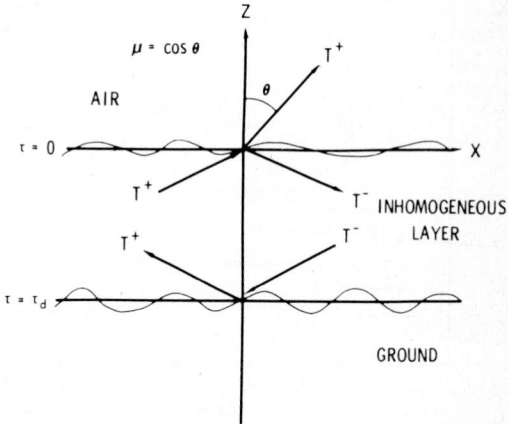

Fig. 4-62. Geometry of the problem of emission from an inhomogeneous, irregular layer.

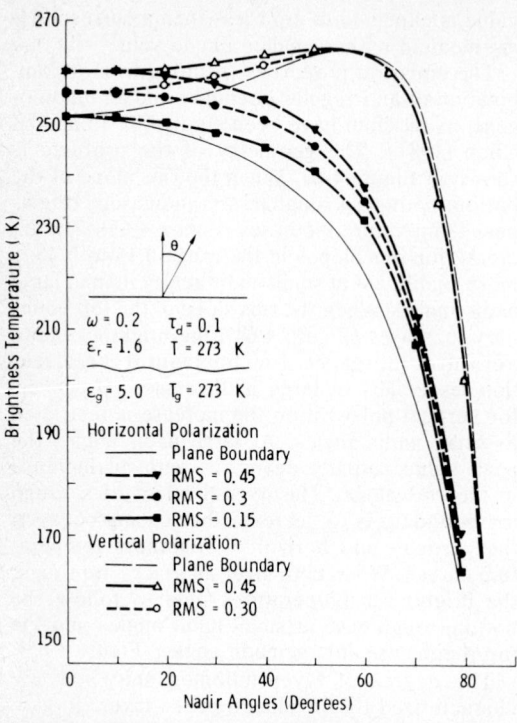

Fig. 4-63. Effect of surface roughness on emission from an inhomogeneous layer with an irregular bottom interface (from Fung and Chen, 1981).

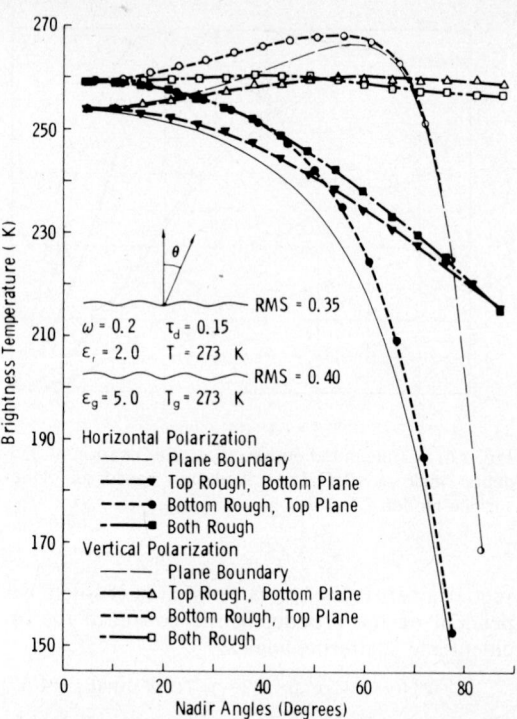

Fig. 4-65. Effect of surface roughness on emission from an inhomogeneous irregular layer (from Fung and Chen, 1981).

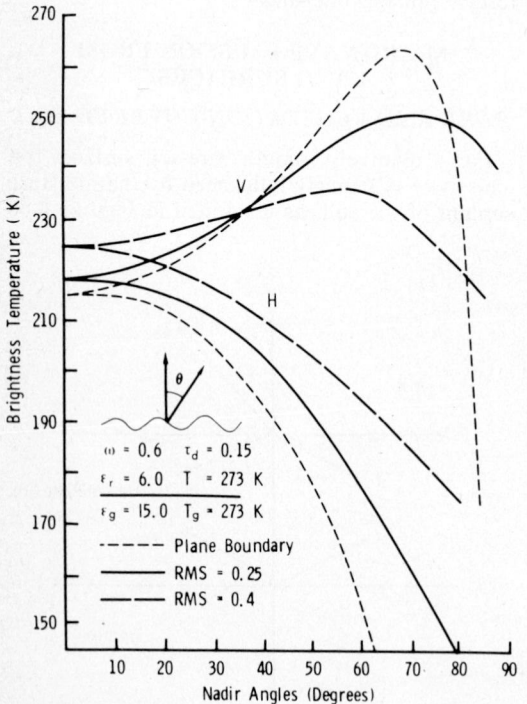

Fig. 4-64. Effect of surface roughness on emission from an inhomogeneous layer with an irregular top interface (from Fung and Chen, 1981).

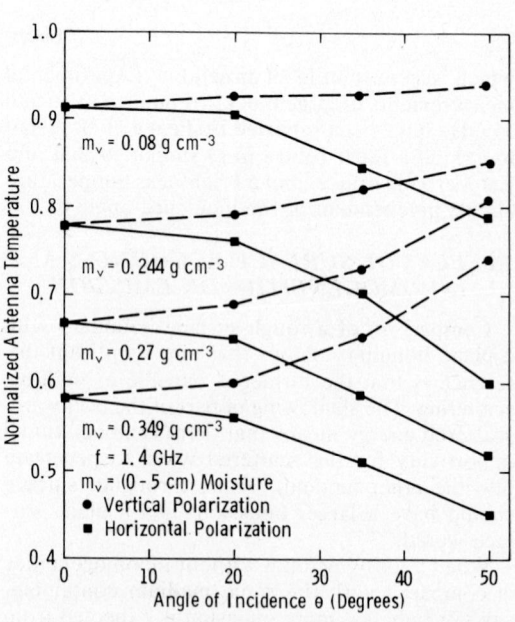

Fig. 4-66. Angular response of normalized antenna temperature for a bare field with a smooth soil surface, for each of four different soil moisture conditions (Newton and Rouse, 1980).

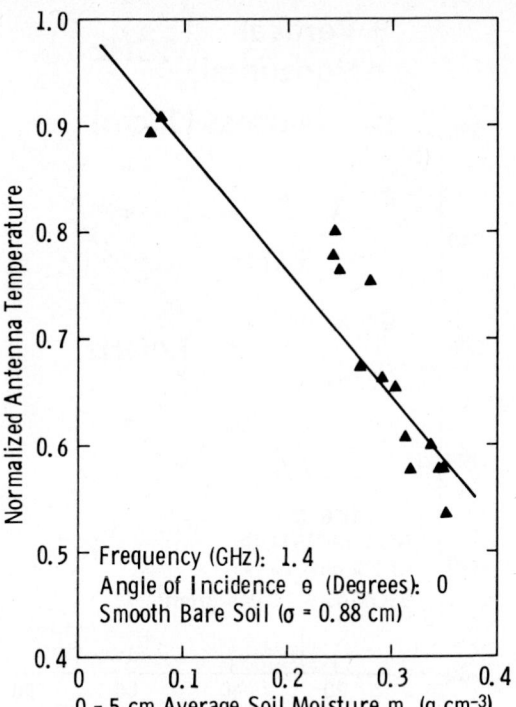

Fig. 4-67. Radiometric response to soil moisture content for a smooth, bare field at nadir (from Newton and Rouse, 1980).

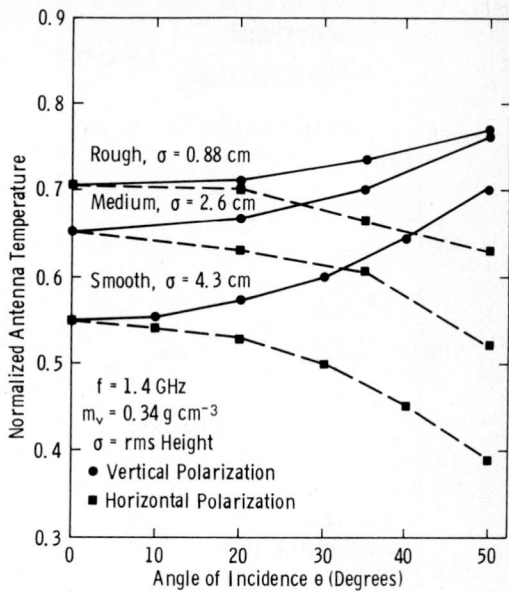

Fig. 4-68. Effect of surface roughness on the microwave emission from bare soil at 1.4 GHz (from Newton and Rouse, 1980).

and 4-67. The data shown in these two figures were measured at 1.4 GHz (λ = 21.4 cm) for a smooth-soil surface characterized by an *rms* height of 0.88 cm (Newton and Rouse, 1980). Figure 4-66 shows the normalized antenna temperature (antenna temperature divided by the ground physical temperature) as a function of θ for four different soil moisture conditions. The linear variation of normalized antenna temperature with volumetric moisture content s, shown in Figure 4-67, corresponds to a sensitivity of about 3.7 K per 0.01 cm³/cm³ for a ground temperature of 300 K. This value of the radiometric sensitivity to soil moisture is for nadir ($\theta = 0°$) observations; at higher angles, the sensitivity is slightly higher for horizontal polarization and lower for vertical polarization. Also, the sensitivity is a function of soil texture, as discussed by Schmugge (1980).

Another parameter that affects the emission from bare soil is surface roughness. Figure 4-68 shows the angular variation of the normalized antenna temperature for three soil surfaces with the same soil moisture content but with different roughness conditions (Newton and Rouse, 1980). Roughness causes the emission to increase as explained in the previous section and the spacing between the horizontal and vertical polarization curves to decrease. These trends are consistent with the theoretical predictions of the model reported by Fung and Chen (1981).

Vegetation cover results in a reduction of the radiometric sensitivity to soil moisture. To illustrate the effects of vegetation cover on the emitted radiation, consider a simple two-layer case consisting of a semi-infinite soil medium with emissivity e_s, covered by an absorbing vegetation layer with a loss factor L_v. Assuming the vegetation layer to be nonscattering, ignoring reflections at the vegetation-air boundary, and assuming that the soil and vegetation have the same physical temperature, the radiative transfer equation leads to:

$$e_v = 1 - \frac{1 - e_s}{L_v^2} \qquad (4\text{-}57)$$

where e_v is the emissivity in the presence of vegetation. If the top layer is lossless (no vegetation present), $L_v = 1$ and $e_v = e_s$, as expected. If, on the other hand, the vegetation layer is very lossy (L_v large), e_v approaches the emissivity of the vegetation layer (assumed to be unity). For a given value of L_v, the sensitivity to soil moisture m is given by

$$\frac{d e_v}{d m} = \frac{1}{L_v^2} \frac{d e_s}{d m}. \qquad (4\text{-}58)$$

Thus, the vegetation reduces the radiometric sensitivity to soil moisture de_s/dm by the factor L_v^2.

Figure 4-69 shows measured values of the brightness temperature T_B at 1.4 and 5 GHz for 10-cm- and 30-cm-tall grass (Wang et al., 1980). For comparison, curves for bare soil also are shown. The presence of the grass cover results in an increase in the level of T_B, and the increase is the largest for the 30-cm grass at 5 GHz. This result is in agreement with the simple model given by

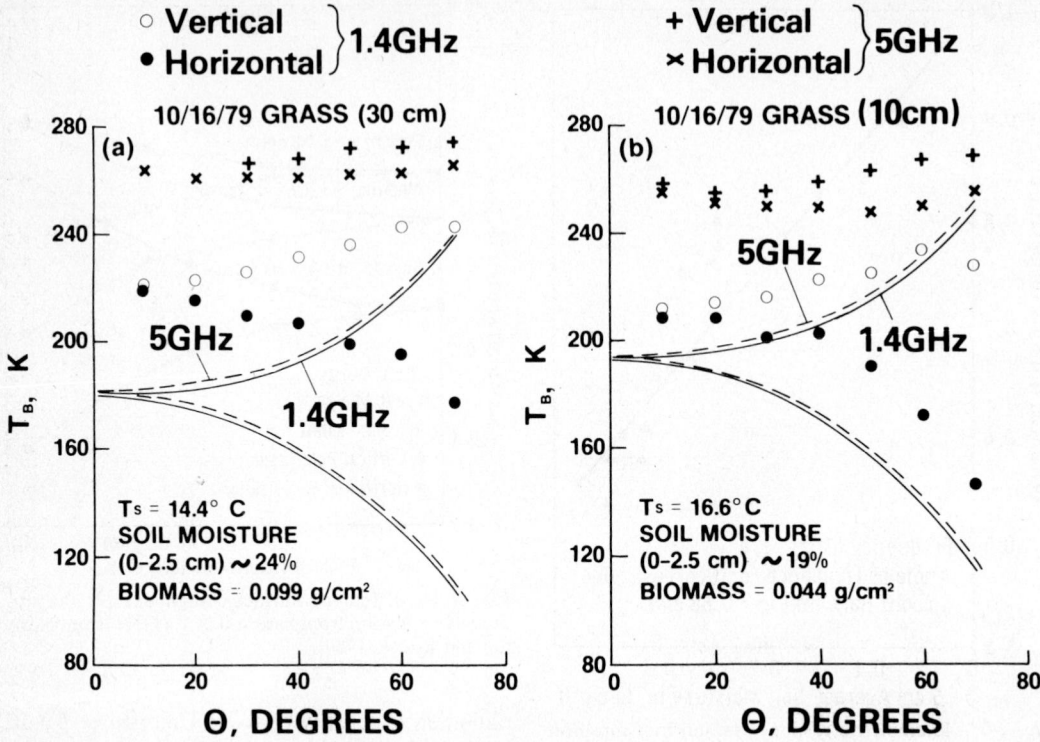

Fig. 4-69. Effect of vegetation cover on the brightness temperature T_B. Measured T_B versus θ for (a) 30-cm-tall grass and (b) 10-cm-tall grass. The smooth curves were calculated to show the contrast with the emission from bare soil. The soil surface temperature is T_S (from Wang et al., 1980).

Eq. 4-55 because L_v increases with both height and frequency,

$$L_v = e^{\alpha\, h\, \sec\, \theta} \qquad (4\text{-}59)$$

where h is the height of the vegetation layer, θ is the angle of incidence, and α is the power attenuation coefficient of the vegetation, which increases with increasing frequency.

Similar measurements made for corn and soybean fields show results similar to those of the 30-cm grass of Figure 4-69 (Wang et al., 1980).

SNOW

Emission from snow-covered ground is similar in some respects to emission from vegetation-covered ground; in both cases, the emission consists of contributions from the ground medium and from the layer above it (vegetation or snow), with the latter acting as attenuator of the ground contribution and as a self-emitter. Also, dry snow has a small loss factor, similar to that of dry vegetation, and wet snow is very lossy, as is healthy green vegetation.

The angular variation of the apparent radiometric temperature of a 27-cm-deep snow layer is shown in Figure 4-70 at 10.7 GHz and 37 GHz for dry- and wet-snow conditions (Stiles and Ulaby, 1980a). The apparent temperature T_{ap} is the same as the brightness temperature at the antenna, T_B, which includes the brightness temperature of the snowpack, T_{BS}, as well as the

scattered temperature T_{SC} (see Eq. 4-45); T_{ap} is used in lieu of T_B to avoid confusion between T_B and T_{BS}. The data of Figure 4-70 indicate that at 10.7 GHz, T_{ap} is weakly sensitive to the change in m_v, the volumetric liquid-water content, while at 37 GHz, a change of m_v (of the top 5-cm layer) from zero to 2 percent caused a drop of the order of 120 K.

The dependence on m_v is illustrated further in Figure 4-71a, which shows T_{ap} as a function of frequency for dry- and wet-snow conditions. The same data were used to compute the emissivity by subtracting T_{SC} and dividing by the snow's physical temperature. The results are shown in Figure 4-71b.

To determine the variation of T_{ap} with snow depth (or water equivalent), experiments were conducted by piling up snow in discrete layers. Figures 4-72 through 4-74 show T_{ap} as a function of water equivalent W (depth × density) for the dry snow at 10.7, 37, and 94 GHz (Ulaby and Stiles, 1980). Upon correcting the data for emissivity, one obtains the results shown in Figure 4-75. At each frequency and angle combination shown, the emission decreases with W until a level is reached representing the emission from a semi-infinite snow layer beyond which the emission remains approximately constant as a function of W. The emissivity of a semi-infinite snow layer is proportional to $(1 - w)$, where w is the scattering albedo of the snow volume. Due to its dependence on the

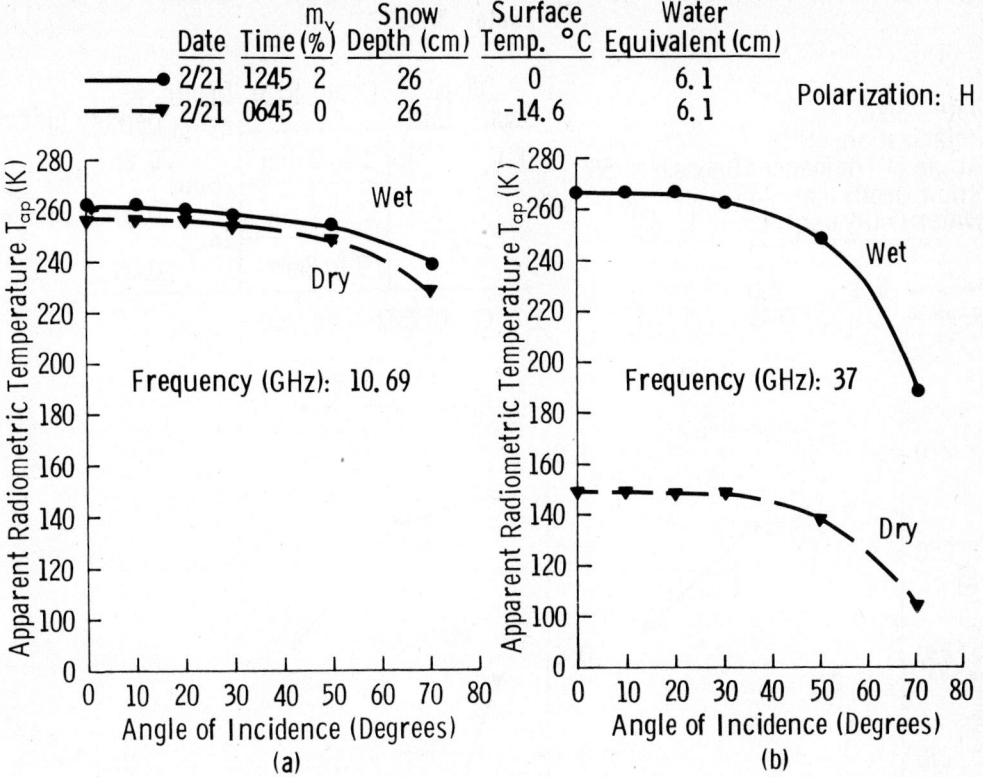

	Date	Time	m_v (%)	Snow Depth (cm)	Surface Temp. °C	Water Equivalent (cm)
●——●	2/21	1245	2	26	0	6.1
▼--→	2/21	0645	0	26	-14.6	6.1

Polarization: H

Fig. 4-70. Angular response of the measured apparent temperature for dry- and wet-snow conditions at (a) 10.7 GHz and (b) 37 GHz (from Stiles and Ulaby, 1980a).

size of the ice crystals (in wavelength units), the albedo increases with increasing frequency. Hence, the emissivity of the semi-infinite snow layer decreases with increasing frequency, as illustrated in Figure 4-75. As expected, the depth approximating semi-infinite conditions, which corresponds to total masking of the ground contribution, decreases with angle of incidence due to the corresponding increase in the slant depth of the snow layer, and decreases with frequency due to the fact that the attenuation of snow increases with increasing frequency.

The material described above is intended to provide an overview of the radiometric behavior of snow. Although the presentation concentrates on the dependence of snow emission on snow water equivalent and liquid-water content, other snow parameters may also be important, including ice-crystal size distribution, surface roughness, the presence of ice layers in the snowpack, and the wetness of the underlying ground surface. A review of results obtained prior to 1975 is available in Chapter 4 of the first edition of the *Manual of Remote Sensing*. Investigations conducted since 1975 are reported in articles by Shiue et al. (1978), Tiuri (1981), Rango et al. (1979), Hofer and Mätzler (1980), Stiles and Ulaby (1980a), Ulaby and Stiles (1980), and Fung and Chen (1981).

SEA ICE

As mentioned in previous sections, sea ice is a mixture of ice, salt, air bubbles, and liquid brine. These inhomogeneities are responsible for the darkening effect (lower brightness temperature) on emission. It is known that first-year sea ice is very lossy and contains inhomogeneities that are very small in physical size, while the multi-year ice is only slightly lossy and contains much larger inhomogeneities (mainly air bubbles). As a result, emission from the first-year sea ice is expected to be significantly higher than that of the multi-year. Measured brightness temperatures by Poe et al. (1974) are shown in Figure 4-76.

In an attempt to provide a finer distinction between ice types, Tooma et al. (1975) categorized sea ice into five types: (1) rough first-year ice, (2) smooth first-year ice, (3) second-year ice, (4) multi-year ice, and (5) new ice or open water. Dual-frequency data were gathered (19.34 and 31 GHz) and plotted as average brightness temperature versus temperature difference. Figure 4-77 shows the locations of the first four types. The new ice or open-water category is at 130–150 K, a substantially lower range of average brightness temperature. The relative level between the ice types is consistent with the theoretical predictions

Date: 3/24/77

Date: 3/24/77
Polarization: HH
Angle of Incidence (Degrees): 50
Snow Depth (cm): 45
Water Equivalent (cm): 12.6

m_v (%)	Time
━━●━━ 3.3	1200
━━▼━━ 0	0645

	Temp.		Grain Size	Depth	
	Max.	Min.		45 cm	Density (g/cm^3)
	0° C	-7° C	1 to 2 mm		0.26
				26 cm	
			2 to 4 mm		0.31
				14 cm	
			3 to 8mm		0.21
	0.2° C	0° C Thawed Soil			

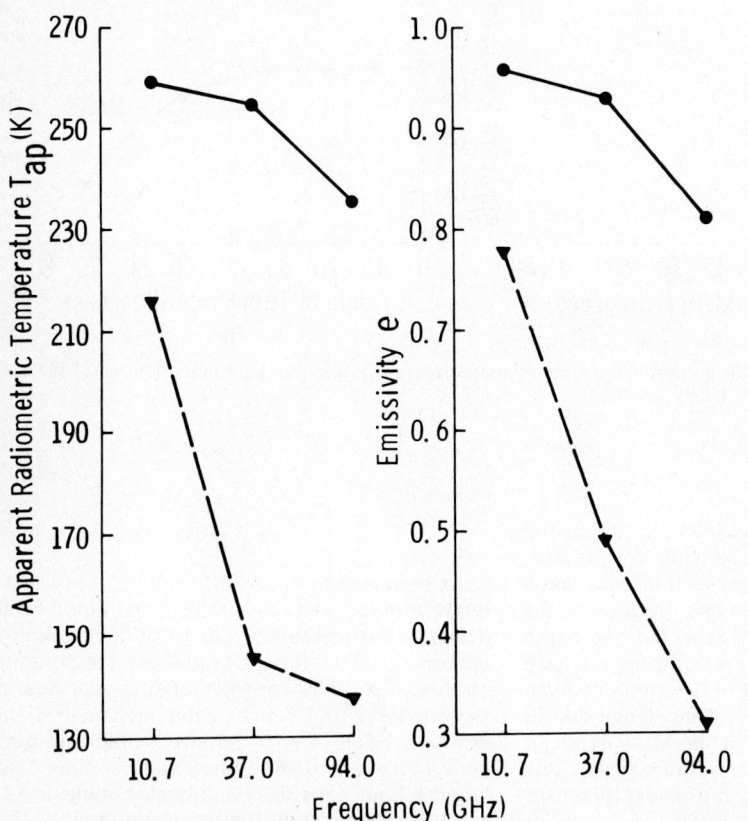

Fig. 4-71. Spectral response of (a) T_{ap} and (b) emissivity at 50° angle of incidence for wet and dry snow (from Stiles and Ulaby, 1980a).

of temperature changes due to surface roughness and volume inhomogeneities reported by Fung and Chen (1981).

A recent sea-ice science report (1980) from NASA/Goddard has categorized sea ice into four types (1) new ice without snow cover (10 cm), (2) young ice without snow cover (10–30 cm), (3) first-year ice (30–200 cm), and (4) multi-year ice (>200 cm). It is shown (Figure 4-78) that, in general, three types can be isolated. At about 30 GHz, all four types have different emissivities.

MICROWAVE RADIOMETRIC CHARACTERISTICS OF OCEANS

The emission of the ocean's surface depends primarily upon its surface roughness, permittivity, and foam cover. The complex permittivity of sea water is itself a function of sea-water temperature, salinity, and incident frequency, while surface roughness and foam cover are dominated by the wind condition. The generation and geometry of sea waves and foam under high wind conditions are very complex and difficult to analyze. However, it has been observed that emission from the sea surface increases with the wind speed (Hollinger, 1971; Wagner and Lynch, 1972) and foam

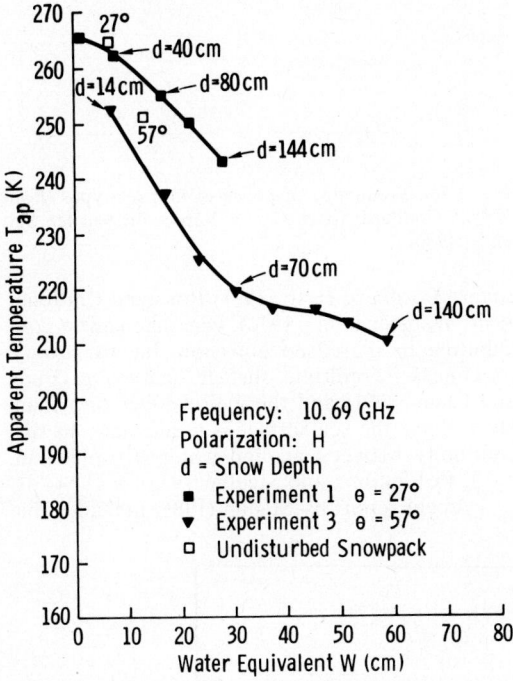

Fig. 4-72. Radiometric apparent temperature response to snow water equivalent at 10.7 GHz (Ulaby and Stiles, 1980).

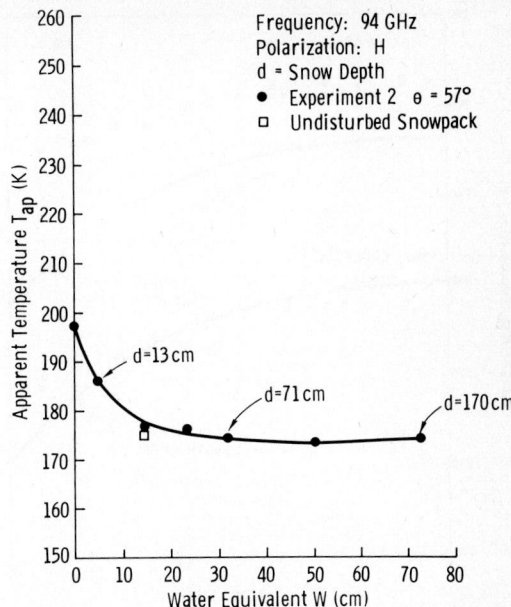

Fig. 4-74. Radiometric apparent temperature response to snow water equivalent at 94 GHz (Ulaby and Stiles, 1980).

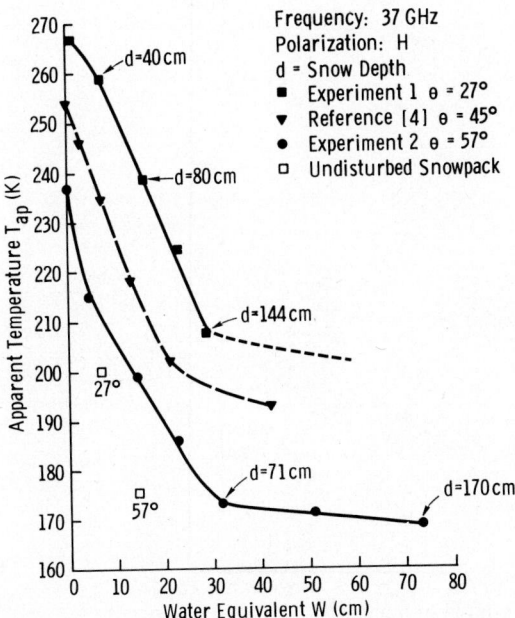

Fig. 4-73. Radiometric apparent temperature response to snow water equivalent at 37 GHz (Ulaby and Stiles, 1980).

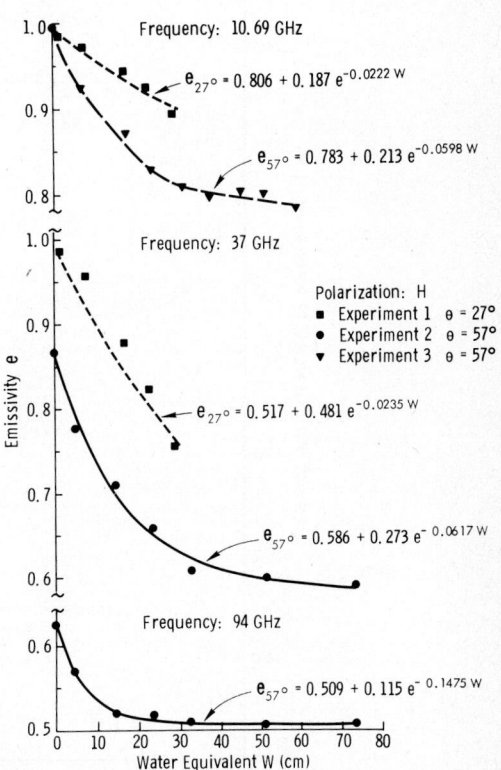

Fig. 4-75. Measured radiometric emissivity response to dry snow water equivalent at 10.7 GHz, 37 GHz, and 94 GHz. Average snow density 42 g cm^{-3} (from Ulaby and Stiles, 1980).

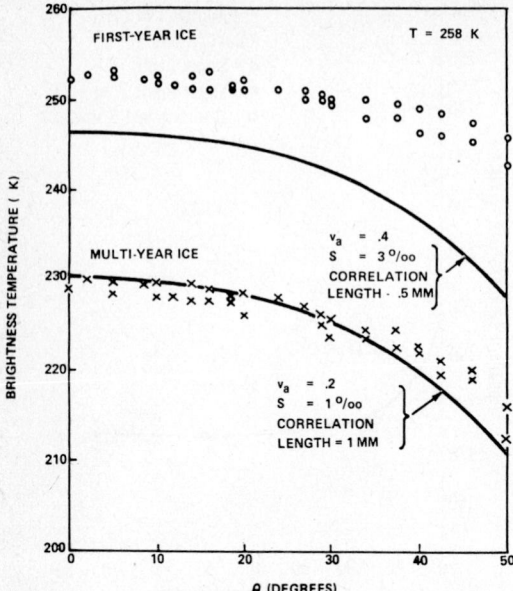

Fig. 4-76. Brightness temperature at 19.4 GHz as a function of angle from nadir (March 1971 data; Poe et al., 1974).

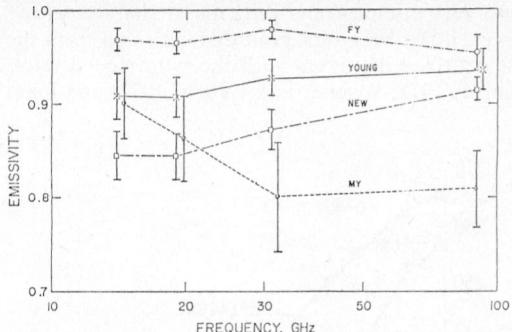

Fig. 4-78. Frequency response of four ice-types (from NASA Goddard Space Center Report on Sea Ice Science, 1980).

cover (Nordberg et al., 1971; Ross and Cardone, 1974; Webster et al., 1976). One mechanism contributing to increased emission due to surface roughness is multiple surface scattering (Fung and Chen, 1981) and the major effect of a foam layer above the sea surface is to alleviate the discontinuity between air and water (Droppleman, 1970; Rosenkranz and Staelin, 1972).

A recent report by Staelin (1981) indicates that

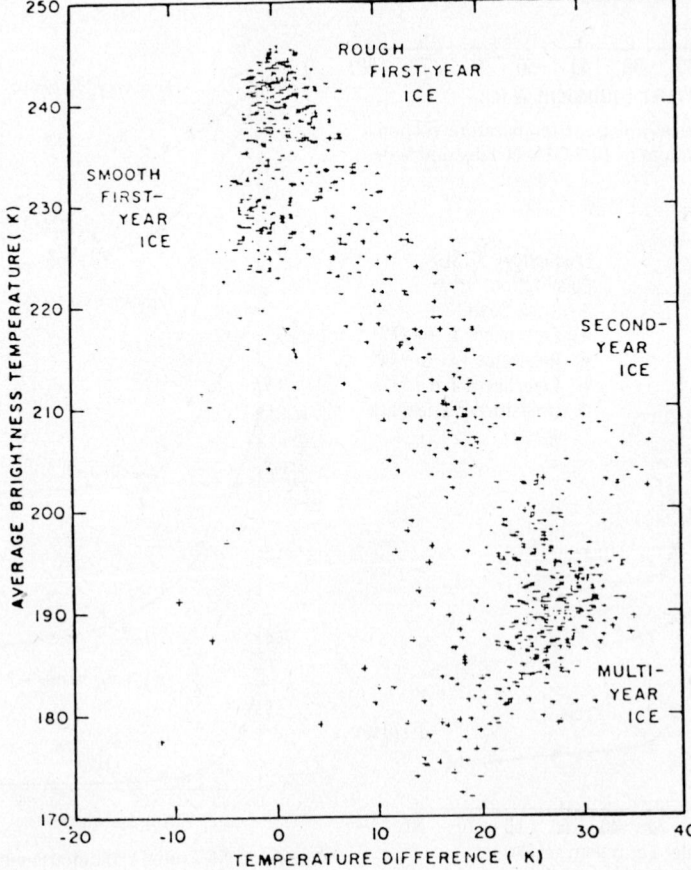

Fig. 4-77. Plot of average brightness temperature versus temperature difference for 22-km section (from NASA/ GSFC Report, Sea Ice, 1980).

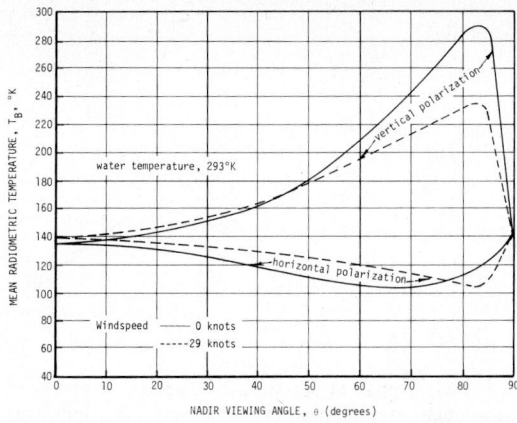

Fig. 4-79. Theoretical antenna temperatures for sea state as functions of nadir viewing angle and wind velocity.

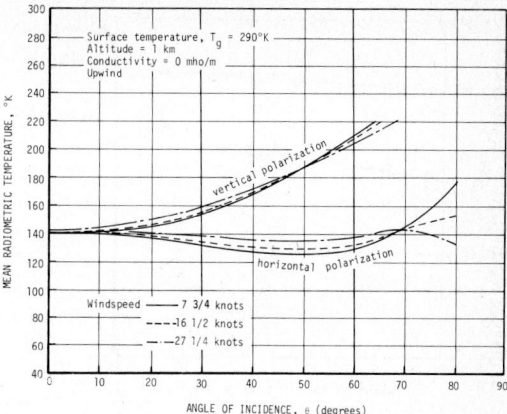

Fig. 4-80. Radiometric temperature invariant points for sea state as functions of angle of incidence and wind velocity.

passive microwave techniques might be useful for the sensing of (1) sea surface, (2) sea-surface temperature, and (3) surface salinity. Hollinger and Mennella (1973) have indicated that the thickness of an oil spill having a depth of more than 0.05 mm on an ocean surface can be determined by multifrequency radiometric observations. In what follows, the principles involved in the recovery of the above-stated parameters are discussed in terms of the brightness temperature characteristics of the ocean surface. It should be kept in mind that observations made at satellite altitudes may be contaminated by atmospheric effects (Wilheit, 1978), which are discussed in Chapter 5.

EFFECT OF SURFACE ROUGHNESS

A common approach to investigating changes in the microwave radiometric response to changing sea states is to plot the antenna temperature against observation angle, as shown in Figure 4-79. This method allows investigators to examine the effects of polarization and changing frequencies on signal amplitudes for various sea states. This simple diagram (Wagner and Lynch, 1972) shows several interesting points that are of value in describing sea states by using remote methods.

The most obvious change is the decrease in the spacing between the vertically and horizontally polarized emissions as the sea becomes rougher (wind of ~40 knots), a result of the fact that the rougher sea surface becomes more diffuse. For a totally diffuse surface, it appears that there should be little sensitivity to polarization and viewing angle. In addition, there is an increase in radiometric temperature for both polarizations at near-nadir angles under rougher sea conditions (or higher wind speed).

One of the most valuable features derived from an angular curve is the invariant point (shown on Figure 4-80 at $\theta = 53°$ and about 188 K). The microwave temperature, which is proportional to

surface temperature, remains constant regardless of sea roughness, which is a function of wind speed. Wagner and Lynch (1972) found that the invariant point was present for all wind speeds to 40 knots, using an expansion of Stogryn's theory, which included multiple scattering and wave-shadowing effects. The invariant point has value in the sense that a simple vertically polarized microwave radiometer viewing the ocean surface will only be subject to change if the surface temperature and atmospheric absorption change also, regardless of sea state. However, should the choice of 53° lead to unacceptable resolution size, nadir viewing may be used along with an empirical wind-speed correction (Blume et al., 1977). Such a scheme permits surface-temperature measurements to have a 1-K accuracy.

EFFECT OF OCEAN FOAM

The effect of ocean foam on microwave radiometric measurements has received considerable attention (Nordberg et al., 1969; Webster et al., 1976; Ross and Cardone, 1974; Droppleman, 1970; Stogryn, 1972; Rosenkranz and Staelin, 1972), and is considered to be an important parameter because of the extensive white-cap generation that accompanies sea chop. Several experiments (Nordberg et al., 1969; Williams, 1971) have shown that foam does indeed cause an increase in microwave surface emissivity. This is probably due to an impedance-matching effect. Ross and Cardone (1974) have shown that the percent of foam cover over ocean is related to wind speed. As a result, it is possible to use radiometric measurements to infer wind speed via the effect of foam cover (Webster et al., 1976; Rosenkranz et al., 1978). A brightness temperature versus wind speed curve at a wavelength of 0.8 cm is shown in Figure 4-81. For nadir observations, Webster et al. (1976) have also generated the slope of the

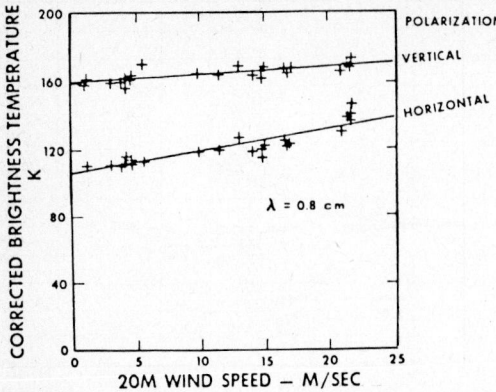

Fig. 4-81. Example of the variation of microwave brightness temperature versus surface wind speed for a wavelength of 0.8 cm, horizontal polarization, at a viewing angle of 38° from nadir (from Webster et al., 1976).

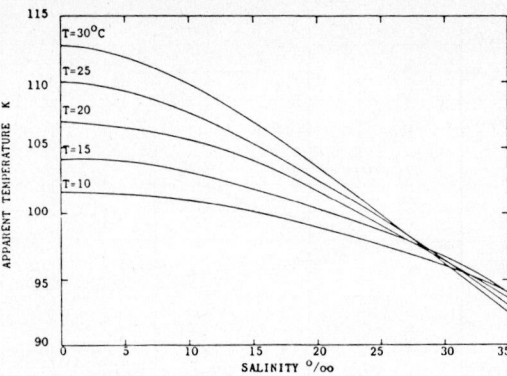

Fig. 4-83. Apparent sea-surface temperature at 21-cm wavelength as a function of salinity and water temperature: 9.3° incidence angle, vertical polarization, plane sea surface (from Thomann, 1976).

brightness temperature versus the wind speed curve as a function of frequency. This is shown in Figure 4-82.

EFFECT OF SALINITY

As shown in Figure 4-83, the radiometric temperature at L-band is sensitive to change in salinity. The sensitivity is significant for salinities above 10 $^0/_{00}$ and water temperatures above 20°C. By taking the sea-water temperature measurements with an infrared radiometer along with a radiometric temperature measurement at 21-cm wavelength, Thomann (1976) has determined the salinity of sea water to an accuracy of about 2 $^0/_{00}$. Since Thomann used a foam-free, plane sea-surface model, he indicated that errors due to foam and surface roughness should be corrected. Other sources of error are galactic radiation and solar

radiation reflected from the sea surface, and instrument offsets.

EFFECT OF OIL SLICK

As in the case of foam, the presence of an oil slick over the sea surface acts as a matching layer between free space and the sea, thus enhancing the brightness temperature (Hollinger and Mennella, 1973). As the thickness of the oil film increases, the apparent temperature at first increases and then passes through alternating maxima and minima due to the standing-wave pattern set up by the sea surface. The maxima and minima occur at successive integral multiples one quarter of the observational wavelength in the oil (Figure 4-84). A clear illustration of this is shown in Figure 4-84, which shows theory and measured data collected under laboratory conditions (Hollinger, 1974). Through the use of two or more fre-

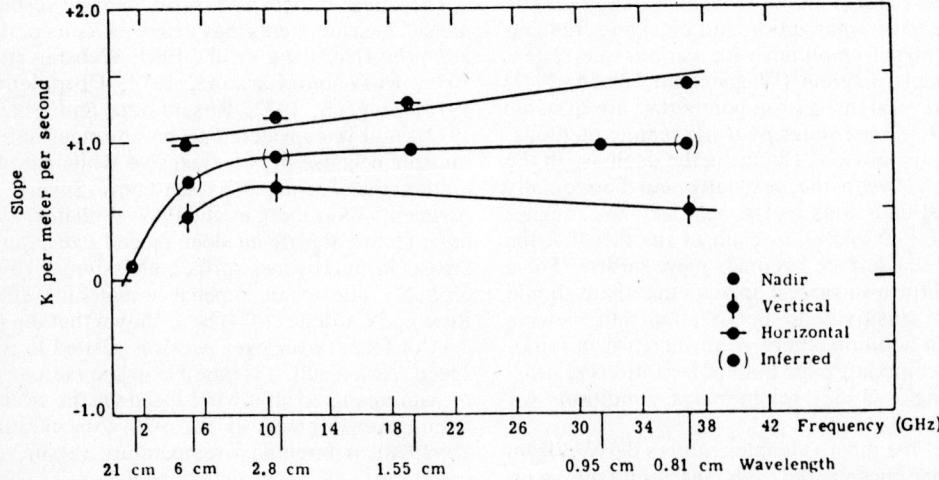

Fig. 4-82. Relationship between brightness temperature and wind speed versus electromagnetic frequency (from Webster et al., 1976).

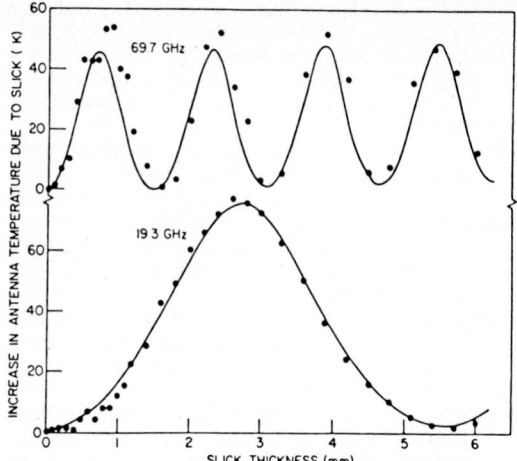

Fig. 4-84. Increased brightness temperature versus oil-slick thickness at 69.75 and 19.34 GHz (after Hollinger, 1974).

quencies, thickness ambiguities introduced by the oscillations can be removed and the film thickness can be determined for a wide range of thicknesses.

REFERENCES

Beckmann, P., and A. Spizzichino, 1963, The Scattering of Electromagnetic Waves from Rough Surfaces; The MacMillan Co., New York.

Blume, H-J. C., A. W. Love, M. J. Van Melle, and W. W. Ho, 1977, Radiometric observations of sea temperature at 2.65 GHz over the Chesapeake Bay; IEEE Trans. Antennas Propag., vol. AP-25, pp. 121–128.

Brekhovskikh, L. M., 1960, Waves in Layered Media; Academic Press, New York, p. 228.

Bush, T., F. Ulaby, T. Metzler, and H. Stiles, 1976, Seasonal variations of the microwave scattering properties of the deciduous trees as measured in the 1-18 GHz spectral range; Remote Sensing Laboratory Technical Report 177-60, University of Kansas, Lawrence, Kansas.

Campbell, M. J., and J. Ulrichs, 1969, Electrical properties of rocks and their significance for lunar radar observations; J. Geophys. Res., vol. 74, pp. 5867–5881.

Carlson, N. L., 1967, Dielectric constant of vegetation at 8.5 GHz; Ohio State University ElectroScience Laboratory Technical Report 1903-5, Columbus, Ohio.

Carswell, A. I., 1965, Microwave scattering measurements in the Rayleigh region using a focused-beam system; Canad. J. Phys., vol. 43, pp. 962–977.

Chan, H. L., and A. K. Fung, 1977, A theory of sea scatter at large incident angles; J. Geophys. Res., vol. 82, no. 24, pp. 3439–3444.

Colbeck, S. A., 1979 (December), Grain clusters in wet snow; J. Colloid & Interface Sci., vol. 72, no. 3.

Cosgriff, R. L., W. H. Peake, and R. C. Taylor, 1960, Terrain scattering properties for sensor system design: Terrain Handbook 2; Eng. Exp. Sta. Bull. 181, Ohio State University Antenna Laboratory, Columbus, Ohio.

Cumming, W., 1952, The dielectric properties of ice and snow at 3.2 cm; J. Appl. Phys., vol. 23, pp. 768–773.

Daley, J. C., 1973, Wind dependence of radar sea return; J. Geophys. Res., vol. 78, no. 33, pp. 7823–7833.

Debye, P., 1929, Polar Molecules, Dover Press, New York.

de Loor, G. P., 1968, Dielectric properties of heterogeneous mixtures containing water; J. Microwave Power, vol. 3, pp. 67–73.

Dellwig, L. F., 1969, An evaluation of multifrequency radar imagery of the Pisgah Crater area, California; Modern Geol., vol. 1, pp. 65–73.

Droppleman, J. D., 1970, Apparent microwave emissivity of sea foam; J. Geophys. Res., vol. 75.

Fung, A. K., 1967, Theory of cross-polarized power returned from a random surface; Appl. Sci. Res., vol. 18, pp. 50–60.

Fung, A. K., 1968, Mechanisms of polarized and depolarized scattering from a rough dielectric surface; J. Franklin Institute, vol. 285, no. 2, pp. 125–133.

Fung, A. K., 1969, Scattering and depolarization of electromagnetic waves by a horizontally weakly inhomogeneous medium; Appl. Sci. Res., vol. 20, pp. 368–380.

Fung, A. K., and M. F. Chen, 1981, Emission from an inhomogeneous layer with irregular interfaces; Radio Sci., vol. 16, no. 3, pp. 289–298.

Fung, A. K., and H. J. Eom, 1979, Multiple scattering and depolarization by a randomly rough Kirchhoff surface; Remote Sensing Laboratory Technical Report 369-4, University of Kansas, Lawrence, Ks. Also appears in IEEE Trans. Propag., vol. 29, no. 3, pp. 463–471, May 1981.

Fung, A. K., and H. J. Eom, 1981(a) Note on the Kirchhoff rough surface solution in backscattering; Radio Sci., vol. 16, no. 3, pp. 299–302.

Fung, A. K., and H. J. Eom, 1981(b), A theory of wave scattering from an inhomogenous layer with an irregular interface; IEEE Trans. Antennas Propag., vol. 29, no. 6, pp. 899–910.

Fung, A. K., and H. S. Fung, 1977, Application of first-order renormalization method to scattering from a vegetation-like half-space; IEEE Trans. Geosci. Electr., vol. 15, no. 4, pp. 189–195.

Fung, A. K., and F. T. Ulaby, 1978, A scatter model for leafy vegetation; IEEE Trans. Geosci. Electr., vol. 16, no. 4, pp. 281–286.

Glen, J. W., and P. G. Paren, 1975, The electrical properties of snow and ice; J. Glaciol., vol. 15, pp. 15–38.

Grant, C. R., and B. S. Yaplee, 1957, Backscattering from water and land at centimeter and millimeter wavelengths; Proc. IRE, vol. 45, pp. 975–982.

Guinard, N. W., and J. C. Daley, 1970, An experimental study of a sea-clutter model; Proc. IRE, vol. 58, no. 4, pp. 543–550.

Hoekstra, P., and P. Cappillino, 1971, Dielectric properties of sea and sodium chloride ice at UHF and microwave frequencies; J. Geophys. Res., vol. 76, pp. 4922–4931.

Hofer, R., and C. Mätzler, 1980 (January), Investigations on snow parameters by radiometry in the 3- to 60-mm wavelength region; J. Geophys. Res., vol. 85, no. C1, pp. 453–460.

Hollinger, J. P., 1971, Passive microwave measurements of sea surface roughness; IEEE Trans. Geosci. Electr., vol. 9, no. 3.

Hollinger, J. P., and R. A. Mennella, 1973, Oil spills:

measurements of their distributions and volumes by multifrequency microwave radiometry; Science, vol. 181, pp. 54–56.

Jones, W. L., L. C. Schroeder, and J. L. Mitchell, 1977, Aircraft measurements of the microwave scattering signature of the ocean; IEEE Trans. Antennas Propag., vol. 25, no. 1, pp. 52–61.

Ishimaru, A., 1978, Wave Propagation and Scattering in Random Media; Vol. 2, Academic Press, New York.

Kalmykov, A. I., and V. V. Pustovoytenko, 1976, On polarization features of radio signals scattered from the sea surface at small grazing angles; J. Geophys. Res., vol. 81, pp. 1960–1964.

Katzin, M., 1957, On the mechanisms of radar sea clutter; Poc. IEEE, vol. 45, pp. 44–54.

Kerr, D. E., 1951, Propagation of Short Radio Waves; McGraw-Hill Book Co., pp. 457, 608.

Krishen K., N. Ulahos, O. Brandt, and G. Graybeal, 1971, Results of scatterometer systems analysis for NASA/MSC earth observations sensor evaluation; Proc. 7th SRSE.

Lane, J., and J. Saxton, 1952, Dielectric dispersion in pure polar liquids at very high radio frequencies. III. The effect of electrolytes in solution; Proc. Roy. Soc., vol. 214 A, pp. 531–545.

Leader, J. C., 1975, Polarization dependence in EM scattering from Rayleigh scatterers embedded in a dielectric slab; J. Appl. Phys., vol. 46, no. 10, pp. 4371–4391.

Linlor, W. I., 1980, Permittivity and attenuation of wet snow between 4 and 12 GHz; J. Appl. Phys., vol. 51.

Long, M. W., 1974, On a two-scatter theory of sea echoes; IEEE Trans. Antennas Propag., vol. 22, no. 5, pp. 667–672.

Lundien, J. R., 1971, Terrain analysis by electromagnetic means; Technical Report 3-693, U.S. Army Engineer Waterways Exp. Sta., Vicksburg, Miss.

MacDonald, H. C., and W. P. Waite, 1971, Soil moisture detection with imaging radars; Water Resources Research, vol. 7, no. 1, pp. 100–110.

Meier, M. F., and A. T. Edgerton, 1971 (May), Microwave emission from snow—a progress report; Proc. 7th Intl. Symp. Rem. Sens. Env., Univ. Michigan, Ann Arbor, Mich.

Meyer, W., and W. M. Schilz, 1981, Feasibility study of density-independent moisture measurement with microwaves; IEEE Trans. Micr. Theory and Tech., vol. MIT-29, no. 7.

Moore, R. K., and L. F. Dellwig, 1966 (September), Terrain discrimination by radar image polarization comparison; Proc. IEEE, vol. 54, no. 9, pp. 1213–1214.

Moore, R. K., and A. K. Fung, 1979, Radar determination of winds at sea; Proc. IEEE, vol. 67, no. 11, pp. 1504–1521.

Moore, R. K., A. K. Fung, G. J. Dome, and I. J. Birrer, 1978, Estimate of oceanic surface wind speed and direction using orthogonal beam scatterometer measurements and comparison of recent sea theories; NASA Contractor Report 158908, Langley Research Center, Hampton, Va.

Moore, R. K., and W. J. Pierson, 1971, Worldwide oceanic wind and wave predictions using a satellite radar-radiometer; J. Hydronautics, vol. 5, no. 2, pp. 52–62.

Morain, S. A., and D. S. Simonett, 1967, K-band radar in vegetation mapping; Photogram. Eng., vol. 33, pp. 730–740.

NASA Goddard Space Flight Center Report, 1980, Sea ice science, NASA/GSFC Report, Greenbelt, Md.

Newton, R. W., and J. W. Rouse, Jr., 1980, Microwave radiometer measurements of soil moisture content; IEEE Trans. Antennas Propag., vol. AP-28, pp. 680–686.

Nordberg, E., J. Conaway, and P. Thaddeus, 1969, Microwave observations of sea state from aircraft; Quart. J. R. Meteorol. Soc., vol. 95, pp. 408–413.

Nordberg, W., J. Conaway, D. B. Ross, and T. Wilheit, 1971, Measurements of microwave emission from a foam-covered wind-driven sea, J. Atmos. Sci., vol. 28, pp. 429–435.

Onstott, R. G., R. K. Moore, and W. F. Weeks, 1979 (July), Surface-based scatterometer results of Arctic sea ice; IEEE Trans. Geosci. Electr., vol. GE-17, no. 3, pp. 78–85.

Parashar, S. K., A. K. Fung, and R. K. Moore, 1978, A theory of wave scatter from an inhomogeneous medium with a slightly rough boundary and its application to sea ice; Rem. Sens. Env., vol. 7, pp. 37–50.

Paris, J. F., 1969 (January), Microwave radiometry and its application to marine meteorology and oceanography; Texas A&M Univ., Dept. of Oceanography.

Peake, W. H., 1959 (December), Interaction of electromagnetic waves with some natural surfaces; IEEE Trans. Antennas Propag., vol. 9, pp. 324–329.

Peake, W. H., R. L. Riegler, and C. H. Schulz, 1966 (April), The mutual interpretation of active and passive microwave sensor outputs, Proc. 4th Intl. Symp. Rem. Sens. Env., Univ. Michigan, Ann Arbor.

Poe, G. et al., 1974, Study of microwave emission properties of sea ice; Final Report 1804FR-1, Aerojet Electrosystems Co., Azusa, Ca.

Rango, A., A. T. C. Chang, and J. L. Foster, 1979 (February), The utilization of spaceborne microwave radiometers for monitoring snowpack properties; J. Nordic Hydrol., vol. 10, pp. 25–40.

Robertson, S. D., 1947, Targets for microwave radar navigation; Bell Syst. Tech. J., vol. 26, pp. 852–869.

Rosenkranz, P. W., and D. H. Staelin, 1972 (November), Microwave emissivity of ocean foam and its effects on nadiral radiometric measurements; J. Geophys. Res., vol. 77, no. 33.

Rosenkranz, P. W., D. H. Staelin, and N. C. Goody, 1978, Typhoon June (1975) viewed by a scanning microwave spectrometer; J. Geophys. Res., vol. 83, pp. 1857–1868.

Ross, D. B., and V. J. Cardone, 1974, Observations of oceanic whitecaps and their relation to remote measurements of surface wind speed; J. Geophys. Res., vol. 79, pp. 444–452.

Ross, R. A., 1966, Radar cross section of rectangular flat plates as a function of aspect angle; Trans. IEEE Antennas Propag., vol. AP-14, no. 3, pp. 329–335.

Schmugge, T. J., 1980, Effect of texture on microwave emission from soils; IEEE Trans. Geosci. Electr., vol. GE-18, pp. 353–361.

Shiue, J. C., A. T. C. Chang, H. Boyne, and D. Ellerbruch, 1978, Remote sensing of snowpack with microwave radiometers for hydrologic applications; Proc. 12th Intl. Symp. Rem. Sens. Env., ERIM, Ann Arbor, Mich.

Skolnik, M. I., 1970, Radar Handbook; McGraw-Hill Book Co., pp. 27–36.

Staelin, D. H., 1981 (July), Passive microwave techniques for geophysical sensing of the earth from satellites; IEEE Antennas Propag., vol. 29, no. 4, pp. 683–687.

Stiles, W. H., and F. T. Ulaby, 1980(a) (February), The active and passive microwave response to snow parameters, Part I: wetness; J. Geophys. Res., vol. 85, no. C2, pp. 1037–1044.

Stiles, W. H., and F. T. Ulaby, 1980(b), Microwave remote sensing of snowpacks; NASA Contractor Report 3263, Goddard Space Flight Center, Greenbelt, Md.

Stiles, W. H., F. T. Ulaby, A. K. Fung, and A. Aslam, 1981, Radar spectral observations of snow; 1981 International Geoscience and Remote Sensing Symposium (IGARSS '81) Digest, Vol. I, pp. 654–668, June 8–10, Washington, D.C.

Stogryn, A., 1967 (March), The apparent temperature of the sea at microwave frequencies; Trans. IEEE Antennas Propag., vol. 15.

Stogryn, A., 1972 (March), The emissivity of sea foam at microwave frequencies; J. Geophys. Res., vol. 77, no. 9.

Sweeney, B. D., and S. C. Colbeck, 1974, Measurements of the dielectric properties of wet snow using a microwave technique; Cold Regions Research and Engineering Laboratory Report, Hanover, N.H.

Tan, H. S., and A. K. Fung, 1979, A first-order theory on wave depolarization by a geometrically anisotropic random medium; Radio Sci., vol. 14, no. 3, pp. 377–386.

Thomann, G. C., 1976, Experimental results of the remote sensing of sea-surface salinity at 21 cm wavelength, IEEE Trans. Geosci. Electr., vol. 14, pp. 198–214.

Tinga, W. R., W. A. G. Voss, and D. F. Blossey, 1973, Generalized approach to multiphase dielectric mixture theory; J. Appl. Phys., vol. 44, pp. 3897–3902.

Tiuri, Martti, 1981, Microwave emission signatures of snow in Finland; 1981 International Geoscience and Remote Sensing Symposium (IGARSS '81) Digest, Vol. I, pp. 670–675, June 8–10, Washington, D.C.

Tiuri, M., K. Jokela, and S. Heikkila, 1980, Microwave instrument for accurate moisture and density measurement of timber; J. Microwave Power, vol. 15, no. 4, pp. 251–254.

Tobarias, J., P. Saguet, and J. Chilo, 1978, Determination of the water content of snow from the study of electromagnetic wave propagation in the snow cover; J. Glaciol., vol. 20, pp. 585–592.

Tooma, S. G., R. A. Mennella, J. P. Hollinger, and R. D. Ketchum, 1975, Comparison of sea ice type identification between airborne dual frequency passive microwave radiometry and standard laser/infrared techniques; J. Glaciol., vol. 15, no. 73, pp. 225–239.

Tsang, L., and J-A. Kong, 1978, Radiative transfer theory for active remote sensing of half-space random media; Radio Sci., vol. 13, no. 5, pp. 763–773.

Ufintsev, P. Ya., 1962, Diffraction of plane electromagnetic waves by a thin cylindrical conductor; Radioteknica Elektronika, vol. 7, p. 260.

Ulaby, F. T., 1980 (June), Microwave response of vegetation; 23rd Ann. Conf. of Committee on Space Res. (COSPAR), Adv. Space Res., vol. I, pp. 55–70.

Ulaby, F. T., and J. E. Bare, 1979 (November), Look direction modulation function of the radar backscattering coefficient of agricultural fields; Photogram. Eng. Rem. Sens., vol. 45, pp. 1495–1506.

Ulaby, F. T., P. P. Batlivala, and M. C. Dobson, 1978, Microwave backscatter dependence on surface roughness, soil moisture, and soil texture: Part I—bare soil; IEEE Trans. Geosci. Electr., vol. 16, no. 4, pp. 286–295.

Ulaby, F. T., F. Kouyate, and A. K. Fung, 1981a, A backscatter model for a randomly perturbed periodic surface; 1981 International Geoscience and Remote Sensing Symposium (IGARSS '81) Digest, Vol. II, pp. 1280–1293, June 8–10, Washington, D.C.

Ulaby, F. T., R. K. Moore, and A. K. Fung, 1981b, (October), Microwave Remote Sensing, Vol. I—Microwave Remote Sensing: Fundamentals and Radiometry; Advanced Book Program, Addison-Wesley Pub. Co., Reading, Mass.

Ulaby, F. T., R. K. Moore, and A. K. Fung, 1982, Microwave Remote Sensing, Vol. II—Radar Remote Sensing and Surface Scattering and Emission Theory; Advanced Book Program, Addison-Wesley, Reading, Mass.

Ulaby, F. T., R. K. Moore, and A. K. Fung, 1983, Microwave Remote Sensing, Vol. III—Volume Scattering and Emission Theory, Advanced Systems, and Applications; Advanced Book Program, Addison-Wesley Pub. Co., Reading, Mass.

Ulaby, F. T., and W. H. Stiles, 1980 (February), The active and passive microwave response to snow parameters, Part II: water equivalent of dry snow; J. Geophys. Res., vol. 85, no. C2, pp. 1045–1049.

Valenzuela, G., 1967, Depolarization of EM waves by slightly rough surfaces; IEEE Trans. Antennas Propag., vol. 15, no. 4, pp. 552–557.

Vant, M. R., R. B. Gray, R. O. Ramseier, and V. Makios, 1974, Dielectric properties of fresh and sea ice at 10 and 35 GHz; J. Appl. Phys., vol. 45, pp. 4712–4717.

Vant, M. R., R. O. Ramseier, and V. Makios, 1978, The complex-dielectric constant of sea ice at frequencies in the range 0.1–40 GHz; J. Appl. Phys., vol. 49, pp. 1264–1280.

Wagner, R. J., and P. J. Lynch, 1972 (March), Analytical studies and emission by the sea surface; Contract N000 14-71-C-0240, NR 387-051/1-14-17, 414, ONR.

Waite, W. P., and H. C. MacDonald, 1970, Snowfield mapping with K-band radar; J. Rem. Sens. Environ., vol. 1, pp. 143–150.

Waite, W. P., and H. C. MacDonald, 1971 (July), Vegetation penetration with K-band imaging radar; Trans. IEEE Geosci. Electr., vol. 9, no. 3, pp. 147–155.

Wang, J. R., 1980, The dielectric properties of soil-water mixtures at microwave frequencies; Radio Sci., vol. 15, pp. 977–985.

Wang, J. R., and T. J. Schmugge, 1980, An empirical model for the complex dielectric permittivity of soils as a function of water content; IEEE Trans. Geosci. Rem. Sens., vol. 18, pp. 288–295.

Wang, J. R., J. C. Shiue, and J. E. McMurtrey III, 1980 (October), Microwave remote sensing of soil moisture content over bare and vegetated fields; Geophys. Res. Let., vol. 7, no. 10, pp. 801–804.

Webster, W. J., T. T. Wilheit, D. B. Ross, and P. Gloersen, 1976 (June), Spectral characteristics of the microwave emission from a wind-driven, foam-covered sea; J. Geophys. Res., vol. 81, no. 18, pp. 3095–3099.

Wetzel, L., 1977, A model for sea backscatter intermittency at extreme grazing angles; Radio Sci., vol. 12, no. 5, pp. 749–756.

Wilheit, T. T., 1978, A review of applications of microwave radiometry to oceanography, Boundary-Layer Meteorol., vol. 13, pp. 227–293.

Williams, G. F., 1971, Microwave emissivity measurements of bubbles and foam; IEEE Trans. Geosci. Electr., vol. 9, no. 4, pp. 221–224.

Wright, J. W., 1968, A new model for sea clutter; Trans. IEEE Antennas Propag., vol. 16, no. 2, pp. 217–223.

Wu, S. T., and A. K. Fung, 1972, A noncoherent model for microwave emissions and backscattering from the sea surface; J. Geophys. Res., vol. 77, no. 30, pp. 5917–5929.

Wu, S. T., and A. K. Fung, 1973 (November), A theory of microwave apparent temperature over the ocean, NASA Contractor Report CR-2329, Remote Sensing Laboratory Technical Report 186-7, University of Kansas, Lawrence, Ks.

CHAPTER 5

Interaction Mechanisms Within The Atmosphere

Author-Editor: MOUSTAFA T. CHAHINE

Contributing Authors: DANIEL J. McCLEESE, PHILIP W. ROSENKRANZ and DAVID H. STAELIN

GENERAL CONTENTS: Atmospheric characteristics; characteristics of solar radiant energy; radiant interaction with atmospheres; particulates; optical thickness; downwelling radiation at the ground; upwelling radiation at the top of the atmosphere; turbulence; atmospheric interactions between EM radiation and atmospheric gases; atmospheric interactions between EM radiation, haze, clouds, and precipitation particles; propagation of EM radiation in the earth's atmosphere. Fundamental techniques for remote sensing of the atmosphere and earth.

NOMENCLATURE

To conserve and eliminate repetition in text and references, the following symbols, units and names have been used in this chapter.

Symbol	Definition	Units
B	Planck function	$W\ cm^{-1}(cm^{-1})^{-1}sr^{-1}$
B	turbidity coefficient	—
c	speed of light in vacuum, 3×10^{10} cm sec^{-1}	
C	contrast	—
d	distance	cm
d_s	penetration distance and scattering depth	cm
D	water vapor density	$g\ cm^{-3}$
E	spectral irradiance	$W\ m^{-2}\ Å^{-1}$
E_D	skylight irradiance	$W\ m^{-2}\ Å^{-1}$
E_G	total spectral irradiance or global flux	$W\ m^{-2}\ Å^{-1}$
E_S	direct sunlight irradiance	$W\ m^{-2}\ Å^{-1}$
e	partial pressure of water vapor	mb
e	gas species	—
f	line shape factor	$(cm^{-1})^{-1}$
g	acceleration of gravity	cm sec^{-2}
G	radiance obscured by fractional cloud cover	$W\ cm^{-2}(cm^{-1})^{-1}sr^{-1}$
H	scale height	km
h	Planck's constant	J sec
H^A	scale height of large particles	km
H^G	scale height of gas	km
H_h	solar irradiance	$W\ cm^{-2}(cm^{-1})^{-1}$
h	solar elevation	deg.
H	height of satellite	km
I_ν	intensity	$W\ cm^{-2}(cm^{-1})^{-1}sr^{-1}$
I	observed clear-column radiance	$W\ cm^{-2}(cm^{-1})^{-1}sr^{-1}$
I_a	atmospheric emission	$W\ cm^{-2}(cm^{-1})^{-1}sr^{-1}$
I_d	reflected atmospheric downward flux	$W\ cm^{-2}(cm^{-1})^{-1}sr^{-1}$
I_h	reflected solar flux	$W\ cm^{-2}(cm^{-1})^{-1}sr^{-1}$
I_s	surface emission	$W\ cm^{-2}(cm^{-1})^{-1}sr^{-1}$
I'	atmospheric downward radiation	$W\ cm^{-2}(cm^{-1})^{-1}sr^{-1}$
I	current	amperes
i	$\sqrt{-1}$	—
I	outgoing radiance	$W\ cm^{-2}(cm^{-1})^{-1}sr^{-1}$

Symbol	Definition	Units
$\bar{I}$	outgoing radiance	$W\ cm^{-2}(cm^{-1})^{-1}sr^{-1}$
$\bar{I}_k$	observed radiance	$W\ cm^{-2}(cm^{-1})^{-1}sr^{-1}$
I	clear-column radiance	$W\ cm^{-2}(cm^{-1})^{-1}sr^{-1}$
I	observed solar radiance	$W\ cm^{-2}(cm^{-1})^{-1}sr^{-1}$
I_o	solar radiation outside the atmosphere	$W\ cm^{-2}(cm^{-1})^{-1}sr^{-1}$
J_ν	scattering source function	$W\ cm^{-2}(cm^{-1})^{-1}sr^{-1}$
J	quantum number of total angular momentum vector	—
k	Boltzmann's constant, 1.381×10^{-23} J K^{-1}	
k_ν	spectral line shape parameter-gas absorption coefficient	
$\bar{k}$	mean line strength	$cm^{-1}(g\ cm^{-2})^{-1}$
L	radiance of light	$W\ cm^{-2}(cm^{-1})^{-1}sr^{-1}$
L_2	atmospherically scattered upwelling radiance	$W\ cm^{-2}(cm^{-1})^{-1}sr^{-1}$
L_3	surface interacting upwelling radiance	$W\ cm^{-2}(cm^{-1})^{-1}sr^{-1}$
$L_{3,1}$	radiance of direct sunlight reflected from ocean surface, eventually emerging in zenith direction after at least one scattering by atmosphere	$W\ cm^{-2}(cm^{-1})^{-1}sr^{-1}$
$L_{3,2}$	radiance of skylight reflected from surface of smooth ocean, eventually emerging from top of atmosphere	$W\ cm^{-2}(cm^{-1})^{-1}sr^{-1}$
$L_{3,3}$	radiance of light reflected at least twice from surface before escaping	$W\ cm^{-2}(cm^{-1})^{-1}sr^{-1}$
L^a	airlight radiance	$W\ cm^{-2}(cm^{-1})^{-1}sr^{-1}$
L^s	radiance transmitted from surrounding	$W\ cm^{-2}(cm^{-1})^{-1}sr^{-1}$
L^t	radiance in direction to target	$W\ cm^{-2}(cm^{-1})^{-1}sr^{-1}$
M	total mass	g
m	average molecular weight of atmosphere = 28.97	—
m_o	atomic mass unit = 1.660 $\times 10^{-27}$kg	
M	spectral radiant emittance	$W\ cm^{-2}(cm^{-1})^{-1}sr^{-1}$
M_2	atmospherically scattered upwelling light	$W\ cm^{-2}(cm^{-1})^{-1}sr^{-1}$

Symbol	Definition	Units	Symbol	Definition	Units
M_3	surface interacting upwelling light	W cm⁻²(cm⁻¹)⁻¹sr⁻¹	$W_{H_2O,UP}$	weighting function for water vapor looking up from surface	cm⁻¹
M	molecular mass of gas	g	x	path length	km
m	absorber amount	g cm⁻²	x_j	single-drop cross section [j = S(scattering), E(extinction), A(absorption)]	cm⁻²
m	mass density of ensemble	g m⁻³	X_j	particle cross section	cm²
m	mass density of liquid water	g m⁻³	X_E	extinction cross section	cm²
m	mass density of cloud	g m⁻³	X_S	scattering cross section	cm²
m	magnetic dipole	amperes − m⁻²	y	contrast transmittance	—
m	optical air mass	—	z	height level (variable)	km
n	number density, molecules	m⁻³	z_o	surface of planet	km
N	total number of particles per unit volume	cm⁻³	z_j	peak value of kernel $K(\nu_j,z)$	km
n	index of refraction	—	z_c	height of cloud cover	km
N_i	population of energy at level i	cm⁻³	z'	height between z and H	km
N	quantum number describing state of quantization that may take on odd values only		α	half-width of spectral line	cm⁻¹
N	refractivity	—	β	volume extinction coefficient for particulates	m⁻¹
N_k	fractional cloud cover	—	γ_ν	extinction coefficient	cm⁻¹
p	pressure	N m⁻²	Γ	gamma function	—
p	scattering phase function	sr⁻¹	γ_D	Doppler line width	cm⁻¹
p	electric dipole moment	coulomb-m	γ_L	Lorentz line width	cm⁻¹
p	atmospheric pressure	mb	γ_A	absorption coefficient	cm⁻¹
Q_{ext}	nondimensional extinction efficiency factor	—	γ_E	extinction coefficient	cm⁻¹
Q_i	energy of level i	J	γ_S	scattering coefficient	cm⁻¹
q	charge on a particle	coulomb	γ_{ij}	contribution to total absorption at frequency ν by molecular transition from energy state i to energy state j	cm⁻¹
q	drop size parameter	—			
q	constant mixing ratio of absorbing gases	g/g			
r	radius	μm	γ_{cloud}	extinction coefficient applicable to clouds	cm⁻¹
R	universal gas constant, 8.3143J K⁻¹ mol⁻¹		γ	total absorption temperature	K
r_c	mode radius of distribution- drop radius at which number distribution function is maximum	μm	γ_{O_2}	absorption coefficient for oxygen	cm⁻¹
$R^{(n)}$	residuals	—	γ_{H_2O}	absorption coefficient of water vapor	cm⁻¹
r	amount of scatterer	m³	δ	mean spacing of spectral lines	cm⁻¹
s	distance	m	ϵ_s	surface emissivity	—
S_ρ	monochromatic solar constant	cm	ϵ_o	static dielectric constant	farad − m⁻¹
S	line strength	cm⁻¹(atm cm)⁻¹	ϵ_∞	optical dielectric constant	farad − m⁻¹
S_{band}	band strength	cm⁻¹(atm cm)⁻¹	Θ	zenith angle	deg.
S_i	line strength at ν_i	cm⁻¹(atm cm)⁻¹	λ	wavelength	μm
S	quantum number describing spin angular momentum of unpaired electrons	—	λ_0	Debye relaxation wavelength	μm
T	temperature	K	μ	cosine of angle of incidence ray	deg.
T	fraction of transmitted monochromatic energy	—	μ'	cosine of angle of emergent ray	deg.
T	turbidity factor	—	μ_{ij}	dipole matrix element	coulomb-m
T	temperature	K	ν	wavenumber	cm⁻¹
T^a	atmospheric transmission	—	ν	frequency of radiation	cm⁻¹
t	mean time between collisions of absorbing gas	sec	ν_0	frequency of line center	cm⁻¹
			ξ_j	efficiency factor	—
T_m	temperature of energy scattered in all directions	K	ξ_A	absorption efficiency factor	—
T_{sc}	temperature of energy scattered in all directions	K	ξ_E	extinction efficiency factor	—
			ξ_S	scattering efficiency factor	—
T_B	observed brightness temperature	K	ρ	mass density	g cm⁻³
			ρ	ground reflectance	—
T_{air}	local atmosphere temperature	K	ρ	partial spectral reflectance	—
			ρ_s	directional surface reflectivity	—
v	speed of propagation in air	cm sec⁻¹	ρ	density of gas	g cm⁻²
w_i	statistical weighting factor for degenerate space quantization and nuclear spin	—	ρ_e	concentration of absorbing gases	g cm⁻²
			σ	extinction cross section of one particle	cm⁻²
$W_{O_2,DN}$	weighting function for oxygen looking down through atmosphere	cm⁻¹	σ	transmittance	—
			τ	optical thickness	—
			τ^A	particulate optical thickness	—

Symbol	Definition	Units
τ^G	scattering optical thickness of atmospheric permanent gases	—
τ_1	specific optical path length	—
τ_1	optical thickness	—
τ	atmospheric spectral transmittance between surface and instrument	—
τ'	transmittance of entire atmosphere clear column traversed by incident solar flux	—
τ_ν	monochromatic optical depth	—
τ	opacity	—
τ_r	scattering coefficient for air molecules	—
τ_z	absorption coefficient for ozone	—
τ_{max}	zenith opacity	—
Φ	lifetime of molecule in excited state	sec
Φ	azimuthal angle	deg.
Φ'	emergent angle	deg.
Φ	instrument function	cm
Ψ_ν	emission coefficient	W cm^{-3}(cm^{-1})$^{-1}$sr^{-1}
Ψ^A	thermal emission	W cm^{-3}(cm^{-1})$^{-1}$sr^{-1}
Ψ^S	energy scattered into direct of interest	W cm^{-3}(cm^{-1})$^{-1}$sr^{-1}
Ψ	polar angle of inclination from vertical axis	radians
ω	single scattering albedo	—
Ω	solid angle incident beam	

INTRODUCTION

This chapter begins with a definition of the important atmospheric parameters needed to develop the interaction mechanisms within the atmosphere and between the atmosphere and the surface. The various types of interaction mechanisms are then formulated as a function of wavelength covering the range from the visible to the infrared and the microwave. The last part of this chapter investigates typical numerical and observations techniques employed to account for the various atmospheric and surface interaction processes. The reference list at the end of this chapter is intended to be used as a supplement to, but not a substitute for, the various derivations given in the main text.

ATMOSPHERIC EFFECTS ON REMOTE SENSING

The upwelling electromagnetic radiation (EMR) from earth is a function of the physical and chemical states of the surface and the atmosphere. Thus, in principle, it should be possible to recover information about the physical and chemical structure of the surface and the atmosphere from analysis of the upwelling EMR. However, the problem in analyzing such data lies in finding ways to uncouple the interactions of the surface radiation from the atmospheric radiation in order to retrieve the true values of each unknown parameter separately. The following sections deal with the problem of atmospheric interactions in the visible, infrared and microwave regions of the spectrum, with particular emphasis on the effects of such interactions on remote sensing.

ATMOSPHERIC CHARACTERISTICS

INTERACTION MECHANISMS

EMR interacts with the atmosphere in two ways: the energy can be either scattered or absorbed. In order to clarify these two concepts, imagine a nearly parallel and monochromatic beam of radiant energy entering, at a constant rate, a thin slab of the atmosphere containing atoms, molecules, and particulates. Radiant energy in the beam is attenuated (reduced in flux density) and emerges from the slab at a reduced rate, because the slab absorbs and scatters energy. The scattered photons have the same energy as the incident photons and emerge from the slab in all directions. The absorbed energy is either reemitted as photons of a different energy, or it increases the internal energy of the slab.

All particles in the atmosphere—atomic, molecular, and large particles, scatter EM energy, although an insufficient number of atoms are present to be of significance. Molecular scattering is important in the near-ultraviolet and visible spectra, but is negligible at wavelengths beyond 1 μm; essentially all molecular scattering is accounted for by N_2 and O_2. The dry atmosphere contains 78% of N_2, 21% of O_2, and 1% of A; percentages which remain fixed throughout the lowest 90 km of the atmosphere (Valley, 1965).

Density

The density of the dry atmospheric gas decreases approximately at an exponential rate with respect to height, which is measured from sea level. The total density can be represented nearly by the expression

$$n(z) = n(O) \exp(-z/H), \qquad (5\text{-}1)$$

where:

n = number density, molecules m^{-3},
H = scale height,
z = height level (variable)
$O = 0$ height = mean sea level.

The number density is also related to the pressure $p(z)$ and temperature $T(z)$ by means of the perfect gas law:

$$n = \frac{P}{kT} \qquad (5\text{-}2)$$

where:

T = temperature in degrees Kelvin,
p = pressure in $N\,m^{-2}$, N = force in newtons
k = Boltzmann's constant,
1.381×10^{-23} J K^{-1}

The number and mass densities at sea level are given in Table 5-1 for the U.S. Standard Atmo-

TABLE 5-1

Density of Standard Atmospheres

Model	Temperature (K)	Pressure (N m^{-2})	Number Density (m^{-3})	Mass Density (kg m^{-3})
U.S. Standard Atmosphere at level $z = 0$ (sea level) and dry	288	1.013×10^5	2.55×10^{25}	1.225
Dry atmosphere at standard temperature and pressure	273	1.013×10^5	2.69×10^{25}	1.298

sphere (1962) including those for standard pressure and temperature. Since number densities in gaseous absorption studies are frequently reduced to the conditions of standard temperature and pressure, the value $n = 2.69 \times 10^{25}$ m^{-3} is used when solving Eq. 5-1.

Scale Height

The scale height (H) *will be defined for this chapter in terms of the total mass of dry air in a vertical column of the atmosphere.* The average mass of one molecule of the dry atmosphere equals the average molecular weight times the atomic mass unit. The mass density is then

$$p(z) = n(z)mm_o, \qquad (5-3)$$

where:

$\rho(z)$ = mass density,
m = average molecular weight of atmosphere = 28.97,
m_o = atomic mass unit = 1.660×10^{-27} kg.

The total mass $M(z_1, z_2)$ in a vertical column of unit cross-sectional area and between a lower (z_1) and an upper level (z_2) can be expressed as

$$M(z_1,z_2) = \left[p(z_1) - p(z_2) \right]/g(z), \qquad (5-4)$$

where:

g = acceleration of gravity at some level z between levels z_1 and z_2.

The average total mass from sea level to the top of the atmosphere is

$$M(0,\infty) = \frac{1.013 \times 10^5 \ N \ m^{-2}}{9.81 \ m \ s^{-2}} \qquad (5-5)$$

$$= 1.034 \times 10^4 \ kg \ m^{-2}$$

where:

M = average total mass.

The sea-level value of g at a latitude of 45° has been used in Eq. 5-5 without compromising accuracy.

The average mass at sea level varies over the earth's surface by about one percent, and the standard deviation of the total mass at an arbitrary place is also about the same value (Valley, 1965). Hence, the total mass of the dry atmosphere in a

vertical column above sea level can be considered constant when computing radiation effects for remote sensing.

The average scale height is related to the total mass by substituting Eq. 5-3 in Eq. 5-1, and integrating from sea level to the top of the atmosphere; then

$$M(0,\infty) = \rho(0)H. \qquad (5-6)$$

The numerical value of the scale height of the dry atmosphere is obtained by substituting the mass density at standard temperature and pressure (Table 5-1) and Eq. 5-5 in Eq. 5-6 to obtain H = 8.0 km. If this value of H is used in Eq. 5-1 to calculate the density, the relative deviation of this density from the density of the U.S. Standard Atmosphere for the same height is less than 20 percent, when the height is less than 20 km. The chief reason for the difference is that an exponential decrease in density (Eq. 5-1) is strictly valid only for an isothermal atmosphere, and the earth's atmosphere deviates from being isothermal by as much as ± 20 percent. The temperature of the U.S. Standard Atmosphere, for example, decreases at the rate of 6.5 K km^{-1} from the ground to a height of 11 km, and is constant from 11 to 25 km (Valley, 1965). The details of several other models of the atmosphere are given in *U.S. Standard Atmosphere Supplements* (1966).

Atmospheric Pressure

The atmospheric pressure also decreases approximately exponentially with respect to height. Both the pressure and density decrease about 10 percent for each height increase of 16 km or 10 miles. Only 1 percent of the atmospheric mass lies above 32 km; therefore, the effects of the atmosphere above 32 km can be neglected. The effective atmosphere is a thin skin lying on the surface of the earth, with a relative thickness analogous to the skin on an apple.

Because of the thinness of the atmosphere, models of it that are used for theoretical and computational studies are usually represented as plane, parallel slabs of material. Such an approximation is reasonably accurate, unless the line of sight is near the horizon (Collins et al., 1972).

COMPOSITION

Gases

Of the atmospheric gases, water vapor, carbon dioxide and ozone dominate the interactions with electromagnetic radiation. Water vapor and carbon dioxide have strong absorption and emission features in the infrared and microwave spectral region and are important in determining the energy budget of the earth's atmosphere and surface. Ozone largely obscures the lower atmosphere and surface to ultraviolet radiation from the sun and has a strong band of spectral lines in the infrared. Absorption of ultraviolet radiation by ozone is a major source of heating in the upper atmosphere.

Water-vapor concentration is highly variable, and greatest in the lower atmospheric layers. In order to simplify computations where high accuracy is not required, the water vapor density is assumed to decrease exponentially with height; the scale height is approximately 2.5 km. The sea-level density of water vapor varies from 10^{-2} g m^{-3} in very cold, dry climates, to a maximum recorded value of 30 g m^{-3} in hot humid regions (Valley, 1965). The average surface concentration is about 10 g m^{-3}; thus the average total mass of water vapor in a vertical column is 2.5 g cm^{-2}. This mass is only 0.3 percent of the total atmospheric mass. Additional climatological data on water vapor are given by Greaves et al. (1971), and by Gaut and Reifenstein (1971).

Water vapor and carbon dioxide are the principal causes of the near-infrared absorption bands, which become more intense further into the infrared spectrum. Water vapor is most important radiatively in the troposphere. Significant absorption bands of water occur at 2.66 μm, 2.74 μm, 6.25 μm and the broad rotation band extending from near 20 μm far into the microwave region.

Carbon dioxide is substantially uniformly mixed from the surface up to about 100 km. Very strong spectral bands of CO_2 occur at 4.3 μm and 15 μm. The atmosphere is virtually opaque at these wavelengths except at very high spectral resolution.

Ozone causes a short-wave cutoff in solar radiation reaching the surface at 0.3 μm due to the strong absorption in the Hartley-Huggins bands lying between 0.2 μm and 0.3 μm as described in Heicklen (1976). Ozone also has a broad, weak band centered at 0.5 μm, and strong bands at 9.6 μm and 14 μm. Ozone is concentrated in a layer between 20 and 50 km above the surface.

The major atmospheric constituents, nitrogen and oxygen, are not radiatively active molecules in the visible or infrared. Oxygen is a source of heating in the earth's thermosphere (near 100 km) due to an absorption band in the far ultraviolet. Absorption for a vertical atmospheric path by the major gaseous absorbers is shown in Figure 5-1.

Trace gases absorb radiant energy to a minor

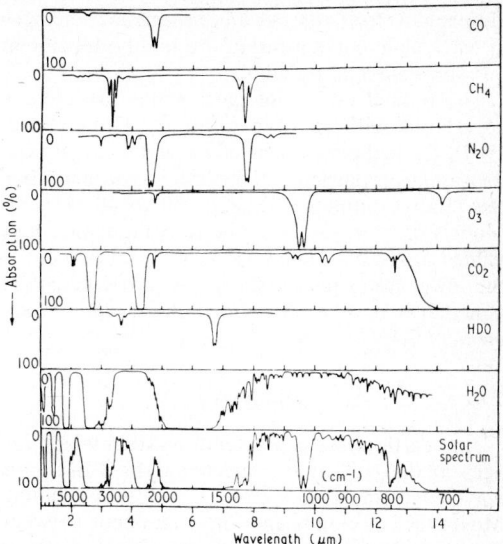

Fig. 5-1. Absorption in the infrared for a vertical atmospheric path by a variety of constituents; (J. H. Shaw, 1970).

extent in the atmospheric windows; particulates behave similarly. The surface temperature of the earth is primarily influenced by water (in the form of both vapor and clouds), whereas the temperature structure of the stratosphere results mainly from absorption by O_2 and (to a lesser extent) CO_2.

Information on the infrared absorption spectra of the trace gases may be found in recent compilations by McClatchey et al. (1973) and Rothman et al. (1978). A discussion of the measurement of ozone concentrations and their variabilities with altitude, latitude, and season is given in NASA Reference Publication 1010 edited by R. Hudson (1977). The detailed radiative balance between the surface of the earth and the atmospheric regions is discussed in CIAP Monograph 4.

Particulates

The atmospheric particles that are much larger than molecules can be divided into two classes: (1) The omnipresent haze or dust particles called *particulates;* and (2) clouds of liquid and solid water. The radii of the particulates of optical importance lie between 0.1 and 10 μm. The radii of the cloud particles are larger and vary from 1 to 100 μm.

The optical effects of particulates are discussed in detail in later paragraphs but a few general comments about their distribution will be made here. They are concentrated in the few lowest kilometers of the atmosphere. The optical effects of particulates are small above a few kilometers, the horizontal visibility being a rough measure of the extent of the optical effects. Because particulates are so small, they remain suspended in the atmosphere for days. As a consequence, changes in particulate concentration are highly dependent on their dispersal by winds.

The optical effects of particulates are closely correlated with relative humidity (Rozenberg, 1967). Optical effects caused by water-vapor condensation on particulates become apparent when the relative humidity exceeds 30 to 40 percent. Models of average atmospheric composition give values of relative humidity of 60 to 80 percent in the lower atmosphere where the particulates are concentrated (*U.S. Standard Atmosphere Supplements*, 1966).

CLOUDS

Any earth-observing satellite system which images in the part of the spectrum short of 1 mm wavelength must take cloud cover into account. Most types of clouds and fogs that occur between the ground and a higher observational platform completely obscure the ground for that part of the spectrum from 0.3 to 3 μm. A few types of clouds, such as thin cirrus, have a small optical thickness and are semitransparent.

Coverage

Statistics of cloud coverage are used for planning various survey missions. *Cloud coverage is defined as the fraction of a horizontal surface that is covered by the vertical projection of the clouds*. For example, a coverage of 0.0 means that no clouds are present and that the sky is clear, 0.6 means that 60 percent of the sky is covered by clouds and 1.0 means that the sky is totally overcast with clouds. Coverages derived from satellite data are sometimes different from those measured from the ground (Glaser et al., 1968), since the satellite coverage depends on the cloud type, on the resolution, and on the spectral band of the sensor. Martin and Liley (1971) have computed statistical relations between cloud coverages observed at the ground and from satellites.

Cloud statistics have to be assembled carefully for mission planning. The statistics should be valid for the region of interest, the season of the year, and even the time of day. The mean cloud coverage can change quickly in a short distance, as it does during the summer between the cloudy California coast and 100 to 200 km inland, where few clouds occur. The monsoon regions have a dramatic change of clouds between seasons. The daytime cloud-coverage for the southeastern United States is about 0.12 higher than the nighttime coverage (Martin and Liley, 1971).

Dependence on Field of View

The cloud statistics that are utilized for mission planning depend on the size of the field of view of the sensor being used, as explained by Greaves et al. (1970). To demonstrate this, the author describes the frequency functions of cloud coverage for a point area and for the entire world. A point is either covered by at least one cloud, or it is not. As a result, the frequency function is strongly peaked at 0.0 and 1.0 coverage, and is zero between these two points. The average cloud coverage for the entire world, on the other hand, is 0.4 (Sherr et al. 1968). It is never observed to be completely clear, not completely overcast. In this case the frequency function is strongly peaked at 0.4. The frequency function of coverage for an area of intermediate size can be completely different from the two values just given.

The effects of cloud cover on two hypothetical satellite-survey missions are discussed by Martin and Liley (1971), assuming the field of view for both missions to be a square with sides of 185 km. The area is located in the United States, east of the Rocky Mountains, and north of latitude 38°N. The time is 1000 local standard time in July. The satellite passes over the region of interest at periods exceeding a few days, which is the approximate time for the cloud statistics to become uncorrelated.

The success of the first mission requires at least one observation of the entire field of view without cloud obscuration. The second mission permits viewing of whatever cloud-free area exists. On successive satellite passes new cloud-free areas may appear. A mosaic of the field of view is assembled. Basic statistics, as derived for cloud coverage, demonstrate a probability of only 0.1 that the entire field of view will be clear at one time. As a consequent, seven passes are required to get a 50 percent probability of success for the first mission, and 28 passes for 90 percent probability of success.

Probabilities for the second type of mission were computed by Monte Carlo methods (Greaves et al., 1970); he found that 90 percent of the field of view can be seen in seven passes with a confidence of 94 percent. Almost the entire field of view was seen by mosaicking with high confidence in seven passes, whereas the entire field of view would be seen with only 50 percent probability in the same number of passes, as previously mentioned.

Recently Goetz (1979) generated cloud cover probabilities for the U.S. using Landsat data collected over a period of six years. Cloud-cover probabilities were calculated by dividing the number of Landsat observations yielding up to 10 percent cloud cover by the total number of observations at that point. Figure 5-2 shows a contour plot for the U.S. of probabilities of 0 to 10 percent cloud cover corresponding for small fields of view of 185 × 185 km.

From the results shown in Figure 5-2 Goetz generated statistics on the number of observations required to obtain cloud-free measurements for a given confidence factor. Figure 5-3 shows contour plots of the 75-percent probability for viewing the U.S. with a 0−10 percent cloud cover.

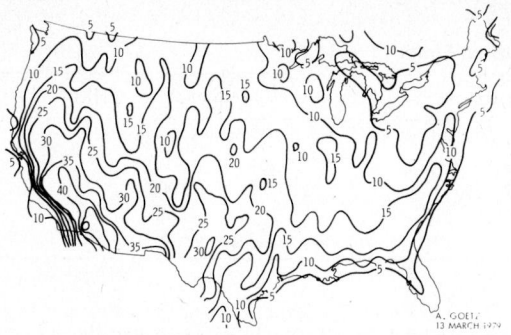

Fig. 5-2. Percent probability of 0 to 10% cloud cover for any observation; (Goetz 1979).

CHARACTERISTICS OF SOLAR RADIANT ENERGY

SPECTRAL IRRADIANCE

The sun is the strongest source of radiant energy for the spectral band of 0.3 to 3 μm, other sources being much weaker in the same spectral range. The full moon is the next strongest source, and the radiant energy from it is about 10^{-6} of that from the sun.

In the infrared the primary source of radiant energy is thermal emission from the earth's surface and atmosphere. Figure 5-4 shows the blackbody curves for solar and terrestrial temperatures. Virtually all the solar radiation is at wavelengths $\lambda < 4.5$ μm and all the terrestrial radiation at $\lambda > 4.5$ μm. For convenience, Table 5-2 gives some examples of spectral position in the units used in this chapter.

Figure 5-5 shows the spectral characteristics of sunlight, the upper continuous curve giving the coarse spectral irradiance above the earth's atmosphere. The maximum irradiance occurs at 0.47 μm. About one-fourth of the sun's energy falls in the spectral band of $\lambda < 0.47$ μm and 46 percent of the total energy falls in the visible band from 0.40 to 0.76 μm. Measurements of the solar irradiance with a fine resolution of 10^{-5} μm from a satellite would reveal numerous absorption lines which are caused by absorption in the sun's atmosphere.

Fig. 5-3. Number of observations required for 75% probability of 0 to 10% cloud cover; (Goetz 1979).

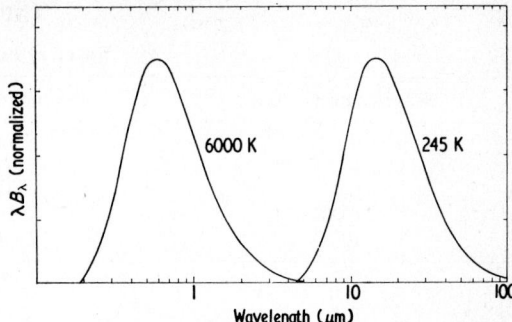

Fig. 5-4. Blackbody curves for solar and terrestrial temperatures. Solar radiation absorbed by the earth-atmosphere system is, on average, balanced by emission of infrared radiation to space; (Houghton and Taylor 1973).

Additional information on solar energy can be found in the NASA report *Solar Electromagnetic Radiation* (1971b).

The dashed curve on Figure 5-5 represents the irradiance of a blackbody at $T = 5,900$ K. The lowest curve represents the irradiance of the direct sunlight at sea level, the difference between the two continuous curves representing attenuation by scattering and absorption. The stippled portion represents loss by absorption only. Ozone absorption causes a cutoff at 0.3 μm and also appears weakly in the red part of the spectrum. Except for strong, narrow-band oxygen absorption at 0.76 μm, the near infrared absorption is dominated by H_2O and CO_2, as shown by Figure 5-5.

Figure 5-6 shows the emission spectrum of the Sahara desert as measured from a space platform. In the $800-1000$ cm^{-1} region the observed temperatures are very nearly the same as the hot surface temperatures. This "window" in the atmosphere is in a spectral region of small atmospheric absorption. The 320 K blackbody has a maximum energy near 600 cm^{-1}. The strong band of CO_2 near 700 cm^{-1} markedly reduces the depth to which the measuring instrument can "see" into the atmosphere. The equivalent brightness temperature at 700 cm^{-1} is just below 220 K and most of the emission originates from the stratosphere. Absorption by ozone near 1000 cm^{-1} occurs at altitudes near 30 km. Since ozone is not opaque in this spectral region, emission from lower levels is also observed. At longer wavelengths water vapor in the troposphere dominates the radiation field.

SOLAR CONSTANT

The rate at which the total solar radiant energy flows across a unit area normal to the direction of propagation and located at the mean distance of the earth from the sun is called the solar constant. The value of this constant is shown in Table 5-3. Variations are correlated with sun-spot numbers. About 35 percent of the sunlight is reflected from the earth, its atmosphere and clouds; about 17 percent is absorbed in the atmosphere; and about

TABLE 5-2

Spectral Location: Units

Wavelength (λ, μm)	Wavenumber (ν, cm^{-1})	Frequency (ν, GHz)
0.1	100,000	3×10^6
10	1,000	3×10^4
100	100	3000
1000	10	300
10^5	0.1	3

47 percent reaches the surface of the earth and is absorbed there (Kondratyev, 1969). Recent satellite measurements give a lower albedo of the earth's atmosphere system of 0.31 (Jacobowitz et al., 1979). If no clouds are present in the sky and θ is the solar zenith angle, the irradiance of the earth's surface by solar energy is roughly proportionate to the $\cos\theta_0$ kW m^{-2}.

RADIANT ENERGY INTERACTIONS WITH ATMOSPHERES

This section begins with a formulation of the radiative transfer equation followed by a listing of references that give the theory of the interaction of radiant energy with atmospheres. Methods of computing radiation quantities are outlined. Because of the importance of particulates, their physical and optical characteristics are discussed in detail. The dependence of the illumination of an object of interest on both the direct sunlight and the diffuse scattered light is discussed. One section is devoted to the characteristics of the downwelling radiant energy at the ground, and the final section presents the characteristics of radiant energy leaving the top of the atmosphere. Data are not given here for intermediate levels, but Dave and Furukawa (1966) have issued computed tables of such data. All of the radiation data here presented apply to cloud-free atmospheres and to monochromatic radiation. The convention for

specifying monochromaticity is to use the adjective "spectral" before the noun that names the parameter. Since all radiant energy parameters that are discussed here refer to monochromatic energy, the term spectral will be omitted.

RADIATIVE TRANSFER IN A SCATTERING AND ABSORBING MEDIUM

The fundamental equation describing the propagation of EMR in the presence of interactions with the medium through which it passes is the *equation of radiative transfer*. At a given point in the medium, which is characterized by an extinction coefficient γ_ν and an emission coefficient Ψ_ν the elemental change in intensity $I_\nu(\Theta,\Phi)$ as it traverses a distance ds in the direction (Θ,Φ) is given by Chandrasekhar (1950) as

$$\frac{d I_\nu(\Theta,\Phi)}{ds} = -\gamma_\nu I_\nu(\Theta,\Phi) + \Psi_\nu(\Theta,\Phi). \quad (5\text{-}7)$$

The first term on the right-hand side of Eq. 5-7 represents the loss of radiation by both absorption and scattering, and the second represents the total gained by thermal emission and scattering into the direction (Θ,Φ).

The emission coefficient may be expanded to the form

$$\Psi_\nu(\Theta,\Phi) = \Psi_\nu^A(\Theta,\Phi) + \Psi_\nu^S(\Theta,\Phi) \quad (5\text{-}8)$$

where:

Ψ_ν^A refers to the process of thermal emission,

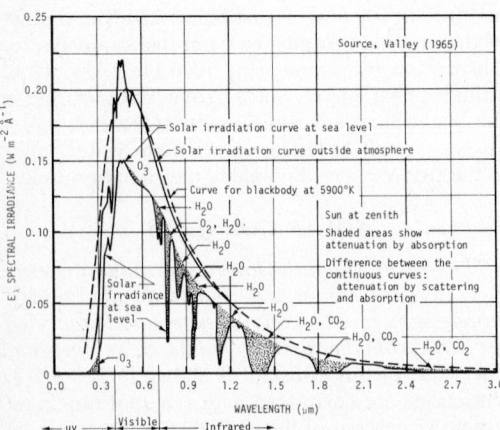

Fig. 5-5. Spectral irradiance (E_λ) of the sunlight before and after it passes through the earth's clear atmosphere.

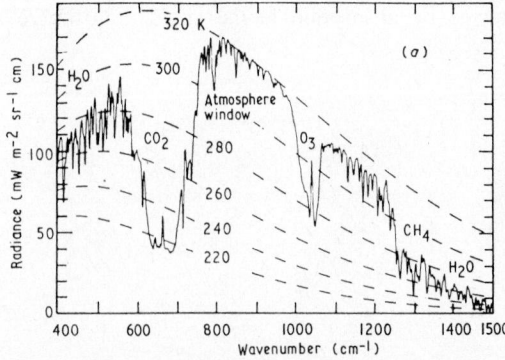

Fig. 5-6. Thermal emission spectrum recorded by IRIS D on Nimbus 4 over the Sahara. Radiances of blackbodies at several temperatures are superimposed; (Hanel et al. 1971).

TABLE 5-3

Summary of Solar Irradiance Observations by Electrically Self Calibrated Cavity Pyrheliometers in Flight Experiments

Experimentor	Platform	Year	Altitude (km)	Solar Constant At 1 A.U. (W/m²)	S.I. Uncertainty (W/m²)	Reference
Kruger	aircraft	1967	12	1372	>34	Thekaekara (1969)
Kendall	aircraft	1968	12	1373	>34	Kendall (1973)
Willson	balloon	1968	25	1366	31	Willson (1971)
Willson	balloon	1968	25	1373	26	Willson (1971)
Willson	balloon	1969	36	1369	<14	Willson (1973)
Plamondon	spacecraft	1969	>100	1362	14	Plamondon (1969)
Plamondon	spacecraft	1969		1364	14	Plamondon (1969)
Willson	rocket	1976		1368	<7	Duncan (1977)
Willson	rocket	1976		1364	<7	Duncan (1977)
Willson	rocket	1978		1373	<7	Willson (1979)
Hickey	spacecraft	1978		1377	<7	Stowe (1979)
Willson	spacecraft	1980		1368	<7	Willson (1980)

Ψ_ν^S refers to the energy scattered into the direction of interest.

Ψ_ν^A can be written for a medium in thermal equilibrium using Kirchhoff's law as

$$\Psi_\nu^A(\Theta,\Phi) = \gamma_\nu^A \, B_\nu(T) \qquad (5\text{-}9)$$

where:

γ_ν^A extinction coefficient,
$B_\nu(T)$ Planck function, given by

$$B_\nu(T) = \frac{2h\nu^3}{c^2} \frac{1}{\exp(h\nu/kT) - 1} \qquad (5\text{-}10)$$

where:

h = Planck's constant,
ν = frequency of radiation,
c = speed of light in a vacuum,
k = Boltzmann's constant, and
T = absolute temperature of the medium.

The scattering term may be written in terms of a scattering function such that

$$\Psi_\nu^S(\Theta,\Phi) = \gamma_\nu^S \, J_\nu(\Theta,\Phi), \qquad (5\text{-}11)$$

where:

$J_\nu(\Theta,\Phi)$ is a function dependent upon the scattering characteristics of the medium.

Making use of the extinction coefficient

$$\gamma_\nu = \gamma_\nu^S + \gamma_\nu^A \qquad (5\text{-}12)$$

the single scattering albedo

$$\omega = \gamma_\nu / \gamma_\nu \qquad (5\text{-}13)$$

and the elemental opacity

$$d\tau = -\gamma_\nu ds, \qquad (5\text{-}14)$$

Eq. 5-7 can be written as

$$\frac{dI_\nu(\Theta,\Phi)}{d\tau} = I_\nu(\Theta,\Phi) - (1 - \omega) B_\nu(T) - \omega J_\nu(\Theta,\Phi).$$

$$(5\text{-}15)$$

THEORY OF ATMOSPHERIC RADIANT ENERGY INTERACTION

Theoretical analyses of the scattering of light by the earth's atmosphere are difficult. They will not be discussed in detail, but basic references will be given which will permit more specific study of these interactions. A treatment of the fundamentals of radiative transfer was given by Milne (1930), while later advances were discussed by Ambartsumyan (1958), Kourganoff (1963), and Sobolev (1963). Chandrasekhar (1950) provided an important stimulus to radiation studies, showing how to account for and compute the polarization properties of scattered light. Some of the difficulties in applying theory to the earth's atmosphere were shown by Kondratyev (1969). These references, for the most part, analyze the physics of radiative transfer in a rigorous mathematical fashion. This treatment, however, is limited in scope since only relatively simple atmospheric models will have a tractable solution. Models characterizing realistic atmospheric conditions such as inhomogeneous layering and strong forward scattering by particulates are best handled by computerized numerical techniques.

Much effort within the last few years has gone into researching a variety of numerical methods for treating radiative transfer in cloudy atmospheres. Two of the more popular techniques are mentioned below; an excellent review of the field is given by Hansen and Travis (1974).

Difficulties of Computation

The principal difficulty in computing the values of such radiation parameters as irradiance, spec-

tral radiance, and degree of polarization is in accounting for light that is scattered more than once. As sunlight filters through the atmosphere, about one-half of the incident photons (depending on the wavelength and solar zenith angle) are scattered, a fraction of them more than once. In order to compute reasonably accurate numerical data, accounting has to be made of the multiple scattering (Dave, 1964).

The strong anisotropy of particulate scattering has been difficult to incorporate into computations. It is only quite recently that the sphericity of the earth and its atmosphere has been accounted for by Monte Carlo techniques (Collins et al. 1972), where spherical effects become increasingly evident as the line of sight approaches the horizon. The computation of polarization characteristics of scattered light introduces new problems, since three additional parameters (degree of polarization, inclination of the plane of polarization, and ellipticity) have to be accounted for. Although Dave and Gazdag (1970) seemingly demonstrate that omission of polarization effects can cause errors as large as 23 percent in radiances computed for model atmospheres that do not consider particulates as components, Hansen (1971) shows that the error is less than 1 percent as compared with a model atmosphere containing particulates.

Models of Earth Atmosphere Systems

Because of the complexities in computing numerical values for radiation parameters, models of the earth-atmosphere system are usually quite simple. The atmosphere is typically assumed to be a plane—specifically, a parallel slab that simulates the earth's gaseous atmosphere and the particulates in it. The optical properties of the slab vary only in the vertical direction. Such a model cannot account for horizontal variations caused by scattered clouds or by the ground; these variations have been investigated by Monte Carlo techniques.

The ground, which is the lower boundary of the atmosphere, is usually assumed to reflect radiation according to Lambert's Law. Current numerical techniques, however, can allow for surfaces with quite general scattering properties when such a complexity is justified.

METHODS OF SOLUTION

Several methods have been used to compute numerical solutions for the radiance and polarization parameters. Chandrasekhar's method (1950), which accounted for multiple scattering and polarization, was used to calculate extensive tables (Coulson et al., 1960) of the radiance and polarization of light emerging from the top and the base of a Rayleigh homogeneous model atmosphere. An atmosphere which excludes large particles and absorbing gases is called a Rayleigh atmosphere, since a small portion of it scatters light according to Rayleigh's law. Chandrasekhar's method, because of its difficulty and inability to account for large-particle scattering, has largely been replaced by iterative methods. Dave and Furukawa (1966) calculated the radiance, polarization parameters, and irradiance as functions of height for a model atmosphere that contained N_2, O_2, A, and O_3, but excluded particulates. Herman (1971) added particulates to the same model. Russian scientists have published tables of radiances and irradiances which accounted for both gaseous and particulate scattering (Atroshenko et al., 1963; Feigelson et al., 1960).

Monte Carlo Method

The Monte Carlo technique has been exploited by Plass and Kattawar (1968–1971) to calculate the radiation parameters for many models of clear, hazy, and cloudy atmospheres, and of an atmosphere-ocean system. The method is based on the technique of following photons through an atmosphere and having them interact with scatterers in a probabilistic sense. This method has sufficient potential flexibility to take into account horizontal variations in the atmosphere and ground, and the location of non-solar sources such as lasers. Collins et al., (1972) extended the Monte Carlo method to compute radiation parameters for models of spherical atmospheres.

Doubling and Additive Method

The doubling and addition-of-layers method (Twomey et al., 1966; Grant and Hunt, 1968; Hansen, 1969) has been used extensively in the last few years in modeling the atmospheres of Venus, Mars, Jupiter and Saturn, in addition to that of the earth. In essence, the addition-of-layers method is an algorithm for determining the radiative transfer properties of an inhomogeneous layer composed of two layers each of whose transfer properties are known. When the two layers are identical, the algorithm is known as the doubling method. These algorithms complement the Monte Carlo technique in that they are easier to use and, in general, are faster computationally, but are limited at the present time to plane-parallel layer geometry.

PARTICULATES

The shape, index of refraction, size, and concentration of particulates have to be known in principle to compute their effect on the transfer of radiant energy. In general, these properties are not known accurately, although the accuracy with which they need be specified depends on another source of error, their spatial variability. Current plane-parallel models of the atmosphere allow for no horizontal variation; only that the total number of particulates depends on height.

Particulates are removed from the atmosphere by three methods according to the *Report of the Study of Man's Impact on Climate* (Matthews,

1971): (1) they can fall out to the ground; sedimentation is effective only for particles larger than 1 μm radius; (2) they can impact foliage, ground-based structures, and the ground, where they are removed from the atmosphere if they stick; and (3) particles are removed by cloud-forming and precipitation processes.

The lifetime of particles that are found in the lowest few kilometers of the atmosphere seems to be about one week. During that time these particles can travel thousands of kilometers, coagulate among themselves and mix with particles from various sources. As a result, the chemical composition and shapes of the particulates are complex.

PHYSICAL CHARACTERISTICS

The sources of particulates are given in Table 5-4, which separates natural and man-made sources. In both cases the particulates can be ejected directly into the atmosphere or they can be formed in the atmosphere from trace gases; the latter process is believed to be dominant. The range of values indicated for each source indicates the uncertainty, which may be an order of magnitude, in their estimates. Only one estimate is given for sea salt, which originates in the spray thrown up by the wave action of the seas. Except for sea salt, the remaining particulates have a land source. The extreme values of the subtotals for the man-made contribution indicate that such contribution is between 7 and 43 percent of the total; most of the particulates are, therefore, of natural origin. Man's contributions are most noticeable and significant near the centers of industrial activity and where farmland is cleared by burning.

Relation of Humidity

About 50 percent of the particulates are water soluble and cause condensation of water vapor (Matthews, 1971). Condensation changes the optical properties of particulates, and therefore of the atmosphere, but it is important only when the relative humidity exceeds 35 percent by one estimate (Rozenberg, 1967; also see Winkler and Junge, 1971) or 60 percent by another estimate (Pueschel et al., 1969). As the relative humidity increases, water condenses on some of the particulates. They increase in size, their indices of refraction change, and usually the extinction efficiency and the amount of light scattered increase. For example, when the relative humidity increases from 70 to 95 percent near the ground, the horizontal visibility decreases by a factor of two to six (Junge, 1963). In general though, the effect of relative humidity on attenuation and scattering is insufficiently understood. Hanel et al. (1971); Hanel (1976).

Size Distribution Functions

Because of the complex characteristics of particulates, it is somewhat surprising that their size-distribution functions are rather similar and

TABLE 5-4

Estimates of Particles Smaller Than 20-μm Radius Emitted into or Formed in the Atmosphere
(units of 10^9 kg/year)*

Natural Particulates	Estimated Particulates
Soil and rock debris†	100 — 500
Forest fires and slash-burning debris†	3 — 150
Sea salt	300
Volcanic debris	25 — 150
Particles formed from gaseous emissions:	
Sulfate from H_2S	130 — 200
Ammonium salts from NH_3	80 — 270
Nitrate from NO_x	60 — 430
Hydrocarbons from plant exudations	75 — 200
Total Natural Particulates	773 — 2200
Man Made Particulates	
Particles (direct emissions)	10 — 90
Particles formed from gaseous emissions:	
Sulfate from SO_2	135 — 200
Nitrate from NO_x	30 — 35
Hydrocarbons	15 — 90
Total Man-made Particulates	185 — 415
Total Particulates	958 — 2615

* Source, *Report of the Study of Man's Impact on Climate* (Matthews, 1971).
† Includes unknown amounts of indirect man-made contributions.

uniform. On a global scale, the aerosol[1] state of the atmosphere within approximately 10 km of the ground can be separated into three classes: (1) the background state, where the atmosphere is extremely clean or the visibility is very good; (2) maritime, and (3) continental.

Typical, average size-distribution functions for these three states are shown in Figure 5-7. The ordinate gives the number of particles in a cubic centimeter of the atmosphere for a unit change in the decadic logarithm of the radius. For example, the Continental function shows that there are about 10^4 particles per cubic centimeter, the radii of which lie between 0.01 and 0.1 μm. The background function indicates that the number of particles decreases rapidly below 0.1 μm radius; it is not considered that the atmosphere contains particles greater than 100 μm in radius.

The maritime particulate distribution differs from the background particulate distribution by having significantly more particles between 0.5 and 20 μm radii. The continental distribution is the largest, especially below 0.1 μm radius.

Little is known about the number of particles below 0.1 μm radius. However, the dashed curves in Figure 5-7 indicate their possible numbers, if small particles are produced within the atmosphere. The distribution functions are different near sources, and may show rapid fluctuations with respect to size. The particulate concentration may be an order of magnitude larger in urban environments than that shown for continental atmospheres, which is representative of rural concentrations in clean atmospheres. It is now becoming evident that the aerosol size distribution is typically bimodal with a fine-particle mode near 0.3 μm and a coarse-particle mode in the range $5-50$ μm, depending on the specific location. A third mode in the range 0.01 to 0.1 μm is also sometimes present near combustion sources (Bloch and Seinfeld, 1979).

The three boxes near the bottom of Figure 5-7 give additional information about particulates. The vertical height of each box is proportional to the contribution of particulates of the corresponding radius. The lowest box indicates that only particles with radii between 0.06 and 100 μm contribute to the mass; the masses for particles having radii between 0.1 and 1 μm and between 1 and 10 μm are equal. The middle box indicates which particles are effective condensation nuclei for cloud formation. The upper box indicates that only the particles between 0.06 and 10 μm radii have an effect on the optical state of the atmosphere. Although there are numerous particles below 0.1 μm radius, each interacts weakly with visible and infrared radiation. On the other hand, as the particles increase in size, they interact more strongly with light, but the number of large parti-

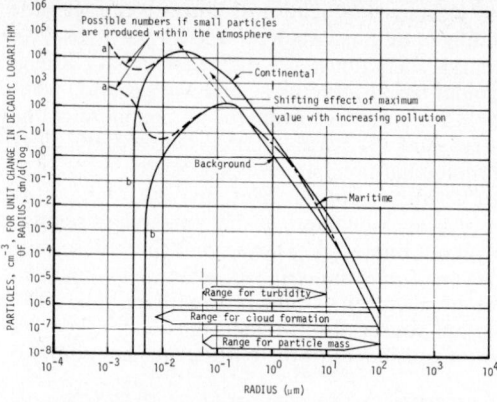

Fig. 5-7. Typical size-distribution functions of particulates. Modified from the *Report of the Study of Man's Impact on Climate* (Matthews, 1971).

cles decreases so rapidly that they become ineffective at about 10 μm radius.

The optically effective particles are concentrated near the surface of the earth and decrease rapidly with height. About 60 percent of the effective particles lie within 1 km of the ground. The size-distribution function is roughly independent of height, but measured changes are quite apparent (Blifford and Ringer, 1969).

EFFECT OF PARTICULATES ON REMOTE SENSING

In order to compute the effect of particulates on remote sensing, several simplifying assumptions have to be made. As indicated on Figure 5-7, the most effective particles optically are those with a diameter of the order of a wavelength of light. The computation of scattering characteristics of irregular shaped particles of such size is not feasible; therefore, the particles are assumed to be spheres. Rolland and Gagne (1970), Hodkinson (1963), Zerull and Giese (1974), and Pinnick et al. (1976) give measured data on the consequences of such as assumption.

In addition to the size of the particle, knowledge of the *index of refraction*, which is a complex number, is required for calculations. Both the real and imaginary parts of the index of refraction depend on the fraction of the particulate that consists of water. As the fraction of water increases, the real part approaches 1.33 and the imaginary part approaches a negligibly small value at visible wavelengths. For irregularly shaped particles, the index should be represented by a matrix that describes the optical activity (bi-refringence, dichroism, . . .) of the particles (van de Hulst, 1957; Fymat and Vasudevan, 1975). However, the entries in this matrix are poorly known.

Size Distribution Functions

Two forms of the size distribution function are commonly used, the Deirmendjian (1969) and

---

[1] Aerosols — Liquids and solids in the atmosphere, the liquids in turn containing absorbed gases.

Junge (1963). Let us consider first the *Deirmendjian distribution function*.

Deirmendjian distribution function

Let $dn(r,z)/dr$ represent the number density of particles per unit of radius (r) at a height z above sea level. Deirmendjian (1969) has introduced the following expression for dn/dr:

$$dn(r,z)/dr = a(z)r^A \exp(-br^B), \; 0 \leqslant r \leqslant \infty.$$
$$(5\text{-}16)$$

The height dependence $a(z)$ in Eq. 5-16 is given by

$$a(z) = \frac{Bb(A + 1)/B}{[(A + 1)/B]} N(z),$$

where:

    Γ = gamma function,
    $N(z)$ = total number of particles per unit volume.

Deirmendjian (1969) shows that the function dn/dr can represent particle sizes in different types of haze and even in clouds by varying the parameters A, B, and b.

Junge distribution function

Another common size-distribution function, which represents the continental particulates of Figure 5-7 in the size range of optical effectiveness is given by Junge (1963),

$$dn(r,z)/dr = N(z)(\nu - 1)r_{min}^{\nu-1} r^{-\nu}, \quad (5\text{-}17)$$

where:

    $r_{min} \leqslant r \leqslant r_{max}, \; r_{min} << r_{max}$
    N = total number density
    ν = variable parameter.

The minimum and maximum values of r can be taken as $r_{min} = 0.03 \; \mu m$ and $r_{max} = 10 \; \mu m$. The exponent ν usually lies between 3.5 and 4.5 (Junge, 1963) and is taken as 4 in Figure 5-7. The total number density (N) depends critically on the smallest size of the particles measures. However, published values of N do not always give the range of particle sizes for which it applies.

The size distribution function of Eq. 5-17 is usually referred to as the *Junge distribution*. It has been used with varying degrees of success to explain the optical characteristics of particles found in continental regions. Bullrich (1964) has used it extensively. Chayanova and Shifrin (1968) suggest that the exponent $\nu = 4$ or 5, if the surface visibility is less than or more than, respectively, 10 km. Ivanov (1968) states that the Junge distribution is quite satisfactory, except that the exponent has a somewhat lower value of $\nu = 2.5$. However, the Junge distribution is not always accurate. Other optical data given by Ivanov et al. (1968, p. 47) are the basis for questioning the utility of the Junge distribution, since they show that it is incorrect more than one-half of the time. Rosenberg (1967)

says that the Junge distribution represents the average size-distribution of continental particulates, but in general is not applicable to any specific place and time. It can be shown that the different size-distribution functions used in atmospheric studies, including those in Eqs. 5-16 and 5-17, are but different solutions of the celebrated K. Pearson's differential equation of statistical theory which, thus, provides a convenient unifying representation (Fymat, 1977).

Height Distribution of Density

The height distribution of the number density of large particles is frequently approximated by an exponential relation (Laktionov, 1964; Elterman, 1970):

$$N(z) = N(0) \exp(-z/H^A) \quad (5\text{-}18)$$

where:

    H^A = scale height of large particles $\sim$ 1 km.

Equation 5-18 does not give the correct distribution of large particles above 5 km. However, these particles contribute only 20 percent to the total large-particle optical thickness from the ground to the top of the atmosphere (Elterman, 1968). Blifford and Ringer (1969) have measured particle sizes to a height of 10 km. They show size-distribution functions changing with respect to height, and no exponential decrease of N.

PARTICULATE SCATTERING CHARACTERISTICS

The standard reference on the scattering of EM energy by spheres is the work by van de Hulst (1957); supplementary information is given by Kerker (1969). The large aerosol particles that are optically important occur in the range of sizes 0.05 $< r <$ 2.5 μm (Bullrich, 1964). Their intermediate size makes them inappropriate for application of the simple laws of Rayleigh scattering for small particles and of geometric optics for large-scale particles in the visible- and near-infrared. Hence, the complex Mie theory is used in the accurate computations. Additional references on Mie computations are found in Bullrich (1964), Dave (1969, 1970), and Deirmendjian (1969).

Volume Extinction Coefficient

One example of the use of the Mie theory is in the computation of the *volume extinction coefficient* for particulates. *The extinction* (σ) *of a single particle has the dimensions of area, and represents the cross-sectional area of an incident beam that interacts with the particle.* Another way of considering the extinction cross-section is to assume that the ratio of σ to the cross-sectional area of a beam of light will interact with the particle. The extinction cross-section of one particle of radius r is expressible in terms of a factor Q_{ext}, which is usually computed from the Mie theory (van de Hulst, 1957):

$$\sigma(\lambda,r) = \pi r^2 \, Q_{ext} \, (\lambda,r), \qquad (5\text{-}19)$$

where:

σ = extinction cross-section of one particle
Q_{ext} = nondimensional efficiency factor.

The values of σ are given in Figure 5-8. For example, $\sigma = 6.7 \times 10^{-6}$ cm^2 for a sphere of radius $r = 10$ μm, which has an index of refraction relative to its surroundings[2] of n = 1.54. Since the imaginary part of the index is zero, all of the extinction is caused by scattering. In this example, the sphere is large relative to the wavelength. Its extinction cross-section is approximately twice its geometric cross-sectional area.

The volume extinction-coefficient is obtained by adding up the extinction cross-sections of the individual particles (σ) within the volume. If there are N particulates per unit volume, all of which have the same cross-section, then their volume extinction-coefficient (β^A), or extinction cross-section per unit volume, is given by

$$\beta^A(\lambda) = N \, \sigma(\lambda), \qquad (5\text{-}20)$$

where:

β^A = volume extinction-coefficient, where A is taken to signify particulates
N = particulates per unit volume.

The dimensions of β^A are inverse length. Since the atmospheric particulates vary in size, the volume extinction-coefficient is obtained by an integration:

$$\beta^A(\lambda) = \int_o^\infty \frac{dn}{dr} (r) \, \sigma(\lambda,r) \, dr. \qquad (5\text{-}21)$$

The integrand of Eq. 5-21 divided by β^A is the *probability density-function* for extinction by particles of size r. This function is illustrated in Figure 5-8 for a Junge distribution (Eq. 5-17, $\nu = 3$) that extends from radius 0.01 μm to 10 μm. Although the smallest particles are the most numerous, the probability of extinction by them is small because their cross-sections are so small. On the other hand, the cross-section for the largest particles is large, but the probability of extinction by the largest particles is small because there are so few of them.

Angular Distribution of Scattered Light

The angular distribution of scattered light is described by means of a phase function. It gives the probability that light will be scattered into a unit solid angle. If no absorption occurs, the integral of the phase function over 4π steradians equals one (Chandrasekhar, 1950). The phase function is assumed to depend only on the polar angle between

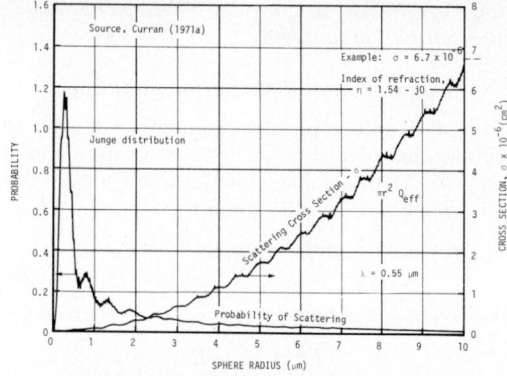

Fig. 5-8. Scattering cross-section and probability of scattering as a function of radius for a model of continental particulates.

the directions of the incident and scattered pencils of light. Light scattered near to the direction of propagation is loosely referred to as *forward scattering*. Scattering by particles of radii comparable to or greater than the wavelengths is strongly peaked about the forward direction. Such strong anisotropy is common for the atmosphere.

Scattering Phase Functions

The effect of particulates on the scattering phase functions of the atmospheric aerosol, including both particulates and gases, is shown in Figure 5-9. These measured functions are averaged for 10 different classes of atmospheric turbidity. The increasing number labels on the curves indicate decreasing visibility, or increasing turbidity, and hence increasing influence of the particulates. Curve 1(a) represents the average phase functions measured at a mountain observatory, at 3.2 km above sea level with average visibility of 220 km and with dry air. The phase function is very nearly that of a Rayleigh atmosphere (curve 1), which excludes particulate scattering. As the particulate influence increases, the phase function

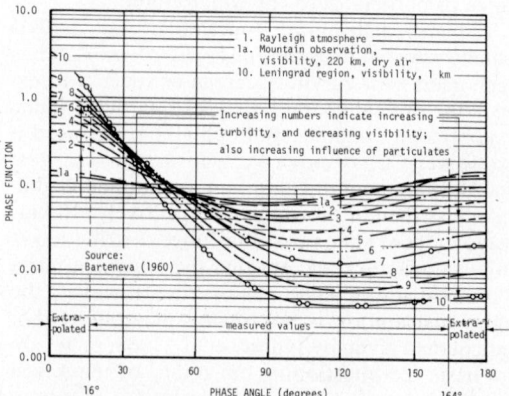

Fig. 5-9. Average measured phase functions for different states of the atmosphere.

---

[2] n is the usual designation of the index of refraction.

becomes more asymmetrical, and the forward scattering relative to backward scattering increases. The largest asymmetry was measured in the Leningrad region when the visibility was 1 km (curve 10).

The data of Figure 5-9 were measured only for the range of angles $16°-164°$, and were extrapolated over the remaining range of angles. Chayanova and Shifrin (1968) have computed the phase functions for $0°-16°$ and $164°-180°$. They find that extrapolated values in Figure 5-9 are valid for curves 1 to 5, but are too low for curves $5-10$.

OPTICAL THICKNESS

LAMBERT-BOUGUER LAW OF TRANSMISSION

A fundamental parameter for atmospheric optics is the optical thickness, which can be introduced by means of the *Lambert-Bouguer law of transmission*. As a beam of radiant energy propagates from a source at x_1 to a receiver at x_2, there is a loss of energy because of scattering and absorption. The fraction of monochromatic energy that is transmitted (T) along a homogeneous path is given by the Lambert-Bouguer law

$$T(\lambda, x_2 - x_1) = \exp\left[-\beta(\lambda)(x_2 - x_1)\right]. \quad (5-22)$$

where:

T = fraction of transmitted monochromatic energy.
β = volume extinction-coefficient.
x = path length, a variable.

β, at sea level, is of the order of 0.1 km^{-1}, in which case 37 percent of the energy is directly transmitted in a horizontal distance of 10 km.

It is convenient to replace the product βx by one parameter (τ) called the optical thickness: $\tau = \beta x$. In general β is not constant along the path. Then the optical thickness is calculated by an integral

$$\tau(\lambda, x_2 - x_1) = \int_{x_1}^{x_2} \beta(\lambda, x)\, dx. \quad (5-23)$$

The optical thickness along a slant path from the top of the atmosphere will be derived with the aid of Figure 5-10. The slant path is inclined at the polar angle Ψ from the vertical axis. Then $dx = -sec\ \Psi\ dz$. When this is substituted in Eq. 5-23, the optical thickness along the slant path between two heights z_1 and z_2 is

$$\tau(\lambda, z_2 - z_1) = sec\ \Psi \int_{z_1}^{z_2} \beta(\lambda, z)\, dz, \quad (5-24)$$

where:

z = height variable,
Ψ = polar angle of inclination from vertical, and where the implied assumption is that the extinction coefficient does not change in the horizontal direction.

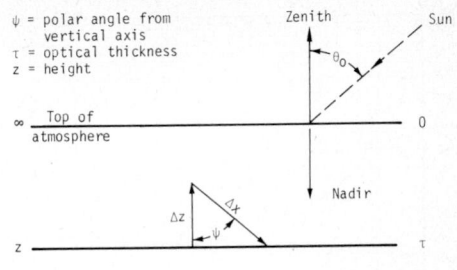

Fig. 5-10. Geometric relationships within the atmosphere, showing optical thickness coordinates.

The optical thickness along the vertical direction

$$\tau(\lambda, z) = \int_{z}^{\infty} \beta(\lambda, z)\, dz \quad (5-25)$$

is called the *normal optical thickness* and is the quantity that will be referred to in the following discussions.

Normal Optical Thickness

The normal optical thickness equals the sum of the separate optical thicknesses of all the attenuating constituents and each optical thickness is a function of λ and z:

$$\tau(\lambda, z) = \tau^G + \tau^{O_3} + \tau^{H_2O} + \tau^A, \quad (5-26)$$

where:

τ^G = scattering optical thickness of permanent gases, argon, nitrogen, oxygen, and carbon dioxide.
τ^A = particulate optical thickness (see Eq. 5-28).

Data that are given by Penndorf (1957) can be used to obtain the formula for τ^G in the 2 to 20 μm spectral region for argon, nitrogen and oxygen:

$$\tau^G(\lambda, z) = 1.07 \times 10^{-3}\ \lambda^{-4.10}\ H^G\ exp(-z/7.99) \quad (5-27)$$

where:

H^G = scale height of gas in km
λ = wavelength μm
z = height in km.

The optical thicknesses of other absorbing gases are omitted, since surface remote-sensing procedures utilize relatively non-absorbing spectral bands. Eq. 5-27, plotted on Figure 5-11, indicates that an atmosphere without particulates would scatter blue light much more strongly then red light. The optical thickness of ozone is also shown in Figure 5-11. If the wavelength is greater than 0.35 μm, ozone absorption is small enough to be neglected except in accurate computations.

Water Vapor Optical Thickness

The optical thickness of water vapor is negligible for $\lambda < 0.7$ μm, but must be accounted for at

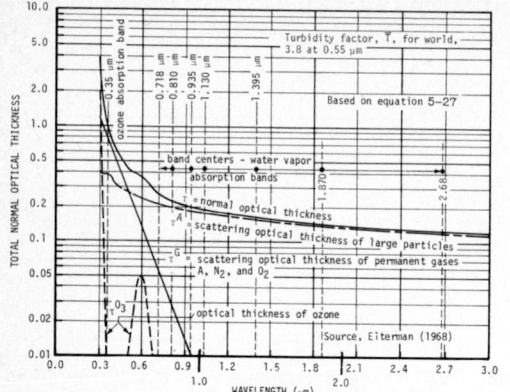

Fig. 5-11. Total normal optical thickness of several constituents of atmosphere.

longer wavelengths. The water-vapor bands and their strengths in the near infrared are given in Table 5-5, where the vertical transmission is calculated for a slab of water vapor equivalent in mass to liquid water of 1-cm depth. This amount of water vapor in a vertical column of the atmosphere frequently occurs in mid-latitudes. Water-vapor bands in the infrared are numerous and complex. Kyle (1975) has provided an excellent summary of the atmospheric transmission from 1 to 2600 cm^{-1} for altitudes above 54, 45, 40, 30, 14 and 4 km. The tables and figures include data in this spectral interval for H_2O, CO_2, O_3, N_2O, CO and CH_4.

Particulate Optical Thickness

The particulate optical thickness (τ^A) above a height z can be calculated from β^A, the extinction cross-section per unit volume for particulates (Eq. 5-21), as follows:

$$\tau^A(\lambda,z) = \int_z^\infty \beta^A(\lambda,d)\,dz \qquad (5\text{-}28)$$

TABLE 5-5

Near Infrared Water Vapor Absorption Bands*

The Total Amount of Precipitable Water Along a Vertical Atmospheric Path is 1 g cm^{-2}

Band Center, μm (absorption maximum)	Band Limits μm	Vertical Transmission
0.718	0.70 − 0.74	0.969
0.810	0.79 − 0.84	0.960
0.935	0.86 − 0.99	0.87
1.130	1.03 − 1.23	0.86
1.395	1.24 − 1.53	0.58
1.870	1.53 − 2.10	0.68
2.68	2.52 − 2.84	

* Source, Kondratyev (1969).

where, as before, the superscript A stands for particulate.

It can be shown that if the physical characteristics are independent of height and the number density decreases exponentially, the β^A is also an exponential function:

$$\beta^A(\lambda,z) = \beta^A(\lambda,0)\exp(-z/H^A), \qquad (5\text{-}29)$$

where:

H^A = scale height of particulates in km.

Measured data show that $\beta^A(\lambda,0)$ can be sometimes approximated (Junge, 1963; Rozenberg, 1967; Irvine and Peterson, 1970) by

$$\beta^A(\lambda,0) = \alpha'\,\lambda^{-b}. \qquad (5\text{-}30)$$

Substitution of Eqs. 5-29 and 5-30 in 5-28 yields this expression for the large particle optical thickness:

$$\tau^A(\lambda,z) = \alpha'\,\lambda^{-b}H^A\exp(-z/H^A). \qquad (5\text{-}31)$$

The exponent b usually lies between 0 and 2; a value of 1.3 is often used for continental regions (Robinson, 1966). The constant α' varies by an order of magnitude, when λ is expressed in micrometers; and $H^A = 1.2$ km, Kondratyev (1969), Elterman (1968), and McClatchey (1970) also discuss computations of τ^A. These references are also excellent sources of data on atmospheric optical thickness. Flowers (1969) gives statistical data on the optical thickness of the atmosphere for the United States. Elterman (1970) gives relations between the surface horizontal visibility and the optical thickness.

REFERENCE STANDARD FOR ATMOSPHERIC PARTICULATES

Elterman (1968) has constructed a model of atmospheric particulates, which is used as a reference standard. The values of τ^A for this model are given in Figure 5-11. The slope of the τ^A curve is much less than that of the τ^G curve. Below 0.4 μm, gaseous attenuation is dominant, but the particulate attenuation becomes much greater in the near infrared. Except in the ozone absorption band for $\lambda < 0.35$ μm and in the water-vapor bands for $\lambda > 0.7$ μm, as indicated on Figure 5-11, atmospheric attenuation at visible wavelengths is caused chiefly by scattering. The much stronger scattering in the short wavelengths accounts for the blue color of the sky.

The variance of the normal optical thickness of Eq. 5-26 is

$$\mathrm{var}(\tau) = \mathrm{var}(\tau^G) + \mathrm{var}(\tau^{O_3}) + \mathrm{var}(\tau^{H_2O}) + \mathrm{var}(\tau^A), \qquad (5\text{-}32)$$

if each of the variables can be considered as independent. These variances are in Table 5-6 for $\lambda = 0.55$ μm. Ninety-six percent of the variance is caused by particulates. The other constituents have negligible effect on the variance of the normal optical thickness for the visible spectrum.

TABLE 5-6

Variances of Total Normal Optical Thickness of the Extinction Constituents of a Cloud-Free Atmosphere, at $\lambda = 0.55\ \mu$m

Constituent		Optical Thickness τ^i	τ^i hr /τ^G	Variance (τ^i)
Water vapor,	τ^{H_2O}	0.003	0.03	$(3 \times 10^{-6})^2$
Ozone,	τ^{O_3}	0.03	0.3	$(1 \times 10^{-2})^2$
Permanent Gases,	τ^G	0.10	1.0	$(1 \times 10^{-3})^2$
Particulates,	τ^A	0.17	1.7	$(5 \times 10^{-2})^2$
				$(5.1 \times 10^{-2})^2$

However, the variance of the optical thickness of water vapor becomes important in the infrared.

TURBIDITY FACTOR AS MEASURE OF OPTICAL THICKNESS

Another measure of the atmospheric transparency is given by a *turbidity factor*, (T), of which there are many. One of them is the ratio of the total optical thickness τ to τ^G, which is essentially constant:

$$T(\lambda, z) = \frac{\tau(\lambda, z)}{\tau^G(\lambda, z)} . \qquad (5-33)$$

The average sea-level value of $\overline{T}$ suggested for the world, is

$$\overline{T} = 3 \text{ for } 0.55\ \mu\text{m}$$

(Shifrin and Shubova, 1964). In Figure 5-11,

$$\overline{T} = 3.8 \ (0.55\ \mu\text{m})$$

for the Elterman model. $\overline{T}$ increases with increasing wavelength from about 2 at 0.4 μm to 8 at 0.75 μm. Consequently, the optical thicknesses of the atmosphere and of the large particles are comparable in the blue spectrum; but τ^A is much greater than τ^G in the red spectrum, as has been indicated previously on Figure 5-11. The standard deviation of $\overline{T}$ is one-half at 0.55 μm (Shifrin and Shubova, 1964).

DOWNWELLING RADIATION AT THE GROUND

SPECTRAL IRRADIANCE OF SUNLIGHT AND SKYLIGHT

Sunlight reaching the ground is conveniently separated into two components: (1) direct sunlight, which passes directly through the atmosphere without being absorbed or scattered; and (2) skylight, or the diffuse light, which is light that has been scattered at least once.

The total spectral irradiance (E_G) on a horizontal surface of the ground equals the sum of the irradiances of the direct sunlight (E_S) and of the skylight (E_D):

$$E_G(\tau_1) = E_S(\tau_1) + E_D(\tau_1), \qquad (5-34)$$

where:

E_G = total spectral irradiance; frequently called the *global flux*.
E_S = direct sunlight irradiance.
E_D = skylight irradiance.

The spectral irradiance of the direct sunlight can be easily calculated from the following expression:

$$E_S(\tau_1) = S_\lambda \cos \Theta_o \exp(-\tau_1 \sec \Theta_o), \qquad (5-35)$$

where:

τ_1 = specific optical path length,
S_λ = monochromatic solar constant, as given in Figure 5-5,
Θ_o = zenith angle of the sun.

The computed ratio of the irradiance of direct sunlight to the global irradiance at the ground is shown as a function of the optical path length $(\tau_1 \sec \Theta_o)$ in Figure 5-12. The computations apply to a model of a nonabsorbing Rayleigh atmosphere and ground reflectance of less than 0.15. *The reflectance of a surface is defined as the ratio of the upward to downward fluxes of radiant energy at the surface or to the ratio of irradiance to radiant emittance.*

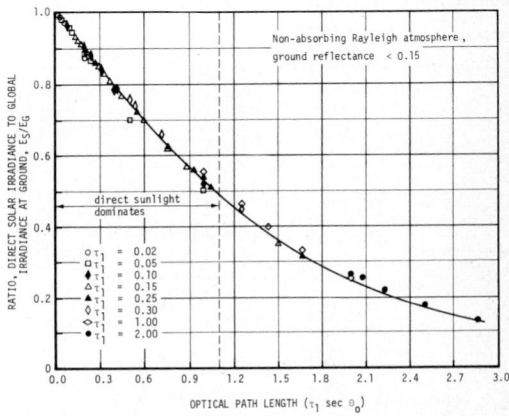

Fig. 5-12. Ratio of the direct solar irradiance to the global irradiance.

Figure 5-12 shows that if the optical path length is less than 1.1, the direct sunlight dominates. For an example of the relative importance of direct sunlight and skylight, assume a representative value of optical thickness of $\tau_1 = 0.50$ and a solar zenith angle of $\Theta_o = 45°$. Then $\tau_1 \sec \Theta_o = 0.7$; Figure 5-12 shows that the direct sunlight contributes 65 percent of the global irradiance. The curve is slightly steeper for an atmosphere with particulates; when the reduced solar flux contributes one-half of the irradiance, then $\tau_1 \sec \Theta_o = 0.96$. As the ground reflectance increases, the curve will become somewhat steeper, since the diffuse component of irradiance increases.

SPECTRAL COMPOSITION OF SUNLIGHT AND SKYLIGHT

The data of Figure 5-12 can be used to show that there is a difference between the spectral composition of the skylight and of the direct sunlight at the surface. At an arbitrary solar zenith angle, the skylight is bluer than the sunlight, since the ratio of the irradiances of direct sunlight and global flux decreases with decreasing wavelength (increasing τ_1). Furthermore, the relative contribution of the sunlight to the illumination of the object space increases with solar zenith angle.

Average measured irradiances of the ground by the entire solar spectrum on clear days are shown in Figure 5-13. Curve E_D represents the irradiance of the skylight, curve E the direct sunlight on a plane perpendicular to the solar rays, and curve E_G the global irradiance. E_G is related to E_D and E by

$$E_G = E_D + E \cos \Theta_o. \qquad (5\text{-}36)$$

The noon solar flux (E) is about 25 percent lower in June than in December because the sun is 7 percent farther from the earth; the solar zenith angle is larger, which accounts for a 5 percent decrease; and the turbidity is higher. The larger turbidity causes more scattered light and higher diffuse irradiance (E_D). The ratios of the diffuse and global irradiances are 0.20 and 0.43 at noon-

time during December and June, respectively. These ratios increase with increasing solar zenith angle.

PATTERN OF SKY RADIANCE

The distribution of the diffuse radiant energy across the sky must be considered for some remote sensing experiments because this light, when reflected from the ground, can be one of the principal components of the upwelling radiation. One experiment where this can occur is in the remote measurement of ocean pollution through the observation of color. These parameters ordinarily are measured outside of the bright glitter pattern caused by the reflection of the direct sunlight.

The light received outside of this pattern consists of light scattered from below the sea surface, in addition to skylight reflected from the surface. The radiance of the reflected skylight depends on both the radiance of the skylight incident on the ocean and the reflection coefficient.

The general pattern of the radiance of the sky is shown for red and blue light in Figure 5-14, as measured for a clear and clean atmosphere in a desert region of the State of New Mexico. Points of the sky are given in polar coordinates. The sun is located at a zenith angle of $\Theta_o = 46°$ and an azimuth angle of $0°$, and the ground reflectance is roughly 0.25. The radiance for each color is normalized to the value of 100 at the azimuth angle of $180°$ and at some zenith angle, and the general radiance patterns are similar for both colors. The sky is very bright near the sun, in a region called the *aureole;* this is caused by the strong forward scattering of direct sunlight by particulates. The sky is darkest in the quadrant opposite the sun; the horizons are relatively bright. The sky radiance is different for a polluted atmosphere (Coulson, 1971); it is also dependent on the ground reflectance. The upwelling radiation into the base of the atmosphere can be considered as a secondary source of illumination of the atmosphere.

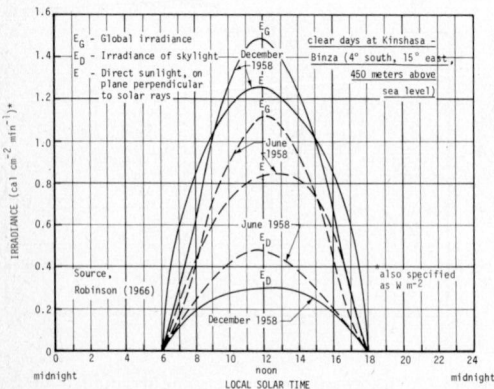

Fig. 5-13. Average daily measured variation of ground irradiance on clear days by entire solar spectrum.

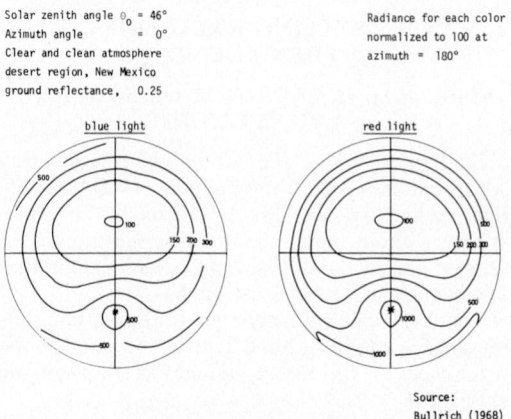

Fig. 5-14. Measured relative radiance of the sky for blue and red light.

SCATTERING BACK FROM UPWELLING RADIATION

The computed fraction of light that enters the base of a nonabsorbing molecular atmosphere and is scattered back down to the ground is shown as a function of the total normal optical thickness in Figure 5-15. Two curves are shown: one for the case where a rough ground reflects light diffusely according to Lambert's law and the other curve for a case where a smooth surface reflects light specularly according to Fresnel's law. The solar zenith angle is $\Theta_o = 60°$. The reflectance of the atmosphere is independent of the solar zenith angle for a Lambert surface, and for a Fresnel surface increases slowly with Θ_o for $\Theta_o < 60°$ and rapidly for $\Theta_o > 60°$. Figure 5-15 shows that the reflectance of the atmosphere is relatively insensitive to extreme variations in the character of the ground reflectance.

Approximately 30 percent of the upwelling light that enters the base of the atmosphere is scattered back down to the ground for the visible spectrum ($\tau_q = 0.4$). If the reflectance of the ground is small, for example, 5 percent as for oceans, then the atmosphere is weakly illuminated from below; about 30 percent of this illumination is reflected back to the ground for the visible spectrum, but augments the global irradiance by only two percent. On the other hand, in snow-covered regions where the reflectance can be as high as 90 percent, one-fourth of the global irradiance on the ground consists of light reflected from the snow and then scattered back down to the ground again.

Computed skylight radiances can be found in a number of references, among which are: Atroshenko (1963), Bullrich (1964), Coulson et al. (1960), Dave and Furukawa (1966), Feigelson et al. (1960), and Plass and Kattawar (1968–1971). A small amount of measured radiance data is given by Bullrich (1968), Kondratyev (1969), and Coulson (1971). Figure 5-15 represents data from Kondratyev relating to solar irradiance at the earth for different optical thicknesses.

UPWELLING RADIATION AT THE TOP OF THE ATMOSPHERE

Variations in the radiant energy leaving the top of the atmosphere depend on the solar zenith angle, principally because the solar irradiance varies with $\cos \Theta_o$, the reflectance of the ground, the particulates in the atmosphere, and also with the water-vapor content, especially in the near-infrared water absorption bands. The dependence of the upwelling energy on these parameters will be demonstrated by a discussion of the radiant emittance of the earth-atmosphere system; although the effects of clouds will continue to be neglected.

Since remote sensing is usually done by measuring radiances, an examination of their characteristics follows. Polarization is beginning to be used in remote sensing, and it will be mentioned briefly.

SHORTWAVE RADIATION

The total *spectral radiant emittance M* of the upwelling light from the earth-atmosphere system can be expressed as the sum of two components: The first is the light that interacts with the ground or water, and eventually emerges from the top of the atmosphere (M_3); and the second component (M_2) never interacts with the surface, but is scattered just by the atmosphere.

Information about the object space is contained in the M_3-component. The M_2-component limits the amount of information that can be extracted from the measured radiation. The sum of the two components gives the total radiant emittance:

$$M = M_2 + M_3 \qquad (5\text{-}37)$$

Relative and Absolute Values

The relative values of these radiant emittances are given in Figure 5-16 for a model atmosphere that is nonabsorbing and scatters light according to Rayleigh's law. Although particulate effects are excluded from this model, it can be used to illus-

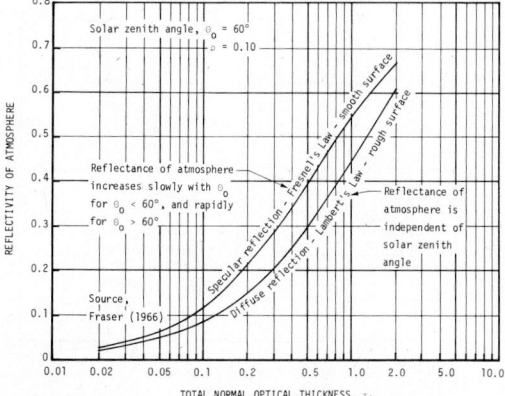

Fig. 5-15. Reflectance of a Rayleigh atmosphere for illumination from below.

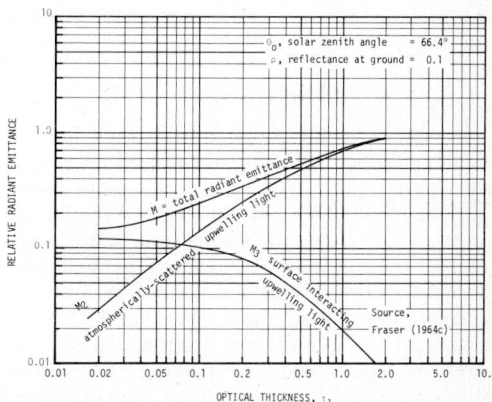

Fig. 5-16. Computed radiant emittance at the top of a Rayleigh atmosphere as a function of optical thickness.

trate the essential features of the radiation field. The computed radiant emittances in this section, and radiances in the following section, are relative values but can be converted to absolute values by multiplying them by the monochromatic solar constant S_λ and dividing by π. In Figure 5-16 the solar zenith is $\Theta_o = 66.4°$ and the ground reflectance is $\rho = 0.1$, which is a low value that can be associated with oceans. The radiant emittance of just the airlight (M_2) vanishes if the atmosphere disappears ($\tau_1 \to 0$), and at the other extreme of an infinitely thick atmosphere equals the solar irradiance, $\pi \cos(66.4°) = 1.26$. In the latter case all incident solar radiation is scattered outwards by the atmosphere, since no radiant energy reaches the ground. The radiant emittance of the light reflected from the ground (M_3) exceeds the radiant emittance of the airlight (M_2) only for small optical thicknesses, where $\tau_1 < 0.08$.

Effect of Ground Reflectance

If the ground reflectance is increased above the value of 0.1 used in Figure 5-16, the relative contribution of the light reflected from the ground will also increase. As a good approximation, the total spectral radiant emittance is linearly proportional to the reflectance of the ground:

$$M = M_2 + \alpha\rho \qquad (5\text{-}38)$$

where:

α is independent of ρ, but depends to some extent on the solar angle and particulates.

For observations of objects on the ground from a large distance in the visible spectrum, the unwanted light scattered by the atmosphere can greatly exceed that returned from a surface of low reflectance, where $\rho < 0.1$.

Effect of Surface Roughness

The radiant emittance of an earth-atmosphere system is only weakly dependent on the roughness of the surface as long as its reflectance stays constant (Fraser, 1964c). However, the roughness does have a significant effect on the radiance of the upwelling light, as will be shown in later paragraphs.

Effect of Solar Zenith Angle

The effect of the solar zenith angle on the radiant emittance at the top of the model Rayleigh atmosphere and smooth ocean is shown in Figure 5-17. The incident solar irradiance is $\pi \cos \Theta_o$ and decreases by a factor of 5 as the solar zenith angle increases from 0° to 78.5°. However, at small optical thicknesses the radiant emittance increases as the solar zenith angle increases when $\Theta_o > 37°$. This occurs because the thin atmosphere has only a negligible effect, and the reflectance of the solar radiation from the smooth surface increases faster than the incident solar energy decreases. The surface reflectance is less important when the at-

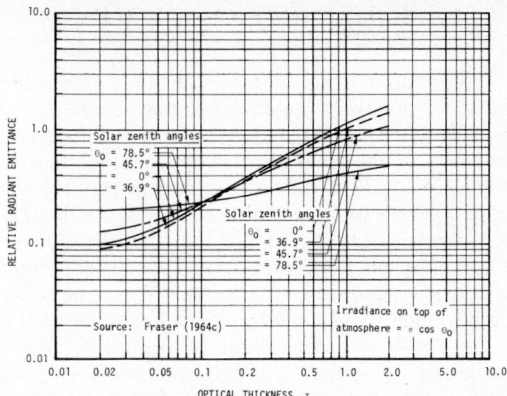

Fig. 5-17. Computed radiant emittance at the top of a model Rayleigh atmosphere and smooth surface.

mosphere is optically thick—the radiant emittance is then more nearly proportional to the irradiance of incident sunlight.

Effect of Particulates

The effect of particulates on radiant emittance is not know accurately, since the absorption characteristics of many solid and liquid particles have yet to be determined. The computed effect is given in Figure 5-18 for the model of an earth-atmosphere that contains both the permanent gases and nonabsorbing continental particulates, the size-distribution function of which is given in Figure 5-7. The optical thicknesses as of the gaseous components are $\tau^G = 0.205$ and 0.101 for $\lambda = 0.46 \ \mu m$ and $0.54 \ \mu m$, respectively. The solar zenith angle is $\Theta_o = 40.5°$, the incidence solar irradiance is 2.39, and the ground has the low reflectance of $\rho = 0.03$. The radiant emittance increases with increasing particulate optical thickness. However the particulates are not as efficient in scattering light outwards as are the atmospheric gases, because the phase function for scattering

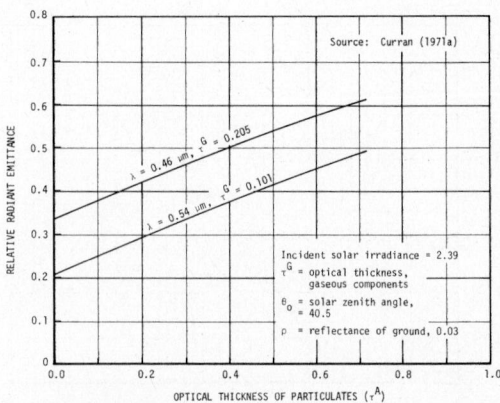

Fig. 5-18. Computed radiant emittance of an earth-atmosphere system as a function of the optical thickness of the contained particulates.

the direct sunlight in the upward directions is smaller for particulates than for the gases (Figure 5-9).

For example, the radiant emittance of the up-welling light from an atmosphere, the particulate optical thickness of which equals that of just the gases, is only 25 percent greater than the radiant emittance from a clear atmosphere without the particulates. As a result, it can be expected that the characteristics of the light scattered outwards will be dominated by Rayleigh scattering, if the optical thickness of the gas and of the particulates are comparable. Since particulates scatter sunlight much more strongly in the infrared spectrum, the characteristics of the upwelling infrared radiation in the atmospheric windows are dominated by particulate scattering.

The relative *radiance L* at the nadir of an ocean-atmospheric model is given in Figure 5-19 for the same model used for the computations of the radiant emittance, as given in Figure 5-16. The total radiance is

$$L = L_2 + L_3, \qquad (5\text{-}39)$$

where:

L = radiance of light (radiant emittance per unit solid angle) reflected from the surface, and emergent from top of the atmosphere,

L_2, L_3 = subscripts having the same meaning as in Figure 5-16 and Eq. 5-37.

Upwelling light through the surface is neglected in this computation, the radiant emittance of this component being comparable to the radiant emit-

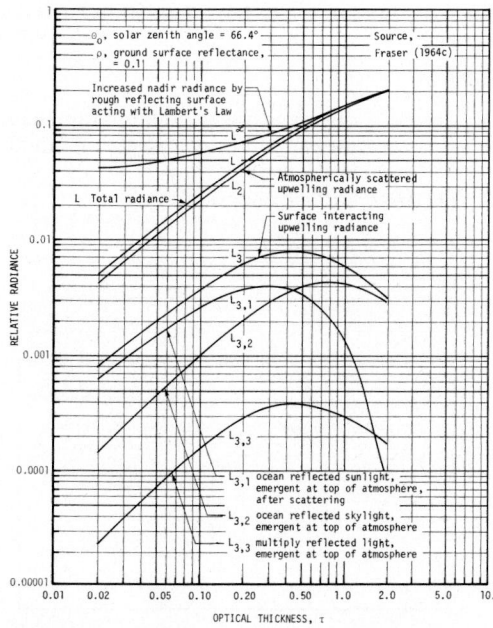

Fig. 5-19. Computed relative radiance of the nadir of a model Rayleigh atmosphere and smooth ocean.

tance of the reflected light (Davis, 1941). The important point to be learned from Figure 5-19 is that the radiance of the airlight (L_2) in the visible spectrum is many times greater than the radiance of the light reflected from the sea.

Effect of Surface Roughness

The roughness of a surface has a strong effect on the radiance of an earth-atmosphere system. The radiances identified by double subscripts ($L_{3,1}$; $L_{3,2}$; $L_{3,3}$) in Figure 5-19 refer to light reflected from a smooth ocean which eventually escapes from the top of the atmosphere. This reflected light is separated into three components:

$$L_3 = L_{3,1} + L_{3,2} + L_{3,3}, \qquad (5\text{-}40)$$

where:

$L_{3,1}$ = the radiance of direct sunlight reflected from the ocean surface, eventually emerging in the zenith direction after at least one scattering by the atmosphere;

$L_{3,2}$ = the radiance of skylight reflected from the surface of a smooth ocean, eventually emerging from the top of the atmosphere;

$L_{3,3}$ = the radiance of light reflected at least twice from the surface before escaping (this radiance is negligible).

Effect of Direct Reflected Sunlight

Again using Figure 5-19, the direct reflected sunlight that is not scattered emerges from the atmosphere at a nadir angle of 66.4°, which is the same value as the solar zenith angle. If the ocean surface is rough, the inclined facets of the sea will occasionally reflect the direct sunlight toward the zenith direction. As a result, a larger fraction of the reflected light will escape in the zenith direction. The curve labeled L indicates how much the nadir radiance is increased by a rough surface that reflects light according to Lambert's law. In this case the total radiance is, to a good approximation, linearly proportional to the reflectance,

$$L = L_2 + c\rho \qquad (5\text{-}41)$$

where:

c depends on solar zenith angle Θ_o and an optical thickness τ_1.

Effect of Particulates

The radiance of a model of an earth-atmosphere containing particulates is given in Figure 5-20; this is the same model referred to in the discussion of Figure 5-18. The absolute value of radiance is obtained from Figure 5-20 by multiplying the values shown by the monochromatic solar constant (S_λ). The ground is assumed to reflect light according to Lambert's law with a reflectance of $\rho = 0.03$. The radiance at the top of the atmosphere is nearly symmetric about the nadir. The radiance of the

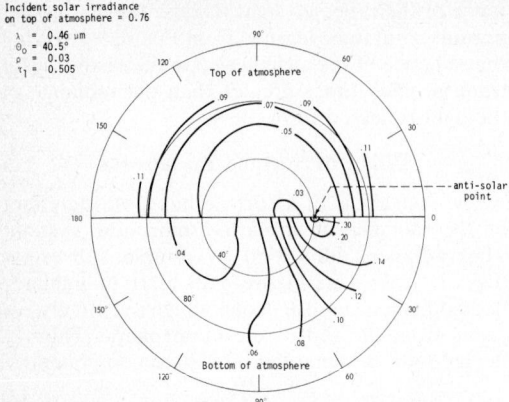

Incident solar irradiance
on top of atmosphere = 0.76

λ = 0.46 μm
Θ_0 = 40.5°
ρ = 0.03
τ_1 = 0.505

Fig. 5-20. Computed relative radiance at the base and at the top of an atmosphere containing particulates.

light incident on the ground, which is given in the lower half of the figure, lacks this symmetry. The radiances at the base and at the top of the atmosphere are thus quite different.

Effect of Solar Zenith Angle

The radiance at the top of the atmosphere is weakest near the antisolar point, which is indicated on Figure 5-20. It is significant that for remote sensing the unwanted airlight is relatively weak near the nadir. The radiance increases slowly with increasing nadir angle until the nadir angle reaches 60°. In order to minimize the unwanted contribution of airlight, the rim of the earth near the horizon should be avoided in remote sensing scanning.

The computations just discussed were based on plane-parallel models; these are good models until line-of-sight approaches the horizon (Collins et al., 1972). Then a spherical model is a more accurate representation.

Effect of Polarization

Some of the unwanted airlight can be removed under certain conditions where it is highly polarized. The principle is the same as using a polarizer with a camera to enhance the contrasts in a distant outdoor scene. If the scattered airlight within the field of view of a sensor is completely plane polarized, an analyzer in the system (such as a piece of polaroid film) can prevent this light from reaching the detector. On the other hand, if the scattered airlight and light reflected from the object space are unpolarized, an analyzer is not helpful, and will only reduce the energy transmitted to the detector by one-half. Discussions of the characteristics of polarization of airlight are given by Coulson (1966) and by Plass and Kattawar (1970). Additional information on polarization will be found in Chapter 3.

Measured Radiation over Ocean Surfaces

Radiances measured over the ocean from a satellite agreed well with computed values, when the field of view seemed to be either completely free of clouds or entirely covered by clouds (Fowler, et al., 1971). An example of nadir radiances measured over a common area of the ocean at heights of 0.92 km and 14.9 km is given in Figure 5-21.

The radiance of the blue light is five times larger at the upper height than at the lower height, while the radiance of the red light increases three times with height. The measurements at 0.92 km indicate the approximate radiance of light that has interacted with the ocean. About three-fourths of this light is transmitted to 14.9 km; as a consequence, the radiance of just the airlight is about four times the radiance at 0.92 km for blue light. Additional measured radiances for land and cloud surfaces in the spectral band of 0.4 to 2.4 μm are given in the NASA report *Earth Albedo And Emitted Radiation* (NASA, 1971a).

LONGWAVE RADIATION

Because the atmosphere is not completely transparent even in the least absorbing regions of the infrared spectrum, the outgoing spectral radiance of the earth will be influenced not only by the earth's surface but also by the composition and thermal structure of the atmosphere. The observed radiance will be further modified by the presence of clouds and by scattered solar radiation. It is therefore necessary to take all of these factors into account if accurate, reliable, surface parameters are to be obtained.

Elements of Surface Radiance Measurements

The total spectral radiance $I(\nu,\Theta)$ observed at frequency ν and zenith angle Θ can be expressed in terms of four main components:

$$I(\nu,\Theta) = I_s(\nu,\Theta) + I_a(\nu,\Theta) + I_d(\nu,\Theta) + I_h(\nu,\Theta)$$
$$(5\text{-}42)$$

where

$\qquad I(\nu,\Theta)$ = observed clear-column radiance, measured in the absence of clouds in the field of view,

$\qquad I_s(\nu,\Theta)$ = surface emission,

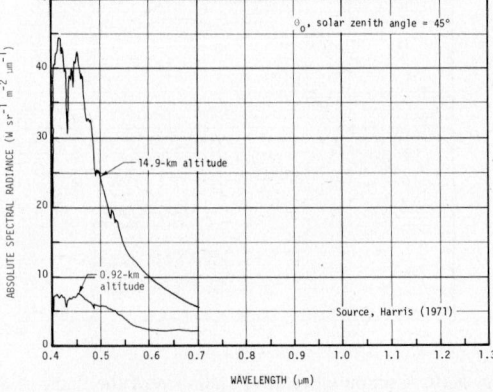

Fig. 5-21. Nadir radiances measured over the ocean at two heights (Hovis, 1971).

$I_a(\nu,\Theta)$ = atmospheric emission,
$I_d(\nu,\Theta)$ = reflected atmospheric downward flux,
$I_h(\nu,\Theta)$ = reflected solar flux.

The geometry associated with observations made when the instrument is viewing the earth at a zenith angle Θ is shown in Figure 5-22. The surface is assumed to be horizontal so that the local normal is parallel to the vertical axis. In general the outgoing radiance has both zenith and azimuth angular dependence; in this section the azimuthal dependence will be suppressed. The cloud effect will not be discussed in this section because we will assume that the instrument field-of-view (FOV) is sufficiently small to allow for more frequent occurrence of cloud-free observations. The effect of atmospheric scattering is not strong at long wavelengths and for our purposes can be neglected at wavelengths larger than ~ 8 μm.

Surface Emission

The emitted radiance reaching the instrument from the surface can be written as

$$I_s(\nu,\Theta) = \epsilon_s(\nu,\Theta)\, B(\nu,T_s)\, \tau(\nu,\Theta,z_s) \quad (5\text{-}43)$$

where $\epsilon_s(\nu,\Theta)$ is the surface emissivity, and $\tau(\nu,\Theta,z_s)$ is atmospheric spectral transmittance between the surface and the instruments. The Planck function $B(\nu,T)$ at frequency ν and temperature T is given by

$$B_\nu(T) = \frac{2h\nu^3}{c^2} \frac{1}{\exp(h\nu/kT) - 1} \quad (5\text{-}44)$$

as in Eq. 5-10.

Atmospheric Emission

The atmospheric emission for a plane, parallel, and homogeneous atmosphere in local thermodynamic equilibrium can be expressed as

$$I(\nu,\Theta) = \int_{z_0}^{\bar{z}} B[\nu,T(z)]\, \frac{\partial \tau(\nu,\Theta,z)}{\partial z}\, dz \quad (5\text{-}45)$$

where $T(z)$ is the vertical atmospheric temperature profile as a function of height, z, and $\tau(\nu,\Theta,z)$ is the clear-column atmospheric transmittance between height z and the observing instrument at $\bar{z}$.

For most atmospheric conditions the use of the trapezoidal rule is adequate for evaluating $I(\nu,\Theta)$.

Reflected Atmospheric Downward-Flux

The reflected thermal downward-flux originates from the atmosphere above the surface. In general it is seen from Figure 5-23 that the surface element dA receives radiant energy through an elementary beam of solid angle $d\Omega_i$ from a direction (Θ_i,Φ_i). The radiant energy reflected into the solid angle $d\Omega_r$ in the direction (Θ_r,Φ_r) comes from all directions above the surface. This radiant energy will be attenuated by the atmosphere when it traverses the atmosphere layers between the surface and the observing system. At the observation point the reflected thermal downward-flux can be expressed in its general form as

$$I_d(\nu,\Theta_r,\Phi_r) = \tau(\nu,z_s,\Theta_r,\Phi_r) \int_{FOV} dA \int d\Omega_i$$

$$I'(\nu,\Theta_i,\Phi_i)\, \rho(\nu,\Theta_i,\Phi_i,\Theta_r,\Phi_r)\, \cos\Theta_i \quad (5\text{-}46)$$

where $I'(\nu,\Theta_i,\Phi_i)$ is the atmospheric downward-radiation in the direction (Θ_i,Φ_i), $d\Omega_i = \sin\Theta_i\, d\Theta_i\, d\Phi_i$, and $\rho(\nu,\Theta_i,\Phi_i,\Theta_r,\Phi_r)$ is the partial spectral reflectance. The expression for I_d can be simplified by assuming an optically thin isotropic atmosphere where

$$I'(\nu,\Theta_i,\Phi_i) = I_a(\nu,0)/\cos\Theta_i.$$

In this case $I_d(\nu,\Theta)$ can be written as

$$I_d(\nu,\Theta) = \rho_s(\nu,\Theta)\, \hat{I}_a(\nu)\, \tau(\nu,\Theta,p). \quad (5\text{-}47)$$

with $\hat{I}_a(\nu) = 2\, I_a(\nu,0)$.

For a Lambertian surface the directional surface reflectivity is $\rho_s(\nu,\Theta)$ is related to $\epsilon_s(\nu,\Theta)$ by

$$\rho_s(\nu,\Theta) = 1 - \epsilon_s(\nu,\Theta) \quad (5\text{-}48)$$

Reflection of Solar Flux

During daytime observations an additional term caused by the scattering of solar flux from the surface, $I_h(\nu,\Theta)$ should be included. The sun can be considered as a source subtending a small solid

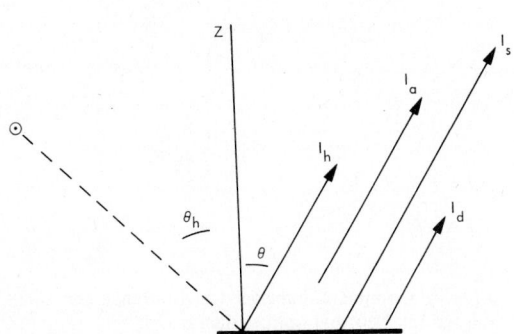

Fig. 5-22. Components of surface emission.

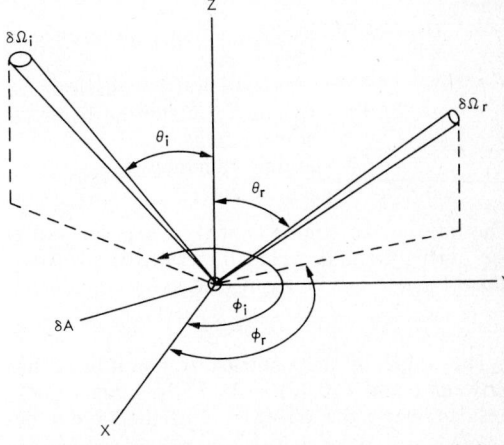

Fig. 5-23. Geometry of incident and reflected radiation.

angle $d\Omega_h$ in the direction Θ_h, and the reflected solar flux measured by the instrument can be written as

$$I_h(\nu,\Theta) = H_h(\nu) \cos \Theta_h \, \tau'(\nu,\Theta_h,p_s) \, \rho'_s(\nu) \, \tau(\nu,\Theta,p_s)$$

$$(5\text{-}49)$$

where the solar irradiance for normal incidence outside the atmosphere at $T_h \approx 6000$ K is

$$H_h(\nu) = 2.16 \, \pi \, 10^{-5} \, B(\nu,T_h)$$

and

$$\rho'_s(\nu) = \rho'_s(\nu,\Theta_i,\Phi_i,\Theta_r,\Phi_r) \neq 1 - \epsilon_s(\nu)$$

where Θ_h is the direction of the sunbeam, and $\tau'(\nu,\Theta_h,p_s)$ is the transmittance of the entire atmospheric clear column traversed by the incident solar flux.

CONTRAST AND CONTRAST TRANSMITTANCE

A useful parameter for measuring the quality of an image is the apparent contrast between an object or target and its surrounding terrain. For the purpose of this chapter, contrast (C) will be defined as the difference between radiance in the directions to the target (L^t) and to its surrounding terrain (L^s) divided by L^s, thus,

$$C = (L^t - L^s)/L^s, \qquad (5\text{-}50)$$

At a finite distance from the object space, the radiant energy emanating from the object is attenuated by the atmosphere, but is also augmented by the radiance of the intervening column of air (L^a). The airlight radiances in the direction to the target and to its nearby surroundings are essentially the same, when the target is small. Hence, the airlight radiances cancel when taking the difference ($L^t - L^s$) at the observer. The apparent contrast then equals the difference at the object space ($L^t_o - L^s_o$) multiplied by the atmospheric transmission, and divided by the radiance transmitted from the surroundings plus the airlight radiance:

$$C = (L^t_o - L^s_o)T^s_o/(L^t_o \, T^a + L^a), \qquad (5\text{-}51)$$

where:

$(L^t_o - L^s_o)$ = airlight radiance difference at object space

T^a = atmospheric transmission

$L^s_o \, T^a$ = radiance transmitted from the surroundings

L^a = airlight radiance

subscript o = object space.

The ratio of the contrasts at the observer and at the object space is called the *contrast transmittance* (y) and can be expressed as

$$y = \left[1 + L^a/(L^s_o \, T^a)\right]^{-1}. \qquad (5\text{-}52)$$

The value of the contrast transmittance lies between 0 and 1 ($0 < y < 1$). If the optical thickness between the observer and the object becomes very large, the transmission approaches zero. Then the contrast transmittance is negligibly small. On the other hand, the contrast transmittance is up to one if the amount of the atmosphere between the object and observer vanishes; then the transmission $T^a = 1$, and the airlight radiance $L^a = 0$. Unfortunately, the contrast transmittance depends on more than just the optical state of the ground surrounding the target. The contrast transmittance increases with increasing radiance of the ground.

Effect of Airlight on Contrast Transmittance

The computed effect of airlight on contrast transmission through the entire atmosphere is shown in Figure 5-24 for an optically thin atmosphere (in which $\tau_1 = 0.02$), which is the approximate lower limit of the optical thickness in the near-infrared spectrum, and for $\tau_1 = 1.0$, which is a value encountered for the near-ultraviolet. The data of Figure 5-7 are based on an earth-atmosphere model of terrain reflecting light according to Lambert's law with reflectance of $\rho = 0.25$ and of a Rayleigh atmosphere. Then the data on the left apply to a wavelength of $\lambda = 0.809 \, \mu m$ and on the right apply to $\lambda = 0.312 \, \mu m$. The maximum contrast transmittances are $y = 0.99$ and $y = 0.23$ for the lower and higher optical thickness, respectively. Both maxima occur at about the same zenith angle in the principal plane, which contains the observer, nadir, and sun. The contrast transmittance for the large optical thickness approaches zero at a zenith angle of $\Theta = 81°$. In general one can expect that the contrast transmittances in an arbitrary direction of observation will increase with decreasing thickness.

Effect of Model Atmospheres on Contrast Transmittance

The computed contrast transmittance at the top of different model atmospheres will be given. Lambert's law of reflection is assumed for the

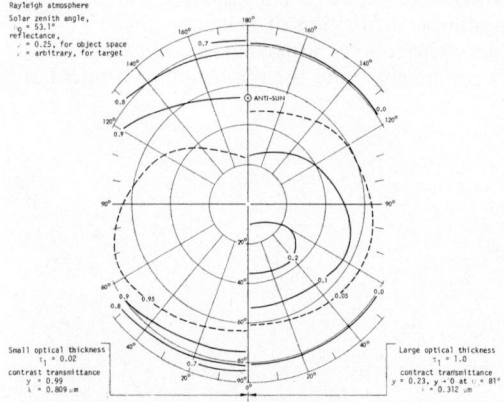

Fig. 5-24. Computed contrast transmittance for small and large total normal optical thickness.

object space, except for the target. The light reflected from the small target can be of arbitrary radiance and polarization. The contrast transmittance in the principal plane is shown in Figure 5-25 for the case in which the terrain surrounding the target has a reflectance $\rho = 0.25$ and for $\lambda = 0.625$ μm. The dashed lines give the contrast transmittance (y). The contrast transmittance is small near the horizons because of both the high radiance of the airlight and the low radiance of the light transmitted from the target. The highest dashed curve gives y for an atmosphere free of particles and shows y for $\lambda = 0.625$ μm and $\rho = 0.25$. The contrast transmittance decreases for the model atmospheres containing continental and Los Angeles-type particulates.

Effect of Polarization on Contrast Transmittance

Because the light leaving the atmosphere is polarized, the contrast transmittance can be enhanced by extinguishing part of the airlight with an analyzer in an optical receiver. The solid lines (y_ℓ) of Figure 5-25 are calculated for the case in which the transmission plane of a perfect analyzer (one that transmits no light through a plane perpendicular to the transmission plane) in the receiver optical system is parallel to the principal plane. Use of an analyzer increases the contrast transmittance, except near the antisolar point, where the polarization of the light is small.

Effect of Height on Contrast Transmittance

Examples of the measured apparent contrast as a function of height are given in Figures 5-26a, b, c, and d. The intrinsic contrast is not given, but may be 0.99. If so, the values of apparent contrast in these figures are only one percent larger than the contrast transmittance (y). The apparent contrast decreases more with height at the shortest wavelength and least with the longest wavelength. Figures 5-26a, b, c, and d reflect the general conclusions from a total of 27 experiments: the appar-

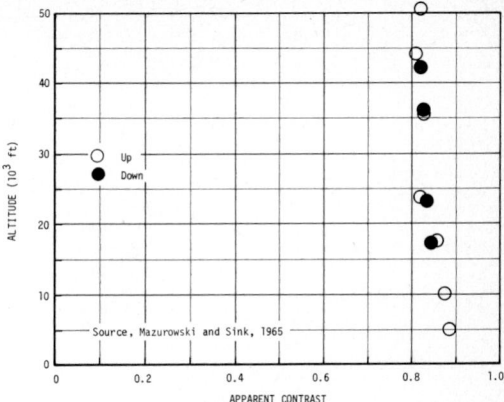

Fig. 5-26a. Measured apparent contrasts at various altitudes with red filters.

ent contrast shows a small spectral dependence below 10,000 ft; at an arbitrary height above 10,000 ft the apparent contrast increases with increasing wavelength. The greatest loss of contrast occurs in the lowest 34,000 ft (Mazurowski and Sink, 1965).

Duntley et al. (1964) have investigated a method of measuring the contrast transmittance from the ground. The ground-based measurements agree well with high-altitude measurements.

TURBULENCE

The effects of atmospheric turbulence on light are apparent to the most casual observer. If, for example, on a hot day, his line of sight to an object passes close to a hot surface such as a highway, the object appears to move and to be distorted. Another common manifestation of turbulence is the twinkling of stars. Telescopic observations show, even though the observer knows there is no such movement, that a star appears to change position by about 10 sec of arc, and that the apparent radiance fluctuates.

The computed effect of the turbulent atmo-

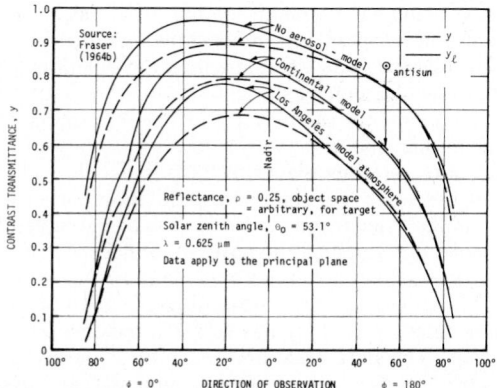

Fig. 5-25. Computed contrast transmittance for three model atmospheres.

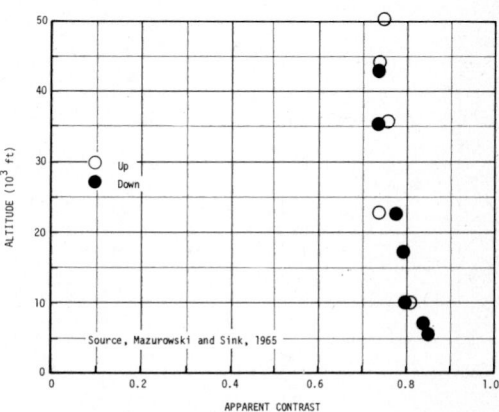

Fig. 5-26b. Measured apparent contrasts at various altitudes with green filters.

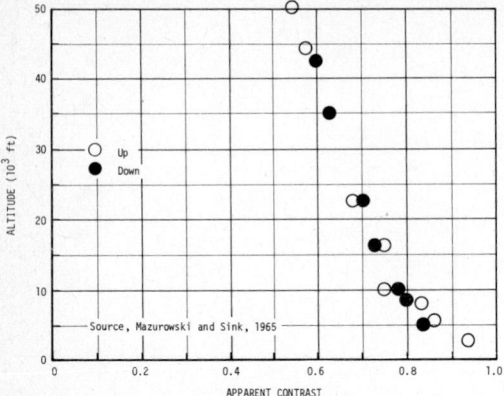

Fig. 5-26c. Measured apparent contrasts at various altitudes with blue filters.

sphere on light propagating upwards shows that this effect is negligibly small for remote sensing (Farhat and DeCou, 1969; Fried, 1966a,b; Lee and Harp, 1969; Tatarski, 1961; Weiner, 1967). The smallest resolvable size on the ground increases with sensor height up to a height of about 50 km, and then remains constant. Above a height of 50 km the smallest resolvable distance is of the order of centimeters, which is much less than that required in earth resource programs.

ATMOSPHERIC RADIATIVE TRANSFER PROCESSES IN THE INFRARED

The infrared spectral interval is widely used in remote sensing of the physical state of the surface and atmosphere. "Atmospheric windows," regions in the infrared spectrum in which attenuation due to atmospheric gases is minimal, permit observations of surface thermal emission from space. Radiative characteristics of surface materials can, in some cases, provide unique identification of soil types and ground cover. Vital to meteorological research is the measurement of sea surface and land brightness temperatures in these

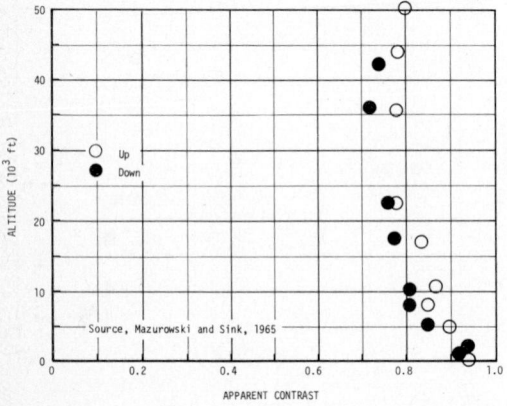

Fig. 5-26d. Measured apparent contrasts at various altitudes with no filter.

infrared window regions. The uniformly mixed gas, carbon dioxide, is used extensively for sounding the vertical and horizontal temperature field of the atmosphere from the surface to near 100 km using a strong spectral band of CO_2 to 15 μm. Spectral features of other gases, notably water vapor at 6.7 μm, have been observed to determine the composition of the atmosphere. Measurements of cloud height, amount, and type are rapidly developing applications of remote sensing thermal radiation.

The mechanisms of gaseous line-, band- and continuum-absorption and emission as well as the radiative properties of cloud particles in the infrared are discussed below.

LINE ABSORPTION

Transfer of infrared EMR through the atmosphere involves complex interactions with a variety of molecular species. Interactions of radiation with molecules may produce motion, or molecules possessing motion may emit radiation. Water vapor, carbon dioxide and ozone dominate the processes of infrared absorption and emission in the earth's atmosphere. The interaction of molecular vibration and rotation of these polyatomic molecules with radiation is considered here.

Spectral lines have their origins in transitions which occur between the energy levels of a molecule. Energy can be absorbed if the transition is from a lower to a higher energy level, and emitted if the transition is from a higher to a lower energy level. The energies and spectral positions of line features depend on the resonances due to electrons, vibrating atoms in a molecule, and rotating molecules. These resonances depend on the physical structure of the molecule itself. In rotation, the three principal moments of inertia of the molecule determine the characteristics of the spectrum. Four types of rotating molecule may be distinguished. If all three principal moments of inertia are different, the molecule is called an asymmetric top (H_2O is a molecule of this type): if at least two of these are equal, it is called a symmetric top. Of this type, the spherical top has all three principal moments of inertia equal at the center of mass (CH_4 is an example) and the linear molecule (such as CO_2) in which the positions of the atoms lie along a straight line and thus has two equal moments of inertia and one of negligible magnitude. No energy transitions are allowed in the pure rotation spectrum for a molecule which possesses no permanent dipole moment. The symmetric linear molecule, CO_2, is an example of a molecule which has no pure rotational spectrum. For this same reason the most abundant of atmospheric gases, the homonuclear, diatomic molecules, N_2 and O_2, have no infrared absorption bands. The pure rotation spectrum of the asymmetric top molecule, H_2O, has an apparently disordered structure as a result of the greatly differing moments of inertia. The problem of finding the

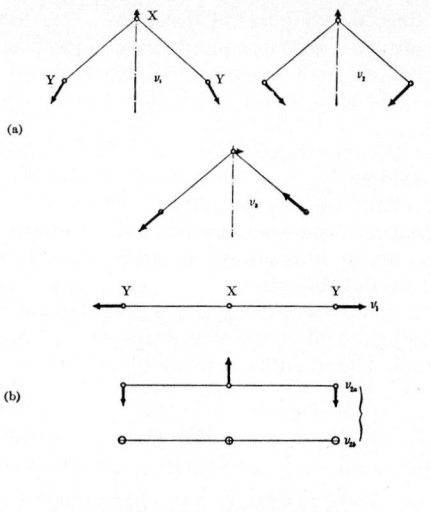

Fig. 5-27. Normal vibrations of (a) bent and (b) linear XY$_2$ molecules. H$_2$O is an example of a non-linear XY$_2$ molecule; CO$_2$ is an example of a linear molecule; (Herzberg 1945).

energy levels of such molecules is a difficult one and no explicit formula can be found. However, by considering the asymmetric molecule to be an intermediate case of the two types of symmetric top molecules the energy levels may be approximated and the line positions computed. Tabulations of measurements of the rotation band of H$_2$O may be found in McClatchey et al. (1973).

A non-linear molecule consisting of N atoms has 3N degrees of freedom. Translational motion accounts for three degrees of freedom and rotation for an additional three. There remain 3N-6 coordinates to describe the motion of nuclei relative to axes with their origin at the center of mass (linear molecules possess 3N-5 degrees of freedom). The normal modes of vibration of a

triatomic non-linear molecule, H$_2$O, and a triatomic linear molecule, CO$_2$, are shown in Figure 5-27.

The ν_1 vibration of CO$_2$ involves no change of dipole moment and is therefore inactive in the infrared spectrum. The ν_2 band of CO$_2$ is centered at 15 μm as a result of bending-type vibration. This band has been intensely studied as a result of its extensive use in the remote sounding of atmospheric temperature. The atmosphere is opaque over a broad region about the band center. The asymmetric stretch mode ν_3 of CO$_2$ produces a strong band near 4.3 μm.

While the rigid rotator and harmonic oscillator models of molecules provide insight into the main features of the infrared spectrum, it is necessary to consider the interaction between rotational and vibrational motion. This interaction arises from the fact that the effective moment of inertia of a molecule is different for different vibrational states because of the change in the mean square value of the internuclear distance R (see Houghton and Smith, 1966). Branching of the band structure, for example the 15 μm band of CO$_2$ (Figure 5-28), is evidence of rotation-vibration interaction. The rotational transitions appear as fine structure near the frequency of the vibration transition.

The principal complication in infrared radiative transfer calculations for the atmosphere is the broadening of spectral lines. Spectral lines in a rotation-vibration band do not have zero widths. The natural width of a line is a consequence of the finite lifetime of the molecule in an excited state. This width cannot be less than

$$\frac{\Delta \nu}{2} \approx \frac{1}{4 \pi \Phi}$$

where Φ is the lifetime of the excited state. At normal atmospheric temperatures this width is negligible when compared with the Doppler width

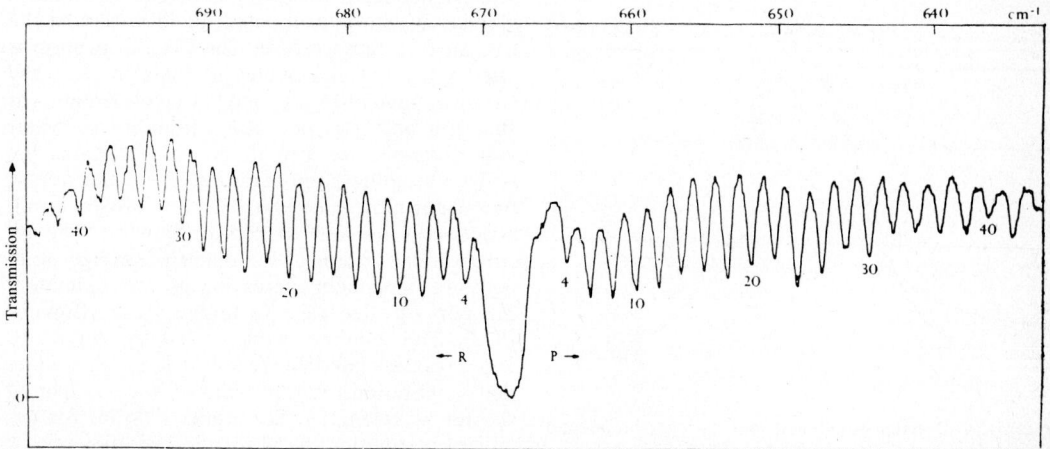

Fig. 5-28. The ν_2 band at 15 μm. The additional structure in the wings of the P-branches is due to the isotope ^{13}C^{16}O$_2$, (Plyler 1960).

of the line. The thermal motion of the molecules of a gas during emission or absorption of radiation gives rise to a Doppler shift of the radiation frequency. The resultant shape of the spectral line is given by the formula:

$$k\nu_o = \frac{S}{\gamma_D \sqrt{\pi}} \exp\left\{ - \left(\frac{\nu - \nu_o}{\gamma_D}\right)\right\} \quad (5\text{-}53)$$

where:

$$S = (\text{the line strength}) = \int_0^\infty k_\nu \, d\nu \text{ with some}$$

temperature dependence,

ν = frequency,
ν_o = the frequency of the line center,
γ_D = (the Doppler line width)
$$= \frac{\nu_o}{c}\left(\frac{2\pi R T}{M}\right)^{1/2},$$

c = velocity of light in the atmosphere,
T = temperature of the absorbing or emitting gas,
R = the universal gas constant,
M = molecular mass of the gas.

Doppler broadening produces a Gaussian profile.

A further source of line broadening arises from pressure-induced collisions and mutual interactions between molecules. The shape of collision-broadened (Lorentz) lines is given by

$$k_{\nu_L} = \frac{S}{\pi} \frac{\gamma_L}{(\nu - \nu_o)^2 + \gamma_L^2} \quad (5\text{-}54)$$

where γ_L (Lorentz half-width) = $(1/(2\pi t))$, where t is the mean time between collisions of the absorbing gas, and S, ν, ν_o are defined as before. The mean time between collisions of the molecules in the atmosphere is inversely proportional to the pressure p, hence

$$\gamma_L = p/p_o \, \gamma_L \, (p_o) \quad (5\text{-}55)$$

A line of Doppler shape is more intense at the line center and weaker in the wings than a Lorent-

zian line; (see Figure 5-29) hence, a line that is fully absorbed at the center (referred to as black at line center) will increase in absorption in the wings due to collision rather than to Doppler effects.

The Lorentz width dominates at high pressures, for example, $\gamma_L \approx \gamma_D$ near 30-km altitude for the 15-μm band of CO_2. In the upper atmosphere the infrared spectral lines are of Doppler shape and consequently temperature dependent rather than pressure dependent.

The band strength S_{band} is a useful quantity in assessing the influence of an absorber in a spectral interval. The quantity is given by

$$S_{band} = \int_{band} k_\nu \, d\nu \quad (5\text{-}56)$$

CONTINUUM ABSORPTION

Background absorption or emission which occurs between spectral lines and bands is of major importance when viewing into the lower atmosphere from a spacecraft or aircraft platform. Atmospheric windows are, in fact, contaminated by interactions between the atmosphere and the radiation field. The most common source of opacity is to be found in the slowly decaying wings of distant spectral lines. Far-wing absorption frequently does not follow the Lorentzian profile but, instead, falls off more slowly and may thus affect window regions. Also, broad spectral regions are never completely free of weak lines which may significantly reduce the transparency of the window.

The two most useful intervals in the infrared which are substantially transparent are at 3.5–4.1 μm and 10.5–12.5 μm. The 11 μm window is of particular importance since it is centered near the energy peak of a black body at typical surface temperatures. Carbon dioxide and water vapor contribute to the opacity in this atmospheric window. The effect of CO_2 is generally much smaller than that of H_2O, and since CO_2 is uniformly mixed and has a relatively constant distribution globally it can be accounted for in retrieving the true surface temperature. The CO_2 absorption is due to the extreme wings of lines in the very strong ν_2 band at 15 μm, and to two weak bands at 10.4 μm and 9.4 μm. The variability of water vapor complicates any attempt to determine the surface brightness temperature. The water dimer, two water molecules joined by a hydrogen bond, is active in the 10.5- to 12.5-μm interval. The absorption coefficient of the dimer is strongly temperature dependent increasing by approximately 2% per 1°C decrease in temperature (Bignell, 1970). The window is bracketed by the water vapor rotation band on the long wave edge, and by the ν_2 vibration-rotation band of water vapor at shorter wavelengths. See Roberts (1976) for details of computing the absorption coefficients for the dimer and line wings for water vapor in this interval. Rozenberg (1977) reports findings which

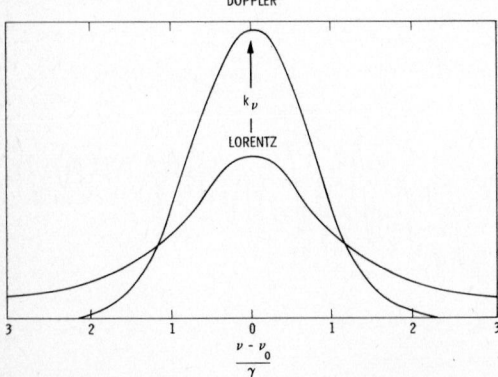

Fig. 5-29. Doppler-broadened and Lorentz-broadened line shapes are compared for two lines with the same strengths, line centers, pressure and temperature, such that $\gamma_L = \gamma_D$ (Houghton, 1968).

suggest that submicron aerosols may be a major source of opacity in the window under apparently high visibility conditions. Contributions from all sources of attenuation may decrease the observed surface temperature as seen by an aircraft- or spacecraft-borne instrument by as much as 10°C for typical atmospheric conditions.

In the 3.5- to 4.1-μm atmospheric window, collision-induced absorption by nitrogen is significant (Farmer and Houghton, 1966), as is CO_2 line absorption. The contribution of opacity from water vapor is due to wings of distant lines. This window is less affected by gaseous absorption than the 10.5- to 12.5-μm window, but the black body energy is substantially lower for surface emission, and solar reflection from the surface may be non-negligible.

To understand Eq. 5-7 more fully, the contributions due to gaseous absorption, particle absorption, and particle scattering may be considered separately. The extinction coefficient γ_ν is defined as

$$\gamma_\nu^A = \beta_\nu + k_\nu \rho \qquad (5\text{-}57)$$

where:

β_ν = particle extinction-coefficient due to absorption within the particles,
k_ν = gas absorption coefficient,
ρ = density of the gas.

For the special case in which there are no particles in the medium we find the extinction coefficient

$$\nu_\nu = k_\nu \rho$$

the single scattering albedo

$$\omega = 0$$

the elemental opacity

$$d\tau = -k_\nu \, \rho \, ds$$

and, finally, the transfer equation to be

$$dI_\nu(\Theta,\Phi) = -k_\nu \rho \, I_\nu(\Theta,\Phi)ds + k_\nu \rho \, B_\nu(T)ds \qquad (5\text{-}58)$$

or

$$\frac{dI_\nu(\Theta,\Phi)}{ds} = -k_\nu \rho \left[I_\nu(\Theta,\Phi) - B_\nu(T) \right] \qquad (5\text{-}59)$$

The quantity $exp\text{-}[\int_{path} k_\nu \, ds]$ is the transmission over the path. It is the polychromatic transmission through an atmosphere free of particles which we will consider first.

The mean transmission in the spectral interval (ν_1, ν_2) for a vertical path, limited by the altitudes z_1 and z_2, in an atmosphere containing only gas is given by

$$\overline{T}(z_1, z_2) = \frac{1}{\nu_1 - \nu_2} \int_{\nu_1}^{\nu_2} exp\left\{ - \int_{z_1}^{z_2} k_\nu(z) \, \rho \, dz \right\} d\nu \qquad (5\text{-}60)$$

where ν is the frequency, z is the altitude and $k_\nu(z)$

is the absorption coefficient at frequency ν of the absorbing gas whose density is ρ. In general, $k_\nu(z)\rho$ is the sum of the absorption coefficient and density products of the spectral lines of all gases in the atmosphere at ν. Eq. 5-60 cannot be evaluated analytically. A number of numerical techniques have been suggested for handling the complex integration over frequency and over altitude (see for example Scott, 1974). These techniques, referred to as "line-by-line methods" permit computations of radiative transfer in realistic atmospheres using compilations of line data for atmospheric gases (McClatchey et al., 1973). The accuracy of the model calculations is largely determined by the sophistication of the transfer algorithms used, which are in turn dictated by the computing power available, and the quality of the line data for the gases in the spectral interval of interest. The data base is generally adequate for the major gaseous species for most applications throughout the infrared.

A very different approach to the method of determining the transmission functions of gases is to replace the integration over frequency in Eq. 5-60 with a model of the characteristics of all the lines within a spectral interval.

This approach is most often used when the computation time for line-by-line calculations is prohibitive, and when the requirements on accuracy are not severe. The most widely used band model is the statistical model due to Goody (1952) which assumes the spectral lines to be randomly distributed in the band and the line intensities to have an exponential distribution. The mean transmission is expressed in the statistical model as an integration over a single line

$$\overline{T} = exp\left\{ -1/\delta \int_{-\infty}^{\infty} \frac{\tau_\nu}{1 + \tau_\nu} \, d\nu \right\} \qquad (5\text{-}61)$$

where δ is the mean line spacing and τ_ν is the monochromatic optical depth. Using a mean line strength k for the spectral interval, and a Lorentz shape with half-width α, the mean transmission becomes

$$\overline{T} \approx exp\left\{ -\frac{km}{\delta}\left(1 + \frac{km}{\pi \, \alpha} \right)^{-1/2} \right\} \qquad (5\text{-}62)$$

over a path at constant pressure and absorber amount m. The band parameters are defined by

$$k/\delta = \frac{1}{\nu_1 - \nu_2} \sum_i S_i \qquad (5\text{-}63)$$

and

$$k/\pi\delta = \left(\frac{k}{\delta} \right)^2 \left(\frac{2 \sum_i \sqrt{S_i \alpha_i}}{\nu_1 - \nu_2} \right)^{-2} \qquad (5\text{-}64)$$

where, S_i and α_i are the line strengths and half-widths, respectively, at ν_i within the spectral interval $\nu_1 - \nu_2$.

The Lorentz line shape is not valid when the

pressure varies as it does in the atmosphere along the path followed by the radiation. The Curtis-Godson approximation avoids the need for integration over the optical path for each ν necessary to define the correct line shape. It does so by assuming the line to be Lorentz and defining the half-width as

$$\bar{\alpha} = \frac{\int \alpha \, dm}{\int dm} \ . \qquad (5\text{-}65)$$

The absorber amount along the inhomogeneous path is given by $\int dm$.

An example of the use of the Goody band model in atmospheric radiative transfer calculations can be found in Rodgers and Walshaw (1966). Care must be taken in applying band models to ensure the validity of the assumptions in the model by performing representative line by line calculations for comparison. In considering more than one gas in a spectral interval the frequency-averaged transmissions may be multiplied if the line positions of the gases are not correlated.

In the case where the atmosphere contains clouds or aerosols the scattering-source function term in Eq. 5-15 becomes important. The scattering term is

$$J_\nu(\Theta,\Phi) = 1/4\pi \int_0^{4\pi} I_\nu(\Theta,\Phi) \, p(\cos\Theta) \, d\Omega \qquad (5\text{-}66)$$

where:

$$\Omega = \text{solid angle of incident beam}$$
$$p(\cos\Theta) = \text{the scattering phase function,}$$
which describes the angular distribution of the scattered beam

and,

$$\cos\Theta = \mu'\mu + (1 - \mu^2)^{1/2} (1 - \mu'^2)^{1/2} \cos(\Phi',\Phi) \qquad (5\text{-}67)$$

where:

$$\mu = \text{cosine of the angle of the incident ray}$$

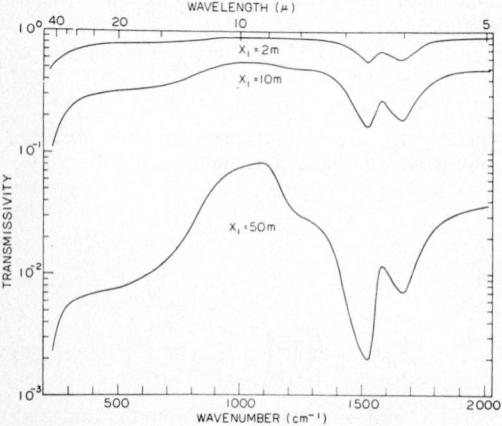

Fig. 5-30. Spectral transmissivity of clouds of various thicknesses versus wavenumber or wavelength (Yamamoto et al., 1970).

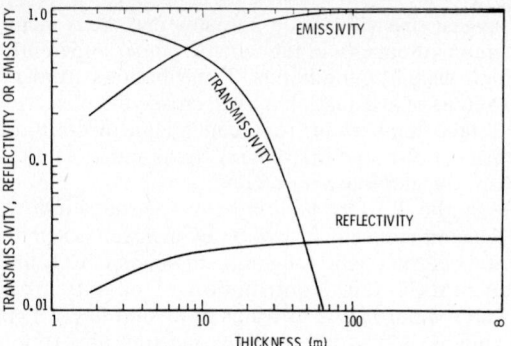

Fig. 5-31. Emissivity, reflectivity and transmissivity of clouds for the infrared region from 5-50 μ versus cloud thickness in meters (Yamamoto et al., 1970).

$$\mu' = \text{cosine of the angle of the emergent ray;}$$
$$\Phi \text{ and } \Phi' \text{ are the azimuthal and emergent angles, respectively}$$

Mie (1908) presented the solution of Maxwell's equations necessary to determine $p(\cos\Theta)$ for spheres of arbitrary size. Van de Hulst (1957) gives a clear and detailed account of the theoretical and practical aspects of scattering from single particles. Hansen and Travis (1974) give an excellent review of the computational methods available for treating multiple scattering in planetary atmospheres.

Figure 5-30 shows the spectral transmissivity of altostratus clouds of various thicknesses having a water droplet distribution with minimum, maximum and mode diameters of 2, 26 and 9 μm, respectively. This figure is a result of computations due to Yamamoto et al. (1970), which includes the radiative processes due to both cloud droplets and water vapor. Note that the transmissivity decreases with increasing cloud thickness and that the transmissivity is smaller at 6.3 μm (1580 cm^{-1}) in the water vapor band region, and in the far infrared where droplet and water vapor rotation-band absorption are large. The mean emissivity, reflectivity and transmissivity of this cloud type over the entire infrared spectral region from 5-50 μm are shown in Figure 5-31. The mean emissivity and reflectivity tend to 0.97 and 0.03, respectively, as the cloud thickness approaches infinity.

The radiative characteristics of clouds in the infrared depend strongly on the particle-size distribution, number-density and composition (water or ice). Irvine and Pollack (1968) have tabulated the infrared optical properties of water and ice spheres.

SCATTERING, EXTINCTION AND ABSORPTION BY CLOUDS

The extent to which a particular cloud type effects the properties of EM radiation within a spatially localized region depends upon the scattering, extinction, and absorption coefficients, given by:

$$\gamma_j(n,\lambda) = \int_r dn(r)/dr \; x_j(n,\lambda,r)dr \quad (5\text{-}68)$$

with

$$N = \int_r dn(r)/dn \quad (5\text{-}69)$$

where:

x_j = single-drop cross-section ($j = S$, A, E; S = scattering; A = absorption; E = extinction),
$dn(r)/dr$ = number of drops per cm³ in the radius range from r to $r + dr$,
N = total number density at the point in the medium.

In the Rayleigh limit, the single-drop cross-sections for extinction and absorption are dependent only upon the mass of the droplet. Hence, Eq. 5-68 takes the simple form

$$\gamma_E \simeq \gamma_A = mf(\lambda,T), \quad (5\text{-}70)$$

where:

m = mass density of ensemble in g m⁻³,
f = a function of the wavelength and temperature.

Models of Cloud-Cover Conditions

A total of 19 cloud models have been taken from Reifenstein and Gaut (1971), representing both fair-weather and rain-bearing cloud conditions, and including commonly occurring categories of low, middle, and high stratus, and cumulus clouds. The properties of these models are given in Table 5-7. Each model is characterized by one or more horizontal layers for which the composition (water cloud, ice cloud, precipitation, etc.), the mass density, and three parameters describing the drop-size distribution, are specified. In Table 5-7, column 1 contains a reference number for each cloud model, and column 2 lists the most descriptive name of the models. The base height and top height are specified in meters in columns 3 and 4; the separate specification of a base- and top-height for each layer allows for the possibility of layers which are physically separated or stratified in height. Columns 6-8 contain the parameters characterizing the cloud-drop size distribution, $dn(r)/dr$.

There are many possible choices for the distribution of dn/dr. For this study, however, the analytical distribution of Deirmendjian (1964) was used. His work enables a given cloud layer to be characterized for cloud-drop size distribution by four parameters: the mass density m of liquid water in gm⁻³ (see column 5); mode radius of distribution, r_c; and two shape parameters, C_1 and C_2. Deirmendjian's suggested distribution, which

is identical to Eq. 5-16, except for a change in notation, is expressed as

$$dn(r)/dr = A\, r^{C_1} \exp\left[-Br^{C_2}\right] \; (cm^{-3}\,\mu m^{-1}), \quad (5\text{-}71)$$

which includes, in addition to the above listing, two scale parameters, A and B, where

$$A = \frac{mC_2 B^{\left[\frac{C_1+4}{C_2}\right]}}{\frac{4}{3}\pi \times 10^6 \Gamma\left[\frac{C_1+4}{C_2}\right]} \quad (5\text{-}72)$$

$$B = \frac{C_1}{C_2\, r_c^{C_2}} \quad (5\text{-}73)$$

where

Γ = gamma function
m = mass density of the cloud, in g m⁻³,

and where B is a function of

r_c = drop radius at which the number density distribution function is maximum, as tabulated in column 6 of Table 5-7.

Both A and B are also functions of the shape parameters,
C_1 (see column 7, Table 5-7), and
C_2 (see column 8, Table 5-7).

The three ice-cloud models represent cirrus clouds at three predominantly different altitude ranges, corresponding to the three different climatological conditions of the "standard" atmospheres. The information necessary to construct the models has been taken largely from Valley (1965), Fletcher (1966), and Mason (1957); data of Laws and Parsons (1943), as tabulated by Crane (1966), have been used for the rain-bearing layers. For modeling purposes, the ice layers have been assumed to be made up of ice spheres. The mode radius of 40 μm was based on the work of Blau et al. (1966), with the shape parameters designed to include the effect of random-oriented crystal faces of 100 μm or more. The properties of haze, included as Model 20-4, have been taken from Deirmendjian (1964). The catalog does not intend to be inclusive, but instead covers a range of cloud conditions of interest in the study of remote-sensing applications.

Drop-Size Distributions

Two drop-size distributions representing extremes of cloud and precipitation conditions have been modeled using Eq. 5-71, and are shown in Figure 5-32. The model labeled (CU) has a mode radius of 10 μm and average condensed water density of 0.01 g m⁻³. It represents a very thin fair-weather cumulus. The cumulonimbus model (CB + rain) has a mode radius of 400 μm. The density used (8.0 g m⁻³) is near the upper limit found in the atmosphere, according to Weickman

TABLE 5-7

Properties of Standard Cloud Models

1	2	3	4	5	6	7	8	9	10
		Cloud Height		Mass Density H_2O	Mode radius of distribution	Shape Parameters		Principal composition of each cloud layer	Source Reference
Reference Cloud Number	Descriptive Cloud Name	Base (m)	Top (m)	liq. ($g\ m^{-3}$)	r_c (μm)	C_1	C_2	Comp.	(See below)
1-A-1	CIRROSTRATUS, ARCTIC, 12-18 kft	4000.	6000.	0.10	40.0	6.0	0.5	ICE	1,4
1-M-1	CIRROSTRATUS, MID-LAT., 15-21 kft	5000.	7000.	0.10	40.0	6.0	0.5	ICE	1,4
1-T-1	CIRROSTRATUS, TROPICAL, 18-24 kft	6000.	8000.	0.10	40.0	6.0	0.5	ICE	1,4
10-1	ALTOCUMULUS 8000-9650 ft	2400.	2900.	0.15	10.0	6.0	0.5	WATER	1,2,3,
14-1	ALTOSTRATUS 8000-9650 ft	2400.	2900.	0.15	10.0	6.0	1.0	WATER	1,2,3
20-1	LOW-LYING STRATUS 500-2000 ft	150.	650.	0.25	10.0	6.0	1.0	WATER	1,2,3
20-2	LOW-LYING STRATUS 1500-3000 ft	500.	1000.	0.25	10.0	6.0	1.0	WATER	1,2,3
20-3	FOG LAYER, GROUND to 150 ft	0.	50.	.15	20.0	7.0	2.0	WATER	3
20-4	HAZE, HEAVY	0.	1500.	10^{-3}	0.05	1.0	0.5	WATER	1,5
21-1	DRIZZLE, 0.2 mm/hr	0.	500.	1.00	20.0	6.0	0.5	RAIN	6
		500.	1000.	2.00	10.0	6.0	0.5	WATER	
		1000.	1500.	1.00	10.0	6.0	0.5	WATER	
21-2	STEADY RAIN, 3 mm/hr	0.	150.	0.20	200.0	5.0	0.5	RAIN	6
		150.	500.	1.00	10.0	6.0	0.5	WATER	
		500.	1000.	2.00	10.0	6.0	0.5	WATER	
		1000.	1500.	1.00	10.0	6.0	0.5	WATER	
21-3	STEADY RAIN, 15mm/hr	0.	300.	1.00	200.0	5.0	0.5	RAIN	6
		300.	1000.	2.00	10.0	6.0	0.5	WATER	
		1000.	2000.	3.00	10.0	6.0	0.5	WATER	
		2000.	4000.	2.00	10.0	6.0	0.5	WATER	
22-1	STRATOCUMULUS 1000-2000 ft	330.	660.	0.25	10.0	6.0	0.5	WATER	1,2,3
22-2	STRATOCUMULUS 2000-4000 ft	660.	1320.	0.25	10.0	6.0	0.5	WATER	1,2,3

25-1	FAIR WEATHER CU. 1500-6000 ft	500.	1000.	0.50	10.0	6.0	0.5	WATER	1,2,3
		1000.	1500.	1.00	10.0	6.0	0.5	WATER	
		1500.	2000.	0.50	10.0	6.0	0.5	WATER	
25-2	CUMULUS WITH RAIN 2.4 mm/hr	0.	500.	0.10	400.0	5.0	0.5	RAIN	3,6
		500.	1000.	1.00	20.0	6.0	0.2	WATER	
		1000.	3000.	2.00	20.0	6.0	0.2	WATER	
25-3	CUMULUS WITH RAIN 12 mm/hr	0.	400.	0.50	400.0	5.0	0.5	RAIN	3,6
		400.	1000.	2.00	20.0	6.0	0.2	WATER	
		1000.	4000.	4.00	10.0	6.0	0.2	WATER	
25-4	CUMULUS CONGESTUS, 3000-9000 ft	1000.	1200.	0.30	10.0	6.0	0.5	WATER	3
		1200.	1600.	0.50	15.0	5.0	0.4	WATER	
		1600.	2000.	0.80	20.0	5.0	0.3	WATER	
		2000.	2500.	1.00	20.0	5.0	0.3	WATER	
		2500.	3000.	0.50	20.0	5.0	0.3	WATER	
26-1	CUMULONIMBUS W. RAIN 150 mm/hr	0.	300.	6.30	400.0	5.0	0.2	RAIN	2,3,6
		300.	1000.	7.00	20.0	6.0	0.2	WATER	
		1000.	4000.	8.0	10.0	6.0	0.2	WATER	
		4000.	6000.	4.00	10.0	6.0	0.2	WATER	
		6000.	8000.	3.00	10.0	6.0	0.2	WATER	
		8000.	10000.	0.20	40.0	6.0	0.5	ICE	

REFERENCES:
1. Valley (1965)
2. Fletcher (1966)
3. Mason (1957)
4. Blau, Espinola, and Reifenstein (1966)
5. Deirmendjian (1964)
6. Crane (1966), tabulating data of Laws and Parsons (1943)

Definitions of Principal Compositions

water = cloud with liquid water, 100% humidity, and a continuous drop-size distribution
ice   = cloud layer primarily of ice crystals
rain = cloud layer primarily of precipitation, <100% humidity, drop-size distribution concentrated at radii > 100 μm.

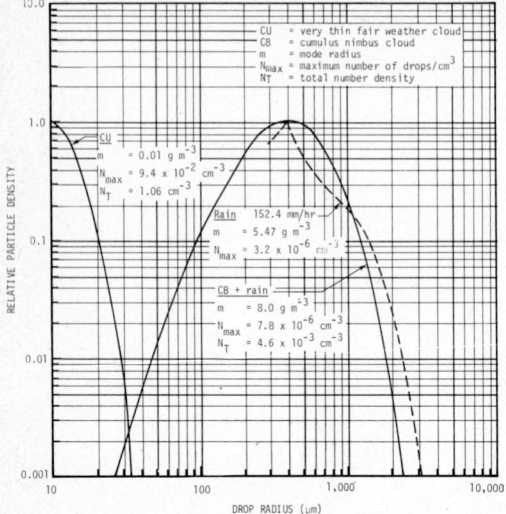

Fig. 5-32. Drop-size distributions for three model clouds, normalized to maximum value.

(1953). Figure 5-32 also shows, for comparison, a distribution of Crane (1966) based on Laws' (1943) data, corresponding to a rain rate of 152.4 mm hr^{-1}.

ATMOSPHERIC RADIATIVE TRANSFER PROCESSES IN THE MICROWAVE

The EMR reflected or radiated from a ground surface experiences interacton with the atmosphere as it propagates to the remote sensor, located in an aircraft or satellite. As the radiation interacts with the atmosphere, the information it carries from the surface can be altered or degraded considerably. The extent of this interaction is a function of the atmosphere and its composition, and the wavelength, or band of radiation.

Similarly, as for the visible or infrared spectrum, microwave radiation reflected, scattered, or radiated by the atmosphere can be quite beneficial. Satellite applications for profiling water vapor, temperature, and water content of the atmosphere are being studied by many researchers.

The microwave part of the EM spectrum is unique when related to the physical properties of the earth's atmosphere. Within the spectral range from 1 to 300 GHz, two important transitions take place: (1) the influence of atmospheric gases upon propagated energy increases from very weak to very strong, and (2) cloud and rain particles undergo a transition from very weak absorbers and scatterers to very significant contributions to the EM environment.

These properties enable the microwave spectrum to be used for several important but quite different purposes. At the long wavelength end, large numbers of communications- and radar-systems operate. Recent advances in microwave

technology and the theory of remote sensing provide new and exciting means for using the emission from the earth and atmosphere to study and identify certain important properties of each.

Microwave instruments have already successfully measured the vertical temperature structure and water-vapor distribution in the atmosphere, and many interesting features of the earth's surface correlate with microwave spectral measurements. An important part of the future is tied together with the unique relationship that millimeter and centimeter waves have to clouds and precipitation.

The degree to which water droplets interact with electromagnetic energy is related to the ratio of the droplet circumference to the wavelength of the radiation. Where this ratio approaches unity and larger, the interaction is important and complex. At infrared wavelengths, most clouds contain droplets larger than the wavelength; thus the ability of infrared radiation to penetrate to the surface, or to regions below the cloud level, is greatly diminished. Since a large fraction of the earth is covered at any one moment by clouds, the same large fraction of the earth is denied to infrared surveillance.

However, because microwaves cover the transition region between weak and strong interaction with clouds and rain, the influence that clouds have upon observation can be controlled to a degree. The value of using this transition region is threefold: (1) the effects of clouds can be minimized for purposes such as surface-to-space communication and observations of the surface from space; (2) the clouds themselves can be emphasized to extract information about their water content and extent; and, (3) measurements of atmospheric properties other than cloud parameters can be obtained even in the presence of clouds.

For the above reasons and others, interest in microwaves has quickened in the areas of remote sensing.

One can conveniently separate the important interactions between EMR and atmospheric gases into two parts: those dealing with the interactions of individual molecules and the ambient fields; and those that are mostly related to the macroscopic properties of the gases.

The former category includes the absorption and emission by molecules which result in spectral lines. The latter category contains the interactions that depend upon the bulk dielectric properties of the gases; they give rise to refraction, ducting, scintillation, and trophospheric scattering. Some of these subject areas are explored herein.

SPECTRAL LINE ABSORPTION

A number of molecules that are found in the atmosphere either as natural species or contaminants exhibit microwave spectral features (see Barrett, 1963 and Waters, 1976). Some of these are oxygen (O_2), water vapor (H_2O), ozone (O_3), nitrous oxide (N_2O), nitrogen dioxide (NO_2), carbon monoxide (CO), sulfur dioxide (SO_2), and the CℓO radical. The last six are observable, if at all, only in the stratosphere where the lines are narrow (on the order of MHz). In terms of atmospheric effects on propagation, water vapor and oxygen are overwhelmingly more important than any other gases. To understand spectral line absorption in the atmosphere, the contributions of oxygen and water vapor must, therefore, be understood. To place these in perspective, however, it is valuable to first briefly explore the factors which enter into the origin, strength, and width of rotational spectral lines, as discussed by Townes et al. (1955) and Herzberg (1945).

Pure Rotational Spectral Lines

Several forms of energy are associated with gas molecules, among which are the following: the kinetic of translation, the energy associated with rotation, the energy of vibration, and the energy associated with orbital electrons.

The last three forms are quantized and interact with the ambient radiation field. Each form can be considered as an approximately independent domain of energy.

In the case of oxygen and water vapor, the spectral lines which appear in the microwave region all arise from transitions between rotational energy levels only; the lines are therefore called *pure rotational spectral lines*. The contributing molecules are in the electronic ground state, and the lowest vibrational energy state. Schematic representations of the oxygen and water-vapor molecules and their rotational modes are shown in Figure 5-33.

The frequency at which spectral lines appear is directly related to the energy difference between the initial and final energy states of a transition, and is approximately given by the Bohr relationship,

$$\nu_{ij} = |Q_j - Q_i|/h, \qquad (5\text{-}74)$$

where:

ν_{ij} = resonant frequency of the transition between energy levels,
i, j = energy levels,
Q_i, Q_j = energy values,
h = Planck's constant.

The Absorption Coefficient

The absorption coefficient for radiation is found from quantum mechanical considerations to be dependent upon at least five factors: (1) frequency, (2) intrinsic line strength, (3) square of the dipole moment strength, (4) population of molecules that can participate in a transition, and (5) line shape factor.

A rapid insight can be had by referring to the general mathematical model, or expression for the

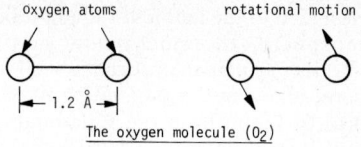

The oxygen molecule (O_2)

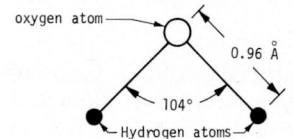

The water vapor molecule (H_2O)

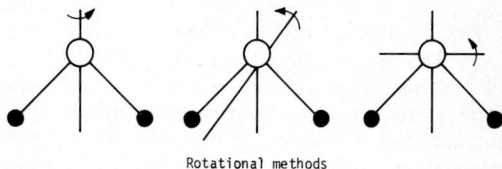

Rotational methods

Fig. 5-33. Schematic representation of the oxygen and water-vapor molecules and their rotational modes.

absorption coefficient of a spectral line in the microwave region:

$$\gamma_{ij} = \frac{8\pi^3\nu}{3hc}|f(\nu_{ij},\nu)|\,(N_i|\mu_{ij}|^2 - N_j|\mu_{ji}|^2), \quad (5\text{-}75)$$

where:

γ_{ij} = contribution to the total absorption at frequency ν by the molecular transition from energy state i to energy state j,
ν = frequency,
c = speed of light,
ν_{ij} = frequency of resonance,
$f(\nu_{ij},\nu)$ = line shape factor,
N_i and N_j = populations of the energy levels i and j,
$|\mu_{ij}|^2$ and $|\mu_{ji}|^2$ = squares of the dipole matrix elements derived from quantum mechanical analysis of the transition.

Equation 5-75 brings out the dependence of the absorption coefficient upon the populations of the two energy levels. Transitions occur from one energy level to the other and absorption takes place only when

$$|\mu_{ij}|^2 N_i > |\mu_{ji}|^2 N_j.$$

This condition is met when i represents the state of lower energy, and when thermodynamic equilibrium exists. When the reverse condition prevails, i.e., i is the state of lower energy and

$$|\mu_{ij}|^2 N_i < |\mu_{ji}|^2 N_j$$

energy is added to the ambient radiation field. For the atmosphere, thermodynamic equilibrium exists for the principal absorbers well into the mesosphere, allowing the population of levels (N_i and N_j) to be found from the Boltzmann theory. The population of any given level is found to be proportional to

$$w_i \exp(-Q_i/kT),$$

where:

w_i = statistical weighting factor for degenerate space quantization and nuclear spin
Q_i = energy of level i
k = Boltzmann's constant
T = absolute temperature

The resonant character of the spectrum is described by the line shape factor $f(\nu_{ij}, \nu)$ in Eq. 5-75. This function is related (in a semi-classical picture) to the Fourier transform of the time autocorrelation function of the molecule's dipole moment. In all but the very highest regions of the atmosphere, it is the effect of collisions between molecules that determines $f(\nu_{ij}, \nu)$. Collisional, or pressure, broadening of spectral lines results from disturbance by collisions of the molecule's natural interaction with the electromagnetic field (Van Vleck and Weisskopf, 1945; Gross, 1955). The frequency of such disturbances, and thus the amount of broadening, is proportional to the density of the atmosphere. It is only at very low densities (≥ 70 km altitude) that another mechanism, Doppler or thermal broadening, is significant for microwave lines. Doppler broadening is proportional to the line frequency times the square root of the temperature.

The detailed behavior of the function $f(\nu_{ij}, \nu)$ in the case of collisional broadening may depend on the presence of other nearby lines in the spectrum of the molecule. The reason for such *interference* between spectral lines is that if, for example, a collision changes the rotational state of the molecule, the rotational quantum number may (depending on the shape of the intermolecular potential function) be more likely to change by a small amount than by a large amount. Thus the state of the molecule after the collision will be correlated with its state before the collision. If lines associated with those two states overlap, then the collision will be less effective in disrupting the molecule's coupling with the electromagnetic field than if there were no overlap. Such an effect is observable in the microwave spectrum of oxygen at atmospheric pressures. Gordon (1966, 1967), Rosenkranz et al. (1975), and Lam (1977) discuss methods of calculation of spectral line interference effects.

Oxygen

The oxygen molecule is a diatomic molecule

with no permanent electrical dipole moment.[3] However, two of the orbital electrons are unpaired, resulting in permanent magnetic dipole moment.[4] The molecule rotates end-over-end with quantized angular momentum.

Knowledge of these quantized states, as obtained from the study of quantum statistics, provides information about the absorption spectra of oxygen. The quantum number N describing the state of quantization may take on odd values only. The spin angular momentum of the unpaired electrons is described by a quantum number $S = 1$. The two momenta combine to form a total angular momentum vector J, which is the vectorial sum of N and S. For every N state there are three J states in which

$$J = N - 1 \qquad (6\text{-}76)$$

$$J = N \qquad (6\text{-}77)$$

$$J = N + 1 \qquad (6\text{-}78)$$

where:

J = quantum number or total angular momentum vector.

Each of these J states has a slightly differing energy from the other two. For a given N, the three energy levels associated with J are known collectively as *rho-type triplet*. See Van Vleck (1947a).

Transitions within a given N quantum state are allowed from

$$J = N \text{ to } J = N + 1$$

and from

$$J = N \text{ to } J = N - 1$$

and lead to a series of spectral lines which are important between 50 and 70 GHz, with the peak very near 60 GHz. A single line also occurs at 118.7505 GHz. The magnetic dipole fine-structure transitions are weak when compared to typical electrical dipole transitions (as with water vapor). However, in the atmosphere, the total path length through oxygen is so great that a very intense absorption band is produced in the region of 0.5-cm wavelength. In addition to the two transitions listed above, the oxygen molecule can absorb energy in the states $J = N \pm 1$ since these two states have nonzero diagonal dipole matrix elements μ_{ii}. This *nonresonant* absorption is very

---

[3] Dipole moment—the product of one of the electrically or magnetically charged particles of opposite sign of the dipole unit by the distance separating the two dipolar charges. The vector length between the two infinitesimal charges of $+q$ and $-q$ is defined as d; then the electric dipole moment may be defined as $p = q\,d$. The definition of multipoles follows along the same lines.

[4] A current I circulating about a path, enclosing a differential vector area dS defines a magnetic dipole, or $m = IDS$.

weak, but gives rise to a measurable low frequency "tail" in the oxygen spectrum. Because of this tail, oxygen dominates the clear-air absorption spectrum at low frequencies.

To visualize better the spectral features of oxygen, Figures 5-34 and 5-35 have been prepared as representative of absorption for a mixing ratio of oxygen and nitrogen found at sea level. Figure 5-34 is presented as an overview for the spectrum over the entire range from 1 to 300 GHz, at a constant temperature and at two illustrative pressures. At the most intense point of the overlapping spectral-line absorption near 60 GHz, the peak value for the stated conditions of temperature and for 1,000-mb of pressure reaches 17 db km^{-1}. The maximum in the satellite line at 118.8 GHz is 1.5 db km^{-1}. At frequencies above 140 GHz, oxygen absorption is weak, and in the atmosphere is overshadowed in importance by water vapor.

At 100-mb pressure and 173 K, the spectrum at low frequencies has been reduced by almost two orders of magnitude. At the peak of the resonance region some decrease in intensity is evident, but less than one order of magnitude. However, detailed analysis in the vicinity of 50 to 70 GHz would show individual spectral lines well-resolved.

Figure 5-35, an enlarged and more detailed presentation of the 60-GHz spike shown on Figure 5-34, is included to portray graphically the effect of temperature variation on the spectral region

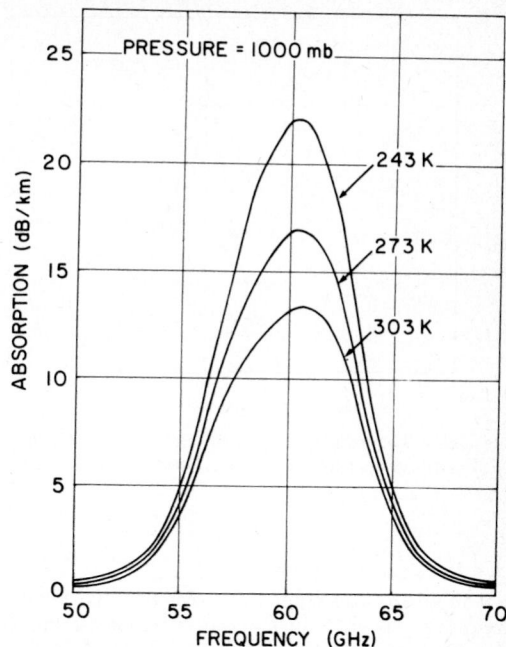

Fig. 5-35. Absorption spectrum of atmospheric oxygen in the spectral region near 60 GHz, as a function of temperature at constant pressure.

near 60 GHz. Since the absorption coefficient is proportional to T^{-3} (where T = absolute temperature in K), a very striking increase in total absorption occurs when the temperature is lowered from 303K to 243K.

More extensive calculations of oxygen absorption and comparison with measurements can be found in papers by Liebe (1977) and Liebe and Hopponen (1977).

Water Vapor

The molecular configuration and size of the water-vapor molecule have already been shown in Figure 5-33. The principal axes of inertia are also shown, about which the rotation of the molecule is most easily described. Since the moments of inertia are all unequal, the water molecule is called an *asymmetric rotor* or *asymmetric top molecule*. The energy levels that occur can be identified by the notation J_τ, where the total angular momentum quantum-number is J, and τ is an auxiliary integer which can run from $-J$ to $+J$ and describes the order of energy levels for a given J state. In this notation, the lowest frequency transition for water vapor, at 22.2 GHz, occurs between the 5_{-1} and 6_{-5} energy states. The next line of water vapor, at 183.3 GHz, is the result of transitions between the 2_{+2} and the 3_{-2} states.

Because of the irregularity of energy levels, spectral lines also occur irregularly and without the semi-periodicity found in the spectra of symmetric rotors and diatomic molecules.

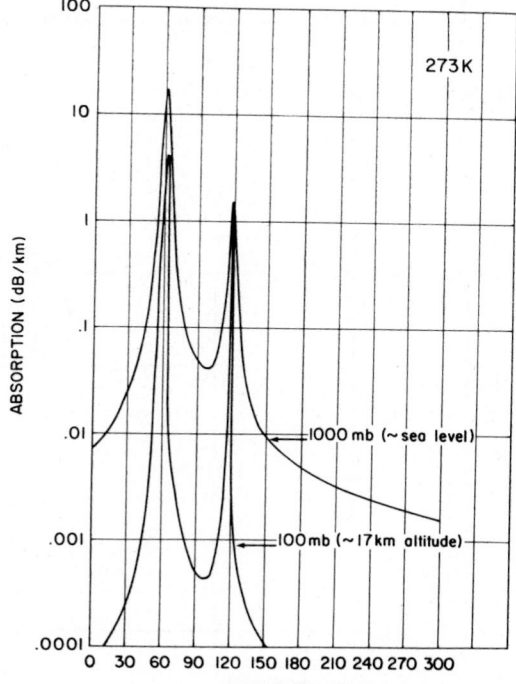

Fig. 5-34. Absorption spectrum of atmospheric oxygen from 1 to 300 GHz for two pressures at constant temperature.

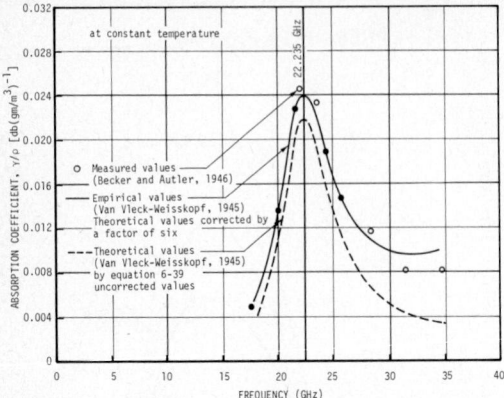

Fig. 5-36. Theoretical, empirical, and measured values of absorption near the 22.235 GHz spectral line of water vapor.

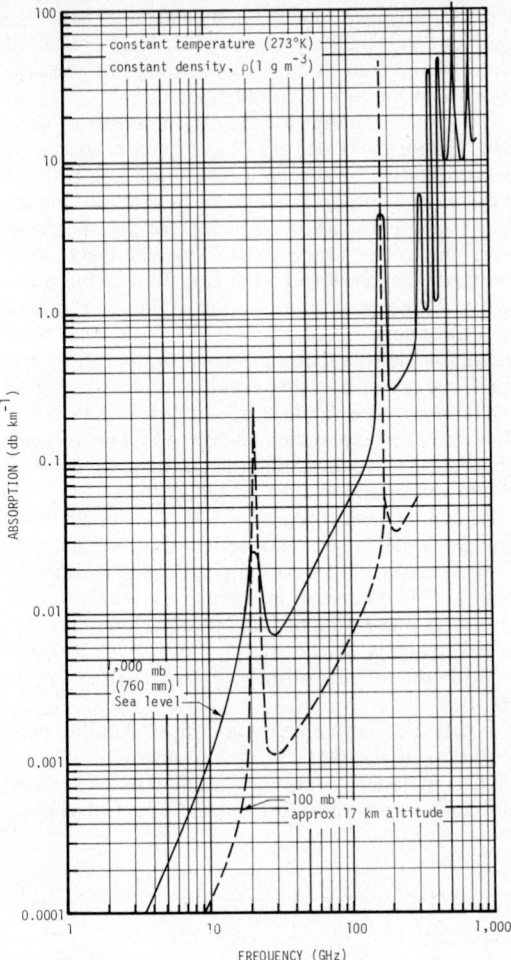

Fig. 5-37. Absorption spectrum of water vapor from 1 to 300 GHz for two pressures, at constant pressure.

The absorption coefficients for the two lowest frequency spectral lines of water vapor are based upon an expansion of the model of Eq. 5-75, but with certain empirical corrections to force the expressions to produce results which fit experimental measurements. See Van Vleck (1947b), Van Vleck and Weisskopf (1945), Becker and Autler (1946), Gaut (1967), Frenkel (1966), Waters (1976), and Liebe and Gimmestad (1978).

It appears that the theoretical expression predicts that absorption due to the submillimeter and infrared lines of water-vapor diminishes away from resonance more rapidly than actually occurs. This effect is illustrated in Figure 5-36. The lower curve depicts the 22.235 GHz water-vapor line as it is computed by Eq. 5-52 and applied unmodified to water vapor, and the upper curve represents the same line but with the contributions from the wings of all other lines increased by a factor of six. The open circles are experimental data points.

In order to gain some insight into the behavior of water-vapor absorption, Figures 5-37 and 5-38 have been prepared. Figure 5-37 gives the curve for absorption under the conditions of air of 273 K, 1,000 and 100 mb of pressure, and 1 g m^{-3} of water vapor. The most striking feature of the curves is the increase in absorption on resonance as the pressure is reduced. This change occurs because absorption (from the same amount of water vapor) is spread over a narrower frequency band.

Water-vapor absorption varies over five orders of magnitude between 3 and 300 GHz. At lower frequencies it is quite negligible for most purposes.

Figure 5-38 provides some feel for the variation of absorption near the 22-GHz line as a function of temperature. Unlike oxygen, the peak absorption changes little. However, the line is broadened considerably at low temperature.

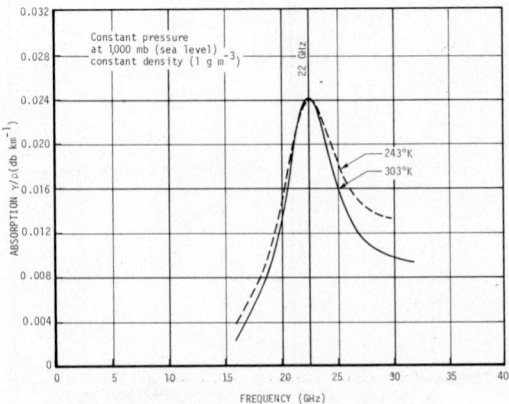

Fig. 5-38. Absorption spectrum of water vapor near the 32-GHz line, as a function of temperature at constant pressure.

THE DIELECTRIC PROPERTIES OF MOIST AIR

The index of refraction of moist air is defined as

$$n = c/v.$$

where:

c = speed of light in a vacuum,
v = speed of propagation in air.

It is greater than unity because of two effects: (1) charges within molecules are distorted in the presence of an external EM field; and (2) those molecules with a permanent dipole moment try to align themselves with respect to the impressed field.

The velocity of the propagating wave is less where the density of gas molecules is greater. Near the surface of the earth, EM waves travel in the neighborhood of 100 km s^{-1} slower than waves in empty space.

A working equation to calculate the index of refraction in a mixture of oxygen, water vapor, and CO_2 for frequencies less than 200 GHz is given by Bean et al. (1966) as

$$N = (n - 1) \times 10^6 = 77.6 \frac{p}{T} + 3.73 \times 10^5 \frac{e}{T^2},$$

where:

N = refractivity,
n = index of refraction,
p = the total atmospheric pressure in mb,
T = the absolute temperature,
e = the partial pressure of water vapor in mb.

Many of the effects of refraction and other related propagation phenomena can be qualitatively understood through use of this equation.

Under average conditions, p, T, and e decrease with height in such manner that N decreases roughly exponentially with height. The propagation of a nonvertical traveling wavefront is, therefore, tilted away from the vertical and back toward the earth.

Conditions of such magnitude can occur in the distributions of T and e that the former increases with height in a given interval and the latter decreases in the same vertical interval. Equation 5-79 shows that both of these trends tend to reduce the value of N. It often occurs in stable meteorological conditions that the curvature or bending of a wave front is greater than the curvature of the earth itself. The wave is then returned to the surface of the earth and trapping occurs. These stable layers have been known to exist over hundreds and sometimes thousands of miles. They are the source of many anomalously long radar-ranging reports and unusual radio propagation.

Scintillation occurs in the atmosphere when the index of refraction, over regions not large with respect to the wave front being monitored, changes rapidly in time. The intensity of an incoming signal, perhaps from a satellite source, will be observed to fluctuate randomly. The effect varies with atmospheric conditions.

More important, and more difficult to analyze than scintillation, are the scattering effects of the inhomogeneities in the atmosphere caused by turbulence, stable layers, and the density and compositional changes between adjacent air masses. These inhomogeneities are found on all size scales and throughout the entire atmosphere. They have a mean activity over which are impressed large local variations in density and, therefore, in electrical properties.

The reason for the importance of the atmospheric inhomogeneities lies in their ability to scatter enough energy so that narrow beams of radiation directed toward the horizon can be detected hundreds of kilometers below the horizon.

INTERACTIONS BETWEEN MICROWAVE RADIATION AND HAZE, CLOUDS AND PRECIPITATION PARTICLES

The interaction of microwave radiation with atmospheric gases is the result of absorption and emission by transitions between energy levels of individual molecules. This section will therefore consider the effect of material particles—i.e., aggregates of molecules—upon the propagation of EM energy. Specifically, the atmospheric particles of interest of fall into three categories: water droplets; ice particles; and dust particles.

The effect of the first is dominant under normal conditions, and for simplicity, present discussions are therefore confined to spherical water droplets the radii of which lie in the range from 0.1 μm to 0.5 cm. This range includes haze, clouds, and extreme cases of precipitation (Barnett, 1963, and Bean et al. 1966).

An EM wave incident upon a material particle causes an interaction with the free and bound charges present within the material by subjecting them to a force due to the incident electric field. The charges, moving under the action of this force, generate new fields which propagate within the material or are carried away as outgoing radiation. The resulting EM field outside the particle consists of the vector superposition of the *scattered* field. The result of this interaction is a redistribution of the incident energy in such a manner as to attenuate the forward traveling wave. This attenuation, or "extinction" actually consists of two parts: that which reappears as scattered radiation; and that which is absorbed because the moving charges within the medium experience damping forces which transfer their mechanical energy to the surrounding medium. In addition, the medium itself emits thermal radiation that is characteristic of its equilibrium temperature.

The effect of a single water droplet upon the

external field depends critically upon the ratio of the drop radius

$$r \text{ to } \frac{\lambda}{|n|}$$

where:

λ = (free space) wavelength of incident radiation and

n = the (complex) index of refraction of the droplet.

In the so-called *long-wavelength limit*, this ratio is small, and the EM field is effectively uniform throughout the droplet. The motion of the constituent charges is, therefore, coherent throughout the droplet and the reradiated field is that of an induced dipole oscillating at the frequency of the incident field. This limit is referred to as the *Rayleigh-scattering limit* due to the λ^{-4} dependence of the power radiated by an oscillating dipole.

The *short-wavelength* or *geometrical limit*, for which $r > \lambda$, lies at the opposite extreme. In this case, the motion of charges within the droplet is coherent only along a wavefront of the exciting radiation, and the secondary fields are propagated within the medium in such manner as to interfere destructively except at these points. The laws of reflection and refraction apply, and a ray of incident radiation is decomposed at each surface of the droplet into reflected and refracted rays which then constitute the *scattered* radiation.

Between these two extremes, the long- and short-wavelength limits, the EM fields are best regarded as superpositions of partial waves which represent, physically, the modes of excitation of the dielectric sphere. Thus, in the simplest case, corresponding to the long-wavelength limit, only the lowest order mode—the induced dipole—is significant. As the radius of the sphere is increased, successively higher multipoles become significant.

Knowledge of the interaction of a single particle makes possible the study of large-scale ensembles since, for a random distribution of scatterers, there are no coherent phase relationships between the individual scatter fields. Hence, if the "shadowing" of one particle by another may be neglected, the total extinction due to the ensemble is the algebraic sum of the contributions due to the individual particles. For this reason, an effective analysis of the effects of haze, clouds, and precipitation consists of these steps: (1) evaluation of the properties of individual droplets at a given wavelength (for which the index of refraction is known) as a function of drop radius; (2) determination of unit-volume (intensive) radiative transfer properties of an assumed drop-size distribution; and (3) determination of geometry-dependent (extensive) effects using the equation of radiative transfer.

INTERACTION OF INDIVIDUAL WATER DROPLETS WITH MICROWAVE RADIATION

The rigorous solution for the diffraction of a plane monochromatic wave by a homogeneous dielectric sphere of arbitrary radius was first obtained by Mie (1908), followed shortly by Debye (1909).

The Mie efficiency factors for scattering, extinction, and absorption are defined as the ratio of the actual cross-section to the geometrical cross-section:

$$\xi_j(n,q) = X_j(n,q)/\pi r^2, \qquad (5\text{-}80)$$

where:

j = $S, A, E,$
S = scattering
A = absorption
E = extinction
X_j = appropriate cross-section
r = radius of the drop,
ξ_j = the corresponding efficiency factor, which is a function of: the complex index of refraction, n, and the dimensionless drop-size parameter:

$$q = 2\pi r/\lambda. \qquad (5\text{-}81)$$

The efficiency factors are obtained by solving the microwave field equations for a dielectric sphere of complex index of refraction, with the solution taking the form of an expansion in multipoles:

$$\xi_S = \frac{2}{q^2} \sum_{\ell=1}^{\infty} (2\ell + 1) \left[|a_\ell(n,q)|^2 + |b_\ell(n,q)|^2 \right]$$

$$(5\text{-}82)$$

$$\xi_E = \frac{2}{q^2} \sum_{\ell=1}^{\infty} (2\ell + 1) \, \text{Re}\left[a_\ell(n,q) + b_\ell(n,q) \right]$$

$$(5\text{-}83)$$

$$\xi_A = \xi_E - \xi_S, \qquad (5\text{-}84)$$

where the coefficients a and b in the Mie series are determined by the boundary conditions at the surface of the sphere (see Stratton, 1941).

The complex index of refraction, n, must be known or assumed for computation of the single-drop interaction with the EM field. At microwave frequencies, the most significant contribution to the possible polarization of water molecules, and hence the index of refraction of liquid water, is that due to dipole rotation. The contributions of ionic and electronic interactions occur in the infrared and optical regions of the spectrum, respectively. For this reason, the complex index of refraction is given accurately, for the microwave range, by the Debye formula (Debye, 1929),

$$n^2 = \frac{\epsilon_o - \epsilon_\infty}{1 + i(\lambda_o/\lambda)} + \epsilon_\infty, \qquad (5\text{-}85)$$

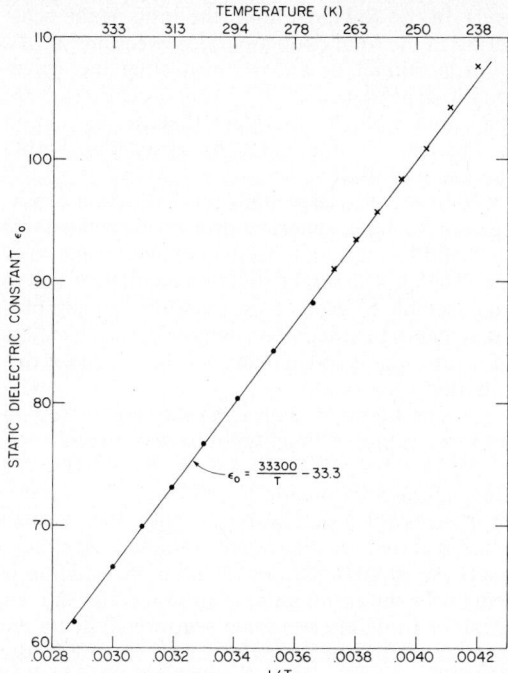

$$\epsilon_o = \frac{33300}{T} - 33.3$$

Fig. 5-39. Static dielectric constant of pure water versus reciprocal temperature. Measurements · — (Collie et al., 1948), x — (Hasted and Shahidi 1976).

where:

ϵ_o and ϵ_∞ = static and optical dielectric constants,

λ_o = Debye relaxation wavelength, resulting from damping effects in the liquid,

i = $\sqrt{-1}$.

The parameters ϵ_o and λ_o depend on temperature and are plotted in Figures 5-39 and 5-40. The relaxation wavelength was determined by Collie et al. (1948) by measuring the dielectric constant at three frequencies and fitting the Debye function to the data. The parameter ϵ_∞ is difficult to determine from microwave measurements and was therefore fixed at a value of 5.5.

The temperature range of interest for atmospheric propagation is approximately from 235 to 315 K. The lower limit is less than 273 K because not all cloud droplet nuclei are suitable for sublimation and supercooled clouds are therefore encountered at temperatures as low as −38°C (Aufm Kampe and Weichmann, 1957). On the other hand, when clouds are frozen, both the real and imaginary parts of the index of refraction of the ice crystals are much smaller than for liquid water. Consequently, ice clouds (cirrus) have for most purposes negligible effect on microwave propagation.

Mie efficiency factors for extinction and scattering are shown for three frequencies: (3, 30, and

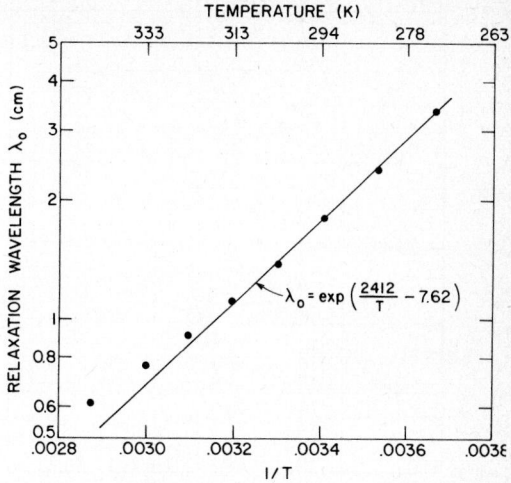

Fig. 5-40. Relaxation wavelength of pure water versus reciprocal temperature. Measurements (Collie et al. 1948).

300 GHz) in Figures 5-41, 5-42, and 5-43. The curves at each frequency are presented as a function of drop radius, which is proportional to the drop-size parameter q. The long-wavelength or Rayleigh limit corresponds to $q < 1$, for which the efficiency factors ξ_S and ξ_E take the form:

$$\xi_S = \frac{8q^4}{3}|K|^2 + \dots \quad (5\text{-}86)$$

$$\xi_E = 4q\,Im(-K) + \frac{8q^4}{3}|K|^2 + \dots \quad (5\text{-}87)$$

$$\xi_A = 4q\,Im(-K) + \dots , \quad (5\text{-}88)$$

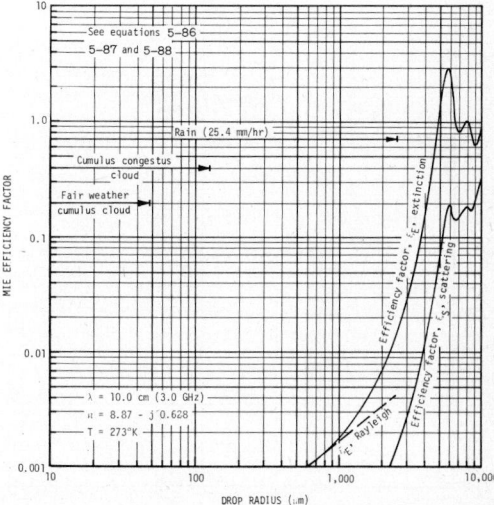

Fig. 5-41. Mie efficiency factors for scattering and extinction as a function of drop radius, at 3.0 GHz.

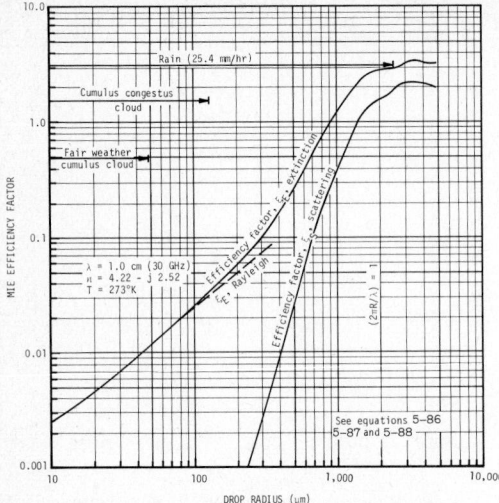

Fig. 5-42. Mie efficiency factors for scattering and extinction as functions of drop radius, at 30 GHz.

where:

$$K = \frac{n^2 - 1}{n^2 + 2} \quad . \qquad (5-89)$$

From Figures 5-41, 5-42, and 5-43, it is clear that there exists a range of drop sizes at each frequency for which the extinction factor ξ_E is linear in the drop radius. That is, there is a range for which the extinction is given accurately by the Rayleigh limit (Equations 5-86, 5-87, and 5-88). At 3 GHz, the approximation is valid for all drop sizes less than about 1,000 μm, and at 30 and 300 GHz, the departure from the approximation occurs for drop sizes of about 100 and 40 μm respectively. Several conclusions may be drawn from these results:

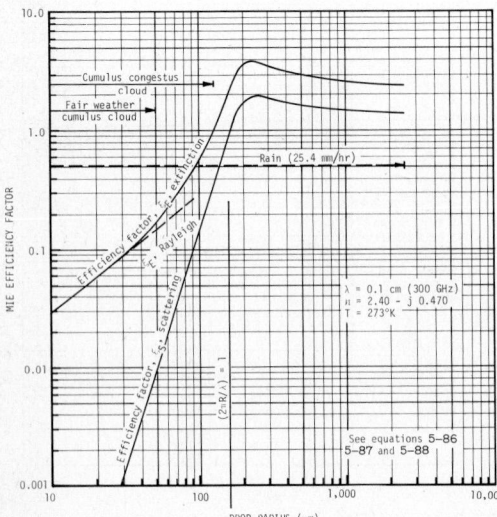

Fig. 5-43. Mie efficiency factors for scattering and extinction as functions of drop radius, at 300 GHz.

(1) In the Rayleigh limit, the ratio of the scattering to the total extinction cross-sections X_S/X_E is proportional to $L(r/\lambda)^3$, neglecting the wavelength dependence of K. From Equations 5-86, 5-87, and 5-88, it is evident that the extinction in this limit is dominated by absorption within the material itself.

(2) In the Rayleigh limit the extinction cross-section X_E for a spherical droplet is proportional to r^3 and hence simply to the volume (or mass) of the droplet. The total extinction coefficient due to an ensemble of droplets is, therefore, simply proportional to the total mass density of the ensemble of drops, and is independent of the drop-size distribution.

(3) The limits of drop sizes normally assumed for representative cloud types have been indicated in Figures 5-41, 5-42, and 5-43. At 3 GHz, the Rayleigh approximation is valid up to the largest drop sizes for all cloud types, with only a small error in extinction made in the rain-bearing cloud.

(4) At 30 GHz, the Rayleigh approximation is valid over the entire range of drop radii for the fair weather cumulus, but some scattering effects are apprarent for the cumulus congestus. For the rain-bearing cloud, the approximation is clearly invalid, and a serious error results from the neglect of scattering.

(5) At 300 GHz, the upper limit of the frequency range considered here, the Rayleigh approximation is adequate only for the fair-weather cumulus.

On the basis of single drop interactions with microwave radiation, it is clear that water droplets affect the entire microwave region from 1 to 300 GHz.

UNIT VOLUME RADIATIVE TRANSFER PROPERTIES OF CLOUDS

The function f (see Equation 5-70) has been determined by Goldstein (1951) and Staelin (1966) with the result that the extinction coefficient applicable to clouds in the Rayleigh limit and for frequencies less than 80 GHz is given by

$$\gamma_{cloud} = \frac{m \cdot 10^{[0.0122\ (291-T)-6]}}{\lambda^2} \ (neper\ cm^{-1}).$$

$$(5-90)$$

Neper (napier) is a unit used to express the scalar ratio of two currents or two voltages, the number of nepers being the natural logarithm of such ratio; the reduction of a value $1/e$ from its intial value, where $e = 2.7183$, the natural base of the natural logarithm.

Outside the Rayleigh regime, the extinction and scattering coefficients must be determined by formal solution Equations 5-82, 5-83, 5-84, and 5-68, using the full Mie theory, and adopting an appropriate analytic form for the drop-size distributions $dn(r)/dr$. See equation 5-71.

The results of computations for two of the mod-

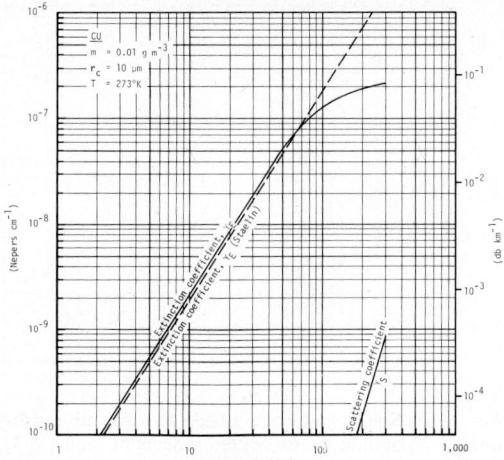

Fig. 5-44. Unit-volume extinction and scattering coefficients computed as functions of frequency for the model cumulus cloud.

els in Figure 5-32 are shown in Figures 5-44 and 5-45 for the fair-weather cumulus and cumulonimbus. In each case the extinction and scattering coefficients are shown in comparison with the extinction coefficient given by the Staelin approximation (Eq. 5-90).

For the fair-weather cumulus, the agreement between the approximate and exact values for γ_E is close for frequencies less than about 80 GHz. Above this frequency, the approximation breaks down due to the departure from the simple λ^{-2} dependence. In the case of the rain-bearing cumulonimbus, the Staelin formula substantially underestimates the extinction coefficient due to the departure from the Rayleigh limit, as noted earlier.

The question of absorption and scattering by

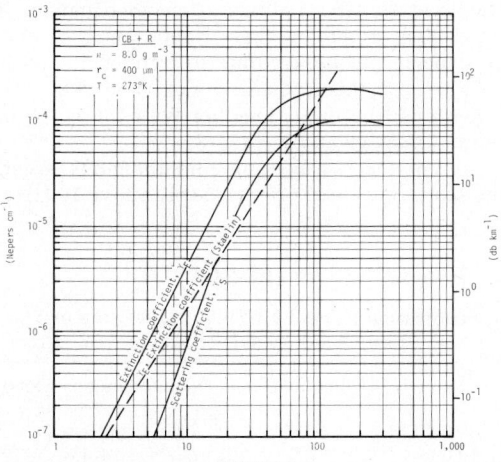

Fig. 5-45. Unit-volume extinction and scattering coefficients computed as a function of frequency for the model cumulonimbus cloud.

solid particles in the form of ice and snow is discussed in Goldstein (1951). In the discussion, it is implied that under normal considerations attenuation by dry ice crystals and hailstones is considerably less than by liquid water of similar mass, at least at frequencies less than perhaps 50 GHz. However, it was found by Kerker et al. (1951) that at 1- and 3-cm wavelengths melting ice crystals could have twice the cross-section for extinction that completely melted drops of similar mass would exhibit.

PROPAGATION OF MICROWAVE RADIATION IN THE EARTH'S ATMOSPHERE

The mechanics of various interactions between microwaves and gases, liquids and solids found in the atmosphere have now been reviewed. In this section, these properties will be combined with models of the atmospheric environment to help establish a picture of the manner in which the atmosphere modifies and contributes to propagation.

At microwave frequencies, where $hv << kT$ and the Rayleigh-Jeans approximation can be made, Eq. 5-15 may be integrated to give

$$T_B(\Theta,\Phi,0) = T_B(\Theta,\Phi,\tau) \exp(-\tau)$$

$$+ \int_0^\tau T_{eff}(\Theta,\Phi,\tau') \exp(-\tau')d\tau', \quad (5\text{-}91)$$

in which

$$T_{eff}(\Theta,\Phi,\tau') = \left[1 - \omega(\tau')\right] T_m(\tau')$$

$$+ \omega(\tau') T_{sc}(\Theta,\Phi,\tau'). \quad (5\text{-}92)$$

$T_m(\tau')$ is the radiation temperature associated with the medium, equal to the kinetic temperature when thermodynamic equilibrium exists. $T_{sc}(\Theta,\Phi,\tau')$ is a temperature associated with the energy scattered from all directions into the direction (Θ,Φ). In the special case, $\omega \to 0$ corresponding to a scatter-free medium, Eq. 5-91 reduces to the familiar form with the effective temperature equal to the medium temperature. The brightness temperature may be accurately estimated if the absorption coefficient and temperature are given for all points along the line of sight.

Figure 5-46 is an example of radiative transfer calculations done by Tsang *et al.* (1977) for the case of a 1-km-thick layer of rain over an ocean background. The drop-size distribution was identical to that listed in Table 5-7 for the cumulus with rain 12 mm/hr (0−400 m). At 30 GHz, Figure 5-46 shows that the brightness temperature at any given angle is reduced, compared to a calculation in which scattering is not considered (shown by the dashed lines). This result is due to scattering of upwelling radiation out of the line of sight and the lesser intensity of downwelling radiation which may be scattered into the line of sight. The effect is enhanced at larger angles and for vertical

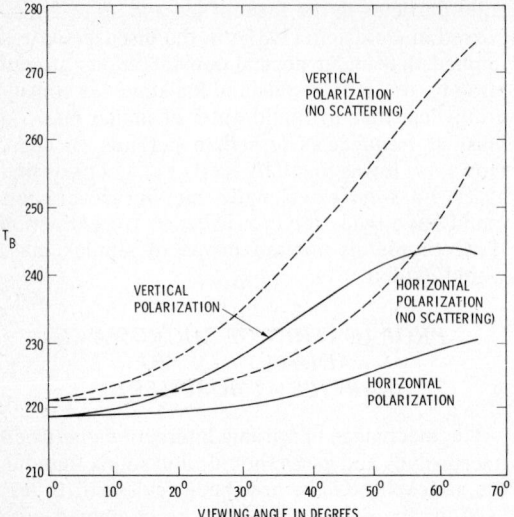

Fig. 5-46. Brightness temperature at 30 GHz of a 1-km-thick rain layer over ocean vs. viewing angle from nadir (Tsang et al., 1977).

polarization because the emission in the absence of scattering is greater.

The solution of Eq. 5-91 for even the simplest cases in which scattering exists is formidable due to the fact that every point in the medium depends upon every other point, through Eq. 5-92. Certain simplifying assumptions can be made, however, which enable one to estimate the validity of assuming a scatter-free medium.

A necessary condition for neglect of the effects of scattering is that the effective mean free path for a scattering event must be large compared to the scale length of the medium itself. One may formalize this criterion by noting that an error of approximately 0.1 neper in the opacity τ (due to neglect of scattering terms) leads to an error of approximately 10 percent in the evaluation of the attenuation of the medium. The error $\Delta\tau$ in τ made in traversing a distance d in a uniform medium is:

$$|\Delta\tau| = (\gamma_E - \gamma_A)d = \gamma_S d = \omega\,\tau \qquad (5\text{-}93)$$

in which γ_E is the extinction coefficient. Hence an error of 0.1 neper occurs in a *penetration distance* d_S such that

$$\gamma_S d_S = 0.1$$

The dependence upon the scattering albedo ω is clear from Eqs. 5-91 and 5-92 which when combined lead to an infinite series of integrals in successive orders of ω with terms linear in ω representing single scattering, ω^2 double scattering, and so on. Hence an error of approximately ω is made in the determination of T_B due to omission of (single) scattering; approximatley ω^2, due to consideration of single scattering but neglect of multiple scattering; and so forth.

With these criteria in mind, it is apparent that the validity of omission of scattering terms from

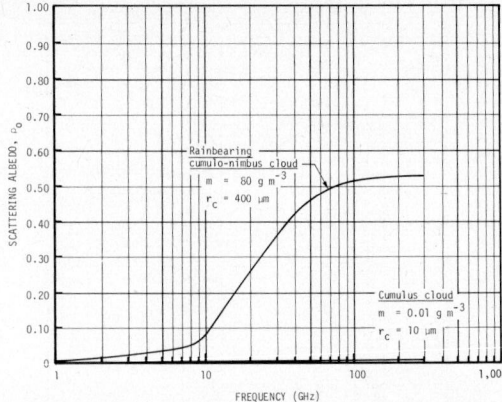

Fig. 5-47. Single-scattering albedo as a function of frequency computed for two cloud models of Figure 5-32.

the radiative transferr equation depends upon the scattering coefficient γ_S and the scattering albedo ω, as determined from a full Mie-scattering treatment of the medium under study. A sufficient condition for omission of scattering terms is that at every point along the path ω is small, and that the *scattering depth* d_S is everywhere large compared to the physical dimensions involved.

Figure 5-47 depicts the single scattering for two of the three cloud models of Figure 5-32 as a function of frequency, and the penetration depths for single scattering (single-scattering albedos) are given for selected frequencies in Table 5-8.

The results of these computations may be summarized as:

(1) The scattering albedo for the fair-weather cumulus cloud is negligible over the entire range from 1- to 300-GHz; the penetration depth is everywhere orders of magnitude greater than the physical dimensions of reasonable cloud structures, even at 300 GHz. As a result, the Rayleigh approximation may be used, together with the simple scatter-free radiative transfer equation. If Eq. 5-90 is used for γ, the γ^{-2} approximation may be expected to be accurate to frequencies of 80 GHz. Above this frequency, Eqs. 5-86, 5-87, and 5-88 should be used, together with the index of refraction as given by Eq. 5-85.

(2) For the rain-bearing cumulonimbus cloud, the scattering albedo rises sharply above 10 GHz.

TABLE 5-8

Penetration Depths for Single Scattering in The Two Cloud Models of Figure 5-32

Frequency	Cumulus Cloud	Cumulonimbus Cloud
3.0 GHz	$>10^{10}$ km	100 km
10.0	$>10^3$	1.43
30.0	$>10^6$	0.04
100.0	7100	0.01
300.0	1280	0.01

At the same time, the penetration depth for scattering is less than 1 km for all frequencies greater than about 10 GHz, whereas buildups of 3 km are not uncommon for clouds of this type. The full Mie theoretical treatment must be used, at least from frequencies above 3 GHz, for the scattering and extinction coefficients. Above 10 GHz, the effects of single scattering must be taken into account in the radiative transfer equation. Above 30 GHz, the effects of multiple scattering become important, making an accurate estimate of the effect of such a cloud much more complicated.

REMOTE SENSING IN THE INFRARED

To this point, the discussion has emphasized the understanding and the delimiting of the interaction between the atmosphere and surface radiation. The purpose of this section is to present some results which bear upon the use of upwelling radiations as probes of the natural environment, and to discuss the feasible additional areas of application of passive remote sensing. Where possible, the shortcomings of techniques and unsolved problems will be shown. Many of these applications are covered by Staelin (1969) and Chahine (1977).

Any of the physical interactions between the radiation field and the environment offer potential means for probing that part of the environment which is involved in the interaction. The problem is to find interactions that are strong enough to be observable and that are also sensitive and isolated enough from other contaminating signals so that their interpretation is unambiguous.

A number of parts of the EM spectrum do meet these three criteria when applied to certain parameters of the atmosphere that have been shown to be measurable from space. They are: tropospheric water vapor, atmospheric temperature, and cloud parameters. Figure 5-48 gives the regions of the EM spectrum that have been most commonly used for earth observation.

The following sections are presented to illustrate briefly the development of infrared and microwave inversion techniques. Additional details are reported by Chahine (1977), Waters (1968),

Staelin (1969), and Smith (1976) and a comprehensive and detailed treatment of this subject is given by Twomey (1977).

MATHEMATICAL FORMULATION

The formulation of remote sounding problems in radiative transfer leads often to nonlinear integral equations of the form

$$I(\nu) = B\left[\nu, g(z_o)\right] C(\nu, z_o)$$
$$+ \int_{z_o}^{\bar{z}} B\left[\nu, g(z)\right] K(\nu, z)\, dz. \quad (5\text{-}94)$$

In the microwave cases, however, the resulting integral equations are linear in $g(z)$ of the form

$$I(\nu) = g(z_o) C(\nu, z_o) + \int_{z_o}^{\bar{z}} g(z) K(\nu, z)\, dz. \quad (5\text{-}95)$$

In Eqs. 5-94 and 5-95, $C(\nu, z_o)$ and $K(\nu, z)$ describe specific radiative transfer processes such as absorption, emission or scattering in the atmosphere. In many problems of remote sounding this kernel $K(\nu, z)$ reaches its maximum peak at different values of z for different values of ν. $I(\nu)$ is a given function, usually measured at a discrete number of observations $\tilde{I}(\nu_j)$, and $g(z)$ is the distribution of an atmospheric parameter to be determined. Thus, the inverse solution here reduces to finding a function $g(z)$ such that, when it is substituted into Eqs. 5-94 or 5-95, will yield values of $I(\nu_j)$ equal to the corresponding measurements $\tilde{I}(\nu_j)$, with

$$\tilde{I}(\nu_j) - I(\nu_j) = 0,$$

for all the given values of ν_j.

REMOTE SENSING OF TEMPERATURE PROFILES

In the problem of remote sounding of atmospheric temperature profiles, the following equation occurs

$$I(\nu) = B\left[\nu, T(z_o)\right] \sigma(\nu, z_o)$$
$$+ \int_{z_o}^{\bar{z}} B\left[\nu, T(z)\right] K(\nu, z)\, dz. \quad (5\text{-}96)$$

Eq. (5-96) is the integral form of the radiative transfer equation for a plane parallel homogeneous and nonscattering atmosphere in local thermodynamic equilibrium. $I(\nu)$ is the outgoing radiance measured at a vertical distance $\bar{z}$ from the surface z_o of a planet within a narrow solid angle around the local vertical axis, z. B is the Planck function explicitly given by

$$B_\nu(T) = \frac{2h\nu^3}{c^2} \frac{1}{\exp(h\nu/kT) - 1} \quad (5\text{-}97)$$

as in Eq. 5-10. $\sigma(\nu, z)$ is the transmittance of a

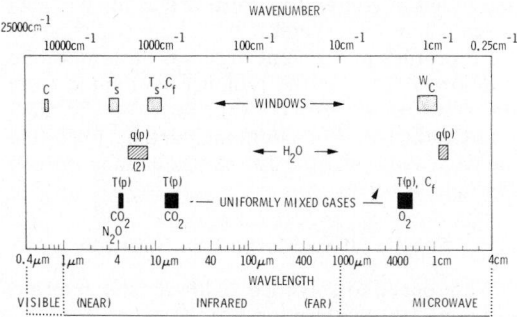

WAVENUMBER

25000cm^{-1}

10000cm^{-1} 1000cm^{-1} 100cm^{-1} 10cm^{-1} 1cm^{-1} 0.25cm^{-1}

C T_s $T_s \cdot C_f$ W_c
0   □   □   ←— WINDOWS —→   □

q(p)   q(p)
▨   ←— H_2O —→   ▨
(2)

T(p)   T(p)   T(p), C_f
■   ■ ·—— UNIFORMLY MIXED GASES —— ◀   ■
CO_2 CO_2 O_2
N_2O

0.4μm   1μm   4   10μm   40   100μm   400   1000μm   4000   1cm   4cm
WAVELENGTH

VISIBLE (NEAR)   INFRARED   (FAR)   MICROWAVE

Fig. 5-48. Spectral regions found useful for atmospheric remote sensing (Smith, 1972).

column of absorbers between levels $\bar{z}$ and z and is defined for monochromatic observations as

$$\sigma(\nu,z) = \exp\left[- \int_{z}^{\bar{z}} \rho_s(z') \, k(\nu,z') \, dz' \right]$$

(5-98)

where $k(\nu,z)$ is the absorption coefficient at ν due to all lines, and can be represented for a Lorentz profile by the equation

$$k(\nu,p) = \sum_{i} \frac{S_i(T)}{\pi} \frac{\alpha_i'(T,p)}{(\nu - \nu_i)^2 + \alpha_i'^2(T,p)}$$

where S_i is the strength, α_i' is the half-width of the line and ν_i is the frequency at peak intensity of the line. The density profile of an absorbing gas e is given by $\rho_e(z')$. The kernel $K(\nu, z, \rho_e \ldots)$ is defined as

$$K(\nu,z) = \frac{\partial\sigma(\nu,z)}{\partial z} .$$

(5-100)

The pressure $p(z)$ is related to z through the hydrostatic equation

$$dp = - \rho(z) \, g \, dz$$

(5-101)

where $\rho(z)$ is the density profile of the entire atmosphere.

In practical observations, measurements of $I(\nu)$ are made at a discrete number of frequencies ν_j centered within a finite band $\Delta\nu$ with

$$I(\nu_j) = \int_{\nu''}^{\nu'} \Phi(\nu_j,\nu) \, I(\nu) \, d\nu$$

(5-102)

where $\Phi(\nu_j,\nu)$ is the instrument function.

From a practical point of view it is advisable (but not necessary for the present method) to substitute Eq. 5-96 into Eq. 5-102 and define the transmittance $\sigma(\nu_j, z)$ in the interval $\Delta\nu$ as

$$\sigma(\nu_j,z) = \int_{\nu''}^{\nu'} \Phi(\nu_j,\nu) \, \sigma(\nu_j,z) \, d\nu.$$

(5-103)

Since $B(\nu,T)$ is a smooth function in the interval $\Delta\nu$ we take

$$B(\nu,T) = B(\nu_j,T)$$

and rewrite Eq. 5-102 as

$$I(\nu_j) = B\left[\nu_j, \, T(z_j)\right] \sigma(\nu_j, z_o)$$

$$+ \int_{z_o}^{z} B\left[\nu_j, \, T(z)\right] K(\nu_j,z) \, dz.$$

(5-104)

From a given set of values of $\tilde{I}(\nu_j)$, $j = 1, 2 \ldots J$, we want to determine the temperature profile $T(z_j)$, assuming that $\rho_e(z)$, $\sigma(\nu_j,z)$ and $K(\nu_j,z)$ are known. In this problem, the selection of the set ν_j and the determination of $T(z_j)$ are strongly related and form the basis for the method of inverse-solution of Eq. 5-104.

By selecting a set of frequencies ν_j with varying degrees of atmospheric attenuation such that

$\sigma(\nu_1,z_o) \geq \sigma(\nu_2,z), \ldots \geq \sigma(\nu_j,z)$ we can generate a set of kernels $K(\nu_j,z)$ such that for each value of ν_j the kernel possesses a maximum at a different value of z_j, as shown in Figure 5-49.

The Mapping Transformation

Equation 5-104 is a nonlinear integral equation with fixed limits which may be viewed as a nonlinear transformation from $T(z)$ to $I(\nu)$ as in

$$\tilde{I}(\nu) = N \, T(z).$$

To obtain $T(z)$ we need to perform an inverse transformation as in

$$T(z) = N^{-1} \, \tilde{I}(\nu).$$

Figure 5-49 strongly suggests a mapping transformation from the ν axis to the z axis: since kernels $K(\nu_j,z)$ are strongly decaying functions, variations of $T(z)$ around z_i will affect the values of $I(\nu_j)$ very strongly, while variations of $T(z)$ at values of $z \ll z_j$ and $z \gg z_j$ do not affect $I(\nu_j)$ appreciably. Hence, we propose to map ν_j into z_j where z_j corresponds to the peak value of the kernel $K(\nu_j,z)$. Mathematically, we derive the transformation

$$\nu_j = \nu(z_j)$$

(5-105)

from the solution of the equation

$$\frac{\partial K(\nu_j,z)}{\partial z} = 0$$

(5-106)

for $j = 1, 2, 3 \ldots J$. Note, that z and ρ are related through the hydrostatic equation given in Eq. 5-101. Equation 5-106 can be used immediately to map the set of J points on the ν axis into a set of J points on the z axis.

But in order to map the I axis into the T axis we need a relationship between $I(\nu_j)$ and $T(z_j)$. Chahine (1968, 1970) applied the mean value theorem to Eq. 5-104 and derived the following relaxation equation

$$\frac{B\left[\nu_j, T^{(n+1)} (z_j)\right]}{B\left[\nu_j, T^{(n)} (z_j)\right]} \approx \frac{\tilde{I}(\nu_j)}{I^{(n)} (\nu_j)}$$

(5-107)

Equation 5-107 relates changes in the outgoing radiance for one frequency ν_j with changes in the Planck function at one level z_j. Equation 5-107 is expressed in an iterative form useful for our purposes where $T^{(n)} (z)$ and $T^{(n+1)} (z)$ are two temperature profiles at different orders n of an iterative solution. $I^{(n)} (\nu_j)$ is the radiance computed from Eq. 5-104 for a given $T^{(n)} (z)$ and $I(\nu)$ is the measured radiance. For additional details regarding the derivation of Eq. 5-107 see Equations (6)−(9) in Chahine (1970).

The Iterative Method Of Solution

We proceed to solve Eq. 5-104 for the determination of $T(z)$ by iteration as follows:

Assume a set of measured radiances, $\tilde{I}(\nu_j)$ is given for $j = 1, 2, \ldots J$.

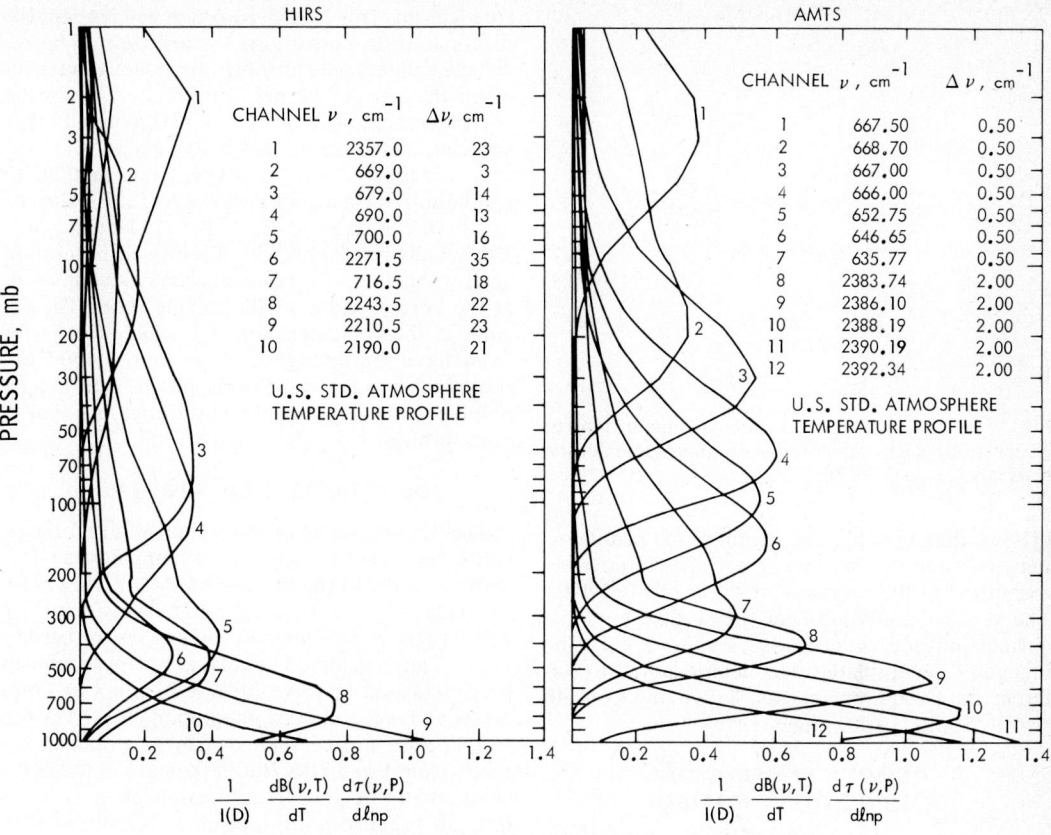

Fig. 5-49. Weighting functions for infrared temperature sounders.

1. Make an initial guess ($n = 0$) for $T^{(n)}(z)$.
2. Substitute $T^{(n)}(z)$ into Eq. 5-104 and evaluate the corresponding $I^{(n)}(\nu_j) j = 1, 2, \ldots J$ using in this case a linear interpolation formula.
3. Check the residuals

$$R^{(n)} = \frac{1}{J}\left[\sum_{j=1}^{J}\left(\frac{I^{(n)}(\nu_j) - \tilde{I}^{(n)}(\nu_j)}{\tilde{I}^{(n)}(\nu_j)}\right)^2\right]^{1/2},$$

(5-108)

for each frequency and in an *rms* sense. If $R^{(n)}$ is small, then $T^{(n)}(z)$ is a solution. If not,
4. Obtain a new guess

$$T^{(n+1)}(z_j) = \alpha_j^{(n)} T^{(n)}(z_j)$$

where the scaling factors are obtained from Eqs. 5-97 and 5-107 as

$$\langle\Delta T^{(n)}\rangle_{av} = \frac{1}{J}\sum_{j=1}^{J}\left| T^{(n)}(z_j) - T^{(n-1)}(z_j) \right|.$$

(5-110)

The iteration is terminated when $\langle\Delta T^{(n)}\rangle_{av}$ is less than some prescribed value, say 0.1 C.
5. Go to step 2 and repeat until step 3 or Eq. 5-110 is satisfied. It is possible however to use more than one measurement to recover the solution at one point z_j. This can be done by applying weighted scaling factors as

$$T^{(n+1)}(z_j) = \bar{\alpha}_j^{(n)} T^{(n)}(z_j),$$

where each $\bar{\alpha}_i^{(n)}$ is obtained as a weighted average of more than one sounding frequency as discussed in Chahine (1974).

We note here that in case the weighting func-

$$\alpha_j^{(n)} = \frac{b\nu_j/T^{(n)}(z_j)}{\ln\left\{1 - \left[1 - \exp\left(b\nu_j/T^{(n)}(z_j)\right)\right] I^{(n)}(\nu_j)/\tilde{I}^{(n)}(\nu_j)\right\}}$$

(5-109)

A second criterion to establish convergence of the iterative solution was applied here. The criterion is obtained by obseving the rate of convergence of the temperature profile with respect to itself as

tions are broad it becomes difficult to resolve small details in the profile even in those regions where the corresponding peaks are narrowly spaced. We note also that in general the resolution

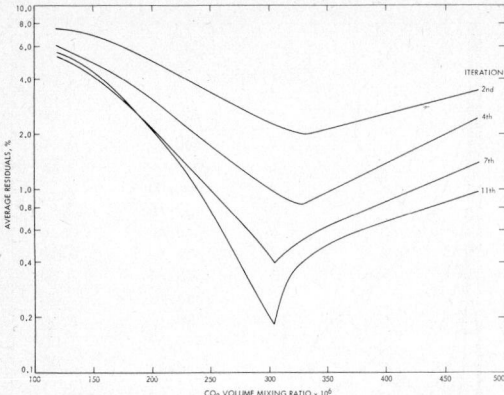

Fig. 5-50. Determination of the constant mixing ratio of an absorbing gas by the criterion of minimization of the residuals (Chahine, 1970).

of small detail in only one region of the profile, to the exclusion of the rest, is not always possible regardless of the accuracy of the measured data. The vertical resolution near the surface for most current infrared sounders is close to 2 km. The mapping transformation has been adapted to different data requirements as shown by Conrath (1970), Smith (1970), Shaw (1970).

REMOTE SENSING OF COMPOSITION PROFILES

The dependence of the relative transfer equation on the concentration of absorbing gases $\rho_e(z)$ appears in the kernel of the equation as shown in Eqs. 5-98, 5-100 and 5-104. If the mixing-ratio profile is a constant q_e the dependence of $K(\nu,z)$ on q_e is simple and the retrieval of q_e can be obtained by a minimization process. However, if the mixing ratio is a function of z the unknown $\rho_e(z)$ will appear as a *functional* in the integral equation, and the retrieval of $\rho_e(z)$ in this case becomes more difficult. We will examine these two cases in this section.

Case of The Constant Mixing Ratio

The property of the residuals given in Chahine (1970) to derive the constant mixing ratio q of ab-

troduces an error in $K(\nu_j, p)$ which will prevent the residuals from converging to near zero. The residuals will reach an absolute minimum value only when the correct kernel is used, i.e., when the correct mixing ratio is known, assuming all other sources of error to be relatively small.

As an illustration, we applied this method to synthetic data assuming, however, the value of q_{CO_2} (which is equal to 462×10^{-3}) to be unknown. The results in Figure 5-50 of the twelfth iteration clearly show that the residuals have one minimum at the correct value of the mixing ratio. The results of the third iteration, for which $T(z)$ is far from having converged, show that a good approximation to the value of the mixing ratio can be obtained with just a rough knowledge of temperature profile.

Case of the Variable Mixing Ratio

The determination of the composition profile ρ_e can be obtained by applying a mapping transformation similar to the one used for $T(p)$. However, the relaxation equation required to transform the I axis into the ρ_e axis may, in some cases, be hard to express analytically. The relaxation approach can be generalized to solve for any function or functional under the sign of integration. If $g(z_j)$ is the temperature profile then α_j can be obtained directly from Eq. 5-107. But in the case of the composition profile the determination of α_j is more difficult because $\rho_e(z)$ appears as a *functional* in the kernel,

$$K(\nu_j,z, \langle \rho_e(z) \rangle, \ldots)$$
$$= \frac{\partial \sigma(\nu_j,z, \langle \rho_e(z) \rangle, \ldots)}{\partial z}, \qquad (5\text{-}111)$$

since σ and K depend on the distribution of $\rho_e(z)$ between $\bar{z}$ and z. The notation $\langle \rho_e(z) \rangle$ indicates that the transmittance and the kernel are functionals of $\rho_e(z)$.

The General Approach

To determine $\rho_e(z)$, let us first integrate Eq. 5-104 by parts and write the result as in Chahine (1972)

$$I(\nu_j) = B[\nu_j, T(\bar{z})] - \int_{z_0}^{\bar{z}} \sigma[\nu_j, z, \langle \rho_e(z) \rangle, \ldots] \frac{\partial B}{\partial z} \, dz . \qquad (5\text{-}112)$$

sorbing gases, such as the constant mixing ratio of carbon dioxide, could be used to determine the constant mixing ratio for any gas. It was shown in Chahine (1970) that an error in the value of the mixing ratio q_{CO_2} used in computing the kernel in-

To determine $\rho_e(z_j)$ from a given set of radiance measurements, assuming that the temperature profile is known and is not isothermal, we map the ν_j axis into the z_j axis then make an initial guess $\rho_e^{(n)}(z_j)$ and solve the equation

$$I(\nu_j) - B[\nu_j, T(\bar{z})] = \int_{z_0}^{\bar{z}} \sigma[\nu_j, z, \langle \alpha_j^{(n)} \rho_e^{(n)}(z_j) \rangle, \ldots] \frac{\partial B}{\partial z} \, dz \qquad (5\text{-}113)$$

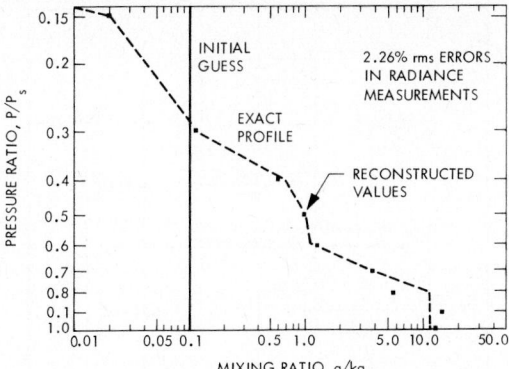

Fig. 5-51. Determination of the variable water vapor profile from synthetic data with random noise (Chahine, 1972).

to obtain a set of scaling factors $\alpha_j^{(n)}$, for $j = 1, 2 \ldots J$. We generate the next iteration through the relaxation equation

$$\rho_e^{(n+1)}(z_j) = \alpha_j^{(n)} \rho_e^{(n)}(z_i). \quad (5\text{-}114)$$

This iteration process is repeated until each value of the scaling constants approaches unity, which is equivalent to satisfying the residuals. This relaxation method of solution leads to accurate determination of composition profiles without any *a priori* information for the expected solution as shown in Figure 5-51.

The relaxation method can be applied in conjunction with any interpolation formula. The extent of interpolation is dictated by the quadrature requirements, and by the need to optimize the quality of solutions obtained from a finite set of sounding frequencies. Additional details on this subject can be found in Chahine (1972) and Chahine (1974).

Approximate Relaxation Equation

The determination of $\alpha_j^{(n)}$ can be rather time-consuming because it requires reevaluation of σ many times. Two approximate relaxation equations have been derived in equations (13) and (17) of Chahine (1972). We give here one approximation to the relaxation equation

$$\sqrt{\alpha_j^{(n)}} = 1 - \frac{\bar{I}(\nu_j) - I^{(n)}(\nu_j)}{\int_{z_0}^{\bar{z}} \sigma^{(n)} \ln \sigma^{(n)} \frac{\partial B}{\partial z} dz} > 0,$$

where σ and B are functions of both ν and z. Eq. 5-115 proved to be very useful particulary when σ is close to unity.

CORRECTIONS FOR THE EFFECTS OF CLOUDS

Equation 5-104 is derived for the case of plane parallel, *homogeneous* and *nonscattering* atmospheres. This is an ideal case which does not usu-

ally apply to observations in the presence of clouds or other horizontal inhomogeneities. In this section we treat the problem of remote sounding of cloudy atmospheres for the determination of the "clear column" vertical profiles, i.e., the vertical temperature profiles in the clear portions of the field of view. We will separate the problem into two parts. The first part deals with the simple case of a single cloud layer or a single degree of horizontal inhomogeneity and the second part deals with the general case of multiple cloud layers. The treatment of this problem in this section will be brief and will describe only the results which have been published in Chahine (1975, 1977), Chahine et al. (1977), and Aumann (1976).

Observations in the Presence of a Single Cloud Layer

We consider two adjacent fields of view having different fractional cloud covers at the same height z_c. We express the outgoing radiance $\tilde{I}_1(\nu)$ and $\tilde{I}_2(\nu)$ from the first and second fields of view as

$$\tilde{I}_1(\nu) = I_1(\nu)^{\text{clear}} - N_1\, G(\nu, z_c, \ldots)$$
$$\tilde{I}_2(\nu) = I_2(\nu)^{\text{clear}} - N_2\, G(\nu, z_c, \ldots) \quad (5\text{-}116)$$

Equation 5-116 makes no assumption about the radiative transfer properties of clouds. It simply states that the observed radiance $\tilde{I}_k(\nu)$ is equal to the clear-column radiance which would have been measured in the absence of clouds minus the radiance G "obscured" by the presence of a fractional cloud cover N_k. G is unknown and depends on ν, z_c and the spectral properties of clouds.

If the two fields of view are small and contiguous we can assume that

$$\bar{I}(\nu) = I_1(\nu)^{\text{clear}} = I_2(\nu)^{\text{clear}}$$

and substitute into Eq. 5-104 to eliminate G and get

$$\bar{I}(\nu) = \tilde{I}_1(\nu) + \eta_1[\tilde{I}_1(\nu) - \tilde{I}_2(\nu)] \quad (5\text{-}117)$$

where $\eta = $ unknown $= (N_1)/(N_2 - N_1)$. Thus if η is known we can reconstruct the clear-column radiance according to Eq. 5-117 for any frequency which saw the same field of view.

According to Eq. 5-117, η can be determined from a knowledge of $I(\nu)$ at any frequency, say ν', as

$$\eta = \frac{I(\nu') = \tilde{I}_1(\nu')}{\tilde{I}_1(\nu') - \tilde{I}_2(\nu')} \quad (5\text{-}118)$$

However, the exact value of the clear-column radiance $I(\nu')$ is actually unknown because $I(\nu')$ itself depends on $T(z)$. Thus the determination of η and $T(z)$ should be carried out simultaneously. The selection of ν' is discussed in Chahine (1974). The use of an atmospheric sounding frequency from the microwave region of the spectrum is very efficient for determining η, particularly since microwave observations are not affected by most types of clouds.

Determination of the Amounts and Heights of Clouds

For the case of black clouds the term $G(v, z_c, \ldots)$ in Eq. 5-116 can be expressed analytically as

$$G(v,z_c) = -B[v,T(z_c)]\,\tau(v,z_c)$$
$$+ B[v,T(z_o)]\,\tau(v,z_o)$$
$$+ \int_{z_o}^{\bar{z}} B[v,T(z)]\,K(v,z)\,dz, \quad (5\text{-}119)$$

where z_c is the cloud-top height. By substituting Eq. 5-119 into Eq. 5-116 and using Eq. 5-104 we can express the measured radiance in any field of view $\bar{I}_k(v)$ as an integral function of the clear-column temperature profile $T(z)$, the cloud-top height z_c and the fractional cloud cover N, according to equation (1) in Chahine (1975) as

$$\bar{I}(v) = N\Big\{ B[v,T(z_c)]\,\sigma(v,z_c)$$
$$+ \int_{z_c}^{\bar{z}} B[v,T(z)]\,K(v,z)\,dz \Big\}$$
$$+ (1-N)\Big\{ B[v,T(z_o)]\,\sigma(v,z_o) \quad (5\text{-}120)$$
$$+ \int_{z_o}^{\bar{z}} B[v,T(z)]\,K(v,z)\,dz \Big\}.$$

The value of the fractional cloud cover N and the height z_c of clouds can be determined when the radiative transfer properties of the clouds in $G(v,z_c,\ldots)$ are given.

For the case of black clouds Chahine (1970) showed that by substituting Eq. 5-120 into Eq. 5-116 and by taking the ratio for two different sounding frequencies v_1 and v_2 we get

$$\frac{I^{(n)}(v_1) - \bar{I}_k(v_1)}{I^{(n)}(v_2) - \bar{I}_k(v_2)} = \frac{G^{(n)}(v_1,z_c)}{G^{(n)}(v_2,z_c)} \quad (5\text{-}121)$$

where Eq. 5-121 is now a function of one unknown z_c, assuming that $T^{(n)}(z)$ has been determined as described earlier. The solution for z_c can be obtained by minimization techniques as described in Chahine (1974). The corresponding value of N_k is obtained directly from equation (35) as

$$N_k = \frac{I^{(n)}(v_1) - \bar{I}_k(v_1)}{G(v,z_c)}. \quad (5\text{-}122)$$

Observations in the Presence of Multiple Cloud Formations

The method described earlier for the elimination of the effects of a single layer of clouds can be extended to multiple cloud formations. Derivation of the required equations can be found in Chahine (1977).

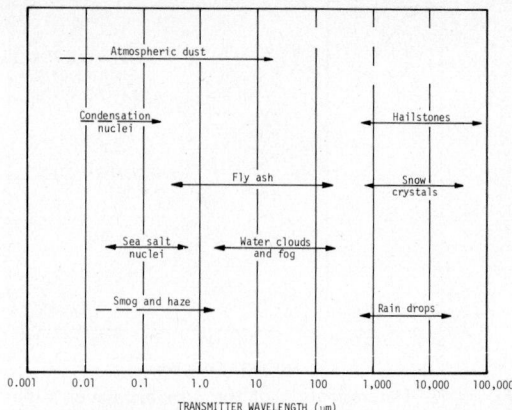

Fig. 5-52. Relation of radiation wavelengths to diameters of atmospheric particles.

CORRECTIONS FOR ATMOSPHERIC PARTICULATES

Particles suspended in the atmosphere modify the solar and terrestrial radiation fluxes. Scattering, absorption and emission by those particulates (here we consider aerosols other than condensates in the form of clouds) may be important in remote sensing for wavelengths up to 100 μm and longer. The relation of wavelength to diameter of the major types of atmospheric particles is shown in Figure 5-52.

Optical Properties of Particulates

The *volume extinction absorption coefficient* is a useful parameter in determining the effect of particulates on the radiation fluxes. The extinction of a monochromatic beam propagating in a scattering medium is given by

$$dE(\lambda) = -\beta_{ext}(\lambda,n)E(\lambda)dr, \quad (5\text{-}123)$$

where:

$E(\lambda)$ = irradiance of beam defined as the energy per unit bandwidth, at wavelength λ, transmitted per second through a unit area normal to the propagating direction;

dr = amount of scatterer in a volume of unit cross-sectional area and length dl;

$\beta_{ext}(\lambda,n)$ = volume extinction absorption coefficient (per unit ext length) for scatterers with index of refraction

$$n = m_1 + im_2. \quad (5\text{-}124)$$

Long and Rensch (1970) have calculated β_{ext} for a Junge distribution with a particulate concentration of 10^9 m^{-3} and relative humidity of less than 90 percent. Results are shown in Figure 5-53. The variable parameter v in the Junge distribution Eq. 5-17 usually lies between 3.5 and 4.5 (Junge, 1963). The distribution for $v = 4$ is shown in Figure 5-7.

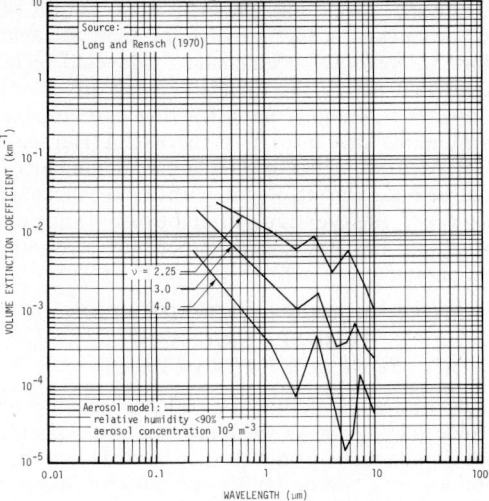

Fig. 5-53. Volume extinction coefficient versus wavelength for aerosol model with relative humidity <90% and aerosol concentration of 10^9m^{-3}.

Particulate Composition

Air pollutants can occur in the form of gases, solid particles, or liquid aerosols, forms which can exist either separately or in combinations. Gaseous pollutants constitute about 90 pecent of the total mass emitted to the atmosphere while particulates comprise the rest. Given suitable conditions, many of the primary pollutants will participate in reactions in the atmosphere that produce secondary pollutants, such as photochemical smog. The most important pollutant types are: (1) Particulates, including carbon fly ash, lead, zinc oxide, and arsenic; (2) nitrogen compounds, including NO, NO_2, NO_3, HNO_2, and HNO_3; (3) sulfur compounds, including SO_2, SO_3, and H_2SO_3; and (4) oxygen compounds, in particular O_3.

In addition, hydrogen chloride and other halogen compounds, as well as ammonia and peroxylacetyl nitrate (PAN) are important pollutants to be measured and monitored.

Remote Sensing in the Presence of Particles

Aerosol particles, including pollutant particles, are not equally distributed in size. The carrying capacity of the atmosphere is far greater for smaller particles than for larger ones. Typical size-distribution plots were shown in Figure 5-7.

Scattering intensity is a function of the particulate number concentration, size distribution, shape, and absorption-emission characteristics. Pollutant particulates are generally passive scatterers. Because of their relative by high numberdensities and random shapes they are sometimes assumed to exhibit no shape characteristics in the scattering process. An ensemble of randomshaped particles would then be equivalent to the same number of spherical-shaped particles of some mean diameter. Recent work by Pollack and Cuzzi (1978) allows some irregular-shaped scat-

terers to be approximated by spheres if the scattering characteristics of the spheres are appropriately scaled. At a wavelength of 1 μm (β_{ext} being 0.004 km^{-1}), an approximate signal change of ~1 percent in 2 km can be expected. At 10 μm, the signal change is approximately an order of magnitude less. The exact value is, of course, dependent on the particle number density and size distributions along the line of sight.

The importance of aerosols in modifying visible radiation fluxes is indicated by a factor known as the turbidity coefficient. The turbidity coefficient B is the decadic extinction coefficient at a wavelength of 0.50 μm defined (following Flowers, 1969) as

$$I/I_o = 10^{-(\tau_r + \tau_z + B)m} \qquad (5\text{-}125)$$

where:

I = observed solar radiance at $\lambda = 0.50$ μm, adjusted to the mean sun-earth distance;

I_o = solar radiation outside the atmosphere at $\lambda = 0.50$ μm at mean sun-earth distance;

τ_r = scattering coefficient for air molecules; for air mass 1 at sea level ($\tau_z = 0.0634$);

τ_z = absorption coefficient for ozone for air mass 1 at sea level ($\tau = 0.004$);

m = optical air mass

$\qquad m \simeq \csc(h); \; 10° < h \leqslant 90°$

where

$$h = \text{solar elevation.}$$

Turbidity data for the continental U.S. are shown in Figure 5-54 where turbidity, as defined here, is due exclusively to aerosol particles. From Eq. 5-125, the transmission to the earth's surface of solar radiation in an aerosol-free atmosphere ($B = 0$, $m = 1$) is $I/I_o = 0.86$. For the highest urban turbidity in Figure 5-54, B = 0.22, the radiation reaching the surface falls to 0.75. The *total transmission* in a vertical path down to the earth's surface and back out to space is then 0.56. At these levels turbidity is controlled by sulfate aerosols in the boundary layer. Clearly, any attempt to measure surface reflectivity from a space platform is dependent on knowledge of the aerosol content of

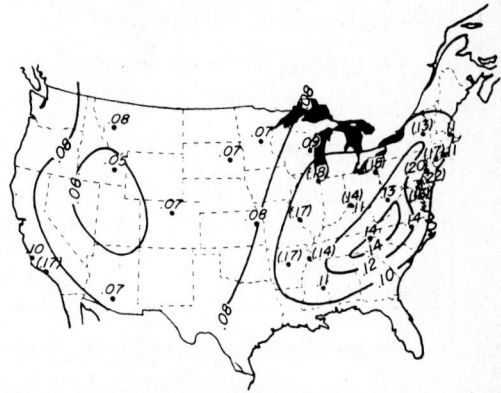

Fig. 5-54. Annual average rural turbidity, B, in the United States, 1961–1966 (Flowers, 1969). Urban values are in parentheses [Unit 10^2 (air mass)$^{-1}$].

the atmospheric column through which the measurement is made.

At longer wavelengths aerosols may significantly attenuate the upwelling radiation from the surface. Rozenberg (1977) has suggested that submicron particles composed of sulfur compounds may be responsible for appreciable extinction of thermal radiation in the 8-12 μm atmospheric window. In the $\lambda \geqslant 8$ μm region, scattering by aerosols becomes negligible and absorption dominates. For $\lambda \geqslant 12$ μm aerosol extinction is small except, perhaps, for conditions of extremely high dust-loading.

The small number densities per unit volume for aerosols make multiple scattering processes negligibly small throughout the visible and infrared regions. This considerably simplifies the computational effort involved in correcting observed radiation fluxes for aerosol effects. The problem remains, however, of determining the vertical

distribution, sizes and chemical composition of the particulates in the line of sight. Considerable effort has recently been made to develop techniques that permit the remote sensing of aerosol parameters. Methods for retrieving particle size distributions have been most successful (Herman, 1977). Measurements of total column abundances of aerosols have also met with some success; however, the vertical distribution function throughout the atmosphere is not now a retrievable parameter.

SIMULTANEOUS DETERMINATION OF ATMOSPHERIC AND SURFACE PARAMETERS

The optimum method for remote sensing of surface parameters is to obtain the atmospheric data simultaneously with surface data. The results obtained by Chahine (1977) are shown in Figure 5-55 and were obtained from radiance data mea-

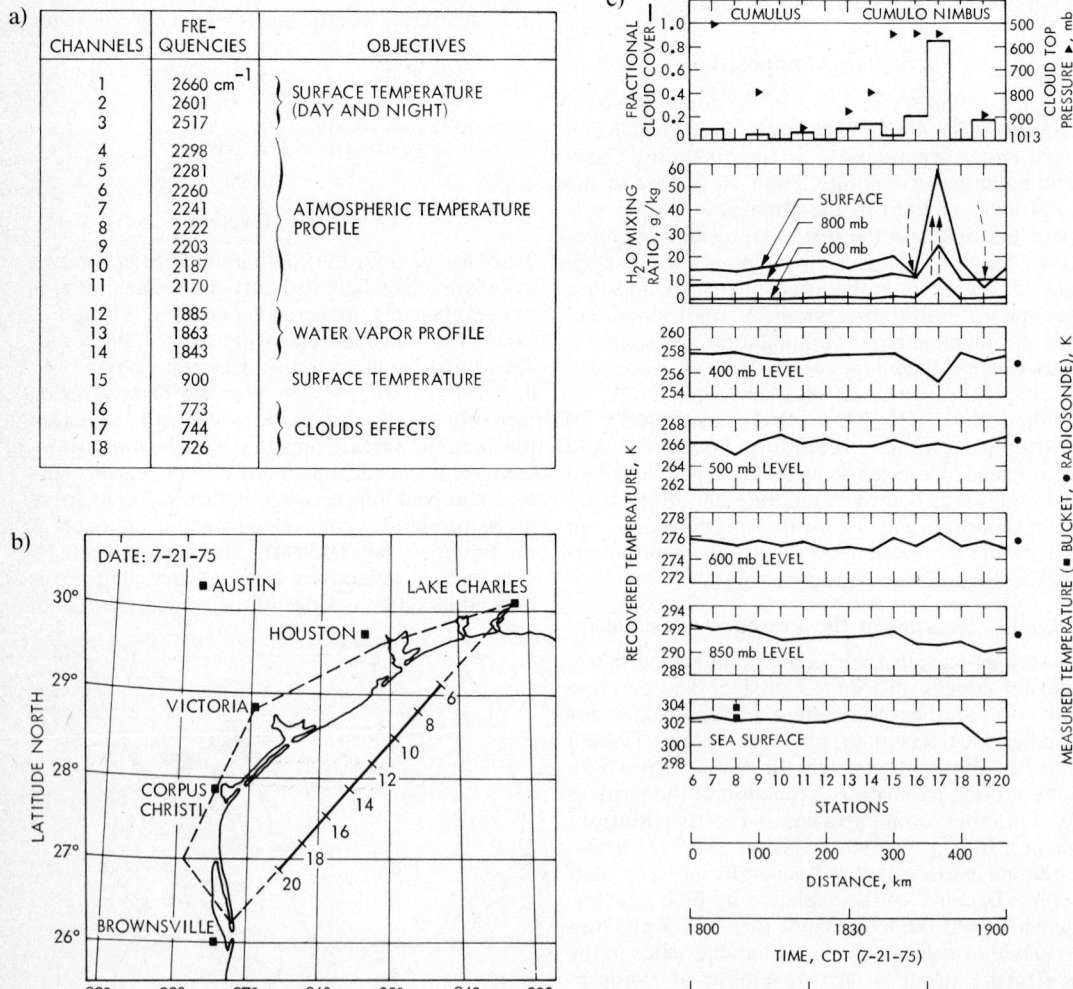

Fig. 5-55. (a) Set of sounding frequencies used and their main functions. (b) Flight path and location of stations shown in (c). (c) Surface, atmospheric and cloud parameters recovered from simultaneous multispectral measurements taken from aircraft (Chahine, 1977b).

sured simultaneously in 4.3-μm and 15-μm CO_2 bands for temperature profiling, the 6.3-μm band for determining the water vapor mixing ratio profiles, and the 11-μm and 3.7-μm window regions for surface temperatures.

REMOTE SENSING IN THE MICROWAVE

ATMOSPHERIC ABSORPTION AND EMISSION FROM 1 TO 300 GHz

The spectra so far presented were computed for homogeneous conditions in the respective gases. However, it is of greater interest to know representative values for the total absorption through the atmosphere and the values of emission spectra that result.

The scatter-free equation of radiative transfer can be written at some frequency for observations looking at the zenith as (see Eq. 5-73):

$$T_B = T_{sky}\exp(-\tau_{max}) \qquad (5\text{-}126)$$

$$+ \int_0^\infty T_{air} \, \gamma(z) \, \exp\left[- \int_0^z \gamma(z') \, dz' \right] dz,$$

where:

T_B = observed brightness temperature,
T_{air} = local atmosphere temperature,
γ = total absorption temperature
z = height.

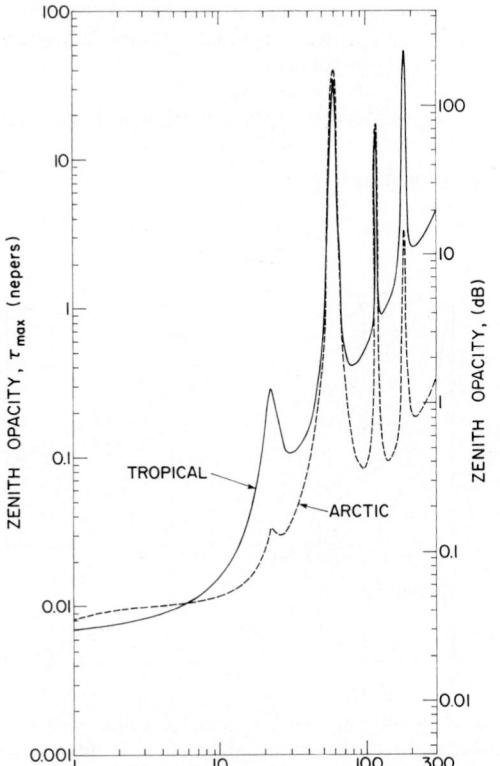

Fig. 5-56. Zenith opacity computed as a function of frequency for the Standard Atmosphere 1962, Tropical and Arctic Supplements.

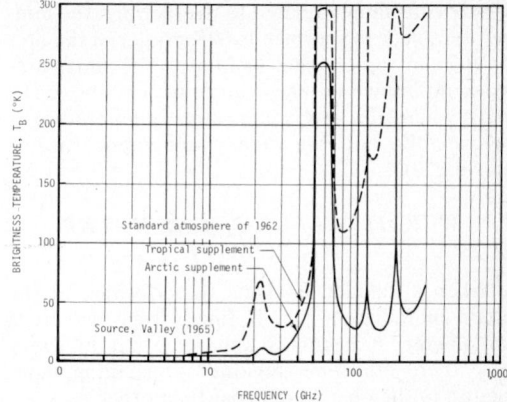

Fig. 5-57. Emission spectrum as seen from the earth's surface computed as a function of frequency for the Standard Atmosphere 1962 Tropical and Arctic Supplements, (Valley, 1965).

The zenith opacity τ_{max} is expressible as (see Eq. 5-14)

$$\tau_{max} = \int_0^\infty \gamma(z) \, dz . \qquad (5\text{-}127)$$

For the "Standard Atmosphere 1962, Tropical and Arctic Supplements" (see Valley, 1965) both τ_{max} and T_B were computed over the range of 1 to 300 GHz; they are presented in Figures 5-56 and 5-57. The differences between the distributions of temperature and water vapor in the two atmospheres were significant. The Tropical Supplemental Atmosphere contained 4.03 g cm^{-2} of precipitable water in vapor form while the Arctic Supplemental Atmosphere held only 0.21 g cm^{-2}. The surface temperature in the tropical case was 300K; it was 249K for the arctic case. These atmospheres are representative of extremes found over the earth and have resulted in extremes for absorption and brightness temperature spectra.

Figure 5-56 has several interesting features. At the lowest frequencies the cold arctic temperature drives the oxygen absorption above that expected in the warmer tropical environment. The difference in water vapor is so great that the water-vapor absorption at both resonances is reduced by an order of magnitude from the moist to the dry atmosphere. The absorption in the window between the 118 GHz line of oxygen and the 183 GHz line of water vapor is more affected by the differences in total water than in the windows that are lower and higher in frequency.

Figure 5-57 depicts the emission spectrum as seen from the surface of the earth. Again the difference in water vapor causes the most striking effect. In the tropical atmosphere the brightness temperature on the 22.235 GHz resonance is 72 K, while in the arctic it has been reduced to approximately 12K. Near the very opaque lines of both oxygen and water vapor, the differences in brightness temperatures between the two atmospheres are simply related to the fact that the temperature near the surface is 60K cooler in the arctic than in the tropical example. The most dramatic difference in temperature occurs above 200 GHz where the low water-vapor temperature is below 50K, and the high water-vapor case is above 270K.

WEIGHTING FUNCTIONS IN THE MICROWAVE

When a detailed analysis is performed of the origin of radiation that is finally recorded by a radiometer, it becomes apparent that in every case there are some regions contributing more energy to the observed signal than other regions. The relative contribution from a given layer to the total is often found to be quite stable for a given frequency, showing only mild variations from one climatological region to another. In order to better understand these patterns of contributions from the atmosphere, various so-called *weighting functions* have been defined. They aid in understanding the origin of spectra, and judging the feasibility of remote sensing of atmospheric parameters. Figures 5-58, 5-59, 5-60, and 5-61 present several varieties of weighting function and related curves.

Figure 5-58 is a weighting function for emission by oxygen as viewed from above the atmosphere.

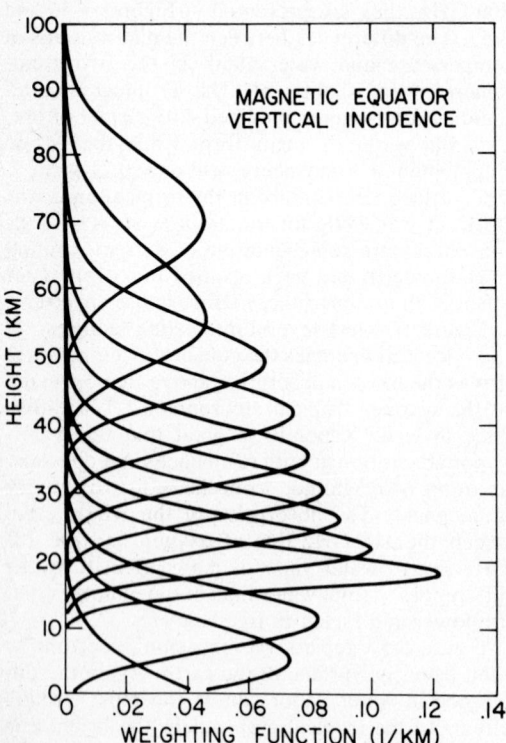

Fig. 5-58. Temperature weighting functions as a function of altitude above the surface for observations from space which view the nadir. The curves are based upon Eq. 5-128 and are the results of emission by oxygen.

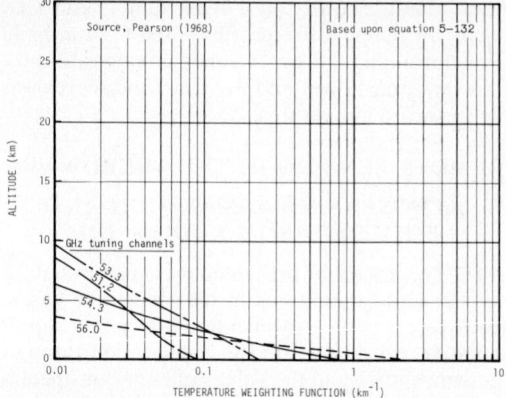

Fig. 5-59. Unnormalized weighting functions for temperature as functions of height above the surface for observations from the surface looking at the zenith. The curves are based upon Eq. 5-132 and are the results of emission by oxygen.

The weighting function is defined as follows (see Lenoir, 1968):

$$W_{O_2,DN}(\nu,z) = \gamma_{O_2}(\nu,z)\exp\left[-\tau(\nu,z)\right], \quad (5\text{-}128)$$

where

$W_{O_2,DN}$ = weighting function for oxygen looking down (DN) through the atmosphere, and is a measure of the contribution that the temperature at height z will make to the brightness temperature observed,

ν = frequency,

$\gamma_{O_2}(\nu,z)$ = absorption coefficient for oxygen at ν and z,

τ is defined as

$$\tau(\nu,z) = \int_0^H \gamma_{O_2}(\nu,z')\,dz', \quad (5\text{-}129)$$

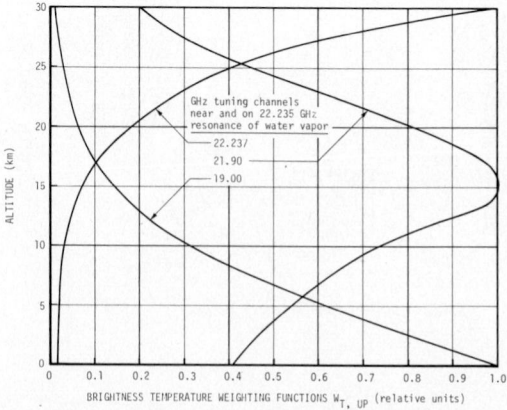

Fig. 5-60. Normalized weighting function curves for water-vapor density in the atmosphere at three representative frequencies near and on the 22.235 GHz resonance of water vapor. The curves are derived for brightness temperature measurements from the surface of the earth.

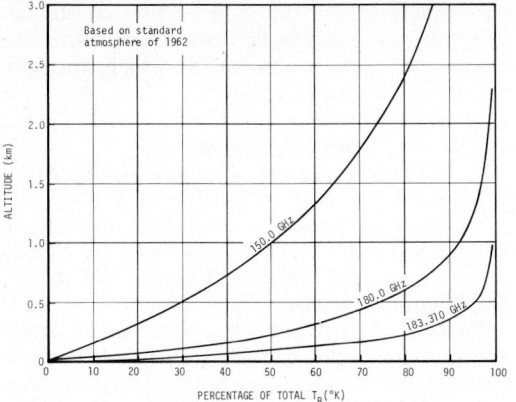

Fig. 5-61. The altitude below which originates a given percentage of power normally incident upon the terrestrial surface, for various frequencies near the 183-GHz water-vapor resonance.

where:

H = height of the satellite
z' = a variable of height between z and H.

The definition of the weighting function is such that the observed brightness temperature spectrum is

$$T_B(\nu) = \int_0^H T_{\text{air}}(z') \, W_{O_2,DN}(\nu,z') \, dz. \tag{5-130}$$

From the shape of the curves in Figure 5-58, it is clear, for example, that most of the energy received by the channel tuned to 60.82 GHz will be from the region near 18 km. These weighting functions—in addition to aiding in the understanding of spectra—also form the basis for effectively utilizing microwave spectral data to probe the atmospheric temperature structure from a satellite.

The origin of the bell-shaped curves in Figure 5-58 lies with the very opaque nature of the spectrum near the oxygen complex of lines and the exponential distribution of atmospheric pressure. If one successively adds layers of air starting from the top of the atmosphere he finds that at first, because each deeper layer has more oxygen than the last, it contributes more energy to the observing radiometer. Finally, however, the thickness of the atmosphere above the layer being sampled is so great that more radiation is lost by increased absorption than is gained from the emission of the new layer. The contributions of succeeding layers then diminish rapidly until the surface is encountered, or until the opacity is so high that negligible energy penetrates the intervening layers.

The bell-shaped curves of Figure 5-58 contrast strongly with the weighting functions for oxygen emission looking up from the surface, as shown in Figure 5-59 (see Pearson, 1968). Now the contri-

bution to the measured brightness temperature is greatest close to the surface and diminishes very rapidly with increased altitude. The scale heights of the weighting function curves decrease as more highly absorbing parts of the spectrum are reached, as might be expected.

The opacity of a fixed amount of water vapor near the 22.235 GHz line varies significantly from level to level in the atmosphere. However, the sensitivity varies only slightly over the range of atmospheric temperatures; therefore, it can be written for water vapor opacity or for the temperature contribution by each layer to the total observed brightness temperature. In the case of water vapor opacity, the weighting function takes the form

$$W_{H_2O} \, \gamma(\nu,z) = \frac{\gamma_{H_2O}(\nu,z)}{D(z)} \tag{5-131}$$

where

γ = absorption,
H_2O = coefficient of water vapor,
D = water-vapor density.

For the brightness temperature observed at the surface, the weighting function becomes

$$W_{H_2O,UP}(\nu,z) = \frac{\gamma_{H_2O}(\nu,z)T_{air}\exp(-\tau)}{D(z)}$$
$$\tag{5-132}$$

where

$$\tau = \int_0^z \gamma_{H_2O}(z') \, dz', \tag{5-133}$$

and where

$W_{H_2O,UP}(\nu,z)$ = weighting function for water vapor looking up from the surface.

Because the atmosphere is so transparent in this region, $\exp(-\tau)$ is almost equal to unity, and the dominating variable in Eq. 5-132 becomes γ_{H_2O}. Therefore, the shapes of the weighting functions from Eqs. 5-131 and 5-132 are very similar. The weighting functions for three reprsentative frequencies derived from Eq. 5-132 are shown in Figure 5-60.

The characteristic features are: (1) on resonance (22.235 GHz) the sensitivity increases inversely as the pressure $1/P$; (2) somewhat off resonance there exists a maximum sensitivity to water vapor at an altitude that is highly frequency-dependent; and, (3) in the further wings of the line, the sensitivity decreases as the total pressure decreases.

Near the 183 GHz water-vapor line, the opacity is so high that weighting functions looking up and down through the atmosphere would have shapes similar to those shown in Figures 5-58 and 5-59 for oxygen. However, they would change drastically as the water vapor in the atmosphere changed.

The curves shown in Figure 5-45 can be used instead of weighting functions. It can be seen that at resonance (183.310 GHz) more than 80 percent of the energy received arrives from the lowest 250 meters in the standard atmosphere.

If measurements are desired which have greater vertical resolution, or if measurements are to be made of trace atmospheric constituents in the altitude region 10–150 km, then a limb-scanning geometry is generally best (Staelin, 1977). A satellite viewing the limb of the earth has great sensitivity to the atmosphere at that altitude where the ray path is tangent to a constant pressure surface; the ray averages over a volume extending horizontally ~300 km. The temperature-weighting function corresponding to this geometry is illustrated in Figure 5-62 for a ray tangent at 12-km altitude; its effective width is ~2–3 km. If the atmosphere is opaque along this ray, the high resolution benefit is lost, as shown for a frequency only 0.5 GHz from the highly opaque 118 GHz oxygen resonance.

Because of the ~300 km path available for limb scans, compared to ~8 km for nadir views, the sensitivity to trace constituents can be increased by more than a factor of 30; no such satellites have flown yet, but they are being proposed. Trace constituents that might be observed include: H_2O, O_2, CO, NO, N_2O, HNO_3, NO_2, ClO, H_2O_2, OH, O, and others (Waters, 1976; Waters and Wofsy, 1978). Already the profiles of O_3 (Shimabukuro et al., 1977) and CO (Waters, 1976) have been observed from the ground. Such observations are generally restricted to gases in the stratosphere or

mesosphere where the reduced pressure-broadening permits them to be distinguished from the broad H_2O and O_2 resonances which normally dominate the spectrum.

OBSERVATIONS OF ATMOSPHERIC PHENOMENA

The general principles of passive microwave remote sensing have been reviewed by Staelin (1969), Tomiyasu (1974), and Staelin and Rosenkranz, editors (1978).

Passive microwave observations of atmospheric temperature profiles are possible because atmospheric oxygen has both a very uniform mixing ratio and also opaque resonances near 60 and 118 GHz. These properties of oxygen result in temperature-weighting functions (defined in Eq. 5-130) such as those illustrated in Figure 5-58. In general, the radiation emitted to space at any wavelength is proportional to the average temperature in a layer defined by the weighting function, which typically is ~8–12 km thick. Multifrequency passive microwave radiometers can thus sense the average air temperatures at several altitudes simultaneously; mathematical retrieval techniques can then yield accurate estimates of the full atmospheric temperature profile.

The first microwave instrument in space to measure temperature profiles was the *Nimbus-E Microwave Spectrometer* (NEMS), which was launched into a polar orbit in 1972 (Staelin, 1973); it carried five radiometers which viewed nadir with ~200 km resolution and 0.1–0.2K sensitivity, and it operated at 22.234, 31.4, 53.65, 54.9, and 58.8 GHz. This experiment was preceded by aircraft tests of the same instrument which proved the technique and revealed that there is remarkably little fine structure in atmospheric temperature fields when averages of ~8 km altitude are observed (Rosenkranz, 1972).

The first techniques used to retrieve these temperature profiles were linear minimum-square-error estimators based upon statistically representative *a priori* sets of data; the *rms* temperature retrieval errors were ~2K (Waters et al., 1975). It was found that only large precipitating clouds or mountains, both of which can be opaque at altitudes above ~2–3 km, noticeably perturbed the results (Staelin, 1975).

The temperature-profile accuracy is worst (2–5K, *rms*) near the terrestrial surface and tropopause because both regions often exhibit sharp discontinuities in the temperature profile which cannot be resolved by the vertical resolution of the weighting functions. Another source of error arises because local climates may differ from the *a priori* average, and the result is a small location-dependent bias. Both of these problems have been reduced somewhat by the use of a Kalman filter retrieval technique (Ledsham and Staelin, 1978) which provides ~30 percent improvement in retrieval accuracies.

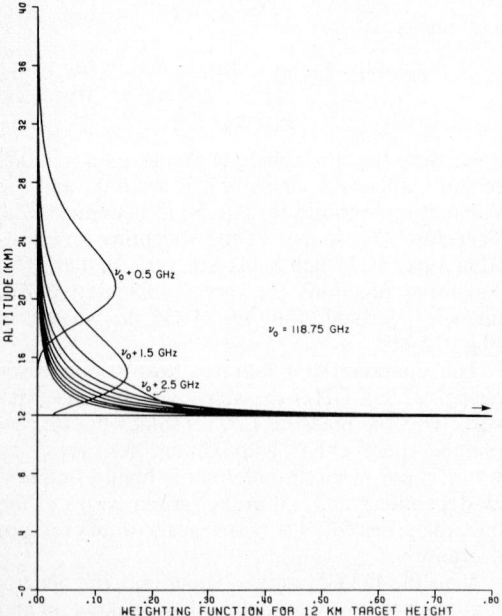

Fig. 5-62. Temperature weighting functions for an infinitesimal pencil-beam tangent to the terrestrial atmosphere at 12 km altitude; the frequencies are on the wings of the strong 118 GHz oxygen resonance.

More useful data were obtained by the *N*imbus-6 *S*canning *M*icrowave *S*pectrometer (SCAMS), launched in June, 1975, which imaged the earth continuously until the spring of 1976 with ~150–300 km resolution (Staelin, 1977). Observations at only three oxygen-band frequencies (52.85, 53.85, 55.45 GHz) were sufficient to retrieve temperature profiles for the range 100–1000 mbar. The three images in Figure 5-63 represent the retrieved average temperatures of the layers 1000–500, 500–250, and 250–100 mbar for a single orbit which extends from the north pole to the south pole and back. The width of the strip is ~2400 km and the width of each temperature contour is 2K. Several dynamic weather systems are evident; note the differences with altitude. The multi-contoured feature in the center of the image is the spurious result of the 4 km elevation over central Antarctica. Systems similar to this now operate on the United States Tiros-N and Block 5D operational weather satellites.

It is also possible to measure atmospheric temperature profiles from the ground, as suggested by the weighting functions in Figure 5-59. The altitude resolution is greatest near the surface and the accuracy varies from a fraction of a degree at the surface to a few degrees at 10 km altitude (Westwater, 1975; Miner, 1972; Westwater and Decker, 1977).

The first passive microwave spectrometer to monitor atmospheric water vapor and liquid water was a four-channel instrument on Cosmos-243, which operated for a couple of weeks after launch in 1968 (Basharinov, 1969). It operated at wavelengths between 8 mm and 8 cm and observed water vapor, clouds, and precipitation in emission against the cold background provided by the reflective oceans. The general principles of sensing atmospheric water have been discussed by Staelin (1966), and data from several satellites have been published (Obukhov and Tatarskaya, 1969; Gorelik et al., 1975; Dombkovskaya et al., 1976; Dombkovskaya et al., 1977; Staelin, 1976, Rosenkranz, 1978; Wilheit, 1976, Allison, 1974; Wilheit, 1977, and Lipes et al., 1979).

In Figure 5-64 are enhanced images of global

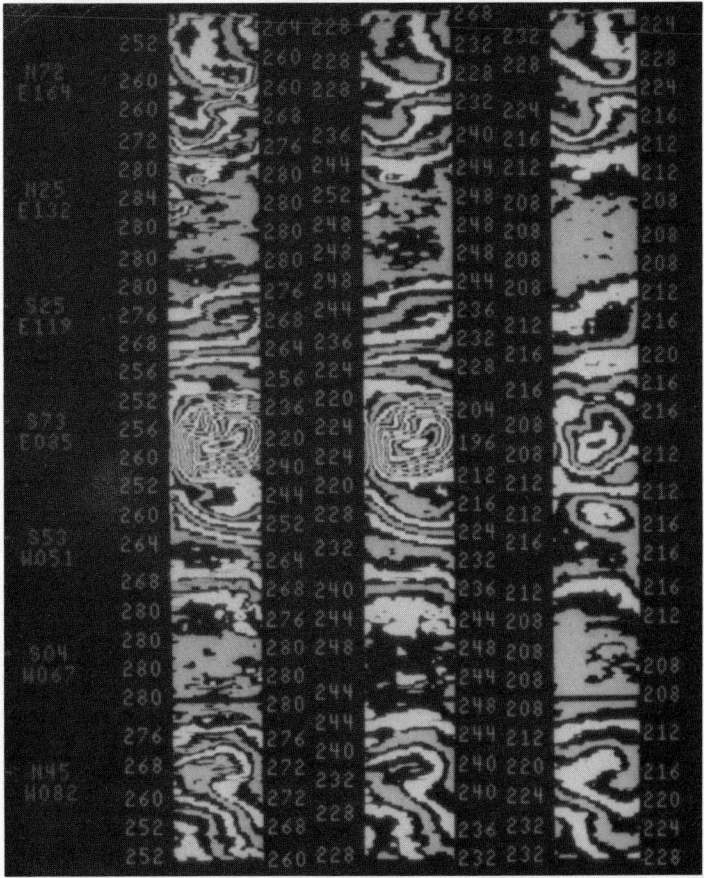

Fig. 5-63. Retrieved average temperatures for the layers 1000-5000 mb, 500-250 mb, and 250-100 mb (left to right), as obtained from SCAMS 25 May 1976. The images are ~2400 km wide and circle the globe once; Antarctica is in the center of the images and the equator crossings are at longitudes 126° East and 68° West in the top and bottom portions of the images, respectively. The contours are 2 K wide, and the incompletely calibrated equatorial layer temperatures here are ~280 K, 248 K, and 208 K, respectively.

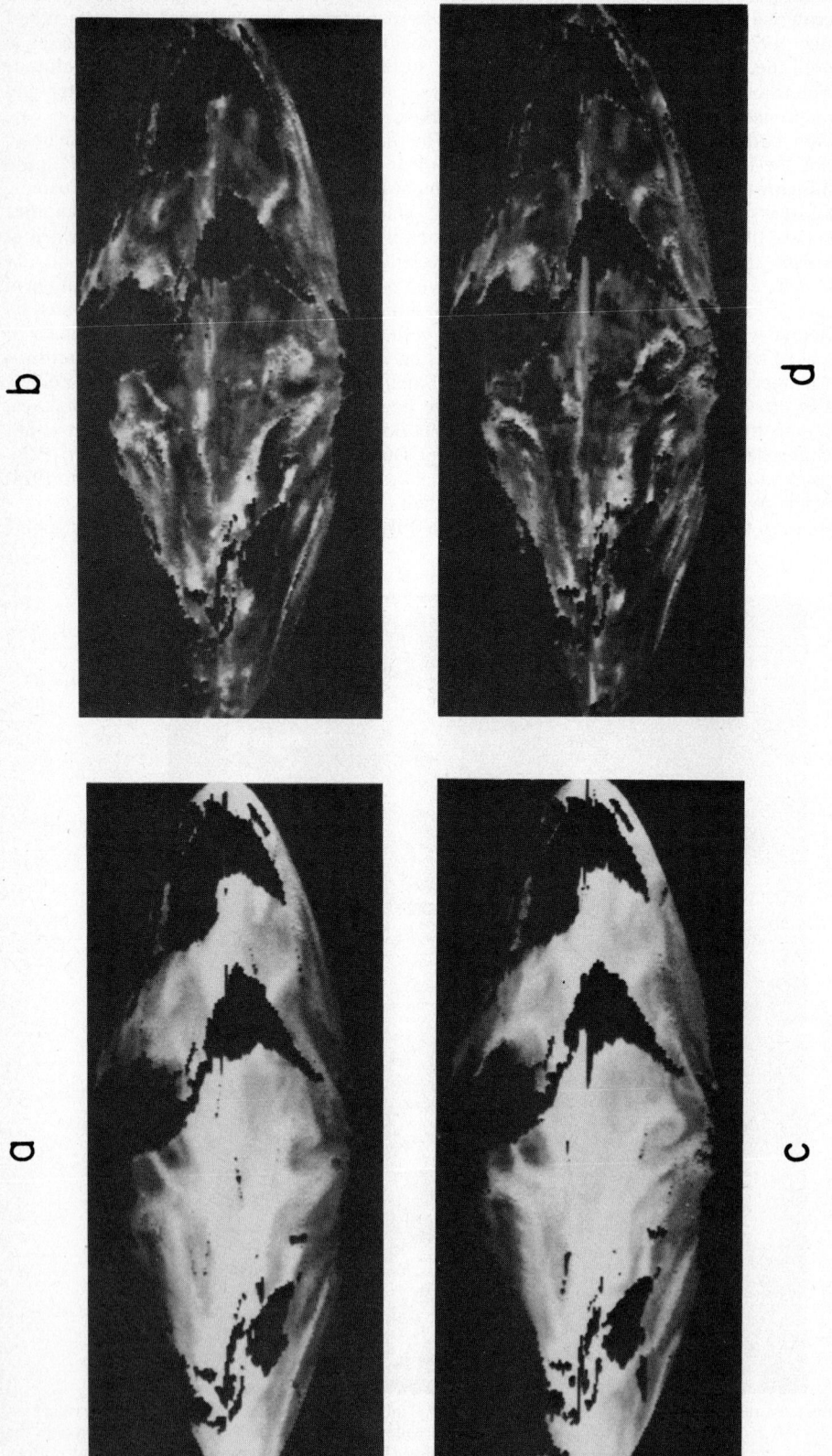

Fig. 5-64. Retrieved water vapor (a and c) and liquid water (b and d) maps observed by SCAMS on 22 January 1976 (a and b) 23 January 1976 (c and d). Darker areas over ocean are drier; black areas are land or missing data.

water vapor and precipitation obtained by SCAMS over 12-hour periods on January 22 and 23, 1976. The intertropical convergence zone is evident in the precipitation map, and many interesting features of tropospheric water vapor are visible. The evolution of the major frontal system is quite marked over this 24-hour interval.

Figure 5-65 displays two infrared images and a 1.55 cm wavelength image of a typhoon located midway between India and Antarctica (see the ESMR image). The boundaries between air masses of different moisture content are clearly visible. At the top of the figure the water vapor

and liquid-water estimates obtained from NEMS in the central 200 km of these maps are plotted, and suggest that these two parameters are statistically nearly orthogonal. For example, the water-vapor retrievals exhibit no perturbation near the precipitation spike at 40°S.

In the same figure the brightness temperature for 54.9 GHz is plotted near the eye of the storm. The eye is approximately 3 K warmer than is the periphery due to the low pressures there and the dynamics of the storm; the horizontal temperature gradient is approximately proportional to the vertical gradient in wind. The effect has been studied

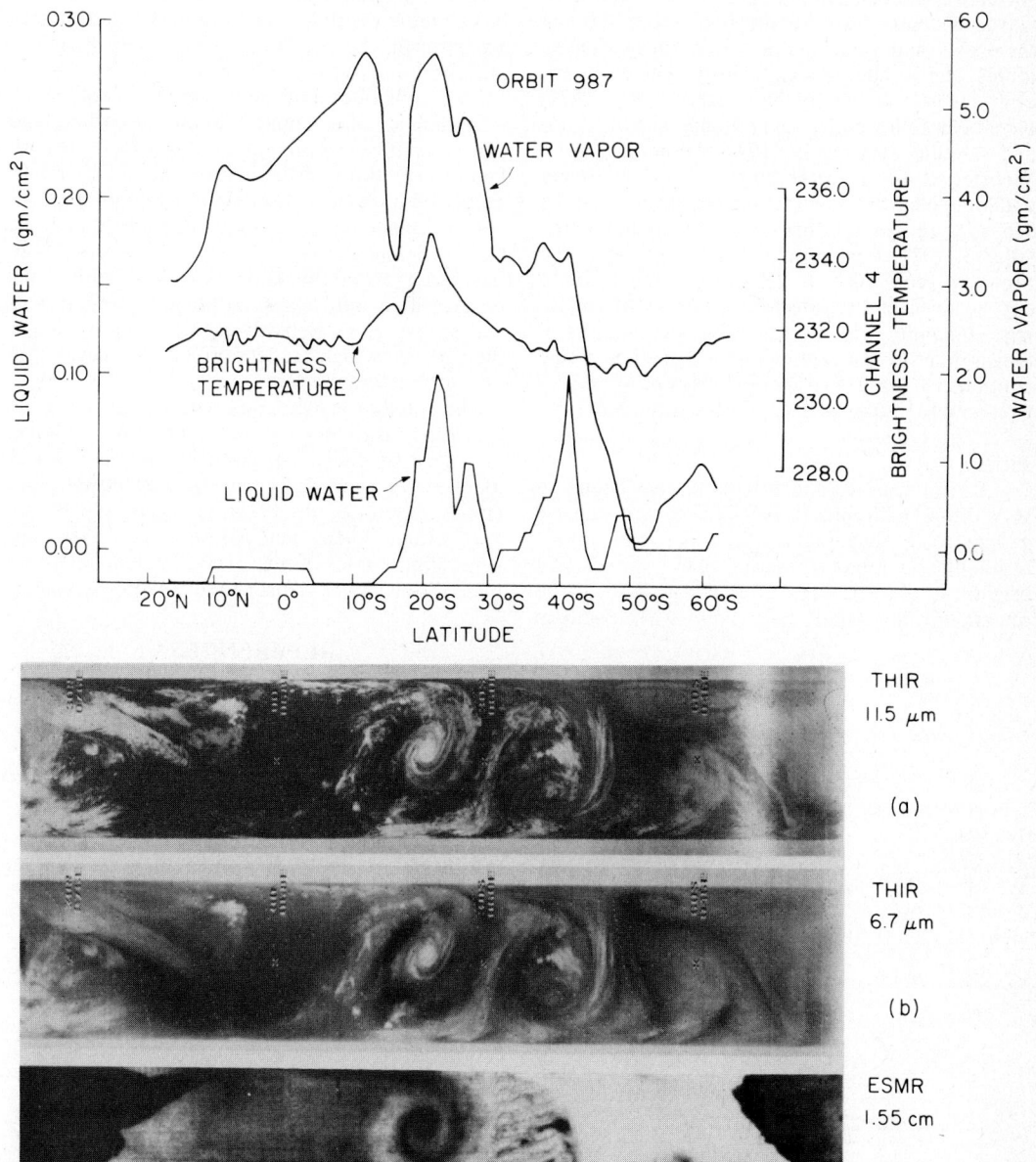

Fig. 5-65. Tropical cyclone viewed at night by the Nimbus-5 NEMS, THIR, and ESMR instruments on 22 February 1973 south of India near 62° longitude. (Staelin et al. (1976)).

in several hurricanes by Grody et al. (1979) and Kidder, (1978).

The dynamic relation between geostrophic wind and temperature can be expressed in terms of wind weighting functions; the horizontal gradient in brightness temperature is proportional to the integral over altitude of the wind profile and the wind weighting function. Such weighting functions, computed by Grody et al. (1978), appear in Figure 5-66. They are proportional to the derivative with respect to altitude of the temperature weighting function for that frequency.

Winds over ocean can also be deduced from the roughness of the sea. The roughness and the associated foam can readily be sensed with microwaves because the normally high reflectivity of the ocean (and associated low brightness temperatures) can be altered significantly. The early aircraft measurements by Nordberg, (1969, 1971) have been followed by several analyses of *in situ* and satellite data (Swift, 1974; Thomann, 1976; Webster et al., 1976; Hollinger, 1971; Hollinger and Mennella, 1973; and Rosenkranz, 1978). Even sea surface temperatures have been measured (Blume, 1977; Basharinov et al., 1969), but the accuracy required (~1–2K) strains the technology because of the interfering effects of roughness, humidity, etc. Fortunately, wavelengths of longer than several centimeters are much less susceptible to these extraneous sources of error. Conversely, atmospheric observation must respect the emission characteristics of the ocean surface.

Snow, sea ice, and firn have also been observed to exhibit microwave signatures which are geophysically meaningful and which can impact atmospheric measurements. The microwave brightness temperatures of these media can be remarkably low (130K near 1 cm wavelength in Antarctica) due to internal scattering which occurs many centimeters below the surface. The theory of internal scattering has been addressed by Stogryn (1970), Gurvich et al. (1973), England (1974, Tsang and Kong (1976), Fisher (1977), and others. Observations of sea ice and the increased levels of scattering as the ice ages and brine departs have been reported by Campbell et al. (1975), Kunzi et al. (1976), and others. Firn and snow over land also exhibit significant scattering which depends largely upon the temperature history of the snow and the grain size distribution. Because of this dependence, snow accumulation rates over Greenland and Antarctica can be mapped, and snow depths over other land masses have been crudely estimated (Gloersen and Salomonson, 1975; Chang et al., 1976; Kunzi et al., 1976; and Zwally, 1977).

Rain cells over land often appear as cold spots at millimeter wavelengths, as noted earlier, and can be confused with water-soaked land. Theory and observations of the microwave signatures of moist soil have been discussed by Schmugge et al. (1974), Eagleman and Lin (1976), Njoku and Kong (1977), Schmugge et al. (1977), and others. It appears that terrain that is not heavily vegetated can be classified into a few significant categories of soil moisture, particularly at wavelengths longer than 20 cm, which are less affected by vegetation and surface granulation.

The satellite instruments required for such observations are similar to the microwave radiometers used by radio astronomers. Satellite instruments have been described by Basharinov et al. (1969), Staelin et al. (1973), Gloersen and Barath (1977), and Staelin and Rosenkranz (1978). The general principles of microwave radiometers have been discussed by Tiuri (Kraus, 1966).

REFERENCES

Allison, L. J., E. B. Rodgers, T. T. Wilheit, and R. W. Fett, 1974, Tropical cyclone rainfall as measured by the Nimbus 5 electrically scanning microwave radiometer; Bull. Amer. Meteor. Soc., vol. 55, pp. 1074–1089.

Ambartsumyan, V. A., 1958, Theoretical Astrophysics; Pergamon Press.

Atroshenko, V. S., E. M. Feigelson, K. S. Glazova, and M. S. Malkevich, 1963, Calculation of the Brightness of Light, Part 2; New York, Consultants Bureau.

Aufm Kampe, H. J., and H. K. Weickmann, 1957, Physics of Clouds; Meteor. Monographs, vol. 2, p. 182.

Aumann, H. H., and M. T. Chahine, 1976, An infrared multidetector spectrometer for remote sensing of temperature profiles in the presence of clouds; Appl. Optics, vol. 15, pp. 2091–2093.

Barnett, A. H., 1963, Microwave spectra lines as probes of planetary atmospheres; Mem. Soc. Roy. Sci. Liege, vol. 8, p. 197.

Barteneva, O. D., 1960, Scattering functions of light in the atmospheric boundary layer; Izy. Geophys. Ser., pp. 1237–1244 in English translation.

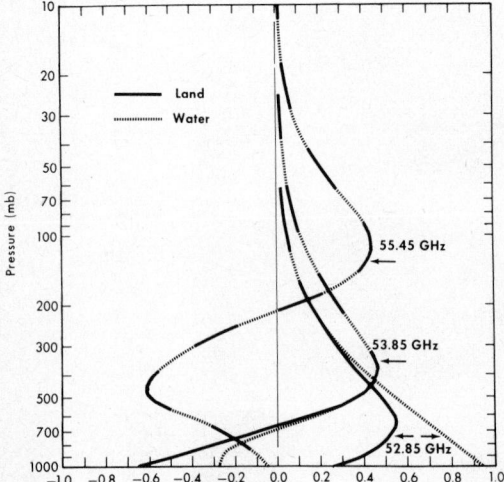

Fig. 5-66. Wind weighting functions calculated for nadir observations over land and sea at the SCAMS frequencies near the 5-mm weavelength oxygen absorption band. (Grody Staelin and Rosenkranz, 1978).

Basharinov, E., A. S. Gurvich, and S. T. Yegorov, 1969, Determination of geophysical parameters according to the measurements of thermal microwave radiation on the artificial satellite Cosmos-243; Doklady AN SSSR, vol. 188, no. 6, p. 1273.

Bean, B. R., and E. G. Dutton, 1966, Radio Meteorology; NBS Monograph 92.

Becker, G. E., and S. H. Autler, 1946, Water vapor absorption of electromagnetic radiation in the centimeter wave-length range; Phys. Rev., vol. 70, p. 300.

Bignell, K. J., 1970, The water vapour infrared continuum; Quart. J. R. Met. Soc., vol. 96, p. 390.

Blau, H. H., R. P. Espinola, and E. C. Reifenstein, III, 1966, Near infrared scattering by sunlit terrestrial clouds; Appl. Optics, vol. 5, no. 4, p. 555.

Blifford, I. H., Jr., and L. D. Ringer, 1969, The size and number distribution of aerosols in the continental troposphere; J. Atmos. Sci., vol. 26, pp. 716–726.

Bloch, F. N., and J. H. Seinfield, 1979, Scientific Research Objectives in Tropospheric Pollution; NASA Tech. Rept. (in press).

Blume, H. C., A. W. Love, J. J. Van Melle, and W. W. Ho, 1977, Radiometric observations of sea temperature at 2.65 GHz over the Chesapeake Bay; IEEE Trans. Ant. Propag., vol. AP-25, p. 121.

Born, M., and E. Wolf, 1964, Principles of Optics; Pergamon, Oxford.

Bullrich, K., 1964, Advances in Geophysics; vol. 10: (H. E. Landsberg and J. von Mieghem, eds.) pp. 101–260, Academic Press.

Bullrich, K., W. G. Blattner, T. Conley, R. Eiden, G. Hanel, K. Heger, and W. Norwak, 1968, Research on Atmospheric Optical Radiation Transmission; Meteorologisch-Geophysikalisches Institut der Johannes Gutenberg-Universitat, Mainz, Germany.

Campbell, W. J., W. F. Weeks, R. O. Ramseier, and P. Gloersen, 1975, Geophysical studies of floating ice by remote sensing; J. Glaciol., vol. 15, pp. 305–328.

Chahine, M. T., 1968, Determination of the temperature profile in an atmosphere from its outgoing radiance; J. Opt. Soc. Amer., vol. 58, pp. 1634–1637.

Chahine, M. T., 1970, Inverse problems in radiative transfer: determination of atmospheric parameters; J. Atmos. Sci., vol. 27, pp. 960–967.

Chahine, M. T., 1972, A general relaxation method for inverse solution of the full radiative transfer equation; J. Atmos. Sci., vol. 29, pp. 741–747.

Chahine, M. T., 1974, Remote sounding of cloudy atmospheres. I. The single cloud layer; J. Atmos. Sci., vol. 31, pp. 233–243.

Chahine, M. T., 1975, An analytical transformation for remote sensing of clear-column atmospheric temperature profiles; J. Atmos. Sci., vol. 32, pp. 1946–1952.

Chahine, M. T., 1977, Remote sounding of cloudy atmospheres. II. Multiple cloud formations; J. Atmos. Sci., vol. 34, p. 744.

Chahine, M. T., H. H. Aumann and F. W. Taylor, 1977, Remote sounding of cloudy atmospheres. III. Experimental verifications; J. Atmos. Sci., vol. 34, p. 758.

Chandrasekhar, S., 1950, Radiative Transfer; Oxford Univ. Press.

Chang, T. C., P. Gloersen, T. Schmugge, T. T. Wilheit, and H. J. Zwally, 1976, Microwave emission from snow and glacier ice; J. Glaciol., vol. 16, pp. 23–29.

Chayanova, E. A., and K. S. Shifrin, 1968, The scattering indicatrix of the atmospheric boundary layer; Izv., Atms. and Oceanic Phys., vol. 4, no. 2, pp. 133–134.

CIAP Climatic Impact Assessment Program, The natural and radiatively perturbed troposphere; Monograph 4, DOT-TST-75-64.

Collie, C. H., D. M. Ritson, and J. B. Hasted, 1948, The dielectric properties of water and heavy water; Proc. Phys. Soc. (London), vol. 60, p. 145.

Collins, D. G., W. G. Blattner, M. B. Wells, and H. G. Horak, 1972, Backward Monte Carlo calculations of the polarization characteristics of the Radiation emerging from spherical-shell atmospheres; Appl. Optics, vol. 11, p. 2684.

Conrath, B. J., R. A. Hanel, V. G. Kunde, and C. Parbhakara, 1970, Infrared interferometer experiment on Nimbus 3; J. Geophys. Res., vol. 30, pp. 5831–5857.

Coulson, K. L., 1966, Effects of reflection properties of natural surfaces in aerial reconnaissance; Appl. Optics, vol. 5, pp. 905–917.

Coulson, K. L., 1971, On the solar radiation field in a polluted atmosphere; J. Quant. Spectrosc. Radiat. Transfer, vol. 11, pp. 739–755.

Coulson, K. L., J. V. Dave, and Z. Sekera, 1960, Tables Related to radiation Emerging From a Planetary Atmosphere with Rayleigh Scattering; Los Angeles, Univ. of Calif. Press.

Crane, R. K., 1966, Microwave scattering parameters for New England rain; MIT Lincoln Laboratory, Tech. Report 426.

Dave, J. V., 1964, Importance of higher order scatter in a molecular atmosphere; J. Opt. Soc. Amer., vol. 54, pp. 307–315.

Dave, J. V., 1969, Effect of coarseness of the integration increment on the calculation of the radiation scattered by polydispersed aerosols; Appl. Optics, vol. 8, pp. 1161–1167.

Dave, J. V., 1970, Intensity and polarization of the radiation emerging from a plane-parallel atmosphere containing monodispersed aerosols; Appl. Optics, vol. 9, pp. 2673–2684.

Dave, J. V., and P. M. Furukawa, 1966, Scattered Radiation in the Ozone Absorption Bands at Selected Levels of a Terrestrial, Rayleigh Atmosphere; Boston, Am. Meteor. Soc.

Dave, J. V., and J. Gazdag, 1970, A modified Fourier transform method for multiple scattering calculations in plane parallel Mie atmosphere; Appl. Optics, vol. 9, pp. 1457–1466.

Davis, F. J., 1941, Surface loss of solar and sky radiation by inland lakes; Wisconsin Acad. Sci., Arts, and Letters, vol. 33, pp. 83–93.

Debye, P., 1909, Ann. Phys., vol. 30, p. 57.

Debye, P., 1929, Polar Molecules; Chemical Catalogue Co., New York.

Deirmendjian, D., 1964, Scattering and polarization properties of water clouds and hazes in the visible and infrared; Appl. Optics, vol. 3, p. 187.

Deirmendjian, D., 1969, Electromagnetic Scattering on Spherical Polydispersions; New York, Elsevier Publ. Co.

Dombkovskaya, E. P., V. V. Ozerkina, and I. S. Skuratova, 1976, Meteorological interpretation of microwave polarization measurements aboard the meteor satellite in the 0.8 cm region; Space Research XVI—Akademie-Verlag, Berlin, pp. 149–153.

Dombkovskaya, E. P., A. E. Salomonovich, S. V. Solomonov, and A. S. Khaikin, 1977, Studies of submillimeter radiation of tropical cloud clusters from

Cosmos-669; Physika Atmospheri i Oceana,vol. 13; no. 4, pp. 391–398.

Duncan, C. H., et al., 1977, Rocket calibration of the Nimbus 6 solar constant measurements; J. Appl. Optics, vol. 16, p. 2690.

Duntley, S. Q., R. W. Johnson and J. I. Gordon, 1964, Ground-based measurements of earth-to-space beam transmittance, path radiance, and contrast transmittance; Wright Patterson Air Force Base.

Eagleman, J. R., and W. C. Lin, 1976, Remote sensing of soil moisture by a 21 cm passive radiometer; J. of Geophys. Res., vol. 81, pp. 3660–3666.

Elterman, L., 1968, UV, visible, and IR attenuation for altitudes to 50 km, 1968; AFCRL, Environ. Res. Paper 285, AFCRL-68-0153.

Elterman, L., 1970, Relationships between vertical attenuation and surface meteorological range; Appl. Optics, vol. 9, pp. 1804–1810.

England, A. W., 1974, Thermal microwave emission from a half-space containing scatterers; Radio Sci., vol. 9, pp. 447–454.

Farhat, N. H., and A. B. De Cour, 1969, Relation between wave structure function looking up and looking down through the atmosphere; J. Opt. Soc. Amer., vol. 59, pp. 1489–1490.

Farmer, C. B. and J. T. Houghton, 1966, Collision-induced absorption in earth's atmosphere; Nature, vol. 209, p. 1341.

Feigelson, E. M., M. S. Malkevich, S. Ya. Kogan, T. D. Koronatova, K. S. Glazova, and M. A. Kuznetsova, 1960, Calculation of the Brightness of Light–Part 1; New York, Consultants Bureau, Inc.

Fisher, A. D., 1977, A model for microwave intensity propagation in an inhomogeneous medium; IEEE Trans. Antennas Propag., AP-25, pp. 876–882.

Fletcher, N. H., 1966, The Physics of Rainclouds; Cambridge University Press.

Flowers, E. C., R. A. McCormick, and K. R. Kurfis, 1969, Atmospheric turbidity over the United States, 1961–1966; J. Appl. Meteor., vol. 8, pp. 955–962.

Fowler, W. B., E. L. Reed, and J. E. Blamont, 1971, Bidirectional reflectance of the moonlit earth; Appl. Optics, vol. 12, pp. 1657–2660.

Fraser, R. S., 1964(a), Computed intensity and polarization of light scattered outwards from the Earth and an overlying aerosol; J. Opt. Soc. Amer., vol. 54, pp. 157–168.

Fraser, R. S., 1964(b), Apparent contrast of objects on the Earth's surface as seen from above the Earth's atmosphere; J. Opt. Soc. Amer., vol. 54, pp. 289–300.

Fraser, R. S., 1964(c) Theoretical Investigation, the Scattering of Light by a Planetary Atmosphere; TRW Space Tech. Lab., Redondo Beach, California.

Fraser, R. S., 1966, Theoretical investigation, the scattering of light by a planetary atmosphere; TRW, Inc., Redondo Beach, California.

Frenkel, L., and D. Woods, 1966, The microwave absorption by H_2O and its mixtures with other gases between 100 and 300 GHz; Proc. IEEE, vol. 54, p. 4.

Fried, D. L., 1966(a), Optical resolution through a randomly inhomogeneous medium for very long and very short exposures; J. Opt. Soc. Amer., vol. 56, pp. 1372–1379.

Fried, D. L., 1966(b), Limiting resolution looking down through the atmosphere; J. Opt. Soc. Amer., vol. 56, pp. 1380–1384.

Fried, D. L. 1967; J. Opt. Soc. Amer., vol. 57, p. 175.

Fymat, A. L., 1977, Remote sensing of environmental particulate pollutants: optical methods for determinations of size distribution and complex refractive index; Proc. 4th Joint Conf. Sensing of Environmental Pollutants, Ed. V. E. Derr, pp. 719–723.

Fymat, A. L., and R. Vasudevan, 1975, A new approach to radiative transfer theory using Jones' vectors; Astrophysics and Space Science, vol. 38, pp. 9–124.

Gaut, N. E., 1967, Studies of Atmospheric Water Vapor by Means of Passive Microwave Techniques; Ph.D. Dissertation, Dept. of Meteorology, M.I.T., also Tech. Rpt. 457, M.I.T. Res. Lab. for Electronics.

Gaut, N. E. and E. C. Reifenstein, III, 1971, Interaction Model of Microwave Energy and Atmospheric Variables; Environ. Res. and Tech., Inc., Waltham, Mass., for sale by CFSTI, $3.00.

Glaser, A. H., J. C. Barnes, and D. W. Beran, 1968, Apollo landmarks sighting: An application of computer simulation to a problem in applied meteorology; J. Appl. Meteor., vol. 7, pp. 768–779.

Gloersen, P. and F. Barath, 1977, A scanning multichannel microwave radiometer for Nimbus-G and Seasat-A; IEEE J. of Oceanic Engineering, vol. OE-2, no. 2, pp. 172–178.

Gloersen, P. and V. V. Salomonson, 1975, Satellites—new global observing techniques for ice and snow; J. Glaciol. vol. 15, no. 73, pp. 373–389.

Goetz, A. F. H., 1979, Preliminary Stereosat mission description; Jet Propulson Laboratory Report No. 720–33.

Goldstein, H., 1951, in Propagation of Short Radio Waves; (D. E. Kerr, ed.), McGraw-Hill.

Goody, R. M., 1952, A statistical model for water-vapour absorption; Quart. J. R. Met. Soc., vol. 78, p. 165.

Goody, R. M., 1964, Atmospheric Radiation; New York, Oxford Univ. Press.

Gordon, R. G.,, 1966, Semiclassical theory of spectra and relaxation in molecular gases; J. Chem. Phys., vol. 45, pp. 1649–1655.

Gordon, R. G., 1967, On the pressure broadening of molecular multiplet spectra; J. Chem. Phys., vol. 46, pp. 448–455.

Gorelik, A. G., E. P. Dombkovskaya, V. V. Ozerkina, V. I. Semiletov, I. S. Skooratoa, and A. V. Frolov, 1975, Microwave polarization measurements from the "Meteor" satellite; Meteorologia i Hydrologia, no. 7, p. 36.

Grant, I. P., and G. E. Hunt, 1968, Solution of the radiative transfer problems using the S_n method; Mon. Not. Royal Astron. Soc., vol, 141, p. 27.

Greaves, J. R., P. E. Sherr, and A. H. Glaser, 1970, Cloud cover statistics and their use in the planning of remote sensing missions; Remote Sensing of Environ., vol. 1, pp. 95–101.

Greaves, J. R., D. B. Spiegler, and J. H. Wiland, 1971, Development of a Global Cloud Model for Simulating Earth-viewing Space Missions; Allied Res. Associates, Inc., Concord, Mass.

Grody, N. C., 1978, Microwave radiometry applied to synoptic meteorology and climatology; in Staelin and Rosenkranz, op. cit., p. 6-1 to 6-40.

Grody, N. C., C. M. Hayden, W. C. C. Shieu, P. W. Rosenkranz, and D. H. Staelin, 1979, Typhoon June winds estimated from scanning microwave spectrometer measurements at 55.45 GHz; J. Geophys. Res. vol. 84, no. C7, pp. 3689–3695.

Gross, E. P., 1955, Shape of collision-broadened spectral lines; Phys. Rev., vol. 7, p. 891.

Gurvich, A. S., B. I. Kalinin, and D. T. Matveyev, 1973, Influence of the internal structure of glaciers on their thermal radio emission; Atmos. Oceanic Phys., vol. 9, no. 12, pp. 1247–1256.

Hanel, G., 1971, New results concerning the dependence of visibility on relative humidity and their significance in a model for visibility forecast; Beitrage zuer Physik der Atmosphare, vol. 44, pp. 137–167.

Hanel, G., 1976, The single scattering albedo of atmospheric aerosol particles as a function of relative humidity; J. of Atmos. Sci., vol. 33, p. 1120–1124.

Hanel, R. A., B. Schlachman, D. Rogers, D. Vanous, 1971, Nimbus–4 Michelson interferometer; Appl. Optics, vol. 10, p. 1376.

Hansen, J. E., 1969; Ap. J., vol. 155, p. 565.

Hansen, J. E., 1971, Multiple scattering of polarized light in planetary atmospheres, Part I, The doubling method; J. Atmos. Sci., vol. 28, pp. 120–125.

Hansen, J. E., and L. D. Travis, 1974, Light scattering in planetary atmospheres; Space Sci. Rev., vol. 16, p. 527.

Harris, F. S., 1969, Tellus, Particle characteristics from light scattering measurements; vol. 21, p. 223.

Hasted, J. B. and M. Shahidi, 1976, The low frequency dielectric constant of supercooled water; Nature, vol. 262, pp. 777–778.

Heicklen, J., 1976, Atmospheric Chemistry; Academic Press, New York, pp. 35–36.

Herman, B. M., 1977, Application of Modified Twomey Techniques to Invert Lidon Angular Scatterer and Solar Extinction Data for Determining Aerosol Size Distributions; Univ. Arizona, Inst. Atmospheric Phys.

Herman, B. M., S. R. Browning, and R. J. Curran, 1971, The effects of atmospheric aerosols on scattered sunlight; J. Atmos. Sci., vol. 28, pp. 419–428.

Herzberg, G., 1945, Infrared and Raman spectra of polyatomic molecules; in Molecular Spectra and Molecular Structure, Van Nostrand. vol. 2.

Hodkinson, J. R., 1963, in Electromagnetic Scattering (M. Kerker, ed.); Pergamon Press, pp. 87–100.

Holland, A. C., and G. Gagne, 1970, The scattering of light by polydispersed systems of irregular particles; Appl. Optics, vol. 9, pp. 1113–1121.

Hollinger, J. P., 1971, Passive microwave measurements of sea surface temperature; IEEE Trans. Geoscience Elect., vol. GE-9, p. 165.

Hollinger, J. P., and R. A. Mennella, 1973, Oil spills: measurements of their distributions and volumes by multifrequency microwave radiometry; Science, vol. 181, p. 54.

Houghton, J. T., 1968, The Earth's Upper Atmosphere—Radiation and Radiative Transfer; European Space Research Organization, SP-31.

Houghton, J. T., S. D. Smith, 1966, Infra-red Physics; Oxford University Press.

Houghton, J. T., and F. W. Taylor, 1973, Remote sounding from artificial satellites and space probes of the atmospheres of the Earth and the planets; Rep. Prog. Phys., vol. 36, pp. 827–919.

Hovis, W. A., Jr., 1971; private communication.

Hudson, R. D., ed., 1977, Chlorofluromethanes and the Stratosphere; NASA Reference Publication 1010.

Irvine, W. M., and J. B. Pollack, 1968, Infrared optical properties of water and ice spheres; Icarus, vol. 8, p. 324.

Irvine, W. M., and F. W. Peterson, 1970, Observations of atmospheric extinction from 0.315 to 1.06 microns; J. Atmos. Sci., vol. 27, pp. 62–69.

Ivanov, A. L., G. Sh. Livshits, V. Ya. Pavlov, B. T. Tashenkov, and Ya. A. Teyfel, 1968, Light Scattering in the Atmosphere, Part 2, Nauka Press, Alma Ata. USSR: Transl. NASA TT F–553; for sale by CFSTI, $3.00.

Jacobowitz, H., W. L. Smith, H. B. Howell, F. W. Nagle, and J. R. Hickey, 1979, The first 18 months of planetary radiation budget measurements from the Nimbus 6 ERB experiment; J. Atmos. Sci., vol. 36, p. 501.

Junge, C. E., 1963, Air Chemistry and Radioactivity; Academic Press.

Kattawar, G. W., and G. N. Plass, 1968, Radiance and polarization of multiple scattered light from haze and clouds; Appl. Optics, vol. 7, pp. 1519–1527.

Kendall, J. M., Sr., 1973, Factors affecting accuracy of radiometer measurements and results of a measurement of the solar constant; paper presented at the Smithsonian Symposium on Solar Radiation: Smithson. Inst., Rockville, MD.

Kerker, M., N. P. Langlben, ad R. L. S. Gunn, 1951, Scattering of microwaves by a melting spherical ice particle; J. Meteor., vol. 8, p. 424.

Kerker, M., 1969, The Scattering of Light and other Electromagnetic Radiation; Academic Press.

Kidder, S. O., W. M. Gray, and T. H. Vonder Haar, 1978, Estimating tropical cyclonic central pressure and outer winds from satellite microwave data; Mon. Weather Rev., vol. 106, pp. 1458–1464.

Kondratyev, K. Ya., 1969, Radiation in the Atmosphere; Academic Press.

Kourganoff, V., 1963, Basic Methods in Transfer Problems; New York, Dover Publications, Inc.

Kraus, J. D., 1966, Radio Astronomy; McGraw-Hill Book Co., New York.

Kunzi, K. F., A. D. Fisher, D. H. Staelin, and J. W. Waters, 1976, Snow and ice surfaces measured by the Nimbus 5 microwave spectrometer; J. Geophys. Res., vol. 81, pp. 4965–4980.

Kyle, T. G., 1975, Atlas of Computed Infrared Atmospheric Absorption Spectra; National Center for Atmospheric Research, TN/STR-112.

Laktionov, A. G., 1964, On the relation of light scattering in the free atmosphere to vertical distribution of aerosol particle concentration; Izv. Geophs. Ser., no. 6. pp. 579–581, in Engl. transl.

Lam, K. S., 1977, Application of pressure broadening theory to the calculation of atmospheric oxygen and water vapor microwave absorption; J. Quant. Spectrosc. Radiat. Transf., vol. 17, pp. 351–383.

Laws, J. O., and D. A. Parsons, 1943, The relation of rain drop-size to intensity; AGU Trans, vol. 24, p. 452.

Ledsham, W. H., and D. H. Staelin, 1978, An extended Kalman-Bucy filter for atmospheric temperature profile retrieval with a passive microwave sounder; J. App. Meteorol. vol. 17, pp. 1023–1033.

Lee, R. W., and J. C. Harp, 1969, Weak scattering in random media, with applications to remote probing; Proc. IEEE, vol. 57, pp. 37–406.

Lenoir, W. B., 1968, Microwave spectrum of molecular oxygen in the mesosphere; J. Geoph. Res., vol. 73, p. 361.

Liebe, H. J., and G. G. Gimmestad, 1978, Calculation of clear air EHF refractivity; Radio Sci., vol. 13, pp. 245–251.

Liebe, H. J., G. G. Gimmestad, and J. D. Hopponen, 1977, Atmospheric oxygen microwave spectrum experiment versus theory; IEEE Trans. Antenn. Propag., vol. AP-25, pp. 327–335.

Liebe, H. J., and J. D. Hopponen, 1977, Variability EHF air refractivity with respect to temperature, pressure, and frequency; IEEE Trans. Antenn. Propag., vol. AP-25, pp. 336–345.

Lipes, R. G., R. L. Bernstein, V. J. Cardone, K. B. Katsaros, E. G. Njoku, A. L. Riley, D. B. Ross, C. T. Swift, F. J. Wentz, 1979, Seasat scanning multichannel microwave radiometer—Results of the Gulf of Alaska Workshop; Science, vol. 204 (4400), p. 1415.

Long, R. K., and D. B. Rensch, 1970, Comparative studies of extinction and back scattering by aeroosls; Appl. Optics, vol. 9, p. 1563.

Martin, D. C., and B. Liley, 1971, ERTS Cloud Cover Study; N. Amer. Rockwell Space Div., Downey, California.

Mason, B. J., 1957, The Physics of Clouds; Oxford University Press.

Mastenbrook, H. J., and D. R. Purdy, 1969, Vertical Distribution of Water Vapor over Washington, D. C., During 1966 and 1967; NRL, Report 6891.

Matthews, W. H., 1971, Report of the Study of Man's Impact on Climate; MIT Press.

Mazurowski, J. J., and D. R. Sink, 1965, Attenuation of photographic contrast by the atmosphere; J. Opt. Soc. Amer., vol. 55, pp. 26–30.

McClatchey, R. A., W. S. Benedict, S. A. Clough, D. E. Burch, R. F. Calfee, K. Fox, L. S. Ruthman, and J. S. Garing, 1973, AFCRL Atmosphere Absorption Line Parameters Compilation; AFCRL-TR-73-0096.

McClatchey, R. A., R. W. Fenn, J. E. A. Selby, J. S. Garing, and F. E. Volz, 1970, Optical Properties of the Atmosphere (revised); AFCRL.

Meeks, M. L., and A. E. Lilley, 1963, The microwave spectrum of oxygen in the Earth's atmosphere; J. Geoph. Res., vol. 68, p. 1683.

Mie, G.,1908; Ann. Phys., vol. 25, p. 377.

Milne, E. A., 1930; in Handbuch Der Astrophysik, vol. 3, Part 1, Chap. 2; also found in Selected Papers on the Transfer of Radiation, ed. by D. H. Menzel; New York, Dover Publications, Inc., pp. 77–269.

Miner, G. F., D. D. Thornton, W. J. Welch, 1972, The inference of atmospheric temperature profiles from ground-based measurements of microwave emission from atmospheric oxygen; J. Geophys. Res., vol. 77, p. 975.

NASA, 1971(a), Earth Albedo and Emitted Radiation; for sale by NTIS, $3.00.

NASA, 1971(b), Solar Electromagnetic Radiation; NASA SP-8005 for sale by NTIS, $3.00.

Njoku, E. G., and J. A. Kong, 1977, Theory of passive microwave remote sensing of near-surface soil moisture; J. Geophys. Res., vol. 82, pp. 3108–3118.

Nordberg, W., J. Conaway, and P. Thaddeus, 1969, Microwave observations of sea state from aircraft; Quart. J. R. Met. Soc., vol. 95, p. 408.

Nordberg, W., J. Conaway, D. B. Ross, and T. Wilheit, 1971, Measurements of microwave emission from a foam-covered, wind-driven sea; J. Atmos. Sci., vol. 28, p. 429.

Obukhov, A. M. and M. S. Tatarskaya, 1969, Field of integral atmospheric water vapor over the southern hemisphere from measurements of thermal microwave radiation on the "Cosmos-243" satellite; Meteorologia i Hydrologia, no. 11, pp. 36–39.

Palmer, E., 1970, Aerosol-particle parameters from light scattering data. Session II paper at the 3rd LRC.

Pearson, M. D., and W. T. Kreiss, 1968, Ground based microwave radiometry for recovery of average temperature profiles of the atmosphere; Boeing Sci. Res. Lab. Tech. Report, D1-82-0781.

Penndorf, R., 1957, Tables of the refractive index for standard air and the Raleigh scattering coefficient for the spectral region between 0.2 and 20.0 μ and their application to atmospheric optics; J. Opt. Soc. Amer., vol. 47, pp. 176–182.

Pinnick, R. G., R. E. Carroll and R. J. Hofman, 1976, Polarized light scattered from monodisperse randomly oriented nonspherical aerosol particle measurements; Appl. Optics, vol. 15, pp. 384–393.

Plamondon, J. A., 1969, Thermal control flux monitor; Space Programs Su. 37–59, p. 162, Jet Propulsion Lab., Pasadena, Calif.

Plass, G. N., and G. W. Kattawar, 1968, Calculations of reflected and transmitted radiance for Earth's atmosphere; Appl. Optics, vol. 7, pp. 1129–1148.

Plass, G. N., and G. W. Kattawar, 1969(a), Radiative transfer in an atmosphere-ocean system; Appl. Optics, vol. 8, pp. 455–466.

Plass, G. N., and G. W. Kattawar, 1969(b), Effect of changes in complex part of the refractive index on polarization of light scattered from haze and clouds; Appl. Optics, vol. 8, pp. 2489–2495.

Plass, G. N., and G. W. Kattawar, 1970, Polarization of the radiation reflected and transmitted by the Earth's atmosphere; Appl. Optics, vol. 9, pp. 1122–1130.

Plass, G. N., and G. W. Kattawar, 1971, Comment on the scattering of light by polydispersed systems of irregular particles; Appl. Optics, vol. 10, pp. 1172–1173.

Plyler, E. K., et al. 1960, Vibration-rotation structure in absorption bands for the calibration of spectrometers from 2 to 16 microns; J. Res. National Bur. Stand. A64, p. 29.

Pollack, J. B., and J. N. Cuzzi, 1978; Proc. Third Conf. Atmos. Radiation, p. 20.

Pueschel, R. F., R. J. Charlson, and N. C. Ahlquist, 1969, On the anomalous deliquescence of sea-spray aerosols; J. Appl. Meteor., vol. 8, pp. 995–998.

Reifenstein, E. C. III, and Gaut, N. E., 1971, Microwave Properties of Clouds in the Spectral Range 30–40 GHz; Tech. Rep. No. 12, Environ. Res. and Tech., Inc.

Robinson, N., (ed.), 1966, Solar Radiation; New York, Elsevier Publishing Co.

Rodgers, C. D., and C. D. Walshaw, 1966, Computation of infra-red cooling rate in planetary atmospheres; Quart. J. R. Met. Soc., vol. 92, p. 67.

Rosenkranz, P. W., 1975, Shape of the 5 mm oxygen band in the atmosphere; IEEE Trans. Antenn. Propag., vol. AP-23, pp. 498–506.

Rosenkranz, P. W., F. T. Barath, J. C. Blinn, E. J. Johnston, W. B. Lenoir, D. H. Staelin, and J. W. Waters, 1972, Microwave radiometric measurements of atmospheric temperature and water from an aircraft; J. Geophys. Res., vol. 77, pp. 5833–5844.

Rosenkranz, P. W., D. H. Staelin, and N. C. Grody, 1978, Typhoon June (1975) viewed by a scanning microwave spectrometer; J. Geophys. Res., vol. 83, pp. 1857–1868.

Rothman, L. S., S. A. Clough, R. A. McClatchey, L. G. Young, D. Snider, and A. Goldman, 1978: AFGL trace gas compilation; Appl. Optics, vol. 17, p. 507.

Rozenberg, G. V., 1967, The properties of an atmospheric aerosol from optical data; Izv. Atms. and Oceanic Phys., vol. 3, pp. 545–551, in Engl. Transl.

Rozenberg, G. V., Yu. S. Georigiyevskiy, U. N.

Kapastin, Yu. S. Lyuboutseva, A. P. Orlou, S. M. Pirogou, A. I. Chauro, and A. KH. Shukurou, 1977, Submicron aerosol fraction and light absorption in the $8-12$ μm transparency window; Izvestiya Atms. and Oceanic Physics, vol. 13, p. 815.

Schmugge, T. J., P. Gloersen, T. Wilheit, and F. Geiger, 1974, Remote sensing of soil moisture with microwave radiometers; J. of Geophys. Res., vol. 79, pp. 317–323.

Schmugge, T. J., J. M. Meneely, A. Rango, and R. Neff, 1977, Satellite microwave observations of soil moisture variations; Water Resources Bulletin, 13, p. 265.

Scott, N. A., 1974, A direct method of computation of the transmission function of an inhomogeneous gaseous medium—I. description of the method; J. Q. S. R. T., vol. 14, p. 691.

Shaw, J. H., 1970, Determination of the earth's surface temperature from remote spectral radiance observations near 2600 cm^{-1}; J. Atmos. Sci., vol. 27, p. 950.

Shaw, J. H., M. T. Chahine, C. B. Farmer, L. D. Kaplan, R. A. McClatchey, and P. W. Schaper, 1970, Atmospheric and surface properties from spectral radiance observations in the 4.3 micron region; J. Atmos. Sci., vol. 27, pp. 773–780.

Sherr, P. E., A. H. Glaser, J. C. Barnes, and J. H. Willard, 1968, World-wide Cloud Cover Distributions for Use in Computer Simulations; Concord, Mass., Allied Research Associates, Inc.

Shifrin, K. S., and G. L. Shubova, 1964, The statistical characteristics of the vertical transparency of the atmosphere; Izv. Geophs. Ser., no. 2, pp. 161–163, in Engl. transl.

Shimabukuro, F. I., P. L. Smith, W. J. Wilson, 1977, Estimation of the daytime and nighttime distribution of atmospheric ozone from ground-based mm-wavelength measurements; J. App. Meteor., vol. 16, p. 929.

Smith, W. L., 1970, Iterative solution of the radiative transfer equation for the temperature and absorbing gas profile of an atmosphere; Appl. Opt., vol. 9, pp. 1993–1999.

Smith, W. L., 1972: Bull. Am. Meteor. Soc., vol. 53, no. 11, p. 1074.

Smith, W. L., and H. M. Woolf, 1976, The use of eigenvectors of statistical covariance matrices for interpreting satellite sounding radiometer observations; J. Atmos. Sci., vol. 33, pp. 1127–1140.

Sobolev, V. V., 1963, A Treatise on Radiative Transfer: Van Nostrand Company.

Staelin, D. H., 1966, Measurements and interpretations of the microwave spectrum of the terrestrial atmosphere near 1-centimeter wavelength; J. Geoph. Res., vol. 71, p. 2875.

Staelin, D. H., 1969, Passive remote sensing at microwave wavelengths; Proc. IEEE, vol. 57, p. 427.

Staelin, D. H., 1977, Inversion of passive microwave remote sensing data from satellites, in Inversion Methods in Atmospheric Remote Sounding; Adarsh Deepak, Ed., Academic Press, New York, New York, pp. 361–394.

Staelin, D. H., A. H. Barrett, J. W. Waters, F. T. Barath, E. J. Johnston, P. W. Rosenkranz, N. E. Gaut, and W. B. Lenoir, 1973, Microwave spectrometer on the Nimbus 5 satellite: meteorological and geophysical data; Science, vol. 182, pp. 1339–1341.

Staelin, D. H., A. L. Cassel, K. F. Kunzi, R. L. Pettyjohn, R. K. L. Poon, P. W. Rosenkranz, and J. W. Waters, 1975, Microwave atmospheric temperature sounding: effects of clouds on the Nimbus 5 satellite data; J. Atmos. Sci., vol. 32, pp. 1970–1976.

Staelin, D. H., K. F. Kunzi, R. L. Pettyjohn, R. K. L. Poon, and R. W. Wilcox, 1976, Remote sensing of atmospheric water vapor and liquid water with the Nimbus 5 microwave spectrometer; J. Appl. Meteor., vol. 15, pp. 1204–1214.

Staelin, D. H., P. W. Rosenkranz, F. T. Barath, E. J. Johnston, and J. W. Waters, 1977, Microwave spectroscopic imagery of the earth; Science, vol. 197, pp. 991–993.

Staelin, D. H., and P. W. Rosenkranz (eds.), 1978, High resolution passive microwave satellites; Mass. Inst. of Tech., Res. Lab of Electronics.

Stogryn, A., 1970, The brightness temperature of a vertically structured medium; Radio Science, vol. 5, pp. 1397–1406.

Stowe, L., et al., 1979, Radiometric performance of the Nimbus 7 Earth radiation budget experiment; NASA Technology Improvement Workshop, U. of MD, College Park, MD.

Stratton, J., 1941, Electromagnetic Theory; McGraw-Hill.

Swift, C. T., 1974, Microwave radiometer measurements of the Cape Cod Canal; Radio Sci., vol. 9, p. 641.

Tatarski, V. I., 1961, Wave propagation in a turbulent medium; McGraw-Hill.

Thekaekara, M. P., R. Kruger, and C. H. Duncan, 1969, Solar irradiance measurements from a research aircraft; J. Appl. Optics, vol. 8, p. 1713.

Thomann, G. C., 1976, Experimental results of the remote sensing of sea-surface salinity at 21-cm wavelength; IEEE trans., Geosci., Electronics, vol. GE-14, p. 198.

Tiffany, W. B., 1968; Appl. Opt., vol. 7, p. 67.

Tomiyasu, K., 1974, Remote sensing of the earth by microwaves; Proc. IEEE, vol. 62, p. 86.

Townes, C. H., and A. L. Schawlow, 1955, Microwave spectroscopy; McGraw-Hill.

Tsang, L., and J. A. Kong, 1976, Thermal microwave emission from half-space random media; Radio Sci., vol. 11, pp. 599–609.

Tsang, L., J. A. Kong, E. Njoku, D. H. Staelin, and J. W. Waters, 1977, Theory for microwave thermal emission from a layer of cloud or rain; IEEE Trans. Antennas and Propag. vol. AP-25, p. 650.

Twomey, S., 1970; J. Atmos. Sci., vol. 27, p. 515.

Twomey, S., 1977, Introduction to the Mathematics of Inversion in Remote Sensing and Indirect Measurements; Elsevier Scientific Publishing Company.

Twomey, S., H. Jacobowitz, and H. B. Howell, 1966, Matrix methods for multiple-scattering problems; J. Atmos. Sci., vol. 23, pp. 289–296.

U.S. Standard Atmosphere, 1962, National Aeronautics and Space Administration, U.S. Air Force: U.S. Printing Office.

U.S. Standard Atmosphere Supplements, 1966, National Aeronautics and Space Administration, U.S. Air Force: U.S. Printing Office.

Van Vleck, J. H., 1947(a), The absorption of microwaves by oxygen; Phys. Rev., vol. 71, p. 413.

Van Vleck, J. H., 1947(b), The absorption of microwaves by uncondensed water vapor; Phys. Rev., vol. 71, p. 425.

Van Vleck, J. H., and V. F. Weisskopf, 1945, On the shape of collison broadened lines; Rev. Mod. Phys., vol. 17, nos. 2 and 3, p. 227.

Waters, J. W., 1976, Absorption and emission by atmospheric gases, in Methods of Experimental Physics; vol. 12: Astrophysics, Part B (M. L. Meeks, ed.) Academic Press, New York.

Waters, J. W., K. F. Kunzi, R. L. Pettyjohn, R. K. L. Poon, and D. H. Staelin, 1975, Remote sensing of atmospheric temperature profiles with the Nimbus 5 microwave spectrometer; J. Atmos. Sci., vol. 32, no. 10, pp. 1953–1969.

Waters, J. W., and D. H. Staelin, 1968, Statistical inversion of radiometric data; MIT Res. Lab of Electronics, Cambridge, Mass., Quart. Progr. Rep. 39.

Waters, J. W., and S. C. Wofsy, 1978, Applications of high resolution passive microwave satellite systems to the stratosphere, mesosphere, and lower thermosphere; in Staelin and Rosenkranz, op, cit., pp. 7-1 to 7-69.

Waters, J. W., W. J. Wilson, and F. I. Shimabukuro, 1976, Microwave measurement of mesospheric carbon monoxide; Science, vol. 191, p. 1174.

Webster, W. J., T. T. Wilheit, D. B. Ross, and P. Gloersen, 1976, Spectral characteristics of the microwave emission from a wind-driven foam-covered sea; J. Geophys. Res., vol. 81, p. 3095.

Weickman, H., and H. J. Aufm Kampe, 1953, Physical properties of cumulus clouds; J. Meteor., vol. 10, p. 204.

Weiner, M. W., 1967, Atmospheric turbulence in optical surveillance systems; Appl. Optics, vol. 6, p. 1984.

Westwater, E. R., J. B. Snider, and A. V. Carlson, 1975, Experimental determination of temperature profiles by ground-based microwave radiometry; J. Appl. Meteor., vol. 14, p. 524.

Westwater, E. R., and M. T. Decker, 1977, Application of statistical inversion to gound-based microwave remote sensing of temperature and water vapor profiles; in Inversion Methods in Atmospheric Remote Sounding, A. Deepak, Ed.: Academic Press, New York, pp. 395–425.

White, P. G., 1969, in Second annual Earth resources aircraft program status review, Sept. 16–18, 1969; NASA/MSFC, Houston, Texas, vol. 3, sec. 50.

Wilheit, T. T., J. S. Theon, W. E. Shenk. L. J. Allison, and E. B. Rodgers, 1976, Meteorological interpretations of the images from the Nimbus 5 electrically scanned microwave radiometer; J. App. Meteor., vol. 15, pp. 166–172.

Wilheit, T. T., A. T. C. Chang, M. S. U. Rao, E. B. Rodgers, and J. S. Theon, 1977, A satellite technique for quantitatively mapping rainfall rates over the oceans; J. App. Meteor., vol. 16, pp. 551–560.

Willson, R. C., 1971, Active cavity radiometric scale, international pyrheliometric scale, and solar constant; J. Geophys. Res., vol. 76, p. 4325.

Willson, R. C., 1973, New radiometric techniques and solar constant measurements; Solar Energy, vol. 14, p. 203.

Willson, R. C., C. H. Duncan, J. Geist, 1979, Direct measurement of solar luminosity variation; Science, vol. 207, p. 177.

Willson, R. C., and H. S. Hudson, 1980, Variations of solar irradiance; Astrophysical Journal Letters (accepted for publication).

Winkler, P., and C. E. Junge, 1971, Comments on anomalous deliquescence of sea spray aerosols; J. Appl. Meteorol., vol. 10, pp. 159–163.

Yamamoto, G., M. Tanaka, and S. Asano, 1970, Radiative transfer in water clouds in the infrared region; J. Atmos. Sci., vol. 27, p. 282.

Zerull, R., and H. Giese, 1974, in Planets, Stars and Nebulae Studied with Photopolarimetry; T. Ghrels, ed., University of Arizona Press.

Zwally, H. J., 1977, Microwave emissivity and accumulation rate of polar firn; J. Glaciol., vol. 18, pp. 195–215.

CHAPTER 6

Photographic Systems For Remote Sensing

Author-Editor: PHILIP N. SLATER

Contributing Authors: FREDERICK J. DOYLE, NORMAN L. FRITZ, ROY WELCH

GENERAL CONTENTS: Camera spectroradiometry; the optics of aerial cameras; atmospheric effects; spectral signatures; aerial photographic film; aerial black-and-white film; aerial color film; filters; antivignetting filters; spectral filters; polarization filters; cameras; conventional cameras; multiband cameras; references.

NOMENCLATURE

To conserve and eliminate repetition in text and references, the following symbols, units and names have been used in this chapter.

Symbol	Units	Name
AFS	(lux-s)$^{-1}$	aerial film speed
AWAR	cy mm^{-1}	area weighted average resolution
C_D	—	differential contrast
C_L	—	logarithmic contrast
C_R	—	contrast ratio
D	—	photographic density
D	m	diameter of entrance pupil
dA	m^2	elemental area
DS	—	density scale
E	W m^{-2}	irradiance
E_V	lux	illuminance
E_0	W m^{-2}	solar irradiance at top of earth's atmosphere
f	m	focal length
G	—	granularity
H	lux-s, m cd s, erg cm^{-2}, or J m^{-2}	exposure
I	W sr^{-1}	radiant intensity
K	K	absolute temperature (kelvins)
L	W m^{-2} sr^{-1}	radiance
L_V	nit	luminance
M	—	modulation
mrad	—	milliradian
MTF	—	modulation transfer function
N	—	f-number
nm	nm	nanometer (10^{-9} m)
Q	J	radiant energy
s	s	second
t	s	time
v/h	s^{-1}	velocity-to-height ratio
$V(\lambda)$	—	photopic spectral luminous efficiency
α	—	absorptance
γ	—	slope of characteristic curve
θ	degree	angle of incidence
λ	μm or nm	wavelength
μm	μm	micrometre (10^{-6} m)
ν_c	cycles mm^{-1}	cutoff frequency
ρ	—	reflectance
$\sigma(D)$	—	rms deviation in density
τ_A	—	atmospheric transmittance
τ_0	—	optical transmittance
Φ	W	radiant flux
Φ_V	lm	luminous flux
ω	sr	solid angle

INTRODUCTION

Unlike the allied fields of photogrammetry and aerial reconnaissance, very few texts are available on the subject of photographic remote sensing. For this reason, the choice has been made to emphasize in this chapter the fundamentals of photographic remote sensing in terms of basic concepts and techniques, camera system considerations, and camera systems themselves. For space reasons alone, this emphasis has been at the expense of coverage of camera subsystem engineering and instrumentation, and of many aspects of photographic and physical optics. Descriptions of stabilized camera mounts, shutters, viewfinders, intervalometers, v/h sensors, FMC mechanisms, data annotators, etc., have been omitted; so have discussions of the theory of the photographic process, the physical properties of aerial films, optical design, and details of photographic image evaluation.

This chapter does include consideration of the spectroradiometry of the camera. The fundamental concepts can be found in standard optics texts, but such texts do not explain the complicated case of high-altitude photography which is essential in this context. Until recently, the internationally accepted terms, symbols, and units have not been generally used. The main reasons, however, for presenting the material in this much detail are twofold: First, the material is basic to photographic remote sensing, and some aspects are not fully appreciated or properly understood by many involved in this field. Second, application of the equations developed allows ground scene radiance or spectral signature information to be derived from remotely sensed data. The determination of spectral signatures adds a new dimension to the well developed photointerpretive skills of identification and mensuration, and the photogrammetric techniques developed for topographic map compilation.

The factors which influence the measurement of ground scene radiance through the atmosphere are discussed in later sections of this chapter. Aspects of camera optics, photographic films, and photographic remote sensing systems important to the user are considered, again with emphasis on quantitative ground scene radiance measurement.

FUNDAMENTAL CONSIDERATIONS

CAMERA SPECTRORADIOMETRY

Cameras are used in remote sensing for reconnaissance, cartography, and the determination of spectral signatures. The detailed spectroradiometric properties of the camera are of concern only for determining spectral signatures. Conversely, correct exposure and reasonably uniform spectral irradiance across the format are the only important radiometric considerations for reconnaissance and cartography.

The fundamental concepts described in this chapter also appear in Chapters *3*, *10*, and *11*. The repetition is designed to be of benefit to the reader as the emphasis and development of these important concepts vary according to the theme of the individual chapters.

In the following discussion[1] the spectroradiometric properties of aerial cameras will be derived from basic radiometric and geometrical optics considerations. The internationally accepted radiometric and photometric terms and symbols used here are those introduced in Chapter *3* and described in USA Standard RP-16 (1967). Radiometric quantities encountered in photography are defined in Table 6-1.

Spectroradiometric Theory

Those involved with remote sensing are concerned with determining surface radiance by measuring the radiant flux, Φ, emerging from a given portion of a surface; for a point source, this has to do only with the radiant flux per unit solid angle. The concept of solid angle measurement has been presented in Chapter 3, together with the definition of steradian. To review, if a point source of radiation is located at S, the center of a sphere, the irradiance at the surface of the sphere will decrease as the square of r, the radius of the sphere; this is the well known inverse square law. *Generally, in photographic remote sensing, the user is concerned with extended surfaces (sources).* It will be shown later in this chapter that the inverse square law does not apply in this

[1] Where confusion may arise, radiometric quantities are differentiated from photometric quantities by use of subscripts *e* and *v*, respectively. The subscript λ indicates per wavelength interval (μm). The symbol λ in parentheses indicates a function of wavelength. Subscripts r and i indicate reflected and incident quantities, respectively. Unless otherwise noted, subscript O will denote the object, and I will indicate the image. Differential expressions for radiometric quantities, e.g. $d^2\Phi$ and $d\Phi$, will be used in the following text whenever the dependencies in the above definitions are implied.

case, and that the irradiance of the image is proportional to the squared ratio of aperture diameter to focal length of the camera lens.

In the particular case of photography, that is, for wavelengths less than about 1 μm, this emerging flux is the product of the direct and atmospherically scattered solar flux incident at the surface and the reflectance, ρ, of the surface, as shown in Figure 6-1. As discussed in Chapter *3*, and illustrated by Figure 6-2, if the radiance at P caused by reflected incident radiation is L, in W sr^{-1} m^{-2}, then the radiant flux, $d^2\Phi$, emerging from P may be expressed as the product of the radiance L; the projection of the area normal to the direction of interest as given by $dA \cos \theta$; and the solid cone angle $d\omega$ subtended at P as

$$d^2\Phi = L \cos \theta \, dA \, d\omega. \qquad (6\text{-}1)$$

Eq. 6-1 is one form in which radiance is expressed in Table 6-1.

Before considering the application of Eq. 6-1 to aerial photography, the relationship between the radiance of a ground surface and the irradiance of the surface due to the sun and the sky must be determined.

First, it is necessary to describe the reflection properties of the surface. The measured reflectance value of a surface for a given wavelength range will depend on the geometry of the arrangement of radiation source, surface, and detector; the polarization of the irradiance; and the spectral distribution of the irradiance at the surface. For the present, these factors will be neglected, and the assumption will be made that the irradiance at the surface is unpolarized, and that the measurement is made monochromatically or over a narrow wavelength range. A further assumption will consider that the surface is a perfectly diffuse reflector, a condition that is approximated by, for example, a snow or cloud layer, or a matte- or frosted-glass surface.

Many natural terrain features roughly approximate diffuse reflectors (Table 6-5); however, water surfaces and man-made objects are not diffuse reflectors and cause a distribution of reflected radiant flux that lies *between* those shown in Figure 6-3b and c. These nondiffuse reflectors show a peak in their reflectance distributions, the so-called *specular reflectance component*, when the angles of incidence and reflectance lie in the same plane and are equal. *Lambert defined a perfectly diffuse surface, hence the commonly designated Lambertian surface, as one for which the radiance, L, is constant for any angle of reflectance, θ, to the surface normal.* Thus, the radiant flux, $d^2\phi$, in watts, from a Lambertian surface of area dA and radiance L given by Eq. 6-1 shows a maximum of $L \, dA \, d\omega$ along the normal to the surface. The locus of $d^2\phi$ falls along the circumference of a circle of diameter $L \, dA \, d\omega$ and is zero for $\theta = 90°$. In the usual three-dimensional case, the distribution of radiant flux from the surface is spherical.

TABLE 6-1

Definition of Radiometric Quantities

Quantity	Symbol	Definition	Units	Unit Symbol
Radiant energy	Q		joule (erg)	J
Radiant flux	Φ	$\dfrac{dQ}{dt}$	watt (erg/s)	W erg s^{-1}
Irradiance	E	$\dfrac{d\Phi}{dA}$	watt/square meter	W m^{-2}
Radiant intensity	I	$\dfrac{d\Phi}{d\omega}$	watt/steradian	W sr^{-1}
Radiance	L	$\dfrac{d^2\Phi}{d\omega\,dA\,\cos\theta}$ or $\dfrac{dI}{dA\,\cos\theta}$	watt/steradian/square meter	W sr^{-1} m^{-2}
Reflectance	ρ	$\dfrac{\Phi_r}{\Phi_i}$		

Now assume that the elemental Lambertian surface dA in Figure 6-4 is irradiated by E in W m^{-2}, and that the radiant flux reflected from dA in any direction θ to the surface normal is given by Eq. 6-1. Because the elementary solid angle $d\omega$ subtended by the annulus at dA (Fig. 6-4) is given by

$$d\omega = \frac{2\pi r \sin\theta\,(r\,d\theta)}{r^2}, \qquad (6\text{-}2)$$

the total radiant flux in watts reflected into the hemisphere is given by

$$d\Phi_h = 2\pi \int_0^{\pi/2} L\,dA\,\cos\theta\,\sin\theta\,d\theta = \pi L\,dA, \qquad (6\text{-}3)$$

where:

the subscript h stands for total hemispherical. The ratio of the total reflected radiant flux to the radiant flux incident at the surface, $d\phi_i = E\,dA$, defines the *diffuse* or *hemispherical reflectance* of the surface, or

$$d\Phi_h/d\Phi_i = \rho = \pi L/E. \qquad (6\text{-}4)$$

So far, the spectral distribution of the flux reflected by and incident at the surface has not been considered. The ratio of these quantities is proportional to the hemispherical spectral reflectance, $\rho(\lambda)$, of the surface. The wavelength dependence is indicated mathematically by

$$\rho(\lambda) = \pi L_\lambda/E_\lambda. \qquad (6\text{-}5)$$

Thus the spectral distribution of the surface radiance, L_λ, (W m^{-2} sr^{-1} λ^{-1}) depends on both

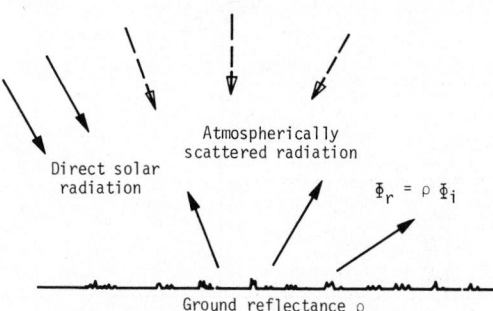

Fig. 6-1. Reflection at a surface where Φ_i is the sum of direct and atmospherically scattered solar radiation incident on the surface and Φ_r is the total flux reflected into the hemisphere above the surface.

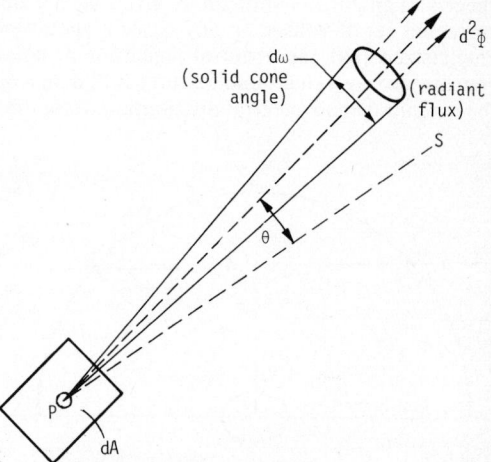

Fig. 6-2. Geometry for determining radiant flux from elemental surface area, dA, where PS is normal to dA.

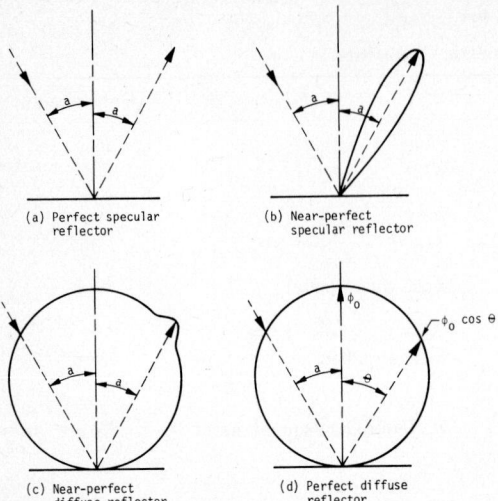

(a) Perfect specular reflector

(b) Near-perfect specular reflector

(c) Near-perfect diffuse reflector

(d) Perfect diffuse reflector

Fig. 6-3. Distribution of reflected radiant flux from different types of reflecting surfaces.

the spectral reflectance of the surface $\rho(\lambda)$ and the spectral distribution of the incident irradiation E_λ (W m^{-2} λ^{-1}).

As shown in Chapter 4, spectral reflectance and quantities related to it are used extensively in remote sensing to aid in characterizing a ground feature. For example, when a multiband camera is used, it may photograph the same ground feature in four spectral bands each 0.1 μm wide. Sensitometry and densitometry of the photography will yield values for the irradiance of the image of the ground feature incident on the film in each of these four wavelength bands.

These spectral irradiance values, used in conjunction with the spectroradiometric calibration of the camera, and when corrected for atmospheric effects, correspond to the spectral radiance values for the ground feature, and are often referred to as its *spectral signature*. It should be noted that the spectral signature is sometimes given as the uncorrected set of values. In any case, it should be emphasized that the spectral signature is not a constant for a given ground feature. It depends on the magnitude and spectral distribution of the flux

incident on the ground feature and the geometrical relationship between the solar angle and the camera angle to the surface. Such factors as the polarizing properties of the ground feature further increase the variability of the spectral signature. A ratioing technique is often used to reduce variations because of the above factors.

Applications to Aerial Photography

The general case in aerial photography will now be considered where the optical axis of the camera is tilted at an angle β to the vertical (Fig. 6-5). A flat earth is assumed and R is the distance from the lens of focal length f and diameter D to the ground feature of area dA_O (the subscript O denotes the object; the subscript I the image) at a field angle of α. There is an angle θ between the vertical at the object and the ray incident at the lens at field angle α.

Eq. 6-1 can be rewritten in a form that gives the spectral radiant flux, $d\Phi_\lambda$, intercepted by the lens from the ground feature of spectral radiance L_λ. Thus, by reference to Figure 6-5, and with the solid angle approximation referred to in Chapter 3, which will be used throughout this chapter

$$d\Phi_\lambda = L_\lambda (dA_O \cos \theta) \frac{\pi D^2}{4} \frac{\cos \alpha}{R^2}. \quad (6\text{-}6)$$

Now, if the lens is lossless, this intercepted flux will be incident over the area of the image dA_I. Eq. 6-1 can then be applied to both object and image, making the product of the projected area of the object and the solid angle subtended by the lens at the object equal to the product of the projected area of the image and the solid angle subtended by the lens at the image, or

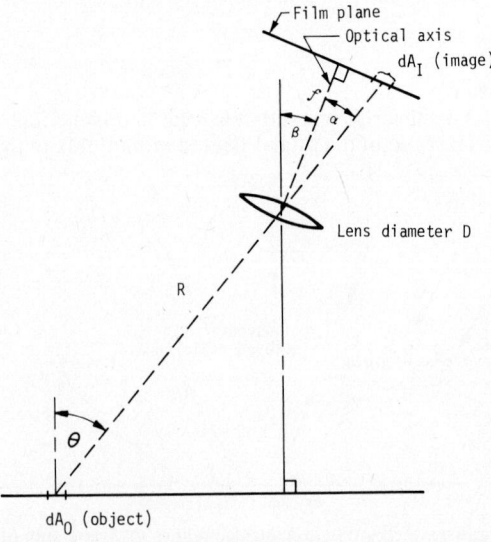

Fig. 6-5. Diagram of the general case in aerial photography of a tilted camera photographing an off-axis ground feature.

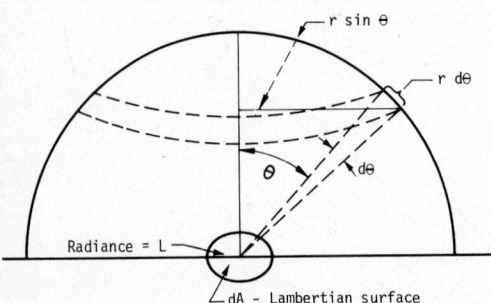

Fig. 6-4. Geometry of hemispherical interception of radiant flux from a Lambertian surface dA of radiance L.

$$dA_I \frac{\cos^3 \alpha}{f^2} = dA_O \frac{\cos \theta}{R^2}. \qquad (6\text{-}7)$$

Therefore, the spectral radiant flux incident over the area of the image is given by

$$d\Phi_\lambda = \frac{\pi}{4} L_\lambda dA_I \left(\frac{D}{f}\right)^2 \cos^4 \alpha. \qquad (6\text{-}8)$$

An additional contribution to the spectral radiant flux over the image area is due to sky spectral radiance $L_{\lambda s}$. When $L_{\lambda s}$ is included, Eq. 6-8 becomes

$$d\Phi_\lambda = \frac{\pi}{4} dA_I \left(\frac{D}{f}\right)^2 \cos^4 \alpha (L_\lambda + L_{\lambda s}). \qquad (6\text{-}9)$$

Now, for a Lambertian feature, the relationship between L_λ, the spectral radiance, and E_λ, the spectral irradiance due to the Sun and sky, was found to be given by Eq. 6-5:

$$L_\lambda = \rho(\lambda) \frac{E_\lambda}{\pi}, \qquad (6\text{-}5)$$
repeated

where:

$\rho(\lambda)$ = spectrally dependent diffuse reflectance of the ground feature.

In addition, the spectral transmittance of the atmosphere $\tau_A(\lambda)$ and the lens-filter combination $\tau_0(\lambda)$ can be included in Eq. 6-9:

$$d\Phi_\lambda = \frac{dA_I \cos^4 \alpha \left[E_\lambda \rho(\lambda)\tau_A(\lambda) + \pi L_{\lambda s}\right] \tau_0(\lambda)}{4N^2}, \qquad (6\text{-}10)$$

where:

$N = f/D$, and is referred to as the relative aperture of the lens.

The numerical value of the ratio is the f-number of the lens. Thus, if the effective focal length is twice the effective diameter of the lens, the relative aperture is written as $f/2$ and the f-number is 2.

If there is interest in a particular wavelength range $\Delta\lambda$ (from λ_1 to λ_2), the contributions of the wavelength-dependent factors can be summed in Eq. 6-10. The total spectral radiant flux incident over the image is thus given in radiometric quantities as

$$d\phi_{e\Delta\lambda} = \frac{(dA_I \cos^4 \alpha)}{4N^2} \int_{\lambda_1}^{\lambda_2} \left[E_{e\lambda}\rho(\lambda)\tau_A(\lambda)\right.$$
$$\left. + \pi L_{e\lambda s}\right]\tau_0(\lambda)\, d\lambda, \qquad (6\text{-}11)$$

and the spectral irradiance of the image E_I, in the wavelength range of interest, is obtained from

$$E_{eI\Delta\lambda} = \frac{\cos^4 \alpha}{4N^2} \int_{\lambda_1}^{\lambda_2} \left[E_{e\lambda}\rho(\lambda)\tau_A(\lambda)\right.$$
$$\left. + \pi L_{e\lambda s}\right]\tau_0(\lambda)\, d\lambda. \qquad (6\text{-}12)$$

This is the basic expression, which will be reconsidered later in this section.

The reader is referred to Chapter 4 for another discussion of this subject.

More Detailed Considerations

With the inclusion in Eq. 6-12 of $\cos^4 \alpha$, $\tau_0(\lambda)$, and N^2, one may be misled into thinking that all camera-dependent factors that can affect $E_{eI\Delta\lambda}$ have been taken into account. In practice, the effects of stray light, vignetting, and departure from the fourth power of cosine α may all be significant. Therefore, a more general form will be discussed for the spectral irradiance at the image plane.

Eq. 6-12 shows that *the spectral irradiance at the image plane is proportional to the product of the spectral transmittance of the camera and the spectral radiant flux incident at the camera.* Some of the flux that is not transmitted is absorbed by the lens elements, and some is reflected out of the camera. Of greater importance is the loss in spectral transmittance due to scattering by surface imperfections, such as dust and scratches, and microscopic inhomogeneities within the lens elements. In addition, there are reflections at and multiple reflections from lens surfaces, camera film, mounting fixtures, diaphragm, and shutter surfaces and edges. These sources of stray light or flare give rise to additional spectral irradiance $E_{\lambda f}$ over the image surface, which reduces the contrast in the image and is commonly referred to as *veiling glare.* In addition to the above sources, $E_{\lambda f}$ depends on lens design, coating and fabrication, f-number, field position, and wavelength range.

As an example, assume 5%, a typical value of $E_{\lambda f}$ for veiling glare; that is, a perfectly black object against an extended surround of uniform radiance will be imaged with an irradiance of 5% of the surround. Then the *camera flare factor* (which is defined as the ratio of the object contrast to image contrast) will be 6 for an object contrast of 100:1[2] and 1.1 for an object contrast of 2:1. Atmospheric scattering, L_s, usually causes scene contrasts to be 2:1 or less, so aerial camera flare factors are often close to unity. Nevertheless, the fact that the additional spectral irradiance is even a few percent of the high irradiance values in the image must be accounted for in accurate spectroradiometry.

Vignetting (defined later in this section) is field-angle and f-number dependent. Vignetting will be denoted here by $K_N(\alpha)$. The $\cos^4 \alpha$ relationship discussed later in this section can be modified; for more general application it will be replaced by $\cos^n\alpha$, where n usually lies between 2.5 and 4. A more general form for the spectral

---

[2] In this case the object is not perfectly black, since it has a value of 1% of the surround. The image contains an additional 5% irradiance, giving a ratio of object to image contrast of 6. Because the surround is much larger in area than the object, the loss from the extended surround is assumed to be negligible. The same argument holds for the 2:1 case.

irradiance in the image plane is then, substituting in Eq. 6-12,

$$E_{eI\Delta\lambda} = \frac{K_N(\alpha)\cos^n\alpha}{4N^2}$$

$$\int_{\lambda_1}^{\lambda_2}\left[E_{e\lambda}p(\lambda)\tau_A(\lambda) + \pi L_{e\lambda s}\right]\tau_0(\lambda)\,d\lambda + \int_{\lambda_1}^{\lambda_2} E_{e\lambda f}\,d\lambda.$$

$$(6\text{-}13)$$

This general expression is included as a caution against oversimplifying camera spectroradiometric calculations. However, for the remainder of this chapter Eq. 6-12 will be usually referred to, as a matter of convenience. It is important to understand that Eq. 6-12 should be expanded in practice to include as many of the additional factors in Eq. 6-13 as are significant to the case under consideration.

Use of Photometric Quantities

Should the limited case of luminous radiation be of interest, then photometric terms and units may be used, and Eq. 6-11 can be written as

$$d\Phi_v = \frac{dA_I\cos^4\alpha}{4N^2}(E_v\rho\tau_A + \pi L_v)\tau_0, \quad (6\text{-}14)$$

where:

Φ_v = luminous flux in lumens (ℓm),
E_v = illuminance in ℓm/unit area,
L_v = luminance[3] in candela (cd)/unit area of the skylight intercepted by the lens,
ρ = the ground object reflectance, and
τ_0 and τ_A = the transmittances of the optics and atmosphere, respectively.

Also,

$$E_{vI} = \frac{\cos^4\alpha}{4N^2}(E_v\rho\tau_A + \pi L_v)\tau_0, \quad (6\text{-}15)$$

where

E_{vI} = illuminance of the image in ℓm/unit area.

Eq. 6-15 can be rewritten as

$$E_{vI}t = \frac{t\cos^4\alpha}{4N^2}(E_v\rho\tau_A + \pi L_v)\tau_0, \quad (6\text{-}16)$$

where[4]

E_{vI} is in lux,
t is the exposure time in seconds,
L_v is in[4] candela/m².

[3] The International System (SI) unit of luminance is the candela (cd), where 60 candelas are defined as the luminous intensity per square centimeter of a blackbody radiator at the temperature of solidification of platinum, 2,046 K. (For precise equality, 58.9 candelas are equivalent to one old international candle.)

[4] The lux is the SI unit of illuminance, and is equal to 1 lumen/m². It was previously referred to as the "meter-candle." The candela/m² is the SI unit of luminance, also referred to as the "nit."

In the field of radiation measurement there are two different measurement concepts and sets of terms and units based on the use of the watt (radiometry) and the lumen (photometry). The confusion that results is compounded by the adherence of some scientists to the ft-lb-sec system of units. Efforts to standardize the use of radiometric quantities expressed in metric units have thus far been only partially successful. Important data continue to be reported in other forms to large extent because the sensitivities of electro-optical detectors and photographic films are still expressed in photometric terms by many of the manufacturers. Eq. 6-16 may also be rewritten with the use of appropriate conversion factors as

$$E_{vI}t = \frac{2.7t\cos^4\alpha}{N^2}(E_v\rho\tau_A + L_v)\tau_0, \quad (6\text{-}17)$$

where in this case there is a mixture of units with
$E_{vI}t$ in lux-s,
E_v in foot candles, and
L_v in foot-lamberts.

This form of the equation is included because E_v and L_v are frequently given in these units, and film manufacturers typically express film speed in terms of reciprocal lux-seconds.

Radiometric to Photometric Conversions

It is important to be able to convert from photometric to radiometric units, or from radiometric to photometric units. It is necessary to use values of the relative luminosity response for the eye to make this conversion. As shown in Figure 6-6, the bell-shaped curve for this parameter peaks at 0.555 μm for the light-adapted eye. The Commission Internationale de l'Eclairage (CIE) has defined the values listed in Table 6-2 as the response of the standard eye. At 0.555 μm, 1 W = 683 ℓm, thus

$$E_v = 683\int_{\lambda_1}^{\lambda_2} V(\lambda)E_{e\lambda}\,d\lambda, \quad (6\text{-}18)$$

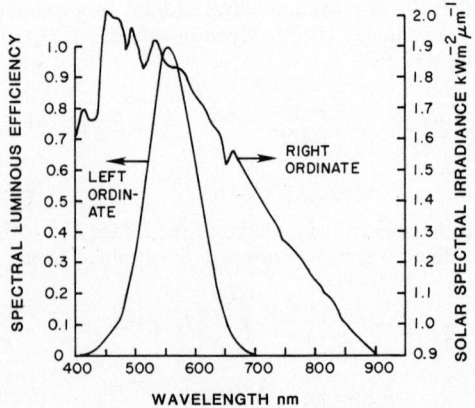

Fig. 6-6. Spectral luminous efficiency of the eye and the solar irradiance on an area normal to the sun and above the atmosphere, Labs and Neckel (1973).

TABLE 6-2

Spectral Luminous Efficiency for the Eye Relative to 1.00 at 0.555 μm and Solar Irradiance at Sea Level* on Area Normal to sun in W m^{-2} μm^{-1}, where $E_0 = 1322$ W m^{-2}

Wavelength (μm)	Efficiency	W m^{-2} μm^{-1}	Wavelength (μm)	Efficiency	W m^{-2} μm^{-1}
0.40	0.0004	470	0.60	0.631	1,167
0.41	0.0012	672	0.61	0.503	1,168
0.42	0.0040	733	0.62	0.381	1,165
0.43	0.0116	787	0.63	0.265	1,176
0.44	0.023	911	0.64	0.175	1.175
0.45	0.038	1,006	0.65	0.107	1,173
0.46	0.060	1,080	0.66	0.061	1,166
0.47	0.091	1,138	0.67	0.032	1,160
0.48	0.139	1,183	0.68	0.017	1,149
0.49	0.208	1,210	0.69	0.0082	978
0.50	0.323	1,215	0.70	0.0041	1,108
0.51	0.503	1,206	0.71	0.0021	1,070
0.52	0.710	1,199	0.72	0.00105	832
0.53	0.862	1,188	0.73	0.00052	965
0.54	0.954	1,198	0.74	0.00025	1,041
0.55	0.995	1,190	0.75	0.00012	867
0.56	0.995	1,182	0.76	0.00006	566
0.57	0.952	1,178			
0.58	0.870	1,168			
0.59	0.757	1,161			

* For air mass of 2, i.e., solar zenith angle = 60°.
[These solar irradiance values and other useful data can be found in Valley (1965)]
For zero air mass, see data presented in Figure 6-6.

where:

E_v = illuminance in ℓm/unit area,
$V(\lambda)$ = photopic spectral luminous efficiency,
λ_1 to λ_2 = wavelength range of interest, and
$E_{e\lambda}$ = known spectral irradiance distribution (W m^{-2}).

The method of making such a conversion may be illustrated by an example: the determination of the relationship between watts and lumens in the wavelength range of 0.59 to 0.72 μm for the irradiance values as listed in Table 6-2. This is typical of the kind of calculations that are made in multiband photography work.

Using the trapezoidal rule, the area under the spectral solar irradiance curve plotted in Figure 6-6 from Table 6-2 is found to be 147 W m^{-2} in the specified wavelength range. Then, as Eq. 6-18 indicates, the product of the two functions $V(\lambda)$ and $E_{e\lambda}$ has to be found over the wavelength range λ_1 to λ_2. Again, using the trapezoidal rule, the area under the curve resulting from the plot of this product against wavelength is found to be 30 W m^{-2} in the specified wavelength range. These are effective values of W m^{-2}, and it is interesting to note how the decreasing sensitivity of the eye in the red spectral area reduces the actual W m^{-2} available by a factor of about five in the conversion to effective W m^{-2}. The product of 30 W m^{-2} × 683 ℓm W^{-1} gives 20,490 ℓm m^{-2}, and thus, under the specified irradiance conditions, 1 W is equal to roughly 140 ℓm in the wavelength range of 0.59 to 0.72 μm.

Comments Concerning Basic Equation for Image Spectral Irradiance

There are several important comments to be made regarding Eq. 6-12, which is the basic expression relating image spectral irradiance to camera radiometry, ground feature irradiance and reflectance, and atmospheric radiance and transmittance.

First, the equation shows that the image irradiance is inversely proportional to the square of the relative aperture. That is, in order to halve the image irradiance, the f-number should be increased by a multiplying factor of $(2)^{1/2}$. For this reason the diaphragm markings or stop openings for a lens are marked in a geometrical progression, using the geometrical ratio of $(2)^{1/2}$. (Where the lowest f-number does not coincide with the standard progression, it is usually quoted separately as 1.9, 2.0, 2.8, 4.0, etc.) The accuracy of stop markings is usually $\pm(2)^{1/6}$ of the f-number, a third of a stop, or $\pm12\%$ of the f-number. Thus the amount of light transmitted by a lossless lens may vary by $\pm24\%$ from the value corresponding to the stop marking. In the case of complex lenses consisting of many air-to-glass interfaces or elements that absorb an appreciable amount of the radiation to be recorded, the f-number is not an accurate measure of the radiation gathering power or speed of the lens. The T-number is then often used, which, for a circular aperture, is given by

$$T\text{-}number = \frac{f\text{-}number}{\sqrt{\tau_0}}, \qquad (6\text{-}19)$$

where

τ_0 = transmittance of the optics.

The error in the T-stop markings of a lens should not exceed one-tenth of a stop, or $\pm 7\%$ of the axial image irradiance. N, rather than the T-number, is used in Eq. 6-12 because τ_0 in the general case is a function of wavelength. A further use of N is in the calculation of the depth of focus of a lens; this will be discussed in the next section of this chapter.

Secondly, using Eq. 6-12 and ignoring atmospheric effects, the spectral irradiance of the image is independent of the distance of the camera lens to the ground object. Since the radiance collected by the lens from a ground object decreases as the square of the altitude, so does the area of the image of the object formed by the lens. Thus the ratio of the radiance collected to the area of the image is independent of altitude, and the irradiance of the image is constant as long as the image is of a uniform extended object.

The third comment to be made concerns the fact that, according to Eq. 6-12, and again ignoring atmospheric effects, the irradiance of the image decreases as $\cos^4$ of the field angle α. This applies to a simple lens and one in which no vignetting takes place. Some multi-element wide-angle lenses have been designed so that the falloff of irradiance follows a relationship of approximately $\cos^3 \alpha$. This is achieved by introducing coma[5] into the entrance pupil of the lens to enlarge the area of the entrance pupil as the field angle is increased. Such an improvement is obviously valuable because, for a 90° lens with $\cos^4 \alpha$ falloff on a standard mapping camera, the irradiance at the corner of the frame is a factor of four less, or two f-stops greater, than that at the center of the frame. Even with a $\cos^3 \alpha$ falloff, the change in irradiance is more than can be tolerated by a reversal color film where the exposure latitude is about ± 0.5 f-stop. To overcome this problem, an antivignetting filter is placed in front of the lens.

Vignetting is the progressive reduction in the cross-sectional area of a beam of light passing through a lens as the field angle is increased. The reduction is usually caused by the lens mounts within the lens. Thus, vignetting for a given lens is f-number dependent, becoming smaller at larger f-numbers. In contrast, $\cos^4 \alpha$ falloff is independent of f-number, and therefore a single antivignetting filter can be used to render the image irradiance uniform for a lens not afflicted with vignetting.

Reference to Eq. 6-13 will serve to underscore the dependence of spectral signatures on the basic spectroradiometric conditions pertaining to aerial photography. It should be noted that the same expression holds for any kind of camera, whether the image is recorded photographically or electro-optically.

In order to carry out accurate calculations based on this equation, the spectroradiometric response of the camera must be determined in the laboratory, allowance must be made for departures from the assumed Lambertian characteristics of the scene reflectance and atmospheric scattering, and close attention must be paid to the film sensitometry.

Finally, the exposure of the film in the camera has to be considered. The radiant energy, Q, in ergs, which exposes a given area A of the film, is a function of exposure time, t, as well as E, the image irradiance, since

$$Q = E \times t \times A. \qquad (6\text{-}20)$$

Therefore, the exposure time has to be known accurately, that is, the average effective open time of the shutter corresponding to the nominal marked value should be known, as should the repeatability of the shutter.

THE OPTICS OF AERIAL CAMERAS

There are several comprehensive basic and advanced texts dealing with general optical concepts (Longhurst, 1973; Welford, 1962; Kingslake, 1965; Born and Wolf, 1959); there are also a number that deal specifically with aerial photography (American Society of Photogrammetry, 1966, 1968 and 1981; Brock 1967, and 1970; Brock et al, 1965; Jensen, 1968). This chapter will describe, primarily, those optical factors that are a measure of, or affect, the image-forming performance of the camera, and that might be determined by, or be useful to, the camera user and data interpreter. Discussions of geometrical optics, aberration theory, and optical design methods are not within the scope of this manual. Details of optical transfer function theory and measurements can be found in the references listed above and will be treated only briefly. Other references can be found in Chapter *13*. Topics that will be covered are (1) contrast; (2) resolving power and area weighted average resolution (AWAR); (3) modulation transfer functions and threshold modulation; (4) scale and change of resolution with field angle; (5) depth of focus; and (6) distortion and calibrated focal length.

Contrast

The contrast of an object or image is frequently referred to both qualitatively as high- or low-contrast, and as an abbreviation of contrast ratio or contrast difference. Unfortunately, several expressions are in common use to describe contrast and are sometimes confused in their use.

Figure 6-7 will aid in understanding and defining these terms. Assuming that the distribution of image irradiance is as shown, and that E_{max} and E_{min} are the maximum and minimum image ir-

---

[5] Coma is a monochromatic aberration that causes an off-axis object point to be imaged as an unsymmetrical comet-shaped blur.

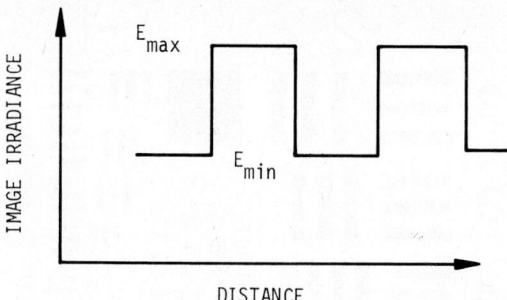

Fig. 6-7. Assumed distribution of irradiance in image plane.

radiances, respectively, then the generally accepted terms are:

Contrast ratio	$= C_R = E_{max}/E_{min}$	(6-21)
Differential contrast[6]	$= C_D$	(6-22)
	$= (E_{max} - E_{min})/E_{min}$	
Logarithmic contrast	$= C_L$	(6-23)
	$= \log_{10} (E_{max}/E_{min})$	
Modulation	$= M$	(6-24)
	$= \dfrac{(E_{max} - E_{min})}{(E_{max} + E_{min})}.$	

Of these, contrast ratio (usually abbreviated to contrast) and modulation are the terms most widely used in aerial photography, and they will be used exclusively throughout this chapter.

Note: when $E_{min} = 0$, M is unity and C_R is infinity; when $E_{min} = E_{max}$, M is zero and C_R is unity (not zero as often loosely stated).

All the above expressions are interrelated in this manner:

$$C_R = \frac{1 + M}{1 - M} = C_D + 1 = 10^{C_L} \qquad (6\text{-}25)$$

$$C_D = \frac{2M}{1 - M} = C_R - 1 = 10^{C_L} - 1 \quad (6\text{-}26)$$

$$C_L = \log_{10} \frac{1 + M}{1 - M} = \log_{10} (C_D + 1) = \log_{10} C_R$$
$$(6\text{-}27)$$

$$M = \frac{C_D}{C_D + 2} = \frac{C_R - 1}{C_R + 1} = \frac{10^{C_L} - 1}{10^{C_L} + 1}. \quad (6\text{-}28)$$

Resolving Power and Area Weighted Average Resolution

The *limit of resolution* of an optical system is reached when, according to a given criterion, the system is judged to be just able to separate the elements of a well defined test object; e.g., a dou-

ble star in astronomy, the lines of a grating in microscopy, and the bars of a bar target in photography. *In photography, the resolving power of the system is the reciprocal of the center-to-center separation of the bars (lines) in the image of the target at the resolution limit.* Usually the resolving power of a system or its component parts such as the lens, the film, or an electro-optical device are stated in lines per mm[7] or cycles per mm. A convenient way to express the resolving power of an image-forming system is in angular measure, e.g., mrad/cycle, from which the lines/m in object space or lines/mm in image space can be determined with equal facility. Resolving power and resolution are often used interchangeably, although this distinction should be made: *Resolving power applies to a system or a component of a system, whereas resolution applies to the image produced.*

Discussion of image formation by a perfect or aberration-free lens can be found in the standard optics texts previously mentioned. Suffice it to say here that a point source is imaged by a perfect lens as an *Airy pattern.* The size of the central disk in the pattern (the disk contains 84% of the irradiance in the pattern) depends on the wavelength and numerical aperture, or *f*-number, of the image-forming cone according to the nomogram and equations in Figure 6-8. The profile of the Airy pattern or the point spread function for an aberration-free lens with a circular aperture is shown in Figure 13-103 in Chapter 13.

Lenses of moderate *f*-number can approach perfection if the design field of view is narrow. Thus microscope objectives working on, or nearly on-axis, and some panoramic camera lenses of *f*/5 and covering a 10° field, can be designed to be almost aberration-free. Wide-angle aerial camera lens designs show substantial amounts of aberration, and Figure 6-8 cannot be applied to these lenses.

Methods of Determining and Measuring Resolving Power

The resolving power of a system can be determined by a variety of methods and against several different criteria. The geometry, radiometry, and image recording and processing have to be noted in all cases to give the resolving power value any real significance. A standard method is usually followed to shorten the description of the measuring technique and the processing used in determining the resolving power of a camera, and to allow for ready comparison with other cameras. The most frequently used standard method in the United States is described in Military Standard

---

[6] Also known as universal contrast (Duntley et al., 1964).

[7] For brevity, this chapter will refer to lines/mm rather than line pairs/mm or optical line pairs/mm. Caution: Do not equate lines/mm as used in this chapter with lines/mm as used in TV engineering; 2 TV lines/mm = 1 line/mm as considered herein, and in photographic usage.

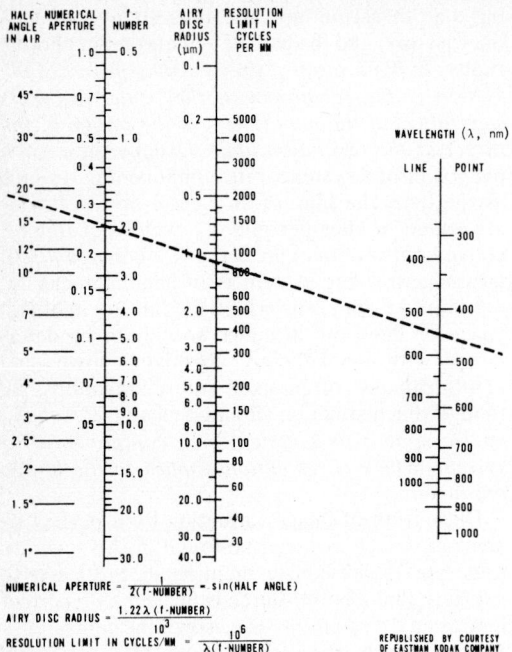

NUMERICAL APERTURE $= \frac{1}{2(f\text{-NUMBER})} = \sin(\text{HALF ANGLE})$

AIRY DISC RADIUS $= \frac{1.22\lambda(f\text{-NUMBER})}{10^3}$

RESOLUTION LIMIT IN CYCLES/MM $= \frac{10^6}{\lambda(f\text{-NUMBER})}$

REPUBLISHED BY COURTESY OF EASTMAN KODAK COMPANY

Fig. 6-8. Theoretical limits of resolution and radius of Airy disk for a perfect lens.

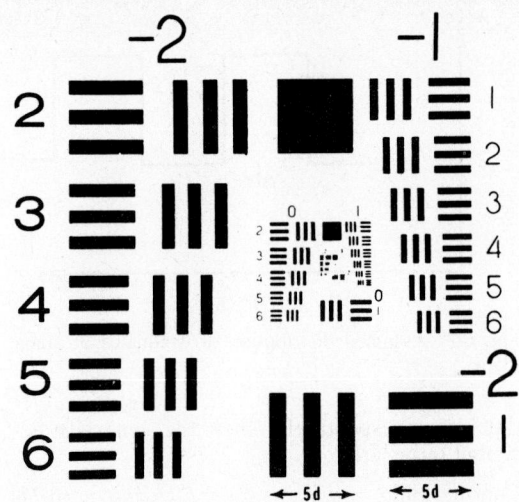

Fig. 6-9. The United States Air Force 1951 resolving-power test target. The reciprocal of *2d* is the frequency of the target in lines/mm and the reciprocal of *d* is the frequency in TV lines/mm.

150A (U.S. Govt., 1963); this method is identical in all important respects to the method described by the American National Standards Institute (1969). These standards, which define the use of a three-bar target with elements of various sizes, have been in wide use in this country for about the past thirty years, except by the National Bureau of Standards, which, until quite recently, frequently used a different design of test target. In other countries a variety of different target configurations are in use, e.g., annuli, two bar, multiple bar, and various simple geometrical shapes. Three-dimensional test targets have also been devised in an attempt to simulate the real world.

Test Targets.

The standard test target consists of a series of patterns decreasing in size according to some root of two, usually $(2)^{1/6}$. The standard target is shown in Figure 6-9, and consists of two patterns (two sets of lines) at right angles to each other drawn in the proportions indicated. For testing aerial cameras or aerial camera lenses, transmission targets are used, in which the bars have a greater transmittance than the surround. The density difference between bar and surround is usually chosen to be greater than 2.0 (high-contrast ratios greater than 100:1); equal to 0.8 ± 0.05 (medium-contrast ratio of 6.3:1); or equal to 0.2 ± 0.05 (low-contrast ratio of 1.6:1).

Measuring Resolving Power of a System.

The standard procedure for measuring the resolving power of a camera or lens is as follows: (1)

The standard target is placed in the focal plane of a well-corrected collimator. (2) The camera or lens under test is mounted on a nodal slide so that the lens can be rotated to different field angles, while the image position remains conveniently stationary. (3) The target is irradiated by simulated clear-day radiation from a quartz iodide source or a tungsten source with color compensating filter, i.e., from a blackbody source of approximately 6000 K. (4) Exposures are made at various field angles, usually in multiples of 1.25°, to provide at least five equal increments across the semifield of the lens. (5) A series of exposures is made to determine the combination of exposure time and focal setting yielding the highest average resolution over the entire picture area. The camera is tested with the film that will be used in flight, and the spectral and antivignetting filters specified for use with the camera should be employed during the test, or misleading data will result. The same strictures apply to the test of a lens not mounted in a camera, although in this case there is sometimes an interest in determining the resolving power of the lens alone. In this case, very high resolving power film can be used, and its effect on the measured resolution can be neglected. (6) When the correct exposure and focal settings have been determined, a series of exposures is made at these settings-along the two diagonals of the field (additional azimuths are sometimes used) at selected angular increments. (7) The frame is carefully processed, and the conditions and gamma of the processing are noted. (8) The photographic images of the target are then examined under a binocular or stereoscopic microscope.[8] At least two experienced readers

[8] Hooker (1970) compares three stereoscopic microscopes suitable for photointerpretation work. He reports the square wave responses of the microscopes at four

should read the images and an average should be taken. The *numerical aperture* (NA) of the objective, the overall magnification of the microscope, and the level of irradiance of the image should be carefully selected.

Selwyn (1954) has shown that it is useful to choose the overall magnification that approximately equals the numerical value of the limiting resolution. To many this corresponds to too high a magnification; consequently, a good choice is to use the lowest magnification and image irradiance that will provide the maximum confidence, ease, and convenience in reading. Resolution readings are difficult to make, and their accuracy depends on the training of the reader, his degree of concentration, and his visual acuity, which deteriorates rapidly with fatigue. This fatigue is accelerated if the image irradiance is held at too high a level.

Is the Target Element Resolved?

The decision as to whether a marginally resolved target element should be called resolved or not resolved is a difficult subjective matter. Military Standard 150A, previously referenced, states that an element is resolved if the reader is able to count the correct lines in the recorded image, over the entire length of the lines in the correct orientation, subject to the provision that no coarser element is unresolved. As a useful check against over-optimism, it is worthwhile for the reader to question whether, with a greater than 50:50 confidence level, he could identify the target as comprising three separate straight bars if he did not have prior knowledge of the configuration of the target.

The subjective and statistical nature of resolving power should not be overlooked, nor should the fact that the $(2)^{1/6}$ increments in the size of the standard elements in a target correspond to a change of 12% in the spatial frequency. The reader may often be uncertain over a range of two or three target elements.

Area Weighted Average Resolution

Reference was made above to the highest average resolution over the entire picture area; this is another way of referring to the *area weighted average resolution (AWAR)*, a convenient single average value used to describe the resolving power of a lens-film combination. To determine the AWAR, the picture format is divided into concentric annular zones, the boundaries of which are determined from the angles midway between successive test angles. For the outer zones, which extend partially beyond the format, only the area within the format is used to determine the weighting ratio. The resolution obtained at any

given test angle is multiplied by the ratio of the area of the zone for that angle to the total area of the picture format. The ratio of areas for a lens with a 45° semifield angle and a square picture format at 5° intervals is shown in Figure 6-10 to indicate the heavy weighting of midfield angles in comparison with near-zero and extreme field angles.[9] Often, and especially for nearly diffraction-limited[10] optical systems, the average solution of 0.7 of full field is a close approximation to the AWAR for a square format.

The geometrical mean of the tangential and the radial resolutions is determined for each field angle. The AWAR is then calculated from

$$AWAR = \sum \frac{A_i}{A} \sqrt{R_i T_i}, \qquad (6-29)$$

where:

A = total area of the picture format over which the summation is made,

A_i = area of a particular zone,

R_i = average radial resolving power in zone A (or the radial resolving power at the midpoint of the zone),

T_i = average tangential resolving power in zone A (or the tangential resolving power at the midpoint of the zone).

Now by definition R and T are the resolving powers for *radial* and *tangential lines*, respectively, and are therefore at right angles to the direction in which the resolution is actually measured. The R and T values measured using a collimator method have to be corrected to correspond to an aerial camera with a vertical axis that is used to photograph an assumed flat earth. The correction for the tangential resolution value in lines/mm is a factor of $\cos^2\theta$, and for the radial resolution it is $\cos\theta$, where θ is the field angle. (This applies if the image is recorded on film in the camera focal plane. If the image is recorded on film in a plane orthogonal to that of the collimator, as is frequently done in the nodal slide test, then the corrections are $\cos^3\theta$ and $\cos\theta$ for tangential and radial resolution, respectively.) These corrections are made if the number of lines/mm is determined from the spatial frequency of the target element and a knowledge of the demagnification of the collimator camera combination; they are not made

[9] The values in Figure 6-10 emphasize the fact that a single on-axis check of resolution can be misleading because the focal plane for highest AWAR usually does not coincide with the focal plane for the highest axial resolution. Sometimes the focal plane for the highest AWAR yields an axial resolution that is lower than the resolution in that focal plane at intermediate field angles.

[10] *Diffraction-limited* refers to a lens that produces an image with a maximum irradiance which is $\geqslant 0.8$ times the maximum irradiance in an Airy disk formed by a perfect or aberration-free lens of the same numerical aperture and at the same wavelength (Born and Wolf, 1959). Thus a diffraction-limited lens can possess small amounts of aberrations; an aberration-free lens is diffraction-limited.

field angles for each of four magnifications and compares the performance of the microscopes for use in both search and detailed examination modes of operation.

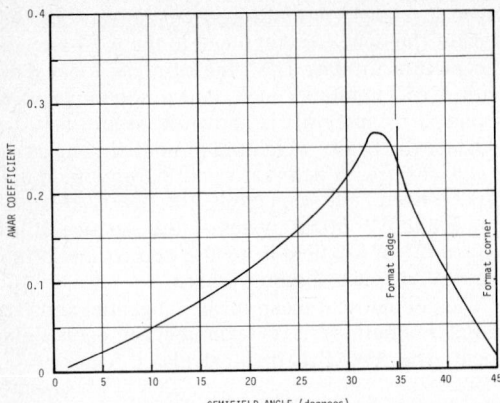

Fig. 6-10. AWAR coefficient as function of semifield angle for a square format with 90° full diagonal. The AWAR coefficients were calculated at semifield angles at the midpoints of 5° annuli. The value for the central zone was calculated for a semifield angle of 1.25°.

if the spatial frequency is determined in the *image plane* from a microdensitometer scan, for example.

Modulation Transfer Function and Threshold Modulation

At present the standardized method to report the spatial image-forming characteristics of an optical system is in terms of resolving power. The measuring of resolving power requires close attention to detail, or erroneous values will result. For most remote-sensing purposes a knowledge of the resolving power of the camera is sufficient to provide the data user a benchmark for reference.

The *modulation transfer function* (MTF) provides a more complete description of the image-forming properties of a lens than does the resolving power. However, in photographic remote sensing the emphasis is on the spectroradiometric analysis and residual distortion characteristics of the photography; the detailed relationship between MTF and information content has not been established, although suggestions have been made on how this could be accomplished (Slater and Schowengerdt, 1973).

The MTF and film threshold modulation curves are of value to the optical designer for predicting the three-bar resolving power of a given design. The camera operator can also use such data to predict how the photographic resolution of his camera will change with a change in film type. It is a simple matter to investigate the influence of target contrast and the effects of image motions of linear, sinusoidal and other types on the resolution. Atmospheric turbulence effects can also be taken into account using the subsequent Eq. 6-40. Accounts of these techniques can be found elsewhere (Brock, 1970; Brock et al, 1965; Scott et al, 1965; Welch, 1971), such details not being within the scope of this manual.

Modulation Transfer Function

A brief description of the MTF of an optical system and the threshold modulation is pertinent, since these characteristics are often referred to when describing a camera system.

Any object can be considered to consist of a spectrum of spatial frequencies of various amplitudes and phases. The simplest case is a sine-wave distribution of amplitude. Another is a series of square-wave distributions of amplitude, which can be represented as a sine wave having the same frequency, ν, as the square wave (called the fundamental), and a series of odd harmonics having the frequencies 3ν, 5ν, 7ν, etc., and amplitudes 1/3, 1/5, and 1/7, etc., the amplitude of the fundamental. When a square wave or a bar pattern is imaged by an optical system, the edges of the bars become rounded owing to the reduction by the system of the amplitude of the higher harmonics. The lens acts as a low-pass spatial frequency filter, and the MTF, as a function of a transfer factor, $\tau(\nu)$, and a given spatial frequency, ν, is given by the ratio

$$\tau(\nu) = \frac{M_I(\nu)}{M_O(\nu)}, \qquad (6\text{-}30)$$

where

$$\nu = \text{spatial frequency}$$
$$M_I(\nu) \text{ and } M_O(\nu) = \text{modulation of the image and object as a function of } \nu.$$

The transfer function curve, Figure 6-11, is a plot of $\tau(\nu)$ against ν, and is shown for an aberration-free lens on-axis, with a sinusoidal and square wave input modulation. The abscissa is normalized by setting

$$\nu_c = 1/\lambda N = 1, \qquad (6\text{-}31)$$

where:

$$\nu_c = \text{spatial frequency cutoff of aberration-free lens}$$
$$N = f\text{-number}$$
$$\lambda = \text{effective wavelength.}$$

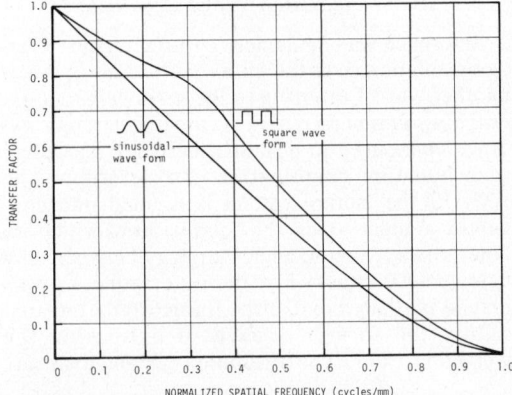

Fig. 6-11. Transfer function curves for an on-axis aberration-free lens for sinusoidal and square-wave input functions.

It is more convenient in practice to plot the MTF in log-log form (Fig. 6-12). Then, with assumptions of linearity, the MTF's of the various components of a system can be cascaded (multiplied) by adding the ordinate values to obtain the total system MTF (Scott et al, 1965).

Use of the MTF

The MTF is sometimes misinterpreted, and the cutoff frequency is equated to the size of the smallest object resolvable. To use the MTF to determine whether an image is resolved, the image has to be separated into its various component spatial frequencies; the modulations at these frequencies are reduced according to the MTF, and the modified spatial frequency spectrum is transformed to obtain the image. Then, based on a given criterion, the modulation in the image is judged to be greater than, equal to, or less than that required to resolve the image. It should be noted in passing that the smaller the image, the greater its bandwidth, and as the MTF becomes flatter through this bandwidth, the resemblance of image to object will become closer.

An example of the use of MTF in the analysis of photographic imagery is given later in this chapter in the discussion of the photography obtained with the Earth Terrain Camera.

Film Threshold Modulation Curve

In practice, threshold modulation (TM) values are used to relate a modulation transfer curve to the resolution limit of a system. For example, the normal eye can usually resolve image detail having a modulation greater than about 0.05 if there is adequate magnification. This fact can be used to qualify visually estimated resolution data for lenses or films. For example, the aerial image produced by a lens or the processed image on film is viewed under adequate magnification using a well-corrected microscope, and the resolution limit of the lens-eye or film-eye combination corresponds to a modulation level of about 0.05. *For a camera system a TM curve can be established for the film, and the spatial frequency at the intersection point of the TM curve with the MTF of the lens is the resolution limit of the camera system* (Scott et al, 1965).

A TM curve is determined experimentally by exposing on film the image formed by a microscope objective of three-bar targets with various modulations; the exposure values are chosen to yield the maximum resolution at each modulation. The processed film is then examined under magnification to determine visually the resolution limit at each modulation. The modulation is that of the image impressed on the film and includes the reduction in modulation that occurs due to the microscope objective. These modulation values can then be plotted against the corresponding resolution limits to construct the threshold modulation or aerial image modulation curve for the film. A detailed account of the production of TM curves has been presented by Lauroesch et al (1970), and some curves are shown in the subsequent Figure 6-35.

Scale and Change of Resolution with Field Angle

Scale

Scale indicates the correspondence of a distance between two points as measured on a map to the actual separation of the two points. Thus we may have a map drawn to a scale of 1:10,000, where 1 mm on the map represents an actual distance of 10 m.

Scale is used in aerial photography in a similar fashion where the demagnification produced is the scale factor. *For vertical high-altitude photography, the scale factor is simply the ratio of altitude to focal length.* Unfortunately, scale and ground resolution are used interchangeably by some engaged in photographic remote sensing. This probably relates to the earlier days of mapping photography when the finest detail recorded on the original film could be seen by the unaided eye. Then indeed the altitude or scale could directly indicate the ground resolution to be expected. Nowadays the finest detail recorded on the film in the camera can be observed only under magnification. The use of magnification corresponds in effect to a change in scale. A statement that a certain terrain feature can be observed with photography taken at a certain scale (ratio of altitude to focal length) is by itself meaningless. Reference has to be made to the image-forming properties of the camera or the additional magnification required. These and related factors are discussed in detail by Welch (1972). The perceptual quality and geometric cartographic quality of ERTS-1 imagery have been related to the maximum printing scale for the imagery by Colvocoresses and McEwen (1973).

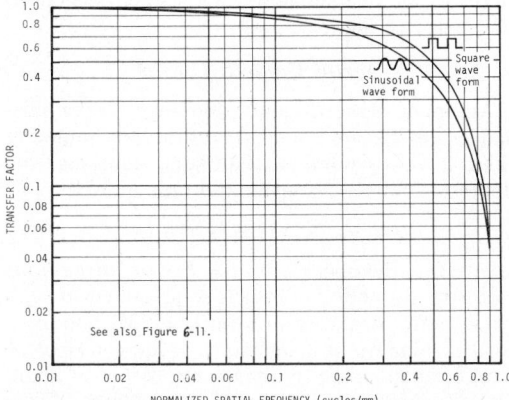

Fig. 6-12. Plot of curves in Figure 6-11 in log-log form.

Change of Resolution with Field Angle

In the succeeding section on atmospheric effects, it will be seen that as the angle off-nadir increases, the sky radiance likewise increases. As a consequence, the number of lines per meter resolvable on the ground decreases. The decrease in resolution is compounded by two other factors.

(1) *The resolving power of the camera lens decreases with an increase in field angle.* Typically, as the field angle increases, the aberrations of a lens system increase. This generality cannot be extended further because even lenses of the same class (such as Tessar or double Gauss) have different balances of their residual aberrations, depending on the choice and skill of the designer, the time devoted to the design, and fabrication tolerances. However, it is worthwhile noting that, in the case of a hypothetical aberration-free wide-angle lens, the resolving power of the lens decreases as the field angle increases, simply because of the geometry.

Looking back at the exit pupil of a simple lens from an off-axis point P (Fig. 6-13), the pupil will appear elliptical with minor axis $D \cos \theta$, where D is the pupil diameter and θ is the off-axis angle. This ellipticity reduces the resolving power for tangential lines by $\cos \theta$. Now the focal length is increased by $\sec \theta$, so the cone angle of the image-forming light (which for a given wavelength and for an aberration-free lens governs the resolving power) is reduced. The projection onto the camera focal plane accounts for a third $\cos \theta$ factor. Thus

$$R_T = R_A \cos^3 \theta$$

and　　　　　　　　　　　　　　　　　　(6-32)

$$R_R = R_A \cos \theta$$

where:
　R_T, R_R and R_A　refer to tangential, radial, and axial resolving powers, respectively.

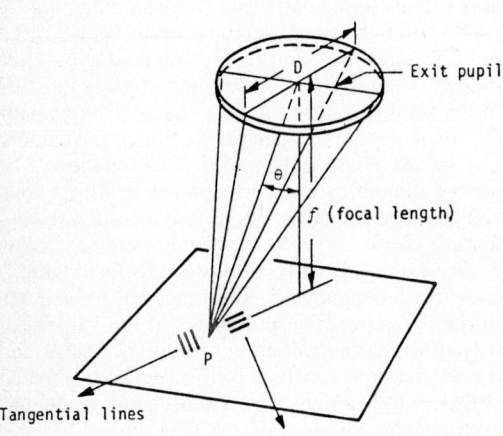

Tangential lines

Radial lines

Fig. 6-13. Geometry of image formation off-axis.

Note that these cosine factors are inherent in the geometry of the hypothetical lens considered. They should not be confused with the cosine correction factors used when target frequencies are read off on a collimator test of a lens.

(2) *The finite thickness of the photographic emulsion is a second factor tending to reduce the resolving power at wide angles for a flat film surface.* In the absence of scattering, a ray at a field angle of 60° passes through the emulsion at an angle of about 40°. For a 10-μm thick emulsion, an infinitely thin tangential line image incident at the top of the emulsion will appear as a line about 8 μm thick when the emulsion is processed and viewed at normal incidence.

Depth of Focus

The depth of focus of an optical system can most simply be defined as the distance through which a screen, normal to the optical axis, can be moved to either side of focus before the image on the screen begins to appear to be out of focus. In all cases it depends on the f-number of the lens, and for a camera system on the resolving power of the lens in relation to that of the film.

Diffraction-Limited Lens

In the case of a diffraction-limited lens, Rayleigh's criterion states that the maximum departure of δd from the wavefront centered on the Gaussian or paraxial image point should be no greater than $\lambda/4$, where λ is the effective wavelength of the image-forming radiation. Then

$$\delta d_{df} = \pm 2 \, \lambda \times (f\text{-number})^2, \qquad (6\text{-}33)$$

where
　df = diffraction-limited.

At the extreme ranges of this tolerance for a diffraction-limited lens, the modulation at half the cutoff frequency drops from 0.4 to 0.3; whereas at one-tenth the cutoff frequency, the modulation drops from 0.88 to 0.82. Thus a high-resolving-power film capable of resolving the modulation at half the cutoff frequency of the lens will be more sensitive to the $\lambda/4$ tolerance than film just capable of resolving the modulation at one-tenth the cutoff frequency of the lens.

Aerial Camera Lenses

Mapping cameras and wide-angle reconnaissance cameras are not nearly diffraction-limited at their lowest f-number. For such cameras, the depth of focus δd is given approximately by

$$\delta d_{(aerial)} = \pm D \times (f\text{-number}) \qquad (6\text{-}34)$$

where D is the diameter of the *disk of least confusion,* the formation of which is described in such basic optics texts as Longhurst (1973) and Welford (1962). As an example, a value for D of 20 μm, corresponding to a resolving power of about 50 cycles/mm, gives for an $f/5$ lens a depth of focus $\delta d = \pm 100 \, \mu$m.

Eqs. 6-33 and 6-34 provide useful guides to the depth of focus tolerance, although usually the change with defocus in the MTF curve of the lens design is used to predict quantitatively the effect of the tolerance on the resolution obtainable. For most lenses the best focal setting is a compromise because the surface of best focus is not plane.

Figure 6-14 shows some results obtained by Mayo (1968) for a high-quality reconnaissance lens. It is noticeable that from on-axis to 10° off-axis, the focal shift is 100 μm. Also, because the resolving power is less at 10°, the value for D in Eq. 6-34 is larger, and thus δd is larger. The shape of the 0° and 10° curves is evidence of the correspondingly greater focus tolerance at 10°. In some frame cameras the vacuum platen is contoured to compensate for field curvature and to afford maximum resolution over the field.

Distortion and Calibrated Focal Length

Radial Distortion

A lens is afflicted with distortion if, for example, a square grid pattern is imaged as a pincushion or barrel-shaped pattern. These are the two characteristic radial distortion patterns for a flat image plane in the case of a lens that has been accurately assembled. Distortion, like compensated field curvature as described in the previous paragraph, is not an aberration that reduces the resolution of the image; its effect is to displace an image point from the position where it would accurately represent the geometry of the object. A more complete account can be found in basic optics texts (Longhurst, 1973; Welford, 1962), and detailed reports are available on distortion measurement from the National Bureau of Standards (Washer, 1965).

For mapping purposes, the distortion introduced by the lens should be as small as possible and accurately known. As we will see in a later section on "Mapping Cameras," the distortion over the 23-cm-square format of a modern

150-mm focal length mapping lens is close to the limit of accuracy of distortion measurement; that is, it is approximately ± 2 μm.

Calibrated Focal Length

Distortion occurs because of a departure in the image position from the product $f \tan\theta$, where θ is the semifield angle, and f is the focal length of the lens that yields the highest AWAR (Area Weighted Average Resolution). Generally this focal length, f, does not balance the distortion across the format. In making measurements from mapping photography, a focal length f_c is selected to scale the measurements to ground distances and at the same time to redistribute the distortion across the image plane. (Several different values of f_c could be used to compensate accurately for the distortion, but this is too tedious an operation for general mapping work.) *The symbol f_c is referred to as the calibrated focal length, and is usually chosen to equate the maximum values of positive and negative distortion across the diagonal of the format.* The choice, however, is arbitrary, and for some mapping applications the distortion has been balanced "across the flats," i.e., along a line connecting the midpoints of opposite sides of the format.

When discussing calibrated focal length, it must be remembered that this refers to a fictitious focal length, which is assumed in scaling measurements from photography taken at another focal length. For a nominal 150-mm focal-length lens, the calibrated focal length is usually no more than about 1 mm different from the effective focal length.

Tangential Distortion

Radial distortion, a true optical aberration, is distinct from tangential distortion, which is due to errors of tilt and decentering introduced during fabrication and assembly of the lens. For a modern mapping lens, tangential distortion should not amount to more than about 3 or 4 μm at any point across the format.

Calibration Facilities

The measurement of radial and tangential distortion to the precision required for modern mapping lenses is a difficult and time-consuming undertaking. The best approach is to use a collimator bank such as is set up at the National Research Council (Carman, 1969); the U.S. Geological Survey (Karren, 1968; Tayman, 1974; Washer, 1965); and Ogden Air Force Base, Utah. At these facilities distortion testing is conducted on a routine basis, and the subtle systematic errors introduced by the instrumentation have been determined. Other methods, such as rotating the lens on a nodal slide in front of a collimator, photographing a star field (Beccasio, 1971; Fritz and Schmid, 1974), or photographing the grid

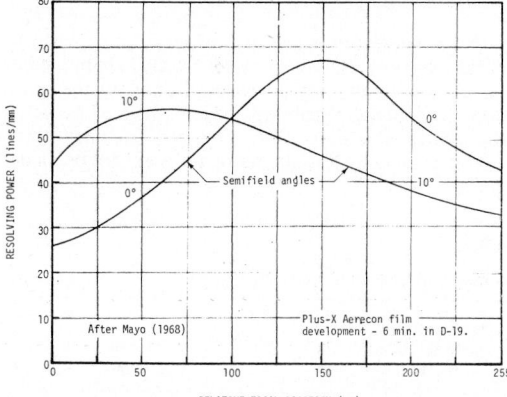

Fig. 6-14. Resolving power as a function of focal position for a 46-cm $f/4.0$ Bell and Howell lens. High contrast target illuminated by 2-ms xenon strobe.

pattern from crossed diffraction gratings (Voggenthaler, 1972), are used from time to time. With care these methods yield accuracies commensurate with the collimator bank approach, but there is no question that measurements of this nature are most efficiently conducted at facilities where such measurements are made on a routine basis.

ATMOSPHERIC EFFECTS

The effect of the atmosphere on aerial and space photography of the earth can be divided into the four general categories of scattering, absorption, refraction, and turbulence. Of these, scattering is the dominant effect in the great majority of situations.

Atmospheric scattering will be treated here only in terms of its gross effects on the contrast and spectral signatures of ground objects as viewed from altitude. A fuller account has been given in Chapter 5. Radiation from the Sun incident on the surface of the Earth is scattered in its passage through the atmosphere. Thus shadows on the Earth's surface are weaker than those on the lunar surface, simply because the Earth is surrounded by a scattering atmosphere. In the Earth's atmosphere, part of the incident radiation is backscattered, and part of the radiation reflected from the ground scene is forward scattered. The net effect of multiple scattering is that the components scattered into the camera reduce the contrast of the image as they do not contain information regarding the object; a reduction in image contrast reduces the spatial resolution of the photography.

Scattering

The scattering that occurs in the atmosphere can be described in terms of a combination of scattering from particles in three broad size ranges (Table 6-3). Even at its clearest, the atmosphere does not approximate a Rayleigh atmosphere. It has been reported (Curcio, 1961) that a clear real atmosphere scatters according to a $\lambda^{-0.7}$ to $\lambda^{-2.0}$ law, as shown in the cross-hatched portion of Figure 6-15. A clear atmosphere with a large component of nonselective scattering typically shows 50% more scattering in the blue spectral areas than in the red. Gross differences occur depending on the humidity and dust content; thus, a tropical maritime atmosphere will scatter more than a continental polar atmosphere, and the scattering

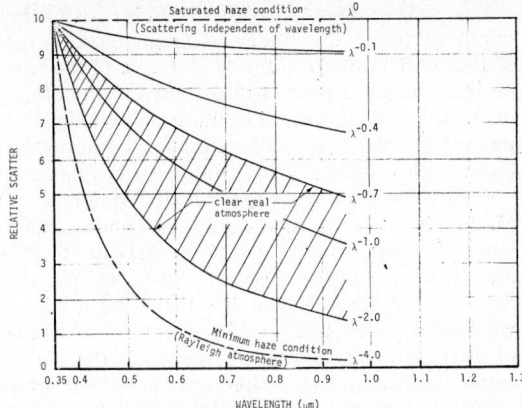

Fig. 6-15. Relative scatter as a function of wavelength for various magnitudes of atmospheric haze.

in a continental desert atmosphere will depend strongly on the dust content of the winds in the lower atmosphere. The wavelength dependence of atmospheric scattering accounts for the blue of the sky and the red of the Sun when seen through a long atmospheric path.

Scattering causes the atmosphere to have a radiance of its own. The effect of atmospheric radiance on aerial photography, as might be expected, is a function of many variables—camera altitude; nature, concentration and size-distribution of atmospheric aerosols; solar altitude; spectral sensitivity range of the camera; angle of view from nadir and its azimuth with respect to the Sun; and polarization.

An increase in the camera altitude or atmospheric haze causes an increase in the atmospheric radiance incident at the camera. Results of high-altitude photography (Fig. 6-16) in which a filter was used over the camera to remove blue radiation, show that when the solar altitude decreases from 90° (Sun immediately overhead), the atmospheric luminance[11] increases as the result of the larger effective scattering volume caused by the longer path of solar radiation through the at-

[11] It is necessary to use photometric terms when reporting work in which the results are given in photometric units without detailed spectral data. In the above case the spectral distribution of the scattered radiation, the spectral sensitivity of the film, etc., are unspecified, so the conversion to radiometric units cannot be made.

TABLE 6-3

Atmospheric Scattering Processes

Scattering Process	Wavelength Dependence	Approximate Particle Size (in λ)	Kind of Particles
Rayleigh	λ^{-4}	$\leqslant 1$	Air molecules
Mie	λ^{0} to λ^{-4}	0.1 to 10	Smoke, fumes, haze
Nonselective	λ^{0}	>10	Dust, fog, cloud

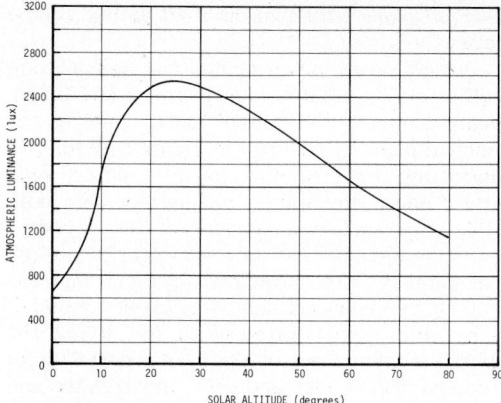

Fig. 6-16. Atmospheric luminance from high-altitude photography with a minus-blue filter as a function of solar altitude angle.

mosphere. The atmospheric luminance at a solar altitude angle in the neighborhood of 20° to 30° reaches a maximum of about twice the value reached at a solar altitude of 90°. For smaller angles, the atmospheric luminance decreases because increasing absorption in the lengthening atmospheric path begins to be the governing factor.

In looking vertically down through the whole atmosphere, the degree of visibility is often categorized in the broad regions listed in Table 6-4, in which cd refers to candela.

For comparison, this is about the same range of atmospheric luminances as encountered for a clear atmosphere at an altitude of 6 km when the off-nadir angle varies from 0° to 90° in the direction into the Sun and the solar altitude is 50° (Fig. 6-18b).

The spectral sensitivity range of the camera is an important consideration because of the increase in scattering that occurs as the wavelength of the radiation decreases (Fig. 6-15). For this reason, minus-blue filters are used in front of the lens for black-and-white high-altitude photography, and, depending on the haze conditions, a choice is made of partially attenuating blue filters for use in front of the lens for high-altitude color photography, unless the filter is included as an integral part of the film.

Boileau (as reported by Duntley et al, 1964) determined atmospheric luminance as a function of altitude from airborne photometers at about mid-

day, with solar altitude of 48.5°, on February 28, 1956, over a region south of Crestview, Florida. The day was clear (cloudless), but with pronounced haze in the first 1.2-km altitude. Evidence is seen for the presence of a haze layer in the plot of altitude versus transmittance shown in Figure 6-17. The atmospheric path luminance can be determined from Boileau's data by using the following relationship (Duntley et al, 1957),

$$L_P(z,\theta,\phi) = L_A(z,\theta,\phi) - L_O(z,\theta,\phi)\tau(z,\theta),$$
$$(6\text{-}35)$$

where:

z = altitude,
θ = angle to nadir; actually, Duntley refers to θ as the zenith angle, but Coulson's use of nadir angle is preferred (Coulson, 1966),
ϕ = azimuth angle,
$L_P(z,\theta,\phi)$ = path luminance,
$L_A(z,\theta,\phi)$ = apparent luminance,
$L_O(z,\theta,\phi)$ = inherent luminance,
$\tau(z,\theta)$ = beam transmittance.

Using this relationship, plots were made, as shown in Figures 6-18a, b, and c, of the polar distributions of atmospheric path luminances at altitudes of 1.5, 6, and 18 km (the latter is an extrapolation also described by Duntley et al, 1957). These polar distributions show the marked asymmetry of atmospheric luminance in the azimuth into the Sun ($\phi = 0°/180°$).

Polarization

No quantitative measurements seem to have been reported on polarization effects produced by atmospheric scattering for the downward-looking case. Coulson (1966) has made predictions based on theoretical studies and laboratory measurements that show the polarization dependence in terms of the variation of *contrast transmission coefficient* with viewing angle to the nadir in the

TABLE 6-4

Categories of Atmospheric Visibility

Visibility	Atmospheric Luminance
Very clear	2500 cd m^{-2}
Light haze	3500 cd m^{-2}
Medium haze	5000 cd m^{-2}
Heavy haze	7000 cd m^{-2}

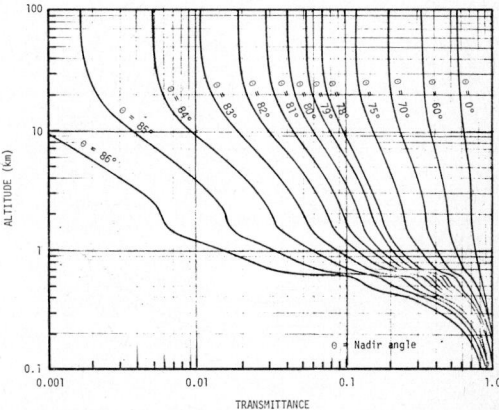

Fig. 6-17. Atmospheric beam transmittance for nadir angles shown, based on data by Boileau, as reported by Duntley et al. (1964).

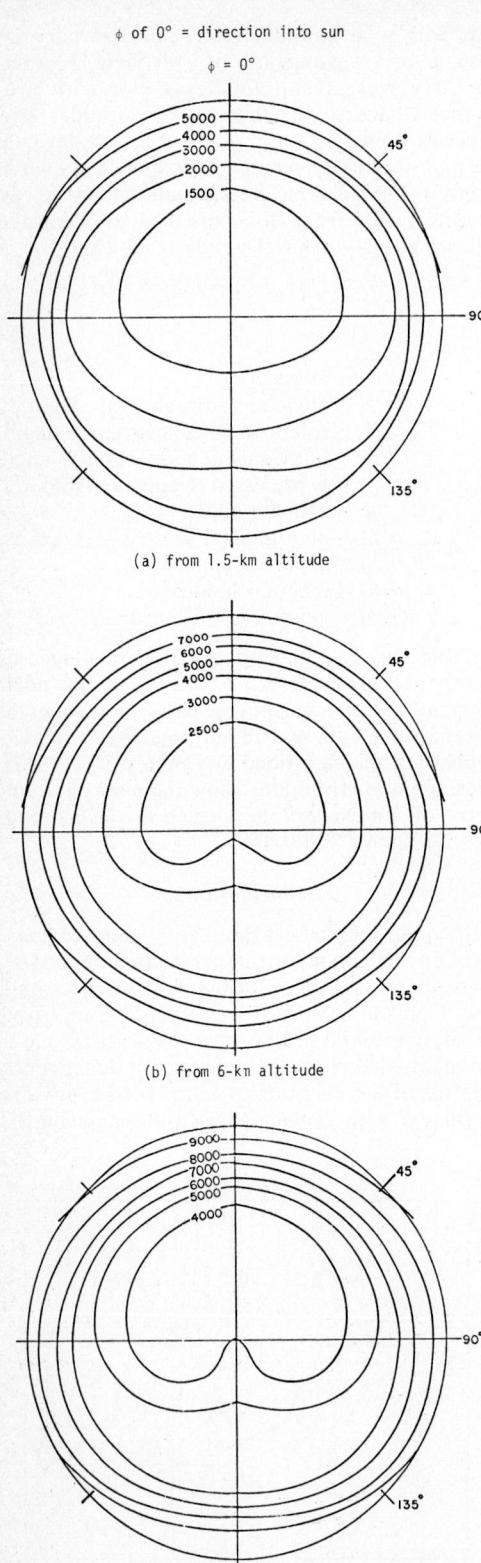

φ of 0° = direction into sun

φ = 0°

(a) from 1.5-km altitude

(b) from 6-km altitude

(c) from 18-km altitude

Fig. 6-18. Polar luminance (cd/m²) distribution, looking downward from various altitudes, from data by Boileau, reported by Duntley et al. (1957).

principal plane[12] for two different ground reflectances.

Coulson's data, covering contrast transmission coefficients for total radiant intensity and for intensity components parallel and normal to the principal plane, are shown in Figure 6-19 for two contrasting areas of reflectance—a desert sand surface and a green grass turf surface. For both reflectance surfaces considered, the use of a polarizing filter to pass only the intensity component parallel to the principal plane can increase contrast by maximum factors of about 1.5 and 3, respectively, compared with not using the polarizing filter. Coulson does not report data for azimuths of φ = +90° and −90°, in which the tendency should be for the improvement in the contrast transmission factor to increase with increasing angles to nadir. The drawback to the use of a polarizing filter to improve contrast is that the polarizing filter reduces the amount of radiation from the ground object typically by a factor of about three because of the combined effect of polarization and absorption by the filter. There is a gain in aerial image contrast, but there is an accompanying increase in the resolving power of the camera only if the longer exposure time required does not cause an offsetting amount of image smear.

In summary, the use of a polarizing filter to improve the resolution of high-altitude photography does not seem to have been completely explored. Likewise, the use of a polarizing filter as an aid to discriminating between or identifying terrain features does not seem to have been adequately exploited. Although the work of Coulson, Curran (1979), Solomon (1981) and Walraven (1977) indicate that the use of a polarization-spectral signature may be a useful refinement to multispectral remote sensing.

Contrast Reduction

As indicated earlier, the effect of atmospheric luminance is to increase the illuminance of the camera image plane and, at the same time, to reduce the contrast of the image. Thus, the contrast ratio and modulation of the scene are given by combining Eq. 6-15 with Eq. 6-21, and then with Eq. 6-24, to yield

$$C = \frac{E\rho_{max}\tau_A + \pi L}{E\rho_{min}\tau_A + \pi L} \qquad (6\text{-}36)$$

and

$$M = \frac{\rho_{max} - \rho_{min}}{\rho_{max} + \rho_{min} + \dfrac{2\pi L}{E\tau_A}}, \qquad (6\text{-}37)$$

where:

E = illuminance, in lux, due to Sun and sky,

[12] The principal plane is the vertical plane containing the sun, the ground target, and the observer; it is in the φ = 0°/180° azimuth where φ = 0° represents the direction into the sun.

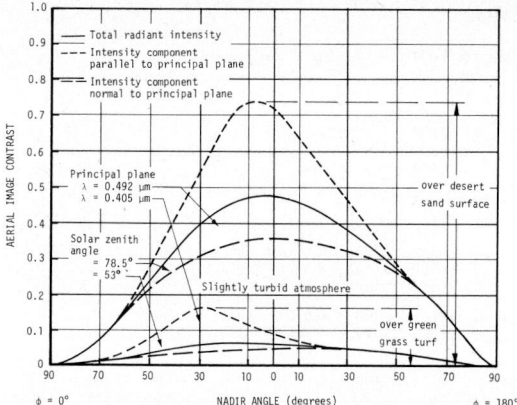

Fig. 6-19. Polarization dependence in terms of contrast transmission coefficients for two different reflectances and a slightly turbid atmosphere, from Coulson (1966).

ρ_{max} = reflectance of scene highlight areas,
ρ_{min} = reflectance of scene lowlight areas,
τ_A = atmospheric transmittance,
L = atmospheric luminance, in cd m^{-2}.

To appreciate the effect of the atmospheric scattering on apparent ground scene contrast, the contrast values encountered with and without an atmosphere are determined. First, consider the list (Table 6-5) of directional luminous reflectances as compiled by Gordon and reported by Duntley et al (1964) and amplified by Gordon and Church (1966).

Although it is seen that high contrast ratios can occur, it is also obvious from Table 6-5 that the majority of contrast ratios are less than 5:1. This table does not take into account the effect of shadow. Shadow increases contrast values in a scene, and can be particularly useful in the photointerpretation of manmade objects. Shadow is also an aid in geological studies, and low solar altitudes are often requested in photogeology; however, in agriculture, forestry, etc., the presence of shadow can be disadvantageous.

As an example then, let us take a ground contrast ratio of 5:1 between say, rock at 0.2 reflectance, and forest at 0.04 reflectance with a solar altitude of about 50°. By reference to Figure 6-20, it is seen that this solar altitude, E, corresponds to an illuminance value of 8.6×10^4 lux. Figure 6-16 shows that the atmospheric luminance for this solar altitude is about 2,000 cd m^{-2}. If an atmospheric transmittance coefficient of 0.8 is assumed, then Eq. 6-36 gives a constant ratio, C, of

$$C = \frac{8.6 \times 10^4 \times 0.8 \times 0.2 + \pi \times 2 \times 10^3}{8.6 \times 10^4 \times 0.8 \times 0.04 + \pi \times 2 \times 10^3} = 2.2.$$

The effect of atmospheric luminance in this example was to reduce the ground scene contrast from 5:1 to 2.2:1.

Aerial image modulation, as a function of ground target modulation, has been determined by Geltmacher (1964) for four average ground reflectances for the cases of a continental polar air mass and a maritime tropical air mass. Figure 6-21 shows these data, with the polar air mass in solid lines, and the tropical air mass in dashed lines.

Now, with reference to the data on Figures 6-20 and 6-16, which show solar horizontal plane illuminance and atmospheric luminance values as functions of solar altitude, the decrease in scene contrast due to the atmosphere can be determined as a function of solar altitude. Figure 6-22 illustrates the results of such a calculation where a constant lowlight reflectance value of 0.04, and ground scene contrast values of 2.5:1, 5:1, 10:1, and 20:1, have been assumed.

Referring again to Boileau's data as plotted in Figures 6-17 and 6-18b, the aerial image contrast, as a function of angle off nadir, was plotted for ground contrasts of 2.5:1, 5:1, 10:1, and 20:1, assuming that the solar horizontal plane illuminance at a solar angle of 48.5° is 8×10^4 lux, and that the background reflectance is constant at 0.04. These curves, for a camera altitude of 6 km, are shown for the $\phi = -90°/90°$ azimuth in Figure 6-23 in solid lines, and for the $\phi = 0°/180°$ azimuth, as dashed lines in the same figure. The asymmetry in the $\phi = 0°/180°$ azimuth is again noticeable.

Two general characteristics of atmospheric contrast reduction are worthy of note: first, the effect of atmospheric luminance is to reduce nonlinearly the apparent contrast of the scene; and secondly, referring to Eq. 6-36, it is seen that, for a given scene contrast and atmospheric luminance, as the lowlight reflectance is decreased, the resulting atmospheric contrast reduction is correspondingly increased.

Comparatively little work has been done to determine the wavelength dependence of atmospheric contrast reduction as a function of altitude for "average" real atmospheres. Mazurowski and Walker (1962) and Mazurowski et al (1963) report on measurements made up to altitudes of about 10 km, and some of their results are summarized in Figure 6-24.

Atmospheric Refraction and Ground Positional Errors

A correction has to be made to the position of the image of ground objects in precise mapping applications because of refraction through the atmosphere. The higher the altitude and the greater the angle off nadir, the larger the correction.

Schut (1969) has presented tables and equations that allow the photogrammetric error due to atmospheric refraction to be calculated as a function of camera altitude, ground elevation, and angle off nadir. An example of the magnitude of the photogrammetric error is given in Table 6-6. It is interesting that the error introduced by the tilt of the camera off nadir is small compared with the error introduced by atmospheric refraction. At an altitude of 6 km, the effect of omitting the positional error is smaller than the above tilt error because of the curvature of the earth. As a practical point,

TABLE 6-5

Directional Luminous Reflectance of Terrain Backgrounds

Description	Sun zenith angle	Azimuth of the path of sight relative to the sun	Nadir angle of view 0°	15°	30°	45°5	60°	75°	80°	85°
1. Pine trees, small, uniformly spaced. Data are for unresolved terrain over which atmospheric data given were collected.	41.5°	0°	0.0333	0.0241	0.0214	0.0214	0.0261	0.0379	0.0463	0.0859
		45		0.0222	0.0202	0.0194	0.0210	0.0303	0.0387	0.0549
		90		0.0315	0.0311	0.0317	0.0317	0.0337	0.0387	0.0463
		135		0.0335	0.0382	0.0392	0.0387	0.0438	0.0463	0.0572
		180		0.0402	0.0444	0.0578	0.0640	0.0711	0.0758	0.0825
2. Grass, thick, rather long, pale green, dormant, dryish, little ground showing.[a]	41.5	0	0.088	0.081	0.076	0.077	0.088	0.094	0.096	0.094
		180		0.098	0.119	0.146	0.150	0.153	0.153	0.160
3. Asphalt, oily, with dust film blown onto oil.[a]	42.0	0	0.061	0.057	0.058	0.060	0.068	0.090	0.104	0.127
		180		0.067	0.080	0.101	0.090	0.086	0.086	0.088
4. "White" concrete, aged.[a]	42.2	0	0.266	0.263	0.254	0.254	0.266	0.298	0.320	0.374
		180		0.289	0.313	0.343	0.367	0.350	0.343	0.320
5. Calm water, intimate optical depth.[b]	41.5	0	0.0222	0.0234	0.0297		0.0569	0.139	0.267	0.461
		45		0.0230	0.0240	0.0272	0.0357	0.107	0.199	0.325
		90		0.0221	0.0222	0.0234	0.0293	0.0711	0.121	0.214
		135		0.0213	0.0212	0.0220	0.0270	0.0665	0.113	0.203
		180		0.0214	0.0212	0.0216	0.0267	0.0718	0.125	0.254
6. Grass, lush green, closely mowed thick lawn.[c]	40.4	0	0.100	0.096	0.098	0.108	0.120	0.149	0.168	
	39.6	90		0.103	0.110	0.121	0.138	0.159	0.168	
	39.6	135		0.107	0.125	0.148	0.166	0.178	0.178	
	39.9	180		0.109	0.109	0.119	0.122	0.125	0.125	
7. Macadam, washed off and scrubbed.[c]	48.5	0	0.113	0.115	0.119	0.128	0.148	0.194	0.229	
	60.1	90		0.110	0.109	0.116	0.122	0.139	0.147	
	46.0	180		0.126	0.141	0.156	0.166	0.172	0.176	
8. Dirt, hard packed, yellowish.[c]	53.2	0	0.243	0.230	0.229	0.239	0.252	0.300	0.330	
	56.5	90		0.243	0.258	0.260	0.276	0.300	0.304	
	51.1	180		0.272	0.313	0.370	0.422	0.432	0.434	
9. Mixed green forest, deciduous (oak) and evergreen (pine).[d]	39.0°	0°	0.0360	0.0325	0.0291	0.0205	0.0205	0.0342		
	37.0	180		0.0410	0.0493	0.0493	0.0820	0.263		
10. Pine forest.[d]	33.5	0	0.0385	0.0385	0.0308	0.0246	0.0246	0.0200		
11. Grass, dry meadow, dense, mid-summer.[e]	45	0	0.0955	0.0897	0.0960	0.0952	0.108	0.129		
	45	90		0.0778	0.0890	0.101	0.111	0.130		
	45	180		0.116	0.131	0.143	0.153	0.170		
	45	270		0.107	0.121	0.134	0.137	0.132		
12. Ilyas, sparse and dry, yellowish grass on sand at end of summer.[e]	40	0	0.231		0.320		0.342	0.356		
	40	90			0.163		0.176	0.198		
	40	180			0.295		0.353	0.359		
	40	270			0.262		0.237	0.229		
13. Sand dunes, sharply expressed micro-relief, dry.[e]	40	0	0.288		0.183		0.337	0.353		
	40	90			0.284		0.329	0.306		
	40	180			0.246		0.259	0.276		
	40	270			0.278		0.410	0.281		
14. Podsol, ploughed, moist.[e]	50	0	0.0600	0.0680	0.0646		0.0555			
	50	90		0.0662	0.0953	0.0715	0.0614	0.0761		
	50	270		0.149	(0.180)[f]		0.168	0.168	(0.180)[f]	
15. Pasture meadow at end of summer (K No. 75-77)[e]	45	0	0.0822							
	45	90					0.0882			
	45	180					0.0968			
16. Meadow (with clover and timothy)—dense growth, with flowers, mid-summer (K No. 82-84)[e]	45	90					0.135	0.154[g]		0.143
17. Meadow—with crow foot, dense grass with abundant flowers (K No. 89-91)[e]	45	90					0.108	0.105[g]		0.133
18. Dry meadow—sparse low grass (K No. 135-137)[e]	45	90	0.0845				0.0879	0.120		
19. Dry meadow—more dense low grass (K No. 138-140)[e]	45	0	0.0796							
	45	90					0.109			
	45	180					0.144			
20. Oats—with spikes (K No. 186-188)[e]	45	90					0.0866	0.130[g]		0.168
21. Millet—ripening (K No. 198-200)[e]	45	0					0.0716			
	45	90					0.0593			
	45	180					0.114			
22. Wheat—before harvesting (K No. 201-206)[e]	50	0					0.0895	0.274[g]		0.371
	50	180					0.111	0.292[g]		0.385
23. Wheat—in the flowering period (K No. 207-211)[e]	50	0					0.0702	0.122[g]		0.268
	50	90						0.0863[g]		0.0980
24. Wheat-after mowing (K No. 212-213)[e]	50	90					0.133[g]			0.152

TABLE 6-5 (Continued)

Description	Sun zenith angle	Azimuth of the path of sight relative to the sun	Nadir angle of view 0°	15°	30°	45°5	60°	75°	80°	85°
25. Barley–spiked (K No. 224-226)[e]	45	90				0.113	0.155[g]			0.144
26. Sand dunes–with sharply expressed microrelief, dry, no shadows (K No. 248-254)[c]	40	90	0.250				0.370		0.375	0.345
	40	270					0.222		0.243	0.233
27. Soil–podsol–ploughed, dry (K No. 287-289)[e]	45	0				0.105				
	45	90				0.189				
	45	180				0.179				
28. Soil–sandy loam, ploughed, moist (K No. 295-298)[e]	45	0				0.203				
	45	90				0.112				
	45	180				0.212				
	45	270				0.180				
29. Soil–sandy loam, ploughed, dry (K No. 299-301)[e]	45	0	0.153			0.159				
	45	90				0.159				
30. Soil, black earth, rich, ploughed, wet (K No. 304-306)[e]		0	0.0204			0.0574				
		180				0.0614				
31. Soil, black earth, rich, ploughed, dry (K No. 307-310)[e]	45	0	0.0291			0.0339				
	45	90				0.0403				
	45	180				0.0466				
32. Dirt–flat desert road freshly bulldozed to remove encroaching sage[h]	42	90	0.226	0.229	0.229	0.234	0.247	0.261	0.271	0.270
33. Oiled dirt–flat desert road, oiled, with a light covering of dust[h]	42	90	0.100	0.102	0.102	0.104	0.109	0.118	0.121	0.124

[a] These terrains were measured on the ground by means of a goniophotometer beneath, and during the collection of the data by Duntley, 1964.

[b] Computed from equations by Duntley (1964) for the lighting condition prevailing for items 1 and 2 in this table.

[c] Data taken with a goniophotometer, 10 October 1956.

[d] Data taken with a photoelectric telephotometer from a helicopter at 300 ft. (91.4-m) altitude, mountain forested area near Julian, California, 23 September 1959.

[e] Luminous directional reflectance for terrains 11 through 31 (Brown, 1952) were computed from spectrophotometrid data by Krinov (1953) using C.I.E. Illuminant B. Disparity between data for azimuths 90° and 270° "is explained apparently by the direction of shallow furrows in relation to the sun."

[f] Parentheses indicate estimates based on incomplete spectral data.

[g] The zenith angle of the path of sight is 115°.

[h] Data taken with a goniophotometer at Naval Ordinance Test Station, China Lake, California, on 16 July 1962. The spectral reflectance of a sample of the dirt was measured with a Hardy spectrophotometer. Using C.I.E. Illuminant B, chromaticity coordinates were $x = 0.370$, $y = 0.361$, $z = 0.269$; dominant wavelength $= 0.58 \ \mu m$, and excitation purity $= 10\%$.

Schut notes that the positional errors introduced by a real atmosphere can vary by 10% from those calculated from a standard atmosphere. However, because of the small size of the refraction errors involved, the uncertainty of 10% is a negligible quantity.

Colvocoresses (1970) examined the effect of tilt on the accuracy of mapping from space with reference to a small field camera such as the return beam vidicon cameras used on Landsats 1 and 2. He showed that by properly scaling and rectifying the imagery, the errors caused by the Earth's curvature could be reduced to 50 m at a distance from nadir of 130 km and an altitude of 900 km (this

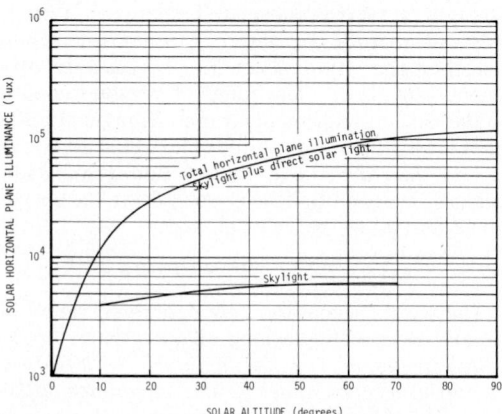

Fig. 6-20. Solar horizontal-plane illuminance for sky light alone and solar plus sky light as a function of solar altitude (Brown, 1952; Duntley, et al., 1964).

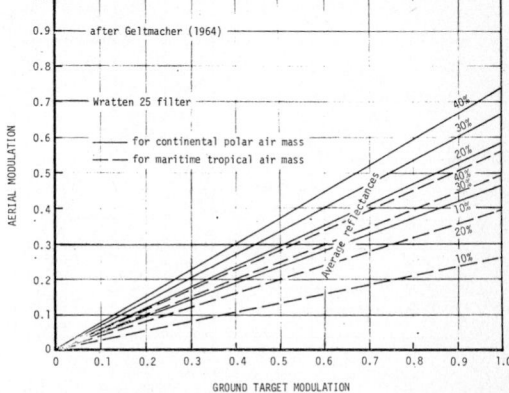

Fig. 6-21. Aerial modulation above 9 km as a function of ground target modulation for continental polar and maritime tropical air mass (Geltmacher, 1964).

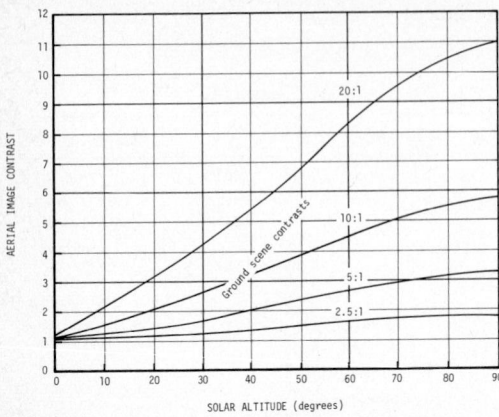

Fig. 6-22. Aerial image contrast as a function of solar altitude for the ground contrasts indicated.

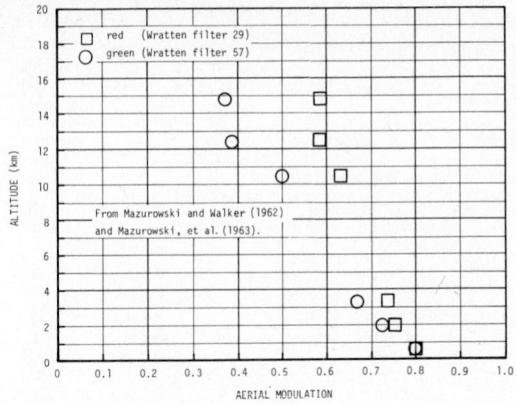

Fig. 6-24. Contrast reduction in the red and green spectral bands as a function of altitude.

does not include camera distortion errors). He also showed that with an angle of 1° to the vertical, a point in the corner of a 16° diagonal frame will be differentially displaced, in relationship to the center of the photography, by about 450 m. However, if the angle to the vertical is *known,* then optical or mechanical rectification can be performed on the imagery. Colvocoresses states that if the angle is known to 0.14°, then the displacement error will amount to about ±50 m, when coupled with the residual curvature effect in rectification.

Effect of Atmospheric Turbulence on Spatial Resolution of Aerial Cameras

Atmospheric turbulence lowers the spatial resolution of aerial cameras and becomes more significant at higher altitudes until its effect asymptotically decreases as the altitude increases above 10 km. The decrease of spatial resolution is more pronounced when there is a greater degree

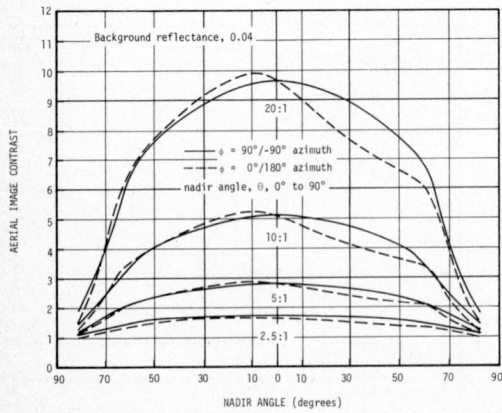

Fig. 6-23. Aerial image contrast reduction for azimuths of $\phi = 90°/-90°$ and $0°/180°$, and for nadir angles of $\theta = 0°$ to 90° as functions of object contrast for fixed background reflectance of 0.04.

of turbulence in the atmosphere. For low angular resolving power systems, for example, mapping cameras with angular resolving power of the order of 20 mrad/cycle, the effect of atmospheric turbulence is negligible. For systems of high angular resolving power, say 0.2 mrad/cycle, the effect of atmospheric turbulence may reduce the spatial resolution by an order of magnitude.

Hufnagel (as reported by Scott et al, 1965) has derived the following expression for the reduction in image modulation caused by atmospheric turbulence

$$<\tau(\nu)> = \tau_0(\nu) \exp \left[-(2\pi\sigma f\nu)^2 \right], \quad (6\text{-}38)$$

where

$<\tau(\nu)>$ = average MTF,
ν = spatial frequency in the image plane in cycles/mm,
f = focal length (in millimeters),
σ = seeing strength factor of atmosphere.

This is based on the use of a system of MTF $\tau_0(\nu)$ to photograph through an atmosphere of seeing strength factor, σ. The value for the strength factor σ is chosen to be between 0.15, or 0.9 rad/altitude in meters, depending upon whether the atmosphere is photographically "good" or "bad." This range of σ values applies to the case of nadir photography from an altitude of at least 10 km. For altitudes less than 10 km, σ is less than the above values. For an off-nadir angle θ, σ is multiplied by $\sec^{1/2}\theta$ for the image plane normal to the viewing angle.

SPECTRAL SIGNATURES

The word "signature," by definition, refers in general to a distinguishing characteristic. *In remote sensing, a "spectral signature" of a feature comprises a set of values for the reflectance or the radiance of the feature, where each value corresponds to the reflectance or radiance averaged over a different, well defined, wavelength interval.* Generally, because of limitations in sensor

TABLE 6-6

Radial Error Due to Atmospheric Refraction
(after Schut, 1969)

Angular distance from camera axis (degrees)	Wide-angle camera f = 152.4 mm Altitude = 6 km		Super-wide-angle camera f = 88.2 mm Altitude = 6 km	
	Tilt 0° (μm)	Tilt 2° (μm max)	Tilt 0° (μm)	Tilt 2° (μm max)
9	− 1.5	− 1.8	− 0.9	− 1.0
18	− 3.2	− 3.6	− 1.9	− 2.1
27	− 5.7	− 6.2	− 3.3	− 3.6
36	− 9.9	−10.7	− 5.7	− 6.2
45	−17.9	−19.2	−10.4	−11.1
59	—	—	−32.5	−35.2

f = camera focal length

sensitivity, the quantity of available film, or (in the case of electro-optical sensors) the data transmission bandwidth, the spectral width of the various wavelength intervals is large, about 0.05 to 0.2 μm in the photographic spectral range. Introductory or basic material on the concept of signatures begins with Chapter 3. Examples for ground materials are found in Chapter 4, and special attention to the atmosphere in Chapter 5.

In the laboratory, a spectral signature (although there it is seldom referred to as such) is obtained from measurements using a spectrophotometer. The measurements give the distribution of reflectance of a material as a function of wavelength compared to that of a diffuse white surface. The spectral resolution of the measurements is generally about 0.001 μm. In comparison consider a case in photographic remote sensing, where the spectral resolution is about 0.1 μm, as defined by the width of the spectral filters used. If four such bands are used, then four discrete spectral intervals, rather than 400 in the case of the spectrophotometer, cover a spectral region of 0.4 μm. Obviously this hundredfold decrease in spectral resolution makes it much harder to discriminate between materials having roughly the same spectral-reflectance properties, in other words, the spectral signature obtained is often not unique. Furthermore, the optical properties of the intervening atmosphere, the photographic geometry used, etc., modify the spectral irradiances recorded by the camera.

In order for the spectral reflectance or spectral radiance of a feature to be determined by a remote sensor, the effect of the atmosphere has to be accounted for, and the spectroradiometric calibration of the camera has to be invoked. The spectral signature, given in terms of *spectral radiance,* is not constant for a given feature because it depends on the varying spectral irradiance due to Sun and sky. A less variable measure is that of *spectral reflectance* as it takes into account variations in spectral irradiance. As discussed previously in this chapter, there are still variations with angle of view and irradiance for a spectral signa-

ture quoted in terms of spectral reflectance values, because of the non-Lambertian properties of the surface and the polarization dependence of the surface and intervening atmosphere. The following discussion describes how the effects of varying atmospheric factors can be minimized or corrected. The correction of spectral signatures for atmospheric effects is discussed by Slater (1980).

AERIAL PHOTOGRAPHIC FILM

AERIAL BLACK-AND-WHITE FILM

This section will consider the performance characteristics of aerial photographic films, with the exception of emulsion and processing chemistry, and the theory of latent image formation, which will not be treated here. The reader is referred to Carroll (1966) for partial summaries of these topics; to Todd and Zakia (1969) for a discussion of sensitometry and tone reproduction; to James and Higgins (1960) and Sturge (1977) for instructional texts; to the SPSE Handbook of Photographic Science and Engineering (1973) and to Mees and James (1966) and James (1977) for a complete treatment of the theory of the photographic process; and to Dainty and Shaw (1974) for discussions on emulsion modeling and photographic image recording. Technical data on Kodak aerial films and plates and the physical properties of Kodak aerial films can be found in publications of Eastman Kodak Company (1972a and b). Useful additional data are published from time to time in Eastman Kodak *Tech-Bits* (1963 to date).

Characteristic Curve

The characteristic curve of processed film is a plot of optical densities against the logarithm of the corresponding exposures, where exposure, H[13], is the product of irradiance and the time during which the irradiance is incident on the

[13] Note that H is now the internationally accepted symbol for exposure; the symbol E is used only for irradiance.

emulsion surface. The idea of presenting the photographic input-output relationship in this form is due to Hurter and Driffield in 1890, and the characteristic curve, or *D*-log *H* curve, is often called the *H* and *D* curve as a consequence. They defined density, *D,* as the logarithm of the opacity, *o,* where opacity is the reciprocal of the transmittance, τ,

$$D = \log_{10} o = -\log_{10} \tau. \qquad (6\text{-}39)$$

A typical characteristic curve is shown in Figure 6-25 for Panatomic-X aerial film. Note the captioned qualifications that must be recorded to give the curve meaning. Figure 6-26 shows plots of τ against *H* and *D* against *H* for the same film and conditions as in Figure 6-25. These linear plots can be useful in the special case of image tube film cameras where a few quanta are recorded, or in other applications, such as photoelectric scanners, where the eye is not involved, and interest is primarily in exposure levels at the threshold of the film.

The various parts of the characteristic curve in Figure 6-25 are commonly referred to in the following manner: The region A to B is the *base-plus-fog density level*. The base usually has a density in the range of 0.01 to 0.02 density units; the fog density is the result of chemical reduction of unexposed grains by the developer. (Note: Photographic processing simply exploits the different rates at which silver halide grains are reduced; thus grains not exposed to radiation are reduced, although at a much slower rate than those grains exposed to radiation.) The region from B to C is referred to as the *toe of the curve* and is the region where, for the processing conditions, the exposure is at or just above the threshold, and the density is distinguishable from the base-plug-fog density. From C to F the curve approximates a straight line. For many aerial films the straight-line portion starts at a density of about 0.8. Several workers have shown that a low-contrast

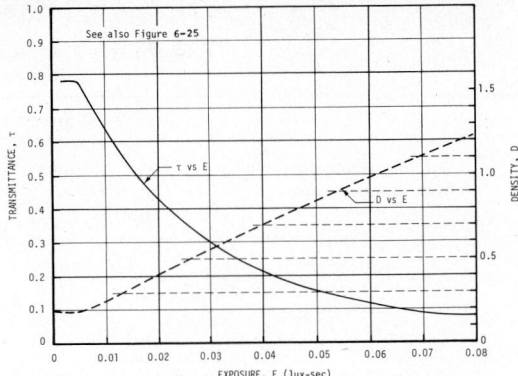

Fig. 6-26. Curve drawn in Figure 6-25 replotted as transmittance-versus-exposure and density-versus-exposure.

image will be recorded at its maximum resolution when the average density is between 0.8 and 1.2. Below this density range the image has reduced contrast, and above this range the increasing granularity[14] reduces the resolution of the image.

The end of the straight line portion of the curve at F typically lies at a density of between 2.5 and 3.5, although some films such as Plus-X have an approximately straight line range between 0.8 and 6.0. The projection of CF onto the abscissa is referred to as the *latitude of the film*. The region from F to G is the *shoulder of the curve,* and the dashed line above G corresponds to D_{max} for the film.

The use of the term *characteristic curve* is misleading, because it gives the impression that a single curve relates density and the logarithm of exposure for a given emulsion. In fact, a large number of factors influence the shape and position of the curve (Slater 1980)—the processing conditions are among the most important. Even though the importance of the physical and chemical conditions may be recognized, it should not be overlooked that density and exposure themselves need to be carefully defined.

Gamma and Contrast Stretch

The slope of the straight-line portion of the characteristic curve is referred to as the *gamma of the processed film* and, by definition, is given by

$$\gamma = \tan \theta = \frac{\Delta D}{\Delta \log H}. \qquad (6\text{-}40)$$

For most aerial films, the manufacturers' recommended processing conditions and the corresponding speeds quoted for the films give gammas in the neighborhood of two. It should be noted that a gamma greater than unity increases the contrast of the recorded image. For medium- and

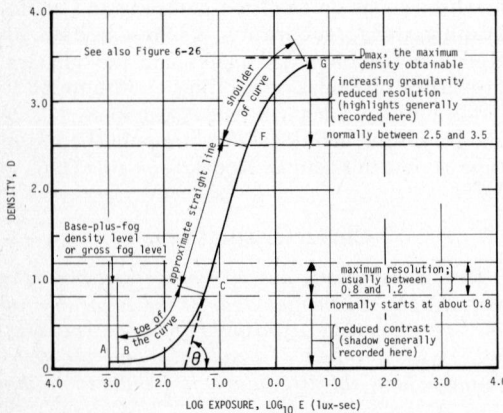

Fig. 6-25. Characteristic curve for Panatomic-X aerial film (daylight illumination, developed for 5 min in D-19 at 20° C in a sensitometric machine).

[14] Granularity is roughly proportional to the square root of density.

high-altitude aerial photography, a gamma of two or three is usually desirable.

Contrast stretch, a well-known procedure in digital image processing, can be implemented photographically using high gamma duplicating film. In some applications, extremely high gamma duplicating film is useful. Such film has an essentially binary response from clear to high density without intermediate gray tones. An interesting analysis of a photographic transparency can be made by contact printing or enlarging the original transparency onto high gamma duplicating film. If this is done on a number of pieces of film, each being given a different exposure, the analyst will be able to discriminate very subtle contrast differences, at different density levels in the original photograph, that would not otherwise be detectable.

Density

When defining density, it is necessary to describe the spectral distribution of the radiation used in the measurement and the incident and collection cone angles. There are basically three different geometries for density measurements: *Specular* – when the incident and collection cone angles approach zero; *diffuse* – when the incident cone angle approaches 0° and the collection cone angle approaches 90° (hemispherical), or vice versa; and *double diffuse* – when the incident and collection angles both approach 90°.

A density measured specularly has a higher value than the same density measured diffusely, simply because part of the transmitted beam is scattered so as not to fall on the detector. The ratio of specular to diffuse density is referred to as *Callier's Q*. For a density of about 1, Callier's Q can vary between about 1.1 and 1.5, depending on the processing and measuring conditions and the emulsion type.

Determination of Characteristic Curve

To determine the characteristic curve, the film must be irradiated with a series of known, different exposures that do not overlap. There are two simple ways to do this: One is to keep the level of irradiance fixed and to vary the exposure time; the other is to keep the exposure time fixed and vary the level of irradiance, usually by the use of a calibrated step tablet of density.

Unfortunately, these two approaches do not always give the same result. This is referred to as *reciprocity law failure;* the discrepancy is small enough to be neglected in general for most modern aerial films for exposure time from 1 to 100 ms. However, a check should be made for precise work, or before extrapolating beyond the above exposure time range.

It is also necessary to be careful regarding the color temperature of the irradiance. In aerial photography it is usual to expose the film in a sensitometer, in which a light source with an *effective color temperature* of approximately 6000 K is used for an exposure time of 10 msec. The color temperature of 6000 K is chosen to match the average color temperature of clear daylight irradiance. Although it is convenient to refer to the well-established color temperature of 6000 K, there is some question as to its practical value. For example, for the photography of predominantly agricultural crops or oceanographic areas, it may be advantageous to change the color temperature of the sensitometry to match more closely the spectral distribution of the radiant flux reflected from the surface being photographed.

Listing of Films

It is appropriate at this point to list the black-and-white aerial films commonly used in remote sensing, and to include with them the aerial color films to be discussed later. A general conclusion can be drawn regarding photographic films, partially from studying this list in Table 6-7: *the slower the speed of the film, the higher are the resolving power and the gamma, while the granularity and the latitude are less.*

Some of the terms not so far mentioned (e.g., film speed and granularity) will now be defined, and then emulsion MTF and threshold modulation curves will be discussed.

Other Basic Parameters

Film Speed

Photographic speed values are simply a convenient way of comparing the average sensitivities of different photographic emulsions. Because there are many distinctly different applications for photography, there are several different systems for measuring and using photographic speeds. There are both absolute speed systems and relative speed systems. The relative speed systems are usually used when an exposure measurement is difficult to make, as in the photography of an oscilloscope trace. This discussion will be restricted to *aerial film speed* (AFS).

It is important to realize that there is no such thing as a conversion factor between one speed system and another. Furthermore, without specifying the so-called *speed point* for a film, it is not possible to determine the speed of the film, or to compare the speeds of two films. To illustrate these statements, let us try to determine the speeds of two emulsions with the characteristic curves A and B, as shown in Figure 6-27. It is seen first that the curves have different contrasts or gammas, and that they intersect at a point 0.6 density unit above gross fog. At a speed point 0.1 density unit above gross fog, film A is four times the speed of film B. At 0.6 density unit above gross fog the speeds are equal, and at 1.0 density unit above gross fog, film B is twice as fast as film A. So the choice of film depends on whether a light or heavy density image is required.

AFS is defined as the reciprocal of two-thirds

TABLE 6-7

Aerial Photographic Films

Film Name	Film Number	Nominal Base Thickness (μm)	Aerial Film Speed (AFS)	Developer or processes	Resolving Power, lines/mm target-object contrast 1000:1	1.6:1	Diffuse RMS granularity[f]
Plus-X Aerographic	2402	100	200	D-19	100	50	19
Tri-X Aerographic	2403	100	650[c]	D-19	80	25	33
Double-X Aerographic	2405	100	320[c]	DK-50	100	50	26
Panatomic-X Aerocon	3410	60	40	D-19	250	80	13
Plux-X Aerocon	3411	60	200	D-19	125	50	19
High-Definition Aerial	3414	60	8	D-19	630	250	9
	1414	40	8	D-19	630	250	9
Panatomic X Aerocon II	3411	60	40	D-19	400	160	9
Infrared Aerographic[a]	2424	100	400	D-19	80	40	30
Aerochrome Infrared[b]	2443	100	40	EA-5	63	32	17
	3443	60	40	EA-5	63	32	17
High Definition	SO-127	100	6[c]	EA-5	160	50	9
Aerochrome Infrared[b]	SO-131	60	6[c]	EA-5	160	50	9
	SO-130	40	6[c]	EA-5	160	50	9
Aerocolor Negative	2445	100	100[c]	Aero-Neg. color proc.	80	40	13
Ektachrome MS Aerographic	2448	100	32[c]	EA-4	80	40	12
Aerial Color	SO-242	60	6	ME-4	200	100	11
	SO-255	40	6	(modified)	200	100	11
Ektachrome EF Aerographic	SO-397	100	64[c]	EA-4	80	40	13
	SO-154	60	64[c]	EA-4	80	40	13
Anscochrome	D/200	100	90[d]	AR-1C	100	50	25
Anscochrome	D/500	100	230[d]	AR-1C	80	40	45

[a] Speeds are for no filter.

[b] Speeds are for a Wrattennumber 12 filter.

[c] Effective film speeds are relative speeds, found by comparison to films whose speeds are determined according to ANSI Standard PH2.34, 1969.

[d] At time of compilation the manufacturer had not released an AFS for this film.

[e] Haze filters such as HF-3, HF-4, and HF-5 may be used without changing exposure.

[f] The RMS-Diffuse granularity value represents 1000 times the standard deviation in diffuse visual density produced by scanning a uniformly exposed and developed sample with an f/2.0 system of 48-μm scanning spot size.

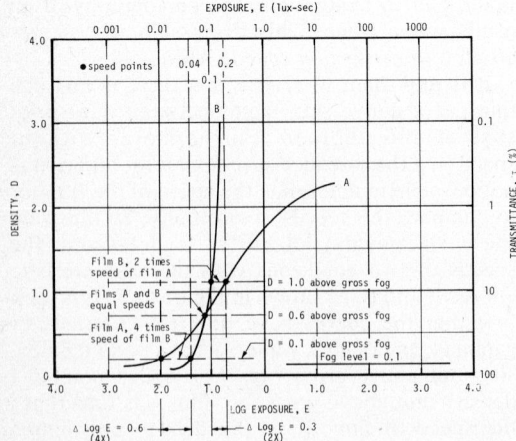

Fig. 6-27. The manner by which a film with a characteristic curve of A can be judged to be faster or slower than another with curve B, depending on the selection of a speed point.

the exposure (in lux-s) at the point on the toe of the characteristic curve having a density 0.3 above gross fog. It applies only to black-and-white films. The speeds of aerial color films are relative film speeds found by comparison with black-and-white aerial films.

Film speed is greatly dependent on both the pre-exposure history of the film and the processing conditions. Prefogging the film increases the speed, and "forcing" the development of the film also increases the speed. Factors of two or three can sometimes be gained in these ways at the expense of decreased resolving power and increased granularity and minimum density values. The speed values listed in Table 6-7 can be used as fairly conservative data; that is, faster film speeds for these films can be assumed under special circumstances.

For publication, film speed values are usually rounded off to a multiple of $(2)^{1/3}$ (about 1.25). It should be remarked that the speed systems described are all arithmetic; thus an emulsion with a speed of 100 is twice as fast and requires half the exposure of a film of speed 50.

Spectral Sensitivity

The spectral sensitivity of most aerial black-and-white films shows what is often referred to as an extended red panchromatic response. That is, the films not only are sensitive throughout the visible spectrum, but also show a peak at about 0.69 μm before falling to about a tenth this peak value at 0.72 μm. As we can see from Figures 6-28, 6-29, 6-30, and 6-31, where the spectral sensitivity is shown as a plot of the logarithm of the sensitivity S[15] against wavelength, the curves show a dip in some cases (e.g., for 3414 in Fig. 6-28) at about 0.45 μm. The general shape of the curve for 3414 is advantageous in aerial photography because it is preferred to use red and green radiation for imaging, rather than blue, which includes a high proportion of atmospherically scattered radiation that reduces the overall contrast of the image. (The sensitivity of 3414 in the range of 0.40 to 0.45 μm can be easily removed by use of a minus-blue filter.) Thus the earliest emulsions, which were blue-sensitive only, and orthochromatic emulsions (blue and green sensitive) are not as suitable as panchromatic or extended red panchromatic emulsions for aerial photography.

Filter Factor

Filters are employed in aerial photography to remove unwanted blue radiation (minus-blue); other spectral filters are used in multiband photography. When using filters, it is necessary either to increase the exposure time or to decrease the f-number in order to compensate for the radiation removed by the filter. The factor by which the exposure with filter is greater than the exposure without filter is known as the *filter factor*.

The filter factor can be determined by a mathematical calculation similar to that following Eq. 6-18, as follows: First, a curve is plotted which is the product of the spectral transmittance curve of the lens, the spectral distribution curve of the unfiltered (usually daylight) irradiance, and the spectral sensitivity curve of the film. The area A_1 under the resulting curve is proportional to the exposure time, which is known by calculation or experiment to be t_1. Second, this curve is multiplied by the spectral transmittance curve of the filter, and the area A_2 under the resulting curve is again determined. Then the ratio A_1/A_2 is the filter factor, and the required exposure time t_2 is given by

$$t_2 = \frac{A_1}{A_2} t_1. \qquad (6\text{-}41)$$

[15] *S is the reciprocal of exposure in erg cm^{-2} to produce a density of 1.0 above gross fog, where 1 W = 10^7 erg s^{-1}.*

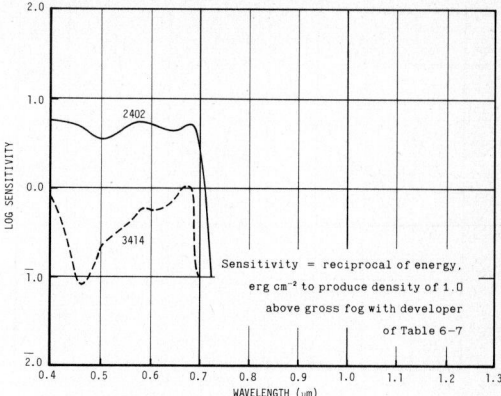

Fig. 6-28. Spectral sensitivity curves for aerial black-and-white films 2402 and 3414.

Granularity

The effect of granularity, which is due to variations in the concentration of silver grains in a processed emulsion, is to cause fluctuations in the amount of light transmitted through the scanning aperture of a microdensitometer. The visual impression due to granularity is graininess—a qualitative descriptor.

In order to obtain a measurement of granularity, a length of the film, for which the granularity is to be found, is uniformly exposed and processed to a density of 1.0. The film is then scanned with a microdensitometer having a scanning aperture 48 μm in diameter. The standard deviation $\sigma(D)$ in the density variations about a density of 1.0 can then be determined from successive readings of density taken from the scan at equal distance intervals greater than d, the width of the scanning aperture in the direction of scan. The most convenient equation to use is

$$\sigma(D) = \left[\frac{n \sum D_i^2 - (\sum D_i)^2}{n(n-1)} \right]^{1/2}, \qquad (6\text{-}42)$$

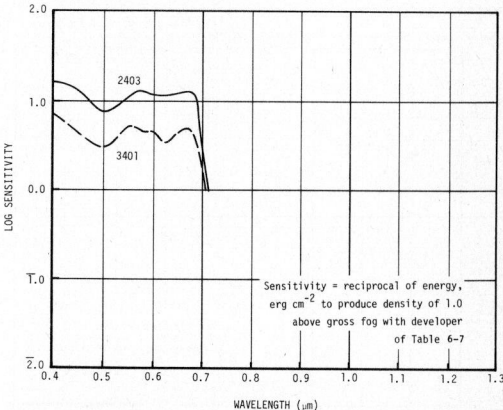

Fig. 6-29. Spectral sensitivity curves for aerial black-and-white films 2403 and 3401.

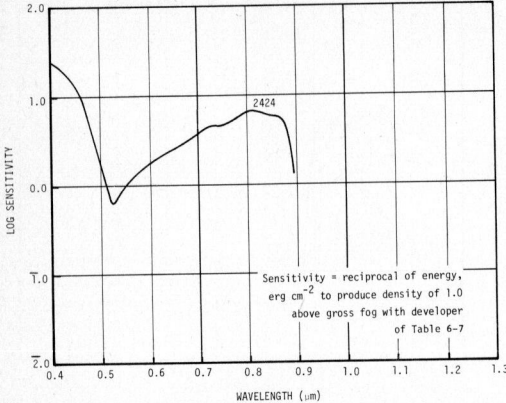

Fig. 6-30. Spectral sensitivity curves for infrared aerial black-and-white film 2424.

where:

D_i = an individual density reading,

n = the number of readings, about 1,000 for good precision.

The *standard deviation* or *granularity*, $\sigma(D)$, obtained can then be used to determine *Selwyn granularity*, G, from

$$G = \sigma(D)A^{1/2}, \qquad (6\text{-}43)$$

where:

A = area of the scanning aperture.

Knowing G, the rms granularity for other scanning aperture areas can be found. It should be noted that, for convenience, the rms granularity values listed in Table 6-7 have been multiplied by a factor of one thousand. Generally, granularity measurements with a 48-μm aperture, when determined in terms of diffuse visual density, will indicate the magnitude of the graininess sensation produced by viewing the diffusely illuminated sample with 12X monocular magnification.

The granularity of a given film depends on the type of developer used, the extent of develop-

ment, and the average density of the developed image, as well as on the size distribution and concentration of the silver-halide crystals. As remarked earlier, granularity increases with density; thus overexposure or overdevelopment should be avoided, and the mean density of low-contrast details should not be much greater than 1.0.

In many cases only tests can determine which is the better combination of emulsion and developer to use to minimize granularity for a specific aerial photography mission. The choice is between the use of a high-speed, coarse-grained emulsion with a so-called fine-grain developer, and a lower-speed, finer-grained emulsion with a standard developer.

A knowledge of the granularity of emulsions is important when comparing them as information storage media. Detailed treatments of this subject, which is outside the scope of this manual, can be found in Higgins (1966) and Jones (1958). Jones states that a useful figure of merit, M, for an emulsion can be determined from

$$M = \frac{K \times film\ speed \times \gamma}{granularity}, \qquad (6\text{-}44)$$

where:

K = a constant

γ = the slope of the straight-line portion of the D-log H curve.

It is often of interest to know the number of *density levels*, N, that can be distinguished in the presence of grain noise for a given emulsion. To make this determination, the *density scale*, DS, must be known. The density scale is the density range between minimum and maximum densities for the emulsion. A measure of the rms variation of density, σ, about several mean densities within the range DS, is also needed. It is necessary to make a choice of the number, K, of standard deviations by which adjacent density levels should be separated—a quantity that is related to the confidence of distinguishing between adjacent levels. The number, N, of available levels is given approximately by

$$N \simeq \frac{DS}{2\ K\ \bar{\sigma}}, \qquad (6\text{-}45)$$

where:

K = number of standard deviations to separate adjacent density levels,

$\bar{\sigma}$ = average value for the σ's for the various mean density levels.

The value of N for film number 3414 scanned by an aperture of 25 μm is about 50 for a K value of unity. The information capacities of four films and high-density magnetic tape have been compared by Slater (1980).

Modulation Transfer Function of Emulsions

It is well established that when a sinusoidal distribution of radiance is imaged by a linear optical

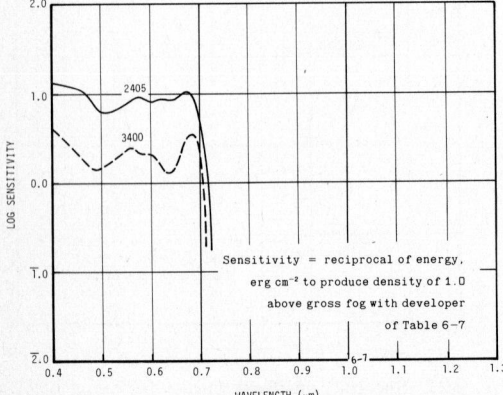

Fig. 6-31. Spectral sensitivity curves for aerial black-and-white films 2405 and 3400.

system or recorded by a detector having a linear response, the resulting output will be an image having a sinusoidal distribution of irradiance. Thus, if it is required to determine the MTF of an emulsion, a sinusoidal variation of radiance at several spatial frequencies can be imaged onto the emulsion. The ratio of image to object modulation as a function of spatial frequency is referred to as the *modulation transfer function* (MTF) of the emulsion.

Mathematically, the process can be explained simply by describing the sinusoidal distribution of radiance $L(x)$ for the object as

$$L(x) = L_0 + L_1 \cos (2\pi \nu x), \qquad (6\text{-}46)$$

where:

L_0 = constant background radiance,
L_1 = amplitude of radiance variation,
ν = spatial frequency of the modulation,
x = distance.

From the definition of modulation quoted earlier in this chapter, it can be seen that the modulation, M_O, of the object is given by

$$M_O = \frac{L_1}{L_0}. \qquad (6\text{-}47)$$

The MTF of an emulsion at spatial frequency, ν, is given as $\tau(\nu)$; thus

$$\tau(\nu) = \frac{M_E(\nu)}{M_O(\nu)}, \qquad (6\text{-}48)$$

where:

$M_O(\nu)$ = modulation of the aerial sinusoidal distribution as it is incident on the emulsion,
$M_E(\nu)$ = modulation of the sinusoidal distribution within the emulsion during the exposure.

Obviously, by varying ν a complete plot can be obtained of the emulsion transfer function as a function of spatial frequency.

The main difficulty encountered in accurately determining the MTF of an emulsion is the preparation of sufficiently good sinusoidal targets, particularly for spatial frequencies greater than about 100 cycles/mm. Two types of test objects are often used—one is film having a sinusoidal transmittance, and the other is a sinusoidal mask. In the latter case the mask is either imaged by a cylindrical lens, or the target or emulsion is moved in a direction normal to the frequency axis of the mask. Several organizations manufacture low-frequency targets of the former type and provide with the target the results of analyses of microdensitometer scans to indicate the unwanted higher harmonic content of the distribution.

Experimentally, the test object is imaged by a high-quality camera. For very high frequencies, a microscope objective is used with the target at the long conjugate and the emulsion at the short conjugate. A slit can be scanned across the image in

exactly the same plane as the emulsion is exposed. With a photomultiplier tube behind the slit, the modulation $M_O(\nu)$ in Eq. 6-48 is determined. This accounts for any reduction in the modulation of the target introduced by the reducing lens. The exposed emulsion is processed and also scanned with the microdensitometer and, by reference to an exposure step tablet (sometimes known as a gray scale, or tablet), the modulation on the developed emulsion is expressed in terms of exposure to give $M_E(\nu)$. This procedure is then repeated for other frequencies to obtain the complete MTF for the film (see Chapter 8).

In practice, the emulsion MTF depends, generally to a small extent, on several experimental factors. A wide cone of radiation incident on an emulsion from a fast imaging system will cause a spreading of the image, particularly for thick emulsions. Fortunately, emulsion thicknesses are about 5 μm for high-resolving-power black-and-white aerial films; however, for color films the total emulsion thickness is about 25 μm, and this effect is significant. Adjacency effects tend to enhance edge sharpness, in particular at low and intermediate frequences when the incident modulation is high. This accounts for several emulsions showing transfer functions greater than unity at frequencies in the neighborhood of 5 to 10 cycles/mm. The effect expectedly is dependent on developer and agitation conditions. Several emulsions show a transfer function of less than unity at zero spatial frequency. This can be attributed to halation effects[16] and can be reduced if the film base is dyed or a dyed backing is used. (The dye is removed during processing.)

It should be noted that the MTF of an emulsion does not include the effects of granularity, which are averaged out by the use of a long slit in the scanning microdensitometer. There is evidence that *grain noise is white,* that is, *the density deviations around the mean density level remain constant as a function of spatial frequency.* Thus an operational resolution limit for an emulsion may be defined when scanned by a small spot in a microdensitometer. This limit occurs at the point where the modulation of the developed image is comparable to the grain noise.

The MTF curves of some black-and-white aerial films are shown in Figures 6-32, 6-33, and 6-34.

Threshold Modulation Curves

An earlier discussion concerned the experimental determination of threshold modulation curves, and their use in predicting the three-bar resolving power of aerial cameras, provided the MTF of the camera lens is known. Figure 6-35 (Keene, 1972) shows threshold modulation curves

---

[16] Halation effects are caused by the image irradiance reflected off the film base exposing the emulsion around the focused image.

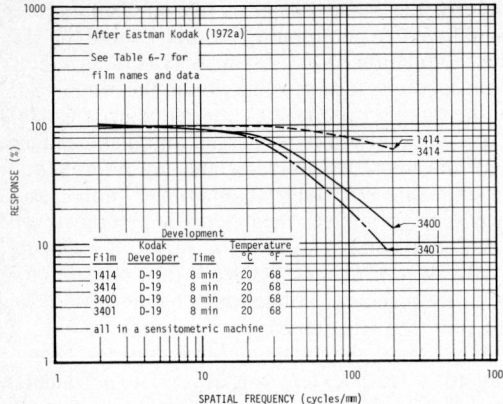

Fig. 6-32. MTF curves for aerial films 1414, 3400, 3401, and 3414.

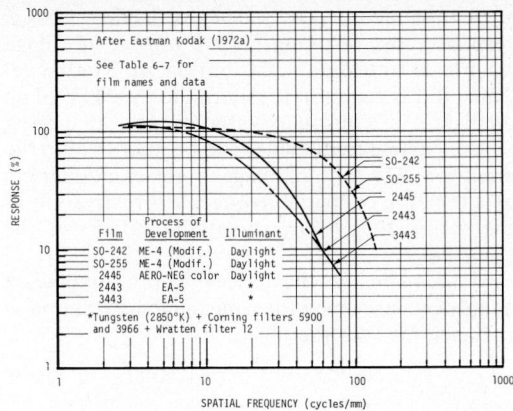

Fig. 6-34. MTF curves for aerial films SO-242, SO-255, 2445, 2443, 3443.

correspondingly to the equations listed in Table 6-8 where M is the modulation at spatial frequency ν.

In quoting these data, it must be emphasized that they are probably not reliable to better than ±10% because of the combination of photographic and human-vision processes involved in their determination. In addition, it should be noted that the threshold curves for color films are determined in terms of conventional black-and-white targets. There is no indication therefore of how the position of the threshold modulation curve will change with wavelength; for some of the recent color aerial films this change may be quite large.

Film as a Radiometric Detector

The superiority of film as a readily available, easy to use, high-resolution-image recording and storage medium is limited to applications involving the visual display and interpretation of recorded imagery. When film is to be used as a radiometer, that is, when meaningful density or transmission measurements are required of re-

corded images, many problems arise owing to the nonlinear photographic process, film and processing variations, and other causes.

Film and processing variations and the logarithmic relationship between density and exposure can be largely accounted for by exposing a gray step scale (using a sensitometer or a uniformly exposed gray wedge of known density steps) alongside each frame of photography. Densities recorded on the film can thereby be related to the exposure level required to produce them *regardless of the actual sensitivity of the batch of film used and the processing conditions* (type of developer and time and temperature of development, etc.). This is true in an *absolute* sense only if a sensitometer is used that provides a known irradiance level for a known exposure time and when the spectral distribution of irradiance for the

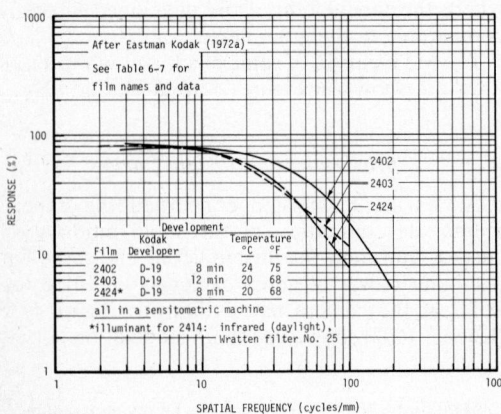

Fig. 6-33. MTF curves for aerial films 2402, 2403, and 2424.

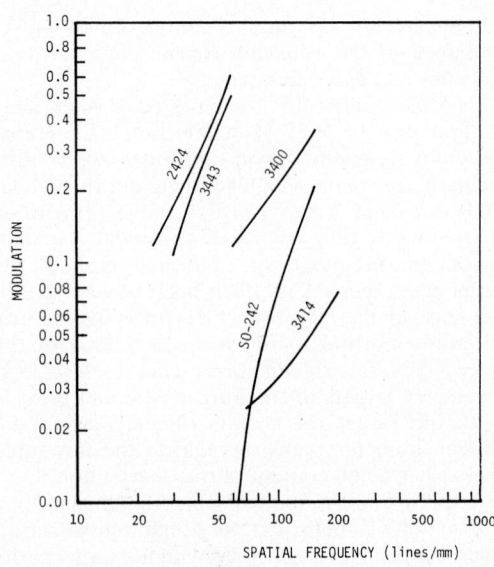

Fig. 6-35. Threshold modulation curves for five aerial films.

TABLE 6-8

Threshold Modulation Equations for Various Films

Film Name	Number	Threshold Modulation Values	Applicable Ranges for v (cycle/mm)
Infrared Aerographic	2424	$M = 0.022 + 160 \times 10^{-6} v^2$	25–60
Aerochrome Infrared	3443	$M = -0.027 + 146 \times 10^{-6} v^2$	30–60
Panatomic-X Aerial	3400	$M = 0.067 + 14.6 \times 10^{-6} v^2$	60–150
High Definition Aerial	3414	$M = 0.017 + 1.47 \times 10^{-6} v^2$	70–200
Aerial Color	SO-242	$M = -0.039 + 11.6 \times 10^{-6} v^2$	65–150

step wedge is the same as that of the image irradiance. In a *relative* measurement, the density step levels need not be related to known exposure levels in ergs cm^{-2} or m.cd.s, but the spectral distributions of the flux exposing the gray scale and the image should be the same. In practice the condition of matched spectral distributions is usually not met. The distribution of irradiance used to produce the step wedge is that due to a 6000 K source, which is a reasonable approximation to average daylight conditions, but the spectral variation in the reflectance (radiance) of the object being imaged can give rise to errors in radiometric measurements of its image if the spectral sensitivity of the film is not flat over the entire spectral range used in the photography. Fortunately, by its nature, multispectral photography makes use of spectral intervals about 100 nm wide, and most black-and-white films are nearly spectrally flat over such intervals so errors due to this effect are small.

Errors introduced during the development process are in some cases impossible to account for; they are mentioned here to alert the photographic analyst as to when these insidious effects may be influencing his data. These errors are generically referred to as developer effects and can be classified under three headings: adjacency effects, the Eberhard effect, and the Kostinsky effect. All of these effects result because, in a heavily exposed area of film, the developer used to reduce silver halide to silver becomes less effective in its action, or depleted, compared to developer in proximity to unexposed or lightly exposed, areas of film.

The adjacency effect occurs when the film image of a sharp edge (a high-low step in irradiance) is developed. Developer depletion causes the density far to the left of the edge in Figure 6-36 to be low. As the edge is approached from the left, the density increases owing to the increasing concentration of undepleted developer associated with the unexposed region to the right of the edge. The fringe region is caused by the presence of a higher concentration of depleted developer in that region than farther to the right. The effect of a border and fringe around the image is analogous to edge enhancement in digital processing. It can be advantageous in photointerpretive work, but of course microdensity measurements in the edge region are in error. Depending on the processing conditions, the edge region can be up to about a millimeter in width.

In the case of the development of a heavily exposed line image, the adjacency effect occurs at both edges of the line. As shown in Figure 6-37, as the line decreases in width its density increases. This is referred to as the Eberhard effect, and it indicates the extreme difficulty in making accurate density measurements of small areas, e.g., < 1 mm², because the density values then become image-area dependent. To emphasize this point, Figure 6-38 shows two characteristic curves for the same piece of film, processed under the same conditions. The lower curve is the macro- and the upper curve the micro-characteristic curve. Actually, as Figure 6-37 implies, there is a family of such curves as a function of micro-image size.

The third developer effect is the Kostinsky effect; it concerns the lateral displacement of two images rather than errors in their density values. If the film images are heavy exposures and close together, then the unexposed region between the two will be less developed than the unexposed region surrounding them. As seen in Figure 6-39, a spurious increase in image separation is thereby produced.

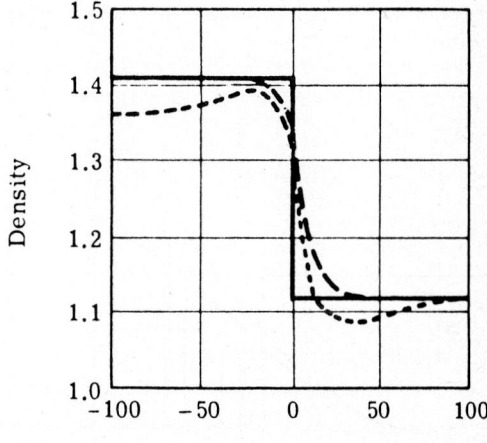

Fig. 6-36. The adjacency effect.—Ideal edge response; − − Edge image without adjacency effect; - - - Edge image with adjacency effect.

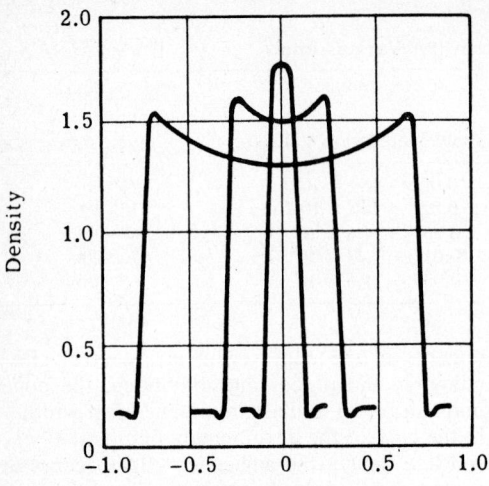

Distance, millimeters

Fig. 6-37. The Eberhard effect.

Fig. 6-39. Image without (a) and with (b) the Kostinsky effect.

AERIAL COLOR FILMS

Additive and Subtractive Presentation of Color[17]

There are two primary systems for obtaining a color photographic rendition of a scene: additive and subtractive. The principles of the additive system are illustrated in Figure 6-40, where three colored lights (red, green, and blue) are shown to be partly overlapped. In addition to the three primary colors started with, it can be seen that combinations of these colors produce other colors. For example, the combination of nearly equal intensities of red and green light produces yellow; red and blue produce magenta; and blue and green produce cyan. By varying the individual intensities of these lights, one can obtain a wide gamut of colors. Where equal amounts of all three lights superimpose, the resulting area appears white.

A way of making a color photograph with additive color photography is to make three black-and-white photographs of the same scene through red, green, and blue filters, and to develop and print the film to give positive images. When these films are projected in superposition through the filters used to make the original photographs, a very good color reproduction can be obtained.

Additive color photography is the system used for multispectral aerial photography. It is a very flexible system in that the photographs can be made through filters selected to maximize the spectral information content of the scene photographed, rather than to obtain maximum color fidelity. It can also be used to obtain information in spectral regions to which the human eye is insensitive, such as the infrared.

The system is also flexible in that the projection filters can be completely different from those used in the camera. Under these circumstances a false-color rendition will result, but greater detail-enhancement often can be attained. The primary drawback of the additive system is its complexity, not only in the number of steps required, but also in the complexity and precision of the equipment needed to produce high definition in the resulting photographs.

Additive color photography is not easily incor-

Development effects are more pronounced the greater the image contrast. They can be reduced by (a) the choice of an active developer, e.g. D-19, and (b) brisk agitation of the film in the developer. In respect to the latter, brush development or the use of a processing machine that sprays the developer onto the film is advantageous. In addition to these steps to reduce developer effects, the agitation should be random and such as to provide repeatable results.

In summary, the photographic analyst should recall that at a gamma of unity, an uncertainty in density of 0.04 corresponds to an uncertainty in exposure of 10%. The above discussion, together with an appreciation of variability of shutter speeds and environmental conditions and the limits on the accuracy of optical system radiometric calibration, should make it clear that *a radiometric measurement uncertainty of less than 10% is extremely hard to achieve in photographic remote sensing.*

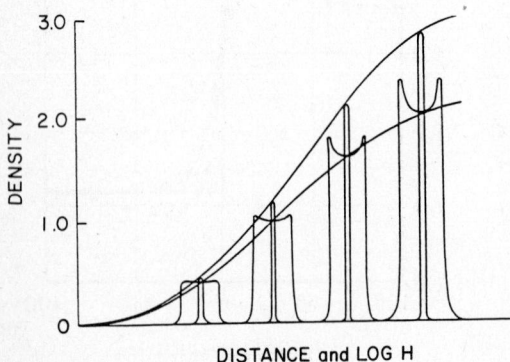

DISTANCE and LOG H

Fig. 6-38. Macro- and micro-image characteristic curves.

[17] The subject of color specification was covered in the first edition of the *Manual of Remote Sensing* (Slater, 1975) and in the *Manual of Color Aerial Photography* (Rib, 1968). The reader is referred to these sources for further information on these subjects.

porated into practical photographic film systems, so most color photography uses what is known as the subtractive system. When white light is available for viewing the photographs, the subtractive photographic system allows the use of three dyes, each of which controls the light of only one primary region of the visible spectrum. These dyes are cyan, magenta, and yellow. The cyan dye transmits freely in the blue and green primary regions but absorbs (or controls by its concentration) the transmission in the red primary region. In a similar manner the magenta dye transmits in the blue and red regions and absorbs or controls the green, and the yellow dye transmits in the red and green regions and controls the blue. The principles of the subtractive system are shown in Figure 6-41, where the three colored filters are shown partially overlapped. A wide gamut of colors can be obtained with these three dyes; for example, if the magenta and yellow dyes are superimposed, the magenta absorbs the green light, the yellow absorbs the blue light, and the combination will appear red. Likewise, combinations of yellow and cyan will appear green, combinations of cyan and magenta will appear blue, and variations in the concentrations of the dyes will vary the colors of the combinations. Because the three dyes can be incorporated in three layers on a single film base, the combination of which is sometimes referred to as an *integral tripack*, the camera function for making a color photograph is as simple as that for making a black-and-white photograph, and much easier than that required for additive color photography.

Color Films

Figure 6-42 shows a schematic cross-section of the usual arrangement of the layers of a color film. Other configurations are used, and the most important of these will be described later. The top layer is blue sensitive and its exposure controls the formation of the yellow dye. The middle layer is green sensitive and forms the magenta dye, and the bottom layer is red sensitive and forms the cyan dye. Since all the layers are inherently sensitive to blue radiation, the yellow filter is used to prevent this radiation from exposing the green- and the red-sensitive layers. The basic principles and detailed descriptions can be found in Fritz (1969, 1971, 1974) and Eastman Kodak Company (1971, 1972a).

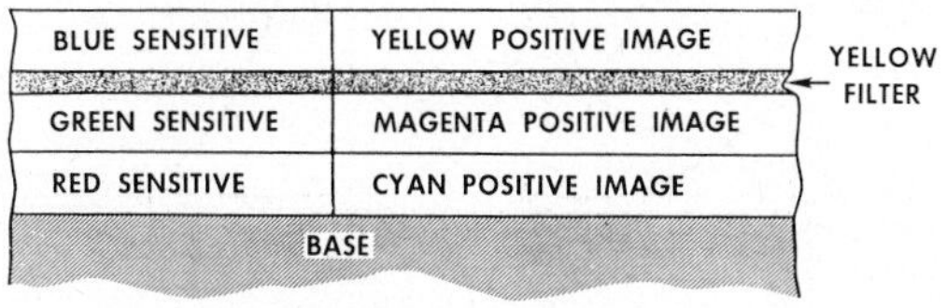

BLUE SENSITIVE	YELLOW POSITIVE IMAGE	YELLOW
		← FILTER
GREEN SENSITIVE	MAGENTA POSITIVE IMAGE	
RED SENSITIVE	CYAN POSITIVE IMAGE	
BASE		

Fig. 6-42. Schematic representation of the layers of a normal-color film.

There are basically two different methods of processing color films: reversal and negative. Films are generally made for only one type of processing. With reversal processing, a positive photograph is produced; that is, light areas reproduce as light, dark areas as dark, and the colors closely represent those of the scene photographed. With negative processing, light areas reproduce as dark, dark areas as light, and the colors are complementary to those of the scene photographed. When a photograph or print is made of this negative on another negative material, a reversal to the complementary colors will again be obtained, and the final colors will again correspond closely to those of the original scene photographed.

Figure 6-43 will help to explain the basic principles of color photography. At the top of the figure is a representation of a scene being photographed which consists of areas that are white, black, red, green, and blue. The white provides radiation in all three primary regions and thus exposes all three layers of the film. There is no radiation from the black; consequently, none of the film layers is exposed. The red light passes through the top layer, the yellow filter, and the middle layer, (which is not sensitive to red light,) and then exposes the bottom layer. The green radiation passes through the top layer, which is not sensitive to it, through the yellow filter, and exposes the middle layer. The remaining energy passes through the bottom layer, which is not sensitive to green. The blue radiation exposes the top layer and is prevented from reaching the two bottom layers by the yellow filter.

There are two processes currently being used to provide an image-forming dye in each of the film layers; these are the dye-formation and the dye-bleach processes. The former is used for both negative and reversal systems. The latter is used only in the reversal.

For reversal development in the dye-forming process, the exposed film is first immersed in a negative black-and-white developer where the exposed silver halide grains are reduced to metallic silver. The remaining silver halide is then made developable by exposing it with a suitable light, or this step can be accomplished by a chemical reaction in the color developer. When the film is placed in the color developer, the newly exposed silver halide is reduced to silver, and at the same time a dye is formed in each of the layers, its concentration being proportional to the amount of silver halide that is reduced. Thus, a positive image is formed in which the dye concentration is inversely proportional to the amount of the original exposure. The film is then bleached, fixed, and washed, which removes the silver, leaving only the dye image. If white light is allowed to pass through this film, then as can be seen in the first column of the left half of Figure 6-43, since there is no dye, the resulting transmittance is white. In the second column, the dyes absorb in all three primary regions, allowing no radiation to pass, and thus black results. As mentioned previously, the combination of yellow and magenta (column 3) transmits only red radiation, yellow and cyan (column 4) transmit green, and magenta and cyan (column 5) transmit blue; the final reproduction matches closely the colors of the original scene photographed.

To develop the film as a negative, it is initially put into the color developer, where the dyes are formed as the exposed silver halide is developed. The silver and silver halide are then removed, leaving again a dye image. In the right half of Figure 6-43, it can be seen that in the first column all three dyes are formed and the reproduction color is black; in the second column no dyes are formed and the reproduction is white; and the individual cyan, magenta, and yellow dyes are formed as a result of exposure by the red, green, and blue light. Thus, each color formed in the negative system is the complement of the color that was originally photographed.

In the dye-bleach system, the dyes are initially present in all three layers along with the silver halide. After exposure, the film is first placed in a black-and-white developer, as in the reversal dye-forming system, where the exposed silver halide is reduced to form a negative silver image. The film is then put into a dye-bleach solution, where the dye is selectively reduced in proportion to the amount of silver present. The film is then put through the usual silver bleaching and fixing solutions, leaving the positive dye images in which the concentration is inversely proportional to the amount of the original exposure.

It should be noted that the foregoing descriptions depict color photography as a sort of "go, no-go" system. The film is depicted as having sensitivity in a spectral region or as having no sensitivity. The dyes are shown to transmit or to block a spectral region completely. Actually, the system is not as ideal as that. Each of the layers has some sensitivity in regions other than that of its primary sensitivity, and each of the dyes has absorptions in regions other than that of its primary absorption. Also, correct exposure of most objects will result in some exposure of all three layers, which provides the transmissions in the three primary regions that will combine to form the gamut of colors possible with color photography.

Figure 6-44 shows the actual spectral sensitivities of the three layers of a typical color reversal camera film, and Figure 6-45 shows the absorptions of the dyes that are formed. Similar sensitivities and dyes are used for color negative camera films. The dyes, although primarily absorbing in one spectral region, do have absorption in the other regions; and they do not absorb uniformly over any spectral region. When these three dyes are combined, as is shown by the upper curve, the visual neutral formed is not a spectral

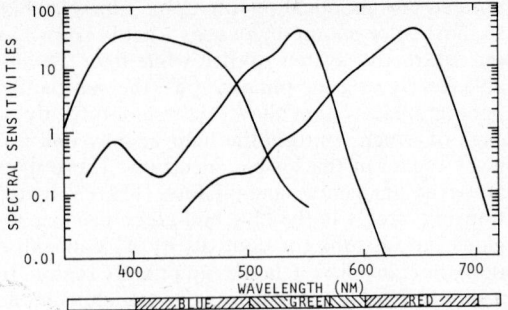

Fig. 6-44. Spectral sensitivities of the three layers of a typical normal-color reversal film.

neutral; that is, it does not have the same density at all wavelengths, even though it appears neutral to the human eye. Similarly, combinations of the three dyes will form colors that visually match those of an object photographed, but the spectral densities will not be identical. Therefore it is generally not possible to derive the spectral reflectance of an object from spectral measurements of its photographic image.

False-color Films

The relationship between the sensitivity of a layer, and the dye formed in that layer, does not have to be the same as has been described for films which produce a normal-color rendition. Any combination of sensitivity and dye may be used in each layer and, if different from the above, will produce a false-color rendition. The most noted variants of this type are the infrared-sensitive color films in which one of the layers is made sensitive to the infrared spectral region and the other layer or layers have their sensitivities in the visual spectral region. These films may be made for either the positive or negative mode of

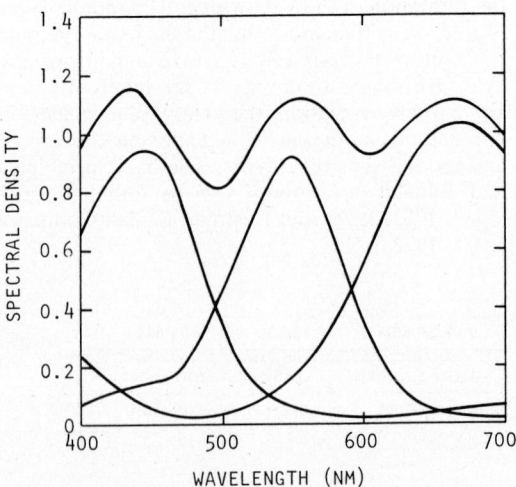

Fig. 6-45. Spectral densities of a unit neutral gray and its yellow, magenta, and cyan dye components for a typical color reversal film.

processing. Basic principles and detailed descriptions can be found in Tarkington and Sorem (1963) and Fritz (1967).

Many films giving a false-color rendition have been made for special applications, but the most important ones and the only ones that will be described here are the three-layer films in which one layer is sensitive to the infrared spectral region. This type of film was originally made for camouflage detection but has since found many applications in remote sensing and earth resources studies.

Figure 6-46 shows the basic operational characteristics of the normal-speed version of this false-color film compared to that of a normal-color film, to illustrate both the similarities and dissimilarities.

If the spectral region to which photographic films are sensitive is divided into the ultraviolet, blue, green, red, and infrared spectral regions, it will be remembered that the normal-color film is sensitive to the blue, green, and red spectral regions. Associated with these sensitivities are the yellow, magenta, and cyan dyes, respectively, which, after processing, combine to produce the colors blue, green, and red, which closely match those of the original scene photographed.

With the infrared-sensitive color film known as Infrared Ektachrome, or Aero Ektachrome Infrared, all three layers are sensitive in the blue spectral region; the individual layers are also sensitive to green, red, and infrared radiation. A yellow filter is always used over the camera lens to prevent blue light from reaching the film, the result is a film-filter combination that has effectively three layers, sensitive, respectively, to the green, red, and infrared regions. Again, the same three dyes—yellow, magenta, and cyan—are associated with these sensitivities; and it can be noted that they are used in the same order with respect to increasing wavelength, but their sensitivities have

been shifted by one "block" toward longer wavelengths. As with normal-color film, combinations of the three dyes form blue, green, and red colors; but now the blue has resulted from a green exposure, the green from a red exposure, and the red from an infrared exposure. Figure 6-47 shows the actual spectral sensitivities of the three layers of such a film, plotted with the transmittance of the yellow filter.

High-Definition Films

In addition to the main classifications of color aerial films as negative, reversal, or false-color, they may also be subdivided into categories relating to their definition characteristics, or conversely, their speed. Divided this way they fall into two distinct categories; normal-speed films that include a majority of the products available, and the slow, high-definition films for use in high-performance camera systems at very high altitudes (Moser and Fritz, 1975). The high-definition films are currently available only for reversal processing and have speeds roughly 1/10 to 1/6 that of normal-speed films. Thus they require fast camera systems and stable platforms.

For a normal-color, high-definition film, the layer arrangement is different from that of the usual color films. The arrangement is shown in Figure 6-48 and can be compared to the normal layer arrangement shown in Figure 6-42. The reasons for this arrangement are as follows: The apparent impression of sharpness in almost any color photograph is primarily dependent on the characteristics of the magenta-forming layer. If this layer has high definition, then the whole picture will generally appear sharp. In the usual film structure, the magenta-forming layer is in the middle, and light reaching it must pass through the layer above. Because any emulsion layer slightly diffuses the light passing through it, the magenta

Spectral region	Ultraviolet	Blue	Green	Red	Infrared
Normal reversal color film sensitivities		Blue	Green	Red	
Color of dye layers		Yellow	Magenta	Cyan	
Resulting color in photograph		Blue	Green	Red	

Infrared reversal color film sensitivities		Blue	Green	Red	Infrared
Sensitivities with yellow filter			Green	Red	Infrared
Color of dye layers			Yellow	Magenta	Cyan
Resulting color in photograph			Blue	Green	Red

Fig. 6-46. Principles of operation of normal-color film and infrared-sensitive color film.

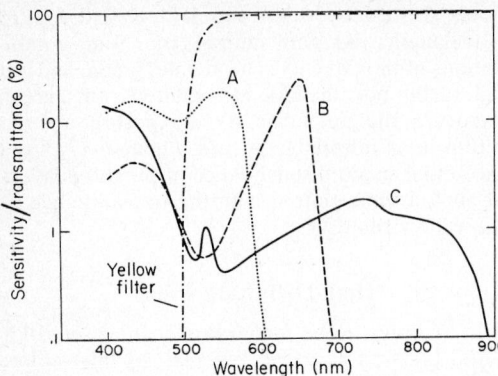

Fig. 6-47. Spectral sensitivity curves for a color-infrared film and the spectral transmittance of the yellow filter through which it is exposed. A is the yellow-forming layer, B is the magenta-forming layer, and C is the cyan-forming layer.

image is not as sharp as it would be without the layer above it. For this reason, one of the important features of this film that helps to provide its extremely high definition is that this green-sensitive layer has been placed above the other two. However, as with the normal-speed film, a yellow filter is needed to prevent blue-light exposure of the cyan- and magenta-forming layers and thus must be above all three emulsion layers. The blue-light density of this filter is sufficiently high to prevent exposure of the red- and green-sensitive layers, but still permits transmission of enough radiation to expose the extremely sensitive yellow-forming layer on the bottom. Since this layer contributes little to the visual impression of sharpness, a very fast, coarse-grained emulsion is used. The high definition provided by this film is due partly to the layer arrangement and partly to the fact that the magenta- and cyan-forming layers have very thin, slow-speed, fine-grained emulsions that result in improved optical qualities. A resolving power of 200 lines/mm for a test object contrast of 1000:1 and 100 lines/mm for a test object contrast of 1.6:1 along with a diffuse RMS granularity of 9 can be obtained in a film of this type. This compares to values of about 80, 40, and 12, respectively, for the normal-speed films.

There is also a high-definition version of the infrared-sensitive color film that has the same speed as the above film, and it is intended for the same types of applications. This film has several significant improvements over the normal-speed

product. The most important is the improved spectral sensitivity of the infrared-sensitive, cyan-forming layer. Figure 6-49 shows the spectral sensitivity curves for this type of film; these can be compared to those of Figure 6-47 for the normal-speed product. The sensitivity range of the infrared-sensitive layer is narrower with a peak at about 800 nm, and a very low sensitivity in the red spectral region. Thus, better color differentiation can be attained because of the decrease in dilution of the infrared exposure by red light.

The yellow filter that has always been required in the camera lens system for the infrared-sensitive color films is not needed with the high-definition film. A yellow filter is still required to prevent blue radiation from exposing the inherent blue sensitivities of all three layers, but it is coated as part of the film and thus is not needed separately. This arrangement provides better optical characteristics than if the filter were over the lens; it assures that the correct filter is used and also that the filter is always present.

An important improvement in these films is that they have keeping stability that is similar to that of most normal-color films. Thus, the color balance remains relatively constant with age, and special handling requirements are no more stringent than those for regular aerial color films.

Color Film Sensitometry

Basically, the sensitometry of color films parallels that of black-and-white films in the requirements for precise exposure intensity and time, and repeatable processing. However, the analysis of the resulting images is complicated by the fact that each of the three principal layers of the film produces its own sensitometric curve, and these layers are generally nonseparably overlapping. Because color films are used in a variety of ways, there are a number of types of density measurements that are made on them. Two general conditions must be met: the geometric and the spectral characteristics of the densitometer must conform to those of the intended use of the film. Color film densitometry can be broadly categorized into two types, integral and analytical. Density measurements of the three layers, taken to-

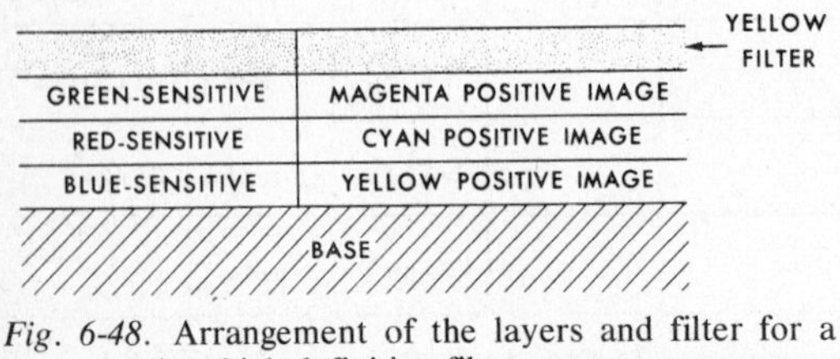

	YELLOW FILTER
GREEN-SENSITIVE	MAGENTA POSITIVE IMAGE
RED-SENSITIVE	CYAN POSITIVE IMAGE
BLUE-SENSITIVE	YELLOW POSITIVE IMAGE
BASE	

Fig. 6-48. Arrangement of the layers and filter for a normal-color, high-definition film.

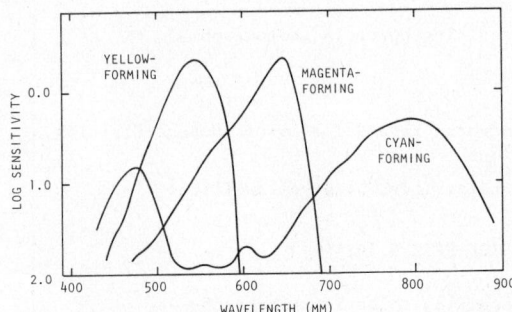

Fig. 6-49. Spectral sensitivity curves for a high-definition, infrared-sensitive color film.

gether, are known as *integral densities*. For example, the density of a color film image measured with red illumination would be a red integral density. Densities of the individual layers are called *analytical densities* because they represent an analysis into the individual components of the film.

There are several types of integral densities, each corresponding to a particular use that will be made of the film. Figure 6-50 shows the characteristic curves of a typical color reversal film measured by integral densitometry using red, green, and blue filters.

Color *printing densities* are determined for films from which prints will be made, to show the effective densities as "seen" by the print material. Red, green, and blue filters are designed for the densitometer in such a way that its response is the same as that of the three layers of the print film. This is an oversimplification of printing density, because filters are standardized and used for many products, and the response of a given layer of print material is somewhat dependent on the development of the other layers. These effects can be corrected for by the use of linear equations relating the densitometer response to the actual film response. Figure 6-51 shows the characteristic curves relating printing density to log exposure for a typical aerial color negative film.

Another type of integral density is *colorimetric density*, which is used to evaluate photographic images that will be viewed directly. For this, the densitometer filters are made to evaluate the densities of the dyes in terms of the visual response to the transmitted light. The measurements are in terms of the negative logarithm to the base 10 of the tristimulus values.[18] Because the $\bar{y}$ function is identical to the luminous efficiency function, the density values through this filter are known as *visual densities*.

Spectral integral densities are the densities of a dye or combination of dyes as a function of wavelength, and are measured with a spectrophotometer or with narrow-band interference filters. The top curve of Figure 6-45 represents the spectral integral densities of the combination of dyes shown. For some applications it is not necessary to determine the whole curve, and measurements are made only at or near the peak densities of the three dyes.

For some applications, such as in the manufacture of film or the photometry of terrain features in aerial photographs, it is necessary to know the response of the individual film layers, that is, the densities of the cyan, magenta, and yellow dyes. These are known as *analytical densities*, and as with integral densities, there are several types. They cannot be measured directly with spectrally selective filters because, as is shown in Figure 6-45, each of the dyes has some absorption in other than its principal region of absorption, and

[18] See previous footnote (17).

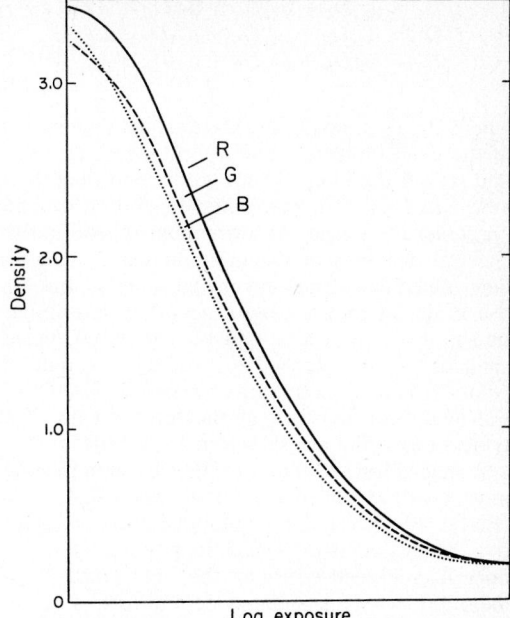

Fig. 6-50. Characteristic curves of a typical color reversal film.

thus measurement in any spectral region includes absorption of all three dyes.

It has been found that if the *spectral analytical densities,* or more simply, the spectral density characteristics of the individual dyes, are known, it is possible to calculate the analytical densities of a given combination of dyes from their integral densities by means of simple linear equations. These equations are of the form:

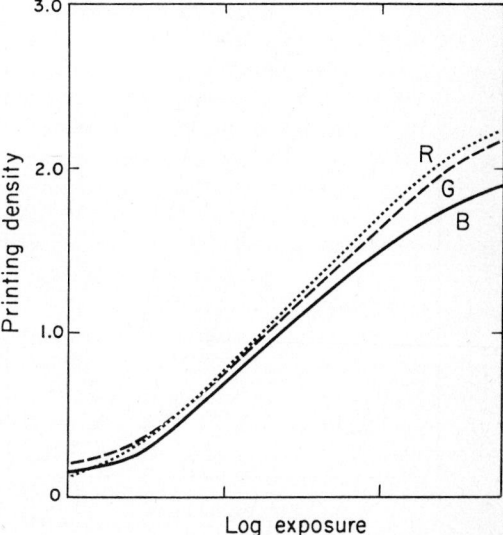

Fig. 6-51. Characteristic curves of a typical aerial color negative film.

$$D_C = a_{11}D_R + a_{12}D_G + a_{13}D_B + a_{14}$$
$$D_M = a_{21}D_R + a_{22}D_G + a_{23}D_B + a_{24}$$
$$D_Y = a_{31}D_R + a_{32}D_G + a_{33}D_B + a_{34}$$
$$(6\text{-}49)$$

where D_C, D_M, and D_Y, are the analytical densities of the cyan, magenta, and yellow dyes; D_R, D_G, and D_B are the integral red, green, and blue densities: and a_{MN} are constants depending on the particular dyes used. As a first approximation the spectral densities of the individual dyes can be determined from single-layer film samples, but for use in densitometry a more accurate determination is made by a least-squares procedure using samples of film exposed to produce a gamut of colors around a central gray exposure. The spectral analytical densities of the three layers of a typical color film are shown in Figure 6-45.

A special form of analytical density, and the one most used, is called *equivalent neutral density* (END). The END of one component of a subtractive dye system is the visual density it would produce if combined with just the right amounts of the other two dyes of the system to form a visual neutral. It is essentially a normalization of the analytical densities so they will be equal at the neutral point. Figure 6-52 shows the characteristic curves of a typical infrared-sensitive color film determined in the terms of equivalent neutral analytical densities. With a normal-color film, the three sensitometric curves are almost superimposed, while with this film the cyan-forming layer is about one stop slower than the other two layers. The reason for the decreased speed of the cyan-forming layer can be explained by considering the high infrared reflectance of most foliage. If the cyan layer were as fast as the other layers and the camera exposure was such that the other two layers were exposed correctly, the infrared exposure would be recorded on the toe of the cyan curve, and any variation of infrared exposure would produce negligible cyan density change. Small differences would not be detected.

Print and Duplicating Films

The terms "print" and "duplicating" are often used interchangeably, although to be precise, a print material is negative-working and is used to make prints from camera negatives. A duplicating film is a reversal product and is used to make *duplicates* of either camera negatives or positives.

To record as much image detail as possible, print and duplicating films must have considerably higher definition than the camera films being printed. There are always some information losses in any printing operation, but they can be minimized with high definition print or duplicating film and the optimization of other system elements such as lens quality, focus, and contrast. High-definition films tend to be slow, but this is of little consequence because longer exposures can be used in the printing operation than are available for the original photography. These films also have different spectral sensitivities than camera films. Camera films are used to photograph a natural scene with daylight illumination, while print and duplicating films are essentially used to photograph other photographs (the camera originals) with artificial illumination. For this reason their spectral sensitivities must correlate with the spectral densities of the dyes in the camera film, rather than with actual scene reflectances. Figure 6-53 shows the spectral sensitivities of a typical color duplicating film. It may be compared to Figure 6-44 to show the differences from a camera film and to Figure 6-45 to show the correlation with the camera film dye densities.

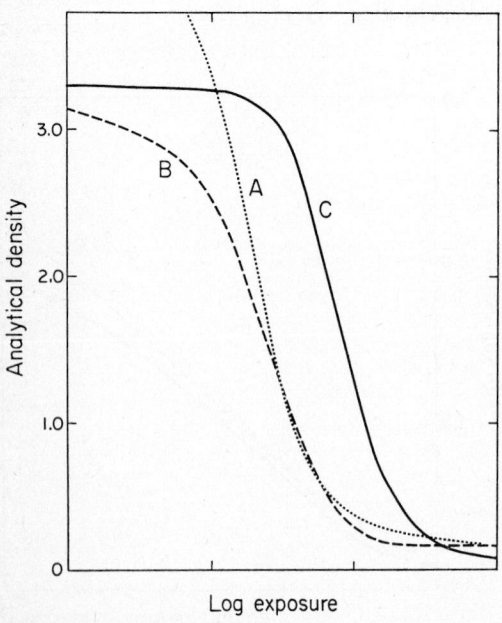

Fig. 6-52. Characteristic curves of a typical infrared-sensitive color film. A is the yellow layer, B is the magenta layer, and C is the cyan layer.

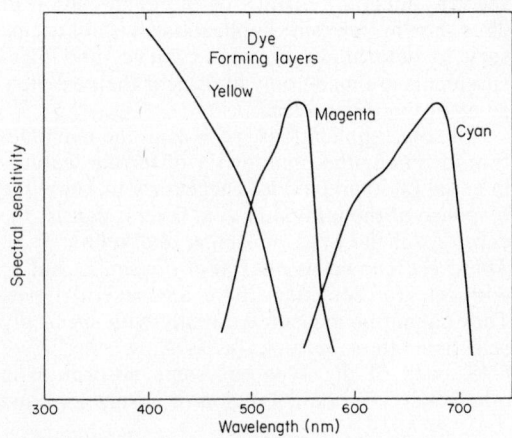

Fig. 6-53. Spectral sensitivities of a typical color duplicating film.

The color print and duplicating films have somewhat different sensitometric characteristics from each other. The print films need high contrast in order to compensate for the low contrast of the camera negative film, and also for the somewhat low contrast of the aerial scene. The duplicating films generally need low contrast (gamma 1.0 or slightly higher) to provide tone reproduction characteristics similar to those of the original camera films. Also, the low contrast allows multiple-generation duplicates to be made without a great contrast buildup. Both printing and duplicating films should have matched contrasts in the three layers to provide uniform color balance over the exposure range. It is not necessary to have matched speeds for the three layers of either film because of the adjustments possible with filters when making the prints. Duplication films are available in both the dye-formation and dye-bleach systems.

Reflection print materials are available as either negative or reversal products for making prints from camera negatives or positives, respectively. They have an opaque base that is usually paper but that may be made of pigmented acetate or polyester materials. Again, the reversal type is available with both the dye-formation and dye-bleach systems. These materials have characteristics similar to those of the print and duplicating films in that the spectral sensitivities must correlate with the camera film dyes, the negative materials must have high contrast, the reversal must have low contrast, and the speeds of the three layers need not be equal.

Film Thickness and Spool Capacity

For high-altitude and space photography it is often required that film weight and size be minimized. These requirements can be satisfied

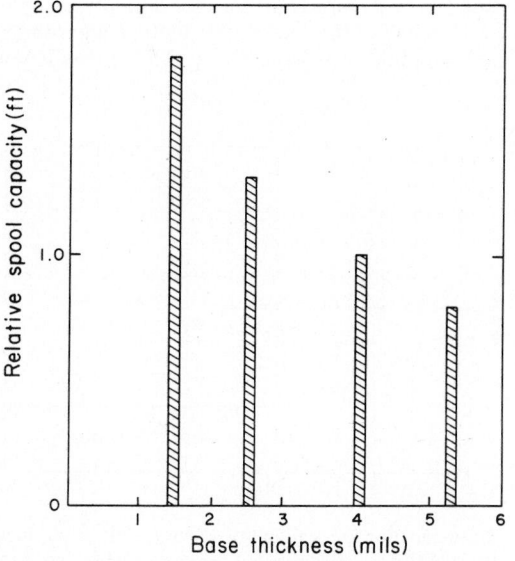

Fig. 6-54. Effect of base thickness on spool capacity.

by a reduction of format size, but if this is not accompanied by an increase in system definition, a loss in information content will result. An obvious method of increasing film payload is to reduce the thickness of the film, thereby increasing spool capacity proportionally. Emulsion thickness can be reduced slightly, and thinner emulsions have the advantage of providing increased sharpness. This method is more effective with pelloid-backed films because the curl-balancing gelatin backing can also be reduced proportionally. However, reduction of the film base thickness produces the greatest space saving, as can be seen in Figure 6-54. By changing from 130-μm (5.2-mil) acetate to 100-μm (4-mil) polyester base, spool capacity is increased by about 25%. A further 30% capacity increase can be attained by reducing the thickness to 64 μm (2.5 mils). A few films are available with 38-μm (1.5-mil) base, but the usefulness of this base is limited by manufacturing problems and difficulty with handling and tracking in cameras, processors, and printers.

FILTERS

A wide variety of filters is used in remote sensor photography. Table 6-9 lists typical applications, the optical properties required, and the filter type employed. The literature referenced has covered the subject in much detail, and the reader is encouraged to utilize these sources to the greatest extent possible. This section will present only a general overview of the three categories of antivignetting, spectral, and polarization filters—more detailed data are outside the scope of this discussion.

ANTIVIGNETTING FILTERS

Antivignetting filters are used to correct the cos⁴ falloff in image plane irradiance and vignetting effects (Slater, 1980). They are usually produced by depositing Inconel (a metal alloy) on glass in such a way that the central area of the filter is strongly absorbing and the circumferential region is clear. They are rarely designed to give a uniform irradiance over the whole format of, say, a 90° field camera, for two reasons. First, the loss in *T*-number would be large, the uniformity being gained at the expense of the transmission of the axial and near axial full *f*-number beams. Second, the asymmetrical variations in atmospheric backscatter make such careful corrections unwarranted.

For example, the antivignetting filter for the Kollsman Geocon IV lens holds the irradiance level constant to 1% over 76° of the total 90°. Spectrophotometric data for the 152-mm Wild Aviogon and Universal Aviogon, given by Duddek (1967), present spectral transmittance and relative irradiance curves for lenses. At *f*/8 the relative irradiance for the 152-mm Aviogon with antivignetting filter is about the same as that quoted above for the Geocon IV.

It should be noted that antivignetting filter de-

TABLE 6-9

Filters for Remote Sensing

Application	Optical Properties	Filter Type
Antivignetting	Graded neutral density	Neutral absorption, usually the metal alloy Inconel
Haze reduction for black-and-white film	Sharp spectral cutoff absorbing completely below about 0.4 μm	Selective absorption
Haze reduction for color film	Absorption increases with decreasing wavelength through visible to sharp cutoff, absorbing completely below about 0.4 μm	Selective absorption
Haze reduction for black-and-white and color films	Reduce polarized component of scattered haze light	Polarization filter
Use with color or black-and-white infrared film	Sharp spectral cutoffs absorbing completely below about 0.5 and 0.7 μm, respectively	Selective absorption
Multiband photography	Bandpass, long and short wavelength pass	Selective absorption or interference filter. Short wavelength pass obtainable with interference filter
Color correction for color films	Slowly varying change of transmittance with wavelength	Selective absorption

signs are always tailored for a given lens type. In some cases filters are available for each F stop setting of the lens.

To reduce the number of filters and the number of between-filter reflections, antivignetting filters are often deposited on haze-cutting filters or whatever other absorption filters might be used with the camera. Their use is critical when color reversal film is used with a wide-angle camera because the half-stop tolerance of such film can be exceeded several times due to the effects of $\cos^4$ and vignetting.

SPECTRAL FILTERS

Basic Types

As indicated by Table 6-9, spectral filters can be either absorption or interference filters. Before describing those suitable absorption filters that are readily available (the number and variety of commercially available interference filters are too many to list), some general comments should be made regarding the basic types of filters.

(1) *Absorption filters* are available as long wavelength pass (very efficient[19]) and passband filters (inefficient). They are *not* available as short wavelength pass. (2) *Interference filters* (also known as *thin-film,* or *dielectric* filters), unlike absorption filters, are incidence-angle sensitive.

[19] Efficiency is defined by the ratio of the amount of radiation transmitted in the passband of the filter (as defined by the 50% transmittance points) to the amount of radiation incident on the filter in that passband.

All interference filters show a wavelength shift to the blue with increasing angle of incidence (Fig. 6-55). They can, however, be used with wide-angle lenses in special cases (McKenney and Slater, 1970). (3) The variety of absorption filters is limited with regard to wavelength selectivity. Interference filters have no such restriction. (4) Both absorption and interference passband filters can show substantial offband transmission unless suitably blocked. Long- or short-wavelength pass interference filters suffer from the same drawback. (5) Comparisons of absorption and interference filters from manufacturers' data are complicated because the curves are plotted logarithmically and linearly, respectively. The reason is not

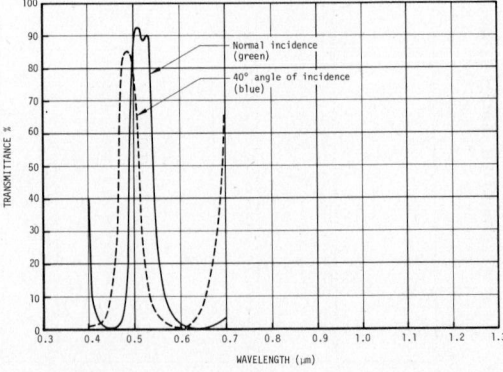

Fig. 6-55. Bandpass filter wavelength shift at 40° incidence. (The wavelength shift becomes significant at incidence angles greater than 15°).

hard to understand when the shapes of the curves are studied. A logarithmic plot for an interference filter would show the relatively high value of transmittance in offband regions. A linear plot for absorption passband filters would emphasize the low-peak transmittance, and the relatively slow rate of change of transmittance with wavelength at the half-power points.

Absorption Filters

Absorption filters are manufactured by Corning Glass Works (1965), Schott and Gen. (1962), and Eastman Kodak Company (1970). Corning and Schott produce colored glass filters, although only Schott glass filters are made of optical quality glass. In Schott filters, the coloration is caused either by simple or complex ions in true solution or by submicroscopic colored crystals in the glass; the correct size of these crystals is obtained by the temperature treatment of the glass.

Most Wratten filters consist of organic dyes mixed in gelatin and lacquered for durability. These gelatin filters can be carefully cemented between plane-parallel plates of optical quality glass and used with systems of high angular resolution. The Tiffen Optical Company markets a large variety of absorbing filters cemented between glass plates.

Glass filters suitable for aerial photographic use must be free of striations, optically flat, and with plane-parallel surfaces. The transmittance values of glass filters can be controlled by optical grinding and polishing to control thickness.

Transmittance Curves for Absorption Filters. – Dobrowolski et al. (1977) reported on a comprehensive study of some 800 commercially available color filters. This is the best single source of data on absorption filters. Spectral-transmittance curves for some of the Wratten filters commonly used in remote sensing are plotted linearly in Figures 6-56 through 6-59. Corresponding logarithmic plots can be found in Kodak Publication B-3 (Eastman Kodak Company, 1970). Figure 6-56 shows the curves for haze-

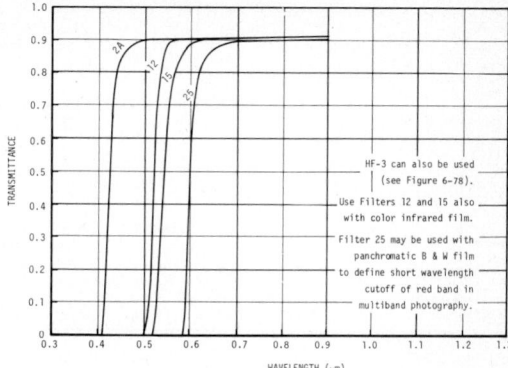

Fig. 6-57. Haze-reduction (minus-blue) filters for use with black-and-white film.

cutting filters for color aerial photography, types HF-3, HF-4, and HF-5. Combinations of type HF-3 with type HF-4 or HF-5 are used to provide the proper color balance in the blue. Type HF-3 with HF-5 is used when there is a strongly blue-scattering atmosphere between camera and ground, and HF-3 is used with HF-4 at other times. Haze reduction for black-and-white films can also be provided by HF-3.

Other typical filters used for the purpose are 2A, 12, 15, and 25, as shown in Figure 6-57. Filters 12 and 15 are also used with color infrared film. Filter 25 is frequently used with panchromatic black-and-white film to define the short wavelength cutoff of the red band in multiband photography.

The blue and green bands in multiband photography can conveniently be obtained by using filters 47 and 57A, as shown in Figure 6-58. Filter 89B, also shown in Figure 6-58, is usually used to define the short-wavelength cutoff in black-and-white infrared photography.

Color Compensating Filters

Color compensating filters affect the color balance in color photography by absorbing in the red,

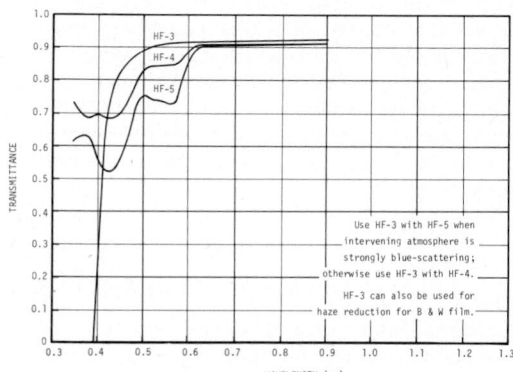

Fig. 6-56. Haze-cutting filters for color aerial photography.

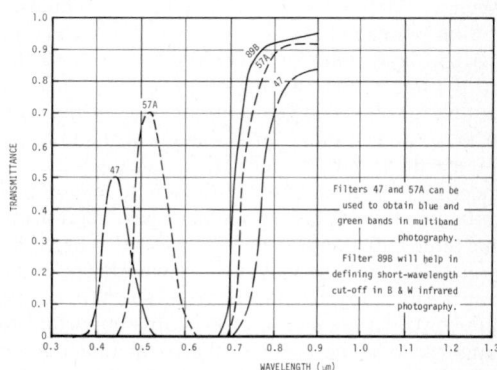

Fig. 6-58. Filters commonly used in multiband photography.

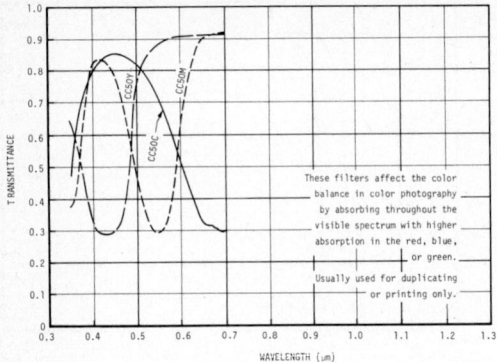

Fig. 6-59. Cyan, magenta, and yellow color-compensating filters.

blue, or green. As can be seen from Figure 6-59, they absorb throughout the visible spectrum, with higher absorptions in the red, blue, or green regions. They are referred to as cyan, yellow, and magenta; or minus red, minus blue, and minus green, respectively. Sets of filters of different absorptions are available. Figure 6-56 shows the spectral transmittance curves for the most highly absorbing in each set.

Color compensating filters are usually used during duplication or printing; they are rarely used in aerial photography in taking the photography, although some advantages have been found under certain conditions in color infrared photography in modifying the color balance of the film.

Interference Filters

Interference filters, which work on the principle of reflecting the unwanted radiation and transmitting the required radiation, are manufactured by several companies in the United States. Basically, the long- or short-pass filters comprise quarter-wave optical-path-thickness layers of alternately high (η_H) and low (η_L) refractive-index dielectric materials. These layers are usually coupled to the substrate with special matching layers. Such multilayer stacks can be of many different designs. Three computed by Baumeister (Defense Supply Agency, 1962) are shown in Figure 6-60. The broken line in each case shows the transmittance of an eight-layer filter on a glass substrate, and the solid lines show the envelope for an infinite number of layers of the same design. In the captions, t_H and t_L stand for geometrical thicknesses for the high and low refractive index layers, respectively, and λ_0 can be chosen to be of any value, the high or low wavelength pass sections of the filter can therefore be positioned where desired. Note that the values of transmittance in the stop bands (shaded portions) go to zero as the number of layers increase. It is also important to point out that the ripple in the passbands can be reduced by the use of suitable matching layers between the stack and air and substrate.

Figure 6-61 is a useful reference for defining the characteristics of an interference filter.

As can be seen from Figure 6-60, the passbands of this type of interference filter are wider than usually employed in photographic remote sensing. A more suitable form of interference filter for remote sensing has been described (McKenney and Slater, 1970). The design is complicated, and no attempt has so far been made to construct one. An interference filter construction that can provide passbands in the approximate range of 2.5×10^{-2} to 1×10^{-4} μm follows that of the well known Fabry-Perot etalon, except that the air space between the two high-reflectance surfaces is replaced by a dielectric. For the highest efficiency bandpass filters, the high-reflectance layers are the dielectric stacks described earlier. When the bandpass is wide, aluminum films can be used instead of the dielectric stacks.

In general, all forms of interference filters show unwanted transmittances at higher and lower wavelengths than those of interest. Higher wavelength transmittances can only be eliminated by an additional interference filter (or filters) with suitably located stopbands. Unwanted lower wavelength transmittances, although they can be removed by another interference filter (or filters), are usually easier to remove by absorption long wavelength pass filters. Filters used to remove unwanted transmittances are referred to as *blocking filters.*

POLARIZATION FILTERS

A polarization filter can be used to *penetrate* haze, thereby significantly increasing the contrast, and thereby the resolution of aerial photographic imagery. It has also been suggested that a polarization spectral signature could be a useful additional means of discriminating ground scenes. Polarization effects are sensitive to the geometry of the photography; thus solar altitude, nadir, and azimuthal angles are all important factors.

Little information seems to be available on the optical quality of polarization filters. Some forms of filter will definitely degrade the spatial resolution of the camera. However, there are other forms that consist of a thin polarizing film (approximately 100 μm), which can be coated on an optical quality glass substrate. Although no specific data are available, it would seem likely that such filters could be made of a quality commensurate with that of the highest quality absorption glass filters.

CAMERAS
CONVENTIONAL CAMERAS

The salient features of conventional types of aerial cameras—mapping frame cameras, frame reconnaissance cameras, panoramic cameras, and strip cameras—were described in the first edition

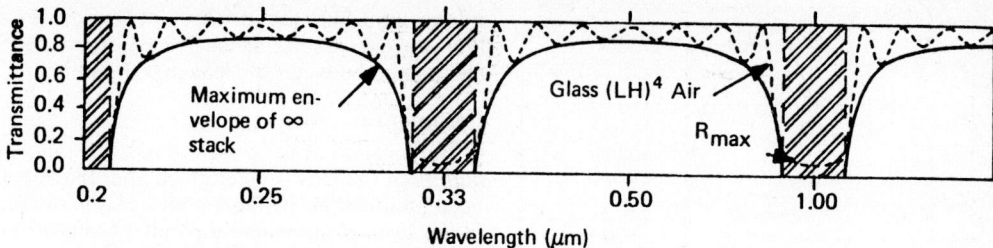

(a) Computed spectral transmittance of an eight-layer quarterwave stack (---) and its envelope of maximum transmittance (solid line) $n_H t_H + n_L t_L = \lambda_0/2$.

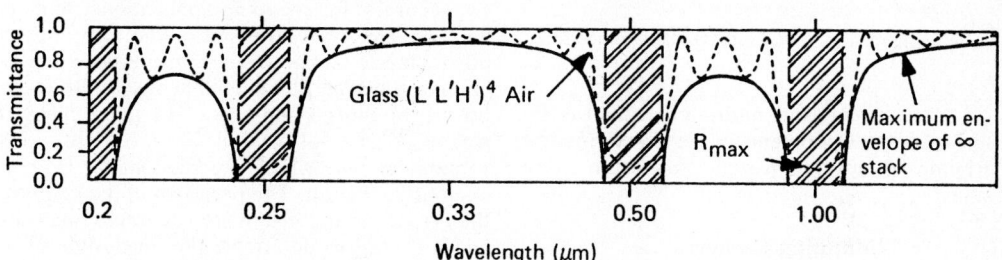

(b) Computed spectral transmittance of an eight-layer 2:1 stack (---) and its envelope of maximum transmittance (solid line). $2n_L t_L + n_H t_H = \lambda_0/2$.

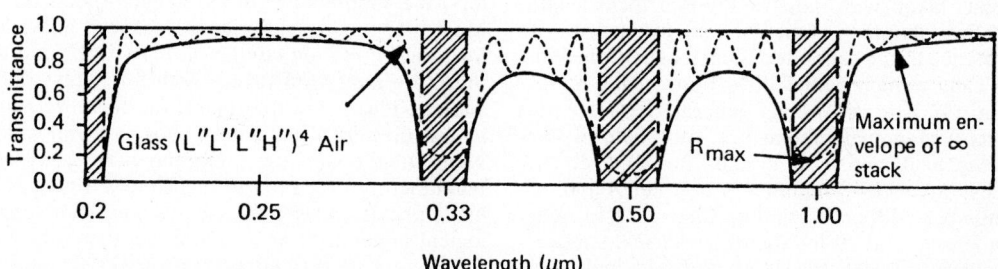

(c) Computed spectral transmittance of an eight-layer 3:1 stack (---) and its envelope of maximum transmittance (solid line). $3n_L t_L + n_H t_H = \lambda_0/2$.

Fig. 6-60. Spectral transmittance curves for three different multilayer designs, from Baumeister in *Mil-Hdbk-141*, Defense Supply Agency (1962).

of the *Manual of Remote Sensing* (Slater, 1975). Tabulations of most of the widely used cameras in the United States (worldwide in some categories) were included, with listings of the operational characteristics commonly needed in the selection of a camera for a particular mission.

A well-illustrated description of "air survey" cameras can be found in *Atlas of Photogrammetric Instruments* (Cimerman and Tomasegovic,

1970); other useful listings have been made by Data Corporation (1965 and 1967); McDonnell-Douglas Corporation (1979) and the *Manual of Photogrammetry* (American Society of Photogrammetry, 1981). Because of the detailed coverage provided in these references to reconnaissance frame and panoramic cameras, only such cameras associated with Earth resources space missions will be described here. Emphasis is

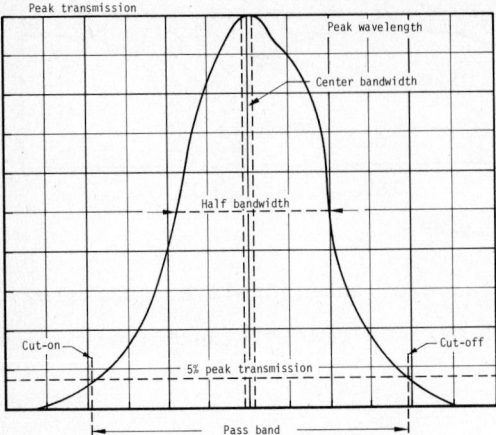

Fig. 6-61. Characteristics of typical interference filter.

placed on mapping and multiband cameras including recent developments in these systems and descriptions of systems planned for future space missions.

Mapping Cameras

Mapping cameras (often referred to as metric cameras or cartographic cameras) are the simplest type of aerial camera to describe because all are of basically the same general configuration. The distinctive feature of the mapping camera is its high degree of distortion correction. Aerial photographs taken with modern 150-mm focal length, 90° field mapping cameras, and measured with reference to the calibrated focal length of the lens, typically exhibit radial distortion values of less than ±10 μm. Additional general characteristics of most mapping cameras are: total field of view across the diagonal: 90° or 120°; format size: 23 × 23 cm (9 × 9 in); f-number is between 4 and 6.3; T-number with antivignetting filter is 10 for modern lenses, and 20 for the older lenses; shutter—intralens; fiducials for location of principal points of format are an integral part of lens cone; reseau and fiducial markers are exposed on film simultaneously with exposure of ground scene; platen—vacuum; and extensive flight and camera data are recorded alongside each frame.

Lens Cone Assembly

Mapping cameras differ in design from frame reconnaissance cameras primarily in the lens-cone assembly. The inner cone of a mapping camera is a square frame on which the fiducial markers are attached to the lens-mount assembly with Invar supporting members. An outer cone, surrounding the inner cone, protects the inner cone and shields extraneous light from the film plane. The alignment of the lens in the inner cone and the position of the fiducials are established with high precision. The film, which is held flat on a vacuum platen, is brought into contact with the square frame of the inner lens cone prior to exposure, and thus a very precise geometrical relationship is maintained between the lens and the film.

Collimation

The location of the *principal point of autocollimation* of the lens is determined after the lens has been mounted in the inner cone. This is the point in the focal plane where a parallel beam of light, perpendicular to the film plane, is brought to a focus by the lens. The fiducial marks are then adjusted so that straight lines joining opposite fiducials intersect at the principal point of autocollimation.

The relative positions of image points across the photograph have to be determined to the highest accuracy for mapping applications, and thus any stretching or shrinking of the film after exposure has to be known. For this purpose a reseau is imaged on the film simultaneously with the exposure of the ground scene. In most cases the reseau consists of several fine cross slits in the vacuum platen, which are illuminated from the back of the platen. The positions of the centers of the crosses in the platen are accurately measured and are used to determine the magnitude of any dimensional change in the film.

Lenses

The emphasis on wide field and freedom from distortion results in lenses for mapping cameras of moderate f-number and, with antivignetting filters, the T-number is in the neighborhood of 10. Most mapping cameras do not have a provision for *forward motion compensation* (FMC) and are therefore restricted to use with fairly fast films, such as Plus-X. Furthermore, the flight velocity, v, and altitude, h, have to be chosen to yield a low value for the v/h ratio. Mapping camera systems manufactured recently in the United States have made provision for FMC—a considerable technological advance. For example the Fairchild KC-6A camera, described by Norton (1968), utilizes film-plane FMC. Figure 6-62 shows a cutaway view of the Geocon IV lens used with the system, and Figure 6-63 shows the FMC fiducials and other special features incorporated in many modern cartographic cameras.

Viewfinder

The camera is only part of the overall airborne system used for taking aerial mapping photographs. There is also a viewfinder which, in addition to showing the camera operator the scene being photographed, determines the v/h ratio. An intervalometer can then be set to give the desired percentage of forward overlap for stereoscopic purposes.

Resolving Power of Mapping Cameras

The wide field and freedom from distortion required of a mapping camera lens are achieved at

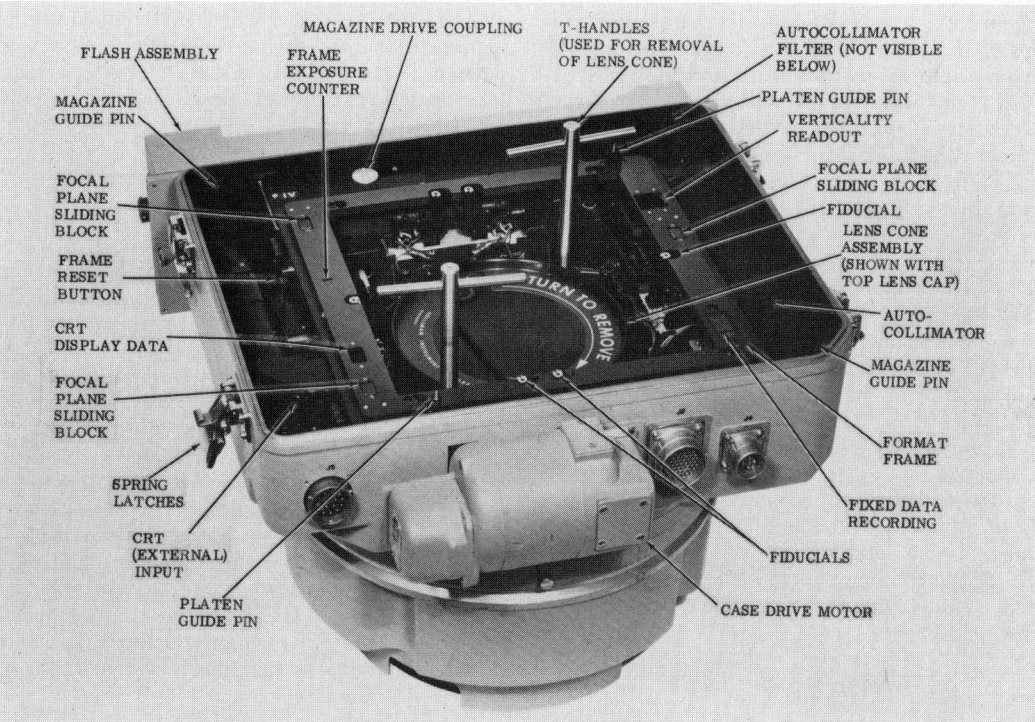

Fig. 6-62. Cutaway view of the Geocon IV mapping lens.

Fig. 6-63. View of the KA-6A camera showing FMC fiducials and other special features.

TABLE 6-10

Mapping—Frame Aerial Cameras

nera	F-639	KC-1A	KC-1B	KC-4	KC-6A	T-11 & T-11B	Model VI
nufacturer	Fairchild	Fairchild	Fairchild	Fairchild	Fairchild	Fairchild	Galileo Corp. of Amer.
nary Use	med/hi alt mapping	med/hi alt mapping	med/hi alt mapping	med/hi alt mapping	med/hi alt mapping	med/hi alt mapping	mapping
mat (cm)	23 × 23	23 × 23	23 × 23	23 × 23	23 × 23	23 × 23	23 × 23
n Length (m)	215	120	120	120	120/240	120	110
e Thickness (μm)	thin	standard	standard	standard	standard/thin	standard	NA
mes/roll	770	450	450	450	470/940	450	NA
gazine	integral	integral	integral	integral	integral	integral	interchangeable
le Time (sec)	2.0 maximum	2.5	2.5	2.5	2.0 maximum	3.0	5.0
C Type	moving platen	none	none	none	moving platen	none	NA
C Rate (mm/sec)	1.3–30.5	NA	NA	NA	2.5–20.3	NA	NA
C Error	NA	NA	NA	NA	NA	NA	NA
'AR, 1000:1 (ℓp/mm)	55	20	20	38	45*	20	NA
n	Pan-X	8401	8401	8401	5401	8401	NA
cessing	NA	NA	NA	NA	NA	NA	NA
e of Measurement	laboratory dynamic	dynamic, inflight	dynamic, inflight	dynamic, inflight	laboratory dynamic	dynamic, inflight	NA
npatible Lenses	Geocon I	Planigon II	Planigon II	Geocon I	Geocon IV B**	Metrogon II	Orthogon
nufacturer(s)	Fairchild	B&L, Curtis, Goerz	Goerz	Fairchild, J.G. Baker	Kollsman	B&L	Officine Galileo S.p.A.
al Length (mm)	150	150	150	150	150	150	
o. Range	5.6	6.3–11	6.3–11	5.6–11	5.0–11.2	6.3–11	5.6–11
velength Range (μm)	0.4–0.7	0.4–0.7	0.4–0.7	0.4–0.7	0.45–0.7	0.4–0.7	NA
ers	W12, W25, anti-vignetting	clear or yellow anti-vignetting	clear or yellow anti-vignetting	yellow anti-vignetting	yellow anti-vignetting	yellow anti-vignetting	NA
tortion (μm) radial							
-tangential)	<10	NA	<10	<10	±8R,<8T	NA	±20
i be Used W/Normal or Film?	yes	yes	yes	yes	yes	yes	yes
i be Used W/IR or Film?	NA	NA	NA	NA	yes	NA	yes
inge of Focus quired for IR or Film?	NA	NA	NA	NA	no	NA	no
gular Coverage	74° × 74°	74° × 74°	74° × 74°	74° × 74°	74° × 74°	74° × 74°	74° × 74°
itter Speed (sec)	1/25–1/900	1/10–1/500	1/10–1/500	1/25–1/500	1/50–1/800	1/10–1/500	1/125–1/300
itter Type	intralens Rapidyne	intralens Rapidyne	intralens Rapidyne	intralens Rapidyne	intralens Rapidyne	intralens Rapidyne	NA
ight (kg), w/o n & Mount	45 w/film	43 w/film	43 w/film	43 w/film	66 w/film	39 w/film	36
unt(s)	hard, w/vibration isolators	A-8, -11, -25, -28, -32, ART-21, -25	A-8, -11, -25, -28, -32, ART-21, -25	A-28, ART-25, Model 6	LC-7A	A-8, -11, -25, -28, -32, ART-21, -25	special
a Annotation	counter, clock, card, etc.	counter, clock, card, etc.	counter, clock card, etc.	counter, clock card, etc.	CRT, counter, fixed data	counter, clock, card, etc.	NA
rvalometer	yes	yes	yes	yes	yes	yes	yes
posure Control	auto (AEI 6, 20, 64)	manual	manual	manual	auto (AEI 20–2,000)	manual	NA
narks	platen reseau	no reseau	no reseau	platen & peripheral reseu	platen reseau	no reseau	reseau optional

* minimum req'd by specification
** high contrast AWAR, lens only, is 60 lp/mm on Pan-X process in D-19 w/γ = 1.4

the cost not only of *T*-number, but also of resolving power. The high-contrast axial resolving power of modern mapping cameras tested with operational films is found to be about 70 cycles/mm, whereas the AWAR is less than 60 cycles/mm, sometimes as low as 40 cycles/mm. These values compare unfavorably with those pertaining to reconnaissance cameras of the same focal length but smaller field in which AWAR values greater than 100 cycles/mm are common.

The results of resolving power tests of seven cartographic cameras used with various color films have been reported by Tayman et al (1968). Welch and Halliday (1973) have reported the imaging characteristics of a Wild RC8 camera with a Universal Aviogon 152-mm-focal-length lens and a Zeiss RMK A 30/23 with a Topar lens of 306-mm focal length. The evaluations were made in terms of modulation transfer function, resolving power, and the detectability and measurability of small detail. In addition, the limiting capabilities of the cameras were determined by analyzing laboratory and aerial photographs and taking into account such factors as image motion, film type, and target contrast.

Mapping Camera Listing

Table 6-10 is a listing of most modern mapping cameras in worldwide use. Parameters and criteria listed for these and other tabulated cameras are in accordance with data supplied by the manufacturers and do not represent results of independent tests. Values for film length, weight, focal length, etc., should be considered as nominal.

The Large Format Camera

The Large Format Camera is a high performance mapping system to be flown in an early Space Shuttle flight. The camera, described in detail by Doyle (1979) and Mollberg (1979) has a 305-mm focal length and a 230-mm × 460-mm format—hence its name.

The camera system is shown in Figure 6-64, and the parameters of interest are listed in Tables 6-10 and 6-11. The system is flown with the long dimension of the format in the flight direction in order to obtain the necessary stereo coverage for topographic map compilation.

The predicted ground resolutions (in meters/cycle) for a variety of different films and as a function of orbital attitude are shown in Figure 6-65. For altitudes less than 300 km, the higher resolving power films will provide ground resolutions comparable to the Earth Terrain Camera, described in the next section.

The positioning capability of the Large Format Camera is shown in Figure 6-66. If measurements

TABLE 6-10 (Continued)

MRB 15/2323	MRB 9/2323	MRB 11 5/1818	RC-10			RMK A 8.5/23
Jena	Jena	Jena	Wild Heerbrugg			Carl Zeiss
med/hi alt mapping	med/hi alt mapping	med/hi alt mapping	general mapping			aerial survey
23 × 23	23 × 23	18 × 18	23 × 23			23 × 23
120–150	120–150	120–150	150			120
NA	NA	NA	100			160
450	450	540 maximum	580			470
interchangeable w/MRB 9	interchangeable w/MRB 15	NA	cassettes			NA
2.0	2.0	NA	1.5			2-180
NA	NA	NA	none			NA
NA	NA	NA	NA			NA
NA	NA	NA	NA			NA
NA	NA	NA	55°*		NA	NA
NA	NA	NA	NA	NA	NA	NA
NA	NA	NA	NA	NA	NA	NA
NA	NA	static	static	NA	NA	NA
Lamegon PI	Super Lamegon	Lamegon	Universal Aviogon	Universal Aviogon II	Super Aviogon II	S-Pleogon A
Jenoptic Jena GmbH	Jenoptic Jean GmbH	Jenoptic Jena GmbH	Wild Heerbrugg	Wild Heerbrugg	Wild Heerbrugg	Carl Zeiss
150	90	115	150	150	90	85
4.5–8	5.6–11	4.0–8.0	5.6–22	4	5.6–22	4.–8.
NA	NA	NA	0.45–0.85	0.4–0.8	0.45–0.85	0.4–0.9
anti-vignetting	anti-vignetting	yellow, orange	yellow & red (1.4 A.V.)	NA	yellow & red (2.2 A.V.)	Zeiss KL, KLF, A, B, C, D, E, F, G, H, I, K
±6R, <5T	±7	±8R, <5T	±10R, <5T	<4	±10R, <5T	7
yes	yes	NA	yes	yes	yes	yes
yes	yes	NA	yes	NA	yes	yes
no	no	NA	no	no	no	no
74° × 74°	104° × 104°	90° × 90°	74° × 74°	74° × 74°	104° × 104°	107° × 107°
1/100–1/1,000	1/50–1/500	1/100–1/1,000	1/100–1/1,000	1/100–1/1,000	1/100–1/1,000	1/50–1/500
rotating disc	rotating disc	rotating disc	rotary	rotary	rotary	Aerotop rotating disc
63	52	37	78	NA	75	60
Universal MRB-A	Universal MRB-A	Universal MRB-A	PAV-10	NA	NA	ASV shock & vibration absorbing auxiliary
clock, altitude, gray scale, etc.	clock, altitude, gray scale, etc.	clock, altitude, gray scale, etc.	clock, altitude, counter, etc.	clock, altitude, counter, etc.	clock, altitude, counter, etc.	
yes	yes	yes	yes	yes	yes	IRU
optional	optional	optional	manual, remote	manual, remote	manual, remote	auto w/EMI 2
peripheral reseau	peripheral reseau	peripheral reseau	no reseau	no reseau	no reseau	no reseau

* measured to NBS specifications, lens only

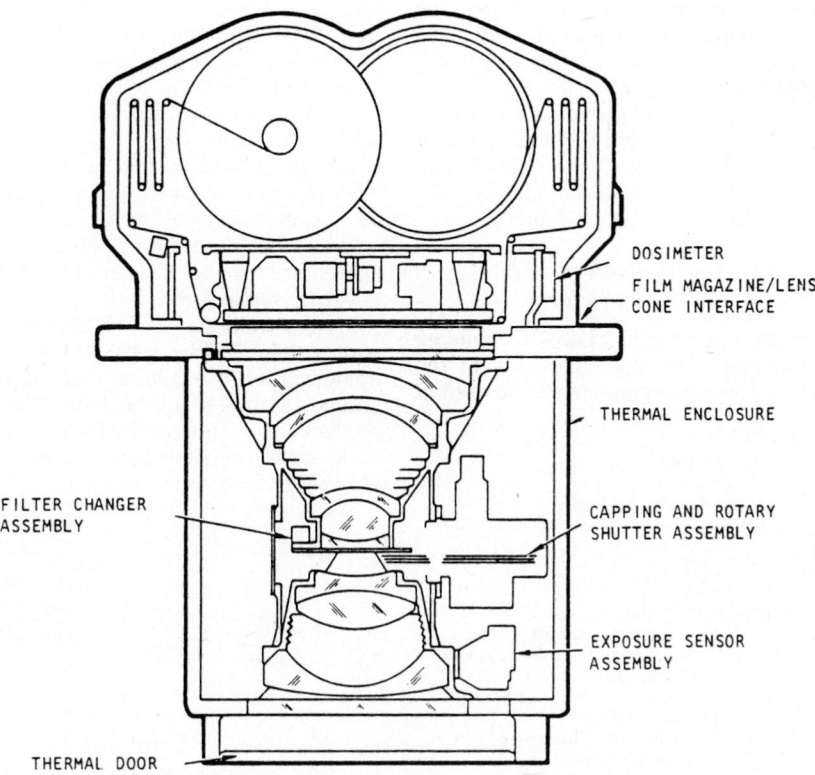

Fig. 6-64. The Large Format Camera (courtesy ITEK Corp.).

TABLE 6-10 (Continued)

Camera	RMK A 15/23	RMK A 21/23	RMK A 30/23	RMK A 60/23	Lunar Mapper
Manufacturer	Carl Zeiss	Carl Zeiss	Carl Zeiss	Carl Zeiss	Fairchild
Primary Use	aerial survey	aerial survey	aerial survey	aerial survey	lunar mapping
Format (cm)	23 × 23	23 × 23	23 × 23	23 × 23	11.4 × 11.4
Film Length (m)	120	120	120	120	460
Base thickness (μm)	160	160	160	160	thin
Frames/Roll	470	470	470	470	3,000
Magazine	NA	NA	NA	NA	external cassettes
Cycle Time (sec)	2–180	2–180	2–180	2–180	NA
FMC Type	NA	NA	NA	NA	moving platen
FMC Rate (mm/sec)	NA	NA	NA	NA	10–40 mrad/sec
FMC Error	NA	NA	NA	NA	NA
AWAR, 1000:1 (ℓp/mm)	51*	NA	NA	NA	NA
Film	Aviphot Pan 30 PE	NA	NA	NA	NA
Processing	Perutin	NA	NA	NA	NA
Type of Measurement	NA	NA	NA	NA	NA
Compatible Lenses	Pleogon A	Toparon A	Topar A	Telikon A	
Manufacturer(s)	Carl Zeiss	Carl Zeiss	Carl Zeiss	Carl Zeiss	Fairchild
Focal Length (mm)	150	210	305	610	80
f/No. Range	5.6–11.	5.6–11.	5.6–11.	6.3–12.5	NA
Wavelength Range (μm)	0.4–0.9	0.4–0.9	0.4–0.9	0.4–0.9	0.4–0.7
Filters	Zeiss KL, KLF, A, B, C, D, E, F, G, H, I, K	Zeiss KL, KLF, A, B, C, D, E, F, G, H, I, K	Zeiss KL, KLF, A, B, C, D, E, F, G, H, I, K	Zeiss, KL, KLF, A, B, C, D, E, F, G, H, I, K	NA
Distortion (μm) (R-radial T-tangential)	5	4	3	50	NA
Can be Used W/Normal Color Film?	yes	yes	yes		
Can be Used W/IR Color Film?	yes	yes	yes	yes	yes
Change of Focus Required for IR Color Film?	no	no	no	yes	NA
Angular Coverage	74° × 74°	57° × 57°	41° × 41°	no	NA
Shutter Speed (sec)	1/100–1/1,000	1/100–1/1,000	1/100–1/1,000	23° × 23°	74° × 74°
Shutter Type	Aerotop rotating disc	Aerotop rotating disc	Aerotop rotating disc	1/100–1/1,000 Aerotop rotating disc	1/15–1/240 rotary shutter
Weight (kg), w/o Film & Mount	57	55	57	54	NA
Mount(s)	AS shock & vibration absorbing	ASII, ASIII shock & vibration absorbing	ASII, III shock & vibration absorbing	ASI, III shock & vibration absorbing	special
Data Annotation	auxiliary	auxiliary	auxiliary	auxiliary	time, altitude, exposure
Intervalometer	IRU	IRU	IRU	IRU	NA
Exposure Control	auto	auto w/EMI 2	auto w/EMI 2	auto w/EMI 2	auto
Remarks	optional reseau with Pleogon AR lens	no reseau	optional reseau with Topar AR lens	no reseau	reseau

*typical resolution at 1.6:1
contrast is 27 (ℓp/mm)

System consists of mapping
and Stellar camera. Film
flattened by glass plates

of 15-μm accuracy are made off the imagery, the corresponding ground position accuracy will be 15 m, which is appropriate for map compilation at a scale of 1:50,000.

The relative elevation accuracy as a function of altitude for three base-to-height ratios is shown in Figure 6-67. Note that an accuracy of about 10 m at an altitude of 300 km should be adequate for compiling contours at 30-m vertical intervals.

The Metric Camera Experiment for Spacelab-1

The European Space Agency (ESA) is building modular components for the cargo bay of the NASA Shuttle. These components will be used in the joint ESA-NASA mission designated as Spacelab-1. This will be a Shuttle sortie mission of one week's duration at an altitude of 250 km and an inclination of 57° to be launched in the spring of 1983. Among many other experiments from both NASA and ESA, Spacelab-1 will carry an Earth-observing camera system mounted to an optical window in the manned module. The camera will be a standard Zeiss aerial camera type RMK A30/23, with three film magazines, modified for operation in space. The parameters for this camera are given in Table 6-10.

From an altitude of 250 km, the scale of the photography will be 1/820,000, the coverage per frame will be 190 × 190 km, and the area weighted ground resolution will be 20 m. The photographs produced should be useful for map compilation at scales up to 1:50,000 and for many geoscience applications of photointerpretation.

TABLE 6-11

Large Format Camera

Lens:	
Focal length	305 mm
f number	6.0
Chromatic correction	0.4–0.9 μm
Distortion	15 μm max over format 10 μm over central 230 × 230 mm
Format	230 × 460 mm
Exposure range	3 to 24 ms
FMC range	10 to 45 mrad s^{-1}
Reseau	50-μm-diameter illuminated apertures at 50-mm intervals
Magazine capacity	1200 m of film for 2400 frames

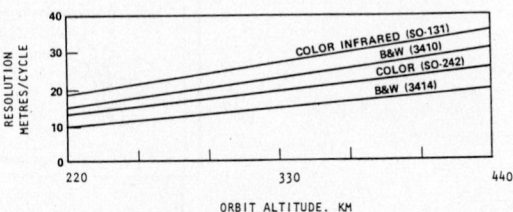

Fig. 6-65. Ground resolutions of the Large Format Camera as a function of altitude, from Doyle (1979).

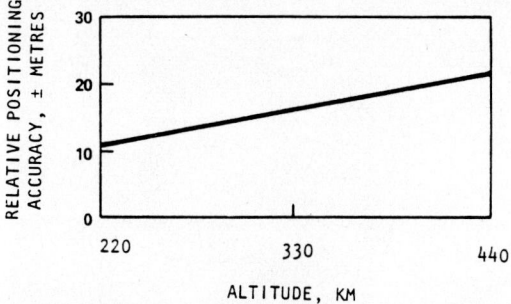

Fig. 6-66. Relative ground position accuracy as a function of altitude for the Large Format Camera, from Doyle (1979).

The Spacelab-1 camera is Phase A of an overall ESA camera development program called ATLAS being conducted by the Federal Republic of Germany. Phase B, presently under discussion, will consider several options:

a. Add image motion compensation (IMC) to the RMK A30/23 camera and mount it in a pressurized container on the external Shuttle pallet.

b. Increase the focal length to 60 cm, add IMC, increase the film magazine capacity, and mount on external pallet.

c. Develop a new camera with 60-cm focal length, IMC, and 11.5-cm × 23-cm format for operation in the manned module.

Phase C of the ATLAS program will investigate the operation of these cameras on free-flying spacecraft launched and serviced by the Shuttle.

Earth Terrain Camera System

The Earth Terrain Camera (ETC) consisted of a 46-cm focal length lens covering a film format 11.25 cm square. The ETC used on Skylab was referred to as the S190B experiment and was provided by Actron Industries, Inc., under contract to NASA. ETC was basically a high-performance frame reconnaissance camera with rocking camera forward motion compensation and a lens chromatically corrected over the extended wavelength range 0.40 to 0.90 μm.

An annotated view of the camera is shown in Figure 6-68. The outer lens barrel, which was attached to a window in the spacecraft, provided the mount for the camera. The forward motion compensation (FMC) mechanism, working against the stationary camera mount, rocked the lens cone, body assembly, and magazine about the lens-cone pivot shown in the figure. The bidirectional focal plane shutter yielded speeds of 1/100, 1/140, and 1/200 s. The forward motion compensation had a variable rate of from 0 to 25 mrad/s. The cycle rate could be varied from 0 to 25 frames/min in an automatic mode and could operate in a single-frame mode.

Predicted resolving powers for the system as a function of film and filter are listed in Table 6-12.

The ETC closely resembles the lunar topographic camera listed in Table 6-10, other operational details can be found there. The main difference is that the lens for the ETC is color corrected for a much broader wavelength range than is the lens for the lunar topographic camera.

Skylab photographs were recorded on Eastman Kodak reconnaissance films types 3414, SO-242, and SO-131 at a scale of approximately 1:946,000. Of these films, 3414 and SO-242 are high-resolution panchromatic and color films, whereas SO-131 is a color-infrared film with greatly improved image structure properties as compared to the commercially available type 3443. Significantly, the duplicating films 2430 and 2447 are inferior to the original emulsions.

The resolving power capabilities of the ETC system under laboratory conditions are summarized in Table 6-13 for first-generation (original) photographs recorded on 3414, SO-242, and 3443 films (Gimlett, 1975a). Although these resolution values demonstrate the inherent capabilities of the camera system, they are not a realistic measure of the quality of the second-generation photographic products distributed to investigators.

System performance and second-generation photo quality were assessed (Welch, 1976) using the MTF analysis techniques described in the following paragraphs. Referring to Figure 6-69, the appropriate component MTF's were cascaded to obtain the predicted system MTF's given in Figure 6-70. These predicted MTF's represent theoretical performance levels that might be achieved in a laboratory environment and are indicative of optimum quality second-generation photographs. Uncompensated image motion, vacuum failure, incomplete contact between the original and duplicating materials during the reproduction process, or any of a number of other factors can degrade photographs recorded and duplicated under operational conditions.

MTF's for selected second-generation ETC photographs provided by NASA were next determined from microdensitometer edge traces performed across field boundaries and aircraft runway patterns with a Joyce Loebl Mark III CS microdensitometer (2 × 125 μm effective slit). These edge traces were then converted to system MTF's using procedures previously described by

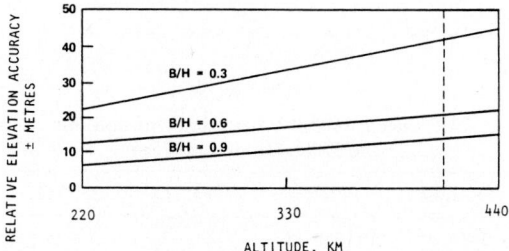

Fig. 6-67. Relative elevation accuracy as a function of altitude for the Large Format Camera, from Doyle (1979).

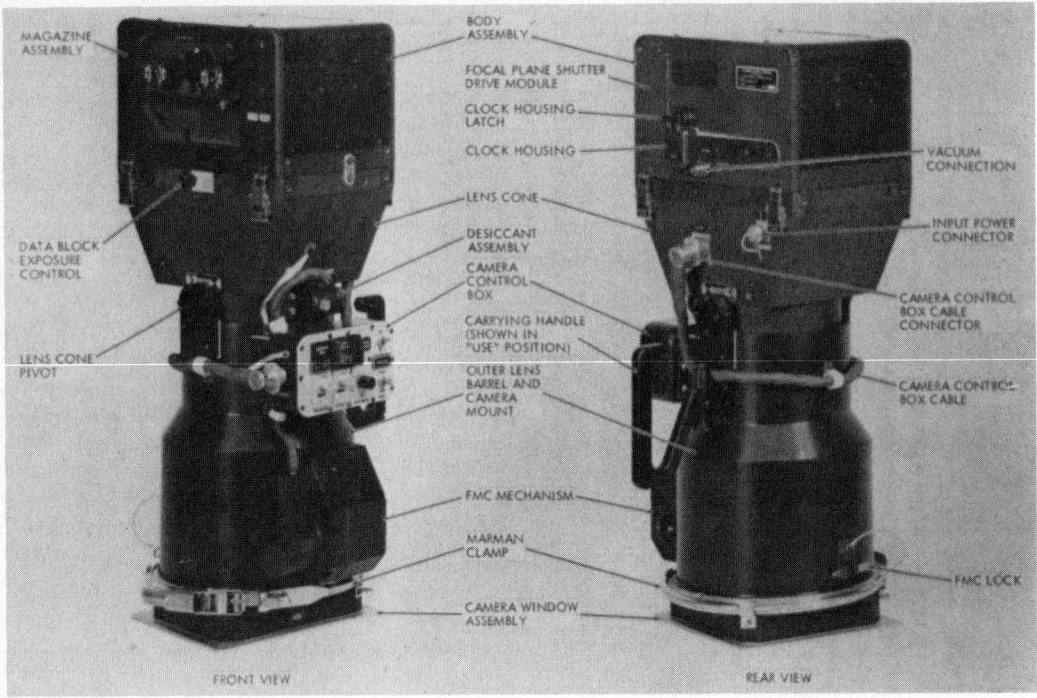

Fig. 6-68. Earth Terrain Camera system (courtesy of NASA and Actron Industries, Inc.).

Welch (1971). Sensitometric data for both the original and duplicating films were obtained from NASA (1973a,b,; 1974a).

The mean MTF's derived from these analyses are shown in Figure 6-71, and a comparison of these measured operational MTF's with the predicted curves in Figure 6-70 indicates correspondence (in response) to within approximately 10 per cent for 3414/2430; 25 per cent for SO-242/2447; and 2 per cent for SO-131/2447. In each of these comparisons the measured MTF's are reduced from the predicted curves by the given percentages, a pattern also noted in the earlier analyses of the S-190A MPF photographs. With the exception of the SO-242/2447 photographs, the correspondence between measured and predicted MTF's is excellent and confirms that ETC operational system performance was about as expected.

The quality of the second-generation photographs specified in terms of low-contrast resolution values is of particular interest to photogram-

metrists. In the absence of imaged targets, resolution estimates for a target contrast ratio of 1.6:1 at the camera lens were obtained by translating the measured MTF's to a response of 23 per cent (which is equivalent to a contrast ratio of 1.6:1) on the log-log paper and accepting the indicated maximum spatial frequency in the threshold modulation range of 5 to 10 per cent. A small upward adjustment was made for the color photographs because color differences as well as modulations contribute to the recorded resolution. The general procedure is illustrated in Figure 6-72, and the estimated resolution values for the ETC photographs are given in Table 6-14. Based on the estimated resolution values, ground resolutions of 15 to 30 m are representative for ETC photographs, as compared to 60 to 145 m for MPF (S-190A) images.

Suitable scales for photomaps produced from the Skylab ETC photographs were objectively estimated, again with the aid of the MTF's (Welch,

TABLE 6-12

Predicted Resolving Power Values in Lines/mm for
The Earth Terrain Camera

Film Type	SO-242	3443	3400	3414
Filter	None	W-12	W-12	W-12
AWAR 1000:1 contrast	100	44	75	180
AWAR 2:1 contrast	65	25	50	100

TABLE 6-13

Measured AWAR Resolution (lpr/mm) for
the ETC (Gimlett, 1975a)

EK Film	Filter	1000:1 TOC	2:1 TOC
3414	Wratten 12	206	143
SO-242	None	137	127
3443	Wratten 12	56	33

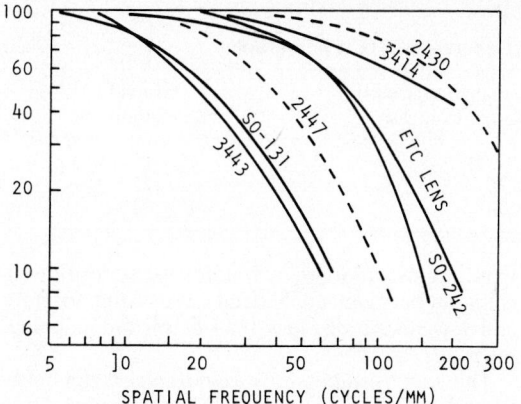

Fig. 6-69. MTFs for the ETC lens (Gimlett, 1975b) and the Eastman Kodak films employed to record and duplicate ETC photographs.

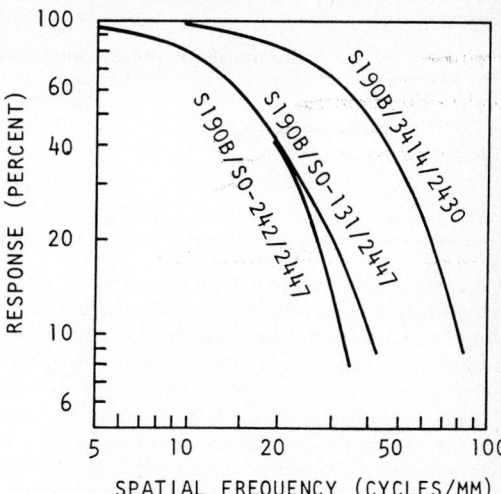

Fig. 6-71. Average measured MTFs for properly exposed second-generation ETC photographs.

1972). In Figure 6-73 the measured image MTF's and the MTF of the human eye are plotted. Based on this figure, the measured MTF's for the panchromatic and color/color infrared photographs must be translated to the left by factors of approximately 20× and 10× respectively in order to correspond to the MTF for the eye. Since these translation factors may be regarded as enlargement ratios at which the image will begin to appear blurred, maximum scales for "sharp" photomaps are limited to *approximately* 1:50,000 scale for products prepared from the black-and-white photographs and 1:100,000 scale if the color/color infrared photographs are employed. These objectively determined enlargement ratios generally have been confirmed by U.S. Geological Survey experiments involving the production of photomaps from ETC photographs.

MULTIBAND CAMERAS

The multiband camera plays a leading role among remote-sensor instruments. For this reason this section will provide an updated and detailed description of most of the systems that have been developed for aircraft and spacecraft platforms. An earlier review of this subject has been made by Slater (1972).

Requirements and Tolerances

In multiband photography, a camera records a number of images of a scene through several spectral filters. The imagery is ordinarily converted into black-and-white positive transparencies, which are illuminated through a second set of spectral filters and brought into register to form an additive color display. In other instances the various images are scanned, and the outputs are operated on either simultaneously or sequentially, and in analog or digital processors.

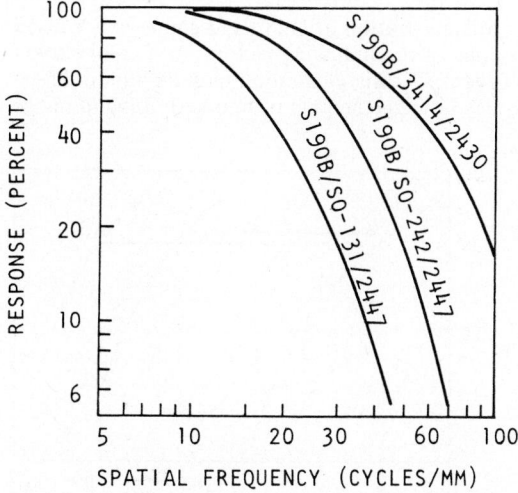

Fig. 6-70. Predicted MTFs for second-generation ETC photographs obtained by cascading the lens MTF with the appropriate MTFs for the original and duplicating films.

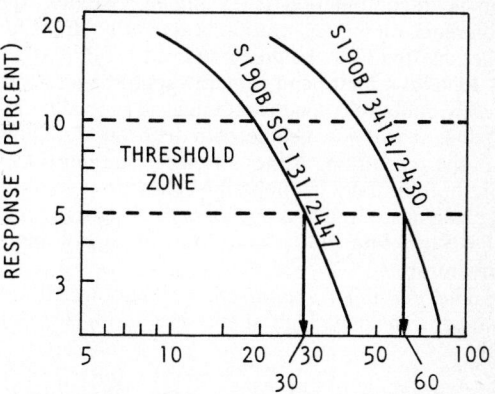

Fig. 6-72. Resolution prediction for low-contrast (1.6:1) targets.

TABLE 6-14

Resolution Estimates for Second Generation ETC Photographs

Film/Duplicating Film	Resolution Estiamtes (lpr/mm) for 1.6:1 Target Contrast	Ground Resolution, m
3414/2430	60–70	15
SO-242/2447	35	25
SO-131/2447	30	30

In any event, multiband photography is concerned with a comparison of signals, the registration of which governs the spatial resolution of the recombined imagery. *The spatial resolution of the recombined imagery is termed multiband spatial resolution, and is defined as the limiting resolution for which spectral fidelity is maintained in a multiband recombination.* This limit is the result of *color* fringing due to misregistration, and/or spatial resolution differences between multiband images. According to this definition, the multiband spatial resolution of any perfectly registered set of imagery is the spatial resolution of the lowest resolution band of imagery. Also, for a set of imagery of identical spatial resolution, R, the multiband spatial resolution, R_M, resulting when one or more images is misregistered, is approximated by

$$R_M \simeq R/(1 + 10^{-3} Rd), \qquad (6\text{-}50)$$

where:

  d = distance of misregistration, in μm.

The exact form of this expression will change with the shape of the *spread function*[21] of the system; however, for practical purposes the approximation is sufficient. In fact, the accuracy of measurement of registration error is commensurate with the approximation. The definition of multiband spatial resolution will be expanded upon in succeeding paragraphs. However, it is worthwhile noting at this point that the above definition of multiband resolution is based solely upon geometrical considerations, thus it applies to any machine analysis of the data where the images are scanned in high resolution. When a multiband image recombination is examined visually, the tolerance on misregistration is generally larger for the blue than for the green and red bands.

Ideally, a multiband camera should be an accurately calibrated spectroradiometric recording instrument with low geometrical distortion and high spatial resolution. Although these characteristics are not mutually incompatible, they do constitute a challenge to the lens and camera designer and to the filter, film, and electro-optical sensor manufacturer.

The *geometric requirements* for multiband cameras include: high spatial resolution in each band; registration to a fraction of a resolution element between each band across the format; and low distortion, less than 5 μm for mapping purposes.

The *spectroradiometric requirements* for multiband cameras include: uniform spectral irradiance across format; sharply defined spectral sensitivity in each band; and shutter repeatability.

Resolution

In spite of the special problems created by the need to have precisely registerable images from different spectral bands, it is possible and practical for a set of highly corrected, matched $f/2.8$ lenses covering fields of 25° to be designed, each covering a different 0.1-μm wavelength range in the visible and photographic infrared regions of the spectrum. For application to the currently available electro-optical sensors, such as vidicons and solid-state arrays with resolving powers of less than 100 cycles/mm, a well-corrected $f/2.8$ optical system is more than adequate. Film comes closer to exploiting the potential of high-performance designs; for example, Eastman Kodak's high-definition film No. 3414 can yield a resolution on the order of 400 cycles/mm with an $f/2.8$ diffraction-limited lens for a high-contrast target.

Registration

Misregistration in multiband cameras is caused by one of the following factors, or a combination of them. (1) Misregistration may be due to differences in image heights from one multiband image

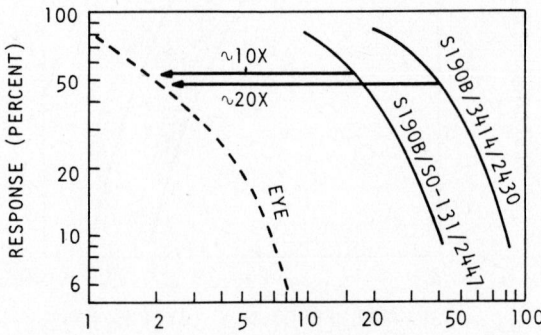

Fig. 6-73. MTD analyses indicate that ETC high-resolution panchromatic and color/color-infrared photographs may be enlarged by factors of approximately 20X and 10X before they will appear blurred.

---

[21] The spread function of a system is the image irradiance distribution of a point source object, such as a star.

TABLE 6-15

Multiband Camera Spatial Resolution as a Function of Misregistration

Resolution, cycles/mm		Maximum misregistration in terms of −		Curve number when applicable (Figures 6-74 and 6-75)
On each film	Of misregistered film combination	Resolution elements	μm	
10	5	1	100	1
10	7	0.5	50	2
10	8	0.25	25	3
20	10	1	50	2
20	13	0.5	25	3
20	16	0.25	12.5	4
40	20	1	25	3
40	27	0.5	12.5	4
40	32	0.25	6.3	5
60	30	1	16.7	—
60	40	0.5	8.9	—
60	48	0.25	4.1	—
80	40	1	12.5	4
80	53	0.5	6.3	5
80	64	0.25	3.1	6
100	50	1	10	—
100	67	0.5	5	—
100	80	0.25	2.5	—
120	60	1	8.3	—
120	80	0.5	4.2	—
120	96	0.25	2.1	—
140	70	1	7.1	—
140	93	0.5	3.6	—
140	112	0.25	1.8	7

to another because of aberrations or assembly errors in the image-forming optics, errors such as differential asymmetrical distortion, chromatic variation in focal length, chromatic distortion, or lateral chromatic aberration. (2) Boresighting error due to the lack of parallelism of the axes of the various optical channels gives rise to a difference in *keystoning* between the various images. (3) Nonsynchronization of midpoints of the shutter exposures gives rise to induced boresighting errors, depending on the attitude rates of the camera carrier. (4) Misregistration can also be caused by a lack of flatness of the film across the frame, and (or) by (5) differential film distortion because of tension, temperature, or humidity differences between films in the cameras or processing equipment.

Based upon use of Eq. 6-50, the multiband spatial resolutions, as calculated for various amounts of misregistration, are listed in Table 6-15.

The table applies to the pair of photographs in a multiband combination that are most out of register. The addition of photographs in better register than these will not affect the multiband resolution as defined here. It was assumed in Table 6-15 that both photographs of the pair have the same resolution.

Table 6-15 can be used in conjunction with appropriate graphs to determine the relationship between misregistration tolerance, the focal length difference, and the boresighting error. If each optical channel has the same distortion characteristics, then the misregistration Δx due to a focal length difference Δf is given by

$$\Delta x = \Delta f \tan\alpha, \qquad (6\text{-}51)$$

where:

α = semifield angle.

Curves of Δf vs α for several values of Δx are shown in Figure 6-74.

As an example, consider two lenses of semifield angle 45°, which, at this angle, yield a tangential resolution[22] on film of 40 cycles/mm. Then, assume that no less than 27 cycles/mm tangential multiband resolution are needed at this field angle from photographs taken simultaneously by the two lenses. Table 6-15 shows that the maximum misregistration is 0.5 resolution elements or 12.5 μm, and that curve 4 is the corresponding curve on Figure 6-74. The tolerance on focal-length match is found, from this curve, to be 12.5 μm at 45°. This is a tight tolerance; however, it should be remembered that the depth of focus tolerance will be in the range of ±50 μm. Therefore, assembly errors amounting to 112.5 μm difference in focal length between two lenses can be tolerated if precision refocusing is allowed.

[22] See definition of tangential and radial resolution following Eq. 6-29, and shown in Figure 6-13.

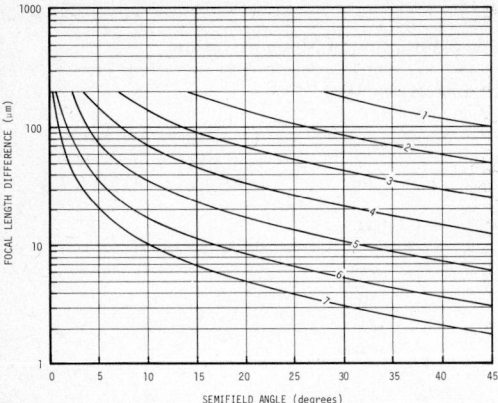

Fig. 6-74. Multiband camera band-to-band misregistration as function of focal length difference related to semifield angle, *see* Table 6-15.

Misregistration, Δx, due to boresighting error, θ, is given by

$$\Delta x = f\,\theta\,\tan^2\alpha, \qquad (6\text{-}52)$$

where:

f = focal length of two identical lenses,
α = semifield angle.

This relationship is plotted for several values of Δx in Figure 6-75 for $f = 150$ mm. Again using the requirements in the above example, curve 4 of Figure 6-75 shows us that for $\Delta x = 12.5\ \mu m$, the tolerable boresighting error is 0.083 mrad, or 17 arc sec. It can also be predicted from this figure that, if this tolerance is held, there will be less than 9% decrease in multiband resolution if the tangential resolution should increase to 80 cycles/mm at a semifield angle of 26°.

Uniformity of Image Plane Irradiance

Uniformity in image-plane spectral irradiance is desirable in multiband photography to ensure high

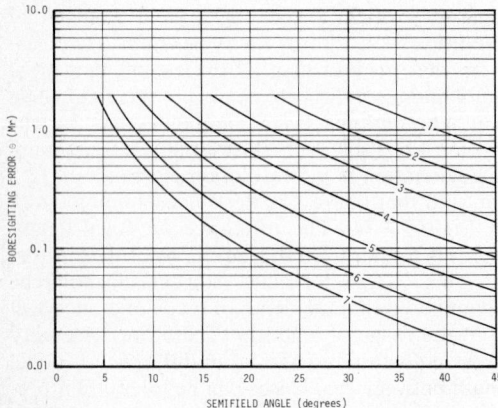

Fig. 6-75. Multiband camera band-to-band misregistration as a function of boresighting error related to semifield angle, *see* Table 6-15.

tonal fidelity across each frame. The increased atmospheric path associated with angles off axis will modify the image-plane irradiance. This varies according to conditions, and it is preferable to use a system of good uniformity, which, for wide-angle lenses, necessitates the use of an antivignetting filter.

In several instances the exposure times in the various channels of a multiband camera array have been adjusted by changing the f-number rather than the shutter time. This is an incorrect procedure because the uniformity in image plane irradiance is strongly affected by the f-number for most lenses. The average $f/2.8$ photographic lens, for example, exhibits vignetting at the low f-numbers, and it is not until the lens is stopped down to about $f/8$ that the image-plane irradiance assumes a falloff of approximately $\cos^4\alpha$. In many multiband applications it may be worthwhile to suffer some image motion by operating at a modest f-number on all lenses rather than to avoid image motion by using the lower f-numbers on some of the lenses.

Shutter Repeatability

One of the worst problems encountered in quantitative multiband work is shutter repeatability. Focal-plane and leaf shutters, by their action, not only can affect uniformity of image plane irradiance, but can give different exposures as the result of temperature and pressure. These changes, unfortunately, are usually not the same from shutter to shutter, and therefore taking the ratio of the data from the various channels does not correct the error. The best solution is to use a continuously rotating disk-type shutter that is intrinsically more repeatable than the others and in addition is highly efficient.

With reference to processing, it can be stated that in a precision processing laboratory, variations across the width or along a length of several meters of film should not exceed 0.02 density unit, corresponding to 5% in exposure level. The tolerances on image-plane uniformity of irradiance and shutter repeatability should be set with this value as a guide.

Some of the special camera systems that have been used or proposed for multiband work were described by Slater (1975). Parametric characteristics of some of these special multispectral cameras are tabulated in Table 6-16.

Space Multiband Cameras

The Camera for the S065 Experiment

The multiband camera used in the S065 experiment was flown on the Apollo 9 flight of March 1969. It consisted of an array of four Hasselblad cameras (Fig. 6-76) attached during flight to the hatch window of the Command Module for earth photography. Experiment S065 took the first multiband pictures of Earth from space in the NASA

TABLE 6-16

Multispectral Aerial Cameras

	Nine lens	Model 11	Mark I	Aero I	MPF
Camera	Nine lens	Model 11	Mark I	Aero I	MPF
Manufacturer	Itek	Spectral data	I²S	Dol Products, Inc.	Itek
Primary use	multispectral	multispectral	multispectral	multispectral	multispectral
Format (cm)	5.7 × 5.7	8.9 × 8.9	8.9 × 8.9	5.7 × 5.7	5.7 × 5.7
Film length (m)	76	76	76	120	30
Base thickness (μm)	100	100	100	100	100
Frames/roll	325	188	300	470	NA
Magazine	A9B modified	A-5A or A-9	A5A	A-9	cassette
Cycle time (sec)	1.25–1.75	2.0	2.0	2.0	2.0+
FMC type	moving film	moving film	NA	NA	rocking mount
FMC rate (mm/sec)	2.5–125	2.5–700	NA	NA	15.8 mrad/sec
FMC error	NA	NA	NA	NA	<10%
AWAR, 1000:1 (ℓ p/mm)	NA	45*	NA	NA	varies w/band
Film	NA	2424	NA	NA	NA
Processing	NA	NA	NA	NA	NA
Type of measurement		laboratory	NA	NA	NA
Compatible lenses Manufacturer(s)	Xenotar Schneider & Leitz	Xenotar Schneider	Xenotar Schneider	Xenotar Schneider	MPF Itek
Focal length (mm)	152 mm	150, 100	150, 100	150, 100	150
f/No. range	2.4–16	2.8–16	2.8–16	2.8–16	2.8–16
Wavelength range (nm)	9 bands	4 bands	4 bands	4 band	500–900 (4 bands) 499–700, 500–900
Filters	Wratten	forty available	400–470 W47B, 470–590 W58, 590–690 W25, 740 + W88*		special
Can be used w/normal color film?	yes	yes	yes	yes	yes (1 band)
Can be used w/IR color film?	yes	yes	yes	yes	yes (1 band)
Changes of focus requires for IR color film?	NA	NA	NA	NA	no
Angular coverage	21° 14 × 21° 14	varies w/focal length	varies w/focal length	varies w/focal length	21° 12' × 21° 12'
Shutter speed (sec)	1/30–1/120	1/25–1/133 or 1/150–1/350 1/350–1/800	1/150–1/350 (A), 1/350–1/800 (B)	1/150–1/350 (A), 1/350–1/800 (B)	0.0025, 0.005, 0.01
Shutter type	focal plane	focal plane (2 types: A,B)	focal plane (2 types: A,B)	focal plane (2 types: A,B)	rotary, intralens
Weight (kg), w/o film and mount	23	43	26	57	57
Mount		A-8, -11, -11A, -27, -27A	A-8, -11, -11A, -27, -27	A-8, -11, 11A, -27, -27A & NR1	
Data annotation	Band no. & flight detector	NA	NA	external	card, digital
Intervalometer		yes	external	NA	
Exposure control		NA	NA		auto
Remarks	Fiducials	*Peak	modified K-22 Camera w/4 lens 24-cm	modified K-22	grid reseau 7.0-cm film 6 lenses

Fig. 6-76. The four-Hasselblad-camera array used in the S065 experiment on Apollo 9.

program. The equipment was simple, and the components were not specially selected to give matched multiband data. Nevertheless, the results did show promise for the future of multiband photography from space; much credit belongs to those at NASA/MSC who took advantage of the opportunity afforded by Apollo 9.

The camera's historic significance and the fact that it represents a typical multiband camera assembled from off-the-shelf components make it worthwhile to list some of the camera data and performance figures (Table 6-17). In the table, resolution data [included for red (DD camera) and black-and-white infrared (CC camera)] are based on laboratory measurements of high-contrast standard Air Force targets. Ground-resolution figures, derived from these laboratory measurements, and assuming an average altitude of 200 km, are also summarized in this table.

The uniformity of image plane irradiance as a function of field angle and f-number (Cuneo, 1970) is shown in Figure 6-77. Further data on the S065 multiband camera can be found in Keenan et al (1970).

S190A Skylab Multiband Camera

The multiband camera system, referred to as experiment 190A, was part of the Earth Resources Experiment Package on Skylab, which was launched in May 1973 with an orbital inclination of 50°. A photograph of the system is shown in Figure 6-78, from which can be seen the six lenses and the bar around which the six cameras rotate to provide forward motion compensation.

The camera covered very similar bands to those of the Landsat Multispectral Scanner System: 0.5 to 0.6 nm, 0.7 to 0.8 nm, and 0.8 to 0.9 nm. In addition, color infrared film was used in the fifth camera and normal color film was used in the sixth camera. As seen from Table 6-18, the ground resolved distance varied in the range from 39 to 114 m/cycle. It should be noted that, except for the sixth camera, the performances of the individual cameras were unnecessarily limited by the resolving power of the film used. Higher resolving power films were available for use and the forward motion compensation was adequate to allow them to be used with advantage.

Details of the performance of the aircraft version of the S190A are provided in Table 6-16, and a detailed comparison of the S190A and the Zeiss JENA MKF-6 systems have been made by Slater (1980), Womble (1979), and Zickler (1978).

Forkey and Womble (1972) have described the unique design of the Petzval lenses used in the six cameras, and Kenney and Demel (1975) have dis-

TABLE 6-17

Four-Camera (Hasselblad) Array Used in SO65 Experiment, Apollo 9

CRITERIA									RESOLUTION					
									Semi field angle					
Camera	Wratten Filter No.	"Color"	Film Type (Eastman Kodak)	Focal Length* (mm)	f-Number*	Shutter Speed (sec)*	Focus Setting (ft)	Target Contrast	0°	7.5°	15°	22.5°	AWAR ℓp/mm	Ground Resolution m (ft)
AA	15	Yellow	SO180 IR Color 3400	80	8	1/250	50							
BB	58	Green		80	4	1/125	∞							
CC (Radial) (Tangential)	89B	Deep Red	Panatomic-X SO246 IR B&W	80	16	1/250	30	>100:1	37 / 37	36 / 35	28 / 31	30 / 26	31 / 31	80 (270) / 80 (270)
								low (est)					20	125 (400)
DD (Radial) (Tangential)	25A	Red	3400 Panatomic-X	80	4	1/250	∞	>100:1	67 / 67	75 / 69	41 / 53	40 / 38	51 / 51	50 (160) / 50 (160)
*Nominal								low (1.6:1) (est)					36	70 (230)

REGISTRATION ERRORS (center to corner of format)

Type of Error	Typical Magnitude of Error	Resulting Image Height Error, μm
Chromatic variation in focal length	500 μm	250
Chromatic variation in distortion	12 μm	12
Filter wedge	3 arc min	20
Film flatness	100 μm	50
Boresighting	1°	300

TABLE 6-18

Lens $6	Film[b] $ 030,___	Spectral Range (μm)	Image Resolution (lines mm^{-1})				
			Design Spec.	Paper Design	Lab Test	Test	GRD[c]
1	3400	0.5−0.6	56	60	56	55	52
2	3400	0.6−0.7	58	73	69	62	46
3	2424	0.7−0.8	24	29	28	25	114
4	2424	0.8−0.9	24	29	28	25	114
5	3443	0.5−0.88	24	35	30	30	95
6	SO-242	0.4−0.7	49[d]	85	75	73	39

[a] Data from Womble and Breen (1974).
[b] All films are Eastman Kodak products.
[c] Ground resolved distance in meters per line pair for 435-km altitude.
[d] Specified on film type SO-121.

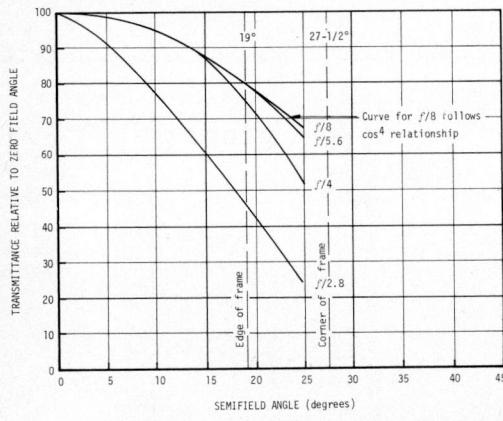

Fig. 6-77. Lens transmittance as a function of field angle for 80-mm Zeiss planar lens (used on the S065 experiment multiband camera) at indicated f-number. (Note that the curve for $f/8$ follows the cos^4 relationship.)

Fig. 6-78. The S190A camera used on Skylab (courtesy ITEK Corp.).

TABLE 6-19

Characteristics of Zeiss JENA MKF-6 Camera

Number of lens	6
f-number minimum	4.0
f-number adjustment	½-stop increments from $f4$ to $f/13.5$
Focal length	125 mm
Field of view	40.5°
Film width	70 mm
Film format	56 mm × 80 mm
Base thickness	90 or 180 μm
Image distortion	±3 μm
Registration	<2 μm boresighting <2 μm focal-length match
Image-plane irradiance	$\cos^4$ to 38° field angle
Boresighting	±10 arc sec
Forward-motion compensation	16.8–39.8 mrad s^{-1}, error ±0.8 mrad s^{-1}
Shutter speeds	7, 10, 14, 20, 28, 40, 56 ms
Spectral intervals	0.46–0.50, 0.52–0.56, 0.58–0.62, 0.64–0.68, 0.70–0.74, 0.79–0.89 μm
Shutter synchronization	<2 ms
Weight of camera body	73 kg
Power consumption	~100 W
Dimensions for camera body	640 × 620 × 226 mm³

cussed the very successful operation of the system in the Skylab program.

Zeiss JENA MKF-6 Multispectral Cameras

The Zeiss JENA MKF-6 camera is similar in many respects to the S190A camera just de-scribed. The main differences between the two systems are that the S190A system is f/2.8 covering a 30° total field and the MKF-6 is f/4 covering a 40.5° total field. Zickler (1978) has given a detailed description of the system, in which he compares its performance to that of the S190A. Womble (1979) has prepared a rebuttal to some of the comments made in this paper.

Figure 6-79 is a photograph of the MKF-6 system that was orbited on Soyuz 22. Unfortunately, there are no detailed resolution data available for the MKF-6 camera; the only datum available is a resolution of 160 lines/mm in the visible for high-contrast targets on Soviet film type T-18(a). A list of the camera parameters is presented in Table 6-19.

REFERENCES

American National Standards Institute, 1969, Method for determining the resolving power of photographic materials: ANSI-PH2. 33-1969, ANSI.

American Society of Photogrammetry, 1966, Manual of Photogrammetry, Third ed.

American Society of Photogrammetry, 1968, Manual of color aerial photography.

American Society of Photogrammetry, 1981, Manual of Photogrammetry, Fourth ed.

Beccasio, A. D., 1971, Stellar calibration of the lunar mapping camera system; SPIE Journal, vol. 5, no. 3, pp. 77–82.

Born, Max, and Emil Wolf, 1959, Principles of optics: London, Pergamon Press.

Brock, G. C., 1967, The physical aspects of aerial photography: New York, Dover Publications.

Brock, G. C., 1970, Image evaluation for aerial photography: An appraisal of current techniques: London, Focal Press.

Brock, G. C., D. I. Harvey, R. J. Kohler, and E. P. Myskowski, 1965, Photographic considerations for aerospace: Itek Corporation, Lexington, Mass.

Brown, D. R., 1952, Natural illumination charts; U.S. Navy Report No. 374-1.

Carman, P. D., 1969, Camera calibration laboratory at the National Research Council; Photogram. Eng., vol. 35, pp. 372–376.

Carroll, B. H., 1966, Sensitivity of silver halide emulsions with development in Photographic Systems for Engineers: F. M. Brown, H. J. Hall, and J. Kosar, eds., SPSE.

Cimerman, V. J., and Z. Tomasegovic, 1970, Atlas of photogrammetric instruments: New York, Elsevier Pub. Co.

Colvocoresses, A. P., 1970, ERTS-A satellite imagery: Photogram. Eng., vol. 36, pp. 555–560.

Colvocoresses, A. P., and R. B. McEwen, 1973, EROS cartographic progress; Photogram. Eng., vol. 39, pp. 1303–1309.

Corning Glass Works, 1965, Color glass filters: Publication CF-3, Corning, New York.

Coulson, K. L., 1966, Effects of reflection properties of natural surfaces in aerial photography: Appl. Opt., vol. 5, pp. 905–917.

Cuneo, W. J., Jr., 1970, Atmospheric limitations on the field of view in multiband aerial photography: Technical Report 60. OSC.

Curcio, J. A., 1961, Evaluation of atmospheric aerosol particle size distribution from scattering measure-

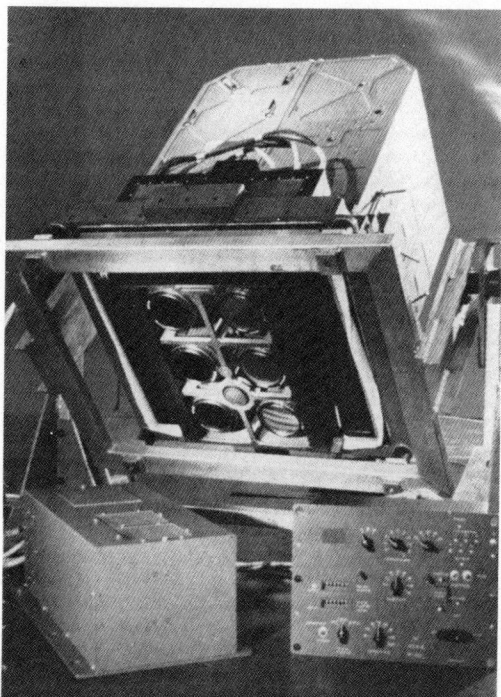

Fig. 6-79. Photograph of the Zeiss JENA MKF-6 system.

ment in the visible and infrared: J. Opt. Soc. Am., vol. 51, pp. 548–551.

Curran, P. J., 1979, The use of polarized panchromatic and false color infrared film in the monitoring of soil surface moisture: Remote Sens. Environ. vol. 8, pp. 249–266.

Dainty, J. C., and Shaw, R., 1974, Image Science. Academic Press, New York.

Data Corporation, 1965, Airborne photographic equipment, Vols. I, II, and III: Report RC013200 for Recon Central, WPAFB (and supplements, Report RC076575).

Data Corporation, 1967, Aerial camera lenses: Contract AF33C65D-14443 for Recon Central, WPAFB.

Defense Supply Agency, 1962, Optical design: MIL-HDBK-141, Washington, D.C.

Dobrowolski, J. A., Marsh, G. E., Charbonneau, D. G. Eng, J., and Josephy, P. D., 1977, "Colored Filter Glasses: an Intercomparison of Glasses Made by Different Manufacturers," Appl. Opt. vol. 16, p. 1491.

Doyle, F. J., 1979, A large format camera for Shuttle: Photogram. Eng., vol. 45, p. 73.

Duddek, M., 1967, Practical experiences with aerial color photography; Photogram. Eng., vol. 33, pp. 1117–1125.

Duntley, S. Q., A. R. Boileau, and R. W. Preisendorfer, 1957, Image transmission to the troposphere; J. Opt. Soc. Am., vol. 47, pp. 499–506.

Duntley, S. Q., J. I. Gordon, J. H. Taylor, C. T. White, A. R. Boileau, J. E. Tyler, R. W. Austin, and J. L. Harris, 1964, Visibility: Appl. Opt., vol. 3, pp. 549–598.

Eastman Kodak Company, 1963 (to date), Tech-bits; Rochester, New York.

Eastman Kodak Company, 1970, Kodak Wratten filters for scientific and technical use: Kodak Publication B-3, Rochester, New York.

Eastman Kodak Company, 1971, Kodak data for aerial photography, ed. 3, Kodak Publication M-29, CAT 151-3381, Rochester, New York.

Eastman Kodak Company, 1972a, Kodak aerial films and photographic plates: Kodak Publication M-61, CAT 143-S001, Rochester, New York.

Eastman Kodak Company, 1972b, Physical properties of Kodak aerial films: Kodak Publication M-62, CAT 150-0321, Rochester, New York.

Forkey, R. E., and D. A. Womble, 1972, Unique lens design of multi-spectral photographic cameras: Presented for Commission I, XII Congress; ISP, Ottawa, Canada, July-August.

Fritz, Norman L., 1967, "Optimum Methods for Using Infrared-Sensitive Color Films," Photogram. Eng., vol. 33: pp. 1128–1138.

Fritz, Norman L., 1969, "Basic Principles of Color Photography and Color Aerial Photography," Proceedings of the Workshop on Aerial Color Photography in the Plant Sciences, pp. 1–15.

Fritz, Norman L., 1971, "New Color Films for Aerial Photography," Proceedings of the Third Biennial Workshop on Color Aerial Photography in the Plant Sciences, pp. 44–68.

Fritz, Norman L., 1974, "Available Color Aerial Photographic Materials," Photogram. Eng., 40: 1423–1425.

Fritz, L. W., and H. H. Schmid, 1974, Stellar calibration of the orbigon lens: Photogram. Eng., vol. 40, p. 101–111.

Geltmacher, H. E., 1964, Contrast conditions for the evaluation of aerial photographic images: United States Air Force Avionics Laboratory, WPAFB, AL-TRR-64-232.

Gimlett, J. I., 1975a, Lens for the Skylab S-190B Camera, MDC U0055, Actron, Monrovia, California.

Gordon, Jacqueline I., and Peggy V. Church, 1966, Sky luminances and the directional luminous reflectances of objects and backgrounds for a moderately high sun: Appl. Opt., vol. 5, pp. 793–801.

Higgins, G. C., 1966. Information capacity of photographic materials; in Photographic Systems for Engineers, F. M. Brown, H. J. Hall, and J. Koser, eds., SPSE, pp. 167–207.

Hooker, R. B., 1970, A comparison of the square wave response of three microscopes commonly used in photointerpretation: Technical Report 53, OSC.

James, T. H., 1977, Theory of the Photographic Process, Fourth Ed.: MacMillan, New York.

James, T. H., and G. C. Higgins, 1960, Fundamentals of photographic theory: New York, Morgan and Morgan.

Jensen, Niels, 1968, Optical and photographic reconnaissance systems: New York, Wiley.

Jones, R. C., 1958. On the quantum efficiency of photographic negatives: Phot. Sci. Eng., vol. 2, no. 2, pp. 57–65.

Karren, R. J., 1968, Camera calibration by multicollimator method: Photogram. Eng., vol. 34, pp. 706–719.

Keenan, P. B., R. A. Schowengerdt, and P. N. Slater, 1970, Interim post-flight calibration report on Apollo 9 multiband photography experiment S065: Tech. Memo 2, OSC.

Keene, G. T., 1972, Eastman Kodak, private communication to the author.

Kenny, G. P., and Demel, K. J., 1975, Skylab Program, Earth Resources Experiment Package, Sensor Performance Evaluation, Final Report, vol. 1, (S190A), NASA-CR-144563.

Kingslake, Rudolph, ed., 1965, Applied optics and optical engineering: New York, Academic Press.

Krinov, F. L., 1953, Spectral reflectance properties of natural formations; Aero Methods Laboratory, Acad. Sci. USSR, 1947; G. Belkov, transl., Tech. Trans. TT-439, NRC.

Labs, D. and H. Neckel, 1973, Proc. of Symp. on Solar Radiation: Smithsonian Institution, Washington, D.C.

Lauroesch, T. J., G. G. Fulmer, J. R. Edinger, G. T. Keene, and T. F. Kerwich, 1970, Threshold modulation curves for photographic films: Appl. Opt., vol. 9, pp. 875–887.

Longhurst, R. L., 1973, Geometrical and physical optics, Third Ed., New York, Wiley.

Mayo, J. W., 1968, Photographic resolving power of aerial reconnaissance lenses as a function of target modulation: Thesis, Univ. of Arizona, Tucson.

Mazurowski, M. J., and J. E. Walker, 1962, Project Photorek II; A study of aerial photographic contrast attenuation by the atmosphere, prepared by CAL, for U.S. Air Force Avionics Laboratory, WPAFB, CAL Report No. VF-1478-P-2.

Mazurowski, M. J., F. B. Silvestro, and J. D. Rinaldo, 1963, A study of photographic contrast attenuation by the atmosphere: prepared by CAL for U.S. Air Force Avionics Laboratory, WPAFB ASD-TDR-63-541.

McDonnel Douglas Corporation, 1979, Reconnaissance reference manual: prepared for Naval Air Systems Command by McDon. Doug. Reconnaissance Laboratory, St. Louis, Missouri.

McKenney, D. B., and P. N. Slater, 1970, Design and use of interference passband filters with wide-angle lenses for multispectral photography: Appl. Opt., vol. 9, p. 2435–2440.

Mees, C. E. K., and T. H. James, 1966, Theory of the photographic process: MacMillan.

Mollberg, B. H., 1981, Performance characteristics of the Orbiter camera payload system's large format camera (LFC); Proc. SPIE, vol. 278, pp. 66–72.

Moser, James S., and Fritz, Norman L., 1975, "High-Definition Color Films for Terrain Photography," *Photographic Science and Engineering* 19: No. 4, 243–246.

NASA, 1973a, *SL/2 Sensitometric Data Package,* July.

———, 1973b, *SL/3 Sensitometric Data Package,* November.

———, 1974a, *SL/4 Sensitometric Data Package,* June.

Norton, C. L., 1968, Aerial cameras for color; Photogram. Eng., vol. 34, p. 36–42.

Rib, H. T., 1968, pp. 12–24 in American Society of Photogrammetry, 1968, Manual of color aerial photography.

Schott & Gen., 1962, Color filter glass: Schott and Gen., Mainz, 365e, (U.S. Rep.: Fish-Schurman Corp., 70 Portman Rd., New Rochelle, N.Y.).

Schut, G. H., 1969, Photogrammetric refraction: Photogram. Eng., vol. 35, p. 79–86.

Scott, R. M. et al, 1965, The practical application of modulation transfer functions: Phot. Sci. Eng., vol. 9, p. 237–264.

Selwyn, E. W. H., 1954, Theory of resolving power, Chapter 16, *in* Optical Image Evaluation: NBS, Circ. 526.

Slater, P. N., 1972, Multiband cameras; Photogram. Eng., vol. 38, p. 543–555.

Slater, P. N., 1975, Photographic Systems for Remote Sensing, in Manual of Remote Sensing (R. G. Reeves, ed.), Chapter 6, 235. Amer. Soc. of Photogrammetry, Falls Church, Va.

Slater, P. N., 1980, Remote Sensing: Optics and Optical Systems. Addison-Wesley, Reading, Mass.

Slater, P. N., and R. A. Schowengerdt, 1972 (February), The specification of sensor performance for Earth resources studies: Photogram. Eng., vol. 39, p. 197–201.

Solomon, J. E., 1981, Polarization Imaging: Appl. Opt., vol. 20, pp. 1537–1544.

SPSE (Society of Photographic Scientists and Engineers), 1973, SPSE handbook of photographic science and engineering: Thomas Woolief, Jr., ed., New York, Wiley.

Sturge, J. M., ed., 1977, Neblette's Handbook of Photography and Reprography Materials, Processes and Systems. Seventh Ed., Van Nostrand Reinhold Co., New York.

Tarkington, Raife C., and Sorem, Allan L., "Color and False-Color Films for Aerial Photography," *Photogrammetric Engineering* 29: 88–95, (1963).

Tayman, W. P., 1974, Calibration of lenses and cameras at the U.S. Geological Survey: Proc. of 40th Ann. Meeting, pp. 455–460.

Tayman, W. P., W. V. Hull, and F. E. Washer, 1968 Use of color film in location of plane of best average definition; *in* Proc. of 34th Ann. Meeting, ASP, pp. 229–294.

Thekaekara, M. P., R. Kruger, and C. H. Duncan, 1969 (August), Solar irradiance measurements from a research aircraft: Appl. Opt., vol. 8, no. 8, pp. 1713–1732.

Todd, H. N., and R. D. Zakia, 1969, Photographic sensitometry, the study of tone reproduction: New York, Morgan and Morgan.

USA Standard RP-16, 1967 (Ausugt 16), Nomenclature and definitions for illuminating engineers; USAS Z 7.1—1967, USA Standards Institute.

U.S. Govt. Printing Office, 1963 Military Standard 150A (revised): Military Standard Photographic Lenses; Washington, D.C.

Valley, S. L., ed., 1965, Handbook of geophysics and space environment: New York, McGraw-Hill, pp. 16–9.

Voggenthaler, J. A., 1972, An artificial star-field camera calibrator: Presented at 12th Congress of ISP, Commission I.

Walraven, R., 1977, Polarization imagery: SPIE vol. 112, pp. 164–167.

Washer, F. E., 1965, Laboratory manual on precise optical measurements: NBS Circ. 8904, 8913, 8946, 8971.

Welch, R., 1971, Modulation transfer functions: Photogram. Eng., vol. 37, pp. 247–259.

Welch, R., 1972, Quality and applications of aerospace imagery: Photogram. Eng, vol. 38, pp. 379–398.

Welch, R., 1976, Skylab S-190B ETC photo quality, Photogram. Eng., vol. 42, pp. 1057–1060.

Welch, R., and J. Halliday, 1973, Imaging characteristics of photogrammetric camera systems: Photogrammetria, vol. 29, pp. 1–43.

Welford, W. T., 1962, Geometrical optics, Vol. I, Second ed.: Amsterdam, North-Holland Pub. Co.

Womble, D. A. (1979). Optimizing Optical Features of Multispectral Space Cameras. Presented at 45th Annual Meeting of the Amer. Soc. of Photogram., Washington, D.C.

Womble, D. A., and Breen, D. L. (1974), Lens Considerations for Multiband Cameras. Presented at 14th Annual Meeting of Photographic Scientists and Engineers, Fall Symposium, Washington, D.C.

Zickler, A. (1978). Design and Technical Parameters of the MKF-6 Multispectral Camera and the MSP-4 Multispectral Projector. Int. Symp. on Remote Sensing of Environment, Manila, Philippines.

CHAPTER 7

Electro-Optical Non-Imaging Sensors

Authors: B. F. ROBINSON and D. P. DeWITT

GENERAL CONTENTS: Radiometric measurement systems; the measurement equation; field-of-view relationships; detectors and their spectral responses; referencing; electronic signal processing; signal-to-noise ratio; radiometric instruments; broadband infrared; broadband visible/near-infrared; direct solar radiation; radiation thermometers; multiband radiometers; spectroradiometers; photo detection mechanisms; thermal detection mechanisms; detector noise; figures of merit; references.

NOMENCLATURE

To conserve and eliminate repetition in text and references, the following symbols, units and names have been used in this chapter.

Symbol	SI Units	Name
A	—	gain
A	m^2	area
A	$cm^3 \cdot s^{-1}$	radiative combination coefficient
$B, \Delta f$	Hz	electrical bandwidth
C_f	μF	capacitance
C_z	v	second term of Fourier Series Expansion
d	—	differential operator
D	W^{-1}	detectivity
D	m	diameter
D^*	$cm\ Hz^{1/2}W^{-1}$	dee-star, figure of merit
D^{**}	$cm\ Hz^{1/2}W^{-1}$	dee-double-star, figure of merit
e	C	electronic charge
f	m	focal length
f	$m^{-3}\ s^{-1}$	effective photons per unit volume per unit time
f	Hz	frequency
F	—	noise figure (decimal)
F_{dB}	dB	noise figure (decibel)
g_m	μS	transconductance
G	—	gain
h	m	separation between target and radiometer
i	A	noise current
I	A	current
k	$1.380 \times 10^{-23} JK^{-1}$	Boltzmann's constant
k	—	index
K	—	transfer coefficient, transfer function
l	m	length
L'	$W\ m^{-2}sr^{-1}$	radiance
L'_λ	$W\ m^{-2}sr^{-1}\mu m^{-1}$	spectral radiance
$L'_{\lambda 1}$	$W\ m^{-2}\ sr^{-1}$	in-band radiance, centered about λ_1
n	m^{-3}	electron density
N_r	m^{-3}	density of recombination centers
NEP	W	noise equivalent power
p	m^{-3}	hole density
r	m	radial spherical coordinate
$\mathcal{R}$	$V\ W^{-1}$	responsivity
R	ohms or Ω	resistance
R	m	radial coordinate
R_f	$k\Omega$	transimpedance
S_n	m^2	cross-sectional area for recombination process

Symbol	SI Units	Name
t	s	time
T	—	transmittance
T	s	period of periodic signal
T	K	temperature
T_λ	K	spectral radiance temperature
U_n	$m \cdot s$	thermal velocity
v	V	noise voltage
V	V	signal voltage
w	m	width
Δ	—	finite-difference operator
ϵ	—	emissivity
ϵ	$F \cdot m^{-1}$	permittivity
η	—	quantum efficiency
θ	° or rad	zenith angle, spherical coordinate
θ_f	° or rad	angular field-of-view (full angle)
λ	μm	denotes spectral concentration, wavelength
λ_1	μm	denotes spectral or inband condition at λ_1
μ	$m^2 \cdot s^{-1} \cdot V^{-1}$	electron mobility
σ	$(ohm \cdot cm)^{-1}$ or $(ohm \cdot m)^{-1}$	conductivity
ϕ	° or rad	azimuth angle, spherical coordinate
Φ	W	radiant flux
ω	sr	solid angle
ω	radian	angular frequency
τ	s	time constant

Subscripts

a	aperture stop, amplifier
b, B	bias
c	cavity
d	detector, dark condition
f	feedback, field-of-view (full angle)
fov	field-of-view
G	gate
L	radiance, load
n	noise component
o	reference condition
r	reference source
s	system, systematic noise
t	target, transfer
Φ	radiant power
λ	spectral quantity
BP	bandpass
HP	high pass
LP	low pass
in	input
out	output
P	parallel, preamplifier

Subscripts

dB	decibel
eq	equivalent
x	chopping frequency
v	voltage
pp	peak to peak
rms	root mean square
*	random noise

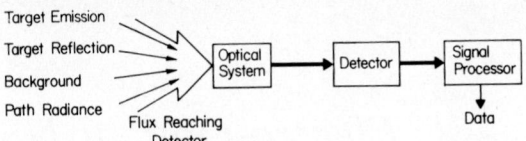

Fig. 7-1. General configuration of the radiometric measuring system for measuring reflected and emitted flux from a target.

INTRODUCTION

The purpose of this chapter is to provide an overview of nonimaging instruments that measure electromagnetic radiation using optical techniques. The instruments are nonimaging in the sense that they do not produce a picture, but rather, integrate over time, space, and wavelength to produce a spectral curve, a set of numbers, or a single number that characterizes the electromagnetic power that is emitted from, reflected by, and/or transmitted through a surface or region of space.

The fundamental application of nonimaging devices is the measurement of *optical quantities* such as radiant flux, irradiance, and radiance, which describe the intensity of the radiation field or the *optical properties* of a surface or a region of space. These quantities are defined in Chapter 2 of this edition and conventional/direct measurement is discussed in Grum and Becherer (1979) and Nicodemus (1976).

In addition to the direct measurement of optical properties and quantities, nonimaging devices are used to determine, *indirectly*, parameters such as the biological properties of crop canopies (Bauer, et al., 1981), the percent moisture in grain samples and the quality of protein (Birth, 1975). Other fields of application include measurement of (a) the temperature of surfaces, Nutter (1974); (b) the arterial blood oxygen content, Merrick and Hayes (1976); and (c) the serum bilirubin concentration in neonates, Hannemann, et al. (1979). The number of such applications is rapidly growing.

While the electro-optical systems that are used for these direct and indirect measurements have distinctive features, they also share many common characteristics. The diagram in Figure 7-1 illustrates a general measurement system comprised of three major elements wherein radiant flux reflected or emitted from the target is sensed. The elements are: the *optical system* consisting of lenses, mirrors, apertures, modulators, and dispersion devices; the *detector* providing an electrical signal proportional to the irradiance on its active surface; and the *signal processor* performing specified functions on the electrical signal to provide the desired output data. Such a system for measuring radiant flux is referred to as a *radiometer*.

RADIOMETRIC MEASUREMENT SYSTEM

The major performance characteristics of a radiometer are introduced along with the mea-

surement equation which relates the irradiance on the detector to the signal output.

THE MEASUREMENT EQUATION

For a general description of radiometer performance, three characteristics can be defined. The radiometer *responsivity, R,* can be defined as the change in output voltage, ΔV, divided by the change in incident flux (input) on the detector, $\Delta \Phi$, and is expressed as

$$\mathfrak{R}_\Phi = \frac{\Delta V}{\Delta \Phi} \left[V \cdot W^{-1} \right]. \qquad (7\text{-}1)$$

The second characteristic is the *detectivity, D,* which can be defined as the reciprocal of the noise-equivalent flux incident on the detector or as the responsivity divided by the noise voltage, v_n, (the *rms* noise fluctuation of the output). That is,

$$D = \frac{\mathfrak{R}_\Phi}{v_n} \left[W^{-1} \right]. \qquad (7\text{-}2)$$

The third characteristic identifies the level of incident radiation corresponding to the zero output reading of the instrument; this is referred to as the *reference radiation, Φ_0.*

Through use of the above characteristics, the output of a radiometer written in terms of the input radiant flux and other characteristics is referred to as the *measurement equation* and has the form

$$V = \mathfrak{R}_\Phi(\Phi - \Phi_0) + v_n \left[V \right]. \qquad (7\text{-}3)$$

The radiant flux reaching the detector is determined by the field-of-view (FOV) of the radiometer which is dependent upon the construction features of the radiometer. The *spectral response* of the radiometer is largely determined by the nature of the detector and, of course, any dispersion devices present in the optical system. The manner in which *signal processing* is accomplished will influence the signal-to-noise ratio (S/N). These three major aspects of radiometric measurement systems are discussed in the following sections.

FIELD-OF-VIEW RELATIONSHIPS

The *field-of-view* (FOV) is the solid angle through which radiant flux is accepted by the radiometer and depends upon the optical configuration of the radiometer.

The Simple Radiometer

The *simple radiometer* is schematically represented in Figure 7-2 and the field-of-view is determined by the position of the entrance aperture stop, A_a, relative to the detector area, A_d, which serves as the field stop.

If the detector is small such that $r_d/f << r_a/f$, then the plane angle, θ_f is approximated by

$$\theta_f = 2\arctan\left(\frac{r_a}{f}\right) [rad] \cdot \qquad (7\text{-}4)$$

The angle, θ_f, is referred to as the *angular* field-of-view. All the rays of radiation originating from the target which are contained within the solid angle, ω_{fov}, will be accepted by the radiometer and reach the detector. In terms of the plane angle, θ_f, the field-of-view can be written as

$$\omega_{fov} = \frac{\pi}{4} \tan^2\left(\frac{\theta_f}{2}\right) [sr] \cdot \qquad (7\text{-}5)$$

Typically, remote sensing field radiometers have 15° fields-of-view (θ_f) such that $\omega_{fov} = 0.0564$ sr. From the geometry of Figure 7.2, the target diameter, D_t, and separation (elevation) distance, h, can be expressed as

$$D_t = 2h \tan\left(\frac{\theta_f}{2}\right) [m] \qquad (7\text{-}6)$$

and the *target* area, A_t, is

$$A_t = \pi h^2 \tan^2\left(\frac{\theta_f}{2}\right) [m^2] \cdot \qquad (7\text{-}7)$$

Selected values of the target diameter for selected solutions for 1° and 15° FOV radiometers are presented in Table 7-1.

TABLE 7-1
Target Diameter (m) as a Function of Elevation Above a Target Surface for 1° and 15° Fields-of-View

Field-of-View	Elevation (m)			
	1	2	4	8
1°	0.018	0.035	0.070	0.14
15°	0.54	1.08	2.14	4.29

The flux reaching the detector can be derived in a direct manner if the radiance from the target is isotropic and uniform. It follows that the radiance, L', intercepted by the detector is that of the target. From the geometry of the situation as shown in Figure 7-3, the flux (or irradiation) incident on the detector is

$$\Phi = \int_{fov} L' \cdot dA_d \cdot \cos\theta \cdot d\omega \; [W] \quad (7\text{-}8)$$

where L' represents the uniform, isotropic incident radiance from the target. Recognizing that, for distant points in the field-of-view, the solid angle subtended by the detector is small and that the solid angle $d\omega$ in the spherical coordinate system is $\sin\theta \cdot d\theta \cdot d\phi$, the flux is

$$\Phi = L' \cdot A_d \int_0^{2\pi} \int_0^{\theta_f/2} \cos\theta \cdot \sin\theta \cdot d\theta \cdot d\phi \quad (7\text{-}9)$$

$$\Phi = \pi L' \cdot A_d \sin^2\left(\frac{\theta_f}{2}\right) \cdot \qquad (7\text{-}10)$$

It is important to recognize that this simple result is a consequence of the target radiance being uniform and isotropic.

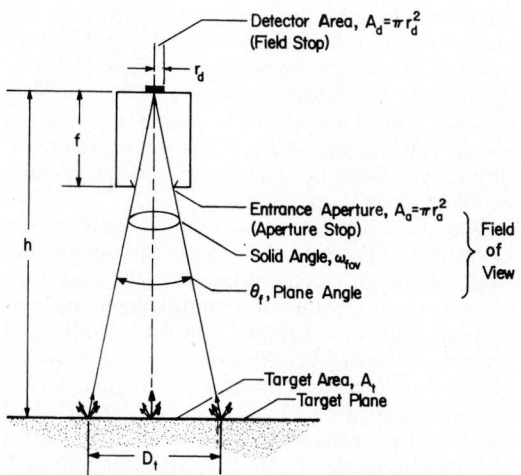

Fig. 7-2. Defining the field-of-view for the simple radiometer.

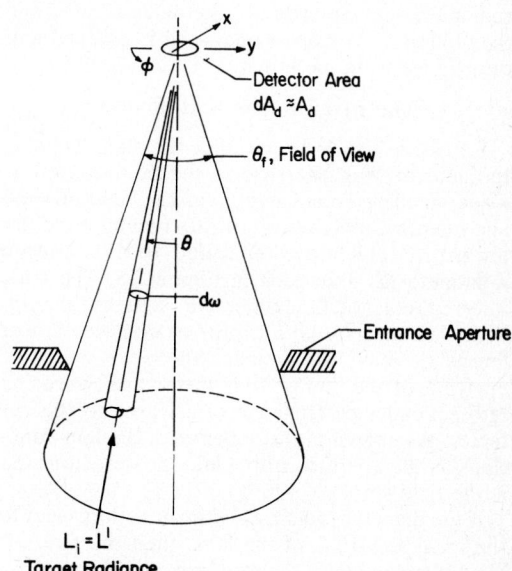

Fig. 7-3. Radiant flux accepted by the simple radiometer in terms of target radiance and field-of-view.

The radiant flux accepted by the radiometer and reaching the detector may be expressed in terms of target-plane parameters. Consider the differential area, dA_t, of the target plane shown in Figure 7-4, noting carefully the positions of the entrance aperture and field stops. The flux reaching the detector is of the form

$$\Phi = \int_{fov} L'(R, \theta_t, \phi) \cdot \cos\theta_t \cdot dA_t \cdot \omega_t \qquad (7\text{-}11)$$

where L' is the directional radiance in θ_t direction of the target plane at the coordinate R, ϕ and ω_t is the small solid angle subtended by the detector. The differential target area dA_t, can be written as

$$dA_t = R \cdot d\phi \cdot dR \qquad (7\text{-}12)$$

$$dA_t = (h \cdot \tan\theta_t) \cdot d\theta \cdot (h \cdot \sec^2\theta_t \cdot d\theta_t) \qquad (7\text{-}13)$$

The solid angle, ω_t, has the form

$$\omega_t = \frac{A_d \cos\theta_t}{(h/\cos\theta_t)^2} . \qquad (7\text{-}14)$$

Substituting Eqs. 7-13 and 7-14 into Eq. 7-11 and identifying the integration limits, the flux relation is

$$\Phi = \int_0^{2\pi} \int_0^{\theta_f/2} L'(R, \theta_t, \theta) \cdot \cos\theta_t \cdot \sin\theta_t d\theta_t \cdot d\phi. \qquad (7\text{-}15)$$

The integral can be readily evaluated if two conditions are met: (1) the target plane fills the field-of-view and (2) the target radiance is isotropic and uniform over the target area. If these conditions are fulfilled, the flux relation is identical to Eq. 7-10,

$$\Phi = \pi L' A_d \sin^2 \left[\frac{\theta_f}{2} \right] . \qquad (7\text{-}16)$$

That is, the flux is dependent upon the two radiometer characteristics—the detector area and the field-of-view plane angle—and is independent of distance from the target.

The Fixed-Focus Radiometer

Equation 7-16 indicates that the flux reaching the detector will decrease as the field-of-view is made smaller. To achieve a smaller field-of-view and maintain a high level of flux reaching the detector, a lens may be added to the simple radiometer as illustrated in Figure 7-5. The lens, referred to as the field objective (actually, the optical element may be a mirror or combination of lenses or mirrors), replaces the open entrance aperture of the simple radiometer and serves to redirect radiance from the target area to the detector. As shown in the schematic, the lens functions as the aperture stop while the detector area is the field stop.

If the detector radius, r_d, is small with respect to the focal length, f, of the lens, then the field-of-view plane angle of the radiometer will be

$$\theta_f = \tan^{-1} \left[\frac{r_d}{f} \right] . \qquad (7\text{-}17)$$

For a distant extended source, $h >> (r_a/r_d)f$, the radiant flux incident on the detector can be written as

$$\Phi = A_a \int_{\omega_d} L' \cdot \cos\theta \cdot d\omega \qquad (7\text{-}18)$$

where L' is the radiance from the target that is redirected by the lossless lens onto the detector and A_a is the area of the lens of radius, r_a. Integrating in the same manner as for the simple radiometer, the flux incident on the detector from a distant extended source of uniform, isotropic radiance, L', is

$$\Phi = A_a \int_0^{2\pi} \int_0^{\theta_f/2} L' \cdot \sin\theta \cdot \cos\theta \cdot d\theta \cdot d\phi$$

$$\Phi = \pi L' A_a \sin^2 \left[\frac{\theta_f}{2} \right] . \qquad (7\text{-}19)$$

Note again that the flux is dependent upon the two radiometer characteristics—the lens area and the field-of-view plane angle, and is independent of distance from the target.

The two types of radiometers are compared in Figure 7-6. It is useful to note that if $A_{a,1} = A_{a,2}$; $f_1 = f_2$, and $A_{d,1} = A_{d,2}$, then the fluxes from Eqs. 7-16 and 7-19 expressed as

$$\Phi_1 = \pi L' A_{d,1} \sin^2 \left[\frac{\theta_{f,1}}{2} \right] .$$

$$\Phi_2 = \pi L' A_{a,2} \sin^2 \left[\frac{\theta_{f,2}}{2} \right] .$$

will be approximately equal for fields-of-view that are less than 15°.

DETECTORS AND SPECTRAL RESPONSE

Common Detectors

The function of a detector of optical radiation is to transform radiant power into an electrical voltage signal. Ideally, the flux incident upon the detector and the voltage output signal would closely conform to a unique linear function. This is nearly the case for the detectors considered below. That is, if the optical and electrical processing systems are properly designed and if the device temperature is held constant, then *linear* instruments may be made using those detectors.

Detectors are classified on the basis of the physical mechanisms that cause the conversion from radiant to electrical energy. The two main classes are quantum (or photon) detectors and thermal detectors. Thermal detectors produce an electrical signal because the temperature is changed by the incident optical radiation. Quantum detectors produce an electrical signal when the mobility, or number of free charge carriers, is changed by incident photons. The last four sections of this chapter review detection mechanisms, detector noise, and figures-of-merit. Limperis and Mudar (1979) and Jacobs (1978) provide

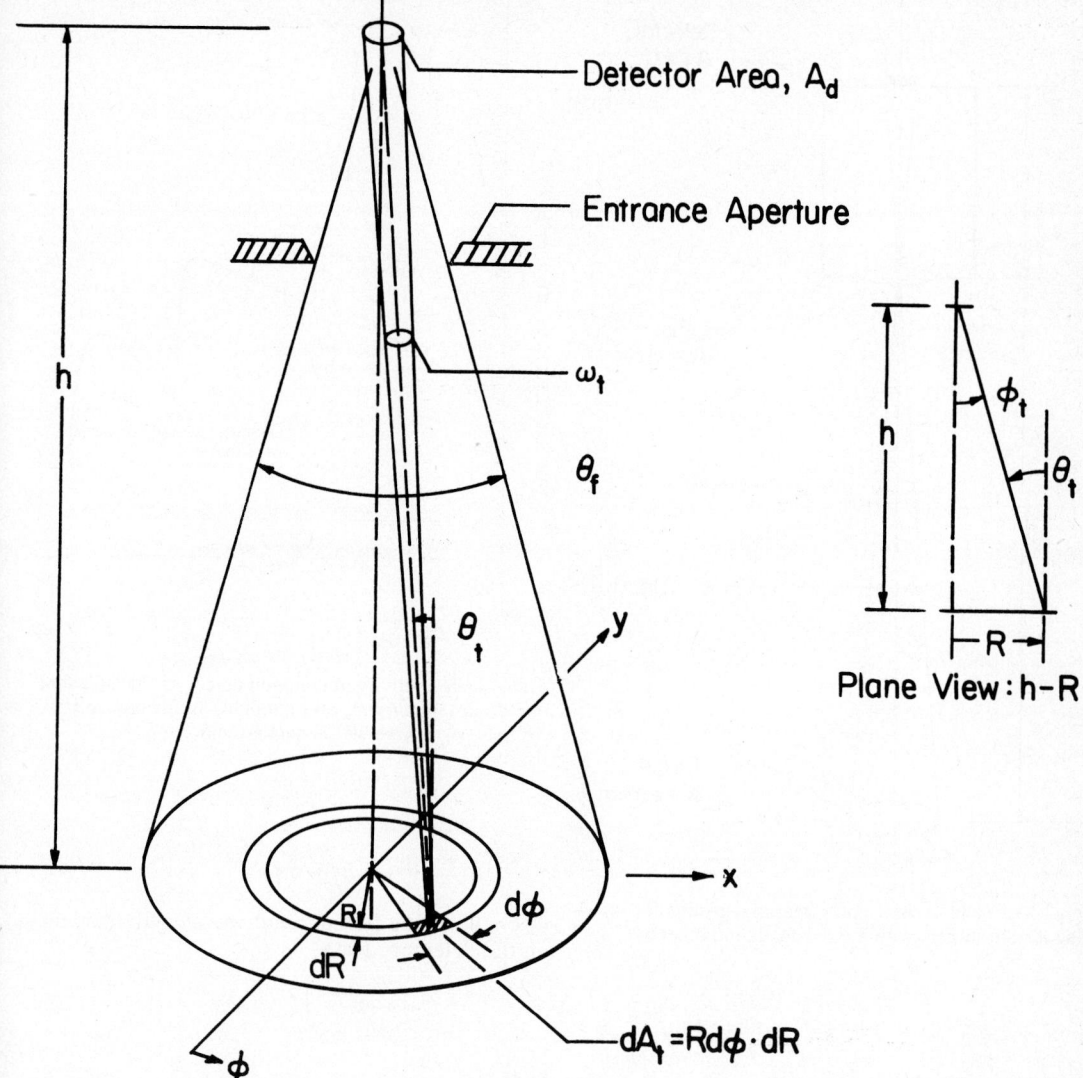

Fig. 7-4. Radiant flux accepted by the simple radiometer in terms of target parameters and the field-of-view.

a thorough discussion of the mechanisms of modern detectors. In addition, current and past editions of the *Optical Industry Systems Purchasing Director* (1981) contain useful technical discussions and lists of available detectors.

A comparison of the relative performance and spectral coverage of eight commonly used detectors is given in Figure 7-7 and Table 7-2. The figure-of-merit, D^*, is characteristic of a detector material rather than an individual detector (discussed in the section on figures of merit). The curves represent available detector performance, but not the "state of the art" for these materials. The detectors may be operated at different chopping frequencies and temperatures, which may significantly alter performance. Since the dominant noise mechanism in photomultiplier tubes is

shot noise (which is signal dependent), D^* is seldom used to describe photomultiplier-tube performance and less frequently used for comparison with other types of detectors.

Spectral Selection

For the simple radiometer of Figure 7-3, if the spectral radiance at each point (R,ϕ) of the target plane is $L_\lambda'(R,\theta_t,\phi)$, then the spectral flux reaching the detector is

$$\Phi_\lambda = A_d \int_{fov} L_\lambda'(R,\theta_t,s) \cdot \sin\theta_t \cdot \cos\theta_t \cdot d\theta_t \cdot d\phi.$$
$$[W \cdot \mu m^{-1}] \qquad (7\text{-}20)$$

If the spectral radiance of the target plane is uniform and isotropic,

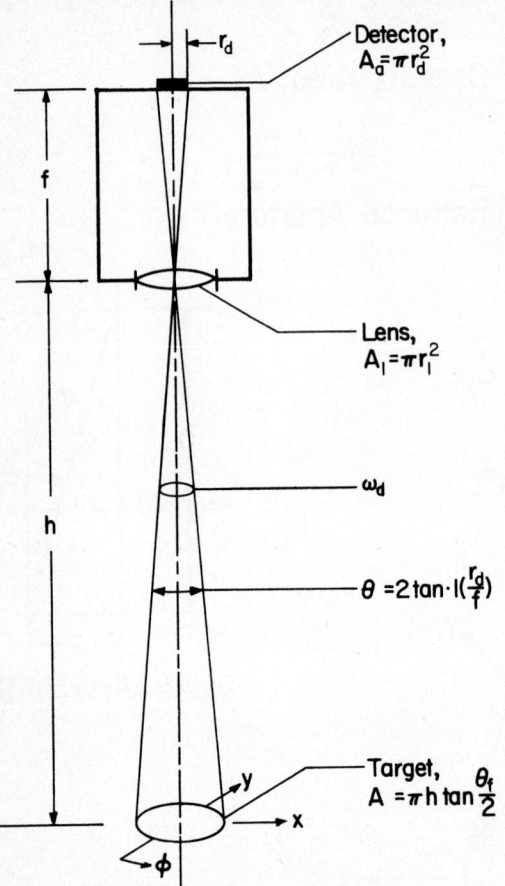

Fig. 7-5. Field-of-view and target dimensions for viewing distant targets with a fixed-focus radiometer.

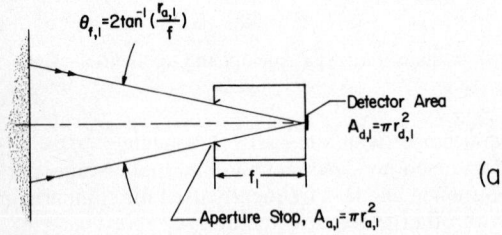

(a)

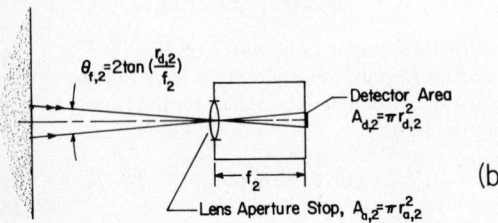

(b)

Fig. 7-6. Comparison of radiometer fields-of-view for the (a) simple and (b) fixed-focus radiometers. Note that when $f_1 = f_2$, $A_{d,1} = A_{d,2}$, $A_{a,1} = A_{a,2}$ then $\Phi_1 = \Phi_2$.

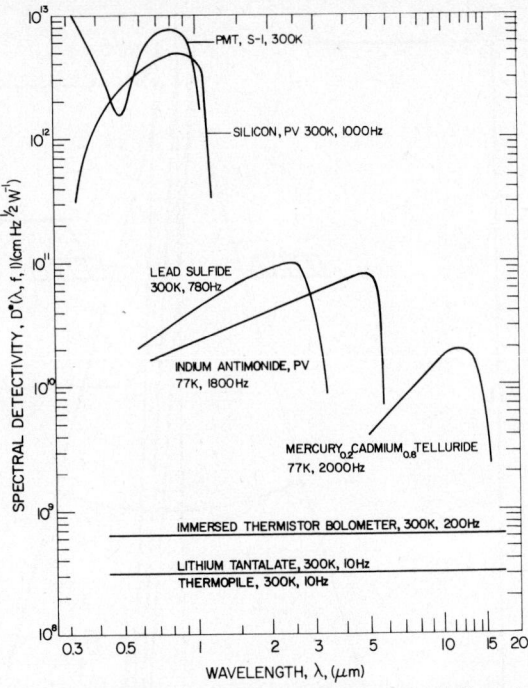

Fig. 7-7. Detectivity of common detectors for indicated operating temperature and chopping frequency and for 2π field-of-view to 300 K background.

$$\Phi_\lambda = \pi L_\lambda' \cdot A_d \cdot \sin^2 \left| \frac{\theta_f}{2} \right| \qquad (7\text{-}21)$$

where Φ_λ is the flux per unit wavelength incident on the detector. Then

$$\Phi = \int_0^\infty \Phi_\lambda d\lambda \qquad (7\text{-}22)$$

and

$$\Phi = \pi A_d \sin^2 \left| \frac{\theta_f}{2} \right| \int_0^\infty L_\lambda' d\lambda \quad [W] \qquad (7\text{-}23)$$

where Φ, the total flux reaching the detector, is obtained by integration over all wavelengths.

To measure the spectral distribution of radiation, elements such as gratings, prisms, interference filters, colored glasses, and combinations of these elements are used to select the spectral radiation which is passed by the optical system to the detector. Conceptually, the simplest means of defining an optical passband is through the use of an interference filter (see Slater, 1980).

Figures 7-8 and 7-9 show an interference filter inserted in front of the simple radiometer and its effect on the spectral flux reaching the detector. With the filter in place,

$$\Phi = \int_0^\infty \Phi_\lambda \cdot T(\lambda) \cdot d\lambda \quad [W] \qquad (7\text{-}24)$$

TABLE 7-2
Characteristics of Eight Common Detectors

Detector	Response Time (μsec)	Operating Range (K)	Linearity (%)	Stability % per °C	Remarks
Photomultiplier Tube S-1 surface	0.001	200 to 323 See Remarks	1	0.5	Fragile and intolerant of humidity and vibration. Long warm-up time. Significantly higher sensitivity is available. Dark current and responsivity are functions of recent history of input radiation signal. Several mechanisms for nonlinearity and noise are present.
Silicon Photodiode	PC, 0.005 PV, 1	220 to 400	0.1	0.1	Rugged and tolerant of vibration and abuse. Photoconductive mode is used where speed is important. Photovoltaic mode is used where linearity and low noise are important. Widely used in laboratory and field instruments.
Lead Sulfide	5000 @ 77 k	77 to 350	0.1<	>4	Rugged and tolerant of vibration and abuse. Usually cooled to maintain constant device temperature and thus may require special handling. May be operated at ambient with reduced D*. When device temperature is not held constant, special techniques must be used to compensate for temperature-dependent changes in responsivity. Widely used in laboratory and field instruments due to 1-μm to 4-μm performance.
Indium Antimonide	PV, 1 PC (77K), 10 PC (300K) 10^4	30 to 350	0.1<	>5.6	Rugged and tolerant of vibration and abuse. May be operated at 300 K with reduced D*; however, responsivity is strongly temperature dependent.
Mercury (1-x) Cadmium (x) Telluride	PV(77 K); 0.01 PV(300 K); 0.01 PC(77 K), 1 PC(300 K), 0.4	77 to 323	0.1<	>6	Same as Indium Antimonide.
Pyroelectric	See Remarks	263	1<	0.1<	Nanosecond response available. Normally operated between 10 Hz and 2 KHz. Linearity stated is for operation as a sensor from 10.4 μm to 12.5 μm for targets from 253 K to 343 K. 2% linearity to 50 m W/cm². Reported to be nearly free from microphonics. Curie Temperature: 700–800 C.
Thermopile	See Remarks	253 to 353	0.1	0.1<	Response time varies, with construction, from 2 ms to 80 ms. Modern deposited-film-type units are rugged and more tolerant of abuse than their wire-junction ancestors.
Thermistor Bolometer	1000 to 20,000	238 343	1.0	0.5<	Suitable for operation in environmental extremes. Tolerant of vibration and shock. Precision instruments usually entail immersion of detector in a temperature controlled lens and housing.

where Φ is the total filtered flux reaching the detector and $T(\lambda)$ is the transmittance of the filter as a function of wavelength.

Recalling Eq. 7-7 and temporarily setting $\Phi_0 = 0$ the voltage response of the detector to the filtered flux is

$$V = G \int_0^\infty \mathcal{R}_\Phi(\lambda) \cdot \Phi_\lambda \cdot T(\lambda) \cdot d\lambda \quad [V] \quad (7\text{-}25)$$

where $\mathcal{R}_\Phi(\lambda)$ is the responsivity of the detector as a function of wavelength and G is the gain of the electronics.

If $T(\lambda_1)$, is the transmittance of an ideal filter at λ_1, with narrow passband, $\Delta\lambda$, and if $\mathcal{R}_\Phi(\lambda_1)$ and Φ_λ are relatively smooth functions over $\Delta\lambda$, as shown in Figure 7-10, then

$$V_1 = \Phi_{\lambda 1} \cdot \Delta\lambda \cdot G \cdot \mathcal{R}_\Phi(\lambda_1) \cdot T(\lambda_1) \quad [V] \quad (7\text{-}26)$$

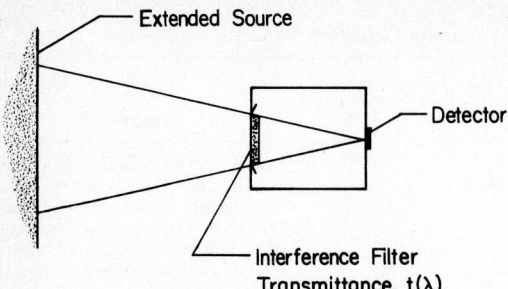

Fig. 7-8. Simple radiometer with interference filter.

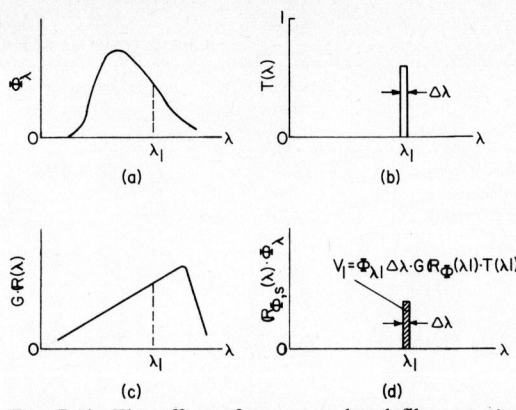

where V_1 is the system output response and $\Phi_{\lambda 1} \cdot \mathcal{R}_\Phi(\lambda_1)$, and $T(\lambda_1)$ are averages over the passband, $\Delta\lambda$. Then the voltage response has the form

$$V_1 = \Phi_1 \cdot \mathcal{R}_{\Phi,s}(\lambda_1) \qquad (7\text{-}27)$$

where $\mathcal{R}_{\Phi,s}(\lambda_1) = G \cdot \mathcal{R}_\Phi(\lambda_1) \cdot T(\lambda_1)$ is the system flux responsivity of the passband and Φ_1 is the inband flux ($\Phi_{\lambda 1} \cdot \Delta\lambda$). Using Eq. 7-21,

$$V_1 = L_1 \cdot A_d \cdot \pi \sin^2\left[\frac{\theta_f}{2}\right] \cdot \mathcal{R}_{\Phi,s}(\lambda_1)$$

$$V_1 = L_1' \cdot \mathcal{R}_{L,s}(\lambda_1) \qquad (7\text{-}28)$$

where $\mathcal{R}_{L,s}(\lambda_1)$ is the radiance responsivity of the system and L_1' is the in-band radiance of the source ($L_\lambda' \cdot \Delta\lambda$).

The importance of careful interpretation in the comparison of optical quantities measured by different instruments and the importance of the choice of appropriate passbands are illustrated in Figure 7-11 where two ideal filters of different passbands are centered at λ_c. Figures 7-11(b) and 7-11(c) show that two sources having different spectral distributions may yield the same total flux at the detector. Figures 7-11(c) and 7-11(f) show that significantly different responses may be mea-

Fig. 7-10. The effect of a narrow-band filter on the spectral flux reaching the detector: (a) Φ_λ, the spectral flux incident on the detector before filtering $[W \cdot \mu m^{-1}]$, (b) $T(\lambda)$, the transmittance of the ideal filter, (c) $G \cdot \mathcal{R}_\Phi(\lambda)$, the product of the electronic gain of the system and the flux responsivity of the detector $[V \cdot W^{-1}]$, (d) $\mathcal{R}_{\Phi,s} \cdot \Phi_2$ the spectral voltage response of the system including the filter $[V \cdot \mu m^{-1}]$.

sured for the two sources when a different bandwidth is used. Figures 7-11(b) and 7-11(e) show that the measured flux may be different for the same source; this effect must be considered when comparing optical quantities measured by instruments having continuously variable spectral coverage since the spectral passbands of the instruments may be different due to the use of different spectral selection devices.

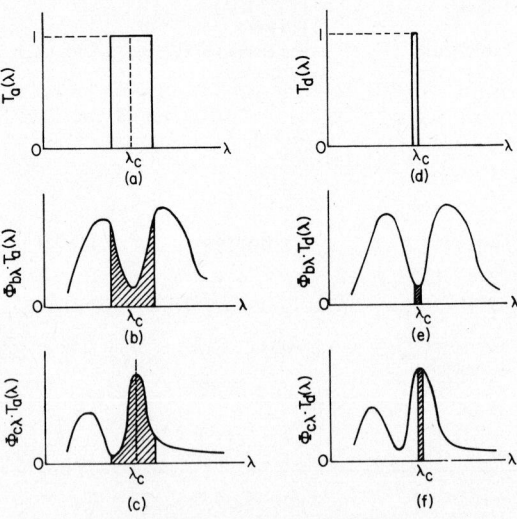

Fig. 7-11. The effect of filter bandwidth: (a) $T_a(\lambda)$, the transmittance of a relatively wide ideal filter, A, (b) $\Phi_{b,\lambda} \cdot T_a(\lambda)$, the spectral flux incident on the detector from a blackbody target after filter A, (c) $\Phi_{c\lambda} \cdot T_a(\lambda)$, the spectral flux incident on the detector from target C after filter A, (d) $T_d(\lambda)$, the transmittance of a relatively normal ideal filter, D, (e) $\Phi_{b,\lambda} \cdot T_d(\lambda)$, the spectral flux incident on the detector from a blackbody target after filter D, (f) $\Phi_{c,\lambda} \cdot T_d(\lambda)$, the spectral flux incident on the detector from target C after filter D.

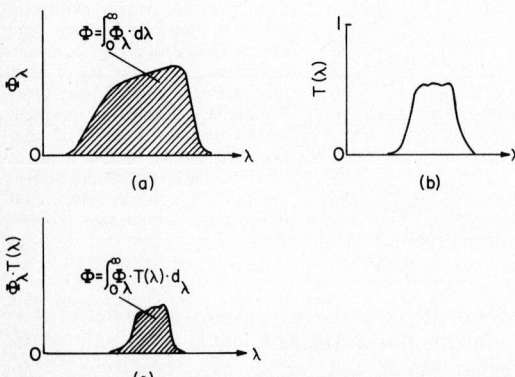

Fig. 7-9. The effect of a wide-band filter on the spectral flux reaching the detector: (a) Φ_λ, the spectral flux incident on the detector before filtering $[W \cdot \mu m^{-1}]$, (b) $T(\lambda)$, the transmittance of the filter, (c) $\Phi_\lambda \cdot T(\lambda)$, the spectral flux incident on the detector after filtering $[W \cdot \mu m^{-1}]$.

REFERENCING

For the case of the simple radiometer shown in Figure 7-8, the flux reaches the detector from two sources: the target and the optical cavity. The voltage response to these sources can be represented as

$$V = \int_0^\infty \int_0^{2\pi} \int_0^{\theta_f/2} L_\lambda'(R,\theta_t,\phi)\cdot \mathfrak{R}_{L,s}(\lambda)\cdot \cos\theta_t\cdot\sin\theta_t\cdot d\theta_t\cdot d\theta\cdot d\lambda$$

$$+ \int_0^\infty \int_0^{2\pi} \int_{\theta_f/2}^{\pi/2} L_{\lambda,c}(\theta,\phi)\cdot \mathfrak{R}_{L,c}(\lambda)\cdot \cos\theta\cdot\sin\theta\cdot d\theta\cdot d\phi\cdot d\lambda$$

$$+ \; V_{offset} \quad [V] \tag{7-29}$$

where the second term represents the optical cavity, which is assumed to be perfectly absorbing ($\epsilon = 1$). Thus the radiance of the cavity is due to its temperature. The third term represents the offset voltage of the amplifying electronics. Equation 7-29 is a linear function of L_λ'. That is, if L_λ' is uniformly attenuated at each wavelength, V will vary as a linear function of the attenuation. For a simple narrow-band radiometer

$$V = L_{\lambda 1}'\cdot \pi\sin^2\left[\frac{\theta_f}{2}\right]\cdot \mathfrak{R}_{L,s}(\lambda)$$

$$+ L_{c1}\cdot \pi\cos^2\left[\frac{\theta_f}{2}\right]\cdot \mathfrak{R}_{L,c}(\lambda)$$

$$+ V_0 \quad [V] \tag{7-30}$$

where $L_{\lambda 1}'$ and L_{c1} are the inband radiances of the target and cavity and $\mathfrak{R}_{L,s}(\lambda)$ and $\mathfrak{R}_{L,c}(\lambda)$ are the responsivities of the target and cavity, respectively. Note that the radiance of the cavity is unfiltered. From Eq. 7-30 it can be seen that two levels of $L_{\lambda 1}'$ need to be measured to determine the linear relation between $L_{\lambda 1}'$ and V. The techniques for producing these levels vary with the type of radiometer.

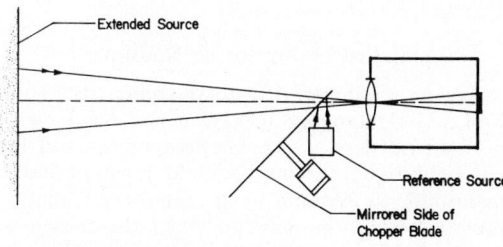

Fig. 7-12. A simple chopping scheme with reference source.

Referencing for a dc Radiometer

A simple narrow-band radiometer with a detector that does not respond to radiation emitted by ambient temperature surfaces, (i.e. one for which $\mathfrak{R}_{\Phi,s}(\lambda) = 0$ beyond 3 μm) may be covered with a device that blocks all incoming radiation.

This is *dark-level* referencing. When the cover is in place, the first two terms in Eq. 7-30 vanish and

$$V - V_{dark} = L_{\lambda 1}\cdot \pi\sin^2\left[\frac{\Phi_f}{2}\right]\cdot \mathfrak{R}_{L,s}(\lambda), \tag{7-31}$$

since $V_{dark} = V_{offset}$. An important assumption is that the detector does not respond to the filtered thermal radiation from the cover. In order to determine the value of the product of factors to the right of $L_{\lambda 1}'$, several values of known target radiance should be measured. Since, in general, the radiance responsivity of the system is a function of detector temperature, frequent *in situ* reference to a target of known radiance and to a dark-level reference is required for accurate absolute-radiance measurements. This is good practice even for systems with detectors maintained at constant temperature.

Referencing for an ac Radiometer

If a chopper is installed in front of the simple narrow-band radiometer as shown in Figure 7-12, then the field-of-view of the detector can be alternately filled with the reference source and the target. The voltage response of the system is indicated in Figure 7-13. The ac signal due to the variation of the detector output voltage is measured and the dc value is ignored. The output of the system is then proportional to the peak-to-peak variation. Then, from Eq. 7-30, the output voltage, V_1, is

$$V_1 = L_{\lambda 1}'\pi\sin^2\left[\frac{\theta_f}{2}\right]\mathfrak{R}_{L,s}(\lambda_1)$$

$$- L_r\pi\sin^2\left[\frac{\vartheta_f}{f}\right]\mathfrak{R}_{L,s}(\lambda_1)$$

$$+ V_{offset} \quad [V] \tag{7-32}$$

where $L_{\lambda 1}'$ and L_r are the in-band radiances of the target and reference, respectively, and V_{offset} is the offset voltage of the amplifying electronics.

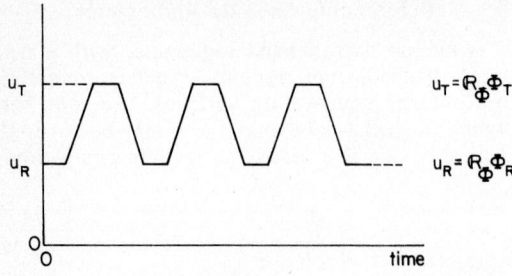

Fig. 7-13. Typical variation of detector output voltage.

V_{offset} may be determined by blocking the input to the cavity so that no signal variation is introduced to the detector. This is analogous to measuring the dark-level reference; however, in this case, the detector may be responsive to thermal radiation. In some cases, the reference radiation is known and measurement of a single additional reference target of known radiance enables the absolute measurement of radiance. In other cases, all that is known of the reference is that it is constant. In these cases, two targets of known radiance are required for the making of absolute measurements.

ELECTRONIC SIGNAL PROCESSING

The function of signal processing is to convert the detector output signal to an electrical signal which may be read from a meter, permanently recorded, or used to drive other devices. Figure 7-14 diagrams the common signal-processing systems that are discussed in the following sections.

Signal Processing for dc Radiometers

The dc system is illustrated in Figure 7-14a. The detector-preamplifier module serves to amplify the signal to a level suitable for processing by an amplifier which determines the system frequency response, provides the required output impedance, and amplifies the signal to the desired voltage.

The *photovoltaic* silicon diode coupled with a *current mode* operational amplifier is shown in Figure 7-15a. While frequently used as a preamplifier, this system may be used to perform all the functions of Figure 7-14a. For example, a signal-processing system may be designed to provide a signal output of about 0.8 V when used in a simple passband radiometer ($8_f = 15°$; $\Delta\lambda = (0.76-0.90)\mu m$; $T = 0.5$; $A_d = 0.032$ cm^2) to view a horizontal, 100-percent reflecting surface irradiated by sun and sky at midday. From Eq. 7-23,

$$\Phi_1 = L'_{\lambda 1} \cdot \Delta\lambda \cdot A_d \cdot \pi \sin^2\left[\frac{\theta_f}{2}\right]$$

$$\Phi_1 = \frac{0.075}{\pi} \times 0.14 \times 0.032 \times \pi \times (0.017) \quad (7\text{-}33)$$

$$\Phi_1 = 5.7 \times 10^{-6} \ W$$

where Φ_1 is the in-band flux on the detector (without filter) and where the average spectral irradiance over the passband is 0.075 W·cm^{-2}·sr^{-1}·μm^{-1}. Then, for $R_f = 800$ kiloohm the output voltage is

$$V_1 = \Phi_1 \cdot \mathcal{R}_{\Phi,s}(\lambda_1)$$

$$V_1 = \Phi_1 \cdot T(\lambda_1) \cdot G \cdot \mathcal{R}_{\Phi}(\lambda_1)$$

$$V_1 = \Phi_1 \cdot T(\lambda_1) \cdot (-R_f) \cdot \mathcal{R}_{\Phi}(\lambda_1)$$

$$V_1 = (5.7 \times 10^{-6})W \cdot 0.5(-8 \times 10^5\Omega) \cdot 0.35 A W^{-1}$$

$$V_1 = -0.8 \ V$$

where, in this case, G is the transimpedance, $-R_f$, and 0.35 $A \cdot W^{-1}$ is the typical responsivity of the photovoltaic silicon PIN diode detector operated with the current mode amplifier. If a 0.01-μF feedback capacitor, C_f, is used, the time constant of the system ($R_f \cdot C_f$) will be 8 milliseconds and the low-pass corner frequency ($\frac{1}{2}\pi\ R_f C_f$) will be about 20 Hz. Operation of a single-stage system at high gains or wide bandwidths requires a thorough analysis of the amplifier and detector impedance characteristics.

A *photoconductive* detector with *voltage mode* amplification is illustrated in Figure 7-15b. In this case, the transistor amplifier has a voltage gain of approximately $R_S/(R_S + 1/g_m)$ and an output impedance of $R_S/(1 + g_m R_S)$ where g_m is the transconductance of the field-effect transistor. Since a typical value of g_m is 2000 microsiemens when R_S is 33 kiloohms, the voltage gain of the preamplifier is about 0.99 and the output impedance is approximately 500 ohms. Thus, this configuration is used to provide a high impedance load for the detector-bias resistor combination and deliver the detector signal with little loss to the following amplifier stage.

The common emitter, common collector, and common base configurations of bipolar transistors are also used for preamplifiers. In addition, common-source configurations are used for field-effect transistors. The transistor and configuration depend on the application and are of critical importance in determining the noise performance of the entire system. Sloan (1979) discusses the important characteristics of transistor types and circuit configurations. In general, for detector-bias resistor combinations less than 500 ohms, bipolar transistors provide better noise performance; above 10 kiloohms, field-effect transistors are usually preferred.

Signal Processing for ac Radiometers

The ac radiometer systems shown in Figure 7-14 may be designed to have nearly black chopper surfaces ($\epsilon(\lambda) \approx 1$). The temperatures of the cavity and chopper may be held constant and/or measured. In addition to the reference radiation provided when the detector views the choppers, an improved signal-to-noise ratio may be obtained by operating the amplifiers and detector at higher

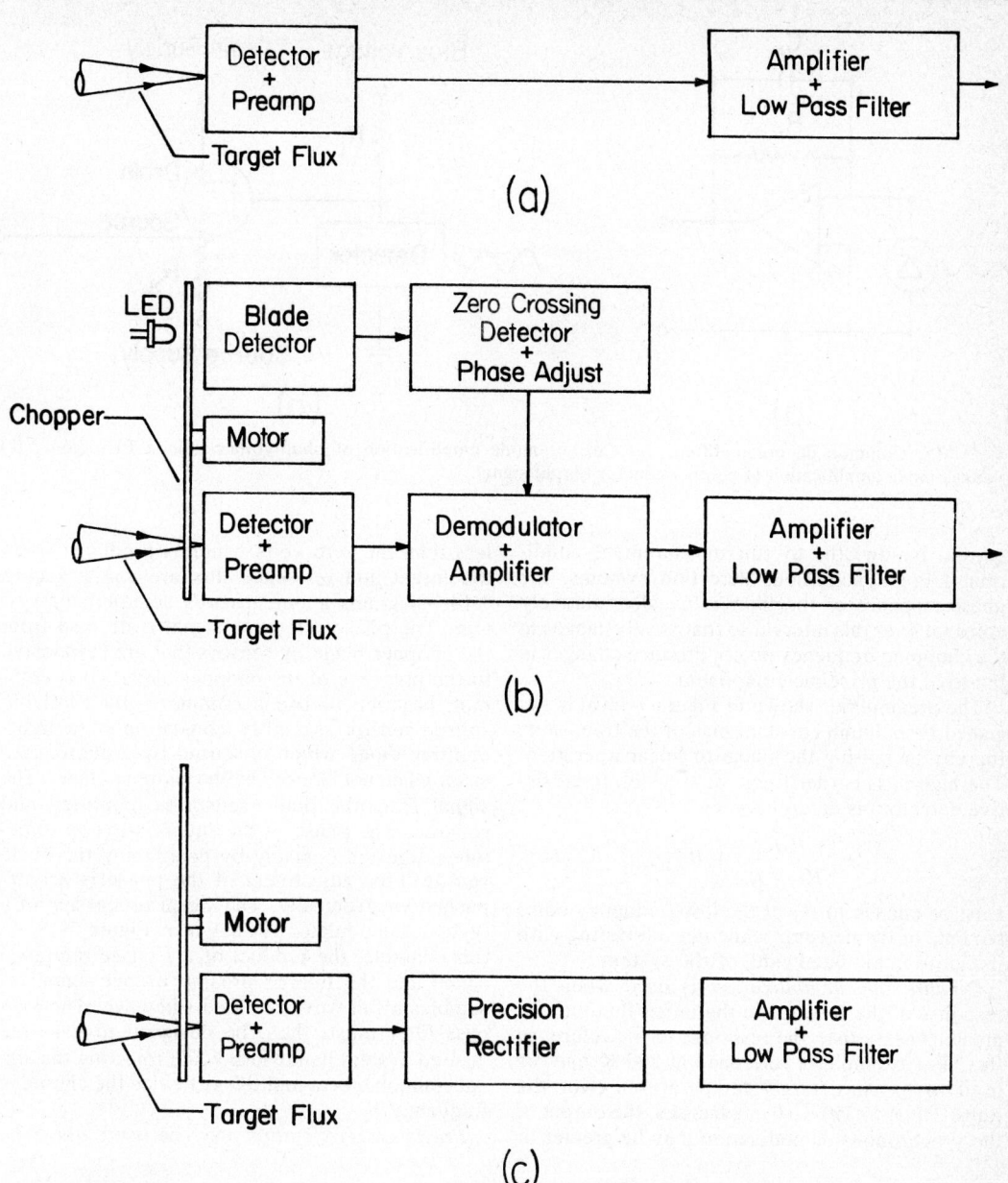

Fig. 7-14. Common signal-processing systems: (a) dc radiometer, (b) ac radiometer with synchronous demodulation, (c) ac radiometer with precision rectification.

frequencies, which avoids the low-frequency $1/f$ noise generated by both devices.

In many cases the detector of an ac radiometer is direct-coupled to the preamplifier (as shown in Figure 7-16a). The following stage is usually a bandpass filter formed by an amplifier and the coupling circuitry. The electronic filter is chosen to reject all but the principal component of the detector signal while maintaining a bandwidth that does not restrict the maximum system bandwidth. Figure 7-17 shows the effect of a bandpass filter applied to a preamplifier output signal for a 200-Hz chopping frequency. Rejection of the low-frequency components of the detector signal is important in most systems, particularly in spectroradiometers which automatically scan spectral signals and which must stabilize quickly to follow only the peak-to-peak amplitude changes produced by the changing spectral flux of the signal and the reference. The filter is reasonably flat over the range of the system bandwidths on either side of the principal component. (System bandwidth is usually determined by the low-pass filter at the output which may be adjustable for

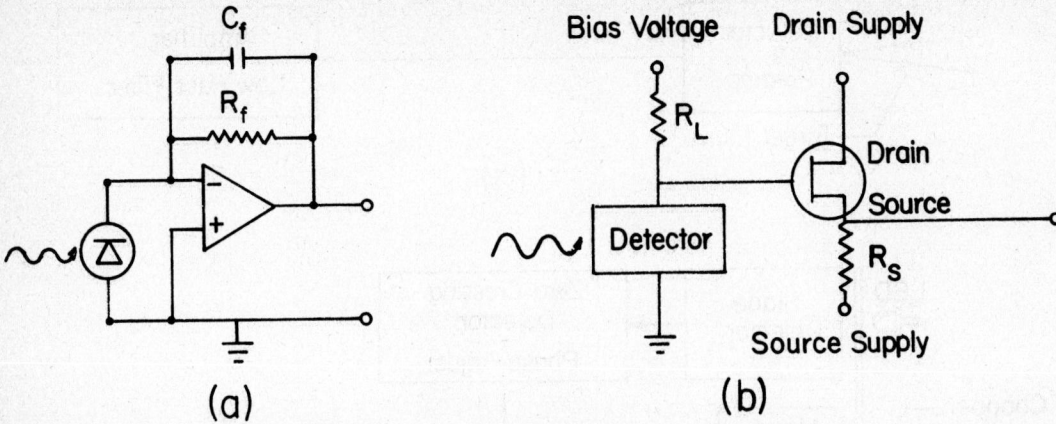

Fig. 7-15. Common dc preamplifiers: (a) Current mode amplification of photovoltaic silicon PIN diode, (b) voltage-mode amplification of photoconductor output signal.

several bandwidths to suit measurement conditions.) For synchronous detection systems, the phase response of the filter is usually relatively constant over this interval so that small changes in the chopping frequency do not produce changes in phase of the principle component.

The preamplifier shown in Figure 7-16(b) is designed to maintain constant bias of the transistor, thereby increasing the range of linear operation. The high-pass corner frequency, which for resistive detectors is determined by

$$f_c \approx \left\{ 2\pi \left[\frac{R_L R_d}{R_L + R_d} + R_G \right] C_G \right\}^{-1} \qquad (7\text{-}35)$$

must be chosen to reject the low-frequency components of the detector while not interfering with the information bandwidth of the system.

Synchronous demodulation is used when the response of the detector to the target flux may be greater or less than the response to the reference flux. For example, a reference at 300 K may be used to measure target temperatures over the range 250 K to 350 K. In such cases, the output of the synchronous demodulation may be greater or less than the zero volts which is produced when the target and reference flux are equal. Figure 7-14b diagrams a synchronous demodulator system. The phase reference signal is derived from the chopper blade by sensors that are responsive to the presence of the chopper blade. (It is common practice to use a commercially available source-sensor assembly consisting of a light-emitting diode which is sensed by a photo transistor when not blocked by the chopper blade.) The signal from the blade sensor is amplified and squared. The phase of the square wave is sometimes adjusted coarsely by positioning the blade sensor. Fine adjustment of the phase is accomplished electronically. The signal processing of a typical demodulator is shown in Figure 7-18. In this example, the product of the phase reference signal and the filtered preamp output signal resembles a full wave rectified sinewave. The low-pass filter must, then, be designed to pass the desired system bandwidth while rejecting the signal components at multiples of twice the chopping frequency.

Precision rectification may be used when the

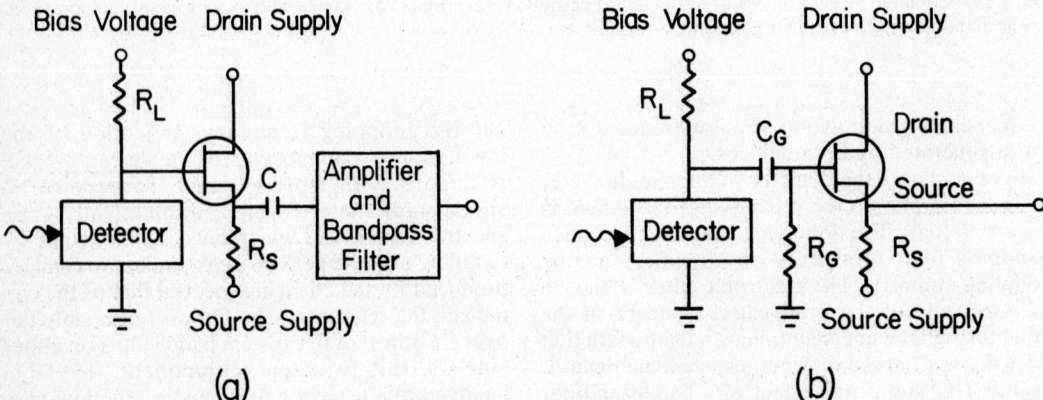

Fig. 7-16. Common ac preamplifiers: (a) ac coupling to preamplifier output, (b) ac coupling to preamplifier input.

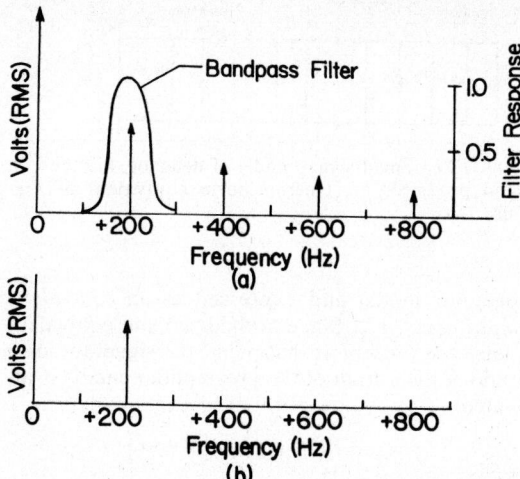

Fig. 7-17. Frequency-domain representation of chopped signals: (a) unfiltered, showing the bandpass filter response, (b) filtered signal.

response of the detector to the target flux is always greater than or equal to (or always less than or equal to) the response to the reference flux. The precision rectifier performs, with high linearity, the absolute-value function on the filtered preamp signal. The advantage of the precision rectifier is that it is entirely electronic and is insensitive to phase changes caused by changes in detector temperature and small changes in chopping frequency. The low-pass filter is designed as for the synchronous demodulator.

SIGNAL-TO-NOISE-RATIO

Noise may be present in the radiant flux arriving at the detector. This *external noise* may be generated in the processing of the optical signal by mechanical vibrations of the optical components, inadequate spectral filtering, inadequate definition of the required field of view, random scattering of stray light (dependent on the radiance distribution of the target) and other random or systematic sources of unwanted variation of the flux incident on the detector. In addition, electrical noise may be generated in the system (including the detector) by electromagnetic fields originating from chopper motors, local radio stations, switching power supplies, etc. Electrical noise can also be generated by the mechanical vibration of signal cables (triboelectric) and crystals (piezoelectric). Most of the unwanted flux variations and electrical-signal variations can be eliminated through proper design and construction. However, *quantum noise* generated by the random arrival of photons from a "constant" source cannot be eliminated (see Chapter 8). The root-mean-square sum of the quantum noise and the residual external noise will be denoted as v_{ext}.

The *internal noise*, due to the detector and signal processing, will be denoted: v_{int}. Assuming these sources are uncorrelated, the system output noise can be expressed:

$$v_{sys} = \sqrt{(v_{ext})^2 + (v_{int})^2} \qquad (7\text{-}36)$$

This section will discuss the quantum noise and internal noise of radiometers. The major sources of internal noise are the detector, the detector-bias circuit, the preamplifier, and, in chopped systems, the nonrandom demodulation residuals.

RANDOM NOISE

In many cases, the detector-bias-preamplifier circuit may be modeled as in Figure 7-19 when the noise sources have been omitted. For these cases,

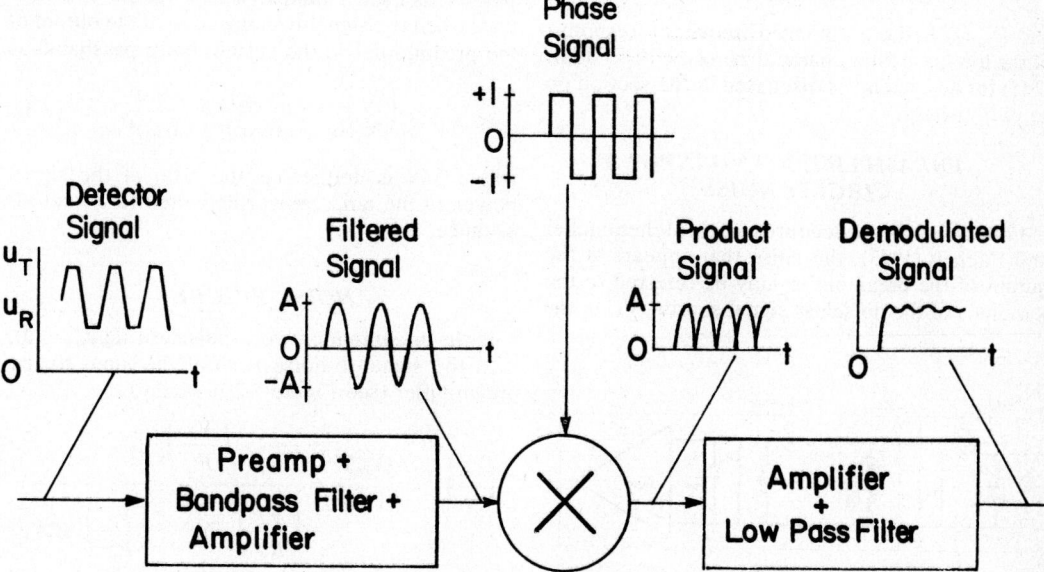

Fig. 7-18. Synchronous detector system with time-domain representation of electrical signals.

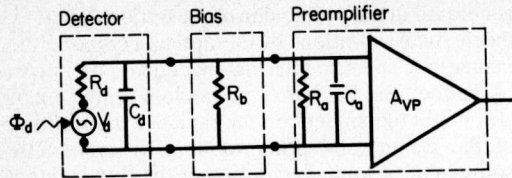

Fig. 7-19. Noiseless small-signal model of detector, bias circuit and preamplifier.

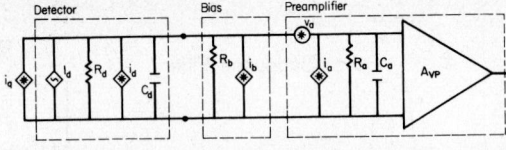

Fig. 7-21. Small-signal model of detector, bias circuit, and preamplifier. Current form equivalent of Figure 7-20.

the voltage across the detector, $V_d(t)$, and the current through the detector for the Norton equivalent circuit, $I_d(t)$, are related to the detector input flux signal, $\Phi_d(t)$, by:

$$V_d(t) = I_{d(t)}R_d = \Phi_d(t) \cdot R_{\Phi,d} \quad [V] \quad (7\text{-}37)$$

where R_d is the resistance of the detector and R_Φ is the voltage responsivity of the detector to input flux. Then, the mean-square-values of the signals are

$$V_d^2 = I_d^2 \cdot R_d^2 = \Phi_d^2 R_{\Phi,d}^2 \quad (7\text{-}38)$$

For ac systems, the signals are assumed to have mean values of zero.

Figures 7-20 and 7-21 include the noise sources of the detector, bias resistor, and preamplifier where the preamplifier is assumed to have a constant gain, A_{VP}, and a low-pass bandwidth wider than the system requirements. Characteristics of the noise sources are discussed in the examples below.

Noise sources in the following are assumed to be evaluated in the system noise passbands:

$$v^2 = \int_0^\infty \overline{v^2(f)}|T^*(f)|^2 df \quad (7\text{-}39)$$

where

$$T^*(f) = H_{LP}(f) \text{ for } dc \text{ systems}$$

and $H_{LP}(f)$ is the normalized frequency response of the low-pass filter characterizing the dc system. $T^*(f)$ for ac systems is discussed in the section on ac radiometers.

PREAMPLIFIER AND INPUT CIRCUIT NOISE

Following the procedures of Motchenbacher and Fitchen (1973), the noise that appears at the output of the preamplifier may be referred to the terminals of the noiseless signal source, V_s, in the

detector model and expressed as an *equivalent input noise*, v_{eq}. Since both signal and equivalent noise are present at that point, the signal-to-noise ratio at the output of the preamplifier can be evaluated:

$$v_{out} = |K_t| \cdot v_{eq} \quad [V \text{ rms}]$$

and

$$V_{out} = |K_t| \cdot V_d \quad [V \text{ rms}] \quad (7\text{-}40)$$

where, for simplicity, $V_d(t)$ has been assumed to be a constant or sinusoidal signal and where V_{out} and v_{out} are the root-mean-square signal and noise voltages, respectively, at the output of the preamplifier. K_t is the magnitude of the voltage transfer function from the input source to the output of the preamplifier. For the system modeled by Figures 7-20 and 7-21 the mean square of the equivalent input voltage will have the form:

$$v_{eq}^2 = \left[(i_q)^2 + (i_d)^2 + (i_a)^2\right]|Z_P|^2 + v_a^2 \quad (7\text{-}41)$$

where i_q is the qu: um noise; i_d is the noise current of the detector, v_d/R_d; where i_b is the noise current of the bias resistor, v_b/R_b; and where i_a is the input noise current of the preamplifier. Z_P is the parallel combination of R_d, $(1/j\omega c_d)$, R_b, R_a, and $(1/j\omega C_a)$ shown in Figure 7-21. Since the noise sources are evaluated in the system passbands (see example below for ac radiometers), then the signal-to-noise ratio at the output of the preamplifier in the system noise passbands is

$$\left(\frac{S}{N}\right)_{out} = \frac{(V_{out})^2}{(v_{out})^2} = \frac{(V_d)^2}{(v_{eq})^2} \quad (7\text{-}42)$$

where S/N is defined as the ratio of the signal power to the noise power delivered to a load resistance.

NOISE FIGURE

If the preamplifier were noiseless ($i_a, v_a = 0$), then the signal-to-noise ratio at the input to the preamplifier (see Figure 7-20) would be

$$\left(\frac{S}{N}\right)_{in} = \frac{(V_{in})^2}{(v_{in})^2} = \frac{\left(\frac{V_d}{R_d}\right)^2 \cdot |Z_P|^2}{\left(\frac{v_d}{R_d}\right)^2 |Z_P|^2 + \left(\frac{v_b}{R_b}\right)^2 |Z_P|^2} \quad (7\text{-}43)$$

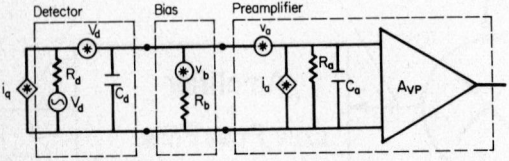

Fig. 7-20. Small-signal model of detector, bias circuit and preamplifier.

$$\left(\frac{S}{N}\right)_{in} = \frac{(v_d^2)^2}{(v_d)^2 + \left(\dfrac{R_d}{R_b}\right)^2 (v_b)^2} \quad (7\text{-}44)$$

where V_{in} and v_{in} are the rms signal and noise voltages, respectively, delivered to the preamplifier input terminals as measured in the system noise passbands.

The signal-to-noise ratio at the output of the preamplifier has been defined in Eq. 7-42 above. Then the preamplifier may be characterized as decreasing the signal-to-noise ratio at its input (as well as increasing the magnitude of the signal at its output).

The *noise figure, F,* of the preamplifier is defined by

$$(S/N)_{out} = \frac{(S/N)_{in}}{F} \quad (7\text{-}45)$$

where both the input and output noises are evaluated in the system noise passbands. Since the largest possible value of $(S/N)_{out}$ is desired, designers strive to make F as small as possible; however, F is always greater than unity. It should be noted that the constant internal noise of the photoconductive photodiode (the shot noise due to the dark current) is increased greatly by the external noise (the photon noise due to the average signal current) and that the input signal-to-noise ratio does not increase as the square of the input current.

Manufacturers' specifications usually express F in decibels:

$$F_{dB} = 10 \log_{10} F \quad (7\text{-}46)$$

so that conversion to decimal form is required. Modern preamplifiers, which are designed to amplify the signals from specific detector types, have *wideband noise figures, $F_{P,dB}$* which range from 0.5 dB to about 6 dB with typical performances of 3 dB for the electrical passband of the amplifier (e.g., Santa Barbara Research Center Bulletin Brochure, 1979). The decimal value of the wideband noise figure ranges from 1.1 to 4.0 with a typical value of 2.0. Thus, if the decimal noise figure, F_P of the preamplifier is 2.0, then the signal-to-noise ratio is reduced by a factor of 2 by the preamplifier:

$$(S/N)_{out} = (S/N)_{in}/F_P \quad (7\text{-}47)$$

and, then,

$$\frac{V_{out}}{v_{out}} = \frac{1}{\sqrt{2}} \frac{V_{in}}{v_{in}} \cdot \quad (7\text{-}48)$$

In this case, the preamplifier has increased the rms noise-to-signal ratio by a factor of $\sqrt{2}$.

Properly designed, the amplification stages following the preamplifier have little effect on the noise figure, F_S of the preamplifier–amplifier system. This is apparent from the classic work by Friis (1944), as evidenced by the equation

$$F_S = F_P + \frac{F_A - 1}{A_P} \quad (7\text{-}49)$$

where F_A is the noise figure of the stage following the preamplifier and A_P is the available power gain of the preamplifier which may easily have values ranging from 10^3 to more than 10^6.

SIGNAL-TO-NOISE RATIO AC RADIOMETER

The noise sources may be evaluated in the system noise passbands (Eq. 7-39 and below), referred to the system input, and compared with the effective mean-square input signal to determine the signal-to-noise ratio at the output of the system. Nonrandom demodulation noise may be considered in the presence of the random noise.

Figure 7-24 shows a system for processing the signals in a radiometer. To evaluate the noise sources in the system noise passbands the effect of the bandpass filter and the product with the square wave on the spectral density of the noise delivered to the low-pass filter may be considered.

SYSTEM FILTERS

The bandpass amplitude response shown in Figure 7-22 eliminates system drift due to temperature-induced changes in detector resistance and preamplifier offsets; furthermore, even a single-pole high-pass filter will easily dominate $1/f$ noise near zero Hertz. The transfer function $|K_t(f)|$ may be included in the bandpass amplifier (for example, see Eq. 7-42) and may be balanced by a low-frequency high-pass filter so that the net frequency response is relatively flat at the chopping frequency (200 Hz). Additional filtering may be designed to maintain relative flatness at the chopping frequency for both the amplitude and phase characteristics.

The low-pass filters shown in Figure 7-23 are chosen to respond to signal variations in the desired bandwidth (0-20 Hz) and to reject the harmonic components of the rectified signal.

SYSTEM NOISE PASSBANDS

Figure 7-24 shows a system for processing the signal from a biased photoconductor. The spectral density of the noise at the output by the preamplifier is

$$v_P^2(f) = \left\{ \overline{i_q^2} + \overline{i_d^2} + \overline{i_b^2} + \overline{i_a^2} \left[\frac{C_d + C_c}{C_c} \right]^2 \left[1 + \left(\frac{f_i}{f} \right)^2 \right] \right.$$

$$\left. + \frac{\overline{v_a^2}}{R_1^2} \left[1 + \left(\frac{f}{f_1} \right)^2 \right] \right\} |K_t(f)|^2 \quad (7\text{-}50)$$

where i_d^2, etc., are the spectral densities that are functions of frequency and where $f_i = 1/(2\pi R_1(C_d + C_c))$; $f_1 = 1/(2\pi R_1 C_d)$; $R_1 = R_d \| R_b$; and

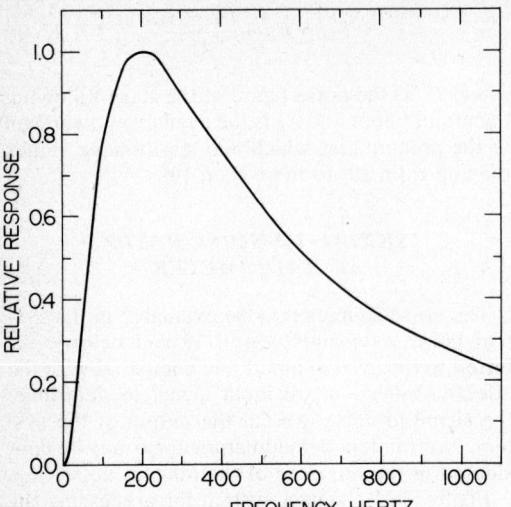

Fig. 7-22. Plot of the normalized transfer function given by:

$$|H_{BP}(f)| = \cfrac{1}{\sqrt{1 + \left(\cfrac{f}{900} - \cfrac{44.4}{f}\right)^2}}$$

$$\cdot \cfrac{1}{\sqrt{1 + \left(\cfrac{f}{250}\right)^2}} \cdot \cfrac{1}{\sqrt{1 + \left(\cfrac{160}{f}\right)^2}},$$

where the first factor is due to the detector input circuitry.

$$K_t(\omega) = \frac{R_P \tau_c A_v}{\tau_a + \tau_1 + \tau_c} \cdot$$

$$(7\text{-}51)$$

$$\cfrac{1}{1 + j\left[\cfrac{\omega(\tau_1\tau_a + \tau_a\tau_{1c} + \tau_1\tau_{ac})}{(\tau_a + \tau_1 + \tau_c)} - \cfrac{1}{\omega(\tau_a + \tau_1 + \tau_c)}\right]}$$

where

$\tau_1 = R_1C_d$; $\tau_a = R_aC_a$; $\tau_c = [R_1 + R_a]C_c$; $\tau_{1c} = R_1C_c$; $\tau_{ac} = R_aC_c$. R_P is $R_d\|R_b\|R_a$. Then $K_t(\omega)$ may be written

$$K_t(f) = \frac{R_P\tau_cA_v}{\tau_a + \tau_a + \tau_c} \cdot \cfrac{1}{1 + j\left[\cfrac{f}{f_{LP}} - \cfrac{f_{HP}}{f}\right]} \qquad (7\text{-}52)$$

where f_{LP} and f_{HP} are determined from Eq. 7-51. To design for a real value at the chopping frequency which will yield relatively flat phase and amplitude response:

$$\omega_X = \frac{1}{\sqrt{R_1R_a(C_dC_a + C_aC_c + C_dC_c)}} \cdot \qquad (7\text{-}53)$$

If the amplifier capacitance can be neglected

$$\omega_X = \frac{1}{\sqrt{R_1C_d \cdot R_aC_c}} \qquad (7\text{-}54)$$

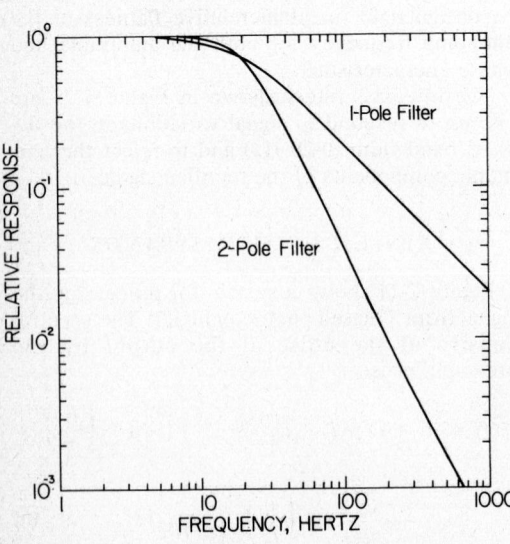

Fig. 7-23. Normalized transfer function, $|H_{LP}(f)|$ for the N pole low-pass filter at the system output. $|H_{LP}(f)| = [1 + [f/20]^{2N}]^{-1/2}$.

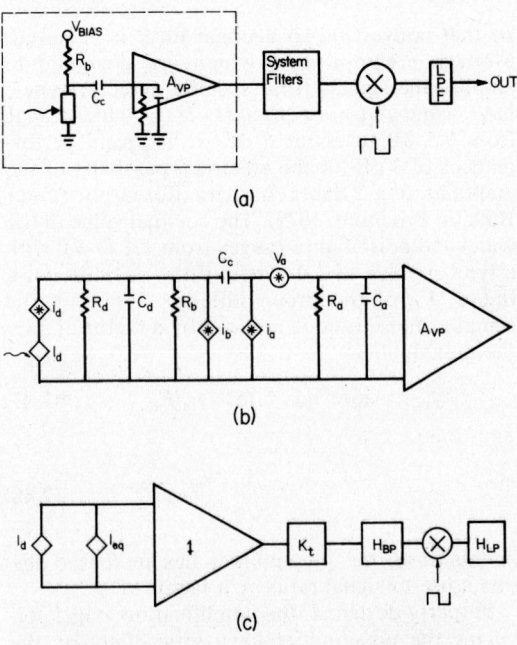

Fig. 7-24. ac system for a biased photoconductor: (a) block diagram, (b) model, (c) equivalent input model.

Equation (7-50) may be written

$$\overline{v_P^2(f)} = \overline{i_{eq}^2(f)} \cdot \left[\frac{R_P \tau_c A_v}{\tau_a + \tau_1 + \tau_c}\right]^2 \cdot \frac{1}{1 + \left[\dfrac{f}{f_{LP}} - \dfrac{f_{HP}}{f}\right]^2} \tag{7-55}$$

When subsequent filtering and amplification are applied, the spectral density at the input to the multiplier is

$$\overline{v_{BP}^2} = \overline{i_{eq}^2(f)} \cdot \left[\frac{R_p \tau_c A_v}{\tau_a + \tau_1 + \tau_c}\right]^2 \cdot K^2 |H_{BP}(f)|^2 \tag{7-56}$$

where $H_{BP}(f)$ is the normalized frequency response of the composite bandpass filter and K is the normalization constant.

The spectral density at the output of the multiplier may be determined, but it is simpler to consider the portion of the noise that will fall within the low-pass output filter after multiplication by the square wave.

The Fourier series representation of the square wave is

$$X(t) = \sum_{\substack{k=1 \\ \text{Kodd}}}^{\infty} \frac{4}{k\pi} \sin(k2\pi f_x t) \tag{7-57}$$

where f_x is the chopping frequency. Then the mean-square noise in the low-pass filter, $H_{LP}(f)$, may be represented from the relation

$$v_{\text{out}}^2 \approx \int_0^{\infty} \frac{1}{2} \sum_{\substack{k=1 \\ \text{kodd}}}^{\infty} \overline{v_{BP}^2(f)} \cdot \left[\frac{4}{k\pi}\right]^2 \cdot |H_{LP}(f - kf_x)|^2 df \tag{7-58}$$

where the factor of ½ arises from the mean-square values of the sinusoids. Then, if B_n is the equivalent noise bandwidth of the low-pass filter,

Figure 7-25 shows v_{out}^2 for the first two noise passbands for the detector noise. The mean-square noise in the system noise passbands will, in this case, be closely estimated by the first term in Eq. 7-59:

$$v_{\text{out}}^2 \approx \left[\frac{1}{2}\right] \cdot \left[\frac{4}{\pi}\right]^2 \cdot K_{vs}^2 \cdot \overline{i_{eq}^2(f_x)} \cdot 2B_n \tag{7-61a}$$

$$\approx \frac{8}{\pi^2} v_{BP}^2 \tag{7-61b}$$

which follows from Eq. 7-56.

EQUIVALENT INPUT SIGNAL

The signal input to the multiplier resembles the trapezoidal wave shown in Figure 7-26. The Fourier series representation of the trapezoid (assuming $|H_{BP}(f)|$ is an ideal high-pass filter) is

$$V(t) = \frac{V_{pp}}{2} \cdot \frac{2}{\pi^2} \cdot \frac{T}{t_1} \sum_{\substack{k=1 \\ \text{kodd}}}^{\infty} \frac{\sin(2k\pi t_1/T)}{k^2} \sin[k(2\pi f_x t)]$$

where V_{pp} is the peak-to-peak value of $V(t)$ and the fraction of time spent on the slope is $4t_1/T$. The slope of the trapezoid is due to the fact that the blade requires a finite time to move from the edge of the aperture (which is finite) to the position where the aperture is completely filled with the reference source (see Hudson, 1969).

$$v_{\text{out}}^2 \approx \left(\frac{1}{2}\right) \cdot \left(\frac{4}{\pi}\right)^2 \cdot K_{vs}^2 \cdot \sum_{\substack{k=1 \\ \text{Kodd}}}^{\infty} \frac{\overline{i_{eq}^2(k \cdot f_x)} \cdot |H_{BP}(k \cdot f_x)|^2 \cdot 2B_n}{k^2} \tag{7-59}$$

where the summation is the evaluation of the noise in the *system noise passbands* at $k \cdot fx$ and from Eq. 7-56:

$$K_{vs} = \frac{R_P \tau_c A_v}{\tau_a + \tau_1 + \tau_c} \cdot K \tag{7-60}$$

Equation (7-63) gives the spectrum of the multiplier output signal for two cases: (a) The bandpass filter is a perfect high-pass filter (see Figure 7-24C), and (b) the bandpass filter is a narrow-band filter centered on the chopping frequency.

$$V_x(t) = \frac{V_{pp}}{2}\left[\left(1 - 2\frac{t_1}{T}\right) - \frac{1}{\pi^2}\cdot\frac{T}{t_1}\sum_{k=1}^{\infty}\frac{\sin^2\left(2k\pi\frac{t_1}{T}\right)}{k^2}\cos(2k(2\pi f_x)t)\right] \tag{7-63a}$$

$$V_x(t) = \frac{V_{pp}}{2}\frac{2}{\pi^2}\cdot\frac{T}{t_1}\cdot\sin\left[2\pi\left(\frac{t_1}{T}\right)\right]$$

$$\times\left[\frac{2}{\pi} - \frac{4}{\pi}\sum_{k=1}^{\infty}\frac{1}{(4k^2 - 1)}\cos[2k(2\pi f_x)t]\right] \tag{7-63b}$$

The first term of each expression is the average value that will appear at the output of the low-pass filter. For intermediate filtering cases, the contributions of the filtered Fourier components (see Figure 7-27) may be multiplied term-for-term with the square wave to determine the average value:

$$V_x(t) = \frac{V_{pp}}{2}\cdot\frac{1}{2}\sum_{\substack{k=1\\k\,odd}}^{\infty}\frac{8}{\pi^3}\cdot\frac{T}{t_1}\frac{\sin\left(2k\pi\frac{t_1}{T}\right)}{k^3}\left|H_{BP}(kf_x)\right| \tag{7-64}$$

However, assuming that only the fundamental component of the trapezoid is rectified, the average value obtained from Eq. 7-36b will be accurate to within a few percent in many cases (see Figure 7-28a).

In these cases, the mean-square voltage of the fundamental component of $V_{BP}(t)$ is used to compute the signal-to-noise ratio at the input to the multiplier. The signal may be expressed:

$$V_{BP}(t) = \frac{V_{dpp}}{2}\cdot\frac{2}{\pi^2}\frac{T}{t_1}\sin\left(2\pi\frac{t_1}{t}\right)\sin(2\pi f_x t)\cdot\frac{K_t(f_x)}{R_d}\cdot K \tag{7-65}$$

Then the mean-square value of the signal at the input to the multiplier is

$$V_{BP}^2 = (V_{dpp})^2\cdot K_x^2(t_1/T)\cdot K_t^2(f_x)\cdot K^2 \tag{7-66}$$

where

$$K_x(t_1/T) = \frac{\sqrt{2}}{2}\cdot\frac{1}{\pi^2}\frac{T}{t_1}\cdot\sin\left[2\pi\frac{t_1}{T}\right] \tag{7-67}$$

is the *chopping factor* which decreases monotonically from 0.45 to 0.29 for $t_1/T = 0$ (square wave) to $t_1/T = \frac{1}{4}$ (triangular wave), respectively.

From Eq. 7-63b, the mean square of the voltage at the output of the low-pass filter is

Using Eq. 7-61b, the signal-to-noise ratio at the output may be written

$$\frac{V_{out}^2}{v_{out}^2} = \frac{\left[\frac{8}{\pi^2}\right]\cdot V_{BP}^2}{\left[\frac{8}{\pi^2}\right]\cdot v_{BP}^2} = \frac{V_{BP}^2}{v_{BP}^2} \tag{7-69}$$

which states that, with the assumptions of a bandpass filter with modest high-frequency attenuation centered on the chopping frequency and a steep low-pass filter following rectification, the signal-to-noise ratio due to random noise is essentially unchanged by synchronous demodulation.

SIGNAL-TO-NOISE RATIO

Random Noise

Given Eq. 7-69, the signal-to-noise ratio at the output due to random noise may be written:

$$\left[\frac{S}{N}\right]_{out(*)} = \frac{V_{BP}^2}{v_{BP}^2} = \frac{\frac{(V_{dpp})^2}{(R_d)^2}\cdot K_x^2(t_1/T)\cdot K_t^2(f_x)\cdot K^2}{i_{eq}^2\cdot K_t^2(f_x)\cdot K^2} \tag{7-70}$$

where the mean-square signal at the input is evaluated with respect to the chopping factor and the mean-square noise is evaluated in the system bandwidth. This may be rewritten

$$\left(\frac{S}{N}\right)_{out(\circ)} = \frac{V_{dpp}^2\cdot K_x^2\left[\frac{t_1}{T}\right]}{v_{eq}^2} \tag{7-71}$$

where $v_{eq}^2 = i_{eq}^2\cdot R_d^2$.

$$V_{out}^2 = [V_{dpp}]^2\cdot[\sqrt{2}\cdot K_x(t_1/T)]^2\cdot\left|\frac{K_t(f_x)}{R_d}\right|^2\cdot K^2\cdot\left[\frac{2}{\pi}\right]^2$$

$$V_{out}^2 = \frac{8}{\pi^2}[V_{BP}^2] \tag{7-68}$$

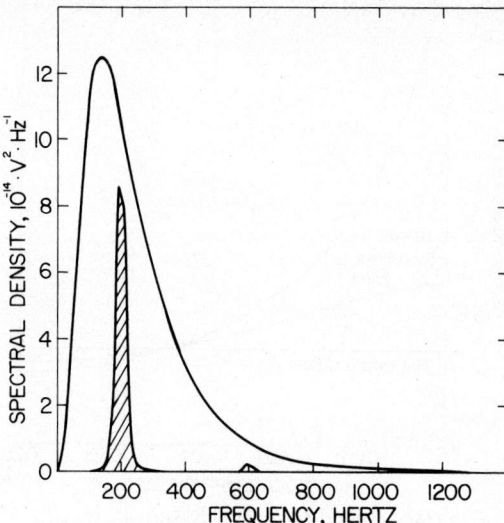

Fig. 7-25. Spectral density of the detector noise at the output of the bandpass filter. The sum of the shaded areas is the mean-square voltage in the system bandwidth.

Nonrandom Noise

The terms of Eqs. 7-63a and b are at even multiples of the chopping frequency. The sum of these terms, following attenuation by the low-pass filter, is a nonrandom source of variation which is added to the mean value and is manifest at the system output as noise. This is the sum of the *demodulation residuals*. Since the amplitude of the higher-order residuals is usually significantly smaller than the residual at $2f_x$, and since the effect of the residuals is based on the sum of the squares of their rms values, only the residual at $2f_x$ is considered. Then, for the ideal high-pass and bandpass filters, the amplitudes of the rectified signals at $2f_x$ (before the low-pass filter) are, respectively:

$$C_{2HP} = - \frac{V_{pp}}{2} \cdot \frac{T}{t_1} \cdot \frac{1}{\pi^2} \cdot \left[\sin\left(2\pi \frac{t_1}{T}\right) \right]^2$$
(7-72)

$$C_{2NBP} = - \frac{V_{pp}}{2} \cdot \frac{T}{t_1} \cdot \frac{8}{3} \cdot \frac{1}{\pi^3} \cdot \sin\left(2\pi \frac{t_1}{T}\right)$$
(7-73)

The amplitude of the rectified signal at $2f_x$ for the typical bandpass filter may be obtained from the product of the square wave with the bandpass-filtered trapezoid by selecting the terms that contain $\cos(2f_x t)$:

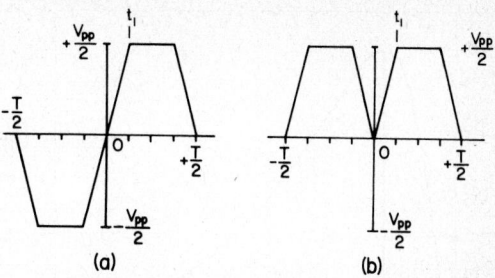

(a)   (b)

Fig. 7-26. Trapezoidal waveform ($t_1/T = 1/8$): (a) signal, (b) rectified signal.

where

$$a = (2\pi t_1/T).$$

The magnitudes of C_2 are given in Figure 7-28(b) as a function of the signal *shape factor*, t_1/T, for the three types of filters where the filter shown in Figure 7-22 is used as the typical bandpass filter.

The magnitudes of the components of three rectified signals (at $2f_x$) are about the same except for the ideal high-pass filtered trapezoid which represents a limiting case. The signal-to-noise ratio for a fixed t_1/T depends on the extent to which the $2f_x$ component is attenuated by the low-pass filter:

$$\left(\frac{S}{N}\right)_{out(s)} = \frac{(\text{Average Value})^2}{\frac{1}{2}(C_2)^2 | H_{LP}(2f_x)|^2}$$
(7-75)

which is *a constant* for fixed t_1/T. For the case $t_1/T = 0.125$ and for the typical bandpass filter, the signal-to-noise ratio can be determined from Figures 7-28(a,b) and 7-23 to be 1.2×10^3 and 490×10^3 for the single- and double-pole filters, respectively. These are power-signal-to-noise ratios, however, which correspond to rms-noise-to-signal ratios of about 2.8% and 0.14%, respectively. In this case, the filters are inadequate for many applications.

To complete the compute-signal-to-noise ratio, assuming the output of the multiplier approximates a rectified sinewave, it can be shown using Eq. (7-63b) that

$$\left(\frac{S}{N}\right)_{out(s)} = \frac{4.5}{| H_{LP}(2f_x)|^2}$$
(7-76)

which is not a function of t_1/T. The other cases may be written explicitly and are shown in Figure 7-28c. For each of these cases, for a given system, the signal-to-noise ratio due to the component of the rectified signal at $2f_x$ does not depend on the signal level.

$$C_{2BP} = \left[- \frac{V_{pp}}{2} \cdot \frac{4}{\pi^3} \cdot \frac{T}{t_1} \right] \cdot$$

$$\left\{ \sin(a) \cdot | H(f_x)| - \sum_{\substack{k=1 \\ k\,odd}}^{\infty} \left[\frac{\sin(ka) \cdot | H(kf_x)|}{k^2 \cdot (k+2)} + \frac{\sin[(k+2)a] \cdot |H[(k+2)f_x]|}{(k+2)^2 \cdot k} \right] \right\}$$
(7-74)

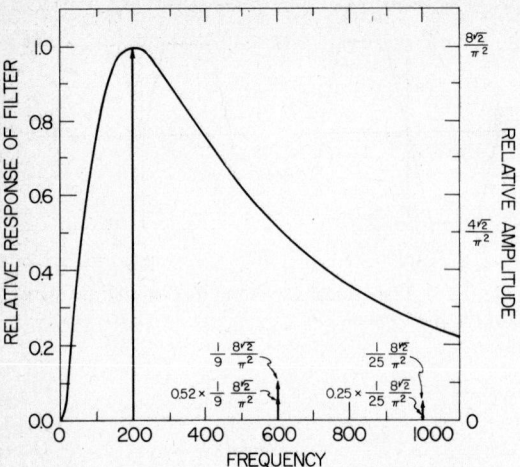

Fig. 7-27. Relative amplitudes of the Fourier components of a trapezoidal wave ($t_1/T = 1/8$) before and after filtering.

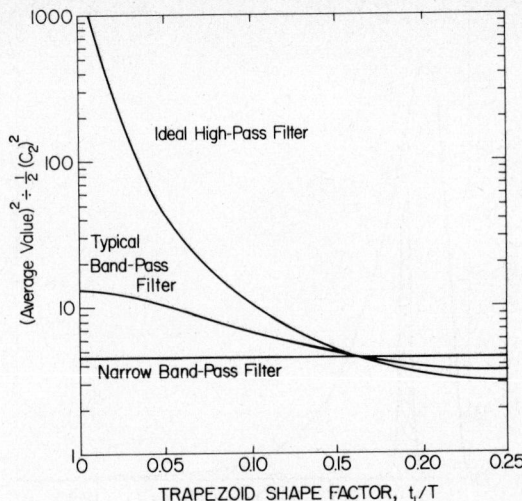

Fig. 7-28c. Signal-to-Noise Ratio: $(S/N)_{out(s)} = $
[(Average Value)$^2/\frac{1}{2}(C_2)^2$] $\div |H_{LP}(2f_x)|^2$.

COMBINED RANDOM AND NONRANDOM NOISE

Recalling Eq. 7-71 and adding the mean squares of the random and coherent noise:

$$\left(\frac{S}{N}\right)_{out} = \frac{V_{dpp}^2 \cdot K_x^2(t/T)}{v_{eq}^2(f_x)\cdot 2B_n + v_s^2} \qquad (7\text{-}77)$$

where v_s^2 is the equivalent input noise due to demodulation residuals:

$$v_s^2 = V_{dpp}^2 \cdot K_x^2 \left(\frac{t}{T}\right) \cdot \left(\frac{S}{N}\right)_{out(s)}^{-1} \qquad (7\text{-}78)$$

For the system shown in Figure 7-24: $R_b = 1.34 \times 10^6\ \Omega$; $R_d = 10^6\ \Omega$; $C_d = 354\ pF$; $C_a = 3\ pF$; $R_a = 6.29 \times 10^6\ \Omega$. Then $C_c = 490\ pF$; $f_{HP} = 44.4$Hz;

$f_{LP} = 900$ Hz and the maximum of the bandpass filter is set to 200 Hz. The shape factor of the trapezoidal wave is chosen to be 0.125.

Then, from Eq. 726-17, the mean-square value of the equivalent input random noise current will be

$$i_{eq}^2 = i_q^2 + i_d^2 + i_b^2 + i_{aeq}^2 + \frac{v_{aeq}^2}{(R_1)^2} \qquad (7\text{-}79)$$

where the individual mean-square values are evaluated in the system noise passbands.

The spectral density of the quantum noise is flat:

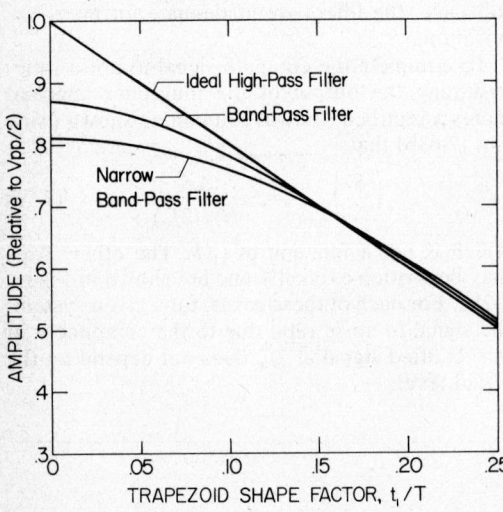

Fig. 7-28a. Average values for rectified trapezoidal signals for three bandpass filters, $|H_{BP}(f)|$, as shown in Figure 7-24c.

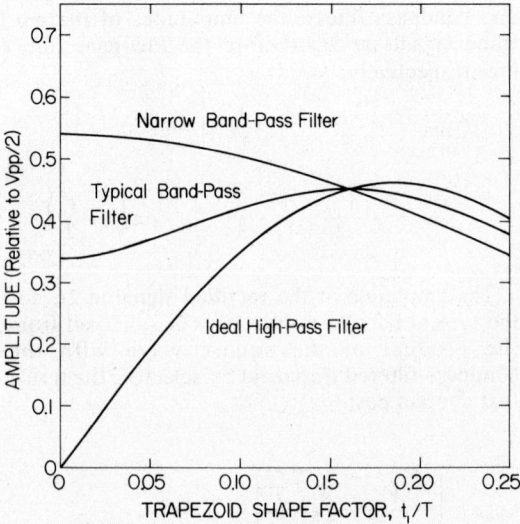

Fig. 7-28b. Amplitude of the rectified signal at $2fx$ for the three bandpass filters, $|H_{BP}(f)|$, as shown in Figure 7-24c.

$$i_q^2(f) = 2eI\left[\frac{A^2}{Hz}\right] \qquad (7\text{-}80)$$

where e is the unit electronic charge, 1.60×10^{-19} Coulomb, and I is the average current due to signal photons. *Assuming* the reference flux to be nearly zero

$$i_d^2 = \frac{v_d^2}{R_d^2} = \frac{1}{R_d^2}$$

$$\times\left[\overline{v_d^2(f_x)}|H_{BP}(f_x)|^2 \cdot 2B_n + \frac{1}{9}\left[\overline{v_d^2(f_x)}|H_{BP}(3f_x)|^2 \cdot 2B_n\right] + \cdots\right]$$

$$i_d^2 = 10^{-12}\left[9.50 \times 10^{-14} + \frac{1}{9}\cdot 0.86 \times 10^{-14} + \cdots\right]\cdot 2B_n$$

$$i_d^2 = 4.26 \times 10^{-24} \qquad (7\text{-}85)$$

$$\overline{i_q^2(f)} = 2e\left(\frac{V_{dpp}/2}{R_d}\right). \qquad (7\text{-}81)$$

Referring to Eq. 7-61 and the bandpass filter (Figure 7-22), the mean-square quantum noise current is

$$i_q^2 = 2e\left(\frac{V_{dpp}/2}{R_d}\right)\cdot\left(1 + \frac{1}{9}|H_{BP}(3f_x)|^2 + \frac{1}{25}|H_{BP}(5f_x)|^2 + \cdots\right)\cdot 2B_n$$

$$= e\cdot\frac{V_{dpp}}{R_d}\cdot[1.03]\cdot 2B_n$$

$$= 7.32 \times 10^{-24}\cdot V_{dpp}\left[A^2\right] \qquad (7\text{-}82)$$

where the equivalent-noise bandwidth B_n of the two-pole Butterworth filter is taken to be 22.2 Hz.

The spectral density of the detector noise may be obtained directly from the data sheets of some manufacturers, or since the spectral density of the noise is independent of the wavelength of the signal flux, from the manufacturer's curves:

$$\overline{v_d^2(f)} = \frac{A_d|\mathcal{R}_\Phi(\lambda_{max},f)|}{D^*(\lambda_{max},f)}$$

$$= \frac{A_d\,\mathcal{R}_\Phi(\lambda_{max},f_{max})\left[1 + \left(\frac{f}{f_d}\right)^2\right]^{-1}}{D^*(\lambda_{max},f)}$$

$$(7\text{-}83)$$

For the lead sulfide photo conductor $\sqrt{A_d}\,\mathcal{R}_\Phi$ is a constant of the material dependent on temperature and manufacture; $\mathcal{R}_\Phi(\lambda_{max},f_{max}) = \mathcal{R}_\Phi(\lambda_{max},0)$ in the case of the resistive detector; $f_d = \frac{1}{2}\pi R_d C_d$ which may be estimated from the manufacturer's curves for responsivity. Figures 7-29a through d show typical curves for these quantities and Figure 7-29d compares with that given by the Santa Barbara Research Center (Brochure 67) as quoted by Jacobs (1978).

The spectral density of the noise at the output of the bandpass filter is given in Figure 7-25

$$v_{dBP}^2(f) = \overline{v_d^2(f)}\cdot|H_{BP}(f)|^2 \qquad (7\text{-}84)$$

Then the mean square equivalent detector noise current is

where the noise equivalent bandwidth of the low-pass filter, B_n, is again taken to be 22.2 Hz.

The spectral density of the noise in the bias resistor is flat. The mean-square value is determined as above for the white quantum noise:

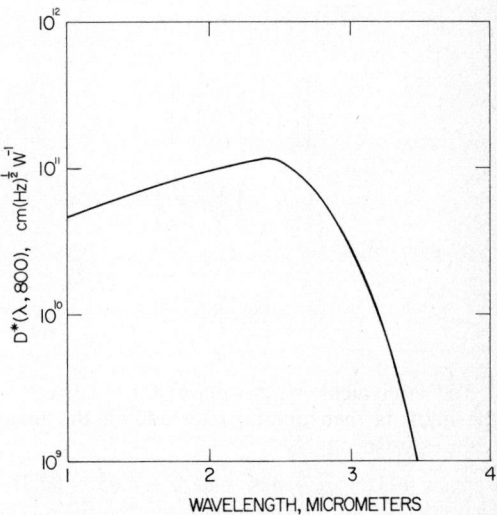

Fig. 7-29a. Detector properties: (a) detectivity vs wavelength, (b) detectivity vs frequency, (c) responsivity vs frequency, (d) spectral density of detector noise.

$$i_b^2 = \frac{4 \cdot k \cdot T \cdot 2B_n}{R_b} \cdot [1.03] \qquad (7\text{-}86)$$

$$= 6.12 \times 10^{-25} \quad [\text{A}^2]$$

where k is Boltzmann's constant, $1.38 \times 10^{-23} \ [\text{W} \cdot \text{s} \cdot \text{K}^{-1}]$; T is the resistor temperature, 298 K; and R_b is the resistance of the bias resistor $[\Omega]$.

The spectral density of the equivalent noise of the current source at the input of the preamplifier is shaped by a transfer factor

$$\overline{i_{aeq}^2}(f) = \overline{i_a^2} \cdot \left[\frac{C_d + C_c}{C_c}\right]^2 \left[1 + \left(\frac{f_i}{f}\right)^2\right] \qquad (7\text{-}87)$$

where $\overline{i_a^2}$ is the value of the spectral density of the source and the transfer is defined for Equation 726-17. The equivalent mean-square noise is

$$i_{aeq}^2 = \left[\frac{C_d + C_c}{C_c}\right]^2 \overline{i_a^2} \cdot \left[\left[1 \times \left(\frac{f_i}{f_x}\right)^2\right]\right.$$

$$\left. + \left[1 + \left(\frac{f_i}{3f_x}\right)^2\right] \cdot \frac{1}{9} \cdot |H_{BP}(3f_x)|^2 + \cdots \right] 2B_n$$

$$i_{aeq}^2 = [2.97] \cdot \overline{i_a^2} \cdot [3.75] \cdot [44.4]$$

$$= 4.95 \times 10^{-26} \quad [\text{A}^2] \qquad (7\text{-}88)$$

where the spectral density of the source is given by the manufacturer to be $10^{-28} [\text{A}^2 \cdot \text{Hz}^{-1}]$. The spectral density of the input equivalent of the voltage source of the preamplifier is also shaped by a transfer factor. In this case the noise is not white but according to Figure 7-30 it is essentially flat beyond 200 Hz. Therefore,

$$\frac{v_{aeq}^2}{R_1^2} = \frac{\overline{v_a^2}(f_x^2)}{R_1^2}$$

$$\cdot \left\{\left[1 + \left(\frac{f_x}{f_1}\right)^2\right]\right.$$

$$\left. + \left[1 + \left(\frac{3f_x}{f_1}\right)^2\right] \cdot \frac{1}{9} \cdot |H_{BP}(3f_x)|^2 + \cdots\right\} \cdot 2B_n$$

$$= [1.22 \times 10^{-26}] \cdot [1.12] \cdot [44.4]$$

$$= 6.07 \times 10^{-25} \quad [\text{A}^2] \qquad (7\text{-}89)$$

The equivalent mean-square noise current at the input is then obtained by adding the mean squares, respectively

$$i_{eq}^2 = (732\,V_{dpp} + 426 + 61.2 + 4.95 + 60.7)$$

$$\times 10^{-26} \quad [A^2] \qquad (7\text{-}90)$$

where, in this case, the dominant noise source is the detector.

V_{dpp} and, subsequently, the signal-to-noise ratio

for random noise, are determined by further specification of the system and the radiance of the target. Since the reference flux is assumed to be zero

$$V_{dpp} = \Phi_{dpp} \cdot \mathcal{R}_\Phi(\lambda, 0) \qquad (7\text{-}91)$$

If the system is assumed to have a 15° FOV, a 1-mm square detector, a perfect filter from 1.15 μm to 1.30 μm with in-band transmittance of 60%, and if the target is a level barium-sulfate diffuser irradiated on a clear day with a solar angle of 45°, then

$$\Phi_{dpp} = \pi (L_\lambda' \cdot \Delta\lambda) \cdot A_d \cdot \sin\left(\frac{\theta_f}{2}\right) \cdot T(\lambda)$$

$$= \pi [15.5] \cdot 10^{-6} \cdot [0.017] \cdot [0.6]$$

$$= 4.97 \times 10^{-7} \quad [\text{W}] \qquad (7\text{-}92)$$

where the in-band radiance of the target is approximately 16.8 watts $\cdot \text{m}^{-2} \cdot \text{sr}^{-1}$. Then, from Figure 7-29a, it can be determined that

$$\mathcal{R}_\Phi \ (1.225 \ \mu\text{m}, 0 \ \text{Hz}) = 1.02 \times 10^5. \ [V \cdot W^{-1}] \qquad (7\text{-}93)$$

Then V_{dpp} is approximately 0.05 volts.

From Eq. 7-72, for $t_1/T = 1/8$, the signal to noise ratio for random noise is

$$\left(\frac{S}{N}\right)_{out(*)} \approx 7 \times 10^{13}. \qquad (7\text{-}94)$$

The equivalent mean-square voltage at the input due to the demodulation residuals may be determined from Eq. 7-78:

$$v_s^2 = 4.43 \times 10^{-10} \ [V^2] \qquad (7\text{-}95)$$

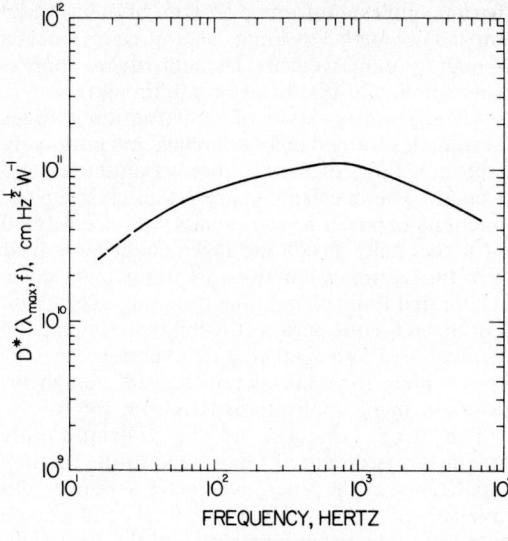

Fig. 7-29b. D* vs frequency.

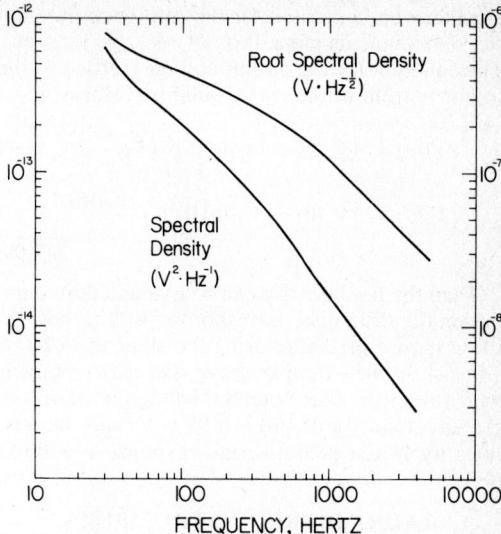

Fig. 7-29d. Spectral densities at λ_{max}.

The signal-to-noise ratio due to the demodulation residuals is determined from Figures 7-28c and 7-23.

$$\left(\frac{S}{N}\right)_{out(s)} = \frac{5.8}{(0.0025)^2} = 9.28 \times 10^5 \quad (7\text{-}96)$$

Then from Eq. 7-77

$$\left(\frac{S}{N}\right)_{out} = 9.3 \times 10^5 \quad (7\text{-}97)$$

which indicates that, in this case, the demodulation residuals determine the signal-to-noise ratio which will remain essentially constant over the dynamic range. It should be noted that the rms noise-to-signal ratio will be about 0.1 percent.

LINEARITY

Although the detector (properly operated with regard to heat sinking, bias voltage, and incident signal levels) is inherently linear, the loaded detector circuit is not. Assuming that all nonlinearity in the system is due to the detector bias circuit, the nonlinearity has the form

$$\frac{dV}{dV_d} = \frac{1}{R}\frac{dV_d}{d\Phi} = \frac{R_b}{(R_b + R_d)} \cdot \left[1 - \frac{\Re \cdot \Phi}{V_B}\right]^{-2} \quad (7\text{-}98)$$

where R_b is the bias resistor; R_d is the dark resistance of the detector; V_B is the bias voltage; and $\Re$ is the responsivity (which is a function of V_B and R_b) for the specified bias conditions. The slope of the curve of the output voltage gradually decreases as input flux increases. The change in

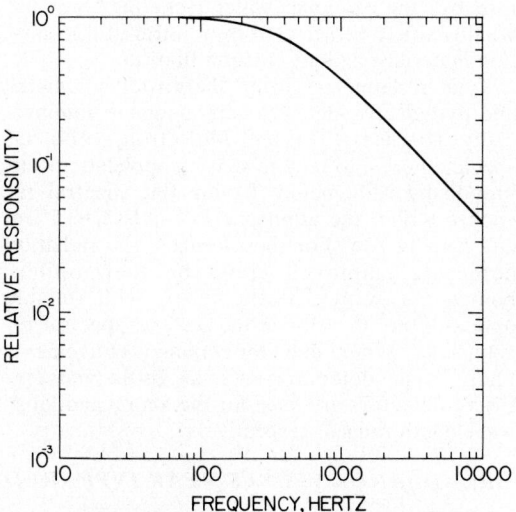

Fig. 7-29c. Relative responsivity vs frequency.

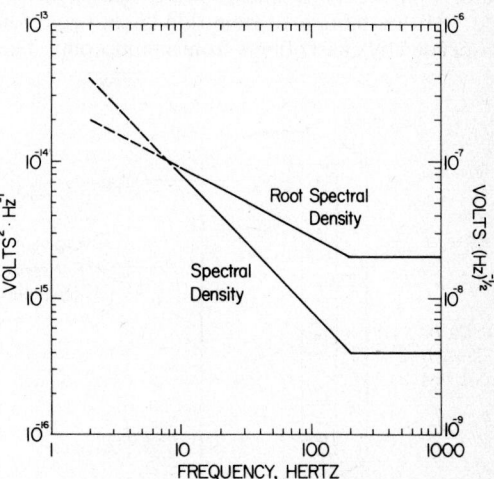

Fig. 7-30. Spectral density of preamplifier input voltage noise source.

slope can be computed for the reference flux and the maximum incident flux; $\Re \cdot \Phi_{ref} = V_{ref}$, etc. The equation for the output voltage (settled to the top or bottom of the trapezoidal waveform) is

$$V(\Phi) = \left[V_d(0) - \Re \cdot \Phi \right] \left[1 - \frac{\Re \cdot \Phi}{V_B} \right]^{-1}$$

$$V(\Phi) = \left[V_d(0) - V_d(\Phi) \right] \left[1 - \frac{V_d(\Phi)}{V_B} \right]^{-1}$$

$$(7\text{-}99)$$

Then the nonlinearity can be evaluated by comparing the difference in response, $V(\Phi \text{ target})$, to the response predicted using the slope at $V(\Phi \text{ reference})$. In the example above, the bias voltage is 93.6 volts (the dark detector voltage is 40 volts), the target voltage $V_d(\Phi)$ is 0.05 volts and the nonlinearity in the peak-to-peak response is -0.053 percent.

RADIOMETRIC INSTRUMENTS

Commercial radiometers are available with a wide variety of features for diverse applications. For convenience in describing instrumentation, five major classes are identified according to the manner in which the spectral performance is determined except for the class of radiometers referred to as radiation thermometers. Only the more common types of instruments particularly useful for field research will be discussed. More general and extensive treatments of radiometers can be found elsewhere (Zissis, 1979; Hudson, 1969; Grum and Becherer, 1979).

BROADBAND-INFRARED

For measurements of the total radiance from a target, the radiometer requires a detector with wideband, flat spectral response over the infrared region. Thermocouple or thermopile detectors with optically blackened receiver surfaces (special black paints, soot, carbon, or gold black) are well-suited for such applications. Typical foil receivers, in circular or linear form as shown in Figure 7-31, are fabricated from thin (5 μm) gold foil suspended by quartz fibers from support pins. The

thermocouples are formed by wire of diameter 25 μm and of length 3 to 4 mm. Thermocouple metals frequently employed are bismuth-silver, copper-constantan, and bismuth-bismuth/tin alloys.

An alternative style of construction utilizes thermopiles formed by overlapping, evaporatively deposited films of metals such as antimony and bismuth. The overlapping areas that make up the junctions exposed to the radiant flux are formed on a thermally insulating layer covering a heat sink; the reference junctions are formed where the evaporated films contact the heat sink. The evaporation technique permits flexibility in shaping the receiver area and arranging the junctions, as well as providing for construction rugged enough for field and space applications (Hudson, 1969).

The time constant of the thermocouple radiometer is directly proportional to the thermal capacitance of the junction/receiver assembly and inversely proportional to the thermal conductance between the sensing junctions and the heat sink. Foil-type receivers may have time constants as large as several seconds, while evaporative-film-type receivers with low thermal capacitance and high thermal conductance paths may have time constants in the 4- to 50-ms range. The responsivity may be expressed as the time constant divided by the thermal capacitance, indicating that it is possible to trade time of response for responsivity.

Most thermopile sensing elements (typically 12 to 16 junctions) are mounted within housings that are either evacuated or filled with air or inert gas. The former will generally have an increased responsivity, increased time constant, and steadier zero reading since the thermocouple is not as well-coupled to its surroundings. Air-type radiometers are suitable for radiant flux densities less than 0.01 to 0.02 W/m². Vacuum-type radiometers may be limited in exposure to 1000 W/m²; for higher flux levels, it is frequently necessary to stabilize the radiometer housing temperature by, for example, water jacketing. Suitable windows may be selected from infrared transmitting materials such as calcium fluoride.

Total radiometers using thermistor-bolometer and pyroelectric detectors are available commercially, (Barnes, 1979; and Molectron, 1979). Of special interest to remote sensing applications are broadband radiometers having flat spectral response within the atmospheric bands 3 to 5 μm and 8 to 14 μm. For these ranges, the radiation detectors employed could be thermopiles, bolometers or pyroelectric types, with suitable optical filters to achieve the desired spectral responsivity. Where detector cooling is convenient, photon-type detectors such as In-Sb and Hg-Cd-Te detectors are used for the short- and long-wavelength ranges, respectively.

BROADBAND—VISIBLE/NEAR INFRARED

The primary application of broadband radiometers with responsivity in the visible–near-infrared

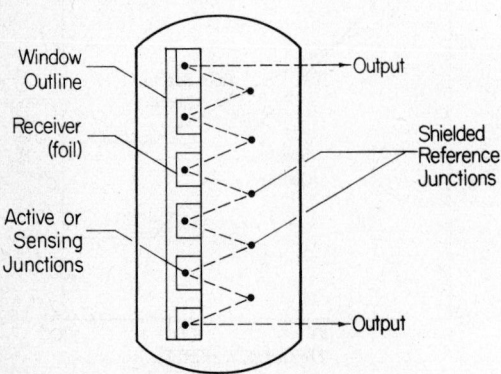

Fig. 7-31. General features of a total radiation thermopile detector.

wavelength spectrum is measurement of radiant fluxes originating from solar insolation. Radiometers generally employ one of three detector types—photomultipliers, photocells, or silicon photodiodes—with a matching optical filter to achieve the desired spectral responsivity. During the past five years, substantial advances in silicon photodiode technology have brought a wide array of instrumentation for total (solar), photosynthetic-action-radiation (PAR), and photopic measurements. Following discussion of these radiometers, instrumentation for direct solar radiation using thermopiles is described.

Total Response

The typical silicon photodiode has spectral detectivity as was shown in Figure 7-5 and a typical spectral responsivity as represented by the dashed line in Figure 7-32. A flat spectral response for a silicon-type detector radiometer—as illustrated by the solid line of this latter figure—can be obtained by using subtractive filtering consisting of multiple elements of five or six colored glasses.

Such a radiometer is inexpensive and provides a convenient means for measuring solar irradiance. For this measurement, the sensing surface is usually fitted with a diffusing transmitter designed to assure uniform response (cosine corrected) from all hemispherical directions. Note from Figure 7-32 that uniform response is limited to the region 450 to 950 nm which includes approximately 50% of the solar flux with 12% and 38% beyond the lower and upper limits, respectively. Using proper calibration procedures, it is possible to develop

correlations algorithms that will permit direct reading of total solar irradiance with accuracies of ±5 percent. Users of such instruments should carefully examine the basis for achieving direct-reading capability and recognize that the accuracy may be dependent upon the nature of the daylight conditions.

Photopic Response

A radiometer with a detector/filter combination having the spectral response of the human eye, is referred to as a photometer. The detector/filter combination may be achieved using a silicon detector as illustrated in Figure 7-33 or a photomultiplier. Photometers based upon the latter type of detector have been commercially available for many years.

The output of a photometer is a measure of the photometric quantities luminance or illuminance which are analogous to the radiometric quantities radiance or irradiance, respectively. The radiometric (energy, e) unit is radiant flux, $\Phi_e[W]$, and is related to the visual (v) quantity, luminous flux, Φ_v, as

$$\Phi_v = K_m \int_0^\infty V_\lambda \cdot \Phi_{\lambda,e} d\lambda$$

where Φ_v has the units of lumens [lm], V_λ is the relative, spectral luminous efficiency for the CIE-standard photometric observer and K_m is the maximum, luminous efficacy, 685 lm/W. The basic photometric unit is the candela (cd) defined as the luminous intensity (lm/sr or cd) from a blackbody having an area of 1/600,000 m² at the freezing point of platinum. Further definitions of

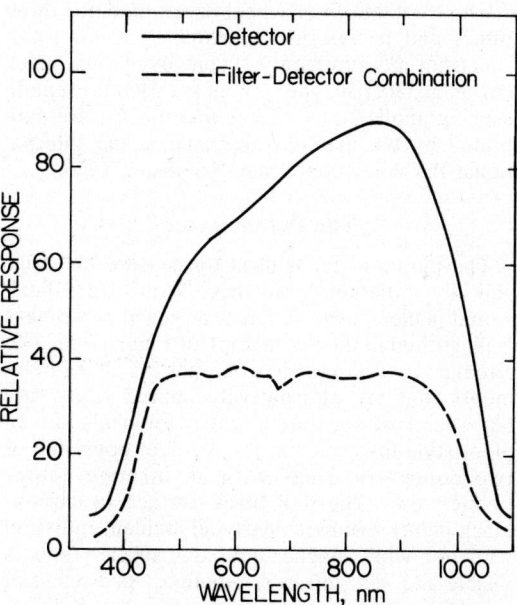

Fig. 7-32. Combining spectral response of a silicon photodiode detector with a multiple-element, colored-glass subtractive filter to obtain a spectrally flat, broadband radiometer.

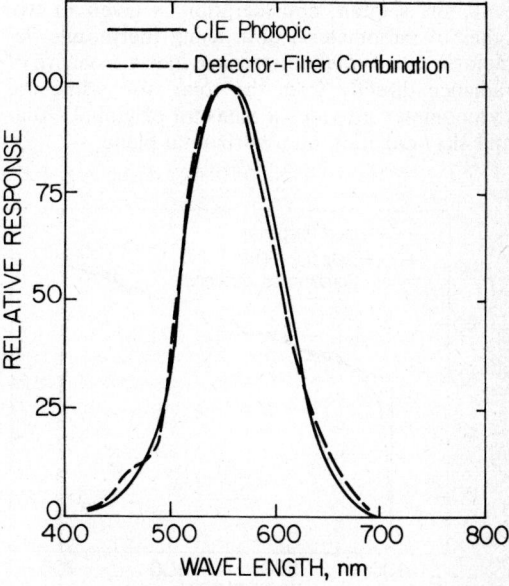

Fig. 7-33. Comparison of relative spectral responsivities of the human eye (CIE photopic) and combination detector/filter photometer.

photometric quantities, their units and interrelations, are presented in standard texts and numerous reference books.

Photosynthetically Active Radiation

Plants use light in the 400–700 nm waveband for photosyntheses. It has been shown that a simple relationship exists between the number of plant molecules changed photochemically and the number of photons absorbed within the prescribed band and that the relationship is independent of photon energy. On this physical basis, the photosynthetically active radiation (PAR) is measured as the number of photons within the PAR spectrum incident per unit time on a unit surface area. The unit of einstein (E) is defined as Avogadro's number of photons. Hence it follows that the photon flux fluence rate can be expressed as

$$1\,\mu E/s \cdot m^2 = 6.02 \times 10^{17}\ photon/s \cdot m^2$$

Alternatively, the einstein is sometimes designated to represent the quantity of radiant energy in Avogadro's number of photons [Li-Cor, 1980].

It follows that since the energy of a photon is $h\nu$ or hc/λ, an energy detector having uniform response to photon fluence must have the spectral responsivity shown as the solid line in Figure 7-34. That is, the ideal spectral responsivity of the radiometer will decrease linearly with wavelength. Such a response can be achieved by using a silicon detector/filter combination as represented by the dotted curve of the figure. In addition to colored glasses, the filter system requires the use of an interference filter to achieve the sharp cutoff at the upper wavelength limit.

DIRECT SOLAR RADIATION

In this section, consideration is given to two types of radiometers, both using thermopile detectors. The pyrheliometer provides a measure of radiance directly from the solar disc while the pyranometer provides a measure of global (solar and sky) radiation on a horizontal plane.

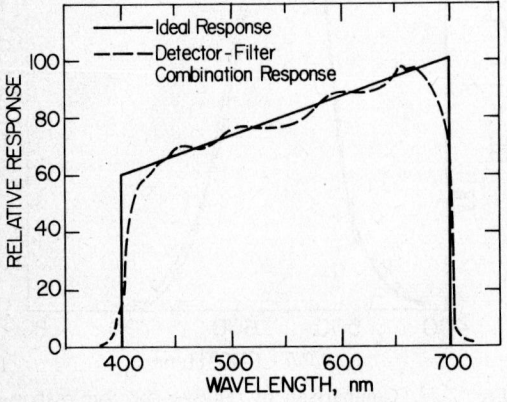

Fig. 7-34. Radiometer response required for measuring photosynthetically active radiation (PAR).

The Pyrheliometer

The function of the pyrheliometer is to measure the directional radiance from the sun within a prescribed bandwidth. Because of the difficulty of sighting an instrument with a small aperture adapted to the solar disc, most instruments are constructed to include a part of the circumsolar radiation. To reduce scattered radiation, the receiver (thermopile) is located at the bottom of a tube which is accurately aimed at the solar disk. The geometry of the tube is, of course, very important and many designs have been critically analyzed over the years (Eppley, 1980a). The normal-incidence pyrheliometer shown in the photograph in Figure 7-35 illustrates typical construction consisting of three major parts: thermopile assembly, tube, and rotatable filter wheel. The sensing element is a wire-wound-type thermopile with a thermistor temperature-compensating circuit to account for ambient temperature changes.

The thermopile is mounted at the base of the tube, constructed from brass, having an aperture-to-length ratio of 1 to 10 and subtending an angle of 5° 43'40". The inside of the tube is suitably diaphragmed and blackened while the outside is chromium plated. The tube is filled with dry air at atmospheric pressure and sealed at the viewing end by a removable crystal-quartz window of 1 mm thickness. Two flanges, one at either end of the pyrheliometer tube, are provided with a sighting arrangement for aiming the instrument directly at the sun. To permit continuous measurement of the solar radiation, the pyrheliometer can be equipped with an equatorial mount which is electrically driven and geared to solar time.

The rotatable filter wheel accommodates three filters and leaves one aperture clear for total-spectrum measurements. Typically, filters which cut-on at 530, 630, and 695 nm are used for remote sensing applications. These instruments are calibrated by the manufacturer against the International Pyrheliometer Scale (Robinson, 1966).

The Pyranometer

The pyranometer is used to measure the solar and sky radiation (irradiance, W/m²) on a horizontal plane. Construction of a typical instrument is illustrated in the photograph of Figure 7-36. The circular receiver consists of eight pie-shaped elements that are alternatively coated white and black, and are separated by thermal insulation. In older-style instruments, the receiver consisted of two concentric rings with an inactive center (white) disk. The dull black surface (a carbon-black paint) absorbes nearly all incident radiation while the white magnesium oxide surface reflects visible and near-infrared radiation. As a result of the different absorptivities, a temperature difference between the surfaces will occur and is measured by a multijunction thermopile. Both surfaces have similar spectral absorptivities for

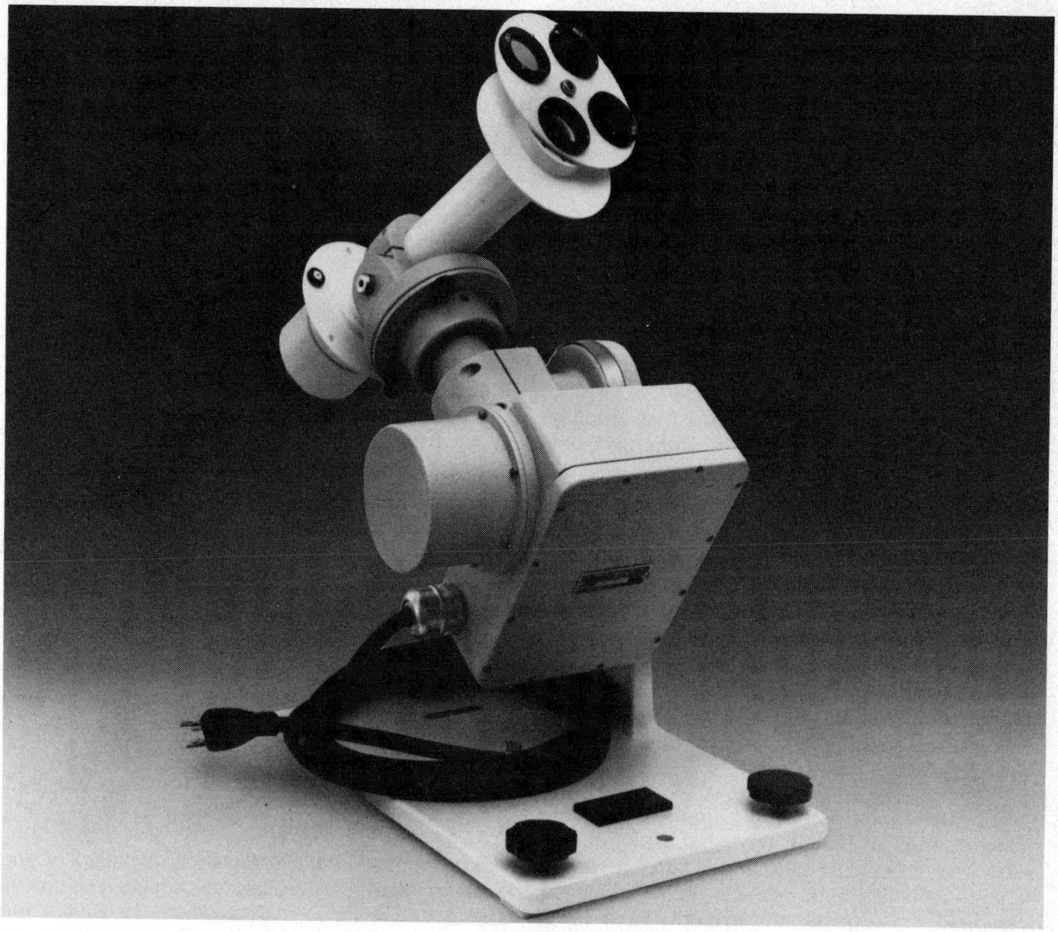

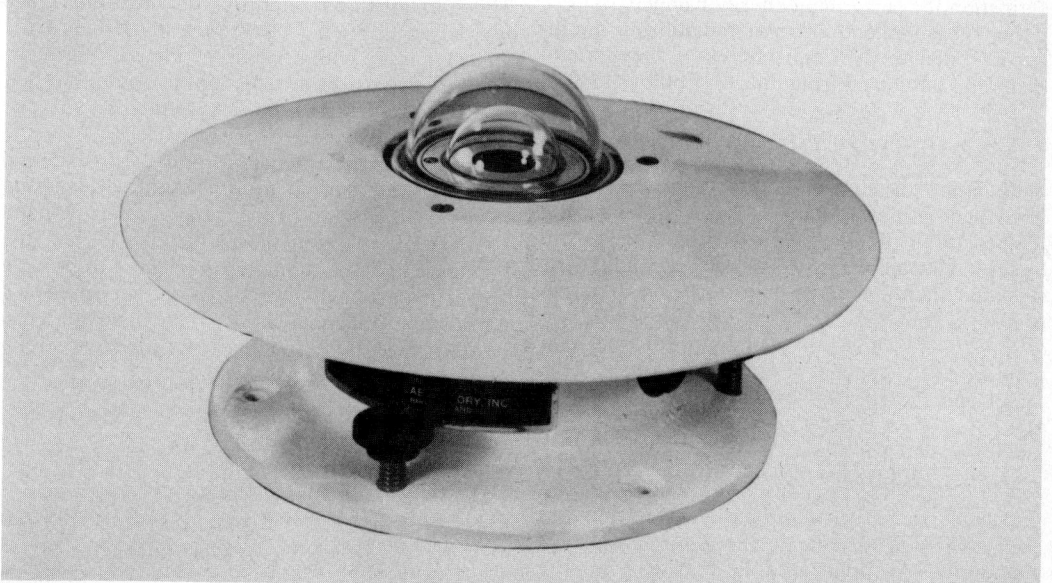

Fig. 7-35. Photograph of a normal-incidence pyrheliometer for measuring direct solar radiation (courtesy of Eppley Laboratories, Newport, RI).

Fig. 7-36. Photograph of a pyranometer for measuring global radiation (courtesy of Eppley Laboratories, Newport, RI).

long-wavelength radiation such that emission from the hemispherical glass bulb does not appreciably affect the temperature of either surface.

The spectral response of the pyranometer is determined by the characteristics of the removable ground and polished concentric spheres. For *total* irradiance measurements, the glass (Schott WG7) has high transparency over the range 285 nm to 2.8 μm. For measurement of band irradiances, the outer hemisphere is made of glasses with long wavelength cut-off characteristics.

With clear glass hemispheres, the instrument has a sensitivity of approximately 9 μV/(W/m^2) and a temperature dependence of less than 1% for the ambient temperature range -20 to $+40°$C. Orientation has no effect on performance although the unit must be horizontal for meaningful results to be obtained.

A further development of the total pyranometer is the *pyrgeometer* for the measurement of undirectional terrestrial long-wavelength radiation. The pyrgeometer is similar in construction having the same type of receiver and temperature-compensating circuitry. However, two important features are different. First, the radiation emitted by the receiver (detector) is subtracted from the output signal which is proportional to the incident flux. This is achieved by a circuit using a thermistor to sense detector temperature which then introduces the necessary subtractive signal. Second, by using KRS-5 rather than glass hemispheres, long-wavelength terrestial radiation can reach the detector. The inner surface has a vacuum-deposited interference filter giving the pyrgeometer a transmission window of 4 to 50 μm. The pyrgeometer is useful for measuring long-wavelength radiation in the downward and upwelling directions. The difference provides information for performing energy balances.

For measurements of solar radiation in the ultraviolet range, the hemispheres of the pyranometer are fabricated from quartz (Eppley, 1980b).

RADIATION THERMOMETERS

Radiation thermometers are radiometers that have been calibrated to correctly indicate the temperatures of blackbodies. When viewing a target having an emissivity of less than unity, the radiation thermometer will indicate an apparent temperature—referred to as the *radiance temperature*—which is lower than the true temperature and is dependent on the emissivity of the target and the spectral bandpass of the radiometer.

Radiance Temperature

Consider a *narrow-band* radiometer that has been calibrated to indicate the temperature of a blackbody. As illustrated in Figure 7-37, the radiant flux accepted by the radiometer when viewing the blackbody will be proportional to the blackbody radiance $L_{\lambda,b}(\lambda,T)$. When viewing the target, which is at the same temperature, the radiant flux accepted by the radiometer will be proportional to $\epsilon_\lambda L_{\lambda,b}(\lambda,T)$ where ϵ_λ is the spectral emissivity of the target. This radiant flux can also be expressed as being proportional to an equivalent blackbody radiance, $L_{\lambda,b}(\lambda,T_\lambda)$ at an apparent temperature, T_λ. This equivalence can be expressed as

$$L_{\lambda,b}(\lambda,T_\lambda) = \epsilon_\lambda(\lambda) \cdot L_{\lambda,b}(\lambda,T) \qquad (7\text{-}100)$$

where the left- and right-hand sides of the relation denote the equivalent blackbody radiance and actual radiance from the target, respectively.

The apparent temperature, T_λ, is referred to as the *spectral radiance temperature* and represents the temperature of a blackbody having the same spectral radiance as the target with a specified spectral emissivity. It follows that the spectral-radiance temperature of a target will be less than or equal to the true temperature depending upon whether the target spectral emissivity is less than or equal to unity.

When the blackbody spectral radiance can be expressed by Wien's approximation to the Planck distribution, a simple relationship between spectral radiance temperature, true temperature, and emissivity can be obtained. Substituting Wien's law into Eq. 7-100 and cancelling like terms, we find that

$$\exp\left[-\frac{C_2}{\lambda T_\lambda}\right] = \epsilon_\lambda \cdot \exp\left[-\frac{C_2}{\lambda T}\right] \cdot (7\text{-}101)$$

Taking natural logarithms of both sides of the equation and rearranging, the sought relation is obtained

$$\frac{1}{T_\lambda} = \frac{1}{T} - \frac{\lambda}{C_2}\ln\epsilon_\lambda. \qquad (7\text{-}102)$$

This relationship is appropriate for $\lambda T < 2898$ μm$\cdot$K since Wien's approximation will provide values that are within 1% to the Planck values. To be treated as spectral radiation, it is required that the bandpass ($\Delta\lambda$) be less than 1% of the wavelength (λ).

For measurements at normal environmental temperatures, there is insufficient target emission to permit operation of inexpensive, narrow-band or spectral radiation thermometers. Hence, the bandpass of the radiometer is selected to provide an adequate signal. The confusion that arises as a consequence of considering a finite bandwidth can be illustrated by extending the equivalence relation of Eq. 7-100 which becomes

$$\int_0^\infty \mathcal{R}_{L,s}(\lambda) \cdot L_{\lambda,b}(\lambda,T_\lambda)d\lambda$$

$$= \int_0^\infty \epsilon(\lambda) \cdot \mathcal{R}_{L,s}(\lambda) \cdot L_{\lambda,b}(\lambda,T)d\lambda \quad (7\text{-}103)$$

It follows that the *in-band radiance temperature*, T_r, is the equivalent temperature of a blackbody having the same in-band radiance as that of the

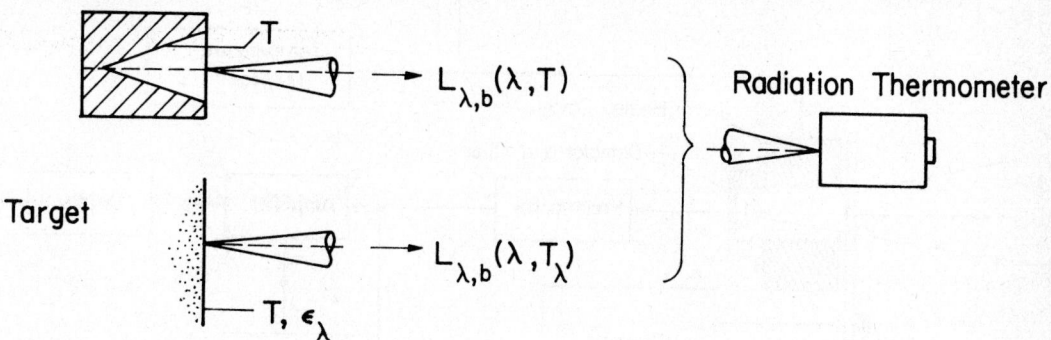

Fig. 7-37. Radiation thermometer viewing blackbody and target at same temperature indicates spectral radiance temperature, T_λ, of the target.

target over the spectral bandpass of the radiometer. It also follows that comparison of radiance temperatures from targets having different emissivity spectra could be misleading when attempting to infer the true temperatures of the targets. For the special case when $\mathcal{R}_{L,s}(\lambda)$ is uniform and constant over the total wavelength range, the integration of the blackbody spectral radiance can readily be performed to give total exitance and using the Stefan-Boltzmann law, Eq. 7-103 reduces to

$$T_t = \epsilon_t^{1/4} \cdot T \qquad (7\text{-}104)$$

where T_t is referred to as the *total radiance temperature*. In general, such a measurement method would not be used in remote sensing work since the presence of atmospheric absorption bands (water vapor and carbon dioxide) restricts sensing to selected windows.

General Characteristics

For practical remote-sensing applications, target temperatures will generally fall in the range -30 to $100°C$. Because of the low spectral self-exitance of bodies at these temperatures, it is necessary to employ wideband radiometers for temperature measurement. Most frequently, to avoid atmospheric absorption bands wideband radiometers operating in the 3- to 5-μm and 9- to 14-μm spectral ranges are used.

Recall from the treatment of the measurement equation, in the section on radiometric measurement systems, that three general characteristics of a radiometer were identified. For the situation where the radiometer is used as a radiation thermometer, these characteristics can be made with more specificity. First, the responsivity needs to be written in terms of the change in signal output for a change in target temperature. This parameter, more frequently expressed as *sensitivity,* gives an indication of the capability of the radiation thermometer to sense a prescribed temperature change. Second, the detectivity can be redefined to indicate the *noise equivalent temperature;*

that is, the signal rms noise fluctuation expressed in terms of temperature. Third, the *reference radiation* is usually introduced by a direct or indirect comparison with a surface or source at a fixed or measurable temperature. These characteristics will be evident for the several instruments discussed in the following section.

Instrument Features

Three instruments will be described as examples of the types of radiation thermometers useful for remote-sensing studies. The examples illustrate the features of modulation of the target flux, methods for obtaining appropriate reference radiation, and target emissivity compensation.

The precision radiation thermometer shown schematically in Figure 7-38 features a mechanical chopper for modulation of the radiant flux to the detector. This permits use of synchronous rectification and ac amplification, eliminating drift associated with dc amplifiers. The chopper also is important to the method for obtaining the reference radiation signal. The infrared detector is located at the base of the optical cavity which is heated to a precisely controlled temperature. The rotating highly polished chopper passing over the open end of the cavity interrupts the radiant flux from the target and allows the detector to alternately sense target flux and reflected radiation from the cavity. The difference between the target flux and the reference radiation from the cavity at a fixed, known temperature is converted by the detector into an electrical signal related to target temperature.

Typical of this type of radiation thermometer is the Barnes Model PRT-5 available with 2° and 20° fields-of-view giving target spot sizes (diameter) of 0.76 m and 3.0 m, respectively, at a distance of 3 m. Using a cassegrain optical system reduces the field of view to 0.14° and results in a typical target spot of 0.76 m at 300 m separation. This version of the instrument would be most useful for high altitude studies. An immersed thermistor bolometer (0.05 × 0.05 mm) is the detector and

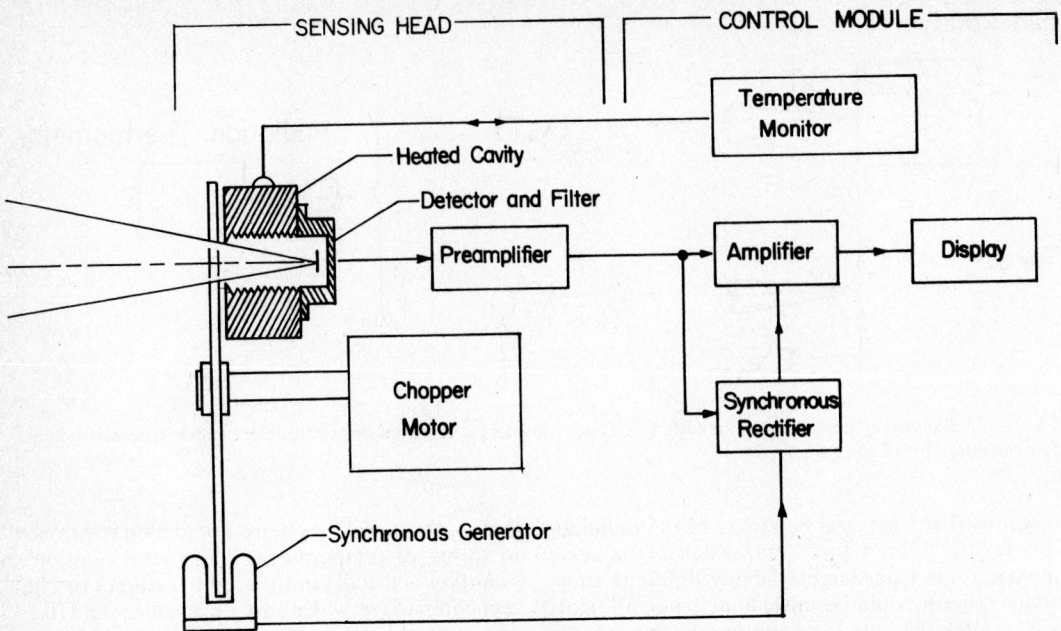

Fig. 7-38. Radiation thermometer featuring modulation of target flux and a reference heated cavity.

the spectral bandpass is determined by one of several available filters, the standard being for the 8- to 14-μm range. The noise equivalent temperature is 0.005°C and absolute system accuracies of ±1°C are possible.

An unchopped or dc-type radiation thermometer is represented schematically in Figure 7-39. The selection of the optical filter and type of optics will depend upon the desired wavelength passband. For near-ambient atmospheric conditions, generally the passbands will be within the atmospheric windows and as wide as possible in order to increase the radiant flux incident upon

the detector. The typical detector arrangement utilizes two bead-thermistor bolometers in a bridge with one detector exposed to the incident flux and the other shielded. The output from the bridge circuit is fed to a high-gain, low-drift dc amplifier which provides a signal proportional to the incident flux. Through proper calibration, a meter indicates the target radiance temperature.

The zero reference point for this type of thermometer is obtained by placing a shutter (typically, a trigger operation if manual or by a shutter drive if automatic) over the active thermistor and adjusting the bridge circuit output to a zero volt-

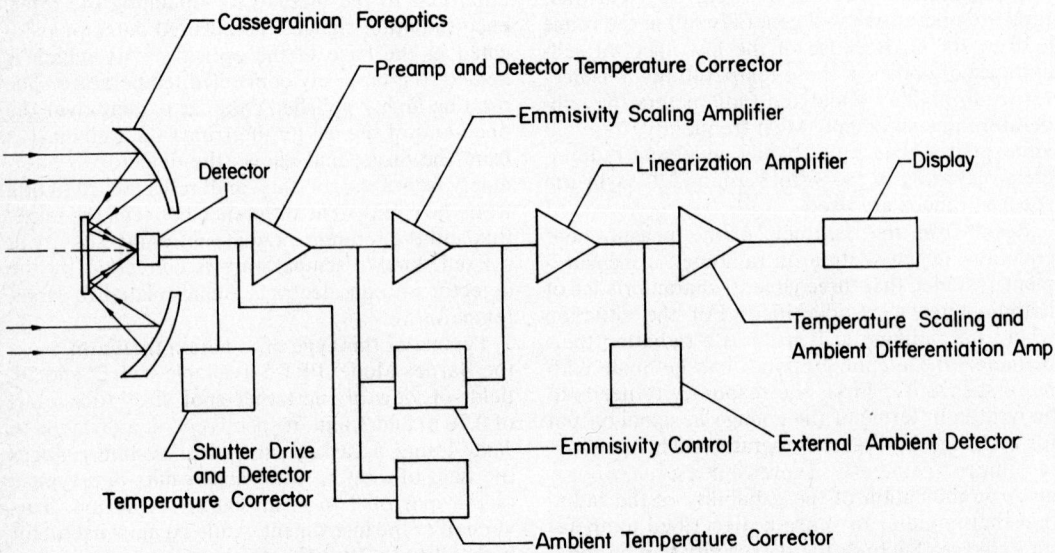

Fig. 7-39. Unchopped-type radiation thermometer featuring target emissivity compensator.

age. This manual zero adjustment compensates for changes in the detector characteristics with use and ambient temperature. Generally internal compensation is also provided in the bridge circuit to allow the instrument to be used over a wide span of environmental temperatures.

As illustrated in Figure 7-39, compensation is achieved by externally adjusting the gain of the dc amplifier. The relationship between the gain or emissivity and radiant flux is linear; however, the relationship for temperature would be nonlinear and hence the gain scale should be calibrated against a blackbody. Another feature of this radiation thermometer is the external ambient detector, which provides a measure of the ambient air temperature. This signal and that from the target are supplied to a differential amplifier which then provides a measure of the target and air temperatures. Jackson, et al. (1977) have shown that this is an important parameter in evaluating crop stress.

The schematic of Figure 7-39 is that of the Telatemp Corporation, Model AG-42 infrared thermometer designed especially for agricultural applications. The scale range is -30 to $+100°$ C with a resolution and noise-effective temperature of $\pm0.1°$ C; the accuracy is $\pm0.5°$ C. The spectral response is for the range 8 to 14 μm and has an emissivity compensation range of 0.1 to 0.99; at a 2-m viewing distance, the target spot size is 10 cm (spot size is viewing distance/20). Another radiation thermometer having many of these features is the Wahl Corporation Model HSA-120, for the range -20 to $60°$ C with an accuracy of $\pm0.6°$ C. This instrument is intended for use on small target areas (typically, 6 mm spot size at 10 cm distance) with working distances less than 1 m.

MULTIBAND RADIOMETERS

The use of a single optical and electronic signal processing system with several optical filters that may be manually positioned to select a desired spectral band, provides an economical multiband radiometer that is suited for a variety of applications where speed of operation is not a factor. Several of these units are described by Zissis (1979).

Multiband radiometers are also made by combining several single-band radiometers into a compact package. Typically, each radiometer has a separate signal-processing system and the individual radiometer outputs are available simultaneously in analog or digital form. The advantage of these instruments is their ability to quickly measure optical quantities and properties from desired portions of the optical spectrum without mechanical and/or electronic adjustment.

Either approach enables the measurement of spectral information with a minimum of data handling, data processing, and data storage requirements (Robinson and Biehl, 1979).

DC Multiband Radiometers

Applications for dc multiband radiometers in remote-sensing field research require the flexibility for rapid operation with spectral bandwidths to 50 nm and, occasionally, narrower. The most reliable dc multiband radiometers are limited to the spectral range of the silicon detector because high-speed detectors that are adequately sensitive beyond 1 micrometer are not sufficiently stable for direct-coupled operation.

The multiband radiometers most widely used in remote-sensing field research have been equipped to simultaneously measure in spectral bands that approximate the Landsat bands (0.5 $-$ 0.6, 0.6 $-$ 0.7, 0.7 $-$ 0.8, and 0.8 $-$ 1.1 micrometers). Some units allow the user to conveniently insert custom spectral filter sets and also provide a filter set for the first four Thematic Mapper bands (0.45 $-$ 0.52, 0.52 $-$ 0.60, 0.63 $-$ 0.69, and 0.76 $-$ 0.90 micrometers) as described by Exotech (1980).

Some units feature fields-of-view (and a sighting telescope) that are coaligned to within tenths of a degree. Fields of view most suited to remote-sensing field research are characterized by a relatively constant response for points within the field and a steep drop in response (to zero) for points beyond the stated field-of-view. The acceptance cone determined by the 50-percent response isophote is defined to be the field-of-view. Circular fields-of-view of 15° (accepting rays from within 7.5° of the optical axis) are typical for measurements from near surface platforms and square or circular fields-of-view of 1° are typical for measurements from light aircraft.

The stability and noise performance of the direct-coupled silicon detector enables the selection of a gain setting that can be maintained in field operations without sacrificing measurements precision (Exotech, 1980 and Matra Optique, 1980).

AC Multiband Radiometers

At this time, there is one commercially available unit designed for field research in remote sensing. This unit has eight modular optical-electronics units centered on a 10.6-cm circle. A front-mounted chopper modulates the target flux entering the optical modules. The back side of the chopper is black and serves as a reference. Seven channels may be operated from 0.4 to 2.7 μm using silicon or lead-sulfide detectors for the appropriate range. The remaining channel is normally devoted to the range of its lithium-tantalate detector. The temperature of the chopper is monitored to provide a reference for this channel.

The standard filter set matches the Thematic Mapper bands (0.45 $-$ 0.52, 0.52 $-$ 0.60, 0.63 $-$ 0.76 $-$ 0.90, 1.55 $-$ 1.75, 2.08 $-$ 2.35, and 10.4 $-$ 12.5 μm) with an additional band (1.15 $-$ 1.30 μm) (Barnes Engineering Company, 1981).

The modular nature of the optical-electronic

units enables field installation of optical filters and of the 1° and 15° field-of-view defining devices. In addition, complete optical-electronic modules may be exchanged in the field. A telescopic sight may be mounted on the unit and centrally sighted with the sharply defined fields of view which are coaligned to within 0.3° (see Figure 7-40).

The gain of each channel may be adjusted to fill the dynamic range of 0 to 5 volts with signals suitable for 12-bit conversion. This eliminates the need for gain changing for most field operations (see Figure 7-40(b)).

The frequency response of each channel may be field-adjusted for operation from stationary platforms, helicopters, and light aircraft. Temperature compensation of the lead-sulfide detectors provides accurate operation for measurement of reflectance factors under field conditions (Robinson, Buckley, and Burgess, 1981).

SPECTRORADIOMETERS

Spectroradiometers are distinguished from multiband radiometers because they measure flux in much narrower spectral bands. In the spectroradiometer the simple filter is replaced with a dispersion device that separates optical radiation into its spectral components which are directed to a detector. A thorough treatment of dispersion devices and a complete directory of laboratory spectroradiometers are given by Zissis (1979), and Slater (1980).

The scope of this section is to briefly describe the general features of the major types of spectroradiometers and their applicability to remote-sensing field research.

In general, the process of sequentially directing the dispersed radiation onto a detector (a spectral scan) takes time. If the target is fixed and the radiance distribution across the target is fixed, then the source of the radiation is roughly the same for each step in delivering the dispersed radiation to the detector. However, if the target (or the spectroradiometer) is moving, the radiance distribution is changing and care must be taken in the interpretation of the result of a given spectral scan. For example, if a spectroradiometer that makes one spectral scan per second is mounted on a helicopter that is moving at 24 meters/sec. at an altitude of 120 meters over a cornfield that is 400 meters long in the direction of travel, about 16 complete spectra will be made as the helicopter traverses the field. If the field is well-sampled by the 16 scans, their average spectrum and standard deviation spectrum can provide valuable information regarding the spectral reflectance of the field. However, it is likely that individual spectra will have no particular meaning.

Filter-Wheel Spectroradiometers

The segmented filter-wheel spectroradiometer is similar to the simple radiometer in that a number of filters, arranged in a circle, are rotated one by one into the position of the single filter. The advantages of this approach include the potential for a large number of spectral bands that can be precisely specified, and the simplicity of the optical and electronic processing systems. The main disadvantage is the time required for data acquisition for a spectral scan. Careful mechanical and electronic design are necessary if a chopper

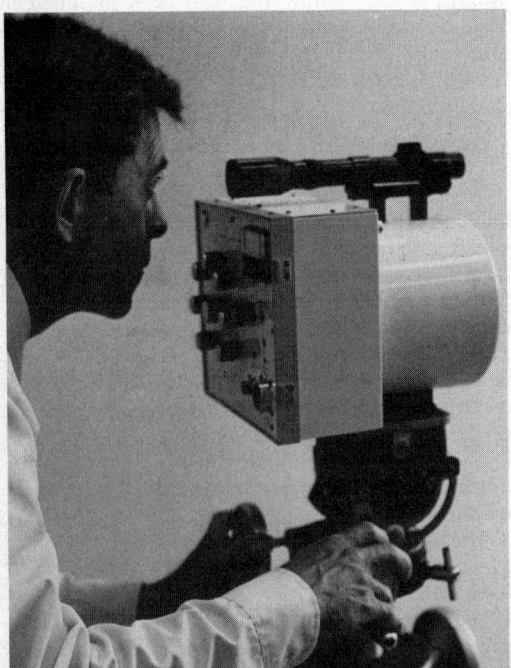

Fig. 7-40. Barnes Model 12-1000 multiband radiometer (a) shown with operator and (b) control panel.

system is required and if the filters are rapidly sequenced.

The operation of the *circular variable filter* (CVF) is similar to that of the segmented type. The CVF is an interference filter fabricated in the form of a thin, circular disc. The filter is designed so that the center of the optical passband is linearly related to angular position. These devices have been used in remote-sensing field research because they facilitate the construction of simple, rugged spectroradiometers having appropriate spectral resolution. Advantages and disadvantages are the same as for the segmented units. Circular variable-filter spectroradiometers that operate from 0.4 to 14 μm have been widely used in the remote-sensing field research described by Robinson, Bauer, Biehl, and Silva (1978). Several devices are described by Zissis (1979) and the CVF is discussed by Wolfe (1979).

Prism Spectroradiometers

In these units, a prism and other optical surfaces are moved in concert to cause the dispersion pattern of the prism to move over an exit slit (which masks all but a narrow band of the dispersed radiation) to produce a spectral scan. The advantage of simple prism approaches lies in their mechanical simplicity. However, if the unit must cover the spectral range from 0.4 to 2.4 μm the spectral resolution of simple designs is not adequate at longer wavelengths. More complex units are not well-suited to portable operation in the field environment.

Grating Spectroradiometers

Traditional grating instruments have produced a high resolution spectral scan by moving the dispersed radiation over an exit slit. Grating spectroradiometers having spectral coverage from 0.4 to 2.4 μm (in two ranges) have been used in field environments. Due to the requirement for order-sorting filters (See Grum and Becherer, 1979) and mechanical complexity, a spectral scan requires about one minute to complete and the exchange of spectral ranging systems requires several minutes.

In recent years, a number of grating instruments that project the dispersed radiation onto a monolithically fabricated array of silicon detectors (typically 256) have been developed. The spectral range of these *rapid scanning* instruments is typically 400 to 1020 nm and care must be taken to insure proper order sorting. The advantage of these instruments is that they have no moving parts and the output of each of the detectors can be scanned at rates from 15 to 60 times per second. Data storage requires from 2 to 60 seconds. The limited spectral range of the silicon detector is a disadvantage for some applications; however, presently available arrays of lead sulfide and pyroelectric detectors may lead to the development of commercially available rapid-scanning spectroradiometers that cover the range of optical radiation.

Fourier Transform Spectroradiometers

This class of instruments applies variations of the Michelson interferometer to produce an interferogram by rapid back-and-forth translation of the "adjustable" mirror.

The changing intensity of the central fringe of the interference pattern is sensed by a detector. The output signal from the detector is usually averaged over many of the mirror translations to produce the interferogram.

For constant monochromatic input flux of wavelength, λ, the interference pattern oscillates between dark and light at a rate $f = v/\lambda$ where v is the velocity of the mirror. The amplitude of the interferogram is proportional to the amplitude of the input flux and the frequency of the interferogram is proportional to the wave number ($\nu = 1/\lambda$) of the monochromatic source. For sources having complicated spectral intensity, the interferogram is the superposition of the sinusoids that would result from each component of the source. To obtain a calibrated spectrum, the Fourier transform of the time-averaged interferogram is taken. The transform results in a spectrum that is a function of wave number and that may be easily converted to a function of wavelength and calibrated with respect to standards of spectral intensity.

Advantages of the Fourier transform spectroradiometer are high sensitivity and spectral resolution. Successful field instruments have usually been limited to the spectral range above 1 μm due to the tight mechanical tolerances required for proper operation. However, one unit was used in the mid 1970s for field measurements from 0.4 to 2.4 μm. (Robinson, et al., 1978).

DETECTION MECHANISMS, NOISE, AND FIGURES OF MERIT

PHOTON DETECTION MECHANISMS[1]

Photoemissive Detectors

In a photoemissive material, incident photons collide with electrons, imparting energy to them. If the energy is sufficient to overcome the work-function barrier, the electron is emitted from the surface, and becomes a part of the signal current. The *work function* is the energy difference between a free electron in a vacuum and one within the material. An energy-level diagram for the vacuum interface is shown in Figure 7-41, the work function being measured in electron volts. The *threshold wavelength* is the wavelength at which the photon energy equals the work function.[2]

---

[1] This section on Photon Detection Mechanisms is taken from Volume 1, of the first edition of the Manual of Remote Sensing (1975), Chapter 7, Electro-Optical Remote Sensors with Related Optical Sensors, L. Ralph Baker and R. MacDonald Scott II, co-authors—editors. For alternate approaches, see Grum and Becherer (1979) and Kingston (1978).

[2] $Q\lambda_0 = 1.2397$, where the function, Q, is in electron volts, and λ_0 is in μm.

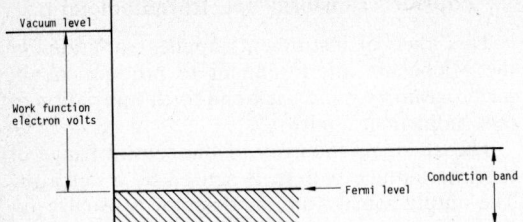

Fig. 7-41. Energy-level diagram for vacuum interface.

If one were limited to pure metals, for which the work function is a few electron volts, then the detector thresholds would not extend toward the longer wavelengths, or past the blue region of the visible spectrum. However, mixtures of the alkali metals have led to the development of photocathodes that operate well into the near infrared.

Efficiency Aspects

Even when an electron is elevated above the work-function barrier, the probability of its emission may still be very low; the elevated electron is kinetically hot and will excite internal acoustic vibrations. Each excitation removes a little of the electron's energy so that, within a path length of about 0.02 μm, its energy has been reduced below the work-function barrier and it cannot escape. This restricts the photoemissive activity to a region very near the surface and makes the quantum efficiency of photoemissive detectors rather low (in the order of 10–20% at maximum).

Quantum efficiency is defined as the ratio of emitted electrons to incident photons. For materials with low work functions the quantum efficiency is greatly reduced—in materials where the response extends to wavelengths of 1 μm, only one out of every thousand incident photons results in an emitted electron.

Some dramatic improvements in infrared quantum efficiencies have recently been achieved by combining photoconductors with such photoemissive materials as Cs_2O. In these materials, the long-wavelength cutoff is determined, not by the work function of the photoemissive material but by the bandgap of the photoconductor. Typical examples are Cs_2O on gallium arsenide, or gallium indium arsenide.

Dark Current

The performance of photoemissive detectors is usually limited by dark current. Dark current is the designation for a phenomenon in which electrons within the material have energies that are determined by the thermal-energy distribution curve at the ambient temperature; some of them have energies high enough to overcome the work function, and are therefore emitted spontaneously.

As the work function is lowered to give longer wavelength sensitivity, the dark current rises, be-

cause at a fixed temperature there are now more electrons with energies above the barrier. If the temperature is lowered, the dark current decreases—for multialkali surfaces, it drops by a factor of two for every 5° decrease in temperature down to 0°C, and at a lower rate down to −160°C. At this level, the dark-current emission from the silver-Cs_2O photocathode is about 1 electron cm^{-2} s^{-1}.

Photoconductive Detectors

In photoconductive detectors, the operating mechanism is the production of a free carrier by an absorbed photon. Photoconductivity induced by energy-gap transitions, in which the photon excites an electron across a bandgap, is called *intrinsic photoconductivity* to distinguish it from the effect due to impurities. Impurities added to materials (such as mercury or copper in germanium) form an energy level within the bandgap, inducing the occurrence of photoconductivity at longer wavelengths (with the photon producing an electron by ionizing the impurity level). This is called *extrinsic photoconductivity*.

In an insulator (where the number of free carriers in the conduction band is insignificant at the operating temperature), the rate of production of free carriers by absorbed photons is

$$\frac{dn}{dt} = f - Anp \qquad (7\text{-}105)$$

where
t = time,
n = electron density,
p = hole density,
f = number of "effective" photons absorbed per unit time,
A = radiative recombination coefficient characteristic of the material.

The net charge throughout the material must be zero; therefore, n must equal p. The steady-state solution of Eq. 7-105 then becomes

$$n_0 = (f/A)^{1/2} . \qquad (7\text{-}106)$$

Current

The conductivity σ is

$$\sigma = n_0 e \mu \qquad (7\text{-}107)$$

where
e = electronic charge,
μ = electron mobility.

For simplicity, it has been assumed that the hole mobility is very much lower than the electron mobility, so that holes do not contribute substantially to the current. The photocurrent for a detector of unit cross-section and length l is

$$I = (n_0 e \mu V)/l \qquad (7\text{-}108)$$

where V is the voltage applied in direction l. By substitution from Eq. 7-106 this equation can be recast into the useful form

$$\frac{I}{e} = fl \left[\frac{V\mu}{L^2(fA)^{1/2}} \right]. \qquad (7\text{-}109)$$

This shows that the current is not directly proportional to the radiation but to its square root. It is also evident that if the quantity within the parentheses is greater than unity, the detector will show a gain. This concept will be examined in more detail later.

The time t_d required for a charge carrier to cross a length l is

$$t_d = l/\mu E = l^2/\mu V \qquad (7\text{-}110)$$

where E is the electric field.

Gain

From Eq. 7-105, it can be shown that the time $t_{1/2}$ required for the charge density to drop to one-half its steady-state value, if the irradiation ceases, is given by

$$t_{1/2} = (fA)^{-1/2}. \qquad (7\text{-}111)$$

The term in parentheses in Eq. 7-109 can be called the gain, G. Then, substituting from Eqs. 7-110 and 7-111, the gain can be expressed as

$$t_{1/2}/t_d \qquad (7\text{-}112)$$

which demonstrates that large values of G require slow response or long recombination times.

Recombination Centers

In practice it will be found that the current in actual detectors is proportional to the radiation rather than to its square root; this is due to the fact that the effective time constant is not the time $t_{1/2}$ referred to above, but is determined by the existence of *recombination centers* and *traps* in the material. A *recombination center* is a region within the crystal, such as dislocation, that will capture an electron and hold it long enough for it to recombine with a hole.

If the density of the recombination center is N_r and the illumination level is sufficiently low that the photon-generated charge densities are small with respect to N_r, then the lifetime of a charge carrier is only that length of time it requires to find a recombination point. This is

$$t = (N_r u_n S_n)^{-1} \qquad (7\text{-}113)$$

where:

u_n = thermal velocity of charges,
S_n = cross-section for recombination process.

This lifetime replaces $t_{1/2}$ in Eq. 7-112, and results in a relation in which the current is directly proportional to the radiation.

Traps

A *trap* is an energy level lying slightly below the conduction band (for electrons) or slightly above the valence band (for holes), with an action that is essentially the temporary capture of charges, since the "depth" of traps is sufficiently small that the charges are quickly released as a result of thermal excitation. The effect is to increase the detector's time constant without increasing its sensitivity.

Traps are similar to recombination centers, but have lower holding powers, so there is a higher probability that the captured electrons will be thermally excited back into the conduction band.

Under steady-state conditions, as many charges are being released from traps as are being captured, so that the charge density is nearly the same either with or without the traps. At the same time, the rise time for charge density is increased because traps must be filled, and the discharge time is also increased by charges released from traps.

Photodiodes

Ordinarily, photoconductors have an objectionable amount of dark current arising from thermal ionization of impurity centers. These centers are present to some degree even in the purest available material; they can be controlled to some degree by cooling. Alternatively, the photoconducting material can be formed into a semiconducting diode so that the internal fields will now drastically reduce these unwanted currents. The advantage of this solution is that it exploits the impurities to reduce the dark current. The semiconducting junction is actually the interface between two different materials, each with a sufficient amount of impurity that the normally resistive materials become conductors through either "hole" mobility for p-type[3] materials, or through electron mobility for n-type.[4] A schematic diagram of such a diode is shown in Figure 7-42. As is characteristic of such junctions, there is an electric field directed from the n-region in which negatively-charged materials are the predominant current carriers, and the p-region, in which positively charged "holes" (or unoccupied energy levels) predominate. This electric field inhibits the diffusion of electrons and therefore the leakage current. If the internal field is cancelled by applying an external or bias field in the opposite, or p-to-n direction, then there is no restriction to charge flow. This external field, (contrary to our application) is known as forward bias because it allows current to pass through the diode in its direction. In most detector applications, the internal field is reinforced by applying a reverse bias in order to further enhance the detector properties.

---

[3] A p-type semiconductor is a semiconductor material that, in the electrically quiescent condition, contains positively-charged hole carriers in excess of free electrons.

[4] An n-type semiconductor is a semiconductor material in which negatively-charged free electrons are present in excess of free holes, in the elastically quiescent condition.

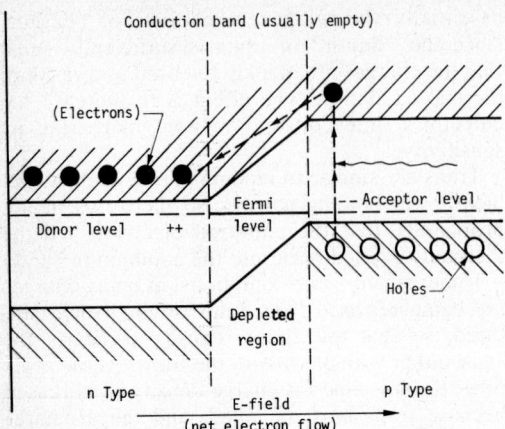

Fig. 7-42. Energy-level diagram for junction solid-state diode.

Junction Mechanism

As indicated in the diagram of Figure 7-42, the junction in its quiescent state has three distinct regions: a highly conductive region (the *p*-material), with mobile positive charges (holes); a depleted region; and another highly conductive region (the *n*-material), with mobile negative charges (electrons).

The depleted region arises because, at the interface, free holes in the *p* region can diffuse into the *n* region where they annihilate the free electrons, leaving behind fixed ions. These ions create the *n* to *p* electric field which finally inhibits further charge flow. The volume containing the fixed charge is known as the depletion region because all the charge carriers which would normally be found here have either been annihilated or swept back into their respective regions. This depleted region now acts like the insulator of a capacitor because it is surrounded by two highly conductive surfaces. The size of this capacitor determines the frequency response of the back bias diode (see Ranish, 1971; Parks, 1966).

Photovoltaic Mode

If this junction is illuminated by photons of sufficient energy to raise electrons into the conduction band in the *p* region, such an excitation creates an electron-hole pair. If the electron also diffuses into the depletion region, the electronic field accelerates it across this region into the *n* region, where it becomes just another member of the free electron population. The work done appears at the terminal of the device as a voltage, which of course appears only when the junction is illuminated. This is the photovoltaic mode. There is no leakage current in this mode because there is no external bias; the response is dependent on load and varies from a logarithmic quantity for very large resistances to linear for zero values. The time constant is rather long because the

changes in field strength which constitute the signal are compensated only by diffusion.

Reverse Bias Diode—(Photoconductive Mode)

The other form of operation is to apply an external voltage to the diode, which (as was pointed out earlier) is in the same direction as the internal field.

The voltage increases the junction field and therefore the depleted volume. The detection process is the same as before except that there is a higher probability that the electron-hole pair will be produced in the depletion region where the effect on the external circuit is faster; there is also a much lower probability of recombination. Once the charges cross the depletion region, the electric field should change, but is constrained by the external bias; equilibrium is reestablished by a current flow in the circuit. Time constants as short as 10^{-11} seconds are typical of this mode of operation. Figure 7-43 shows a circuit schematic for this mode.

Absorption of photons in the depletion region is improved if an insulating layer is provided between the two materials of the normal junction to produce a *p-i-n* (*PIN*) type photodiode. The insulator increases the thickness of the depletion region, making the diode more sensitive.

The principal disadvantage of the reverse bias mode of operation is the presence of dark noise, which results from the thermal excitation of electrons into the conduction band. A typical dark current at room temperature is 10^{-12} amperes for a reverse bias of a hundred volts.

Time Constants

The time constants of photodiodes are determined by their internal capacitance, C, expressed as

$$C = \epsilon A / d \qquad (7\text{-}114)$$

where ϵ is the permittivity of the material, A is the area, and d is the width of the depletion region. The thickness of this region is usually in the order of tens of micrometers and the relative permittivity of silicon is approximately 12. Thus, for any reasonable area the capacitance is large. For an area of one square cm, it can exceed 100 picofarads. The application of higher external fields increases the thickness of the depletion region and lowers the capacitance.

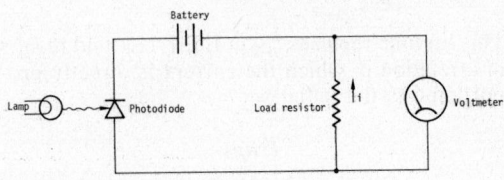

Fig. 7-43. Circuit schematic for reverse-bias photodiode.

THERMAL DETECTION MECHANISMS

Limpiris and Mudar (1979) provide an excellent treatment of the theory of the thermal detectors discussed below.

Thermopile Detectors

A junction of two dissimilar metals forms a *thermocouple* (Seebeck effect). As discussed above (Radiometric Instruments), a thermocouple detector can be constructed by attaching a small piece of blackened foil to a junction formed with short thin wires. The thermocouple and the foil receiving-surface are held in place by materials having high electrical and thermal resistance. The wires may be attached to a heat sink to form the reference junction. The sensitivity of single-element thermocouple detectors is usually inadequate for field measurements.

When more than one thermocouple is used, the device is called a thermopile. A modern thermopile is described by Barnes (1976): "The thermopile consists of six junction pairs formed from vacuum-deposited thin films of dissimilar conductors. One junction of each pair is a reference, in thermal contact with a heat sink. The other is the measurement junction. The six junctions comprise the active 1 × 1 mm area. The measurement junctions are blackened, thermally isolated from the heat sink, and are free to absorb (or emit) radiation. The resulting temperature rise (or fall) causes a voltage to appear across the junction leads. The voltage is proportional to the temperature difference, and the polarity depends on whether the measurement junctions are hotter or colder than the reference junctions. The junction pairs are connected in series to increase the thermopile output voltage to 15 volts/watt for incident radiation. One effect of the reference junctions is the elimination of any temperature effect from −20 to 40°C.

Thermistor Bolometers

Thermistors are semiconductors formed from mixed metal oxides. The resistance of these oxides decreases with temperature due to the increased number of charge carriers in the semiconductor material.

A thermistor bolometer is made with a matched pair of thermistors (sintered wafers of manganese, cobalt, and nickel oxides). One wafer serves as the sensing element for incident radiation and the other, shielded from incident radiation, serves as a reference. (See Figure 7-44). The shielded wafer compensates for ambient temperature changes allowing operation over a wide dynamic range without cooling. For best accuracies, the temperature of the bolometer is usually held constant. Use of an immersion lens, as shown in Figure 7-44, provides increased D* and a heat sink for the radiation sensing element.

Pyroelectric Detectors

When a pyroelectric crystal absorbs incident radiation causing its temperature to rise, the charge distribution in the crystal changes. To construct a pyroelectric detector, a crystal of lithium tantalate (triglycine sulfate, strontium barium niobate, and other materials are also used) is grown and sliced into thin wafers. The wafers are placed between two electrodes to form a capacitor which changes with temperature.

Radiation incident upon the crystal raises its temperature causing a change in the voltage across the capacitor-like element. Since the detector requires a high impedance amplifier, a field-effect transistor is frequently mounted on the wafer and connected directly to the electrodes of the crystal as a source follower. The crystal is blackened to achieve a nearly flat response.

Pyroelectric detectors are somewhat piezoelectric and care must be taken in designing the crystal mount. In addition, they become depoled if cycled through the curie temperature of the material. They may be restored (re-poled) to useful operation by temperature and/or voltage cycling. The curie temperature of DTGS (deuterated triglycine sulfate) is 45°C and the curie temperature of lithium tantalate is 700–800°C.

DETECTOR NOISE[5]

The overall sensitivity of detectors is governed by the responsivity and the detector noise. The output signal from a detector is continuous with time, and noise is that variation in the output level that cannot be predicted. Noise then limits the accuracy with which a signal can be measured at a given observation and determines the minimum detectable signal.

Determination of the limiting noise level as a function of the contribution of the various noise sources within a detector is a complicated exercise in solid-state physics. But, for simplicity, and to specify performance, this problem has been overcome by manufacturers through the introduction of the following figures of merit: noise-equivalent power (NEP), detectivity (D*), and the signal-to-noise ratio (S/N). These will be considered in detail in succeeding paragraphs.

The magnitude of the signal is usually easy to determine; noise, however, has a statistical character and can only be discussed on a probabilistic basis. For purposes of discussion, noise in electro-optical devices can be classified into the three categories of photon noise, Johnson noise, and device-dependent noise.

---

[5] This section, Detector Noise, is taken from Volume 1 of the first edition of the Manual of Remote Sensing (1975), Chapter 7, Electro-optical Remote Sensors with Related Optical Sensors, L. Ralph Baker and R. MacDonald Scott II, co-authors−editors.

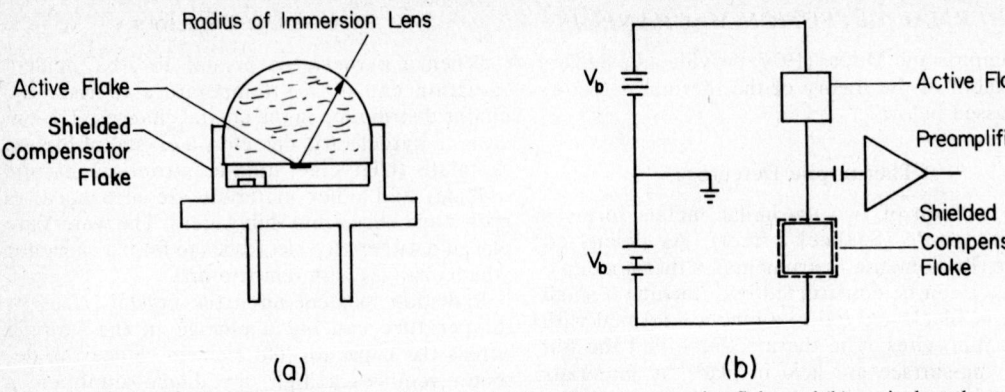

Fig. 7-44. Immersed thermistor bolometer showing (a) active and compensating flake and (b) equivalent thermistor bridge circuit bias V_b.

Photon Noise

The signal consists of a stream of photons. These photons, however, do not arrive at a uniform rate, but at a rate that is subject to random fluctuation. Statistically, if the average number of independent random events occurring in a time period t is n_o, then the mean-square deviation in this average over a large number of such intervals is also equal to n_o. If N is the actual number, then this mean-square deviation, N^2, is

$$N^2 = \overline{(n - n_o)^2} = n_o \qquad (7\text{-}115)$$

and thus, for the noise, we have

$$n = n_o^{1/2} \qquad (7\text{-}116)$$

This kind of noise arises from the energy received in the signal itself, and also from the background. For example, in an imaging situation, the signal is represented by the differences in intensity over the scene; the noise, as defined above, is derived from the average intensity over the whole scene. Somewhat different situations sometimes arise, and it is important to identify the true sources of this kind of noise.

This noise is called *photon noise* or, more commonly, *shot noise*. It is properly computed at the output of a detector, since it applies to the response of the detector to the incoming photons, as well as to the local photons themselves. If the quantum efficiency is less than unity, or if multiplicative events (such as secondary electron emission) take place, these factors must also be included.

Equation (7-116) gives the number of noise events in a sampling time interval, t. To eliminate this time interval and treat the situation on a continuous (or average) basis, one should first apply the Nyquist theorem,[6] which gives the maximum information bandwidth as $t/2$, and then multiply the quantity under the radical by $2B$, where B is

[6] The nyquist theorem states that two samples per cycle will completely define a band-limited signal; i.e., the sampling rate must be twice the highest frequency component of the signal.

the bandwidth in Hertz. If the units of current are used in place of the average number of electrons, the resultant is the familiar expression for shot noise in terms of output noise current:

$$\overline{i_n} = 2(eiB)^{1/2} \qquad (7\text{-}117)$$

where

e = electron charge,
i = average shot noise producing output current,
B = bandwidth.

This is not a rigorous statistical approach in cases where there are several noise-producing mechanisms (such as photon conversion, secondary electron emission, etc.) but it is nearly always adequate for performance computations.

Johnson Noise

A major source of noise in all detection devices is thermal or Johnson noise (also called Nyquist noise). This is the result of fluctuations of charge carriers within the device, and can be related to the radiation of a blackbody. It is usually expressed in a form that treats the detection device as a resistance, R. In terms of current, to make the expression comparable with Eq. 7-117

$$i_n = \left[(4kTB)/R\right]^{1/2} \qquad (7\text{-}118)$$

where:

k = Boltzmann's constant, 1.380×10^{-23} J·K^{-1}
T = absolute temperature, K

$$(7\text{-}119)$$

Device-Dependent Noise

In addition to shot noise and Johnson noise, which are always present in electronic devices, other noise sources, which are unique to the particular form of device being used, are usually present. Among such sources that are quite common in electro-optical devices are current or $1/f$ noise, and *generation-recombination noise*.

$1/f$ noise is present in nearly all photoconductive devices; it is inversely proportional to the operating frequency, and usually follows the form

$$\overline{i_n^2} = Ci_o^2B/fA_d \qquad (7\text{-}119)$$

where:

C = a constant, which varies with the detector under consideration,

i_o = rms current associated with noise,

f = operating frequency.

The physical mechanism giving rise to this noise is not understood; there appears to be some relationship with the surface condition and with the contact regions, but this does not tell the whole story. This noise factor is applied on the basis of experimental measurements of the particular detectors being used. Even the exponent of the current is not a fixed value.

Generation-recombination noise

Generation-recombination noise (G-R) in photoconductors arises from fluctuations in the member of electrons in the conduction band. Even in the absence of signals, this number is being reduced by random recombinations with vacant valence sites. A mathematical statement of G-R noise is given by

$$\overline{i_n}^2 = \frac{2\tau\overline{i^2}B}{N\left[1 + (\omega\tau)^2\right]} \qquad (7\text{-}120)$$

where:

τ = recombination time,

ω = angular frequency of modulation,

N = average charge density.

FIGURES OF MERIT[7]

As previously indicated, three significant measures of performance are currently considered important in comparing electro-optical transducers: the quantum efficiency, usually symbolized by η; the noise-equivalent power, or NEP; and the detectivity, D^* (pronounced dee-star).

It should be emphasized that such figures of merit as NEP and D^* can serve only as rough guides to the relative value of different detectors. The circumstances of use are of great importance, and careful signal-to-noise computations, using all available information on the particular situation, are necessary to make a firm selection of the optimum detector.

Quantum Efficiency

Quantum efficiency (η) may be defined as the number of electrons emitted for each incident photon in a photoemissive surface, or the number of charge carriers per incident photon in a photoconductor. By definition, a quantum efficiency of one would exist if one photon released one electron. Quantum efficiency is always less than

[7] This section, Figures of Merit, is taken from Volume 1, of the first edition of the Manual of Remote Sensing (1975), Chapter 7, Electro-Optical Remote Sensors With Related Optical Sensors, L. Ralph Baker and R. MacDonald Scott II, co-authors—editors.

unity, however. Multiplicative processes, such as a secondary electron emission, are always considered separately, and are called *gain*.

The principal significance of the quantum efficiency is its effect on shot noise. As stated, the output current, for a given average input signal, is proportional to the quantum efficiency, and the shot noise becomes proportional to the square root of the quantum efficiency. Therefore, the S/N ratio for shot noise becomes directly proportional to the quantum efficiency, and thus is reduced significantly for detectors with low quantum efficiency. Typical values of η for photoemissive surfaces are 10 to 20% in the visible spectrum, but falling off to less than 1/10 of one percent at the long-wavelength limit of photoemissivity around 1 μm. The quantum efficiency of photoconductors and photodiodes, on the other hand, is typically very near to unity.

Noise-Equivalent Power

A useful general expression, applicable to all detectors, is *noise-equivalent power, NEP*. It is a measure of minimum detectable power, or the detector noise, expressed in terms of input power. It specifies the power required to generate an S/N ratio of one, where detector noise is usually referred to a 1-Hz bandwidth. The NEP is given, in its simplest form, by the expression

$$NEP = \Phi_d v_n/V_s \qquad (7\text{-}121)$$

where:

Φ_d = peak power (W) of modulated radiation incident on the detector

V_s = peak signal voltage developed by the detector

v_n = rms noise voltage developed by the detector as referred to a 1-Hz bandwidth.

Noise-equivalent power may be defined as the rms value of sinusoidally modulated radiant input power that will result in an rms output signal voltage equal to the rms noise voltage from the detector.

In the interpretation of the NEP value, careful attention must be given to the conditions under which it is measured, which should be stated. The NEP value will normally be given for a particular wavelength, modulation frequency, detector area, or temperature, and (in some cases) a particular field of view; a change in any one of these factors can alter the value considerably. Typically, infrared detectors are measured with a 500-K blackbody as the source, a 900-Hz chopping frequency, and a 1-Hz reference bandwidth. For these criteria the identification would be given as NEP 500 K, 900, 1. Since the noise in solid-state detectors almost always varies with the square root of the detector area and the square root of the bandwidth, the NEP (in general) varies with wavelength, and its value as a function of wavelength is frequently shown as NEP_λ.

Two other terms, responsivity and quantum efficiency, are implicit in the expressions of Eq. 7-119, which could have been expressed by

$$NEP = V_n/\Re \qquad (7\text{-}122)$$

where

$\Re$ = responsivity $[V \cdot W^{-1}]$.

The *spectral noise equivalent power*[8] may be written:

$$NEP_\lambda = \frac{v_n}{\Re(\lambda)} \; [W] \qquad (7\text{-}123)$$

where $\Re(\lambda)$ is the responsivity as a function of wavelength of the signal flux.

Blackbody noise equivalent power is defined as:

$$NEP_{bb} = \frac{v_n}{\Re_{bb}} \; [W] \qquad (7\text{-}124)$$

where $\Re_{bb}$ is the responsivity of the detector for radiation from a blackbody source at a specific temperature.

Detectivity

This measure of detector performance that increases with improved performance was proposed by Jones (1952):

$$D = \frac{1}{NEP} \; [W^{-1}]. \qquad (7\text{-}125)$$

The *detectivity*, D, depends on the wavelength (or spectral distribution) of the incident radiation, the temperature of the detector, the area (and shape) of the detector, bias conditions, the chopping frequency and the equivalent-noise bandwidth of the measuring circuit.

To obtain a figure of merit which is, in many cases, independent of detector area and noise bandwidth, Jones (1959) proposed the *normalized spectral detectivity*. which is written:

$$D^*(\lambda, f) = \frac{\sqrt{A_d \, \Delta f}}{NEP_\lambda}$$

$$= D(\lambda)\sqrt{A_d \, \Delta f} \; [cm \; Hz^{1/2} \; W^{-1}] \qquad (7\text{-}126)$$

where A_d is the area of the detector $[cm^2]$, $\Delta f \, [Hz]$ is the equivalent-noise bandwidth, λ is the wavelength of the incident radiation, and f is the chopping frequency.

The advantage of D^* as a figure of merit is that it applies to many detectors and always applies to background-limited detectors. The main assumption is that NEP_λ is proportional to the square root of detector area. Hudson (1969) reports that this relationship may be in error for detectors whose length is more than five times the width. Kingston (1978) derives D^* in terms of quantum efficiency and background irradiance.

An ideal *background-limited detector* is free from internal noise. *Background noise* is the

quantum noise (shot noise) due to the average irradiance on the detector from its surrounding cavity and, if the target does not fill the field-of-view of the detector, from the background for the target. For detector specifications, it is standard practice to report the background temperature and the field-of-view for the background (298 K and 2π steradians are typical) with the curves for D^*.

To improve instrument performance, a *cold stop* is sometimes used to reduce the quantum noise due to the background irradiance from the optical cavity and restrict the field-of-view to the target. The cavity surrounding the detector and the field stop are cooled to a temperature such that the background noise is negligible and the radiation incident on the detector is from the optical system in front of the cavity. To normalize D^* for test systems using a cold stop to limit the field of view, Jones (1960) defined a new figure of merit, D^{**} (pronounced dee-double-star):

$$D^{**} = \sqrt{\frac{\Omega}{\pi}} D^* \qquad (7\text{-}127)$$

where Ω is the effective weighted solid angle of the cold stop. If the detector is uniformly responsive to rays from all angles and approaches circular symmetry and if the solid angle determined by the detector and the cold stop approximates a circular cone, then

$$D^{**} = D^* \sin\left(\frac{\theta_f}{2}\right)$$

where θ_f is the full angle of the cone.

Since detector performance is strongly affected by each of the operating conditions mentioned above, it is difficult to compare detectors using test results made under different conditions. Limpiris and Mudar (1979) provide a thorough treatment of needed test data and calculations to aid in the comparison of detector performances.

REFERENCES

Barnes Engineering Co., 1976, Bulletin 2-412; 30 Commerce Rd., Stamford, CT 06904.

Barnes Engineering Co., 1979, Bulletin 12-880; 30 Commerce Rd., Stamford, CT 06904.

Barnes Engineering Co., 1981, Bulletin 12-1000; 30 Commerce Rd., Stamford, CT 06904.

Birth, Gerald S., 1975, Electromagnetic Radiation, Chapter 9, Instrumentation and Measurement for Environmental Sciences; Z. A. Henry, ed., Am. Soc. Agric. Engineers Special Pub. SP-0375, St. Joseph, MI.

Bauer, Marvin E., C. S. T. Daughtry, and V. C. Vanderbilt, 1981 (September), Spectral-agronomic relationships of corn, soybean, and wheat canopies; Proc. Colloq. on Signatures of Remotely Sensed Objects, ISP, Avignon, France.

Eppley Laboratory, Inc., 1980a, Model NIP, Normal incidence pyrheliometer; Newport, RI.

Eppley Laboratory, Inc., 1980b, Model PSP, Precision spectral pyranometer; Newport, RI.

Exotech, Inc., 1980, Model 100-AX Hand-Held

---

[8] All following material is added to text of First Edition.

Radiometer; 1200 Quince Orchard Blvd., Gaithersburg, MD 20760.

Friis, H. F., 1944, Noise figures of radio receivers; Proc. IRE, vol. 32, no. 7, pp. 419–422.

Grum, F., and R. J. Becherer, 1979, Optical Radiation Measurements, Volume I—Radiometry; Academic Press.

Hanneman, R. E., D. P. DeWitt, E. J. Hanley, R. L. Schreiner, and P. Bonderman, 1979, Determination of serum bilirubin by skin reflectance: effect of pigmentation; Pediatric Res., vol. 13, p. 1326.

Hudson, R. D., Jr., 1969, Infrared System Engineering; John Wiley and Sons, New York.

Jackson, R. D., R. J. Reginato, and S. B. Idso, 1977, Wheat temperature: practical tool for evaluating water requirements; Water Resources Res., vol. 13, pp. 651–656.

Jacobs, S. F., 1978, Nonimaging Devices, Chapter 4, Handbook of Optics; W. G. Driscoll and W. Vaughn, eds., OSA, McGraw-Hill Book Company.

Jones, R. C., 1952, Detectivity, the reciprocal of noise equivalent input radiation; Nature, vol. 170, pp. 473–478.

Jones, R. C., 1959, Phenomenological description of the response and detecting ability of radiation detectors; Proc. Inst. Radio Engineers, p. 1495.

Jones, R. C., 1960, Proposal of the detectivity, D** for detectors limited by radiation noise; J. Opt. Soc. Am., vol. 50, p. 1058.

Kingston, R. H., 1978, Detection of Optical and Infrared Radiation; Springer-Verlag, New York.

Li-Cor, Inc., 1980, Brochure DS-1180; Lincoln, NB.

Limpiris, T., and J. Mudar, 1979, Detectors, Chapter 11, The Infrared Handbook; W. L. Wolfe and G. J. Zissis, eds., ERIM, Ann Arbor, MI.

Matra Optique, 1980, Terrain Radiometer; 93, Avenue Victor-Hugo, 92502 Ruell-Malmaison, Cedex, France.

Merrick, Edwin B., and Thomas J. Hayes, 1976 (October), Continuous, noninvasive measurements of arterial blood oxygen levels; Hewlett-Packard J., Palo Alto, CA 94304.

Molectron Corporation, 1979, Bulletin PR-200; 177 N. Wolfe Rd., Sunnyvale, CA 94086.

Motchenbacher, C. D., and F. C. Fitchen, 1973, Low-Noise Electronic Design; John Wiley and Sons, New York.

Nicodemus, F. E. (Ed.), 1976, Self-Study Manual on Optical Radiation Measurements: Part 1—Concepts; Chapters 1 to 3, Natl. Bur. Standards (U.S.), Tech. Note 910-1, 93 pp.

Nutter, G. D., 1974, Radiation thermometry—a review; J. Appl. Measurements, vol. 1, p. 83.

Optical Industry Systems Purchasing Directory, 1981, The Optical Publishing Co., P.O. Box 1146, Berkshire Common, Pittsfield, MA 01201.

Parks, M. S., 1966, The story of microelectronics; N. Amer. Aviat., Autonetics Division, Anaheim, CA, Publication, pp. 127–137.

Ranish, M. R., and Isna Hayashi, 1971 (July), A new class of diode lasers; Sci. Amer., pp. 32–40.

Robinson, N., 1966, Solar Radiation; Elsevier Publishing Company, New York.

Robinson, B. F., M. E. Bauer, L. L. Biehl, and L. F. Silva, 1978 (July), The design and implementation of a multiple instrument field experiment to relate the physical properties of crops and soils to their multispectral reflectance; Proc. Intl. Symp. for Observation and Inventory of Earth Resources and Environment, ISP, Freiburg, West Germany.

Robinson, B. F., and L. L. Biehl, 1979, Calibration procedures for measurement of reflectance factor in remote sensing field research; Proc. 23rd Annual Intl. Tech. Symp., SPIE, Bellington, Washington, vol. 196.

Robinson, B. F., R. F. Buckley, and J. A. Burgess, 1981, Performance evaluation and calibration of a modular multiband radiometer for remote sensing field research; Proc. 25th Annual Intl. Tech. Symp., SPIE, Bellington, Washington, vol. 308.

Santa Barbara Research Center, 1979, Bulletin 19, 75 Coromar Drive, Goleta, CA 93017.

Slater, P. N., 1980, Remote Sensing: Optics and Optical Systems; Addison-Wesley Publishing Co., Reading, MA.

Sloan, William W., 1979, Detector associated electronics, Chapter 16, The Infrared Handbook; W. L. Wolfe and G. J. Zissis, eds., ERIM, Ann Arbor, MI.

Wolfe, William L., 1979, Optical materials, Chapter 7, The Infrared Handbook; W. L. Wolfe and C. J. Zissis, eds., ERIM, Ann Arbor, MI.

Zissis, G. J., 1979, Radiometry, Chapter 20, The Infrared Handbook; W. L. Wolfe and G. J. Zissis, eds., ERIM, Ann Arbor, MI.

CHAPTER 8

Electro-Optical Imaging Sensors

Author: VIRGINIA T. NORWOOD and JACK C. LANSING, Jr.

GENERAL CONTENTS: Styles of sensors; Frame sensors; Pushbroom sensors; Mechanical scanners; Selecting a sensor design; Design procedure; Resolution and optical transfer function; Establishing signal-to-noise requirements; Other sources of radiometric error; Separation of spectral energy; Detectors and imaging devices; Noise sources; Figures of merit; Radiometric calibration; Amplitude calibration; Spectral calibration; Spatial calibration; Transfer methods; Calibration categories; Calibration errors; References.

NOMENCLATURE

To conserve and eliminate repetition in text and references, the following symbols, units and names have been used in this chapter.

A_d	detector area
A_{eff}	effective aperture area
BLIP	background-limited infrared photodetector
c	velocity of light
c	diameter of blur circle containing designated percentage of energy
c	capacitance
CCD	charge-coupled device
C_T	thermal capacity
d	differential operator
D^*	detector figure of merit
dB	decibel
e	charge on an electron, 1.6×10^{-19} coulomb
E	energy of a photon
ECPR	electrically calibrated pyroelectric radiometer
f	electrical frequency
fe	clock frequency
f^*	3 dB frequency of D*
f	focal length
$F(\theta)$	electrical transfer function
FET	field-effect transistor
G	conductance; photoconductive gain
h	Planck's constant
H	height above the scene
H_o	solar irradiance in spectral band
I	current
I_{dc}	direct current
IFOV	instantaneous field of view
i	an index
i_n	Johnson noise current
i_n	current noise, amplifier input device
i_{nd}	dielectric loss noise current
i_{ngr}	generation-recombination noise
i_{ns}	shot noise current
j	imaginary number
k	Boltzman's constant; various constants adopted for temporary convenience
K	thermal conductivity
L	radiance
L_a	radiance scattered during transit
L_{sky}	radiance scattered from atmosphere
m	number of IFOVs across scene
M_i	pyroelectric figure of merit
MTF	modulation transfer function
n	number of detectors in spectral band; number of transfers, number of measurements
n-type	majority carrier is electron
N	number of detector junctions; number of photons; number of free carriers

N_λ	number of photons of wavelength λ per unit time
$NE\Delta L$	noise equivalent increment of radiance
NEP	noise equivalent power
p-type	majority carrier is hole
P	power
P_o	peak power
Q_s	output charge of a CCD
Q_{tr}	charge transfer per gate in a CCD
r	distance to scene
R	electrical resistance
R_T	thermal resistance
$\bar{S}$	detector responsivity
$S(\alpha)$	intensity across image
$S_{tr}(f)$	transfer noise spectral density
S/N	peak-to-peak signal voltage to rms noise voltage
t	time
T	temperature; time between samples; transmittance through one atmosphere
V	voltage; velocity of sensor-track along scene
V_o, V_M	voltages, true and measured
W	width of scene across track
$\bar{x}$	mean of measurement set
x_i	one measurement of set
z	zenith angle of sun
α	angle subtended by IFOV; Seebeck coefficient
δ	dielectric loss coefficient
Δ	increment; difference
ϵ	error; fractional charge remaining
ϵ_r	dielectric constant
η	quantum efficiency
θ	field of view across scene-width W
λ	wavelength
ρ	reflectance; density
Σ	summation
σ	standard deviation; electrical conductivity
τ	time; time constant
τ_o	optical transmittance
Φ_b	background photon flux density
ω	angular frequency
Ω	solid angle to be imaged per unit time

INTRODUCTION

Imaging sensors have application for the production of detailed earth images from aircraft, synoptic earth resources images gathered in several spectral bands, cartography, synoptic and mesoscale meteorology from satellites, planetary exploration and military reconnaissance from aircraft or satellites. Photographic cameras (Chapter 6) have served some of these functions in the past; however, this chapter is concerned with electro-optical sensors in which electrical signals can be transmitted from the remote station via radio.

Detectors can be fabricated to be sensitive to radiation at any part of the spectrum, from ultraviolet through the region of reflected sunlight and well into the self-emitting or long infrared region. Particular detectors can be installed in sensors to collect images in several spectral bands simultaneously. The designer now has a diverse range of detectors that can be used in combination with different types of optical systems, with and without mechanical scanners. The selection of the best sensor for an application must proceed from requirements that will be found in Volume *II*.

Each discipline will have preferences in resolution, spectral bands, viewing geometry, time of day, degree of sensitivity and radiometric accuracy. In the following pages the strengths and weaknesses of different kinds of sensors and the characteristics of available photo-sensitive devices are described.

Performance qualities that must be considered in the design include signal-to-noise ratio, modulation transfer function, geometric fidelity, and calibration. These engineering qualities are defined and, where applicable, are translated into terms having meaning for the user of the end-product.

STYLES OF SENSORS

All sensor types gather energy from the scene with an optics aperture of area, A_{eff}, and focus each spot onto a photo sensitive surface that is sensitive to a designated part of the spectrum. The size of the spot is determined either by the size of the detector or, in the case of an electron beam tube, by target structure or scan beam geometry. The electrical signal level that corresponds to the radiance from that spot is properly 'tagged' by processing circuitry and is either stored or transmitted to a location for processing later into a replica of the original scene.

Sensors can be separated roughly into three categories—frame sensors, linear array or pushbroom sensors, and mechanical scanners. The frame sensor is self sufficient in that no scanning motion is needed to image an area. The linear array scans in one direction by electronic readout and requires an auxiliary motion or tilt to scan in the other direction. Mechanical scanners, including a variety of double scanning or one-directional scanning sensors, scan from left to right but they employ external motion for the other direction. Table 8-1 shows the principal characteristics of the three sensor types written from the viewpoint of a moving platform. The moving platform has interest because the platform's motion imposes

TABLE 8-1

Relative Advantages of Basic Sensor Types

	Frame	Pushbroom (Linear Arrays)	Mechanical Scanner (Object Plane)
Optics Required	Wide angle, both dimensions	Wide angle, one dimension	On-axis
Sensitivity	Longest dwell-time	Next longest dwell-time	Most restricted dwell-time
Stereo Viewing	Good	Good	Adequate
Effect of Platform Instability	Least susceptible	Less susceptible	Most susceptible
Multispectral Use	Poor	Registration restricted by array* accuracy. Restricted space. Cooling load high.	Most suitable
IR Capability	Very limited*	Cooling load high	Cooling load modest
Radiometric Calibration	Very difficult	Difficult*	Least difficult
Geometric Accuracy	Poor for electron-beam devices. Detector arrays have good potential.	Limited by array technology	Very high precision possible (1μ rad) (represents mature technology)
Number of Resolution Elements Covered in Scene	Limited by array size and optics*	Limited by array length and optics*	Unlimited

*Improvement is expected with technology advances

constraints on time available for imaging; these constraints are more easily met by some sensors than by others. Another class of sensors that requires wide angle optics has been called image plane scanners and is described below, although not included in the table.

The designer must assemble requirements in terms of number of spectral bands, signal sensitivity, angular resolution, number of elements (instantaneous fields of view*) that are to be imaged per unit time, what part of the needed scanning is provided by the platform, and what coverage is left for the sensor to provide. Both aircraft and low-orbiting spacecraft cause a forward motion so that the sensor need only extend the image from side to side as the scene advances underneath. When the scene and the sensor platform are relatively stationary, it is necessary to substitute an additional scan motion either in the sensor itself or by tilting the platform. One extreme is exemplified by a class of meteorological scanners in geostationary orbit in which detectors at the focus of telescopes are scanned across the earth by spinning the entire spacecraft. A fold mirror is advanced by a small angle on each spin so that the scan swath is moved systematically in the orthogonal direction to complete the frame. Frame cameras permit both directions to be scanned electronically but are not suited to multispectral imaging.

Generally speaking, electro-optical imaging sensors can be divided into two main categories—mechanical and electronic. The following discussion describes these two categories as though both were pure types of sensors. In reality, however, few sensors are purely mechanical or purely electronic. The degree of mix is determined, in any given instance, by the sensor requirements and by state of the technology. Table 8-1 is a summary of the major characteristics of sensors in these two categories. Hybrid designs will exhibit various mixtures of these characteristics in rough proportion to the mixture.

FRAME SENSOR

The earliest frame sensors grew out of the television industry and provided the basis for the vidicons that were adapted for imaging the moon and Mars. In such sensors, photosensitive tubes with electron-beam scanning are located at the focus of wide-angle optics. Sensors mounted on moving platforms are equipped with shutters to prevent image blur and to control exposure time.

Photosensitive surfaces previously were restricted to the visible region of the spectrum which meant that (1) small-sized optics could be used without encountering diffraction problems, and (2) exposure times could be sufficiently long

* The term pixels has sometimes been used synonymously with IFOV; however, we reserve the term pixel for a sample interval of encoded signal.

to insure the receiving of a good signal. The number of resolution elements remained at 500 to 1000 on a side until development of the return-beam vidicon (RBV) which is capable of over 3000 elements on a side. The RBV has been used very successfully as a high-resolution, multiband, earth-mapping device on Landsat. Optics are designed to be compatible with an angular resolution of 33 microradians.

Two-dimensional arrays of detectors are improving to the point where they can be expected to supplant tubes. Detector arrays will offer a frame format with the advantages of extended spectral coverage, long life, and elimination of high voltage. Work is continuing to make arrays with detectors that are more nearly contiguous in both dimensions and to reduce the number of separate preamplifiers by shifting signals out of the entire array simultaneously for immediate multiplexing. The arrays have a potential for geometric accuracy which is superior to electron-beam devices. These sensors are best suited for imaging applications in single spectral bands with modest requirements on radiometric accuracy.

PUSHBROOM SENSOR

This sensor uses a wide-angle optics system in which all of the scene across θ degrees is imaged on a detector array at one time. In the simplest type of array, the signal from each detector is amplified separately and the individual output signals are usually sampled sequentially to provide a serial representation of the line for transmission from a remote location. As the platform moves along the track, successive lines are imaged by the array and sampled by the multiplexer for transmission. More advanced arrays and processors permit the signals to be shifted out in parallel and processed sequentially.

The size of the element imaged at the scene is equal to the product of the distance to the scene times the detector dimension, divided by the focal length of the optical system. For general discussion it is more convenient to use the angular instantaneous field of view, α, in place of d/f.

The time that elapses between the imaging of two successive lines can be as long as the platform takes to move the distance in the scene that is subtended by α. This dwell time is relatively long compared to that of the mechanical scanner that is described later and represents the primary advantage of the pushbroom sensor. It will be shown that a long dwell time leads to low noise in the signal. The other attractive feature of the pushbroom sensor is that no moving parts are required when it is operated from a moving platform.

The primary disadvantage is that a very large number of detectors is involved which means that calibration is difficult and signal level correction presents a real data processing burden. The pushbroom technology is new and it can be expected that these disadvantages can be reduced in importance with future array development and sophisticated signal processing.

MECHANICAL SCANNERS

The entire range of ingenious scanning devices that has been used or proposed for particular applications is too voluminous and specialized for inclusion here. Object space scanners can use on-axis optics or telescopes with a flat scan mirror located in front of the collecting aperture. These have been used extensively in space for meteorology and earth resources data collection. Other scanning methods, in which the scan mirror is located further into the optical train, require wide-angle optics and are described under the heading of near-image scanners.

Object Space Scanners

Figure 8-1 shows a simple telescope, preceded by a flat tilting mirror that scans across the scene and causes each segment of the scene to be swept across the detector in a time determined by the mirror rate. The nature of the reflection is such that the mirror need scan only $\theta/2$ degrees in order to image θ degrees of scene. The scan rate must be set so that each line is finished in the time during which the platform advances by a distance equivalent to one unit in the scene intercepted by the field stop angle α.

In comparing the two systems, it will be found that the dwell time for the scanner is reduced, compared with that of the pushbroom, by at least a factor of θ/α. Often this causes so much noise that additional lines of detectors are added along the track. By the imaging of k more lines with each mirror scan it is possible to extend the time per scan by a factor of k, thereby reducing the noise bandwidth by k.

Chapter 12 has detailed descriptions of the Landsat MSS and Thematic Mapper which illustrate the highly developed state of mechanical scanning sensors.

Near-Image Plane Scanners

Mechanical scan mirrors can also be located within the optical focusing path which has led to the term image plan scanning in distinction to object space scanning. The term near-image plane is used here because the scan device must be inserted in front of the image plane if the image is to slide across a detector; therefore, the converging beam must be interrupted before it reaches the focal point and then be refocused. The optical system must focus the entire scene simultaneously instead of merely the segment of the scene that is being scanned at any given instant. Figure 8-2 shows a design that uses this principle. Generally, refractive correcting elements are used to conserve space so that the designs have spectral bandwith ratios limited to 2-to-1 for resolution finer than about 80 microradians.

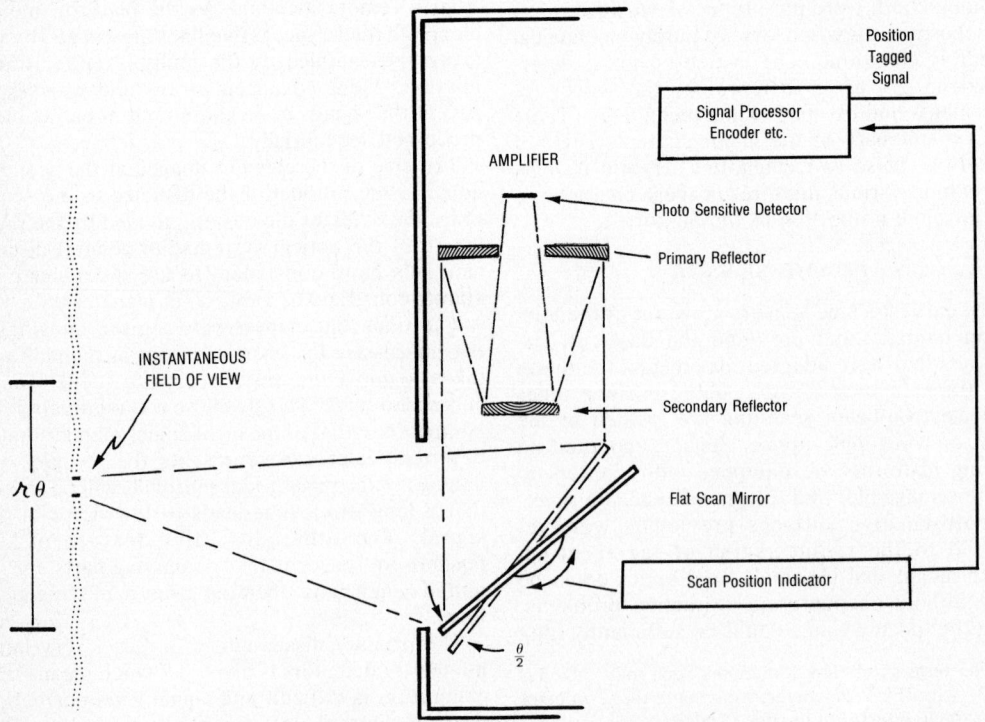

Fig. 8-1. Elements of an object space scanner.

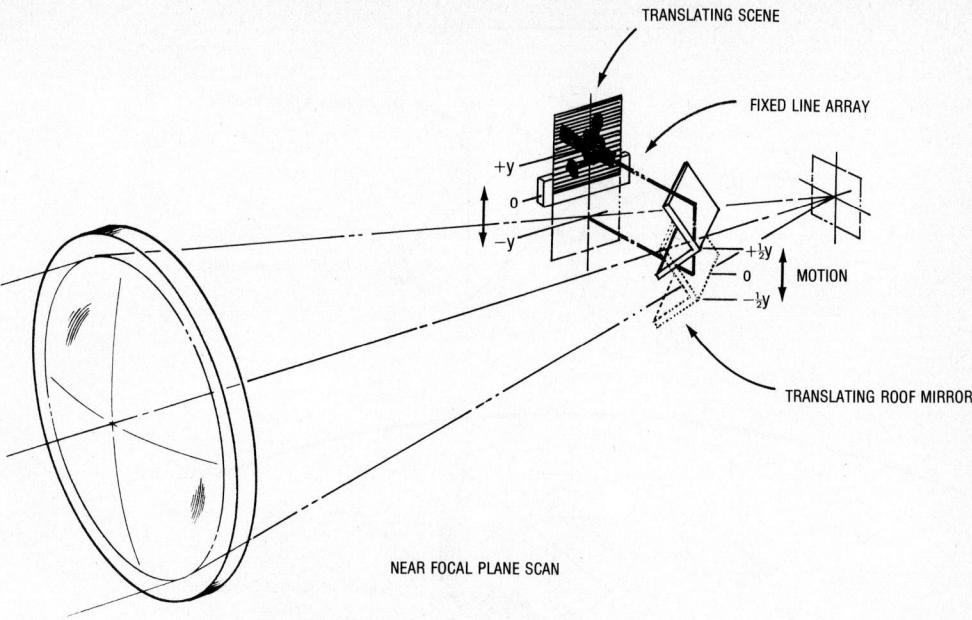

Fig. 8-2. Example of near-focal-plane scanning.

Considerable care must be taken to avoid changes in resolution quality with scan angle. One method of scanning within the focusing path that circumvents this problem is the conical scanner. The scanned strip is annular, hence each scan position uses the circular optics symmetrically. Figure 8-3 from Lowe and Beilock describes a design that was used in Skylab and was preceded by many successful aircraft-borne multispectral imagers. A specially designed conical scanning counterpart should be used for reconstructing the circular strips into an image at the processing site.

A serious drawback to all near-image plane scanners is that only a portion of the collecting aperture is used at one instant so that very large optics are needed for high sensitivity data collection. The method is well-suited to applications that require precision scanning over small fields of view.

A hybrid sensor uses the internal scanner to make a detailed image of a small area and has an auxiliary scan capability that is used to shift the sensor field of view to another area for another detailed look. The auxiliary scan can be accomplished either by an external gimbal system or by flat mirrors that are placed in front of the sensor aperture. The requirements for monitoring severe storms from space suggest the use of a scanning system of this type.

SELECTING A SENSOR DESIGN

DESIGN PROCEDURE

The following is a brief description of one approach for roughing-in the design of a sensor. Details that may be needed in order to quantify various aspects of the design are treated in the sections following. A description of the design process is necessarily serial while, in fact, some parts are carried out in parallel with the final design being arrived at only after several iterations.

Minimum Optics Aperture

The first lower bounding for the size of the collecting aperture is determined by applying a diffraction-limiting criterion in terms of the IFOV desired at each spectral wavelength. The collecting aperture must be sufficiently large that the blur caused by diffraction will not dominate the resolution that could be realized by the field-stop alone. Diffraction gives a first cut on aperture size and subsequent noise calculations may show that an even larger size is needed in order to collect more signal energy.

Figure 8-4 shows the relationship between telescope diameter and diffraction blur size for blur circles containing 90 percent of the energy. Three representative spectral wavelengths are given and others may be interpolated linearly.

The size of the diffraction blur circle relative to the usable IFOV can be arrived at only after considering certain other system factors that degrade the total blur and the modulation-transfer-function requirements. As an example, suppose an IFOV of 40 μ radians were required for a spectral band at 1 μm and also assume that 16 μ radians has been allocated for diffraction effect. A telescope size of 41 cm would be indicated. The same criterion would appear to make the 41-cm telescope unsuitable for use at 11.5 μm unless a much larger IFOV of 460 μ radians were adopted. When the diffraction is placed in context with other factors that contribute to image blur, one realizes that it is

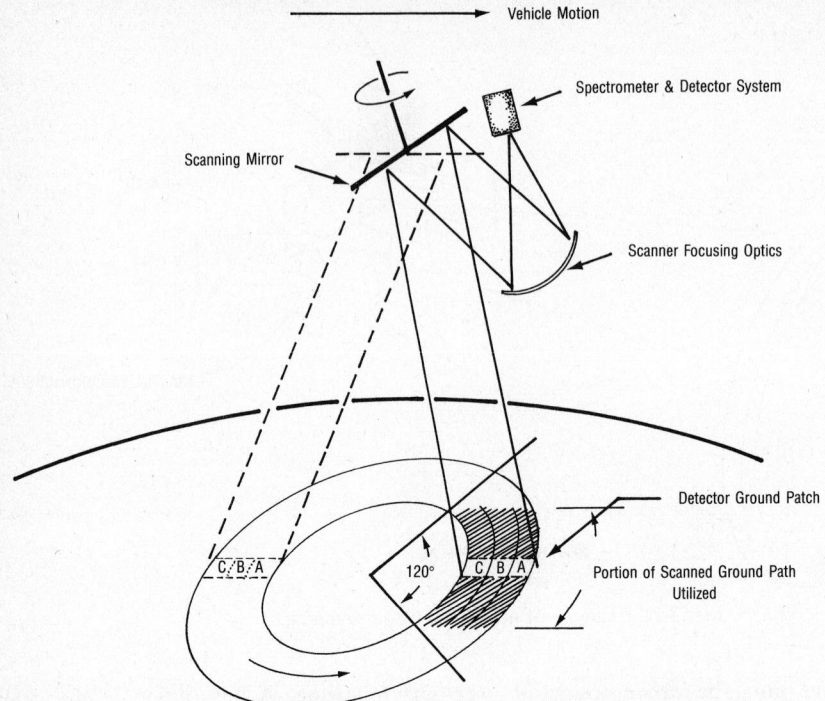

Fig. 8-3. Conical or small-circle scanning technique.

probably not necessary to increase the IFOV so drastically because the optics figure blur is relatively smaller at long wavelengths and the allocation for diffraction might be permitted to equal the IFOV, which would permit the field stop to be set at 160 μ radians.

The example and graphed values are based on a simple primary mirror with a small secondary mirror obscuration effect. Although these values can be used for preliminary estimates, each type of optics requires particular analysis for an accurate assessment of performance. Other factors that influence the choice of aperture follow in the section on determination of MTF.

Number of Detectors Required

Having in mind an approximate size for the collecting aperture, the system designer can next calculate the number of detectors that will be needed to collect imaging data at the required rate and meet the sensitivity requirements. The quantitative effect of the number of detectors is shown in a later section. Again, the moving platform provides an example that shows how additional detectors permit reduction in noise bandwidth when the time-to-image factor is restricted. Assume that the scene is moving past at a rate of n resolution elements per second and that the sensor must view or scan *m* elements across the track in specified spectral bands. If a single detector scanned the scene serially, it would be necessary to scan fast enough to image *nm* resolution ele-

ments per second which translates roughly into a bandwidth of *nm*/2 cycles per second. By adding more detectors along the track, it is possible to scan more slowly and reduce the noise bandwidth.

The designer next reviews the application requirements for noise equivalent radiance increment, NEΔR, in order to identify the spectral band that is critical in terms of state-of-the-art detector capability, spectral energy, etc. The equations given in the signal-to-noise section are used to calculate the smallest dwell time, or largest bandwidth, that can be tolerated.

Comparing the required dwell time with $(nm)^{-1}$ will reveal how many detectors must be used for the most critical band. Each geometry and type of sensor involves different detector configurations, but in all cases the principles are similar.

Options in Optical Systems

The angle that must be imaged across-track, *m* IFOVs, has an impact on the type of optics used. Figure 8-5 shows limitations on IFOV imposed by field-of-view constraints for the best examples of different kinds of optical systems. The analysis was limited to reflecting elements because electro-optical sensor applications usually involve wide spectral bands and maximum performance for the weight, although refractive elements may still be used for particular bands after spectral separation.

For each type of optics, the graph shows the 80

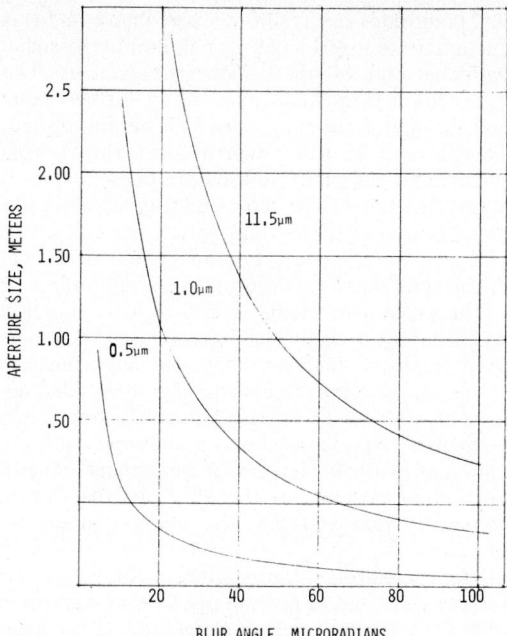

Fig. 8-4. Aperture size required to limit to indicated diffraction blur angle.

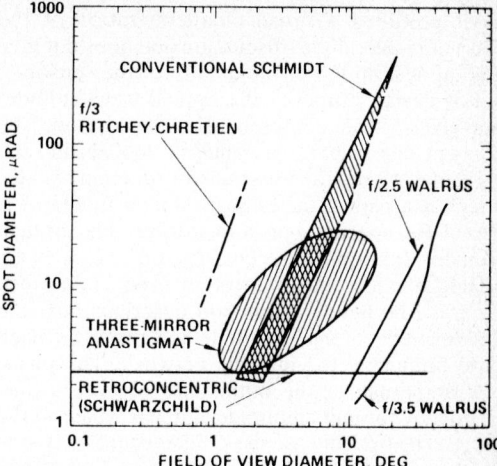

Fig. 8-5. Optical blur diameter vs. usable field-of-view for reflecting systems.

percent blur circle that will be encountered when the indicated field of view is used. As in the diffraction case, the fraction of the IFOV that can be permitted is flexible. A criterion that has been used for earth imaging is to limit the optics blur circle to one third of the IFOV.

Most of the optics designs have had circular focused areas; but the advent of long detector arrays used in the pushbroom arrangement has motivated designers to extend the coverage in one direction by relaxing the requirements in the other direction. The curves "WALRUS" that have been marked in Figure 8-5 represent such a design.

Examination of this graph will show that either wide-angle sensors cannot perform for the application under study, or wide-angle sensors can be considered and all options are open.

Settling the Parameters

Any of the foregoing considerations may result in the elimination of some sensor type. The optics expert may decide that the field of view, aperture size, and weight constraints are not compatible with the use of a wide-angle optics system. Alternatively, a wide-angle pushbroom sensor might be an obvious choice if the number of detectors in the signal-to-noise calculations turned out to be in the order of m, (the number of elements required across the scene), and if the spectral band could be accommodated. More often, the numbers and sizes fall within the equivocal range and the designer analyzes preliminary designs for several types of sensors.

The scanner design involves several more selection processes. If the number of detectors required per band is small, it may still be advantageous to use a larger number and reduce the scan rates. Reduced scan rates ease vibration effects; however the number of lines imaged per scan is restricted by overlap from scan to scan at the edge of the image. Platform motion may impose another restriction on slow scanning.

The pushbroom sensor will be attractive where low noise is needed and scan angles are modest. The primary difficulties with pushbroom sensors result from the large number of detectors involved. For multi-spectral sensing, the problems of spectral separation, cooling, calibration and data processing may be prohibitive.

Frame sensors offer advantages for panchromatic imaging and will be excellent for cartographic mapping as two-dimensional detector arrays are improved.

RESOLUTION AND OPTICAL TRANSFER FUNCTION

Definitions

Resolution originally was defined by astronomers as the minimum angular spacing between two point-sources that could be distinguished or resolved by a telescope. The definition is very specialized for stars against dark backgrounds and is not so useful when applied to image problems in which the sensor should distinguish a range of radiances and a variety of shapes. The term resolution has an intuitive appeal and several practitioners have suggested definitions that would be more applicable to imagers. One example is to state the closest spacing that will be discernible for a periodic pattern of black-and-white bars in the scene when the image is processed. Others use resolution to mean the dimension of the instantaneous field of view. In either case, resolution is most useful as a broad characterization of the class of imager under discussion and does not give a good description of the quality of the sensor.

For design purposes the optical transfer function gives a more comprehensive description. The concept has value as an analytic tool during design, and the modulation transfer function (MTF), which is a part of the optical transfer function, is useful for specification and testing. The optical transfer function (see Overington, 1976) is an analog of the transfer function used in network theory. The network function describes how the network affects the amplitude and phase for each pure frequency fed into the network. The phase and amplitude at the output are plotted against frequency for unit amplitude and zero phase at the input. The optical case uses the concept of spatial frequency introduced by means of special targets placed in the sensor view and the output is the processed image of the test targets. Each target might consist of a series of equally spaced dark and light bars in which the intensity transition from dark to light follows a sine function. The angular spacing between two peaks defines one spatial wavelength and the inverse gives the spatial frequency. If a series of targets is made with the same intensity of dark and light but with different spacing or spatial frequency for each, it would be found that the contrast between peaks and troughs in the processed image becomes smaller as the frequency increases. A plot of the ratio of peak-to-trough in the output versus spatial frequency gives the MTF which is the amplitude portion of the optical transfer function. The curve is normalized to the intensity in the image when viewing a very large area of the dark value and a similarly large area of the light value.

Figure 8-6 shows MTF curves for constituent parts that make up a scanning sensor designed for earth resources. The field stop shows the effect of scanning the instantaneous field of view across the bars. When the angular dimension of a dark and light bar just equals that of the IFOV, the contrast at the output is reduced to zero. Any response to bars representing frequencies beyond

this point does not result in a true image and it is customary to insert an electrical filter in the signal path that cuts off these higher frequencies. The information from the scene has no further value and the high-frequency noise can be diminished. The three-pole Butterworth filter shown was selected for simplicity and smooth phase response across the band. The third curve gives the combined effects of diffraction, optics figure, thermal changes, misalignment tolerances and distortion of the scan mirror caused by reversing forces.

The phase part of the optical transfer function has been lost in the measurement described and is used mostly in analysis. Phase has importance in estimating aliasing effects and also in detailed design of wide-angle optics systems having many elements. Experience shows that simple multiplication of MTFs for individual components gives a very good estimate of the MTF for the overall sensor system and that the phase can be neglected.

The measurement procedure with sinusoidal targets was used to give an operational definition of MTF but is not easily implemented. Other measurements have been developed for particular circumstances. When the entire sensor system is assembled it is convenient to use sets of uniform bars that are alternately opaque and transparent. A bar set is placed in the far field, or more usually in a collimator that simulates a far field. A controlled radiance source is placed behind the bars and sometimes the bars are located on wheels that spin to simulate scanning. A sufficient number of different bar spacings is used to approximate the range of frequencies that is pertinent to the sensor use. The uniform bar structure no longer represents a pure frequency, so it is necessary to extract that portion of the response that is caused by the fundamental frequency either by harmonic analysis or by using a narrow-band filter in the signal channel.

Another method uses the fact that the MTF is the inverse Fourier transform of the image produced by a line source or point source for two-dimensional functions. This method has applicability to optical systems before the detectors are installed. The image at the output of a sensor system that views the line or point is spread into a band or into a blur circle. MTF can be derived by measuring the intensity across the blurred image $S(\alpha)$ and calculating the inverse Fourier transform.

Blur Circle and Equivalent MTF

The blur dimension also offers a useful way of combining many contributors, which have negligible effects individually, to the transfer function. Blur circles were used in Figure 8-5 to describe the optical quality of different types of systems and information for many elements of the composite curve of Figure 8-6 initially were found in terms of blue circle. The MTF was derived by

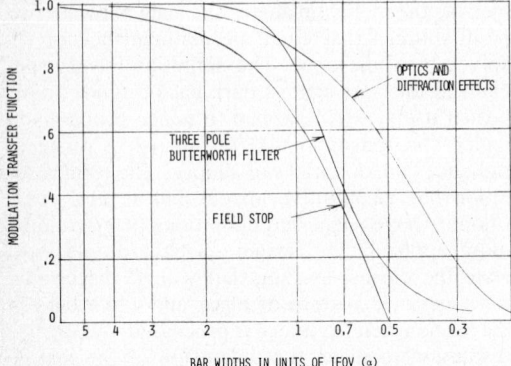

Fig. 8-6. Typical MTF curves for sensor constituents.

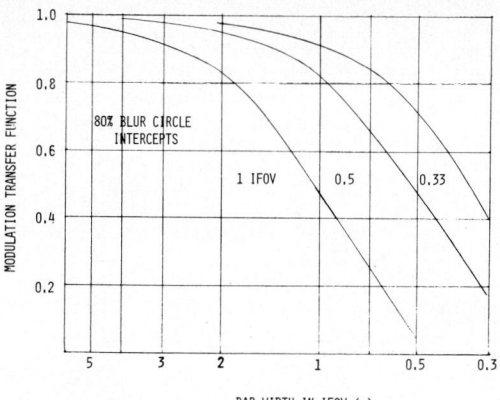

Fig. 8-7. Typical MTF curves for representative 80% blur diameters in terms of IFOV.

applying an inverse Fourier transform to a numerical description of the intensity across the blur circle. Examination of measured blur circles for this type of optical system showed that a Gaussian curve represented an excellent fit. Figure 8-7 shows translation to MTF for three diameters that contain 80 percent of the energy. Unity on the abscissa is the frequency corresponding to test bars that have a width equal to the IFOV and the point 0.5 represents the frequency where no contrast occurs, i.e., a bar pair equals one IFOV. The parametric curves show the effect of blur circles that have widths that are equal to one-half, one-third, and one-fourth of the IFOV. Although these blur-circle diameters include 80 percent of the energy, it should be mentioned that there is little uniformity in stating the cutoff point for energy contained in a blur circle so that the diameter definition should be checked carefully when making evaluations.

Budgeting MTF

Following is an approach for estimating and controlling the modulation transfer function that has evolved in the course of sensor system design. Subsequent measurements that have been applied to the finished system both in the laboratory and from scene analysis after launch have shown excellent agreement with the values predicted using the method.

The *chain* of MTF contributors has been grouped according to the boxes shown below, with the analytic form of the MTF written above each in terms of the angular IFOV, α, and variable θ which is the angle intercepted by one

$\dfrac{\sin \pi \alpha / 2\theta}{\pi \alpha / 2\theta}$	$F(\theta)$	$e^{-.456(c/\theta)^2}$
Field Stop	Electrical Filter	Diffraction Plus All Optics Tolerances

sinusoidal test bar.

The first two terms are major contributors. Practical tolerances on nominal dimensions for detector masks and optical focal length lead to an uncertainty in α of +5 percent which has been translated into a tolerance that must be allowed in the MTF value. The tolerance on filters is set at ± 12 percent for the combined effect of initial circuit elements, temperature drift and aging. This is a very tight tolerance and has consequences on matching detectors for striping effects as well as MTF. The filter shown in the example is known as the "three-pole Butterworth."

MTF for the rest of the contributors was derived from the inverse Fourier transform of the combined 90 percent blur circle and these are listed for a sensor that uses a flat scan mirror in front of a 16-inch Ritchey-Chretien telescope and of 43 microradians. A fold mirror plus a scan-line corrector are in the optical path and detector arrays for each spectral band are located at the focus. MTF for the three cooled detector bands is different from that of the first four bands described here.

Table 8-2 gives the budget for nominal values and worst case tolerance values of elements that make up the blur-circle contribution. A major contributor here is the diffraction effect which varies across the spectral region and is fixed by aperture size. The other terms result from a variety of tolerance effects including distortion of the 21-inch scan mirror caused by scan reversal. The total expected blur circle is calculated by summing all of the first mode distortion (optical power) that can occur simultaneously and root-sum squaring the power term with the others which will have a random variation. The result is then transformed into an MTF curve. Figure 8-8 shows six possibilities that could occur. Various combinations of wide tolerance and diffraction for the different spectral bands can be found by multiplying a curve selected from the top three with a curve selected from the lower two.

Sampling MTF

Another important MTF factor can result from sampling the sensor signals. This has not been included in the sensor budget because the effect is dependent on the number of samples taken in each dwell time and also on the technique used in reconstructing the data. The simplest picture reproduction uses a zero order hold in which each sample interval, T, is represented by a square block of width T and amplitude is the value of the sample. This introduces an additional transfer function of $\sin u/u$ where $u = \pi T/2\theta$. When one sample per IFOV is used, the function has the same form shown for the field stop in Figure 8-6.

More sophisticated processing described in Chapter 17 can reduce the sampling effect significantly. Another means of making the sampling less damaging is to increase the samples, which decreases T. Increased loading of data links

TABLE 8-2
Blur Circle Budget

Contributor		Allocation, μrads	
		Nominal	+ Tol.
Telescope—figure	0.2λ to 0.3λ	7	4
thermal defocus	$0.001''$	7	
misalignment	$0.001''$ to $0.002''$	4	4
thermal gradient	$2°$ to $3°$	7	3.5
Scan Mirror—figure	λ	5	5
thermal mode 1			
	0.25 to $0.32\ \mu m$	4.3	5
dynamic mode 1			
dynamic mode 2	$0.9\ \mu m$	1.4	
Scan Line Corrector & Fold Mirror		3	1.5
SCL + (Fold Mirror)			
Diffraction	$1\ \mu m - 16\ \mu rads$		
	$0.5\ \mu m - 8\ \mu rads$		
Total 90 percent Blue Circle Diameter at 1 micrometer		22	28
Total at $0.5\ \mu m$		17	24

makes this solution less attractive than one that is based upon data-processing improvement.

ESTABLISHING SIGNAL-TO-NOISE REQUIREMENTS

In many instances the user wishes to measure certain spectral reflectance values of scene materials because these characterize the materials regardless of viewing geometry, intensity of the illumination source, or intervening atmosphere. Details of atmospheric effects are described in Chapter 5. Researchers in the field are concerned with the complicated problem of deducing an accurate value for reflectance, given radiance measurements from a sensor combined with ancillary data of atmospheric measurements that might be available. The task for the sensor designer is not so sensitive because it is only necessary to determine the worst-case effect on sensor requirements.

The radiance level intercepted by a sensor detector while viewing a ground patch having reflectance ρ is, in part, a function of ρ but also has a term caused by scattered light which does not relate to ρ and is analogous to "dark noise" or background. The effect of this offset is to raise the requirement for S/N performance of the instrument. The sketch in Figure 8-9 shows sun irradiance in a spectral band, H_o, traveling directly through the atmosphere to the ground patch. At this point the solar energy has been reduced by $\cos z\ T^{\sec z}$ which represents transmittance T through one atmosphere adjusted for slant distance, $\sec z$ and by $\cos z$, the flux intercepted by the patch. In addition, some energy has been scattered from all directions onto the patch. Both of these are reflected up to the sensor through one more atmosphere and diminished again by T. The sensor has been positioned directly above the patch for simplicity and further geometric modifi-

cations must be made for non-normal viewing situations. Energy at the sensor aperture has been translated to radiance, L. The additional term, L_a, is caused by many types of scattering from molecules and aerosols. The radiance at the sensor has the general form shown in the sketch. Tables for H_o are available (Elterman, 1963) H_{sky}, and L_a have been measured for some environments (Fraser, 1974). Signal level from the detector is a linear function of radiance and the noise introduced by the sensor should not exceed the increment of radiance caused by the specified reflectance increment. Setting S/N equal to L/NEΔL in terms of ground reflectance gives,

$$\left(\frac{S}{N}\right) \text{required} = \frac{\rho}{\Delta\rho}$$

$$+ \frac{1}{\Delta\rho \left[\dfrac{L_o}{L_a} \cos z\ T^{\sec z} + 1 + \dfrac{L_{sky}}{L_a} \right]} .$$

The equation must be examined to determine which realistic combination of lighting and atmosphere entails the highest requirements level. The trend of the S/N equation can be seen more readily if one observes that the L_a and L_{sky} are coupled and for the rural midlatitudes, pertaining to crop assessment, L_{sky} is roughly 72 percent of L_a. Factors that increase S/N_{rqd} are: dense atmosphere, which reduces T and increases L_a, and large Z, sun angle, which further reduces the second denominator. There are two trends in these characteristics which tend to balance; that is, the worst lighting occurs in winter and improves as the seasons advance, while the clearest atmosphere occurs in winter and worsens as the seasons advance.

Table 8-3 gives the transfer equations that have been calculated for Thematic Mapper spectral bands relating scene reflectance to radiance ex-

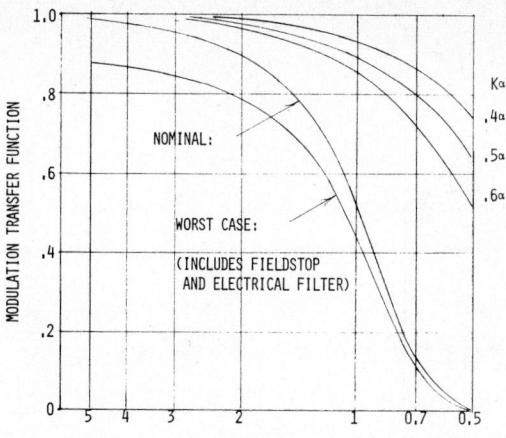

BAR WIDTHS IN UNITS OF IFOV (α)

Fig. 8-8. MTF curves for nominal and worst-case budget values.

pected at the aperture. The sun zenith angles for a 9:15 a.m. orbit were selected to represent highest latitudes that have emergent crops. The dry-air column is included to represent the maximum signal that must be accommodated in the signal processing. Each remote sensing application must be calculated specifically and the effect of atmosphere can be much greater in urban areas where atmospheric contaminants abound (Flowers et al., 1969). On the other hand, reflectances may be higher and coarser incremental distinctions may be tolerated.

The sensor generally has the lowest signal-to-noise performance at low signal levels; therefore the lower end of the required reflectance scale will generally give the critical design point.

It will become apparent, through the use of transfer functions similar to those shown in the table, that the required signal-to-noise ratio must exceed $(\kappa\rho + L_a)/\kappa\,\Delta\rho$ for the specified $\Delta\rho$, the

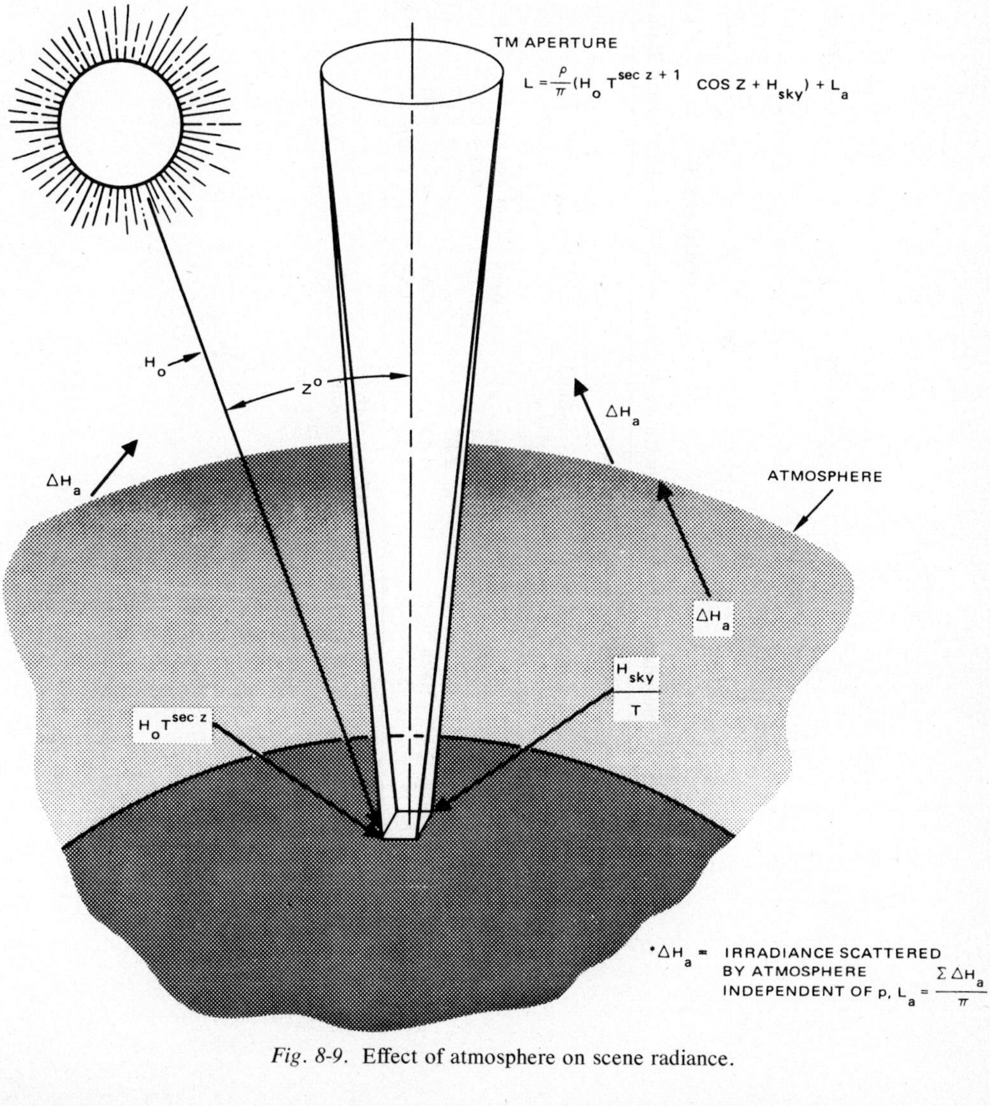

TM APERTURE

$$L = \frac{\rho}{\pi}(H_o\,T^{\sec z + 1}\,\cos z + H_{sky}) + L_a$$

ΔH_a

ATMOSPHERE

ΔH_a

ΔH_a

H_{sky}

T

$H_o T^{\sec z}$

H_o

z^o

*ΔH_a = IRRADIANCE SCATTERED BY ATMOSPHERE INDEPENDENT OF ρ, $L_a = \dfrac{\Sigma\,\Delta H_a}{\pi}$

Fig. 8-9. Effect of atmosphere on scene radiance.

TABLE 8-3

Summary of Transfer Equations: Radiance in Thematic Mapper Bands $= \kappa\rho + L_a$ mw·cm^{-2}·ster^{-1}

| Band | Solar Energy, No Atmosphere | | Spring (Fall) | | Summer | | Rayleigh, Dry Air | |
	Solar Irradiance mw·cm^{-2}·μ^{-1}	Radiance in Band mw·cm^{-2}·ster^{-1}	$Z = 49°$* T		$Z = 43°$** T		$Z = 34°$*** T	
0.45–0.52	207.4	4.62	0.68	$1.28\rho + 0.20$	0.64	$0.60\rho + 0.33$	0.86	$2.96\rho + 0.3$
0.52–0.60	178.3	4.54	0.72	$1.44\rho + 0.19$	0.68	$1.73\rho + 0.27$	0.92	$3.32\rho + 0.26$
0.63–0.69	148.3	2.83	0.79	$1.07\rho + 0.07$	0.75	$1.27\rho + 0.11$	0.96	$2.20\rho + 0.08$
0.76–0.90	108.8	4.85	0.83	$2.0\rho + 0.08$	0.78	$2.23\rho + 0.08$	0.98	$3.80\rho + 0.07$
1.55–1.75	22.3	1.42	0.89	0.69ρ	0.84	0.77ρ	1.0	1.20ρ

* Sun angle in April at 42° Latitude for 9:15 a.m. orbit.
** Sun angle in June at 60° Latitude for 9:15 a.m. orbit.
*** Highest sun angle encountered in 9:15 a.m. orbit.

lowest required ρ and the viewing conditions that cause the highest ratio of L_a/k.

SIGNAL-TO-NOISE PERFORMANCE

System Equations

The signal-to-noise ratio for the sensor is calculated at a point where the signal has been amplified enough to dominate subsequent amplifier noise, usually at the output of the preamplifier. Other sources of noise in the system occur from sampling and binary encoding, data links, and signal processing for end-use. These factors can be combined later with sensor noise for a full system evaluation.

The signal level for a generalized sensor is,

$$S = L \bar{S} \, \tau_0 \, A_{eff} \, \alpha^2$$

where L is radiance in band in watts ster^{-1} cm^{-2}, $\bar{S}$ is detector responsivity in amp/watt (or volt/watt), τ_0 is optical transmission including reflectances, lens attenuation and filter, A_{eff} is the effective aperture corrected for obscuration and scan variation, α^2 is angular extent of field stop in steradians. The only factor that has electrical frequency dependency is the detector responsivity.

Noise, on the other hand, is strongly dependent on the electrical frequency, and the form is a composite of several curves which vary with the detector style as is shown in the section entitled "Combined Noise for System Evaluation." For this discussion, a simplified form is given that permits the electrical bandwidth, Δf, to be factored. Isolation of bandwidth is important for comparing sensor types since the effect of bandwidth produces the primary difference among them. The simplified form can be written as

$$N = \sqrt{2eS + I_0^2} \, (\Delta f)^{1/2}$$

where e is the charge on an electron viz 1.6 $\times$ 10^{-19} coulomb. The first term arises from the statistical uncertainty of photon arrival that varies with signal level and the other terms are independent of signal.

The approximation that allows bandwidth to be factored simplifies manipulation of system parameters. After parameters have been selected, an accurate estimate of performance will require that the actual noise characteristics be combined with the electrical bandpass shape and integrated to determine the *rms* noise current (or voltage).

Determining The Number of Detectors

To make this determination, one must combine all of the signal and nose factors. This is done by lumping all of the terms as $\bar{X}$ that pertain to optical losses, electrical filters, spectral filters and scan efficiency using the relation

$$\frac{S}{N} = \frac{\bar{X} L \bar{S} A_{eff} \, \alpha^2}{\Delta f^{1/2} \left[I_0^2 + 2e \, \bar{X} L \bar{S} A_{eff} \, \alpha^2 \right]^{1/2}} \cdot$$

Electrical bandwidth, Δf, in inverse seconds, depends on the number of detectors used in a spectral band, the solid angle Ω that must be imaged per second, the instantaneous field of view for one detector and, for a scanner, a scanning efficiency factor. These factors are shown in the equation

$$\Delta f = \frac{\Omega/\text{sec}}{2\alpha^2 \, k_{eff}} \, ,$$

where n is the number of detectors. If the solid angle per second is not constant, then the electrical bandwidth must be sufficiently large to accommodate the fastest rate of solid angle, or $d\Omega/dt$.

As an example, consider a vehicle moving at velocity V units per second at an altitude H above the earth and with a requirement to image a swath W across the track. One must transform V and W at distance H into radians and find the number of elements that must be imaged per second, $VW/H^2\alpha^2$. The least bandwidth required to accommodate the number of elements will be half the elements per second with additional allowances for scan efficiency, electrical filter shapes, etc. For most imaging applications a calculation will usually show that the signal gathered using a reasonable aperture will not exceed the noise in this single detector bandwidth with sufficient margin for good imaging. The noise, and hence the bandwidth, can only be reduced by imaging with several detectors at once. This can be accomplished by scanning several detectors across the swath at once or by sweeping a fixed array along the path. Another option is offered through the use of imaging charge-coupled devices arranged along the scan direction in a time-delay integration mode. When this option is used, however, the noise calculations do not fit easily into the form of equations given here.

In general, an increase in the number of detectors decreases the bandwidth linearly and the *rms* noise is decreased as the square root of n. Sensor selection can then be reduced to the balance between optics size and number of detectors inherent in the relationship,

$$A_{eff} \sqrt{n} \geq \text{constant times required } S/N.$$

The designer now begins a balance between optics size needed for A_{eff} and the number of detectors that can be supported in terms of calibration, reliability, testing and signal processing. A fixed array is clearly indicated if the required n is in the same order as the number of resolution elements across the track and if an array of width n can be accommodated by reasonable optics. Often the number n is found to be an order of magnitude lower than W/H in which case a scan hybrid with the attendant small angle optics offers an attractive alternative.

A calculation of noise for the entire system must include sources that follow the sensor, such as signal encoding or data-link noise for analog transmission, tape recorder effects, and data processing. Discussions of these will be found in Chapter *16*. It is hoped that these added noise factors can be kept low compared with those of the sensor, but it may be necessary to perform a further iteration on the sensor sizing if a final calculation of the complete system reveals excessive noise.

Selection of Electrical Filter

For most applications the ideal filter would have uniform phase response and unity transmittance across the band of interest and zero transmittance everywhere else. Another approach would be to require the filter to compensate for other elements that make up the MTF. Neither of these ideals is possible although either could be approximated with an enormous penalty in circuitry. The primary reason for avoiding the use of many circuit elements is that all detector channels should be as nearly identical as possible, which is not consistent with the use of large numbers of components and with aging effects. Other factors may well be the reliability, cost and weight penalties.

Once the degree of complexity that can be tolerated has been established the filter design must be slanted according to the stress of requirements. In order to enhance the MTF it is desirable to shift the cut-off to higher frequencies; whereas curtailing noise is best served by moving the cut-off to lower frequencies. A very sharp fall-off is useful for reducing noise and also aliasing signals; however, very sharp fall-off may cause undesirable overshoot when the scene contains an abrupt transition. Table 8-4 shows trends in noise, bandwidth, MTF, overshoot and aliasing for a series of Butterworth-Thompson filters. The *m*

parameter has the effect of varying the filter from the original Butterworth ($m = o$), which has maximally flat amplitude response with poor phase flatness to the Bessel filter ($m = 1$) which has maximally flat delay with poor amplitude flatness. The intermediate values of *m* combine good qualities of the two gradations for two, three and four pole networks which are identified under "order." For generality, the bandwidths are normalized to the half-power which might be set to correspond to a spatial frequency of $(2\alpha)^{-1}$. The actual positioning of the roll-off frequency is an option of the designer for the MTF and noise compromise mentioned above. The advantage of four poles over three poles has been found to be marginal for most sets of requirements.

The inclination toward simplicity could be reversed when printed circuits of high quality can be fabricated for the frequencies and circuit constants required.

OTHER SOURCES OF RADIOMETRIC ERROR

Electrical noise is emphasized in previous sections and in detector descriptions that follow because of the strong influence on the type and size of sensor that will be used. Several other factors contribute to radiometric error and special attention may be required to keep these from degrading performance.

Two factors that have general applicability are described below. Others that should be reviewed for applicability for particular designs are: polarization bias, electrical and radiant crosstalk, geometric distortion and registration between spectral bands. Every testing program should include provisions for detecting power supply noise, which usually has systematic characteristics in

TABLE 8-4

Butterworth Filter Characteristics

Filter Type	Order	Overshoot, %	Noise Bandwidth Factor	Rise/Fall Times 1%, μsec	Frequency with Respect to 3 dB Bandwidth For MTF =			Percent Aliasing Noise (Scene—6 dB/Octave)	
					0.94	0.89	0.79	1 Sample/ RFOV	1.5 Samples/ RFOV
m = 0.2	2	3.05	1.11	6.9/5.7	0.22	0.66	0.87	8.5	5.0
m = 0.4	2	1.89	1.12	6.9/4.6	0.17	0.66	0.84	11.1	6.7
m = 1.0	2	0.43	1.15	6.9					
m = 0	3	8.15	1.05	7.2/14.4	0.48	0.81	0.93	3.3	1.5
m = 0.3	3	4.75	1.04	7.2/5.1	0.23	0.72	0.87	4.4	2.1
m = 0.4	3	3.87	1.04	7.2/4.8	0.18	0.67	0.87	4.9	2.3
m = 1.0	3	0.75	1.07	6.9	0.14	0.60	0.84	7.5	3.8
m = 0.4	4	5.04	1.00	7.5/4.5	0.17	0.66	0.87		
m = 0.5	4	4.01	1.00	7.5/4.2	0.17	0.66	0.87	3.2	1.2
m = 0.6	4	3.13	1.01	7.5/3.6	0.17	0.66	0.85		
m = 1.0	4	0.84	1.05	7.5	0.14	0.60	0.84	5.7	2.7

distinction to the random nature of the sources listed here. Calibration is a fundamental factor in radiometric accuracy and has been described in a separate section at the end of Chapter 8.

Square Wave Response Error

MTF is usually associated with picture clarity; however, a radiometric error called square-wave response error arises from the contaminating effect of one field of the scene on an adjacent field. This error varies with the field size, which relates to spatial frequency, and with the difference in radiance levels between fields.

Figure 8-10 shows the square-wave response required to reduce errors to levels shown on the abscissa when viewing checkerboards of alternating radiance levels. Because the error is greater for larger differences between adjacent radiances, it is important to determine the range of radiances that must be identified. The two curves given correspond to ratios of ±30 percent and ±15 percent which were found to represent the one sigma variation of growing materials in the 0.7- to 1.0-μm and 0.5- to 0.7-μm parts of the spectrum. This equation represented by the curves is,

$$error\ ratio = 0.5\ [(1-m^2)\ (k^2 - 1)$$

where m is the value for the square-wave response function for bar charts, k is the ratio of the upper one-sigma radiance to the mean, and k^{-1} is the ratio of the lower one-sigma radiance to the mean. The error calculated in this way represents an *rms* error for the agricultural scene and may be root-sum squared with other random noise. If this error is to be kept on a level with other error sources, it will be necessary to control the square-wave response at frequencies corresponding to the minimum size of field to be measured. This implies that fields must have dimensions larger than the IFOV by factors of 7 to 10 in order to ensure errors below 1 to 2 percent.

Scattering

Scattering of radiance within the sensor also contributes to noise. Good optics-design practice requires efforts to ensure that baffles and support structures will not cause radiant energy to be scattered into the detectors. This can pose a difficult problem over wide spectral bands, and especially at long wavelengths where cooled elements may be involved.

Reflecting surfaces must be selected and fabricated to satisfy the scattering requirements and, after installation, strict cleanliness must be maintained to avoid buildup of contaminants that could scatter energy.

The permissible level of scattering should be allocated along with system noise and square-wave response error at a level that is compatible with the radiance accuracy required. The error caused by a scattering coefficient ϵ is similar to the square-wave response error in that it increases with the spread between the radiance being measured and the average radiance entering the aperture. If the true radiance L_o filled the aperture during calibration, the voltage measured will be

$$V_o = K\ L_o + K\epsilon\ L_o,$$

where K is the system transfer function. When a spot of radiance L_o is surrounded by a scene of L_o $(1 + \Delta)$, the measured voltage becomes,

$$V_m = K\ L_o + K\epsilon\ L_o\ (1 + \Delta).$$

The difference of $(V_m - V_o)$ or scattering error will be $\epsilon\Delta$. If $\pm\Delta$ represents the one-sigma variation in the expected radiance from the scene, then the error calculated from this expression for scattering error will represent an *rms* noise and may be root-sum squared with the other random errors.

SEPARATION OF SPECTRAL ENERGY

The many uses of multispectral images are described in Volume *II* and requirements have become quite specific. In the early days experimenters requested as many narrow bands as could be obtained. Consequently, aircraft radiometers made use of gratings and other dispersive elements similar to those developed for laboratory spectroscopy. As the areas to be imaged became larger and more distant it became necessary to increase the energy and simplify the sensors by settling on fewer and broader bands, which coincides with increased user sophistication and the need to reduce data volume.

Spectral bands are most efficiently separated by means of dichroic mirrors or by spectral bandpass filters on detectors that are slightly displaced from each other along the path of image motion. The first method is exemplified by the Skylab multispectral scanner and the spatial method by the Thematic Mapper (Chapter *12*). Thematic Mapper users can tolerate misalignments of only 10 percent of IFOV for bands used for agriculture classification. The Landsat MSS (as described in Chapter *12*) uses a variation of spatial separation in that optical fibers are arranged in an array at the focus and energy is carried back to detectors on

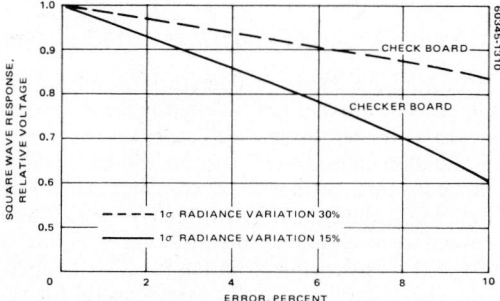

Fig. 8-10. Square-wave response required to ensure error level for checkerboard patterns.

which filters are mounted. Energy for the long IR band is taken from an adjacent region in the focused area and directed through a filter along another optical train to the cold finger of a radiative cooler (90 K) where two detectors are situated. The visible bands served by optical fibers have a measured maximum misregistration of 20 percent of an IFOV. The error is repeatable and is caused by optical fiber misalignment plus varying scan-rates across the scan.

In either method, careful attention must be given to assuring good registration between spectral bands. Repeatable errors in registration will be caused by irregularities in detector arrays and by alignment imperfections in the optical elements when different spectral bands are split along separate paths. Such errors can be calibrated and corrected in processing; however, the corrections necessitate interpolation and resampling of data and great care should be exercised to minimize these errors in the sensor. A similar type of correctable registration error arises from using spatial/spectral separation with a sensor having a systematic scan variation. Data sampling offers an interesting method of correcting this at the source. Usually a multispectral sensor is sampled with timing that catches the different spectral bands as each images the same ground spot. When the samples are processed it is possible to compare the samples that are displaced by a predetermined number for each spectral band without interpolating. Such a scheme is disturbed if the scan rate varies and the sampling rate is constant. By controlling the sampling rate with a position pick-off from the scan mechanism, one can sample in equal angle increments instead of equal time increments, after which the timing can be made uniform for transmission purposes by using a reasonably short in-line storage buffer. Once again, the samples that are separated by a predetermined number correspond to the same spot imaged in different bands.

Spectral bands having remotely located detectors, such as the cryogenically cooled band described for MSS, always present difficult registration problems. Studies comparing refractive and reflective optics trains show that reflective methods have lower attenuation and more predictable thermal change. Two methods which may become more attractive as technology advances are IR optical fibers and heat pipes that conduct heat from the detectors to the remotely located cryogenic cooler.

Temperature change and vibration are serious causes of registration errors as well as general optical defocusing and misalignment. After all of the best materials practices are studied, the designer may elect to use active temperature control to insure stability. Invar and ceramic structure for reflecting surfaces may add so much weight that a temperature control system is preferable. The relative advantages of the two approaches will shift either way as composite materials are improved and temperature systems become more sophisticated.

A novel spectral sorting method utilizes a laser as a local oscillator in a direct analog to a heterodyne receiver. The desired segment of the spectrum is filtered from the mixed signal and processed. This belongs to a new category of active spectral separators including acoustic filters that offer the possibility of variable spectral bands. Details of amplitude uniformity, conversion efficiency, reliability and packaging remain to be worked out.

DETECTORS AND IMAGE DEVICES

A critical step in remote sensing is the transduction of scene radiant energy into a form suitable for processing and transmission, generally an electrical waveform. This detector interface is of paramount importance, as it is generally here that the overall system sensitivity and the dynamic range are established. Detectors for scanning or direct imaging (framing) represent the two major categories of transducers.

In the scanning approach, performance can be evaluated in terms of an individual detector (which may be a member of an array), as the transducer element, transforming the energy received as the scene image moves by into a single, time-varying electrical quantity. On the other hand, the imaging device or area sensor is briefly sensitive, capturing a snapshot of the scene in a spatial distribution of some electrical property, which is then scanned within the device to obtain an electrical waveform.

The fundamentals of detectors and their technology are treated in detail in Keyes, 1977. Fundamental analyses of electro-optical sensors are found in Holter, 1963; and Kruse, 1962. *The Infrared Handbook,* Wolfe, 1978, is comprehensive in basic concepts, materials and techniques. A special issue of IEEE Trans. on Electron Devices, 1980, dealing with infrared materials, devices and applications, has wide coverage in these areas.

As an example of the selection of detector types, Table 8-5 shows the considerations that were involved when choices were being made for the Thematic Mapper, representative of the state of the art in 1976.

DETECTORS

The principal types of detectors useful for remote sensing are thermal and photon (or quantum) detectors. In the thermal detector, absorbed signal radiation causes a temperature increase which changes some property of the detector. For example, a thermocouple, which is a junction between two dissimilar metals, will generate a voltage that is dependent on the junction temperature. An example which has been useful for remote sensing is the thermopile, which consists of a number of thermocouples connected in series.

TABLE 8-5

Thematic Mapper Detector Selection

Wavelength Region	Detector Type	Advantage	Disadvantage	Remarks
		Uncooled		
Short, <1 μm	Photomultiplier tube	Compact, low noise electron amplifier.	High voltage power supplies. Gain variations. Fiber optics required.	
	Si photodiode	High quantum efficiency Linearity Stable responsivity. Focal plane array available.	Requires high impedance close spaced preamp circuitry for ultimate performance.	
		Cooled		
Medium, 1 μm to 3 μm	Ge photodiode		Fabrication techniques not well developed. Required input circuits on cold stage.	Ge responsivity dropping rapidly at 1.7 μm. Transparent at longer λ
	Ge photoconductor		Fabrication techniques need some development.	
	InSb photodiode	Thoroughly developed, exceeds D* requirement.	Requiring input circuit on cold stage. Require blocking filter for longer λ	
	InAs photodiode	Higher temperature operation.	Fabrication techniques need development.	
	HgCdTe photodiode	Higher temperature operation.	Fabrication techniques need development.	
Long, 10 to 13 μm	HgCdTe photoconductor	Thoroughly developed, exceeds D* requirement.		
	PbSnTe photodiode		Available cooling is marginal, bandwidth problem.	

In the photon detector, an absorbed photon of the signal energy will activate an electron, affecting the electrical characteristics of the detector. An example is the photoconductor, made of a semiconductor material in which the release of an electron increases the electrical conductivity.

In general, the photon detectors show faster response because of the more direct process involved. In some applications a thermal detector has an advantage over a photon detector because the latter would require elaborate refrigeration means. A design problem unique to the thermal detector is the mounting of the sensitive element, which must be thermally isolated to allow its temperature to change, while being securely mounted to avoid microphonic effects. The thermal detector also responds to photons from the scene, and the voltage generated depends on the energy of the photon, which is

$$E = hc/\lambda$$

where E is the energy of a photon, h is Planck's constant, c is the speed of light and λ is the wavelength. A thermal detector with constant responsivity to power (energy rate) per unit of wavelength then requires a proportionately greater number of photons at longer wavelengths than at shorter for the same response. Also, this equation shows that photon detectors with equal responsivity to photons of different wavelengths will have greater responsivity to power of longer wavelengths. This accounts for the slope on the left of the peak responsivity (based on power) in Figure 8-12.

The speed of response in quantum-type detectors is a function of various recombination mechanisms, dielectric properties, and transit-time effects. Many detectors have several time constants. Regardless of the mechanism involved, most manufacturers tend to quote a time constant, τ, as the time it takes for the output to reach within $1/e$, or to within 63 percent of its steady-state value after the light is turned on or off. If the detector is modulated at a frequency of $1/(2\pi\tau)$, the output amplitude will be down 3 dB from its low-frequency amplitude. This degraded response is undesirable for good imaging. Hence, the upper frequency of the video bandwidth should be restricted to a lower value.

A sampling of the available detectors is shown in Table 8-6 and in Figures 8-11, 8-12, 8-13 and 8-14, with quantum efficiency and D^* shown as a function of wavelength or temperature. The meaning of these quantities is discussed in later sections.

Photon Detectors

Photoconductor and Photodiode

The photoconductive and photodiode detectors are made of semiconductor materials, in which most of the electrons are bound in the crystal lattice. The quantum of energy for each photon may be imparted to an electron to break it free to move through the lattice. The hole left behind in the lattice can also appear to move. After a lifetime, which is a characteristic of the material, electrons and holes recombine. A pure semiconductor will require a particular minimum photon energy, and therefore a maximum wavelength, to release an electron. Traces of impurities in semiconductors can be used to change the crystal-lattice such that some bonds may be weaker so that electrons are more easily freed (n-type semiconductor) or such that electron acceptors are present, and holes are more easily produced (p-type semiconductor). The semiconductors having the trace impurities will respond to a longer wavelength than the pure material, and are known as extrinsic. The pure materials are called intrinsic. The photoconductive type may be single crystal or may consist of multiple crystallites.

Photodiode detectors, which may be of the same material as some photoconductive detectors, each contain a junction, generally between p- and n-type material in a single crystal. Such a structure will form an electric field across the junction region which will separate charges to cause an external voltage to appear when photons create electron-hole pairs in the region. The current-voltage characteristics are as shown in Figure 8-15. In addition to the p-n junction type, the useful photodiodes include p-i-n, with a region of intrinsic material between the p and the n which increases the region with electric field. The Schottky barrier photodiode has a metal-semiconductor interface, where electrons are emitted into the semiconductor when excited by a photon. The metal may be thin enough for the photon to enter from that side, or entry may be through the semiconductor if it is transparent.

The energy-level diagram is useful for understanding detectors and other electronic devices. Grove (1967) treats this topic in some depth. In the diagram, distance is shown horizontally and energy vertically. Figure 8-16 is an example of a junction diode, with the upward direction corresponding to greater electron energy. The unshaded central band is the forbidden gap, where energy states are not available in the pure material.

The Fermi level is that at which the probability of an electron occupying an energy state is one-half. The nearer this level is to the conduction band, the greater the probability of finding electrons in that band. In the figure, an electron is shown jumping the gap upon the absorption of a photon.

The photodiode can be operated with a reverse bias voltage to increase the size of the region near the junction where an electric field is present, to give a larger volume from which electrons or holes will be swept across the junction. This is useful if the photon absorption coefficient of the semiconductor is low so that a long path in the electric field region is utilized to collect the carriers generated. Reverse bias also reduces the junction capacitance.

Operation of a photodiode with reverse bias is called the photoconductive mode, not to be confused with a photoconductor. The photovoltaic or zero bias mode of operation may be used to minimize noise generated by the photodiode or to avoid heating of the focal plane area.

It is convenient to think of the photodiode as a current generator because the responsivity, expressed as a current change per unit of incident radiant power, is relatively independent of voltage

TABLE 8-6

Significant Characteristics of Some of the More Common Electro-Optical Remote-Sensing Detectors

Name	Available Forms	Spectral Range	Figure of Merit	Time Constant	Operating Temperature	Remarks
			Quantum Efficiency			
PHOTOCATHODES Solar Blind	Rb_2 Te	UV atmo. cutoff to 0.325 μm	0.08 @ 0.25 μm	$<1 \times 10^{-9}$ sec	Ambient	Not sensitive to solar radiation transmitted through atmosphere.
	Cs_2 Te	UV atmo. cutoff to 0.375 μm	0.08 @ 0.25 μm	$<1 \times 10^{-9}$ sec	Ambient	
Visible	Cs-Sb	UV window cutoff to 0.650 μm	0.182 @ 0.35 μm	$<1 \times 10^{-9}$ sec	Ambient	Chemicals used to make S-4, S-5, S-11, S-13, S-19 detector surfaces.
	Ag-Bi-O-Cs	0.320- 0.680 μm	0.055 @ 0.45 μm	$<1 \times 10^{-9}$ sec	Ambient	S-10 material.
	K-Cs-Sb	UV window cutoff to 0.65 μm	0.312 @ 0.38 μm	$<1 \times 10^{-9}$ sec	Ambient	Bi-alkali
	Ga-As-P	UV window cutoff to 0.75 μm	0.21 @ 0.30 μm	$<1 \times 10^{-9}$ sec	Ambient	Periodic Table III-V Type
Near-IR	Na-K-Cs-Sb	UV window cutoff to 0.93 μm	0.239 @ 0.42 μm	$<1 \times 10^{-9}$ sec	Ambient	Multialkali Type, S-20, S-25 Detectors. Varying mix or alkali metal will usually increase long wavelength response and leakage noise.
	Ag-O-Cs	0.45 to 1.10 μm	0.0043 @ 0.80 μm	$<1 \times 10^{-9}$ sec	Ambient	Decreasing maximum quantum efficiency. S-1 photocathodes. Longest wavelength response.
	Ga-As	UV window cutoff to 0.90 μm	0.47 @ 0.30 μm	$<1 \times 10^{-9}$ sec	Ambient	Periodic Table III-V Type absolute. Sensitivity very uniform over its range.
	Ga_{1-x} In_xAs	UV window cutoff to 1.10 μm	0.13 @ 0.40 μm	$<1 \times 10^{-9}$ sec	Ambient	Periodic Table III-V Type with variable red cutoff. The more In the material contains the longer the red cutoff. These tubes have slightly better response than the S-1 and much lower dark leakage.

TABLE 8-6 (Continued)

Significant Characteristics of Some of the More Common Electro-Optical Remote-Sensing Detectors

Name	Available Forms	Spectral Range	Figure of Merit	Time Constant	Operating Temperature	Remarks
PHOTOCONDUCTORS Visible			Responsivity			
	Sb_2S_3	0.38 to 0.70 μm	0.11 amp W^{-1}	<1/30 sec	Ambient	Photosurface for vidicon television cameras.
	PbO	0.38 to 0.70 μm	0.12 amp W^{-1}	>1/30 sec	Ambient	Photosurface for vidicon television cameras.
	ASOS Antimony Sulfide Oxysulfide	0.45 to 0.76 μm	3 amp W^{-1}	2 sec	Ambient	Photosurface for very high resolution. Television tubes (the Return Beam Vidicon).
	Silicon	0.30 to 1.20 μm	6 amp W^{-1} @ 0.80 μm	10^{-9} sec	Ambient	Used in most commercial photodiodes.
			D*(cm Hz $\frac{1}{2}$W^{-1})			
Near-IR	PbS	0.50 to 3 μm	1×10^{11} 6×10^{11} 1.5×10^{11} @ 2 μm	1 msec	Ambient 196 K 77 K	
	InAs	1 to 3 μm	3×10^9 2×10^9 4×10^9	1 μsec	Ambient 196 K 77 K	Photodiodes in photovoltaic mode.
	InSb	1 to 5 μm	5×10^{10} @ 5 μm	1 μsec	77 K	Photodiode used in photovoltaic mode.
	Ge(Au) Gold-doped germanium	1 to 10.6 μm	5×10^9	<1 nsec	77 K	
Long Wavelength IR(Intrinsic)	Hg$_{1-x}$Cd$_x$Te Mercury cadmium telluride	1 to 5 μm 3 to 8 μm 8 to 14 μm	2×10^{11} 8×10^{10} 1×10^{10}	~500 nsec	77 K	x = 0.40 x = 0.30 The properties of this material depend on composition, so that no general specification can be stated. Material is important because it allows operation in the 8- to 14-μm region with only liquid nitrogen cooling.

Category	Material	Wavelength	D^*	Time constant	Temperature	Comments
	$Pb_{(1-x)} Sn_x Te$ lead tin telluride	8 to 14 μm	10^{10}	~500 sec		The properties of this material vary with composition. In the various wavelength regions its detectivity is the same as HgCdTe. Effort in this material continues. Its use in detectors exhibits greater uniformity than HgCdTe.
Long Wavelength IR (Extrinsic)	Ge(Hg) mercury-doped germanium	2 to 14 μm	3×10^{10}	<1 nsec.	5 K	Infrared photon actually ionizes an impurity level.
	Ge(Cd) cadmium-doped germanium	2 to 23 μm	$>10^{10}$	<1 nsec.	5 K	
	Ge(Zn) zinc-doped germanium	2 to 40 μm	$>10^{10}$	<1 nsec.	5 K	
	Silicon doped with As (arsenic) B (boron) Ga (gallium) Sb (tin)	2 to 19 μm and beyond, depending on dopant	$>10^{10}$	<1 nsec.	<5 K	Silicon makes a better host than germanium because of lower index of refraction and dielectric constant.
	GaAs doped with Ge (germanium) and others	2 to 19 μm depending on dopant	$>10^{10}$		<5 K	Very new materials should be better than silicon for the same reasons silicon is better than germanium.
Longer Wavelength IR (Extrinsic)	Ge:Au	1 to 10 μm	2×10^{10}	65 K to 90 K		Name convention is (host material): (dopant).
	Ge:Hg	5 to 14 μm	3×10^{10}	35 K		
	Ge:Cd	10 to 23 μm	2×10^{10}	25 K		
	Ge:Cu	10 to 30 μm	3×10^{10}	15 K		
	Ge:Zn	15 to 40 μm	1×10^{10}	7 K		
	Si:In	3 to 7 μm	8×10^{10}	60 K		
	Si:Ga	10 to 16 μm	3×10^{10}	30 K		
	Si:As	14 to 24 μm	3×10^{10}	30 K		
	Si:P	10 to 29 μm	3×10^{10}	30 K		

** Below this line, the value of D^* is given for a 60° FOV.

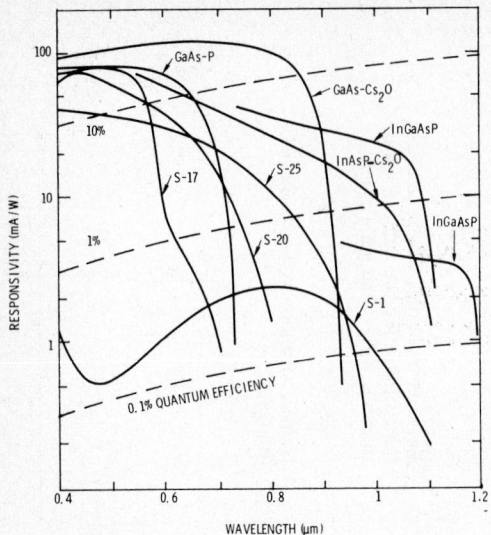

Fig. 8-11. Responsivity vs. wavelength for several representative photocathodes (after Keyes, 1977; courtesy of Varian, LSE Division).

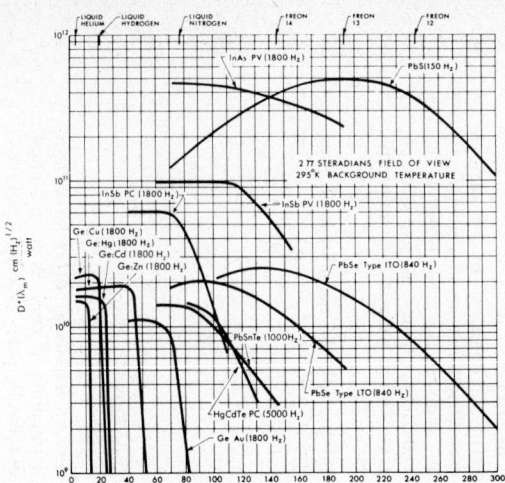

Fig. 8-13. Dependence of detectivity on operating temperature for typical infrared detectors. (Courtesy of Santa Barbara Research Center).

and of radiant power magnitude. The signal current produced by a photodiode detector is

$$I = \eta e N_\lambda$$

where η is the quantum efficiency, e the electronic charge and N_λ the number of photons of wavelength λ absorbed per unit time. The photoconductor signal current at low frequencies is

$$I = \eta e N_\lambda G$$

where G is the photoconductive gain, which is the ratio of the lifetime of a majority carrier to the transit time between electrodes. This gain does not generally represent a performance advantage over photodiodes when noise is also considered.

Photoemissive

In the photoemissive detector, an especially designed surface (a photocathode) is mounted in a

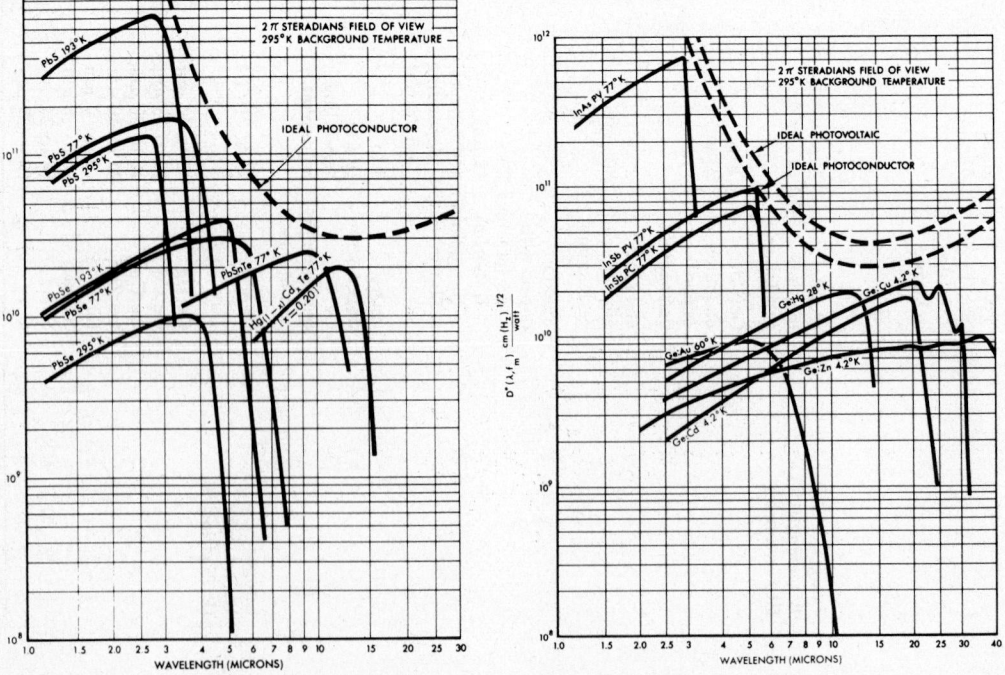

Fig. 8-12. Detectivity (D^*) for representative quantum detectors in the infrared. (Courtesy of Santa Barbara Research Center).

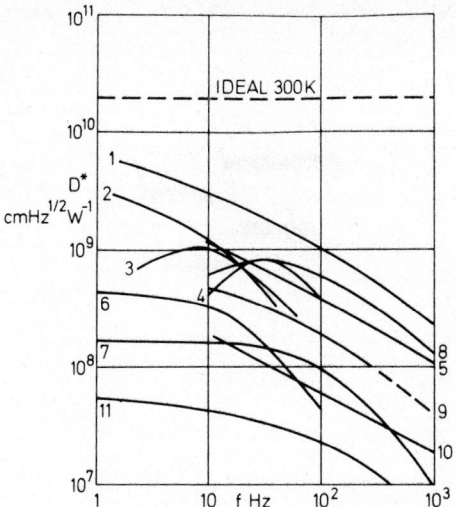

Fig. 8-14. Performance of uncooled thermal detectors. 1, 4, 5, 8, 9, and 10 are pyroelectric detectors; 2 and 6 are thermopiles; 3 is a Golay cell (employs gas expansion); 7 and 11 are bolometers (after Keyes, 1977).

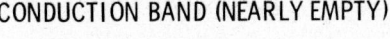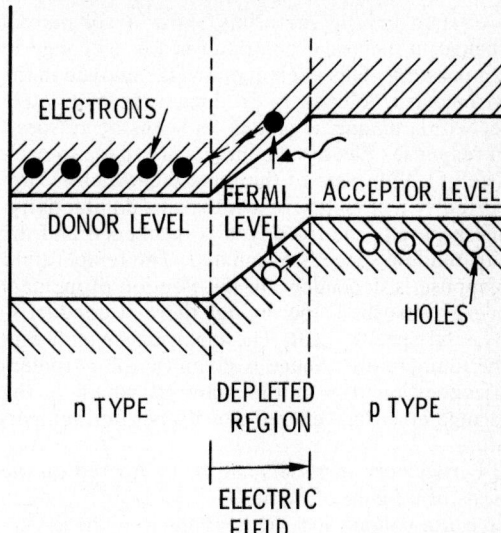

Fig. 8-16. Energy-level diagram for junction solid-state diode.

vacuum and exposed to signal photons. The energy of a photon, if large enough, can enable an electron to leave the surface. A nearby conductor, maintained at a positive voltage relative to the photocathode, collects the electron to produce a signal current. The available emitting surfaces will release electrons upon absorbing photons of wavelengths up to a little greater than 1 μm.

An efficient form of amplification generally used in photoemissive devices is the electron multiplier. The surface encountered by an electron leaving the photodiode is made of a material that, when struck by an electron of sufficient energy, will yield several electrons in the secondary emission process. A number of surfaces are arranged and supplied with accelerating voltages to repeat this process so that a large number of electrons (10^6 for example) will issue for each one released from the photocathode. The advantage of the electron multiplier over conventional amplifiers is that the initial electron current need not flow through a resistor which causes thermal

noise. The electron multiplier requires high voltage, on the order of one kilovolt, which requires special care for high altitude operation.

Another mode of using the photoemissive detector electron-multiplier, or photomultiplier, involves counting the groups of electrons produced by each individual electron leaving the photocathode. First, thresholding is used to reject small groups of electrons which arise within the multiplier portion of the device. Then the remaining groups are counted, thus eliminating effects of variation in the size of the groups. This mode is limited to signal levels where the count rate is not impractically high. For example, it could not be used for earth resources imaging in the order of 50 KHz.

Thermal Detectors

All thermal detectors follow the heat balance equation

$$P = C_T \frac{d(\Delta T)}{dt} + \Delta T/R_T$$

where P is the power absorbed by the thermal mass associated with the detection element, C_T is the thermal capacity of the mass, ΔT is the temperature difference of the mass relative to the heat sink, and R_T is the thermal resistance between the mass and the sink. If P has sinusoidal component $P_o e^{jwt}$, the equation can be solved for ΔT:

$$\Delta T = R_T P_o e^{jwt}/(1 + jwR_T C_T).$$

The maximum R_T occurs when radiation is the only coupling, in which case a limiting noise is that of the radiation.

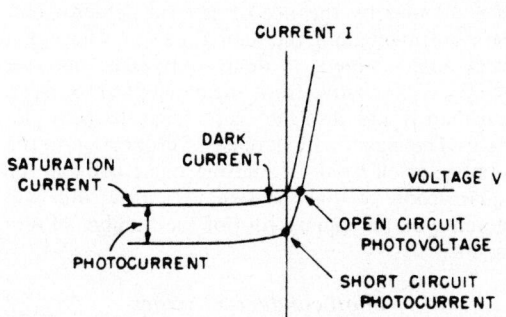

Fig. 8-15. Current-voltage characteristics of a photodiode (after Keyes, 1977).

Pyroelectric

Certain crystal materials show a substantial change in electrical polarization for a change in temperature. This phenomenon is employed in the pyroelectric detector to produce a thermal detector with a unique advantage in terms of its speed of response. Electrical contacts are made to each face of a thin plate of the material, which is supported by a mounting of low thermal conductivity, and generally the front face is blackened and the surrounding space is evacuated. The temperature response is dependent on the fraction of incident energy absorbed, the crystal thermal conductivity, and specific heat. The electrical response to the temperature change is given by the pyroelectric coefficient p of the material which is the change in surface charge density per unit temperature.

Pyroelectric materials can be compared on the basis of a figure of merit $M_i = p(c\rho\epsilon_r^{1/2})^{-1}$ where $c\rho$ is the volume specific heat and ϵ_r is the dielectric constant. A practical limitation in the use of this detector is microphonic susceptibility, since pyroelectric materials are also piezoelectric. A unique characteristic arises from the fact that, if the temperature of the detector is not changing, charge leakage will lead to electrical neutrality in which case there is no response at zero frequency.

Thermopile

The individual thermocouple element of a thermopile generates a voltage at the junction of two dissimilar materials characterized by the Seebeck coefficient, $\alpha = dV/dT$, which is the limiting value of $\Delta V/\Delta T$, based upon the voltage and temperature differences between two junctions of the dissimilar materials. Alternate junctions of a number of series-connected thermocouples are collected on the detector element, while the other junctions are joined to heat sinks. The signal will be $N\Delta V$ for N detector junctions. The thermal conductivity K of the path between detector junctions and heat sink junctions should be minimized to keep the responsivity high. On the other hand, the electrical conductivity σ should be high. A figure of merit for materials has the form $\alpha^2\sigma/K$. Antimony and bismuth form a couple which performs well as a thermocouple. High-sensitivity thermopiles with a short enough time constant for remote sensing applications are made of thin film elements supported on a film substrate, as shown in Figure 8-17.

Arrays of Detectors

It is feasible, in principle, to scan an image with a single electro-optical detector and many remote-sensor systems have been built to operate in this way. More often, the available time to dwell on each scene element is so brief that the required bandwidth becomes very high. This difficulty is at least partially overcome by the use of detector

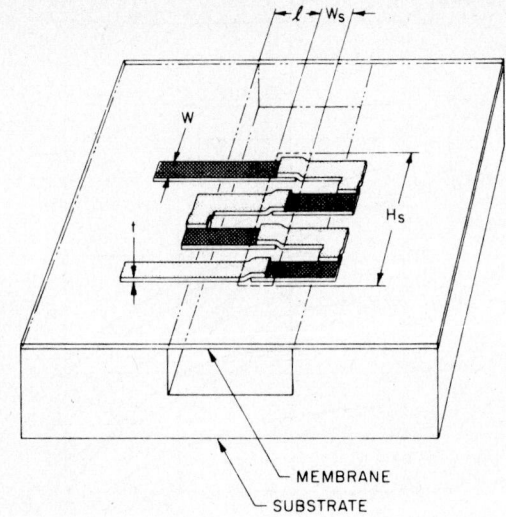

Fig. 8-17. Evaporated-thermopile schematic (after Stevens, 1970).

arrays—arrangements of many detectors which operate in parallel. These arrays may be one- or two-dimensional in extent. The scanning sensor often uses a one-dimensional array across the scan direction. Two-dimensional arrays are used for framing sensors, or for scanning sensors used in the time delay and integration mode discussed below.

There are different approaches in constructing arrays: (1) stacking individual detectors on a passive substrate, (2) partially cutting a single crystal into individual pedestals or (3) delineating individual detectors in a common single crystal, using photolithographic techniques. The present availability of detector arrays that incorporate switching networks on the same substrate represents an important contribution in this field by modern integrated circuit technology. The technology in this field is advancing at a rapid rate spurred by the numerous advantages offered by such detectors, particularly with the charge-coupled devices discussed later in this chapter.

Time Delay and Integration

A signal-to-noise ratio advantage may be gained in a scanner by the use of several detector elements arrayed along the scan direction. The signal from each element is delayed by the time the image takes to move from one element to the next, and then is added to the signal from the next element. The signal will increase in proportion to the number of elements, while the noise adds as the square root, so that the signal-to-noise ratio will increase as the square root of the number of elements.

Photoconductor Arrays

Arrays of photoconductors, including the first two forms of physical construction listed above, have been available for some time and have been used chiefly in the intermediate and far-infrared areas of the spectrum. Linear arrays are more common, though some two-dimensional arrays have been used. Since the information emerges from the array in parallel form, and usually must be converted to serial form, it is necessary to provide intermediate information-storage together with signal multiplexing capability.

Some major problems associated with arrays are nonuniformity of response, limitations on resolution, and crosstalk. The response of individual detectors may vary throughout the array to such an extent that calibration of individual detectors and compensation of the output signals will be necessary for radiometric systems.

Photodiode Arrays

In addition to the operating mode in which each element in the array is connected to an amplifier stage, the photodiodes can be operated in a signal integrating mode, and thus serve as their own information storage medium. They can also be integrally coupled with solid-state switches to provide the necessary multiplexing for readout.

To understand the integration mode, it should be remembered that the photodiode junction is essentially a capacitor. If reverse bias is applied to such a junction when it is dark, the capacitor will be charged; if the bias is removed, the charge will remain. This charge, however, can be neutralized by photon-produced electrons or holes, and the amount of charge neutralized will be directly proportional to the number of photon-produced carriers. In operation, the capacitor is charged by applying reverse bias for a very short period. The bias is removed, and the incident radiation partially neutralizes the charge by the absorption of photons over an exposure interval. The junction is then recharged and the number of photons converted is measured by the current needed to restore the original charge, which constitutes the scene information. Figure 8-18 shows a schematic circuit for such an array. Each diode is connected through a battery to a resistor and a switch. A clock pulse closes the switches sequentially for an interval during which the battery recharges each diode to a set voltage level. The current that flows is measured by means of the voltage developed across the resistor. Typically, the switches consist of a series of transistors built on the same substrate as the detector array, and the resistor is the input to a video amplifier. If there are 100 diodes

in the array, then the integration time is at least 99 intervals.

The limitations imposed by nonuniformity, crosstalk, and resolution constraints apply to photodiode arrays as well as to other forms. In addition, the arrays exhibit switching noise that results because part of the switching transient is transferred to the circuit through the diode capacitance.

Charge-coupled devices

The basic concept of the charge-coupled device (CCD) consists of storing electrons (or holes) in potential wells created at the surface of a doped semiconductor, and moving this charge, as a packet, across the surface by translating the potential minima. Séquin (1975) gives a comprehensive treatment of the CCD. The charges may be developed by absorption of photons, or they may be injected from another source such as an infrared-sensitive photodiode. The devices allow signal handling with high packing density.

As shown in Figure 8-19, a CCD consists of a doped semiconductor covered with an insulating layer; electrodes are provided on the outside of this layer. A voltage on the electrode drives the surface of the semiconductor into depletion of majority charge carriers; minority carriers that are produced in this region by the absorption of photons are held in this location by the voltage on the electrode. If the voltage is switched from one electrode to the next, the minority carriers move sequentially from one electrode to the next until they are delivered to an output at the end of the row. The CCD can be used in a "pushbroom" arrangement in which case it serves to facilitate signal processing and to reduce amplifiers or it can be arranged along the scan direction in a time-delay and integration mode.

These devices are relatively simple to produce and are therefore economical. The efficiency of the charge transfer—which limits the total number of transfers that can be effectively made—is quite high. The devices can be operated at MHz rates and they can be made very small physically. Most CCDs have been made of silicon, but charge transfer action has been demonstrated in other detector materials such as indium antimonide and mercury cadmium telluride.

Thermal Arrays

Thermal arrays can be constructed as arrays of individual elements which must be thermally isolated from each other. Linear arrays of thermocouples or bolometers* were among the earliest instruments used in remote sensing, but the inherently long time-constant restricts their use. This increase in dwell-time may be incompatible with motion rates of the platform.

* Bolometers use the effect of resistance changing with the temperature.

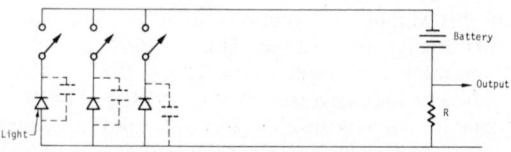

Fig. 8-18. Circuit schematic for photodiode array.

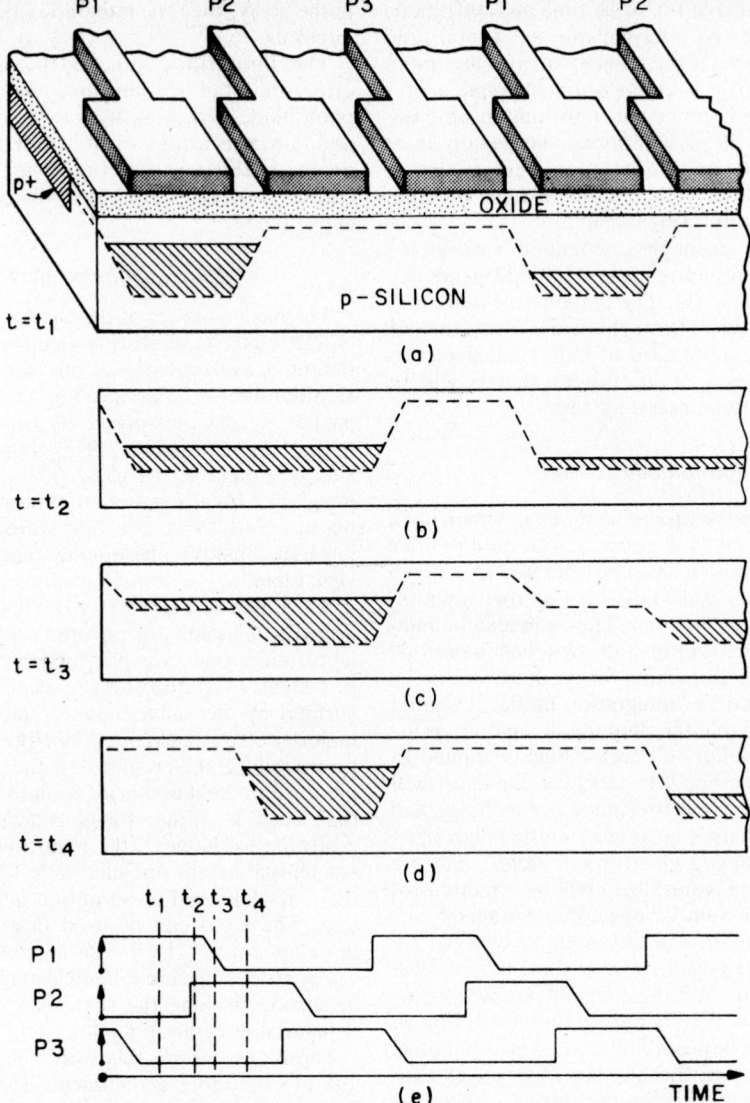

Fig. 8-19. (a) Schematic rendering of a 3-phase n-channel CCD with the charge carrying potential wells shown in the cross-section of the silicon substrate. (b,c,d) Potential wells shown at subsequent time intervals illustrating the transfer of charge. (e) The corresponding time slots marked in the diagram of the waveforms. (After Séquin, 1975. Copyright (1975) Bell Telephone Laboratories. Reprinted by permission.).

NOISE SOURCES

The selection of a detector for a particular mission will depend upon its efficiency in energy conversion, as discussed above, and in addition upon the effects imposed by the noise sources that limit signal accuracy. These sources are not all within the detector, but some of the external sources are influenced by detector characteristics, such as its impedance and its requirement for bias supply. The relative importance of the various sources depends upon the electronic frequency range used by the system.

Most of the noises discussed here are expressed in current terms, but can as well be used as voltages, multiplying by the circuit resistance.

Quantum (also Photon or Radiation) Noise

The irreducible minimum noise is due to the quantum nature of the signal being measured. The number of photons per second arriving at a detector will have randomness on arrival such that the uncertainty or noise is proportional to the square root of the magnitude, an approximation generally valid for the remote sensing applications covered in this Manual. Corrections might become necessary for far-infrared (see Holter, 1962, p. 249) or extremely weak signals (see Keyes, 1977, p. 48) or coherent radiation (see Wolfe, 1978, p. 11-39). A detector such as the charge-coupled device, which stores charge during a time interval $\Delta\tau$ after which it is transferred away, will have an uncertainty in the charge quantity of $(\eta N \Delta \tau)^{1/2}$, where η is the quantum efficiency and N is the number of photons. In other detectors, producing a current $I = \eta e N$ in response to the arriving photons, the noise current is

$$i = (2eI\Delta f)^{1/2}$$

where e is the electronic charge and Δf is the electronic bandwidth.

Background Noise

The preceding equations apply to irradiance to which a detector is sensitive, thus extraneous irradiance arriving with the desired signal will also cause noise. Sources can be unwanted material in the field of view or portions of the scanner itself. The latter is usually most important in long-wavelength sensors and may be minimized by cold shields and cold filters.

Shot Noise

Shot noise consists of current pulses caused by the discrete nature of the electrons in a current flow across a surface or junction. The noise current,

$$i_{ns} = (2eI\Delta f)^{1/2},$$

occurs in the junction regions of diodes at crystallite boundaries such as those in the thin-film lead-salt detectors, and in vacuum tubes. If the current I is due only to signal photons, the shot noise is the quantum noise previously described.

Generation-Recombination Noise

In a semiconductor, the generation of current carriers occurs in a random manner, followed by recombination after a lifetime τ. The resulting generation-recombination (G-R) noise may be expressed as

$$i_{ngr} = I_{dc} \left| \frac{2\tau}{N(1 + \omega^2\tau^2)} \right|^{(1/2)}$$

Where N is the total number of free carriers. Carriers may be generated thermally or by background or signal radiation, but the signal-generated and often the background-generated noises are not included under this label. For a photodiode, carriers generated and swept across the junction join the much more numerous majority carriers, so that the recombination noise is dominated with the result that the 2 is no longer needed in the expression above.

Johnson (also Nyquist or Thermal) Noise

Johnson noise, present in any resistive element and often a major noise source, is caused by the random motion of charge carriers. Its magnitude is

$$i_n = (4kT \Delta f/R)^{1/2}$$

where k is Boltzmann's constant, $(1.38 \times 10^{-23} J/K)$, T is the absolute temperature, and R is the resistance.

Dielectric Loss Noise

In a capacitance, there are energy losses which can be thought of as originating in a conductance.

$$G = 2\pi f C \tan \delta$$

where f is the frequency, C the capacitance, and $\tan \delta$ the dissipation factor (also power factor or loss tangent) of the dielectric material. Except for vacuum or air, the value of $\tan \delta$ is on the order of 0.001 to 0.01. This conductance generates a noise

$$i_{nd} = (4kTG\Delta f)^{1/2}$$

which is included by some writers in the Johnson noise category. Dielectric loss noise is important for the pyroelectric detector and can be the dominant noise source for other detectors in high-impedance, wide-bandwidth systems.

Temperature Noise

Thermal detectors have a temperature fluctuation,

$$\overline{\Delta T^2} = \frac{4kT^2/R}{(1/R_T{}^2 + \omega^2 C_T{}^2)}$$

which is due to the randomness of thermal exchange with the surroundings. The minimum possible noise occurs if the exchange is radiative only, and while not descriptive of actual systems, is used as an ultimate limit for comparison.

$1/f$ (Also Modulation, Contact, or Current) Noise

$1/f$ (one over f) noise is marked by the characteristic that the noise power is nearly inversely proportional to the frequency and directly proportional to the square of current. The vagueness refers to the fact that exponents on the frequency and current terms can vary from unity and two, respectively, for different detectors, or for different lots of the same detectors.

The mechanisms of this noise are not understood, but there is evidence linking at least part of it to surface states of the semiconductor, the nature of the contacts, and carrier traps. The magnitude is determined by measurement and can become dominant below frequencies on the order of 10 to 1000 Hz depending on the detector and the application.

Pattern Noise

Arrays of detectors will have some differences in responsivity for the various elements which can be thought of as a spatial distribution of responsivity along the array known as pattern noise.

CCD Noise Sources

In addition to those described above, the following may be found in CCDs:

Transfer Noise

At transfer, the packets of charge in a CCD will leave behind some small fraction ϵ and pick up a fraction left by the preceding packet. The fluctuations in these fractions are a noise source causing a mean square fluctuation in the charge packet after n transfers of

$$\overline{\Delta Q_s^2} = 2n\ \overline{\Delta Q_{tr}^2} \text{ for } n\epsilon << 1$$

where ΔQ_{tr}^2 is the total mean square fluctuation for each transfer process. The correlation of fluctuations in adjacent charge packets leads to a noise spectrum in which higher frequencies are predominant:

$$S_{tr}(f) = \frac{4nf_c}{e}\ \overline{\Delta Q_{tr}^2}\ (1 - \cos 2\pi f/f_c)$$

where f_c is the clock frequency and e the electron charge. Trapping of the charges can contribute to the fluctuations in the transfer.

Injection Noise

If the detector element is separate from the charge transfer register, noise can be added by the process of inserting the detector charge into the CCD.

Readout Noise

The charge in the CCD is sensed by a sampling process in which the resetting between samples adds noise. A technique called correlated double sampling, which uses the difference caused by the signal, may be used to avoid the reset noise. A different approach not using resetting is the distributed floating-gate amplifier which uses repeated sampling of the same charge packet as it moves past a number of gates.

Clock Feed-Through Noise

The clock pulses used to perform insertion, transfer and readout may couple through to the signal channel.

Input Circuit Noise

In any practical use of a detector, the early stages of amplification must be considered as potential limitations on performance. Several of the noise sources already mentioned also pertain to the input circuit. These include Johnson noise, which is that of the combined resistance of detector, bias resistor and any other shunting resistance of the input circuit. The dielectric loss noise includes all of the capacitances of the input circuit, each with its appropriate loss factor.

The preamplifier input device generates noise of several sorts: shot noise at junctions, Johnson noise in its resistances, and $1/f$ noise. These may be conveniently handled in system performance budgets in the form of the summary parameters $\bar{e}_n$ and $\bar{i}_n$, or voltage and current noise, in units of $V/$

$\sqrt{Hz}$ and $A/\sqrt{Hz}$ respectively, which are affected by frequency, temperature, and electrical operating point of the device. These two noises enter performance calculations as a voltage generator in series with the input signal and a current generator in parallel.

The noise of circuitry following the input stage must be evaluated for possible influence, when the signal out of the first stage is not large in comparison.

Combined Noise for System Evaluation

Understanding of the noise performance of a system is assisted by graphing the noise contributors as a function of frequency. For example, Figure 8-20 shows the estimated noise sources of the silicon photodiode channels of the Thematic Mapper. The signal photon noise which is dependent on signal level is not shown here. In this circuit, a field-effect transistor (FET) is the input device of a current mode amplifier with R_f as the feedback resistor of 10^9 ohms. This resistor with the input capacitance of 5 pF would cause the signal to decrease with frequency above 32 Hz, but for the feedback of the amplifier. This feedback amplifier action causes the FET input-voltage noise to increase with frequency. The decreasing response above 50 KHz is caused by the electronic filter, designed to limit noise at frequencies unnecessary for signal response. This form of graph is useful for optimization studies because any of the curves can be shifted vertically to find the relative effect of a change in that noise source. Also, a graph of experimentally determined total noise may allow identification of a principal contributor, if it is one with a characteristic slope. It should be noted when using the graph that the noise as measured by a meter or viewed on an oscilloscope is the result of integration, over linear frequency, of the square of the values appearing in the graph. Thus, the higher frequencies and the higher amplitudes are of greater relative effect than suggested by this graph.

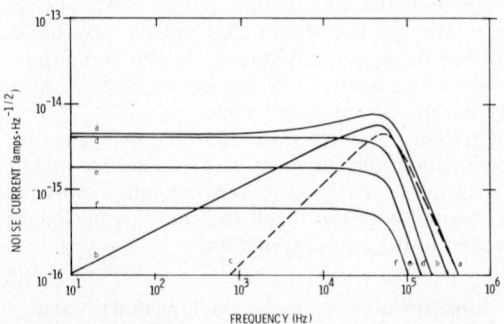

Fig. 8-20. Example of system noise estimates, showing noise components: (a) Total noise, integrated over frequency, 2.04 pA; (b) dielectric loss noise, 1.33 pA; (c) amplifier voltage noise, 1.05 pA; (d) thermal noise, 0.93 pA; (e) dark current shot noise, 0.41 pA; (f) amplifier current noise, 0.14 pA.

In overall system evaluation all types of noise sources must be taken into account, such as electrical pickup, crosstalk between channels, and microphonics. The latter is generally caused by the vibrational variation of a capacitance that has voltage applied. Mechanical refrigeration of the detector or scanning mechanisms is a possible source of vibration.

FIGURES OF MERIT

Figures of merit have been developed to enable useful comparisons to be made in selecting the best detector for a particular mission. The mission's needs will consist of the detection of a minimum quantity of power in some wavelength region with a necessary time resolution. In comparisons of detectors based on a figure of merit, the designer must not neglect aspects of detector performance not included in the figure, such as:

- Uniformity of response over the sensitive area.
- Linearity of signal vs. radiant power.
- Stability of detector characteristics as a function of time, temperature and other environmental factors.

Quantum Efficiency

Quantum efficiency (η) may be defined as the number of electrons emitted for each incident photon in a photoemissive surface, or energized per incident photon in a photoconductor. The ideal quantum efficiency of one means that one photon releases one electron. Multiplicative processes, such as a secondary electron emission, are considered separately, and are called gain.

As stated, the output current, for a given average input signal, is proportional to the quantum efficiency, and the shot noise is proportional to the square root of the quantum efficiency. Therefore, the S/N ratio, for shot noise, becomes directly proportional to the square root of quantum efficiency. Typical values of η for photoemissive surfaces are 10 to 20 percent in the visible spectrum (see Chapter 7), with a long wavelength limit of photoemissivity around 1 μm. The quantum efficiency of photoconductors and photodiodes, on the other hand, is typically very near to unity.

Noise Equivalent Power

Noise equivalent power (NEP), a basic concept for performance, is defined as the amount of signal radiant power (in watts rms) that will give a signal output equal to the noise output. The conditions of the measurement must be specified, such as the frequency response of the amplifying and measuring electronics and the optical spectral content of the signal radiance. In certain cases, for example in system design before the electronic bandwidth is established, it may be convenient to define NEP in units of $W \cdot Hz^{-1/2}$. NEP is also applied to complete systems, as well as to detectors.

D* (D-star)

D^*, a normalized figure of merit derived from the NEP, is widely used, especially for infrared detectors, although its measurement conditions must be noted. It may be defined as

$$D^* = \sqrt{A_d}\, \Delta f / NEP$$

where A_d is the detector area in cm^2 and Δf is the electronic bandwidth. Convention includes certain designations following the symbol. For example, D^* (10 μm, 1000 Hz) indicates the measurement condition of 10 μm wavelength radiant power chopped at 1000 Hz, while D^* (500 K, 500 Hz) indicates that the radiant power is that of a 500 K blackbody chopped at 500 Hz. An upper limit on the value of D^* is that imposed by background radiation and referred to as the BLIP D^* (background limited infrared photodetector).

Some caution is necessary in the use of D^* because the inherent assumptions in its definition may not apply. The predominating noise often does not depend on detector area, for example, and another invalidating factor occurs when the noise frequency-spectrum is not flat. In such cases D^* does not have the invariance quality that is needed for a figure of merit. When D^* is applicable, the equation shows the value of using optics of lower f number, which reduce the detector area necessary for a given IFOV.

RA Product

The sensor is background limited (Keyes, 1977) if two terms in the photoconductor equations have the relationship,

$$RA >> \frac{1}{\eta G^2}\, \frac{kT}{e^2 \Phi_b}$$

where η = quantum efficiency, G = photoconductor gain (1 for photodiode), k = Boltzmann's constant, T = temperature, e = electronic charge, Φ_b = background photon flux density. The RA product is a material property which the detector fabricator strives to maximize. In system design, after selection of the detector area the corresponding resistance is calculated and is combined with other resistances to determine the total Johnson noise.

Response Time

The speed of response of a detector has importance for some systems and has also been used as a figure of merit. The time constant of the responsivity is often cited for this purpose and is appropriate if the predominant noise is uniform with frequency and if the responsivity time-dependence conforms to that of a simple resistor-capacitor circuit. A more complex time-dependence requires more detailed description than can be expressed in a single figure of merit. Another manifestation of response time is the upper frequency at which D^* has decreased by 3 dB. This point is termed f^* (f-star) and is used in the figure of merit, $D^* f^*$.

RADIOMETRIC CALIBRATION

The word "calibration" is ultimately derived from the Arabic "q̄alib" meaning form or mold, with the connection that the size of a mold for shot must conform to the diameter (or caliber) of the gun barrel in which the shot is to be used. The present meaning is the determination of the quantitative relation of an instrument's response to the input values it is intended to measure. The values of prime consideration for electro-optical remote sensing are radiant amplitude, spectral location, and spatial response. The calibration process entails comparison with standards and also involves more portable means of maintaining calibration, as necessary in instrument checkout and environmental testing (e.g. in orbit or flight). Wyatt (1978) covers the fundamentals of the subject while "A Self-Study Manual on Optical Radiation Measurements," (National Bureau of Standards 1976, 1978) contains a detailed and tutorial development of the concepts used in calibration.

AMPLITUDE CALIBRATION

Amplitude calibration involves measurement of the proportionality factor between a change in radiant input and the corresponding change in output units of the system, such as digital numbers. This factor is known as gain. Also, the relation must be known between some fixed reference point on the input scale and a corresponding point on the output scale, a relation called offset. In general, the logical fixed points are zero radiant energy input and digital zero output.

Zero radiant energy is achieved by a shutter at shorter wavelengths, but at longer wavelengths, where ambient temperature emission gives appreciable signal, the reference must be cold enough to radiate negligible energy. In orbit this can be achieved by pointing the instrument at dark space, while in ground calibration a cold, high-emissivity source can be used.

Pointing at space may be impractical, as it is for many earth-imaging satellites, in which case the reference point must be some positive amount of radiant energy which is assigned an output value. Offset is then measured between the assigned output value and the value measured when the selected radiant energy is presented.

The radiant energy observed for optical remote sensing tends to fall into one of two spectral regions, that of reflected sunlight at shorter wavelengths or that of thermal self-emission at longer wavelengths. Clearly the ultimate standard for the first region is the sun itself, while for the second, cavity radiators closely approximating ideal blackbodies are used as standards, and for both regions, standard detectors are available wherein radiant power is alternately compared to electrical power. The last method, known as an electrically calibrated pyroelectric radiometer, or ECPR, is described by Doyle (1976).

Sources

Sun

High-accuracy measurement of the solar energy spectrum from the ground has been difficult because of the variability of the earth's atmosphere. A summary of available information appears in White (1977). Measurement from beyond the atmosphere has been limited by the need for comparison standards of laboratory quality. Actually, the need for accurate knowledge of this spectrum is also mitigated by the atmosphere, which causes variation in the reflected energy sensed from space, as detailed in an earlier section. This variation cannot be reduced by improving the accuracy of solar measurements.

Use of the sun as a reference will insure that relative spectral reflectance measured from orbit can be independent of any change in solar spectral distribution. However, there is some doubt that changes significant to remote sensing imagers occur except at ultraviolet wavelengths.

Blackbodies

Any object of unit emissivity and known, uniform temperature is a blackbody and will have a known spectral radiance calculable from the Planck equation. Cavity radiators close to this ideal are readily available and are appropriate standards except for shorter wavelengths where the required temperatures can lead to rapid oxidation when such a source is used in air. The blackbody radiation also has the convenient properties of being perfectly diffuse and unpolarized. Small, lightweight sources that approximate blackbodies can be incorporated in instruments for checking calibration.

Tungsten Lamps

At shorter wavelengths tungsten lamps are used as standards. Coiled filament quartz-halogen lamps are sources of spectral irradiance and ribbon-filament lamps can produce known spectral radiance. Calibration for certain types of these lamps can be obtained from NBS.

Miniature lamps are useful for on-board maintenance of calibration with low power consumption. In essence, a laboratory standard calibration of the instrument is monitored for change by reference to the miniature lamp.

SPECTRAL CALIBRATION

Spectral calibration is the process of measuring an instrument's response as a function of frequency (wavelength) of the input radiation. The input source or sources must produce radiation in discrete intervals. The spectral purity and position of the calibration sources must be compatible with the objectives of the calibration being conducted. In calibrating the spectral response of an instrument, one must realize that among the features requiring special attention are the slope of the response at a band edge and the need to suppress response outside the band. For example, a band intended to measure earth scenes in the red region should avoid response in the near-infrared where there is a sudden and strong increase in foliage reflectance.

A peculiarity of spectral filters is that they often have leakage at wavelengths longer than the desired passband while easily blocking shorter wavelengths. This can be a problem in an instrument; conversely, a calibration technique can employ part of the peculiarity to detect long wavelength leakage by using a filter which blocks all wavelengths within, and shorter than, the passband while passing those longer.

A different technique for measuring any long wavelength leakage involves varying the temperature of a blackbody source and noting if the resulting shift in spectral location of the blackbody radiance is transmitted through the filter. Figure 8-21, after Wyatt (1978), shows data from such a procedure when leakage is present.

Sources

As primary standards, a laboratory monochromator or spectrometer can use low-pressure, rare-gas lamps or absorbing media where the position of emission or absorption lines is precisely known. Chapters 2 and 7 of *The Infrared Handbook* (1978) show tabulated and graphical information on many of these. A multichannel line scanner might use a laboratory monochromator for its wavelength calibration.

Sources for instrument calibration can use narrow-band filters, the spectral response of each of which has been measured with a laboratory spectrometer, to isolate a portion of the spectrum. A blackbody or tungsten lamp of known spectral characteristics, as previously described, can be combined with the filter to form a calibrated source at a desired spectral location.

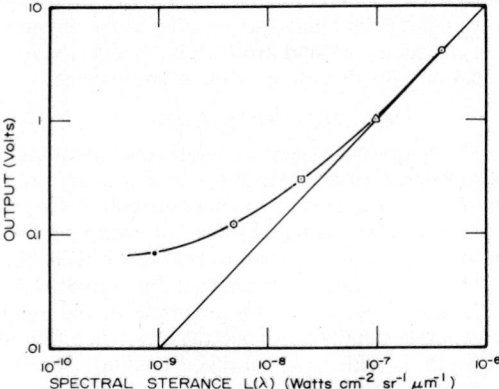

Fig. 8-21. An example of long-wavelength filter leakage for various temperatures: 1200 K (⊙), 1000 K (△), 800 K (☐), 700 K (⊙), and 600 K (●). After Wyatt, 1978.

SPATIAL CALIBRATION

Spatial calibration involves evaluating the valid response to the scene and the spurious response arising from scattering and crosstalk. The scene-feature response may be characterized as response to a bar chart, often transformed to give a modulation transfer function (MTF), as described earlier in this chapter. The spurious response generally is expressed as a graph of the relative response vs. the angle off axis. Several such graphs may be required for different directions off axis.

The bar charts used differ for scanners or linear arrays and for frame sensors. The former can be presented with a series of patterns of equal bright and dark bars, each pattern having different bar size. For the frame sensor, several test patterns have been designed with graduated bar sets, distributed within the frame in order to test the full area.

TRANSFER METHODS

Even with the most accurate of standards available, the problem of transferring the energy of the standard to the instrument remains. For example, the instrument's optical aperture may be much larger than can be covered by any available standard or the standard may emit much more energy than the instrument can accept. A calibration source or calibrator must be matched to the characteristics of the instrument to be calibrated, and a detector used to compare the calibrator output to that from a standard. If the output levels of the two sources are much different, the linearity of the detector should be verified. Bar-chart patterns may be included in the calibrator.

For example, the calibrator for the Thematic Mapper consists of a collimator of larger diameter and smaller central obscuration than that of the instrument, mounted on a precision rotating table and having a number of patterns and sources that can be selected at the focus by a multiposition folding mirror. The rotating table provides coverage of the full scan-angle range of the instrument.

Vacuum chamber tests of the instrument with its calibrator are necessary for instruments which are to be used in space. Discrepancies have been found between calibration tests that have been conducted in ambient and in vacuum. The difference may arise because moisture absorbed during ambient exposure is desorbed in vacuum, causing changes in spectral influence of some optical surfaces such as thin film layers and antireflective coatings. Both the test equipment and the sensor are susceptible so that assignment of error is complicated. Further investigation is needed for a good understanding of this effect.

If a calibrator and standard source have outputs that exceed the linear range of the measuring detector, other steps may be needed to make the comparison. One such step is to change the magnitude of the radiant input geometrically by stops or distance. In this way, the comparison system response can be measured over the appropriate range with a precision limited only by the accuracy of the geometric measurements.

After the linearity has been established, comparison of ribbon filament lamps of known spectral radiance with the large-area sources can proceed by the direct substitution method (see Figure 8–22). The standard lamp filament is imaged onto the monochromator entrance slit in the known area and a reading is taken. The standard is moved out of the way and the large-area source replaces it. The signal for this source is then measured. The standard is then put back into place and another measurement taken to ensure time stability.

The procedure is then repeated for as many wavelength points and spatial areas on the large-area source as required.

Integrating Spheres

The integrating sphere with its multiple diffuse reflections is useful in calibration because of its tendency to average out variations in energy with angle of arrival or variations in polarization of the energy. Fig 8–23 is an example where the lamp output is transformed such that the 10-inch diameter exit-window is filled uniformly, within 1 percent for any half-inch area. This design was used in the NASA/MSC-24-band line scanner.

An integrating sphere source provided by Goddard Space Flight Center for calibration of the Landsat Multispectral Scanner uses 12 internally mounted lamps to furnish a number of stable levels of output. A similar sphere is described in McCulloch (1969) and a variation of the design used for the Thematic Mapper contains mixed sizes of lamps to provide finer graduations in the output.

CALIBRATION CATEGORIES

There are two kinds of calibration, absolute and relative. Absolute calibration implies knowledge of the object in fundamental units, such as spectral radiance, spectral irradiance, power etc., and requires measurements in which the comparison is made with maintained standards. Often one does not need to know the absolute magnitude of

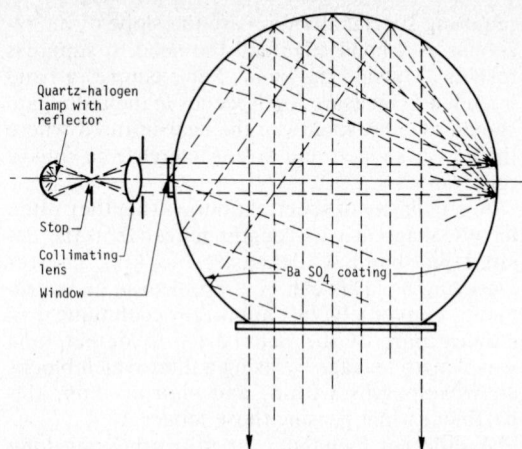

Fig. 8-23. Incomplete integrating sphere calibration technique, geometry.

the radiation from a scene but rather the relative magnitude of the radiance of one scene element with respect to another, or the relative radiance of the same scene element observed at different times, or the magnitude of the spectral radiance in one wavelength band with respect to another. Thus stability and repeatability with time are much more important than traceability to a known source. In the thermal region, for example, it is more important to know that the effluent from the paper mill is heating the river by 2.7°C within 1,000 ft of its discharge point than to pinpoint the effluent temperature as being 29.6°C. In this case, the differential temperature is more important than the absolute temperature.

Absolute calibration for each resource will not be meaningful until methods are devised to measure the atmosphere over each scene and to extract reflectances as described earlier. Nevertheless, efforts should be continued to maintain a calibration history with respect to known standards so that data from different sensors can be used interchangeably. In this way, when the atmospheric problem is solved, it will be possible to have a large data bank, to greatly reduce the need for collecting ground truth data, and to have a capability for detecting subtle scene changes.

CALIBRATION ERRORS

The steps necessary to relate the operational performance of an instrument to a primary standard are many, and each will contribute some uncertainty. The estimated errors of each step may be summed to arrive at an overall error estimate.

The uncertainties in calibrations supplied by NBS are indicated by the example of the 1000-watt lamp standard of spectral irradiance (Saunders, 1977, with mimeographed update). The uncertainty is stated to be 1.2 percent from 450 nm to 1600 nm, increasing with shorter wavelength to 2.6 percent at 250 nm. *The Infrared Handbook* (1978) contains extensive listings of other standards, with accuracies.

Transfer of calibration from a standard to a

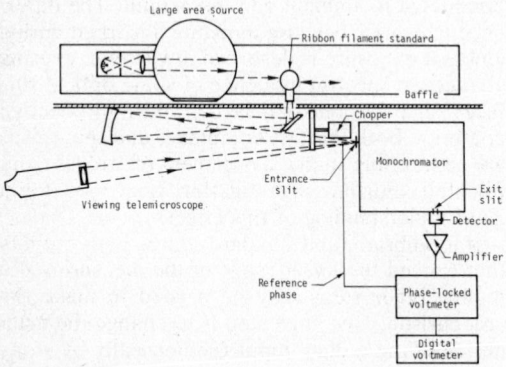

Fig. 8-22. Comparison of known spectral sources (ribbon filament lamps) with large-area sources by transfer calibration technique.

TABLE 8-7

Absolute Accuracy Error Budget Estimated for Thematic Mapper

Accuracy of NBS primary standard	±1%
Transfer from primary to integrating sphere standard	±3%
Transfer from integrating sphere to internal calibrator	±1.5%
Estimating gains and offsets	±0.25%
Detector linearity	±0.1%
Electronics linearity	±0.1%
DC restore accuracy	±0.1%
Internal calibrator stability through launch	±3%
Internal calibrator stability through life	±0.4%
Vacuum shift	±5%
TOTAL (root-sum-square)	±6.8%

calibrator having characteristics matched to the final instrument requires close attention to certain systematic error sources. These include linearity in the transfer detector and geometric differences between the paths to the detector from the standard and from the calibrator.

At each step of calibration transfer, there is the possibility of some change during the time between one step and the next, the effect of which may be mitigated by the use of several observations distributed in time.

On repeated measurement of any of the calibration data, the systematic or random nature of changes should be noted. After the systematic effects have been removed, the different observations may be combined and the error evaluated by calculating the standard deviation of a sample:

$$\sigma = \left[\frac{1}{n-1} \left(\sum_{i=1}^{n} x_i^2 - n\bar{x}^2 \right) \right]^{1/2}$$

where

n = number of measurements
x_i = a measurement
$\bar{x}$ = mean of the measurements

The ratio of $\sigma/\bar{x}$ will influence the size of n required for a significant estimate of the radiance.

An example of an error estimate for a sensor system is shown in Table 8-7.

REFERENCES

Arams, F. R., (ed.), 1973, Infrared-to-millimeter wavelength detectors; Artech House.

Barbe, D. F., (ed.), 1980, Charge-coupled devices; vol. 38 of Topics in Applied Physics, Springer-Verlag.

Bedford, R. E., 1960, Can. J. Phys., vol. 38, p. 1256.

Bode, D. E., 1980, Infrared detectors in applied optics and optical engineering; vol. VI, Kingslake and Thomson, eds., Academic Press.

Boyce, P. B., 1977, Low light level detectors for astronomy; Science, vol. 198, pp. 145–148.

Coltman, J. W., 1954 (June), The specification of imaging properties by response to a sine wave input; J. Opt. Soc. Am., vol. 44, no. 6, pp. 468–471.

Doyle, W. M., et al., 1976 (November/December), Implementation of a system of optical calibration based on pyroelectric radiometry; Opt. Eng., vol. 15, no. 6, p. 541.

Elterman, L., 1963 (July), A method of a clear standard atmosphere for alternation in the visible region and infrared windows; AFCRL 63–675.

Flowers, E. D., R. A. McCormick, and K. R. Kurtis, 1969 (December), Atmospheric turbidity over the United States 1961–1966; J. Appl. Meteorol., vol. 8.

Fraser, R. D., 1974 (August), Computed atmospheric corrections for satellite data; SPIE Proc., vol. 51, pp. 64–72.

Goddard Space Flight Center, Advanced scanners and imaging systems for earth observations; NASA SP-335, Greenbelt, Md.

Grove, A. S., 1967, Physics and Technology of Semiconductor Devices; Wiley and Sons.

Holter, M. R., et al., 1962, Fundamentals of Infrared Technology; Macmillan Co.

IEEE Trans. on Electron. Devices, 1980 (January), vol. ED-27, no. 1.

Jones, R. C., 1959, Noise in radiation detectors; Proc. IRE, vol. 47, p. 1481.

Keyes, R. J., (ed.), 1977, Optical and infrared detectors; vol. 19 of Topics in Applied Physics, Springer-Verlag.

Kruse, P. W., et al., 1962, Elements of Infrared Technology: Generation, Transmission and Detection; Wiley and Sons.

Liu, S. T., and D. Long, 1978, Pyroelectric detectors and materials; Proc. IEEE, vol. 66, no. 14.

Lowe, D. C., and M. M. Beilock, 1970, Imaging with optical mechanical scanners from earth-orbiting spacecraft; AIAA Paper No. 70–314, Annapolis, Md.

Marshall, D. E., et al., 1978 (June), Improved detectivity of pyroelectric detectors; NASA Contractor Report 158917.

McCulloch, A. W., et al, 1969, Evaluation and calibration of some energy sources for the visible and near infrared regions of the electromagnetic spectrum; NASA X-622-69-195, Greenbelt, Md.

Overington, I., 1976, Vision and Acquisition; Crane, Russak and Co.

Putley, E. H., 1970, The pyroelectric detector; Semiconductors and Semimetals, Willardson and Beer, eds., vol. 5, New York Academic Press, pp. 259–285.

Reine, M. B., and R. M. Broudy, 1977, A review of HgCdTe infrared detector technology; Proc. SPIE, vol. 125, no. 62.

Saunders, R. D., and J. B. Shumaker, 1977, Optical radiation measurements: the 1973 NBS scale of spectral irradiance; NBS Tech. Note, USGPO, pp. 594–613.

Séquin, C. H., and M. F. Tompsett, 1975, Charge transfer devices; in Suppl. 8 for Advances in Electronics and Electron Physics, L. Marton, ed., Academic Press.

Stevens, N. B., 1970, Radiation thermopiles; vol. 5 of Semiconductors and Semimetals, Academic Press.

Stillman, G. E., et al., 1977, Far-infrared photoconductivity in high-purity gas; vol. 12 of Semiconductors and Semimetals Willardson and Beer, eds., Academic Press.

White, D. R., (ed.), 1977, The Solar Output and its Variation; Colorado Assoc. Univ. Press.

Wolfe, W. L., and G. J. Zissis, eds., 1978, The Infrared Handbook; USGPO.

Wyatt, C. L., 1978, Radiometric Calibration: Theory and Methods; Academic Press.

CHAPTER 9

Radar Fundamentals and Scatterometers

Author, Editor: RICHARD K. MOORE

GENERAL CONTENTS: Resolution concepts: angle discrimination, range discrimination; speed measurement; ambiguity and resolution; sensitivity, amplitude measurement, fading, and noise; calibration of radars; scatterometer system descriptions; scatterometer types (pulse, FM, velocity-angle and ground-based), references.

NOMENCLATURE

To conserve and eliminate repetition in text and references, the following symbols, units and names have been used in this chapter.

Symbol	SI Units	Name
$1_x, 1_y, 1_z$	—	unit vectors in the x,y,z directions
A	m²	area
A_r	m²	effective area of receiving antenna
$A_r(f)$	s⁻²	Fourier transform of $a(t)$
a	Hz s⁻¹	rate of change of frequency
$a_r(t)$	s⁻¹	receiver impulse response
B	Hz	bandwidth
B_e	Hz	effective bandwidth
B_r	Hz	resolution bandwidth
c	m s⁻¹	speed of electromagnetic waves in space
$c(\tau)$	V² s	range ambiguity
$c(\tau')$	—	correlation function for received and reference pulses
D	m	spacing between receivers, physical-aperture length
E	V m⁻¹	electric field strength
F	—	noise factor or noise figure
f_D	Hz	Doppler frequency
f_{Do}	Hz	maximum Doppler frequency
f_R	Hz	received frequency
f_T	Hz	transmitted frequency for FM radar
f_s	Hz	output (signal) frequency in FM radar
G	—	antenna gain, power gain of receiver
G_r	—	gain of receiving antenna
G_t	—	gain of transmitting antenna
h	m	height of radar above ground
I	A	current
$I(t,R)$	m⁻²	illumination integral
$K_{(fD)}$	V² s	Doppler frequency ambiguity function
L	m	total distance traveled in averaging for scatterometer
ℓ	m	dipole length
M	—	number of resolution cells in averaging length
N, N_i	—	number of independent samples
N_{in}	W	noise input power at 290 K
N_n, N_r	—	number of independent samples of noise, of received signal
N_{out}	W	noise power out of receiver
P	W m⁻²	Poynting vector
$P(f)$	W Hz⁻¹	input frequency
$P_o(f)$	W Hz⁻¹	output frequency
$P(\ )$	*	probability density function
$p(t)$	W	pulse power
R	m	slant range
$R_d(\tau)$	V² s	autocorrelation function of detected voltage
$R_d(\tau)$	V² s	autocovariance for total V_d
S	—	signal-to-noise ratio in terms of input (used when no subscript is required)
S_i, S_o	—	signal-to-noise ratio at input, output
T	s	averaging period
T	s	travel time for radar signal
T_a	K	effective antenna temperature
T_e	K	effective temperature of input
T_e	s	equivalent time duration of Doppler bandwidth
T_o	K	reference temperature (290 K)
T_o, T_p	s	pulse repetition period
t	s	time
$U(\ )$	—	Fourier transform of $u(t)$
$\mathbf{u}, u, u_R$	m s⁻¹	velocity, speed, radial speed
$U(t)$	V	baseband pulse shape
V	V	voltage
$\overline{V_{ac}^2}$	V²	ac component of noise
V_d	V	detected voltage
V_e	V	envelope voltage
V_n	V	noise voltage
V_r	V	received voltage
V_s	V	signal voltage
$v_t(t)$	V	transmitter pulse voltage
$v_{tr}(t)$	V	transmitter pulse voltage after receiver
W	W	power
W_e	W	envelope power
W_n, W_r, W_s, W_t	W	noise power, received, signal, transmitted power
β	rad, degrees	angle from center of beam, antenna beamwidth
β_h	rad, degrees	horizontal or azimuth beamwidth of antenna
β_{int}	rad, degrees	beamwidth for interferometer
γ	rad, degrees	angle of incidence for scatterometer
Δf_D	Hz	equivalent Doppler bandwidth
Δf_s	Hz	bandwidth of filter in FM radar
ΔR	m	slant-range resolution
$\Delta \theta$	rad, degrees	width of lobe

Symbol	SI Units	Name
$\Delta\rho$	m	along-track resolution (scatterometer)
$\Delta\tau$	s	equivalent pulsewidth
δ	s	duration of burst of pulses
θ	rad, degrees	angle of incidence, colatitude angle
θ_m	rad, degrees	mean angle of incidence in scatterometer cell
λ	m	wavelength
λ_c	m	carrier wavelength
μ	—	mean value of a distribution
ξ, η	m	coordinates along and normal to isodop
σ	m^2	radar cross-section
σ	—	standard deviation
σ_{av}	—	standard deviation of envelope power for N samples
$\sigma_n, \sigma_r, \sigma_s$	W	standard deviation of noise power, received power, signal power
σ_{rdn}	W	standard deviation of received power after averaging N independent samples
σ_T	W	standard deviation after averaging for T
σ^0	—	scattering coefficient (differential cross-section)
τ	s	delay time or time interval
τ	s	pulse duration
$\chi_{(\tau, f_D)}$		time-frequency ambiguity function
$\omega, \omega_c, \omega_D$	rad s^{-1}	angular frequency, carrier, Doppler
ω_R, ω_T	rad s^{-1}	received, transmitted angular frequency

INTRODUCTION

Microwave sensors can measure through cloud cover and some rain, although the measurement is almost independent of the weather at the lower microwave frequencies. Thus, microwave sensors are unique in their capability to provide timely information under conditions when other sensors are rendered inoperable by the weather.

Microwave responses are functions of the frequency (or wavelength), polarization, and look angle, as are responses in other parts of the spectrum. The microwavelength range is far from "micro," as compared with the visible and infrared ranges; it may be considered to extend approximately from 1-mm to 1-m wavelength. The wavelength ratio between the longest and shortest microwave is therefore greater than that between the longest useful infrared and the shortest visible wavelength. Polarization is often used as a discriminant in the microwave region because microwave antennas used for reception are most easily built with a single polarization direction, whereas most optical detectors are relatively independent of polarization, and special effort must be made to polarize the signal prior to detection. Incident angle variations in observed signals are common to all regions of the spectrum, but of course at each point in the spectrum these varia-

tions are different. The geometry of radar range measurement encourages the use of observational angles well away from the vertical, while the radiometer (much as with the optical and infrared sensors) tends to be used more often in a near-vertical mode.

Active microwave sensors provide their own illumination, while *passive systems* measure radiation originating somewhere other than in the radiometer. Normally the radar source is coincident with the receiving point, so the optical analog would be observation with back-lighting from the sun and no sky-lighting whatever. Most active microwave sensors both detect and measure range, but some measure only amplitude or speed, not range. Nevertheless, the term *radar* (meaning Radio Detecting And Ranging) is used commonly to describe even the nonranging active microwave systems, and it will be used that way here.

This chapter treats radar fundamentals along with scatterometers and spectrometers. Most users will find imaging radar systems of maximum value because of the fine resolution and geometric information available. These are discussed in Chap. 10.

Resolution with microwave sensors normally is poorer than with sensors operating in the optical and infrared region. This happens because the diffraction limit for angular resolution is directly proportional to the wavelength, and the shortest micro-wavelength used is a factor of 100 larger than the wavelength for thermal infrared sensing. Fine-angular-resolution microwave sensors therefore require very large apertures. Radars may achieve large apertures by coherent storage of the observed signal and subsequent synthetic aperture production, but passive microwave sensors are restricted to those apertures that can be physically constructed.

As discussed in Chapters 4 and 5, and elsewhere in the Manual, radars not only work during bad weather, but they also observe phenomena different from those observed in the optical and infrared parts of the spectrum. Thus applications of radar depend both on the all-weather illumination-independent nature of their performance and on the fact that they "see" different things.

The place of radar in geoscience remote sensing has been recognized only recently. Radar is particularly important for time-dependent surveys because of its capability to be flown at the time needed, regardless of environmental or atmospheric conditions. The first major radar survey, in Darien Province, Panama (Viksne, et al., 1969), and parts of the Brazilian (de Moura, 1972) and other more recent tropical mapping projects were areas nearly always covered with clouds, so that aerial photography was almost impossible.

The value of radar for mapping geologic structures was demonstrated during the first radar mapping projects by MacDonald, et al. (1967) and Wing (1971). Radar images have been widely applied to geology, especially in exploration, since these early efforts. The first operational application of radar images to monitoring time-dependent phenomena was in mapping ice on the sea (Loshchilov and Voyevodin, 1972) and in the North American Great Lakes (Gedney, et al., 1975). Radar has been used successfully to map vegetation in various countries, since the early work by Hardy, et al. (1971), but its use in monitoring vegetation growth, particularly that of crops, has been slow to come of age. The ability of radar to produce such maps, however, has been well demonstrated by ground-based experiments (Bush and Ulaby, 1978) and simulations (Li, Ulaby, and Eyton, 1980). The value of radar for detecting oil spills was established by Guinard (1971), and operational systems are in use in several countries. Radar also appears useful, particularly when combined with poorer-resolution microwave radiometry, for mapping soil moisture (Ulaby, et al., 1975). Radar has been used for mapping floods in several disaster areas.

Radar was shown to be useful for monitoring winds on the sea by Claassen, et al. (1972) based on aircraft measurements, and subsequent demonstrations involving Skylab (Young and Moore, 1977) and Seasat (Jones, et al., 1982), with the latter demonstrating the ability to measure wind direction as well as wind speed. Although such systems work through clouds, the use of microwave radiometers to correct for rainfall attenuation depends on the homogeneity of the rainfall or the use of coincident beams and scan patterns for the active and passive systems. The Seasat synthetic-aperture radar (SAR) demonstrated that images of the surface of the sea contain much valuable information (Vesecky, et al., 1982).

A radar scatterometer is a device that measures the scattering properties of the region observed. Any radar that makes an accurate measurement of the strength of the observed signal is therefore a scatterometer. However, most existing radars that produce images are uncalibrated; consequently, most radar scatterometers are not imaging systems. A calibrated imager, however, would also be a scatterometer.

The radar scatterometer has merit of its own in oceanographic applications where its poor resolution is not of great significance. Over land its primary purpose is collecting information which can then be used to design other radars. The altimeter is primarily of use for low altitude profiling, although some altimeter systems have been made that provide additional information, and an altimeter over the sea can be useful for very high accuracy profiling of mean sea level from satellites.

The term *microwave radiometer* is normally applied to passive microwave systems. The signal received by the radiometer is a complex combination of thermally emitted radiation from ground and atmosphere, together with scattered emission originating in the atmosphere and extra-

terrestrially, both in the sun and outside the solar system. Strictly speaking, *a scatterometer is also a radiometer, since it measures a signal radiated by the scatterometer itself and scattered from the target area, but the term radiometer is reserved here for passive devices.*

BASIC RADAR SYSTEM IDEAS

A radar system contains a transmitter which provides energy emitted from an antenna. Energy travels to the target and returns into a receiving antenna; the output of the receiving antenna goes to a receiver which takes the very weak returning signal and increases its strength. The signal from the receiver is related in some way to the transmitted signal and the result of this processing is displayed for the user. Figure 9-1 illustrates such a basic system. Often the same antenna is used for transmitting and receiving. The passive microwave radiometer contains a receiving antenna and receiver much like the radar, but has no transmitter or correlation circuitry.

RESOLUTION CONCEPTS

The proper definition of resolution has to do with the ability to distinguish objects of different contrasts spaced by the resolution distance. In microwave work a less precise definition is customarily used. *For microwaves the resolution is usually meant to be the half-power response width of the measuring system.* This is somewhat analogous to the TV line width, but not exactly so. The half-power width of the response is easier to describe than true resolution because it does not involve the contrast of the target. The true resolution for targets of any character can be derived from the actual response and estimated from the half-power width that we call *resolution* in dealing with microwave sensors.

Figure 9-2 illustrates the kinds of resolution possible with microwave sensors. The active system can distinguish objects by angular range and speed resolution, whereas the passive system is restricted to the angular measurement alone.

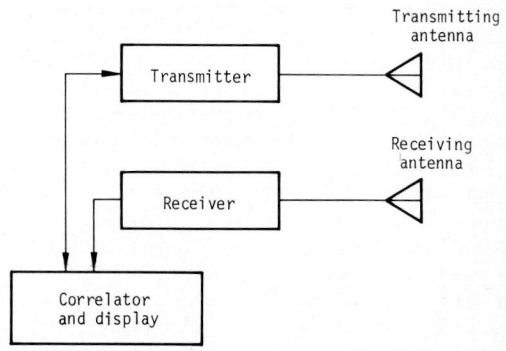

Fig. 9-1. Elements of a radar system. Radiometer lacks transmitter.

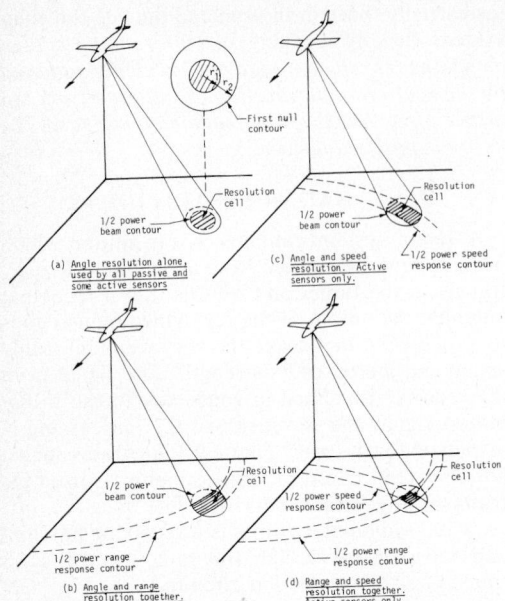

Fig. 9-2. Resolution techniques used in microwave sensing.

Figure 9-2a shows an aircraft flying horizontally with an approximately circular antenna beam pointed at an angle to the side. The cross-hatched area represents the "resolution cell;" that is, the area contained within the half-power contour on the ground if the antenna were used for transmitting. The radiation intensity along the contour is one-half the maximum intensity found at a point within the contour. The size of the resolution cell is determined by the angular beamwidth of the antenna. Both passive and active systems can use this kind of *angle resolution*.

In Figure 9-2b the resolution cell for the antenna used in Figure 9-2a has been reduced by using the range measurement capability of a radar. The resolution in range is defined by the half-power width of the range response. For a particular delay time and a plane ground surface, resolution with range alone would be a ring bounded by concentric circles, the distance of which from the radar would represent the half-power range responses of the radar. The range resolution can be readily made finer than resolution due to antenna beamwidth because the physical limitations on the size of an antenna that set the minimum resolution cell for angle are not present for range measurements.

By comparing the frequency of the transmitted signal with the received frequency, the Doppler frequency shift of the received signal can be observed. This Doppler shift is directly proportional to the relative speed of the aircraft and the point on the ground. Figure 9-2c shows resolution in speed measurement (really resolution in Doppler frequency measurement) combining with angular resolution to reduce the size of the resolution cell. For the horizontally moving aircraft, contours of

constant relative speed are hyperbolas, in which the resolved region on the ground corresponds with the half-power width of the response in Doppler frequency and consequently in relative speed. The figure shows the combination of such a resolution with that due to the angle measurement. As with range measurement, the size of the resolution cell is significantly smaller, but the direction for the fine resolution is different from that for the range measurement.

If angle measurement is not used, but range and speed measurements are combined, the radar resolution cell can be much smaller than would be feasible with the angular measurement due to an antenna beamwidth. This is illustrated in Figure 9-2d. A radar using this technique is called a *synthetic aperture radar*. The technique is also sometimes described as *Doppler beam sharpening*.

Range measurement and Doppler beam sharpening are possible with a radar because the designer can correlate the received signal with the illumination signal. Since the radiometer measures signals emitted by the target or by the atmosphere or extra-terrestrial sources, this correlation is essentially impossible for a passive system. Consequently the fine resolution indicated in Figure 9-2d must be degraded to the coarse resolution of Figure 9-2a if we are to take advantage of other properties of the passive microwave radiometer.

ANGLE DISCRIMINATION

Discrimination in angle by the use of narrow-beam antennas is common to many radars and all microwave radiometers. The limitations of antenna beamwidth are determined by the same diffraction phenomena that limit the resolution obtainable in the visible and infrared regions with lenses and reflectors. In fact, both lenses and reflectors are used in the microwave region.

Elementary Radiators

Antennas frequently consist of many elements a half-wavelength long or smaller. The fundamental element is the *Hertzian dipole*—electric current flowing for a short distance (Fig. 9-3). The Hertzian dipole and short dipole antennas radiate maximum intensity perpendicular to the direction of current flow and do not radiate along the cur-

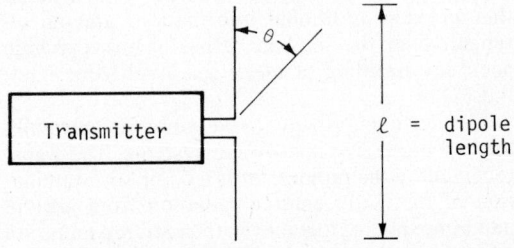

Fig. 9-3. Elementary radiating antenna (dipole).

rent axis. The strength of the electric (E) or magnetic (H) field radiated is

$$E = \alpha_E \frac{I\ell \, \sin \theta}{R} \qquad (9\text{-}1)$$

$$H = \alpha_H \frac{I\ell \, \sin \theta}{R} \qquad (9\text{-}2)$$

where:

$\alpha_{E,H}$ = scale factors,
I = current,
ℓ = dipole length,
R = distance to observation point,
θ = angle from one end of the dipole, measured in a plane containing the dipole, 0° to 180°.

The magnetic field of a dipole at any point in space is tangential to a circle through that point perpendicular to and centered on the current axis. The electric field is perpendicular both to the magnetic field and to a radius from the center of the antenna. It is, therefore, tangential to a circle centered on the dipole and containing the dipole within the plane of the circle. Thus the electric field vector has a component parallel to the dipole and a component toward or away from the axis of current; and the *polarization* of the radiation is in the direction of the electric field vector. These points are illustrated in Figure 9-4.

Power is radiated away from an antenna and the power per unit area in the radiated field is the product of the electric and magnetic fields. The expression for power per unit area at any point in space is therefore proportional to the square of the field expression and is given by

$$P = EH = \alpha_E \alpha_H \frac{(I\ell)^2 \, \sin^2 \theta}{R^2}. \qquad (9\text{-}3)$$

Figure 9-5 illustrates the power per unit area at a fixed radius. (If the radius is one, this is power per unit solid angle.) The short dipole is so nondirectional that it can only be used for angular discrimination *against* a particular direction by orienting the current toward the point where no radiation is desired.

When an antenna is used for receiving, its

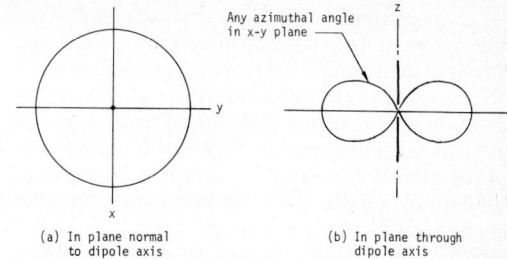

(a) In plane normal to dipole axis

(b) In plane through dipole axis

Fig. 9-5. Distribution of intensity radiated by short dipole.

angular discrimination is essentially the same as for transmitting; that is, its response to signals coming from particular directions is reduced from the maximum response by the same proportion that the signal emitting from the transmitting antenna in that direction is reduced from the signal emitted in the direction of maximum radiation. Hence the patterns of Figure 9-5 may be thought of either as transmitting patterns or as receiving patterns. Pointing the current vector (which means pointing the wire in a short antenna) toward the direction of an incoming signal reduces the response of the antenna to zero. This technique is used in direction finding where the antenna is rotated until the signal is zero and the direction noted. It is not very useful for radar, however.

A loop antenna, the circumference of which is small compared with the wavelength, behaves like a dipole perpendicular to the plane of the loop, except that the role of electric and magnetic fields is interchanged. Such an antenna is therefore thought of as a *magnetic dipole*. The use of loops for direction finding is well known. A directional antenna for radar could be made of a large number of loop elements, but seldom is.

When a slot is cut in a metallic sheet separating a region where uniform electromagnetic fields exist from a region otherwise shielded from them, a signal may leak through the slot. A rectangular slot about a half-wavelength long, and much narrower than it is long, distributes its radiation into the otherwise shielded region like an antenna element, although the directivity of a single slot would be entirely inadequate for most radar purposes. Figure 9-6 illustrates this point.

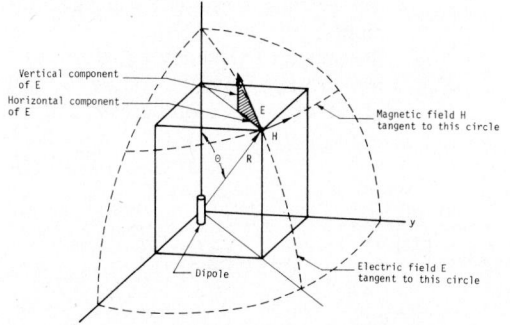

Fig. 9-4. Fields radiated by a dipole.

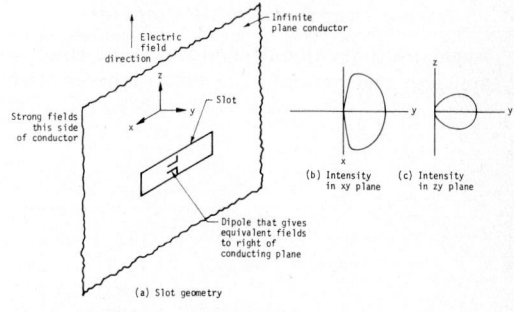

(b) Intensity in xy plane

(c) Intensity in zy plane

(a) Slot geometry

Fig. 9-6. Radiation from a slot.

If an electromagnetic (EM) wave carried in an enclosed guide is to be transmitted into space, the end of the guide may be left open to space. A more effective antenna is achieved if the end is flared out to form a *horn*. These general types of antennas are illustrated in Figure 9-7. Proper design of a horn permits the energy to be concentrated into a narrow beam, so that a horn antenna may be used for angular discrimination. Typically, however, a horn large enough to accomplish this purpose for radar is so long that it is infeasible for aircraft or spacecraft usage. If a relatively large resolution cell is required, a horn antenna may be the answer because, for a large resolution cell, a wide beamwidth can be used and a wide-beamwidth horn may be of reasonable size.

Arrays, Reflectors, and Lenses

Most radars and many radiometers used in remote sensing have antennas consisting of *arrays* of dipoles, slots or horns. At long radar wavelengths, dipoles are fairly uncommon; at shorter wavelengths slots are more widely used. (Such arrays produce narrow beams as discussed below.) The most significant result, however, is that the beamwidth is inversely proportional to the number of wavelengths across the array in the appropriate direction. Thus a narrow horizontal beamwidth requires an array with a horizontal dimension of many wavelengths. A narrow vertical beamwidth requires an array with a vertical dimension of many wavelengths. These points are illustrated in Figure 9-8.

The EM field set up across the aperture of an array by its separate elements may also be set up across a similar aperture by an appropriately shaped reflector accepting a signal coming from a single source; when this happens the radiation is directed in the same manner as for the array. Reflector antennas are commonly used for scanning radars. The energy from the transmitter is usually provided to the reflector by a horn, but sometimes a dipole or array of dipoles or slots may be used to illuminate the reflector. Figure 9-9 shows a reflector and a lens antenna. The arrangement for providing illumination of the lens from the transmitter or for collecting the signal focused by the lens is similar to that for a reflector, as indicated in the figure.

Angle Discrimination Parameters

Angle discrimination depends upon the antenna beamwidth and pattern. The limitations of such

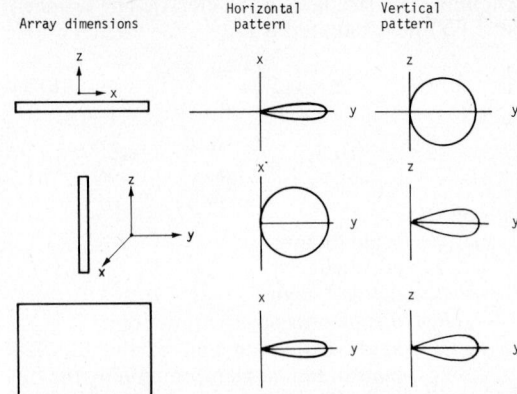

Fig. 9-8. Beam patterns are narrow for direction where array is large.

patterns can only be understood if the method by which the pattern is derived is also understood. The simplest example is a two-element microwave interferometer.

Consider the interferometer shown in concept in Figure 9-10 used as a receiving antenna. Each element is assumed to be able to receive independently of the angle of the incoming radiation, although such isotropic reception is physically impossible and only used as a mathematical convenience. If the interferometer elements are short dipoles, the pattern of each is sufficiently non-directive so that results with the isotropic element are approximately the same; indeed they are identical in the plane perpendicular to the dipoles (Fig. 9-5).

The voltage at a receiver connected to the two interferometer elements by equal-length transmission lines is

$$V = V_a + V_b$$

$$= V_{ao} e^{j\left(\omega t - \frac{2\pi R_a}{\lambda}\right)} + V_{bo} e^{j\left(\omega t - \frac{2\pi R_b}{\lambda}\right)},$$

$$(9-4)$$

where

V = voltage at receiver,
$V_{a,b}$ = the voltages received by the individual elements,
$V_{ao,bo}$ = the voltage amplitudes at the elements,
ω = the angular frequency equal to $2\pi f$,
$R_{a,b}$ = the distance from the source to each of the elements (source assumed far

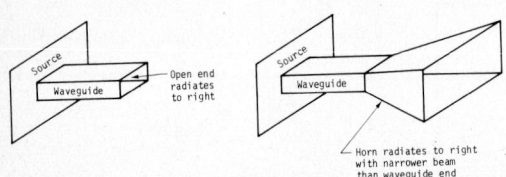

Fig. 9-7. Radiation from open-end waveguide and horn.

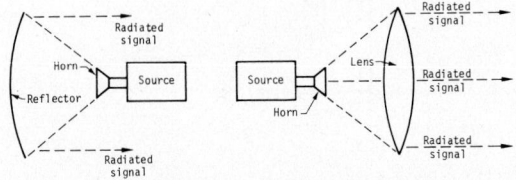

Fig. 9-9. Radiation from reflector and lens antenna.

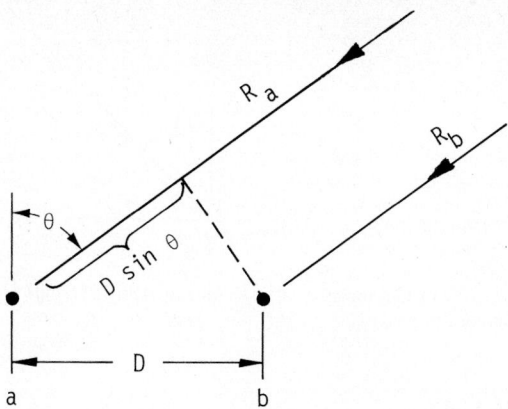

Fig. 9-10. Geometry of interferometer derivation.

enough away so the lines indicated on the diagram for R_a and R_b are parallel),

λ = the wavelength of the received radiation.

Normally the interferometer would be set up so that for a distant source the amplitudes received at the two elements would be equal. Thus

$$V_{ao} = V_{bo} = V_o. \qquad (9\text{-}5)$$

Substituting this in Eq. 9-4, we obtain

$$V = V_o e^{j\omega t}\left(e^{-j\frac{2\pi R_a}{\lambda}} + e^{-j\frac{2\pi R_b}{\lambda}}\right). \qquad (9\text{-}6)$$

The distance from the interferometer to the source may be conveniently expressed by the distance from the center of the interferometer, R. That is, it is convenient to use

$$R = \frac{R_a + R_b}{2}. \qquad (9\text{-}7)$$

In these terms, and using the other parameters shown in the figure,

$$R_a = R + \frac{D \sin \theta}{2}, \quad R_b = R - \frac{D \sin \theta}{2}. \qquad (9\text{-}8)$$

This substitution permits rewriting Eq. 9-6 as

$$V = V_o\, e^{j\left(\omega t - \frac{2\pi R}{\lambda}\right)}$$

$$\left(e^{-j\frac{\pi D \sin \theta}{\lambda}} + e^{+j\frac{\pi D \sin \theta}{\lambda}}\right). \qquad (9\text{-}9)$$

Thus the magnitude of the received voltage can be seen to be

$$|V| = 2V_o\left|\cos\left(\frac{\pi D \sin \theta}{\lambda}\right)\right|. \qquad (9\text{-}10)$$

This is the shape, as plotted in Figure 9-11, of the reception pattern for the interferometer, or the radiation pattern if the interferometer is to be used for transmission. Clearly the return is a maximum on a line perpendicular to the axis of the interferometer ($\theta = 0$), but is of the same magnitude at a number of other angles. Each peak is called a *lobe*.

Characteristics of Lobes

The width of the lobes may be established by observing that $|V| = 0$, when

$$\frac{\pi D}{\lambda} \sin \theta_n = (2n + 1)\frac{\pi}{2}, \qquad (9\text{-}11)$$

so that

$$\sin \theta_n = \frac{2n + 1}{2}\left(\frac{\lambda}{D}\right). \qquad (9\text{-}12)$$

whence the width can be established from

$$\sin \theta_n - \sin \theta_{n-1} = \frac{\lambda}{D}. \qquad (9\text{-}13)$$

For narrow lobes the first few lobes satisfy the criterion

$$\theta_n \ll 1, \qquad (9\text{-}14)$$

so that the width of these lobes is

$$\Delta\theta \approx \frac{\lambda}{D}. \qquad (9\text{-}15)$$

This very significant result states that the width of the lobe, and consequently the resolution, is inversely proportional to the size of the interferometer measured in wavelengths. Thus a one-milliradian lobewidth would require an interferometric baseline a thousand wavelengths long. Although this would only be a meter at a wavelength of 1 mm, it would be 30 m at a more common wavelength of 3 cm; and a kilometer at the possible wavelength of a meter.

For antennas that have only a single major lobe (and normally these are required for remote sensing) the beamwidth is considerably wider than for the interferometer lobe. The pattern of such an antenna is shown, together with the interferometer patterns, in Figure 9-12.

To illustrate what happens with actual antennas, consider an antenna completely filling an aperture of length D. Such an antenna could be a reflector, a lens, or a horn, and approximately the same result would be obtained with an array consisting of many elements a half-wavelength apart or closer. The voltage received by such an antenna is given by

$$|V| = V_1 \int_{-D/2}^{D/2} f(x)\, e^{-j\left(\frac{2\pi x \sin \theta}{\lambda}\right)} dx, \qquad (9\text{-}16)$$

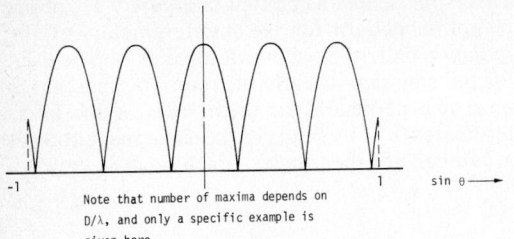

Note that number of maxima depends on D/λ, and only a specific example is given here.

Fig. 9-11. Received voltage on interferometer.

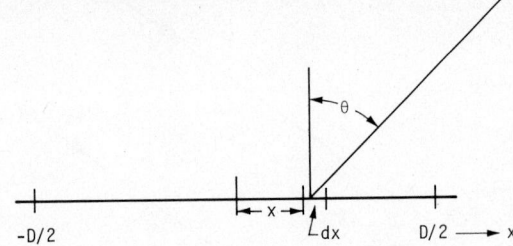

Fig. 9-13. Geometry of a single-line aperture of length D along the x axis.

where:

> $V_1 dx$ = voltage received for a length dx in which the weighting function $f(x)$ is unity.

The geometry of this antenna is shown in Figure 9-13. Equation 9-16 is a general formula. If the antenna extends in two directions, the resulting voltage is the product of an expression like that on the right in Eq. 9-16 for each dimension.

Consider as an example the simple case where the weighting of the reception from each part of the aperture is the same; that is, the aperture has uniform weighting and

$$f(x) = 1. \qquad (9\text{-}17)$$

The resulting expression for the voltage may be expressed using the variable

$$u = \frac{\pi D \sin \theta}{\lambda} \qquad (9\text{-}18)$$

as

$$|V| = V_1 D \left(\frac{\sin u}{u} \right). \qquad (9\text{-}19)$$

This well known pattern is plotted in Figure 9-14. The width of the main lobe may be determined from noting

$$\begin{aligned} V &= V_1 D & u &= 0 \\ &= 0 & u &= n\pi \; (n \neq 0), \end{aligned} \qquad (9\text{-}20)$$

so that

$$\Delta\theta = 2\lambda/D. \qquad (9\text{-}21)$$

Clearly the main lobe for the uniformly weighted aperture is twice as wide as that for the interferometer where reception is only at each end of the aperture. In exchange for this wider lobe, the signals on all other lobes have been reduced. The first *side lobe* has a level 13.2 dB below the main lobe, and successive side lobes are smaller.

Antenna Gain and Directivity

The gain and directivity of an antenna are defined in terms of power rather than voltage. Hence

$$G(\theta) \text{ is proportional to } |V|^2 \qquad (9\text{-}22)$$

where:

> G = gain of antenna.

For uniform weighting across the aperture, this means that the gain is

$$G_u(\theta) \text{ is proportional to} \left(\frac{\sin u}{u} \right)^2. \qquad (9\text{-}23)$$

Beamwidth

The beamwidth is defined at the halfpower points. For the interferometer this means that for the point half a beamwidth ($\beta/2$) to the side

$$\cos^2 u = 1/2. \qquad (9\text{-}24)$$

Consequently this point occurs where

$$u = \pi/4,$$

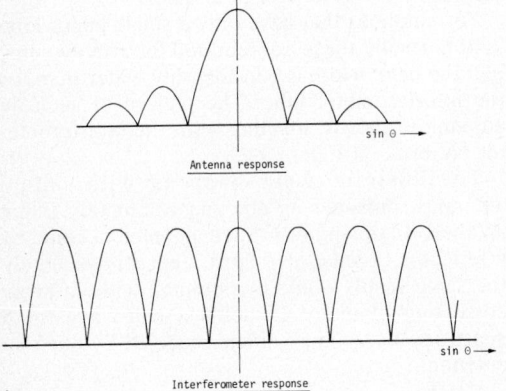

Antenna response

Interferometer response

Fig. 9-12. Comparison of interferometer and antenna responses.

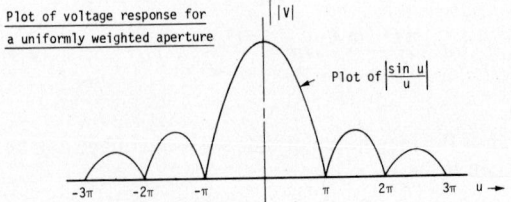

Plot of voltage response for a uniformly weighted aperture

Plot of $\left| \frac{\sin u}{u} \right|$

Fig. 9-14. Plot of voltage response for a uniformly weighted aperture.

and the beamwidth for the interferometer is given by

$$\beta_{int} = \lambda/2D. \qquad (9\text{-}25)$$

For uniform weighting in the antenna, the half beamwidth ($\beta/2$) occurs where

$$\left(\frac{\sin u}{u}\right)^2 = 1/2,$$

which results in a beamwidth of

$$\beta_u \approx 0.88\lambda/D. \qquad (9\text{-}26)$$

Sidelobe Level and Beamwidth Tradeoff

This widening of the beam results in a larger, or poorer, angular resolution, and a larger resolution cell on the ground for a microwave remote sensor than for an interferometer. A tradeoff always exists between sidelobe level and beamwidth. A major problem in antenna design is to achieve suitable beamwidth with sufficiently small sidelobes, since the strength of the sidelobes can be quite important for microwave remote sensors. For a radar looking at a strong isolated target, the signal may be strong enough to be seen not only at the main lobe location, but at several sidelobe locations. Since one normally assumes that a target is located in the main lobe, the other responses give false indications of the location of the object. For the microwave radiometer, signals picked up on minor lobes may come from sources at much higher temperatures than the source at which the main lobe points. Consequently they may increase the observed temperature by a larger fraction than indicated by the actual size of the side lobe.

Table 9-1 gives four examples of the tradeoff between beamwidth for the main lobe and strength of the signal received on a side lobe (Sherman, 1970). Note that the linear taper gives a sidelobe level 13 dB lower than uniform weighting, but this is at the expense of almost a 50% degradation in beamwidth (and resolution). More efficient tapers can be designed, but lower sidelobe levels always mean wider main beams.

Antenna technology has been highly developed, but no reasonable way has been found to circumvent the fundamental idea that the angular resolution is of the order of λ/D. The antennas used in airborne remote sensing radars are quite simple by comparison with those for scanning the airspace from ground or ship. Antennas for microwave radiometers may be more complicated because of more complex scanning and more stringent sidelobe requirements.

Typical remote sensing radar antennas require narrow beamwidths in the horizontal dimension and wide beamwidths in the vertical. Early mechanically scanned radars, and some of the sidelooking radars as well, used parabolic reflecting antennas. Although some such antennas used circular reflectors, those with finer resolution used horizontal parabolic cylinders as reflectors. Thus, the vertical beam was formed by the parabolic surface and the horizontal beam by a long "feed" structure at the focus of the parabola.

Slot arrays are normally used for remote sensing radar antennas. Each horizontal row of slots is cut in the wall of a horizontal waveguide, and several such waveguides are stacked vertically, with appropriate excitation amplitude and phase taper on the aperture (to achieve the desired vertical pattern) provided by the structure connecting the antenna waveguides with the radar.

Lens antennas are seldom used in airborne applications, for the need to locate a feed or pickup at the focal point of the lens usually would require a structure too clumsy to use on an aircraft.

RANGE DISCRIMINATION

Discrimination in range is used by most radars, but is not possible with passive microwave systems. Range measurement or discrimination depends upon measuring the finite time required for a signal to go from the radar to the target and back, or upon discriminating between the times required for travel to and from different targets.

The radar signal is an EM wave and travels through the atmosphere or space with the velocity characteristic of such a wave. Usually this is the same as the velocity of light, although very minor differences occur in the atmosphere because of differences in the refractive properties at optical

TABLE 9-1

Beamwidth-Sidelobe Tradeoff

f(x)	Weighting	½-Power Width (radians)	First Sidelobe Level (dB relative to main beam)
	Interferometer	0.5 λ/D	0
	Uniform	0.88 λ/D	−13.2
	Linear taper	1.28 λ/D	−26.4
	Cos² taper	1.45 λ/D	−32.0

Source—Sherman (1970)

and microwave frequencies. For most purposes the value 3×10^8 m s^{-1} is adequate for describing the performance of a radar system. If extremely accurate range measurements are needed, the values contained in appropriate handbooks may be used and modified as need be because of the effects of the atmosphere. The effects of atmospheric variation, however, represent only a few parts per hundred thousand in the velocity.

The time delay associated with a wave traveling from radar to target and back is that for a wave traveling *twice* the *range* to the target. That is, time is given by

$$T = 2R/c, \qquad (9\text{-}27)$$

where:
R = range
c = mean velocity of the wave.

If a very short duration pulse is transmitted to a target at distance R, and the transmitted pulse is displayed together with an amplified version of the received signal on an oscilloscope, the result is as shown in Figure 9-15. That is, a very small target causes a reproduction of the transmitted pulse at a time delay T following the transmission.

The extremely short pulse is not feasible, although it may be approached by suitable choice among the various wave forms that can be used for radar transmission. The technique of range measurement always requires identification of the transmitted wave and comparison with its replica received from the target. A common transmitted wave form is a short pulse. Sometimes a continuous signal is varied in frequency and the relationship between the transmitted and received frequency at any instant can be used as a measure of range. Sometimes these two techniques are combined; sometimes still other wave forms are used, including even noise modulation. Here we discuss briefly the concepts of pulse modulation, frequency modulation, and their combination in pulse-compression systems. (Modulation concepts have been introduced in Chapter 15). Initially only returns from isolated point targets are treated, although in fact, radars that are used for remote sensing of the earth work with extended rather than with point targets.

Pulse Modulation

First consider ordinary pulse modulation. The ideal pulse would be of zero duration, but such a pulse would require infinite amplitude to achieve the finite energy required for overcoming noise,

besides which it would be impossible to transmit through any kind of realistic equipment. Consequently pulses that are actually used are often rectangular in shape. The rectangular pulse is nearly always used as a model for simplified analysis, but it can only be approximated in practice. Therefore consider here a pulse of somewhat more general shape.

If a pulse $p_T(t)$ is transmitted, the received pulse $p_R(t)$ is given by

$$p_R(t) = Ap_T(t - T) = Ap_T(t - 2R/c), \quad (9\text{-}28)$$

where:
A = a factor containing all the radar parameters, including those of the system and of the path through the space, as well as the target;
T = the time delay described above.

It can be seen that the received pulse is a replica of the transmitted pulse except for the amplitude difference given by A, but it is delayed in time by the interval required for transmission to and from the target.

How do we compare the received with the transmitted pulse to determine the time delay? The simplest method and that used in the early radars, as well as in many present day radars, is illustrated in Figure 9-16. The pulse generator that supplies the transmitter with modulation also supplies a trigger to an oscilloscope. The sweep on the oscilloscope starts simultaneously with the transmitted pulse. The signal out of the receiver is coupled to the vertical deflection circuit of the oscilloscope so that the received signal deflects the display. Thus, as soon as the pulse returns from the target, a *pip* appears on the oscilloscope screen. The time delay can be measured simply by relating to time delay the distance from the start of the sweep to the initial appearance of the pip and knowing the sweep rate for the oscilloscope. If a more accurate measurement is desired, the start of the sweep may be delayed for a known period of time, and the scale on the oscilloscope expanded to facilitate measurement. Such a display, known as an *A-scope,* was commonly the only one found on the early radars. Modern radars occasionally use an A-scope display as an auxiliary, but customarily combine it with some other, more readily interpreted, form of display, as will be discussed later.

Frequency Modulation

Frequency-modulated (FM) radars are almost as old as pulse-modulated radars, the first

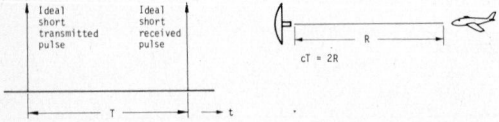

Fig. 9-15. Radar return from point target at range R for very short transmitted pulse.

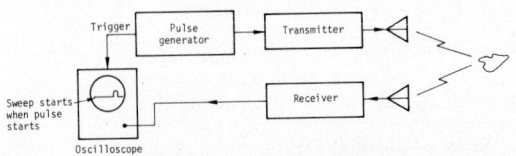

Fig. 9-16. Simple pulse radar.

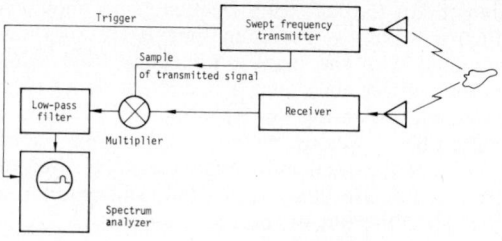

Fig. 9-17. Simple FM radar.

$$f_T = f_0 + at, \qquad (9-29)$$

where:

f_T = transmitted frequency,
f_0 = initial frequency,
a = rate of change of frequency.

The frequency received is that which was transmitted earlier at time $(t - T)$; that is,

$$f_R = f_T(t - T) = f_0 + a(t - T), \qquad (9-30)$$

where

f_R = received frequency.

The input to the multiplier is the product of two sinusoidal signals, one at the transmitted frequency and one at the received frequency. For simplicity, it will be assumed that both may be represented by equal-amplitude cosines without phase terms. The result would be the same in terms of frequency, although more complex algebraically, if amplitude and phase differences were both considered. The output of the multiplier is given by

$$V_m = \cos \omega_T t \cos \omega_R t$$

$$= \frac{1}{2} \left[\cos(\omega_T + \omega_R)t + \cos (\omega_T - \omega_R)t \right], \qquad (9-31)$$

where the ω's are the angular frequencies associated with the frequencies f_T and f_R.

The low-pass filter removes the sum term which is at approximately double the transmitted frequency, so that the signal going to the spectrum analyzer is

$$V'_m = \frac{1}{2} \cos(\omega_T - \omega_R)t. \qquad (9-32)$$

As a result, the spectrum analyzer displays a frequency f'_m given by

$$f'_m = f_T - f_R$$

$$= f_0 + at - f_0 - at + aT = aT = \frac{2aR}{c}. \qquad (9-33)$$

Thus a single frequency appears on the spectrum analyzer for the point target, just as the single line appeared for the point target on the oscilloscope with the ideal short pulse. The scale factor $(2a/c)$ serves the same purpose in relating the range R to frequency that the scale factor $(2/c)$ does for the pulse radar.

For real FM radars, as indicated by Figure 9-18b, the sweep cannot extend over an infinite frequency range, just as we cannot use a zero duration pulse for the pulse radar. The effect of the finite range of sweeping, like that of the finite pulse duration, is to stretch out the response from the point target on the spectrum analyzer. The equivalent duration of the response on the spectrum analyzer measured in terms of range is at best approximately $1/B$ where B is the bandwidth of the sweep. In fact, the range measurement at best will be spread out in time over an interval

frequency-modulated altimeter having been described in the literature in 1939. Figure 9-17 shows the simple FM radar. The transmitter for the FM radar operates continuously, but its frequency is swept over a predetermined range. The block diagram indicates a trigger associated with the starting of the sweep. The received signal is compared in a multiplier with a sample of the transmitted signal. The output of the multiplier contains signals at approximately twice the transmitted and received signal frequencies as well as the difference frequency. The low-pass filter eliminates the double-frequency term and the output can then be displayed on a spectrum analyzer that serves the same purpose as the oscilloscope does for the pulse radar.

To better understand the operation of the FM radar, consider Figure 9-18, which describes what happens over an ideal portion of the sweep for the FM radar in (a) and over a more realistic situation in (b). At this point only (a) will be analyzed. The variable transmitted frequency plotted there may be considered to consist of an initial frequency and a continuously increasing one, as given by

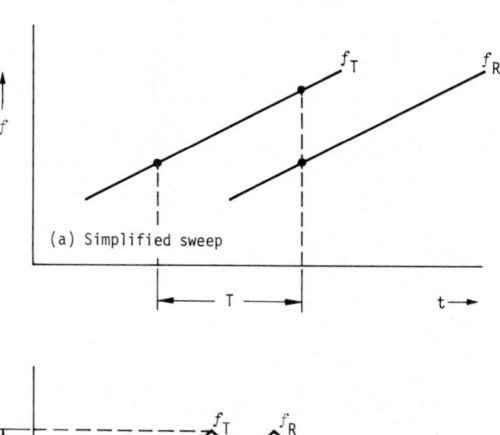

Fig. 9-18. Frequency relations for single FM sweep.

approximately $1/B$, regardless of what type of modulation is used to achieve the bandwidth B.

Pulse-Compression Range Measurement

Achieving a really fine range resolution with a pulse radar requires transmission of a very short pulse. If the pulse is thus very short, it must also be very powerful so that it contains enough energy to overcome the noise. The technique of pulse compression permits use of longer, lower amplitude pulses to achieve the kind of resolution otherwise possible only with very short, high-amplitude pulses.

The most common pulse compression technique is *chirp*, in which the techniques of frequency modulation and pulse modulation are combined. A pulse is transmitted, and its frequency is modulated. Upon reception the resolution obtainable is determined by the bandwidth over which the frequency has been swept, rather than by the pulse duration. In fact, the usual method for processing the received signal converts the long frequency-modulated pulse into a shorter pulse from which the frequency modulation has been removed.

The technique of chirp radar is illustrated in an idealized manner in Figure 9-19 (from Cook and Bernfield, 1967). Here in (a) and (b) the frequency and amplitude of the transmitted signal are shown at the left. The transmitted pulse has a duration τ and is swept over a band from frequency f_1 to f_2 with the total bandwidth being B. After the usual time delay T, a received signal from a point target comes in as a replica of the transmitted pulse, as shown. To take advantage of the chirp nature of the modulation, the received signal is passed through a filter, the time delay of which is depen-

dent upon the incoming frequency, as shown in Figure 9-19c. A minimum time delay t_{d2} is associated with the highest frequency (that transmitted at the end), and a maximum delay t_{d1} is associated with the lowest frequency (that transmitted first). Figures 9-19a and b illustrate the result: that the received signals from the different frequencies are delayed just the right amount so that all come out of the filter together. Since all the energy of the long pulse is concentrated in a very short time, the amplitude at the output is much higher than the received pulse amplitude, as indicated.

The illustration in Figure 9-19 would tend to indicate that the output pulse is of zero duration, but this of course is not possible. In fact, if the transmitted pulse is constant in amplitude and varies only in frequency, as indicated in Figure 9-19 and 9-20a, the output of the filter takes the form of a pulse shaped in the form of $(\sin x)/x$ [Fig. 9-20b]. The effective width of this pulse τ_p is the reciprocal of the transmitted bandwidth, just as with the FM radar. The amplitude of the pulse is increased by the square root of the time-bandwidth product $B\tau$, as shown. This is as would be expected, since the energy in the "de-chirped" signal should be the same as that at the input to the filter.

The $(\sin x)/x$ form of the output wave leads to some possible ambiguity as to the position of a given target; that is, instead of the target appearing at only one position, it appears with reduced amplitude at each peak of this function. For some purposes this is not important. On the other hand, if the *side lobes* of this wave form are sufficiently large to be troublesome, techniques similar to those used to reduce side lobes for antennas may

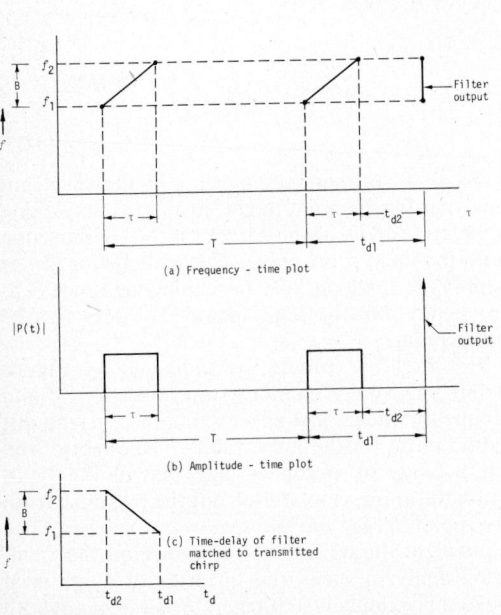

Fig. 9-19. Idealized chirp radar performance.

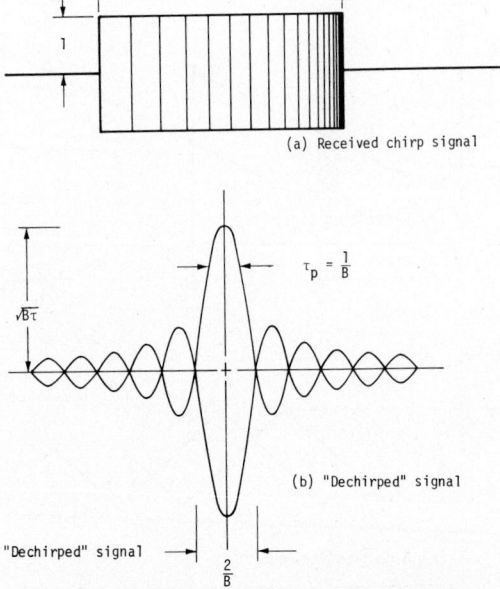

Fig. 9-20. Realistic output waveform for chirp radar.

also be used here. For instance, the transmitted pulse would not have to be of constant amplitude; and even if it were, a filter could be used that would have the effect of reducing the amplitude of the secondary responses. Techniques for generating the chirp signal and for suppressing the undesired responses are discussed in numerous references.

The chirp wave form indicated is not the only one that can be used for pulse compression. Various modern radars use binary codes, usually with each bit during the transmitted pulse having the same amplitude, but with the phase shifting by 180° between one and zero of the binary numbers. Suitable binary codes can achieve results quite similar to those of the FM chirp system. Details of these techniques can be found in appropriate references.

SPEED MEASUREMENT

The Doppler Principle

The possibility of measuring speed or relative velocity with a radar is based upon the Doppler effect. Because of this effect, the frequency returned to the radar is different from the frequency transmitted by the radar, and the difference is proportional to the relative velocity of the target and radar. The acoustic Doppler effect is, of course, familiar to everyone because of the change in frequency of moving single-tone sources such as train whistles.

As previously discussed in this Manual, *the cause of the Doppler frequency shift is a change of the phase of the signal due to motion through the wave pattern in space that adds to the normal phase change due to transmission of a sinusoidal wave.* The instantaneous phase of a signal returned to a radar, neglecting phase shifts in the reflection process, is given by

$$\text{Instantaneous Phase} = \phi = \omega_c \left(t - \frac{2R}{c} \right)$$

$$= \omega_c t - 4\pi R/\lambda_c. \quad (9\text{-}34)$$

Here ω_c is the angular carrier frequency of the radar, and λ_c is the associated wavelength. If there is no motion, the angular frequency, which is the rate of change of phase, is

$$\omega = \frac{d\phi}{dt} = \omega_c. \quad (9\text{-}35)$$

When motion is present, both terms in Eq. 9-34 contribute to the rate of phase change; that is, we have

$$\omega = \frac{d\phi}{dt} = \omega_c - \left(\frac{4\pi}{\lambda_c} \right) \frac{dR}{dt} = \omega_c - 4\pi u_R/\lambda_c. \quad (9\text{-}36)$$

Expressing this in terms of the frequency rather than the angular frequency, it becomes

$$f = f_c - 2u_R/\lambda_c. \quad (9\text{-}37)$$

The second term in Eq. 9-37 is the Doppler frequency f_D. That is

$$f = f_c + f_D; \quad f_D = -2u_R/\lambda_c. \quad (9\text{-}38)$$

In these equations, u_R is the relative velocity, or rate of change of R. The factor of 2 which appears in the Eqs. 9-34, 9-37, and 9-38 is because of the roundtrip distance involved for radar. For Doppler effect in a one-way transmission, this would not be present.

For increasing R, the Doppler effect may be thought to happen because both the wave and vehicle are traveling in the same direction, so that the vehicle tends to a minor extent to "keep up with" the travel of the wave. Consequently it takes slightly longer for a single cycle of the wave to go past the vehicle than it would if the vehicle were standing still. Furthermore, during retransmission by the target the time for a cycle to go past the receiver is longer, since the latter parts of the cycle must travel farther than the earlier parts.

Most airborne remote sensors are used with horizontal flight, so it is instructive to consider the Doppler frequencies that are generated for linear motion parallel to a plane surface such as the Earth. Figure 9-21 shows the geometry. The radar is considered to be at a point h above the origin of a rectangular coordinate system (in the z direction). The vehicle is assumed to be moving in the x direction and the target point is at some point on the $z = 0$ surface, but not necessarily on either axis. Adopting the convention $\mathbf{1}_i$ as the unit vector in the direction i, the velocity is given by

$$\mathbf{u} = \mathbf{1}_x u, \quad (9\text{-}39)$$

and the vector distance from radar to target is given by

$$\mathbf{R} = \mathbf{1}_x x + \mathbf{1}_y y - \mathbf{1}_z h. \quad (9\text{-}40)$$

The relative velocity is the component of u in the R direction, that is, it is the rate of change of R and given by

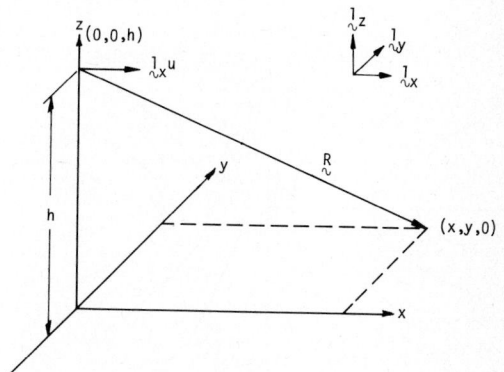

Fig. 9-21. Geometry for Doppler frequency shift calculation.

$$u_R = \frac{dR}{dt} = \frac{\mathbf{u} \cdot \mathbf{R}}{R}. \qquad (9\text{-}41)$$

$$u_R = \frac{ux}{R}.$$

Combining this with Eq. 9-38 gives the Doppler frequency

$$f_D = \frac{2ux}{\lambda_c R}. \qquad (9\text{-}42)$$

From this equation it can be seen that the Doppler frequency is zero along the y-axis where $x = 0$; that is, on a line perpendicular to the flight line, there is no Doppler frequency shift. The maximum Doppler frequency for a given x is along the x-axis; that is, directly ahead of the radar, for this is where R is minimum for a given value of x. Similarly, there is a maximum negative value along the x-axis behind the vehicle. Thus the Doppler frequency for a given depression angle varies from a maximum directly ahead of the radar through zero directly to the side of the radar flight path to a negative maximum directly behind the radar.

This equation can be rearranged, using the rectangular coordinate expression for R, into the form of an equation for a constant Doppler frequency (*isodop*) on the ground:

$$h^2 = x^2 \left[\left(\frac{2u}{\lambda_c f_D} \right)^2 - 1 \right] - y^2. \qquad (9\text{-}43)$$

Note that the quantity in parentheses within the square bracket is greater than one, for the maximum value for f_D is the value it would take on if the radar were going directly toward the target, which would be $2u/\lambda_c$. The contours of constant Doppler frequency are shown in Figure 9-22. These are the contours indicated in Figures 9-2c and 9-2d where resolution on the ground was set by the distance between two isodops separated by the amount associated with the filter bandwidth for Doppler measurement.

Clearly these isodops are approximately parallel to the iso-range circles ahead of and behind the

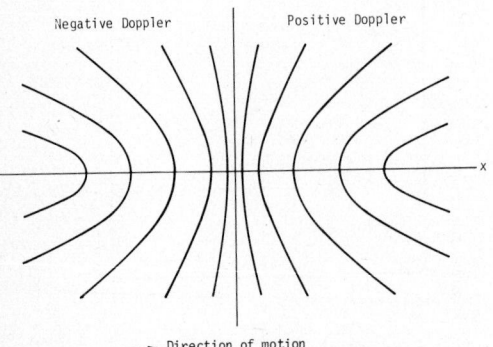

Negative Doppler   Positive Doppler

→ x

→ Direction of motion

Fig. 9-22. Contours of constant Doppler frequency for horizontal aircraft flight.

radar, but are orthogonal to the iso-range circles to the side. Consequently the best resolution for a combined range and speed measurement is attained at the side.

Continuous-wave Doppler Radars

Speed-measurement radars using continuous wave (CW) (unmodulated) transmission have several special uses: the familiar speed measuring by police, the speed (actually velocity) measurement used in a Doppler navigator, various personnel detectors used for military and industrial security, and a radar scatterometer in which the separation of different targets measured on the ground is obtained by filtering out the appropriate Doppler frequencies. For remote sensing, the scatterometer and the Doppler navigator are probably of greatest importance.

The block diagram for a CW-Doppler system can be quite simple, like that of Figure 9-17 for the simple FM radar. The Doppler system is identical except that the transmitter is not to be swept; consequently no trigger from it is necessary.

The Doppler navigator uses three or four separate narrow beams to determine different components of the velocity. A typical version uses two beams pointed forward, one on each side, and two beams pointed aft. If the vehicle is moving in the direction in which it is pointed, the Doppler frequencies for the two forward beams are the same, as they are also for the two aft beams. If, however, the vehicle is crabbing so that actual motion is not in the same direction as the heading, one of the forward beams will receive a higher Doppler frequency than the other; and from the difference the amount of the crab angle can be determined. Of course, if the radar is ascending or descending, the Doppler frequencies are modified accordingly. Such a system uses some kind of device to determine the effective Doppler frequency at the center of each of its beams, rather than the spectrum analyzer shown in Figure 9-17.

Pulse Doppler Systems

Pulse Doppler radars are relatively complex. The simple CW Doppler system uses a sample of the transmitted signal for a reference against which the Doppler frequency can be measured. With the pulse Doppler system, however, this is not possible because reception occurs only while the transmission is not taking place. Thus a more complex system is needed. Typical pulse radars not using Doppler effect use transmitters that are oscillators at the carrier frequency. For example, a common transmitter is a magnetron oscillator. With pulse Doppler systems either a stable microwave source is required, or the magnetron or other less stable transmitter must be used to phase-lock a stable source at a lower frequency.

Figure 9-23 shows a simplified diagram for one type of pulse Doppler radar. The stable microwave source typically is a multiplier chain

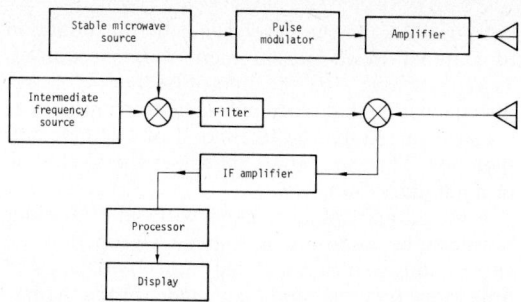

Fig. 9-23. Simplified diagram of a pulse-Doppler radar.

starting with a crystal oscillator in the 10-MHz region, although it might be a carrier-frequency oscillator that is locked in phase with a lower-frequency crystal oscillator. Since the source must operate during transmission to help provide a local oscillator for the superheterodyne receiver, it must be followed by a pulse modulator and an amplifier. The continuously operating source itself must be at a low power level so that a signal from it does not couple directly into the receiver. The pulse modulator therefore must be very effective in cutting off the drive signal to the amplifier so that no residual signal will be amplified to a level that will cause trouble in the receiver. The local oscillator for the receiver is obtained by mixing a sample of the stable carrier frequency with a stable source at the intermediate frequency (IF), thus obtaining either a signal the frequency of which is above the carrier frequency by the amount of the intermediate frequency, or one below the carrier frequency by that amount. The filter removes the unwanted sum or difference frequency so that only one frequency is selected to mix with the incoming Doppler-shifted signal. The output of this mixer is at the intermediate frequency, except for the Doppler shift, and it goes on to processor and display.

Figure 9-24 shows one example of the kind of processor that might be used with a pulse-Doppler radar. The output of the IF amplifier is selected for each range element by a range gate, as shown. To obtain the frequency associated with each range element and consequently the relative velocity with regard to the target at that range, an intermediate frequency source provides a signal

mixed with the output of the range gate. The spectrum analyzer then determines the resulting difference frequency from which the relative velocity can be established. If both positive and negative Doppler frequencies are expected and they must be separated, the source will be at a different frequency than the intermediate frequency, so that the negative Doppler frequencies are not folded on top of the positive frequencies.

Synthetic aperture radars are pulsed-Doppler systems. The processing for a synthetic-aperture radar is different from that for other pulse-Doppler radars, and is discussed later in this chapter in considerably more detail.

FM Radar with Doppler Shift

When speed measurement is combined with range measurement using FM, problems can arise because the indications of range and of speed are frequencies. Figure 9-25 shows what happens with a sawtooth modulation in frequency. Here the dashed line shows the received signal if no Doppler shift is involved; the solid line with the label f_R shows the frequency that is actually received. Assuming a positive Doppler shift (decreasing range), the difference frequency measured by the FM radar during the upward transmitter frequency swing is Δf_1, which is less than would be measured without the Doppler effect. During the downward sweep, the measured frequency is Δf_2, which is greater than would be measured without the Doppler effect.

If a nonsymmetrical modulation wave form had been used for the situation of Figure 9-25, the result would be a net shift in the return frequency due to the Doppler effect which could not be distinguished from the shift due to range. With the symmetrical sawtooth modulation shown, however, the processor can average the results during the upward and downward frequency sweeps to obtain the range measurement and can take the difference between them to determine the Doppler frequency. This technique is widely used with radar altimeters.

AMBIGUITY AND RESOLUTION

The concept of ambiguity is important in measuring both distance and speed with a radar. It could also be important in angular measurement if the antenna should have more than one major beam, so that signals could come in from each of

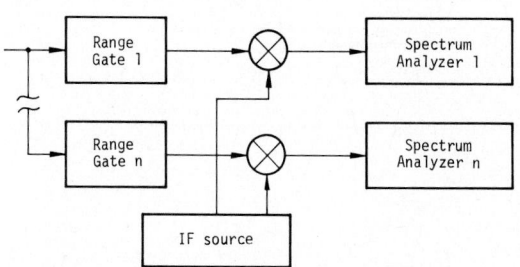

Fig. 9-24. One type of pulse-Doppler processor.

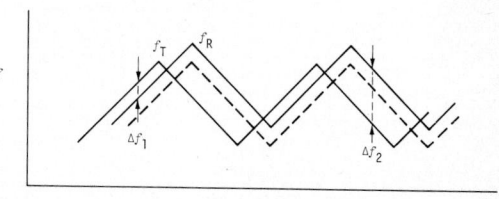

Fig. 9-25. Effect of Doppler shift on FM radar.

two or more angles and not be identified as to direction. The concept of ambiguity is discussed here first in elementary terms; then a concept known as the *ambiguity function* (due originally to P. M. Woodward) is developed. This function not only displays the ambiguous characteristics, but also the combined resolution characteristics in distance and speed.

Range Measurement Ambiguity

To illustrate how range measurement ambiguity can occur, consider Figure 9-26. The top line illustrates a sequence of transmitted pulses separated by time T_0. Normally it is assumed for a radar either that the signals from distances causing greater time delays than T_0 are too small because the distance is too great, or that they are reduced in size by the vertical beamwidth of the antenna. When this is true, a return such as p_{R1} can be observed, but returns at delays like that for p_{R2} and p_{R3} are suppressed by the range or the antenna beam. If neither range nor antenna can suppress signals from these distances, the results shown on the figure can occur for pulses reflected from different objects when radiation from pulse A strikes them. The ambiguity involved for p_{R2} is that the pulse shown, which is actually due to A with time delay $T_0 + T_1$, can easily be mistaken for a signal from a much closer target caused by the transmission of pulse B; that is, the display will show the return at a time T_1 after the start of time measurement, which assumes that this pulse is due to B. Similarly the pulse shown as p_{R3} can be assumed to have started from pulse C. If it is possible for the pulses shown to be due to A, then the ranges associated with time delays T_1, $T_0 + T_1$, and $2T_0 + T_1$ are said to be ambiguous; that is, we cannot tell these ranges apart.

This kind of ambiguity seldom occurred in the early radars, for the interpulse period T_0 was normally made very long compared to the range of the radar. Occasionally abnormal propagation conditions made ambiguous returns possible, but under normal propagation conditions they were

not possible. On the other hand, the introduction of Doppler-speed-measurement radars required, as seen below, that the interpulse period be reduced (that is, that more pulses be transmitted per second) to reduce the likelihood of Doppler ambiguities. This, of course, increases the likelihood of ambiguities in range.

Ambiguities can also exist with an FM radar because the duration of a given sweep may be short enough so that a signal return is delayed in time more than one complete sweep; thus the frequency difference would be ambiguous. This is not a common situation with FM radars, however, because the sweep duration is usually so long that signals with such long time delays are very unlikely.

Speed Measurement Ambiguity

Ambiguities in speed measurement may also occur, as illustrated in Figure 9-27. The top line of this Figure shows the positions of received pulses from a target at a fixed range. If the target is stationary, the frequency received by a CW radar will also be stationary, and the beat frequency output of a CW Doppler radar will be zero. On the other hand, if the target is moving, the frequency received by a CW radar is different from that received from the stationary target, and the difference frequency comes out of the detector of the Doppler radar producing a wave form like that shown in the second line. With a pulse Doppler radar, we may consider that each pulse provides a sample of this wave form; that is, the output of the first pulse will be large and positive, the output of the second pulse large and negative, the third

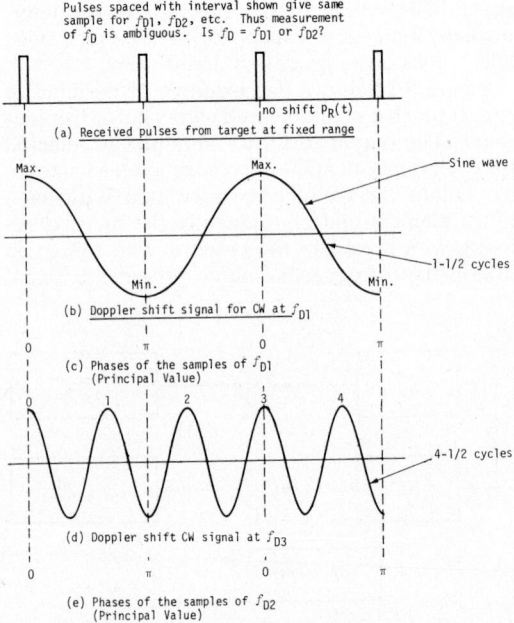

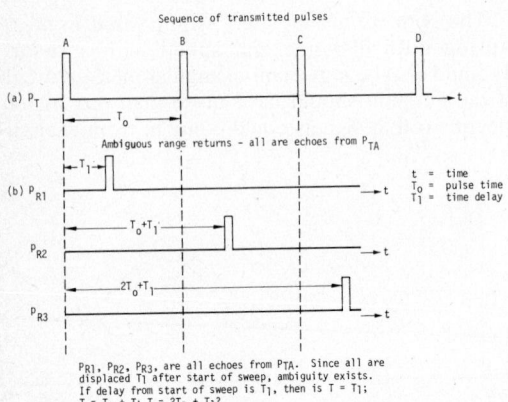

Fig. 9-26. Ambiguous range returns.

Fig. 9-27. Ambiguous Doppler returns.

pulse large and positive, etc. Thus the pulses sample this signal at the interval shown in the third line; that is, 0 and π radians for each cycle. The fourth line shows a higher Doppler-shift output as it would appear for a CW radar; this Doppler shift is three times that of f_{D1} above. The third harmonic is chosen to pictorially show the phenomenon; however, any odd harmonic is applicable. The pulse again samples at the appropriate points, and the samples are the same as for the signal above. In this case, however, the Doppler frequency being sampled is three times as great, so that the second pulse samples the second minimum after the maximum sampled by the first pulse, whereas for f_{D1} the second pulse samples the first minimum. Thus from the output of the pulse Doppler radar, one cannot tell whether f_{D1} or f_{D3} is the Doppler frequency. In fact one cannot tell f_{D1} from any of its odd harmonics. There is thus an ambiguity in speed between the speed appropriate to a Doppler shift f_{D1} and speeds three, five, and other odd integer multiples of the basic measurement speed.

Clearly, transmitting pulses three times as often would permit proper sampling of f_{D3}, and if this were the highest expected velocity, the tripling of the sampling rate seems to be the right thing to do. However, tripling the sampling rate makes range ambiguities (as illustrated in Figure 9-26) more likely, so a trade-off must be made. Some pulse-Doppler radars have very high repetition rates, which means considerable range ambiguity but good speed measurement. In others the range measurement is unambiguous, but the speed measurement is so ambiguous that in effect the radar can only tell whether the target is moving or stationary and cannot really measure its speed!

To understand the ambiguity function and its relation to resolution, the use of a correlation detector must be considered. As a matter of fact, it can be shown that, for some purposes, optimum S/N ratio occurs when a correlation detector or the equivalent matched filter is used. With such a detector the ambiguity function can be used to describe resolution as well as ambiguity. The best resolution is obtained by causing the ambiguity function (which is the output of the correlation detector) to fall to zero as quickly as possible when either range or speed is changed, whichever is the most important quantity to measure.

Consider a transmitted pulse that is a sinusoid modulated by a complex function $u(t)$

$$p_T^R(t) = Re\ u(t)e^{j\omega_c t} = Re\ p(t). \quad (9\text{-}44)$$

Here the real pulse p_T^R is given by the real part of the complex function shown. When this pulse is transmitted, the complex received pulse is given by

$$p_R(t) = u(t - T)e^{j(\omega_c - \omega_D)\ (t-T)}$$
$$= p(t - T)e^{j\omega_D(t-T)}, \quad (9\text{-}45)$$

where both the time delay T and the Doppler angular frequency ω_D are shown.

Range-only Ambiguity Function

To illustrate the use of the ambiguity function, consider first the case for a stationary target; that is, there is no Doppler frequency shift.

In the correlation detector the received signal is compared with a delayed sample of transmitted signal by multiplying the two together and integrating the product over a time long compared with the pulse duration. If the two signals (received and delayed transmitter sample) have the same delay time, they coincide and the output is the square of the input. If, however, they do not overlap, the product is zero everywhere and there is no output. When the complex form of the signal is used, the correlation function is made by comparing the complex conjugate of the delayed signal with the received signal.

Now consider the case where the Doppler frequency is zero. A correlation function can be written in the form

$$\hat{c}(\tau') = \int_{-\infty}^{\infty} u(t - T)e^{j\omega_c(t-T)}u^*(t - \tau')e^{-j\omega_c(t-\tau')}dt$$

$$= \int_{-\infty}^{\infty} u(t - T)u^*(t - \tau')e^{-j\omega_c(T-\tau')}\ dt, \quad (9\text{-}46)$$

where

$\hat{c}(\tau')$ = correlation function
τ' = delay experienced by the transmitter sample
T = delay experienced by the received signal.

Clearly the optimum condition is when τ' is the same as T. For convenience in setting up the integral, define two new variables by

$$\tau = t' - T$$
$$y = t - T. \quad (9\text{-}47)$$

Here

τ = deviation of delay for transmitter sample from what it should be,
y = deviation of position in the product wave form from the time delay for arrival of the pulse.

Using these definitions, and modifying the definition of the correlation function to eliminate the phase effect of the time delay, the simpler form obtained is

$$c(\tau) = \int_{-\infty}^{\infty} u(y)u^*(y - \tau)dy, \quad (9\text{-}48)$$

where:

$c(\tau)$ = range ambiguity function.

This form is like a self-convolution and results in some broadening of the shape of the pulse. Figure 9-28 shows examples of this correlation function, which we define as the *range-only ambiguity function*. The first illustration is for a very short

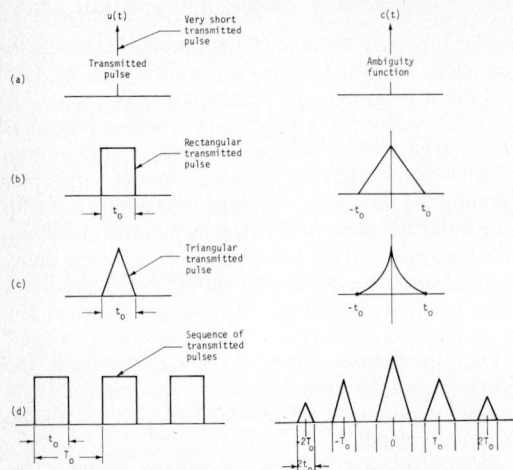

Fig. 9-28. Some transmitted pulses and their range-ambiguity functions: correlation detector outputs compared with relative delay of received and reference pulses.

transmitted impulse which remains the same in the ambiguity function. A rectangular transmitted pulse gives a triangular impulse, the base of which is twice as wide as that of the rectangle. A triangular transmitted pulse results in a parabolic ambiguity function, again with twice the total base width. When a sequence of pulses is transmitted rather than a single pulse, the ambiguity function also contains a sequence. Each "lobe" in the ambiguity function corresponds with an ambiguous time delay; that is, if the correlation is set not at the correct value, but rather at a value different from it by the amount of the interpulse period, the output of the correlation detector will be the same. This can happen either because of an erroneous setting of the time delay, or because two signals being received simultaneously have time delays separated by the amount T_o.

Clearly, if the range ambiguity function $c(\tau)$ is approximately equal to the maximum value occurring where $\tau = 0$, a target separated by this amount from the reference target cannot be distinguished. On the other hand, if $c(\tau)$ is sufficiently small, a target with a spacing τ from the reference target can be easily and unambiguously distinguished.

To obtain an equivalent width for resolution for different pulse shapes, introduce the concept of the equivalent time width $\Delta\tau$ as defined by

$$c^2(0)\Delta\tau = \int_{-\infty}^{\infty} |c(\tau)|^2 \, d\tau. \qquad (9\text{-}49)$$

Thus the width is given by

$$\Delta\tau = \frac{\int_{-\infty}^{\infty} |c(\tau)|^2 \, d\tau}{c^2(0)}, \qquad (9\text{-}50)$$

which is a weighted mean square time duration for the pulse ambiguity function.

This equivalent time resolution can be expressed in terms of frequency by noting that the Fourier transform of the ambiguity function is simply the squared magnitude of the Fourier transform of the transmitted wave form:

$$\mathcal{F}\left[c(\tau)\right] = |U(\omega)|^2, \qquad (9\text{-}51)$$

where:
$\mathcal{F}$ = Fourier transform of ambiguity function
U = Fourier transform of the modulation function.

Using this result and *Parseval's theorem*[1] the equivalent time duration can be expressed by

$$\Delta\tau = \frac{2\pi \int_{-\infty}^{\infty} |U(\omega)|^4 \, d\omega}{\int_{-\infty}^{\infty} |U(\omega)|^2 \, d\omega} = \frac{1}{B_e}, \qquad (9\text{-}52)$$

where it is also given in terms of an effective bandwidth B_e.

This equivalent duration (or the related bandwidth) can be used as a descriptor for the range resolution capability of the radar by making use of the relationship between time and distance:

$$\frac{c}{2}\Delta\tau = \Delta R. \qquad (9\text{-}53)$$

Speed-only Ambiguity Function

An analogous ambiguity function can be generated in the frequency domain for Doppler frequency measurement, and consequently for speed measurement. Consider the Fourier transform of the transmitted wave form:

$$\mathcal{F}\left[p_T(t)\right] = P_T(f) = U(f - f_c). \qquad (9\text{-}54)$$

where:
f_c = carrier frequency
p_T = transmitted waveform.

Note that the Fourier transform U of the modulation function is translated to the carrier frequency f_c. The Fourier transform of the received function is given by

$$\mathcal{F}\left[p_R(t)\right] = U(f - f_c - f_D)e^{j\omega_c T}, \qquad (9\text{-}55)$$

where:
f_D = Doppler frequency,

and where both the phase shift due to time delay and the frequency shift due to speed are indicated. Again one may correlate with a sample of the transmitted frequency function, but this time cor-

---

[1] Parseval's theorem—Total energy is the same when expressed as integral of time function squared or frequency spectrum squared.

relating in frequency. The result, neglecting the phase factor associated with the time delay, is

$$k(f_D) = \int_{-\infty}^{\infty} U(f - f_c - f_D)\, U^* (f - f_c)\, df, \tag{9-56}$$

where:

$k(f_D)$ = Doppler frequency ambiguity function.

This can be shown equivalent to, in terms of the time function,

$$k(f_D) = \int_{-\infty}^{\infty} |u(t)|^2\, e^{j2\pi f_D t}\, dt. \tag{9-57}$$

Clearly Eq. 9-56 is quite analogous with Eq. 9-48, except that 9-56 is in the frequency domain! Both represent a correlation approach to the measurement resolution problem.

In a similar manner to that used for the range resolution, an equivalent bandwidth for Doppler frequency measurement can be defined by

$$\Delta f_D k^2 (0) = \int_{-\infty}^{\infty} |k(f_D)|^2\, df_D. \tag{9-58}$$

Using this definition and Eq. 9-57 together with Parseval's theorem, we find that the equivalent bandwidth is

$$\Delta f_D = \frac{\int_{-\infty}^{\infty} |u(t)|^4\, dt}{\left[\int_{-\infty}^{\infty} |u(t)|^2\, dt\right]^2} = \frac{1}{T_e} \tag{9-59}$$

where here it has been also described in terms of an equivalent time duration T_e.

Converting this from a bandwidth to a velocity resolution is readily achieved by using the simple relationship between them:

$$\Delta u = \frac{\lambda_c \Delta f_D}{2}. \tag{9-60}$$

Time-Frequency Ambiguity Function

The ambiguity functions defined above for time delay alone or for frequency shift alone can be combined, since one is a time correlation and the other a frequency correlation, and the two are related. The ambiguity function may take on either form, as shown by

$$\chi(\tau, f_D) = \int_{-\infty}^{\infty} u(y)u^*(y - \tau)e^{j2\pi f_D y}\, dy \tag{9-61a}$$

$$\chi(\tau, f_D) = \int_{-\infty}^{\infty} U^*(f - f_c)U(f - f_c - f_D)e^{j2\pi f\tau}\, df \tag{9-61b}$$

This is the standard definition introduced by Woodward for the ambiguity function. Although in principle the function is quite simple, calculating it for some wave forms can be difficult.

It can be seen that Eqs. 9-61a and b reduce to either of the separate ambiguity functions when the appropriate variable is set equal to zero. Thus

$$\chi(\tau, 0) = c(\tau) \tag{9-62a}$$

$$\chi(0, f_D) = k(f_D). \tag{9-62b}$$

The ambiguity function has been calculated for numerous waveforms. Some simple results presented by Berkowitz (1965) are shown in Figures 9-29, 9-30, 9-31, and 9-32.

Figure 9-29 shows the ambiguity function for short and long pulses with Gaussian (or normal) shaped envelopes (the rectangular envelope results in a $(\sin x)/x$ shape, the minor lobes of which are difficult to plot in this fashion). The short pulse is narrow in the time domain, which means that the resolution in that direction, and consequently in range, is good; but it is wide in the frequency domain, which means that the resolution in Doppler frequency or velocity is poor. This is to be expected, since the short pulse itself occupies a wide range of frequencies and measurements of relatively small shifts in frequency are difficult. On the other hand, the long pulse (which gives poor range resolution) permits better Doppler frequency resolution because its spectrum is narrow to begin with, so that a shift in its spectrum may be more readily detected.

Figure 9-30 shows the ambiguity function for the long pulse with dimensions included. This pulse is of duration T, and for the Gaussian shape its equivalent width in the time domain is just T. The resulting spectral width causes an equivalent resolution width in the frequency domain that is the reciprocal of the pulse duration. This is approximately equal to the bandwidth occupied by the pulse; it is therefore expected that a signal shifted by more than this bandwidth would have its shift readily detectable.

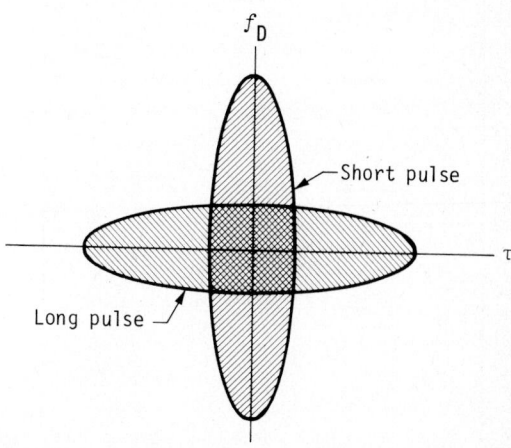

Fig. 9-29. Ambiguity functions of monochromatic pulses (Gaussian envelope).

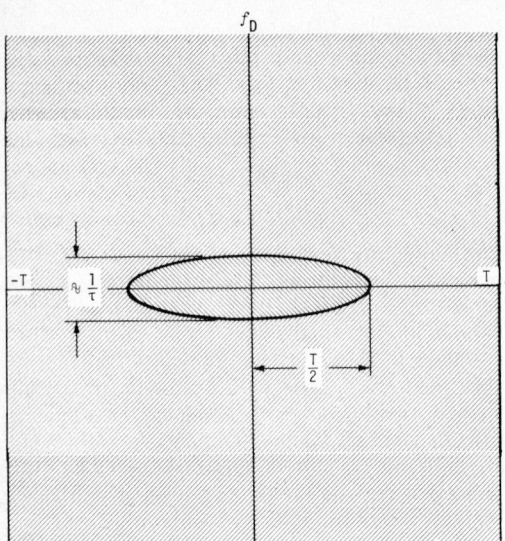

Fig. 9-30. Ambiguity function of an RF pulse of duration T.

The pulse compression "chirp" wave form commonly used (and described previously) results in an ambiguity function, as shown in Figure 9-31. The best resolution would be obtained for targets travelling with such velocities that they could be represented along a line from upper left to lower right, since the ambiguity function would be narrowest in that direction. The poorest resolution would be obtained for signals separated in the other direction; that is, such that increasing time delay corresponds with increasing positive Doppler shift. As can be seen, the effective range (delay time) resolution for zero Doppler shift is the reciprocal of the bandwidth over which the signal is swept and is independent of the actual duration of the pulse. Similarly, for signals at the

same distance ($\tau = 0$), the resolution in frequency is just the reciprocal of the pulse duration, as with the unchirped pulse. Total width of the ambiguity function in the time dimension is the same as for the unswept pulse of the same duration, and total extent in the frequency dimension is the bandwidth occupied by the sweep. For many purposes, however, the narrower resolutions along the axes are more appropriate, and it is for this reason that the swept FM pulse is used.

Figure 9-32 shows the ambiguity function for a burst of pulses, each of which has singly a circular ambiguity spike. The alignment of pulses along the τ axis is comparable with the alignment shown in Figure 9-28d, and the interpretation is the same; that is, ambiguities occur at time intervals corresponding with the interpulse spacing. Similarly, the ambiguities in speed measurement are analogous with the results expected from the analysis of Figure 9-27. That is to say, a frequency the period of which is a submultiple of the interpulse interval (repetition period) cannot be distinguished, as was seen in Figure 9-27. The new concept introduced by use of the ambiguity diagram is the effect of combining the range and Doppler ambiguities to produce other combined ambiguities off the two axes.

An interesting property of the ambiguity function is that the volume under the ambiguity surface described by the amplitudes above the τf_D plane (Fig. 9-32) is a constant. This means that a single spike, if it is relatively narrow, will be quite tall, so that the distinctions between targets very close to the reference target and others at greater distances will be completely clear. This distinction may not be as clear if the volume is made up of relatively small spikes with considerable "in-between" structure.

Ambiguity functions for radars looking at the ground can be modified tò show them in ground coordinates because both the time delay and the Doppler frequency shift are associated with coordinates on the ground, as shown in Figure 9-2. Such illustrations for particular radar cases can be

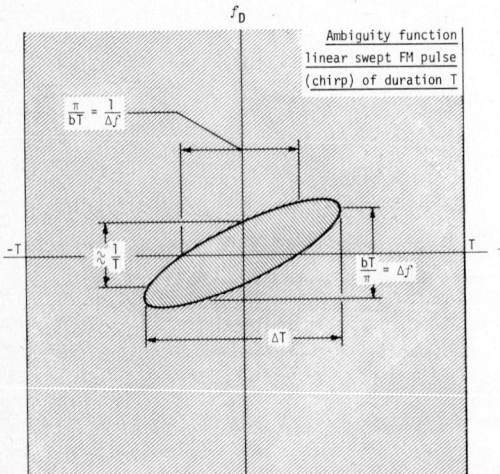

Fig. 9-31. Ambiguity function of a linear swept FM pulse (chirp) of duration T.

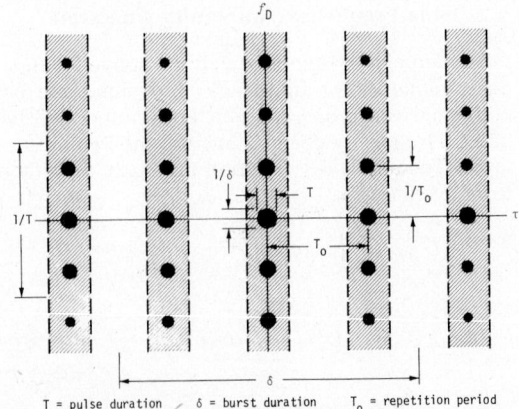

Fig. 9-32. Ambiguity diagram of a burst of pulses.

helpful in checking to see that a signal design accomplishes the desired purpose. The subject of signal design and ambiguity functions is treated at length in numerous textbooks and will not be treated here in further detail.

SENSITIVITY, AMPLITUDE MEASUREMENT, FADING AND NOISE

The *sensitivity* of either a radar or a passive microwave radiometer is determined by the noise level present and the ability of the receiver to extract the received signal from that noise. The signal-to-noise ratio (*S/N*) required for simply detecting the presence or absence of a target may be different from that required for making a precise measurement of its amplitude or range or speed. In addition, the *S/N* required for measuring the amplitude of the return from an area-extensive target may be different from that for a point target, because averaging may be possible over returns from several presumed identical elements in the area-extensive target, but only one element is available for the point target.

The measurement of amplitude, speed, direction or distance may be described in terms of both *precision* and *accuracy*. Frequently these words are assumed to mean the same thing, but in fact their meanings are quite distinct. Usually the precision by which a radar can distinguish one target from another is considerably better than the accuracy with which it can establish the desired parameter of the target. Thus precision measures the fluctuation between one measurement and another made from targets of the same characteristics.

Precision is determined by the stability of the radar, the characteristics of the target, fading of the signal, and the noise level involved. Accuracy, on the other hand, is a measure of the ability to determine the absolute size of the quantity measured.

To make an accurate amplitude measurement the gain of the antennas used must be known accurately, as well as the characteristics of the system itself. Antenna gain measurements are difficult to make, so that a system may be very stable with very low noise and have resultant excellent precision, yet still have poor measurement accuracy because of a *bias error* caused by lack of knowledge of the antenna characteristics.

Fading of signals is very important in establishing both the precision of measurement and the accuracy. Signals from most real targets fluctuate wildly as the target aspect changes very slightly, an amplitude fluctuation which is called *fading*. Radars measuring the angle of arrival of a signal also are subject to random fluctuations in this angle caused by slight changes in target aspect; such fluctuations are called *scintillation*. A noiseless system may still make imprecise measurements for a particular target because the characteristic of the target required is the average

value of the fading signal (or scintillating angle), and the system design does not permit sufficient averaging to obtain a good measure of the mean.

Consideration will be given in this section to the strength of the signal received by a radar (the radar equation), some of the characteristics of noise and fading, and the precision of measurement of amplitude. No attempt will be made here to discuss the precision of measurement of range, angle, or speed.

The Radar Equation

The equation describing the power received by a radar is called the *radar equation.*[2]

$$W_r = \frac{W_t G_t A_r \sigma}{(4\pi R^2)^2} .$$ (9-63)

where:

W_t = transmitted power,
G_t = gain of the transmitting antenna in the direction of the target,
R = distance from transmitter to target,
σ = effective backscatter area of the target,
A_r = effective receiving area of the receiving antenna.

Most radars use either the same or identical antennas for transmitting and receiving. Consequently for such a system we may combine the gain factors, using the relationship between gain and receiving aperture

$$G = G_t = G_r = \frac{4\pi A_r}{\lambda^2} .$$ (9-64)

The received power is then given by

$$W_r = \frac{W_t G^2 \lambda^2 \sigma}{(4\pi)^3 R^4} .$$ (9-65)

An additional loss factor is sometimes included to take into account atmospheric effects which have been neglected here.

Radar equation modified for ground-sensing

Radars remotely sense areas on the ground in which each resolution cell contains many separate scatterers. Thus a somewhat modified form of the radar equation is appropriate for ground-sensing purposes. A commonly used model for scatter from ground targets assumes that the positions of the separate components of a resolution cell are sufficiently randomized that the power received from each may be added to the power from all the others without consideration of phase. Clearly this would not be an appropriate model if a pulse from a radar had such good resolution in range and angle that a single object represented essentially all or even more than all of one resolution

---

[2] The radar equation has previously been introduced in Chapter 4, beginning at Eq. 4-1. This discussion herein, while perhaps repetitive, is included for completeness and unity of thought.

cell. Thus a radar with a resolution cell less than the size of a single automobile would receive an echo from an automobile in which phase differences would have to be considered. In fact, the automobile would be considered a single target with a scattering cross-section having characteristics much like that of an antenna pattern.

For that large class of targets on the ground in which phase may be ignored, the average signal returned may be expressed as

$$W_r = \frac{\lambda^2}{(4\pi)^3} \sum_{i=1}^{N} \frac{W_{ti} G_i^2 \sigma_i}{R_i^4}, \qquad (9\text{-}66)$$

where:

the summation is over the N scatterers within the resolvable cell,

the subscript i points out that the illuminating power, antenna gain, and distance will be different for each of the scatterers.

If the cell is assumed to contain a large number of such scatterers, the summation may be replaced by an integral, using an average value of the scattering cross-section per unit area rather than the actual scattering cross-section associated with each individual area element. Such a radar equation is

$$W_r = \int_{\substack{\text{Scattering} \\ \text{Area}}} \frac{W_t G^2 \, \lambda^2 \, \sigma^0 \, dA}{(4\pi)^3 \, R^4} \qquad (9\text{-}67)$$

where:

dA = a differential element of ground area.

σ^0 = differential scattering cross-section (cross-section per unit area).

This form of the radar equation is commonly used in discussing the performance of radars for remotely sensing the ground, including mapping radars, Doppler navigators, and scatterometers. However, in certain cases, as indicated in the above discussion, Eq. 9-67 must be replaced by Eq. 9-65 if the cell is small and a large target is contained within it.

Approximate radar equation

A special case of great importance is a radar for which all parameters in Eq. 9-67 are essentially unchanged from one part of the resolution cell to the other. When this is true, the integral of Eq. 9-67 may be replaced simply by the product of the factors within it and the area A:

$$W_r = \frac{W_t G^2 \, \lambda^2 \, \sigma^0 \, A}{(4\pi)^3 \, R^4}. \qquad (9\text{-}68)$$

Many radar analyses assume this condition to prevail even though the approximations may be relatively imperfect. Figure 9-33 illustrates the geometry for a pulse radar for which this approximation may be applied. The illuminated area A in this case is given by

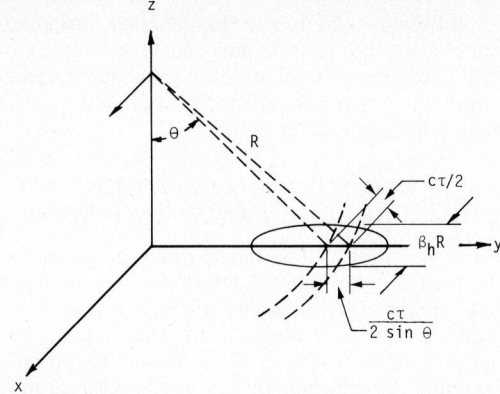

Fig. 9-33. Geometry for approximate received signal calculation.

$$A = (\beta_h R) \frac{c\tau}{2 \sin \theta}. \qquad (9\text{-}69)$$

where:

β_h is the effective beamwidth of the antenna, so that

$\beta_h R$ is the dimension of the area transverse to the radius vector,

The other factor gives the ground resolution cell size in the direction away from the radar for a pulse of duration τ,

where:

c is the velocity of light,

θ is the angle of incidence.

Substituting this expression into equation 9-68 gives a familiar form of the radar equation:

$$W_r = \frac{W_t G^2 \, \lambda^2 \, \sigma^0 \beta_h c\tau}{2(4\pi)^3 \, R^3 \sin \theta}. \qquad (9\text{-}70)$$

This formula may be modified in numerous ways, since β_h and G are not independent, and it is often convenient to express distance in terms of height rather than slant range. No attempt is made here to enumerate here the various forms of the equation.

Signal-to-Noise Ratio

If noise were not present, the received power would need to be calculated to determine the amount of gain required in the amplifier of the receiver, but the signal could always be amplified sufficiently to obtain an adequate measurement.

Of course, this is not the case, because noise does exist in the receiver, and in fact noise is received by the antenna, so the transmitter power required must be sufficient to overcome the noise and achieve the desired signal-to-noise ratio. The noise power available in a 1-Hz bandwidth is the product of Boltzmann's constant k, and the absolute temperature, T. The noise power is proportional to the bandwidth, B, so that the total power received is kTB.

The noise factor of a receiver describes the additional noise applied to the signal by the receiver. If the input is assumed to be at a reference temperature of 290 K, a perfect receiver would have a noise factor of one, for it would introduce no additional noise beyond that unavoidable at the input. Real receivers, of course, always introduce some extra noise. Mathematically the noise factor F is given by

$$F = \frac{N_{out}}{GN_{in}} = \frac{N_{out}}{GkT_oB} = \frac{S_i}{S_o}, \qquad (9\text{-}71)$$

where

F = noise factor
N_{out} = noise output power,
N_{in} = noise input power at the reference temperature 290 K (T_o),
G = gain of the receiver, and
B = bandwidth,
S_i = S/N at input,
S_o = S/N at output,

Thus the equation states that the noise factor is the noise measured at the output divided by the noise that would be measured if the receiver added no additional noise. It also states that the noise factor is the S/N S_i at the input divided by the S/N S_o at the output, since the signal is assumed to be the same at input and output except for the gain factor. We may therefore think of the noise factor as describing the decrease in S/N caused by the presence of the receiver.

The received power is expressed in terms of the radar parameters in Eq. 9-70. In terms of the noise and receiver parameters, it is

$$W_r = kT_oBFS, \qquad (9\text{-}72)$$

where:

$S = $ S/N expressed in terms of the input level, taking into account noise added by the receiver.

When the input noise level is significantly less than the reference level, the received power may be more conveniently expressed in terms of the combination of the receiver noise and the noise received through the antenna:

$$W_r = \left[kT_oB(F - 1) + kT_aB \right] S. \qquad (9\text{-}73)$$

Here the first term gives the contribution of the receiver. The second term, involving the effective temperature seen by the antenna T_a, is the noise signal received; this latter is the total signal received by a microwave radiometer, but ordinarily radar signals are significantly larger. When it is desired to consider the receiver properties in this fashion, a more convenient form is

$$W_r = kT_eBS, \qquad (9\text{-}74)$$

where:

T_e = effective temperature of the input, including; both the antenna temperature and the contribution due to the receiver.

Since S/N is the most significant factor describing the received radar signal, the radar equation of 9-71 or the other forms shown in Eqs. 9-65, 9-66, 9-67, 9-68, and 9-70 should be modified by replacing the received power by one of the expressions of Eq. 9-72, 9-73, or 9-74. The resulting expression, using 9-74, is

$$S = \frac{W_tG_tA_r\sigma}{(4\pi)^2\,R^4kT_eB}. \qquad (9\text{-}75)$$

Clearly S/N can be increased by increasing transmitter power, receiving aperture or transmitter gain, and by decreasing the distance R or the bandwidth B. Furthermore a receiver should be designed to have as low a noise figure, and consequently also as low an effective temperature T_e, as possible.

Required S/N

If an adequate desired value of S/N is known, the other parameters (except σ) can be adjusted, but the S/N required varies greatly depending upon the problem. For example, in communication systems the S/N required for telephone work is of the order of 20-30 decibels (dB). On the other hand, if a digital signal is communicated, only 12-18 dB is required, whereas a TV network requires ratio of about 50 dB.

Radars designed to achieve detection of targets, particularly automatic detection, require ratios of 12-16 dB. The probability of a successful detection changes very rapidly over this relatively small range of S/N. Therefore a ratio much less than 12 dB for most simple types of processing results in high probability of false detection or of missed targets, whereas a ratio greater than 16 dB is seldom required. A radiometer, on the other hand, because of the nature of its measurement, frequently can use S/N much less than -10 dB. Fine resolution imaging radars need only 4 or 5 dB because the fading of the signal makes greater ratios meaningless.

The signal-to-noise ratio required for a particular application depends upon the required probability of error or, in the case of television or some radar pictures, on permissible visual degradation of the output. The value of S/N required at the output may be quite different from that required at the input, for the radar may use various sophisticated processing techniques to improve S/N. A relatively unsophisticated but very effective technique is averaging of great numbers of independent return signals. If the average is over a large enough number of independent signals, the distinction between the average noise and the average signal is made easy. This is the reason that radiometers and some scatterometers can make accurate measurements even when the signal is many dB less than the noise. On the other hand, such processing techniques are not always available. For instance, it may not be possible to average many independent returns because the return is simply not observable for a long enough period to receive a large number of independent signals. Sometimes it is possible to improve this situation by using a wider bandwidth, but this requires a radar with more transmitter power.

The S/N shown in Eq. 9-75 is that at the input to the receiver prior to processing. Thus if the output S/N after processing is considerably better than the input ratio, the required input ratio is accordingly reduced. Therefore any calculation using Eq. 9-75 to establish required transmitter power or antenna parameters, or the permissible range, should first involve translation of the post-processing required S/N into the preprocessing S/N used in equation 9-75.

Noise and Fading Statistics
Gaussian Noise

The noise associated with radar work is customarily described as Gaussian. Statistics of Gaussian noise and the comparable statistics of the fading signal will be discussed here. Noise due to atmospherics and interference from other transmitters is usually not Gaussian and its statistics are different. Since these cases are seldom important in radar, they are not discussed here.

Gaussian noise has an instantaneous voltage best described by a Gaussian or *normal* distribution; that is, the probability of instantaneous voltage, $p(V)$, is

$$p(V) = \frac{1}{\sqrt{2\pi}\,\sigma}\, e^{-V^2/2\sigma^2}, \qquad (9-76)$$

where:
σ is the standard deviation of the noise voltage;[3]

that is, the mean-square voltage is equal to the variance of the distribution:

$$\bar{V}^2 = \sigma^2. \qquad (9-77)$$

This zero-mean distribution is appropriate for the instantaneous voltage which has equal probability of being positive and negative. In many radar applications, however, there is more concern for the amplitude of a relatively narrow-band signal rather than for its instantaneous value. Thus the radar may operate at a frequency of 10^{10} Hz, whereas its bandwidth seldom exceeds 10^8 Hz, and usually is considerably less than that. The result is that the *carrier* at about 10^{10} Hz is modulated with an *envelope* with a highest frequency as much as 10^7 or 10^8 Hz. Hence the statistics of this envelope are very important in radar, as well as in communications which also use relatively narrow bandwidth.

---

[3] σ has a history of representing the standard deviation and as such is not changed herein, even though σ is used equally often for the radar cross-section.

Rayleigh Distribution

The statistics for the amplitude of a narrow-band Gaussian noise or signal are governed by the *Rayleigh distribution,* originally derived by Lord Rayleigh in the 1870's. The Rayleigh distribution for power is also recognizable as a chi-squared distribution with two degrees of freedom. The model for the voltage in the narrow band situation is

$$V = V_e \cos(\omega t - \phi), \qquad (9-78)$$

where:
V_e is the envelope voltage,
ϕ is the phase.

The Rayleigh distribution for V_e and the accompanying uniform distribution for the phase are

$$p(V_e) = \frac{V_e}{\sigma^2}\, e^{-V_e^2/2\sigma^2}; \; p(\phi) = \frac{1}{2\pi}. \quad (9-79)$$

The Rayleigh distribution does not have zero mean; in fact, V_e is only defined for positive values. The average value of the voltage can be expressed in terms of the standard deviation of the noise:

$$\bar{V}_e = \left(\frac{\pi}{2}\right)^{1/2} \sigma. \qquad (9-80)$$

Sometimes a Rayleigh distribution is considered as a steady voltage with a superimposed random alternating voltage. The value of this effective fluctuation power is simply the mean-square envelope voltage, minus the square of the mean (assuming unit resistance). The mean-square voltage $\bar{V}_e^2$ is

$$\bar{V}_e^2 = 2\sigma^2, \qquad (9-81)$$

or

$$\bar{V}_{ac}^2 = \bar{V}_e^2 - \bar{V}_e^2 = \left(2 - \frac{\pi}{2}\right)\sigma^2 = 0.429\sigma^2.$$

Frequently the ac power is considered to be a noise, and the dc level to be a signal, so that the effective S/N for the Rayleigh distribution would be the ratio $\bar{V}_e^2/\bar{V}_{ac}^2$ of 3.66. This is 5.6 dB.

Noise Models

One model for noise assumes it to be the summation of signals from a large number of randomly phased oscillators at slightly differing frequencies. Thus the instantaneous voltage may be expressed by

$$V = \sum_{i=0}^{\infty} a_i \cos \omega_i t + b_i \sin \omega_i t, \qquad (9-83)$$

where a_i and b_i are independent variables distributed with a Gaussian distribution.
Carrying of the summation to infinity assumes that each component is infinitesimally small. The expression for the voltage may also be written in

terms of a single summation involving phase angle as

$$V = \sum_{i=0}^{\infty} V_i \cos(\omega_i t + \phi_i). \qquad (9\text{-}84)$$

Here the distribution for phase is uniform, and that for voltage is Rayleigh. Since the a's and b's are independent, the power involved is

$$W = \sum_{i=0}^{\infty} a_i^2 + b_i^2 = \sum_{i=0}^{\infty} V_i^2. \qquad (9\text{-}85)$$

One reason for using this model is that it permits development of the noise spectrum in terms of the voltages. Thus the power contained within an interval of width $\Delta\omega$ is given by

$$\lim_{n \to \infty} P(\omega_j)\Delta\omega = \sum_{i=j}^{n} V_i^2 \quad \omega_j < \omega_i < (\omega_j + \Delta\omega), \qquad (9\text{-}86)$$

where $P(\omega_j)$ is the power density of the noise at the frequency ω_j.

The summation is carried on over the values of i starting with j, and going up to the value appropriate to the upper end of the interval of width $\Delta\omega$. Of course, as one carries the model to its ultimate, the number of oscillators within such a finite band is considered to be infinite; hence the limit shown. It can be shown, however, that the statistics for a signal like this are essentially the same as for an infinite sum whenever $n - j$ is greater than about 10, the only deviation occurring then far out in the tails of the distribution. The number of randomly phased oscillators needed for the model to be effective is therefore relatively small.

If the received signal is from an unfading target, it is a single carrier frequency. Thus a common problem in measuring the amplitude of such a signal is that of measuring the amplitude of a sine wave immersed in random noise. The distribution for such a signal is called a Rice distribution (Rice, 1944, 1946). Consideration will be given here to the single case where S/N is relatively large, in which case the relatively complicated Rice distribution reduces to a Gaussian distribution centered about the mean value of the envelope voltage for the sinusoidal signal:

$$p(V) = \frac{1}{(2\pi \bar{W}_n)^{1/2}} e^{-(V - V_s)^2/2 \, \bar{W}_n}. \qquad (9\text{-}87)$$

Here

V_s = envelope amplitude for the sine wave,
W_n = average noise power.

The properties of the normal distribution are such that the value for the voltage will lie within plus or minus one standard deviation 68% of the time; that is, the voltage lies in the range $(V_s \pm \sigma)$ 68% of the time. For this situation S/N is

$$S = \frac{V_s^2}{\bar{W}_n}, \qquad (9\text{-}88)$$

so that

$$V_s\left(1 - \frac{1}{\sqrt{S}}\right) < V < V_s\left(1 + \frac{1}{\sqrt{S}}\right) \text{ 68\% of time} \qquad (9\text{-}89)$$

If we are interested in the range occupied by the voltage 90% of the time, it is, using the standard Gaussian expression,

$$V_s\left(1 - \frac{1.645}{\sqrt{S}}\right) < V < V_s\left(1 + \frac{1.645}{\sqrt{S}}\right). \qquad (9\text{-}90)$$

Phase Interference as a Cause of Fading

Fading of a radar signal comes about because of phase interference phenomena. Since phase interference phenomena are also the cause of antenna patterns, the analysis is similar to the earlier review on arrays, and the concept may be best illustrated by an interferometerlike, simplified radar target.

Figure 9-34a shows a pair of radar targets a and b separated by a distance D. It is assumed that these targets are illuminated by a radar at considerable distance and nearly perpendicular to the line joining them. The radar is assumed to be moving along a path parallel to that line for the purpose of the illustration. It is further assumed that there is no phase shift upon backscatter by the two targets, so that the signals received back at the radar are identical and experience only the phase shift associated with the travel distance.

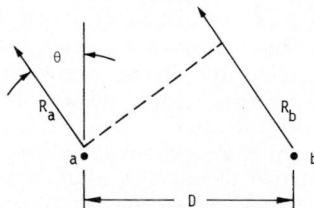

(a) Interferometer target geometry

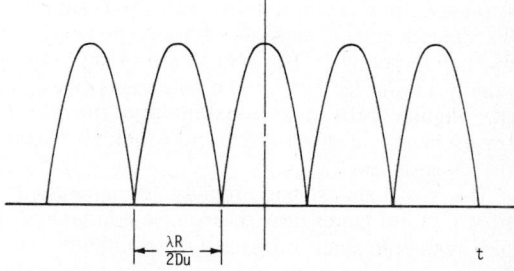

(b) Fading pattern for interferometer target

Fig. 9-34. Simplified target fading.

The equivalent to Eq. 9-6 of this chapter is

$$V = V_0\, e^{j\omega t}(e^{j4\pi R_a/\lambda} + e^{-j4\pi R_b/\lambda}). \qquad (9\text{-}91)$$

Here the phase shift that was used in the earlier section has been doubled because the signal travels both out and back along the distance R_a and R_b. The voltage received back at the radar is that given in Eq. 9-91. Its magnitude is

$$|V| = 2V_0\left|\cos\left(\frac{2\pi D\,\sin\theta}{\lambda}\right)\right|. \qquad (9\text{-}92)$$

Using the same approach as used for Eq. 9-15 on antenna arrays, the beamwidth for the target is found to be

$$\Delta\theta \approx \frac{\lambda}{2D}. \qquad (9\text{-}93)$$

Note that this beamwidth is half that for the antenna of the same dimension, the factor of two being caused by the doubling of the phase shift associated with the return path for the radar target.

Since it is assumed that the radar is at a point nearly normal to the line joining a and b, the assumption can also be made that the angle and its sine are about equal. If x is the displacement from the position directly opposite the target and if the radar travels at a velocity u,

$$\sin\theta \approx \theta \approx \frac{x}{R} = \frac{ut}{R}. \qquad (9\text{-}94)$$

Consequently Eq. 9-92 may be expressed in terms of the time variation as

$$|V| = 2V_0\left|\cos\left(\frac{2\pi Dut}{\lambda R}\right)\right|. \qquad (9\text{-}95)$$

The result shown in Figure 9-34b is the fading pattern associated with this double target. Instead of a constant amplitude received signal, as a single target would give, the signal amplitude varies cosinusoidally between zero and a peak value, with the period $(\lambda R/2Du)$.

This variation is greatly simplified because of the fact that two identical scatterers were assumed rather than a larger number of scatterers with different amplitudes and phase shifts; nevertheless it illustrates important principles of fading. The signal will fade more rapidly if the wavelength is reduced because the smaller wavelength means finer lobes for a given aperture for the target (D). The signal also fades more rapidly at shorter ranges. This happens because the angular pattern is constant and the radar travels across a given angle more quickly when the range is less.

The signal fades more rapidly for a larger D (that is, for a larger target), because a large aperture results in finer lobes in the backscatter pattern, just as a large aperture gives finer resolution with an antenna. The signal also fades more rapidly if the radar moves more rapidly, because the travel through the lobe structure occurs in a

shorter period of time. All of these principles apply to fading from more complex targets, as well as to fading from this simple target.

Fading in Terms of Doppler Frequencies

Instead of fading being treated as a travel through lobes of a pattern, it can be described in terms of the effect of multiple Doppler frequencies being received. Consider the model for a ground target shown in Figure 9-35a. Various contours of constant Doppler frequency are shown within the illuminated area on the ground. At least one target is present on each contour, each target having a different Doppler frequency shift. Consequently the received instantaneous voltage from the ith target is

$$V_i = A_i\cos(\omega_c + \omega_{Di})t. \qquad (9\text{-}96)$$

To show that this is identical with the result obtained from the pattern approach, consider the same situation in Figure 9-35b that was illustrated in Figure 9-34a. Here targets a and b are located a distance $D/2$ ahead and behind the perpendicular or zero-Doppler line. The expression for the instantaneous voltage obtained by summing the contributions indicated in Eq. 9-96 is

$$V = V_0\big[\cos(\omega_c + \omega_{Da})\,t + \cos(\omega_c + \omega_{Db})\,t\big]. \qquad (9\text{-}97)$$

Using the appropriate trigonometric identity, this becomes

$$V = 2V_0\cos\left(\omega_c + \frac{\omega_{Da} + \omega_{Db}}{2}\right)t$$
$$\times\cos\left(\frac{\omega_{Da} - \omega_{Db}}{2}\right)t \qquad (9\text{-}98)$$

Note here that the first factor is at a frequency close to, but in general somewhat removed from, the carrier frequency, and its envelope is described by the product of the amplitude and the second factor. Thus the amplitude is

$$|V| = 2V_0\cos\left(\frac{\omega_{Da} - \omega_{Db}}{2}\right)t. \qquad (9\text{-}99)$$

For the examples shown, the Doppler frequency for target a is the same as that for target b except for sign; so

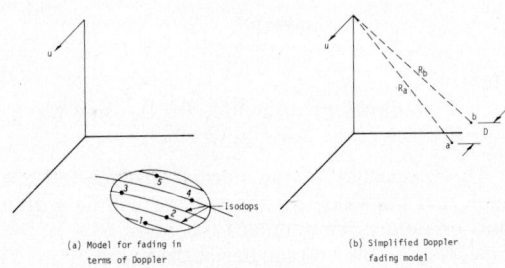

(a) Model for fading in terms of Doppler

(b) Simplified Doppler fading model

Fig. 9-35. Fading in terms of point-target Doppler shift.

$$f_{Da} = \frac{2u_a}{\lambda} = \left(\frac{2u}{\lambda}\right)\frac{(D/2)}{R} = \frac{uD}{\lambda R} = -f_{Db}.$$

(9-100)

Substituting these results into Eq. 9-98 gives for the voltage observed by the radar the same result as in Eq. 9-95; that is

$$|V| = 2V_0 \left| \cos\left(\frac{2\pi Dut}{\lambda R}\right) \right|.$$

(9-95)

(repeated)

Consequently one may consider the fading either as being motion of the radar through a pattern in space like an antenna pattern, or as being due to combinations of different Doppler frequencies returning from the different target elements observed simultaneously.

Returns from Complex Targets

When the target is more complex than the simple two-element model of Figure 9-35b, resembling Figure 9-35a instead, the result is a summation of terms like that of Eq. 9-96. Each of these is a signal at a different frequency and it also is reasonable to assume uniformly-distributed phases. For this reason the model for the Doppler signal return is just like the model describing noise as the sum of a large number of signals at different frequencies described by Eqs. 9-83 and 9-84. Since both signals (noise and fading) are described by the same model, the statistics developed for the noise may also be applied to fading. Consequently the instantaneous voltage may be treated as a Gaussian process, and the envelope for a narrow-band situation (and the Doppler frequency is always small enough to assure a narrow band) is Rayleigh distributed. Therefore the fading signal is like a noise, the bandwidth of which is determined by the maximum spread of the Doppler frequencies observed. For pulse radars the Doppler bandwidth is usually much less than the bandwidth resulting from the short pulse so that, in effect, the pulses sample the Doppler signal as described in connection with pulse-Doppler radars.

The fading model described by Figure 9-35a is the basis for calculation of the spectrum of the fading signal; that is, a good way to calculate the fading spectrum is to assume that the targets along each incremental strip between two isodops (lines of equal Doppler frequency) contribute a spectral component for the frequency increment associated with the isodops. Figure 9-36 shows a general incremental ground map associated with calculations using this model. The distance along an isodop is given by ξ. The distance normal to the isodops is described by η. Thus an increment of area which can be used in the radar equation lies between contours of constant ξ and constant η that are relatively close together.

The differential power associated with differen-

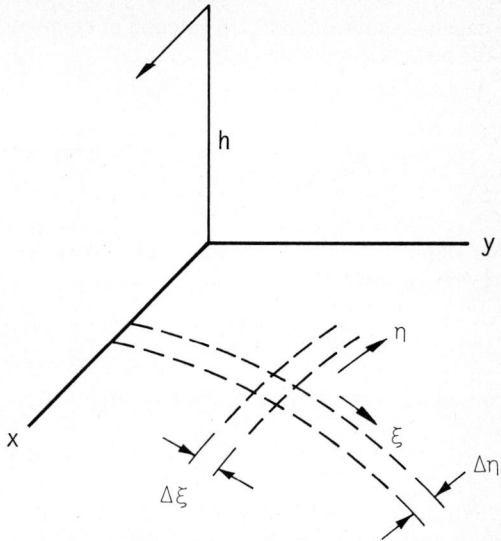

Fig. 9-36. Contours of constant Doppler shift and orthogonal coordinates.

tial width strips may be obtained by writing Eq. 9-68 in the special form of

$$dW_r = \frac{W_t G^2 \lambda^2 \; \sigma^o d\xi d\eta}{(4\pi)^3 \; R^4}.$$

(9-101)

Here the differential element of area is given by $d\xi d\eta$. To calculate the power density associated with a given frequency, the total power lying in the strip of width $d\eta$ is calculated by integrating over ξ. That is, the power density $P(f_D)$ becomes

$$P(f_D)df_D = \frac{\lambda^2 \; d\eta}{(4\pi)^3} \int \frac{W_t G^2 \; \sigma^o \; d\xi}{R^4}$$

$$= \left| \frac{\lambda^2}{(4\pi)^3}\left(\frac{d\eta}{df_D}\right) \int \frac{W_t G^2 \; \sigma^o \; d\xi}{R^4} \right| df_D.$$

(9-102)

Here we first express the power coming from the incremental width $d\eta$ and then convert the incremental ground distance to the increment in Doppler frequency. These calculations can be relatively simple for some geometries, but the orthogonal coordinates (ξ,η) can be difficult to describe analytically for certain types of motion or ground geometry.

Fading Spectrum–SLAR with pulse transmitter

As an example of the application of this technique, consider the relatively simple case of a side-looking airborne radar (SLAR) using pulse transmission. The basic situation is described by the geometry of Figure 9-33. Here the area is, of course, described by Eq. 9-69 and the return power by Eq. 9-70. For this case the contour of constant Doppler is essentially parallel to the y-axis, so the normal to it is parallel to the x-axis. That is,

$$d\xi = dy; \qquad d\eta = dx.$$

(9-103)

Using these substitutions, the integral of equation 9-102 becomes approximately

$$\int \frac{W_t G^2\, \sigma^0 d\xi}{R^4} = \frac{W_t G^2\, \sigma^0}{R^4} \int dx$$

$$\approx \frac{W_t G^2\, \sigma^0}{R^4}\left(\frac{c\tau}{2\sin\theta}\right). \tag{9-104}$$

The Doppler frequency for this special case was previously derived:

$$f_D = \frac{2ux}{\lambda_c R}. \tag{9-42}$$
$$\text{(repeated)}$$

Consequently the derivative shown in Eq. 9-102 is

$$\frac{d\eta}{df_D} = \frac{dx}{df_D} \approx \frac{R\lambda_c}{2u}. \tag{9-105}$$

Substituting this into Eq. 9-102, the power density in the Doppler spectrum becomes

$$P(f_D) \approx \frac{\lambda_c^3 c\tau G^2\, W_t\sigma^0}{4(4\pi)^3\, uR^3\sin\theta}. \tag{9-106}$$

The variation in this power density might be caused by a pulse shape, but the model for the radar assumes the pulse shape to be rectangular. The variations on R will be negligible across the area illuminated by the antenna, but the gain of the antenna itself is not constant, and this gain determines the shape of the received spectral density curve.

If the antenna used for the SLAR has uniform illumination, the form of the gain has previously been given in Eqs. 9-18 and 9-19. Here it is expressed in terms of an effective beamwidth β_h as

$$G = G_0 \left[\frac{\sin\left(\dfrac{\beta\pi}{\beta_h}\right)}{\left(\dfrac{\beta\pi}{\beta_h}\right)} \right]^2. \tag{9-107}$$

Note that for this situation

$$\beta_h \approx \frac{\lambda_c}{D}. \tag{9-108}$$

To apply this to calculate the Doppler spectrum, it is necessary to relate a point in the beam at an angle β from the y-axis in terms of the x coordinates. This is easily done by noting that

$$\beta = \frac{x}{R} = \frac{\lambda_c f_D}{2u} = \frac{f_D}{f_{Do}}. \tag{9-109}$$

Here β has been expressed in terms of the Doppler frequency f_D at the angle β divided by the maximum Doppler frequency that would occur if the radar were going directly toward the point.

[4] The subscript c is conveniently introduced again to indicate that the average carrier wavelength applies as before, or λ_c.

Substituting this into the expression for the gain, we find the gain expressed in terms of Doppler frequency

$$G = G_0 \left[\frac{\sin\left(\dfrac{f_D\pi}{f_{Do}\beta_h}\right)}{\left(\dfrac{f_D\pi}{f_{Do}\beta_h}\right)} \right]^2. \tag{9-110}$$

This expression may be substituted into that for the power density, Eq. 9-106, yielding

$$P(f_D) = \frac{\lambda_c^3 c\tau G_0^2\, W_t\sigma^0}{4(4\pi)^3\, uR^3\sin\theta} \left[\frac{\sin\left(\dfrac{f_D\pi}{f_{Do}\beta_h}\right)}{\left(\dfrac{f_D\pi}{f_{Do}\beta_h}\right)} \right]^4. \tag{9-111}$$

The geometry for this analysis is shown in Figure 9-37, and the resulting Doppler spectrum in Figure 9-38.

The expression in the denominator of the argument of the sine function can be rewritten in the form

$$f_{Do}\,\beta_h = \left(\frac{2u}{\lambda_c}\right)\frac{\lambda_c}{D} = \frac{2u}{D}. \tag{9-112}$$

It is an interesting fact that the Doppler spectrum is largely independent of wavelength. If the beamwidth is wider because of a larger wavelength, the Doppler frequency is higher for that wavelength than it would be with the narrow beam, but the Doppler frequency for a given velocity and angle decreases with wavelength in the same ratio as the beamwidth increases, so the two cancel out, and the width of the Doppler spectrum is independent of wavelength. It may then be stated that *the Doppler spectral width for a given radar depends only on the speed of the aircraft and on the physical size of the aperture*, not on the size of the aperture in wavelengths, as one might suspect. This has interesting implications for the synthetic-aperture radar discussed later.

The Doppler frequencies involved are relatively low compared with the bandwidth of the pulse radar, so that the pulse does, in effect, sample the Doppler signal as in other pulse-Doppler radars.

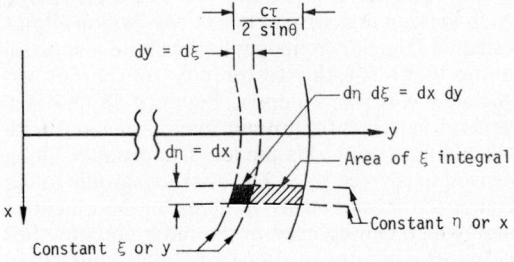

Fig. 9-37. Geometry for SLAR Doppler spectrum computation.

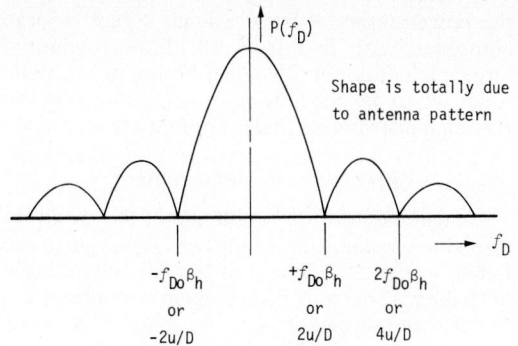

Fig. 9-38. Doppler spectrum for SLAR example.

For example, the spacing between nulls for the antenna pattern described is $4u/D$. For an antenna 4 m long and an aircraft velocity of 200 m/sec, this gives a bandwidth of 200 Hz. This bandwidth is quite small relative to the carrier frequencies used in radar. With an antenna of this size, a real aperture radar would certainly have a carrier frequency of the order of 10^{10} Hz, and often the carrier frequency is 3.5×10^{10} Hz. The bandwidth for a slant-range resolution of 15 m is of the order of 10^7 Hz, so the Doppler bandwidth is only 0.002% of this! Since synthetic-aperture radars utilize the Doppler frequency shifts in their processing, and these shifts are such a miniscule fraction of the carrier frequency, the stability required for synthetic-aperture systems is high indeed, as will be discussed later.

The Doppler bandwidth for the side-looking radar is quite small, but the Doppler bandwidth for a radar looking in any other direction is also small when compared with the carrier frequency. For instance, the value of $2u/\lambda$ for a spacecraft in low earth orbit is, for 3-cm wavelength, only 2.5×10^5 Hz, which is certainly low compared with any conceivable carrier frequency. The bandwidth for any reasonable radar system will be constrained to less than this maximum value because the antenna, or pulse duration, will not permit the full range of possible Doppler frequencies.

Effect of Detector on Spectrum

The kind of detector used in a radar affects the post-detection spectrum of fading. A Doppler radar (including a side-looking synthetic-aperture radar) actually beats the signal down to a carrier frequency of zero or some low value, rather than detecting it. On the other hand, a real aperture SLAR, a plan-position indicator (PPI) radar, and many other kinds of radars simply detect the output of a relatively high intermediate-frequency amplifier by either a square-law device, a linear rectifier, or a peak detector.

The output spectrum from a linear rectifier or peak detector is relatively complicated, although the linear rectifier output has been studied (Rice, 1946). The square-law detector lends itself to relatively simple analysis, and it is often used.

Since the output of a square-law detector is proportional to the square of the time domain signal, it can be represented in the frequency domain by a convolution integral:

$$P_o(f) = \int_{-\infty}^{\infty} P(f - y)P(y)\, dy, \quad (9\text{-}113)$$

where

P_o = output frequency spectrum,
P = input spectrum, including both positive and negative frequencies.

The *convolution process* changes the shape of the predetection spectrum. The output spectrum, centered on zero frequency, is twice as wide as the predetection spectrum. An example that will be discussed widely in connection with precision of measurement is shown in Figure 9-39, which assumes a uniform bandwidth B centered about an apparent carrier frequency f_c. The width of the spectrum is very small compared with the carrier frequency itself. Convolution changes the rectangle for the predetection spectrum into a triangle for the post-detection spectrum. The total width of the post-detection spectrum (positive and negative frequencies) is $2B$, whereas the predetection width was B. If the predetection noise is then thought of as a carrier plus sidebands, the detector output contains frequencies twice as high as the highest sideband. Physically this may be understood in terms of the multiple-oscillator model for fading signal or noise. Beat notes are formed between the highest and lowest frequency components, as well as between all more closely spaced pairs of components. The number of closely spaced components is larger than the

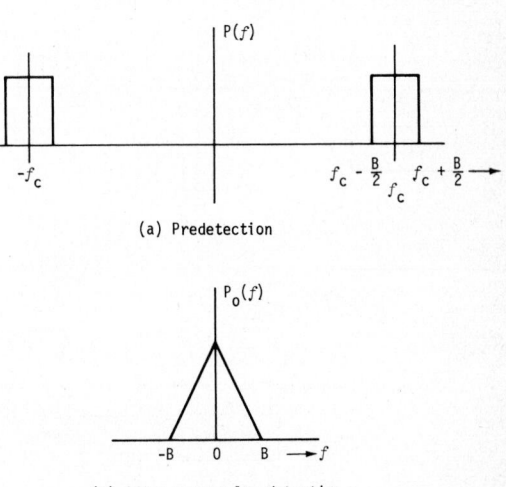

(a) Predetection

(b) After square-law detection

Fig. 9-39. Spectrum modification by detection.

number widely spaced; hence the triangular shape.

Since ordinary detection gives no information on the center frequency of the Doppler band f_c, *coherent detection* must be used if the Doppler frequency is to be measured. This can be achieved by two different schemes (discussed in more detail in appropriate sections dealing with the equipment). In one case, shown in Figure 9-40a, the signal is mixed with a local oscillator to move it down to a low-frequency carrier that permits the most negative Doppler shift to remain at a positive frequency. Thus both positive and negative Doppler frequencies are preserved for analysis. The other scheme involves beating the signal down to zero, but having two separate channels: one mixed with a given sample of the local oscillator, and the other mixed with a local oscillator sample shifted 90° in phase, as in Figure 9-40b. By proper recombination of the outputs of these quadrature channels, the positive and negative Doppler frequencies can be distinguished.

The quadrature detection scheme is not needed for the Doppler navigator because its antenna beam assures that only positive or negative Doppler shift will be present for any measurement, not both. With a sidelooking synthetic aperture radar, however, one of these two schemes is necessary because both positive and negative frequencies are present, and separation is needed. A fan-beam scatterometer may also require this kind of separation, although a pencil-beam scatterometer could get by with ordinary square-law detection.

Continuation on Statistics

Fading has the same statistical properties as noise, and consequently all the developments made previously for noise also apply for fading. Sometimes, however, the target consists not of a large number of small elements, but rather of one or two large elements plus some smaller ones. If the power returned from a single target is large compared with that from all of the remaining target elements, the Rice distribution (Rice, 1946) is needed to describe the statistics, rather than the Rayleigh distribution that is appropriate for noise.

Precision of Measurement

Measurement of the amplitude of the signal returned to a radar requires further discussion of the fading statistics. Because of the difficulty caused by fading, some sort of averaging is required if a precise measurement is to be made. The measurement precision for several cases will be discussed in this section.

Exponential Distribution

The instantaneous distribution for the fading signal voltage is given by Eq. 9-76, and for the envelope of a narrow-band signal by Eq. 9-79. After square-law detection, the detected voltage is proportional to the square of the input voltage envelope, and consequently to the input power. Here the proportionality constants are neglected, and the detected voltage is simply stated as

$$V_d \propto V^2 = W,$$

where:
V_d = detected voltage
V = input voltage
W = input power.

Thus the probability density function for the voltage after square-law detection is just that for the instantaneous received power given by

$$p(V_d)dV_d = \frac{e^{-V_d/2\sigma^2}}{2\sigma^2}dV_d$$

$$= \frac{e^{-V_d/\overline{W}_e}}{\overline{W}_e}dV_d, \tag{9-114}$$

where the subscript e designates the envelope parameters.

This exponential distribution is the previously discussed chi-squared distribution with two degrees of freedom. Clearly this distribution has a large standard deviation and consequently the amount of fading makes a measurement of the mean value difficult. Recall, for example,

$$\sigma = \left(\frac{2}{\pi}\right)^{1/2} \overline{V}_e, \tag{9-115}$$

and

$$\overline{V}_e^2 = 2\sigma^2 = \overline{W}_e. \tag{9-116}$$

Properties of the Rayleigh and Exponential Distributions

The Rayleigh distribution is plotted in Figure 9-41, where both the mean envelope and mean-square envelope voltage are indicated on the dB

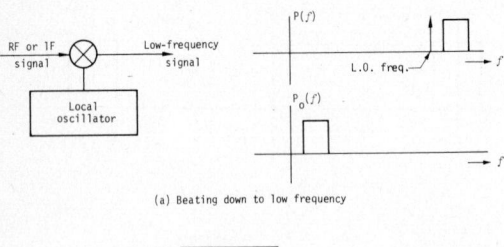

(a) Beating down to low frequency

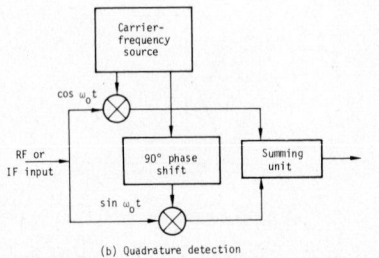

(b) Quadrature detection

Fig. 9-40. Methods for coherent detection.

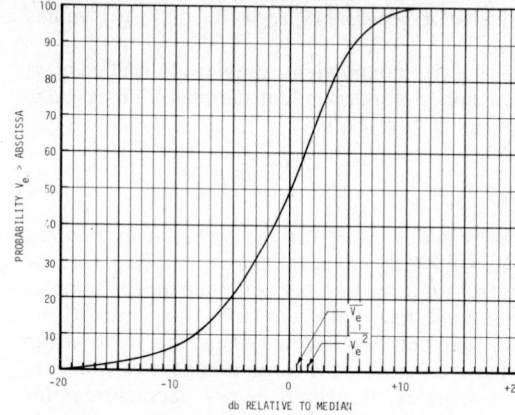

Fig. 9-41. Probability of signals exceeding level.

abscissa. Note: the value exceeded for 95% of the time is 0.254 $\bar{V}_e$; the value exceeded for 5% of the time is 1.96 V_e. Thus, for 90% of the time the signal varies over a range from 5.8 dB above the mean envelope level to 11.9 dB below the mean—a total range of 17.7 dB! Mathematically we may express this as

$$P\left[V_e(dB) < \bar{V}_e(db) - 11.9 \text{ dB}\right]$$
$$= 0.05 = P\left[V_e < 0.254 \bar{V}_e\right]$$
$$P\left[V_e(dB) < \bar{V}_e(dB) + 5.8 \text{ dB}\right] \quad (9\text{-}117)$$
$$= 0.95 = P\left[V_e < 1.95 \bar{V}_e\right].$$

Similarly, if comparison is made with the mean-square voltage of the envelope, which is the mean voltage after square-law detection, it is seen that the range is the same. However, the signal goes a shorter distance above it, and a longer distance below it for the 95% range, as indicated by

$$P\left[V_d(dB) < \bar{V}_d(dB) - 12.9 \text{ dB}\right]$$
$$= 0.05 = P\left[V_d < 0.051 \bar{V}_d\right]$$
$$P\left[V_d(dB) < \bar{V}_d(dB) + 4.8 \text{ dB}\right] \quad (9\text{-}118)$$
$$= 0.95 = P\left[V_d < 3.0 \bar{V}_d\right].$$

Obviously, the mean of a signal with this much fading is indeed difficult to measure accurately. A given measurement of V_d has only 90% probability of lying within a ratio of 60:1, which is certainly a poor measurement. This situation can be proven by integration, smoothing, or filtering. All of these words refer to the same operation, although perhaps indicating different means of performing it. Straight integration represents a form of low-pass filtering, and either filtering or integration serves to smooth the output by removing its higher frequency components.

The easiest case to consider is that for which each measured voltage is statistically independent of every other measured voltage. If the integration process represents straight-forward averaging, the average output voltage is given by

$$V_{av} = \frac{1}{N}\sum_{i=1}^{N} V_{di}, \quad (9\text{-}119)$$

where:
    N is the number of samples integrated,
    V_{di} is the detected output for the ith sample.

If the detector is a square-law device, each V_{di} is described by the exponential distribution, a chi-square distribution with two degrees of freedom. Each element of the sum adds two degrees of freedom to the chi-square distribution so that V_{av} has $2N$ degrees of freedom. The chi-square distribution is given for different numbers of degrees of freedom in both analytical and tabular form in standard references on statistics.

The Gaussian distribution is representative of the fading signal and of the noise. Here the expected average voltage after square-law detection is given by

$$E\left\{V_{av}\right\} = 2\sigma^2 = \bar{W}_e, \quad (9\text{-}120)$$

where:
    E specifies expectation
    σ = standard deviation of the original instantaneous input voltage.
    $\bar{W}_e$ = average envelope power.

For the chi-square distribution, the standard deviation for the average is given by

$$\sigma_{av} = \frac{\bar{W}_e}{N^{1/2}} \quad (9\text{-}121)$$

This illustrates the basic problem with averaging: the standard deviation only improves as the square root of the number of independent samples; the result is that many independent samples are required to achieve a small standard deviation.

The range of fading for the distribution with different numbers of independent samples can be obtained from a chi-squared distribution table, and is shown in Figure 9-42 for the 90% range. The range of fading decreases rapidly as a few samples

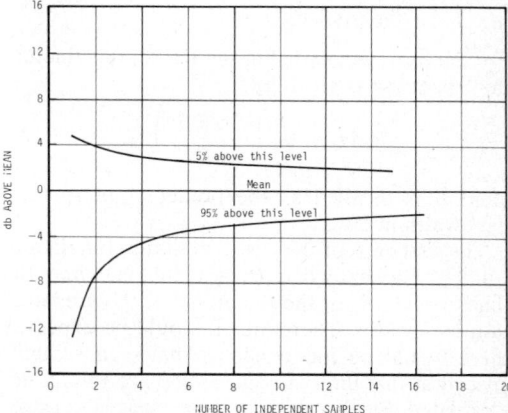

Fig. 9-42. Fading range after square-law detection and integration.

are averaged, but much more slowly as the number of samples increases. Integrating ten samples still leaves a fading range that exceeds 4.5 dB; that is, the measurement is uncertain over a ratio of three to one! For values beyond about ten independent samples, the approximation of the chi-square distribution by a Gaussian distribution (as discussed previously) is valid because the central limit theorem applies.

Spacing Between Independent Samples

The discussion above assumes knowledge of the number of available independent samples. The determination of the number of independent samples existing in an average of a given number of pulses or a given length of time is a problem of considerable importance. With fading signals or noise, this involves knowledge of the spectrum or *autocorrelation function*. In general, *samples are independent* when the autocorrelation function is *zero* or nearly zero, and are dependent for times too short for this decrease of the autocorrelation function to have occurred.

Some idea of the spacing between independent samples may be achieved by examining the autocorrelation function for an appropriate example. In this case, consider the example of Figure 9-39; that is, a square-law detector, the input of which is a rectangular narrow-band spectrum, the output being a triangular spectrum centered on zero. The autocorrelation function for this process is defined as

$$R_d(\tau) = \lim_{T \to \infty} \int_0^T V_d(t) V_d(t + \tau) dt$$
$$[9\text{-}122(a)]$$

$$R_{df} = R_d(\tau) - \overline{V_d}^2 = R_d(\tau) - \overline{W_e}^2.$$
$$[9\text{-}122(b)]$$

where:

R_d = autocorrelation function for the total post-detection voltage V_d,

R_{df} = autocorrelation function for the fluctuating component or autocovariance for the total V_d.

For the example quoted in the figure, this fluctuation spectrum is given by

$$R_{df}(\tau) = \overline{W_e}^2 \left(\frac{\sin \pi B \tau}{\pi B \tau} \right)^2 . \qquad (9\text{-}123)$$

Here B is recalled as the predetection spectral bandwidth in Hz.

The first zero of the autocorrelation function of Eq. 9-123 occurs where $B\tau = 1$; that is, where the time is $\tau = 1/B$, or the reciprocal of the predetection bandwidth. Therefore, it would be expected there would be independence between samples spaced at this interval, since they are totally decorrelated. Furthermore, samples spaced at larger intervals are relatively independent because the autocorrelation function has a small value, al-

though not necessarily zero, for all larger values of τ.

Since the independent samples are spaced by τ, the number of independent samples N observed in a period T is given by

$$N = \frac{T}{\tau} = BT.$$

This result is exact for the situation of samples spaced by $1/B$, but it is approximately correct in general for large numbers of independent samples for a great many different kinds of spectra. However, one can find spectra where this relationship is incorrect, even for large numbers of independent samples. It usually is not very correct for small numbers of independent samples; except, of course, if they are spaced exactly right.

Although many radar systems use pulses or other sampling schemes, this analysis is easiest to perform if we consider an uninterrupted fading signal such as would return to a radar that emitted a continuous wave. If this is the case, an integration is carried on for a period T. Rice (1946) and others have shown that the variance is given by

$$\sigma_T^2 = \frac{2}{T} \int_0^T \left(1 - \frac{x}{T} \right) R_{df}(x) \, dx. \quad (9\text{-}125)$$

Of course, to determine the actual variance, the correlation function associated with the particular problem must be calculated.

This equation may be used to determine an equivalent number of independent samples, so that the statistical techniques and tables appropriate to discrete sampling from independent populations may be used; the curve of Figure 9-42 is appropriate for this use. It is found that σ_T is a standard deviation after integration for time T. It is therefore equated from Eq. 9-125 with the value of the standard deviation of the average from Eq. 9-121

$$\sigma_T^2 = \sigma_{av}^2 = \frac{\overline{W_e}^2}{N} = \frac{2}{T} \int_0^T \left(1 - \frac{x}{T} \right) R_{df}(x) \, dx.$$
$$(9\text{-}126)$$

This may be solved for N, obtaining

$$N = \frac{\overline{W_e}^2 T}{2 \int_0^T \left(1 - \frac{x}{T} \right) R_{df}(x) \, dx} \qquad (9\text{-}127)$$

This expression is quite general. Since R_{df} includes $\overline{W_e}^2$ as a factor, clearly the remaining part of the denominator of Eq. 9-127 represents an equivalent time for a single independent sample. If the correlation function were to decay very slowly (only low frequencies being present), the equivalent time between samples would be governed in large part by the actual duration of the integration, and it would turn out to be about one. If, on the other hand, the decay of R_{df} is fast compared with

T, the duration of the integration has little to do with the effective time interval between independent samples.

For the example of Fig. 9-39, the relative standard deviation is approximately equal to $1/(BT)^{1/2}$, as long as BT is larger than about ten. For smaller values the approximation made in equation 9-129 is not correct. Waite (1970) has calculated the variance reduction for the case of averaging in frequencies, and his curve has been adapted here in Figure 9-43, which shows both the relative standard deviation and its reciprocal, the equivalent number of independent samples. An interesting result is that for $BT = 1$ there is an equivalent independent sampling of 1.52, not 1.00 as would be expected. This is because some decorrelation takes place before the time reaches $1/B$. Of course, if samples were only obtained once every $1/B$ this advantage would not be achieved. The advantage remains out to about $BT = 5$, at which point the equivalent number of independent samples is only about 5.25. For larger values of BT, the equivalent number of independent samples is about equal to the BT product.

Analysis of Pulsed Fading Signal Return. The preceding analysis has dealt with the fading signal return for a continuous transmitted wave. If pulses are transmitted, they have the effect of sampling the continuous-wave signal. In fact, with a typical pulse radar, the signal that would be obtained by sampling each successive pulse return at the same delay after transmission is a sample from a separate fading process for each delay, since each time delay corresponds with a different position on the ground. Because the pulses are customarily very short compared to the total duration of the interpulse period, the sampling usually may be considered instantaneous as far as the fading is concerned. If the sampling rate is at least $2B$ the analysis that applies for the CW signal also applies for the pulse signal. For a lower sampling rate, the samples are more independent, but they do not cause a completely accurate reproduction of the initial spectrum. The calculation of the exact independence depends upon consideration of *aliasing* in the sampling process, a subject discussed in detail by Blackman and Tukey (1958).

For a very high pulse-repetition frequency (prf), the samples may be quite highly correlated. Sometimes the mistake is made of assuming that every pulse return is independent of every other one. This is certainly *not* true for a *high* prf. That is,

$$N = BT < \frac{T}{T_p}, \qquad (9\text{-}128)$$

where:

T_p is the interpulse (repetition) period.

On the other hand, if the pulse repetition rate is low enough, each sample may be almost independent; that is,

$$N \approx \frac{T}{T_p}, \qquad T_p > \frac{1}{B}. \qquad (9\text{-}129)$$

That this is not necessarily so is indicated by the example of Figure 9-44. Here a quite narrow post-detection spectrum is shown, although it does contain some components all the way out to B. The effective bandwidth of this spectrum, however, is relatively narrow and the repetition period would have to be quite long indeed if each sample were to be independent for this case.

Precision of Signal and Noise Measurement Together

The fading characteristics described above have to do with measurement of fading signals or of noise alone, or indeed of a combination of a fading signal with noise. In any real measurement, both signal and noise are present and the combination must be considered, unless S/N is sufficiently high that it can be neglected. With a fading signal, S/N need not be extremely high, for the effect of the fading is so great that the additional fluctuation

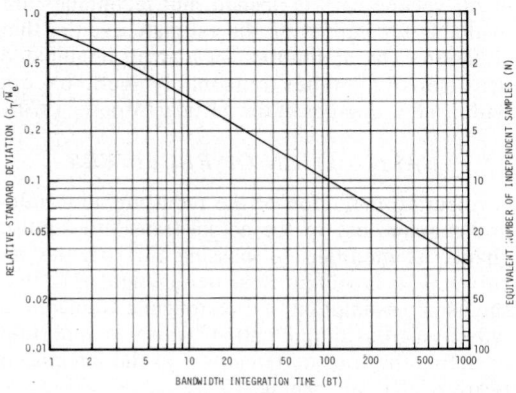

Fig. 9-43. Effect of integration after square-law detection on fading signal or noise with rectangular narrowband predetection spectrum.

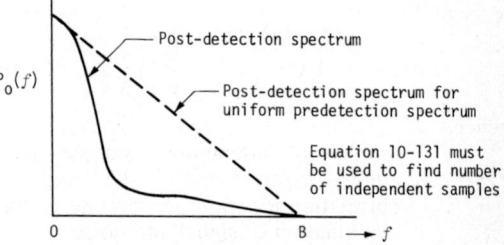

Fig. 9-44. Post-detection spectrum where Eq. 9-129 does not apply.

due to noise is unimportant unless the noise is almost as big as the signal.

The voltage received consists of a signal and noise voltage added together:

$$V_r = V_s + V_n, \qquad (9\text{-}130)$$

where:

subscripts r, s, and n stand for received, signal, and noise, respectively.

Because of the random nature of the signals, not only do the voltages add, but so do the powers; that is

$$W_r = W_s + W_n = W_s \left(1 + \frac{1}{S}\right). \qquad (9\text{-}131)$$

where:

S represents the signal-to-noise ratio.

The standard deviation for the received signal after square-law detection is the same as for the exponential distribution, and is equal to the mean received power; that is,

$$\sigma_{rd} = \bar{W}_r = \bar{W}_s \left(1 + \frac{1}{S}\right), \qquad (9\text{-}132)$$

After integration of N independent samples, the standard deviation, now given as σ_{rdn}, has been reduced by $N^{1/2}$; that is,

$$\sigma_{rdn} = \frac{\bar{W}_s}{N^{1/2}} \left(1 + \frac{1}{S}\right). \qquad (9\text{-}133)$$

Measurement of the signal involves estimating the signal amplitude from the received amplitude and, if it can be separately measured, the noise amplitude; that is,

$$\hat{\bar{W}}_s = \hat{\bar{W}}_r - \hat{\bar{W}}_n, \qquad (9\text{-}134)$$

where:

$\hat{\bar{W}}$ = a measured estimate of the mean value of the power.

The variance for the difference of two random variables is simply the sum of the variances for the separate variables:

$$\sigma_{\hat{s}}^2 = \sigma_{\hat{r}}^2 + \sigma_{\hat{n}}^2, \qquad (9\text{-}135)$$

where:

$\hat{s}$ is a subscript identifying the measured estimate of signal.

Thus the variance on the estimate of the signal is given by

$$\sigma_{\hat{s}}^2 = \frac{\bar{W}_s^2}{N_r}\left(1 + \frac{1}{S}\right)^2 + \frac{\bar{W}_s^2}{N_n}\left(\frac{1}{S}\right)^2, \qquad (9\text{-}136)$$

where:

N_r = number of independent samples calculated using equation 9-127, and the correlation function appropriate to the combination of signal and noise,

N_n = number of independent samples calculated using the autocorrelation function appropriate to noise alone.

Of course, the separate noise measurement may involve integration for a different time, which also affects N_n.

The implications of separate measurement of signal and noise are discussed under the individual discussions of scatterometry in this chapter, imaging radars in Chapter 10, and radiometers in Chapter 11. The concept is the basis of all radio telescopes and microwave radiometers, but it has also been used for scatterometers, and it influences the needed value of the S/N ratio for imaging radars.

CALIBRATION OF RADARS

Radar calibration is necessary if quantitative measurements of amplitude are to be made. Many remote sensing radars are uncalibrated, for they produce images with the prime purpose of determining geometric information about the context of various items on the ground. For such purposes as mapping geologic structure, identifying road networks, etc., this technique of uncalibrated radar imaging is quite adequate. If, however, quantitative identification of materials on the ground is to be made on the basis of return signal amplitude at one or more frequencies and polarizations, calibration is essential. Although scatterometers are by definition calibrated, imaging radars will be calibrated more frequently in the future.

To examine the calibration problem, refer back to the radar Eq. 9-67, and rewrite it as follows:

$$\bar{W}_r(t) = \int \frac{W_t G^2 \, \lambda^2 \, \sigma^\circ \, dA}{(4\pi)^3 \, R^4}, \qquad (9\text{-}137)$$

Note that the integral here is appropriate only for the average power, since the instantaneous value fluctuates as described above; in fact, the differential scattering coefficient σ° is an average quantity. This quantity is the descriptor of the ground that must be established by a calibrated radar.

The illuminated area is often small enough so that we may treat σ° as constant across the region of integration. If σ° cannot be considered constant, special care is needed, but techniques are available for improving the estimate even in that situation. The assumption of a small illuminated area usually applies reasonably well, except within a few degrees of the vertical (Moore, 1970).

CALIBRATION OF RECEIVERS

Accurate definition of the radar equation integral requires use of the actual transmitted pulse shape, as modified by passing through the receiver. This is seldom close to rectangular, so that the usual assumption of a rectangular transmitted pulse is a gross simplification. Thus it is important to define the transmitter power in the integral in terms of the pulse shape:

$$W_t = W_{to} p(t - 2R/c). \qquad (9\text{-}138)$$

Here W_{to} is usually taken to be the peak value of the power, although it could be taken as an effective average power across a specified pulse duration.

The effect of the receiver on the shape of the effective transmitted pulse may be obtained by considering transmission of the transmitted voltage pulse through the receiver. The output that results is

$$v_{tr}(t) = \int_0^t v_t(t-x)a_r(x)\,dx, \qquad (9\text{-}139)$$

where:

$v_{tr}(t)$ = receiver-modified transmitter voltage,

$v_t(t)$ = actual transmitted voltage,

$a_r(t)$ = receiver impulse response.

In the frequency domain this relationship is

$$V_{tr}(f) = V_t(f)\,A_r(f), \qquad (9\text{-}140)$$

where:

$V_{tr}(f)$ and $V_r(f)$ = Fourier transforms of v_{tr} and v_t,

A_r = Fourier transform of a_r and is the transfer function of the receiver.

V_{tr} may be obtained by measuring the output of the receiver when an attenuated transmitted pulse is passed through it, or it may be calculated from Eqs. 9-139 or -140 if the quantities are measured separately. If the receiver bandpass is set by a criterion for optimum S/N, the output voltage peak is somewhat reduced from that for a very wideband receiver that would reproduce the transmitter pulse better, and this difference must be considered in the measurement; i.e., measuring the shape of v_{tr} is not enough.

When $v_{tr}(t)$ has been determined, it may be used to establish $W_t(t)$, since

$$[v_{tr}(t)]^2 = W_t(t). \qquad (9\text{-}141)$$

Use of the voltage superposition integral for passage of the pulse through the receiver, and of the power superposition integral of Eq. 9-71 based on it, can be demonstrated to be proper for determining the mean return for a fading signal because of the averaging out of cross-product terms. When the definition of Eq. 9-138 is used, Eq. 9-67 may be solved for the differential scattering coefficient σ^0, obtaining

$$\sigma^0(t) = \left\lfloor \frac{(4\pi)^3}{\lambda^2} \right\rfloor \times \left\lfloor \frac{\bar{W}_r(t)}{W_{to}} \right\rfloor \frac{1}{\int \dfrac{p(t-2R/c)G^2\,dA}{R^3}}.$$

$$(9\text{-}142)$$

The first bracket contains a quantity that is constant for a particular radar situation. The second bracket is the ratio of the mean received power at a particular delay time (and consequently an apparent location) to the effective transmitter power amplitude. The third factor contains the illumination integral that accounts for the effects of

nonuniform illumination caused by both pulse and gain of the antenna, as well as the (usually negligible) variation of range across the area associated with illumination. Eq. 9-142 may therefore be rewritten as

$$\sigma^0(t) = \left\lfloor \frac{(4\pi)^3}{\lambda^2} \right\rfloor \left\lfloor \frac{\bar{W}_r(t)}{W_{to}} \right\rfloor \frac{1}{I(t,\mathbf{R})}, \qquad (9\text{-}143)$$

where I, the illumination integral, is a function of delay time from the start of the transmitted pulse, as well as the position of the target relative to the radar. Therefore it is stated in terms of a vector $\mathbf{R}$ rather than simply in terms of the range itself, for $\mathbf{R}$ includes appropriate angles relative to boresight.

The *illumination integral* may be calculated for each flight situation and delay time. In fact, it may be used to determine an equivalent illuminated area that would exist if the pulse were rectangular and of amplitude W_{to}, and the antenna gain were constant across a specified beam width and zero elsewhere. Details of this calculation differ with different problems and are therefore not included here.

Accurate determination of the antenna gain is a difficult problem. Typical antenna measurements are accurate at best to a few tenths of a dB and often are even less accurate away from the peak of the pattern. Antenna pattern measurements often consist only of the *principal plane cuts;* that is, a measurement in a plane containing the electric vector (*E-plane cut*) and one in a plane containing the magnetic vector (*H-plane cut*). These measurements are inadequate for quantitative measurements of scattering coefficient; the entire pattern of the antenna, at least in the vicinity of the main beam, should be measured.

The factor containing the mean received power and the transmitted effective power is a ratio. Although the ratio may be obtained by separate measurements of received power and transmitter power, it also may be obtained by allowing a sample of the transmitted power to pass through the receiver so that the ratio may be measured directly at the receiver output. If these samples are obtained frequently enough, the effect of fluctuations in receiver performance may be largely eliminated. Thus long term variations in receiver gain or bandwidth are unimportant for the measurement of σ^0. This is an important result since fluctuations in gain and bandwidth do occur, but need not effect the accuracy of the measurement if the system is properly designed. In fact, the receiver gain need not be measured at all, although most radar measurement programs do in fact include determination of receiver gain as a cross-check on the other measures of performance.

One important factor that must be considered in all such measurements is potential or actual nonlinearities in the receiver performance. Sources of nonlinearity include saturation of both amplifiers and detectors, and low-signal-level nonlinearities in detectors. Thus although the ratio of received-to-transmitted power can be obtained without reference to receiver gain by passing a sample of the transmitted power through the receiver, the transmitter sample should be at approximately the same level as the signal received through the antenna. Since the received radar signal may vary over a range of tens of dB, this means the sample of the transmitter should also be provided at several levels so that a comparison can be made in an amplitude region; appropriate to any received signal.

TECHNIQUES FOR CW RADAR CALIBRATION

CW Radar Calibration by Receiver Input Switching

Figures 9-45 and 9-46 show two techniques that can be used with a continuous wave system. In Figure 9-45 the calibration signal is separated from the received signal in time because the receiver input is switched from the receiving antenna to a transmitter sample path. Clearly this switching must occur often enough so that drift in the performance of the receiver is not important. During the time that the receiver is being calibrated, no signal may be obtained from the antenna.

The transmitter sample path in Figure 9-45 requires a directional coupler that takes a small fraction of the output signal and leads it toward the receiver. Since the output signal from the directional coupler is much stronger than the received signal, and since the sample level should be comparable with that received through the antenna, a calibrated attenuator is included in the path from the directional coupler to the receiver. If this attenuator is adjustable, the level of the sample can be set to be comparable with the level of the received signal.

Clearly the attenuation in the path taken by the sample must be accurately established, usually by a laboratory measurement. Since passive components are used, this attenuation should not change significantly over a reasonably long time interval, so that the measurement need not be made in the field. However, the effects of temperature on the attenuation should be established. This effect may be calibrated in the laboratory; but if it is significant, the temperature must be monitored during the actual radar measurement so that the effect may be removed.

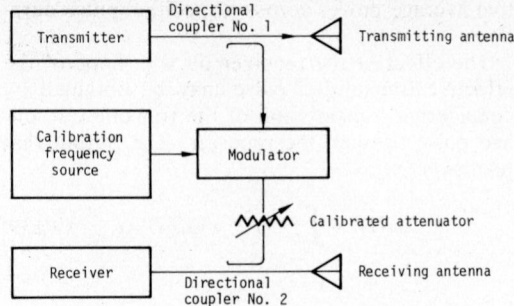

Fig. 9-46. Frequency separation of calibrate signal.

CW Radar Calibration by Modulation Technique

Sometimes the CW system is designed to exclude the carrier frequency and only permit Doppler-shifted signals to pass. In this case the calibration path must include some kind of modulator or frequency shifter so that the sample signal comes through at a frequency different from the carrier. One approach to this is to use the modulation technique shown in Figure 9-46.

In Figure 9-46, frequency separation of the calibration signal is used to permit simultaneous measurement of received signal and calibration sample of the transmitter signal. This is achieved by modulating the sample of the transmitter signal at a frequency high enough so that the lowest modulation sideband is just outside the passband of the Doppler frequency shift for the signal. Thus both may pass through the receiver simultaneously, and they may be separated at the output by filtering. When this method is used, the relative frequency response of the receiver must be accurately established because the desired signal passes through the receiver at a different frequency than does the calibration signal. Fortunately the frequency response is usually fixed mostly by passive elements that are less likely to drift over a short term than are the active elements that determine the gain of the receiver. The gain may drift, but the fact that both calibration and received signals are amplified by the same drifting gain means that the ratio is unaffected.

The technique shown can produce a calibration signal the amplitude of which is largely independent of the amplitude of the calibration frequency source. Thus this amplitude need not be carefully measured. The technique used is to have the modulator serve as a switch. Ideally it would switch from zero attenuation for the "on" condition to infinite attenuation for the "off" condition. Real modulators using PIN diodes or ferrite switches achieve "on" attenuation of a few tenths of a dB and "off" attenuations of 30 dB or more. Thus this is a reasonable approximation to the ideal switch, but the magnitude of the attenuation in the "on" condition must be carefully mea-

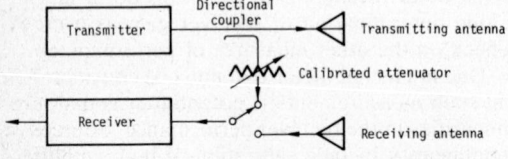

Fig. 9-45. Time separation of calibrate signal.

sured, since this is part of the attenuation through the calibration path. This square-wave modulation produces multiple sidebands, but only the lowest need be monitored.

Techniques for Pulse Radar calibration

Calibration of pulse radars is somewhat more complicated than calibration of CW radars. The problem is that the receiver is not on and active during the transmission period for the pulse radar, so that the method of Figure 9-46 cannot be used; and the transmitter power output is often so great that the receiver cannot be activated even with the use of an attenuator, as with the method of Figure 9-45. This method would be feasible if, as shown, separate transmitting and receiving antennas were used, but attenuation through the usual single-antenna isolation system is insufficient.

Calibration of Pulse Radar Transmitters

A method that has been used for many pulse radar measurements involves separate determination of the transmitter output characteristics and the receiver characteristics. When this is done, the ratio of received-to-transmitted power in Eq. 9-142 must be obtained from separate measurements rather than as a single measurement, as described for CW radars. Figure 9-47 illustrates the kinds of measurements required for this method. The transmitted pulse must be sampled as with any other calibration scheme. In this technique two kinds of measurement are made on the transmitted pulse itself, and in addition its repetition rate must be accurately determined. The pulse average power is measured using a power meter, and the pulse shape is established by displaying a detected sample of the pulse on an oscilloscope. The repetition rate may be determined by counting the pulses out of this same detector.

The power meter works on the average power; i.e., power meters depend upon heating of temperature-variable resistance in which the thermal time constant is long compared with the duration of an individual pulse, so the measure-

ment must be made by averaging over a large number of pulses. Since the average power so obtained must be related to the peak power in the transmitter pulse, the characteristics of the pulse must be carefully measured and used in this conversion. Although the pulse shape may be continually monitored, the average power is often measured before and after an experiment rather than during the experiment. The presumption is made that stability of the output of the pulse sampling detector indicates that the power level was constant during the experiment so that the power meter need not be used continuously. The nonlinearities and transfer properties of the sampling system must be accurately measured.

Calibration of Pulse Radar Receivers

To determine the received power for the power ratio, a pulse is transmitted through the receiver from a signal generator. The shape of this pulse should be comparable with that of the transmitter so that the receiver passband has similar effect on the calibration pulse and the actual received pulse. The calibration pulse may be provided during each transmission period by turning on the calibrator during the sensitive period of the receiver shortly before the transmitter is activated. Appropriate triggering signals are provided by the synchronizer.

The calibration pulse should be used at an amplitude comparable with the expected mean value of the return signal; but, because of fluctuation in the return signal, the level of the calibration pulse must be varied before and after the experiment to determine the nonlinear properties of the receiver. The ratio calculated from these separate measurements is seldom as accurate as that determined by an actual ratio measurement.

The gain of a pulse system is often designed to vary with delay time, so that signals arriving at the detector are of comparable levels even though they come from different ranges and angles of incidence. The calibration obtained by pulsing the signal generator immediately prior to activating the transmitter only provides information on a part of this characteristic of gain vs time. Consequently periodic calibrations must be arranged in which the transmitter is disabled, so that the signal generator pulse may be provided at time delays corresponding to actual ranges to be observed by the radar. A set of calibration curves for different ranges will assist in determining the effects of the nonlinearities of the receiver for different preset gain values. The measurement at the time of each transmitted pulse, however, must necessarily be kept outside the time interval for reception. Thus some uncertainty exists about variation in the circuits controlling the change in receiver gain with time delay, even if the receiver gain remains constant at the long delay associated with the calibration pulse.

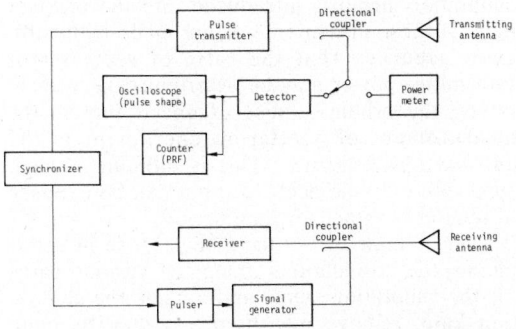

Fig. 9-47. Receiver calibration by delayed sample of transmitted pulse; separate measurement method.

Calibration of Pulse Radar Receivers by Delayed Sample

An attempt to overcome the problems associated with separate measurement of transmitter and receiver characteristics uses a delayed sample of the transmitted pulse to calibrate the receiver. Figure 9-48 illustrates this method. This should be as good a technique as the methods available for the CW radar, but problems exist because suitable delay elements are not available. Delaying the microwave signal for tens of microseconds would be desirable, but delay lines this long are very bulky if their loss is to be kept small. Possibly a super-conducting cryogenic delay line could be used for this purpose, but the inconvenience of working at liquid-helium temperatures makes this method less than desirable.

Figure 9-48 shows two alternate versions of this method. In Figure 9-48a, delay is straightforward, and it is presumed the attenuation is sufficient so that no additional attenuation is required. The other method, shown in Figure 9-48b, uses a section of short-circuited wave guide or transmission line arranged so that the signal bounces back and forth along this line. The result is a stair-step wave form with the first steps of high amplitude and succeeding steps of successively smaller amplitudes because of the attenuation in the short-circuited line section and mismatches. This method would provide a complete calibration of the transmitter pulse passing through the receiver at different levels to take into account the nonlinear properties of the receiver, but it, too, suffers because of the bulkiness of any suitable low-loss delay line.

Modern surface-acoustic-wave delay lines might be used for this type of calibration, since they are compact and can achieve appropriate delays with reasonable attenuation. On the other hand, such surface wave devices introduce additional complications because the accuracy and stability of the transduction from EM to mechanical waves and back must be taken into account.

Calibration of Pulse Radars by Continuous Source

Another method that can be used for calibrating pulse radars by the ratio method is shown in Figure 9-49. Although the transmitter pulse only lasts

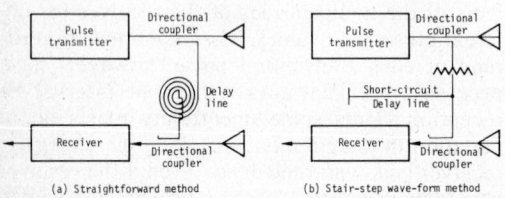

Fig. 9-48. Receiver calibration by delayed sample of transmitted pulse; pulse-delay method.

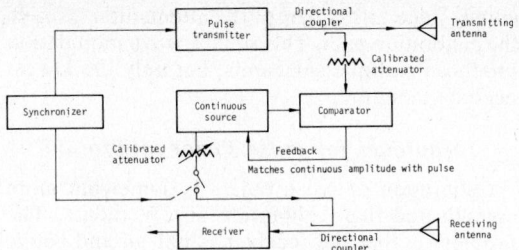

Fig. 9-49. Calibration of pulse radars.

a very short time, a continuous source is adjusted in amplitude so that its amplitude is directly proportional to the peak transmitter pulse amplitude.

The output of the continuous source is available for calibration purposes in the same way that the output of the transmitter is available for the CW radar. Of course, such a signal cannot be continuously supplied to the receiver, for it would interfere with the desired signal. Consequently a switch is necessary to provide a calibration sample at an appropriate time, as in Figure 9-49. The pulse out of this switch should closely parallel the transmitter pulse in shape, so the effect of the receiver transfer function is the same on both.

To calibrate the receiver for different gains used at different ranges, the incoming signals may be interrupted at appropriate intervals, as they are for the CW radar of Figure 9-45. During the interruption, the synchronizer may be directed to provide pulses at different delays and the calibrated attenuator stepped at each delay to give appropriate calibrations.

ACCURACY OF MEASUREMENT

All of the calibration techniques require careful laboratory measurement to determine the attenuation in the path taken by the calibration signal and to determine the attenuation outside that path. This extra path is from the point where the sampling directional coupler attaches to the transmitter signal path on out to the transmitting antenna, and from the point where the signal is received by the antenna into the point at which the calibration signal is introduced into the receiver path. If these measurements are made with sufficient accuracy that the ratio of receiver-to-transmitter power may be determined very accurately, the primary cause of uncertainty in the measurement of scattering coefficient is the *antenna characteristic*. This is difficult to measure with extreme accuracy, but can be exposed to remain constant for long periods.

Even if some errors have been made in determining the attenuations along the various paths for the calibration signal and outside the calibration loop, relative measurements may be quite precise, provided all elements in the transmitter sample loop and between the sampling path and the antennas are fixed and passive. Introduction of microwave waveguide rotary joints between the sampling point and the antennas may cause significant variable errors in that path which will not be detected. The best arrangement is to mount the sampling directional coupler on a portion of the feed system fixed to the antenna, and to connect the calibration pulse to a receiver input circuit also attached rigidly to the antenna.

When antennas are mounted in radomes, additional difficult-to-calibrate problems may be introduced. If the antenna is fixed in the radome, the effect of the radome is constant provided the surface is dry. However, moisture on the surface of the radome may cause reflections that can upset the calibration to some extent. When the antenna rotates within the radome, the complications are more severe, for the effect of radome is different at each angle. This effect may be calibrated, but only with difficulty.

NOISE MEASUREMENT AND REMOVAL

The methods described above for calibrating the receiver can be used when the signal is much stronger than the noise, and can also be used with care when the received signal is comparable with the noise (Moore and Ulaby, 1969). For best results when noise and signal are comparable, measurements should be made not only of the received signal and a sample of the transmitted signal, but also of the received noise during a time when neither transmitter sample nor received signal is present. The noise observed by the radar under these conditions is a combination of that produced by the input circuitry itself and that received by the antenna. The latter would be a signal for a microwave radiometer, but for a radar it constitutes part of the noise. It is, of course, variable because this variability is the quantity measured with the radiometer. Nevertheless, the variability over land is relatively small compared to the variability of the radar signal. The percent variability of the sum of received and receiver noise is even smaller because the received noise is usually less than the receiver noise by a significant amount. The received noise varies by about 3 dB when the radar goes from land to water, although this represents less than a 3-dB variation for the total noise.

If the noise is measured separately, the estimate of both received fading signal and calibration signal may be made using methods similar to those discussed previously. That is, the signal may be extracted from the sum of signal and noise by subtracting the noise if sufficient averaging has taken place to reduce the variance to an acceptable value. Radiometers always use this method, but radars may use it, too, and achieve significantly improved scattering coefficient measurement for weak signals. The variance for this measurement is given by Eq. 9-136. Frequently the noise measurement may be made with enough independent samples to make the second term in

this equation negligible, even though the first term remains important, because of the relatively narrow bandwidth of fading.

Figures 9-50 and 9-51 show this technique for removing noise. In Figure 9-50 the basic *Dicke radiometer* is indicated. For this radiometer, the receiver input is alternately switched between the antenna and a calibrated noise source kept at a carefully controlled temperature. The output is synchronously switched between two integrators; one picks up the signal and receiver noise, and the other picks up the noise source output and receiver noise. The subtractor at the output gives an estimate of the difference between the signal coming in the antenna and the level of the calibrated noise source. The noise introduced by the receiver therefore cancels out except for the variance indicated in Eq. 9-136. Details of various techniques that are modifications and improvements of this basic system are given in Chapter 11.

Figure 9-51 shows application of the same technique to a radar. Here the two integrators are also found; actually a pair of integrators is required for each range interval to be observed. The top integrator is used during the time the signal comes in from the desired range element and the bottom one is used during a time when no signal is present. Provided the gain of the receiver is the same for both ranges, the subtraction is straightforward, in accordance with Eq. 9-136. The timing sequence for this situation is shown in the figure. If, on the other hand, the receiver characteristics are varied with range, the receiver may have to be actuated twice for each transmitter pulse: once during which the received signal is present and the gain varying circuit operates, and once during which only noise is received and the same gain-variation and range-sampling circuitry used.

The noise removal technique described here is seldom used with radars because S/N with radar is usually so large it is not necessary. Sometimes when these techniques are used with radars, more complicated estimation devices are used than simply integration and subtraction. Nevertheless this technique is a valid one and has been used on radiometer-scatterometer combination system (RADSCAT) instruments developed for NASA, as well as on the Seasat scatterometer.

SCATTEROMETERS

Radar scatterometers permit more detailed observation of radar scattering behavior than radar

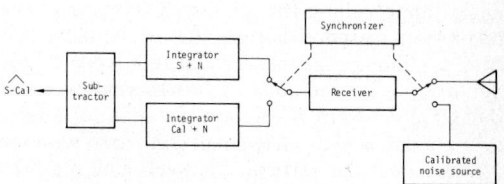

Fig. 9-50. Receiver noise removal; basic Dicke radiometer.

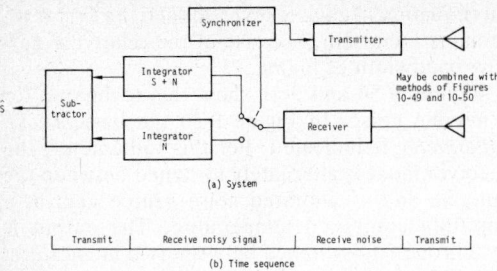

(a) System

| Transmit | Receive noisy signal | Receive noise | Transmit |

(b) Time sequence

Fig. 9-51. Receiver noise removal; radiometer technique for radar.

imagers in the same way that infrared spectrometers permit more detailed observation than infrared imagers. There is, however, a penalty paid for this information gain: degraded resolution and reduced areal coverage. Radar scatterometers are often used to measure the variation of scattering coefficient with incident angle for radar signals. Some scatterometers, like some other radar systems, also may be used to study the effect of polarization and wavelength variations.

Airborne scatterometers frequently use a fan-beam antenna to allow simultaneous observations at different incident angles. The same techniques can be used on spacecraft systems, although the need for higher antenna gain to allow lower transmitter power may force the spacecraft system from the *fan-beam configuration* discussed here to a *pencil-beam configuration*. The basic pencil-beam system is quite simple, but data can be obtained from only one angle at a time. Ground-based scatterometers can be used in research and as auxiliary sources of information for airborne systems. Although the fan-beam technique can be used with ground-based instruments, most use the pencil-beam approach for simplicity.

Radar scatterometers, like other radar systems, can take advantage of both range and velocity measurements to discriminate areas on the ground by combinations of antenna pattern and range, antenna pattern and velocity, or range and velocity measurement. Furthermore, the methods for measuring range and velocity are varied. Some of the general concepts for such systems are discussed here.

A range-angle and a velocity-angle *fan-beam* scatterometer system are discussed in detail, using somewhat idealized assumptions. Resolution along the flight track varies from quite a large distance at the vertical to quite a small one at angles approaching the grazing. Conversely, the resolution distance across the track increases from a minimum at the vertical relatively slowly out to about 60° and quite rapidly thereafter. Even though along-track resolution distance for angles near grazing is extremely small compared with the distance near the vertical, the averaging required to obtain an accurate measurement requires an *effective* resolution distance almost as large as required near the vertical.

Use of the fan beam system requires processing of the data after flight because observations are made simultaneously from different terrain elements at different incident angles, and the observations for a particular terrain element must be collated later. After this has been done, a curve of scattering coefficient vs angle may be plotted for each terrain element.

SCATTEROMETER SYSTEM DESCRIPTIONS

A scatterometer measures the scattering or reflective properties of surfaces (and sometimes of volumes); a radar scatterometer measures reflection and scattering of waves generated by the radar itself. Since for most radar sets the transmitter and receiver are in the same place, most scatterometers measure signals returned to the source. Most signals are returned to the receiver by a scattering process; only occasionally does specular reflection contribute significantly to the radar return. Hence, the term *scatterometer* is more appropriate than the term *reflectometer*.

Many scatterometers measure the variation of differential scattering cross section of a surface with angle of incidence. The angle of incidence may vary from zero (normal incidence) to 90° (grazing incidence), although few scatterometer systems operate over the entire angular range. Airborne scatterometers of the type discussed here may operate near normal incidence, but they have difficulty at grazing incidence because at these angles the range is so great that the signals received are too small.

Some scatterometers transmit and/or receive with different polarizations. For example, a scatterometer may transmit horizontal polarization, and receive both vertical and horizontal polarization, or it may transmit first one and then the other polarization, receiving both with each transmission. Circular polarization may also be used.

Most scatterometers use only a single carrier frequency. However, measurement of scattering properties over a wide range of frequencies provides additional information, and a ground-based system that does this is described.

An ordinary imaging radar measures scattering coefficient if it is calibrated properly, but it can do so for any point in the image at only the particular angle of incidence with which that point is illuminated. Most scatterometers obtain more information than this by measuring the scattering properties at various angles.

The scatterometer, like other radar sets, contains the basic elements previously shown in Figure 9-1. A transmitter sends a signal to the surface being studied, from which the signal is scattered back to the receiver. Some sort of synchronization is maintained between transmitter and receiver; in fact, a copy of the transmitted signal may be delayed and compared in the receiver with the received signal. The receiver output is processed to determine the scattering coefficient at a particular point and the result is displayed.

Radar scatterometers, since they provide their own illumination, can take advantage of knowledge about the illumination to discriminate among the returning signals. Systems dependent upon self-emission from the surface or upon scattering of incident radiation from other sources do not have this advantage. Figure 9-2 has previously shown, in a general manner, the various broad concepts of radar determination of the area sensed. The methods of discrimination are discussed in the text accompanying Figure 9-2.

Range-Angle Systems

Figure 9-2b illustrates the combination of range and angle discrimination. Such a technique may be used either for ground-based or for airborne/spaceborne systems. A radar system may achieve range discrimination by any scheme permitting comparison of a received signal with a delayed version of a modulation signal—with the amount of delay determining the range from which the received signal will be selected. Transmitting a pulse of short duration is the most common method. If a delayed sample of the pulse is used to turn on the receiver at a later time, the output of the receiver is the signal scattered from the distance for which the round trip signal travel time in space is equal to the delay.

A typical *range-angle system* uses an antenna with a fan-shaped beam that is narrow in the direction transverse to the flight path and wide along the flight path. Figure 9-52 shows a contour on the ground of points seen with the same gain by the antenna; this may be treated as the limit of the illuminated ground area. Sections of the circles of constant range are shown within this contour. Thus area A is selected if the distance between the two isorange contours is the length of a transmitted pulse and the internal pulse delay is the same as for a signal returned from the outer contour. Use of a different pulse delay would select an area between another pair of isorange contours.

The isorange contours can also be set by frequency modulating the radar or by using other types of modulation, such as quasi-random binary phase modulation.

With any *range measurement system*, the block diagram in Figure 9-53 is applicable. Modulation

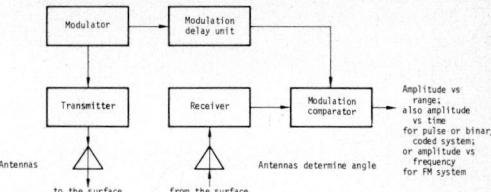

Fig. 9-53. Angle-range-measurement scatterometer system.

in amplitude, frequency, or phase is applied to a transmitter. The output of the transmitter is fed to an antenna which radiates toward the ground, while a sample of the modulator output is simultaneously sent to a delay unit. The signal coming from the transmitter antenna goes to the ground and is returned to the receiver antenna. The output of the receiver and the output of the modulation delay unit go to a modulation comparator. The output of this comparator is based on signals scattered from the range for which the delay unit is set; receiver outputs from other ranges do not pass from comparator to recorder. Thus, if only a single delay is used, the output of the modulation comparator is an amplitude corresponding to a certain range. Normally the modulation delay unit may have numerous possible delays, so the output of the modulation comparator is a set of information concerning the relationship of amplitude and range, with the number of range points being determined by the number of separate delays.

With a *pulse or binary-phase-coded system*, range and time are directly related, so the output of this system also is in the form of amplitude vs time. With an *FM system*, range and frequency are related, so the output is in the form of amplitude vs frequency.

Although separate antennas are shown, many systems use the same antenna for both transmitting and receiving.

Velocity-Angle Systems

The principle illustrated in Figure 9-2c permits separation of ground elements within the angular confines of a beam by measurement of velocity. Such systems are used only from aircraft and spacecraft. Figure 9-54 shows the ground map of this process. Here, again, a fan-beam antenna has been assumed. The area within the isogain contour shown is considered illuminated; the area outside the isogain contour is considered dark. Area A is further restricted by a filter separating out a particular range of Doppler frequencies.

Figure 9-55 shows a block diagram of such a system. The fundamental velocity-angle system does not use modulation; hence, the modulation block that was present in Figure 9-53 does not appear in Figure 9-55. The carrier frequency may be delayed, or the carrier frequency being transmitted at the time the signal is received may be used in the phase comparator. The output of the phase comparator is a spectrum of Doppler fre-

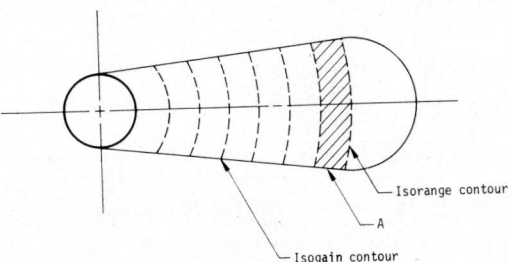

Fig. 9-52. Angle-range-measurement scatterometer illuminated region.

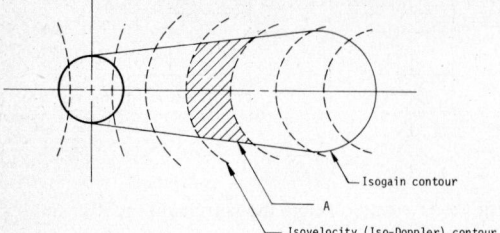

Fig. 9-54. Angle-velocity-measurement scatterometer illuminated region.

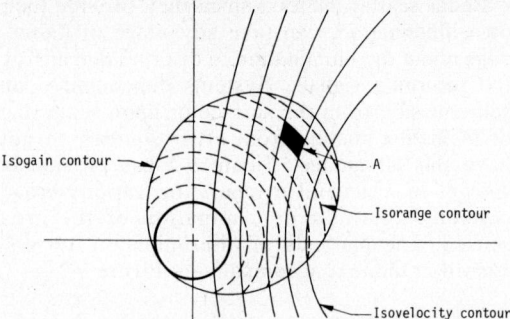

Fig. 9-56. Range-velocity-measurement scatterometer illuminated area.

quencies. Since each frequency is associated with a given relative speed, this output is a measure of amplitude *vs* relative speed. Separating the various frequencies with filters permits equivalent separation of the areas on the ground.

Range-Velocity Systems

Simultaneous range and velocity measurements may be explained by illustration of combined isovelocity contours and isorange contours, as indicated in Figure 9-56. Of course, some sort of antenna is also present, so an isogain contour is shown to limit the other contours. The area *A*, however, is not determined by the isogain contour, but is bounded by two isovelocity contours and two isorange contours. It may be determined by any of the range measurement schemes in combination with Doppler frequency measurements.

Figure 9-57 shows a simplified block diagram of a range-velocity system. Here the modulation is shown feeding the transmitter and a sample of the modulated transmitter output is fed to the delay unit. This modulated output contains the phase information necessary for Doppler measurement and the modulation information necessary for range measurement. The transmitter output goes to the antenna, is radiated to the ground, and then re-radiated to the receiving antenna. A phase and modulation comparator is placed at the output of the receiver. This comparator output gives the amplitude of the return as a function of range (usually expressed as time), and as a function of relative speed (usually expressed as Doppler frequency). By a combination of time and frequency filtering, therefore, the typical area *A* of Figure 9-56 can be separated from the composite return that contains many such areas.

A real system of this kind may have several internal paths between transmitter and receiver. The modulation and phase reference may, for example, be separate from both transmitter and receiver but connected to both. Nevertheless, the principle illustrated in Figure 9-57 still applies.

Systems such as this are commonly used for synthetic-aperture side-looking airborne radars. Most scatterometers use forward-looking systems, but a side-looking scatterometer based on this principle has utility in a meteorological spacecraft for oceanographic observation.

SYSTEM CHOICE

For many uses, a scatterometer should obtain as much as possible of the curve of scattering coefficient as a function of the angle of incidence. If this curve is to be measured for a particular segment of ground, it is necessary that this segment be illuminated successively from different angles at the scatterometer is moved overhead. If the range of angles is to include the vertical, the scatterometer must be flown directly over the illuminated segment. If the scatterometer is being flown in a straight line, this requires that the illuminated area be centered along the flight track. Figure 9-58 illustrates this point. Here a pulsed (or other range-measuring) scatterometer is traveling over a plane surface. At angle θ_4 a first observation is made of a particular segment of ground. Later, as the radar advances, the same segment will be seen at an angle θ_3. Still later it will be seen at angle θ_2, later yet at angle θ_1, and finally the radar will go directly over the segment. If the

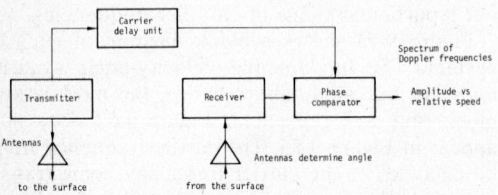

Fig. 9-55. Angle-velocity-measurement scatterometer system.

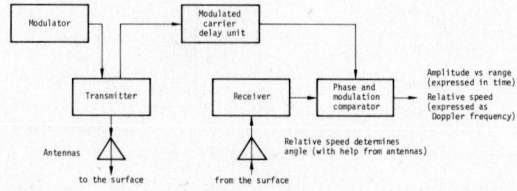

Fig. 9-57. Range-velocity-measurement scatterometer system.

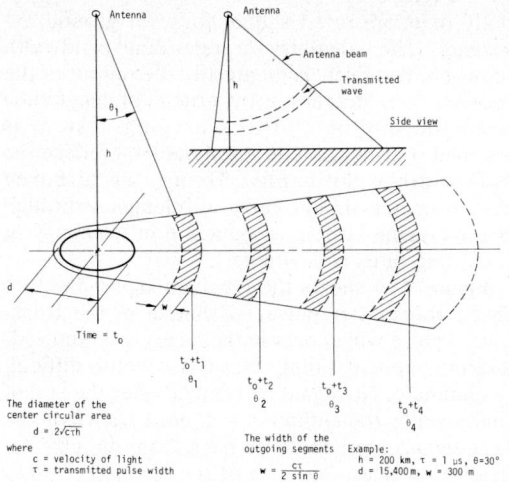

The diameter of the
center circular area

$$d = 2\sqrt{c\tau h}$$

where

c = velocity of light
τ = transmitted pulse width

The width of the
outgoing segments

$$w = \frac{c\tau}{2 \sin \theta}$$

Example:

h = 200 km, τ = 1 μs, θ=30°
d = 15,400 m, w = 300 m

Fig. 9-58. Scatterometer-illuminated area change with increasing time for fan-shaped antenna beam.

antenna pattern extends out the back, the sequence is then repeated in reverse order. If the antenna beam had been pointed in any other direction, it would not have been possible to obtain the vertical incidence record.

Since antenna pattern is the only good way to limit an area to small distances either side of the flight track, a fan-beam antenna illuminating a strip on the ground, as shown in Figure 9-58, is the most useful for scatterometry. A single narrow beam could be used; it could be pointed ahead at a particular segment of ground and kept pointed at that segment as the radar travels toward and over the segment. Such a system only permits observation of selected points rather than of a continuous track. Furthermore, it calls for accurate tracking of the same point on the ground by an antenna beam. Hence, for most purposes, it is inferior to the fan beam.

When a fan-beam system has been selected, the choice must be made between range and velocity measurement. As indicated in later sections, each has its advantages. Discrimination between small angles near the vertical is difficult with range measurement. The Doppler measurement, on the other hand, complicates matters away from the vertical. Detailed studies of system choices give answers that depend on particular problems being attacked, and are beyond the scope of this discussion.

If a ranging system is chosen, a choice must be made between a pulse system and a frequency-modulated system or other CW or nearly CW ranging system. For many purposes, the high peak powers of the pulse system cause problems. Furthermore, the fast pulse circuits necessary to separate ranges with a pulse system can become rather complicated. On the other hand, the FM system depends on isolation between transmitter and receiver antennas if long-range operation is

contemplated. Such isolation is easy to achieve at relatively short wavelengths on large vehicles, but difficult on longer wavelengths and smaller vehicles.

If a velocity-measurement system is to be used, a continuous-wave system is possible, provided antenna isolation can be achieved. If this is not possible, some sort of an interrupted-CW (ICW) system must be used so that the transmitter is turned off during the time the signal is being received. The CW-Doppler system is certainly the easiest system to build.

It is often desirable to measure range for other purposes. For example one may wish to use the same system for an altimeter and a scatterometer. Here, neither the CW system nor the ICW system will work. However, velocity measurement can be made in a suitably designed ranging system, for the Doppler shift exists regardless of the type of modulation used. Such Doppler measurements may or may not be worth the trouble, even where other conditions permit them.

A side-looking system using range and velocity measurement can get by with a considerably smaller antenna than any other system. For this reason, the use of this method may be desirable any time only one angle of incidence is required, especially at longer wavelengths.

PRECISION OF MEASUREMENT IN THE PRESENCE OF NOISE

The scatterometer must average a large number of independent fading signals together to achieve a precise measurement. Since both these fading signals and noise have Gaussian statistics, measurements of signal alone, signal plus noise, and noise alone are subject to the same considerations of precision. This was recognized early in radio astronomy, where extremely large numbers of independent samples are measured, with the result that the mean signal can be extracted from the measurement even if it is tens of decibels weaker than the receiver noise. The same technique is used in radiometers that look at the earth as in those radiometers that look at the sky (radio telescopes). The general concept was described previously in the chapter, under the subheading *Fading.*

The scatterometer usually measures many fewer independent samples than the radiometer because the Doppler bandwidth is measured in Hz to kHz rather than in hundreds of MHz. Nevertheless some airborne and spacecraft scatterometers gather enough independent samples per measurement so that this technique can be used (Moore and Ulaby, 1969).

A scatterometer that measures large numbers of independent samples in the presence of noise can reduce the effect of the noise by making an independent measurement of the noise. This may be done by measuring noise in the same bandwidth as the scattering during a time when no signal is being received, or by measuring noise simultaneously with signal reception but in an adjacent bandwidth segment where no signal is present. The Skylab RADSCAT instrument made such measurements on a time-sharing basis (Hanley, 1972).

The number of independent samples may be different for signal-plus-noise measurements and noise-only measurements because of different integration times, or because of different bandwidths, or both. In certain cases the bandwidth of the receiver used for the signal is significantly larger than the bandwidth of the signal itself. In this situation the calculation of the number of independent samples for the combined signal is more complicated than obtaining a time-bandwidth product. The problem in this case is that the spectrum of the signal combined with noise is not flat. The techniques of handling this particular situation were described by Fischer (1972).

PULSE SCATTEROMETERS

The operation of the pulsed scatterometer using a fan-beam was illustrated in Figure 9-58. Here the illuminated ground segments corresponding to angles from the vertical out to θ_4 are indicated. The radiated pulse first illuminates the circle at the vertical and then illuminates successive rings. Only a portion of each ring is illuminated because of the angular limitation caused by the antenna beam. The width of the illuminated ring is determined by the pulse duration of the radar.

A scatterometer system to be used for this type of operation is presented in Figure 9-59. The timing system provides pulses which turn the transmitter on for the pulse duration τ. It also provides pulses to the gating system at times delayed from the start of the transmitter pulse by the right amount to permit observation at vertical incidence and at, successively, θ_1, θ_2, θ_3, and θ_4. In the notation of Figure 9-58, this means the delays are t_0, $t_0 + t_1$, $t_0 + t_2$, and so on. The transmitted signal goes to the ground and returns through the receiver to the gating system. Those portions of the received signal that are gated through to the output go to separate averaging systems, and the average output corresponding to each of the angles is sent to the data processing unit for collation.

Some elements shown in the sketch need not be present in the flight system. The receiver output could be recorded or telemetered, together with timing information, for ground processing. The gated outputs could be either recorded sequen-

tially or telemetered sequentially for ground averaging. The telemetry or recording bandwidth required, however, is greatest at the output of the receiver and decreases by orders of magnitude until the output of the averaging system is reached. Further processing should not reduce the bandwidth significantly. Hence, an airborne scatterometer should carry all elements through the averaging system if reduction in telemetry or recording rates is desirable.

Figure 9-60 shows the received signal due to a single transmitter pulse. A sample of the transmitted pulse will appear in the receiver if required, and may appear simply because it is too difficult to eliminate. No signal is received after the end of this sample transmitter pulse until t_0, when the first signals start to come back from directly beneath the radar. Because of the geometry, times corresponding to equally spaced angles are spaced unequally, so that the gate times are closer together near the first return (t_1 is close to t_0) and are further apart for later returns (as shown in the diagram).

The observed signal, like that of Figure 9-61, is jagged in appearance. The signals of Figure 9-61 would have appeared spread out much as those of Figure 9-60a if a fan-beam had been used. With a transmitter pulse duration τ, a new independently fading signal appears for every interval τ in the received pulse–provided, of course, that the receiver bandwidth is sufficient. Since each point on the pulse corresponds to a randomly fading signal, and since σ^0 is an average of such signals, the return amplitudes at the various gate outputs for individual pulses are not in themselves meaningful but must be averaged. Thus, if the entire pulse is averaged, one would obtain the signal shown in Figure 9-60b. The system of Figure 9-59, however, will not provide such a continuous pulse, but rather will simply provide the amplitudes at the indicated times. A curve could, of course, be drawn through these points to produce the illustration of Figure 9-60b.

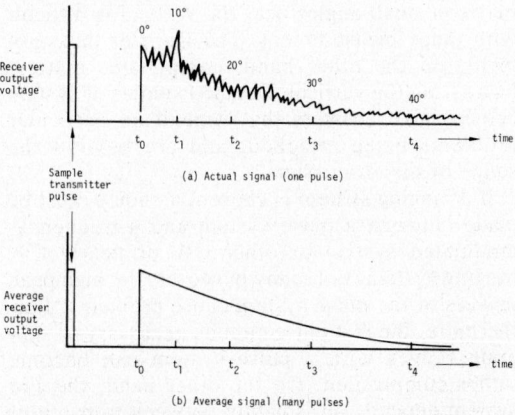

(a) Actual signal (one pulse)

(b) Average signal (many pulses)

Fig. 9-60. Received pulses—fan-beam scatterometer.

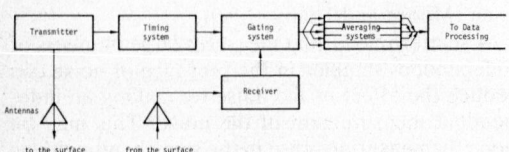

Fig. 9-59. Pulse-range-measurement system.

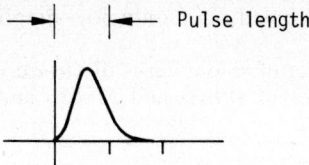

Transmitted pulse shape (not to scale)

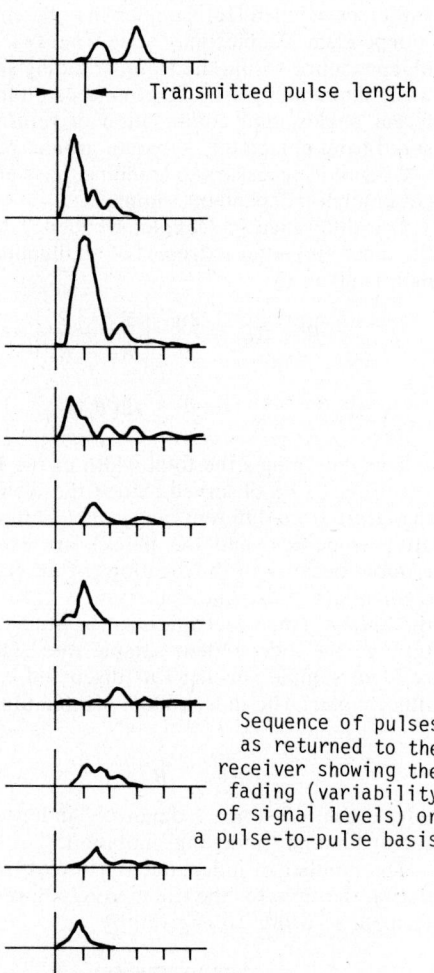

Transmitted pulse length

Sequence of pulses as returned to the receiver showing the fading (variability of signal levels) on a pulse-to-pulse basis

Fig. 9-61. Fading for pulse system.

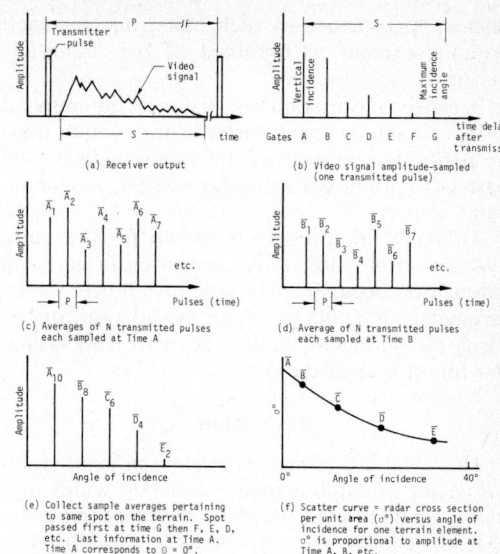

Fig. 9-62. Scatterometer data processing.

Processing

Figure 9-62 indicates the processing required to obtain curves of scattering coefficients as functions of angles of incidence. Figure 9-62a shows the location of the video signal between two transmitted pulses. The time between pulses is P, the total duration of the signal is S. Figure 9-62b gives the output of the gating system for a single transmitted pulse and for 7 time gates. Gate A, at one extreme, is for vertical incidence, while gate G, at the other extreme, is for the maximum inci-

dent angle. Random variations about the mean curve can be seen.

The averaging system acts separately on each of the gate outputs. Thus Figure 9-62c indicates its action on the output of gate A. $\bar{A}_1$, the average output for N pulses, is the average for the first ground segment to pass under the radar at normal incidence. A_2 is the average for the second element to pass under the radar (vertical incidence). The other averages are also indicated up through the 7th element. Figure 9-62d indicates the same sort of averages for gate B. $\bar{B}_1$ is the average return from a segment observed at the angle corresponding to B at the time when the radar is directly over the segment giving $\bar{A}_1$.

Since the outputs of the individual averaging circuits for any particular time correspond to different segments of ground, collation must be accomplished in the data processing. That is, $\bar{A}_1$ does not correspond to the same segment of ground as either $\bar{B}_1$ or $\bar{C}_1$. Preparation of a curve of scattering coefficient vs angle for a particular section of ground must therefore wait until all the returns from that particular section of ground have been obtained and collated. Figure 9-62e indicates this. If the fan beam points ahead, a section of ground is first observed at the maximum incident angle. For this curve, we assume the first observation is in gate E and at time 2. As the radar advances toward the target segment, this segment is observed at the angle corresponding to gate D and the later time 4. Later, (at time 6) it is observed at gate C, still later (time 8) at gate B, and finally the radar passes directly over the ground segment at time 10 so that the output of gate A is appropriate. When these samples are all selected in the appropriate sequence and plotted as a function of angle of incidence, part (e) of Figure 9-62 results. A sequence of such plots, one for each ground segment, is obtained as the radar advances.

Because of variations in distance, illuminated area, and antenna gain, further computation must be performed on each of the values indicated in part (e) to produce a scattering coefficient as indicated in part (f).

Although this process is shown for the pulse system, outputs like those appearing in parts (b) through (f) occur for any type of fan-beam scatterometer—FM, binary phase modulation, or velocity measurement, so that the processing is similar for all system types.

Resolution

The ideal range-measurement fan-beam scatterometer would illuminate a constant-width strip along the flight path. Bands of constant thickness extending across this strip would be established at uniform angular intervals by the range measuring system. Figure 9-63a illustrates this. Idealized antenna patterns with constant gain over a specified angular width and zero gain outside of that are frequently used in discussing radar and other systems. Figure 9-63b shows the pattern of ground illumination caused by such an ideal antenna pattern with a fan beam.

In reality, of course, neither the ideal ground illumination nor the ideal fan beam can be achieved. Figure 9-63c presents a more realistic ground illumination with a fan beam. The antenna is pointed with its maximum in the direction indicated for the maximum gain contour. The pattern, instead of having straight isogain lines on the ground, has a rounded contour. Furthermore, the gain falls off gradually rather than dropping suddenly to zero. Figure 9-63c shows cross-sections of the pattern of Figure 9-63c. The dashed lines indicate the idealized rectangular patterns shown in Figure 9-63b and the solid lines are the actual patterns. Thus, one can sketch contours of constant gain on the ground as in Figure 9-63c. It is customary to use the half-power contour to describe the illumination of a particular antenna, although for some purposes the 1/10 power contour might be more appropriate. In the discussion that follows a pattern similar to the pattern of Figure

9-63b is assumed, but only for simplicity of analysis.

The subject of resolution is discussed in detail in Moore, et al (1968) and Moore and Waite (1969).

Fading-Rate Limitations

Averaging to get meaningful values for σ^o requires the signal to be observed long enough that sufficient independent samples may be collected. Independent samples may arise from two causes: independence within the Doppler fading spectrum while illuminating a single ground patch from different angles, and combination of returns measured from completely separate ground patches.

Often it is necessary to combine these effects to get enough independent samples.

The difference in Doppler frequency between the inner and outer extremes of an illuminated region is given by

$$B = \Delta f_D = \frac{2u}{\lambda} \cdot \left| \frac{R_2}{R_2} - \frac{R_1}{R_1} \right|$$

$$= \frac{2u}{\lambda}(\sin\theta_2 - \sin\theta_1). \qquad (9-144)$$

This determines the total width of the Doppler spectrum to be observed. Since the elements in the return from different ranges have different relative velocities, and the phases are essentially random because of the locations of the scattering elements, a noise-like spectrum is produced by the fading. This spectrum must be studied to determine the independent sample rate. The independent sample spacing was discussed earlier in this chapter. The independent sample time is

$$t_I = \frac{1}{B}, \qquad (9-145)$$

where the subscript i denotes "independent," rather than an "indexing" notation.

The number of independent samples occurring during the time for the radar to pass a resolution element of width $\Delta\rho$ is given by

$$N_i = \frac{\text{time to traverse } \Delta\rho}{\text{time/sample}}. \qquad (9-146)$$

Thus,

$$N_i = \frac{\Delta\rho/u}{t_i} = \frac{\Delta\rho\,\Delta f_D}{u}. \qquad (9-147)$$

This can be expressed as

$$N_i = \frac{2(\Delta\rho)^2\cos^3\theta_m}{\lambda h} \qquad (9-148)$$

in terms of $\Delta\rho$, or as

$$N_i = \frac{(c\tau)^2\cot^2\theta_m\cos\theta_m}{2\lambda h}. \qquad (9-149)$$

in terms of τ, where θ_m is the mean value of θ for the cell.

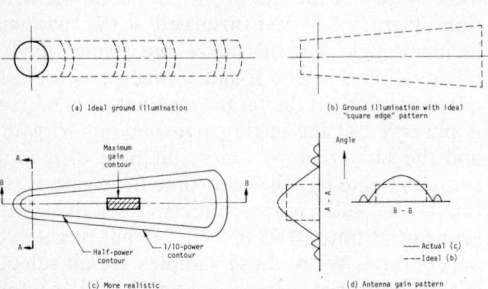

Fig. 9-63. Scatterometer ground illumination.

Eqs. 9-148 and -149 show that the number of independent samples per ground segment is greater nearer the vertical rather than nearer the grazing angle, even if the width of the segment is the same. This comes about because the difference in angle between the extremes of the segment is smaller for angles nearer grazing, so the difference in Doppler frequencies is smaller. Furthermore, since the Doppler frequencies are proportional to the sine of the angle, and since the sine curve is flatter near $\pi/2$, the reduction in number of independent samples is magnified near grazing incidence. Since $\Delta\rho$ itself is smaller near grazing incidence, Eq. 9-149 shows an even greater variation.

To see what this means, consider its influence on an example. Assume a pulse system at 2000 m altitude with cross-track beamwidth of 0.05 radians. Let the wavelength be 10 cm. With this assumption, equation 9-149 yields 600 independent samples at vertical, 6 independent samples at 30°, and only 1 independent sample at 60°. From this it is obvious that at the larger angles several resolution lengths ($\Delta\rho$) must be used to obtain an adequate sample. For example, if the 245-m $\Delta\rho$ for the vertical is used at 30° and 60°, the number of independent samples changes to 600, 47 and 4.

The number of independent samples from the combination of several samples in passing across one resolution cell (described above) and examining returns from several independent resolution cells may be expressed by multiplying the results of Eqs. 9-148 or -149 by the ratio of total distance traversed to $\Delta\rho$. Thus, Eq. 9-148 becomes

$$N_i = \frac{2(\Delta\rho)^2 \cos^3 \theta_m}{\lambda h}\left(\frac{L}{\Delta\rho}\right) = \frac{2L\,\Delta\rho\,\cos^3 \theta_m}{\lambda h}$$

(9-150)

where L is the total distance travelled during calculation of an average.

Eq. 9-149 then becomes

$$N_i = \frac{Lc\tau \cos^2\theta_m \cot \theta_m}{\lambda h}.$$

(9-151)

Figure 9-64 illustrates this.

The number of independent samples required depends upon the precision desired. Of course, if the distance required to obtain enough independent samples is greater than the distance over which a terrain segment is homogeneous, a poorer precision must be accepted. Figure 9-42 has shown the relation between precision and independent sample number.

Dynamic-Range Problems

The dynamic range required for the receiver in a radar scatterometer operating between vertical and 60° is so great that it is extremely difficult to achieve with a linear system. With a logarithmic system, it is not difficult. The worst situation for dynamic range is the return from relatively smooth targets, such as calm seas.

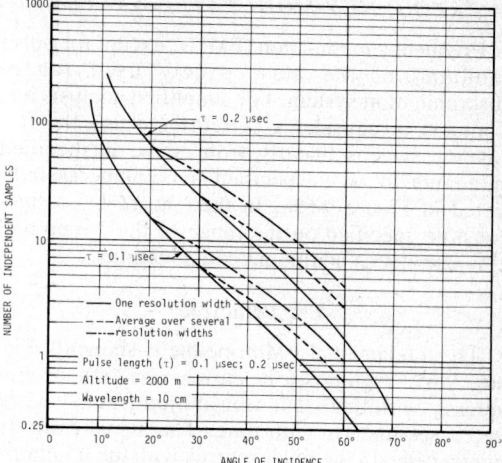

Fig. 9-64. Number of independent samples-examples.

The scattering coefficient σ^o for a smooth sea may vary from 30 dB at vertical incidence to less than −30 dB at 60°. (See Chapter 4). Thus, even if all signals returned were of the mean value, a dynamic range of 60 dB would be required for a receiver covering this total range of angles. Over land the situation is not so bad; +10 dB is a relatively high value for σ^o over land, although occasionally it gets higher. An area with such value at the vertical would probably have a σ^o no lower than −25 dB at 60°. Thus, a 35 dB range is indicated.

The dynamic range of the system must take into account not only variation in σ^o, but also the fading about this value. The 5% to 95% range for a Rayleigh distribution is 18 dB (+5.8 to −11.9 dB). Thus, the receiver must be able to accommodate from +35.8 to −41.0 dB if it is to cover that range for the smooth sea example, or from +15.8 to −36.9 dB for the land example. The total ranges indicated are respectively 78 and 53 dB.

The wide dynamic range can be accommodated in various ways. Possible solutions include: Use of logarithmic IF amplifier, variation of receiver gain with range in accordance with some standard curve, stepping the gain of the receiver for different ranges, and use of separate receiver channels for different angular regions.

Scatterometers that operate over a fairly narrow range of angles do not have dynamic range problems to the same extent as those operating over a wide range of angles; but a scatterometer operating near the vertical has to handle a relatively wide dynamic range, even if it only goes out to about 20°, since the curves of σ^o vs θ are often quite steep near the vertical.

Janza (1963) thoroughly analyzed a pulse system used for near-vertical measurements. The NRL 4-frequency pencil-beam system is described by Guinard and Daley (1970).

FREQUENCY-MODULATED SYSTEMS

Frequency modulation (FM) is, except for pulse modulation, the most widely used range-discrimination system. For simplified analysis it is common to consider a system in which the frequency sweeps linearly from some unspecified minimum to an unspecified maximum, as indicated in Figure 9-65a. In fact, however, a limit must be specified on the range of the sweep and this introduces additional complications.

FM Principle

The fundamental FM principle is shown in Figure 9-65. Figure 9-65a shows variation of frequency with time. The transmitted frequency, f_t, increases linearly with time. The signal from directly beneath the radar returns with the minimum time delay t_h. It is, therefore, at the frequency that was being transmitted t_h seconds earlier. The difference between that frequency and the one being transmitted when it is received is f_s. The received signal f_s may also take on other values for signals returned from larger ranges. Thus the dashed lines in the figure show the longer time delay associated with longer ranges and the frequency differences for them. The four f_s lines show the received signals from four different ranges.

Figure 9-65b shows the spectrum of received

signals. The maximum signal, of course, is received from the shortest range and has the lowest f_s. Signals received from longer ranges have larger frequencies associated with them. This curve should be compared with the analogous pulse shape of Figure 9-60.

Figure 9-66a shows the way transmitted and received frequencies vary for a target at a single range with a saw-tooth modulation. The single frequency corresponding to one element of the spectrum of Figure 9-65b is the horizontal line in Figure 9-66a. Of course, additional frequencies are introduced by the cross-over between increasing and decreasing frequency modulation.

Because the Doppler effect causes a frequency shift, its effect on FM System performance exceeds its effect on pulse system performance since the range and velocity are both measured by frequencies in the FM system. Figure 9-66b shows what happens when the Doppler frequency is positive (again for a single target). The average frequency of the plot at the bottom is the same as without Doppler frequency, but on individual sweeps the signal frequency is shifted either up or down. If the Doppler frequency is relatively small, this merely amounts to broadening of a given spectral line. If it is large, separate filters must be used for the upper and lower frequencies. In fact, it is possible for the Doppler frequency to be larger than the signal frequency, in which case the signal frequency appears as a modulation on a Doppler frequency sub-carrier.

FM System Design

The basic FM scatterometer system is shown in Figure 9-67. The mixer combines a sample of the transmitted signal with the signal being received. Its output is the difference frequency f_s. This is amplified and passed through separate filters corresponding to the different ranges and, consequently, different angles. The filter outputs are averaged and either recorded or displayed.

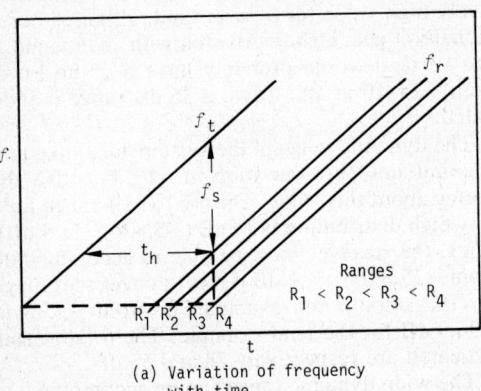

(a) Variation of frequency with time

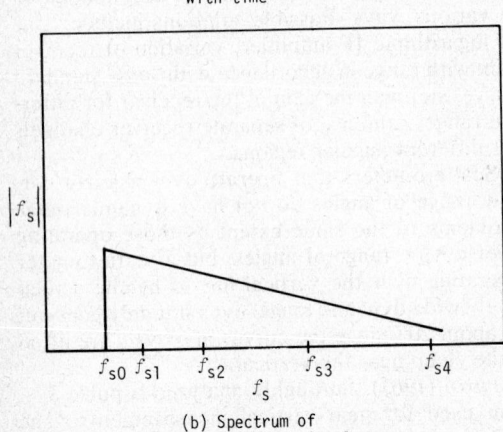

(b) Spectrum of received signals

Fig. 9-65. Basic FM principles.

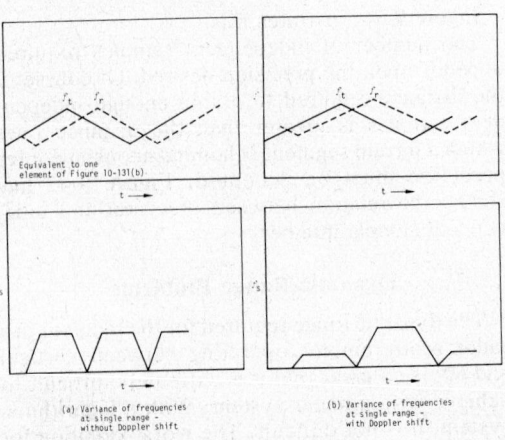

(a) Variance of frequencies at single range – without Doppler shift

(b) Variance of frequencies at single range – with Doppler shift

Fig. 9-66. Sawtooth frequency modulation waveforms—single target.

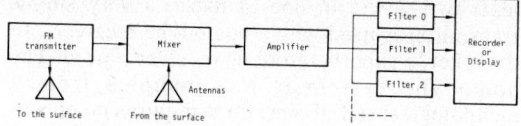

Fig. 9-67. Basic FM scatterometer system.

Numerous variations are possible in this system. For example, a superheterodyne system can be constructed, and the filtering might be done either at the intermediate frequency or after a second mixer. The S/N performance can be improved by introducing a low-noise amplifier between the receiving antenna and the mixer. Various modulation schemes other than the saw-tooth form of Figure 9-66 are also possible. The fundamental relations governing the system of Figure 9-65 are

$$f_t = f_c + at$$

$$f_r = f_c + a\left(t - \frac{2R}{c}\right) \qquad (9\text{-}152)$$

$$f_s = \frac{2Ra}{c},$$

where:
  a = modulation rate
  R = slant range.

If the width of a filter is Δf_s, this means that it is related to a range difference and to a resolution width by

$$\Delta f_s = \frac{2a\Delta R}{c} = \frac{2a\Delta\rho\,\sin\theta_m}{c}. \qquad (9\text{-}153)$$

where θ_m is the mean angle of incidence, ΔR is the slant-range resolution, and $\Delta\rho$ is the ground-range resolution.

The equivalent pulse width is, of course, given by $\Delta R = \dfrac{c\tau}{2}$. Hence,

$$\Delta f_s = a\tau. \qquad (9\text{-}154)$$

By using this equivalent pulse length, the geometric computations for a pulse scatterometer can be used to determine resolution variations.

The effect of finite sweep bandwidth and of Doppler shift complicates these analyses so much that this effect will not be treated in detail here. The finite repetitive sweep changes the continuous spectrum of Figure 9-65b into a discrete spectrum and tends to smear it somewhat, the amount of smearing depending upon the sweep rate and total width. Various systems, primarily for use against single targets rather than against the ground, have been developed to take advantage of the spectral lines by using narrow filters to enhance S/N.

Averaging of Samples

Averaging of independent samples comes about differently with the FM system than with the pulse system. In the pulse system Doppler shift from one end to the other of the ground segment observed is the only means for obtaining the various elements of the fading pattern and, consequently, of the Rayleigh distribution. With the FM system, the sweep in frequency may pass through peaks and nulls of the diffraction pattern of the scattering ground surface, just as motion causes travel through peaks and nulls and results in the Doppler shift. Hence, some averaging may occur in a single sweep and the time between independent samples may be determined by the width of the filter that sets ground resolution rather than by a Doppler effect. Calculations shown here neglect the Doppler effect, on the assumption that a is selected in such a manner as to make Doppler shift negligible. This is possible with relatively low speed aircraft, but with jet aircraft and spacecraft, a more complicated analysis must be made.

Considering the same autocorrelation analysis as before, the time for independent samples is the period corresponding to the spectral width of the filter, provided that square-law detection follows the filter output. That is,

$$t_i = \frac{1}{\Delta f_s} = \frac{1}{a\tau}. \qquad (9\text{-}155)$$

Using this relation in Eq. 9-146, the expression for the number of independent samples in the idealized FM case is

$$N_i = \frac{\Delta f_s \Delta\rho}{u} = \frac{2m(\Delta\rho)^2\,\sin\theta_m}{cu}. \qquad (9\text{-}156)$$

In fact, the idealized situation does not occur, although it may be approximated if the frequency deviation is great enough. In Figure 9-66, consider the duration of the single sweep to be T. The total frequency shift B is

$$B = aT. \qquad (9\text{-}157)$$

If the time between independent samples is the same as calculated before (and it may not be if it is too small compared with T) the total number of independent samples obtained during one sweep is given by

$$N_i = \frac{T}{t_i} = a\tau T = B\tau. \qquad (9\text{-}158)$$

If the deviation is large enough and the sweep is slow enough, the above analysis holds. If, however, the deviation is sufficiently small that the product $B\tau$ is close to unity, the spectrum of the periodic waves is vastly different from that for the aperiodic wave of Figure 9-65a. Furthermore, the effect of Doppler can be quite severe. If the Doppler frequencies are such that the time between independent samples due to Doppler fading is considerably greater than that for the FM system, the effect of the Doppler fading is simply the same as in the pulse system, but with N_i of Eq. 9-158 as a multiplier. This implies that the Doppler frequency shifts are only a small fraction of the bandwidth of the filter used in the FM system. If the Doppler frequency shift is a significant fraction of this filter bandwidth, not only the analysis but the system must be changed.

The application of the above principles can be illustrated with an example. Let us assume 100 MHz deviation and a sweep duration of 10^{-2} sec. Assume $\tau = 10^{-7}$ seconds, $u = 200$ m/sec, $\lambda = 0.1$ m. Applying Eq. 9-158, we find that $N_i = 10$; that is, in one sweep there are 10 independent samples because of the deviation in frequency. This method can be used provided the time for independent samples due to Doppler frequency is long compared with the time for independent samples due to sweeping. This is, indeed, the case as the time between independent samples for sweeping is 1.0 ms, while for the Doppler shift it is 7.25 ms. If frequency deviation of only 10 MHz had been assumed instead of 100 MHz, the situation would have been greatly different.

VELOCITY-ANGLE SYSTEMS

The following sections treat in detail the fan-beam velocity-angle scatterometer illustrated previously in Figures 9-54 and 9-55. A system of this type offers two distinct advantages:

First, data may be collected both fore and aft of the vehicle by broadening the along-track coverage of the antenna pattern. The fore and aft signals may be separated by individually filtering positive and negative frequency shifts.

Second, the sampling of the return at different angles may be accomplished in the frequency domain by simple filtering rather than by sampling in time as required in range measuring systems. This offers a decided improvement in complexity of the preprocessing circuitry.

The CW-Doppler scatterometer used in velocity-angle systems may be implemented in at least two ways. In one of these, a very simple transmitter is used with a homodyne receiver.[5] In the other, a superheterodyne receiver is used. The homodyne system is less complex from a technological standpoint, for it requires much less transmitter stability. On the other hand, the homodyne system cannot be used for zero Doppler frequency, so it cannot permit measurement at vertical incidence for horizontal flight.

CW-Doppler Homodyne System

Figure 9-68 illustrates the CW-Doppler homodyne-receiver system as implemented for a fore-and-aft beam. The quadrature channel shown is required for separation of positive and negative Doppler frequencies, so a system with a beam pointing only ahead of or only behind the aircraft need not have this complexity. The basic idea is that most of the transmitter signal goes out on the transmitting antenna, while a sample is fed to the receiver for use as a local oscillator (phase reference). The signal arriving on the separate receiving antenna is mixed with the transmitter sample to produce an output at the Doppler frequency. Before filtering, it must be amplified, as indicated. The amplifier used ordinarily has a gain-vs-frequency characteristic of such nature that the weaker signals from larger incidence angles are amplified more than the stronger signals from near the vertical; this is analogous to the beam-shaping and STC circuitry used in imaging radars. This output could be filtered directly if only positive or only negative Doppler frequencies were present.

When both positive and negative Doppler fre-

---

[5] Homodyne reception—Also called zero-beat reception; a system of reception using a sample of the carrier frequency voltage as a local-oscillator signal.

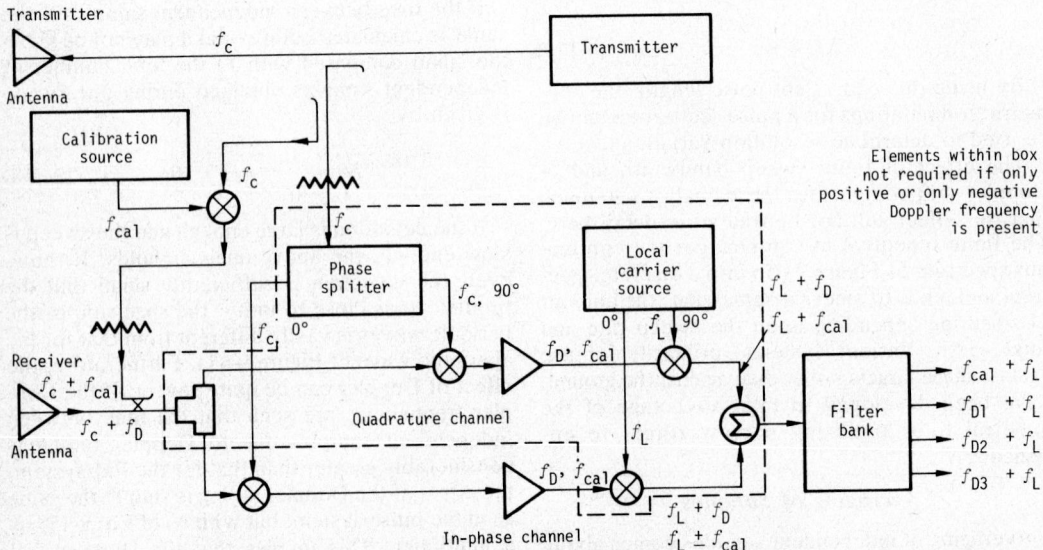

Fig. 9-68. Quadrature homodyne CW-Doppler scatterometer.

quencies are present, some means must be found to distinguish them, since they come out of the 0° mixer superimposed. The technique used is to separate the local signal into two components 90° apart in phase. One output is proportional to the cosine of the incoming signal and the other to its sine. Positive and negative Doppler frequencies are in phase in the cosine of the signal and out-of-phase in the sine of the signal, so a proper combination of the two outputs allows identification of amplitudes of positive and negative Doppler frequencies separately. The technique illustrated here not only combines the two signals, but uses them to construct a replica of the incoming Doppler spectrum centered about a new locally generated carrier frequency. This frequency is chosen so that the separate components corresponding to different incident angles may be filtered easily.

Calibration is accomplished by modulating an audio tone of known amplitude onto a sample of the transmitted carrier, which is then coupled into the receiver. For continuous calibration, this frequency should be outside the Doppler band expected, but calibration within the Doppler band may be achieved when not operating by feeding the transmitter to a dummy load rather than to its antenna, and using calibration frequencies that fall within the Doppler band. The most stable calibration technique uses a calibration signal strong enough to saturate the calibration modulator; in this case the amplitude is determined only by the saturated loss of the modulator, which can be kept quite constant.

Recording may take place at the outputs of the amplifiers, with the remodulation and filtering taking place after playback of the recording, or the entire system may be airborne. This operation may be done in the analog manner shown, or the equivalent operations may be performed in a computer on the digitized outputs of the in-phase and quadrature channels.

Superheterodyne CW-Doppler System

Figure 9-69 shows the superheterodyne CW-Doppler system. Here the intermediate frequency is achieved by offsetting the local-oscillator frequency fed to the signal mixer. The diagram of Figure 9-69 is simpler than that of Figure 9-68 be-

cause no quadrature channel is shown, but the implementation is more difficult because the frequency sources must be very stable. In fact, the IF would have to be quite high for the filter indicated to separate carrier and IF signals, and this might cause problems in the subsequent Doppler filtering. Thus, a block diagram showing all the necessary elements would be much more complicated than the simple one shown for illustrative purposes. The techniques used here, however, are the same as those used in synthetic aperture and moving-target-indicator radars, so they are well developed even though they are complex.

Geometric Considerations

Figure 9-70 details the geometry of an idealized fan-beam scatterometer capable of gathering data both fore and aft of the aircraft. The fan-beam antenna is pointed directly downward with the long axis of the antenna transverse to the velocity vector and parallel to the horizontal ground plane. The cross-track beamwidth of the antenna is angle AOC and the along-track beamwidth is angle LOI. The intersection of this idealized square-beam pattern with the ground is given by $KLMCJIHA$. Let us define the width of the beam on the ground as b and b_0 as the width at vertical incidence (AC). Using these definitions and the geometry indicated in the figure we find

$$b_0 = h\beta, \qquad (9\text{-}159)$$

$$b = \frac{h}{\cos \gamma} \beta = \frac{b_0}{\cos \gamma}, \qquad (9\text{-}160)$$

or

$$\frac{b}{b_0} = \frac{1}{\cos \gamma}. \qquad (9\text{-}161)$$

The variation in width of the illuminated area with angle of incidence is shown in Figure 9-71. It is apparent that the increase in width beyond about ±60° is so rapid that attempts to use the scatterometer beyond this incident angle will result in excessively wide ground segments.

The resolution distance along the flight line is determined by the bandwidth of the Doppler filter. Assuming that the fan-beam pattern of the antenna is sufficiently narrow, the angle of inci-

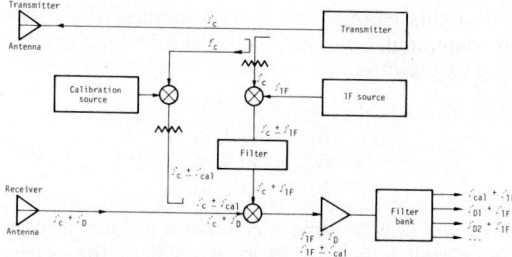

Fig. 9-69. Superheterodyne CW-Doppler scatterometer.

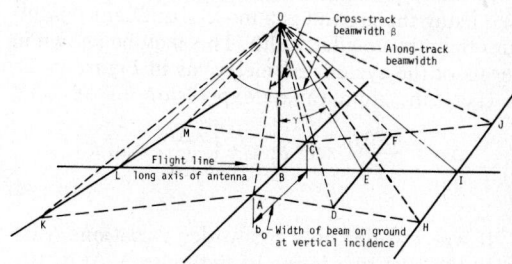

Fig. 9-70. Fan beam geometry.

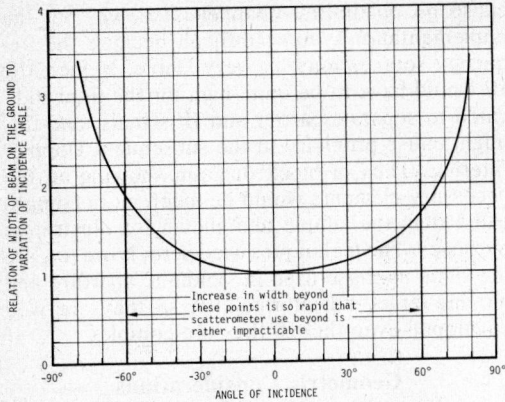

Fig. 9-71. Fan-beam illumination width.

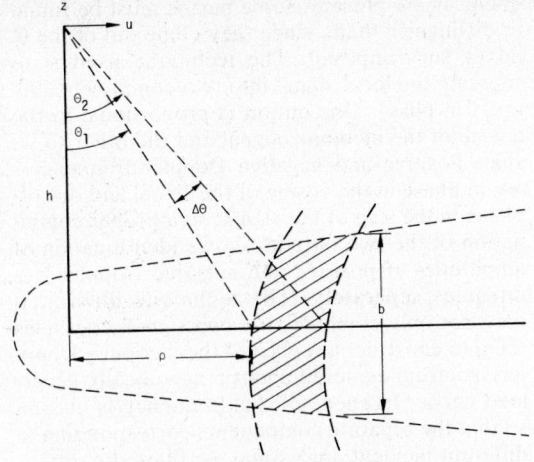

Fig. 9-72. Geometry of angular resolution.

dence may be approximated for the entire area with the angle along the flight track. Using this assumption, the Doppler frequency and filter bandwidth are given by

$$f_D = \frac{2u}{\lambda} \sin \theta, \qquad (9\text{-}162)$$

$$\Delta f_D = \frac{2u}{\lambda} (\sin \theta_2 - \sin \theta_1). \qquad (9\text{-}163)$$

The geometry for angular resolution is illustrated in Figure 9-72.

Defining the incident angle in terms of the average angle and the angular difference, we have

$$\theta_m = \frac{\theta_1 + \theta_2}{2}, \qquad (9\text{-}164)$$

$$\Delta \theta = \theta_2 - \theta_1. \qquad (9\text{-}165)$$

The Doppler bandwidth may then be expressed as

$$\Delta f_D = \frac{2u}{\lambda} \left[\sin\left(\theta_m + \frac{\Delta\theta}{2}\right) - \sin\left(\theta_m - \frac{\Delta\theta}{2}\right) \right]. \qquad (9\text{-}166)$$

The expression for angular width is then

$$\sin \frac{\Delta\theta}{2} = \frac{\lambda \Delta f_D}{4u} \sec \theta_m. \qquad (9\text{-}167)$$

For small $\Delta\theta$, Eq. 9-167 may be approximated by

$$\Delta\theta \approx \frac{\lambda \Delta f_D}{2u} \sec \theta_m. \qquad (9\text{-}168)$$

In many instances the length of the resolution cell along the ground is a more significant parameter than the angular width. This may be shown in terms of the system geometry, as in Figure 9-72. Using the approximate expression for $\Delta\theta$

$$\Delta\rho = \frac{h\lambda \Delta f_D}{2u} \sec^3 \theta \left[1 + \left(\frac{\lambda \Delta f_D}{2u}\right) \tan \theta \right]. \qquad (9\text{-}169)$$

If we consider only first-order variations, [use only the first two terms of the $\tan(\theta + \Delta\theta)$], Eq. 9-169 reduces to

$$\Delta\rho \approx \frac{h\lambda \Delta f_D}{2u} \sec^3 \theta. \qquad (9\text{-}170)$$

The expressions for the width (the dimension transverse to the flight line) of the resolution element, as given by Eq. 9-160, may be combined with the expression for the length along the track given by Eq. 9-170 to define an approximate rectangular resolvable area. An example serves to illustrate the parameters that a practical scatterometer might take:

Assume for the purposes of illustration a scatterometer with the parameters approximately those of the Ryan (1967) 13.3-GHz system used in the NASA Earth Resources Survey Aircraft Program. Summarizing,

$$\beta = 2.5°$$
$$h = 1,000 \text{ m}$$
$$u = 100 \text{ m s}^{-1}$$
$$\lambda = 2.25 \text{ cm}.$$

The Ryan system utilizes the fan-beam geometry shown in Figure 9-70, and collects data fore and aft of the aircraft. The along-track beamwidth is ±60° from nadir. Using Eq. 9-160, we have

$$b = \frac{b_0}{\cos \theta} \qquad (9\text{-}171)$$

For this example $b_0 = 43.6$ meters. For other incident angles we then have the following resolution cell widths:

$$b_{5°} = 43.8 \text{ m}$$
$$b_{15°} = 45.1 \text{ m}$$
$$b_{30°} = 50.4 \text{ m}$$
$$b_{60°} = 87.2 \text{ m}.$$

Let us now assume it is desired to have a resolution cell length of 50 m at each of the above incident angles. The corresponding bandwidths are then:

$$\Delta f_{D5°} = 439 \text{ Hz}$$
$$\Delta f_{D15°} = 398 \text{ Hz}$$
$$\Delta f_{D30°} = 288 \text{ Hz}$$
$$\Delta f_{D60°} = 55 \text{ Hz}.$$

With the above figures go the following angular widths:

$$\Delta\theta_{5°} = 2.85°$$
$$\Delta\theta_{15°} = 2.66°$$
$$\Delta\theta_{30°} = 2.15°$$
$$\Delta\theta_{60°} = 0.71°.$$

From the angular widths for a constant range resolution we see that the approximation of the mean angle as the angle of incidence for the entire cell becomes better as the angle of incidence is increased. This effect is not, however, entirely to our benefit, as shown in the next section dealing with the independence of samples with a resolution cell.

FADING RATE LIMITATIONS AND THE RADAR UNCERTAINTY PRINCIPLE

Averaging to get meaningful values of σ^0 requires the signal to be observed long enough that sufficient independent samples may be collected. As mentioned before, independent samples may arise from two causes: Independence within the Doppler fading spectrum while illuminating a single ground patch from different angles, and combination of returns measured from completely separate ground patches.

To see what this means for the CW-Doppler system, consider its influence on the NASA/Ryan example quoted previously. The number of independent samples for each of the incidence angles used previously is:

$$N_{i5°} = 219 \text{ independent samples}$$
$$N_{i15°} = 199 \text{ independent samples}$$
$$N_{i30°} = 144 \text{ independent samples}$$
$$N_{i60°} = 28 \text{ independent samples.}$$

This situation can be considerably worse if a lower aircraft altitude is assumed, with consequently smaller resolution cell length. If an altitude of 300 meters is assumed, with a resolution cell equal to the cell width at $\theta = 30°$ ($\Delta\rho \approx 15$ m), the following figures are obtained for the number of independent samples:

$$N_{i5°} = 67 \text{ independent samples}$$
$$N_{i15°} = 61 \text{ independent samples}$$
$$N_{i30°} = 44 \text{ independent samples}$$
$$N_{i60°} = 8.5 \text{ independent samples.}$$

The total number of independent samples based on returns from several independent resolutions cells is the product of N_i from Eq. 9-150 and the ratio of total distance traversed to $\Delta\rho$. Thus Eq. 9-150 becomes

$$N_i = \frac{2(\Delta\rho)^2 \cos^3 \theta_m}{\lambda h}\left(\frac{L}{\Delta\rho}\right) \qquad (9\text{-}150)$$
$$\text{repeated}$$

$$= \frac{2L\Delta\rho \cos^3\theta_m}{\lambda h},$$

where L is the total distance travelled during calculation of an average.

As may be seen from Eq. 9-151, the number of independent samples is proportional to the product of resolution cell length and total distance travelled. Thus, use of a larger resolution cell $\Delta\rho$ can result in a decreased actual resolution distance L, for the maximum value of N_i is achieved if $\Delta\rho = L$. Conversely, for a fixed N_i, L is minimized by making $L\Delta\rho = L^2$. In this case, Eq. 9-150 applies. Due to the decrease in sample independence with increased angle, it may well be necessary to vary the resolution cell length with incidence angle to insure adequate sampling.

The problem of independent sample size is most important, and limits the resolution cell length achievable for a given accuracy. Hence, for a given standard deviation of the measurement, Eqs. 9-150 and -151 may be combined to give

$$\frac{\sigma}{\mu} = \left[\frac{\lambda h}{2L(\Delta\rho)^2 \cos^3 \theta_m}\right]^{1/2}, \qquad (9\text{-}172)$$

where σ/μ is the relative standard deviation and μ is the normalizing average.

This may be written as a sort of *uncertainty principle:*

$$\Delta\rho\left(\frac{\sigma}{\mu}\right) = \left[\frac{\lambda h}{2 \cos^3\theta_m}\right]^{1/2}. \qquad (9\text{-}173)$$

If better resolution is required, $\Delta\rho$ is reduced; but (σ/μ), an indication of the relative error of the measurement, becomes larger. The situation is best (minimum error-resolution product) at vertical, where $\theta_m = 0$. The error-resolution product is 2.8 times larger at 60°. This equation assumes $N_i \geq 20$, and the relationship, although similar in trend, is more complex for smaller N_i.

If several resolution cells determined by filter width are combined, Eq. 9-151 must be used instead of -150, giving

$$\frac{\sigma}{\mu} = \left(\frac{\lambda h}{2L\Delta\rho\cos^3\theta_m}\right)^{1/2}, \qquad (9\text{-}174)$$

or

$$(L\Delta\rho)^{1/2}\left(\frac{\sigma}{\mu}\right) = \left(\frac{\lambda h}{2 \cos^3 \theta_m}\right)^{1/2}; \qquad (9\text{-}175)$$

if $L/\Delta\rho = M$,

$$M^{1/2}(\Delta\rho)\left(\frac{\sigma}{\mu}\right) = \left(\frac{\lambda h}{2 \cos^3 \theta_m}\right)^{1/2} \qquad (9\text{-}176)$$

Thus, increasing L by averaging returns from shorter resolution cells gives an improvement in precision proportional to the square root of the number of cells averaged, whereas increasing L by increasing $\Delta\rho$ gives increased precision directly proportional to the increase in $\Delta\rho$ (and thus in L).

GROUND-BASED SCATTEROMETERS

Scatterometers operating at short-range from platforms such as buildings, bridges, and towers, or from trucks or low-flying helicopters, can be used for fundamental studies and to provide so-called "ground truth" (basic data from known test sites where ground parameters are monitored simultaneously with the received scatterometer signals) for airborne systems. They are particularly useful for studying the variation of radar signals with time as ground conditions such as moisture and vegetation growth change. They also provide the opportunity to obtain multifrequency or continuous-spectrum responses that would be much more difficult with systems carried on fast aircraft.

Many ground-based scatterometers have been mounted on fixed platforms, so they are restricted in the range of objects to be examined to those within view of the fixed locations. Some systems, however, are mobile to permit examination of a wider variety of objects. Fixed platform systems have been used to examine returns from the ocean; some have been mounted on towers or ships at sea, and others have been mounted at high points along coast lines. The latter are of course restricted to relatively shallow grazing angles except in the surf zone. Fixed platform systems for examining land targets have been mounted on tall buildings and bridges. In the Netherlands 200-meter-high TV towers with equipment rooms near the top have been used successfully.

Mobile equipment necessarily is restricted to relatively low elevations above the ground. "Cherry picker" booms of 20- to 30-meter height, however, are relatively common in construction work so that these may be purchased at reasonable cost. A boom that high, however, is not truly mobile because the truck must be parked and stabilizing legs lowered to the ground before the boom can be raised to its full height. Helicopters have also been used at 10–30-meter altitudes in hovering or very slow moving modes for this purpose.

System Considerations

Methods for controlling the observed area with ground based systems are the same as for airborne systems not using velocity discrimination; the difficulty of driving at constant speed with an extended boom makes velocity (Doppler frequency) discrimination almost impossible for these systems. Near the vertical, ground-based scatterometers normally use beam-width limitation of the illuminated area, but at angles of incidence exceeding about 45° the difficulty of constraining the observed area with vertical beamwidth usually leads to use of a range-measuring system. Of course, if a fan beam is used, the range measurement provides radial discrimination even near the vertical.

System design is different for short-range systems mounted on truck booms and bridges from that for longer-range systems primarily used for near-grazing measurements. The short-range systems must be able to achieve isolation of transmitted and received signals when the time delays involved are under 0.1 μs whereas the longer range systems do not have this constraint. Accordingly, short-range systems often use separate antennas for transmitting and receiving. This is possible because the antennas need not be large; in fact, they usually must be relatively small to prevent confining the observed area too much. On the other hand, the longer-range systems must use larger antennas if the observed area dimension normal to the line of sight is to be constrained to a reasonably small size, and consequently they normally use single large antennas and pulse transmission.

With separate transmitting and receiving antennas an important part of the system, calibration is accomplished by aligning the beams so that they overlap. This requires that the antenna pair be mounted on a structure that may be placed on the antenna range, adjusted there, and removed to the field without loss of adjustment. The effective observation area is determined by the product of the patterns of transmitting and receiving antennas. A convenient arrangement that makes alignment less critical is use of antennas of different beamwidths so that the narrower beam may be "embedded" in the larger beam.

Longer-range systems are customarily made by modifying standard ground, ship, or aircraft pulse radars. The modification consists of providing a calibration method and some kind of pulse sampling and recording equipment.

Short-range scatterometers may be constructed by using simple CW transmitters and appropriate simple receivers, if the antenna pattern is used to establish the observed area. If the transmitter is strong enough, the receiver may simply be a diode detector followed by a dc amplifier, but this type of receiver is so insensitive that a superheterodyne is usually preferred. This type of scatterometer may be assembled from a simple klystron or solid-state oscillator and (for example) an antenna-pattern-range receiver.

In such a straight-CW system it is almost impossible to eliminate leakage from the transmitter directly into the receiver. The system used in measurements at Ohio State University (Cosgriff, et al, 1960) avoids this problem by requiring motion between radar and sensed area; the motion is achieved either by driving the truck-mounted system along a homogeneous area or by moving the target past the radar in a cart. Because of the motion, the received signal is Doppler-shifted away from the transmitted carrier; the resulting radar output is ac, whereas the leakage signal is detected as dc so that it may be readily removed. This technique, of course, is not possible for in-place targets and booms that are so high the truck must be parked.

Use of range gating eliminates the leakage

problem if the gating can adequately discriminate against the leakage signal. Pulse transmissions can be used for this purpose, but the pulses must be very short. For example, a 50-ns pulse cannot be used at all for distances less than 7.5 meters. Unfortunately, reflections within the transmitter often make the pulse extend to many times its half-power width. Although these reflected signals may be quite weak, they may be strong enough to leak into the receiver and interfere with the received signal. Thus, great care must be taken in developing a pulse system for this purpose.

Frequency modulation has two advantages: the leakage problem is easier to control, and the peak power transmitted is no larger than the average power, so the transmitter need not be capable of high power levels. Commercial signal generators often have adequate power output levels for FM transmission, and they usually also have provisions for frequency modulation. Thus, frequency modulation seems to be preferable for short-range scatterometers.

Independent Samples for Precision

A problem with ground-based scatterometers, as with other types of radar systems, is obtaining enough independent samples of the (more-or-less) Rayleigh distributed signal to permit an accurate measurement. This problem is particularly severe for the ground-based system, for motion past a relatively large target that provides significant numbers of independent samples for airborne systems is not so easy for the ground-based system. The Ohio State University truck-mounted system has been driven far enough past target areas to obtain reasonable numbers of independent samples, but this option is not available for systems that must operate in fixed locations. Use of a helicopter eliminates this problem.

Fig. 9-73. The MAS systems acquiring data on wheat stubble; the MAS 1-8 GHz system is in the background.

One solution is to observe several different areas having essentially the same characteristics and with the same angle of incidence, or nearly the same. This can sometimes be done by scanning the antenna beam, and sometimes by using relatively fine range resolution and combining signals from several different range cells.

For a target consisting essentially of vegetation, motion of the target itself is caused by the wind, so that the number of independent samples may be made large enough by simply performing a time average for a sufficient period of time. This technique must be used with caution, however, for many targets simply do not move enough to make much fading. A test can be made for the validity of this technique by observing the probability distribution function of the returns; if the range approaches the 18 dB expected for a Rayleigh distribution, time averaging may be used safely; if not, some other method must be used.

Sweeping in frequency has the potential of providing many independent samples. A certain bandwidth is required for range resolution, and any excess bandwidth used can reduce the fading effect (and thus improve precision of measurement). The range resolution for the scatterometer may be set by the antenna beam, or it may be set by the ranging system (pulse length or FM filter bandwidth). The resolution bandwidth equation is

$$B_r = 150/(\text{Slant range resolution}) \text{ MHz.}$$
$$(9\text{-}177)$$

Thus, if the slant-range resolution is 1 meter, the use of 600 MHz bandwidth gives 4 independent samples. With 3 meters for slant-range resolution, 12 samples are available with a 600-MHz bandwidth. An FM system can attain this bandwidth easily by simply sweeping over this wide a band. If a matched filter were used at the output, the full resolution associated with the bandwidth would be achieved; but if a wider filter were used, the desired averaging would be achieved (at the expense of the resolution). Thus, in the example, the FM filter would be 4 times as wide as required to match the 600-MHz sweep width for a 1-meter resolution and 12 times as wide for 3-meter resolution.

In the above discussion slant-range, rather than ground-range, resolution was considered. The two are similar near grazing, but near vertical they may be quite different. In fact, at vertical incidence the vertical extent of the target may have more influence in determining the effective slant resolution than does a distance on the ground. This is particularly true with vegetation.

Calibration

Calibration of ground-based systems is sometimes easier than for airborne systems. Because of the relatively short range, a sample of the transmitted signal may be provided to the receiver through a delay line, the length of which is comparable to the actual range being investigated. Since

Fig. 9-74. Closeup photograph of the 8-35 GHz MAS system's antennas and RF sections.

TABLE 9-2

MAS 8-18/35 Nominal System Specifications

	MAS 8-18	35 GHz Channel
Type	FM-CW	FM-CW
Modulating Waveform	Triangular	Triangular
Frequency Range	8-18 GHz	35.6 GHz
FW Sweep: Δf	800 MHz	800 MHz
Transmitter Power	10 dBm	1 dBm
Intermediate Frequency	50 kHz	50 kHz
IF Bandwidth	9.4 kHz	9.4 kHz
Antennas		
Height over Ground	19 m	19 m
Type	46 cm Reflector	Scalar Horn
Feeds	Quad-Ridged Horn	—
Polarization		
Capabilities	*HH, HV, VV*	*HH, HV, VV*
		RR, RL, LL
Beamwidth	4° at 8.6 GHz to	3°
	2.5° at 17.0 GHz	
Incidence Angle Range	0° (nadir) − 80°	0° (nadir) − 80°
Calibration		
Internal	Signal Injection	Signal Injection
	(delay line)	(delay line)
External	Luneberg Lens	Luneberg Lens
	Reflector	Reflector

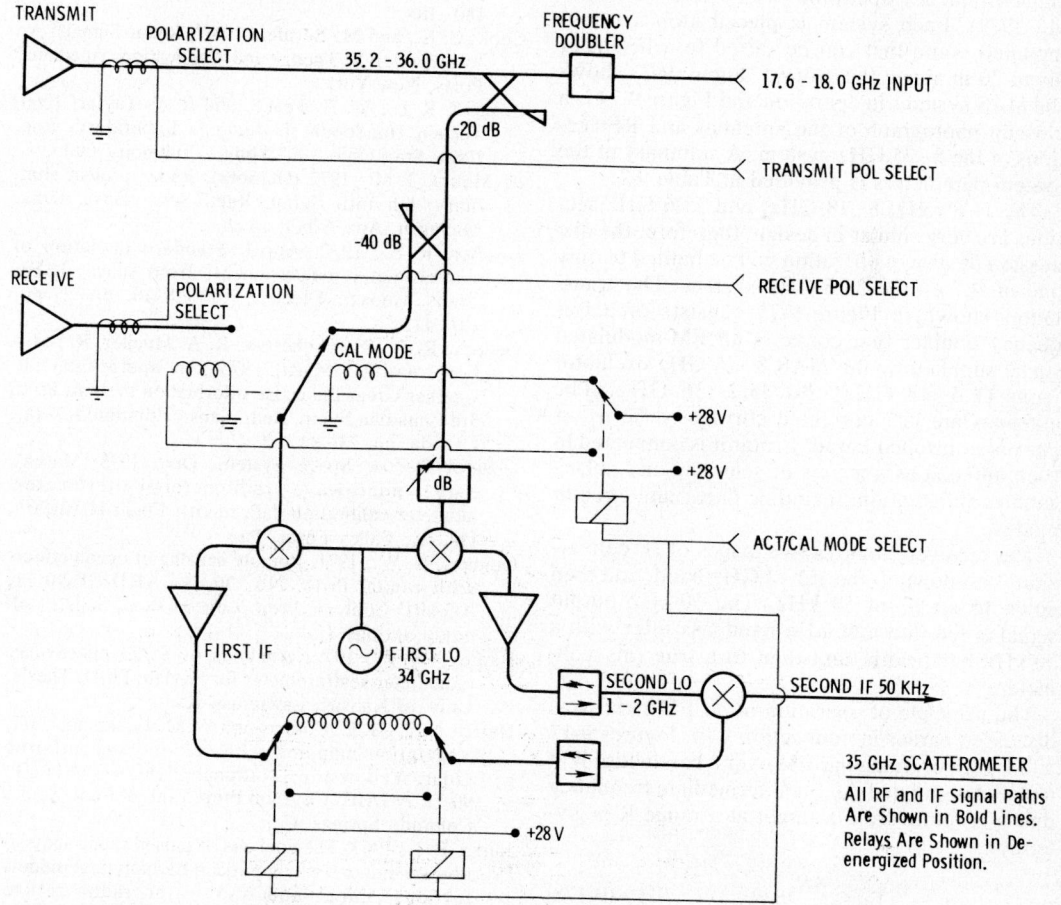

Fig. 9-75. Overall schematic of the 35-GHz radar module. _____

the attenuation of a delay line remains fixed, this provides an excellent and stable calibration, although it does not include the effect of the antennas.

Both spheres and corner reflectors have been used as standard targets for calibration purposes. Corner reflectors give much stronger returns than spheres of the same size, but they must be made very carefully if they are to be used as absolute standards. Spherical Luneberg lens reflectors can be obtained commercially with gains of the order of 20 dB above the scattering cross-section of a metallic sphere of the same size. These Luneberg sphere reflectors have a wider uniform pattern than a corner reflector, and are thus quite useful for calibration of scatterometers.

Example of FM-CW Scatterometer-Spectrometer

The microwave active spectrometers (MAS) developed by the University of Kansas are good examples of ground-based, swept-frequency scatterometers. These are two independent systems: one covers the 1–8 GHz frequency band, and the other covers the 8–18 GHz band plus single-frequency operation at 35.6 GHz (Ulaby et al., 1979). Each system is placed atop a truck-mounted boom that can be raised to a height of about 20 m above the ground. Figure 9-73 shows the MAS systems in operation and Figure 9-74 is a closeup photograph of the antennas and RF sections of the 8–35 GHz system. A summary of the system parameters is provided in Table 9-2.

The 1–8 GHz, 8–18 GHz, and 35.6 GHz sections are very similar in design; therefore, the discussion of system operation will be limited to only one of these—the 35.6-GHz system. The transmitter shown in Figure 9-75 consists of a frequency doubler that converts an FM-modulated signal supplied by the MAS 8–18 GHz oscillator from 17.6–18 GHz into 35.2–36 GHz. The antennas are lens-corrected corrugated horns. A current-controlled Faraday rotator is connected to each antenna as a means of selecting any polarization configuration, including linear and circular modes.

The receiver employs two stages of IF conversion: first down to the 1.2–2 GHz band, and then down to an IF of 50 kHz. The 50-kHz output signal is fed into a 60-kHz bandpass filter with a 10-kHz bandwidth, and then to a true rms voltmeter.

The principle of operation of an FM radar was discussed earlier in connection with Figures 9-17 and 9-18. For triangular FM over a bandwidth B at a modulation rate f_{mod}, the intermediate frequency due to a return from a target at a range R is given by

$$f_{IR} = \frac{4R}{c} B f_{mod}. \qquad (9\text{-}178)$$

In the MAS system, B is kept constant and f_{mod} is electronically tuned (and recorded) so that f_{IF} is always equal to 50 kHz. The modulation frequency is formed by peaking the output of the rms voltmeter. Hence, in addition to measuring the power backscattered from the ground area illuminated by the antenna beam, the MAS system measures the range R to that area as well.

Internal calibration of the system's stability is performed through the use of a 130-ns delay line at the first IF of 1.6 GHz. External calibration is accomplished using a Luneberg-lens reflector of known radar cross-section.

REFERENCES

Berkowitz, R. S., 1965, Modern Radar; John Wiley and Sons, pp. 207–211.

Blackman, R. B., and J. W. Tukey, 1958, The Measurement of Power Spectra; Dover Publ., New York.

Bush, T. F., and F. T. Ulaby, 1978, An evaluation of radar as a crop classifier; Rem. Sens. Env., vol. 7, pp. 15–36.

Claassen, J. P., H. S. Fung, R. K. Moore, and W. J. Pierson, Jr., 1972, Radar sea return and the RADSCAT satellite anemometer; 1972 IEEE Intl. Conf. Engineering in the Ocean Environment (IEEE Pub. 72 CHO 660-10CC), Newport, R. I., pp. 180–185.

Cook, C. E., and M. Bernfeld, 1967, Radar Signals: An Introduction to Theory and Application; Academic Press, New York.

Cosgriff, R. L., W. H. Peake, and R. C. Taylor, 1960, Terrain Handbook II; Antenna Laboratory, Eng. Expt. Sta., Ohio State Univ., Columbus, Ohio.

de Moura, J. M., 1972 (October), Radar project summary; 8th Intl. Symp. Rem. Sens. Env., Univ. Michigan, Ann Arbor, Mich.

Fischer, R. E., 1972 (April), Standard deviation of scatterometer measurements from space; IEEE Trans. Geosci. Electr., vol. GE-10, no. 2, pp. 107–113.

Gedney, R. T., R. J. Schertler, R. A. Mueller, R. J. Jirberg, and H. Mark, 1975, An operational all-weather Great Lakes ice information system; Proc. 3rd Canadian Symp. Rem. Sens., Edmonton, Alta., Canada, pp. 73–82.

General Electric Space Systems Div., 1973 (March), S-193 microwave radiometer/scatterometer/altimeter calibration data report; Flight Hardware; vol. IA, Valley Forge, Pa.

Guinard, N. W., 1971, Remote sensing of ocean effects with radar; Proc. No. 90, AGARD-CP-90-71, AGARD Conf. on Prop. Lim. of Rem. Sens., Colorado Springs, Colo.

Hanley, W. R., 1972, Analysis of S-193 microwave radiometer/scatterometer for Skylab; Ph.D. Thesis, Univ. of Kansas, Lawrence, Ks.

Hardy, N. E., J. C. Coiner, and W. O. Lockman, 1971, Vegetation mapping with side-looking airborne radar: Yellowstone National Park; AGARD-CP-90-71, AGARD Conf. on Prop. Lim. of Rem. Sens., Colorado Springs, Colo.

Janza, F. J., 1963, The analysis of pulsed radar acquisition system and a comparison of analytical models for describing land and water radar return phenomena; Sandia Corp. Monograph SCR-533, Ph.D. Thesis, Univ. of New Mexico.

Jones, W. L., L. C. Schroeder, D. H. Boggs, E. M. Bracalente, R. A. Brown, W. J. Pierson, F. J. Wentz, and G. J. Dome, 1982, The Seasat-A satellite scatterometer: the geophysical evaluation of remotely sensed wind vector; J. Geophys. Res., in press.

Li, R. Y., F. T. Ulaby, and J. R. Eyton, 1980 (June), Crop classification with a Landsat-radar sensor combination; Symp. Machine Proc. of Remotely Sensed Data, Purdue Univ., W. Lafayette, Ind.

Loshchilov, V. S., and V. A. Voyevodin, 1972, Determination of elements of ice cap drift and movement of ice edges by means of the airborne side-scan radar "Toros"; Problemy Arktiki i Antarktiki, vol. 40, pp. 23–30.

MacDonald, H. C., P. A. Brennan, and L. F. Dellwig, 1967, Geologic evaluation by radar of NASA sedimentary test site; Photogram. Eng., vol. 2, pp. 179–193.

Moore, R. K., 1970, Ground echo; Chapter 25, Radar Handbook (M. I. Skolnik, editor), McGraw-Hill.

Moore, R. K., and F. T. Ulaby, 1969, The radar-radiometer; Proc. IEEE vol. 57, pp. 587–590.

Moore, R. K., and W. P. Waite, 1969, Radar scatterometry; Remote Sensing Laboratory Technical Report 118-115, Univ. of Kansas, Lawrence, Ks.

Moore, R. K., W. P. Waite, J. R. Lundien, and H. W. Masenthin, 1968 (April), Radar scatterometer data analysis techniques; Proc. 5th Intl. Symp. Rem. Sens. Env., Univ. of Michigan, Ann Arbor, Mich, pp. 765–780.

Rice, S. O., 1944 and 1946, Mathematical analysis of random noise; Bell Syst. Tech. J., vol. 22, pp. 282–332; vol. 23, pp. 46–156.

Ryan Aeronautical Co., 1967 (September), Scatterometer data analysis program, final report; No. 57667-2, NASA Contract No. NAS 9-6059, San Diego, Calif.

Sherman, J. W., III, 1970, Aperture-antenna analysis; Radar Handbook (M. I. Skolnik, editor), McGraw-Hill Book Co., pp. 9–25.

Ulaby, F. T., P. P. Batlivala, and M. C. Dobson, 1978, Microwave backscatter dependence on surface roughness, soil moisture, and soil texture: Part 1—bare soil; IEEE Trans. Geosci. Electr., vol. GE-16, pp. 286–295.

Ulaby, F. T., G. A. Bradley, and M. C. Dobson, 1979, Microwave backscatter dependence on surface roughness, soil moisture, and soil texture: Part 2—vegetation-covered soil; IEEE Trans. Geosci. Electr., vol. GE-17, pp. 33–40.

Ulaby, F. T., W. H. Stiles, D. Brunfeldt, and E. Wilson, 1979 (May), 1–35 GHz microwave scatterometer; IEEE/MTT-S Intl. Microwave Symp., Orlando, Fla.

Vesecky, J. H., and R. H. Stewart, 1982, The observation of ocean surface phenomena using imagery from the Seasat synthetic-aperture radar—an assessment; J. Geophys. Res., in press.

Viksne, A. T., C. Liston, and C. D. Sapp, 1969, SLR reconnaissance of Panama; Photogram. Eng., vol. 36, pp. 253–259.

Wing, R. S., 1971, Structural analysis from radar imagery; Eastern Panamanian Isthmus; Mod. Geol., vol. 2, pp. 1–21.

Young, J. D., and R. K. Moore, 1977, Active microwave measurement from space of sea-surface winds; IEEE J. Oceanic Eng., vol. OE-2, pp. 309–317.

CHAPTER 10

Imaging Radar Systems

Author-Editor: RICHARD K. MOORE

Contributing Authors: L. J. CHASTANT, L. PORCELLO, AND J. STEVENSON

GENERAL CONTENTS: PPI and B-scan radar; real-aperture SLAR; basic idea; systems and equipment; speckle; profiling; distortions; synthetic-aperture radar (SAR); review, SAR concept; basic configurations; error sources; SAR signal processing; concept summation; SAR processor descriptions; speckle reduction in SAR; multipolarization and multispectral SAR; representative SAR systems and results; references.

NOMENCLATURE

To conserve and eliminate repetition in text and references, the following symbols, units and names have been used in this chapter.

Symbol	SI Units	Name
A	m^2	surface area
A/D	—	analog-to-digital converter
A_e	m^2	effective aperture of antenna
B	Hz	bandwidth
B_D	Hz	Doppler bandwidth
B_{Df}	Hz	Doppler filter bandwidth
B_r	Hz	bandwidth associated with range resolution
c	$m\,s^{-1}$	speed of electromagnetic waves in space
D	m	antenna aperture length
F	—	noise figure of receiver
f_D	Hz	Doppler frequency
f_{LO}	Hz	local oscillator frequency
f_s	Hz	pulse repetition frequency
G	—	antenna gain
H	m	height of aperture
$H(f)$	$V\,Hz^{-1}$	transfer function
HH	—	horizontal transmit–horizontal receive polarization
HV	—	horizontal transmit–vertical receive polarization
I	—	in-phase channel, interpretability
k	$J\,K^{-1}$	Boltzmann's constant, 1.38×10^{-23} $J\,K^{-1}$
k	m^{-1}	wave number
L	m	length available for synthetic aperture
L_p	m	longest possible synthetic aperture
M	—	number of range channels in processor
N	—	number of pulses combined in synthetic aperture
N	s^{-1}	number of resolution elements per second
N	—	number of subcells
N_i	—	number of independent samples
N_o	$W\,Hz^{-1}$	noise spectral density
$n(t)$	V	noise voltage
prf	Hz	pulse repetition frequency
$P_t(f)$	$W\,Hz^{-1}$	spectral density of transmitted signal
$p(t)$	V	baseband transmitted signal voltage
Q	J	energy
Q	—	quadrature channel
R	m	range
R'	m	true slant range (including error)
$R_c(f)$	—	target autocorrelation function in frequency
R_g	m	ground range
r_a	m	along-track resolution
r_{amin}	m	fully focused SAR resolution
r_{ar}	m	along-track resolution for real aperture
r_g	—	gray-level volume
r_R	m	slant-range resolution
r_t	s	time resolution
r_y	m	cross-track resolution
$(S/N)_0$	—	signal-to-noise ratio at output
$S(f)$	$V\,Hz^{-1}$	signal voltage spectral density
T	s	time interval
T_r	K	receiver temperature
u	$m\,s^{-1}$	speed of radar
V	m^2	resolution volume
V	V	voltage
VH	—	vertical transmit–horizontal receive polarization
VV	—	vertical transmit–vertical receive polarization
V_r	V	received voltage
W	W	power
β'	rad, degrees	beamwidth of conventional array
β''	rad, degrees	beamwidth of synthetic array
β_h	rad, degrees	along-track (horizontal) beamwidth
γ	—	modified scattering coefficient (per unit projected area)
ΔR	m	swathwidth for radar
ΔT	s	interpulse period
ϵ_R	m	range error
ϵ_y	m	cross-track positional error
ϵ_z	m	vertical positional error
θ	rad, degrees	angle of incidence (relative to vertical)
λ	m	wavelength
σ	m^2	radar cross-section of target, standard deviation
$\sigma°$	—	scattering coefficient (differential scattering cross-section)
τ	s	pulse length
$\Phi(f)$	rad, degrees	phase of signal spectrum
ϕ	rad, degrees	phase shift, azimuth angle
ψ	rad, degrees	grazing angle (relative to horizontal)
ω	rad, degrees	angular frequency

RADAR IMAGING SYSTEMS

Radar systems used in remote sensing usually produce images of some kind. The most common imaging radar systems are the *side-looking air-*

borne radars (SLAR's) that produce continuous strip images. In these, the scan in the direction perpendicular to the aircraft flight line is obtained by a ranging type measurement, and the scan along the flight line is obtained by synchronizing motion of a film with motion of the aircraft. The original imaging radars, however, were built with mechanically-scanned antennas producing either the *B-scan image* (in which severe distortions exist) or the *plan position indicator,* the PPI (in which the distortion is reduced).

A brief discussion of B-scan and PPI airborne radars is included primarily for historical value, but the majority of the effort is devoted to describing *real-aperture* SLAR's and *synthetic-aperture* SLAR's that can achieve better along-track resolutions than real-aperture systems.

PPI- AND B-SCAN AIRBORNE RADAR

The original PPI- and B-scan airborne radars were developed to aid in navigation of aircraft. The idea was to present a picture to a pilot, or in some cases a bombardier, that allowed him to navigate to a known spot on the earth by providing a continuously updated "map" of the ground. Because the pilot could do most of his navigation by noting sharp boundaries between strongly contrasting surfaces such as water and land or by locating large targets such as major buildings and cities, little attempt was made to achieve a good gray-scale rendition on the early airborne radars; and, indeed, PPI- and B-scan radars still usually present displays that are essentially binary in intensity.

The B-scan is applied to a mechanically- or electronically-scanned antenna covering a sector ahead of the aircraft. The antenna scans back and forth about the flight line through a restricted sector the width of which seldom exceeds about 90° (±45°). The received signal is used to intensity modulate the electron beam on a cathode ray tube so a strong target appears bright on the screen.

Figure 10-1 illustrates the layout of the B-scan display. The range dimension is displayed vertically and the azimuth angle is the horizontal scale. At the left-hand edge of the display is a vertical line consisting of signals from all observable ranges obtained when the antenna is at its farthest rotation to the left. The vertical line in the center of the display corresponds to the signals from different ranges directly ahead of the aircraft.

Clearly this display is a highly distorted map, for the angle between two objects a given distance apart transverse to the flight line is large at near range but small at far range. Figure 10-2 illustrates this point. Actual locations of three pairs of target objects separated in the transverse direction by the same distance are shown at different ranges in Figure 10-2a. Figure 20-2b shows the way they appear on the B-scan display. This distortion is unacceptable for mapping and is usually unacceptable for use by pilots, so that most modern sector-scan radars use a variation of the PPI display (Fig. 10-3c) rather than the B scan.

The PPI was developed during World War II to permit production of an image without the distortion inherent in the angle-range B-scan. Figure 10-3a illustrates the original idea. An antenna rotates through 360° beneath an aircraft, thereby producing an image of all the terrain around and beneath the aircraft. The scan on the CRT does the same. Each pulse transmitted causes a radial line to be displayed on the circular screen in the same direction as the pointing angle for the antenna associated with that pulse. Thus with a flat terrain and a true-ground-range display, this gives a planimetric view of the terrain. For a stationary platform carrying the antenna and an antenna large enough for adequate azimuth resolution, this would seem to be the best type of presentation one could use for mapping. However, both the effect of aircraft motion and the large size of the required rotating structure make this approach less desirable than the side-looking approach, which uses a fixed antenna.

A modification of the PPI can be used with the forward sector scan for which the B-scan display was originally developed. Figure 10-3b and c illustrates the way in which this scan removes the inherent distortion of the B scan.

Because it takes a finite time for the antenna to rotate and the aircraft has moved forward during this time, a unique distortion is present with the

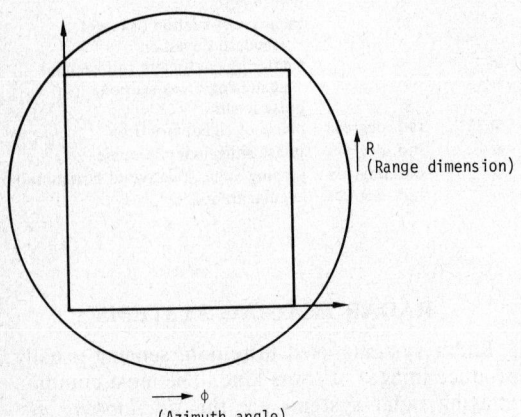

Fig. 10-1. Layout of B-scan display.

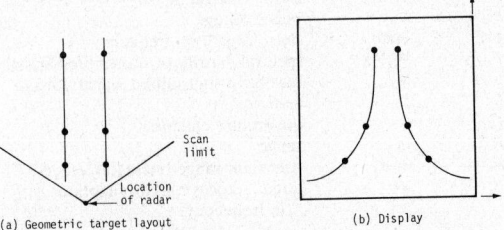

Fig. 10-2. Azimuth distortion on B-scan display.

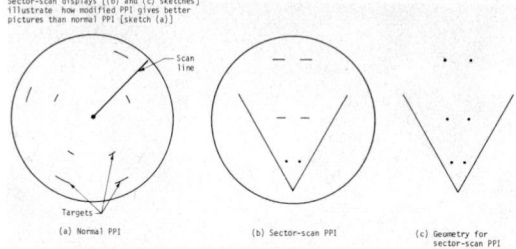

Fig. 10-3. Plan-position indicator (PPI) displays.

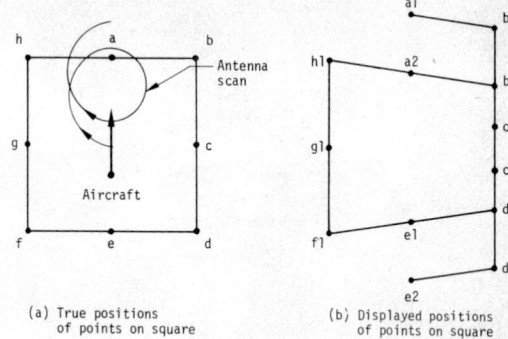

Fig. 10-4. Distortion of PPI due to motion.

PPI radar. The best way to illustrate this kind of distortion is to show the effect of imaging a square as the aircraft moves through the center of the square. Figure 10-4 illustrates this for 1.5 scans of the PPI radar. The corners of the square and the center of each side are labeled by letters in Figure 10-4a. In Figure 10-4b the number following each of these letters represents the scan number. Assume that the scan starts with point a and moves in a clockwise direction. By the time point b is imaged, the aircraft has moved forward somewhat, so point b is at a shorter range. Point c is somewhat ahead of where one would expect it because the aircraft has not reached the center of the square at the time it images point c. Following around the square, it is seen that point f appears farther away and point g is imaged behind point c, because by the time the antenna reaches that angle the aircraft has moved ahead of the center of the square. Coming around to a again, a2 appears much closer to the center of the display, because by this time the aircraft has advanced the distance between a1 and a2. Thus the square never closes and the distortion indicated in the illustration continues. It would be possible to remove this distortion by proper compensation of the PPI display, but it would be considerably more complicated than simply rotating the display with the antenna. For this reason, if no other, the PPI display would be somewhat undesirable for most mapping purposes.

The horizontal aperture that can be rotated is clearly limited by available space. For normal aircraft a diameter of about 2 m is the maximum, although special aircraft have been built to carry antennas with rotatable apertures up to 15 m in the horizontal dimension. Since the beamwidth is inversely proportional to the aperture, the azimuth resolution of such an antenna cannot be very fine because of the size limitation. For example, at 10 GHz with an aperture of 1 or 2 m, the beamwidth is 3° or 1½° which represents a resolution at a distance of only 4 km of 120 to 60 m. Even at 35 GHz the beamwidth for an aperture of this size is only 0.5 to 0.25° with a resolution at 4 km of 34 to 17 m. These resolutions might be adequate, but the 4 km slant range certainly is not very adequate!

Since B-scan and PPI radars are seldom used for remote sensing, they are not discussed further

here. For those who need more information on such systems, standard textbooks (for example, Skolnik, 1980) present more information, as does the *Radar Handbook* (Skolnik, 1970).

REAL APERTURE SLAR

Real-aperture side-looking radar (SLAR) is primarily used as a remote-sensing and surveillance tool. The term "real aperture" is used in contrast to "synthetic aperture" because the along-track resolution is determined by the actual length of the antenna aperture in the real-aperture radar (RAR)[1] whereas it is determined by signal processing equivalent to a longer "synthetic aperture" in the SAR. Real-aperture radar systems can be much simpler than SAR systems, and, when properly designed and operated, can provide images with quality more than adequate for many mapping applications. In its usual form the RAR signals representing scatter from the ground elements are recorded on a continuous strip of film; thus, the images are comparable with those from a strip camera or an optical-IR scanner. Uninterrupted image strips with consistent ground and brightness scales can be mosaicked for a broad overview of a large area.

SLAR was first developed in the early 1950's when advances in antenna design, centimeter wavelength microwave components, and recording techniques made possible high-resolution imagery which was useful for military reconnaissance. Applied research by earth scientists through the 1960's developed the imaging radar into a valuable remote sensing tool, and SLAR surveys using RAR became available on a commercial basis in 1969. SLAR has been used as a means of rapid survey and reconnaissance of areas where timely detection of terrain features require its unique geometry and wavelength, and

---

[1] In the discussion that follows, the term RAR is used for items referring specifically to real-aperture systems, while the term SLAR is used for items that may apply either to RAR or SAR systems.

Fig. 10-5. Imagery obtained from APQ-97 SLAR developed by Westinghouse for U.S. Army Electronics Command—RAR imagery.

in regions where prevailing cloud cover restricted the use of other remote sensors.

Examples of RAR imagery, shown in Figures 10-5, 10-6, and 10-7, illustrate its equal applicability to areas of extreme ruggedness and developed metropolitan centers. Imagery can be interpreted equally well in areas of geomorphology, geology, land use, forestry, natural vegetation, and other fields. A later section gives the details of an operational system.

BASIC IDEA OF SLAR

SLAR is an airborne sensor for displaying the radar backscatter characteristics of the earth's surface in the form of a strip map or picture of a selected area. Microwave energy reflected from the terrain is proportionally converted into light energy and recorded on photographic film as a function of distance along the aircraft track and distance from the aircraft track. The recorded density of the image will vary from point-to-point in accordance with surface roughness, water content, and other surface parameters, to form an image that can be interpreted in terms of the topo-

graphical and man-made features of the terrain. Alternatively, the signal may be recorded on magnetic tape, digitized, and presented on a digital image display. In this form, it may also be corrected for distortions and subjected to pattern recognition and image-enhancement algorithms for special purposes.

The SLAR picture is not a *snapshot,* but is obtained by scanning the terrain, as illustrated in Figure 10-8. In RAR, an antenna with a long horizontal aperture is mounted along the side of the aircraft. The antenna directs microwave energy into a narrow fanshaped beam, which defines a narrow path or line across the terrain strip that is approximately normal to the flight track. The antenna is fed with a pulse of microwave energy, which propagates at the speed of light within the beam and successively illuminates points along the line. The radar returns scattered back from targets at different ranges are separated in time at the radar receiver. Usually synchronized intensity-modulated light spot scans a line across photographic film to record target returns at their scaled slant range or ground range distance. After each line of video return is recorded, another

Fig. 10-6. Imagery obtained from APQ-97 SLAR developed by Westinghouse for U.S. Army Electronics Command—RAR imagery.

pulse is transmitted to obtain a new scan. The strip image is produced by advancing the film past the scanning line proportionally with the aircraft ground speed. Other recording and storage methods ultimately result in a similar product.

SLAR systems use either a slant-range or a "true-ground-range" presentation for the across-track coordinate. *Slant range* is the radial distance from the antenna to the target. *Ground range* is the horizontal distance from the aircraft ground track to the target. The geometry is shown in Figure 10-9.

Slant range, *R,* is the natural radar coordinate, and is obtained directly from the time, *t,* required for the pulse to propagate out and back at the speed of light, *c.* A constant recorder sweep rate will accurately reproduce slant range separation of targets. Slant range presentations have been used together with angle measurements for computation of accurate ground range and terrain elevation.

A ground range (R_g) sweep waveform is computed from slant range and altitude (h_0) above a flat terrain surface, by the simple expression

$$R_g = (R^2 - h_0^2)^{1/2}. \qquad (10\text{-}1)$$

The sweep is faster in the near range and asymptotically approaches the slant-range rate as range increases. A "true-ground-range" sweep only approximates the true horizontal across-track separation, since terrain relief will cause deviations from the design altitude. Because the horizontal shape and match is more accurate than

on a slant range presentation, the ground range presentation is preferred for preparation of mosaics and stereo viewing of image strips in areas of modest relief.

Range-swath and angular coverage of SLAR systems can vary greatly for different applications. A system may map a narrow strip of terrain either at long offset range or at the full range of angles from directly beneath the aircraft to very small grazing angles at the horizon. SLAR systems that are applied to earth resource mapping use a geometry in between these extremes to achieve the desired performance. The angular coverage usually excludes the angles near normal incidence and zero grazing to avoid distortions and shadowing, as discussed in detail in a later section. Range coverage is based on achieving resolution and the high S/N required to map terrain features of interest.

Resolution

Resolution is closely associated with the dimensions of the surface from which signal is simultaneously scattered back toward the radar. A small resolved area is achieved by transmitting a pulse of short duration, τ, within a narrow azimuth beamwidth, β_h. The return received at the antenna at a time, t, after transmission can be a slant range increment from R to $R + c\tau/2$ within the beam, as shown in Figure 10-9. Resolution is defined by the intersection of this pulse packet and the surface. Surface features, intercepted at a grazing angle, ψ

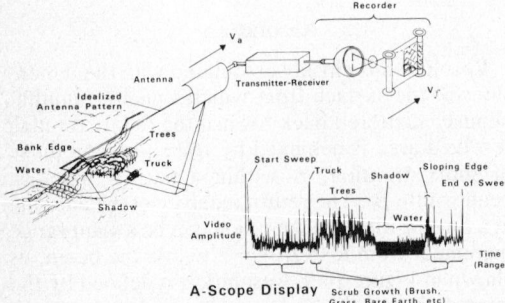

Fig. 10-7. Imagery obtained from APQ-97 SLAR developed by Westinghouse for U.S. Army Electronics Command—RAR imagery.

or incidence angle θ, can be resolved if they are separated by a distance,

$$r_y = \frac{c\tau}{2 \cos \psi} = \frac{c\tau}{2 \sin \theta} \qquad (10\text{-}2)$$

in the range direction, and a distance,

$$r_a = R\beta_h. \qquad (10\text{-}3)$$

in the azimuth or along-track direction. Range resolution will become poorer in the near range as

the spherical shell defined by the pulse becomes oblique to the surface. Along-track resolution will become poorer with range since the beam's angular width is constant.

The narrow azimuth beamwidth required for fine resolution is achieved by radiating a short wavelength of microwave energy. The azimuth beamwidth limit is obtained from antenna or diffraction theory. The beamwidth, β, in radians, is on the order of the reciprocal of the antenna length (D) in wavelengths λ. For an active microwave sensor, the round-trip power pattern is the square of the one-way antenna gain pattern and the ground cell resolution in the azimuthal direction is approximately

$$r_a \simeq 0.7R \ \lambda/D. \qquad (10\text{-}3a)$$

(see Chapter 9 for more details).

Since antenna length is physically limited by the aircraft, most real-aperture SLAR use radar wavelengths between 3 and 0.8 centimeters to achieve beamwidths in the order of milliradians, as required for high-resolution radar mapping. For radars using noncoherent radiation, resolution can be improved by a factor of two over the diffraction

Fig. 10-8. Real aperture SLAR (RAR) technique.

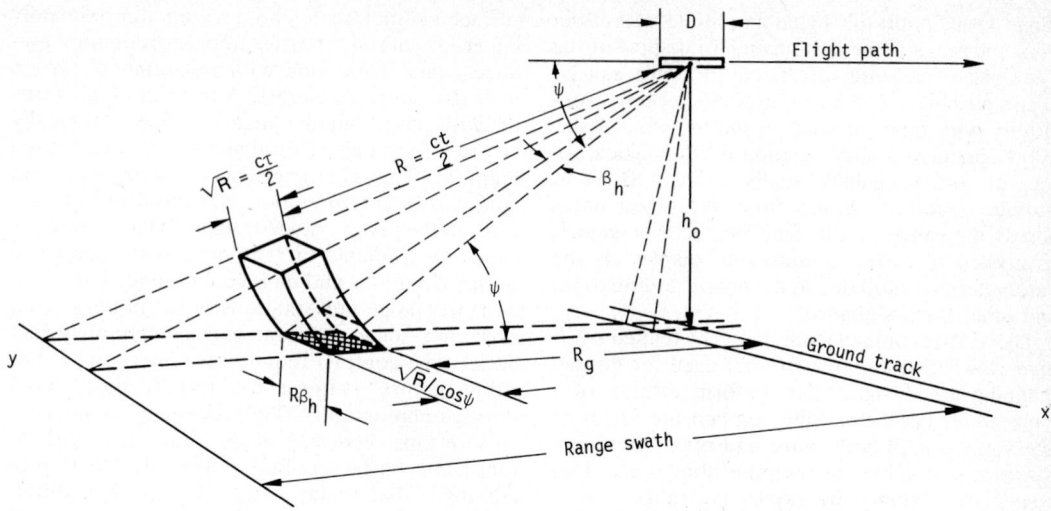

Fig. 10-9. RAR geometry and resolved surface area.

limit by using a technique called *monopulse resolution improvement*. This technique is discussed in succeeding paragraphs, under the heading of "Antenna" in "Systems and Equipment."

SLAR resolution is a measure of the ability to distinguish between closely spaced objects on the radar imagery. Although fundamentally related to pulse duration and antenna length, it includes effects of the receiver bandwidth and the recorder spot size. A typical resolution profile for present-day, real-aperture SLAR is shown in Figure 10-10. Spatial resolution is usually determined from either the half-power width of a point-target response or the threshold distance at which two point targets against a low-level background can be separated on the imagery. The detection and separation of more subtle surface patterns or complex targets are determined by dynamic range, linearity and uniformity of the displayed target reflectivity, and the degree of fading or speckle, as well as spatial resolution.

Displayed Reflectivity

Terrain reflectivity must be displayed in a uniform and proportional manner for meaningful interpretation. Two backscatter coefficients are normally used to describe radar returns from area targets:

σ^0 = scattering cross section per unit surface area; and

γ = scattering cross section per unit projected area; being the area normal to the direction of propagation through which the surface area is illuminated.

The relationship among the measures of cross section is

$$\sigma = \sigma^0 A = \gamma (A \sin \psi) \qquad (10\text{-}4)$$

where A is the surface area resolved by the pulse packet.

To achieve a display adequate to detect features of interest, the design of SLAR is concerned with the uniform display of like surfaces throughout the image strip and the display of target intensity; the target intensity is proportional to the strength of the target cross-section. The range of target cross-sections which can be so displayed is termed the *dynamic range* of the radar system.

Power received at the radar is sensitive to both target cross-section and range. The radar equation previously given (Eq. 9-63) describes the relationship between transmitted power and received power. Signal power from a small target with a constant cross section decreases inversely as the fourth power of the range. Because of this range sensitivity, the system gain must be varied with

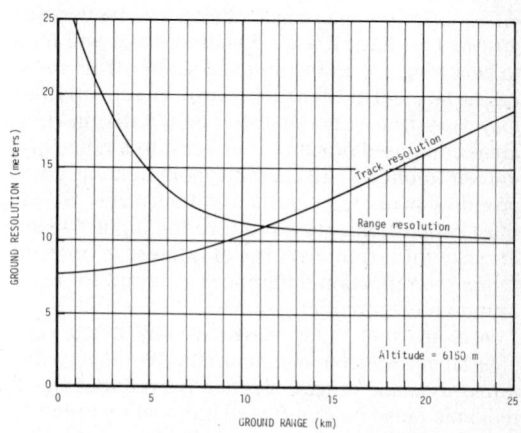

Fig. 10-10. Typical resolution profile for high-resolution, real-aperture SLAR.

range to maintain the signal amplitude within the dynamic range of the system and for display of the appropriate measure of reflectivity. This can be accomplished either by controlling receiver sensitivity with time, or shaping the antenna illumination pattern or gain function in the vertical direction. Antenna gain is usually varied in SLAR to provide a uniform return from like target areas across the range swath, and the receiver gain is controlled to obtain a consistent display as the parameters of altitude, atmospheric attenuation, and other factors change.

The vertical antenna gain patterns (called *modified cosecant-squared patterns*) used for ground mapping are designed for uniform display of a backscatter coefficient and compensate for average variations in both range and resolution area. Requirements for the angular distribution of power are obtained by expressing range dependence in terms of angle and a constant height above flat terrain. The gain patterns are

$$G(\psi) = csc^{3/2} \psi \cos^{1/2} \psi \qquad (10\text{-}5)$$

for constant σ^o, and

$$G(\psi) = \csc^2 \psi \cos^{1/2} \psi \qquad (10\text{-}6)$$

for constant γ.

Since neither σ^o nor γ is typically constant, these patterns are far from perfect. For a more complete discussion of this problem, see Ulaby, et al. (1982).

Nevertheless, these antenna patterns remove the first-order geometric dependence of the radar return. In hilly and mountainous terrain, the resolved surface area and range varies with terrain slope and elevation. These deviations from design assumptions cause a relief-dependent shading within the radar image.

The range of maximum signal to minimum signal, or input dynamic range, is very wide and varies with type of terrain and the targets of interest. One reason for this is that common construction practices produce surfaces that are smooth compared to the wavelength. The directivity of reflections from man-made objects results in radar returns that have a wide dynamic range and tend to be stronger than the diffused scatter from natural surfaces. The dynamic range of reflectivities may vary from a hundred-to-one (20 dB) in rural areas composed of natural surfaces, to a range of a million-to-one (60 dB) in metropolitan areas composed of man-made and natural surfaces. SLAR must proportionately reproduce the input dynamic range for terrain patterns of interest in order to retain the reflection differences required for resolution and recognition.

Real-aperture radar (RAR) usually has a variable signal transfer characteristic to display different dynamic ranges of reflectivity. Input signal dynamic range is compressed into and recorded as a 20-dB dynamic range of brightness on the film (typically, the maximum range that can be effectively utilized visually). The desired detectable surface characteristics produce an approximately linear-logarithmic transfer to proportionately produce signal variations with minimum distortion over the range of interest. A transfer characteristic with 25-dB input dynamic range is typically selected for mapping rural areas or natural terrain features. This characteristic achieves sufficient contrast for interpretation of natural terrain patterns in the image. A wider input dynamic range is generally available to reproduce both man-made return variations and their background. The wider dynamic range presentations result in lower contrast and some apparent loss in natural terrain detail. The contrast required for interpretation of subtle natural patterns can usually be restored through photographic film processing techniques.

With tape recording of the signal the dynamic range can be larger. Subsequent electronic processing, either analog or digital, may allow different portions of the dynamic range to be emphasized for display of the desired target characteristics in different parts of an image—or even in the same part.

SYSTEMS AND EQUIPMENT

A photograph of typical RAR equipment is shown in Figure 10-11. The total airborne package for this radar weighs 191 kg. The antenna in the background is about 4 m long and 20 cm high and is mounted along the sides of the aircraft. Following the antenna, there are three boxes that contain the transmitter-receiver, recorder, and power supplies. In the foreground are a monitor scope and a control panel that allow the operator to select the radar mapping geometry and to adjust the recorded range of signal amplitudes.

The major functional blocks of RAR systems are the transmitter, receiver, antenna, and recorder. These functions and common implementation techniques are outlined in the following discussion and indicated in Figure 10-12. In addition,

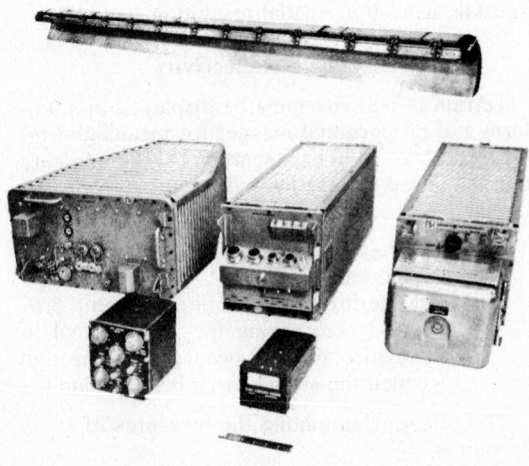

Fig. 10-11. Typical side-looking radar components.

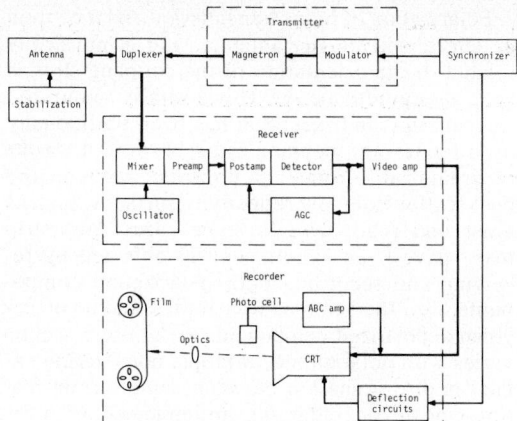

Fig. 10-12. Block diagram of typical RAR.

typical systems contain a synchronizer and duplexer. The synchronizer is the system clock and controls the time for pulse transmission, recorder sweep, and switching functions. The duplexer is a switch that connects the antenna to the transmitter during transmission and to the receiver during the recording time.

Although this block diagram represents the most common system, some of the more modern systems replace the recorder block with a video slowdown unit and either an analog recorder or a digitizer and digital recorder. The final recording on the ground may then be either with a unit like that shown in the recorder block of Figure 10-12 or with a digital image-processing system. Other configurations are also possible. For more discussion see Ulaby, et al. (1982).

Transmitter

The transmitter consists of a modulator that forms the pulse waveform and includes one or more microwave tubes that generate the required power. A single oscillator tube, the *magnetron,* which is used in most noncoherent radars, is capable of generating short pulses of high microwave power. Pulse duration can be as short as 40 nanoseconds, and peak power can be in the order of several hundred kilowatts.

The transmission is in the form of repetitive pulses of energy. The interpulse period is sufficiently long to record the radar return from the range swath of interest and to avoid recording target returns from the farther ambiguous ranges during the following interpulse period. The pulse waveform can either be an ordinary short pulse or a longer dispersed waveform that can be compressed into a short pulse in the receiver. However, the short pulse is most commonly used in RAR.

The frequency of transmission for RAR is typically between 10 and 35 GHz (Figs. 10-5 through

10-7 were produced from a 35-GHz real-aperture system). Lower frequency transmission results in a wider beamwidth and poorer along-track resolution for practical antenna size. Development of SLAR systems at higher frequencies than those mentioned above is discouraged by power and weather difficulties. Peak power is more difficult to generate, and higher atmospheric attenuation and resolution result in greater required power for the same range. More complicated pulse-compression techniques would be required to achieve adequate power. In addition, radar rapidly becomes sensitive to water particles as their size approaches a half wavelength.

Receiver

The receiver modifies the signals from the radar antenna through amplification, filtering, and detection to produce useful signals for recording. Techniques are chosen to meet sensitivity, frequency-selectivity, and dynamic-range requirements for high fidelity, and reproduction of the terrain backscatter characteristics.

A typical receiver consists of a local oscillator, mixer, preamplifier, post-amplifier, detector, and video amplifier. The signal from the antenna is mixed with the local oscillator signal and reduced to an intermediate frequency. A low-noise mixer and preamplifier are used to achieve required sensitivity. The preamplifier boosts the weak signal so the noise figure in following stages need not be as low. The post-amplifier is an *automatic gain controlled* (AGC) amplifier that matches the range of signals from the preamplifier to the detector amplitude characteristic. The detected signal is biased and amplified in the video amplifier to achieve proper levels for recording.

The receiver usually compensates for the difference between input dynamic range of signals and the range of film densities that can be interpreted. The required compression can be accomplished by several techniques, including the *linear-logarithmic (lin-log) detector* and *gain switching.*

The *lin-log detector characteristic* compresses the amplified input signal levels into a suitable range of video levels for recording.

Gain switching uses a linear detector characteristic with the receiver output signal range matched to the recorder, but receiver gain is varied from pulse to pulse to form a system lin-log transfer characteristic by superposition of different input signal ranges on the photographic film. The recorded input dynamic range typically varies between 25 and 50 dB, and is selectable from the control panel in some RAR systems. A variable gain-control reference is used to center target returns of interest within the recorded dynamic range.

The receiver accepts all the signals from the antenna. These include reflected transmitted signals, as well as the signals from other natural and man-made radiations. A selective frequency characteristic must be provided to discriminate against unwanted signals and noise generated by electronic components. Optimum discrimination is obtained by matching frequency characteristics of the amplifier before detection to the characteristics of the transmitted pulse. A close approximation of the matched filter is implemented in most receivers. For a short pulse, the optimum receiver bandwidth is given by the reciprocal of the pulsewidth.

Antenna

In RAR, the antenna is designed to obtain a narrow azimuth pattern for resolution and a wider elevation pattern for uniform return from like surfaces. The antenna also controls polarization and is an important factor in determining power requirements and the operational range of the system.

The size and shape of the antenna are directly related to the directional gain requirements for ground mapping. The horizontal length of the antenna is made as long as is practical to achieve narrow azimuth beam; typically, in the neighborhood of 4.5 m. The antenna height is determined by system requirements to achieve a modified cosecant-squared vertical pattern. The vertical beamwidth is wide enough that the required vertical aperture seldom exceeds 30 cm.

The shaped cylindrical reflector (which is fed by a linear waveguide array) and the planar waveguide array are the antenna types most frequently used. The shape of the gain patterns is controlled by reflector shape and by varying the phase and amplitude radiated at the slots in the array. In some ground-mapping radars the horizontal aperture is divided in half and provisions are made to receive the sum and difference patterns that are typically used for monopulse angle tracking. The *sum pattern* is the same pattern that would be obtained with the total aperture; the *difference pattern* has two lobes or beams within the mainbeam of the sum pattern and can be non-coherently subtracted from the sum pattern to improve azimuth resolution. This technique is called *monopulse resolution improvement*.

SLAR antennas are designed for mounting along the side of or beneath an aircraft. Variation in aircraft roll angle changes the illumination of a target and causes an undesired brightness modulation in the radar imagery if the elevation beam is not stabilized.

Beam stabilization can be accomplished electrically by varying phase across the vertical aperture or mechanically by stabilizing the physical antenna. Most SLAR systems have mechanically roll-stabilized antennas. Pitch and yaw motion produce negligible intensity modulation, but affect geometric accuracy. These motions can be compensated either by stabilizing the antenna or by rotating the recorder sweep.

Polarization of radar transmission and reception is determined by the antenna, and is physically related to the orientation of the coupling slots in the waveguide array. Horizontally polarized transmission and reception has been traditionally used for terrain-mapping radar because it results in greater difference or contrast between the backscatter from low reflectivity surfaces, such as grass and roadways. In some cases, two strip maps have been simultaneously obtained by receiving and recording both polarization components, i.e., the same as transmitted and its orthogonal depolarized component. Radar cross section varies with polarization. Multiple polarization enhances discrimination between some terrain features; however, radar system implementation for multiple polarization requires greater antenna complexity, additional receiving and recorder channels, and more transmitter power to map the smaller depolarized target cross-sections.

Recorder

The recorder produces a photographic record or map of the radar return. The recorder typically includes a *cathode ray tube* (CRT), an optical system, deflection circuits, film drive, and associated circuits and power supplies. The functions of the recorder are: to transform the detected video signal into light signals, which expose the photographic film; to deflect the CRT spot in proportion to range and antenna angle; and to drive the film at a rate proportional to aircraft ground speed.

The recorder is an important factor in determining overall system dynamic range, resolution, and accuracy.

The radar signal is recorded on photographic film through the CRT and optical system. The video signal from the receiver is applied to the grid of the CRT to intensity-modulate the range sweep. A voltage bias is added to the video signal to obtain a linear modulation transfer between signal power and light intensity. An *automatic brightness control* (ABC) is used to ensure a consistent transfer and the detection of low-level signals. The ABC circuitry and photo-cell sense and maintain a minimum, constant background brightness when no signal is received. The stabilized intensity-modulated sweep is projected on photograph film through an optical lens system. A diaphragm or neutral density filter is inserted in the optical path to change the light intensity in proportion to sweep rate and ground speed, and to maintain consistent film exposure.

Recorder resolution is an important factor in determining the tradeoff between overall system resolution and the range swath covered. A present-day CRT with a 13-cm tube face has a spot size of about 0.02 millimeters measured at the half intensity points. Pulse resolution and recorder resolution are typically adjusted to be approximately equal for the maximum coverage.

Azimuth resolution will usually be limited by the recorder at near range and the antenna beamwidth at far range. Allowing the recorder to limit the along-track resolution can lead to increased speckle and reduced interpretability of the image as discussed below.

Matching of the transfer characteristics of the various elements of the receiver and recorder is the key to obtaining high-quality images. This is discussed in more detail by Ulaby, et al. (1982).

SPECKLE AND ITS REDUCTION

Most images produced by radars are more speckled in appearance than comparable pictures produced optically, as a result of fading of the monochromatic signal used by the radar. This fading is averaged out for the optical signal unless it is produced by a laser. Photographs made with laser light also exhibit the speckle phenomenon.

Speckle Reduction by Spatial Averaging

Reduction of the speckle effect due to fading can be achieved by averaging a sufficient number of independent samples for each resolution cell. The averaging must take place incoherently; that is, it must be post-detection averaging. The process is the same as that discussed in detail in Chapter 9 under *Precision of Measurement,* and the causes of fading are the same as those discussed in that section. For the RAR, the fading situation can be described in relatively simple terms using concepts easily developed. The predetection Doppler bandwidth B for a rectangular beam of width β_h is given by

$$B = \frac{2u}{\lambda} \beta_h, \qquad (10\text{-}7)$$

where u is the velocity.

Since the resolution cell width is $\beta_h R$, the time to pass each resolution cell is

$$T = \frac{\beta_h R}{u} \qquad (10\text{-}8)$$

Consequently the time-bandwidth product (BT) is given by

$$BT = \left(\frac{\beta_h R}{u}\right)\left(\frac{2u}{\lambda}\right)\beta_h = \frac{2\beta_h^2 R}{\lambda}. \qquad (10\text{-}9)$$

In the above-referenced section, the relation of the time-bandwidth product to the number of independent samples in a fading signal was also discussed.

Time-bandwidth products for the RAR can be better understood if expressed in terms of resolution and aperture size. Since the azimuth resolution is

$$\beta_h R = r_a, \qquad (10\text{-}10)$$

and the beamwidth is approximately

$$\beta_h \approx \frac{\lambda}{D}; \qquad (10\text{-}11)$$

the time-bandwidth product of equation 10-9 is

$$BT = \frac{2r_a}{D}. \qquad (10\text{-}12)$$

If we assume a rectangular beamwidth, the resulting Doppler spectrum is rectangular and has this time-bandwidth product. This is the same as the earlier example of Chapter 9 illustrated in Figure 9-38. Eq. 9-124 says that the number of independent samples is just equal to the time-bandwidth product for this example. Consequently, the number of independent samples, N_i, where the subscript i specifies independent, is

$$N_i = BT = \frac{2r_a}{D}. \qquad (10\text{-}13)$$

Thus there is one independent sample for each half-antenna length in the resolution cell; that is, the number of independent samples is a ratio of the azimuth resolution to half the length of the antenna. This approximate result can be made more exact by using the spectrum associated with the actual antenna beam, as discussed following Eq. 9-124. Eq. 10-13 also applies to SAR.

Speckle Reduction by Use of Excess Bandwidth

Speckle may be reduced by increasing the time-bandwidth product; that is, by use of excess bandwidth since the time is fixed. This section will describe two major methods of accomplishing this: Under Principle A, a much finer range resolution is made than is otherwise required, and several of these cells are averaged together; and under Principle B, the reduction is accomplished by averaging in frequency. Both of these principles have identical effects.

The resolution cell for Principle A is illustrated in Figure 10-13. Here the entire square is the desired resolution cell, but the range resolution has been improved by a factor of 6 so that there are 6 sub-cells within the desired resolution cell. To improve the range resolution to this degree, additional bandwidth is required beyond that needed for the square cell. One way to obtain this additional bandwidth is to use a short pulse or the

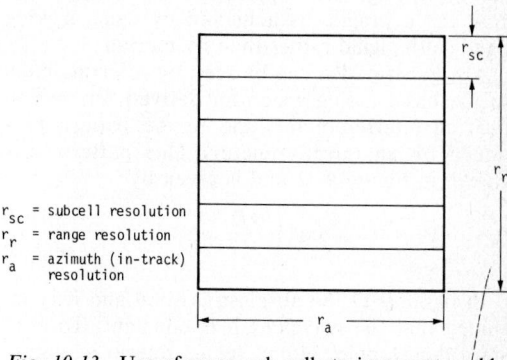

r_{sc} = subcell resolution
r_r = range resolution
r_a = azimuth (in-track) resolution

Fig. 10-13. Use of range sub-cells to improve speckle.

equivalent chirp waveform. After detection, all of the signals from the sub-cells are added together to produce the reduced-fading average. Let B_r be the bandwidth associated with the basic resolution cell. Then for the geometry of Figure 10-9, we have

$$Br \approx \frac{1}{\tau_r} = \frac{c}{2r_y \sin \theta}. \qquad (10\text{-}14)$$

The time duration of the pulse for the square resolution cell τ_r is given in terms of the sub-cell pulse duration τ by

$$\tau_r = N\tau, \qquad (10\text{-}15)$$

where N = number of sub-cells within the larger cell.

Consequently the bandwidth required for the smaller sub-cells is given in terms of the bandwidth for the basic resolution cell by

$$B = \frac{1}{\tau} = \frac{N}{\tau_r} = NB_r. \qquad (10\text{-}16)$$

Thus N is the excess-bandwidth ratio, as well as the number of sub-cells within the larger cell. In other words, to obtain the N independent samples, there would have to be a bandwidth N times greater than would be required without the use of sub-cells.

Principle B refers to averaging in frequency. Laser light is monochromatic, and wide-band light gives smoother pictures because of frequency averaging. The problem of frequency averaging was studied by Waite (1970) and by Thomann (1970). If we use a random scatterer model for the spectrum within a resolution cell, the variance for the received power is given by

$$\sigma_{wr}^2 = \int_{-\infty}^{\infty} R_c^2 (f) |P_t (f)|^2 df, \qquad (10\text{-}17)$$

where:

$R_c(f)$ = target autocorrelation function in frequency,

$P_t(f)$ = spectrum of the transmitted signal.

This result is directly analogous to that associated with Doppler-frequency averaging due to motion through the Doppler spectrum, but in this case the averaging is achieved by using a wide bandwidth signal rather than by motion.

The general idea can be seen by referring back to the basic fading spectrum derived for motion past an interferometer, and to the pattern produced by an interferometer. This pattern was shown in Figure 9-11 and is given by

$$\cdot |V| = 2V_o \left| \cos \left(\frac{\pi D}{\lambda} \sin \theta \right) \right|, \qquad \begin{array}{l}(9\text{-}10)\\ \text{(repeated)}\end{array}$$

In figure 9-11 the abscissa is $\sin \theta$ and it is assumed that the wavelength is constant. To consider the Doppler effect, let us rewrite equation 9-10 as

$$|V| = 2V_o \left| \cos \left(\frac{\pi f D}{c} \sin \theta \right) \right|, \qquad (10\text{-}18)$$

where the frequency is explicitly called out.

Clearly the pattern of Figure 9-11 could have been expressed with an abscissa of f rather than $\sin \theta$ if f had been the variable and $\sin \theta$ the constant. Thus we can see that that signal returned from such an interferometer is periodic in frequency, so that transmitting a wide band of frequencies permits averaging over the periodicities. This concept was used in deriving Eq. 10-17.

If the spectrum across a rectangular-pulse resolution cell is uniform in frequency, as in the example of Fig. 9-38, and a uniform spectrum of width B is transmitted for a time τ, Eq. 10-17 becomes

$$\frac{1}{N_i} = \frac{\sigma_{wr}^2}{\overline{W}_r^2} = \frac{2}{\pi B \tau} \int_0^{\pi B \tau} \frac{\sin^2 x}{x^2} \left(1 - \frac{x}{\pi B \tau} \right) dx. \qquad (10\text{-}19)$$

Note that this equation is directly analogous to Eq. 9-125 for averaging of fading by motion through the Doppler spectrum. Figure 9-43 shows the improvement in the variance achieved by increasing the bandwidth above that required simply to transmit the pulse of duration β. For time-bandwidth products exceeding about 5, the result is essentially the same as that obtained by the simplified analysis of Principle A.

Thus speckle in an image can be removed either by transmitting shorter pulses than used in the final image and post-detection summing of the finer image elements, or by any other means for transmitting the same amount of excess bandwidth. *The amount of averaging achieved (the number of independent samples) is equal to the ratio of the bandwidth transmitted to that required by the basic resolution cell.*

This reduction of speckle may be combined with the reduction inherent in the real-aperture SLAR, as indicated by Eqs. 10-12 and -13; that is, *when excess bandwidth is transmitted, the number of independent samples is the product of the time-bandwidth product obtained inherently by by length of the aperture and that obtained by the excess bandwidth.* Images obtained with time-bandwidth products of 50-100 have the same speckle-free appearance as photographic images. Even a relatively small time-bandwidth product improves the appearance of an image greatly as contrasted to the appearance with only a single independent sample per resolution cell, such as would be obtained at the focal point of an antenna on a RAR or by a fully-focused synthetic aperture.

For visual interpretation of images one can describe the improvement in terms of a gray-scale resolution r_g that can be used in a resolution volume V that replaces the pixel area A as a measure of the quality of the image (Moore, 1979). Thus, this volume is

$$V = r_a r_y r_g \qquad (10\text{-}20)$$

where r_a = along-track resolution,
$\quad\quad r_y$ = cross-track resolution,
and $\quad r_g$ = gray-scale resolution.

The interpretability of the image was found to be related to V approximately by

$$I = I_o e^{-(V/V_o)} \quad\quad (10\text{-}21)$$

where V_o is the characteristic volume for a particular class of surface. The gray-scale resolution is given by

$$r_g = \frac{(\text{signal exceeded 10\% of the time})}{(\text{signal exceeded 90\% of the time})}.$$
$$(10\text{-}22)$$

Since the value of the gray-scale resolution decreases dramatically for an increase in N over the range from $N = 1$ to about $N = 4$, the interpretability also increases rapidly in this region of N. In fact, the interpretability is better for $N = 9$ and azimuth resolution of 90 meters than for $N = 1$ and azimuth resolution of 10 meters for the kinds of interpretations made in the referenced study. Hence, the degradation of interpretability for a RAR, experienced with the degradation in along-track resolution with range indicated by Eq. 10-10, is not as great as might be supposed. This result is discussed in more detail in Ulaby, et al. (1982).

PROFILING

Terrain profiles, suitable for contour mapping, can be obtained simultaneously with SLAR imagery by using the radar interferometer technique. *Interferometer techniques are used throughout the spectrum to measure path lengths in fractions of wavelength by sensing the phase of the interference between EM waves.* Stereo methods may also be used.

The interferometer used for terrain profiling is made up of two parallel antennas that are vertically separated by a fixed distance in a direction approximately normal to the illuminated swath. The antenna separation achieves a path difference that is sensitive to target depression angle. The radar pulse is transmitted from one antenna for proper terrain illumination and received in both antennas for normal terrain mapping and phase comparison. The phase comparison of energy scattered from resolved points across the range swath thus provides a measure of their vertical angular direction from the aircraft.

The SLAR interferometer geometry is shown in Figure 10-14. A wave scattered from a terrain point reaches the lower antenna over a shorter path than the upper antenna. The difference in path length is the distance

$$X = d \sin \psi, \quad\quad (10\text{-}23)$$

where:

d = interferometer separation
ψ = angle from the interferometer boresight.

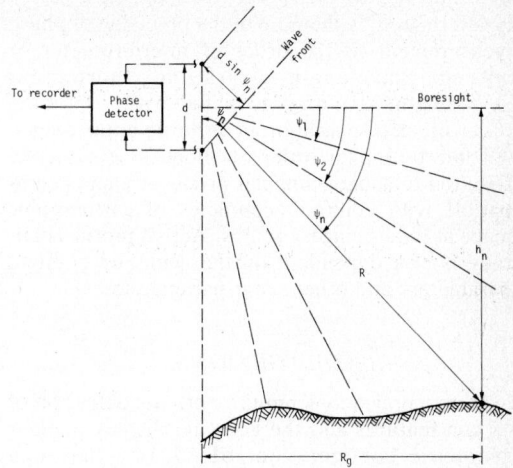

Fig. 10-14. Interferometer geometry.

The phase relationship between waves repeats whenever the path difference is an integral number of wavelengths and a set of angles is defined by a particular phase. These are given by the relationship

$$\frac{d}{\lambda} \sin \psi_n = n + \frac{\phi}{2}; \quad\quad 0 \leq \phi < 2\pi, \quad\quad (10\text{-}24)$$

where:

n = number of wavelengths
ψ_n = set of angles
ϕ = phase.

A convenient method of recording interferometric data on photographic film is to generate and record a pulse whenever a phase event, such as cancellation of waves or a phase equal to π radians, occurs. A series of lines corresponding to the angles δ_n are recorded on a slant range display. The range R, or position of the lines on the display, will change with terrain elevation. The terrain profiles,

$$h_n = R \sin \psi_n, \quad\quad (10\text{-}25)$$

at the ground range distance,

$$R_g = R \sin (90° - \psi_n) = R \sin \theta_n \quad\quad (10\text{-}26)$$

can be computed for each line. Contours can be constructed by interpolation between profiles. The density of profiles can be increased by detecting other phases. For example, alternate dark and light lines corresponding respectively, to *reinforcement* $(\phi = 0)$ and *cancellation* $(\phi = \pi)$, have been used to record two sets of profiles. In general, phase can be continuously measured for each resolved point in the map, and the density of profiles depends on detection, measuring, and recording techniques used.

The interferometer antenna separation in wavelength is directly related to profile accuracy. The precision of phase measurement is usually limited by S/N and accuracy of the detection techniques; the precision of angular measurement is determined by the ratio of the precision of phase measurement to the order of interferometer or antenna wavelength separation. Short radar wavelengths are used to achieve the optimum order of such separation in airborne applications. An interferometer order of about 30 is typically required to achieve angular precision that is compatible with contour accuracies of cartographic maps at radar imagery scales. Actual profile accuracy is also dependent on the accuracy of flight parameters and other radar parameters.

DISTORTIONS

Many applications require both the detection of terrain features and the accurate display of their geometric configuration. SLAR is a line-scan technique and does not have a fixed two-dimensional geometry that guarantees consistent reproduction of shape. The SLAR system must receive inputs from navigation sensors to stabilize the image for accurate measurements. Aircraft motion that exceeds nominal stabilization limits and navigation system errors will cause angular distortion and scale nonuniformity, errors which will vary with time. In addition, there is the common sensor problem of correctly translating, or mapping, a three-dimensional surface into two dimensions. This limits the accuracy of planimetric measurements in the range direction because of the uncertainty and variability of terrain relief.

Range Distortions

SLAR measures slant range, the distance from the antenna to resolved terrain points within the antenna beam. At shallow grazing angles, the slant range separation approaches the ground separation, and the far range imagery has the expression of a vertical view. Near normal incidence, the range circles intercept the terrain at oblique angles. The near-range imagery is similar to an oblique camera view since the target position is more sensitive to relief, and the horizontal separation of targets is compressed. This compression,

shown graphically in Figure 10-15, is called *slant range distortion*.

A more accurate measure of target horizontal separation can be obtained by calculating ground position from slant range and nominal aircraft altitude above terrain. This approximate method is used to generate a *recorder sweep function* for the "true-ground-range" SLAR presentation. Targets at a higher elevation than the datum at the design altitude will be at a closer range and thus displaced toward the radar. The relief displacement, ΔR, is shown graphically in Figure 10-16 and is approximately

$$\Delta R_g = h \tan \psi, \qquad (10\text{-}27)$$

where:

h = target elevation
ψ = depression angle.

Terrain slopes toward the radar will be foreshortened and slopes away from the radar will be elongated on a ground-range presentation. The displacement and foreshortening decreases at shallow depression angles.

Terrain layover is a unique distortion for range measuring systems. Two targets horizontally separated and at different elevations may be simultaneously illuminated and their return mapped into a single resolution cell. Excessive layover occurs when large-scale terrain slopes become normal to the direction of propagation. In mountainous terrain, the effective angular coverage is limited by the confusion of terrain layover in the near range and the shadowing of terrain features in the far range. Both the angle from the vertical and the angle from the horizontal should be less than the large scale slope angle to avoid the loss of terrain detail. Optimum depression angles for different parts of the world have been mapped by Mac-Donald and Waite (1971).

Aircraft Motion

Geometric distortions, introduced by aircraft motion, may be roughly classified according to effect as *short-term* and *long-term*.

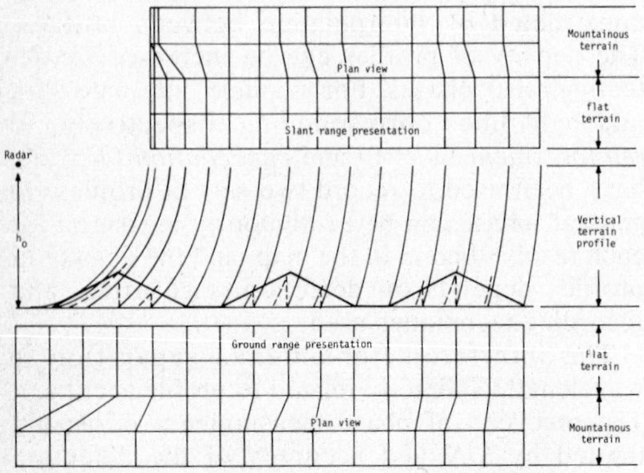

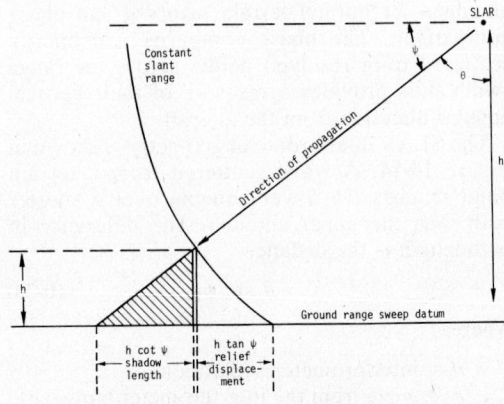

Fig. 10-15. Slant range and ground range display comparison.

Fig. 10-16. SLAR shadow and relief displacement.

Short-term distortions result from a rapid change in the aircraft position or attitude and are evident as an obvious anomaly or smear in a single image strip. These often result in a resolution loss.

Long-term distortions are more closely associated with accuracy and become apparent when matching a SLAR image strip to a map or another image strip. Distortions resulting from relatively slow anomalies, such as variance of aircraft flight about a straight ground track, show up as rectangular patterns mapped into a curvilinear geometry. Unwanted beam scanning due to aircraft attitude changes and velocity changes cause alternate expansion and compression in along-track scale.

The isolated effects of yaw, pitch, and crosstrack deviations on mapping a square grid are shown in Figure 10-17. Actual distortion in a SLAR map is a composite of these effects.

Geometric distortion requirements for SLAR are determined by application. Targets of interest may be within a curved path and a flexible geometry is desirable. The application of SLAR to large-area surveys requires that both long- and short-term distortions be minimized. The variable SLAR geometry is stabilized to a geographic coordinate frame. A straight geographic path, such as a rhumb line, or a great circle and its orthogonal reference, is computed from aircraft navigation sensor data and the image is stabilized to the reference path. Stabilization may be accomplished in several manners: By physically controlling the aircraft, through the autopilot, to follow the computed path, with simultaneous control to assure that the antenna points in the reference direction; and by displacing and rotating the image in the recorder to compensate for deviations in aircraft position and altitude.

A combination of physical control and recorder compensation techniques is often used.

Long-term distortion is determined by navigation sensor accuracy, film drive accuracy, recorder sweep accuracy, and angular alignment, with other factors. The limiting factor is usually navigation accuracy. Scale uniformity within one percent and angular distortion within 10 milliradians is obtained with present-day navigation techniques if relief displacement is neglected.

Navigation system accuracy is not always achieved on a short-term basis. The ideal image stabilization technique would reduce aircraft motion effects to within a resolution cell, thereby eliminating objectionable loss of resolution and local distortion in the image strip. Stabilization techniques are usually designed to achieve this precision under normal atmospheric and environmental condition. Transient conditions will cause occasional local distortions. A large wind shear may drive the aircraft at a rate that cannot be physically compensated for, or an electrical power transient may cause error in the antenna or recorder trace control signal. Image uniformity is usually restored rather quickly after a short-term distortion.

Although these complicated navigational and stabilization efforts are important for large-scale surveys and for any mapping effort where precise location of ground points is required, many applications of SLAR do not require such precautions. If the general nature of the terrain is known, along with suitable landmarks, changes may be detected in, for example, vegetation or flooding. For these applications, the antenna may be hard-mounted to the aircraft and suitable images may be obtained as long as the aircraft is flown in relatively stable air. In the Soviet Union, geologic mapping is accomplished in this way by flying at night.

SYNTHETIC-APERTURE RADAR (SAR)

The earliest open publications describing the remote sensing potential of SLAR centered about the use of imagery generated by the Westinghouse AN/APQ-56 and AN/APQ-97 RAR's. Radars of this type are basically illumination, energy sensing, and display systems which do not rely heavily upon signal-processing concepts and which, in turn, have the virtues of simplicity of design and a relatively modest cost of implementation.

Unfortunately, for reasons to be discussed, noncoherent SLAR's do not lend themselves to applications in which fine-resolution radar imagery must be generated from data collected at long range. While this is not a serious problem for many kinds of aircraft surveys, it impacts seriously on the use of orbiting spacecraft as platforms for remote-sensing imaging radars. This limitation is severe enough for single-wavelength systems operating at wavelengths below about 3 cm, but it is even more severe for longer-wavelength systems and multiple-wavelength systems. The resolution attainable with a feasible antenna at the 25-cm wavelength of the Seasat SAR and the Shuttle SIR-A radars would have been too poor to be of much value even from aircraft—and these systems were flown in space.

The solution to this apparent impasse rests in a

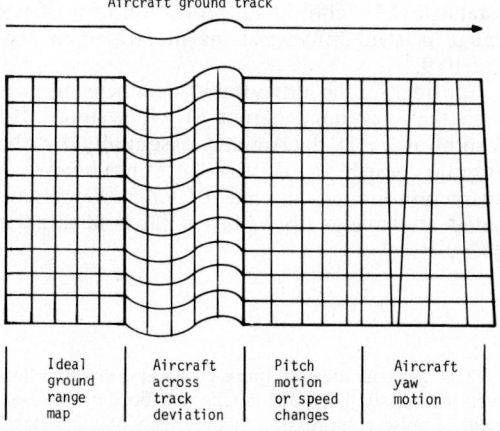

Aircraft ground track

| Ideal ground range map | Aircraft across track deviation | Pitch motion or speed changes | Aircraft yaw motion |

Fig. 10-17. SLAR map of square grid with uncompensated aircraft motion.

second class of SLAR systems known as *synthetic-aperture radar* (SAR) systems, and often referred to as *coherent* SLAR's. This type of radar permits fine-resolution radar imagery to be generated at long operating ranges by the use of signal processing techniques, and can be provided with a multispectral capability in a manner compatible with projected user requirements.

The purpose of this part of Chapter 10 is to describe the SAR concept, to outline some fundamental parametric relationships in SAR systems, to describe typical SAR configurations at the block-diagram level, to discuss signal processing requirements and techniques associated with SAR systems, and to include some illustrative examples of SAR imagery.

Although the mathematical analyses of SAR systems can become rather complex, an explicit attempt is made to keep this text as non-mathematical as possible and, instead, to stress the intuitive concepts which underly the detailed theory. The reader interested in the mathematical basis for SAR performance is referred to an excellent text by Harger (1970). A variety of journal articles and tutorial treatments have also been published in this area [Brown (1967), Brown and Porcello (1969), and Cutrona (1970)]. A more complete treatment at an intermediate level is given by Ulaby, et al (1982).

REVIEW OF SLAR CONCEPTS

The previous section of this chapter, *Real-Aperture SLAR*, described in detail the basic configuration and performance of noncoherent SLAR's. For the system geometry shown in Figure 10-9, it was observed that the radar system generated an image of the microwave reflectivity distribution of a terrain strip which parallels the flight line of the aircraft. This two-dimensional image is a *projection* of the three-dimensional reflecting surface, portions of which may be shadowed by elevated terrain lying between the aircraft and the shadowed region. The resolution of the image is determined in the cross-track, or y-coordinate, by the range-resolving capability of the radar; and in the along-track, or x-coordinate, by the physical width of the antenna beam in that direction.

For flat terrain, image resolution in the y- (or "ground-range") dimension is provided by pulse-ranging techniques. This topic was discussed in Chapter 9 where it was shown that

$$T = 2R/c. \tag{9-27}$$

This relation can be used to show that slant-range resolution r_R is given by

$$r_R = c\tau/2 \tag{10-28}$$

Eq. 10-28 relates the slant-range resolution r_R of a short-pulse radar system to the pulse length τ. When, as in Figure 10-9, the system views a horizontal terrain element at depression angle ψ, the resolving capability r_y in the horizontal (y) coordinate is also a function of ψ.

$$r_y \approx \frac{c\tau}{2} \sec \psi. \tag{10-29}$$

An alternate form of this relationship is frequently seen. It was indicated in (10-16) that a radar system transmitting and receiving a pulse of duration τ must have an overall system bandwidth of

$$B \approx \frac{1}{\tau} \text{ Hz}, \tag{10-30}$$

where τ is in seconds, in order to preserve this pulse width at the output of the radar receiver. If the system transmits a pulse of duration τ, but the overall receiver has system bandwidth $B < 1/\tau$, then the received pulse width is degraded to approximately $1/B$, with the result that

$$r_R \approx \frac{c}{2B}, r_y \approx \frac{c}{2B} \sec \psi \tag{10-31}$$

Furthermore, it is possible to generate transmitter waveforms having duration τ and bandwidth B in such manner that $\tau B \gg 1$. These *large time-bandwidth product* waveforms have the property of compressibility into short pulses of time duration $1/B$, much shorter than that of the original waveform.

These waveforms constitute the basis for the *pulse-compression radars* used for a large variety of applications.[2] This technique is important in systems which require fine range resolution r_R, and are limited in peak transmitter power capability W_{peak}, but which require a large amount of transmitted energy,

$$Q = W_{peak} \times \text{pulse duration (joules per pulse)}. \tag{10-32}$$

Pulse compression features can be, and are, incorporated into SLAR systems, with the end result that the bandwidth B and not the pulse length τ essentially controls r_R. In this case the more general Eq. 10-31 then describes the y-dimension performance. The use of ranging techniques to establish the y-coordinate projection of a SLAR image is straightforward, as illustrated in Figure 10-9.

Considering the along-track (x) dimension, it is seen that, for noncoherent SLAR systems, the resolution r_a is determined essentially by the angular width β_h of the antenna beam. A diffraction-limited antenna of length D, illuminated at wavelength λ, has a far-field angular width of

---

[2] The general idea of pulse-compression radar has been previously discussed in Chap. 9. For further treatment of pulse-compression theory, see Cook and Bernfeld (1967).

$$\beta_h \approx \frac{\lambda}{D} \quad \text{(radians)}^3 \qquad (10\text{-}33)$$

There is therefore, for this antenna, an along-track linear width of approximately $R\beta_h$, provided that $\beta_h \ll 1$ radian. Therefore, as illustrated in Figure 10-9,

$$r_a \approx \frac{\lambda R^4}{D}. \qquad (10\text{-}34)$$

Along-track resolution deteriorates with increasing range R, with increasing wavelength λ, and with decreasing antenna length D. In applications where a given level of resolution r_a is required, the ratio (λ/D) is made as small as possible, subject to other system constraints. The resulting limit on maximum operable slant range of the system,

$$R_{\max} = \frac{D}{\lambda} r_a, \qquad (10\text{-}35)$$

is simply accepted.

This approach is frustrated by several considerations: The λ-dependence of the terrain return may, for some applications, force (from scientific consideration) the use of long wavelengths; as λ decreases, such atmospheric effects as backscatter from precipitation become important and may inhibit all-weather operation; the achievement of diffraction-limited performance from an airborne antenna with angular beamwidth of order 10^{-3} radians or finer, is a formidable and expensive undertaking at any radar wavelength; and at some of the longer wavelengths which may be of interest for earth-resource survey, even an antenna of length

$$D \approx 100\lambda$$

may be longer than the aircraft to be used as its platform.

Therefore, although arbitrarily fine resolution may be realized in the crosstrack dimension simply by increasing the radar bandwidth, no analogous option is open in noncoherent SLAR (RAR) systems with respect to along-track performance.

THE SYNTHETIC-APERTURE CONCEPT

The concept of synthetic-aperture radar may be thought of in several ways, two of which are discussed here in some detail. The first of these is that of a synthesized antenna array, and the second is that of a Doppler beam sharpener.

[3] This expression is an approximation to Eq. 9-26, which gave the beamwidth for a uniformly illuminated aperture. With the usual tapered aperture, Eq. 10-28 gives a better approximation to the true beamwidth.

[4] This is different from Eq. 10-3a because the latter was for the two-way beamwidth for a nearly uniform illumination of the aperture whereas this expression is for the one-way beamwidth of a more tapered illumination.

To illustrate the idea of a synthetic aperture, consider Figure 10-18. An antenna array can be focused at a point T as illustrated in Figure 10-18(a) by connecting the elements of the array to the receiver through transmission lines that make the total time delay (and therefore phase shift) from T to the receiver the same along all paths. This means that the signals from a target at T add up at the receiver in phase, whereas those from a target at another location add up out of phase. Since the time delay along the shorter path of length R_3 is less than that along the longer paths such as R_5, the length of the transmission line for element 3 L_3 must be longer than those for the other elements. The same principle is used with a convex lens, for which the lower-velocity path through the center of the lens is longer than the lower-velocity paths through the outer edges. With the synthetic array, the positions representing array elements are occupied successively in time and the information as to the amplitude and phase of the received signal stored in memory (frequently on film). The round-trip phase shifts $2kR_1$ through $2kR_5$ must be compensated for in the same way that the different transmission-line lengths in Figure 10-18(a) compensate. With the synthetic-aperture system, this compensation is done during the processing, as indicated in Figure 10-18(c). There the phase shifters introduce appropriate shifts ϕ_1 through ϕ_5 so that all signals from T are added in phase, whereas signals from other locations are not properly corrected and do not add in phase.

The total length available for the synthetic aperture is the distance over which the target point T is illuminated by the radar beam, as seen in Figure 10-19. Thus, it is simply L, the total width of the real-aperture beam on the ground. This is the same width that would be the along-track resolution r_a for a RAR with the same beam. It is given by

$$L = \beta_h R. \qquad (10\text{-}36)$$

As the radar-bearing aircraft proceeds along its trajectory, it begins to illuminate the object when the radar is at along-track location $(x_0 - L/2)$. The target then remains illuminated as the radar passes abreast of it at the location x equivalent to x_0, and finally ceases to be illuminated when

$$x > x_0 + L/2.$$

Consider this approach, for the moment ignoring the ranging process and assuming that the target field has structure at only a single slant range R_0. Under these conditions, it is possible, in principle, to combine observations from the moving antenna (assuming suitable signal storage and retrieval) to *synthesize* an array of length $L \approx \lambda R/D$. In a *conventional passive system* (such as an optical telescope), an angular resolution of

$$\beta' \approx \lambda/L \qquad (10\text{-}37)$$

would be realized from a real or synthesized

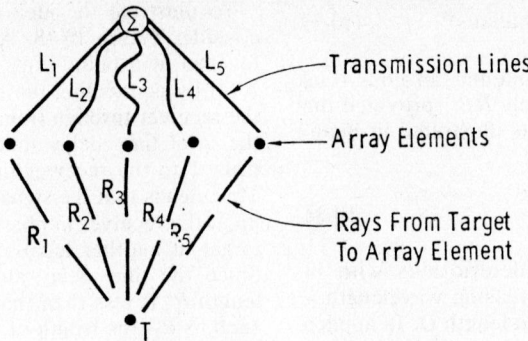

(a) Real-Aperture Equivalent Array Focused on Target T; $L_1 + R_1 = L_2 + R_2 = L_3 + R_3 = L_4 + R_4 = L_5 + R_5$

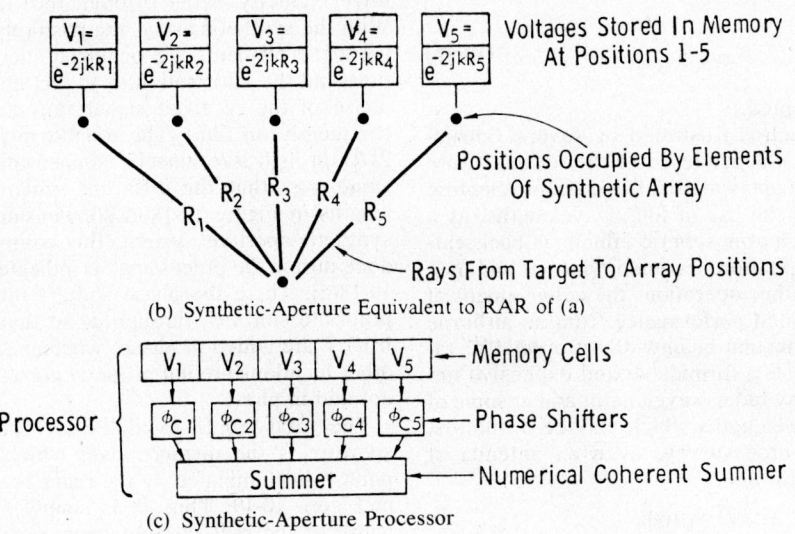

(b) Synthetic-Aperture Equivalent to RAR of (a)

(c) Synthetic-Aperture Processor

Fig. 10-18. Concept of synthetic aperture (Ulaby et al., 1982).

aperture of length L. *Because the radar system is active and incorporates two-way signal propagation,* it can be shown that an array of the form thus described can provide an effective angular resolution, β'', of

$$\beta'' = \frac{\lambda}{2L} \qquad (10\text{-}38)$$

provided that certain system errors are kept within specified tolerances.

The linear along-track resolution at slant range R then becomes

$$r_a = R\beta''. \qquad (10\text{-}39)$$

Recalling from Eq. 10-36 that $L \approx R\beta_h \approx R(\lambda/D)$, this reduces to

$$r_a \approx \frac{D}{2}. \qquad (10\text{-}40)$$

This approximate and empirical result repre-

sents a useful rule-of-thumb for the preliminary design of SAR systems. More precise analyses lead to essentially the same expression. Thus, the theoretically achievable resolution r_a is independent of the operating range R; r_a is also independent of the operating wavelength λ; and the resolution r_a is actually improved when the physical antenna employed by the SAR system is reduced.

These properties, especially the first two, constitute the basic reasons SAR systems are attractive for long-range radar remote-sensing applications. However, they are deceptive in the following sense: Although r_a is independent of range R, the complexity and accuracy requirements imposed on the radar, the storage element, and the processor increase with range for fixed values of r_a and wavelength; likewise, the complexity and accuracy requirements increase with wavelength for fixed values of r_a and range; and furthermore, the radiated RF power requirement grows sharply as D is decreased, thereby leading to a tradeoff

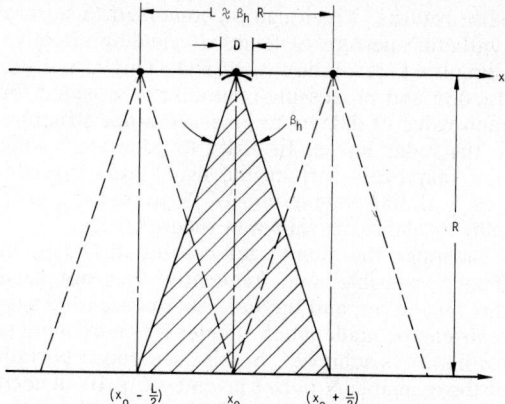

Fig. 10-19. Visibility of x_0 over distance L.

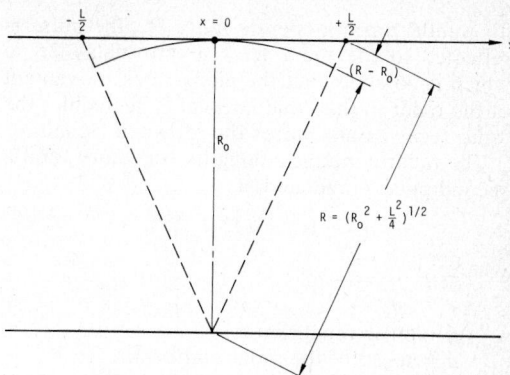

Fig. 10-20. Variation of R with aperture length L.

between resolution and image S/N at long operating ranges. These problems are treated in greater detail later.

Unfocused and Focused SAR Systems

It is well known that *long antennas operating at relatively short ranges must be "focused" in much the same manner as optical systems.* As for optical systems, the Rayleigh criterion provides an estimate of the relationship between array length and the need to focus. This situation is illustrated for a SAR system in Figure 10-20.

Suppose a SAR system operating at a specific wavelength observes a reflector at $x = 0$ and $R = R_0$ while the radar moves through a distance L centered at $x = 0$. At the positions where x lies within $\pm L/2$, the range to the target is given by

$$R = (R_0^2 + x^2)^{1/2}$$

$$\approx R_0 + \frac{x^2}{2R_0} \text{ for } L \ll R_0. \qquad (10\text{-}41)$$

Therefore, as the aircraft passes the target, the range decreases in a quadratic manner from $(R_0 + L^2/8R_0)$ to R_0, and then increases again to $(R_0 + L^2/8R_0)$.

The quadratic phase shift induced by this geometric change, referenced to the phase at $x = 0$, is given by

$$\Delta\phi = \frac{4\pi}{\lambda} \frac{L^2}{8R_0}, \qquad (10\text{-}42)$$

where two-way propagation is taken into account. But according to the Rayleigh criterion, this phase error will be of little consequence if $|\Delta\phi| < \pi/2$; therefore, the error is small if

$$\frac{4\pi}{\lambda} \frac{L^2}{8R_0} < \frac{\pi}{2}, \qquad (10\text{-}43)$$

which can be written

$$L < (\lambda R_0)^{1/2}. \qquad (10\text{-}44)$$

As long as L obeys inequality 10-44, the quadratic phase shift need not be taken into account; that is, it is not necessary to focus a synthetic array of length $L < (\lambda R_0)^{1/2}$. However, if $L \geq (\lambda R_0)^{1/2}$, the array must be focused by inserting a compensating quadratic phase, as shown in Fig. 10-18.

The implications on performance are straightforwardly calculated. Since, for a synthetic array,

$$r_a \approx \frac{\lambda}{2L} R, \qquad (10\text{-}45)$$

it is observed that the best performance available in an unfocused SAR system is given by

$$r_a \approx \frac{1}{2} (\lambda R_0)^{1/2}. \qquad (10\text{-}46)$$

Of course, if the synthetic array is focused, the limitation reverts to $r_x \approx D/2$.

When this resolution is attained with an unfocused SAR, the length of the synthetic aperture is twice the pixel length, and this can complicate the processor more than one might wish to—after all, the primary purpose of using an unfocused rather than a focused SAR is simplicity. The length of the aperture and of the pixel are the same if we allow

$$r_a = (\lambda R_0/2)^{1/2}. \qquad (10\text{-}47)$$

Thus, for the occasional cases where unfocused systems are used, the resolution of Eq. 10-47 is more common than that of Eq. 10-46.

Cross-Track Resolution in SAR

The synthetic-array concept suggested by Figure 10-18 can be implemented to provide along-track resolution r_a of approximately $D/2$. A slight modification of the concept permits the incorporation of a pulse-ranging feature. This is the mechanism by which resolution is achieved in the cross-track or y-dimension. In the simplest case, the radar uses a short-pulse transmission; the following sequence of events then takes place:

The radar radiates a short pulse at location

$$x_1 = (x_0 - L/2);$$

for a reflector at moderate range R_1, the pulse is reflected to the radar after a time delay $2R_1/c$ which is so short that the along-track movement of the radar in that time interval is negligible; the radar receives and stores the reflected signals.

The radar continues along its trajectory, and a second pulse is radiated at

$$x_2 = (x_o - L/2 + \Delta x),$$

where:

$\Delta x = u/f_s$
f_s = pulse repetition rate
u = speed of the radar platform.

The reflected signal from x_2 is likewise stored.

The process is repeated for $x_3, \ldots, x_N$ where these values are spaced a distance Δx apart. The stored return from each radiated short pulse contains ranging information which can be used to provide resolution in the y-dimension. This structure is preserved in the data-storage process. A 1:1 correspondence exists between pulse delay time and slant range for each reflected pulse. In order to preserve this structure, a two-dimensional signal storage matrix of the form shown in Figure 10-21 is used. In this matrix, one dimension is assigned to the propagation delay time for a single pulse, while the second is assigned to the pulse sequence. The complex amplitude (i.e. phase as well as magnitude) of the reflected signal is stored at the appropriate coordinate locations in the matrix.

Now, it is recalled that, for each pulse radiated by the radar, the return after propagation delay time τ_1 is associated with a target at range

$$R_1 = c\tau_1/2.$$

If the storage matrix is interrogated at $\tau = \tau_1$, the reflected signals associated with all targets at range R_1 are recovered, on a pulse-by-pulse basis. In accordance with the earlier discussion, a target at range R_1 gives rise to

$$N = \frac{L}{u} f_s + 1 \qquad (10\text{-}48)$$

pulse returns, which can be combined to form a synthetic aperture of length L yielding effective along-track resolution $r_a \approx D/2$. This signal extraction and processing operation is repeated for each value of delay τ to preserve range structure in the radar image. In order to generate a long strip map of the terrain, the signal processor carries a sliding summation of N pulses for each value of delay, as shown in Figure 10-22.

Although the along-track resolution $D/2$ is, in theory, possible, one frequently does not need this resolution, and the radar and processing systems can be made much simpler if a more modest resolution is achieved. In this case only a portion of the available N pulses given by Eq. 10-48 need be used, so the storage of Figure 10-22 can be smaller.

To a first approximation, it is possible to treat the *range performance* and the *along-track performance* of a SAR system relatively independently, and this will be done. The justification for this decomposition into a *range channel* and an *azimuth channel* is treated by Harger (1970), together with a discussion of higher order interaction effects. When this decomposition is adopted, range performance can be calculated from the form of the range pulse, the transfer function of the radar's range channel, and the elevation pattern of the physical antenna used by the SAR system. The along-track or azimuth performance can be calculated from the along-track pattern of the antenna, the radar pulse-repetition frequency (prf), and the overall transfer function of the azimuth channel. In each case, the overall transfer function includes the transfer functions of the signal storage device and signal processor, and incorporates the effects of errors present in the two channels.

The Doppler Approach to SAR

The original synthetic-aperture radar was called a Doppler beam sharpener by its inventor, Carl Wiley. Thus, the Doppler approach to SAR antedates the SAR approach itself. The general

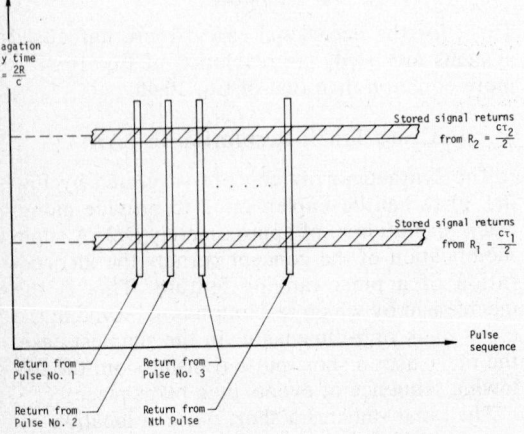

Fig. 10-21. Storage matrix for reflected signals.

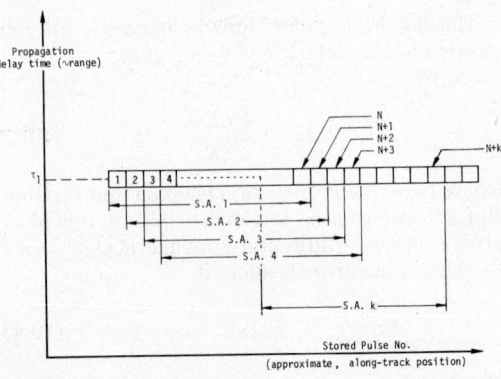

Fig. 10-22. Successive synthetic apertures generated at slant range $R_1 = ct/2$.

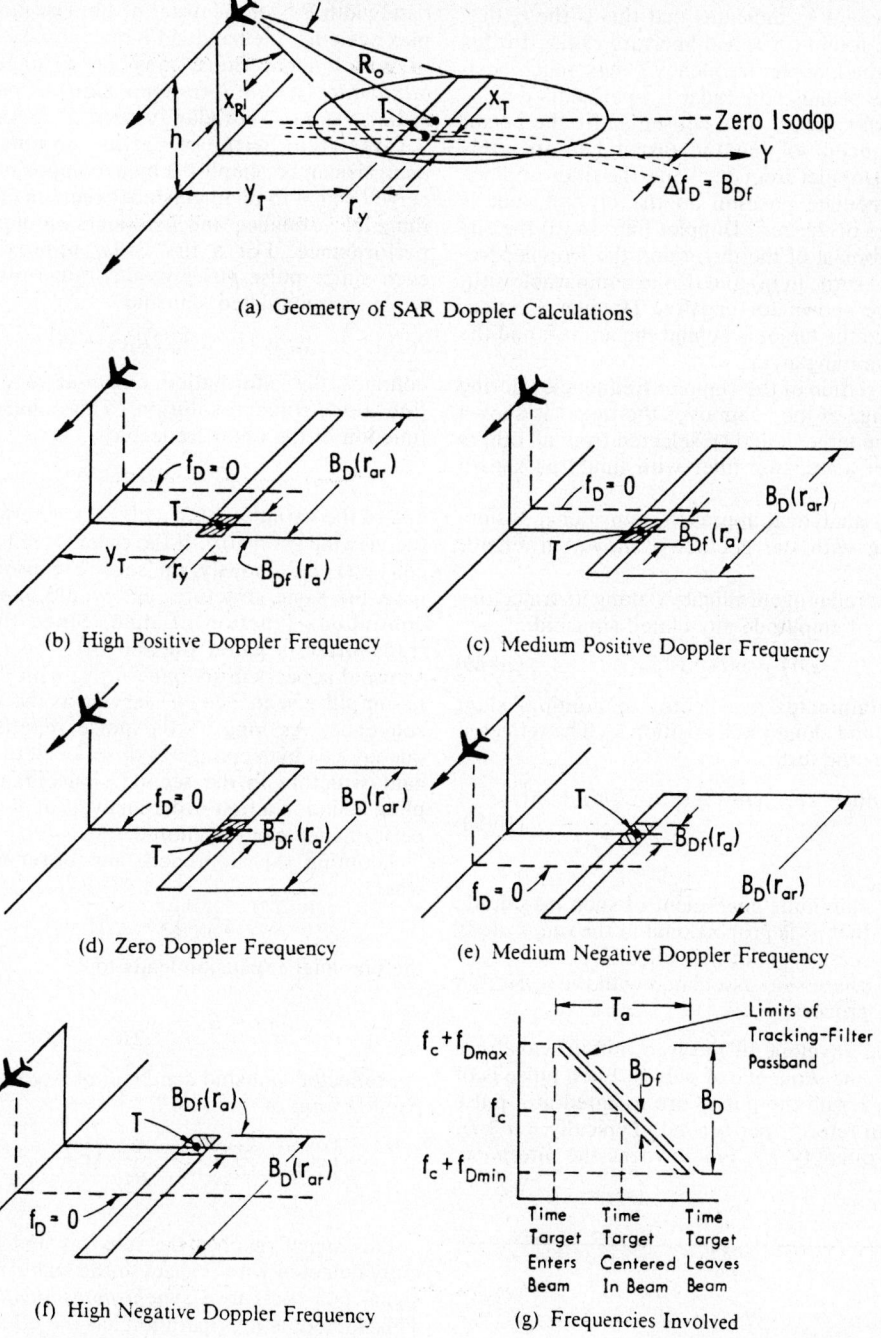

(a) Geometry of SAR Doppler Calculations

(b) High Positive Doppler Frequency

(c) Medium Positive Doppler Frequency

(d) Zero Doppler Frequency

(e) Medium Negative Doppler Frequency

(f) High Negative Doppler Frequency

(g) Frequencies Involved

Fig. 10-23. Doppler beam-sharpening SAR. Concept assumes a tracking narrow-band Doppler filter (Ulaby et al., 1982).

idea is illustrated in Figure 10-23. The basic geometry is illustrated in Figure 9.2(a), where the beam is shown to the side of the aircraft, along with a single range-resolution cell and a pair of contours of constant Doppler frequency (isodops). The target has an x coordinate x_T and the radar an x coordinate x_R. A filter of width $\Delta f_D = B_{Df}$ is used to differentiate the Doppler frequency shift of the target from those of other targets, and the isodops corresponding to the limits of this filter are shown.

In Figures 10-23(b) through (f) are shown the situations as the aircraft advances. In (a) the target has just entered the beam (indicated by the rectangle of width r_y). The Doppler frequencies selected by the filter (and hence the area on the ground) are indicated by the crosshatching. The total width of the rectangle of illuminated area is $r_{ar} = L$ where r_{ar} indicates that this is the r_a that would be found for a real-aperture radar. In this location the Doppler frequency f_D has a high positive value because the radar is approaching it.

In (c) the aircraft, and consequently the beam, has advanced so that the target has an intermediate Doppler frequency and the filter outlines an intermediate position on the ground—but is still ahead of the zero-Doppler line. In (d) the aircraft is abreast of the target and the Doppler frequency is zero. In (e) and (f) the comparable situations are shown for negative Doppler frequencies, since the target is behind the aircraft and the radar is moving away.

The variation of the Doppler frequencies during the passage of the beam over the target is shown in (g). The target could be selected from all others by use of a tracking filter with limits as shown in (g).

For an analytical approach, consider a system operating with the geometry shown in Figure 10-24.

Let the radar at coordinate x along its trajectory transmit an amplitude-modulated sinusoid

$$s_t(t) = p(t) \cos (\omega_0 t), \qquad (10\text{-}49)$$

which illuminates a reflector at nominal slant range R_0 and along-track position x_0. The reflected signal has the form

$$s_r(t) = \alpha p(t - 2R/c)\cos \left[\omega_0(t - 2R/c) + \phi\right], \qquad (10\text{-}50)$$

where:

α = amplitude coefficient of such magnitude that α^2 is proportional to the *radar cross section* (σ) of the reflector
ϕ = phase shift associated with the reflection process.

Let the envelope of the transmitted waveform be a periodic sequence of pulses. Each pulse is of duration τ, and the pulses are radiated at a pulse repetition rate, f_s, per second. Typically $\tau \ll 1/f_s$, and the quantity $1/f_s$ is known as the *interpulse period*. In the case of a *short pulse* system, the RF bandwidth B is of the order of $1/\tau$. For more complex systems, the bandwidth may now be $B \gg 1/\tau$, in which case there may be a *large time-bandwidth-product* transmission as part of a *pulse-compression* radar system.

Consistent with an earlier comment, the analysis can be simplified by decomposing the received signal to distinguish between its effects on range performance and its effects on along-track performance. For a first-order approximation, each single pulse $p(t)$ provides range resolution, while the modulated sinusoid

$$\cos \left[\omega_0 (t - 2R/c) + \phi\right]$$

contains the information essential to obtaining fine along-track resolution. This sinusoid is a function of the radar frequency

$$f_0 = \omega_0/2\pi, \qquad (10\text{-}51)$$

and of the variation of R with x, as determined by the viewing geometry. If the radar were to radiate $\cos (\omega_0 t)$ continuously, the second sinusoid would have the same structure, and would appear as a continuous function of time. Since the radar transmitter is gated on and off, the reflected sinusoid appears in *sampled* form, with the transmitter pulse sequence $p(t)$ serving as the sampling sequence. As long as the pulse repetition frequency f_s is high enough to observe all the significant structure in the second sinusoid, the sampling action in a first-order analysis of along-track performance can be ignored.

Adopting this approach, and observing that, when

$$R_0^2 \gg |x - x_0|^2, \qquad (10\text{-}52)$$

the binomial expansion leads to

$$R \approx R_0 + \frac{(x - x_0)^2}{2R_0}, \qquad (10\text{-}53)$$

the reflected sinusoid can be expressed in the expanded form of

$$\cos \left| \omega_0 t - \frac{2\omega_0 R_0}{c} - \frac{\omega_0}{R_0 c}(x - x_0)^2 + \phi \right|. \qquad (10\text{-}54)$$

This signal reaches the receiver and is coherently detected with respect to the stable reference signal $\cos (\omega_0 t)$ in a *synchronous demodulator*. This demodulation operation has the effect shown in Figure 10-25.

Therefore, the output of the synchronous demodulator in the receiver chain has the form

$$\cos \left| \left(\phi - \frac{2\omega_0 R_0}{c} \right) - \frac{\omega_0}{R_0 c}(x - x_0)^2 \right|, \qquad (10\text{-}55)$$

where

the first term is a time-invariant phase
the second term, a quadratic in $(x - x_0)$, varies

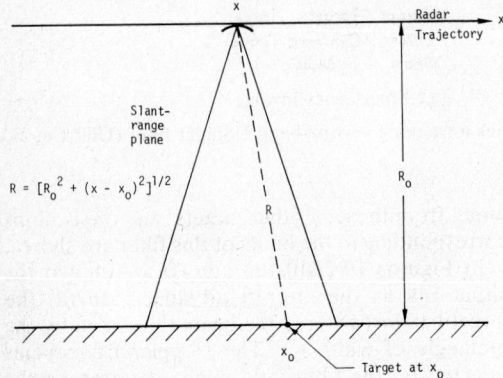

Fig. 10-24. Range variation with $(x - x_0)$.

lead to spurious target returns (or azimuth ambiguities) in the SAR image. For ambiguity-free operation, the sampling rate must be in conformance with Eq. 10-87.

Assume that it is desired to design for limiting performance $r_a \approx D/2$. If Eqs. 10-86 and 10-88 are combined,

$$\frac{u}{r_a} < f_s < \frac{c}{2\Delta R} \;. \tag{10-89}$$

This is a condition which can be met if (and only if)

$$\frac{\Delta R}{r_a} < \frac{c}{2u} \;. \tag{10-90}$$

The significance of this relationship is that, given a vehicle velocity u, no ambiguities will appear in either dimension if (and only if) the slant-range coverage and the along-track resolution r_a requirements conform to relationship 10-90. If this constraint is violated in a single-channel SAR system, ambiguities begin to appear.

When operating requirements do not permit relationship 10-90 to be met, then more complex systems based on multichannel concepts or scanning are necessary.

The ambiguity situation may therefore be briefly summarized: This situation is not a serious problem with aircraft systems, for it is difficult for the ratio of swath width ΔR to resolution r_a to be as large as the other side of Eq. 10-90. For spacecraft systems, however, the velocity u is so large that the problem poses a severe limitation. This problem may be solved at the expense of resolution by scanning the beam to different elevations (Moore, 1981).

First, the number of resolution elements imaged per second by a SAR system is given by

$$N = \left(\frac{\Delta R}{r_R} \right)\left(\frac{u}{r_a} \right). \tag{10-91}$$

Recalling that $r_R \approx c/2B$, simple substitution into the relationship of Eq. 10-91 shows that

$$N < B, \tag{10-92}$$

this is to say, the number of resolution elements per second which can be imaged by a single-channel system, operating at its ambiguous limit, approximates B, the RF bandwidth of the system.

Secondly, the sampling requirements have been treated with the assumption that the Doppler spectrum falls sharply to zero at $f_D = \pm u/D$, and the ranging requirements have been based on the assumption that the elevation pattern of the antenna falls sharply to zero for $y < y_1$ and $y > y_2$. These patterns are never sharply bounded, and do have some spill-over energy in these outer regions. While more accurate constraints have been formulated by Harger (1965), the constraints of the treatment herein do have value, though primarily for order-of-magnitude estimates.

Third, although the ambiguity problem has been discussed with special relationship to the SAR context, it has equal applicability to noncoherent SLAR systems. The range constraint of Eq. 10-86 is identical in both cases. Doppler concepts do not apply to the RAR situation; however, there is an implied requirement to sample the scene in the along-track dimension, and to thus satisfy inequality 10-88. In practice, noncoherent systems are not plagued by the ambiguity problem only because other limitations—in particular their fundamental inability to realize fine resolution at long ranges—apply before the ambiguity problem becomes apparent.

BASIC SAR CONFIGURATIONS

The basic block-diagram configuration of a SAR system follows from the array concepts and signal processing concepts outlined in the previous section.

A transmitter is required to generate a periodic RF waveform, which may be a simple sequence of short pulses, or may exhibit a more complicated structure;

An antenna is required to radiate the RF waveform as an EM wave, and to receive the reflected wave;

A receiver is required to detect and amplify the reflected signal emerging from the antenna;

Since relative phase relationships are critical in establishing the performance of antenna arrays, a common stable phase reference must be supplied to the transmitter waveform generator and to the detector in the receiver;

A signal storage matrix is required which is capable of preserving, for each reflected pulse, the time-delay structure which will eventually provide range resolution, and which is capable of storing a sequence of reflected pulses sufficiently great to permit adequately long synthetic arrays to be formed; a signal processor is required that can accept signals from the storage matrix and effect the required synthetic-array processing without loss of range-dimension structure; a display device is needed to present the output of the signal processor to the user; finally, a timing system is needed to synchronize the functions of all the other system elements.

These basic elements are interconnected as shown in Figure 10-33. Typically, all elements of

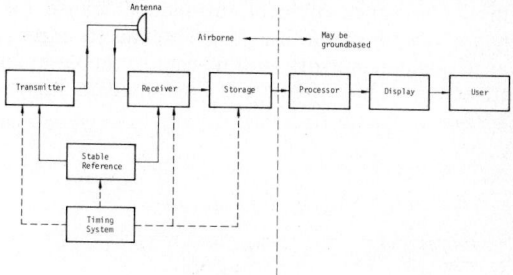

Fig. 10-33. Basic SAR functional diagram.

an airborne SAR system, up to and including the storage device, are in the aircraft, while the processor, display, and user are groundbased. The stored signal simply is transported to the processor after the aircraft lands. When required, it is possible, in principle, to implement an airborne processor/display to provide near-real-time data to an in-flight user, or to insert a data link between the radar-bearing vehicle and the ground to relay either processed or unprocessed data to the user.

ERROR SOURCES IN SAR SYSTEMS

A wide variety of error sources is present in SAR systems, and the resulting errors collectively limit systems performance.

Additive random noise is inserted at the receiver input and at each of several other points within a SAR system, and ultimately propagates through the receiver, storage, processor, and display systems to limit the S/N of the output image, as well as the usable dynamic range of the system.

Phase and/or amplitude errors across the signal spectrum, introduced in the transmitter, receiver, storage subsystem, and processor, effectively limit the range resolution r_R, and upset the control of range-dimension sidelobe structure.

Antenna elevation pointing errors lead to improper illumination across the swath to be imaged, and in some cases can result in range ambiguities.

Likewise, improper azimuth pointing results in an improperly illuminated Doppler spectrum which, in turn, can lead to striations of the imagery and, in severe cases, to azimuth ambiguities.

The designer of a SAR system normally attempts to control the contributing error sources in accordance with an *error budget,* and the effects of these errors can be analyzed in a straightforward manner.

One rather serious error source, however, which in practice is difficult to control, is the spurious cross-track motion of the radar-bearing vehicle itself. Recall that, in formulating the SAR concept, it was assumed that the airborne antenna moved in the x-direction along a perfectly straight trajectory at a perfectly uniform rate, u. In actual flight, however, an aircraft continually experiences three-axis rotations about its center of mass as well as three-dimensional spurious translations with respect to the nominal straightline flight path. These effects, small in smooth air but increasing in the presence of local turbulence, cause the *phase center* of the airborne antenna to experience positional errors with respect to the nominal trajectory. This situation is illustrated in Figure 10-34.

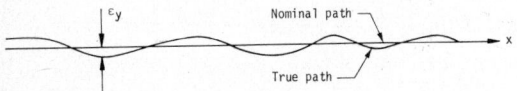

Fig. 10-34. Spurious vs. true flight path.

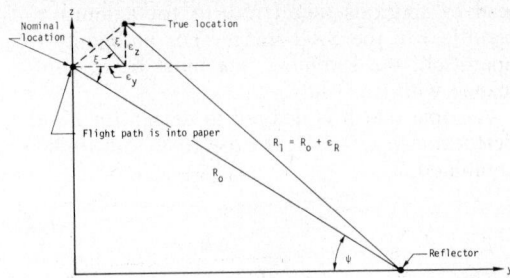

Fig. 10-35. Range error vs. cross-track displacement.

Figure 10-35 shows a reflector at slant range R_0 and depression angle ψ from the nominal trajectory. In the presence of a radar cross-track (horizontal) positional error, ϵ_y, and a vertical positional error ϵ_z, the true slant range R to the reflector is given by

$$R' = R_0 + \epsilon_z \sin \psi - \epsilon_u \cos \psi. \quad (10\text{-}93)$$

Therefore, the *range error*

$$\epsilon_R = R' = R_0 \quad (10\text{-}94)$$

induces a *round-trip phase error* of

$$\phi_{\epsilon_R} = \frac{4\pi\epsilon_R}{\lambda} \quad (10\text{-}95)$$

into the Doppler history of the target. Substituting,

$$\phi_{\epsilon_R} = \frac{4\pi}{\lambda}(\epsilon_z \sin \psi - \epsilon_y \cos \psi). \quad (10\text{-}96)$$

For an airborne side-looking radar, the short-term errors in along-track position lead to phase errors which are usually very small in comparison with the cross-track-induced errors, and these will be ignored for the purpose of this treatment.

In a *conventional antenna array,* performance will be close to the diffraction limit, although with marginal sidelobe control, if the element positions are controlled accurately enough that phase errors are held to approximately $\pm45°$.

In the *synthetic-aperture case,* a phase error limit of $\pm45°$ requires, because of two-way propagation effects, that the ϵ_R must be held to $\pm\lambda/16$ across the aperture. At X-band, with $\lambda \approx 3.2$ cm, a flight path accuracy of approximately ±2 mm is required over apertures which may be as much as tens or hundreds of meters in length. Aircraft seldom, if ever, fly in this manner.

The method normally used to circumvent this difficulty is to incorporate a set of accelerometers at or close to the phase center of the physical antenna. The two accelerations $\ddot{y}$ and $\ddot{z}$ can be monitored continuously, and integrated twice to yield ϵ_y and ϵ_z, respectively. These can then be combined to estimate ϵ_R for a known depression angle. Finally, a phase correction

$$\phi_c = -\phi_{\epsilon_R}$$

can be inserted into the radar signal chain to compensate for the spurious motion.

In practice, an *inertial motion-compensation system* designed to carry out this function may be a reasonably complex and costly device, depending upon the degree of correction it is required to achieve.

Although the concept of using accelerometers is straightforward, it is necessary to orient the accelerometers accurately to avoid domination of the z-channel by gravitational acceleration. Furthermore, since the platform must remain locally horizontal as the aircraft changes latitude and longitude, navigational data are required. The resulting complexity of the system is beyond the scope of this treatment.

Figure 10-36 shows a block diagram of a SAR system modified to accept motion compensation signals.

These problems can be mitigated in part by sacrificing some resolution. If the synthetic aperture need not be so long, the stability requirements on the aircraft need not be so great. Thus, in moderately stable air, a resolution of 10 m or so at a range of 10 km to 20 km is not difficult to achieve with minimum error compensation. On the other hand, achieving a resolution of 2.5 m at 50 km requires using all of the compensation techniques discussed near the limits of feasibility. These "partially focused" SARs are discussed in considerable detail in Ulaby, et al. (1982).

Another potentially serious source of error arises as a result of atmospheric turbulence. The SAR theory outlined earlier assumed that the SAR signals propagated through free space, but is easily modified to account for atmospheres with constant or slowly varying refractive indices. However, the refractive index changes randomly in a turbulent atmosphere with position and with time. This causes a phase error across the synthetic aperture which, in turn, degrades sidelobe control in azimuth and ultimately limits attainable azimuth resolution. Fortunately, these effects are small (Porcello, 1970) for SAR systems of interest for remote sensing applications, and may be ignored for the purposes of this general presentation.

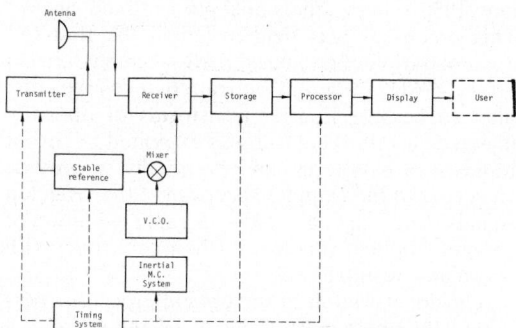

Fig. 10-36. SAR configuration with motion compensation.

SAR SIGNAL PROCESSING

As previously observed, SAR systems employ signal processing to convert stored radar signals into fine-resolution radar images. Once an image or a family of images has been generated in this manner, it may be subjected to further processing to achieve image enhancement, shape recognition, spectral recognition, or other analogous results, in much the same manner as may the output of a conventional photographic system or a multi-spectral scanner.

To avoid semantic difficulties in this treatment, the term *processor* will be used to describe the signal processor required to generate either the radar image itself or a variation of that image (such as a hologram or a digital tape).

Any enhancement or recognition devices which operate on the output of the processor are teamed *post-processors,* and are not treated in this chapter.

CONCEPT SUMMATION

When the SAR concept was first introduced, it was pointed out that the fundamental aperture synthesis process is carried out by storing and summing reflected radar signals observed at the sequence of positions,

$$(x_1, x_2, \ldots x_n).$$

When aperture focusing was introduced, the process was modified to permit insertion of a quadratic phase shift

$$\phi(x) = \alpha x^2$$

across the aperture, prior to the summation; in this case, the coefficient α was a function of the slant range R of the reflector and of the wavelength λ of the radar. It was also observed that this operation needed to be implemented for each and every resolvable range element in the terrain strip to be imaged by the SAR system; that is, a multichannel phase-weighted summation of stored signals is required, with each range element assigned to a given channel in the processor.

The alternate approach involved the *Doppler history* associated with a reflector at slant range R. It called for a tracking filter or a *matched filter,* with a transfer function which was the complex conjugate of the Fourier transform of the Doppler history. One such matched filter is required for each resolvable range element.

Although it is beyond the scope of this manual, it can be shown quite readily that these two approaches impose identical mathematical requirements on the processing, and merely represent two different intuitive approaches to describing the processor in physical terms.

SAR PROCESSOR DESCRIPTIONS

The original processor for the first Doppler beam sharpener was an unfocused one involving a fixed filter rather than a tracking filter like that

shown in Figure 10-23(g). For many years thereafter focused SARs used optical processors exclusively. Today both optical and electronic processors are in use. Optical processors have the advantage that processing is fast, whereas many of
the digital electronic processors must operate relatively slowly because the amount of equipment
required would otherwise be too great. For modest resolutions, however, real-time electronic
processors are available. Optical processors have
disadvantages in that they are difficult to adjust
compared with the electronic processors, and
they are less flexible because of the need to acquire new lenses to accommodate certain changes
in system performance.

Here we can only treat processing briefly. For
more details the reader should consult the references given on optical processors and Ulaby, et
al. (1982), (and the references therein) for electronic processors.

During the early development of SAR systems,
it became evident that processors for unfocused
SAR systems (that is, those that did not require
range-dependent quadratic phase corrections)
could be implemented using recirculating delay
lines, Doppler filters, or other analogous networks.

Optical Processors

In an attempt to produce a processor for a focused SAR system, attention was turned to the
potential offered by optical data-processing channels. A cathode ray tube (CRT) was used to display the reflected radar signal after coherent demodulation.

For use by an optical processor, the *signal film*
is produced in the aircraft or spacecraft in the
form illustrated in Figure 10-21. The production
method for this film is similar to that used to produce the *image film* in the RAR, but the signal film
is different, as discussed below.

Since the signal emerging from the synchronous
demodulator is bipolar, it is necessary to modulate
the electron beam about some bias value to preserve the bipolar nature of the intensity pattern on
the face of the CRT. The optical intensity pattern
on the face of the CRT is recorded on a photographic film, in the manner illustrated by Figure
10-89, and stored in the format of Figure 10-21.

Following this process, the *x*-coordinate of the
film corresponds to the along-track coordinate in
radar space, while the *y*-coordinate on the film
corresponds to the slant-range coordinate, also in
radar space. The data from two reflectors at
ranges R_1 and R_2 are stored in the regions shown
in Figure 10-21. Figure 10-37 is an example of a
typical SAR data film generated in this manner.
This *signal film* is *not* an image like that of Figure
10-6 because of the synchronous demodulation
used with the synthetic aperture system.

The data storage format shown in Figure 10-21
leads directly to the following optical signal-
processing conceptual procedure:

Fig. 10-37. Typical SAR data recorded on a CRT/film
recorder.

Illuminate the film at normal incidence with a
collimated coherent light beam, across which the
optical phase is uniform.

At coordinate position y_1 corresponding to
range R_1, optically insert a quadratic phase correction αx^2, where the coefficient α is chosen to
"focus" the synthetic aperture at R_1.

Sum across the *x*-direction for fixed position y_1,
using data from an aperture of sufficient length L
to yield the desired resolution r_a.

Treating each resolvable value of y as a "channel," repeat the process for the remaining positions $y_2, \ldots, y_n$; with the quadratic correction
appropriate to the range R_j represented by each
value of y_j. The final output of each channel is
recorded on an "output" film (*image film*).

Because of the nature of the optical process, all
of these summations for the length of the apertures and for the total swathwidth take place simultaneously. The light for each range is just focused at a different point on the film from that for
other ranges and the light from each element in the
synthetic aperture arrives at the same point simultaneously and adds coherently to those from
all the other points.

A coherent optical processor with a conical lens
to effect the range-dependent quadratic-phase
correction, and cylindrical lenses to effect multichannel operation, was first assembled in the
late 1950's using the concepts outlined above.
This processor was field tested in the first full
demonstration of a focused SAR system (Cutrona
et al., 1961). Coherent optical channels can display successive Fourier transforms of optically
injected signals. This fact was exploited to permit
removal of bias terms and eventually to provide
an access to the Doppler spectrum of the reflected
signals (Cutrona et al., 1966). A rapid evolution of
coherent optical processors then occurred, with
two major results:

A minor variation of the optical processor permitted the short-pulse radar to be replaced by a
pulse-compression radar using linear frequency
modulation (that is, "chirp" radar), with minimal
increase in processor complexity. In effect, the

Fig. 10-38. Precision Optical Processor (courtesy, ERIM).

optical system was capable of first "compressing," in the range direction, a linearly frequency modulated pulse, and then effectively operating as a multichannel processor to complete the synthetic-array processing. In practice, processing in the range and the along-track dimensions occurs simultaneously (Leith, 1968).

A major change in the processor configuration permitted the elimination of the conical lens, an unwieldy element which was difficult to fabricate. The new processor configuration employed cylindrical and spherical telescopes in modes known as *tilted plane* and *tilted cylinder* processors (Kozma et al., 1972). These configurations represent current optical systems used for the processing of SAR data. The details of operation can be found in the references.

Figure 10-38 shows a laboratory-quality *Preci-* *sion Optical Processor* (POP) at the Environmental Research Institute of Michigan. A typical radar image generated by the POP processor from a data record similar to that of Figure 10-37 is shown in Figure 10-39.

In a conventional optical processor, such as the POP, an optical rendition of the fine-resolution radar image is displayed at the output of the optical system, where it may be either observed visually or recorded in some manner. Because the radar wavelengths employed in SAR systems are several orders of magnitude longer than optical wavelengths, radar images of terrain may exhibit a highly specular character, especially when cultural objects are present in the scene. One consequence of this fact is that the optically projected radar image, prior to recording, exhibits a very large dynamic range, frequently as much as 60 dB or more. Conventional photographic films can display approximately 20 to 30-dB dynamic range in a local scene; hence, these films tend to severely limit the dynamic range of the radar output images if they are used as recording media. Because of the linear processing required, the logarithmic transfer functions described for RAR cannot be used with SAR.

A method for circumventing this problem is available in the form of the *holographic viewer*. A well-known property of an optical hologram is that it is capable of producing an optical reconstruction with a very large dynamic range. If the

Fig. 10-39. SAR image generated by Precision Optical Processor.

output of an optical processor in a SAR system is used to generate a hologram of the radar scene, rather than to expose an output film, a simple holographic viewing device can later be used, at the convenience of the user of the data, to examine the scene with its full dynamic range. A processor is changed from an output-film-recording mode to a hologram-generating mode by making a minor change in configuration and readjusting the focus of the processor. The same optical elements which constitute a processor are used, in a highly altered configuration, to generate the holograms. In fact, when the data are provided to the user in holographic rather than image form, he accrues an additional benefit in the sense that certain errors in the SAR data can be compensated by refocusing the holographic viewer.

Coherent optical processors are widely accepted as processors for focused SAR systems applied to remote sensing. Unfortunately, the user frequently needs to subject the SAR system output image, or family of images, to further processing for shape–and/or spectral-recognition purposes. At present, this requires conversion of the optical image formed at the output of the processor to electronic form. To carry out the conversion, the output film of a processor is replaced with a photosensitive detector, the output of which may either be digitized or preserved as an analog signal, to meet the requirements of the user and the post-processor employed. A suitable transducer for this purpose is the image dissector (Chapter 8). This device, which combines the ability to scan a scene with approximately 1,000 by 1,000 elements and an inherent dynamic range approaching 40 dB, has adequate sensitivity to respond to the illumination levels available in a coherent optical processor. This processor uses a moderate-intensity CW laser as a primary source.

Figure 10-40 presents a pair of radar scenes which were generated on the POP processor using data from the ERIM multispectral radar; digitized using the dissector/digitizer; and finally printed out in a conventional digital display.

Fig. 10-40. Pair of digitized radar images.

Electronic Processing

Electronic processing of SAR signals may take many forms, only a few of which will be treated here, and these only in a cursory fashion. For a more complete discussion, refer to Ulaby et al. (1982). Most electronic processors for focused systems are known as azimuth-sequential because they first separate the different range cells and then operate sequentially on the results for a particular range. This is similar to reading out the data in one of the horizontal lines of Figure 10-21 and operating on it. Range-sequential processors may also be constructed for some applications. In these processors, the azimuth processing is done for all range elements before range gating. These processors are not discussed further here.

A basic block diagram for an azimuth-sequential processor is shown in Figure 10-41. The signal is assumed to have been "chirped" prior to transmission so that range compression is necessary to separate the cross-track pixels; this operation may be performed in analog fashion with a surface-acoustic-wave (SAW) device, or it may be performed digitally. The next step, range gating, is the selection of one of the sequences of Figure 10-21. Azimuth prefiltering is an optional operation that can be used to reduce the size of the corner-turning memory. This is the equivalent of the film process indicated in Figure 10-21; that is, the signal for each pulse transmitted enters the memory corresponding to the recording of one of the vertical lines on the film of Figure 10-21, and the signal is read out, after all the pulses for a synthetic aperture have been stored, in a direction corresponding to the horizontal lines in Figure 10-21. Hence the corner-turning memory is a matrix that is loaded along its columns and read out along its rows if the columns correspond to range and the rows to azimuth. The azimuth-compression unit performs the synthetic-aperture addition (or the Doppler filtering). The signal amplitude is then detected and stored. The subject of multi-look additions is addressed later.

The other blocks of Figure 10-41 relate to compensation for motion and attitude errors. The signal itself may be used in a system called a "clutterlock" to determine the amount of correction that must be applied. Range migration occurs when the antenna does not point directly to the side so that a given point on the ground is not at the same range for all points in the synthetic aperture. When the Doppler centroid and the degree of range migration are known, the processor can compensate for them.

Doppler Processing

When Doppler filtering is used in the azimuth processor, one would need to have a bank of tracking filters operating if the process illustrated in Figure 10-42 were not used. Since tracking filters are more difficult than fixed filters to imple-

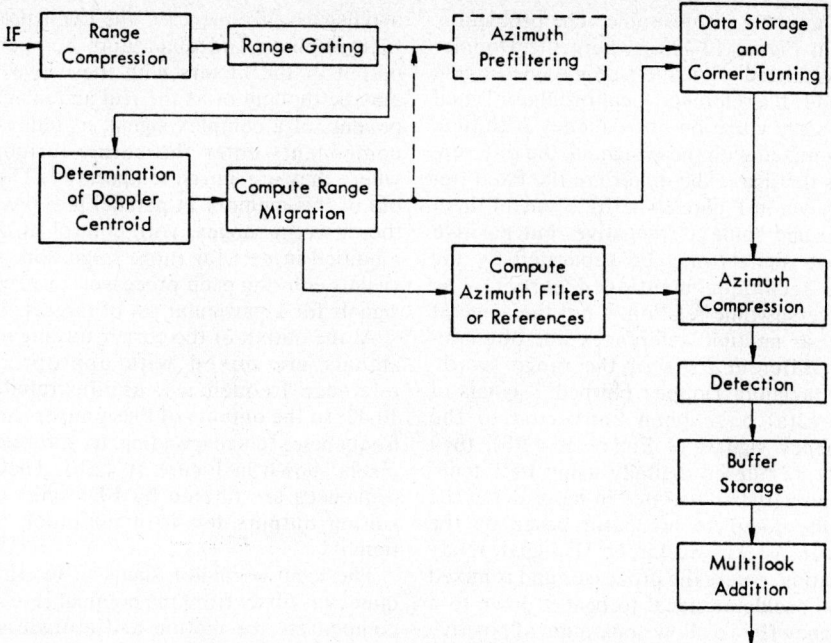

Fig. 10-41. Block diagram for range-gated (azimuth sequential) electronic processor for SAR. Note 1: First step may be sampling and "prf buffering" to reduce overall processing bandwidth. Note 2: A/D conversion is prior to the first digital operation (Ulaby et al., 1982).

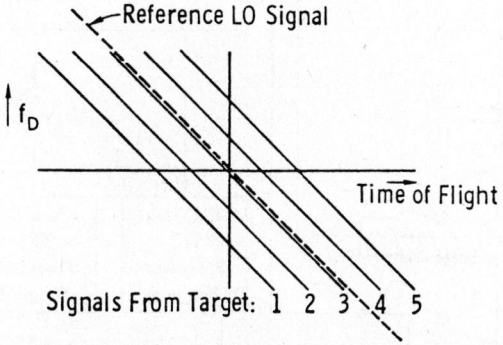

(a) Basic Problem: Signals are shown for five targets at same range. Radar passes abreast of target at time target Doppler frequency is zero.

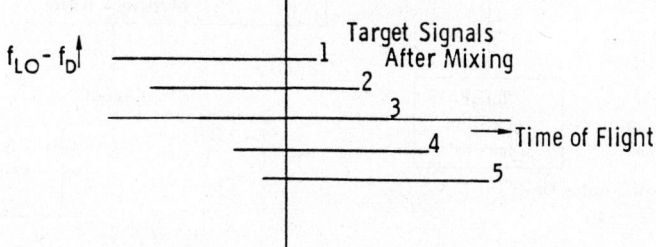

(b) Situation after mixing with a swept LO. LO frequency is same as that of target 3 in (a), except that it sweeps a wider band so it can start when signal 1 starts and end when signal 5 ends.

Fig. 10-42. Doppler shifts for multiple targets (Ulaby et al., 1982).

ment, such a process is desirable. The problem is illustrated in Figure 10-42(a) where the Doppler histories associated with five side-by-side targets are illustrated. If a reference local-oscillator signal having the same variation of frequency with time as shown is mixed with these signals, the different frequencies that leave the mixer are the fixed frequencies shown in Figure 10-42(b). Some of them are positive and some are negative, but positive and negative signals may be separated by the quadrature technique mentioned earlier. The frequency-versus-time relation is not the same at all ranges, so multiple references are often required for different parts of the range swath. When the "azimuth Doppler-chirped" signals of Figure 10-42(a) have been converted to the fixed-frequency signals of Figure 10-42(b), they may readily be filtered digitally using FFT techniques. Figure 10-43 illustrates in more detail the nature of the complete processor based on the swept references. The dechirped IF signal, ready for range gating, enters the processor and is mixed with a local oscillator signal to beat it down to a zero-frequency IF. To allow separation of positive

and negative frequencies, the LO is provided with both in-phase and quadrature components. The output of the mixers with these two LO inputs may be thought of as the real and imaginary components of a complex signal, as indicated. These components enter the corner-turning memory where they are stored temporarily. They are read out of this memory in parallel into processors for the different ranges, with control of the readout modified in part for range migration, so that the signals entering each processor correspond to the signals for a particular set of targets.

At the output of the corner-turning memory the signals are mixed with appropriate swept-reference frequencies, as illustrated in Figure 10-42; so the outputs of these mixers are the fixed frequencies corresponding to different azimuth pixels shown in Figure 10-42(b). These complex sequences are filtered by FFT units and the resulting outputs fed to a multilook buffer (optional).

The local oscillator signal at intermediate frequency is offset from the nominal IF, as shown, to compensate for motion and attitude errors. The

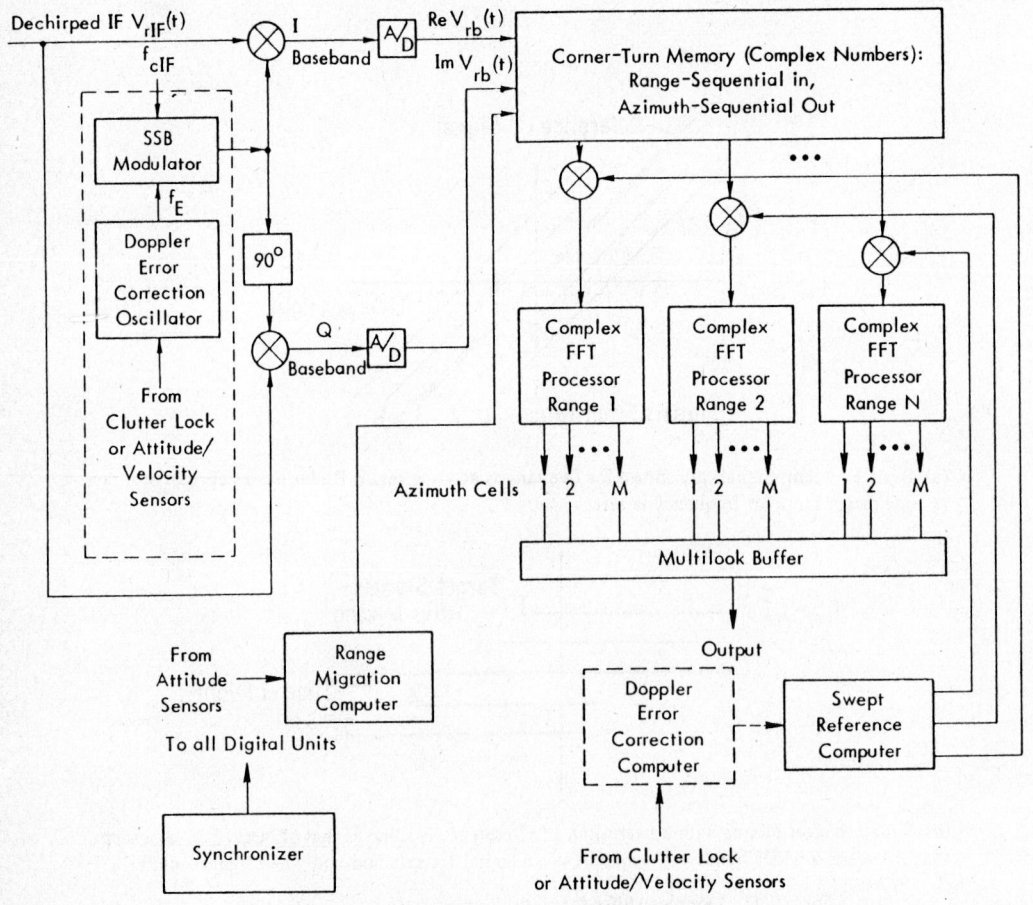

Fig. 10-43. Block diagram of zero-offset digital processor. N range cells; M azimuth cells in L_p (Ulaby et al., 1982).

dashed lines indicate that this correction could also be made in the swept-reference signals rather than in the IF local oscillator.

Matched-Filter Processing

One way to accomplish the matched-filter processing discussed above is shown in Figure 10-44. Matched filtering can also be thought of as a correlation process between the received signal and a reference function, as shown in Figure 10-44. However, if this were implemented as shown, one would need reference functions for all of the azimuth pixels being processed simultaneously. This problem can be overcome by a frequency-domain approach. The input signal $V_s(t)$ is Fourier transformed into $V_s(f)$. Similarly, the reference function is transformed into $V_r^*(f)$. The complex conjugate is indicated because the correlation operation is equivalent to a convolution in the time domain. The two frequency-domain functions are multiplied together and the inverse transform performed on the product. The resulting output $V_o(t)$ is a time-sequential record of the outputs corresponding to different azimuth positions.

SPECKLE REDUCTION IN SAR

Speckle is a more severe problem with SAR than it is with RAR, where it was first discussed. Ordinary SAR-processing gives only a single sample of each pixel, so the speckle is described by the exponential distribution for the intensity of the fading signal, as discussed in Chapter 9. The RAR usually has a resolution greater than $D/2$, so Eq. 10-13 shows that it averages several independent samples for each pixel; the only exception occurs when the spot on the recorder is too small to permit overlap of these samples and consequent summation by the film. Because of the approximate 20-dB dynamic range of film, the 18-dB 90-percent range of fading makes the SAR's fully focused image appear to be a binary speckled image. To see shades of gray, one must stand back

from the image in order to let one's eye perform the averaging for several pixels.

The solution to this problem with SAR is to average several fully focused pixels together or to use multiple-look images with poorer resolution. The expression of Eq. 10-13 is valid for SAR as well as for RAR:

$$N_i = 2r_a/D \qquad (10\text{-}13)$$

Since for SAR the fully focused resolution is

$$r_{amin} = D/2, \qquad (10\text{-}97)$$

as can be seen from Eq. 10-40 and Eq. 10-67 (which were for the fully focused situation), if one selects a value for r_a greater than $D/2$, one can attain more independent looks at the target,

$$N_i = r_a/r_{amin}. \qquad (10\text{-}98)$$

This can be achieved by using a shorter aperture than

$$L = \beta_h R$$

in the expression for the azimuth resolution

$$r_a = \frac{\lambda}{2L} R. \qquad (10\text{-}45)$$

Figure 10-45 illustrates this. In Figure 10-45(a), an example is shown for an antenna 4 m long. The best possible resolution is 2 m ($D/2$), but only one independent sample is available at this resolution. With a 4-m resolution two samples are available, with an 8-m resolution four samples, and with a 16-m resolution eight samples. How do we achieve these independent samples ("looks")? Figure 10-45(b) shows one way—simply process for a 2-m resolution and average four pixels together in the final image. This requires four apertures of length L_p, where L_p is the longest possible aperture, $\beta_h R$.

Processing for the finest possible resolution is difficult, uses more equipment in electronic processing, and imposes more stringent requirements on the radar, navigation equipment, and optical processor than does processing for a poorer resolution. A solution without these negative factors is indicated in Figure 10-45(c). The possible aperture is divided into four subapertures, each of length $L_p/4$. For the pixel shown, these apertures are processed sequentially, with each "squinted" the appropriate amount. With this technique, the amount of processing required is reduced and the advantage of averaging four looks is achieved.

The amount of speckle reduction was discussed in terms of the gray-scale resolution r_g defined in Eq. 10-22 and used in establishing the "resolution volume"

$$V = r_y r_a r_g \qquad (10\text{-}20)$$

that relates to interpretability by

$$I = I_o e^{-V/V_o}. \qquad (10\text{-}21)$$

Although one can see intuitively that reducing

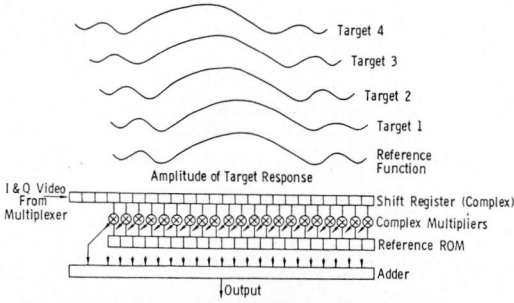

Fig. 10-44. Correlation processing: moving signal past fixed reference. Target 1 signal shown correlated. All signals simultaneously present. Single range cell components shown (Ulaby et al., 1982).

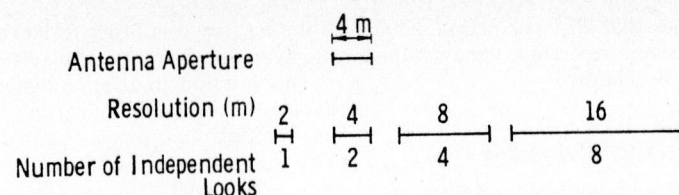

(a) Example of 4-m Antenna Showing Potential Number of Independent Looks

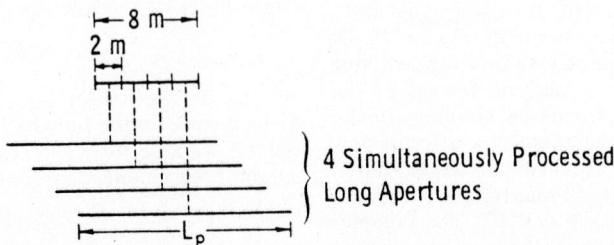

(b) Obtaining 4 Samples by Processing to Finest Possible Resolution Pulses Used Per Aperture, N_p. Total Processing Steps Proportional to $4N_p$.

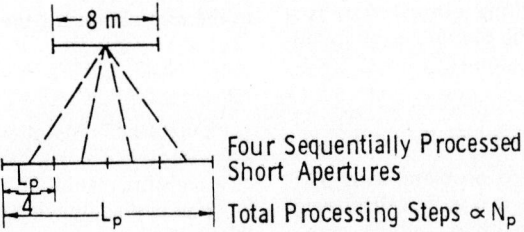

(c) Obtaining 4 Samples by Processing to Final Resolution—Pulses Used Per Aperture $N_p/4$

Fig. 10-45. Obtaining independent looks with SAR (Ulaby et al., 1982).

the speckle will improve the picture, one cannot intuitively determine the tradeoff between this improvement and the degradation of resolution. Moore (1979) examined this tradeoff and showed that one can establish a figure of merit for the visual-interpretation task in his experiment, and that this figure of merit shows that the optimum use of the aperture from an interpretability standpoint is attained by dividing it into three parts and averaging the resulting three independent samples. Since the cross-track resolution does not enter into this discussion, the F of M is simply the ratio of the resolution volume V to r_y times r_{amin}. That is, it is given by

$$F \text{ of } M = V/r_y r_{amin} = \frac{r_a}{r_{amin}} r_g(N) = N r_g(N)$$

This function is plotted in Figure 10-46. The best F of M has the lowest value, so one can see that two or three samples would be equally good for interpretation, but the average of three samples, having less speckle, would be more attractive. One could degrade the resolution by a factor of nine and average nine samples and still achieve the same interpretability as the fully focused SAR—but with a great simplification in the pro-

cessing. This summation is performed in the multilook buffer indicated in the block diagrams of Figures 10-41 and 10-44.

Note that this F of M and the gray-scale resolution were derived for a visual interpretation task. If machine interpretation were to be used, or if

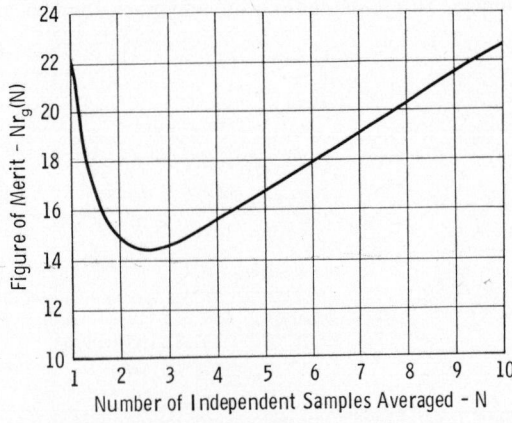

Fig. 10-46. Figure of merit for processing different numbers of samples with fixed antenna length (from Moore, 1979).

one needed to obtain precise measurements of the scattering coefficient over some area, the requirements might be different.

MULTIPOLARIZATION AND MULTISPECTRAL SAR

Multispectral sensing in the visible and infrared regions of the spectrum, followed by pattern recognition based in part on spectral signatures, has proven to be a useful concept for earth resource surveys. Although those physical mechanisms which lead to a rich spectral signature structure in the visible and near-visible regions are quite different from the mechanisms which provide spectral signature structure at microwavelengths, investigations of the potential of multispectral imaging radars show that similar benefits accrue.

Single Wavelength Multipolarization SLAR

The simplest extension of the single-channel SLAR operates at one nominal wavelength, but has a multipolarization capability. Most SLAR systems built to date have utilized linearly polarized airborne antennas usually configured to radiate with horizontal (H) polarization. For H-polarization transmission, it has been determined that various classes of terrain *depolarize* the reflected signal to varying degrees, with the consequence that the reflected signal observed at the radar receiving antenna has both vertically V- and H-polarized components. For scattering from a rough surface, the amount of depolarization is a function of the surface roughness as well as of the dielectric constant and the conductivity of the surface.

If a vertically polarized receiving antenna and a second receiver channel are added, it is possible to simultaneously collect the cross-polarized (HV) return and the parallel-polarized (HH) return from the terrain strip under observation.

In a high prf system, for which

$$f_s \geqslant 2u/D,$$

it is possible to switch a single receiver channel back and forth between horizontally and vertically polarized receiving antennas (while transmitting H-polarization), and to separately store the HH and HV returns for later processing into HH and HV images. Alternately, V-polarization could be transmitted, and both VV and VH returns could be collected.

At a higher level of complexity, it is possible to switch between H- and V-transmissions on alternate pulses, and to receive both H- and V-returns for each transmission. In this manner, four images depicting the HH, HV, VH, and VV returns are obtainable. Only three of these are expected to be independent since, as a result of reciprocity,

$$HV \approx VH.$$

In a somewhat different approach, it is possible to radiate a circularly polarized electromagnetic wave, and to receive circularly polarized returns of both senses.

Any of these polarization approaches makes it possible to generate two or three *independent* images and to use the polarization signatures of the terrain to aid in the interpretation of the data. The polarization signature can be examined by individual resolution cells if desired, or may alternately be examined for average properties over large extents.

A more general form of SLAR provides for the transmission of radar signals at two or more distinct operating wavelengths, sufficiently far apart that terrain reflectivities at the various wavelengths will display systematic differences. Such a multifrequency feature can be combined with the multipolarization feature described in the previous paragraph. A multispectral SLAR system with M wavelengths and N independent polarizations per wavelength would then yield (MN) independent images of the terrain under scrutiny, opening the door to interpretation techniques which capitalize upon spectral and polarization signatures as well as spatial structure. If time multiplexing of channels is used in the design of such a system, it must be arranged in such a manner that, for each wavelength and each polarization, the effective sampling rate of the respective Doppler signals is adequate to avoid aliasing of the Doppler spectrum (for SAR but not RAR).

SOME REPRESENTATIVE SAR SYSTEMS AND RESULTS

A variety of SAR systems capable of generating radar images useful for remote sensing applications has come into existence since the late 1950's. These range from highly experimental one-of-a-kind systems, such as those assembled by the Environmental Research Institute of Michigan ("ERIM," formerly the Willow Run Laboratories of the University of Michigan) and operated by both ERIM and the Canada Centre for Remote Sensing, to military operational systems exemplified by the AN/APQ-102, fabricated by Goodyear Aerospace Corporation. More recently a 25-cm wavelength system was flown on Seasat, and in slightly modified form, on shuttle. A few of these SAR systems are described very briefly in this section, and representative image samples are shown.

The AN/APQ-102

The AN/APQ-102 radar system is an X-band (3-cm) SAR using a horizontally polarized antenna for both transmission and reception of the radar signal. This operational system was fabricated in production quantities and was installed in the RF4C aircraft. It relies upon a CRT-film type photographic recorder in the aircraft, and employs a coherent optical processor to convert the photo-

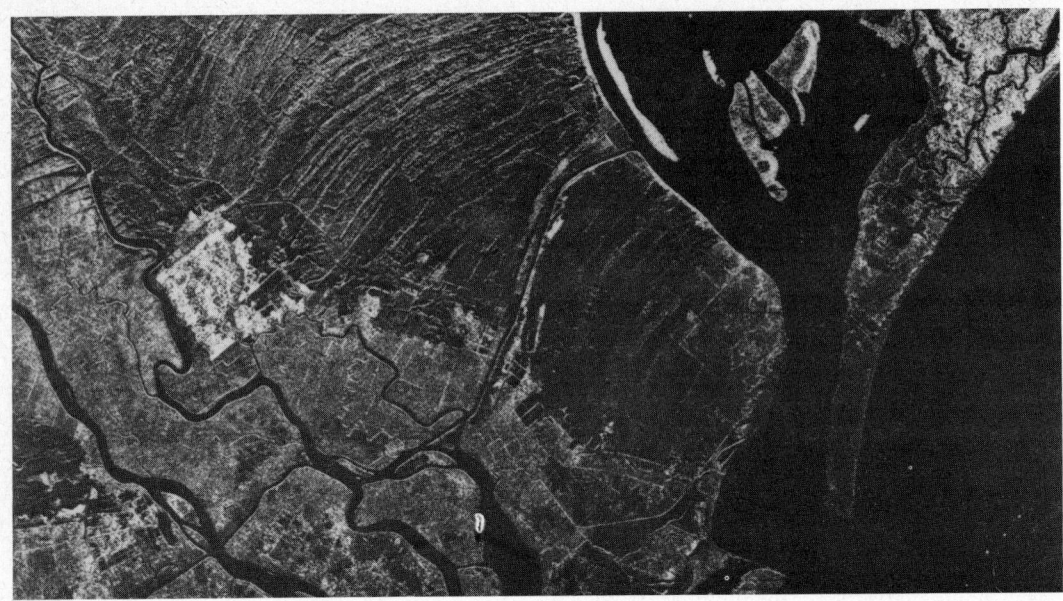

Fig. 10-47. AN/APQ 102 imagery (courtesy, Goodyear Aerospace Corp.).

graphically stored radar signals into an output image which is recorded on a second photographic film.

Large areas of the continental United States have been imaged with this system and imagery with nominal resolution on the order of 10 meters in both range and azimuth has been provided to geoscientists for preliminary analysis. Figure 10-47 is an example of AN/APQ-102 radar imagery (courtesy Goodyear Aerospace Corporation).

The Goodyear Electronic Mapping System

The Goodyear Electronic Mapping System (GEMS), a variation on the AN/APQ-102 radar

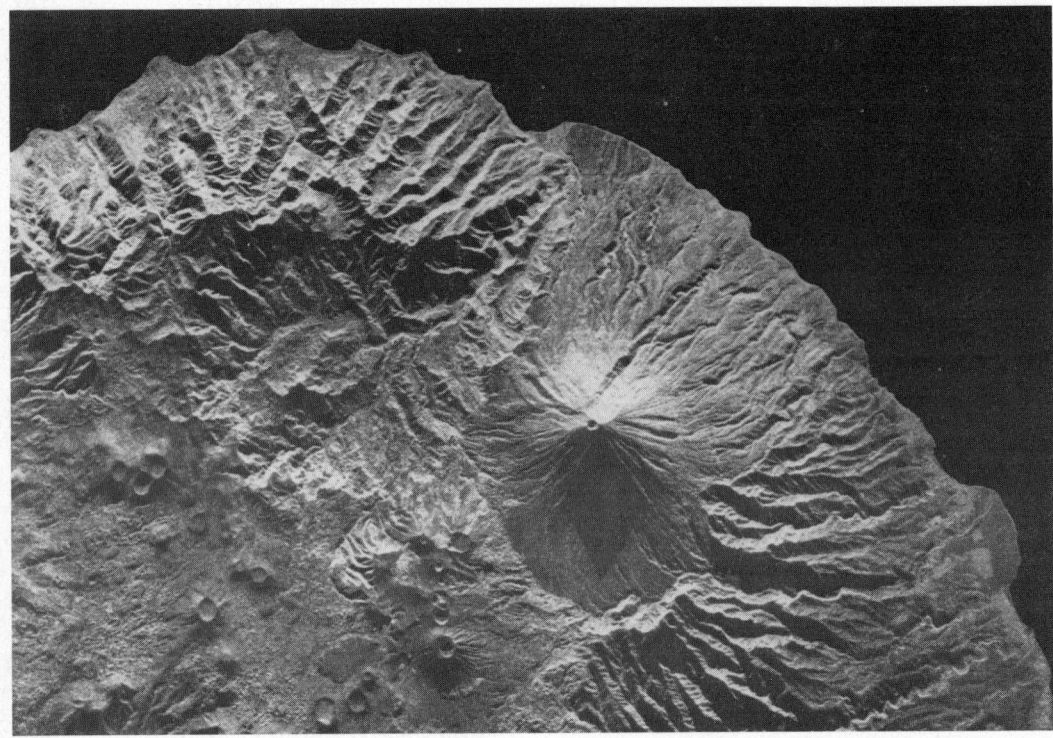

Fig. 10-48. GEMS imagery of a volcano (courtesy, Goodyear Aerospace Corp.).

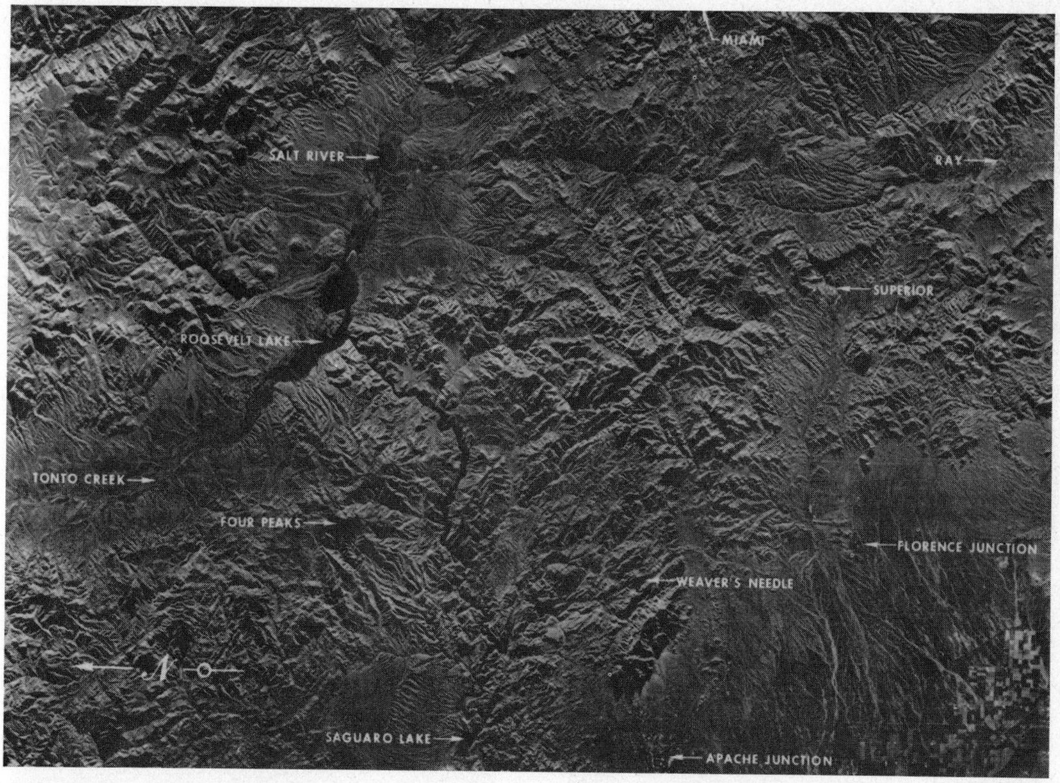

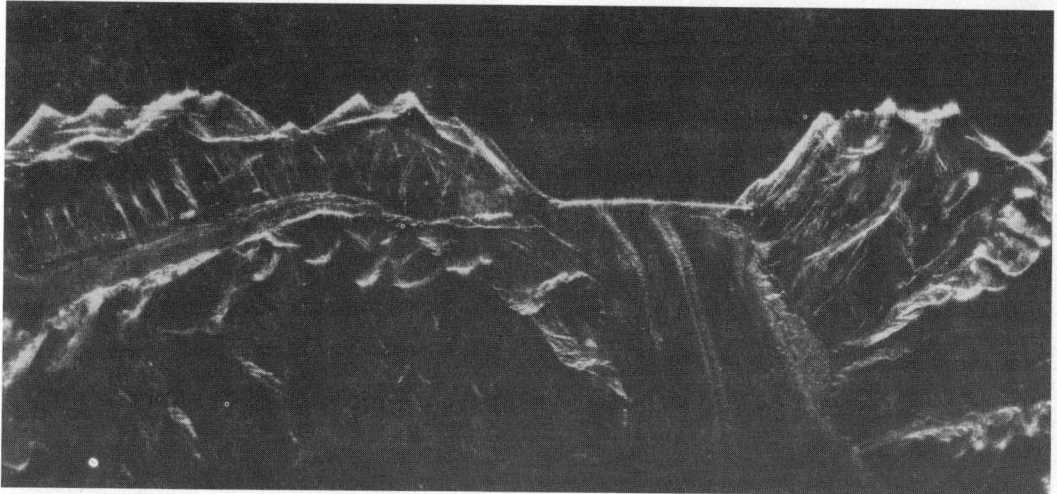

Fig. 10-49. GEMS imagery of Arizona Salt River Storage system, (courtesy, Goodyear Aerospace Corp.).

system, was built by Goodyear Aerospace Corporation and installed in a Caravelle aircraft. This system has engaged in earth resource survey missions in many parts of the world. The radar antenna is horizontally polarized as for the APQ-102; the system relies upon a CRT-film recorder, and employs a coherent optical data processor. The resolution of the imagery generated by this system is on the general order of 10 meters in both range and azimuth.

Figures 10-48 and 10-49 are examples of X-band (3-cm) imagery generated by the GEMS radar (courtesy Goodyear Aerospace Corporation).

The JPL L-Band Radar

The Jet Propulsion Laboratory of Pasadena, California, has fabricated a lightweight L-band (25-cm) SAR system which has been flown on a variety of earth resource survey missions, largely in Arctic and sub-Arctic regions. For most tests,

Fig. 10-50a. JPL L-band radar image, Kaskawulsh Glacier, H-H polarization; note profiling due to steep incidence angle (courtesy, JPL).

Fig. 10-50b. JPL L-band radar image, Kaskawulsh Glacier, H-V polarization, (courtesy, JPL).

this system has been operated on the NASA CV-990 aircraft. The system can transmit with either horizontal or vertical polarization, and is capable of receiving both horizontally and vertically polarized reflected signals. CRT/film recording and optical processing are used to generate the final radar image.

Figure 10-50a and b are typical examples of a pair of parallel- and cross-polarized radar images generated by the JPL L-band system. Nominal

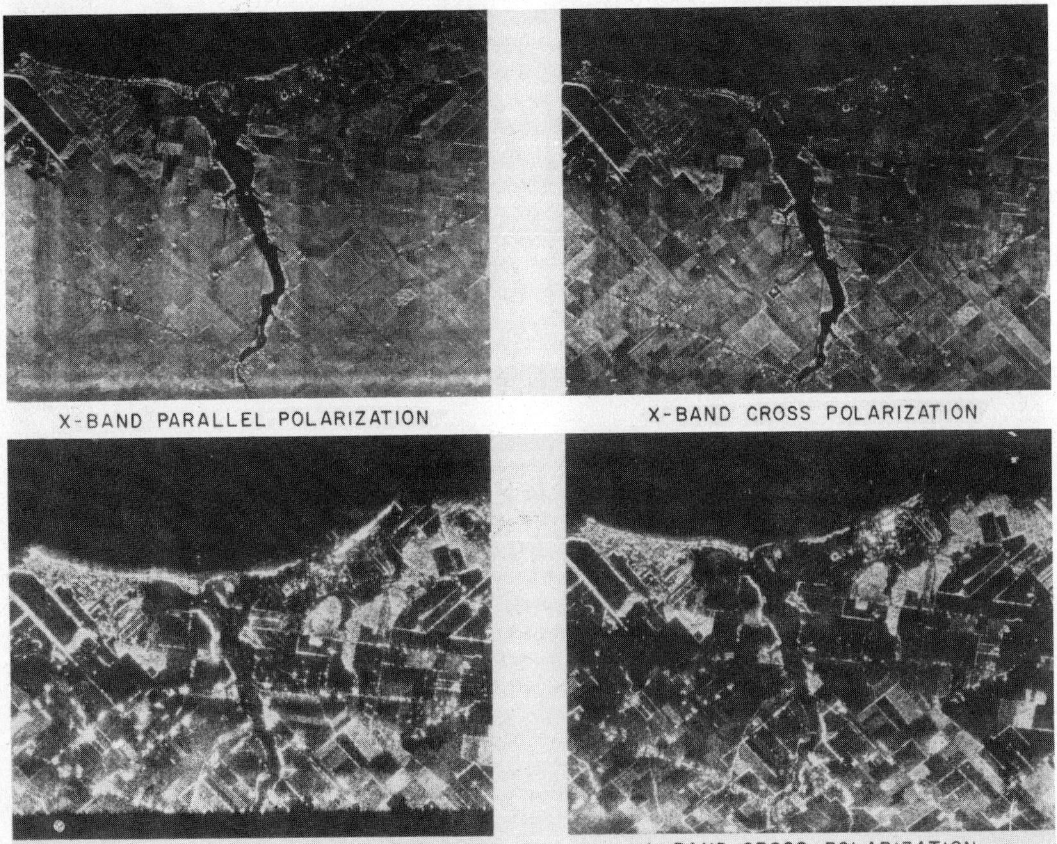

X-BAND PARALLEL POLARIZATION

X-BAND CROSS POLARIZATION

L-BAND PARALLEL POLARIZATION

L-BAND CROSS POLARIZATION

Fig. 10-51. Simultaneous X-L Imagery Enrico Fermi Power Plant, Monroe, Michigan (April 5, 1973, courtesy ERIM).

IMAGING RADAR SYSTEMS 469

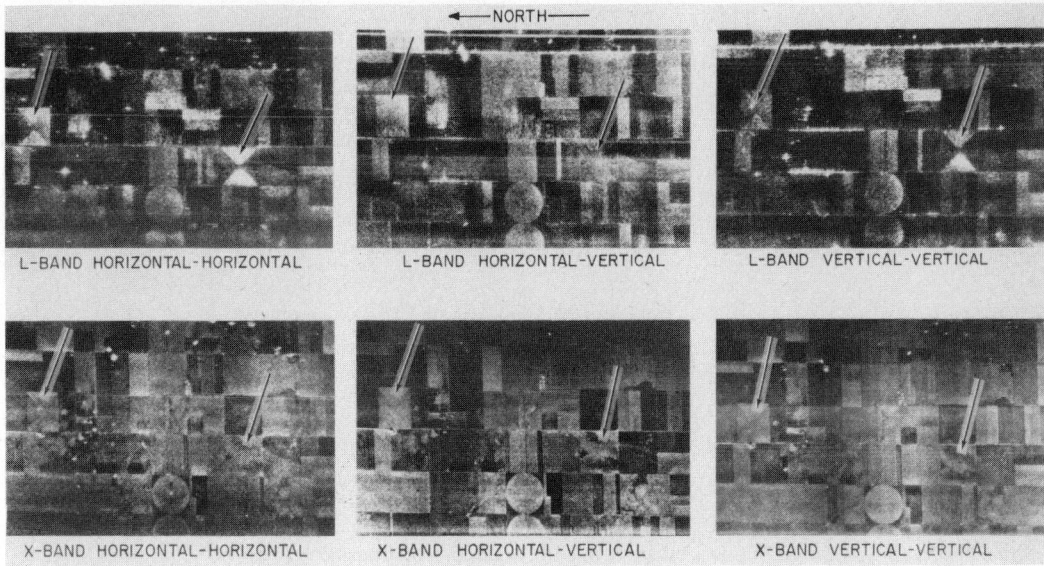

L-BAND HORIZONTAL-HORIZONTAL   L-BAND HORIZONTAL-VERTICAL   L-BAND VERTICAL-VERTICAL

X-BAND HORIZONTAL-HORIZONTAL   X-BAND HORIZONTAL-VERTICAL   X-BAND VERTICAL-VERTICAL

Fig. 10-52. Two-wavelength, three-polarization imagery, agricultural test site, Garden City, Kansas.

Fig. 10-53. VHF SAR image from Apollo 17.

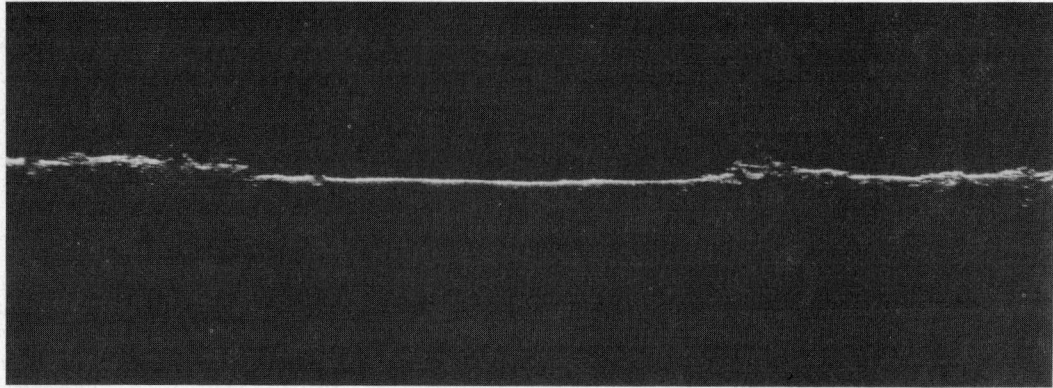

Fig. 10-54. HF SAR image from Apollo 17.

resolution of the radar is on the order of 20 meters.

The ERIM/CCRS Multispectral Radar System

It has been shown above that a SAR system can be configured with multiple channels to provide a multifrequency, multipolarization capability. This type of capability is a prerequisite to implementing a radar remote sensing system using spectral recognition techniques. The ERIM multispectral SAR system, installed in a Convair 580 aircraft, is capable of operating with six channels to generate, essentially simultaneously, two X-band

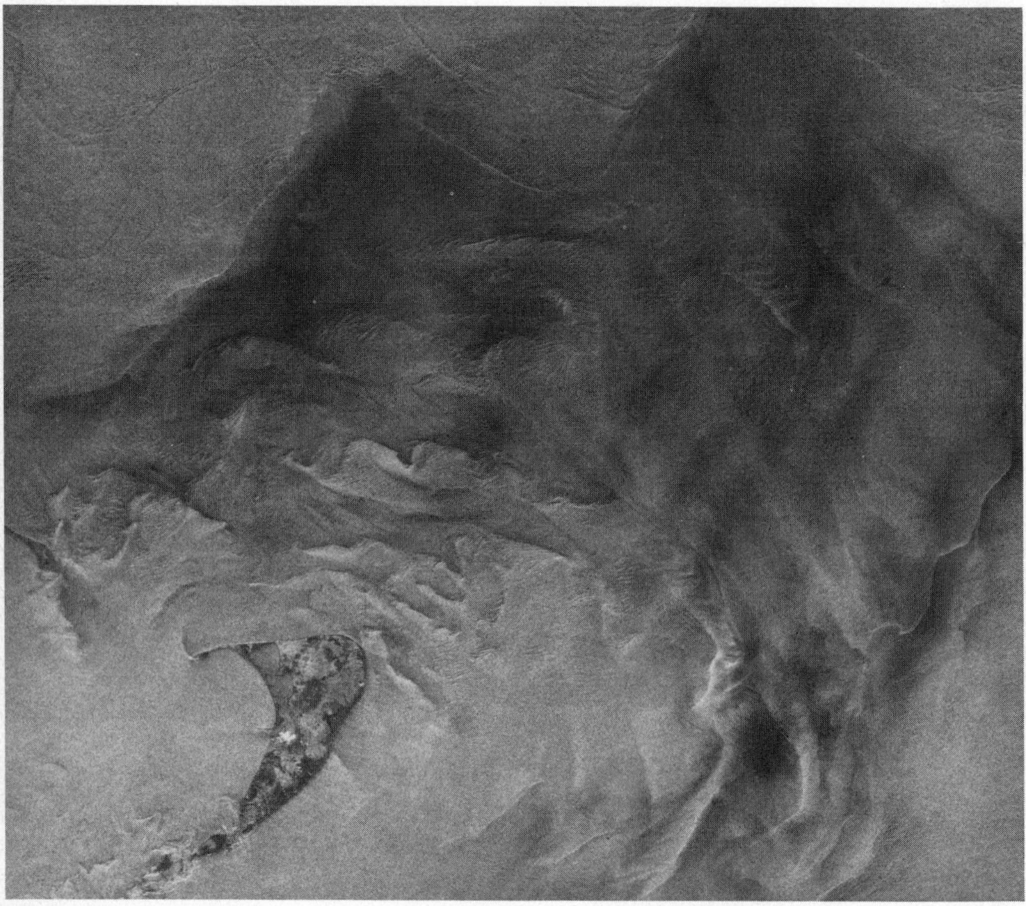

Fig. 10-55. Seasat SAR image of ocean surrounding Nantucket Island. The structure shown indicates the way radar views the surface expression of the undersea topography and upwelling (image courtesy of JPL, Pasadena, Calif.).

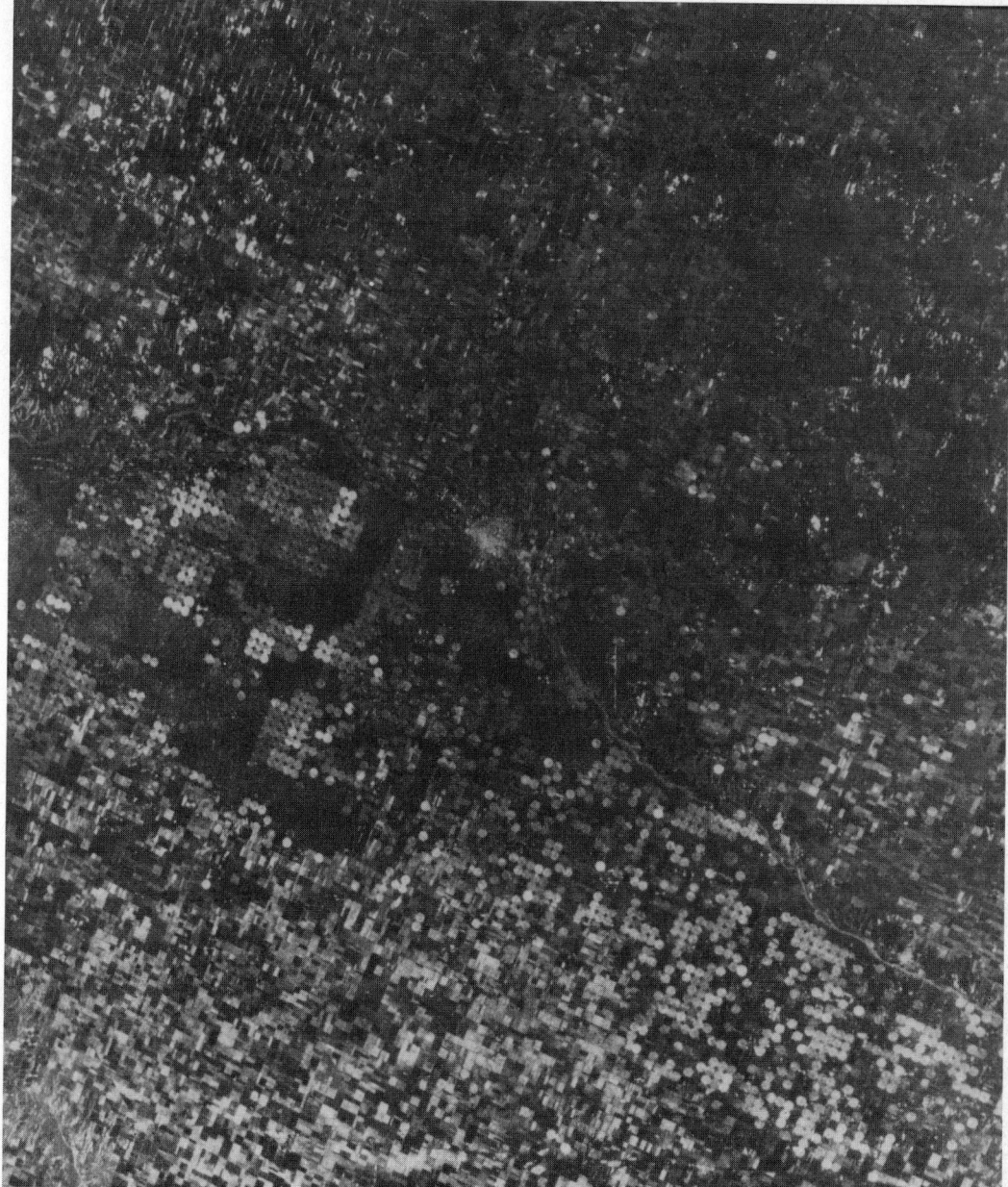

Fig. 10-56. Seasat image showing an area of southwest Kansas, centered approximately on Garden City, Kansas. The image is dominated by circular and rectangular patterns common in the agricultural areas of the Great Plains (image courtesy of JPL, Pasadena, Calif.).

(3-cm), two L-band (24-cm) and two C-band (5.5 cm) images of a terrain strip. It is possible at each wavelength to illuminate with either horizontal or vertical polarization, and to receive both the parallel- and the cross-polarized return.

Figure 10-51 shows a set of four-channel images simultaneously generated by the ERIM system along the Lake Erie shoreline in southeastern Michigan. The transmitted signal was horizontally polarized on each wavelength, with both received polarizations displayed on each.

Figure 10-52 shows a set of six images, generated earlier by the ERIM system over the Garden City, Kansas, Agricultural Test Site. In this family of images, a terrain region is displayed at each of the two wavelengths, with *HH*, *HV*, and *VV* polarization. The first letter of this symbology denotes transmitted-signal polarization; the second letter stands for received polarization. These images were not generated simultaneously, but were produced from four successive passes of the radar, each pass producing a single-wavelength,

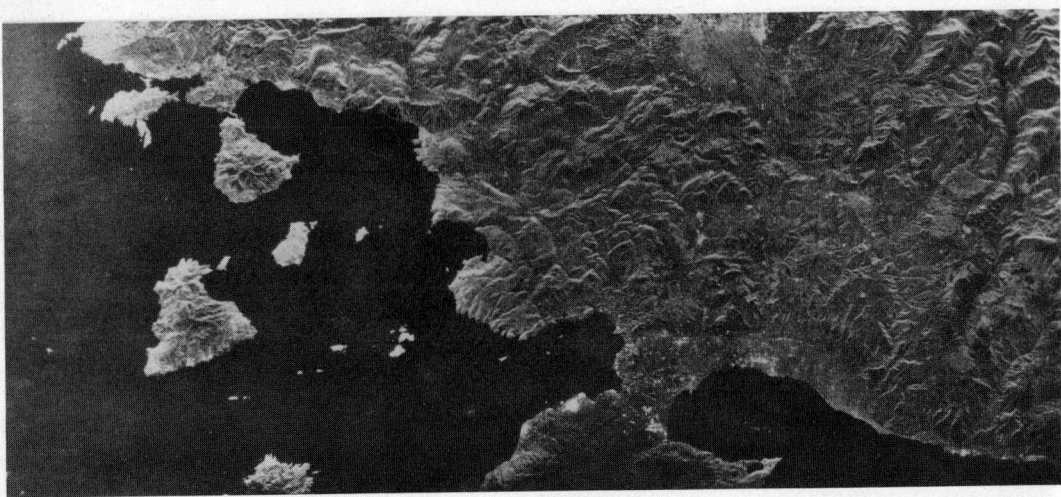

Fig. 10-57. Image showing northern Peloponnesia and part of southern Greece, acquired by JPL's Shuttle Imaging Radar-A (SIR-A) experiment aboard the NASA Space Shuttle Columbia on Nov. 12-14, 1981. The Corinthian Canal is visible as a bright line cutting across the narrow Corinth Isthmus (upper center). The black area to the right is the smooth water of the Aegean Sea; on the left, the Gulf of Corinth. Islands to the right, starting at the top, are Salamis, Aegina, and Angistrion, and the Peninsula of Methana. Southwest of the canal on the gulf coast is the city of Corinth, appearing as bright, white spots. The image covers an area approximately 50 × 100 km (image courtesy of JPL, Pasadena, Calif.).

dual-polarization view. All the imagery shown has a nominal resolution of about 10 meters in both slant-range and azimuth.

The Apollo Lunar Sounder System

Apollo 17 carried with it a three-wavelength SAR system designed to probe the surface and subsurface geologic structure of the moon, from lunar orbit. This instrument, designated as the S-209 experiment package, was the first multi-spectral SAR system to go into space. The constraints of the lunar mission precluded operating this system in earth orbit, although a modified prototype of the S-209 package was flight tested in a KC-135 aircraft prior to the lunar flight. The system was capable of operating simultaneously

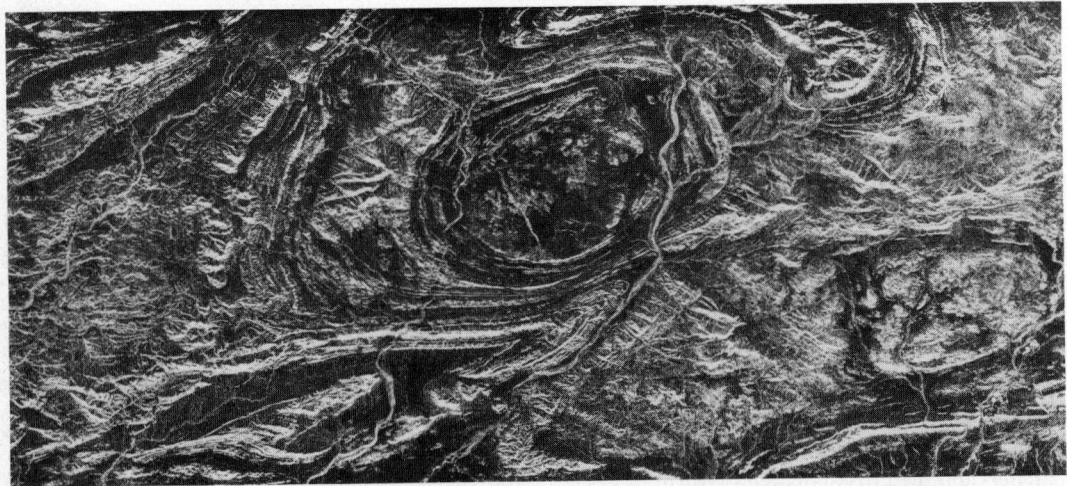

Fig. 10-58. This image, showing the Hamersley Mountain Range in western Australia was acquired by the Shuttle Imaging Radar-A (SIR-A) on Nov. 12-14, 1981. A circular pattern of eroded folds surrounds a prominent granite dome, remnants of a volcanic past, seen in the center of the image. The area displays subtle topography and sparse vegetation. Volcanic rocks and sediments exposed at the surface are ancient (1.5 billion years old). The image covers a 50- × 100-km area of ancient volcanic terrain potentially rich in metal deposits. The Hardey River is seen running vertically to the right of the center circular dome, and the small town of Paraburdoo appears as a patch of tiny, bright rectangles in the lower right corner (image courtesy of JPL, Pasadena, Calif.).

Fig. 10-59. SIR-A image of northern New Jersey and the New York City Burroughs of Richmond (Staten Island) and Brooklyn, acquired by the NASA Space Shuttle Columbia during its Nov. 12-14, 1981 flight. The high returns seen as bright shapes in the upper right-hand corner are, from left to right, Brooklyn, with Coney Island directly below; and the Rockaway and Long Beach areas. The two small bright dots above and to the left of Brooklyn are returns from the Verrazano-Narrows Bridge. Staten Island appears in the upper middle region of the image, and Sandy Hook is visible as the bright land-spit across the Bay from and directly below Brooklyn (image courtesy of JPL, Pasadena, Calif.).

at wavelengths of 20 and 60 meters, or alternatively as a single-wavelength, two-meter SAR system, with a single linear polarization being available at each wavelength. The system was capable of generating images with nominal slant-range and azimuth resolutions r_R and r_a approximating 10λ at each of the three wavelengths. Since the primary emphasis of the S-209 package was on subsurface sounding, with a secondary emphasis on lunar profiling, the system was operated at a steep depression angle not conducive to specific terrain imaging.

The S-209 system relied upon a CRT-film recorder carried on board the spacecraft. The film cassette was recovered by an astronaut by means of EVA during the trans-Earth portion of the Apollo 17 mission. A coherent optical processor was used to generate image transparencies and image holograms. In addition, radar images were digitized, using an image dissector, for later analysis by members of the S-209 experiment team.

Figures 10-53 and 10-54 are a pair of images generated by the Apollo Lunar Sounder during the lunar orbit portion of the Apollo 17 mission. Analysis of S-209 imagery is discussed further in Chapter 17, *Data Management and Information Systems.*

Seasat and Shuttle SARs

An L-band (24-cm wavelength) SAR was flown on the short-lived Seasat spacecraft in 1978, and a modified SAR was flown on the Space Shuttle in 1981. These systems were outgrowths of the JPL L-band radar mentioned above. The Seasat SAR had a nominal resolution of 25 × 25 m, with four independent looks averaged, although it could also be used in a fully focused mode with resolution of 6.25 × 25 m. The system operated at a steep incidence angle of 20° from the vertical, with coverage from about 17.5° to 22.5°. The most striking feature of the system itself was the 3- × 12-m antenna, such a large structure being necessary because of the long wavelength and ambiguity constraints. With this antenna and incidence angle, the system could image a swath 100 km wide to the side of the spacecraft.

The Seasat signals were telemetered to earth for processing. Both optical and digital processing were used at various places in the world. The results over both ocean and land were spectacular, as indicated by the examples shown in Figures 10-55 and 10-56.

The modified SAR flown on the Shuttle was pointed to a 47° angle of incidence and was intended primarily for geological studies. At this angle, the swathwidth is significantly narrower than at the 20° angle used on Seasat. Examples of images from this radar are shown in Figures 10-58 and 10-59.

REFERENCES

Brown, W. M., 1967 (March), Synthetic aperture radar; IEEE Trans. Aerosp. Electron. Syst., vol. AES-3, pp. 217–229.

Brown, W. M., and L. J. Porcello, 1969 (September),

An introduction to synthetic-aperture radar; IEEE Spectrum, pp. 52–62.

Cook, C. E., and M. Bernfeld, 1967, Radar Signals: An Introduction to Theory and Application; Academic Press, New York.

Cutrona, L. J., 1970, Synthetic-aperture radar; Chapter 23, Radar Handbook (M. I. Skolnik, ed.), McGraw-Hill, New York.

Cutrona, L. J., et al., 1961 (April), A high-resolution radar combat-surveillance system; IRE Trans. on Milit. Electron., pp. 127–131.

Cutrona, L. J., et al., 1966, On the application of coherent optical processing techniques to synthetic-aperture radar; Proc. IEEE, vol. 54, pp. 1026–1032.

Harger, R. O., 1965, An optimum design of ambiguity function, antenna pattern, and signal for side-looking radars; IEEE Trans. Milit. Electron., vol. MIL-9, pp. 264–278.

Harger, R. O., 1970, Synthetic Aperture Radar Systems: Theory and Design; Academic Press, New York.

Kozma, A., E. N. Leith, and N. G. Massey, 1972, Tilted-plane optical processor; Appl. Opt., vol. 11, pp. 1766–1777.

Leith, E. N., 1968, Optical processing techniques for simultaneous pulse compression and beam sharpening; IEEE Trans. Aerosp. Electron. Syst., vol. AES-4, pp. 879–885.

MacDonald, H. C., and W. P. Waite, 1971, Optimum radar depression angles for geological analysis; Mod. Geol., vol. 2, pp. 179–193.

Moore, R. K., 1979 (September), Tradeoff between picture element dimensions and non-coherent averaging in side-looking airborne radar; IEEE Trans. Aerosp. Electron. Syst., vol. AES-15, no. 5, pp. 697–708.

Moore, R. K., J. P. Claassen, and Y. H. Lin, 1981, Scanning spaceborne synthetic aperture radar with integrated radiometer; IEEE Trans. Aerosp. Electron. Syst., vol. AES-17, no. 3, pp. 410–421.

Porcello, L. J., 1970, Turbulence-induced phase errors in synthetic-aperture radars; IEEE Trans. Aerosp. Electron. Syst., vol. AES-6, pp. 636–644.

Skolnik, M. I., ed., 1970, Radar Handbook; McGraw-Hill, New York.

Skolnik, M. I., 1980, Introduction to Radar Systems; 2nd ed., McGraw-Hill, New York.

Thomann, G. C., 1970, Panchromatic illumination for radar; acoustic simulation of panchromatic radar; Ph.D. Thesis, Univ. of Kansas; also CRES Tech. Rept. 177-11, Univ. of Kansas, Lawrence, KS.

Tomiyasu, K., 1981 (April), Conceptual performance of a satellite-borne, wide-swath synthetic aperture radar; IEEE Trans. Geosci. Rem. Sens., vol. GRS-19, no. 2, pp. 108–116.

Ulaby, F. T., R. K. Moore, and A. K. Fung, 1982, Microwave Remote Sensing: Active and Passive, Vol. II, Radar remote sensing and surface scattering and emission theory; Addison-Wesley, Reading, Mass.

Waite, W. P., 1970, Broad-spectrum electromagnetic backscatter; Ph.D. Thesis, Univ. of Kansas; also CRES Tech. Rept. 133-17, Univ. of Kansas, Lawrence, KS.

Wiley, C. A., 1967, personal communication.

CHAPTER 11

Passive Microwave Radiometry

Authors: FAWWAZ T. ULABY and KEITH R. CARVER

GENERAL CONTENTS: Antenna fundamentals; Radiative transfer fundamentals; Radiometer receivers; Basic receiver operation; Modulation techniques; Noise sources; Advances in receiver performance; Radiometer calibration; Power transfer through microwave radiometer networks; Receiver calibration; Radiometer radiative transfer equation; Antenna loss correction factors; Antenna loss calibration techniques; Imaging considerations; Examples of radiometric recordings.

NOMENCLATURE

To conserve and eliminate repetition in text and references, the following symbols, units and names have been used in this chapter.

Symbol	SI Units	Name
A	—	argon
A_e	m²	maximum effective aperture
$A_e(\theta,\phi)$	m²	effective aperture function
b	V	intercept
B	Hz	bandwidth
B_{bb}	W m⁻² Hz⁻¹ sr⁻¹	brightness of blackbody
$B(\theta,\phi)$	W m⁻² Hz⁻¹ sr⁻¹	brightness
$BWFN$	rad	beamwidth between first nulls
c	3 × 10⁸ m s⁻¹	speed of light
C_d	V W⁻¹	square-law-detector constant
d_c	m	correlation length
$d\Omega$	sr	unit solid angle
D	m	aperture diameter
D	—	directivity
D_{iso}	—	directivity of isotropic antenna
$D(\theta,\phi)$	—	directivity function
e_v, e_h	—	polarized emissivities
$e(\theta,\phi)$	—	emissivity
E	V m⁻¹	electric-field intensity
E_a	—	antenna emission coefficient
E_w	—	emission coefficient of network
ENR	—	excess noise ratio
f	Hz	frequency
f_e	Hz	pulse generator frequency during calibration
f_I	Hz	input RF frequency
f_{LO}	Hz	local oscillator frequency
f_s	Hz	switch frequency
f_{sc}	Hz	scanning frequency
F_c	—	coupling factor
g	—	voltage gain
g_{agc}	—	voltage gain of agc synchronous detector
g_{sig}	—	voltage gain of signal detector in receiver

Symbol	SI Units	Name
G	—	antenna gain (maximum)
G_1, G_2	—	gain of receiver first stages
G_{IF}	—	gain of IF amplifier
G_M	—	conversion loss of mixer
G_{RF}	—	gain of RF amplifier
G_S	V T⁻¹	system gain constant
$G(\theta,\phi)$	—	antenna gain pattern
h	6.63 × 10⁻³⁴ J s	Planck's constant
H	m	radiometer antenna height
$HPBW°$	degree	half-power beamwidth
I_r	A	terminal current
k	1.38 × 10⁻²³ J K⁻¹	Boltzmann's constant
L	—	loss factor
L_A	—	antenna loss factor
L_{TL}	—	transmission-line loss factor
L_w	—	network attenuation factor
m	V K⁻¹	slope
m_{G1}, m_{2L}	—	mismatch factor
M	—	receiver figure of merit
ML	—	mismatch loss
N_2	—	nitrogen
P	—	fractional beam overlap
P	W	power
P_1	W	power input to port 1
P_{av}	W	average power
P_{bb}	W	power from hohlraum
P_{i1}	W	incident power on port 1
P_{i2}	W	incident power on load
P_{in}	W	input power
P_I	W	input noise power
P_L	W	power absorbed by load
P_L'	W	load power due only to noise from generator
P_L''	W	load power due only to noise from connecting network
$P_n(dB)$	—	normalized power pattern (dB)
$P_n(\theta,\phi)$	—	normalized power pattern

475

Symbol	SI Units	Name
P_N	W	noise power generated by connecting network
P_O	W	output noise power
P_{ohmic}	W	ohmic loss power
P_r	W	received power
P_r, max	W	maximum received power
P_{rad}	W	radiated power
P_{r1}	W	reflected power from port 1
P_{r2}	W	reflected power from load
R	m	distance
R	Ω	resistance
R_r	—	power reflection coefficient
R_v, R_h	—	polarized Fresnel power reflection coefficients
S	W m^{-2}	power density
S_{ij}	—	S-parameters
T	K	temperature
T_A	K	antenna temperature
T_a	K	antenna structure temperature
T_{ac}	K	antenna structure temperature during calibration
T_B'	K	integrated brightness temperature
T_{BM}	K	average main-beam temperature
T_{BS}	K	average side-lobe temperature
$T_B(\theta,\phi)$	K	brightness temperature
T_c	K	brightness temperature of cryoload
T_E	K	equivalent noise temperature
T_G	K	available noise temperature from generator
T_H	K	available noise temperature from hot load
T_I	K	temperature of matched resistor
T_{IF}	K	IF amplifier noise temperature
T_{IN}	K	input noise temperature to receiver
T_{IN}^c	K	input noise temperature to receiver from cold load
T_{IN}^h	K	input noise temperature to receiver from hot load
T_M	K	mixer noise temperature
T_N	K	added noise temperature
T_o	K	ambient temperature
T_{REC}	K	receiver noise temperature
T_{REC}'	K	receiver noise temperature at antenna terminals
T_{REF}	K	reference noise temperature
T_{RF}	K	RF amplifier noise temperature

Symbol	SI Units	Name
T_{SE}	K	self-emission temperature
T_{SKY}	K	zenith sky temperature
T_{SYS}	K	system noise temperature
T_{SYS}'	K	system noise temperature at antenna terminals
T_w	K	waveguide structure temperature
T_x	K	temperature of receiver first stage
u	m s^{-1}	velocity of radiometer platform
v_p	m s^{-1}	phase velocity
V_{agc}	V	receiver output voltage of agc synchronous detector
V_c	V	cold-load receiver voltage
V_d	V	square-law-detector voltage
V_{dac}	V	fluctuating component of V_d
V_h	V	hot-load receiver voltage
V_o	V	receiver output voltage
V_{out}	V	receiver dc output voltage
$\bar{V}_{sig}$	V	receiver output voltage of signal synchronous detector
$VSWR$	—	voltage standing-wave ratio
Z_L	Ω	load impedance
Z_o, Z_{o1}, Z_{o2}	Ω	characteristic impedance
α	Np m^{-1}	voltage attenuation constant
β	rad	antenna beamwidth
β	rad m^{-1}	phase constant
Γ_1, Γ_2	—	voltage reflection coefficients into loaded network ports
Γ_G	—	generator voltage reflection coefficient
Γ_L	—	load voltage reflection coefficient
Δ	K	temperature correction term
Δf	Hz	bandwidth
Δ_{LA}	—	error in antenna loss
ΔP_o	W	noise power contributed by noisy device
ΔT	K	total sensitivity of receiver
ΔT_a	K	error in antenna structure temperature
$\Delta T_B'$	K	error in integrated brightness temperature
ΔT_G	K	receiver radiometric temperature uncertainty due to gain fluctuations
ΔT_{min}	K	minimum detectable temperature

Symbol	SI Units	Name
ΔT_N		receiver radiometric temperature uncertainty due to noise fluctuations
ΔT_{SKY}	K	error in sky temperature
ΔX	m	resolution distance
ϵ_M	—	main-beam efficiency
η	—	antenna efficiency
η_{12}	—	network efficiency
θ_o	rad	incidence angle
(θ,ϕ)	rad	spherical angle coordinates
(θ_o,ϕ_o)	rad	antenna pointing direction
θ_s	rad	scan angle
λ	m	wavelength
ρ	—	voltage reflection coefficient
σ	1.80×10^{-8} W m^{-2} K^{-4}	Stefan-Boltzmann constant
σ_d	V	standard deviation of square-law-detector voltage
σ_{ii}	—	like-polarized bistatic scattering coefficient
σ_{ij}	—	cross-polarized surface bistatic scattering coefficient
τ	s	integration time
τ_{12}	—	available transmission factor
τ_s	s	switching period
Y_{GL}	—	transmission coefficient
Ω_A	sr (steradian)	beam solid angle
Ω_M	sr	main-lobe solid angle
Ω_s	sr	side-lobe solid angle

INTRODUCTION

Since the launch of a four-band microwave radiometer on board the USSR Satellite Cosmos 243 in 1968, over 13 multi-frequency or imaging microwave radiometers have flown in space for earth observations. For satellite-borne radiometer antennas with dimensions of the order of meters, the spatial resolution typically is of the order of tens of kilometers. Hence, the use of satellite microwave radiometers has so far been limited primarily to atmospheric and oceanographic applications and a few land applications where the scene parameters of interest are characterized by low spatial-frequency variations. Several studies are being conducted, however, to determine the feasibility of placing large antenna structures in space. When such structures become available, probably in the 1990s, the use of satellite microwave radiometers will undoubtedly expand to a wider range of land applications. Some of these applications are feasible with today's technology from aircraft platforms. Examples include the monitoring of soil moisture content, floodplain delineation, and canal seepage; these applications are pursued in the USSR by a semi-operational aircraft program (Basharinov et al., 1979 a, b;

Shutko, 1981) which uses multi-frequency radiometers for detecting the presence of water at different depths beneath the soil surface.

Applications of microwave radiometric remote sensing are discussed in Volume II of this Manual. The purpose of this chapter is to provide background material on the fundamental aspects of microwave radiometric measurements and the antennas, receivers, and radiometer systems used for that purpose. Following an introductory review of antenna fundamentals, the theory of radiative transfer is used to derive relations between the emission and scattering properties of a surface or volume and the power received by a radiometer receiver. Basic operation of a microwave radiometer is discussed and the advantages and disadvantages of different receiver configurations are outlined. Techniques used for calibrating radiometers are treated in some detail, and trade-off considerations between radiometric, spectral, and spatial resolutions of imaging systems are presented.

ANTENNA FUNDAMENTALS

A receiving antenna is a structure that serves as a region of transition between an incident free-space wave and a guided wave on a transmission line or a waveguide. Practical microwave radiometer antennas usually take the form of paraboloidal dish reflectors, horns, or arrays of dipoles, although many other antenna structures are sometimes used. Since antenna nomenclature is used to describe radiative transfer in microwave radiometry, this section will briefly summarize some basic terminology.

ANTENNA PATTERNS

Consider an antenna receiving energy transmitted by a point source at a large distance[1] R, as shown in Figure 11-1. If the transmitted power is constant, then as the angles (θ,ϕ) are varied, the received power $P_r(\theta,\phi)$ will also vary. A polar or rectangular graph of this function is known as the *antenna power pattern* and normally will be composed of a *main lobe, sidelobes* and *backlobes*, as shown in Figure 11-2. The polar graph shown is a typical constant ϕ cut of $P(\theta)|\phi = 0$; for other values of ϕ, $P(\theta)$ may have a different shape. For the horn antenna shown in Figure 11-1, the x-z plane cut ($\phi = 0°$) and y-z plane cut ($\phi = 90°$) patterns are known as the *principal-plane patterns*. The width of the main lobe at the half-power point is known as the *half-power beamwidth* (HPBW) or -3 dB beamwidth. The normalized power pattern $p_n(\theta)$ shown has been adjusted to a maximum of unity and is dimensionless. Antenna engineers usually express the power pattern on a logarithmic scale, i.e.,

$$P_n(\text{dB}) = 10 \log p_n(\theta,\phi) \qquad (11\text{-}1)$$

---

[1] Conventionally, $R \cong 2D^2/\lambda$ where D is the maximum aperture dimension of the antenna.

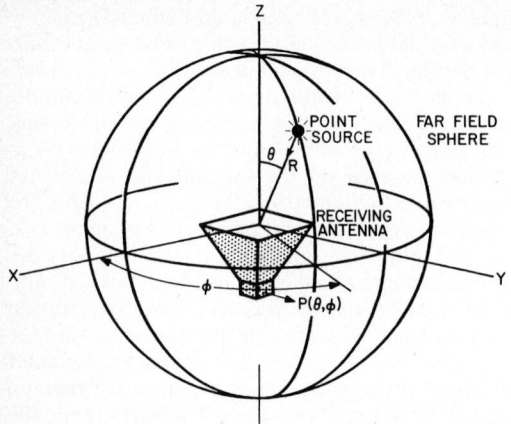

Fig. 11-1. Geometry for describing the receive antenna power pattern.

where

$$\Omega_M = \iint_{\substack{main \\ beam}} p_n(\theta,\phi)d\Omega \qquad (11\text{-}4)$$

is the *main lobe solid angle*. Well-designed radiometer antennas have main-beam efficiencies greater than 95%.

RADIOMETER ANTENNA POLARIZATION

The polarization of a receiving antenna is the same as that of the electric field transmitted by the antenna. For example, a dish antenna fed by a dipole parallel to the earth's surface would transmit a horizontally polarized E vector. The same dish acting as a receiving antenna is said to be horizontally polarized.

GAIN, DIRECTIVITY AND EFFECTIVE APERTURE

From the reciprocity theorem, the receiving and transmitting directional patterns of an antenna are identical.[2] It is thus convenient to define the *directivity* of a transmitting antenna and then use the same quantity for the receiving case. Referring to Figure 11-1, imagine now that the horn antenna is transmitting and that the point source is now used as an isotropic receiving antenna, with received power $P_r(\theta,\phi)$. As the receiving antenna is moved over the far-field sphere, there will be a location (θ_m, ϕ_m) on the main lobe peak where the received power is maximum. As it continues to move over the entire 4π solid angle a spatially average power

$$P_{av} = \frac{1}{4\pi} \iint_{4\pi} P_r(\theta,\phi)d\Omega$$

can also be determined.

The *directivity* is defined as:

$$D = \frac{maximum\ power}{average\ power}\bigg|\ R = constant$$

$$= \frac{P_{r,\ max}}{\displaystyle\iint_{4\pi} P_r(\theta,\phi)d\Omega} \qquad (11\text{-}5)$$

Since $p_n(\theta,\phi) = P_r(\theta,\phi)/P_{r,max}$, Eq. 11-5 can be rewritten as:

$$D = \frac{4\pi}{\displaystyle\iint_{4\pi} p_n(\theta,\phi)d\Omega} \qquad (11\text{-}6)$$

From Eq. 11-2 the denominator of Eq. 11-6 can be recognized as the *beam solid angle* Ω_A, i.e.,

$$D = \frac{4\pi}{\Omega_A} \qquad (11\text{-}7)$$

Thus, the maximum is 0 dB and the highest sidelobe is at 10 log 0.2 = −7 dB for the pattern shown; well-designed radiometer antennas have sidelobe levels of −20 dB or lower. Although it is desirable that all the received energy enter only through the main lobe, in practice it is not possible to reduce the sidelobe and backlobe levels to zero. The *beam solid angle* Ω_A is defined as

$$\Omega_A = \iint_{4\pi} p_n(\theta,\phi)\ d\Omega \qquad (11\text{-}2)$$

where in spherical coordinates $d\Omega = \sin\theta\ d\theta\ d\phi$ and the integration is for θ from 0 to π and ϕ from 0 to 2π. For an isotropic antenna $\Omega_A = 4\pi$, but for highly directive antennas, Ω_A will be much less than this. Another quantity of interest is the *main-beam efficiency* ϵ_M which is defined as:

$$\epsilon_M = \frac{\Omega_M}{\Omega_A} \qquad (11\text{-}3)$$

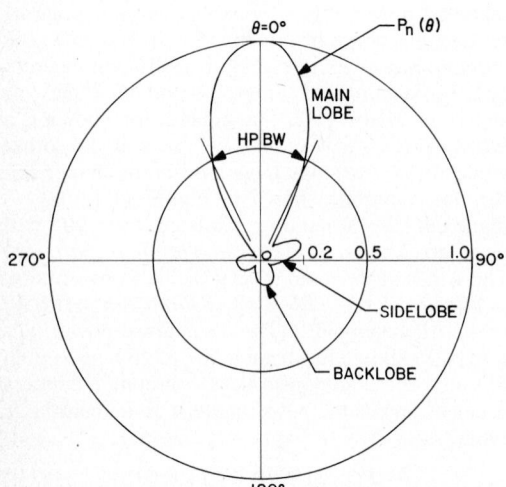

Fig. 11-2. Polar graph of typical normalized antenna power pattern for horn antenna.

[2] The only exception to this occurs when the medium is nonreciprocal, such as when ferrites or other anisotropic materials are used in the antenna.

For an isotropic antenna, $\Omega_A = 4\pi$ so that $D_{iso} = 1$ (0 dB). For antennas with narrow main beams (pencil beams) and low sidelobes, an approximate formula for the directivity is:

$$D \cong \frac{34,000}{(HPBW^\circ)_{xz} \, (HPBW^\circ)_{yz}} \quad (11\text{-}8)$$

where the two denominator terms are the principal-plane beamwidths in degrees. Equation 11-8 is not exact and may be in error by about ± 1 dB, however, it is useful for an estimation of the directivity. It should be noted from Eq. 11-6 that the directivity depends only on the pattern *shape* and is not dependent on antenna loss or efficiency.

The *gain* of a transmitting antenna involves both the pattern shape and the antenna loss, and is a measure of antenna performance relative to another reference antenna. Referring again to Figure 11-1, and letting the test antenna transmit, imagine that the isotropic receiving antenna (of identical polarization) is moved on the far-field sphere to position (θ_m, ϕ_m) where the received power is maximum, $P_{r,\max}^{\text{test}}$. Next, the original test antenna is replaced by a lossless reference antenna of identical polarization and the input power to this reference antenna is kept the same as before. The isotropic receiving antenna is again moved until the power is maximum, $P_{r,\max}^{\text{ref}}$. The *gain* of the test antenna is then:

$$G = \frac{P_{r,\max}^{\text{test}}}{P_{r,\max}^{\text{ref}}} . \quad (11\text{-}9)$$

The reference antenna is usually either a dipole (at VHF or UHF) or a pyramidal horn (at microwave or millimeter frequencies); these are antennas whose gain with respect to isotropic is known very accurately. Unless otherwise stated, the gain G is taken with respect to a hypothetical lossless isotropic antenna of identical polarization.

For practical antennas, not all of the input power appears as radiated power because of ohmic losses (Joule heating) in the antenna structure. The *antenna efficiency* η is defined as:

$$\eta = \frac{P_{\text{rad}}}{P_{\text{in}}} = \frac{P_{\text{rad}}}{P_{\text{ohmic}} + P_{\text{rad}}} \quad (11\text{-}10)$$

and is a dimensionless quantity ranging from 1 (for a lossless antenna) to 0 (for an infinitely lossy antenna). The reciprocal of the antenna efficiency is defined as the antenna loss L_A, i.e.,

$$L_A = \frac{1}{\eta} . \quad (11\text{-}11)$$

It can be shown that

$$G = \eta D = \frac{D}{L_A} \quad (11\text{-}12)$$

where

G = antenna gain over isotropic ($0 \le G < \infty$)
D = antenna directivity ($1 \le D < \infty$)
η = antenna efficiency ($0 \le \eta \le 1$)
L_A = antenna loss ($1 \le L_A < \infty$)

Thus, the gain of an antenna can never exceed its directivity.

The gain G discussed above is sometimes referred to as the *maximum gain*. A *gain pattern* $G(\theta, \phi)$ is then defined as:

$$G(\theta, \phi) = G \, p_n(\theta, \phi) \quad (11\text{-}13)$$

Consider an antenna receiving energy from an incident wave of power density S ($W \, m^{-2}$) and oriented so as to produce the maximum available power P_{av} at the terminals. The *effective aperture* A_e is defined as:

$$A_e = \frac{P_{av}}{S} \quad (11\text{-}14)$$

and is an area which is typically from $40\% - 80\%$ of the physical aperture of the antenna. It may be shown that the *effective aperture* of a receiving antenna and the *gain* (while transmitting) are related by:

$$G = \frac{4\pi A_e}{\lambda^2} \quad (11\text{-}15)$$

where both G and A_e are at their maximum values. As in Eq. 11-13, an effective aperture function $A_e(\theta, \phi)$ may be defined

$$A_e(\theta, \phi) = A_e \, p_n(\theta, \phi). \quad (11\text{-}16)$$

The beamwidth, sidelobe level, gain, and related attributes for an antenna are determined by its aperture width relative to a wavelength, and its aperture distribution. Although a complete mathematical description of these relations for all antennas is beyond the scope of this chapter, it is possible to present simple formulas for two commonly used radiometer antennas. Table 11-1 presents formulas for the half-power beamwidth, gain, and highest sidelobe level for an optimum rectangular horn and for a typical paraboloidal dish reflector with an aperture distribution tapering from 0 dB at the center to -10 dB at the dish edge. The formulas given are not exact, but are useful as a rule of thumb.

INPUT IMPEDANCE AND MISMATCH

The input impedance to an antenna is frequency-dependent and has both a resistive and reactive part. Radiometer antennas with coaxial connectors are usually designed for a nominal 50 Ω input impedance, although there may be appreciable variations from this over the radiometer bandwidth. At frequencies above about 7 GHz, waveguides are customarily used in order to reduce losses. In either case, there is maximum transfer of power from the receiving antenna to the receiver input if the input impedance to the receiver is the complex conjugate of the input impedance to the antenna. Normally, the input impedance to the receiver is purely resistive and is equal to the characteristic resistance R_0 of the feedline between the antenna and the receiver. The mono-

TABLE 11-1

Formulas for Horn and Dish Antennas

Antenna Type	Aperture Sketch	Gain	Half-power Beamwidth (deg).	Sidelobe Level (dB)
Rectangular horn (optimum)		$G = \dfrac{4\pi A_e}{\lambda}$	$\text{HPBW}° = 56\dfrac{\lambda}{b}$ (E-plane)	-13 dB (E-plane)
		$A_e = 0.5ab$	$\text{HPBW}° = 67\dfrac{\lambda}{a}$ (H-plane)	< -30 dB (H-plane)
Paraboloid (dish) with -10 dB edge taper in E plane		$G = \dfrac{4\pi A_e}{\lambda}$	$\text{HPBW}° = 72\dfrac{\lambda}{D}$ (E-plane)	-23 dB (E-plane)
		$A_e = 0.6A_p$ $A_p = \pi\left(\dfrac{D}{2}\right)^2$	$\text{HPBW}° = 86\dfrac{\lambda}{D}$ (H-plane)	-30 dB (H-plane)

chromatic *reflection coefficient* looking into the antenna from the feedline is given by

$$\rho = \frac{Z_A - R_0}{Z_A + R_0} \qquad (11\text{-}17)$$

and the monochromatic *mismatch loss ML* is defined by

$$ML = \frac{Power\ available\ from\ the\ antenna}{Power\ delivered\ by\ the\ antenna}. \qquad (11\text{-}18)$$

Therefore,

$$ML = \frac{1}{1 - |\rho|^2}. \qquad (11\text{-}19)$$

When the antenna is matched to the feedline $ML = 1$ and all the power available from the antenna is delivered to the receiver (assuming a lossless feedline). If $Z_A \neq R_0$, $ML > 1$, and the power input to the receiver is less than the available power. This loss is in addition to any ohmic (Joule heating) losses in the antenna structure itself.

The discussion above is for single-frequency situations. In a subsequent section, the effect of both ohmic loss and mismatch loss on radiometer performance will be discussed.

RADIATIVE TRANSFER FUNDAMENTALS

A microwave radiometer consists of an antenna, a wide-band receiver and a recording device; it is used to measure thermally generated electromagnetic radiation from extended scenes. In this section, basic physical mechanisms and mathematical relationships describing the transfer of thermal radiation are given.

Material bodies such as the earth's surface emit weak electromagnetic waves as the result of thermally induced random motion of electrons and protons. Often the direction of these charged

particles is basically random, so that the polarization of the emitted wavelets is also random. The net radiation is then said to be *randomly polarized*. In addition, such thermally generated noise waves have all frequency components.

BRIGHTNESS CONCEPTS

Referring to Figure 11-3, consider the noise power dP from a thermal radiator which is incident in a bandwidth df on an area dA through a solid angle $d\Omega$, expressed by [Kraus, 1966]

$$dP = B(\theta,\phi)\,\cos\theta\;d\Omega\;dA\;df \qquad (11\text{-}20)$$

where

dP = infinitesimal noise power (W),

B = brightness of thermal source (W m^{-2} Hz^{-1} sr^{-1}),

$d\Omega = \sin\theta\;d\theta\;d\phi$ = infinitesimal solid angle (sr),

dA = infinitesimal plate area (m^2),

df = infinitesimal bandwidth (Hz),

(θ,ϕ) = spherical coordinates of solid angle.

The $\cos\theta$ factor in Eq. 11-20 accounts for the projection of dA onto the plane normal to $d\Omega$, so that $\cos\theta\;dA$ may be viewed as an infinitesimal effective receiving aperture. If the thermal radiation incident from direction (θ,ϕ) is of uniform brightness over the extent of the plate area A, we may write

$$dP = A\,B(\theta,\phi)\,\cos\theta\;d\Omega\;df. \qquad (11\text{-}21)$$

The *brightness B* is seen to be a measure of the thermal noise power received per unit bandwidth, per unit area, and per unit solid angle.

Referring to Figure 11-4, the plate is now replaced by an effective receiving aperture $A_e(\theta,\phi)$ of a radiometer antenna, and the noise power dP is associated with the available power at the antenna terminals. Since all antennas are completely polarized, it follows that only half of the incident

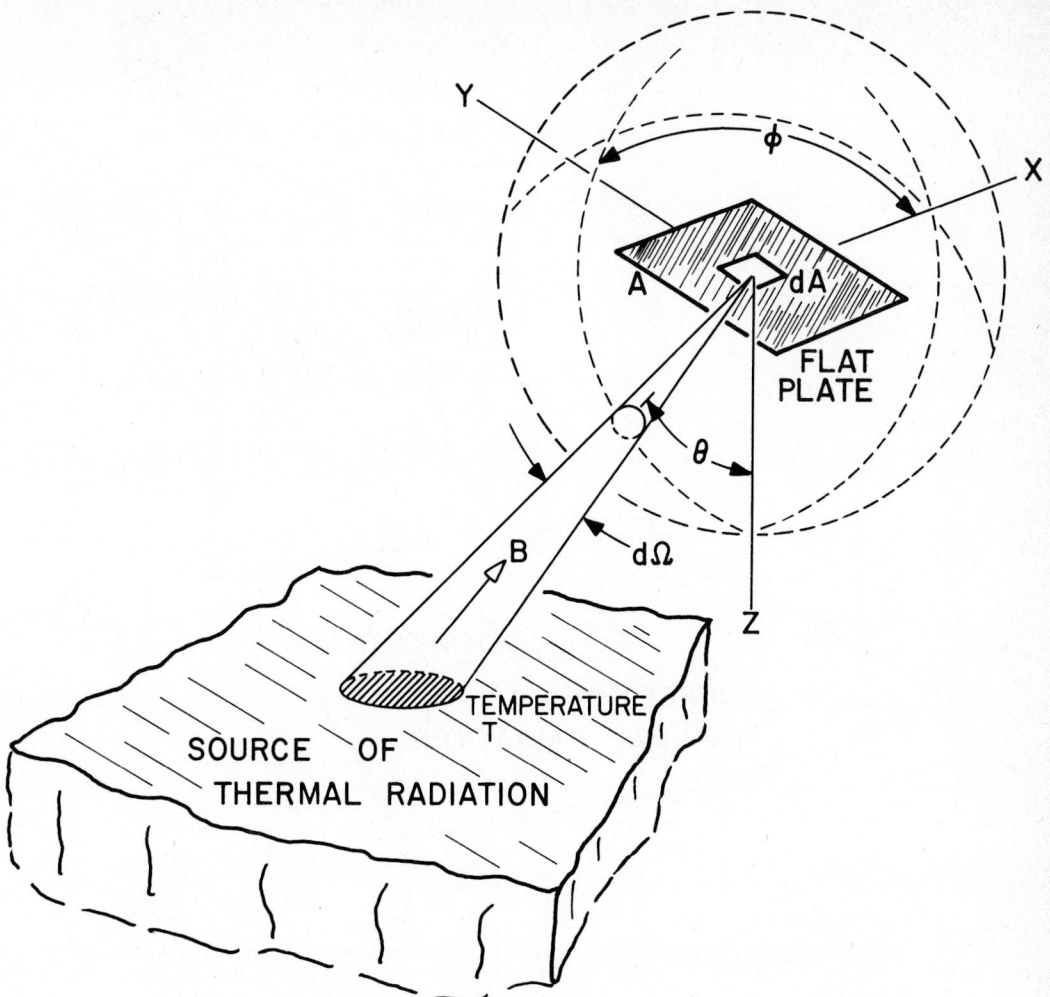

Fig. 11-3. Geometry for radiation incident on a flat plate from a thermal source.

power is available at the terminals of the antenna. Thus, Eq. 11-21 is replaced with

$$dP = \frac{1}{2} A_e(\theta,\phi)\ B(\theta,\phi)\ d\Omega\ df. \quad (11\text{-}22)$$

The total noise power available at the antenna terminals over a bandwidth Δf and from all sources of thermal radiation is given by

$$P = \frac{1}{2} \int_{f}^{f+\Delta f} \iint_{4\pi} A_e(\theta,\phi)\ B(\theta,\phi)\ d\Omega\ df. \quad (11\text{-}23)$$

The available noise power is seen to be dependent on the radiometer frequency f, the radiometer bandwidth Δf, the radiometer antenna effective aperture A_e, and the brightness B of the thermal source. Utilizing the relation of Eq. 11-15 between gain and effective aperture, Eq. 11-23 can be rewritten as

$$P = \frac{1}{2} \int_{f}^{f+\Delta f} \frac{\lambda^2}{4\pi} \iint_{4\pi} G(\theta,\phi)\ B(\theta,\phi)\ d\Omega\ df. \quad (11\text{-}24)$$

The observed brightness of a thermal radiator depends in general on its temperature, the observation frequency, the material composition of the source, its geometrical shape, and the polarization of the observing instrument. Thus, passive microwave radiometers operating at a fixed frequency and polarization are used to *remotely sense* the material composition of a thermal source (e.g., terrain) as well as its three-dimensional geometrical shape (e.g., surface roughness or flatness). Since the radiometer measures the noise power given by Eq. 11-24, it is essential in remote sensing (1) to know the antenna gain function $G(\theta,\phi)$ and (2) to have accurate mathematical models which relate the observed brightness $B(\theta,\phi)$ to the material composition and geometry of the thermal source.

BLACKBODY RADIATION

The term *blackbody* is given to a theoretical object that is an idealized perfect absorber of electromagnetic energy at all frequencies and also

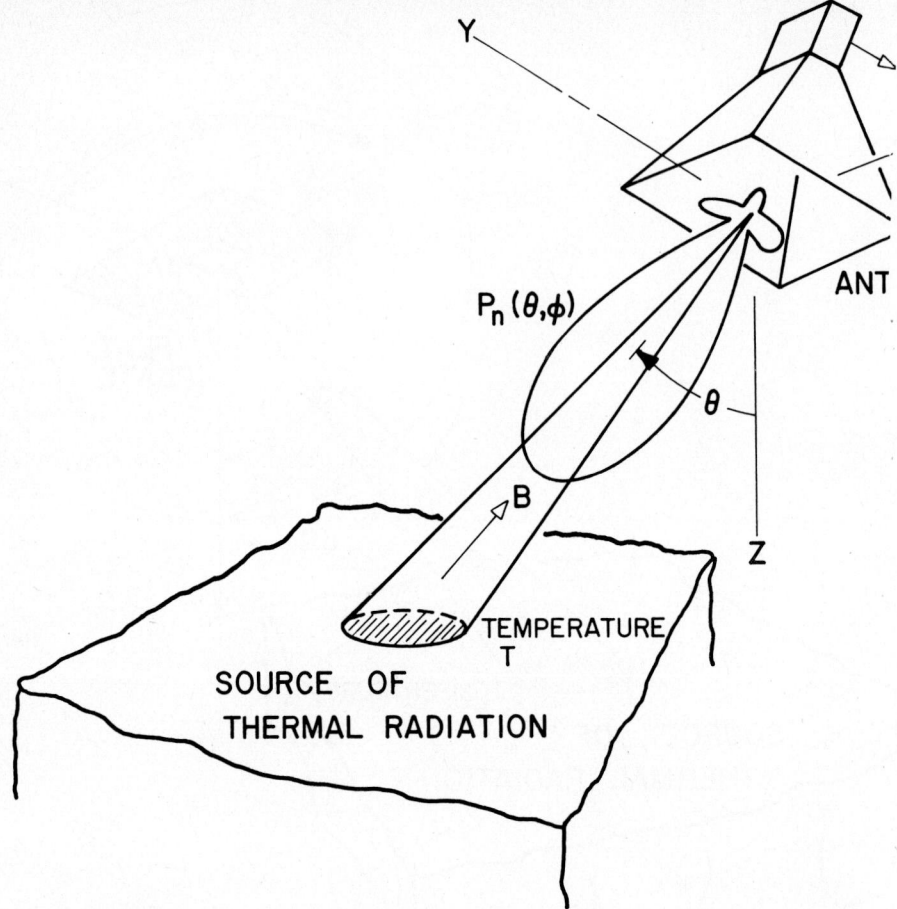

Fig. 11-4. Geometry for radiation incident on an antenna from a thermal source.

is a perfect radiator of electromagnetic energy at all frequencies. It is the simplest thermal radiator whose brightness can be mathematically described. The opposite of a blackbody is a perfect reflector, which absorbs no incident radiation. Although no perfect blackbodies are known to exist in nature, many objects behave approximately as blackbodies over a portion of the electromagnetic spectrum. For example, a *hohlraum* (German for hollow enclosure) is an absorber-lined enclosure maintained at an isothermal temperature T and functions as a blackbody, producing interior thermal radiation which is dependent only on the frequency and the temperature, independently of coordinate location within the box.

The brightness of a perfect blackbody is described as follows by Planck's Law, first published in 1901 by the German physicist Max Planck:

$$B_{bb} = \frac{2hf^3}{c^2} \frac{1}{e^{hf/kT} - 1} \qquad (11\text{-}25)$$

where

B_{bb} = brightness of blackbody (W m^{-2} Hz^{-1} sr^{-1})

h = Planck's constant = 6.63×10^{-34} J s

k = Boltzmann's constant = 1.38×10^{-23} J K^{-1}

c = speed of light = 3×10^8 m s^{-1}

T = blackbody temperature (K)

F = frequency (Hz)

A graph of B_{bb} (f) from Eq. 11-25 is shown in Figure 11-5 for blackbody temperatures of 30 K and 300 K. It is seen that, for this range of temperatures, the peak brightness occurs in the infrared portion of the spectrum and that in the radio spectrum the log B curve becomes a straight line. This can be shown from Eq. 11-25 by noting that for $T < 300$ K, $hf/kT << 1$ for frequencies less than about 300 GHz. Thus, the denominator term $e^{hf/kT} - 1 \cong hf/kT$ so that Eq. 11-25 reduces to

$$B_{bb} = \frac{2kT}{\lambda^2} \qquad (11\text{-}26)$$

which is known as the *Rayleigh-Jeans Law,* a special case of Planck's Law holding only in the radio spectrum. This relationship, published in 1900 by Lord Rayleigh and Sir James Jeans, was derived on the basis of classical theory which holds that a thermal radiator has a continuum of energy

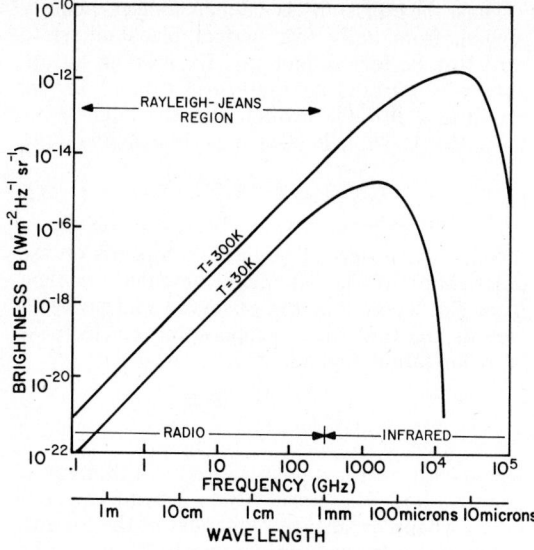

Fig. 11-5. Planck's Law brightness curves for black-bodies at $T = 30$ K; $T = 300$ K.

levels. Planck's Law is based on quantum theory, which holds that thermally induced energy transitions are discrete, resulting in the emission of discrete energy quanta.

Returning momentarily to Planck's Law, Eq. 11-25, and Figure 11-5 for blackbodies with temperatures comparable to ambient (200 K-300 K), the peak brightness occurs in the thermal-IR portion of the spectrum. The *total spectral brightness* B_T is found by integrating B_{bb} over all frequencies from 0 to ∞. The result is the *Stefan-Boltzmann relation*

$$B_T = \sigma T^4 \qquad (11\text{-}27)$$

where

B_T = total spectral brightness (W m^{-2} sr^{-1})
T = blackbody temperature (K)
σ = Stefan-Boltzmann constant = 1.80×10^{-8} W m^{-2} K^{-4}.

This relation shows that the spectrally integrated brightness is proportional to the fourth power of the blackbody temperature. Moreover, almost all the contribution to the total spectral brightness of blackbodies at these temperatures comes from the thermal-IR portion of the spectrum, thus accounting for the frequent assertion that wide-band thermal infrared radiometers are sensitive to the *fourth power* of the temperature in degrees Kelvin. It will be shown later that microwave radiometers, by contrast, are sensitive to the *first power* of the temperature.

THERMAL NOISE FROM A RESISTOR

A resistor at temperature T can generate thermal noise as a result of the random motion of electrons within it. At radio frequencies, the time-average noise power available in a band-

width Δf is given by the following formula, first published in 1928 by H. Nyquist of the American Telephone and Telegraph Company:

$$P = kT \, \Delta f \qquad (11\text{-}28)$$

where

P = time-average noise power (W)
k = Boltzmann's constant = 1.38×10^{-23} J K^{-1}
T = resistor temperature (K)
Δf = bandwidth (Hz).

Such noise is often called Johnson noise after J. B. Johnson of Bell Labs, who was one of the first to study the statistical nature of thermally generated noise in resistors and vacuum tubes. It is important to note that the noise power computed from Eq. 11-28 is the *maximum available power* which would be delivered to a matched load. If there is a mismatch between the resistor and the load (detector), the actual power delivered will be less than this.

THERMAL NOISE FROM AN ANTENNA

The input resistance of a lossless transmitting antenna is known as its *radiation resistance R*, an equivalent resistance which for an input current I_t would dissipate as heat the same amount of power $|I_t|^2 R$ which is actually radiated by the antenna. Conversely, the radiation resistance R for the same antenna acting as a receiver with terminal current I_r flowing into a matched load, represents the equivalent resistance which would absorb the received power $|I_r|^2 R$.

Consider a lossless receiving antenna of radiation resistance R inside a perfectly absorbing/emitting hohlraum (blackbody) at temperature T, as shown in Figure 11-6. The thermal radiation is uniform throughout the hohlraum so that the brightness is isotropic and given by Eq. 11-26. Substituting Eq. 11-26 into Eq. 11-24, the noise power available from the antenna is found to be

$$P_{bb} = \frac{1}{2}(2kT) \int_f^{f+\Delta f} \frac{1}{4\pi} \iint_{4\pi} G(\theta,\phi) \, d\Omega \, df. \qquad (11\text{-}29)$$

If it is further assumed that the antenna gain pattern varies negligibly over the bandwidth Δf, then Eq. 11-29 becomes

$$P_{bb} = kT\Delta f \frac{1}{4\pi} \iint_{4\pi} G(\theta,\phi) \, d\Omega. \qquad (11\text{-}30)$$

It can be shown that for lossless antennas the value of the integral in Eq. 11-30 is 4π (regardless of the shape of the gain pattern) so that Eq. 11-30 reduces to

$$P_{bb} = k T \, \Delta f \qquad (11\text{-}31)$$

which is seen to be identical in form to the Nyquist formula, Eq. 11-28. This means that the available noise power from a lossless antenna surrounded by an isothermal blackbody is the same as that

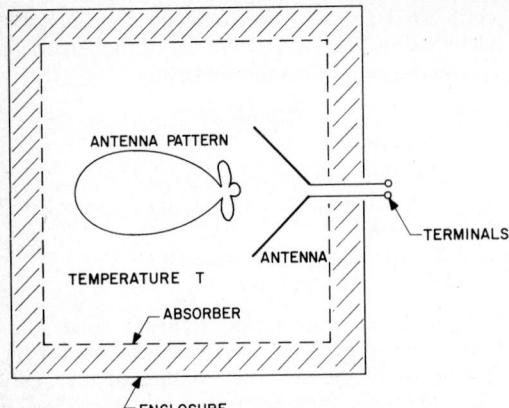

Fig. 11-6. Antenna inside hohlraum at temperature T.

available from a resistor at temperature T having a value R equal to the antenna radiation resistance. Furthermore, the simple form of Eq. 11-31 gives an easy method for interchangeable use of *temperature* and *noise power*. Even though the antenna structure temperature in the hohlraum is also T, the measured noise power Eq. 11-31 is associated with the background blackbody temperature T and does not originate from the antenna structure itself.

ANTENNA TEMPERATURE, BRIGHTNESS TEMPERATURE AND EMISSIVITY

In the preceding sections, formulas were developed for the brightness and observed noise power of blackbodies. In reality, no objects behave as perfect blackbodies; at a given temperature, a real body produces less thermally radiated electromagnetic energy than a blackbody at the same temperature. Moreover, the observed brightness of such graybodies depends on both direction and polarization, as well as on their material composition and shape.

The noise power measured by an antenna observing a thermally radiating background with brightness $B(\theta,\phi)$ can be related to an *antenna temperature* T_A given by

$$P = k\,T_a\,\Delta f \qquad (11\text{-}32)$$

which is recognized as the Nyquist formula. This antenna temperature is therefore an equivalent noise temperature equal to the absolute temperature of a resistor which gives the available noise power P.

The *emissivity* of a thermal radiator is defined as

$$e(\theta,\phi) = \frac{B(\theta,\phi;T)}{B_{bb}(T)} \qquad (11\text{-}33)$$

i.e., it is the ratio of the power radiated through a unit solid angle $d\Omega(\theta,\phi)$ by a unit surface area of an object at temperature T, to that radiated through the same $d\Omega(\theta,\phi)$ by a unit area of a blackbody at the same temperature T.

Thus, the emissivity is a dimensionless quantity ranging from unity (for perfect blackbodies) to zero (for perfect reflectors). By relating an apparent *brightness temperature* $T_B(\theta,\phi)$ to the brightness $B(\theta,\phi)$ through the Rayleigh-Jeans Law, Eq. 11-26, it is clear that we can also write

$$e(\theta,\phi) = \frac{T_B(\theta,\phi)}{T}. \qquad (11\text{-}34)$$

Since the observed brightness depends on the polarization of the radiometer antenna it is clear from Eq. 11-34 that the observed emissivity as well as the brightness temperature will likewise be polarization-dependent, i.e.,

$$e_i(\theta,\phi) = \frac{T_{Bi}(\theta,\phi)}{T} \qquad (11\text{-}35)$$

where $i = v$ (vertical polarization) or h (horizontal polarization). The emissivity is dependent solely on the shape or surface roughness of the thermal radiator and its material composition. It is independent of the material temperature unless the material composition is temperature sensitive. For example, an infinite, perfectly flat surface bounding the half-space of a homogeneous medium has polarized emissivities given by

$$e_{\substack{v\\h}}(\theta,\phi) = 1 - R_{\substack{v\\h}}(\theta,\phi) \qquad (11\text{-}36)$$

where R_v is the Fresnel power reflection coefficient for vertical polarization and R_h is the Fresnel reflection coefficient for horizontal polarization. A more comprehensive treatment of the emissivity is given in Chapter 4.

Since $B(\theta,\phi) = 2\,k\,T_B(\theta,\phi)/\lambda^2$, Eq. 11-24 becomes

$$P = \frac{k}{4\pi}\int_f^{f+\Delta f}\iint_{4\pi} T_B(\theta,\phi)\,G(\theta,\phi)\,d\Omega\,df. \qquad (11\text{-}37)$$

Assuming as before that the antenna gain pattern varies negligibly over the bandwidth Δf,

$$P = k\Delta f\frac{1}{4\pi}\iint_{4\pi} T_B(\theta,\phi)\,G(\theta,\phi)\,d\Omega.$$

Comparing Eq. 11-38 to Eq. 11-32 it is clear that

$$T_A = \frac{1}{4\pi}\iint_{4\pi} T_B(\theta,\phi)\,G(\theta,\phi)\,d\Omega. \qquad (11\text{-}39)$$

This relation is of fundamental importance in microwave radiometry, and describes the observed antenna temperature T_A as a gain-weighted sum of the individual brightness temperatures from each direction.

A primary scientific objective of microwave remote sensing is to determine the emissivity $e(\theta,\phi)$ of the terrain, sea or atmosphere, and from this to deduce secondary information about surface roughness or material composition. By separately measuring the surface temperature (e.g., with an infrared radiometer) the emissivity $e(\theta,\phi)$ can be determined from Eq. 11-34 if the brightness temperature $T_B(\theta,\phi)$ can be measured. Unfortunately, it is not always possible to measure $T_B(\theta,\phi)$ with spatial precision because of the smoothing effect of the antenna gain pattern on the true brightness temperature distribution $T_B(\theta,\phi)$, as described in Eq. 11-39. This situation is illustrated in Figure 11-7 for a terrain brightness temperature $T_B(\theta,\phi)$ incident on an antenna with gain pattern $G(\theta,\phi)$. If the antenna is scanned in angle, then Eq. 11-39 is replaced by

$$T_A(\theta_0, \phi_0) = \frac{1}{4\pi} \iint_{4\pi} T_B(\theta,\phi) \, G(\theta-\theta_0, \phi-\phi_0) \, d\Omega \tag{11-40}$$

where (θ_0, ϕ_0) are the coordinates of the main-beam peak. As a simple example, consider a lossless antenna whose pattern is independent of θ and which scans a brightness distribution $T_B(\phi)$. In this case, Eq. 11-40 is replaced by

$$T_A(\phi_0) = \frac{1}{2\pi} \int_0^{2\pi} T_B(\phi) \, G(\phi-\phi_0) \, d\phi$$

$$= \frac{D}{2\pi} \int_0^{2\pi} T_B(\phi) \, p_n(\phi-\phi_0) \, d\phi. \tag{11-41}$$

This expression for the antenna temperature is a convolution integral, i.e., the normalized power pattern $p_n(\phi-\phi_0)$ is convolved with the brightness temperature $T_B(\phi)$. It follows from Eq. 11-6 that, for this case,

$$D = \frac{2\pi}{\int_0^{2\pi} p_n(\phi) \, d\phi}. \tag{11-42}$$

Therefore Eq. 11-41 becomes

$$T_A(\phi_0) = \frac{\int_0^{2\pi} T_B(\phi) \, p_n(\phi-\phi_0) \, d\phi}{\int_0^{2\pi} p_n(\phi) \, d\phi}. \tag{11-43}$$

It is clear from Eq. 11-43 that the observed antenna temperature $T_A(\phi_0)$ is a weighted value which is dependent on the antenna beamwidth. This is illustrated in Figure 11-8 for a T_B pulse function of width $\Delta\phi$ and height 50 K superimposed on a 100-K background. In the top graph, a narrow-beam antenna having a beamwidth between first nulls (BWFN) equal to half the source-width (BWFN = $\Delta\phi/2$) scans across the pulse source and produces an observed $T_A(\phi_0)$ antenna temperature distribution which is a slightly rounded version of the true distribution. In the bottom graphs, a wide beam with BWFN = 1.5 $\Delta\phi$ scans across the same pulse source, and produces an observed $T_A(\phi_0)$ distribution which is about 20% wider than the original and with about one-half the true increase in temperature. It is apparent from this example that in order for a radiometer antenna to clearly resolve an object of width $\Delta\phi$, the antenna pattern should have a BWFN smaller than $\Delta\phi$. One commonly used criterion is that BWFN $\lesssim \Delta\phi$ for a continuous source distribution of width $\Delta\phi$ and BWFN $\lesssim 2 \Delta\phi$ for two point-sources separated by $\Delta\phi$, this last condition being known as the *Rayleigh criterion*.

The preceding simple example used one-dimensional source- and antenna-pattern distributions. We now return to the more general two-dimen-

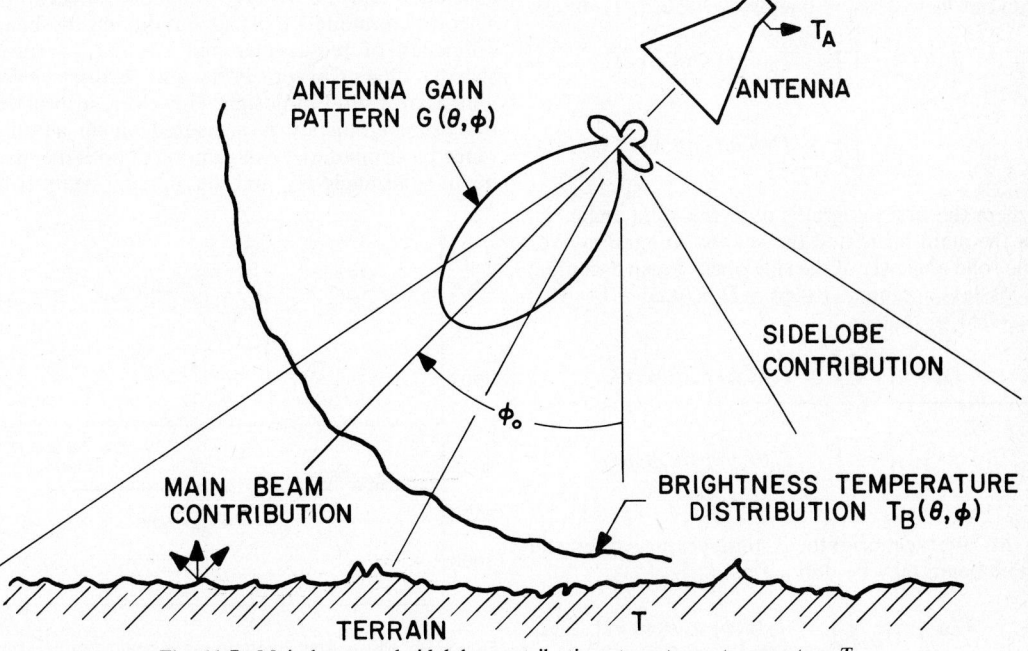

Fig. 11-7. Main-beam and sidelobe contributions to antenna temperature T_A.

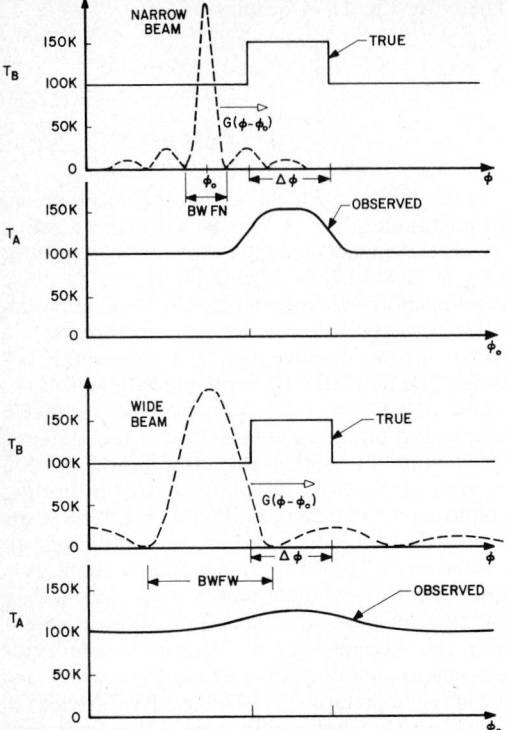

Fig. 11-8. True brightness temperature distribution T_B and observed antenna temperature distribution T_A for narrow-beam and wide-beam antennas.

sional case described by Eq. 11-40. It is of interest to know what fraction of the total antenna temperature is contributed by the main beam and what remaining fraction is from the sidelobes. We can put $\theta_0 = 0$, $\phi_0 = 0$ and rewrite Eq. 11-40 as

$$T_A = \frac{1}{4\pi} \iint_{\Omega_M} T_B(\theta,\phi)\, G(\theta,\phi)\, d\Omega$$

$$+ \frac{1}{4\pi} \iint_{\Omega_S} T_B(\theta,\phi)\, G(\theta,\phi)\, d\Omega \quad (11\text{-}44)$$

where the first integral is over the solid angle Ω_M of the main beam and the second integral is over the solid angle Ω_S of the sidelobes. Again assuming a lossless antenna, $G(\theta,\phi) = D\, p_n(\theta,\phi) = (4\pi/\Omega_A)\, p_n(\theta,\phi)$ so that

$$T_A = \frac{1}{\Omega_A} \iint_{\Omega_M} T_B(\theta,\phi)\, p_n(\theta,\phi)\, d\Omega$$

$$+ \frac{1}{\Omega_A} \iint_{\Omega_S} T_B(\theta,\phi)\, p_n(\theta,\phi)\, d\Omega. \quad (11\text{-}45)$$

An average brightness temperature over the main beam may be defined as

$$T_{BM} = \frac{1}{\Omega_M} \iint_{\Omega_M} T_B(\theta,\phi)\, p_n(\theta,\phi)\, d\Omega \quad (11\text{-}46)$$

and an average brightness temperature over the sidelobes may be defined as

$$T_{BS} = \frac{1}{\Omega_S} \iint_{\Omega_S} T_B(\theta,\phi)\, p_n(\theta,\phi)\, d\Omega. \quad (11\text{-}47)$$

Therefore Eq. 11-45 may be rewritten as

$$T_A = \frac{\Omega_M}{\Omega_A} T_{BM} + \frac{\Omega_S}{\Omega_A} T_{BS}. \quad (11\text{-}48)$$

Noting that $\Omega_S = \Omega_S - \Omega_M$ and recalling the definition Eq. 11-3 for main-beam efficiency,

$$T_A = \epsilon_M T_{BM} + (1 - \epsilon_M) T_{BS}. \quad (11\text{-}49)$$

In a practical situation it is desirable that the second term be much smaller than the first, i.e., that the sidelobe level be kept very low. This is shown in Figure 11-9 where the top graph of $T_B(\phi)$ shows a baseline varying slightly about 290 K except for a dip to about 170 K in the region of the antenna main beam. The near-in sidelobes see the colder temperature levels, thus dropping the average sidelobe temperature to 260 K. The main-beam average temperature is 180 K. Assuming a main-beam efficiency of 95%.

$$\begin{aligned} T_A &= 0.95 \times 180 + .05 \times 260 \\ &= 171\text{ K} + 13\text{ K} \\ &= 184\text{ K}. \end{aligned}$$

Therefore, the sidelobe contribution causes the antenna temperature of 184 K to be 4 K higher than the average temperature over the main beam. Since it is this T_{BM} value which is the desired measurement, the actual measurement of 184 K is high by about 2%. In a practical situation, it is desirable to accurately assess both the main-beam efficiency and the average sidelobe temperature. Theoretical models exist for predicting the beam efficiency of rectangular and circular apertures (Nash, 1964; Carver, Potts and Widner, 1971) which are useful in design. However, in practice the beam efficiency is measured on an antenna range by computerized integration of both the main beam solid angle Ω_M and the antenna beam solid

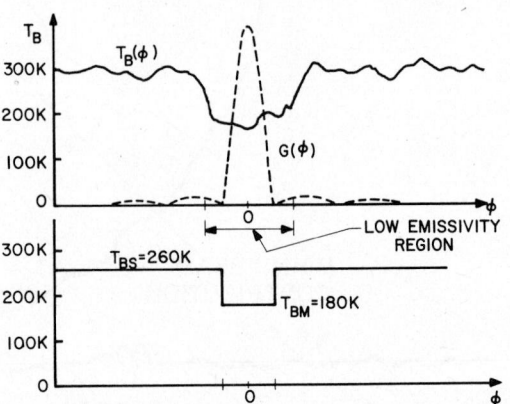

Fig. 11-9. Illustrating average main-beam temperature T_{BM} and average sidelobe temperature T_{BS}.

angle Ω_A. The average sidelobe temperature is often estimated from *à priori* knowledge of surface emissivity and temperature profiles.

ATMOSPHERIC RADIATIVE TRANSFER

The atmosphere of a planet will absorb and emit according to its density profile and also according to the energy transition resonant frequencies of its constituent molecules. For the earth's atmosphere, absorption and emission at microwave frequencies are dependent on both H_2O and O_2 molecules, and become particularly significant above about 10 GHz. The *absorption coefficient* α (Np m^{-1}) of the atmosphere depends on both the height above the surface and the frequency. Its value at any one frequency depends on the water vapor content, the content of free oxygen, and on the effects of pressure-broadening by the atmosphere. Therefore, the absorption coefficient can change with atmospheric turbulence. Under conditions of horizontal stratification in the atmosphere, the one-way attenuation or loss between the surface and height H (see Figure 11-10) may be calculated from a solution to the fundamental equation of radiative transfer [Chandrasekhar, 1960],

$$L_a = e^{\tau(O,H)\,\sec\,\theta_0} \qquad (11\text{-}50)$$

where $\tau(z_1,z_2)$ is the *optical thickness* of the atmosphere between heights z_1 and z_2, i.e.,

$$\tau(z_1,z_2) = \int_{z_1}^{z_2} \alpha(z)\,dz. \qquad (11\text{-}51)$$

This loss factor represents the reduction in received power P_r in comparison to the power P_{ro} received if the atmosphere were not present, i.e., $L_a = P_{ro}/P_r$. The loss factor for the entire atmosphere under ICAO standard conditions is shown in Figure 11-11 for values of θ_0 ranging from 0° to 90°. The increased loss at 90° is due to the greater thickness of the atmosphere at this aspect angle. The peak at 22 GHz is due to pressure-broadened absorption by atmospheric water vapor and the

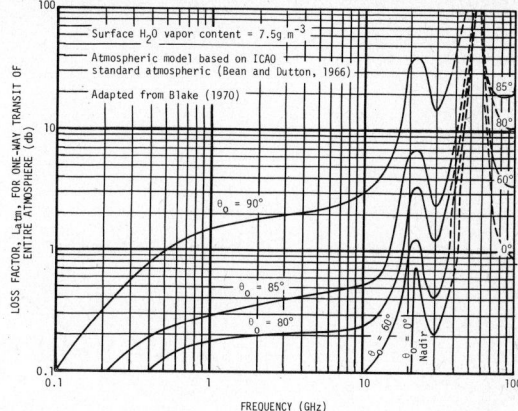

Fig. *11-11*. Absorption loss for one-way transit of entire atmosphere, as a function of frequency, and at various incidence angles.

peak from 50−70 GHz is due to absorption by free O_2 molecules.

Figure 11-12 indicates the principal contributors to the apparent brightness temperature seen by an airborne or spaceborne radiometer. Downwelling radiation T_D through and from the atmosphere scatters from the surface, producing an equivalent scattered temperature at the surface T_{SC}. Added to this is the surface brightness temperature T_S. Both of these components are attenuated by the atmospheric loss L_A before they reach the antenna. Finally, the upwelling radiation from the atmosphere is T_U. The equation of radiative transfer for this situation is:

$$T_{B_i}(\theta_0) = \frac{T_{SC_i} + T_{S_i}}{L_a} + T_U \qquad (11\text{-}52)$$

where the subscript i stands for h (horizontal) or v (vertical) polarization. Since the upwelling atmospheric radiation is randomly polarized, no polarization subscript is needed. If the absolute air temperature is denoted as $t_a(z)$, the upwelling temperature T_U may be calculated as

$$T_U = \sec\,\theta_0 \int_0^H t_a(z)\,\alpha(z)\exp\left[-\tau(z,H)\sec\,\theta_0\right]dz.$$
$$(11\text{-}53)$$

This is shown in Figure 11-13 as a function of frequency for several angles θ_0, and where it is assumed that the radiometer is at satellite altitude, i.e., $H\rightarrow\infty$. For frequencies below about 1 GHz, noise from our galaxy can become greater than noise generated within the atmosphere. However, since this noise enters the sidelobes which would ordinarily be 20−30 dB below the main-beam peak, the effective added galactic noise at these lower frequencies is usually 2 to 3 orders of magnitude smaller than the dashed lines shown for the galactic center and the galactic pole.

A similar graph can be used to calculate the downwelling temperature T_D, which is given by

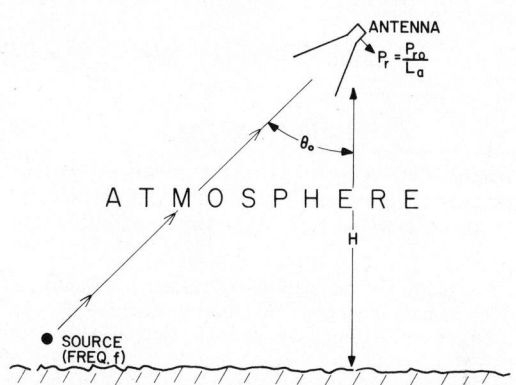

Fig. *11-10*. Geometry for demonstrating atmospheric loss factor L_a.

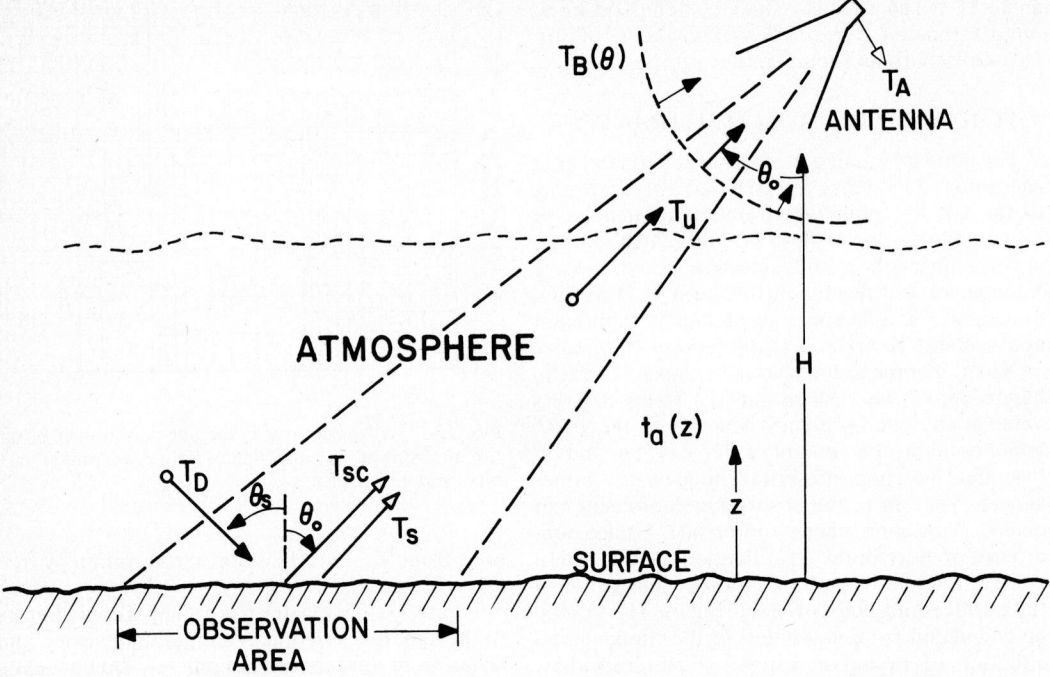

Fig. 11-12. Illustrating the principal contributors to the apparent brightness temperature $T_B(\theta_0)$.

$$T_D(\theta_s)$$

$$= \sec \theta_S \int_0^\infty t_a(z)\, \alpha(z)\, \exp\left[-\tau(0,z)\sec\theta_s\right]\,dz \tag{11-54}$$

which is seen as similar in form to Eq. 11-53. This expression includes *only* contributions from the atmosphere and ignores galactic, solar and man-made sources of noise, which are negligible above about 1 GHz. The surface brightness temperature T_S is polarization dependent and is given by

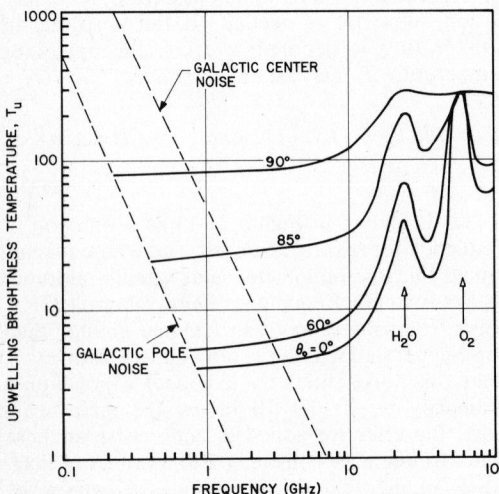

Fig. 11-13. Calculated upwelling atmospheric brightness temperature T_U based on standard atmosphere, for the entire atmosphere.

$$T_{Si} = e_i(\theta_0)\, t_S \tag{11-55}$$

where $e_i(\theta_0)$ is the polarized emissivity in the direction θ_0 and t_S is the absolute thermometric temperature of the surface observation area.[3]

As discussed in Chapter 4, the scattered radiation is due to the atmospheric downwelling radiation $T_D(\theta_s)$ from all angles θ_s which scatters partially in the direction θ_0,

$$T_{SC_i}(\theta_0) = \frac{1}{4\pi\cos\theta} \iint_{2\pi} T_D(\theta_S, \phi_s)$$

$$\left[\sigma_{ii}^0(O,S) + \sigma_{ij}^0(O,S)\right] d\Omega_S \tag{11-56}$$

where σ_{ii}^0 and σ_{ij}^0 are the surface bistatic scattering coefficients, as defined in Chapter 4. If the surface is specular (electrically flat), Eq. 11-56 reduces to

$$T_{SC_i}(\theta_0) = T_D(\theta_0)\left[1 - e_i(\theta_0)\right] \tag{11-57}$$

or

$$T_{SC_i}(\theta_0) = R_i\, T_D(\theta_0) \tag{11-58}$$

where R_i ($i = v$ or h) is the polarized Fresnel power reflection coefficient for a flat surface.

As an example, consider the apparent bright-

[3] Actually the temperature t_S varies with depth and more refined models are required to describe the contribution to T_S from each layer. In Eq. 11-55 the temperature t_S may be taken as that of the first layer of thickness equal to the skin depth, i.e., $\delta(cm) = 1.6/\sqrt{f\sigma}$ where f is the frequency (GHz) and σ is the conductivity (mhos/m).

ness temperature T_B seen by both vertically and horizontally polarized channels of a dual-polarized K_u-band (22 GHz) radiometer which scans in angle from $\theta_0 = 0°$ to $80°$. We will assume that the radiometer is satellite-borne and is viewing a 20°C specular sea surface. The polarized emissivities of calm sea water at this frequency with $\epsilon_r = 19 - j\,36$ may be calculated using the Fresnel reflection coefficients (cf. Chapter 4) and are summarized in the first two columns of Table 11-2. The atmospheric loss for a standard ICAO atmosphere is found from Figure 11-11 and is listed in the third column. The upwelling temperature of the atmosphere is found from Figure 11-13 and is listed in the fourth column. The polarized surface brightness temperatures T_S are calculated from Eq. 11-55 using the listed emissivities and $t_S = 293$ K. The surface scattered temperatures T_{SC} are calculated from Eq. 11-57 and noting that $T_D(\theta_0) = T_U(\theta_0)$.

It is clear from the temperatures listed in Table 11-2 that the major contributor to the apparent brightness temperature is the surface brightness temperature, except for large angles where the horizontally polarized term T_{sh}/L_a is comparable to the surface-scattered temperature for this polarization. A graph of the apparent brightness temperature $T_B(\theta_0)$ and the true surface brightness temperature $T_S(\theta_0)$ for each polarization is shown in Figure 11-14.

This example is oversimplified, but illustrates the masking effect of the atmosphere on the observed brightness temperature at K_u-band of the sea. In reality, the sea will not be specular at 22 GHz since this would require an rms smoothness of 1 mm or less. The emissivities of Table 11-2 should be replaced by those of a rough surface, as discussed in Chapter 4. Moreover, the emissivities will fluctuate with time and location within the antenna-beam observation cell.

RADIOMETER RECEIVERS

According to Eq. 11-32, the noise power available at the terminals of a lossless antenna is given by $P = kT_A\,\Delta f$, where the antenna temperature

T_A represents the integrated power incident upon the antenna from all directions, as in Eq. 11-39. A typical radiometer receiver is capable of detecting changes in T_A in the order of 1 K or less. For a receiver bandwidth $\Delta f = 1$ MHz, this corresponds to a change in P of $\Delta P = k\,\Delta T_a\,\Delta f \cong 1.3 \times 10^{-17}$ watts, which is several orders of magnitude smaller than the internally generated noise power of the receiver itself. How, then, does a radiometer accomplish such a high degree of measurement precision? To answer this question, we first need to briefly review the standard methods used for characterizing the noisiness of a device or system, and then proceed to discuss the operation of a microwave radiometer receiver.

NOISE REPRESENTATION

This section provides a brief review of the methods used for characterizing the noise generated by devices and systems. In all cases, the devices are assumed to be matched at their input and output terminals. The effects of impedance mismatches are considered in a future section.

The device (or system) shown in Figure 11-15a has a bandwidth B^4 and power gain G. The input noise power P_I is equivalent to the thermal noise that would be generated by a matched resistor of temperature T_I such that

$$P_I = k\,T_I\,B. \qquad (11\text{-}59)$$

The output noise power P_O is given by

$$P_O = G\,k\,T_I\,B + \Delta P_O \qquad (11\text{-}60)$$

where ΔP_O is the noise contribution generated by the noisy device. The *equivalent noise temperature* T_E of the device is defined on the basis of the equivalence of Figures 11-15a and 11-15b, with respect to the input and output noise powers. The noisy device of Figure 11-15a has been replaced in Figure 11-15b by a combination of a noise-free device and a fictitious noise source of temperature T_E whose magnitude is given by:

[4] In discussions of thermal noise considerations of devices, the symbol B, rather than Δf, is commonly used for the bandwidth.

TABLE 11-2

Contribution to K_u-band Apparent Brightness Temperature of Specular Sea

Angle	Emissivity		Atmos. Loss	Upwell. Temp.	Surface Brightness Temperature		Surface Scattered Temperature	
θ_0	e_v	e_h	L_a	T_U	T_{Sv}	T_{Sh}	$T_{SC,v}$	$T_{SC,h}$
0°	.415	.415	1.174	6.0 K	121.6 K	121.6 K	3.5 K	3.5 K
10°	.420	.411	1.179	6.3	123.1	120.4	3.6	3.7
20°	.435	.396	1.185	6.6	127.5	116.0	3.7	4.0
30°	.462	.372	1.192	7.0	135.4	109.0	3.8	4.4
40°	.504	.377	1.202	7.5	147.7	98.7	3.7	5.0
50°	.566	.292	1.230	8.2	165.8	85.6	3.6	5.8
60°	.657	.235	1.312	9.0	192.5	68.9	3.1	6.9
70°	.788	.168	1.549	13.0	230.9	49.2	2.8	10.8
80°	.923	.089	2.089	19.0	270.4	26.1	1.5	17.3

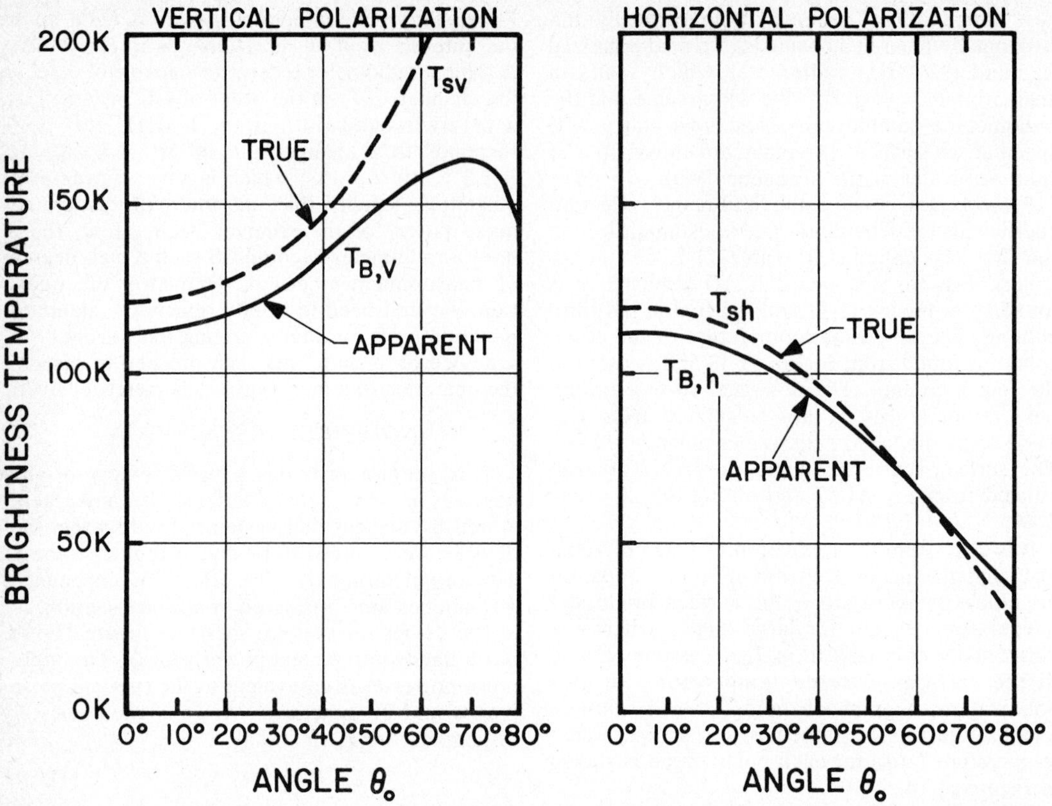

VERTICAL POLARIZATION

HORIZONTAL POLARIZATION

Fig. 11-14. Calculated surface (true) and apparent brightness temperatures of specular sea at K_u-band from satellite altitude.

$$T_E = \frac{\Delta P_O}{GkB} \qquad (11\text{-}61)$$

It should be noted that the equivalent noise temperature T_E is referred to the input terminals of the device.

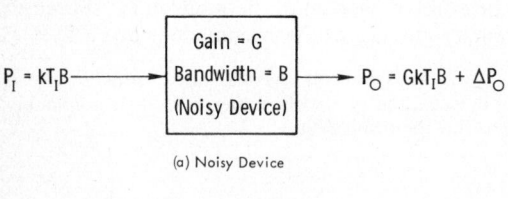

(a) Noisy Device

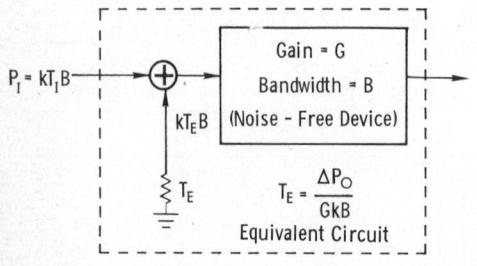

(b) Equivalent Representation

Fig. 11-15. The noisy device shown in (a) can be represented in terms of an equivalent circuit consisting of a noise-free device and a fictitious noise source added at the input of the device.

Noise Temperature of Cascaded System

The system shown in Figure 11-16a consists of two devices in cascade, with parameters G_1, T_{E1} for the first device and G_2, T_{E2} for the second. Equivalent representation in terms of noise-free devices is shown in Figure 11-16b for a bandwidth B. The output noise power is given by

$$P_O = k\,B\,\left[G_1\,G_2\,(T_1 + T_{E1}) + G_2\,T_{E2}\right].$$
$$(11\text{-}62)$$

The same noise power appears at the output of the equivalent system shown in Figure 11-16c, which is characterized by a power gain $G_1\,G_2$ and an equivalent (input) noise temperature T_E given by

$$T_E = T_{E1} + \frac{T_{E2}}{G_1}. \qquad (11\text{-}63)$$

If the above relation is extended to a linear system of N subsystems in cascade, the equivalent noise temperature can be shown to be given by

$$T_E = T_{E1} \;+\; \frac{T_{E2}}{G_1}$$
$$+\; \frac{T_{E3}}{G_1\,G_2}$$
$$+\; \cdots \;+\; \frac{T_{EN}}{G_1\,G_2 \ldots G_{N-1}}. \qquad (11\text{-}64)$$

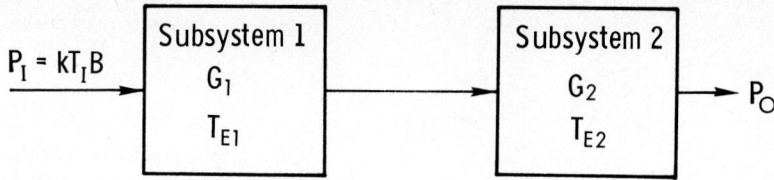

(a) Two Noisy Subsystems in Cascade

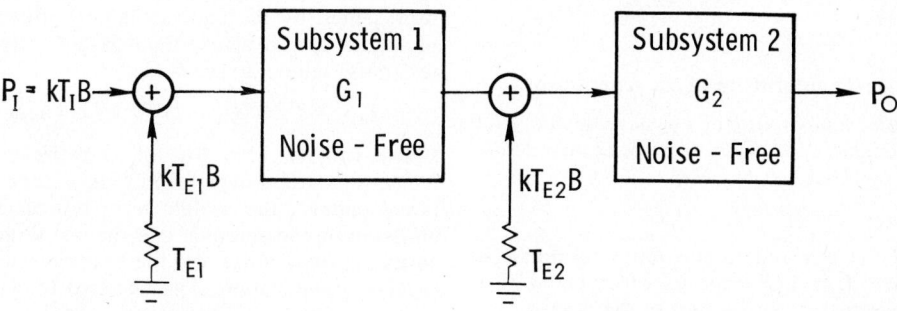

(b) Equivalent Representation of (a) in Terms of Noise-Free Subsystems

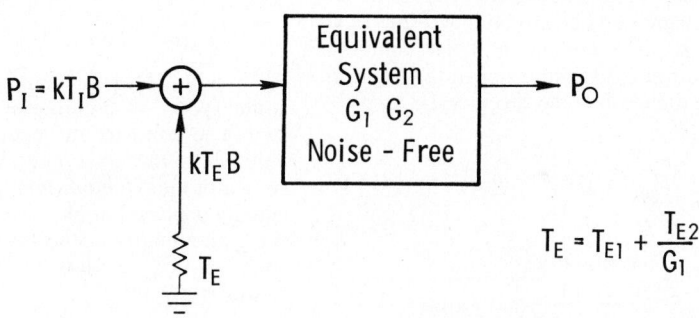

$$T_E = T_{E1} + \frac{T_{E2}}{G_1}$$

(c) Equivalent Single - System Representation of (b)

Fig. 11-16. Effective input noise temperature of two systems in cascade.

Thus, the equivalent noise temperature and gain of the first subsystem, T_{E1} and G_1, exercise the greatest influence on the noisiness of the overall system. For the superheterodyne receiver shown in Figure 11-17, the receiver noise temperature (referred to its input terminals) is given by:

$$T_{REC} = T_{RF} + \frac{T_M}{G_{RF}} + \frac{T_{IF}}{G_{RF}\,G_M} + \dots$$

(11-65)

where the subscripts *RF, M* and *IF* denote *RF* (radio-frequency) amplifier, mixer, and *IF* (intermediate-frequency) amplifier, respectively. To illustrate the significance of the various terms in Eq., 11-65, let us consider the following example of a moderately low-noise receiver: $T_{RF} = 300$ K,

$G_{RF} = (1000 \, (= 30 \text{ dB}), \, T_M = 300$ K, $G_M = 0.2$, $T_{IF} = 100$ K, $G_{IF} = 1000$. Inserting the above values into Eq. 11-65 results in

$$T_{REC} = \left(300 + \frac{300}{1000}\right.$$
$$\left. + \frac{100}{1000 \times 0.2} + \dots\right) \text{ K}$$
$$= (300 + 0.3 + 0.5 + \dots) \text{ K}$$
$$\cong 300 \text{ K}$$
$$= T_{RF}.$$

The above result, $T_{REC} \cong T_{RF}$, comes about from the fact that G_{RF} is large and it appears in the denominator of all terms succeeding the first term.

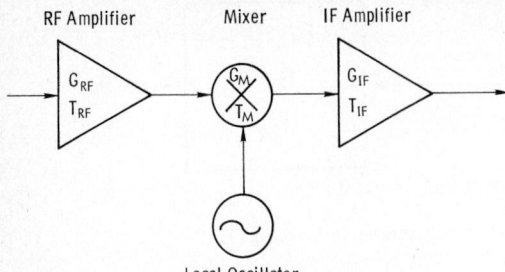

Fig. 11-17. Front-end section of superheterodyne receiver.

Noise Temperature of an Attenuator

The noise temperature of a passive device, such as an attenuator, a transmission line or an antenna, is given by (Ulaby et al., 1981):

$$T_E = (L - 1) T_O \qquad (11-66)$$

where T_E is referred to the input terminals of the device, $L = 1/G$ is the loss factor and T_O is the thermometric temperature of the device.

Figure 11-18a shows a receiver connected to an antenna via a transmission line of loss L_{TL}. The receiver has an input noise temperature T_{REC}, the antenna has a loss factor L_A (= $1/\eta$, where η is the antenna radiation efficiency), and the antenna and transmission line are both at a (physical) temperature T_a. The equivalent noise temperature at the antenna/transmission-line junction, looking in the direction of the receivers, is given by

$$T'_{REC} = T_{TL} + \frac{T_{REC}}{G_{TL}} \qquad (11-67)$$

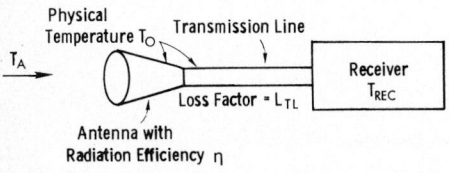

(a) Antenna Connected to Receiver

(b) Noise - Free Equivalent

Fig. 11-18. The configuration shown in (a) is represented by a noise-free equivalent with all sources of noise referenced to the antenna input in the form of T_{SYS}.

where T_{TL} is the transmission line noise temperature and $G_{TL} = 1/L_{TL}$ is its gain. Using Eq. 11-66 for T_{TL}, we have

$$T'_{REC} = (L_{TL} - 1) T_a + L_{TL} T_{REC}. \qquad (11-68)$$

We may extend the above formulation a step further to incorporate the thermal noise generated by the antenna structure. The antenna integrates radiation incident upon it, which was characterized by an antenna temperature T_A in a previous section, and generates thermal noise of its own if not lossless. The system of Figure 11-18a may be represented by the equivalent noise-free system of Figure 11-18b whose input is given by a system noise temperature, T_{SYS},

$$T_{SYS} = T_A + (L_A - 1) T_a + L_A T'_{REC}. \qquad (11-69)$$

The first term in Eq. 11-69 represents the incident radiation, weighted by the antenna directional pattern, the second term represents self-emission by the antenna, and the last term represents thermal noise by the transmission-line/receiver combination. The last two terms represent noise at fictitious antenna terminals to the left of the antenna. Upon inserting Eq. 11-68 into Eq. 11-69, we have:

$$T_{SYS} = T_A + (L_A - 1) T_a + \left(1 - \frac{1}{L_{TL}}\right)$$
$$L_A T_a + L_A L_{TL} T_{REC}. \qquad (11-70)$$

As will be discussed in the next sections, some types of radiometer receivers employ a switch to compare the energy received from the antenna to the noise energy supplied by a noise source of known noise temperature. Such a switch usually is placed as close to the antenna terminals (i.e., antenna/transmission-line junction) as possible. At this junction, the total equivalent noise power is given by

$$\begin{aligned} T'_{SYS} &= \frac{1}{L_A} T_{SYS} \\ &= \frac{T_A}{L_A} + \left(1 - \frac{1}{L_A}\right) T_a + T'_{REC} \\ &\qquad (11-71) \\ &= T'_A + T'_{REC} \end{aligned}$$

where

$$T'_A \triangleq \frac{T_A}{L_A} + \left(1 - \frac{1}{L_A}\right) T_a. \qquad (11-72)$$

BASIC RECEIVER OPERATION

One of the simplest receiver configurations used in microwave radiometry is the *total-power* receiver. This configuration will be used in this section to discuss the basic operation of a microwave radiometer and to define its figures of merit.

The total-power radiometer shown in Figure 11-19 consists of an antenna connected to a superheterodyne receiver of bandwidth B and total

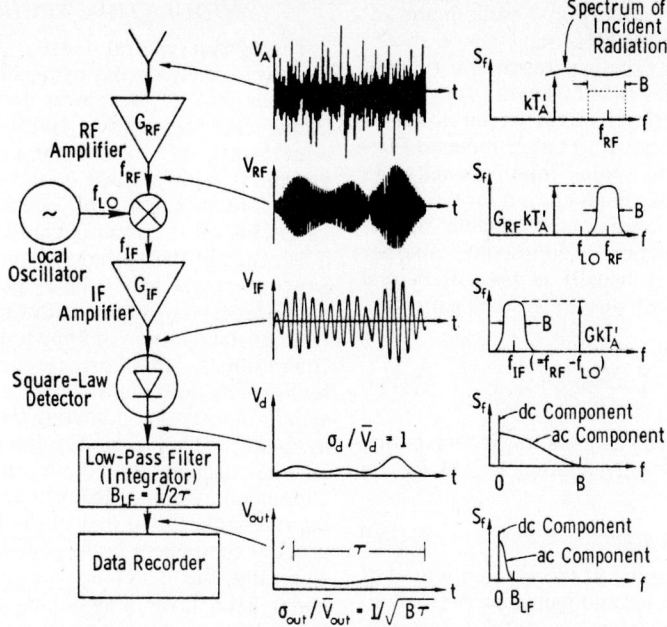

G = Power gain of predetection section (between RF amplifier input and IF amplifier output).

S_f = Power spectral density, W Hz^{-1}

Fig. 11-19. Total-power radiometer with a superheterodyne receiver. The signal voltage and corresponding spectrum are shown at various stages (from Ulaby et al., 1981).

power gain G, followed by a square-law detector and a low-pass filter. Also shown are the voltage waveforms and corresponding power spectra at several points between the radio-frequency (RF) input and the final output. Using the noise equivalent concepts of the previous section, the receiver may be regarded as noise-free by referring its noise to its input terminals. At the antenna terminals, the total available equivalent noise power is T_{SYS} as defined by Eq. 11-71.

The output of the square-law detector, $V_d(t)$, consists of a *dc* component, $\bar{V}_d$, and a fluctuating component $V_{dac}(t)$,

$$V_d(t) = \bar{V}_d + V_{dac}(t).\qquad(11\text{-}73)$$

The voltage of the input noise usually is characterized by Gaussian statistics, which after square-law detection results in (Ulaby et al., 1981),

$$\bar{V}_d = C_d\,G\,k\,B\,T'_{SYS}$$
$$= C_d\,G\,K\,B\,(T'_A + T'_{REC})\qquad(11\text{-}74)$$

and

$$\frac{\sigma_d}{\bar{V}_d} = 1\qquad(11\text{-}75)$$

where C_d is the square-law detector constant and σ_d is the standard deviation of $V_d(t)$ (or the *rms* value of the *ac* component, $V_{dac}(t)$). The quantity

δ_d represents the uncertainty associated with the measurement of $\bar{V}_d$, which in turn is a measure of the uncertainty associated with the measurement of the antenna temperature T'_A. To reduce this large degree of uncertainty to an acceptable level, a low-pass filter (integrator) is used to eliminate the high-frequency variations in the noise spectrum. If the integration time (not time-constant) of the low-pass filter is τ and if its voltage gain is g_{LF}, its *dc* output voltage is given by

$$\bar{V}_{out} = g_{LF}\,C_d\,G\,k\,B\,T'_{SYS}\qquad(11\text{-}76)$$

and the standard-deviation-to-mean ratio of $V_{out}(t)$ is

$$\frac{\sigma_{out}}{\bar{V}_{out}} = \frac{\sigma_d}{\bar{V}_d}\cdot\frac{1}{\sqrt{B\tau}} = \frac{1}{\sqrt{B\tau}}.\,(11\text{-}77)$$

In other words, the uncertainty is reduced by the square root of the time-b   product, $\sqrt{B\tau}$. If all receiver paramete   on-stant, the combination of E leads to

$$\Delta T_N = \frac{T'_{SYS}}{\sqrt{B\tau}} =$$

where ΔT_N is the stan This quantity is also ' sensitivity or radiom script N in ΔT_N den

494

MANUAL OF REMOTE SENSING

due to noise fluctuations. For a radiometer receiver with $T'_{SYS} = 1000$ MHz, and $\tau = 1$ s, $\Delta T_N = 0.1$ K. This is an excellent resolution relative to the range of values that T_A covers for natural earth surfaces, which extends between about 50 K and 320 K, and relative to target-induced fluctuations in T_A due to spatial inhomogeneity. In practice, however, the values chosen for the above parameters may not always be realizable, as discussed in a later section. Additionally, another source of uncertainty usually is present in real receivers, namely that due to system gain fluctuations,

$$\Delta T_G = \left(\frac{\Delta G_S}{G_S}\right) T'_{SYS} \qquad (11\text{-}79)$$

where G_S is a system gain factor incorporating all the system parameters relating $\bar{V}_{out}$ to T'_{SYS} in Eq. 11-76,

$$G_S = g_{LF} \, C_d \, G \, k \, B \qquad (11\text{-}80)$$

and ΔG_S is the *rms* value of the *ac* component of G_S. Assuming the noise and gain uncertainties to be statistically independent, the total uncertainty is given by

$$\Delta T = \left[(\Delta T_N)^2 + (\Delta T_G)^2\right]^{1/2} \qquad (11\text{-}81)$$

$$= \left[\frac{1}{B\tau} + \left(\frac{\Delta G_S^2}{G_S}\right)\right]^{1/2} (T'_A + T'_{REC}).$$

The majority of gain variations are attributed to the stability of the receiver front-end components such as the RF amplifier and mixer-IF amplifier assembly. Typically, $(\Delta G_S/G_S)$ is of the order of 10^{-3} to 10^{-2}, which for $T'_{SYS} = 1000$ K, results in ΔT_G between 1 K and 10 K. This uncertainty due to gain variations is one to two orders of magnitude larger than the 0.1-K noise-generated uncertainty mentioned earlier. Several receiver configurations have been conceived to reduce the magnitude of ΔT_G, or eliminate it altogether. This is accomplished through the use of modulation techniques, as discussed in the next section.

MODULATION TECHNIQUES

The power spectral density of the system gain variation usually decreases rapidly with frequency, with the bulk of the power density contained at frequencies below a few Hertz. That is, the gain uncertainty ΔT_G is caused primarily by slow variations with periods of the order of 0.1 s or longer. In principle, the effects of system gain variations on the input signal could be removed if a record of the variations were available. Such a record may be generated by periodically measuring the receiver output, with the input connected to a constant source of known noise temperature. Operationally, this correction procedure can be realized by modulating the receiver input and synchronously demodulating the output as shown in Figure 11-20. The modulation consists of switching the receiver input between the antenna and a temperature-controlled matched load at a switching rate higher than that of the highest significant spectral component in the gain variation spectrum. Typically, the switching rate f_s is between 50 Hz and 1 kHz. Over a switching period τ_s ($= 1/f_s$), the system gain G_S essentially is constant, and therefore has the same value for the half-cycle during which the switch is connected to the antenna and the next half-cycle during which it is connected to the constant load.

The function of the synchronous demodulator is to change the polarity of the detected voltage in synchronism with the position of the input switch. When this is properly done the final output of the integrator is proportional to the difference between the detected voltage due to the energy received from the antenna and the detected voltage due to energy from the constant load. Since each of these two detected voltages contains the same receiver noise component, the subtraction action of the demodulator ends up cancelling out this component in the final *dc* output:

$$\bar{V}_{out} = \tfrac{1}{2} G_S (T'_A - T_{REF}) \qquad (11\text{-}82)$$

where T_{REF} is the noise temperature of the reference load. The ½-factor is due to the fact

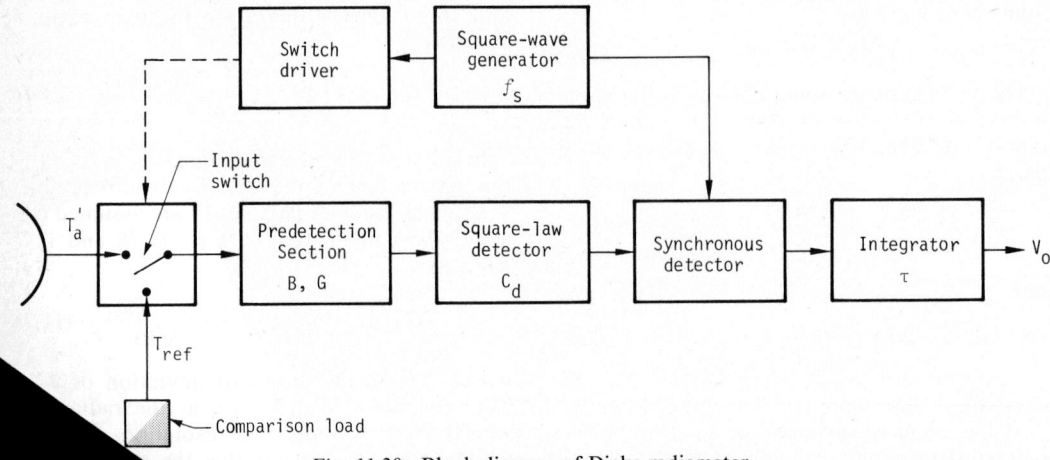

Fig. 11-20. Block diagram of Dicke radiometer.

that T_A' and T_{REF} are each observed for one-half of the time.

The modulation radiometer is more commonly known as the comparison-, switched-, or Dicke-radiometer, the latter being in reference to R. H. Dicke who was the first to introduce the modulation/demodulation technique to microwave radiometry (Dicke, 1946).

The radiometric resolution associated with the above modulation scheme can be shown to be given by (Ulaby et al., 1981):

$$\Delta T = \left[\frac{2(T_A' + T_{REC}')^2 + 2(T_{REF} + T_{REC}')^2}{B\tau} + \left(\frac{\Delta G_S}{G_S} \right)^2 (T_A' - T_{REF})^2 \right]^{1/2} \quad (11\text{-}83)$$

Figure 11-21 compares ΔT as given by the above expression with ΔT of the total-power radiometer (Eq. 11-81), both plotted as a function of the antenna noise temperature T_A'. For the values of $\Delta G_S/G_S$, T_{REF} and T_{REC}' specified in the figure, ΔT of the modulated radiometer always is smaller than that of the total-power radiometer, and the difference between the two is most significant for high values of T_A'.

Close inspection of Eq. 11-83 would show that as a function of T_A', ΔT is minimum when $T_A' = T_{REF}$. For this case, Eq. 11-83 reduces to:

$$\Delta T = 2 \frac{(T_{REF} + T_{REC}')}{\sqrt{B\tau}} , \text{ for } T_A' = T_{REF}, \quad (11\text{-}84)$$

which is independent of $(\Delta G_S/G_S)$. This is known as the "balanced" condition because, according to Eq. 11-82, the output voltage is zero.

Null-Balancing Techniques

To take advantage of the balanced condition, several feedback schemes have been proposed for maintaining the radiometer receiver in this condition on a continuous basis. These schemes may be divided into the following groups:

(a) Reference-Channel Control Technique: The reference temperature T_{REF} is made to track T_A' on a continuous basis through the use of an external feedback loop (Figure 11-22a) that maintains the output voltage in the null condition. Since $T_A' = T_{REF}$ always, T_A' is determined by recording the control voltage V_c, having established a relationship between V_c and T_{REF} through calibration. For the configuration shown in Figure 11-22,

$$T_{REF} = T_N' \quad (11\text{-}85)$$

$$T_N' = \frac{T_N}{L} + \left(1 - \frac{1}{L} \right) T_O \quad (11\text{-}86)$$

where T_N is the noise temperature of the noise source and L and T_O are the loss factor and ambient temperature of the voltage-controlled variable attenuator.

(b) Antenna-Channel Control Technique: The null condition is realized by feeding the appropriate amount of noise into the antenna channel (Figure 11-23) so that the total, T_A'', is equal to T_{REF}, the latter being maintained constant. That is,

$$T_A'' = T_{REF} \quad (11\text{-}87)$$

$$T_A'' = \left(1 - \frac{1}{F_c} \right) T_A' + \frac{T_N'}{F_c} \quad (11\text{-}88)$$

where F_c is the coupling factor of the directional coupler and T_N' is given by Eq. 11-86.

The noise injected through the directional coupler may be provided on a continuous basis or in the form of short pulses with high power levels such that the average noise level satisfies Eqs. 11-87 and 11-88. A description of the pulsed noise-injection scheme is available in the paper by Hardy et al. (1974).

IF Gain Modulation

The receiver output can be nulled by square-wave modulating the gain of the IF amplifier in synchronism with the input switch such that $T_A' G_1 = T_{REF} G_2$ where G_1 and G_2 are the power gains of the IF amplifier during the half-cycles when the input switch is connected to the antenna and to the reference source, respectively. If G_1 (or G_2) is maintained constant, and the relationship between G_2 (or G_1) and the feedback control voltage V_c is known, then T_A' can be determined from V_c. The IF gain-modulation technique is not very popular in microwave radiometry due to several drawbacks (Ulaby et al., 1981) when compared with the techniques discussed earlier.

Automatic-Gain-Control Techniques

Traditional receivers use automatic-gain-control (AGC) to maintain the output voltage at a constant level. Continuous AGC is inapplicable to radiometers because the AGC in effect eliminates all variations, including those due to T_A'. Hence, the sampled AGC technique was introduced by Seling (1964) to monitor the receiver's gain during only the half-cycles of the square-wave modulation when the receiver is connected to the reference load (maintain at a constant temperature). In this case, the AGC feedback loop responds to system gain variations and not to T_A'. One of the limitations of this technique is that the AGC loop cannot differentiate between system

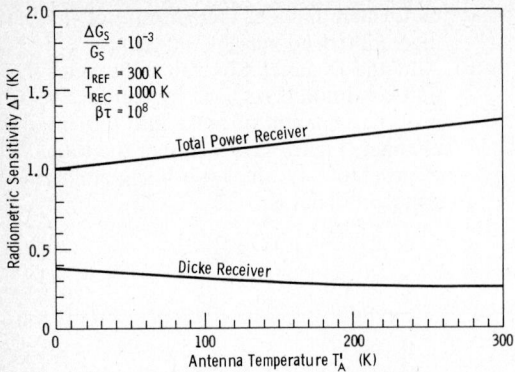

$$\frac{\Delta G_S}{G_S} = 10^{-3}$$

$T_{REF} = 300$ K
$T_{REC} = 1000$ K
$\beta\tau = 10^8$

Total Power Receiver

Dicke Receiver

Fig. 11-21. Radiometric sensitivity of total-power and Dicke receivers as a function of antenna temperature.

gain variations and slow variations in the receiver noise temperature.

A different form of sampled AGC was introduced by Hach (1968) who proposed the use of two constant-temperature loads and two synchronous detectors. His configuration, shown in Figure 11-24, uses an input switch and a reference switch with the switching frequency of the input switch being exactly twice that of the reference switch. The receiver has two outputs, V_{sig} at the output of the signal synchronous detector, and V_{agc} at the output of the AGC synchronous detector. The dc values of V_{sig} and V_{agc} can be shown to be given by (Hach, 1968; Ulaby et al., 1981):

$$\bar{V}_{sig} = G' g_{sig} (2 T_A' - T_1 - T_2) \quad (11\text{-}89)$$

$$\bar{V}_{agc} = G' g_{agc} (T_2 - T_1) \quad (11\text{-}90)$$

where G' is a system gain factor, g_{sig} and g_{agc} are the voltage gains of the signal and AGC synchronous detectors, respectively, and T_1 and T_2 are the constant reference temperatures.

The Hach radiometer may be used with or without feedback. In the absence of feedback, it is necessary to measure V_{sig} and V_{agc}, whose ratio is independent of G', and from which T_A' is obtained via

$$T_A' = \frac{1}{2}\left[\left(\frac{V_{sig}}{V_{agc}}\right)\left(\frac{g_{agc}}{g_{sig}}\right)(T_2 - T_1) + (T_2 + T_1)\right].$$

$$(11\text{-}91)$$

Alternatively, $\bar{V}_{agc}$, which is independent of T_A' and T_{REC}', may be measured experimentally and set equal to a constant reference voltage V_R. Variations in $(V_{agc} - V_R)$ are then due to variations in G', which can be compensated for by the feedback loop shown in Figure 11-24. In this case T_A' is given by the same expression given above except for V_{agc} being replaced by V_R.

The Hach receiver and the Dicke balanced receiver have two important features in common; these are (a) insensitivity to system gain variations, and (b) insensitivity to receiver noise variations. The major difference between the two types of receivers is that the Dicke balanced receiver relies on controlling the noise output of RF devices while the Hach radiometer relies on controlling the gain of a video amplifier (as well as the use of two synchronous demodulators). This is an attractive feature of the Hach radiometer because it is generally easier to control and calibrate video frequency devices than it is to control and calibrate microwave devices. On

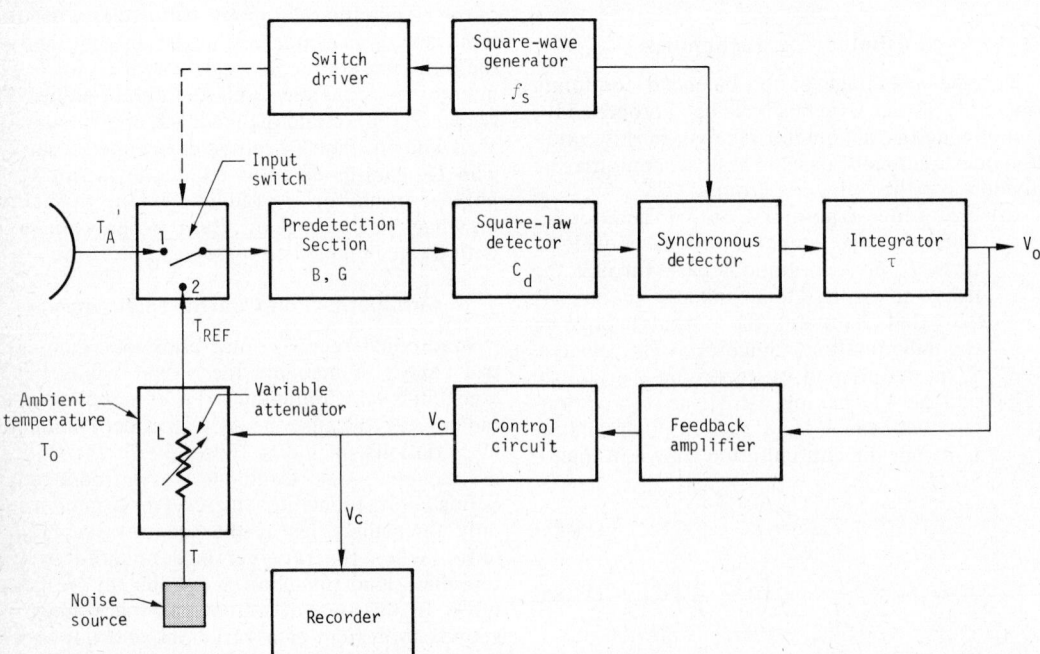

Fig. 11-22. Dicke radiometer using feedback to control the reference temperature T_{REF} to achieve balance.

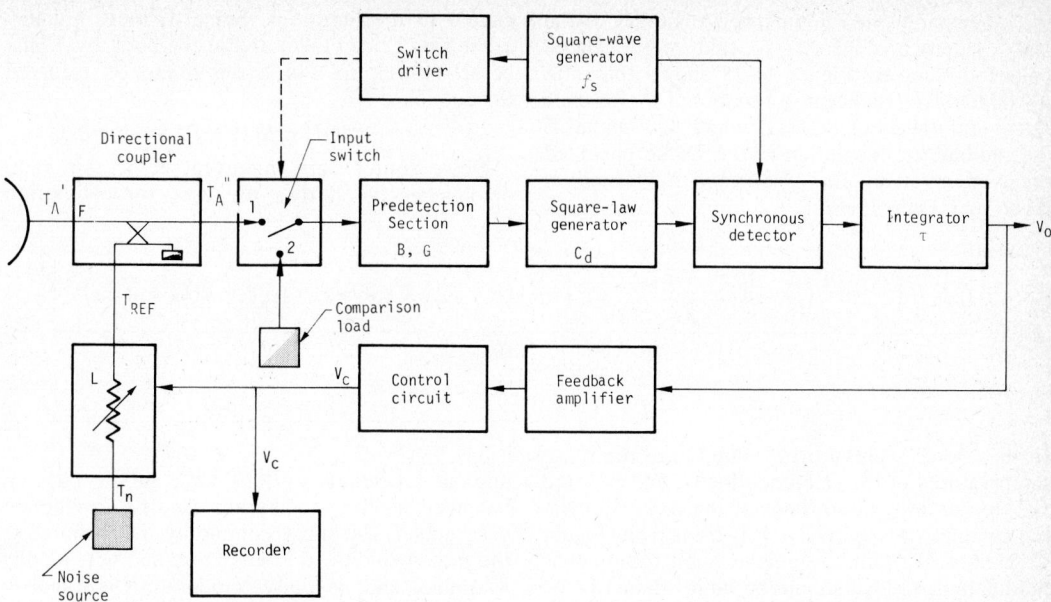

Fig. 11-23. Dicke radiometer using feedback to control the level of the injected noise temperature T_n' to achieve balance.

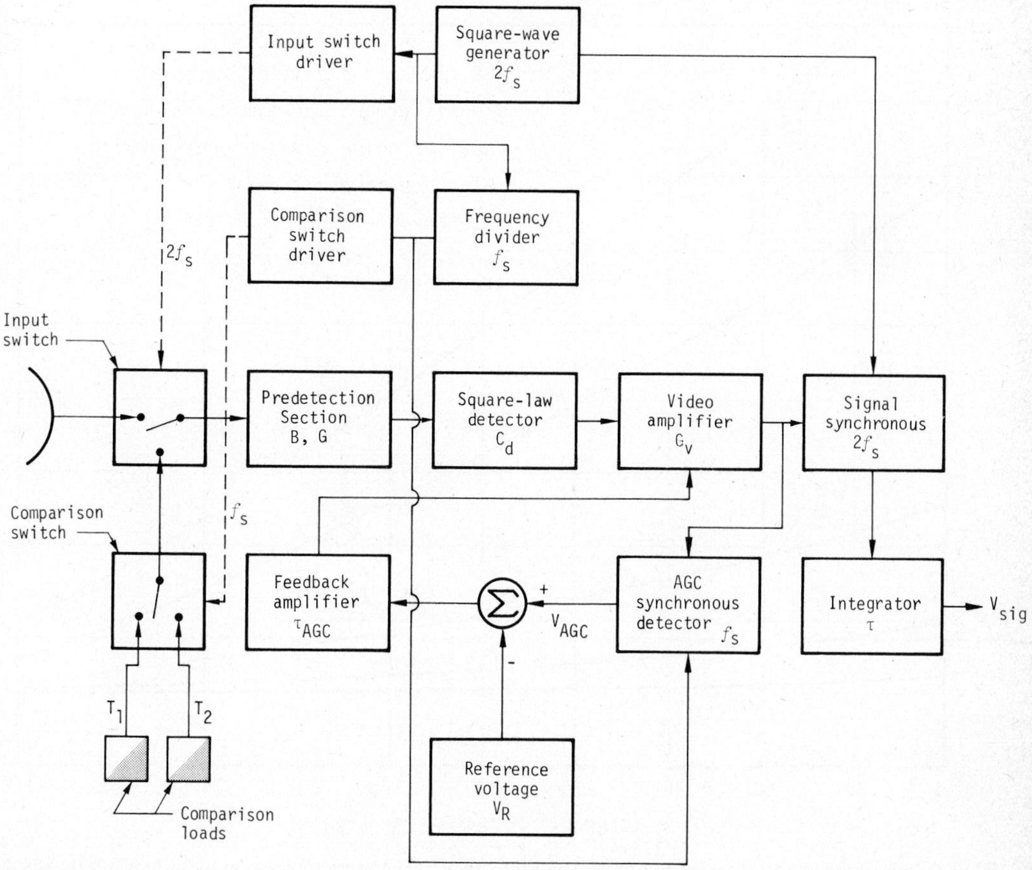

Fig. 11-24. Hach radiometer.

the other hand, the radiometric sensitivity of the Hach receiver is inferior to that of the Dicke balanced-receiver. Figure 11-25 shows the ratio ΔT (Hach)/ΔT (Dicke) as a functional T_A' for each of several values of τ_{agc}/τ_{sig}, where ΔT (Dicke) is the radiometric resolution of the Dicke balanced receiver, given by Eq. 11-84, and ΔT (Hach) is given by (Hach, 1968):

$$\Delta T = \frac{\left\{ \left[1 + \left(\frac{1}{1 + \tau_{agc}/\tau_{sig}} \right) \left(\frac{T_2 + T_1 - 2T_A'}{T_2 - T_1} \right)^2 \right] \left[(T_2 + T_{REC}')^2 + (T_1 + T_{REC}')^2 + 2(T_A' + T_{REC}')^2 \right] \right\}^{1/2}}{\sqrt{B} \, \tau_{sig}}$$

(11-92)

In the above expression, T_1 and T_2 are the noise temperatures of the reference loads, and τ_{agc} and τ_{sig} are the integration times of the AGC and signal channels, respectively. It is clear from Figure 11-25 that in order to achieve good radiometric modulation, τ_{agc}/τ_{sig} should be larger than 10.

Other Receiver Configurations

In addition to the commonly used receiver configurations described above, several other types of special-purpose receivers also have been re-

ported in the literature. Some of these are discussed by Price (1976) and in the book by Ulaby et al. (1981), to which the reader is referred for details.

NOISE SOURCES

Noise sources are important devices in radiometer receivers; they are used for calibration and as reference sources and, in the case of balanced receivers, they are used to provide excess noise. The most commonly used source is the matched load. It is easy to construct, readily available, and its noise temperature is equal to its physicial temperature. The matched load often is referred to as a passive noise source.

Active noise sources usually deliver noise power levels in excess of kT_0B. The *excess noise ratio*, ENR, is defined as

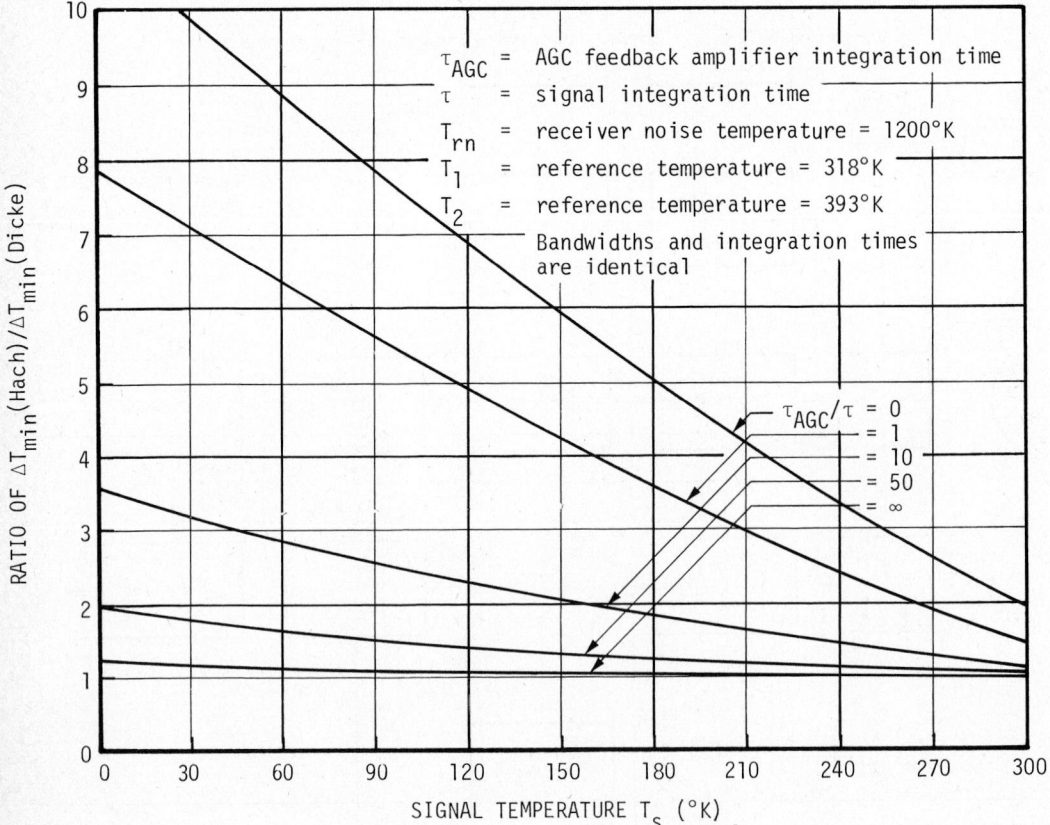

Fig. 11-25. Ratio of the temperature resolution of a two-reference temperature radiometer (with AGC) to the resolution of a conventional balanced-Dicke radiometer.

$$ENR = \frac{kB(T_N - T_O)}{kT_O}$$

$$= \frac{T_N}{T_O} - 1 \qquad (11\text{-}93)$$

where T_N is the output noise temperature of the device and T_O is its physical temperature. Active noise sources include the gas-discharge tube and several solid-state devices, most notably the avalanche diode.

As will be discussed later, internal calibration of a radiometer receiver usually necessitates the use of a matched "cold" calibration source with a noise temperature in the 50–100 K range. Traditionally, this has been done by physically cooling a matched load to the desired temperature by immersing it in a cryogenic fluid, such as liquid nitrogen. Recently, however, Frater and Williams (1981) demonstrated that a field-effect transistor (FET) can be made to deliver a noise temperature that is much lower than its physical temperature; at 1.4 GHz, a noise temperature as low as 50 K was measured at room temperature. With such a device, which its inventors call a COLDFET, cryogenic cooking can be avoided. Of course, the concept of *excess* noise temperature becomes meaningless in this case since $T_N < T_O$, and therefore ENR is negative.

ADVANCES IN RECEIVER PERFORMANCE

One of the most important performance parameters of a radiometer receiver is its noise temperature T_{REC}. Until the early 1970s, typical values of T_{REC} for uncooled receivers ranged between 500 K and 1000 K for frequencies below 10 GHz, and higher than this for higher frequencies. As an example, the receiver noise temperature of the 13.9 GHz radiometer flown on Skylab in 1973 was 1195 K (see Table 11-3). Advances in microwave-integrated-circuit technology have led to significant improvements in the noise performance and stability of microwave receivers. Figure 11-26 shows a schematic of an MIC that includes the front-end RF components of the 2.65-GHz radiometer shown in block-diagram form in Figure 11-27 from Hardy et al. (1974). The authors state that the MIC, shown in the hatched constant-temperature enclosure of Figure 11-27, was "fabricated in microstripline on substrates of high-purity alumina and placed on opposing broad walls of a short length of rectangular waveguide of standard dimensions, 2.84 by 1.34 in. Coupling to the waveguide is by means of probes." Using this configuration, the receiver noise temperature was reported to be only 70 K.

In most receivers, the overall noise performance is dictated by the noise temperature of the RF amplifier or by that of the mixer-preamp if no RF amplifier is used. In the receiver shown in Figure 11-27, for example, 50 K of the 70-K receiver noise temperature was contributed by the RF

TABLE 11-3

Summary of the Radiometer Section of Skylab's RADSCAT. (General Electric, 1973).

Antenna:	
Antenna Type	Parabolic Reflector
Reflector Diameter	114 cm
Half-Power Beamwidth	1.5°
Beam Efficiency	90%
Antenna Loss	0.3 dB
Polarization	Horizontal and Vertical
Radiometer:	
Center Frequency	13.9 GHz
Predetection Half-Power Bandwidth	210 MHz
Receiver Noise Temperature	1195 K
Comparison Temperatures	318 K and 393 K
Switching Frequency	996 Hz
Video Amplifier Gain	42 dB ± 4 dB
AGC Integration Time-Constant	0.9 s
ΔT_{min} for:	
32 ms integration time	1.15 K
128 ms integration time	0.6 K
256 ms integration time	0.425 K

parametric amplifier. In general, radiometers operating at frequencies below about 10 GHz usually are configured as single-sideband receivers with RF amplifiers, while millimeter wave (30 ≥ GHz) radiometers are double-sideband receivers with a mixer-preamp at the front end. In the in-between frequency range between 10 and 30 GHz, both configurations are used, depending on other receiver specifications. The major reason for this dichotomy is that low-noise solid state and un-

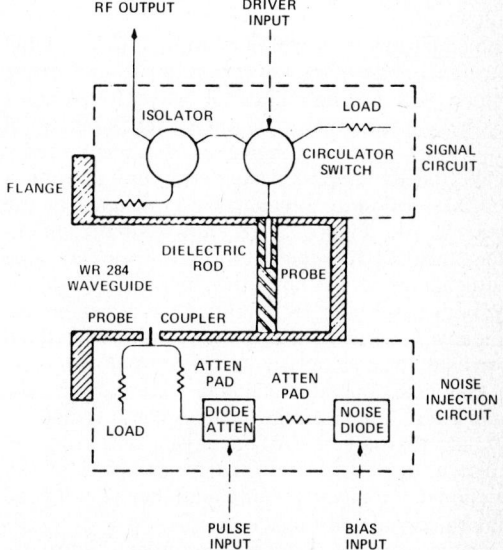

Fig. 11-26. Schematic of Microwave Integrated Circuit [MIC] (from Hardy et al., 1974).

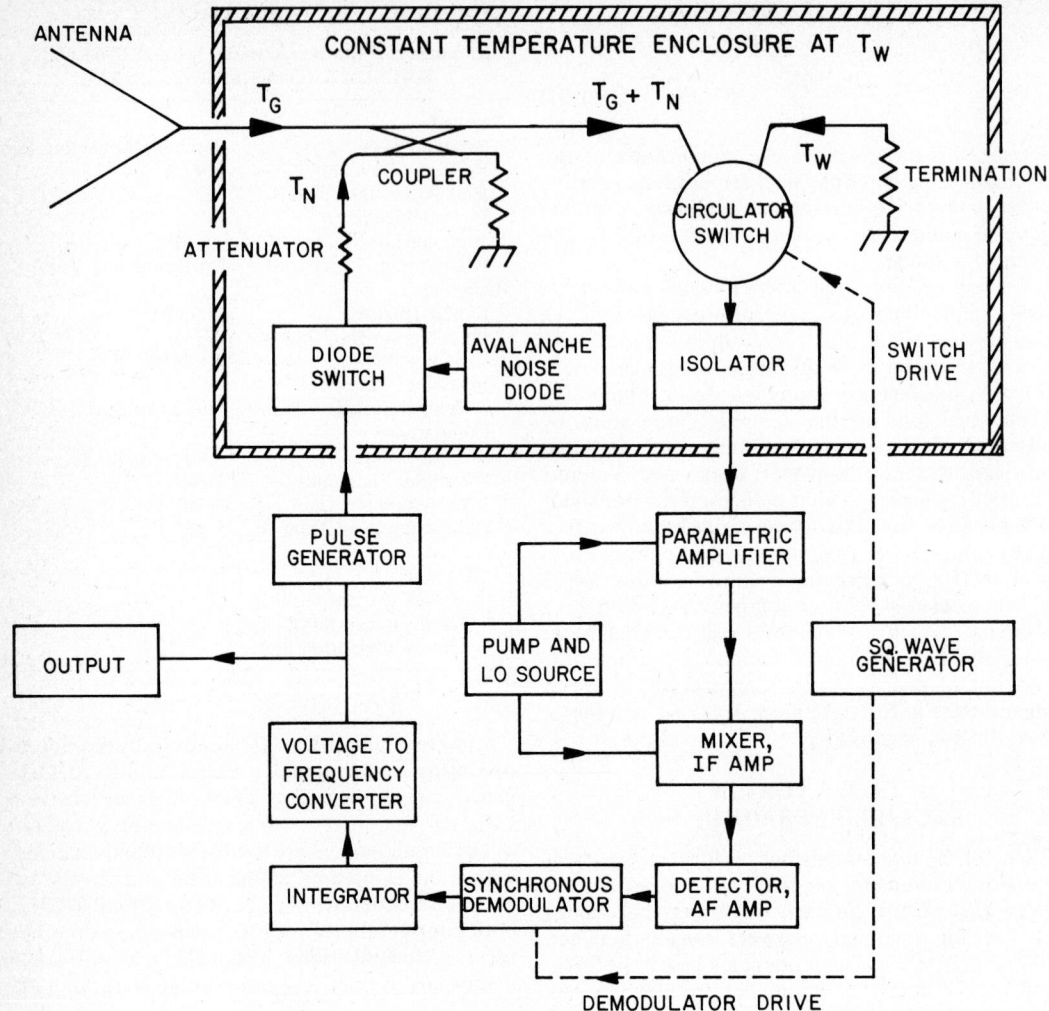

Fig. 11-27. Simplified block-diagram of feedback-mode radiometer (after Hardy et al., 1974).

cooled parametric amplifiers are readily available for frequencies in the lower part of the microwave region, but not at millimeter wave frequencies at the present time. The rapid development of solid state amplifiers, however, will undoubtedly make the RF amplifier front-end configuration a feasible option at millimeter wavelengths in the near future. Figure 11-28 shows simple block-diagrams of two superheterodyne receivers, one with an input RF amplifier, and one without. In both cases, the IF amplifier has a center frequency f_{IF} and a bandwidth B, and the local oscillator frequency is f_{LO}. The output spectrum of the mixer contains bands centered around the sum and difference of the local oscillator frequency f_{LO} and the input RF frequency f_I. The sum frequency, $f_{LO} + f_I$, is rejected by the preamplifier section of the mixer-preamp, and therefore it need not be considered further.

In the absence of an RF amplifier (Figure 11-28a), two RF bands exist that, when mixed with the LO signal, result in the IF spectrum of the IF amplifier. Each of these RF bands is of bandwidth B and their center frequencies are:

$$f_{I1} = f_{LO} + f_{IF} \qquad (11\text{-}94a)$$

and

$$f_{I2} = f_{LO} - f_{IF} \qquad (11\text{-}94b)$$

as illustrated in Figure 11-28a. Thus, the IF signal contains energy from a total RF bandwidth of $2B$.

The reader may recall that the equivalent noise temperature T_E of a device is defined according to the equivalency of Figures 11-15a and b. If a matched resistor is placed at the input of the mixer-preamp of Figure 11-28a, the output noise power is given by:

$$P_O = 2G\,K\,T_I\,B + \Delta P_O \qquad (11\text{-}95)$$

where T_I is the resistor temperature, G is the gain and ΔP_O is the internally generated noise power of the mixer-preamp. The equivalent noise temperature T_E is defined such that if a fictitious

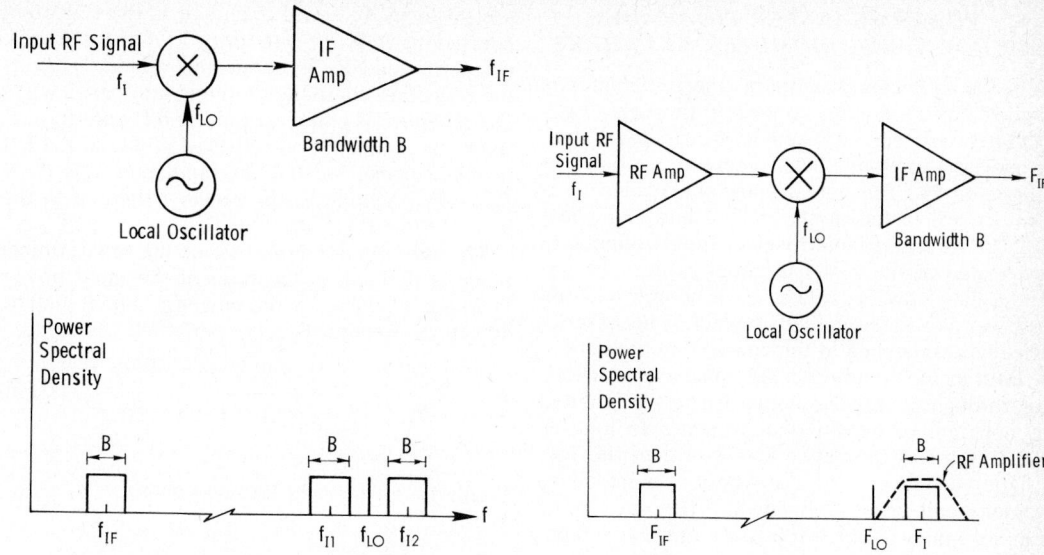

Fig. 11-28. Comparison of spectra for (a) double-sideband and (b) single-sideband receivers.

broadband noise source of temperature T_E is added to the input of the device, the same output power P_O is generated with the device assumed to be noiseless.

Thus,

$$P_O = 2G\,K\,T_I\,B + 2G\,K\,T_E\,B. \quad (11\text{-}96)$$

Relating Eq. 11-95 to Eq. 11-96, we obtain

$$T_E\,(\text{DSB}) = \frac{\Delta P_O}{2\,G\,K\,B} \quad (11\text{-}97)$$

where DSB stands for double-sideband. If, on the other hand, a filter or RF amplifier were used (as in Figure 11-28b) to limit the available input noise to only one of the two RF bands (centered about f_{I1} and f_{I2}), the total RF bandwidth would have been B rather than $2B$. In that case we would obtain the following expression for the equivalent noise temperature of the mixer-preamp:

$$T_E\,(\text{SSB}) = \frac{\Delta P_O}{G\,K\,B} \quad (11\text{-}98)$$

where SSB stands for single-sideband. Comparison of Eqs. 11-97 and 11-98 leads to:

$$T_E\,(\text{DSB}) = \tfrac{1}{2}\,T_E\,(\text{SSB}) \quad (11\text{-}99)$$

The receiver noise temperatures for the two configurations shown in Figure 11-26 are:

$$T_{REC}\,(\text{DSB}) = T_M\,(\text{DSB}) + \frac{T_{IF}}{GM} \; ; (\text{DSB})$$

$$T_{REC}\,(\text{SSB}) = T_{RF}\,(\text{SSB}) + \frac{T_M\,(\text{SSB})}{G_{RF}}$$

$$+\ \frac{T_{IF}}{G_M\,G_{RF}} \; ; (\text{SSB}).$$

In the 1–10 GHz frequency range, RF amplifiers with $T_{RF} < 300$ K and $G_{RF} > 100$ are readily available, thereby leading to SSB receiver noise temperatures of the order of 300 K or less. At millimeter-wave frequencies, however, such RF amplifiers are not available, especially at frequencies above 100 GHz. Mixer-preamp configurations with T_M (DSB) of the order of 300–500 K are available at frequencies as high as 170 GHz; and when such devices are used in radiometer receivers, overall receiver noise temperatures, T_{REC} (DSB), of the order of 500–900 K are obtained (Cardiasmenos, 1980; Vowindel et al., 1980).

RADIOMETER CALIBRATION

A radiometer is said to be calibrated when an accurate relationship has been established between the receiver output voltage V_0 and the antenna integrated absolute brightness temperature. This process normally involves two steps: (1) calibration of the receiver, in which a linear relationship is found between V_0 and the input temperature T_{IN} to the receiver first stage, and (2) finding a relationship between T_{IN} and the integrated brightness temperature T_A. In the latter step, compensation is made for the attenuation and mismatch losses suffered in the antenna and connecting network (e.g., waveguide) as well as thermal noise emitted by those structures. In a practical situation, establishing an accurate radiometer calibration which holds over long periods of time can be a demanding, difficult task. This section discusses the mathematical basis and practical techniques of radiometer calibration.

POWER TRANSFER THROUGH
MICROWAVE RADIOMETER NETWORKS

In the following discussion, the radiometer is modeled as an antenna connected through a two-port network (e.g., waveguide, transmission line, or isolator) to a wide-band receiver, as shown in Figure 11-29. The antenna and waveguide connectors are mated at the $x = 0$ terminal plane and the waveguide and receiver input connectors are mated at the $x = l$ terminal plane. The antenna may now be viewed as a source and the receiver as a load, neither of which is necessarily impedance-matched to the connecting network.

Both incident and reflected noise waves traveling through microwave networks may be treated as coherent or quasi-monochromatic so long as the length l of the network is less than the correlation distance $d_c = v_p/\Delta f$ where v_p is the phase velocity and Δf is the bandwidth (Carver, 1977). For example, if $\Delta f = 100$ MHz and $v_p = 0.6c$, $d_c = 1.8$ m, which is considerably longer than most waveguides or transmission lines used in practical radiometers. The assumption of coherence allows the use of conventional microwave network theory, as presented, for example, in the excellent treatment by Kerns and Beatty (1967). A brief summary of a few useful network equations is given here for a reader assumed to be familiar with the concept of network S-parameters.

Consider a two-port network (e.g., transmission line, attenuator, isolator, etc.) characterized by the matrix $[S_{ij}]$, as shown in Figure 11-29. The characteristic impedances of the input and output waveguides are denoted as Z_{o1} and Z_{o2} respectively and are assumed to be both positive and real. The source and load voltage reflection coefficients

are given by Γ_G and Γ_L, and Γ_1 is the reflection coefficient looking into port 1 of the network when the load Z_L is connected. Similarly, Γ_2 is the reflection coefficient looking into port 2 when the generator is connected. P_{i1} and P_{r1} are respectively the incident and reflected power at the left terminal plane, so that the input power is $P_1 = P_{i1} - P_{r1}$. Similarly, the power delivered to the load is $P_L' = P_{i2} - P_{r2}$.

The *mismatch factor* m_{G1} at the left terminal plane is defined as the ratio of the input power P_1 to the available source power P_{i1} and it may be shown (Kerns and Beatty, 1967) that

$$m_{G1} = \frac{P_1}{P_{i1}} = \frac{(1 - |\Gamma_G|^2)(1 - |\Gamma_1|^2)}{|1 - \Gamma_G \Gamma_1|^2}. \quad (11\text{-}100)$$

Similarly, at the right terminal plane,

$$m_{2L} = \frac{P_L'}{P_{i2}} = \frac{(1 - |\Gamma_2|^2)(1 - |\Gamma_L|^2)}{|1 - \Gamma_2 \Gamma_L|^2}. \quad (11\text{-}101)$$

The *efficiency* η_{12} of the network is defined as the ratio of the load power P_L' to the input power P_1 and it may be shown that

$$\eta_{12} = \frac{P_L'}{P_1} = \frac{Z_{o1} |S_{21}|^2 (1 - |\Gamma_L|^2)}{Z_{o2} |1 - S_{22} \Gamma_L|^2 (1 - |\Gamma_1|^2)}. \quad (11\text{-}102)$$

The *available transmission factor* τ_{12} is defined as the ratio of the available power P_{i2} at the output port to the available power P_{i1} from the source (Otoshi, 1968), i.e.,

$$\tau_{12} = \frac{P_{i2}}{P_{i1}} = \frac{P_L' m_{G1}}{m_{2L} P_1} = \frac{\eta_{12} m_{G1}}{m_{2L}}. \quad (11\text{-}103)$$

Substitution of Eqs. 11-100, 11-101 and 11-102 into Eq. 11-103 yields

$$\tau_{12} = \frac{1}{L_w} \left[\frac{(1 - |\Gamma_G|^2)(1 - |S_{11}|^2)}{|1 - S_{11} \Gamma_G|^2 (1 - |\Gamma_2|^2)} \right] \quad (11\text{-}104)$$

where L_w is the network attenuation factor (≥ 1) given by (Otoshi, 1968)

$$L_w = \frac{Z_{o2} (1 - |S_{11}|^2)}{Z_{o1} |S_{21}|^2}. \quad (11\text{-}105)$$

Two additional useful relations are

$$\Gamma_1 = S_{11} + \frac{S_{12} S_{21} \Gamma_L}{1 - S_{22} \Gamma_L} \quad (11\text{-}106)$$

$$\Gamma_2 = S_{22} + \frac{S_{21} S_{12} \Gamma_G}{1 - S_{11} \Gamma_G}. \quad (11\text{-}107)$$

If the network is an ideal uniform waveguide or transmission line, $S_{11} = S_{22} = 0$ and $S_{12} = S_{21} =$

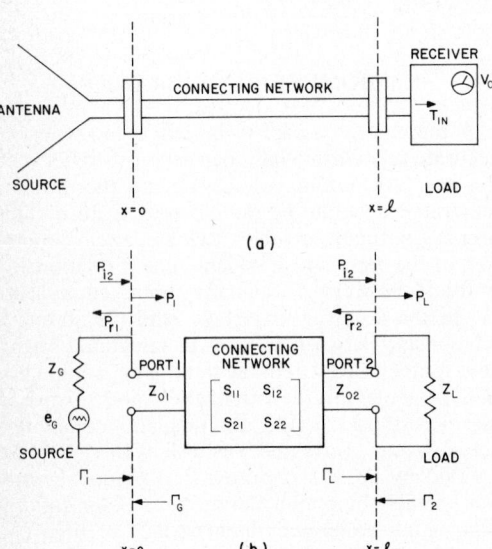

Fig. 11-29. (a) Physical model of microwave radiometer; (b) network model of microwave radiometer.

$e^{-\gamma l}$ where $\gamma = \alpha + j\beta$ is the complex propagation constant. In this case, $\Gamma_1 = \Gamma_L e^{-2\gamma l}$ and $\Gamma_2 = \Gamma_L e^{-2\gamma l}$. For an ideal isolator with phase shift ϕ, $S_{11} = S_{12} = S_{22} = 0$ and $S_{21} = e^{j\phi}$. Thus, for an ideal isolator, $\Gamma_1 = \Gamma_2 = 0$.

Eqs. 11-100 to 11-107 can be used to describe the transfer of noise power through a microwave radiometer by use of the Nyquist formula, Eq. 11-26. However, it should be remembered that the "sources" of noise power may not be only those network elements conventionally thought of as sources, and that any resistive elements that dissipate power will also act as emitters of noise power. Thus, if the antenna of Figure 11-29 is replaced by a commercial noise source of available temperature T_G, then $P_{i1} = k\, T_G\, \Delta f$. The noise power absorbed by the receiver (load) is $P_L = k\, T_{IN}\, \Delta f$. Although most of T_{IN} is due to T_G (after attenuation and mismatch by the connecting network), an additional fraction is due to noise originating in resistive elements of the connecting network as well as lossy elements in the first stage of the receiver.

RECEIVER CALIBRATION

Radiometer receiver calibration requires that a linear relation be established between the output voltage or counts and the input noise power or temperature, i.e.,

$$V_o = m\, T_{IN} + b \qquad (11\text{-}108)$$

where m is the slope and b is the intercept. By the use of two external noise generators, i.e., a hot load and a cold load as shown in Figure 11-30, two widely separated known input temperatures can be generated so that two equations in two unknowns are obtained:

$$V_c = m\, T_{IN}^c + b \qquad (11\text{-}109)$$

$$V_h = m\, T_{IN}^h + b \qquad (11\text{-}110)$$

with the solution

$$m = \frac{V_c - V_h}{T_{IN}^c - T_{IN}^h} \qquad (11\text{-}111)$$

$$b = \frac{V_h\, T_{IN}^c - V_c\, T_{IN}^h}{T_{IN}^c - T_{IN}^h}. \qquad (11\text{-}112)$$

Commercially available calibration noise sources for coaxial transmission-line systems make use of precision 50-ohm loads held at a constant known temperature; the hot load is usually maintained at about 370 K and the cold 50-ohm load is placed in a cryogenic flask cooled by liquid nitrogen at its boiling temperature. Figure 11-31 is a photograph of one such commercially available, dual-temperature calibration noise source which uses 14-mm precision coaxial connectors. Because of impedance mismatch between the noise source and the receiver, it is possible for the input temperature T_{IN} to differ significantly from the manufacturer's stated available noise temperature T_G of the hot and cold loads. Waveguide noise sources also use nominally matched loads and are also subject to mismatch effects.

Assuming that a direct connection is made between the noise source and the receiver input, so that $\eta_{12} = 1$ and $\Gamma_1 = \Gamma_L$, Eqs. 11-100 and 11-102 reveal that

$$T_{IN} = \frac{(1 - |\Gamma_G|^2)\,(1 - |\Gamma_L|^2)}{|1 - \Gamma_G \Gamma_L|^2}\, T_G + T_{SE}$$

$$(11\text{-}113)$$

where T_G is the available noise temperature (equal to T_h or T_c) which would be delivered to a matched load. The second term in Eq. 11-113 accounts for the self-emitted noise from the receiver first stage at temperature T_x which is transmitted

Fig. 11-31. Commercially available dual-temperature noise source with 14-mm coaxial connector terminals. The top of the cold-load cryoflask is shown extending through the top of the unit.

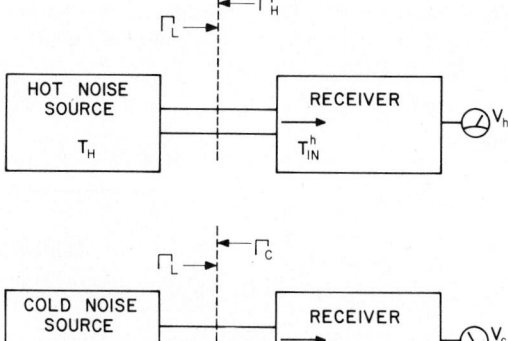

Fig. 11-30. Connection of hot- and cold-load noise sources to radiometer receiver.

by the radiometer and then partly reflected back into the receiver. According to Eq. 11-113, T_{SE} is the measured input temperature when a short circuit is placed on the input. T_{SE} is related to T_x by

$$T_{SE} = (1 - m_{L2})\, T_x$$

$$= \left\{ 1 - \frac{(1 - |\Gamma_L|^2)\,(1 - |\Gamma_G|^2)}{|1 - \Gamma_G\,\Gamma_L|^2} \right\} T_x.$$

Evaluation of the denominator of Eqs. 11-113 and 11-114 requires a knowledge of the phases of the reflection coefficients as well as their amplitudes. However, since the reflection coefficients usually have a small amplitude (less than 0.1), the input temperature may be calculated for maximum and minimum values by letting the denominator become $(1 \pm |\Gamma_G||\Gamma_L|)^2$ so that Eq. 11-113 becomes

$$T_{IN}^{\pm} = T_x + \frac{(T_G - T_x)\,(1 - |\Gamma_G|^2)\,(1 - |\Gamma_L|^2)}{\left[1 \mp |\Gamma_G||\Gamma_L|\right]^2}.$$

$$(11\text{-}115)$$

If both noise generator and receiver are matched, then $T_{IN} = T_G$, independently of T_x. As an example, suppose that the VSWR looking into the hot load is 1.25 so that $|\Gamma_H| = .1111$, and the VSWR looking into the receiver is 1.09 so that $|\Gamma_L| = .0431$. Assume further that the hot load temperature is $T_H = 325$ K and the receiver front-end ambient temperature is $T_x = 300$ K. Then, from Eq. 11-115

$$T_{IN}^h = \begin{cases} 324.41 \text{ K (minimum)} \\ 324.88 \text{ K (maximum)} \end{cases}.$$

This means that there is a systematic uncertainty of 0.47 K in the actual input temperature of the hot load due to the unknown phases of the reflection coefficients. This uncertainty can be reduced further by lowering the hot load VSWR through the use of a low-loss tuner or by the use of a network analyzer to measure the phases of the reflection coefficients.

It is clear from Eqs. 11-111 and 11-112 that uncertainties in the hot- and cold-input temperatures lead to corresponding uncertainty in the calibration slope m and intercept b values. In addition to this, the receiver gain may slowly drift over a period of hours, days, or even weeks so that it may be necessary to calibrate the receiver at periodic intervals. In some cases, it is possible to establish a relationship between the slope and intercept values and the receiver box temperature.

RADIOMETER RADIATIVE TRANSFER EQUATION

The transmission line or waveguide network that connects the radiometer antenna to the receiver may attenuate the antenna temperature, add to this its own emitted temperature, and through mismatch it may cause a self-emission

temperature from the receiver first stage. In a practical radiometer, these effects are usually significant enough to deserve careful attention during the calibration of the instrument. In this section, we obtain a relationship for the total noise power delivered to the receiver which incorporates all of these effects.

Referring to Figure 11-29, we may replace the combination of the original noise source plus the connecting network wtih an equivalent noise source with available temperature T_e, i.e., $P_{i2} = k\, T_e\, \Delta f$ where P_{i2} is the same available noise power as before and Γ_2 is also the same as before. Thus, the noise power P_L' absorbed by the receiver due *only* to sources to the left of the $x = l$ terminal plane is (Wells, Daywitt and Miller, 1964)

$$P_L' = m_{2L}\, k\, T_e\, \Delta f = m_{G1}\, \eta_{12}\, k\, T_G\, \Delta f + P_n$$

$$(11\text{-}116)$$

where the first term on the right-hand side is due to noise from the original generator and the second term is the additional noise generated by lossy elements of the connecting network. When the original noise source, network, and receiver are in thermal equilibrium at temperature T_w ($=T_G$ $=T_e$), then the power absorbed by the network must be re-emitted in an equal amount. Therefore, from Eq. 11-116,

$$P_n = k\, T_w\, \Delta f\, (m_{2L} - m_{G1}\, \eta_{12}) \quad (11\text{-}117)$$

so that in general (Otoshi, 1968),

$$P_L' = k\, \Delta f\, m_{G1}\, \eta_{12}\, T_G + (m_{2L} - m_{G1}\, \eta_{12})\, T_w.$$

$$(11\text{-}118)$$

This assumes that the entire connecting network is at an isothermal temperature T_w. The noise power P_L'' absorbed by the receiver as a result of the reflection of its own self-emission is found from Eq. 11-114 as

$$P_L'' = k\, \Delta f\, (1 - m_{L2})\, T_x. \quad (11\text{-}119)$$

Since P_L' and P_L'' are independent noise sources, they may be added so that

$$P_L = P_L' + P_L'' = k\, T_{IN}\, \Delta f \quad (11\text{-}120)$$

where $T_{IN} = m_{2L}\, T_e$. Therefore, the receiver input temperature is given by the fundamental radiative transfer equation

$$T_{IN} = Y_{GL}\, T_G + E_w\, T_w + R_r\, T_x \quad (11\text{-}121)$$

where

$$Y_{GL} = m_{G1}\eta_{12} = \frac{Z_{o1}}{Z_{o2}}\, \frac{|S_{21}|^2\,(1 - |\Gamma_G|^2)\,(1 - |\Gamma_L|^2)}{|1 - \Gamma_G\,\Gamma_1|^2|1 - S_{22}\,\Gamma_L|^2}$$

$$(11\text{-}122)$$

$$E_w = \frac{(1 - |\Gamma_2|^2)\,(1 - |\Gamma_L|^2)}{|1 - \Gamma_2\,\Gamma_L|^2} - Y_{GL}$$

$$(11\text{-}123)$$

$$R_r = 1 - \frac{(1 - |\Gamma_L|^2)\,(1 - |\Gamma_2|^2)}{|1 - \Gamma_L\Gamma_2|^2}. \quad (11\text{-}124)$$

The term Y_{GL} is known as the *transmission coefficient* $(0 < Y_{GL} \le 1)$, the term E_w as the *emission coefficient* $(0 < E_w \le 1)$, and R_r as the *power reflection coefficient* $(0 < R_r \le 1)$. Furthermore, it is seen from Eqs. 11-122, 11-123 and 11-124 that

$$Y_{GL} + E_w + R_r = 1. \qquad (11\text{-}125)$$

Therefore, if the noise sources, connecting network and receiver are in thermal equilibrium $(T_g = T_w = T_x = T_o)$, then from Eq. 11-121, $T_{IN} = T_O$. Eq. 11-125 may therefore be viewed as an extension of Kirchhoff's Law.

One connecting network of particular interest is a reflectionless but lossy uniform waveguide or transmission line, for which $S_{11} = S_{22} = 0$, $S_{12} = S_{21} = e^{-\gamma l}$, and $Z_{o1} = Z_{o2} = Z_o$. In this case,

$$Y_{GL} = \frac{1}{L_w} \frac{(1 - |\Gamma_G|^2)(1 - |\Gamma_L|^2)}{|1 - \Gamma_G \Gamma_L e^{-2\gamma l}|^2}$$

$$E_w = \left(1 - \frac{1}{L_w}\right) \frac{(1 + |\Gamma_G|^2/L_w)(1 - |\Gamma_L|^2)}{|1 - \Gamma_G \Gamma_L e^{-2\gamma l}|^2}$$
$$(11\text{-}127)$$

$$R_r = 1 - \frac{(1 - |\Gamma_G|^2/L_w^2)(1 - |\Gamma_L|^2)}{|1 - \Gamma_G \Gamma_L e^{-2\gamma l}|^2}$$
$$(11\text{-}128)$$

where

$$L_w = e^{2\alpha l} \qquad (11\text{-}129)$$

is the dissipative attenuation factor. There are several limiting mismatch situations which may be encountered in practice.

CASE II: *Negligible Generator Mismatch* $(\Gamma_G = 0)$

For this case,

$$Y_{GL} = \frac{1}{L_w}(1 - |\Gamma_G|^2) \qquad (11\text{-}130)$$

$$E_w = \left(1 - \frac{1}{L_w}\right)\left(1 + \frac{|\Gamma_G|^2}{L_w}\right) \qquad (11\text{-}131)$$

$$R_r = \frac{|\Gamma_G|^2}{L_w^2}. \qquad (11\text{-}132)$$

CASE II: *Negligible Generator Mismatch* $(\Gamma_G = 0)$

For this case,

$$Y_{GL} = \frac{1}{L_w}(1 - |\Gamma_L|^2) \qquad (11\text{-}133)$$

$$E_w = \left(1 - \frac{1}{L_w}\right)(1 - |\Gamma_L|^2) \qquad (11\text{-}134)$$

$$R_r = |\Gamma_L|^2. \qquad (11\text{-}135)$$

CASE III: *No Mismatches* $(\Gamma_G = \Gamma_L = 0)$

For this case, the coefficients reduce to

$$Y_{GL} = \frac{1}{L_w} = e^{-2\alpha l} \qquad (11\text{-}136)$$

$$E_w = 1 - \frac{1}{L_w} = 1 - e^{-2\alpha l} \qquad (11\text{-}137)$$

$$R_r = 0 \qquad (11\text{-}138)$$

so that in this simple situation,

$$T_{IN} = \frac{T_G}{L_w} + \left(1 - \frac{1}{L_w}\right)T_w \qquad (11\text{-}139)$$

which is the formula seen in many articles on radiometer calibration, and is equivalent to Eq. 11-68.

However, although this is by far easier to use than the previous more complicated formulas, it often introduces calibration errors which are unacceptable in practical operation. Nonetheless, eq. 11-139 does yield a simple relationship for finding the dissipative loss factor $e^{2\alpha l}$, i.e.,

$$L_w = \frac{T_G - T_w}{T_{IN} - T_w}. \qquad (11\text{-}140)$$

In order to use this relation to find L_w, it is necessary to insure that the reflection coefficients are negligible (e.g., by the use of low-loss tuners) and that the transmission line or waveguide is kept at a constant temperature over its length. If these constraints are met, then only three measurements are necessary in order to find the dissipative loss factor: (1) the available noise temperature from the generator, T_G; (2) the input temperature to the receiver, T_{IN}; and (3) the thermometric temperature of the waveguide or line, T_w.

In the geometry discussed in this section, the generator reflection coefficient is associated with the input to the antenna. The magnitude of Γ_G can be found from the antenna VSWR through the use of $|\Gamma_G| = (\text{VSWR} - 1)/(\text{VSWR} + 1)$. However, it is more difficult to measure the receiver reflection coefficient, since the power levels associated with microwave test equipment are normally much higher than the operating range of the receiver. As a result, an attempt to directly measure the input reflection coefficient by the application of milliwatt-level signals will at best give an erroneous reading and at worst burn out the switch or RF amplifier. Commercially available radiometer receivers are usually furnished with specifications that includes the maximum input VSWR, which can then be used to estimate the worst-case reflection coefficient of the load or receiver.

In order to illustrate the importance of the preceding formulas for waveguide correction factors to the absolute calibration of a radiometer, an example using typical operating factors will be given. Consider a radiometer whose available antenna noise temperature is $T_G = 150$ K from a parabolic dish antenna with input VSWR = 1.20, i.e., $\Gamma_G = .0909$. The radiometer receiver has a maximum input VSWR = 1.10, i.e., $\Gamma_L = .0476$, and an available self-emission noise temperature of $T_o = 290$ K which was determined by placing a short circuit at the receiver input. Connecting the antenna and the receiver is a length of transmission line whose temperature is 290 K ($=T_w$). In order to illustrate the importance of maintaining very low line-loss in achieving absolute calibration, we will consider two dissipative loss factors of 0.5 dB ($L_w = 1.1220$) and 0.1 dB ($L_w = 1.0233$). Since we do not know the phase of the reflection coefficients, we will use $|1 - \Gamma_G \Gamma_L e^{-2\gamma l}|^2 \rightarrow [1 \pm |\Gamma_G||\Gamma_L|/L_w]^2$ in the denominator of Eqs. 11-126, 11-127 and 11-128 for an upper and lower bound. The results of this are shown in Table 11-4.

This means that the system with a 0.5-dB dissipative loss in the transmission line would have a systematic uncertainty of 1.90 K in the input temperature; with a 0.1-dB dissipative loss, this uncertainty would be increased to 2.30 K as a result of decreased padding on the generator reflection coefficient. In a practical situation, the radiometer would register some input temperature T_{IN} and from this, Eq. 11-121 would be used to solve for the apparent antenna (generator) noise temperature. As an example, suppose that the 0.1-dB loss transmission line was used and that the radiometer input temperature is $T_{IN} = 140.00$ K. Then, by applying the correction factors of Table 11-4 and using Eq. 11-121, the maximum estimated antenna temperature would be 136.19 K, and the minimum T_G would be 133.56 K, an uncertainty of 2.63 K.

ANTENNA LOSS CORRECTION FACTORS

Practical radiometer antennas such as horns or paraboloidal dishes are constructed from aluminum, copper, brass, etc., and usually have low heat-losses, although these effects are not necessarily negligible when calibrating the radiometer system. When an electromagnetic wave is incident on an antenna, a portion of the energy is scattered or re-radiated by the antenna, another portion is lost as heat within the structure, and the remainder is delivered to the load. The antenna loss factor L_A was defined in Eq. 11-11 and its effect on the antenna temperature T_A was implicitly included in Eq. 11-39 within the antenna gain, i.e.,

$$T_A = \frac{1}{4\pi L_A} \iint_{4\pi} T_B(\theta,\phi) D(\theta,\phi) \, d\Omega = \frac{T_B'}{L_A}$$

where T_B' is the integrated brightness temperature and D is the antenna directivity function. Therefore, the antenna temperature is inversely proportional to the antenna loss.

Due to thermal emission within the antenna structure, an additional term must be added to T_A in order to account for the total noise temperature available at the antenna terminals. The available thermal noise power due to emission within the antenna structure depends on the current distribution on the antenna in much the same way as that within a waveguide or transmission line, except that it is much more difficult to describe mathematically since the structure is neither of one-dimensional nor of uniform cross-section. There are no explicit solutions available to date that treat antennas with any general thermometric temperature distribution. One approximation to this solution may be obtained by assuming that an antenna emission coefficient E_a exists in a form similar to that for a waveguide (Eq. 11-127) except that $\Gamma_G = 0$ and Γ_L is replaced by Γ_l, i.e.,

$$T_G = \frac{T_B'}{L_A} + \left(1 - \frac{1}{L_A}\right) T_a \quad (11\text{-}142)$$

where T_A is given by Eq. 11-141, T_a is the antenna structure thermometric temperature, and $E_a = 1 - 1/L_A$ is the antenna emission coefficient.

If the antenna reflection coefficient is negligible ($\Gamma_G = 0$), then the receiver input temperature is found from Eq. 11-121 to be

$$T_{IN} = \frac{1 - R_r}{L_w} T_G + (1 - R_r)\left(1 - \frac{1}{L_w}\right) T_w.$$
$$+ R_r T_x \quad (11\text{-}143)$$

Substitution of Eq. 11-142 into Eq. 11-143 gives a simple relationship between the receiver input temperature and the integrated brightness temperature:

$$T_{IN} = \frac{1 - R_r}{L_A L_w} T_B' + \frac{(1 - R_r)}{L_w}\left(1 - \frac{1}{L_A}\right) T_a$$
$$+ (1 - R_r)\left(1 - \frac{1}{L_w}\right) T_w + R_r T_x. \quad (11\text{-}144)$$

This is a radiative transfer equation which applies to a radiometer with an isothermal antenna having uniformly distributed loss and which has

TABLE 11-4

Calculated Correction Factors

Factors	Dissipative Loss L_w = 0.5 dB		Dissipative Loss L_w = 0.1 dB	
Y_{GL}	.88874 (max)	.87514 (min)	.97519 (max)	.95884 (min)
E_w	.11013 (max)	.10845 (min)	.02310 (max)	.02271 (min)
R_r	.00113 (min)	.01641 (max)	.00170 (min)	.01845 (min)
$T_G Y_{GL}$	133.31 K (max)	131.27 K (max)	146.28 K (max)	143.83 K (min)
$T_w E_w$	31.94 (max)	31.45 (min)	6.70 K (max)	6.59 K (min)
$T_x R_r$	0.33 K (min)	4.76 K (max)	0.49 K (min)	5.35 K (max)
T_{IN}	165.58 K (min)	167.48 K (max)	153.47 K (max)	155.77 K (min)

a reflectionless connection to an isothermal wave-guide with uniformly distributed loss; receiver self-emission is accounted for in the last term. The dissipative loss in waveguide is normally very small; for example, a 1-meter length of copper waveguide operating at 2.4 GHz has a loss factor of $L_w = 1.004$, or .017 dB. Dissipative loss in coaxial transmission line is higher, especially at frequencies above 1 GHz. However, if the wave-guide loss is negligible ($L_w = 1$), Eq. 11-144 becomes

$$T_{IN} = \frac{1 - R_r}{L_A} T'_B + (1 - R_r)\left(1 - \frac{1}{L_A}\right)T_a.$$
$$+ R_r T_x \qquad (11\text{-}145)$$

Since a well-calibrated receiver actually measures T_{IN}, a more useful form of Eq. 11-144 is

$$T'_B = \frac{L_A L_w}{1 - R_r}\left[T_{IN} - R_r T_x\right] - L_A\left[T_a\right.$$
$$\left. + (L_w - 1)T_w\right], \qquad (11\text{-}145a)$$

which can then be used to determine the integrated brightness temperature T'_B for the special case where the antenna reflection coefficient Γ_G is negligible and the intervening waveguide is re-flectionless ($S_{11} = S_{22} = 0$).

ANTENNA LOSS CALIBRATION TECHNIQUES

If the radiometer antenna views an isothermal background at temperature T_B, then $T'_B = T_B$ and

Eq. 11-142 may be used to solve for the antenna loss factor L_A:

$$L_A = \frac{T_B - T_a}{T_G - T_a}. \qquad (11\text{-}146)$$

The use of Eq. 11-146 to find the antenna loss requires three assumptions, viz., that:
1. The background radiation T_B is sufficiently isothermal and is much less than T_a.
2. The antenna structure is at a uniform tem-perature T_a.
3. The available antenna (generator) noise tem-perature T_G can be determined accurately enough from the receiver input temperature T_{IN}, by applying the correction factors of Eq. 11-121.

Two techniques have been used to present a cold isothermal background T_B to the radiometer antenna, namely the *bucket technique* and the *cryoload technique*.

Bucket Technique

In this method, which is well-adapted to phys-ically large radiometer antennas, the background zenith sky temperature T_{SKY} is used as the cold temperature source and the relatively hot noise radiation from the surrounding terrain is blocked out by placing the radiometer inside a large metal bucket, as shown in Figure 11-32. Since the sky noise varies only slightly from its zenith value over ±40°, the assumption of an isothermal sky is good for antenna beamwidths less than about

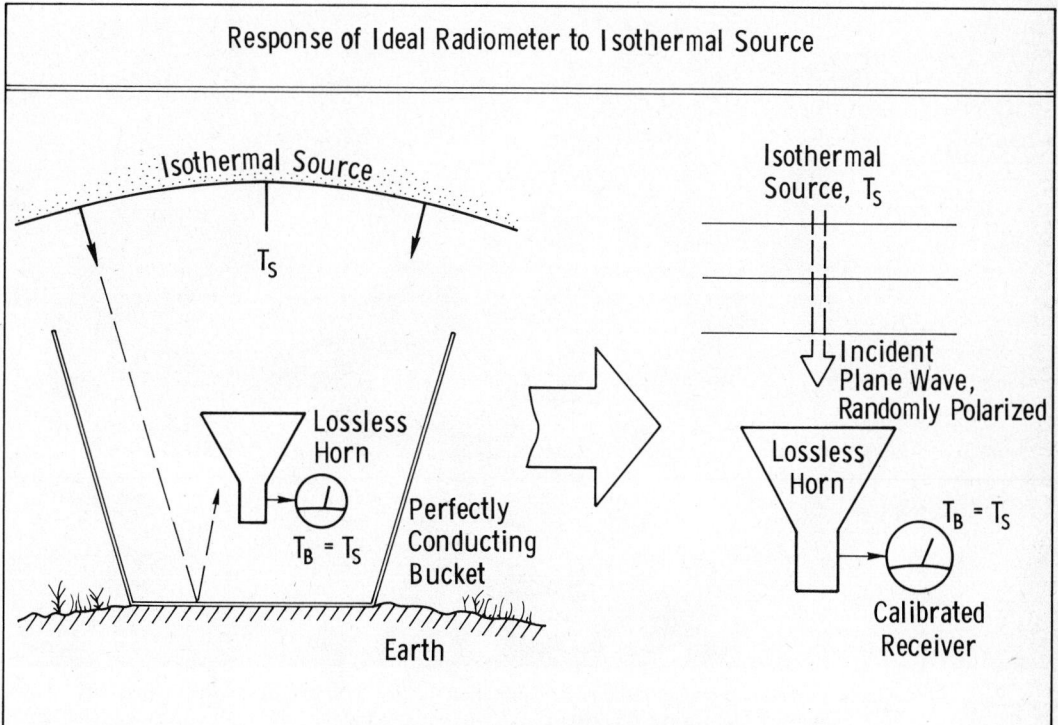

Fig. 11-32. Illustrating the principles in the bucket method of calibrating microwave radiometer antennas.

35°. The reflection of this incident thermal noise from the bucket walls serves to bathe the antenna in an isothermal noise level equal to the zenith sky noise temperature. At frequencies from below 1 GHz to above 60 GHz, the sky temperature may be calculated by assuming a universe background blackbody temperature of 2.7 K and calculating the intervening absorption and emission for a spherically stratified earth atmosphere of pressure-broadened water and oxygen molecular constituents. In order to meet the assumed conditions of horizontal atmospheric layering, it is necessary that the bucket be located at altitudes over 1200 m above mean sea level and that measurements be taken during clear-sky conditions and with low humidity. A computer program (SKYTEMP) is available for calculation of the zenith sky temperature (Paris, 1971) which uses atmospheric height profiles of pressure, air temperature, and relative humidity as an input. These may be obtained from radiosonde data taken at the same time of the loss calibration measurement and in the same location as the bucket. Figure 11-33 is a photograph of a radiometer calibration bucket facility on A-Mountain (1460-m elevation) near New Mexico State University in Las Cruces. The bucket is 4.7 m in height, with a square opening 10.9 m on a side; the base is 3 m on a side. It is a wooden structure with

aluminum-foil-covered plywood sides on an aluminum-covered concrete floor. The emissivity of the bucket is less than 0.001. An extensive report on this facility and an analysis of its performance are available (Carver, 1975).

By assuming that systematic errors in T_{SKY}, T_a and T_G are independent, a most probable error in the measured loss L_A may be computed by

$$\Delta L_A = \left[\left(\frac{\Delta T_{SKY}}{T_{SKY} - T_a} \right)^2 + \left(\frac{T_{SKY} - T_G}{(T_G - T_a)^2} \Delta T_a \right)^2 \right.$$
$$\left. + \left(\frac{T_{SKY} - T_a}{(T_G - T_a)^2} \Delta T_G \right)^2 \right]^{1/2}. \qquad (11\text{-}147)$$

In addition to being frequency-dependent (see Figure 11-13), T_{SKY} is also sensitive to changes in atmospheric water vapor, particularly at 22 GHz where the water molecule exhibits an absorption peak. Typical systematic errors to be expected in T_{SKY} under clear-sky conditions can be computed (Paris, 1975) and are listed in Table 11-5.

Systematic errors in T_a are usually less than ± 0.5 K when high-quality thermistors are used. Systematic errors in T_G were discussed earlier and under good design conditions will be about ±2 K.

As an example, consider a 22-GHz radiometer with $T_a = 290$ K, $T_{SKY} = 15$ K, and a measured $T_G = 21$ K. From Eq. 11-146, $L_A = 1.0223$. If it is further assumed that $\Delta T_{SKY} = 3$ K, $\Delta T_a = 0.5$ K,

Fig. 11-33. Photograph of aluminum bucket radiometer calibration facility at New Mexico State University, showing the airborne MultiFrequency Microwave Radiometer system (without radome) during test. Three scalar horns for K_u, K, and K_a bands are visible along with the L-band printed-circuit planar array antenna.

TABLE 11-5
Computer Errors in T_{SKY} Temperatures

Frequency (GHz)	ΔT_{SKY} (K)
1.414 GHz	0.3 K
10.000	0.4
18.000	0.8
22.000	3.0
37.000	1.8

and $\Delta T_G = 2$ K, Eq. 11-147 gives a most probable error in the antenna loss of $\Delta L_A = .0133$, which means that the actual antenna loss lies in the probable range 1.0090 to 1.0356.

In view of the proliferation of errors, it is easy to lose sight of the objective, which is to estimate the error in the scene brightness temperature T_B'. Using Eq. 11-142, the most probable error in T_B is

$$\Delta T_B' = \{[(1 - L_A)\Delta T_a]^2 + [(T_G - T_a)\Delta L_A]^2$$
$$+ [L_A \Delta T_G]^2\}^{1/2}. \qquad (11\text{-}148)$$

To continue the numerical example from above, let us assume that the 22-GHz radiometer is now used to view a scene of integrated brightness temperature whose nominal value is $T_B' = 150$ K. Using $T_a = 290$ K and $L_A = 1.0223$, Eq. 11-142 predicts that $T_G = 153.05$ K. Further assuming that $\Delta T_a = 0.5$ K, $\Delta L_A = .0133$ and $\Delta T_G = 2$ K, Eq. 11-149 yields

$$\Delta T_B' = (.0001 + 3.313 + 4.186)^{1/2} = 2.74 \text{ K}$$

This probable error in the scene brightness temperature originates primarily from uncertainties in the 22-GHz sky temperature and to systematic errors in the available generator temperature T_G due to mismatch and efficiency factor errors.

Cryoload Technique

If the basic radiative transfer equations, Eqs. 11-121 and 11-125, are combined, it is seen that

$$T_{IN} = (1 - E_w - R_r) T_G + E_w T_w + R_r T_x \qquad (11\text{-}149)$$

Thus, the emission and mismatch losses associated with E_w and R_r give rise to an apparent temperature increase of T_{IN} over T_G, given by

$$T_{IN} - T_G = E_w (T_w - T_G) + R_r (T_x - T_G) \qquad (11\text{-}150)$$

For example, an S-Band radiometer viewing the sea will have T_G of about 100 K, whereas both T_x and T_w might each be 300 K, so that $T_{IN} - T_G \cong 200 (E_w + R_r)$ (Hardy, Gray and Love, 1974). If an absolute accuracy of 0.1 K is desired, then E_w and R_r must remain constant to about 0.001 dB, a condition that cannot be met in practice because (1) there is no independent method of measuring losses to this precision, and (2) transmission-line connector stability and repeatability cannot be relied upon to 0.001 dB.

It is noted from Eq. 11-15 that no correction at all is necessary if both T_w and T_x are equal to T_G, so that the stability of E_w and R_r is no longer a problem. A feedback-mode Dicke radiometer which achieves this condition has been designed by Hardy, Gray and Love (1974) (and was shown earlier in Figure 11-26). This design allows the addition of a noise temperature T_N so that the sum $T_G + T_N$ is brought up to the level of T_w, the temperature of the enclosure. Variable noise injection is provided by a constant pulse-width pulsed noise source so that T_N is proportional to the pulse repetition rate. The radiometer output is an *ac* signal at the modulation frequency, which then produces a null when $T_G + T_N = T_w$, the enclosure temperature. It is possible to establish an extremely linear relationship between the pulse frequency and the average injected noise power, which then leads to the following expression for the input temperature T_{IN} to the parametric amplifier input:

$$T_{IN} = \frac{f}{f_c} T_c + \left(1 - \frac{f}{f_c}\right) T_w + \Delta \qquad (11\text{-}151)$$

where

f = frequency of pulse generator,
f_c = frequency of pulse generator during calibration,
T_c = brightness temperature of noise radiation incident on antenna during calibration (from cryoload),

and where Δ is a small correction term (typically, less than 0.2 K) accounting for losses in the antenna, given by

$$\Delta = (L_A - 1)\left[(T_w - T_a) - \frac{f}{f_c}(T_w - T_{ac})\right] \qquad (11\text{-}152)$$

In Eq. 11-152, L_A is the antenna loss, Eq. 11-11, T_a is the antenna structure temperature and T_{ac} is the antenna structure temperature during calibration. The brightness temperature T_c can be very accurately established by using a cryogenically cooled, microwave-absorbing, pyramidal panel which subtends the entire antenna aperture, as shown in Figure 11-34 (Hardy, 1973). The porous microwave absorber of low reflection coefficient is fitted with a non-porous low-density polyurethane foam cap which is shaped to fit closely over the absorber's pyramidal surface. When the absorber is then filled with either liquid nitrogen or liquid argon, the cap forces the boiling cryogen to conform to the shape of the absorber. Small vent holes above each absorber peak allow the escape of the gas produced by the boiling cryogen and ensure that all of the absorber is saturated. A flexible aluminum shield between the horn and the aluminum cryoload container ensures that all of the energy entering the horn originates in the cooled absorber. The brightness temperature T_c thus produced deviates less than 0.1 K from the boiling points of nitrogen and argon which are respectively

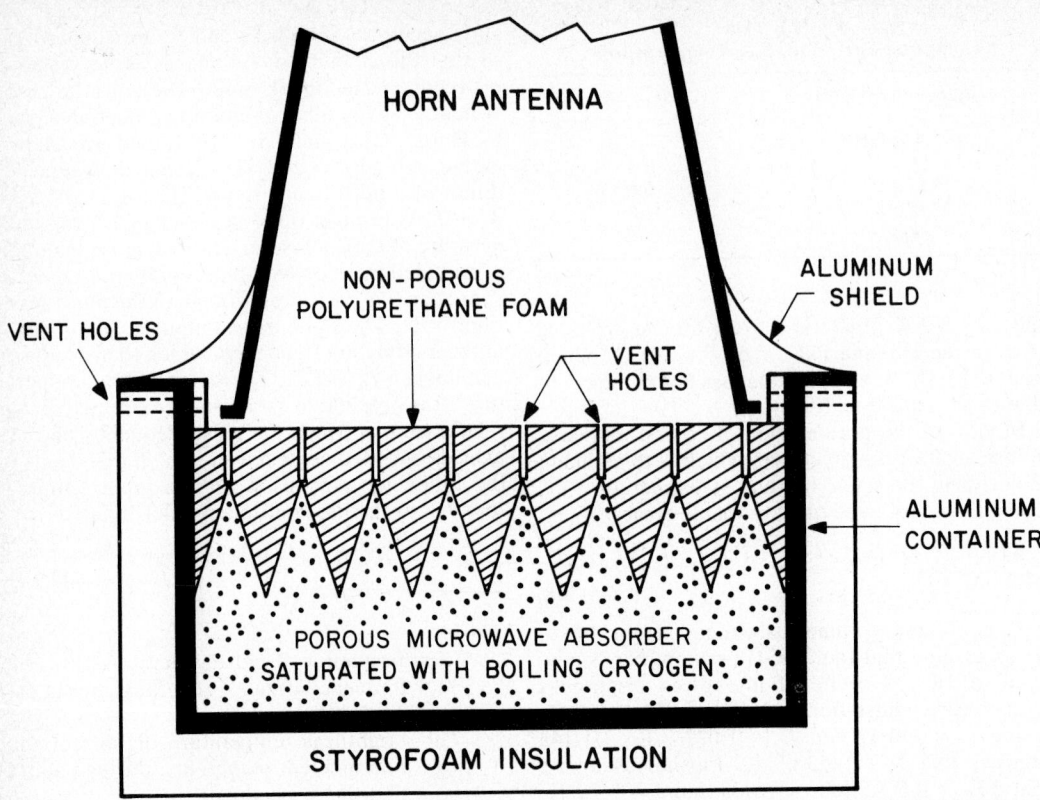

Fig. 11-34. Construction of cryoload for calibration of radiometers with horn antennas (after Hardy, 1973).

$$T_c(N_2) = 77.36 + .011\ (P - 760)\ (K)$$
$$(11\text{-}153)$$

$$T_c(A) = 87.28 + .010\ (P - 760)\ (K)$$
$$(11\text{-}154)$$

where P is the barometric pressure in millimeters of mercury. Figure 11-35 is a photograph of L-band and S-band microwave radiometers in a laboratory cryoload calibration mounting arrangement at NASA Langley Research Center. These radiometers have been used by NASA for airborne measurement of the emissivity of the sea, and have been demonstrated to have absolute accuracies approaching 0.1 K for brightness temperatures around 100 K (Hardy, Gray and Love, 1974). The S-band radiometer has been calibrated at intervals of 3½ years, and has been shown to have an 0.7-K *rms* calibration repeatability, with an average temperature deviation (bias error) of 0.03 K (Blume, 1977).

It is clear that the cryoload technique has considerably better inherent accuracy than the bucket technique, when applied to horn antennas without large radomes. However, no cryoloads have yet been constructed which are large enough to adequately subtend the apertures of larger radiometer antennas systems. These may include large reflector antennas (e.g., the Large Area Mapping Microwave Radiometer which uses a 4.6-m diam-

eter spaceborne offset-fed reflector antenna) or airborne radiometer systems using large nose radomes which can add appreciable dissipative and mismatch noise effects. For these larger systems, the bucket technique is preferred.

IMAGING CONSIDERATIONS

Angular resolution of a microwave radiometer may be defined as the halfpower beamwidth of its antenna main-beam. As indicated earlier in this chapter, a beamwidth of λ/L radians is associated with antennas, where λ = wavelength in meters, and L = antenna length, also in meters. Actual beamwidth values may vary by about a factor of two larger or smaller, as shown in Table 11-1. In the case of circularly symmetrical antennas, such as parabolic reflectors, L is the antenna diameter and the resulting main beam has a circular cross-section. In practice, however, if the antenna feed is linearly polarized, the main-beam cross-section may have a slight degree of ellipticity in its shape.

To obtain a radiometric (sometimes called thermal) image of a scene of interest, it is necessary to scan with the antenna main-beam. If the radiometer is stationary relative to the scene, line-by-line scanning is usually used, while with moving airborne platforms, scanning only in the cross-track dimension may suffice. Antenna scanning, or beamsteering, can be accomplished either mechanically or electronically. Mechanical beam-

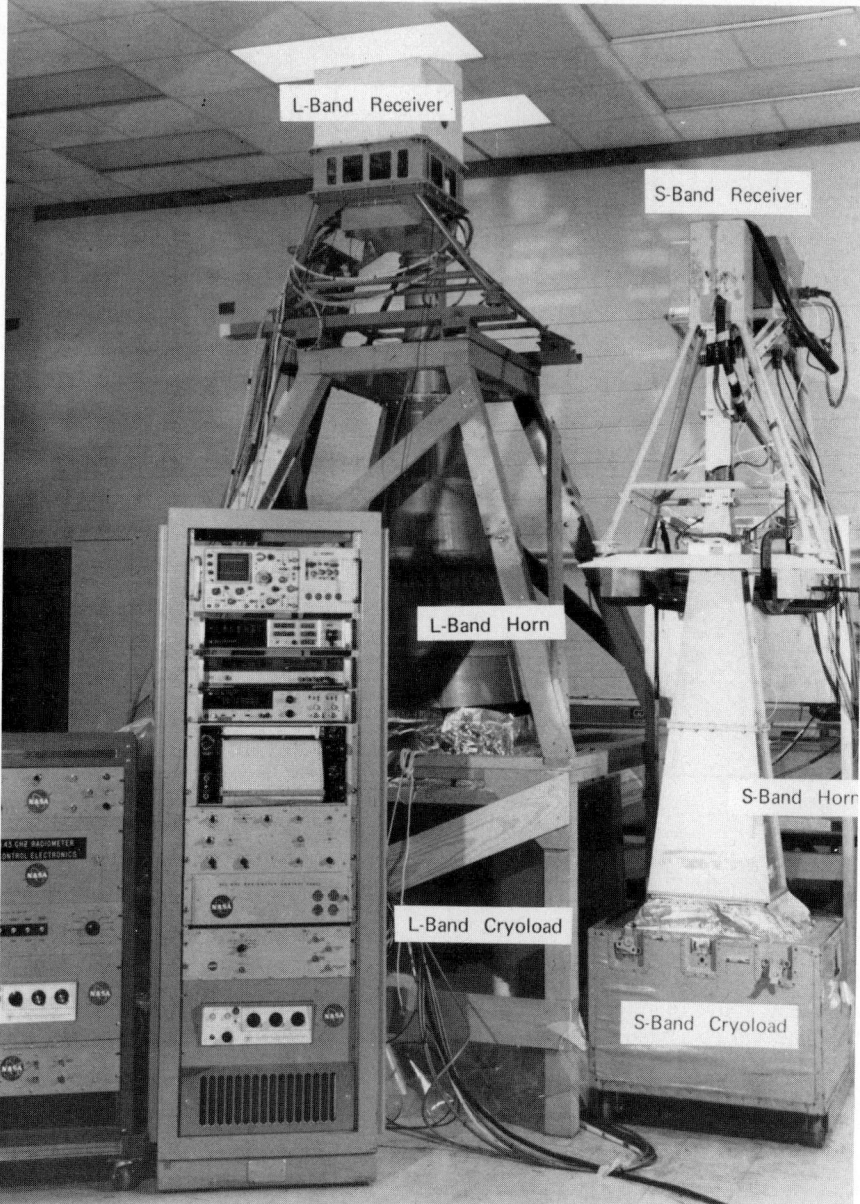

Fig. 11-35. Photograph of S-band and L-band radiometers with associated cryoloads (courtesy of NASA/Langley Research Center).

steering involves changing the pointing direction of the antenna axis by moving (in angle) the entire antenna or its feed.

Electronic beamsteering is achieved by using a planar-phased array consisting of many radiating elements (usually waveguide slots, microstriplines, or small horns). The main beam can be steered in both dimensions by electronically controlling the relative phase of each element (Cheston and Frank, 1980). In an operational system, a computer is required to perform the steering computations. If the antenna is to scan in only one dimension, it is sufficient to control the relative phase in that dimension. This is illustrated in Figure 11-36, in which electronic phase shifters are used to control the phase of each individual feed line. The two most significant advantages of a phased-array antenna are its noninertial beam-steering capability and the high speed at which scanning can be accomplished. This type of antenna does have drawbacks, when compared to a conventional antenna, such as a parabolic dish of the same dimensions: complexity, losses in the phaseshifters, and high cost and weight.

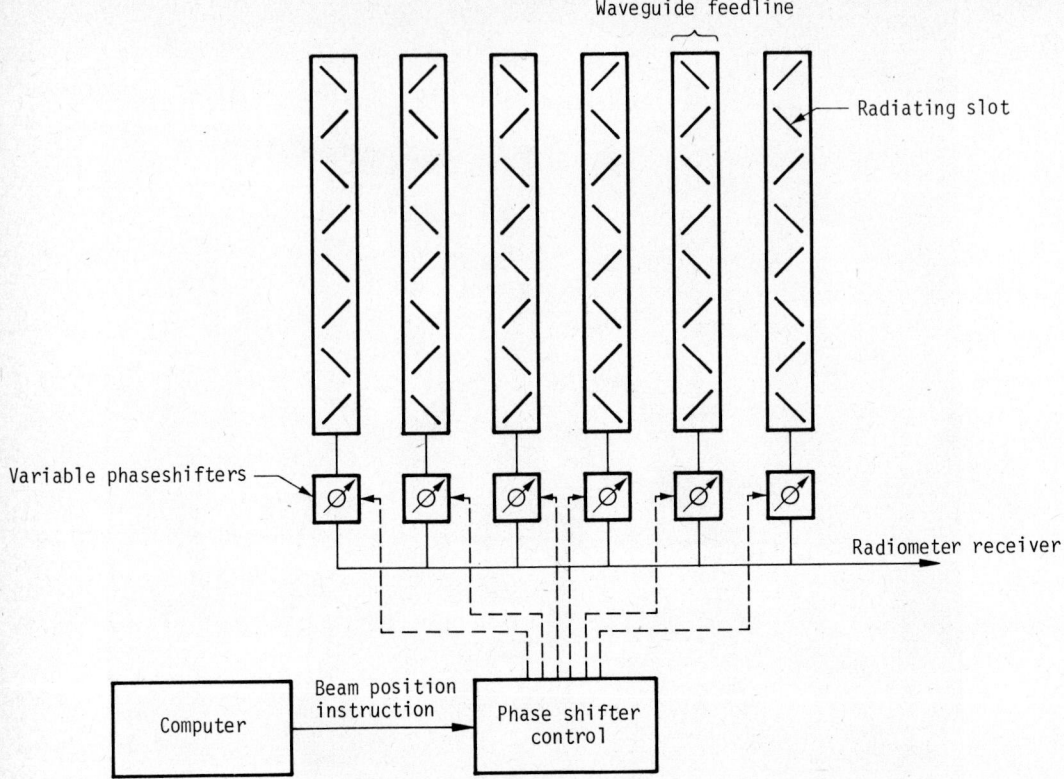

Fig. 11-36. Phased-array antenna.

MOVING PLATFORM PARAMETERS

Consider the situation depicted in Figure 11-37. A mapping radiometer is shown scanning over a scan angle θ_S transverse to the flight direction from an altitude H above the ground. The movement of the platform at a velocity u is used to accomplish a line-by-line scanning format. The size of the ground resolution cell is a function of the antenna beamwidth β, the height H, and the incidence angle θ_0. At nadir, the diameter of the resolution cell is simply H.

For a given set of platform parameters, the scanning frequency f_{sc} (beamwidths per second) can be expressed in terms of H, u, β, θ_S, and the desired beam overlap from scan to scan P (≤ 1) as follows:

$$f_{sc} = \frac{u\theta_S}{PH\beta^2}, \qquad (11\text{-}155)$$

which was adapted from McGillem and Seling (1963). As the antenna beam scans in the cross-track dimension at a constant angular speed, the dwell time spent over each point within a resolution cell is:

$$\tau = \frac{1}{f_{sc}}. \qquad (11\text{-}156)$$

Hence, τ is the available integration time per resolution cell.

To relate the flight parameters to the radiometric resolution $\Delta T_{\min}$ of the measured output, a receiver figure of merit M independent of τ is usually defined in the form:

$$\Delta T_{\min} = \frac{M}{\sqrt{\tau}}. \qquad (11\text{-}157)$$

The above expression is valid for most of the radiometer receiver configurations considered in the previous sections if the effects of gain variation are negligible. Combining Eqs. 11-155 to 11-157 leads to:

$$\Delta T_{\min} = \frac{M}{\beta}\left(\frac{u\theta_S}{PH}\right)^{1/2}. \qquad (11\text{-}158)$$

Near to the nadir, the resolution distance is

$$\Delta x = \beta H. \qquad (11\text{-}159)$$

By eliminating β from Eq. 11-159, the radiometric resolution $\Delta T_{\min}$ can be expressed in terms of the spatial resolution Δx as follows:

$$\Delta T_{\min} = \frac{M}{\Delta x}(uH\theta_S)^{1/2} \qquad (11\text{-}160)$$

where P is set equal to 1 (contiguous mapping).

It is clear that for a given set of flight parameters u and H, and a scanning angle θ_S, radiometric and spatial resolutions are reciprocally related. This result is not surprising, since decreasing

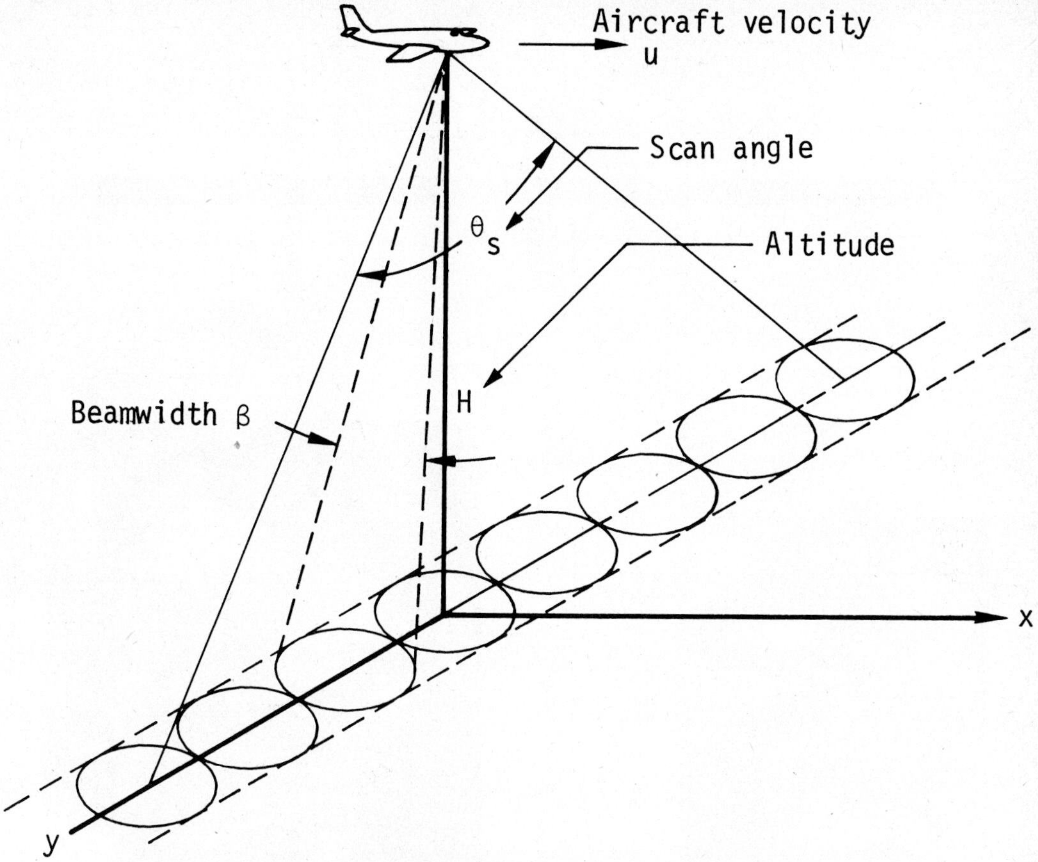

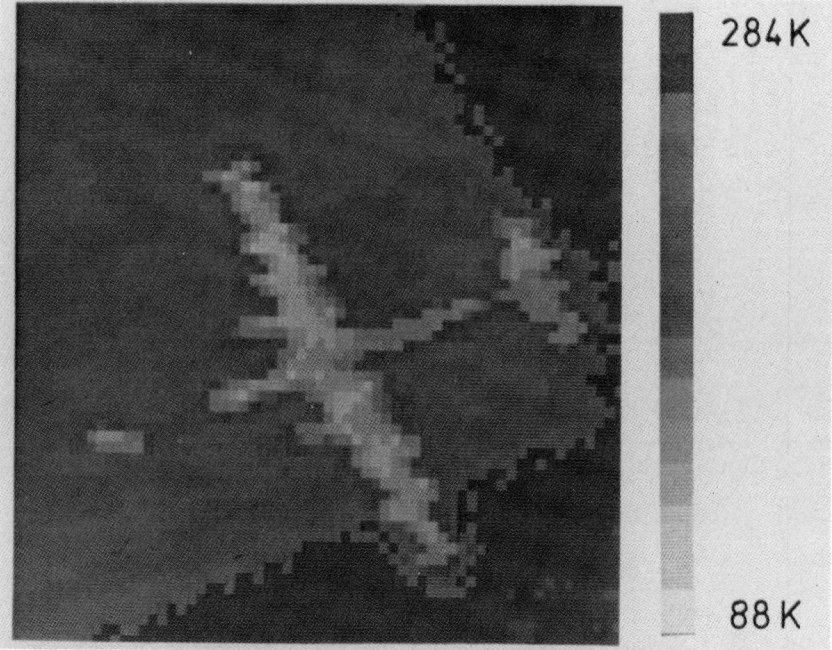

Fig. 11-37. Geometry of airborne scanning microwave radiometer.

Fig. 11-38. Image of a parked twin-engine aircraft on a concrete platform. The image area of 45 m × 45 m was scanned within 1 s. For the color reproduction of this figure, see Color Figure 11-38.

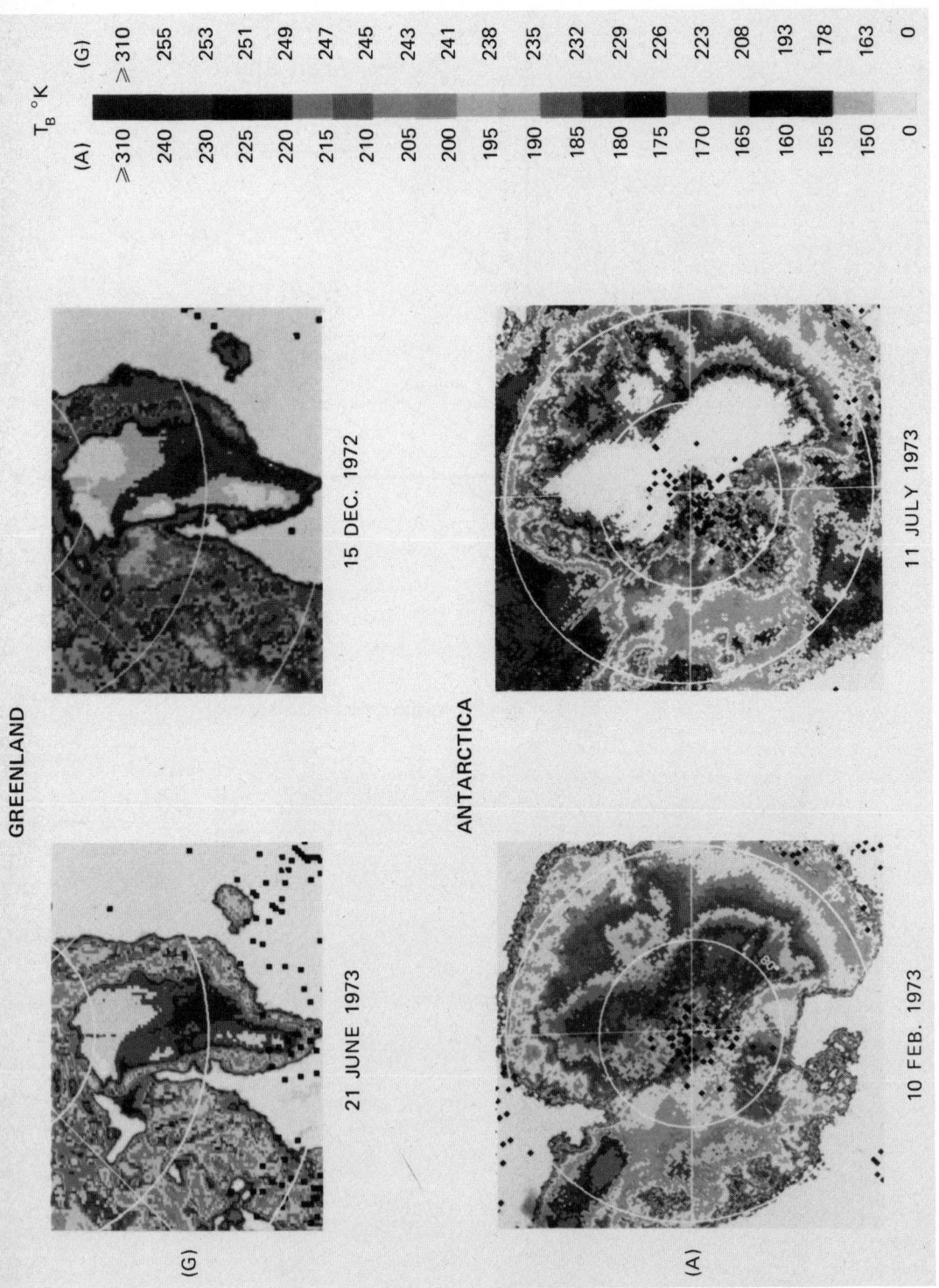

Fig. 11-39. Seasonal variations of the radiometric temperatures over the continental ice sheets in Greenland and Antarctica (courtesy of NASA/

the resolution distance (or beamwidth) results in a shorter dwell time, which corresponds to a larger temperature uncertainty, and vice versa.

Examples of Radiometric Recordings

Because the spatial resolution of a microwave radiometer is determined by the antenna beamwidth and the range to the scene, most imaging systems have been designed to operate at millimeter wavelengths. The image shown in Figure 11-38* shows a twin-engine aircraft parked on a concrete platform and a small vehicle parked in front of it. The image was produced by a scanning 90-GHz radiometer flown at an altitude of 85 m above ground. The angular resolution of the system is about 1°, which resulted in a spatial resolution of about 1.6 m (Vowinkel et al., 1981). As expected, metallic targets exhibit low brightness temperatures that are due primarily to reflected sky radiation, in contrast to the high emissivity concrete and meadow (brown color) backgrounds.

Scanning microwave radiometers have been flown on several satellites, including Nimbus 5, 6 and 7, and Seasat. Detailed descriptions of some of these systems are given in Chapter 13. Figure 11-39* shows typical examples of images produced from satellite altitudes; in this case the seasonal variations of the radiometric temperatures over the continental ice sheets in Greenland and Antarctica were recorded by the 1.55-cm wavelength Electrically Scanning Microwave Radiomenter (ESMR) flown on Nimbus 5 (Gloersen et al., 1973).

REFERENCES

Basharinov, A. E., I. M. Butenko, E. I. Reutov, and A. M. Shutko, 1979a, The use of microwave radiometry for the operational mapping of soil moisture, in Space Photography and Thematic Mapping: A Method for Processing Multi-channel Photography; NASA Goddard Tech. Translation, pp. 154–159.

Basharinov, A. E., M. S. Krylova, A. I. Maslov, and A. M. Shutko, 1979b, Remote sensing of subsurface soil moisture by means of microwave radiometers; Water Resources (English), vol. 5, no. 4, pp. 538–542.

Blume, H. J. C., 1977 (October), Noise calibration repeatability of an airborne third generation radiometer; IEEE Trans. Microwave Theory & Tech., MTT-25, no. 10, pp. 852–855.

Cardiasmenos, A. G., 1980, Practical MICs ready for millimeter receivers; Micro. Sys. News, vol. 10, no. 8, pp. 37–51.

Carver, K. R., 1975 (May), Antenna and radome loss measurements for MFMR and PMIS; Physical Science Laboratory Tech. Report PA00817, New Mexico State University, Las Cruces.

Carver, K. R., 1977 (May–June), Radiometric recognition of coherence; Radio Science, vol. 12, no. 3, pp. 371–379.

Carver, K. R., B. Potts, and R. Widner, 1971 (September), Beam and aperture efficiencies vs. sidelobe level; Digest IEEE 1971 Int'l. Symp. Antennas & Propag., Los Angeles.

Chandrasekhar, 1960, Radiative Transfer, Dover Press, New York.

Cheston, T., and J. Frank, 1970, Array antennas; Chapter 11, Radar Handbook (M. I. Skolnik, ed.), McGraw-Hill.

Dicke, R. H., 1946 (July), The measurement of thermal radiation at microwave frequencies; Rev. Sci. Instr., vol. 17, pp. 268–275.

Frater, R. H., and D. R. Williams, 1981, An active "cold" noise source; IEEE Trans. on Microwave Theory & Tech., MTT-29, pp. 344–347.

Gloersen, P., W. Nordberg, T. J. Schmugge, T. T. Wilheit, and W. J. Campbell, 1973a, Microwave signatures of first-year and multiyear sea ice; J. Geophys. Res., vol. 78, p. 3564.

Hach, Johann-Peter, 1968 (September), A very sensitive airborne microwave radiometer using two reference temperatures; IEEE Trans. on Microwave Theory & Tech., MTT-16, no. 9, pp. 629–636.

Hardy, W. N., 1973 (March), Precision temperature reference for microwave radiometry; IEEE Trans. on Microwave Theory & Tech., MTT-21, pp. 149–150.

Hardy, W. N., K. W. Gray, and A. W. Love, 1974 (April), An S-band radiometer design with high absolute precision; IEEE Trans. on Microwave Theory & Tech., MTT-22, pp. 382–390.

Kerns, D. M., and R. W. Beatty, 1967, Basic theory of waveguide junctions and introductory microwave network analysis; Pergamon Press, Oxford.

Kraus, J. D., 1966, Radio Astronomy, McGraw-Hill Book Co., New York.

McGillem, C. D., and T. V. Seling, 1963 (October), Influence of system parameters on airborne microwave radiometer design; IEEE Trans. on Military Electronics, pp. 296–302.

Nash, R. T., 1964 (December), Beam efficiency limitations of large antennas; IEEE Trans. Antennas & Propag., AP-12, pp. 918–923.

Otoshi, T. Y., 1968 (September), The effect of mismatched components on microwave noise-temperature calibrations; IEEE Trans. on Microwave Theory & Tech., MTT-16, pp. 675–686.

Paris, J. F., 1971, A program for computing the brightness temperature of a clear atmosphere from radiosonde data; Tech. Report LEC/HASD 6490.21.068, Lockheed Electronics Co., Inc., Aerospace Systems Div., Houston, Texas.

Paris, J. F., 1975 (February), Sky brightness temperature error analysis for microwave sensor calibration; Job Order 7-415, Lockheed Electronics Co., Inc., Aerospace Systems Div., Houston, Texas.

Price, R. M., 1976, Radiometer fundamentals, section 3.1 in Astrophys., Part B: Radio Telescopes; (M. L. Meeks, ed.), Academic Press, New York.

Seling, Theodore V., 1964 (September), The application of automatic gain control to microwave radiometers;

---

* In addition to appearing within Chapter 11, Figures 11-38 and 11-39 appear as color plates in the color section of this book.

IEEE Trans. Antennas and Propag., vol. AP-12, pp. 636–639.

Shutko, A. M., 1981, Microwave radiometry of lands under natural and artificial moistening; URSI Symposium on Signature Problems in Microwave Remote Sensing, University of Kansas, Lawrence, Kansas.

Ulaby, F. T., R. K. Moore, and A. K. Fung, 1981, Microwave Remote Sensing: Active and Passive, vol. 1, Addison-Wesley Pub. Co., Reading, MA.

Vowindel, F., J. K. Peltonen, W. Reinert, K. Gruner, and B. Aumiller, 1981 (June), Airborne imaging system using a cryogenic 90-GHz receiver; IEEE Trans. on Microwave Theory and Tech., MTT-29, pp. 535-541.

Wells, J. S., W. C. Daywitt, and C. K. S. Miller, 1964 (March), Measurement of effective temperatures of microwave noise sources; IEEE Trans. Instrumentation & Measurement, IM-13, pp. 17–28.

CHAPTER 12

Landsat Satellites

Authors: STANLEY C. FREDEN AND FREDERICK GORDON, JR.

GENERAL CONTENTS: LANDSAT SPACECRAFT: Observatory configuration; Subsystems; Landsat payload. LAUNCH AND ORBITAL OPERATIONS: Landsat launches; Landsat orbit; Worldwide reference system. GROUND SYSTEMS: Ground communication stations; Operations control center; NASA data processing facility; Image processing facility; EROS Data Center. LANDSAT-D SATELLITE: Instrument systems; Flight segment; Thematic mapper; Ground processing; Transportable ground station; Assessment system. LANDSAT-D BECOMES LANDSAT-4.

INTRODUCTION

The series of satellites now known as the Landsats evolved in concept from the photographic observations of early Mercury and Gemini orbital flights. The data from those manned flights indicated the practicality of observing from space orbit what is broadly referred to as earth resources. These observations and the thoughts they generated led to the NASA satellite program that developed the Earth Resources Technology Satellite-1 (ERTS-1), which is now called Landsat 1.[1]

The imagery from these early orbital flights, being primarily photographic, led the initial thinking about a sensor system for a dedicated "earth resources" satellite to be of a framed nature (i.e., a series of sequentially acquired total scene images). In 1967, a NASA study team looked into the problem of what sensor system should be used on the satellite. The recommendations were in line with the framing device concept. Their recommendation for the primary sensor system was a television-type system, specifically the return-beam vidicon (RBV), which sequentially acquired framed images that would be telemetered to ground stations. A non-NASA government organization that was interested in the potential use of the data from the satellite was the United States Geological Survey (USGS), whose interest areas were primarily geology, cartography, and related subjects. The USGS visualized the sensor system as a three-camera effort with one TV camera filtered in visible red and a second one in the near (reflective) infrared. Two of the camera systems should have a resolution of about 100 meters. The third camera should have higher resolution (about 25 m) and could be panchromatic. NASA decided on a three-camera, three-spectral band RBV. The bands chosen were green (0.475 to 0.575 μm), red (0.580 to 0.680 μm), and near infrared (0.698 to 0.830 μm) at a resolution of about 80 meters.

NASA also received input from other potential users including the United States Department of Agriculture (USDA). USDA's experience resulting from the use of color and infrared aerophotogrammetry indicated the need for spectral data in agricultural studies. A USDA-university team, headed by Dr. A. Park, determined that although the RBV could provide spatial resolution, it would not be satisfactory in a spectral sense. The team thought that the spectral requirements would probably be best satisfied by a multispectral scanner (MSS) system. Experience with the Michigan multispectral scanner had shown the practicality of a flight version of such an instrument. The result was the decision to include a multispectral scanner (MSS) system on the ERTS to provide the more precise spectral data needed primarily for the anticipated uses of the agricultural community. The spectral bands recommended were the result of the various experiences with color infrared imagery and scanners. The original recommendations were on band centers only and were 550, 650, and 750 nanometers. The exception was the fourth band in which the recommendation was "whatever can be acquired in the 800-nanometer plus region." The actual MSS bandwidths in the visible and reflective infrared regions (0.5 to 0.6, 0.6 to 0.7, 0.7 to 0.8, and 0.8 to 1.1 μm) were the result of instrument decisions with regard to sensitivity and other technological factors. The decision to add the MSS to the payload turned out to be an excellent one because the high quality of the MSS data proved to be the key to Landsat's success.

The orbit chosen for the ERTS (Landsat) was the best compromise in terms of the type of coverage desired, the constraints of the sensor system, and the dictates of the laws of orbital mechanics. The type of orbit desired was circular (for near-constant resolution and scale), cyclically repetitive (for periodic observation of the same sites), and sun-synchronous (for fairly constant illumination). The result was an 18-day repetitive

---

[1] The change in name was announced on January 13, 1975, just before launch of Landsat 2 (January 22, 1975).

cycle at a nominal altitude of about 913 km. Associated with the sun-synchronous aspect of the orbit was one selectable parameter, local sun time at a given latitude of overpass of the satellite. The original time considered, primarily on the basis of the requirements of geomorphologists, was an equatorial crossing that was either early morning or late afternoon to provide the long-shadow profiles that would assist in topographical determination. The potential users that preferred maximum illumination, such as agriculturists, protested this choice and opted for a near high-noon overpass. To satisfy a diverse constituency, a compromise was indicated, which meant mid-morning or mid-afternoon. The final choice was mid-morning (i.e., about 9:30 a.m. local time). The corresponding mid-afternoon orbit was ruled out because of the nature of cloud cover in most terrestrial areas. Afternoons generally tend to be more cloudy than mornings. These decisions concerning the spacecraft system needs were finalized in about 1968. The result of all these discussions, and the realities of physical and fiscal constraints, was a NASA-planned program for a series of earth resources satellites, the first of which would be launched in early 1972.

The Landsat total system, like any satellite remote-sensing system, consists of two main parts: a satellite system and a ground system. The former is basically the acquirer of the data of interest; the latter is the processor of the data into a form that makes them usable to those who will convert these data into information, such as the agriculturist, the land-use specialist and the geologist. The total system, as graphically shown in Figure 12-1, would include the spacecraft, its payload and subsystems for data acquisition, and the means of telemetering them to earth, as well as the total ground-handling system to receive, process, and distribute the data. There were initially three U.S. ground stations capable of receiving the data and communicating with the satellite and one Canadian ground station that could receive data. After 1972, the number of receiving sites increased to include other non-U.S. locations that covered much of the world's land mass (Figure 12-33).

At this writing, the original Landsat period is in its twilight. Landsats 1 and 2 are no longer functioning. By the time of publication of this manual, this initial Landsat era (i.e., Landsats 1, 2, 3) will probably be over, and the new generation of Landsat-D satellites initiated. However, the period from July 23, 1972 through the final demise has been one of great significance to the world of satellite remote sensing. During this period, the three Landsats acquired MSS scenes (unique in time but repetitive in area) of several tens of bil-

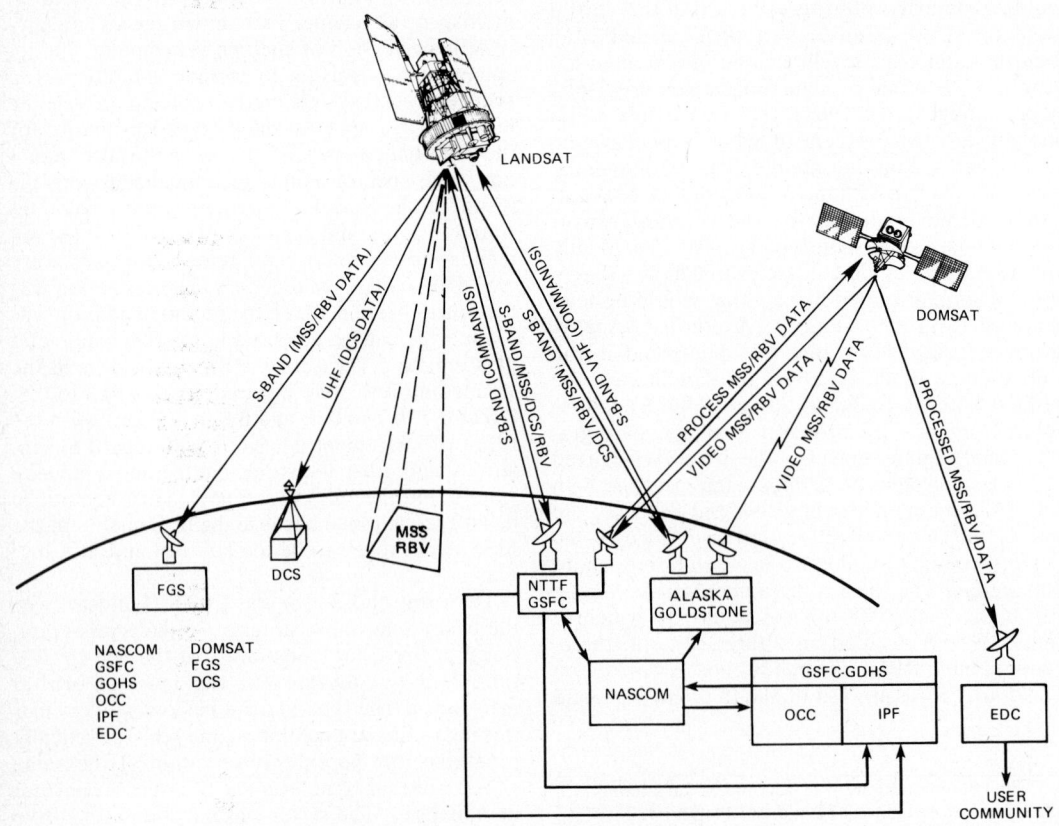

Fig. 12-1. Landsat total communication and processing.

lions of square kilometers of the earth's surface. The Landsat 3 RBV is in the process of producing a scene catalog of all the world's land masses. RBV coverage has amounted to about 8 billion square kilometers. Multiple earth resources applications, in such discipline areas as agriculture, mineral and petroleum exploration, land use, water resources, and forestry, have been developed using satellite data. Tables 12-1 and 12-2 present a performance history of these three Landsat satellites from 1972 through May 1982. To paraphrase a noted British statesman, "Never have so much data been generated to benefit so many by so few sensor systems."

LANDSAT SPACECRAFT

OBSERVATORY: GENERAL CONFIGURATION

The observatory used for ERTS (Landsat) is essentially the same vehicle that has been used for the Nimbus series of weather satellites, and several of the spacecraft functional subsystems were carryovers or modified carryovers from the Nimbus series. The Nimbus observatory model was chosen because of its proven success in a somewhat similar orbit (near-earth, near-polar) and could be fairly easily adapted to the ERTS requirements. These spacecraft had been in orbit since 1964 and, by early 1969, there were three successful spacecraft in orbit. The vehicle chosen to launch ERTS was a thrust-augmented Delta, which had not been used with any of the Nimbus satellites in orbit at that time. Although Deltas have been in use since 1960, previous models did not have sufficient lift capability to put a spacecraft of the size of ERTS (Landsat) into a polar orbit at the altitudes of interest (~915 km).

The Landsat observatory (Figure 12-2) is approximately 3 m (10 ft) high and 1.5 m (5 ft) in diameter. The attendant solar paddles have an extended width of about 4 m (13 ft). The overall weight of the observatory is about 950 kg (2,100 lbs). The Landsat observatory has the function of providing the "environment" in which the payload can operate, the power to permit it to perform, the means of communicating the sensor-acquired data and spacecraft status to the ground stations, and a capability of receiving and acting upon commands relative to spacecraft control and operation. The "environment" can be considered as both structural and physical and includes such factors as the framing mount for the payload and functional support systems, the means of maintaining the thermal levels of the payload within allowable limits, and the ability to maintain the orbital location of the spacecraft so that the sensors look at their targets from an acceptable and known perspective. The torso of the observatory is the sensory ring (Figure 12-3). It contains and/or mounts most of the observatory subsystems that functionally support the payload. The differences between the Landsat 1 and 2 sensory package and that of Landsat 3 are apparent in this figure.

TABLE 12-1

Scenes Acquired by Landsat Spacecraft

	Landsat 1 Terminated on 1/6/78	Landsat 2 as of 12/27/80	Landsat 3 as of 12/27/80
MSS: Total scenes acquired by U.S. and foreign stations and in U.S. inventory	149,611	161,534	70,885
Real-Time Data	102,648	77,602	31,599
Recorded Data	46,963	67,259	39,243
Pakistan			
(Transportable NASA ground station for acquisition of Lacie data. Now removed.)	0	13,337	0
Italy and Sweden (European Space Agency (ESA))	0	1,844	0
Brazil	0	1,492	43
Total scenes acquired by foreign ground stations and not in U.S. inventory	122,146	293,275	142,395
Canada	55,178	121,856	36,425
Brazil	33,881	39,970	26,336
Italy and Sweden (ESA)	33,087	95,671	39,988
Iran (closed on January 8, 1979)	0	1,085	4,158
Japan	0	8,608	8,543
India	0	15,896	18,762
Australia	0	8,087	8,183
Argentina	0	774	0
S. Africa	0	1,328	0

9.4 billion square nautical miles imaged (all spacecraft): ~32 billion square km.
Average number of images received per day (all spacecraft): 305.

TABLE 12-2

Landsat Operating Experience (as of November 12, 1979 and December 27, 1980)

	Launch Date	Days in Orbit	Orbits	MSS Hours	RBV Hours	WBVTR-1 Hours*	WBVTR-2 Hours*
Landsat 1[1]	7/23/72	2,003[2]	27,930[2]	2,806[2]	14[2]	733[6]	5[3]
Landsat 2	1/22/75	2,170	30,206	4,656	31	122[4]	1,176
Landsat 3	3/05/78	1,032	14,339	2,397	1,168	265[5]	699

[1] Operations of MSS and RBV terminated on January 16, 1978.
[2] Numbers at time of January 16, 1978, termination of sensor operations.
[3] Failed on Orbit 148.
[4] Failed on April 5, 1978.
[5] Recorder malfunction on June 3, 1979, precludes use with MSS data (recorder is fully operable with RBV data input).
[6] Failed on July 2, 1974.

* Does not include ground test time of recorders.

The observatory subsystems designed to help meet these support functions include:

- Orbit adjust subsystem (OAS)
- Attitude control subsystem (ACS)
- Attitude measurement subsystem (AMS)
- Power subsystem (PS)
- Thermal control subsystem
- Telemetry, storage, and command subsystems

The last item includes the various transmission and receiving telecommunication subsystems, the command electronics, and the various tape recorders. These observations and their payloads were designed for a nominal 1-year operational life. Landsat 1 operated for 5½ years with only minor malfunctions except for the failure of the wideband tape recorders, one of which failed weeks after launch and the second of which lasted almost 1,000 hours, about twice its design lifetime.

Landsat 2 operated for 7 years. At the time of this writing, Landsat 3 has been operating for 4 years; obtaining somewhat restricted data because of a late line start problem and an intermittent multiplexer problem that put it out of service temporarily for a few months.

SUBSYSTEMS

Orbit Adjust Subsystem

The OAS may be considered as having two functions:

a. To apply corrections for inserting the spacecraft into "best" orbit after the launch vehicle has placed it in a nominal orbit.
b. To maintain and/or reestablish the orbit during the lifetime of the satellite.

The first function is necessary because the launch insertion orbit of a satellite (particularly one requiring a circular, repetitive, sun-synchronous orbit) is seldom perfect and must be corrected for factors such as in-plane injection errors. The second function is to reestablish the desired orbit because it is perturbed by sources such as environmental drag, nonperfect inclination/period relationship, etc. (This will be discussed more completely in the section entitled "Landsat Orbit.")

The Landsat OAS uses a monopropellent, hydrazine-fueled propulsion system (Figure 12-4). There are three rocket-engine assemblies: two oriented on the pitch axis and one on the roll axis. The OAS module is mounted on the upper plane of the sensory ring. The orientation of the rocket engines is such as to have the thrust vector in each pass through the center of mass of the spacecraft. The thrust chamber in each engine contains a catalyst (Shell 405) that causes the hydrazine to decompose into ammonia, hydrogen, and nitrogen gases at about 1000°C. The expulsion of these gases through the nozzle produces the thrust. The

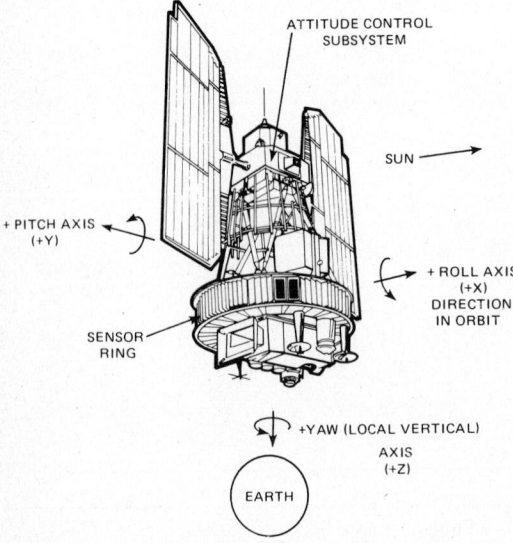

Fig. 12-2. Landsat spacecraft.

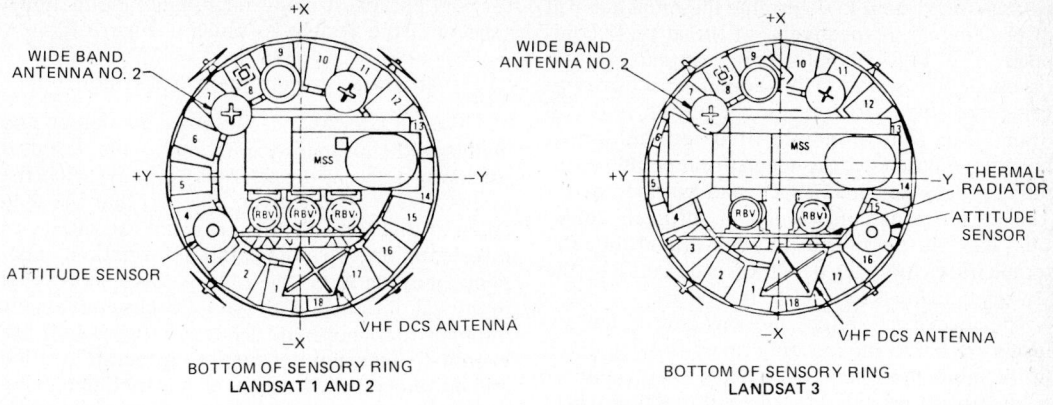

Fig. 12-3. Sensory ring component location.

OAS is initially fueled with 30.4 kg of hydrazine, which in the case of Landsat 1 lasted for over 5 years. During that period, there were 51 "burns": 3 for initial orbit correction, 11 for orbit maintenance at various times, and in late 1976 to early 1977, 37 burns to bring about the major orbital relocation that changed the Landsat 1/Landsat 2 relationship from 9-day/9-day to 6-day/12-day (see section on Landsat Orbit).

Attitude Control Subsystem

The attainment of usable data by the Landsat sensor system is heavily dependent on the proper

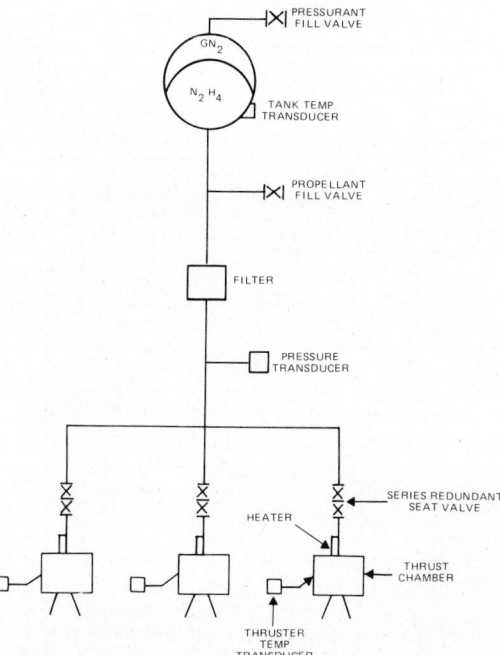

Fig. 12-4. Orbit adjust subsystem block diagram.

orientation of the spacecraft relative to its target. Data accuracy and precision are dependent on the direction and stability of the spacecraft. There is fairly tight tolerance for angular variation of the spacecraft's vertical axis to the local geoid vertical or nadir line. The specifications for the Attitude Control System (ACS) require that the angular displacement for pitch be less than $\pm 0.7°$ and that the maximum angular displacement for roll be $\pm 1.0°$. The associated stability, as indicated by an angular rate of change, must be such that this rate of change shall be no greater than $0.04°s^{-1}$. In addition to considering the displacement from the vertical (roll and pitch), it is important to maintain the rotational orientation of the spacecraft around the line of the local vertical so that the sensor system is orthogonal to the ground trace of the spacecraft. This maintenance of yaw stability is specified to be also less than $\pm 1.0°$. Data acquired with off-nadir or rotational variation within these tolerances can be corrected for in the ground data processing.

In addition to payload orientation, another critical spacecraft subsystem that requires an independent attitude control is the solar paddles. To achieve the maximum energy output from this solar-cell assemblage, their active plane surfaces must be kept as close to perpendicular to the sun-spacecraft vector as practicable. Maintaining proper orientation of both the payload and the solar paddles is the function of the ACS.

The ACS is a physically separate unit mounted near the top of the observatory and structurally connected to the sensory ring by a truss assembly (Figure 12-2). To perform its function, the ACS needs two capabilities: (1) to detect the orientation of the systems relative to pertinent references, and (2) to make changes in orientation to ensure optimum performance of these systems.

The pitch and roll orientations are detected by two horizon scanners pointing fore and aft and parallel to the roll axis. Detection of the displacement and rotational orientation of the horizon permits the system to determine the pitch and roll displacements, respectively, in the attitude computer. The ACS computer then communicates to the corrective systems what action is necessary to bring the spacecraft to correct pitch and roll orientation. The principle of determination of horizon is based on the viewed temperature difference between cold space and warmer earth. The presence of cold clouds in sufficient quantities can cause problems in pitch-roll attitude determination. Any yaw deviation is detected by the yaw gyro compass.

The normal roll, pitch, and yaw control mechanisms are based on speeding up or slowing down of the momentum reaction wheels. Yaw damping is effectively provided by the roll reaction wheel through dynamic coupling of roll-yaw motions. Momentum acquired by these wheels is unloaded by the ACS gas reaction jets.[2] Jet firings normally occur several times per orbit. Since the normal yaw control is exclusively through the reaction wheel, malfunction of that wheel causes loss of yaw orientation. In the case of Landsat 2 in late 1979, the yaw reaction wheel froze. This temporarily ended the ability to acquire data until the wheel became unstuck in mid-1980.

Like the OAS, the ACS has two functions: (1) orientation during initial acquisition after injection into orbit, and (2) maintenance of proper orientation for the remainder of the satellite life. The roll and pitch attitude control operations are the same

[2] Magnetic torquing is also used for unloading the momentum acquired by the reaction wheels. This technique has effectively prolonged the life of the ACS gas indefinitely.

for both phases. A block diagram of the functional aspects of the system is shown in Figure 12-5.

Attitude Measurement Subsystem

Although the ACS is designed to detect and maintain the proper orientation of the Landsat sensor systems, it is necessary to know what the orientation is at any given time, so that the data taken at that time can be corrected for variations in orientation. To obtain this information, independent of the ACS, the Landsat has an AMS that in no way interacts with the ACS but conveys its findings to the ground system. These data are eventually incorporated on the spacecraft location and attitude tape (SLAT), which is used in making systematic corrections to data during ground processing. The AMS detects only the pitch and roll attitudes. The AMS uses a germanium lens/Irtan disc/thermocouple optics detector arrangement, which is directed earthward from the lower plane of the sensory ring. It detects any misorientation of the sensor system's optical planes (which are identical with the lower plane of the sensory ring) to the local vertical. The germanium lens and the Irtan disc materially act as a bandpass filter system for thermal infrared (14 to 16 μm) radiation from the earth's atmosphere and differentiate it from the spatial background surrounding it. The germanium lens focuses its passed radiation on the plane of the Irtan disc, which has its face divided into four fields. The patterns of exposure among the different areas result in differential outputs that are interpreted by the ground system into pitch and roll locational information to within ±0.07°.

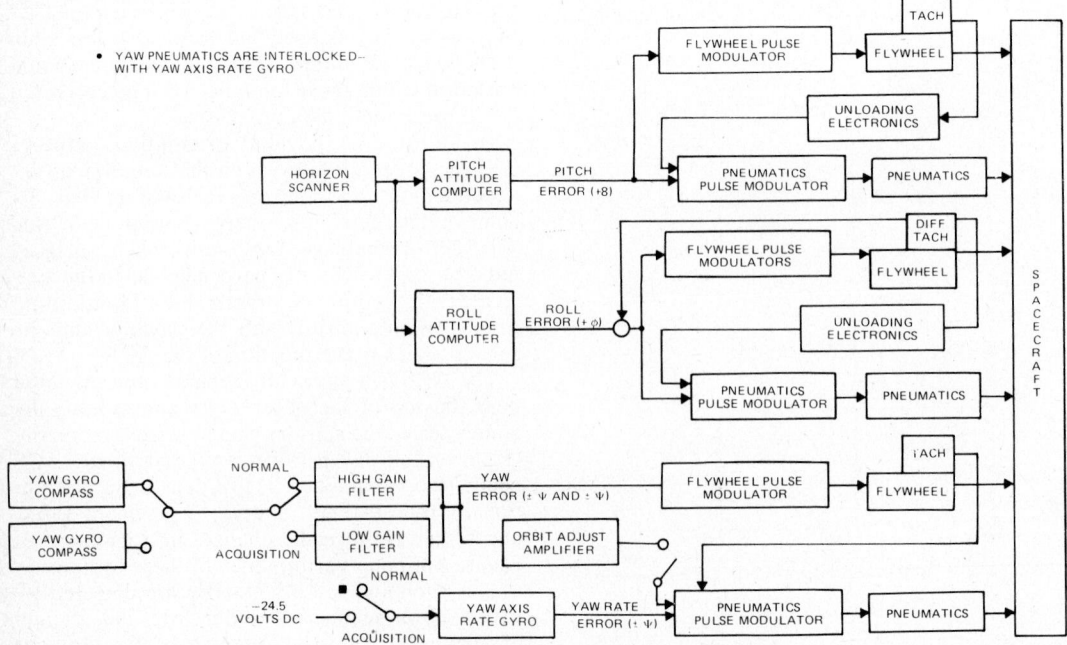

Fig. 12-5. Attitude control system functional diagram.

Power Subsystems

The power subsystem of the observatory has the function of generating, storing, and distributing the electrical power required by all the spacecraft subsystems. The electrical energy is generated by the solar cells mounted on the solar paddles, each of which is 2.36 m (93 in) high and 0.97 m (38.5 in) wide. Energy storage and supply are accomplished through the batteries in conjunction with the solar paddles. The solar cells are phosphorus-doped (N on P) cells mounted in nine individual circuit arrangements on each paddle. Two of these circuits on each paddle are monitored by telemetry. The solar paddle temperature is also telemetered. These paddles are capable of delivering approximately 560 w of power at launch, but this degrades somewhat with time because of microparticle and space radiation damage to the solar cells.

During payload operation of the satellite, 480 w are required for real-time operation and 521 w for remote operation. During sunlit periods, the electrical power subsystem can deliver 980 w of regulated power (−24 v) for short periods. This is done by load-sharing between the solar panels and the 8-storage modules, each module consisting of 23 nickel-cadmium storage cells, rated at 4.5 ampere-hour capacity. When surplus solar-cell power is available, it is used, in a regulated regime, to recharge the storage batteries. Current maxima and voltage/temperature relationships are used to control the charge sessions. Excess power from the solar panel is dissipated in shunt loads that are under the control of the power control module. Auxiliary loads can be switched in on command to avoid any buildup of excessive heat in the battery modules. The various power switching and regulator modules are used to control the flow of power to the payload and the various support subsystems. In addition to the onboard programmed command of power systems, the capability is available to command it through telemetered instructions from ground stations.

Thermal Control Subsystem

The thermal control subsystem provides a controlled environment of 20° ± 10°C for most vehicle and sensor components. The ACS is controlled to a temperature of 25° ± 10°C, and the orbit adjust subsystem is controlled to 27° ± 22°C. Thermal control is accomplished by passive radiators, insulation, and coatings, by semipassive shutters, and by active heaters. Shutters are located on each of the peripheral compartments on the sensory ring and are actuated by two-phase, fluid-filled bellows assemblies. These assemblies are clamped tightly to heat-dissipating components and position the shutter blades to the proper heat-rejection levels. Heaters are bonded at various locations in the sensory ring to prevent temperatures from falling below minimum levels during extended periods of low equipment-duty cycles. The heaters are energized selectively by ground command when the temperature levels at these locations fall below predetermined values. The upper and lower surfaces of the sensory ring are insulated to reduce heat transfer through these areas. External structural and radiating surfaces are coated to provide the required values of emissivity and absorptivity. Passive radiators coated with low-absorptivity, high-emissivity finishes are used to assist the shutters in ejecting the heat from the sensory ring. Radiators are provided for the RBVs, the MSS, the wideband video tape recorders (WBVTRs), and the narrowband tape recorders (NBTRs).

Communications and Data-Handling Subsystems

These subsystems are responsible for all the information flow in the spacecraft, including the incoming and outgoing telemetry, information storage, the data flow from sensor systems to antenna, and all communications systems and links on the spacecraft. These subsystems are divided into two general categories: (1) the wideband subsystem and (2) the narrowband subsystem.

Wideband Telemetry Subsystem

The wideband telemetry subsystem accepts and processes data from the RBV, the MSS, and both WBVTRs, and transmits them to the ground receiving stations as illustrated in Figure 12-1.

The subsystem consists of two S-band frequency modulated (FM) transmitters (which can be operated at either a 10-w or 20-w normal output mode) and associated filters, antennas, and signal-conditioning equipment. As shown in Figure 12-6, the subsystem permits transmission from any two data sources simultaneously, either real-time or recorded, over either of the two downlinks (one data source each). Commandable power-level-tracking wave-tube (TWT) amplifiers and shaped beam antennas provide maximum fidelity of the sensor data at a minimum power. Cross-strapping for dual mode operation (either of two data sources) with a single TWT amplifier is available for redundancy in the event of equipment malfunctions.

Narrowband Telemetry, Tracking, and Command Subsystem

The telemetry, tracking, and command subsystem collects and transmits satellite and sensor housekeeping data to the ground stations, provides tracking aids, receives commands from NASA's Spaceflight Tracking and Data Network (STDN), and implements these commands on board the satellite. In addition, it provides the link for transmitting the data collection system (DCS) data.

The total of 912 telemetry points (576 analog, 16 ten-bit digital words, and 320 one-bit binary

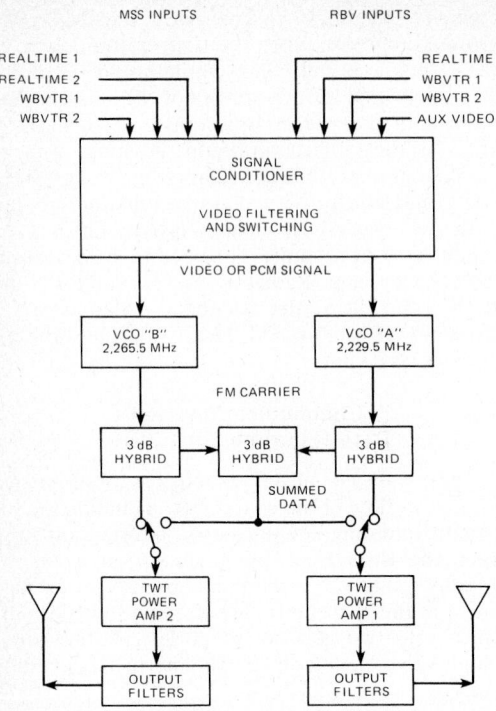

Fig. 12-6. Wideband telemetry subsystem functional block diagram.

words) can be sampled at rates between 1 per 16 seconds and 5 times per second. The data are pulse-code modulated (PCM) and can then be transmitted in real-time either over the VHF or the unified S-band (USB) links at a 1-kbps rate (Figure 12-7). Up to 210 minutes of data can be stored on each of the two NBTRs for subsequent playback at 24 kbps. Analog data have 8-bit accuracy or one part in 256.

The USB equipment has the capability to

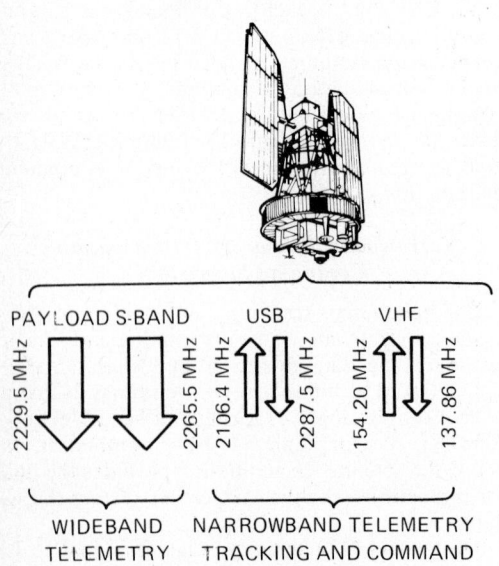

Fig. 12-7. Landsat communication links.

transmit on separate subcarriers real-time telemetry (768 kHz), playback data (597 kHz), DCS data (1.024 MHz), and pseudo-random ranging information simultaneously over the same 2287.5-MHz carrier. The playback data can be derived from either of the NBTR's or either of the auxiliary tracks of the WBVTRs. Only real-time or playback data (from either of the WBVTRs) can be transmitted at one time over the 137.86-MHz VHF equipment. All three of the Landsat receiving stations in the United States normally use the USB downlink.

Commanding can be performed either by the VHF link at 154.20 MHz or by the USB link at 2106.4 MHz into redundant sets of receivers on the satellite. These commands can be any of the 512 possible commands executable by the command/clock or any of the 8 commands recognizable by the command integrator unit. A total of 30 command/clock commands can be "stored" for execution outside the range of the ground stations. All remote payload operations are performed using stored commands.

LANDSAT PAYLOAD

The Landsat satellites carry an MSS and an RBV camera system. WBVTRs store the outputs of these systems when the satellites are out of range of a U.S. ground receiving station. The satellites also carry the space-borne relay component of a DCS. The payload is schematically indicated in Figure 12-8 for Landsats 1 and 2. Landsat 3 does not have the third RBV camera, and a dummy read signal of 1.6 MHz is substituted in the video stream when the cameras are sequenced.

Multispectral Scanner

The Landsat multispectral scanners are line scanning devices that continually scan the earth in a 185-km (100-n.mi.) swath nominally perpendicular to the Landsat orbital track. Scanning is accomplished in the cross-track direction by an oscillating mirror; six lines are scanned simultaneously in each of the four spectral bands for each mirror sweep. On Landsat 3, a fifth band was scanned, two lines per mirror sweep. The forward motion of the satellite provides the along-track progression of the scan lines (Figure 12-9).

The Landsat 1 and 2 MSS responded to earth-reflected sunlight in four spectral bands, whereas the Landsat 3 MSS carries an additional band that responded to thermal infrared radiation (Table 12-3).

The analog signals produced by the MSS detectors are digitized and formatted into a 15-megabit data stream for onboard recording and/or transmission to an earth receiving station (Figure 12-10). During subsequent data processing at the NASA Image Processing Facility (IPF), the MSS data are transformed into framed imagery. The areal coverage of Landsat 1 and 2 MSS data is approximately the same as that of the Landsat 1 and 2 RBV framing cameras.

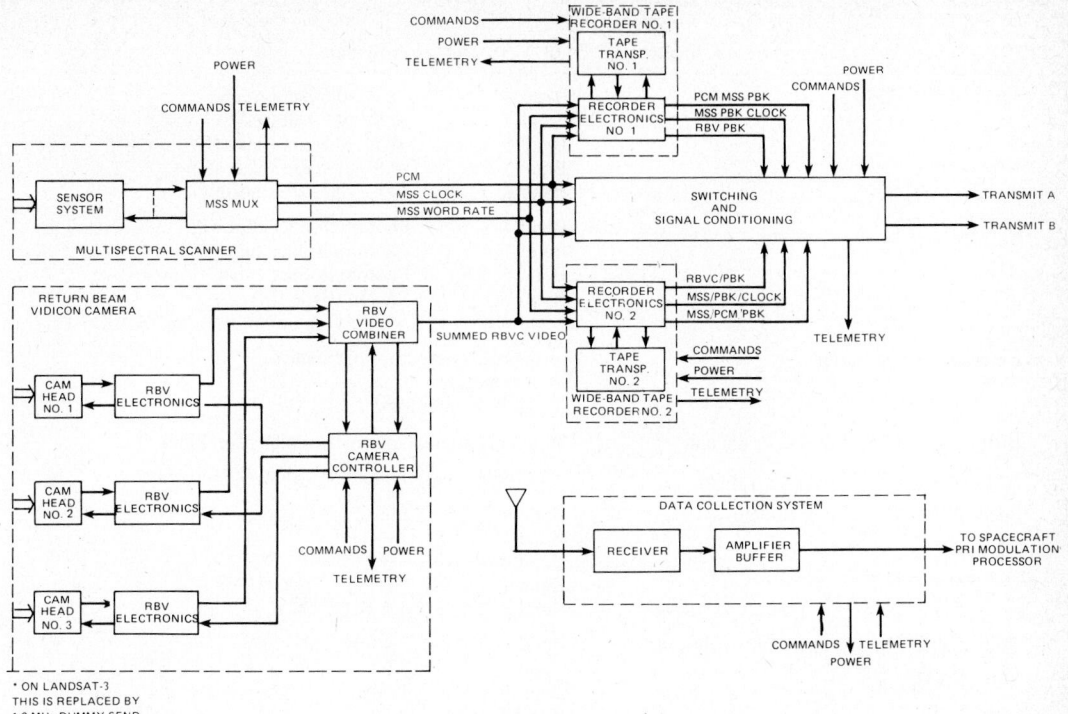

Fig. 12-8. Payload subsystem.

MSS Detector System

The oscillating mirror shown in Figure 12-11 creates the swath pattern as shown in Figure 12-9. While it is sweeping out this 185-km wide swath, the mirror continually focuses the ground image into its optical focal plane. Arranged in that optical plane is a 4 × 6 array of the receiving ends of 24 light pipes (fiber optics). Each light pipe leads to a filter-detector arrangement that responds to the radiance focused on it. This 4 × 6 matrix consists of the four reflective optical bands (0.5 to 0.6,

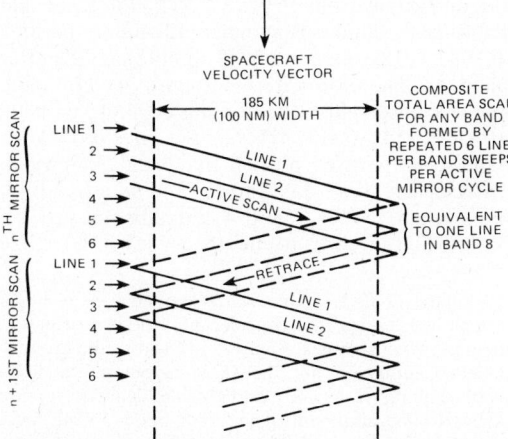

Fig. 12-9. Ground scan pattern for a single MSS detector.

0.6 to 0.7, 0.7 to 0.8, and 0.8 to 1.1 μm) with six lines for each band. The thermal infrared band, which has only one-third of the linear resolution of the other bands, uses germanium optics and two detectors.

In conjunction with the MSS optical system, the dimensions of the individual fiber ends provide an angular instantaneous field of view (IFOV) of 0.086 mrad, which, at the nominal altitude of the Landsat satellite ($\sim$913 km), results in a nominal IFOV of 79 m × 79 m. The mirror oscillates at a frequency of 13.61 Hz (73.5 ms/cycle) and is only actively scanning as the imaged ground is swept from west to east. The active scan time is 33 ms. The scan is nominally sweeping across the terrain at 5.612 m μs^{-1}. The electrical output of each of the 25 detector output channels is sampled every 9.958 μs. This means that the 24 reflective detectors plus either one (alternately) of the thermal IR or a comparable signal must be individually sampled during that period (25 measurements), which is 0.3983 μs per measurement. When each individual detector is again sampled after a period of 9.958 μs, the detector's view of the ground has moved 56 m. Because the IFOV is 79 m, a given detector is still looking at about 11.5 m of the previous pixel and 11.5 m of the next pixel. This amounts to about a 40-percent oversampling along the scan. Therefore, the effective sampled pixel size is 56 m by 79 m. The layout of the fiber optics in the focal plane is shown in equivalent ground distances in Figure 12-12.

TABLE 12-3

Characteristics of MSS Detector System

Spectral Response	Band 4	0.5– 0.6 μm (green)
	Band 5	0.6– 0.7 μm (red)
	Band 6	0.7– 0.8 μm (near IR)
	Band 7	0.8– 1.1 μm (near IR)
	Band 8*	10.4–12.6 μm (thermal IR)
Detector Type	Band 4	Photomultiplier tube (PMT)
	Band 5	Photomultiplier tube
	Band 6	Photomultiplier tube
	Band 7	Silicon diode (Si)
	Band 8*	Mercury-cadmium-telluride (Hy-Cd-Te)
Detector Configuration	(4 bands, 6 detectors per band)	
	(4 × 6 matrix)	
	(band 8, 2 detectors)	
Instantaneous Field of View, angular	86 microradians (258 microradians for Band 8)	
Instantaneous Field of View, linear	~79 meters (237 meters for Band 8), typical	
Spectral Band Signal to Noise	Band 4	113
	Band 5	86
	Band 6	72
	Band 7	123
NEΔT	Band 8	1.5 K @ 300 K
Radiance level to produce full-scale output	Band 4	2.48 nw cm^{-2} Sr^{-1}
	Band 5	2.00
	Band 6	1.76
	Band 7	4.60
Dynamic Range	Band 8	260–340 K

* Landsat 3 only.

Because of the finite sampling time, the same area on the ground as represented by a pixel is not, for a given sweep line, sampled at the same time, or even in the same minor frame[3] for each band. Table 12-4 gives the actual delay times across the four bands for a given pixel location in a given sweep. This is depicted in Figure 12-13 for one line of a sweep. If the band[4] detector for that line is viewing pixel P_o at time T_o, bands 5, 6, and 7 are viewing other ground areas when sampled in the same minor frame. The fiber optics layout is such that when the time sequence differ-

entials are taken into consideration, each detector is viewing an area two ground pixels to the west of the band immediately preceding it on its line. Note that there is one sample time difference (0.3983 μs) between bands 4 and 5 in a given line so that the equivalent static gap between them on the focal plane is reduced by about 2 m (0.3983 ms × 5.612 m μs^{-1}), or from 35 to 33 m. The center-to-center spacing of observed pixels is not (79 m −2) + 33+ (79-2) = 112 m (2 cross-scan pixels). If the arithmetic for the other bands is done on the same line, a two-pixel differential for each band is indicated. Noting from Figures 12-12 and 12-13 and Table 12-4, band 6 is sampled 12 sample periods (0.3983 × 12) and band 7, 13 sample periods after band 4.[5] Therefore, it is not until six complete minor frames plus 13 sampling periods, 64.9 μs [((6 × 9.958) + (13 × 0.3983) = 64.9 μs)], after band 4 has viewed pixel P that band 7 views it. Thus, when these data are aligned as a band-by-band areal matrix, band 4 must be moved six pixels in the time frame to achieve the same

[3] A minor frame is a complete single-sampling sequence, 25 samples (9.950 μs). A major frame is a complete sampling scan or one active sweep.

[4] The MSS bands are numbered 4 through 8 with increasing wavelength. The original Landsat sensing system was the three-band RBV plus the MSS, and the RBV bands were numbered 1 through 3.

[5] Figure 12-13 is a time/space sequence of a single 4-band line, taking into consideration the sensor movement between samples. Figure 12-12 shows the ground-distance equivalent for the static layout in the focal plane. Figures 12-12 and 12-13 give an idealistic version of line-to-line abutment of the fiber optics in the focal plane. Actual spacing differences are such that there is some variation from the ideal spatial IFOV. (See References 2, 3, and 4 for further discussion of this subject.)

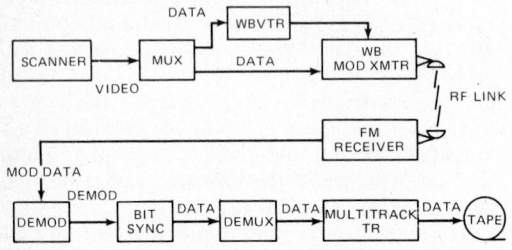

Fig. 12-10. Generalized MSS data flow.

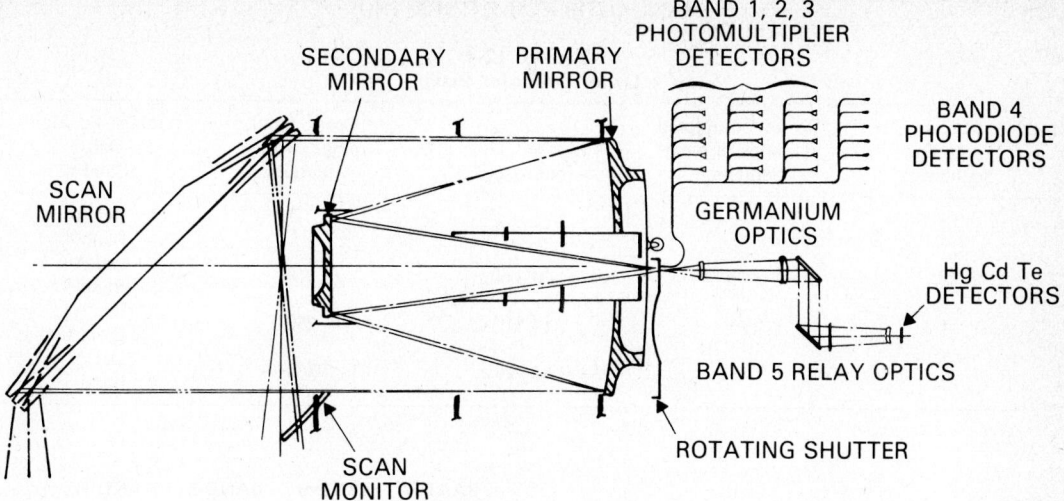

Fig. 12-11. Landsat Multispectral Scanner (MSS) optics schematic.

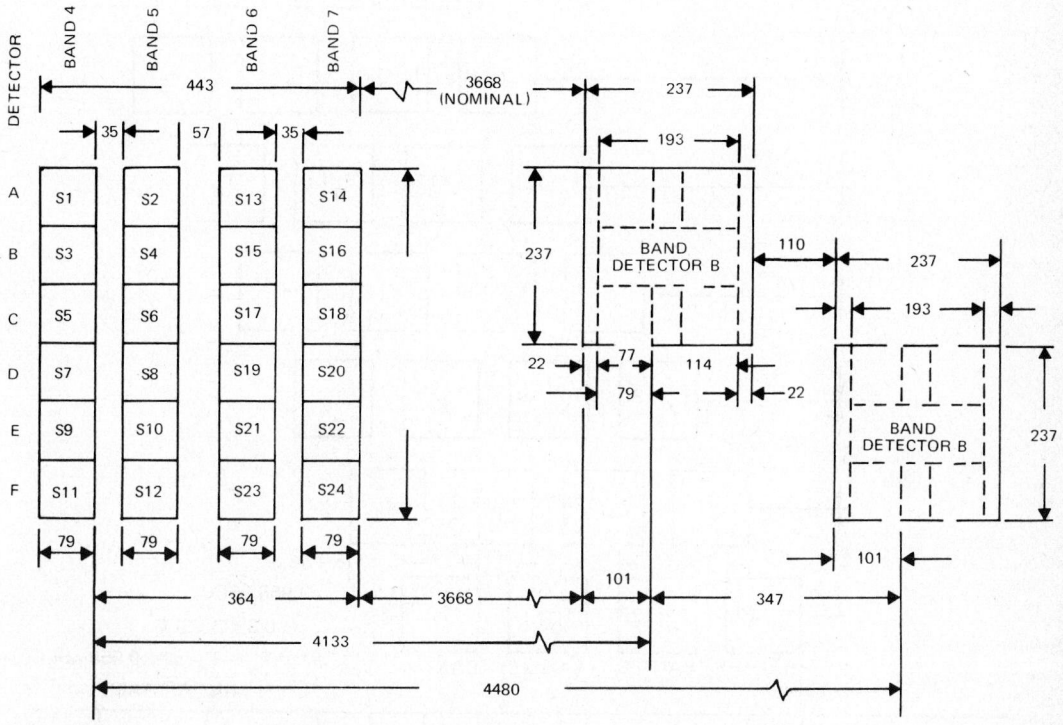

NO. OF SPACECRAFT BYTES = METERS
5.612 METERS/μs 0.38932

NO. OF IN-BAND BYTES = METERS
5.612 METERS/μs 9.785

25 SPACECRAFT BYTES = ONE IN-BAND BYTE

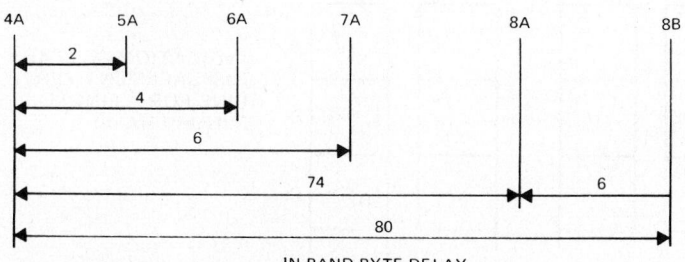

IN-BAND BYTE DELAY

Fig. 12-12. Fiber optics layout (equivalent ground distances).

TABLE 12-4
Delay Times for Pixels

No. of Complete Sampling Sequences	No. of Completed Samples in Next Sequence	Elapsed Time (microsec)	Cross-Track Image Motion (meters)	Image Position Del X Band n
0	0	0	0	Del A Band 4
2	1	20.314	114	Del A Band 5
4	12	44.812	250	Del A Band 6
6	13	64.926	364	Del A Band 7

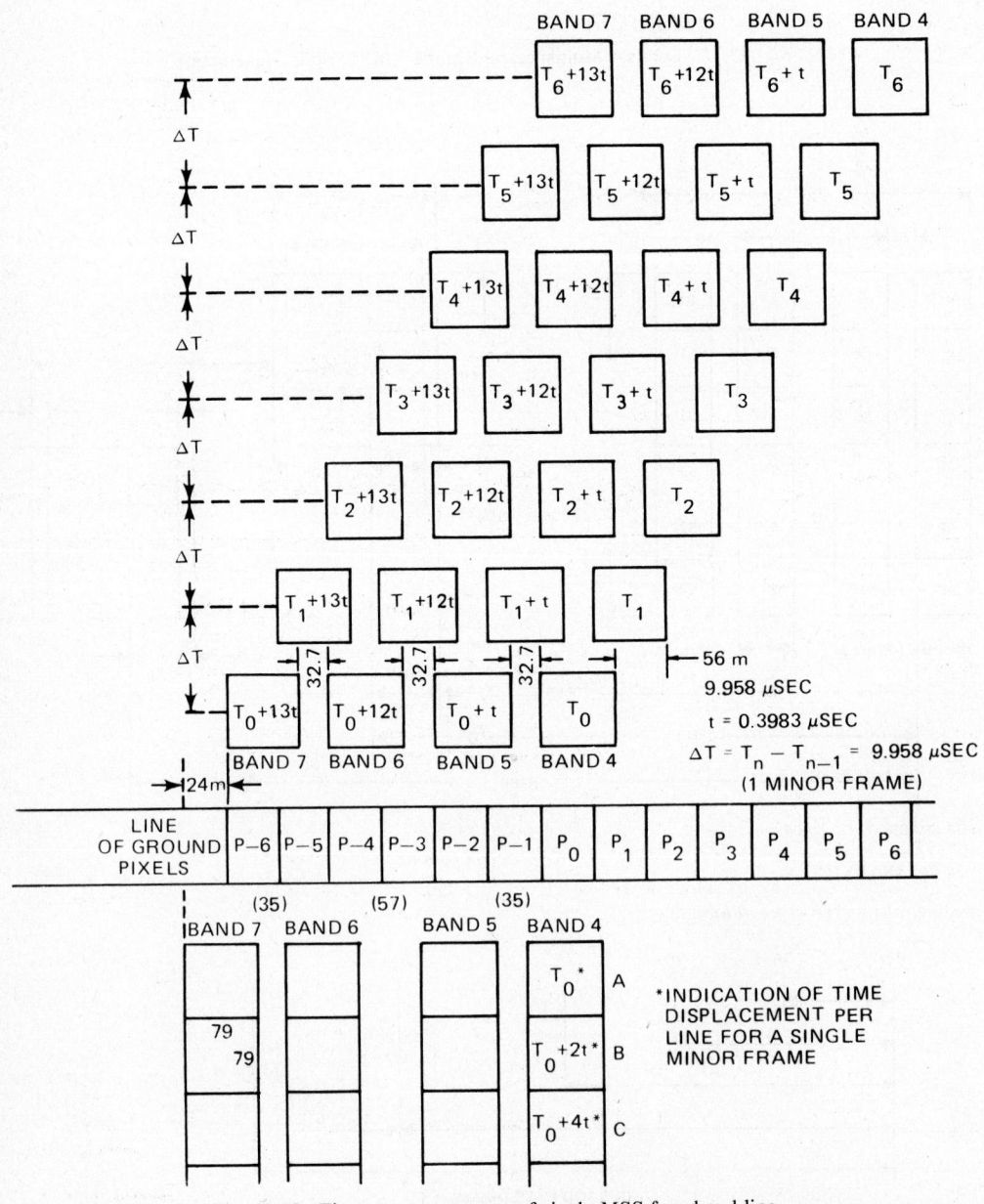

Fig. 12-13. Time-space sequence of single MSS four-band line.

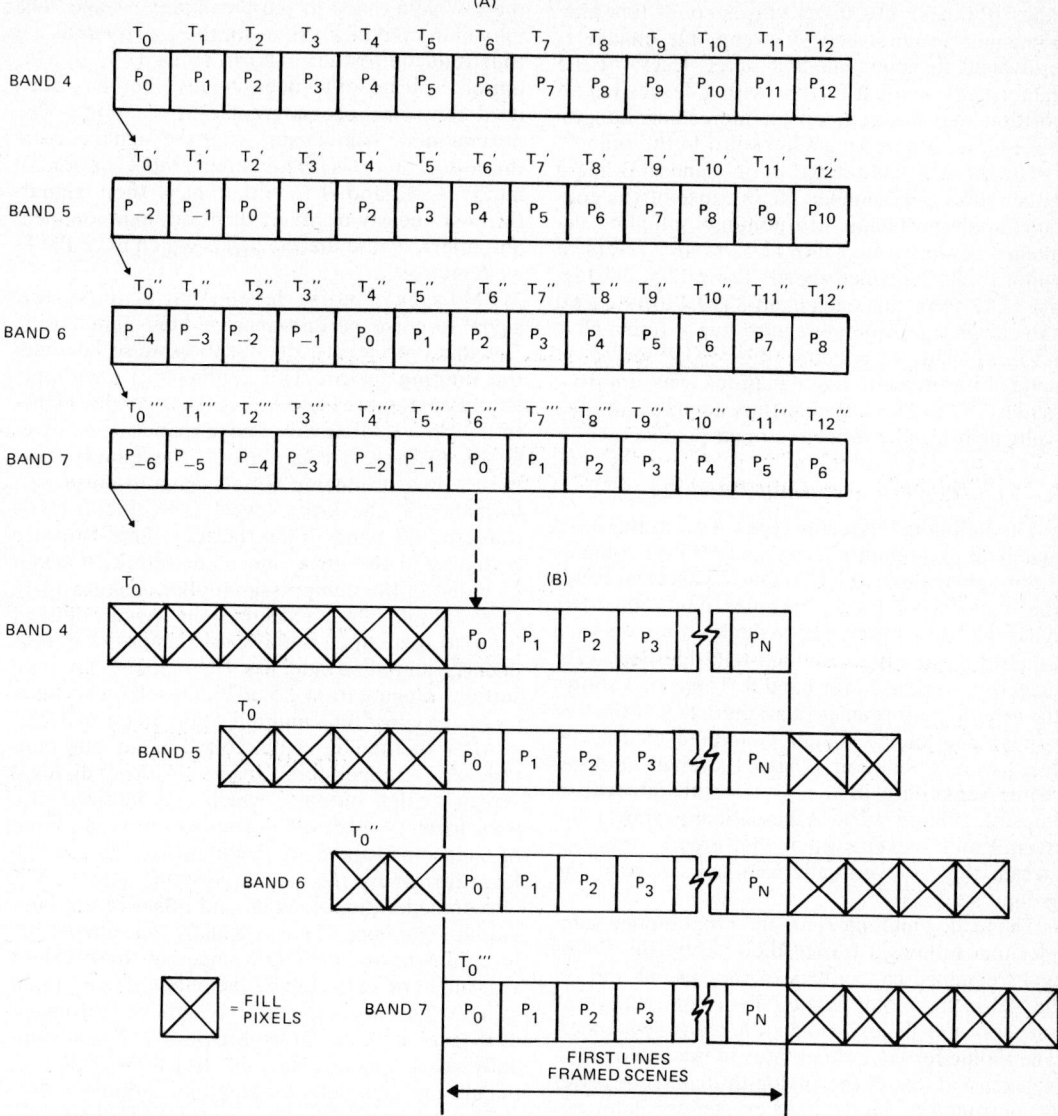

Fig. 12-14. Image-forming realignment of four MSS bands to account for actual time-space relationships of acquisition sequence.

ground matrix as band 7, with the two-pixel increment holding, proceeding from band to band (Figure 12-14).

There is also a line-to-line locational differential for a given band (see sampling sequence of Figure 12-12). Note that in any given band column, there is a difference of two sampling times between the lines, or that there is space equivalent of 10 sampling times, $(10 \times 0.3983 \times 5.612) = 22.4$ *meters,* between lines A and F for any given band. The band-to-band (2 pixel integral) difference was taken care of with the pixel-interleave format in computer-compatible tapes (CCT's) (X-tapes) generated in the original (Landsats 1 and 2) NASA Data Processing Facility (NDPF), but not so for the line-to-line correction. The original NDPF

photographic production system (using the electron beam recorder (EBR) and the associated control computer) did correct for the line-to-line misregistration as well. (In the geometrically corrected tapes (P-tapes) now available, the resampling makes all necessary geometry corrections.)

As each successive sweep moves along the orbital path, the earth rotation is such as to slightly displace each six-line sweep westward from the previous one. Over the distance of 185 km of orbit path in a scene, there is about 12.5 km effective movement to the west of a given pixel column. This must also be accounted for in aligning pixels. This correction is made in the CCT generation of the "raw" tapes.

The generation of the geometrically corrected

tapes (P-tapes) also takes into account this phenomenon. From sweep to sweep, the change is equivalent to about one-half pixel. Users of the data usually account for this with a deskewing algorithm that corrects on an increasing integral pixel basis, progressing southward in the scene.

The thermal band (band 8) on Landsat 3 had a rather short-performance life because of gas contamination problems. Its nominal sampling sequence is shown in Table 12-5. MNFS refers to minor frame sequence signal. Table 12-5 and Figure 12-12 show that each band-8 (10.4 to 12.6 μm) detector is sampled every third minor frame (3 $\times$ 9.958 s). This is because the IFOV for these optics-filter-detector combinations was approximately 237 m (3 $\times$ 79) and thus acquires an area equivalent to nine reflective-band pixels.

Onboard MSS Data Handling

The individual detector types used in the MSS are small photomultiplier tubes (PMTs) for bands 4 through 6. Beyond 0.8 μm in the spectral band, the response of PMTs becomes nonusable. For band 7 (0.8 to 1.1 μm), silicon diodes were used. On Landsat 3, mercury-cadmium-telluride (HgCdTe) detectors were used for band 8. Table 12-3 shows the general performance characteristics of the five bands. The analog signals generated by the detectors are a function of signal strength and detector sensitivity in the wavelength interval of interest. Figure 12-15 schematically shows the overall multiplexer system that creates the data stream that is eventually beamed down to the ground stations.

The analog multiplexer is the programmed sampler that follows a format dictated by the timing and frequency capabilities of the system and reflected in the minor frame (single sampling of all sensors) and major frame (full scan) sequences. The timing for the commutator of the analog multiplexer and that of the analog-to-digital converter is controlled by an onboard crystal-oscillator circuit. The analog commutator's output is a serial stream of data. Before the analog-to-digital conversion by the comparator occurs, there are two signal-conditioning options in the analog chain. The first is the commandable option of either high gain or low gain in bands 4 and 5. The high gain provides three times the voltage amplification of

the low-gain mode in the preamplifier stage. The maximum radiance values for these two bands are thus reduced to those shown in Table 12-6. This option is used only occasionally, but has been used for some oceanographic applications. The normal mode is low gain. After the signal is commutated, there is the commandable option of bands 4, 5, and 6 to either pass their signals through linearly or divert them through nonlinear amplifiers. These are the bands which have PMTs as detectors.

PMTs have noise levels that limit system signal-to-noise performance at high light levels. For low light levels, quantization noise becomes the limiting factor. The nonlinear (logarithmic) amplifier compresses the signal from the higher radiation signal levels and expands the lower levels to more closely match the quantization noise, thus producing better signal-to-noise performance at the lower levels. The silicon diode detector of band 7 has better signal-to-noise matching in the linear mode; therefore, it is not included in the compression option. Figure 12-16 shows graphically the result of the various options prior to digitizing. Band 8 on Landsat 3 is also linearly amplified and has 8 commandable gain settings ranging from 2.5 to 10.5, each higher than its predecessor by a multiplication factor of 1.22.

After the analog signal conditioning, the multiplexer's comparator digitizes all the individual sensor analog outputs, which are digitized into 6-bit formats (64 levels). The compressed output of bands 4, 5, and 6 are decompressed into a 128-level format by the ground system, which takes into consideration the gain and offset of the individual amplifiers: (1) as originally determined before launch, and (2) any changes in these values that might be indicated by the onboard calibration system. This decompression is achieved through a 128[6]-level look-up table. Figure 12-17 schematically shows this look-up table transformation as it occurs for each detector-amplifier output.

The calibration data are acquired between active scan periods (during sweepback) on alternate scans. Provision for such acquisition is controlled by a shutter mechanism that shuts off the ground input during this period and exposes the focal plane fiber optics to calibration lamps. These calibration data are added at the end of the data stream for each line scan for each detector. Also in the end-sequence for each line, there is indicated the number of actual pixels (samplings) taken for that line. This will be used in the ground processing for adjusting line length to a standard for a given scene. Each data output line also has a preamble and time code added that permits scan-to-scan synchronization for each detector. Figure 12-18 shows an example of the outline of a typical bit stream that would be recorded at a re-

TABLE 12-5

Minor Frame (MSS) Sequence

25th Channel	Data Channels (1 through 24)
MNFS	1, 2, 3 24
Detector A, Band 8	1, 2, 3 24
Detector B, Band 8	1, 2, 3 24
MNFS	1, 2, 3 24
Detector A, Band 8	1, 2, 3 24
Detector B, Band 8	1, 2, 3 24

[6] There are still only 64 possible levels for each detector output, but they are expanded into a 0- to 127-level scale via a look-up table for each detector.

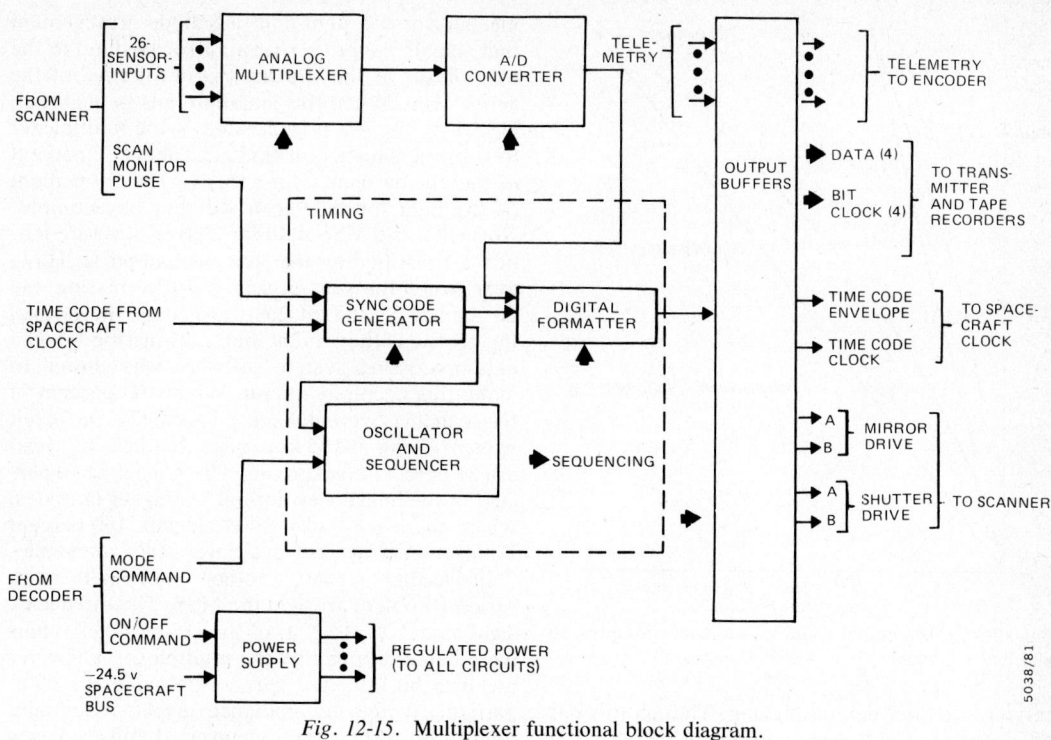

Fig. 12-15. Multiplexer functional block diagram.

TABLE 12-6

Maximum Radiance Values

Spacecraft Item	Landsats 1 and 2			Landsat 3
	Camera 1	Camera 2	Camera 3	
Nominal Spectral Band (micrometers)	0.475−0.575 Blue-Green	0.580−0.680 Yellow-Red	0.698−0.830 Red-IR	0.505−0.750 Panchromatic
Abbreviated Band Reference	Blue	Yellow	Red	—
Signal-to-Noise Ratio (at 100% highlight) Aperture Correction Out	33 dB	33 dB	30 dB	33 dB
Exposure Time (milliseconds)				
No. 1	4.0	4.8	6.4	2.4
No. 2	5.6	6.4	7.2	4.0
No. 3	8.0	8.8	8.8	5.6
No. 4	12.0	12.0	12.0	8.8
No. 5	16.0	16.0	16.0	12.0
Number of Cameras	3			2
Ground Coverage/Camera (km)	185 × 185 (100 n.mi.)			98 × 98 (n.mi.)
Ground Coverage/Frame	(100 × 100 n.mi.)			183 × 98 (98.5 × 53 n.mi.)
Ground Resolution (meters)	79			24
Erase/Prepare/Expose Cycle	Simultaneous			Staggered
Horizontal Scale Rate (lines/sec)	1250			1250
Number of Scan Lines (active)	4125			4125
Cycle Time (sec)	25			12.5
Nominal Lens EFL (mm)	125			236
Prepare Time/Camera (sec)	14			8
Subsystem Average Power (watts)	173.7			132.9
Subsystem Weight (1b)	203			167

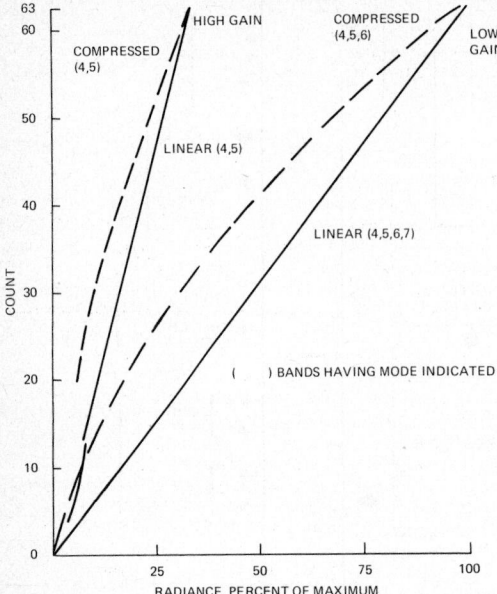

Fig. 12-16. MSS output count vs. radiance, compressed and linear modes.

ceiving site after demultiplexing. Though this has the appearance in the figure of an analog output, it is actually a digital bit stream. In the 33 ms of active scan time, nominally about 3300 pixels are acquired per line/band/sweep. The ground data processing tolerates a range from 2650 to 3480 pixels per line within which it can make adjustments.

Landsat 3 developed a line-start problem in early 1979 that gradually got worse until almost 100 percent of the lines failed to start at the western edge of the swath. The normal line-start

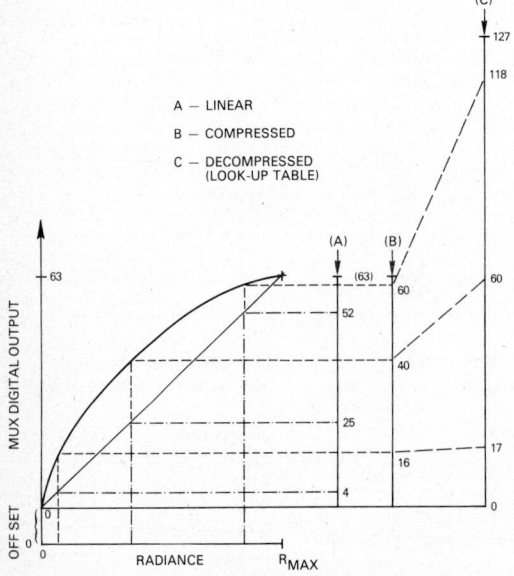

Fig. 12-17. Radiance input vs. digital output of MSS, reflecting ground processing decompression option.

mechanism is a light-emitting diode arrangement that signals the proper scan mirror position to the multiplexer at the starting (western) edge of the active scan. When this situation fails to occur, an electrical impulse is generated in the multiplexer to start the data acquisition after about 30 percent of the mirror scan. Thus, the western 30 percent of the data for that scan will not be acquired. Normally, the MSS data for a given scan are left- or west-justified by the start-of-scan pulse. However, when the start-of-scan pulse is missing, the data must be shifted eastward to fit the normal image (i.e., they must be right-justified). The data-processing system software was altered to make this accommodation. When 100 percent of the scan lines were not started normally, the result was a 185-km × 185 km image that had the western 30 percent blacked out. The remaining 70 percent of the image was normal. At higher latitudes, where there was substantial sidelap, 100-percent Landsat coverage per cycle was still achievable.

In December 1980, another functionally more serious problem arose in the MSS. The three least significant bits (1, 2, 4) of the six-bit pixel values were being dropped by the multiplexer whenever the fifth bit (16) was zero. This made the data, particularly for low radiance levels, essentially nonusable. When this occurred, Landsat 3 was turned off. However, in March 1981, this problem miraculously disappeared, and the full-digit (0 to 63) readout was once more available. The Landsat 3 was then returned to limited service, still having the line-start problem.

Return-Beam Vidicon

As indicated in the introduction, the RBV was the basic sensor system first considered for the ERTS (Landsat) payload. It was considered at the time of conception of this satellite series to be representative of the state of the art. Being a framed-image device, the RBV would acquire scenes sequentially along the path of the orbit. The time sequence of RBV scene acquisitions, originally every 25 s, became the driver for the framing sequence of the MSS. For Landsats 1 and 2, the intent was to get spectral as well as spatial data with the RBV. In these spacecraft, a three-camera RBV system was used. Each camera was focused on the same area (Figure 12-19a) with each camera having a spectral filter that would permit optical transmission of only one of the bandwidths of interest (Figures 12-20a and b). The spectral regions roughly approximate the regions covered by infrared photographic films, reflecting both the system's conceptual prehistory as mentioned in the introduction, and the spectral limitations of the vidicon tube whose photosensitive surface has a long-wavelength sensitivity limit of about 0.9 μm, which is similar to that of infrared photographic film.

The RBVs of Landsats 1 and 2 are mainly of historical interest since they were little used in

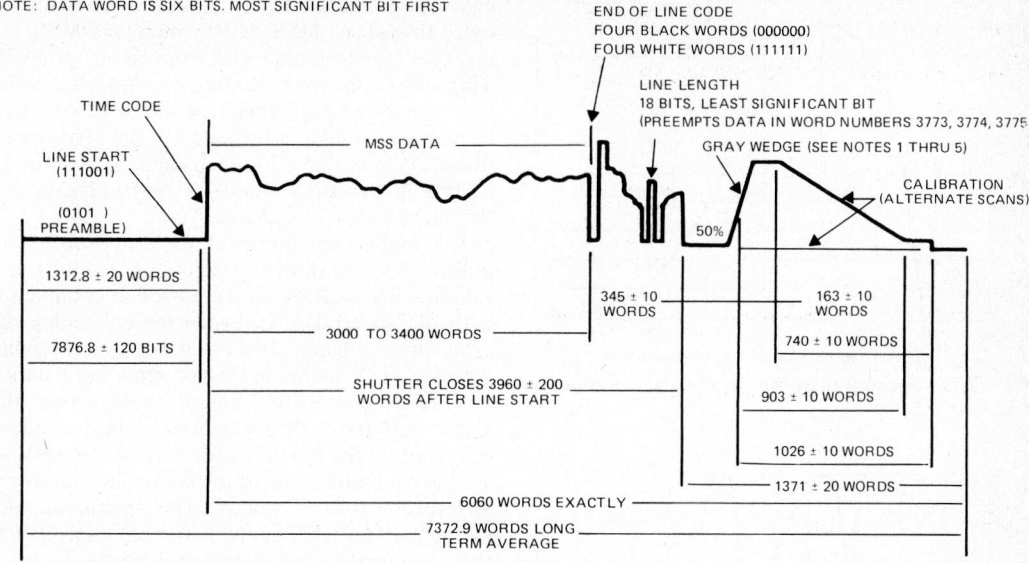

NOTE: DATA WORD IS SIX BITS. MOST SIGNIFICANT BIT FIRST

END OF LINE CODE
FOUR BLACK WORDS (000000)
FOUR WHITE WORDS (111111)

TIME CODE

LINE LENGTH
18 BITS, LEAST SIGNIFICANT BIT
(PREEMPTS DATA IN WORD NUMBERS 3773, 3774, 3775)

MSS DATA

GRAY WEDGE (SEE NOTES 1 THRU 5)

LINE START
(111001)

CALIBRATION
(ALTERNATE SCANS)

(0101   )
PREAMBLE)

50%

1312.8 ± 20 WORDS

345 ± 10
WORDS

163 ± 10
WORDS

7876.8 ± 120 BITS

3000 TO 3400 WORDS

740 ± 10 WORDS

SHUTTER CLOSES 3960 ± 200
WORDS AFTER LINE START

903 ± 10 WORDS

1026 ± 10 WORDS

1371 ± 20 WORDS

6060 WORDS EXACTLY

7372.9 WORDS LONG
TERM AVERAGE

NOTES:
1. GRAY WEDGE RISE TIME APPROXIMATELY 20 WORDS
2. FIRST WORD AFTER LINE START IS WORD NUMBER 0
3. PREAMBLE LENGTH IN BITS WILL NOT BE DIVISIBLE BY 6 EXCEPT BY CHANCE
4. DECOMMUTATED SPACECRAFT TIME CODE IS FOUND IN WORDS 0 AND 1
5. THE GRAY WEDGE AND BLANKING OF ALTERNATE SCANS IS REPLACED BY
   BAND 8 HOT AND COLD REFERENCE LEVELS IN CHANNEL 25

Fig. 12-18. Multispectral scanner tape data format from Ampex tape recorder FR1928 (one track-typical).

Landsat 1 and 2 operations because of the spectacular performance of the MSS. Comparative characteristics are shown in Table 12-6 for the Landsat 1, 2, and 3 RBVs. An indication of the early emphasis on the spatial qualities of the RBV is provided by the fact that the original geometric precision processing system for the ground data-handling system at the Goddard Space Flight Center (GSFC) was based on using the EBR-generated data from the RBV as input.

With Landsat 3, there was a change in utilization philosophy. Landsat-1 and -2 results had

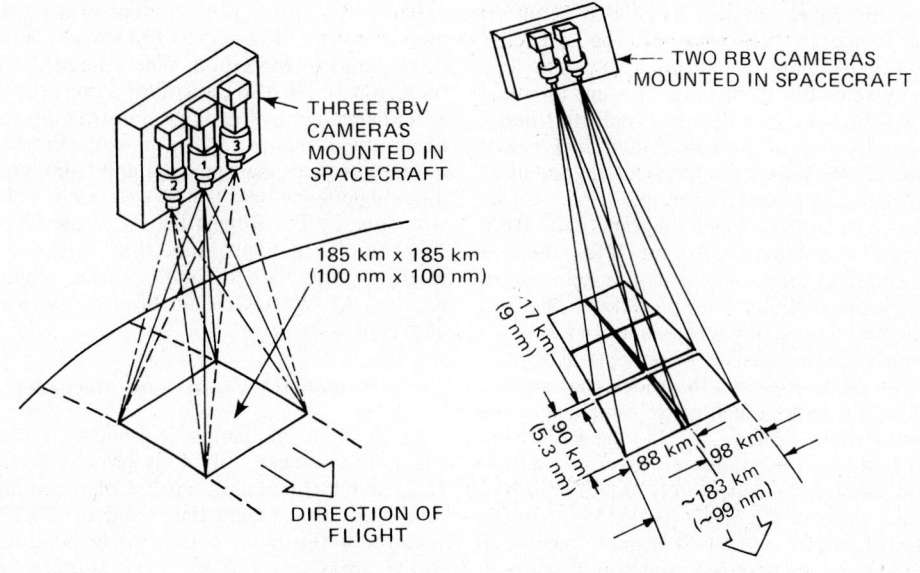

THREE RBV
CAMERAS
MOUNTED IN
SPACECRAFT

TWO RBV CAMERAS
MOUNTED IN SPACECRAFT

185 km x 185 km
(100 nm x 100 nm)

17 km
(9 nm)

90 km
(5.3 nm)

88 km   98 km

~183 km
(~99 nm)

DIRECTION OF
FLIGHT

a. LANDSAT 1 AND 2

b. LANDSAT 3

Fig. 12-19. RBV scanning patterns.

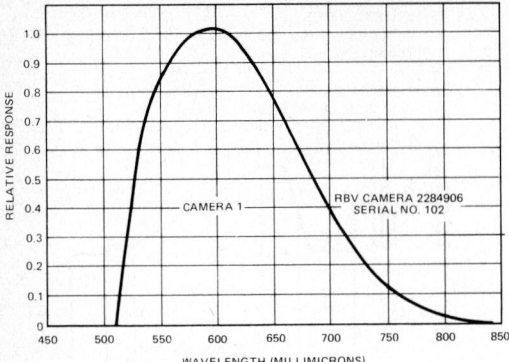

Fig. 12-20. Spectral response, RBV camera system on Landsats 1-2.

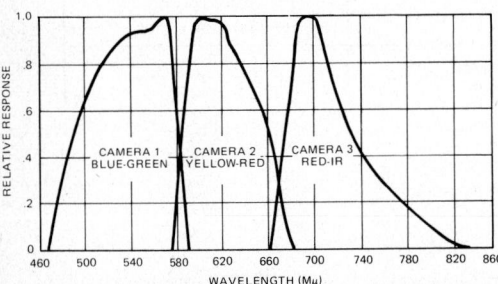

Fig. 12-20. Spectral response, RBV camera system on Landsat 3.

proved the spectral and radiometric superiority of the MSS; therefore, it was decided to use the RBV in the area of its main feature strength—within the Landsat system concept—spatial resolution. The original resolution could be doubled by some fairly simple changes in the optical system (e.g., doubling the focal length). This also required minimal changes in the electronics. The Landsat-3 RBV is a two-camera, panchromatic system. The vidicon systems for all the satellites are identical except for the spectral filters in Landsats 1 and 2 and a modification of the lens system in Landsat 3. Figure 12-19b shows the physical alignment of the two-camera Landsat 3 system.

Figure 12-21 shows a block diagram of the RBV spacecraft subsystem. Unlike the MSS, there is no conversion of the analog sensor system output to a digital signal aboard the spacecraft. The data are telemetered in analog form via an FM-FM signal. Each camera consists of an optical lens, a shutter, an RBV sensor, a thermoelectric cooler, deflection and focus coils, erase lamps, and the sensor electronics (Figure 12-22). The RBV cameras on Landsat 3 are mounted to view side-by-side and aligned to collectively cover approximately the same swath width as the MSS. In the direction of orbital path, two frames from each RBV camera are required to equal the frame size of the MSS. Therefore, there is a 4:1 area coverage ratio of an MSS scene to an RBV single camera frame. The optical resolution is the inverse

equivalent, and the RBV has a linear resolution of twice that of the MSS, RBV ~40 m and MSS ~80 m. The panchromatic spectrum (0.505 to 0.750 μm) is the same for both cameras. Since the linear field of view of each camera is about 98 km, there is nominal sidelap of about 13 km. The image frame cycle is 12.5 s for each camera, as opposed to 25 s in Landsats 1 and 2. Figure 12-23 shows this comparison of camera cycles.

The higher resolution of the Landsat-3 RBV produced a new dimensionality concept for Landsat data. These RBV data are used in conjunction with the spectral MSS bands for enhancing discrimination. Figure 12-24 is a portion of a standard rectified MSS scene, in the conventional bands 4, 5, 7 color composite. Figure 12-25 shows this scene, with the RBV data merged. The delineation provided by the greater resolution of the RBV assists in outlining some of the manmade features in this forest area of Maine. The spatial location factor with an RBV scene is further improved by the continued use of reseau grid marks.

In April 1978, a satellite was launched whose sensor system was primarily devoted to the acquisition of thermal infrared data. This satellite project was called the Heat Capacity Mapping Mission (HCMM). Its sensor system was called the Heat Capacity Mapping Radiometer (HCMR) and took data in two optical bands; visible panchromatic (0.5-1.1 micrometers) and thermal infrared (10.5-12.5 micrometers), at a linear resolution of about 600 meters. It is not the intention to discuss this program in detail here. For that information we suggest the *HCMM Anthology* by Short and Stuart (see reference). However, just as the RBV and MSS data were merged to bring out additional surface information, so the HCMM thermal infrared data were merged with the MSS data to enhance the information content of a scene. Figure 12-26 is an MSS scene of Morocco, using only three bands of MSS data. When the MSS data are merged with HCMM data of the same area there is an enhancement of geologic information that can be obtained. Figure 12-27 is an image in which this merge has been incorporated and from which additional geologic data can be extracted. This work was done by Dr. Rupert Haydn of the University of Munich, in a program that was sponsored jointly by the NASA/Goddard Space Flight Center and the West German Ministry of Research and Technology.

Wideband Video Tape Recorder

At the time of launch of Landsat 1 (July 23, 1972), there were only four ground stations (3 U.S. and 1 Canadian) capable of receiving real-time data. That meant that Landsat (ERTS) data acquired in this mode could only be acquired over global areas covered by these stations: mainly North America. Any data that were to be gathered over the rest of the globe would have to be acquired and stored on board the spacecraft until

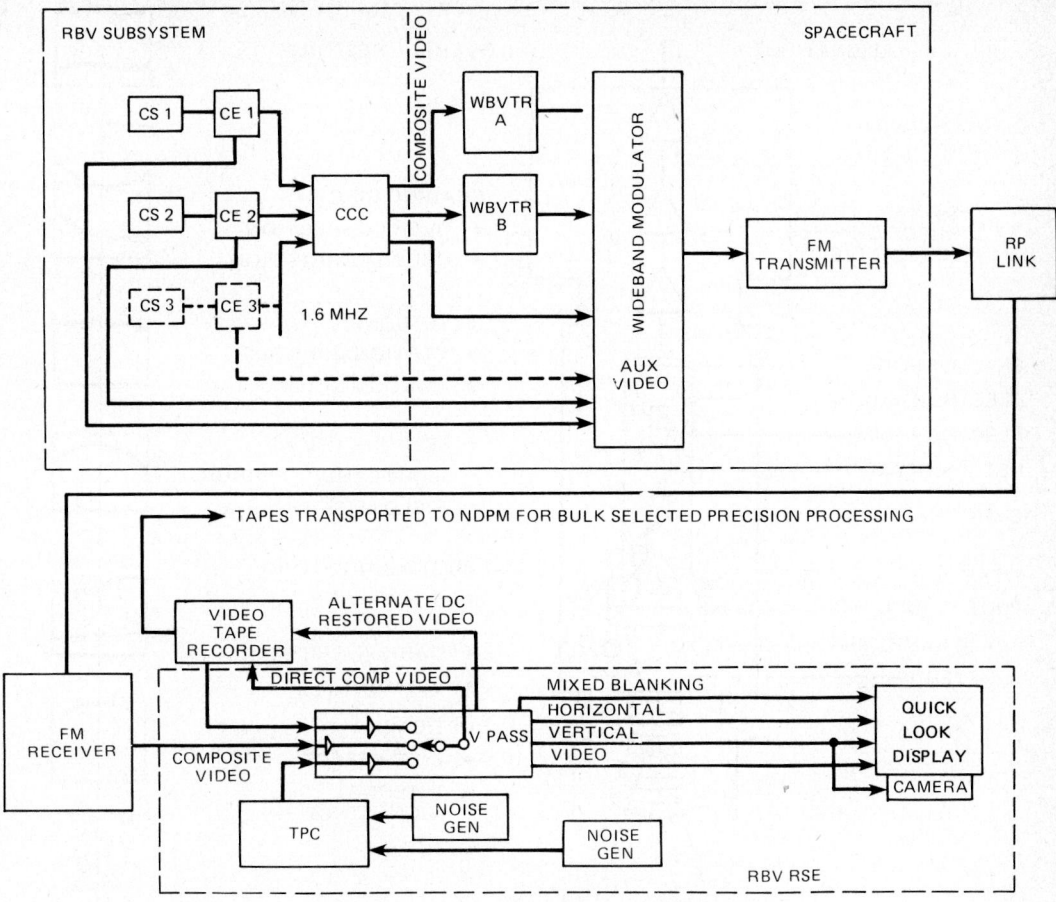

Fig. 12-21. RBV subsystem block diagram.

they could be played back at one of the U.S. sites. The wideband video tape recorders (WBVTRs) stored these data on board the spacecraft for subsequent transmission to earth. Over the years since 1972, several new ground stations have been built around the world to receive Landsat data. The need for onboard storage or some alternative method still exists at the time of this writing, however, because there are still large areas of the globe not covered by ground stations.

Each WBVTR is capable of recording 30 minutes of either 3.2-MHz video analog data from the RBV or 15-Mbps digital data from the MSS multiplexer. The RBV analog data are transformed into the FM domain by video circuitry in the WBVTR. Either or both WBVTRs aboard a given Landsat can be commanded to record RBV or MSS data (Figure 12-28a). The four-headed wheel shown in Figure 12-28b records across a 2-inch-wide tape. Both recording and playback are at a tape speed of 30.5 cm/sec (12 in./sec) and are executed in the same direction.

The history of tape recorders aboard unmanned spacecraft has not established these systems as among the most reliable components. Landsat's history in this matter is mixed, but generally on the positive side. On Landsat 1, the WBVTR-2 lasted about 10 days, but the WBVTR-1 lasted until July 1974 (over 2 years), 733 operational hours. On the Landsat 2 WBVTR-1, a single head (of the four) failed after 5 months, but the recorder was usable for RBV recording until a second head failed in January 1977 (2 years after launch). The second WBVTR on Landsat 2 failed at a later date after almost 1,200 hours of use. On Landsat 3, the WBVTR-1 became nonusable for MSS after 165 hours of operation (June 1979), but continued to be usable for RBV. Landsat 3 WBVTR-2 continues to be fully operational at the time of this writing.

Data Collection System

The DCSs on the Landsats, while considered part of the payload, are somewhat of an anomaly. Although they assist in sensing various parameters associated with "earth resources," the DCSs are not sensor systems, but communication/relay systems. Each system has two major components: the relay system aboard the spacecraft and the data collection platforms (DCPs) located on the ground. Figure 12-29 shows the general scheme

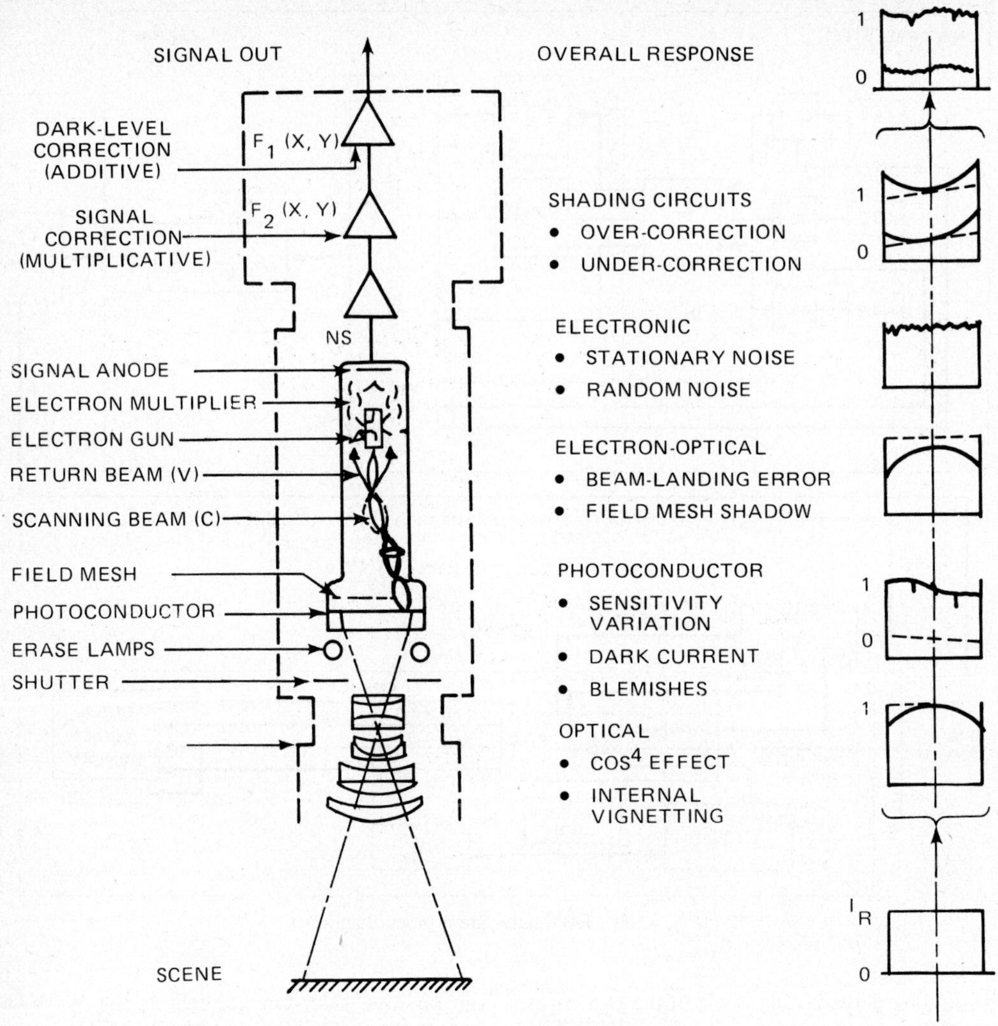

SIGNAL OUT
OVERALL RESPONSE
DARK-LEVEL CORRECTION (ADDITIVE)
F_1 (X, Y)
SIGNAL CORRECTION (MULTIPLICATIVE)
F_2 (X, Y)
SHADING CIRCUITS
• OVER-CORRECTION
• UNDER-CORRECTION
NS
ELECTRONIC
• STATIONARY NOISE
• RANDOM NOISE
SIGNAL ANODE
ELECTRON MULTIPLIER
ELECTRON GUN
RETURN BEAM (V)
SCANNING BEAM (C)
ELECTRON-OPTICAL
• BEAM-LANDING ERROR
• FIELD MESH SHADOW
FIELD MESH
PHOTOCONDUCTOR
PHOTOCONDUCTOR
• SENSITIVITY VARIATION
• DARK CURRENT
• BLEMISHES
ERASE LAMPS
SHUTTER
OPTICAL
• COS⁴ EFFECT
• INTERNAL VIGNETTING
SCENE
I_R
1
0

Fig. 12-22. RBV camera subsystem and radiometric error sources.

for the acquisition, relaying, and recording of data via the DCS. All data must be acquired in real time because the DCS has no onboard tape recorders.

There must also be a mutuality of coverage among the sending location, the satellite, and the receiving location in order for a message to be received (i.e., the satellite must have both the transmitting station and the receiving station in view (Figure 12-30a)). Figure 12-30b shows what

REV ON
EXPOSE 3 CAMERAS
50 SECOND WARMUP
1
2
3
1
LANDSAT 1/2
14.5 SEC
25 SEC
9 SEC
12.5 SEC
12.5 SEC
EXPOSE CAMERA 1
EXPOSE CAMERA 2
1.6 MHz DUMMY READ
50 SECOND WARMUP
LANDSAT 3

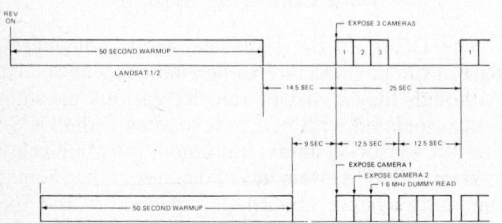

Fig. 12-23. Comparison of two- and three-camera timing.

the area of coverage might be for a single Landsat orbit over the United States.

Initially, there were only the three U.S. ground receiving stations (Fairbanks, Alaska; Goldstone, California; and Greenbelt, Maryland) that had the capability for receiving the DCS data. Later dedicated downlinks were added, such as those of the Army Corps of Engineers at Waltham, Massachusetts. DCS receiving capability was also added to the NASA Santiago, Chile station. The most extensive network of Landsat DCPs was that of the Canadian Department of Environment, which was used mainly for water-flow monitoring throughout Canada. Much of the present activity in remote site sensing of earth resources has been switched from the Landsat satellites to those of the NASA/NOAA, Geostationary Operational Environmental Satellites (GOES), because of their operational nature.

The sensors used in the DCS system are user selected, but must interface with the electronics of the DCP. These have included many types of

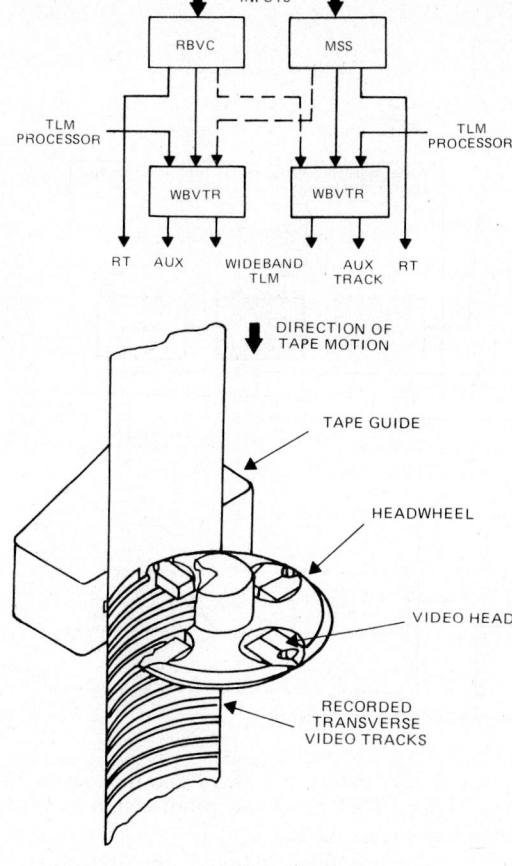

OPTICAL INPUTS

RBVC   MSS

TLM PROCESSOR

TLM PROCESSOR

WBVTR   WBVTR

RT   AUX   WIDEBAND TLM   AUX TRACK   RT

DIRECTION OF TAPE MOTION

TAPE GUIDE

HEADWHEEL

VIDEO HEAD

RECORDED TRANSVERSE VIDEO TRACKS

RECORDING HEAD STOPS FOR WBVTR

Fig. 12-28. *Top:* WBVTR system block diagram. *Bottom:* Recording head stops for WBVTR.

water gauges, temperature detectors, moisture and snow measuring instruments, geologic instrumentation, etc. Figures 12-31a and 12-31b show the DCP electronic block diagram and sketch of the package. The DCP will accept analog, serial-digital, or parallel-digital input data, as well as combinations of these inputs. Eight analog inputs or 64 bits of digital input can be accepted. Combined inputs are selected by individual analog inputs and groups of 8 bits of digital input up to a total equivalent of 64 bits. Selection of the type of input is made by the switch positions on the front panel of the platforms. For all types of inputs, the nominal signal amplitude range is from 0 to +5 Vdc. The source impedance must be less than 10,000 ohms resistance and less than 1000 picofarad capacitive. Input impedance is greater than 1 MΩ. For analog inputs, the analog-to-digital converter converts the normal signal range of 0 to +5 Vdc into 8 bits of binary with a resolution of 19.53 mv per bit; conversion error is less than 1 percent of full scale, including quantization.

Serial-digital data (of up to 64 bits) are accepted as a single input. An enable command device and

a 2.5-kHz clock are supplied to enable the transfer of the serial-digital data. Up to 64 parallel-digital bits can be accepted by the DCP. These parallel bits are sampled in 8-bit groups in sequence during a 68-ms period corresponding to the entire platform "on" time (warm-up and message transmission). A data gate is provided during this period. Before transmission, each DCP message is encoded to produce a 190-bit message output. A message is sent every 90 to 180 s, depending on the setting of the time-selection switch on the front panel of the DCP. Access by the satellite is random. Up to ~1,000 DCPs could be handled by one satellite with minimal loss of data.

LAUNCH AND ORBITAL OPERATIONS

LANDSAT LAUNCHES

As with all U.S. near-polar orbital satellites, the Landsats 1, 2, and 3 were launched from the NASA Western Test Range (WTR) at Vandenberg AFB, California. The rocket-vehicle type used for these Landsat launches was the McDonnell-Douglas Delta-900 series. This is a two-stage rocket that is thrust-augmented with a modified Thor booster, using nine Thiokol solid-propellant strap-on engines. The launch dates for the three Landsats were July 23, 1972; January 22, 1975; and March 5, 1978. All three satellites achieved orbits fairly close to nominal at initial orbital injection and required little orbital adjustment to acquire the best orbit. The operating orbital parameters of the three satellites are shown in Table 12-7.

LANDSAT ORBIT

As mentioned in the introduction, the orbit chosen for the Landsat series was a near circular, repetitive, sun-synchronous orbit with an altitude of about 913 km at the equator and with a descending node at about 9:30 a.m. local sun time. This orbit results in a cycle that repeats itself every 18 days. There are 251 orbits in this cycle. The longitudinal distance between successive orbits is 25.82°. The orbital coverage creeps westward by one orbit swath each day, providing adjacent coverage on consecutive days. The angular longitudinal distance between adjacent swaths is 1.434° and, in the 18-day cycle period, fills in the space between the successive orbits. Figure 12-32 shows the pattern of coverage as a function of day and swath. There would be approximately 14 of these coverage patterns around the earth (actually 13.944 orbits) per day. Figure 12-33 shows the orbital coverage for a single day.

After Landsat 2 was launched, the plan was for Landsats 1 and 2 to have a half-cycle relationship to each other (i.e., to be separated in time by 9 days to essentially double the data acquisition capability over a given area). Subsequently, Landsats 2 and 3 would have this same relationship. Figure 12-32 indicates this situation for all

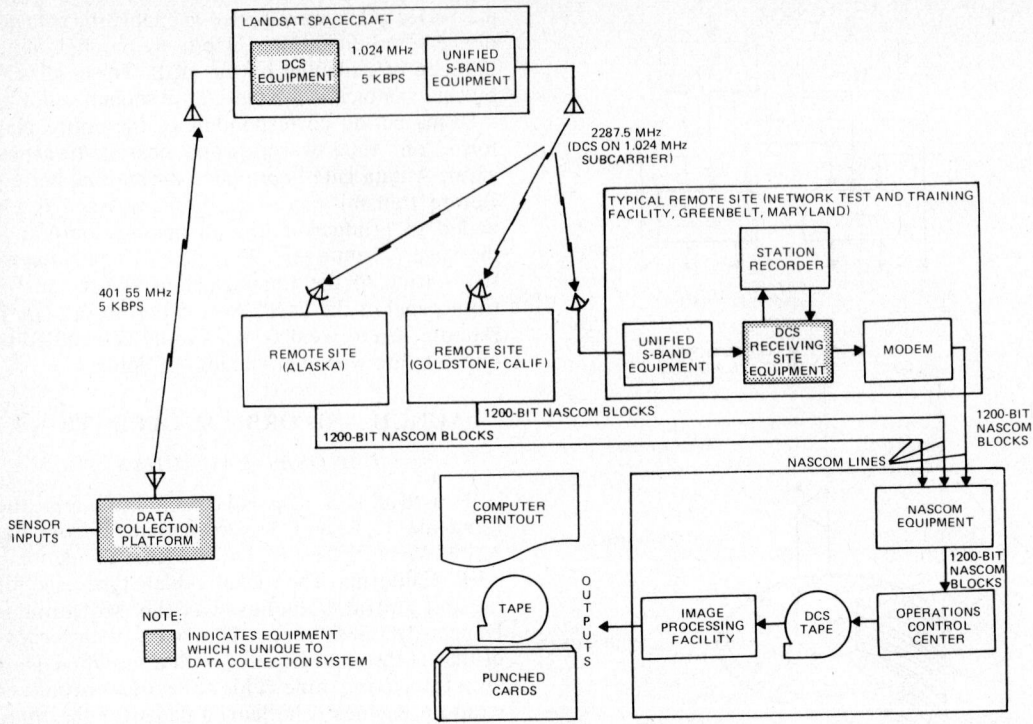

Fig. 12-29. Data collection system (DCS) block diagram.

three satellites, including the 1974 and 1975 change in the Landsat-1 and -2 relationship described below.

The concept of a circular orbit that is constantly repetitive and sun-synchronous is an ideal. The orbit, even if initially close to the ideal, is continually perturbed by various forces like the oblateness of the earth, small difference in the period-orbital angle relationships, sun and moon gravitational effects, atmospheric drag, etc. The result is that the orbit drifts at a slightly different rate than that desired to remain perfectly sun-synchronous and therefore coverage repetition patterns are approximate. This is anticipated and every so often the OAS is used to reestablish the desired orbit. The allowable cross-track drift in the orbital track, before correction was required, was specified at ± 18 km (10 n. mi.) at the equator. Figure 12-35a shows the corrective actions that were taken for Landsats 1 and 2 for most of each satellite's lifetime. As indicated in the figure, the initial adjustment to go from the launch injection orbit to the nomimal orbit is not shown. The major orbital change in 1974 and 1975, (referred to previously, in the Landsat 1 orbit-adjust history) is also shown in Figure 12-31b. By the summer of 1974, the relative drifts of Landsats 1 and 2 had caused a situation in which their relative positions and times were such that neither could be corrected by the usual orbit adjust routines. It was decided to relocate the Landsat 1 orbit so that, instead of having a 9-day/9-day situation, the adjusted relationship was 6-day/12-day coverage. Landsat 2 covered a swath 6 days before Landsat 1 and then again 12 days later. This coverage relationship is also shown in Figure 12-32.

The approximate distance between corresponding orbit paths on consecutive days at the equator is 159 km. The swath of the Landsat sensor systems if 185 km wide. This results in an equatorial sidelap between adjacent swaths of about 14 percent. Because the orbit paths converge with increasing latitude, this sidelap increases with latitude (Table 12-8). This increases the probability of cloud-free Landsat coverage on adjacent days, which is particularly useful in many areas.

WORLDWIDE REFERENCE SYSTEM

The repeatability of the Landsat orbit has allowed the development of the framed world concept. This concept is implemented in the worldwide reference system (WRS), Each orbit within a cycle is designated as a path. Along these paths, the individual nominal sensor frame-centers designate the rows of the WRS. The center point for the reference row is taken as the equator (0° latitude). Since Landsat sensors are daylight oriented, the descending node is this reference point. There are about 119 potential daylight scenes per orbit. The scene whose center is on the equator has been designated row 60. The frame immediately to the north of this, in the same path, is designated 59. This scene will have its center at the distance equivalent to 25 *seconds* of spacecraft nadir ground trace time. This number-

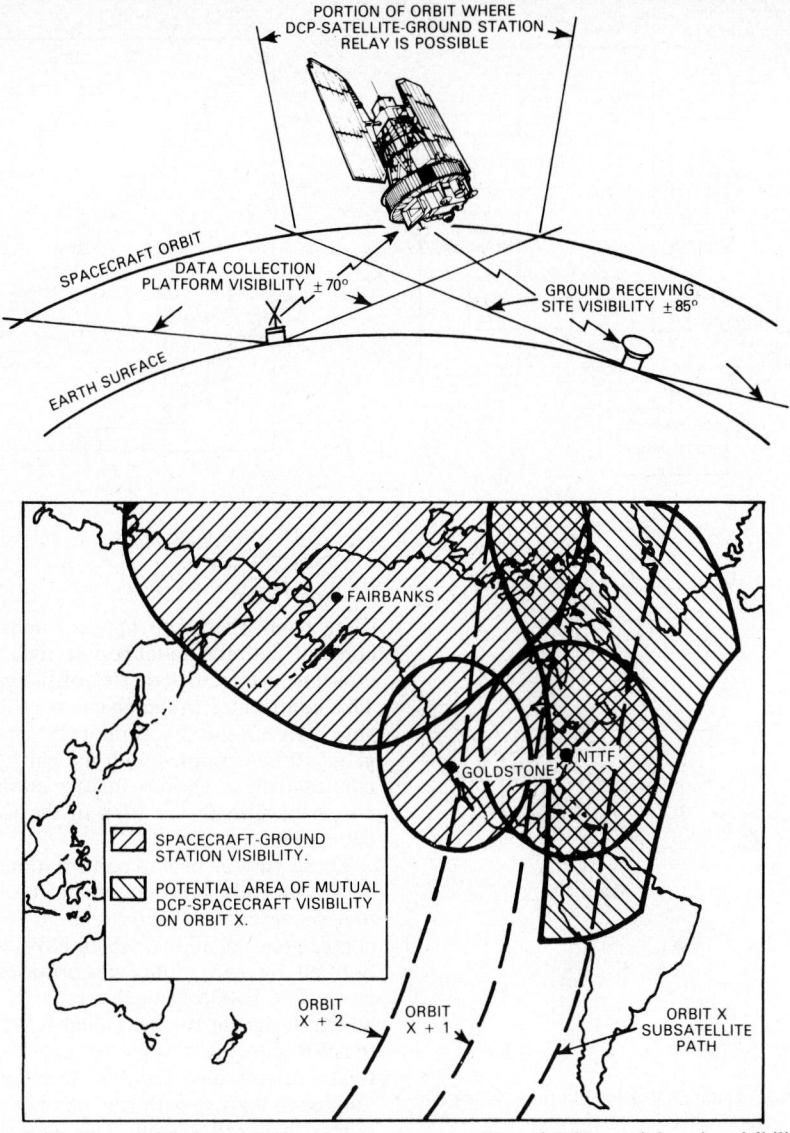

Fig. 12-30. Top: DCS data relay geometry. *Bottom:* Mutual DCP-receiving site visibility.

ing is continued northward to frame 1, whose center is located at 80°01′12″ N latitude. Southward, in the direction of the daylight course of the satellite, the numbers increase until frame 119 is reached at 80°10′12″ S latitude. The location of the ends of the daylight frames is at approximately 81° latitude because this is the extent of the satellite's global range as determined by the inclination of the orbital plane. If the night passes of the ascending node are included for thermal infrared acquisitions, the orbital path around the earth has 248 frames.

The paths correspond to the orbits. The WRS has 251 paths corresponding to the number of Landsat orbits required to cover the earth in one 18-day cycle. Path 1 is that orbit whose frames first include some of the mainland of eastern North America. As the earth is rotating from west

to east under the satellite, these paths are numbered progressively westward until path 251 is reached. Path 252 will correspond to path 1, and the cycle will start over again.

In order to have a sun-synchronous orbit, the angle between the normal to the plane of the satellite orbit and the sun-earth line must be maintained constant. This means that the Landsat orbital plane precession must be at the same rate as the rotation rate of the earth about the sun (Figure 12-35). This can be attained by achieving the proper relationship between the radius (or the period) of the nearly circular orbit and the satellite orbit inclination angle. However, the ideal orbital altitude—inclination angle relationship is not exactly achieved; therefore, the exact precession relationship necessary for an ideal sunsynchronous orbit is not achieved. Orbital perturbations

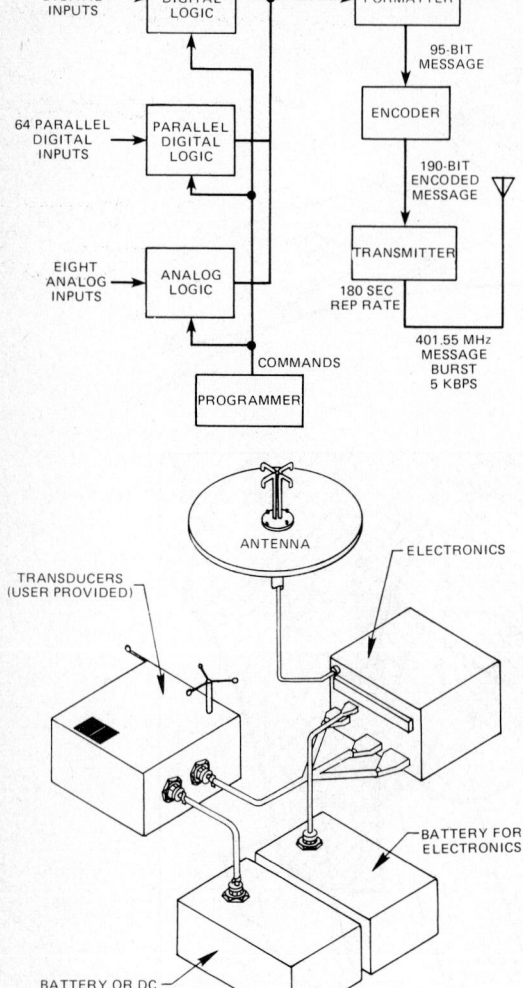

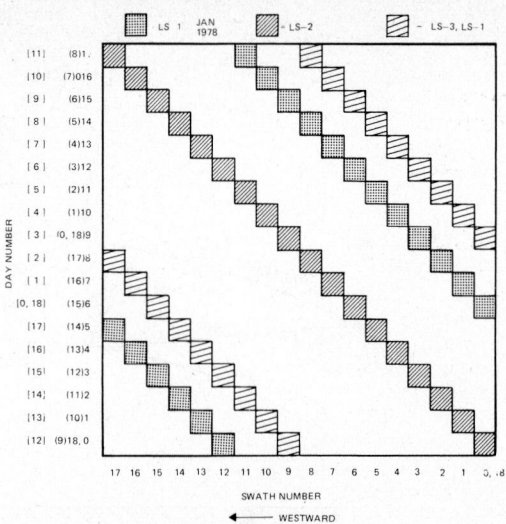

Fig. 12-32. Day/swath-number pattern of coverage for Landsats 1, 2, and 3.

Fig. 12-31. Top: Data collection platform block diagram. Bottom: Data collection platform.

(time from apogee to apogee) does not exactly coincide with the nodal period (time from equator crossing to equator crossing of the ground trace). This contributes to the changing altitude relationship between the spacecraft orbit and the nominal geoid. When coupled with the earth's natural oblateness, these various factors contribute to the observed altitude variation of the Landsats from 880 to 930 km.

These various orbital perturbations and the satellite's attitude variation result in a scatter of actual scene center points relative to the nominal center point locations, which have been uniquely defined by geographic coordinates on a fixed earth. The EROS Data Center (EDC) computers, which designate the individual Landsat scenes in the WRS, handle this by an algorithm that effectively selects any Landsat scene center as belonging to a given path-row number if its distance is less than 73 km from the nominal center. The usual scene designation by path-row is a pair of three-digit numbers (e.g., 121 to 042). The WRS is the same for Landsats 1, 2 and 3, and a scene can be uniquely defined by specifying the cycle number or date of pass for the given satellite and the path-row of interest. Figure 12-36 shows the

also change this relationship continually. The slight difference in precession rate between the ideal and the actual causes a small drift in local time of equatorial crossing. The orbit also is not exactly circular; therefore, the anomalistic period

TABLE 12-7

Orbital Parameters of Landsat

Orbital Parameter	Satellite		
	1	2	3
Semimajor Axis (km)	7285.438	7285.989	7285.776
Inclination (deg)	99.906	99.210	99.117
Period (min)	103.143	103.155	103.150
Eccentricity	0.001070	0.001019	0.001330
Time of Descending Node Equatorial Crossing (local time)	8:50 a.m.	9:08 a.m.	9:31 a.m.

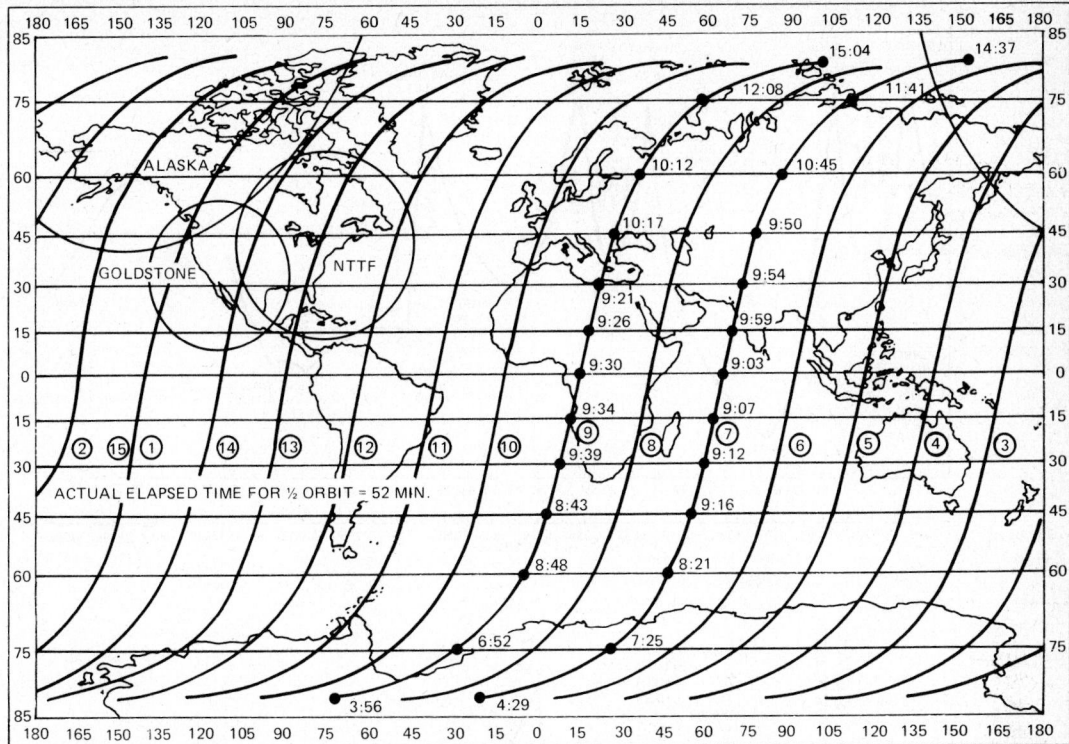

Fig. 12-33. Typical Landsat daily ground trace (daylight passes only), showing local time variations within an orbit.

framed world center concept for a portion of the United States.

GROUND SYSTEMS

GROUND COMMUNICATION STATIONS

The ground communication stations are the radiotelemetry links between the Landsat satellites and the earth. They fall into two general categories: (1) those which have the sole function of receiving sensor and attitude data from the satellites, and (2) those which, in addition to receiving sensor and attitude data, can receive satellite housekeeping data and transmit commands. The former are essentially all the non-U.S. stations, whereas the latter are the three U.S. sites located at Greenbelt, Maryland; Goldstone, California; and Fairbanks, Alaska.

Figure 12-37 shows the real-time coverage area of the various receiving sites throughout the world. Table 12-1 gives an indication of the amount of data that were acquired by the three Landsats up to May 1982.

In 1979, the use of commercial communications satellites was incorporated as a data link between the Alaska and California receiving sites and the ground data-handling system (GDHS) at GSFC. It is the GDHS that preprocesses and/or processes the Landsat data prior to product generation at the United States Department of the Interior (USDI) EROS Data Center at Sioux Falls, South Dakota (see below). Prior to this, the tapes of all data

acquired at these sites were sent to GSFC by mail. The data acquired by the non-U.S. receiving sites are processed under the auspices and/or in the facilities of the corresponding countries and are distributed through their own centers. Table 12-9 shows the chronology of the non-U.S. receiving sites.

Figure 12-34 also shows that, even after the activation of the various non-U.S. receiving sites, much of the land mass of the world remains outside ground-station acquisition areas. These non-real-time areas require that any Landsat data acquired of them be tape-recorded on board the spacecraft. When this is done, the data are dumped at Fairbanks or GSFC, usually during a nighttime pass.

The receiving site role changed somewhat with the use of the Domsat commercial communications satellite. Prior to this, preprocessing of the serial sensor data received from the satellite was performed. At the receiving site, it was parallel-recorded on a 28-channel tape unit, with one detector output per channel, and later mailed to the NASA Data Processing Facility (NDPF) at GSFC. The data received at the Goddard/Greenbelt site were hardwired into the Operations Control Center (OCC) and preprocessed there as at the other U.S. sites. These preprocessed data were hand-carried to the NDPF.

Since the initiation of Domsat, the data from all three U.S. sites go into the Domsat Interface Facility at GSFC, where they are formatted for input

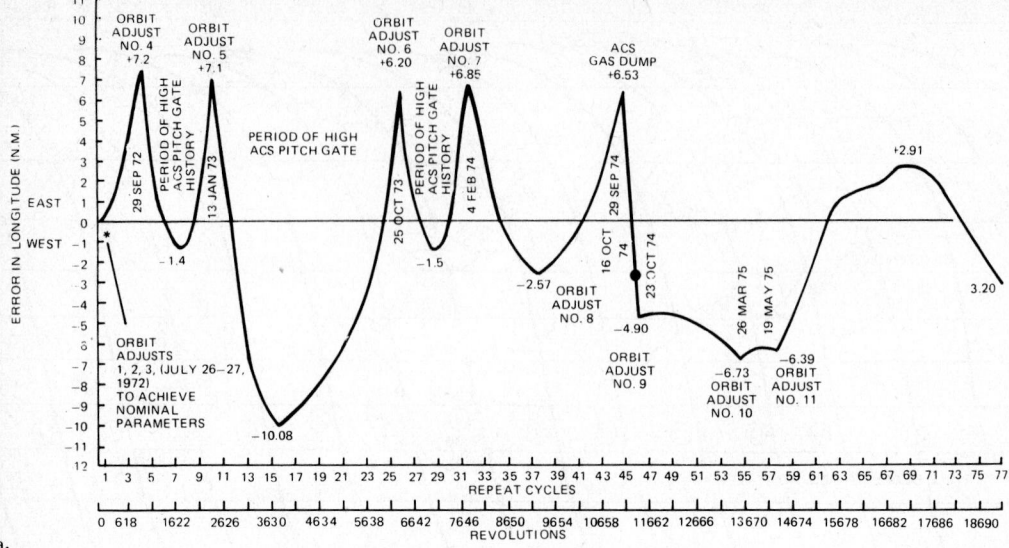

a.

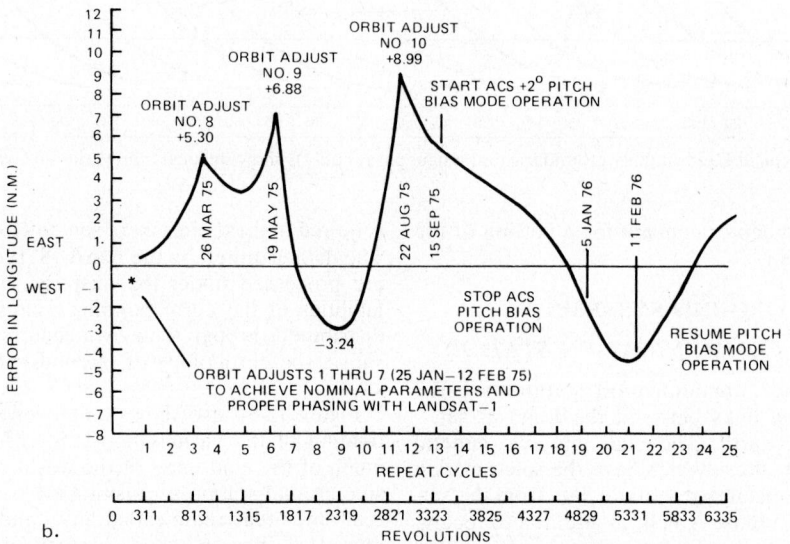

b.

Fig. 12-34. Cross-track drift and orbit adjustments (a. Landsat; b. Landsat 2).

into the Image Processing Facility (IPF), the successor to the NDPF.

OPERATIONS CONTROL CENTER

The OCC located at GSFC is the heart of all the ground activities associated with the Landsat system. The spacecraft functional operations and data acquisition and transmission are controlled and monitored at this point in the system. The OCC computers process all the housekeeping data associated with spacecraft and payload performance and generate the commands transmitted to the spacecraft. The system data, which generally are transmitted via S-band but which can use VHF, and the commands to the spacecraft can be

communicated to any of the three U.S. antenna sites from the OCC via the NASCOM network. Prior to the initiation of the Domsat relay system (Figure 12-38), the MSS/RBV data received at the Network Test and Training Facility (NTTF) located at GSFC were hardwired into the OCC and recorded and preprocessed there. These data now go directly to the GSFC Domsat Interface Facility (DIF) for these operations. The OCC also receives (via NASCOM) and records the DCP data collected at the remote sites. In the OCC, there are control and display consoles that interact with the computers. These permit direct monitoring and interaction between the satellite systems and the operators. The essential elements of the OCC are shown in Figure 12.39.

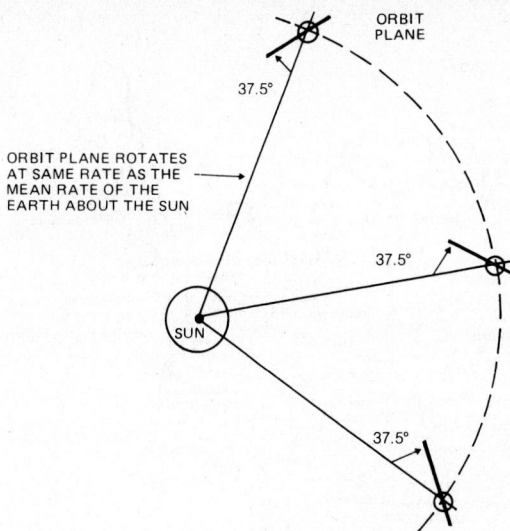

Fig. 12-35. Motion of orbit plane in sun-synchronous orbit.

NASA DATA PROCESSING FACILITY

Until the all-digital IPF, presently at GSFC, became operational, all the Landsat-1, -2 and -3 data were processed by its predecessor, the NDPF. This had as its archival output a 70-mm film product that was created by an EBR from the incoming sensor data. In parallel to the film product, an X-format CCT could be generated. These tapes were created upon request (i.e., one was not created for every scene acquired by the MSS or RBV). The archival 70-mm film was forwarded to the USGS EROS Data Center (EDC) at Sioux Falls, South Dakota, for product production. When the EDC received a first-time order for a CCT, a CCT master would be generated by the NDPF and forwarded to the EDC for archiving, copying, and distribution. Subsequent requests for that CCT scene would then be made from the tapes in the EDC library.

TABLE 12-8

Sidelap—as a Function of Latitude

Latitude (deg)	Image Sidelap (%)
0	14.0
10	15.4
20	19.1
30	25.6
40	34.1
50	44.8
60	57.0
70	70.6
80	85.0

IMAGE PROCESSING FACILITY

All MSS data acquired by the Landsats beginning on February 1, 1979, are being processed by the IPF. Since September 1980, the RBV data are also being processed in the digital mode through the IPF. Prior to that, the Landsat-3 RBV data had been processed through the NDPF using the film mode via the EBR.

The IPF has two parallel computer systems that are interdependent but do not interact directly. The main system is the IPF itself that processes the sensor data as received from the DIF. The second system is the information and product control system (IPCS) that keeps track of the data being processed and initiates work requests on what should be processed. Figure 12-40 shows the data flow for the IPF.

The MSS videotapes (MVTs) and the RBV videotapes (RVTs) are preprocessed in the respective MSS processor (MPP) or RBV processor (RPP) subsystems. When the MPP or RPP functions are completed, a performance record (MPPT, RPPT) is generated by the MPP or RPP and entered into the IPCS. A comparable performance record is sent after each step of the IPF operation. The MPP organizes the data into major (single scan) and minor (single data sampling set) frames and associates with each data segment the appropriate header information. These are fed into a high-density digital recorder, and the resultant tape (HDT-FM) is the input to the Master Data Processor (MDP). A similar operation occurs in the case of the RBV data, resulting in the production of HDT-FR and RPPT tapes. Since the RVT data are in analog form, the RPP must also perform the analog-to-digital conversion into the serial-digital format required by the MDP. Both RPP and MPP have the capability of displaying the scene being processed so that a cloud-cover assessment can be made by the operator. Similarly, both subsystems have the ability to make a quality assessment of the lines in each channel and, if a line is unacceptable, the subsystem can substitute an adjacent line for that line. The output of either the RPP or the MPP can be used to create a latent 70-mm image by the quick-look film recorder (QLFR). This image can be produced for quality-control purposes or to supply a user's need for quick-look data. The QLFR produces an image up to about the equivalent of a full MSS scene size (3600 pixels/scan line by 2340 lines/band), upon which neither radiometric nor geometric corrections have been made.

Master Data Processor

All the corrective processing of the raw data as reformatted in the MPP or RPP is performed in the MDP. The tapes that result from the processing will be the archival products that are sent to the EROS Data Center for user product production. Note that except for an ancillary side-look at the

Fig. 12-36. Worldwide Reference System (WRS) index map of the United States.

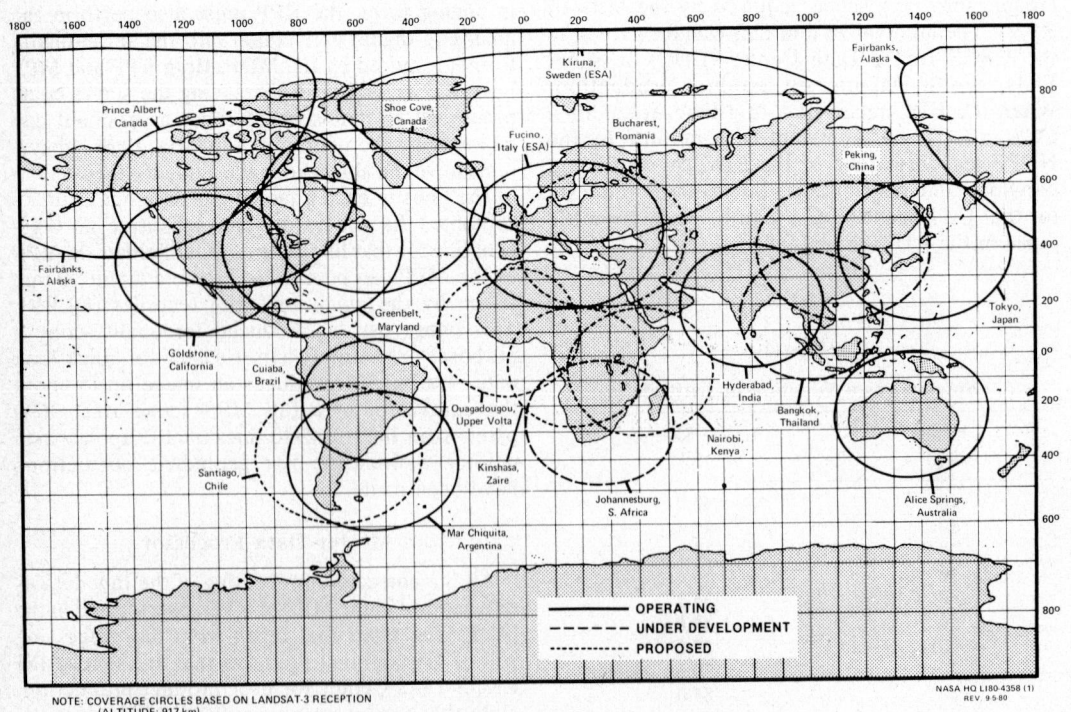

Fig. 12-37. Current and probable Landsat ground stations.

TABLE 12-9

Ground Station Chronology

Country	Date of Initial Agreement	Date Station Contract Signed	Date Station Begins Receiving Data	Date Station Begins Data Processing
Argentina	1976	1978	1980	1980
Australia	1978	1978	1980	1980
Brazil	1973	1973	1973	1974
Canada				
(Prince Albert)	1972	1971	1972	1972
(Shoe Cove)			1977	1977
Chile	1976	—	—	—
India	1978	1978	1979	1980
Iran	1974	1976	1978	1978
Italy (ESA)	1974	1974	1975	1976
Japan	1978	1978	1978	1979
Sweden (ESA)	1978	1977	1978	1979
Zaire	1975	—	—	—

latent image from the QLFR, this entire processing procedure is in the digital domain. This is the essence of the difference between the NDPF and the IPF operation. The prime product of the NDPF has been an analog/photographic product from the electron beam recorder, a 70-mm film positive for each band, whereas the IPF prime product is a high-density digital tape.

The inputs to the MDP are the HDT-FM and/or HDT-FR from the preprocessors, ancillary data tape (ADT), order information, and pertinent ground control point (GCP) data. The ADT is produced by the IPCS from input of the spacecraft location and attitude tape (SLAT) and is used by the MDP for framing and annotating the sensor data and for providing data for the calculations of the geometric break points that are used to interpolate the data for the geometric correction process. The ground control points that can be used in the geometric correction process are stored in the GCP library, which is part of the MDP. GCP updated information can be added, as needed. The

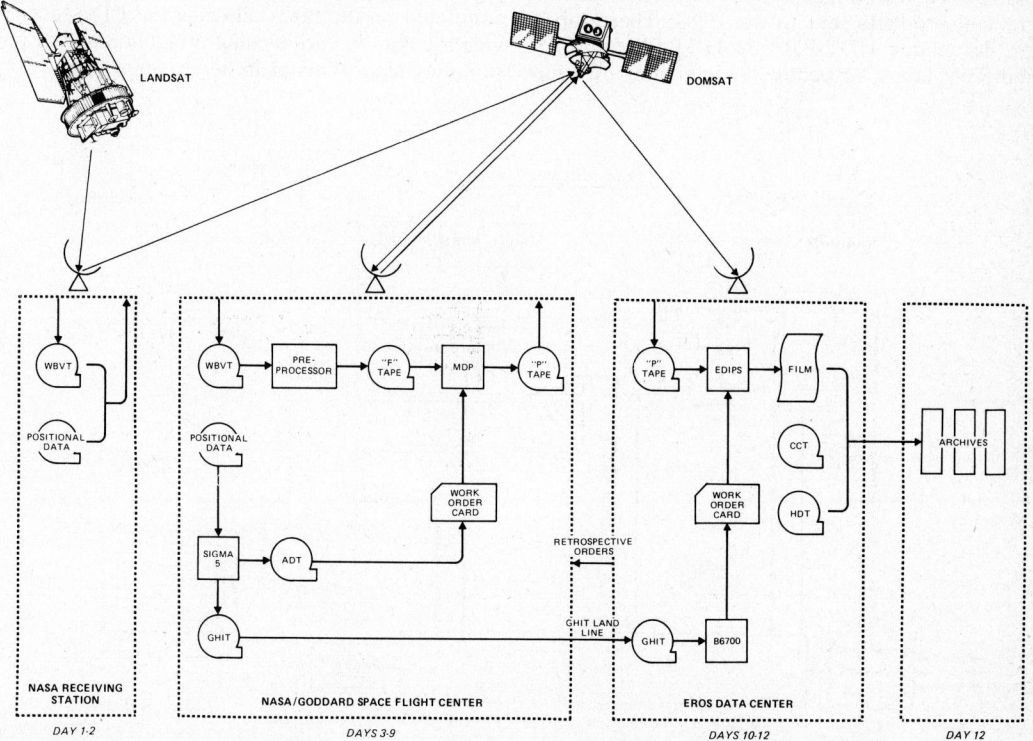

Fig. 12-38. Landsat communication and ground processing pattern after initiation of Domsat relay.

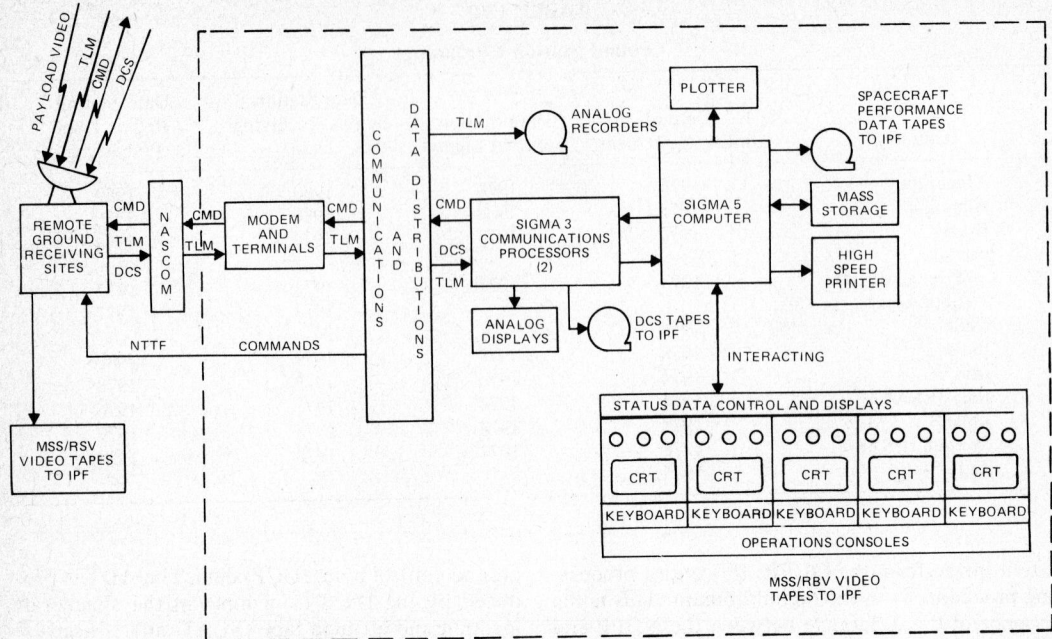

Fig. 12-39. Major elements of operations control center.

initial objective is to have a GCP library that contains pertinent GCPs for 860 preselected Landsat MSS scenes.

Since the advent of the MDP on February 1, 1979 for the MSS and on September 1, 1980 for the RBV, the standard outputs of the MDP have been the fully corrected high-density tapes that are the archival products sent to the EDC. These tapes are designated HDT-PM and HDT-PR for MSS and RBV tapes, respectively. Beginning on June 1, 1981, and September 1, 1981, respectively the IPF has been supplying HDT-AM and HDT-AR tapes to the EDC for product generation. These are, respectively, MSS and RBV high-density digital tapes on which the geometric corrections have been calculated using GCPs, but are not yet applied to the data. The correction functions are annotated on the tapes allowing the EDC to provide the data in various map-projection and/or resampling algorithms or in uncorrected form.

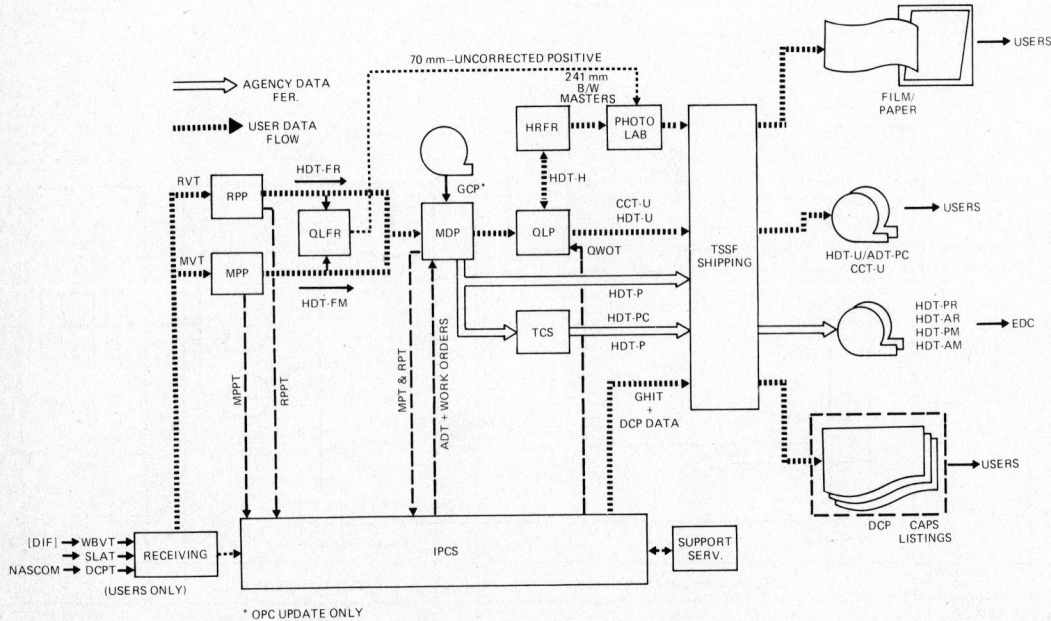

Fig. 12-40. User data flow in the IPF.

When sufficient numbers of GCPs cannot be obtained, an image will be system-corrected based only on the spacecraft's AMS and the orbital ephemeris data. In this case, the accuracy of these data depends on the performance of the AMS and the validity of the ephemeris data. Table 12-12 shows the expected differences in accuracy between geodetically corrected and uncorrected data.[7]

The HDT-P tapes have a band sequential (BSQ) format. The resampling algorithms in the geometric correction process are based on cubic convolution techniques. The standard map projection being used at the present time is the hotline oblique mercator (HOM), a close approximation to the space oblique mercator (SOM). It is planned eventually to convert to the SOM projection.

Quick-Look Processor

The HDT output of the MDP can be used as input to a quick-look processor (QLP) if any need is indicated. If a quick-look work order tape (QWOT) is generated by the IPCS in response to a user or operator request, the QLP will generate the specified product. These products may be in HDT or CCT format. The HDTs produced include HDT-U, which include only the specific HDT scene(s) required by the requestor. If the request is for a scene in CCT format, the QLP will reformat the input HDT data.

High-Resolution Film Recorder

An additional output of the QLP is an HDT-H that is formatted for input into the high-resolution film recorder (HRFR). The HRFR is a laser beam recorder that accepts the BSQ-formatted HDT-H and produces 241-mm black-and-white pictures from each band. The bandwidth (BW) output for each band of the scene may be sent to the GSFC Photographic Processing Laboratory (PPL) for production of black-and-white (BW) transparencies or prints, color products, etc. The PPL will also process the 70-mm imagery produced by the QLFR, if required.

---

[7] At this writing, GCPs only exist for MSS data. Thus, RBV data are not geodetically corrected.

Shipment to EDC

The GSFC Tape Staging and Storage Facility (TSSF) collects the output from the MDP and the IPCS for shipment to EDC. The MDP output is the archival HDTs that are transmitted to EDC via Domsat. The IPCs, working from information on the MPTs and RPTs, generate the necessary shipping documents (labels, etc., when required) and the Goddard HDT inventory tape (GHIT), which is sent to EDC for use with the associated HDT's. The DCP data that are received and taped at the OCC are processed in the IPCS into punched cards and/or listings, which are sent to the users through the TSSF.

EROS DATA CENTER

The USDI EROS Data Center (EDC) at Sioux Falls, South Dakota, is the facility that produces all the Landsat user-oriented products and passes the data into the public domain. All Landsat data recorded by the U.S. ground stations are available to the general public at EDC, provided they are processable. The Landsat data are considered to be in the public domain once they are received and catalogued by EDC. Until the summer of 1979, the output products from GSFC were mailed to EDC. At that time, the Domsat link was initiated, and now the standard procedure is to send the data through this satellite to EDC. Backup tapes may be mailed to EDC, if requested. The computer system that EDC uses to produce the Landsat data products is the EDC digital image processing system (EDIPS). The standard input to EDIPS is now the HDT-A. The output products include film and digital tapes. The conversion of the incoming HDT-A tapes into corrected (P) 241-mm film products is also performed for all scenes. Digital products, either CCTs or HDTs, are made on customer demand. Two types of further data refinement beyond the standard radiometric and geometric corrections applied to the data are available from the EDIPS. These apply haze removal and edge enhancement algorithms. In early EDIPS operation, a contrast-enhancement algorithm was also applied, but was eliminated because of the negative effects that resulted from it in subsequent temporal comparisons of scenes.

LANDSAT-D SATELLITES[8]

GENERAL

The Landsat-D and -D' satellites are the bridge between the old and the new in Earth resources satellite systems. Each has as a payload a multispectral scanner (MSS) and a thematic mapper

[8] For the remainder of this chapter the terms Landsat-D and Landsat-4 are used somewhat interchangeably. This reflects the fact that part of the material was written before, and part after, the successful launch of Landsat-D on July 16, 1982, at which time its name automatically was changed to Landsat-4.

TABLE 12-10

Geometric Accuracy of HDT-Ps

With "Good" GCPs	Without "Good" GCPs
• At least 25 well-distributed GCPs Geodetic accuracy of the Landsat image—≤1.0 pixel (57 m)	• Geodetic accuracy of the Landsat image— 60 pixels (3.5 km)
• 8 to 24 CCPs—10 pixels	
• 1 to 7 GCPs—20 pixels	

(TM). The MSS is included in the Landsat-D and -D' sensor packages to provide continuity of the data type that has been provided since 1972 by Landsats 1, 2, and 3. The TM is a more advanced scanner, having more spectral, radiometric, and geometric sensitivity than its predecessors. Landsat-D is scheduled for launch in late 1982. Landsat-D' will be stored in a launch-ready condition and will be used when Landsat-D fails.

LANDSAT-D INSTRUMENT SYSTEMS

The MSS's on the Landsat-D and -D' are identical in spectral-band location and width with the four reflective bands of Landsats 1, 2, and 3: 0.5 to 0.6, 0.6 to 0.7, 0.7 to 0.8, and 0.8 to 1.1 μm. The TMs will have seven bands, three of which are in the spectral range of the MSS (green through reflective infrared). The new spectral coverage regions are the blue and the mid-infrared ranges. The blue band will be 0.45 to 0.52 μm. The mid-infrared (short wave) bands will be 1.55 to 1.75 and 2.08 to 2.35 μm. The three bands that approximate the region of the four MSS bands are 0.52 to 0.60 μm (green), 0.63 to 0.69 μm (red), and 0.76 to 0.90 μm (reflective infrared). The green and red bands are narrower than their MSS predecessors, primarily to improve the sensitivity to spectral changes that result from agricultural phenomena in this region. The reflective IR band is narrower than the combined bands of the MSS in this region, having its center in a region of maximum sensitivity to plant vigor. The seventh band is the thermal infrared band (nominal 10.4 to 12.5 μm), which is similar to the band that was on Landsat 3. (Actual Landsat-D TM prelaunch-calibrated bandwidth was 10.42 to 11.61 μm.)

The new blue band has been included to expand the use of satellite data in the field of bathymetry and to assist in stress evaluation for agricultural crops. The mid-infrared bands will enable agriculturalists to better study water-stress problems in crops and will help geologists to better distinguish between rock classes. The 1.55- to 1.75-μm band will also improve the ability to discriminate between snow and cloud coverage. Table 12-11 synopsizes the functions and requirements of the instrument.

The linear geometric resolution of the TM is about 2.6 times that of the MSS. The IFOV of the TM is a 30-m × 30-m pixel, as compared with the 79-m × 79-m pixel of the MSS of the original Landsats. (The MSS on the Landsat-D MSS will have an IFOV of 82m × 82m). The TM's higher resolution will permit observation of smaller areal segments (e.g., smaller agricultural fields). Table 12-12 shows the comparative characteristics of the two sensor systems aboard Landsat-D and -D'.

The greater radiometric sensitivity of the TM is achieved by going to a digitizing scheme that has a sensitivity of one part in 256, 8-bit quantization in the analog-to-digital conversion process. The MSS uses 6-bit quantization, a sensitivity of one part in 64. This finer radiometric detectability in the TM should permit observation of smaller changes in radiometric magnitudes in a given spectral band and also provides a greater sensitivity to changes in relationships between bands.

The TMs greater number of spectral bands, higher radiometric sensitivity and the more resolution elements (pixels) per unit ground area, all contribute to a much greater data bit rate. This results in the need for a much higher communication frequency to telemeter the TM data. The TM data rate, after digitizing, will be 84.9 Mbps. The S-band microwave carrier used on Landsats 1, 2, and 3 cannot handle this rate. Therefore, in direct communication to ground stations, X-band (8.025 to 8.4 GHz) will be used for transmitting the TM and MSS data, which collectively come to 100 Mbps. The MSS data will also be independently

TABLE 12-11

Thematic Mapper Functions and Requirements

Bands	Range (μm)	Radiometric Resolution	Principal Applications
1	0.45–0.52	0.8% NE$\Delta\rho$	Coastal Water Mapping, Soil/Vegetation Differentiation, Deciduous/Coniferous Differentiation
2	0.52–0.60	0.5% NE$\Delta\rho$	Green Reflectance by Healthy Vegetation
3	0.63–0.69	0.5% NE$\Delta\rho$	Chlorophyll Absorption for Plant Species Differentiation
4	0.76–0.90	0.5% NE$\Delta\rho$	Biomass Surveys, Water Body Delineation
5	1.55–1.75	1.0% NE$\Delta\rho$	Vegetation Moisture Measurement, Snow/Cloud Differentiation
6	10.4–12.5	0.5 K NEΔT	Plant Heat Stress Management, Other Thermal Mapping
7	2.08–2.35	2.4% NE$\Delta\rho$	Hydrothermal Mapping

Absolute Radiometric Accuracy	10%	Scan Profile Repeatability	<6 m
Band-to-Band Radiometric Precision	2%	Along-Track Overlap/Underlap	<6 m
Channel-to-Channel Radiometric Precision	$< \dfrac{\text{RMS Noise}}{4}$	Swath Width	185 km
Band-to-Band Registration	<6 m		

TABLE 12-12

Radiometer Characteristics

	Thematic Mapper (TM)		Multispectral Scanner Subsystem (MSS)	
	Micrometers	Radiometric Sensitivity (NE$\Delta\rho$)	Micrometers	Radiometric Sensitivity (NE$\Delta\rho$)
Spectral Band 1	0.45− 0.52	0.8%	0.5−0.6	0.57%
Spectral Band 2	0.52− 0.60	0.5%	0.6−0.7	0.57%
Spectral Band 3	0.63− 0.69	0.5%	0.7−0.8	0.65%
Spectral Band 4	0.76− 0.90	0.5%	0.8−1.1	0.70%
Spectral Band 5	1.55− 1.75	1.0%		
Spectral Band 6	10.40−12.50	0.5 K (NEΔT)		
Spectral Band 7	2.08− 2.35	2.4%		
Ground IFOV		30 m (Bands 1−6) 120 m (Band 7)	82 m (Bands 1−4)	
Date Rate		85 Mbps	15 Mbps	
Quantization Levels		256	64	
Weight		258 kg	68 kg	
Size		1.1 × 0.7 × 2.0 m	0.35 × 0.4 × 0.9 m	
Power		332 watts	50 watts	

transmittable in S-band. Thus, the existing ground stations, unless modified, will be able to receive MSS data, but not TM data.

To achieve worldwide coverage, independent of the use of tape recorders or foreign ground stations, NASA will use the NASA Tracking and Data Relay Satellite System (TDRSS). This two-satellite geosynchronous system will receive sensor data from the Landsat-D and retransmit them to the TDRSS receiving site at White Sands, New Mexico. Then, these data will be again retransmitted to GSFC via the Domsat communications satellite for processing in the GSFC Landsat-D ground-processing system. The Ku-microwave band (∼15 GHz) will be used for data transmission between the Landsat-D and the TDRSS satellites. Figures 12-41 and 12-42 show the general data communication patterns and major elements of the Landsat-D system, respectively. Figure 12-43 shows the possible coverage by ground stations for the MSS data. The coverage of Landsat-D for both sensor systems, using the TRDSS, is the entire world, except for (1) the zone of exclusion shown over India and the Indian Ocean, and (2) circumpolar areas having a latitude of greater than 81°. The TDRSS satellites will be geosynchronously located above the equator at 41 W and 171 W longitudes.

The present plan for launching the Landsat-D into the nominal 705-km orbit will result in the coverage pattern shown in Figure 12-44. This orbit will have a repeat cycle of 16 days. The swath width for both sensors will be 185 km. The sun-synchronous, repetitive orbit will cross the equator on its descending orbit at about 9:45 a.m. local sun time, which is consistent with the local times of the previous Landsats. The orbit is lower than that of Landsats 1, 2, and 3 that were at a nominal altitude of 913 km. This new orbit was dictated primarily by the desire to make the Landsat-D and its successors compatible with the concept of retrievability by the Space Shuttle. The lower orbit also eases the problems associated with achieving higher resolution.

LANDSAT-D FLIGHT SEGMENT

The Landsat-D flight segment consists of the spacecraft and its payload. The spacecraft is a Multimission Module Spacecraft (MMS), which is larger than those used in the previous Landsat series. It will have all the functional operations of supplying power, providing attitude and altitude control, and will contain all the communications and control systems. The attitude control will be much more stringent than previously required, dictated by the higher performance specifications of the sensor system. The pointing accuracy is specified as 0.01° (1 standard deviation) and the stability is $10^{-6°}$ s^{-1} (1 standard deviation). These may be compared with the values on previous Landsats, which were 0.7° and 0.04° s^{-1}, respectively. The general configuration of Landsat-D flight segment is shown in Figure 12-45. The flight segment weight is 1996 kg (4400 lb) as compared with the 953 kg (2100 lb) of Landsat 3. The high-gain antenna shown protruding from the spacecraft is the microwave link with the TDRSS and must be capable of pointing to the geostationary orbital locations of the TDRSS. This is achieved by the double-gimbaled mounting that connects the antenna to the mast. The average end-of-life power capacity of the solar panels, which are essentially cantilevered to one side, is 1497 watts during the 65 minutes in sunlight.

THEMATIC MAPPER

The general performance characteristics of the TM were described previously. However, how

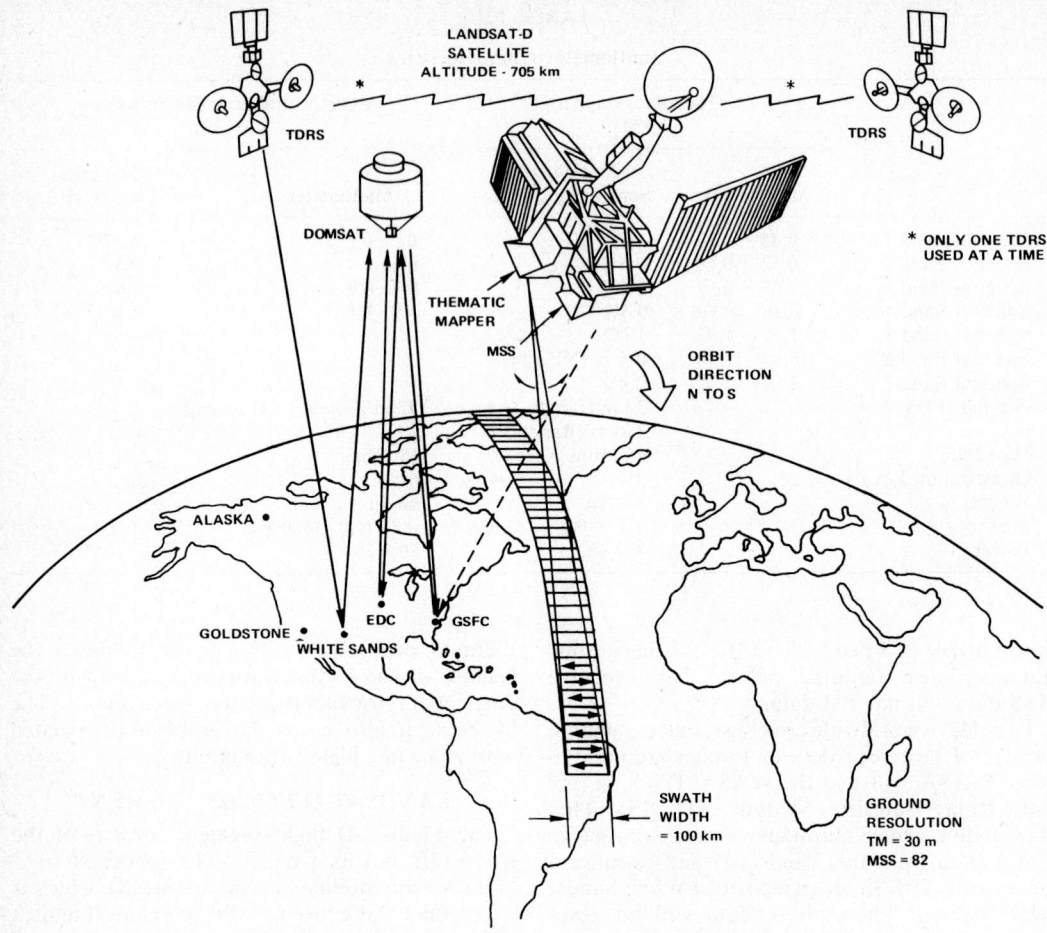

Fig. 12-41. General data communication pattern.

these performance parameters are implemented relative to the more complex performance requirements of this scanner system is of special interest. The general optical field-stop, focal-plane concept used in the MSS is still retained in the TM, but with a number of innovative features warranted by the needs of the more complex system. Figure 12-46 shows the general layout of the TM optical system. A unique feature of this system is the introduction of a second set of moving mirrors called the scan-line correctors into the optical path between the scan mirror and the detector focal plane. To get the necessary signal strength so that an acceptable signal-to-noise detector performance can be achieved, it is necessary to slow down the sweep rate of the scanning mirror (6.999 Hz as compared with 13.67 Hz for the MSS). This is necessitated by the smaller IFOV of each detector (42 μrad for the TM versus 86 μrad for the MSS), the larger number of detectors that must be sampled in a minor frame, and a large number of complex interrelated system features. The MSS has a 6 × 4 matrix of detectors, whereas the TM has 16 × 6 reflective band detectors plus 4 thermal detectors, com-

prising a 100-detector matrix. The actual difference in minor frame rate counts is 102 for TM versus 25 for the MSS.

To meet the general system requirements, both directions of the scan mirror sweep must be active in the TM (the MSS is active only on the west-to-east sweep in the descending or daylight orbital path). Figure 12-47 (a) shows a simplified version of what would happen if only the scan mirror motion were used to reflect the ground data into the sensor. Part of the sweep would have redundant data (overlap) and part would miss areas entirely (underlap). If, however, the individual detectors in the focal plane are effectively field stops which only interrupt part of the total image that is available to the scan mirror aperture, then it is possible to effectively move these field-stop areas around to eliminate the redundancies and fill in the gaps. The scan-line corrector mirrors are there for this purpose. Figure 12-47(b) shows the action of these two parallel mirrors in the vertical plane as they rotate in the optical path between the scan mirror and the focal plane. As the scan mirror sweeps west to east, or vice versa (Figure 12-48), the scan-line corrector mirror effectively moves the

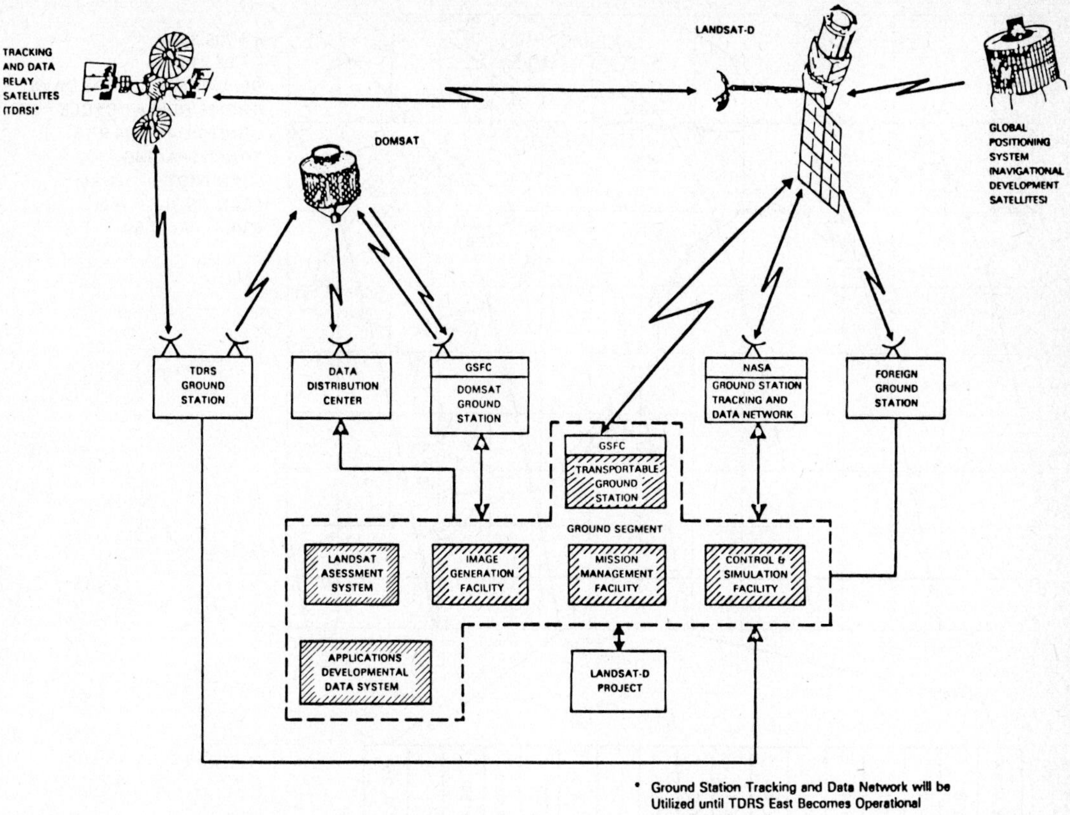

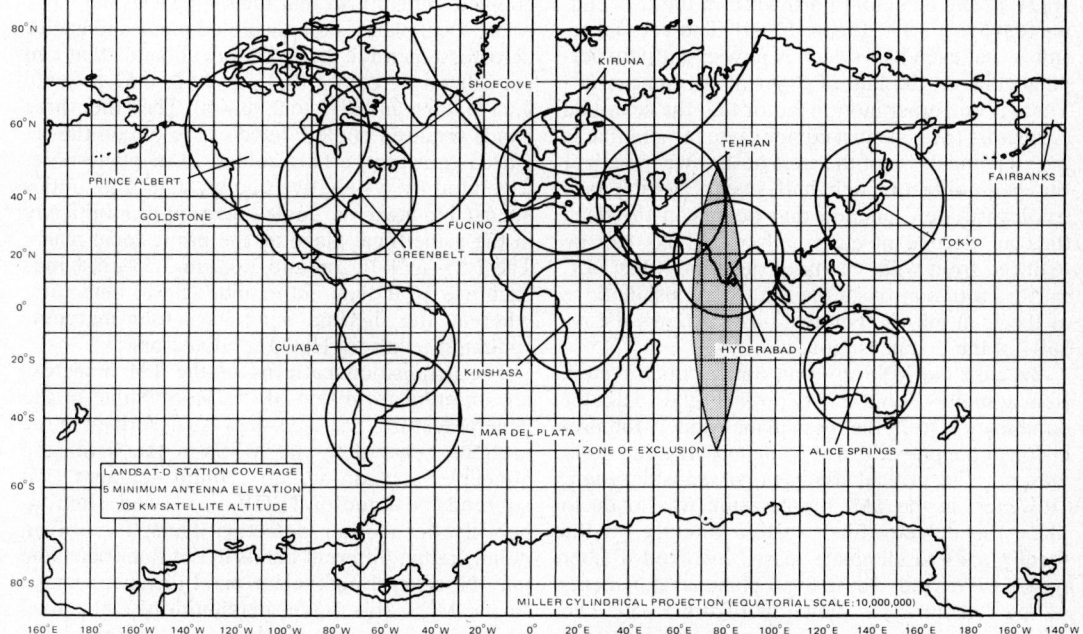

Fig. 12-42. Major elements of the Landsat-4 system.

Fig. 12-43. Landsat-4 ground receiving station coverage.

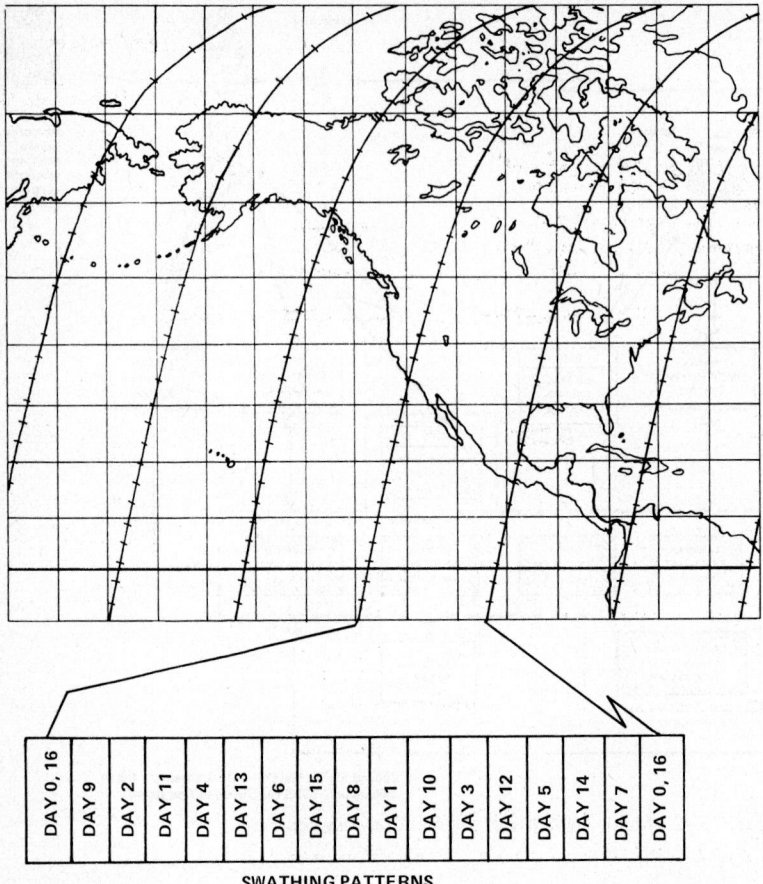

h = 705.3 km
i = 98.210°
REPEAT PERIOD = 16 DAYS
ORBITS/REPEAT CYCLE = 233
ORBITS/DAY = 14 9/16
TRACE SPACING = 172
SCAN WIDTH = 185 km
SCAN ANGLE = 14.9°
OVERLAP = 7.6%

DAY 0, 16 | DAY 9 | DAY 2 | DAY 11 | DAY 4 | DAY 13 | DAY 6 | DAY 15 | DAY 8 | DAY 1 | DAY 10 | DAY 3 | DAY 12 | DAY 5 | DAY 14 | DAY 7 | DAY 0, 16

SWATHING PATTERNS

Fig. 12-44. Landsat-4 orbit ground trace pattern.

IFOV of the detectors northward in the descending (daylight) orbit (Figure 12-49). The vertical result is that each half-sweep is now essentially perpendicular to the line of flight of the spacecraft. There is one other motion factor that the scan-line correctors (SLCs) must compensate for and that is the stepping down of the start of the sweep path at the beginning of each half-sweep, which is a southward step function that occurs in the dead time at the end of each half-sweep. As can be deduced from what has been described, the SLC mirrors rotate at twice the rate of the oscillations of the scan mirror. Figure 12-47(c) shows the results of the compensation.

Because the MSS used an optical fiber system that actually isolated the optical filter/detector combination from the focal plane proper, the precision of the physical arrangement of the detectors was not the critical dimensional consideration. However, in the TM, the decision was made to place the detector/filter system directly on the focal planes to eliminate losses involved if fiber optics were used. These focal plane arrangements of the detectors are shown in Figure 12-50. The effective focal plane arrangement shown in Figure 12-51 is actually a combination of two focal planes

coupled optically so that they appear as one. The prime focal plane is the uncooled plane and has 16 detectors each for the four optical bands that can use silicon detectors (0.45 to 0.52, 0.52 to 0.60, 0.63 to 0.69, and 0.76 to 0.90 μm). The other three bands required cooled detectors to obtain the required sensitivity. This cooled focal plane is refrigerated by a radiative cooler, and relay optics are used to make the cooled plane appear optically in the same focal plane as the prime focal plane. The 1.55- to 1.75-μm and 2.08- to 2.35-μm bands each use 16 indium-antimonide (InSb) detectors, whereas the thermal band uses four mercury-cadimum-telluride (HgCdTe) detectors.

The acquisition patterns of the TM detectors are different from that of the MSS. Some of the general features are the same. As described in the sections concerning the MSS, in the field stop concept of scanning, no ground segment that subtends a detector IFOV is viewed simultaneously be another detector. Thus, in a given minor frame, different bands in the same scan line may be several equivalent pixel distances apart. In the MSS, this was a maximum of eight pixel distances. Figure 12-51 shows that between the first and last of the reflective bands on the TM,

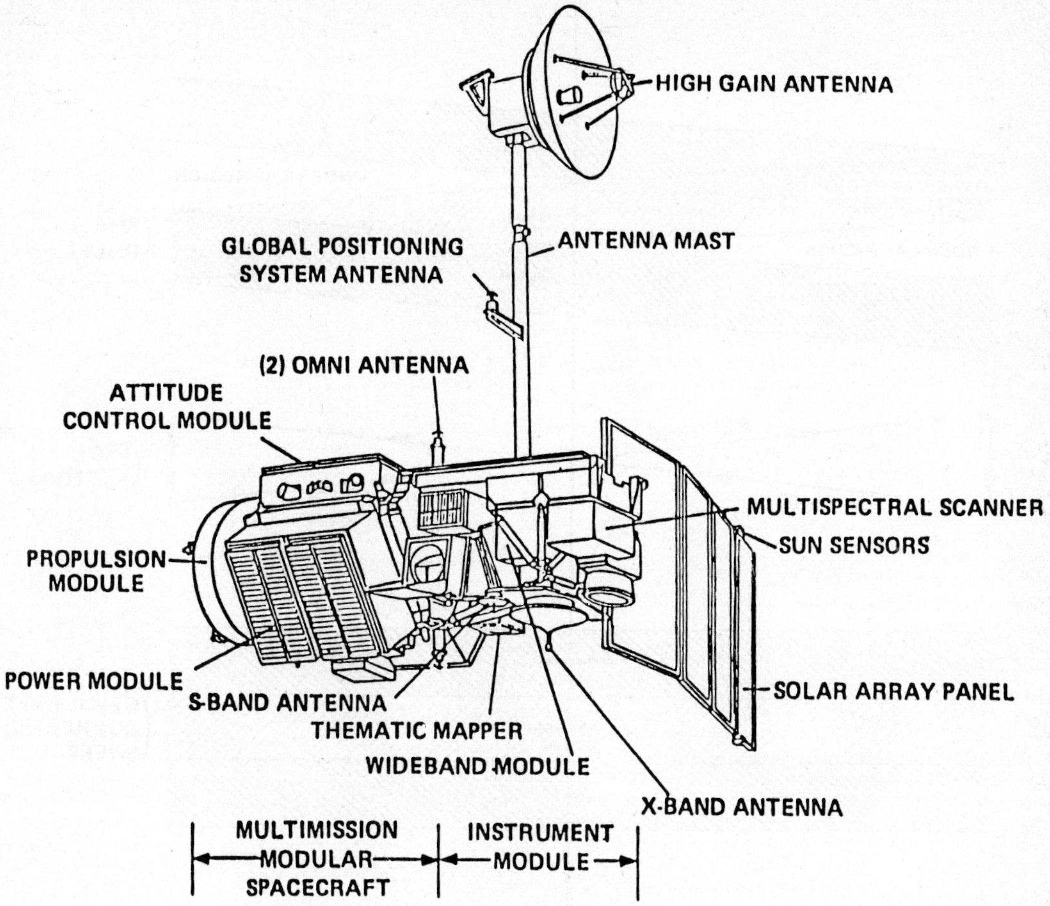

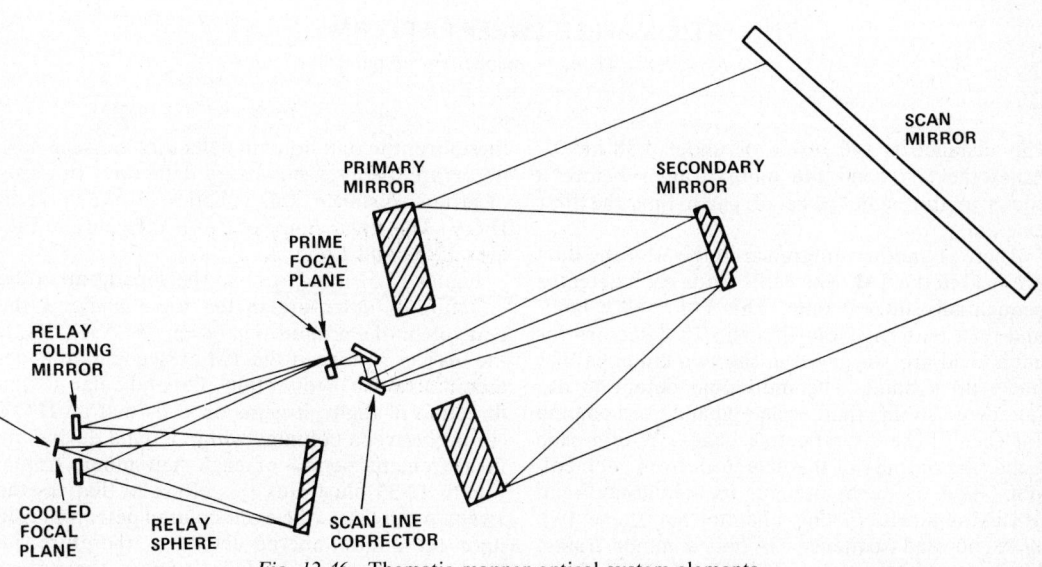

HIGH GAIN ANTENNA

GLOBAL POSITIONING
SYSTEM ANTENNA

ANTENNA MAST

(2) OMNI ANTENNA

ATTITUDE
CONTROL MODULE

MULTISPECTRAL SCANNER

SUN SENSORS

PROPULSION
MODULE

POWER MODULE

S-BAND ANTENNA

THEMATIC MAPPER

WIDEBAND MODULE

X-BAND ANTENNA

SOLAR ARRAY PANEL

MULTIMISSION
MODULAR
SPACECRAFT

INSTRUMENT
MODULE

Fig. 12-45. General configuration of Landsat-4 flight segment.

SCAN
MIRROR

PRIMARY
MIRROR

SECONDARY
MIRROR

PRIME
FOCAL
PLANE

RELAY
FOLDING
MIRROR

COOLED
FOCAL
PLANE

RELAY
SPHERE

SCAN LINE
CORRECTOR

Fig. 12-46. Thematic mapper optical system elements.

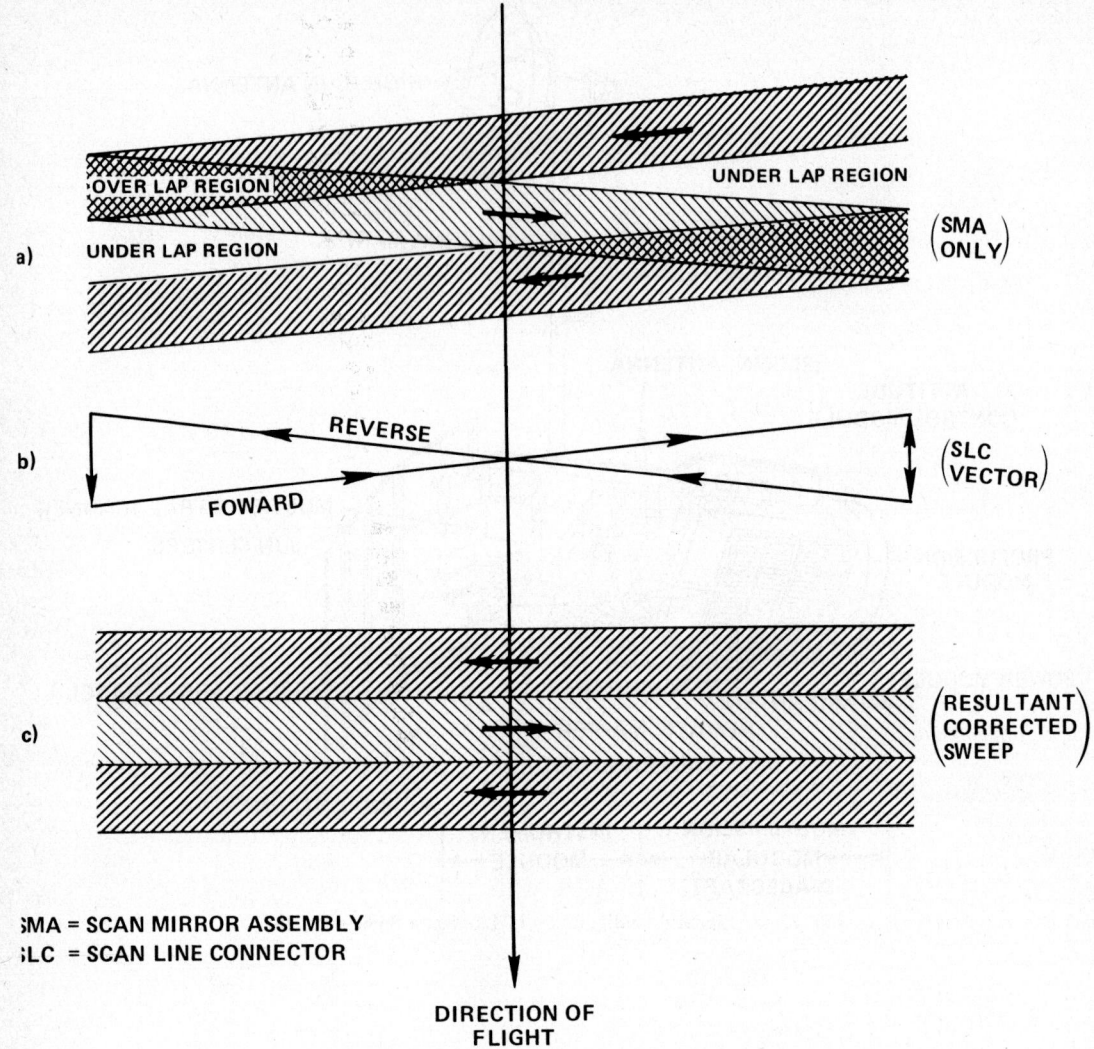

a)

OVER LAP REGION

UNDER LAP REGION

UNDER LAP REGION

$\begin{pmatrix} \text{SMA} \\ \text{ONLY} \end{pmatrix}$

b)

REVERSE

FOWARD

$\begin{pmatrix} \text{SLC} \\ \text{VECTOR} \end{pmatrix}$

c)

$\begin{pmatrix} \text{RESULTANT} \\ \text{CORRECTED} \\ \text{SWEEP} \end{pmatrix}$

SMA = SCAN MIRROR ASSEMBLY
SLC = SCAN LINE CONNECTOR

**DIRECTION OF
FLIGHT**

THEMATIC MAPPER SWEEP PATTERN

Fig. 12-47. Thematic mapper sweep pattern.

this distance is 146 pixels or about 4.38 km. It takes the last band 146 minor frames before it looks at the same piece of earth that the first one did.

There is another difference in the way the data are read in the TM. The MSS reads each detector sequentially in real time. The TM uses a hold-and-read pattern. Note that the 16 detectors for each band are staggered in the two columns that make up a band. The individual detectors are numbered so that there is an odd and even column for each of the six reflective bands. Within each band, the outputs of the detectors from each column (odd or even) are put in a hold-and-read status separately. The phasing for these two hold-and-read sequences is half a minor frame. The time duration of a minor frame is 9.611 μs;

therefore, the odd-column detectors are read 4.81 μs before the even-numbered detectors. In terms of ground distance, this is about 15 m or half an IFOV. The significance of this is indicated in Figures 12-52 and 12-53.

Figure 12-52 indicates that the separation of the columns of detectors in the focal plane is the equivalent of a ground separation of 2.5 IFOVs. If the optical system of the TM is sweeping this detector array in a direction perpendicular to the direction of flight, it gains or loses half an IFOV (15 m) between columns within a band during the measurement period of each half minor frame. Figure 12-53 illustrates this effect in that, as the sweep progresses, the even-column detectors gain upon the odd-numbered detectors; therefore, instead of their being 2.5 IFOVs apart, there is an

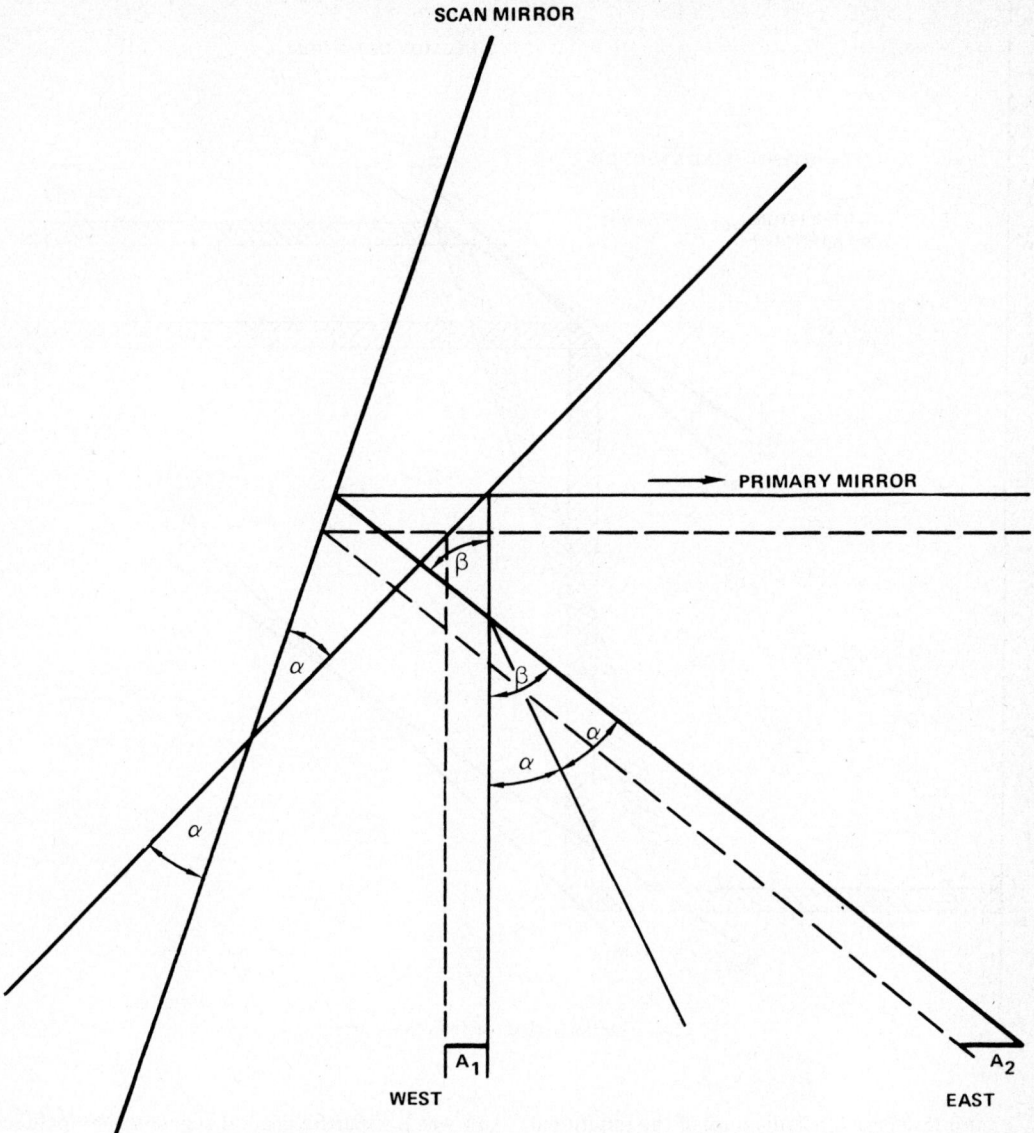

Fig. 12-48. Scan mirror (alone) sweep pattern.

effective separation of only two IFOVs. The converse is true in the east-to-west sweep so that the half IFOV distance is lost and the resulting separation is three IFOVs. In reconstruction of the TM data into an image, the construction of the image matrix must take this into consideration. Also, because the data timeline is serial, the reverse sweep data must be turned around relative to the forward sweep data to maintain spatial relationships in the image. Since the MSS only uses the data in a sweep in one direction, each sweep of six lines could be left-justified with regard to line start. The TM data must be right- and left-justified with each alternate sweep, however.

The block diagram of the multiplexer (Figure 12-54) and the minor frame sampling pattern (Figure 12-55) can be used to demonstrate how the reading arrangement is accomplished. There is a track-and-hold analog multiplexer for each band that has its own analog-to-digital converter. The exception to this is that band 5 shares its analog-to-digital converter with the four detectors of the thermal infrared band. One detector of the thermal band is read each minor frame and has its own time slot (Figure 12-55). The outputs of the individual analog-to-digital converters are serialized through a single digital multiplexer in an 84.90×10^6 bits-per-second stream that is subsequently transmitted by either Ku- or X-band microwave, depending on which options have been selected.

GROUND PROCESSING

Table 12-13 indicates the requirements for the ground segment.

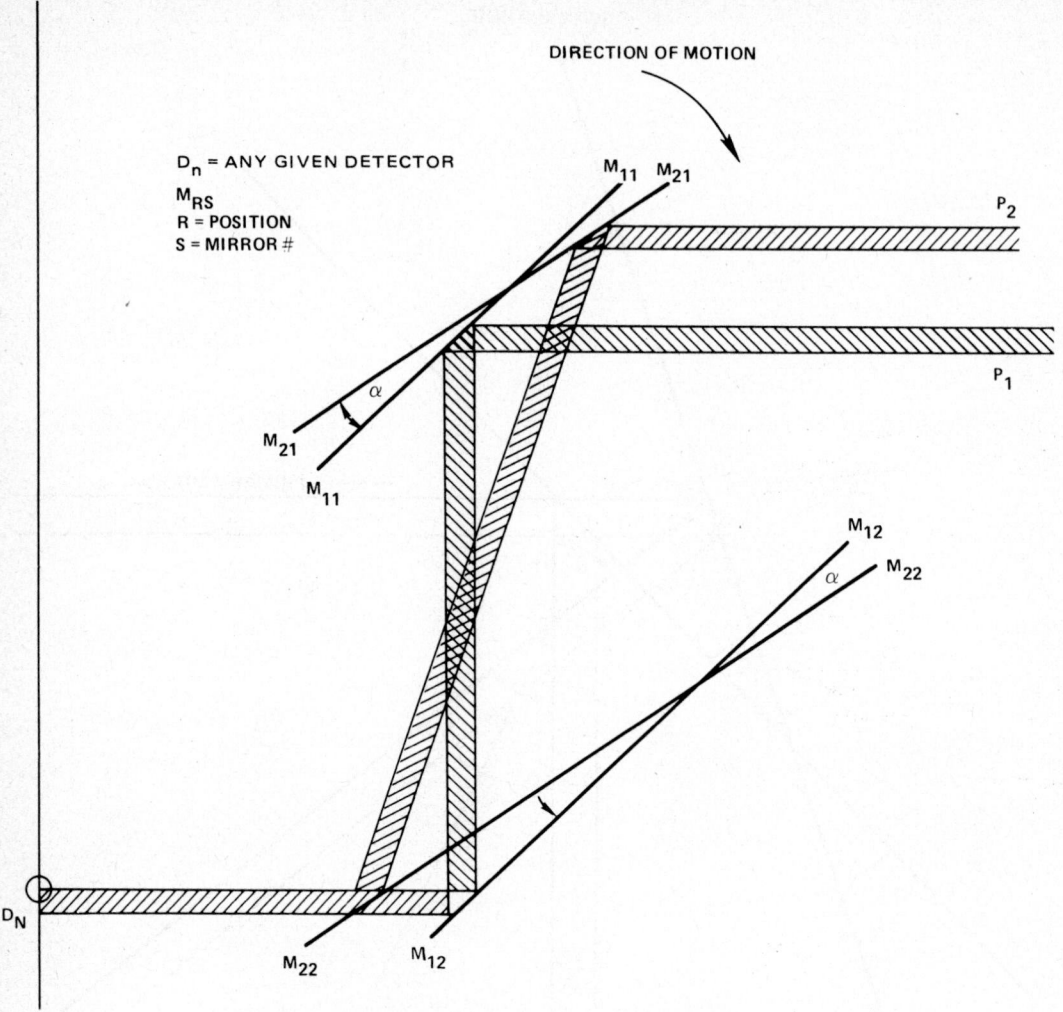

Fig. 12-49. Optical effect of scan line corrector.

Figure 12-56 is a block diagram of the functional units of the ground segment. The ground segment and the transportable ground station (TGS) are located at GSFC. The overall relationship of the ground segment to the total system is shown in Figure 12-57. The Landsat-D management and operational structure is more complex than the ground data processing system described for the original Landsats. This is not only because the system is more complex but also because its structure reflects many of the lessons learned from operating the earlier Landsat ground systems.

The original specification for Landsat-D (1977) called for a ground system arrangement that was similar to what had been used with the original Landsat ground processing activity; i.e., a two-faceted system consisting of an OCC and a data management system (see previous sections). However, in 1979, it was decided to take advantage of a common segment data base, and the re-

sult was a tripartite ground segment that included a Control and Simulation Facility (CSF), a Mission Management Facility (MMF), and an Image Generation Facility (IGF). The functional relationship of the new system to the old is shown in Figure 12-58. In early 1980, a high-level government decision was made to make the National Oceanic and Atmospheric Administration (NOAA) responsible for the operational aspects of earth-resources satellites. With regard to the Landsat-D satellites, this was defined as NASA's relinquishing its operational responsibility to NOAA as soon as the total system was successfully functional. This administrative decision made it necessary to subdivide the MMF and IGF operations into those specifically related to the MSS and those related to the TM. This was because of the different time lines for the readiness of the data processing for the two different sensor systems. The fully operational MSS was scheduled to be taken over by NOAA 6 months after the

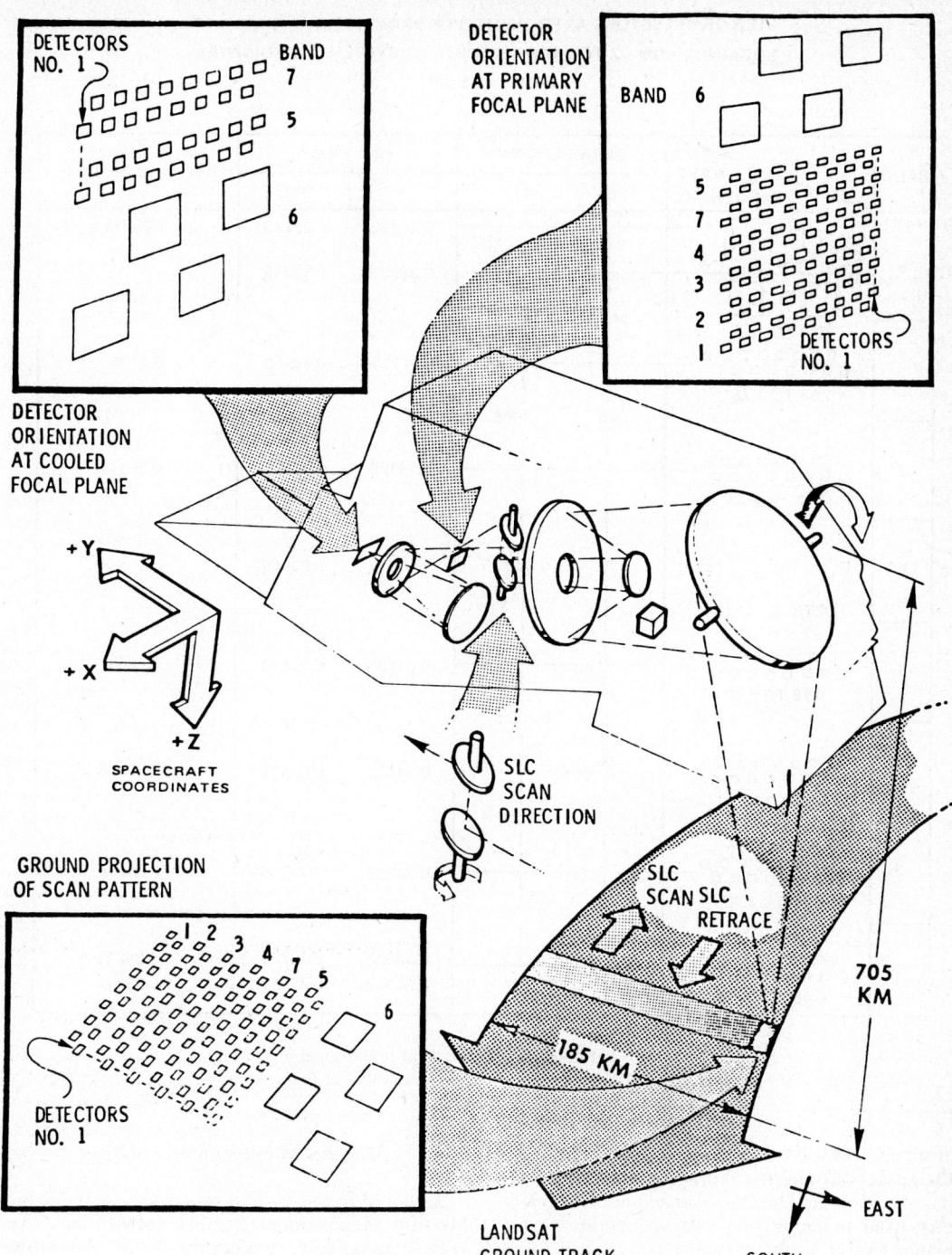

Fig. 12-50. Detector array projections on ground track.

launch of Landsat-D. Phasing of the TM processing was extended over the next 2 years, although the TM data would be acquired by the spacecraft, limited only by the constraints of the ground receiving stations and the availability of the TDRSS. The independent MSS and TM systems are indicated in Figure 12-52. The ultimate product types and their quantities that will be produced by the

Landsat-D and -D′ ground segment are shown in Table 12-14.

Control and Simulation Facility

The Control and Simulation Facility (CSF) has those functions of the old OCC-related planning and scheduling of the satellite (flight segment) op-

VIEW OF DETECTORS AS PROJECTED ON PRIME FOCAL PLANE
LOOKING FROM -Z TOWARDS +Z (OPTICAL SYSTEM COORDINATES)
1FOV = 42.5 μrad

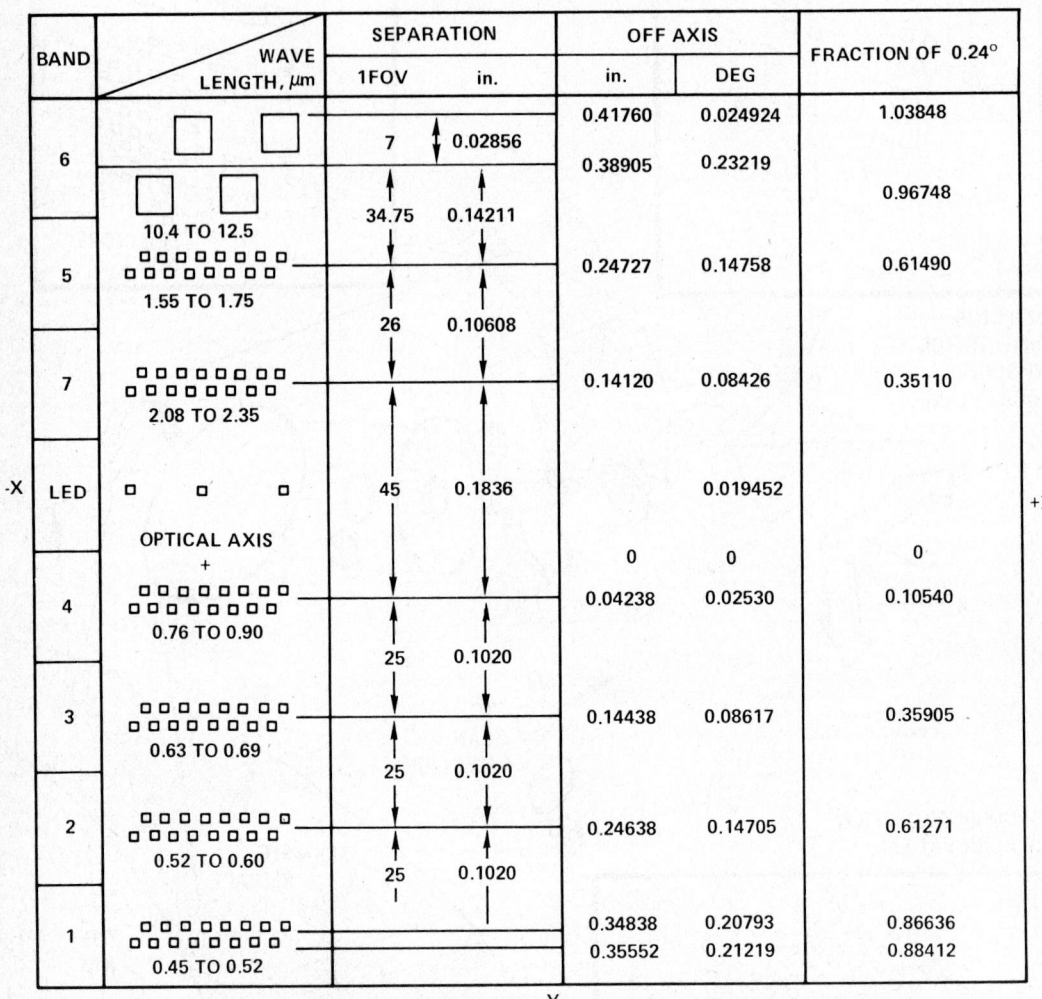

Fig. 12-51. TM detector projection at prime focal plane.

erations. These include command and control of the spacecraft and the reprogramming of the on-board computer. The CSF also acquires the telemetry used in monitoring the spacecraft performance as well as the ancillary data used in image processing, such as attitude monitoring. In addition to these control and command functions, the CSF has the capacity for simulation of flight-segment operation for test and training purposes and has a built-in capability for self-test. Its primary function is to coordinate and direct TM and MSS operations for the acquisition of image data and delivery of these data to the Image Generation Facility. The CSF also interfaces with the Network Control Center to schedule and coordinate the TDRSS support activities.

Mission Management Facility

As noted in the ground processing section, the Mission Management Facility (MMF) has two subsections: one subsection is for the MSS, MMF-M; the other for the TM, MMF-T. The MMF consists of the hardware, software, operations and procedures for providing user-request processing, image-production management, ground control-point library generation, inventory control, and ground-segment general management. The main difference between the MMF-M and MMF-T is that the MMF-M is the single interface with the CSF for acceptance of all acquisition data requests, both MSS and TM. The TM data are transferred from the MMF-M to the

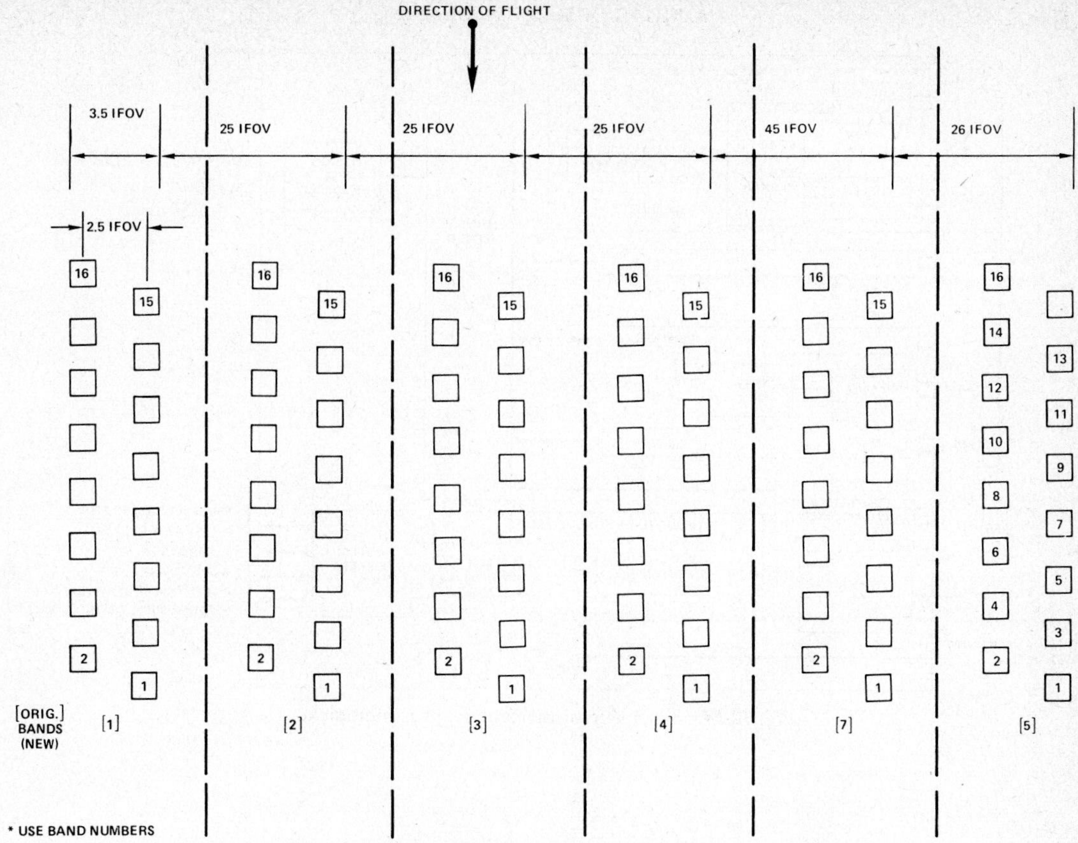

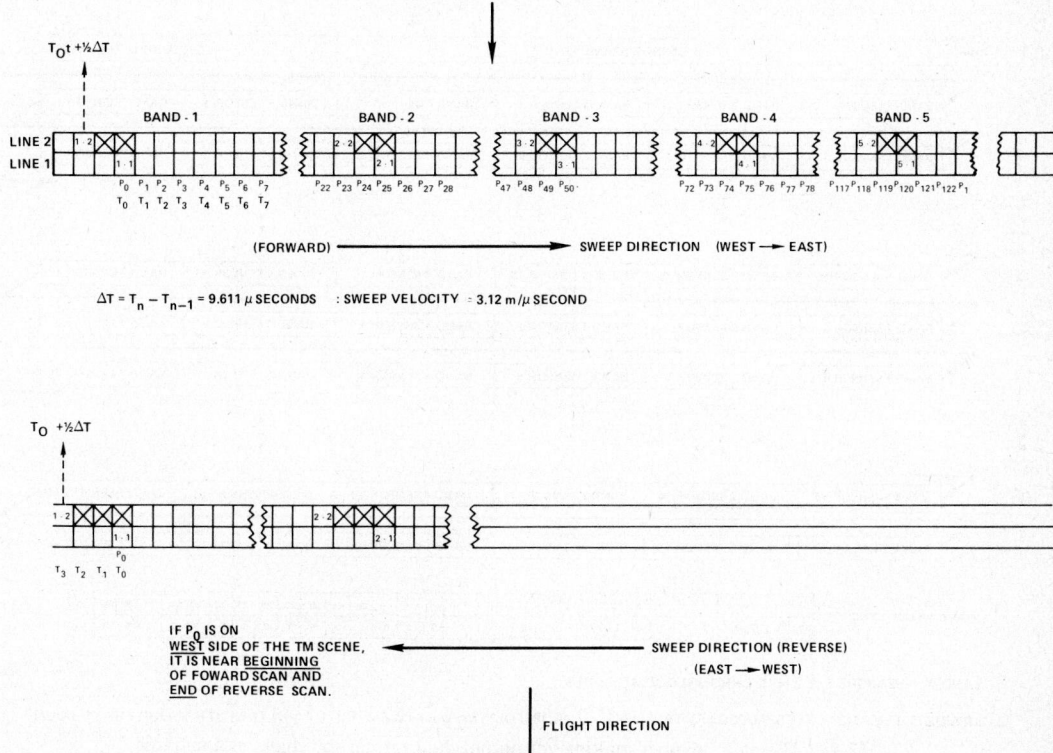

DIRECTION OF FLIGHT

3.5 IFOV | 25 IFOV | 25 IFOV | 25 IFOV | 45 IFOV | 26 IFOV

2.5 IFOV

[ORIG.] BANDS (NEW)   [1]   [2]   [3]   [4]   [7]   [5]

* USE BAND NUMBERS

Fig. 12-52. Detector layout in combined (effective) focal plane of thematic mapper.

FLIGHT DIRECTION

$T_Ot + \frac{1}{2}\Delta T$

BAND - 1   BAND - 2   BAND - 3   BAND - 4   BAND - 5

LINE 2
LINE 1

P_0 P_1 P_2 P_3 P_4 P_5 P_6 P_7 P_{22} P_{23} P_{24} P_{25} P_{26} P_{27} P_{28} P_{47} P_{48} P_{49} P_{50} P_{72} P_{73} P_{74} P_{75} P_{76} P_{77} P_{78} P_{117} P_{118} P_{119} P_{120} P_{121} P_{122} P_1

T_0 T_1 T_2 T_3 T_4 T_5 T_6 T_7

(FORWARD) ——————→ SWEEP DIRECTION (WEST → EAST)

$\Delta T = T_n - T_{n-1} = 9.611 \mu$ SECONDS : SWEEP VELOCITY = 3.12 m/μ SECOND

$T_O + \frac{1}{2}\Delta T$

P_0

T_3 T_2 T_1 T_0

IF P_0 IS ON
WEST SIDE OF THE TM SCENE,   ←——————   SWEEP DIRECTION (REVERSE)
IT IS NEAR BEGINNING   (EAST → WEST)
OF FOWARD SCAN AND
END OF REVERSE SCAN.

FLIGHT DIRECTION

Fig. 12-53. Space-time relationships of pixels in single minor frame.

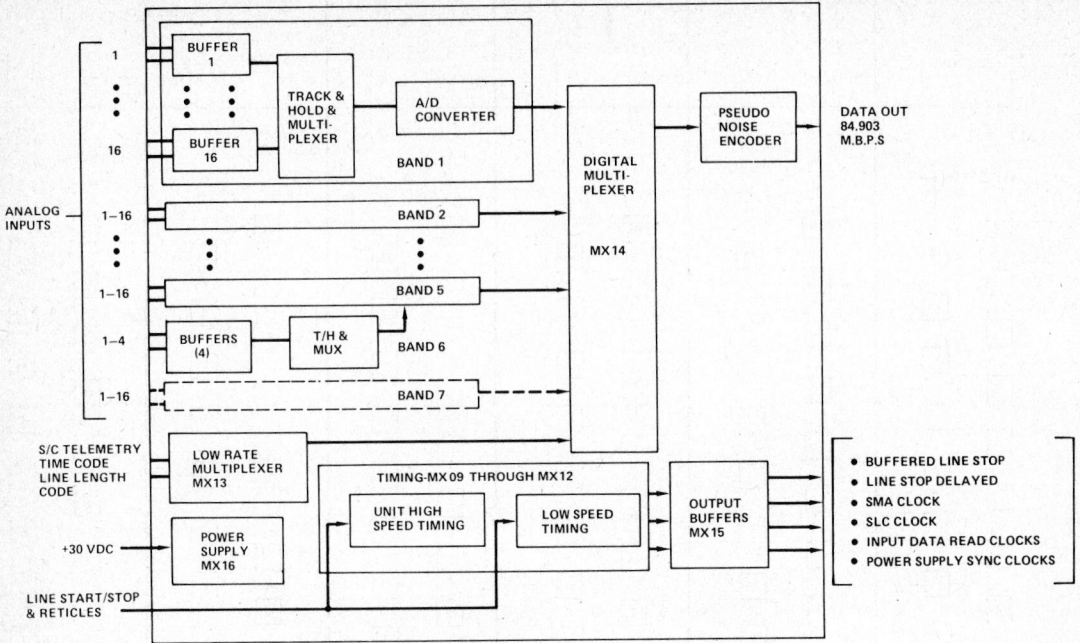

Fig. 12-54. Schematic of thematic mapper multiplexer.

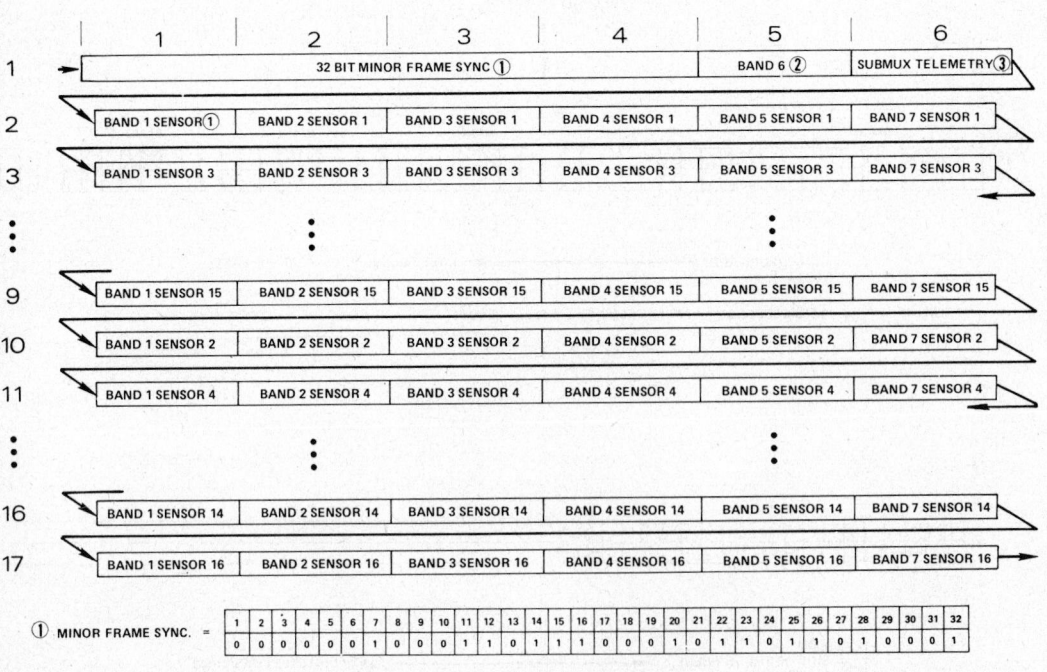

① MINOR FRAME SYNC. =

1	2	3	4	5	6	7	8	9	10	11	12	13	14	15	16	17	18	19	20	21	22	23	24	25	26	27	28	29	30	31	32
0	0	0	0	0	0	1	0	0	0	1	1	0	1	1	1	0	0	0	1	0	1	1	0	1	1	0	1	0	0	0	1

② BAND 6 = SENSORS 1, 3, 2, 4 IN CHRONOLOGICAL ORDER.

③ SUBMUX TELEMETRY = 15 SPACECRAFT TELEMETRY WORDS, FOLLOWED BY ONE WORD OF MULTIPLEXER MAJOR FRAME COUNT.

Fig. 12-55. Thematic mapper multiplexer output data format for single minor frame.

TABLE 12-13

**Landsat-D/D′ Ground-Segment Image
Performance Requirements**

Function/Operation	Performance Objective
Turnaround	48 hours from collection of raw sensor data to generation of archival products
Radiometric error correction (relative interdetector)	1 quantum level over full range
Geometric error correction (nominal conditions with ground central points (GCPs)	0.5 sensor pixel (90 percent of the time) (with sufficient correlatable GCPs)
Temporal registration error	0.3 sensor pixel (90 percent of time) (with sufficient correlatable control points)
Map projections supported	Space Oblique Mercator Universal Transverse Mercator/polar stereographic
Resampling algorithms supported	Cubic convolution Nearest neighbor

MMF-T by a switched disk. Otherwise, the performance requirements for both MMFs are identical for the assigned sensor system. These MMF functions are as follows:

- Processes requests for data acquisition.
- Provides candidate scene acquisition lists.
- Plans and schedules operations.
- Accounts for telemetry data acquisition.
- Accounts for image data acquisition.
- Schedules archival processing.
- Maintains archival data base and produces image catalogs.

- Processes requests for product generation.
- Schedules product generation.
- Maintains ground-segment supplies inventory.
- Tracks ground-segment problems.
- Provides verification and self-test capability.
- Provides management reports.

Image Generation Facility

It is through the Image Generation Facility (IGF) that the sensor data are processed into user products. When it was decided to update the design of the ground segment in 1980, the components of the IGF became essentially autonomous and were raised to the facility level. The facility level operations were then the Data Receive, Record, and Transmit System (DRRTS), the MSS Image Processing System (MIPS), and the TM Image Processing System (TIPS).

Data Receive, Record, and Transmit System

The Data Receive, Record, and Transmit System (DRRTS) acquires MSS and TM data directly from the TGS or through the Domsat Interface Facility (DIF). The former is by direct electronic hookup into the DRRTS. The latter is by a tape that is recorded at the DIF from a transmission from the White Sands, New Mexico, receiving site of the TDRSS. The DRRTS generates a data file correlating IRIG-A time on the edge track of a high-density tape with the spacecraft time that is embedded with the image data. This directory file is transferred to the MMF-M to identify the MSS image data on a particular R-tape. The DRRTS also does error checking on the image data, monitors the quality of the data entering and being recorded, and reports the quality to the MMF-M.

In addition, the DRRTS responds to requests from the MMF-M and transmits processed data

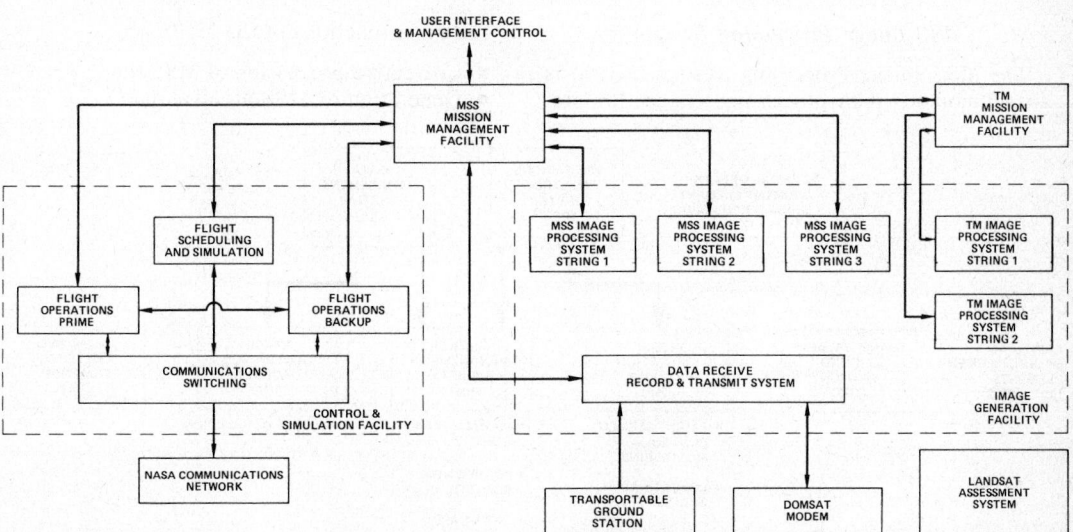

Fig. 12-56. Landsat-4 ground segment.

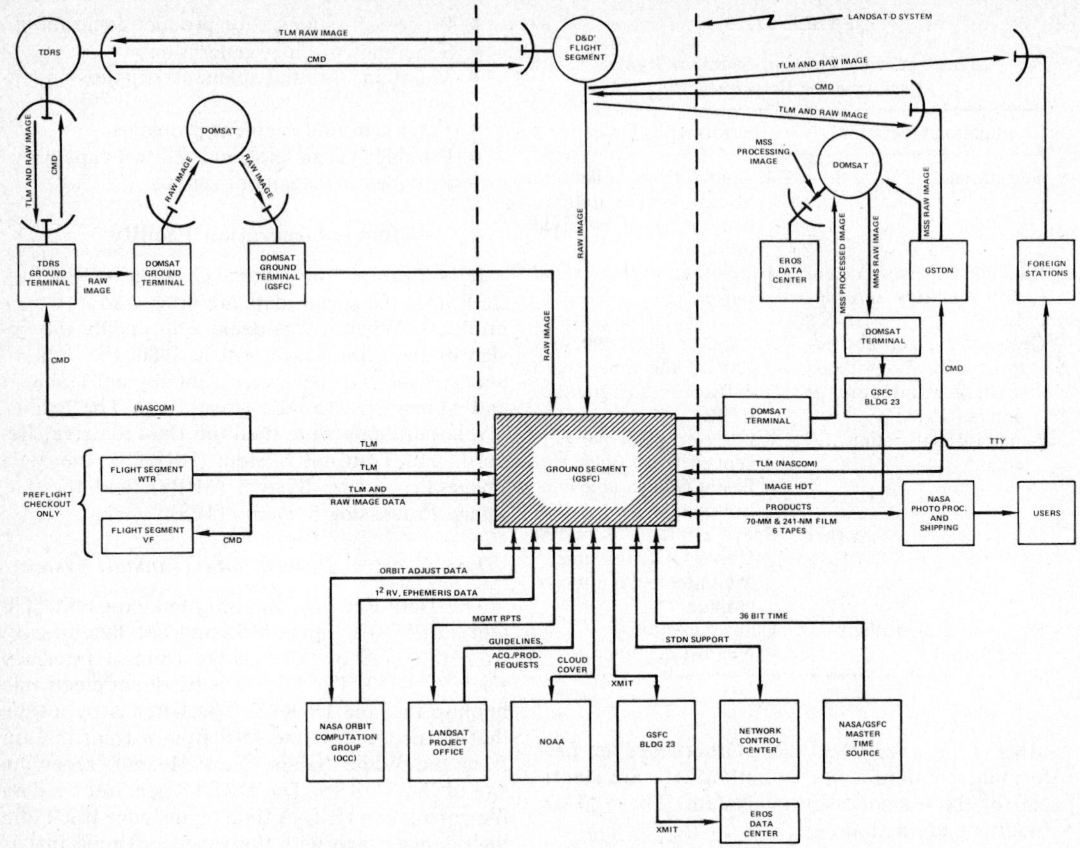

Fig. 12-57. Landsat-4 system and associated elements.

via Domsat to EDC at Sioux Falls, South Dakota. If the Domsat link fails, the DRRTS is capable of processing a 14-track data tape for shipment to EDC. Whenever a transmission is completed, the DRRTS notifies the MMF-M of the action by an electronic feedback file. The DRRTS can transmit up to 200 MSS scenes per day via Domsat.

MSS Image Processing System

The MSS Image Processing System (MIPS) is an autonomous data processing system for pre-

paring the MSS imagery for eventual distribution by EDC. As noted in a previous section, the plan is to turn over to NOAA this part of the Landsat-D ground segment, along with the MMF-M, within 6 months after the launch of Landsat-D. Therefore, it must be essentially independent of the part of the IGF that will process the TM data.

The main functions of the MIPS are:

- Corrective processing of MSS data
- Generation of MSS archival data

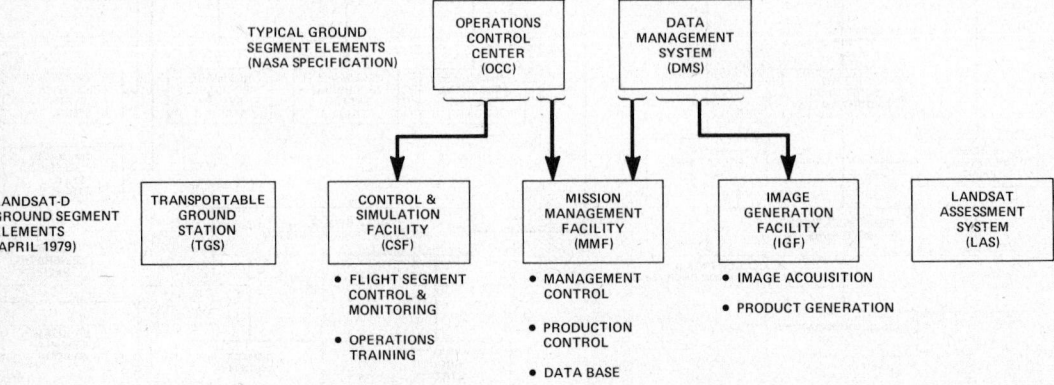

Fig. 12-58. Evolution of Landsat-4 ground segment major item terminology.

TABLE 12-14

Summary of Ground System Products from Landsat D/D′ GSFC Ground Segment

Instrument	Product	Quantity* Scenes/Day	When Available
MSS	MSS A Tape (HDT) (User Product) MSS 70-mm film (QC Product) (One band QC Product)	200	a. For first 3 months operating 2 lines, 2 shifts will produce 125 scenes. b. After first 3 months operating 3 lines, 2 shifts will produce 200 scenes. c. Turn over operational system to NOAA, 200 scenes/day at Landsat-D launch plus 6 months.
	MSS CCT (A or P) (QC Product) MSS 241-mm film (QC Product)	2 4	a. At launch of Landsat-D a. At launch of Landsat-D (2 scenes/day) b. Launch plus 90 days (4 scenes/day)
TM	Thematic Mapper	1	a. Provide developmental TM processing using selected *a priori* jitter algorithms for 1 scene/day at the launch of Landsat-D. Film, A and P tapes, and CCTs will be provided.
	TM A Tape (HDT) (User Product) 70-mm film (QC Product)	12 to 100	a. During the TM R&D period (which starts 1 year after the launch of Landsat-D), capability of processing 12 scenes/day with *a priori* knowledge of flight segment disturbances (jitter) b. One year after start of TM R&D period, 12 scenes/day shall be demonstrated operationally using *a posteriori* knowledge of jitter. c. Six months later, the system capability will be increased to 100 scenes/day and use *a posteriori* knowledge (in orbit measurements) of jitter. The system will be turned over to NOAA at this time.
	TM P Tape (HDT) (User Product) 241-mm film (QC and User Product)	12 to 50	a. During the Tm R&D period (which starts 1 year after the launch of Landsat-D), capability of processing 12 scenes/day with *a priori* knowledge of flight segment disturbances (jitter). b. One year after start of TM R&D period, 12 scenes/day shall be demonstrated operationally using *a posteriori* knowledge of jitter. c. Six months later, the system capabilities will be increased to 50 scenes/day and use *a posteriori* knowledge (in orbit measurements) of jitter. The system will be turned over to NOAA at this time.
	TM CCT (A or P) (User Product)	2 to 10	a. By April 1984, 2 scenes/day will be demonstrated. b. Turn over operational system to NOAA, 10 scenes/day in January 1985.

* A scene/day is defined as a ground system output with a 48-hour turnaround averaged over a 10-day period.

- Evaluation of product generation
- Construction and maintenance of ground control point library

The principal function is the processing of the MSS data as the data come to the MIPS through the DRRTS and the MMF-M. In the processing of the MSS data proper, the associated peripheral data must be validated and prepared. This includes attitude data, ephemeris data, etc., when made available by the CSF. After these data are processed, they are used for correlation with the imagery data to make up the worldwide reference system (WRS) scenes. In addition, MIPS generates the scene correction data, including scene center times and locations, band-line adjustment coefficients, scene IDs, and any departures from nominal of the attitude and ephemeris data. If there is a request from the MMF-M, the MIPS will perform the correction process on the particular scene. As indicated in Figure 12-56, there are three independent MIPS strings, each capable of performing the foregoing jobs.

TM Image Processing System

The third semiautonomous part of the Image Generation Facility is the TM Image Processing System (TIPS). This sub-facility does for the TM data exactly what the MIPS does for the MSS data. Its relationship to the MMF-T is also the same as the MIPS is to the MMF-M. However, the timeline for readiness is quite different. It will not begin archive generation until July 1983, and then only 12 scenes per day. It will not attain its full capacity until January 1985. Then, it will process 50 scenes per day to high density geometrically corrected tapes (HDT-PT). The image data on the HDT-PT will be radiometrically corrected to ± 1 quantum level, geodetically accurate to ± 0.5 pixel (90 percent of the time), and will have a temporal registration accuracy of 0.3 pixel (90 percent of the time).

One unique function that is performed in the TIPS during the archiving operation is automatic cloud-cover assessment. Inclusion of the mid-infrared band in the 1.55- to 1.75-μm region permits the preparation of an algorithm that allows the system to discriminate between clouds and snow or sand. Table 12-15 generally indicates the elimination technique used for the assessment process. Each scene is subsampled on a 128×128 pixel grid basis during this automatic cloud-cover assessment.

Transportable Ground Station

The Transportable Ground Station (TGS) is a relocatable receiving site that will be initially located at GSFC. It is capable of receiving, directly from the Landsat-D satellite, both X- and S-band transmissions. The X-band telemetry is for both the MSS and TM imagery data, as well as TM data alone. The S-band frequency provides MSS data and spacecraft status telemetry data. Figure 12-59 shows the approximate coverage of the TGS when located at GSFC. In the absence of TDRSS availability, this will represent the only TM cov-

erage of the continental United States during the period of initial launch of Landsat-D.

Landsat-D Assessment System

The Landsat-D Assessment System (LAS) is an off-line facility whose prime function is to evaluate the performance of the Landsat-D systems, both as to meeting technical requirements and to ensuring improved utility of the data. This latter aspect is particularly oriented toward the TM data. However, because the TIPS will not be available at time of Lansat-D launch in 1982, LAS will, in conjunction with the GSFC Application Developmental Data System (ADDS), provide early access to TM data at a rate of one TM scene per day. This is particularly important for evaluating and rectifying spacecraft effects on the data because of anticipated, but not completely quantifiable, flight-segment disturbances (jitter). The LAS facility has independent hardware and software systems for performing its analytic and production functions, including interactive display systems and computers. The sized-processing input is a maximum of 5×10^8 pixels of TM data per day through typical corrective and analytic (classification) functions. For MSS data, the daily aggregate for the same-type functions is 1×10^8 pixels.

To help characterize the capability of the Landsat-D imaging systems, qualified researchers and users from the scientific and technical community are being used for performing certain assessments. Initial orientation of this evaluative work is toward image quality in general, as opposed to the latter work that will be oriented toward specific discipline utility. The TM image quality assessment will be interested in performance parameters such as scan mirror velocity profile, the effect of the scancorrector mirrors, band-to-band registration, flight-segment inter-component disturbance factors (jitter), calibration, temporal and geodetic registration, etc. If anomalies are perceived early enough in the LAS operation, it will be possible to apply corrective methods to the TIPS before it becomes fully operational.

In the realm of MSS data, LAS will be interested not only in the *per se* performance of the Landsat-D MSS system but also in comparing its data performance with the previous Landsats, es-

TABLE 12-15

Classification Characteristics for Automatic Cloud-Cover Assessment

TM Band	Bandwidth (μm)	Threshold	Identifies
4 (near-infrared)	0.76 to 0.90	Rad ≥ 2.5 mW/cm²-sr	Cloud, snow, sand
5 (mid-infrared)	1.55 to 1.75	Rad ≥ 0.4 mW/cm²-sr	Cloud, sand
6 (infrared)	10.4 to 12.5	Temp $< 5°$C	Cloud

Fig. 12-59. Approximate transportable ground station land-mass coverage.

pecially in consideration of the orbital differences (705 km vs. 913 km). Through these comparisons the current users of Landsat MSS data will learn whether Landsat D assures them of a continuity of such data.

LANDSAT-D BECOMES LANDSAT-4

The Landsat-D satellite was successfully launched from the NASA Western Test Range at Vandenberg AFB at 16:59 GMT on July 16, 1982. When the launch was declared a success, i.e., the spacecraft was injected into a nominally acceptable orbit, the mission became Landsat-4. The initial orbital parameters were somewhat off those required for the desired circular, repetitive, sun-synchronous orbit. Through the use of a combination of natural orbital drift and the orbit adjust system of the spacecraft, however, the satellite was injected into its desired World Reference System (WRS) orbit on July 29, 1982. This WRS orbit has a 16-day repetitive cycle, 233 orbits per cycle and an orbit skip factor of 7, as discussed in an earlier part of this chapter. The period from July 30 to August 10, 1982, was called Cycle zero and much of the preliminary testing of the spacecraft systems and the payloads went on during this period. Cycle 1 was initiated on August 11 and the regular collection of sensor data for eventual release into the public domain started in this cycle, although there were occasional problems in the data acquisition as is to be expected in the shakedown period of any new system.

The satellite's two sensor systems, the Multispectral Scanner (MSS) and the Thematic Mapper (TM), generally have worked well from the outset. The MSS was activated on day 4 after launch, after the preliminary spacecraft system checks were made. On day 5, the first four bands of the TM were turned on. These were the bands with the silicon detectors that were mounted on the uncooled focal plane (0.45−0.52, 0.52−0.60, 0.63−0.69 and 0.76−0.90 micrometers). The data from the two sensor systems were examined and proved to be satisfactory. When the 4-band TM data that were taken over the Detroit-Toledo-western Lake Erie area were seen (this was the first good cloud free scene taken), it became apparent that the TM was going to be a great success with its increased resolution. Details could be seen on blowups to single pixel resolution, that had never been achieved previously in the Landsat program, e.g., cul-de-sacs and side streets could be observed in housing subdivisions. This scene is shown in false color in Figure 12-60, but as a full scene, and thus these features mentioned may not be appreciated at this scale. Therefore attention is invited to the enlargement of a portion of this area, in both natural color and false color, as seen in the frontispiece section of this volume on page v. The Landsat-4 MSS data proved to be of comparable quality to the MSS data of its predecessors, Landsats-1, -2 and -3, as illustrated by the left half of the figure that appears on page vi of the frontispiece section.

After an outgassing period that lasted about one month, the TM bands on the cooled focal place were activated. These were band 5 (InSb detector

Fig. 12-61. Landsat-4 Thematic Mapper imagery from band 1 (0.45–0.52 μm) of the Dyersberg, Tennessee quadrangle. This scene includes a portion of the Mississippi Valley near the area of the Tennessee, Missouri, Arkansas and Kentucky borders.

material; 1.55–1.75 micrometers), band 7 (InSb detector material; 2.08–2.35 micrometers) and band 6 (HgCdTe detector material; 10.4–12.5* micrometers). The first two of these are mid-infrared bands, still in the reflective region; band 6, however, is in the thermal infrared or emissive region. The performance of the TM in these three regions met the same standards as those of the first four bands.

One of the first cloud-free 7-band TM scenes acquired was in the Mississippi valley, near where Arkansas, Tennessee, Missouri and Kentucky come together. The Application Branch of the US Geological Survey at the EROS Data Center in Sioux Falls took a segment of this scene, identified by EDC as the Dyersberg, Tennessee, TM map quad and geometrically corrected it, using ground control points and resampling it to a 20-meter pixel size. The results are shown in color figures 12-61 through 12-71. Figures 12-61 through 12-67 show the single-band black-and-white images for bands 1, 2, 3, 4, 5, 7 and 6, respectively. In these one can see how different features are emphasized in each of the bands. These black-and-white images were produced from the digital data using the EROS Data Center laser beam recorder. Figures 12-68 through 12-71 show some of the color products that can be produced from these single-bands used in various combinations of three and varying

the color filters for each band. Figure 12-68 is essentially the equivalent of the false color imagery used conventionally with the MSS data in the past, that is, using TM bands 2 (green), 3 (red), and 4 (near infrared), with a blue filter for band 2, a green filter for band 3 and a red filter for band 4.[9] Figure 12-69 replaces the green band 2 with the blue band 1, but uses the same filter combination. Only subtle differences can be noticed, such as slight difference in the color of the water in the Mississippi River. Figure 12-70 represents the first opportunity we have had with Landsat data to approach a true color picture, i.e., one composed of the spectral bands which can be sensed by the human eye, band 1 (blue), band 2 (green), band 3 (red). This approaches what a human observer might see if he were riding along in the Landsat-4 and gazing down. The general "blueness" can be in part attributed to normal Rayleigh scattering phenomena. Such scattering occurs both to the incident sunlight, as it passes down through the

---

[9] It will be noted that the band numbers are coded differently in those pictures produced by the EROS Data Center than those which might indicate the filters used. This is because in producing the color products, EDC used a subtractive reversed color film process. While the process that was used produces blue, green and red coloring in the final film product, the corresponding colors achieved during the process are yellow, magenta and cyan. The color pictures produced at GSFC show the band numbers in the colors of the filters, blue, green and red.

---

* Actual prelaunch calibration indicated an upper edge of 11.7 micrometers for this band.

Fig. 12-62. Landsat-4 Thematic Mapper imagery from band 2 (0.52−0.60 μm) of the area shown in Figure 12-61.

Fig. 12-63. Landsat-4 Thematic Mapper imagery from band 3 (0.63−0.69 μm) of the area shown in Figure 12-61.

Fig. 12-64. Landsat-4 Thematic Mapper imagery from band 4 (0.76−0.90 μm) of the area shown in Figure 12-61.

Fig. 12-65. Landsat-4 Thematic Mapper imagery from band 5 (1.55−1.75 μm) of the area shown in Figure 12-61.

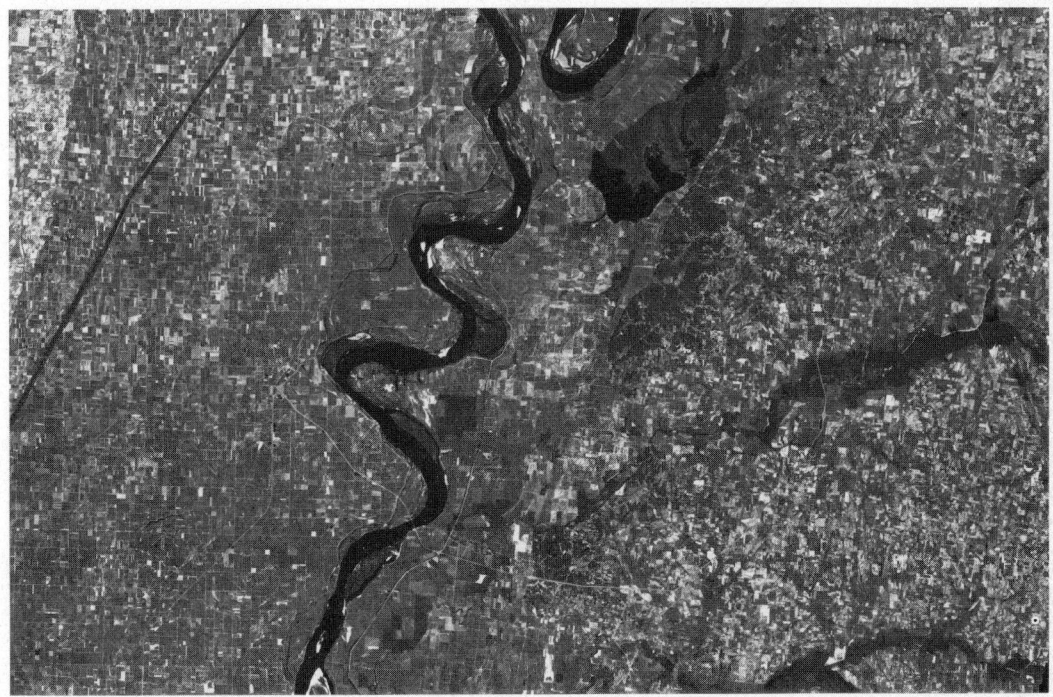

Fig. 12-66. Landsat-4 Thematic Mapper imagery from band 7 (2.08–2.35 μm) of the area shown in Figure 12-61.

Fig. 12-67. Landsat-4 Thematic Mapper imagery from band 6 (10.4–11.6 μm), (thermal infrared) of the area shown in Figure 12-61.

atmosphere, and to the reflected light on its return journey to the sensor. Figure 12-71 represents a combination of bands which heretofore have not been seen in Landsat data, band 1 (blue), band 3 (red), and band 5 (mid infrared). No attempt is made here at interpretation of the advantages of this expanded capability. Color figures 12-72 and 12-73 are full TM (185 Km × 185 Km) scenes of the area from which Figures 12-61 through 12-71 are subsets. These were produced at Goddard Space Flight Center from geometrically corrected CCTs that were worked into an Optronic Beam Recorder. Here again we have the conventional blue, green and red filters used in Figure 12-72 for bands 2, 3 and 4 (false color) and in Figure 12-73 for bands 1, 2 and 3. These products are considered nonstandard as they use an experimental linear stretch technique that is being tried out to determine its acceptability.

During the early days of acquisition, one scene of prime interest to those working with the Landsat-4 system and also to many users, both scientific and those public relations oriented, was the scene that included the Nation's Capitol—Washington, DC. However, the vagaries of the weather were such that it was not until November 2, 1982, that a clear image of this area was acquired, after five previous unsuccessful attempts. Color figures 12-74 and 12-75 are the false (bands 2, 3, 4) and true (bands 1, 2, 3) versions of this full scene. Looking at this image at this size and scale it does not appear too different from its MSS counterparts. However, as one begins to look at subsets of this scene, the differences in detail become apparent. Even looking at 1/16 of the scene at the same size as the full scene (Color figures 12-76 and 12-77) one can begin to see detail that was not apparent in the MSS scenes. If we take a magnifying glass to look at ground features such as the Pentagon and Washington National Airport, one can see details that were never apparent at the resolution of the MSS. Even the shadow of the Washington Monument can be seen in the false color picture (which incidently acts as a sundial and gives an indication of the time of day when the image was taken). But if one is really desirous of appreciating the difference in resolution quality between the TM and the MSS, it is necessary to zoom into a region where single pixels are discernible in the MSS and look at the equivalent TM image of this area. The two figures that appear on page vi of the frontispiece to this volume do just this for the area that covers both the federal core of Washington, DC, and that part of Virginia which includes the Pentagon and National Airport. Here again note the detail in the various wings of the Pentagon in the TM image and the runway detail at the airport. Neither of these features is readily identifiable in the MSS data. Various other comparisons are left to the reader.[10]

No attempt is made here to evaluate the scientific advantages of the use of TM data in analyses for various disciplines such as agriculture, geology, hydrology, etc. What we do wish to indicate is that in Landsat-4 the potential for acquiring ever more useful remotely sensed data from space has been demonstrated. As such data are used by specialists in various applications, this expanded utility will become more apparent. Landsat-4 also points toward a still brighter future in this field, because, as impressive as the data from TM are, they are only a promise of what the future holds in this field of remote sensing from space—for the benefit of all mankind.

REFERENCES

Blanchard, L. E., and O. Weinstein, 1979, Design challenges of the thematic mapper; Machine Proc. Rem. Sens. Data Symp., Purdue Univ., West Lafayette, In.

Engel, J. L., 1980 (October), Thematic mapper—interim report on anticipated performance; Hughes Aircraft Corp., Santa Barbara, Calif.

Hughes Aircraft Corp., 1972, Multispectral scanner system for ERTS; HS324-5214, Santa Barbara, Calif.

Lansing, J. C., and R. W. Cline, 1975 (July/August), The four- and five-band multispectral scanner for Landsat; Opt. Eng., vol. 14, no. 14.

NASA Contract Report NAS5-24200, 1978 (June), Thematic Mapper; vols. I–V, Hughes Aircraft Corp., Santa Barbara, Calif.

NASA Contract Report NAS5-25300, June, 1981, Landsat-D ground segment design description; General Electric Co., Lanham, Md.

Parkosh, A., and E. P. Beyer, 1981 (June), Landsat-D thematic mapper image resampling for scan geometry correction; 1981 Machine Proc. Rem. Sens. Data Symp., Purdue Univ., West Lafayette, In.

Salomonson, V. V., P. L. Smith, A. Park, W. C. Webb, and T. J. Lynch, 1979, An overview of progress and design and implementation of Landsat-D systems; Machine Proc. Rem. Sens. Data Symp., Purdue Univ., West Lafayette, In.

Salomonson, V. V., D. L. Williams, and J. L. Barker, 1981 (May), Information expeditions from Landsat-D; 15th Intl. Symp. Rem. Sens. Env., Ann Arbor, Mich.

Short, W. M., L. M. Stuart, 1982, The HCMM (Heat Capacity Mapping Mission) Anthology. (To be published by Govt. Printing Office.)

Slater, P. N., 1974 (November), A reexamination of the Landsat MSS; Photogram. Eng. & Rem. Sens., vol. XLV, no. 11, pp. 1479–1485.

Taranik, J. V., 1978 (January), Characteristics of the Landsat multispectral data system; USGS Open-File Report 78-187, Sioux Falls, S.D.

Thomas, V. L., 1975 (November), Generation and physical characteristics of Landsat 1 and 2 MSS computer compatible tapes; GSFC Report X-563-75-233, Greenbelt, Md.

USGS, 1979, Landsat Data User's Handbook; Revised Edition, USGS Branch of Distribution, Arlington, Va.

Weinstein, O., L. R. Linstrom, and J. C. Brewer, 1978, Development of thematic mapper; GSFC Internal Document, Greenbelt, Md.

Williams, D. L., and V. V. Salomonson, 1979, Data acquisition and projected applications of the observations from Landsat-D; ASP Conf., Sioux Falls, S.D.

---

[10] See Color Figure 12-78 for an example of "synthetic stereo" as applied to Landsat imagery.

CHAPTER 13

Microwave and Infrared Satellite Remote Sensors

Author-Editor: CHARLES ELACHI

Contributing Authors: ALEXANDER GOETZ, ROLANDO JORDAN, ANNE KAHLE, and ENI NJOKU.

GENERAL CONTENTS: ACTIVE MICROWAVE SENSORS: Synthetic aperture radar; Spaceborne SAR; Scatterometers; Altimeters; Seasat sensors. PASSIVE MICROWAVE SENSORS; INFRARED SENSORS; Heat Capacity Mapping Mission; Shuttle Multispectral Infrared Radiometer, Surface thermal emission; multispectral infrared sensing of rocks.

NOMENCLATURE

To conserve and eliminate repetition in text and references, the following symbols, units and names have been used in this chapter.

Symbol	SI Units	Name
PRF	pulse/sec	pulse repetition frequency
T	sec	interpulse period
X	m	slant swath width
C	m/sec	speed of light
h	m	satellite altitude
θ	degrees	look angle
ϕ, ψ	degrees	beam width in range, azimuth
W, L	m	antenna width, length
λ	m	wavelength
V	m/sec	satellite velocity
v	m/sec	equatorial earth rotation velocity
γ	degree	incidence angle
r_r, r_z	m	range and azimuth resolution
R_c	m	range curvature
$\mathscr{L}$	m	azimuth footprint
B	Hz	bandwidth
T_A, T_R	°K	antenna, receiver temperature
τ	sec	integration time

INTRODUCTION

The scattered and emitted electromagnetic radiation from natural surfaces contains information about the chemical and physical properties of these surfaces. This radiation propagates through the earth's atmosphere and is partially transmitted to outer space. Spaceborne sensors detect some of this radiation and measure its intensity and, in some cases, its phase and polarization. These measurements, in imaging or non-imaging format, are then used to derive information about the surface properties.

In the visible and near infrared regions of the spectrum, the energy source is the sun which emits mainly in those spectral regions. The emitted radiation from the earth's surface is mainly in the middle (or thermal) infrared and the microwave region of the spectrum. In the lower microwave and radio-frequency regions the energy source is usually man-made (i.e., radar sensors).

The electromagnetic spectrum is not fully accessible for remote sensing from space. The earth's atmosphere and ionosphere are opaque to a number of bands in the spectrum (Figure 13-1). In the high frequency ranges (UV to upper microwave), absorption and scattering by the atmospheric constituents lead to the presence of opaque bands. In the lower frequency region, the ionosphere limits the transmission to frequencies above the plasma frequency (i.e., about 10 MHz).

The interaction of electromagnetic radiation with natural surfaces and the earth's atmosphere is discussed in Chapters 3, 4 and 5. In this chapter we discuss the principles of microwave and infrared satellite remote sensors and the types of information which they provide. Active microwave sensors (synthetic aperture radar, scatterometer, altimeter and sounder) are discussed as well as passive sensors and infrared sensors. Landsat sensors and meteorological satellite sensors are not included in this chapter. They are the subject of Chapters 12 and 14 respectively.

ACTIVE MICROWAVE SENSORS

The term "active microwave sensors" refers to the sensors that generate the microwave energy used to illuminate a surface being sensed. These sensors usually transmit a train of microwave pulses, detect the returned echoes, and process them to derive information about the surface or subsurface properties. Because they generate their own illumination, these sensors can operate any time of the day or night, and the illumination geometry is selectable and controllable by the experimenter. All the active microwave sensors flown on satellites operated in the frequency range below K-band. Therefore their operation was not affected by cloud cover or weather conditions (except in extreme weather conditions). In effect, these sensors can be considered as all-time, all-weather systems.

In this section, we discuss only the active microwave sensors which have been flown on satellites. They include the imaging synthetic aperture radar (SAR) which was flown on Seasat, the scatterometers which were flown on Skylab and Sea-

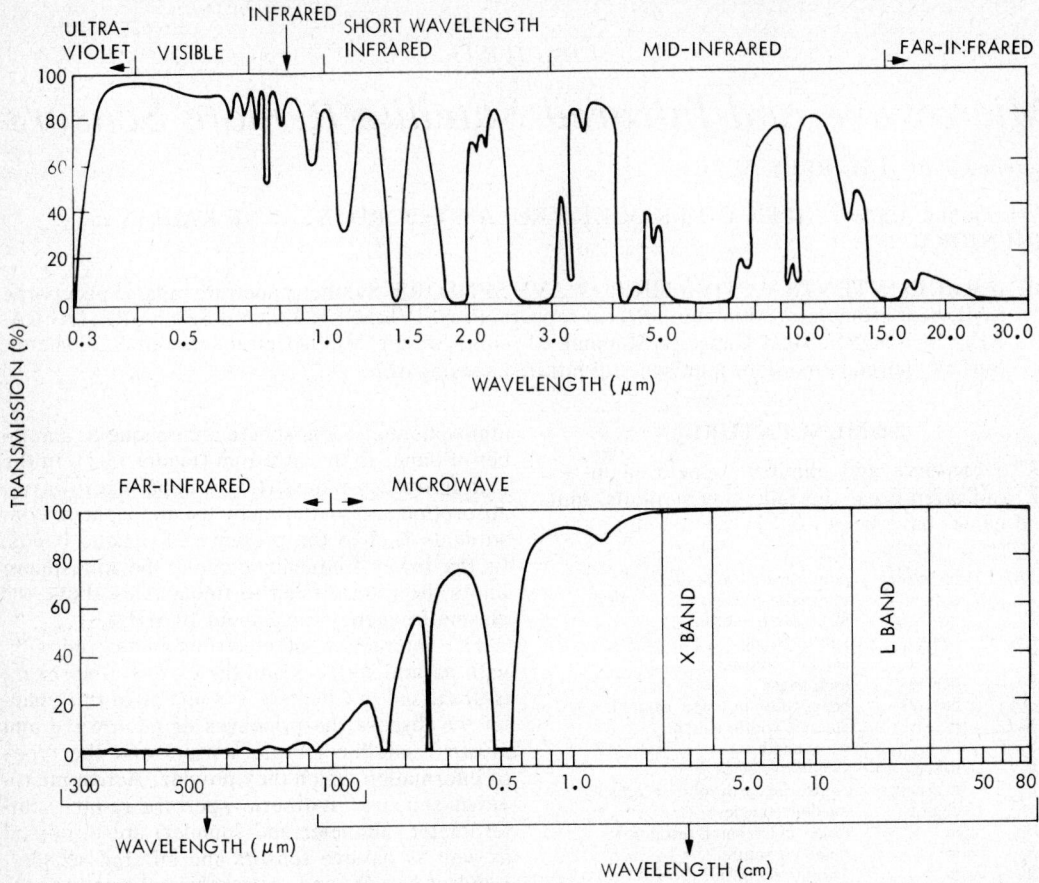

Fig. 13-1. Electromagnetic spectral transmission through the earth's atmosphere at zenith with the named spectral regions outlined.

sat, the altimeters which were flown on GEOS-C and Seasat, the synthetic aperture sounder which was flown on Apollo 17, and the Pioneer Venus radar mapper.

Synthetic Aperture Radar (SAR)

The imaging SAR sensor on the Seasat satellite which was launched in June 1978, and operated for 100 days, provided for the first time synoptic high resolution radar images of the earth's surface from an orbiting platform. The successful development and operation of this complex instrument opened a new dimension in our capability to observe, monitor and study the earth's surface. It extended the capability of the Landsat- and Nimbus-type sensors by providing high-resolution surface images in the microwave region of large areas in North America, Western Europe, the North Pacific and North Atlantic.

The SAR uses the reflected echo to generate a high-resolution image of the surface. The radar image tone is a two-dimensional representation of the surface backscatter at the frequency of operation. The surface interaction mechanisms are discussed in Chapter 4. In this chapter, we briefly

discuss the principle of SAR image formation with particular emphasis on the unique problems related to spaceborne platforms. We then present in detail the characteristics of the Seasat sensor and data handling systems and examples of the Seasat data and their interpretation.

SAR Image Formation

In the synthetic aperture technique, the Doppler information in the returned echo is used simultaneously with the time delay information to generate a high-resolution image of the surface being illuminated by the radar. The radar usually "looks" to one side of the moving platform (to eliminate right-left ambiguities) and perpendicular to its line of motion. It transmits a short pulse of coherent electromagnetic energy toward the surface. Points equidistant from the radar are located on successive concentric spheres. The intersection of these spheres with the surface gives a series of concentric circles centered at the nadir point (see Figure 13-2). The backscatter echoes from objects along a certain circle will have a well-defined time delay but different Doppler characteristics.

Points distributed on coaxial cones, with the

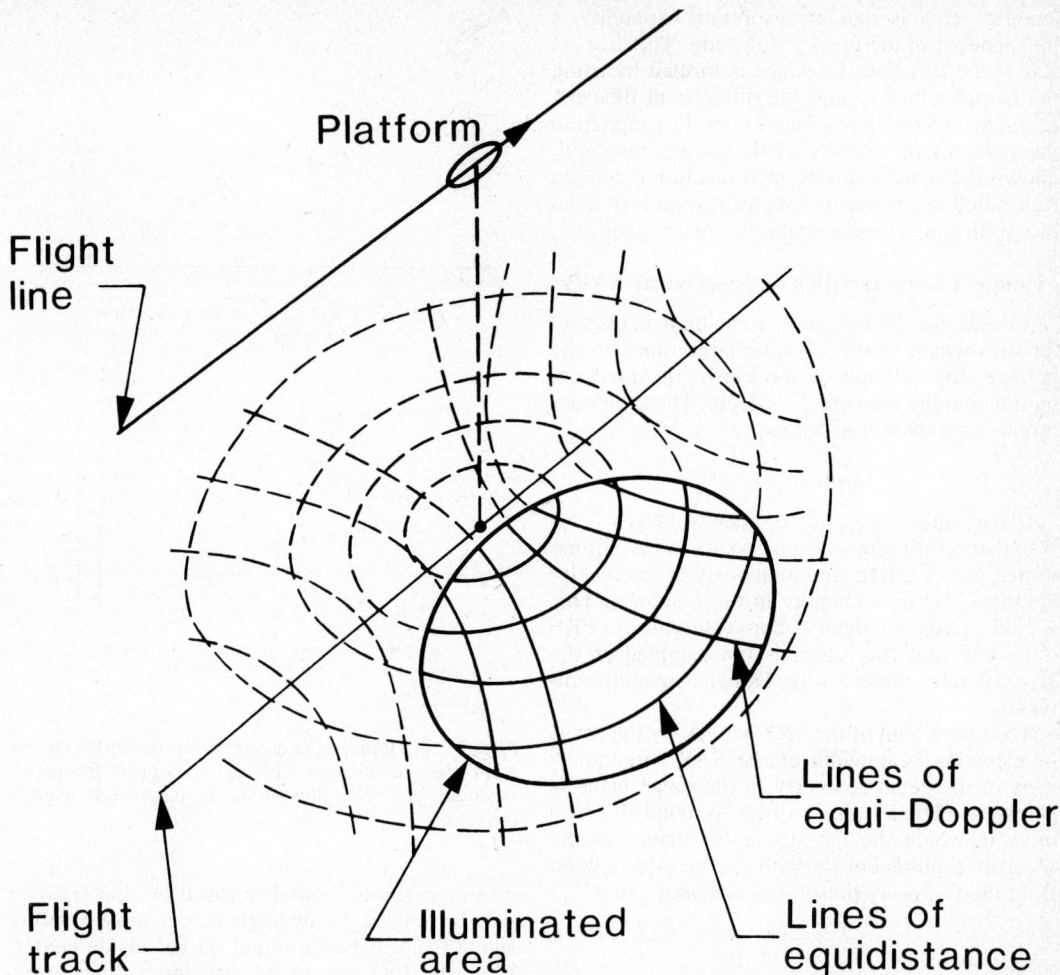

Fig. 13-2. Synthetic aperture radar geometry. The constant time-delay and constant Doppler lines form the radar imaging coordinate system. The backscatter return from each cell on the surface can be uniquely determined by filtering from the total return the energy in the appropriate time-delay bin and Doppler shift bin.

flight line as the axis and the radar as the apex, provide identical Doppler shifts of the returned echo but different delays. The intersection of these cones with the surface gives a family of hyperbolas (Figure 13-2). Objects on a specific hyperbola will provide equi-Doppler returns. Thus, if the time delay and Doppler information in the returned echoes are processed simultaneously, the surface can be divided into a coordinate system of concentric circles and coaxial hyperbolas (Figure 13-2), and each point on the surface can be uniquely identified by a specific time delay and specific Doppler. The brightness that is assigned to a specific pixel (picture resolution element) in the radar image is proportional to the echo energy contained in the time delay bin and Doppler bin which corresponds to the equivalent point on the surface being imaged. The resolution capability of the imaging system is thus dependent on the measurement accuracy of the differential time delay and differential Doppler (or phase) between two neighboring points on the surface.

In actuality, the situation is somewhat more complicated. The radar transmits a pulsed signal, which is necessary to obtain the time delay information. To obtain the Doppler information unambiguously, the echoes from many successive pulses are required with a pulse repetition frequency which meets the Nyquist sampling criterion. Thus, as the moving platform passes over a certain region, the received echoes contain a complete Doppler history and range-change history for each point on the surface that is being illuminated. These complete histories are then processed to identify uniquely each point on the surface and to generate the image. This is why a very large number of operations is required to generate each pixel in the image. Such is not the case in optical sensors. A simplified comparison is that the radar sensor generates the equivalent of a hologram of the surface, and further processing is required to obtain the image. This processing can be done either optically or digitally.

One unique feature about the synthetic aperture

imaging radar is that its resolution capability is independent of the platform altitude. This is a result of the fact that the image is formed by using the Doppler history and the differential time delays, none of which is a function of the range from the radar to the surface. This unique capability allows the acquisition of high-resolution images from satellite altitude as long as the received echo has sufficient strength above the noise level.

Unique Characteristics of Spaceborne SARs

Spaceborne SARs have some unique characteristics which result from the large range to the surface, the rotation of the earth, the platform motion and the ionospheric effects. These characteristics are discussed below.

Ambiguities

If the pulse repetition frequency (PRF) is so high that return signals from two successive transmitted pulses arrive simultaneously at the receiver, there will be ambiguity in the response. This is called range ambiguity. Conversely, if the PRF is so low that the return is not sampled at the Nyquist rate, there will be Doppler azimuth ambiguity.

The upper limit of the PRF is fixed by the range or elevation beamwidth of the SAR antenna. A view of the beam geometry in the range plane is shown in Figure 13-3. In order to avoid the situation in which the far-edge echo arrives at the receiver simultaneously with the near-edge echo from the following pulse, it is necessary that

$$T > 2X/C \qquad (13\text{-}1)$$

i.e.,

$$T > \frac{2h}{C}\frac{\tan\theta}{\cos\theta} \qquad \phi = \frac{2\lambda h}{CW}\sin\theta/\cos^2\theta \qquad (13\text{-}2)$$

or

$$PRF < CW\cos^2\theta/2\lambda h\ \sin\theta \qquad (13\text{-}3)$$

where T is the time between successive pulses, λ is the radar wavelength, W is the antenna width, θ is the look angle, h is the platform height and ϕ is the beamwidth (see Figure 13-3a). The above equation assumes that ϕ is small and the pulse length is much smaller than T; the earth's surface curvature is neglected.

The lower limit of the PRF is imposed by the requirement that the PRF must equal or exceed the maximum Doppler shift of the return signals. The transmitted signal has spectral components separated in frequency from the carrier determined by the Doppler shift from the targets illuminated by the antenna beam. Targets at the center of the broadside beam will return signals with zero Doppler shift. Targets ahead of broadside center are characterized by positive Doppler and those behind by negative Doppler frequencies. If the return of a target is shifted in frequen-

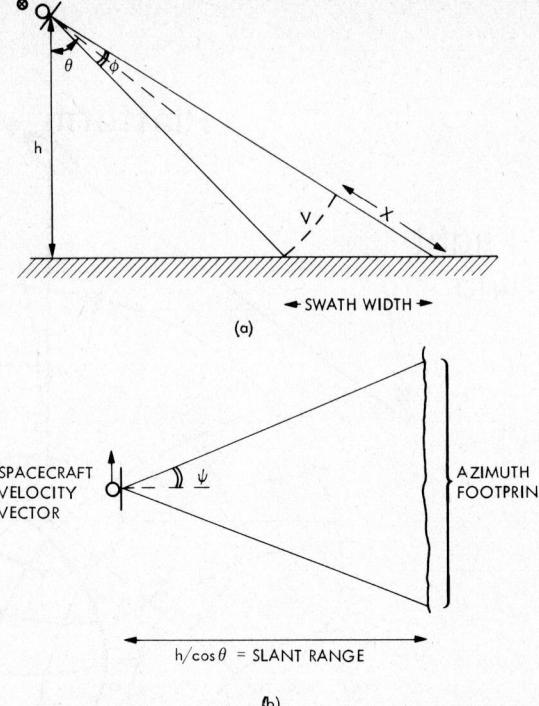

Fig. 13-3. (a) Radar range geometry illustrating the parameters used in the text. (b) Radar Doppler frequency (azimuth) geometry illustrating the parameters used in the text.

cy by an amount equal to the PRF, the receiver will be unable to distinguish the pulsed return signal from that of a target on broadside center. Thus the PRF has to be sufficiently high to exceed the maximum Doppler shift of targets located at the beam edge. The Doppler shift f_D due to a radial velocity V_r is given by

$$f_D = \frac{2V_r}{C}f \qquad (13\text{-}4)$$

where f is the radar frequency and C is the light velocity. By referring to Figure 13-3b, $V_r = V \sin \psi \simeq V\psi$. The angle ψ to the first null of the one-way radar illumination beam is $\psi = \lambda/L$. Thus the lower limit of the PRF is

$$PRF_{low} = f_D = \frac{2V}{L} \qquad (13\text{-}5)$$

The two limitations derived above give

$$\frac{CW\cos^2\theta}{2\lambda h\sin\theta} > PRF > 2V/L \qquad (13\text{-}6)$$

which can be rewritten as:

$$WL > \frac{4V\lambda h\sin\theta}{C\cos^2\theta} \qquad (13\text{-}7)$$

This condition imposes a minimum size on the antenna area (WL), which is required to avoid ambiguities. Eq. 13-7 should be used only in preliminary calculations. In actual design, the earth's curvature should be taken into account.

Earth Rotation Effects

The satellite orbit plane is fixed in space. Targets on the earth's surface move with a velocity which is dependent on their latitude. This additional Doppler shift must be compensated for during processing. The Doppler frequency shift f_d caused by the rotating earth is given by [Tomiyasu, 1978]:

$$f_d = (2f/C) \, v \sin\gamma \sin A \, \cos(\pi-\beta) \quad (13\text{-}8)$$

where:

v = equatorial earth rotation velocity = 463 m/sec

A = arc distance from the pole (see Figure 13-4)

γ = incidence angle

β is defined in Figure 13-4.

f_d varies from a maximum value at the equator to zero at the maximum latitude. The effect of the earth rotation can be compensated for in three ways: (i) Rotating the antenna beam in yaw about the satellite nadir to include the zero Doppler frequency direction. Because the rotation angle varies with latitude, this requires that the satellite attitude be continuously adjusted. (ii) Continuously adjusting the receiver local oscillator frequency with respect to the transmitted frequency to cancel the effect of the Doppler from the earth rotation. (iii) Subtracting the Doppler due to the earth rotation from the received signal during processing. The last approach was the one used by the Seasat SAR. A detailed discussion of this approach is given later in the section on processing.

Orbit Eccentricity Effects

Orbit eccentricity causes an altitude change and displaces the image in the track direction. With descending altitude, the velocity vector will dip below the local horizon. The zero Doppler frequency direction is perpendicular to the satellite velocity vector so that the image will shift rearward relative to the subsatellite point [Harger 1970, Tomiyasu, 1978]. The image shift is given by:

$$s = \frac{h}{v} \frac{dh}{dt} \cos\theta \quad (13\text{-}9)$$

Similar to the earth-rotation effect, compensation for altitude change can be accomplished by rotating in yaw the antenna boresight axis about the local satellite nadir, by trimming the receiver local oscillator frequency or during processing. It should be noted that since the orbit perigee will migrate, the orbital position for maximum altitude change rate will migrate.

Ionospheric Effects

The ionosphere is composed of layers of free electrons that exhibit an index of refraction of less than unity. The ionosphere affects the electromagnetic wave phase velocity and, hence, phase coherency, as well as the wave polarization vector because of the earth's magnetic field. For a microwave beam traversing the ionosphere, the quantity of interest is the total columnar electron content N along the path given in units of electrons per square meter of cross-section. At midlatitudes, N typically varies diurnally between $10^{16} \, e/m^2$ and $10^{17} \, e/m^2$. Above equatorial regions and at high latitudes, N is typically $10^{18} \, e/m^2$. During magnetic storms N can be as high as $10^{19} \, e/m^2$ [Aarons, 1977]. With an N of $10^{19} \, e/m^2$ and a radar frequency above 1 GHz, calculations show that the Faraday rotation and the amount of dispersion are negligible. However, at lower frequencies, these effects have to be taken into account.

Amplitude and phase scintillations are caused by ionospheric irregularities [Aarons et al., 1971; Crane, 1977]. The phase scintillations could affect the formation of the synthetic aperture. The phase granularity can be calculated from time-delay formulations given by Burns and Fremouw [1970]. If this granularity along a synthetic aperture length leads to a phase shift in excess of a small fraction of a radian, it will result in a degradation of the image quality.

Range Curvature

During integration along the synthetic aperture length, the target range may change in excess of the range resolution. This effect is called range curvature [Leith, 1977]. If R_o is the nearest range

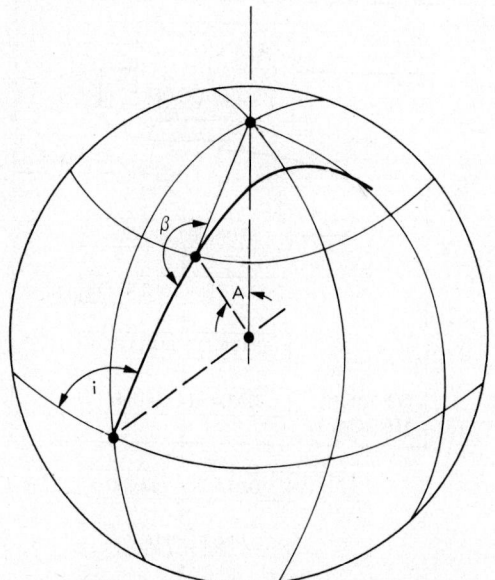

Fig. 13-4. Orbital geometry showing the parameters used in the text.

to the target and R_m the maximum range at the edge of the synthetic aperture, then

$$R_m = \sqrt{R_o^2 + \left(\frac{V\zeta}{2}\right)^2} = R_o + \frac{1}{8}\frac{(V\zeta)^2}{R_o} \quad (13\text{-}10)$$

where ζ is the integration time ($\zeta = \lambda R_o/2Vr_z$), and r_z is the azimuth resolution. The range curvature is given by:

$$R_C = R_o - R_m = \frac{1}{8}\frac{(V\zeta)^2}{R_o} = \frac{\lambda^2 R_o}{32 r_z^2} \quad (13\text{-}11)$$

Relative to the range slant resolution r_r, then

$$\frac{R_c}{r_r} = \frac{\lambda^2 R_o}{32 r_r r_z^2} \quad (13\text{-}12)$$

For the Seasat A SAR system, $r_r = r_z = 7$ meters, $\lambda = 0.25$ m and $R_o = 850$ km. The range curvature is equal to:

$$\frac{R_c}{r_r} = 4.3$$

This implies that a point migrates through 4.3 resolution elements during the formation of the synthetic aperture. This effect must be taken into account during processing.

Seasat SAR Sensor System

The Seasat synthetic aperture radar (SAR) was the first imaging SAR system used as a scientific sensor from earth orbit. It was launched in June 1978 and operated for 100 days, at which time the satellite failed in orbit as a result of a massive short circuit in the electrical system. During its orbital operations, the SAR provided a unique and voluminous set of data concerning the earth's sur-

face. It generated radar images with a 100-km swath at a resolution of 25 meters. The total area imaged was about 126 million km².

The Seasat SAR system, shown in Figure 13-5, consists of a planar array antenna, a transmitter/receiver sensor, an analog data link, a data formatter, a high density digital recorder and a processing subsystem. All these subsystems are described in this section except for the data processing subsystem, which is discussed in the next section. This description is mostly based on the papers by Jordan [1980] and Wu [1980]. The characteristics of the Seasat SAR are summarized in Table 13-1.

Antenna:

The Seasat-A antenna system consists of a 10.74 m × 2.16 m phased array system deployed after orbit insertion. This deployed antenna is configured to fly with the long dimension along the spacecraft velocity vector and bore-sighted at an angle of 20.5° from the nadir direction in elevation (cone) and 90° from the nominal spacecraft velocity vector (clock). The antenna dimensions (10.74 m × 2.16 m) are dictated by a desire to limit range and Doppler ambiguities to acceptable low levels (Eq. 13-1 to 13-7). At a nominal 20.5° look angle from nadir, a total beamwidth in elevation of 6.2° is required to illuminate a 100-km swath on the earth's surface from an 800km-high orbit. Thus, the antenna cross-track dimension is 2.16 m to limit the radiation to these sets of angles. The area illuminated on the surface of the earth is from 240 km to 340 km to the right of the subspacecraft point. The antenna elements in elevation are weighted in illumination to limit sidelobes in the

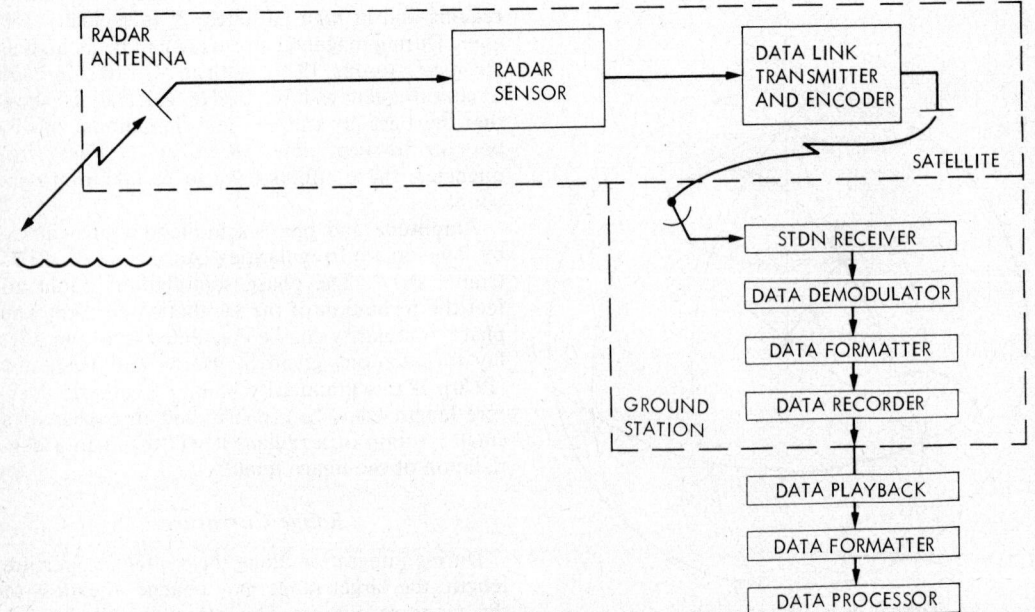

Fig. 13-5. Simplified functional diagram of the Seasat SAR system.

TABLE 13-1

Seasat-A SAR System Characteristics

Satellite altitude	800 km
Wavelength	0.235 m
Theoretical resolution	25 m × 25 m on the surface at 4 looks
Contrast ratio	9 dB
Swath width	100 km
RF bandwidth	19 MHz
Transmit pulse length	33.4 μsec
Pulse repetition range	1463 pps to 1640 pps
Transmit time bandwidth product	634
Radar transmitter peak power	1000 W
Telemetry transmitter power	5 W
Sensitivity time control range	9 dB
Data recorder bit rate	110 Mbits/sec (5 bits per word)
Radar DC power	500 W
Radar Antenna dimensions	10.7 m by 2.16 m
Radar antenna gain	35 dB
Telemetry antenna	Quadrafilar helix
Telemetry transmit antenna gain	5 dB
Telemetry ground station antenna	9 m parabolic antenna
Gain	35 dB at SAR sensor electronics
VSWR	1.5 (±11 MHz bandwidth)
Power	1500 W peak

cross-track direction. The minimun antenna long-track length is limited by a desire to keep azimuth sampling ambiguities to an acceptably low level while the maximum alongtrack length is determined by the requirement to illuminate a sufficiently large patch of terrain to allow processing of the data to four looks.

These two requirements limit the antenna length along the velocity vector to between 10.5 and 14 meters. The 10.74-m antenna length was dictated by the available volume within the satellite shroud. The level of integrated ambiguities in azimuth was estimated to be between −18 dB and −24 dB, depending on the PRF and processor bandwidth.

The Antenna Subsystem consists of eight mechanically deployed, electrically coupled, flat microstrip panels. This array is shown in Figure 13-6 in both stowed and deployed configurations. The performance and design requirements placed on this subsystem are contained in Table 13-2. The construction of the microstrip is depicted in Figure 13-7 and the deployment and extension mechanization is shown in Figure 13-8.

Radar Sensor Description

The radar sensor provides the antenna with a series of high-power coherent pulses of energy at L-band, and amplifies the weak return echoes which are received by the antenna. The radar sensor consists of four subassemblies: transmitter, receiver, logic and control, and power converter. A diagram of the sensor is shown in Figure 13-9, and the principal sensor parameters are tabulated in Table 13-3.

To obtain an adequate signal-to-noise ratio from a system whose range resolution is 25 m on the surface and which utilizes a solid-state transmitting device, it is necessary to use a long transmitted pulse and pulse-compression techniques to

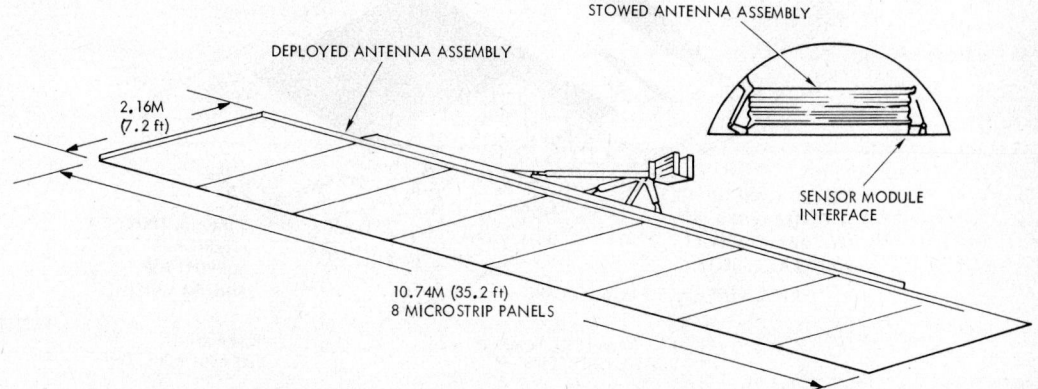

Fig. 13-6. Seasat SAR antenna shown in its stowed and deployed configuration.

TABLE 13-2

SAR Antenna Performance

Beamwidth	
Azimuth	1.73° at −4 dB
Cone angle (range)	6.2° ± 0.1° between +33 dB far side and +31 dB near side
Polarization	Horizontal linear of 99% purity
Frequency	1.275 GHz
Sidelobes	−12.5 dB
Phase errors	
Quadratic	±20° over ±0.5° from beam center
Higher order	±2° over ±0.5° from beam center
Beam pointing (nominal)	90° clock angle (AZ)
	20.5° cone angle (EL)
	0° rotation
Beam alignment	±0.2°, 3σ clock and cone
goal (alignment between	(determined to ±0.1°, 3σ)
electrical boresight and	±0.5°, 3σ rotation
antenna reference axes)	(determined to ±0.2°, 3σ)

reduce the peak power requirement. The output of the transmitter assembly is, as a result, a linearly swept frequency-modulated pulse (or chirp) with a 634:1 compression ratio. It is generated in a surface acoustic wave (SAW) device located in the chirp generator subassembly of the transmitter assembly. The output of the transmitter is coupled to the Antenna Subsystem through an output combiner. A portion of the output (leakage) is also impressed on the receiver input where a load is placed in the circuit each time the transmitter operates. This prevents the leakage pulse from burning out the input stage of the receiver.

Echo returns are coupled into the receiver assembly through the output network in the transmitter. Because the echo's intensity is expected to vary in proportion to the variation of antenna gain with angle, a sensitivity time control (STC)

has been incorporated in the receiver. The STC action, initiated by satellite stored commands, linearly decreases the receiver gain by 9 dB during the first half of the return echo period, and then increases the gain until the end of the echo has been received. The application of the STC results in a nearly uniform signal (echo) return for a uniform scattering field, and, as a result, the dynamic range required to transmit the resultant data to earth is reduced by 9 dB.

The sensor receiver output is sent to the data link along with timing and frequency references derived from the SAR system local oscillator (STALO). The STALO generates a very stable (in frequency) signal at a nominal frequency f_s, where $f_s = 91.059$ MHz. A portion of this signal is delivered to the multiplier assembly. Another portion is used to derive both square-wave clock

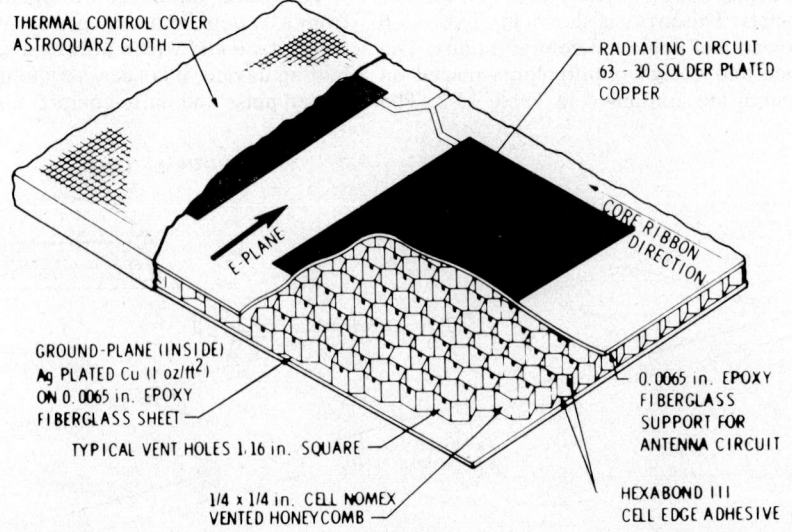

Fig. 13-7. Construction of the microstrip honeycomb panels that were used in the Seasat antenna.

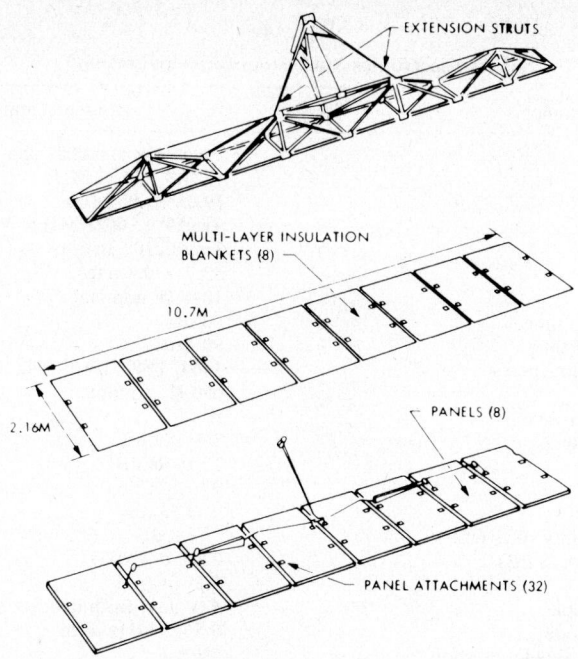

Fig. 13-8. Seasat SAR antenna mechanical components and assemblies.

and sine-wave signals at $f_s/3$, which are used in synchronizing other Sensor Electronics Subsystem functions. The frequency multiplier assembly provides signals at $3 f_s$, $9 f_s$, and $18 f_s$. The $3 f_s$ and $18 f_s$ signals are delivered to the chirp generator where, along with a portion of the STALO signals, f_s, they are used to generate the linear FM pulse (chirp) signal at the frequency of $14 f_s$. The signal at $9 f_s$ and a portion of the signal at f_s are delivered directly to the Data-Link Subsystem along with a signal derived from the PRF event which divides the interpulse interval into 4096 sectors.

The remaining two assemblies in the Sensor Electronics Subsystem are the logic and control and the power converter. They provide the primary electrical interface with the satellite. The logic and control assembly receives commands from the satellite, decodes them and causes the Sensor Electronics Subsystem to assume one of a number of operating modes. In addition, the logic and control interfaces between the satellite and

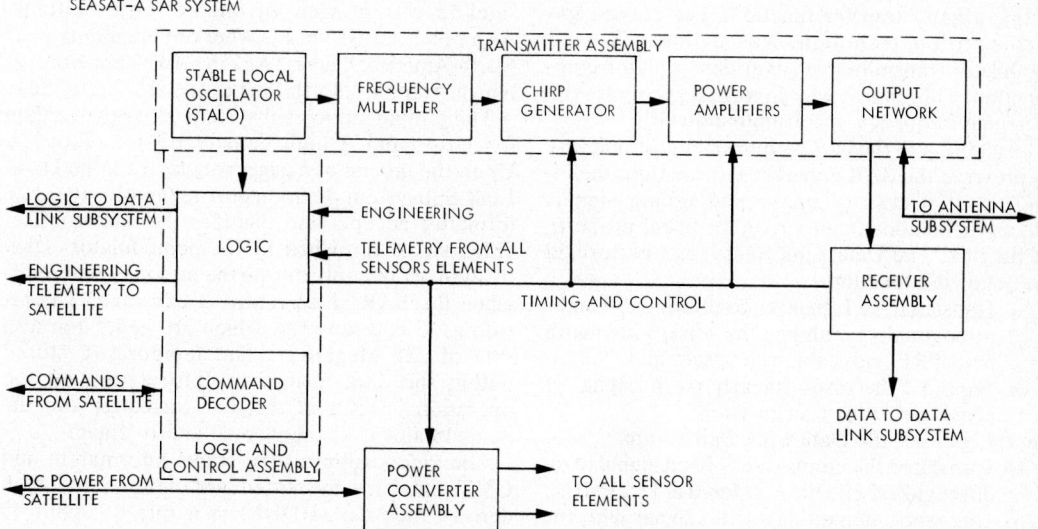

Fig. 13-9. Block diagram of the Seasat SAR sensor.

TABLE 13-3

Sensor Electronics Subsystem Implementation

Electronics	Implementation
Modes	Four: off, standby, operate transmit
Center frequency	1274.8 ± 0.31 MHz
Bandwidth	$19.05 \pm .05$ MHz
STALO frequency	$91.059 \pm .0022$ MHz
STALO stability	3 parts/10^{10} in 5 ms
Pulse width	$33.9 \pm 0.8 \,\mu$sec
Peak power	1000 W nominal
Pulse envelope rise time	90 nsec
Pulse envelope fall time	90 nsec
Pulse repetition frequencies	1464, 1540, 1580, 1647 $\pm$.1PPS
Noise temperature	650 K, nominal
Receiver gain control steps	8
Gain control step spacing	3 ± 0.3 dB
Gain control range	77 to 98 dB
STC gain variation	9
Receiver gain flatness	± 0.33 dB
Receiver gain stability ($0 - 60°$C)	± 1.0 dB
Receiver bandwidth (3 dB)	22 ± 0.2 MHz
	$5.3°$ overall
Receiver phase ripple	$4.0°$ dev. from quadratic
Transmitter FM slope	0.5622 MHz/μsec
Transmitted pulse envelope droop	5%
Transmitter amplitude response	±0.05 rms
Transmitter phase response	$3°$ rms

the intra-subsystem engineering telemetry. The power converter assembly provides the stable, isolated power required by all the SAR Subsystems.

Data-Link Subsystem:

The purpose of the Data-Link Subsystem is to transmit the radar echo to the ground for digitization, storage, and subsequent processing. The link also maintains the phase and time references necessary to the processing function, thus providing a unity transfer function. The chosen implementation technique was a linear S-band modulator/transmitter/receiver/demodulator combination. This choice was governed chiefly by the available frequency spectrum/bandwidth.

In addition to the basic requirement for linearity to preserve the SAR coherent information, the inclusion of necessary phase and timing signals placed an additional burden on the linear property of the link. The Data-Link Subsystem performed the following functions:
  a) Translated the L-band echo return to S-band,
  b) orthogonally combined the offset video with both PRF and stable reference, and
  c) amplified the result linearly (with negligible phase error) for transmission:
On the ground, the Data-Link Subsystem
  a) Translated the composite S-band signal to an offset video frequency centered at 11.25 MHz,
  b) coherently demodulated the signal with the spacecraft local oscillator,
  c) removed the link-induced Doppler from the composite signal, and

  d) reconstructed the PRF and STALO signals from the video.

Ground Station Subsystem

The Seasat SAR system required unique equipment at the ground receiving station. Only three U.S. stations (Fairbanks, Goldstone and Merritt Island), one Canadian station (Shoe Cove) and one British station (Oak Hangar) were equipped to receive the SAR data. The SAR operations were limited only to the time periods when the satellite was in view of one of those stations. As a result, radar images were obtained only over North America, Central America, Western Europe, North Atlantic, North Pacific and the Arctic Sea.

The ground station subsystem consists of a data formatter and a high density digital recorder. Upon the receipt of a trigger signal from the Data-Link Subsystem demodulator assembly, the data formatter accepts and digitizes the analog offset video signal furnished by the demodulator. Digitization occurs only during the period ($\sim$300 μsec) when the SAR video return is expected. The resulting 13,680 samples, which are generated at a rate of 227 Megabits/s, are temporarily stored within the data formatter. Information on the operational status of the data formatter and the demodulator is also collected and retained.

The video samples, the status information and GMT time are formatted and sent to the high density recorder (HDDR) at a rate of about 13 Mbps. The HDDR records this high-rate stream on one-inch width magnetic tape. Recording uses 40 (of 42) parallel tracks on the tape at a record-

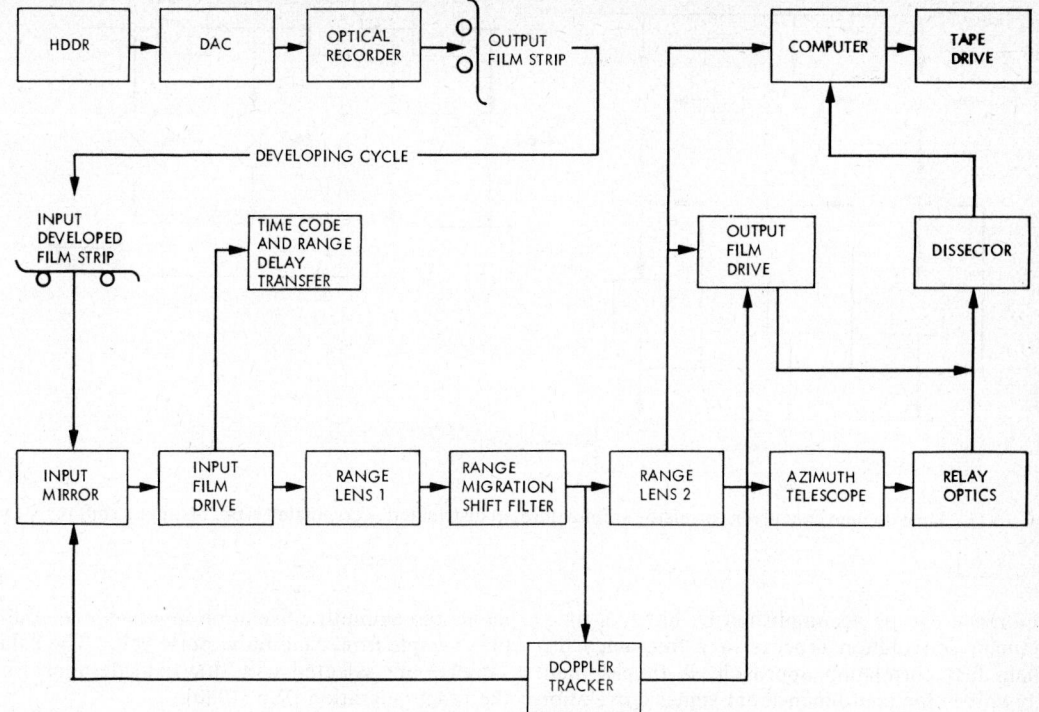

Fig. 13-10. Block diagram of the Seasat SAR optical data processor.

ing speed of 150 ips. Parity is included on each track and timing information is carried on one of the remaining tracks.

Optical Processing of SAR Data

The SAR sensor provides a record of the received echoes. This record is called the signal record. The signal data are then processed in a correlator to generate the synthetic aperture which provide the final high resolution image. Data correlation can be achieved either optically or digitally. Here we will emphasize the functions which are unique to spaceborne systems. The Seasat SAR optical correlator is used as an example. Digital correlation is discussed in the next section.

A block diagram of the Seasat SAR optical data processing subsystem is shown in Figure 13-10. The signal film is a record of the echo history. It is generated by playing back the signal tape into a video generator followed by an optical recorder. The signal film also contains two channels where the time code and range delay are optically written. The film drive speed is slaved to predicted spacecraft relative ground speed at swath center. Before correlation, the azimuth Doppler is determined from targets in the data or from altitude information. The Doppler tracking system steers the illumination beam via the input mirror so that the Doppler phase histories are centered in the optical system. The range migration correction optics are moved in azimuth

in synchronism with the input mirror so that the proper correction is maintained.

The azimuth telescope is automatically adjusted to account for range shifts in the digitization window. Demagnification is adjusted for each quarter swath so that an aspect ratio of unity is achieved at each swath center. Relay optics include a magnification lens to adjust the image scale factor to 500,000:1. The output film drive is tilted above the optical axis in synchronism with the input mirror movements to correct for azimuth data skew caused by Doppler beam steering. The time code is detected from the signal film and transferred to the image film. Time code and film speed are adjusted to account for Doppler steering.

Digital Processing of SAR Data

The concepts of digital processing of SAR data can be divided in two major approaches. In one approach the target's time-phase history is treated as an impulse response, and the image is generated by correlating a reference waveform with the radar return. In the second approach, the SAR data processing is handled as the signal processing from a linear phased array. The ultimate long synthetic aperture is formed by several stages of array processing, each forming a longer aperture from the outputs of shorter apertures. In this approach, pulse compression is accomplished without the generation of the implicit target response waveform.

The correlation between the signal data and the

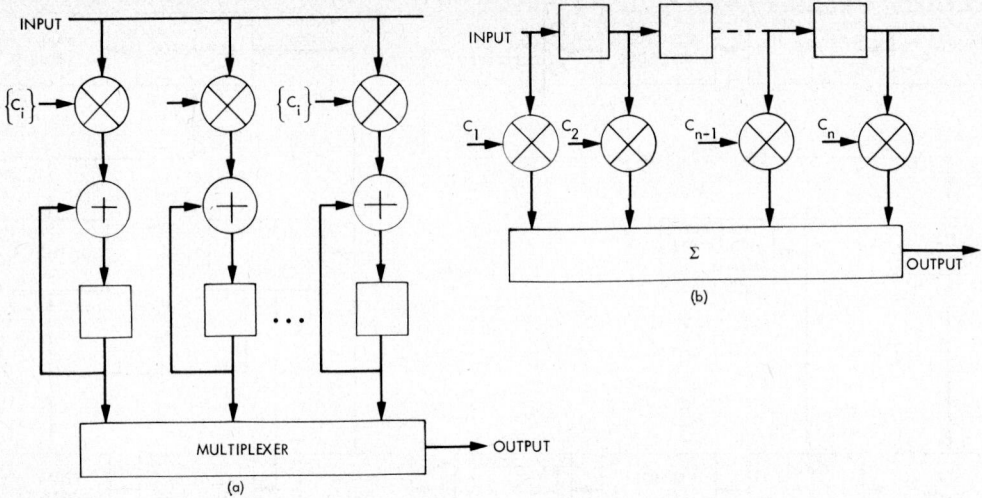

Fig. 13-11. Time domain convolver correlator architecture, a) distributed accumulated type, b) tapped shift register type.

reference can be accomplished by either a time-domain convolution process or a frequency domain fast correlation approach. A time-domain convolver for one-dimensional signal correlation can use a distributed accumulator or a taped shift register (Figure 13-11). In the simplest arrange-

ment, the azimuth correlation involves one complex sample from each radar pulse echo. The data samples are selected with due consideration for the range migration [Wu, 1980].

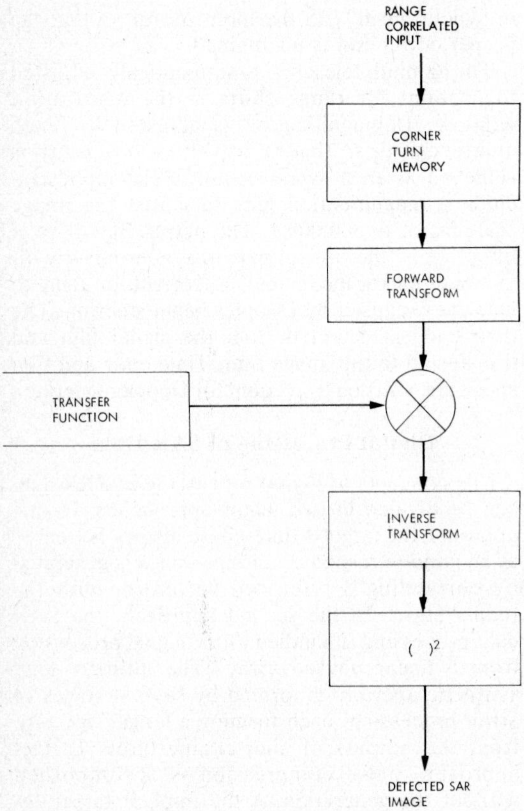

The fast Fourier transform (FFT) is often used to perform fast correlation operations. The correlation is performed over a block of data in range dimension. A range "corner-turn" operation is needed in such an approach (Figure 13-12).

In the array formation approach, at each stage of the processing a longer synthetic aperture is formed from shorter ones. The idea is similar to phased array radar signal processing, and the fast Fourier transform is commonly employed in each state of processing. The data processing is divided into two stages. In the first stage, processing is applied to successive radar echo data at a constant range delay. The output of this stage consists of subbeams within the physical beamwidth. By sifting the input echo array along the sensor path, successive sets of "subarrays" are formed. From the output of those subarrays one can locate the subbeams which provide registered coverage on the same target footprint. The extent of this area under a subbeam is much smaller than the original radar footprint. By applying another stage of array processing the resolution can be further improved. Details of this multistage approach are given by Martinson [1975] and Van de Lindt [1977].

A specific digital SAR processing approach is characterized by arithmetic complexity and control function complexity. The arithmetic complexity depends on a number of factors, the predominant ones being the throughput rate, the pulse compression ratio and the number of looks. The arithmetic processing rate can be measured by the product of the pixel rate at the output of the processor and the number of arithmetic operations required to produce each pixel. The pulse

Fig. 13-12. Frequency domain convolver for SAR azimuth correlator.

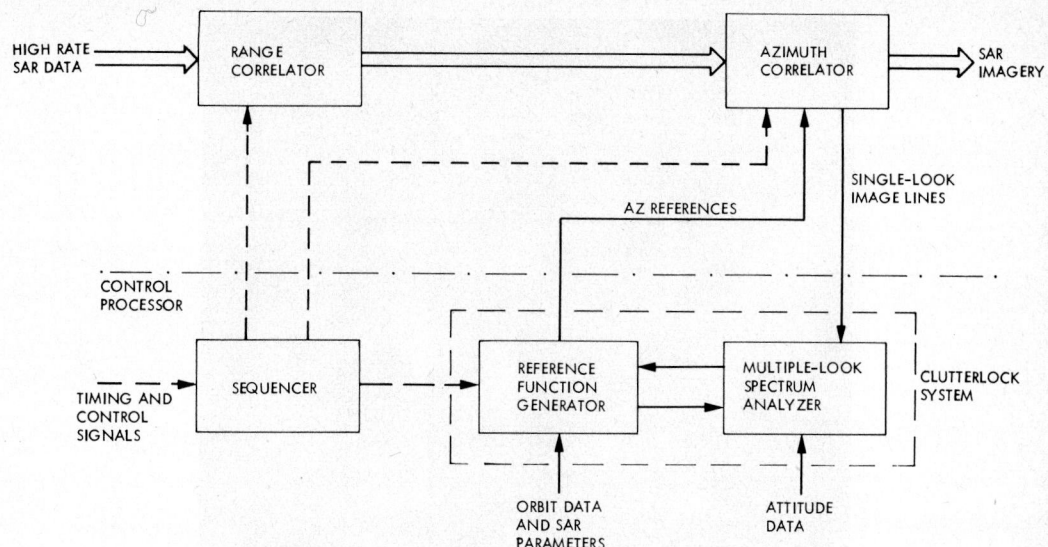

Fig. 13-13. Real-time SAR processor control block-diagram.

compression factor of SAR azimuth correlation is equal to the ratio of the synthetic aperture length to the corresponding azimuth resolution. This is equal to the number of complex arithmetic operations K_t

$$K_t = \frac{\mathscr{L}}{r_a} = \lambda R/2(r_a)^2 \qquad (13\text{-}13)$$

where $\mathscr{L}$ is the azimuth footprinted on the surface. For a near optimal SAR system, the echo window is nearly equal to the separation between pulses and r_a is equal to half of the physical aperture. The pixel rate P is very close to the bandwidth B of the radar transmitted pulses. Thus

$$P \simeq B = \frac{C}{2r_r \sin\theta} \qquad (13\text{-}14)$$

The arithmetic complexity measure A_T is then given by

$$A_T = \frac{\lambda RC}{4r_r (r_a)^2 \sin\theta} \qquad (13\text{-}15)$$

For the frequency domain processing approach using the FFT multistage processing, the number of arithmetic operations per pixel is approximated by

$$K_f = \log_2 K_t \qquad (13\text{-}16)$$

and the complexity measure A_f is given by

$$A_f = \frac{C}{2r_r \sin\theta} \, \log_2 \left| \frac{\lambda R}{2r_a^2} \right| \qquad (13\text{-}17)$$

Equations 13-15 and 13-17 are useful for a quick assessment of the arithmetic complexity of the processor for various SAR systems.

The control function complexity is not easily quantifiable. In general, the control complexity increases as the system performance requirements (resolution, sidelobe levels, etc.) become more stringent. The SAR processor must include a controller which determines accurately the parameters required for processing. These include exact description of the range history between the radar and the target, which requires knowledge of the orbital and attitude characteristics as well as surface curvature and velocity at the point of observation. Figure 13-13 summarizes the control functions. Predictions of SAR Doppler processing parameters can be obtained from the spacecraft altitude and state vectors provided by navigation sensors. Refinement of the SAR Doppler center frequency can be performed through a feedback loop. Rapid updating of the azimuth correlation reference functions and range migration compensation coefficient increases the complexity of the overall processor.

SAR Data Analysis and Application

The tone of the radar image is a representation of the surface backscatter cross-section, which in turn is a function of the surface slope, surface roughness at the scale of the radar wavelength, and the complex dielectric constant. The interaction mechanisms of microwaves with natural surfaces is discussed in detail in Chapter 4.

The interpretation of the radar images is based on two types of information, (i) geometric patterns, forms and shapes, and (ii) image tone and texture. Examples of the first are lineaments, folds, domes, drainage patterns, cultivation fields, ocean waves, current boundaries, etc. These patterns, forms and shapes are interpreted in a way similar to that used with regular photography. Figures 13-14 and 13-15 show a Landsat and Seasat SAR image of the Los Angeles Basin and the Western Mojave region in California. Many of the major geologic, geographic and urban features are easily recognizable on both images. The San

Fig. 13-14. Seasat SAR image of the Los Angeles Basin (lower part of the image) and the western Mojave Desert (upper part of the image).

Andreas and Garlock faults are visible because of the topographic variation across the fault strike which is reflected as an abrupt change in the image tone and texture. Numerous other faults are visible due to topographic alignments. In the urban Los Angeles area, the major highways are visible as dark-toned linear features. The very bright regions correspond to townships where the streets are almost parallel to the orbital track. This leads to the cardinal effect where strong returns are due to appropriately oriented corner reflectors formed by the streets' surface and the buildings' walls. In Long Beach harbor, large vessels are visible as bright point features.

Image tone and texture are primarily a function of the surface roughness and subresolution small-scale topography, the surface complex dielectric constant, and surface variations on the scale of few resolution elements. The dielectric constant variation is most useful in the study of vegetated and moist surfaces. The tonal and textural data on the radar image provide new information that is not available with optical or infrared photography. Their interpretation requires an understanding of the interaction of microwaves with natural surfaces [Schaber, et al., 1980, Blom and Elachi, 1981]. Figure 13-16 shows a Seasat SAR image and a Landsat image of cultivation fields in the Imperial Valley, California. The tonal variations in the SAR image are not directly correlated with the tonal variations in the Landsat image. This shows that the surface properties have dif-

Fig. 13-15. Landsat image of the same area covered in Figure 13-14.

ferent effects on the response of each sensor. Work is presently ongoing on how to use these two sets of data in a complementary way to improve the discrimination and identification of crops.

Figure 13-17 is a Seasat SAR image of Jamaica. The area in the image is completely covered by vegetation. Variations in the image texture are clearly visible. These variations correlate well with the surface geology. The fine texture corresponds to limestone areas where the karst topography is well developed. The coarse texture corresponds to areas with igneous or metamorphic rocks. The topographic texture is visible on the radar image because of the strong sensitivity of the backscatter intensity to variations in the surface slope.

Spaceborne radar data have applications in a number of fields. In the following, we present some examples of Seasat SAR images to illustrate some of these potential applications.

Geologic Mapping

The radar backscatter cross-section is very sensitive to the surface slope, particularly at small (less than 30°) and large (larger than 60°) incidence angles. At intermediate angles, the backscatter is mostly sensitive to the surface roughness. Thus the radar sensor is most useful for the study of patterns and features that are expressed as changes of slope or roughness. Lineaments, faults, fractures and contacts are usually expressed on the surface as sharp changes in surface topography, morphology or cover [Ford, 1980, Sabins et al., 1980]. These features affect

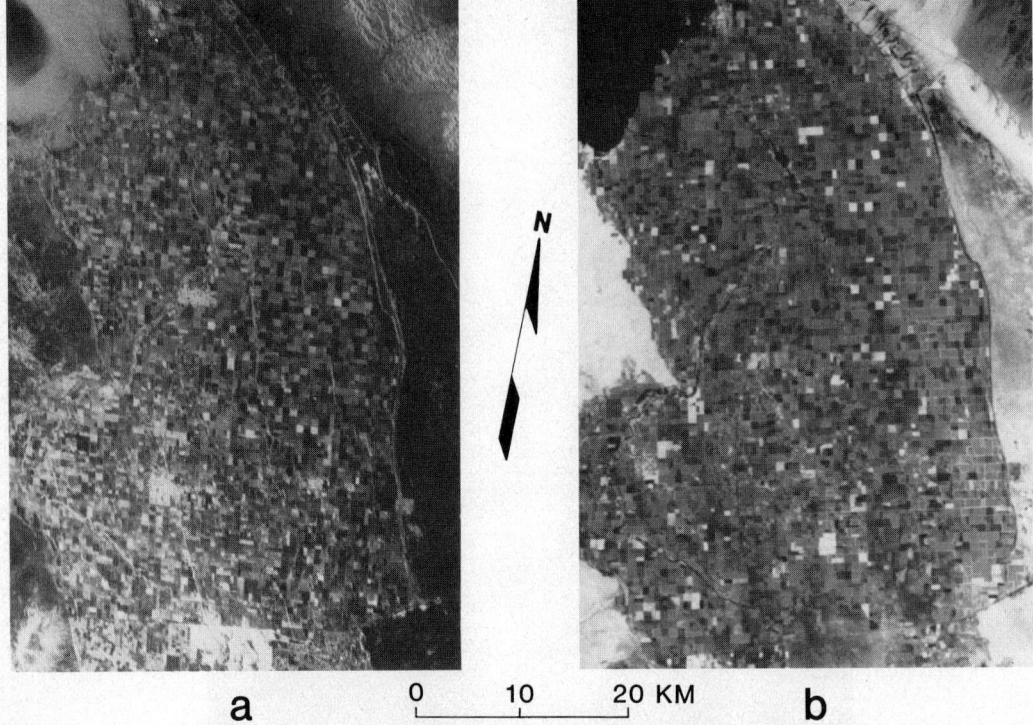

a   0   10   20 KM   b

Fig. 13-16. Radar and Landsat image of the Imperial Valley in southern California. The checkerboard pattern is the cultivation fields. At the time these images were obtained (July 1978), the cultivation in the fields was mainly alfalfa and cotton. Many fields were fallow. The Salton Sea is in the upper left corner of the images.

the radar scattering and are observed on the radar images as a tonal or textural change. In Figure 13-14, the San Andreas and Garlock faults are clearly visible as a change in the tone and the texture. Such changes are due to the sharp topographic change across these faults.

Topographic structural features such as domes, anticlines, synclines, cinder cones and folds are observed because of the high sensitivity of the radar return to the slope change. Figure 13-18 shows folded terrain near Harrisburg, Pennsylvania. The pattern of "noses" formed by the mountains and valleys represents plunging anticlinal and synclinal structures. Regional drainage crosses the structural strikes forming gaps as is the case with the Susquehanna and Juniata rivers. The heavily forested mountain slopes exhibit fine smooth texture, while the valleys have a wide range of tones and textures due to the urban areas and fields.

Canyons, such as the Grand Canyon in Arizona, represent the extreme in topographic variation. In this case, shadowing plays an important role in identifying surface features [Elachi and Farr, 1980]. Figure 13-19 shows an image of the Grand Canyon. The hard rock formations are observed as dark bands in the canyon wall because of the shadow due to their vertical profile.

Hydrologic Applications

Figure 13-20 is a radar image of the Clearfield, Pennsylvania region. The drainage patterns in that

region are observed because of the local topographic change at the edge of the river segments. The radar image can be used to map the drainage network and determine its geometric characteristics. The classification of the different drainage patterns observed in a radar image is identical to what is done in photo interpretation [Ray, 1960]. The drainage network provides information about the hydrologic characteristics of the basin and the nature of the underlying rocks. Dense drainage patterns usually occur in relatively soft rocks. Drainage pattern density decreases over harder rocks. In Figure 13-20, the transition between the regions of high and low drainage pattern density corresponds to the contact region between the Catskill Formation (relatively soft red-to-brown shale and sandstone) and the Mississipian Pocono Group (relatively resistant conglomerate and sandstone with some shale).

Crop Discrimination

Figure 13-16 shows radar and Landsat images of the Imperial Valley, California. The images were acquired almost simultaneously. The agricultural field patterns are clearly depicted in the radar image. The different fields have a wide range of tones. The relationship between the image tone and the field characteristics (crop type and stage, soil moisture, row direction, vegetation cover density, etc.) is not well known. Much research is still necessary to determine to what extent imaging radars can, by themselves or in conjunction

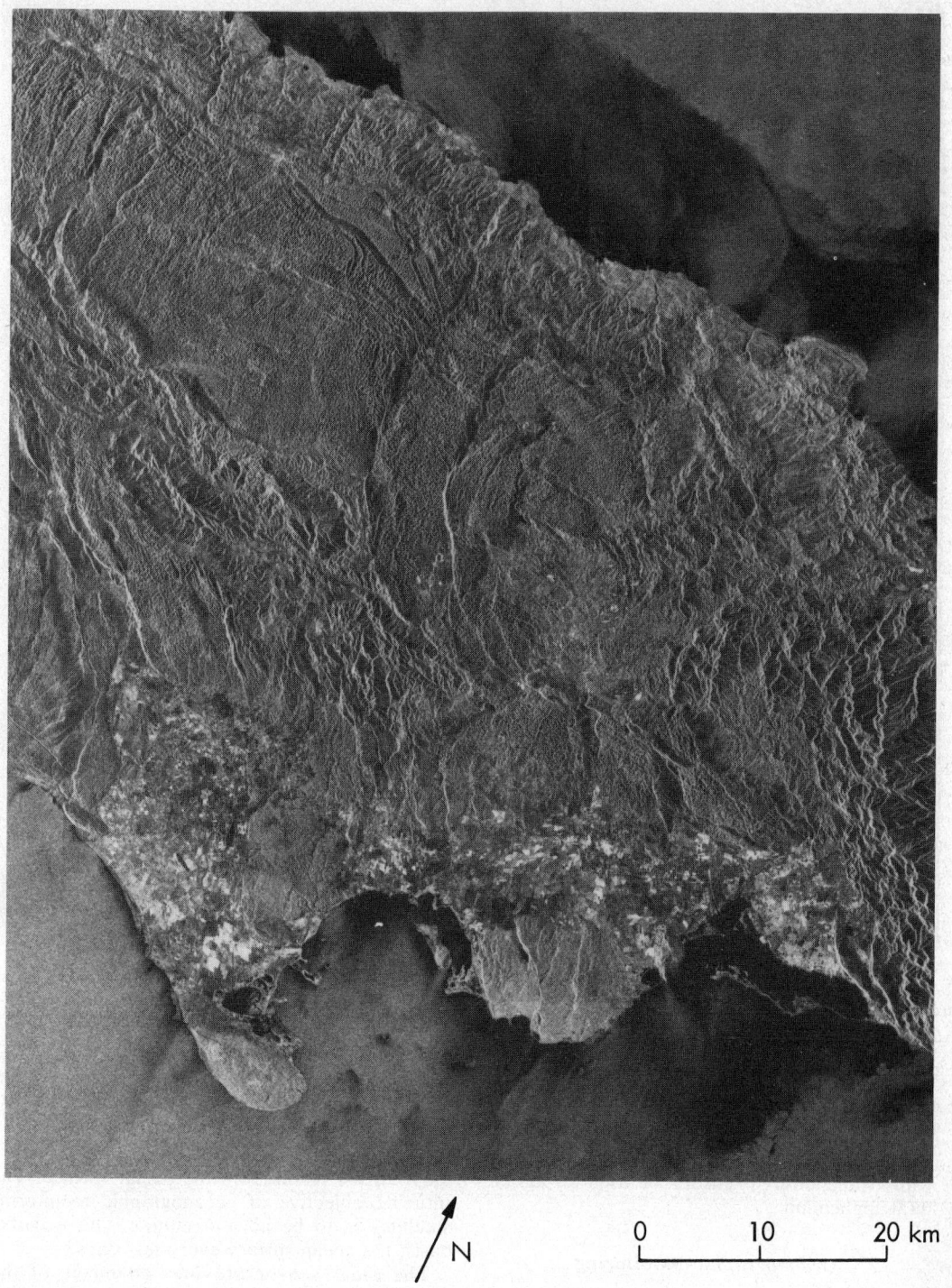

N

0   10   20 km

Fig. 13-17. Radar images of the Island of Jamaica, West Indies. Kingston and the surrounding agricultural areas stand out in the flat alluvial southern region. Much of the island consists of younger forest-covered limestone easily distinguished because of the well-developed karst topography. The eastern region is a Cretaceous igneous and metamorphic complex. Faults are visible in an east-west trend and northwest-to-southeast trend.

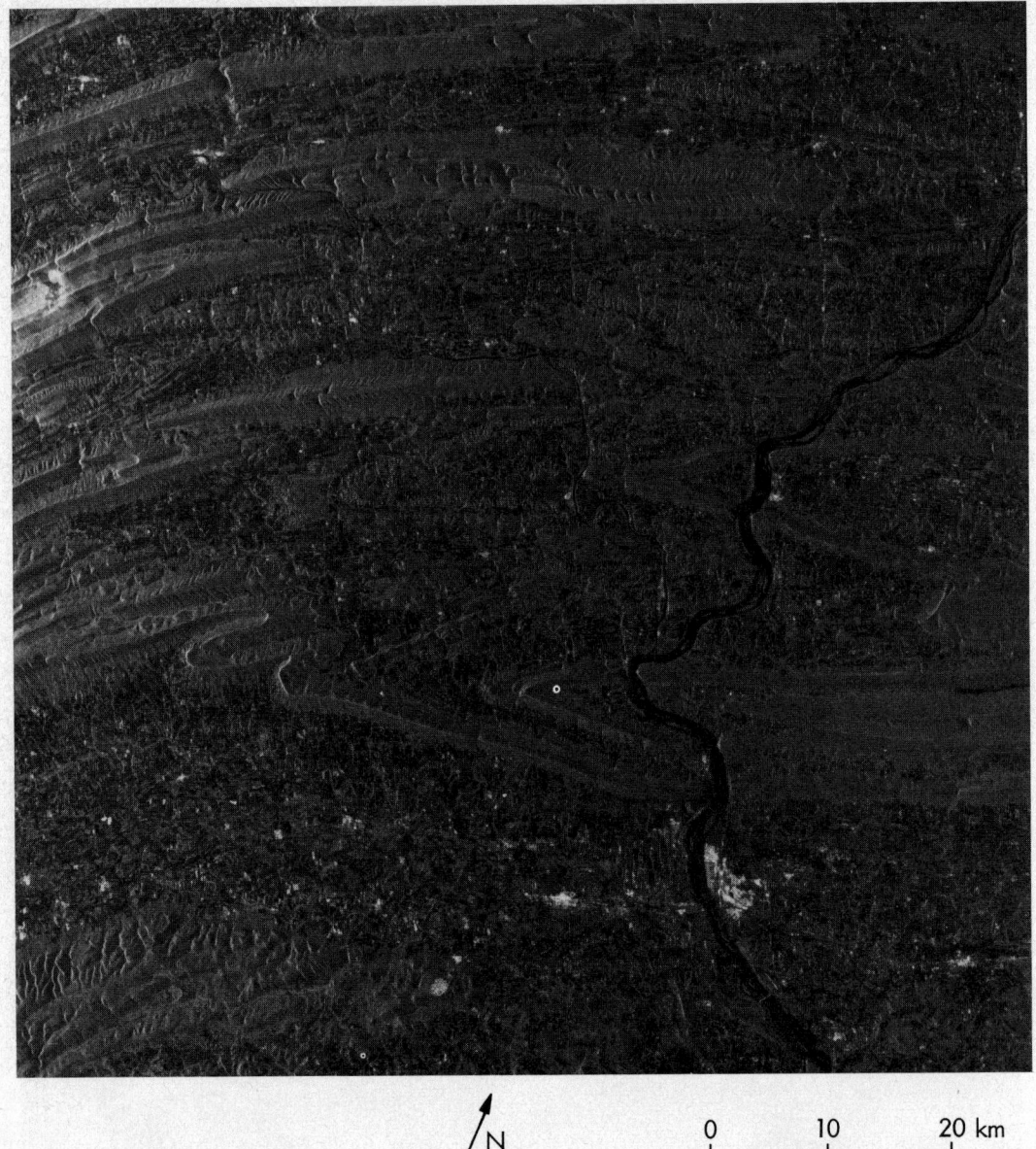

N   0   10   20 km

Fig. 13-18. This image depicts the folded Appalachian Mountains in east-central Pennsylvania. The pattern of "noses" formed by the mountains and valleys on the image represents plunging anticlinal and synclinal structures. Parts of the City of Harrisburg show a very bright tone region on the bank of the Susquehanna River (dark tone).

with visible/infrared sensors, be used for agricultural application.

Oceanographic Monitoring

In oceanographic applications, the imaging radar sensor has a unique and essential characteristic: the capability of obtaining high-resolution surface images independent of the cloud cover and at any time of the day or night. This capability is essential because of the dynamic nature of al-most all the features on the ocean surface. The ultimate objective of oceanographic monitoring satellites is to be able to monitor, on a global basis, the ocean surface every few days.

The radar sensor provides an image of the ocean surface backscatter characteristics. The backscatter is controlled by the small-scale surface topography (short gravity and capillary waves which scatter the radar energy by the Bragg scattering mechanisms) and the local surface tilt, which is due to the presence of large surface

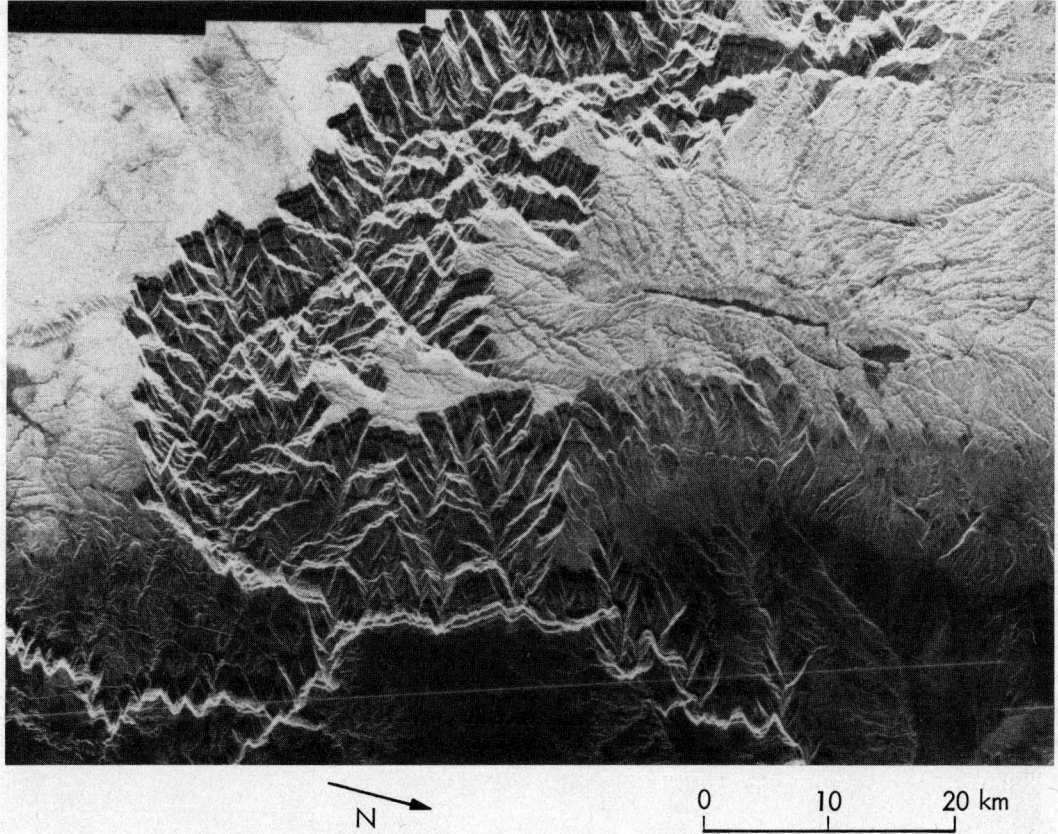

N →   0   10   20 km

Fig. 13-19. Radar image of the Grand Canyon region. Grand Canyon Village is at upper left. The dark strips in the canyon wall correspond to vertical cliffs that are in shadow.

waves or swells. The radar imaging mechanism is also dependent on the surface motion due to swells or currents. This motion leads to a Doppler shift which is reflected in the image formation [Elachi and Brown, 1977; Alpers and Rufenach, 1979; Valenzuela, 1979]. Thus, the SAR is capable of imaging surface and near-surface phenomena that affect the surface roughness directly or indirectly. These phenomena include surface waves, internal waves, currents, weather fronts, subsurface topography in shallow water, eddies, etc.

Figure 13-21 shows a radar image of ocean waves near Shetland Island, England. The waves are visible as a periodic change in the image tone. Wave refraction and diffraction by the islands are clearly visible, particularly around the Foula Island and the southern tip of Shetland Island. Such an image can be used to determine the wavelength and direction of the wave pattern in the region.

Figure 13-22 shows internal wave patterns which are observed because of their effect on the surface small scale roughness [Elachi and Apel, 1976]. The image is in the Gulf of California. The internal waves are observed as packets of convex strips with spacing becoming shorter toward the center of curvature.

Figure 13-23 shows the radar image of the Nantucket Island region. Most of the curvilinear features and periodic strips reflect closely the bottom topography in the shallow waters of that region. It is thought that the bottom topography has a direct effect on the surface and near surface current, which in turn affects the surface roughness. The radar, in turn, provides an image of the variation of the surface roughness. Therefore, the image is an indirect reflection of the bottom topography.

Polar Ice Mapping

Figure 13-24 shows two Seasat radar images taken three days apart of the same area in the Beaufort Sea. Ice floes are clearly visible on both images. Appreciable change and displacement did occur over the period of three days. Some of the ice floes moved in excess of 30 km. Others broke up into smaller floes. The streaking observed in the October 1 image is probably a result of surface wind.

Scatterometers

Scatterometers have been flown on two space platforms: Skylab [Moore and Fung, 1973] and

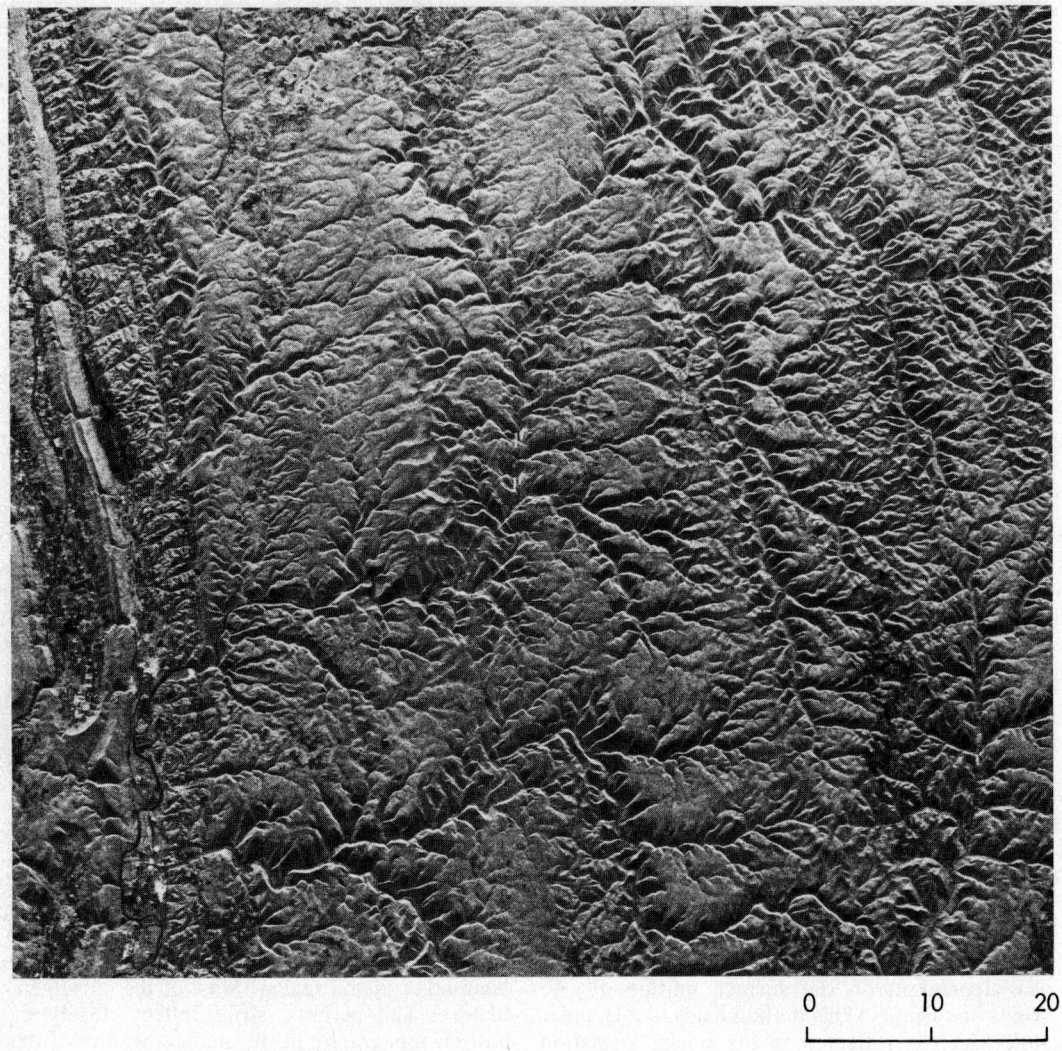

0   10   20

Fig. 13-20. Drainage patterns in the Lock Haven region, Pennsylvania. The drainage density is dense in the relatively soft rock region (right) and light in the relatively hard rock region (left).

Seasat-A [Grentham et al., 1977; Jones et al., 1979; Johnson et al., 1980; Bracalante et al., 1980]. The Seasat-A Satellite Scatterometer (SASS) will be discussed here in some detail.

A scatterometer provides the backscatter cross-section as a function of incidence angle for the area under observation. In the case of the SASS, the main interest was in measuring the ocean surface backscatter as a means to derive the surface wind vector. The physical basis for this technique is that the strength of the radar backscatter is proportional to the amplitude of the surface capillary waves (Bragg scattering), which in turn is related to the wind speed near the surface. Moreover, the radar backscatter is anisotropic, allowing the wind direction to be derived from backscatter measurements at different azimuth angles. A historical perspective of the development in using scatterometers for ocean

surface wind measurement is given by Barrick and Swift (1980).

The SASS Sensor Description

The SASS sensor operated at a frequency of 14.6 GHz (i.e., wavelength of 2 cm). It incorporated four dual-polarized fan beam antennas which produced an X-shaped pattern of illumination on the surface (Figure 13-25). The fixed fan beam antenna design was combined with electronic Doppler filtering to achieve the desired swath width and resolution cell size. Twelve Doppler filters were used to subdivide the antenna footprint electronically into resolution cells approximately 50 km on the side. The total swath covered was 750 km, with the incidence angle ranging from 25° to 65° from vertical. Three additional Doppler cells provided measurement near

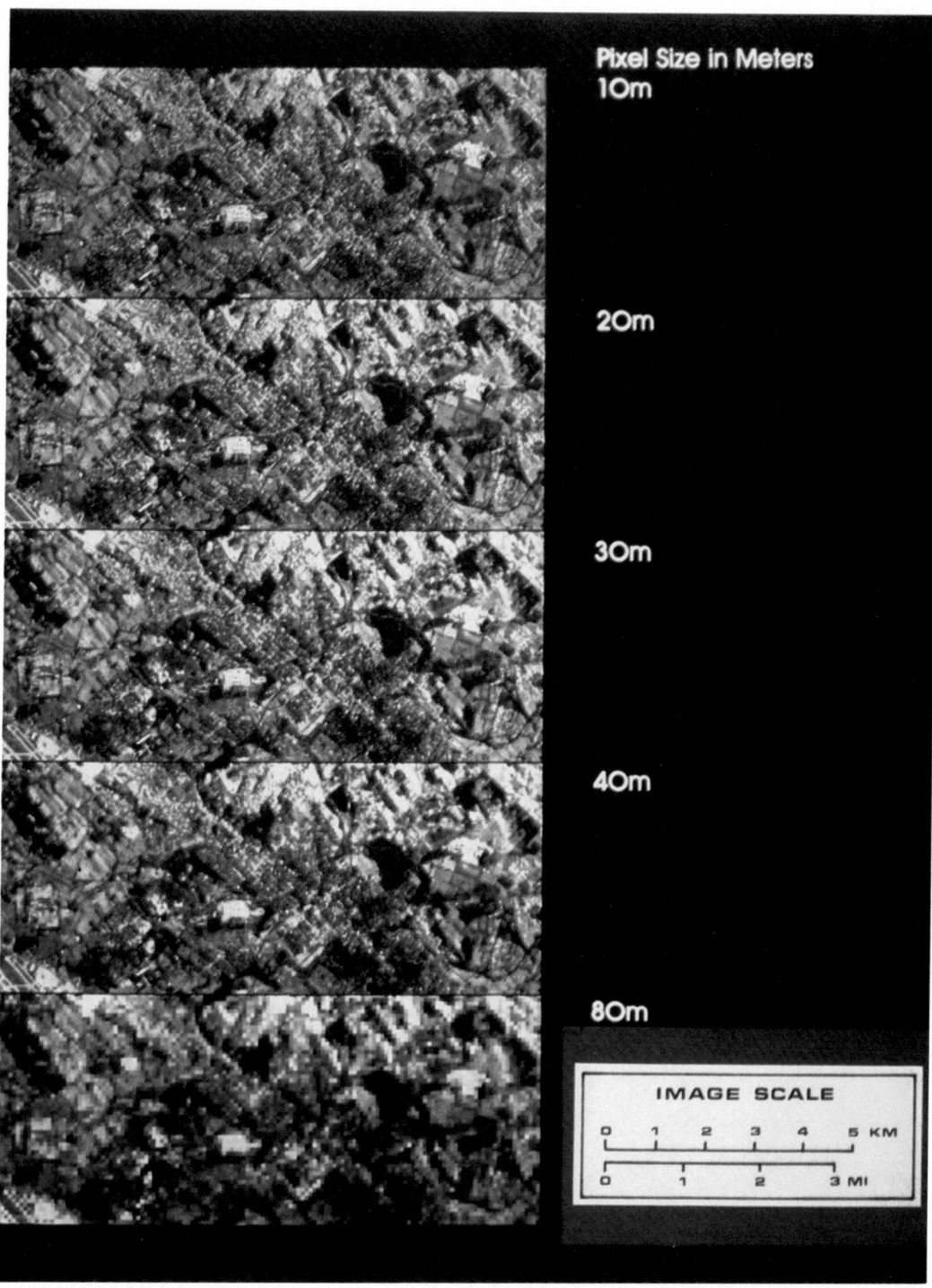

Fig. 1-7. Simulation of varying spatial resolutions for an urban test site at Marysville, Tennessee. By using the airport runway located in the lower left corner in each of the images, one can see how changes in spatial resolution can affect interpretability.

Fig. 1-12a,b. Two presentations of a Landsat image, August, 1973, of Swakopmund and the Namib Desert. Fig. 1-12a is a three-band eigenimage constructed from combinations of the original four spectral bands of Landsat. Greater detail can be seen in the off-shore area where sediments were carried by the current near Walvis Bay. Inland geologic features, such as marble beds and sand flats, are also easy to plot. (courtesy Charles Sheffield, Earth Satellite Corporation, and Sidgwick and Jackson Limited 1981)

Fig. 1-12b. Comparison with figure on facing page shows lack of much relevant detail on this figure. (Color Figure 1-11 has been deferred to next page in order to facilitate comparison of Figures 1-12a and 1-12b).

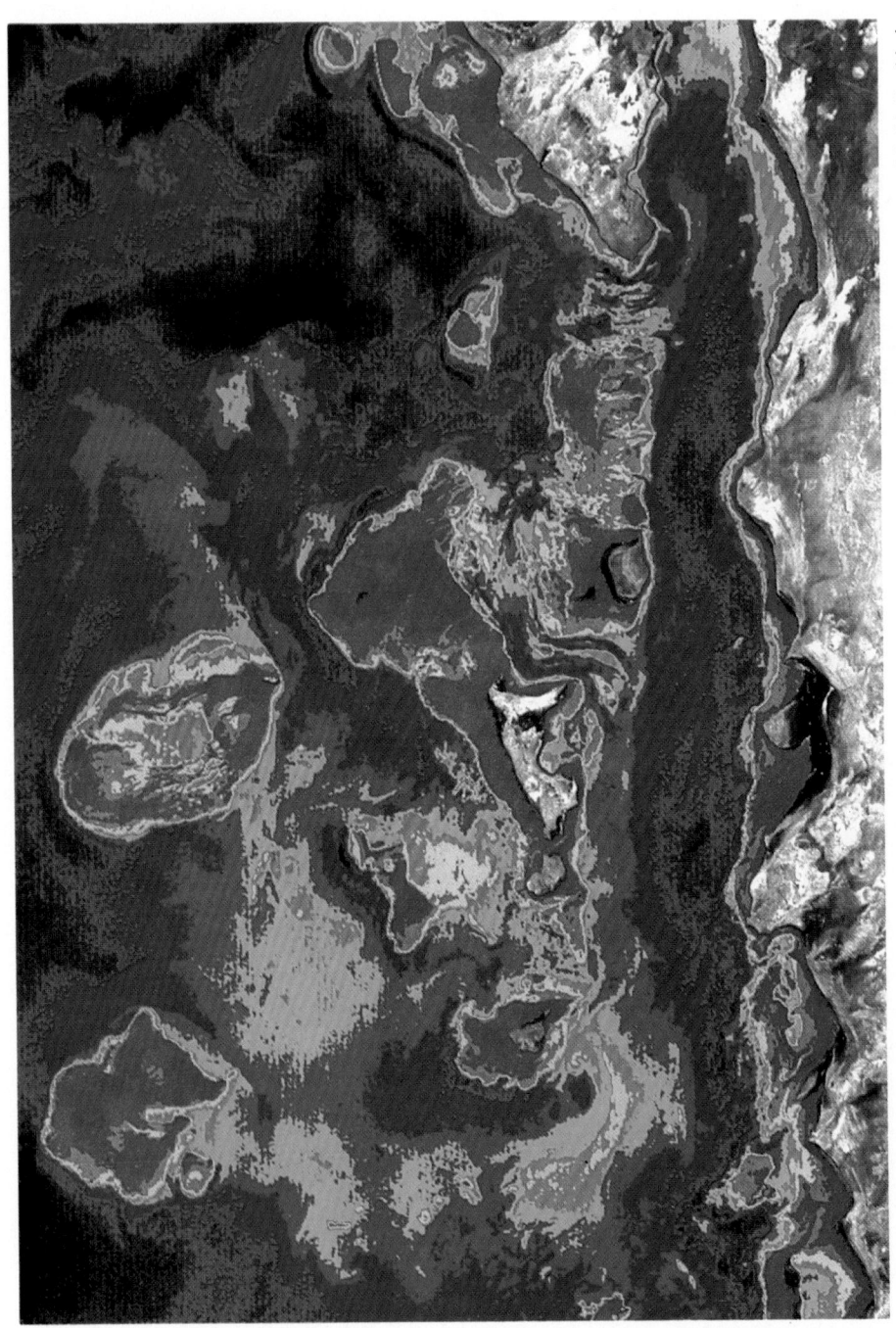

Fig. 1-11. Landsat image, October 1978, of shoreline and off-shore area of the United Arab Emirates, about 30 miles west of Abu Dhabi. In this image water has been selectively enhanced to show variation in depth, while areas of land remain in standard color. (courtesy Charles Sheffield, Earth Satellite Corporation, and Sidgwick and Jackson Limited, 1981)

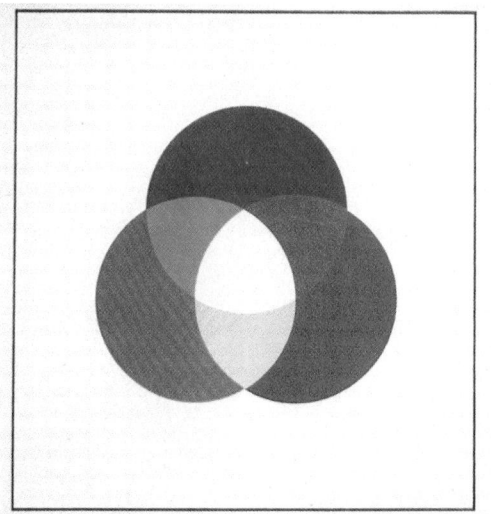

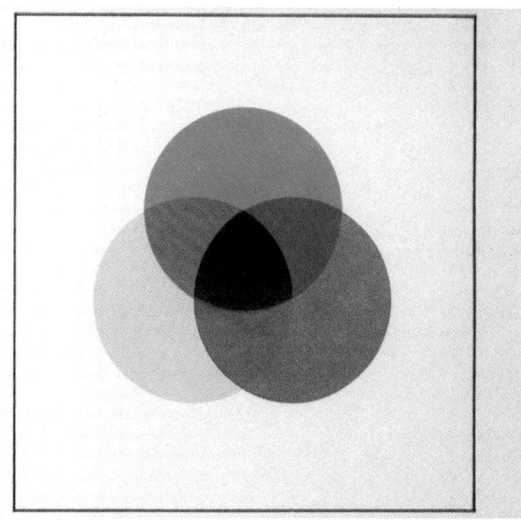

Fig. 6-40. Production of color by the additive process. In this case, the color circles are as seen when projected onto a colorless screen.

Fig. 6-41. Production of color by the subtractive process. The circles represent colored filters viewed in transmission.

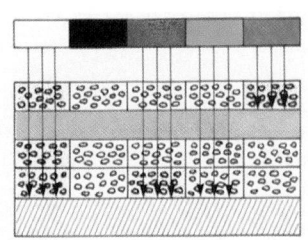

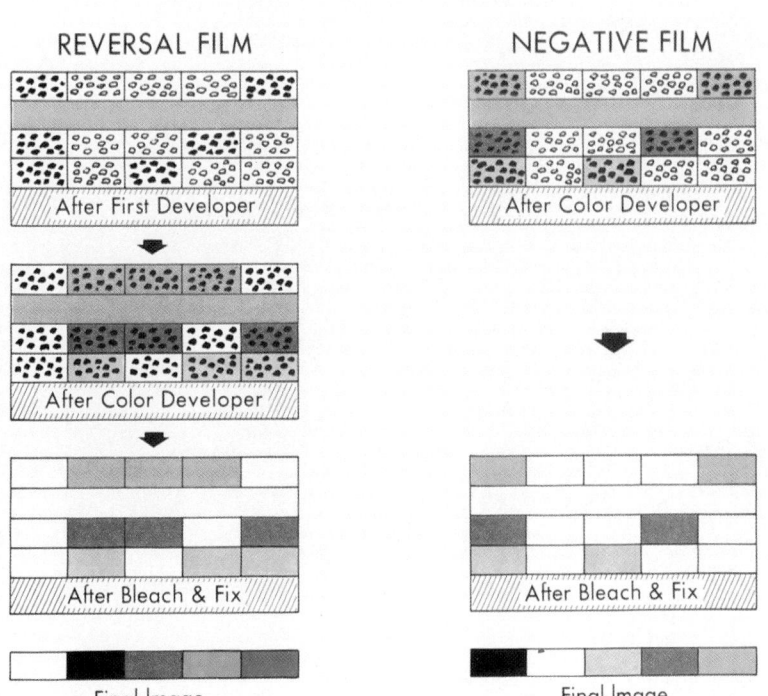

Fig. 6-43. Reproduction of colors by a reversal color film and a negative color film.

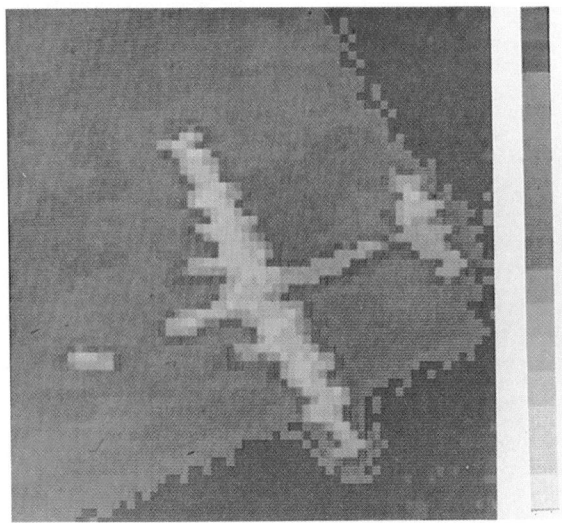

Fig. 11-38. Image of a parked twin-engine aircraft on a concrete platform. The image area of 45 m × 45 m was scanned within 1 s.

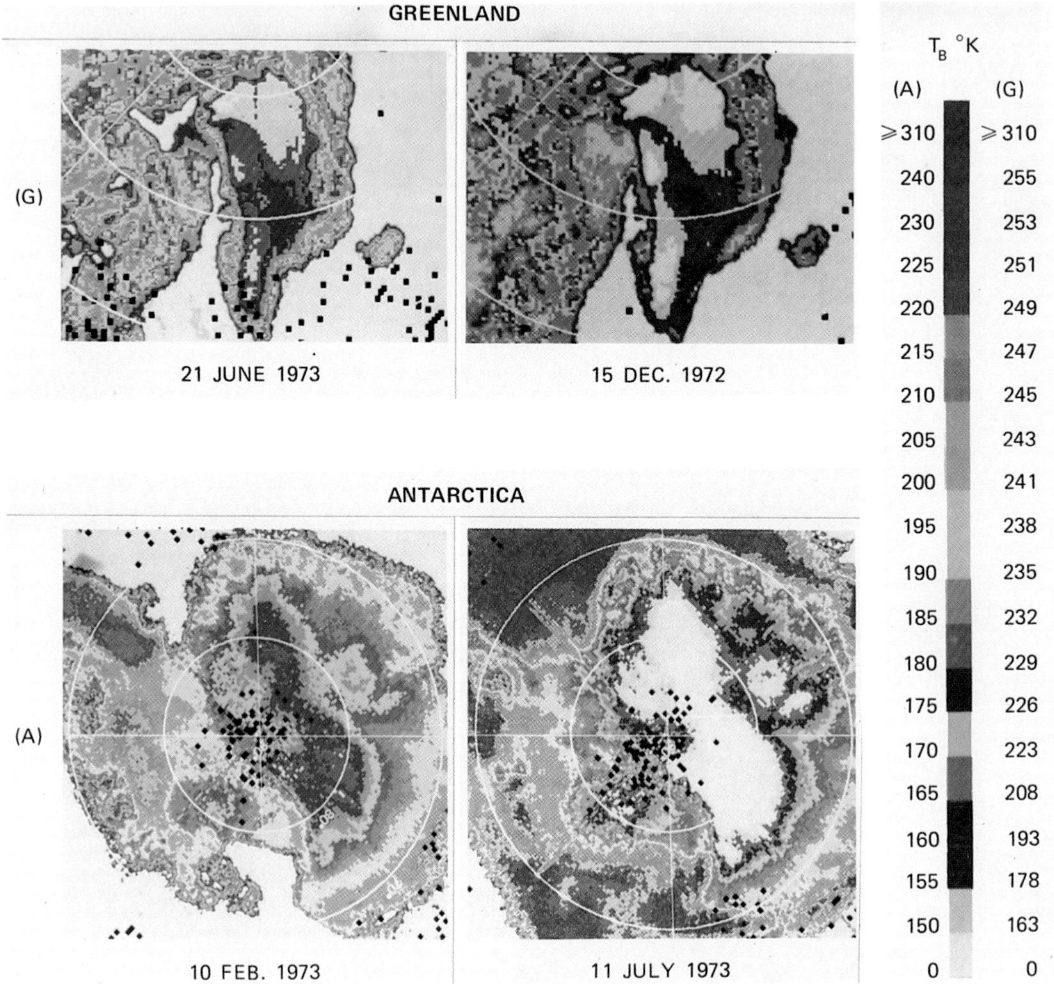

GREENLAND

21 JUNE 1973

15 DEC. 1972

T_B °K

(A)	(G)
≥310	≥310
240	255
230	253
225	251
220	249
215	247
210	245
205	243
200	241
195	238
190	235
185	232
180	229
175	226
170	223
165	208
160	193
155	178
150	163
0	0

ANTARCTICA

(A)

10 FEB. 1973

11 JULY 1973

Fig. 11-39. Seasonal variations of the radiometric temperatures over the continental ice sheets in Greenland and Antarctica (courtesy of NASA/GSFC).

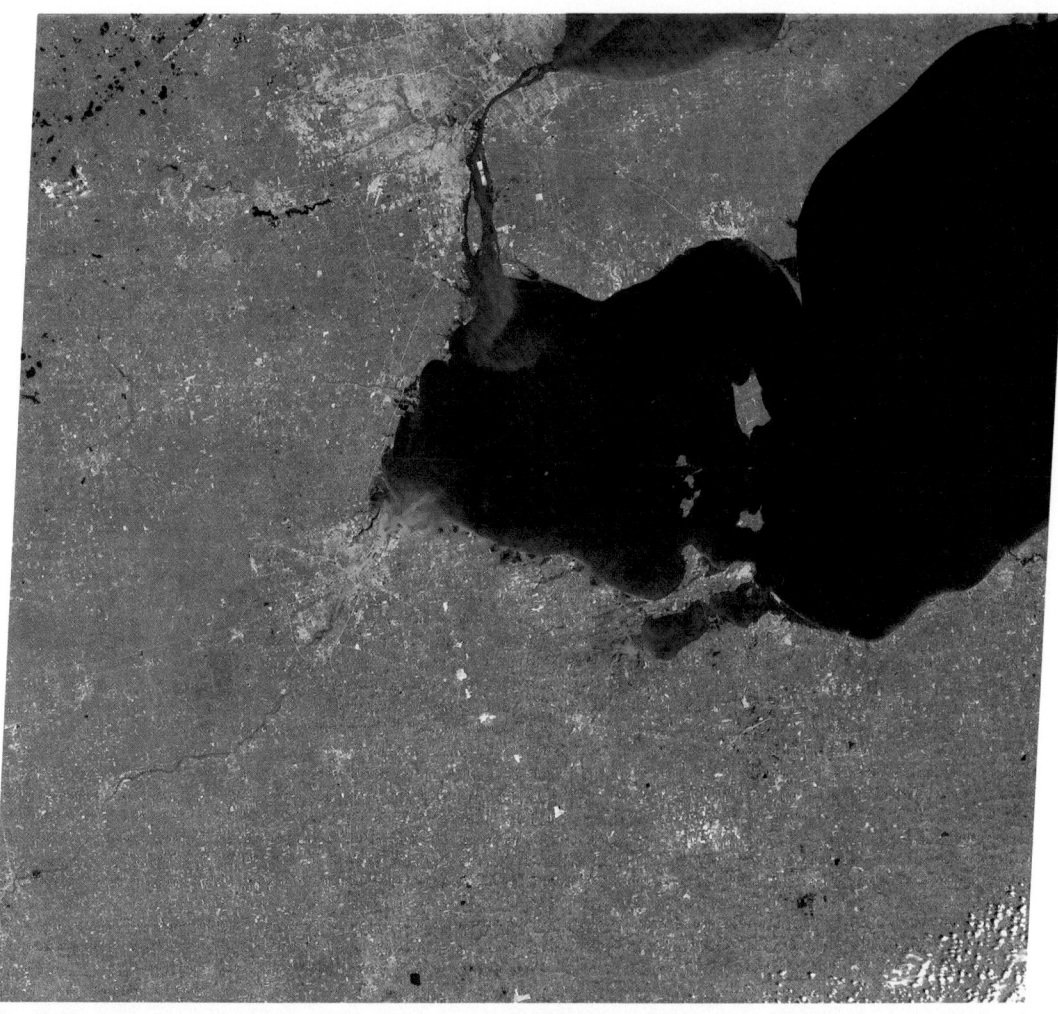

Fig. 12-60. Early Landsat-4 Thematic Mapper scene of the Detroit and Toledo areas in false color using TM bands 2, 3 and 4. (Color Figures 12-24, 12-25, 12-26 and 12-27 have been deferred to facilitate comparisons.)

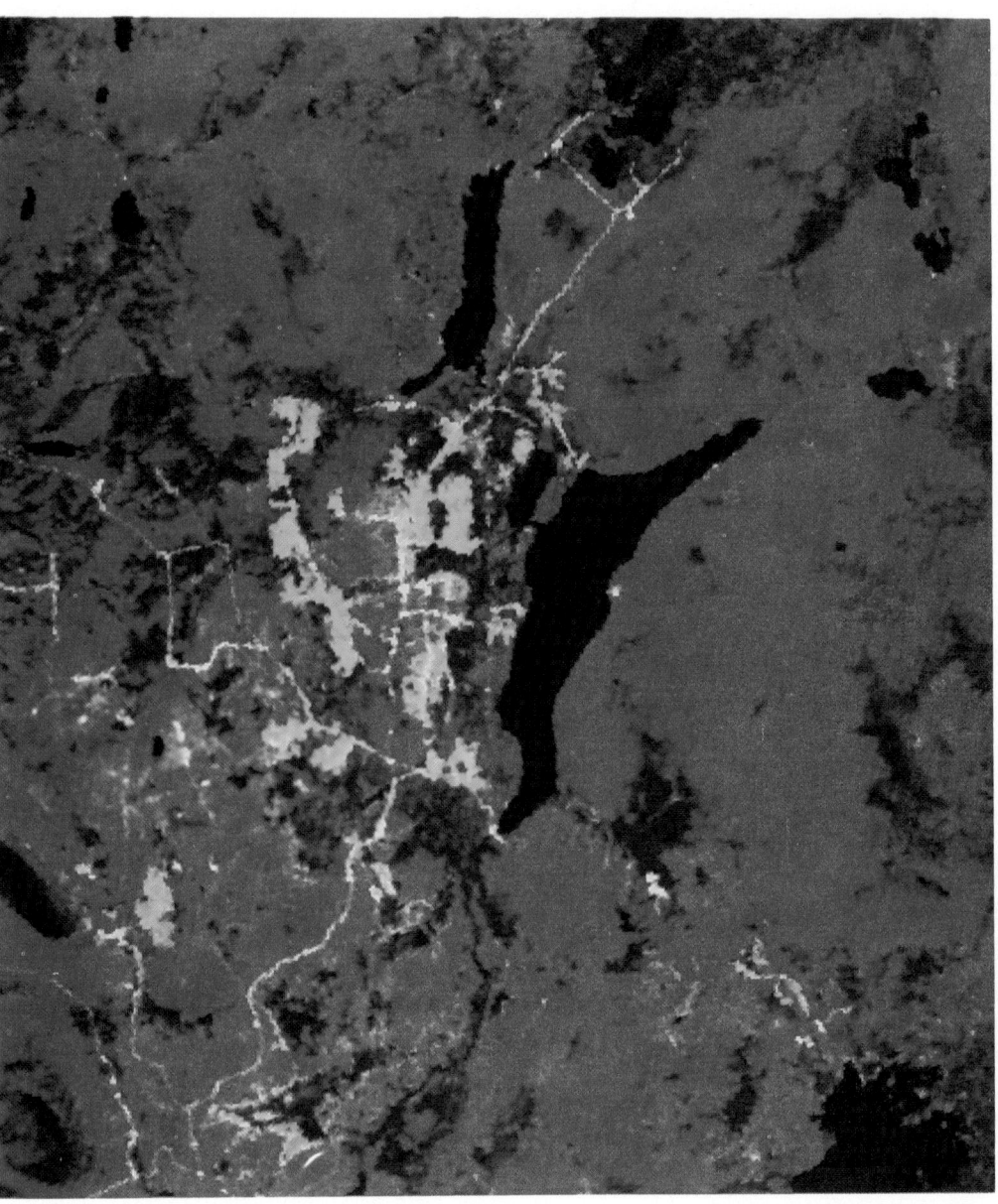

Fig. 12-24. Landsat MSS false color composite of a forestation study area near Moosehead Lake in Maine, U.S.A. (courtesy of M. Imhoff, ERRSAC, NASA/GSFC)

Fig. 12-25. This is same area as Figure 12-24 with the Landsat RBV data merged with the MSS data. The increased sharpness of detail can be noted in the road system and classified areas. (Courtesy of M. Imhoff, ERRSAC, NASA/GSFC)

Fig. 12-26. Landsat MSS standard false color composite of an area of the Anti-Atlas Mountains in Morocco. (Courtesy of NASA/GSFC)

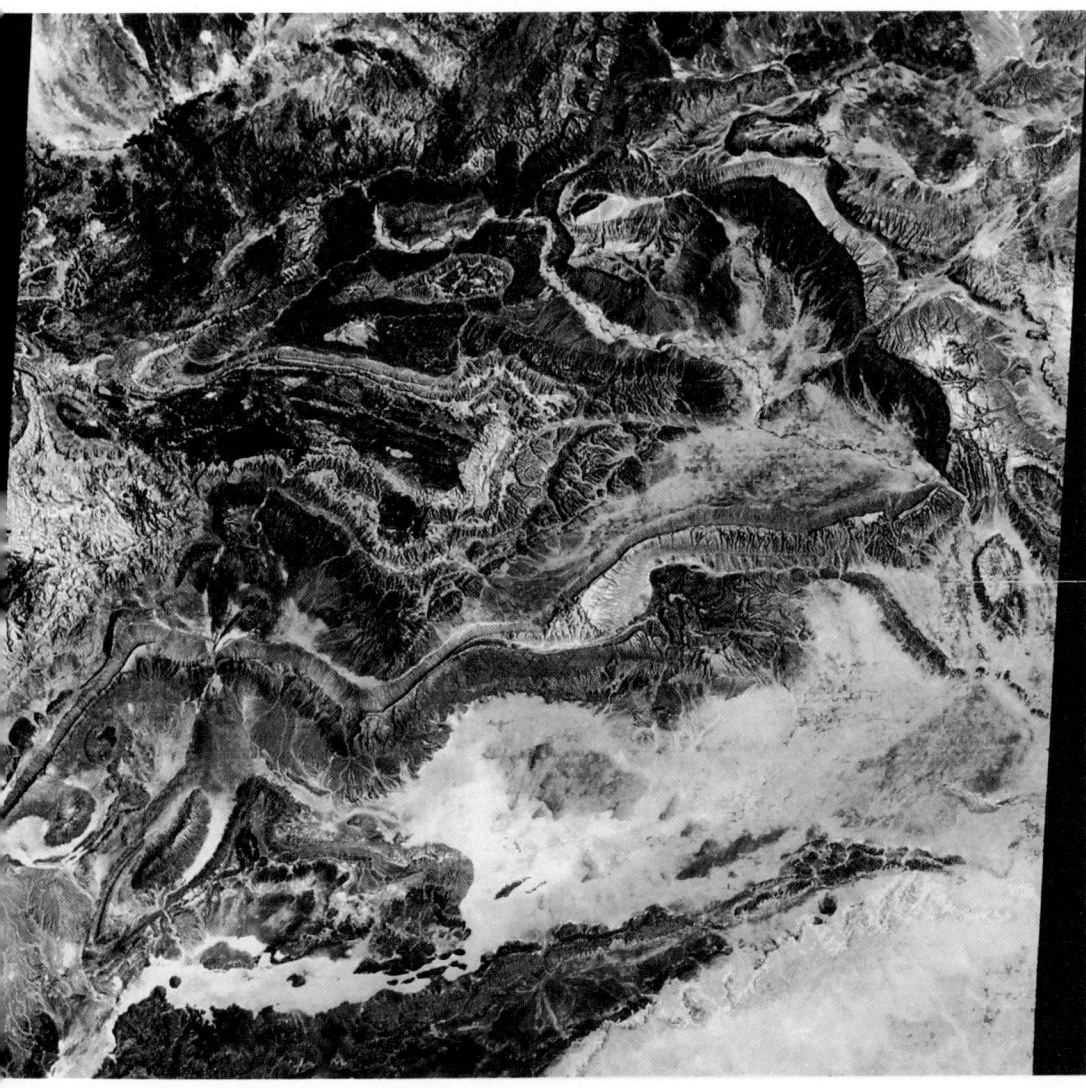

Fig. 12-27. Landsat MSS data have been merged with thermal infrared data from the Heat Capacity Mapping Mission satellite acquired at night. This shows an enhancement of mineral signature data in the same region in Figure 12-26. (Courtesy of NASA/GSFC)

Fig. 12-68. Landsat-4 Thematic Mapper false color composite imagery of the Dyersberg, Tennessee quadrangle. This scene includes a portion of the Mississippi Valley near the area of the Tennessee, Missouri, Arkansas and Kentucky borders. This imagery was produced by using bands 2, 3 and 4 with blue, green and red filters respectively. This composite uses transparencies of figures 12-62, 12-63 and 12-64 as the successively exposed images.

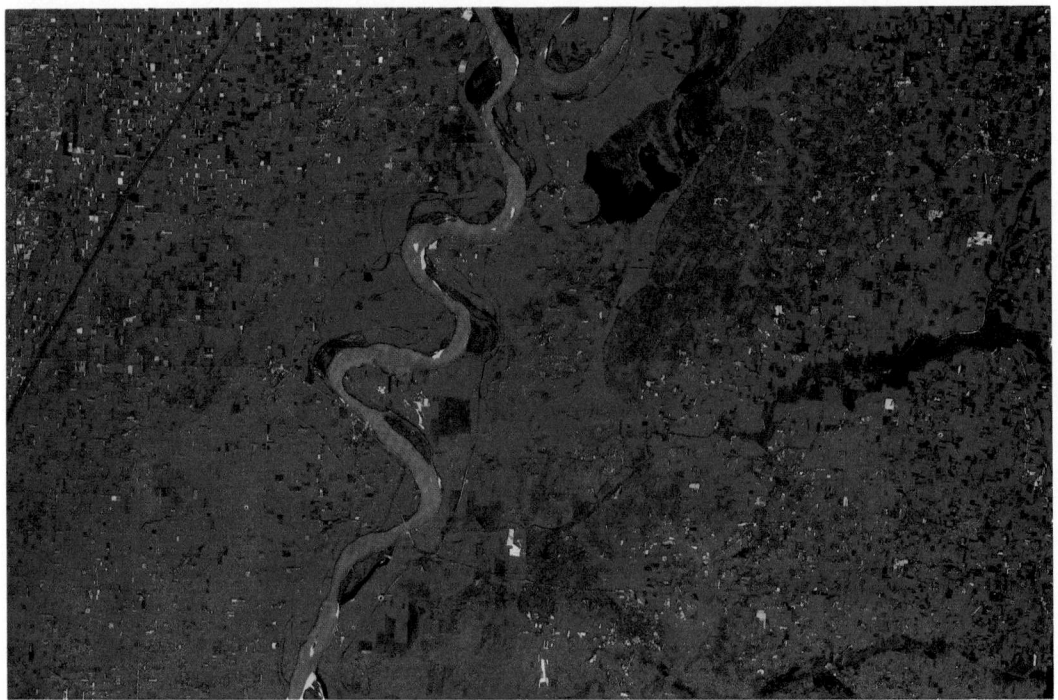

Fig. 12-69. Landsat-4 Thematic Mapper false color composite imagery of the area shown in Figure 12-68. This imagery was produced by using bands 1, 3 and 4 with blue, green and red filters, respectively. In this case the composite uses transparencies of figures 12-61, 12-63 and 12-64.

Fig. 12-70. Landsat-4 Thematic Mapper "true" color composite imagery of the area shown in Figure 12-68. This imagery was produced by using bands 1, 2 and 3 with filters that directly correspond to the regions of color sensitivity of their respective detectors—i.e. by using a blue filter for band 1, a green filter for band 2, and a red filter for band 3, in conjunction with the transparencies of figures 12-61, 12-62 and 12-63.

Fig. 12-71. Landsat-4 Thematic Mapper false color composite imagery of the area shown in Figure 12-68. This imagery was produced by using band 1 with a blue filter, band 3 with a green filter and band 5 with a red filter. Transparencies of figures 12-61, 12-63 and 12-65 were used.

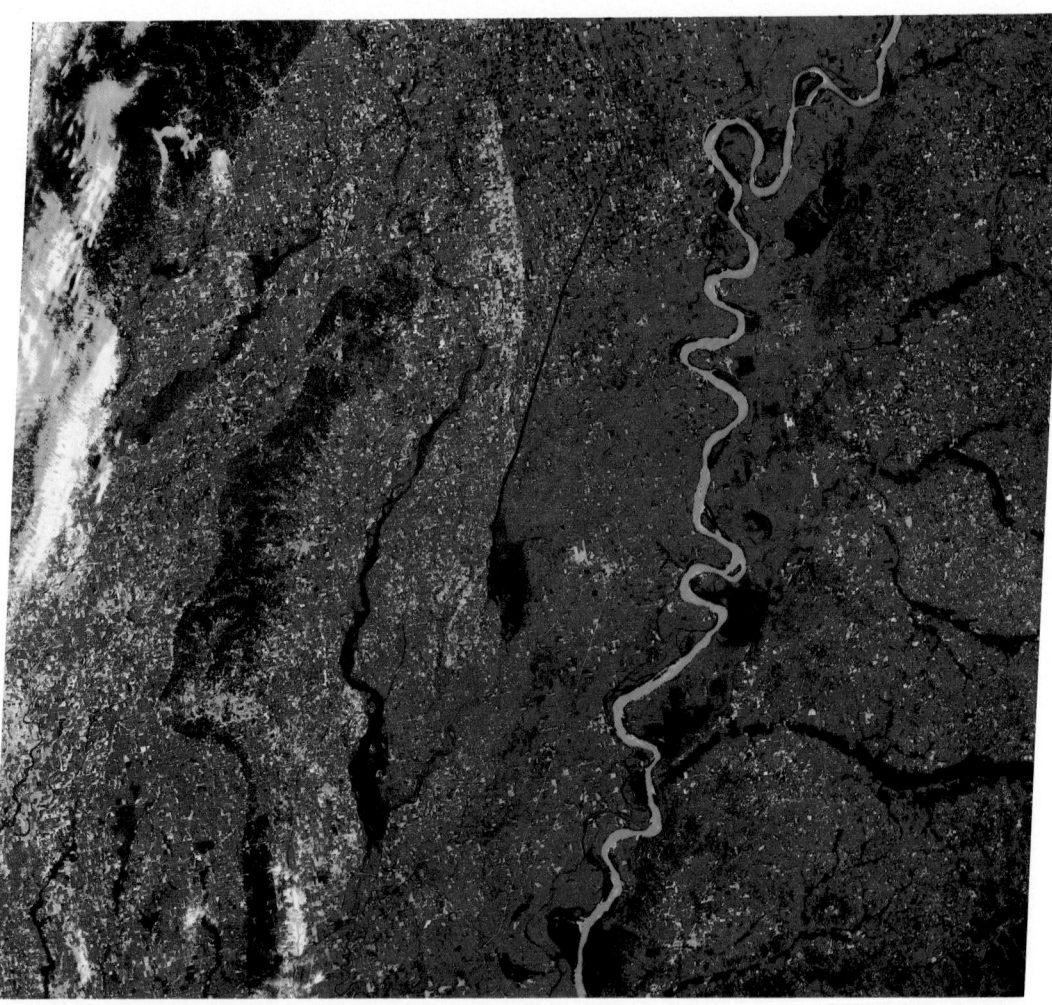

Fig. 12-72. Landsat-4 Thematic Mapper full scene (185 Km × 185 Km) of which figures 12-61 through 12-71 are subsets. This image is in conventional false color—using bands 2, 3 and 4. The individual bands were linearly stretched, using a nonstandard experimental algorithm.

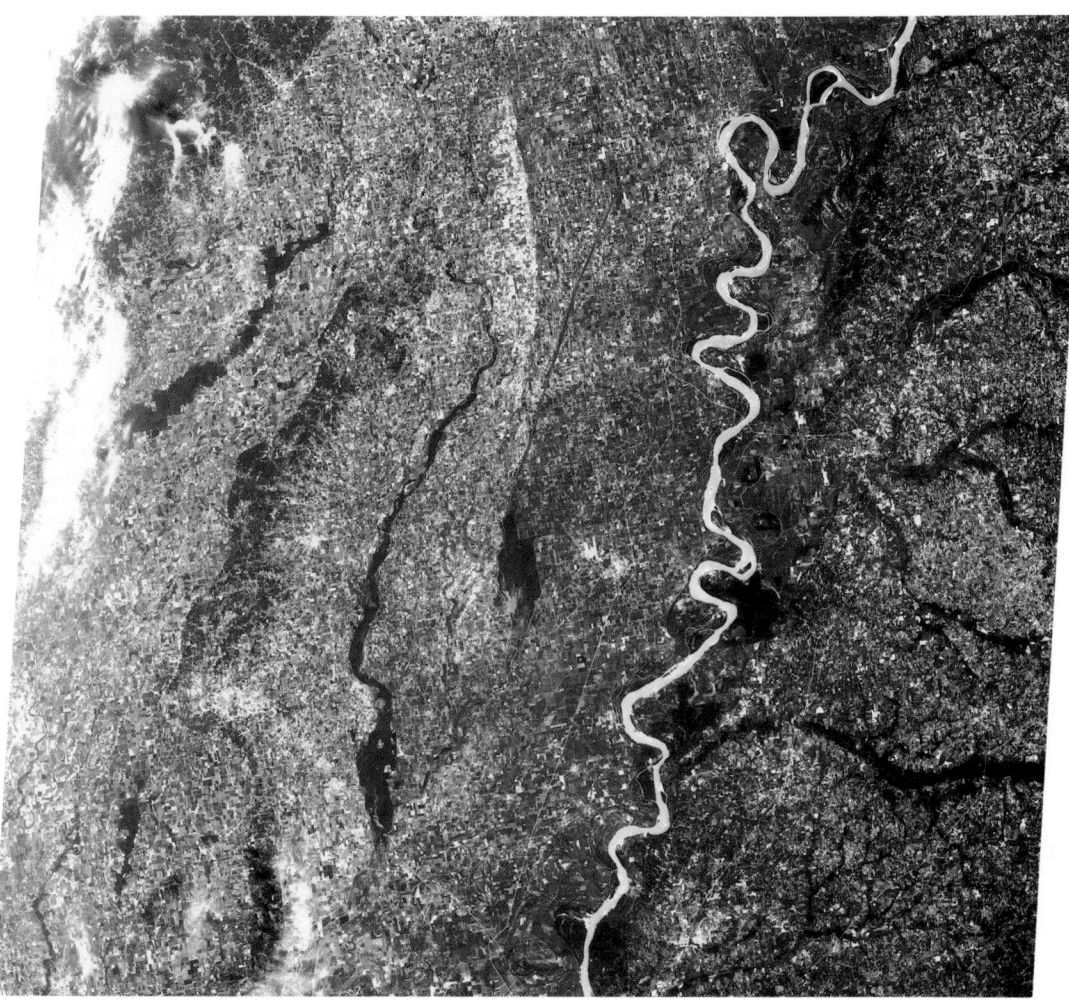

Fig. 12-73. Landsat-4 Thematic Mapper full scene (185 Km × 185 Km) of which figures 12-61 through 12-71 are subsets. This image uses bands 1, 2 and 3 with filters corresponding to the color regions of these bands (blue, green and red, respectively). The linear stretches used tend to distort the color balance; hence the picture does not represent an optically true perspective.

Fig. 12-74. Landsat-4 Thematic Mapper false color scene of the Washington, DC, Baltimore, Chesapeake Bay area, produced by using TM bands 2, 3 and 4, with blue, green and red filters, respectively. This is a full scene (185 Km × 185 Km).

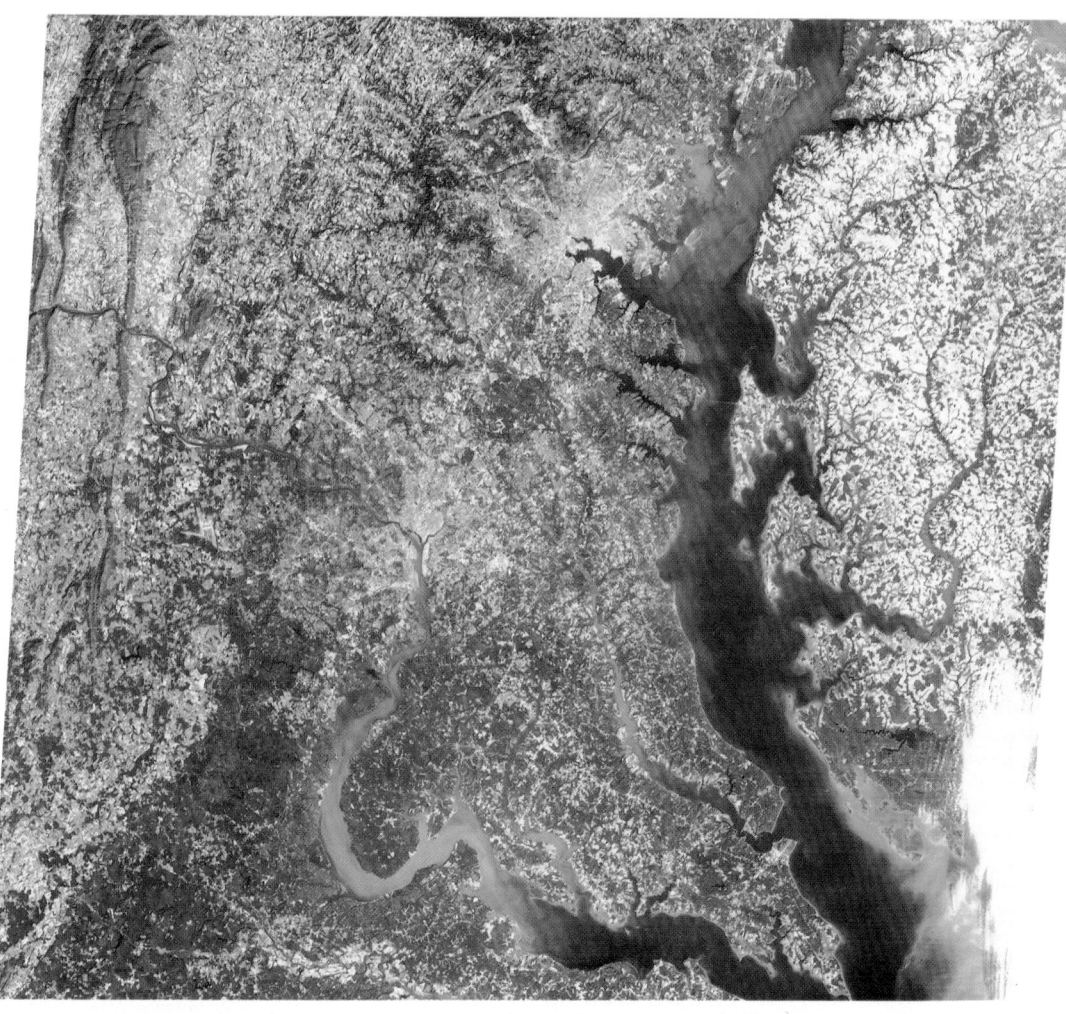

Fig. 12-75. Landsat-4 Thematic Mapper full scene (185 Km × 185 Km) of the area shown in Figure 12-74. This imagery was produced by using bands 1, 2 and 3 with the corresponding filters of blue, green and red, respectively. This is an essentially "true" color image of an autumn day (November 2, 1982) in this area.

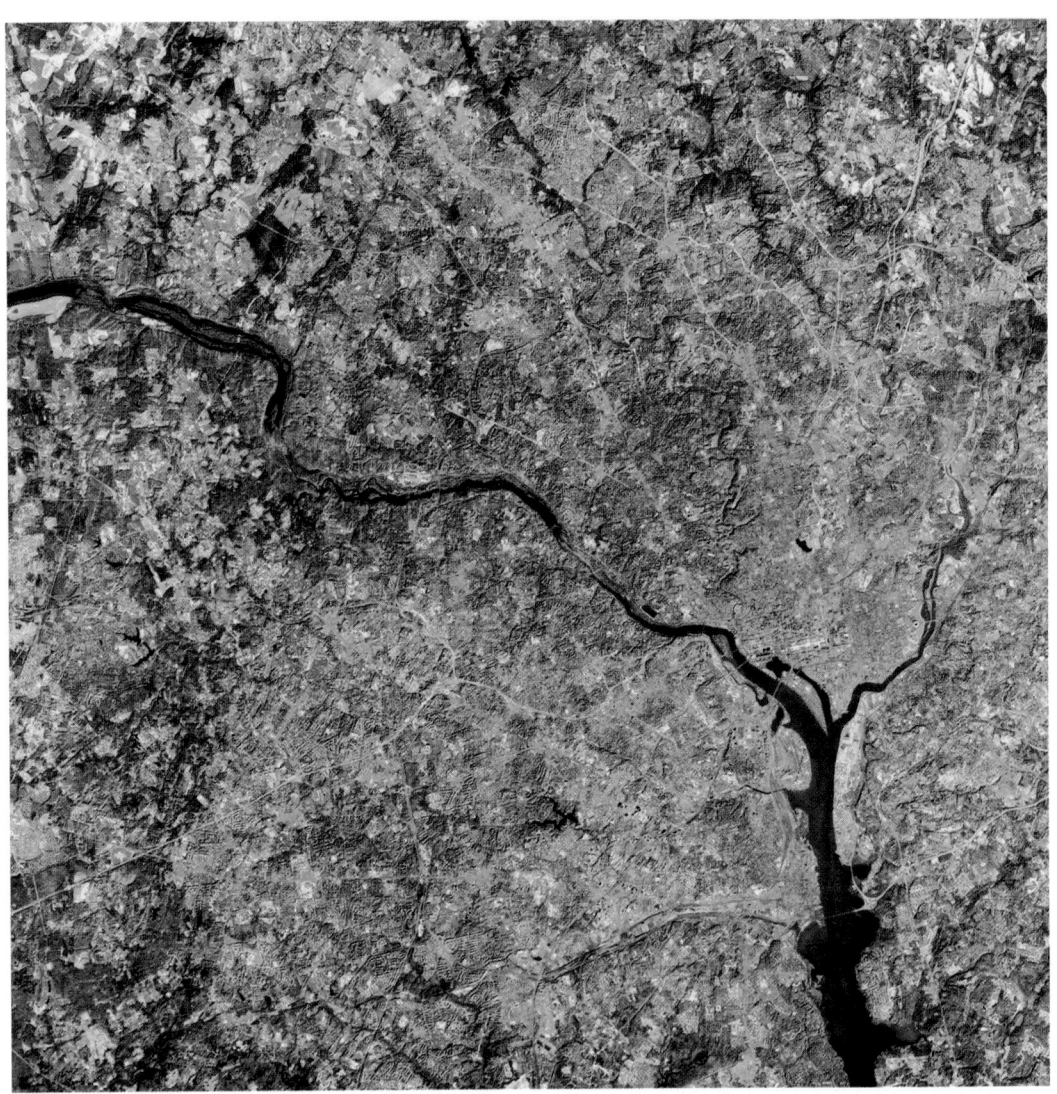

Fig. 12-76. This image is a 1/16 subset of the area of figure 12-74. The color conditions are the same as in figure 12-74.

Fig. 12-77. This image is a 1/16 subset of the area of figure 12-75. The color conditions are the same as in figure 12-75.

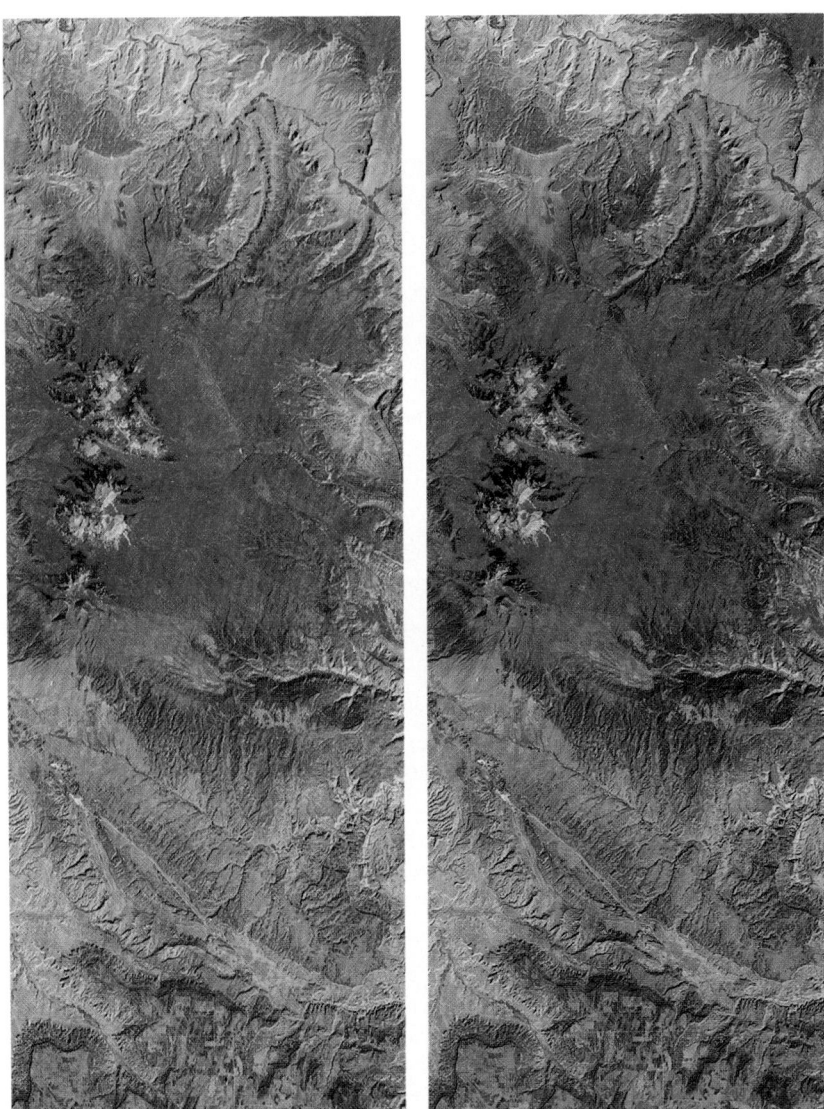

Fig. 12-78. A decorrelated Landsat MSS (4, 5, and 7) image of a portion of the Paradox Basin, Utah, has been merged with Digital Terrain Data to produce this synthetic stereo imagery at a scale of 1:500,000. The MSS data are differentially displaced east or west in the resampled DTD scan format by an amount equal to the elevation times 0.6 (the base-to-height ratio). This product reveals a variety of lithologic, structural, and botanical phenomena not readily apparent on other enhancement products. The Mancos shale appears as light blue over a large area in the north and outcrops throughout the image. Stereo viewing reveals an anticline-syncline-anticline sequence proceeding southward from the LaSal Mountains. The eye-shaped anticlinal structure in the southern third of the image is the Lisbon Valley anticline, notable for the Mancos outcrops along the southern limb and the bleached formations at the northwestern end associated with known underlying hydrocarbon reservoirs. Across the LaSal massif are NE-SW and NW-SE trending fault traces, enhanced by the stereo effect. (Prepared by Earth Satellite Corporation for Japan Petroleum Exploration Company, Ltd. and Earth Resources Satellite Data Analysis Center.)

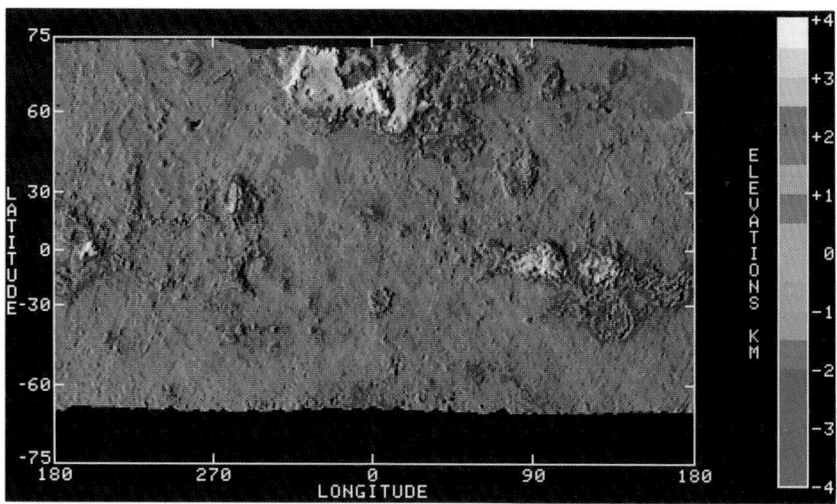

Fig. 13-36. Global relief of Venus in Mercator projection. The colors represent constant levels as shown on the right. Shaded relief enhancement has been superimposed to show three-dimensional effect. Color Figures 13-47a and 13-47b see next page.

Fig. 13-63. Color-ratio composite NASA aircraft image of the Cuprite, Nevada, mining district. The following band ratios and color combinations are used: 1.6/0.48 as green, 1.6/2.2 as red, and 0.6/1.0 as blue. Red and magenta indicate intense absorption centered near 2.2 μm due to the presence of kaolinite and alunite in opalized rocks; green, limonitic rocks; dark blue to black, spectrally flat rhyolite flows and tuffs and highly silicified, goethite-coated rocks; yellowish green, limonite altered rocks with kaolinite.

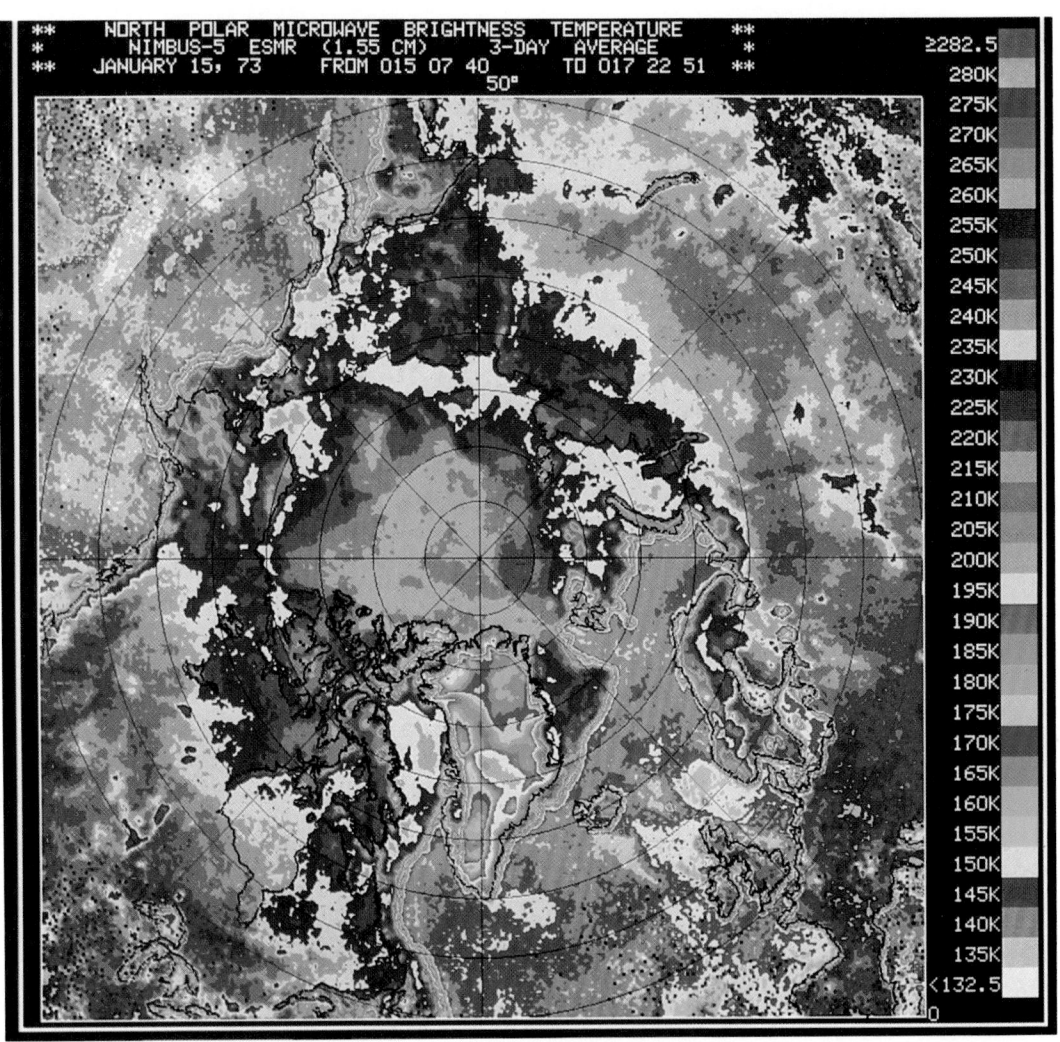

Fig. 13-47. Images of north polar region generated from brightness temperatures measured by the Nimbus 5 ESMR: (a) winter (January 1973), (b) (on next page) later summer (September 1973).

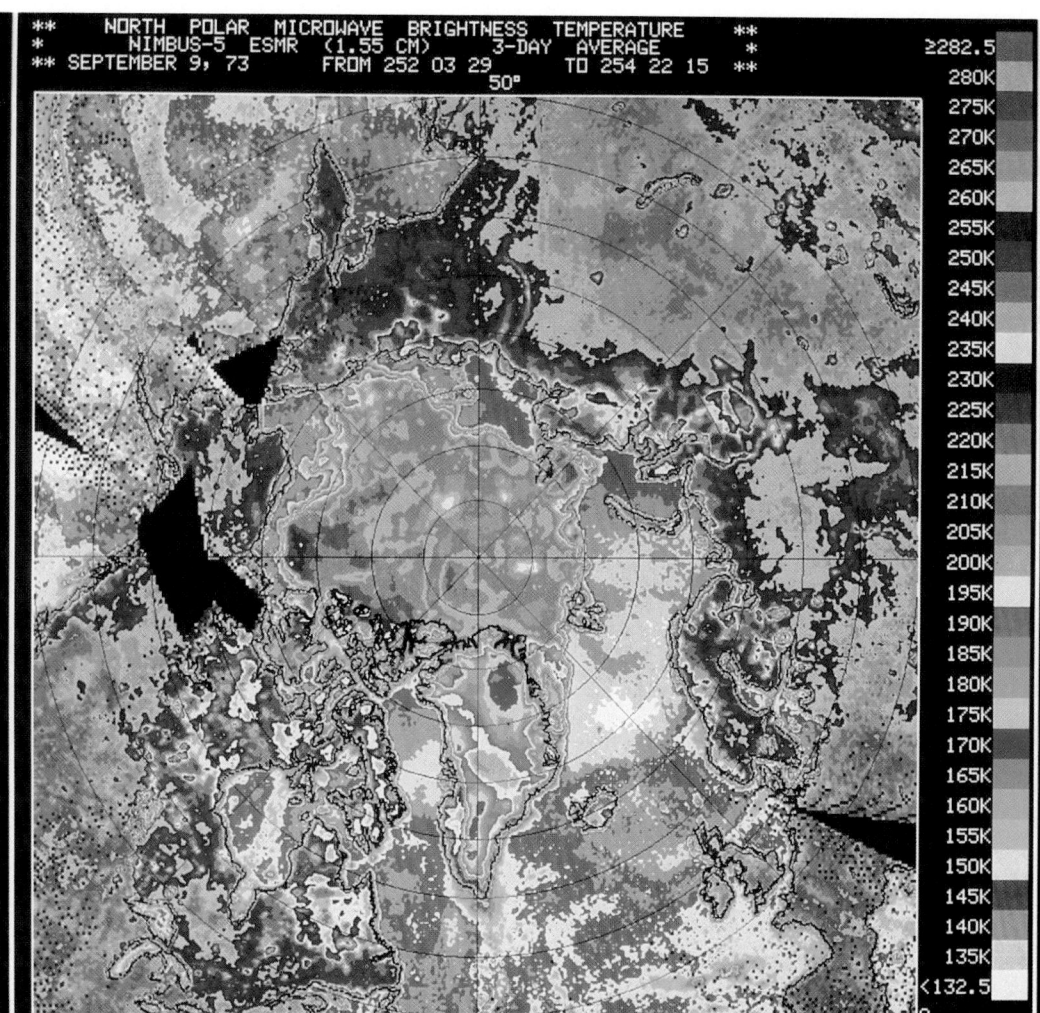

** NORTH POLAR MICROWAVE BRIGHTNESS TEMPERATURE **
* NIMBUS-5 ESMR (1.55 CM) 3-DAY AVERAGE *
** SEPTEMBER 9, 73 FROM 252 03 29 TO 254 22 15 **
50°

≥282.5
280K
275K
270K
265K
260K
255K
250K
245K
240K
235K
230K
225K
220K
215K
210K
205K
200K
195K
190K
185K
180K
175K
170K
165K
160K
155K
150K
145K
140K
135K
<132.5

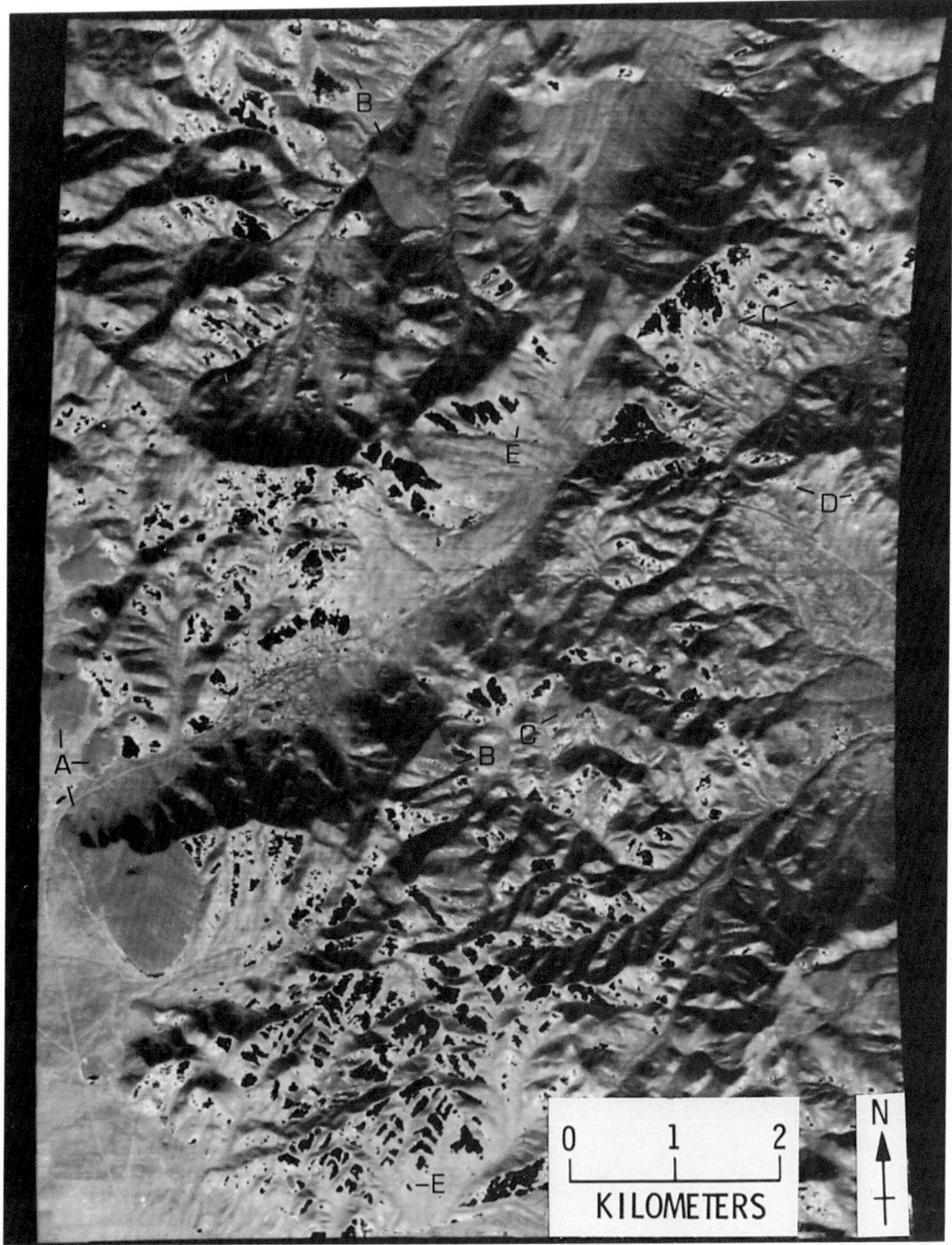

Fig. 13-67. Principal component color-composite image of the central part of the East Tintic Mountains, Utah, made from multispectral mid-infrared NASA aircraft data. A. quartz-rich rocks; B. interlayered quartz-rich and carbonate rocks; C. silicified rocks; D. quartz latite and quartz monzonite; E. latite and monzonite; and F. areas that exceeded the thermal response range of the scanner.

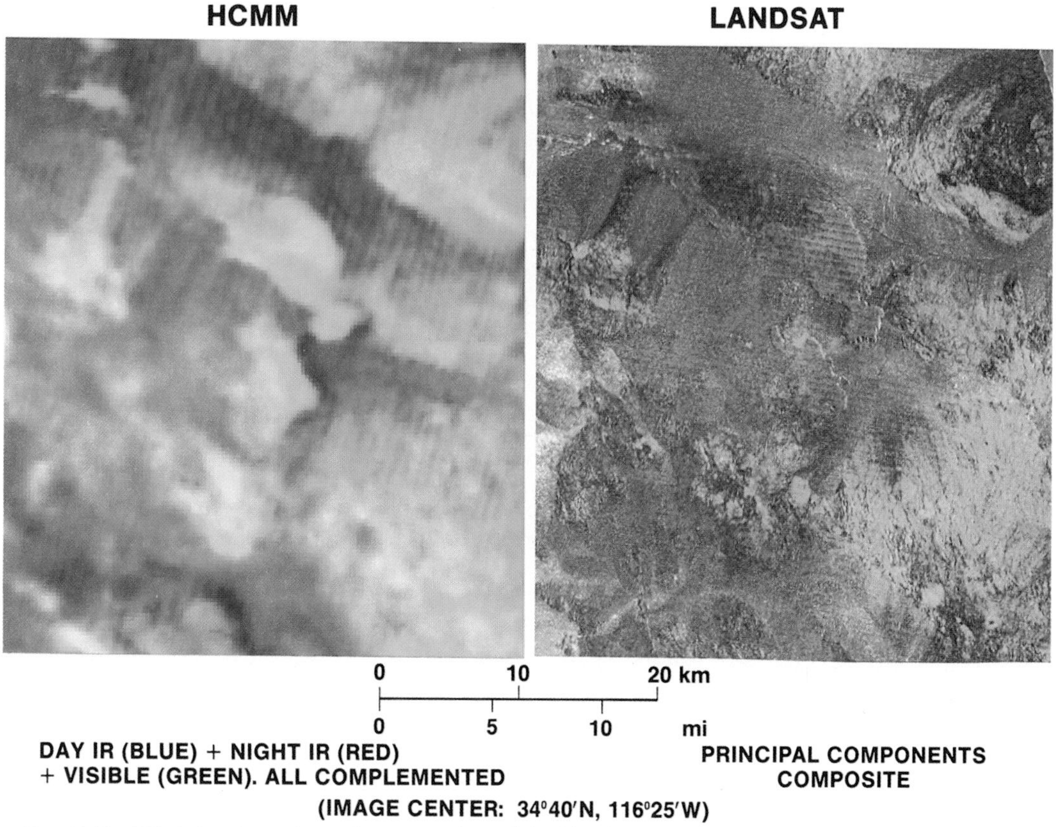

HCMM   **LANDSAT**

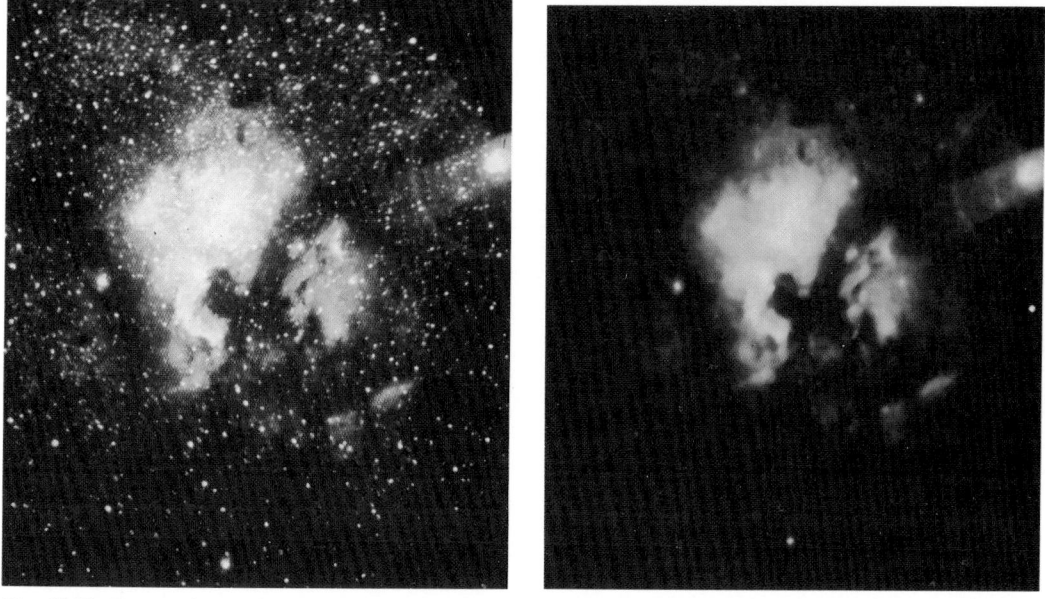

0   10   20 km
|---------------|---------------|

0   5   10   mi
|---------------|---------------|

DAY IR (BLUE) + NIGHT IR (RED)   PRINCIPAL COMPONENTS
+ VISIBLE (GREEN). ALL COMPLEMENTED   COMPOSITE
(IMAGE CENTER: 34°40′N, 116°25′W)

Fig. 13-79. HCMM and Landsat color composites of the Pisgah Lava Flow area, Mojave Desert, California.

Fig. 17-70. Removal of spike noise with a two-dimensional (5 × 5 pixel) median window filter (Klinglesmith, Goddard SFC).

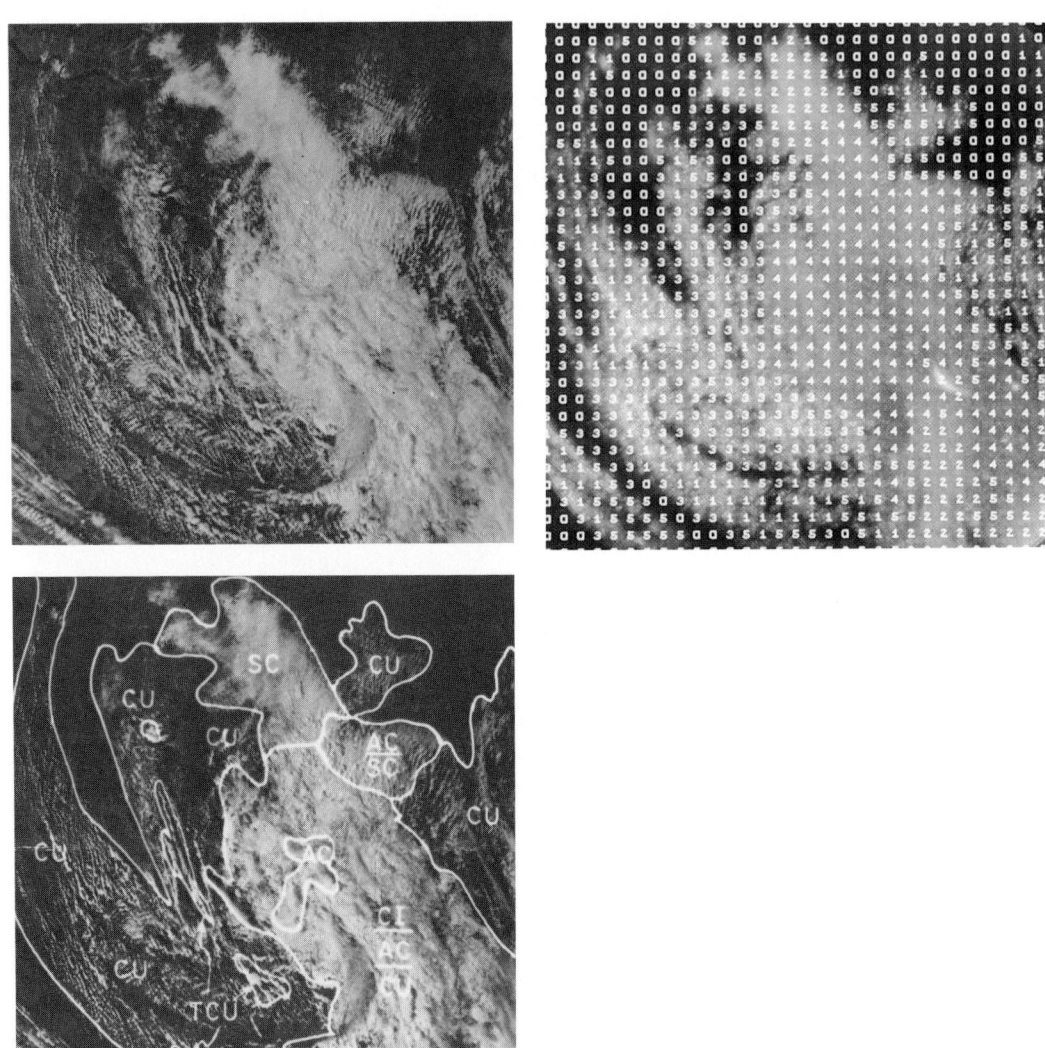

Fig. 17-83. Polychromatic processing for cloud classification. (a) (top left) Original satellite image. (b) (top right) Optical system output with cluster analysis classification results superimposed. (c) (bottom left) Photointerpreter classification results.

Fig. 21-30. Mosaic of eight Landsat Scenes Over Montana & Wyoming.

Fig. 24-4. Color aerial photograph of the discharge of the Cuyahoga River into Lake Erie at Cleveland, Ohio. Color variations in water body are related to variations in suspended-sediment load (USGS, 1966).

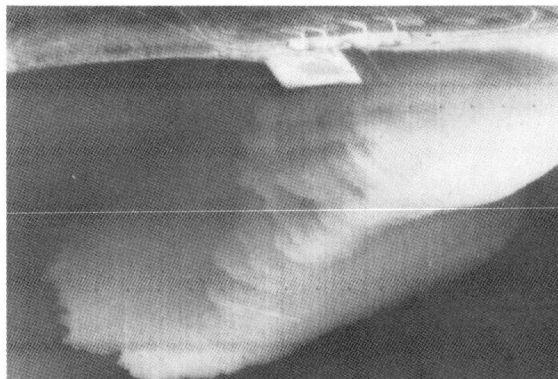

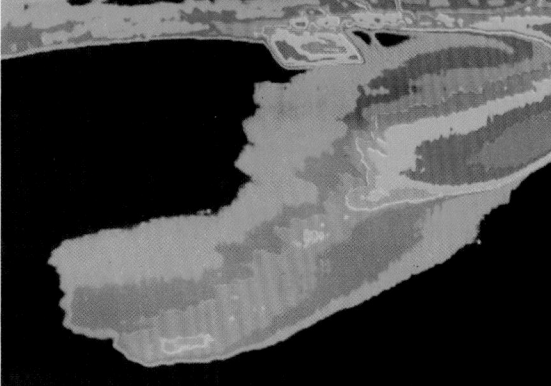

Fig. 24-5. See caption on facing page.

Fig. 24-69. High-oblique aerial photograph of Santa Barbara, California (Courtesy Pacific Western Aerial Surveys).

Fig. 24-5. Color presentation of density slicing for power-plant effluents discharged into Lake Michigan. Thermal infrared image (top) displays surface-water temperatures as shades of grey; warmer water is shown as lighter tones. The color example (below) is prepared by converting differences in optical density of the black-and-white thermal image to color zones that represent varying temperatures. With the addition of surface-temperature points measured simultaneously with the aerial infrared coverage, isothermal boundaries can be drawn and surface extent and local area of specific temperature levels estimated. This information has been useful in surveying major thermal discharges from power plants and refineries.

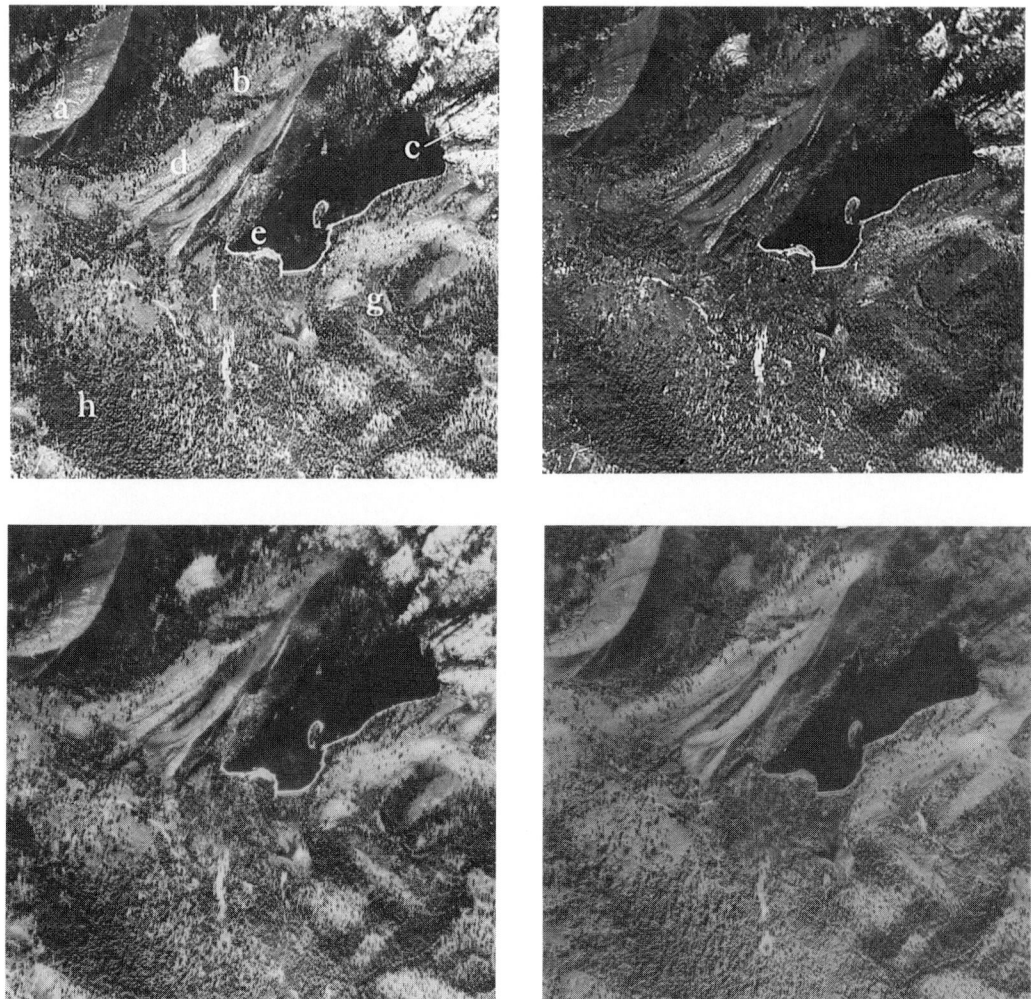

Fig. 24-74. Aerial photographs of a wildland area near Silver Lake in northeastern California, taken on 29 October 1966 at a scale of approximately 1:22,000; a = hardwood stand; b = forage; c = granite area; d = soil boundary; e = lake and underwater detail; f = dead and dying trees; and (g-h) = different species of conifers (Thorley, 1968). Film-filter combinations are as follows: *top left,* Ektachrome Aero, Wratten HF-3; *top right,* Ektachrome Infrared Aero, Wratten 12; *lower left,* Plus x Aerographic, Wratten 61; *lower right,* Infrared Aerographic, Wratten 89.

Fig. 24-82A. Three multiband photos used to reconstitute the "natural" color display in Figure 24-82B. The area on each photo is a 7- × 7-in. section of the 9- × 9-inch format (Orr, 1968).

Fig. 24-82B. (*lower right*) Color image made from tricolor projection of the multiband photos in Figure 24-82A (Orr, 1968).

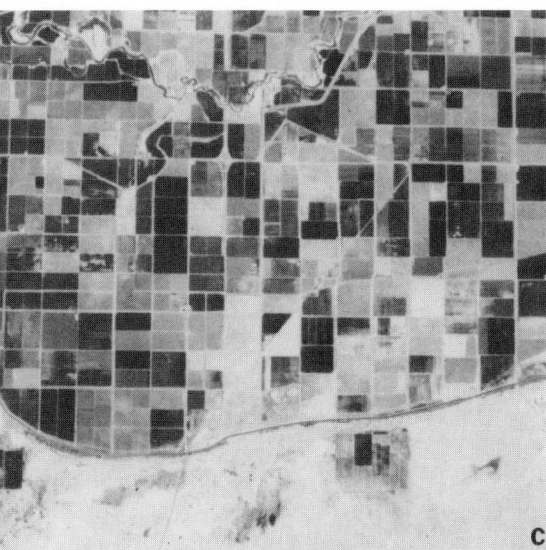 

Fig. 24-84. Color multidate optical image-enhancement of a portion of the Imperial Valley. Three panchromatic "highflight" photographs, taken in March (A), April (B), and May (C), 1969, of the Imperial Valley were optically combined into a single color-composite image. The objective was to produce a color composite in which each important earth-resource feature would be uniquely color coded. The different colors represented on this image are indicative of variations in temporal cropping cycles, patterns, and practices (NASA and University of California, Berkeley and Riverside).

Fig. 25-39. Low sun angle Landsat band-7 and Seasat false color digital composite. (Chavez and Sanchez, 1982). Scale approximately 1:500,000.

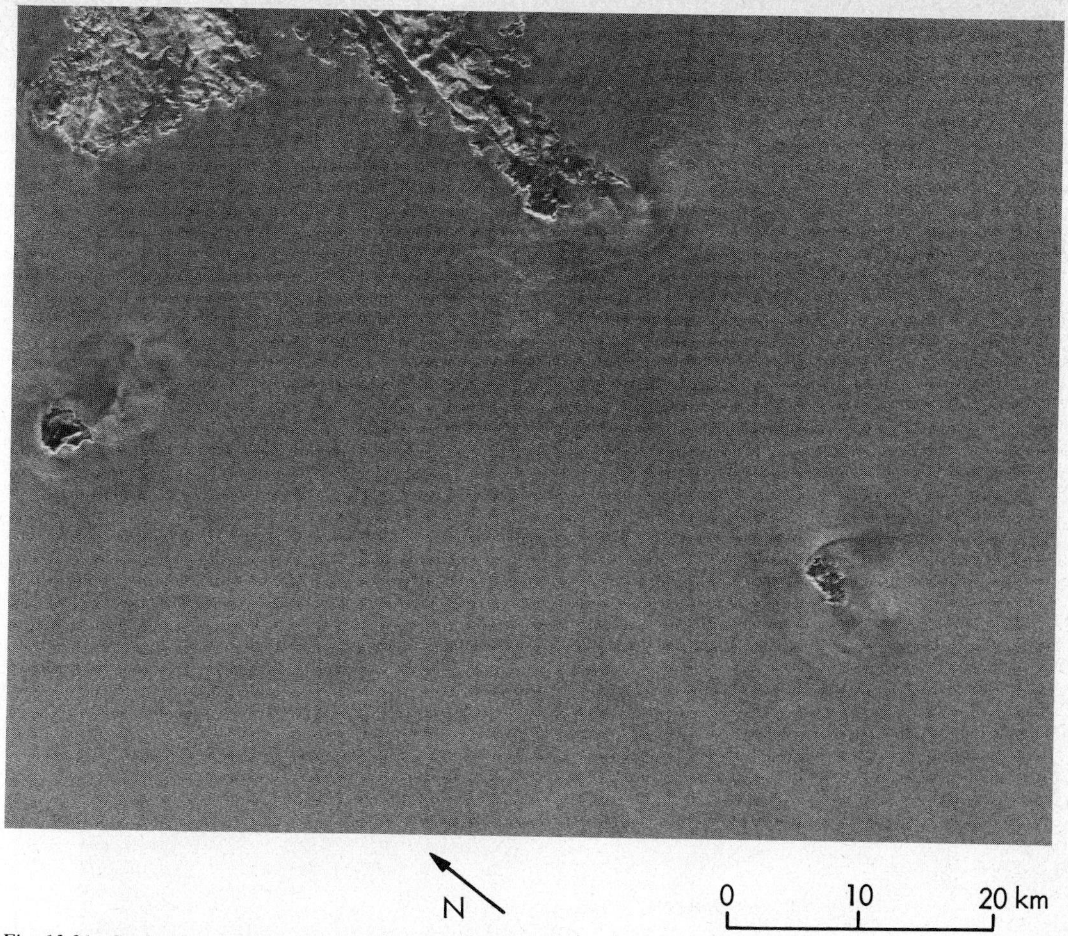

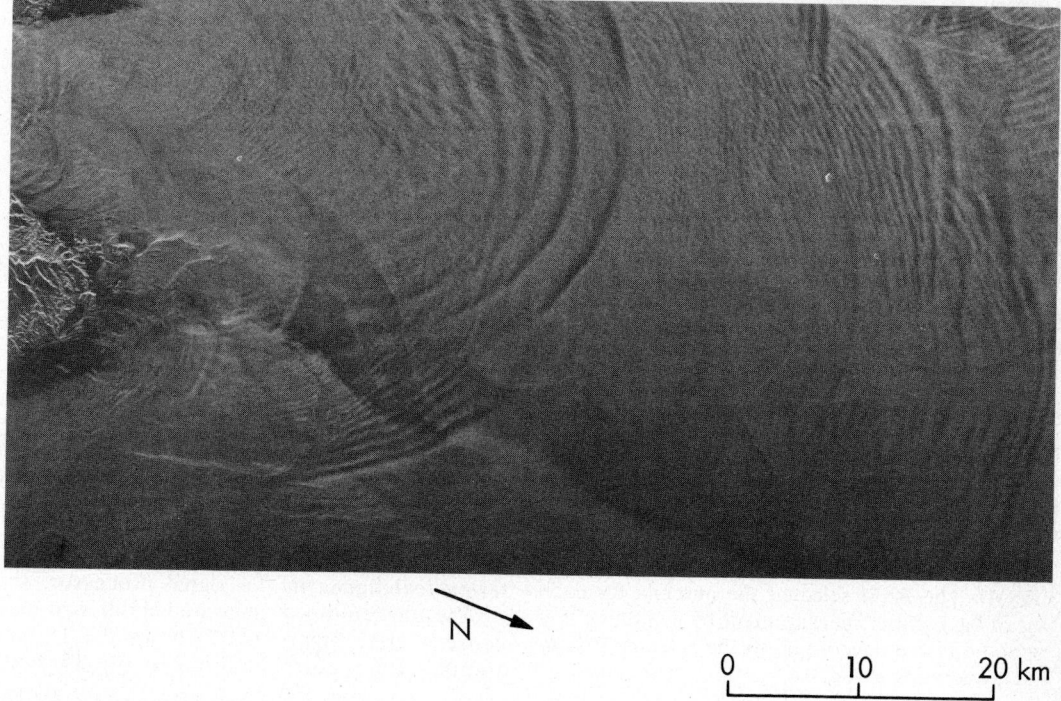

Fig. 13-21. Surface ocean waves observed near the southern tip of the Shetland Islands in the Norwegian Sea. The two small islands are Foule (northern) and Fair Isle (southern). The swell had a wavelength of about 300 meters. The image was taken at about 8 a.m. GMT on September 15, 1978.

Fig. 13-22. Radar image over the Gulf of California just north of Angel de la Guarde Island. Numerous internal wave trains are observed as large, curvilinear strips of dark, bright, or dark followed by bright tones.

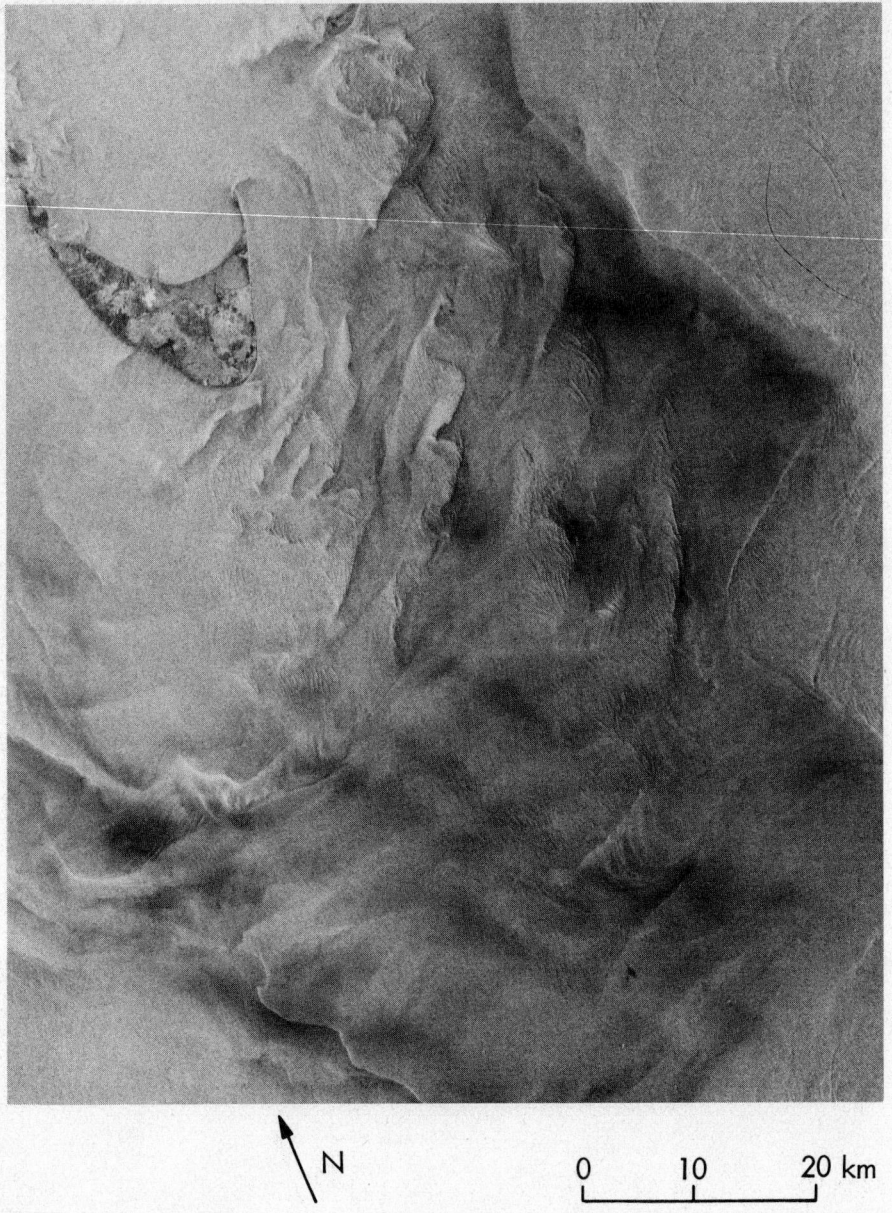

N

0   10   20 km

Fig. 13-23. Radar image of the Nantucket Island region. The island is visible in the upper left corner of the image. Most of the other patterns observed on the ocean surface reflect the bottom topography in the shallow region.

the satellite track at incidence angles of 0°, 4° and 8°.

A simplified diagram of the sensor is shown in Figure 13-26. A frequency synthesizer provided an excitation signal at 14.6 GHz to the transmitter and reference signals to the receiver. The TWT amplified the signal to a 100-*watt* peak level before transmission to the antenna switching matrix (ASM). The ASM selected the antenna for each set of backscatter measurement by switching in a periodic fashion according to the selected instrument operating mode.

In all operating modes, 15 backscatter measurements were made every 1.89 s with an antenna switching cycle completed every 7.56 s. This timing was designed to provide measurements which were located on a 50-km spacing in the along-track direction. The noise source provided a periodic receiver gain calibration every 250 seconds. Through the use of range gating and Doppler filtering techniques in the signal processor, the backscatter return was measured at different incidence angles. Figure 13-27 shows the 15 iso-Doppler lines corresponding to the 15 scat-

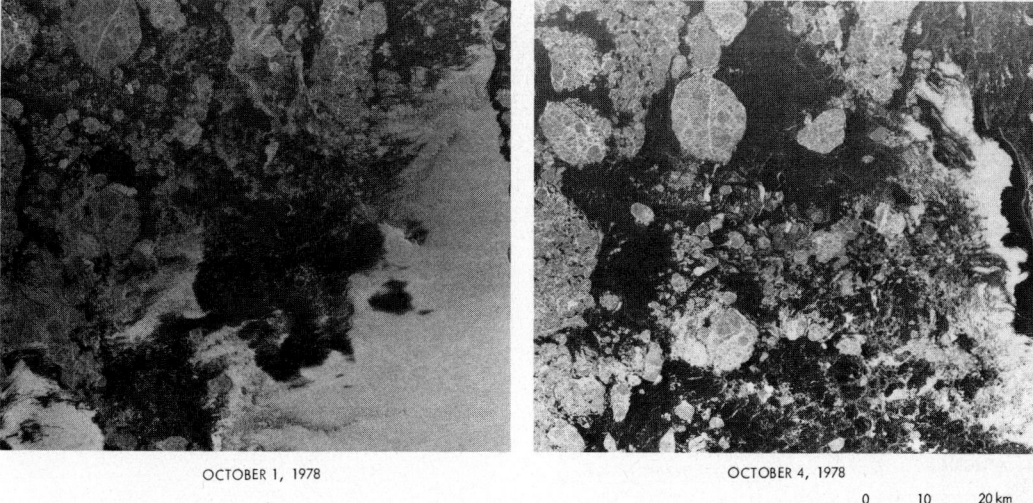

OCTOBER 1, 1978   OCTOBER 4, 1978

0   10   20 km

Fig. 13-24. Floating polar ice off the western coast of Banks Island. The two images were taken three days apart (October 1 and October 4, 1978) early in the morning. Some of the floes moved up to 30 km during the three-day period. The streaking effect visible in the October 1 image is probably a result of local wind.

terometer Doppler cell center frequencies. The Doppler lines have a symmetry axis that is rotated relative to the spacecraft subtrack. This is due to the earth's rotation.

The intersection of the antenna beam pattern and Doppler lines determines the resolution cell size and location. The cell size is determined by the Doppler filter noise bandwidth and the antenna beamwidth.

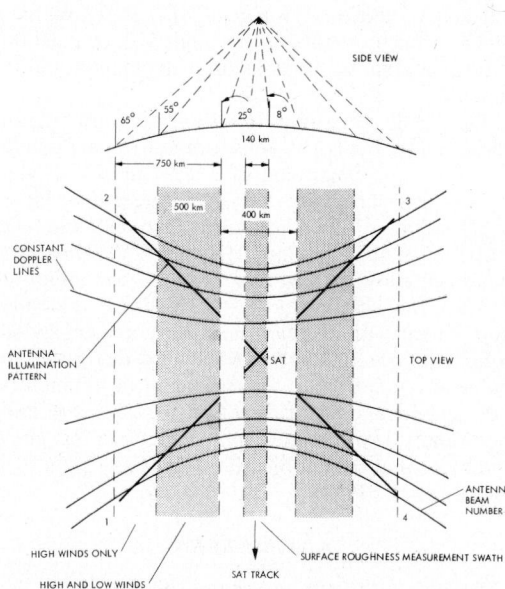

Fig. 13-25. SASS measurement geometry. The antennas' footprints are arranged in an X-pattern such that each point in the swath is observed twice as the satellite passes by it [Johnson et al., 1980].

In the receiver, a square-law detector and gated integrator are used to sample the reflected power from an antenna 64 times during a 1.84 s measurement period. The first three samples are used to place the receiver-processor in the most appropriate of four possible gain states, thereby putting the received signal level in the linear portion of the square-law detector. The remaining 61 samples are then processed using the selected gain state. The mean values of the 61 integrated voltage levels are then entered into the data stream for each of the 15 Doppler cells. The 1.84 s measurement interval is repeated continuously, but a different antenna is activated for each consecutive sampling period.

Examples of SASS Data

Measurements of the backscatter cross-section σ over the Amazon rain forest were used to determine the relative biases between the four dual-polarized antennas. The σ of the Amazon forest is believed to be isotropic and insensitive to polarization because of the extremely heavy foliation and random orientation of the scatterers involved. Figure 13-28 shows an example of the SASS measurement during an orbital pass (Rev. 952) over South America, in which data returned from the Amazon forest as well as from other regions are identified. Figure 13-29 shows σ versus incident angle θ for several orbit passes encompassing the different beam-polarization combinations; the data are representative of all rain forest data examined.

Three SASS Doppler filters were designed to provide measurement near nadir (0°, 4° and 8°). Examples of these data are shown in Figure 13-30 for a pass over Typhoon Carmen (Aug. 14, 1978).

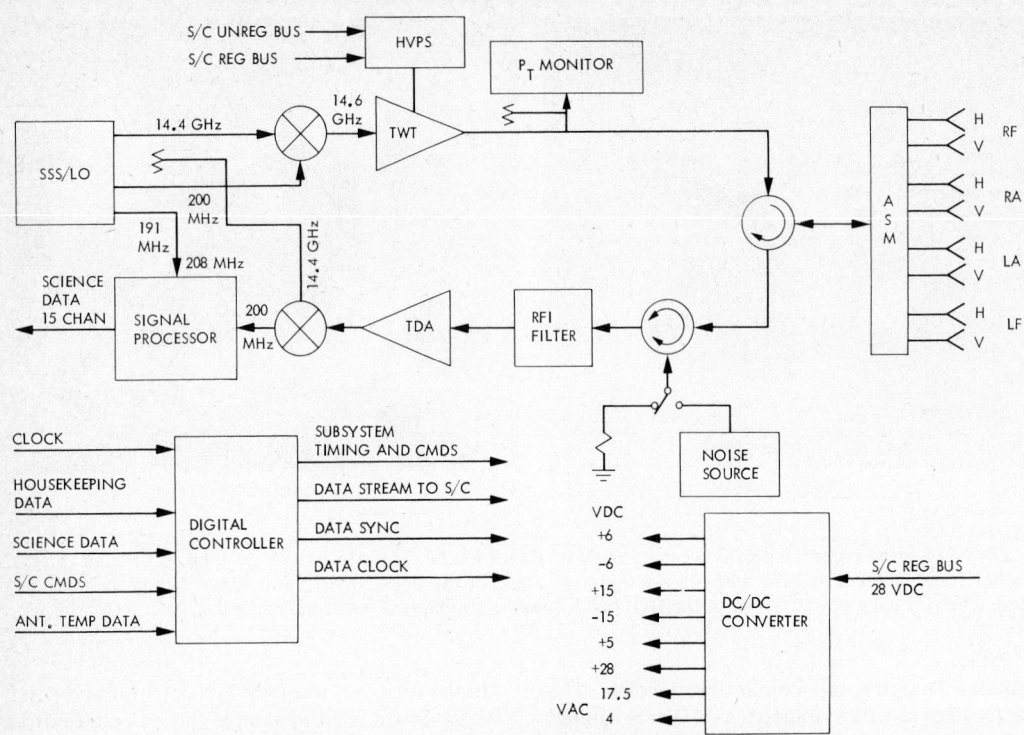

Fig. 13-26. SASS block diagram [Johnson et al., 1980].

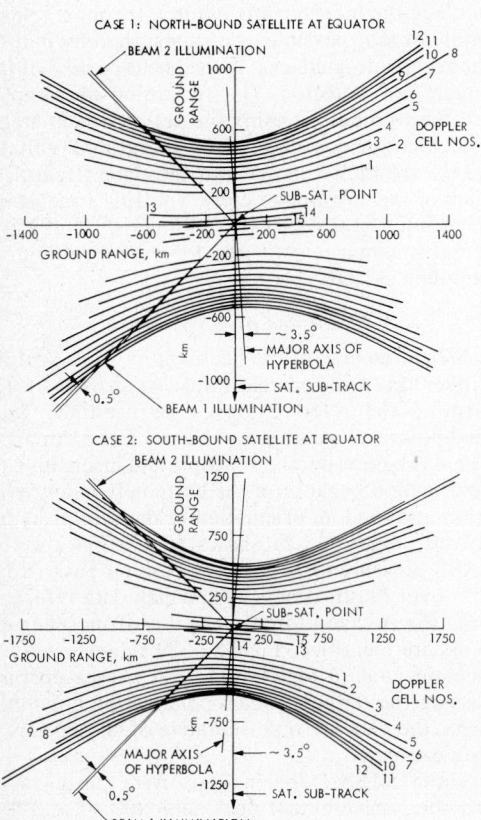

Fig. 13-27. SASS iso-Doppler lines projected on the surface. The tilt of the symmetry axis is a result of the earth's rotation. This tilt angle varies as a function of latitude [Bracalante et al., 1980].

The nadir return varies by more than 10 dB as a result of wind-speed variation, while the return at 8° varies by less than 4 dB. Aircraft experiments have shown that the values of σ associated with angles around 13° are relatively independent of ocean surface roughness over a wide range of surface conditions [Jones et al., 1977; Wentz, 1978]. Measurements at this angle will be used on future satellite systems as an instrument stability indicator.

During the three-month period of operations of the satellite, the SASS sensor obtained data which will form the beginning of a data library of the back-scatter cross-section $\sigma°$ of different surfaces as a function of incidence angle (0° to 65°) and as a function of polarization, HH and VV. This data base will allow an assessment of the capability of spaceborne scatterometers to measure, globally and in nearly all weather, near surface wind speed over the ocean. The SASS was designed to achieve a wind-speed measurement of ±2 meters per second and wind direction to ±20°. Preliminary analysis of the data over large-area test sites indicates that those objectives may have been met [Jones et al., 1974].

ALTIMETERS

Satellite altimetry has been used mostly for the study of ocean surface topography. Accurate measurements of the ocean surface mean level contribute to the detection and measurement of ocean currents, tides and storm surges, and to the accurate mapping of underwater features. Altimeters can also measure the surface wave

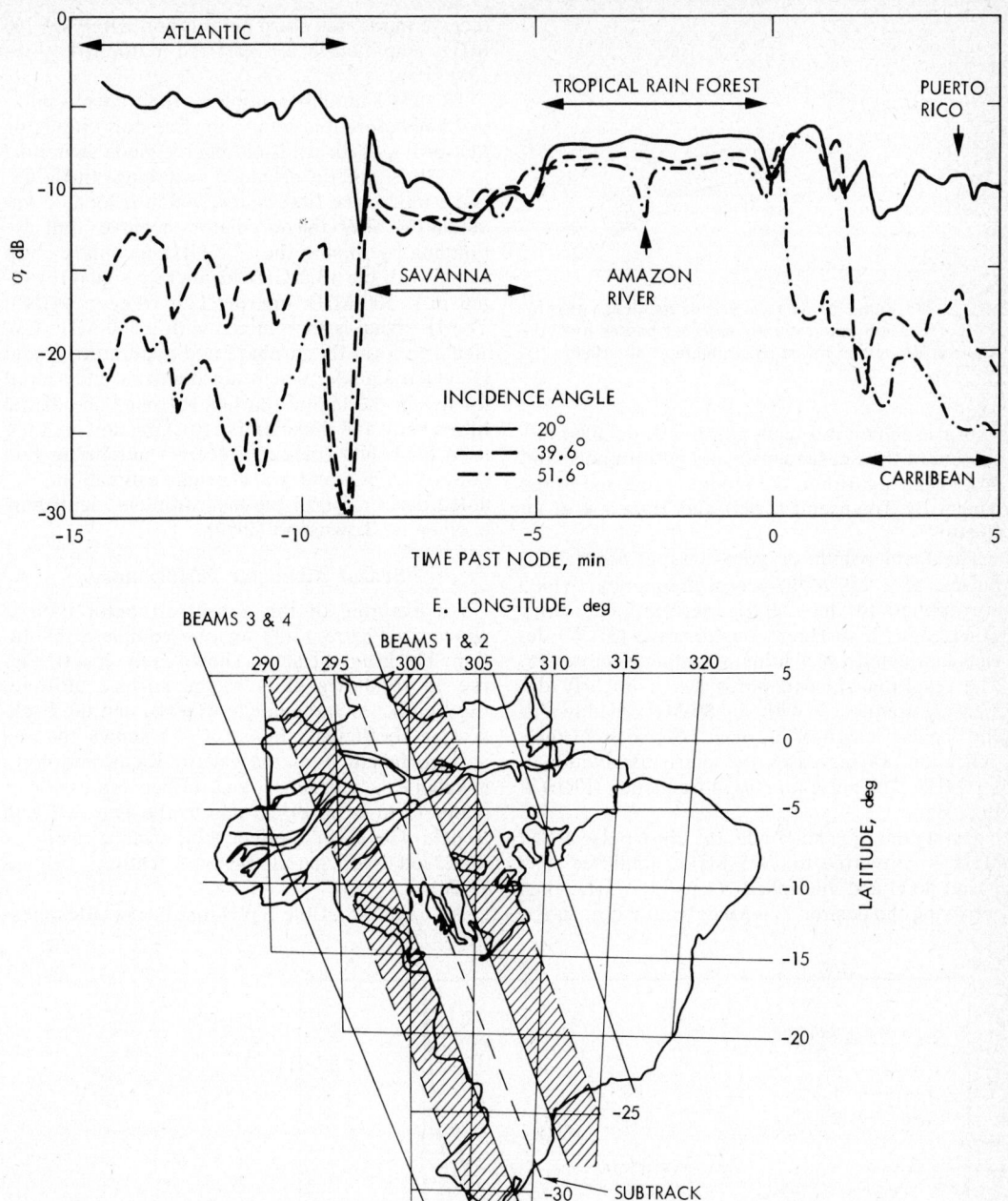

Fig. 13-28. Backscatter data over South America during Rev. 952. The main geographical regions are pointed out. The swath-track is shown in the insert. The different curves correspond to different incidence angles. The polarization is VV [Bracalante et al., 1980].

height. Altimeters have been flown in earth orbit on Skylab, GEOS-3, and Seasat. The Seasat altimeter, which is the latest and most advanced sensor, achieved an altitude measurement precision of less than 10 cm and a significant wave height accuracy of ±0.5 meters. Only the Seasat altimeter is discussed here in detail for illustration purposes.

A large part of the Venusian surface was also mapped with the Pioneer Venus Orbiter Radar Mapper. Altimetry data were acquired for the area

between latitudes of 74°N. and 63°S. The PVO altimeter is discussed in this section. Altimetry measurement was also obtained with the Apollo 17 lunar sounder (ALSE) during two lunar orbits. The ALSE System is discussed in the next section.

Seasat Altimeter Description

The Seasat altimeter operated at a frequency of 13.5 GHz using a 1-m-diameter horn-fed parabolic dish antenna. A simplified block diagram of the

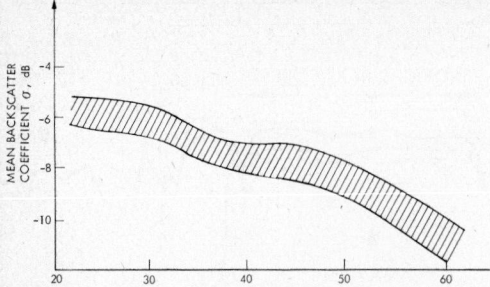

Fig. 13-29. Range of mean σ versus incident angle for VV polarization taken during different passes over the Amazon River rain forest [Bracalante et al., 1980].

sensor is shown in Figure 13-31. For detailed discussion of the Seasat sensor, its performance and processing algorithm, the reader is referred to the papers by Townsend (1980) and Hancock et al. (1980).

The local oscillator generates 12.5-nsec impulses, at a 250-MHz center frequency, which are applied to the chirp generator. The chirp generator is a surface acoustic wave (SAW) device fabricated on a lithium tantalate substrate. The resulting chirped pulse has a linearly decreasing frequency with an 80-MHz bandwidth and a pulse length of 3.2 μsec. Subsequent multiplication ×4 increases the pulse bandwidth to 320 MHz. The pulse repetition frequency (PRF) is 1020 Hz.

During the transmit mode, the chirp pulse at 250 MHz is converted to 3375 MHz, amplified to a 1-watt level and multiplied ×4 to 13.5 GHz, thus achieving the desired 320-MHz bandwidth. In the receive mode, the chirp pulse is converted to 3250 MHz, amplified to 0.1 watt and multiplied ×4 to 13.0 GHz.

The TWT amplifier amplifies the transmit pulse to 2 kW before it is sent to the five-port circulator that provides for transmit/receive mode switching as well as calibration mode switching. In the receive mode, the first mixer, which is located immediately after the circulator, achieves full deramping by mixing the 13.5-GHz incoming chirp signal with the 13.0-GHz local chirp signal resulting in a 500-MHz intermediate frequency (IF). The IF signal is then mixed with a 500-MHz CW signal to form the in-phase and quadrature (I and Q) video signals, which are digitized and stored for use in the digital filtering scheme. The digital filters bank and the adaptive tracking unit are then used for height tracking, receiver automatic gain control (AGC) and wave height estimation. Detailed description of the measurement algorithms is given by Townsend (1980).

Seasat Altimeter Performance

An example of the Seasat altimeter data is shown in Figure 13-32 for one complete revolution. The ground track is shown in the insert. Figure 13-32 shows plots of the surface altitude, the significant wave height (SWH), and the backscatter coefficient. Figure 13-33 shows the sea surface height over the Puerto Rican and Venezuelan Coast trenches. Sea surface depression of about 15 m is observed due to the Puerto Rican trench. The reverse effect (i.e., surface rise) was observed over sea-mount-type features (Figure 13-34).

An example of the SWH and backscatter mea-

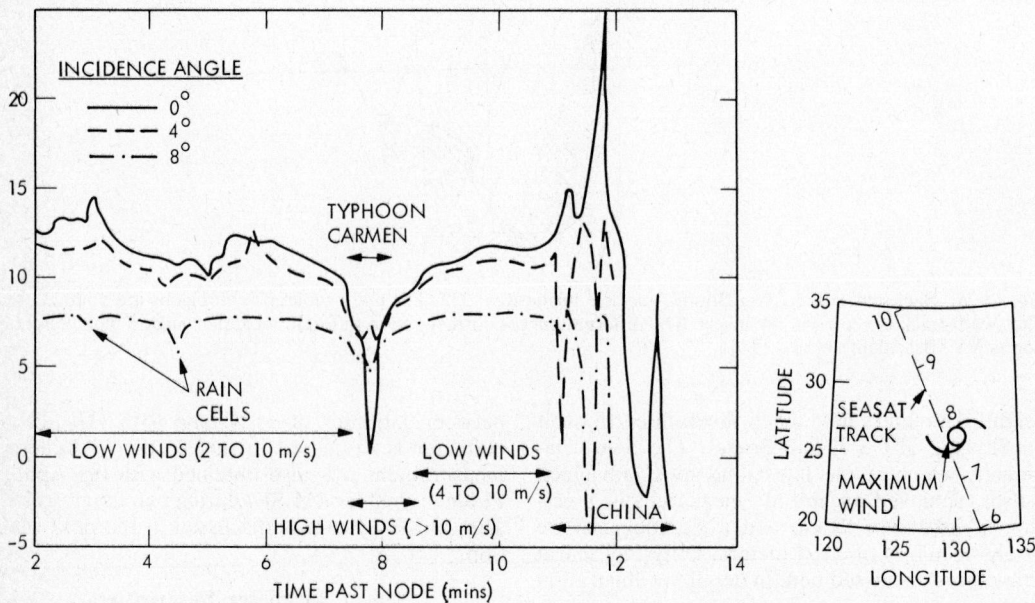

Fig. 13-30. σ data near Typhoon Carmen for near-vertical incidence angles [Bracalante et al., 1980].

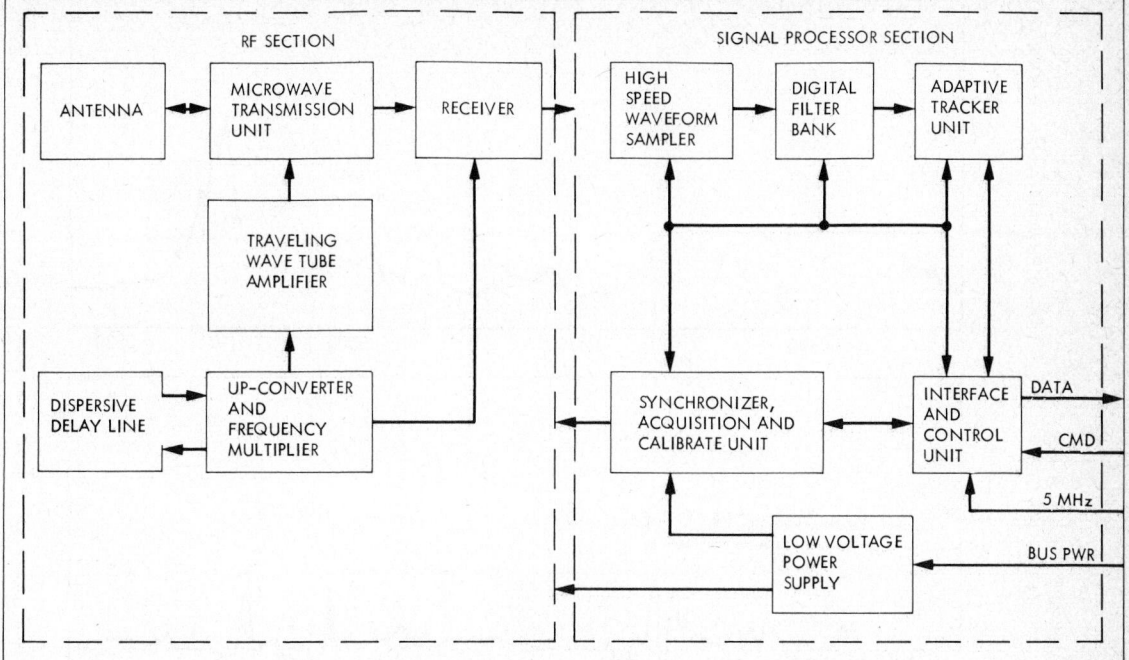

Fig. 13-31. Seasat altimeter major functional elements [Townsend, 1980].

surement is shown in Figure 13-35. It corresponds to an orbital track which came within 100 km of the center of Hurricane Fico between Hawaii and the west coast of the U.S. on July 16, 1978. The peak SWH observed was about 10 m at the point of closest approach to the hurricane center.

Pioneer Venus Orbiter Radar Mapper

The prime objective of the PVO radar was to obtain altimetry measurements over a large percentage of the Venusian surface. The Pioneer Venus Orbiter was put in orbit around Venus in December, 1978. Altimetry data were acquired between latitudes 76°N and 63°S. The altimeter footprint varied from about 7 km to about 100 km depending on the altitude of the spacecraft, which was in a highly elliptical orbit. The height measurement accuracy was about 200 meters.

The PVO radar also acquired images with a 30 km resolution between latitudes 65°N and 15°S. A detailed description of the sensor is given by Pettengill (1977) and Pettengill et al. (1980). In this section, a brief summary of the instrument characteristics is given.

The PVO radar has an operating frequency of 1.757 GHz (i.e., wavelength of 17 cm) and uses a 38-cm dish which is mounted on the spinning spacecraft (spin period about 12 seconds). The transmitted peak power is 20 watts. The transmitted signal is modulated with a 55-element, bilevel phase PN code. The instrument operates in two distinct modes: altimetry and imaging. The altimetry mode operates at altitudes below 4700 km. Estimates of the echo's absolute time delay, its time dispersion and its intensity are obtained simultaneously; the latter two quantitites are used to yield information on local surface properties. The imaging mode operates only when the spacecraft is below 550 km and when the radar antenna is on one side or the other of the radar as the spacecraft rotates around its axis. In this "side-looking" mode the echo time-delay and Doppler characteristics are used to generate a map of the surface radar backscatter at various angles between 30° and 58°.

Example of PVO Data

Figure 13-36☆* is a false color rendition of the topography of Venus as derived from the PVO altimetry data. This map is cast in a standard Mercator projection and covers the majority of the Venusian surface except for the polar regions. The two main "high" regions are the Ishtar Terra and the Aphrodite Terra, the latter one being spatially larger than South America. Except for the prominent elevated features, the planet is quite flat. The highest point observed has an altitude of 11.1 km relative to a median radius of 6051.2 km. The lowest point observed is at −1.9 km.

---

* Any reference to a figure marked with a star (☆) indicates that the figure referred to does not appear within this chapter. It appears as part of the color section within this book.

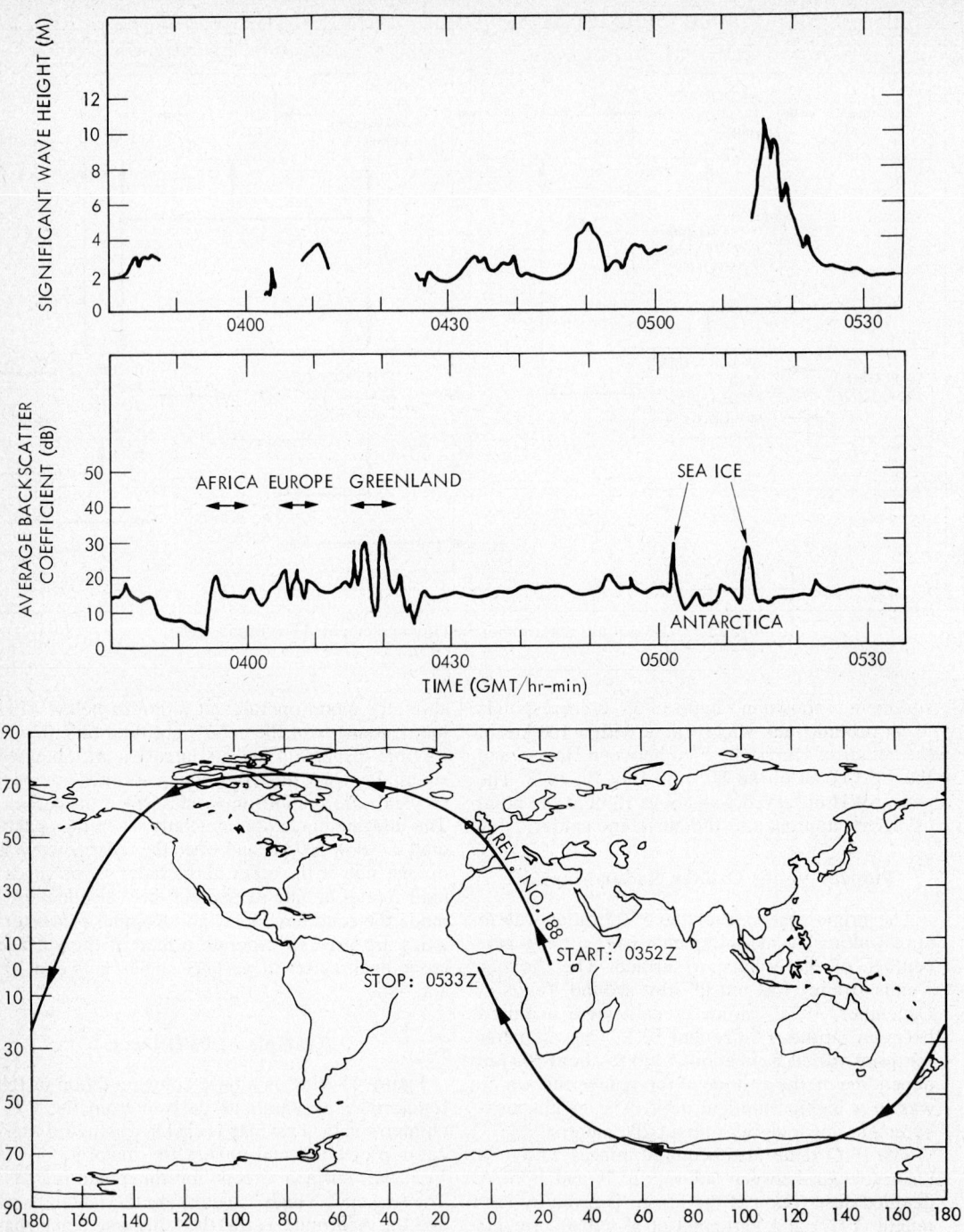

Fig. 13-32. Seasat altimeter measurement of wave height and backscatter coefficient taken during Rev. 188. The orbit track is shown in the insert [Townsend, 1980].

SOUNDERS

The first synthetic aperture radar flown on a satellite was the Apollo 17 sounder which orbited the moon in December 1972. The objective of the Apollo 17 lunar sounder experiment (ALSE) was to detect and map subsurface features and discontinuities. To accomplish this, a three-wavelength synthetic aperture system was developed. It also obtained surface profiles (altimetry) and surface images. A major impediment to the detection of subsurface features is the strong re-

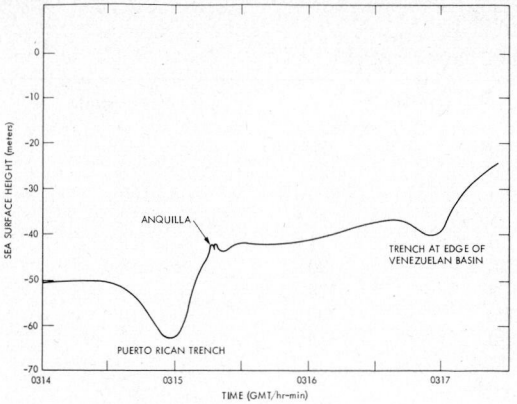

Fig. 13-33. Sea-surface height over a trench-type feature [Townsend, 1980].

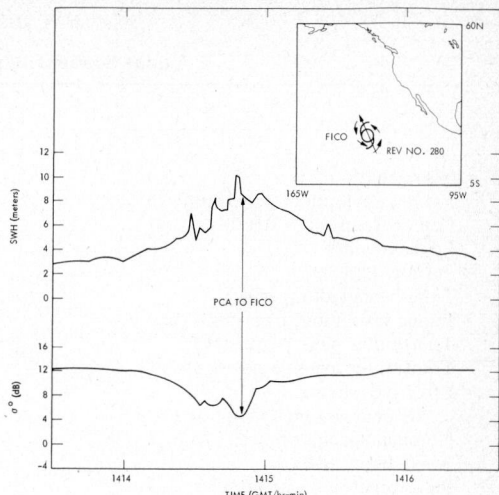

Fig. 13-35. Significant wave height and backscatter measurement over Hurricane Fico [Townsend, 1980].

turn from the surface beneath the spacecraft against which the weak subsurface returns must be observed. The strength of these subsurface returns depends primarily on the attenuation characteristics of the lunar material to be observed and the relative dielectric constant of the discontinuity. On the basis of measurements of lunar material loss and estimates of subsurface reflectivities, the three operating frequencies were 5, 15 and 150 MHz.

In order to obtain depth resolution, pulse compression techniques were used. At each frequency, a system bandwidth equal to 10% of the center frequency was selected with a time bandwidth product of 128. This depth resolution was inversely proportional to frequency. The along-track resolution for each of the three systems was equal to five times the wavelength and was determined by the amount of data integration possible during the generation of the synthetic aperture. The ALSE system as shown in Figure 13-37 consisted of two distinct coherent radar systems, one operating at two frequencies in the

HF band, the other operating at VHF. Each radar system had a separate antenna but shared a common data-storage device—an optical recorder. The HF subsystem radiated the two HF signals at 5 and 15 MHz on alternate pulses via a center-fed dipole antenna. The VHF system radiated using an eight-element Yagi antenna. The principal parameters for the ALSE radar subsystem are shown in Table 13-4.

The instrument and optical recorder were mounted in the Scientific Instrument Module (SIM) bay of the Command Service Module (CSM), and the antennas on the aft surface of the CSM between the SIM bay and the Service Propulsion System engine.

Radar operation, as well as HF antenna deployment and retraction, was controlled by the

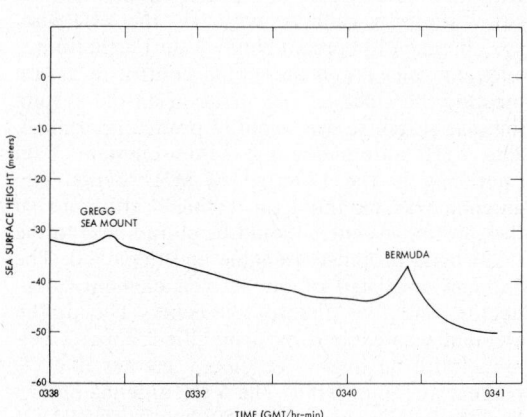

Fig. 13-34. Sea-surface height over a mount-type feature [Townsend, 1980].

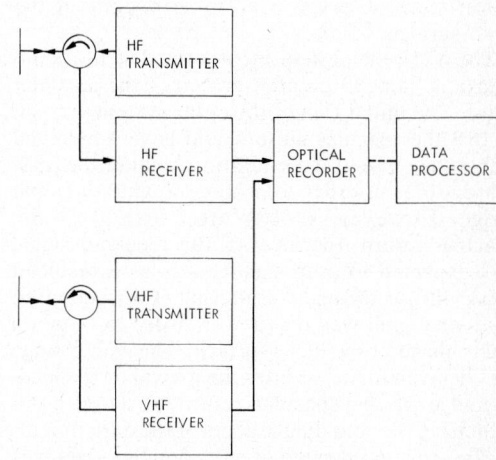

Fig. 13-37. Simplified block diagram of the ALSE radar.

TABLE 13-4

Lunar Sounder System Characteristics

	HF Mode		VHF Mode
	HF1	HF2	
Wavelength (m)	60	20	2
Estimated depth of penetration (n)[a]	1300	800	160
Center frequency (MHz)	5.256	15.8	158
RF bandwidth (MHz)	0.5333	1.6	16.0
Pulsewidth (μs)	240	80	8.0
Time-beamwidth product	128	128	128
Range resolution, free space (m)	300	100	10
Transmitter peak power (W)	130	118	95
Transmitter average power (W)	12.4	3.7	1.5
Effective antenna gain (dB one-way including efficiency)	−0.8	−0.7	+7.3
Nominal specular power (dBm)	−53	−57	−73
Noise figure (dB)	11.4	11.4	10.0
Pulse repetition rate (s^{-1})	397	397	1984
Recording duration (μs)[b]	600	600	70
AGC gain range (dB)	12.1	12.1	13.9
Echo tracker	No	No	Yes

[a] Based on model analysis.
[b] Interlaced on HF1 and HF.

astronauts in the Command Module. Since no internal access from the Command Module to the optical recorder was available, the photographic film on which the recorder stored the radar data was retrieved by an astronaut during an EVA; this was carried out toward the end of the Apollo 17 mission, during the trans-earth phase.

Automatic gain control was incorporated at both HF and VHF. To avoid undue complexity in the data processing, the gain setting was allowed to change no more often than once every 30 seconds and then only in discrete steps. The gain could be varied by up to 12 dB at each frequency. During the flight of Apollo 17, the power return did follow predicted specular values on the average; however, significant variations in return power were observed over some regions of the lunar surface.

The VHF subsystem incorporated a 12-dB increase in gain, 13 μs after arrival of the specular pulse. The initial gain of the radar system was set so that the specular pulse would have a received amplitude close to the saturation point of the optical recorder in order to achieve a wide dynamic range. However, shortly after arrival of the specular return, the level of the received signal was expected to drop significantly as a result of weak diffuse off-nadir scattering. Consequently, additional gain was required in order to obtain a radar image of the lunar surface. The gain change was delayed for 13 μs after the arrival of the leading edge of the specular return to avoid compromising the sounding aspects of the experiment. By this time no portion of any sounding data was expected to be present. The additional gain was limited to 12 dB because of system noise.

The desired maximum radiation gain at the HF frequencies was achieved with two retractable, centerfed, 10.4-m elements driven by a suitable matching network. Selection of the antenna length was based on 1) pattern studies obtained from a computer simulation of a wire grid model of the spacecraft and antenna elements and 2) impedance properties of the various antenna element lengths simulated. The simulated radiation system models were verified by pattern and impedance measurements on a 1/10-scale model of the spacecraft and antenna elements. Overall antenna radiation efficiencies of 54 and 32 percent were obtained in the 5- and 15-MHz bands, respectively.

The VHF antenna was designed to have a maximum gain on the right side of the spacecraft, with the one-way gain in the nadir direction 3 dB below the peak gain. As with HF, the VHF subsystem had a 10 percent bandwidth. Furthermore, adequate phase and amplitude control to assure meeting the sidelobe specification on the system impulse response was again of prime importance. The VHF antenna was a seven-element Yagi operating in the 150- to 166-MHz band. The antenna was mounted on a hinged structure so that, during ascent, it could be stored next to the CSM main propulsion engine and deployed. The antenna consisted of one driven element, a reflector, and five director elements. The driven element was excited by a modified Roberts balance, with an overall efficiency greater than 90 percent at center-band. The VHF antenna had an E-plane 3-dB beamwidth of 50° and H-plane 3-dB beamwidth of 60°.

Subsequent to landing, the film was developed

and processed optically and digitally. Coherent optical processing techniques were used to generate range and azimuth compressed images on photographic film, optical holograms and digital records on computer compatible tape.

The ALSE sensor obtained data over two complete lunar orbits. The surface profile data were used to study the internal geophysical properties of the Moon (Kobrick, 1976).

FUTURE SYSTEMS

As a result of the data acquired with the active microwave sensors flown on Seasat, the interest in the use of these sensors has increased appreciably. A number of advanced sensors are planned for development and flight in the 1980s. In the area of the SAR, the main thrust is on the development of flexible multiparameter (multifrequency, polarization and incidence angle) experimental sensors to be used on the Shuttle. The objective is to understand the effect of the different radar parameters on the image information content and on determining the most favorable combination of parameters for specific applications. Experimental free-flying SARs are also planned particularly for application requiring long-term monitoring capability such as ice dynamics, soil moisture measurement, vegetation monitoring and ocean observations. One major challenge is to determined how to use the data from multiparameter radars in conjunction with IR and visible sensors data to acquire maximum information about the surface. A SAR system will also be put in orbit around Venus in 1988 to map the cloud-hidden surface of our sister planet.

Spaceborne scatterometers and altimeters are being considered for use in a quasi-operational system for ocean surface monitoring. An advanced altimeter is also planned for the TOPEX (Topography Experiment) mission, which has the objective of detailed mapping of the ocean-surface topography.

PASSIVE MICROWAVE SENSORS

The term "passive microwave sensors" refers to sensors that measure emitted microwave energy from the earth's surface and atmosphere. This information can then be used to generate a two-dimensional image of the surface microwave emission. In this section, the development and progress of passive microwave remote sensing from space will be discussed, and a description given of the microwave radiometer systems launched prior to 1980. In addition, the technology advances and data-utilization approaches necessary to meet the growing user needs in remote sensing will be discussed. For a background on principles of microwave radiometry and microwave interaction mechanisms, the reader is referred to the discussions in Chapters 4 and 11.

BACKGROUND

Historical Development

Subsequent to the early ground-based passive microwave observations of the atmosphere by Dicke at al. (1946), it was realized that the use of radiometers in space for microwave observations of the earth's emitted radiation was a valuable means for studying the global properties of the earth's atmosphere and surface. The first test, however, of a spacecraft microwave radiometer was on the Mariner 2 Venus flyby mission, during which the spacecraft passed within 34,350 km of the surface of Venus on 14 December 1962. The microwave radiometer aboard the spacecraft successfully measured emission from the planet at frequencies of 15.8 GHz and 22.2 GHz on three scans of the planetary disk (Barath et al., 1964). Results showed that the planetary emission was characterized by limb-darkening, and confirmed the high surface-temperature of Venus.

The application of microwave radiometry from space to remote sensing of the earth began with observations from the Russian satellite Cosmos 243, which was launched into an elliptical orbit around the earth on 23 September, 1968. Measurements were made at frequencies of 3.5, 8.8, 22.2, and 37.5 GHz in a nonscanning, nadir-viewing mode, from which estimates of atmospheric water vapor and liquid water were made (Basharinov et al., 1971). This was followed by a series of earth-viewing radiometers on both U.S. and Russian satellites. The most extensive observations to date have been made by radiometers on the Nimbus series of satellites. Microwave radiometers launched prior to 1980 are listed in Table 13-5 with brief descriptions of their characteristics. These radiometers, together with a well-established program of supporting research and aircraft radiometer flights, have served to demonstrate the growing potential of microwave radiometry for operational remote sensing from space.

In progressing toward the development of operational spaceborne remote sensors, two kinds of systems have evolved. Atmospheric sounders provide information about vertical profiles of temperature and molecular constituents in the atmosphere by making measurements near the molecular resonance frequencies. Surface sensors on the other hand operate primarily at window frequencies, where atmospheric absorption is low and surface features can be imaged or measured quantitatively. For most of these surface sensors the microwave instruments have a distinct advantage over their counterparts at infrared or visible wavelengths in that the presence of most cloud cover does not significantly degrade the measurement accuracy. Thus, microwave measurements from orbit provide a global, day or night, nearly all-weather remote sensing capability. These advantages, together with the unique

TABLE 13-5

History of Microwave Radiometry on Spacecraft (Adapted from Staelin and Rosenkranz, 1978)

Year of Launch	Spacecraft	Instrument Acronym	Frequencies (GHz)	Antenna Type	Swath Width of Scan (km)	Smallest Resolution Element (km)	Principal Parameters Measured or Inferred
1962	Mariner 2 (Venus flyby)	—	15.8, 22.2	Mechanically scanned parabola.	Planetary	1300	Limb darkening of planetary emission
1968	Cosmos 243	—	3.5, 8.8	Nadir-viewing parabola	—	13	Atmosphere: Water vapor content, liquid water content Surface: Sea temperature, sea ice concentration
1970	Cosmos 384	—	22.2, 37				
1972	Nimbus 5	ESMR	19.3	Electrically scanned array	3000	25	Atmosphere: Rain rate Surface: Sea ice concentration, ice classification
		NEMS	22.2, 31.4, 53.6, 54.9, 58.8	Five lens-loaded horns, Nadir-viewing	—	200	Atmosphere: Temperature profile, water vapor content, liquid vapor content Surface: Ice classification, snow cover
1973	Skylab	S-193	13.9	Mechanically scanned parabola	180	16	Surface: winds, precipitation
		S-194	1.4	Nadir-viewing array	—	115	Surface: soil moisture
1974	Meteor	—	37	Dual polarization 35° from nadir	—	—	Atmosphere: liquid water content
1975	Nimbus 6	ESMR	37	Electrically scanned array, dual-polarization	1300	20 × 43	Same as Nimbus 5 ESMR
		SCAMS	22.2, 31.6, 52.8, 53.8, 55.4	Three rotating hyperbolic mirrors	2700	150	Same as Nimbus 5 NEMS
1978	Block 5D	SSM/T	50.5, 53.2, 54.3, 54.9, 58.4, 58.8, 59.4	Single rotating mirror	1600	175	Atmosphere: Temperature profile
1978	Tiros N (2 satellites)	MSU	50.3, 53.7, 55.0, 57.9	Dual rotating mirrors	2300	110	Atmosphere: Temperature profile
1978	Seasat	SMMR	6.6, 10.7, 18, 21, 37	Offset-fed oscillating parabola, dual-polarization	600	18 × 28	Atmosphere: Water vapor content, liquid water content, rain rate Surface: Sea temperature, wind speed, sea-ice concentration, ice classification, snow cover, soil moisture
1978	Nimbus 7		18, 21, 37		800	22 × 35	

signatures of some surface and atmospheric features at microwave wavelengths, have encouraged the development of spacecraft passive microwave sensors in recent years.

Atmospheric Sounding

One of the earliest meteorological applications of spaceborne radiometry was in global atmospheric temperature sounding. The mixing ratio of oxygen in the atmosphere is quite uniform and time-invariant; thus measurements of atmospheric emission at frequencies around the oxygen resonance complex (centered near 60 GHz) are proportional to atmospheric temperature at altitude levels defined by temperature weighting functions (Meeks and Lilley, 1963; Lenoir, 1968). By choosing frequencies whose weighting functions peak at different altitudes, retrieval techniques can be used to estimate vertical temperature profiles by inversion of the microwave data (Waters and Staelin, 1968; Staelin, 1969).

The Nimbus-5 Microwave Spectrometer (NEMS) was the first instrument of this type, with three oxygen channels for temperature sounding. Since the weighting functions were weakly dependent on atmospheric water content and surface reflectivity, two additional channels were added, one on the water vapor absorption line (22.2 GHz) and one at a window frequency (31.4 GHz). Results from NEMS have been reported by Waters et al., 1975; and by Staelin et al., 1975, 1976. These results indicated temperature profile rms retrieval accuracies of ~2°C in the region of the troposphere and integrated water vapor accuracies of ~0.4 gm/cm². They also showed these retrievals to be relatively unaffected by cloud cover. A second instrument of this type, the Scanning Microwave Spectrometer (SCAMS), was flown on Nimbus 6. SCAMS provided global maps of atmospheric temperature structure and further demonstrated the feasibility of the technique for studying meteorological phenomena (Grody and Pellegrino, 1977; Rosenkranz et al., 1978). The success of these two experimental sensors led to the development of microwave temperature sounders for operational use in synoptic meteorology. One of these is the four-channel Microwave Sounding Unit (MSU) launched on the TIROS-N satellite series. Another is a seven-channel microwave temperature sounder (SSM/T) operating on the Defense Meteorological Satellite Program (DMSP) Block 5D satellite. Operational temperature sounders will not be treated in this chapter but are discussed in Chapter 14 on Meteorological Satellites.

Techniques are presently being studied to use resonances other than the oxygen 60-GHz complex for atmospheric temperature and composition sounding. The oxygen line at 118 GHz shows promise for temperature sounding from geosynchronous orbit, since the higher frequency reduces the antenna size required to achieve satisfactory spatial resolution from that altitude. The 183-GHz water vapor line is sensitive to changes in atmospheric water vapor gradients above about 2 km. Thus, measurements around this frequency may be used to study the vertical distribution of water vapor. Measurements of temperature and water vapor profiles from geosynchronous orbit are necessary for future operational severe-storm monitoring from space (Wilheit et al., 1978).

Another approach to monitoring atmospheric constituents, particularly in the upper atmosphere, is the limb-sounding technique. This technique can provide substantially improved sensitivities and vertical resolution over downward-looking sytems. Ozone in particular, among other molecular species, is known to have a marked effect on the earth's environment, and there is considerable incentive for measuring its spatial and temporal distribution. Limb-sounding radiometers operating at microwave and millimeter wavelengths are currently under development, and will enable global measurements of these species to be made (Waters and Wofsy, 1978).

Surface Sensing

Surface sensing applications can be divided into three main categories: Ocean, Ice, and Land. In general, atmospheric window frequencies are used to study the surface phenomena. However, measurements at these frequencies are affected to some extent by tropospheric water vapor, clouds and rainfall. Hence most surface sensing radiometer systems include frequency channels sensitive to atmospheric water vapor and liquid water to measure global distributions of these parameters and to correct for their effects on the measurement of surface parameters. Microwave radiation from atmospheric liquid water and surface materials is by the mechanism of nonresonant thermal emission; thus the precise frequencies chosen to observe these parameters are usually noncritical. Frequencies are generally chosen to maximize sensitivity to particular parameters under different environmental conditions and to operate in regions of the spectrum free of interference from ground-based or earth-to-space radars and communication links.

Radiometers on the Cosmos satellites, and the two low-frequency channels on NEMS and SCAMs, made the first measurements of atmospheric water vapor and cloud liquid water over the oceans (Basharinov et al., 1971; Staelin et al., 1976). These measurements were possible due to the contrast between the warm radiative temperature of atmospheric water and the cold radiative background of the ocean. Against the warm background radiation over land, these measurements cannot be made. The radiometers also measured sea-ice concentration, ice-type, and snow cover (Kunzi et al., 1976). The sea-ice measurements were substantially improved by measurements from the 19.35-GHz Electrically Scanning Microwave Radiometer (ESMR) on Nimbus 5 and the 37-GHz ESMR on Nimbus 6. The ESMRs provided medium resolution (~35 km) images of the polar ice, containing information on the variations of the ice packs on time scales ranging from several days to seasons (Zwally and Gloersen, 1976). The technique uses the large differences in emissivity and polarization properties of the various ice types, and of the neighboring sea water. Potential uses of such data for ship navigation and offshore engineering in polar regions and for climatological studies appear promising.

The ESMR instruments have also been used for global measurement of rainfall over the oceans. Rain has a stronger effect on brightness temperature, versus nonprecipitating liquid water, due to having larger droplet sizes than those found in nonraining clouds, and due to the greater total water content in rain clouds. While quantitative measurements are difficult at present due to the complicated modeling problems involved, precipitation areas in the vicinity of storm systems have been identified, and relative global distributions of precipitation have been obtained (Wilheit et al., 1976, 1977).

During 1973 and 1974 two microwave radiometers at 13.9 and 1.4 GHz were operated aboard the orbiting Skylab space station. Three short-duration missions during this period permitted earth observations by the two sensors, designated S193 and S194, respectively, between latitudes of ±50°. Real-time adjustments in operating modes were made possible by communication between ground investigators and crew members on the spacecraft. The S193 sensor was a combined radiometer/scatterometer/altimeter instrument, enabling comparisons to be made between passive and active microwave responses to a variety of land and ocean surfaces (Sobti and Moore, 1976). The S194 sensor made coarse spatial resolution brightness temperature measurements, which have shown correlations with large-scale phenomena, such as soil moisture over the central and western portion of the U.S. (Eagleman and Lin, 1976).

The latest in the series of spaceborne surface sensing radiometers is the Scanning Multichannel Microwave Radiometer (SMMR). This instrument was launched on both the Nimbus 7 and Seasat satellites in 1978. The instrument has been described in detail by Gloersen and Barath, 1977, and Njoku et al., 1980. The additional capabilities of the SMMR over previous radiometers are primarily in the measurement of sea-surface temperature and wind speed. However, the five-frequency dual-polarization instrument also measures atmospheric water vapor and liquid water, rain rate, sea-ice concentration, ice type, snow cover, and may provide information on soil moisture. The effect of surface temperature on the emissivity and brightness temperature of the ocean is well understood (Klein and Swift, 1977; Blume et al., 1977). Wind speed, however, affects the surface emissivity only indirectly through the generation of ocean waves and foam. The complexity of this indirect relationship has made it difficult to develop precise models of brightness temperature dependence on wind speed (Nordberg et al., 1971; Hollinger, 1971; Webster et al., 1976; Wilheit 1979). Techniques for inverting data from the ten brightness temperature channels to obtain the ocean and atmospheric parameters, and determinations of the accuracies of the retrieved parameters are still undergoing evaluation. Initial results of these evaluations have been encouraging, and indicate that sea-surface temperature and wind speed rms accuracies of 1.2°C and 2 m/s, respectively, are achievable with the SMMR (Lipes et al., 1979; Hofer et al., 1981). The multispectral nature of the measurements will also improve the ability to discriminate ice types and to determine snow accumulation rates and snowpack properties over land.

The success of the Seasat oceanographic satellite mission, which operated a number of active microwave and infrared sensors in addition to the SMMR, has generated interest within NASA, NOAA and the military for the future development of operational ocean-monitoring satellites, in which passive microwave sensors will undoubtedly play a major role. Operational ice-monitoring satellites with microwave radiometers aboard can also be expected to receive impetus in the years ahead. The outlook for operational spaceborne radiometers in earth-resource and land applications is less clear. Although spacecraft radiometer experiments for the land applications will most likely be continued in the coming years, much research and experimental analysis remains to be done to understand the complex terrain-electromagnetic interactions before the microwave data can become of operational use.

SPACEBORNE RADIOMETER SYSTEM CONCEPTS

Spaceborne radiometer systems may have a variety of characteristics related to their specific applications. However, there are certain basic features common to most systems which can be broadly classified into radiometer system features and features related to the spacecraft platform. These general features will be discussed here, prior to a more detailed discussion of individual sensors in the next section.

Radiometer System Description

Radiometer systems for microwave remote sensing typically consist of three basic subsystems: (1) an antenna and scan subsystem, which receives incoming radiation from specified beam-pointing directions; (2) a radiometer receiver and electronics subsystem, which detects and amplifies the received radiation within a specified frequency band; and (3) a data and control subsystem, which provides timing and sequencing signals for the antenna and radiometer subsystems, and performs digitizing, multiplexing and formatting functions on the radiometric and housekeeping data to form the output digital data stream. In spaceborne applications, the microwave radiometer may be only one of a number of different sensor types on the spacecraft. Thus, the radiometer output is typically interfaced with data from the other sensors and transmitted to ground via the spacecraft data system. On the ground, data from each sensor are separated and combined with ancillary spacecraft data such as ephemeris and timing data to form the basic sensor data record. For the microwave radiometer, the sensor data record forms the input to three stages of data processing: (1) radiometer calibrations, in which the radiometer digital output voltages are converted into units of antenna temperature; (2) antenna pattern corrections, in which effects of radiation entering the antenna sidelobes are accounted for; and (3) geophysical parameters retrieval, in which brightness temperatures measured by the radiometer system are interpreted in terms of the corresponding surface or atmospheric features. Stages (1) and (2) are usually considered as instrument-related stages in the data processing, with brightness temperature as the final processed sensor output, whereas stage (3) involves modeling of the surface and atmospheric emission characteristics and the use of various data inversion techniques. For some applications, not all of these data processing steps are necessary. For example, brightness temperature images of the polar regions are sufficient to show the variations in extent and type of sea ice, by virtue of the large brightness temperature contrasts between ice and water and between the different ice types.

To provide global coverage with adequate spatial resolution from earth-viewing remote sensors in orbit, scanning antennas are required. Microwave antennas may be scanned either mechanically as in rotating reflectors, or electrically as in phased array antennas. Whichever type is used, scanning is generally in the cross-track dimension, and coverage in the other dimension is provided by the spacecraft's motion (the situation is somewhat different at geosynchronous altitude, where the spacecraft remains fixed over a given point on the earth). One particular scan geometry which has become popular for both scanned reflector antennas and phased arrays is the conical scan, in which the beam is offset at a fixed angle from nadir and scanned about the vertical (nadir) axis. The beam thus sweeps out the surface of a cone. If the full 360° of a scan is utilized, double coverage fore and aft of the spacecraft is obtained. The main advantage of this type of scan is that the angle of incidence of the antenna beam at the earth's surface is constant, independent of scan position. This significantly increases the accuracy with which the brightness temperature data can be interpreted in terms of surface parameters.

For spaceborne applications most experience to date has been achieved using radiometers of the Dicke-switched superheterodyne type. An example of this type of system is shown in Figure 13-38. Prior to the receiver portion of the radiometer is a network of switches. The first of these determines which of a number of singals are to be fed into the receiver, either a signal from the vertical or horizontal antenna ports, or from one of the two calibration targets. Measurements of the calibration targets are used later in the data processing to calibrate the radiation received via the antenna in units of antenna temperature. The Dicke switch switches periodically at a high rate between the

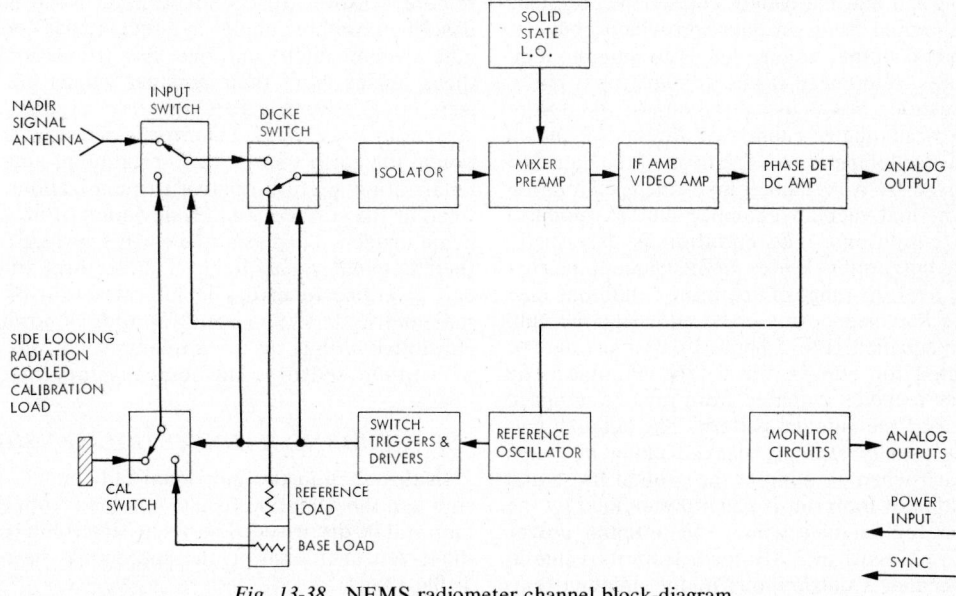

Fig. 13-38. NEMS radiometer channel block-diagram.

incoming signal and a known reference source. A synchronous detector at the other end of the receiver demodulates the signal. By these high-frequency comparisons of the signal with a known reference source, the effect of low-frequency gain variations in the amplifiers and other active devices is minimized. The local oscillator (LO) frequency is at the center of the desired radiometer passband and is mixed with the input signal to produce an intermediate-frequency (IF) signal which is then amplified and detected. In this system the critical noise component is the mixer, and considerable effort is made to keep it as low-noise as possible to improve the receiver sensitivity. After demodulation, the signal is integrated and sampled, and output to the data system. The receiver parameters that determine the temperature sensitivity (ΔT_{rms}) of the radiometer are the receiver noise temperature T_R, the receiver predetection bandwidth B, and the integration time τ. For a Dicke radiometer with square-wave demoulation the sensitivity can be expressed as:

$$\Delta T_{rms} = \frac{2(T_A + T_R)}{B\tau},$$

where T_A is the antenna temperature. As available integration times become smaller in advanced systems designed for high surface spatial resolution, total-power radiometers become advantageous, eliminating the Dicke switch and demodulators, and in theory increasing the available sensitivity. In addition, the development of low-noise rf amplifiers may eliminate the requirement for local oscillator/mixer heterodyne systems at the lower microwave frequencies. Other radiometer systems than described above have been used in the past for various applications with varying degrees of success. Some of these radiometer types are described in Chapter 11.

The instrument data processing usually occurs in the ground segment and consists of data calibrations and antenna pattern corrections. The data are calibrated using an equation which converts radiometer output voltage levels to antenna temperatures, referenced to the antenna input ports. This equation makes use of radiometric data from the two radiometer calibration targets of known microwave temperature. The form of the equation is derived from prelaunch instrument calibration in a thermal-vacuum chamber which simulates space conditions. The equation is essentially linear, but nonlinearities in instrument performance over the range of operating conditions may require nonlinear terms to be added to the calibration equation. The calibrated data must also be corrected for effects due to the antenna. The antenna receives radiation from regions of space defined by the antenna pattern. The antenna pattern is usually strongly peaked along its beam axis, and when pointing at the ground its spatial resolution or footprint is generally defined by the angular region over which the antenna power pattern is less than 3 dB down from its value at beam-center. Contributions to the antenna temperature from regions outside the footprint region are accounted for by inversion of the antenna pattern equation. This inversion procedure rapidly becomes complex as increasing accuracy is sought in deriving brightness temperatures for the 3-dB footprints. A tradeoff usually occurs between data processing complexity and brightness temperature accuracy. The situation can be improved considerably by the design of antennas with high beam efficiency and low sidelobes.

The remaining data processing consists of interpreting the brightness temperatures in terms of the desired geophysical variables. This involves physical models of microwave emission from the earth's atmosphere and surface, as described in Chapter 4, and the use of various data inversion techniques. Since these techniques depend on the particular application, further discussion will be delayed until the description of specific spacecraft sensors in the next section.

Spacecraft Considerations

Microwave radiometers are typically one of several sensors flown together on a spacecraft to form a combined remote sensing payload. Certain aspects of the spacecraft or satellite mission design can influence the utility of the radiometer data. These are: (1) orbit design, (2) attitude control, and (3) data processing and transmission network.

For the non-geosynchronous, circular earth orbits most commonly used in microwave remote sensing, the orbital parameters of importance are altitude, inclination, and the related ground trace pattern. The orbital altitude and the sensor scan geometry together determine the swath width on the earth's surface. The orbit inclination determines the latitude range of coverage. The altitude and inclination together determine the nodal precession rate and the ground trace pattern. The spacecraft attitude is also an important factor for remote sensors. Spacecraft attitude is normally described by three angles in a rectangular coordinate system: pitch, roll, and yaw. Deviations of these angles from their nominal values are referred to as attitude errors and affect a radiometer system in three ways: (1) error in footprint location at the earth's surface; (2) rotation of antenna polarization vectors relative to polarizations defined at the surface; and (3) deviation of antenna beam incidence angles at the earth's surface from their nominal value. Each of these three effects can give rise to errors in interpretation of the radiometric data. Spacecraft attitude is normally controlled within specified tolerances by the use of attitude sensors and compensatory mechanisms.

SPACEBORNE RADIOMETRIC SENSORS

In this section the individual radiometric sensors and the scientific results obtained from their data will be discussed. The discussion is limited to those sensors flown on the spacecrafts listed in Table 13-6.

TABLE 13-6

Earth Orbiting Spacecraft with Passive Microwave Sensors

Satellite Characteristics	Nimbus 5	Skylab Earth Resources Experimental Package (EREP)	Nimbus-6	Seasat	Nimbus 7
Launch date	December, 1972	May, 1973	June, 1975	June, 1978	October, 1978
Main Program Objectives	Development of measurement techniques for atmospheric processes related to meteorology and general circulation	Development of visible, infrared, and microwave sensors, and data evaluation techniques for earth resource applications from space.	To sound the atmosphere using advanced techniques extending measurement capabilities demonstrated on previous Nimbus satellites.	Proof-of-concept mission to demonstrate: (1) techniques for global monitoring of oceanographic phenomena; (2) key features of an operational ocean monitoring system.	To conduct experiments in the pollution, oceanographic and meteorological disciplines. Refine atmospheric measurement capabilities demonstrated on previous Nimbus satellites.
Sensor Payload	Temperature Humidity Infrared Radiometer, Surface Composition Mapping Radiometer (SCMR), *Electrically Scanning Microwave Radiometer (ESMR)*, Infrared Temperature Profile Radiometer (ITPR), Selective Chopper Radiometer (SCR), *Nimbus-E Microwave Spectrometer (NEMS)*	Multispectral Photographic Cameras (S190A), Earth Terrain Camera (S190B), Infrared Spectrometer (S191), Multispectral-Scanner (S192), *Microwave Radiometer/Scatterometer and Altimeter (S193)*, *L-band Radiometer (S194)*	THIR, *ESMR*, High Resolution Infrared Radiation Sounder (HIRS), *Scanning Microwave Spectrometer (SCAMS)*, Earth Radiation Budget (ERB), Limb Radiance Inversion Radiometer (LRIR), Pressure Modulator Radiometer (PMR)	Radar Altimeter (ALT), Seasat-A Satellite Scatterometer (SASS), Synthetic Aperture Radar (SAR), *Scanning Multichannel Microwave Radiometer (SMMR)*, Visible and Infrared Radiometer (VIRR)	THIR, *SMMR*, ERB, Coastal Zone Color Scanner (CZCS), Limb Infrared Monitor of the Stratosphere (LIMS), Stratospheric Aerosol Measurement (SAM 11), Stratospheric and Mesospheric Sounder (SAMS), Solar Backscatter Ultraviolet and Total Ozone Mapping Spectrometer (SBUV/TOMS)
Orbit Characteristics: Altitude	Circular Sun-Synchronous 1100 km	Circular 438 km	Circular Sun-Synchronous 1100 km	Circular 800 km	Circular, Sun-Synchronous 955 km
Inclination	99°	50°	99°	108°	99.2°
Attitude Stabilization: Pitch	±1.0°	±2.0°	±0.5°	±0.5°	±0.7°
Roll	±0.5°	±2.0°	±0.5°	±0.5°	±1.0°
Yaw	±0.5°	±2.0°	±1.0°	±0.5°	±1.0°

NEMS (Nimbus-5)

The Nimbus-5 Microwave Spectrometer (NEMS) was the first microwave temperature sounder flown in space (Staelin et al., 1972). The instrument can be considered as composed of two parts: (1) three channels at center frequencies 53.65, 54.9 and 58.8 GHz for measuring primarily the atmospheric temperature profile; (2) two channels at 22.235 and 31.4 GHz, sensitive over the oceans to water vapor and liquid water, and indicating emissivity and surface temperature over land. Although these two parts could operate separately, performance could be improved by using data from all channels simultaneously in the profile and parameter estimations. A photograph of the NEMS instrument is shown in Figure 13-39.

The instrument comprised four main functional units as shown in the functional block diagram (Figure 13-40). These were: (1) the H_2O radiometer unit, (2) the O_2 radiometer unit, (3) the data/ programmer unit, and (4) the power supply unit. A fifth unit, the bench checkout equipment (BCE), was a ground-support and calibration unit.

The H_2O radiometer unit consisted of channels R_1 (22.235 GHz) and R_2 (31.4 GHz), which were completely independent. Each was fed by a lens-loaded circular horn, nadir-viewing, with a 3-dB beamwidth of 10°. This provided a surface footprint of about 190-km diameter. The two radiometers shared a space radiator to cool their calibration loads, and the ambient temperature reference and base loads of both channels were physically linked to ensure that they were at the same temperature. The O_2 radiometer unit consisted of channels R_3, R_4 and R_5 (53.65, 54.9 and 58.8 GHz), which were also independently operating radiometers. They shared a common space radiator for their cooled calibration loads and had their ambient reference and base loads physically linked.

The five radiometer channels were functionally identical and may be described by reference to Figure 13-38. Each was of the Dicke-switched, super-heterodyne type, with a two-point calibration system. The input and calibration switches allowed a selection of inputs to the instrument from either the antenna, the space-cooled calibration load, or instrument ambient-temperature base load. The receiver portion of the channel is a conventional superheterodyne system isolated from the Dicke switch to minimize effects due to impedance modulation. The signal was routed to a low-noise balanced mixer with a solid-state local oscillator. An intermediate frequency (IF) amplifier, video amplifier, phase detector and *dc* amplifier completed the radiometer channel, delivering a *dc* voltage proportional to the difference

Fig. 13-39. Photograph of NEMS instrument, with O_2 and H_2O channels in separate units, and power supply.

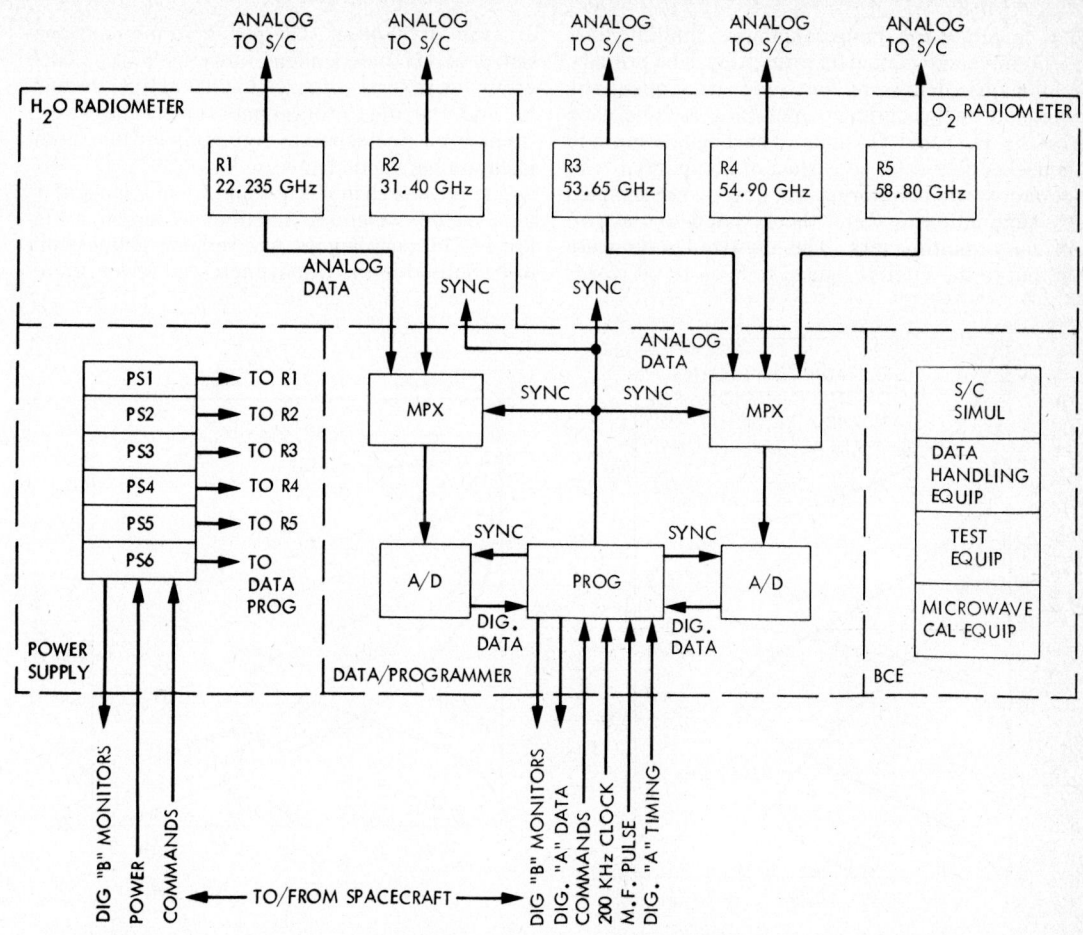

Fig. 13-40. NEMS system block-diagram.

between the Dicke reference load and the signal presented to the input port of the Dicke switch. The *dc* output passed to the data/programmer unit, where it was processed prior to entering the data stream as prime data. It was also passed through a buffer circuit, and output as an analog

signal for "quick look" and prime data backup capability. The instrument characteristics are summarized in Table 13-7.

The data/programmer unit provided timing and control signals (synchronized from the spacecraft's 200-Hz clock) for antenna and load switch-

TABLE 13-7

NEMS Instrument Characteristics

Characteristics	R_1	R_2	R_3	R_4	R_5
Local oscillator frequency (GHz)	22.235	31.4	53.65	54.9	58.8
RF bandwidth (MHz)	250	250	250	250	250
Integration time (sec)	2	2	2	2	2
Sensitivity, ΔT_{rms} (K)	0.24	0.23	0.29	0.29	0.24
Dynamic range (K)	100−325	100−325	100−325	100−325	100−325
Absolute accuracy (K) (longterm)	2	2	2	2	2
IF frequency range (MHz)	10−110	10−110	10−110	10−110	10−110
Antenna beamwidth (deg)	10	10	10	10	10
Antenna beam efficiency (%)	95	94	94	93	95

ing, integrate and dump commands, multiplexing, A/D conversion, and data formatting. The primary radiometric data were integrated and sequentially sampled by two primary multiplexers, one each for the H_2O and O_2 units. The engineering and housekeeping data, consisting of component temperatures and monitoring functions, were sampled by submultiplexers and then passed to the two primary multiplexers. The digitized data were output to the satellite data system as 10-bit words

at a rate of 50 *bits/s*. The power supply unit consisted of six independent power supplies, each providing several voltages, to channels R_1 through R_5 and the data/programmer unit. The power supplies carried separate commons for maximum isolation between channels.

The NEMS channels 3, 4 and 5 had atmospheric temperature weighting functions as shown in Figure 13-41a, enabling the atmospheric temperature to be sensed in the troposphere and lower strato-

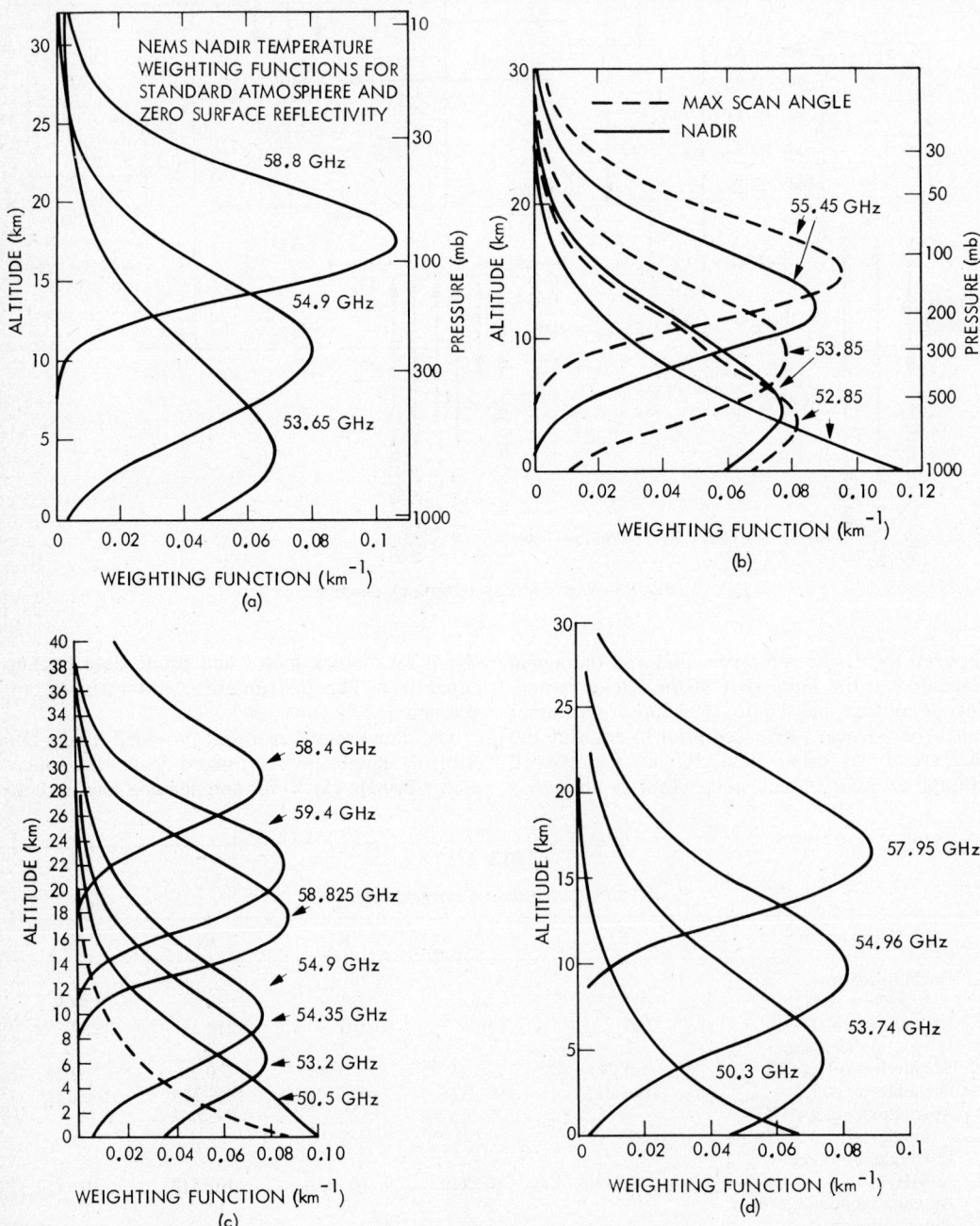

Fig. 13-41. Atmospheric temperature weighting functions, viewing the earth from space: (a) NEMS (zero surface reflectivity), (b) SCAMS (calm sea background), (c) SSM/T (calm sea background), (d) MSU (calm sea background).

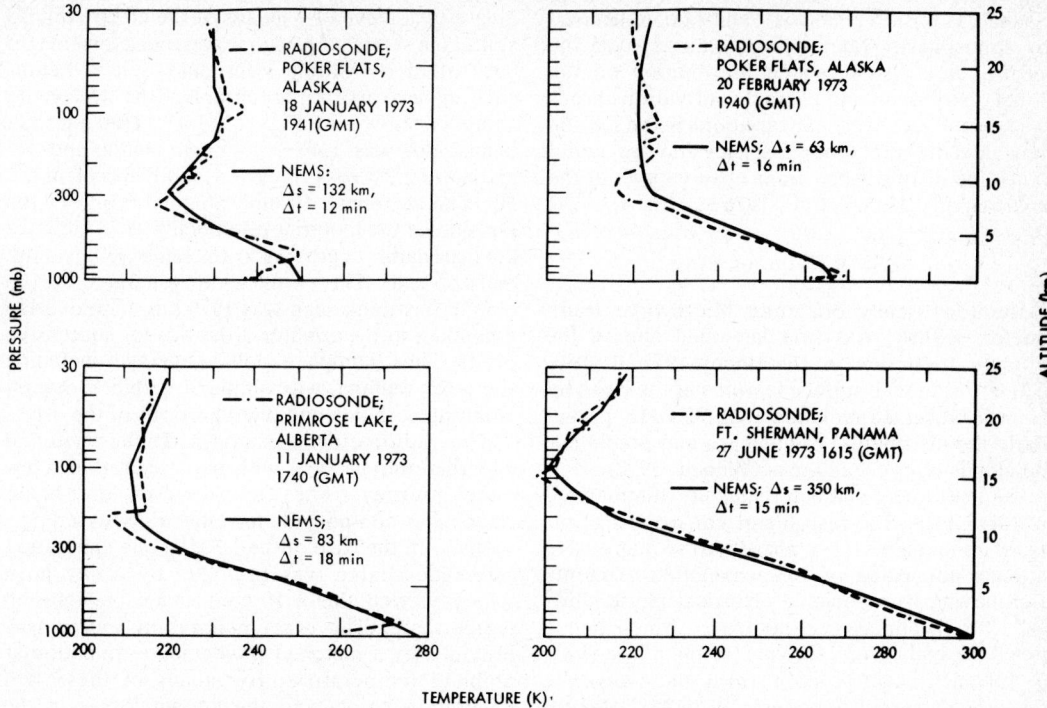

Fig. 13-42. Comparisons of individual temperature soundings by NEMS and coincident radiosondes. The spatial difference Δs and time difference Δt of the NEMS sounding relative to the radiosonde are indicated for each sounding (Waters et al., 1975).

sphere. To determine the atmospheric temperature profile at a number of discrete points using data from the three brightness temperature channels, a regression approach was used, incorporating *à priori* statistics from radiosonde data and atmospheric model calculations. Results from NEMS using this approach are illustrated in Figure 13-42 (Waters et al., 1975). These show the NEMS-retrieved profiles compared with radiosonde data in four different situations. The errors are largest where sharp changes in profile occur, showing the limitations in vertical resolution imposed by the width of the weighting functions. Comparisons using the most reliable NMC temperature profiles indicate that the NEMS *rms* temperature accuracy averaged over the profile is about 2°C. The effects of clouds on the profile retrieval accuracy were investigated by Staelin et al. (1975a). Clouds were observed to affect less than 0.5 percent of the temperature profile soundings. Most such effects occurred in the intertropical convergence zone (ITCZ) and altered the inferred temperature profile by less than a few degrees centigrade. Furthermore, these large clouds could be detected by NEMS channels 1 and 2 and thus accounted for in the data analysis.

The NEMS channels 1 and 2 were used for determining columnar atmospheric water vapor and liquid water abundances using simple regression approaches (Staelin et al., 1976; Grody, 1976). The accuracy of NEMS water vapor estimates was tested by comparison with radiosondes

launched from ships and islands. The results are shown in Figure 13-43 and indicate an overall *rms* accuracy of about 0.4 g/cm². Accuracy for the NEMS liquid water estimates was difficult to determine experimentally since there were no reliable sources of comparison data. Theoretical *rms* accuracy was estimated at 0.01 g/cm².

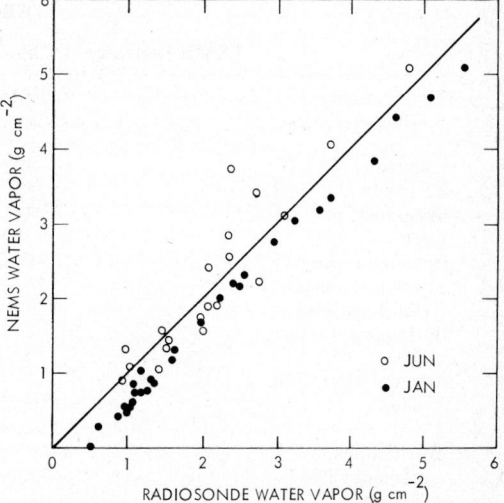

Fig. 13-43. Comparisons of NEMS water-vapor retrievals with radiosonde water-vapor measurements (Staelin et al., 1976).

Whereas NEMS channels 1 and 2 could be used for atmospheric water determinations over the oceans, they also provided information on the global distribution and character of various types of snow and ice. Seasonal variations in the ice and snow, and different types of sea ice and firn, could clearly be distinguished from observations in the polar regions (Kunzi et al., 1976).

ESMR (Nimbus 5)

The Electrically Scanning Microwave Radiometer (ESMR) was first launched aboard the Nimbus 5 satellite in December 1972. Its objective was to map surface features of the earth by means of received radiation at 19.35 GHz, particularly the distribution of polar ice and precipitating clouds over ocean areas (Wilheit, 1972).

The instrument characteristics are summarized in Table 13-8. The instrument consisted of four major components: (1) a phased array microwave antenna consisting of 103 waveguide elements each having its associated electrical phase shifters; (2) a beam-steering computer which determined the coil current for each of the phase shifters for each beam position; (3) a microwave receiver with center frequency of 19.35 GHz; (4) timing control and power circuits. The instrument is shown diagrammatically on the Nimbus-5 spacecraft in Figure 13-44. The block diagram is similar to one channel of that shown for the Nimbus-6 ESMR (see later).

The aperture area of the phased array antenna was 83.3 cm × 85.5 cm, with polarization linear and parallel to the spacecraft velocity vector. The antenna beam was scanned cross-track (i.e., perpendicular to the spacecraft velocity vector) in 78 steps, ±50° from nadir, every four seconds. Scanning was achieved by means of the electric phase shifters in series with each waveguide element and controlled by timing commands via a beam-steering network. A photograph of the waveguide array is shown in Figure 13-45. The antenna beamwidth was 1.4° × 1.4° near nadir, and degraded to 2.2° cross-track × 1.4° downtrack at the 50° scan extremes. From a nominal orbit of 1100 km altitude the footprint resolution was 25 km × 25 km near nadir, degrading to 160 km × 45 km at the ends of scan. The swath width generated by the ±50° cross-track scan was 1570 km, with overlap extending to the equator crossings for successive orbits. Thus, complete global coverage, including the polar regions, was obtained within 12 hours, counting both day and night portions of the orbit.

The radiometer was of the Dicke-switched superheterodyne type with two calibration reference sources. The functions of the basic radiometer components have been discussed previously. In the case of the ESMR, one calibration reference source was provided by a sky horn which viewed the 3 K cold space background temperature. The other calibration source was provided by a matched waveguide termination at ambient temperature. By means of these two calibration points and the known losses in the waveguide and switch components, the radiometer/antenna system could be calibrated to a designed long-term absolute accuracy of 2 K in antenna temperature. The temperature sensitivity of the instrument (noise △T) was also approximately 2 K. The instrument featured a stepped automatic gain control (AGC) which controlled the gain of the postdetection amplifier and optimized the use of the 10-bit A/D converter over the instrument dynamic range. The instrument had three outputs. The main data output was a

TABLE 13-8
ESMR Instrument Characteristics (Nimbus 5 and 6)

Characteristics	ESMR (Nimbus 5)	ESMR (Nimbus 6)
Frequency (GHz)	19.35	37
RF bandwidth (MHz)	250	250
Integration time (msec)	47	
Sensitivity, ΔT_{rms} (K)	1.5	
Dynamic range (K)	50–330	
Absolute accuracy (K) (longterm)	2	
IF frequency range (MHz)	5–125	
Antenna beamwidth (deg)	1.4 × 1.4 (nadir position) 1.4 × 2.2 (scan extreme)	1.0 × 0.95 (scan center) 0.85 × 1.17 (scan extreme)
Antenna beam – efficiency (%)	90 to 92.7	
Polarization isolation (dB)	−25 to −36	

NIMBUS 5

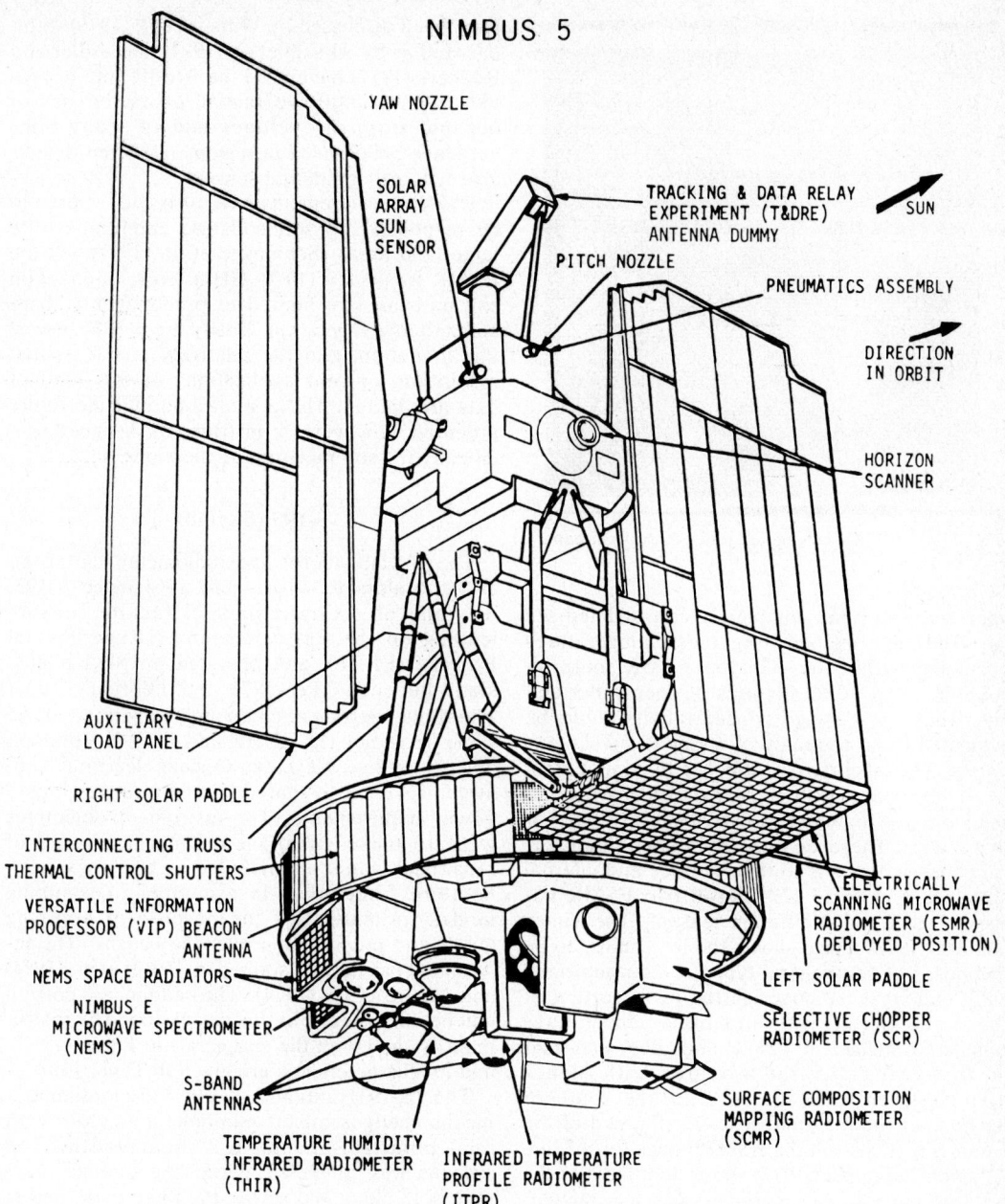

Fig. 13-44. Nimbus 5 satellite, launched December 1972.

digital bit stream at a rate of 200 *bits/s,* comprising the radiometric data multiplexed with AGC step number and housekeeping data necessary for the data processing. Other outputs were a set of command relay and frame identification status words, and a set of analog voltages of engineering interest, sampled once per 16 seconds. The layout of the ESMR radiometer and electronics is shown in Figure 13-46.

To account for effects of antenna sidelobes a simple one-dimensional correction was applied to the antenna temperatures. This was sufficient due to the extremely low antenna sidelobe levels in the plane parallel to the waveguide elements. A matrix operation on the 78 antenna temperatures in each scan performed the necessary corrections for each beam position.

The ESMR capability for providing images of the polar icecaps, through clouds, and both day and night, has found important applications in polar research and marine operations. There is a large contrast in brightness temperature between

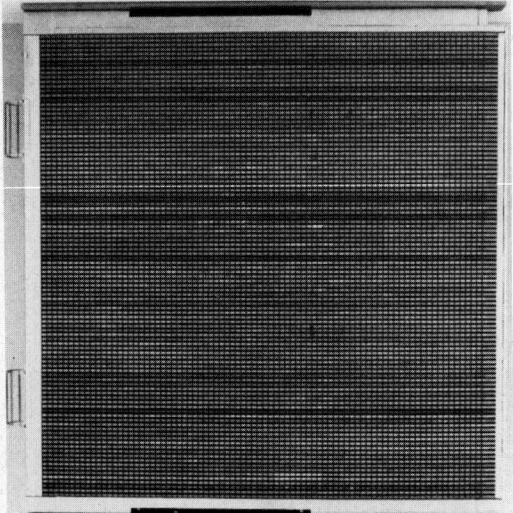

Fig. 13-45. ESMR waveguide phased array antenna.

open sea water and completely consolidated sea ice. Thus intermediate brightness temperatures occurring within the ~35 km ESMR footprint could be ascribed to mixtures of open water and ice, whose percentage concentrations could be estimated to an accuracy of about 15 percentage points. This accuracy is affected to some extent by the existence of three radiometrically distinct sea-ice types: first year, multiyear, and first-year thin sea ice. These ice types each have characteristic emissivities due to their surface and internal structure. Figure 13-47☆ shows two ESMR images of the arctic at different seasons. The winter (January 1973) image shows the ice canopy to be composed of two general types of ice: the principally multiyear ice covering the main portion of the Arctic Ocean, with brightness temperatures ranging from 209 K to 223 K, and either first-year or first-year/multiyear mixtures with higher brightness temperatures covering the southern portions of the marginal seas. By the end of the summer melt season the ESMR image for the late summer (September 1973) shows that most of the ice that covered the marginal seas has melted. Detailed examination of year-round images such as these has provided a wealth of information on the complex morphology and dynamics in the arctic and antarctic regions (Gloersen et al., 1974; Gloersen and Salomonson, 1975; Campbell et al., 1976; Zwally and Gloersen, 1976; Gloersen et al., 1978).

Over the oceans the ESMR has been shown to be capable of detecting precipitation areas. These appear in the ESMR images as regions of high brightness temperature caused by the larger droplet sizes and total water content in precipitating clouds (Wilheit et al., 1977). Meteorological applications of these images compared to images from infrared and visible sensors

have been discussed by Wilheit et al. (1976). Similar studies by Allison et al. (1974), and Adler and Rodgers (1977) have used the ESMR data to provide semi-quantitative rainfall characteristics of oceanic tropical cyclones and to study convergence phenomena in regions not often detectable with other satellite sensors.

ESMR measurements have also been shown to be responsive to soil moisture conditions over large land areas (Schmugge et al., 1977). At the ESMR frequency (19.35 GHz), correlations with soil moisture are limited to predominantly bare soil and low vegetation density areas. Because of this limitation, and the relatively coarse spatial resolution, current applications of this kind of data are limited. These studies do indicate future potential, however, for instruments designed specifically for soil moisture applications.

S-193 (Skylab)

The Skylab microwave radiometer/scatterometer and altimeter instrument, designated S-193, was one of six experimental remote sensors making up the Earth Resources Experimental Package (EREP), and operated on Skylab missions between May 1973 and February 1974. The three instrument components operated at the same frequency (13.9 GHz) and shared a common antenna and scan system. Certain electronic subsystems were also shared to minimize weight and power requirements. The instrument objectives were to make simultaneous measurements of radar backscatter and radiometric brightness temperature from orbit, in a number of scanning modes, primarily for the purpose of studying winds and precipitation over the oceans. The instrument has been described by Potter et al. (1974) and Moore et al. (1974). The radiometer portion will be emphasized in this discussion. The instrument is shown on the spacecraft in Figure 13-48, and its characteristics are given in Table 13-9.

The (S-193) antenna was a 115-cm-diameter mechanically scanned parabolic reflector, with dual polarization and a 2° beamwidth. The antenna was gimballed, permitting antenna scanning both along and across the flight path, and at the nadir position provided a circular footprint of 16-km diameter. Drive commands were received from an antenna controller which could follow any of four preset scan modes. A block diagram of the system is shown in Figure 13-49. The signal from the antenna was preamplified using a tunnel-diode amplifier, and the IF signal generated by the mixer was split between the radiometer and scatterometer receivers. After detection and integration the output was digitized and recorded on an EREP 10 kb/s tape recorder channel. The calibration source box contained two calibration sources, used for temperature reference and to set the automatic gain control of the radiometer receiver. Three choices of receiver integra-

Fig. 13-46. ESMR radiometer and scan system electronics.

tion times were available in an effort to accommodate the different scan modes of the antenna. Complete calibration of the system including the antenna was not performed prior to launch. Instead, the antenna insertion loss and the system transfer function were computed in orbit, based on measurements with the antenna pointed towards cold space of assumed microwave temperature 2.7 K.

Analysis of the radiometric data was directed towards studying correlations between the active and passive data over a variety of surface types. These correlations were studied using data measured at antenna scan angles between 0° and 48° from nadir. Lowest correlations were found between active and passive data at vertical polarization and 30° nadir angle (Sobti and Moore, 1976). Effects of clouds and weather, however, probably

significantly affected the results, making conclusions about surface effects difficult. Studies of the effects of soil moisture on the radiometer response were reported by Moore et al. (1975). Since soil moisture surface measurements were difficult to obtain, a related antecedent precipitation index (API) was used instead. Although a trend was evident towards lower emissivities with higher API, the correlation was relatively low, probably due to the many assumptions in the API model and the significant effects of surface roughness and vegetation at 13.9 GHz. Another study by Ulaby et al. (1975) compared data from the S-193, S-194 (1.4 GHz), and ESMR (19.35 GHz) sensors, over the Utah Great Salt Lake Desert. Within this region, large decreases in brightness temperature and increases in backscattering coefficients were observed, as a result of the smooth,

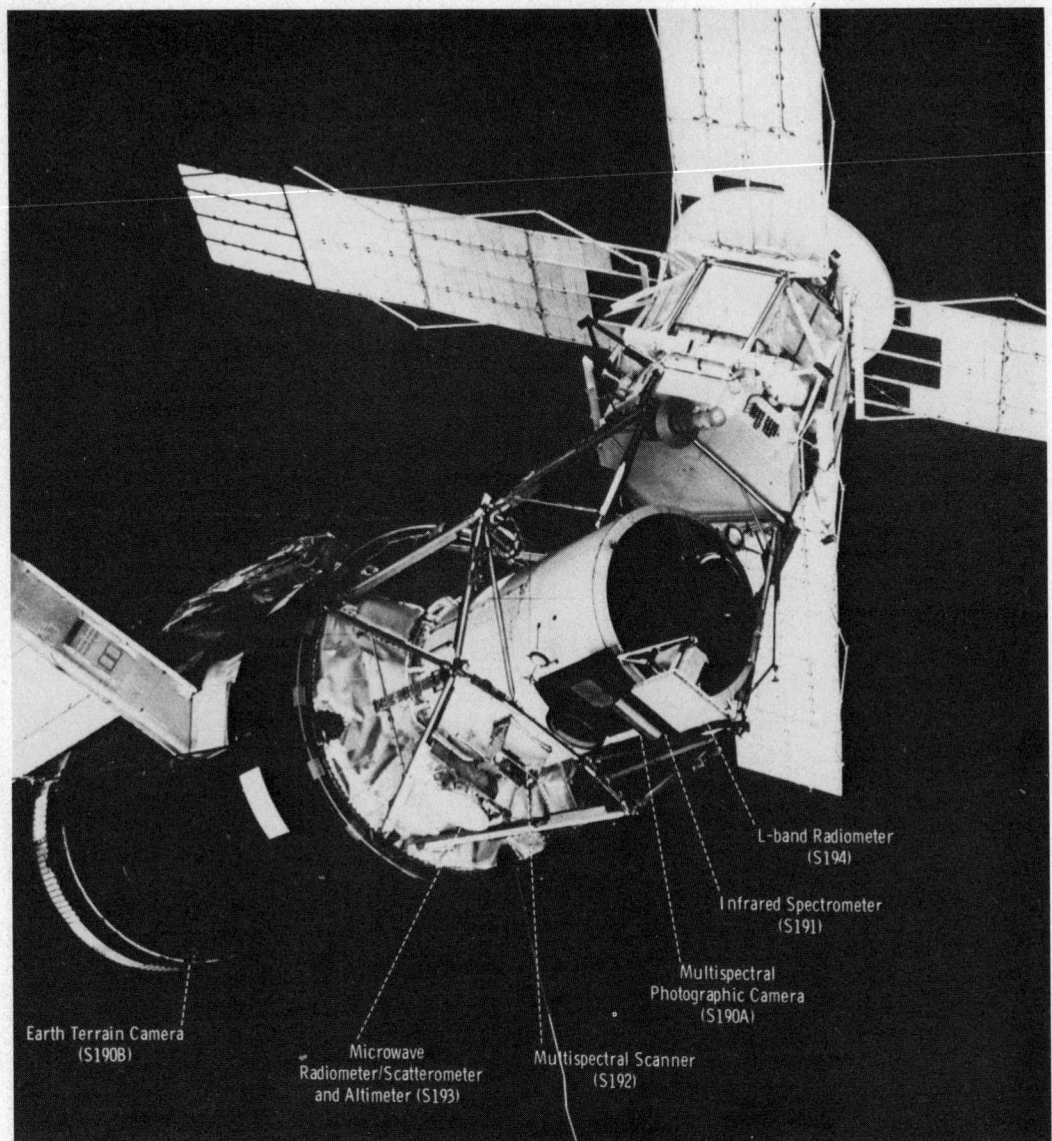

Fig. 13-48. Skylab, launched in May 1973, showing location of EREP sensors on the multiple docking adaptor.

TABLE 13-9

Skylab S193 and S194 Instrument Characteristics

Characteristics	S-193	S-194
Frequency (GHz)	13.9	1.4
RF bandwidth (MHz)	200	27
Integration time (msec)	256, 128, or 32	
Sensitivity, ΔT_{rms} (K)	0.4, 0.6, or 1.2	~0.85
Dynamic range (K)	50–350	0–350
Absolute accuracy (K) (longterm)	~2	~2
Antenna beamwidth (deg)	2	15
Antenna beam efficiency (%)	90	
Polarization isolation (dB)	−16.3	

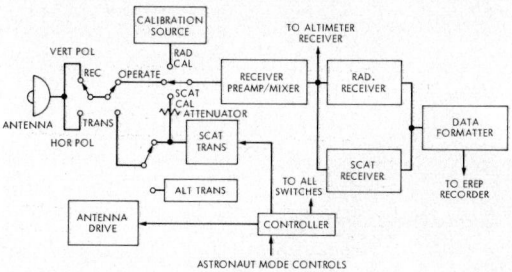

Fig. 13-49. Skylab S-193 System block-diagram [Moore et al., 1974].

bare nature of the surface and possible effects of subsurface brine sediments.

S-194 (Skylab)

The S-194 was an L-band (1.4 GHz) radiometer system flown as part of the EREP remote sensing payload on Skylab. The objective was to measure radiation from the earth at this frequency, with application to earth resources and in particular soil moisture. The instrument consisted of a fixed, nadir-viewing, phased array antenna, and receiver, calibration, and data subsystems (Potter et al., 1974). The antenna beamwidth was 15°, providing a surface footprint of 115 km diameter. The radiometer was a modified Dicke type, with tuned RF receiver and gain modulation. Other characteristics are given in Table 13-9. The instrument is shown on the spacecraft in Figure 13-48.

In addition to comparisons with S-193 and ESMR data (described above under S-193), the data from S-194 were studied for correlations with soil moisture by Eagleman and Lin (1976). Due to the large footprint size, water balance models were used to extrapolate point measurements of surface soil moisture for comparison with the spacecraft data. The observed correlations with soil moisture in the top 2.5 cm were quite high. However, the data points used were not strictly independent since they represented footprints with significant amounts of overlap. Studies by

McFarland (1976) using the antecedent precipitation index showed similar results.

SCAMS (Nimbus-6)

The Scanning Microwave Spectrometer (SCAMS) on Nimbus 6 had similar objectives to the NEMS instrument, i.e., to determine atmospheric temperature profiles and, over ocean surfaces, the abundance of water vapor and liquid water in the atmosphere (Staelin et al., 1975b). Whereas NEMS observed atmospheric parameters at nadir, along the subsatellite track, SCAMS operated by scanning to either side of this track to produce maps of these parameters with nearly full earth coverage. SCAMS was intended to provide a unique global data set for research and trial operational use in meteorological applications, and for use in the Global Atmospheric Research Program (GARP).

The SCAMS instrument characteristics are given in Table 13-10. A photograph of the instrument mounted in its assembly fixture is shown in Figure 13-50. There is one antenna each for the two H_2O channels and a single antenna for the three O_2 channels. Scanning was achieved by stepping a reflector in front of each antenna. Each reflector, inclined at 45° to its antenna, scanned the antenna beam in a left-handed sense with respect to the spacecraft velocity vector, and a complete 360° scan took 16 seconds. Thirteen earth-data samples, each separated by a 7.2° scan step, were recorded between ±43.2° from nadir. Three seconds at the end of each earth scan were reserved for reflector rotation and system calibration, which took place by 1-second views of cold space and of an instrument blackbody source. The antenna beamwidths of 7.5° gave ground resolutions of 145 km at nadir and 220 km × 360 km at the edges of the scan (43.2° from nadir). The ±43.2° scan covered a swath width of 2400 km, which provided nearly full-earth coverage every 12 hours. The polarization vector rotated as the beam scanned. At nadir, polarization was parallel to the satellite velocity vector for all channels except channel 5 (55.45 GHz), for which it was perpendicular.

TABLE 13-10

SCAMS Instrument Characteristics

Characteristics	1	2	3	4	5
Frequency (GHz)	22.235	31.65	52.85	53.85	55.45
RF bandwidth (MHz)	220	220	220	220	220
Integration time (sec)	0.95	0.95	0.95	0.95	0.95
Sensitivity, ΔT_{rms} (K)	0.2	0.2	0.6	0.5	0.5
Dynamic range (K)	0−350	0−350	0−350	0−350	0−350
Absolute accuracy (K) (longterm)	2	2	2	2	2
IF frequency range (MHz)	10−110	10−110	10−110	10−110	10−110
Antenna beamwidth (deg)	7.5	7.5	7.5	7.5	7.5
Antenna beam efficiency (%)	97.3	97.3	96.9	97.1	97.4

Fig. 13-50. Photograph of SCAMS instrument.

The functional operation of the SCAMS was similar to that described for the NEMS instrument (Figure 13-40, with the addition of a scan subsystem). The 22.2- and 31.6-GHz channels shared no components in common, whereas the three O_2 channels shared a common antenna and calibration target. A channelizing filter separated the frequencies of the two O_2 channels with like polarizations; the third O_2 frequency was separated by a polarization transducer. The radiometers were of the Dicke-switched superheterodyne type and, from the Dicke switch back, were identical in operation to those used on the NEMS, (figure 13-38). The scan subsystem consisted of the scan controller, which controlled the scan stepping and timing sequences; the scan driver with motor, drive electronics and position readouts; and the scan mechanism with shafts, pulleys, belts and bearings. The system was designed so that alignment and synchronization of the three antenna beams was achieved to within $\pm 0.2°$. The radiometer outputs were sequentially integrated and sampled, digitized in 10-bit words, and with engineering and housekeeping data were output as a "Digital A" stream into the spacecraft data system. Other backup radiometer analog signals

and key indications of radiometer state were fed into a "Digital B" data system. The use of external calibration targets viewed by the antennas was an improvement in the calibration design. By this means a more accurate system calibration was achieved for the orbital radiometric data. After calibration in the ground processing, the data were deconvolved to remove the antenna sidelobe contributions. The high beam efficiencies of the SCAMS antennas ($\geq 97\%$) enabled this to be done fairly simply.

Weighting functions for the three temperature-sounding channels are shown in Figure 13-41b. These provided theoretical retrieval accuracies comparable to NEMS. SCAMS-derived temperature profiles were compared with radiosonde temperatures for several European frontal systems by Grody and Pellegrino (1977). Figure 13-51a shows a cloud image from the NOAA-4 VHRR for a case-study region over Europe, with two adjacent SCAMS grid patterns for 26 and 27 January, 1976. A 700-mb temperature analysis derived from available radiosonde reports is shown in Figure 13-51b, revealing a deep trough with strong temperature gradients. Figure 13-51c shows the 700-mb SCAMS result in which the po-

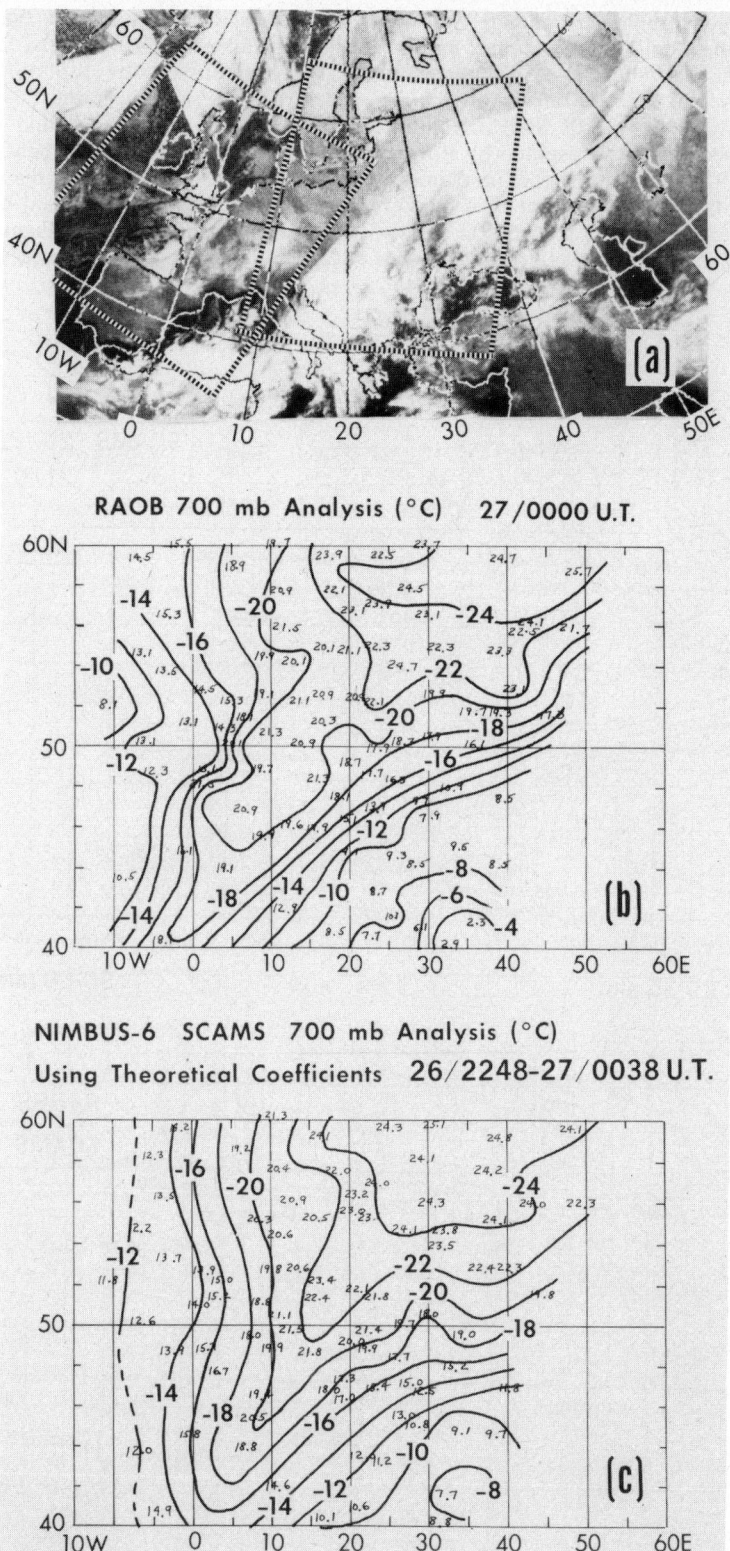

Fig. 13-51. Comparisons between SCAMS 700-mb retrieved temperature measurements and radiosonde data, January 26–27, 1976: (a) NOAA-4 VHRR satellite image showing cloud features and SCAMS grid coverage, (b) analysis of 0000 UT radiosonde temperature, (c) analysis of 2248-0038 UT SCAMS retrieved temperatures (Grody and Pellegrino, 1977).

sitions of the major trough and ridge are well defined, although the amplitudes are somewhat un-
derestimated. The *rms* temperature accuracies at
this level for the SCAMS retrievals are about
2 K. A more stringent test of SCAMS cloud-
penetrating capabilities was presented by Rosenkranz et al. (1978), in which the instrument
was used to study the development of Typhoon
June over the Pacific. Results showed the possibility of deriving cloud liquid water content and

sea surface winds around the typhoon from the
22 GHz and 31 GHz channels. The 55.45 GHz
channel displayed the typhoon's warm core
temperature structure at the 200-mb level. Improved spatial resolution would have increased
the impact of these measurements. Kidder et al.
(1978) have also investigated the use of SCAMS
data for estimation of tropical cyclone central
pressure and outer winds. Figure 13-52 shows
SCAMS-derived columnar water vapor contours

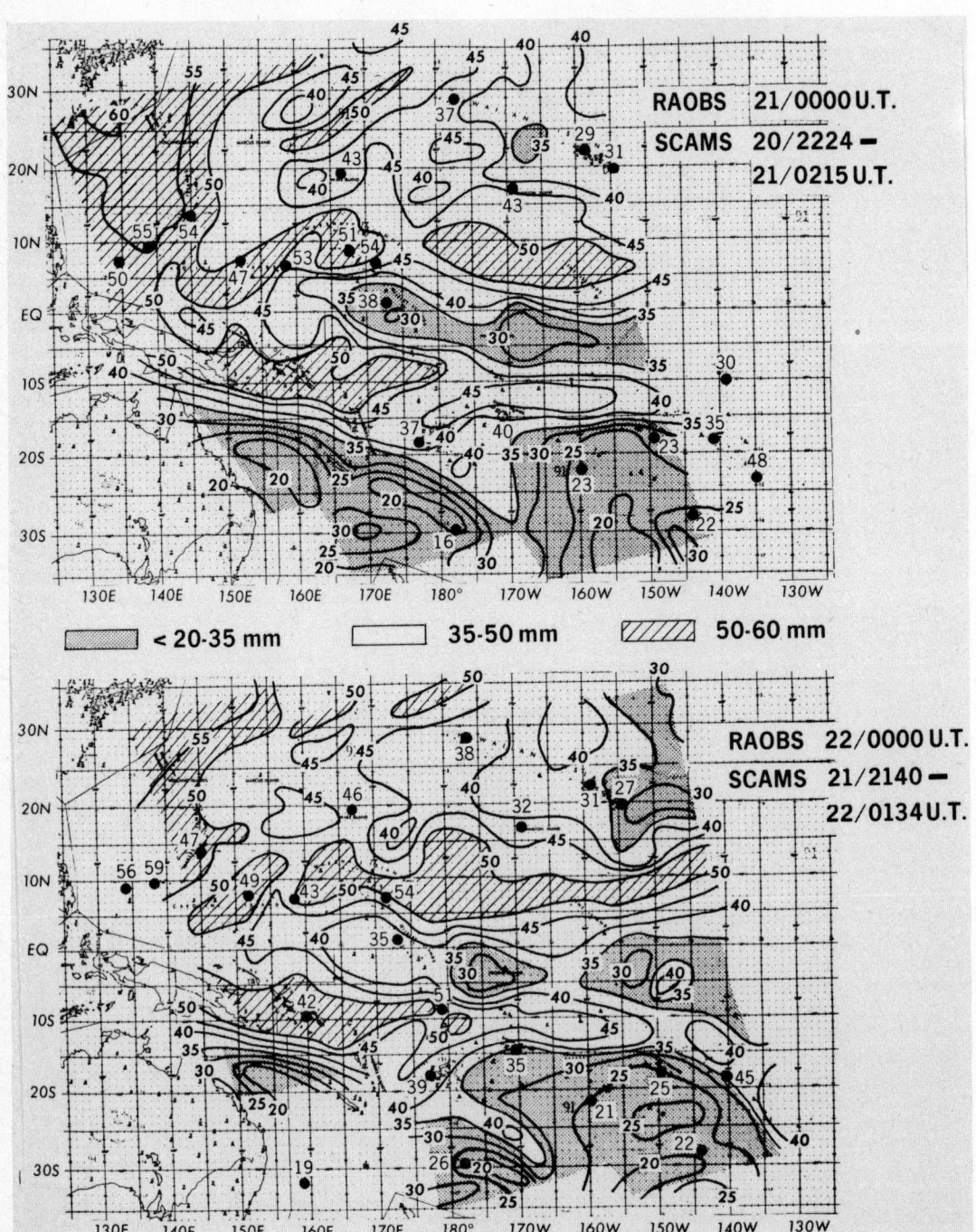

Fig. 13-52. Comparisons between SCAMS retrieved precipitable water (mm) and radiosonde measurements, August 20–22, 1975 (Grody et al., 1980).

in the western Pacific, compared to local radiosonde observations. Knowledge of these large variations in water vapor within the tropical environment is of utmost importance in understanding the development of tropical cyclones and cloud clusters (Grody et al., 1980).

ESMR (Nimbus-6)

The successful operation of the Electrically Scanning Microwave Radiometer (ESMR) on Nimbus 5 laid the basis for a second ESMR which was launched on the Nimbus-6 satellite in June 1975 (Wilheit, 1975). The operating wavelength for the Nimbus-6 ESMR (ESMR-6) was 37 GHz (as opposed to 19.35 GHz for ESMR-5). This frequency change had the effect of tripling the instrument's sensitivity to water droplets while keeping its sensitivity to water vapor essentially the same, thus making it easier to distinguish light rain areas from areas of high water vapor. A secondary effect was to roughly double the contrast between first-year and multiyear ice, although some increase in ambiguity of ice concentration determination occurred. Snowfield mapping capability was also expected to improve.

The ESMR-6 measured both horizontally and vertically polarized radiation at the 37-GHz frequency, as opposed to horizontal only for ESMR-5. The polarization information facilitated quantitative interpretation of the radiometric measurements, and provided information on winds over the oceans. The ESMR-6 characteristics are summarized in Table 13-8.

The antenna beam of the ESMR-6 scanned ahead of the spacecraft along a conical surface, with a constant angle of 45° with respect to the antenna axis (Figure 13-53). The beam scanned in azimuth ±35° about the forward direction in 71 steps. The 3-dB beamwidth varied from 0.95° × 1.0° at scan center to 1.17° × 0.85° at the scan extremes. The antenna axis was tipped forward 5° from the vertical axis of the spacecraft. The variation of incidence angle across the scan was about ±0.6°. The 3-dB footprint on the earth's surface had dimensions approximately 20 km × 43 km. The scan pattern resulted in a swath width of 1300 km, which allowed polar coverage up to latitudes of 85.7°.

The radiometer section of the instrument was similar in operation to that of ESMR-5. However, there were two separate receivers, one for each polarization, with common Gunn diode local oscillators (see Figure 13-54). Radiation of both polarizations was received and separated in the phased array antenna before being fed into the radiometers. The output digital bit stream comprised vertical and horizontal radiometer data with associated thermistor and other housekeeping data, at a rate of approximately 300 *bit/s*.

Calibration was achieved by making use of data from two calibration references sources, a sky horn and ambient matched termination. These reference points, together with the known losses

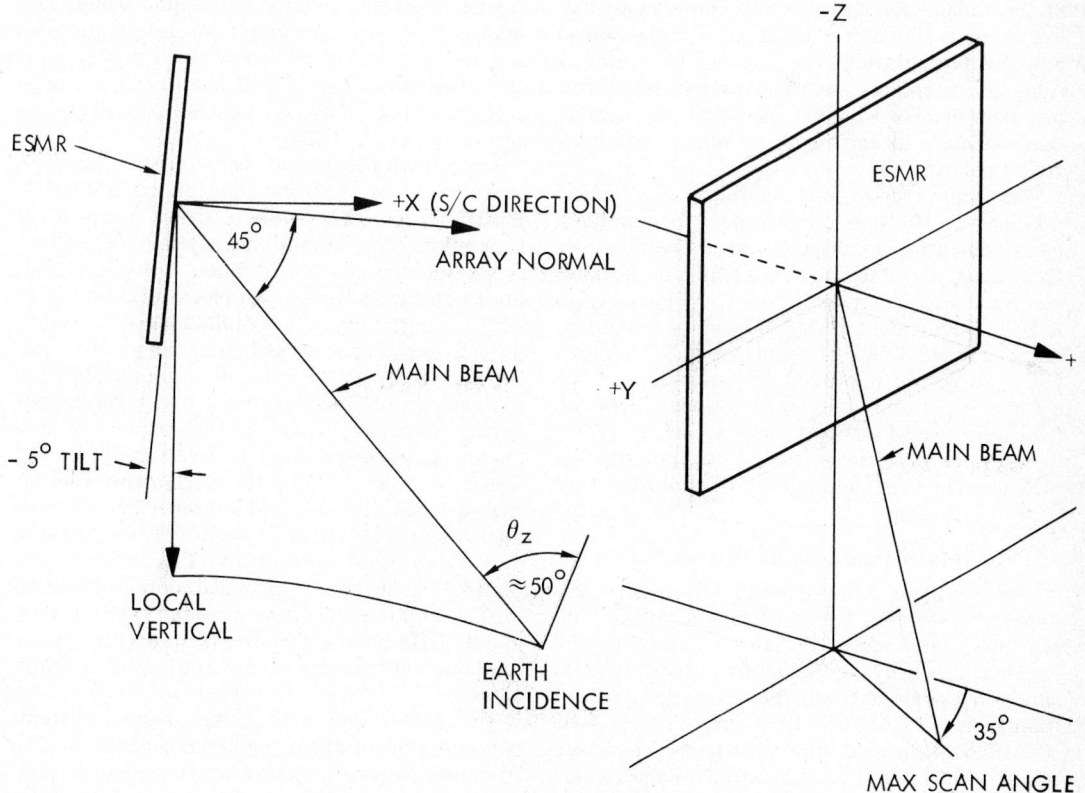

Fig. 13-53. ESMR antenna scan geometry (Nimbus 6) (Wilheit, 1975).

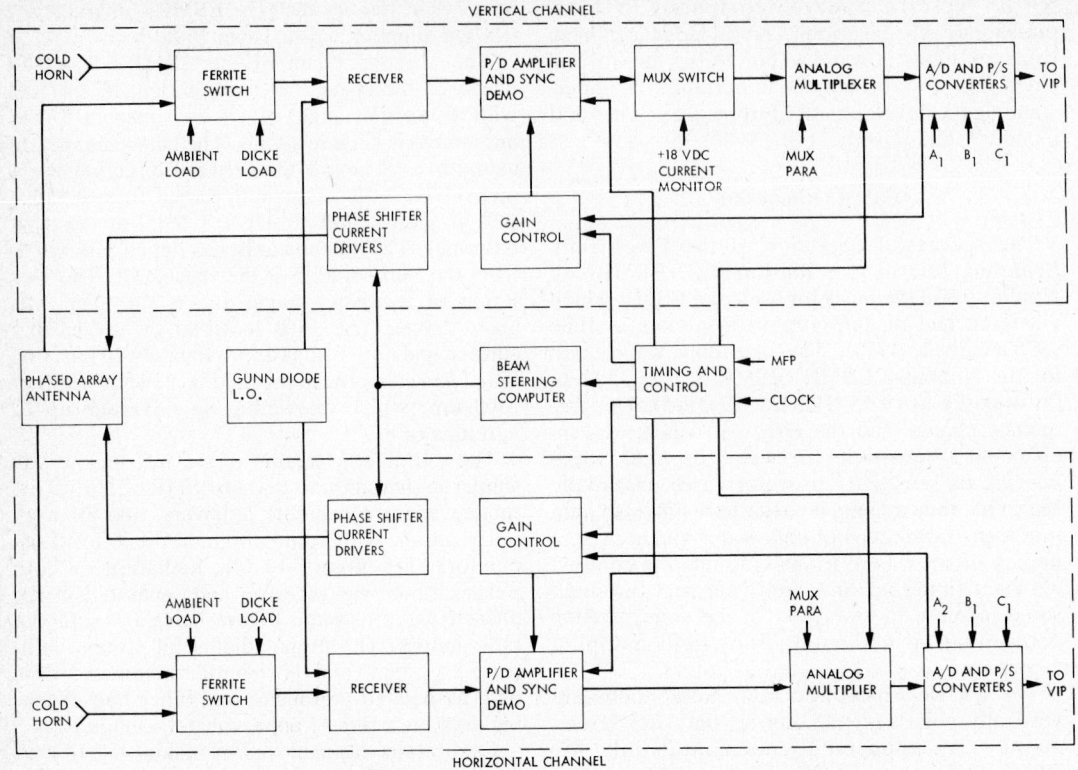

Fig. 13-54. Simplified system block-diagram of Nimbus 6 ESMR (Wilheit, 1975).

of the radiometer and antenna components, allowed the radiometer outputs to be converted to antenna temperatures. A scheme for antenna sidelobe corrections was used, similar to but simpler than that for ESMR-5. An additional correction was made for mixing of the vertical and horizontal polarizations.

The dual-polarization capability of ESMR-6 was used to study its effectiveness for improved measurements of wind speed and rainfall rate over the oceans, in addition to continuation of the polar ice studies begun with ESMR-5. Wilheit (1970) analyzed coincident ESMR-6 brightness temperatures and wind measurements from NOAA data buoys to get an improved understanding of the effect of ocean emissivity on wind speed. ESMR-6 data were used by Weinman and Guetter (1977) to show the potential for using dual-polarization measurements for determining rainfall over land as well as ocean.

SMMR (Seasat and Nimbus-7)

The Scanning Multichannel Microwave Radiometer (SMMR), launched on the Seasat and Nimbus-7 satellites in 1978, is a five-frequency imaging microwave radiometer system. It measures dual-polarized radiation from the earth at frequencies of 6.6, 10.7, 18, 21 and 37 GHz. The SMMR data provide information on sea-surface temperature, wind speed, atmospheric water

vapor, cloud and precipitating liquid water, and sea-ice type and concentration. Information on soil moisture and terrain snow cover properties is also obtainable. The SMMR instrument has been described in detail by Gloersen and Barath (1978), and Njoku et al., 1980a).

The SMMR instrument characteristics are given in Table 13-11. A front view of the SMMR is shown in Figure 13-55. There are six independent Dicke-type superheterodyne radiometers, fed by a single antenna and a calibration subsystem. Figure 13-56 shows the system block diagram. At 37 GHz, two radiometers simultaneously measure the horizontal and vertical components of the received signal. At the other four frequencies the radiometers alternate between two polarizations during successive scans. In this manner 10 data channels, corresponding to five dual-polarized signals, are provided by the instrument. The ferrite switches, isolator, and reference and ambient loads are packaged as a single unit for low-loss and isothermal operation. The mixers are Shottky-barrier diode, balanced, double-sideband mixers with integral IF preamplifiers having a 10- to 110-MHz bandwidth. The local oscillators are fundamental-frequency, cavity-stabilized Gunn diodes.

The SMMR has a scanning antenna system, consisting of an offset parabolic reflector with a 79-cm-diameter collecting aperture and a mul-

TABLE 13-11

SMMR Instrument Characteristics (nominal)

Characteristics	1	2	3	4	5
Frequency (GHz)	6.6	10.69	18	21	37
RF bandwidth (MHz)	220	220	220	220	220
Integration time (msec)	126	62	62	62	30
Sensitivity, ΔT_{rms} (K)	0.9	0.9	1.2	1.5	1.5
Dynamic range (K)	10−330	10−330	10−330	10−330	10−330
Absolute accuracy (K) (longterm)	2	2	2	2	2
IF frequency range (MHz)	10−110	10−110	10−110	10−110	10−110
Antenna beamwidth (deg)	4.2	2.6	1.6	1.4	0.8
Antenna beam efficiency (%)	87	87	87	87	87

tifrequency feed assembly. The antenna reflector is mechanically scanned about a vertical axis, with a sinusoidally varying velocity, over a ±25° azimuth angle range. The antenna beam is offset 42° from nadir, thus the beam sweeps out the surface of a cone and provides a constant incidence angle at the earth's surface. Calibration is achieved by alternately switching in a "cold horn" viewing deep space and a "calibration load" at instrument ambient temperature, at the scan extremes. The reflector is supported by a hexapod structure attached to a ring surrounding the feed horn. Both reflector and pods are made from graphite-epoxy; the reflector is aluminized and the pods are wrapped with aluminum foil for thermal and electrical reasons. The multifrequency feed horn is a ring-loaded corrugated conical horn with a sequence of waveguide tapers, resonators and orthomode transducers at the throat, to which are coupled the 10 output ports. The cold calibration horns are similar in design to the feed horn. One horn serves the 6.6- and 10.7-GHz channels, another the 18- and 21-GHz channels, and a third the 37-GHz channels. These horns are scaled to provide equal beamwidths of 15°. The antenna hexapod supporting ring is driven back and forth sinusoidally by a friction-clutch-equiped motor shaft via a double-cogged belt of polyurethane impregnated Dacron-filament construction. A separate belt from the antenna ring drives two redundant 12-bit optical shaft angle encoders for antenna position readout. Angular momentum compensation is accomplished by loading the motor rotor, which acts as a flywheel, with the appropriate mass.

A data-programmer unit in the electronics assembly provides the timing, multiplexing and synchronization signals, contains A/D converters, multiplexers and shift registers, and provides formatting and buffering functions between the instrument and the spacecraft digital systems.

The Seasat SMMR data processing flow is shown schematically in Figure 13-57. The SMMR digital bit stream is extracted from the satellite data stream in the early stages of ground processing to form the SMMR Sensor Data Record (SDR). This contains the SMMR radiometric, en-

gineering and housekeeping data, in addition to satellite and footprint location data. The SDR is input to the calibration processing, which converts the digital radiometer output data into antenna temperatures. This is done by the standard two-point calibration technique, using radiometric data from the sky horn and internal ambient load, engineering data consisting of monitored component temperatures, and the known system component losses (Swanson and Riley, 1980). The antenna temperatures are then input to the antenna pattern correction algorithms, which reformat the data, compensate for antenna sidelobe and cross-polarization effects, and perform a number of other operations such as corrections for spacecraft attitude and data quality flagging. These antenna pattern corrections are performed in a rather detailed manner for the SMMR, since the beam efficiencies are low, and the brightness temperatures are required to high accuracy, particularly for the 6.6- and 10.7-GHz frequencies (Njoku et al., 1980b). The brightness temperature data are then passed on to the final stage of data processing in which the geophysical parameters are derived.

The ocean-parameter retrieval algorithms are based on physical models of the effects of surface temperature and wind speed on brightness temperature (Wilheit, 1978). The 6.6-GHz frequency was chosen for sensitivity to sea-surface temperature. At 19.69 GHz there is increased sensitivity to wind speed with little comtamination by atmospheric effects. The 18-, 21- and 37-GHz frequencies collectively provide information on atmospheric water vapor, cloud liquid water, and rainfall, and make corrections for these effects at the lower frequencies. Dual polarization is used at all frequencies to provide additional discriminatory capability. In practice, combinations of particular brightness temperatures are used in the derivation of each parameter, as determined by regression analyses or other retrieval techniques. The dual-polarization, multifrequency capability is also used for improved discrimination of ice types and measurement of snow properties.

Initial results of comparisons between Seasat SMMR measurements and data from other

Fig. 13-55. Photograph of SMMR instrument in its handling fixture.

sources have been described by Lipes et al. (1979) and Hofer et al. (1981). Figure 13-58 shows the results of comparing the Seasat SMMR and conventional sea surface temperature (SST) measurements in the Pacific. The *rms* discrepancy between the two data sets is approximately 1.2 K.

An 8-days' average SMMR SST map is shown in Figure 13-59a, with a conventionally derived map (from ship observations) shown in Figure 13-59b for comparison. A scatterplot of SMMR wind-speed measurements versus winds derived form surface observations in the North East Pacific is

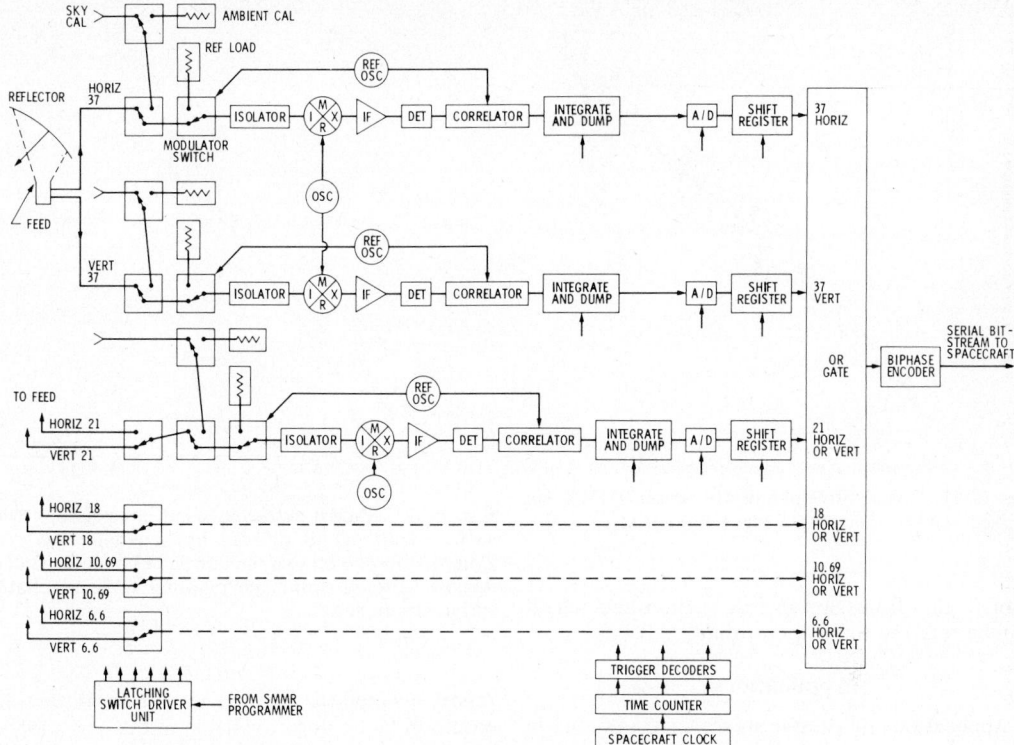

Fig. 13-56. SMMR instrument functional block-diagram.

shown in Figure 13-60. The *rms* discrepancy is approximately 2 m/s which is close to the expected accuracy of the surface measurements themselves. Continuing analysis of SMMR data from both Seasat and Nimbus-7 satellites will enable studies of a global nature to be performed in the areas of oceanography, weather, and climate.

FUTURE DEVELOPMENTS

During the past decade passive microwave remote sensing from space has made dramatic progress. In several application areas spaceborne microwave sensors have demonstrated their potential as key elements of global monitoring systems. This potential will quite possibly be realized during the 1980s, and new applications will be ad-

vanced, as our technology and data utilization capabilities continue to develop. As the applications proliferate and new sensor types evolve, it becomes clear that some coordination is required to design efficient sensors that can satisfy broad groups of satellite data users in a way that is economic and yet satisfies the various foreseeable government- and commercial-program objectives. In an attempt to gain insight into some of these issues, a panel comprising instrument scientists, researchers and operational data users was assembled by NASA in 1977 to review the current state-of-the-art, and to make recommendations on directions which future system development should take. This Application Review Panel (ARP) produced a final report (Staelin and Rosenkranz eds., 1978) whose contents are highly rele-

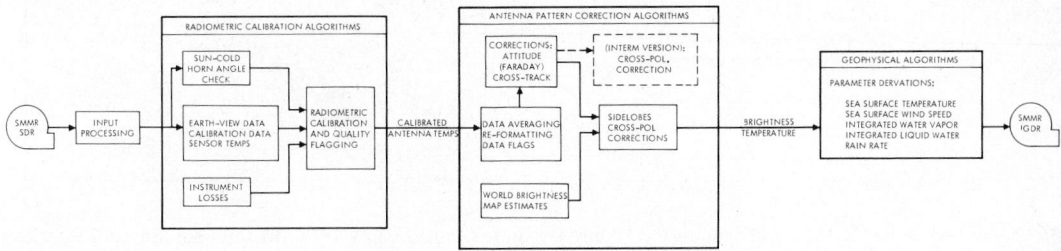

Fig. 13-57. Seasat SMMR data-processing flow diagram.

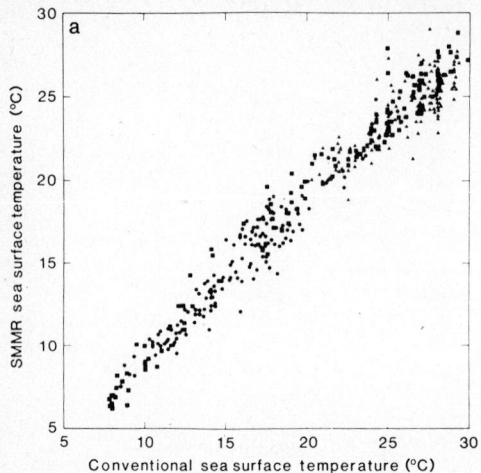

Fig. 13-58. Comparative plot of 418 Seasat SMMR and conventional SST measurements (Hofer et al., 1981).

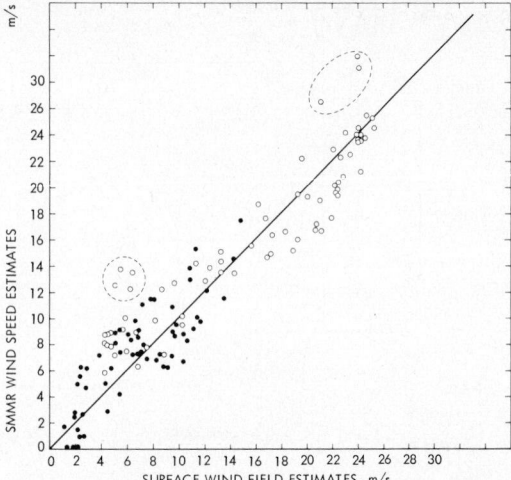

Fig. 13-60. SMMR retrieved wind comparisons with surface truth winds derived by kinematic analysis. Comparisons are for two satellite passes over the Gulf of Alaska (circled points correspond to less reliable surface-truth data).

vant to the discussion in this section and which will be referred to as appropriate.

Applications

Applications of passive microwave systems in space may be divided into two broad groups, those that require imaging of the earth and atmosphere beneath the satellite, and those that require viewing of the earth's atmosphere at the planetary limb. The primary earth-imaging applications are listed in Table 13-12 (adapted from the ARP report). The spatial resolutions, receiver sensitivities, and revisit times needed to satisfy these applications with respect to the various parameters of interest are also listed. In order to assess the ability of a single system to jointly satisfy several of these application requirements, the ARP

report selected the baseline system for discussion based on technology available in the early 1980s. The baseline system had a 4-meter mechanically scanned antenna compatible with Shuttle capability. The spatial resolution and sensitivity achievable with this system for each of the listed applications are also given in Table 13-12 for a platform altitude of 900 km.

For synoptic meteorology and climatology applications, passive microwave sensors are beginning to be exploited operationally to supplement the traditional passive sensors in the infrared and visible regions of the spectrum. Examples are the microwave temperature sounders: MSU on

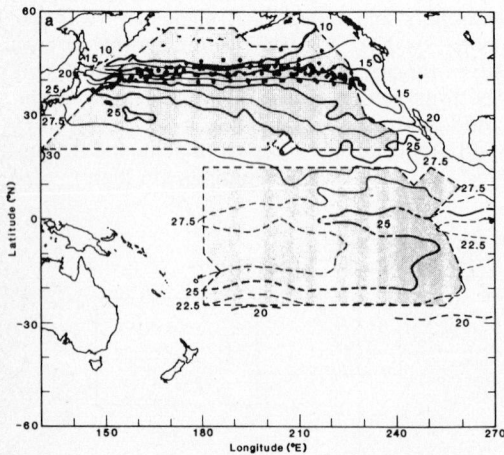

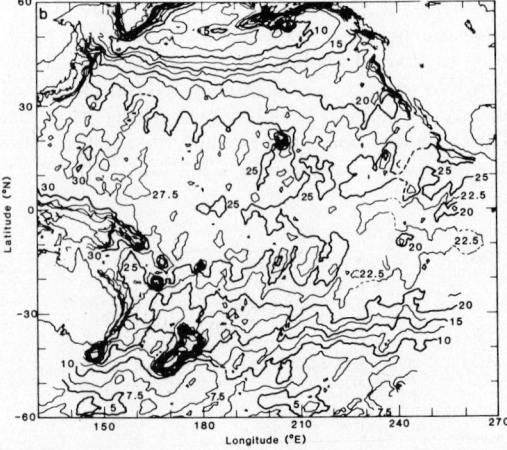

Fig. 13-59. (a) Seasat SMMR SST map of the Pacific Ocean for 16 to 23 July 1978. (b) National Marine Fisheries Service (NMFS) monthly coverage SST map for July 1978. Dashed contours indicate sparse data; circles indicate location of XBT measurements included in scatterplot of Figure 13-58.

TABLE 13-12

Applications and Observable Parameters for High Resolution Passive Microwave Satellites
(Adapted from Staelin and Rosenkranz, 1978)

Observable Parameters	Desired Specifications			Spot Size (km) at Critical Frequency	ΔT_{rms} at Desired Resolution	[Spot Size(+)]/ [Desired Size]
	Spot Size (km)	ΔT_{rms}	Revisit Time (hr)			
SYNOPTIC METEOROLOGY & CLIMATOLOGY						
Temperature Profile	50	0.3	3−12	3.3	0.3	0.06
Water Vapor Profile	15	0.5	3−12	8.6	0.6	0.6
Water Vapor Profile (Non-Tropical)	15	0.5	3−12	1.0	1.0	0.06
Liquid Water Abundance, Rain Rate	2−10	2	3−12	4.9	1.3	0.5
Sea Surface Temperature	50	0.3	6−36	28	0.1	0.6
Sea Surface Wind (Magnitude)	50	1	3−12	17	0.1	0.3
Sea Surface Wind (No Precipitation)	50	1	3−12	4.9	0.2	0.1
SEVERE STORMS						
Temperature Profile	3−30	0.3	1−6	3.3	0.6	0.11
Water Vapor Profile	3−15	0.5	1−6	8.6	0.6	0.4
Water Vapor Profile (Non-Tropical)	3−15	0.5	1−6	1.0	1.0	0.06
Liquid Water Abundance, Rain Rate	2−10	2	1−6	4.9	1.3	0.5
Sea Surface Temperature	3−30	0.3	2−6	28	0.2	0.9
Sea Surface Wind (Magnitude)	3−30	1	1−6	16.9	0.2	0.6
Sea Surface Wind (No Precipitation)	3−30	1	1−6	4.9	0.4	0.2
OCEAN SURFACE						
Surface Wind Velocity	2−50	1	3−12	16.9	0.1	0.3
Surface Wind Velocity (No Precipitation)	10−50	1	3−12	4.9	0.2	0.1
Sea Surface Temperature	1−50	0.3	6−36	28	0.1	0.6
Salinity	0.5−10	0.3	3−12	130	0.5	13.0
Oil Slicks, etc. (No Precipitation)	0.5	0.3	1−12	4.9	25	9.8
LAND PARAMETERS						
Soil Moisture (Low Frequency)	3−25	1	6−36	130	0.2	5.2
Soil Moisture (Higher Frequency)	3−25	1	6−36	28	0.2	1.1
Snow Cover and Type, Frozen Ground	3−25	1	6−12	4.9	0.5	0.2
SEA AND LAND ICE						
Sea Ice Concentration	1−5	2	3−12	4.9	2.5	1.0
Sea Ice Type	1−5	1	6−12	4.9	2.5	1.0
Land Ice Properties (e.g., Firn)	10−50	1	12−36	4.9	0.2	0.1

TIROS-N, and SSM/T on DMSP Block 5. In addition to atmospheric temperature measurements or pressure-level thickness determinations, the oxygen channel brightness temperature gradients are also useful in deriving wind fields at different levels in the atmosphere, particularly for geostrophic winds. To date, the temperature sounding instruments have had horizontal resolutions in excess of 100 km, making them useful only for observing synoptic scale meterological systems. Such sensors would be of even greater value if instruments having smaller fields of view were available to probe cloud-shielded sub-synoptic and mesoscale weather situations. Within the tropical environment, knowledge of the large variations in water vapor are of great importance in understanding the development of tropical cyclones and cloud clusters. Most numerical prediction models require at least two vertical layers of water vapor amounts in the atmosphere rather than the total (integrated) amounts derived from previous microwave instruments. Utilization of the 183-GHz water vapor line in addition to the line at 22.2 GHz may improve the ability to derive water vapor profile information. For operational use in these applications, satellite data with complete global coverage approximately every six hours is desirable.

In severe storm applications the time scales of interest are short (~1 hr) and the spatial scales are small (~10 km); thus the use of geostationary platforms and higher frequencies such as the

MANUAL OF REMOTE SENSING

TABLE 13-12 Continued

| Observable Parameters | [ΔT_{rms} (Available)] / [ΔT_{rms} (Required)] | SMMR | | | | | | | TIROS-N | | |
		1.4	3.0	6.6	10.7	18	21	37	55	90	183
SYNOPTIC METEOROLOGY & CLIMATOLOGY											
Temperature Profile	1.1							•	⊕		
Water Vapor Profile	1.2					+	⊕	+			+
Water Vapor Profile (Non-Tropical)	6.0									+	⊕
Liquid Water Abundance, Rain Rate	0.6			•	•	•	+	+	⊕	•	+
Sea Surface Temperature	0.4	•	+	⊕	•	+	•	+			
Sea Surface Wind (Magnitude)	0.1	+	•	+	⊕			+			
Sea Surface Wind (No Precipitation)	0.2					+	+	⊕			
SEVERE STORMS											
Temperature Profile	1.9							•	⊕		
Water Vapor Profile	1.2					+	⊕	+			+
Water Vapor Profile (Non-Tropical)	6.0									+	⊕
Liquid Water Abundance, Rain Rate	0.6			•	•	•	+	+	⊕	•	+
Sea Surface Temperature	0.6	•	+	⊕	•	+	•	+			
Sea Surface Wind (Magnitude)	0.2	•	•	+	⊕			+			
Sea Surface Wind (No Precipitation)	0.4					+	+	⊕		+	
OCEAN SURFACE											
Surface Wind Velocity	0.1	+	•	+	⊕			+			
Surface Wind Velocity (No Precipitation)	0.2					+	+	⊕		+	
Sea Surface Temperature	0.4	•	•	⊕	•	+	•	+			
Salinity	1.6	⊕	•	+	•	+		+			
Oil Slicks, etc. (No Precipitation)	84			•	•	+		+	⊕		
LAND PARAMETERS											
Soil Moisture (Low Frequency)	0.2	⊕		+							
Soil Moisture (Higher Frequency)	0.2	•		⊕					•		
Snow Cover and Type, Frozen Ground	0.5			•	+	+		⊕		+	
SEA AND LAND ICE											
Sea Ice Concentration	1.3					+		⊕			•
Sea Ice Type	2.5			•	•	+	+	⊕			•
Land Ice Properties (e.g., Firn)	0.2			•	•	+	+	⊕			•

\+ Frequencies of primary importance
• Frequencies of secondary importance
⊕ "Critical" frequency that correspond to the spatial resolution listed

118-GHz O_2 line and the 183-GHz H_2O line become desirable. Key measurement parameters, in addition to atmospheric temperature profile and water vapor, are precipitation and sea-surface winds. Precipitation has been observed over the oceans using ESMR data at 19.35 and 37 GHz, although the measurements suffer from the complexity of the modeling problem and the coarse spatial resolution with respect to dimensions of typical rain cells. The use of frequencies above the 60-GHz oxygen complex shows promise for more quantitative rain measurements over both ocean and land areas. Studies of hurricanes and tropical cyclones using microwave data from Nimbus 5 and 6 sensors have shown much potential in storm forecasting studies and in understanding storm development. These applications will be much improved by sensors with better time- and spatial-resolution.

Ocean remote-sensing applications using microwave radiometers are in an early stage of development. The primary measurement requirements are for surface temperature, surface winds and salinity. Oil-slick detection and pollution monitoring are further possiblities. The applications can be broadly classified into two groups. Open-ocean phenomena typically have large-scale features on the order of tens of kilometers with variabilities on time scales of a few days. These can be studied using wide-area observational coverage by spacecraft sensors at medium spatial resolutions (10–50 km). The results can be applied to medium- to long-range weather forecasting, and large-scale modeling of ocean dynamics and air-sea interactions. Coastal applications on the other hand focus on localized currents and smaller-scale eddies which generally require spatial resolutions of less than 5 km, and on salinity and pollution monitoring, typically requiring resolutions of less than one kilometer. The need for low microwave frequencies (1 to 10 GHz) in ocean-surface measurements, combined with the need for high spatial resolution in some of the applications, is a driving force towards the development of larger space antennas. The SMMR instruments on Seasat and Nimbus 7 have provided the first steps in ocean microwave sensing from orbit, and more advanced sensors are in the planning stages.

Sea and land ice applications have received a promising start from measurements made by NEMS, SCAMS and the ESMR's on Nimbus 5 and 6, and more recently by the SMMR on Nimbus 7 and Seasat. The principal measurement goals are the determination of temporal and spatial variations of sea-ice concentration and type, and horizontal and vertical temperature distributions and snow accumulation rates on the ice sheets of Greenland and Antarctica. These measurements are important in advancing the understanding of ice dynamics, improving maritime operations and ship navigation in polar regions, and in climate modeling studies. For these applications, improved spatial resolutions of less than 5 km would aid in the location of specific large leads in the ice pack and enable more accurate ice concentration determinations to be made.

Passive microwave sensing of land parameters, such as soil moisture and snowpack water equivalent, is far from maturity. This is mainly due to the variable nature of the observed phenomena and the poor spatial resolutions of current space sensors for making these measurements. In water resources applications, soil moisture information can be applied in hydrological models for predicting runoff and flood conditions. Similarly snowpack measurements can be used to estimate spring runoff and melt water. Both applications require data with resolution on the order of less than 3 km. For agricultural applications, where soil moisture information is used in crop-yield forecasting models and large-area resource planning, spatial resolutions of about 10 km are desirable. For studies of the effects of soil moisture and snow cover on climate, somewhat larger spatial resolution is acceptable. Soil moisture is best measured using low frequencies in the 1- to 3-GHz range, where depth penetration on the order of centimeters can be achieved, and confusing effects of surface vegetation and roughness are minimized. For snowpack monitoring a multifrequency approach is most promising, using several frequencies in the range 3 to 90 GHz.

Microwave sensing of the earth's upper atmosphere (stratosphere, mesosphere and lower thermosphere) is a relatively new application. Global measurements in the upper atmosphere of the abundances of several molecular species, and measurements of temperature, winds, magnetic field, and reference pressure levels, will assist in understanding the complex chemistry and transport processes of that region. Limb-sounding techniques are best suited to these measurements since they provide substantially improved sensitivities and vertical resolutions over downward-looking experiments. With such techniques several important measurements can be made using existing technology. The extension of microwave technology to submillimeter wavelengths will allow many more important measurements to be made.

Technology Advancement

The improvement in characteristics of the various spacecraft sensors described in previous sections, and the expanding applications and requirements discussed above, lead naturally to a discussion of the technological capabilities for meeting these requirements in the 1980s. In addition, as pointed out in the ARP report (Staelin and Rosenkranz, eds., 1978), there are substantial economic benefits to be gained in satisfying a large number of applications by using a few versatile system designs. This is especially true of applications that require improved spatial resolutions and hence larger and more costly antennas.

The most critical elements in passive microwave remote sensing systems are the receivers, antennas, and data systems. The state-of-the-art for ground-based receivers is summarized in Figure 13-61, using system noise temperature as the quality indicator. Most of the receiver specifications can be duplicated in space, although those that employ cryogenics may prove difficult. Uncooled receivers generally fall near the line marked "useful" in the figure. It appears that

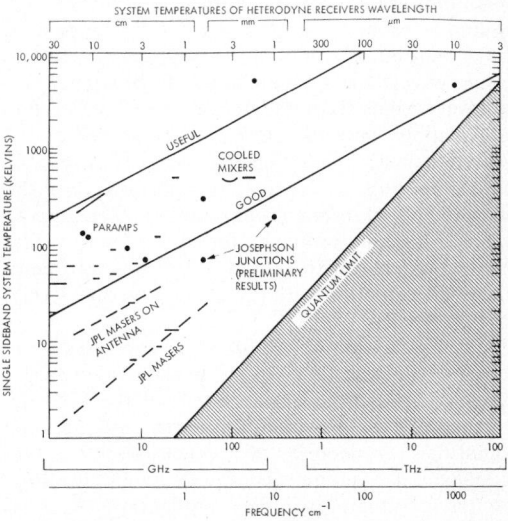

Fig. 13-61. State-of-the-art single sideband system noise temperatures (K) (Staelin and Rosenkranz, 1978).

present receiver technology is adequate to meet a majority of the applications discussed in this section; the exceptions will be discussed later. Microwave antennas have been built for ground-based systems that could satisfy all the spatial resolution requirements from orbit discussed above. However, these antennas are unsuitable for remote sensing from space since they would need rapid scanning either mechanically or by electrical beam-steering. Antennas that require unfurling or space fabrication, due to packaging size limits for Shuttle launch, generally cost substantially more and involve higher risk. A compromise for the near term is the use of a 4-meter offset fed parabolic reflector antenna which could be stowed for launch within the Shuttle payload bay, and when deployed in space could be scanned mechanically at a rate of about 60 rpm. This type of antenna was used as a baseline for deriving the performance characteristics presented in Table 13-12. An off-axis feed location was chosen to avoid blockage and to provide high beam efficiency. The antenna would scan about a vertical axis causing the beam to sweep out a conical surface, with a constant incidence angle at its intersection with the earth. For the purpose of the calculations in Table 13-12 an incidence angle of 55° was assumed, the sum of the receiver noise temperature and the antenna temperature was assumed to be $(400 + 20\nu)$ K, where ν is the radiometer frequency in GHz, and the receiver bandwidth for each channel was assumed to be 100 MHz. These numbers may vary in actual systems but are good baseline approximations. The table also suggests a range of frequencies which could be used to serve the listed applications.

By studying Table 13-12 it can be seen that the critical frequency spot size is significantly greater than the maximum desired spot size only for the salinity, oil slick, and soil moisture measurements. To meet the requirements of these applications much larger antenna systems are required. A number of other parameters appear to have marginal resolutions, but useful information could nevertheless be obtained. High spatial resolution implies short integration times, and certain applications may therefore require receiver sensitivities that are unobtainable. With a foreseeable future doubling of receiver sensitivities, only the water vapor determinations using 183 GHz appear to have serious problems for lack of sensitivity. Cryogenic receivers could solve this problem. Oil-slick detection with full earth coverage would not be feasible.

The discussion above suggests four classifications of systems for future passive microwave sensing from orbit, (1) a system of the kind described above, providing full earth coverage and satisfying the majority of applications; (2) large, specialized antenna systems designed for low-frequency, high-resolution applications such as salinity and soil moisture; (3) geosynchronous at-

mospheric sounders operating around the O_2 and H_2O resonance frequencies; and (4) limb-scanning systems for upper atmospheric studies. Development of these systems will depend largely on the priorities assigned to the various applications.

INFRARED SENSORS

The Landsat visible and near-infrared sensors are discussed in Chapter 12. In this section, we discuss remote sensors that probe the surface at the infrared atmospheric window regions between 1.0 and 3.0 μm, between 3 and 5 μm, and between 8 and 15 μm (Figure 13-1).

BACKGROUND

The short-wavelength infrared region, 1 to 3 μm, provides more diagnostic spectral information about the composition of minerals and rocks than the visible and near-infrared regions. Absorption bands at 1.4 and 1.9 μm are caused by the bound and unbound water contained in surface materials. Unfortunately, these bands (Figure 13-62) coincide with strong atmospheric water bands (Figure 13-1) that render the atmosphere opaque for all but the highest topographic elevations.

The region around 1.6 μm exhibits the highest reflectance for most rocks because it is nearly midway between the ultraviolet-visible iron absorption bands and a strong fundamental OH^- vibration at 2.74 μm. In particular, altered rocks containing clay with or without a short-wavelength Fe^{3+} absorption, exhibit a pronounced peak reflectance (Figure 13-62), often greater than 70 percent, at 1.6 μm (Rowan et al., 1977). Evaluations of field spectra and multispectral aircraft images show that bandpasses centered near 1.6 and 2.2 μm are especially useful for mapping altered rocks, because OH^--bearing minerals, commonly contained in these rocks, give rise to a relatively sharp absorption band near 2.2 μm and a decrease in reflectance associated with the 2.74 μm OH^- absorption. The 2.2-μm band is either absent or weak in the spectra for most unaltered regional lithologic units in arid and semiarid environments (Figure 13-62) (Goetz and Rowan, 1981).

Color-ratio composite images made from National Aeronautics and Space Administration (NASA) aircraft data that include bandpasses centered at 2.2 and 1.6 μm illustrate that bleached, nonlimonitic altered rocks in the Cuprite, Nevada, mining district, which were not identified in Landsat MSS images, are correctly categorized (Figure 13-63☆) (Abrams et al., 1977). Colors in the image can be related to the spectral characteristics of the rocks in the district (Figure 13-64).

In an identical color-ratio composite of the Coaldale, Nevada, area, limonitic unaltered tuffs were readily identified as having limonite but no

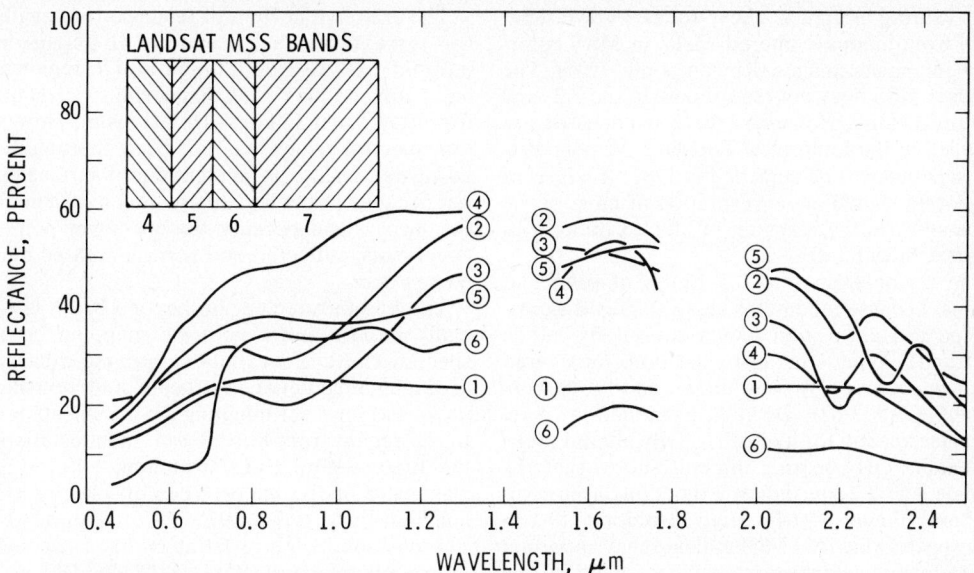

Fig. 13-62. Field-acquired reflectance spectra: 1. unaltered tuff fragments and soil; 2. argillized andesite fragments; 3. silicified dacite; 4. opaline tuff; 5. tan marble; 6. ponderosa pine. The gaps at 1.4 and 1.9 μm are the result of atmospheric water absorption.

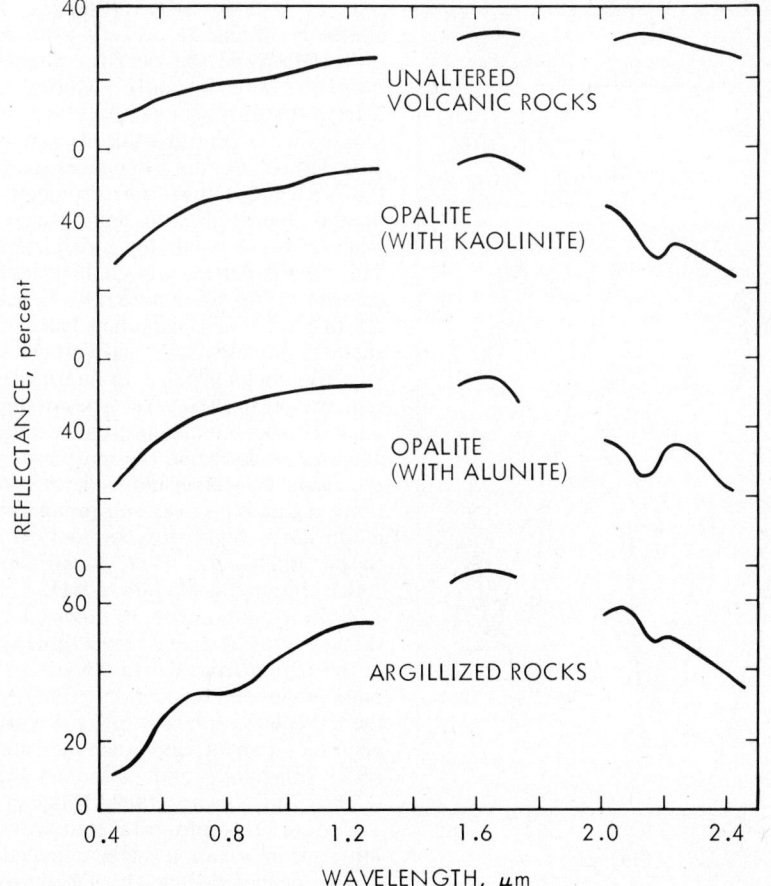

Fig. 13-64. Representative spectral reflectance curves of altered and unaltered rocks in the Cuprite, Nevada, mining district (Abrams et al., 1977).

OH^--bearing minerals. These rocks were insepa-rable from limonitic altered rocks in MSS color-ratio composite images (Rowan et al., 1977). The Landsat MSS does not record data in the 2.2- and 1.6-μm regions. However, these bandpasses are included in the Landsat D Thematic Mapper sys-tem scheduled to be launched in 1982. Analysis of these data should allow resolution of most of the ambiguities in the mapping of altered rocks with Landsat MSS images.

The region from 2 to 2.5 μm is of particular interest because it contains sharp, highly diagnos-tic spectral absorption bands caused by lattice overtone bending-stretching vibrations for layered silicates such as clays and micas, and for carbon-ates (Figure 13-62). Detailed examination of re-flectance spectra for hydrothermally altered rocks containing OH^--bearing mineral shows that the minima near 2.2 μm shift as a function of the com-position (Figure 13-64). High-resolution labora-tory spectra (Figure 13-65) indicate that important mineralogical information can be gained if one exploits this wavelength region (Hunt and Ashley, 1979). In order to test this approach to remote mineralogical determination, the Shuttle Multi-spectral Infrared Radiometer (SMIRR) was con-structed to be flown on space-shuttle flights. (Goetz, 1981).

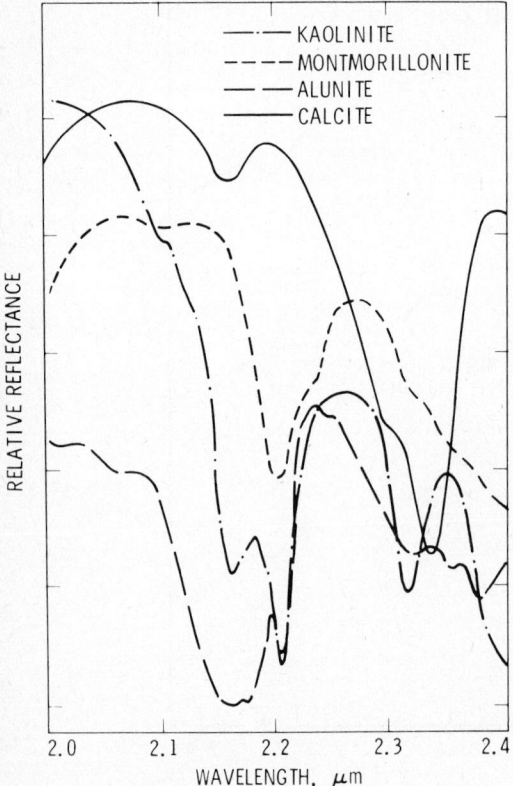

Fig. 13-65. High-resolution laboratory reflectance spectra in the region from 2.0 to 2.4 μm for some typical hydrothermal alteration minerals (Hunt and Ashley, 1974).

The emissive portion of the spectrum available for terrestrial observation, called the mid-infrared, extends from 3 to 15 μm. The region from 3 to 5 μm contains weak diagnostic spectral bands for nitrates and sulfates (Hovis, 1966). However, this spectral region has not been thoroughly in-vestigated for use as an identifier of surface mate-rial because of the low energy flux available from the surface and because the crossover point be-tween solar-reflected and surface-emitted energy occurs here.

The mid-infrared region beyond 8 μm is espe-cially important for geologic mapping because spectral emittance variations provide a basis for distinguishing between silicate and nonsilicate rocks and for discriminating among silicate rocks. In the region from 8 to 14 μm, manifestations of the fundamental Si-O stretching vibration are diagnostic of the major types of silicates (Hunt and Salisbury, 1974, 1975, 1976) (Figure 13-66). The position of the reststrahlen bands or regions of metallic-like reflection are dependent on the extent of interconnection of the Si-O tetrahedra comprising the crystal lattice. The end-members are represented by quartz, with complete sharing of the oxygen molecules, and olivine, which con-sists of isolated SiO_4 tetrahedra. The spectral emittance of silicate rocks is especially sensitive to variations in the quartz content (Kahle and Rowan, 1980). Recently acquired multispectral MIR (8-13 μm) scanner data of the East Tintic Mountains in central Utah, as obtained from the now defunct Bendix 24-channel scanner flown on the NASA C-130 aircraft provided an unprece-dented opportunity to demonstrate the impor-tance of these bands for geological applications. This area is particularly challenging for lithologic determinations by remote sensing because a vari-ety of rock types, including Paleozoic quartzite, shale, carbonate rocks, and Tertiary extrusive and intrusive rocks of felsic to intermediate composi-tion, are present. Several types of hydrothermally altered rocks are also present, with argillized and silicified rocks being the most widespread (Mor-ris, 1964a, b; Morris and Lovering, 1979). In addi-tion, vegetation cover and topographic relief are moderately high (Rowan and Abrams, 1978; Siegal and Goetz, 1977; Morris and Lovering, 1961; Milton and Madura, 1981).

A principal component color-composite image (Figure 13-67☆) derived from three spectral bands in the region from 8.3 to 9.8 μm exhibits red to pink, green, and blue colors primarily. In general, the red colors represent rocks in which quartz is a major component, and green indicates nonsilicate rocks (limestone and dolomite) and vegetation (Kahle and Rowan, 1980; Kahle, et al., 1980). These results indicate that very important lithologic information can be obtained in bandpass-es that do not include the 9.6-μm ozone absorp-tion band (Figure 13-1). Ozone has been consid-ered to be a limiting factor in multispectral mid-infrared imaging from orbit.

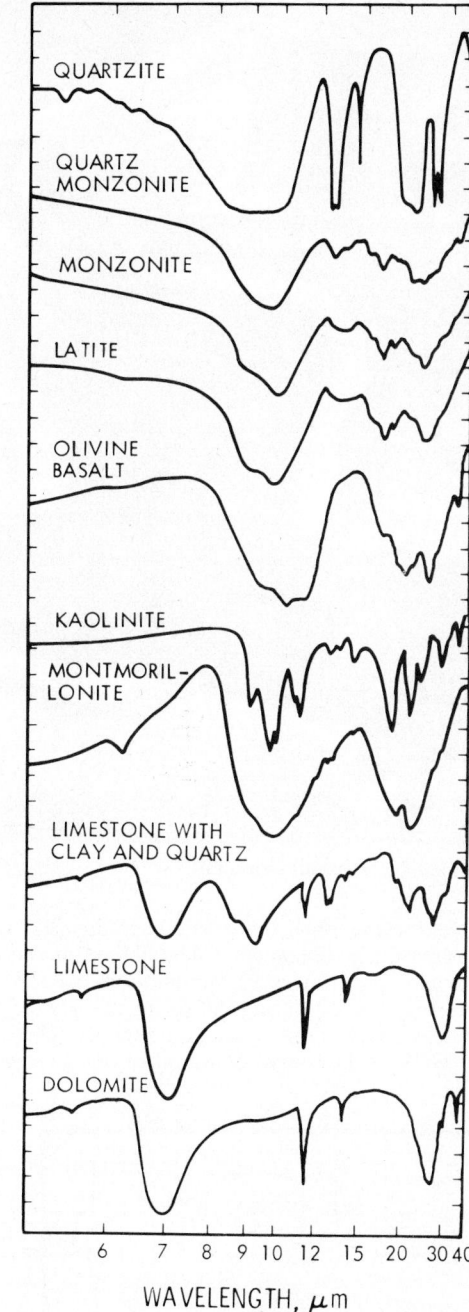

RELATIVE TRANSMISSION

QUARTZITE

QUARTZ
MONZONITE

MONZONITE

LATITE

OLIVINE
BASALT

KAOLINITE

MONTMORIL-
LONITE

LIMESTONE WITH
CLAY AND QUARTZ

LIMESTONE

DOLOMITE

6   7   8   9 10 12   15   20 30 40

WAVELENGTH, μm

Fig. 13-66. Transmission spectra of some common silicates [Hunt and Salisbury, 1975]. Regions of low transmission are associated with reststrahlen bands and are equivalent to regions of low emittance.

In addition to the information available from multispectral data in the middle infrared, the surface temperature can be determined from measurement of the total radiance in a broad middle-infrared wavelength band, typically from 8 to 14 μm or from 10.5 to 12.5 μm. Some information concerning body properties as opposed to surface

properties can be obtained if one analyzes the changes in surface temperature that are induced by diurnal solar heating. This property, called thermal inertia, is defined as $(K\rho c)^{1/2}$, where K is the thermal conductivity, ρ is the density, and c is the specific heat (Kahle, 1977; Gillespie and Kahle, 1977; Kahle, et al., 1976; Watson, 1973, 1975). This method allows measurement to a depth of about 10 cm or less.

Thus, the infrared region of the spectrum holds great potential for future remote sensing. Most of the infrared satellite systems flown to date have either been of a research nature, or are designed primarily for meteorological applications. In this section are described the two recent research satellite systems, the Heat Capacity Mapping Mission (HCMM) satellite and the Shuttle Multispectral Infrared Radiometer (SMIRR). These are discussed in the context of geological applications, but they obviously will have many other applications, as described in Volume II. The SMIRR is designed to measure multispectral data in moderate and narrow bands in the short wavelength infrared. The HCMM satellite was flown to measure surface temperatures, using a 10.5- to 12.5-μm band, at the daily maximum and minimum of surface heating, in order to infer values of surface thermal inertia.

HCMM SYSTEM

Mission Description

The HCMM (Figure 13-68) was the first of a planned series of Applications Explorer Missions that involve the placement of small, dedicated spacecraft in special orbits to satisfy mission-unique data acquisition requirements. The HCMM supported scientific investigations to determine the feasibility of using thermal infrared temperature measurements of the earth's surface within a 12-hour interval, at times when the temperature variation was at maximum to determine thermal inertia.

A mission duration of one year was planned. HCMM was launched on April 26, 1978, into a nearly sun-synchronous 620-km circular orbit inclined 97.6 degrees (retrograde) to the equator. This orbit was achieved by the Scout launch vehicle and the spacecraft on-board propulsion system. Local times of equator crossings were approximately 2 p.m. (ascending node) and 2 a.m. (descending node). At Northern Hemisphere mid-latitudes, the crossing times were 1:30 p.m. and 2:30 a.m. Southern Hemisphere mid-latitude crossing times were 2:30 p.m. and 1:30 a.m. The HCMM orbit covered every area of the earth's surface between the latitudes of 85 degrees north and 85 degrees south at least once during the day and once during the night within a 16-day interval. Both night and day passes over selected areas within a 12-hour period were repeated at 16-day intervals. The areas between 22° and 33° latitude (north and south) received only 36-hour coverage

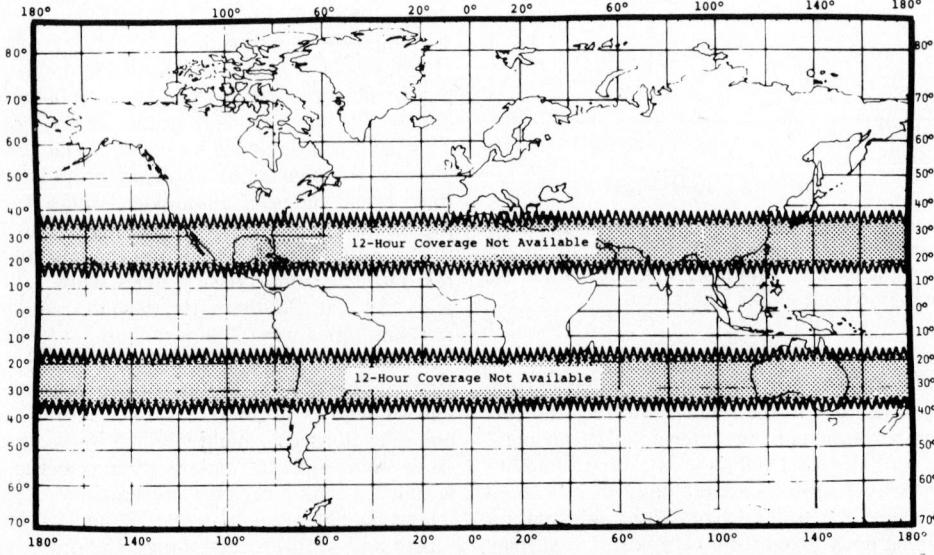

Fig. 13-68. HCMM (Heat Capacity Mapping Mission) spacecraft.

(Figure 13-69). The typical coverage pattern for a combination 12-hour night/day pass across the US is shown in Figure 13-70.

On February 21–23, 1980, the HCMM orbit was lowered from 620-km altitude to 540-km altitude representing a 16-day, 241-orbit repeat cycle, rather than the previous 16-day, 237-orbit repeat cycle. Because of degradation of the batteries, nighttime data acquisition became extremely limited during the spring of 1979, and all satellite operations ceased on Sept. 30, 1980.

HCMM data were collected in real-time when

Fig. 13-69. Geographical extent of HCMM 12-hour night/day coverage.

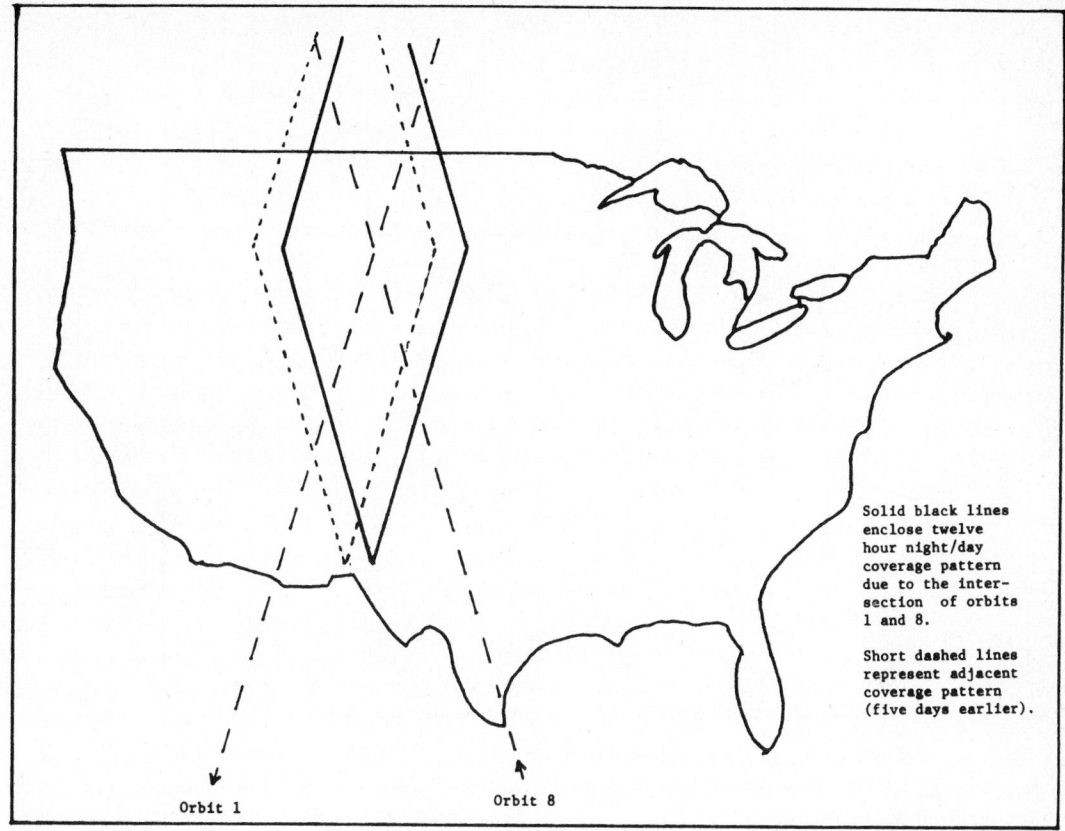

Solid black lines
enclose twelve
hour night/day
coverage pattern
due to the inter-
section of orbits
1 and 8.

Short dashed lines
represent adjacent
coverage pattern
(five days earlier).

Orbit 1

Orbit 8

Fig. 13-70. HCMM coverage pattern.

the satellite was within reception range of NASA receiving stations. Idealized tracking station acquisition circles for the stations that received HCMM data are shown in Figure 13-71. Actual station coverage varied slightly depending upon antenna characteristics and local terrain. HCMM data were processed by NASA at the Goddard Space Flight Center (GSFC). These data were used to produce geometrically corrected and calibrated images of radiance or equivalent black-body temperature, registered images of daytime and nighttime temperature difference and, where applicable, apparent thermal inertia images, which combine daytime reflected energy with temperature difference (Goddard Space Flight Center, 1978).

Spacecraft Description

The HCMM basic spacecraft was made up of two distinct modules: an instrument module, containing the HCMR and its unique supporting gear, and a base module, containing the necessary data-handling, power, communications, command, and attitude control subsystems required to support the instrument module.

The HCMR instrument was a modified spare Surface Composition Mapping Radiometer similar to that flown on Nimbus 5. The HCMR had a geometric instantaneous field-of-view (IFOV) of 0.83 milliradians. The instrument was a two-

channel scanning radiometer. One spectral channel covered the reflectance band from 0.5 to 1.1 micrometers, while the other channel viewed the thermal infrared band between 10.5 and 12.5 micrometers. The two channels thus provided measurements of reflected solar and emitted thermal energy, respectively. Measurement accuracy in the two channels was limited by the analog telemetry system to a Noise Equivalent Radiance (NER) of 0.2 mw/cm^2 in the 0.5 to 1.1 micrometer channel, and Noise Equivalent Temperature Difference (NEΔT) of 0.4 K at 280 K in the 10.5 to 12.5 micrometer channel when NASA Spaceflight Tracking and Data Network (STDN) stations equipped with 9-meter receiving antennas were used to acquire the data. Accuracy was greatest for zenith tracking and degraded somewhat when the satellite was low on the horizon. Exclusive of the telemetry system, the thermal channel sensor itself had a measured NEΔT of 0.3 K at 280 K.

From the nominal orbit altitude of 620 km, the spatial resolution of the infrared channel was approximately 600 m × 600 m at nadir, and the resolution in the reflectance channel was 500 m × 500 m. These values were masked by data processing, which generated registered data at a 481.5-meter pixel size. Registration between channels was to 0.2 resolution elements. The swath of data coverage along the track was approximately 716-km wide. Table 13-13 summarizes the major charac-

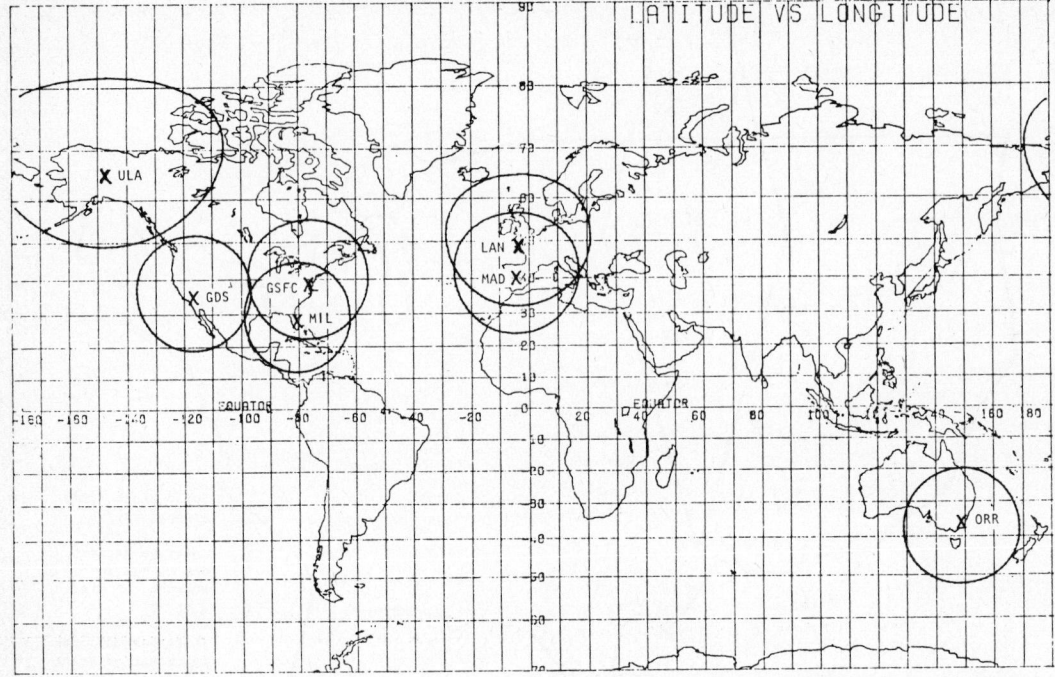

Fig. 13-71. HCMM data acquisition stations. The circles are the ten-degree horizon for each station.

teristics of the HCMR. Figure 13-72 is a simplified block-diagram of the instrument.

The HCMR was composed of four major subassemblies mounted in a common housing. These were:

- Scan Drive Subassembly
- Optics Subassembly
- Electronic Subassembly
- Radiant Cooler Subassembly

The scan drive subassembly provided cross-track scanning of the IFOV with reference to the satellite ground track. The optics subassembly provided energy collection and spectral definition of the two channels. The electronic subassembly contained signal amplifiers and provided the telemetry interface; it conditioned the analog sensor data so that it was compatible with the HCMR data system and it accepted the data from the detectors and multiplexed them for transmission to receiving stations. The radiant cooler subassembly was designed to cool the thermal detector to a controlled temperature of 115 K.

The optics subassembly (Figure 13-73) was a

TABLE 13-13

Heat Capacity Mapping Radiometer Summary Data Sheet

Orbital altitude = 620 km
Angular resolution = 0.83 milliradians
Resolution = 0.6 km × 0.6 km at nadir (infrared)
           0.5 km × 0.5 km at nadir (visible)
Scan angle = 60 degrees (full angle)
Scan rate = 14 revolutions/sec
Sample rate = 1.19 samples/resolution element at nadir
Sampling interval = 9.2 μs
Swath width = 716 km
Information bandwidth = 53 KHz/channel
Thermal channel = 10.5 to 12.5 micrometers; NEΔT = 0.4°K at 280°K
Usable range = 260 to 340 K
Visible channel = 0.55 to 1.1 micrometers; SNR = 10 at ~1% albedo
Dynamic range = 0 to 100% albedo
Scan mirror = 45 degree elliptical flat
Nominal telescope optics diameter = 20 cm
Calibration = Infrared: View of space, seven-step staircase electronic calibration, and blackbody
                      calibration once each scan.
           Visible: Pre-flight calibration assumed valid.

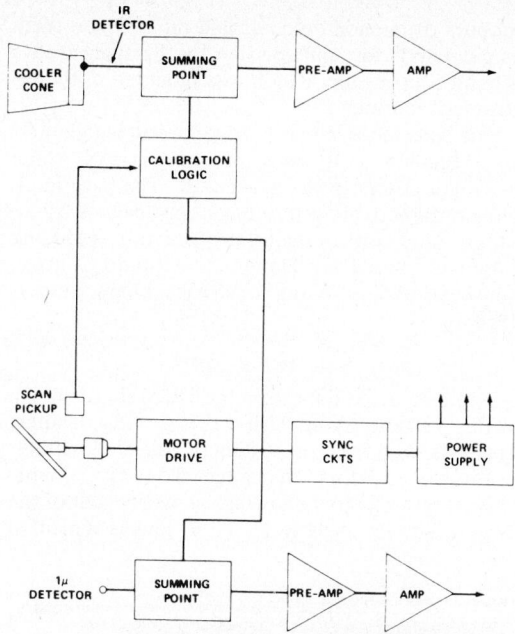

Fig. 13-72. HCMM sensor block-diagram.

cm) diameter. Spectral separation was provided by a dichroic beam splitter positioned in the collimated beam from the secondary mirror which acted as a folding mirror for the 10.5- to 12.5-micrometer band and transmitted energy at shorter wavelengths.

The reflectance channel optics consisted of a long wavelength (greater than 0.55 μm) pass interference filter, focusing optics and an uncooled silicon photo-diode. The long wavelength cut-off of the silicon detector limited the bandpass to wavelengths of less than 1.1 μm. The sensitive area of the detector was approximately 0.15 mm square.

The thermal infrared beam was focused onto the mercury-cadmium-telluride detector using a germanium lens. Final focusing and spectral trimming was accomplished by a germanium aplanat located at the detector.

The electronic subassembly and its operation were as follows: The detectors produced a small electrical signal that was proportional to the difference in radiant energy between the scene and space. The electrical signals from the detectors were amplified in each video amplifier to a level required for processing. Each amplifier contained a low-noise preamplifier, video filter, and post-amplifier. A space clamping technique was used to establish the *dc* zero level once every rotation of the scanner by clamping the output to zero

catadioptric collector with an afocal reflecting telescope. The telescope was a modified Dall-Kirkham configuration which reduced the optical beam from an 8-inch (20.32 cm) to a 1-inch (2.54

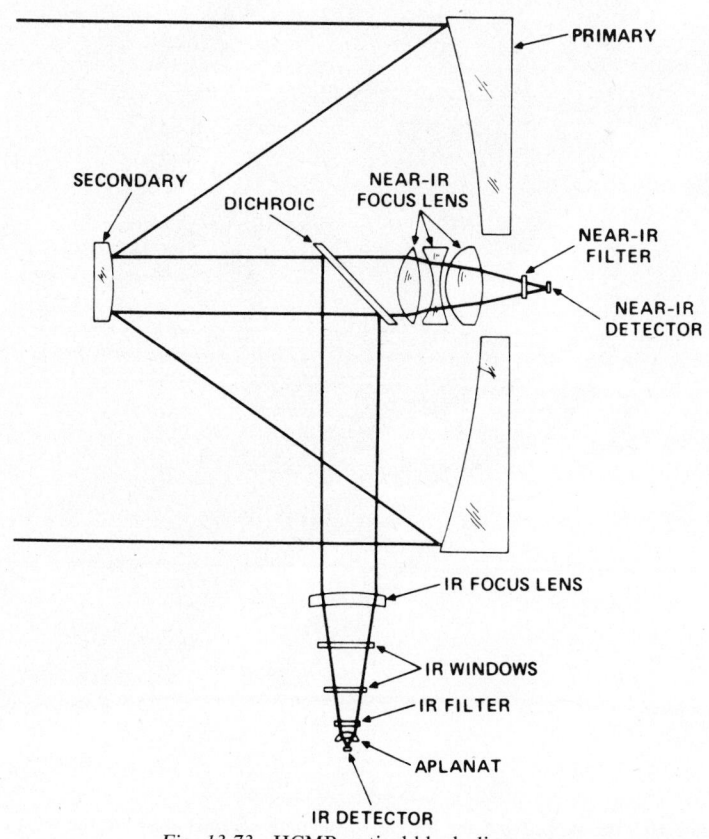

Fig. 13-73. HCMR optical block-diagram.

when viewing cold space, and holding this level for the duration of the scan. The overall video amplifier gain was such that the highest energy scene (340 K or unit reflectivity) produced a 6-volt output signal. Calibration signals consisting of a six-step staircase waveform were inserted on every scan line at the amplifier input, as well as at the amplifier output, to provide constant calibration and complete assessment of the amplifier performance. At the amplifier output, synchronizing pulses were gated in, along with the output calibration to make up the composite video. Output buffer amplifiers with unit gain and low output impedance were used for the output interface to the data system.

The voltage calibration circuitry consisted of an accurate, stable, digital-to-analog converter that generated a staircase of six 1-volt steps for insertion at the amplifier input and output.

A data signal multiplexer accepted the analog outputs from each detector and multiplexed them as sidebands for transmission by the spacecraft's S-band transmitter, which was included in the instrument module.

The base module was hexagonal and weighed 97 kg. Attached to it were the solar arrays which provided power to the spacecraft. The base module contained all the subsystems necessary to support and control the total spacecraft, including Command and Data Handling, Attitude Control, and Orbit Adjust (Goddard Space Flight Center, 1978).

Selected Results

Some selected data from the HCMM are shown in this section, along with a few geological inferences derived from these data (Kahle et al., 1981).

Figure 13-74 is an image from the visible wavelength channel showing the west coast of the United States, on May 31, 1978. This is typical of

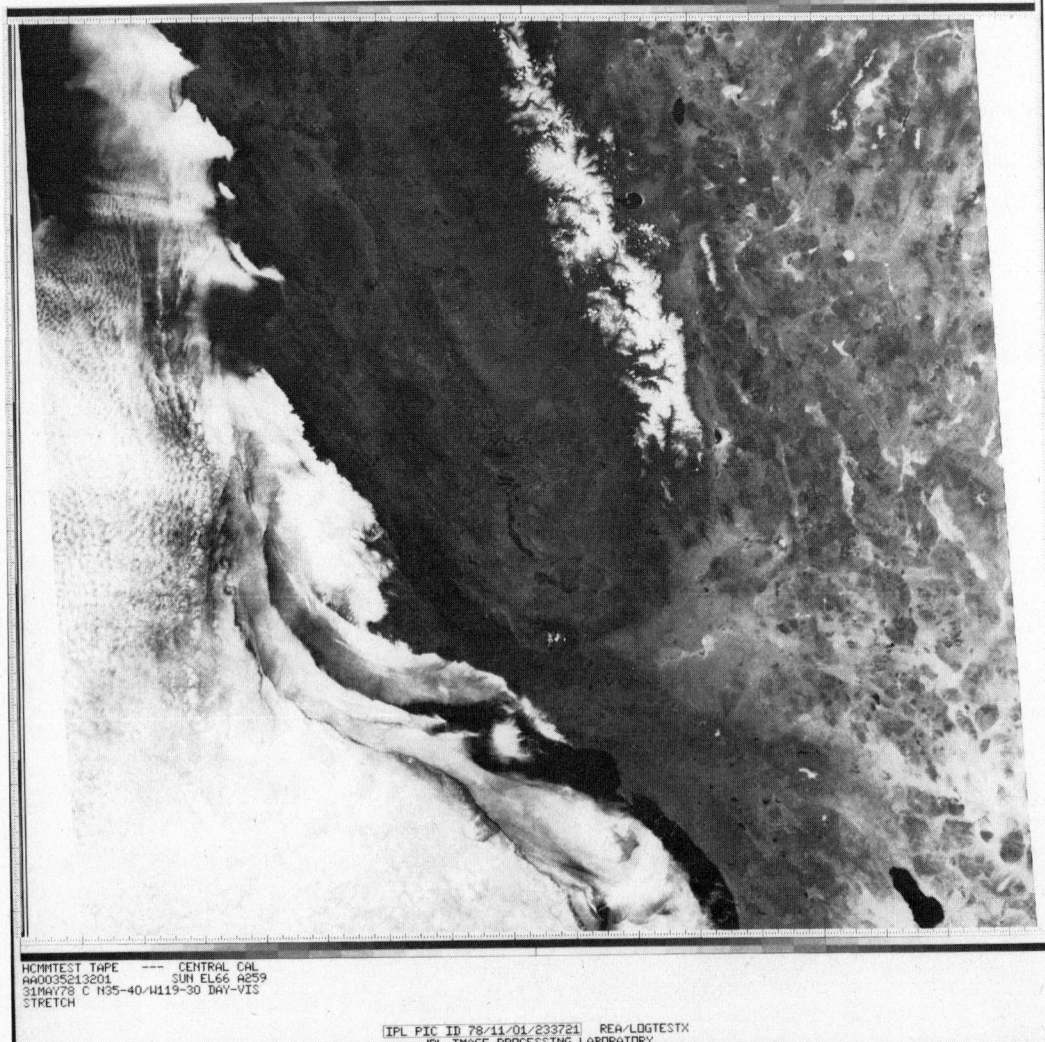

HCMMTEST TAPE --- CENTRAL CAL
AA0035213201   SUN EL66 A259
31MAY78 C N35-40/W119-30 DAY-VIS
STRETCH

IPL PIC ID 78/11/01/233721 REA/LOGTESTX
JPL IMAGE PROCESSING LABORATORY

Fig. 13-74. Full-frame HCMM image from the visible wavelength channel showing the west coast of the United States on May 31, 1978.

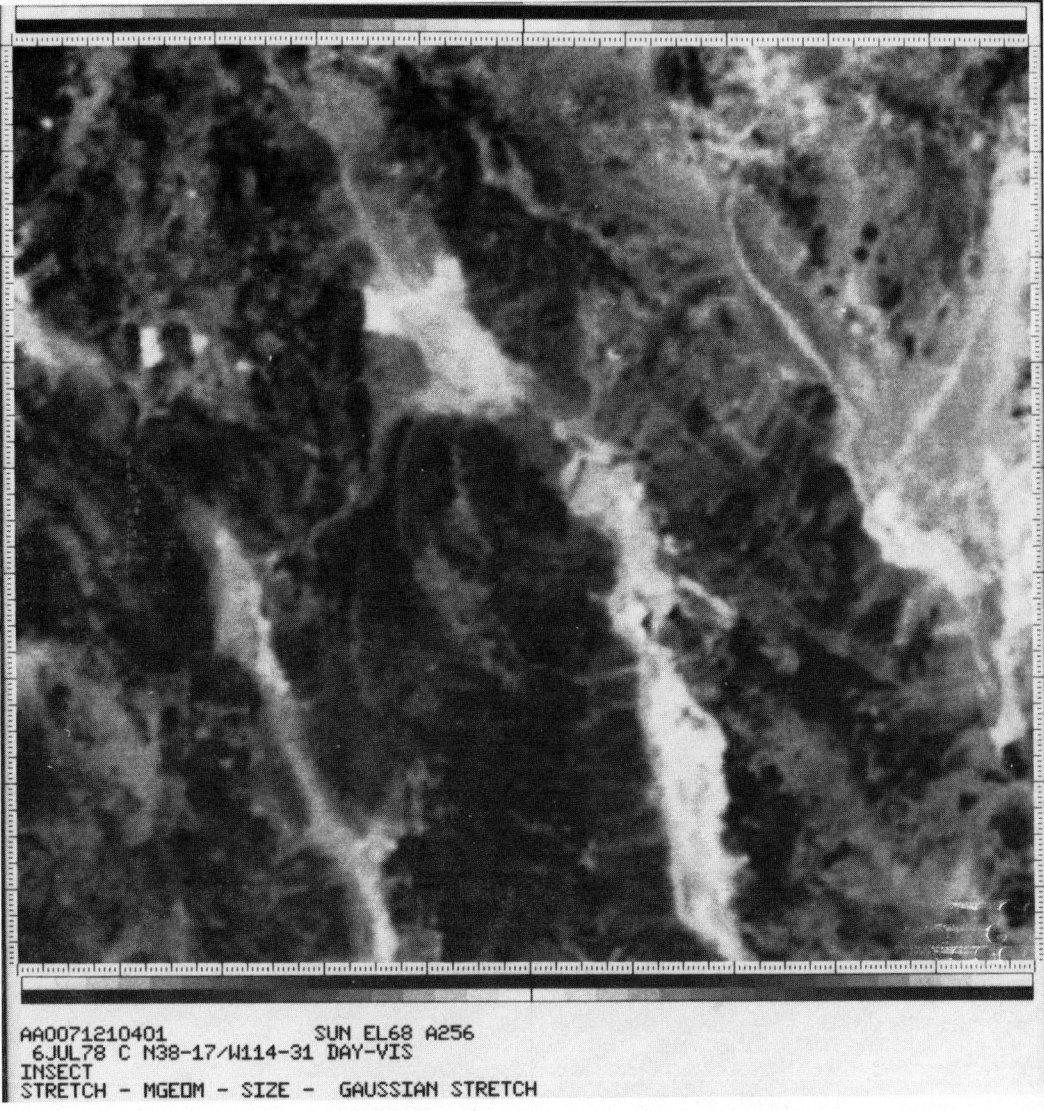

AA0071210401   SUN EL68 A256
6JUL78 C N38-17/W114-31 DAY-VIS
INSECT
STRETCH - MGEOM - SIZE - GAUSSIAN STRETCH

Fig. 13-75. Portion of an HCMM day-visible image, Death Valley, California, July 6, 1978.

the data on one tape, representing a portion along the path of one satellite pass. Figures 13-75, 13-76 and 13-77 show day visible, day IR, and night IR data for a small subsection of one image, at Death Valley, Calif. Figure 13-78 is a thermal inertia image derived from the visible and IR data using the Jet Propulsion Laboratory algorithm (Kahle et al, 1981). Bright areas have high thermal inertia and dark areas have low thermal inertia. The brightest features in the scene correspond to areas underlain by dolomite, limestone, quartzite, or granite. These rock types stand out distinctly on the thermal inertia image and are quite easy to map. The darkest features correspond to areas of young alluvium in the mountain valleys while the floor of Death Valley is bright to medium gray; the former areas are those that are probably more moist than the other areas.

The thermal inertia image also allows accurate delineation of the bedrock-alluvium contact. This differentiation is often difficult to make on visible or reflected IR images, particularly when moderate to low spatial resolution prevents the use of textural information.

In order to examine the interrelationship of the three HCMM data types, several color composite images were produced by using various combinations of data types as components of a color additive triplet. One such image for the Pisgah Lava Flow area in the Mojave Desert, Calif. is shown in Figure 13-79☆ (left). This image is a combination of the day IR, night IR, and visible, displayed as blue, red, and green respectively. All three components were complemented, so areas that are cold or have low albedo appear as strongly colored in the image. For comparison, a Landsat

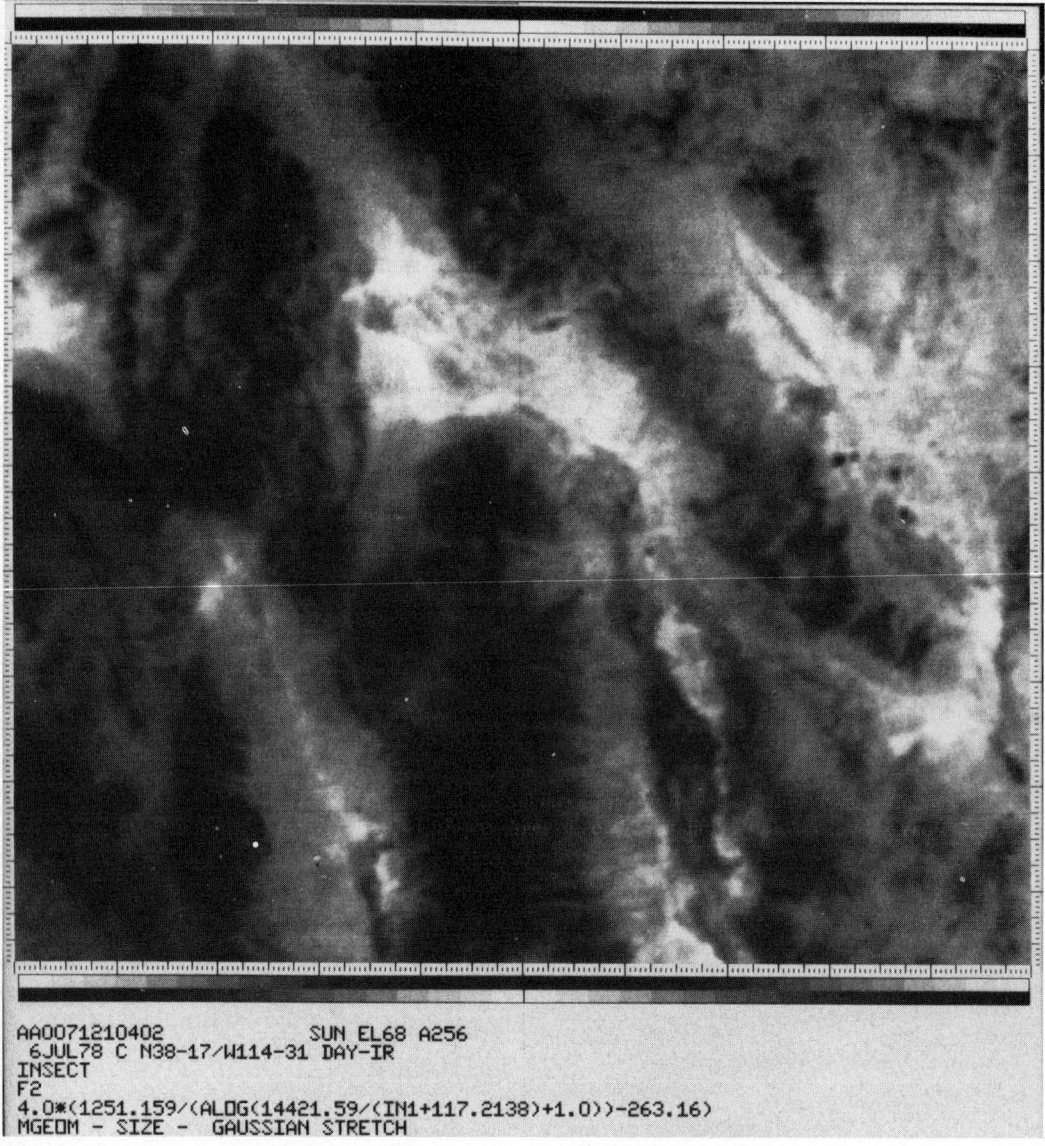

```
AA0071210402              SUN EL68 A256
 6JUL78 C N38-17/W114-31 DAY-IR
INSECT
F2
4.0*(1251.159/(ALOG(14421.59/(IN1+117.2138)+1.0))-263.16)
MGEOM - SIZE -  GAUSSIAN STRETCH
```

Fig. 13-76. Portion of an HCMM day-IR image, Death Valley, California, July 6, 1978.

scene of the same area was computer processed; of several enhancements produced, the best, a principal-components transformed image (Blodget et al., 1978), is shown in Figure 13-79☆ (right). Interpretation maps of each are presented in Figure 13-80.

The HCMM image presents a wide variety of different colors for bedrock areas, whereas most of these areas are similarly colored on the Landsat image. Many of the rocks are quite flat spectrally in the Landsat wavelength region, and so cannot be separated. These same rocks, however, appear to have different thermal properties, and so can be separated using HCMM data.

On the HCMM image, basalts are green. The *aa* flows at Pisgah and west of Pisgah are yellow, and easily separable. They are colder at night than the

pahoehoe flows due to their lower thermal inertia. Granitic rocks are generally orange; the biotite quartz monzonites in the south are somewhat redder. The dacite west of Pisgah is the same orange color; compositionally this rock is similar to the granites. The sand dunes in the northwest corner are red (cold at night, warm in the day, high albedo) as is Lavic Lake playa. The other extrusive rocks are orange-yellow. The varied colors of the alluvial areas reflect the variation in their source-rock composition.

On the Landsat image the bedrock areas are distinguishable from the alluvial areas. However, very few separations can be made between the various rock types. The major improvement that Landsat data show over HCMM data in this particular, spectrally flat, area is the higher spatial res-

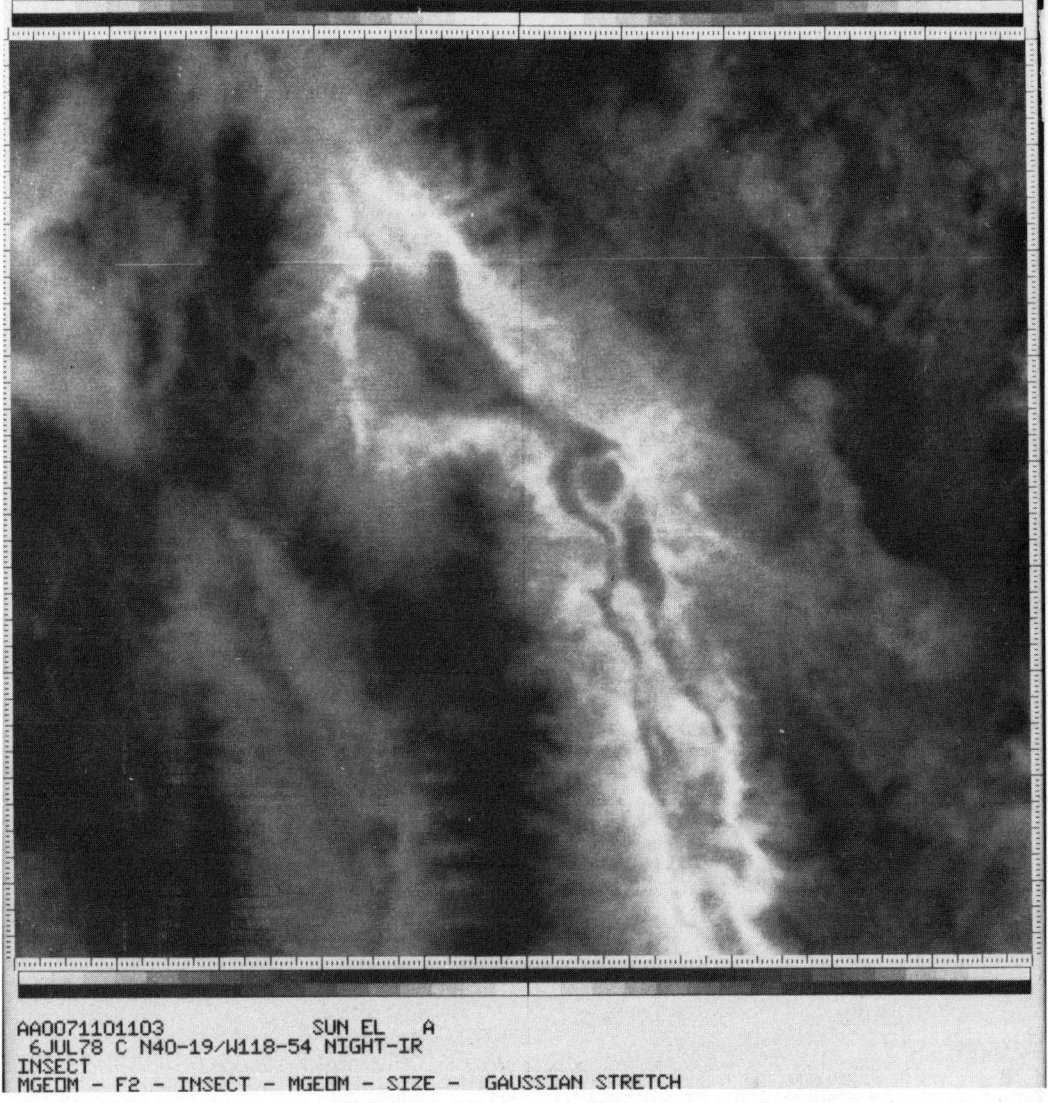

AA0071101103   SUN EL   A
 6JUL78 C N40-19/W118-54 NIGHT-IR
INSECT
MGEOM - F2 - INSECT - MGEOM - SIZE - GAUSSIAN STRETCH

Fig. 13-77. Portion of an HCMM night-IR image, Death Valley, California, July 6, 1978.

olution which permits mapping of smaller features. Very little other information can be interpreted from the Landsat image.

SMIRR SYSTEM

Mission Description

The Space Shuttle, on its second flight, carried the first science and applications payload scheduled by the Space Transportation System. This payload, called OSTA-1, was developed by NASA's Office of Space and Terrestrial Applications (OSTA) to provide an early demonstration of the Space Shuttle's research capabilities.

During its time in orbit, the Shuttle assumed an earth-viewing orientation, thus accommodating the experiments of the OSTA-1 payload. In this attitude the Shuttle's payload bay faces the earth on a line perpendicular to the earth's surface.

SMIRR Description

The Shuttle Multispectral Infrared Radiometer (SMIRR) of the OSTA-1 package will evaluate 10 bands in the 0.5- to 2.5-μm range to determine their effectiveness in identifying geological units when the data are gathered from orbit.

The SMIRR data will be correlated with field spectrometer data to determine whether the data gathered on the ground over a small area are sufficient to specify bands for future orbiting multispectral scanners designed for geological mapping. The experiment will assess the variability in reflectance signature of similar geological units in different climatic environments. The SMIRR experiment will also assess the effect of variable atmospheric absorption on the quality of the data.

Although not a multispectral mapping instrument, SMIRR is an important precursor to sub-

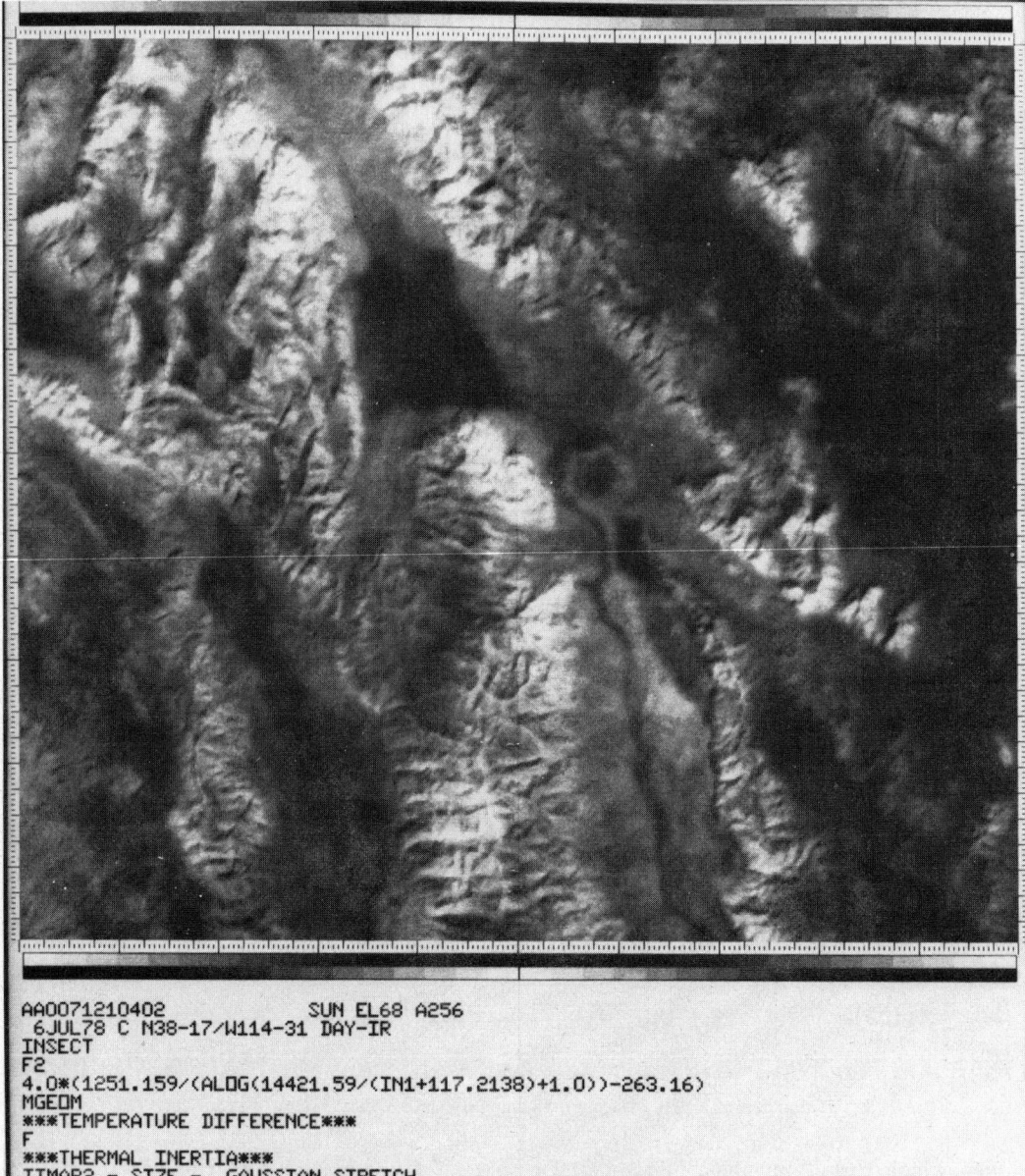

```
AA0071210402              SUN EL68 A256
 6JUL78 C N38-17/W114-31 DAY-IR
INSECT
F2
4.0*(1251.159/(ALOG(14421.59/(IN1+117.2138)+1.0))-263.16)
MGEOM
***TEMPERATURE DIFFERENCE***
F
***THERMAL INERTIA***
TIMAP2 - SIZE -  GAUSSIAN STRETCH
```

Fig. 13-78. Thermal inertia image of Death Valley, California, derived from the HCMM visible- and IR-data.

sequent missions of that type. The SMIRR is equipped with boresighted film cameras and provides visual and infrared measurements of the ground-track radiance in 10 selected bands. The film images are used to correlate the spectral signatures with the surface morphology and from that to maps. The utility of the various spectral bands for detection and identification of mineralogic units can then be determined.

The SMIRR shown in Figure 13-81 consists of a 1371-mm focal length Cassegrain telescope followed by a rotating filter and chopper wheel and a thermo-electrically cooled detector. A mercury-cadmium-telluride photoconductor with 3.5-μm cutoff wavelength is used for all spectral channels.

The instrument electronic sample both signal and dark reference sections of the filter wheel, and subtract the instrument background. Two 16-mm single-frame cameras, boresighted on the radiometer FOV, alternately acquire black-and-white images of the ground track for geographic reference.

The spectral bands shown in Table 13-14 were selected to provide discrimination among a number of mineral types.

Calibration and Rock Sample Tests

Laboratory calibration of SMIRR was carried out using a light cannon, consisting of a hemisphere illuminated with four quartz-iodide lamps

HCMM   LANDSAT

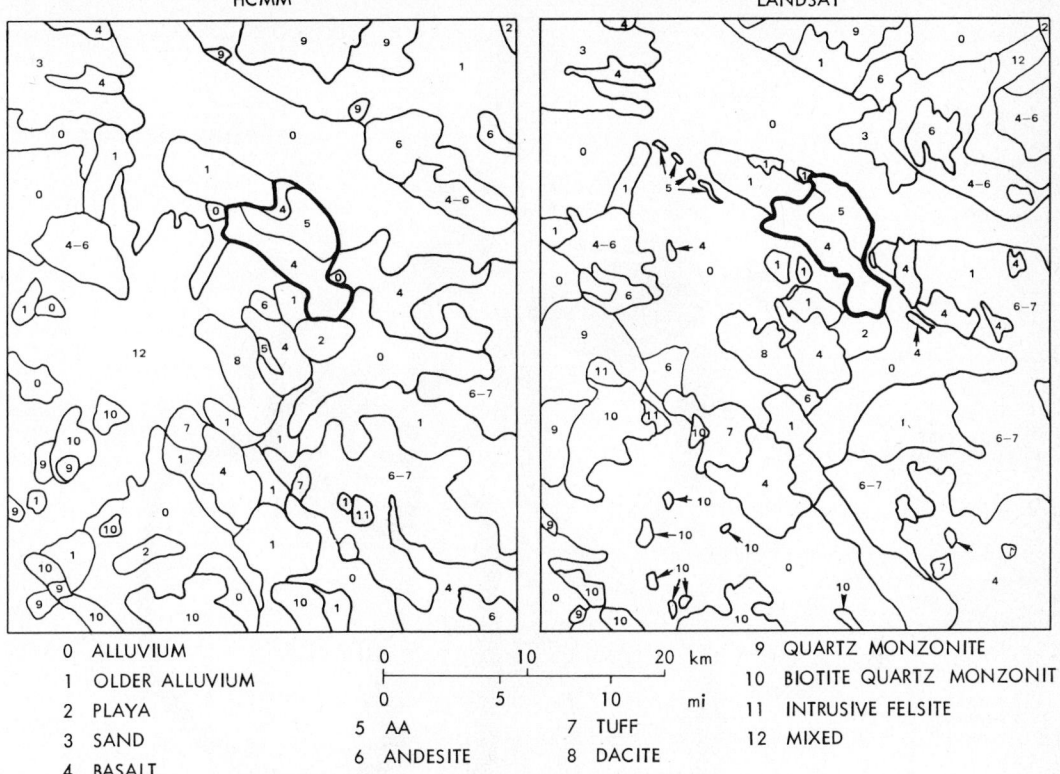

0	ALLUVIUM			9	QUARTZ MONZONITE
1	OLDER ALLUVIUM	0	10 20 km	10	BIOTITE QUARTZ MONZONIT
2	PLAYA	0	5 10 mi	11	INTRUSIVE FELSITE
3	SAND	5	AA	7	TUFF	12	MIXED
4	BASALT	6	ANDESITE	8	DACITE		

Fig. 13-80. Interpretation maps for Figure 13-79.

connected to a cone having a ground-glass screen at its base (Wellman and Goetz, 1980). The intersection of the cone and the hemisphere contains an iris used to control the radiance at the ground-glass plate. An absolute calibration of the light cannon was made using a spectrometer system at the Johnson Space Center.

Relative calibration under simulated measurement conditions was carried out outside the laboratory using the sun as a source. A number of rock and mineral specimens, as well as a standard white material having a flat response throughout the wavelength region of interest, were arrayed in front of the SMIRR telescope. Partial results of these measurements are shown in Figure 13-82.

The values of the sequential spectral band ratios 2.1 through 2.35 μm are plotted. The data are first normalized to a standard (Fiberfrax, a white ceramic wool is used). Ratios are used because absolute measurements of brightness have little meaning in comparing one area on the earth's surface to another. The radiance measured at the spacecraft has atmospheric components as well as a large dependency on the terrain observed. The brightness variations associated with terrain slope are much greater than those associated with spectral reflectance properties of the materials. Therefore, ratios are used to, in effect, normalize the terrain effect.

Statistical data handling techniques are also used to overcome the high correlation of spectral band data. The errors in Figure 13-82 associated with measurement of the ratios due to system noise are approximately ±0.03 for very similar materials such as kaolinite, illite, and montmorillonite, —all clays. The variation in the ratios is still significant enough to allow direct identification. The outdoor calibrations will be very important in reducing data obtained from the spacecraft. Additional data were collected by placing the SMIRR in an aircraft.

Results from Aircraft Tests

The SMIRR was mounted in the JPL Queen-Air aircraft, and data were collected from 10 test sites in Nevada, California, and Arizona. The aircraft was flown approximately 2000 m above the terrain at an air speed of approximately 93 m/s (180 knots). The resulting instantaneous-field-of-view (IFOV) diameter was approximately 1 m, but the aircraft speed was sufficiently slow so that oversampling was obtained. In fact, the ratio of velocity to altitude for the Shuttle is approximately 19 and for the aircraft test approximately 20.

The position of the FOV of SMIRR during the the aircraft flights was determined by the boresighted 16-mm framing camera. However, because of the roll instability of the aircraft, it was not possible to extrapolate between frames and determine the exact ground location. The situation made data interpretation more difficult, but it was not a limiting factor.

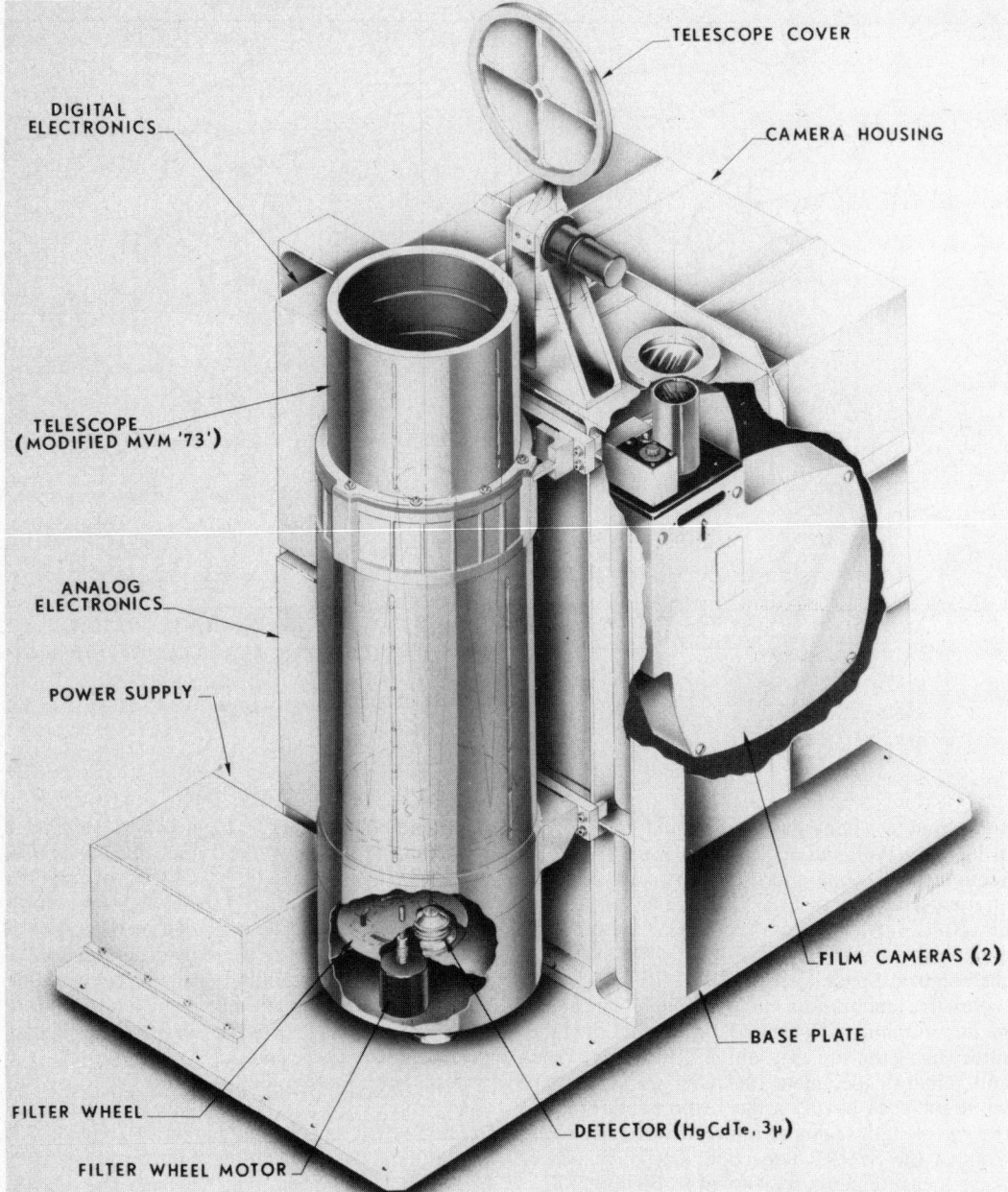

Fig. 13-81. Shuttle Multispectral Infrared Reflectance Radiometer (SMIRR).

Figure 13-83 shows a plot of two spectral band ratios during a 70-s period of a test site overflight. For purposes of comparison of these data to potential data from the Shuttle, only every 100th point was plotted. The test site contains two major rock types, marble (a calcium carbonate), and granodiorite (an igneous silicate rock containing a high percentage of quartz and feldspar). In the exposed areas, both marble and granodiorite have a reddish appearance. The ratio 0.5/0.6, equivalent to the ratio between Landsat bands 4 and 5, shows little difference between the marble and the granodiorite responses. But the ratio 2.22/2.35 is

considerably higher in marble than it is in the granodiorite. 2.35 μm is the position of the absorption band for calcium carbonate.

Figure 13-84 shows a canonical analysis of the same data shown in Figure 13-83 in which all the sequential ratios are used. The axes correspond to linear combinations of the ratio values that yield the greatest separation between the two components, marble and granodiorite. The U-statistic is related to the weight applied to a particular ratio in developing the canonical variables. This figure shows that the ratio 0.6/1.05 is the most important for separating the marble from the granodiorite.

TABLE 13-14

Spectral Bands for the SMIRR

Channel	Center, μm	Half-Power Bandwidth, μm
1	$0.5 \pm .02$	0.1
2	$0.6 \pm .02$	0.1
3	$1.05 \pm .02$	0.1
4	$1.2 \pm .02$	0.1
5	$1.6 \pm .02$	0.1
6	$2.1 \pm .02$	0.1
7	$2.17 \pm .005$	0.02
8	$2.20 \pm .005$	0.02
9	$2.22 \pm .005$	0.02
10	$2.35 \pm .015$	0.06

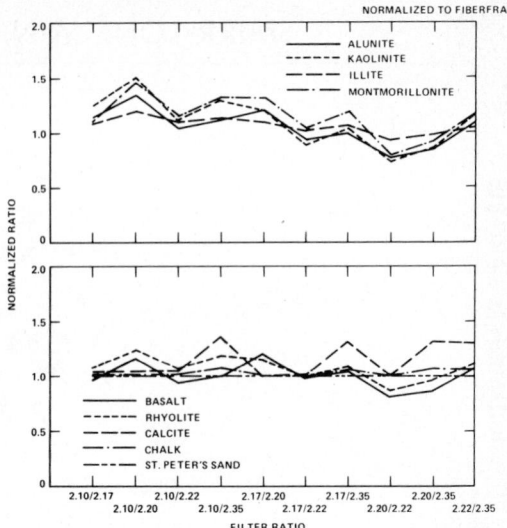

Fig. 13-82. SMIRR calibration results using natural materials in sunlight (curves normalized to the spectrally flat standard of Fiberfrax).

The reason is that the ratio 0.6/1.05 is very sensitive to the quantity of vegetation in the scene. More vegetation is found on the granodiorite than on the marble, therefore, the statistical analysis shows a heavier weighting for 0.6/1.05 than it did for the 2.22/2.35 ratio which is a direct measure of the rock type differences. *In situ* spectral reflectance measurement as well as laboratory spectral photometry substantiate the results taken by SMIRR from an aircraft (Wellman and Goetz, 1980).

FUTURE SYSTEMS

Landsat D

As discussed in a previous section the new Landsat satellites include short wavelength infrared bands centered at 1.6 and 2.2 μm. This instrument is further described in Chapter 12.

Thermal Infrared Multispectral Scanner (TIMS)

Subsequent to the analysis of the multispectral infrared scanner data described earlier, NASA initiated construction by Daedalus of a new Thermal Infrared Multispectral Scanner (TIMS) for airborne use. This instrument, completed in the spring of 1982, includes six bands between 8 and 12 μm: 8.2–8.6, 8.6–9.0, 9.0–9.4, 9.4–10.2, 10.2–11.2, and 11.2–12.2 μm. It has a 2.5-mrad IFOV, an 86° scan angle, selectable scan rates and V/H, and internal blackbody reference sources. The instrument is considered to be a forerunner of a possible orbiting system, after sufficient aircraft data have been analyzed to demonstrate the utility and requirements for such data.

Advanced HCMM

A possible follow-on to HCMM, the Advanced HCMM is under consideration. Such a future system would retain the mission objectives of day and night coverage within a 12-hour period, with measurements near times of maximum and minimum surface temperatures. Improvements over HCMM would include greater spatial resolu-

tion, greater temperature resolution, possible inclusion of a second IR band to allow for atmospheric corrections, and better calibration.

Imaging Spectrometer (IS)

Looking beyond the Thematic Mapper, it is apparent that the emerging technology of solid-state array imaging sensors frequently referred to as multispectral, linear arrays (MLA), offers significant performance advantages over scanning systems. Each picture element in an MLA pushbroom imager may integrate for up to one line time, whereas scanning detectors must be sampled at each pixel time. Thus the MLA sensor has an enormous sensitivity advantage because of its longer integration time per sample.

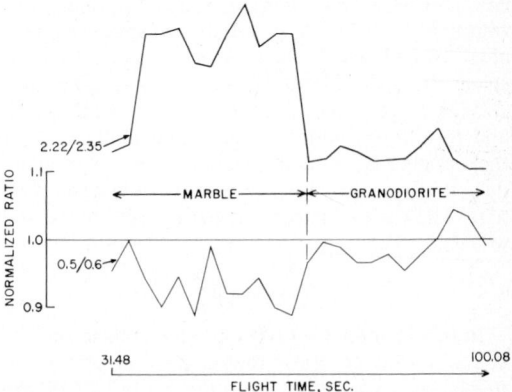

Fig. 13-83. Two SMIRR band ratios plotted during 70-s segment from an overflight of Lone Mountain Nevada (data sampled every 100 points for display; the ratio 2.22/2.35 is sensitive to the CO_3-absorption at 2.35μm).

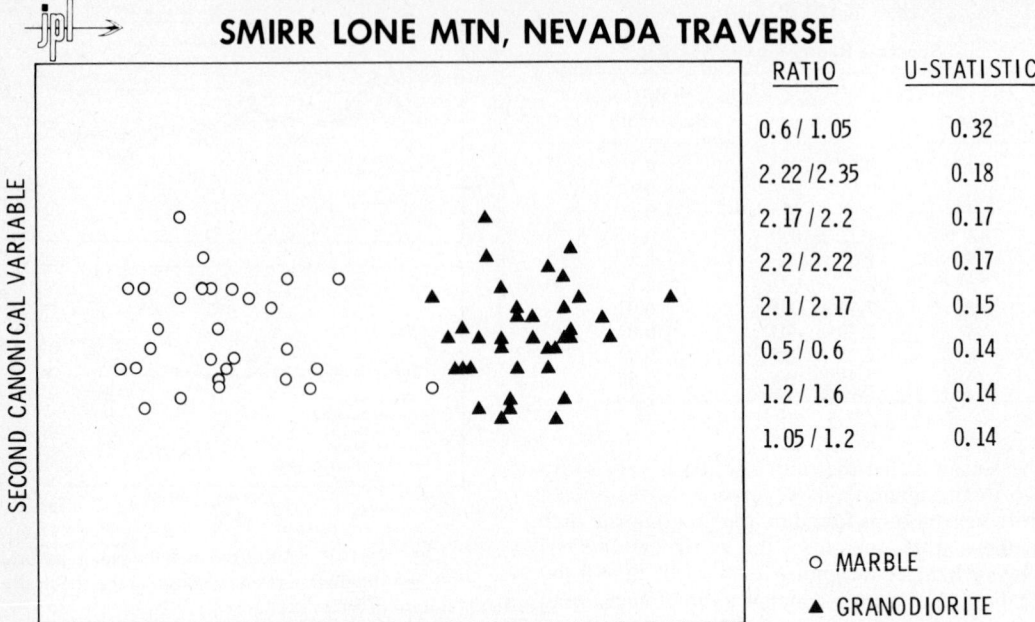

SMIRR LONE MTN, NEVADA TRAVERSE

RATIO	U-STATISTIC
0.6 / 1.05	0.32
2.22 / 2.35	0.18
2.17 / 2.2	0.17
2.2 / 2.22	0.17
2.1 / 2.17	0.15
0.5 / 0.6	0.14
1.2 / 1.6	0.14
1.05 / 1.2	0.14

○ MARBLE

▲ GRANODIORITE

SECOND CANONICAL VARIABLE

FIRST CANONICAL VARIABLE

Fig. 13-84. Canonical analysis of all the ratios obtained in the overflight segment in Figure 13-83 (axes are linear combinations of the ratio values for ratios shown in the legend).

The solid-state array sensor does not require a moving scan mirror in order to acquire an image. As a result, the MLA sensor can be expected to exhibit longer operating life than a scanner. In addition, the fixed geometry afforded by the detector arrays results in high geometric accuracies in the line direction which will simplify the image reconstruction and processing tasks.

NASA has established an MLA program to examine the requirements for future systems, to develop the technologies required for their implementation, and to demonstrate an advanced system from earth orbit within this decade.

The Imaging Spectrometer (IS), utilizes a spectrometer section to separate the spectral channels and area array detectors to acquire images simultaneously in a large number of spectral channels. Onboard data processing is used to construct the spectral bands desired for a given observation, to select among several swath-width and resolution modes, to radiometrically correct the imaging data, and to compress the imaging data to a rate compatible with the telemetry system. The major instrument parameters are listed in Table 13-15.

SUMMARY

In this chapter we reviewed spaceborne remote sensors that cover a major part of the electromagnetic spectrum, from the visible to the microwave region. The type of information acquired about the surface is highly dependent on the spectral band of observation. Microwave active sensors are mainly sensitive to the physical and electrical properties of the surface. Microwave passive and middle infrared sensors are sensitive to the surface thermal properties. Multispectral short and middle infrared sensors are sensitive to the chemical composition of the surface. All sensors are sensitive to the surface topography. No single sensor can provide a complete description of the surface properties. The main challenge in the future is to use the appropriate combination of sensors to acquire sufficient information to characterize the physical, chemical and thermal properties of surfaces—especially the surface of the earth. There still remains much virgin territory in spaceborne remote sensing that will require new instruments and techniques. Some of these new developments include:

1) Global topographic mapping with scanning laser altimeters, and stereoscopic image coverage with visible and radar sensors to aid in landform interpretation; also high resolution (few centimeters) radar altimeters for ocean topography mapping.

2) High spectral resolution imaging in the 2- to 2.5-μ regions, and exploitation of the 8- to 14-μ region through multispectral scanners and active laser spectral reflectance measurement.

3) Multifrequency and multipolarization radar systems for increased surface units and cover discrimination.

4) Combined imaging radars and scatterometers for high-resolution wind field mapping and use of coherent radar for ocean currents mapping.

TABLE 15

Imaging Spectrometer Design Parameters

Field of View	15 degrees
Swath Width at 700 km Altitude	185 degrees
Instantaneous Field-of-View	15 m, 30 m, 90 m
Spectral Coverage	
Visible	0.4 to 1.0 μm
SWIR	1.0 to 2.5 μm
Spectral Resolution	
Visible	32 channels in 0.4 to 1.0 μm
SWIR	32 channels in 1.0 to 2.5 μm
Spectral Bands	Any 6 non-overlapping portions
	of the covered spectral range
Optics Clear Aperture	52 cm
Individual Detector Size	21 μm $\times$ 21 μm
Individual Array Size	64 $\times$ 204 elements
Total Arrays Required	60 VNIR + 60 SWIR
Optically Active Focal Plane	
Visible	12240 $\times$ 32 pixels
SWIR (Operational)	12240 $\times$ 32 pixels
Focal Plane Temperature	170 K
Data Encoding	8 bits/pixel
Raw Data Rate	264 Mbps
Compression Ratio	3:1 Maximum

5) Active imaging systems in the near- and short-wavelength infrared regions for making observations using very narrow spectral bands.

REFERENCES

Aarons, J. (ed.), 1973, Total electron content and scintillation studies of the ionosphere; AGARDograph no. 166.

Aarons, J., H. E. Whitney, and R. S. Allen, 1971, Global morphology of ionosphere, scintillations; Proc. IEEE, vol. 59, pp. 159–172.

Abrams, Michael J., Roger P. Ashley, Lawrence C. Rowan, Alexander F. H. Goetz, and Anne B. Kahle, 1977, Mapping of hydrothermal alteration in the Cuprite Mining District, Nevada, using aircraft scanner imagery for the 0.46–2.36 μm spectral region; Geology, vol. 5, p. 713.

Alder, R. F., and E. B. Rodgers, 1977, Satellite-observed latent heat release in a tropical cyclone; Mon. Weather Rev., vol. 105, p. 956.

Allison, J. J., E. B. Rodgers, T. T. Wilheit, and R. W. Fett, 1974, Tropical cyclone rainfall as measured by the Nimbus 5 electrically scanning microwave radiometer; Bull. Amer. Met. Soc., vol. 55, no. 9.

Alpers, W. R., and C. L. Rufenach, 1979, The effect of orbital motions on synthetic aperture radar imagery of ocean waves; IEEE Trans. Ant. Prop., vol. AP-27, pp. 685–690.

Barath, F. T., A. H. Barrett, J. Copeland, D. E. Jones, and A. E. Lilley, 1964, Mariner 2 microwave radiometer experiment and results; Astron. J., vol. 69, no. 1.

Barrick, D. E., and C. T. Swift, 1980, The Seasat microwave instruments in historical perspective; IEEE J. Oceanic Eng., vol. OE-5, pp. 74–80.

Basharinov, A. E., A. S. Gurvich, S. T. Yegorov, A. A. Kurskaya, D. T. Matveyev, and A. M. Shutko, 1971, The results of microwave sounding of the earth's surface according to experimental data from the satellite Cosmos 243; Space and Research, vol. 11, Akademie-Verlag, Berlin.

Blodget, H. W., F. J. Gunther, and M. H. Pdwysocki, 1978, Discrimination of rock classes and alteration products in southwestern Saudi Arabia with computer-enhanced Landsat data; NASA Technical Paper 1327, Scientific and Technical Information Office, 40 pp.

Blom R., and C. Elachi, 1981, Spaceborne and airborne imaging radar observation of sand dunes; J. Geophys. Res., vol. 86, p. 3061.

Blume, H. C., B. M. Kendall, and J. C. Fedors, 1977, Measurements of ocean temperature and salinity via microwave radiometry; Boundary Layer Meteorology, vol. 13, p. 295.

Bracalante, E. M., D. H. Boggs, W. L. Grantham, and J. L. Sweet, 1980, The SASS scattering coefficient algorithm; IEEE J. Oceanic Eng., vol. OE-5, pp. 145–153.

Burns, A. A., and E. J. Fremouw, 1970, A real time correction technique for "transionospheric ranging error;" IEEE Trans. Ant. Prop., vol. AP-28.

Campbell, W. J., P. Gloersen, W. J. Webster, T. T. Wilheit, and R. O. Ramseier, 1976, Beaufort sea ice zones as delineated by microwave imagery; J. Geophys. Res., vol. 81, no. 5.

Crane, R. K., 1977, Ionospheric scintillations; Proc. IEEE, vol. 65, pp. 180–199.

Dicke, R. H., R. Beringer, R. L. Kyhl, and A. B. Vane, 1946, Atmospheric absorption measurements with a microwave radiometer; Phys. Rev., vol. 70, p. 340.

Eagleman, J. R., and W. C. Lin, 1976, Remote sensing of soil moisture by a 21-cm passive radiometer; J. Geophys. Res., vol. 81, no. 21.

Elachi, C., 1980, Spaceborne imaging radar: geologic and oceanographic applications; Science, vol. 209, pp. 1073–1082.

Elachi, C., and J. Apel, 1976, Internal wave observations made with an airborne synthetic aperture imaging radar; Geophys. Res. Lett., vol. 3, pp. 647–650.

Elachi, C., and T. Farr, 1980, Observation of the Grand

Canyon wall structure with an airborne imaging radar; Remote Sensing Env., vol. 9, p. 171.

Elachi, C., and W. E. Brown, 1977, Models of radar imaging of the ocean surface waves; IEEE Trans. Ant. Prop., vol. AP-25, pp. 84–95.

Ford, J., 1980, Seasat orbital radar imagery for geologic mapping: Tennessee-Kentucky-Virginia; Am. Assoc. Petrol. Geol., vol. 66, p. 2064.

Gillespie, A. R., and A. B. Kahle, 1977, The construction and interpretation of a digital thermal inertia image; Photogram. Eng. and Rem. Sens., vol. 43, p. 983.

Gloersen, P., and F. T. Barath, 1977, A scanning multichannel microwave radiometer for Nimbus-G and Seasat-A; IEEE J. Oceanic Eng., vol. OE-2, no. 2.

Gloersen, P., and V. V. Salomonson, 1975, Satellites— new global observing techniques for ice and snow; J. Glaciol., vol. 15, no. 73.

Gloersen, P., H. J. Zwally, A. T. C. Chang, D. K. Hall, W. J. Campbell, and R. O. Ramseier, 1978, Time-dependence of sea-ice concentration and multi-year ice fraction in the Arctic basin; Boundary Layer Meteorology, vol. 13, p. 339.

Gloersen, P., T. T. Wilheit, T. C. Chang, W. Nordberg, and W. J. Campbell, 1974, Microwave maps of the polar ice of the earth; Bull. Amer. Met. Soc., vol. 55, no. 12.

Goddard Space Flight Center, NASA, 1978, Heat Capacity Mapping Mission (HCMM) Data User's Handbook for Applications Explorer Mission-A (AEM).

Goetz, A. F. H., 1981, Shuttle multispectral infrared radiometer (SMIRR) experiment description, Jet Propulsion Laboratory, Pasadena, Calif., in press.

Goetz, A. F. H., and L. C. Rowan, 1981, Geologic remote sensing; Science, vol. 211, p. 781.

Grantham, W. L., E. M. Bracalante, W. L. Jones, and J. W. Johnson, 1977, The Seasat-A satellite scatterometer; IEEE J. Oceanic Eng., vol. OE-2, p. 200.

Grody, N. C., 1976, Remote sensing of atmospheric water content from satellites using microwave radiometry; IEEE Trans. Ant. Prop.; vol. AP-24, no. 2.

Grody, N. C., A. Gruber and W. C. Chen, 1980, Atmospheric water content over the tropical Pacific derived from the Nimbus-6 scanning microwave spectrometer; J. Appl. Meteorol., vol. 19, p. 986.

Grody, N. C., and P. P. Pellegrino, 1977, Synoptic-scale studies using the Nimbus-G scanning microwave spectrometer; J. Appl. Meteorol., vol. 16, p. 816.

Hancock, D. W., R. J. Forsythe, and T. Lorell, 1980, Seasat altimeter sensor file algorithms; IEEE J. Oceanic Eng., vol. OE-5, pp. 93–99.

Harger, R. O., 1980, Synthetic aperture radar systems, theory and design, New York, Academic Press.

Hofer R., E. G. Njoku, and J. W. Waters, 1981, Microwave radiometric measurements of sea surface temperature from the Seasat satellite: first results; Science, vol. 212, p. 1385.

Hollinger, J. P., 1971, Passive microwave measurements of sea surface roughness; IEEE Trans. Geosci. Electron., vol. GE-9, p. 165.

Hovis, W. A., 1966, Infrared spectral reflectance of some common minerals; Applied Optics, vol. 5, p. 245.

Hunt, G. R., and J. W. Salisbury, 1974, Mid-infrared spectral behavior of igneous rocks; U.S. Air Force Cambridge Research Laboratories Report AFCRL-TR-74-0625, Environmental Research Papers, no. 496.

Hunt, G. R., and J. W. Salisbury, 1975, Mid-infrared spectral behavior of sedimentary rocks; U.S. Air Force Cambridge Research Laboratories Report AFCRL-TR-75-0356, Environmental Research Papers, no. 520.

Hunt, G. R., and J. W. Salisbury, 1976, Mid-infrared spectral behavior of metamorphic rocks; U.S. Air Force Cambridge Research Laboratories Report AFCRL-TR-76-0003, Environmental Research Papers, no. 543.

Hunt, G. R., and R. P. Ashley, 1974, Spectra of altered rocks in the visible and near infrared; Economic Geology, vol. 74, p. 1613.

Johnson, J. W., L. A. Williams, E. M. Bracalante, F. B. Beck, and W. L. Grantham, 1980, Seasat-A satellite scatterometer instrument evaluation; IEEE J. Oceanic Eng., vol. OE-5, pp. 138–144.

Jones, W. L., et al., 1979, Seasat scatterometer: results of the Gulf of Alaska workshop; Science, vol. 206, pp. 1413–1415.

Jordan, R. L., 1980, The Seasat-A synthetic aperture radar system; IEEE J. Oceanic Eng., vol. OE-5, pp. 154–163.

Kahle, A. B., 1977, A simple thermal model of the earth's surface for geological mapping by remote sensing; J. Geophys. Res., vol. 82, p. 1673.

Kahle, A. B., A. R. Gillespie, and A. F. H. Goetz, 1976, Thermal inertia imaging: a new geologic mapping tool; Geophys. Res. Lett., vol. 3, p. 26.

Kahle, A. B., and L. C. Rowan, 1980, Evaluation of multispectral middle infrared aircraft images for lithologic mapping in the East Tintic Mountains, Utah; Geology, vol. 8, p. 234.

Kahle, A. B., D. P. Madura, and J. M. Soha, 1980, Middle infrared multispectral aircraft scanner data: analysis for geological applications; Applied Optics, vol. 19, p. 2279.

Kahle, A. B., J. P. Schieldge, M. J. Abrams, R. E. Alley, and C. J. LeVine, 1981, Geologic application of thermal inertia imaging using HCMM data; HCMM Final Report, JPL Publication No. 81-55.

Kidder, S. Q., W. M. Gray, and T. H. Vonder Haar, 1978, Estimating tropical cyclone central pressure and outer winds from satellite microwave data; Mon. Weather Rev., vol. 106, p. 1458.

King, J. C., 1975, Quantization and symmetry in periodic coverage patterns with applications to earth observation; AAS Astrodynamics Specialist Conf., Nassau, Bahamas, July 28.

Klein, L. A., and C. T. Swift, 1977, An improved model for the dielectric constant of sea water at microwave frequencies; IEEE Trans. Ant. Prop., vol. AP-25, p. 104.

Kobrick, M., 1976, Random processes as a cause of the lunar asymmetry; The Moon, vol. 15, p. 83.

Kunzi, K. F., A. D. Fisher, D. H. Staelin, and J. W. Waters, 1976, Snow and ice surfaces measured by the Nimbus-5 microwave spectrometer; J. Geophys. Res., vol. 81, no. 27.

Leith, E. N., 1977, Complex spatial filters for image deconvolution; Proc. IEEE, vol. 65, pp. 18–28.

Lenoir, W. B., 1968, Microwave spectrum of molecular oxygen in the mesosphere; J. Geophys. Res., vol. 73, p. 361.

Lipes, R. G., R. L. Bernstein, V. J. Cardone, K. G. Katsaros, E. G. Njoku, A. L. Riley, D. B. Ross, C. T. Swift, and F. J. Wentz, 1979, Seasat scanning multi-channel microwave radiometer: results of the Gulf of Alaska workshop; Science, vol. 204, no. 4400.

Martinson, L., 1975, A programmable digital processor for airborne radar; IEEE 1975 Int'l. Radar Conf. Record, pp. 186–191.

Masursky, H., E. Eliason, P. G. Ford, G. E. McGill, G. H. Pettengill, G. G. Schaber, and G. Schubert, 1980, Pioneer Venus radar results: geology from images and altimetry; J. Geophys. Res., vol. 85, pp. 8232–8260.

McFarland, M. J., 1976, The correlation of Skylab L-band brightness temperatures with antecendent precipitation; Proc. NASA Earth Resources Symp., NASA TMX-58168, p. 2243.

Meeks, M. L., and A. E. Lilley, 1963, The microwave spectrum of oxygen in the earth's atmosphere; J. Geophys. Res., vol. 68, p. 1683.

Milton, N. M., and D. Madura, 1981, Vegetation distribution of the central part of the East Tintic Mountains, Utah; U.S. Geological Survey Miscellaneous Field Studies Map 1195.

Moore, R. K., and A. K. Fung, 1979, Radar determination of winds at sea; Proc. IEEE, vol. 67, pp. 1504–1521.

Moore, R. K., F. T. Ulaby, and A. Sobti, 1975, The influence of soil moisture on the microwave response from terrain as seen from orbit; Proc. 10th Int'l. Symp. on Rem. Sens. Env., Ann Arbor, Mich.

Moore, R. K., J. P. Claassen, A. C. Cook, D. L. Fayman, J. C. Holtzman, A. Sobti, W. E. Spencer, F. T. Ulaby, J. D. Young, W. J. Pierson, V. J. Cardone, J. Hayes, W. Spring, R. J. Kern, and N. M. Hatcher, 1974, Simultaneous active and passive microwave responses of the earth—the Skylab Radscat experiment; Proc. 9th Int'l. Symp. on Rem. Sens. Env., Ann Arbor, Mich., p. 189.

Morris, H. T., 1964a, Geology of the Eureka quadrangle, Utah and Juab Counties, Utah; U.S. Geological Survey Bulletin, vol. 1142-K, 29 pp.

Morris, H. T., 1964b, Geology of the Tintic Junction quadrangle, Tooele, Juab and Utah Counties, Utah; U.S. Geological Survey Bulletin, vol. 1142-L, 23 pp.

Morris, H. T., and T. S. Lovering, 1961, Statigraphy of the East Tintic Mountains, Utah; U.S. Geological Survey Professional Paper 361, 145 pp.

Morris, H. T., and T. S. Lovering, 1979, General geology and mines of the East Tintic mining district, Utah and Juab Counties, Utah; U.S. Geological Survey Professional Paper 1024, 203 pp.

Njoku, E. G., E. J. Christensen, and R. E. Cofield, 1980, The Seasat scanning multichannel microwave radiometer (SMMR): antenna pattern corrections—development and implementation; IEEE J. Oceanic Eng., vol. OE-5, p. 125.

Njoku, E. G., J. M. Stacey, and F. T. Barath, 1980, The Seasat scanning multichannel microwave radiometer (SMMR): instrument description and performance; IEEE J. Oceanic Eng., vol. OE-5, p. 100.

Nordberg, W., J. Conaway, D. B. Ross, and T. Wilheit, 1971, Measurements of microwave emission from a foam-covered, wind-driven sea; J. Atm. Sci., vol. 28, p. 429.

Pettengill, G. H., 1971, Orbiter radar mapper instrument, Pioneer Venus experiment description; Space Sci. Rev., vol. 20, pp. 512–515.

Pettengill, G. H., E. Eliason, P. Ford, G. B. Loriot, H. Masursky, and G. E. McGill, 1980, Pioneer Venus radar results: altimetry and surface properties; J. of Geophys. Res., vol. 85, pp. 8261–8270.

Potter, A. E., C. K. Williams, A. L. Grandfield, K. J. Demel, M. C. Trichel, T. L. Barnett, R. D. Juday, W. E. Hensley, N. M. Hatcher, W. E. McAllum, J. T. McGoogan, J. C. Jones, O. N. Brandt, J. G. Braithwaite, R. H. McLaughlin, R. Collins, W. H. Peake, and R. K. Moore, 1974, Summary of flight performance of the Skylab earth resources experimental package (EREP); Proc. 9th Int'l. Symp. on Rem. Sens. Env., Ann Arbor, Mich., p. 1803.

Ray, R. G., 1960, Aerial photographs in geologic interpretation and mapping, USGS Professional Paper 373.

Rosenkranz, P. W., D. H. Staelin, and N. C. Grody, 1978, Typhoon June (1975) viewed by a scanning microwave spectrometer; J. Geophys. Res., vol. 83, no. C4.

Rowan, L. C., and M. J. Abrams, 1978, Evaluation of Landsat multispectral scanner images for mapping altered rocks in the East Tintic Mountains, Utah; U.S. Geological Survey Open-File Report 78-736, p. 73.

Rowan, L. C., A. F. H. Goetz, and R. P. Ashley, 1977, Discrimination of hydrothermally altered and unaltered rocks in visible and near-infrared multispectral images; Geophysics, vol. 42, p. 522.

Sabins, F. F., R. Blom, and C. Elachi, 1980, Seasat radar image of the San Andreas Fault, California; Amer. Assoc. of Petrol. Geol., vol. 64, p. 614.

Schaber, G. G., C. Elachi, and T. Farr, 1980, Remote sensing data of SP lava flow and vicinity in north central Arizona; Rem. Sensing Env., vol. 9, p. 169.

Schmugge, T. J., J. M. Meneely, A. Rango, and R. Neff, 1977, Satellite microwave observations of soil moisture variations; Water Resources Bull., vol. 13, no. 2.

Siegal, Barry S., and Alexander, F. H. Goetz, 1977, Effect of vegetation on rock and soil type discrimination; Photogram. Eng. and Rem. Sens., vol. 43, p. 191.

Sobti, A., and R. K. Moore, 1976, Correlation between microwave scattering and emission from land and sea; IEEE Trans. Geosci. Electron., vol. GE-14, no. 2.

Staelin, D. H., 1969, Passive remote sensing at microwave wavelengths; Proc. IEEE, vol. 57, no. 4.

Staelin, D. H., A. L. Cassel, K. F. Kunzi, R. L. Pettyjohn, R. K. L. Poon, P. W. Rosenkranz, and J. W. Waters, 1975a, Microwave atmospheric temperature sounding: effects of clouds on the Nimbus 5 satellite data; J. Atm. Sci., vol. 32, p. 1970.

Staelin, D. H., A. H. Barrett, P. W. Rosenkranz, F. T. Barath, E. J. Johnston, J. W. Waters, A. Wouters, and W. B. Lenoir, 1975b, The scanning microwave spectrometer (SCAMS) experiment; The Nimbus 6 User's Guide, NASA/Goddard Space Flight Center, Greenbelt, Md.

Staelin, D. H., and P. W. Rosenkranz, (eds.), 1978, High resolution passive microwave satellites; Applications Review Panel Final Report, MIT Research Lab. of Electronics, Cambridge, Mass.

Staelin, D. H., F. T. Barath, J. C. Blinn, and E. J. Johnston, 1972, The Nimbus-E microwave spectrometer (NEMS) experiment; The Nimbus 5 User's Guide, NASA/Goddard Space Flight Center, Greenbelt, Md.

Staelin, D. H., K. F. Kunzi, R. L. Pettyjohn, R. K. L. Poon, R. W. Wilcox, and J. W. Waters, 1976, Remote sensing of atmospheric water vapor and liquid water with the Nimbus-5 microwave spectrometer; J. Appl. Meteorol., vol. 15, no. 11.

Swanson, P. N., and A. L. Riley, 1980, The Seasat

scanning multichannel micro radiometer (SMMR): radiometric calibration algorithm development and performance; IEEE J. Oceanic Eng., vol. OE-5, p. 116.

Tomiyasu, K., 1978, Tutorial review of synthetic aperture radar with applications to imaging of the ocean surface; Proc. IEEE, vol. 66, no. 5, pp. 563–583.

Townsend, W. F., 1980, An initial assessment of the performance achieved by the Seasat radar altimeter; IEEE J. Oceanic Eng., vol. OE-5, pp. 80–92.

Ulaby, F. T., L. F. Dellwig, and T. Schmugge, 1975, Satellite microwave observations of the Utah Great Salt Lake Desert; Radio Science, vol. 10, no. 11.

Valenzuela, G., 1980, An asymptotic formulation for SAR images of the dynamical ocean surface; Radio Science, vol. 15, pp. 105–114.

Van de Lindt, W. J., 1977, Digital technique for generating synthetic aperture radar images; IBM J. Res. and Develop., vol. 21, pp. 415–432.

Waters, J. W., and D. H. Staelin, 1968, Statistical inversion of radiometric data; Quart. Prog. Report no. 89, MIT Research Lab. of Electronics, Cambridge, Mass.

Waters, J. W., and S. C. Wofsy, 1978, Applications of high resolution passive microwave satellite systems to the stratosphere, mesosphere and lower thermosphere; High Resolution Passive Microwave Satellites, Staelin and Rosenkranz, eds., MIT Research Lab. of Electronics, Cambridge, Mass.

Waters, J. W., K. F. Kunzi, R. L. Pettyjohn, R. K. L. Poon, and D. H. Staelin, 1975, Remote sensing of atmospheric temperature profiles with the Nimbus 5 microwave spectrometer; J. Atm. Sci., vol. 32, no. 10.

Watson, K., 1973, Periodic heating of layer over a semi-infinite solid; J. Geophys. Res., vol. 78, p. 5904.

Watson, K., 1975, Geologic applications of thermal infrared images; Proc. IEEE, vol. 63, p. 128.

Webster, W. J., T. T. Wilheit, D. B. Ross, and P. Gloersen, 1976, Special characteristics of the microwave emission from a wind-driven, foam-covered sea; J. Geophys. Res., vol. 81, p. 3095.

Weinman, J. A. and P. J. Guetter, 1977, Determination of rainfall distribution from microwave radiation

measured by the Nimbus-6 ESMR; J. Appl. Meteorol., vol. 16, p. 437.

Wellman, J. B., and A. F. H. Goetz, 1980, Experiments in infrared multispectral mapping of earth resources; Proc. AIAA Sensor Systems for the 80's Conference, Colorado Springs, Colorado, December 2–4.

Wilheit, T., 1972, The electrically scanning microwave radiometer (ESMR) experiment; the Nimbus 5 User's Guide, NASA/Goddard Space Flight Center, Greenbelt, Md.

Wilheit, T., 1975, The electrically scanning microwave radiometer (ESMR) experiment; The Nimbus 6 User's Guide, NASA/Goddard Space Flight Center, Greenbelt, Md.

Wilheit, T., 1978, A review of application of microwave radiometry to oceanography; Boundary Layer Meteorology, vol. 13, p. 277.

Wilheit, T., 1979, A model for the microwave emissivity of the ocean's surface as a function of wind speed; IEEE Trans. Geosci. Elect., vol. GE-17, p. 244.

Wilheit, T., 1979, The effect of wind on the microwave emission from the ocean's surface at 37 GHz; J. Geophys. Res., vol. 84, p. 4921.

Wilheit, T., A. T. C. Chang, M. S. V. Rao, E. B. Rodgers, and J. S. Theon, 1977, A satellite technique for quantitatively mapping rainfall rates over the oceans; J. Appl. Meteorol., vol. 16, p. 551.

Wilheit, T., J. S. Theon, W. E. Shenk, L. J. Allison, and E. B. Rogers, 1976, Meteorological interpretations of the images from the Nimbus-5 electrically scanned microwave radiometer; J. Appl. Meteorol., vol. 15, p. 166.

Wilheit, T., R. R. Adler, R. Burpe, R. Sheets, W. E. Shenk, and P. W. Rosenkranz, 1978, Monitoring of severe storms; High Resolution Passive Microwave Satellites, Staelin and Rosenkranz, eds., MIT Research Lab. of Electronics, Cambridge, Mass.

Wu, C., 1980, Considerations on real-time processing of spaceborne synthetic aperture radar data; SPIE 24th Int'l. Symp., San Diego, August 1.

Zwally, H. J., and P. Gloersen, 1976, Passive microwave images of the polar regions and research applications; Polar Record, vol. 18, no. 116.

CHAPTER 14

Meteorological Satellites

Author-Editors: LEWIS J. ALLISON and ABRAHAM SCHNAPF

Contributing Authors: BERNARD C. DIESEN, III, PHILIP S. MARTIN, ARTHUR SCHWALB, WILLIAM R. BANDEEN

GENERAL CONTENTS: Polar Orbiting and Low Earth-Orbiting Satellites: TIROS; ESSA; ITOS; TIROS-N; Advanced TIROS-N; NIMBUS. The Defense Meteorological Satellite Program: Evolution; Block 5D instruments; Special sensors. Geostationary Meteorological Satellites: ATS; SMS/GOES: GOES D, E, and F. The Earth Radiation Budget Experiment. International Weather Satellites. Oceanographic Satellites: Seasat-A. Space Shuttle. References.

INTRODUCTION

This chapter presents an overview of the meteorological satellite programs that have been evolving from 1958 to the present and reviews plans for the future meteorological and environmental satellite systems that are scheduled to be placed into service in the 1980s. The development of the TIROS family of weather satellites, including TIROS, ESSA, ITOS/NOAA, and the present TIROS-N (the third-generation operational system) is summarized (Schnapf, 1979). The contribution of the Nimbus and ATS technology satellites to the development of the operational polar-orbiting and geostationary satellites is discussed. Included are descriptions of both the TIROS-N and the DMSP payloads currently under development to assure a continued and orderly growth of these systems into the 1980s.

POLAR ORBITING AND LOW EARTH-ORBITING SATELLITES

TIROS

The TIROS (Television and Infrared Observation Satellite) system and its successor, TOS (TIROS Operational System), the ITOS (Improved TIROS Operational System) system, and TIROS-N/NOAA, the current operational system, have been the principal global operational meteorological satellite systems for the United States over the past 22 years. Table 14-1 highlights the launch dates, orbits, and payloads for the U.S. weather satellites. Figure 14-1 depicts the performance in orbit for each of the TIROS/ESSA/ITOS/NOAA series of satellites (Schnapf, 1980). These systems matured from a research and development program, marked by the successful mission of TIROS-1 in April 1960 (Allison and Neil, 1962). A semi-operational system soon evolved in which nine additional TIROS satellites were successfully launched in the period from 1960 to 1965. Each TIROS satellite carried a pair of miniature television cameras and in approximately half of the missions a scanning infrared radiometer and an earth radiation budget instrument were included with the instrument complement.

ESSA

The commitment to provide routine daily worldwide observations without interruption in data was fulfilled by the introduction of the TIROS Operational System (TOS) in February 1966. This system employed a pair of ESSA (Environmental Science Services Administration) satellites, each configured for its specific mission. Through their on-board data storage systems, the odd-numbered satellites (ESSA 1, 3, 5, 7, 9) provided global weather data to the U.S. Department of Commerce's CDA (Command and Data Acquisition) stations in Wallops Island, Va., and Fairbanks, Alaska, and then relayed data to the National Environmental Satellite Service at Suitland, Maryland, for processing and forwarding to the major forecasting centers of the United States and to nations overseas. The even-numbered group of satellites (ESSA 2, 4, 6, 8) provided direct real-time readout of their APT (Automatic Picture Transmission) television pictures to simple stations located around the world (Schnapf, 1980). Nine ESSA satellites were successfully launched between 1966 and 1969. One of them, ESSA-8, remained in operation until March 1976. Larger television cameras (2.54 cm vidicon) developed for the Nimbus satellite program were adapted for use on the ESSA series, providing a significant increase in the quality of the cloud cover pictures over that obtained from the earlier TIROS cameras which used a 1.27 cm vidicon (Schwalb and Gross, 1969).

ITOS

The second decade of meteorological satellites was introduced by the successful orbiting on

TABLE 14-1
U.S. ENVIRONMENTAL SATELLITE PROGRAMS

Name	Launched	Period (Min)	Perigee (km)	Apogee (km)	Inclination (Deg)	Remarks
TIROS I	01APR60	99.2	796	867	48.3	1 TV-WA and 1 TV-NA
TIROS II	23NOV60	98.3	717	837	48.5	1 TV-WA, 1 TV-NA, passive & active IR scan
TIROS III	12JUL61	100.4	854	937	47.8	2 TV-WA, HB, IR, IRP
TIROS IV	08FEB62	100.4	817	972	48.3	1 TV-WA, IR, IRP, HB
TIROS V	19JUN62	100.5	680	1119	58.1	1 TV-WA, 1 TV-MA
TIROS VI	18SEP62	98.7	783	822	58.2	1 TV-WA, 1 TV-MA
TIROS VII	19JUN63	97.4	713	743	58.2	2 TV-WA, IR, ion probe, HB
TIROS VIII	21DEC63	99.3	796	878	58.5	1st APT TV direct readout & 1 TV-WA
Nimbus I	28AUG64	98.3	487	1106	98.6	3 AVCS, 1 APT, HRIR "3-axis" stabilization
TIROS IX	22JAN65	119.2	806	2967	96.4	First "wheel"; 2 TV-WA global coverage
TIROS X	02JUL65	100.6	848	957	98.6	Sun synchronous, 2 TV-WA
ESSA 1	03FEB66	100.2	800	965	97.9	1st operational system, 2 TV-WA, FPR
ESSA 2	28FEB66	113.3	1561	1639	101.0	2 APT, global operational APT
Nimbus II	15MAY66	108.1	1248	1354	100.3	3 AVCS, HRIR, MRIR
ESSA 3	02OCT66	114.5	1593	1709	101.0	2 AVCS, FPR
ATS I	06DEC66	24 hr	41,257	42,447	0.2	Spin scan camera
ESSA 4	26JAN67	113.4	1522	1656	102.0	2 APT
ESSA 5	20APR67	113.5	1556	1635	101.9	2 AVCS, FPR
ATS III	05NOV67	24 hr	41,166	41,222	0.4	Color spin scan camera
ESSA 6	10NOV67	114.8	1622	1713	102.1	2 APT TV
ESSA 7	16AUG68	114.9	1646	1691	101.7	2 AVCS, FPR, S-Band
ESSA 8	15DEC68	114.7	1622	1682	101.8	2 APT TV
ESSA 9	26FEB69	115.3	1637	1730	101.9	2 AVCS, FPR, S-Band
Nimbus III	14APR69	107.3	1232	1302	101.1	SIRS A, IRIS, MRIR, IDCS, MUSE, IRLS
ITOS 1	23JAN70	115.1	1648	1700	102.0	2 APT, 2 AVCS, 2 SR, FPR, 3-axis stabilization
Nimbus IV	15APR70	107.1	1200	1280	99.9	SIRS B, IRIS, SCR, THIR, BUV, FWS, IDCS, IRLS, MUSE
NOAA 1	11DEC70	114.8	1422	1472	102.0	2 APT, 2 AVCS, 2 SR, FPR
NOAA 2	15OCT72	114.9	1451	1458	98.6	2 VHRR, 2 VTPR, 2 SR, SPM
Nimbus 5	11DEC72	107.1	1093	1105	99.9	SCMR, ITPR, NEMS, ESMR, THIR
NOAA 3	06NOV73	116.1	1502	1512	101.9	2 VHRR, 2 VTPR, 2 SR, SPM
SMS 1	17MAY74	1436.4	35,605	35,975	0.6	VISSR, DCS, WEFAX, SEM
NOAA 4	15NOV74	101.6	1447	1461	114.9	2 VHRR, 2 VTPR, 2 SR, SPM
SMS 2	06FEB75	1436.5	35,482	36,103	0.4	VISSR, DCS, WEFAX, SEM
Nimbus 6	12JUN75	107.4	1101	1115	99.9	ERB, ESMR, HIRS, LRIR, T&DR, SCAMS, TWERLE, PMR

APT	Automatic Picture Transmission TV	NEMS	Nimbus E Microwave Spectrometer
AVCS	Advanced Vidicon Camera System (1" Vidicon)	PMR	Pressure Modulated Radiometer
AVHRR	Advanced Very High Resolution Radiometer	SAM-II	Stratospheric Aerosol Measurement-II
BUV	Backscatter Ultraviolet Spectrometer	SAMS	Stratospheric and Mesospheric Sounder
CZCS	Coastal Zone Color Scanner	SBUV	Solar Backscatter Ultraviolet Spectrometer
DCS	Data Collection System	SCAMS	Scanning Microwave Spectrometer
ERB	Earth Radiation Budget	SCMR	Surface Composition Mapping Radiometer
ESMR	Electronic Scanning Microwave Radiometer	SCR	Selective Chopper Radiometer
FPR	Flat Plate Radiometer	SEM	Solar Environmental Monitor
FWS	Filter Wedge Spectrometer	SIRS	Satellite Infrared Spectrometer
HB	Heat Budget Instrument	SMMR	Scanning Multichannel Microwave Radiometer
HEPAD	High Energy Proton and Alpha Particle Detector	SPM	Solar Proton Monitor
HIRS	High Resolution Infrared Sounder	SR	Scanning Radiometer
HRIR	High Resolution Infrared Radiometer	SSU	Stratospheric Sounding Unit
IDCS	Image Dissector Camera System	T&DR	Tracking and Data Relay
IR	Infrared - 5 Channel Scanner	THIR	Temperature Humidity Infrared Radiometer
IRIS	Infrared Interferometer Spectrometer	TOMS	Total Ozone Mapping Spectrometer
IRLS	Interrogation, Recording and Location Subsystem	TV	Television Cameras (½" Vidicon)
IRP	Infrared Passive		NA Narrow Angle - 12°
ITPR	Infrared Temperature Profile Radiometer		MA Medium Angle - 78°
LIMS	Limb Infrared Monitoring of the Stratosphere		WA Wide Angle - 104°
LRIR	Limb Radiance Infrared Radiometer	TWERLE	Tropical Wind Energy Reference Equipment
MEPED	Medium Energy Proton and Electron Detector	VAS	VISSIR and Atmosphere Sounder
MRIR	Medium Resolution Infrared Radiometer	VHRR	Very High Resolution Radiometer
MSU	Microwave Scanner Unit	VISSR	Visible Infrared Spin-Scan Radiometer
MUSE	Monitor of Ultraviolet Solar Energy	VTPR	Vertical Temperature Profile Radiometer
		WEFAX	Weather Facsimile

TABLE 14-1 (Continued)
U.S. ENVIRONMENTAL SATELLITE PROGRAMS

Name	Launched	Period (Min)	Perigee (km)	Apogee (km)	Inclination (Deg)	Remarks
GOES 1	16OCT75	1436.2	35,728	35,847	0.8	VISSR, DCS, WEFAX, SEM
NOAA 5	29JUL76	116.2	1504	1518	102.1	2 VHRR, 2 VTPR, 2 SR, SPM
GOES 2	16JUN77	1436.1	35,600	36,200	0.5	VISSR, DCS, WEFAX, SEM
GOES 3	15JUN78	1436.1	35,600	36,200	0.5	VISSR, DCS, WEFAX, SEM
TIROS-N	13OCT 78	98.92	849	864	102.3	AVHRR, HIRS-2, SSU, MSU, HEPAD, MEPED
Nimbus 7	24OCT78	99.28	943	955	104.09	LIMS, SAMS, SAM-II, SBUV/TOMS, ERB, SMMR, THIR, CZCS
NOAA-6	27JUN79	101.26	807.5	823	98.74	AVHRR, HIRS-2, SSU, MSU, HEPAD, MEPED
GOES-4	9SEPT80	1436.1	35,600	35,600	0.5	VAS, DCS, SEM,WEFAX
GOES-5	15MAY81	1436.1	35,600	35,600	0.5	VAS, DCS, SEM, WEFAX
NOAA-7	23JUN81	101.92	852	869	98.9	AVHRR, HIRS-2, SSU, MSU, HEPAD, MEPED

January 23, 1970, of ITOS-1,[1] the second-generation operational weather satellite. This satellite dramatically surpassed the capabilities of the predecessor ESSA satellites, moving rapidly closer to the objectives of the U.S. National Operational Meteorological Satellite System. ITOS-1 provided a single spacecraft the combined capability of two ESSA spacecraft—the direct readout APT system, and the global stored images of the AVCS system. Additionally, ITOS-1 provided, for the first time, day-and-night radiometric data in real time, as well as stored data, for later payback. Global observation of the earth's cloud cover was provided every 12 hours with the single ITOS spacecraft as compared to every 24 hours with two of the ESSA satellites. ITOS-1 was equipped with a flat plate radiometer for earth radiation measurements. A second ITOS spacecraft, NOAA-1 (ITOS-A), was launched on December 11, 1970.

As the ITOS system evolved to become the ITOS-D system, the flexibility inherent in the spacecraft design permitted a broader and more sophisticated array of environmental sensors to be carried, with only minor changes to the spacecraft. This new sensor complement provided day-and-night imaging by means of Very High Resolution Radiometers (VHRR's) and medium resolution Scanning Radiometers (SR's) (Conlan, 1973). It included Vertical Temperature Profile Radiometers (VTPR's) for temperature soundings of the atmosphere and a Solar Proton Monitor (SPM) for measurements of proton and electron flux. Six spacecraft (ITOS-D, E-2, F, G, H, and I) were planned for the ITOS-D series. NOAA-2 (ITOS-D), the first satellite in this series, was successfully launched on October 15, 1972. Three additional satellites of this type (NOAA-3, NOAA-4,

and NOAA-5) were placed into orbit in 1973, 1974, and 1976, respectively (Fortuna and Hambrick, 1974). Due to the longevity experienced in orbit by the ITOS/NOAA satellites, ITOS-E-2 and -I launches were cancelled. The ITOS system, as it matured, brought closer the realization of the goals of the U.S. National Operational System.

The ITOS satellite system evolved from the proven technology of the TIROS and ESSA spacecraft. Many devices and techniques employed on the earlier series were enhanced, and the enhanced versions were used on the ITOS spacecraft. This orderly evolution permitted growth from a spin-stabilized spacecraft to a three-axis, stabilized, earth-oriented, despun platform.

The principal objectives of this growth pattern during the evolution from an R&D satellite to a global operational system were improved performance, the provision for increased quality and more frequent acquisition of meteorological data, and more timely dissemination of the processed data to the users. The evolving system had to be compatible with the global ground network of local receiving stations as well as the two principal command- and data-acquisition sites. Finally, the operational system had to be cost-effective and have the capacity for future growth.

TIROS-N

The third-generation operational polar-orbiting environmental satellite system, designated TIROS-N, completed development and was placed into operational service in 1978. Eight spacecraft in this series will provide global observational service from 1978 through 1986 (Schnapf, 1980). This new series has a new complement of data-gathering instruments. One of these instruments, the Advanced Very High Resolution Radiometer (AVHRR), increases the amount of radiometric information for more accurate sea-surface temperature mapping and identification of snow and sea ice, in addition to day and night imaging in the visible and infrared bands. Other in-

---

[1] This spacecraft was originally designated TIROS-M. After being placed into orbit, it was redesignated ITOS-1. Subsequent spacecraft in this series were named NOAA-1 NOAA-2, etc. by the National Ocean and Atmospheric Administration, the successor to ESSA as operator of the system.

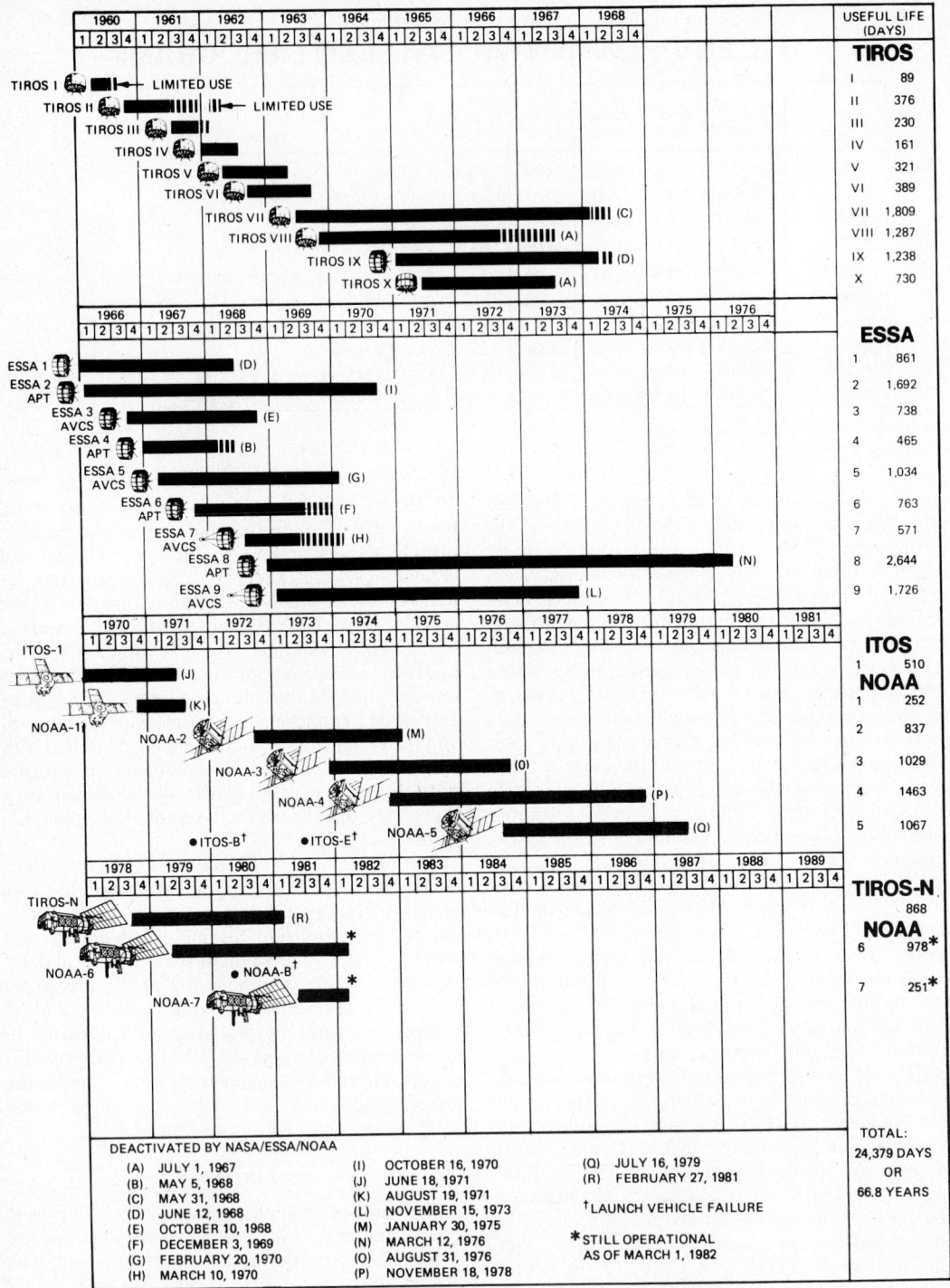

Fig. 14-1. TIROS/ESSA/ITOS/NOAA program summary.

struments, contained in a subsystem known as the TIROS Operational Vertical Sounder (TOVS), provide improved vertical sounding of the atmosphere. These instruments are the High Resolution Infrared Radiation Sounder (HIRS/2), the Stratospheric Sounding Unit (SSU), and the Mi-crowave Sounding Unit (MSU). A Data Collection System (DCS) receives environmental data from fixed or moving platforms such as buoys or balloons and retains them for transmission to the ground stations. A Solar Environmental Monitor is included to measure proton, electron, and alpha

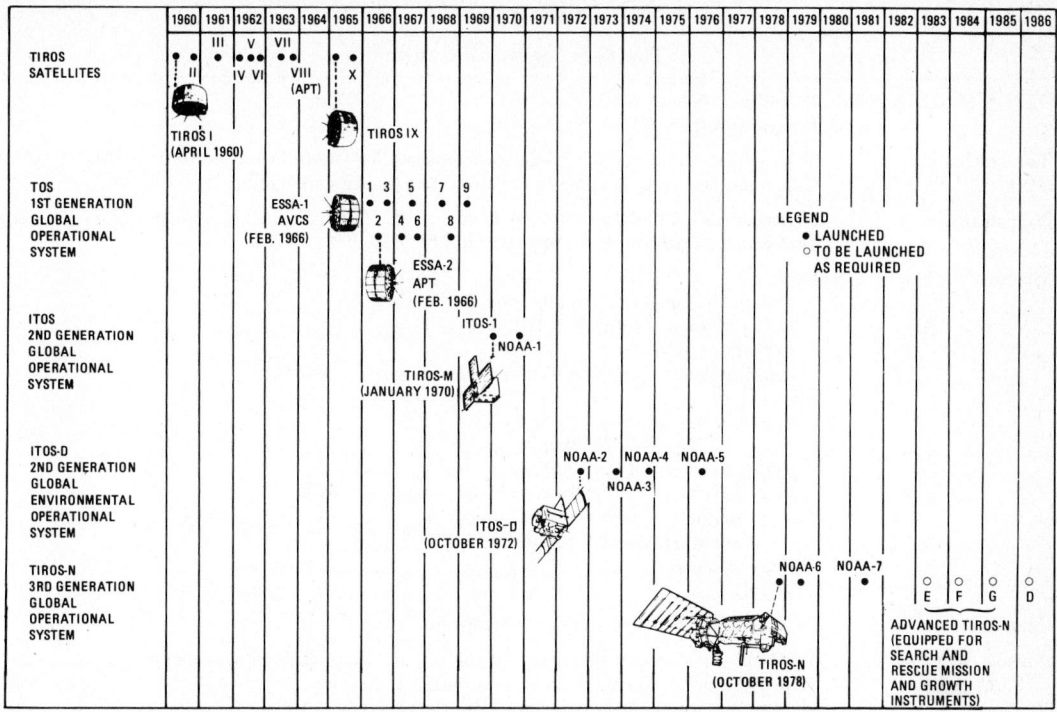

Fig. 14-2. Evolution of TIROS/ESSA/ITOS/NOAA meteorological satellites.

particle densities for solar disturbance prediction (Hussey, 1979). Figure 14-2 depicts the evolution of the TIROS-ESSA-ITOS-NOAA family of satellites.

The TIROS-ESSA-ITOS-NOAA spacecraft series was designed and built by RCA Astro-Electronics under the technical management of the National Aeronautics and Space Administration, Goddard Space Flight Center, and procured (operational series) and operated by the U.S. Department of Commerce; National Oceanic and Atmospheric Administration.

The following section contains a more technical description of the TIROS-N system. Refer to Table 14-2 for historical data and Figure 14-3 for a depiction of the TIROS-N system.

Tiros-N Instruments

Advanced Very High Resolution Radiometer (AVHRR)

The Advanced Very High Resolution Radiometer (AVHRR) for TIROS-N (Schwalb, 1978) and four follow-on satellites will be four-channel scanning radiometers, sensitive to visible/near IR and infrared (IR) radiation. The instrument channelization (Table 14-3) has been chosen to permit multispectral analyses which are expected to provide improved determination of hydrologic, oceanographic, and meteorological parameters. The visible (0.6 μm) and visible/near IR (0.9 μm) channels are used to discern clouds, land-water boundaries, snow and ice extent and

when the data from the two channels are compared, an indication of ice/snow melt inception. The IR window channels are used to measure cloud distribution and to determine temperature of the radiating surface (cloud or surface). Data from the two IR channels will be incorporated into the computation of sea surface temperature. By using these two data sets, it is possible to remove an ambiguity introduced by clouds filling a portion of the field-of-view. On NOAA-7 and on later spacecraft in the series, a third IR channel was added to provide the capability for removing radiant contributions from water vapor when determining surface temperatures. Prior to inclusion of this third channel, corrections for water-vapor contributions will be based on statistical means using climatological estimates of water-vapor content (Schnapf, 1980).

TIROS Operational Vertical Sounder (TOVS)

The TIROS Operational Vertical Sounder (TOVS) system consists of three separate and independent instruments, the data from which may be combined for computation of atmospheric temperature profiles. The three instruments are:

a. The High Resolution Infrared Radiation Sounder (HIRS/2)
b. The Stratospheric Sounding Unit (SSU)
c. The Microwave Sounding Unit (MSU)

The TOVS has been designed so that the acquired data will permit calculation of (1) temperature profiles from the surface to 10 mb, (2)

TABLE 14-2

TIROS-N Spacecraft System

TIROS-N/NOAA A-G:	Protoflight, NASA funded Follow-on flights (7) are NOAA funded
Ground Stations:	NOAA funded and operated—Satellite Operational Control Center: Suitland, MD, CDA—Command and Data Acquisition: Fairbanks, Alaska and Wallops Is., Va.
Launch:	NOAA/NASA funded; USAF managed (Atlas E/T vehicles). Launched from Western Test Range, Lompoc, California on October 13, 1978.
Characteristics:	Orbit 　near-polar, sun-synchronous 　833 or 870 km (450 n. mi.; 470 n. mi.) 　102 minute period 　morning descending or afternoon ascending Physical 　weight—1421 kg (3,127 lbs)—lift off 　　　　　783 kg (1,725 lbs)—in orbit 　payload weight—194 kg (427 lbs) 　size—length—3.71 meters (146 inches)—non-deployed 　　　　　diameter—1.88 meters (74 inches) 　　　　　solar array—11.6 square meters (125 sq. ft.) Design Lifetime 　2 years
Manufacturer:	RCA, Hightstown, New Jersey (except for sensors)
Special Services:	Realtime transmission of sensor data to a wide range of users worldwide. High Resolution Picture Transmission (HRPT) Service Automatic Picture Transmission (APT) Service Direct Sounder Transmission (DST) Service Data Collection/Transmission

water-vapor content at three levels of the atmosphere, and (3) total ozone content.

High Resolution Infrared Radiation Sounder (HIRS/2)

The High Resolution Infrared Radiation Sounder (HIRS/2) is an adaptation of the HIRS/1 instrument designed for and flown on the NIMBUS 6 satellite. The instrument, being built by the Aerospace/Optical Division of ITT, measures incident radiation in 20 spectral regions of the IR spectrum, including both longwave (15 μm) and shortwave (4.3 μm) regions.

The HIRS/2 utilizes a 15-cm (6-in) diameter optical system to gather emitted energy from the earth's atmosphere. The instantaneous field of view (IFOV) of all the channels is stepped across the satellite track by use of a rotating mirror. This cross-track scan, combined with the satellite's motion in orbit, provides coverage of a major portion of the earth's surface.

The energy received by the telescope is separated by a dichroic beam-splitter into longwave (above 6.4 μm) energy and shortwave (below 6.4 μm) energy, controlled by field stops and passed through bandpass filters and relay optics to the detectors. In the shortwave path, a second dichroic beam-splitter transmits the visible channel to its detector. Essential parameters of the instrument are shown in Table 14-4. Primary system components include:

a. Scan system
b. Optics, including filter wheel
c. Radiant cooler and detectors
d. Electronics and data handling
e. Mechanisms

Stratospheric Sounding Unit (SSU)

The Stratospheric Sounding Unit (SSU) is supplied by the United Kingdom Meteorological Office. It employs a selective absorption technique to make measurements in three channels. The principles of operation are based on the selective chopper radiometer flown on Nimbus 4 and 5, and the Pressure Modulator Radiometer (PMR) flown on Nimbus 6. Basic characteristics are shown in Table 14-5.

The SSU makes use of the pressure modulation technique to measure radiation emitted from carbon dioxide at the top of the earth's atmosphere. A cell of CO_2 gas in the instrument's optical path has its pressure changed (at about a 40-Hz rate) in a cyclic manner. The spectral characteristics of the channel and, therefore, the height of the weighting function is then determined by the pressure in the cell during the period of integration. By using three cells filled at different pressures, weighting functions peaking at three different heights can be obtained. The primary objective of

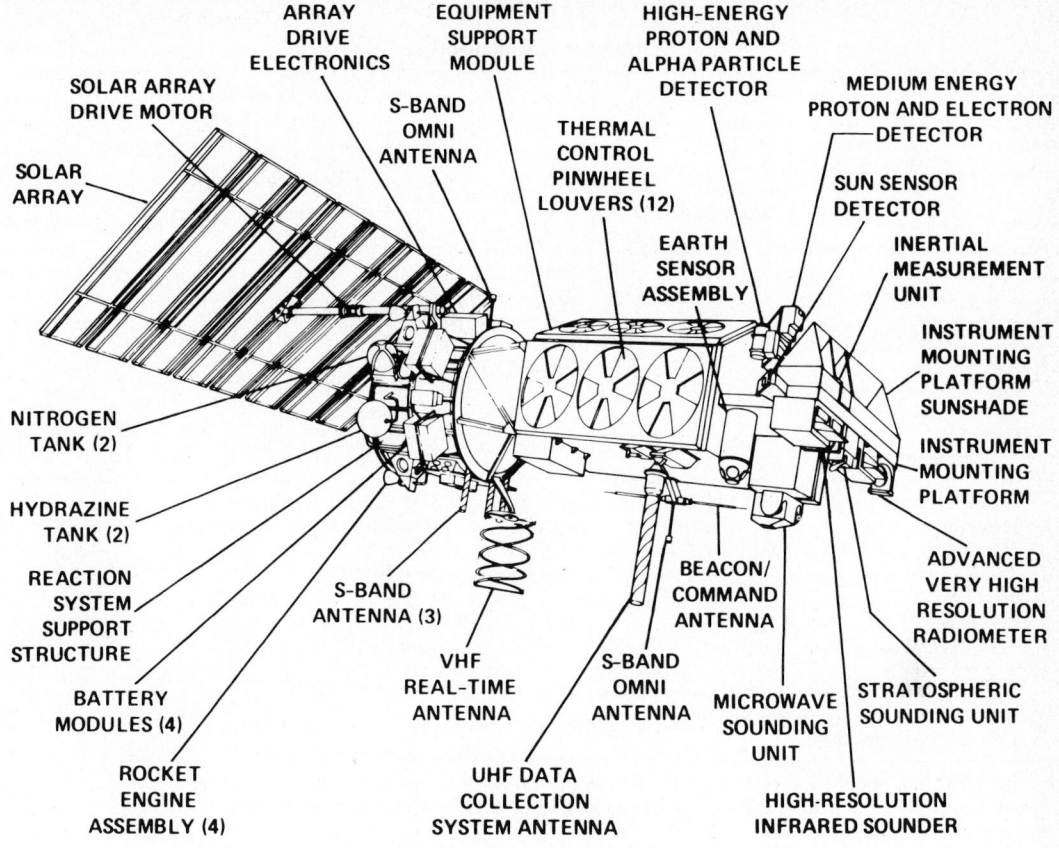

SOLAR ARRAY DRIVE MOTOR

SOLAR ARRAY

ARRAY DRIVE ELECTRONICS

EQUIPMENT SUPPORT MODULE

S-BAND OMNI ANTENNA

HIGH-ENERGY PROTON AND ALPHA PARTICLE DETECTOR

THERMAL CONTROL PINWHEEL LOUVERS (12)

MEDIUM ENERGY PROTON AND ELECTRON DETECTOR

SUN SENSOR DETECTOR

EARTH SENSOR ASSEMBLY

INERTIAL MEASUREMENT UNIT

INSTRUMENT MOUNTING PLATFORM SUNSHADE

INSTRUMENT MOUNTING PLATFORM

NITROGEN TANK (2)

HYDRAZINE TANK (2)

REACTION SYSTEM SUPPORT STRUCTURE

S-BAND ANTENNA (3)

BEACON/ COMMAND ANTENNA

ADVANCED VERY HIGH RESOLUTION RADIOMETER

BATTERY MODULES (4)

VHF REAL-TIME ANTENNA

S-BAND OMNI ANTENNA

MICROWAVE SOUNDING UNIT

STRATOSPHERIC SOUNDING UNIT

ROCKET ENGINE ASSEMBLY (4)

UHF DATA COLLECTION SYSTEM ANTENNA

HIGH-RESOLUTION INFRARED SOUNDER

Fig. 14-3. TIROS-N spacecraft.

the instrument is to obtain data from which stratospheric (25–50 km) temperature profiles can be determined. This instrument will be used in conjunction with the HIRS/2 and MSU to determine temperature profiles from the surface to the 50-km level.

Instrument Operation

The single primary telescope with its 10° IFOV is step scanned perpendicular to the subpoint track. Each scan line is composed of eight individual 4.0-second steps and requires a total of 32 seconds, including mirror retrace.

The SSU uncooled pyroelectric detectors integrate the radiance in each channel for 3.6 seconds during each step. The integrated output signal level is sampled eight times during this period. Quantization is to 12-bit precision.

Microwave Sounding Unit (MSU)

The Microwave Sounding Unit (MSU) is an adaptation of the Scanning Microwave Spectrometer (SCAMS) experiment flown on the Nimbus 6 satellite. The instrument, which is built by the Jet Propulsion Laboratory of the California Institute of Technology, is a four-

channel Dicke radiometer making passive measurements in four regions of the 5.5-mm oxygen region. The frequencies are shown in Table 14-6, which lists the instrument parameters.

The instrument has two scanning reflector antenna systems, orthomode transducers, four Dicke superheterodyne receivers, a data programmer, and power supplies.

The antennas scan ±47.7° either side of nadir in 11 steps. The beamwidth of the antennas is 7.5° (half-power point), resulting in a ground resolution at the subpoint of 109 km.

Data Collection System (DCS)

The Data Collection and Location System (DCS) for TIROS-N was designed, built, and furnished by the Centre National d'Études Spatiales (CNES) of France, who refer to it as the ARGOS Data Collection and Location System. The ARGOS provides a means for obtaining environmental (e.g., temperature, pressure, altitude) data, and earth location from fixed or moving platforms. Location information, where necessary, may be computed by differential Doppler techniques using data obtained from the measurement of platform carrier-frequency as re-

TABLE 14-3

AVHRR Channelization

Protoflight Instrument (1)	*Four-Channel Flight Instruments (4)*
1.* 0.55 − 0.90 μm	1. 0.55 − 0.68 μm
2. 0.725− 1.10 μm	2. 0.725− 1.10 μm
3. 3.55 − 3.93 μm	3. 3.55 − 3.93 μm
4. 10.5 −11.5 μm	4. 10.5 −11.5 μm
5. Channel 4 data repeated	5. Channel 4 data repeated

AVHRR/2—Five Channel Instruments (3)
1. 0.58 − 0.68 μm
2. 0.725− 1.10 μm
3. 3.55 − 3.93 μm
4. 10.3 −11.3 μm
5. 11.5 −12.5 μm

	Channels				
Characteristics	1	2	3	4	5
Spectral Range (μm)	0.58−0.68	0.725−1.1	3.55−3.93	10.3−11.3	11.5−12.5
Detector	Silicon	Silicon	InSb	HgCdTe	HgCdTe
Resolution (km)	1.1	1.1	1.1	1.1	1.1
IFOV (mrad)	1.3	1.3	1.3	1.3	1.3
NETD @ 300 K	—	—	0.12	0.12	0.12
S/N 0.5% albedo	>3:1	>3:1	—	—	—
MTF (IFOV/single bar)	0.3	0.3	0.3	0.3	0.3

Optics: 8-inch diameter a focal cassegrainian telescope
Scanner: 360-rpm hysteresis synchronous motor
Cooler: 2-stage passive

* In-orbit data obtained after completion of the protoflight instrument has shown the necessity of eliminating spectral overlap with channel 2 if snow-cover areal extent is to be accurately measured.

ceived on the satellite. When several measurements are received during a given contact with a platform, location can be determined. The environmental data messages sent by the platform will vary in length depending on the type of platform and its purpose. The ARGOS (DCS) system consists of three major components:

a. Fixed or free-floating platforms

b. On-board instrument and spacecraft communication and storage

c. Data acquisition and processing center

Platforms

The terrestrial platforms may be developed by the user to meet his particular needs so long as it

TABLE 14-4

HIRS/2 System Parameters

Parameter	Value
Calibration	Stable Blackbodies (2) and Space Background
Cross-Track Scan	±49.5° (±1120 km)
Scan Time	6.4 Seconds
Number of Channels	20
Number of Steps	56
Optical FOV	1.25°
Step Angle	1.8°
Step Time	100 Milliseconds
Ground IFOV (Nadir)	17.4 km Diameter
Ground IFOV (End of Scan)	58.5 km Cross-Track by 29.9 km Along-Track
Distance Between IFOV's	42 km Along-Track
Data Rate	2880 Bits/Second
Dectors: Long Wave	HgCdTe
Short Wave	InSb
Visible	Silicon

TABLE 14-5

SSU Characteristics

Channel Number	Central Wave No. (cm^{-1})	Cell Pressure (mb)	Pressure of Weighting Function Peak	
			mb	km
1	668	100	15	29
2	668	35	5	37
3	668	10	1.5	45

Calibration	Stable Blackbody and Space
Angular Field-of-View	10°
Ground IFOV—Nadir	147.3 km
Number of Earth Views/Line	8
Time Interval Between Steps	4 Seconds
Total Scan Angle	±40° from Nadir
Scan Time	32 Seconds
Data Rate	480 Bits/Second
Detector	TGS Pyroelectric

TABLE 14-6

MSU Instrument Parameters

Characteristics	Value				Tolerance
	CH 1	CH 2	CH 3	CH 4	
Frequency (GHz)	50.3	53.74	54.96	57.05	±20 MHz
RF Bandwidth (MHz)	220	220	220	220	Maximum
NEΔT K	0.3	0.3	0.3	0.3	Maximum
Antenna Beam* Efficiency	>90%	>90%	>90%	>90%	
Dynamic Range K	0−350	0−350	0−350	0−350	
Calibration	Hot Reference Body and Space Background Each Scan Cycle				
Cross-Track Scan Angle	±47.35°				
Scan Time	25.6 sec				
Number of Steps	11				
Step Angle	9.47°				
Step Time	1.84 sec				
Angular Resolution	7.5° (3dB)				
Ground IFOV (Nadir)	109 km				
Data Rate	320 bps				

* >95% Obtained.

meets the interface criteria defined by CNES. Before being accepted for entry into the system, the platform design must be certified as meeting these criteria. By international agreement, entry into the system is limited to platforms requiring location service or for those situated in polar regions out of the range of the DCS on geostationary satellites. General platform criteria are shown in Table 14-7.

On-board Instrument

The on-board instrument is designed to receive the incoming platform data, demodulate the incoming signal, and measure both the frequency and relative time of occurrence of each transmission. The on-board system consists of three modules: the power supply and command interface units, the signal processor, the redundant receiver and search units, and an antenna.

Platform signals are received by the receiver search unit at 401.65 MHz. Since it is possible to acquire more than one simultaneous transmission, four processing channels [called Data Recovery Units (DRU)] operate in parallel. Each DRU consists of a phase lock loop, a bit synchronizer, Doppler counter, and a data formatter. After measurement of the Doppler frequency, the sensor data are formatted with other internally generated data and the output transferred to a buffer interface with the spacecraft telemetry information processor (TIP). The DCS output data rate is controlled to 720 bits per second.

Space Environment Monitor (SEM)

The Space Environment Monitor (SEM) has been designed and built by the Ford Aerospace and Communication Corporation. The instrument measures solar proton, alpha particle, and elec-

TABLE 14-7

ARGOS Platform Characteristics

Carrier Frequency	401.650 MHz
Aging (During Life)	±2 kHz
Short Term Stability (100 ms)	1:10⁹ (Platform Requiring Location)
	1:10⁸ (Platform Not Requiring Location)
Medium Term Stability (20 min)	:0.2 Hz/min (Requiring Location)
Long Term Stability (2 hr)	:±400 Hz
Power Out: 34.8 dBm (3W) Nominal	
Range During transmission (Stability)	:0.5 dB
Antenna: Vertical Linear Polarization	
Message Length: 360 ms to 920 ms	
Repetition Period for Message:	40−60 sec (Requiring Location)
	60−200 sec (Not Requiring Location)
Data Sensors: 4−32 Eight-Bit Sensors	
for Environmental Data	
Total Number of Platforms:	4,000 global
	450 Within View

tron flux density, energy spectrum, and the total particulate energy disposition at satellite altitude.

The three components of the SEM are:
a. Total Energy Detector (TED)
b. Medium Energy Proton and Electron Detector (MEPED)
c. High Energy Proton and Alpha Detector (HEPAD)

This instrument is a follow-on to the Solar Proton Monitor (SPM) flown on the ITOS series of NOAA satellites. The new instrument modifies the SPM capabilities and adds the monitoring of high energy protons and alpha flux. The package also includes a monitor of total energy deposition into the upper atmosphere. The instrument will augment the measurements already being made by NOAA's Geostationary Operational Environmental Satellite (GOES).

Total Energy Detector (TED)

The TED uses a curved plate analyzer and channeltron detector to determine the intensity of particles in the energy bands from 0.3 KeV to 20 KeV. Four curved plate analyzers (two measuring electrons, two protons) measure incoming particles reaching the instrument.

Medium Energy Proton and Electron Detector (MEPED)

The MEPED senses protons, electrons, and ions with energies from 30 KeV to greater than 60 MeV. This instrument is comprised of four directional solid-state detector telescopes and one omnidirectional sensor. All five components use solid-state nuclear detectors. Outputs from the detectors are connected to a signal analyzer which senses and logically selects those events which exceed specific threshold values.

High Energy Proton-Alpha Detector (HEPAD)

The HEPAD senses protons and alphas from about 370 MeV to greatr than 850 MeV. The instrument is essentially a Cerenkov detector. The Cerenkov crystal is installed within a telescope in association with two solid-state detectors; the telescope is shielded to establish the instrument's field-of-view.

THE ADVANCED TIROS-N SPACECRAFT

The last three spacecraft in the TIROS-N/ NOAA A-G series of spacecraft have been designated the Advanced TIROS-N (ATN) spacecraft configuration. The ATN (Figure 14-4) mission not only must satisfy the current TIROS-N mission payload but must additionally carry new payload instruments: the Search and Rescue (SAR) repeater, the SAR 406 MHz processor, and SAR Antenna; the Solar Backscatter Ultra-Violet (SBUV) instrument; and the Earth Radiation Budget Sensing System (ERBSS) non-scanner and the ERBBS scanner (Schnapf, 1980).

NOAA-E, the first in the ATN series, will be equipped with the Standard TIROS-N payload plus the Search and Rescue payload. NOAA-F and -G will contain SAR, SBUV, and the ERBSS instruments.

To accommodate these added payloads, the TIROS-N spacecraft has been elongated by 50-cm. and the solar array power has been increased from 1260 to 1470 watts. The overall lift-off weight of the ATN spacecraft is 270 kg greater than TIROS-N.

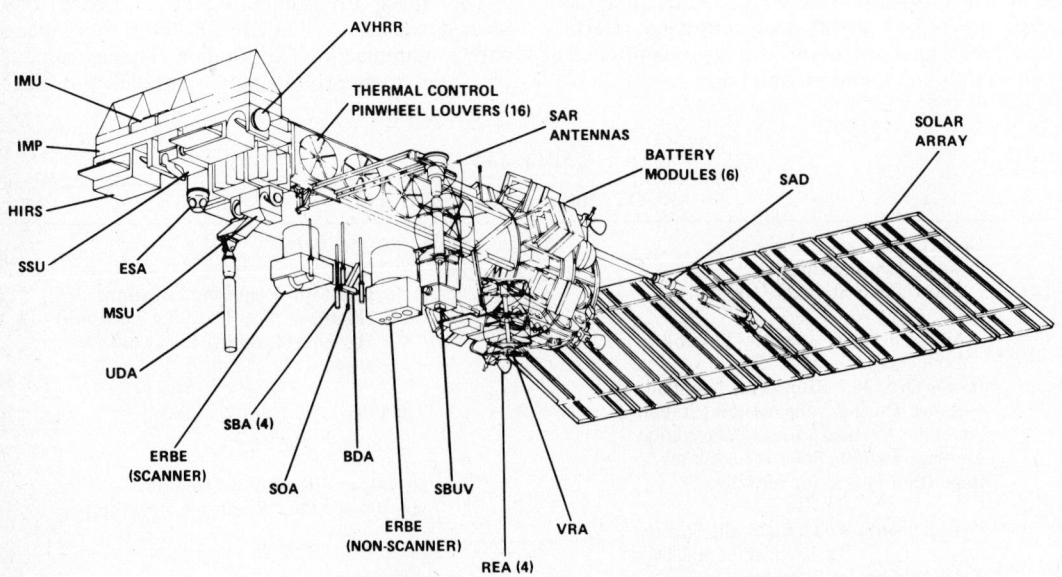

Fig. 14-4. Advanced TIROS-N spacecraft.

Search and Rescue Mission

The SAR mission is a joint International experiment with USA, Canada, France, and USSR to demonstrate the feasibility of the timely detection and location of downed aircraft or marine vessels in distress and the transmission of the data to the emergency rescue forces.

The SAR system will detect signals from aircraft Emergency Locator Transmitters (ELT's) and from Emergency Position Identification Radio Beacons (EPIRB's) on ships. Government regulations require the general aviation to be equipped with ELT's. ELT's and EPIRB's transmit distress signals on 121.5 and 243 MHz automatically. A new and improved ELT operating on 406 MHz is to be introduced during this experiment. In addition to the repeater mode of operation, the 406 MHz signals will be processed by the satellite's 406 MHz processor and will be capable of providing location information of the distressed vehicle.

Solar Backscatter Ultraviolet Instrument (SBUV/2)

The SBUV/2 instrument provides ultraviolet (UV) measurements which can be used to determine the total-ozone and vertical-ozone distribution in the stratosphere and provide a detailed UV spectrum of either the sun or the solar radiation reflected from the earth. It has two operational modes when viewing the earth and the same two calibration modes when viewing the sun. The SBUV/2 makes obsevations along the local vertical. Characteristics of the SBUV/2 are shown in Table 14-8.

The first operational mode consists of observation at 12 fixed wavelengths, between approximately 250 nanometers (nm) and 340 nm, in the ozone-absorption band. The spectral scan mode is achieved by a stepped-position grating in a double monochromator. The 12 fixed wavelengths can be changed while in orbit by uplinking new position information in order to coincide with the UV experiments orbited during the 1970's. A cloud-cover radiometer views the same field as the monochromator to assist in data processing; its spectral band is centered at 380 nm. The second operational mode is a constant scan, between 160 nm and 400 nm, with the same monochromator. Selection between these modes is made by ground command. The SBUV/2 is always in one of the operational modes except during calibration. The calibration mode can consist of the two modes above except that a diffuser is used to direct radiation from the sun into the instruments aperture. Switching to the calibration mode is controlled by ground command.

The output of the SBUV/2 is digitized to a precision of 12 bits for inclusion in the GAC data and is processed by NOAA/NESS.

Earth Radiation Budget Sensor System (ERBSS)

The Earth Radiation Budget Sensor System to be included on the Advanced TIROS-N spacecraft consists of two instruments that will make measurements to determine both the portions of the solar irradiation on the earth that is reflected and the portion that is reradiated from the earth and its atmosphere. The two instruments, the ERBE-Scanner (ERBE-S) and the ERBE-Non-Scanner (ERBE-NS) utilize different methods to make the total radiation measurements.

ERBE-NS

The ERBE-NS makes the measurement by viewing the irradiance coming from the total earth's disk as seen from the satellite. A second set of detectors measures a smaller portion of the earth's disk. The radiation is measured in (1) the total spectral region from ultraviolet through the infrared and (2) the solar region. Characteristics of the ERBE-NS are given in Table 14-9.

ERBE-S

The ERBE-S measures the irradiance in a small instantaneous field-of-view which is mechanically

TABLE 14-8

SBUV/2 Characteristics

Characteristics	Spectrometer	Cloud-Cover Radiometer
Spectral Range (nm)	160 to 400	380
Detector Type	Photomultiplier	Photodiode
Dynamic Range (mW/m^2-sr-A)		
o Maximum	40	40
o Minimum	2.5×10^{-4}	0.125
Ground Resolution—		
Nadir Only (km)	160×160	160×160
Spectral Scan Period (sec)	32	N/A

TABLE 14-9

ERBE-NS Characteristics

Characteristic	Channels				
	1	2	3	4	SCM
Spectral Region (μm)	0.2 to 50	0.2 to 5	0.2 to 50	0.2 to 5	0.2 to 50
Detector Type	Active Cavity Radiometer				
Resolution Earth View					
Angle (deg)	136		65		N/A
Equivalent Earth	6907		1093		N/A
Coverage (km)	(Horizon-to-Horizon)				
Solar Calibration	18		18		18

scanned from horizon to horizon. Three spectral regions, short-wave (0.2 μm to 5 μm), near infrared (0.7 μm to 3 μm), and total (0.2 μm to 50 μm), are measured. The characteristics of the ERBE-S are given in Table 14-10. To assure measurement accuracy, the calibration of both instruments is made by viewing the sun periodically. On the ground, the ERBE-related information is stripped from the Global Area Coverage (GAC) data by NOAA/NESS and forwarded to NASA's Langley Research Center for processing to a scientific product. The ERBE and the other climate experiment (the SBUV) are low-rate instruments, so their data are also available to local users via S-band and VHF links.

NIMBUS

The Nimbus satellite program was initiated by the National Aeronautics and Space Administration in the early 1960s to develop an observational system capable of meeting the research and development needs of the nation's atmospheric and earth scientists (Schnapf, 1979).

The general objectives of the program were: (1) to develop advanced passive radiometric and spectrometric sensors for daily global surveillance of the earth's atmosphere and thereby provide a data base for long-range weather forecasting; (2) to develop and evaluate new active and passive sensors for sounding the earth's atmosphere and

mapping surface characteristics (Press and Huston, 1968); (3) to develop advanced space technology and ground techniques for meteorological and other earth-observational spacecraft; (4) to develop new techniques and knowledge useful for the exploration of other planetary atmospheres; and (5) to participate in global observation programs (World Weather Watch) by expanding daily global weather observation capability (National Research Council, 1978).

The Nimbus System was designated to be (1) the test-bed for advanced instruments for the future operational TIROS polar-orbiting satellites and (2) the research system for remote sensing and data collection. The Nimbus spacecraft system was developed under NASA/GSFC management, with the General Electric Company as the spacecraft integration contractor. RCA, Hughes, ITT, Texas Instruments, and a number of other companies and universities provided sensors and data processing and storage equipment for the Nimbus series. The project has matured to become the nation's principal satellite program for remote-sensing research. Each new satellite in the Nimbus series has represented significant growth in sophistication, complexity, weight, capability, and performance.

A total of seven Nimbus spacecraft were successfully placed into orbit from 1964 through 1978 (Staff Members, 1965). The final spacecraft, Nimbus 7, was launched in November 1978. This

TABLE 14-10

ERBE-S Characteristics

Characteristic	Channels		
	1	2	3
Spectral Region (μm)	0.2 to 50	5.0 to 50	0.2 to 3.5
Detector Type	Thermistor Bolometer	Thermistor Bolometer	Thermistor Bolometer
Accuracy (W/m²)	7.0	7.0	7.0
Ground Resolution at Nadir (km)	83.4 × 72.8		
Scan Field-of-View	Horizon-to-Horizon		
Scan Period (sec)	8		
Samples/Scan Line	62		

spacecraft was instrumented with sensors to monitor the atmospheric pollutants, oceanography, and weather and climate (Figure 14-5) (Staff Members, Goddard Space Flight Center 1976, Vostreys and Horowitz, 1979). The payload consisted of eight instruments:

1. Scanning Multichannel Microwave Radiometer (SMMR)—Measures radiances in five wavelengths and ten channels to extract information on sea surface roughness and winds, sea surface temperature, cloud liquid-water content total water-vapor content, precipitation (mean droplet size), soil moisture, snow cover, and sea ice.

2. Stratospheric and Mesospheric Sounder (SAMS)—Measures vertical concentrations of H_2O, N_2O, methane (CH_4), carbon monoxide (CO), and nitric oxide (NO); measures temperature of stratosphere to ~90 km and trace constituents.

3. Solar-Backscattered Ultraviolet/Total Ozone Mapping System (SBUV/TOMS)—Measures direct and backscattered solar UV to extract information on variations of solar ir-

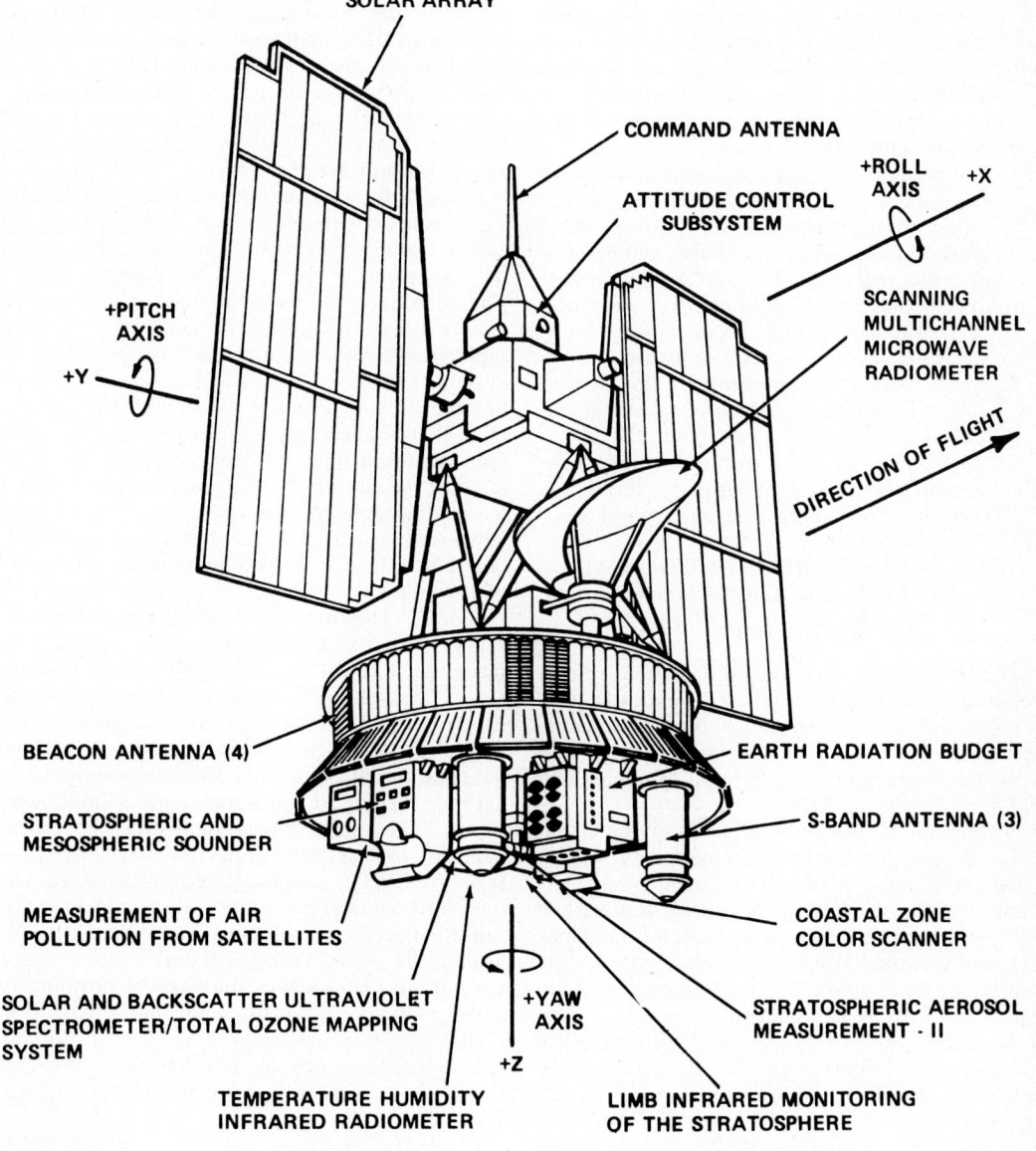

NIMBUS

Fig. 14-5. Nimbus-7 spacecraft.

radiance, vertical distribution of ozone and total ozone on a global basis.

4. Earth Radiation Budget (ERB)—Measures short- and longwave upwelling radiances and fluxes and direct solar irradiance to extract information on the solar constant, earth albedo, emitted longwave radiation, and the anisotropy of the outgoing radiation.

5. Coastal Zone Color Scanner (CZCS)—Measures chlorophyll concentration, sediment distribution, gelbstoff (yellow substance) concentration as a salinity indicator, and temperature of coastal waters and open ocean.

6. Stratospheric Aerosol Measurement II Experiment (SAM II)—Measures the concentration and optical properties of stratospheric aerosols as a function of altitude, latitude, and longitude. Tropospheric aerosols can be mapped also if no clouds are present in the IFOV.

7. Temperature-Humidity Infrared Radiometer Experiment (THIR)—Measures the infrared radiation from the earth in two spectral bands (11 and 6.7 μm) both day and night to provide pictures of cloud cover; three-dimensional maps of cloud cover; temperature maps of cloud-, land-, and ocean-surfaces; and atmospheric moisture.

8. Limb Infrared Monitoring of the Stratosphere Experiment (LIMS)—Makes a global survey of selected gases from the upper troposphere to the lower mesosphere. Inversion techniques are used to derive gas concentrations and temperature profiles.

THE DEFENSE METEOROLOGICAL SATELLITE PROGRAM (DMSP)

EVOLUTION

DMSP satellites provide both tactical (direct readout) and strategic (stored readout) data transmission capabilities. Data from the spacecraft have been routinely transmitted directly to U.S. Air Force and Navy ground terminals and Navy carriers since 1971.

Sixteen air transportable data receiving terminals are now located in the Philippines, Spain, Guam, Okinawa, Alaska, the Canal Zone, Germany, Korea, the Kwajalein Missile Range, and the continental U.S. to provide tactical commanders with real-time DMSP data. The Navy is currently operating direct data readout sites at Rota, Spain and San Diego, California.

In March 1969, DOD approved a joint-service development effort. As a result, a feasibility model shipboard receiving terminal was developed by the Air Force with assistance from Navy personnel attached to the DMSP System Program Office (SPO). This terminal was installed aboard the USS Constellation and proved extremely effective in several deployments to Southeast Asia. A prototype station has since been placed in oper-

ation on the USS Kennedy and production units are in operation on the USS Independence, the USS Kitty Hawk, the USS Midway, and the USS Enterprise. All CV-class carriers are slated for eventual receipt of this equipment.

In December 1972, DMSP data were declassified and made available to the civil/scientific community through the National Oceanic and Atmospheric Administration (NOAA) (Nichols, 1975a; 1976; Brandli, 1976).

In September 1976, the first Block 5D-1 satellite was launched. Although anomalies precluded collection of operational data immediately after launch, the satellite was restored to nominal operational condition and provided meteorological data of unmatched quality. The second, third, and fourth Block 5D satellites have subsequently been launched and placed in operation (Figure 14-6).

Since 1965, there have been three major spacecraft models flown: Block 4, Versions A & B; Block 5, Versions A, B, and C; and Block 5D-1 and its current 5D-2.

Block 4. Block 4 employed a pair of VIDICON cameras to acquire television pictures showing the earth's cloud cover and some terrain features as they appeared in the visible wave-length region. The resolution of these pictures was approximately 1.5 nautical miles at nadir, but degraded rapidly toward the picture edge. A supplementary system to roughly measure albedo was also incorporated on later Block 4 spacecraft. This system of 16 thermopile sensors, known as the "C" system, acquired data on energy emitted by large areas of the earth in two selected IR intervals: 0.4 to 4.0 micrometers (energy from reflected sunlight) and 8.0 to 12.0 micrometers (energy self-emitted by the earth). Resolution was on the order of 100 nautical miles.

Block 5. The first Block 5 was launched in February 1970. Block 5 versions A, B, and C, replaced the VIDICON cameras with a new primary sensor known as the Sensor Aerospace Vehicle Electronics (AVE) Package (SAP) to gather visual and infrared data at improved resolutions. Visual data and IR data were collected at one-third nautical mile resolution and at two-nautical-mile resolution. The one-third nautical mile data were available to U.S. Air Force Global Weather Central (AFGWC), Omaha, Nebraska, while the one-third nautical mile data were routinely transmitted directly to AF and Navy tactical sites around the globe. Versions B and C incorporated various special sensors for vertical profiling of atmospheric temperature, for measuring precipitating electron activity at spacecraft altitude, for atmospheric density profiling, etc. Many of these sensor packages have been improved for Block 5D.

Block 5D. The first Block 5D-1 was launched in September 1976. Although anomalies precluded collection of operational data immediately after launch, the satellite was restored to nominal operational condition and provided meteorological

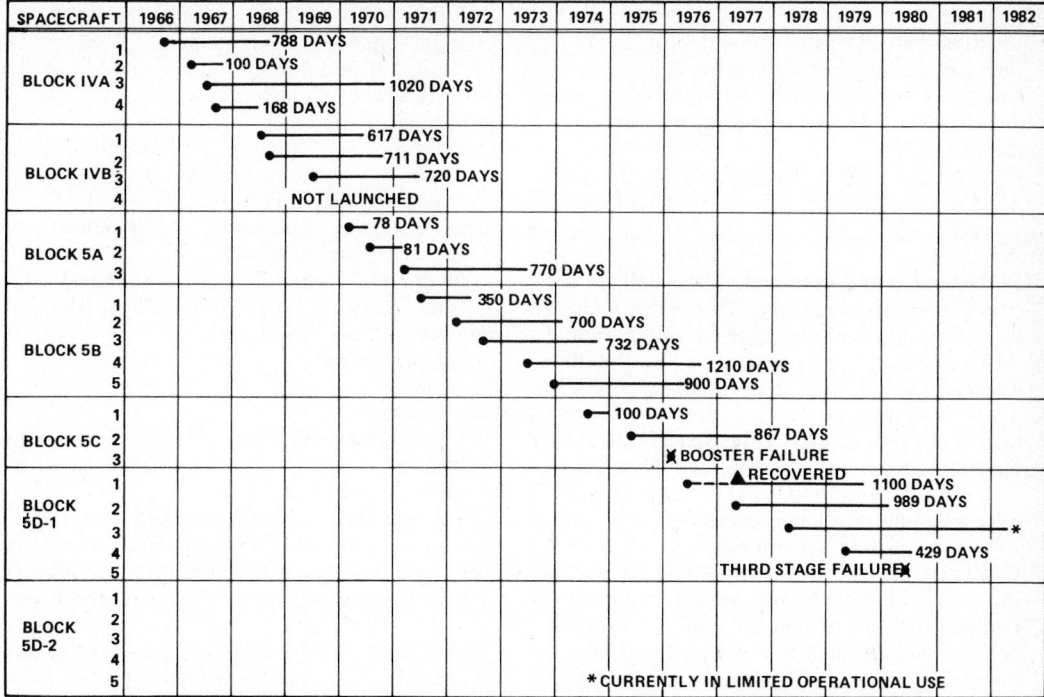

Fig. 14-6. DMSP performance in space.

data of extremely high quality. The second, third, and fourth Block 5D satellites have subsequently been launched and placed in operation. The 5D version included a new primary sensor, the Operational Linescan System (OLS), which provided improved resolution of 0.3 nautical miles for both visual and IR fine data and 1.5 nautical mile for smoothed data. The biggest improvement over the SAP was that the OLS increased resolution uniformity along the scan line. At 450 nautical miles from nadir, the SAP smooth data experienced a degraded resolution of 13 nautical miles while the OLS maintained a resolution of two nautical miles or better for smoothed data.

The following section will describe the Block 5D-1 and -2 instruments and system in more detail (Nichols, 1975b):

DMSP Block 5D-1:
DOD Meteorological Satellite Program satellites 5D-1 replaced the 5C models.
Ground Support:
DOD funded and operated. The satellites are commanded and controlled from sites located at Loring AFB, Maine (Site 2) and Fairchild AFB, Washington (Site 1) and the Satellite Control Facility Hawaiian Tracking Station. These stations also receive stored data from tape recorders on board the spacecraft. The data are relayed to AFGWC at Offutt AFB, NE (Site 3) and FNOC at Monterey, CA, over a communication satellite link.

The program's Command and Control Center (CCC) is at Offutt AFB, Nebraska (Site 5). The 4000th Aerospace Applications Group (SAC) is responsible for the on-orbit commanding through Sites 1 and 2 and the orbital telemetry analysis performed at the CCC.
Launch:
The first Model 5D-1 was launched on September 11, 1976 on a Thor (LV-2F) booster by Aerospace Defense Command's 10th Aerospace Defense Squadron at Vandenberg AFB, California.
Characteristics:
Orbit: near polar circular sun-synchronous
Altitude: 833 km (450 ± 10 nm)
Inclination: 98.7 ± 0.15 degrees
Period: 101 minute
Sun Angle: 0–95 degrees
Physical:

weight	468 kg	(1032 lbs)
		(on orbit)
payload weight	136 kg	(300 lbs)
size–length	5.89 m	(232 inches)
w/solar array extended		
diameter	1.21 m	(48 inches)
solar array	8.92 sq m	(96 sq ft)
power	300 watts	

Design Lifetime
2 years

Manufacturer:
RCA, Hightstown, New Jersey (except for sensors)

Special Services:

Realtime Direct Digital transmission (DDT) of fine visual and smoothed infrared imagery or smoothed visual and fine infrared imagery to Air Force and Navy tactical sites worldwide.

DMSP Block 5D-2:

DOD Meteorological Satellite Program Satellite 5D-2 replaces the 5D-1 models. DOD funded. (Figure 14-7)

Ground Support:

DOD funded and operated. The satellites are commanded and controlled from sites located at Loring AFB, Maine (Site 2) and Fairchild AFB, Washington (Site 1) and the Satellite Control Facility Hawaiian Tracking Station. These stations also receive stored data from tape recorders on board the spacecraft. This data is relayed to AFGWC at Offutt AFB, NE (Site 3) and FNOC at Monterey, CA, over a communication satellite link.

The programs's Command and Control Center (CCC) is at Offutt AFB, Nebraska (Site 5). The 4000th Aerospace Applications Group (SAC) is responsible for the on-orbit commanding through Sites 1 and 2 and the orbital telemetry analysis performed at the CCC.

Launch:

The first Model 5D-2 is to be launched in 1982 on an Atlas (E/F) booster at Vandenberg AFB, California.

Characteristics:

Orbit: near polar circular sun-synchronous altitude

Altitude: 833 km (450 ± 10 nm)
Inclination: 98.7 ± 0.15 degrees
Period: 101 minutes
Sun Angle: 0–95 degrees

Physical:

weight	698 kg	(1540 lbs)
	(on orbit)	
payload weight	159 kg	(350 lbs)
size–length	6.39 m	(250 inches)
w/solar panel extended		
diameter	1.21 m	(48 inches)
solar array	11.15 sq m	(120 sq ft)
power	400 watts	

Design Lifetime

3 years

Manufacturer:

RCA Hightstown, New Jersey (except for sensors)

Special Services:

Realtime Direct Digital Transmission (DDT) of fine visual and smoothed infrared imagery or smoothed visual and fine infrared imagery to Air Force and Navy tactical sites worldwide.

The primary objective of the 5D-2 development is to increase on-orbit life through improved reliability. The 5D-2 design for reliability improvement is based on "Functional Module Redundancy." The 5D-2 system is subdivided into functional modules which are made redundant and independently controllable as to use of primary or backup. This approach allows switching around failed units until both modules of a required function

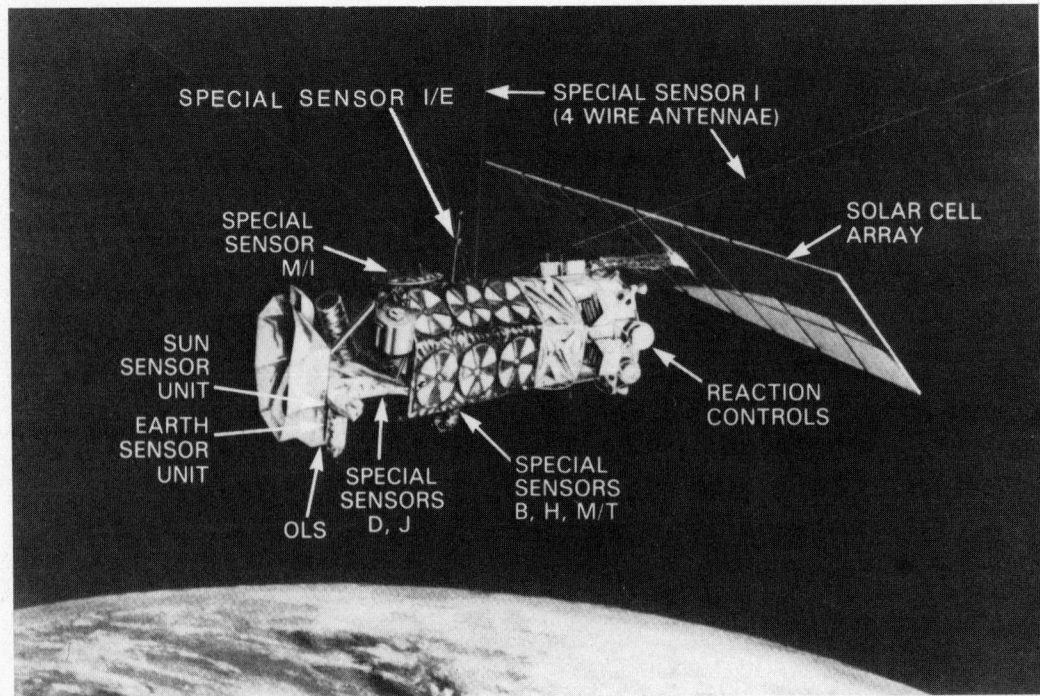

Fig. 14-7. DMSP 5D-2 in orbit.

have failed. In addition to the "Functional Module Redundancy" design feature, the quality of parts, materials and processes will be significantly upgraded. Other objectives are increasing command and control performance, improving producibility, and providing minor performance improvements.

The 5D-1 Sensor Processing System (SPS) has incurred major alterations. The power supply has been removed and, in the 5D-2 configuration, exists as a separate Power Supply Unit (CPSU) with redundant supplies. The 5D-1 SPS analog circuit boards have been moved to the PSU box, leaving the SPS a pure digital subassembly. All SPS functional blocks—memory, processor, I/O and SDS and SDF formatters are redundant.

The RDT and SSP formatters and processors in the 5D-2 configuration are redundant and are located in the Special Sensor Processing Unit (SPU).

The 5D-2 configuration includes three encrypters and four tape recorders; an addition of one each. The 5D-1 encryption interface box has been eliminated. The Output Switching Unit (OSU) is fully redundant.

BLOCK 5D INSTRUMENTS

Operational Linescan System (OLS)

The OLS is the primary data acquisition system on the Block 5D satellite. This system gathers and outputs in real time or stores multi-orbit day and night visual and infrared spectrum data from earth scenes for transmission to ground stations with appropriate calibration, indexing, and auxiliary signals. Data are collected, stored, and transmitted in fine (0.3 nm) or smoothed (1.5 nm) resolution.

Fine Resolution Visual Data

The visual daytime response of the OLS is in the spectral range of 0.4 to 1.1 microns to provide maximum contrast between earth, sea, and clouds in the image field. The visual fine resolution (0.3 nm × 0.3 nm) is provided for day scenes only.

Smooth Resolution Visual Data

The smooth resolution (1.5 nm × 1.5 nm) visual data are provided across a dynamic range from full sunlight down to quarter moonlight. Nighttime visual data are provided from a photomultipler tube operated in the same spectral range as the visual fine data and is energized automatically as the radiance decreases. Daytime, smooth-resolution data are derived from fine-mode data by analog and digital data processing by the OLS. Five fine-mode resolution cells are averaged along the scan line to produce a series of 0.3 nm × 1.5 nm cells. Then five such cells are digitally averaged along the track to produce a single smooth resolution cell, 1.5 nm × 1.5 nm.

Fine Resolution Infrared

The OLS infrared detector is a segmented tri-metal (HgCdTe) detector operating at approximately 105 K with a spectral response of 8.0 to 13.0 microns* to provide optimal detection of both water and ice-crystal clouds. The sensor output is normalized to equivalent blackbody temperature of the radiating object such that the sensor output voltage is a linear function of scene temperature. The tri-metal detector is accurate to within 1 k *rms* across the temperature range 210–310 K.

Smooth Resolution Infrared

The smooth resolution data are obtained from fine resolution infrared data in the same manner as described for the smooth resolution visual data. Fine-mode visual and infrared data are gathered through the same optics and are digitally identified; the smooth-mode data are also digitally identified. Thus corresponding visual and infrared data cells maintain a unique one-to-one location correspondence throughout the data processing chain.

OLS Data Processing

The OLS data processing subsystem performs command, control, data manipulation, storage, and management functions of the instrument. All data are processed, stored, and transmitted in digital format. Special Sensor data and OLS telemetry data are merged by the OLS into the smooth-data stored format.

OLS Data Transmission

S-band transmitters are provided for data transmission. Two may be operated simultaneously for stored data playback. A third S-band transmitter is dedicated to transmission of Direct Digital Transmission (DDT) to tactical sites world wide at 1.024 Mbps. DDT data are normally encrypted.

The OLS system including the Sensor Subsystem, Signal Processing System, and the Recording Subsystem weighs about 290 lbs and requires 170 watts of power.

The 5D-2 OLS System has the same performance requirements as the 5D-1 System as previously described. The basic premise of the 5D-2 OLS is to increase overall system reliability by providing functional redundant circuitry, although some changes, such as increased IR digitization (from 7 to 8 bits), and improved sensitivity to low temperature values (190–210 K) have been incorporated.

---

* With F-4, the OLS IR Spectral Band was changed from 8–13 μm to 10.5–12.6 μm to improve the sea surface temperature resolution.

Characteristics of the 5D-1 Operational Linescan System (OLS)

Meteorological data collected in visible and infrared spectra

Visible data collected as 0.3 nm × 0.3 nm during day and 1.5 nm × 1.5 nm at night

Infrared data collected as 0.3 nm × 0.3 nm at all times

Oscillating scanner collects data in both directions along 1600 nm swath

Near constant resolution as a function of scan angle

Three digital tape recorders for data storage

Each recorder can store 20 minutes of interleaved visual and infrared 0.3 nm × 0.3 nm data

Analog filtering and digital averaging are used to smooth data to 1.5 nm × 1.5 nm for on-board global storage

Each recorder can store 400 minutes of interleaved visual and infrared 1.5 nm × 1.5 nm data

Telemetry and special meteorological sensor data are included within the primary smoothed data stream

Real-time encrypted transmission of 0.3 nm and 1.5 nm data

SPECIAL SENSORS

Special Sensor (SSH)—A Humidity, Temperature, and Ozone Sounder

This instrument is an infrared multispectral sounder for humidity, temperature and ozone. Vertical profiles of temperature and humidity and a single measurement of ozone were derived from radiances sensed from vertical and slant paths lying under and to ±48° of the sub-satellite track. The SSH is a 16-channel sensor with one channel (1020-cm^{-1}) in the 9.8 micrometer ozone absorption band, one channel (835-cm^{-1}) in the 12-micrometer atmospheric window, six channels (747-, 725-, 708-, 695-, 677-, 668-cm^{-1}) in the 15-micrometer CO_2 absorption band, and eight channels (from 535-cm^{-1} to 353-cm^{-1}) in the 22- to 30-micrometer rotational water-vapor absorption band. The instrument consists of an optical system, detector and associated electronics, and a scanning mirror. The scanning mirror was stepped across the satellite subtrack, allowing the SSH to view 25 separate columns of the atmosphere every 32s over a cross-track ground swath of 2000 km. While the scanning mirror is stopped at a scene station, the channel filters are sequenced through the field of view. The surface resolution is approximately 39 km at nadir. Radiance data are transformed into temperature water vapor and ozone profiles by a mathematical inversion technique.

A Cassegrain objective forms a 2.7° FOV centered on an axis parallel to the flight path. A step-rotating diagonal scanning mirror scans the FOV

according to a pre-established scan program in a plane and normal to the flight path.

The SSH instrument weighs 31 pounds and consumes 13 watts of power.

Spacecraft F-1 through F-4 carried the SSH-1 as described. SSH units starting with F-5 are designated SSH-2. The SSH-2 is a modified SSH-1. The 1020-cm^{-1} ozone channel was deleted. The 835-cm^{-1} atmospheric window channel was replaced by the 898-cm^{-1} channel. In the 15-μm CO_2 absorption band, the 725-cm^{-1} channel was deleted while the 731- and 799-cm^{-1} channels were added. In the 22- to 30-μm rotational water-vapor absorption band, the 347- and 355-cm^{-1} channels were deleted and replaced by a 497-cm^{-1} channel. A 2700-cm^{-1} window channel was also added. The total number of channels remained at 16, while the channel selection was changed to improve the SSH instrument sensitivity.

Special Sensor M/T (SSM/T)—A Passive Microwave Temperature Sounder

The SSM/T is a seven-channel scanning passive microwave radiometer which measures radiation in the 50–60 GHz frequency region to provide data for temperature profiling from the earth's surface to above 30 km. The SSM/T scans in synchronization with the SSH (an infrared temperature and humidity sounder). The microwave sounder complements the infrared sounder by providing temperature sounding over previously inaccessible cloudy regions of the globe. Temperature profiles to higher altitudes than was previously possible with infrared sensors alone are also provided.

The SSM/T operates in the 50–60 GHz absorption band of molecular oxygen. Since the mixing ratio of oxygen is essentially constant in the atmosphere, the contribution of any layer of the atmosphere to the total signal of a given frequency received by a radiometer flown on a spacecraft is primarily a function of the temperature and density of the layer and the amount of absorption in the atmosphere above the layer. By choosing frequencies with different absorption coefficients on the wing of the O_2 absorption band, a series of weighting functions peaking at preselected atmospheric heights may be obtained. Frequencies were chosen to obtain seven channels with weighting functions peaking at altitudes ranging from the surface to above 30 km. The surface channel at approximately 50 GHz was chosen to permit removal of background terrain contributions to the channels which peak in the lower atmosphere. Other frequencies were chosen (Table 14–11) to peak at altitudes that produce an optimum temperature profile.

The Multi-channel, Single Antenna

Radiometer scans across the nadir track on seven scan positions and two calibration positions (cold sky and 300 K). The dwell time for the

TABLE 14-11

SSM/T Channel Specifications

Number	Peak Height (km)	Frequency (GHz)	NETD* (K)
1	0	50.5	0.3
2	2 ± 2	53.2	0.3
3	6 ± 2	54.35	0.3
4	10 ± 2	54.9	0.3
5	16 ± 2	58.825	0.3
6	22 ± 3	59.4	0.4
7	30 ± 3	58.4	0.5

* Noise Equivalent Temperature Difference.

crosstrack and calibration positions is 2.7s each. The total scan period is 32s. The instrument has an instantaneous field of view of 12° and scans ± 36° from nadir.

The SSM/T weighs 25 pounds (11 kg) and consumes 14 watts of power.

Special Sensor B (SSB)—Gamma Detector

The SSB or the SSB/O sensors are gamma radiation measurement sensors which were carried on each 5D-1 spacecraft except F-4.

The SSB sensor system consists of a four-detector array, each detector being positioned 30° from the vertical. Each detector is basically a scintillator disc coupled to a photomultiplier tube. The scintillator is surrounded by a tantalum ring shield to give a directional characteristic. The SSB also contains a piggyback electronics (PBE-2) package which is an optical system for gathering background information. It contains three selectable silicon sensors.

The SSB/O is sensitive to X-rays in the energy range of approximately 1.5 KeV to more than 100 KeV. It uses large-area proportional counter and cadmium telluride (CdTe) solid-state detectors to provide excellent spectral resolution and high detection efficiency. By virtue of its extended low energy response, the SSB/O is able to identify signals associated with Bremsstrahlung generated by energetic electrons precipitating in the upper atmosphere.

Special Sensor J* (SSJ*)—Space Radiation Dosimeter

The SSJ* measured the radiation dose in silicon under aluminum shielding of four thicknesses representative of the Block 5D DMSP spacecraft. The dosimeter was launched aboard the first Block 5D satellite (F-1).

The interest in such data was engendered by the discovery of the "softness" of current Complementary Metal Oxide Semiconductors (CMOS) circuitry to radiation (10^4 rads results in failure) coupled with the present substantial uncertainty and controversy concerning the natural energetic electron dose received by a satellite in a low polar orbit. Direct measurement of this dose, correlated with performance of on-board circuitry, was therefore of crucial importance in providing data for planning of future missions (particularly where use of CMOS components is envisioned).

The dosimeter used technology proven by flight aboard many USAF and NASA satellites. The instrument consisted of four separate single-detector units, each using small cubical lithium drifted silicon detectors to perform two major functions. First, it directly measured the ionization in the silicon cube caused by natural radiation (the radiation dose). Second, the dosimeter served as an electron-proton spectrometer, thus yielding the fluxes of energetic electrons and protons encountered in the DMSP orbit as a function of time.

A space radiation dosimeter is scheduled to fly on spacecraft F-7. Although different in manufacture, its operational objectives remain essentially the same as the 5D-1 dosimeter carried on F-1.

Special Sensor D (SSD)—Atmospheric Density Sensor

The Atmospheric Density Sensor (SSD) provides a measure of major atmospheric constituents (nitrogen, oxygen and ozone) in the earth's thermosphere (100 to 250 km in altitude) by making earth-limb observations of the ultraviolet radiation from this atmospheric region. The sensor measures the radiation emitted in the ultraviolet spectral region from excitation of molecular nitrogen by impinging solar radiation. The intensity of the emitted radiation is proportional to the excitation rate and the number of molecules at any given altitude. Funneltrons and a photomultiplier tube will detect the radiation after it passes through a collimator which provides a 0.1 × 4.0° field-of-view. The SSD is mechanically driven to scan vertically through the earth's limb in about 30 seconds. This 18-pound instrument provides approximately 50 sets of density profiles on the daylight portion of each orbit.

The SSD is currently being evaluated aboard the DMSP F-4 satellite launched in June 1979. This is an experimental sensor with no current Program Office plans to integrate it into subsequent 5D spacecraft.

Special Sensors J/3 and J/4 (SSJ/3 and SSJ/4)—Precipitating Electron Spectrometers

A small, lightweight (3 lb) (1.4 kg) sensor, the SSJ counts ambient electrons with energies ranging from 60eV to 20KeV. Utilizing a time-sequenced variable electrostatic field to deflect the particles toward the channeltron detector, the sensor determines the number of electrons having energies within certain sub-ranges of the 60eV to 20KeV spectrum.

The SSJ/3 and SSJ/4 are electron spectrometers that detect and analyze low-energy electrons precipitating into the atmosphere, producing the auroral display.

The SSJ/4 units for Block 5D-2 will incorporate the capability to measure precipitating protons in the 50eV to 20KeV energy range as well as electrons measured on 5D-1. It is expected that this change will give new information relative to magnetic substorm activity.

Special Sensor C
(SSC)—Snow/Cloud Discriminator

The Snow/Cloud Sensor was an experimental unit used in conjunction with the OLS sensor on spacecraft F-4. Simultaneous in-orbit use of these two sensors proved the proposition that snow/cloud scene discrimination can be obtained through the combination of near-IR (1.6 micrometer wavelength) sensor data and OLS L-Channel (visual information. The success in discrimination of snow/cloud data during the life of the F-4 instrumentation was about the 90 percent rate.

The Snow/Cloud Sensor is a "push-broom" scan radiometer that will depend upon the orbital velocity of the 5D spacecraft to provide the along-track scan and a linear array of 48 detector elements at the image plane of a wide-angle lens to provide a 40.2° cross-track scan. The sensor depends upon reflected solar energy in the 1.51- to 1.63-micrometer spectral band for its input signal.

Special Sensor M/I (SSM/I)—Microwave
Environmental Sensor System

The SSM/I is currently undergoing system development for integration aboard later Block 5D-2 vehicles. The sensor is a passive microwave imager that scans in seven channels. It will operate at four frequencies, three with both vertical and horizontal polarization (at frequencies of 19.35, 37.0, and 85.5 GHz), and one with only vertical polarization (at a frequency of 22.23 GHz).

The SSM/I will be designed to scan cross-track (with respect to the satellite's velocity vector) in a conical scan pattern, and will gather data over an approximate 1400 km swath width. The data gathered will provide information that will be used to determine the location, extent, and intensity of precipitation. In addition, the data collected will provide for a determination of soil moisture content, wind speed over the ocean, and the morphology and extent of sea ice.

Special Sensor I/P (SSI/P)—Passive
Ionospheric Monitor

The SSI/P is a high-frequency (HF) receiver which passively monitors the ionospheric noise breakthrough frequency used in ionospheric forecasting. Noise from man-made and natural sources below the ionospheric F2 layer peak can be detected by an HF topside receiver only at frequencies above the F2 critical frequency (foF2). By sweeping through the HF spectrum,

the SSI/P can monitor the frequency of the noise breakthrough. This frequency agrees to within 1 MHz of the measured foF2 above 85 percent of the time over mid-latitude land masses. The foF2 is a prime parameter used in constructing electron density profiles used in forecasting the state-of-the-ionosphere.

Special Sensor I/E (SSI/E)—Ionospheric
Plasma Monitor

The SSI/E is an electron- and ion-probe designed to measure the average ion mass, satellite potential, electron density, and electron and ion temperatures at the spacecraft altitude. The SSI/E provides a measurement of the electron density and parameters for the calculation of the plasma scale height at the satellite altitude. The latter information is a prime input to the production of existing vertical electron density models.

The SSI/E to be carried on Block 5D-2 satellites will have the additional capability of making scintillation measurements at orbital altitudes. This instrument will be identified as SSI/ES.

GEOSTATIONARY METEOROLOGICAL
SATELLITES

ATS, APPLICATIONS
TECHNOLOGY SATELLITE

The increased launch-vehicle capabilities available during the middle 1960s permitted satellites to be placed at geostationary altitudes and thus provided atmospheric scientists with a new dimension in observations, namely: continuous observations from a single platform of almost one-third of the earth's surface. A NASA research program involving geostationary satellites was implemented in the Applications Technology Satellite (ATS) series. Although primarily designed to demonstrate communications satellite technology, several of the ATS series carried high-resolution cameras for atmospheric observation.

On December 7, 1966, ATS-1 was placed into geostationary orbit. One function of this technology satellite was to demonstrate the capability of providing a picture of the western hemisphere every 20 minutes through the use of a spin-scan camera. Useful data were provided from approximately 55°N to 55°S latitude. The ability to receive sequential photographs of the same area improved the possibility of early detection of severe storms and tornadoes, and provided real-time data with respect to cloud and frontal movements.

A second technology satellite, ATS-3, was launched November 1967. This satellite, using a multispectral spin-scan camera, returned the first color images of almost the full earth disc. Copies of these pictures have been used for many applications in addition to meteorology. ATS-1 and ATS-3 were developed by NASA GSFC, with

Hughes Aircraft as the prime contractor (Suomi and Vonder Haar, 1969).

The performance history of the Nimbus and ATS technology satellites is shown in Figure 14-8.

SMS/GOES (OPERATIONAL GEOSTATIONARY SATELLITE)

The successful application of atmospheric observations from geostationary altitudes led to NASA's development of a satellite designed specifically for that purpose. This satellite, the SMS/GOES, was designed and integrated by the Western Development Laboratories of Ford Aerospace and Communications Corporation. NASA's prototype Synchronous Meteorological Satellite, SMS-1, was successfully launched in May 1974. Placed over the equator at 45°W longitude, it provided continuous near-hemispheric coverage. The principal instrument for SMS is a 16-inch aperture telescope for visible and infrared scanning. Built by the Santa Barbara Research Center and called VISSR (Visible and Infrared Spin-Scan Radiometer), this sensor permits day and night observation of clouds and the determination of temperatures, cloud heights, and wind fields (Johnson, 1979).

The SMS also relays data received from remotely located data collection platforms such as river gauges, ocean buoys, ships, balloons, and aircraft. Its space environmental monitor (consisting of an X-ray sensor, an energetic particle sensor, and a magnetometer) detects unusual solar activity, such as flares, and measures the flow of electron and proton energy and the changes in the geomagnetic field. Observation and forecasting of atmospheric phenomena not specifically related to meterology are thus possible on an operational basis (Corbell et al., 1976).

Four additional satellites of the SMS design have been launched: SMS-2 on February 6, 1975;

the first operational version, GOES-1 (Geostationary Operational Environmental Satellite), on October 16, 1975; GOES-2 on June 16, 1977; and GOES-3 in June 1978. These operational satellites are owned and operated by NOAA. The SMS/GOES satellite history is depicted in Figure 14-9.

The SMS/GOES satellites have been maneuvered to various stations to optimize the data and to support special tasks. The following is the disposition of SMS/GOES as of the spring of 1982.

SMS-1 is at 130°W longitude and GOES-1 at 126°W. GOES-2 is located at 107°W and GOES-3 at 90°W. GOES-2 is being used to relay weather data to this hemisphere and to assist lesser-developed countries in the receipt of processed satellite data. GOES-4 at 135° West and GOES-5 at 75° West are the two primary West Coast and East Coast operational GOES satellites, respectively.

GOES D, E, AND F

Improvements to the GOES system were introduced with GOES-D, E and F, which contain the VAS (Visible Infrared Spin-Scan Radiometer Atmospheric Sounder). The first two in this series, GOES-4 and -5, (formerly GOES D & E) were launched on September 9, 1980 and May 15, 1981. These spacecraft were built by Hughes Aircraft. This is an advanced version of the Visible Infrared Spin-Scan Radiometer (VISSR) developed for worldwide geostationary meterological satellite systems. The VISSR is a dual-band (visible and infrared) spin-scan imaging device utilized for day and night, two-dimensional, cloud-cover pictures. The VAS retains the VISSR dual-band imaging function. However, the infrared channel capabilities have been expanded using a more complex detector configuration together with selectable narrow-band optical filters. The additional spectral bands provided are sensitive to the

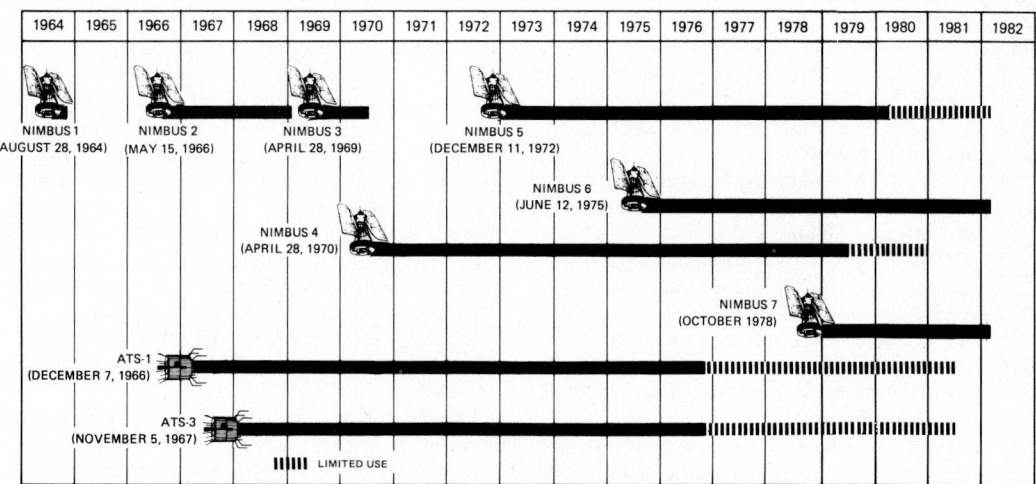

Fig. 14-8. Meteorological technology satellites: performance history.

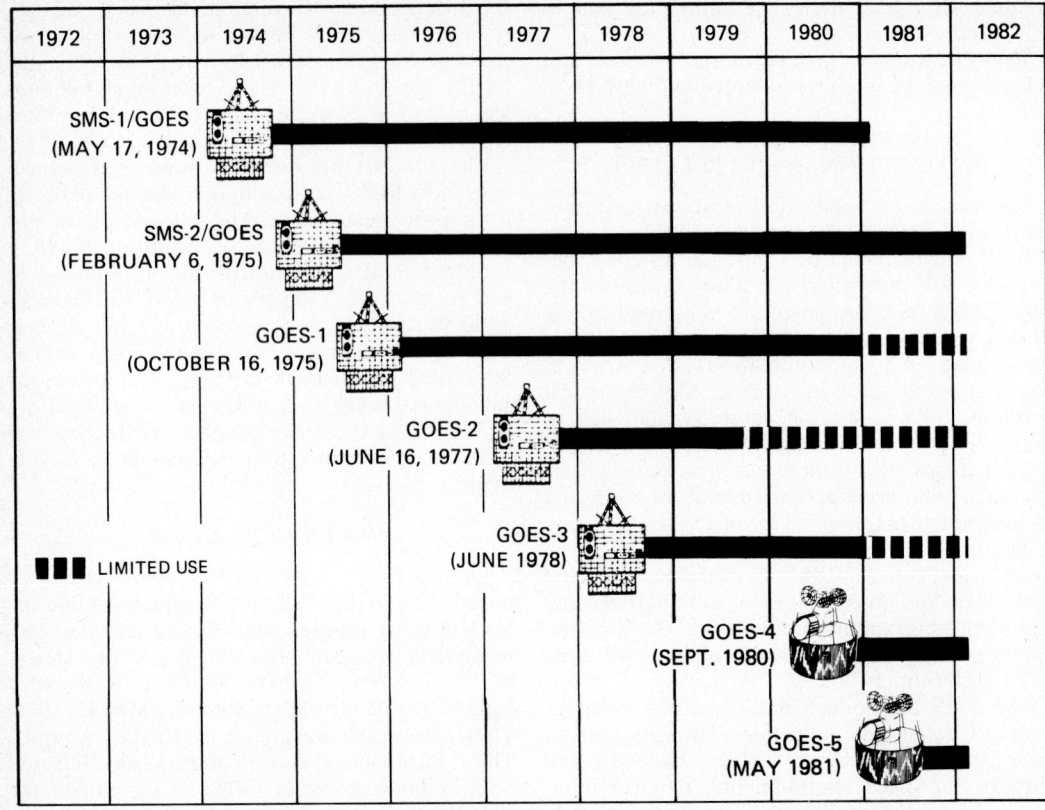

| 1972 | 1973 | 1974 | 1975 | 1976 | 1977 | 1978 | 1979 | 1980 | 1981 | 1982 |

Fig. 14-9. Geostationary operational environmental satellites, performance history.

effects of atmospheric constituents which makes it possible to determine not only the surface and cloud-top temperatures, as in VISSR, but also the three-dimensional structure of the atmospheric temperature and water-vapor distribution.

The VAS system consists of a scanner which contains a telescope assembly, and a separate electronic module. The telescope assembly is a 40.6-cm (16.0-inch) diameter aperture optical system with an object-space scan mirror for accomplishing the N-S step scan (Staff Members, 1980(a)). The complete scanner contains:

1. A position-controlled object-space scan mirror
2. A Ritchey-Chretien primary collecting optics subassembly
3. A set of secondary optics, optical filters, detectors and preamplifiers for the visible and infrared channels
4. A two-stage radiation cooler for passively cooling the infrared channel detectors
5. A separate optical focus drive assembly for the visible and the infrared channel
6. Two (a primary and a redundant) scan mirror drive motor and optical position angle encoder subassemblies
7. An optical subassembly with reduced size aperture to permit an inflight radiometric check-of-calibration of the visible channels on the sun

8. A temperature monitored, heatable, calibration blackbody and motor drive/shutter subassembly to permit an inflight radiometric calibration of the infrared channels
9. Eighteen scanner temperature sensors
10. A scan mirror stow (electromagnet latch) subassembly
11. High voltage power supplies for the visible channel detectors
12. A temperature-controlled twelve-position spectral filter wheel assembly for selecting the infrared optical passband.

The VAS has six infrared detectors. Two have a sub-satellite point resolution of 6.9 km (IGFOV of 0.192 mr × 0.192 mr) and are used primarily for imaging. Four have a resolution of 13.8 km (IGFOV of 0.384 mr × 0.384 mr) and are used for sounding information. The two small infrared channel detectors are mercury-cadmium-telluride (HgCdTe) long-wavelength detectors. Two of the large infrared channel detectors are HgCdTe. The other two are indium antimonide (InSb).

Although there are six VAS infrared detectors, only two will be in use during any satellite spin period.

The following are the three VAS operating modes:

1. *VISSR:* The sensor operates in a normal cloud-mapping mode.
2. *Multispectral Imaging:* This mode provides

the normal VISSR cloud-mapping function. In addition, it supplies data in any two additional spectral bands selected with a spatial resolution of 13.8 km. This mode of operation takes advantage of the condition that the VAS infrared imaging detectors (small HgCdTe) are offset one scan line in the N-S plane. Using the data from these detectors simultaneously produces a complete infrared map when they are operated every other scan line. This allows using the larger detectors during half of the imaging/scanning sequence period to obtain additional spectral information.

3. *Dwell Sounding:* Up to twelve spectral filters covering the range 680 cm^{-1} (14.7 μm) through 2535 cm^{-1} (3.9 μm) can be positioned into the optical train while the scan mirror is on a single N-S scan line. In addition, the filter wheel can be programmed so that each spectral band (filter) can dwell on a single scan line for from 0 to 255 spacecraft spins. Either the 6.9 km or 13.8 km resolution detectors can be selected for the seven filter positions operating in the spectral region 703 cm^{-1} (14.2 μm) through 1490 cm^{-1} (6.7 μm). For the remaining five spectral bands the 13.8 km resolution detectors are used. Selectable frame sector size, position and scan direction are the same as in the Multispectral Imaging (MSI) mode of operation (Montgomery and Endres, 1977).

In some of the spectral regions, multiple line data are required to improve the signal-to-noise ratio of the sounding data. The total number of satellite spins at the same N-S scan line position required to obtain the desired sounding data for all spectral bands is called the VAS "spin budget." The VAS "spin budget" for sounding an N-S swath having a 30 km × 30 km resolution is approximately 157 spins. Therefore, with the appropriate interlacing of scan lines formed by the detector FOV pairs, the time required to accomplish sounding over a 0.65° N-S swath (400 km) with a resolution of 30 km × 30 km will be 23 minutes.

THE EARTH RADIATION BUDGET EXPERIMENT (ERBE)

The requirements for the ERBE were specified to overcome the previous problems of existing radiation budget observations (Curran, 1980). The experiment incorporates three sets of radiometers on different satellite platforms. NOAA-F and -G, the last two satellites, are the sun-synchronous TIROS-N series satellites with differing equatorial crossing times and the third satellite is the medium inclination (i ≃ 46) forthcoming Earth Radiation Budget Satellite (ERBS). This satellite is to be Shuttle launched. The medium inclination orbit can be accommodated with a Shuttle launch from the Kennedy Space Center in Florida. It is anticipated that a payload will be scheduled to

fully utilize the Shuttle capabilities permitting the launch of other free-flying satellites or onboard (Spacelab) experiments that are compatible with this orbital inclination. The free flyer will be lifted from the pallet by the articulated arm. While still attached, the entire system including instruments and spacecraft will be checked. Once free of the environment of the Shuttle, on-board thrusters will propel the ERBS spacecraft to the nominal 60 km altitude orbit.

The 600-km orbit and the 46° inclination will provide an orbital precession rate relative to the earth-sun vector of more than 180° per month. This precession rate will permit a minimum of one observation per hour-angle per month for each 1000 km × 1000 km area in the tropics and much better sampling at mid-latitudes. The combination of ERBS and the near-polar orbiting satellites will provide a large number of samples uniformly distributed in space and time, permitting higher precision and less-biased estimates of the monthly averaged radiation budget components.

The instrumentation to be used for the radiation budget observations consists of two parts: a fixed field-of-view section and a scanner section. An artist's concept of the instrumentation on the ERBS is illustrated in Figure 14-10. The fixed field-of-view section has five radiometers, four earth-viewing and one shuttered sun-viewing. The earth-viewing radiometers consist of pairs of broad-band (0.2 to 50 + μm wavelength) and shortwave (0.2 to 5 μm wavelength) detectors. The two spectral bands are also represented by two different instantaneous fields of view, one of which views the entire earth from limb to limb while the other has a restricted field of view with a "foot print" of approximately 1000 km.

Both of these radiometers have cosine law detectors. The wide field-of-view radiometers, because of their cosine response and nadir viewing direction, measure the flux of upwelling radiation from the observable earth at satellite altitude. These flux measurements will be used in accurately determining the global radiation energy balance and its variation with season. These observations of the fluxes at satellite altitude are the only earth flux observations which do not require knowledge of the angular nature of the radiation leaving the earth. To meet the scientific requirements of the ERBS, transformation procedures must be defined to change the observations into the scientifically usable product.

The solar observations will be made with a cavity radiometer, based on the technology of the Solar Maximum Mission. This type of detector will provide periodic observations of the sun to extend the period of solar observations and will also serve as a secondary standard through simultaneous solar observations with the cavity radiometer and all the earth-viewing detectors.

The separate scanning instrument consists of shortwave, longwave and broadband channels which scan perpendicular to the velocity vector of

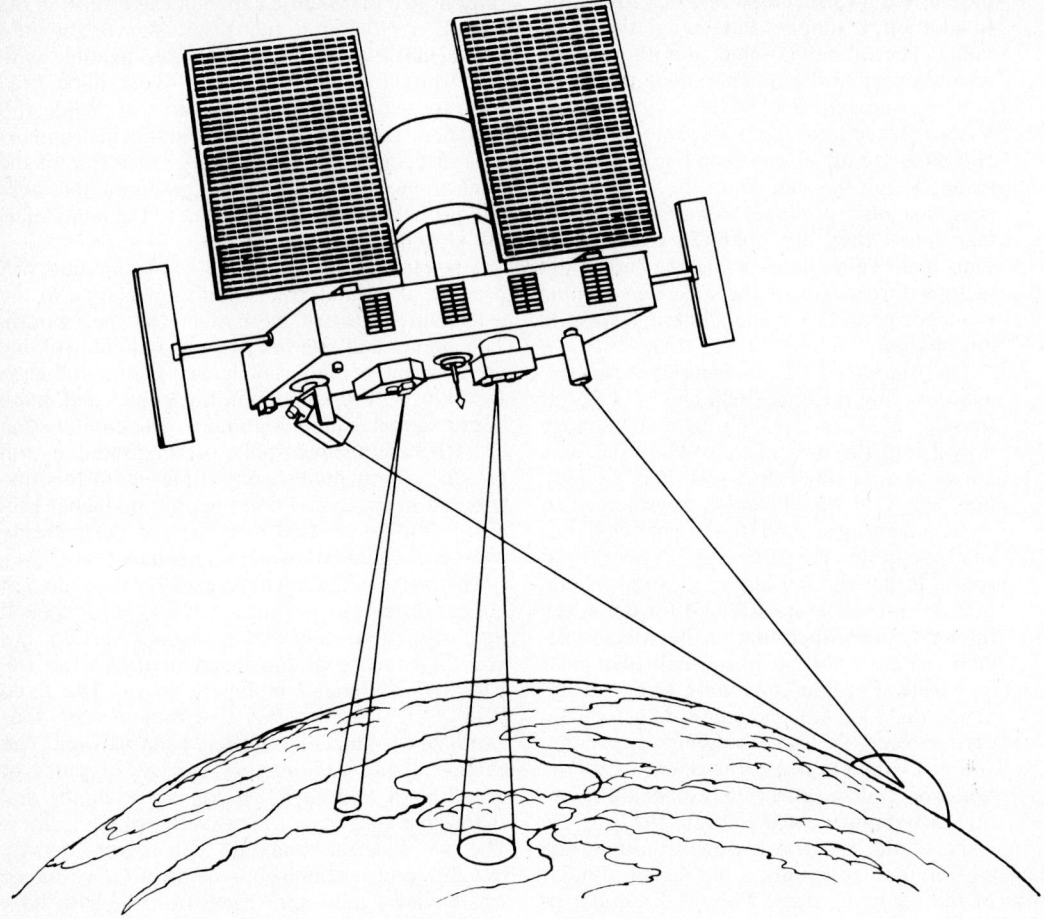

Fig. 14-10. Artist's concept of the ERBS.

the spacecraft. The instantaneous field of view of each of the scanning channels is 3°. Calibration of these channels will take place both through observation of an internal blackbody source and observations of a diffuser plate exposed to direct sunlight. The specifications of the ERBE instrumentation were developed with the cooperation of a number of scientists at NASA, NOAA/NESS and several universities. The implementation of the scientific objectives of the ERBS will be carried forward by a team of scientists recently selected through an Announcement of Opportunity procedure.

The ERBS will carry two instruments in addition to the ERBE scanner and nonscanner. One instrument is the SAGE-II (Stratospheric Aerosol and Gas Experiment) instrument which will provide observations of the stratospheric aerosols, ozone, and nitrogen dioxide. Both ozone and stratospheric aerosols are thought to have an effect on the earth's radiation budget and in turn, are thought to be affected by the magnitudes of the shortwave and longwave fluxes passing through the atmosphere. This is particularly true following major volcanic eruptions which inject particulates and gases into the stratosphere. The ERBS instruments and the SAGE-II instrument will be used in climate studies of the interrelationships among the earth's radiation budget, stratospheric aerosols and ozone. The second additional instrument on ERBS in the Halogen Occultation Experiment (HALOE) instrument. This instrument is part of a study of environmental quality and flies on ERBS as a satellite of opportunity.

INTERNATIONAL WEATHER SATELLITES

As part of the international cooperation and participation within the World Meteorological Organization, the United States, Japan, the European Space Agency (ESA), and the USSR have placed satellites in orbit which supported the First GARP Global Experiment (FGGE) during 1978–79 (Schnapf, 1980). A summary of international weather satellite evolution is shown in Figure 14-11. Japan's geostationary satellite (Himawari-1) was launched on July 14, 1977 and positioned over the western Pacific Ocean. The ESA's geostationary satellite (METEOSAT) was

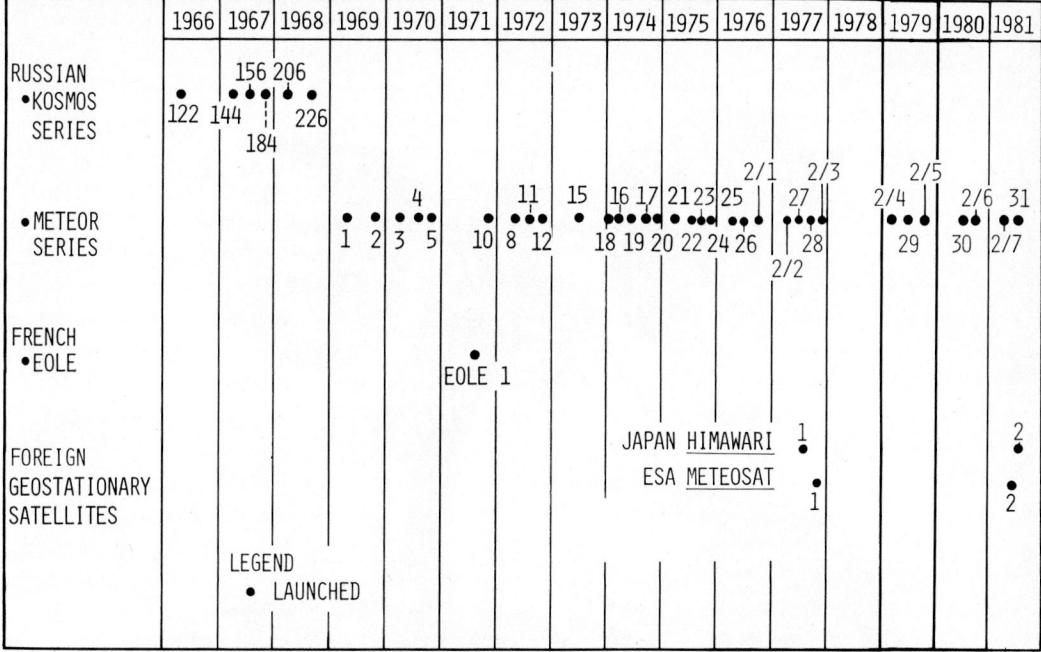

Fig. 14-11. Evolution of international weather satellites.

launched on November 23, 1977 and positioned over the eastern Atlantic Ocean. Both were launched by the U.S., each aboard a NASA Delta launch vehicle from Cape Canaveral, Florida. Himawari-2 and Meteosat-2 were launched on August 10, 1981 and in June, 1981, respectively, as replacement satellites. The Himawari-1 and the Meteosat are shown in Figures 14-12 and 14-13, respectively. More complete descriptions of these two geostationary satellites and the *USSR* Meteor system may be found in articles by Hirai et al., 1975; Morgan, 1978; and Diesen, 1978, respectively.

OCEANOGRAPHIC SATELLITES

The oceans, which cover more than two thirds of the earth's surface, have important influences on our weather and climate and are a vital natural resource needing further exploration (McClain, 1980).

SEASAT-A

Seasat-A (Figure 14-14) was launched in June 1978 and carried instruments fully dedicated to oceanic prediction (NOAA Staff Members, 1977). The sensor applications are as follows:

1. *Altimeter*—A nadir-looking instrument that measures the displacement between the satellite and the ocean surface to a processed accuracy of 10 cm every 18 km and an rms roughness of that surface to 10 cm.
2. *Radar*—A 100-km, swath-width image of the ocean surface with a 25-m spatial resolution viewed every 18 days.
3. *Scatterometer*—Low to intermediate surface wind velocity determined over a swath-width of 1200 km, providing global coverage (±75° latitude) every 36 h on a 100-km grid basis.
4. *Microwave Radiometer*—Ice boundaries and leads, atmospheric water vapor, sea surface temperature, and intermediate to high wind speeds are provided over a swath-width of 1000 km every 36 h. Spatial resolution of ice features is 25 km, and 125 km for sea surface temperature (Gloerson and Barath, 1977).
5. *Visible and IR Radiometer*—Provide a 7-km spatial resolution imagery for feature identification for the microwave data.

SPACE SHUTTLE

From early in the 1980s, and at least throughout the following decade, the Shuttle will be NASA's transportation system for access to space. Beyond that, the joint venture with European countries in developing the Spacelab will provide a functioning manned laboratory facility in space. These two programs will play vital roles in the development of future remote sensors (Allison et al., 1978). The four functional categories that these programs will provide are as follows: (*a*) calibration of instruments on operational weather satellites, (*b*) direct monitoring of slowly changing earth/atmosphere parameters, (*c*) development and demonstration of new remote sensors, and (*d*) special experiments impossible or impractical by other means.

The difficulties inherent *to* calibration and cor-

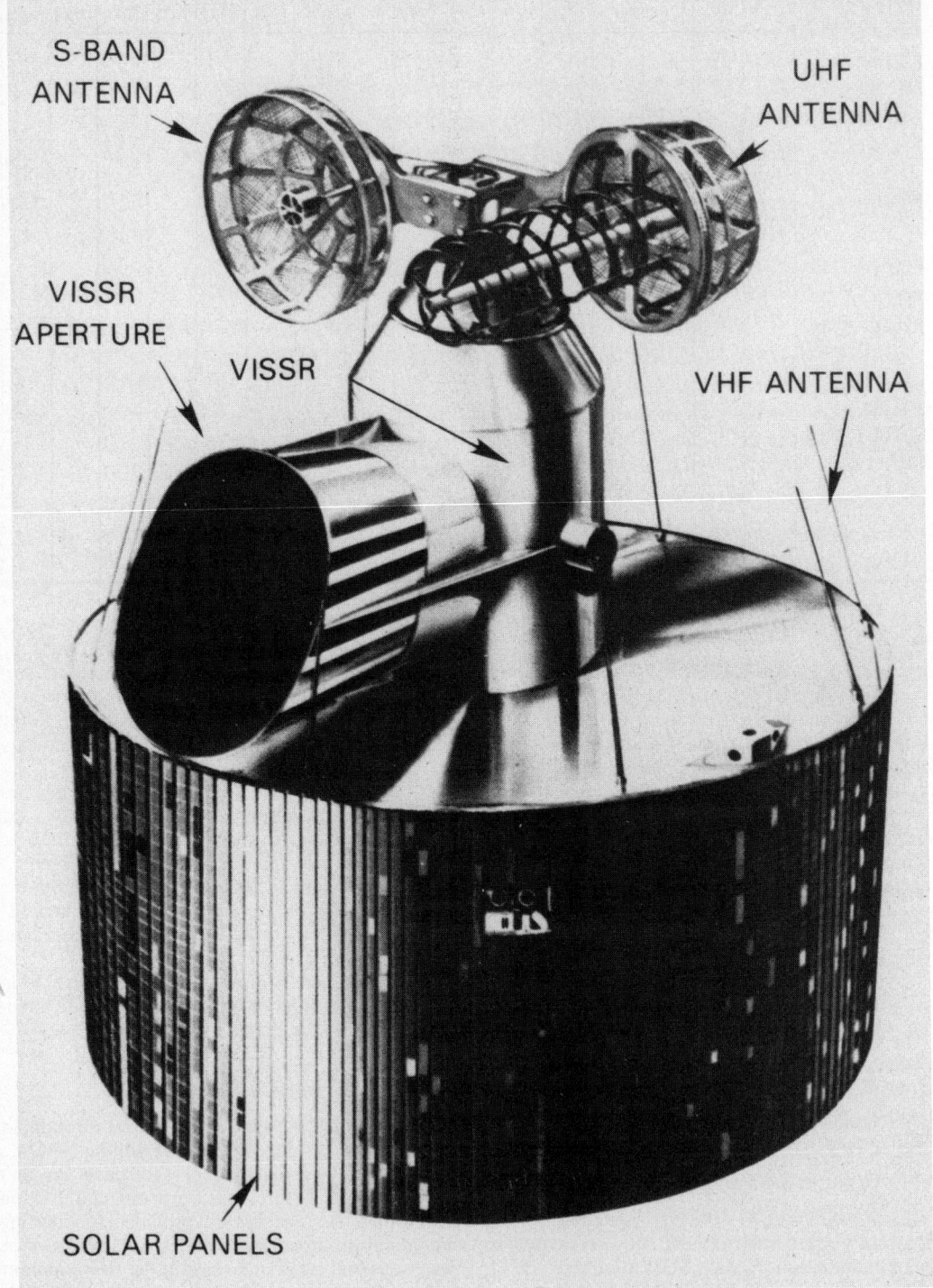

Fig. 14-12. Geostationary meteorological satellite (GMS) (Himawari).

rection for long-term drift of satellite instruments have, in the past, presented manifold problems. First, logistic problems associated with deployment of enough ground systems and interrelating their independent errors imposed severe lim-

itations. Second, ground instruments do not generally measure the same quantities as satellite instruments. For example, a satellite radiometer measures a brightness temperature, T_B, integrated over the entire field of view, while ground-truth

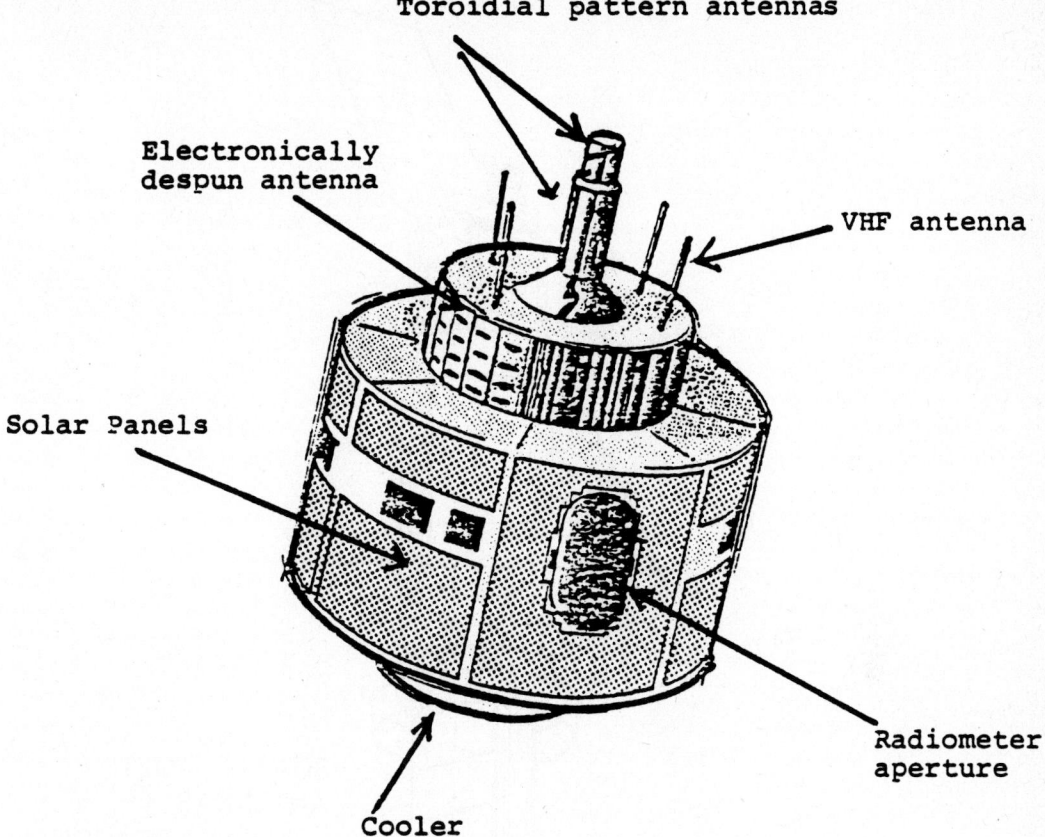

Fig. 14-13. METEOSAT external appearance.

data are restricted usually to point measurements. Thus, calibrations of satellite instruments over their several years of useful lifetimes and the ability to interrelate succeeding satellites, are crucial requirements for the emerging U.S. Climate Program. The Shuttle can fulfill these requirements by periodic flights of instruments, calibrated against standards before and after flight, that can provide comparative measurements at points of orbit conjunction with the satellites when identical scenes are in view.

Experience with the Nimbus-6 ERB instrument substantiates the desirability of continuing total solar flux measurements at six-month intervals. Such measurements will provide an independent and highly calibrated set of data by themselves, but will also serve to interrelate the total solar flux measurements of the ERBE instrumentation. Other parameters that can be directly monitored by the Shuttle and Spacelab include a number of the well-mixed atmospheric constituents, ozone, and tropospheric aerosols. For example, annual measurements are probably adequate for CO_2, CFMs, N_2O, NO_x, and CH_4.

Spacelab will also provide the needed facilities to develop, test, and demonstrate new remote sensors. These include:

1. *Sea Surface Temperature*—Improvements that appear possible include increased signal-to-noise ratios and additional wavelengths to reduce errors due to atmospheric water vapor and clouds.
2. *Vertical Temperature Profile*—Accuracy of passive sounding below 20 km altitude can be improved with measurements at 4 μm and with bandwidths of 2 wave numbers. Active Lidar promises even higher accuracy along with an order-of-magnitude improved vertical resolution.
3. *Winds*—Lidar techniques may be able to measure wind speed and direction in cloud-free regions.
4. *Stratospheric Aerosols*—Stellar sources provide a large number of occultations not limited to the ecliptic plane, so that reasonably good coverage of stratospheric aerosols is possible.
5. *Tropospheric Aerosols*—At present, no satellite technique exists for measuring tropospheric aerosols, but a pulsed lidar system is believed to hold promise.

As plans are formulated for the development of these instruments (as well as any others that may yet be identified) to the point where space flight

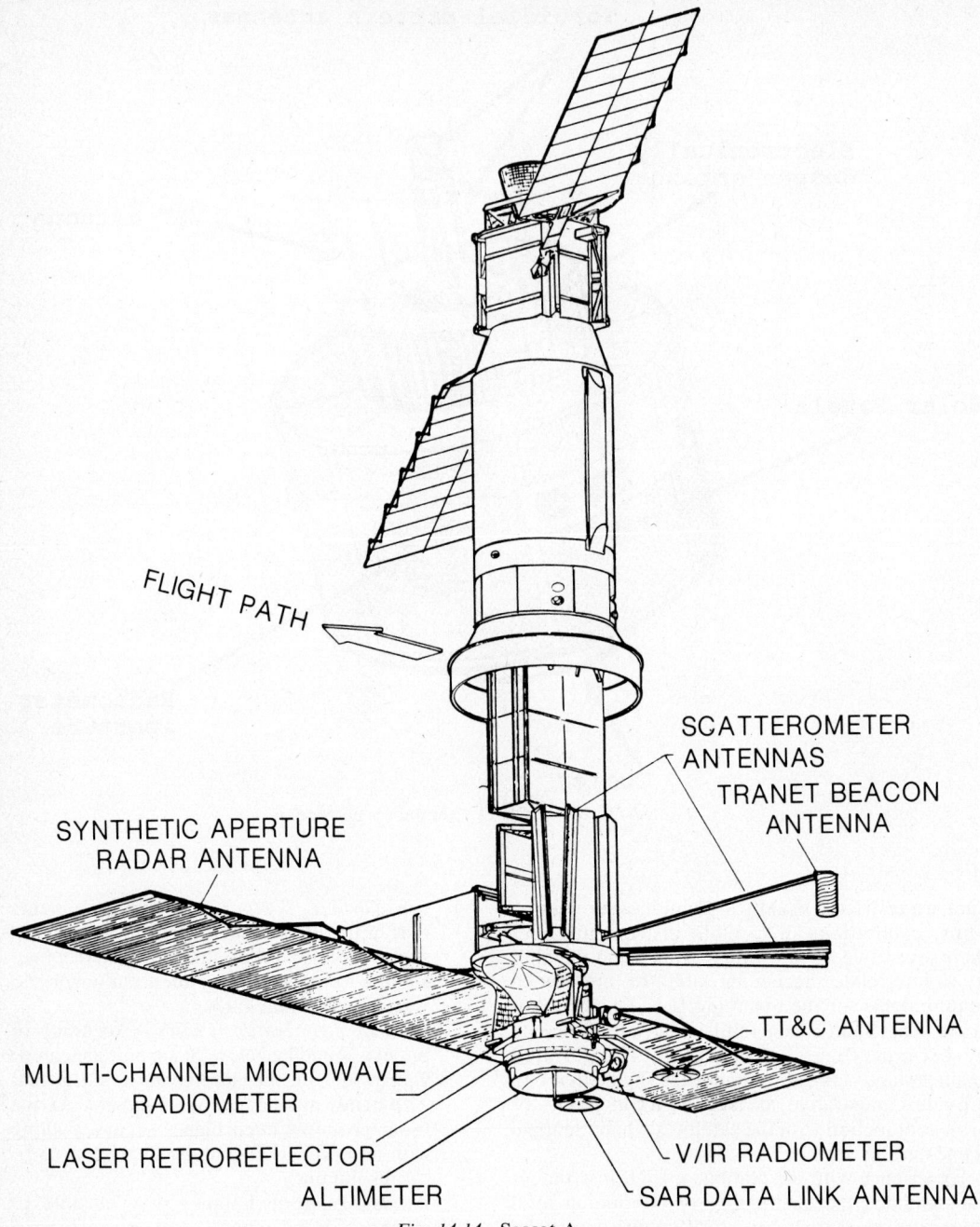

FLIGHT PATH

SCATTEROMETER
ANTENNAS
TRANET BEACON
ANTENNA

SYNTHETIC APERTURE
RADAR ANTENNA

TT&C ANTENNA

MULTI-CHANNEL MICROWAVE
RADIOMETER

LASER RETROREFLECTOR

V/IR RADIOMETER

ALTIMETER

SAR DATA LINK ANTENNA

Fig. 14-14. Seasat-A.

development and testing can be visualized, the advantages that Spacelab can provide will be capitalized upon to the maximum extent possible.

There are two particularly important special experiments in the area of remote sensing that are especially suited to Spacelab's capabilities. These are extended cloud physics and radiation studies and precipitation over land and water. In the case of the extended cloud and radiation studies a group of instruments, including two separate radiometers and a lidar system, are required. In the case of the precipitation studies, a microwave radiometer, a visible-band radiometer, an IR radiometer, and a 10-cm radar are required. A number of short-duration flights are adequate to meet the objectives of both of these studies and the low-orbiting Spacelab offers a unique capability for their accomplishment.

REFERENCES

Allison, L. J., and E. A. Neil, 1962, Final report on the TIROS 1 meteorological satellite system; NASA Technical Report R-131, Goddard Space Flight Center, Greenbelt, Md. 20771.

Allison, L. J., R. Wexler, C. R. Laughlin, and W. R. Bandeen, 1978, Remote sensing of the atmosphere from environmental satellites; Special Technical Publication 653, American Society for Testing and Materials, Philadelphia, Pa., pp. 83–155.

Brandli, H. W., 1976, Satellite Meteorology AWS-TR-76-264; Hq. Air Weather Service, U.S. Air Force, Scott AFB, Illinois, p. 187.

Conlan, E. F., 1973, Operational products from the ITOS scanning radiometer data; NOAA Technical Memo NESS 52, U.S. Dept. of Commerce, Washington, D.C.

Corbell, R. P., C. J. Callahan, and W. J. Kotsch, 1976, The GOES/SMS User's Guide; U.S. Dept. of Commerce, NOAA, NESS, Washington, D.C., p. 118.

Curran, R. J., 1980, Satellite measurements of earth radiation budget for climate applications; Space Shuttle: Dawn of an Era, vol. 41, Advances in Aeronautical Sciences, American Astronautical Society, San Diego, Calif.

Diesen, B. C., III, and D. L. Reinke, 1978, Soviet meteor satellite imagery; Bull. Amer. Meteorol. Soc., vol. 59, no. 7, pp. 804–807.

Fortuna, J. J., and L. N. Hambrick, 1974, The operation of the NOAA polar satellite system; NOAA Technical Memo NESS 60, U.S. Dept. of Commerce, NOAA, NESS, Washington, D.C.

Gloersen, P., and F. T. Barath, 1977, A scanning multichannel microwave radiometer for Nimbus-G and Seasat-A, IEEE J. Oceanic Eng., vol. OE-2, no. 2, pp. 172–178.

Hirai, M., K. Watanabe, H. Tsuru, M. Miyaki, and M. Kimura, 1975, Development of geostationary meteorological satellite (GMS) of Japan; Proc. of the Eleventh International Symposium on Space Technology and Science, Tokyo, Japan, pp. 461–465.

Hussey, W. J., 1979, The TIROS-N/NOAA operational satellite system; U.S. Dept. of Commerce, NOAA, NESS, Washington, D.C., p. 35.

Johnson, D. S., 1979, NOAA's operational satellite service, C³A, Signal, Armed Forces Communications and Electronics Assn., Falls Church, Va.

McClain, E. P., 1980, Passive radiometry of the ocean from space—an overview; Boundary Layer Meteorology, vol. 18, D. Reidel Pub. Co., Dordrecht, Holland, pp. 7–24.

Montgomery, H., and D. Endres, 1977, Survey of dwell sounding for VISSR atmospheric sounder (VAS); NASA X-942-77-157, Goddard Space Flight Center, Greenbelt, Md., p. 85.

Morgan, J., 1978, Introduction to the meteosat system—Issue 1; MDMD/MET/JM-bd/833 European Space Operations Centre, Darmstadt, Germany.

National Research Council, 1978, The global weather experiment—perspectives on its implementation and explorations; Report of the FGGE Advisory Panel, National Academy of Sciences, Washington, D.C., p. 104.

Nichols, D. A., 1975(a), DMSP Block−4 compendium; Space Division/YDE, U.S. Air Force, Air Weather Service, Los Angeles, Calif.

Nichols, D. A., 1975(b), DMSP Block−5D compilation; Space Division/YDE, U.S. Air Force, Air Weather Service, Los Angeles, Calif.

Nichols, D. A., 1976, DMSP Block−5A, B, C compendium; Space Division/YDE, U.S. Air Force, Air Weather Service, Los Angeles, Calif.

NOAA Staff Members, 1977, NOAA program development plan for SEASAT-A research and applications; U.S. Dept. of Commerce, NOAA, NESS, Washington, D.C.

Press, H., and W. B. Huston, 1968, NIMBUS, a progress report; Astronautics and Aeronautics, March 1963, pp. 56–65.

Schnapf, A., 1979, Evolution of the operational satellite service; 1958–1984; NOAA-A Colloquia, Washington, D.C.

Schnapf, A., 1980, Twenty years of weather satellites: where we have been and where we are going; 17th Annual Space Congress, Cocoa Beach, Fla.

Schwalb, A., and J. Gross, 1969, Vidicon data limitations; ESSA Technical Memo and NESC Technical Memo 17, U.S. Dept. of Commerce, ESSA, NESC, Washington, D.C.

Schwalb, A., 1978, The TIROS-N/NOAA A-G satellite series; NOAA Technical Memo NESS 95, U.S. Dept. of Commerce, NOAA, NESS, Washington, D.C., p. 75.

Staff Members, Goddard Space Flight Center, 1965, Observations from the Nimbus 1 meteorological satellite; NASA SP-89, Goddard Space Flight Center, Greenbelt, Md. 20771, p. 90.

Staff Members, Goddard Space Flight Center, 1976, Nimbus G, Nimbus observation processing system, (NOPS) Design Study Report; Goddard Space Flight Center, Greenbelt, Md. 20771.

Staff Members, 1979(a), Concept for a Block 6 meteorological satellite system technical requirements document; Headquarters SAMSO/DMSP, U.S. Air Force, Air Weather Service, Los Angeles, Calif.

Staff Members, 1979(b), ICEX, Ice and Climate Experiment; Goddard Space Flight Center, Greenbelt, Md. 20771, p. 10.

Staff Members, 1980(a), Visible infrared spin-scan radiometer atmospheric sounder system description; Santa Barbara Research Center, Goleta, Calif.

Staff Members, 1980(b), National Oceanic Satellite System (NOSS), Algorithm Development Plan; Goddard Space Flight Center, Greenbelt, Md. 20771.

Suomi, V. E., and T. H. Vonder Haar, 1969, Geosynchronous meteorological satellite; J. of Spacecraft and Rockets, vol. 6, no. 3, pp. 342–344.

Vostreys, R. W., and R. Horowitz, 1979, Report on active and planned spacecraft and experiments; NSSDC/WDC-A-R&S 79-03, National Space Science Data Center, Goddard Space Flight Center, Greenbelt, Md. 20771.

CHAPTER 15

Communication and Data Transmission Systems

AUTHOR: JOHN A. BECKMAN

GENERAL CONTENTS: The communication process; information input; radio circuits; modulation function; receiving; the radio optical spectrum; optical communication; communication system design; free-space propagation; atmospheric effects; lenses and antennas; optical resolution; noise; system power budget; references.

NOMENCLATURE

Symbol	SI Units	Name
A	m²	effective aperture, or area, of an antenna
A_r	m²	effective area of a receiving antenna
α	dB	attenuation
B	Hz	bandwidth
c	m s⁻¹	velocity of light in vacuum
C	bits s⁻¹	channel capacity
C	W	carrier power level of an EM signal
D (also d)	m	diameter of a circular antenna
dB	—	decibels
EM	—	electromagnetic wave or signal (radio or light radiation)
f	Hz	frequency
f_c	Hz	carrier frequency
f_m	Hz	modulating frequency
φ	rad	phase angle
G	—	gain (power) of an antenna
G_t	—	transmitter antenna power gain
G_r	—	receiver antenna power gain
λ	m	wavelength
m_f	—	modulation index in FM (frequency modulation)
NF	dB	noise figure of a receiver
nif	—	noise improvement factor
P	W m⁻²	power density of an EM wavefront
P_r	W m⁻²	same as P, at the receiving antenna
R	m, (km, miles)	range, or distance from transmitter to receiver
RF	—	radio frequency (frequencies)
T	K	temperature
t	s	time
W_r	W	power intercepted by receiving antenna
W_t	W	transmitted power
ERP	W	effective radiated power
θ	rad	angle of resolution of a lens or antenna

INTRODUCTION

Radio, or radio-frequency (RF), communication provides a vital link between remote sensors on aircraft and satellites and ground-based control facilities. Remote sensor data are collected and transmitted to central facilities for processing, interpretation, and storage. Various modulation methods are utilized to transfer these data with minimum error and degradation, and to command necessary control functions on the space platform. Radio communication links have made unmanned satellite remote sensor systems, such as Voyager I, operationally feasible. Basic theory and pertinent aspects of RF links are reviewed for remotely-sensed imagery.

The objective is twofold: (1) to bridge the variances in terminology and technology between specialists in the optical/visual and radio/electronic fields, and (2) to acquaint the reader of this manual, who is familiar with various aspects of the reconnaissance field, with the basic principles of radio communications as they apply to image transmission.

The *unifying factor* between the historically diverse fields is their common fundamental description in terms of the same electromagnetic (EM) phenomena.

The most direct method for getting these data to the central point is physical transport of the information, such as return to the ground of exposed film and other raw data, with no consequent degradation of the information content due to any intervening process. The earliest and most successful examples of physical transfer of acquired information include balloons, aircraft and the Discoverer series of reconnaissance satellites; in the Discoverer series, capsules bearing the recorded information on film and magnetic tape are re-entered into the atmosphere and parachuted to aircraft waiting to recover them with huge nets!

There are disadvantages to physical transport of the information on a routine basis. The vehicle may not return to its base, or a re-entry capsule may be lost. Additionally, the information garnered may be of urgent portent, and therefore the delay inherent to physical transport may be unacceptable. Furthermore, some space vehicles are intended for indefinite life in orbit, and therefore cannot afford to eject capsules. Direct physical transport of recorded data therefore has only limited application; the practical alternative is transmission via radio communication.

Radio transmission may be performed immediately, in real time, as the data are sensed at the remote vehicle, or it may be delayed until the sensing vehicle is in the best position with respect to a ground receiving station. The approach largely depends upon the flight path or orbit of the remote sensing vehicle. A delay in transmission (from real time) implies storage and then transmission of the data, all to be accomplished within the remote vehicle, most probably upon a command from a ground control station. This technique is frequently employed with weather observation (meteorological) and several other satellite systems (Mueller, 1964).

The imagery obtained with the satellite sensors is scanned by various methods in order to place the information in a series analog form. This analog signal (or its digital equivalent) then modulates an RF carrier. The modulation methods are reviewed in the following sections.

The chapter deals with basic considerations found in the transmission of imagery and information from a space platform to the ground station.

Other signals must also be communicated between the sensing platform and its base station. These may include command and control of the platform's course or orbit and the platform's sensing functions, such as quite simply turning cameras on and off. Perhaps these must be performed in a secure fashion, so that unauthorized individuals cannot acquire control over the remote sensors.

The remote platform may also transmit to its base station such internal conditions as temperature, power supply, and other *housekeeping* data; it may also acknowledge receipt of command and control instructions.

THE COMMUNICATION PROCESS

The communication process must accomplish these functions: (1) Accept an electrical representation of the image or other data; (2) prepare it for transmission, (3) convey it by an EM wave from the transmitting to the receiving station, and (4) restore the original information at the receiving terminal.

This broad view includes a multiplicity of individual, intermediary links along the chain from input to output. The step-by-step advance of information is illustrated by the typical signal flow diagram shown in Figure 15-1.

INFORMATION INPUT

Communication systems have been designed to convey all types of information, from the slowest telegraph to the very high content of multiple channels grouped together. The discussions herein will emphasize image transmission, although similar techniques and equipment are found in the various communication services.

The information input has its inherent frequency spectrum, commonly called the *baseband*. For imagery, it arises from scanning, sampling, and quantization processes described in this manual.

The need for understanding the techniques of transferring information by the communication link is becoming increasingly important because data collection platforms are resorting to ever more sophisticated technology which provides a far greater volume of collected data (e.g., greater

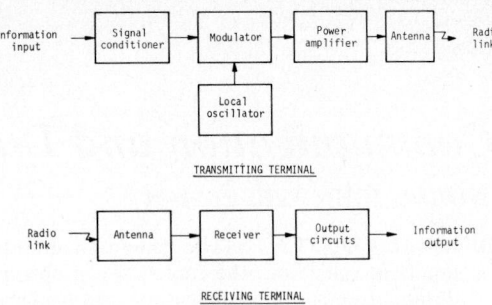

Fig. 15-1. Transmit—receive functions.

resolution, color rendition, etc.). Thus advances in sensing techniques inexorably lead to a greater burden on transmission of the additional data which have been acquired. The basic steps of communication remain the same as shown in "Signal Flow," although the implementation of each function may change as the communication art advances to accommodate the greater information input from the sensors (Sci. Amer., 1972).

The circuits which adapt one or more information inputs to the requirements of the *modulator* are termed, collectively, the *signal conditioner*. It includes matching voltages and impedances, and switching among inputs, or combining (multiplexing) several inputs, and other features which will give a more versatile transmitting instrument. The term is used most commonly in describing spacecraft operation.

RADIO CIRCUITS

Next, the baseband frequencies are "translated" by modulation, a non-linear circuit operation, to the radio frequency required for transmission from remote sensing platform to base terminal.

Therefore, all of the discussion which follows in this Chapter will be concentrated upon RF.

Local Oscillator

The *local oscillator* (LO) determines the operating frequency of the system. It comprises a high-stability oscillator with attendant buffers or amplifiers to isolate the oscillator itself from any outside disturbances which might affect its stability.

This point of high stability is important in reducing congestion, or preventing one station channel from wandering onto another adjacent channel, and also in permitting minimum bandwidth at the receiving terminal, which does not have to accommodate transmitter frequency uncertainties.

MODULATION FUNCTION

The modulation function, which occurs in the modulator, impresses the information onto the

LO output. Modulators are nonlinear circuits (their instantaneous output voltage is not directly proportional to their input).

When the local oscillator is modulated by the information, a new set of frequency components is generated. This is characteristic of passing any signal or signals through a nonlinear circuit, as is readily demonstrated mathematically.

In the new composite signal, the LO is now the *carrier.* Surrounding it are representations of the baseband, the *sidebands.* They are symmetrically disposed as mirror images about the carrier, with an *upper sideband* above the carrier frequency, and a *lower sideband* below.

The sideband structure may become quite complex and extend far beyond the minimum possible *bandwidth,* which is the maximum baseband frequency component. The total bandwidth depends upon the type of modulation.

Modulation techniques for optical sources were described in Chapter 7, Electro-Optical Non-Imaging Sensors. In a typical case, the RF source will be replaced by an optical source, the power amplifier deleted, and a lens substituted for the antenna.

The most important modulation methods for transmitting images from space platforms are described in the following sections.

Amplitude Modulation

Amplitude modulation (AM) is conceptually the simplest form of modulation, in which the amplitude of the carrier is varied as the instantaneous amplitude of the information. If the carrier frequency is f_c, and the modulating signal (the information) is f_m, the modulator output frequencies and: f_c, $f_c + f_m$, and $f_c - f_m$.

Thus the spectrum occupied is twice that of f_m alone. It should be emphasized that in imagery f_m is not a discrete, single frequency, but a spectrum (the baseband). Therefore the total spectrum involved is composed of f_c, $f_c +$ (baseband spectrum) and $f_c -$ (baseband spectrum).

An illustration of an AM signal is given in Figure 15-2 (Terman, 1955), which shows that the modulated signal bears the *waveshape* of the modulating signal. This characteristic is exclusive to AM (by definition), and it makes for the easiest

reception (*demodulation*), since a very simple circuit will extract the information content.

AM has several disadvantages, such as: (1) The upper and lower sidebands duplicate each other, so the bandwidth is twice that necessary for the baseband information content; (2) the carrier power, very substantial, is wasted because it conveys no information; (3) power amplification of an AM signal requires an inefficient linear amplifier to maintain its waveshape; (4) fading in the propagation link appears as spurious AM, so it must be compensated against lest it appear as information at the system output; (5) and the signal-to-noise ratio at the output of the demodulator may be worse than the input *S/N.*

AM is infrequently used for image transmission except for one of its variants termed "vestigial sideband," which is the television image standard.

Frequency Modulation

In *frequency modulation* (FM), the output frequency is varied, but the amplitude remains constant; the frequency swings from its normal f_c both above the carrier and below in response to the information (signal amplitudes of the input). The waveform of FM is shown in Figure 15-3. Terman (1955). It will be noted that the frequency increases with the amplitude of the information input; the opposite convention also may be used.

FM waveform is of misleading simplicity, because analysis of its frequency spectrum involves a sine function, f_c, of another sine function, f_m; this leads to a very complicated spectrum which must be expressed by Bessel functions. However, a useful property of FM is given by the *modulation index, m_f,* where

$$m_f = \frac{\text{frequency deviation}}{\text{modulation frequency}} = \Delta f / f_m \quad (15\text{-}1)$$

When m_f is near to 1, the resultant spectrum is very similar to AM. There are "higher order" sidebands in addition, at multiples of $\pm f_m$ from the carrier; they decrease in amplitude as they depart, and appear at $\pm 2f_m$, $\pm 3f_m$, etc. In this case of $m_f =$

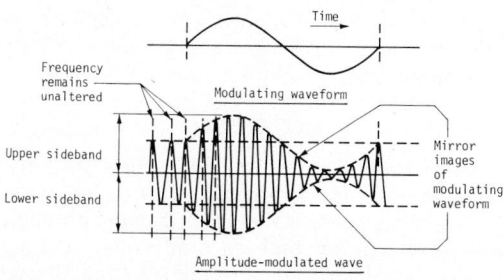

Fig. 15-2. Amplitude modulation (AM) signal form.

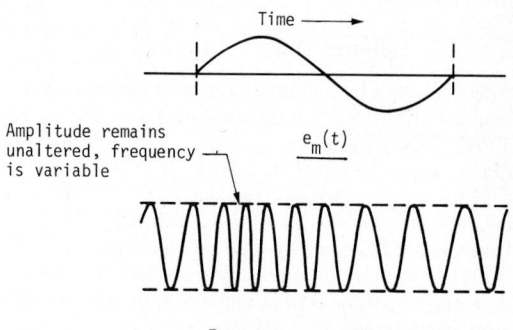

Fig. 15-3. Frequency modulation (FM) signal form.

1, only the $2f_m$ component is important, being about 0.2 of the f_m sideband in its amplitude.

The question naturally arises: if the spectra for AM and FM are so similar, why does AM vary so in amplitude while FM remains constant in amplitude? The answer lies in the relative phases of the upper and lower sidebands and the carrier. It turns out that FM is loaded with such subtleties; Terman (1955).

The most valuable feature of FM is realized when the modulation index m_f is larger than unity, approximately 3 to 5 or more. This is because there is a *noise improvement factor*, nif, which is approximately

$$nif = 3 \ m_f^2. \qquad (15\text{-}2)$$

Thus the square factor becomes very significant as m_f is increased. Note that it is one or greater as m_f equals or exceeds 0.6. The equation must be applied carefully to practical cases, for there are a number of qualifications which must be met to insure its validity.

The improvement factor is exploited effectively in high-fidelity FM broadcasting, where m_f is 5 or greater; Taub and Schilling (1971).

Phase Modulation

Phase modulation (PM) may be best considered from the nature of the modulating signal; that is, whether it is analog or digital.

For an analog information input, PM and FM exhibit very similar characteristics. Their waveforms have the same appearance, any difference being primarily the resultant of any signal conditioning which may have been applied. This is because of the intimate relationship between frequency and phase, where frequency, f, may be defined as the time derivative of phase, ϕ, as

$$f = d\phi/dt, \qquad (15\text{-}3)$$

where:

f = *frequency, in hertz*
ϕ = *phase angle, in radians*
t = *time, in seconds*

The above equation must be applied with great care, for various complexities underlie its simple formulation.

Digital Data Modulation

Modulation by a digital data information source is quite different from the analog cases illustrated above. Many varieties have been used; some primary ones are shown in Figure 15-4.

Here a digital stream of binary "1" and "0" bits is shown, with the resulting modulated carriers from AM, FM, and PM. The carrier phase for FM is inverted for a "0" as compared to a "1". Such a concept is widely used in modern digital transmission systems.

In the constant effort to pack more information into a given bandwidth, the two-level input (0 and

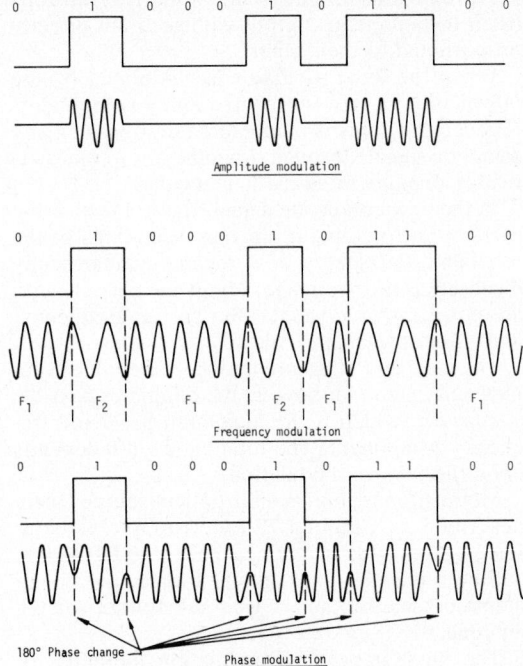

Fig. 15-4. Modulation methods for digital data.

1) may be converted to a multi-level equivalent. In general, n bits can be put into a given time slot (called a baud) by describing the n bits by 2^n levels. Thus, an 8-level signal describes 3 bits.

Power Amplifier

The power amplifier (PA) raises the modulator output to the power level required to meet system requirements. For example, a typical modulator may have an output of 10 milliwatts, but 10 watts to the antenna may be required. Then the PA power gain must be 1,000. It may also include frequency-multiplier circuits to raise the modulator output to a higher portion of the spectrum. In this case, all frequency and phase components of the spectrum are multiplied, together with the carrier frequency.

Frequency-selective filters are included in the PA to preclude spurious transmission on other than the desired frequency. When a PA must provide linear amplification, as for AM, it inevitably operates at a low efficiency as perhaps 30%, as contrasted with 70% when using nonlinear PA's for FM and PM.

RECEIVING

As shown in Figure 15-1, the process of receiving the data or imagery begins with the antenna.

This being a terminal station, it most probably will have a large high-gain antenna, thereby maximizing the signal power level from the remote satellite or aircraft.

The receiver design will indeed utilize the most modern techniques to reject unwanted inputs, such as adjacent frequencies. It will be of a low-noise design, as will be described below.

Output Circuit

To complete this chain of step-by-step functions, the communication process concludes with reception at the base station, and then preparation of the received signal for interpretation and analysis.

Output circuits may be the inverse of the signal conditioning circuits at the transmitting station. They serve to convert the particular receiver outputs to the format required by the succeeding stages, beyond the communication system itself.

THE RADIO/OPTICAL SPECTRUM

In recent decades, both optical and radio activities have expanded greatly, so that their technologies are now approaching each other in the overall spectrum. This undefined area between optical and radio techniques is constantly shrinking, and may now lie somewhere between 300 GHz to 3 THz (1 mm to 100 μm), an expanse unbounded from either direction.

The techniques of optics and radio have already drawn upon each other in the laboratory. It can be expected that this trend will not only continue, but that it will be accelerated. Optics and radio may each continue from their own bases, but each will show increasing reliance upon the other discipline.

HISTORICAL PERSPECTIVE

The optical world has had an essentially unlimited frequency band in which to expand both upward and downward, and thereby to permit the discovery and exploitation of entirely new capabilities.

On the other hand, radio communication has an insatiable appetite for frequency, but not for wavelength. The entire spectrum from 100 kHz to 0 Hz can accommodate only 30 voice channels, although the wavelength is increasing indefinitely.

Transmission of a given amount of information at a certain rate demands a particular bandwidth, which assumes that more and more hertz are available as communications requirements expand. The definition of information as a product of bandwidth and time is a result of communication theory, which is a basic restraint upon the utilization and availability of communication channels.

Therefore radio developed into ever-higher frequencies, from initial emergency maritime usage, to standard broadcast, then to long-distance short waves, followed by the impetus of radar and navigation for shorter wavelengths and higher frequencies (to accomplish precision location-fixing), and now to operate in space systems.

Frequency Descriptors

Because of the particular techniques most useful and available, optical technology has progressed with wavelength as its basic descriptor; conversely, radio technology has developed with frequency as its more usual unit. Optical technology rarely uses frequency in describing an EM wave. Frequency predominates in radio practice.

The one generally accepted breakdown of the radio spectrum into bands is based upon a decade division of the total region usually considered as radio frequencies. Its wide acceptance has been achieved through adoption by the International Telecommunications Union. These bands are shown in Table 15-1.

Detailed allocation of the radio frequency spectrum is accomplished through international agreements. In this way, very good exploitation of the radio bands is established to avoid undue interference among the radio services. The most recent of these allocations was determined at the 1979 Geneva plenary session of the World Administrative Radio Conference.

There are exceptions when the radio community does express itself in wavelength:

Two terms are widely used: (1) *Microwaves*, a term generally used to denote wavelengths shorter than, roughly, 10 cm, or 3 GHz, but greater than 10 mm or 30 GHz and (2) *millimeter waves,* the term used for wavelengths shorter than 10 mm, or 30 GHz.

TABLE 15-1
Frequency Band Designation of International Telecommunications Union

Abbrev.	Frequency, f	Wavelength, λ	Name
VLF	3– 30 kHz	100– 10 km	Very Low Frequency
LF	30– 300 kHz	10– 1 km	Low Frequency
MF	300–3,000 kHz	1,000–100 m	Medium Frequency
HF	3– 30 MHz	100– 10 m	High Frequency
VHF	30– 300 MHz	10– 1 m	Very-High Frequency
UHF	300–3,000 MHz	1,000–100 mm	Ultra-High Frequency
SHF	3– 30 GHz	100– 10 mm	Super-High Frequency
EHF	30– 300 GHz	10– 1 mm	Extremely High Frequency

OPTICAL COMMUNICATION

There is no clear demarcation between radio and optical waves, but the optical may be considered to be shorter than the far-infrared (1 mm). As to modulation, be it analog or digital information, (AM) is used almost exclusively, for simplicity. FM may be received through use of optical filters (to provide frequency/wavelength discrimination) as the sensing element. PM is very difficult to accomplish at optical wavelengths because phase coherence at such extreme frequencies is impractical.

The greatest disadvantage, of course, is the vulnerability of optical wavelengths to the weather vagaries of the atmosphere.

An incidental advantage of optical communication as contrasted to radio is that no operating or station licenses are required from the Federal Communications Commission (U.S.) for use of optical wavelengths, with consequent elimination of the serious problems of time, expense, and frequency coordination when obtaining a license. The feature of the optical spectrum may therefore be very attractive, especially for line-of-sight special and private services.

Spectrum Utilization

The broad radio spectrum from VLF to EHF has widely disparate propagation characteristics, equipment requirements, and historical development. Therefore, through evolution and regulations, particular services have gravitated (or have been legislated) to the portions of the spectrum most suited to their needs and capabilities. The paragraphs which follow will describe this growth of spectrum utilization, with emphasis on communication of imagery. It will be seen that as frequency increases, the environment for image transmission becomes better. Still, it is necessary to survey the whole radio spectrum in order to place all capabilities and limitations into proper perspective.

Very Low/Low Frequencies

The Very Low and Low (VLF/LF) bands (3 kHz to 300 kHz) are not used for image transmission because of their severe bandwidth limitations.

Medium Frequencies

The Medium Frequency (MF) band (300 kHz to 3 MHz) is similar to the VLF/LF band in its propagation. It is dominated by the AM standard broadcast band, which extends from 0.54 to 1.60 MHz in the United States. There are maritime and special services above and below the broadcast band. Its upper portion (2–3 MHz) has been used for low-quality imagery, such as weather maps.

High Frequencies

Propagation at the High Frequency (HF) band (3 to 30 MHz), via the ionosphere, is very effective. This is the *Shortwave Band*, which provides long-range performance with simple equipment and antennas. It serves a great variety of communications users for all types of information, but unfortunately has some disadvantages, which are: (1) the great frequency congestion caused by far too many users for the spectrum available, thereby creating severe interference, and (2) the changeable nature of the ionosphere, which exhibits its greatest and most rapid variations at HF, and which often requires changes in frequency to maintain service.

The capability inherent to HF thus produces excessive interference, as undesired signals may occur because of a change in the ionosphere, or by another signal suddenly appearing as a result of a change in its operating frequency. These problems are only moderately controlled by choice of frequencies and by directive antennas. HF is the lowest frequency band in which antenna gain becomes significant; 10 and even 20 dB are readily achievable. Antenna efficiences are usually high, over 90 percent.

Propagation often goes by two or more paths as reflections from transmitter to ionosphere to ground to ionosphere, etc., take place before receipt at antenna. In addition to fading, signals may be *stretched* or *smeared* up to 2 to 4 msec, which causes *ghosts* to appear in images. Multipath transmission, combined with lack of frequencies, has limited image transmission to low-speed service for news photos and weather maps.

In summary, ionospheric propagation is the rule in the HF spectrum, and therefore communications are subject to a variety of severe natural and manmade disturbances.

Very High Frequencies

At the Very High Frequency (VHF) band (30 to 300 MHz) and above, the ionosphere no longer has significant effect, since ionization is ordinarily insufficient to reflect these frequencies, and transmission may be considered as *line-of-sight*. However, VHF and other radio systems are less restrictive than truly optical ones, and discrepancies from line-of-sight can be accommodated. One approximation is a four-thirds earth diameter, which describes propagation over an earth which is more nearly flat, because of the effects of atmospheric refraction, yielding greater transmission ranges than might be expected because of the true geometry of the link.

An exception should be noted on ionospheric operation: there is useful medium distance service (1,000 to 2,000 km) called *ionospheric scatter*, up to about 60 MHz, based upon ionization found at the lower levels of the ionosphere. Another mode is from the ionized trails produced by meteors entering the upper atmosphere, leaving a transitory (1 second) propagation path; there is limited practicality to this mode because of the large but random incidence of meteors. There are also temporary paths called *sporadic E*, from temporary ionization in the E layer (100 to 120 km) of the ionosphere.

VHF is a medium in which many services are accommodated under very stable conditions; broadcasting and general services share VHF.

Image transmission is a major occupant of this portion of the spectrum. In the U.S., television channels 2 through 13 are heavy users of the spectrum, using a total of 72 MHz, with 6 MHz allocated to each channel. More than 4 MHz is used for image information, while the rest of the 6 MHz is used for audio and guardbands to protect adjacent channels, because of its type of modulation.

Another domestic consumer, FM audio broadcasting, occupies the spectrum from 88 to 108 MHz in the United States.

Ultra High Frequencies

The primary points of the discussion above on VHF also apply to the Ultra High Frequency (UHF) band (300 to 3,000 MHz, or 0.3-3 GHz).

Above the HF band, antenna gain becomes significant and practical; as frequency is raised through the VHF, UHF, and higher bands, the antenna characteristics play an increasingly important role. Not only is more power gain available, but side lobes become smaller to reduce unwanted pickup of spurious signals and noise. *UHF is perhaps the most benign and stable segment of the spectrum.* Its position in the spectrum places it above the area of erratic ionospheric effects, but below the frequencies where local weather has significant effects.

Television is a major occupant of the UHF band. Channels 14 through 83 are assigned 470 to 890 MHz in the United States.

Super High Frequencies

The general trends beginning with VHF and UHF continue as frequency increases into the Super High Frequency (SHF) band (3 to 30 GHz), or *microwaves*.

In the SHF band, frequency congestion in point-to-point service is effectively reduced through careful frequency and geographical planning. This point-to-point relay service is used very heavily in image (television) communication. It is based largely upon the very narrow beamwidths available from antennas of one to three meters aperture. The community is extensively utilizing this band for the many communications, TV and scientific orbiting satellites.

Extremely High Frequencies

The same comments as for SHF apply more emphatically as frequencies/wavelengths in EHF (30 to 300 GHz) approach the optical region. The wavelengths covered (10 to 1 mm) give rise to the term *millimeter waves* for this band.

EHF is more susceptible to atmospheric effects than are lower frequencies. A very severe molecular absorption occurs at about 60 GHz (5 mm); absorption there is essentially total.

Even this outage can be put to an advantage, however. There are occasions when interception of, or interference with military communications must be prevented. This comes about naturally in the case of satellite-to-satellite links by using these absorption frequencies. The signals are immune from ground-based *intercept* or *jamming*. This service is called *Data Relay Satellite,* utilizing high-altitude satellites for relaying information around the earth, as planned for certain advanced imagery systems. A secondary advantage is that there is no reflection from the earth as an interfering multipath signal. EHF is still a highly experimental band for which fundamental propagation data are being gathered. Its great potential lies in the vast expanse of frequencies available for broadband services.

COMMUNICATION SYSTEM DESIGN

The design of an image communication system must include the considerations presented above, in the communication process and functions, as well as spectrum characteristics. However, these are rather broad criteria.

Therefore, this section emphasizes the more particular factors bearing upon the design of a remote-sensor communication system. The major headings that follow are:

- Free-Space Propagation
- Atmospheric Effects
- Lenses and Antennas
- Noise
- System Power Budget

FREE-SPACE PROPAGATION

This description of EM propagation for communications will begin with the simplest case, that of transmission through a vacuum, and upon this basis the more complex cases encountered in practice will be developed.

The propagation characteristics of EM waves in a vacuum are independent of frequency and wavelength. Since this is true, it is especially unfortunate that optical and electronic technologies have so completely evolved into separate ways of describing the same basic phenomenon.

Decibel Notation

In discussing EM wave transmission, one encounters extremely large numbers. A shorthand is therefore necessary and expedient in expressing equations and the magnitudes of the variables (power, gain, frequency, range) involved. As an example consider the shorthand decibel notation applied to power.

Conversion for *power* ratios is often expressed

in *decibels*, (dB), defined as

$$dB = 10 \log W_1/W_2, \qquad (15\text{-}4)$$

where the powers W_1 and W_2 are to be compared with each other. As with many physical phenomena, the ratios become very unwieldy unless converted into logarithmic equivalents, such as decibels.

The notation of Eq. 15-4 is illustrated by tabulating the decibel equivalents of several ratios of power comparison:

W_1/W_2	dB
1	0
2	3
10	10
100	20
1,000	30
1,000,000	60
0.5	− 3
0.1	−10

In the examples above, W_2 might be considered as the base to which W_1 is compared. Sometimes certain levels of W_2 are especially useful, such as the watt and milliwatt. These may be expressed as dBW and dBm, respectively. Thus:

$$1 \text{ watt} = \quad 0 \text{ dBW} = 30 \text{ dBm}$$
$$1 \text{ kW} = 10^3 \text{ watts} = \quad 30 \text{ dBW} = 60 \text{ dBm}$$
$$1 \text{ mW} = 10^{-3} \text{ watt} = -30 \text{ dBW} = \quad 0 \text{ dBm}$$

In this manner, very large ratios can be accommodated in an economical expression.

Transmission Path Loss

The utility of this decibel notation will be seen in the simple expression given by Eq. 15-12 obtained for calculating transmission path loss, as discussed below.

Distance Effect

The geometry of a transmitting and associated receiving station determines some of the salient characteristics of radio and optical propagation, especially attenuation and path delay. All propagation, especially attenuation and path delay. All propagation effects are deduced from the nature of an *isotropic* transmitting source of power W_t watts. Since such a nondirectional source uniformly illuminates the surface of a sphere centered about this point, and since the area of this sphere is $4\pi R^2$, where R is the radius in meters, the flux or flow at any point on the sphere is

$$P = W_t/4\pi R^2, \qquad (15\text{-}5)$$

where:

P = *power density, watts per square meter.*

That is, the received power varies inversely as the square of the transmission distance. Note that this does not represent any real loss of radiated power, such as loss in some absorptive medium, but is just the case that the total power is being spread over a larger *surface* as distance increases. Note that this loss is independent of frequency.

Incident Power Flux

Now an apparent violation of this rule will be introduced. That is, wavelength or frequency will enter into the calculations of space loss, in seeming contradiction of basic principles. Actually there will not be such a violation as appears on the surface. Rather, a great convenience can be effected toward calculation of path loss when wavelength or frequency is used as a working tool in practical cases.

To develop this concept, first consider the sequence of primary components and functions which comprise radio transmission: The transmitter, of power out W_t watts at frequency f Hz (wavelength λ meters); the transmitting antenna, of power gain G_t; transmission of the EM wave through space, with resulting path attenuation over the distance R, in meters; the receiving antenna, of power gain G_r and effective area A_r, in meters^2, with intercepted power W_r, in watts.

The incident power flux at the receiving antenna, P_r in watts per square meter, is

$$P_r = \frac{W_t G_t}{4\pi R^2}, \qquad (15\text{-}6)$$

the power intercepted by the receiving antenna, W_r watts, is

$$W_r = P_r A_r = \frac{W_t G_t A_r}{4\pi R^2}, \qquad (15\text{-}7)$$

now the effective area of the receiving antenna can be replaced by its equivalent in power gain G_r and wavelength λ,

$$W_r = \frac{W_t G_t}{4\pi R^2} \times \frac{G_r \lambda^2}{4\pi}. \qquad (15\text{-}8)$$

The equation above is converted into frequency through the relation $\lambda = c/f$ to yield

$$W_r = \frac{W_t G_t G_r c^2}{(4\pi)^2 \, R^2 \, f^2}. \qquad (15\text{-}9)$$

Power Ratio

The feature of primary importance here is the ratio of the transmitted-to-received power:

$$\frac{W_t}{W_r} = \frac{(4\pi)^2 \, f^2 \, R^2}{G_t G_r c^2}. \qquad (15\text{-}10)$$

This ratio can be considered the free-space loss between transmitting and receiving antennas. The terms G_t and G_r have been carried along to this point to illustrate the practical case in which one or both may be large, and therefore most important to overall system performance. Taking G_t and G_r equal to one, a more compact and useful expression can be developed for path loss computations:

$$W_t/W_r = 1757\ R^2 f^2. \qquad (15\text{-}11)$$

For greater convenience, the equation can be converted to decibel notation, using α as the symbol for free-space attenuation:

$$\alpha = 32.4 + 20 \log R + 20 \log f. \qquad (15\text{-}12)$$

where

α is in dB, and is free-space attenuation,
R is in km,
f is in MHz.

This is the most useful expression for calculating free-space propagation loss (attenuation). It is readily converted to range R in either statute or nautical miles, as:

$$\alpha\ (statute\ miles) = 36.6 + 20 \log R_{s.\mathrm{mi.}}$$
$$+ 20 \log f \qquad (15\text{-}13)$$

or,

$$\alpha\ (nautical\ miles) = 37.8 + 20 \log R_{n.\mathrm{mi.}}$$
$$+ 20 \log g. \qquad (15\text{-}14)$$

To convert from MHz to GHz, add 60 dB to the constant, thus changing to 92.4, 96.6, and 97.8 dB

for kilometers, statute miles, and nautical miles, respectively.

The determination of α from range, frequency, and antenna gains represents the first step in developing the *power budget* for a communication link, a concept which will be discussed below.

Figure 15-5, Sams (1968), is a nomogram for easy determination of path loss. The dashed line shows the case for 48 km and 5 GHz: the loss is 140 dB.

Reception

The receiving antenna is a significant part of an Earth terminal. In practice, the primary limiting factor is cost as higher performance is sought.

The ground stations of satellite-to-earth communications systems incorporate very large-aperture antennas, up to 30 meters or more in diameter, and therefore of high gain and narrow beamwidth (Dawson, 1962). To follow a traversing satellite (as contrasted to a so-called geostationary one) such antennas are often aimed by a computer-generated program to follow the particular orbit.

Under the topic of "noise," discussed below, it will be seen that the receiving antenna is a major factor in controlling system noise. The signals from the receiving antenna are applied to the receiver, which includes highly-selective filters to reject extraneous frequencies. Here they are am-

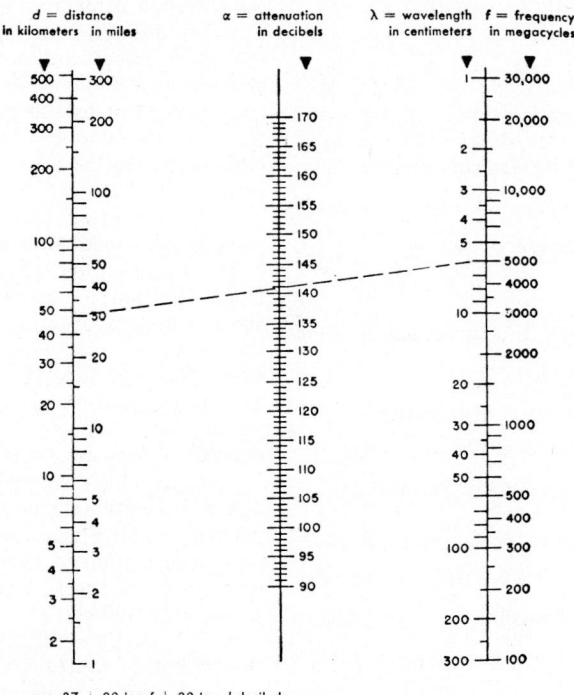

$\alpha = 37 + 20 \log f + 20 \log d$ decibels

Example shown: distance 30 miles, frequency 5000 megacycles; attenuation = 141 decibels

Fig. 15-5. Nomogram showing the relationship of free-space path attenuation (α) to distance and wavelength/frequency.

plified many times to compensate for the severe transmission path loss. The receiver *demodulates* the received signal; that is, it strips the information content from the radio-frequency input. Under desired working conditions, the output is essentially the same as the input to the modulator at the transmitting station. As with the receiving antenna, the receiver's noise characteristics are of primary importance.

Effective Radiated Power

In remote sensing systems, the distant platform is operating under severe constraints in size, weight, and prime power. Thus the system design must set a balance between PA output and antenna gain.

For satellite sensing station, it is advantageous to increase PA output to the 1 to 10 watt level (depending on frequency), instead of concentrating on antenna gain.

A widely-used measure of the transmitting process is *effective radiated power*, (ERP). *This is the power output from the transmitter, as augmented by the gain of the transmitting antenna;* there must be deducted some relatively small amounts for transmission losses between the PA and the antenna, and for misalignment or other circuit degradation.

For a typical case, if the PA output is 5 watts (7 dBW = 37 dBm), losses are 1 dB, and transmitting antenna gain is +3 dB, then:

$$\text{ERP} = 37 - 1 + 3 = +39 \text{ dBm.}$$

Path Delay

Path delay, or propagation time t, is due to the finite velocity of EM waves, described below as c, with a value of 3×10^8 meters per second. Thus, path delay is given by

$$t_{\text{sec}} = \frac{R}{c} = \frac{\text{distance, meters}}{3 \times 10^8} \ . \quad (15\text{-}15)$$

Some practical space EM propagation time examples are:

Sun-to-Earth: $t = \dfrac{93 \times 10^6 \times 1.61 \times 10^3}{3 \times 10^8}$

$= 500$ seconds or 8.3 minutes,

Earth-to-Moon: $t = \dfrac{2.4 \times 10^5 \times 1.61 \times 10^3}{3 \times 10^8}$

$= 1.3$ second,

where:

$1.61 \times 10^3 = $ *conversion factor, miles to meters,*

$93 \times 10^6 = $ *earth-sun distance of* 93,000,000 *miles,*

$2.4 \times 10^5 = $ *earth-moon distance of* 240,000 *miles.*

The importance of these path delays to remote sensing is that feedback control around the entire loop from earth-to-subject-to-earth must be a slow process for the longer communication links. As dramatically illustrated in the Apollo moon explorations, about 2½ seconds, plus human and machine reaction time, is required to sense and control moon-situated sensors through a feedback loop.

ATMOSPHERIC EFFECTS

The various *components of the atmosphere* cause profound effects on the signals transmitted through it, from the lowest radio frequencies to the shortest optical wavelengths. Atmosphere, in this context, is considered in its broadest sense, that of the total gaseous envelope surrounding the planet; therefore it encompasses the troposphere, stratosphere, ionosphere, exosphere, etc. As this subject is of great scope and complexity, only limited descriptions of the phenomena are given here. However, some significant points are briefly introduced:

1. The level of signals received, as calculated for propagation in a vacuum, is the most optimistic level, and any atmospheric effects can only degrade performance below the free-space figure.

2. Atmospheric conditions can be very dynamic in most parts of the world, and therefore transmission is far from being at a static level.

3. To spacecraft or high-altitude vehicles, atmospheric effects are most pronounced with passage through greater density and distance. Thus, transmission at low angular elevations (close to the horizon) is doubly affected by the atmospheric conditions. On the other hand, vertical transmission suffers the least losses.

4. Transmission loss increases with shorter wavelengths under severe weather conditions.

The Earth's atmosphere comprises mostly nitrogen, oxygen, oxides of carbon, inert gases, and small amounts of other constituents. Each of these may have its particular effects on transmission, often as a function of frequency. They may exhibit molecular absorption, or they may dissociate into ions and free electrons; this is the essence of ionospheric properties.

The Sun's high-energy radiation on the Earth's upper atmosphere causes formation of ionized particles and free electrons, the process of *ionization*. These ionized particles are layers (designated D, E, F_1 and F_2 layers) at altitudes between about 50 to 400 km.

As the irradiation from the sun varies, so does the extent of ionization. Very large variations occur, based upon diurnal, annual, and 11-year sunspot cycles, as well as from auroral and solar storm conditions. Geographical latitude is a factor, as is the geomagnetic field which affects EM propagation through the medium.

The usual nature and function of the ionosphere is that of a gigantic refractor or reflector. Up to some maximum frequency for each path, the wave will be returned from a spacebound course back to the Earth's surface, but higher frequencies are not so returned. The maximum frequency is usually around 30 to 50 MHz, depending upon a host of factors.

Thus, the ionosphere does not act as a shield around the Earth for higher frequencies, but it does at lower ones. This is fortunate, since nearly all of the remote sensing applications utilize frequencies above 100 MHz in the EM spectrum.

Atmospheric transmittance to the surface is likewise highly variable with wavelength. Figure 15-6 illustrates the concept of atmospheric *windows*, or points of high transmittance factor as a function of wavelength. For further details see Chapter 5.

Scatter is a term often applied to nontypical modes of propagation which can extend transmission distance by acting as reflectors or refractors. *Ionospheric scatter* is based on patches of ionization found at the lower levels of the ionosphere. *Tropospheric scatter* is based upon anomalous air masses; they can provide a propagation mode by refraction through or around their volumes.

Fading

More than one effective path of communication frequently exists between two points; this is called *multi-path transmission*. It may be the result of the complex modes of ionospheric transmission, in which, for example, a simultaneous set of two and three path (or *hop*) reflections between sky and ground may occur. The high directivity of the usual antenna reduces the probability of multipath transmission of microwaves at high elevations, as used in space systems. However, low (essentially zero) elevation angles are used in point-to-point

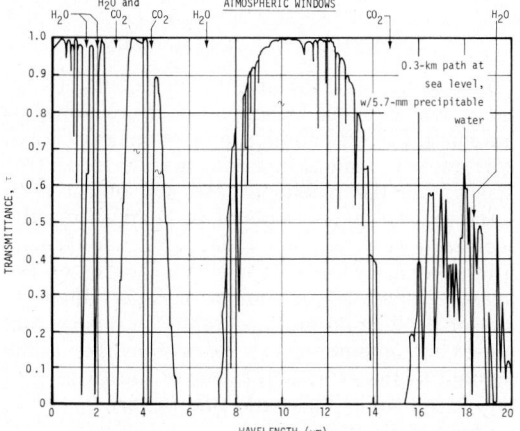

Fig. 15-6. Atmospheric transmittance as a function of wavelength.

services, and reflection from the ground or other obstacles creates multipath. It is also caused by many other physical configurations, such as the transitory passage of aircraft.

Multipath signals, which consist of a pair or more of signals, have unknown relative amplitudes and phases between them because the geometry is uncontrolled. These ratios change in time with atmospheric and ionospheric conditions. The composite of two or more signals of this nature is one which can change widely, when sensed at the receiving terminal; nearly complete cancellation sometimes results. When their relative phases first aid and then oppose each other, as they do with a varying transmission medium, a *cyclic* change in their composite amplitude results: this phenomenon is known as *fading*.

Two other multipath effects may be significant: (1) The time *duration* of signals can be increased up to the amount of their difference in path delay; this may be troublesome to digital data transmission; and (2) analysis of the combination of signals arriving at separate time reveals that the path is frequency sensitive; depressions from a flat frequency response will appear (non-uniform response to frequency).

Attenuation

Atmospheric effects exhibit both fixed and variable *attenuation factors*. The fixed effects are the result of molecular absorption, while the variable effects relate to local weather conditions. Atmospheric attenuation is greatest at low-elevation angles (paths close to the horizon), essentially because weather vagaries are more pronounced. In addition, the path through the atmosphere is longer traversing a greater amount of dense air than for high-elevation angles (toward the vertical) which suffer less from both distance and density of the atmosphere.

Both of these atmospheric attenuations (fixed molecular and variable weather) are in contrast to "attenuation" in free space, which is solely caused by the spreading of an EM wave over a larger volume with increased distance. There is no energy loss in the free-space case, but there is energy loss in the cases of atmospheric attenuation.

Molecular Absorption

The primary losses caused by atmospheric components are related to the electric dipole moment of the water-vapor molecule, and to the magnetic dipole moment of the oxygen molecule. Other molecular losses are relatively insignificant.

Figure 15-7 illustrates that molecular attenuation is very frequency-dependent for these two absorption components. There is a strong trend to greater attenuation with increasing frequency. Near 60 GHz a complete loss of communications may occur in many practical cases.

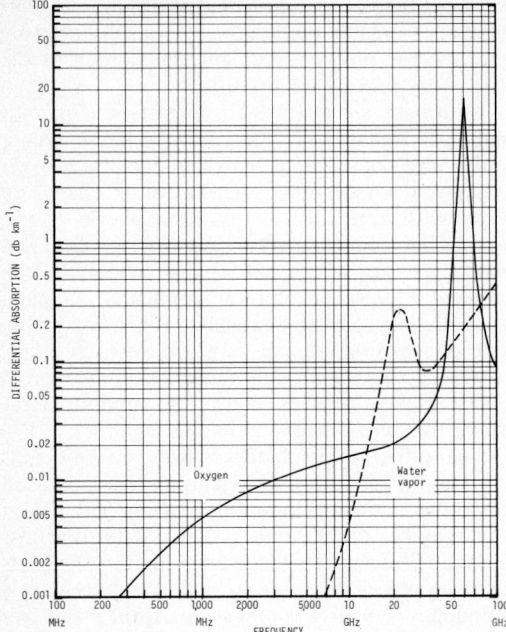

Fig. 15-7. Water vapor and oxygen molecular attenuation in atmosphere.

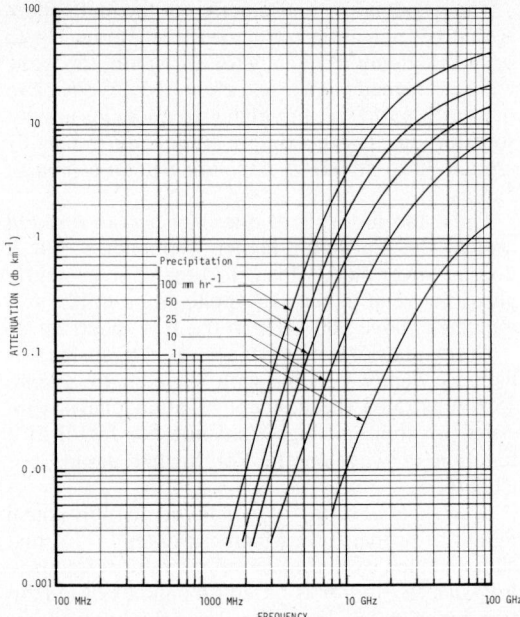

Fig. 15-8. Precipitation attenuation in atmosphere.

Precipitation

Unlike the fixed loss caused by molecular absorption, precipitation losses are wholly dependent upon the ambient weather conditions. The loss mechanism operates because small drops of water scatter the energy which impinges upon them. The result is a diminution of signal power in the desired direction. Although not precipitation, fog in various degrees of density causes the same scattering phenomenon, and somewhat similar losses as with precipitation.

The losses due to precipitation are shown in Figure 15-8. Attenuation also increases with frequency as for molecular absorption, but the resonant peaks are absent.

The system design must allow for the severe attenuation which may occur during variable weather conditions. The incidence of fog, rain, and other atmospheric phenomena throughout the world is highly changeable, and therefore no fixed margin or reserve can be stated. However, the safety factor in system power level for a station in Alaska will vary considerably from one in the Sahara Desert.

Optical Attenuation

Severe molecular absorption also occurs in the optical band. Weather effects are apparent to the casual observer and usually have more severe results in the optical band than for even the shortest radio waves.

Refraction

The density of the atmosphere, and therefore its index of refraction, η, normally decreases with altitude, so EM waves tend to *bend* to conform to the earth's surface. Total conformation would be equivalent to a flat earth. The amount of bending is somewhere between a flat and a true earth in its apparent propagation, depending upon local conditions. An intermediate value, often used as a compromise, is four-thirds of the earth's true radius. This permits calculations of values which are closer to the values actually encountered. It should be noted that other factors may be more appropriate; for example, temperature inversions may reverse the trend to longer paths due to refraction.

LENSES AND ANTENNAS

One of the most important components of a remote sensor is the collecting aperture (lens, or antenna). Lenses in optics, and antennas in radio, perform very similar functions in their respective spectra. In the receiving case, the object is to intercept the largest amount of energy from the incident EM wave. This is accomplished through a large-area lens or mirror which focuses on the sensor a much larger amount of energy than the sensor itself could have collected. In this respect lenses and antennas achieve *power* gain, G, in the amount of the ratio of the lens or mirror area to the sensor area. This ratio often exceeds 10^{10} in optics and 10^4 in radio.

The same considerations apply in the transmitting case, in which the purpose is to collimate a wide-angle source into a narrow, intense beam.

The conditions of transmitting and receiving, and of optics and radio, have similar fundamental derivations. Thus even the equations which describe their performance have the same appearance, although different nomenclature will be encountered.

However, there is a basic difference between optical lenses or mirrors and radio antennas. That is, while the optical component accepts EM radiation and emits EM in some desired geometrical pattern, and the antenna also provides geometrical shaping, the antenna additionally acts as a *transducer.* As contrasted with a lens or mirror, which has the relatively simple function of EM radiation in—EM radiation out, an antenna system makes a conversion between an electrical signal and a radiated EM wave. The electrical signal is at the same frequency, and has the same information content as the EM wave it produces. Depending upon the nature of the antenna, the efficiency of the conversion from electrical to EM radiation may range from a few percent to 90 or greater percent; for the types of antennas considered here, the efficiency is usually 50 to 60 percent.

Despite this basic difference, both optical and radio components can receive similar treatment in systems design. While each is treated in turn, it will be seen that even the same equations apply, with some modifications.

OPTICAL RESOLUTION

For purposes of this discussion, the resolution of an optical system is defined as the minimum angle between two point sources at which these can be separated, i.e., recognized as a pair rather than one source.[1] The classical resolution equation from optics is

$$\theta = 1.22\lambda/d \qquad (15\text{-}16)$$

where:

θ = *resolution in radians*
λ = *wavelength,*
d = *diameter of a circular aperture, in the same units as λ.*

(One radian is $360°/2\pi = 57.3°$, or 206,265 seconds of arc.)

This point of *seconds of arc* is emphasized because it illustrates the tremendous angular selectivity yielded by even moderate apertures at visible wavelengths. For example, an optical objective with a diameter of 30 mm has a possible resolution of 10 seconds of arc, while the 200-inch Palomar (Hale) telescope can relatively resolve 0.02 seconds of arc.

Determination of resolution is often a subjective

[1] For a more complete discussion, see Chapter 6.

matter, however, so that the factor of 1.22 in Eq. 15-16 may diminish to about 1.0 in practical applications.

The comments on lens performance given above are probably well known to the reader, but they have been presented to provide ease of comparison with the radio beamwidth analog, which follows.

ANTENNAS

There is a basic difference between optical lenses or mirrors, and radio antennas. That is, while the optical component accepts EM radiation and emits EM in some desired geometrical pattern, and the antenna also provides geometrical shaping, the antenna additionally acts as a *transducer.* As contrasted with a lens or mirror, which has the relatively simple function of EM radiation in—EM radiation out, an antenna system makes a conversion between an electrical signal and a radiated EM wave. The electrical signal is at the same frequency, and has the same information content as the EM wave it produces. Depending upon the nature of the antenna, the efficiency of the conversion from electrical to EM radiation may range from a few percent to 90 or greater percent; for the types of antennas considered here, the efficiency is usually 50 to 60 percent.

Despite this basic difference, both optical and radio components can receive similar treatment in systems design. While each is treated in turn, it will be seen that even the same equations apply, with some modifications.

Beamwidth

A concept very similar to optical resolution is the definition of the beamwidth of an antenna. *This is the angular separation between the two points, one on each side of the main beam, where the antenna response is one-half the power of the maximum on-axis power.*

Here another unifying feature occurs between optics and radio, because a similar equation applies for beamwidth as for resolution,

$$\theta = \lambda/D, \qquad (15\text{-}17)$$

where:

θ = *beamwidth*
D = *diameter of antenna aperture.*

One of the primary practical differences between optical and radio practical considerations—the great difference in beamwidths—is now apparent.

This may be illustrated by the example of a typical parabolic antenna which is 2.5 m in diameter. At a wavelength of 5 cm, its beamwidth is

$$\theta = 0.05/2.5 = 0.02 \ radian, \ or \ 1.2 \ degrees.$$

By comparison with the optical field, an optical mirror of this size would reduce this beam further by a factor of 10,000 using visible light waves.

Equally important, spurious responses (side lobes) would be correspondingly compressed around the main beam.

Gain

In remote sensing, within the radio spectrum, antenna gain becomes a key consideration. *The gain of an antenna, meaning power gain over an isotropic point source, and designated G, is fundamentally related to its beamwidth.* This is because gain is realized through concentration of total power in the preferred direction. Necessarily, power must be correspondingly diminished in other directions, in favor of the intended direction. Integrated over the entire sphere about the radiator, these losses equal the gain in the preferred direction (ignoring inefficiencies).

The gain of standard antenna configurations is derived in various texts; the calculations are based upon the effective cross-section of a receiving antenna. The gain of a transmitting antenna is taken to be the same as its receiving counterpart, based upon the *theory of reciprocity,* which states that the roles of the antenna, either in transmission or reception, can be interchanged, with its characteristics remaining the same (Blake, 1966).

Parabolic Reflectors

The parabolic reflector type antenna, the most widely used configuration for the microwave re-gion of the spectrum (where λ is around 3 cm) because of its versatility and economy, forms the basis for the following discussion.

The arrangement of this type of antenna is similar to that of a reflecting telescope. But instead of an optical element, a small antenna of minimal gain is placed at the focus to accept the energy intercepted by the parabola in receiving, or to "illuminate" the parabola in transmitting. While the terms *parabola* or *parabolic antenna* are commonly used with no loss of understanding, the reflector surface is of course a paraboloid (parabola of revolution on its symmetric axis).

The gain of a parabolic antenna is

$$G = 0.54 \, (\pi D/\lambda)^2. \qquad (15\text{-}18)$$

The factor of 0.54 is generally accepted in the field as the average value of the efficiency of such an antenna; true values of efficiency vary from about 0.4 to 0.6. Figure 15-9 presents a nomogram for solution of the gain of a parabolic reflector, based on use of the 0.54 factor (see Sams, 1963). The example given on the graph by the dashed line shows that a parabola of 1.8 m has a power gain of 32 decibels at 3 GHz.

One of the highest-gain antennas ever designed and built operates at 35.2 GHz (Dawson, 1962). Its diameter is 8.5 m, and it has a measured gain of 67.4 dB. The theoretical maximum gain for an antenna of this size and frequency is 70 dB. This discrepancy of 2.6 dB corresponds to only 55 percent efficiency.

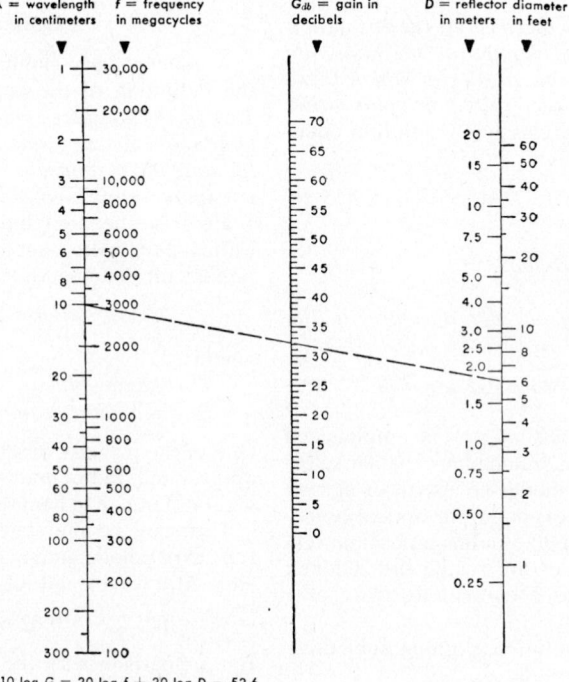

$$10 \log G = 20 \log f + 20 \log D - 52.6$$

Example shown: Frequency 3000 megacycles, diameter 6 feet; gain = 32 decibels

Fig. 15-9. Nomogram showing relationship of the gain of a parabolic reflector to wavelength/frequency and reflector diameter.

By comparison with the optical regime, the 200-inch (5.1 m) Palomar telescope has a theoretical maximum gain of 150 dB, and may realize about 148 dB at 0.5 μm. Thus, the optical instrument considered to be one of the best in the world has a power gain advantage over one of the best radio examples of some 80 dB, or a factor of increase of 10^8 (one-hundred million).

Polarization

Unlike most optical sources and receivers, the radio antenna usually exhibits pronounced *polarization. The plane of polarization of the antenna is defined as that of the electric vector of the EM wave.*

It is necessary that the transmit and receive antennas have nearly similar polarizations for maximum power transfer. The intercepted field strength is proportional to the cosine of the angle by which they differ; therefore, the power is proportional to the cosine squared; thus there is no response when the antennas are at right angles or cross polarized. Some polarization error is acceptable, since the incident electric field strength varies only as the cosine of polarization misalignment. In practice there is always some residual signal, but it may be a signal as much as 40 dB below the signal transmitted and received when there is correct alignment between antennas.

The situation is readily controlled in point-to-point communication between earth terminals, but it becomes serious when the relative polarizations are random, as in the case of an unstabilized satellite transmitting to a ground station.

The solution to this condition is to have either the transmit or the receive stations operate with both of the two orthogonal polarizations, usually termed horizontal and vertical. If the transmit antenna emits both polarizations, the receive antenna needs to accept only one, or vice versa. It must be observed that in this case only half of the transmitted power is useful, since the other polarization is ignored; this results in a loss of 3 dB (half-power). The loss can be overcome by the use of a special technique utilizing *circular polarization*. Because of its specialized nature, the reader is referred to Kraus (1950).

NOISE

To this point the discussion has been devoted only to conveying the input information to a receiving station. This section will introduce the concept of *noise*, which will *compete* against the desired signal and thereby degrade the quality of communications. See also Chapters 7 and 9.

Noise is an all-pervasive phenomenon which is inherent to any electrical circuit, and is also introduced from external sources.

Normally, the absolute magnitudes of signal and noise are not important, but rather their relative values, called the signal-to-noise ratio (S/N), expressed as a power ratio or the equivalent decibel figure. The term $(S + N)/N$ is often seen in the literature. Under poor receiving conditions with $S = N$, or $S/N = 0$ dB, then $(S + N)/N$ is 3 dB greater than S/N. But under more typical and useful conditions where S is 100 times N ($S/N = 20$ dB) then S/N and $(S + N)/N$ are closely identical. $(S + N)/N$ is sometimes more useful in calculations.

Every communication system requires some minimum S/N at its receiver, depending upon the type of modulation, bandwidth and the quality required at the signal output.

There is a theoretical limit by which the transfer of information is limited by noise (*Shannon's Law*). It states that the capacity of a channel is:

$$C = B \log_2 (S + N)/N, \qquad (15\text{-}20)$$

where:

$C =$ *the channel capacity in bits/second*
$B =$ *bandwidth in hertz*
$\log_2 =$ *logarithm to the base 2.*

Typical operational communications are poorer than this theoretical maximum value by some 3 to 10 dB in the factor $(S + N)/N$.

The literature will often show a different usage of the symbol C. There it is used to denote the carrier power of a transmitted signal as received at a distant point. Therefore, it appears as C/N, the equivalent of S/N, (carrier-to-noise, or signal-to-noise ratio, respectively).

Thermal Noise

Any electrical conductor is a source of noise output power, given by (Munford and Scheibe, 1968),

$$W = kTB, \qquad (15\text{-}21)$$

where:

W *is in watts*
k *is Boltzmann's constant,* 1.38×10^{-23} *($J\,K^{-1}$)*
T *is absolute temperature, (K)*
B *is bandwidth in hertz.*

This noise is termed thermal, white, or Johnson (after a pioneer investigator).

The noise power is caused by the random motion of free electrons because of thermal agitation. It produces a voltage at the open ends of any conductor, including resistors. The power is independent of the value of the resistance, so the voltage varies as the square of the resistance. Noise itself is not bandwidth limited.

The temperature emcountered in practice is an *effective* one, T_e, as will be described below. It is frequently compared to a generally accepted reference, the *standard noise temperature,*

$$T_0 = 290 \; K \; (17°C). \qquad (15\text{-}22)$$

The source noise power at 290 K is -204 dB with respect to 1 watt per hertz of bandwidth, or 4×10^{-21} watt per hertz. This is a very small amount of power, but so is the signal level at the receiving antenna. Also, for a wide bandwidth system of 10 MHz, the noise level of -204 dBw is increased by 70 dB to -134 dBw.

Receiver Noise

The effective noise temperature of receivers has been brought down from more than 1,000 K some years ago to much lower values, approaching to within a few degrees of absolute zero. The two major advances which have brought about this notable improvement are the *parametric amplifier* and the *Maser,* (*M*icrowave *A*mplification by *S*timulated *E*mission of *R*adiation).

Parametric amplifiers may have temperatures from 90 to 150 K; when refrigerated with liquid helium, the value is as low as 9 K at 4 GHz. Some of these amplifiers are likely to suffer from spurious frequency responses in addition to the desired signal, which will raise the equivalent temperature by introducing additional noise.

The maser, the precursor of the optical laser, can give the best performance, frequently 5 to 15 K.

One receiver measure of merit is its *noise figure, (NF)*. It is the ratio of its actual noise output power as compared to an ideal receiver at the same temperature and bandwidth. The ratio is always greater than one for a realizable equipment; the larger the noise figure, the poorer is the receiver.

Some confusion has arisen through the years concerning the exact use and meaning of the definition of noise figure. Consequently, it has now been largely supplemented by the term *noise temperature* at higher frequencies, especially for space systems. The relation between these two terms may be expressed as:

$$NF = 1 + T_e/T_o, \qquad (15\text{-}23)$$

where:
    T_e = effective noise temperature of the receiver.

It shows that NF can never be less than 1, no matter how small T_e becomes. The terms *noise figure* and *noise factor* are both found in the literature, and they are synonomous when correctly applied.

Because noise figure is a power ratio, it is often expressed in decibel notation. As an example, if a receiver has an effective noise temperature of 870 K,

$$NF = 1 + \frac{870}{290} = 4, \text{ or } 6 \text{ dB.}$$

Each component of a receiving system adds noise and thereby raises the system temperature. This additional noise increment, in degrees, is proportional to the amount of loss it contributes and to the temperature of the component. For example, a transmission line may be of a significant temperature, 300 K, but if it is 99% efficient

(with a loss of 1%) it will add approximately 3 K to the overall system temperature. This would be very serious to a system with very low noise, but inconsequential to a poor one with higher integral noise.

External Noise

Noise from the antenna is a major contributor to overall system noise. The noise is not primarily caused by the antenna itself, but by noise sources sensed by the antenna beamwidth, from sky or earth. The noise intercepted by the antenna varies widely with frequency, and it comes from a variety of sources, including galactic, ionospheric, and tropospheric sources. In addition, minor antenna lobes pick up noise from the surrounding earth.

A composite of the several contributors is shown in Figure 15-10, Blake (1966). It illustrates that the noise temperature decreases rapidly from 0.1 to about 1 GHz.

The considerable differences in temperature depicted for a given frequency depend upon antenna orientation. At the lower frequencies (100 to 1,000 MHz), the center part of the galaxy causes the higher temperatures. For higher frequencies (10 to 100 GHz), the higher temperatures are present at low elevation angles.

In addition to the noises from natural causes as discussed above, manmade signals, called *interference,* will also occur, either from accidental or deliberate jamming causes.

The incidence of interference is somewhat haphazard at the lower frequencies (below 30 MHz) because of irregular ionospheric propagation. At much higher (microwave) frequencies transmission paths are much more stable, and through planning, licensing, and international

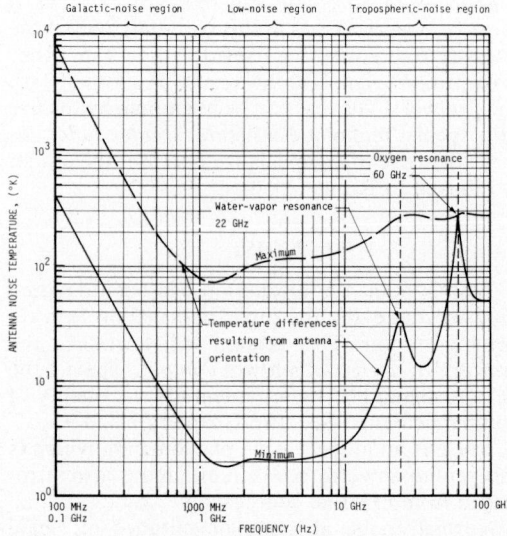

Fig. 15-10. Irreducible noise temperature of an ideal antenna.

agreements, interference is generally well controlled. A major technical factor which assists in the reduction of interference is the trend toward maximum use of highly-directional (very narrow beamwidth) microwave antennas. These antennas optimally reject signals from other than the desired direction.

Man-made Noise

In addition to the noises from natural causes as discussed above, manmade signals, called *interference,* will also occur, either from accidental or deliberate jamming causes.

The incidence of interference is somewhat haphazard at the lower frequencies (below 30 MHz) because of irregular ionospheric propagation. At much higher (microwave) frequencies transmission paths are much more stable, and through planning, licensing, and international agreements, interference is generally well controlled. A major technical factor which assists in the reduction of interference is the trend toward maximum use of highly-directional (very narrow beamwidth) microwave antennas. These antennas optimally reject signals from other than the desired direction.

Optical Noise

The noise of the detector in an optical communication system must be expressed in a much more complicated form than that shown above for a radio receiver. The several types of optical detectors (photovoltaic, photoconductive, and others), depending upon their physical prinicples of operation, require different formulations to describe their performance. Each is chosen for its low noise and sensitivity across the system's EM spectrum. The noise of amplifiers which normally follow the detector is inconsequential as compared to the detector itself.

The performance data for optical detectors are too specialized and complex for inclusion here, but are to be found in Chapter 7, Electro-Optical Remote Sensors with Related Optical Scanners. (See IEEE Proc., Oct., 1970; RCA Handbook, 1968).

SYSTEM POWER BUDGET

The many factors influencing an overall system require judicious compromises and trade-offs among sometimes conflicting objectives and constraints.

Methods of combining the several components of a system in their individual details to describe total performance from input to output is illustrated by a pertinent example in the following section. The example emphasizes the concept of a power budget, which applies to any communication system, but is most often used with space-ground communication.

A power budget is an accounting of the positive and negative contributions of the many steps in the communication process.

TABLE 15-2
Representative Space-to-Ground Link Power Budget. Power Output, 10 Watts. Reference Power, 0 dbm (1 m watt) (−30 dbw).

Satellite Output		
Power output (10 watts)	+40 dBm	
Miscellaneous losses	− 1 dB	
Transmitting antenna gain	+ 3 dB	
Effective radiated power		+ 42 dBm
Transmission Path		
Path loss (500 km, 2.14 GHz)	−153 dB	
Received power		−111 dBm
Receiving Terminal		
Receiving antenna gain	+34 dB	
Miscellaneous losses	− 1 dB	
Receiver noise figure	− 3 dB	
Receiving gain		+ 30 dB
Receiver Useful Signal Level		− 81 dBm
Noise Power (1 MHz Bandwidth)		−114 dBm
Resultant System S/N Ratio		+ 33 dB

The balance sheet gives the output signal-to-noise ratio, which confirms or denies the validity of the system, and it also pinpoints where improvements should be made.

To illustrate the procedure, a power budget for a representative space-to-ground link is given in Table 15-2 for a nominal power output of 10 watts. The more abstruse losses are excluded, but would be important additions to the design and planning of an actual system. In this power budget, the power levels proceed as shown with the flow of the signal. For simplicity, no safety margins have been added for such important factors as low-elevation propagation, antenna ambient noise levels, weather effects, or equipment degradation and detuning with time.

The 33 db output signal-to-noise ratio is more than adequate for any *digital* transmission technique. However, it would yield a noisy image in an *analog* system, thus it must be raised to the 45 to 50 dB level.

A valuable feature of the power budget is in guiding the designer toward realistic improvements. Here it might be out of the question to raise transmit output power or antenna gain, since they impinge upon all aspects of spacecraft design. At the ground terminal, the noise figure cannot be improved more than 2 dB. A higher gain antenna is a direct solution, albeit an expensive one.

Thus some other approaches are required if this is to be a viable design. Two methods could be employed: (1) Reduce bandwidth by slowing the information rate, i.e., by taking more time to scan each image frame, or reduce image resolution; and (2) investigate various modulation methods to improve output S/N, such as wideband FM to exploit its noise-improvement-factor. Perhaps, too much is being demanded of the equipment available. If so, some reduction in system performance must be accepted.

In conclusion, the power budget is the culmination of many factors, but it is only one criterion of a transmission system. It is a necessary but not a sufficient condition for the overall successful design of the Image communication system.

REFERENCES

American Society of Photogrammetry, 1966, Manual of Photogrammetry, third ed.

American Society of Photogrammetry, 1975, Manual of Remote Sensing, first ed.

Aviation Week and Space Technology, 1971 (23 August), Special issue on communications satellites.

Aviation Week and Space Technology, 1972 (31 July), Special report: Earth Resources Technology Satellites.

Bennett, W. R., and J. R. Davey, 1965, Data transmission; McGraw-Hill, New York.

Blake, L. V., 1966, Antennas; John Wiley and Sons, New York.

Dawson, J. W., 1962 (June), A 28 ft liquid-spun radio reflector for millimeter wavelengths; Proc. IRE, v. 50, p. 1541.

Dynair Electronics, Inc., 1968, Video transmission techniques; San Diego.

Electronics Magazine, Sept. 25, 1972.

Greenwood, T., 1973, Reconnaissance and area control; Scientific American (February), v. 228, no. 2, p. 14–25.

Hamsher, D. H., 1967, Communication system engineering handbook; McGraw-Hill, New York.

Jaffe, L., 1966, Communications in space; Holt, Rinehart and Winston, New York.

Jamieson, J. A. et al., 1963, Infrared physics and engineering; McGraw-Hill.

Jensen, N., 1968, Optical and photographic reconnaissance systems; John Wiley and Sons, New York.

Kraus, J. D., 1950, Antennas; McGraw-Hill, New York.

Kruse, P. W., L. D. McGlauchin, and R. B. McQuistan, 1962, Elements of infrared technology; John Wiley and Sons, New York.

Martin, J. 1969, Telecommunications and the computer; Prentice Hall, Englewood Cliffs, N.J.

Mueller, C. E., E. R. Spangler, 1964, Communication satellites; John Wiley and Sons, New York.

Mumford, W. W., and E. H. Scheibe, 1968, Noise performance factors in communication systems; Horizon House-Microwave Inc., Dedham, Mass.

Pandelides, J., et al., undated, the ERTS wideband image communication system; NASA, Goddard Space Flight Center.

Institute of Electrical and Electronic Engineers, 1970 (October), Proc., Special issue on optical communication; pp. 1407–1466.

RCA Commercial Engineering, 1968, Electro-optics handbook, EOH-10; Harrison, N.J.

Sams, H. W., and Company (ITT), 1968, Reference data for radio engineers, fifth edition; Indianapolis, Indiana.

Scientific American Magazine, 1972 (September), Special issue on communications.

Taub, H., and D. L. Schilling, 1971, Principles of communication systems; McGraw-Hill, New York.

Terman, F. E., 1955, Electronic and radio engineering; McGraw-Hill, New York.

CHAPTER 16

Orbital Mechanics for Remote Sensing

Co-authors: KENNETH I. DUCK and JOSEPH C. KING

GENERAL CONTENTS: Discussion of circular orbits; orbital coverage; elliptical and other conic orbits; impulsive and continuous thrusting maneuvers; propulsion systems; maintenance of remote sensing orbits; geosynchronous orbits; sun synchronous orbits; references.

NOMENCLATURE

To conserve and eliminate repetition in text and references, the following symbols, units and names have been used in this chapter.

Symbol	Units	Definition
A	m²	satellite cross section area
a	m	orbit semi-major axis
a_R, a_s, a_w	ms⁻²	radial, circumferential, normal acceleration components
D	s	mean solar day (86400 sec)
D_n	s	nodal day
D_s	s	sidereal day (86164.09 sec)
E	rad	orbit eccentric anomaly
ε	m²s⁻²	orbit energy per unit mass
e	—	orbit eccentricity (dimensionless)
F_s	N	circumferential thrust component
$F_\perp$	N	normal thrust component
G	—	constraint equation in minimization problem (dimensionless)
g	ms⁻²	acceleration of gravity (9.80665m/sec²)
H	rad	eccentric anomaly, hyperbolic orbits
$h_{\odot,\Omega}$	hrs	solar hour angle
h	m²s⁻¹	orbit angular momentum per unit mass
h_s	m	orbit altitude
I_{sp}	s	propellant specific impulse
i	°	orbit inclination (degrees)
J_2	—	earth oblateness coefficient (dimensionless)
J_{22}	—	earth triaxiality coefficient (dimensionless)
M	rad	orbit mean anomaly
m	kg	instantaneous satellite mass
m_o	kg	initial satellite mass
m_p	kg	propellant mass
N	day	coverage repeat period
n	s⁻¹	orbit mean motion (rad/sec)
p	m	orbit semi-latus rectum
Q	—	ratio of orbital angular rate $(\dot{\theta})$ to earth rotation rate relative to orbit plane $(\dot{\Lambda}_n)$ (dimensionless)
q	day⁻¹	orbit revolutions per day
R	—	orbital revolutions per repeat period (dimensionless)
r	m	radius from center of earth to satellite
$r_\oplus$	m	radius of earth (6,378,160m)
r'	km	r in kilometers
S	°	longitudinal interval between consecutive crossings of the same parallel if latitude in the same direction of nadir trace (degrees)
s	°	longitudinal interval between adjacent crossings of the same parallel of latitude in the same direction of nadir trace (degrees)
s'	km	ground track shift from nominal at equator
T	s	orbit period
T_n	s	orbit plane rotation period
t	s	instantaneous time
δt	s	solar time change per nodal day (sec)
U		performance function to be minimized
v	ms⁻¹	instantaneous orbit speed
v'	kms⁻¹	instantaneous orbit speed
v_c	ms⁻¹	circular orbital speed
$\alpha_\odot$	hrs	solar right ascension
β	°	local azimuth (geographic) of orbit (degrees)
γ	°	azimuth of viewing line-of-sight relative to orbit plane (degrees)
δ	°	off-nadir viewing angle (degrees)
ϵ	°	obliquity of the ecliptic (degrees)
η	—	Lagrange multiplier (dimensionless)
θ	rad	orbit true anomaly
Λ	°	earth angular displacement relative to vernal equinox (degrees)
Λ_n	°	earth angular displacement relative to orbit plane (degrees)
λ	°	longitude
λ_D	°	longitudinal displacement of viewpoint from ascending node (degrees)
λ_D'	°	longitudinal displacement of viewpoint from ascending node in nonrotating model (degrees)
$(\Delta\lambda)_R$	°	rotational (earth and orbit plane) contribution to satellite longitude change (degrees)
λ_s	°	longitudinal displacement of satellite from ascending node (degrees)
λ_s'	°	longitudinal displacement of satellite from ascending node in nonrotating model (degrees)
μ	m³s⁻²	earth gravitational constant (3.98601 × 10¹⁴ m³/sec²)
ν	—	trace shift number (dimensionless)
ν_a	—	apparent trace shift number (dimensionless)
ξ	°	angular arc over which thrust is applied (degrees)
ρ	kgkm⁻¹	atmospheric density
τ	day	time

Symbol	Units	Definition
γ	—	direction of vernal equinox (dimensionless)
Φ	°	total angle in orbit measured from ascending node (degrees)
ϕ	°	latitude (degrees)
ϕ_D	°	latitude of viewpoint (degrees)
ϕ_D	°	latitude of viewpoint (degrees)
ϕ_s	°	latitude of satellite (degrees)
ψ	°	angle subtended at earth center by viewing line of sight (degrees)
Ω	°	orbit longitude of ascending node (degrees)
$\Omega_\odot$	year^{-1}	rotation rate of orbit plane relative to sun (revs/year)
ω	°	orbit argument of perigee (degrees)

Prefix		
Δ	—	increment or change (depends upon parameter being measured)

INTRODUCTION

The intent of this chapter is not to provide an all-encompassing treatment of orbital mechanics. Rather, those aspects that apply to remote sensing applications have been selected and are discussed. The primary emphasis is on circular orbits as most remote sensing applications require such orbits. However, for completeness, a brief discussion of elliptic and hyperbolic orbits is provided. In addition, there are discussions of geosynchronous and sun-synchronous orbits that are used for remote sensing applications and of maneuvers, propulsion systems performance, and orbit maintenance.

THE CIRCULAR ORBIT

It is helpful to consider first the circular orbit, rather than elliptic or other conic forms, because the circular orbit is the simplest and most basic type. It is also the most useful for earth-oriented remote sensing because of the uniform viewing that it provides. Geometric relationships, orbit selection, and control are all simpler with circular orbits. Hence the following discussion will refer specifically to circular orbits, although many of the relationships developed are more broadly applicable.

BASIC DESCRIPTION

The circular orbit is specified most simply in polar coordinates (r, θ):

$$r = r_\oplus + h_s \qquad (16\text{-}1)$$

$$\theta = \dot{\theta} t \qquad (16\text{-}2)$$

where

$r_\oplus$ = radius of earth
h_s = orbital altitude
t = time

Note that $r_\oplus$, h_s, r, and $\dot{\theta}$ are all constants. $\dot{\theta}$ is determined directly from the free-fall speed v_c,

i.e., the speed at which the centripetal acceleration $v_c^2/r = r\dot{\theta}^2$ equals the acceleration of gravity (μ/r^2). Thus the angular speed in circular orbit is

$$\dot{\theta} = \sqrt{\mu/r^3} \qquad (16\text{-}3)$$

The corresponding orbital period is $T = 2\pi/\dot{\theta}$, or

$$T = 2\pi\sqrt{r^3/\mu} \qquad (16\text{-}4)$$

ORIENTATION

The orientation in space of a geocentric orbit is usually specified in relation to the earth's equatorial plane and the vernal equinox (γ), as shown in Figure 16-1. The vernal equinox in this sense is the direction in space defined by the earth-sun line at the moment of the familiar vernal equinox, the time of day-night equality about March 21. More specifically, it is the direction of the sun from earth at that moment, when the earth's equatorial plane passes southward through the sun's center (the sun appears to cross the celestial equator moving northward).

The angle between the orbit plane and the equatorial plane is specified by the inclination angle i. The orientation of the node line (intersection of orbit and equatorial planes) is given by the longitude of node angle Ω. It is measured relative to the vernal equinox γ.

The orientation of the orbit relative to the sun is also important in many applications, particularly where solar illumination is needed for sensing. For a given orbital inclination, sun-angle control must be achieved by placing the node line (specifying Ω) at the desired angle to the sun's hour circle. This procedure amounts to specifying the hour angle of the sun $h_\odot, \Omega$, as seen from the orbit node.

In a typical example, illustrated in Figure 16-1, a "2 pm ascending node" is 2 hours or 30° east of the sun's hour circle,[1] or at the ascending node, the hour angle of the sun is 2 hours. In Figure 16-1, the right ascension of the sun is placed at 3 hours (45°) which, with the desired 2 pm (30°) ascending node, places the required nodal longitude (Ω) at 45° + 30° = 75°.

GEOSYNCHRONOUS ORBITS

An important special case in the circular orbit class is the geosynchronous type, frequently used in communications, meteorology, and other applications. The geosynchronous property is obtained by placing the satellite at an altitude such that it revolves at an angular rate equal to the earth rotation rate. In terms of the periods, this condition is

$$T = D_s = 2\pi\sqrt{(r_\oplus + h_s)^3/\mu}$$

$$h_{24} = \left[\mu(D_s/2\pi)^2\right]^{1/3} - r_\oplus \qquad (16\text{-}5)$$

With $\mu = 3.98603 \times 10^{14}$ m^3/sec^2, $D_s = 86,164.1$ sec, and $r_\oplus = 6,378,160$ m, eq. 16-5 gives $h_{24} =$

---

[1] Daily values of the sun's right ascension $\alpha_\odot$ are tabulated in Nautical Almanac Office 1982 (p. C4 ff) along with approximate formulae for computing $\alpha_\odot$ continuously (p. C20).

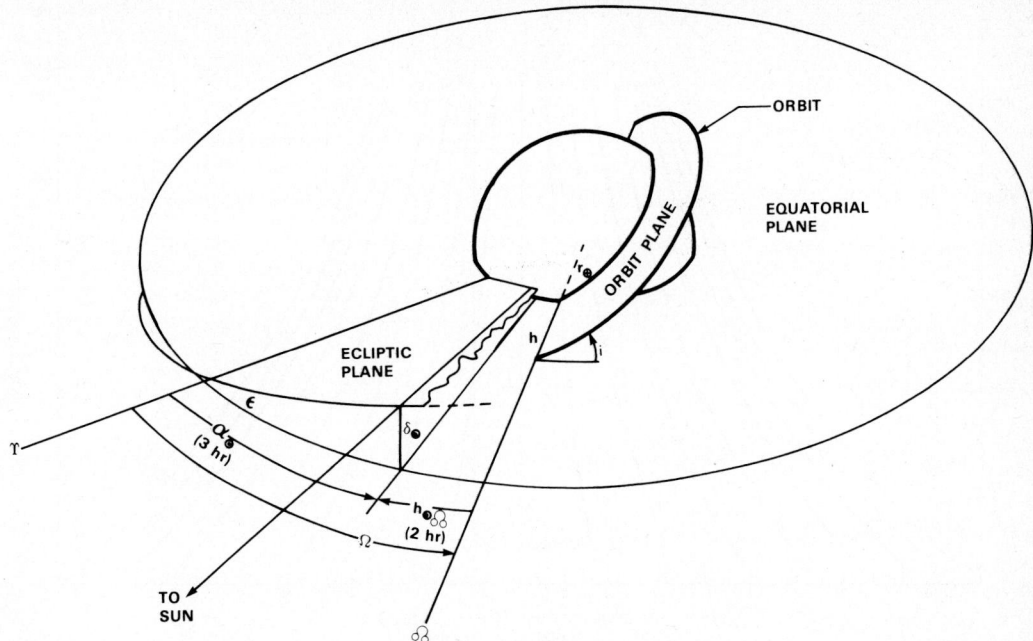

Fig. 16-1. Satellite orbit orientation.

35,786,103 m. (D_s is the sidereal day, the earth's rotation period relative to the vernal equinox.)

The geosynchronous orbit maintains the satellite over a narrow longitude band, producing a figure-8 shaped trace of latitude range $2i$. If $i = 0$, the trace shrinks to a point (in the simplified model—some deviation occurs in practice), and the orbit is called geostationary.

Requirements for periodic orbit adjustment depend on the particular parameters of the application, such as required placement accuracy and duration of service. Adjustment methods are discussed later in the chapter.

ORBIT PRECESSION

The largest variations in orbit orientation are usually due to precession, the rotation of the orbit plane about the polar axis. Precession, which is caused primarily by the earth's oblateness, is usually described as the rate of change of the nodal longitude Ω. It can be approximated by

$$\dot{\Omega} = -\tfrac{3}{2} J_2 \, r_\oplus^2 \sqrt{\mu} (r_\oplus + h_s)^{-7/2} \cos i \quad (16\text{-}6)$$

Earth oblateness enters as the dimensionless number $J_2 = 0.00108263$,[2] the coefficient of the second zonal harmonic of the geopotential.

Equation 16-6 indicates that the nodal rate $\dot{\Omega}$ is a function of the two primary orbit parameters altitude (h_s) and inclination (i). The relationship is plotted in Figure 16-2, showing $\dot{\Omega}$ vs h_s, with i as a parameter.

SUN-SYNCHRONOUS ORBITS

The second nodal-rate scale, $(\dot{\Omega})_\odot$, relative to the sun, is included to provide a convenient solar

[2] Nautical Almanac Office 1982, E83.

correlation for applications where solar incidence angle is important. The widely used (e.g., Landsat) sun-synchronous orbits, for example, are seen to have $(\dot{\Omega})_\odot = 0$, as required to maintain the desired constant node-to-sun angle. Note that the sun-relative scale $(\dot{\Omega})_\odot$ differs from the inertial[3] scale $\dot{\Omega}$ by one rev/year, corresponding to the one rev/year apparent motion of the sun.

ASYNCHRONOUS ORBITS AND LIGHTING VARIATIONS

The nodal day D_n is the earth-rotation period relative to the orbit plane or node. It is expressed most directly as its corresponding rotation rate $\dot{\Lambda}_n$.

$$\dot{\Lambda}_n = \dot{\Lambda} - \dot{\Omega} \quad (16\text{-}7)$$

where $\dot{\Lambda}$ is the inertial earth rotation rate. The usual rate-to-period conversion yields the nodal day,

$$D_n = \frac{2\pi}{\dot{\Lambda}_n} = \frac{2\pi}{\dot{\Lambda} - \dot{\Omega}} \quad (16\text{-}8)$$

Dividing eq. 16-7 by 2π yields the relationship between the corresponding periods

$$\frac{1}{D_n} = \frac{1}{D_s} - \frac{1}{T_n} \quad (16\text{-}9)$$

where D_s is the sidereal (inertial) day and T_n is the rotation period of the orbit plane.

The nodal day enters into the analysis of coverage-pattern formation, discussed in the next section. A closely related quantity, scaled with D_n in Figure 16-2, is the solar time change per nodal day, δt. This quantity indicates the change in solar time, and thus the change in local lighting, at a

[3] Relative to space-fixed reference frame (e.g. star background).

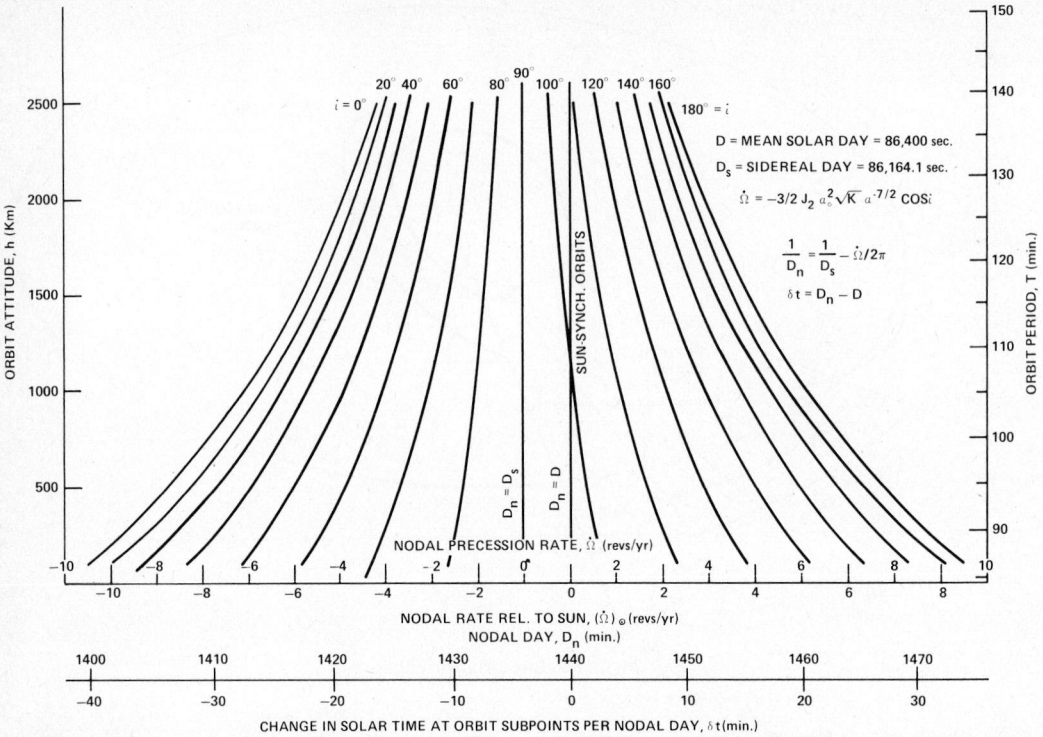

Fig. 16-2. Satellite orbit precession and related viewing parameters.

point on the earth between consecutive passes under the orbit. It is simply the difference between the nodal and solar (D) days:

$$\delta t = D_n - D \qquad (16\text{-}10)$$

Note in Figure 16-2 that, as expected, $\delta t = 0$ when $(\dot{\Omega})_\odot = 0$, i.e., for sun-synchronous orbits.

COVERAGE

The basic function of the satellite in providing coverage—of particular geographic areas on some schedule—indicates the importance of this aspect of remote sensing mission design. Coverage as used here separates naturally into two elements:

1. the nadir trace or ground track of the satellite, determined by the orbit, and
2. the sensor view-area or swath, determined by the orbit and by the field of view and pointing relative to the nadir.

NADIR TRACE

The nadir trace is generated by the combined action of two primary motions:

Orbital (satellite)
Rotational (earth relative to satellite orbit-plane)

The orbital rate ($\dot{\theta}$) is given by equation 16-3. The relative rotational rate ($\dot{\Lambda}_n$) is the difference between the earth's inertial rate ($\dot{\Lambda}$) and the inertial precession rate ($\dot{\Omega}$), given by equation 16-7.

Determination of the nadir trace consists of specifying the geographic coordinates (longitude λ and latitude ϕ) of any point $S'(t)$ on the trace. The analysis is simplified by the fact that the orbital and rotational motions act independently. The latter is a constant-rate rotation about the polar axis, so its contribution is linear (with time) and involves the longitude coordinate only. Specifically, the rotational contribution $(\Delta\lambda)_R$ to the longitude coordinate will be just the relative rotational rate $\dot{\Lambda}_n$ times elapsed time:

$$(\Delta\lambda)_R = \dot{\Lambda}_n t \qquad (16\text{-}11)$$

The "orbital" contribution to longitude is conveniently obtained by analyzing a "nonrotating" model (orbital motion only—no rotation). The resulting "nonrotating" longitude value, added to the rotational term $(\Delta\lambda)_R$, yields the actual longitude value desired. The nonrotating model also yields the actual latitude coordinate directly.

The nonrotating model is sketched in Figure 16-3. The satellite S is in a circular orbit of altitude h_s and inclination i. It moves with the angular speed $\dot{\theta}$, determined by h_s (equation 16-3). At any time t, the satellite will be displaced from the node N by the angle $\theta = \dot{\theta}t$. The coordinates of the nadir point S' are determined from θ by solving the spherical triangle $NS'L_s$[4] to obtain:

$$\lambda'_s = \tan^{-1}(\cos i \tan\theta) \qquad (16\text{-}12)$$

$$\phi_s = \sin^{-1}(\sin i \sin\theta) \qquad (16\text{-}13)$$

[4] The "polar" triangle NS'P can also be used.

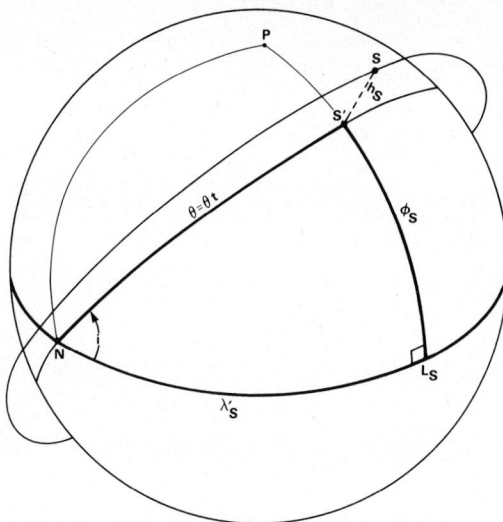

Fig. 16-3. Determination of nadir point

Adding the required rotational contribution to 16-12 as explained above:

$$\lambda_s = \tan^{-1}(\cos i \tan\theta) + \dot{\Lambda}_n t \qquad (16\text{-}14)$$

Equations 16-13 and 16-14 define the actual nadir trace. Equations 16-12 and 16-13 define the "nonrotating" trace, which is actually a great circle.

NADIR TRACE PATTERN FORMATION

As noted above, the nadir trace is generated by the combined action of satellite orbital motion and earth rotation relative to the orbit plane. Without rotation (nonrotating model), the nadir trace resembles a sine wave on the world map, with its great circle path developing a full 360 degree period (longitude). The same path is retraced on succeeding revolutions. Adding rotation causes a steady westward creep of the trace which is proportional to the ratio of the angular rates

$$Q = \dot{\theta}/\dot{\Lambda}_n \qquad (16\text{-}15)$$

This ratio Q is of fundamental importance in determining pattern development and configuration. It is clear, for example, that the rotation-produced creep referred to above accumulates in one satellite revolution to

$$S = 360/Q \qquad \text{degrees longitude} \quad (16\text{-}16)$$

S is often called the "step," i.e. the interval (longitude) between consecutive crossings (in the same direction) of the equator or other parallel of latitude.

The size of the step, or more fundamentally the rate ratio Q, determines the development of the pattern. If Q is an integer R, for example, then R steps equal 360 degrees exactly, and the westward-marching set of equator crossings laid down on the first nodal day is retraced on all subsequent days. Patterns of this type are one-day repeaters.

Multiday repeating patterns are produced when an integral number (R) of steps occupy exactly the same angular interval as another integral number (N) of 360° earth rotations. In these cases, the satellite moves through exactly R revolutions while the earth is making exactly N rotations relative to the orbit node, resulting in return of the ground track to its starting point, followed by indefinite repetition. It is apparent that the revolution ratio (R/N) must be the same as the rate ratio $(Q = \dot{\theta}/\dot{\Lambda}_n)$:

$$R/N = \dot{\theta}/\dot{\Lambda}_n = Q \qquad (16\text{-}17)$$

Equation 16-17 reveals the basic pattern-repetition condition: the rate ratio Q must be a *rational* number (ratio of *integers* R and N). In practice, the achievement of a desired repeating pattern is often a matter of tuning the orbital period to a desired revolution number pair (R, N). This approach becomes clearer when the angular rates in equation 16-17 are converted into periods:

$$T = D_n N/R \qquad (16\text{-}18)$$

Repeating coverage patterns are often advantageous because they permit repetitive viewing of particular geographic areas at fixed, specified intervals. Because of the integer constraint on the revolution numbers N and R, however, the distribution of repeating patterns is discrete rather than continuous, as illustrated in Figure 16-4. That figure shows the complete array of all repeating patterns/orbits within the range of satellite altitudes and pattern repeat-periods plotted. Apparent omissions in the symmetrical array of points in Figure 16-4 occur where the particular N, R integer pair is not relatively prime. Such pairs, therefore represent (redundant) multiples of shorter period cycles (obtained by removing common factors from both N and R).

Figure 16-4 is based on equation 16-18, with $D_n = D$, the mean solar day (i.e., sun-synchronous orbits). For $D_n \neq D$, the abscissa scales of Figure 16-4 yield only approximate values of T and h_s. Correct values are obtained from Figure 16-2 by matching Q values, as required by equation 16-17, between Figure 16-2 (D_n/T) and Figure 16-4 (R/N). Some iteration is ordinarily required. In a typical selection problem, characterized by desired values of altitude (h_s), inclination (i), and cycle period (N), one would enter h_s and i into Figure 16-2, read T and D_n, and compute Q. Multiplying Q by N yields R, to which the nearest integral R value on the specified N ordinate on Figure 16-4 is selected. That value, with N, yields an adjusted Q, to be reconciled in Figure 16-4 by adjusting the original choice of h_s and/or i.

GENERAL PATTERN FEATURES

The discussion of the one-day repeating-trace pattern in the previous section demonstrates that such patterns divide the equator into R equal segments by R ascending or descending equator crossings (subnodal points). It has also been shown by

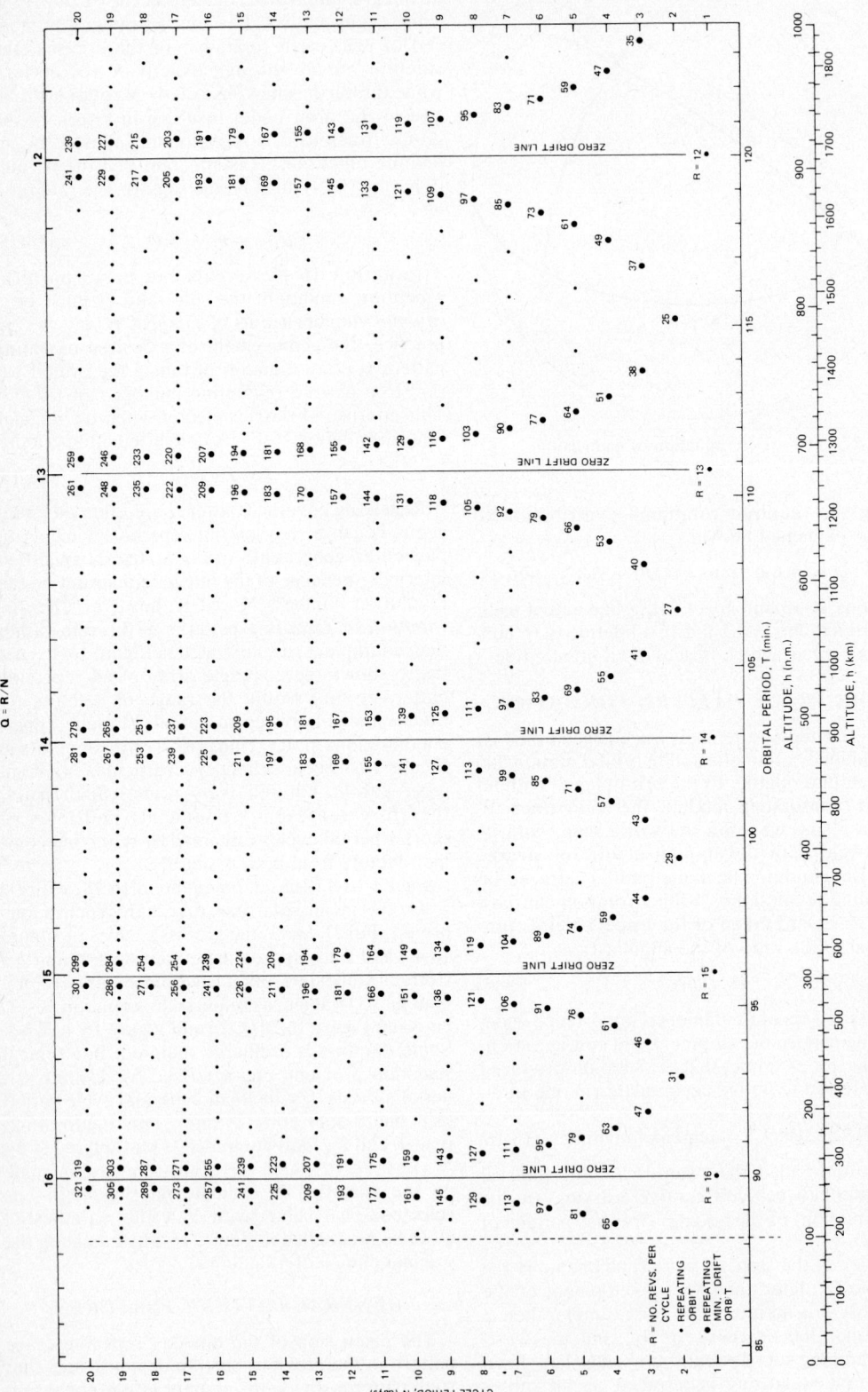

Fig. 16-4. Array of periodic coverage patterns.

King (1976) that the general N-day repeater produces the same type of fixed, R-fold division. The reference develops additional general properties of repeating coverage patterns, which are summarized below:

- The R ascending (or descending) subnodal points are earth-fixed, and they divide the equator into R equal segments $s = 2\pi/R$.
- Each of the fixed subnodal points is intersected once and only once (in one direction) by the nadir trace during each N-day period.
- The interval between consecutive subnodal point crossings is $N \cdot s$.
- The two opposite sets of subnodal points (ascending and descending) coincide if N and R have the same parity and alternate (spaced $s/2$) if N and R have opposite parity.
- All parallels of latitude within the pattern area are divided equally into the angular s-units by each set of trace crossings (ascending and descending).
- All trace intersections occur on a discrete set of meridians that are evenly spaced and quantized by the basic s-unit.
- All trace intersections occur on a discrete set of parallels that are symmetric about the equator and spaced at intervals which decrease with latitude.

PASS SEQUENCING

The previous section refers to the uniform spacing (at intervals $s = 2\pi/R$) of adjacent passes in a complete repeating pattern, and the spacing of *consecutive* passes at Ns intervals. It also implies that the remaining N-1 subnodal positions between a consecutive pair (inherent in the pattern) must be intersected at some time in the cycle.

The time sequencing of these subsequent crossings can be derived from Figure 16-4, from the position of the particular pattern dot (N,R) relative to the integral Q (zero drift) lines. There are N-1 positions, each occupied by a dot or a hole, between any pair of zero drift lines. An apparent shift number ν_a is associated with each position, beginning with zero on the line and counting consecutively to the left (positive, east shifting), and to the right (negative, west shifting). Thus each daily subpattern shifts by ν_a positions each day until all subnodal positions have been intersected at the end of N days. For the one-day repeaters, $\nu_a = 0$; for the "minimum drift" patterns (adjacent passes on consecutive days), $\nu_a = \pm 1$; etc. The complete pattern sequence can be traced by following the daily shift through the N days of the cycle.

The shift number can also be found analytically using the expressions:

$$\nu = R - \left\lfloor \frac{R}{N} \right\rfloor N \qquad (16\text{-}19)$$

$$\nu_a = \begin{cases} \nu & \text{for } \dfrac{R}{N} - \left\lfloor \dfrac{R}{N} \right\rfloor \leq \tfrac{1}{2} \\[2em] \nu - N & \text{for } \dfrac{R}{N} - \left\lfloor \dfrac{R}{N} \right\rfloor > \tfrac{1}{2} \end{cases}$$

$$(16\text{-}20)$$

The brackets [ ] above denote "integral part of." The "analytical" shift number ν is convenient for analysis, whereas the apparent shift number ν_a reflects the direction and amount of shift sensed by an observer.

OFF-NADIR POINTS AND TRACES

The mapping of finite geographic areas viewed from orbit involves off-nadir lines of sight, such as SD in Figure 16-5. SD is specified relative to the orbit plane and the nadir by the azimuth angle γ (orbital frame) and the off-nadir angle δ. As with the nadir trace, the main task in determining off-nadir points and traces is in specifying the geographic coordinates of a general point, in this case the off-nadir point D. This is accomplished by solving the spherical triangle S'DP using the spherical laws of cosines (ϕ_s, $\beta + \gamma$ and ψ known) and sines for ($\lambda_D' - \lambda_s'$) and ϕ_D:

$$\phi_D = \arccos[\sin\phi_s \cos\psi + \cos\phi_s \sin\psi \cos(\beta + \gamma)]$$

$$(16\text{-}21)$$

$$\lambda_D' - \lambda_s' = \arcsin \left[\frac{\sin\psi}{\cos\phi_D} \sin(\beta + \gamma) \right]$$

$$(16\text{-}22)$$

Since $\lambda_D' - \lambda_s' = \lambda_D - \lambda_s$ (viewing geometry unaffected by rotation), equations 16-21, 16-14, and 16-22 yield:

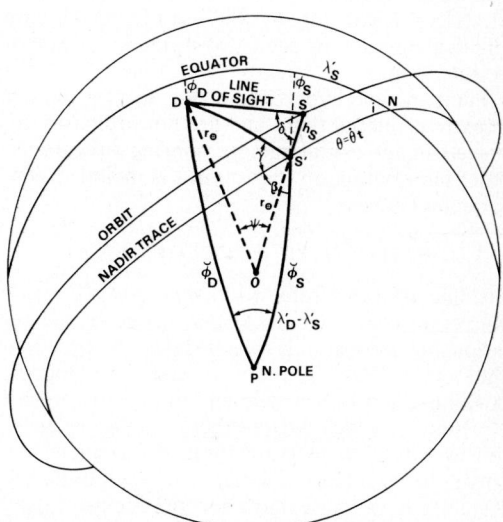

Fig. 16-5. Determination of off-nadir points.

$$\lambda_D = \arctan (\cos i \tan \theta) + \dot{\Lambda}_n t + \arcsin$$

$$\left\lfloor \frac{\sin \psi}{\cos \phi_D} \sin(\beta + \gamma) \right\rfloor \qquad (16\text{-}23)$$

$$\phi_D = \arcsin \left[\sin \phi_s \cos \psi \right.$$

$$\left. + \cos \phi_s \sin \psi \cos(\beta + \gamma) \right] \qquad (16\text{-}24)$$

The orbit azimuth angle β (geographic frame) is obtained from the solution of triangle NS'L$_x$ (Figure 16-3):

$$\beta = \arctan (1/\tan i \cos \theta) \qquad (16\text{-}25)$$

The central angle ψ subtended by the line of sight SD is found by solving the plane triangle OSD in Figure 16-5:

$$\psi = \arcsin \left\lfloor \frac{r_\oplus + h_x}{r_\oplus} \sin \delta \right\rfloor - \delta \qquad (16\text{-}26)$$

With the capability to determine arbitrary view points on the earth, various coverage patterns can be mapped. A simple and useful type results from holding the pointing angles γ and δ constant, producing viewing swath boundaries parallel to the nadir trace. Other special applications involve various programmed pointing sequences.

ELLIPTIC AND OTHER CONIC ORBITS

BACKGROUND

The complete family of conic orbits includes the elliptic, parabolic, and hyperbolic types, as well as the circular orbit (a special case) already discussed. While most operational orbits for terrestrial remote sensing will be nominally circular, it is not possible in practice to establish and maintain an exactly circular orbit. Thus these orbits will be at least slightly elliptical, and may need to be treated accordingly in more precise calculations. In addition, elliptical transfer orbits will be involved in various maneuvers (see section on MANEUVERS). Finally, the use of substantially elliptical operational orbits, while unlikely, cannot be ruled out.

Parabolic and hyperbolic orbits occur primarily in extraterrestrial flight and are not ordinarily involved in terrestrial remote sensing operations. Brief information on these types is included here for completeness.

HISTORICAL DEVELOPMENT

Conic orbits are referred to also as two-body or Keplerian orbits. These terms recognize the relationship between Kepler's laws, deduced in 1609–1618 solely from observations of planetary positions, and the subsequent analytical work of Newton, on which current theory is based. Newton used his law of gravitation to formulate and solve the two-body problem, which confirmed Kepler's laws and established the identity of all two-body orbits as conics. Since Kepler's work was concerned with the planets, his three laws refer specifically to planetary motion:

1. The orbit of each planet is an ellipse with the sun at one of its foci.
2. The radius vector of each planet sweeps over equal areas in equal intervals of time.
3. The squares of the periods of the planets are proportional to the cubes of their mean distances from the sun.

The solution of the two-body problem placed Kepler's observational inferences on a solid theoretical foundation, and permitted mathematical restatement of Kepler's laws as:

$$r = p/(1 + e \cos \theta) \qquad (16\text{-}27)$$

$$h = \sqrt{\mu p} \qquad (16\text{-}28)$$

$$T = 2\pi \sqrt{a^3/\mu} \qquad (16\text{-}29)$$

where

p = $a(1 - e^2)$ = semilatus rectum
e = eccentricity
h = angular momentum
a = semimajor axis

Equation 16-27 is the familiar ellipse equation in polar coordinates from analytic geometry (see Figure 16-6). Note that angular momentum h is constant (conserved). It is also twice the "areal velocity" of Kepler's second law, which must therefore be conserved also, confirming the law. Equation 16-29 corresponds to equation 16-4, which is a special case.

Figure 16-6 gives some of the more useful geometric properties of the ellipse, along with corresponding information on the parabola and hyperbola for comparison and reference.

DYNAMICS OF CONIC ORBITS

Some relationships governing the dynamics of conic orbits are given in Table 16-1. A marked symmetry can be seen throughout the family. A single expression for angular momentum holds for all, and angular momentum and energy are conserved throughout.

The ellipse time-relationship (equation 16-35) is called Kepler's equation. It is used extensively in relating space (r, θ) and time in elliptic orbits. The coefficient of t is called the "mean motion", and the variables M and E, along with θ, are the "anomalies" of celestial mechanics:

n = $\sqrt{\mu/a^3}$ = mean motion
M = nt = mean anomaly
E = eccentric anomaly
θ = true anomaly

The parabolic time relationship (equation 16-37) is called Barker's equation. The hyperbolic time equation (equation 16-38) is analogous to Kepler's equation. Both Kepler's equation and its hyperbolic analog are transcendental and require numerical solutions when the argument is time, but this difficulty can often be avoided by solving in the opposite direction (E or H as argument, solve for t).

ELLIPSE GEOMETRY (0<e<1)

$r = \dfrac{p}{1 + e \cos\Theta}$ (SEE NOTE)

$p = a(1-e^2)$

$b = a\sqrt{1-e^2}$

$r_P = a(1-e)$

$r_A = a(1+e)$

$e = \dfrac{c}{a}$

$c = \sqrt{a^2-b^2}$

$e = \dfrac{r_A-r_P}{a}$

$p = \dfrac{b^2}{a}$

$p = \dfrac{r_P r_A}{a}$

SF + SF' = 2a

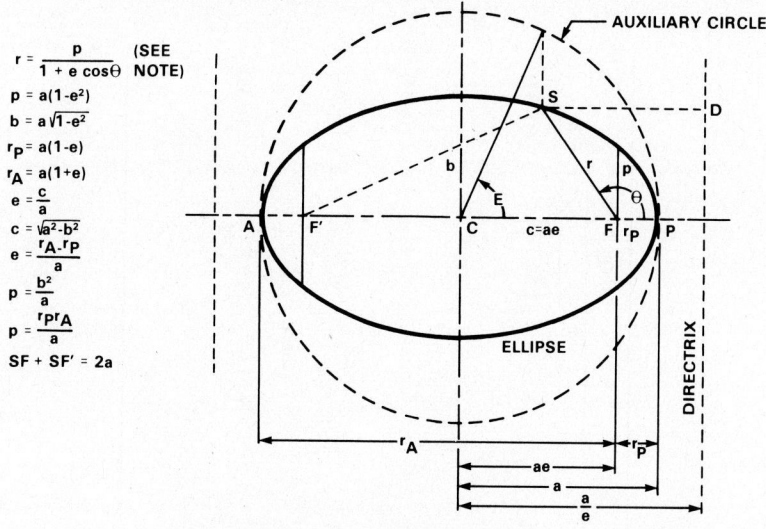

HYPERBOLA GEOMETRY (e>1)

$r = \dfrac{p}{1 + e \cos\Theta}$ (SEE NOTE)

$p = a(e^2-1)$

$b = a\sqrt{e^2-1}$

$r_P = a(e-1)$

$e = \dfrac{c}{a}$

$c = \sqrt{a^2 + b^2}$

$p = \dfrac{b^2}{a}$

$\cos\eta = \dfrac{1}{e}$

SF' - SF = 2a

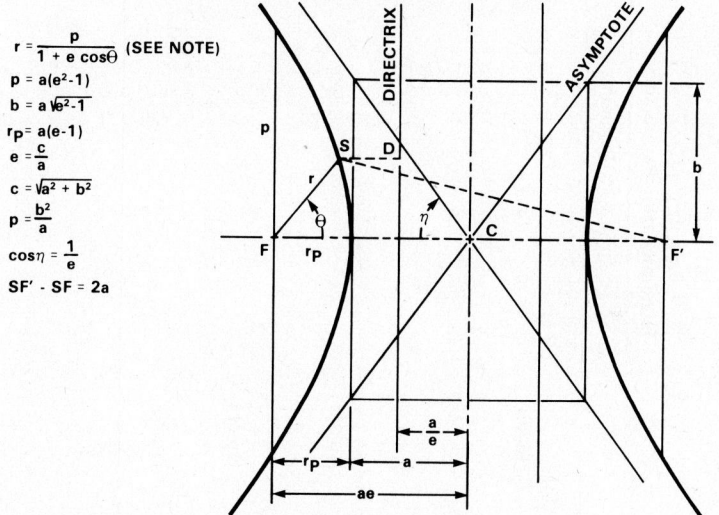

PARABOLA GEOMETRY

$r = \dfrac{p}{1 + \cos\Theta}$ (SEE NOTE)

NOTE:

THIS IS THE BASIC CONIC EQUATION IN POLAR COORDINATES. IT FOLLOWS DIRECTLY FROM THE DEFINITIVE CONIC PROPERTY $\dfrac{r}{SD}$ = e, WHICH DEFINES:

• ELLIPSES WHEN 0<e<1
• PARABOLAS WHEN e = 1
• HYPERBOLAS WHEN e>1

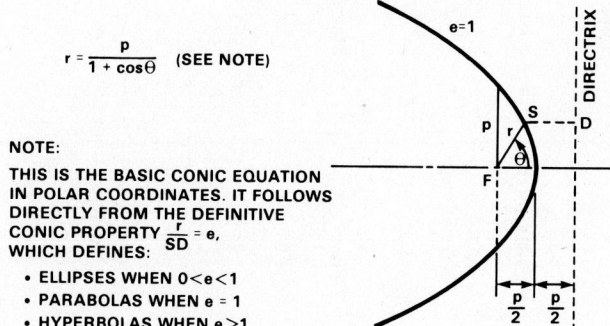

Fig. 16-6. Conic geometry.

TABLE 16-1
Dynamic Orbit Parameters

	Ellipse		Parabola		Hyperbola	
Velocity, v	$v = \sqrt{\mu\left(\frac{2}{r} - \frac{1}{a}\right)}$	(16-30)	$v = \sqrt{\frac{2\mu}{r}}$	(16-31)	$v = \sqrt{\mu\left(\frac{2}{r} + \frac{1}{a}\right)}$	(16-32)
Angular Momentum, h	$h = \sqrt{\mu p}$	(16-33)	$h = \sqrt{\mu p}$		$h = \sqrt{\mu p}$	(16-34)
Energy, ε	$\varepsilon = -\frac{\mu}{2a}$	(16-35)	$\varepsilon = 0$		$\varepsilon = \frac{\mu}{2a}$	
Time, t[5]	$\sqrt{\frac{\mu}{a^3}}\, t = M = E - e\sin E$	(16-36)	$2\sqrt{\frac{\mu}{p^3}}\, t = \tan\frac{\theta}{2} + \frac{1}{3}\tan^3\frac{\theta}{2}$	(16-37)	$\sqrt{\frac{\mu}{a^3}}\, t = e\sin H - H$	(16-38)
where	$E = \cos^{-1}\frac{a - r}{a\,e}$	(16-29)			where	
					$H = \cosh^{-1}\frac{a + r}{ae}$	(16-39)
	$T = 2\pi\sqrt{\frac{a^3}{\mu}}$					

[5] Measured from perifocus

APPLICATIONS TO EARTH SATELLITE ORBITS

In its general form, the two-body problem describes the motion of two bodies of unrestricted mass, revolving about their common center of mass. The problem is somewhat simpler in the earth-satellite case in that the satellite has negligible mass compared to the earth and thus revolves effectively about the earth's center.

The basic mission-analysis problem, as before, is to determine the position of the satellite as a function of time. Again the polar coordinate system is convenient, but now a second parameter is involved, the "argument of perigee", ω. The geometry is shown in Figure 16-7 (cf. Figure 16-1). Because of earth oblateness, ω is not constant but varies gradually, as approximated by

$$\dot{\omega} = (\tfrac{3}{4})\sqrt{\mu/a}\, J_2 (r_\oplus/p)^2 \,(5\cos^2 i - 1) \tag{16-40}$$

Determination of the nadir trace proceeds as with circular orbits (equations 16-13 and 16-14; see also Figure 16-3), except that now $\omega + \theta$ is substituted for θ. Also, $\dot{\theta}$ is no longer constant, because the radial distance r varies, and angular momentum must be conserved (equation 16-28).

A convenient mode of calculation is to choose θ as the argument, calculate r for incremented values of θ, then E and t in sequence (using equations 16-27, 16-36, and 16-35). Note that the nadir point is still determined by θ alone (for a given orbit plane orientation and value of ω). Note also that off-nadir points are determined as with circular orbits, except that h_s $(r - r_\oplus)$ is now a variable.

MANEUVERS

Since most orbits utilized for remote sensing applications are circular, the maneuvers treated here will be constrained to transfer between circular orbits and the adjustment of circular orbits though there are many other types of maneuvers. The equations to estimate the fuel requirements are provided as well as parametric plots that may be used in lieu of the equations.

IMPULSIVE MANEUVERS

The minimum energy transfer between two coplanar circular orbits makes use of what is called the Hohmann Transfer Maneuver. This maneuver consists of impulsively incrementing (or decrementing if downward transfer) the orbit velocity to create an ellipitcal transfer orbit that is tangent to both the initial and final orbits at the perigee and apogee points. When the spacecraft reaches the final orbit in the transfer another impulse is applied to circularize the orbit. The Hohmann maneuver is illustrated for an upward transfer by Figure 16-8. The speed at any point in an elliptical orbit can be determined from the Vis Viva equation

$$v = \sqrt{\mu\left(\frac{2}{r} - \frac{1}{a}\right)} \tag{16-41}$$

For the initial and final circular orbits, the speeds are

$$v_1 = \sqrt{\frac{\mu}{r_1}} \tag{16-42}$$

$$v_2 = \sqrt{\frac{\mu}{r_2}} \tag{16-43}$$

respectively. The transfer orbit has a perigee radius, $r_p = r_1$ and an apogee radius, $r_a = r_2$. The semi-major axis of the transfer orbit is thus

$$a_T = \tfrac{1}{2}(r_1 + r_2) \tag{16-44}$$

The respective perigee and apogee speeds for the transfer orbit are

$$v_{PT} = \sqrt{\mu\left(\frac{2}{r_1} - \frac{1}{a_T}\right)} \tag{16-45}$$

and

$$v_{aT} = \sqrt{\mu\left(\frac{2}{r_2} - \frac{1}{a_T}\right)}. \tag{16-46}$$

The incremental velocities to be applied at perigee and apogee are

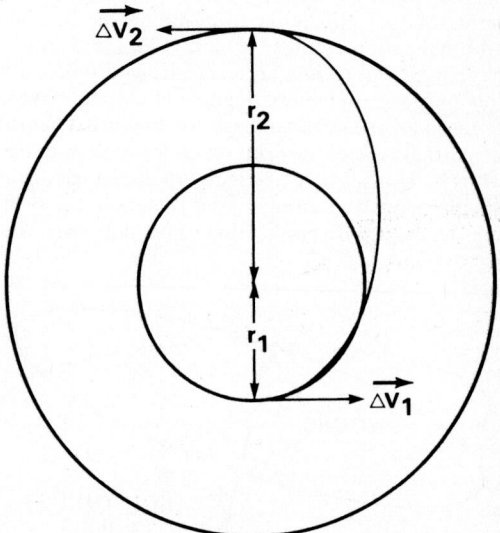

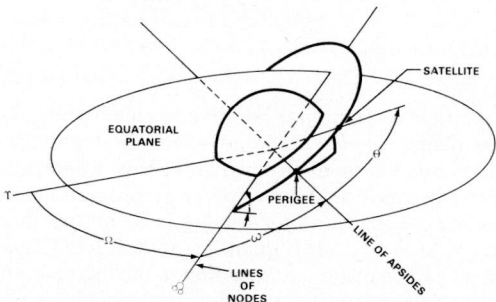

Fig. 16-7. Elliptical orbit orientation.

Fig. 16-8. Hohmann transfer geometry.

$$\Delta v_1 = \sqrt{\mu\left(\frac{2}{r_1} - \frac{1}{a_T}\right)} - \sqrt{\frac{\mu}{r_1}} \quad (16\text{-}47)$$

$$\Delta v_2 = \sqrt{\frac{\mu}{r_2}} - \sqrt{\mu\left(\frac{2}{r_2} - \frac{1}{a_T}\right)} \quad (16\text{-}48)$$

The total incremental velocity change is consequently

$$\Delta v_T = \Delta v_1 + \Delta v_2 \qquad (16\text{-}49)$$

To change the angular orientation of the orbit plane requires the application of a thrust component (or an impulse) normal to the velocity vector. Assuming an earth orbit, if the impulse is applied at the equator crossing, the inclination relative to the equator is changed (Figure 16-9). If the out-of-plane impulse is applied 90° in the orbit away from the equator, the inertial orientation of the orbit plane is changed. This can be visualized physically if the orbit is viewed as a gyro on which a torque is being applied in the same direction as (or diametrically opposed to) the angular momentum vector.

Remembering that the forces being applied are impulses, the burn geometry is depicted by Figure 16-10 where the observer is looking onto the intersection of the orbit planes before and after the maneuver.

Applying the law of cosines to Figure 16-10 yields

$$\Delta v = \sqrt{v_1^2 + v_2^2 - 2v_1v_2 \cos \Delta i} \quad (16\text{-}50)$$

If only the inclination to the reference plane is to be changed, let $v_1 = v_2$ and solving equation 16-50 yields

$$\Delta v_\perp = 2v, \ \sin\left(\frac{\Delta i}{2}\right) \qquad (16\text{-}51)$$

Frequently it is desired to transfer between two orbits that are not coplanar. While the maneuvers can be executed separately, it is advantageous in terms of fuel utilization to combine the altitude and inclination changes. Duck (1968) and others have used the technique to be derived below for this purpose. Remembering the Hohmann Transfer used for the coplanar case, let the thrust vector lie partially out of the orbit plane for each impulse. Thus both an altitude and inclination change occur simultaneously. Letting v_1 and v_3 denote the orbital speeds prior to application of the impulses, and

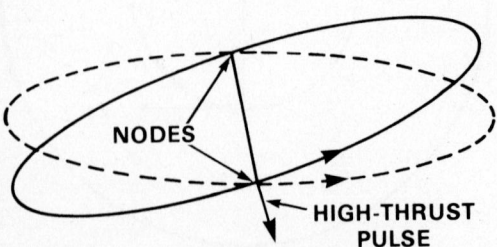

Fig. 16-9. Impulsive inclination change.

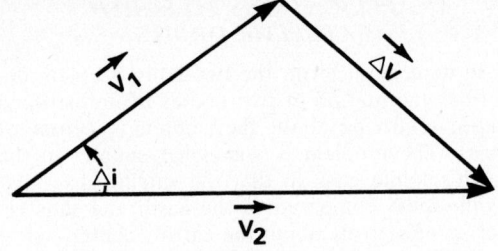

Fig. 16-10. Out-of-plane thrusting geometry.

v_2 and v_4 denote the speeds after the impulses, the velocity increments are:

$$\Delta v_1 = \sqrt{v_1^2 + v_2^2 - 2v_1v_2 \cos\Delta i}, \quad (16\text{-}52)$$

and

$$\Delta v_2 = \sqrt{v_3^2 + v_4^2 - 2v_3v_4 \cos \Delta i_2} \quad (16\text{-}53)$$

The total velocity increment as for the coplanar Hohmann maneuver is:

$$\Delta v_T = \Delta v_1 + \Delta v_2$$

The problem at hand is to minimize Δv_T subject to the constraint:

$$G = \Delta i_1 + \Delta i_2 - \Delta i = 0 \qquad (16\text{-}54)$$

Using the technique of language Multiplier found in a number of applied mathematics texts (e.g. Dettman) it is desired to minimize U where

$$U = \Delta v_T + \eta G \qquad (16\text{-}55)$$

The necessary conditions for a stationary point are:

$$\frac{\partial U}{\partial \Delta i_1} = \frac{\partial U}{\partial \Delta i_2} = \frac{\partial U}{\partial \eta} = 0$$

Taking the partial derivatives, equating them to zero and using algebraic manipulation yields a set of necessary conditions for the minimization of Δv_1, namely:

$$\frac{v_1 v_2}{\Delta v_1} \sin \Delta i_1 = \frac{v_3 v_4}{\Delta v_2} \sin \Delta i_2 \qquad (16\text{-}56)$$

$$\Delta i_1 + \Delta i_2 - \Delta i = 0 \qquad (16\text{-}57)$$

Substituting equations 16-52, 16-53, and 16-57 in 16-56 yields:

$$\frac{v_1 v_2 \sin \Delta i_1}{\sqrt{v_1^2 + v_2^2 - 2v_1v_2 \cos \Delta i_1}}$$

$$= \frac{v_3 v_4 \sin(\Delta i - \Delta i_1)}{\sqrt{v_3^2 + v_4^2 - 2v_3v_4 \cos(\Delta i - \Delta i_1)}}$$

$$(16\text{-}58)$$

Equation 16-58 cannot be solved explicitly for Δi_1, the plane change executed at the first impulse. However, it is easily solved using numerical methods on a digital computer or programmable pocket calculator. After solving 16-58 for Δi_1, then $\Delta i_2 = \Delta i - \Delta i_1$ and equations 16-52, 16-53, and 16-57 can be evaluated to estimate the incremental velocity that must be applied to implement the transfer maneuver.

CONTINUOUS THRUST MANEUVERS

Over the past two decades a significant effort has gone into the development of propulsion systems for maneuvering spacecraft. These devices, which have sufficiently small thrust levels when used with reasonably large spacecraft, result in small accelerations and must be applied over orbital arcs of finite duration. With the advent of the Space Shuttle and its potential for lifting numerous and large payloads into low earth orbit, the mission planner needs to be aware of how to estimate propulsion requirements for low acceleration applications. The continuous thrust analyses to be developed here utilize the Gauss form of the Lagrange Planetary equations of motion which are found in most celestial mechanics texts. This formulation of the equations of motion presents the rate of change of the Keplerian orbital elements in terms of radial, circumferential, and normal acceleration components (a_R, a_s, a_w). In addition, they lend themselves to linearization around circular orbits as well as analytical and numerical averaging over an orbital arc. Equations 16-59 through 16-64 are the Gauss form of the Lagrange equations.

$$\frac{da}{dt} = \frac{2}{n(1 - e^2)^{1/2}} \left(a_R e \sin \theta + \frac{a(1 - e^2)}{r} a_s \right)$$
(16-59)

$$\frac{de}{dt} = \frac{(1 - e^2)^{1/2} \sin \theta \, a_R}{na}$$
$$+ \frac{(1 - e^2)^{1/2}}{ena^2} \left(\frac{a^2(1 - e^2) - r^2}{r} \right) a_s$$
(16-60)

$$\frac{d\Omega}{dt} = \frac{r \sin(\omega + \theta)}{na^2(1 - e^2)^{1/2} \sin i} a_w$$
(16-61)

$$\frac{di}{dt} = \frac{r \cos(\omega + \theta)}{na^2(1 - e^2)^{1/2}} a_w$$
(16-62)

$$\frac{d\omega}{dt} = - \frac{(1 - e^2)^{1/2} \cos \theta}{ane} a_R$$
$$+ \frac{(1 - e^2)^{1/2} \sin \theta}{ane} \left(1 + \frac{r}{a(1 - e^2)} \right) a_s$$
$$- \frac{r \sin(\omega + \theta) \cot i}{a^2 n (1 - e^2)^{1/2}} a_w$$
(16-63)

$$\frac{dM}{dt} = n + \left(\frac{(1 - e^2) \cos \theta}{ane} - \frac{2r}{na^2} \right) a_R$$
$$- \frac{(1 - e^2) \sin \theta}{ane} \left(1 + \frac{r}{a(1 - e^2)} \right) a_s$$
(16-64)

For continuous thrust transfer between two circular orbits, assuming the thrust acceleration is sufficiently small that the instantaneous orbit is near circular and the thrust direction is circumferential, equation (16-59) can be written as:

$$\frac{da}{dt} = \frac{2}{n} \left(\frac{a}{r} \right) a_s$$
(16-65)

Since the orbit is assumed to be near circular, $a \approx r$. Also:

$$n = \sqrt{\frac{\mu}{a^3}}$$
(16-66)

Thus, equation 16-65 becomes:

$$\frac{dr}{dt} = \frac{2}{\sqrt{\frac{\mu}{r^3}}} \left(\frac{F_s}{m} \right)$$
(16-67)

Rearranging and integrating yields:

$$\int_{r_1}^{r_2} \sqrt{\mu} \, r^{-3/2} \, dr = 2 \int_0^t \frac{F_s}{m} \, dt$$
(16-68)

$$2 \left(\sqrt{\frac{\mu}{r_1}} - \sqrt{\frac{\mu}{r_2}} \right) = 2 \int_0^t \frac{F_s \, dt}{m_0 - \dot{m}t}$$
(16-69)

And the fuel flow rate is:

$$\dot{m} = \frac{F_s}{g \, Isp}$$
(16-70)

Substituting equation 16-70 into 16-69 and noting that

$$v_1 = \sqrt{\frac{\mu}{r_1}} \text{ and } v_2 = \sqrt{\frac{\mu}{r_2}}$$

yields:

$$v_1 - v_2 = g \, Isp \, ln \left(\frac{m_0}{m_0 - \dot{m}t} \right)$$
(16-71)

The right side of equation 16-71 is the well known rocket equation which is the velocity change resulting from a rocket engine. The incremental velocity required for continuous coplanar transfer is the difference in the respective orbit speeds.

A slightly different approach will be taken in estimating the effect of continuous thrusting on the orbit inclination. Again assuming a circular orbit, equation 16-62 can be written as:

$$\frac{di}{dt} = \frac{\cos \Phi}{v} \cdot \frac{F_\perp}{m}$$
(16-72)

Noting that the out-of-plane acceleration, a_w, is $F_\perp/m$ and the total angle from the nodal crossing, Φ, is $\omega + \theta$, it is now desired to determine the average inclination change resulting from thrusting over some arc. Note that:

$$\Phi = \Phi_0 + n t$$
(16-73)

$$d\Phi = n \, d t$$
(16-74)

Equation 16-72 now becomes:

$$di = \frac{\cos \Phi}{v} \frac{F_\perp}{m} \frac{d \Phi}{n}$$
(16-75)

but for a circular orbit $v = rn$, thus:

$$di = \frac{r \cos \Phi}{v^2} \frac{F_\perp}{m} d\Phi \qquad (16-76)$$

It will be assumed that the mass of the propellant consumed over one orbit is much less than the total vehicle mass and that the thrusting will be applied symmetric to the nodal crossing with the thrust direction reversed (see Figure 16-11). This latter stipulation is made so as to negate any node rotation that is incurred by thrusting at other than the instant of nodal crossing (see equation 16-61). Let ξ denote the thrust arc on one side of the orbit; then the average inclination change per orbit is:

$$\overline{\Delta i}/\text{orbit} = \frac{2r}{v^2} \frac{F_\perp}{m} \int_{-\xi/2}^{\xi/2} \cos \Phi \, d\Phi \quad (16-77)$$

Integrating yields:

$$\overline{\Delta i}/\text{orbit} = \frac{4r}{v^2} \left(\frac{F_\perp}{m} \right) \sin(\xi/2) \quad (16-78)$$

Note, however, that:

$$\xi = n\Delta t/2 \qquad (16-79)$$

where Δt is the thruster operation-time for the entire orbital pass. Equation 16-78 becomes:

$$\overline{\Delta i}/\text{orbit} = \frac{4r}{v^2} \left(\frac{F_\perp}{m} \right) \sin(\xi/2) \quad (16-80)$$

Remembering that $(r/v) = (1/n)$ and $(\Delta t/2\xi) = (1/n)$, then:

$$\overline{\Delta i}/\text{orbit} = \frac{1}{v} \left(\frac{F_\perp \Delta t}{m} \right) \frac{\sin (\xi/2)}{(\xi/2)} \quad (16-81)$$

It is easily shown that when the mass consumed is $<<$ than the total mass, then $m\Delta v \approx F_\perp \cdot \Delta t$, thus:

$$\overline{\Delta i}/\text{orbit} = \frac{\Delta v}{v} \frac{\sin (\xi/2)}{(\xi/2)} \quad (16-82)$$

Note that equation 16-82, applies to both the impulsive and continuous thrusting cases because as ξ, the thrust arc, approaches zero (an impulse) sin $(\xi/2)/(\xi/2)$ approaches one in the limit. Assuming that the inclination change per orbit is small and the incremental velocities can be summed for a multiple orbit maneuver, then the Δv for inclination change is:

$$\Delta v = \frac{\Delta i \, v \, (\xi/2)}{\sin (\xi/2)} \qquad (16-83)$$

Remembering the discussion given above about a combined altitude and inclination change maneuver using impulses, note that an analogous maneuver exists for the continuous thrust case. The derivation will not be repeated here as it is well treated in Edelbaum's paper. The Δv to transfer between two non coplanar circular orbits is:

$$\Delta v = \sqrt{v_1^2 + v_2^2 - 2v_1 v_2 \cos (\pi \cdot \Delta i/2)}$$
$$(16-84)$$

Edelbaum's paper also provides the optional thrust steering algorithm for the maneuver. It will not be discussed here other than to say that one of the authors has used the algorithm with considerable success in numerically integrating the equations of motion. Figure 16-12 shows the Δv required to transfer from a 450 km circular orbit to a predetermined orbit. This is typical of transferring from a Space Shuttle parking orbit.

ORBIT TRIM MANEUVERS

Frequently in the lifetime of a spacecraft it is necessary to apply corrective trimming of the orbit to remote effects of launch-vehicle guidance and propulsion-system errors as well as results of such environmental perturbations such as atmospheric drag, earth asphericity, solar radiation pressure and third-body gravitational fields. The more significant of these effects will be discussed below for both geosynchronous and sun-synchronous orbits. A number of approximate equations have been developed by Jensen et al. (1962, 1963) and Edelbaum (1961) for trimming an orbit. These formulae are derived using mathematical methods similar to those given above. The derivations are not repeated here, though the most used formulae are provided for convenience. Table 16-2 provides this information which can be used for both high and low thrust cases. The parameters given in Table 16-2 are the more frequently adjusted parameters. Angular position within the orbit and orbit-plane orientation typically are not adjusted. Note that while nodal rotation rate is sometimes adjusted to maintain sun-synchronism, this is usually done with an altitude change. This will be discussed in more detail when station keeping is addressed for sun-synchronous orbits. Figures 16-13 and 16-14 present the incre-

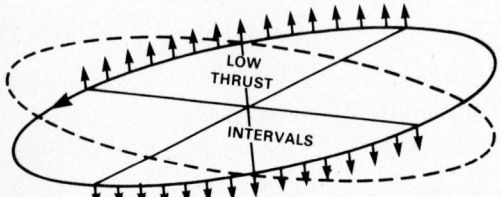

Fig. 16-11. Inclination change using continuous thrust.

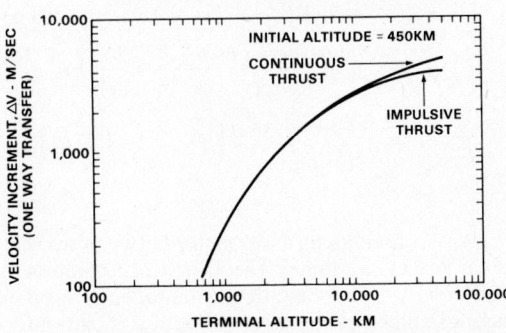

Fig. 16-12. Velocity increment for coplanar transfer.

TABLE 16-2
Orbit Trim Velocity Increments For
Near Circular Orbits

Period

$$\Delta v = \frac{v}{3}\frac{\Delta T}{T} \qquad (16\text{-}85)$$

Radius

$$\Delta v = \frac{v}{2}\frac{\Delta r}{r} \qquad (16\text{-}86)$$

Inclination

$$\Delta v = 2\, v\, \sin\!\left(\frac{\Delta i}{2}\right) \qquad (16\text{-}87)$$

Eccentricity

$$\Delta v = \frac{V}{2}\, e \qquad (16\text{-}88)$$

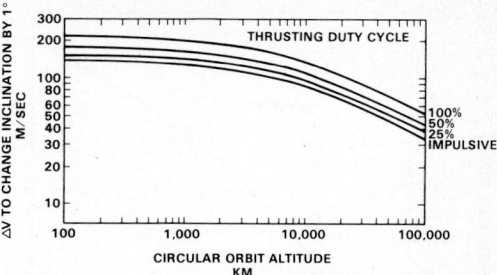

Fig. 16-14. Velocity increment to chagne inclination by 1 degree.

mental velocity to change the altitude (semimajor axis) by one kilometer and the inclination by one degree as a function of altitude. Figures 16-15 and 16-16 give the circular orbit speed and period respectively as a function of altitude. The authors find such plots to be very useful when estimating propulsion requirements.

PROPULSION SYSTEMS

Considerable discussion has been given above to estimating the incremental velocity required for orbital transfer and vernier trimming. The incremental velocity needs to be converted to a propellant mass. The equation used for this conversion is called the "Rocket Equation" and it represents the velocity incurred as a result of integrating the acceleration caused by the rocket engine. The derivation of this equation is found in most texts on space flight (e.g., Jensen et al., 1962, 1963; Nelson and Loft, 1962; Stearns, (1963) and will not be repeated here. The velocity increment resulting from a rocket is:

$$\Delta v = g\, Isp\, ln\!\left(\frac{m_o}{m_o - \dot{m}t}\right) \qquad (16\text{-}89)$$

where g denotes the acceleration of gravity (9.80 m/sec) and Isp the propellant-specific impulse. The ratio of the burnout to initial mass results by rearranging equation 16-89 and is:

$$\frac{m_f}{m_o} = e^{-\Delta v/gIsp} \qquad (16\text{-}90)$$

The propellant mass is thus

$$m_p = m_o(\, - e^{-\Delta v/gIsp}) \qquad (16\text{-}91)$$

A brief discussion of the available spacecraft propulsion-systems is now in order. The treatment is by no means exhaustive, but is intended to provide the potential user with comparisons of performance and relative development state. The typical performance parameter used for comparison is the propellant-specific impulse, i.e., the total impulse delivered per unit mass of propellant. Table 16-3 presents a comparison of several candidate systems.

The cold inert gas system was the earliest form of spacecraft auxiliary propulsion. However, as spacecraft became longer lived and propulsion needs grew, the low specific impulse became a significant penalty. It was desired to develop a hot gas system as the specific impulse is proportional to thrust-chamber temperature in gas systems. The initial hot gas system flown in the early 1960's used hydrogen peroxide (H_2O_2) as propellant, despite its instability, and rapid decomposition which presented a problem with propellant storability.

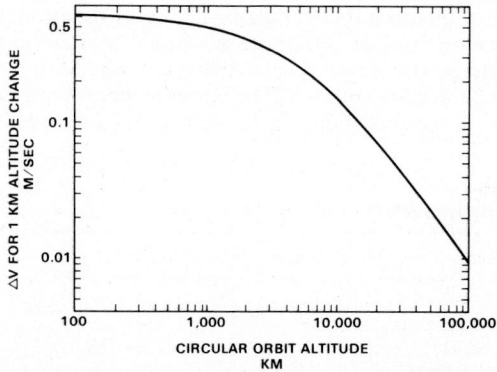

Fig. 16-13. Velocity increment to change altitude by 1 km.

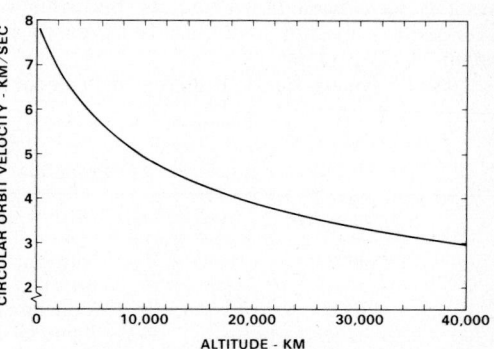

Fig. 16-15. Circular orbit velocity.

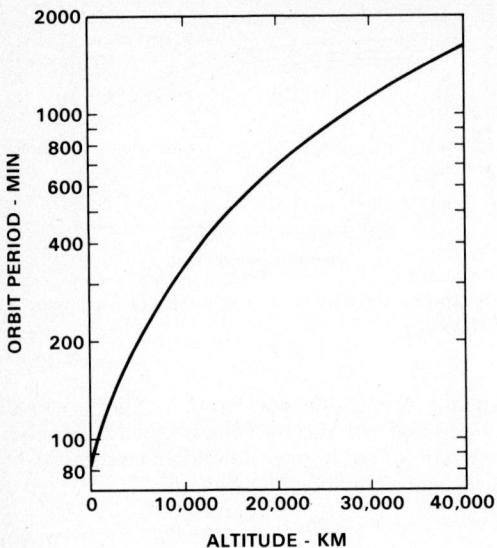

Fig. 16-16. Orbit period.

The system that has become the standard for spacecraft auxiliary propulsion is the monopropellant hydrazine (N_2H_4) system in which the propellant is passed over a catalyst bed where the resulting chemical reaction heats the gas. Such systems have had considerable flight experience over the past 15 years.

With the advent of the Space Shuttle and 7- to 10-year mission lifetimes, a larger propulsion burden is placed on future spacecraft. The most sensible approach in this environment is to have the Shuttle deposit the payload in low earth orbit, check out the payload and then use spacecraft propulsion to adjust the orbit to the desired one. A significantly larger useful payload can be delivered this way since the heavy Shuttle orbiter is not being boosted to high orbits. These factors create a need for higher performance systems.

One system under development is a bipropellant system using nitrogen tetroxide and monomethyl hydrazine (N_2O_4/MMH) as oxidizer and fuel respectively. This concept is a scaled down version (thrust and size) of what is used in launch vehicles (including the shuttle itself). Such systems have been flown and the performance over monopropellant hydrazine is increased by about 20−25 percent.

Another system that will increasingly become

important is the ion system where either cesium or mercury is ionized and then accelerated using an electrostatic field. These systems offer an order of magnitude increase in specific impulse, but at the expense of multiple kilowatt electric-power requirements and very high voltages. Due to the power constraint, thrust levels are limited and orbit maneuvers require significantly more time. In addition, as was shown previously, there is usually a Δv penalty for using low continuous thrust. However, if the specific impulse is sufficiently high it more than compensates. The penalty arises from thrusting at non optimum times. The ion systems, a number of which have been flown experimentally over the past 15 years, are the propulsion systems of the future. Today they are expensive and need more testing. Specific problems are related to large solar arrays operating for long periods of time in the Van Allen belts, radiation-hardening of electronic components and boxes, and long-lifetime demonstration tests.

The reader is referred to the journals of the American Institute of Aeonautics and Astronautics for additional information on these and other propulsion concepts. It is also suggested that the talents of a propulsion specialist be used when configuring a system. The intention here has not been to advocate a particular propulsion-system type or describe all possible candidates. The major system types as well as a brief evolution was given. The final selection must be based on user requirements, reliability, the risks users will accept, system complexity, available electric power and cost.

MAINTENANCE OF REMOTE SENSING ORBITS

For most remote sensing missions of long duration, it is desirable to have an orbit that repeats spatially and temporally with respect to the earth. Environmental disturbances (e.g., atmospheric drag, solar radiation pressure, third body gravitational perturbation, earth asphericity, misaligned thrusters, etc.) tend to drive the orbit away from the desired nominal conditions thus necessitating the use of propulsion devices to maintain these conditions. In addition, sometimes it is desired to alter the orbital coverage by changing the orbital period and/or inclination. Such maneuvers change the phase relationship between the orbit and the earth rotation. The intention here is not to cover perturbations in general as they are well

TABLE 16-3
Propellant Specific Impulse

System Type	Propellant	Isp (sec)
Cold Gas	Nitrogen	65
	Freon	50
Hot Gas	Monopropellant N_2H_4	220−235
	Bipropellant N_2O_4/MMH	290−300
Ion	Hg, Cs	3000−5000

covered in the literature. What are discussed, however, are the major disturbances acting on geosynchronous and sun-synchronous orbits and estimates of fuel required to compensate for them.

GEOSYNCHRONOUS ORBITS

Geosynchronous spacecraft orbit the earth at the same rate as the earth rotation and thus tend to hover over a given point on the equator. This type of orbit is most frequently used for communication satellites, but is also used for meteorological applications.

The primary orbit maneuvers performed on the geosynchronous orbits are station relocation and station keeping. Station relocation refers to changing the longitude over which the satellite hovers while station keeping is maintaining the relative synchronism with the earth in spite of environmental perturbations. The drift of a geosynchronous spacecraft in degrees/day can be approximated by (Isley and Duck, 1974)

$$\omega_D = -540 \frac{\Delta h_s}{(r_\oplus + h_s)} \qquad (16\text{-}92)$$

The negative sign implies a westward drift for a positive altitude increment. Specifically, if the altitude is increased above the geosynchronous value, the orbit period is increased thus requiring longer for one satellite orbit than for one earth rotation and the satellite moves to the west of the reference position on the equator. The formulae relating incremental velocity to altitude change previously derived can be used to estimate the fuel required to relocate the station. Note, however, that at least two engine firings are required, i.e., one to initiate the drift and one to stop it. It should be further noted that the faster one wants to relocate, the greater the cost in fuel expenditure.

The primary environmental disturbances acting on the geosynchronous orbit are earth asphericity, solar and lunar gravity, and solar-radiation pressure. Since solar-radiation pressure is so dependent on the cross-sectional area and orientation of the spacecraft it will not be explored here other than to say that it is significant only for vehicles with large area to mass ratios.

The east-west drift results primarily from the J_{22} tesseral harmonic in the earth's gravity field. This harmonic arises due to the earth being slightly elliptical in the equatorial plane. Figure 16-17 presents the velocity increment that must be applied yearly to offset this effect. If uncorrected, the satellite will librate about the earth's minor axis with a very long period.

The major out-of-plane disturbance on the geosynchronous orbit comes from the gravitational attraction of the sun and moon. Of these two fields, that of the moon exceeds that of the sun by a factor of about three. These forces induce a daily oscillation of the altitude along with very long period oscillation in inclination. The

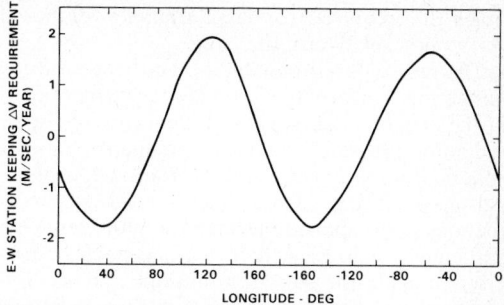

Fig. 16-17. Geosynchronous satellite east-west station-keeping requirement.

latter grows to a value of 14.67° from an initial value of 0° in 26.6 years and then returns back to 0. The mean inclination rate is 0.86°/year. The actual rate of change is a function of the lunar orbit orientation to the earth equatorial-plane. The lunar orbit is inclined to the elliptic plane at 5.15°. Due to orbit regression, its inclination to the earth's equator varies between 18 and 28° and has a period of 18.6 years. Previously, north-south station keeping has not been performed due to the significant fuel requirements (e.g., Δv of 55 meters/second/1° of inclination change). What has been done has been to inject the spacecraft into orbit with a small inclination and proper choice of inertial location of the line of nodes such that the inclination decreases to zero during the first half of the mission and then increases to its initial value during the second half. If more precise north-south control is required, then there is no recourse but to use thrusters.

SUN SYNCHRONOUS ORBITS

The sun-synchronous orbit makes use of a perturbation effect for its very existence. As was mentioned earlier in this chapter, sun synchronism results from choosing an altitude-inclination combination such that the nodal motion caused by the J_2 zonal harmonic (oblateness) equals the mean rate at which the earth orbits the sun. The net result is that the spacecraft always passes over a given patch of earth at the same local time and attempts to remove one of the variables (i.e., time of day) in the multi-temporal observations.

The sun synchronous-orbit plane can be reoriented relative to the sun by changing the altitude, the inclination or both. As a result, the precession rate of the line of nodes is changed and can be estimated as follows

$$\Delta \dot{\Omega} = -\frac{5}{2} \dot{\Omega}_o \left(\frac{\Delta h_s}{r_\oplus + h_s} \right) - \dot{\Omega}_o \tan i \, \Delta i$$

$$(16\text{-}93)$$

where $\dot{\Omega}_o = 0.9856°/day$ and Δi is in radians. It should be noted, however, that the drift rate is rather insensitive to changes in the parameters, meaning that large quantities of propellant or long times are required for large changes of nodal orientation relative to the sun.

The primary environmental disturbances acting on the sun-synchronous orbit are high order terms of the geopotential, solar and lunar gravity, solar radiation pressure and, very frequently, atmospheric drag. For many of the earlier meteorological missions (e.g., Tiros, Nimbus) no orbit corrections were applied because the altitudes were sufficiently high plus there was no need to precisely register the data from multiple passes. Examining time histories of those missions showed the presence of a small amount of drag and a resonance caused by solar gravity (Duck, 1975). This latter effect dominated and caused the orbit-plane orientation to librate about the earth-sun line with a period of 30–50 years. This motion caused a change in time of day of nodal crossing.

The first need for active station keeping of sun-synchronous orbits was for Landsat 1. It is interesting to note, however, that on Landsat, the major disturbance was the unloading of the pitch-axis momentum-wheel by use of thrusters not oriented in a couple. This will not be the case for Landsat D which has a different spacecraft configuration and will be flown about 200 kilometers lower than the previous Landsats.

For the sun-synchronous missions currently envisioned, it appears that atmsopheric drag may be the most significant long term perturbation on ground-track repeatability. The rate of shift of the ground track due to atmospheric drag is approximately

$$\dot{s}' = \frac{240\, q\, \sqrt{r'}\, C_D\, A\rho}{mv'}\tau \qquad (16\text{-}94)$$

where

$r' = r_\oplus + hs$ (km)
q = number of orbits/day
C_D = drag coefficient (2.2–4 typical)
A = cross sectional area (m²)
ρ = atmospheric density kg/km³
τ = time (days)
v' = orbit velocity (km/sec)

Using averaging techniques and the Lagrange equations the altitude change resulting from drag is approximately

$$\Delta r = \frac{-55\sqrt{r'}\, C_D\, A\rho\tau}{m} \qquad (16\text{-}95)$$

The shift in ground track, s', is thus

$$s' = s_o' + \dot{s}'\tau + 120\,\frac{q\sqrt{r'}\, C_D\, A\rho\tau^2}{mv'} \qquad (16\text{-}96)$$

Figure 16-18 gives an estimate of atmospheric density based on the 1976 standard atmosphere. There is now sufficient information to estimate the fuel required to maintain the ground track within specified bounds. What is proposed is use of the "soft" limit-cycle concept as is depicted by Figure 16-19.

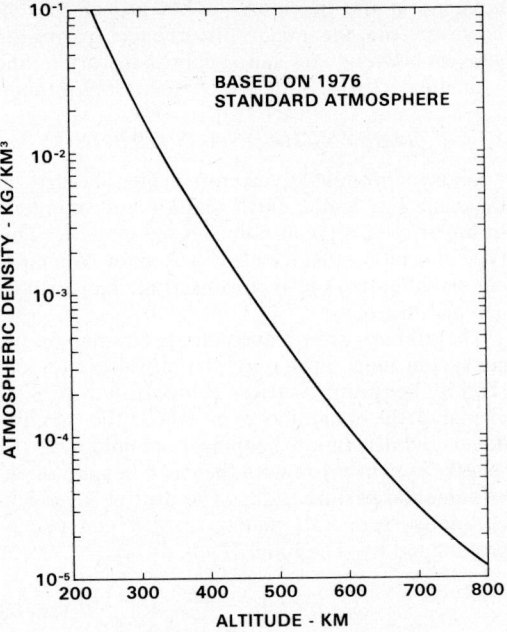

Fig. 16-18. Atmospheric density.

The ground track is allowed to grow until s' reaches the upper (or lower) bound. At this point, the orbit is raised by an amount such that the disturbing force turns the drift around and does not exceed the other bound. Assuming that the orbit is perfectly placed initially, it is desired to estimate the correction interval and ultimately the Δv. From equation (16-96) at point B in Figure 16-19, it can be shown that

$$2s_L' = \frac{120\, q\sqrt{r'}\, C_D\, A\rho\tau^2}{mv'} \qquad (16\text{-}97)$$

Thus, the correction interval in days is

$$\tau_c = \sqrt{\frac{s_L\, m\, v'}{15\, q\sqrt{r'}\, C_D\, A\, \rho}} \qquad (16\text{-}98)$$

The altitude change that must be made at each correction is

$$\Delta r = \frac{55\sqrt{r'}\, C_D\, A\rho\, \tau_c}{m} \qquad (16\text{-}99)$$

and the incremental velocity (Δv) per correction is

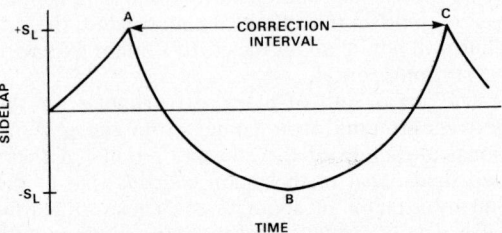

Fig. 16-19. Soft limit cycle.

$$\Delta v = \frac{55 \ v' \ C_D \ A\rho\tau_c}{2\sqrt{r'} \ m} \qquad (16\text{-}100)$$

The yearly incremental velocity is thus

$$\Delta v/\text{year} = 10038 \ \frac{v' \ C_D \ A\rho}{\sqrt{r'} \ m} \qquad (16\text{-}101)$$

REFERENCES

Allione, M. S. et al., 1968, The N-Body problem and special perturbation techniques; Guidance, Flight Mechanics and Trajectory Optimization, vol. VI, NASA CR-1005.

Battin, R. H., 1964, Astronautical Guidance, McGraw-Hill Book Company.

Cooley, J. L., 1972, Orbit selection considerations for Earth Observatory Satellites, Goddard Space Flight Center Report X-551-72-145.

Dettman, J. W., 1962, Mathematical Methods in Physics and Engineering, McGraw Hill Book Company.

Driver, M., 1981, Earth applications orbit analysis for a shuttle-mounted multispectral mapper; American Aeronautics and Astronomics Association Paper 81-182.

Duck, K. I., 1963, An approximation method for determination of propulsion requirements associated with low thrust coplanar transfer in a central force field, Goddard Space Flight Center Report X-623-63-117.

Duck, K. I., 1968, A method for estimating optimum launch vehicle and kickstage performance, Goddard Space Flight Center Report X-734-68-150.

Duck, K. I., 1975, Long period nodal motion of sun-synchronous orbits, Goddard Space Flight Center Report X-726-75-215.

Edelbaum, T. N., 1961, Propulsion requirements for controllable satellites; Am. Rocket Soc. Jour., vol. 31, no. 8, pp. 1079−1089.

Endres, D. L., 1967, Orbit trim propulsion requirements for sun-syncronous satellites; Goddard Space Flight Center Report X-734-67-288.

Endres, D. L., 1967, The influence of ground track constraints upon orbit requirements for sun-synchronous satellites; Goddard Space Flight Center Report X-734-67-598.

Escobal, P. R., 1965, Methods of Orbit Determination, John Wiley and Sons.

Fuchs, A. J., and F. A. Stafella, 1970, ERTS orbit selection analysis, Goddard Space Flight Center Report X-832-70-144.

Herrick, S., 1970, Astrodynamics, Van Nostrand Reinhold.

Hohmann, Walter, 1925, Dire Erreichbarkeit der Himmelskorper (The Attainability of Heavenly Bodies), R. Oldenbourg (Munich-Berlin).

Holcombe, L. B., and D. H. Lee, 1974, Survey of auxiliary propulsion systems for communications satellites, communications satellite technology; Progress in Astronautics, vol. 33, The MIT Press.

Isley, W. C., and K. I. Duck, 1974, Propulsion requirements for communication satellites; Progress in Astronautics and Aeronautics, Vol. 33, The MIT Press.

Jasper, P. E., 1972, Satellite trajectory analysis for remote sensing (STARS); Report From NASA Contract NAS5-11903.

Jensen, J. et al., 1962, Design Guide to Orbital Flight, McGraw-Hill Book Company.

Jensen, J. et al., 1963, Orbital Flight Handbook, NASA Report SP-33.

Kampos, B., 1968, General perturbations theory; guidance, flight mechanics and trajectory optimization, vol. IX, NASA Report CR-1008.

Karrienberg, H. K., E. Levin, and R. D. Luders, 1969, Orbit synthesis; Jour. Aerospace Sciences, vol. XVII, no. 3.

Kaufman, B., and R. Dasenbrock, 1973, Semianalytic theory of long-term behavior of earth and lunar orbiters; Jour. Spacecraft Rockets, vol 10, no. 6 pp. 377−383.

King, J. C., 1971, Swathing patterns of earth-sensing satellites and their control by orbit selection and modification, American Astronautical Society Paper No. 71-353.

King, J. C., 1973, Subnodal points—their location and use as geographic benchmarks for satellite pass registration, Goddard Space Flight Center Report X-591-73-296.

King, J. C., 1975, Concept and analytical basis for REVISTAS—a fast, flexible computer/graphic system for generating periodic satellite coverage patterns, Goddard Space Flight Center Report X-932-75-261.

King, J. C., 1976, Quantization and symmetry in periodic coverage patterns with applications to earth observation; Jour. Aerospace Sciences, vol. XXIV, no. 4.

Macklis, H., R. L. Sackheim, and B. Vogt, 1981, Selection of an optimized integrated propulsion system; Jour. Spacecraft and Rockets, vol. 18, no. 6, pp. 449−505.

Nautical Almanac Office, U.S. Naval Observatory, 1982, Astronomical Almanac, U.S. Government Printing Office.

Nelson, W. C., and E. E. Loft, 1962, Space Mechanics, Prentice Hall Publishers.

NOAA et al., 1976, U.S. Standard Atmosphere, U.S. Government Printing Office.

Pritchard, E. I., 1979, Stnadard mission destination for sun-synchronous satellites study, Aerospace Corporation Report No. ATR-79(7678-01)-1.

Russell, N. H., et al., 1945, Astronomy I. The Solar System, Ginn and Company.

Sackheim, R. L., D. E. Fritz, and H. Macklis, 1980, Performance trends in spacecraft auxiliary propulsion systems; Jour. Spacecraft and Rockets, vol. 17, no. 5, pp. 390−395.

Slater, J. C. and N. H. Frank, 1947, Mechanics, McGraw-Hill Book Company.

Stearns, E. V. B., 1963, Navigation and Guidance in Space, Prentice Hall Publishers.

Thomson, W. T., 1961, Introduction to Space Dynamics, John Wiley and Sons.

Townsend, G. E., and M. B. Tamburro, 1968, The two-body problem; guidance, flight mechanics and trajectory optimization, Vol III, NASA Report CR-1002.

Wagner, C. A., 1974, Effect of resonance-oblateness coupling on a satellite orbit, Goddard Space Flight Center Report X-920-74-334.

Yetman, A. A., 1982, Private communication.

Zendell, A., R. D. Brown, and S. Vincent, 1975, Gravity fields of the solar system, NASA Report SP-8117.

CHAPTER 17

Data Processing and Reprocessing

Author-Editor: FRED C. BILLINGSLEY
Contributing Authors: PAUL E. ANUTA, JAMES L. CARR, CLARE D. MCGILLEM, DARRYL M. SMITH, TIMOTHY C. STRAND

GENERAL CONTENTS: The basic imaging process; image and data conversion; image sampling and reconstruction; information and encoding; geometric modifications and methods of correction; spatial frequency modifications; coherent noise; optical image processing; references.

INTRODUCTION

The traditional method for extracting data from images has been by manual interpretation. The recording of the information thus extracted was of necessity also by time-consuming manual processes subject to the many vagaries of human operations. In recent years, the modern digital computer has assisted the human interpreter by removing camera distortions and other anomalies, emphasizing details, modifying the tonal range, and producing new color displays. Events have shown that the best analyses may be obtained (at least from images originally obtained digitally) by digital extraction of the desired information. Modern image sensors produce digital data to allow this processing.

In particular, computer operations allow operations to be carried out that are infeasible by other methods. These include meaningful calibrations; geometric manipulations, such as map matching and overlay registration; and mathematical operations, such as ratioing. Such preprocessing, however, produces data alterations which must be understood by the user/analyst.

The systems outlined in other chapters are snapshots in time of a rapidly evolving situation; details of the configurations and the specific processing provided will change. The data attributes which drive a given system design, such as timeliness, reasonable cost, or data quantity will dictate systems designs requiring various degrees of preprocessing; there is no way that these can individually be outlined nor discussed in detail. However, since preprocessing removes many users one or more steps away from the raw data, it is critical that users become familiar with the effects of processing on the data.

This chapter will therefore be tutorial on data characteristics as affected by the system and the various preprocessing steps required. This approach is taken because many of the steps are generic and will have been performed on the data prior to access by the analyst, and because many users will wish to perform these same functions on their own data. This chapter concentrates on processing data from sensors involving digital images from the Landsat and GOES satellites. However, many of the processes are generic to all sensors.

Many of the illustrations and much of the text have been adapted from the previous edition of the Manual. The use of these is gratefully appreciated.

THE SYSTEMS MILIEU—AN OVERVIEW

HISTORICAL SITUATION

Experimenters on early NASA earth observation missions used data principally from one instrument, and were responsible for much of the mission design, often the instrument design, and the data analysis (Figure 17-1a). Because of this intimate connection with the data, the experimenter knew the data conditions and had the clear responsibility to assure proper analyses and extraction of the desired scientific information.

Later missions such as Landsat and GOES are beginning to have an operational flavor. These missions serve many users, on a one-to-many plan (Figure 17-1b); while one instrument is the prime data source, these data are used by many experimenters in parallel, each performing his own analyses. Under these conditions, many analysts perform the same generic tasks; they also perform some functions, such as calibration, which can profitably be done by the flight project to service all. To allow communication of results between users and to assure correct data to all, the details of the processing steps employed must be clear to all.

It is becoming evident that many tasks require data input from many sources (e.g., many spacecraft passes, data from multiple spacecrafts, maps, point data, etc.) (Figure 17-2). This places a further burden on the system and on analysts to assure that the various data are made compatible. As secondary users begin to combine instrument data with information derived by others, the understanding of someone else's processing becomes even more important, if correct results are to be obtained.

This also requires that the data from the various sources be prepared in compatible formats and, at least for geographically related data, be registered to each other. Historically this has been done by each user, requiring much parallel development of the required capabilities. Again, a system service

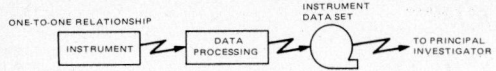

Fig. 17-1a. The early space missions have been built on a one source—one user plan (NASA GSFC).

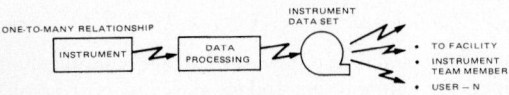

Fig. 17-1b. Recent missions such as Landsat and Seasat serve many users (NASA GSFC).

may be provided in which all users must be aware of the processing performed and its effect on the data received.

Although it is unrealistic to expect sensor data to be obtained in compatible form, it is not unrealistic to expect a coordinated effort to provide data in compatible form to the users (Figure 17-3). Such a coordinated effort is beginning to occur, requiring at the systems level overt attempts to put data from various sources into compatible formats with even more documentation of processing performed, and with commensurate attention to the effects this processing will have on the utility of the data to an ever-widening group of users.

TECHNICAL DATA ATTRIBUTES

The evolving data milieu as outlined above has implications on the types of preprocessing allowed and on the basic characteristics of the data themselves. The generic characteristics desired for all data may be identified as:

Digital—Many analysis processes are capable of greater finesse if the analysis is digital. Therefore, if the data are available digitally, all the preprocessing and archiving should be digital.

Correct—Data rectification, either radiometric, geometric, or otherwise, if perfect, will obviate the need for users to perform this step. However, if not perfect, the effect may be non-reversible. The rectification parameters, whether applied or not, should be available with the data.

Early Generation—Each subsequent operation on data tends to degrade it somewhat. Therefore, the earliest generation data should be archived. It is recognized that, for some missions, results of analysis may also be made available to users.

Easily Analyzable—The ability to apply generic processes during retrieval will provide data in formats easily analyzed and correlated with other data sets.

In the following discussion we have found it useful to define the various conditions (levels) of the data as follows:

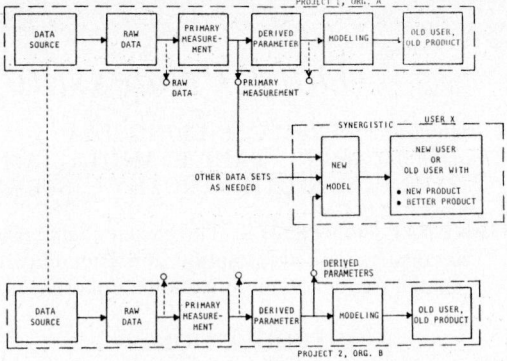

Fig. 17-2. Combined data from several sources is beginning to be used by many users (NASA GSFC).

Level 0   — Reconstructed sensor raw data at full resolution

Level 1A — Reconstructed sensor raw data at full resolution, time-referenced and annotated with ancillary information, with radiometric and geometric calibration coefficients and georeferencing parameters computed and appended but not applied to the Level 0 data. In some cases, individual flight projects may choose to apply radiometric calibrations to the data, so long as this process is reversible.

Level 1B — Level 1A data, with radiometric and geometric coefficients applied; raw image data may be resampled in accordance with user-specified parameters such as:
—resampling (nearest neighbor, cubic, etc.)
—map projection
—pixel size

Level 2 and Beyond—Information derived from Level 1 or other data.

Within the constraints of the attributes as outlined, data preprocessing will typically massage the data at each point (one pixel at a time, as in radiometric calibration), locally (in a neighborhood around a pixel of interest, as in edge enhancements), or globally (the entire image, as in geometric warping for geographic registration).

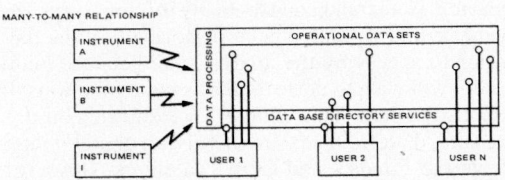

Fig. 17-3. Development of a matrix-like data distribution structure is expediting data delivery and usage (NASA GSFC).

IMPACT OF SENSOR CHARACTERISTICS ON PROCESSING REQUIREMENTS

For remote sensing applications, images are produced by sensors which may be grouped into the following broad classes: photographic, television, mechanical scanners, and linear arrays. Each of these has its own peculiarities:

Photographic—These images contain a very large amount of detail, but this detail is only available if the photographic film is physically returned to the analyst. This is not a trivial limitation, as it is not now cost-effective to return film from long-lasting NASA-satellite remote sensing missions, and therefore film photography in a practical sense is not available from these missions. Photography suffers from the limitation that it can only reproduce data in the visible and near-infrared spectral regions and that the photographic process, especially for color film, is nonlinear and quite variable. Data analysis must be manual or optical unless the film is scanned and converted for digital processing. Camera design is stringent; not a problem for the user, but this may be a restricting factor in the consideration of new missions which might request cameras as sensors. See the discussion in Chapter 6.

Television—Exemplified by the Return Beam Vidicon (RBV) cameras on the Landsat satellites. These can produce data in real time for transmission to Earth, with good resolution and a reasonable noise figure. The RBV on Landsat 3 produced images with about 4000 scan lines (the critical factor, as the ground resolution can be varied by varying the lens focal length). Spectral range is restricted, and multispectral images may be difficult to assemble at full resolution since typically duplicate sensors are used, one for each spectral band. For high-resolution television systems, raster linearity (and stability of raster shape) is a problem, resulting in the need for geometric correction. In common with film systems, the exposure is essentially instantaneous, so that motion during exposure is a secondary effect.

Scanners—Exemplified by the Multispectral Scanner (MSS) and Thematic Mapper (TM) on Landsat, the Coastal Zone Color Scanner (CZCS), and the Heat Capacity Mapping Mission (HCMM) scanner. These utilize a reciprocating mirror to sweep the field of view of a detector-group perpendicular to the spacecraft motion, while utilizing this motion to successively displace the scan lines and so produce a two-dimensional image. The requirements for precision scanning result in a quite complex device, exacerbated as more spectral bands are required. The expectation is that, because the spectral bands are more or less simultaneously imaged, they will be in perfect registration; the extent to which this is true depends on the particular scanner design. Global geometric distortions that arise from earth rotation and spacecraft wobble during the time required for the spacecraft to travel the distance of one frame (typically 20–30 seconds), and the difficulty of producing a precisely linear scan, cause along-scan-line distortions. These may all be modeled, but require appreciable computer time for removal. The spectral range available extends from the ultraviolet to the thermal infrared. In general, the detectors have a very wide linear dynamic range, so that radiometric calibration activities may be minimized. See also the discussion in Chapters 8 and 12.

A second type of scanner uses a rotating prism or mirror system for the crosstrack sweep, producing a conical scan pattern on the ground. This has the advantage of always having a constant path length and angle to the ground, but the circular scan causes added difficulty in image rectification. A third scan system uses a fixed detector on the spacecraft (s/c) and uses the spin of the entire s/c for the crosstrack sweep. See the discussion in Chapter 14.

Linear Array Camera—This is the latest type of sensor to be developed, and is proposed for future NASA missions, for the SPOT (Systeme Probatoire d'Observation de la Terre) system developed by the French, and for future Japanese missions. See also the discussion in Chapter 13. It uses a line of sensors per spectral band aligned perpendicular to the line of s/c motion. The image of this array is "pushed" along the ground in much the same manner as the head of a push broom; hence the nickname "pushbroom sensor." The important characteristics are: fixed and equal detector-to-detector spacing; a long detector-integration time, allowing very high signal/noise and imaging in very narrow spectral bands; light weight and small size compared to a scanner, allowing a pointable instrument. Production image processing will differ from that for a scanner in that the N detectors per scan line per spectral band will require radiometric calibration. Geometric processing will be essentially the same as is required for a mechanical scanner because earth rotation, s/c wobble and panoramic projection effects will still occur. However, there will not be as much along-line detector geometric rectification required.

An advanced version of the "pushbroom" utilizes an area array sensor. The long dimension of the array is oriented crosstrack as in the pushbroom just discussed; the spectrum of each pixel is spread by a dispersing element and imaged across the short dimension. Thus a large number of narrow (typically 0.01 to 0.02 μm) spectral bands are potentially available, which are inherently registered. This sensor is currently in the research stage. (Wellman, 1981)

It should be noted that the above discussions apply to both airborne and spaceborne sensors. For digital processing of photographs, it will be necessary also to convert the images to digital form. Discussion of the details of this process and the required scanner mechanisms may be found in Chapter 6, in Castleman (1979), and Montuori (1980), and Slater (1980). Film scanning, if used, is yet another link in the series of conversions from the true scene to the extracted information.

IMAGE AND DATA CONVERSION

THE BASIC IMAGING PROCESS

For transmission of an image to the ground from a spacecraft (except for film physically returned) and for digital processing of the data, conversion of the continuum image radiance field to a quantized array (in both space and image irradiance) is required. Keeping in mind the various characteristics of the different sensor classes, we may consider that most systems may be characterized as approximately linear, space-invariant, and non-vignetting, at least to the extent of allowing separable processing for each of the three. These assumptions will be true in the local case, if not globally, and are a necessary set of assumptions for tractable analysis. Global characteristics cannot be ignored, however, and cause major problems in geometry which will be treated later.

The following definitions will be found to be useful:

PIXEL—A Picture Element. A single digital measurement. For the Landsat multispectral scanner system (MSS) the pixels are located on the ground on 57 (cross track) × 79 (along track) meter centers.

INSTANTANEOUS FIELD OF VIEW (IFOV)—That scene area contributing to a single digital measurement, as defined by the instrument aperture. For the Landsat MSS, it is an area on the ground of approximately 76 × 76 meters, as defined by a fibre optics light pipe between the focal plane and the detector.

EFFECTIVE RESOLUTION ELEMENT (ERE)—That part of a scene contributing to a single digital measurement. It is primarily defined by the IFOV, but is also affected by the optics, electronics and sample spacing in the instrument. It may be further affected by image-motion smear and the variable effect of the atmosphere. A precise definition must be made to suit a given application, viz.: 1) The spacing required to make two adjacent equal-brightness points just discernible, given a certain background brightness and system noise; 2) The size of an element for which (say) 95% of the energy is recorded by the sensor, given a background of a certain brightness; 3) The square-bar target spacing for a contrast reduction to (say) 5% of the target contrast, under defined atmospheric conditions; 4) The point at which 50% MTF is reached; etc. (see also Welch, 1977; Norton et al. 1977). Recognizing that all of the above are different, although of similar magnitude, and that some are instrument-only-related while others are scene dependent, the Landsat MSS may be said to be a "90 meter ERE" system.

Nature of an Image

In analyzing a digital image system it is found that the point, local and global characteristics of the system may often be treated separately. We will first treat the local analysis, considering that the system is linear to light (a point function), that the system is geometrically distortionless and that there is no vignetting (global). In addition, we will assume that the local characteristics are shift-invariant, that is, they are the same at all points in the image. Figure 17-4 illustrates the general imaging situation in which a brightness element at location x',y' of the scene is imaged into a radiance region at location x,y in the image. The shape of the image region will be defined by the impulse response, or point spread function,

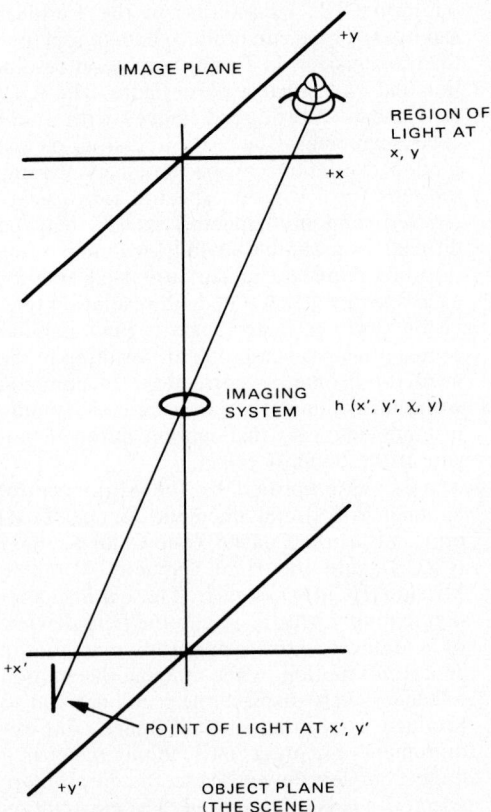

Fig. 17-4. The general imaging situation in which one ground point contributes to a region in the image.

$h(x',y', x,y)$ of the imaging system. The shift-invariance assumption will allow us to consider that h is $h(x,y)$ only.

Thus, in Figure 17-5, a patch of light is imaged at location x,y with an amplitude distribution $h(x,y)$ and we wish to find the total resultant image radiance at point x_0,y_0 caused by the summation of all of the $h(x,y)$'s contributing to the radiance at x_0,y_0. This image radiance is

$$q_{x_0,y_0} = \sum_{x,y} f((x_0 - x), (y_0 - y)) \cdot h(x,y) \quad (17\text{-}1)$$

which, by knowing the relation of the image reference axis to the impulse response reference axis

$$x = x_0 - x$$
$$y = y_0 - y$$

may be written as

$$q_{x_0,y_0} = \sum_{x,y} f(x,y) \cdot h((x_0 - x), (y_0 - y)) \quad (17\text{-}2)$$

Noting that the function $h(x_0 - x)$ is the mirror image of $h(x)$, displaced to the x_0 point on the x axis, the convolution Eq. 17-2 representing the total brightness at x_0,y_0 may be stated as:

To find one point q_{x_0,y_0} in the output image, $h(x,y)$ is rotated 180° and placed at location x_0,y_0 over the input image $f(x,y)$. Each value of h is multiplied by the coincident value of $f(x,y)$ over the range where both functions are non-zero. Then all are added to obtain q_{x_0,y_0}. This function is normally written in abbreviated form as

$$q(x,y) = f(x,y) * h(x,y) \quad (17\text{-}3)$$

when the arguments and origins are clear.

From Eqs. 17-2 and 17-3 it is evident that the measured or observed image is different from the true image as a result of the interaction of the sensor and the true image signal. Methods of estimating the true image from the measured image (called image restoration) have been developed and are useful in a number of remote sensing applications. The performance achievable depends on the nature of the point spread function of the sensor and the character and magnitude of the noise present in the measured image.

The satellite view q of the earth f (Figure 17-6) is affected by at least three impulse response functions: atmosphere h_a, the optics h_o, and the sampling aperture h_s. The resultant image

$$q(x,y) = f * h_a * h_o * h_s$$

will have low contrast, primarily due to the widespread effect of h_a, and loss of detail primarily due to the scanning aperture h_s. The optics generally (although not necessarily) have only a small effect. In the Landsat multispectral scanner an electronic filter following the detector adds one further component in the along-scan-line direction (see the discussion in Chapter 8). The net effect of the various spacecraft components, ignoring the h_a, is to produce a Landsat MSS composite point spread-function of approximately Gaussian cross section with a radius of gyration of about 24 meters. (For a circularly-symmetric Gaussian function having a standard deviation σ, the radius of gyration R is related to σ by $R = (\sigma/\sqrt{2})$.) A more detailed evaluation of the MSS spatial and spectral characteristics is given in Slater (1979). Other effects are discussed below in conjunction with spatial-frequency considerations.

System Point Spread Function

It is helpful to consider the system point spread-function from more than one point of view.

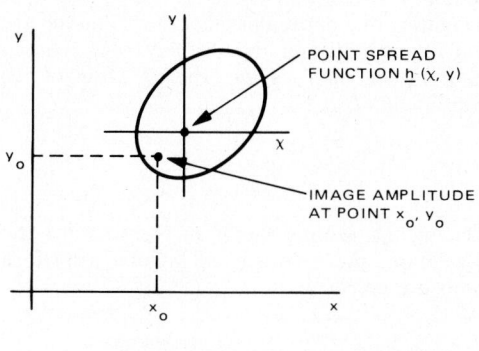

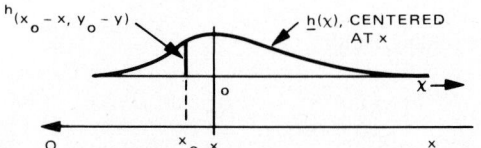

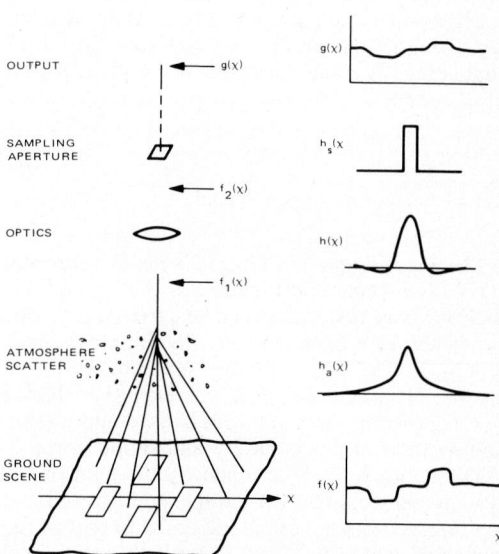

Fig. 17-5. Light from each image impulse contributes to total brightness at x_0, y_0.

Fig. 17-6. The total effective point spread response of a system has contributions from several components.

Generally speaking the PSF consists of a large central lobe and some skirts that extend out to distances many times the width of the central lobe. The PSF results from the interaction of many components such as the lenses, the detector, amplifiers and possibly the recorder. As a consequence of this the net PSF is actually the convolution of many individual functions. Each convolution is a type of smoothing operation and, as a consequence, the PSF of a complex remote sensing system can often be reasonably well approximated by a smooth shape such as a two-dimensional Gaussian pulse. The mathematical representation of such a PSF is

$$h(x_1,y_1) = K \exp\left[\frac{1}{2(1 - \alpha^2)}\left[\frac{x_1^2}{R_{x_1}^2} - \frac{2\alpha}{R_{x_1}R_{y_1}}x_1y_1 + \frac{y_1^2}{R_{y_1}^2}\right]\right]$$

where R_{x_1} is the radius of gyration about the x_1-axis, R_{y_1} is the radius of gyration about the y_1-axis and α is a measure of the coupling between the PSF in the two directions. By an appropriate rotation of coordinates it is possible to decouple the two axes and simplify the expression to

$$h(x,y) = K \exp\left[-\frac{1}{2}\left(\frac{x^2}{R_x^2} + \frac{y^2}{R_y^2}\right)\right] \quad (17-4)$$

The required transformation is

$$\left.\begin{array}{l} x = x_1 \cos\theta - y_1 \sin\theta \\ y = x_1 \sin\theta + y_1 \cos\theta \\ \theta = \frac{1}{2}\tan^{-1}\left|\frac{2\alpha R_{x_1}R_{y_1}}{R_{x_1}^2 - R_{y_1}^2}\right| \\ (R_x)^2 = R_{x_1}^2 \cos^2\theta - R_{y_1}^2 \sin^2\theta \\ (R_y)^2 = R_{x_1}^2 \sin^2\theta + R_{y_1}^2 \sin^2\theta \end{array}\right\} \quad (17-5)$$

Assuming that this rotation and rescaling of the radii of gyration have been carried out, the PSF can then be expressed in the form of Eq. 17-4. This expression is now in the form of what is called a separable point spread-function; i.e., one consisting of the product of two functions each containing only one variable. In the present case this becomes

$$h(x,y) = h_1(x)\cdot h_2(y) = \sqrt{K}\exp^{\dfrac{x^2}{2R_x^2}} \cdot \sqrt{K}\exp^{\dfrac{y^2}{2R_y^2}}$$
$$(17-6)$$

The big advantage of being able to represent $h(x,y)$ as a separable function is that operations such as image restoration can be carried out in one dimension at a time, thus greatly simplifying the processor. This is true for any separable PSF, not just the Gaussian; however, the Gaussian shaped PSF is probably the most widely used approximation because of its versatility and simple form.

The factor K is determined by the gain of the system and would be 1 if a single point were to be reproduced with an amplitude equal to that of the input and would be $(2\pi R_x R_y)^{-1}$ if the gray level were to be preserved by the system. The assumption of no vignetting is associated with K being constant across the entire image. This will be true in scanning systems such as the MSS. However, for systems in which an entire image is formed simultaneously, such as in film camera systems, lens vignetting (brightness falloff away from the center of the image) will cause a decrease in K radially from the center. This decrease is a slowly varying function, allowing its effect to be separated from the PSF function being addressed here. Further, as discussed in Chapter 6, correction for this is often included in the camera itself. If not corrected elsewhere, the vignetting must be digitally corrected before digital analysis, such as multispectral classification, is attempted. As the vignetting is a radial decrease in exposure, it may be modelled as a multiplicative correction.

Spatial Frequency and the Fourier Transform

The process described above, in which the image is considered to be the summation of point spread-functions, each having an amplitude corresponding to its scene radiance, is but one way of decomposing an image. In general, an image $F(x,y)$ may be decomposed into a set of coefficients $L_{m,n}$ representing the amplitudes of the corresponding members of any complete orthogonal basis set $\phi_{m,n}$ of sampling functions (the normal sequential sampling process being represented as a series of delta-function-like apertures). The ordered sequence of these coefficients is known as the transform of the image.

The output transform values $L_{m,n}$ can be considered as the degree of match between a mask function $\phi_{m,n}(x,y)$ [where $\phi(x,y)$ is the spatial weighting of a particular mask $\phi_{m,n}$] and the scene $F(x,y)$, each of total dimension $N \times M$. Thus, the total transform L is an ordered array of these coefficients $L_{m,n}$:

$$L = \sum_{m=1}^{M} \sum_{n=1}^{N} L_{m,n}$$

Placing the scaling factor in the forward transformation, the forward and inverse transformations are given by:

$$L_{m,n} = \frac{1}{K} \sum_{x} \sum_{y} F(x,y) \, \phi_{m,n}(x,y) \quad (17-7)$$

K = scaling factor

$$F(x,y) = \sum_{m=1}^{M} \sum_{n=1}^{N} L_{m,n} \, \phi_{m,n}(x,y) \quad (17-8)$$

A basis set is termed Walsh-like if its elements consist of piecewise constant functions, each spanning the entire image area. The possible set of functions is infinite; one simple example is illustrated in Figure 17-7 for a 4 × 4 array (Andrews, 1976; Andrews, 1979; Larsen, 1976; Arazi, 1975) and discussed further in the section on transform encoding.

The basis function need not be piecewise constant: the Fourier Transformation results when the basis set consists of two series of sinusoids, one in x and one in y, with their spatial scaling being given in cycles/unit distance. This domain is thus referred to as the spatial-frequency domain, and the Fourier transform is discussed here in this context.

The (two dimensional) Fourier transform of a function $f(x,y)$ is defined in the continuum domain as

$$F(u,v) = \mathscr{F}\{f(x,y)\} \qquad (17\text{-}9)$$

$$= \int_{\infty}^{\infty} \int_{\infty}^{\infty} f(x,y) \exp\left[-j2\pi(ux + vy)\right]dxdy$$

The Fourier transform is a linear operator for which an inverse exists, so that the original function (x,y) can be recovered from its transform $F(u,v)$ by the operation

$$f(x,y) = \mathscr{F}^{-1}\{F(u,v)\} \qquad (17\text{-}10)$$

$$= \int_{\infty}^{\infty} \int_{\infty}^{\infty} F(u,v) \exp\left[j2\pi(ux + vy)\right]dudv$$

It can be readily shown that the convolution of two functions in the spatial domain produces an identical result to that obtained by taking the inverse Fourier transform of the product of the Fourier transforms of the two original functions; i.e.

$$f*h = \mathscr{F}^{-1}\{F(u,v)\, H(u,v)\} \qquad (17\text{-}11)$$

This property of the Fourier transform makes it possible to carry out equivalent operations in either the spatial or the frequency domain when they involve shift invariant operations on an image. However, for the large images in particular, two difficulties are often encountered: first, each incoming image point contributes to every Fourier domain point, so that large storage and

clever data handling are required. Second, each point in the Fourier domain contributes to every point in the output image resulting from the inverse transformation. To keep round-off error tolerable, it is necessary to carry many significant figures in the processing operation. For these reasons, and because often the operators are local in nature (i.e., the PSF is non-zero only over a small neighborhood), processing frequently is carried out directly in the spatial domain even though the operation is designed on the basis of frequency-domain considerations.

Reconstruction of the image takes place by superimposing a set of sinusoids corresponding to the sinusoidal sampling set, each with a relative amplitude denoted by the corresponding coefficient (Eq. 17-10). At some suitable high spatial frequency it may be considered that the resulting image content is negligibly small. For undistorted reproduction of the scene the relative amplitudes of the sinusoids must not be disturbed in the imaging or reproduction processes. Thus, the system should have a flat frequency response up to the highest frequency of interest.

One property of Fourier transforms that is of particular importance in changing between spatial and frequency domain processing is that a function, whether it be an image, a filter transfer function or anything else, cannot be of limited extent in both the time domain and the frequency domain. This means that processing an image in the spatial domain with a short length filter is equivalent to processing in the frequency domain with a filter that extends to very high frequencies. It is particularly important to realize, when a particular processing operation is carried out with a truncated version of a more extended design, that this will alter the operation that results and that sometimes the alteration may lead to undesirable results. Some care must also be given to be sure that processing carried out in the discrete domain of the image representation has a desired and meaningful effect when viewed in the continuous domain; i.e., the validity of analogous operations in the discrete and continuous domains must be established.

System Degradations in the Spatial Frequency Domain

The Fourier transform of the system impulse-response function will produce a spatial-frequency plot much as sketched in Figure 17-8a. Similarly, the Fourier transform of the scene will display a spatial-frequency content as sketched in Figure 17-8b. Consideration of the convolution equivalence Eq. 17-11 will show that the spatial-frequency content of the resulting image is the frequency-by-frequency product of the scene content and the impulse-frequency response. Thus, the series of convolutions is converted to a series of frequency-by-frequency multiplications, often an appreciably simpler process.

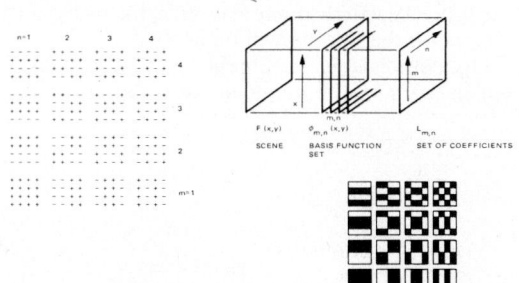

Fig. 17-7. The M = 4, N = 4 ordered array of weights $\phi_{m,n}(x,y)$ for a Hadamard transformation. All element magnitudes = ±1.

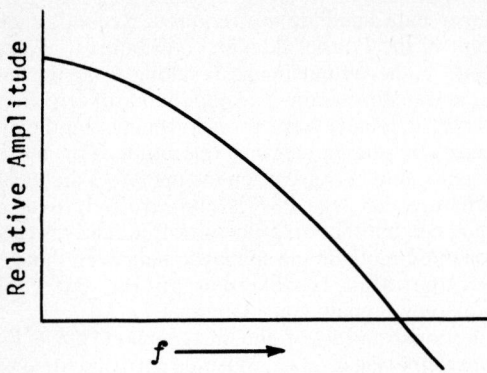

Fig. 17-8a. (left) Typical system spatial frequency response.

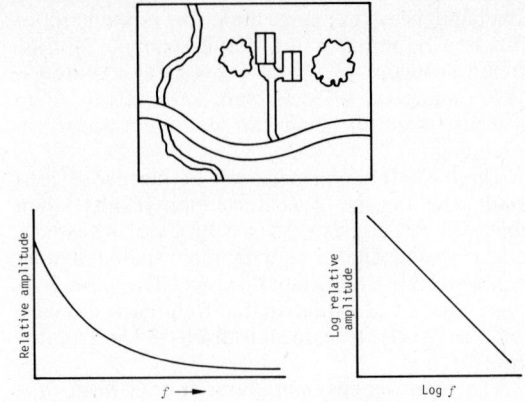

Fig. 17-8b. (right) Typical scene spatial frequency content.

The most common sources of image degradations are (Figure 17-9) contrast reduction by the atmosphere, atmospheric turbulence effects, image quality performance of the optical system, image quality characteristics of the transducer, uncompensated image motions from various causes, geometric distortions, radiometric dynamic range reduction, and noise of all types throughout the system.

The nature of the degradations imposed on the imagery is best described in the spatial-frequency domain by means of the modulation transfer function. This relation plots the modulation transfer (the ratio of output modulation to input modulation) as a function of spatial frequency (cycles or line-pairs per unit length) for an element or a combination of elements of the system. Subject to certain validity constraints, the modulation transfer functions (MTF) of each of the separate system elements may be multiplied together to obtain the MTF for the complete system. The modulation of the output image is obtained by multiplying the overall system MTF by the modulation of the scene or target.

Modulation may be defined as:

$$M = \frac{E_{max} - E_{min}}{E_{max} + E_{min}}$$

where:

E_{min} and E_{max} are the minimum and maximum radiances, respectively, in the scene or in the image, and are related to the contrast ratio by:

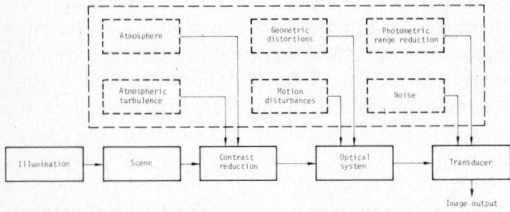

Fig. 17-9. Degradation of an image.

$$M = \frac{C - 1}{C + 1}$$

where the contrast ratio is given by:

$$C = \frac{E_{max}}{E_{min}}.$$

Absorption and scattering by the atmosphere (Chapters 5 and 6) between a scene or target and a remote sensor results in a reduction in contrast of all the elements in the scene. From a detailed point of view, the effect is quite complex, and this may be important in some instances; but in most cases, the effect on system performance is adequately represented by assuming a constant contrast reduction factor which is independent of spatial frequency.

Turbulence effects in the atmosphere—which astronomers refer to as "seeing"—deviate the direction of propagation of the light on a small scale basis. From another point of view, the effect is to distort the incident wave-fronts. Image points are displaced randomly in time and space. Figure 17-10 shows the MTF associated with this type of effect. The abscissa is the normalized spatial frequency (K_o), with unit values equivalent to the reciprocal, or $1/a$, where a is the rms value of the image displacement in an exposure interval. However, in viewing the earth's surface from orbital altitudes with the coarse spatial resolution of the typical earth resources sensors, the turbulence effects on the images will be negligible.

Optical systems are characterized by an MTF which has a spatial frequency cutoff at the value,

$$\frac{1}{F\lambda}$$

where:

$$F = f - number = \frac{focal\ length}{lens\ diameter}$$

λ = wavelength of the light.

Typical curves are shown in Figure 17-11 for an ideal circular aperture, an obstructed aperture,

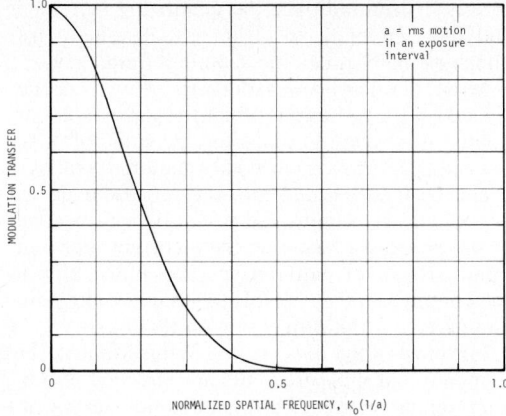

Fig. 17-10. Modulation transfer function for random image motion (normalized to a spatial frequency $K_o = 1/a$ where a equals rms motion in an exposure interval).

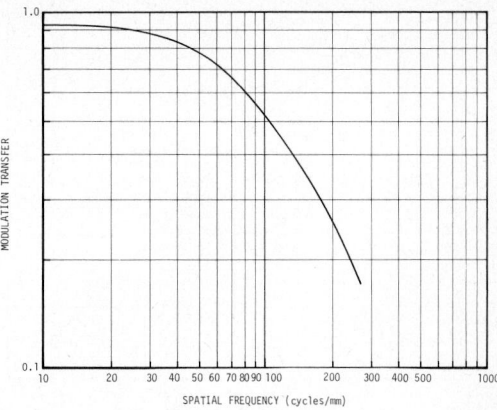

Fig. 17-12. Modulation transfer function for a typical positive duplicating film (EK-2430).

and a circular aperture with a small amount of defocus.

All real optical systems have a greater or lesser amount of residual optical aberrations, and also suffer from small errors in fabrication, assembly, and alignment. These errors result in depressing the MTF curve in the mid-frequency region, in much the same way as obscuration and focus errors.

The transducer in the focal plane of the system will have its own degradation of the image, whether it be a photographic film, a scanning detector, or some other element. Figure 17-12 shows the MTF of a typical photographic film, and Figure 17-13 shows that of a Gaussian scanning spot—typical of a television-type raster scan.

Motions of the image during the exposure time result in a smear which degrades the image. These motions may be linear, such as result from uncompensated forward motion of the vehicle; sinusoidal, such as result from vibrations; or may take other forms. Figure 17-14 shows the MTF's for linear and sinusoidal image motion.

Modification of the spatial-frequency response of the system to reduce the degradations outlined above are discussed in the section on Spatial Frequency Modifications.

System Noise

The presence of noise in the measurement process leads to a measured image that differs from the true scene by the error introduced by the noise. For purposes of analysis noise can be divided into two broad categories: signal-independent noise and signal-dependent noise. For signal-independent noise there is a component added to the image that comes from the sensor, the transmission path or some other source. The image resulting from signal-independent noise degradation can be modelled as

$$f_3(x,y) = f_2(x,y) + n(x,y)$$
$$= f_1(x,y)*h(x,y) + n(x,y) \qquad (17\text{-}12)$$

where $n(x,y)$ is the amplitude of the random noise component. This type of model provides a good representation for detector thermal noise and shot

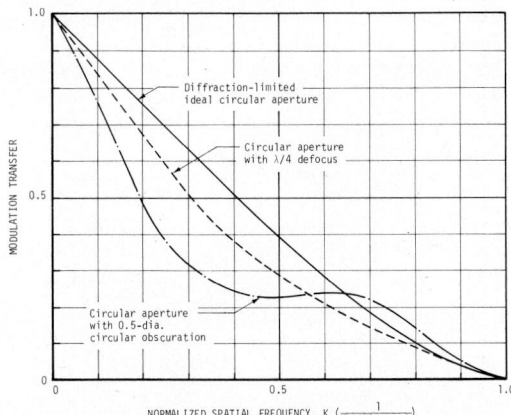

Fig. 17-11. Modulation transfer function for diffraction-limited optical systems.

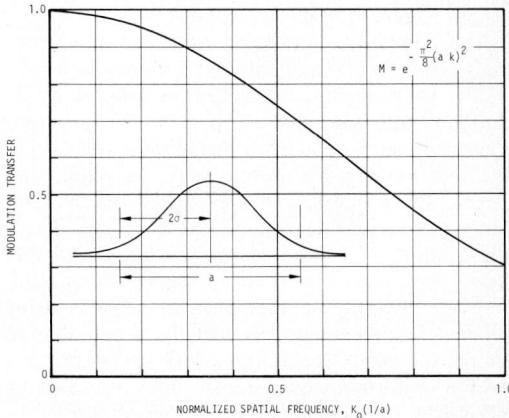

Fig. 17-13. Modulation transfer function for a circular scanning spot with Gaussian transmission function.

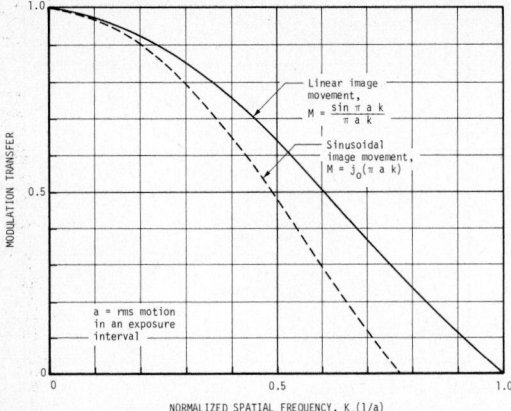

Fig. 17-14. Modulation transfer functions for image motion (normalized to a spatial frequency $K_o = 1/a$ where a equals rms motion in an exposure interval).

noise and for quantization noise occurring in A/D conversion.

For the case of signal-dependent noise, the noise component is of the form $w_1 f$ where w_1 is a random coefficient. For this case the model is

$$
\begin{aligned}
f_3(x,y) &= f_2(x,y) + w_1 f_2(x,y) \\
&= f_2(x,y)\left[1 + w_1\right] \\
&= w \, f_2(x,y)
\end{aligned}
\tag{17-13}
$$

It is clear from Eq. 17-13 why this is called multiplicative noise. This type of noise occurs in photographic recording in which the images are formed from irregular shapes and sizes of grains in the emulsion; it occurs when the transmission medium has a fluctuating absorption coefficient as in the case of turbulence; and it occurs when the signal is used to modulate the intensity of an electron beam used for recording purposes. One approach to dealing with multiplicative noise is to work with the logarithm of the image intensity. This gives

$$
\log f_3(x,y) = \log f_2(x,y) + \log w \tag{17-14}
$$

in which it is seen that the noise is now additive, which is generally more tractable.

The complete characterization of a noise process requires a statistical description which provides the probability density-functions of single and multiple samples taken from the process. Generally such detailed information is not available and some assumptions must be made that permit the processing to be carried out without danger of serious degradation of the image. There are a number of situations in which a reasonably good representation of the noise is known (Billingsley, 1975). For example, for sensors using electronic detectors it is usually acceptable to model the noise as a stationary Gaussian process with a uniform spectral density that increases as 1/f at low frequencies (Scott, 1963; Shlien 1979). For sampled and quantized data the probability density function of the quantization noise is uni-

formly distributed over the quantizing increment and the frequency spectrum is constant to frequencies many times the sampling frequency.

Most picture-processing algorithms are designed on the assumption that the noise is uncorrelated from pixel to pixel and has a uniform frequency spectrum. If more information is available it can be incorporated into the processor design and should give improved performance over that of the processor based on the incorrect representation. However, without specific information to the contrary it is reasonable to assume white noise and carry out the processing on that basis.

For most of the data treated in this Manual, it is assumed that measurements are encoded into binary digital words. The quantization levels are commonly referred to as gray, brightness, or radiance levels, although it is recognized that some other physical quantity may be of concern. Noise can be characterized by its effect on the probability that a given digital sample will be assigned the precise value corresponding to the true scene radiance at that point (Billingsley, 1975). Briefly, the quantizer transforms the magnitude of the signal into a discrete number of steps. In a noiseless system there is no ambiguity in the designation of a particular signal level as a certain digital number (DN). However, in the presence of noise (assumed to be random) it is the signal plus noise which is quantized, and the level of the signal alone is somewhat uncertain from inspection of the digital number. Specifically, there is a finite probability, measured by the relative area of the probability distribution bounded by the quantization step boundaries, of assigning the wrong DN to a given measurement. This is illustrated in Figure 17-15. A derivation and a discussion of the effect of noise were given by Friedman (1965) and the resulting curve is the ± 0 DN curve of Figure 17-16. This curve can be approximated in the range $1 < \beta < 7$ by $\beta \log P = -0.40$. This analysis may be extended to consider the probability of

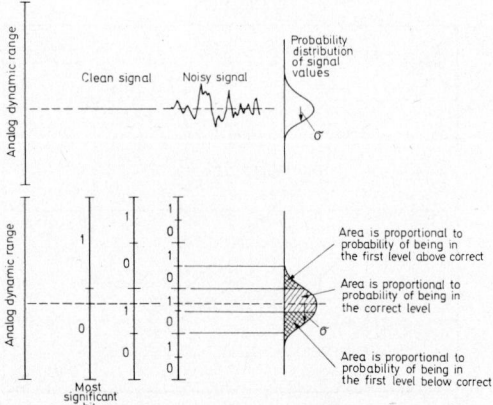

Fig. 17-15. Noise will cause the indicated brightness digital number to have a finite probability of being incorrect.

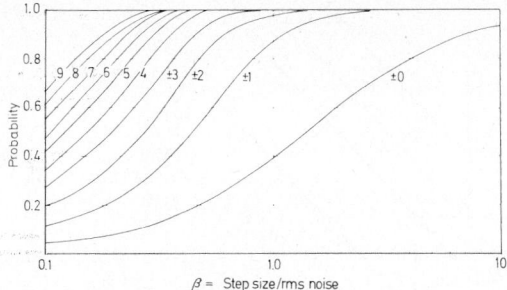

Fig. 17-16. Given a signal uniformly distributed over the quantization intervals. Given a Gaussian noise of value = σ. The curves show the probability of correctly assigning a digital value corresponding to the noise-free signal within ± 0, ± 1, . . . , ± 9 DN (inclusive) as a function of the ratio β = step size/σ. (Billingsley, 1975).

correct digitization to within $\pm L$ digital levels of the correct value (Fultz, 1965). Curves for this are also shown in Figure 17-16.

For the case of multispectral (primarily image) data, this analysis provides a simple model of the effect of noise on context-independent multispectral classifications. A series of samples of a given material is normally taken to be Gaussianly distributed in brightness with a restricted range σ. Classification amounts to determining whether a given pixel is within a certain range, the class size, defined appropriate to the average radiance and its σ for this given material. It can be seen that this class size is precisely analogous to the width of one digitizing step in the previous example, so that the ± 0 curve serves to define the probability that a given sample is within the class interval.

If the noise is uncorrelated from pixel to pixel (approximately true) as $n \times n$ pixels are averaged together, the noise will decrease by n. This, combined with the considerations of the previous paragraph, leads to the set of coupled graphs of Figure 17-17 to estimate the system performance in the presence of noise.

Intensity Quantization Effects

It can be appreciated that, for a given dynamic range, as the number of quantization levels increases, the size of each level decreases. For a constant system-noise performance, this results in a decrease in β and a decrease in the probability of exact assignment of the correct DN to the measurement. However, since the interval defined by one DN is decreasing, the measurement is becoming more precise, and the variance of the quantization noise

$$\sigma^2 = \frac{(\text{step size})^2}{12} \qquad (17\text{-}15)$$

is continually decreasing.

One estimate of the number of useful digitization steps may be obtained by analysis of the effects of the various noise sources on the probability of correct classification. Define the perfect

sensor as having no random noise nor quantization error (i.e., an infinite number of bits). This will define

$$\beta_o = \frac{\text{class size} \cdot n}{\sigma_{\text{scene}}} \qquad (17\text{-}16)$$

and

$$P_o = 10^{-0.4/\beta_o} \qquad (17\text{-}17)$$

For the real sensor, $\beta < \beta_o$ because of the finite σ_{sensor} and $\sigma_{\text{quantization}}$. The new probability of correct classification, P, is related to P_o by:

$$P = P_o^{(\beta_o/\beta)} \qquad (17\text{-}18)$$

The loss in classification accuracy, $\Delta P = P_o - P$, thus depends on P_o and the ratio

$$\frac{\beta_o}{\beta} = \sqrt{1 + \frac{\sigma_2^2 + \sigma_3^2}{\sigma_1^2}} \qquad (17\text{-}19)$$

where:

σ_1 is the scene noise
σ_2 is the sensor random noise
σ_3 is the quantization noise.

A plot of the loss in classification accuracy vs P_o is given in Figure 17-18, for the parameter families β_o/β and σ_2/σ_1. Noise allocation starts with the definition of the desired P_o and ascertaining that the required β_o can be obtained. Definition of the allowed ΔP determines the allowed σ_2/σ_1. An estimation of the scene noise for which the other conditions apply allows the calculation of the total sensor-noise allowed. The final step is to partition this noise between sensor random noise and quantization noise. For example, let P_o = 85 percent. Then $\beta_o = -0.40/\log 0.85 = 5.67$. If we allow a 2 percent loss in accuracy due to the sensor, $\sigma_2/\sigma_1 = 0.6$. If the expected scene for these accuracies has a noise (nonuniformity) $\sigma_1 = 2$ percent, the sensor can have a noise figure $\sigma_2 = 0.06 \ \sigma_1 = 1.2$ percent. If the $NE\Delta\rho$ (random) of the sensor is expected to 1 percent, the allowable $\sigma_3 = \sqrt{1.2^2 - 1^2} = 0.66$ percent. This can be met with 6-bit quantizing, for which $\sigma_3 = 1/64 \ \sqrt{12} = 0.45$ percent.

IMAGE SAMPLING AND RECONSTRUCTION

INFORMATION AND REDUNDANCY

In the following discussion of information systems, the terms *data* and *information* will both be used. From a scientific point of view there is a semantic difference between the two: *data* is a more general term and refers to the numerical results of any set of measurements, regardless of whether or not the measurements have been acquired with a certain purpose in mind; data become *information* after they have been retrieved and processed for a particular use (Tomlinson, 1972).

More precisely stated, "information is an aggregate of facts so organized or a datum so

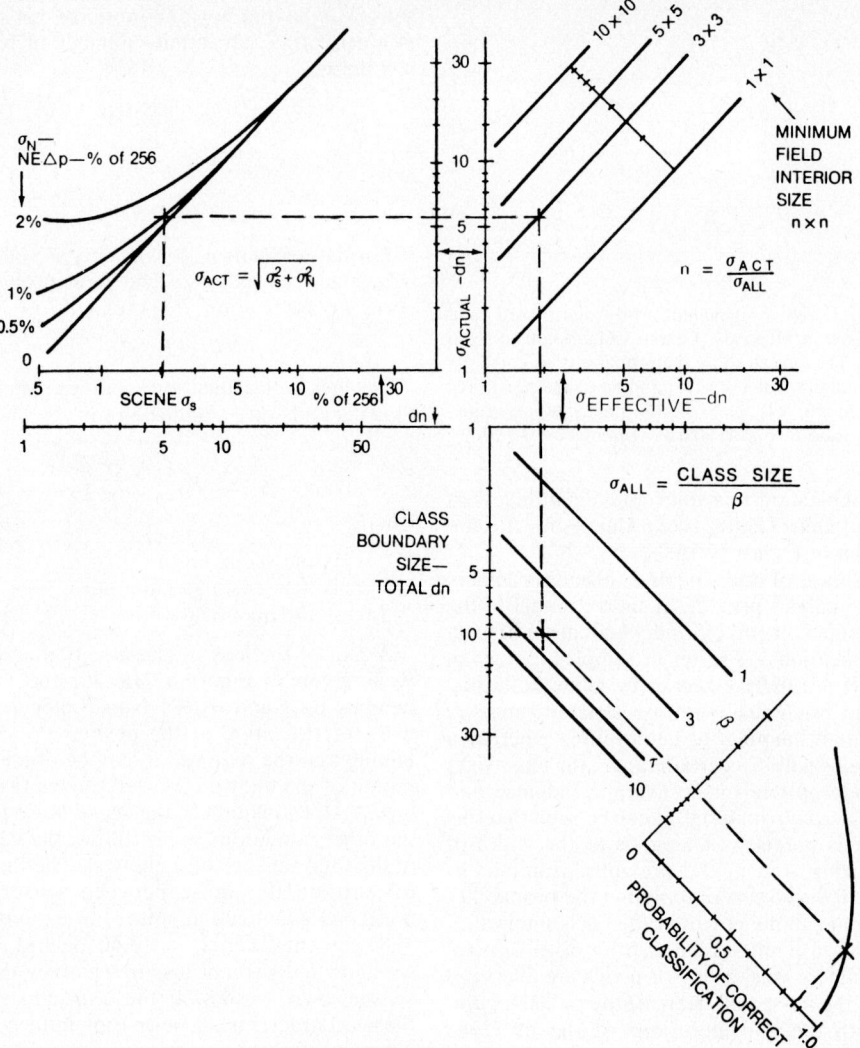

Fig. 17-17. Comparison of noise to classification class size allows estimation of the probability of correct classification (Billingsley, 1982).

utilized as to be knowledge or intelligence. Information is meaningful data, whereas data as such have no intrinsic meaning or significance'' (Rosove, 1967). It will not be possible to make this distinction consistently in this discussion, and the two terms will often be used interchangeably. However, the difference should be kept in mind when it is important.

Information can be treated very much like a physical quantity, such as mass or energy (Tribus and McIrvine, 1971), and a system can be devised which measures the amount of this information. The quantitative treatment of information is known as *information theory*, and was first formulated by Shannon (Shannon and Weaver, 1949).

The following discussion is based mainly on Young (1971).

Suppose that a message consists of E elements and that each element can take on S different states, all equally likely to occur. Then there are

$$M = S^E$$

different messages possible, and this is a measure for the information capacity of a system using such messages.

It has been found convenient to express the amount of information logarithmically, i.e., to take

$$\log{}_b M = E \log{}_b S. \qquad (17\text{-}20)$$

The choice of the base to which the logarithm is taken determines the unit of information. With $b = 2$, the measurement is made in binary units of information. If a message contains just one element (i.e., $E = 1$), which can assume two possible states, then $M = 2$ and $\log_2 M = 1$. This consti-

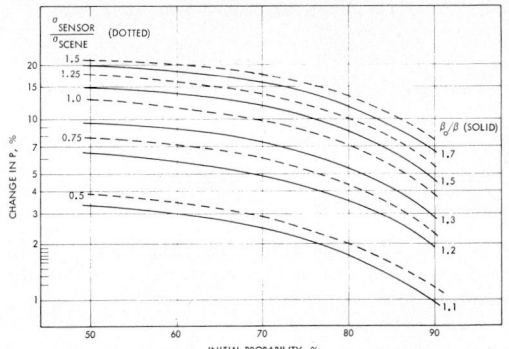

Fig. 17-18. A change in noise causing a change in β will result in a change in the probability of correct classification (Billingsley, 1982).

tutes one *bit* (or binary digit) of information. For a message consisting of two elements ($E = 2$),

$$M = 2^2 = 4 \text{ and } \log_2 M = 2$$

and thus each time a binary digit is added, the amount of information conveyed is doubled.

Note that, in this discussion, the message which is encoded is the scene (for an image) plus any noise added prior to the encoder. The encoder has no knowledge of this noise and treats everything as a true message. The effects of noise have been treated in the discussion related to Figures 17-16 and 17-17.

By considering unequal probabilities, one can find a quantitative expression for the effective amount of information received. First, consider the case of equal probabilities for S possible states. Then the probability of any one state occurring is

$$P = 1/S.$$

The information measure for one element ($E = 1$) can now be rewritten as

$$\log_2 M = \log_2 S = -\log_2(1/S) = -\log_2 P.$$
$$(17\text{-}21)$$

If the probabilities differ for the different states, then a weighted average known as *entropy* can be calculated:

$$H_{avg} = -\sum_{i=1}^{S} P_i \log_2 P_i \text{ bits of information} \quad (17\text{-}22)$$

where

$$H = \text{information content.}$$

This is the average amount of effective information received per elementary message. If a message consists of E elements, then it contains:

$$H = -E \sum_{i=1}^{S} P_i \log_2 P_i \text{ bits of information} \quad (17\text{-}23)$$

Consider the case of a quantized digital picture with gray levels regarded as a set of messages.

The introductory discussion of information and its measurement indicated that the information in a message can be thought of as measured by the degree to which the receiving of it reduces our uncertainty as to what the message will be. Consequently, a highly probable message contains little information, while an improbable message has a high information content. Therefore, if the probability P_i of the ith message is unity, the message carries no information. As P_i goes to zero, the information content of the message becomes arbitrarily great.

If there are N possible messages with respective probabilities

$$P_1 \ldots, P_N,$$

then the average amount of information as given by Eq. 17-22 may be rewritten with N replacing S, as

$$H_{avg} = -\sum_{i=1}^{N} P_i \log_2 P_i \text{ bits.} \quad (17\text{-}24)$$

Extrapolating previous discussions, it may be stated that H is greatest if each message has a probability of $1/N$. The maximum possible average information is therefore

$$H_{\max} = \sum_{i=1}^{N} (1/N) \log_2 N = \log_2 N \text{ bits} \quad (17\text{-}25)$$

To apply these concepts to pictues, a quantized digital picture may be regarded as a set of messages by considering the grey levels of each element as a message. If there are N gray levels, the total amount of information in an $M \times M$ digital picture (the average amount per element times the number of elements) can be as great as

$$H_{tot} = M^2 \log N. \quad (17\text{-}26)$$

The actual available information content then depends on the probabilities with which the gray levels occur.

Sampling Rate

The discussion so far has been concerned with independent messages, independent of their order. For images, and for many one-dimensional signals, the sequence of the messages (the spatial array of the gray tones) is all-important. Further, conventional analysis as above does not allow the recipient to generate a new set of messages (interpolate) based on the set sent. Specifically, the ability to estimate the next message from the last (redundancy) is necessary to be able to estimate those between (interpolate).

The study of information theory discloses a one-to-one correspondence between redundancy and the spatial-frequency content of pictorial data. The maximum information (minimum ability to predict the next message) is represented by an image with a Wiener spectrum (the relative radiant exitance of a scene as a function of the spatial frequency) that is uniform (equally likely) from zero to the upper cutoff frequency. Conventional wisdom for determining the required sample spacing is based on the requirement that all of the spatial frequencies present in the scene be reproducible without change in their relative amplitudes, independent of their possible relative contribution to the scene understanding.

Referring to Figure 17-19, let a band-limited function $f(x)$ be sampled at intervals of Δx with an instantaneous (zero width) δ-function to produce a sampled function

$$f_s(x) = \sum_{K=-\infty}^{\infty} f(K\Delta x)\, \delta\,(x - K\Delta x). \quad (17\text{-}27)$$

If the Fourier transform of $f(x) = F(u)$, the Fourier transform of $f_s(s)$ is:

$$F_s(u) = \sum_{K=-\infty}^{\infty} F(u - K/\Delta x). \quad (17\text{-}28)$$

That is, the frequency components of the original baseband signal are repeated at intervals of $1/\Delta x$, as shown in Figure 17-20. Separation of the original and generated spectral components is possible if the skirts do not overlap; this requires that $1/\Delta x = 2$ baseband bandwidth. Stated in other terms, two samples are required per period of the highest spatial frequency originally present. This is known as the Nyquist sampling rate criterion.

However, data derived from typical remote-sensor images do not have such a Wiener spectrum. The spectrum associated with these scenes tends almost universally to have a characteristic profile, with the low spatial frequencies dominating over the highs (Scott, 1963; Shlien, 1979). The high spatial frequencies, though weak in absolute magnitude, contribute to the image structure or fine detail, whereas the low spatial frequencies contribute to the macro-image structure. Figure 17-8b shows a typical Wiener spectrum of an aerial scene, which implies that the luminance difference between points in a typical scene is not independent of their distance apart. Thus, as we proceed from point to point in a scan of such a picture, the probability for the luminance value of each new point being unchanged is substantially greater than zero (the closer it is to the previous point, the more likely it is to have the same value), and thus the scene has redundancy. This is the

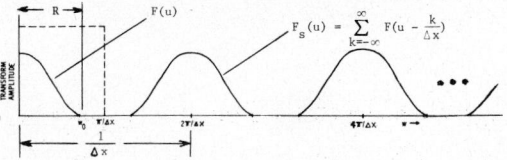

SPATIAL FREQUENCY SPACE OF SAMPLED SIGNAL

Fig. 17-20. Sampling causes generation of a series of new passbands spaced $1/\Delta x$ apart and each with the shape of the baseband spectrum.

consequence of a scene being a continuum with no absolutely sharp edges that might have allowed adjacent points to be independent. This is further illustrated by the joint probability distribution of Figure 17-21. The *joint probability density function* is the relative frequency of occurrence of pairs of luminance values at two points in a scene separated by some specified distance. Figure 17-21 plots this function for the scene shown in Figure 17-22. The vertical axis represents the relative probability of joint occurrence, and the other two axes represent the luminance of the two scene points separated by a specified distance. Linear axes are used.

Note that the presence of most of the data along the diagonal indicates a high degree of correlation between the luminance of points separated by the small distance used (15 cm on the ground). The integral of the joint probability density along either luminance axis will yield the marginal probability density function. As the separation of the points in the scene increases, the joint probability density-function spreads out until there is a noticeable lack of data along the diagonal, signifying that points separated by large distances are most likely not equal.

The progressive dissimilarity of points as the separation increases can be expressed by the autocorrelation function. In the discussion leading to Eq. 17-2, it was noted that one of the functions was reversed in forming the convolution. If this reversal is not performed, the cross-correlation of

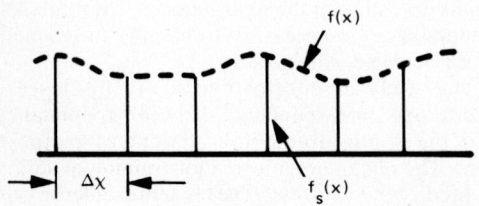

Fig. 17-19. A function $f(x)$ is sampled at a series of points Δx apart to produce a sample sequence $f(k\Delta x)$.

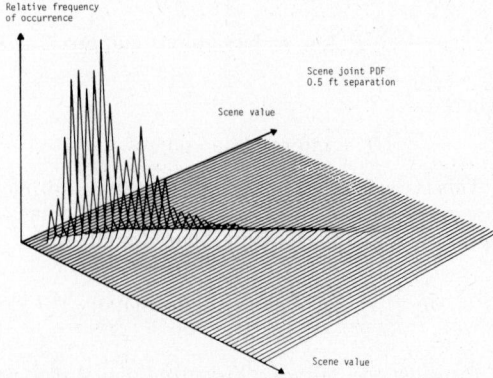

Fig. 17-21. Joint probability distribution function.

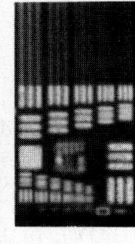

Fig. 17-22. Unsampled original launch site scene.

the two functions is formed (expressed here in one dimension continuous form for simplicity):

$$R_{fg}(a) = f(x)^* g(-x) = \int_{-\infty}^{\infty} f(x) \, g(x + a) dx \tag{17-29}$$

This represents the degree of similarity between the two functions at a given displacement a. If the same function is used for both f and g, the autocorrelation results.

The Fourier transform of the autocorrelation function is the power spectrum:

$$\mathscr{F}\{R_f(x)\} = \mathscr{F}\{f(x)^* f(-x)\} = F(u)F(-u)$$
$$= F(u)F^*(u) = |F(u)|^2 \tag{17-30}$$

These functions will re-appear in subsequent discussions.

EFFECT OF THE SAMPLING APERTURE

The scanning aperture in most systems may be considered to have constant amplitude transmittance over a distance equal to the width of the aperture, normally called the instantaneous field of view (IFOV). As this aperture sweeps across the scene, it will filter the scene spatial frequencies; because it is a rectangular function it will have a sinc x frequency response. Taking the first zero crossing of this function as the system bandwidth, it can be shown that two samples per IFOV distance should be taken for perfect reconstruction. This produces the redundancy required for interpolation. However, because of the general $1/f$ nature of the frequency content of the scene, practice has shown that one sample per IFOV is sufficient for practical purposes. (See also the discussion in Chapter 8.) Redundancy in this case is supplied by the scene, under the assumption that the very high spatial frequencies, having low amplitudes, will produce pixel-to-pixel amplitude variations of negligible consequence. For sensors such as the return beam vidicon (RBV) or, in the case of cathode ray tube or laser scanning of film images, the aperture does not have constant cross section, but, rather, approximates a Gaussian and is circular in cross section. This will form a filter function having a first zero at a period = $1/f$ = half amplitude diameter.

Again, the required sampling distance is two samples per period of this highest frequency.

IMAGE RECONSTRUCTION

Most image manipulation is done under the assumption that the sampling has been done somewhere near the Nyquist rate, with the implication that there is sufficient redundancy between samples to allow interpolation. This assumption is generally satisfied for imaging sensors, but may not be satisfied for other sensors where a series of measurements is made with non-overlapping IFOVs (footprints). Consider an imaging sensor with adjacent, non-overlapping IFOVs and no other smoothing: in the absence of other information, each sample is completely independent and no information as to the distribution of light within either sample is available. In this case, no interpolation is theoretically possible, and the best that can be done is to reconstruct the image with an array of "patches" of brightness, each located to be coincident with the ground position and scaled to the size of its IFOV. The task of image reconstruction is to estimate in some way the continuum scene from which the samples are taken. From this, a smooth representation (image) may be made, and new sample grids established for data registration (interpolation).

The normal, and usually sufficient, assumptions are that although the real samples may be collected independently by the sensor, the scene itself is continuous, and that a sample taken at a location bridging the true samples would have shared the radiances from its neighbors—that is, that interpolation is possible. This assumed continuity assures that the total sensed scene spatial-frequency content after filtering by the sequence of point-spread functions of the sensor is below the Nyquist rate, and that smooth transitions are formed which allow the interpolation.

The estimation of the continuum scene from which the samples are taken may be considered in either the spatial frequency or the image domains. In the former, this amounts to filtering out the frequencies formed by the sampling (Figure 17-20), leaving only the baseband signal. In the latter, this filtering function is expressed as the interpolation function. The usual approach is to shape the interpolation function to produce the desired spatial-frequency filtering. For an image to be faithfully reproduced, all of its baseband spatial-frequency components must be reproduced without relative amplitude distortions, and no others introduced.

Thus, the interpolation filter must have a "flat" freuency response up to the Nyquist frequency, and zero response upward from there. It can be shown that the Fourier transform of a rectangular function is a sin x/x = sinc x function; thus, the series of spatial reconstruction functions to be placed at the physical location of the original samples and which form the interpolation filter are of the sinc x form, each with an amplitude proportional to the original sample at that point. The continuum image thus formed by the superposition of all of the sinc functions (filling in between the samples) is an approximately exact representation of the original scene. It may be reproduced as a smooth image, or itself resampled on a new sampling grid. Because of practical considerations, it will be found that the number of terms required (and hence the number of adjacent samples required), for sinc x to be used for interpolation, is intractably large if ringing is to be avoided. Thus, as discussed in more detail below, truncated approximations to this function have been developed.

INTERPOLATION OR RESAMPLING

The problem of resampling arises whenever pixels are required in an output grid at locations different from where input pixels are located. (See also the discussion in Chap. 21.) Many methods of new pixel value generation are available; however, computation cost considerations tend to influence heavily the choice of method. The simplest method is called 'nearest-neighbor resampling' and it assigns a value (DN) to a new output pixel location (i,j) according to the value $DN_{k,l}$ of the spatially nearest pixel (k,l) in the input array to the precisely desired input location (x,y). The closest pixel (k,l) is found by:

$$k = \text{integer part of } (x + 0.5)$$
$$l = \text{integer part of } (y + 0.5)$$

so that

$$DN_{i,j} = DN_{k,l}. \qquad (17\text{-}31)$$

This method preserves the exact value of pixels in the input data set, and thus introduces no new spectral classes; but it introduces spatial shift errors such that the local geometry may be inaccurate by up to $\sqrt{2}$ of the instantaneous field of view (IFOV), or the size of a pixel on the ground. Worse yet, the pixel from which the gray level is derived shifts suddenly in location from the pixel just before the correct resampling location (x,y) to the pixel just after it. This problem becomes annoying during digital picture comparisons because, while the registration of detail in the two images may be perfect in one location, elsewhere there is misregistration. Since the contribution of a given pixel is constant whenever the output sample is to be drawn from $\pm\frac{1}{2}$ pixel spacing, and zero outside of that range, the average frequency response is *sinc x*, with a first zero at one sample/cycle (Figure 17-23). It thus imperfectly filters the sampling sidebands, and has appreciable attenuation of the high baseband frequencies. The visible effect is the "blocky" appearance of images interpolated by this method. Nevertheless, for many purposes this interpolation is adequate, and may be accomplished with insignificant cost

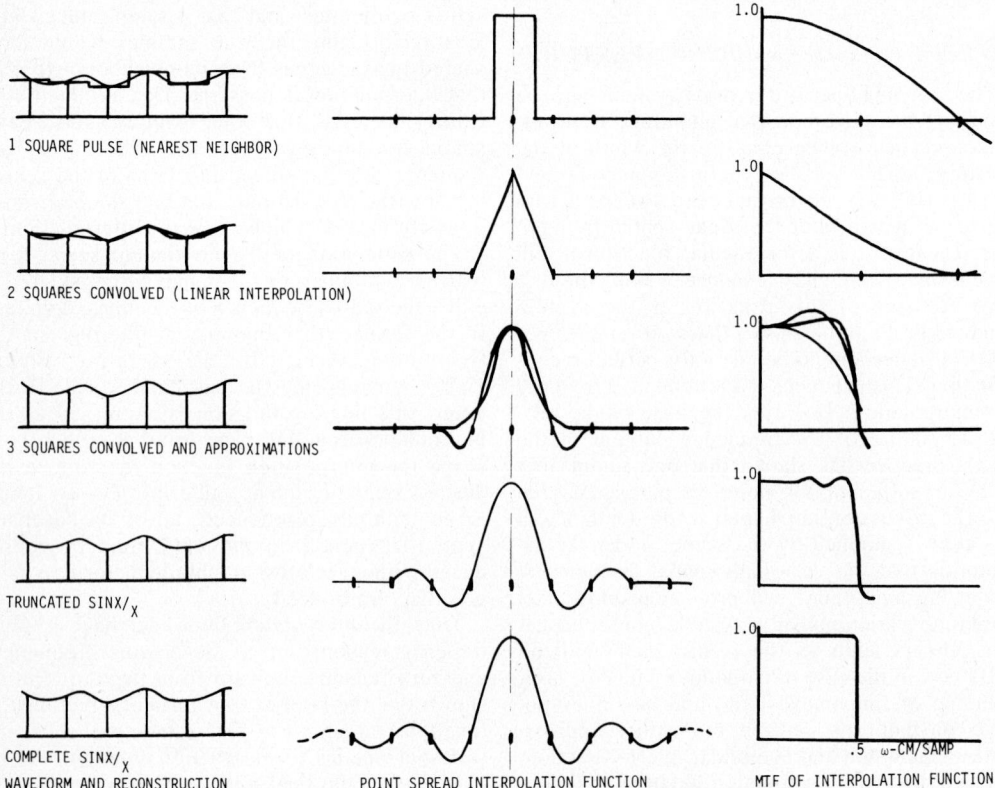

1 SQUARE PULSE (NEAREST NEIGHBOR)

2 SQUARES CONVOLVED (LINEAR INTERPOLATION)

3 SQUARES CONVOLVED AND APPROXIMATIONS

TRUNCATED SINX/$_X$

COMPLETE SINX/$_X$
WAVEFORM AND RECONSTRUCTION

POINT SPREAD INTERPOLATION FUNCTION

.5 ω-CM/SAMP
MTF OF INTERPOLATION FUNCTION

Fig. 17-23. Various interpolation functions and their frequency response curves.

since only address-rounding is needed for each new pixel assignment.

A smoother approximation to the assumption of continuity is obtained when the adjacent pixel is allowed to influence the estimation of between-pixel values. When only the adjacent neighboring pixels are used, only a first-order (bilinear) interpolation is possible. In bilinear interpolation, $DN_{i,j}$ is found by using an interpolation scheme with the four nearest pixels surrounding the resampling location (x,y) to determine the DN at (x,y) (see Figure 17-24). If (x,y) lies between samples k and $k + 1$ and lines ℓ and $\ell + 1$, then the gray level at (x,y) can be found by using:

$$DN_{xy} = (y - \ell)\lfloor(x - k)\,(DN_{\ell,k+1} - DN_{\ell k}) - (x - k)\,(DN_{\ell+1,k+1} - DN_{\ell+1,k})\rfloor$$

which reduces to

$$DN_{xy} = (x - k)\,(y - \ell)\,DN_{\ell,k+1} + DN_{\ell+1,k} - DN_{\ell k} - DN_{\ell+1,k+1}. \tag{17-32}$$

Since the contribution of a given pixel falls off linearly with distance to a distance $\pm$ one pixel spacing, $(\pm a)$, the average frequency response (H) in one dimension can be found by taking the Fourier transform of the triangular convolution kernel (h):

$$h_{bilinear} = \underset{-a\;+a}{\bigwedge} = \left(1 - \frac{x}{a}\right)^a_{x=0} + \left(1 + \frac{x}{a}\right)^0_{x=-a} \tag{17-33a}$$

$$H_{bilinear} = \int_{-a}^{a} he^{-2\pi f_x x} dx \tag{17-33b}$$

where f_x is the frequency in the x direction and a is the sampling interval. Alternatively, $h_{bilinear}$ may be recognized as the convolution of two rect-

angular functions. In either case, the resulting frequency response is found to be $sinc^2 x$, with the first zero at one sample/cycle. Resultant images are much smoother than those from nearest-neighbor interpolation, have about ¼ the mean-squared resampling error (Shlien, 1979) of nearest neighbor, and require appreciably more computer time, primarily because of the four multiplications involved.

Accuracy can be improved further by increasing the number of pixels in the vicinity of the resampling location from the nearest 4 to the nearest 16 (4 × 4 matrix) or more. The additional points offer an opportunity, not available using the simpler methods, to shape the pass band by adjusting the relative contributions (via weighting factors) of the various pixels. Cubic methods have emerged as the most significant higher-order resampling method and several variations exist. Classical cubic polynomial Lagrangian interpolation is the most common and produces a smooth resampled image with good frequency response. Sidelobes can cause overshoot; however, the cubic case is a good compromise between artifact introduction and computation cost. A spline function developed by Riffman (1973) has proven to be quite satisfactory in producing a reasonably shaped passband which provides some high-frequency enhancement:

$$\left.\begin{array}{ll} f_1(x) = 1 - 2x^2 + |x|^3 & 0 < |x| \leq 1 \\ f_2(x) = 4 - 8|x| + 5x^2 - |x|^3 & 1 < |x| \leq 2 \\ f_3(x) = 0 & 2 < |x| \end{array}\right\} \tag{17-34}$$

This function and its passband are sketched in Figure 17-25. It uses ± 2 neighbors for interpolation, has no contribution past 2-pixel spacings, and has continuous first derivatives. It has a

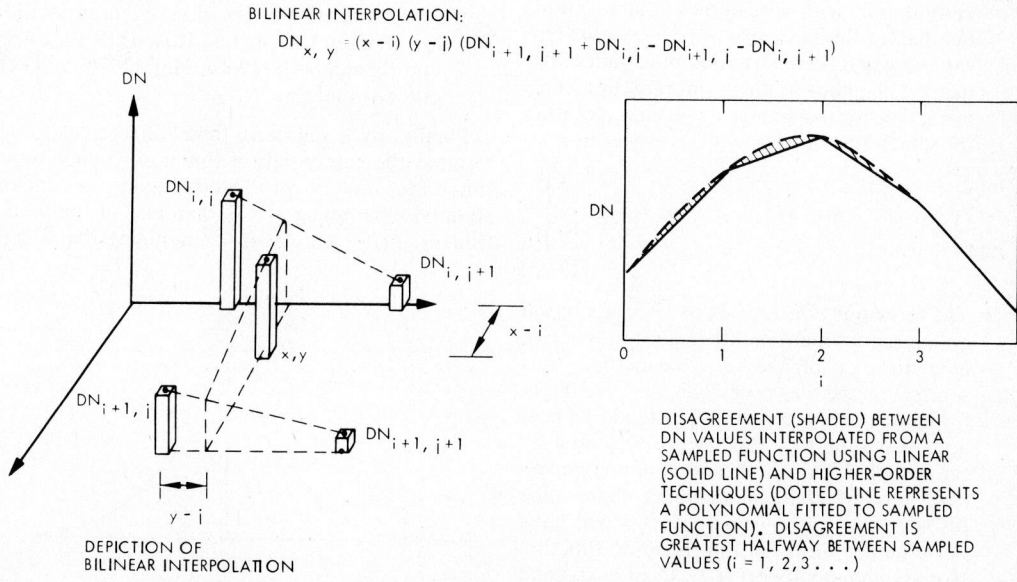

BILINEAR INTERPOLATION:

$$DN_{x,\,y} = (x - i)\,(y - j)\,(DN_{i+1,\,j+1} + DN_{i,\,j} - DN_{i+1,\,j} - DN_{i,\,j+1})$$

DEPICTION OF
BILINEAR INTERPOLATION

DISAGREEMENT (SHADED) BETWEEN DN VALUES INTERPOLATED FROM A SAMPLED FUNCTION USING LINEAR (SOLID LINE) AND HIGHER–ORDER TECHNIQUES (DOTTED LINE REPRESENTS A POLYNOMIAL FITTED TO SAMPLED FUNCTION). DISAGREEMENT IS GREATEST HALFWAY BETWEEN SAMPLED VALUES (i = 1, 2, 3 . . .)

Fig. 17-24. Bilinear interpolation resampling techniques (Gillespie, 1975).

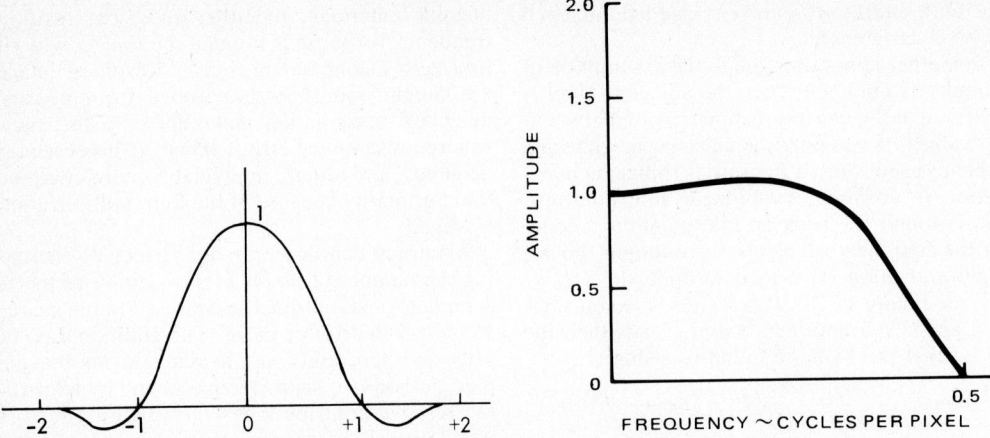

Fig. 17-25. Sketch of an interpolation function using ±2 neighbors.

mean-squared resampling error about ⅓ that of bilinear (Shlien, 1979), but requires four multiplies for each dimension.

Modifications of the cubic case have been designed to minimize undesirable characteristics while maintaining the four-multiply-each dimension per new pixel cost. Two examples will suffice.

1) By reducing the slope at the first zero crossing of the function that is of the type of Eq. 17-34 to minus one-half, a new function (Eq. 17-35) is produced having essentially the same transient rise distance but with less overshoot: (Simon, 1975; Riffman, 1975; Tabor, 1973). This is the only function of the parametric family defined by the slope that is third order convergent. (Keyes, 1981). In addition, Park and Schowengerdt, (1982) have shown that for any band-limited and sufficiently sampled image, parametric cubic convolution with the slope equal to minus one half at the first zero crossing yields less sampling and reconstruction blur than either nearest neighbor or linear interpolation. This is not true for the standard cubic convolution for which the slope = −1.

$$\left.\begin{aligned} f_1(x) &= 1 - 1/2(5x^2 - 3|x^3|) & 0 < |x| \le 1 \\ f_2(x) &= 1/2(4 - 8|x| + 5x^2 - |x|^3)\,1 \le |x| \le 2 \\ f_3(x) &= 0 & 2 \le |x| \end{aligned}\right\}$$

$$(17\text{-}35)$$

2) The functions of Eq. 17-34 or 17-35 are based on the requirement that they go through zero at a distance of 1 pixel from center, thus producing the negative lobe and the high-frequency enhancement. If, instead, the best approximation to a function represented by the samples is desired, and it is recognized that measurements made more than some distance away from a given pixel will have no influence on it, an interpolating function having limited support (i.e., local basis) and with the smoothest interpolation of all func-

tions passing through the same set of points is desired. For equal spacing of the measurements, such a function is the cubic B-spline (Hou and Andrews, 1978), having continuity in the function and its first two derivatives at the knots (the sample points), zero slope at the center and at the second knot, zero amplitude past the second knot, and a summation of contributing overlapping splines equal to unity. Invoking these conditions, the cubic B-spline is found to be (symmetrical around $x = 0$):

$$\left.\begin{aligned} f_1(x) &= 1/6\left[3|x|^3 - 6x^2 + 4\right] & 0 < |x| \le 1 \\ f_2(x) &= 1/6\left[-|x|^3 + 6x^2 - 12|x| + 8\right]1 \le |x| < 2 \\ f_3(x) &= 0 & 2 \le |x| \end{aligned}\right\}$$

A plot of this function is given in Figure 17-26. It can be shown (Peyrovian, 1976) that for sampling near the Nyquist rate, the cubic B-spline is the optimum interpolator. Other interesting properties of spline interpolation are given in LaFata and Rosen (1970), Curry and Schoenberg (1966), Hou (1976) and references therein.

Finally, by using more than ±2 samples to determine the interpolation function, higher order functions may be produced having less mean-squared resampling errors than any of the above (Shlien, 1979). However, recognizing that Land-

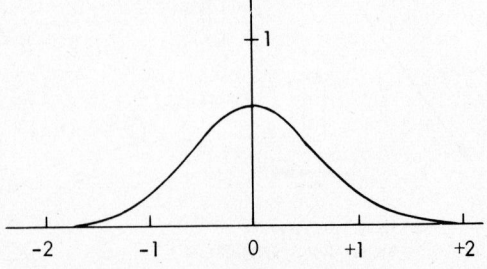

Fig. 17-26. B-spline of degree 3 (cubic B-spline).

sat, for example, has a finite amount of sensor noise, the utility of reproducing this noise with greater fidelity is suspect, and the higher order interpolators have not found widespread use.

INFORMATION AND ENCODING

THE MODEL IS THE MESSAGE

In a previous section of this chapter, the difference between data and information was stated as: Data refers to the results of any set of measurements, regardless of whether the measurements have been made with a certain purpose in mind; information is an aggregate of facts so organized or a datum so utilized as to be knowledge or intelligence. Information is meaningful, whereas data as such have no intrinsic meaning or significance. We may consider that information is generated from data using the analyst's model for that transformation. The model is not included in the data, but is brought from external knowledge by the analyst. Inaccuracies in either the data or the model may invalidate the information so derived. It therefore becomes crucial to verify the model, and to prepare data for correct (as well as efficient) use within the model. Transformations among types of data representation are often nominally reversible; transformations of data to information often substitute a decision (e.g., "that pixel is wheat") for the data, and will often not be reversible.

As an illustration of the importance of using the correct model for interpretation, consider the four cases of Figure 17-27. In the first case, the sample aperture is located half on dark fields and half on light fields; all sample values are the same and at some intermediate value. In the second case, the aperture positions are precisely on the fields; the sample values truly represent the radiances of the fields. In the third case, the aperture/field phasing is about 1/3–2/3; an intermediate set of high and low values results. In the fourth, the aperture registers precisely with a set of fields having intermediate radiances; the same set of measured values as in case three results. There is no unambiguous way to tell which of the sets of values is correct, nor is there any way to ascertain that the samples were from a sequence of uniform fields. The model "The Nyquist sampling rule has been followed and represents true samples of a continuum" is incorrect; the sampling was not according to Nyquist. Each sample in all cases was correct; the problem arises in the interpretation.

As a second example: consider a wheat field in which some scattered pixels do not have the same spectral signature as the rest, perhaps due to a lower canopy cover, and are therefore classified as "not wheat". To a "windshield surveyor" these pixels are wrong, as they are clearly part of the wheat field, but the classifier is also clearly correct. The implicit model being violated is "all wheat has the same signature", leading to "different signature = non-wheat". The per-field and

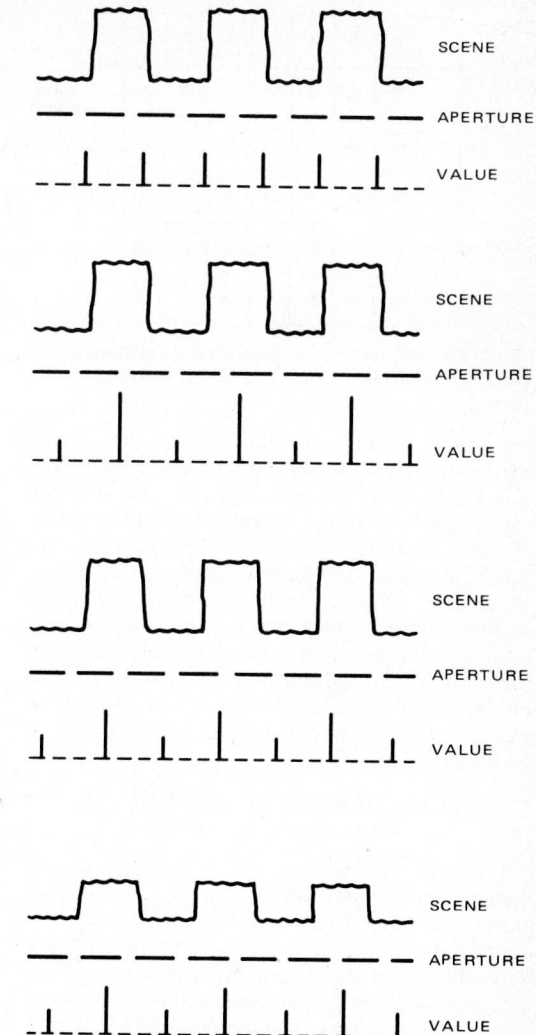

Fig. 17-27. Interpretation of the data may be invalid if the wrong data-to-information model is used.

context-dependent classifiers attempt to correct this type of discrepancy, but may substitute their own errors when they change a true non-wheat inclusion to "wheat" to make a uniform field.

This second example points out the difference between object recognition (the wheat field, which requires several to many pixels for recognition), and individual pixel classification, in which surrounding pixels are irrelevant. In the former case, the additional data in the surrounding pixels may supply additional information to the one being considered, and the shape and context of the entire group (external model parameters) is used for recognition.

As a counter-example, a model may be used to "correct" "faulty" data, or at least faulty display of the data. Consider the situation of Figure 17-28, in which a bright road traverses a darker background, producing the set of sample values shown. Clearly, Nyquist is not being followed,

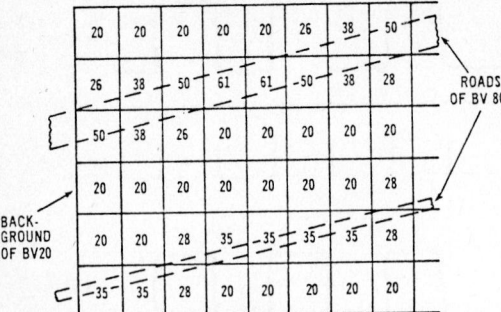

Fig. 17-28. Spatial array of pixels from a portion of one Landsat band with brightness values (BV) shown. (After Taranik)

and interpolation is invalid, so that the display must employ only the original pixels at their measured values, producing a characteristic stairstep pattern. But we rebel; nature, or even the works of man, does not produce these stairsteps. We invoke a model "Nyquist was being followed." This implies pixel-to-pixel redundancy and allows an interpolation which produces a smooth linear road. But this same model implies objects having a size of at least two pixels—the smooth road so produced will be wider than actual, and there is no way to recognize the true radiance and the narrow width of the true road.

CODING, COMPRESSION AND RESTORATION

The data user will recognize that any instrument is an encoder which converts some real-world parameter (such as reflected energy) to a quantized sample set, with more or less fidelity depending upon the instrument design. One of the analyst's tasks is to reconstruct a version of the world, using his understanding (model) of the instrument, and then to do further analysis. One goal of the instrument-coding chain is to supply both the data and the accompanying understanding of the encoding to minimize the possible misuse of the data.

An ever greater proportion of remote sensor imagery is being transmitted in digital form over communication channels. It is necessary to encode it in some form at the transmitting end, and to correspondingly decode it at the receiving end. An ever-increasing amount of digital data is accumulating in various archives, to the point that it is beginning to be impossible to retain it all, at least in its current form. For both the transmission and the archival storage, it is necessary to consider the most efficient way of reducing the amount of data. For many user applications, maximum benefits would result if the data compression could be obtained through extraction of only essential features which can be selectively accessed and interpreted directly without extraneous decoding. Selection of these features will require detailed knowledge of the characteristics of all of the user models which the data are in-

tended to serve to assure that the compression scheme does not hinder the subsequent analysis.

In general, the process of redundancy removal might occur in various stages, at the sensor or later in the storage or communication link. Even for a dedicated application, the user definition of adequate quality is an uncertainty which suggests a need of a capability for options. Therefore, a good data compressor emphasizes flexibility, allowing a high performance rate vs. quality tradeoff over wide ranges of compression ratios and applications. In order to aid the user in supervising the redundancy removal, a data compressor should use features which provide easy understanding of the relationship between the feature quality and the resulting quality of image approximation or of computer analysis. The subject of data encoding cannot be reviewed in depth here; this section seeks only to review the encoding methods that seem most appropriate for transmission or storage of imagery and to compare some of their characteristics. The intent of this section is to acquaint the analyst with encoding schemes which he may encounter or wish to employ and to present their basic characteristics.

In the domain of imagery transmission for remote-sensing applications there are three dominating considerations in the choice of coding methods—minimization of required bandwidth, maximum freedom from the degrading effects of noise, and introducing minimum degradation into the subsequent analyses. This considerably narrows the choice of competing systems.

Those methods which are of most interest in remote sensing are: (1) direct-encoding in the spatial domain: pulse code modulation, delta modulation, differential pulse code modulation, homomorphic differential pulse code modulation, noiseless coding; (2) transform techniques: Fourier, Walsh/Hadamard, cosine, RM2; (3) direct multispectral techniques such as principal component coding and cluster coding.

Each of these is discussed briefly in this section.

DIRECT ENCODING IN THE SPATIAL DOMAIN

Simple Pulse Code Modulation (PCM)

The first method of image-data encoding to be discussed is pulse-code modulation. It is the simplest form, but inadequate as an optimum encoding process. This is revealed by an examination of the statistical properties of scenes in an information theoretic setting. Such an examination also provides guidance for the selection of more efficient encoding methods.

Pulse-code modulation refers to a method wherein the individual picture elements (pixels) are converted to a series of pulses, which are then coded in some manner to indicate the quantization levels. Possible coding includes: pulse-amplitude modulation (PAM), in which the pulses vary in

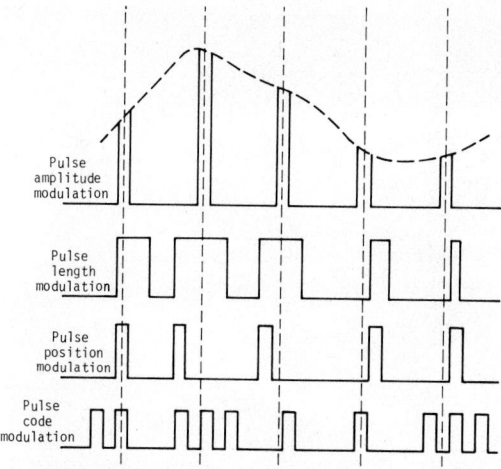

Fig. 17-29. Pulse codes.

For the quantized picture to be acceptable from the viewpoint of information conveyance, it may be necessary to use a large number of quantization levels (gray levels). For the case of an area where the luminance changes slowly, the quantization may result in a series of steps, the boundaries of which appear as conspicuous false contours. These false contours define spurious objects that may compete with or even conceal the real objects in the picture. Figures 17-30 and 17-31 show the effect on the rendition of a scene that is achieved by changing the number of quantization levels. These pictures represent the output of a pulse-code-modulated sampled image system having 4 and 64 quantization levels, in each case spanning approximately the same total range of luminance. These pictures simulate imagery that would be obtained from a PCM transmission system employing vidicon scanning and cathode-ray tube display-equipment. The false contours, which are quite apparent in Figure 17-30 diminish and disappear as the quantization level increases.

amplitude according to the modulation; pulse-length modulation (PLM), in which the pulses vary in duration, or length; pulse-position modulation (PPM), in which the pulses vary in position; and pulse-code modulation (PCM), in which the equal-height pulses form a digital word. The latter (pulse-code modulation) is perhaps most frequently used. The above forms are illustrated in Figure 17-29.

Predictive Methods

These points highlight the basic rate/quality trade-off inherent in choosing the operating parameters of a given encoding system or in selecting between systems. In the case of PCM, the use of only 4 quantization levels allows the

Fig. 17-30. Two-bit quantization of sampled image.

Fig. 17-31. Six-bit quantization of sampled image.

system to communicate at only 2 bits/pixel whereas 64 quantization levels would require 6 bits/pixel. The previous discussion suggests that the quality provided by only 4 quantization levels is unacceptable under most, but not all, situations whereas 64 levels is acceptable. Thus a direct use of PCM alone is generally restricted to applications in which a rate requirement of 6 bits/pixel and up is not a problem. However, the higher quality corresponding to many levels of quantization can generally be achieved by various predictive techniques which take advantage of the highly correlated nature of image data to achieve "data compression."

The general approach is to communicate the difference (error) between a pixel's actual value and its predicted value rather than the original pixels as in basic PCM. Because adjacent pixels tend to be similar, such predictions lead to sample (error) distributions far narrower than for the original pixels. The approach can be as simple as predicting that the next sample will be the same as the last, or it can be quite complex, utilizing many previous samples.

Noiseless Coding

These observations may be used to obtain data compression by several approaches. When very high quality imagery corresponding to many

quantization levels is desired, the most powerful technique is to apply (universal) noiseless coding directly to the prediction errors (Rice and Plaunt 1971, Rice 1979a, Rice 1979b, Rice and Schlutsmeyer 1980).

Data compression will be possible in data sequences in which the data samples are correlated with themselves or with previous information. There is usually some simple transformation which results in a new sequence in which the samples are approximately independent. More important, in the new sequence, the uncertainty in what the sample values will be is greatly reduced. The less uncertainty there is, the greater is the potential for data compression. The first step in the process, therefore, is to perform the correlation-removing operations.

Many decorrelation processes are available; a simple and effective one is to take differences between adjacent samples along a scan line. This produces approximately independent difference samples having zero mean and tending to be tightly distributed. Previous information might be the previous image line; appropriate use of this information generally leads to a similar result but with difference samples more tightly distributed yet.

Given the set of symbols (for example, the difference samples above) resulting from the chosen decorrelation algorithm, for a wide class of problems, their priority ordering is *a priori* known (or at least well approximated). In fact, their order tends to remain the same even though the shape of the probability-distribution function changes dramatically. Specifically, these symbols can generally be relabeled into the positive integers 0, 1, 2, . . . , such that $p_0 \geqslant p_1 \geqslant p_2 \geqslant \ldots$ where $p_i = Pr$ [symbol = i]. Although the actual p_i will tend to vary in practice, this ordering tends to remain valid. These reversible preprocessing operations are summarized in Figure 17-32. If the p_i were fixed, the standard Huffman algorithm to derive an optimum variable-length code for a known (fixed) distribution would yield coding efficiency about as well as could be expected. However, if the data character changes significantly over a span of samples, then the average per sample performance will decrease as the distribution deviates from the design distribution. Performance may be recovered by adaptively changing the coding to suit the changes. Rice (1979) defines a "Basic Compressor" containing four sets of codes among which a choice can be made for each block of samples; the structue of these is such that at least one of them will give performance approaching the maximum possible in the range of entropy between 0.4 and 4.0 bits/pixel. In operation, for a block size of (typically) sixteen coded symbols (for example, sixteen difference samples), the total coding length using each of the four operators is calculated and the shortest chosen. The resulting data string will be a 2-bit ID identifying the code chosen followed by the set of sixteen coded symbols. Additional simple preprocessing functions attached to the Basic Compressor yield combined algorithms which extend efficient performance to any entropy above zero. The main point in the discussion is that this performance can be obtained without prior knowledge of the sample distribution over a wide range of entropies.

The coding allows exact reconstruction of the original data while achieving a significant reduction in the bits/pixel required. Typically, images quantized to 256 levels can be coded (and reconstructed perfectly) with an average of only 4 bits/

pixel (the PCM rate for 16 quantization levels). Certain low-detail images, where the predictive process is more accurate, can even be coded with as few as 2 bits/pixel average. This illustrates that the number of bits used by such noiseless coding is dependent on scene detail creating special buffering requirements in some applications. However, a fixed rate may be assured by adaptively adjusting the quantization as needed (Rice et al., 1979). Nearly flawless image quality can be achieved by this approach at selected rates above 3 bits/pixel. This approach will be used on the Galileo mission to Jupiter.

The related techniques of delta-modulation and DPCM have received considerable attention in the literature and will be described in the following paragraph. The basic concept of these systems is to transmit differences rather than absolute values. Each sample in the image is compared with a predicted value based upon previous samples, and the difference between the actual value and the prediction is transmitted (Davisson, 1969; Rosenfeld, 1969a).

Simple Delta Modulation

This term has come into use to describe a 1-bit delta system. A block diagram of such a system is shown in Figure 17-33. The sampling rate is

$$f_s = 1/T \text{ samples per second.}$$

The quantizer has two output levels, k, where k is called the step size. The feedback loop in the encoder has the transfer function

$$H_1(s) = \frac{\beta D}{1 - \beta D} = \beta D + \beta^2 D^2 + \beta^3 D^3 + \ldots . \quad (17\text{-}37)$$

where:

$\beta < 1$ is a damping factor
D is the delay operator $D = e^{-sT}$.

Thus the output of the feedback loop is the weighted sum of all past inputs delayed by T. At the decoder the transfer function is

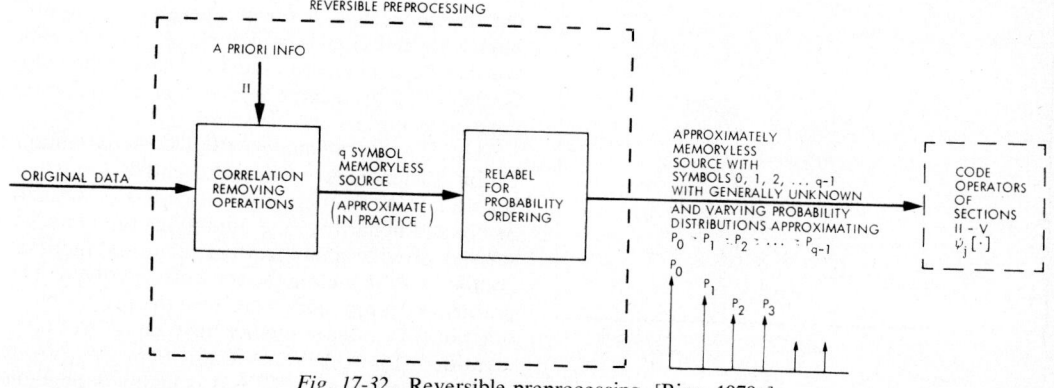

Fig. 17-32. Reversible preprocessing. [Rice, 1979a]

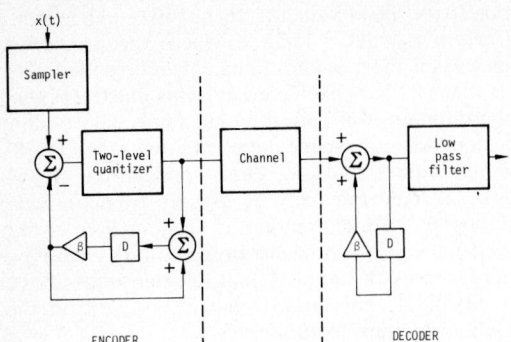

Fig. 17-33. Block diagram of a delta modulation system.

$$H_2(s) = \frac{1}{1 - \beta D} = 1 + \beta D + \beta^2 D^2 + \cdots$$

$$(17\text{-}38)$$

The output of the decoder is the sum of all past inputs at the output of the channel.

If the system is performing well, the samples at the input to the decoder low-pass filter should be identical to those at the output of the sampler. Often they are not, the difficulty being that in regions where the input signal is varying rapidly (the upward slope of the input signal) the delta modulator cannot keep up with it. This phenomenon is called slope overload noise. Another type of noise characteristic of this system is granular noise, resulting from the random 1-bit uncertainty in regions where the input is not changing. These two types of noise are illustrated in Figure 17-34. A third type of noise, characteristic of all systems which rely on changes to the last sample, is error propagation. This causes all samples subsequent to one having error to also be in error (until the system is reset by sending a true value), often resulting in quite visible streaks in the images.

It is clear that, for a given sampling rate f_s, the slope overload noise can be eliminated by increasing the step size, k. The product of the sam-

pling rate and the step size, kf_s, must be greater than the maximum slope of the signal. However, if the step size is increased to combat slope overload noise, the amplitude of the granular noise increases. As might be expected, there is an optimum step size in terms of subjective image quality for a given sampling rate and signal. For low values of the step size, k, the slope overload noise is dominant; for large values of k the granular noise dominates.

Differential Pulse Code Modulation (DPCM)

The form of encoding system most frequently encountered is the one known as differential pulse-code modulation, or DPCM. In this method the differences in output amplitude which are transmitted are quantized into several levels, the quantized value being encoded as a digital word. Full linear quantization followed by noiseless coding leads to the systems discussed above and in Rice and Plaunt (1971), Rice (1979a and 1979b), Rice and Schlutsmeyer (1980), Rice et al. (1979), Rice (1975, 1978b, 1976, 1974a, 1974b, and 1978a), Odenwalder (1974), Liu and Woo (1980), Liu and Lee (1981), and Hooke (1979).

The distinction between these systems and what has come to be known as DPCM is that compression is primarily obtained from non-linear quantization of the predictive error signal as noted below.

In all differential systems, the possibilities of errors (differences between the values encoded at the transmission end and the values decoded at the receiver end) caused by the transmission channel become more numerous, and it is nearly always necessary to insert periodic PCM updates in order to correct these errors. In any delta-modulation system, the final output image persists until it is corrected or until an offsetting error occurs.

In the case of simple delta modulation, the only bit error which can occur is a single bit, which represents one increment of amplitude. These occur at random, and average out to zero. Even in adaptive analog systems, where the magnitude of the error varies, the effect is usually not too serious in the output image. However, when the values are encoded into digital binary words the effect is much more serious. There is no possible binary formulation in which there are not some words where an error of one bit changes the value of the word by a large factor. Errors and error correction are discussed below.

Many analytical models for DCPM communication systems have been presented, but when such a nonlinear encoding system has to actually be created in hardware, assumptions found necessary to provide analytical tractability sometimes conflict with modeling the tradeoffs involved. The problem is compounded because the proper merit functions for picture quality have not as yet been identified (Limb, 1967; O'Neal, 1966).

The performance of DPCM is highly dependent

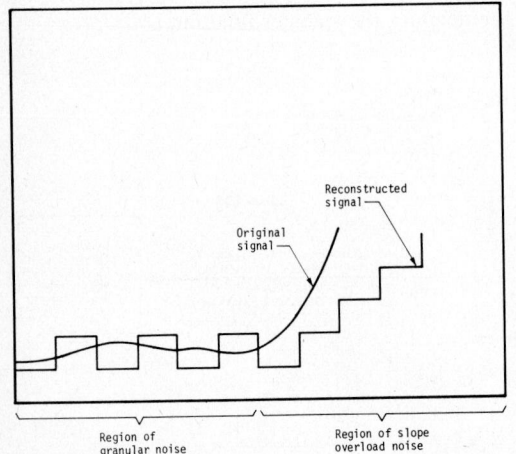

Fig. 17-34. Slope overload and granularity noise.

on sampling rate; even a one-bit delta modulator may perform as well as a two-bit system if the sampling rate is high enough. Spatial sampling is generally accomplished in a one-dimensional fashion as in conventional line-scan systems.

A block diagram of the elements of a DPCM communication system is shown in Figure 17-35. The feedback encoding technique incorporates a nonuniform quantizer in the forward path of the encoder and a unit sample period predictor (labeled unit delay in the figure) in the feedback loop. The function of the predictor is to make an estimate, based on the previously transmitted signal, of what the next sample value will be. The leak, or attenuation factor, depends upon the assumed spatial correlation properties of the scene. The estimate is then subtracted from the actual value of the sample. The difference between the true sample value and its estimate is nonuniformly quantized, encoded, and transmitted over a digital communications channel. The channel output is decoded back into incremental levels by the receiver decoder. The decoded levels are processed by the receiver predictor, which duplicates the action of the source predictor, to obtain the output signal. The receiver predictor is periodically updated to provide tracking between source and receiver. This can be done by setting both source and receiver predictors to zero at the beginning of each line.

Qualitative Analysis

The behavior of the generic form of the DPCM in response to variations in setting the quantizer levels and thresholds can best be described in qualitative terms. Quantization error in all PCM and DPCM systems is defined as the difference between the quantizer input and output values. In basic PCM image-transmission systems, as has been stated, this error is in the form of granularity, and gives rise to contouring artifacts. In DPCM imagery, a quantization error can be classified into three major categories—granularity error, slope-overload error, and combination error.

Granularity error in DPCM, as in PCM encoding, results from the inability of the system to encode a signal that is smaller than the smallest level. Limit cycling of the smallest level is a property of DPCM during periods of constant signal input for quantizers in which the smallest level is not zero. This form of picture noise is objectionable to the observer because of the noticeable checkerboard pattern created in constant radiance areas. Limit cycling can be alleviated by setting the smallest step equal to the rms value of the scanner noise at the encoder input. Scanner noise then prevents constant density regions from occurring and granularity becomes randomized. This effect, or a constraint to be able to transmit the smallest observable signal, may fix the value of the smallest level at the rms value of the noise. Slope overload is the inability of the largest level to accommodate large adjacent differences in the input scene luminance. The result is a staircase pattern of steps taken in an attempt to track transient inputs. Information present in areas where slope overload occurs is lost, as shown in Figure 17-36. The amount of tolerable slope-overload error sets the criterion for establishing the size of the largest level.

Combination error is a mixture of a form of slope overload for a given level and a form of granularity of the larger level (Figure 17-37). For a large ratio between the smaller level and the threshold of the larger level, this error manifests itself primarily as a smaller step slope-overload, and requires the quantizer to take several smaller steps to reach a level just below the threshold of the larger level. During this period of smaller level slope-overload small signals are lost (Figure 17-37). Likewise, if the threshold is just slightly exceeded, the step taken is too large. Several smaller steps are then required to correct the situation. Combination error is minimized by keeping the ratio between adjacent levels moderate.

Threshold values of the quantizer are chosen midway between quantizer levels. In such a quantizer, the error amplitude ranges between plus and minus half a quantum step (Max, 1960).

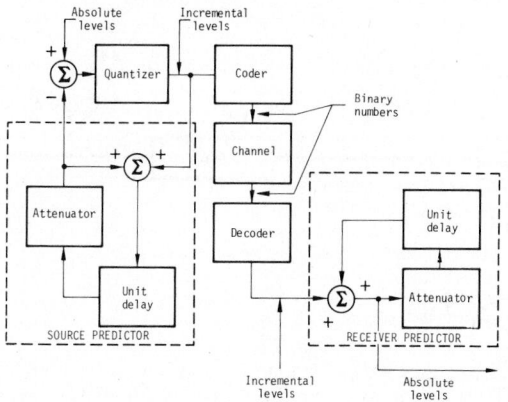

Fig. 17-35. Differential pulse code modulation, block diagram.

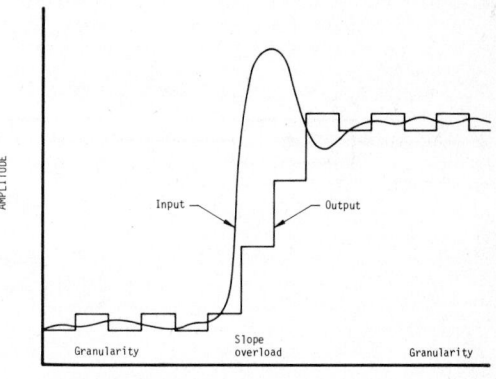

Fig. 17-36. Slope overload and granularity in DPCM.

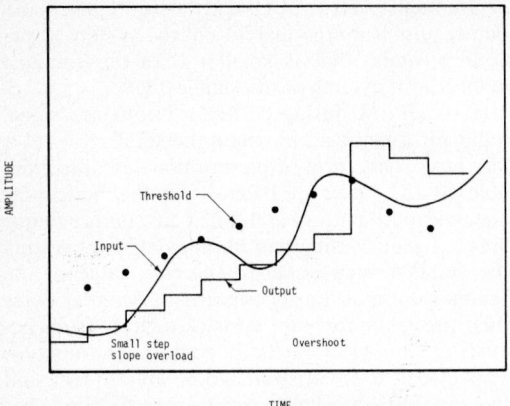

Fig. 17-37. Combination error.

Fig. 17-38. Block diagram of DPCM with PCM update.

An effective approach to reducing the effects of channel errors is to transmit periodically PCM updates that will set the receiver predictor equivalent to that of the transmitter, as illustrated in Figure 17-38. The extent of a bias error at the receiver output may thereby be reduced from a fraction of the picture line to a fraction of the update interval (Arguello et al., 1971). A large body of literature exists on the detailed analyses of DPCM systems, including discussion in the previous edition of this Manual and numerous issues of various Transactions of the IEEE. The reader is encouraged to pursue these for more analytical details.

Homomorphic Coding and Processing

It can be shown that the encoding of photographic density rather than transmittance can improve the subjective performance of an imagery-transmission system which works with a photographic transparency as an input. Figure 17-39 is a block diagram of a nonlinear image-processing system (developed by Stockham, University of Utah) which applies this principle. It can effect a reduction in dynamic range of imagery and simultaneously increase sharpness of detail. It is known as homomorphic processing, which is derived from the mathematical term "homomorphology." This term is defined as a single-valued transformation mapping of input variables in such manner that a singular operation (such as multiplication) is preserved for all variables (Oppenheim et al., 1968; Stockham, 1969).

The rationale with respect to imagery for the system is that images are formed as the product of irradiance and reflectance. Except for the effect of shadows, the irradiance in a typical scene does not vary rapidly over the scene; hence it is composed primarily of low frequency components.

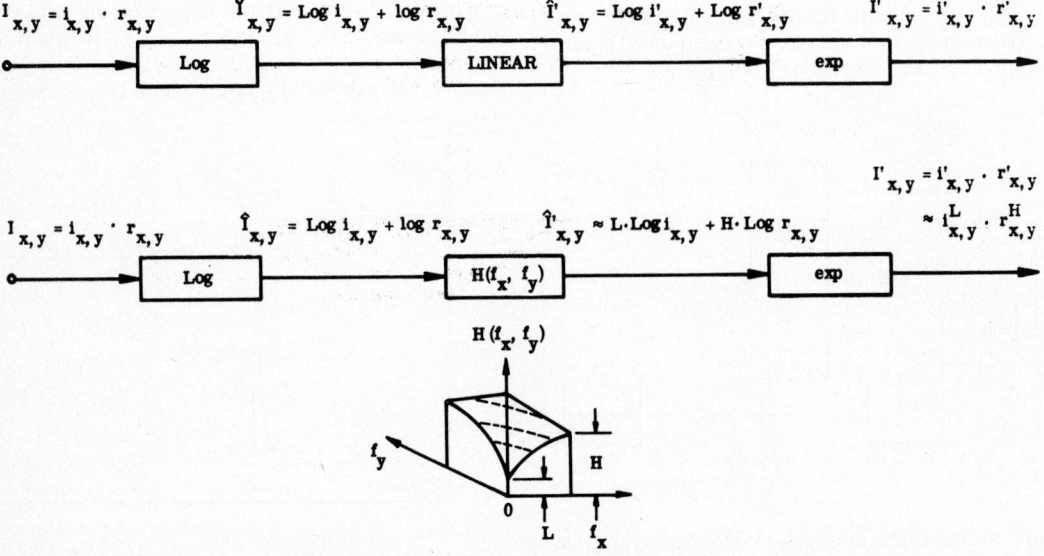

Fig. 17-39. Homomorphic processing of imagery.

The reflectivity, on the other hand, is a property of the objects in the scene, and does change rapidly; hence high frequency components dominate here. The reflectivity represents the actual information in the scene.

If the processor has a response which amplifies high frequencies and attenuates low frequencies, the tendency will be for the output to represent a reinforced reflectance or increased sharpness in detail, combined with an attenuated irradiance or reduction of dynamic range. Extracting the logarithm of the scene values separates the irradiance and reflectance components into two elements of a sum rather than a product.

A bandwidth-reduction scheme, when embedded in a homorphic processor, as shown in Figure 17-40, can be expected to outperform its conventional equivalent. Subjective image quality will be improved because the properties of the human visual tract are then combined with the scene message-source. In a simulation of such a system (Arguello and Sellner, 1971), it was determined that a tapered 4-bit quantizer was required to ensure that sampled imagery would be essentially free from quantization artifacts such as slope overload and granularity noise.

MONOCHROMATIC TRANSFORM ENCODING

Principles of Transform Encoding

Imaging data can be transmitted over a communications link by encoding a linear transformation of the image rather than the actual spatial image samples themselves. The transform coefficients, in general, each represent information which is distributed over the entire scene. For example, the Fourier transform, which exhibits the relative amplitude of the various spatial-frequency components of the scene, contains information from the entire scene at each spatial frequency.

This modified distribution of information has two advantages over direct spatial encoding. One is that it is possible to transmit only a part of the transform and still have information about the whole scene—for example, to transmit a low-resolution picture by transmitting only the low-frequency part of a Fourier transform. If the scene proves to be of interest, the remainder of the transform (if it has been stored in the meantime) can be transmitted to provide the higher resolution information. In the case of a spatial scan of an image, however, the only practical way of accomplishing this would be to scan the input image

separately at two different sampling rates, or to store the entire high resolution image at the source and to first transmit only the low-frequency data derived from this image.

The other advantage of non-adaptive transform techniques is that the reconstructed data are less affected by transmission errors. Any error in the value of a non-adaptive transform coefficient has an effect which is more or less uniformly distributed over the entire scene. Noise, too, is generally seen as a very low-frequency interference in the image and is less offensive to the observer than localized errors in the spatial domain.

As before, if $F(x,y)$ represents the radiance distribution over x and y of a scene, the image can be expressed in terms of a complete set mn of orthogonal functions $\phi_{mn}(x,y)$ as follows:

$$F(x,y) = \sum_{m=1}^{\infty} \sum_{n=1}^{\infty} L_{mn}\phi_{mn}(x,y) \quad (17\text{-}39)$$

where L_{mn} values are constant coefficients given by

$$L_{mn} = \iint F(x,y)\phi_{mn}(x,y)dxdy. \quad (17\text{-}40)$$

When the image is represented by a square array of $N \times N$ discrete samples, then these relationships become

$$F(x,y) = \sum_{m=1}^{N} \sum_{n=1}^{N} L_{mn}\phi_{mn}(x,y); \quad (17\text{-}41)$$

$$L_{mn} = \sum_{m=1}^{N} \sum_{n=1}^{N} F(x,y)\phi_{mn}(x,y). \quad (17\text{-}42)$$

Image coding in the transform or frequency domain is achieved by transmitting the L_{mn} coefficients rather than the spatial samples $F(x,y)$.

Many different schemes for transform coding have been proposed and demonstrated in the technical literature. Each involves a choice of the set of orthogonal functions

$$\lfloor \phi_{mn}(x,y) \rfloor,$$

as well as some method of quantizing and encoding the L_{mn} coefficients. The more familiar of these schemes make use of the Fourier or the Hadamard/Walsh linear transformations (Anderson and Huang, 1971; Andrews, 1970; Pratt et al., 1969; Wintz, 1972; Habibi and Robinson, 1974), and the discrete cosine transformation (Ahmed et al., 1974; Rose et al., 1975; Chen and Smith, 1976). Hybrid techniques combining DPCM with such one-dimensional transforms was discussed by Habibi (1976).

Transform methods have the goal of expressing a block of N data samples in a form which can be more conveniently or more effectively encoded for one reason or another. The fast Fourier transform can be calculated rapidly, and, as is well known, has asymptotically uncorrelated values (Cooley, 1967). An additional advantage is that a frequency weighting can be performed directly on

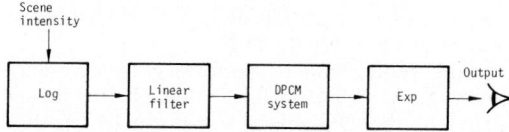

Fig. 17-40. Homomorphic DPCM system.

the transformed frequency values. The Walsh-Fourier or Hadamard transform, which requires no multiplication operation, is computationally still faster. Generalized methods of developing orthogonal function sets are given by Jain (1977) and by Moharir and Varma (1980).

When mean square error is the optimality criterion, the principal components or Karhunen-Loeve transformation is the best and is discussed in some detail below in the context of multispectral coding. However, this technique is generally not used in transform coding image-compression work because of its computational complexity. Haralick and Griswold (1975) describe a fast almost-Karhunen-Loeve transformation for block coding of stationary, isotropic images which approaches closely the performance of the true Karhunen-Loeve while having a fast computational algorithm. But most images are clearly not stationary, so that partitioning of the image into blocks which are closer to stationary will result in better performance, as each block may be optimized independently.

Fourier Transform Encoding

The discrete Fourier transformation is defined by the case where the orthogonal functions are complex exponentials,

$$\phi_{mn}(x,y) = \frac{1}{N} \exp - \frac{2\pi j}{N} (mx + ny). \quad (17\text{-}43)$$

The L_{mn} coefficients defined by Eq. 17-42 become complex and must be represented as a pair of numbers. Several schemes for quantizing these coefficients are identified below.

The first method, demonstrated by Andrews (1970), involves the quantization of all Fourier coefficients in the $N \times N$ Fourier transform of an image. Ideally, the quantization levels should be chosen in such manner that all levels would become equally probable, thereby maximizing the source entropy. However, this requires knowledge of the probability distribution of Fourier coefficients, which varies from image to image. Andrews proposed that the Fourier coefficients should be assumed to be normally distributed, with zero mean and a variance that would decrease as a function of the spatial frequency. Andrews also proposed that this variance function should be Gaussian in shape. His method of Gaussian quantization has given results with good subjective image quality. It yields better results than linear quantization, since the spacing of quantization levels is closer for low-amplitude Fourier coefficients, where there is a higher population of coefficients.

A second quantization scheme, also proposed by Andrews, is called threshold sampling. In this scheme, only the transform samples that exceed a preset threshold (which may be a function of spatial frequency) are transmitted. Small Fourier coefficients are discarded, but it becomes neces-sary to code the positions of the significant samples as well as their values. This position coding is performed with run-length encoding and leads to an asynchronous transmission system.

Another approach to transform quantization and encoding was demonstrated by Anderson and Huang (1971). Their scheme has the image divided into subsections, and transforms and encodes the subsections independently (piecewise transformation). The purpose is to add adaptivity to the method. In each subsection, the L largest Fourier coefficients are quantized and transmitted. Further, the number of quantization levels depends on the standard deviation of the spatial samples in the subsection. Positional information for the L coefficients is given in run-length encoding, leading again to an asynchronous system. The primary advantage is the ability to change the quantization scheme adaptively as a function of spatial location in the image. Busy portions of a scene are assigned more quantization levels than are relatively uniform regions. Anderson and Huang achieved good image quality with typically as few as 1.25 bits per picture element.

Hadamard/Walsh Transform Encoding

A further group of image-coding schemes is based on the use of Walsh functions. These functions have received considerable attention because Walsh transforms are easily implemented by semiconductor technology, since they are binary in form. The Walsh transformation involves only additions and subtractions, and can be computed much more rapidly than Fourier transforms. The Hadamard transforms are two-dimensional extensions of the Walsh transforms, with the orthogonal set $[\phi_{mn}(x,y)]$ defined as,

$$\phi_{mn}(x,y) = \frac{1}{N} (-1)^P \quad (17\text{-}44)$$

where

$$P = \sum_{\ell=1}^{N-1} \left[g_\ell^b(m)x_\ell^b + g_\ell^b(n)y_\ell^b \right], \quad (17\text{-}45)$$

$g_\ell^b(m) = \ell^{\text{th}}$ bit in binary representation of $g(m)$
$x_\ell^b = \ell^{\text{th}}$ bit in binary representation of x
$g_\ell^b(n) = \ell^{\text{th}}$ bit in binary representation of $g(n)$
$y_\ell^b = \ell^{\text{th}}$ bit in binary representation of y.

Whereas a Fourier transform explains a given picture function in terms of sinusoidal waves with varying spatial frequencies, a Hadamard transform can be considered as the description of a picture in terms of square waves of different "sequencies." A sketch of a 4 × 4 Hadamard array is given in Figure 17-7.

Image coding can be performed by encoding either the Walsh transform of each line in an image, or the Hadamard transform of the full two-dimensional image. Andrews (1970) has published results of Hadamard transform coding, using the same Gaussian quantization procedure as described above in quantizing the Hadamard coefficients. Again, the assumptions of normally distributed coefficients resulted in good image quality for a variety of different images. A multitude of others have since used the Hadamard transform in various coding schemes well cited in the literature.

Figures 17-41a and 17-41b show a simulation of this type of algorithm. Figure 17-41a is the log of the magnitude of the Hadamard transform of the original scene shown in Figure 17-22. The reconstruction using the inverse Hadamard transform is shown in Figure 17-41b. Compare Figure 17-41b with the original unmodified scene given in Figure 17-22.

Discrete Cosine Transform Encoding

Jain (1977) introduced a complete family of sinusoidal transforms, in which, in general, each member is a Karhunen-Loeve transform of a non-stationary Markov process. The even discrete cosine transform of Ahmed (1974), Gray (1972), and Shanmugam (1975) is a member of this family whose basis vectors approximate the eigenvectors of a correlation matrix of a first order Markov process. To the extent that images may be approximated by first order Markov properties, this implies that the discrete cosine transform is a better choice of transforms to use than the discrete Fourier transform in image-coding compression work (Haralick, 1976).

The discrete cosine transform (DCT) has a number of applications in image processing (Chen, Smith and Fralick, 1977; Chen and Fralick, 1977). A circular convolution-multiplication relationship has been derived which allows it to be used in linear filtering with some computational advantages (Chen and Fralick, 1977; Brigham, 1974). Compared to other orthogonal transforms

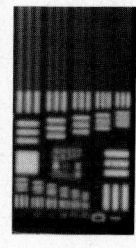

Fig. 17-41b. Hadamard transform of Hadamard transform.

its performance seems to compare most favorably with the optimum Karhunen-Loeve transform for a large number of signal classes (Zelinski and Noll, 1977; Ahmed et al., 1974; Chen and Smith, 1976). In the transforming of image sub-blocks it introduces almost no edge effects. Its energy-compaction property (Figure 17-42) is far superior to the discrete Fourier transform (Chen and Smith, 1976), and for the practical case of images having a correlation coefficient $\rho > 0.5$, is superior to any known transform having a fast computational algorithm.

The discrete cosine transform of a discrete function $x(n)$, $n = 0, 1, \ldots, N - 1$ is defined as

$$F(k) = \frac{2c(k)}{N} \sum_{n=0}^{N-1} x(n) \cos\left[\frac{(2n + 1) k \pi}{2N}\right];$$

$$k = 0, 1, \ldots, N - 1 \qquad (17\text{-}46)$$

and the inverse transform is

$$x(n) = \sum_{k=0}^{N-1} c(k) F(k) \cos\left[\frac{(2n + 1) k \pi}{2N}\right];$$

$$n = 0, 1, \ldots, N - 1 \qquad (17\text{-}47)$$

where

$$c(k) = \frac{1}{\sqrt{2}} \text{ for } k = 0$$

$$= 1 \qquad \text{for } k = 1, 2, \ldots, N - 1.$$

The relationship of this formulation by Ahmed can be shown to be related to the Fourier transform of an even extension of $x(n)$ (Makhoul, 1980).

Let $x(n)$ be a discrete-time signal and $X(\omega)$ its Fourier transform. In one definition the cosine transform of $x(n)$ is the real part of $X(\omega)$. The real part of $X(\omega)$ is also equal to the Fourier transform of the even part of $x(n)$, defined by $x_e(n) = [x(n) + x(-n)]/2$ (see Oppenheim and Shaffer, 1975, for example). Therefore, the cosine transform of $x(n)$ is equal to the Fourier transform of $x_e(n)$. Now, if $x(n)$ is causal, i.e., $x(n) = 0$ for $n < 0$, then $x_e(n)$ and, therefore, the cosine transform uniquely specifies $x(n)$. Therefore, the cosine transform of a causal $x(n)$ can be obtained as the Fourier transform of an even extension of $x(n)$.

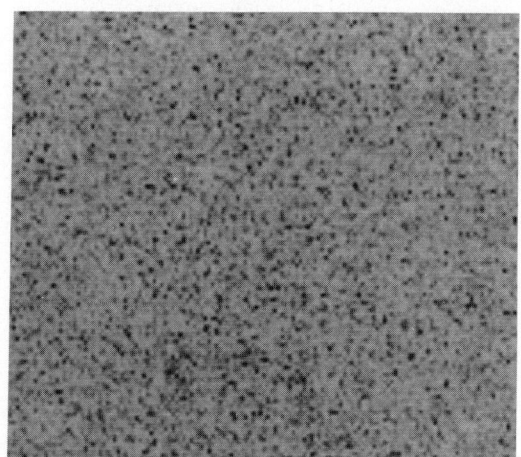

Fig. 17-41a. Logarithm of magnitude of Hadamard transform.

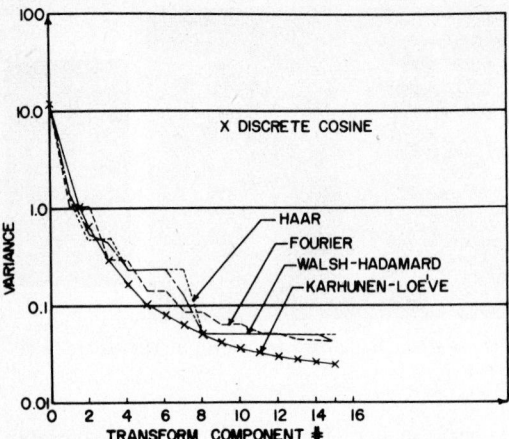

Fig. 17-42. Transform domain variance of various transforms; M = 16, ρ = 0.9 (Ahmed et al., 1974).

It has been shown that the DCT of an N-point signal can be computed using a $2N$-point discrete Fourier transform (DFT) (Ahmed et al., 1974). Chen, et al., (1977) used matrix factorization to derive a special algorithm to compute the *DCT* of a signal with N a power of 2, resulting in a saving of ½ over the previous method when the latter uses the fast Fourier transform (FFT). Haralick (1976) and more recently, Narasimha and Peterson (1978) developed methods that employ an N-point DFT of a reordered version of the signal (where N is assumed to be even), resulting in a similar saving of ½. When N is a power of 2, use of the *FFT* results in a saving comparable to that of Chen et al. However, in the algorithm of Narasimha and Peterson, one can use existing software to compute the *FFT* instead of implementing a special algorithm for the *DCT*. Makhoul (1980) presented a fast transform utilizing the even-extension properties.

A Globally Adaptive, Rate-Controlled Algorithm: RM2

The vast majority of transform-compression efforts have focused on nonadaptive algorithms which provided operation at one or two rates optimized for an assumed stationary data source. However, most real image sources are nonstationary in nature. The early efforts of Anderson and Huang (1971) (and others to follow) recognized this nonstationarity, providing improved performance at the expense of a varying, data-dependent rate and a nontrivial buffering problem. Improved performance was achieved by adjusting coding parameters to match local statistics.

Rice (1974) provided a significant step in performance and user flexibility with the development of a globally adaptive, rate-controlled algorithm called RM2 (Rice, 1975, 1978b, 1976). RM2 allowed a user to arbitrarily select a (fixed) operating rate for each frame, thus providing

complete flexibility to trade-off rate and quality. In addition, improved performance over other adaptive algorithms at all rates was observed in numerous comparisons (Rice, 1976). Global adaptivity was provided by first partitioning an image into subpictures and, before actual coding, allocating the total user-selected bits/image to these subpictures according to activity features related to differential entropy. This allocation procedure could be easily modified to enhance the quality of specific features detected by pattern-recognition algorithms while still maintaining the selected user rate.

The actual coding of subpictures involved the application of universal noiseless coding techniques already noted (Rice, 1979a, 1979b) to the linearly quantized coefficients of cascaded 2 × 2 Hadamard transforms (Rice, 1974a), perhaps the computationally simplest transform used in practice. The noiseless coding effectively provides adaptivity down at the local level. Further subtle improvements in reconstructed quality were obtained by an adaptive inverse transform which utilized surrounding data in a controlled manner to adjust the reconstructed coefficients. The results of applying RM2 to one of a 7-band 5-meter-resolution image of Ventura County, California, is shown in Figure 17-43 for selected rates of 1.33, 1.0, and 0.67 bits/pixel.

The high level of adaptivity, particularly the universal noiseless coding, made the RM2-coded data particularly sensitive to transmission errors. However, this problem was virtually eliminated as a factor for deep space applications by combining it with a concatenated Reed-Solomon/convolutional-Viterbi channel in the form of the Advanced Imaging Communication System (AICS) (Rice, 1974b, 1978a; Odenwalder, 1974; Liu and Woo, 1980; Liu and Lee, 1981).

This same link is now proposed for NASA and ESA standardization both for deep space and high-rate satellite-communication systems and thus its attributes may be assumed for a broad range of applications (Hooke, 1979).

MULTISPECTRAL ENCODING

In some applications of multispectral imaging, extremely good fidelity may be required. In this case the near-optimum universal noiseless-coding techniques noted for monochromic data earlier (Rice and Plaunt, 1971; Rice 1979a, 1979b; Rice and Schlutsmeyer, 1980) are equally applicable here. Some potential for slight per band improvements in rate may be possible. The total flexibility provided by the global rate-control structure of RM2 to adaptively direct bits to regions of specific interest makes it a powerful candidate for achieving high information return at low rates. In addition, two other approaches which directly make use of the multispectral nature of such image data offer some additional and possibly unique benefits: multispectral transform coding and cluster coding.

Fig. 17-43. RM2 applied to a 7-band 5 meter resolution image of Ventura County, California, for 1.33, 1.0, and 0.67 bits/pixel (R. F. Rice, JPL).

Multispectral Transform Encoding

Transform coders perform a sequence of two operations. The first operation is a linear transformation that transforms the set of statistically dependent data elements into a set of more independent coefficients. The second operation is to individually quantize and code each of the coefficients. A variety of linear transformations is available to implement transform coding on MSS data, but the eigenvector transformation is the optimum linear transformation in two senses: the mean square error between the original and reconstructed data is less than for any other linear transformation; also, it eliminates all correlations in the data.

The procedure is as follows:

1) Compute the covariance matrix C of the given data. This matrix is $n \times n$.

2) Obtain the n eigenvectors and associated n eigenvalues of C. The eigenvectors are n-dimensional.
3) Choose the m eigenvectors associated with the m largest eigenvalues of C, where $m < n$.
4) Form an $m \times n$ transformation matrix A' whose rows are the m eigenvectors selected in Step (3).
5) Reduce all original vectors x into a set of y vectors by means of the transformation $y = A' x$.

For $m = n$, the reconstructed data $\hat{x}$ are identical to the original data x; i.e., $\hat{x} = x$. For $m < n$, some error $(\hat{x} - x)$ is incurred. However, the eigenvector transformation results in the least mean square error

$$\epsilon^2 = \left\{ E \| \hat{x} - x \|^2 \right\} \qquad (17\text{-}48)$$

of all linear transformations. The elements y_1, y_2, ... y_n of y are uncorrelated and have variances given by the eigenvalues $\lambda_1, \lambda_2, \ldots \lambda_n$. It can also be shown that the mean square error Eq. 17-48 is given by

$$\epsilon^2 = \sum_{i=1}^{n} \lambda_i - \sum_{i=1}^{m} \lambda_i = \sum_{i=m+1}^{n} \lambda_i \qquad (17\text{-}49)$$

so that retaining only the first m of the n coefficients results in a mean square error given by the sum of the variances (eigenvalues) of the discarded coefficients.

In any particular application, the number of coefficients n that must be retained depends on how fast the eigenvalues $\lambda_1 > \lambda_2 > \lambda_3 > \ldots > \lambda_n$ decrease. If the data samples are not correlated, all n eigenvalues have significant values and we must choose $m = n$ for negligible distortion. On the other hand, if the data samples are highly correlated, then the eigenvalues decrease rapidly so that only the first few have significant values and all but the first few can be discarded. In this case $m << n$ and significant sample compression ratios

$$R_s(\epsilon^2) = n/m \qquad (17\text{-}50)$$

can be obtained with negligible distortion. Hence, the compression ratio that can be achieved depends on the amount of redundancy in the data and the amount of error that can be tolerated.

For correlated data, the m-retained coefficients have different rms values $\sqrt{\lambda_1} > \sqrt{\lambda_2} > \ldots > \sqrt{\lambda_m}$ so that a different number of bits should be used to code each coefficient. The mean square quantization error is minimized by choosing $m_i \sim log\ \lambda_i$. This is called block quantization (Wintz, 1972). If the original N data samples have b bits each, then the bit-compression ratio achieved by the eigenvector transformation and block quantization is

$$R_b(\epsilon)^2 = \frac{bN}{b_1 + b_2 + \ldots + b_n} \qquad (17\text{-}51)$$

Because of the requirement to determine the covariance matrix of the original multispectral data, the computation can be quite lengthy. For this reason, one of two approximations is often employed to minimize the calculation: 1) select a desired number of features (data clusters) in multispectral space, whose separation and optimum coding are important. These features (the training set) can be represented in a number of pixels much less than the original image, so that the covariance matrix of this reduced set can readily be found; 2) use a "standard transform independent of the actual image content; it will be found that the following one will approximately decorrelate most Landsat MSS images:

$$\begin{bmatrix} Y_1 \\ Y_2 \\ Y_3 \\ Y_4 \end{bmatrix} = \begin{bmatrix} G \\ R \\ I1 \\ I2 \end{bmatrix} \begin{bmatrix} 1 & 3 & 1 & 1 \\ 1 & 1 & -1 & -3 \\ 1 & -1 & -1 & 3 \\ 1 & -3 & 1 & -1 \end{bmatrix} \qquad (17\text{-}52)$$

3) use the cosine transformation discussed above. From Figure 17-42 it can be seen that this transformation closely approaches the Karhunen-Loeve in the rate of decrease of the eigenvalues.

Experience to date seems to indicate that three transform bands are required to adequately represent the total variance of a single pass of MSS data. For this reason, no more than about 25 percent data-reduction can be expected by this method. To date there has not been sufficient experience with data with more spectral bands, such as from the Landsat Thematic Mapper or the CZCS, to be able to estimate the performance expected.

Multispectral Feature Extraction Encoding

In the coding schemes outlined above, it is generally difficult to tailor the type of data degradation introduced so as to tend to preserve accuracy of multispectral classification and other computer analyses. This decreases the efficiency of the redundancy reduction allowed. Furthermore, these approaches generally require decompression of the compressed data to their expanded form before selected access or interpretation is possible.

A different approach to data encoding is the extraction and transmission of a series of primitives from the data, from which the data may be approximated during analysis. This technique may be encountered in several forms, such as: 1) Blob (Gupta and Wintz, 1975), in which the attempt is to define homogeneous areas in which the statistical descriptions of the chosen measurement vectors are constant to within defined tolerances, after which the boundaries are encoded and the list of primitives sent for each region; 2) cluster coding (Hilbert, 1977), in which multi-dimensional primitives are extracted from spatially contiguous (but not necessarily homogeneous) regions in the image data, with the transmission of the parameters of the primitive plus a feature map in which each initial measurement vector (pixel) is associated with a derived primitive.

Figure 17-44 shows a model for the general concept of joint feature extraction/data compression (Hilbert, 1977). Let

$$\underline{X}^i \triangleq \left\{ X^i \right\}_{i=1}^{n} \qquad (17\text{-}53)$$

be a sequence of n measurement vectors obtained from a subset of multispectral image data. The entire sequence $\underline{X}^i$ is analyzed to extract features

$$\underline{\psi}^j \triangleq \left\{ \Psi^j \right\}_{j=1}^{m} \qquad (17\text{-}54)$$

for the sequence $\underline{X}^i$. These features could also be considered as primitives, or as a basis for approximating the measurement vectors. The sequence of m features $\underline{\Psi}^j$ provides a description of the characteristics pertaining jointly to the entire sequence of measurement vectors $\underline{X}^i$ and in some applications may be the only output required. In other applications, each measurement vector needs to be approximated in terms of $\underline{\psi}^j$ by the sequence of scalar numbers

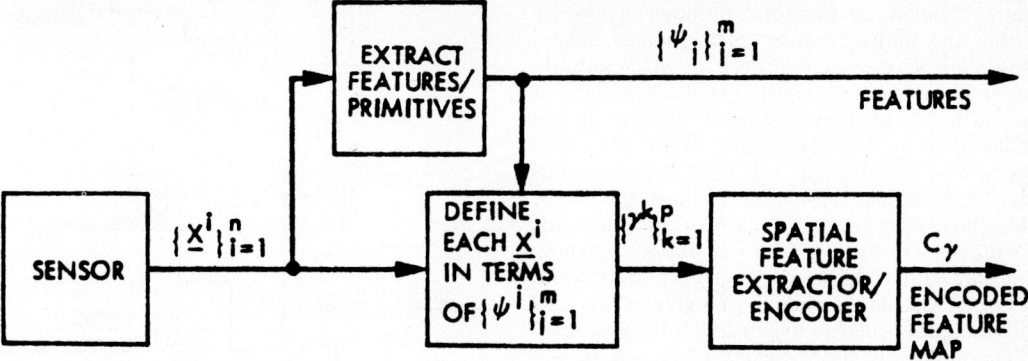

Fig. 17-44. A joint feature extraction/data compression model (Hilbert, 1977).

$$\gamma^k \triangleq \left\{\gamma^k\right\}_{k=1}^{p} \qquad (17\text{-}55)$$

which assigns to each measurement vector an approximation in terms of one of the primitives. The scalar sequence γ^k constitutes a spatial map of the measurement vectors in terms of the primitives, and is called the feature map. Spatial features can be extracted from γ^k, and source encoding can be used to represent efficiently the spatial characteristics of $\underline{X}^i$ through the encoded feature map denoted by $C\gamma$. For each sequence $\underline{X}^i$ of measurement vectors the model basically uses feature extraction concepts to determine a set of features or primitives, and then the model uses data compression techniques to represent efficiently the spatial features in terms of the primitives.

Cluster Compression Algorithm (CCA)
(Hilbert, 1977)

As before let

$$\underline{X}^i \triangleq \left\{X^i\right\}_{i=1}^{n}$$

be a sequence of n d-dimensional spectral irradiance measurements for a spatially local region of a multispectral image. These vectors generally tend to form groups in multispectral irradiance space. A typical plot of $\underline{X}^i$ for two bands in spectral irradiance space is shown in Figure 17-45(a). An intuitively natural set of features to use for representing $\underline{X}^i$ in Figure 17-45(a) are the descriptions for groups of measurement vectors, for example, the groups depicted in Figure 17-45(b–d). Each feature then consists of whatever

set of parameters are used to describe the source distribution for the group of measurement vectors. The mean and covariance are an obvious example of a group description, or feature. The number of features extracted from any sequence of measurement vectors would depend on how accurately $\underline{X}^i$ must be represented, and correspondingly what data rate is acceptable. An example of how two, four, or eight features might be chosen is shown in Figures 17-45(b–d).

Clustering is a means for automatically grouping multidimensional data (Kanal, 1974; Kan, 1972; Anderberg, 1973). The features resulting from clustering a set of measurement vectors are typically a subset of the following: cluster means, cluster variances per band, cluster covariances, number of vectors in each cluster, set of intercluster distances, etc. Anything contributing to the description of the cluster can be considered a feature, even the set of vectors themselves. If the features chosen for transmission include the multispectral means, image reconstruction can consist simply of using the appropriate cluster mean for each vector in that cluster.

One cannot simply obtain the thousands of clusters that will be obtained for a complete image by performing the clustering on the complete image. The large number of clusters could increase the computation time beyond practical limits. However, one can cluster the vectors in very small array sizes (e.g., local sources of, say, 16×16 pixels). Also, it is reasonable to expect a low average number of classes of interest per local

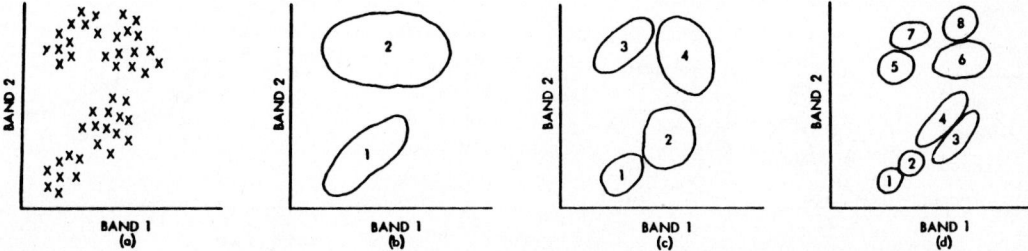

Fig. 17-45. Typical display in two band spectral intensity space for measurement vectors and for groups of measurement vectors used as features. (a) measurement vector $\left\{X^i\right\}_{i=1}^{n}$, (b) two groups, (c) four groups, (d) eight groups (Hilbert, 1977).

source relative to the total number of classes within the global source (total image). Thus, clustering at the local source level is initially easier to implement because there are fewer classes and fewer samples to cluster simultaneously. In addition, as shown in Figure 17-46, the global conditional distributions will have larger spreads than the local distributions due to averaging over many local distributions, each with corresponding various perturbations due to class variance, moisture, atmosphere, sun angle, etc. Therefore, at the local scale the vectors of a given class tend to form tighter clusters which are more separable.

The basic scheme of Figure 17-44 may be expanded to include the sequential clustering of local sources as shown in Figure 17-47. Typical data rates obtained for Landsat MSS data using n = 256 (16 × 16 pixels) per local source, f = 24 bits per feature (i.e., 4 spectral bands × 6 bits per band), m = 8 classes expected maximum per local source, and 4-band data input are 0.1875 bits per pixel per band for the spectral data and 0.75 bits per pixel per band for the spatial map, for a total average data rate of 0.9375 bits per pixel. The compression obtained may be controlled by specifying the number of classes allowed per local source. The mean square error produced decreases as the number of classes increases. An estimation of the MSE obtained is shown in Figure 17-48 for various m, obtained from a test portion of a Landsat MSS scene. Some further compression may be obtained through cascaded clustering, noiseless coding and other techniques (Hilbert, 1977).

The most significant advantage of the Hilbert Cluster Compression Algorithm is the rather significant reduction in computation required to perform supervised multispectral classifications. Such classification can be performed directly on the clusters without the need to "decompress" into an image. The reduction factor in computation over standard techniques is essentially the ratio of the number of original pixels to the number of clusters.

For example, a 1000 × 1000 multispectral image can be represented adequately for most visual applications using only 1000 clusters (obtained by clustering the local clusters). But classifying this image using such clusters yields accuracy equivalent to direct pixel-by-pixel classification. This

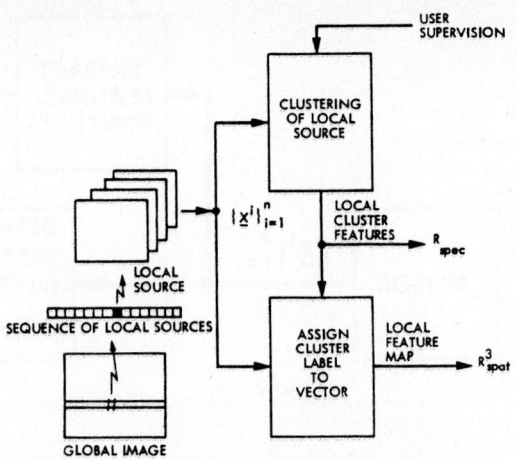

Fig. 17-47. A block diagram of the uncoded CCA (Hilbert, 1977).

result represents a 1000:1 reduction in computation requirements making near real-time minicomputer classification a reality.

COMPARISONS AND CONCLUSIONS

The above discussion has not included all of the possible encoding schemes for digital imagery, nor even all of those which have been exploited in real or conceptual systems. Those explored, however, seem the most promising.

It is found that when PCM is applied to sampled images, quantization artifacts (contouring) arise, which can be made arbitrarily small by increasing the number of quantization levels. PCM distortion at low rates is different than that produced by other differential schemes. Delta modulation (the

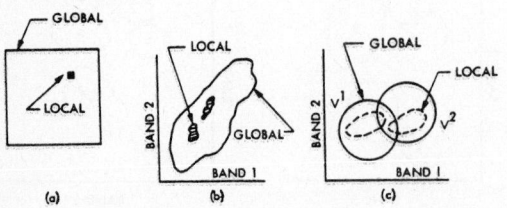

Fig. 17-46. Global vs. local sources (a) spatially, (b) distributions, (c) conditional distributions (Hilbert, 1977).

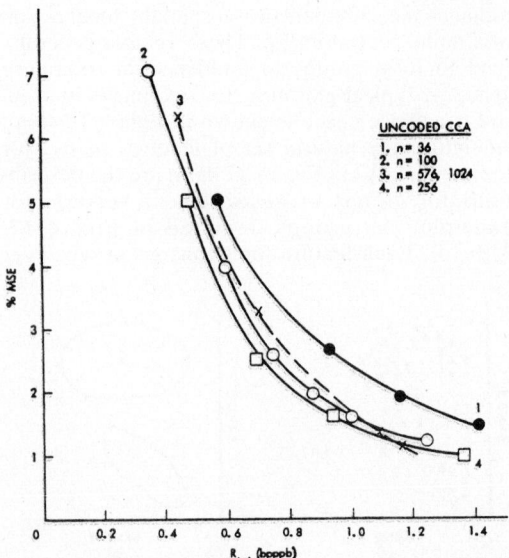

Fig. 17-48. %MSE vs. R_{tot} for various values of n for the uncoded CCA (Hilbert, 1977)

simplest difference scheme) provides the greatest bandwidth reduction, but the poorest spatial- and gray level-fidelity. These shortcomings are tempered somewhat by choosing an adaptive scheme such as HIDM. DPCM offers greater freedom from both spatial and quantization distortions than simple delta modulation, and is more efficient than tapered PCM. Four-bit DPCM approaches 8-bit linear PCM in apparent picture quality. Universal noiseless coding applied to PCM data provides the most efficient approach to very high quality data. Homomorphic processing of any sampled imagery reduces the irradiance dependence (provides dynamic range compression) and allows accessibility for processing of reflectivity. When applied in conjunction with 2-bit DPCM, it significantly reduces quantization artifacting as well.

The use of linear transformations and block quantization appears to offer the potential of efficient encoding of data at low rates. It is also possible to obtain adequate quality for many applications at 1 bit per pixel using adaptive schemes. However, although the subjective encoding achieves a significant decrease in bit rate, the implementation required is substantially more complex than DPCM, for example. The increase in the availability of fast computational algorithms, when coupled with the reduction in cost and size of high-speed integrated circuits, is making implementation of these two-dimensional encoding techniques feasible.

The discrete cosine transform has fast algorithms available and performs data compaction almost as well as the Karhunen-Loeve transform; it also has the advantage over the Karhunen-Loeve that it is deterministic—that is, it does not require data analysis for determination of the transform parameters.

The RM2 adaptive algorithm offers high performance along with broad user flexibility to adjust the user's rate/quality tradeoff to changing requirements. The most sophisticated form of cluster coding appears to offer better compression of multispectral data for a given MSE than do the other schemes (Figure 17-49, Hilbert, 1977) and simpler reconstruction. The key advantage, however, is the possibility of analysis directly in the compressed domain, offering significant computational advantage.

In conclusion, it is apparent that the design of an efficient sampled-imagery digital-communication system must account for the statistical nature of the image itself, and for various types of noise introduced by either the sensor or the quantization process. In addition, the encoding and decoding process should be matched to the channel and its noise characteristics, as well as to the visual tract when the human observer is the final user of the information, and to the analysis procedures to be used.

This field of research is continually developing. Continuing perusal of the publications of the

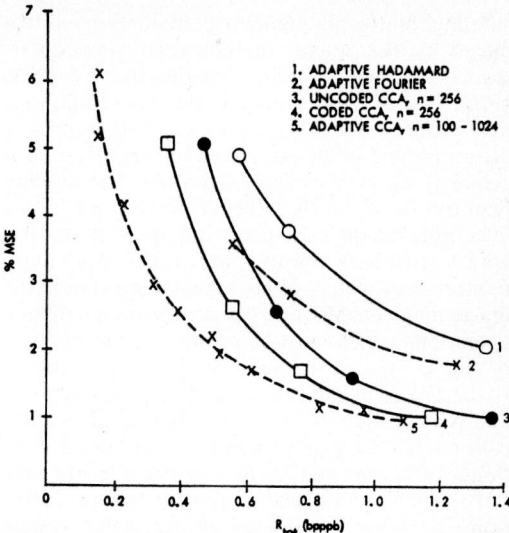

Fig. 17-49. %MSE vs. R_{tot} for various CCA configurations and for adaptive Hadamard and Fourier transforms (Hilbert, 1977).

IEEE (Institute of Electrical and Electronic Engineers), particularly the Transactions on Computers, Communication, and Acoustics, Speech and Signal Processing is recommended. The IEEE Trans. Comp. of November 1977 was completely devoted to Image Bandwidth Compression. In addition, Photogrammetric Engineering and Remote Sensing, and Computer Graphics and Image Processing frequently carry related papers.

THE GLOBAL SITUATION—GEOMETRIC DISTORTIONS

In the early part of this chapter the assumption was made that local processes could be separated from the global, and the chapter to this point has been concerned only with local processes. Let us now turn to some of the generic causes of global (geometric) distortions, and a brief discussion of correction methods. These corrections will be treated in detail in Chapter 21. The discussion will focus primarily on the Landsat spacecraft, as it exemplifies most of the distortions to be encountered.

GEOMETRIC MODIFICATIONS

Sources of Geometric Errors in Remote Sensing Imagery

A wide variety of image geometric distortions can be introduced by both the sensor and the nature of the scene which the sensor is viewing. The sensor system consists of the actual radiation or energy-sensing mechanism plus the platform on which the system is mounted. In many cases the platform introduces greater errors geometrically than the sensor itself. In the case of the Landsat multispectral-scanner system operating on a satellite platform, geometric distortions introduced by the sensor include mirror-sweep irregularities and sampling irregularities (NASA, 1976). The scanning mirror-sweep irregularities may produce a few hundred meters of geometric distortion and other errors in the sensor produce errors of the same or less magnitude. The satellite platform itself, on the other hand, has positional uncertainties (in roll, pitch and yaw primarily) which produce many hundreds and even thousands of meters of positional distortion in the image data set. Most of the geometric correction operations which must be performed on this type of data are due to the platform-data irregularities. In the case of an aircraft scanner-system operating in a vehicle such as a C-47 or other small aircraft, roll, pitch, and yaw variations, plus lateral positional variations due to steering irregularities and cross winds, can cause extremely severe distortions in imagery generated by a scanner system which is mounted rigidly on the aircraft. In the discussion that follows, the case of the aircraft and satellite multispectral scanner will be explored in detail. Many other types of sensors are used in remote sensing, such as cameras, vidicons, and side-looking radar. The geometric distortions from these systems are complex and worthy of more extensive treatment than can be presented here. Discussions of some of these are in Chapters 6, 12, 13, and 14. As scanning sensors are of wide current interest in remote sensing, this discussion is applicable to many sensors.

Landsat Orbit-controlled Systematic Distortions

The Landsat orbit (see also Chapter 12) is almost-polar, and represents in its geometrical aspects many of the mid-altitude (say 200–1000 km altitude) spacecraft having sensor scan-lines perpendicular to the nominal spacecraft path (see, in addition, HCMM, CZCS, and Stereosat as discussed in Chapter 13). It will be used as surrogate for the others in this discussion.

The Landsat MSS is effectively a one-dimensional scanner which scans a set of lines on the ground perpendicular to the spacecraft motion. The spacecraft motion in turn produces a second dimension. Earth rotation during the time of scanning of one frame causes progressive displacement of each scan-line set, resulting in an image skew which must be accounted for. In addition, the horizontal and vertical sample spacings as received are different and, in general, the entire image is rotated with the respect to latitude and longitude. Systematic distortions introduced in Landsat MSS images include skewing and scale distortions, which may be characterized mathematically. The numbers used are for Landsats 1–3.

Skewing is caused by the rotation of the earth during the 28 seconds required for Landsat to scan a single frame (185 km on a side). Typically, this skew is about 3°, but varies with latitude as the angle between the tangential component of the earth and the velocity of Landsat changes:

$$\vec{E}'(\phi) = \vec{E} - \vec{T}(\phi) \qquad (17\text{-}56)$$

where $\vec{E}(\phi)$ is the velocity of Landsat projected onto the rotating earth, $\vec{E}$ is the velocity of Landsat projected onto an inertial earth, and $\vec{T}(\phi)$ is the tangential velocity of the earth at geocentric latitude ϕ (Figure 17-50). The magnitude of E will be represented as $|\vec{E}|$ or simply as E, and is nearly constant:

$$|\vec{E}| = E = \frac{\text{circum earth}}{\text{orbital period}} = \frac{2\pi \text{ x } 6371 \text{ km}}{\sim6000 \text{ sec}}$$
$$= \sim6.5 \text{ km/sec} \qquad (17\text{-}57)$$

Although the contribution of earth velocity in the track direction is a function $|\vec{E}'(\phi)|$, and the scale in the sample direction is a function of the scan-mirror velocity and thus independent of $|\vec{E}|$ or $|\vec{E}'(\phi)|$, the aspect ratio (ratio of meters per line to meters per sample) is constant (for round earth) and independent of latitude, although it changes with spacecraft altitude. An additional complication is introduced because the MSS acquires six lines at a time. Thus, compensation for skew must not disturb the relative geometry within these six-line groups, or swaths; skew calculations must be made on a swath basis. Samples from successive lines in a single channel are acquired one-twelfth of the dwell time apart; consequently, each line within a six-line swath must be offset to the east by one-twelfth sample (or about 5 m) from the preceding line.

Correction for skewing requires two pieces of information: angle of skew and correct aspect ratio. These will be derived for the round earth case. More exact analysis should take the eccentricities of the earth, local elevation and orbit variations into account; this will modify the numerical results slightly, and is important for precise mapping. The spacecraft velocity as projected onto a nonrotating earth is:

$$|\vec{E}| = \frac{2\pi Re}{\tau_{s/c}}$$

The equatorial earth velocity is:

$$|\vec{T}(eq)| = \frac{2\pi Re}{\tau_{day}}$$

The magnitude of $\vec{T}$ will be represented as T. This is reduced at latitude ϕ to:

$$\vec{T}(\phi) = \frac{2\pi Re}{\tau_{day}} \cos \phi$$

Fig. 17-50. Landsat orbit as seen on the downward portion during imaging.

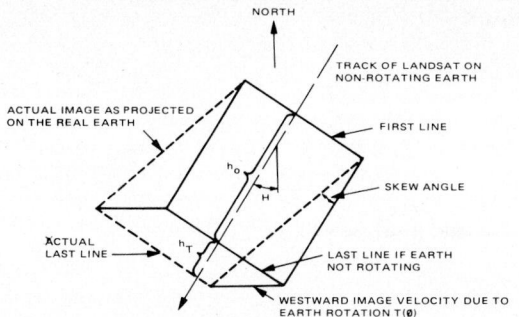

Fig. 17-51. Skewing of the image caused by the earth's rotation. Change in aspect ratio is caused by component of earth velocity in track direction.

During the time required to acquire one frame (about 28 seconds) the earth rotates eastward under the spacecraft orbit, so that each successive line is displaced westward, skewing and stretching the image (Figure 17-51).

In the spherical triangle comprising the equator between the longitude at the spacecraft (s/c) latitude ϕ and the reference longitude, the longitude line at s/c latitude ϕ, and the orbit portion between the equator and latitude ϕ (Figure 17-52):

$$\sin H = \frac{\cos \Psi}{\cos \phi}$$

where Ψ is the orbit minor tilt (see Figure 17-50). Combining the above, the effective heading angle H_{eff} is:

$$H_{eff} = \tan^{-1} \frac{\text{Total } \vec{E}'(\phi)w}{\vec{E}(\phi)s} \qquad (17\text{-}58)$$

$$= \tan^{-1} \frac{\left| \dfrac{\tau_{s/c}}{\tau_{day}} \cos \phi + \dfrac{\cos \Psi}{\cos \phi} \right|}{\cos H = \sqrt{1 - \left(\dfrac{\cos \Psi}{\cos \phi} \right)^2}}$$

This is the angle between the side or centerline of the skewed image and a N-S longitude line, as projected on the real rotating Earth.

The skew can be found by (see Figure 17-52):

$$\tan \text{skew} = \frac{T(\phi) \cos H}{T(\phi) \sin H + E} = \frac{\sqrt{\cos^2\phi - \cos^2\Psi}}{\cos \Psi + \dfrac{\tau_d}{\tau_{s/c}}} \qquad (17\text{-}59)$$

The amount of stretch compared to the picture height (along the track, if projected onto a non-rotating earth) is:

$$\text{stretch} = \frac{ho + h\tau}{ho} = \frac{E + T(\phi) \sin H}{E}$$

$$= 1 + \frac{\dfrac{2\pi Re}{\tau_d} \cos \phi \left(\dfrac{\cos \Psi}{\cos \phi} \right)}{\dfrac{2\pi Re}{\tau_{s/c}}}$$

$$= 1 + \frac{\tau_{s/c}}{\tau_d} \cos \Psi \qquad (17\text{-}60)$$

The actual heading (non-rotating Earth) on the downward part of the orbit will be:

$$\text{Heading} = 180° + H \qquad (17\text{-}61)$$

The length of a siderial day $\tau_d = 86164$ seconds. This and the orbit parameters below are sufficient for the required calculations.

	$\tau_{s/c}$	Ψ	
Landsat 1−3	6196	80.912°	
Landsat D	5940	81.8°	(Anticipated)

Most future Landsat images will be corrected when received. However, this will not be done retrospectively. Hence, the following resultant equations will be useful in correcting the older Landsat 1−3 images:

$$\text{stretch} = 1 + 0.01136 \text{ (average)} \quad (17\text{-}62)$$

$$H = \sin^{-1} \frac{0.15795}{\cos \phi} \qquad (17\text{-}63)$$

$$H_{eff} = \tan^{-1} \frac{0.07191 \cos \phi + \dfrac{0.15795}{\cos \phi}}{\cos H} \qquad (17\text{-}64)$$

$$\text{skew}° = 4.07387 \sqrt{\cos^2\phi - 0.02495} \quad (17\text{-}65)$$

Attitude Distortions

In addition to the systematic orbit-controlled distortions modeled above, other distortions will occur due to platform attitude-variations.

Figure 17-53 shows the six degrees of freedom which are inherent in scanning geometry of all multispectral scanner systems. Three rotational degrees of freedom represent the way the scanning

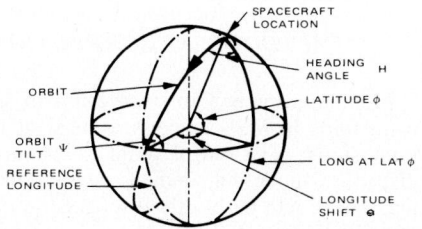

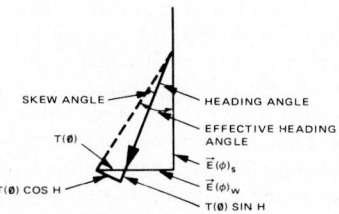

Fig. 17-52. Angular relationships with spacecraft at latitude ϕ.

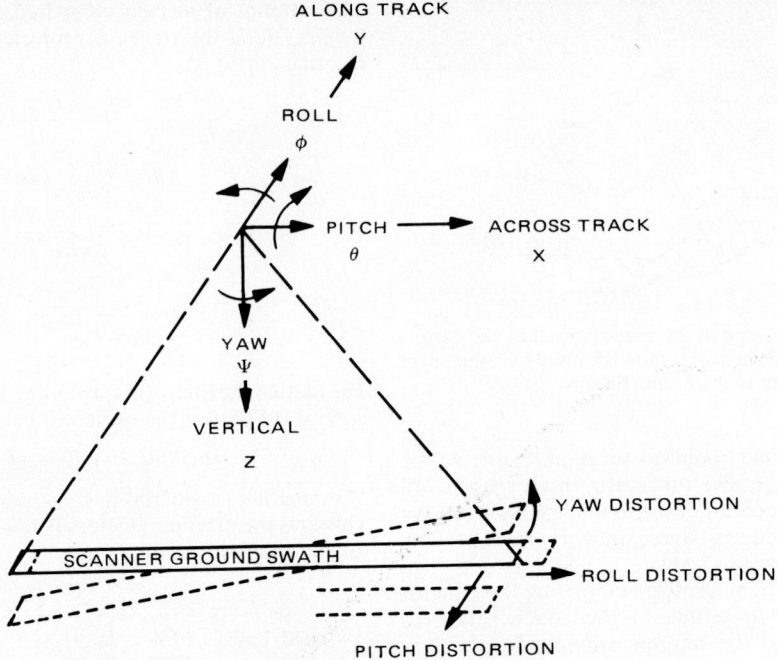

Fig. 17-53. Sources of geometrical distortion due to platform attitude.

sensor can be rotated on three axes with respect to the scene. Rotation along the longitudinal axis of the sensing platform is called "roll." Rotation along a line passing vertically through the sensing platform and intersecting the scene is called "yaw." Rotation on the axis orthogonal to the first two is called "pitch." Similarly, motions in the along-track direction (i.e. the direction in which the sensing platform is moving) are called the "along-track direction, y." Motions to the side are called "cross-track, or x." Motions vertically are called the "vertical, or z."

In the case of the Landsat MSS sensor, all six of these variables affect image geometry; however, the positional or x, y, z quantities can potentially be known with high accuracy via satellite tracking and orbit-determination systems. The angular uncertainties are not as well known and contribute significantly to image distortions. Even with these distortions, the effect is not visible in a typical satellite image without detailed comparison to a good map, and we go to the aircraft systems to find examples for the purpose of illustration.

The illustration in Figure 17-54 shows a very good example of the kind of distortion which can be obtained from a multispectral scanner system. In the left image we see a reconstructed data set obtained from an aircraft multispectral scanner flying over typical rural terrain at an altitude of approximately 2,000 feet. Flying conditions were relatively calm and the lines representing roads and field edges in the scene are reasonably close to what the true positions of these features are on

the earth. In the image on the right, the same area was flown at a later time, when a crosswind condition existed in the area through which the sensing aircraft was flying, resulting in a large yaw angle. Note the distortions in the roads running across the scene; they are tilted in a northeast/southwesterly direction. This skew distortion is characteristic of all multispectral scanner systems rigidly mounted on the vehicle and which view the scene in successive scan lines horizontally across the image while the carrying aircraft moves the sensor forward down a scene, or along track. An even more severe example of this type of distortion is seen in Figure 17-55 in which not only were cross-winds causing skew in the image, but the aircraft ground-track was randomly varying, causing wavy features to be produced where lines in the original image were straight. These same distortions exist in the satellite case but with smaller magnitudes. The distortions listed in Figure 17-56, adapted from the NASA Landsat Users Handbook, summarize the errors found in the MSS sensor.

METHODS OF CORRECTION

Most approaches to correction of geometric distortions in MSS imagery consist of three distinct steps: (1) Control point determination, (2) distortion modeling, and (3) correction processing. These will be briefly discussed in order; See also the more detailed discussion in Chapter 21.

 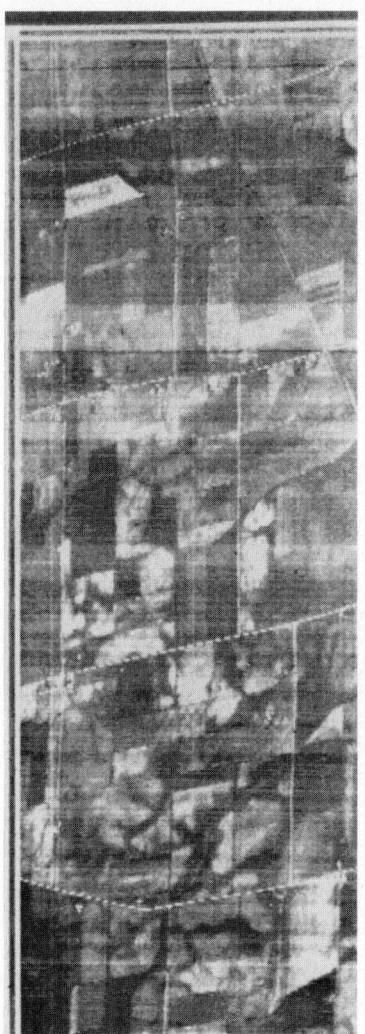

Fig. 17-54. Normal (left) and skewed (right) aircraft multispectral scanner images (left: normal, with no crosswind) (right: strong crosswinds caused craft to fly with large yaw angle). Data from ERIM M-7 system over Northern Indiana, June 1969).

Control Point Location

The basic requirement for geometric-correction processing is the acquisition of control points in the image to be corrected. The coordinates of points in the image and corresponding points in a reference are the inputs to the distortion modeling process by which an image ultimately is corrected. Two categories of reference exist: (1) image reference, and (2) geographic reference. In the first case, the image is to be corrected to match another image, which can be of any shape; and in the second case, a precise correspondence is required between the corrected image and a geographic reference system. In either case, the corresponding points must be found either manually or by computer image-processing techniques. Manual or visual identifi-

cation is a tedious and straightforward process which does not require extensive discussion. Various image reproduction methods may have to be used to make features sufficiently visible for human identification. A discussion of alternative reproduction methods is presented in Appendix B of NASA reference publication 1039 (NASA, 1979).

Automatic methods of control-point location are the subject of extensive research. The usual approach to automatic control-point finding is to compute the correlation coefficient between two images at a large number of relative shifts and choose the location of the maximum value as the misregistration (Anuta, 1970). The equation for the image-to-image correlation, assuming zero means images, is:

$$P_{xy}(k,\,1) = \frac{\sum_i \sum_j X(i,j)\, Y(i+k,\,j+1)}{\left[\sum_i \sum_j X^2(i,j) \sum \sum Y^2(i+k,\,j+1)\right]^{1/2}}$$

(17-66)

The X and Y are the radiance values for the two images being registered and the summation is over a rectangular subimage in X and Y. This method requires a large amount of computation and other more efficient methods have come into use. The difference-correlation function which is of the form:

$$d_{XY}(k,\,l) = \sum_i \sum_j |X(i,j) - Y(i+k,\,j+l)| \quad (17\text{-}67)$$

is more efficient and performs as well as the correlation function in most cases. A form of the difference correlator called the SSDA (Barnea and Silverman, 1972) is implemented in the Master Data Processor (Peavy, 1977) which performs all geometric correction operations on Landsat 1–3 MSS data. Various enhancements are often performed on the subimages before correlation to improve the likelihood of finding the correct misregistration; these are mostly variations of the image-gradient transformation. A common form of the gradient is (Swedlow, McGillem, and Anuta, 1978),

$$G(i,j) = \left[(X(i,j+1) - X(i,j-1))^2 \right.$$
$$\left. + (X(i+1,j) - X(i-1),j))^2 \right]^{1/2} \quad (17\text{-}68)$$

which is the magnitude of the discrete approximation to the two-dimensional gradient vector. Other modifications use thresholded gradients and binary image representations (Nack, 1975). Swedlow, et al. (1978) have demonstrated evidence that all methods perform well when the correlation coefficient ρ between the two images is above 0.5 and that the gradient processors gave marked improvement for the $\rho < 0.5$ cases.

Geometric Distortion Modeling

Two approaches generally can be taken to remove errors in remote sensing imagery. One is to describe the errors with general polynomials or other functions not directly related to the specific distortion sources. The other is to generate mathematical models which characterize the error sources and remove the effects of those errors which can be described. Any residual errors may then be removed using general functions. In the case of the Landsat MSS, several errors are taken out from *a priori* knowledge of the nature of the errors. In the MDP correction system for all Landsat 1–3 hard-copy data implemented at the NASA Goddard Space Flight Center, the following errors are corrected with *a priori* information (NASA, 1979):

Fig. 17-55. Aircraft multispectral scanner imagery (0.58–0.65 μm band) showing severe distortion due to crosswinds. LARS run no. 71054100. Area is immediately west of Crawfordsville, Indiana. Date; August 17, 1971. Altitude: 5000 ft.

Table G.2-3 MSS Geometric Corrections

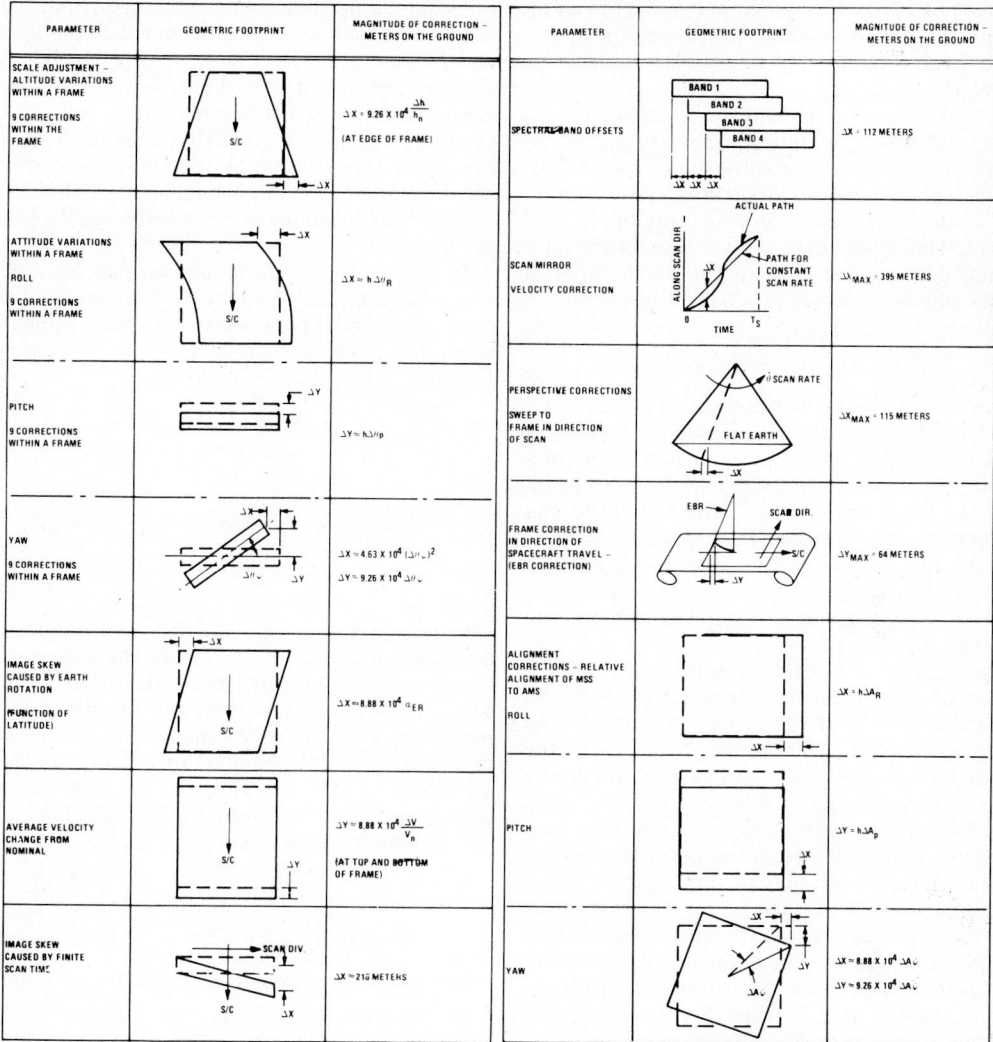

Fig. 17-56. Geometric distortion in Landsat data (courtesy NASA GSFC).

Scan Mirror Errors
Panoramic Distortion
Scan Skew Errors
Earth Rotation Skew
Map Projection Errors
Scan Line-Length Errors
Band-to-Band Offset Errors
Detector Sampling Delay Errors

The effects of spacecraft attitude and position errors are then modeled, using ground-control points. In this process, 20 or more control points are located in each image and on maps and used to model roll, pitch, yaw, and altitude variations over the Landsat-frame scan time. Modeling of the specific distortion sources is preferable but is possible only if the sources can be identified and the required measurements of control-point loca-tions can be made to the precision required for the source-parameter estimation.

Spacecraft Parameter Modeling

With reasonable precision, it is practical to as-sume that the spacecraft attitude and ephemeris may be expressed as low order equations; that is, they are smooth functions. Thus, if suitable data are available at several points in the trajectory, continuity may be invoked to estimate spacecraft attitude and position throughout the trajectory. This estimated parameter set may then be used to model the effects into the image domain to gener-ate the required image-warping function. The re-quired data will typically take the form of orbit location and spacecraft location within the orbit, spacecraft roll, pitch, and yaw, and roll, pitch and yaw derivatives. In addition, any spacecraft vibrations and sensor scanning-parameters must be known. Finally, the effects of projection to the earth, earth rotation, and map projections will be included.

A look-point model may be constructed involving the spacecraft position and attitude, based on the continuity of inertial position, and including such effects as map projection, earth rotation and calibration parameters as previously discussed. This model associates a pixel in the data stream with its location on the earth. Such a model may be regarded as a pair of functions

$$x = x(i,j;p_1,p_2, \ldots p_n)$$
$$y = y(i,j;p_1,p_2, \ldots p_n)$$

$$(17\text{-}69)$$

where x and y are the map coordinates of a pixel, i indexes the scan lines, j indexes the pixel position within a scan line, and $p_1, p_2, \ldots, p_n$ are a set of n parameters required to model the effects to be included. Such parameters might include the spacecraft attitude-angles (roll, pitch, and yaw) and the spacecraft orbital-elements (semimajor axis, eccentricity, inclination, longitude of ascending node, argument of perigee, and perigee passing time).

The accuracy of the look-point calculation is subject to both the intrinsic accuracy of the model and the accuracy of the values assigned to the model parameters. The intrinsic accuracy of the model may always be improved by including higher order effects such as scan mirror non-linearities or deviations of the earth's surface from a sphere. However, the accuracy of the model-parameter values can only be improved by better measurements. In the case of the Landsat series of satellites, the primary source of model parameters would be the telemetry data-stream: the attitude data from the spacecraft attitude-control system, and, if available, the spacecraft-ephemeris data from an on-board model.

In general, the available spacecraft parameters will not be of sufficient accuracy to allow precision projections of the data pixels to the ground. For example, the rated attitude-control system accuracy of Landsat-D ($1\sigma = 0.01°$) at the spacecraft altitude of 705 km will allow a pixel 1σ-error of 120 meters (four TM pixels). Improvement of the model-parameter measurements may be accomplished by comparing the known location of a set of ground-control points with their locations as calculated from the look-point model using nominal or *a priori* parameter values (i.e., the best available set of parameter estimates prior to the comparisons). Such a set of measurements can then be used to statistically estimate a more accurate set of parameter values which, in turn, can be used in the look-point model to produce an image of better geometric quality.

The least-squares estimate to be described is essentially a least-squares fit of the model parameters to the known control-point location data. It will be generally applicable to scenes for which the model parameters may be considered to be constant during the data-collection period. This stipulation does not, however, preclude the use of attitude and ephemeris rates or other higher derivatives, provided that these, in turn, are constant. If it is known that some effects will not be constant (for instance, vibration induced in the sensor of the Landsat-D spacecraft by various moving parts), correction must be made in the expected pixel locations prior to involving the look-point model.

In practice, it is safe to assume that the differences between the true and *a priori* values of the look-point model parameters are small enough to allow the model equations 17-69 to be approximated as affine transformations:

$$x = x(i,j;p_1^{(o)},p_2^{(o)}, \ldots, p_n^{(o)}) + \sum_{k=1}^{n} \mu_{xk}(i,j)\delta_k$$

$$y = y(i,j;p_1^{(o)},p_2^{(o)}, \ldots, p_n^{(o)}) + \sum_{k=1}^{n} \mu_{yk}(i,j)\delta_k$$

$$(17\text{-}70)$$

where $p_1^{(o)},p_2^{(o)}, \ldots, p_n^{(o)}$ are the *a priori* parameter values and $\delta_1,\delta_2, \ldots, \delta_n$ are the differences between the parameters' true and *a priori* values. The μ-coefficients represent the sensitivity of a pixel's location to small changes in the model parameters and may be functions of i and j. They are the partial derivatives associated with the first order terms of a Taylor series expression for x and y evaluated at the nominal parameter point:

$$\mu_{xk} = \left(\frac{\partial x}{\partial p_k}\right) \qquad \mu_{yk} = \left(\frac{\partial y}{\partial p_k}\right)$$
$$\text{with } \delta_1 = \delta_2 = \ldots \delta_n = 0.$$

One example of constructing the model will be given. Let:

y_{sc} = position of the spacecraft along its orbit
H = spacecraft altitude above surface
θ_p = angle of pitch, the *a priori* value
δ_p = deviation of the true and *a priori* positions of θ_p
μ_p = sensitivity in y-position to δ_p

As shown in Figure 17-57, the slant range H_s due to the pitch is $H_s = H/\cos\theta_p$, and the displacement of the pixel ground position is $H \tan \theta_p$. The displacement caused by δ is evaluated at the pitch angle θ_p and found to be $(H/\cos^2\theta_p)\delta_p$. Thus the total expression for the ground position y of the pixel (allowing only y_{sc} and pitch) is:

$$y = y_{sc} + H\tan\theta_p + \frac{H}{\cos^2\theta_p}\delta_p. \quad (17\text{-}71)$$

The more complete expression would involve yaw and recognize the potential error in knowledge of y_{sc} and the effects of changes of altitude.

The affine approximation embodied in Equations 17-70 will result in some small deviations from the true geometric positions, but may be jus-

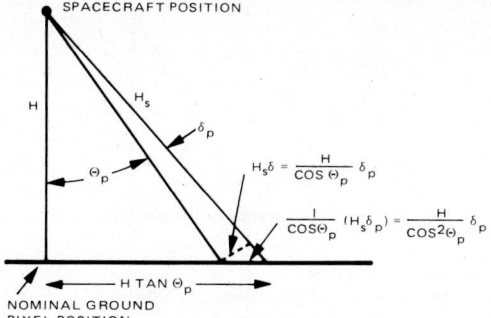

Fig. 17-57. Displacement of pixel ground position due to pitch θ_p and pitch error δ_p.

tified in the practical situation for spacecraft (implying small values of the δs), for which the required geometric accuracy will be obtained. Higher geometric accuracy may be obtained by expanding the model to higher order or by including more distortion sources, provided that the additional required model parameters can be accurately obtained.

The affine transformation-equations are most clearly expressed in matrix notation as

$$\vec{x} = \vec{x}^{(o)} + \mu\vec{\delta}, \qquad (17\text{-}72)$$

where $\vec{x}$ has the components x and y, and the model parameters are regarded as an n-dimensional vector. The goal is to determine $\vec{\delta}$ by comparing a set of nominal control point locations $\vec{x}_1^{(o)}, \vec{x}_2^{(o)}, \ldots, \vec{x}_m^{(o)}$ as they are found in the image formed with $\delta = 0$ with their actual locations $\vec{x}_1, \vec{x}_2, \ldots, \vec{x}_m$ in the actual image. In general, the sensitivity matrix μ is not invertible and therefore the determination of $\vec{\delta}$ is not completely straightforward. In addition, what is measured is the difference between the true and the nominal control-point positions. These measurements may be accomplished by first applying the look-point model using *a priori* parameter values, to create a sub-image (the control point neighborhood image) likely to contain a given control point and centered at the control point's expected position. Next, the control point is located within the sub-image using correlation methods described elsewhere. This determines the offset between the expected position and the actual position.

The offset measurements $\Delta\vec{x}$ are subject to uncertainties which, in turn, are a result of uncertainties in determining the true map coordinates of control points, effects of resampling of the subimages, and the resolution and accuracy of the control point correlation. It will be assumed that the uncertainty in a measurement $\Delta\vec{x}$ can be represented as a zero-mean random variable, $\vec{\xi}$, which has a covariance matrix R, where the covariance matrix is the expectation value of $\vec{\xi}\vec{\xi}^T$ (Wilks, 1962). In the (likely) case in which the R matrix is unknown, that matrix must be assumed, noting that the R's reflect the relative importance

of the various control points. The R's will be required in the estimation of performance. When the matrix-equation 17-72 is applied to a set of m control-point offset measurements, the set of m equations results:

$$\Delta\vec{x}_\ell + \vec{\xi}_\ell = \mu_\ell\vec{\delta} \qquad (17\text{-}73)$$

where $\ell = 1,2, \ldots, $ m.

Letting $\mu = \mu_\ell$ recognizes that the sensitivity matrices may depend on the scan line and also on the pixel position within the scan line at which the control point may be located; furthermore, letting $\vec{\xi} = \vec{\xi}_\ell$ recognizes that the statistics of the uncertainties may be different for every control point.

In making an estimate $\hat{\vec{\delta}}$ of $\vec{\delta}$, the quadratic form

$$J(\hat{\vec{\delta}}) = \frac{1}{2}\sum_{\ell=1}^{m} (\mu_\ell\hat{\vec{\delta}} - \Delta\vec{x})^T R^{-1} (\mu_\ell\hat{\vec{\delta}} - \Delta\vec{x})$$

$$(17\text{-}74)$$

may be taken as a measure of the "badness of fit" of $\hat{\vec{\delta}}$ to the data $\Delta\vec{x}_1, \Delta\vec{x}_2, \ldots, \Delta\vec{x}_m$. $\mu_\ell\hat{\vec{\delta}} - \Delta\vec{x}_\ell$ is the difference between the control-point offset that would result if $\vec{\delta} = \hat{\delta}$, and the measured control-point offset. It is weighted by the inverse covariance matrix and, therefore, components with greater measurement certainty receive the greater weight. Additionally, each term is a positive number by virtue of the positive definite property of covariance matrices (Wilks, 1962, p. 80). Under this definition, the $\hat{\vec{\delta}}$ which best fits the data minimizes J and is said to be the "least squares estimate". It should be noted that if the measurement errors $\vec{\xi}$ are independent and Gaussian, then their joint probability density-function is proportional to $\exp(-J(\vec{\delta}))$. In this case, the least squares estimate which minimizes $J(\vec{\delta})$ maximizes the joint probability density function. Under these conditions the least squares estimate is also the maximum likelihood estimate; that is, the estimate $\hat{\vec{\delta}}$ is such that the measured set of control-point offsets is the most likely set to be measured.

It can be shown that the least squares estimate is a solution of the linear system

$$M^{-1}\hat{\vec{\delta}} = \vec{Y} \text{ where } M^{-1} = \sum_{\ell=1}^{m} \mu_\ell^T R_\ell^{-1}\mu_\ell \text{ and}$$

$$\vec{Y} = \sum_{\ell=1}^{m} \mu_\ell^T R_\ell^{-1}\Delta\vec{x}. \qquad (17\text{-}75)$$

There will be a solution for $\hat{\vec{\delta}}$ as long as every observed $\Delta\vec{x}$ is explainable in terms of the model, that is, if for each $\Delta\vec{x}$, there is a $\vec{\delta}$ such that $\Delta\vec{x} = \mu\vec{\delta}$. However, such a solution is not guaranteed to be unique.

Whether or not a solution is unique (that is, whether the spacecraft model parameters may be correctly obtained from the set of control point measurements) depends upon the number and distribution of control points, and the sensitivity coefficients μ involved. Consider, for example, an MSS-type sensor and a two-parameter model in which the parameters are the spacecraft altitude and its velocity. As shown in Figure 17-58, variations in altitude will cause the image line (across track) to vary. However, there will be little variation in the along-track direction as the velocity changes will be small. No number of control points distributed along the center line (or any line parallel to it) will be able to detect the altitude variation. However, if the change in width can be determined by control points spaced across track, altitude variations can be determined through

$$\frac{W_1}{W_2} = \frac{H}{H + \Delta H}$$

Similarly, changes in velocity will cause length distortions along track. Control points grouped across track will not detect this velocity change, but the ones along track will.

However, consider the situation in Figure 17-59. Here the spacecraft time (corresponding to position along the track) and pitch are allowed in the model. In this case, pitch will also cause an elongation of the scan line, masquerading as an increase in altitude. (Note that a decrease in altitude is unique, as the nadir pointing already produces the shortest scan line for a given altitude.) If the spacecraft time is sufficiently precise, the dislocation of the image line along track can be determined, and the correct cause ascribed to the increased line length.

The parameters for Landsat-D are: $H = 705$ km, Δt accuracy = 20 msec, $W_1 = 185$ km (nominal), and θ_p (1σ) = 0.01°. At the design altitude, the time uncertainty results in 130 meters uncertainty in spacecraft position and the θ_p allows an uncer-

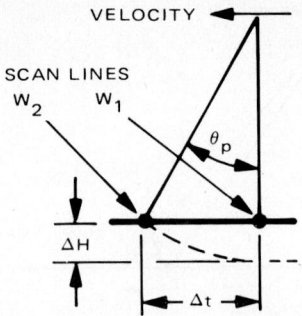

Fig. 17-59. Pitch causes a change in image line length, which could also have been caused by altitude change ΔH.

tainty (1σ) in position of 120 meters. Pitch-variation effect on ground-line scaling is

$$\cos\theta_p = \frac{W_1}{W_2} = \frac{H}{H + \Delta H}$$

so that the 0.01° translates into an image-width increase of 1.00017×, or about 32 meters. However, this could also have been caused by an altitude change of about 123 meters. Unless a timing knowledge better than the 20 msec accuracy specification is available, the cause of the variation cannot be determined.

In the case where one of the components of $\hat{\vec{\delta}}$ is indeterminant, a proper procedure might be to set it to zero, recognizing that the *a priori* estimate of the corresponding parameter value remains our best estimate. However, if there are several components of $\hat{\vec{\delta}}$ which are indeterminant, it is not clear as how to choose a suitable estimate. The one circumstance which obviates the necessity of choosing amongst equally good estimates is when all solutions for $\hat{\vec{\delta}}$ produce the same pixel displacement (i.e., $\mu\hat{\vec{\delta}}$ is the same for all $\hat{\vec{\delta}}$s), in which case any solution for $\hat{\vec{\delta}}$ will produce the same results. For example, if the measurements cannot distinguish time from pitch, no discernable difference in the results will be obtained if all the errors are ascribed to one or the other.

Once a suitable solution for $\hat{\vec{\delta}}$ is obtained, it may be used in the affine transformation-equation 17-72 to better estimate pixel locations, i.e., to generate a corrected image having improved geometric and locational accuracy:

$$\vec{x}^{(1)} = \vec{x}^{(o)} + \mu\hat{\vec{\delta}} \qquad (17\text{-}76)$$

This is the most accurate projection of $x(i,j)$ on to the ground possible with the given model and the set of control points. The pixel locations determined in this manner are designated $\vec{x}^{(1)}$ because they may still deviate from the true locations by a set of residual errors. There are two components to the residual error: those resulting from a difference between the true and estimated spacecraft parameters, and those due to a difference between

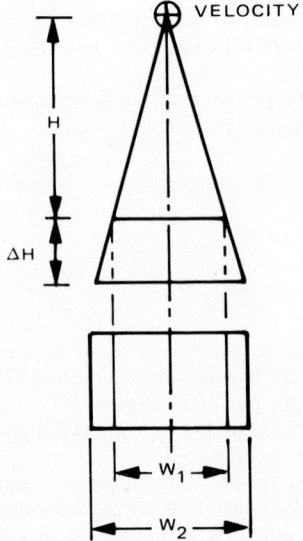

Fig. 17-58. Spacecraft altitude variation ΔH causes a change in image line length.

the model and reality. The first component may be found by subtracting Eq. 17-76 from Eq. 17-72:

$$\vec{x} - \vec{x}^{(1)} = \mu(\vec{\delta} - \hat{\vec{\delta}}). \quad (17\text{-}77)$$

The magnitude of the second component is indicative of the validity of the look-point model and Eq. 17-76. It may be estimated by comparing the actual locations of the control points $\vec{x}_1, \vec{x}_2, \ldots, \vec{x}_n$ with their locations as determined by Eq. 17-76: $\vec{x}_1^{(1)}, \vec{x}_2^{(1)}, \ldots, \vec{x}_n^{(1)}$. The differences between the measured residual errors $\vec{x} - \vec{x}^{(1)}$ and those predicted by Eq. 17-77 are the measures of this component of the residual error. The statistical significance of this component may be assessed by assuming a probability density-function for the measurement errors $\vec{\xi}$. In the case where the measurement errors are Gaussian and independent, it can be shown that the residual errors of the first type are Gaussian distributed with zero mean and covariance $\mu M \mu^T$.

Thus the least-squares estimation technique provides both a method to estimate a pixel's map-coordinates and a method of assessing its positional accuracy in doing so.

Polynomial Modeling

The more general case is one in which an image contains unknown distortions and general polynomials are fit to control points using least squares or minimax error criteria to model the corrections directly in the image domain without explicitly determining the distortion sources. For moderate distortions in a constrained area, a six-parameter affine-transformation works quite well to restore imagery to a geometric reference. This transformation can represent six kinds of distortions in imagery. The first is skew. This could be due to the yaw angle not being 90° to the direction of flight, or due to earth rotation as derived above. The x axis influences the position of points along the y axis when skew is present so that the y dimension after skew is $y' = y + a_{sk}x$ and the x dimension is unaffected, where a_{sk} is a constant proportional to skew. The transformation can also represent a rotation; however, this is a global rotation and not the rotation due to a yaw variation in the scanner system. The third and fourth kinds of variation that the transformation can account for are translation in the x and y directions. The last two variations are scale variations on the x and y axes. When these six operations are combined into a single expression, the equation shown below is obtained:

$$\left.\begin{array}{l} x' = a_0 + a_1x + a_2y \\ y' = b_0 + b_1x + b_2y \end{array}\right\} \quad (17\text{-}78)$$

The six parameters a_i, b_i $(i = 0-2)$ implicitly define the six degrees of freedom, but do not explicitly relate to the degrees of freedom (roll, pitch, yaw, etc.) of the causes of the distortion.

Corrections represented by the above equations are based on the assumption that the scene is a plane with no vertical relief. In the case where

significant topographic variations occur in the area being scanned, distortions due to variations in height of points must be considered. The approach commonly used to rectify imagery with relief distortion is to use the colinearity equations discussed in Mikhail (1973), Ethridge (1976), and Konecny (1979) which have the form:

$$\left.\begin{array}{l} x' = \dfrac{a_{11}(x - x_0) + a_{12}(y - y_0) + a_{13}(z - z_0)}{a_{31}(x - x_0) + a_{32}(y - y_0) + a_{13}(z - z_0)} + x'_c \\[3mm] y' = \dfrac{a_{21}(x - x_0) + a_{22}(y - y_0) + a_{23}(z - z_0)}{a_{31}(x - x_0) + a_{32}(y - y_0) + a_{33}(z - z_0)} + y'_c \end{array}\right\}$$

$$(17\text{-}79)$$

The x, y, z are positions in the output image and the x', y' are corresponding image positions in the input image. The subscripted x's and y's refer to reference coordinates. In the case of satellite-borne scanners, the look angle is generally very small and panoramic and topographic distortions are minimized. For more complex distortions, higher degree polynomials are used; however, the degree is usually limited to five in practice due to the complexity and computational cost of evaluation. A biquadratic function has been successfully used in a general image-registration system (Anuta, 1977), although full-frame Landsat images cannot be handled. Most data sets needed for analysis are significantly less than a frame in this application. For full-frame Landsat MSS correction, the Master Data Processor described in Chapter 12 uses a fifth-degree polynomial to describe all the *a priori* and control-point modeled distortions. This is discussed in more detail in Chapter 21.

Geometric Correction Processing

Once a single high degree polynomial or other complete mathematical description of the distortions is established, correction of the image can proceed. The primary consideration in implementing a correction function is processing efficiency. The question of resampling also arises since new pixels will generally be required from locations between original pixels and some form of interpolation is needed to estimate the new values. Two methods of resampling are common: (1) Nearest neighbor and (2) cubic convolution. These are discussed in detail in the section on interpolation.

The question of efficiency in distortion-function evaluation is important as the degree of representation increases. For a two-dimensional, fifth-degree distortion function, 30 multiplies are required for each evaluation. It is prohibitively costly to evaluate for each pixel in, say, a Landsat frame which has nominally 7.5 million pixels; thus a blocking scheme can be used to linearly interpolate the distortion function. For example, if the fifth-degree function for a Landsat frame would be evaluated only every 300 pixels and linear interpolation used for all pixel locations between, drastic savings in computation would be achieved. In this example, each evaluation of the fifth-degree polynomial for the two dimensions requires 30 multiplies. Thus the 300-pixel block would contain 90,000 pixels and require 2,700,000 multiplies to evaluate at each pixel. If linear interpolation is used in the block, then four multiples per pixel, or 270,000 would be required, plus the 120 for evaluating the fifth-degree function at the four corners.

MAP PROJECTIONS FOR LANDSAT DATA

The basic problem of any map projection is to represent the curved surface of the Earth, a spheroid, in a plane. This must be done while attempting to preserve true terrestrial properties such as areas, shapes, distances, and directions. Since it is impossible to preserve all of these properties simultaneously, it is necessary to select a projection that preserves those desired. Map makers generally consider the characteristic of conformality, that is, correct shape, to be the most important. (Thompson, 1979.)

Conformality retains equal scale locally in all directions and preserves angular relationships. The imagery obtained by Landsat warrants description as conformal, and conformal maps are used to correlate Landsat imagery to ground locations.

A conformal projection can be developed by transferring (projecting) details of the globe to a cylinder that is tangent or secant to a great circle (the Equator, for example). This cylinder can be cut along a meridia and laid flat, thus forming a planar projection surface. Cylindrical conformal projections are named after Mercator, who first conceived them. The regular Mercator projection is most frequently used in navigation. On it, meridians and parallels appear as straight lines crossing at right angles, and distances between parallels increase as latitudes grow higher. The regular Mercator is the only projection on which a rhumb line (a line which crosses successive meridians at constant angle) appears as a straight line.

When the cylinder of the regular Mercator projection is turned 90° about an axis through the Equator and the center of the globe, it becomes a tangent or secant to a meridian. This is the "transverse" case of the Mercator projection (see Figure 17-60). All the properties of the regular Mercator are preserved, except that rhumb lines are no longer straight. In addition, the line of zero distortion is no longer the Equator but the central meridian to which the cylinder is tangent. In the secant case, two lines parallel to a central meridian have zero distortion. This makes the Transverse Mercator projection well suited for mapping a large extent of latitude having a restricted longitude. The system is also suited to a "universal" application in repeated columns (or zones) of longitude.

The Universal Transverse Mercator (UTM) projection is widely used in conjunction with

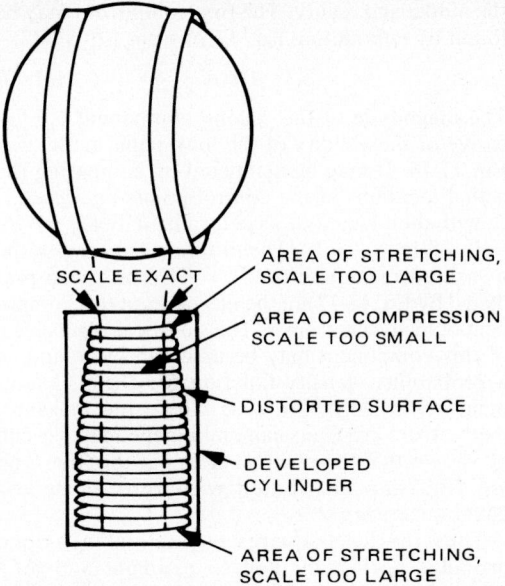

SCALE EXACT

AREA OF STRETCHING, SCALE TOO LARGE

AREA OF COMPRESSION SCALE TOO SMALL

DISTORTED SURFACE

DEVELOPED CYLINDER

AREA OF STRETCHING, SCALE TOO LARGE

Fig. 17-60. The Universal Transverse Mercator projection is based on longitudinal zones represented on a cylinder which has been laid flat.

Landsat imagery. It satisfies the need for conformality, and it is based on north-south strips (60 of them, equally spaced around the Earth), which is roughly how Landsat imagery is acquired. UTM maps are widely available at scales which are appropriate for most Landsat products.

There are certain problems in using UTM projections, however. UTM projections impose scale distortions of up to 1 part in 1,000 (1:1,000). Landsat imagery can theoretically be projected with far less distortion than this because of its geometric fidelity. Also, UTM projections are not suited for use in the polar regions where north-south strips converge. Even in the lower latitudes scenes frequently fall into two or more zones.

To solve these problems, some rather specialized variations on the basic Mercator projection have been devised. These are the Hotine Oblique Mercator (HOM) and the Space Oblique Mercator (SOM) projections (Snyder, 1978). They are both based on north-south strips that are oblique, or inclined at an angle away from, the polar axis, matching, therefore, the ground track of the satellite more nearly. Both are distortion-free, for all practical purposes, since they create a scale difference of no more than 1 part in 10,000 (1:10,000). HOM and SOM projections do not fall into multiple zones; they are suitable for imagery acquired at any latitude, and they are conformal.

Briefly, the HOM projection divides the Earth into five zones of latitude. Within each, oblique strips corresponding to individual Landsat paths are projected onto a plane in such a way that the projection axis approximates the path of the scene centers.

The SOM projection is based on an entirely different concept. The projection plane is modeled on the dynamics of satellite motion. The movements of the satellite platform, the sensors, and the Earth, expressed as functions of time, are used to calculate which latitudes and longitudes on the Earth correspond to locations in the projection plane. Thus, a continuous projection of the entire area of coverage is obtained.

A drawback to using either the HOM or the SOM projection for Landsat imagery is that of projection incompatibility with the UTM system. The reason is scale distortion. Whereas scale differences can be easily adjusted to fit an SOM or HOM image to a UTM grid, it is the change in scale over the area involved that is critical.

The maximum distortion in a UTM projection occurs at the Equator on a zone boundary. If the left side of an SOM (or HOM) image is fitted to a UTM grid at the Equator so that the same ground distance (185 km) is represented in both at the left side, the ground distance at the right side of the SOM (or HOM) projection will vary from that of the UTM by 204 meters. This is the worst case, and the differences decreases with latitude, so that at 45° latitude they are only half as much as at the Equator. Although such distortions are quite small for a single image, a mosaic of several SOM or HOM images could not be expected to fit a UTM grid without measurable error.

For the vast majority of applications, these discrepancies between UTM and SOM/HOM are virtually insignificant and very difficult to detect. However, for digital analysis, one other situation pertains. Because each spacecraft orbit is along a different earth track, each orbit has a different projection cylinder onto which the pixel grid is cast. For this reason, no two Landsat 1-3 MSS frames, even of the same nominal area, will have grids in common, so that some degree of geometric registration and pixel interpolation will always be necessary. For the Landsat-D Thematic Mapper, all repeat views of a given area are interpolated onto a common grid, eliminating this problem. However, even along one orbit, adjacent scenes are slightly rotated with respect to each other, so that digital images are on different grids and so cannot simply be concatenated. It is to be hoped that in the future a common grid system might be developed to allow delivery of digital imagery in truly registered from.

A full account of map projections for use with satellite data will be found in Chapter 21.

LANDSAT-D SYSTEM PROCESSING CONSIDERATIONS

The Landsat-D program represents the culmination of over a decade of experience in developing and operating satellite-based earth-resources data-acquisition and distribution systems. Table 17-1 indicates some of the major features of the Landsat-D program.

Three identical, independent MSS-image data-processing lines have been developed as part of the Landsat-D Ground Segment. The system is batch-oriented as opposed to real-time and therefore is more amenable to modifications should they become necessary. The Landsat-D ground segment is designed to provide up to 200 partially processed MSS scenes per day for transmission to the EROS Data Center (EDC) in Sioux Falls, South Dakota for further dissemination. Data will be available to EDC within 48 hours of acquisition by the ground segment.

TABLE 17-1

Landsat-D Program Features

- Significant Increase in Data Rates and Volume over Earlier Landsats
- 48-Hour Image Processing Turnaround Time to Deliver Data to EDC
- Sub-Pixel Geometric Accuracies:
  —Temporal Registration = 0.3 IFOV to a Reference Image
  —Geodetic Rectification = 0.5 IFOV to a Reference Map
- Multiple Communication Paths for Telemetry, Command and Imagery
  —Tracking and Data Relay Satellite System (TDRSS)
  —Goddard Space Tracking and Data Network (GSTDN)
  —Transportable Ground Station (TGS)
  —Foreign Ground Stations
- New Multi-Mission Spacecraft Used
  —Platform is More Stable than Earlier Landsats
  —Global Positioning System Experiment Provides Improved Ephemeris Data for Geodetic Accuracies
  —Shuttle Compatable for Retrieval
- NASA Turnover to NOAA is to Occur in Stages: MSS First, then TM
- "Lessons Learned" During Landsat 3 are being Implemented as Part of Landsat-D Ground Segment
  —Developed with Operational Philosophy
  —Ability to Fully Evaluate Output Products Internally
  —Built-in Parameter Checks and Check Points Throughout the Processing Cycle
  —Single Prime Contractor for Overall Design, Development, Integration, Test and Operation
- MSS Interfaces with EROS Data Center are Essentially Unchanged

The Landsat-D Ground Processing System will generate all of the geometric correction matrices that are appended to the partially processed data transmitted to the EROS Data Center. The Ground Segment can also apply these corrections to selected quality-assurance output products to ensure that output products are accurate.

The new Thematic Mapper (TM) instrument on board Landsat-D and the corresponding ground processing system will be operated for several years in a research mode with only limited film products and computer-compatible tapes in the public domain. Beginning in July of 1983, up to 12 scenes will be fully processed per day and the additional raw TM data acquired by the spacecraft will be stored on tape. All film and CCT products will be fully corrected both geometrically and radiometrically.

Table 17-2 compares several of the more salient features of the MSS and TM instruments. In January of 1985, TM production capacity is intended to increase to 50 fully processed scenes per day. Due to the relative unavailability of TM data prior to that time, the remainder of this section only discusses Landsat-D MSS image-processing and compares it to earlier Landsat digital processing.

The Landsat-D mission design was constrained to be as transparent as possible for current MSS data users, but several modifications were necessary. Image framing and coverage have been modified, and both geometric accuracy and the radiometric correction accuracy have been improved.

Data Formats

The format of the image data transmitted to EDC from the NASA/NOAA Ground Segment has been constrained to be essentially the same as for Landsat 2 and 3 data. This allows the EDC processing-system, which applies the geometric corrections and generates the user output-products, to remain unchanged. Some additional forms of quality information have been included in previously unused data fields to provide better

management information and knowledge of prior processing quality.

The format of the data on the CCTs supplied to the user from EDC is concurrently being changed to the new International Format. This format is the first step toward the unification of data structure planned to eventually allow various types of data to be used together in digital geographic information systems.

Image Framing and Coverage

The Landsat D orbit is 200km lower than earlier Landsats and provides either 7- or 9-day adjacent-swath overlap of imagery. The 16 day repeat cycle provides 233 orbit paths with 248 potential scenes in each. Each scene covers 185km cross track by 170km in track. The center of each scene is identified by unique latitude and longitude points as part of a Landsat-D World Reference System that is necessarily different from that of earlier Landsats. The sidelap of imagery between adjacent paths is less than with the earlier Landsats.

Geometric Correction

The purpose of geometric correction is to locate image samples such that: 1) the locations are known with respect to a map-grid reference system (geodetic rectification) and 2) each satellite pass over an area may be digitally registered (temporal registration). The Landsat-D system has been designed to provide the following accuracies for both MSS and TM imagery:

geodetic: 0.5 IFOV (90% of the time) to a reference map
temporal: 0.3 IFOV (90% of the time) to a reference image

These values are essentially the same requirements that the Landsat 2 & 3 Image Processing Facility (IPF) was designed to meet. The only known controlled measurement of accuracies (performed using 22 ground-control points by the DOI Geological Survey) indicates that the mean geodetic error is 1.4 pixels with a 2.0-pixel stan-

TABLE 17-2

Landsat-D Instrument Comparison

Measure	Miltispectral Scanner (MSS)	Thematic Mapper (TM)
• Nominal Ground Sample Spacing	60 × 80 m	30 × 30 m
• Swath Width	185 km	185 km
• Spectral Bands (μm)	—	0.45 to 0.52
	0.5 to 0.6	0.52 to 0.60
	0.6 to 0.7	0.63 to 0.69
	0.7 to 0.8	0.76 to 0.90
	0.8 to 1.1	1.55 to 1.75
	—	2.35 to 2.55
	10.4 to 12.6	10.4 to 12.6
• A/D Conversion (Bits Per Pixel)	6	8
• Pixels Per Scene (All Bands)	32 m	300 m

dard deviation. Likewise a registration mean-error of 1.0 pixel with a standard deviation of 0.7 pixels was determined using 23 ground-control points over a pair of scenes analyzed at Goddard Space Flight Center.

There is both good news and bad news with respect to the Landsat-D geometric correction potential. With respect to the spacecraft, the bad news is that MSS design was modified slightly to account for the lower orbital altitude. This provides larger mirror-scan angles, higher scan-mirror velocity and hence potentially greater scan nonlinearities. Also, the spacecraft platform is a new structure and is faced with many sources of vibrational excitation such as the scan mirrors from both the MSS and TM, the solar-array pointing system, and the Tracking and Data Relay Satellite System antenna-pointing system.

The good news is that the spacecraft has a better solar-array control system than earlier Landsats and also has better attitude control and measurement capabilities. The attitude-control system has better pointing accuracy (7 to 1 improvement) and better long term stability. This, combined with the eventual improved real time spacecraft locational knowledge to be provided by the global positioning experiment, will allow better geodetic location to be achieved than at present when ground control points are not available.

On the ground, the improved low frequency (0 to 0.4 Hz) attitude-correction information is used in the MSS image processing. In addition, improved automatic control-point correlation techniques have been incorporated to help provide the geodetic accuracies specified.

Table 17-3 provides a summary of key geometric-correction accuracy expectations and the related caveats.

Radiometric Correction

The Landsat-D radiometric calibration accuracy provides a factor of 2 improvement over the techniques used for Landsat 1, 2 or 3 processing. The Landsat-D calibration algorithm uses both the MSS's internal calibration-lamp data and scene-content information to achieve ±1 digital level (out of 128) accuracy and to eliminate band-to-band striping effects. The algorithms have been evaluated using Landsat 1, 2 and 3 data.

The 6 steps for MSS calibration are defined as follows:

- STEP 1: ACCUMULATE CAL WEDGE SAMPLES AND SCENE HISTOGRAMS
  —4 along-track segments per scene
- STEP 2: PERFORM CAL WEDGE PROCESSING
  —uses 6 samples from CAL Wedge data
- STEP 3: PERFORM HISTOGRAM ADJUSTMENT
  —Cal Wedge data used to determine reference histogram and range of histograms to be processed
- STEP 4: BLEND SCENE SEGMENTS
  —3 sub-segments per segment are used to smooth the transition
- STEP 5: GENERATE RADIOMETRIC CORRECTION LOOK-UP TABLES (RLUTS)
  —Data Decompression and Radiometric Correction are combined bined
- STEP 6: APPLY RLUTS TO DIGITAL IMAGERY AND GENERATE ARCHIVE TAPE

SPATIAL FREQUENCY MODIFICATIONS

In earlier sections of this chapter, image degradation was characterized in terms of cumulative spreading of the impulse-response functions (also termed the point-spread function, PSF) and as deterioration of the spatial frequency of the system. The equivalence of the sequential convolution of system element impulse-response functions and the sequential multiplication of the frequency-response functions was discussed in conjunction with Eq. 17-11. Intentional modifications to the system response, usually performed to compensate for the degradations, or to "restore" the image, may also be treated in dual fashion.

TABLE 17-3

Summary of Geometric Correction Accuracy Expectations

Geodetic Rectification
- 0.5 IFOV (90% of the Time)
- Referenced to Standard Map
- Assumes Accurate Knowledge of Ground Control Point Locations
- Verified Over Areas with Little Topographical Variations

Temporal Registration
- 0.3 IFOV (90% of the Time)
- Based on Instrument Performance That is Equivalent to Actual Performance of MSS's on Landsats 1, 2 and 3
- To a Reference Scene

Specifically, parameters for compensation may be derived in either domain, and the processing carried out in either domain independent of the domain of analysis.

When operations are carried out in the spatial domain, they are performed directly on the original picture function, and the operations carried out by convolution. In order to simplify the notation it is convenient to represent the coordinates as a single vector such as $\underline{x} = (x, y)$ or $\underline{v} = (u, v)$. With this notation, Eq. 17-3 can be written as

$$q(\underline{x}) = f(\underline{x})^* h(\underline{x}) \qquad (17\text{-}80)$$

THE FUNDAMENTAL RESTORATION PROBLEM

The basic problem of image restoration is to recover the original image $f(\underline{x})$, given the blurred image $g(\underline{x})$ and the PSF $h(\underline{x})$, by filtering $g(\underline{x})$ with filter function $p(\underline{x})$. The problem is further complicated by the fact that there is noise present in addition to the blurred image so that the recorded data set $g(\underline{x})$ is actually

$$g(\underline{x}) = q(\underline{x}) + n(\underline{x})$$
$$g(\underline{x}) = f(\underline{x})^* h(\underline{x}) + n(\underline{x}) \qquad (17\text{-}81)$$

The image-restoration model for this purpose is given in Figure 17-61. It is evident from Eq. 17-80 that the restoration problem amounts to solving an integral equation whose kernel is the system PSF. An exact solution is impossible because of the presence of the noise $n(\underline{x})$ which is a sample function from a stochastic process. In general, attempting to directly solve equations of this type is very difficult as they belong to a class of what are called ill-conditioned integral equations. Instead, the procedure is to make an estimate $\hat{f}$ of the true image that satisfies some appropriate criterion of goodness. There are many approaches that have been used for attacking this problem but before considering them it is useful to examine the most obvious one which is the inverse filter. By neglecting the noise it is theoretically possible to recover the original image from the blurred image in the following manner. As discussed previously, convolution in the spatial domain corresponds to multiplication in the spatial frequency domain so that taking the Fourier transform of Eq. 17-80 gives

$$Q(u,v) = F(u,v)\, H(u,v)$$
$$F(u,v) = \frac{Q(u,v)}{H(u,v)}$$
$$f(x,y) = q(x,y)^* \mathscr{F}^{-1}\left\{\frac{1}{H(u,v)}\right\} \qquad (17\text{-}82)$$

The difficulty with this approach is that the inverse transform $\mathscr{F}^{-1}\{1/H(u,v)\}5$ cannot physically exist. This is most easily seen by recognizing that the optical transfer function (OTF) of the system, $H(u,v)$, falls to zero at high frequencies and therefore its reciprocal $1/H(u,v)$ must increase without bound at high frequencies and is therefore not a valid Fourier transform. In order to avoid this difficulty one approach has been to limit the maximum value of $1/H(u,v)$ to some finite value and to truncate the function at some appropriate cutoff frequency. The limiting operation is nonlinear and its effects on distortion and fidelity are difficult to assess. However, the effect of truncation is readily determined by carrying out the operations of Eq. 17-82 with an inverse filter having limited frequency extent as follows:

$$\hat{F}(u,v) = \frac{Q(u,v)}{H(u,v)} \operatorname{rect}\left(\frac{u}{W}\right)\operatorname{rect}\left(\frac{v}{W}\right)$$
$$= q(x,y)^* \mathscr{F}^{-1}\left\{\frac{1}{H(u,v)}\right\}^*$$

$$W \operatorname{sinc}(W\,x)^*\, W \operatorname{sinc}(W\,y)$$

$$\hat{f}(x,y) = f(x,y)^*\, W \operatorname{sinc}(W\,x)^*\, W \operatorname{sinc}(W\,y)$$
$$(17\text{-}83)$$

It is seen that the truncation of the frequency response with the windowing rect functions has led to convolving the true image with the highly oscillatory sinc function which is certain to produce artifacts in the restored image. It is possible to improve performance somewhat by using a smoother roll-off of $1/H(u,v)$ in the spatial-frequency domain but this leads to poorer enhancement. In addition, the noise is increased by filtering operation leading to further uncontrolled deterioration (Nathan, 1968). Shlien (1979) discusses the windowing problem in the context of interpolation.

However, the above procedure has assumed that the noise is zero (by neglecting it). Even the slightest amount of noise will completely ruin the procedure (Frieden, 1979). Basically, this is because the matrix representing the PSF is mainly filled with zeros and small elements near the diagonal, which causes its inverse to have very large elements. Hence, at points where the noise is finite, the error term due to the noise is quite large. Further, the error is largely oscillatory (Phillips, 1962), a type of error generated by the noise during restoration and to which the original samples are invariant. Phillips (1962) derived a smoothing method with impressive results, but the two-dimensional computation is formidable.

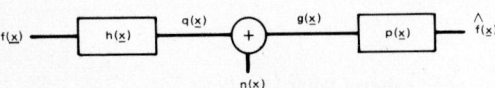

Fig. 17-61a. Restoration model for a noisy image.

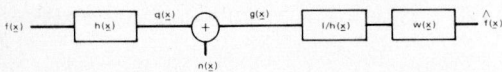

Fig. 17-61b. Restoration filter, decomposed.

LINEAR RESTORATION

Normally image restoration is carried out with a linear processor or filter (Figure 17-61a). This operation can be expressed in the space and frequency domains as

$$\hat{f}(\underline{x}) = p(\underline{x}) * g(\underline{x}) \qquad (17\text{-}84)$$

$$\hat{f}(\underline{v}) = P(\underline{v}) \, G(\underline{v}) \qquad (17\text{-}85)$$

The problem is to select a suitable filter impulse response $p(x)$ or filter transfer function $P(v)$. If the forward impulse response $h(x)$ that is to be compensated has no zeros in the range of interest, $p(x)$ may be decomposed into

$$p(\underline{x}) = \frac{1}{h(\underline{x})} * w(\underline{x}) \qquad (17\text{-}86)$$

where $w(\underline{x})$ is the Wiener compensation function accounting for the noise (Figure 17-61b).

Estimation of the Input Function

It has been shown (Wiener, 1949; Helstrom, 1967; Castleman, 1979; Frieden, 1979) that for uncorrelated signal and noise, if the power spectra S of the signal $\left[S_{ff}(\underline{v})\right]$ and of the noise $\left[S_{nn}(\underline{v})\right]$ are known or can somehow be estimated or modeled, the Wiener filter $w(\underline{x})$ becomes, in frequency space,

$$w(\underline{v}) = \frac{S_{ff}(\underline{v})}{S_{ff}(\underline{v}) + S_{nn}(\underline{v})} \qquad (17\text{-}87)$$

For the concatenated-compensation Wiener filter, the incoming noise spectrum is

$$\frac{N(\underline{v})}{H(\underline{v})}$$

so that the Wiener filter becomes

$$W(\underline{v}) = \frac{S_{ff}(\underline{v})}{S_{ff}(\underline{v}) + \left[\dfrac{N(\underline{v})}{H(\underline{v})}\right]^2} \qquad (17\text{-}88)$$

This is combined with the compensating filter $1/H(\underline{v})$ to give

$$P(\underline{v}) = \frac{1}{H(\underline{v})} \left[\frac{S_{ff}(\underline{v})}{S_{ff}(\underline{v}) + \left[\dfrac{N(\underline{v})}{H(\underline{v})}\right]^2} \right]$$

$$= \frac{H^*(\underline{v})}{|H(\underline{v})|^2 + \dfrac{S_{nn}(\underline{v})}{S_{ff}(\underline{v})}} \qquad (17\text{-}89)$$

Although this filter minimizes the mean square error, in the presence of large values of noise $S_{nn}(\underline{v})$ it may attenuate the signal $S_{ff}(\underline{v})$ too much and produce too blurred an image. Backus and Gilbert (1970) have suggested separately minimizing the output noise and the departure of the restored signal from some windowed estimate of the object. This can be done (Frieden, 1979) by modifying the Wiener compensator of Eq. 16-89 by adding the window $Y(\underline{v})$ and a Wiener gain-function α:

$$P_{BG}(\underline{v}) = \frac{H^*(\underline{v}) \cdot Y(\underline{v})}{|H(\underline{v})|^2 + \alpha \, \dfrac{S_{nn}(\underline{v})}{S_{ff}(\underline{v})}} \qquad (17\text{-}90)$$

Further development of the method, with discussions of windowing techniques which avoid discontinuities at the window edges, is given by Lim (1980).

Although these filters can be implemented in the real domain it is generally not practical to do so because of the large sizes of the matrices involved. Rather, they are typically implemented in the frequency domain using the FFT approximation to the Fourier transform (Tikhonov and Arsenia, 1977; Rosenfeld and Kak, 1976; Andrews and Hunt, 1977).

In spite of these theoretical difficulties, pragmatically successful enhancements have been accomplished. These are typically based on the realization that spatial-frequency degradations (and, therefore, restorations) are local in nature, encompassing the spatial distance over which the PSF has distributed the incoming light from an object impulse. Because of the limited physical extent, convolution filters may be of similar size, typically 3×7 to 7×7 or perhaps even 15×15 pixels. In addition, a general argument may be made that many system PSF's will tend to be Gaussian because of the combination of a number of individual PSF's, and that an optimally sampled system will be sampled somewhere near the Nyquist limit. Surprisingly good restoration results may be obtained with a generalized filter based on these parameters independent of the actual camera used. An example of this is shown in Figure 17-62, in which a Landsat image was enhanced using a filter calibrated for a typical vidicon camera.

Minimization of the PSF Radius

An alternative approach which has the advantage of allowing *a priori* determination of performance is that of compensating the PSF. The procedure is to make use of the linearity of the convolution operator to reformulate the problem as follows. The output (restored) image is given by

$$\begin{aligned} \hat{f}(\underline{x}) &= \left[f(\underline{x}) * h(\underline{x}) + n(\underline{x}) \right] * \rho(\underline{x}) \\ &= f(\underline{x}) * \left[h(\underline{x}) * \rho(\underline{x}) \right] + n(\underline{x}) * \rho(\underline{x}) \\ &= f(\underline{x}) * c(\underline{x}) + n(\underline{x}) * \rho(\underline{x}) \end{aligned} \qquad (17\text{-}91)$$

where $c(\underline{x})$ is the composite point spread function of the system that results from the convolution of the blurring and restoration functions. If there were no noise present the desired form of $c(\underline{x})$ would be a two dimensional impulse as this would give perfect restoration. Actually that would be the inverse filter implementation and is not practical for the reasons given earlier. The procedure used is to make $c(\underline{x})$ as much like an impulse as

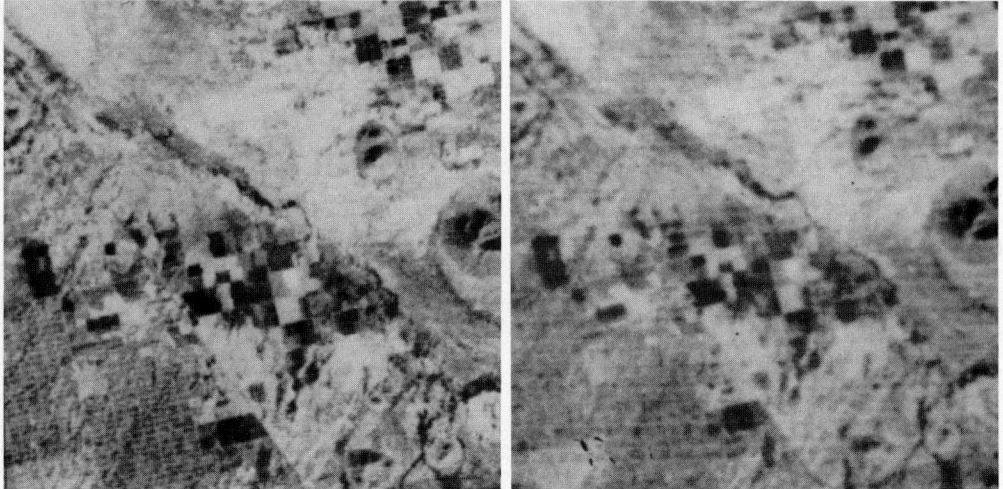

Fig. 17-62. Portion of a Landsat scene showing uncalibrated enhancement (a), left: original (b), right: enhanced image.

possible while satisfying other constraints such as limiting the noise component in the output. The incorporation of a noise constraint also stabilizes the filter. The most frequently used narrowness measure is the radius of gyration of the composite PSF. This is given by

$$R = \frac{\int_{-\infty}^{\infty} |\underline{x}|^2 \, c^2(\underline{x}) \, d\underline{x}}{\int_{-\infty}^{\infty} c^2(\underline{x}) \, d\underline{x}}$$

$$= \frac{\int_{-\infty}^{\infty} |\underline{x}|^2 [\rho(\underline{x})^* h(\underline{x})]^2 \, d\underline{x}}{\int_{-\infty}^{\infty} [\rho(\underline{x})^* h(\underline{x})]^2 \, d\underline{x}} \quad (17\text{-}92)$$

the noise constraint takes the form

$$E_x\{[n(\underline{x})^* \rho(\underline{x})]^2\} \leq \sigma^2$$

$$= \int_{-\infty}^{\infty} \int_{-\infty}^{\infty} R_{nn}(\underline{\alpha} - \underline{\beta}) \, \rho(\underline{\alpha}) \, \rho(\underline{\beta}) \, d\underline{\alpha} d\underline{\beta}$$

$$(17\text{-}93)$$

where $R_{nn}(\Delta \underline{x})$ is the two-dimensional autocorrelation function of the noise. For the continuous variable case, the minimization of Eq. 17-92 subject to the constraint of Eq. 17-93, leads to the requirement for solution of a second-order differential equation in the spatial-frequency domain. Such solutions have been obtained and effective restoration achieved (Pratt, 1978). However, the real utility of this approach is in the case where the image blurring is a function of continuous variables while the processing and image reconstruction employ discrete variables corresponding to sampled and quantized data. Such a representation of the imaging operation is referred to as the

continuous-discrete model and leads to very compact, stable and predictable processors.

Figure 17-63 shows a block diagram of the continuous-discrete model that is a good representation of many actual imaging systems. The samples of $g(\underline{x})$ are given by

$$g(m,n) = \sum_m \sum_n g(x,y) \, \delta(x - m\Delta x) \, \delta(y - n\Delta y)$$

$$= \sum_m \sum_n \left[h(x,y)^* f(x,y) \right.$$

$$\left. + n(x,y) \right] \delta(x - m\Delta x) \, \delta(y - n\Delta y)$$

$$(17\text{-}94)$$

where Δx and Δy are the sampling intervals. The restoration problem now becomes one of selecting a processor that operates on these samples to produce a new set of samples which represent the estimate $\hat{f}(m,n)$ of the image. The design procedure is most easily understood by considering a one-dimensional case. This is also the most widely used procedue and is applicable in all cases where the PSF can be approximated as being separable (e.g., Gaussian) and greatly reduces the processor complexity. For the one-dimensional case Eq. 17-94 becomes

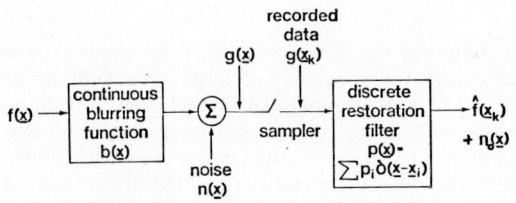

Fig. 17-63. Continuous-discrete model of image restoration.

$$g_s(x) = \sum g(x)\,\delta(k\Delta x) = \sum \left\{\left[h(x)* f(x)\right]\right.$$
$$\left. + n(x)\right\}\delta(x - k\,\Delta x) \qquad (17\text{-}95)$$

It is not necessary that the restoration filter extend over all of the data; in fact, one of the major advantages of this procedure is that the designer can select the length of the processing filter to be used (Nathan, 1968; Arguello, 1972; Frieden, 1979). By keeping the length short one can greatly reduce the processing time and the transients associated with edge effects are minimized. Excellent results can often be obtained with filters containing only a few points. Let it be assumed that the filter will be symmetrical and have $2n + 1$ points. It can then be represented as

$$\rho_s(x) = \sum_{k=-n}^{n} \rho_k\,\delta(x - k\Delta x) \qquad (17\text{-}96)$$

The restored image is given by

$$\hat{f}_s(n) = \left\{ \sum_{k=0}^{M} \left[h(x)* f(x) + n(x)\right]\,\delta(x - k\Delta x) \right\} *$$

$$\sum_{j=-n}^{n} \rho_j\,\delta(x - j\Delta x) = \sum_{k=0}^{M}\sum_{j=-n}^{n}\left[h(x)* f(x)\right.$$

$$\left. + n(x)\right]\rho_j\delta\left[x - (k - j)\,\Delta x\right] \qquad (17\text{-}97)$$

The procedure is now to choose the ρ_j so that the effect of $h(x)$ on $\hat{f}(n)$ will be a minimum. This is again accomplished by minimizing the radius of gyration while constraining the noise in the output to be equal to or less than some specified value. Because only discrete samples of the image will be processed it is possible to formulate the problem in terms of vectors and matrices whose elements reflect the system charactristics. In matrix notation the problem becomes (Riemer and McGillem, 1977; Stuller 1972)

Minimize

$$R^2 = \frac{\rho^T \underline{A}\rho}{\rho^T \underline{B}\rho} \qquad (17\text{-}98)$$

Subject to the constraint

$$\rho^T \underline{N}\rho \leqslant \sigma^2 \qquad (17\text{-}99)$$

where ρ is the vector of samples of the restoration filter and $\underline{A}$, $\underline{B}$ and $\underline{N}$ are square matrices defined by

$$a_{ij} = \int_{-\infty}^{\infty} x^2\, b(x - i\Delta x)\, b(x - j\Delta x)\,dx$$

$$b_{ij} = \int_{-\infty}^{\infty} b(x - i\Delta x)\, b(x - i\Delta x)\,dx$$

$$n_{ij} = E\{n(x - i\Delta x)\, n(x - j\Delta x)\} = R_n\lfloor(i - j)\Delta x\rfloor$$
$$(17\text{-}100)$$

and $R_n(\tau)$ is the autocorrelation function of the noise. The solution of Eqs. 17-98 and 17-99 is ob-

tained using Lagrange multipliers and leads to the following problem. Find $\underline{P}$, λ_1, and λ_2 such that

$$\left.\begin{array}{c} \underline{A}\rho + \lambda_2\underline{N}\rho = 0 \\ \rho^T\underline{B}\rho = 1 \\ \rho^T\underline{N}\rho = \sigma^2 \end{array}\right\} \qquad (17\text{-}101)$$

This is called a generalized eigenvalue problem and requires some type of iterative or search procedure to find the eigenvalues λ_1 and λ_2 and the eigenvector (filter) ρ. Most computer scientific subroutine-packages have programs for carrying out such solutions. For short filters (e.g., less than 15 points) it can generally be assumed that $\lambda_2 = 0$ since such filters cannot have large gain because of their finite length and therefore will not cause major noise degradation. This greatly simplifies the solution and the noise increase produced by the filter can be computed from the relationship

$$\frac{Noise\ Power\ Out}{Noise\ Power\ in} = \frac{\rho^T\underline{N}\rho}{R_n(0)} \qquad (17\text{-}102)$$

Figure 17-64 shows impulse responses of four restoration filters computed for the case of no noise constraint. These restoration filters are for a Gaussian blurring function with a radius of gyration $R = 1$ and a sampling interval of 1. The composite PSFs for these filters ($p * b$) are shown in Figure 17-65 along with the original blurring function. It is evident that substantial improvement is obtained even with short filters. One undesirable feature of the restoration processing is the presence of sidelobes in the composite PSF. This

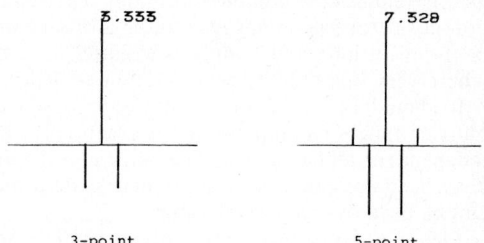

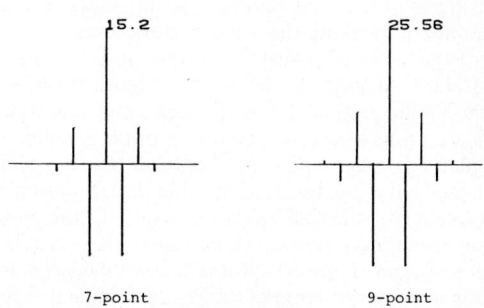

Fig. 17-64. A set of four discrete restoration filters of length 3, 5, 7, and 9 for the Gaussian blurring function of unity radius of gyration and unit sampling interval.

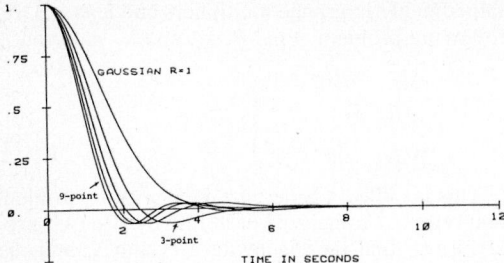

Fig. 17-65. The Composite system functions corresponding to the filters of length 3, 5, 7, and 9 shown in Figure 17-64.

leads to "ringing" or "ghosts" near discontinuities in the image. For filters of this type the sidelobe amplitude is essentially constant at about 7 percent of the peak. It is possible to reduce this by using other criteria than the radius of gyration or by adding additional constraints to the problem (Stuller, 1972; Chu and McGillem, 1978). Frieden (1974) has derived a series of filters in which the competing criteria are the resulting PSF radius vs the height of the first sidelobe. His filter-weights for a 5-point and a 7-point filter are:

5 point 1.0 (center) −0.28798 0.08561

7 point 1.0 (center) −0.45835 0.25478 −0.12151

First sidelobe magnitudes are 0.12 and 0.15, respectively. In addition he shows a 15-point filter requiring only additions. The (integer) weights are 1, (center) 1, 0, 0, −1, −1, 0, 1. In spite of the large size of this filter, the absence of multiplies greatly reduces the computation time. The reduction in sidelobe level always causes a broadening of the main lobe and is only warranted in cases where very high-fidelity reproduction is desired.

It should be noted that restoration based on blurring-function compensation is essentially data independent and that once the filter coefficients have been computed for a particular system they can be used over and over again.

Figure 17-66 is an example of restoration processing. Figure 17-66a is an aerial photograph of Dulles airport blurred with a Gaussian PSF ($R = 1$). Figure 17-66b is a restoration of this image using two one-dimensional 3-point filters applied sequentially along the x and y directions.

Figure 17-67 shows a Landsat 2 image of O'Hare Airport, Chicago. In Figure 17-67a is shown the original data with each image point replicated four times in each direction to produce an enlarged image. Figure 17-67b shows the same image enlarged by 3×4 cubic interpolation to make it smoother and to compensate for the different sampling intervals along- and across-track in this system. Figure 17-67c is a combined restoration and interpolation of the image using a discrete filter of 32×35 points. This filter reduces the composite PSF to 0.65 of its original value and leads to a 17 dB increase in noise. Figure 17-67d is

(a)

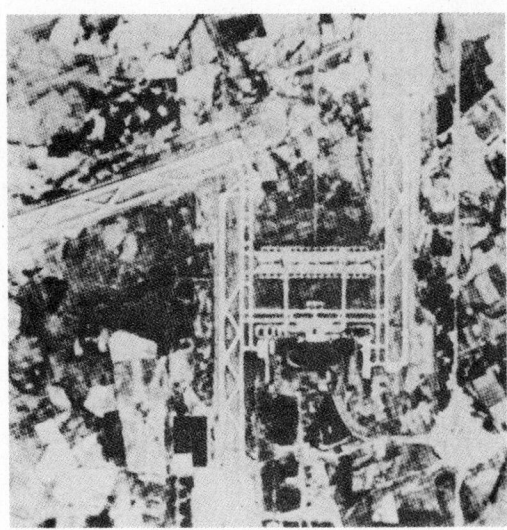

(b)

Fig. 17-66. (a) Gaussian blurred image of Dallas Airport (radius of gyration = 1); (b) restoration of (a) using two one-dimensional 3-point filters.

the image processed with the same size filter but with a 0.45 performance factor and a 22 dB noise increase.

In the case of Landsat 2 imagery it has been found empirically that the PSF can be approximated as being Gaussian in shape with a radius of gyration of about 24 meters (Chu and McGillem, 1978). Since the sampling interval is different in the along-track and across-track directions by a ratio of 1.4:1, the filter must be different in the two directions. Sample values for the along-track and across-track filters that combine cubic interpola-

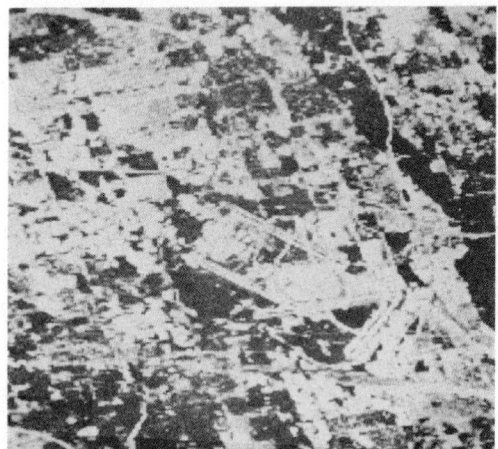

Fig. 17-67a. O'Hare Airport, Chicago, Landsat imagery.

Fig. 17-67c. 3 × 4 Enhancement of Figure 17-67a using an optimal discrete enhancement filter with a performance ratio of 0.65 and noise increase of 17 db.

Fig. 17-67b. 3 × 4 Cubic interpolation of Figure 17-67a.

Fig. 17-67d. 3 × 4 Enhancement of Figure 17-67a using an optimal discrete enhancement filter with a performance ratio of 0.45 and noise increase of 22 db.

tion and enhancement are used to produce Figure 17-67c. The procedure for using the filters is to augment the original Landsat data by adding three zeros between each of the successive across-track points and then convolve the augmented image with the cross-track filter. The new set of points is then augmented by adding two zeros between each of the successive along-track points and convolving with the along-track filter. More efficient algorithms for carrying out this kind of processing can be utilized (McGillem et al., 1975).

NON-LINEAR RESTORATION

To date there have been many examples of uses of linear models in digital environments for inversion purposes. However, almost all results with these models give surprisingly disappointing restorations. One of the underlying reasons for this phenomenon is that considerably more *a priori* knowledge about the imaging process is available than is utilized in the linear models de-

scribed previously. Specifically, the fact that light-sensing devices are energy-sensitive suggests that a constraint of positive object, image, and impulse responses must also be included for our models to be more relevant (Biraud, 1969; Janssen et al., 1970). Initially it might seem a trivial point, because positivity can always be obtained by simply adding enough constant bias to our processing results. However, on closer examination the positivity constraint on image, object, and impulse responses turns out to have a very significant impact on digital image-restoration.

Frieden (1980) has reviewed a number of nonlinear models, and sums up the situation as follows: Since image restoration is an ''ill-conditioned'' problem, inequality constraints have to be built into the output in order to control the severe error-propagation problem. Various forms of *a priori* information may be input into the restoring scheme through the constraints. Depending upon the *a priori* information at hand, an appropriate statistical model for the object may be devised. Each model gives rise to a new restoring principle.

Simple positivity is amenable to a simple-particle model and invocation of maximum degeneracy for the object. These lead to a condition of maximum entropy in the object (Jaynes, 1968; Ables, 1974; Wernecke and D'Addario, 1977; Frieden, 1972).

Upper and lower boundedness may be attained by a "checkerboard" model, wherein each pixel site contains a multiplicity of grain sites for possible allocation of grains. This leads to an entropy-like expression.

A quantum statistical model and the criterion of maximum degeneracy lead to an object representation which is a generalization of maximum entropy. Only in two specific limits does the representation go over into entropy-like forms.

Simply connected objects are amenable to median-window operation in conjunction with any linear restoring algorithm (Frieden, 1976).

Power-spectral objects are well restored if the amplitude data giving rise to them are regarded as having maximum entropy. Then Burg's M E formalism results (Burg, 1976).

Finally, if the object and image are regarded as probability laws on photon position, the object that maximizes information throughput from object to image may be regarded as the restoration. This constitutes a new norm of estimation *per se*, as well as a new norm of restoration (Frieden, 1980.)

There is not sufficient space here to review each of these restorations in detail. Rather, because of the availability of suitable fast algorithms (Huang et al., 1979) which make image enhancement practical, the median-window restoration will be discussed as surrogate for the group. As with all "restoration" processes, with the exception of the $1/H$ process, the spatial-frequency modifications will be in the direction of restoring the MTF of the entire system, but will not be calibrated. In this sense, perhaps all MTF modifications should be considered as enhancements rather than rectifications, a difference which may be of more than semantic importance in some situations.

Median window restoration (Frieden, 1976) is intrinsically a nonlinear operation that, when cyclically used with any linear restoring algorithm, can typically enhance edge gradients by a factor of 5:1 with nearly a complete absence of oscillations or overshoots. Each cycle consists of two basic steps (Figure 17-68): a linear restoration followed by the use of a median-window filter. Median-window filtering is defined as follows: Let an $m \times n$ window be placed over an $m \times n$ area of the picture, centered sequentially at locations i,j of the image $g(i,j)$. At each location, the image value g_{ij} is replaced by the median of the $m \times n$ gray levels lying within the window. In the algorithm reported by Huang (1979), picture elements are replaced sequentially along each line, with the window moving one column per element. The new histogram, from which the median is derived, is formed by discarding the pixels from the left column, and adding the pixels from the newly covered column to the remainder. This type of filter tends to remove extremum data lying within the window, and so removes the oscillations produced by linear restorations. Thus the use of a cyclic process, which asymptotically reduces the difference-image to below a threshold, allows termination. As the window removes noise and oscillations of length shorter than the window, a window of twice the length of the Nyquist sampling interval will remove the oscillations produced by normal linear filtering.

A one-dimensional example will illustrate the effects better than examining an image. In Figure 17-69a (Frieden, 1976) a noiseless edge was MW restored, giving the starred restoration shown in the second row; the top row shows a conventional linear restoration of the same data (shown by dots). In Figure 17-69b 20 percent noise was added to the blurred edge data; the bottom row shows the MW restoration and the top row the linear restoration.

The characteristic of the median-window filter to remove extremum data lying within the window also causes it to remove "spike" noise of width narrower than the filter window. An example of this action is shown in color Figure 17-70, in which the "spike" noise to be removed is the set of smaller stars. The image is of the North American Nebula, taken at Kitt Peak, as part of a study of the broad emission region. The window used 5 × 5 pixels, and successfully removed stars of that size or smaller. Processing was done at the Laboratory for Astronomy and Solar Physics, Goddard Space Flight Center.

COHERENT NOISE

In a typical image-processing application the data to be extracted from the pictures consist of

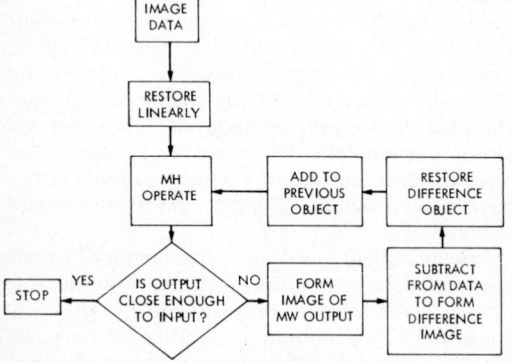

Fig. 17-68. Flow chart of median window restoring MWR algorithm (Frieden, 1976).

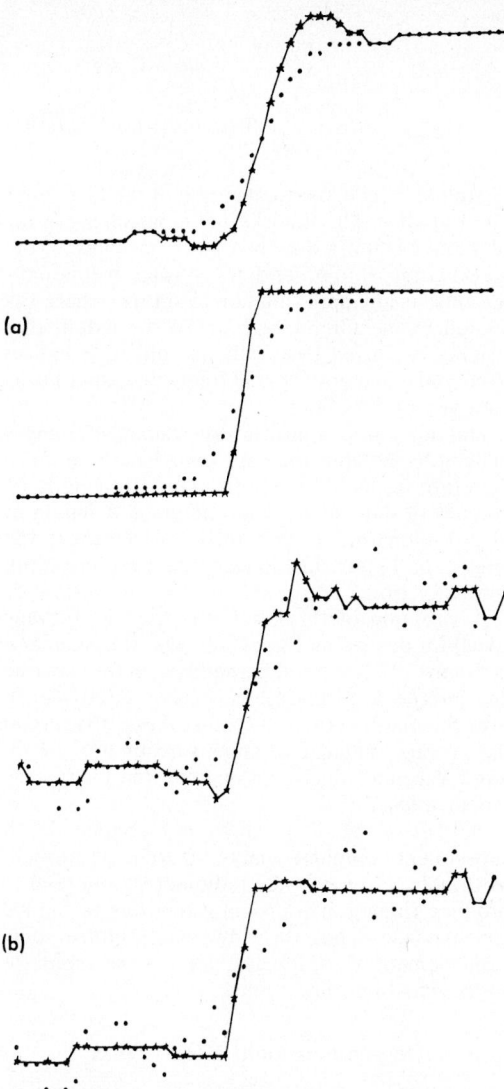

(a)

(b)

Fig. 17-69. (a) A step object restored linearly (top, starred curve) and by the MWR algorithm (bottom, starred curve). Five cycles of MWR were used, with a window length of one Rayleigh resolution length (11 consecutive points). Dots show the image data, which here suffer no noise. (b) As in (a), except with 20% amplitude noise added to the image data (Frieden, 1976).

gree to which it is corrected by the system processing will vary.

In addition, evidence is emerging (Duggin and Ellis, 1980) that differences in the shape of the spectral responses among the set of detectors in a given spectral channel may interact with the reflectivity variations of the scene to give striping that is dependent on the type of material (vegetation, soil, etc.) being viewed. This effect will not be considered in the algorithms discussed here.

Although considerable freedom is available in the order in which rectification processing is performed, there may be some constraints. For example, horizontal banding, such as seen in the MSS data, is relatively simple to remove if it is done first; however, if the images have been photogrammetrically rectified or rotated to match map projections, subsequent striping removal is extremely difficult.

COSMETIC CLEANUP

For Landsat MSS images, it has been found that quite often in the older images every sixth line in the imagery appears much darker than the rest of the image. This suggests that five of the detectors in a band are fairly uniform, but that the sixth is responding in a different way. The first cleanup effort might attempt to remove or correct the data from this designated "bad" detector. This approach is simply cosmetic, to produce imagery that can be examined without too much distraction. The data from the designated detector are replaced by the average of the data from the adjacent detectors above and below on a pixel-by-pixel basis. This degrades the vertical resolution of that line to a certain extent, but nevertheless works quite well in producing an image that is much cleaner than the original.

Before the detector response can be normalized, it must somehow be characterized. One possible method requires finding a number of large, homogeneous regions of different albedo which will be observed by Landsat in the same frame. If the reflectance spectrum at each site can be monitored on the ground while Landsat is overhead, then satisfactory light-transfer characteristics can be devised. However, the scarcity of suitable sites and the presence of atmospheric effects, coupled with the difficulty and expense of simultaneously monitoring their reflectivity, reduces the usefulness of this approach.

Locally, each of the six detectors comprising one Landsat spectral channel can look at a different target; however, over a sufficiently large region, each detector will encounter the same types of scenes in the same proportions. Thus, the probability density function (PDF) and the histogram of radiance occurrences of each detector from the same spectral channel should be similar. Any differences can be attributed to irregularities in the detector response. It follows that manipulating the data from each detector in a way that causes the detector PDFs to resemble each other will re-

the differences in radiance of each pixel as seen in different spectral bands. Since these differences may be small relative to the total radiance range, small errors or non-linearities in the transfer curves may be larger than the data values and thus impair or destroy the ability to analyze the pictures. The Landsat MSS raw data suffer from intensity artifacts due to non-linearities of the radiance to DN transfer characteristics and non-matching of the various detectors, resulting in the 6-line banding so visible in many of the older Landsat images. This type of coherent noise may be expected in any multi-detector system; the deduce radiometric noise. For instance, the means and standard deviation of the six detector PDFs for a given MSS band can be made to agree by applying a different linear-contrast stretch to the data from each detector:

$$DN'_{ijk} = a + bDN_{ijk} \qquad (17\text{-}103)$$

where DN is the gray level at line i sample j of MSS detector K, and a and b are empirically determined constants, different for each of the six detectors. However, some striping still tends to persist after the means and standard deviations of the PDFs agree. Alternatively, a higher order function could be derived, but the observed variability of the detector responses suggest that this complication may not be justified.

Another method used to equalize the histograms of the various detectors is to equalize the cumulative probability-distribution function (CPDF). In this method one detector or the average of several detectors is chosen as the master to which the others are to be matched. The CPDF of each detector and of the master are obtained by placing, in each histogram-radiance cell, the normalized total of all pixels having that radiance or less (normalized to CPDF = 1 at $DN = 255$). Then the matching for each detector is accomplished by table lookup, one input-output table being derived for each detector from the CPDFs.

CONVOLUTION FILTERING

Filtering out the spatial frequencies which produce banding, either in one or in two dimensions, has been the most successful cosmetic destriping technique. This filtering may take on one of three distinct forms, each having its own implications on the computer requirements.

Cosmetic Convolution

The convolution equation 17-80 may be written in discrete form for filtering as:

$$\hat{f}(x,y) = \sum_{u=-a}^{+a} \sum_{v=-b}^{+b} h(u,v) f(x+u, y+v)$$

$$(17\text{-}104)$$

where $\hat{f}(x,y)$ is the pixel value at x, y after filtering; $f(x+u, y+v)$ is the area of the scene surrounding point x, y; $h(u,v)$ is the filter kernel which is centered at x, y. The sum of all the filter weights $h(u, v)$ is generally adjusted to equal unity to avoid shift in the average value of the image radiance. Filtering is seen to require $u \cdot v$ multiplications plus one addition of $u \cdot v$ factors at each pixel point x, y. Unless special measures (such as folding a symmetrical filter) are taken, v lines of the image are required to be in core simultaneously.

The simplest destriping filter is cosmetic and consists in subtracting a one-line low-pass filtered image from the original image; this process is repeated a line at a time. In terms of the previous equation,

$$\hat{f}(x,y) = g(x,y) - \frac{1}{2a+1}$$

$$\sum_{u=-a}^{+a} g(x+u, y) \cdot h(u) \quad (17\text{-}105)$$

where $(2a + 1)$ is the total length of the filter, and $h = 1$ at all u. The filter kernel in which $h = 1$ for all u and v is termed a "box filter" because of the rectangular shape, and its spatial frequency-response is an approximation to sin x/x, where x is related to the filter length u. To the extent that striping is caused by a radiance offset, it has an effect only near zero spatial frequency, and thus is removed by this filter.

This filter also removes slow radiance changes in lengths greater than its own length; it must therefore be long enough to avoid bringing large patches of dark or light to mid-gray. A length of 31-301 elements is generally satisfactory; the longer the better, but longer filters require more computer time.

A variation of this filter was used by Nathan (1968) for destriping Ranger images. It is sketched in Figure 17-71. Although requiring more than one line in core at a time, it may have an advantage over the one-line filter in that it takes into account the average radiance of areas surrounding the filtered line, and thus may degrade the large-scale radiance less.

Neither of the filters above is recommended if subsequent computer analysis is to be performed, as they both destroy the radiometry, and tend to produce ringing at transient edges due to the sin x/x response. They do, however, produce some enhancement of the images, and may be useful for some visual interpretations.

One-Dimensional Notch Filter

The filter is oriented in the vertical (track) direction; the weights are found from the Fourier transformation of a "notch" filter which is fitted to the vertical power spectrum of the image (Figure 17-72). The notches are centered on the frequencies at which noise is observed (0.166, 0.333, and 0.500 cycles/pixel). The shape of the notches is specified to ensure that the product of the filter spectrum and image spectrum shows no noise

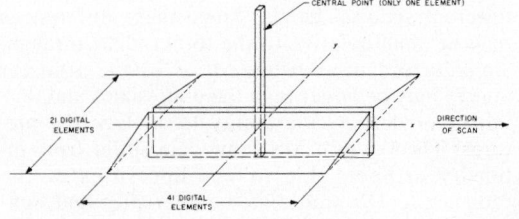

Fig. 17-71. Box filter having tapered cross-line form (Nathan, 1968).

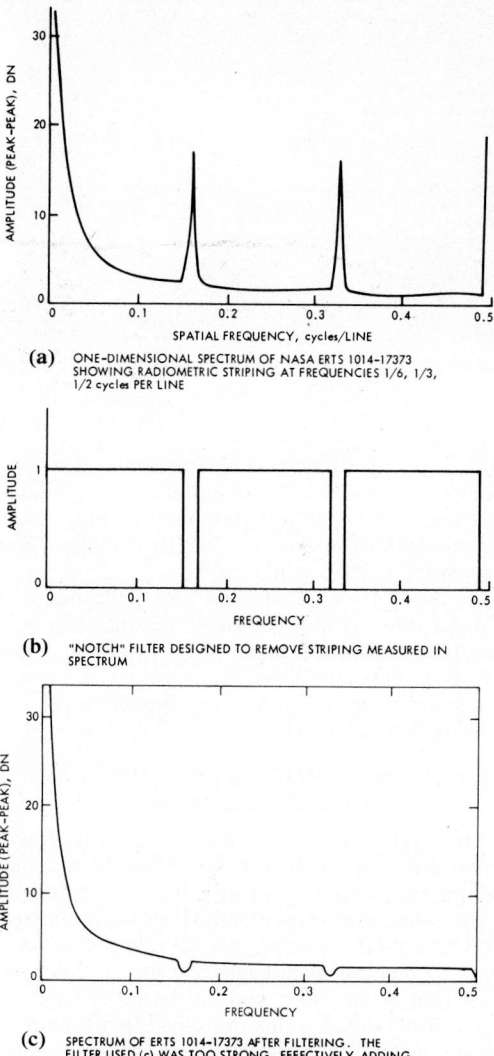

(a) ONE-DIMENSIONAL SPECTRUM OF NASA ERTS 1014-17373 SHOWING RADIOMETRIC STRIPING AT FREQUENCIES 1/6, 1/3, 1/2 cycles PER LINE

(b) "NOTCH" FILTER DESIGNED TO REMOVE STRIPING MEASURED IN SPECTRUM

(c) SPECTRUM OF ERTS 1014-17373 AFTER FILTERING. THE FILTER USED (c) WAS TOO STRONG, EFFECTIVELY ADDING LOW-AMPLITUDE STRIPING OF OPPOSITE SENSE TO IMAGE. THIS MAY BE CORRECTED BY ADJUSTING FILTER AMPLITUDE IN THE "NOTCHES"

Fig. 17-72. Vertical notch filter for line striping removal (a) one-dimensional spectrum showing radiometric striping at frequencies $1/6$, $1/3$, and $1/2$ cycles per line (b) notch filter designed to remove striping measured in the spectrum (c) spectrum after filtering. The filter used (17-72b) was too strong, effectively adding low-amplitude striping of opposite sense to image. This may be corrected by adjusting the filter amplitude in the notches (Gillespie, 1975).

spikes. The spatial frequencies involved in the notches are across lines, so that the filter must be vertical. If the striping were caused solely by offset, the striping components would occur only at zero frequency along lines, requiring filters only one element wide. The gain components of the striping will cause the Fourier-domain components to spread in the line direction, requring a two dimensional filter to remove such spread. However, the one-dimensional form has been

quite successful. Since a convolution filter requires a number of lines in core equal to its v-dimension, the core requirement becomes excessive for Landsat, with its 3240 pixels/line. Therefore it is usually more efficient to rotate the image 90°, and treat the problem as a one-line or few-line filter.

Two-Dimensional Filtering

An example of two-dimensional filtering is shown in Figure 17-73. Figure 17-73a (left) shows a portion of Landsat frame 1024-15071, the average of bands 5 and 6. Figure 17-73b (center), its Fourier transform, shows clearly a group of four energy concentrations on the vertical center axis. These represent the energy involved in the repetitive banding pattern. The clean-up procedure is to set areas of the intensity component surrounding these points to equal zero in the Fourier domain and then to Fourier retransform the resultant back to the real-image domain. The result of doing this is shown in Figure 17-73. This cleanup is one of the most effective methods used. Its chief disadvantage is that, for large image areas, the two-dimensional transform requires much computer time, primarily for the 90° rotation required for the x, y separable-transform process. A final factor to be considered is that the transform is most efficiently handled if the extracted part processed is a multiple of a power of 2 (thus restricting flexibility in the size of the subscene processed).

OPTICAL IMAGE PROCESSING

The human eye provides man with his greatest source of data on the external world. Similarly, in the area of remote sensing, image sensors have proven to be a major tool in data acquisition. The dominance of such image data as aerial and satellite photographs is due in part to historical precedent. Photographic recording was a well developed science long before the concept of remote sensing emerged. Furthermore the early use of photography in reconnaissance led naturally to remote sensing applications. However, there are two other characteristics of photographs and other types of image data that set them apart from other data sources. The first of these is the tremendous amounts of data which can be easily and quickly collected. However, large amounts of data in and of themselves are useless. Thus the second important characteristic of imaging is that the data are presented in a form that is easily digestible for the human user. This fact was especially pertinent in the early days of remote sensing before the digital computer began to have an impact on analysis in remote sensing. The application of computer processing of image data has affected our perspective on the two characteristics of image data just discussed. On the one hand, digital processing has led to an increased awareness of the large amount of raw data contained in

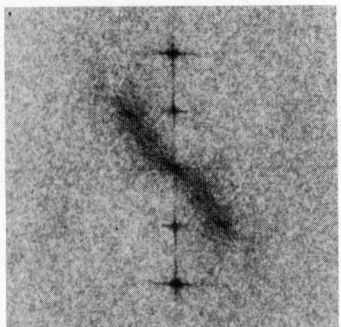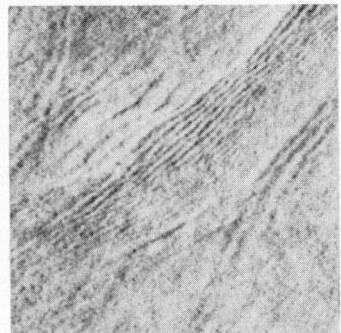

Fig. 17-73. Portion of Landsat frame showing two-dimensional cleanup (a) original (b) Fourier transform (c) cleaned image.

imagery: a 23×23 cm aerial photograph that is being digitized for computer analysis can be scanned with a 25-micrometer aperture producing about 10^{10} picture elements (pixels), each containing on the order of 8 bits of information. While placing greater emphasis on the wealth of data available in imagery, computer analysis has tended to diminish the need for raw data presented in a form suitable for a human analyst. This slight relaxation of the requirements on data format has been overshadowed, however, by quantity of data being generated.

The burgeoning amount of data being collected by modern photographic and electronic sensors has created problems in data processing and data storage. In the search for a solution to these problems attention has been drawn to the same technology that created the problem,—optics. It has been found that optical systems not only present a convenient means of recording data, but they also are naturally suited for processing and storing the recorded imagery.

The factor that makes photographic systems such efficient data collectors is their parallel nature; all image points on a photograph are recorded simultaneously having been imaged in parallel by the optical system. It is thus their innate parallel-processing capability which make optical systems attractive for data-processing applications. There are three types of signal processing which rely upon optics. The first broad area includes those photographic and geometrical optics techniques which operate on the image directly. These are generally point operations such as contrast enhancement, level slicing and color manipulations where the operation at any given point is independent of all other data points. The second type of optical processing consists of systems where the data are transformed and operated on in a non-image plane. This non-image plane is generally related to the original image via a Fourier transform. These techniques have evolved in the last thirty years and comprise the field known as Fourier optics. The final type of optical signal processing involves the application of new optical technologies, such as integrated

optics, to signal-processing problems. The primary concern in this discussion is the area of processing involving Fourier optics. The other techniques will be discussed briefly in conjunction with some of their applications.

In the following section the principles of Fourier optics will be outlined and their application to image processing will be explored. After the basic concepts have been presented various examples of optical processing applied to remote sensing will be described.

FOURIER OPTICS AND COHERENT OPTICAL PROCESSING

The science of Fourier optics represents a new perspective on an old subject. Although one can debate the question of where it had its beginnings, it has come into its own in the past twenty years. This recent efflorescence has been fed on the one hand by technological advances, in particular the invention of the laser, and on the other hand by theoretical contributions from the other relatively young fields of information theory and communication theory. The following paragraphs present a heuristic development of the concepts of optical processing. The reader interested in a more detailed discussion of the theory and the history of Fourier optics is referred to one of the many texts on the subject (e.g., Goodman, 1968 and Gaskill, 1978).

In order to develop the ideas of Fourier optics it is necessary to delineate the initial assumptions. Unless otherwise stated, the formulas will assume essentially monochromatic illumination as would be obtained from a laser. Although one is generally interested in two-dimensional fields, the formulas derived here will be one-dimensional for notation simplicity, the extension to two dimensions being straightforward. Furthermore, we will deal only with the complex amplitude of the electromagnetic field. The complex amplitude, $o(x)$, composed of a magnitude, $a(x)$, and a phase component, $\phi(x)$, is expressed as

$$o(x) = a(x) \exp(j\phi(x)) \qquad (17\text{-}106)$$

where $j = \sqrt{-1}$. By considering scalar fields, the

polarization properties are neglected. The temporal dependence of the electromagnetic wave will not be explicitly included because detectors can only measure the time-averaged intensity, $O(x)$,

$$O(x) = |o(x)|^2 = |a(x) \exp(j\phi(x))|^2 = a^2(x)$$

$$(17\text{-}107)$$

of a wavefront. The fact that detection requires a time-averaged measurement of luminance bears directly on one other fundamental concept, namely, the difference between "coherent" and "incoherent" illumination. If the phase difference between the complex amplitudes of two points is constant in time, the two points are said to be mutually coherent. This will generally be the case for any two points in an image illuminated with a laser source. Two points on a thermal source, such as an incandescent lamp, radiate independently. The relative phase difference between two such points varies randomly in time. When the fields from two mutually coherent points are added, it is the complex amplitudes that are added, thus preserving information on the phase difference between the two components. For two incoherent points there is no fixed phase relationship between the fields. In this case it is the individual luminances which are added to find the total luminance, rather than summing complex amplitudes.

The most important fact of Fourier optics is that a lens can be used to perform a two-dimensional Fourier transform on the complex amplitude in a given plane. To see how this arises, we will refer to Figure 17-74. First consider a monochromatic point source located at x in the front focal plane of lens L. From geometrical optics it is known that this will produce a plane wave behind the lens propagating at an angle

$$\theta = \tan^{-1}(x/f) \qquad (17\text{-}108)$$

with respect to the optical axis. Generally, consideration will be restricted to small values of θ so that

$$\theta \simeq x/f \simeq \sin^{-1}(x/f) \qquad (17\text{-}109)$$

The field in the back focal plane of the lens (the x' plane in Figure 17-74) is a tilted plane/wave

which can be represented as a uniform amplitude with a linear phase factor proportional to $\sin \theta$

$$\text{tilted plane-wave} = \exp(2\pi jx' \sin(\theta)/\lambda)$$
$$\simeq \exp(2\pi jxx'/\lambda f)$$

$$(17\text{-}110)$$

The next step is to consider what happens if there is an extended object $o(x)$ in the front focal plane instead of a single point. Each point x of the input object will produce a plane-wave in the back focal plane with a different tilt angle and with a uniform amplitude proportional to the object amplitude $o(x)$. The total field in the back focal plane is the sum of all these individual tilted plane-waves. In the case of a continuous object this summation becomes an integral over x. Then the field O in the back focal plane is

$$O(x') = \int_{-\infty}^{\infty} o(x) \exp(2\pi jxx'/\lambda f)\, dx$$

$$(17\text{-}111)$$

This assumes that the object is coherently illuminated so that the complex amplitudes are summed rather than intensities. The important result is that $O(x)$, is, aside from a constant factor, equal to the Fourier transform of $o(x)$ (it should also be noted that, in this heuristic derivation, a constant factor which should appear in Eq. 17-111 has been dropped since it is not pertinent to the material discussed here). This Fourier-transform relationship will be used to evaluate all other optical systems in this section.

Many applications of optics to remote sensing are based on the ability to generate the Fourier transform of an image. The measured intensity of this Fourier transform is referred to as the spatial frequency power-spectrum of the image. The term "spatial" frequency refers to the fact that the Fourier transform is taken with respect to a spatial variable, e.g. x, rather than with respect to time, in which case the more familiar concept of temporal frequency is the quantity of interest. Applications which only involve measurements of the power spectrum are called spectrum analysis or Fourier analysis. Examples of such applications will be presented later.

By cascading two Fourier-transforming systems, an imaging system is obtained as shown in Figue 17-75. This follows from the property of Fourier transforms that

$$\mathcal{F}\left[\mathcal{F}\left[f(x)\right]\right] = f(-x) \qquad (17\text{-}112)$$

where $\mathcal{F}$ represents the Fourier transform operation (Bracewell, 1965). The minus sign implies that the system of Figure 17-75 produces an inverted image.

If the object in plane P_1 of the imaging system of Figure 17-75 is illuminated by a monochromatic plane view, the system is referred to as a coherent imaging system. In this case, besides having an image available in plane P_1, the Fourier transform of the object is observable in plane P_2. The great

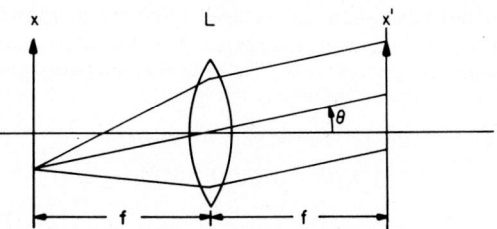

Fig. 17-74. Fourier transform properties of a lens. The amplitudes in the front and rear focal planes of a lens are related by a fourier transform.

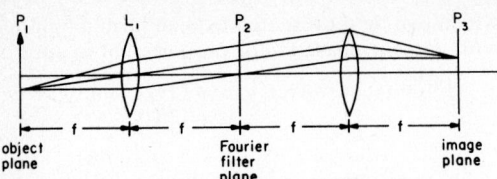

P_1 L_1 P_2 P_3

object plane   Fourier filter plane   image plane

Fig. 17-75. Telecentric imaging system consisting of two consecutive Fourier transform systems.

power of coherent optical processing derives mainly from the capability of manipulating the object by altering the Fourier spectrum of the object. If a transparency with complex amplitude transmittance $H(x)$ is placed in the Fourier transform plane P_2, the image will be altered. This transparency is referred to as a filter for reasons that will become obvious. The system can be easily analyzed starting from the object $o(x)$ in plane P_1. In plane P_2 the complex amplitude before the filter H is the Fourier transform O of the object. After the filter field in plane P_2 is given by the product, $H \cdot O$, of the object Fourier transform and the filter transmittance. The complex amplitude in the output plane is simply the Fourier transform of this product. The Fourier transform of a product is a convolution of the Fourier transforms of the multiplicands. In this case the image contains the convolution of the object with the Fourier transform h of the filter H. Denoting the image as $i(x)$, convolution is

$$i(x) = o(x) * h(x) \equiv \int_{-\infty}^{\infty} o(x') \, h(x - x') \, dx$$

$$(17\text{-}113)$$

The entire imaging system can be represented schematically as shown in Figure 17-76 as a simple linear system. The filter function H is the transfer function of the system and its Fourier transform h is the impulse response, i.e., the image of a point source. The transfer function H can be manipulated and utilized to implement numerous types of systems which will be discussed later, within the applications section.

At this point brief consideration will be given to the implementation of the filter H. If one is only interested in manipulating the amplitude of the Fourier transform, the filter can be generated quite easily by writing the desired attenuation pattern, on a piece of photographic film, for

example. Such amplitude filters are useful particularly in the form of binary masks which transmit only selected portions of the Fourier spectrum. Thus, one can fabricate low-pass, high-pass or bandpass filters which transmit only a specific range of spatial frequencies. In contrast to filters for electronic (temporal) signals, these binary filters can be made with arbitrarily sharp cutoffs since the spatial signals operated on in optics are not restricted by the causality requirements of temporal signals.

On occasion it is necessary to manipulate not only the amplitude of the Fourier transform, but also the phase. Producing a phase filter is problematic since photographic film is generally useful only for controlling amplitude. There are several techniques which have been developed to circumvent this problem. Some techniques involve generating a direct-phase filter, while other techniques encode the phase information in a form that can be readily decoded.

In many cases it is sufficient to have only two phase values, 0 and π. This allows one to produce any real-valued, positive or negative, filter H. Such binary phase filters can be produced by a number of techniques, such as selective etching of a transparent thin film on a transparent substrate. The thickness of the film layer is adjusted so that there is an optical path difference of π between regions where the film is present and regions where it has been etched away. Tsujiuchi (1960) first used this approach for optical image enhancement. A second method has been developed by Chu et. al. (1973) which allows arbitrary complex amplitudes to be implemented. This method involves using two emulsions of a color photographic film; one emulsion modulates only the amplitude of the transmitted light while a second emulsion modulates the phase. This technique is limited by the quality of color photographic emulsions which tend to have relatively low resolution and introduce grain noise and phase noise.

There are many techniques in which the phase is encoded in a filter which was only a positive, real transmittance function. The first of these is a method which evolved from Leith and Upatnieks' (1962) technique of off-axis holography. The idea is to generate a spatial carrier-frequency and modulate this carrier with the desired amplitude and phase signal. Such modulation can be easily achieved optically by recording the interference pattern between the desired complex amplitude function $o(x) = a(x) \exp(j\phi(x))$ and a tilted plane wave $R\exp(j\alpha x)$ which acts as a reference wave. The recorded intensity will be

$$I(x) = |a(x) \exp(jk\phi(x) + R\exp(j\alpha x)|$$
$$= a^2(x) + R^2 + 2Ra(x) \cos(\alpha x - \phi(x))$$

$$(17\text{-}114)$$

The last term in this expression contains both amplitude and phase information. A transparency with an amplitude transmittance $t(x)$ proportional

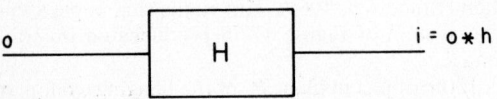

o ———— | **H** | ———— $i = o * h$

Fig. 17-76. Linear system representation of an imaging system. The image i is the convolution of the object o with the impulse response h. The system is characterized by its transfer function H which is the Fourier transform of h.

to the intensity $I(x)$ can be used to regenerate the signal $o(x)$ or its complex conjugate. This is done by illuminating the transparency with the original reference wave to produce

$$V(x) = R\exp(j\alpha x)\, t(x) = R\exp(j\alpha x)\left[a^2(x) + R^2\right.$$
$$\left. + 2Ra(x)\cos(\alpha x - \phi x)\right]$$
$$= R^2 o(x) + R^2 o^*(x)\exp(j^2\alpha x) + R(a^2(x)$$
$$+ R^2)\exp(j\alpha x) \qquad (17\text{-}115)$$

where o^* implies the complex conjugate of $o(x)$. Thus $V(x)$ contains the original object wave $o(x)$ plus its complex conjugate $o^*(x)$ multiplied by a tilted plane-wave plus a third term, riding on a different tilted plane-wave. Due to the tilted plane-wave factors, as these wavefronts propagate they will draw apart from one another. Assuming $o(x)$ has a spatially limited extent, it is always possible to choose α large enough that the terms will be spatially separated after a sufficient propagation distance. Vander Lugt (1964) applied this idea to a filtering problem called matched filtering. The idea is to detect an object $o(x)$ in an input image. This is accomplished by producing a filter which encodes O^* the complex conjugate of the Fourier transform of the object $o(x)$. The simplest way to produce this filter is to record an off-axis hologram of the Fourier transform of the object (see Figure 17-77a). This holographic filter encodes both O and its complex conjugate as shown in Eq. 17-114. If such a filter is placed in the Fourier plane of the filtering system of Figure 17-75, a bright spot will appear in the output plane whenever the object $o(x)$ is in the input. The reason for this is as follows: if $o(x)$ is in the input plane, in the Fourier plane we have O; after the filter this becomes $O \cdot o^* = |O|^2$; Fourier-transforming this gives us the autocorrelation of o,

o^*o. The autocorrelation function is in general a sharply peaked function which can be used as a detection peak in the output plane. The system functions as a correlation detector. Figure 17-77 shows how a matched filter can be used in a filtering system.

An overview of work in spatial filtering has been given by Vander Lugt (1968). Spatial filtering is just one of many coherent optical processing techniques. All of the systems discussed here are linear, space-invariant systems. Considerable work has also been performed to develop systems capable of implementing nonlinear transformations and space-variant operations. These techniques and other coherent optical operations which are not discussed here have been reviewed by Goodman (1977).

INCOHERENT OPTICAL PROCESSING

Although coherent optical processing is a very powerful tool, it is not without its problems. Coherent illumination leads to a grainy noise structure called speckle. Coherent systems are also very sensitive to dust, scratches or other defects in the imaging system that produce artifact noise. In addition to noise problems, coherent filtering systems are very sensitive to misalignment problems. These problems can be significantly reduced or eliminated by using incoherent illumination. As will be seen these advantages are gained at the expense of certain processing capabilities.

In an incoherent system intensities add rather than complex amplitude. From the analysis above, it follows that if the object consists of a single point at x_0 (which can be represented by the Dirac delta function $\delta(x - x_0)$) the image will be the impulse response $h(x)$ centered at x_0 (actually $-x_0$) in the output plane (see Figure 17-78). The intensity of the impulse-response image is referred to as the point spread-function (PSF) of the optical system. Now, if the object is a continuous function, each point in the object produces its corresponding PSF image. The resultant image $I(x)$ is found by adding up (integrating) all of these individual PSF images, each weighted by the intensity

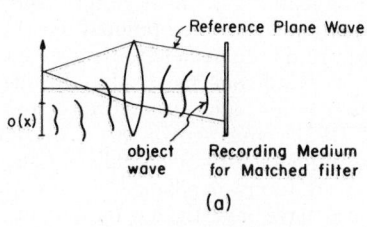

Reference Plane Wave

o(x)

object wave | Recording Medium for Matched filter

(a)

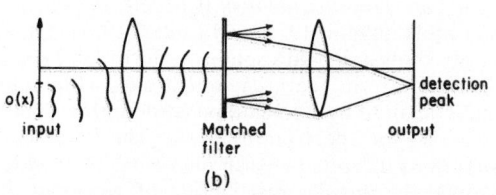

o(x) | input | Matched filter | output | detection peak

(b)

Fig. 17-77. Matched filtering. (a) A matched filter is made by recording a Fourier plane hologram of the object $o(x)$. (b) When the developed holographic matched filter is placed in an imaging system, an autocorrelation peak is detected in the output plane whenever the original object appears in the input plane.

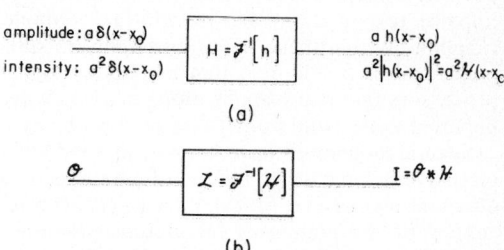

(a)

(b)

Fig. 17-78. Linear system representation of incoherent imaging. (a) Image intensity of a single point object is the point spread function $\mathcal{H}$. (b) For an incoherent image the intensities of the individual points add so that the image is the convolution of the object intensity with the point spread function.

$\mathcal{O}(x)$ of the object point. Refer also to Figure 17-4 and its related discussion. This is expressed as a convolution of the object intensity with the PSF intensity $\mathcal{H}$

$$I(x) = \mathcal{O} * \mathcal{H} = \int_{-\infty}^{\infty} \mathcal{O}(x') \mathcal{H}(x - x')\, dx'$$

$$(17\text{-}116)$$

This implies that an incoherent imaging system is a linear system which operates directly on the image intensity as depicted schematically in Figure 17-78b. This in turn implies that the optical system can be characterized by a function equal to the inverse Fourier transform of the point spread-function intensity $\mathcal{H}$. This function, which will be denoted as $\mathcal{L}$,

$$\mathcal{L} = \mathcal{F}^{-1}[\mathcal{H}] \qquad (17\text{-}117)$$

is called the optical transfer function (OTF). The OTF is in general a complex quantity. In describing the performance of optical systems it is common to use just the magnitude of L, referred to as the modulation transfer function (MTF), since that quantity is easily measured.

The optical transfer function can be expressed in terms of the coherent transfer function H by noting the relationship between the PSF and the coherent impulse response h.

$$\mathcal{L} = \mathcal{F}^{-1}[\mathcal{H}] = \mathcal{F}^{-1}[|h|^2] = \mathcal{H}\mathcal{H}^{\star}. (17\text{-}118)$$

The optical transfer function is seen to be the autocorrelation of the coherent transfer function. The limitation in incoherent processing is that it is not possible to manipulate the transfer function $\mathcal{H}$ directly. It can only be indirectly altered by changing $\mathcal{H}$, the complex amplitude transmittance of the filter. The class of incoherent transfer functions which can be implemented is limited to those realizable as an autocorrelation function. One important example of a filter that does not fall in this class is a low-pass filter, since an autocorrelation function is always non-zero at the origin and in fact takes on its maximum value there. The limitations of the incoherent system can also be appreciated by realizing that the incoherent point-spread function is necessarily a non-negative real quantity since it is an intensity while the coherent impulse response can be an arbitrary complex function. Within these limitations there is still a great deal of freedom so that incoherent optical processors find many applications in areas where coherent noise would otherwise be a problem.

Several techniques have been developed which circumvent some of the inherent limitations of incoherent processors. Monahan et al. (1977) give a review of the numerous incoherent processors relying on geometrical optics which perform operations on bipolar signals. Considerable research has also been carried out on multiple-pupil systems that trade off system bandwidth for the ability to process bipolar and complex signals (e.g., Lohmann and Rhodes, 1978). Rogers (1977) discusses many other techniques, including the use of temporal modulation methods to extend the flexibility of incoherent systems.

REMOTE SENSING APPLICATIONS OF OPTICAL PROCESSING

In the last section it was seen that the power of optical processing rests largely on the ability to perform two-dimensional Fourier transforms of images. The purpose of this section is to show some of the ways in which this capability can be applied to practical problems. Several examples of remote sensing applications are given for the areas of spectral analysis and spatial filtering.

Spectrum Analysis Applications

The simplest operation in coherent optics is to generate the two-dimensional Fourier transform of an object. Despite the development of very efficient algorithms for computing Fourier transforms, digital transformation of high resolution, large format (e.g., 10^8 pixels) imagery is still not practical. For these reasons, spectrum analysis, the evaluation of imagery on the basis of its Fourier power spectrum, is an important area of optical processing.

Spectrum analysis is applicable whenever one is interested in looking for features which have a distinctive spatial-frequency composition. One good example of this is in the analysis of water-surface waves. Figure 17-79 shows a picture of the wave patterns around a small island. Power spectra of different portions of the image clearly show the changing direction and periodicity of the wave patterns.

Spectrum analysis also has obvious applications in geology. Jacobsen et al. (1974) used the technique for determining the azimuthal distribution of linear structural features such as joints, folds, faults and bedding planes. They pointed out that the Fourier spectrum is equivalent to the rose diagram used in studying rock mechanics. Besides applications to macroscopic structural geology, Nyberg et al. (1971) have shown that spectrum analysis is useful for microscopic studies of such things as glacial striations on a limestone surface.

Several groups have investigated the potential use of spectrum analysis in physiographic classification and land-use analysis (Corbett, 1973, Ulaby and McNaughton 1975, and Lukes, 1977). Figure 17-80 shows the diffraction patterns of several distinct terrain features, employing a detector manufactured by Recognition System, Inc. that is designed for spectrum analysis. The latter consists of 64 discrete detector elements, 32 of which sample the angular distribution of power in the Fourier plane and the 32 of which measure the radial distribution of power.

Spectrum analysis has also been applied to problems of image-quality determination and cloud screening (Rotz and Greer, 1974; and

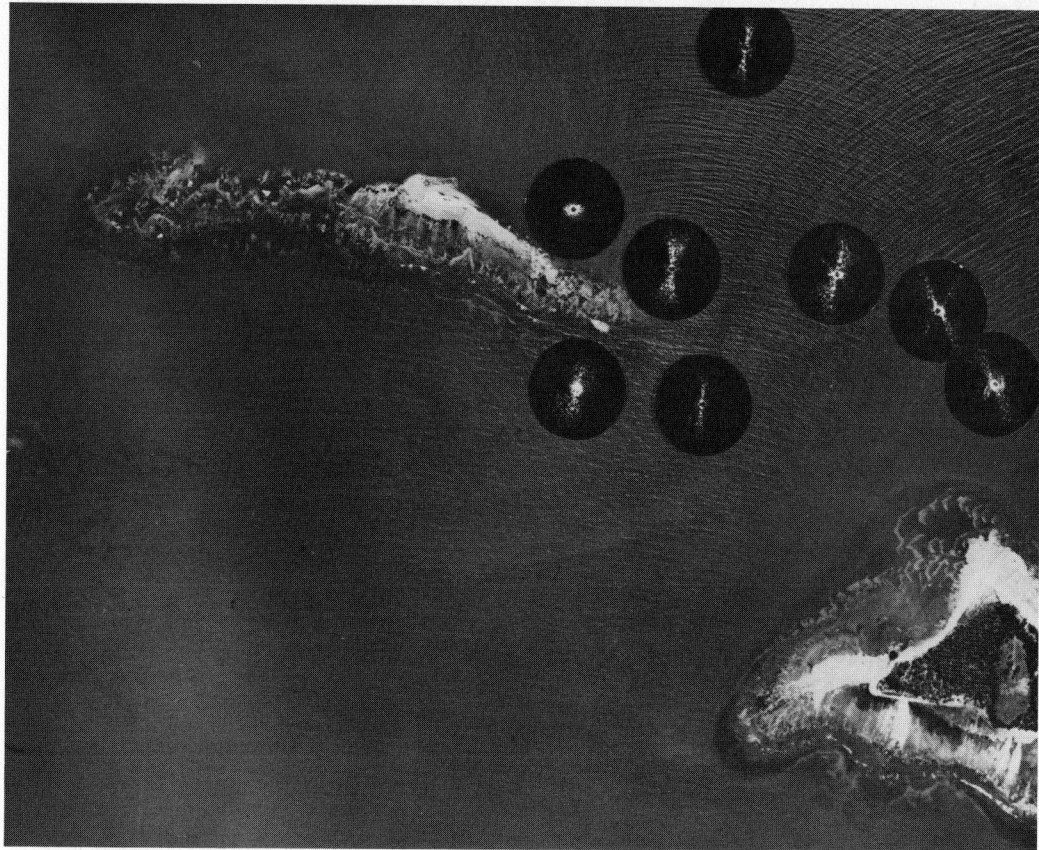

Fig. 17-79. Two-dimensional spectra of a water surface (R. H. Mitchel, Univ. Michigan).

Lukes, 1977) and to cloud classification (Mason, 1977).

Archeology has also found an application for spectrum analysis. Chevallier et al. (1970) showed that spectrum analysis could be used in the search for archeologically significant patterns in aerial photographs. Using this technique, they were able to locate ancient Roman land divisions in France which were largely obliterated by succeeding civilizations.

Spatial Filtering Applications

The spatial filtering concepts which were developed earlier allow one to manipulate an image and to view the results of that manipulation in the image plane.

The simplest forms of spatial filtering are those utilizing binary filters, i.e., masks whose transmittance is either 0 or 1. Such filters are used to select or reject specific portions of the Fourier spectrum. Thus, the periodic mensuration lines found in certain satellite imagery can be eliminated, if desired, by blocking just the spatial-frequency component(s) associated with the grid pattern. An example of using a binary filter to eliminate an unwanted disturbance is shown in Figure 17-81 where a filter has been used to block

the spatial-frequency components associated with the horizontal scan lines in the Orbiter V lunar image (Pincus, 1969). Directionally selective filters have been shown to be useful in the study of glacier crevasse-structure (Bauer et al., 1967) and in studying land features which are obscured by overlying or intermingled features (Harnett et al., 1978; Pincus and Dobrin, 1966). In all these instances spatial filtering made possible the removal of obliterating detail to allow the analysis of the structure of interest.

In some cases useful geological information can be extracted using filters that are easily implemented optically but difficult to obtain with digital techniques. One example of this involves determining the relative ages of regions of the surface of the planet Mercury by measuring the crater density on the surface (Chavel et al., 1980). One simple means of obtaining a measure of crater density is to band-pass-filter images of the surface to essentially detect crater rims and follow that by an averaging process or low-pass filtering. This produces an image intensity proportional to the local crater density. Experiments by Chavel et al. indicated that this operation was not feasible with a single-pass convolution with a small kernel such as could be implemented in hardware or done with

Fig. 17-80. Selected diffraction patterns for terrain patterns imaged in a cartographic aerial photograph (G. E. Lukes, Army ETL).

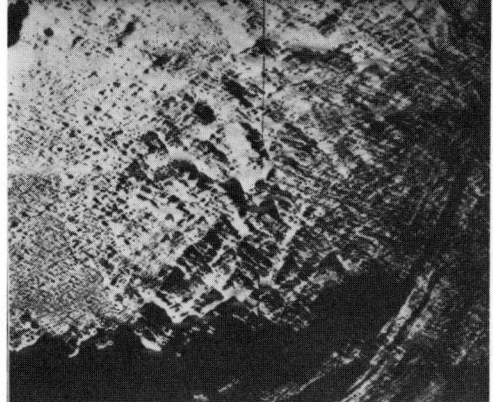

Fig. 17-81. Optical filtering to remove scan lines in a transmitted image. (Top): input image (Aristarcus); (bottom): horizontal scan lines removed by directional filtering (H. J. Pincus, Univ. Wisconsin-Milwaukee).

a minicomputer. They were able to obtain the desired results with an optical system, however, employing an annular ring filter combined with local averaging via a defocus blur. The original image is shown in Figure 17-82 along with the processed image where the intensity indicates the local crater density, or equivalently local surface age.

One variant of spatial filtering with binary masks was a system introduced by Bescos and Strand (1978), to color encode an image such that regions with differing textures appear in different colors. This is achieved by using a spatially-coherent white (polychromatic) light source rather than the monochromatic source associated with strictly coherent processing. Through the use of a color filter in the Fourier plane, different spatial frequencies are coded with different colors. This technique has been utilized to classify clouds in satellite imagery (Mantock et al., 1980) as shown in Color Figure 17-83.

By utilizing nonbinary filters, more complex operations can be achieved. Operations of interest typically involve the search for a particualr object or feature, i.e., a form of pattern recognition

(Balasubramanian and Leighty, 1975). If the object of interest has a well-defined form, such as that of a particular building or a certain type of vehicle, the technique of matched filtering can be utilized (Vander Lugt, 1964). However, in many instances, the object of interest may not have a precise predeterminable form. In such cases matched filtering is impractical. Baures and Duvernoy (1978) have developed a technique of statistical spatial filtering that allows one to design a detection filter that takes into account the statistical nature of the object. They applied this to the detection and analysis of Roman ruins using aerial photographs as their data source.

One very important form of complex spatial filtering is the correlation process. One of the main applications of correlation is the identification of common areas in two aerial photographs. In identifying common areas, the correlation process precisely determines their location and orientation. Thus this procedure is a valuable tool for registering a set of images with respect to one another. Such registration is necessary for aligning stereo pairs of images and for piecing together large mosaic images. Color Figure 17-84 shows an example of the application of correlation detection (Rotz and Greer, 1974). From the process of aligning stereo pairs, it is a straightforward extension to develop a system for obtaining the parallax data needed to generate a topographic map (Krulikoski and Forrest, 1972; Rotz and Greer, 1974; Balasubramanian and Leighty, 1975). One of the major difficulties encountered in correlation studies is the degradation in performance which occurs if there are any geometric or radiance distortions which cause the same object to have a slightly different appearance in two images. These distortions arise from differences in the lighting and viewing angles and differences in scale which often occur in separate photographs of a given area. The problem of recognizing these distortions, and of compensating for them prior to correlation, has been addressed by Casasent and Barniv (1978). In particular, correlators have been developed which are unaffected by low spatial-frequency intensity-variations and are unaffected by scale changes.

Other Techniques

In the past decade, a new field of optics called "integrated optics" has evolved and is beginning to have an impact on optical processing and remote sensing. Analogous to integrated circuitry in electronics, the goal of integrated optics is to build entire optical systems, including sources, modulators, wave guides, equivalents of optical elements, and detectors, on a chip. Although this is still a very experimental area, work has begun on developing an integrated optical processor that could be included in a satellite instrument package (Verber, 1980). The function of the processor would be to screen incoming image data and

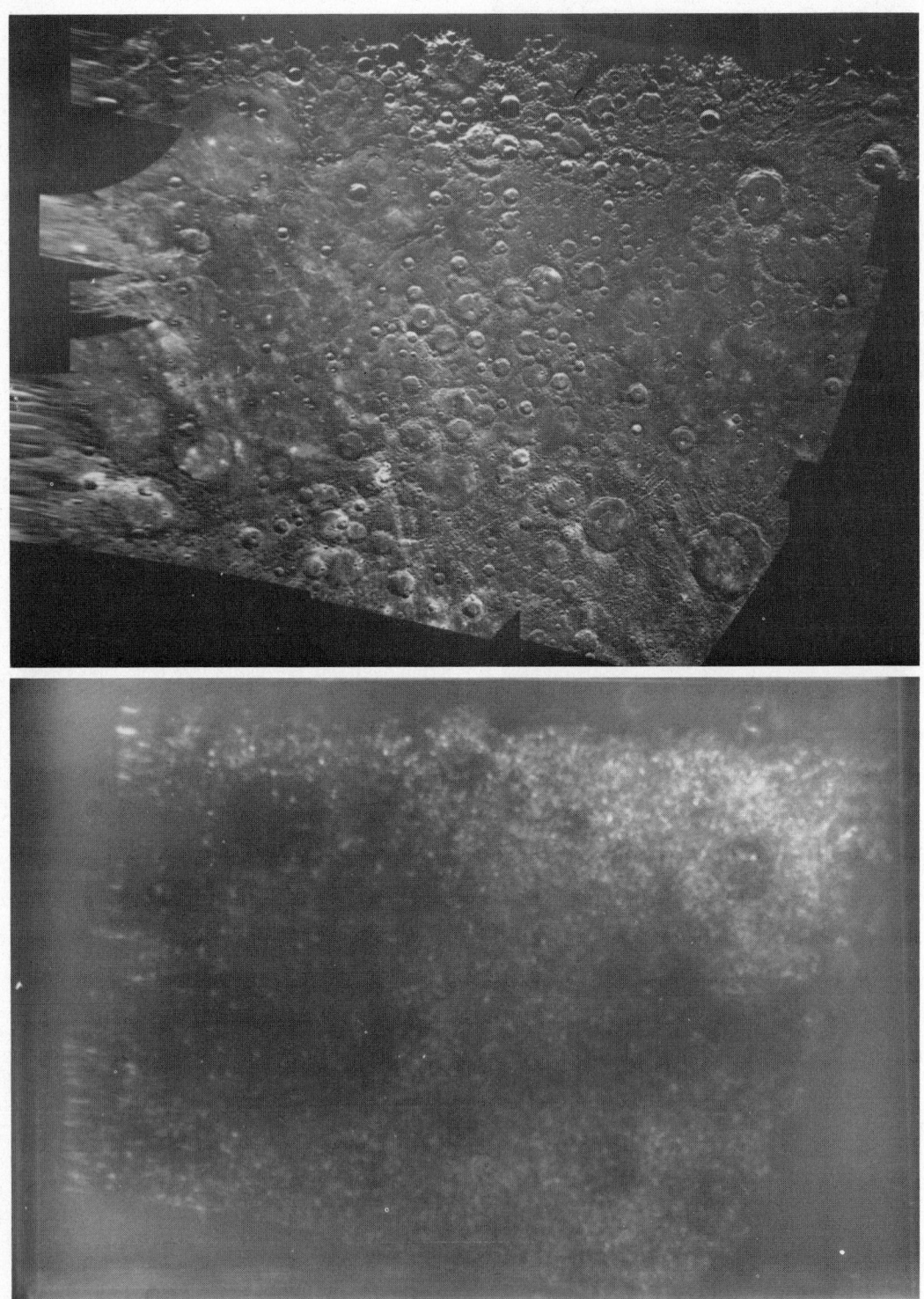

Fig. 17-82. Optical filtering to measure crater densities for surface age determination. (Top): input image of the surface of Mercury; (bottom): filtered image where the intensity relates to the crater density in the original image (P. Chavel, Institut d'Optique, France).

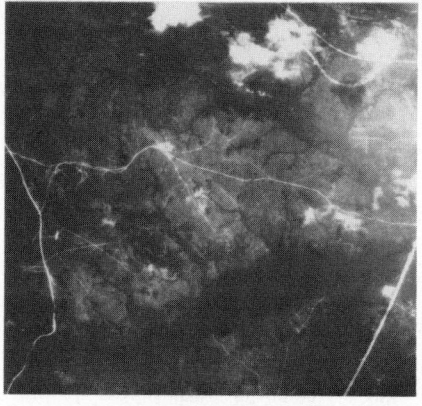
(a) Reference data

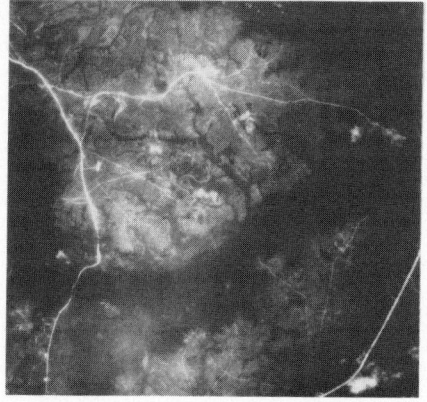
(b) Image to be mapped

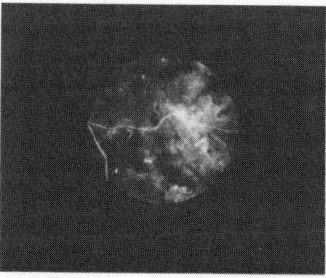
(c) sub-region used for correlation

(d) autocorrelation of reference data

(e) cross-correlation of image and reference data

Fig. 17-84. Example of photographic mapping. (a) Reference data. (b) Image to be mapped. (c) Sub-region used for correlation. (d) Autocorrelation of reference data. (e) Cross-correlation of image and reference data.

sense data that lack any interesting detail. Such data would not be transmitted to the ground station, thus saving power and reducing the amount of processing and storage required on the ground.

In summary, it can be said that optical processing has a wide range of applications in remote sensing. These applications include Fourier spectrum analysis techniques, spatial filtering tech-

niques ranging from edge enhancement to matched filtering detection, and correlation processing for locating objects and registering images. In principle all of these operations can also be done using digital image-processing techniques. However, practical considerations usually make one approach more desirable than the other. The question of when to use optical processing techniques and when to use digital techniques in certain remote sensing applications has been discussed by Colwell and Katibah (1976). Although there is no fixed answer to this question, in general digital processing offers greater flexibility and superior accuracy in processing than optical systems. Optical processors on the other hand have the advantage of being able to process the large amounts of data associated with high resolution imagery in real-time (given suitable input-output devices). Optical systems have the added advantage that they can often be implemented for a relatively low cost compared with digital processing systems.

REFERENCES

Ables, J. G., 1974, Maximum entropy spectral analysis; Astron. and Astrophys. Suppl. 15, p. 383.

Ahmed, N., T. Natarajan, and J. R. Rao, 1974, Discrete cosine transform; IEEE Trans. on Comput., vol. C-25, p. 90.

Anderson, G. B. and T. S. Huang, 1971, Piecewise Fourier transformation for picture bandwidth compression; IEEE Trans. on Comm., vol. COM-19, no. 2, p. 133.

Andrews, H. A., 1976, Two Dimensional Transforms; in Topics in Applied Physics, vol. 6, Picture Processing and Digital Filtering, edited by T. S. Huang, Springer Verlag, Berlin-New York.

Andrews, H. A., 1976, Outer product expansions and their uses in digital image processing; IEEE Trans. on Comput., vol. C-25, Feb. 1976.

Anuta, P. E., 1970, Spatial registration of multispectral and multitemporal digital imagery using fast Fourier transform techniques; IEEE Trans. on Geosci. Elect., vol. GE-8, no. 4, pp. 353–368.

Anuta, P. E., 1977, Computer-assisted analysis techniques for remote sensing data interpolation; Geophysics, vol. 42, no. 3, pp. 468–481.

Anuta, P. E., D. M. Freeman, B. M. Shelly and C. R. Smith, 1978, Synthetic Aperture Radar/Landsat MSS Image Registration Systems Study; Inf. Note 082478, LARS, Purdue Univ., West Lafayette, Ind.

Arguello, R. J. and H. R. Sellner, 1971, Optimal processing bandwidth for optical transfer-function compensation; Photog. Sci. and Eng., vol. 15, no. 4, p. 329.

Arguello, R. J., H. R. Sellener, J. A. Stuller, 1971, The effects of channel error in the differential pulse code modulation transmission of sampled imagery; IEEE Trans. on Comm., vol. COM-19, no. 6, p. 926.

Arguello, R. J., H. R. Sellner, J. A. Stuller, 1972, Transfer function compensation of samples images; IEEE Trans. on Comput., vol. C-21, no. 7, p. 812.

Arazi, B., 1975, Optimal coding scheme for a certain type of Ensemble; IEEE Trans. on Comm., vol. COM-23, no. 7, p. 741.

Backus, G. and F. Gilbert, 1970, Uniqueness in the inversion of inaccurate gross Earth data; Phil. Trans. Royal Soc., (London), vol. 226, ser. A, p. 123.

Balasubramanian, N. and R. D. Leighty, 1975, Coherent optics in mapping; Optical Eng., vol. 14, pp. 211–216.

Barnea, D. I. and H. F. Silverman, 1972, A class of algorithms for fast digital image registration; IEEE Trans. on Comput., vol. C-21, p. 179.

Bauer, A., A. Fontanel, and G. Grau, 1967, The application of optical filtering in coherent light to the study of aerial photographs of Greenland glaciers; Jour. Glaciol., vol. 6, pp. 781–793.

Baures, P. Y. and J. Duveroy, 1978, Statistical and spatial filtering: Application to aerial photographs; Applied Optics, vol. 17, pp. 3395–3401.

Bernstein, R., 1973, Results of Precision Processing of ERTS-1 Image Using Digital Image Processing Techniques; IBM Tech. Rept. FSC-73-0048, IBM Federal Systems Div., Gaithersburg, MD.

Bernstein, R. and H. Silverman, 1978, Digital Techniques for Earth Resource Image Data Processing; IBM Tech. Rept. FSC-71-6017, IBM Federal Systems Div., Gaithersburg, MD., also in Proc. AAIA 8th Annual Mtg., vol. C21, AIAA Paper 71-978.

Bescos, J. and T. C. Strand, 1978, Optical pseudocolor encoding of spatial frequency information; Appl. Optics, vol. 17, pp. 2524–2531.

Billingsley, F. C., 1975, Noise Considerations in Digital Image Processing Hardware; Topics in Applied Physics, vol. 6, Picture Processing and Digital Filtering, edited by T. S. Huang, Springer Verlag, Berlin-New York.

Billingsley, F. C., 1982, Modeling misregistration and related effects on multispectral classification; Photogramm. Eng. and Rem. Sens., vol. 48, no. 3, p. 421.

Biraud, Y., 1969, A new approach for increasing the resolving power by data processing; Astron. and Astrophys., vol. 1, p. 124.

Burg, J. P., 1967, Maximum entropy spectral analysis; presented at 37th Annual Soc. Explor. Geophys. Mtg., Oklahoma City.

Casasent, D. and Y. Barniv, 1978, Coherent optical processing for multisensor image processing in aerial reconnaissance; SPIE Proc., Airborn Recon. II/ Collec. and Exploit. of Recon. Data, pp. 57–63.

Chavel, P., P. Thomas, P. H. Masson, and J. C. Saget, 1980, to be published.

Chen, W. and C. H. Smith, 1976, Adaptive coding of coherent images using cosine transform; Proc. of ICC 76, IEEE cat. #76CH1085-0 CSCB, p. 47–7.

Chen, W. and S. C. Fralick, 1977, Image enhancement using cosine transform filtering; Proc. Symp. on Image Science Mathematics, Navy Postgraduate School, Monterey, CA., C. O. Wilde, ed.

Chen, W., C. H. Smith and S. C. Fralick, 1977, A fast computational algorithm for the discrete cosine transform; IEEE Trans. on Comm., vol COM-25, p. 1004.

Chevallier, R., A. Fontanel, G. Grau, and M. Guy, 1970, Applications of optical filtering to the study of aerial photographs; Photogrammetria, vol. 36, pp. 17–35.

Chu, D. C., J. R. Fienup and J. W. Goodman, 1973, Multi-emulsion on-axis computer generated hologram; Appl. Optics, vol. 12, pp. 1386–1388.

Chu, N. and C. D. McGillem, 1978, Methods and Performance Bounds for Constrained Image Restoration; LARS Tech. Rept. TR 061678, Purdue Univ. West Lafayette, IN.

Chu, N. and C. D. McGillem, 1979, Image restoration filters based on a 0-1 weighting over the domain of PSP; IEEE Trans. Acoustics Speech and Signal Processing, vol. ASSP-27, no. 5, pp. 457–464.

Colwell, R. N. and E. F. Katibah, 1976, Optical vs. electronic enhancement of remote sensing imagery; SPIE Proc. Image Processing, pp. 111–118.

Cooley, J. W., et al., 1967, What is the fast Fourier transform?; IEEE Trans. on Audio and Electroacoustics, vol. AU-15, no. 2.

Corbett, F. J., 1973, Terrain recognition in ERTS-1 imagery by diffraction pattern analysis; Proc. Fall Conf. ASP, p. 431.

Curry, H. B. and I. J. Schoenberg, 1966, On polya frequency functions IV. Fundamental spline functions and their limits; Jour. d'Analyse Mathematique (Jerusalem), vol. 17, p. 71.

Davisson, L. D., 1969, An application of data compression concepts; Proc. Symp. on Info., School of Elec. Eng., Purdue Univ., West Lafayette, IN., vol. 2, p. 519.

Duggin, M. G. and P. J. Ellis, 1980, Limitations on the spectral discrimination of the Landsat MSS; submitted to Photogram. Eng. and Remote Sens.

Ethridge, M. M., and E. M. Mikhail, 1976, Geometric restitution of single coverage aircraft multispectral scanner data; presented at 2nd Pecora Symp., October.

Forrest, R. B., 1971, Geometric Processing of ERTS Images; Bendix Res. Lab. Rept. 71-342, also presented at 1971 ASP-ACSM Convention.

Forrest, R. B., 1973, EOS Mapping Accuracy Study; Bendix Research Lab. Rept. 6750, BRL, Southfield, MI.

Frieden, B. R., 1972, Restoring with maximum likelihood and maximum entrophy; Jour. of the Optical Soc. of Amer., vol. 62, no. 4, p. 511.

Frieden, B. R. 1974, Image restoration by discrete convolution of minimum length; Jour. of the Optical Soc. of Amer., vol. 64, no. 5, p. 682.

Frieden, B. R., 1975, Image Enhancement and Restoration; Topics in Applied Physics, vol. 6, Picture Processing and Digital Filtering, edited by T. S. Huang, Springer Verlag, Berlin-New York.

Frieden, B. R., 1976, A new restoring algorithm for the preferential enhancement of edge gradients; Jour. of the Optical Soc. of Amer., vol. 66, no. 3, p. 280.

Frieden, B. R., 1980, Statistical models for the image restoration problem; Computer Graphics and Image Processing, vol. 12, p. 40.

Friedman, H. D., 1965, On the expected error in the probability of misclassification; Proc. IEEE, vol. 53, p. 658.

Fultz, G. L., 1965, The effect of source noise on quantization accuracy and of PE statistics; JPL Tech. Memo 3341-65-5.

Gillespie, A. R., 1975, A discussion; in Application of ERTS images and image processing to regional geologic problems and geologic mapping in Northern Arizona, A. F. H. Goetz et al., JPL Pub. 32-1597.

Goodman, J. W., 1977, Operations achievable with coherent optical information processing systems; Proc. IEEE, vol. 65, no. 1, pp. 29−38.

Gupta, J. N. and P. A. Wintz, 1974, Multi-Image Modeling; TR-EE 74-24, School of Electrical Engineering, Purdue Univ., Lafayette, Indiana.

Gupta, J. N. and P. A. Wintz, 1975, A boundary finding algorithm and its applications; IEEE Trans. on Circuits and Systems, vol. CAS-22, no. 4, p. 351.

Habibi, A. and P. A. Wintz, 1971, Image coding by linear transformation and block quantization; IEEE Trans. on Comm. Tech., vol. 19, no. 1.

Habibi, A., 1974, Hybrid coding of pictorial data; IEEE Trans. on Comm., vol. COM-22, no. 5.

Habibi, A. and G. S. Robinson, 1974, A survey of digital picture coding; IEEE Computer, vol. 7, p. 22.

Haralick, R. M., 1969a, The Bayesian approach to identification of remotely sensed environment; CRES Tech. Report 133-9.

Haralick, R. M. 1969b, Multi-Image pattern recognition: ideas and results; CRES Tech. Report 133-11.

Haralick, R. M., 1972, Data processing at the University of Kansas; 4th AEPR, Vol. II, Chap. 38, p. 1.

Haralick, R. M. and N. Griswold, 1975, A fast two-dimensional Karhunen-Loeve transform; Vol. 66-Efficient Transmission of Pictorial Information, Soc. of Photo-Optical and Instr. Engrs., Palos Verdes Estates, CA.

Haralick, R. M., 1976, A storage efficient way to implement the discrete cosine transform; IEEE Trans. on Computers, July 1976, p. 764.

Harnett, P. R., G. D. Mountain and M. E. Barnett, 1978, Spatial filtering applied to remote sensing imagery; Optica Acta, Vol. 25, pp. 801−809.

Hein, D. and N. Ahmed, 1978, On a real-time Walsh-Hadamard/cosine transform image processor; IEEE Transactions on Electromagnetic Capability, Vol. EMC-20, pp. 453−457, August 1978.

Helstrom, C. W., 1967, Image restoration by the method of least squares; J. of the Optical Soc. of Amer., vol. 57, p. 297.

Hilbert, E. E., 1977, Cluster compression algorithm—a joint clustering/data compression concept; JPL Publication 77-43, Jet Propulsion Laboratory, Pasadena, CA.

Hooke, A., User-oriented end-to-end transport protocols for real-time distribution of telemetry data from NASA spacecraft; proceeding of the SPIE Spring Symposium, April, 1979.

Hou, H. S., 1976, Least squares image restoration using spline functions; USCIP Report 650, University of S. CA., Los Angeles, CA.

Huang, T. S., G. J. Yang, and G. Y. Tang, 1979, A two-dimensional median filtering algorithm; IEEE Trans. on Acoustics Speech and Sig. Proc., vol. ASSP-27, Feb. 1979.

Hunt, B. R., 1975, Digital image processing; Proc. IEEE, vol. 63, no. 4, p. 693.

Hunt, B. R., 1976, Some remaining mathematical problems in non-linear image restoration; Proc. Symp. on Image Science Mathematics, Navy Postgraduate School, Monterey, CA., C. O. Wilde, ed.

Jacobsen, W. L., L. K. Lepley, and J. D. Gaskill, 1974, Automatic rose diagrams for rock mechanics and structural geology—diffraction patterns; Proc. Ann. Conf. on Rem. Sens. in Arid Lands, University of Arizona, pp. 172−180.

Jain, A. K., 1977, Some new techniques in image processing, image enhancement using cosine transform filtering, Proc. Symp. on Image Science Mathematics, Navy Postgraduate School, Monterey, CA., C. O. Wilde, ed.

Janssen, P. A., R. H. Hunt, E. K. Pyler, 1970, Resolution enhancement of spectra; Journal of the Optical Society of America 60, p. 596.

Jaynes, E. T., 1968, Prior probabilities; IEEE Trans. Systems Science Cybernet, vol. SSC-4, p. 227.

Jones, H. W., D. N. Hein, and S. C. Knauer, 1978, the Karhunen-Loeve, discrete cosine, and related transforms obtained via the Hadamard transform; International Telemetering Conference, ITC '78, vol. XIV, pp. 87−98, November 1978.

Kan, E. P. F., 1972, Data clustering: an overview; Lockheed Electronics Co., Inc., HASD, Houston, TX. Tech Report 640-TR-080, March 1972.

Kanal, L., 1974, Patterns in pattern recognition: 1968−1974; IEEE Trans. Info. Theory, vol. IT-20, Nov. 74, p. 697.

Keyes, R. G. Cubic convolution interpolation for digital image processing, IEEE Trans ASSP-29(6), 1981, p. 1153–1160.

Konecny, G., 1972, Metric problems in remote sensing; invited paper at Commission IV, International Congress of Photogrammetry, Ottawa.

Konecny, G., and W. Schuhr, 1975, Digitale entzenang der Daten von Zeilenabtastern; Bildmessung und Luftbildwesen, pp. 135–143.

Konecny, G., 1979, Methods and possibilities for digital differential rectification; Photogrammetric Engineering and Remote Sensing, vol. 45, no. 6, June 1979, pp. 727–734.

Kratky, V., 1971, Precision processing of ERTS imagery; ASP-ACSM Conference proceedings, fall conference, pp. 481–513, Sept. 1971.

Kratky, V., 1976, Grid-modified polynomial transformation of satellite imagery, remote sensing of environment; vol. 5, pp. 67–74.

Kraus, K., 1978, Rectification of multispectral scanner imagery; Photogrammetric Engineering and Remote Sensing, vol. 44, no. 4, pp. 453–457.

Krulikoski, S. J., Jr., and R. B. Forrest, 1972, Coherent optical terrain-relief determination using a matched filter; Bendix J., vol. 5, pp. 11–18.

LaFata, P., and J. P. Rosen, An interactive display for approximation by linear programming; Comm. of the ACM. vol. 13, no. 11, November 1970.

Landgrebe, D. A., 1976, Final Report T-1309, NASA Contract NAS9-14016, LARS, Purdue University, West Lafayette, IN, May 1976.

Larsen, R. D., and W. R. Madych, 1976, Walsh-like expansions and Hadamard matrices; IEEE Trans. on Acoustics Speech and Signal Processing, vol. AASP-24, no. 1., February 1976, p. 71.

Leith, E. N., and J. Upatnieks, 1962, Reconstructed wavefronts and communication theory; J. Opt. Soc. Amer., vol. 52, no. 10, pp. 1123–1130.

Lim, J. S., 1980, Image restoration by short space spectral subtraction; IEEE Trans. on Acoustics Speech and Signal Processing, vol. ASSP-28 #2, p. 191.

Liu, K. Y., and J. J. Lee, An experimental study of the concatenated Reed-Solomon/Viterbi channel coding system performance and its impact on space communications; JPL Publication 81-58, Jet Propulsion Laboratory, Pasadena, CA, September 1981.

Liu, K. Y., and K. T. Woo, The effects of tracking phase error on the performance of the concatenated Reed-Solomon Biterbi channel coding system; proceedings of the 1980 National Telecommunications Conference, Houston, Texas, November 30, 1980.

Limb, J. O., 1967, Source-receiver econding of television signals; IEEE Special Issue on Redundancy Reduction, vol. 55, March 1967, p. 364.

Lohmann, A. W., and W. T. Rhodes, 1978, Two pupil synthesis of optical transfer function; App. Optics, vol. 7, p. 1141.

Lukes, G. E., 1977, Rapid screening of aerial photography by OPS analysis; SPIE Proc. Data Extraction and Classification from Film, vol. 117, pp. 89–97.

Maharir, P. S. and S. K. Varma, 1980, A new class of ortho-normal transforms; IEEE Trans. Acoustics Speech Signal Processing, to be published.

Makhoul, J., 1980, A fast cosine transform in one and two dimensions; IEEE Trans. Acoustics Speech and Signal Processing, vol. ASSP-28, no. 1.

Mantock, J., A. A. Sawchuk, and T. C. Strand, 1980, Hybrid optical/digital texture analysis; Opt. Eng., vol. 19.

Mason, J. B., 1977, Optical Fourier transform analysis of satellite cloud imagery; SPIE Proc. on Adv. in Laser Tech. for the Atm. Sciences, pp. 77–82.

Max, J., 1960, Quantizing for minimum distortion; IRE Trans. Information Theory, March 1960, p. 1.

McGillem, C. D., T. E. Riemer, and M. Mobasseri, 1975, Resolution enhancement of ERTS imagery; Proc. Conf. on Machine Processing of Remotely Sensed Data, Purdue University, West Lafayette, IN, June 3–5, 1975.

McGlone, J. C., J. R. Baker, and E. M. Mikhail, 1979, Metric information from aircraft multispectral scanner (MSS) data; Proc 46th meeting of Amer Soc of Photogrammetry, Washington, D.C., Paper 19-132, p 274.

Mikhail, E. M., Baker, J. R., 1973, Geometric aspects in digital analysis of multispectral scanner (MSS) data; Info. Note 042473, Laboratory for Applications of Remote Sensing (LARS), Purdue University, West Lafayette, IN. 47907.

Monahan, M. A., K. Bromly and R. P. Bocker, 1977, Incoherent optical correlators; Proc. IEEE, vol. 65, pp. 121–129.

Montuori, J. S., 1980, Image scanner technology; Photogram. Eng. and Rem. Sens., vol. 46, no. 1.

Nack, M. L., 1975, Temporal registration of multispectral digital satellite images using their edge images; AAS/AIAA Astrodynamics Specialist Conference, Nassau, Bahamas, Paper #AAS75-104.

Narasimha, M. J., and A. M. Peterson, 1978, On the computation of the discrete cosine transform; IEEE Trans. on Comm., vol. COM-26, p. 934.

NASA, 1976, Landsat Data Users Handbook; NASA Goddard Space Flight Center Document 76SD4258.

NASA, 1979, Reference Publication 1039, Synthetic Aperture Radar/Landsat MSS Image Registration; NASA Report RP-1039.

Nathan, R., 1968, Picture Enhancement for the Moon, Mars, and Earth; Pictorial Pattern Recognition; G. C. Chen, ed. Thompson, Washington, D.C.

Nathan, R., 1971, Image processing for electron microscopy; enhancement procedures; Advances in optical and electron microscopy; Barer and Cosslett, eds. vol. 4, Academic Press, New York.

Norton, C. L., G. C. Brock, and R. Welch, 1977, Optical and modulation transfer functions; Photogram. Eng. and Rem. Sens., vol. 43, no. 5, pp. 709–719.

Nyberg, S., T. Orhang and H. Svensson, 1971, Optical processing for pattern properties; Photog. Eng., pp. 547–554.

Odenwalder, J. P., 1974, Concatenated Reed-Solomon/Viterbi channel coding for advanced planetary missions: analysis, simulations, and tests; submitted to JPL by Linkabit Corp., San Diego, CA, Final Rept., Contr. 953866.

O'Neal, J. B., 1966, Delta modulation quantizing noise analytical and computer simulation results for Gaussian and television input signals; BSTJ, vol. 45, no. 1, p. 117.

Oppenheim, A. V., R. F. Schafer, and T. G. Stockham, 1968, Non-linear filtering of multiplied and convolved signals; Proc. IEEE, vol. 56, p. 1264.

Otepka, G., 1978, Practical experience in the rectification of MSS images; Photogram. Eng. and Rem. Sens., vol. 44, no. 4, pp. 459–467.

Park, S. K. and R. A. Schowengerdt, 1982. Image reconstruction by parametric cubic convolution. Computer Graphics and Image Processing. (To be published, 1982).

Peavy, B., 1977, High volume digital image

processing—fact or fiction; paper 28/2, Mideon 77, Nov. 8–10, 1977, Chicago, ILL. (IEEE)

Peyrovian, M. J., 1976, Image Restoration by Spline Functions; USCIPI Report 680, Univ. So. CA., Los Angeles, CA.

Phillips, D. L., 1962, A technique for the numerical solution of certain integral equations of the first kind; JACM, vol. 71, pp. 4861–4869.

Pincus, H. J., 1969, The analysis of remote sensing displays by optical diffraction; Proc. 6th Int. Symp. on Remote Sensing of Environment, pp. 261–274.

Pratt, W. K., J. Kane, and H. A. Andrews, 1969, Hadamard transform image coding; IEEE Proc., vol. 57, no. 1, p. 58.

Pratt, W. K., 1973, Bibliography on digital image processing and related topics; USCEE, Report 473, Univ. So. CA.

Ready, P. J. and P. A. Wintz, 1972, Multispectral data compression through transform coding and block quantization; LARS Inf. Note 050572, Purdue Univ., West Lafayette, IN.

Rice, R. F., 1974, RM2: transform operations; Technical Memorandum 33-680, JPL.

Rice, R. F., 1974, Channel Coding and Data Compression System Considerations for Efficient Communication of Planetary Imaging Data; Chapter 4, Technical Memorandum 33-695, JPL.

Rice, R. F., 1975, An advanced imaging communication system for planetary exploration; SPIE Seminar Proceedings, vol. 66, pp. 70–89.

Rice, R. F., 1976, RM2 rms error comparisons; Technical Memorandum 33-804, JPL.

Rice, R. F., 1978, A concept for dynamic control of RPV information system parameters; Proc. 1978 Military Electronics Exposition, Anaheim, CA.

Rice, R. F., 1979a, Some practical universal noiseless coding techniques; JPL Publication 79-22, JPL.

Rice, R. F., 1979b, Practical universal noiseless coding: SPIE Synp. Proc., vol. 207.

Riemer, T. E. and C. D. McGillem, 1977, Optimum constrained image restoration filters; IEEE Trans. Aerospace Eng., vol. 13, no. 2.

Rifman, S. S., 1973, Digital rectification of ERTS multispectral imagery; Proc. Symp. Significant Results Obtained from Earth Resources Technology Satellite-1, NASA Pub. SP-327, pp. 1131–1142.

Rifman, S. S., W. B. Allendoerfer, D. M. McKinnon, K. W. Simon, 1975, Experimental study of digital image processing techniques for ERTS data—task 1 final report; Report no. 26232-6001-RU01, TRW Systems, Redondo Beach, CA.

Rose, J. A., W. K. Pratt, G. S. Robinson, A. Habibi, 1975, Interframe transform coding and predictive coding methods; Proc. ICC 75, IEEE Cat. no. 75CH0971-2GSCB, p. 23.17.

Rosenfeld, A., 1969, Image processing; Space/Aeronautics, p. 48.

Rosenfeld, A., 1972, Picture processing (surveys) computer graphics and image processing; Onward, yearly.

Rotz, F. B. and M. O. Greer, 1974, Photogrammetric and reconnaissance applications of coherent optics; SPIE Proc. Tutorial Seminar and Tech. Utilization Program, pp. 139–148.

Ruth, S. S. and P. J. Kreutzer, 1972, Data compression for large business files; Datamation, vol. 18, no. 19, p. 62.

Sawchuk, A. A., 1972 Space-variant image motion degradation and restoration; Proc. IEEE, vol. 60, July 1972, pp. 854–61.

Scott, R. M., 1965, The practical application of modulation transfer functions; Perkin-Elmer Corporation Electro Optical Division, March 1963.

Shlien, S., May 1979, Geometric correction, registration and resampling of Landsat imagery; Canadian Jour Rem. Sens., V5 No. 1, p. 74.

Simon, K. W., 1975, Digital image reconstruction and resampling for geometric manipulations; IEEE Symp. on Machine Processing of Remotely Sensed Data, June 1975, P. 3A-1.

Slater, P. N., 1979, Re-examination of the Landsat MSS; Photogramm. Eng. and Rem. Sens., vol. 45, no. 10.

Snyder, J. P., 1978, The space oblique mercator projection; Photogramm. Engr. and Rem. Sens., vol. 44, no. 5, pp. 585–596.

Salomonson, V. V., P. L. Smith, A. B. Park, W. C. Webb, and T. J. Lynch, 1980, An Overview of Progress in the Design and Implementation of Landsat-D Systems; IEEE Trans on Geoscience and Remote Sensing, vol. GE-18 #2, p. 138.

Steiner, D., 1974, Digital geometric picture correction using a piecewise zero-order transformation; Rem. Sens. of Env., vol. 3, pp. 261–283.

Stockham, T. G., 1969, Natural image information compression with a quantitative error model; Conf. on Pertinent Concepts in Computer Graphics, Urbana, ILL.

Stockham, T. G., 1972, Image processing in the context of a visual model; Proc. IEEE, vol. 60, p. 828.

Stuller, J. A., 1972, An algebraic approach to image restoration filter design; Computer Graphics and Image Processing, vol. I, no. 2.

Svedlow, M., C. D. McGillem, P. E. Anuta, 1976, Analytical and experimental design and analysis of an optimal processor for image registration; LARS Information Note 090776, Purdue University.

Svedlow, M., C. D. McGillem, P. E. Anuta, 1978, Image registration: similarity measure and preprocessing method comparisons; IEEE Trans on Aerospace and Electronic Systems, vol. AES-14, no. 1, pp. 141–150.

Tabor, J. E., 1973, Evaluation of digitally corrected ERTS images; Third ERTS Symposium, vol. 1, NASA SP-351, p. 1837.

Thompson, M. M., 1979, Maps for America: U.S. Geological Survey, U.S. Gov't Printing Office, No. 024-001-03145-1.

Tomlinson, R. F., 1972, ed., Geographical Data Handling; Int. Geogr. Union Comm. on Data Sensing and Processing for UNESCO/IGU, 2nd Symposium on Geogr. Infor. Systems, Ottawa, Conn.

Tribus, M. and E. C. McIrvine, 1971, Energy and information; Sci. Amer., vol. 225, no. 3, p. 179.

Trussel, H. J., 1980, Relationship between image restoration by the maximum a posteriori method and a maximum entropy method; IEEE Trans. Acoustics Speech and Signal Processing, vol. AASP-28 no. 1, p. 114.

Tsujiuchi, J., 1960, Restitution des images aberrantes par le filtrage des frequences spatiales; Optica Acta, vol. 7, pp. 243–261.

Ulaby, F. T. and J. McNaughton, 1975, Classification of physiography from ERTS imagery; Phtogram. Eng. and Rem. Sens., vol. 41, pp. 1019–1027.

Vander Lugt, A., 1964, Signal detection by complex spatial filtering; IEEE Trans. Info. Theory, vol. IT-10, pp. 139–145.

Vander Lugt, A., 1968, A review of optical data-processing techniques; Optica Acta, vol. 15, pp. 1–33.

Verber, C. M., 1980, An integrated optical circuit for performing vector subtraction; SPIE Proc. Devices and Systems for Optical Signal Processing, vol. 218.

Welch, R., 1977, Progress in the specification and analysis of image quality; Photogram. Eng. and Rem. Sens., vol. 43, no. 6.

Wellman, J. B., Technologies for the multispectral mapping of earth resources, 15th Sym. on Remote Sensing of the Environ., U. Mich., Ann Arbor May 1981.

Wernecke, S. J. and L. R. D'Addario, 1977, Maximum entropy image reconstruction; IEEE Trans. Computers, vol. C-26, p. 351.

Wintz, P. A., 1972, Transform picture coding; Proc. IEEE, vol. 60, p. 809.

Zelinski, R. and P. Noll, 1977, Adaptive transform coding of speech signals; IEEE Trans. Acoust. Speech Signal Proc., vol. AASP-25, p. 299.

BOOKS

Bozic, S. M., 1980, Digital and Kalman Filtering: An Introduction to Discrete Time Filtering and Optimun Linear Estimation; Wiley.

Anderberg, M. R., 1973, Cluster Analysis for Applications; Academic Press, New York.

Andrews, H. A., 1970, Computer Techniques in Image Processing; Academic Press, New York.

Andrews, H. A. and B. R. Hunt, 1977, Digital Image Restoration; Prentice-Hall, Inc., Englewood Cliffs, N.J.

Bennett, W. R., Electrical Noise; 1960, McGraw-Hill Book Co., New York.

Bernstein, R., ed., 1978, Digital Image Processing for Remote Sensing; IEEE Press, New York, IEEE Cat. No. 0-87942-106-1.

Bracewell, R., 1965, The Fourier Transform and its Applications; McGraw Hill, New York.

Brigham, E. O., 1974, The Fast Fourier Transform; Prentice-Hall Englewood Cliffs, N.J.

Castleman, K. R., 1979, Digital Image Processing; Prentice-Hall, Englewood Cliffs, N.J.

Gaskill, J. D., 1978, Linear Systems, Fourier Transforms and Optics. John Wiley and Sons, New York.

Goodman, J. W., 1968, Introduction to Fourier Optics; McGraw-Hill Book Co., New York.

Mikhail, E. M., 1976, Observations and Least Squares; IEP-Dun-Donnelly, New York.

Oppenheim, A. V. and R. W. Shaffer, 1975, Digital Signal Processing; Prentice-Hall, Englewood Cliffs, N.J.

Pratt, W. K., 1978, Digital Image Processing; John Wiley & Sons, New York.

Rabiner, L. A. and B. Gold, 1975, Theory and Applications of Digital Signal Processing; Prentice-Hall, Englewood Cliffs, N.J.

Rogers, G. L. 1977, Noncoherent Optical Processing; John Wiley and Sons, Inc.

Rosenfeld, A. 1969 (a), Picture Processing by Computer; Academic Press, New York.

Rosenfeld, A. and A. Kak, 1976, Digital Picture Processing; Academic Press, New York.

Rosove, P. E., 1967, Developing Computer Based Information Systems; John Wiley & Sons, New York.

Shannon, C. E. and W. Weaver, 1949, The Mathematical Theory of Information; University of Illinois Press, Urbana, Ill.

Schwartz, M., W. Bennett and S. Stein, 1966, Communication Systems and Techniques, Part II Pulse Modulation; McGraw-Hill Book Co., New York.

Slater, P. N., 1980, Remote Sensing Optics and Optical Systems; Addison-Wesley, Reading, MA.

Sorensen, H. W., 1980, Parameter Estimation, Marcel Dekker, Inc., pp. 17–26.

Tikhonov, A. and V. Arsenia, 1977, Solution of Ill Posed Problems; John Wiley & Sons, New York, No. 7.

Wiener, N., 1949, The Extrapolation, Interpolation and Smoothing of Stationary Time Series; John Wiley & Sons, New York.

Wilks, Samuel S., 1962, Mathematical Statistics; Wiley.

Young, J. F., 1971, Information Theory; Butterworth, London.

CHAPTER 18

Pattern Recognition and Classification

Author-Editors: ROBERT M. HARALICK and KING-SUN FU

GENERAL CONTENTS: Basic pattern recognition concepts. Principles of spectral discrimination. Economic consequences of decisions. The Bayes decision rule. The maximum decision rule. The Gaussian assumption. Feature selection. Principle components. Bhattacharyya distance. Syntactic pattern recognition applied to remote sensing problems. Recognition of clouds and shadows. References.

INTRODUCTION

BASIC PATTERN RECOGNITION CONCEPTS

Pattern recognition as used here refers to the automatic machine determination of salient patterns in remotely sensed image data. From the pattern-recognition perspective, the world to be sensed is composed of units defined by the sensor. For digital imaging sensors, as a first approximation, the units can be thought of as small non-overlapping areas on the ground: one such area for each picture element (pixel) in the image. The sensor makes an ordered set of measurements on each unit sensed. The ordered set of measurements is called a measurement vector or measurement pattern. Each value measured in this set is a number proportional to the energy received by the sensor in some band of the electromagnetic spectrum at some specified observation time. The basic pattern-recognition problem is first to automatically and consistently determine the information class or category of each distinct region on the ground using the set of sensor measurement-patterns and second to estimate the error rate for the automatically determined assignments.

Specific examples of pattern recognition for remote sensing applications include determining

1) tree-species composition in a forest
2) hot spots of incipient forest fires
3) natural vegetation cover-types
4) crop types
5) state of health or stressed vegetation
6) percent of sedimentation in a river or lake
7) percent of pollutant in a river or lake
8) geological formation and rock types
9) lineament patterns
10) degree of mineralization
11) number of small objects in a smooth background
12) urban land-use patterns

The automation of these tasks requires a corresponding variety of methods and techniques varying from simple to highly complex. It is the purpose of this chapter to describe the most commonly used techniques.

LITERATURE DEALING WITH PATTERN RECOGNITION CONCEPTS

Books describing the principles of pattern recognition have been written by Sebestyen (1962), Nilsson (1965), Arkadev and Braverman (1966), Fu (1968), Kanal (Ed.) (1968), Watanabe (Ed.) 1969), Mendel and Fu (1970), Fu (Ed.) (1971), Andrews (1972), Fukunaga (1972), Meisel (1972), Patrick (1972), Watanabe (Ed.) (1972), Chen (1973), Duda and Hart (1973), Ullman (1973), Tou and Gonzalez (1974), Batchelor (1974), Young and Calvert (1974), Fu and Whinston (Ed.) (1977), and Batchelor (1978). Some of these books have been reviewed and the reader might be interested in consulting the reviews listed in Table 18-1 before attempting to read any of these books.

Shorter reports and review articles include those by Nagy (1968), Ho and Aggrawala (1968), Fu, Landgrebe, and Phillips (1969), Casy and Nagy (1971), Nagy (1972), Kanal (1972), and Kanal (1974). Reprints of important pattern-recognition articles can be found in Sklansky (1973) and Aggrawala (1977). The May 1979 issue of the **IEEE Proceedings** was a special issue on pattern recognition and image processing. Journal papers on pattern recognition appear in the **IEEE Transaction on Computers, IEEE Transactions on Systems, Man and Cybernetics,** and **IEEE Transaction on Pattern Analysis and Machine Intelligence.** The Pattern Recognition Society publishes a journal called **Pattern Recognition.** Conference papers appear in the **International Joint Conference on Pattern Recognition, The Pattern Recognition and Image Processing Conference, The Purdue Symposium on Machine Processing of Remotely Sensed Data,** and the **Environmental Research Institute of Michigan Remote Sensing of Environment Conferences.**

SUMMARY RELATIVE TO PATTERN RECOGNITION CONCEPTS

To automate pattern recognition, we must define the classes of entities of interest, that is, the kinds of objects between which we must discriminate; we must choose instruments or sensors

TABLE 18-1

Listing of Various Books on Pattern Recognition Where They Have Been Reviewed

AUTHOR(S)	TITLE	WHERE REVIEWED
Harry Andrews	Introduction to Mathematical Techniques in Pattern Recognition; Prentice Hall, New Jersey, 1972, 504 pages.	IEEE Information Theory vol. IT-19, no. 6, November, 1973, p. 831
Richard Duda & Peter Hart	Pattern Classification & Scene Analysis; Wiley, New York, 1973, 482 pages.	IEEE Computer Transactions, vol. C-23, no. 2, February, 1974, p. 223
		IEEE Information Theory, vol. IT-19, no. 6, November, 1973, p. 827−829.
King-Sun Fu	Syntactic Methods in Pattern Recognition; Academic Press, New York, 1974, 397 pages.	IEEE Systems Man Cybernetics, vol. SMC-6, no. 8, August, 1976, p. 590.
Keinosuke Fukunaga	Introduction to Statistical Pattern Recognition; Academic Press, New York, 1972, 382 pages.	IEEE Systems Man Cybernetics, vol. SMC-4, no. 2, March, 1974, p. 238.
		IEEE Information Theory, vol. IT-19, no. 6, November 1973, pp. 829−830.
William Meisel	Computer-Oriented Approaches to Pattern Recognition; Academic Press, New York, 1972, 262 pages.	IEEE Systems Man Cybernetics, vol. SMC-3, no. 2, March, 1973, p. 209.
		IEEE Computer Transactions, vol. C-23, no. 1, January, 1974, p. 112.
		IEEE Computer Transactions, vol. C-22, no. 4, April, 1973, p. 429.
		IEEE Information Theory, vol. IT-19, no. 6, November, 1973, pp. 832−833.
Edward Patrick	Fundamentals of Pattern Recognition; Prentice Hall, New Jersey, 1972, 528 pages.	IEEE Systems Man Cybernetics, vol. SMC-3, no. 5, September, 1973, p. 528.
		IEEE Information Theory, vol. IT-19, no. 6, November, 1973, pp. 830−831.
Julius Tou and Rafael Gonzales	Pattern Recognition Principles, Addison-Wesley; Mass. 1974, 377 pages.	IEEE Systems Man Cybernetics, vol. SMC-6, no. 4, April, 1976, pp. 332−333.
		IEEE Information Theory, vol. IT-22, no. 5, September, 1976, pp. 632-633.
Jullian Ullmann	Pattern Recognition Techniques; Crane-Russak, New York, 1973, 412 pages.	IEEE Computer Transactions, vol. C-23, no. 2, February, 1974, pp. 220−222
		IEEE Information Theory, vol. IT-20, no. 3, May, 1974, p. 400.
Satosi Watanabe (Ed.)	Methodologies of Pattern Recognition; Academic Press, New York, 1969, 579 pages.	IEEE Information Theory, vol. IT-17, no. 5, September, 1971, pp. 633−634.

which can measure the environment in which the objects occur; and we must provide a methodology permitting the recognition of an object in the class of objects of interest from those not in the class of objects of interest. Using this methodology we also must construct a decision rule which will decide what kind of object a particular object is, on the basis of the measurements made from the observed small-area ground patches.

Defining the class of objects of interest might seem to be easy since it is an intrinsic part of the automation need. We will see, however, that it is not so easy since the sensor may not gather sufficient information to allow the discrimination to take place. In these cases we may be forced to define our classes as the more discriminable ones even though they may be of less interest to us. To help us do this we need to employ a clustering process which tells us what the naturally distinguishable classes are given the sensor's data.

Choosing the measuring instruments or sensors and designing a way to preprocess—to standardize, to normalize, and to extract the relevant information in its simplest form from the measurements—so that objects of interest can be simply recognized from those of non-interest (and so that each class or category of objects of interest has a particularly simple description in terms of the preprocessed measurements) are among the most difficult problems in pattern recognition. These problems are called feature-extraction- or preprocessing-problems and are concerned with presenting in some standard form only the simplest and most important information to the decision rule.

Finally, the problem in constructing a decision rule we call the pattern-discrimination problem. It is based on a probability model and it allows us to estimate the error rates of the automatic decision process.

Most pattern recognition of remotely sensed image data is done by processing each pixel's information separately or independently. This means that a category assignment is made to each pixel purely on the basis of its own information. Processing proceeds on a pixel-by-pixel basis over the entire image.

When the pixel's information consists only of the sensor measurement-pattern obtained from one observation time, the measurement pattern is called a multispectral feature-vector and the kind of pattern recognition is called multispectral pattern-recognition. When items of spectral information from more than one observation time for the same ground area are stacked in the same measurement-pattern vector, this kind of pattern recognition is called multispectral-multitemporal pattern-recognition. When the measurement pattern for each pixel contains spectral information from its associated ground area as well as from neighboring ground areas, or when the decision rule which makes category assignments uses the information from a pixel and some of its neighboring pixels, the pattern recognition is called spatial pattern recognition, or spatial-spectral pattern recognition.

PRINCIPLES OF SPECTRAL DISCRIMINATION

In order to understand the pattern-discrimination methodology consider a simplified example. Suppose that there are three types of surface-cover material: vegetation, soil, and water. Suppose further that each of these has a unique spectral response which does not vary with season, atmospheric haze, sun-angle etc. Let these be the responses shown in Figure 18-1. Now select two wavelengths λ_1 and λ_2 for a remote sensor to make some measurements. Then, for each surface-cover category, use wavelengths λ_1 and λ_2 to determine its spectral measurement pattern. Plot these in measurement space as shown in Figure 18-2. Since they obviously plot in areas that are nicely separated from each other we would expect no difficulty in designing a decision rule to recognize these categories. Any time a new measurement pattern needs to be assigned to a category we see if it lies as the point in measurement space associated with vegetation, or soil, or water. If it does, we assign it to the corresponding category. If it doesn't we assign it to an unknown category.

In reality, the spectral response patterns from these surface categories as well as others vary due to natural random variations, systematic seasonal causes, atmospheric haze, etc. There is not a unique measurement pattern associated with each category. Rather, associated with each category is a probability distribution indicating, for any measurement pattern, the relative frequency of occurrence that may arise from a ground area of the given category.

If, using some training data, we plotted five observations of each of three vegetation categories, viz. soybean, corn and wheat, we might obtain the measurement-space plot of Figure 18-3. To assign a new measurement pattern, v, to one of the classes is now not such an easy problem. In essence we must use our training observations to estimate for each new measurement pattern v, the probability that soybeans, corn or wheat is its true category. If we can do that we can associate with each measurement that category having the highest conditional probability given the measurement. In effect, this association partitions measurement space as shown in Figure 18-4. Since our new measurement pattern is in the part of measurement space associated with soybeans the decision rule assigns it to the soybean class.

The procedure by which the measurement space of Figure 18-3 was partitioned is simple. Use the training data for each class to determine the class sample-mean. Then partition the measurement space so that each class has associated with it all the measurement patterns closest to its

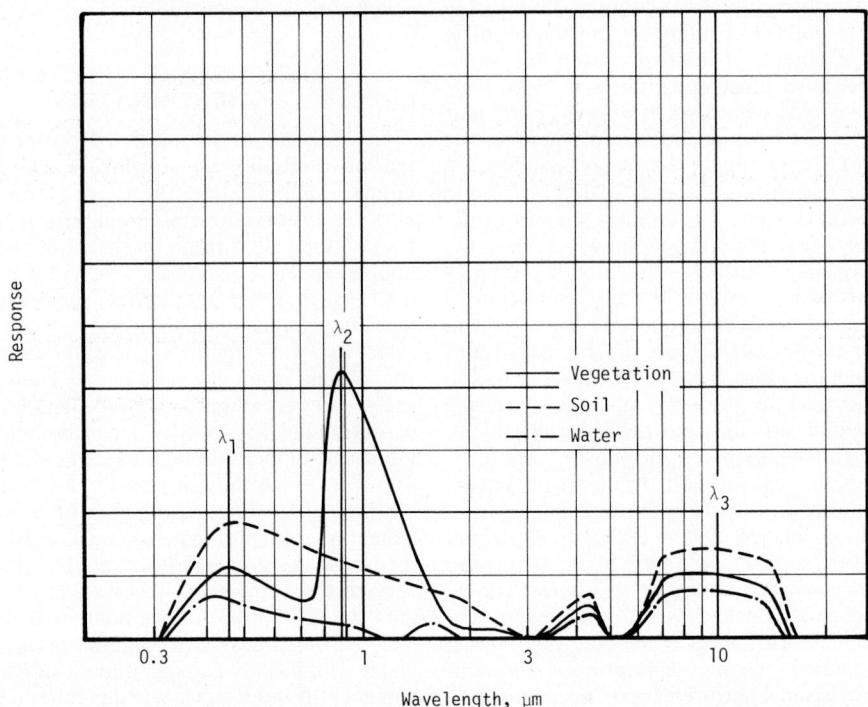

Fig. 18-1. Typical relative response curves for different materials, illustrating the possibility of discrimination by comparison of the curves at different wavelengths. Source: Landgrebe (1972b).

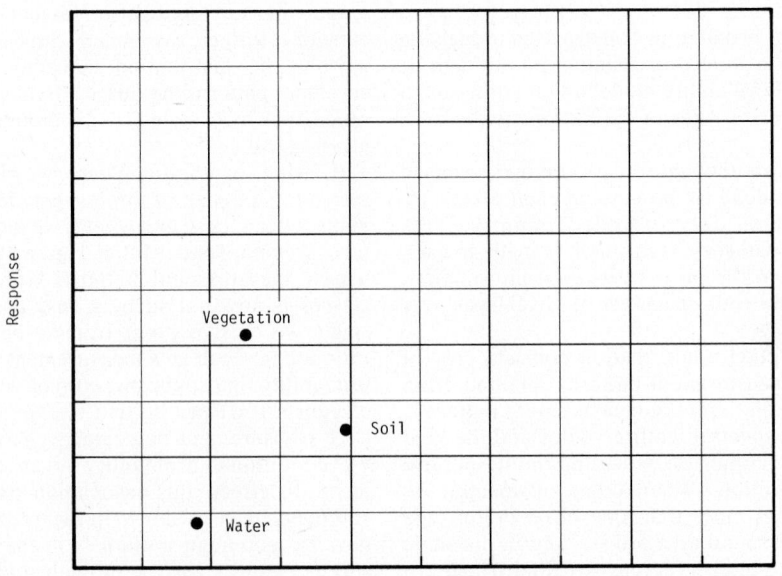

Fig. 18-2. The categories vegetation, soil, and water have distinct responses on wavelengths $\lambda 1$ and $\lambda 2$. Shown in this figure are these categories plotted in a measurement space whose axes are their $\lambda 1$ and $\lambda 2$ responses.

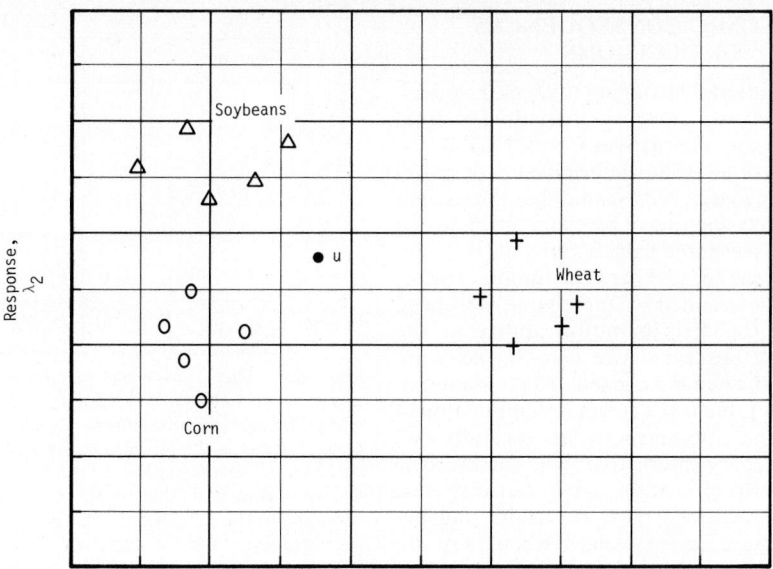

Fig. 18-3. A given material will not always have the exactly same response in a group of samples, but each material tends to cluster together. A typical two-dimensional sampling of three materials is shown. Source: Landgrebe (1972).

sample mean. Unfortunately, without a probability model we cannot say that this procedure is the one that yields the lowest error rate or maximizes any utility function. However, there is a probability model under which this is the appropriate thing to do.

It is the purpose, therefore, of the next sections to develop a probabilistic-decision theoretic-model for pattern discrimination which suggests techniques for decision-rule construction having certain optimal properties which we can measure in terms of utility or economic consequences.

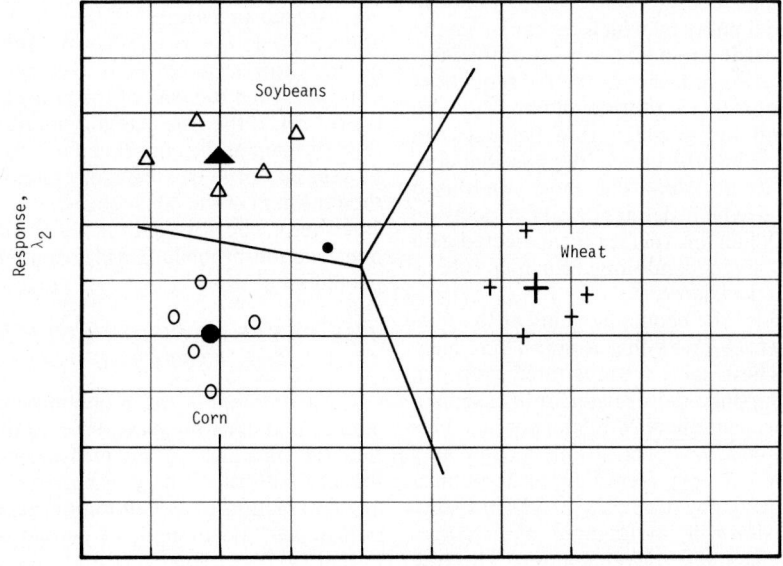

Fig. 18-4. Division of two-dimensional sampling space into domains assigned to different materials. In this case the unknown point u is considered to be soybean because of its location in the sampling space. Source: Landgrebe (1972).

ECONOMIC CONSEQUENCES
OF DECISIONS

For each pattern d belonging to D, $d \epsilon D$, a decision rule f assigns a category alternative c^k from the set of category alternatives $C = \{c^1, \ldots, c^K\}$. The assignment may be deterministic or probabilistic. In any case, we assume that the assignment by the decision rule of category alternative c to a pattern d measured from a unit u carries economic consequences. These economic consequences are determined by the people who need to automate the discrimination ability of the trained human observer. The consequences are generally good when the chosen category alternative c is, in fact, the true category identification of the unit u. The consequences are generally bad when the category alternative c is not the true category identification of the unit u. Because such identification decisions must be made, and because they cause consequences when they are made, we may view the goal of decision-rule construction as the construction of a decision rule which in some sense maximizes the good consequences.

To speak of maximizing good consequences implies that we must have some numerical measure indicating the economic gain or loss of the consequence when the decision rule assigns category c^i to a unit u with measurement d when the true category identification of unit u is category c^j. Let $e(c^j, c^i)$ be the net worth or economic gain of such a consequence. In general, $e(c^i, c^i)$ will be positive signifying a gain for a correct identification, and $e(c^i, c^j)$, for $i \neq j$, will be negative, signifying a loss for an incorrect identification.

In determining a decision rule, we must choose a criterion of optimality by which we can judge the worth of the decision rule on the basis of the various economic gains or losses of the consequences (c^i, c^j). The optimality criterion defines how to judge how well the decision rule balances, in terms of these gains and losses, the possible consequences of its decision. The most often-used criterion is one which defines the best decision rule to be one which maximizes the expected gain under certain given conditions. Such a rule is called a Bayes decision rule.

Let us consider the economic gains of the possible consequences given that a unit u has measurements d. These gains are illustrated simply in Figure 18-5. Suppose the decision rule assigns a unit u having measurements d to category c^i. This assignment, at best, however, is only an educated guess; the true category identification for unit u can actually be any one of $c^1, c^2, \ldots, c^K$. In Figure 18-5 the decision-rule assignment of c^i corresponds to a selection of the ith column. The true category identification of unit u corresponds to a selection of some row. This row, intersected with the ith column, yields an entry which is the economic gain consequence.

The question of concern is how often will the

Fig. 18-5. This shows the economic gains obtained under various alternatives conditioned on the measurement d being made of a unit K. Given that the observed measurement is d, the probability that nature chooses category c^j, corresponding to the jth row, is $P_d(c^j)$. The decision rule will choose some category c^K, corresponding to the Kth column. The result of nature choosing category c^j and the decision rule choosing category c^K is the economic consequence (c^j, c^K).

true category identification of a unit u be category c^j when the unit u has measurement d. We denote by $P_d(c^j)$ the probability of the true category identification of a unit u being in category c^j given that the unit u has measurements d. It is these conditional probabilities which can be estimated from the training data or ground-observation data.

The decision rule has no information regarding the true category identification of any unit. It only knows that the unit gives rise to a pattern d and that it has available estimates of the conditional probabilities $P_d(c^k)$, $k = 1, 2, \ldots, K$. The decision rule must assign the unit to a category, say c^i. This corresponds to a section of the ith column. For this course of action a number of different consequences can occur. If the true category identification is c^1, then the gain of the consequence (c^1, c^i) is $e(c^1, c^i)$. If the true category identification is c^2, then the gain of the consequence (c^2, c^i) is $e(c^2, c^i)$. In general, if the true category identification is c^j, then the gain of the consequence (c^j, c^i) is $e(c^j, c^i)$. The next section discusses a decision-rule construction-procedure which maximizes the expected gain.

THE BAYES DECISION RULE MAXIMIZES
EXPECTED GAIN

Let $f_d(c)$ denote the probability that the decision rule assigns the category c to the unit, given that the unit has pattern measurement d. Since, for any pattern d, there is no reason to suppose any interaction or collaboration between nature, (which may be thought of as choosing the true category identification) and the pattern discriminator, (which may be thought of as employing the decision rule to assign categories) we may assume that nature and the pattern discrimination are statistically independent. Thus, the probability that the unit has measurements d and the decision rule assigned the category c^k to the unit and the true category identification for the unit is c^j may be written as $f_d(c^k)P_d(c^j)P(d)$. Therefore, the expected gain for the decision rule f may be expressed by

$$E[e;f] = \sum_{d\epsilon D} \sum_{j=1}^{K} \sum_{k=1}^{K} e(c^j,c^k)f_d(c^k)P_d(c^j)P(d).$$

$$(18\text{-}1)$$

To see how to find the decision rule which maximizes the expected gain, we rewrite the expression for $E[e;f]$ as

$$E[e;f] = \sum_{d\epsilon D} P(d) \sum_{k=1}^{K} f_d(c^k) \sum_{j=1}^{K} e(c^j,c^k)P_d(c^j).$$

$$(18\text{-}2)$$

$P(d)$, being the probability of measuring pattern d for a unit, is non-negative. Hence $E[e;f]$ will be maximized (maximum taken over all f) if and only if for each $d\epsilon D$ the expected gain given d using f is maximized; that is,

$$E[e|d;f] = \sum_{k=1}^{K} f_d(c^k) \sum_{j=1}^{K}$$

$$e(c^j,c^k)P_d(c^j) \text{ is maximized.} \quad (18\text{-}3)$$

Since $\sum_{k=1}^{K} f_d(c^k) = 1$ and $f_d(c^k) \geq 0$, $k = 1,2,\ldots,K$, it is easy to see that the maximum of the above expression is

$$i = 1,2,\ldots,K \sum_{j=1}^{K} e(c^j,c^i)P_d(c^j) \quad (18\text{-}4)$$

and the decision rule f will certainly achieve this maximum if

$$f_d(c^i) = \begin{Bmatrix} 1, i = k \\ 0, i \neq k \end{Bmatrix} \text{ where } k \text{ is any index such that}$$

$$\sum_{j=1}^{K} e(c^j,c^k)P_d(c^j) \geq \sum_{j=1}^{K}$$

$$e(c^j,c^i)P_d(c^j), i = 1,2,\ldots,K. \quad (18\text{-}5)$$

In this case the optimal decision rule can be deterministic if the index k is unique or it can be either deterministic or probabilistic if k is not unique. Any optimal decision rule is called a Bayes rule.

For example, suppose there are three categories c^1, c^2, and c^3 with conditional probabilities and economic gains for the various alternatives and consequences as shown in Figure 18-6. The optimal decision rule will assign the unit u to category c^3 since the average gain for row 3 is 5/6 which is larger than the average gain for row 1 which is $-1/3$ or for row 2 which is ½.

	c^1	c^2	c^3
$P_d(c^1) = 1/6$ c^1	4	−2	0
$P_d(c^2) - 1/2$ c^2	0	1	0
$P_d(c^3) = 1/3$ c^3	−1	0	3

Fig. 18-6. Illustrates the economic gains for an example problem where the pattern measurements d are made on a unit and there are three possible categories.

BAYES DECISION RULES AND CATEGORY PRIOR PROBABILITIES

It is often the case that the conditional probabilities $P(c/d)$ are not known but that the conditional probabilities $P(d/c)$ of the measurements, given the categories, are known. Fortunately, there is a well known relationship between $P(c/d)$ and $P(d/c)$ which involves the prior probabilities of $P(d)$ and $P(c)$ of the measurements and categories, respectively.

By the definition of conditional probability, we may express $P_d(c)$ by

$$P_d(c) = \frac{P_c(d)P(c)}{P(d)} \quad (18\text{-}6)$$

so that the average gain obtained by the use of decision rule f may be rewritten as

$$E[e;f] = \sum_{d\epsilon D} \sum_{k=1}^{K} \sum_{j=1}^{K} f_d(c^k)e(c^j,c^k)P_{c^j}(d)P(c^j).$$

$$(18\text{-}7)$$

$E[e;f]$ is maximized if and only if for each $d\epsilon D$, the gain conditioned on d,

$$E[e|d;f] = \sum_{k=1}^{K} f_d(c^k) \sum_{j=1}^{K} e(c^j,c^k)P_{c^j}(d)P(c^j)$$

$$(18\text{-}8)$$

is maximized. The maximum value of $E[e|d;f]$ is

$$\sum_{j=1}^{K} e(c^j,c^k)P_{c^j}(d)P(c^j) \quad (18\text{-}9)$$

where k is some index for which

$$\sum_{j=1}^{K} e(c^j,c^k)P_{c^j}(d)P(c^j) \geq \sum_{j=1}^{K}$$

$$e(c^j,c^i)P_{c^j}(d)P(c^j), i = 1,2,\ldots,K. \quad (18\text{-}10)$$

An optimal deterministic decision rule f may therefore be defined by

$$f_d(c^i) = 1 \; i = k \atop \quad\;\; = 0 \; i \neq k \; \Bigg| \; \text{where } k \text{ is any index such that}$$

$$\sum_{j=1}^{K} e(c^j, c^k) P_{c^j}(d) P(c^j) \geq \sum_{j=1}^{K}$$

$$e(c^j, c^i) P_{c^j}(d) P(c^j), \; i = 1, 2, \ldots, K. \quad (18\text{-}11)$$

Note the strong dependence which f has on the category probability $P(c)$. Because of this, any time we define an optimal Bayes decision rule, we must state that it is optimal only relative to the category prior-probability function $P(c)$.

MAXIMIN DECISION RULE

Figure 18-7 illustrates the expected gain of a Bayes decision rule in a two-category classification problem with the identity gain function. Selecting a value of prior probability, the corresponding value of expected gain is the highest expected gain achievable by any decision rule. Therefore, use of any decision rule which is not a Bayes rule is guaranteed to perform below the curve. In particular, if a Bayes rule is used in a new situation where the encountered prior probability function differs from the one employed in the design, then the Bayes rule is not optimal in the new situation.

Recognizing this, a conservative decision-rule designer will attempt to construct a decision rule which maximizes the smallest gain achieved by the decision rule under some encountered prior probability function. It turns out that the expected gain for a decision rule which maximizes the smallest gain as the encountered prior probability function varies has a value equal to the smallest possible Bayes gain (the lowest point on the curve

of Figure 18-7). This value is the same regardless of the actually encountered prior probability functions. This kind of decision rule is called a maximin decision rule since it maximizes the minimum expected gain. In general, it is not a deterministic decision rule and designing a maximin decision rule is equivalent to solving a large linear-programming problem.

THE GAUSSIAN ASSUMPTION

The conditional probability $P_c(d)$ of the data-measurement vector d given the category c plays an essential role in decision rule determination. $P_c(d)$ could be stored as a table. However, because of the large number of possible data vectors, $P_c(d)$ is often represented as a parametric function, the parameters being the category mean measurement μ_c and its covariance matrix Σ_c. The simplest probability density-function having these parameters is the Gaussian one which is defined by

$$P_c(d) = \frac{1}{(2\pi)^{N/2} |\Sigma_c|^{1/2}} e^{-1/2(d - \mu_c)' \; \Sigma_c^{-1}(d - \mu_c)}$$

$$(18\text{-}12)$$

where N is the dimension of the measurement vector d.

In the case of identity-gain function e and equal probabilities for the prior $P(c)$ for each category c, the Bayes decision rule assigns measurement d to that category c minimizing

$$\log |\Sigma_c| + (d - \mu_c)' \; \Sigma_c^{-1}(d - \mu_c) \quad 18\text{-}13$$

This kind of decision rule is sometimes called quadratic or piecewise quadratic because the de-

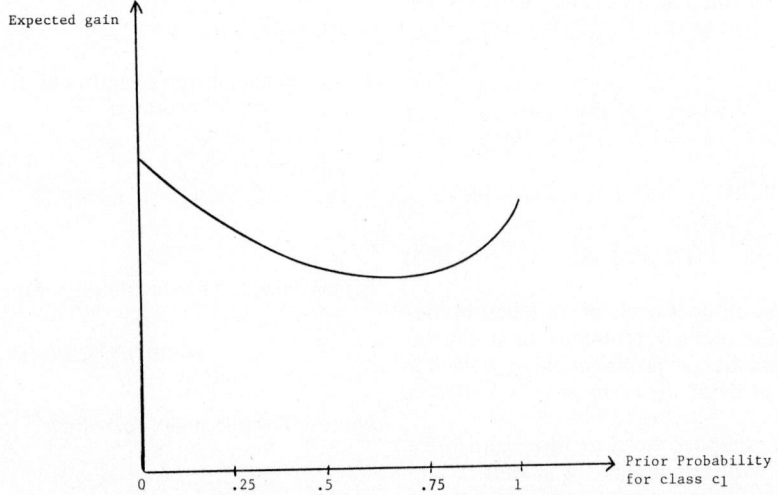

Fig. 18-7. Illustrates how the expected gain of a Bayes decision rule can vary with a change in prior probability for class c_1 in some two class example problem. Notice that as the prior probability for class c_1 becomes 1, the prior certainty reflects itself in an a posteriori certainty which makes the expected gain high. When the prior probability for class c_1 becomes 0, the prior probability for class c_2 becomes 1 and the situation is similar. For class c_1 prior probabilities between 1 and 0, the prior situation is less certain and the expected gain must be less than the end cases. The shape of the function is guaranteed to be convex.

cision boundaries they form in measurement space are piecewise hyperquadratic boundaries.

In case the covariance matrices for all categories are equal, the Bayes decision rule reduces to assigning the measurement d to that category c minimizing the Mahalanobis distance

$$(d - \mu_c)' \sum^{-1} (d - \mu_c) \qquad 18\text{-}14$$

between the measurement vector d and the category mean vector μ_c for category c. This decision does simplify to a linear decision rule. Assign measurement d to that category c maximizing

$$(\mu'\Sigma^1)d - \left(\frac{\mu_c'\Sigma^{-1}\mu_c}{2} \right) \qquad (18\text{-}15)$$

By precomputing the terms in parentheses, the number of multiply and add operations for this decision rule is only $N + 1$ per category, a significant saving over the quadratic rule, especially when the dimensionality N is large.

FEATURE SELECTION

Multitemporal multispectral remotely sensed imagery can produce a ten- or twenty-dimensional data vector for each pixel. The data have inherent redundancies and processing all of the data or storing all of them may not be cost effective. Feature-selection procedures are used to select those dimensions most suitable for processing.

There are two kinds of feature selections depending on whether the classes and their statistics are known or not known. If they are not known, the best feature-selection procedure is called principal components. If they are known, the easiest-to-use feature selection is based on Bhattacharyya distance.

PRINCIPAL COMPONENTS

Principal components is a standard statistical technique for selecting that subspace of given dimension in which the most data variance lies. If $x_1, \ldots, x_N$ are the sample data vectors, μ the sample-mean vector, and $\sum$ the sample covariance-matrix, the best K dimensions in which to project the data would be that K-dimensional subspace spanned by the K eigenvectors of Σ having the largest eigenvalue. Thus if T is a matrix whose K rows are these eigenvectors, the K principal components of $x_1, \ldots, x_N$ is $Tx_1, \ldots Tx_N$, each Tx_N being a K-dimensional vector.

BHATTACHARYYA DISTANCE

The Bhattacharyya distance is a measure of the separability between two classes. For two Gaussian classes having means and covariances μ_1, Σ_1 and μ_2, Σ_2 respectively, the Bhattacharyya distance is given by

$$\frac{1}{8} (\mu_1 - \mu_2)' \left(\frac{\Sigma_2 + \Sigma_2}{2} \right)(\mu_1 - \mu_2)$$

$$+ \frac{1}{2} \ln \frac{\left| \dfrac{\Sigma_1 - \Sigma_2}{2} \right|}{|\Sigma_1|^{1/2}|\Sigma_2|^{1/2}} \qquad (18\text{-}16)$$

To use this distance measure for selecting the best K features from the original N dimensions on an L-class problem, the Bhattacharyya distance needs to be calculated between each of the $L(L - 1)/2$ pairs of classes for each of the $\binom{K}{N}$ possible ways of choosing K features from N dimensions. The K dimensions which are best are those K dimensions whose sum of the Bhattacharyya distances between the $L(L-1)/2$ pairs of classes is highest. The Bhattacharyya distance between a pair of classes for a selection of K dimensions out of N dimensions is calculated using the mean and covariance matrix in the selected K dimensions.

SYNTACTIC PATTERN RECOGNITION APPLIED TO REMOTE SENSING PROBLEMS

GENERAL APPROACH

The approach of using hierarchical structures and grammar rules to describe the structures of pattern has recently received increasing attention (Fu, 1974). This approach is often called the structural or syntactic approach to distinguish it from the decision-theoretic or statistical approach. Practical applications include the description of chromosome images, the recognition of characters, spoken digits, electrocardiograms, and two-dimensional matehmatical expressions, the identification of bubble chamber- and spark chamber-events, and the recognition of fingerprint patterns (Fu, 1978). In the syntactic approach, each pattern is described in terms of its parts, i.e., subpatterns. Each subpattern can again be described in terms of its parts. The simplest subpatterns are called the pattern primitives, and they constitute the basic symbols (the set of terminals) of the pattern language. The description of each primitive can be either deterministic or statistical and the recognition of primitives is often based on the decision-theoretic approach. Each class of patterns is now described by a set of sentences consisting of the primitives, and it can be generated by a pattern grammar. With the above description, it might be said that in the syntactic approach we often use the decision-theoretic approach for primitive recognition; however, the emphasis will be on the use of syntactic rules to describe the structure of patterns (the compositions rules of the primitives and subpatterns).

Multispectral signals measured by Landsat over Marion County (Indianapolis), Indiana were analyzed using clustering analysis. Fourteen clusters were found and the data from the urban

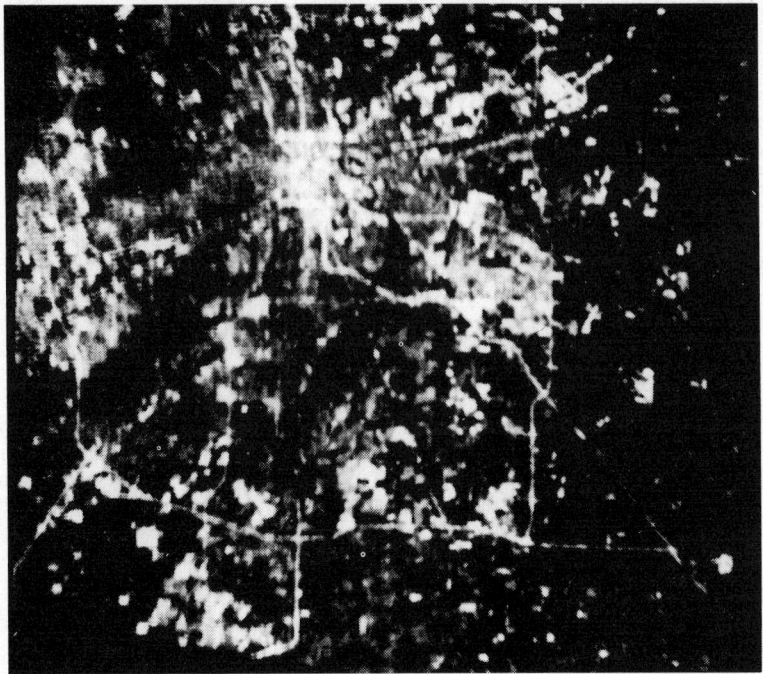

Fig. 18-8. Photograph of Marion County imagery from digital display.

area within the scene were accordingly classified using a Bayes classifier. The result of the Bayes classifier (Figure 18-8) provides the basic pattern primitives. Some manual preprocessing based on these dependencies was used to improve the accuracy of the classification. A hierarchical graph model for these relationships can be constructed as shown in Figure 18-9. Obviously, there are spatial dependencies among the various classes.

The hierarchical graph model shown in Figure 18-9, was constructed directly from observations of the classified data and of aerial photographs of the corresponding region. For simplicity, there are some relationships between the entities in the figure which have not been included; for example, the fact that the SCENE is made up of the EARTH and CLOUD PAIRs (i.e., clouds and shadows). The CLOUD PAIRs obscure the EARTH, so a relation O for "obscures" could be shown linking CLOUD PAIRs to EARTH. Also, if a pair of entities are related, then their descendants are also related. However, these relations are shown only at the level at which they first occur. The form of this diagram is the same as the derivation diagram for a web grammar.

This scene consists, at the highest level, of the EARTH obscured by CLOUD PAIRs. Each CLOUD PAIR consists of a CLOUD and a SHADOW, related by a distance-and-angle R. A CLOUD consists mostly of points classified as clouds (blank) but also points classified as concrete (X) and as suburban (S). This confusion seems to arise because both concrete and clouds are highly reflective. The suburban class is a

mixture of concrete and grass. A SHADOW tends to consist mostly of points classified as shadows (★) but also points classified as commercial (C) and inner city (I). The confusion here seems to occur because the commercial and inner-city classes consist largely of asphalt rooftops with low reflectance.

The EARTH consists of URBAN and RURAL areas. The RURAL area consists of open grassy (O) and wooded (W) areas. The URBAN area consists of the DOWNTOWN area, surrounded by the INNER CITY area, with nearby SUBURBAN areas and a system of HIGHWAYS. The DOWNTOWN area is characterized by the fact it contains the largest concentration of commercial land use. The INNER CITY area surrounds the DOWNTOWN and contains a high concentration of inner city points. The SUBURBAN and HIGHWAY areas are both near the DOWNTOWN and contain mostly suburban-classified points. They are distinguished by the fact that HIGHWAYS occur in linear patterns. This model is now used to guide the analysis of the picture. In essence this analysis is an attempt to verify the model and to make it more specific. Each subentity of the picture and each relationship can now be elaborated and tested separately.

RECOGNITION OF CLOUDS AND SHADOWS

As pointed out in the previous section, clouds and shadows are characterized by the fact that a cloud is a bright area which has associated with it

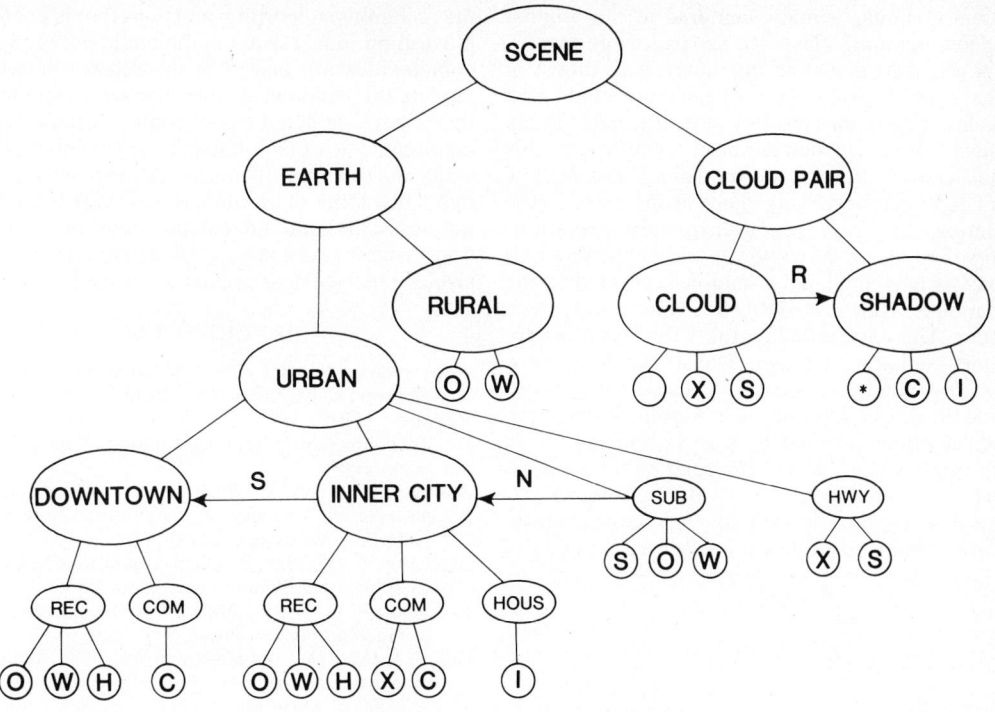

Fig. 18-9. A hierarchical graph model of the scene in Fig. 18-8.

a congruent dark area (the shadow) at a certain specific distance and orientation. It is found that the bright areas are generally classified by the pointwise classifier as clouds, but sometimes they are mistakenly classified as concrete or suburban. The dark areas are usually classified as shadows but sometimes as commercial or inner city. Examples can be observed in Figure 18-8.

If the relationship R between clouds and shadows is not known from some other source such as sun angle and height of clouds, the analysis process must determine R by some type of cross-correlation operation. For one-dimensional patterns, this process of finding similar patterns that are an arbitrary number of symbols away in the pattern can be modelled by a context-sensitive string operation. This is almost certainly also a context-sensitivie problem in a web system for two-dimensional patterns such as clouds and shadows. If the relation R is known from other sources the processing is simplified. Shape matching of a cloud and its partner shadow can be performed to confirm that they actually are pairs. This process is similar to the one-dimensional problem of being able to recognize all strings of the form ww^R (where w^R is w reversed). This is known to be a context-free operation in a string system.

Finally, an even simpler type of recognition would be to check a finite radius around a given cloud point. If there are more cloud points, this verifies the classification of the original point.

Then an equivalent area at a distance given by the relation R can be searched. This process is illustrated in Figure 18-10. Since the radius r of the area searched is finite, the process is essentially finite state.

In the case of the clouds and shadows the simplest possible algorithm was tested first. The picture is scanned left-to-right and top-to-bottom. Whenever a shadow point is encountered, the translater searches a finite window at a distance and angle away given by relation R. If a cloud point is found, the pair qualifies as a cloud-shadow pair and neighboring ★, X, and S points within the window around the cloud are also interpreted as cloud points. Likewise if the pattern qualifies as a pair, neighboring C and I points

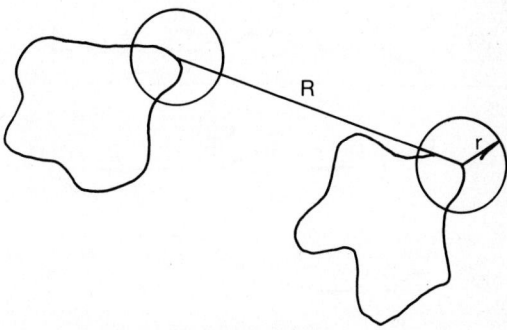

Fig. 18-10. Finite State Recognition of Clouds.

within a similar window centered at the original shadow point are classified as shadow points.

A grammar depicting this analysis is shown in Figure 18-11. Rules (1)–(4) generate cloud and shadow points that are not part of a pair. These points will be regarded as noise, not as true clouds and shadows. Rule (5) shows that a cloud-shadow pair (CP) can occur anywhere in the picture (the relation ''a'' represents ''arbitrary'' relationships). Rule (6) shows that a cloud-shadow pair consists of a cloud and a shadow separated by the relation R. Rules (7)–(10) show how a cloud can occur. The corresponding rules for a shadow are similar and are not shown. Rules (7) and (8) show that any cloud must contain at least one ★ point. Rule (9) shows that once one ★ point is detected, then any point labelled ★, X or S occurring within the window (i.e., within the relation w) is classified as a cloud point. Rule (10) terminates the search when the entire window has been scanned.

This grammar models the essential features of the recognition algorithm and therefore is good for illustration, but it is not complete in detail. Some complexities are buried in the relation w which models the window. A grammar which modelled this window in detail would contain a sequence of counting states that stimulate the scanning of the window. This more detailed grammar would contain a few more nonterminals and rules but would not better illustrate the complexity of the recognition process. As long as the window is of finite size, a regular-linear grammar can be found.

REFERENCES

Aggrawala, A. K., (Ed.), 1977, Machine Recognition of Patterns; IEEE Press, New York.

Andrews, Harry, 1972, Introduction to Mathematical Techniques in Pattern Recognition; Prentice Hall, New Jersey.

Arkadev, A. G., and E. M. Braveman, 1966, Computer and Pattern Recognition; Thompson Book Company Inc., Washington, D.C.

Batchelor, B. G., 1974, Practical Approaches to Pattern Classification; Plenum Press, New York.

Batchelor, B. G., (Ed.), 1978, Pattern Recognition Ideas in Practice; Plenum Press, New York.

Casey, Richard A., and George Nagy, 1971, Advances in pattern recognition; Scientific American, vol. 224, no. 4, April, pp. 56–71.

Chen, C. H., 1973, Statistical Pattern Recognition; Hayden, Washington, D.C.

Duda, Richard, and Peter Hart, 1973, Pattern Classification and Scene Analysis; Wiley, New York.

Fu, K. S., 1968, Sequential Methods in Pattern Recognition and Machine Learning; Academic Press, New York.

Fu, K. S., D. A. Landgrebe, and T. L. Phillips, 1969, Information processing of remotely sensed agricultural data; Proceedings of the IEEE, vol. 57, no. 4, April, pp. 639–653.

Fu, K. S., (Ed.), 1971, Pattern Recognition and Machine Learning; Plenum Press, New York.

Fu, K. S., 1974, Syntactic Methods in Pattern Recognition; Academic Press, New York.

Fu, K. S., and A. B. Whinston, 1977, Pattern Recognition Theory and Application; Noordhoff-Leyden, Netherlands.

Fukunaga, Keinosuke, 1972, Introduction to Statistical Pattern Recognition; Academic Press, New York.

Ho, Y. C., and A. K. Aggrawala, 1968, On pattern classification algorithms introduction and survey; Proceedings of the IEEE, vol. 56, no. 12, December, pp. 2101–2114.

Kanal, L. N. (Ed.), 1968, Pattern Recognition; Thompson Book Company, Washington, D.C.

Kanal, L. N., 1972, Interactive pattern analysis and classification system: a survey and commentary; Proceedings of the IEEE, vol. 60, no. 11, October, pp. 1200–1215.

Kanal, L. N., 1974, Patterns in pattern recognition: 1968–1974; IEEE Transactions on Information Theory, vol. IT–20, no. 6, November, pp. 697–722.

Meisel, William, 1972, Computer-Oriented Approaches to Pattern Recognition; Academic Press, New York.

Mendel, J. M., and K. S. Fu, (Eds.), 1970, Adaptive, Learning and Pattern Recognition Systems: Theory and Applications; Academic Press, New York.

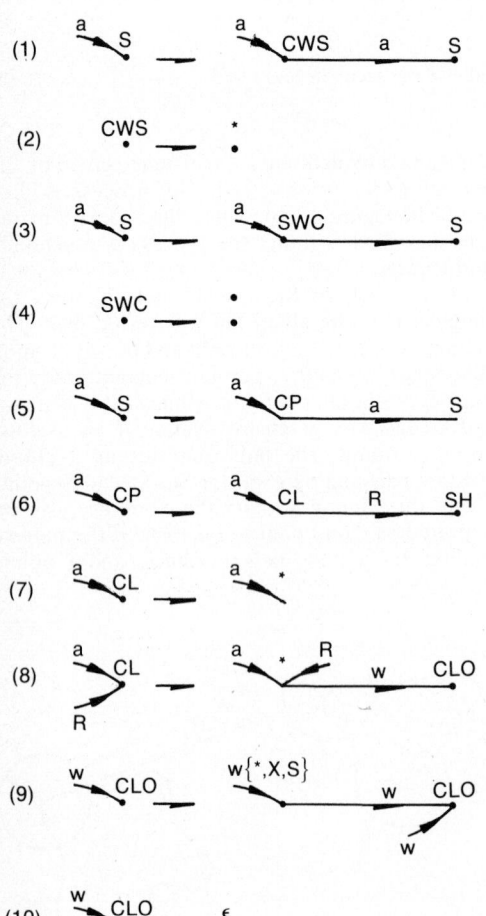

Fig. 18-11. The Cloud/Shadow Grammar.

Nagy, George, 1968, State of the art in pattern recognition; Proceedings of the IEEE, vol. 56, no. 5, May, pp. 836–862.

Nagy, George, 1972, Digital image-processing activities in remote sensing for earth resources; Proceedings of the IEEE, vol. 60, no. 10, October, pp. 1177–1199.

Nilsson, N. J., 1965, Learning Machines; McGraw-Hill, New York.

Patrick, Edward, 1972, Fundamentals of Pattern Recognition; Prentice Hall, New Jersey.

Sebestyen, G. S., 1962, Decision Making Processes in Pattern Recognition; The Macmillan Company, New York.

Sklansky, J., (Ed.), 1973, Pattern Recognition, Introduction and Foundation; Dowden, Hutchinson and Rose Inc., Pennsylvania.

Tou, Julius, and Rafael Gonzalez, 1974, Pattern Recognition Principles; Addison-Wesley, Mass.

Ullman, Jullian, 1973, Pattern Recognition Techniques; Crane-Russak, New York.

Watanabe, Satosi, (Ed.), 1969, Methodologies of Pattern Recognition; Academic Press, New York.

Watanabe, Satosi, 1972, Frontiers of Pattern Recognition; Academic Press, New York.

Young, T. Y., and T. W. Calvert, 1974, Classification, Estimation, and Pattern Recognition; American Elsevier, New York.

CHAPTER 19

Remote Sensing Software Systems

Author-Editor: PETER A. BRACKEN

Contributing Authors: WILLIAM B. GREEN AND ROBERT M. HARALICK

GENERAL CONTENTS: Evolution of Remote Sensing Software Systems; changing needs; software requirements; Design of Remote Sensing Software Systems: current software trends; batch versus interactive processing; user-system interface; software structure. Present Large Scale Computer Software Systems: ORSER; LARSYS; ERIPS; VICAR. Present Minicomputer Software Systems: IDIMS; KANDIDATA; GIPSY; MINI-VICAR; AOIPS. Future Trends. References.

INTRODUCTION

Unlike most aspects of remote sensing, very few texts are available on the subject of remote sensing software systems. For this reason, the emphasis in this chapter has been placed on fundamental considerations for designing remote sensing software systems, on the characteristics of existing large and small scale computer based software systems, and on future trends which will significantly influence the development of remote sensing software during the 1980's.

For the purposes of this chapter, a software system is considered to be a unified collection of computer programs which are designed to perform calculations or data processing tasks in support of remote sensing applications. The classes of computer programs which are used to develop, debug, maintain or execute other computer programs are part of a computer's operating system software and are not discussed extensively in this chapter.

The first section reviews the evolution of remote sensing software and explores changing needs for software capabilities. The second section focuses on current trends in design of remote sensing software.

The remainder of the chapter examines present remote sensing software systems and the impact of technology developments on the design of future systems. Discussions cover software design considerations, design features and capabilities of representative systems, and software transportability. The chapter ends with a discussion of the impact of evolving computer hardware and future trends toward wider software sharing and improved data management system capabilities.

Thanks are extended to Terry Phillips and Jan Owings for supplying the LARSYS and VICAR system figures. The authors are indebted to H. K. Ramapriyan, B. F. Merembeck, J. T. Dalton and F. Gordon for their many suggestions during the preparation of materials on current software systems. The help and extra effort devoted by Grace J. Dodson in typing and preparing the manuscript is gratefully acknowledged.

EVOLUTION OF REMOTE SENSING SOFTWARE SYSTEMS

By 1965, computer aided analysis of multispectral images became feasible. Until then, most remote sensing specialists used photographs from multispectral cameras and manual photo interpretation to recognize scene features and extract required information. The appearance of multi-channel aircraft scanners in the mid 60's, along with rapid advances in digital computer systems technology and pattern recognition analysis techniques, provided the catalysts for rapid development of advanced remote sensing software systems.

Software systems such as the LARSYS, VICAR, ORSER and ERIPS systems, described later in this chapter, evolved flexible capabilities for correcting, registering, enhancing, transforming, and classifying digital image data. Early versions of such systems demonstrated capabilities for handling large volumes of digital image data rapidly and, when used in support of photo interpretation analysis, significantly increased an investigator's ability to extract new information of benefit in understanding and monitoring applications phenomena.

The launch of the first of the Landsat satellites in 1972 and the development of meteorological imaging sensors flown on the Nimbus and Geostationary Operational Environmental Satellites (GOES), with their tremendous data volumes, strongly influenced the continuing trend toward computer-aided digital analysis of remote sensing data.

In response to the need for handling large volumes of image data, remote sensing software systems of the late 60's and early 70's were designed for operation on large scale computer systems. These software systems were tailored for efficient batch-processing operations, and relied on magnetic tape for data storage, and on line printer maps and tabular listings for output products. When using such batch-oriented processing systems, the investigator was only peripherally involved in the actual information-extraction process.

CHANGING NEEDS

As more users became aware of the potential of remote sensing digital-analysis techniques during the '70's, new applications requirements began to emerge. Stronger needs for multi-temporal data analysis, for analyzing combined mutli-source data sets, and for closer participation by the investigator in the information-extraction process significantly impacted the design of remote sensing software systems. The rapid development of image-display terminals with digital refresh memories and the emergence of minicomputers during the '70's, combined with these changing needs, led to the development of the first minicomputer-based interactive remote sensing software systems. The IDIMS, KANDIDATS, AOIPS and Mini-VICAR systems, described later in this chapter, typify the current state of minicomputer-based interactive software systems used in support of remote sensing applications.

The proliferation of data formats and image analysis techniques, and the rapid increase in the number of software packages developed to support remote sensing applications during the last decade have resulted in needs for data format and software standardization. Because of these needs, data-format standards and software-design techniques which emphasize modularity and software transferability are receiving stronger emphasis.

SOFTWARE REQUIREMENTS

In order to be able to meet the general needs of remote sensing applications, software systems must meet a number of requirements concerning the hardware environment of such systems, the environment of the systems programmers who care for and maintain the systems, and the environment of the users whom such systems must ultimately satisfy. All three environments must be expected to change. Technology will introduce new hardware systems. Neophyte users will become experienced users, the experienced users will leave, and new neophyte users will come. Likewise for the system programmers. While all this is going on, users will request new processing operations which must be quickly and easily added to the software by the system programmer. If the total software package is not designed to be easily used, easily learned, easily programmed, easily modified, and easily transportable in this changing environment, then a 100,000 source-statement software system can become a patched-up, unworkable, worthless burden in a matter of a few years.

DESIGN OF REMOTE SENSING SOFTWARE SYSTEMS

Most existing remote sensing software systems, including those described in later sections, provide batch and interactive processing capabilities, an easily used system-user interface, capabilities to perform all of the standard image-processing functions normally required for remote sensing applications, and a thorough set of user and program documentation. These features, in combination with needs for standardization and software transferability, suggest the desirable software system-design characteristics discussed in the next few sections. Useful information on the pitfalls to be avoided when designing software systems is presented by Guinn and Kennedy (1975), and Boehm (1976).

CURRENT SOFTWARE TRENDS

Remote sensing software must resolve some of the most difficult problems confronting large scale application software: handling large amounts of data which can only be accessed a segment at a time with varied and complicated processing tasks that provide heavy processing loads to the computer system. This kind of software tends to be complex even for the simplest of remote sensing processing tasks.

The processing of remotely sensed data is becoming interactive, as documented by Green, (1978). An operator, user, or analyst sits at a console with a means of accessing, preprocessing, feature extracting, classifying, identifying, and displaying the original imagery, training sets, test sets, or classified imagery for subjective evaluation and interaction.

To offer maximum utility in this environment, remote sensing software systems should be designed in a modular fashion and should incorporate other desirable characteristics including:

- System capabilities for interactive and batch processing.
- A flexible user-system interface which incorporates default parameters for all processing options.
- Use of processing session history records to document processing operations.
- Use of standard data formats and file structures for ingesting and storing data files within the system.
- System functions to monitor and control processing operations.
- Use of well defined program interfaces to the system.
- Capabilities for chaining processing tasks into multi-step jobs.
- Capabilities for automatic cataloging of data sets.
- A set of utility programs to handle common service functions.

- Avoiding the use of program overlay structures.
- A rich set of processing programs to meet applications requirements.

In the following sections, these ideas for the structure and design of remote sensing software systems are discussed in more detail. The ideas presented are motivated by systems, such as those described in later sections, that are or have been used in the large scale and mini-computer environment.

BATCH VERSUS INTERACTIVE PROCESSING

Remote sensing software systems should provide capabilities for both batch and interactive processing. Batch processing capabilities are used to execute previously defined processing scenarios which do not require user involvement in the actual processing, and for extensive processing jobs requiring lengthy computations and large volumes of data.

As noted earlier, the user's world is evolving toward interactive rather than batch processing. In interactive computing, the user participates as an active part of the on-line processing cycle. The interactive user's chief concerns are how easy it is to interact with the system to get the job done and how much bookkeeping and remembering the system can do to lessen this burden. For interactive processing operations, user input is typically entered from a keyboard or cathode ray tube (CRT) display terminal, and computer response should occur within 5 seconds.

A major difference between batch processing and interactive processing is what the program must do if it discovers that some user input data are out of range, inconsistent, or incorrect. In batch processing, incorrect input data result in a program abort. In interactive processing, incorrect input data should result in a system request for the user to re-input the required data. To avoid having the user re-input all data to correct one character, interactive processing programs must prompt the user for an item at a time, checking each input for validity.

Interactive output is handled differently from batch output in that it must be kept to a minimum. The user should not be overwhelmed with a full screen or page of output. In addition, for interactive users it is preferable to output pictures and graphs rather than to output tabular data and sentences.

Another aspect of interactive processing that differs from batch processing is that the documentation describing the use of any system feature or command should be stored on-line for interactive access rather than only in external report form. Such interactive documentation is typically implemented through the design of a HELP function which allows the interactive user to access documentation pertinent to a given interest and to request more detailed documentation on any specific item.

USER-SYSTEM INTERFACE

For batch operations, the system design should include a conversational user interface to permit interactive preparation and checkout of program control specifications and processing parameters, and capabilities for executing processing tasks from punched card decks or card image-command files stored on a disk. The system should include capabilities for chaining batch processing tasks into multi-step jobs. The system design should also permit batch users to create procedures which define mutli-task processing scenarios and which can be executed by issuing a single procedure-command statement.

The system interface for the interactive user must be functional for the neophyte as well as the experienced user. This suggests that a variety of different ways of interacting with the system should be allowed, including a command-language interface with conversational prompting for use by the experienced user and a menu-driven interface for use by the inexperienced user.

The neophyte user may only know enough about processing remotely sensed data to select a general rather than a specific processing area. In this case, a menu format which displays, on the console screen, four to twelve different broad image processing categories will usually be enough to get the neophyte underway. After selection of one of the broad areas, a second level menu appears on the screen which shows either a list of related processing categories or a list of actual processing operations. A third level, if necessary, would show a list of four to twelve image processing operations.

The more experienced user will find the menu interaction system tiresome; hence a quicker way is needed to initiate a processing task, preferably a way which will permit the user to type in one command word or process acronym and be done. This emphasis on brevity has consequences. The experienced user may forget the exact spelling or phrasing of a processing name. In this situation, some help is needed such as an alphabetically organized list of the command-name acronyms that can be requested for display on the console screen. A single system command can be implemented to provide this information.

For the user who has an intermediate level of experience, however, an alphabetically organized list of process acronyms will not be enough, for there may be two or more prgrams that might be suitable. For this class of user, a vocabulary command needs to be provided which can explain briefly what any command acronym or abbreviation means.

Finally, there is the case where an experienced or neophyte user is not able to get satisfactory operation for some process because of a misunderstanding of what the process does or of how its parameters may correctly be set up. Here, a complete explanation is needed to define what the process does, legal ways of calling the process into action, what the parameters of the process do, and legal values for the parameters. An explain-command can fulfill this need.

Command Language Syntax

Every remote sensing image-processing operation works on an input image and produces an output image, statistics, or features of the input image. Once the image-processing operation is specified by a menu or command-name procedure, the input and output files or devices must be specified. One natural way for the software system to obtain this information is through a question/answer approach. The question "What is the input image name?" appears on the console screen and the user types in the name, whereupon the software makes sure the file-name syntax is correct and then asks about the output-image name.

The question and answer approach is not the only one. There are a variety of different convenient ways of combining command names with input and output file names into a command string. For example, take the process of quantizing which might have the command name QUANT.

Possible command-string syntaxes include:

QUANT   OUTPUT IMG < INPUT IMG

INPUT IMG > QUANT > OUTPUT IMG

OUTPUT IMG = QUANT (INPUT IMG)

The first syntax is used in the KANDIDATS system at the University of Kansas and can be interpreted as: quantize to an image-named OUTPUT IMG the image-named INPUT IMG. The second syntax, analogous to the engineering diagram showing input, black box, and output, is used in the IDIMS system at the Electromagnetic Systems Laboratory. It can be interpreted as: take the image INPUT IMG and go through the process QUANT to produce the resulting image called OUTPUT IMG. The third syntax, a mathematical functional-form notation, is used in the EIES system at Control Data Corporation and can be interpreted as: the image OUTPUT IMG is a quantized version of the image INPUT IMG.

When the image-processing task does not produce another image but produces statistics about the input image, there is the question of whether these statistics should be displayed on the user console-screen for examination, printed on a line printer for hard copy, or made into a source file to be saved on a mass storage device. Natural variations of the basic command string can easily permit these distinctions to be made. Take, for example, the process of getting a histogram of the grey tones of an image. Let the command name be HIST. The following three command-strings can direct the histogram to the terminal, to the line printer or to the disk pack as a file called TONE HST:

HIST TT < INPUT IMG

HIST LP < INPUT IMG

HIST TONE HST < INPUT IMG

Once the basic combination of input image, output image, and command name have been specified, the image process itself may require additional parameters. The equal-interval quantizing command, for example, will need to know how many quantized levels the user desires for the output image. Other tasks, such as the histogram task, will need to know what portion of the image should be processed. In this case, the user will have to specify a subimage size and location. Because it often is the case that similar kinds of information will be needed for a variety of different commands, the prompting questions that the software provides should be identical in format and style. Certainly, a request for the same information should always appear to a user in the same format.

Processing History Records

Interactive remote sensing image-processing usually proceeds on a step-by-step basis with many intermediate images created and then deleted before a final classified or enhanced image is produced. The intermediate images are often deleted to save mass storage file space. If the user has not kept careful notes on how an output image was produced, it may be difficult to reconstruct accurately the processing steps which were performed on an input image to achieve an output image. Since computers are excellent bookkeepers, the user should expect that the image-processing software will keep processing-history records for each image.

Processing-history records fall in three general categories. The first category consists of some free format information that the user wants associated with the image. Such information could be about the image's origin, associated photography, method of digitization, tape identifiers for the tapes storing copies of the original digital image, associated files of surface-truth information, specific information about targets in the image, and perhaps some information about the purpose of the image processing. What the user desires to put in these records, how much information is to be included, and the format to be used should all be left free. These records should be considered like ASCII source-information with no structure. They should flexibly allow the user to tell any story.

The second category of descriptor record is more structured and accurately tells what processes the image has been through. The input image(s), the process name, the output image, the date and time of the processing, and all the parameters required by the process all should appear in these records. Because the kind and number of parameters vary as a function of the image process, these descriptor records may have to include a data structure to allow easy access by program or user.

The third category of descriptor-record contains either statistical information about the image, which the processing programs create, or surface truth information which the user must input. This information cannot be free-format because it is information that other processing operations must access in order to do their job. Hence, the order in which the pieces of this kind of information appear must be standardized. Examples of surface-truth information include (1) tables which relate a land use type or target name to be associated with some index symbol on a channel of the image and (2) lists giving land-use types, index symbol, and associated polygonal areas on the image of the named land-use type.

A user should be able to examine these descriptor records by commands very much in the same manner that the remote sensing software system is directed to perform a processing task. Descriptor-record operations should include the capability to list all descriptor records, list only the names of the processes the image has undergone in order of their occurrence, list only free-format records, and list only the statistical records. In addition, the user will need to have a way to add unstructured information to the descriptor file. A descriptor-record deletion operation is needed so that the user can correct any misinformation added to the descriptor-record history. However, the user should not be allowed to delete processing history records created by processing programs.

The data base that includes the processing-history descriptor records can be viewed as a companion to the larger digital-image data base. The user should be able to query the descriptor data-base and extract digital images that satisfy the search criteria for display or further processing. This architecture would provide the system user with a powerful image data-base interrogation capability, and enable extraction of the image set that meets a complex user-query specification.

SOFTWARE STRUCTURE

In order to satisfy requirements for easy addition of new processing operations, high portability to new computer-hardware environments, and easily understandable software by the system's programmers, remote sensing software systems must be structured and highly modular. In addition, processing programs and other parts of the system should be implemented in a high-level programming language whenever possible. These and similar ideas are explored in the following sections and by Hamlet and Haralick (1978).

System Implementation Considerations

FORTRAN is the most widely accepted high level language used in scientied applications and serves as a *defacto* standard for implementing remote sensing software. The American National Standard Institute (ANSI) has developed a specification for a machine-independent FORTRAN (see ANSI, 1978) which should be adopted by designers of remote sensing software. In addition, remote sensing software systems should be designed and developed using a standard software methodology such as the top-down, structured-programming methodology discussed by Baker (1972) and described in detail by McGowan and Kelly (1975).

ANSI standard FORTRAN is only partially amenable to the implementation of structured-programming language constructs, such as the DO WHILE and IF THEN ELSE statements. However, a number of FORTRAN preprocessors have been developed to aid in the implementation of structured FORTRAN programs.

Structured FORTRAN preprocessors such as RATFOR, described by Kernigan (1975), translate structured-programming constructs into standard FORTRAN language statements which serve as input to FORTRAN compilers. The use of top-down, structured-programming techniques and ANSI standard FORTRAN helps to assure the development of modular, reliable and easily maintained software and enhances software transferability between computer systems.

The design of remote sensing software systems should incorporate the use of standard data formats and internal data-file structures. Adopting such standards facilitates communications between processing programs, the addition of new programs to the system, and sharing of data and software between systems.

The system structure should include well defined internal and external software interfaces. The internal interfaces should be defined as standard FORTRAN subroutine calls and should provide access to all common system services for use by the processing programs. The external program interfaces can be used to link to data-management systems, spatial/geocoded information-management systems, report generators or other software systems which provide capabilities not implemented within the remote sensing software system.

The system design should include capabilities for automatic cataloging of data sets, interactive catalog search, and data retrieval by data-set attribute (e.g. by time, spatial location, sensor type, data type, etc.).

System designers should avoid use of program

overlay strutures. Overlays are slow and involve the definition of program dependent data areas for inter-module communication. In addition, overlays are highly dependent upon both the machine and the operating system and therefore defeat software transportability. Avoiding overlays is easiest when designing software systems for current large scale and mini-computers which use virtual memory operating systems.

Finally, the remote sensing software system should be structured to include a complete set of utility programs and a rich set of processing programs needed to perform the required image processing and information extraction operations. The utility programs perform common services for data display, input/output, output-product generation, data plotting and other generally required data handling functions. The inclusion of a utility package in the system design frees the processing programs from routine data handling operations and makes the design of new processing programs easier.

Software Control Structure

An effective way of structuring a remote sensing software system is to incorporate a system Executive program in the system design. The Executive serves to localize interfaces to the computer's operating system and to control all processing operations and task sequencing. The Executive should perform common system services in support of the user and the processing programs, including:

—error handling
—dynamic resource allocation
—input/output handling
—interrupt handling
—checkpoint/restart operations
—system bookkeeping
—data file management.

A typical operation sequence within the system would be for the Executive to call a command-string interpreter or menu-prompting routing to accept a user-processing assignment. The menu mode of operation could be initiated by a command through the command-string interpreter. Once a processing option has been selected, input and output file names given, source or destination devices assumed or given, and required processing parameters entered, the Executive calls the selected processing program and passes it the required input parameters. If additional parameters are needed during interactive processing operations, the processing program requests the required services from the Executive. In addition, the processing program calls on the Executive to dynamically allocate resources, to handle input/output operations and for any other generally used system services. When the selected processing task is completed, the processing program writes the required output data sets (using system supplied services) and passes control to the Executive.

During each stage of processing, errors should be prevented from occurring by performing extensive error checks within the system. When errors do occur, the processing program should use Executive services to seek remedial action. The Executive should inform the user of serious errors and prompt the making of corrective input or a new processing assignment as appropriate. In the case of fatal errors during batch-processing operations, the Executive should terminate the offending task with appropriate error messages entered in the task's output file.

System Interrupts

There is another kind of special condition which the software system must be able to handle: user-generated interrupts. Sometimes a user initiates a processing task and then realizes that a legal but incorrect parameter was entered. Rather than wait the required time for the task to finish, the user would like to interrupt the process, return to parameter input interaction again or perhaps even to the Executive level in order to specify a different task. Such user interrupts can be indicated by typing a control key on the terminal.

Interrupt handling is a job carried out by the operating system and may be done differently in every system. It should be possible, however, under every operating system, to have the operating system call a user-designated subroutine after a user interrupt. This subroutine can be part of the software system Executive, and performs any special interrupt-handling operations.

Operating System Interface

The remote sensing software-system Executive program should be designed to provide all interfaces between the system's processing programs and the computer operating system. The Executive should also provide all services generally required by the processing programs including interrupt handling; file level input and output operations; error handling; control, monitoring, and sequencing of processing tasks; and dynamic allocation of memory, tape drives and disk storage space.

For remote sensing software systems that are implemented on computers having sophisticated operating systems, the Executive can accomplish these functions by translating service requests from FORTRAN subroutine calls issued to the Executive by processing programs into operating system service requests. For computers having unsophisticated operating systems, the Executive may have to incorporate modules which perform those service functions needed by the processing programs but not available through the computer's operating system.

In either case, localizing operating system functions and interfaces to a few modules within the Executive provides a high degree of machine independence and software transportability.

PRESENT LARGE SCALE COMPUTER SOFTWARE SYSTEMS

This section describes the capabilities of four example major remote sensing software systems: the ORSER, LARSYS, ERIPS and VICAR systems. Each of these systems was designed to operate on large scale digital computers, and each provides a powerful set of image-processing functions for remote sensing applications. Detailed comparisons of the processing features of the LARSYS and VICAR systems are presented by Carter et al (1977).

ORSER SOFTWARE SYSTEM

The Office for Remote Sensing of Earth Resources (ORSER) at the Pennsylvania State University has developed an operational software system to process and analyze multispectral image data in support of a variety of remote sensing applications. The ORSER software system is used primarily to process Landsat data but also accommodates other remote sensing inputs including Skylab and aircraft data.

The ORSER software executes on an IBM System 370, Model 168 computer and utilizes commercially available equipment for all peripheral devices including remote terminals. Users can access the ORSER system in one of three ways: through low-speed remote job-entry terminals, through local or remote intermediate-speed batch terminals, and through local and remote high-speed dispatch points operated by the Pennsylvania State University Computation Center. ORSER programs are stored in on-line program library files and can be called for execution through any of these entry points.

The ORSER software system is designed to provide rapid turnaround and low processing cost for small processing jobs. Job-execution priority is based on the maximum requested processing time and on the amount of input and output required. Jobs requesting less than one minute of processing time and requiring input and output of 500 or fewer records are usually completed within 2 minutes with results returned to the remote terminal user. Limited amounts of output can be returned to remote terminals for display or printing while sizeable output files are routed to high-speed line printers or stored on magnetic tape. The primary output products available include printed lists; tables and maps; photographic image products; and computer compatible tapes.

Outside investigators can arrange to use the system either locally at ORSER or through remote terminals connected to the ORSER central computer by long distance telephone lines. For remote users, magnetic tapes used for input or output and other bulk output products are shipped to and from the computation center by mail.

ORSER Software Characteristics

The ORSER software system does not include an executive or monitor program to coordinate processing and program execution. Instead, the ORSER software is designed as a collection of interrelated programs which are executed separately. Communication between the ORSER programs is coordinated by maintaining a common data tape format and by passing output data files produced by one program as input data files for a successive program in a processing sequence. All of the ORSER programs are written in FORTRAN IV.

The ORSER software is a magnetic tape-oriented system in which all multispectral image data are processed directly from computer compatible tapes. All of the ORSER programs, with the exception of a data clustering routine, make only one pass through the input data and for the most part operate at tape speed. A special ORSER tape format is used to insure data format compatibility within the system. The ORSER format accomodates all current multispectral scanner data and supports multiple data files on a single tape, as well as multi-tape files. Each ORSER-formatted tape file contains table of contents records which describe the specific subimages within the file. Individual data records contain one complete scan line of the stored image. Response values for individual picture elements within an image scan line are stored as vectors with as many components as there are spectral channels. The response vectors can be expanded to include components which are derived from additional data sources.

The basic ORSER software operates in a batch-processing mode and does not allow user interaction during program execution. Users accessing the software by remote terminal can prepare processing control specifications in an interactive mode prior to submitting an ORSER program for a batch-execution request.

Each ORSER program is executed as a separate computer job. Control specifications for jobs requiring multiple processing tasks cannot be strung together for execution of a single multi-step job. However, the software is designed so that output files created by certain programs serve as input to other programs. Each ORSER program accepts input parameters provided in the user-supplied control specifications, performs the required processing functions on the input data, and prepares output files of the processing results. The programs currently supported within the ORSER system include those described in Table 19-1. More detailed information on the ORSER processing system is presented in McMurtry et al (1977).

TABLE 19-1

Commonly Used ORSER Program (Derived from McMurtry et al, 1977)

Program Name	Program Function
TPINFO	Outputs information about contents of a multispectral image data tape.
SUBSET	Subsets one or more subimage areas of a Landsat data tape into ORSER format.
SUBAIR	Subsets data from aircraft tapes into ORSER format.
SUBOUND	Subsets irregularly bounded areas into ORSER format.
SUBGM	Version of SUBSET which geometrically corrects, rotates, and scales data during subsetting.
MERGE	Merges data from two passes of Landsat over the same area.
SPAN	Spans data from two adjacent panels of the same Landsat scene or concatenates data from two consecutive Landsat scenes of the same pass.
SUBTRAN	Reformats input data to Landsat or ORSER format and allows linear transformation, ratioing, averaging, and data scaling and translation.
HISTRAN	Used prior to SUBTRAN. Generates scaling and translation parameters for SUBTRAN.
REFORMAT	Reformats a channel sequential input tape to ORSER format.
CONSCL	Consolidates data from different tapes onto single ORSER tape.
NMAP	Creates printed grey scale brightness map which shows overall scene features.
UMAP	Maps areas of local spectral uniformity.
STATS	Computes statistics for training areas defined by UMAP or other means.
USTATS	Version of STATS which computes training area statistics using UMAP spectral uniformity criteria.
CLUS	Unsupervised classifier which develops a set of spectral signatures by clustering.
CLASS	Performs supervised classification using Euclidean distance algorithm.
GMCLASS	Version of CLASS which outputs a geometrically corrected, scaled, and rotated classification map of Landsat data.
STCLASS	Produces classification map based on algorithm used in CLASS and generates training area statistics as performed in STATS.
CANAL	Uses canonical analysis techniques to derive an orthogonal transformation which maximizes category separability on as few axes as possible.
PRINCOM	Uses principal component analysis to compute transformation matrix from a set of observations within a user selected data site.
RATIO	Ratios two selected channels and classifies ratioed data based on ratio boundaries derived from training signatures.
PPD	Classifies data using parallelpiped algorithm.
MINDIS	Parametric classifier with linear discriminant function. Classifies on basis of separation distance between spectral categories.
PARAM	Parameteric classifier with quadratic discriminant function which classifies according to the maximum likelihood decision rule.
NPAR	Nonparameteric trainer with linear discriminant function.
NPARMAP	Nonparametric classifier with linear discriminant function.
QUADNPAR	Nonparametric trainer with quadratic discriminant function.
QUADMAP	Nonparameteric classifier with quadratic discriminant function.
MAPCOMP	Performs element by element comparison of two digital classification maps of the same area.
LMAP	Outputs a black and white line map scaled to user specifications.
ORSEROUT	Reads map output of other ORSER programs and prints color maps.
DISPLAY	Reads map output of other ORSER programs and produces a variety of map displays and final map output products

Typical ORSER Processing Scenario

The ORSER software provides substantial capabilities for analyzing and classifying multispectral scanner data. This section describes the processing flow utilized to perform a typical multispectral classification analysis using the ORSER programs.

The user initiates a classification or other analysis task by selecting the area of interest from maps or by selecting images to be processed from catalogs of available data tapes maintained by the ORSER computation center. If the input-image tape required for the analysis is not contained in the ORSER tape library, the user must provide the required tape and must request its inclusion in the tape library.

Having identified the scenes for processing, the

user executes one of the SUBSET programs to select the required subimages and to generate an ORSER-formatted tape containing the selected data. After subsetting, a job is executed using the NMAP program to create a printed grey scale map showing overall scene features. NMAP output is used to locate areas of interest and to verify the presence of scene features.

The UMAP program is next run to map areas of spectral uniformity. UMAP output shows patterns of uniformity and permits the user to specify uniform areas as training sites for supervised classification. Training area signatures and statistics are next obtained by executing the STATS or USTATS programs. Each of these programs computes the multivariate statistics for user-specified training areas. The user can also request grey level histograms for selected spectral channels.

At this point in the process, the user can perform a variety of preprocessing operations on the input data or training-area data by executing the CANAL, PRINCOM, RATIO, HISTRAN, or SUBTRAN programs. When the training areas have been identified, a supervised classification is performed using the CLASS program. The output from CLASS is a character map, with each classification category represented by a unique symbol. These maps are geometrically distorted due to the fixed printing geometry of printer output. However, the user can correct Landsat data to remove these and other distortions by using the SUBGM Program initially or the GMCLASS program (instead of CLASS) to generate geometrically correct character maps.

Final map output products and displays are produced by executing the DISPLAY and LMAP programs. The user also has the option of deriving training areas and associated statistics by executing the CLUS program and the STCLASS program prior to running CLASS. The entire classification analysis requires the separate execution of from 8 to 12 ORSER programs to derive the final classification results.

Improved User Interface

Use of the ORSER software requires a minimal knowledge of IBM 360/370 Job Control Language, familiarity with the ORSER program input parameter specifications, and a knowledge of remote terminal commands. A new conversational translation package has been developed for the ORSER software to eliminate the need for learning these details. This package is called the ORSER Complete Conversational User Language Translator (OCCULT). OCCULT includes the ORSER software and provides three levels of interaction based on the user's experience. The system allows the user to prepare control specifications for executing ORSER programs by interacting with the computer in simple English. At any point during the interaction, the user can obtain needed information by typing HELP. OCCULT validates

the users free-format responses and provides opportunities for the user to correct errors. The system also maintains a catalog of the user's data sets.

OCCULT is modular in design and allows easy inclusion of new programs. The system is available for use on a national commercial time-sharing network, COMNET, operated by the Computer Network Corporation. The OCCULT software is also available from NASA's software-dissemination facility COSMIC, at the University of Georgia, Athens, Georgia, 30602. The OCCULT system capabilities are described by Ramapriyan (1979).

Advantages of the ORSER Software

The ORSER programs provide substantial multispectral analysis capabilities. The software is well documented and debugged and is easily adapted to include new processing routines and utility programs. The system, particularly with the OCCULT enhancement, is very easy to learn and to use. It provides rapid turnaround and low processing cost for tasks requiring processing of small image areas.

LARSYS SOFTWARE SYSTEM

In 1964, the Laboratory for Applications of Remote Sensing (LARS), at Purdue University, initiated development of a multispectral data-analysis system, called LARSYS. The LARSYS software system has evolved into one of the most heavily used remote sensing software systems and supports a variety of applications in hydrology, agriculture, geology, land use and forestry.

The latest version of LARSYS was released in 1975 with revisions continuing through 1980. A thorough user documentation-package is available for LARSYS with recent revisions published periodically in the LARS SCAN LINE Bulletins. A validated, well debugged version of LARSYS is available through COSMIC, with complete user and program documentation.

LARSYS Processing System Features

The LARSYS software system executes on an IBM 3031 computer under the Virtual Machine/ Cambridge Monitor (VM/CMS 370) operating system. Commercially available equipment is used for all peripheral devices including remote terminals and image-display stations. Users can access LARSYS through remote job entry stations, CRT and typewriter remote terminals, and local image-display stations. All of the LARSYS programs are stored in on-line program libraries and can be called for interactive time sharing or batch execution by local or remote users.

The LARSYS software is designed to provide rapid turnaround and low processing costs for interactive and batch-processing jobs. Interactive users experience rapid turnaround of a few seconds while compiling job-control specifications and processing parameters. Response times for actual processing jobs vary depending on the process requested, the volume of data to be processed, and overall system activity. Longer processing jobs are usually submitted for batch processing; such jobs experience turnaround times varying from less than one half hour to 24 hours depending on job requirements. Limited amounts of output can be routed to remote terminals. Sizeable output files are directed to magnetic tapes, disk storage units, or high-speed line printers. The primary output products produced include printed lists, tables and maps; photographic image products; and computer-compatible tapes.

Outside investigators can arrange to use the system either locally at LARS or by remote terminal. Remote terminals and remote job-entry stations are connected to the LARS computer by long-distance telephone lines. The LARS facility provides an extensive magnetic data tape library containing satellite and aircraft imagery and output results of prior analysis runs. The LARS staff provides support services to help users in applying LARSYS to their needs, and an extensive set of training materials, including audio tape courses and documented examples of LARSYS applications, is available.

LARSYS Software Characteristics

The LARSYS software system does not include an executive or supervisor program to control processing and program execution. Instead, the LARSYS software is designed as a collection of interrelated programs which can be executed separately or as multi-task jobs.

Communication is maintained between the LARSYS programs by using common data-tape and disk-file formats. The magnetic tape and disk data sets generated by certain LARSYS programs are used as input files for other programs during a processing sequence.

LARSYS accepts input image data on magnetic tapes in a special LARSYS format or in the Johnson Space Center (JSC) Universal Tape format. The LARSYS format includes an identification record and data records which contain one scan line of data from each spectral channel of the stored image. Additional channels of ancillary data can be included on LARSYS tapes. The system supports multi-file tape handling operations, and utility programs are available for converting non-standard multispectral scanner tape formats to the LARSYS format.

The LARSYS programs use magnetic tapes and disk units to store intermediate results and working data sets during processing operations. Final output tapes and data available in the tape library are cataloged automatically by LARSYS.

The LARSYS software system consists of two parts: The standard LARSYS and a developmental system, called LARSYSDV. The standard LARSYS is well documented and contains

TABLE 19-2

Standard LARSYS Processors

Processor	Function
PICTUREPRINT	Produces printed histograms and maps of image data.
IMAGEDISPLAY	Same as PICTUREPRINT but output is displayed on TV display unit.
CLUSTER	Clusters multispectral image data and prints cluster map.
STATISTICS	Calculates training area statistics for use in classification.
SEPARABILITY	Calculates transformed divergence between classes.
CLASSIFYPOINTS	Performs maximum likelihood classification of individual picture elements.
SAMPLECLASSIFY	Classifies selected fields and prints classification map.
PRINTRESULTS	Uses classification results. Prints classification map and tabulates number of pixels in each class.
IDPRINT	Prints header information from image data tape.
DUPLICATERUN	Copies image data tapes.
TRANSFERDATA	Prints actual data values on input image tape for specified channels.
COPYRESULTS	Copies classification results file from disk or tape to another tape.
LISTRESULTS	Reads classification-results tape and prints header records.
PUNCHSTATISTICS	Punches a copy of the statistics file stored on a classification-results tape.
LINEGRAPH	Graphs a line of multispectral image data on line printer.
COLUMGRAPH	Graphs a column of multispectral image data on line printer.
HISTOGRAM	Generates histograms of image data.
GRAPHHISTOGRAM	Prints histograms produced by PICTUREPRINT, IMAGEDISPLAY, or HISTOGRAM.

18 major programs, called processors, and a number of related routines. The standard processors are described in Table 19–2 and a block diagram of the standard LARSYS is presented in Figure 19–1. Program Abstract files are stored on the system disk for user access, and document the operation of each standard LARSYS program.

The LARSYSDV software was developed starting in 1975 and includes special versions of standard LARSYS programs and additional new processors.

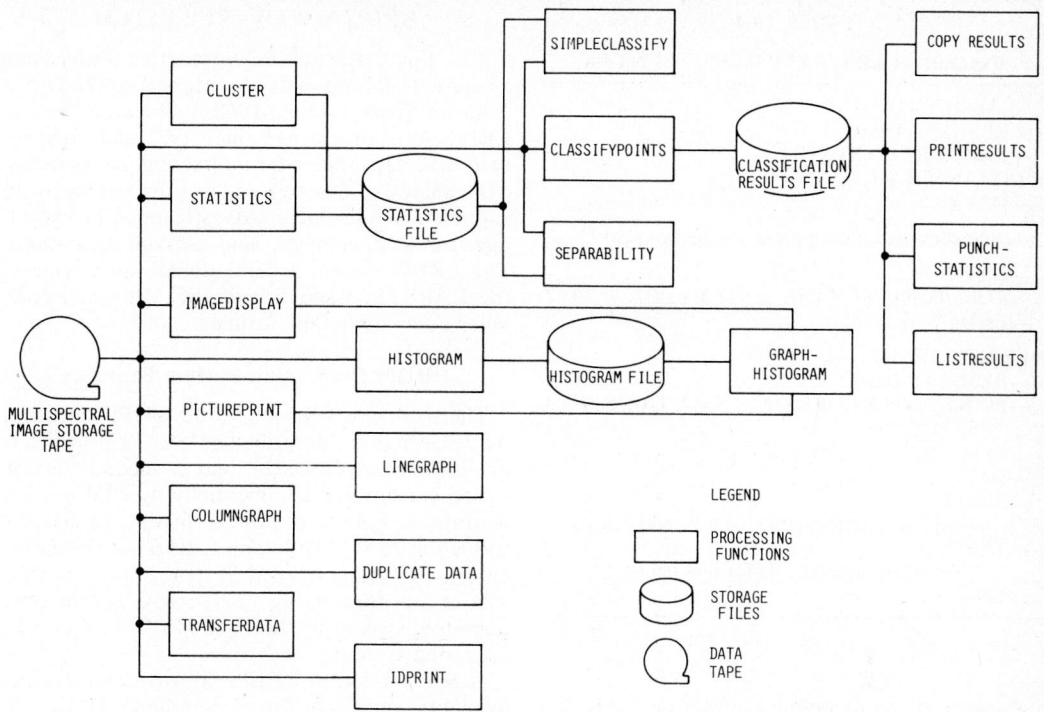

Figure 19-1. LARSYS Processing System. Source: LARS Purdue University.

Program Abstracts and user documentation have not been generated for all of the LARSYSDV programs. However, a description of the control specifications needed to execute LARSYSDV programs is stored on the system disk for user access. Currently, LARSYSDV contains 8 new processors and several modified LARSYS processors which are described in Table 19−3. Capabilities of the LARSYS processors are described in LARS (1979) and by Spencer (1974).

LARSYS Processing Scenario

The user initiates and controls the execution of LARSYS programs by creating card image files or punched card decks which contain the control parameters, processing commands and data needed to execute a LARSYS job. A free-form, keyword and parameter format, with simplified syntax and punctuation, is used to prepare input card decks and card image files.

Extensive error checks are performed during interactive sessions and the system provides opportunities for the user to correct identified errors. For batch users, a CHECKOUT program can be executed to verify LARSYS job control decks prior to submission for processing. An example of a command sequence required to execute a job which performs a multispectral classification and prints the resulting classification map is listed Table 19−4.

The processing scenario for performing a

TABLE 19-3

LARSYSDV Processors

Processor	*Function*
ECHO CLASSIFIER	Groups data into homogeneous cells and classifies each cell as if it were a pixel.
LAYERED CLASSIFIER	Uses tree-structured classification algorithm to classify multispectral image.
MINIMUM DISTANCE CLASSIFIER	Classifies each pixel by using minimum Euclidean distance to class means.
MERGESTATISTICS	Combines multiple statistics decks into one deck.
BROWSE	Searches LARSYS tape catalog and prints list of images satisfying user request.
RATIO	Uses mean vectors of classes in statistics file and calculates and prints ratios of values for selected channels and the sum of each class.
GDATA	Modification of Standard LARSYS PICTUREPRINT processor. Generates output on printer-plotter.
GRESULTS	Modification of standard LARSYS PRINTRESULTS processor. Generates output on printer-plotter.

TABLE 19-4

Example—LARSYS COMMAND SYNTAX

```
* CLASSIFY POINTS
  —RESULTS DISK
  —CARDS READ STATS
  —CHANNELS 1, 3, 4, 6
  —DATA

    (Statistics deck from previous statistics run)

  —DATA
  —RUN (80021500), LINE (1,512,2), COL (1,300,2)
  —END
* PRINT RESULTS
  —RESULTS DISK
  —PRINT OUTLINE (TRAIN, TEST), TRAIN (F,C),
    TEST (F,C,P)
  —SYMBOLS C, S, W, O, G
  —DATA
  —TEST1
    (Field description cards for test field 1.)
  —TEST5
    (Field description cards for test field 5.)
  —END
```

multispectral-classification analysis on LARSYS includes executing:

- The PICTUREPRINT processor to obtain histograms and a printed map of the image area to be classified. This map is used to obtain the image coordinates of training areas and test sites to be used in the classification process.
- The CLUSTER, STATISTICS and SEPARABILITY processors to calculate training-area statistics and to refine training areas prior to classification.
- The CLASSIFYPOINTS or SAMPLE-CLASSIFY processors to classify individual picture elements or specified fields using a maximum-likelihood decision rule.
- The PRINTRESULTS processor to obtain a printed alphanumeric map of classification results.

Photographic products of final classification results can also be obtained.

Advantages of the LARSYS-Software

LARSYS offers access to advanced remote sensing software for low cost. The system has been used extensively and is well documented and debugged. The user-interface language is easy to learn and the system is simple to use, particularly in view of the extensive LARSYS documentation and training aids available. LARSYS is modular and readily accomodates the addition of new programs. The standard LARSYS software, with complete user and program documentation, is available from COSMIC.

ERIPS SOFTWARE SYSTEM

The Earth Resources Interactive Processing System (ERIPS) was developed at NASA's Johnson Space Center (JSC) in Houston, Texas. ERIPS has been in use since 1972 and provides extensive capabilities for interactive multispectral data analysis. The system is used primarily to support earth-resources investigations in hydrology, forestry, geology, land use and agriculture. The ERIPS system accepts digital image data in the LARSYS, Landsat and JSC Universal computer compatible tape formats.

ERIPS Processing System Features

There are two versions of the ERIPS software: the JSC version, (described in JSC, 1979) which is available from COSMIC and a second version which is marketed commercially by IBM's Earth Resources Laboratory in Houston, Texas (described in IBM, 1979). The ERIPS software executes on a variety of IBM System 360 and System 370 computers using either JSC's real-time operating system or the IBM MVT, VS1 or VS2 operating system.

IBM's version of ERIPS utilizes commercially available equipment for all peripheral devices including interactive image-display terminals. Users access the system by means of menus presented at CRT display stations. A user enters processing-control specifications by selecting multiple choice options presented by the system menus using a cursor or by typing in required parameters. Processing results are displayed on alphanumeric TV screens or as color or grey-shade images on image-display devices. The ERIPS software checks user responses to menu, prompts for errors and provides the user opportunities to correct errors as they are detected.

The ERIPS system uses approximately 30 million bytes of disk storage for program libraries and approximately 350 thousand bytes of main processor memory. Additional disk space is used by the system to store working data sets and images. The number of magnetic tape drives required varies from one to four depending on the kinds of processing performed. Response times at the user terminals is dependent on the job mix, the operating system used and the kinds of processing requested by the user. For short jobs, response times from a few seconds to one minute are typical. The primary output products available include printed lists, tables and maps; black-and-white and color image displays; photographic image products; and computer-compatible tapes.

ERIPS Software Characteristics

The ERIPS system is structured around six major software components: a Supervisor program and six software subsystems, called Applications. Figure 19-2 depicts this structure and

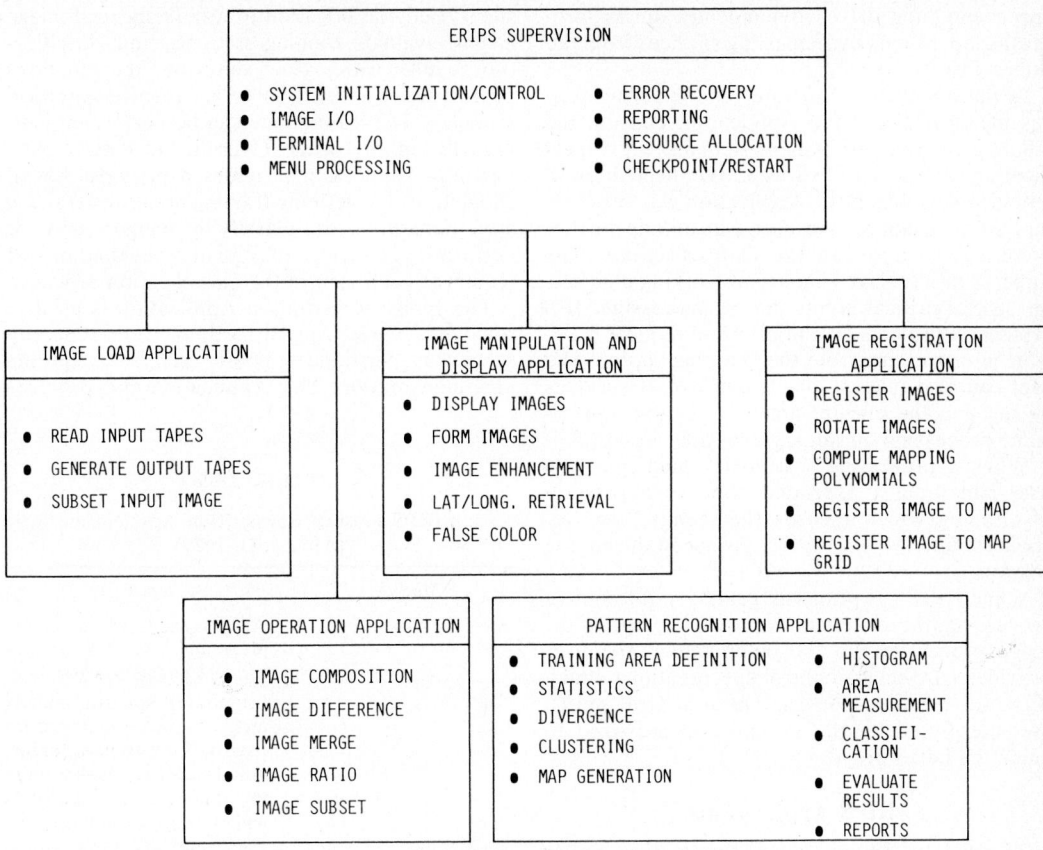

Figure 19-2. ERIPS Software System. After JSC, 1979.

describes the overall functions performed by these software components.

Development efforts for IBM's version of ERIPS were begun in 1975. This system was implemented using top-down and structured-programming design and implementation techniques. The ERIPS software is highly modular and expandable for incorporation of new Applications. The IBM version of ERIPS is programmed in a High Level Assembler Language (60%) which uses structured-programming macro instructions, PL/1 (35%), and Fortran (5%).

The ERIPS Supervisor monitors and controls all activity within the system. The Supervisor isolates the user and the Applications programs from the hardware and the computer's operating system by performing input/output and related services required by the Applications. Localizing common services in the Supervisor provides a high degree of autonomy for the Applications and simplifies their design. It also provides a degree of software transportability since only the Supervisor must be modified when changing operating systems or computer hardware.

The Supervisor initializes the system by loading the required software into memory, preparing re-

quired control blocks, and displaying an Application selection menu from which the user selects the desired processing function. The Supervisor performs all menu processing services, validates user inputs, and passes required parameters to the individual application program. When an application program requires a menu to be displayed, it passes a menu identification number to the Supervisor and waits for menu inputs. The Supervisor displays the requested menu and prompts the user for the required input. After the user enters the needed information, the Supervisor checks the user responses, and passes the input parameters to the requesting applications for further validation and processing.

ERIPS is designed as a disk-storage-oriented processing system in which all image data and other working data sets are loaded onto disk-storage units prior to processing by the Applications. Applications will then display images and read-and-write disk data sets by requesting the Supervisor to perform the required input/output services. A special image-access method is included in the Supervisor to allow high speed transfer of image data to and from disk storage. Applications can also allocate and deallocate disk space and tape drives dynamically during processing operations by requesting services from the Supervisor.

In the event of a software failure during processing operations, the Application causing the failure is removed from the system and the Supervisor allows the user to specify the disposition of any data sets which the Application was processing. After a failure, the user can initiate further processing or terminate the analysis session. The Supervisor also saves important working data sets on disk at critical points during processing. If a system failure occurs, the user can request a Restart in processing from the last checkpoint. The user can also request checkpoint/restart services to suspend the current analysis session and resume processing during a successive session.

When a processing session is completed, the user can request a printed copy of a session-history log which records the menus, user responses, and system reports produced during the session.

While ERIPS is primarily geared to interactive processing, the system does provide an option for batch processing. In the batch mode, the user provides card-image control specifications which serve an menu responses. These control inputs are interpreted by the system and are used to guide the batch process.

ERIPS Applications

The ERIPS Image Load Application reads computer-compatible tapes in the LARSYS, Landsat and JSC Universal formats and stores the data on disk units for further processing by other programs. The Image Load Application provides options for selecting subsets of the input image data to be loaded and for producing reports describing the characteristics and contents of the loaded data. It also provides capabilities for unloading data from disk to magnetic tape. In conjunction with other ERIPS programs, the Image Load Application can input a variety of ancillary data and merge such data with image data to include additional channels of information for processing.

The Image Creation Application can combine two images of the same scene taken at different times into a single multi-temporal image. It is also used to spatially merge or mosaic several images for cartographic analysis and provides capabilities for differencing, ratioing and subsetting images. The Image Manipulation and Display Application is used to display and enhance images. It provides flexible options for false and pseudo-color enhancement, image scrolling, image manipulation and for determining the precise location of any point in a displayed image.

The Pattern Recognition Application consists of twelve modules which contain most of the multi-spectral analysis capabilities in the system. Its primary use is to perform a supervised or un-

supervised classification of a multispectral image and to evaluate training statistics and classification results. Table 19-5 describes the functions performed by the twelve pattern-recognition modules. These functions can be performed individually, in any order. Overall, the Pattern Recognition Application provides a powerful set of capabilities for defining training areas and training area signatures, for classifying images, and for performing a variety of statistical evaluations of results at each step in the classification process.

The Image Registration Application is used to remove a variety of distortions from remotely sensed imagery and supports analyses requiring area mensuration. This Application also provides

TABLE 19-5

ERIPS Pattern Recognition Application
(After JSC, 1979)

Module	Capability
Field Selection	Allows user to define training areas and test sites for subsequent image classification.
Statistics	Calculates spectral means and covariance matrices for user-defined training areas. Allows user to display training signatures and statistics and to redefine training areas.
Divergence	Uses Statistics output to derive divergence measure for defining best set of channels for classification.
Classification	Performs supervised classification based on maximum-likelihood or Bayes decision-rule.
Clustering	Performs unsupervised classification using a single pass or interactive clustering algorithm.
Classification Map	Produces color-image display or printer map of classification or clustering results.
Report	Generates reports which format and display results produced by the above processors.
Probability of Error	Uses analytical model to predict classifier results as a function of training statistics. Allows user to evaluate effects of changing training signatures.
Statistics Modification	Allows user to display and modify training statistics.
Homogeneity Test	Tests for spectral uniformity of classified regions.
Pixel Value	Allows user to view grey-level values for selected picture elements.

capabilities for registering multitemporal data sets and for registering images generated by two entirely different sensors. Images can also be registered to a Universal Transverse Mercator (UTM) or a latitude/longitude map grid.

Advantages of the ERIPS System

The ERIPS software system provides powerful multispectral analysis capabilities for both the batch and interactive system user. The software is modular, well documented and debugged, and can be easily expanded to include new Applications. The menu-driven user-system interface is easy to learn and to use. ERIPS has been used extensively at Johnson Space Center in support of agricultural remote sensing applications including the Large Area Crop Inventory Experiment (LACIE). ERIPS is attracting new users and will continue to evolve new Applications in the future.

VICAR SOFTWARE SYSTEM

The Video Image Communication and Retrieval System, VICAR, has been under development at the Jet Propulsion Laboratory (JPL) for over a decade. JPL's Image Processing Laboratory (IPL) began development of VICAR on an IBM System 360, Model 44 computer in 1966. Several versions of VICAR were developed for that system, and a version that could operate under the OS/MVT operating system on IBM computers was first developed in 1971 and then expanded significantly in 1974–1975 (Jepsen, 1976) for use on an IBM 360/65 computer system. The latest version of VICAR was released in 1979 and is available from COSMIC. VICAR is one of the most widely used image-processing software systems, having been installed at several NASA centers (Goddard Space Flight Center and the Ames Research Laboratory), and at numerous other facilities around the world. VICAR has been used to support the image processing requirements of a wide variety of remote sensing, astronomical, planetary and biomedical applications.

VICAR Processing System Features

The VICAR software executes on IBM 360/370/30XX computers under the standard IBM operating systems and utilizes commercially available equipment for all peripheral devices including remote terminals and image-display stations. Users can access the VICAR system through remote terminals, remote job-entry stations, and local image-display stations for both batch and interactive processing. VICAR programs are stored in on-line program library files and can be called for execution through any of these entry points.

VICAR was originally designed to enable users who are not computer or programming experts to perform complex image-processing tasks by specifying processing functions and parameters that control processing via an easily understood syntactical structure. The main consideration during the evolution of VICAR has been ease of use by an uninitiated user who wants to perform image-processing tasks. Commands are available that set up data-storage devices, invoke particular image-processing programs, enter parameters to control processing by each selected program, and store or display output imagery. A list of available VICAR commands is given in Table 19–6.

VICAR incorporates a general purpose digital-image format, and the applications software is designed to accommodate the VICAR format. A series of logging programs is available to convert imagery from its original form into the standard VICAR format. The logging programs read the digital image in its original format, extract auxiliary information (e.g. position, pointing, time tag, or other information), insert this information into an appropriate location in the VICAR format, and produce a VICAR-formatted image that can then be processed by any of the applications programs. It is thus possible to process imagery acquired by Landsat, synthetic-aperture radar systems, astronomical observing systems, biomedical systems, planetary spacecraft, and other sources with the same set of applications software in the standard VICAR system.

For short processing jobs handling subimage areas of up to 512 lines by 512 picture elements, response times of from a few seconds to one minute are typical, depending on the processing performed. The primary output products available include printed lists, tables and maps; black-and-white and color image displays; photographic image products; and computer-compatible tapes.

VICAR Software Characteristics

The VICAR software system has been under continual evolution and now incorporates several powerful features including: a capability for generating procedures which can invoke multistep jobs to perform sequences of image-processing tasks through a single control statement; an interactive executive, called LIBEXEC, which enables interactive image processing under the Time Sharing Option (TSO) supported by IBM's OS/MVT operating system; and a set of display software that controls image displays on a variety of image display devices during interactive processing. Interfaces to the commercially available MARK IV file management system, developed by the Informatics Corporation, provide semiautomated capability to catalog processed imagery by a series of image attributes that can be stored and interrogated through MARK IV.

The overall structure of the VICAR software system is depicted in Figure 19–3. The system is programmed in FORTRAN IV (50%) and IBM 360/370 Assembler Language (50%). VICAR is designed as a disk-oriented processing system in which image data and other working data sets are

TABLE 19-6

Selected Standard VICAR Commands

RESERVE BLOCK ALLOCATE A	Reserve temporary direct access storage (data sets)
SAVE CATLG	Reserve permanent data sets (imagery) for use in a subsequent job
FIND RELEASE	Access data sets (images) created in a previous job
READ PREAD WRITE PWRITE TAPE PTAPE	Specify device and data formats for tapes
EXEC E	Specify applications program to be used; specify input and output data sets and enter required parameters for that applications program
PARAMS P	Define a symbolic name for a set of parameters used by applications software
LABEL L RELABEL R	Specify label lines (annotation) to be added to an output data set (image)

loaded onto disk storage units prior to processing. The VICAR design incorporates an Executive routine, called VMAST, which monitors and controls activity within the system. VMAST is the only VICAR program that is permanently resident in core storage. The Executive executes VICAR command statements, monitors loading of applications programs into main memory from disk, as required, and passes user-specified parameters from VICAR syntax commands to the applications programs during execution. This design philosophy reduces the amount of memory required for VICAR operations. VMAST also contains programs needed for input/output, data management, system bookeeping and other common functions required by the Applications programs.

The second major component of the system is the VICAR Language Translator, VTRAN. VTRAN is designed to read image-processing control statements generated by users in a simple syntax that requires a minimum of computer knowledge. An example of the command syntax supplied by a user to perform a simple image-processing task is listed in Table 19–7. VTRAN translates the user-generated statements into IBM Job Control Language statements and VICAR program parameter cards in order to set up a standard computer job which can be executed under OS/MVT.

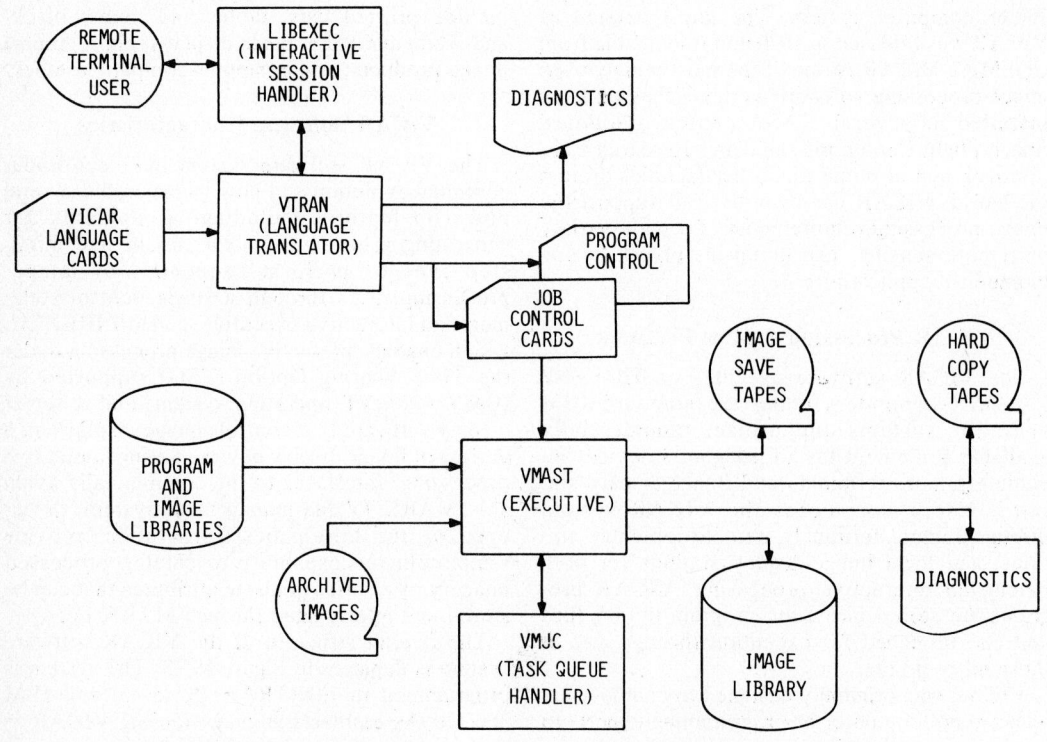

Figure 19-3. VICAR Software System Structure. Source: Owings, 1980.

TABLE 19-7

Typical VICAR Job Setup

VICAR Syntax	Explanation
REGION,160k	Establishes size for OS region during execution
TAPE,TI,ERT602,S,6	Allocates tape drive for Landsat CCT labeled ERT602 and indicates tape format; defines symbolic name TI
TAPE,SCRGRE,GRE,9	Allocates tape drive for output tape to be sent to film recorder
FIND,(JQP.940.DATA.ERA001),WB	Allocates OS data set, defines symbolic name WB for the data set
B,2,1000,1060,*,(A,B)	Allocates 2 blocked disk data-sets that will hold image, up to size 1000 lines by 1060 samples; defines symbolic names A and B for those data sets
E,VERTSLOG,TI,3,A	Reads Landsat CCT file 3; reformats into VICAR format in data set A (*E*xecutes the program VERTSLOG)
L,1,HUMBOLT COUNTY AREA	Adds annotation label to the output image
E,ASTRTCH2,A,B,(GAUSS,GSSG,2.5)	Performs gaussian contrast-stretch on the image; stores output image in data set B (*E*xecutes the program ASTRTCH2)
E,MASK,B,GRE	Writes output image onto film recorder tape for playback
END	Terminates this VICAR job

The third major component of VICAR is its LIBEXEC supervisor. LIBEXEC accepts control specifications from an interactive-terminal user working under IBM's TSO operating system and translates these specifications into VICAR command inputs which are processed by VTRAN. LIBEXEC handles all interfaces with TSO during interactive processing sessions.

The fourth major component of the system is the processing-task queue handler, VMJC. VMJC reads the task queue which is setup in OS Job Control Language by VTRAN and VMAST, and causes each task to be executed in sequence. It transfers control to the IBM operating system for execution, and then regains control after the execution of each application program and proceeds to initiate the next task as generated by VTRAN/VMAST.

The design of LIBEXEC, VTRAN, VMAST and VJMC isolates applications programs from the computer's hardware and operating system and simplifies the design of new processing programs. It also provides a high degree of software transferability since only these programs must be modified when changing operating systems or adding new hardware.

VICAR Applications Programs

The other major component of the VICAR system is its extensive libary of image-processing programs. There are hundreds of applications programs available at JPL, and a subset of this software is available through COSMIC. The material available from COSMIC includes validated software and user documentation that is generally useful to a wide range of image-processing applications. Software that is peculiar to the IPL

hardware and operating system, or to a specific planetary mission, is not normally transported through COSMIC. The COSMIC version of VICAR includes the basic VICAR system and approximately one hundred general-purpose image-processing programs that can be executed in batch mode on large scale IBM computers. Table 19−8 lists the major processing capabilities supported by VICAR applications programs. Seidman and Smith (1979) document with VICAR system capabilities.

TABLE 19-8

VICAR Image Processing Capabilities

VICAR Programs are available to perform the following functions:

- Image Annotation
- Fourier Analysis
- Image Labeling
- Magnification and Reduction
- Geometric Transformations
- Multispectral Classification
- Tiepoint Generation
- Pattern Recognition
- Image Mosaicing
- Polarization Analysis
- Image Registration
- Spatial Filtering
- Image Subsetting
- Image Synthesis
- Color Enhancement
- Tape Mounting/Positioning
- Image Arithmetic
- Transfer Function Compensation
- Contrast Enhancement
- Distortion Removal
- Image Display/Manipulation
- Interactive Image Analysis
- Histogram Manipulation
- Data Management
- Image Statistics
- Texture Analysis
- Feature Selection
- Edge Enhancement

VICAR Extensions

Two major enhancements of the VICAR system have been developed. The first was developed at NASA's Goddard Space Flight Center and is called the Small Interactive Image Processing System (SMIPS). The batch mode of SMIPS/VICAR is essentially identical to that of the VICAR system. Several VICAR enhancements were developed under SMIPS including extensions of the VICAR command language, additional input tape formats, the inclusion of many new image processing programs not contained in the original VICAR system, and interactive processing capabilities. SMIPS is available through COSMIC and is well documented [Moik, (1979a)]. The SMIPS version of VICAR has been widely disseminated and is in use by several corporations engaged in remote sensing applications.

The second major extension of VICAR was developed at the Jet Propulsion Laboratory and is called the Image Based Information System (IBIS). IBIS was designed as a geocoded information system to overlay, correlate, model and analyze multi-source data sets for remote sensing applications. There are over twenty IBIS programs (see Zobrist and Wilczynski 1977) which are available from COSMIC and which execute as standard VICAR processing programs. The functions and capabilities of the major IBIS routines are listed in Table 19−9. The applications and use of the IBIS software are documented in Bryant and Zobrist (1976), and Bryant, et al (1977).

Advantages of the VICAR System

The VICAR system is probably the most extensively used image processing system available. The VICAR command language is easy to learn and to use. The basic system design is uncomplicated and isolates the user and the processing programs from the operating systems. Localization of input/output operations and other common services in the VICAR Executive makes it easy to develop new applications programs. The IBIS extension to VICAR adds significant capabilities for combining, overlaying and analyzing multiple data sets including both image and non-image data. All of these attributes offer significant advantages to prospective users.

PRESENT MINICOMPUTER SOFTWARE SYSTEMS

This section describes the capabilities of five remote sensing software systems: the IDIMS, KANDIDATS, GIPSY, MINI-VICAR, and AOIPS systems. Each of these systems was designed to operate on a minicomputer, and each provides interactive image-processing support for remote sensing applications. Detailed comparisons of the processing features offered by the IDIMS and KANDIDATS systems versus the capabilities offered by the large scale computer-

TABLE 19-9

Selected IBIS Software Modules

Program	Function
VTRACT1	Processes Urban Atlas Census Tract tapes into VICAR disk file of census tracts in lat./long. coordinates.
VTRACTC	Processes Urban Atlas tapes and generates disk file of census tract centroids and tract identification numbers.
POLYREG	Registers and transforms polygon data sets in lat./long. coordinates to image line/sample coordinates.
POLYGEOM	Uses tie-points to accurately remove local distorations and register polygon data set to selected image.
POLYSCRB	Transforms polygon data set to raster image file of polygon borders.
PAINT	Converts image file or polygon borders to multicolored map.
CTRMATCH	Merges multicolor polygon image-file from PAINT with centroid file generated by VTRACTC.
POLYOVLY	Reads PAINT image polygon-file and another image such as a classification map. Generates tabular output file which lists all polygon regions by name and ID number and which gives number of picture elements of each class in a given polygon.
REPORT	Reads POLYOVLY tabular file and generates statistical summaries and reports.

based software systems, described in the previous section, are discussed by Carter et al. (1977). The selection and discussion of these systems are illustrative: There are several dozen small to medium or even large scale systems now operating on mini and microcomputers.

IDIMS SOFTWARE SYSTEM

The Interactive Digital Image Manipulation System, IDIMS, was developed by the Electromagnetic Systems Laboratories (ESL), Incorporated of Sunnyvale, California. The current version of the IDIMS system is marketed by ESL as a turnkey system and includes a complete minicomputer with related peripheral equipment, extensive image processing software, a complete documentation package, and user training. The system has been used successfully for a variety of remote sensing and pattern-recognition applications.

IDIMS Processing System Features

The IDIMS software executes on a Hewlett-Packard 3000 minicomputer under the Multiprogramming Executive operating system. The system can be configured to include local and remote typewriter and CRT display terminals, and local image display stations. A version of the system is available which includes an array processor for rapid bulk-image processing operations. The IDIMS software is stored on disk files and can be called for interactive or batch processing. The system supports multiple batch and interactive users, simultaneously.

Image data are entered into the system from magnetic tapes in the LARSYS, JSC Universal, or Landsat computer compatible tape formats. Images and working data sets are maintained on disk storage units during processing operations. The system also accepts a wide variety of digitized ancillary data including surface-truth information, digitized photographs and maps, topographic data, and related digital information. The output products available include printed lists, tables, plots, and maps; black-and-white and color photographic products and TV image displays; and computer-compatible tapes.

IDIMS Software Characteristics

The IDIMS software was designed as an image analysis tool for the non-programmer user. As described in ESL (1978), it provides capabilities for automated file and disk space management, and for cataloging images and information describing stored images. The system maintains an image-history file containing complete information on the processing history of stored images including all commands, data parameters, input-image names, and processing functions performed on an image.

The user can interact with processing programs during program execution. Job-control specifications and processing parameters are specified by entering system commands in a keyword and parameter format, using simplified punctation. Command language statements can span multiple lines and can be strung together for input to the IDIMS language processor. The system allows users to name images for identification purposes and accepts processing commands according to the following syntax:

Input Image > Function Specification
> Output Image

This format allows the user to specify multiple input images, multiple processing functions, and/or multiple output images in a single command statement. An entire sequence of processing functions can be specified in a single command by listing multiple processing functions. Any intermediate image files required during a multi-step process are created, managed and deleted automatically by the system. The input and output image specifications usually consist of image names which can be up to 35 characters in length. The function specification is the name of a valid system-processing function.

The user can specify language statements in two modes: in menu mode or in command mode. The menu mode is exploited by users not familiar with the system, while the command mode offers advantages for experienced users who do not require the extensive prompting services of the menu processor.

In menu mode, the user interactively specifies Processing, System Control, and Display Control Commands. The menu-mode interface functions as a preprocessor to the IDIMS language processor. The menus are presented to the user on CRT display terminals. The normal menu flow is designed as a tree structure where the system leads the user through successive layers of menus, and prompts him for needed parameters in order to compile control-inputs for the language processor. The menus are stored as disk files and can be easily updated by including new menus or menu entries. The menu processor compiles a history file of the menus chosen, and the parameters and commands specified during an analysis session for later use in documenting or duplicating processing results. The menus also provide function keys which allow the user to randomly select the next processing menu rather than to continue sequentially down a particular branch of the menu tree.

In the command mode, the user enters command statements and parameters directly but can also be prompted by the system to supply parameters interactively during program execution. The command-language processor performs extensive syntax and error checking and allows the interactive user to correct errors as they are detected. In both the menu and command modes, the system provides default values for all required parameters left unspecified by the user.

IDIMS Applications Programs

There are over a 150 processing programs in the current version of IDIMS. Table 19–10 lists the types of processing available. The system allows the user to process full or partial multispectral images by specifying the exact image area and spectral channels to be processed.

IDIMS allows the user to interact with processing programs and to display and generate intermediate and final output products. The system ingests digital image data from magnetic tapes and stores the input data and working image files on disk storge units. The system provides capabilities to correct, filter, enhance and perform Fourier analyses of image data. The user can enhance an input image file and perform radiometric and geometric corrections prior to executing an unsupervised or supervised classification. Training areas, test sites, and related statistics can be generated, refined and displayed during classification operations. The IDIMS software is very flexible and accomodates a wide variety of image-processing and pattern-recognition applications.

TABLE 19-10

General IDIMS Processing and Display Functions

Type	Capability
Utility Functions	Over 25 utility programs to print, copy, subset and manipulate images, histograms and related data.
Arithmetic Functions	Performs simple arithmetic and logical operations or images.
Intensity Transformations	Performs a variety of grey-level transformations on selected image.
Geometric Transformations	Provides capabilities to invert, expand, rescale, rotate, geometrically correct, shift and transpose an image.
Complex Data Manipulation	Provides extensive complex data-manipulation capabilities including autocorrelation, cross correlation, and power-spectrum analyses.
Fourier Transforms	Calculates forward/reverse Fourier transforms of a complex or real valued image.
Filtering	Provides a variety of filtering algorithms including programs for motion-blur compensation, lens defocusing, spatial filtering, etc.
Image Input/Output	Performs a variety of image copying, formatting, mosaicing and output generation tasks.
Multispectral Data Processing	Performs supervised and unsupervised classifications using a flexible set of programs; provides training-area selection, statistics computation, histogram generator and a variety of classification algorithms.
Classification Summary	Performs operations for combining statistics files and comparing or summarizing classification results.
Image Display	Displays, zooms, windows and scrolls images. Overlays images with annotation and ancillary data.

Session-history logs are maintained by the system for optional printing at the conclusion of a user-analysis session. The session-history printout serves to document the entire sequence of operations performed and includes the names of all images processed or generated along with the functions and parameters specified by the user during the entire processing sequence.

Advantages and Disadvantages of the IDIMS Software

The IDIMS software is modular in design and easily accomodates new applications programs. The system automatically controls input/output operations and automatically tracks all input images and working data sets used during processing. This feature frees the user and applications program from routine bookkeeping tasks. The image-history files and session-history logs make it easy to document and duplicate processing sequences performed during analysis sessions.

The IDIMS software has been used extensively and is well debugged and documented. The availability of IDIMS as a turnkey system, with backup support from ESL, makes IDIMS an attractive, cost-effective system for organizations specializing in interactive image-analysis applications and related remote sensing research.

The principal disadvantage of the IDIMS system is also its principal advantage, namely that the hardware and software are sold as a turnkey package. Thus it may be infeasible for small facilities to acquire because of the total system cost involved, although it may be very satisfactory for specialist groups. Similar situations occur with other commercial systems.

KANDIDATS SOFTWARE SYSTEM

The Kansas Digital Image Data System, KANDIDATS, was developed at the Remote Sensing Laboratory of the University of Kansas. KANDIDATS offers a versatile set of processing capabilities for correcting, enhancing, compressing, transforming, clustering, classifying and displaying digital image data. It is designed as a flexible research tool to allow users to formulate and develop processing algorithms which can be applied to large volumes of image data. The system has been used successfully to support a variety of applications in remote sensing including texture analysis, image compression, multispectral classification, and related image analysis tasks.

KANDIDATS Processing System Features

The KANDIDATS software executes on a Digital Equipment Corporation PDP-15 computer operating under DOS. The system includes local CRT display terminals and image TV display stations. It supports both interactive and batch-image processing operations.

Image data are input from magnetic tapes in a variety of formats and are stored internally on disk units in a standard image-file format. Multispectral images are permitted with each channel containing either numeric or symbolic values.

This allows the inclusion of ancillary data as additional image features or data overlays for analysis operations. In addition, image files can be stored in integer, floating point, double integer, double-precision floating-point formats and can be manipulated by standard FORTRAN IV input/output statements for sequential or random access.

A standard image file begins with an identification record which indicates the size of the stored image, its minimum and maximum grey levels, the number of quantization steps from the minimum to maximum grey level, the number of channels, the resolution-element size, the number of symbolic channels, and the data type (i.e. integer, real, etc).

The next section of a standard image file is called the history record section. Here, every processing operation and its parameter set, as applied to the original input image, are recorded in sufficient detail to allow complete recreation of the processing sequence that created the stored image.

The last portion of the image file is the image data. Detailed descriptions of both the image file and the KANDIDATS image-access protocols are found in Haralick et al (1977).

The output products available include printed lists, tables, plots, and maps; black-and-white and color photographic products and TV image displays; and computer-compatible tapes.

KANDIDATS Software Characteristics

KANDIDATS is designed to provide users, having varied degrees of familiarity with computer systems, an easy access to a powerful set of image-processing capabilities. The system accepts free-form command strings entered by a terminal user either directly or in response to system menus and prompts. The user can also create command-string files interactively for later execution of batch-processing jobs. During command-string specification, the system provides default options for all required parameters.

The KANDIDATS command-string interpreter permits user specification of the processing command, input file name(s), and output file name according to the syntax:

COMMAND OUTPUT IMAGE < INPUT
IMAGE(S) (FLAGS)

Use of the flags specification allows the user to specify which of the standard system defaults are to be used. After the command-string interpreter verifies that the image-file names are syntactically correct and the specified command is valid, it passes all the input information to a command-driver subroutine through a labeled common area.

For ease in debugging KANDIDATS software, all programs in the system push their names on a stack in their first execution statement and pop their names off the stack prior to normal exit. The stack thus maintains a record of the subroutines entered, and in the order in which they were entered. If an error occurs in a subroutine, the subroutine exits without popping its name off the stack. When control reaches the user-interaction level, the stack can be dumped to find the trace of subroutines calls. Also, by setting a system flag on the command string every time a subroutine name is pushed onto or popped from the stack, the name of the subroutine entered or exited from can be displayed, thereby affording the user a dynamic trace of subroutine calls.

The system is designed to provide help to the inexperienced user enabling him to execute system commands without prior knowledge of system details. For example, system commands can be executed to obtain detailed explanations of processing-command formats, processing-function requirements, and input/output parameter specifications; and a menu function is provided to simplify user selection and specification of commands.

KANDIDATS Processing Programs

KANDIDATS provides capabilities for performing a myriad of image-processing functions by executing programs contained in one of four subroutine packages. The Utility package, the Cluster package, the Pattern Discrimination package, and the Image Transformation package. Table 19–11 lists selected capabilities of the KANDIDATS program packages. Haralick and Minden (1978) pro-

TABLE 19-11

Selected KANDIDATS Processing Capabilities

Package	Capability
Utility Package	Provides a wide range of utility operations to read, display, subset, create, mosaic, combine, overlay, register, and correct images. Provides capabilities for image arithmetic, ratioing, histogramming, linear combination, content mapping, masking, transforming images and computing image statistics.
Cluster Package	Performs spatial and spectral clustering, edge extraction and textural analysis; is a versatile package.
Pattern Discrimination	Performs feature selection, decision-rule creation, multispectral classification by table-lookup algorithm, and evaluation of classification results.
Image Transformation	Performs image filtering, data compression, image coding, and image transformation operations on subimages.

vide additional information on the capabilities of the KANDIDATS packages.

The image processing capabilities provided by KANDIDATS are flexible and easily modified during processing operations. The processing options offered cover the complete spectrum of image-processing techniques required for remote sensing applications.

Advantages and Disadvantages of the KANDIDATS Software

The KANDIDATS system can be purchased for low cost and offers a complete set of processing programs needed for remote sensing support. The system is modular in design, easy to learn and to use, and is easily enhanced by the addition of new applications programs. The user-system interface accommodates users with a varied computer-systems background and provides effective access to KANDIDATS processing capabilities. The major disadvantage of the system is its relatively high degree of machine dependence, a feature which has been nearly eliminated in the system that developed from KANDIDATS, viz. GIPSY (General Image Processing System).

GIPSY SOFTWARE SYSTEM

Before considering details of the GIPSY software system it is appropriate first to consider portability issues since such issues crucially influenced the design of GIPSY, which from the beginning was intended to have a high degree of portability.

Image-processing applications packages are an important part of the computing workload in remote sensing. Although there are important differences among these package programs, they share the following:

(1) A package is large—the inclusion of hundreds of thousands of lines of code is usual.
(2) A package makes significant use of the service facilities of the operating system underlying it, particularly services like files, memory control, and input-output streams.
(3) A package contains a great deal of special expertise from its field of application—its algorithms are peculiar to that field, and their programming is less difficult than the ideas behind them.

It is the interaction among these features that makes package programs so common and so useful. The package captures what practitioners of a field need and does so in a form that requires far less knowledge of computers, and of the field itself, than that required to program a problem directly. Once the effort has gone into creating a package, there is considerable pressure to use it with many machines, operating systems, and installations. Because so much of the content does not really depend on the computer implementation, but rather on the special algorithms involved, the "conversion" of packages is feasible, and much-attempted. It is unfortunate that sometimes the task is harder than it first appears to be, and weeks stretch into months with the package still not "up" on the new system.

Portability can be achieved. The technique is to encapsulate and standardize all operating-systems service calls. In this way system communication occurs without regard for details that change from system to system, task to task, and problem to problem.

Operating systems all support the needs of packages for file operations, memory and process control, and direct input/output. As mentioned, machine-independence is achieved by obtaining these services only through specified subroutine calls which are the same from machine to machine. This collection of entry points constitutes a kernel that makes all operating systems look the same to the package.

Three forces shape the definition of this operating system interface:

(1) The services must be sufficiently powerful and easy to use.
(2) The kernel must be able to be built around any existing operating system.
(3) The implementation of the kernel for a new system is easy for a local systems expert.

The first two forces tend to increase the size of the kernel; the last limits its size.

In an operating system whose services are nearly the same as those of the interface, the kernel is only a calling-sequence converter, transforming the subroutine calls into monitor calls with some additional error checking added. When the system provides very different services, however, it may take a lot of code to build them out to the proper form.

Transportability is achieved when the application software uses restricted code-calling only on the standardized operating system interface for all operating system services. Moving to a new machine then requires only the implementation of the kernel supporting the standard calls. GIPSY employs a kernel of routines that interface to the peculiar operating system of each machine, providing sophisticated but standard operating system services. This kernel makes the operating system of each computer appear identical to that of all others and does not pose a difficult implementation problem. Above this interface, all GIPSY code is machine-independent.

The General Image Processing System, GIPSY, is under continuing development at Virginia Polytechnic Institute and State University. GIPSY offers image processing capabilities similar to those found in the IDIMS and KANDIDATS minicomputer based systems. It operates on a VAX 11/780 under VMS, and on an IBM 370 under CMS. It thus serves the high end of the minicomputer spectrum as well as larger mainframe systems.

GIPSY is a general interactive image-processing software package designed to be easily used, easily learned, easily modified, and easily transported from computer to computer. It works with single or multiband images which can be in integer or real format. Its capabilities include almost 200 operations to do image filtering, classification, geometric spatial transformations, numeric and symbolic recursive and nonrecursive neighborhood operations, spatial clustering, region-growing and property-file generations. GIPSY is user-friendly and has all its user documentation on-line and available through the GIPSY commands. User interaction with GIPSY is done first with a command string which then is followed by questions and answers. The result of all user interactions on each image is kept permanently as part of the image in history records. This facility minimizes user bookkeeping.

GIPSY Image Processing Capabilities

GIPSY handles single or multi-band images in real or integer mode. Bands of an image can be designated in numeric or symbolic form. Numeric bands contain grey tone values and can be operated on with arithmetic operators. Symbolic bands contain indexes for categorizing names and can be operated on with label propagation or boolean operators.

GIPSY has capabilites for remote sensing classification tasks, filtering, geometric spatial transformations, test image generation, image utility operations, label propagation operations, numeric and symbolic neighborhood operations, single band and multi-band point operations, Hough transforms, clustering, and region property-file creation and manipulation.

Remote sensing classification operations include ground truth and training set input, best band selection, nonparametric Bayes classification, Gaussian classification, and contingency-table computation.

Filtering operations include forward and reverse discrete Fourier Transform, forward and reverse discrete Cosine Transform, rectangular-to-polar and polar-to-rectangular conversion, Gaussian and rectangular-box filter smoothing, and use of arbitrary, specified linear filters in spatial or frequency domain.

Geometric transformations include flipping, skewing, rotating, transposing, arbitrary linear transformations, and geographic-coordinate transformations.

Test images can be generated for standard bar and checkerboard patterns, random noise patterns, and wallpaper-like repeating patterns.

User aids for examining an image include printing numeric or symbolic images with characters suitable for a standard printer, examining image-identification block and descriptor records, examining and editing pixels in the image, and histogramming. Image utilities include appending, blocking, combining, expanding, compressing, and subimaging.

Symbolic label-propagation operations include connected-region shrinking, determining maximally connected components, region filling, shrinking, cleaning, boundary marking, and computing reachability-regions from extrema.

Numeric neighborhood operators include median filter, Laplacian, Roberts gradient, Box filter smoothing, edge detecting, and removal of spot noise.

Point operators include all arithmetic operations between two images, linear combining, absolute difference, all Boolean functions between two symbolic images, computing image norms, inverting an image, taking absolute values, taking logarithms, exponentiating, raising an image to a power, thresholding, and quantizing.

Other operators can create and manipulate the region property file from a segmentated image, compute the region adjacency graph, compute the generalized occurrence matrix, merge regions, classify regions and do minimal spanning tree-clustering with region properties.

GIPSY System Features

GIPSY has a variety of different and convenient ways of combining image operator names with input and output image file names into a command string. For example, take the process of quantizing which might have the command name QUANT. Legal command-string syntaxes in GIPSY include:

QUANT   OUTPUT IMG < INPUT IMG

QUANT   INPUT IMG < OUTPUT IMG

INPUT IMG > QUANT > OUTPUT IMG

OUTPUT IMG = QUANT (INPUT IMG)

When the image-processing task does not produce another image but produces statistics about the input image, there is the need to specify whether these statistics should be put on the user console screen for examination, on a line printer for hard copy, or be made into a source file to be saved on the mass store device. GIPSY allows natural variations of the basic command string which easily permits these distinctions to be made. Take, for example, the process of getting a histogram of the grey tones on an image. Suppose the command name is HIST. The following three GIPSY command strings can direct the histogram to the terminal, to the line printer or to the disk pack as a file called TONE HIST:

HIST TT < INPUT IMG

HIST LP < INPUT IMG

HIST TONE HST < INPUT IMG

Once the basic combination of input image, output image, and command name have been specified, additional processing parameters may be required. The equal-interval quantizing command, for example, will need to know how many quantized levels the user desires for the output image. Other processes such as the histogram command may need to know what range of image values should be matched in the histogram. In this case the user would specify the lower and upper limits by responding to a question.

Because there are many times when the same (or almost the same) processing steps may be run many times, a batch-like mode of operation is implemented in GIPSY. This batch-like mode can save the user from having to enter the same sequence each time. The information (command lines and parameters) can be put into a file with a standard text editor and then executed by the GIPSY RUN operator.

GIPSY History Records

Interactive image processing usually proceeds on a step by step basis with many intermediate images created and then deleted before a final classified or enhanced image is produced. If the user has not kept careful notes about how it was that he produced the final image, all the processing may be lost due to a possible forgetting of what processing steps he had done to his initial image to achieve his final image. To take care of this, GIPSY employs history records which are permanently part of each image file and which indicate everything done to the image from the time of its creation.

GIPSY Software Characteristics

GIPSY employs a monitor to take care of the flow of control. A typical operation sequence would be for the monitor to call the command-string interpreter or menu-prompting routine to accept a user-processing assignment. Once a processing option has been selected, input and output image-file names given, and source of destination devices assumed or given, control is returned to the monitor.

The monitor next calls the process-driver subroutine and passes to it all of the information that it has received from the menu or command-string interpreter. Depending on the task, various parameters may need to be obtained. The process driver can call various character I/O prompting subroutines, each one asking a specific question, to obtain the required information. When all parameters are specified consistently, the process driver allocates the available memory and then calls the processing subroutine specifying all the information it has in the calling argument list.

The processing subroutine initializes proper input and output image files and copies the descriptor records associated with the input image(s) to the descriptor records associated with the output image. Then image processing begins by accessing the image records and calling a number-crunching routine which does the actual processing. On return from the number-crunching routine, the image-processing routine writes the processed-image record to the output-image file and accesses the next image record on the input image. When all records have been processed, the files are closed and control is passed back to the driver which returns control to the monitor.

MINI-VICAR SOFTWARE SYSTEM

The mini-VICAR software system was developed at the JPL's Image Processing Laboratory for execution on Digital Equipment Corporation PDP-11 minicomputers under the RSX-11 M operating system. Mini-VICAR includes a subset of the full VICAR software described earlier in this chapter. The same emphasis on easily understood user syntax and optimized input-output operations for image processing featured in the VICAR system has been carried forward into mini-VICAR.

MINI-VICAR Characteristics

The mini-VICAR system permanently allocates disk storage areas for image files and working data sets for each processing task, thus reducing the overhead involved in dynamic allocation and management of data files. All input/output in mini-VICAR is performed using general operations to provide more efficiency than is available through the standard DEC Files-11 processor. The mini-VICAR programs are implemented in FORTRAN-IV PLUS.

The system is designed in a modular fashion, and modules which interact with the operating system can be isolated and modified to operate under other operating systems or on computers from other manufacturers.

Mini-VICAR tasks can be invoked directly by a user from a terminal, from an indirect command file stored on disk, or from a user-developed applications program. Most tasks include default-parameter options which can be overridden by the user during execution.

The basic mini-VICAR command syntax is very similar to that of the VICAR system. The applications programs currently included in mini-VICAR are listed in Table 19–12. Complete system and user documentation, mini-VICAR source code, and required command files for installing the system are available from COSMIC. System capabilities are fully described in COSMIC (1979).

Mini-VICAR Extensions

A PDP-11 minicomputer version of the IBIS extension to VICAR, as described earlier in this chapter, is also available from COSMIC for use with mini-VICAR. Mini-IBIS includes most of the full IBIS capabilities for geocoded information management and analysis and is described in de-

TABLE 19-12

Mini-VICAR Applications Software

ATT—Add to Tape
DIF—Difference of Two Pictures
DISP—Display Picture on Line Printer
DTD—Disk to Disk Dataset Transfer
DTT—Disk to Tape Dataset Transfer
EXPAND—Increase Picture Size
FILTER—Apply Convolution Box Filter to Picture
FLIP—Invert a Picture
GEN—Generate Picture
GEOM—Image Transformation
INSECT—Insert Sector of One Picture into Another
LAB—Add a Label
LIST—Print Picture Information
MASK—Add Border to Picture
MF2—Mini-VICAR F2
MTT—Magnetic Tape Title
PAT—Print Access Table
PAUSE—Pause
PICREG—Register Picture Interactively
PRO—Protect and Release Datasets
PSUM—Picture Sum
RELAB—Relab a Picture
ROT—90-Degree Rotate
RTT—Rewind Tape
SAR—Segment Averaging Routine
SHRINK—Decrease Picture Size
STA—Sum to Average
STR—Stretch the Contrast of a Picture
TCOPY—Tape Copy
TSCAN—Tape Scan
TTD—Tape to Disk Dataset Transfer
UIC—Set Terminal UIC
UPLAB—Update Label

tail by Zobrist and Bryant (1979) and in COSMIC (1979).

An extension of mini-VICAR which incorporates a powerful system Executive is under development at NASA's Goddard Space Flight Center. This system, although designed to process Landsat-D data, is useful for a wide range of interactive minicomputer-based image-processing applications, and is described by Bracken et al (1979).

AOIPS SOFTWARE SYSTEM

The Atmospheric and Oceanographic Information Processing System (AOIPS) was developed as an interactive, minicomputer-based system at NASA's Goddard Space Flight Center. AOIPS is used primarily for image analysis and information extraction operations in support of a variety of remote sensing applications in the weather, climate, oceanography, and earth resources disciplines. AOIPS development was initiated in 1975 and is expected to continue through the 1980's with the addition of software packages to support interactive analysis of data derived from the Landsat-D program, NASA's Climate Program and related remote sensing investigations.

AOIPS Processing System Features

The AOIPS software executes on Digital Equipment Corporation PDP-11 minicomputers as shown in Figure 19−4. A variety of CRT display terminals and image display stations are attached to the system. As shown in the Figure, Image

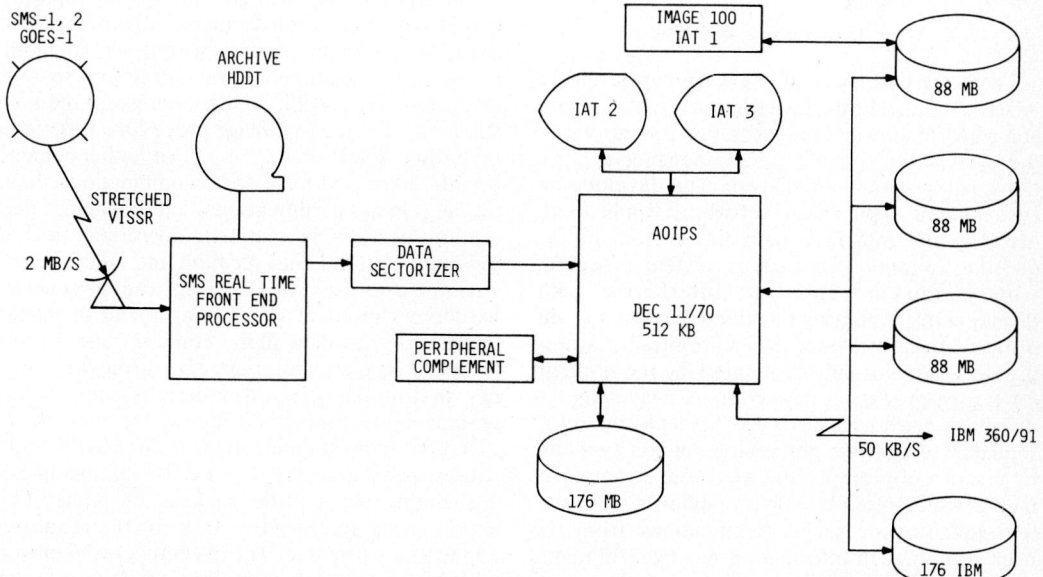

Figure 19-4. AOIPS System Configuration. Courtesy: J. T. Dalton, NASA/Goddard Space Flight Center.

Analysis Terminal 1 is a General Electric Company Image 100 system and Image Analysis Terminals 2 and 3 are state-of-the-art image-processing terminals designed at the Goddard Space Flight Center and built by Hazeltine Corporation. The latter terminals provide powerful hardware capabilities, are fully controlled by the host computer, and are commercially available from Hazeltine. The PDP-11 processors used in AOIPS are interconnected by shared disk units and use both the RSX-11 D and RSX-11 M operating systems. A more detailed description of the system hardware is presented by Bracken et al. (1978a).

The AOIPS is interconnected to a large scale IBM System 360/91 computer. This connection allows AOIPS users to transfer data sets to and from the 360/91 computer and to execute large-scale applications models or the SMIPS/VICAR image-processing system, described earlier.

The system utilizes computer-compatible tapes of multispectral image data generated by satellite and aircraft image scanners as well as digital ancillary data required to perform specific information extraction operations. The system accepts image data in several formats including the Landsat, LARSYS, JSC Universal, SMIPS/VICAR, and NOAA Ingest tape formats. Input-image data and working data sets are stored on disk during processing operations.

The typical output products available include shade prints of image areas; printed lists, tables, plots and maps; color and black-and-white photographic products of images or images overlayed with contours, plots and annotation; color displays of images and images overlayed with a variety of information; and computer-compatible data tapes.

User Interface to AOIPS

There are two types of users supported on the AOIPS: applications investigators who specify and perform information-extraction operations on the system for specific remote sensing applications, and programmer/analysts who develop new systems and applications software. Applications investigators interface with the system by requesting execution of various system processes from menus presented on interactive CRT display-units. Adopting the menu approach as the primary user interface has simplified training problems and virtually eliminated the requirement for a user to learn a new computer language. A discipline-oriented user is led through the entire sequence of possible processing options available by a series of prompts and a layered structure of menus which require simple responses. The system solicits processing specifications from the user until enough information has been obtained to execute the desired process and display processing results.

During an analysis session, the user has several options available for displaying intermediate results, generating output products, and defining sequences of processing operations. Operations may be terminated with the capability to restart in subsequent analysis sessions on the system. The user has the option to start and to save/restart analysis sessions using magnetic tape and/or disk units to store data sets and processing information.

The second type of system user, the programmer/analyst, typically utilizes operating-system facilities and the operating-system commands to code, compile, link, edit, and checkout the software being developed. The programmer/analyst does not usually interact in the menu environment until the final checkout and integration phase in the development of a new applications-software package.

AOIPS Software Characteristics

The AOIPS software has been designed as a modular, interrelated set of processing programs which use common data sets and internal data formats. This feature allows users to access processing capabilities of individual software packages even though these packages may have been developed for dissimilar applications.

The AOIPS system software is designed to provide automatic cataloging and tracking of data sets once the data have been entered into the system. At appropriate places in the information-extraction process, the user selects the desired data sets to be processed from data-selection menus. These menus typically display catalogs of data sets currently resident on system-storage devices and allow the user to input additional data sets and name new data sets for future reference where appropriate.

The system provides capabilities for inputting digital data in several formats; performing standard image preprocessing operations, including image registration, geometric correction, scanner distortion correction, and zooming and reducing subimages; executing image processing functions, including level slicing, contrast enhancement, pseudo color and false color combinations, band ratioing, linear combinations, and histogram generation; performing analytical operations, including multi-spectral classification and special applications analyses; displaying and generating hardcopy output of single images and of images overlayed with data plots, contours and various forms of image annotation; and displaying imagery in time-lapsed, three-dimensional stereographic color displays for interactive analysis.

A user requests execution of an AOIPS applications package by typing a RUN command and designating the specific package by name. This action starts an executive task for the requested applications package. The executive task controls all subsequent menu processing and initiates other system tasks as necessary to perform user-selected functions.

All AOIPS applications packages interface with the user through a hierarchical structure of menus. At any stage of processing, the user is presented with a numbered list (menu) of processing operations available. The user selects an option from the menu by typing in the number corresponding to that option. The result of the selection will be either the display of the next menu level of options or the execution of an applications task to perform a specific function.

AOIPS Application Software

Numerous software packages are available on AOIPS; some are geared to the requirements of a specific application and some are of a general support nature. Figures 19-5 and 19-6 present an overview of the major software packages available. The most heavily used applications packages include:

METPAK—An interactive meteorological analysis system which provides image navigation, registration and other processing capabilities needed to support 4-dimensional cloud-motion studies. METPAK computes cloud heights and the displacement of selected clouds in a series of time-lapsed single images or stereographic image pairs. It generates wind-vector fields and a variety of analyses of the derived wind fields. Results are displayed as color or black-and-white images in time lapsed 2-D and 3-D displays. Images can be overlayed with data plots, contours and annotation. Hardcopy photo and printer products are also provided.

CLASSPAK—An interactive maximum-likelihood multispectral classification package. CLASSPAK accepts up to 24 channels of input data, provides statistics computation, and performs a maximum-likelihood classification using up to 8 selected channels. CLASSPAK provides a highly flexible, interactive environment for selecting, defining, displaying, editing, and combining training sites and training-site statistics. It generates classification maps of up to 25 classes and displays of class correlation and covariance matrices as well as statistical analysis results.

Aircraft Sensor Analysis Package (ASAP)—An interactive software package for displaying, processing, correcting, and enhancing multispectral

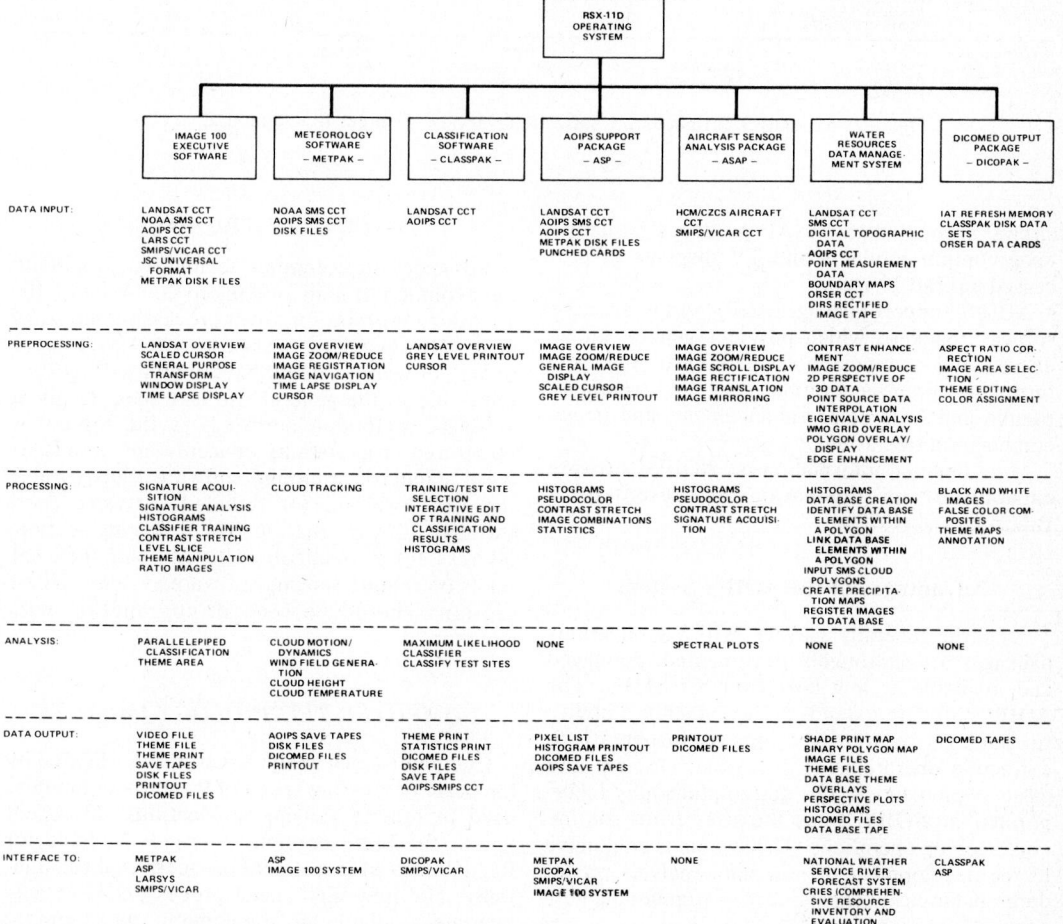

Figure 19-5. AOIPS Terminal 1 Software Overview. Source: Bracken, et al., 1978.

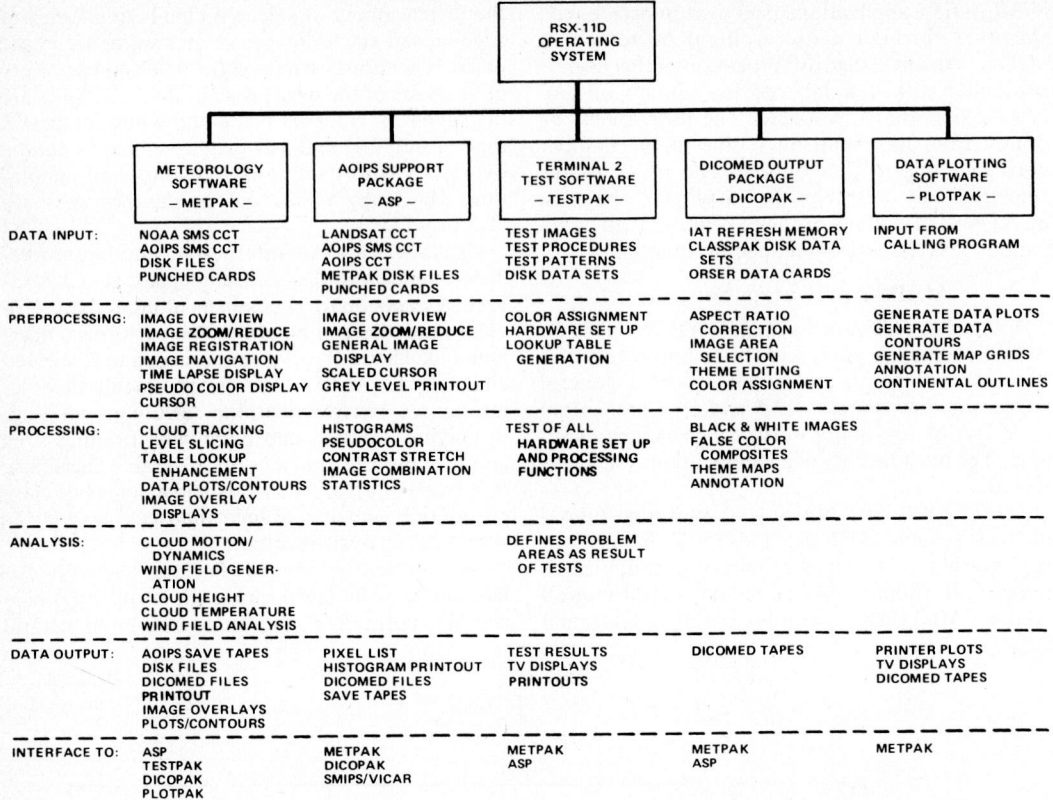

	METEOROLOGY SOFTWARE – METPAK –	AOIPS SUPPORT PACKAGE – ASP –	TERMINAL 2 TEST SOFTWARE – TESTPAK –	DICOMED OUTPUT PACKAGE – DICOPAK –	DATA PLOTTING SOFTWARE – PLOTPAK –
DATA INPUT:	NOAA SMS CCT AOIPS SMS CCT DISK FILES PUNCHED CARDS	LANDSAT CCT AOIPS SMS CCT AOIPS CCT METPAK DISK FILES PUNCHED CARDS	TEST IMAGES TEST PROCEDURES TEST PATTERNS DISK DATA SETS	IAT REFRESH MEMORY CLASSPAK DISK DATA SETS ORSER DATA CARDS	INPUT FROM CALLING PROGRAM
PREPROCESSING:	IMAGE OVERVIEW IMAGE ZOOM/REDUCE IMAGE REGISTRATION IMAGE NAVIGATION TIME LAPSE DISPLAY PSEUDO COLOR DISPLAY CURSOR	IMAGE OVERVIEW IMAGE ZOOM/REDUCE GENERAL IMAGE DISPLAY SCALED CURSOR GREY LEVEL PRINTOUT	COLOR ASSIGNMENT HARDWARE SET UP LOOKUP TABLE GENERATION	ASPECT RATIO CORRECTION IMAGE AREA SELECTION THEME EDITING COLOR ASSIGNMENT	GENERATE DATA PLOTS GENERATE DATA CONTOURS GENERATE MAP GRIDS ANNOTATION CONTINENTAL OUTLINES
PROCESSING:	CLOUD TRACKING LEVEL SLICING TABLE LOOKUP ENHANCEMENT DATA PLOTS/CONTOURS IMAGE OVERLAY DISPLAYS	HISTOGRAMS PSEUDOCOLOR CONTRAST STRETCH IMAGE COMBINATION STATISTICS	TEST OF ALL HARDWARE SET UP AND PROCESSING FUNCTIONS	BLACK & WHITE IMAGES FALSE COLOR COMPOSITES THEME MAPS ANNOTATION	
ANALYSIS:	CLOUD MOTION/ DYNAMICS WIND FIELD GENER- ATION CLOUD HEIGHT CLOUD TEMPERATURE WIND FIELD ANALYSIS		DEFINES PROBLEM AREAS AS RESULT OF TESTS		
DATA OUTPUT:	AOIPS SAVE TAPES DISK FILES DICOMED FILES PRINTOUT IMAGE OVERLAYS PLOTS/CONTOURS	PIXEL LIST HISTOGRAM PRINTOUT DICOMED FILES SAVE TAPES	TEST RESULTS TV DISPLAYS PRINTOUTS	DICOMED TAPES	PRINTER PLOTS TV DISPLAYS DICOMED TAPES
INTERFACE TO:	ASP TESTPAK DICOPAK PLOTPAK	METPAK DICOPAK SMIPS/VICAR	METPAK ASP	METPAK ASP	METPAK

Figure 19-6. AOIPS Terminals 2/3 Software Overview. Source: Bracken, et al., 1978.

aircraft-scanner data. ASAP produces hardcopy image output products and TV displays of processed aircraft imagery.

AOIPS Support Package (ASP)—An interactive software package which provides general image display and manipulation capabilities. ASP performs histogram generation, contrast stretching, pseudo and false color enhancement, and image combination operations.

More detailed information on AOIPS software capabilities and applications is presented by Bracken et al (1978b).

Advantages of the AOIPS System

All of the generally useable AOIPS applications packages are thoroughly documented, debugged and available at low cost from COSMIC. The AOIPS software extends an investigator's ability to perform interactive image information-extraction operations in a flexible, efficient and easily used manner. The design philosophy incorporated in AOIPS frees the user from routine bookkeeping operations. The system is conducive to the development of new data-analysis procedures and is easily expanded to accommodate new applications programs.

FUTURE TRENDS

Advances in computer technology, including the evolution of mini- and micro-computers, offer continued promise for lower processing-hardware costs. However, the costs associated with developing software systems have increased substantially due to the general rise in wages. Evolving software methodologies such as the top-down, structured-programming concepts and trends toward standardization and software transportability offer promise for reduced software costs through software sharing. The following sections address future technology trends and their impacts on remote sensing software systems. These sections should be read in conjunction with Chapter 20.

COMPUTER TECHNOLOGY IMPACTS

Computer-aided analysis was initially limited by the basic processing speed of the early computers used in remote sensing applications. In recent years, the basic computing speed available for fixed dollars of investment has increased substantially, and new high speed processors have also become available to supplement the computational power of general-purpose computer systems. The cost of on-line data storage is also declining, and the storage density of digital storage devices is increasing annually.

For remote sensing applications, future challenges will center around the development of computer-based systems needed to store and interrogate large data bases of image data and ancillary information, to overlay and combine multisource and multitemporal data sets in geocoded data structures (Green, 1978), and to transfer image and related data between remote sites rapidly and efficiently. Several important new technology developments in the computer industry and their impact on future remote sensing software systems are described in the following paragraphs and by Green (1979) and Bracken (1981).

Virtual Memory Systems—Recent large scale computers and minicomputers incorporate virtual-memory address-translation hardware and virtual-memory operating systems. Such systems provide automatic memory-paging features and allow the execution of programs which exceed the physical memory space allocated to them during actual processing. Thus, virtual memory systems eliminate physical memory size as a software-design constraint, and eliminate the need for program-overlay structures. Both of these features will prove of significant benefit in the design of future remote sensing software systems.

High Speed Processors—Large scale, high speed parallel and array processors are available which offer execution speeds in the range of 20 to 100 million instructions per second (MIPS). Several small-scale, low-cost array processors in the 3 to 8 MIPS range have emerged (Wittmayer, 1978) and offer 2 to 3 times the compute power of the largest minicomputers currently available. In addition, special purpose parallel processors will be available in 1983 which offer speeds of 2 to 5 billion operations per second (Strong et al, 1979). These technologies are capable of rapidly processing huge volumes of image data while using complex processing algorithms. To utilize such processors, remote sensing software systems will need to support data transfer to and from these devices at speeds that match their inherent computational speed. Also, new software-development tools will have to be developed to support general purpose programming of these systems.

Mass Storage Systems—Mass storage systems capable of storing large volumes of digital imagery are becoming available. Most of these systems currently rely on high-density magnetic-tape recording, and the future systems will include digital optical-disk-based, read-only, storage devices with very high storage densities (typically 10^{11} bits on each side of an optical disk recording). Future software systems will need to provide data base management routines that are capable of cataloging and searching large image data bases quickly

and efficiently, utilizing the mass storage devices of the future as the basic storage medium.

Communications Capabilities—A variety of high speed digital-data transfer networks, including local fiber-optic networks and domestic satellite communications, will be available within the next decade. The impact on image-processing applications will be that users will have the capability of accessing image data-bases remotely. A remote user will have the data acquired by a variety of imaging sensors available literally at his finger tips. Image-processing software systems will need to accommodate the data-base interrogation and retrieval requirements of remote users, and must provide the tools and protocols for transferring large volumes of image data over high-speed communications interfaces.

Microprocessors—Microprocessors are being used in the development of low cost data-display terminals, remote terminals, smart disk-storage systems, data-acquisition systems, and a host of other special-purpose processing equipment. Future remote sensing software systems will utilize microprocessor software modules to handle a variety of input/output, data manipulation, data reformatting, and other similar data handling functions.

Impact of Evolving Display Hardware

The first digital image-display devices were developed by individual institutions on an *ad hoc* basis. Typically, main memory of a digital computer was utilized as the refresh memory, and a video screen was driven at an appropriate data rate by the computer, which continually transferred digital data from memory through digital to analog converters to the video display. This operation required the full channel capacity of the computer, and processing operations were difficult to perform while an image was being displayed from memory.

The first commercially available image-display systems utilized digital-disk or solid-state memory as the TV display refresh medium, so that the main computer could transfer digital-image data to the local memory in the display device and then proceed to other tasks. The display memory was read at video rates, the conversion to analog format occurred, and the image could be displayed, essentially flicker-free, at video resolution (typically up to 512 × 512-pixel resolution). Later models made it possible to store three or more images in the local display memory and utilize three of the images to form a single color-image display. These displays were generally limited to 512 × 512-resolution because of the high data rate required to refresh color displays at higher resolution and the lack of high-quality color monitors at higher resolution.

Commercially available display hardware began to incorporate local computational capability around 1976, utilizing look-up table capabilities and microprocessors that provided image-manipulation capability locally within the display system. A microprocessor-based system makes it possible to transfer several images in color or monocolor to a local display memory from the host computer, and interactively process and manipulate the displayed imagery using the display resources independent of a host computer. Such systems act as independent image processors with limited capability. Processing functions that are typically available in micro-processor-based display systems include contrast enhancement, image differencing, image ratioing, convolutional filtering (with some limitations on the size of the convolutional filter), and image- and map-data compression/decompression. An example of microprocessor-driven color-image display terminal, available at moderate cost, is described by Dalton and Billingsley (1979).

The reduction in cost of solid state memory has meant that images can be stored in local display memory which accommodates imagery larger than the system-display resolution capability. Such systems provide the capability to roam the large image at full resolution in black-and-white or color, to zoom the image, or to display the full image at reduced resolution, all under user control at the console.

The evolution of image-display systems will have the following implications for remote sensing software systems in the future:

- Future systems will be highly distributed, with one or more intelligent display processors linked to one or more host computers in the same network. There will be an increased emphasis on image-data management software required to access, retrieve, and transfer imagery to a variety of image-display systems located within the same computer network.
- The increasing size of images acquired by digital systems will not be accommodated in the near future by digital image-display systems. Image sizes of over 1024×1024 pixels will be common, but capability for color display of images of this size or larger is still not yet widely available. Software systems, either on host computers or within the microprocessors supporting the display function, will need the capability of acquiring, storing, accessing and displaying large format imagery with a variety of options.
- As more and more processing capability resides within the local display processor, there will be an increased emphasis on user programming of functions on microprocessors that support data-display requirements. Software systems will need to provide development tools for software designed to operate in display-based microprocessors. In addition, the ability to microcode display hardware directly will need to be provided.

Impact of Custom VLSI Technology

Current efforts for Very Large Scale Integration (VLSI) in the electronics industry have resulted in the development of fabrication techniques that make it possible to design and fabricate single chips containing many thousands of transistors. It is now possible to design customized LSI chips that will execute a particular image-processing algorithm at extremely high speed. One recent effort at JPL has been the design of a custom LSI chip that performs edge detection on real-time video data acquired by stereo cameras located on the JPL Robotics vehicle. The chip, when fabricated, will perform edge detection on a full video image (512×480 pixels) in the 1/30 second frame time of the video system.

It is clear that very high speed specialized chips will form the basis for flexible, intelligent image-processing user stations that can operate in stand-alone mode or as remote terminals to central computer and data-base storage facilities. It is also clear that this technology will form the basis for custom high-speed pipeline and/or parallel image-processing systems. The only obstacle is the tedious effort currently required to design a complex chip. Future computer-aided techniques will alleviate that difficulty within the next decade, and the image processors of the future will be able to fabricate their own low cost high speed systems for image-processing applications. The future impact of VLSI technology is discussed in more detail by Mead and Conway (1980).

IMPACT OF EVOLVING SENSORS

Future sensor development for remote sensing applications will include sensor with finer spatial resolution (a larger number of picture elements per image), improved spectral resolution (imagery acquired of the same area through increasing numbers of spectral channels), and an increase in the number and variety of remote sensors (conventional imaging and multispectral scanning systems, synthetic aperture radar systems, visible and infrared radiometers, thermal mapping systems, etc.). Remote sensing will be characterized by continuing large increases in data volume, and extracting the information content from a variety of data types will become a severe challenge in the coming years.

These trends will influence remote sensing software systems of the future significantly. In particular, software systems with the following attributes will be required:

- The software must be flexible enough to handle imagery of varying size.
- Software systems must be modular, so that particular processing sequences can be easily constructed to meet changing user needs.
- Future software systems will need the ability to accept a variety of image data provided

from a large number of missions in a variety of data formats.

- The ability to geometrically register or correlate a large number of disparate data types will be required, in order to perform information-extraction processing that utilizes all data acquired for a particular area, possibly even as a function of time.
- Improved information-extraction techniques that utilize disparate data types will be required.
- Information-extraction procedures will utilize auxiliary data bases of a non-imaging nature in conjunction with a variety of image-data types. Systems will need the ability to correlate existing auxiliary data such as gravity and magnetic data, cultural data bases (e.g. census statistics) and others with remotely sensed imagery to augment the information-extraction process. For example, it may be useful to utilize gravity and magnetic data in conjunction with remotely sensed multispectral imagery, synthetic-aperture radar imagery and remotely sensed thermal data in mineral-exploration applications.

DATA BASE MANAGEMENT SYSTEM TRENDS

As future requirements for handling multisource and multitemporal data sets and a wider variety of data types grow, the need for developing sharable remote sensing data bases and flexible data-base management software will dominate the design of future remote sensing software systems. The remote sensing applications requirements and technology issues faced by designers of future data-management systems are discussed by Billingsley (1979) and Knapp and Szczur (1979). See also discussions in Chapter 22.

Systems such as the IBIS, described in an earlier section, ODYSSEY (see White, 1979), and the many commercially available spatial information management systems offer a first step toward providing the capabilities needed to meet remote sensing data management requirements. In addition, the relational data model, first proposed by Codd (1970), is receiving wider attention for use in remote sensing data management. The relational model offers the user a natural and logically flexible view of his data as data tables. In such models, the ability to create new logical relationships (i.e. data tables) between elements of a data base shows great promise for applications requiring the use of multisource data sets. Recent data-base management systems which incorporate the relational data model are described by Kim (1979), and the requirements for and design of a relational data-base management system for remote sensing data are presented by Moik (1979b). As noted in these papers, relational data-base management systems are particularly suitable for providing

flexible user-system interfaces. Most relational systems incorporate powerful query-language interfaces which allow the user to formulate complex data-access requests using a simple English language-based syntax.

Future data-base management systems for remote sensing applications must also provide capabilities for automatically cataloging data sets, and for browsing, accessing, and retrieving data interactively. Since remote sensing data bases are likely to exist at widely distributed locations, future data-base management systems will have to support catalog access and data retrieval by remote data users.

CONFLICTING SOFTWARE SYSTEM DESIGN GOALS

Earlier sections on the design of remote sensing software systems discussed the need for modularity and simplicity in software systems, explored techniques for enhancing the transportability of software, highlighted the importance of providing flexible user-system interfaces, and noted emerging needs for data-format standardization. Conflicting software-design goals must be faced when attempting to implement systems to meet all of these needs.

For example, seeking software transportability and operating system independence usually results in increased design complexity. Implementing systems in higher-level programming languages and using software methodologies which result in modular designs tends to impose higher system overhead and thus reduces computational efficiency when handling large volumes of data and varied data types. Providing simple user-system interfaces conflicts with goals for accommodating a variety of user-experience levels. Conflicts also arise when attempting to design compact and easily understood data formats which must, at the same time, accommodate a wide variety of data types. These and other design goal conflicts cause serious problems for designers of remote sensing software systems.

As noted by Landgrebe (1976), progress toward solving these conflicts remains more in the future than in the past. Efforts are needed to standardize data formats, data-handling protocols, and data-systems interfaces. Only with such agreements and with continued technology advances, can future systems adequately meet applications requirements, and offer the flexibility, ease of use, transportability, and modularity expected of remote sensing software systems of the 1980's.

ABBREVIATIONS AND ACRONYMS

ACM	Association for Computing Machinery, Inc.
AFIPS	American Federation of Information Processing Societies, Montvale, New Jersey

ANSI	American National Standards Institute, New York, New York
ASP	American Society of Photogrammetry, Falls Church, Virginia
COSMIC	Computer Software Management and Information Center, Athens, Georgia
ESL	Electromagnetic Systems Laboratories, Inc., Sunnyvale, California
ERIM	Environmental Research Institute of Michigan, Ann Arbor, Michigan
ERRSAC	Eastern Regional Remote Sensing Applications Center, NASA/GSFC
IBM	International Business Machines Corp., Houston, Texas
IEEE	Institute of Electrical and Electronic Engineers
JPL	Jet Propulsion Laboratory, California Institute of Technology, Pasadena, California
LARS	Laboratory for Applications of Remote Sensing, Purdue University
NASA	National Aeronautics and Space Administration
NASA/GSFC	Goddard Space Flight Center of NASA
NASA/JSC	Johnson Space Center of NASA
NOAA	National Oceanic and Atmospheric Administration, Washington, D.C.
USGS	United States Geological Survey, Reston, Virginia

REFERENCES

ANSI, 1978, American national standard programming language—Fortran; ANSI Doc. No. X3.9—1978, New York, New York.

Baker, F. T., 1972, Chief programmer team management of production programming; IBM Sys. J., Vol. 11, no. 1.

Billingsley, F. C., 1979, Data base systems for remote sensing—cost and technology considerations; Conf. Proc on Economics of Remote Sensing, JPL.

Boehm, B. W., 1976, Seven basic principles of software engineering; Eng. Coll. Series, NASA/GSFC.

Bracken, P. A., J. T. Dalton, J. J. Quann, and J. B. Billingsley, 1978a, AOIPS—an interactive image processing system; Proc. of the National Computer Conf., vol. 47, pp. 159-171, AFIPS Press, Montvale, New Jersey.

Bracken, P. A., P. H. Van Wie, and J. T. Dalton, 1978b, Computer mapping software on the AOIPS; Proc. First Int. Users Conf. on Computer Mapping Software and Data Bases, Harvard University.

Bracken, P. A., G. G. Knoble, and C. B. Howell, 1979, The design and applications of the Landsat-D assessment system; Proc. 2nd Int. Conf. on Computer Mapping Hardware, Software and Data Bases, Harvard University.

Bracken, P. A., 1981, Earth resource observations data

systems in the 1980s; Jour. Astronautical Sciences, vol. 29, no. 4, pp. 307–319.

Bryant, N. A., A. J. George, and R. Hegdahl, 1977, Tabular data base construction and analysis from thematic classified Landsat imagery of Portland, Oregon; Symp. on Machine Processing of Remotely Sensed Data, LARS.

Bryant, N. A., and A. L. Zobrist, 1976, IBIS: A geographic information system based on digital image processing and image raster datatype; Symp. on Machine Processing of Remotely Sensed Data, LARS.

Carter, V., F. C. Billingsley, and J. Lamar, 1977, Summary tables for selected digital image processing systems; USGS Open File Rep 77-414.

Codd, E. F., 1970, A relational model for large shared data banks; Comm. ACM, vol. 13, no. 6, pp 377-387.

COSMIC, 1979, Mini-VICAR software documentation; Rep. no. NPO-14892, Athens, Georgia.

Dalton, J. T., and J. B. Billingsley, 1979, Interactive mapping software of the domestic information display system; Proc. Fourth Int. Symp. on Computer Assisted Cartography, AUTO-CARTO IV, ASP.

ESL, 1978, IDIMS functional guide, volume 1; Sunnyvale, California.

Green, W. B., 1978, Applications of interactive digital image processing to problems of data registration and correlation; Proc. National Computer Conf, vol. 47, pp. 141-149, AFIPS Press, Montvale, New Jersey.

Green, W. B., 1979, Future trends in image processing software and hardware; Proc. Symp. on Machine Processing of Remotely Sensed Data, IEEE cat. no. 79-CH1430-8-MPRSD.

Green, W. B., et al., 1980, Analysis of multiple imagery at JPL's image processing laboratory; Opt. Eng. Special Issue on Digital Image Processing, vol. 19, issue 2.

Guinn, C., and M. Kennedy, 1975, Avoiding system failure: Approaches to integrity and utility; Contract Rep. 31-109-38-3139, Argonne Nat. Lab.

Hamlet, R. G., and R. M. Haralick, 1978, Transportable package software; Tech. Rep. No. -706, Computer Science Dept., Univ. of Maryland, College Park, Maryland.

Haralick, R. M., et al, 1977, Kandidats image processing system; Remote Sensing Laboratory Report, Univ. of Kansas, Lawrence, Kansas.

Haralick, R. M., and G. J. Minden, 1978, Kandidats—an interactive image processing system; Comp. Graphics and Image Processing, vol. 8, pp. 1-15.

IBM, 1979, ERL/ERIPS system description summary; Houston, Texas.

Jepsen, P. L., 1976, The software/hardware interface for interactive image processing at JPL's image processing laboratory; Proc. of Digital Equipment Users Society.

JPL, 1976, Proc. of the Cal. Tech.—JPL conference on image processing; Pub. SP43-30.

JSC, 1979, ERIPS—LACIE users guide; Contract Report, IBM, Houston, Texas.

Knapp, E. M., and M. R. Szczur, 1980, Spatial data integration requirements for earth resources applications; Proc. of the Second Int. Conf. on Computer Mapping Software and Data Bases, Harvard University, Cambridge, Massachusetts.

Kernigan, B. W., 1975, RATFOR—a preprocessor for a

rational Fortran; Software—Practice and Experience, vol. 5, pp. 395-406.

Kim, W., 1979, Relational data base systems; ACM Computing Surveys, vol. II, no. 3, pp. 185-211.

Landgrebe, D., 1976, Computer based remote sensing technology—a look to the future; Remote Sensing of the Environment, vol. 5, pp. 229-246.

LARS, 1979, LARSYS description memo; Scan Line Bulletin, vol. 5, no. 3, Purdue University.

McGowan, C. L., and J. R. Kelly, 1975, Top-down structured programming techniques; Petrocelli Character Press, New York, New York.

McMurtry, G. J., F. Y. Borden, H. A. Weeden and G. W. Petersen, 1977, the ORSER system for processing and analyzing Landsat and other mss data; Tech. Rep. 7-77, ORSER, Penn. State University.

Mead, C. A., and L. A. Conway, 1980, Introduction to VLSI Systems; Addison Wesley.

Moik, J. G., 1979a, SMIPS/VICAR application program descriptions: NASA/GSFC, Greenbelt, Maryland.

Moik, J. G., 1979b, A data base management system for remote sensing data; Proc. of the First Int. Conf. on Computer Mapping Software and Data Bases, Harvard Univ., Cambridge, Massachusetts.

Owings, J. N., 1980, VICAR system structure; Internal Tech. Doc., NASA/GSFC.

Ramapriyan, H. K., 1979, Occult software announcement; ERRSAC Bul., NASA/GSFC.

Seidman, J. B., and A. Y. Smith, 1979, VICAR image processing system—guide to system use; Pub. 77-37, Revision 1, JPL.

Spencer, P. W., 1974, LARSYS users manual; LARS, Purdue University.

Strong, J. P., et al., 1979, The massively parallel processor and its applications; Proc. 13th Int. Sym. on Remote Sensing of the Environment, ERIM.

White, D., 1979, Odyssey design structure; Mapping Software and Cartographic Data Bases, vol. 2, pp. 207-216, Harvard University, Cambridge, Massachusetts.

Wittmayer, W. R., 1978, Array processor provides high throughout rates; Computer Design, vol. 17, no. 3, pp. 93-100.

Zobrist, A. L., and N. A. Bryant, 1979, Image-based information systems (IBIS)—system guide; Tech. Rep. 900-909, JPL.

Zobrist, A. L., and H. Wilczynski, 1977, Image based information system—software description; JPL.

CHAPTER 20

Digital Hardware

Author/Editor: DAVID NICHOLS

Contributors: JIM FREW, DANNY MARKS, GUY LOHMAN, KEN WALLGREN

General Contents: System components; processors; mass storage of information; data acquisition devices; data display devices; system configurations; technology developments; references.

INTRODUCTION

Remote sensing is moving from reliance on analogue photographic methods to systems which make extensive use of digital technology. Photographic remote sensing applications continue to be important but the use of new remotely sensed data sources requires new tools. Satellite remote sensing and computer technology have been brought together and now constitute the technological underpinnings of modern non-photographic sensing systems.

Digital technology is an important factor in satellite tracking and guidance, ground data systems for sensor and spacecraft testing and integration as well as data telemetry and final product formatting (Stakem, 1981). Digital technology is a necessary part of all aspects of non-photographic remotely sensed data collection systems. Its importance does not cease with the digital product being delivered to the user. The massive amount of information collected in digital form by modern sensor systems dictates digital approaches to utilization as well as collection.

The intent of this chapter is to describe the digital hardware necessary to use remotely sensed data. The focus will be on data applications. Digital hardware elements of the sensors themselves, or any of the developmental or operational support systems, will not be addressed except as related to applications. Basic computer technology and specialized devices peculiar to remote sensing applications are described. Additional information on certain topics or devices may be found in the referenced literature.

Digital data processing is rapidly changing, primarily due to advances in hardware. Costs for constant performance are decreasing; components are becoming faster, smaller, and more data can be stored in a given space. Remote sensing applications benefit from these advances by either adopting hardware developed in response to other requirements, or by stimulating development of specialized devices specifically for remote sensing. Some technologically active areas which are expected to impact remote sensing applications

within the next ten years are described at the end of the chapter.

IMAGE PROCESSING AND REMOTE SENSING

Remote sensing applications generally involve processing data of a two-dimensional (or higher) form. Very large volumes (10^7 bytes or more) are common, requiring special techniques. Data may represent continuous-tone images, abstract forms such as maps, or may simply be discrete geo-located observations of some geophysical or socio-economic variable. Types of processing which may be applied to such data can run the gamut of what is termed scientific data processing. This includes many forms of mathematical, statistical, graphical and data management processing. However, there is one unifying factor in almost all environmental remotely sensed data—the data are spatial. In an image context, a natural assemblage of the data forms a geographical distribution. From the standpoint of a discrete observation, significance is derived both from its absolute location and its location relative to other observations.

Disciplinary areas, (e.g. statistics, geosciences, cartography) contribute methodologies for dealing with spatial data. However, to implement these methods, data must be structured so they can be encoded for computer manipulation. One such structuring paradigm is the raster, or cell-based system. It is highly amenable to computerization and has the added advantage that it is the same form as that provided directly by imaging sensors. Tools for manipulating raster structures are found most highly developed and well-packaged in the area of image processing (Castleman, 1979). Image processing itself includes many contributions from the disciplinary areas. Another structural paradigm for spatial data is the line graph, or vector-based graphics representation. Tools for manipulating this structure are most highly developed in the discipline of computer graphics (see Newman and Sproull, 1973; Foley and Van Dam, 1981). For these reasons, image

processing and computer graphics are considered to drive the primary functional requirements for hardware which aids remote sensing applications.

GENERAL COMPUTER ORGANIZATION

Image processing and computer graphics are by nature, interactive. In a typical scenario, an individual observes the output of an operation, and based on that observation, makes a decision about the subsequent operation. Users require prolonged access to the computer and peripheral devices and dedicated access to CRT monitors, which provide the means of interacting with the pictorial data. Historically, large mainframe computers have not supported interactive computing very well, and it is difficult to attach specialized graphics devices to them. Therefore, most image processing and computer graphics applications were developed and matured on minicomputer systems, specifically designed for interactive processing. Minicomputers continue to provide the basic hardware environment for image processing and will do so for a long time. Micro-computers will begin to play a larger role, but the difference between micro-computers and minicomputers has become more one of packaging than of capability.

Computer organization is a subject of much discussion, particularly because the building blocks—processing elements and memory—are becoming very inexpensive. Most general purpose minicomputers, however, follow a common organizational philosophy. Minicomputers are organized around a central processing unit (CPU) which is attached to memory and peripheral devices by a multi-line connection called a bus. Memory and peripheral devices are connected to the bus by devices called controllers (Figure 20-1).

The central processing unit (Figure 20-2) consists of a control unit supervising data transfer and process initiation activities, and an arithmetic-logical unit (ALU) for binary arithmetic and boolean operations. A set of general purpose registers provides temporary storage for staging data between memory and the CPU functional units. Special purpose registers may exist to provide features such as automatic index incrementing or decrementing. A program counter keeps track of the memory location from which to retrieve the next instruction.

The bus is a set of electrical paths which carry signals for various purposes. Some signals represent data. Others are for control information, such as to notify the CPU that a particular device requires service. Some lines provide the actual power and grounding for the electrical components. Others are used for the device address which specifies the device to which a particular set of bus signals applies.

Peripheral devices are connected to the bus via components called controllers. Many similar devices, such as disk drives, can be serviced by the same controller. The controller determines which physical device is to be addressed, handles all bus-timing, signal synchronization and handshaking, and generates the signals necessary to perform a particular device operation. Memory may look exactly as any other device on the bus or may occupy a separate bus dedicated to CPU-memory transfers.

Bus-structured computer architecture is convenient for building specialized minicomputer-based systems. By making the controller a separate component, all bus-specific electronics can be placed within it, allowing the peripheral devices themselves to be independent of bus designs. Bus designs often vary among manufacturers and sometimes within manufacturers themselves, although industry-wide standards are being developed. Specialized devices can readily be attached to or detached from the bus for particular applications. The existence of a limited number of stan-

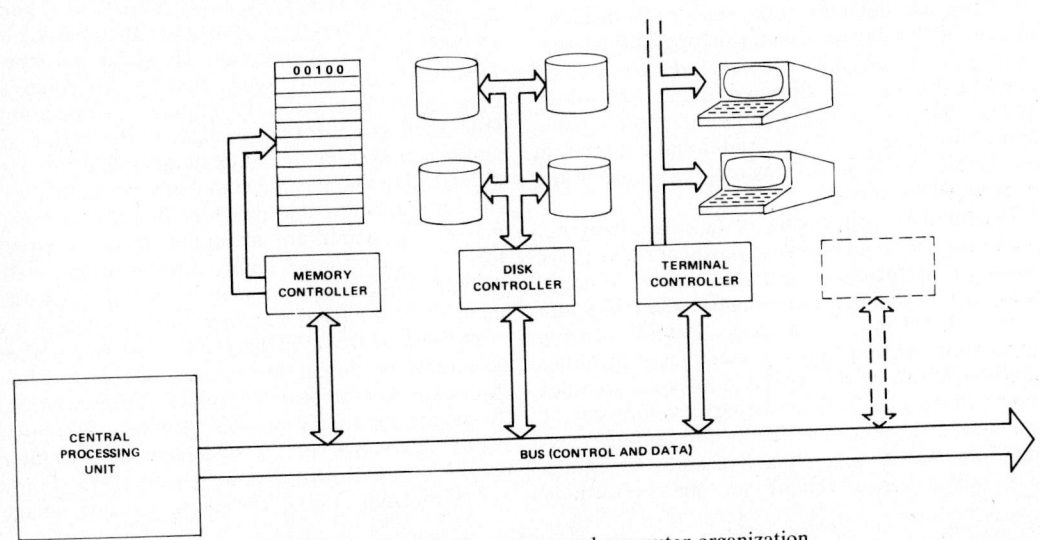

Fig. 20-1. Generalized bus-structured computer organization.

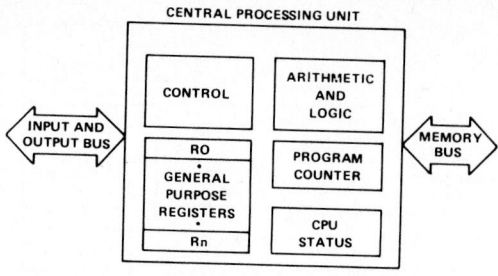

Fig. 20-2. Central Processing Unit organization. (Levy, 1980).

dard buses, and published specifications for the electrical and physical characteristics of the buses, allow manufacturers of specialized devices to build relatively inexpensive interfaces and avoid CPU and manufacturer dependence.

HARDWARE AND SOFTWARE

The distinction between hardware and software is often fuzzy and has become more so with the widespread use of micro-coding (Tanenbaum, 1976) yet the difference is significant. Software consists of instructions which tell the hardware how to do its job. Sometimes the hardware consists of very low level functions (e.g. a logical operation on two bits) or may provide much more complex functions such as vector-to-raster conversion. With the advent of VLSI, carefully selected high level functions will be performed in hardware, as opposed to software. Software, however, will never lose importance, because of its inherent flexibility. Firmware is software which resides in hardware on a relatively permanent basis. That is, instructions are not read from a stored program in main memory, but rather from a program stored in a special memory called ROM (Read Only Memory) which cannot easily be over-written.

SYSTEM COMPONENTS

Image processing systems are made up of a number of discrete components, each of which has a particular function. Integrated hardware systems, complete with software, are available from commercial vendors. These are known as 'turn-key' systems. It is also common for such systems to be assembled and integrated from components supplied to each user directly by the manufacturer. This section is intended to provide some familiarity with the primary system components required for applications in image processing systems.

PROCESSORS

Processors are the brains of any computer system. They provide two basic functions—control and mathematical operations. Control functions perform the housekeeping for data processing; getting data from memory, synchronizing data transfers, etc. Mathematical operations alter data in some fundamental way. Mathematical operations can be either arithmetic or logical. Arithmetic processors can be implemented within the general purpose processor, as a special device connected to the CPU, or as a peripheral device connected to the bus. Separating the arithmetic processor from the CPU allows operations to be performed concurrently in each processor, thus speeding execution times. Arithmetic processors which are optimized for speed generally follow some form of parallel architecture.

General Purpose Processors

The heart and soul of any conventional computer system is the Central Processing Unit (CPU). Since the CPU defines the basic characteristics of general purpose processors, the following discussion on processors will focus on CPUs. CPUs come in many different prices and configurations and offer a variety of speeds and I/O capabilities. Many commercially-available computer systems consist of a CPU plus an assortment of peripheral devices. Alternatively, CPUs may be purchased without any of the supporting hardware (the CPU may consist of a single chip or single printed circuit board) and it is up to the end user to build an integrated computer system. Although processing elements may exist in peripheral devices and controllers, the discussion here addresses CPUs which form the basis for an entire computer system.

The characteristic most descriptive of CPUs is word length. This refers to the size of a datum which can be manipulated by one of the CPU registers (internal memory locations). Word length is what is referred to when a CPU is described as, for example, an "8-bit" processor. Historically, the terms micro-computer, minicomputer, and mainframe have been applied to 8-bit, 16-bit, and 32-bit (or larger) CPUs, respectively. These distinctions have become blurred as CPUs have become increasingly miniaturized. Today, "microcomputer" refers to a computer whose CPU is contained on a single integrated-circuit chip, which in turn is called a micro-processor. 8-bit micro-computers are quite common, and are the basis of most, widely available, personal computer systems. 16-bit micro-computers are now becoming available, and 32-bit micro-computers are in the prototype stages. The distinction between minicomputers and mainframes has become almost entirely one of price.

There are several factors to consider when selecting a general purpose processor. A significant factor in CPU capability is address space. Address space refers to the range of main memory directly accessible by a program. This range places a limit on the size of programs (instructions plus data) the CPU can run. Address space is determined principally by the number of bits in the registers used to form memory location addresses.

An N-bit address is capable of accessing 2^N discrete locations.

Common address space sizes are 16 or 32 bits. The former allows programs to address up to 65,536 memory locations (referred to as 64K). Many image data sets are large enough (a 512 × 512 image contains 262,144 pixels) to require piecemeal processing on such systems. This is not always a problem, but for some applications both program complexity and program execution time can be quite high. Some CPUs, with limited address space, use a technique known as memory mapping (Pohm and Smay, 1981) to connect more physical memory to the CPU than can be directly addressed by any single program (Figure 20-3). The operating system (executive program) can then accomodate more than one user program in the available physical memory. Note, however, that the memory available to any single program is not increased by this procedure.

A 32 bit address space allows over 4 billion locations to be addressed by a single program. While no existing computer system has anywhere near this much actual memory available to it, many systems simulate the non-existent memory by using disk storage. Simulated memory is copied into real memory when it must be examined or modified, a process known as demand paging. Operating systems which perform this physical memory extension (Figure 20-4) are called virtual operating systems (Pohm and Smay, 1981) From the user's point of view it is possible to handle immense data sets within a program. A 512 × 512 image, for example, can be manipulated as an array entirely within memory. Actual physical memory on such systems is typically within the 2 to 8 megabyte range.

The speed at which a CPU operates is largely a function of cycle time. In the strictest sense, this is the time required to execute a single instruction. A single instruction, however, can be quite complex. On most CPUs the time varies according to the particular instruction being executed. What appears to the programmer as a single machine instruction is actually a program consisting of a number of micro-instructions. Each micro-instruction typically executes within one cycle. Within families of CPUs sharing a compatible ar-

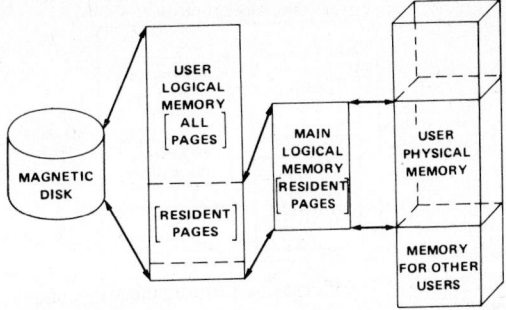

Fig. 20-4. Memory address extension via paging. (Pohm and Smay, 1981).

chitecture, one of the inducements to purchase a more expensive model is a reduction in instruction cycle time.

The effective speed of a CPU is also limited by the cycle time of its main memory. In general, bipolar semiconductor memory is faster than metal-oxide semiconductor (MOS) memory, which is faster than magnetic core memory. Memory speed must be matched to CPU speed so as not to slow down the CPU, or to waste the speed of fast memory.

Current computer architectures use a hierarchy of memory types exhibiting varying speeds. Data and programs are placed in the appropriate memory according to frequency of access (Pohm and Smay, 1981). On systems with large address spaces, the purchase of additional memory is often one of the most reliable ways to improve system performance. The more physical memory that is available, the less time is spent moving portions of program between real and simulated memory.

CPU performance is also affected by the width of its data paths. Data paths are the lines on the bus between the CPU and the outside world over which data move. Data paths also connect the various internal components of the CPU. Width refers to the number of bits that can be transfered simultaneously. A CPU with 16-bit data paths can transfer a 16-bit quantity in a single instruction cycle. Data path-width becomes significant in two circumstances. When transferring large quantities of data to or from a peripheral device, a wider path can transfer more data per unit time. Data path-width can also affect the speed of arithmetic operations. For example, to move a 32-bit word from an internal register to the ALU requires at least two machine cycles on a CPU with 16-bit data paths but only one cycle with a 32-bit path. Typically, a CPU will have data paths as wide as its characteristic word length, although notable exceptions exist.

The principal intrinsic advantage of a large word length is the ability to manipulate large units of data with single instructions. Since larger word lengths are almost always accompanied by increased address space and wider data paths, there

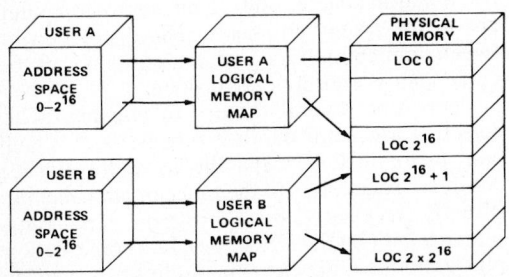

Fig. 20-3. Address space allocation in a memory-mapped system.

is a considerable performance increase as well. For example to retrieve a 32-bit integer number from memory, add 1 to it, and place it back into memory requires 3 cycles in a 32-bit CPU but 12 in an 8-bit CPU.

Arithmetic is commonly one of the most time-consuming operations. Most 8-bit CPUs do not have instructions for performing floating point arithmetic and require several instructions to perform most arithmetic calculations (it is difficult to do any meaningful arithmetic with 8-bit integers). Extensive hardware arithmetic capabilities, including floating point instructions, are usually found in 16 and 32-bit CPUs. Some computers perform floating point operations in a separate processing unit known as a co-processor, which is optimized for the task.

Parallel Processors

Image processing almost always involves operations on very large data sets. Therefore, performance is a salient issue. Performance can be optimized in different ways depending upon system hardware and software architecture. A significant factor in performance for highly repetitive operations on large data sets is the organization of the arithmetic processing elements. By the employment of more than one processing element and overlapping the operations in time (parallel processing) very large increases in speed can result. When parallel computer performance is compared to that of conventional computers, increases by factors of 10 to 1000 times greater are common.

Parallelism can be accomplished by overlapping arithmetic stages and memory access operations on a single data stream (pipelining) or by employing multiple processors assigned to different data streams, all performing the same operation (Karplus and Cohen, 1981). Combinations of the two approaches are also used. Processors can be grouped into four classes (Figure 20-5); (1) single instruction, single data stream (SISD); (2) single instruction, multiple data stream (SIMD); (3) multiple instruction, single data stream (MISD); and (4) multiple instruction, multiple data stream (MIMD).

SISD processors are not parallel processors. Most general purpose computers are SISD machines. Processors of arrays (MISD) are usually termed pipeline or vector processors. Examples of MISD processors are the CRAY 1, the CDC 205, and the Floating Point Systems AP-120. The AP-120 is configured as a peripheral device which can be attached to any one of many minicomputers (Charlesworth, 1981). The other processors mentioned are integrated units within large, expensive mainframe computers.

SIMD and MIMD parallel processors have an internal structure that reproduces the two-dimensionality of the data they most efficiently process. Examples of the SIMD class are the Digital Array Processor (DAP), and the Massively

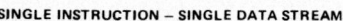

SINGLE INSTRUCTION – SINGLE DATA STREAM

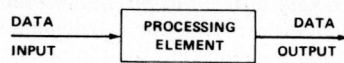

SINGLE INSTRUCTION – MULTIPLE DATA STREAM

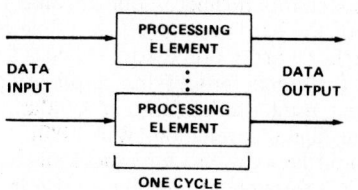

MULTIPLE INSTRUCTION – SINGLE DATA STREAM

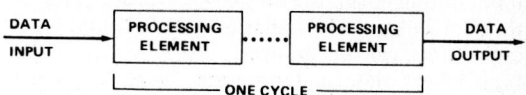

MULTIPLE INSTRUCTION – MULTIPLE DATA STREAM

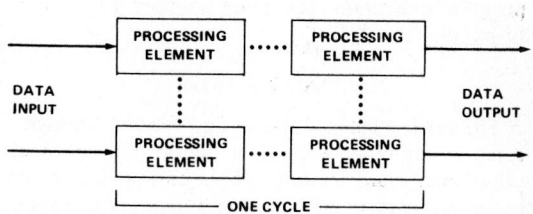

Fig. 20-5. Parallel processor organizations.

Parallel Processor (MPP) (Batcher, 1980). Examples of the MIMD class are ILLIAC IV (Barnes et al., 1968) and the Heterogenous Element Processor (HEP) (Smith, 1981). The interconnection pattern between adjacent processing elements allows the user to efficiently exploit the natural relationships that exist in two-dimensional data. Typically, interconnections are to nearest neighbors as in the MPP and ILLIAC IV.

The number of processing elements in multiple data-stream processors range from 16 in HEP to 16,384 in MPP. As the number increases, the complexity of each element usually decreases. The tradeoff between element complexity and the number of elements is determined by the required system reliability and the class of computing problems that the system is designed to solve.

Each parallel processor type has advantages and disadvantages. Single instruction processors are easy to program because each processing element executes the same operation. The SIMD machine is effective for certain image processing tasks because of its two-dimensional architecture. The MIMD processors are most effective where many related but different calculations must be performed, such as in nuclear physics and crystallography.

Multiple instruction processors, both MISD and MIMD, are very difficult to program because the programmer is responsible for managing not only the instruction sequence but the timing of several disparate concurrent operations. Although high-level language-programming tools are being developed (Brode, 1981) it is still necessary to write many procedures in micro-code to take full advantage of the hardware speed. Multiple instruction peripheral-array processors (MISD) are very popular for signal processing applications and have been used successfully in synthetic aperture radar digital corellation (Wu, 1980).

As would be expected for processors with extremely fast throughout, problems arise in getting data to and from the processor fast enough. The input/output structure of parallel processors is therefore extremely important. Input/output architectures can be grouped into two types; the type where data and program share the same memory, and the type where data and program each have separate memory. The separation of data memory from program-store memory allows data-access operations to overlap in time with the program execution, providing another dimension of parallelism.

MASS STORAGE

The terms 'mass storage' and 'online storage' refer to computer-accessible digital data storage other than main memory. This equipment is an essential part of any digital image-processing system, for either data storage or data processing. Large satellite image datasets have made mass storage the limiting factor in many small processing facilities. The cost of storage is decreasing, but the data volumes of new and proposed satellite products are getting larger, so storage will continue to be a crucial component in image processing.

The two principal categories of mass storage devices are disks and tapes. Both are available in a wide range of capacity, speed, and cost. Disk storage is most appropriate when ease of access is the primary consideration. Disks permit random access and support very fast transfer rates (up to 10^6 bits per second). Magnetic tapes, by contrast, can only be accessed sequentially—data at the end of the type can only be read by reading all of the preceding data. Magnetic tape transfer rates are significantly lower ($10^4 - 10^5$ bits per second). Tape is most appropriate where the data are to be archived or transported in digital form. A reel of magnetic tape is both portable and cheap—far more so on both counts than disk media of comparable capacity.

Disks

A magnetic disk can be made of either rigid metal (hard disk), or flexible plastic (floppy disk). Both types are coated on one or both sides with a magnetizable substance. Information is recorded on the surface of the disk in concentric circles called tracks. This is achieved by spinning the disk on a spindle while a read/write head is positioned at a constant distance from the spindle. The read/write head accesses separate tracks by moving radially across the surface of the disk; a motion called seeking.

Hard disks are often stacked several 'platters' to a spindle in a disk pack. The drive will contain a corresponding stack of read/write heads, one for each disk surface. The heads are moved in unison to a given track position. The corresponding vertically-contiguous set of tracks on the disk pack is called a cylinder (Figure 20-6).

A magnetic disk subsystem consists of one or more disk drives connected to a controller, which interfaces the disks to the rest of the computer system. A controller typically requires that data be read and written in constant sized blocks called sectors, which are integral subdivisions of a track. A typical sector contains 512 bytes of data.

The speed at which data may be recovered from a disk is limited by two mechanical factors—seek time and rotational latency. Seek time is the time required to move the read/write head(s) to the desired track. Latency is the time required for the desired portion of the track (sector) to rotate beneath the head. Access time can be reduced by faster rotation of the disk, to reduce latency, and faster positioning of the heads, which reduces seek time.

Magnetic disk subsystems span a wide price range, with greater price generally buying both larger data capacity and faster access time. The cheapest disk subsystems are built around floppy disks. These are flexible, removable storage media, either 5¼ or 8 inches in diameter, and enclosed in a stiff paper envelope. These media store from 10^5 to 10^6 bytes per disk. Their low capacity and slow access times make them just barely suitable for small-scale image processing. A 512 by 512 image could easily fill up an 8 inch disk, depending on the recording technique used.

The next increment in price and performance from floppy disks was, until quite recently, removable hard disk pack subsystems. These systems are much faster than floppies, and have capacities ranging from 10^6 to 10^8 bytes per pack. Removable packs offer a degree of portability

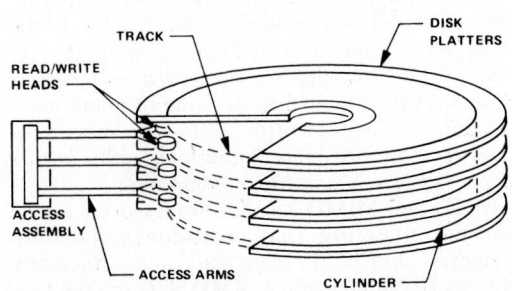

Fig. 20-6. Multi-platter disk drive architecture.

between similar disk subsystems, and are sometimes used as a means of backing-up important datasets. These advantages are mitigated, however, by the high cost of multiple-platter disk packs (typically $1,000) and by the danger of contamination when packs are interchanged.

Removable hard disk subsystems are currently losing favor to non-removable-pack subsystems (Winchester drives) for certain applications. The major reason is the better reliability of non-removable-pack disk drives. These drives are sealed, drastically reducing the likelihood of contaminants finding their way onto the disk surface. Contamination is a legitimate concern when one considers that a high-performance disk may be rotating at 10,000 rpm, with the head only about 10 microns from the disk surface. A single particle of cigarette smoke, resting on the surface of the disk, can strike the head with sufficient force to cause the head to set down on the disk. This is called a head crash, which generally destroys the disk pack and head resulting in irretrievable loss of data.

Because of the sealed, contaminant-free environment, it has become possible to construct smaller and lighter disk heads. The tracks can be placed closer together, and the heads made to fly closer to the disk surface; thus more data can be compressed onto a disk surface. This has led both to very high capacity drives (approaching 10^9 bytes), and to drives as small as floppy disk subsystems with an order of magnitude greater capacity.

Magnetic Tape

Magnetic tape has two principal uses in the handling of digital remote-sensing data: transportation and storage. Its usefulness as a vehicle for transporting digital data between systems derives from its durability and portability, while its usefulness as a storage medium is enhanced by permanence and low cost. Both uses benefit from the physical compactness.

Standard magnetic tape is a narrow (½ inch) strip of thin mylar or other flexible synthetic, one side of which is coated with a magnetic substance. Tape is sold in reels of varying diameters; standard sizes are 7, 8½, and 10½ inches, which contain 600, 1200, and 2400 feet of tape, respectively. A typical 2400 foot reel of tape costs from $15 to $30.

Data are recorded on (or read from) tape by passing the tape over a read/write head as it is wound from one reel to another (Figure 20-7). The read/write head writes data onto the tape in parallel strips called tracks. A byte is written to tape by simultaneously writing each bit to a separate track (Figure 20-8).

In accordance with the ubiquity of 8-bit character codes, almost all magnetic tape hardware reads and writes 9 tracks (the extra track is used for a parity bit). Older hardware may

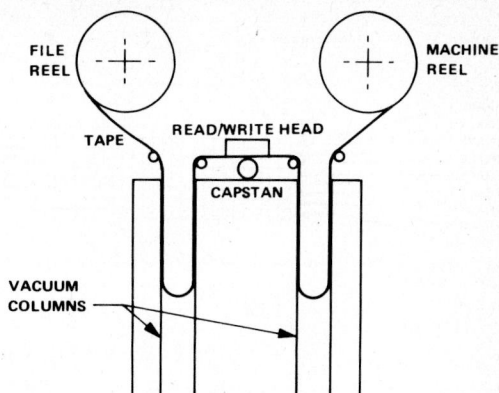

Fig. 20-7. Typical tape drive configuration. (Baer, 1980).

permit reading and writing 7 track tapes as well; but this is not a useful feature and, except in very special circumstances, use of 7 track recording should be avoided.

Two parameters govern the amount of data that can be recorded on a tape of a given length, density and block size. Density refers to the number of bits per unit length of tape that are recorded on a single track. Note that this is the same as bytes per unit length, if all 9 tracks are considered. Density is usually expressed in bits per inch (BPI). The most common tape densities are 800, 1600, and 6250 BPI.

Block size refers to the number of bytes written to the tape in a single read or write operation. This is significant because each write operation leaves a space on the tape after it completes. The space, called an inter-block gap (IBG), is a constant size (usually 0.6 inches) regardless of how big the previous block was (see Figure 20-8). Thus, the smaller the block size, the greater the fraction of the tape that is taken up by the unused space of IBGs.

The interaction of block size and density is illustrated in Table 20-1, which shows the length of tape that would be occupied by 1,048,576 bytes of data (corresponding to a 1024 by 1024 image with 1-byte pixels).

Tape hardware consists of a drive, which contains the mechanical tape transport and the read/write head, and a controller, which contains the electronics needed to interface the drive to the host system. One controller can usually support four or eight drives.

The magnetic tape subsystem may allow the user to select the density at which a tape will be read or written—a useful feature since both 800 BPI and 1600 BPI are commonly used. At this time, the most commonly supported density is 1600 BPI. The high density 6250 BPI tape drives have recently come into widespread use and will quickly become a standard for larger computer systems. The 6250 BPI drive does not cost signifi-

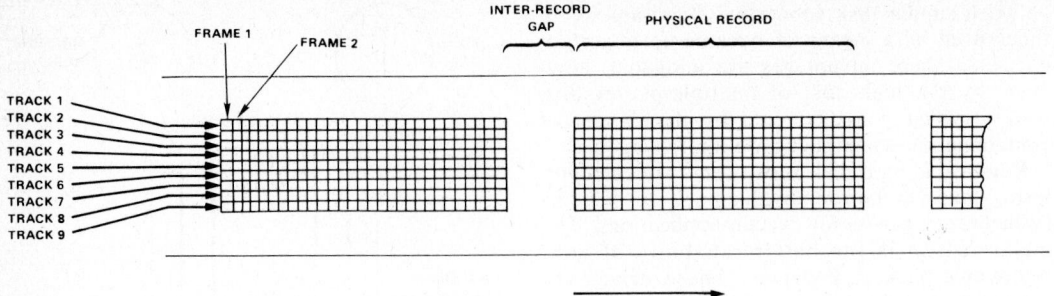

Fig. 20-8. Magnetic tape data format. (Tanenbaum, 1976).

cantly more than lower density drives but significantly reduces the time to load or unload large datasets or perform system back-up.

DATA ACQUISITION DEVICES

Fundamental to remote sensing data processing is the conversion of environmental data to digital form. This process usually takes place out of the control of the remote sensing analyst, as occurs with most satellite data. There is a large body of existing data, however, that must be digitized in order to facilitate further processing or merging with remotely sensed data.

The hardware employed in the digitizing task depends chiefly on whether the data are images or line drawings (e.g., boundaries on a map) and how the data are to be used. Images are most often digitized in raster form; that is, by scanning in a regular fashion and sampling the scan to yield a grid of pixels. Line drawings, on the other hand, are most often digitized in vector form, where a line is approximated by a series of connected straight segments represented by the cartesian coordinates of the end points. Devices which assist in this operation are known as coordinate digitizers.

Scanners

Images are digitized by sampling according to some regular geometric pattern. A rectilinear pattern is normally used and each scan line is known as a raster line. Hardware for image digitization achieves the raster by various combina-

tions of mechanical and electrical means. Devices incorporating a mechanical raster are referred to as scanners, while purely electronic digitizers are called video digitizers.

Scanners come in two basic varieties: flatbed and drum scanners. Both are designed to work with two-dimensional images (transparencies or prints). Flatbed digitizers construct a raster by moving a flat stage to which the image is affixed, past a stationary photodetector. The image is usually illuminated from below although reflected light configurations are available.

Both the image and the photodetector move in drum scanners. The image is affixed to the outside of a cylinder, which is then rotated about its axis. The photodetector remains stationary while sampling a single image line, then moves longitudinally along the drum to the next line position. The light source may be either inside or outside the drum, depending on whether a transparency or opaque image is being scanned. Drum scanning is often a feature of film recorders since the basic mechanisms are the same (Figure 20-9).

Both types of scanners may be used to construct color separations from color imagery. This involves either manual or automatic selection of the appropriate filter for either the light source or the photodetector, and scanning the image for each color channel.

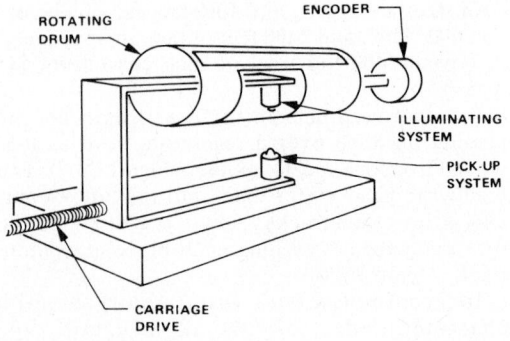

Fig. 20-9. Diagram of a drum-type image scanner. (Source; Optronics International, Inc.)

TABLE 20-1

Relationship of magnetic tape recording density, blocking factors and amount of tape used

	DENSITY (in Bytes Per Inch)		
	800	1600	6250
1024	161 ft	106 ft	66 ft
BLOCK SIZE (in Bytes)			
10240	115 ft	60 ft	20 ft

The spatial resolution of current optical-mechanical scanners is on the order of 25–200 microns; higher resolutions are available at a significant increase in cost. Many scanners allow the user to select from a number of possible spatial resolutions.

Drum scanners are significantly faster than flatbed scanners, with data rates on the order of 10^3 pixels per second. Drum scanners, however, are limited to a rectangular raster, and cannot scan images larger than the surfaces area of the drum. Flatbed scanners can be programmed to move the stage in a variety of non-standard scan patterns (e.g., circular). They can also be made to accept roll film instead of single frames.

Like all precision mechanical devices, scanners are susceptible to wear and abuse, and require more frequent maintenance than most digital components. Degraded mechanical alignment leads to errors in the output such as irregular spacing between scan lines, or misregistered color separations. Most scanners require trained operators to function properly.

Video Digitizers

Video digitizers use the inherent raster of a vidicon (television) camera system to sample an image. Unlike scanners, video digitizers may be used to acquire real-life scenes. There are two principal types of video digitizers, unbuffered and frame grabbers. The difference is mainly one of speed. Both types contain a vidicon camera and a computer interface. The vidicon camera scans an entire scene every 1/60 second. Figure 20-10 shows a video digitizer used at the University of California, Santa Barbara.

As with television, successive pairs of scenes are interlaced. That is, alternate frames contain only the odd scan lines or only the even ones. This reduces the illusion of flicker to humans viewing the image in real time. Therefore, two frame times or 1/30 second are required to obtain complete coverage of a scene.

Vidicon cameras have a spatial resolution of about 559 lines. Sampling at 512 samples per line with 8-bit radiometric resolution, yields a data rate of over 7×10^6 bytes per second, which is above the capacity of most computer channels. Unbuffered video digitizers circumvent this problem by sampling the vidicon output at a much lower rate. One scheme is to sample the first pixel of each scan line of a single frame, then the second, and so on until an entire scene is acquired. With the resolution given above, and accounting for scan interlace, an unbuffered video digitizer would require at least 17 seconds to sample an entire scene (assuming no system-imposed delays). For some applications, this may be an unacceptably long time. The unbuffered approach also requires that the scene to be digitized is static.

Frame-grabbing video digitizers can make use of the same refresh memory technology as digital video displays. It is possible to build memories which can be read or written at video rates. A frame grabber thus consists of a vidicon camera, an analog-to-digital converter, and a digital memory. When the command to digitize is given, the output of the vidicon camera is converted to digital form and written to the memory in real time. Thus stored, the contents of memory can then be examined at leisure. The time to acquire a frame is the same as for the vidicon camera itself, 1/30 second.

Unbuffered video digitizers are more attractive than frame grabbers simply because of their lower cost, although the rapidly falling cost of memory may narrow this gap considerably. Many video display systems, which already contain digital memories, can be fitted with vidicon cameras and A/D converters at a very reasonable cost, resulting in a frame grabber architecture.

The difference between scanners and video digitizers is mainly price and convenience versus performance. Scanners have much better spatial resolution (10^{-5} m) than video digitizers, and better radiometric consistency. Vidicon systems suffer from irregularities in the phosphor coating of the imaging tube as well as poorly controlled source illumination. The optical system in a scanner introduces less geometric distortion than a vidicon tube. However, scanners are much slower, more expensive, and more fragile than video digitizers.

Coordinate Digitizers

Remotely sensed data very seldom constitute the only data required for a particular application. Remotely sensed data must usually be merged with other data to form a more complete data base. A primary source for such ancillary data is in manuscript form as maps or photographic images. Encoding point- or line-data abstracted from these sources is generally done with the aid of a coordinate digitizer. Line networks digitized by raster-oriented devices such as scanners or video digitizers may be processed by various boundary-finding algorithms (Peuquet, 1981), and the linear information extracted. The computational overhead associated with such processing can be considerable. Often it is simpler and faster to enter the data by tracing the lines on a coordinate digitizer. In general, automated methods of line digitizing are suitable only for very high throughput situations. Digitizing tablets also provide a very effective means for controlling the graphics cursor on a volatile display device. A tablet in the $20'' \times 20''$ size range can serve both for cursor control and manuscript data entry.

Coordinate digitizers typically consist of a flat surface on which a manuscript is placed. A cursor is positioned over a point to be digitized, and its coordinates are sent to the host computer. The cursor may be a stylus, in which case it operates much like a pen; or a cross-hair in a transparent

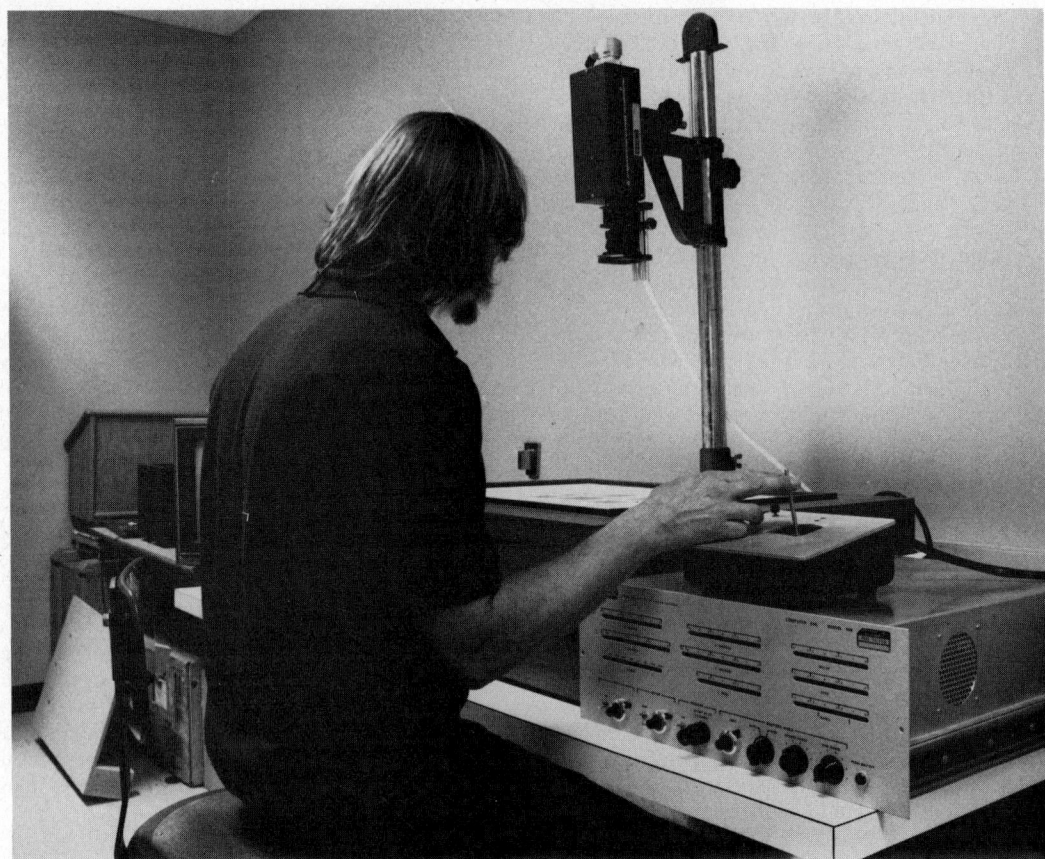

Fig. 20-10. A video digitizer for earth resources analysis. (Source; Univ. of Calif., Santa Barbara, Dept. of Geography)

disk. The cross hair is mounted in a small frame which may contain several buttons for entering status information or special operation codes.

Most commercial coordinate digitizers operate either sonically or electromagnetically. Early digitizers which used electro-mechanical principles are no longer competitive. Sonic digitizers sense the location of a high-frequency sound source located in the cursor. Electro-magnetic digitizers sense a magnetic field in the cursor using a wire mesh embedded in the digitizing surface or placed directly under it. Sonic digitizers are generally cheaper, since no special wire-mesh table is required, but the size of the digitizing surface is still fixed by the range of the microphones. Some people can hear sonic digitizers and find the sound quite irritating. Electro-magnetic digitizers offer greater positional accuracy than sonic digitizers, but the special surfaces add to their cost, especially for the larger tables (up to 3 ft. × 4 ft).

A coordinate can be sent from the digitizer to the host computer under a variety of conditions. A manual trigger (typically, depressing the pen stylus or pressing a button on the cursor assembly) will send a single coordinate pair and a small amount of status information. This is sufficient for

many cartographic applications, where objects are defined by a limited number of straight line endpoints or determinate curves connecting them. This is called point-mode digitizing.

When more complex data (e.g., strip chart recordings) must be digitized, the ability to automatically sample points per given increment of time or motion of the stylus, becomes useful. If the points are generated continuously while the cursor is in proximity to the surface, it is called stream digitizing. If the operator must hold a button or depress a stylus to initiate and continue the point stream, it is termed track digitizing. Automatic digitizing modes should be used with discretion, since they often lead to digitizing far more coordinates than are needed to characterize the input. Most digitizers have hardware time and distance filters so the operator can adjust the sampling rates. Stream operating mode is used when the digitizer is acting as a cursor control device. To prevent an unnecessary burden on the computer, it is critical that the cursor proximity flag be sent to the host by the digitizer controller.

Digitizers can be connected to the host system over standard asynchronous lines, exactly as a common terminal. When in stream mode, a low asynchronous line (300 BPS) will allow about 2 points/second; fast enough for point mode operations but insufficient for cursor tracking or line following.

DATA DISPLAY DEVICES

Remote sensing data analysis requires a great deal of visual interaction. There are many peripheral devices which display data of a graphical or pictorial nature. Computer graphics is an exploding field and new devices are being introduced almost weekly. There are many classifications of these devices possible but one functional factor is of overriding importance—whether or not the display produces a permanent copy. This feature has profound ramifications in the way the device can be used.

Graphical displays are important for viewing imagery and in assessing operations performed on the imagery. Vector-type displays are most important in the display of cartographic information, statistical series and image content summaries. Vector drawing display devices are also important in providing the interface necessary for interactive coordinate-digitizer data entry and editing.

Volatile Displays

Volatile graphics displays produce non-permanent images. This is usually done by constructing the image or line drawing on a cathode-ray tube (CRT). There are two methods for doing this. In the calligraphic method, lines are drawn on the screen by deflecting an electron writing beam from one location to another. In the raster approach, lines are formed by setting a bit in digital memory which represents a map of the screen and refreshing the screen in a regular pattern according to the contents of the map.

Calligraphic Displays

Calligraphic displays were, until recently, the preferred choice for the display of line drawings. Raster displays are now providing strong competition because of rapidly falling memory prices. Calligraphic displays are still preferred, however, for situations where maximum spatial resolution is of paramount concern. Many calligraphic displays provide spatial resolutions on the order of 4000 × 4000 points. Line quality is generally superior to raster displays because there are no jagged diagonal lines, and lines can be drawn very thin.

Storage Tube Display

Direct-view storage tubes (DVSTs) are the simplest type of calligraphic display. Their principal advantage is that no description of the picture needs to be maintained in memory. The picture is "stored" on the screen itself. This is achieved by using two electron guns. A writing gun knocks electrons off the phosphor screen, which initially possesses a strong negative charge. A flood gun continually bombards the screen with lower-energy electrons. These are repelled except in those areas where the writing gun has removed enough electrons to reduce the negative charge. The phosphors in these areas glow under the bombardment of the flood gun. In order to erase an image, the screen is pulsed with a strong positive charge, in effect "writing" to the entire screen and causing a bright flash. The charge is then switched to negative, and the screen may be written again.

The inability to selectively erase a portion of the screen is one of the biggest disadvantages of DVST displays, making them unsuitable for applications where the picture must be rapidly updated. The lack of selective erasure also makes DVSTs less than ideal for text display, since text cannot be scrolled. DVSTs have much lower contrast than stroke-refresh displays and as a result are frequently used under conditions of reduced ambient light. Balanced against these drawbacks are the high spatial resolution achievable with DVSTs (currently up to 4K in either dimension, as opposed to about 2K for stroke-refresh displays). DVST displays are also simpler to program than stroke-refresh displays. Essentially they are "volatile plotters". Indeed, some popular DVSTs and plotters may be interchanged without software modification. The price of a DVST graphics terminal is generally lower than for devices with similar resolution using other construction.

Stroke-Refresh CRTs

Stroke-refresh CRTs are another early form of graphic output device. Whereas in DVSTs the image is stored directly on the screen, in stroke-refresh displays the image is stored in a separate memory as a sequence of instructions. These instructions, called a display list, describe the beam movements needed to draw the picture on the screen, and are similar to the instructions sent to a DVST display. However, the CRT in a stroke-refresh display has no flood gun, and lines traced by the writing gun fade quickly (according to the persistence of the phosphor coating on the screen). The entire picture must therefore be refreshed (i.e., redrawn) by repeatedly executing the instructions in the display list.

Stroke-refresh displays are limited in picture complexity, measured as the number of vectors in the picture. One factor is the size of the memory in which the display list is stored. Another factor is the result of the nonpersistence of the screen phosphors. If there are too many vectors to draw, given that memory exists to store the display list, the amount of time spent between successive refreshes may be long enough to allow the image to fade noticeably. The picture as a whole will then appear to flicker. Some stroke-refresh displays attempt to minimize flicker by using longer-persistence phosphors, or by sorting the display list to minimize beam movement. There remains, however, a limit to picture complexity that does not exist on DVST or raster displays.

The principal application of stroke-refresh displays is dynamic graphics; situations where the displayed picture is subject to continual modification. The picture on a stroke-refresh display may be modified by simply changing the appropriate instruction(s) in the display list. Selective erasure, impossible on a DVST, is achieved by deleting instructions from the display list. An object on the screen may be "moved" by modifying its coordinates in the display program. Since many of the objects in the display list are represented as relative coordinates, moving an object consists of simply changing one coordinate pair.

The choice between DVST and stroke-refresh displays is almost entirely based on intended application. Until recently, price differences may have been a significant factor but with the development of micro-processors and cheap memory, that is no longer true. DVSTs are currently the best choice where an exceedingly complex picture is to be displayed. Stroke-refresh displays are appropriate for applications which require the ability to rapidly and/or selectively modify the picture.

Raster Displays

Raster displays create an image on a cathode-ray tube by successively scanning a regular pattern across the screen, much like a television. If a gray-scale effect is to be achieved, the electron beam is modulated according to some digital value. The entire screen is written every 1/30 or 1/60 of a second, depending upon whether or not the video signal is interlaced. The values (either on-off or scalar) driving the electron beam are stored in digital memory at an address that corresponds (maps) to its location on the screen, and read each time the screen is refreshed. Raster displays are not affected, as stroke-refresh displays are, by the amount of data on the screen at any one time. There is always as much memory in the display map as there are possible locations on the screen.

Raster displays are extremely important for image processing applications because they offer the best overall technology for displaying continuous tone images. Although image display devices are related to the common alpha-numeric CRT terminal, they are much more complex and expensive. As memory prices have dropped, raster graphics devices have become widespread in many applications such as CAD/CAM, where they were previously unsuitable (Linsalata and Scalea, 1981).

Alpha-numeric Displays

Alpha-numeric displays are the simplest raster display devices. Coupled with keyboards, they form the basis for most computer terminals. Memory normally consists of one byte for every character position on the screen. A typical terminal would have 2000 bytes of memory resulting in a 25 line by 80 column display. Each character is displayed as a series of dots formed by the raster beam.

Alpha-numeric terminal usage for image display is generally not feasible, owing to the limited spatial resolution. Some terminals have special character sets which permit crude line graphics. The principal application of alpha-numeric terminals in image processing is in support of the human-computer dialogue which controls the overall functioning of the system.

Raster Graphics Displays

Raster graphics terminals offer considerably higher resolution than alpha-numeric displays, such that true line graphics are possible. The memory of a raster graphics terminal is bit-mapped. That is, each bit in memory corresponds to a single point on the screen. The increase in memory requirements over alpha-numeric displays is significant. Rather than a memory of 2000 bytes, a raster graphics display with a resolution of 512 by 512 points requires over 32000 bytes. Some of the more common alpha-numeric terminals can be retro-fitted with bit-memories, making them into low cost graphics terminals. The typical resolution of a low-cost raster-graphics device is 250 lines by 500 to 600 samples.

Raster-graphics devices are not generally addressed by the programmer in a raster mode. Graphics are not created on the screen by the host computer software writing horizontal line after horizontal line. Instead, the terminals are given instructions to move or draw any place on the screen. The device contains vector-to-raster conversion hardware or firmware which maps the vector into the appropriate locations of the bit plane forming the screen memory. Franklin (1979) provides a discussion of the various algorithms used for the vector-to-raster conversion.

Memory is becoming so cheap that raster displays are beginning to rival stroke-refresh displays in medium resolution capabilities (1000 × 1000). Some companies are even offering hybrid-type (Shaw, 1980) displays which combine the characteristics of each. The fact remains that stroke-refresh is still superior when large parts of a line-drawing must be rapidly modified, as the time required to reconstruct an image in a bit-map memory generally precludes real-time animation rates (see for example, Jackson, 1980).

Raster graphics have the advantage over both stroke-refresh and DVST displays by being able to display solid, as well as line graphics. Many raster-graphics displays exploit this capability by offering firmware which automatically fills in a polygonal area (Pavlidis, 1981) which has been specified by its vertices.

In comparison with DVST displays, raster displays offer selective eraseability, better image

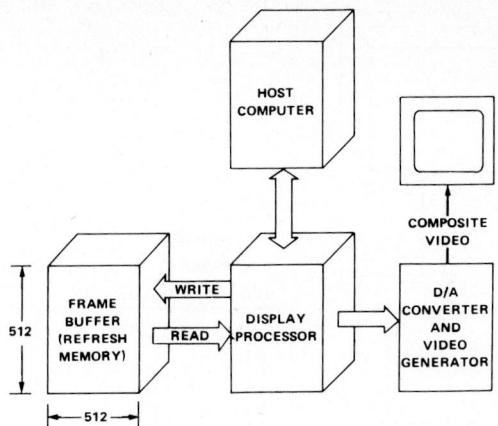

Fig. 20-11. A simple raster display processor architecture.

contrast, and color capability. The DVST is still the terminal of choice, however, if large amounts of monochromatic high resolution line drawings must be displayed at one time and selective erasure is not a prime requirement.

Image Display Processors

Probably the most important peripheral device in an interactive image processing system is the display processor. Certainly it is a device which is unique to image processing and computer graphics. The display processor is used to temporarily store images, create the video signals which drive the display monitor, digitize video signals from video input devices, and peform intensity transformations on the image data at various stages. It can be used to perform complex arithmetic operations on images in real or near-real time. The display processor provides the essential user interface in a pictorial interactive information processing environment.

Image display devices are really quite simple. At the most basic level they consist of memory (called a frame buffer), a D/A converter, and a processor which controls the loading and reading of the data and passes the data to the D/A converter. The processor and memory must be fast enough to read and display an entire frame every 1/30 to 1/60 of a second, (Figure 20-11). A typical memory size (and therefore, output image size) is 512×512. Display processors get more complex as processing is done on the data between these essential elements (Driscoll and Walker, 1981).

If image memory were $512 \times 512 \times 1$ bit, the device would only be able to construct a binary display. Therefore the number of bits used to store the digital value for each pixel is significant. This is often referred to as the frame-buffer depth. A six bit deep frame buffer would be able to display 64 gray levels. Monochrome image displays require a sufficient number of gray levels so that

the transition from one to another appears smooth to the viewer. Empirical studies suggest that at least 5 to 6 bits (32 to 64 gray levels) are required before the difference between adjacent gray levels ceases to be irritating.

Color is incorporated into display processors in one of two ways. In a 'pseudo-color' system the image is stored in a single memory plane. Color is produced by mapping each digital pixel value to a palette value which corresponds to one of many colors. The number of colors displayable at any single moment on the screen is still limited by the memory bit depth. An 8 bit deep memory would be able to display 256 simultaneous colors. The number of colors in the palette is determined by the word length of the look-up table in which re-mapping is performed (Figure 20-12).

True color systems require at least 3 separate memory planes, one for each primary color (Figure 20-13). Output look-up tables are similarly replicated. The processor must also provide the ability to assign a particular memory plane to a given output color. By storing a data set in the processor memory, and manipulating only the look-up tables, one can make extremely fast image-contrast alterations, without changing the actual values in the frame buffer. The ability to switch any plane into a given primary color output allows rapid evaluation of 'false-color' combinations when viewing multi-spectral imagery.

State of the art image display processors can be configured with up to 20 memory planes (Dinwiddie, 1981), each plane being 8 bits deep. Often one of these planes, or a specially designated one, is used for graphics overlay; line drawings and text are superimposed on the image. Extra memories can be mosaicked for storing images larger than 512×512, and can also function as a scratchpad memory when performing arithmetic operations within the display processor.

Display processors which contain more memory planes than can be displayed at one time may allow these memories to be logically merged. For example, a display processor with four 512×512 frame buffers may permit them to be accessed as if they were a single (8-bit deep) 1024 by 1024 buffer (Dinwiddie, 1981). Since this exceeds the resolution of many CRTs, the display-processor hardware will provide "roam" and "zoom" functions, which map arbitrary subsets of this frame onto the screen. If it is desired to view the entire frame, the hardware may allow subsampling (displaying every Nth line and sample). At the other extreme, the hardware may provide replicative zooming, with adjacent screen lines and/or samples being mapped from the same frame buffer location.

Roaming, sub-sampling, and zooming are examples of display-processor functions which invite a high degree of interaction with a human operator. Indeed, the original motivation for the development of video image displays was to provide greater interaction than existed when image

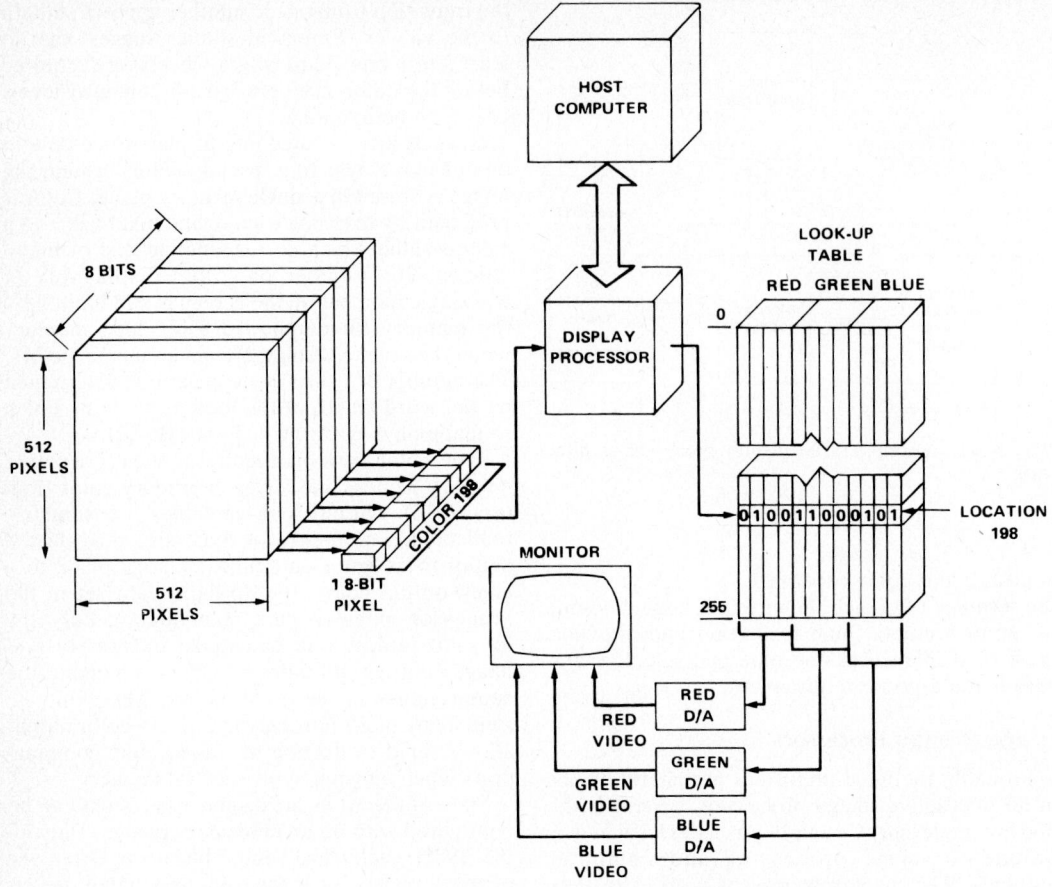

Fig. 20-12. A pseudo-color display processor architecture.

output was limited to hardcopy devices. Human input is most often provided by an alphanumeric terminal connected to the host computer, supplemented by graphic input devices connected directly to the display processor. Graphic input devices allow the operator to specify line and sample coordinates on the screen. The most common devices are joysticks, trackballs, and digitizer tablets.

The term "image display processor" stems partly from the fact that most currently available image display hardware includes some form of arithmetic processing capability. Processing may include modification of the data as they enter the display processor from the host computer, modification of the data in the frame buffer, and modification of output via mapping from the frame buffer to the display. These devices are called video-rate processors because they can perform simple arithmetic operations in a single video frame time of 1/30 second.

The combination of high-rate arithmetic capabilities within the display processor itself, with a feedback mechanism for routing the output back to a memory plane instead of to the video outputs, gives rise to many extremely fast image

processing operations (Driscoll and Walker, 1981). These operations are generally limited only by the fact that image sizes must conform to frame buffer sizes and cannot be arbitrarily sized without considerable software control.

There are currently two basic approaches to providing this processing capability. One approach involves parallel pipeline processing and the other is based on a multiple instruction—single data stream (MISD) processor. To illustrate these two architectures, the Model 70 image computer, manufactured by International Imaging Systems (I^{2}S) and the DeAnza Systems 8500 digital processing system will be discussed.

The I^{2}S video-rate processor (Figure 20-14) is configured as three parallel pipeline processing channels, one for each of red, green, and blue (I^{2}S, 1979). Each channel is fed by one of up to 12 refresh memories or the video digitizer. The outputs of the hardware zoom and split screen are passed to a look-up table, one for each of the 13 potential inputs. This means each of the 12 possible refresh memories can be routed to any one of the three pipeline processing channels. Arithmetic processing includes an adder array, a min/max accumulator, and a constant adder. Additional pro-

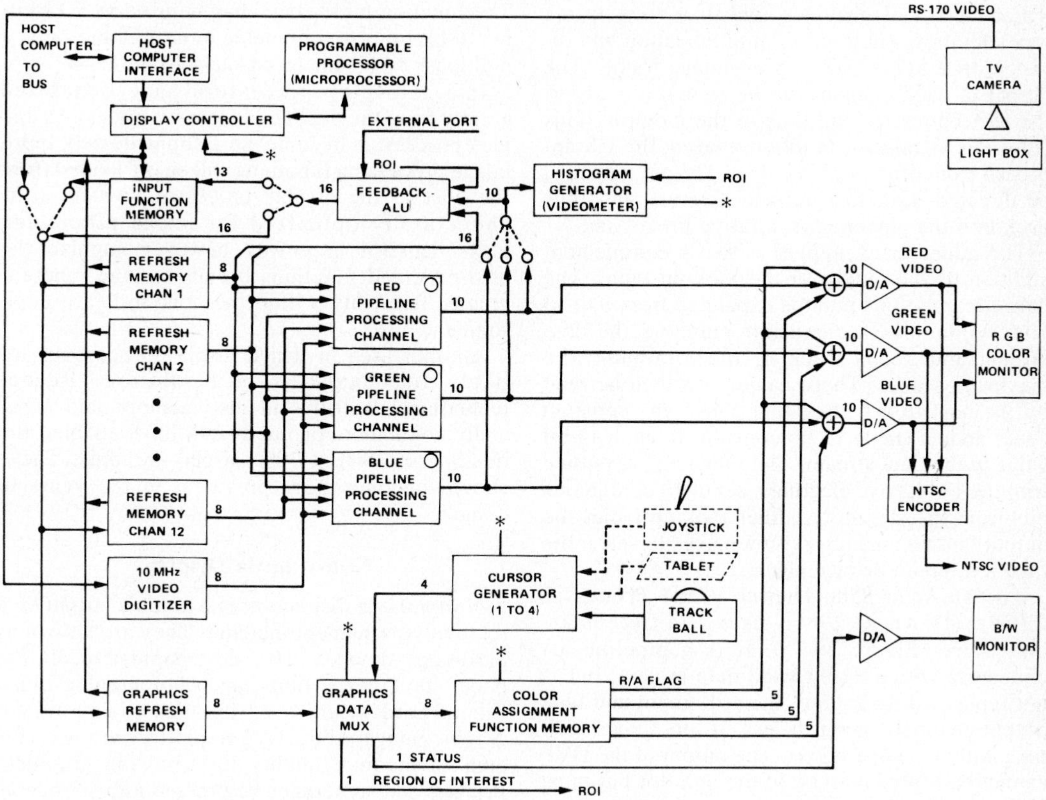

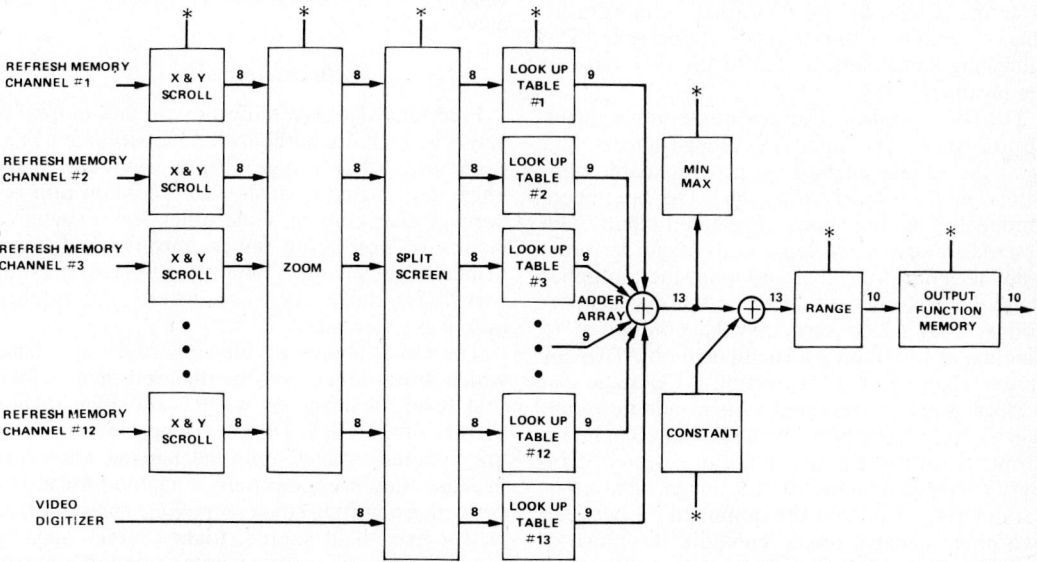

Fig. 20-13. A full-color display processor architecture. (Source; International Imaging Systems, Inc.)

Fig. 20-14. A Parallel-pipeline video-rate display processor. (Source; International Imaging Systems, Inc.)

cessing is provided by a range register and a final look-up table.

Each channel provides real-time (one frame time) addition, subtraction, multiplication and division on a $512 \times 512 \times 8$ bit image frame. The output of each channel can be routed directly to the D/A converter and thus to the monitor. This feature is significant in differentiating the I²S approach from other architectures. Output may also be directed back to a refresh memory and thus back into the pipeline for iterative processing.

The adder array applies a two's complement addition to the output of the look-up table. The data stream at this point is expanded from 8 to 13 bits. A min/max accumulator monitors the data stream and keeps track of the minimum and maximum values. These values can then be read by the host to set other functions. The constant adder adds a single 13 bit constant to each 13-bit value in the data stream. The range registers then trim the 13 bits by selecting a set of 10 contiguous bits from the stream. Another look-up table, the output function memory, is then used to select the final red, green or blue signal.

The DeAnza 8500 Digital Video Processor (DVP), (DeAnza, 1981), uses a different architecture (Figure 20-15). It is a pipeline architecture with a 32 bit wide data stream, but is not replicated for each of the red, green and blue output channels; nevertheless, it too allows real-time arithmetic operations. The output of the DVP cannot be routed directly to the monitor but must follow the feedback loop into one or more of the refresh memories.

The DVP can be fed up to 10 image memories. Two of the image memories can be routed to the control bus where they can be used to control operations within the pipeline. In this fashion separate operations can be performed on arbitrarily shaped regions within a single frame time. The remaining 8 channels are fed to the first stage of the pipeline.

The DVP consists of several consecutive stages of processing. The input select stage selects data from the refresh memories, programmable constants, or from a function table. The multipliers (optionally) multiply pairs of selected inputs. The first ALU (arithmetic-logic unit) stage performs input selection from the outputs of the multiplier stage and performs arithmetic and/or logical operations. This ALU stage may be configured to function as four 8-bit ALUs, two 16-bit ALUs, or as one 32-bit ALU. The second ALU stage can perform similar operations and in addition contains a 36 bit accumulator which can compile information over the entire image or selected, arbitrarily shaped regions of the image. The comparator stage can trim the output to be between two programmable limits, compute the range of output values within a region, generate the pixel by pixel maximum or minimum image, or compare 8-bit, 16-bit, unsigned magnitude or two's complement pixels at the rate of about 76 nanoseconds

per pixel. The shifter stage can modify the output by shifting or rotating the 32, 16, or 8 bit results. The function table stage may be used as a 12-bit-in, 16-bit-out programmable look-up-table or as a multipass 16-bit-in, 16-bit-out look-up-table.

Image display processors have benefitted greatly from the decline in memory prices. As display processors become more sophisticated, more image processing functions will be off-loaded from the host to the display processor—which is architecturally optimized for image processing tasks. This will in no way, however, supplant the host-executed functions because of the inherent greater flexibility within the host software environment.

Another area in which display processors are likely to change is in resolution. Recent technological strides in fast memory and especially color monitor bandwidth have enabled the building of 1000×1000 display systems. These should become more prevalent in the years to come.

Non-volatile Displays

Non-volatile displays create graphic output on a relatively permanent medium. They are known as hard-copy devices. The most popular media are paper, mylar-type films, and photographic films. Film recorders are used for high-quality photographic image products. Pen plotters are used for high to medium quality line drawing graphics. Printers can sometimes be pressed into service as low cost alternatives to both image recording and line drawing devices. All these devices, though, begin with a digital signal. Another class of hard-copy devices, display-screen copiers, are designed to read the analogue video signals driving a display and duplicate the image on a film or paper medium.

Film Recorders

Film recorders are more precise and more expensive, in both purchase and operating costs, than any other image-output devices. Most image-processing systems which contain film recorders also contain some other less expensive means of producing image output, such as a display-screen copier. Film recorders are thus reserved for those occasions where high-fidelity output is essential.

The two families of film recorders are those which write directly on the film and those which first form an image on a CRT and then take a picture of the CRT. Direct-writing film recorders are generally called opto-mechanical recorders because they use a mechanical method for either deflecting a writing beam or moving a piece of film past a fixed light source. Light sources may be monochromatic (e.g. LEDs), panchromatic ('glow tubes') or even lasers. Direct-writing film recorders provide the highest quality film products possible. CRT image-forming recorders gen-

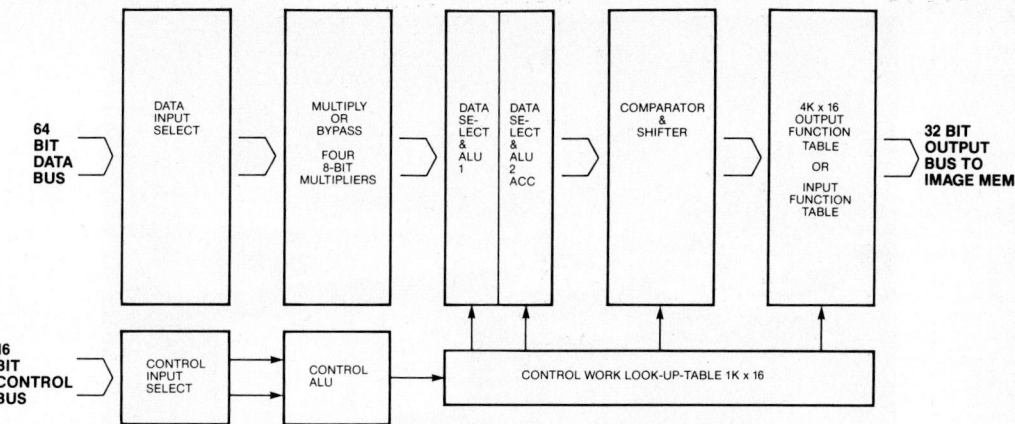

Fig. 20-15. A pipeline video-rate display processor architecture. (Source; DeAnza Systems, Inc.)

erally have much higher throughput capabilities because of less stringent film-positioning requirements, but exhibit reduced geometric fidelity.

In a typical opto-mechanical film-recorder design (Figure 20-16), the film is affixed to the outside of a cylinder, which rotates on its axis inside a light-tight housing. A slit in the housing allows a light source to move longitudinally along the cylinder. An image line is produced by making the light source stationary and modulating its output while the cylinder rotates. Successive lines are written by stepping the light source along the axis of the cylinder. Figure 20-17 shows a film recorder, along with the computer which controls it and a tape drive which is used for data input in a stand-alone operational mode.

Color images may be produced by writing three monochrome images which produce a set of three color-separation negatives. The separation negatives are then used to produce a color print of the scene. More expensive recorders can generate images on color film by exposing the film through three filters. In addition to the filter mechanism this requires a higher-intensity, broader-spectrum light source than a monochrome recorder.

Opto-mechanical film recorders have two major drawbacks. The first is that they are mechanically complex. This translates into greater fragility and more frequent maintenance than all-electronic devices. The second major drawback is that they require darkroom capability to process the film. There are no self-developing films which can be fitted to the curved drum surface. Many facilities cannot easily provide photographic support to their image-processing operations, and/or cannot accept the delays imposed by photographic processing.

CRT image-forming recorders create an output image by modulating the electron beam in a CRT. The beam is deflected according to the raster of the input image, tracing the image on a high-resolution CRT. The film is exposed, a line at a time, by focusing the CRT image onto the film.

Color images are formed by making successive exposures through red, green, and blue filters. Figure 20-18 shows how such a device operates.

Compared with opto-mechanical scanners, CRT image-forming recorders are significantly faster. They accept instant-developing film packs, making them useful for "quick-look" hardcopy. Significant disadvantages are the limited film size (usually 4 by 5 inches), and radiometric and geometric variations introduced by the CRT. Image-forming recorders also have lower spatial resolution than opto-mechanical scanners.

Laser film recorders are just beginning to become available as standard products. Custom laser recorders were built throughout the 1970's for very special high-throughput applications, such as for the EROS Data Center image products and the Jet Propulsion Laboratory planetary-imaging projects. Turek and Walker (1979) describe one such system. These custom devices

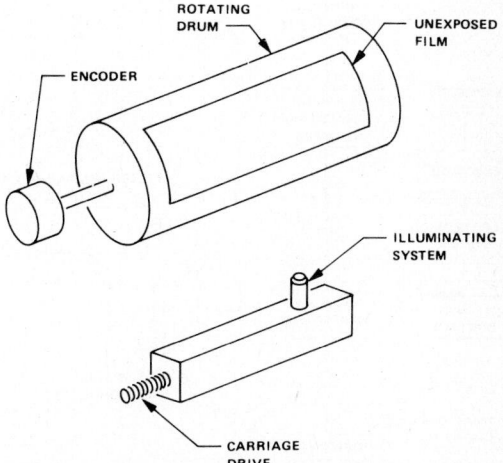

Fig. 20-16. Schematic of a typical drum-type film recorder and associated computer. (Source; Optronics International, Inc.)

Fig. 20-17. A drum-type stand-alone film recorder system. (Source; Optronics International, Inc.)

commanded high prices (several million dollars). The new, commercially available, laser film recorders are much cheaper, costing $100,000 to $200,000 for black-and-white, and 50 per cent higher for color.

Laser light-sources, because of higher energy levels, provide several advantages for film recording. The light beam can be modulated faster, resulting in much higher recording speeds. The spot size can be made smaller, although extremely small spot sizes are of dubious value since they may exceed the resolution of the film. Probably the most significant advantage is that the laser beam has sufficient energy to be split into red, green, and blue components, modulated separately, and recombined, giving the ability to expose a color image in a single pass (Figure 20-19). This is much less time-consuming than separate exposures for each primary color, and eliminates the problem of color-separation misregistration.

Other advantages of lasers derive from their stability. Laser light-sources exhibit much less luminance variation over time than conventional light sources. Conventional light sources take longer to reach operating temperature after being switched on, and exhibit greater fluctuation in color temperature over their life spans, requiring more frequent calibration. Laser light sources also last longer—a typical laser light source has a rated life of 20,000 hours, while a non-coherent glow-tube light-source can typically be expected to last only 60 to 100 hours.

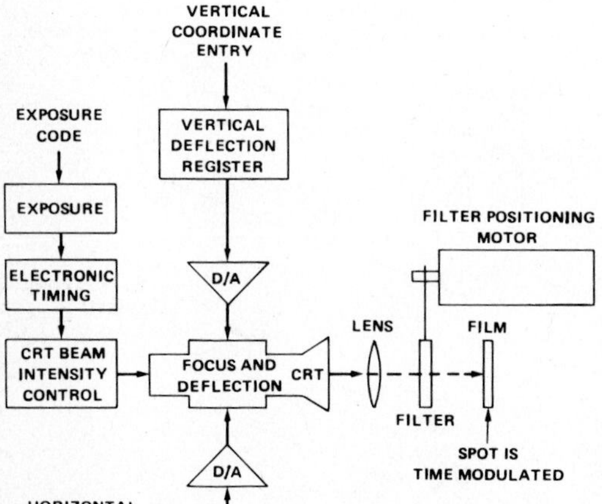

Display Screen Copiers

Display screen copiers are used to obtain hard copy of line drawings or monochrome images from a video display. Such devices are reliable, and require minimal maintenance, though they lack the flexibility and quality of more expensive copiers. Input is a video signal which is used to produce a paper or transparency print. The type of technology employed is dependent upon whether binary (line drawing) or continuous-tone black-and-white or color prints are desired.

Binary display copiers work on either electrostatic or photographic principles. Electrostatic display copiers use the video input signal to control the transfer of an electrostatic charge to a horizontal strip of the output medium. Successive scan lines are charged by moving the paper past a strip of electrodes. The charged paper is then passed through a bath of toner. This process is quite similar to that employed by electrostatic

Fig. 20-18. Schematic of a CRT image-forming film recorder. (Source; Dicomed, Inc.)

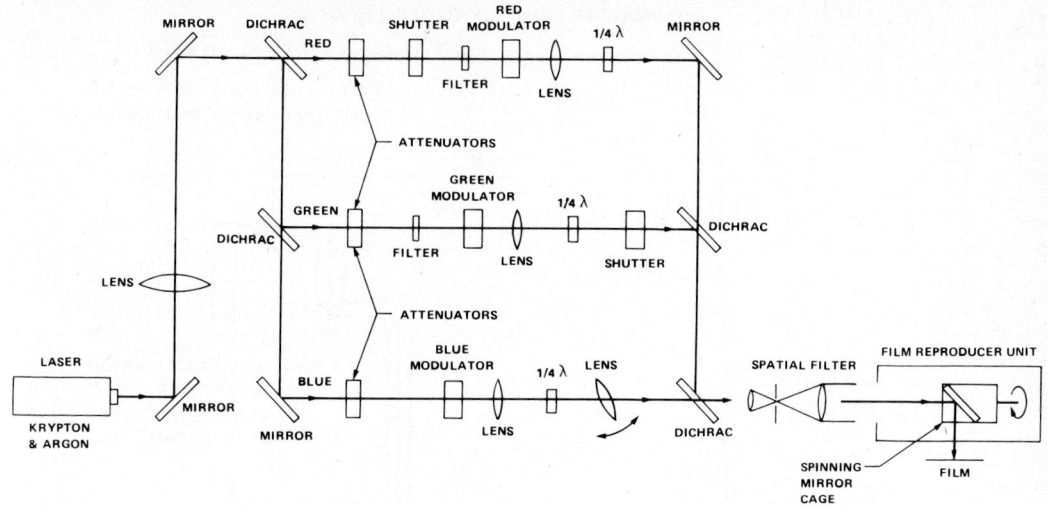

Fig. 20-19. Laser Film Recorder Schematic. (Source; Jet Propulsion Laboratory)

printer-plotters (see Figure 20-20). The difference is that in display copiers the image is produced by analog rather than digital signals.

Photographic display copiers use the input video signal to modulate a linear array of light sources which are directed at a moving piece of photo-sensitized paper. The photochemistry "develops" upon exposure to heat. The advantage of this process is that there is no toner to replenish. The photo-sensitive paper, however, is significantly more expensive than the paper used in an electrostatic process and, since the paper is heat-sensitive, the finished copy must be kept cool or it will fade.

Gray-scale display copiers use a modification of the photographic process employed by binary display copiers. The video signal drives the light sources at a variety of levels, instead of merely switching them on and off. Typically, from 6 to 16 gray levels are achievable by gray-scale copiers. While this is comparable to the gray-level resolution obtainable from dot-matrix or character-overstrike printers, the spatial resolution of gray-scale display copiers is much higher, since the gray levels can vary on a per-dot basis. Gray-scale copiers are thus useful for "quick-look" hard copy of monochrome images, filling a price-performance niche between line printers and film recorders.

Graphics cameras are an ideal way to get good quality film hard-copy directly from a display screen. With the burgeoning color computer-graphics market, graphics camera-technology has begun to come into its own. They are particularly well suited for color applications in which no real paper-based alternative exists.

In a graphics camera the video input is displayed internally on a very high resolution CRT, which is imaged by a relatively conventional camera. Formats of 35 mm, 16 mm cinema, 4 × 5 inch and 8 ×

10 inch Polaroid are generally available. Figure 20-21 is a block diagram of one brand of graphics camera.

Plotters

The principal use of a plotter is to produce hard-copy line drawings. Plotters are used in remote sensing primarily for maps or graphs. Important parameters by which plotters can be evaluated are:

- Resolution (usually expressed in points per unit distance). Typical resolutions are of the order of 100 to 1000 points per inch. The higher the resolution, the smoother the lines that will be drawn.
- Maximum possible size of output drawing. This is especially important for cartographic applications, where large maps may be desired. Price usually increases sharply along with maximum plot size.
- Multi-color capability. Some plotters can draw in more than one color, under computer control, by alternating pens or ink supplies.
- Intelligence. Some plotters have built-in hardware or firmware for generating circles, arcs, characters, and other symbols, or for converting between raster and vector representations of the input data. The presence of such hardware can relieve the host computer of a significant computational burden.

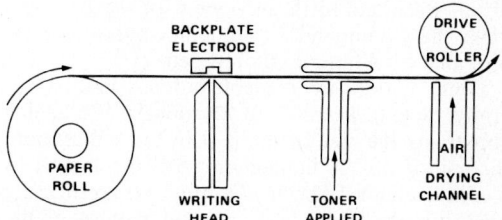

Figs. 20-20. Electrostatic plotter principle of operation. (Source; Versatec, Inc.)

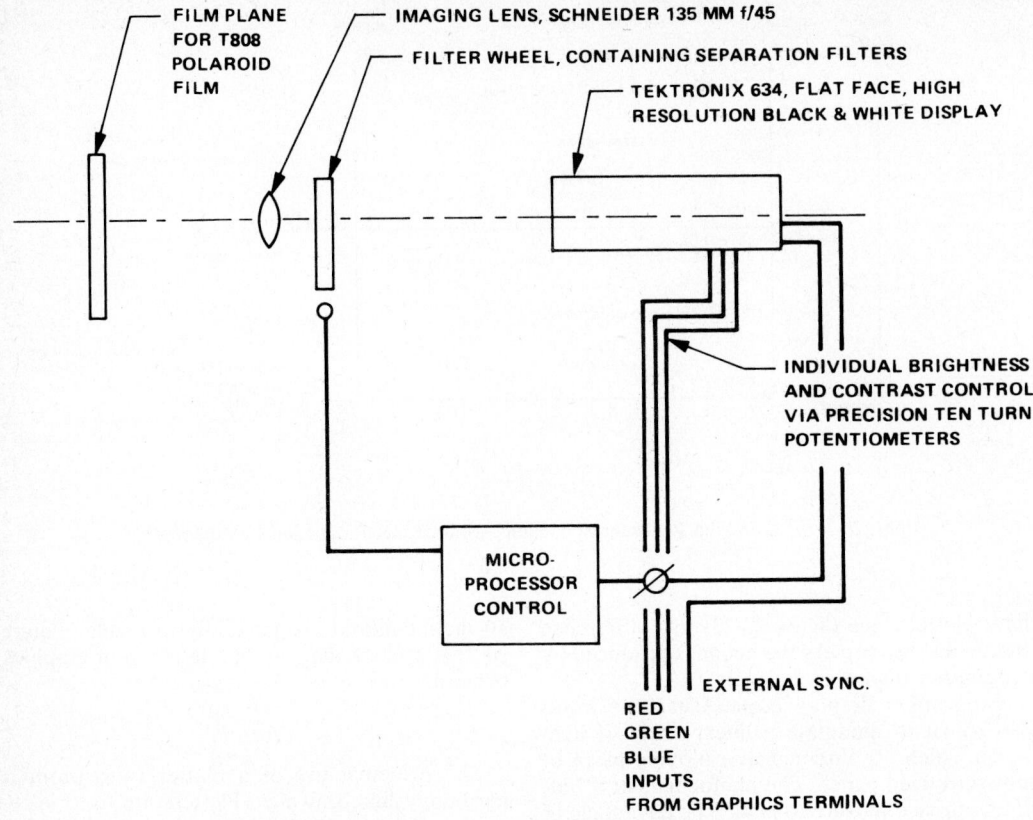

FILM PLANE FOR T808 POLAROID FILM

IMAGING LENS, SCHNEIDER 135 MM f/45

FILTER WHEEL, CONTAINING SEPARATION FILTERS

TEKTRONIX 634, FLAT FACE, HIGH RESOLUTION BLACK & WHITE DISPLAY

INDIVIDUAL BRIGHTNESS AND CONTRAST CONTROL VIA PRECISION TEN TURN POTENTIOMETERS

MICRO-PROCESSOR CONTROL

EXTERNAL SYNC.

RED
GREEN
BLUE
INPUTS
FROM GRAPHICS TERMINALS

Fig. 20-21. Block diagram of the Dunn 631 color camera. (Gruber, 1979)

X-y plotters produce a drawing by moving a pen in relation to a piece of paper. Plotters are generally slow enough that they can be operated using low-speed (and low-cost) asynchronous connections to the host computer. The minimum instruction repertoire of an x-y plotter consists of 'move to location (x, y)', and commands to raise and lower the pen. Electrostatic plotters operate on a raster principle leading to differences in speed and quality.

Drum Plotters

In a drum plotter the output medium is supplied in a continuous roll which is held taut across a sprocketed drum. Forward and reverse rotation of the drum provides the x-component. The pen is suspended on a bar longitudinally across the drum. Back-and-forth movement of the pen provides the y-component. Multicolor output can be provided by different colored pens.

Drum plotters are available in fixed widths generally ranging from 12 to 36 inches. While this constrains the size of the plot in the y-direction, the x-direction is limited only by the length of media available. The media must be specifically made for the width and sprocket pattern of the plotter. Various types of paper and mylar-type films are available for most plotters.

Flatbed Plotters

With flatbed plotters the medium is held on a flat surface by a vacuum, an electrostatic charge, or by mechanical means. The pen is mounted on a carriage which moves in the x-direction, while the pen moves along the carriage in the y-direction. On some plotters, the pen may be replaced by a light source in order to expose photosensitive film, or by a scribe for cutting map overlays on opaque film. Flatbed plotters are available in sizes ranging from 12 by 18 inches to several feet on a side.

Multi-color plotting can be accompanied either by a multiple-pen writing assembly, or by an array of pens along one side of the plotter which are fetched when requested. The latter method allows a wider selection of colors, at the expense of a slight delay whenever the pen is changed.

Some flatbed plotters provide local control of pen movement via a joystick or trackball. In conjunction with the ability to report the pen position to the host computer, this allows the plotter to serve as a coordinate digitizer in a very crude sort of way.

Electrostatic Plotters

Electrostatic plotters differ from line plotters in that individual lines are not constructed by moving a pen or stylus from one end point of a line segment to the other. Instead, a raster principle is used in which, for each line across the paper, all points which should be dark are made black and others remain blank.

The process is actually very much like xerography. Figure 20-20 is a schematic diagram of the electrostatic plotting mechanism. A writing head, which consists of a linear array of writing nibs (usually 100–200 per inch), creates electrostatic dots on electrographic paper. An image is then produced by passing liquid toner over the charged paper. Typically, dual writing heads are used which produce an overlapping dot pattern improving graphic quality.

A significant advantage of electrostatic plotters is their flexibility and speed. They function well as line printers and are often used as direct hardcopy devices for graphics terminals. The high dot resolution also allows continuous-tone monochromatic images to be constructed by adjusting dot density in local areas. The speed of the plotter itself is not affected by data density, and remains constant at around 1 inch per second (36 inches wide and greater will plot more slowly) provided data can be fed to it that rapidly. Since many plots are created as a set of line segments, the graphic elements must be 'rasterized' before they can be plotted, which can be a burden for many computers. Electrostatic plotter manufacturers recognize this problem and provide hardware vector-to-raster conversion, at a significant price.

Line Printers

Line printers are designed primarily for outputting pages of alpha-numeric text. Almost all computer systems contain some form of line printer, its principal use being to produce listings of program source, data files, program documentation and program output. Line printers can also be used for fast, cheap, image output. Although the spatial resolution thus obtained is often marginal, the ubiquity of line printers merits their consideration as "entry-level" imaging devices.

All line printers contain a hardware character-set which maps an input character code into a printed output character, usually occupying a fixed-size resolution cell. Typical line-printer resolutions are 10 or 12 characters per inch horizontally and 6, 8 or 10 characters per inch vertically. Dot matrix line printers may allow the user to access the individual dot positions which comprise the characters, thus offering the opportunity of increased spatial resolution. Freedman and Simpson (1982), provide a summary of the different line-printer technologies and their operational characteristics.

Character-impact printers behave like typewriters. Characters are formed by striking a cast of the character against the paper. The user is limited to those characters supplied with the printer, though special character sets are often available. The spatial resolution of the output is likewise fixed.

The method of generating image output on a character-impact printer is to select a set of characters which, when typed one upon the other and viewed at a reasonable distance, span a range of gray levels. The value of each pixel determines a particular set of characters to be over-printed. Using from 1 to 4 characters per output pixel, it is usually possible to generate at least 8 readily distinguishable gray levels, and using more, a continuous tone image can be approximated.

Images produced by character over-printing are the cheapest possible form of image output, albeit the least desirable in terms of spatial resolution. However, the large pixel size can be advantageous when precise location of a single pixel is desired (such as for finding control points in image-registration applications).

Dot-matrix printers form characters by turning dots on and off within a fixed size matrix, typically 9×7. An input line of character codes yields several consecutive but concurrently written output lines of dot patterns which form a raster image of the desired characters. The actual printing may be by impact or by electrostatic or thermal means. Impact dot-matrix printers typically have resolutions on the order of 50 to 100 dots per inch, while electrostatic printers may have resolutions of 200 dots per inch or more.

Some dot-matrix printers allow over-printing, in which case images may be generated in the same fashion as on a character-impact printer. Another interesting feature of some dot-matrix printers allows individual dot positions to be addressed. By constructing bit patterns to represent gray levels, the user can generate images with much better spatial resolution than if constrained by the standard character set. These printers can also be used as low-quality line plotters by providing vector-to-raster conversion in the host processor.

SYSTEM CONFIGURATIONS

Computers and peripheral devices come in many different levels of capability, quality, and cost. Building or selecting an image-processing system is a matter of matching requirements and budget to the available hardware. Prices for image-processing systems are falling, along with computer systems in general, but it is still expensive to get started as compared with application areas which do not require high-resolution color-graphics displays.

Image processing systems can range in price from $10,000 to well over $1,000,000. To give an indication of how systems might be configured and peripherals matched, three levels of image-processing systems are described. The systems are classified according to low, medium and high cost with each successive system being an order of magnitude more expensive than the previous. These examples are not presented as a shopping list, but rather to suggest the level of capability which can be obtained for a given price.

The cost of a system includes more than its initial purchase price. Maintenance should be expected to cost at least 10 percent of the purchase price per year for the life of the equipment, typically 7–12 years. Larger systems will also need continuous support from system programmers and operators.

SMALL SYSTEMS

A small system costs from $10,000 to $30,000. It is a single-user system, which means that it can be used by only one person at a time. The entire system can be packaged in a desk-top or desk-drawer configuration. No special electrical or environmental conditioning is required. Basic processing power is supplied by an 8-bit or 16-bit micro-processor.

The computer for a small system consists of a chassis to house the circuit boards and power supply, and provide the physical framework and bus connections. To the bus are connected the processor board, memory boards, communications boards, peripheral interface boards, and any other specialized devices or device interfaces. The most popular bus for 8-bit micro-computers is the S-100 bus, which is also a widely accepted standard with many compatible peripherals. Standard buses exist for 16-bit micro-computers but none has established dominance yet. A typical complement of memory would be 32-64K bytes for the 8-bit processors and 256-512K bytes for the 16-bit system.

Mass-storage devices for a small system depart from the traditional 9-track tape drives, which are too large and too costly. Both digital-tape cartridge and floppy disks have been developed to the point where they are capable of handling the volume of data necessary for image-processing applications. Current tape-cartridge technology provides up to 67 megabytes of capacity on a single 600 foot cartridge of 1/4 inch magnetic tape. This is sufficient for storing several Landsat images, especially if data-compression techniques are employed. Disk drives available for micro-computer image processors include floppy disks and small Winchester disks which surpass the floppy disk capacity by an order of magnitude while requiring the same physical space. Mass-storage devices such as these can store several large-area images or numerous smaller ones on-line, along with the necessary system and applications software.

To qualify as an image-processing system, a video display of some sort must be included. Full color frame-buffer systems with high quality monitors would be too expensive and could not easily be interfaced to the micro-processor bus. Small frame buffers for raster graphics have been built for micro-processor oriented buses and these are quite suitable for low cost image applications. A 256 by 256 pixel by 12 bit-deep display has been shown to be quite useful (Wagner, 1981). Color images can then be displayed with 16 different shades in each of the three primary colors. A television, modified for red, green, and blue video input, provides a low cost color monitor. The video display components nevertheless constitute a significant amount of the system cost. Associated with the video display are a cursor generator and a control device, such as a joystick, used for operator input of locational information via the monitor.

In addition to the major components described above, the small system needs a communications controller, a console terminal, and an acoustically-coupled modem for transferring data over telephone lines.

The U.S.G.S. EROS Data Center has developed a micro-computer-based image-analysis system utilizing the technology described above (Wagner, 1981). This system, called the Remote Image Processing System (RIPS), was designed to provide a low-cost, portable image-processing capability for use in smaller field offices and as a support unit to provide digital image-analysis capabilities to survey teams in the field. Figure 20-22 shows the entire system. Software was designed for this unit which permits the use of the device both as a stand-alone processor and as a component of a distributed processing network. Image data-compression techniques and telecommunication protocols were developed (Nady, 1981), to support these units at remote sites. Table 20-2 is a list of components, along with approximate costs, which make up the Remote Image Processing System (Wagner, 1981).

Small systems have limited hardware capabilities because of limited addressable memory and the constricted data and I/O paths. Software capability is limited by hardware and as a result, micro-computer operating systems do not have the range of features available to larger systems. However, much significant processing can be done on such systems, particularly if they are connected to a larger system to provide back-up processing and high-volume data storage (Waltz and Wagner, 1981; Wagner, 1981).

At this time, 16-bit micro-processor-based frame buffer systems are beginning to compete favorably with 8-bit systems. The new 16-bit micro-processors are designed with address spaces, typical up to now only in large mainframe computers. Execution speeds rival the large minicomputers of the recent past. This will enable small systems to be built which do not have nearly the constraints of those listed above. Low price desk-top image-processing systems, which do not need to rely on a powerful host for many tasks, are certainly possible with today's hardware.

MEDIUM SYSTEMS

A medium-sized image-processing system costs between $50,000 and $100,000. It is distinguished from small systems in that it need not rely on a

Fig. 20-22. The Remote Image Processing System (RIPS) workstation. (Source; U.S.G.S. EROS Data Ceter)

host and can perform all major image-processing functions independently. It can support more than one user. Although interactive image-processing tasks are usually limited to one user at a time, other concurrent users can be doing program development, text editing, etc. A medium-sized system is not able to support multiple image-

TABLE 20-2

**Components of the RIPS low-cost
image-processing system**

Component	Approximate Cost
Chassis and Power Supply	$ 1,000
Z80A CPU	400
Memory—52K Bytes	1,200
Floppy Disk Controller	500
Floppy Disk Drive	3,500
Video Frame Buffer	6,000
Color Monitor	1,300
Joystick and Controller	425
Communications Controller	600
Console Terminal	900
1200 Baud Modem	800
	$16,625

processing work stations as can a larger system. Normally built around a 16-bit processor, the medium-sized system is constrained by memory resources and by processor speed.

Certainly there are exceptions to the above generalities. While medium systems make effective stand-alone systems, they can be networked (Donovan and Hoffman, 1981) or included as subsystems in larger, more powerful, computing environments. The 16-bit minicomputers of the 1970s are being replaced by powerful 16-bit micro-computers with greater speed, and memory-addressing capabilities equal to the 32-bit processors. These micro-computers will give rise in the 1980s to medium-cost image-processing systems with greater capability and at a lower cost than current medium-cost systems (Donovan and Hoffman, 1981), once the hardware components are integrated and suitable software adapted.

The basic processing power of a medium-sized system is provided by a 16-bit minicomputer or micro-computer. These processors support floating-point arithmetic, memory-mapping (or extended address space), multi-user operating systems and high-level programming languages. The bus architecture supports high-speed data transfers and facilitates incorporation of a wide variety of peripherals into the system. From 64K bytes to 4 megabytes of memory is provided.

The medium-sized system must be able to function independently. It must also be able to store a significant number of images and be able to accept images from outside sources. For these reasons, mass storage is a critical part of the system design. In recent years, the cost of relatively large-capacity hard-disk systems has dropped to where it is reasonable to include an 80- to 300-megabyte disk drive. This level of disk capacity permits interactive manipulation of multiple images, as well as the application of processing techniques that generate large temporary files.

Magnetic tape is the most cost-effective medium for bulk-data storage, and for the transfer of large amounts of data between installations. A tape subsystem also allows direct processing of large images without requiring intermediate data storage. Tape is also important for performing system back-ups. For the medium-sized system the tape subsystem will include a dual-density (800/1600 BPI) tape controller and drive.

A desirable enhancement for a medium system is the ability to perform floating-point arithmetic efficiently. Some processors include floating-point operations in their basic instruction sets, while others require auxiliary hardware. Either of these can increase the speed and efficiency of floating-point calculations to the level of much larger systems.

The video-display system is built around a medium resolution (512×512 pixels) color-frame buffer (3 channels, 8 bits per channel) with look-up tables for mapping the output of a given channel to the red, green, or blue A/D converter. Additional features include a graphics-overlay channel for color-vector graphics and alphanumeric text, and hardware vector-to-raster generators. The operator interface includes an RGB color monitor and a coordinate-input device (e.g., digitizing tablet, joystick, trackball). As in the small system, the video system is likely to involve a major portion of the cost of the system.

The system also requires a controlling terminal, line printer, and 1200 baud modem to connect to other systems. If multiple users are to be supported, a communications controller and additional terminals are needed. Selecting a dot-addressable matrix printer, instead of an ordinary line printer, provides graphic and low-resolution image-output capability.

Table 20-3 is a list of components and costs for a hypothetical medium-system. The system is fashioned after the MIDAS system described by Donovan and Hoffman (1981). Unlike small systems, environmental conditions such as temperature and humidity are important. The necessity and cost of preparing and maintaining the physical plant required by medium-sized systems is a consideration.

There are many medium-sized turn-key systems

TABLE 20-3

Components of a medium-cost image processing system

Component	Approximate Price U.S. dollars
COMPUTER 16-bit processor including chassis, memory, and all system software	12,000
300 megabyte disk drive	20,000
9-track dual density tape	10,000
Video Display including: 3,8-bit 512 × 512 planes 1,4-bit graphics overlay plane RGB monitor 11″ × 11″ digitizing tablet High speed CPU interface	40,000
CRT terminals (3)	24,000
Dot-matrix line printer	8,000
Miscellaneous communications equipment	2,000
	94,400

available, primarily from video display manufacturers. One such system is shown in Figure 20-23. Turn-key systems may be a convenient way to acquire an integrated system but these systems vary widely in amount of software support— something which should be as carefully evaluated as any hardware feature.

LARGE SYSTEMS

A large system supports a dozen or more concurrent users, several of which may be performing image-processing tasks, and an extensive set of peripheral devices. Such a system is built around a 32-bit processor which allows essentially unlimited addressable storage. Large systems cost from $200,000 to $1,000,000. Still more expensive systems tend to be either highly specialized computers or business-oriented processors which do not easily support scientific computing or specialized peripherals. Large systems are built around minicomputers, though this is not indicative of a lack of processing power. Because of their extensive addressing capability, many operate as virtual machines, treating disk storage as an adjunct to main memory.

The large system supports a variety of operating systems, and a full range of high-level programming languages. As with medium systems, the compatibility of the operating system, programming languages, existing software packages, and hardware is critical to the implementation of the image-processing system. For image processing, it is essential that the system allow interactive programming and control.

Large-system capability can sometimes be acquired by adding peripherals to existing general-purpose computer systems. For most applications and users this is a useful option because it reduces

Figs. 20-23. A medium-priced turn-key image processing system. (Source; DeAnza Systems Inc.)

the cost of achieving image-processing capabilities. Many large image-processing systems are in fact multi-purpose computing environments, where image processing is just one of several often unrelated activities which the system supports. However, because of the considerable resources required by image processing, it is not an additional duty that should be taken lightly.

The large-system CPU is built around a 32-bit processor. The CPU includes cabinet, power supplies, and at least one high-speed bus. It also includes a substantial amount (several megabytes) of memory. The larger the memory available, the more operations can be performed in-core. Hardware options to improve arithmetic performance (e.g., hardware floating-point processors) are also included.

Extensive on-line disk storage is a critical characteristic of the large system. This permits sophisticated operations (e.g., mosaicking, multi-image registration, rotation) on large image datasets. Demand-paged systems also require a significant amount of disk storage for paging under heavy load conditions. For a system with a dozen or more non-concurrent image-processing users, 1000 megabytes of storage may be reasonable, depending on the application mix. For a large system with only a few heavy users, two or

three 300 megabyte drives are sufficient. As the number of users increases, it becomes advantageous to have several smaller disk drives as opposed to one large drive, in order to minimize the contention for access. However, more hardware is both more expensive and more likely to fail.

A magnetic tape subsystem is essential for data storage and transfer. Two or more drives are required to adequately service a multi-user community (e.g., tape duplication, reformatting). Tape drives of 6250 bpi should be used if there is to be heavy data traffic into and out of the system. The higher density drives also expedite backing-up the large disk datasets.

Many image processing calculations are concisely formulated in vector arithmetic but are slowly executed on conventional processors. The budget for a large system may permit the inclusion of an array processor. This device is essentially an attached peripheral processor which can be invoked to perform the same calculation upon each element of an array very quickly. Vector-oriented operations (e.g., FFT, convolution) can be speeded up several orders of magnitude by one of these devices.

The image-display processor for the large system should be carefully considered. Full utilization of the system requires that as many display functions (e.g., lookup-table transforms, histogramming, simple classification, zooming) as possible be downloaded from the host system, allowing it to concentrate on processing without display requirements. The display processor should include additional image buffers (besides the 3 needed to generate a color display) to support flickering between images, or arithmetic operations which require temporary intermediate storage. The display processor should support large-frame buffer configurations (e.g., combining four 512 × 512 images into one 1024 × 1024 image). Split-screen display of multiple images is a desirable feature which allows easy comparison of related images. To take advantage of the investment in memory for some applications, the display processor should allow several different independent image-analysis stations to be attached to it.

Special peripherals supported by a large system vary depending on the intended application. Existing large computing facilities often already support graphics output devices such as plotters and calligraphic displays. If the image volume is large enough, or the requirements justify, a high-quality film scanning and/or recording device is included.

The large system includes a console terminal, a high-speed line printer, and a multiport controller for connecting a dozen or more terminals. The system should have one or more dial-up ports so that users can connect to the system through the telephone lines. A long-distance network interface may be useful for exchanging messages and programs with remote systems, although these links are currently too slow to support frequent exchange of large-image datasets.

Since large systems are frequently built by adding image-processing hardware and software to an existing general-purpose scientific system, the image-processing user often has less control over the final environment than with a small or medium system. In such circumstances, the user should be careful that the target system is sufficiently interactive to support image processing (many large systems are more batch-oriented, with interactive terminal support added as a poorly-functioning afterthought). The existing software environment of the target system should be carefully evaluated since, if it is currently being actively used, it is not likely to change.

Table 20-4 is a list of components, and their approximate cost, making up a typical large image-processing system. The system described is fashioned after the facility at the Computer Systems Laboratory at the University of California, Santa Barbara (Figure 20-24).

TECHNOLOGY DEVELOPMENT

The field of digital hardware is extremely fast-moving, which makes descriptions of such equipment frustrating, exciting and always out of date.

TABLE 20-4

Components of a high-cost image processing system

CPU	
DEC VAX-11/780	160,000
UNIBUS adapters (2)	25,000
4 megabytes MOS memory	12,000
Floating Point Accelerator	25,000
RS-232 asynchronous communications lines (48)	10,000
300 megabyte disk drives with UNIBUS controller (3)	52,000
9 track 800/1600 BPI tape drives (2)	15,000
Array Processor	90,000
Video Display System	82,000
512 × 512 × 8-bit image buffers (8)	
512 × 512 × 4-bit graphics buffer (1)	
Lookup tables	
Real-time histogram	
Digital video-rate processor	
Image/Graphic Hardcopy	
Opto-mechanical film recorder/scanner	85,000
Desktop flatbed plotter	8,000
Electrostatic printer/plotter	11,000
Dot-matrix line printer	11,000
CRT user terminals (20)	16,000
DVST Graphics Terminal	12,000
	$614,000

Image processing and remote sensing applications require tremendous amounts of data to be processed. To handle this load we must continuously be sensitive to emerging technologies so they can be adapted to our particular applications. This section is intended to highlight some of the technological areas which could potentially have a great impact in remote sensing applications. These technologies primarily address issues surrounding the storage, management and processing speed for very large datasets.

The technologies discussed in this section are highly interrelated. VLSI is a key technological area. From it, parallel-processing architectures become more manageable, leading to such things as parallel processors dedicated to data-base management. When optical mass-storage techniques are perfected, the data-base machines will play a crucial role in management of the very large data bases.

VLSI

Very Large Scale Integration (VLSI) is the process of creating digital circuits with many thousands of transitors on a small area (Mead and Conway, 1980). Tobias (1981) defines LSI circuits as consisting of 1000 logic gates per chip and VLSI as an order of magnitude greater. Within three to five years that number is expected to increase by about two orders of magnitude. This fact, along with ever-improving automated pro-

Fig. 20-24. A typical large image processing system based on a "super" minicomputer. (Source; Univ. of Calif., Santa Barbara, Computer Science Laboratory)

duction tools, means that complex functions which do not have an extremely large market appeal (image processing does not have as large a market as automotive applications) can be implemented in hardware. Hardware organization can be specifically optimized for individual applications, unlike general-purpose computers. In this regard, VLSI can be viewed as an extension of software.

Bracken (1980) spelled out the need for improving the processing speed of image data collected from an increasing number of satellites, each with a larger bandwidth than its predecessor. In particular, vast quantities of image data will need to be processed, geometrically projected, mosaicked, manipulated, and merged at rates that far exceed the capacities of present computer systems. The high-throughput problems to which VLSI can be a solution loom mainly for systematic ground processing. However, VLSI products developed for these large applications will eventually become available for smaller, less demanding, configurations.

The ability to do VLSI is based upon the recent development of sophisticated automated design and fabrication techniques. However, not all the pieces are yet in place. Computer languages are currently under development which will eventually allow high level statements to be made which

establish functional criteria and automatically create chip designs. Some of the design effort though, must still be carried out manually in order to perform design-rule checks.

VLSI appears to show promise for applications in the following areas related to remote sensing.

- Filtering
- Parallel processing
- Radar Processing
- Optical disk controllers
- Telemetry decommutation
- Cross-correlation
- Image warping
- Data-base management and machines

PARALLEL PROCESSING

The technologies of VLSI and design automation promise to enable large processors, with a high degree of parallelism, to be built in the future. (Haynes et al., 1982). The potential of achieving 10^{12} or 10^{13} operations per second exists in large parallel systems with a 1K by 1K array of processors. For comparison, a large minicomputer will perform $1\text{-}2 \times 10^6$ operations per second. The availability of VLSI circuits designed specifically for parallel structures will provide the opportunity to build these very high performance two-dimensional processors (Fairbairn, 1982).

As an example of such a processor, it is useful

to consider the Massively Parallel Processor (MPP) currently under development at the Goddard Space Flight Center (Batcher, 1980; Tsoras, 1981). The MPP is a forerunner to potentially larger systems, but the technology found in the MPP may eventually find its way into smaller systems. It is a single-instruction, multiple-data-stream (SIMD) machine consisting of a 128 × 128 array of bit-serial processing elements.

The 128 × 128 array size was determined by reliability concerns. Each processor chip actually contains 8 individual processing elements. The MPP has a separate array-control unit, a data-staging memory buffer and a host computer to manage the user interface. Effective computational rates are given in Table 20-5.

The high computational performance of the MPP can be achieved for computationally-intensive problems or for data-intensive problems where adequate I/O data rates can be supported. The MPP system is designed to have data transfer rates of 160 megabytes per second between the staging buffer and the array unit. The staging buffer can be loaded at 40 megabytes per second. The high computational rate, combined with these high I/O rates will provide an operational throughput rate two to three orders of magnitude higher than conventional computers for image processing applications.

DATA BASE MACHINES

For years, data managers for remote sensing applications have sought to exploit the concepts and features of commercially available data-base management-system (DBMS) software. A DBMS serves as a centralized manager of data bases, controlling all access to data. This reduces data and software redundancy, increases the share-ability of data among users, and facilitates the control of the integrity (correctness) and security of all data. The robust data-structure definition-capabilities of DBMSs augment the limited data definition-capabilities of programming languages such as Fortran, facilitating data structures tailored to flexible access mechanisms (Cardenas, 1979).

Commercially available DBMSs have several limitations in remote sensing applications. Most importantly, DBMS software does not perform well when storing and retrieving large blocks of data, such as images. This problem has been particularly acute with the relatively new and popular relational data base model (Codd, 1970; Date, 1977), which is simple to understand but has to recompute relationships during retrieval rather than store them explicitly in the data base.

Although some relief from these limitations will come from greater awareness by vendors of the requirements of remote sensing applications, significant improvement in performance is most likely to result from data-base machines (DBM). Data-base machines are hardware devices, dedicated and tailored to data-base management functions. Specialization has become cost-effective because of the plummeting cost of mini- and micro-processors, and the increasing proportion of computer time needed to accomplish data storage and retrieval in data-intensive (as opposed to compute-intensive) applications having very large (greater than 10^{10} bits) data bases.

The major cause of complexity and inefficiencies in DBMS software is the mismatch between data-intensive operations and the computer-intensive hardware instructions of today's sequential machines. Architectural alternatives to the sequential machine for general-purpose DBMs may be classified into four types (Su, 1980): cellular-logic systems, back-end computers, associative-memory systems and integrated DBMs. While experimental prototypes of all four exist, only the second and fourth types have been implemented in commercial machines (Rauzino, 1981).

Cellular-logic systems delegate some data-base management functions to intelligent secondary storage devices. When the function of filtering out irrelevant data is delegated, data can be searched associatively; that is by content (values desired) or contexts (neighborhood of the data desired) rather than by physical location. Examples of this type of architecture include the Tektronix INDY research project, the RAP project at the University of Toronto, and the Context Addressed Segment Sequential Memory (CASSM) (Su, 1979).

The key concept of back-end DBMs is to off-load data-base management functions from the host computer to a dedicated processor. These processors are usually general-purpose computers, but special-purpose machines may also be used (Rauzino, 1981). Advantages include: (1) relieving the host of time-consuming data-base manipulations; (2) increased performance through

TABLE 20-5

Execution speeds of the Massively Parallel Processor (Batcher, 1980)

Operations	Execution Speed in MOPS*
Logic Operations	
8-bit	6826
16-bit	3413
Addition of Arrays	
8-bit integers	6553
16-bit integers	3344
32-bit floating point	430
Multiplication of Arrays (Element by Element)	
8-bit integers	1861
16-bit integers	538
32-bit floating point	216

* Millions of Operations Per Second.

functional specialization and parallel processing by the host and back-end machines; (3) increased security and control by making the DBM the only access route to the data in series with the host operating system; (4) several dissimilar hosts can share data through the DBM and be insulated from changes to the data bases or storage devices; and (5) growing and/or distributed data bases can be accommodated by multiple back-ends. The first back-end DBM was the Experimental Data Management System (XDMS) at Bell Laboratories. Other examples are the Data Computer developed by Computer Corporation of America for ARPANET and the Adabas DBM of Software AG (Su, 1980).

High speed associative-memory devices differ from cellular-logic systems in that each bit or word of a separate memory has a processing element. Hence, associative memories are more expensive per unit storage than cellular-logic systems. Pages of data must be transferred from main or secondary storage to the associative memory, where extremely fast content searches can be performed in parallel. The best known example is the STARAN computer system, which contains 32 associative processor arrays, each containing 256 word by 256 bit memory arrays with bi-directional access to a maximum of 256 bits in parallel (Su, 1980).

Integrated DBMs combine a number of specialized processors, including any of the above types of machines. Examples include the Data Base Computer (DBC) project at Ohio State University, the RAP.2 expansion upon RAP at the University of Toronto, INFOPLEX at MIT, and the single-instruction multiple data stream (SIMD) MICRONET at the University of Florida (Su, 1980). The only commercially available DBM of this type is the Intelligent Database Machine (IDM 500) built by Britton-Lee. This machine might also be classified as a dedicated-function back-end machine (Rauzino, 1981) but is itself made up of several general-purpose microprocessors. The key component that makes the IDM 500 unique is an optional auxiliary "data base accelerator", a 10 million instruction per second (MIPS) pipeline processor (Rauzino, 1981) with data management-oriented primitive instructions (Britton-Lee, 1981).

The impact of DBMs for remote sensing applications may be a while in coming. Software AG has demonstrated a 25% average increase in throughput for an off-loaded Adabas versus a host-resident Adabas running the same jobs. The Adabas DBM costs approximately $400,000 (Rauzino, 1981). Britton-Lee claims a standard rate of 2 to 5 data-base transactions (e.g., an update) per second, increasing to 20 to 25 transactions per second using the data-base accelerator. The cost of the IDM 500 ranges from $50,000 to $150,000, depending upon the hardware configuration chosen (Rauzino, 1981; Britton-Lee, 1981). Hawthorn and DeWitt (1982) provide an analytic comparison on the performance of several DBMs

for different types of queries. As competing manufacturers bring these costs down and improve performance it will become more cost-effective for very large data base applications, such as remote sensing, to reap the benefits of DBMS technology and to use the more elegant relational DBMS model.

OPTICAL MASS STORAGE

Optically-coded disks for the video-communication market are now a reality. These disks store a tremendous amount of information in a small area. The laser-driven digital versions record data by burning a pit with the laser, about 1 micron in diameter, in a reflective coating which has been deposited on the disk surface (Bartolini et al., 1978). To read back the information, a light source is directed at the surface and the size and shape of the pit are used to modulate the signal. When such techniques are applied to the writing and recording of binary digital data, densities of 10^{10} to 10^{11} bits are possible (Kenney et al., 1979). For comparison, it would require approximately 25 magnetic tapes, written at 6250 bytes per inch, to store that amount of data. The disks can be written only once, but they can be read as many times as desired. Random access to data is possible.

Digital optical disks do not record a video signal. Rather they record binary digits directly. Video-image recording can tolerate a certain amount of noise in the signal. Digital applications are less tolerant. Digital magnetic media now perform at about a 10^{-12} error rate whereas digital optical disk manufacturers are reporting rates on the order of 10^{-10} to 10^{-11} (SBS, 1980).

Early-on it was thought that for optical disks to compete in the mass storage market they would need to be eraseable. However, it has become apparent that many applications which require quick access to massive volumes of data do not also require updating the data. The data are really of an archival nature and updates do not make sense. The sheer mass of data often makes the logistics of updates infeasible (SBS, 1980) anyway. This situation applies very well to remote sensing applications.

It is in gaining quick access to the large archives of unchanging image data that digital disks will play their most important role. Capacities of magnetic disks are sufficient to handle the day-to-day operations of most image-processing applications. The error rates inherent in optical-disk systems should not bother image processing. Image data have less stringent requirements because the spatial correlation within images allows greater success in error recovery than, for example, text processing.

Optical-disk mass-storage systems would probably consist of "juke box" type configurations in which individual disks would be retrieved and loaded into the player automatically. The mechanism would allow the disk to be retrieved in less than 3 seconds. Ammon (1980) describes such a mass-storage system proposed by RCA which would consist of a unit storing 100 disks, each containing 10^{11} bits, resulting in a 10^{13}-bit system capable of storing approximately 40,000 Landsat frames. By configuring several of these units as peripheral devices to a data-base management computer/controller, another order of magnitude could be obtained.

REFERENCES

Ammon, G. J., 1980, Wideband optical disc data recorder systems; Laser Recording and Info. Handling: Proc. of Symp., San Diego, pp 64−72.

Baer, J. L., 1980, Computer Systems Architecture; Computer Science Press, Potomac, Maryland.

Barnes, G. H., and R. Brain, M. Kato, D. Kuck, D. Slotnick, R. Stokes, 1968, The ILLIAC IV computer; IEEE Trans. on Comp., vol. C-17, pp 746−757.

Bartolini, R. A., A. E. Bell, R. E. Flory, M. Lurie, and F. W. Spong, 1978, Optical disk systems emerge; IEEE Spectrum, Aug., pp 20−28.

Batcher, K. E., 1980, Design of a massively parallel processor; IEEE Trans. on Comp., vol. C-29, no. 9, pp 836−840.

Bracken, P. A., 1980, Earth resource observations data systems in the 1980's; Annual Meeting of American Astronautical Society and AIAA, paper 80−240.

Britton-Lee, Inc., 1981, IDM 500, intelligent database machine, product description; promotional literature, Britton-Lee, Inc., Los Gatos, CA, March.

Brode, B., 1981, Precompilation of Fortran programs to facilitate array processing; IEEE Computer, vol. 14, no. 9, pp 46−51.

Cardenas, Alfonso F., 1979, Data Base Management Systems; Allyn and Bacon, Boston.

Castleman, K. R., 1979, Digital Image Processing; Prentice-Hall, Englewood Cliffs, New Jersey.

Charlesworth, A., 1981, An approach to scientific array processing: the architectural design of the AP-120B/FPS-164 family; IEEE Comp., vol. 14, no. 9, pp 18−27.

Codd, E. F., 1970, Relational model of data for large shared data banks; Comm. ACM, vol. 13, no. 6, pp 377−387.

Date, C. J., 1977, An Introduction to Data Base Systems; Second edition, Addison-Wesley, Reading, MA.

DeAnza Systems Inc., 1981, IP8500 Series Image Processor Product Description Manual; San Jose, CA.

Dinwiddie, K., 1981, Memory organization and high speed processor facilitate unique image display capabilities; Comp. Design, vol. 20, no. 7, pp 115−116.

Donovan, W., and L. Hoffman, 1981, A personal MC68000-based image display and analysis system; Summ. of First EROS/RIPS Workshop, Oct.

Driscoll, T., and C. Walker, 1981, The evolution of image processing technology; Comp. Graph. World, vol. 4, no. 6, pp 29−36.

Fairbairn, D. G., 1982, VLSI: a new frontier for system designers; IEEE Comp., vol. 15, no. 1, pp 87−96.

Foley, J. D., and I. Van Dam, 1981, Fundamentals of Interactive Computer Graphics; Addison-Wesley, Philippines.

Franklin, W. R., 1979, Evaluation of algorithms to display vector plots on raster devices; Comp. Graph and Image Proc., vol. 11, no. 47, pp 377−397.

Freedman, D. and D. Simpson, 1982, Line printers: band and matrix technologies hold the fort; Mini-micro Sys., vol. 15, no. 1, pp 157−170.

Gruber, Leonard, 1979, Color computer graphics and imaging with Polaroid 8 × 10 Polacolor Land film and the Dunn 631 color camera; Harvard Computer Graphics Conf., paper #715.

Hawthorn, P. B., and D. J. DeWitt, 1982, Performance analysis of alternative database machine architectures; IEEE Trans. on Soft. Eng., vol. 8, no. 1, pp 61−75.

Haynes, L. S., R. L. Lam, D. P. Siewiorek, and D. W. Migell, 1982, A survey of highly parallel computing; IEEE Comp., vol. 15, no. 1, pp 9−24.

International Imaging Systems, 1979, Product Descriptions, Model 701 Image Computer and Display Terminal; Sunnyvale, CA.

Jackson, J. N., 1980, Dynamic scan-converted images with a frame buffer display device; Comp. Graph., vol. 14, no. 3, pp 163−169.

Karplus, W. J., and D. Cohen, 1981, Architectural and software issues in the design and application of peripheral array processors; IEEE Comp., vol. 14, no. 9, pp 11−17.

Kenney, G. C., D. Y. D. Lou, R. McFarlane, A. Y. Chan, J. S., Nadon, T. K. Kohla, J. G. Wagner, and F. Zelnike, 1979, An optical disk replaces 25 mag tapes; IEEE Spectrum, Feb., pp 33−38.

Levy, H. M., 1980, Computer Programming and Architecture—The VAX-11; Digital Press, Bedford, Massachusetts.

Linsalata, R., and R. Scalea, 1981, Raster graphics: expanding its frontiers; Comp. Design, vol. 20, no. 7, pp 91−94.

Mead, C. A., and L. A. Conway, 1980, Introduction to VLSI Systems; Addison-Wesley, Reading, MA.

Nady, L., 1981, Preliminary implementation for image telecommunication protocol package; Summ. of First EROS/RIPS Workshop, Oct.

Newman, W. M., and R. F. Sproull, 1973, Principles of Interactive Graphics; McGraw-Hill, New York.

Pavlidis, Theo, 1981, Contour filling in raster graphics; Comp. Graph., vol. 15, no. 3, pp 29−36.

Peuquet, D. J., 1981, An examination of techniques for reformatting digital cartographic data: the raster-to-vector process; Cartographica, vol. 18, no. 1, pp 34−48.

Pohm, A. V., and T. A. Smay, 1981, Computer memory systems; IEEE Comp., vol. 14, no. 10, pp 93−110.

Rauzino, Vincent C., 1981, The looming battle between data base machines and software data base management systems; Computerworld, Jan. 5, pp 8−10.

SBS, 1980, Impact of optical memories (videodiscs) on the computer and image processing industries; Strategic Business Services, Inc., San Jose, CA.

Shaw, P. J., 1980, Architecture for combined vector/raster graphics; Mini-micro Sys., vol. 13, no. 12, pp 107−112.

Smith, B., 1981, Heterogeneous Array Processor; Denelcor, Inc., Prod. Description, Denver, Colorado.

Stakem, P., 1981, One step forward—three steps backup: computing in the U.S. space program; Byte, vol. 6, no. 9, pp 112−144.

Su, Stanley Y. W., 1979, Cellular-Logic devices: concepts and applications; IEEE Comp.; vol. 12, no. 3, pp 11−25.

Su, Stanley Y. W., et al., 1980, Database machines and

some issues of DBMS standards, Proc. of 1980 National Comp. Conf., AFIPS Press, Arlington, VA, pp 191−208.

Tanenbaum, A. S., 1976, Structured Computer Organization; Prentice-Hall, Englewood Cliffs, New Jersey.

Tobias, J. R., 1981, LSI/VLSI building blocks; IEEE Computer, vol. 14, no. 8, pp 83−101.

Tsoras, J., 1981, A massively parallel processor: innovation in high speed processors; AIAA Comp. in Aerospace Conf. III, San Diego.

Turek, J. E., and D. J. Walker, 1979, Large format laser scanner/plotter system; Proc. Auto-Carto IV, vol. II, pp 42−49.

Wagner, H. L., 1981, Communications requirements for remote image processing stations; Summ. of First EROS/RIPS Workshop, Oct.

Waltz, F. A., and H. L. Wagner, 1981, The remote image processing station project; Summ. of First EROS/RIPS Workshop, Oct.

Wu, C., 1980, A digital fast correlation approach to produce Seasat SAR imagery; IEEE Int'l Radar Conf., Convention record, pp 153−160.

CHAPTER 21

Image Geometry and Rectification

Author-Editor: RALPH BERNSTEIN

Contributing Authors: CHARLES COLBY, STEPHEN W. MURPHREY and JOHN P. SNYDER

GENERAL CONTENTS: Image radiometry, geometry and rectification; need for image correction; sources of image radiometry and geometry errors; techniques for image correction; determination of image errors; mathematical techniques used; image processing and manipulation; systems for image correction; cartographic transformations of images; references.

INTRODUCTION

The objective of this chapter is to provide a description of digital image corrections and the techniques for performing such corrections. A background for digital image processing is provided, including a brief description of optical and electro-optical image processing in order to show where the technology came from and to establish the direction toward which digital image processing is heading. This will allow the projection of future developments and needs. The technical discussion will be oriented towards image scanners, such as the NASA Landsat Multispectral Scanner (MSS) which is representative of current imaging technology.

A definition of multispectral images, digital image processing, image radiometry and geometric correction is provided to aid the reader in understanding the concepts and terminology. This is followed by a discussion of the need to correct or preprocess the data, the sources of image errors, techniques for image correction and a definition of the equations used to correct the data. Systems that perform operational digital image processing are described.

Digital image processing experiments on digitized camera data were conducted in the early 1960's (Mach & Gardner, 1962). These experiments involved the production of orthophotos and elevation contours from digital stereo-pair photos. Although these early experiments were successful technically, the computer costs associated with them were prohibitively expensive.

New sensing systems were developed in the late 1960's that used photodiodes and other detectors to convert radiance information into analog voltages, replacing the photographic systems. These sensors were capable of detecting into the thermal infrared region, in addition to the visible and near infrared. These new sensors promoted the use of analog processors to convert the signals into thematic categories, such as agricultural, forest, cultural, hydrological and other major themes. The next step involved the refining and improvement of the algorithms and the implementation of more complex operations on digital computers. Aircraft-acquired sensor data were extensively used to investigate the adequacy of the multispectral classification techniques for agricultural crop discrimination and disease detection (MacDonald, et al, 1975).

In order to increase the accuracy of crop classification, researchers used multitemporal sensor data, i.e., data acquired at different times during the growth of crops. This provided increased discrimination between crops with similar spectral characteristics. Several preprocessing requirements developed from these activities: (1) the need to geometrically register data acquired at different times and with different geometry, (2) the need to convert the data into a geometric projection that would conform to common cartographic projections from which thematic maps could be prepared, and (3) the need to radiometrically calibrate and correct the sensor data. These technical problems were addressed by several investigators with excellent results. Digital algorithms were developed that would automatically locate and accurately register digital ground-control-point features, and would geometrically correct and resample image data to an accuracy of a fraction of a picture element.

Since 1972 the NASA Landsat satellites have provided a tremendous stimulus to digital processing of earth observation data. The significant reduction in the cost of digital technology has made practical tasks which were previously economically prohibitive. The performance of general purpose computers has been doubling each year against fixed costs, and investigators have had the benefit of a significant reduction in costs since the early experiments. For example, in 1954 it would cost $1.26 to perform 100,000 multiplications, compared to $0.01 in 1976. During this time, computer speed has increased from two thousand to over two million multiplications per second.

NASA currently has an operational, high throughput, high accuracy digital image-processing system capable of processing over 100 billion bits of data per day. This system provides users with data that are geometrically and radiometrically corrected and suitably formatted. The system uses a special microprogrammed processor that performs 20 million multiplications per second. Efficient means for the input and output of digital data have been developed. High density digital tape units with 14 to 42 tracks, and with data density of 50 thousand bits per centimeter per track, provide the media for data I/O and storage.

Landsat-D has more spectral bands and higher resolution. This will be followed in the near future by high-resolution linear-array and two-dimensional mosaic detector-array configurations to replace the existing electro-mechanical sensor configurations of today. Charge-coupled or similar device technology will provide intelligence to the sensors and will be used to implement on-board processing. Progress in large-scale integrated circuits, which results from combining sophisticated manufacturing techniques with progress in solid state physics and material science, will continue to improve circuit speed and reduce circuit power, size, and cost. This will allow applications to be considered and implemented that may seem impractical today.

Several technological factors are combining to provide this potential. These include: Sensors with multispectral sensing capabilities, the combination of data from various missions and sensors, the source-digitizing of sensor data, sophisticated information-extraction algorithms that require high speed digital processing, the continually decreasing costs of digital circuitry, and the practical need for the information.

Since the data from current and future imaging sensors are in a digital form, processing the data in a computer provides several benefits. Modern imaging sensors can distinguish 64 to 256 radiance levels; thus the computer is provided with significantly improved dynamic range for image-data analysis and interpretation compared with the human eye, which can discern about 30 levels of intensity on conventional film. Future satellites will provide reliable 8-bit data (256 levels) and some integrating sensors may provide more. Another important advantage is that the computer can implement functions repetitively and with great accuracy. This can greatly reduce the analysts' work and provide improved information.

DEFINITIONS

A *Digital Image* is a sampled and quantized numeric representation of a scene (see Figure 21-1). The scene is spatially partitioned by the sensing device into a regular array of numbers whose values represent the radiance or brightness of the sampled region in one or more spectral bands. In addition to the spatial and brightness dimensions, a digital data set can have spectral and temporal dimensions since the scene can be viewed in many spectral bands, and at different times. Thus a digital image is a multidimensional matrix of numbers that characterizes a scene.

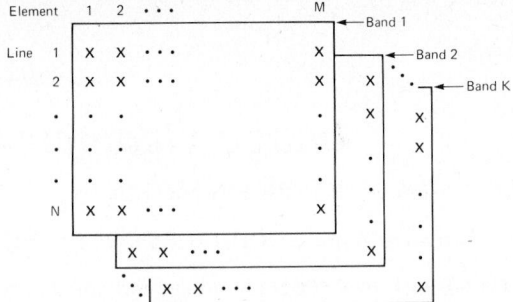

Fig. 21-1. Digitized Multispectral Scanner Images. Each of K spectral bands consist of an M × N array of numbers which represent the image intensity at each picture element (pixel).

A multispectral digital image, I, consisting of K spectral bands can then be represented as K arrays of $M \times N$ elements of non-negative values, or

$$I^{(k)} = (P_{ij}^{(k)}) \qquad (21\text{-}1)$$

where

k = band number = 1, 2, . . . , K
i = line (row) number = 1, 2, . . . , N
j = element (sample or column) number = 1, 2, . . . , M

Each picture element (or pixel) consists of B bits per sample. The total data content in a multispectral image is then $MNKB$ bits.

Digital Image Processing involves the use of a computer to digitally manipulate or operate upon this matrix of numbers for a particular objective. These operations include noise removal, geometric and radiometric correction or modification of the image data, remapping of the image onto a different scale, enhancement of the image, information extraction, data compaction, image display and recording, and image-data manipulation and management. Image processing can be broadly classified into two categories: image correction, and information extraction (e.g., multispectral classification). Image correction implies both geometric and radiometric errors. Though radiometric correction is more directly related to information extraction than it is to cartographic presentation, it is usually performed simultaneously with geometric correction. It involves changing each pixel intensity value to achieve calibration correction or intensity enhancement. Geometric correction or projection modification of an image involves repositioning the sample elements from where they are to where they should be to achieve a desired geometry.

Image Radiometry generally refers to the digital representation of the sensed data. In the case of visible and infrared sensors, the radiometry represents the digitized or numeric representation of the scene radiance or reflectance. The digital numbers are commonly called "pixels", for picture elements.

Image Geometry refers to the projection, scale, and orientation of the image, while *Geometric Correction* refers to the modification of the input geometry to achieve the desired geometry. This usually includes the removal of geometric errors due to the sensor, orientation of the platform, and scene effects. *Resampling* generally means computing the value of new sensor radiance or intensity values between the existing values using one of various interpolation functions.

NEED FOR IMAGE CORRECTION

A major concern in remote sensing information-extraction and data handling is that of ensuring that the imagery has the proper geometric characteristics. The applications that are to be implemented influence the image-processing requirements:

a. the acquired imagery is to be used in applications such as cartography, land use, geography, etc. requiring a high geometric accuracy or fidelity with respect to the scene;

b. the application involves change detection or image enhancement, which involves the subtraction, addition, or combination of multiple and diverse image-data sets. This requires that the images must be correlated or geometrically registered with respect to each other (for example, several images taken simultaneously in different spectral regions, images of the same scene taken at different times, using data from different sensors, etc.). Applications that commonly require this capability include land-use planning, agriculture, hydrology, forestry, and others;

c. the application involves the combination of images over a wide area. This involves mosaicking multiple scenes covering adjacent and overlapping spatial regions. In order to do this correctly, geometric corrections are required in order to assure that common ground features from the diverse data sets are identically positioned.

The radiometry or intensity of the sensor output may be changed to correct the data, to match its output with the display or recording devices, or to merge diverse data sets.

a. Advanced sensors are made from solid-state detectors, photo-multiplier tubes, or charge-coupled devices and produce analog and digital signals. These devices require calibration and data correction to provide the user with reliable sensor data. Thus an important operation is to determine the relationship between the individual detector output and the scene radiance, and to correct the data for these calibration parameters.

b. Image data may also contain relative radiometric or intensity errors. These errors may be caused by sensor defects or improper calibration of the data. For example, the MSS has six detectors for each spectral band that scan the scene in parallel. If one detector's output is greater or less than the others, the image will have "banding" or "striping" characteristics. These effects, if not removed, will not only cause an undesirable cosmetic defect in the image, but will interfere with later computer-implemented information-extraction operations, because of the differing intensities for the same ground radiances.

c. Frequently, the image radiometry is intentionally modified to provide compatibility with the dynamic range of the image display or recording device, or to enhance the image for a particular range of intensities. This is valuable for manually interpreting the data and supporting preprocessing operations such as locating ground control points. The same programs used for calibration, destriping, or enhancement can often be used for this intensity modification operation.

Thus, image geometry and radiometry require correction and modification to compensate for sensor errors and to achieve the proper conditions to support image-data analysis and interpretation.

SOURCES OF IMAGE RADIOMETRY AND GEOMETRY ERRORS

Earth-observation data acquired by on-board spacecraft sensors are affected by a number of electronic, geometric, mechanical, and radiometric distortions that would diminish the accuracy of the extracted information and thereby reduce the utility of the data, if the data were not corrected.

OVERVIEW

Previous methods of correction, using electro-optical processing techniques have had serious limitations. Investigators have shown the superiority of a digital approach over that of electro-optics as a consequence of the former's processing flexibility, fewer required data conversions, and improved accuracy and quality of the information developed.

To correct the image data, the internal and external errors must be determined—they must be either predictable or measurable. Internal errors are due to sensor effects; they are generally systematic (known or predictable) and stationary (essentially constant for all practical purposes), and can be determined from prelaunch calibration measurements. External errors are due to platform perturbations and scene characteristics, which are variable in nature but can be determined from ground control and tracking data. Thus, the information required for correcting data distortion can be obtained (within certain limits of precision).

Figure 21-2 shows the sources of image error. They can be classified into radiometric and geometric error or compensation sources.

```
MSS GEOMETRIC CORRECTION

  Internal:    Mirror Scan Velocity Profile
               Detector Sampling Delay
  External:    Panoramic Distortion
               Scan Skew
               Earth Rotation
               Spacecraft Velocity
               Perspective Geometry
               Attitude (roll, pitch, yaw)
               Altitude
               Desired Projection (eg. UTM)

MSS RADIOMETRIC CORRECTION
  Internal:    Detector Response (Bias and Gain)
               Calibration Source Errors
  External:    Atmospheric Attenuation
               Film Recorder Gamma
```

Fig. 21-2. Image Error Sources and Sensor Data Compensation for the Landsat Multispectral Scanner.

SOURCES OF IMAGE RADIOMETRY ERRORS

Detector: Current and future remote sensors use detectors that convert the sensed radiance into a voltage or digital number (see Chapters 6 through 11). Multiple detectors are used in the sensor systems, and due to variations in individual detector outputs, there will be different outputs for the same ground radiances. Figure 21-3 shows the effect of this error source for two detectors viewing the same radiance area. It is apparent that a different detector gain or bias will cause the detector output to vary.

A/D Converter: An analog-to-digital converter can cause both quantization and signal errors. Because the converter has a finite number of bits, quantization noise with a mean square value of $q^2/12$ will result, where q is the quantization value (least significant bit) of the digitized signal. Selecting a converter with a sufficient number of bits (8 bits or better) will minimize this effect. If the A/D converter has a bias or gain error, it will also degrade the data in a manner similar to a detector gain or bias error.

Atmospheric Effects: The atmosphere introduces both scattering and attenuation effects. Although not an error, as this is the "signal", techniques have been developed to compensate the sensor data for these effects. This usually involves the removal of the first order effects by increasing the gain or contrast in the shorter wavelength bands.

Striping due to Detector Gain and Bias Error

Striping Removed by Calibrating Detectors

Fig. 21-3. Effect of Non-Uniform Detector Response.

SOURCES OF IMAGE GEOMETRY ERRORS

Altitude: Departures of the spacecraft from nominal altitude produce scale distortions in the sensor data. For the MSS, the distortion is along-scan only and varies with time: the magnitude of correction is up to 1.5 km.

Attitude: Nominally, the sensor axis system is maintained with one axis normal to the Earth's surface and another parallel to the spacecraft velocity-vector. As the sensor departs from this attitude, geometric distortion results. For the MSS, the complete satellite attitude time-history must be known in order to compensate for the distortion; up to the magnitude of correction is: pitch 12 km; roll, 12 km; yaw, 2.46 km; pitch rate, 0.93 km; role rate, 0.54 km; and yaw rate, 0.040 km.

Scan-Skew: During the time required for the MSS mirror to complete an active scan, the spacecraft moves along the ground track. Thus, the ground swath scanned is not normal to the ground track but is slightly skewed, which produces cross-scan geometric distortion; the magnitude of correction is 0.082 km.

Velocity: If the spacecraft velocity vector departs from nominal values, the ground track covered by a given number of successive mirror sweeps changes, producing along-track scale distortion: typically the magnitude of correction is 1.5 km.

Earth Rotation: As the MSS mirror completes successive scans, the Earth rotates beneath the sensor. Thus, there is a gradual westward shift of the ground swath being scanned. This causes along-scan distortion: the magnitude of correction is 13.3 km.

Map Projection: For Earth resources use, image data are usually required in a specific map projection. Although map projection does not constitute a geometric error, it does require a geometric transformation of the input data, and this can be accomplished by the same operations and compensate for distortion in the data: the magnitude of correction is 3.7 km along scan and along track (for the continental U.S.).

Sensor Mirror Sweep: The MSS mirror-scanning rate varies nonlinearly across a scan because of imperfections in the electro-mechanical driving mechanism. Since data samples are taken at regular intervals of time, the varying scan rate produces along-scan distortion: the magnitude of correction is 0.37 km.

Panorama: The image ground-area is proportional to the tangent of the scan angle rather than to the angle itself and, since data samples are taken at regular intervals, this produces along-scan distortion: the magnitude of correction is 0.12 km.

Perspective: For most Earth resources applications, the desired Landsat images represent the projection of points on the Earth on a plane tangent to the Earth at the nadir, with all projection lines normal to the plane. For the MSS, this produces only along-scan distortion: the magnitude of correction is 0.08 km.

TECHNIQUES FOR IMAGE CORRECTION

Optical Processing

In the past, images were acquired using conventional cameras and silver-halide film. The films were chemically developed and the images processed using optical and analog techniques. For example, if the scale of the images needed to be changed, an enlarger would be used to enlarge (or reduce) the image to the proper scale. If the image had an aspect error, due to imaging a scene off-nadir, the enlarger could also compensate for this effect by tilting the negative plane and the print plane. This could easily compensate for first-order effects. Higher order geometric distortion effects required more sophisticated electro-optical processing.

Electro-Optical Processing

Analog devices have been used to implement geometric and radiometric correction to image data for decades. The upper part of Figure 21-4 shows the earliest system for geometrically and radiometrically correcting Landsat data. Note that the digital data are first converted into a 70mm film by a computer controlled Electron Beam Recorder (EBR), with partial or systematic corrections applied. When a high precision film or tape product was produced, the 70mm film was rescanned by a CRT, and recorded onto film with another CRT, and geometric and radiometric corrections were made in the recording process. The first CRT system was also used to locate Ground Control Points (GCP) on the basis of a GCP film-chip library. In this system, the final print went through five generations of film processing, and went from digital to analog to film to analog and then to film. The digital data were produced from the scanned film data. Obviously, this process introduced many unnecessary conversions with a resultant degradation of film quality. This process has now been replaced by an all-digital image-processing operation, with a significant improvement in quality and operation.

Digital Processing

Digital image processing is an obvious solution for the correction of sensor data derived from a digital sensor, as it processes the data in the same form (digital) as it is acquired (see lower part of Figure 21-4). The data are read into a digital computer where they are reformatted, geometrically and radiometrically corrected, annotated, and recorded on digital tape for later processing, such as information extraction and data merging. Even the supporting data used for image correction, such as ground control points, are in a digital array form. Thus, there are no unnecessary analog or film conversions in the image-correction pro-

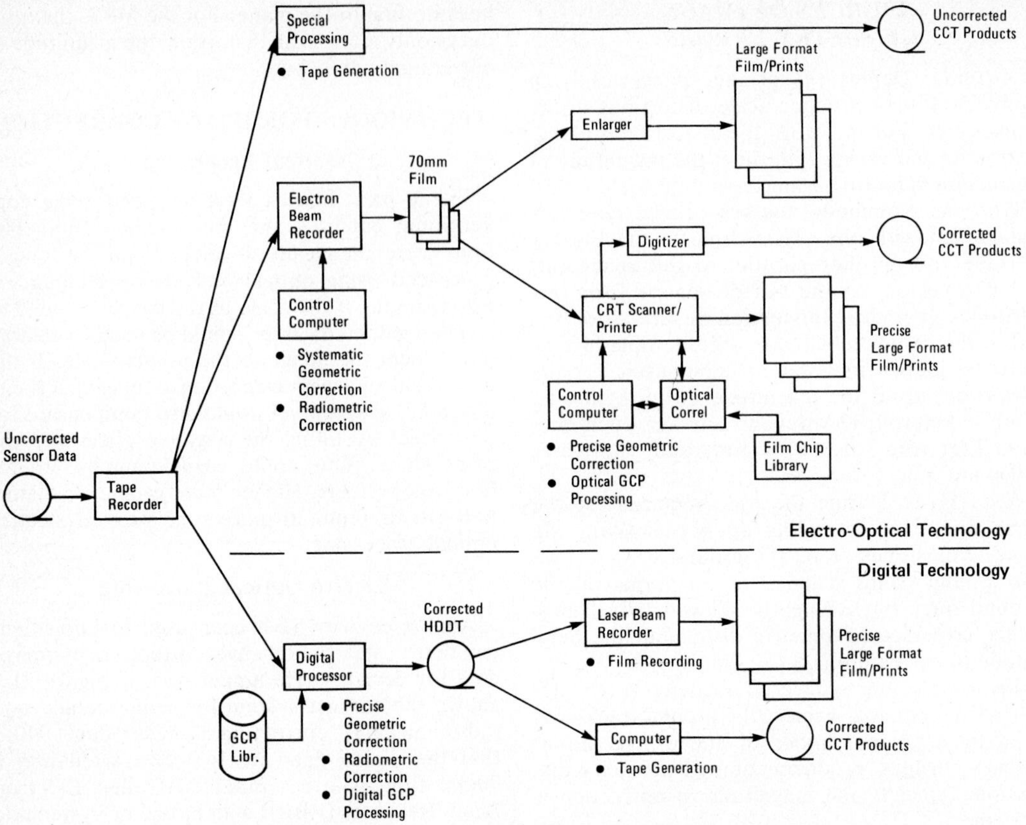

Fig. 21-4. Technologies for Processing Earth Observation Data.

cess, with no loss in resolution and dynamic range.

This approach has now been adopted by NASA and the USGS, and systems for operational image processing now exist for the correction of planetary, celestial, and terrestrial image data.

METHOD TO DETERMINE IMAGE ERRORS

RADIOMETRIC CALIBRATION SOURCES

This section summarizes both the nominal radiometric calibration and correction of the MSS data and the supplemental calibration which can be applied if the nominal methods fail fully to compensate for differences in the detectors.

Nominal Calibration/Correction: The nominal method for calibrating the output of the 24 MSS detectors is described in the section on mathematical techniques following this one and is summarized in Figure 21-5. Basically an on-board incandescent lamp and a variable neutral-density filter generate a calibration wedge which produces points on the input/output curve of each detector. Occasional solar observations are used to correct for changes in the output of the lamp. In this way, gradual changes in the output of each detector can be detected and measured.

The MSS detector outputs are digitized prior to transmission to the ground. Since six-bit quanti-

zation is used, each detector can produce at most 64 discrete values. The calibration data can be used to construct for each detector a table which specifies the "correct" radiometric intensity for each value output by the detector. Radiometric correction can thus be reduced to a simple table look-up operation in which a value output by a given detector is used to extract the "correct" value from the correction table. Although a unique table is required for each of the 24 detectors, the storage to accommodate these tables is quite small by present computer standards.

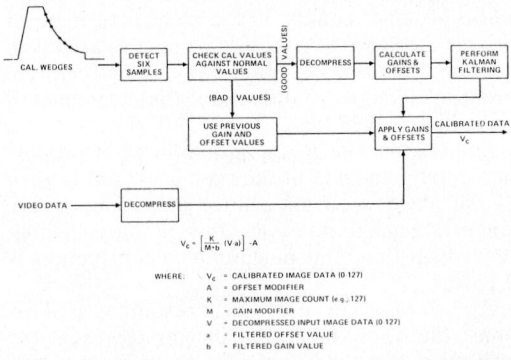

$$V_c = \left[\frac{K}{M \cdot b} (V \cdot a) \right] \cdot A$$

WHERE:
- V_c = CALIBRATED IMAGE DATA (0-127)
- A = OFFSET MODIFIER
- K = MAXIMUM IMAGE COUNT (e.g., 127)
- M = GAIN MODIFIER
- V = DECOMPRESSED INPUT IMAGE DATA (0-127)
- a = FILTERED OFFSET VALUE
- b = FILTERED GAIN VALUE

Fig. 21-5. Radiometric Calibration Algorithm (USGS, 1979).

Supplemental Calibration: These nominal techniques do not always fully compensate for differences in the outputs of the various MSS detectors. When this occurs, the effects are sometimes sufficiently large to produce visible horizontal "stripes" or "banding" in the images. Supplemental calibration can often be used to reduce these effects below the level of visual detectability.

There are several possible approaches to supplemental calibration. One is to note the response of each detector in areas of uniform radiance at different intensity levels. Another approach is to compile histograms of the responses of the detectors over a large number of data samples. In all cases, the objective is to measure the differences in the "corrected" response curves of the detectors in each spectral band. These measurements then can be used to modify the nominal radiometric-correction tables to produce identical corrected response curves for the detectors in each band.

It should be emphasized that supplemental calibration provides only a cosmetic correction. Unlike the nominal calibration techniques, supplemental calibration does not attempt to make the detector outputs correct. It only attempts to make them equal.

GROUND CONTROL POINTS

Because tracking and spacecraft attitude data are not known precisely, Ground Control Points (GCP) are used to obtain external reference information. A GCP is a physical feature detectable in a scene, whose location and elevation are known precisely. Typical GCPs are airports, highway intersections, land-water interfaces, geological and field patterns, etc. A registration operation is used to match a small image (subimage) area containing the GCP with a scene to be corrected and containing the same feature (see Figure 21-6). Typically, the window or GCP dimension $M_p = N_p = 32$, and the search area dimension $N_s = M_s = 128$.

Computationally efficient techniques have been developed to locate GCPs in digital data arrays. Some are based upon spatial domain techniques, and some operate in the frequency domain. The techniques are discussed in detail in the section on mathematical techniques.

RELATIVE CONTROL POINTS

A Relative Control Point (RCP) is a subimage area from a scene that has been previously geometrically corrected, or whose geometry is accurately known. This area or numeric array is used to establish the relative geometry between two or more images, so that these images can be registered to each other. A number of RCP's are selected, based upon the desired accuracy of registration. As with GCPs cross-correlation techniques are used to establish relative registration.

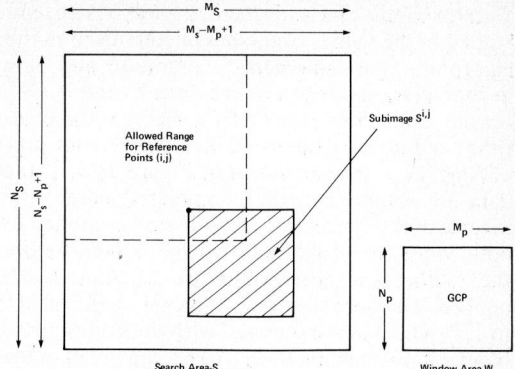

Fig. 21-6. Locating a Ground Control Point in an Image. The ground control point window W is indexed to various locations in the search area S, and a cross-correlation algorithm finds the match.

ATTITUDE/ALTITUDE/POSITION INFORMATION

If the platform attitude (roll, pitch, and yaw), altitude above the ground, and position (latitude and longitude or northing and easting) were known precisely, the external geometry would be known and a mapping function could be directly defined. At the current time, these parameters are not known accurately enough, and only provide an approximate estimate for the correction and mapping parameters. It is anticipated that future satellite and aircraft systems, using the Global Positioning System and star-trackers will be able to provide sufficiently accurate information to minimize or eliminate the use of ground control points.

MATHEMATICAL TECHNIQUES USED

This section discusses digital techniques that have been developed to preprocess (correct) Earth observation data. These techniques have been applied to a variety of problems in addition to those used here as examples. The procedure developed for the correction of satellite imagery is based upon changing the sensor data (intensity) to correct for radiometric errors, and relocating picture elements in order to achieve the proper geometry. Calibration data are used to support the former, and ground control points for the latter. To correct the geometry, it is necessary to compute coefficients of the mathematical models that describe the distortion in the image. After obtaining these coefficients, a mapping function can be synthesized for geometric correction of the sensor data.

RADIOMETRIC MODIFICATION

MSS Radiometric Modification

The radiometric calibration algorithm, shown in Figure 21-5, applies in general to all Landsat MSS data (USGS, 1979). The differences in the radiometric corrections, as they are applied to the various bands of Landsats 1, 2, and 3, are embodied in the values of the parameters used in the algorithms. All radiometric calibration and data decompression operations are done by the NASA Master Data Processor (MDP), wherein the algorithm and look-up tables reside. The flow of processing steps is also shown in Figure 21-5. If the data are acquired in the compressed mode, the inverse of the spacecraft compression is applied to both video and calibration-wedge values before the radiometric corrections are calculated and applied. The decompressed values, ranging from 0 to 127, which are associated with the compressed input values, ranging from 0 to 63, are given in the USGS Users' Manual (USGS, 1979). Description of the elements of the algorithm follows:

$$V_c = \left(\frac{K}{M \cdot b}(V - a)\right) - A \qquad (21\text{-}2)$$

V_c = calibrated pixel value, which will range from 0 to 127 for 7-bit pixels
K = maximum possible value for V_c (127 for 7-bit pixels)
a = the filtered offset value
b = the filtered gain value
V = input pixel-value (from 0 to 127 for 7-bit pixels)
M = multiplicative gain modifier
A = additive offset modifier

M and A are detector-dependent parameters used to control long term drifts in relative detector responses. They are derived by analyzing Landsat data over a period of time to determine the frequency with which these parameters need updating in order to equalize detector gain and offset changes. At launch, $M = 1$ and $A = 0$. The values of a and b are determined as follows.

For band 8 (thermal IR band) a, b are defined as:

$$a = \frac{N_R V_O - N_C V_R}{N_R - N_C}, \qquad b = \left(\frac{V_R - V_O}{N_R - N_C}\right)\Delta N$$

where
N_R = black body reference radiance
N_C = "cold" reference radiance
ΔN = Nmax − Nmin, the range between maximum and minimum radiances of the IR detectors
V_R = corresponding pixel value (relative radiance); for N_R, determined by averaging 6 samples, extracted from alternate mirror sweeps on the input raw data High Density Tape

V_O = corresponding pixel (relative radiance) for N_C

For all other bands, a and b are determined once per scan line using estimates of $\hat{a}$ and $\hat{b}$, made from the calibration data:

$$\left.\begin{array}{l} \hat{a} = \displaystyle\sum_{i=1}^{6} C_i V_i \\[2em] \hat{b} = \displaystyle\sum_{i=1}^{6} D_i V_i \end{array}\right\} \text{ Linear regression} \qquad (21\text{-}4)$$

In these equations V_i is the input value of the calibration wedge word i, and c_i and D_i are regression coefficients which are determined on the basis of prelaunch radiance tests. Then, a and b are filtered for every scan line, n, as follows:

$$b(n) = \begin{cases} \hat{b}(n)\,[\hat{b}(n) - b(n-1)] & , \text{ for } n = 1 \\[1em] b(n-1) + W(n)\,[\hat{b}(n) - b(n-1)] & , \text{ for } n > 1 \end{cases} \qquad (21\text{-}5)$$

$$a(n) = \begin{cases} \hat{a}(n)\,[\hat{a}(n) - a(n-1)] & , \text{ for } n = 1 \\[1em] a(n-1) + W(n)\,[\hat{a}(n) - a(n-1)] & , \text{ for } n > 1 \end{cases} \qquad (21\text{-}6)$$

where
n = sequential number of scan lines per swath;
$\hat{b}(n), \hat{a}(n)$ = estimated values of b and a
$b(n-1), a(n-1)$ = previous filtered value of b and a
$b(n), a(n)$ = new filtered value of b and a
$W(n) = 1/n + 1$ for $n = 1$ to 15
$W(n) = 1/16$ for $n = 16$

Atmospheric Effects Correction

An algorithm has been developed that models the first order effects of atmospheric scattering, or "haze" (USGS, 1979). An off-set grey-level is established for each spectral band. The off-set corresponds to the amount of haze in the bands and is subtracted from each pixel value. The atmospheric effects correction-algorithm is thus defined as follows:

$$I'(i,j) = I(i,j) - \text{Bias} \qquad (21\text{-}7)$$

where
$I(i,j)$ = Input pixel value at line i and sample j
$I'(i,j)$ = Enhanced pixel value at same locations (i,j)

The bias is the amount of offset for each spectral band. The USGS allows the user to specify the bias values if he requests this type of radiometric modification to the sensor data.

AUTOMATIC LOCATION OF GROUND CONTROL POINTS

A number of algorithms have been developed to automatically locate ground control points (GCP's) in an image. The GCP's are used to determine the external geometry of the image to support subsequent image correction. Two techniques will be discussed here: the Sequential Similarity Detection Algorithm (SSDA) and the Fast Fourier Transform (FFT) algorithm.

Overview

The pattern-recognition problem encountered in the geometric correction preprocessing of digital imagery from orbiting sensors, such as the Landsat Multispectral Scanner (MSS), is that of accomplishing the translational registration of a prototype digital-image of a geodetically identified feature (termed a window) with another image of the feature which exists as a congruent subimage in a larger-area digital image (termed a search area).

Figure 21-7 is a schematic drawing which illustrates this terminology for the case of a typical feature, a road intersection. The size of the window area must be determined empirically. Among other factors, it is a function of the area represented by one image sample, the extent of the feature in question, and the size of the region over which potential image scaling and rotational effects can be neglected. The size of the search area is determined by the accuracy of the available ancillary information (e.g. emphemeris and attitude data) which can be used to establish the approximate location of the feature.

The algorithms which are employed to solve this translational registration problem fall into three categories:

a. those which provide a measure of the "difference" between two congruent digital images, and which are the primary pattern matching mechanisms,

b. those which serve to reduce the computational load of the image comparison process, which in practice requires an exhaustive comparison of the window area with all pos-

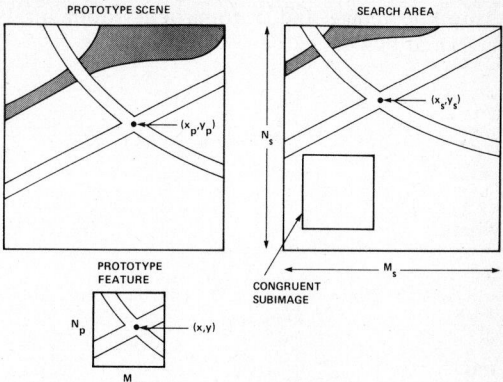

Fig. 21-7. Diagram for Feature Pattern Recognition of Ground Control Points. The road feature derived from the prototype scene is used as a ground control point to locate the same feature in the search area.

Comparison Measures

As is shown in Figure 21-7, the prototype feature is a matrix of N_p rows and M_p columns, with the value of each matrix element determined by the energy reflected from the object area corresponding to the sample. The search area is a matrix of N_s rows and M_s columns with matrix values similarly determined. The search for a translational registration overlay consists of comparing the $N_p M_p$ samples in the window area to the $N_p M_p$ samples in each of the $(N_s - N_p + 1)(M_s - M_p + 1)$ congruent subimages which exist in the search area. Three methods of performing this comparison have been evaluated in the course of developing a practical solution to the translational registration problem: similarity, mean-normalized similarity, and mean-normalized correlation (covariance).

Similarity: Denoting by (N,M) the row and column coordinates within the search area of the upper-left-hand sample of a specific congruent subimage, and by $p(i,j)$ and $s(i,j)$ the values of specific samples from the prototype and search-area images, respectively, the "difference" between the prototype image and the congruent subimage is measured by

$$E_O(N,M) = \sum_{n=1}^{N_p} \sum_{m=1}^{M_p} \left| p(n,m) - s(N + n - 1, M + m - 1) \right| \qquad (21\text{-}8)$$

sible congruent subimages from the search area, and

3. those which are employed to estimate the translational registration coordinates to a precision greater than the granularity of the image sampling lattice.

These classes of algorithms are described separately below.

The particular congruent subimage, with coordinates denoted by (N_O, M_O), which matches the prototype image is recognized by the fact that $E_O(N_O, M_O) = 0$. Note that recognition of a match requires that the corresponding matrix values be identical.

Mean-normalized Similarity: With this comparison measure, the "difference" between the prototype image and a congruent subimage is measured by

$$E_1(N,M) = \sum_{n=1}^{Np}\sum_{m=1}^{Mp} \left| (p(n,m) - \bar{p}) - (s(N+n-1,M+m-1) - \bar{s}_{NM}) \right| \qquad (21\text{-}9)$$

where

$$\bar{p} = \frac{1}{N_pM_p}\sum_{n=1}^{Np}\sum_{m=1}^{Mp} p(n,m)$$

and

$$\bar{s}_{NM} = \frac{1}{N_pM_p}\sum_{n=1}^{Np}\sum_{m=1}^{Mp} s(N+n-1,M+m-1)$$

Again, the particular congruent subimage, with coordinates (N_0,M_0), which matches the prototype image, is recognized by the fact that $E_1(N_0,M_0) = 0$. However, this measure will tolerate an additive bias on the sample values between the prototype image and the congruent subimages, while still generating the unique error sum for the particular congruent subimage which matches the prototype.

Mean-normalized Correlation: For this comparison measure, the "difference" between images is measured by

$$R(N,M) = \frac{\sum_{n=1}^{Np}\sum_{m=1}^{Mp}(p(n,m)-\bar{p})(s(N+n-1,M+m-1)-\bar{s}_{NM})}{\left[\sum_{n=1}^{Np}\sum_{m=1}^{Mp}(p(n,m)-\bar{p})^2\right]^{1/2}\left[\sum_{n=1}^{Np}\sum_{m=1}^{Mp}(s(N+n-1,M+m-1)-\bar{s}_{NM})^2\right]^{1/2}} \qquad (21\text{-}10)$$

where $\bar{p}$ and $\bar{s}_{NM}$ are as defined before. In this case, the particular congruent subimage, with coordinates denoted by (N_0,M_0) that match the prototype image, is recognized by the fact that $R(N_0,M_0) = 1$. This measure will tolerate any spatially-invariant, linear, radiometric relationship between the sample values of the prototype image and the congruent subimage, while still generating the unique value of 1 for the particular congruent subimage which matches the prototype.

The three measures of "difference" have been presented in order of increasing robustness and computational complexity. In a practical application, the chosen comparison measure should be able to tolerate the anticipated variations in the reflected energy from a given feature, as encountered in different images of the feature, while not imposing an unnecessary computational burden. In the case of the images returned by the Landsat MSS, a major radiometric effect is the seasonal variation in ground illumination. The reflected

energy sensed by the MSS thus exhibits a seasonal variation which has been found to render the similarity measure ineffective, and to degrade the performance of the mean-normalized correlation measure. In other applications, where the illumination is more stable, use of the correlation measure might not be required.

In practice, these comparison techniques usually do not yield the expected unique identifiers (i.e. 0 or 1) when the prototype feature-image is compared to the congruent subimage containing the feature. In some cases, visual inspection of the feature images reveals that changes have occurred in the feature or its surroundings. However, even when such changes are not evident, the location of the image samples with respect to the feature can differ in the prototype image and the matching congruent subimage, so that no exactly matching congruent subimage exists. This means that the comparison measure must be applied between the prototype image and each of the congruent subimages from the search area, in order to identify the particular congruent subimage which yields an extremal value of the comparison measure. Furthermore, it also means that there is an inherent uncertainty of ± 0.5 sampling interval in the translational registration coordinates of the matching congruent subimage.

Techniques for Reducing the Computational Requirements of Pattern Matching

While the computational load of pattern matching is small compared with the overall data-processing load encountered in digital preprocessing for the geometric correction of Landsat imagery, it is desirable to take advantage of computational economies where possible. Two approaches to obtaining such economies will be discussed: the sequential similarity-detection algorithm (SSDA) (Barnea and Silverman, 1972) and the use of the Fast Fourier Transform to speed correlation calculations.

Sequential Similarity Detection Algorithms: This class of algorithms was originally proposed as an efficient solution to the problem of translational image registration, with the potential for reducing computational requirements by two orders of magnitude. The essential feature of these algorithms is the concept of an error-threshold function which describes the variation of the comparison measure (e.g. E_1, E_0,) as a function of the number of image samples compared between the prototype feature-image and the congruent subimage containing the feature. An example may help to clarify this concept. In terms of Figure 21-7 and the similarity "difference" measure, one considers the partial error sum

rapid rejection of a significant fraction of those congruent subimages not containing the feature, then the probability of identifying the congruent subimage containing the feature decreased. If, on the other hand, the threshold function was set to provide a high probability of identification of the correct congruent subimage, then the comparison of the prototype image with each of the congruent subimages included most of the possible sample pairs (i.e. $K \simeq N_p M_p$ for all subimages). The sensitivity of the algorithm to threshold adjustment depended on the particular feature considered.

$$E_0{}^K(N,M) = \sum_{k=1}^{K} \left| p(n(k),m(k)) - s(N + n - 1, M + m - 1) \right| \qquad (21\text{-}11)$$

where $n(k)$ and $m(k)$ are tabulations of all possible index pairs for the samples of the prototype image. The existence of a threshold function, $T(K)$, is postulated. $T(K)$ is to have the property that, for the congruent subimage containing the feature, with coordinates (N_0, M_0),

$T(K) \geq E_0{}^K(N_0, M_0)$ for all $K \leq N_p M_p$

while $T(K) < E_0{}^K(N,M)$ for the other congruent subimages for some $K << N_p M_p$. Thus, the function $T(K)$ should provide a mechanism for terminating the comparison of prototype image samples with samples of a congruent subimage, for those congruent subimages which do not contain the feature, when only a small fraction of the total $N_p M_p$ samples have been considered.

Investigations of two realizations of a sequential similarity-detection algorithm were conducted as part of development work in the preprocessing of Landsat MSS data. One realization, employing the similarity measure, proved unable to cope with the radiometric variations encountered in this type of image data. The other realization, which employed the mean-normalized similarity, proved effective enough for an extensive experimental study to be performed. This study used MSS imagery spanning a year interval, from three distinct geographic areas. Over a hundred different features were selected as prototypes, encompassing numerous examples of such general feature types as land-water interfaces, interstate highway intersections, and airfields.

With respect to the sequential similarity-detection algorithm itself, several observations have been made:

a. A satisfactory threshold function could not be found. This function, which had to be empirically determined, required a compromise to be made between successful feature identification and computational speed. If this threshold were adjusted for

b. The mean-normalized similarity measure of this algorithm, combined with the seasonal variation in solar illumination, produced a seasonal variation in the partial error sums, so that seasonal adjustment of the threshold function would be necessary.

c. As originally implemented, the algorithm employed a randomization of the index sets $n(k)$ and $m(k)$ in order to provide, in general, a great deal of "new" information as each term was incorporated in the partial sum. In the absence of any a priori information of the significance of particular samples in the prototype feature image, this is a reasonable procedure which, on the average over a large number of prototype feature images, should improve performance. However, in the case of Landsat preprocessing, the problem was to accomplish rapid identification of specific features, each with its own unique structure, and not well-characterized by random sampling. The best strategy for the sequential sampling proved to be the manual specification of samples (e.g. the boundary samples of a land-water interface).

Thus, in an operational preprocessing system for geometric correction of digital imagery, optimal performance of the pattern-recognition processing with a Sequential Similarity Detection Algorithm should be achieved by tailoring the sequential comparison to the characteristics of specific features, by tailoring the threshold function to the specific features, and by providing a set of threshold functions for each feature in order to account for seasonal differences in solar illumination.

Correlation Using the Fast Fourier Transform: The mean normalized correlation measure previously discussed is the pattern-comparison measure chosen for use in the NASA Master Data Processor (MDP) for Landsat digital image preprocessing. This algorithm was chosen because of its relative insensitivity to the solar illumination variation anticipated for the image data, and because the processing unit employed in this system contains a high-speed arithmetic unit well suited to perform correlation using FFT techniques. In the mean-normalized correlation measure (Eq. 21-10) the term S_{NM} in the numerator can be neglected, since

$$\bar{s}_N \sum_{n=1}^{Np} \sum_{m=1}^{Mp} (p(n,m) - \bar{p}) \equiv 0. \qquad (21\text{-}12)$$

The remaining portion of the numerator is the discrete correlation of the mean-normalized prototype image with the search area. Use of Fast Fourier Transform techniques to efficiently perform this calculation has been discussed extensively (see, for example, Oppenheim and Schafer, 1975). Of the denominator factors, the first depends only on the prototype feature-image, and can thus be pre-computed. In fact, for the bulk of the processing, it can be ignored, since in seeking the maximum of the mean-normalized correlation, it represents a simple scaling multiplier on all the covariance values. The second factor in the denominator can be rewritten as

precise estimate, it is necessary to employ a reconstruction of the analog images which the digital images represent. The justification for such an effort rests in the Sampling Theorem for band-limited signals, which specifies the conditions under which such reconstruction is possible, and the mechanism for such reconstruction, the $sinc(x)$ function. A straightforward calculation can be made to show that the cross correlation of two band-limited functions can be obtained by $sinc(x)$ interpolation of the discrete correlation of sampled representations of the functions.

For the purpose of estimating a precise, two dimensional, translational registration between the prototype image of a feature and its representation in a search area, the first step is to locate the registration coordinates to the precision of the sampling lattice. Then the discrete correlation coefficients in a two-dimensional neighborhood of this are computed. For Landsat MSS data, a five-by-five array of correlation coefficients, centered on the position of the discrete correlation maximum, has been found to be sufficient to define the shape of the correlation function. For typical features employed in Landsat preprocessing, the correlation function tends to be rather broad so that, instead of using $sinc(x)$ as the interpolator for reconstruction of the continuous cor-

$$\left\{ \left[\sum_{n=1}^{Np} \sum_{m=1}^{Mp} s^2(N + n - 1, M + m - 1) \right] - N_p M_p \, \bar{s}_{NM}{}^2 \right\}^{1/2} \qquad (21\text{-}13)$$

Values for the first term in this square-root expression are most efficiently obtained by a direct computation on the image samples in the search area. Values for the second term can be obtained via FFT techniques by convolving the search area itself (whose Fast Fourier Transform must be produced for the numerator calculation) with a unity weight-array of the same size as the prototype image-array. The remainder of the calculation required to locate the maximum of the mean-normalized correlation values is simple arithmetic and comparison using the values generated by the previous transform computations.

Since the prototype features employed by the MDP are 32 samples by 32 samples, and the search areas are 128 samples by 128 samples, the data arrays which are transformed are well into the region where the FFT provides significant computational-speed advantage.

Vernier Translational Registration

As has been observed earlier, the discrete comparison measures can produce an estimate of translational registration to a precision of only ± 0.5 sampling interval. In order to obtain a more

relation surface, the interpolation is performed by the least-squares fitting of a bivariate polynomial to the discrete correlation samples. The final, precise, estimate of the coordinates of registration is then obtained by locating the maximum of this interpolating polynomial.

The registration accuracy which can be obtained by this procedure has been studied extensively by means of simulated imagery, by means of feature images purposely resampled onto offset lattices, and by means of analysis of variance techniques applied to feature clusters in Landsat imagery. In all cases, the estimated accuracy was approximately 0.1 of the sampling interval. This represents a significant improvement over the 0.5 sampling-interval accuracy which can be obtained with discrete calculations alone.

MAPPING FUNCTIONS

Spacecraft roll, pitch, and yaw data are not provided with sufficient accuracy by the satellite attitude-determination system, and the ephemeris data do not provide either altitude or ground position with sufficient accuracy; these parameters must be calculated from knowledge of the GCP locations.

Differences between actual and observed GCP locations are used to determine the coefficients of cubic polynomial time-functions of roll, pitch, yaw and a linear time-function of altitude deviation. The GCPs are first located in the input image (sensor data) and then mapped into the Earth-tangent cylinder using models based on all those errors that can be predicted or determined from tracking data. The tangent-plane projection contains a Cartesian coordinate system with its origin at the center of the image and the x-y plane tangent to the Earth ellipsoid at this origin. The positive x axis is in the direction of the nominal spacecraft ground track; the z axis is oriented away from the center of the Earth; and the y axis completes a right-handed coordinate system. The coordinate convention is shown in Figure 21-8.

RESAMPLING METHODS

After the position of an output picture-element on the input image has been determined, several methods can be used to calculate the intensity value of the output element. For example:

The nearest-neighbor method, which selects the intensity of the closest input element and assigns that value to the output element:

$$I(x,y) = I(u,v). \qquad (21\text{-}14)$$

The bilinear interpolation method, uses four neighboring input values to compute the output intensity by two-dimensional interpolation:

$$I(x,y) = a_1 I(u,v+1) + a_2 I(u,v)$$
$$+ a_3 I(u+1,v+1) + a_4 I(u+1,v) \qquad (21\text{-}15)$$

The cubic convolution method, originally suggested by Rifman and McKinnon (1974), uses 16 neighboring values to compute the output intensity:

$$I(x,y) = \sum_{m,n} a_{m,n} I(u+m,v+n), \quad -1 \leqslant m,n \leqslant 2.$$

$$(21\text{-}16)$$

Resampling of MSS image data can be used to eliminate spatial discontinuities due to nonsimultaneous detector sampling and geometric image-correction operations.

Figure 21-9 shows an MSS subimage, extracted from a central California scene (Figure 21-10), before and after use of the three resampling algorithms (Bernstein, 1976). Discontinuities in the input data have been eliminated by the bilinear interpolation and cubic convolution methods. Some high-frequency loss, due to low-pass filtering of the image data, can be noted in the corrected sub-image for which bilinear interpolation was used.

IMAGE ENHANCEMENT TO SUPPORT PREPROCESSING

Many image-enhancement transformations have been developed to increase the information content of images to be presented for visual interpretation. Since some of these enhancements may usefully precede and enhance the image registration and rectification steps, they are important also for our purposes in correcting image geometry. Several techniques will be discussed. They include contrast stretching, filtering, and edge enhancement. Change detection, while it is more of an information enhancement than an image enhancement technique, is also discussed, since it too is closely tied to geometric rectification.

CONTRAST STRETCHING

A number of algorithms have been developed to perform contrast stretching to enhance digital imagery. These algorithms can be used both to obtain increased contrast over a certain range of intensities while preserving the remaining intensities, and to match image intensities to the characteristics of displays and recording devices. The discussion which follows describes one of these algorithms, a particularly versatile technique based on the cumulative distribution of the radiometric intensities in a scene. For conciseness, the discussion is presented in terms of a 256 gray-level image. The algorithm is performed as follows:

a. Integrate the histogram (p_i for $i = 1, \dots,$ 256) to form a cumulative distribution of pixel brightness

$$q_i = \sum_{j=1}^{i} p_j.$$

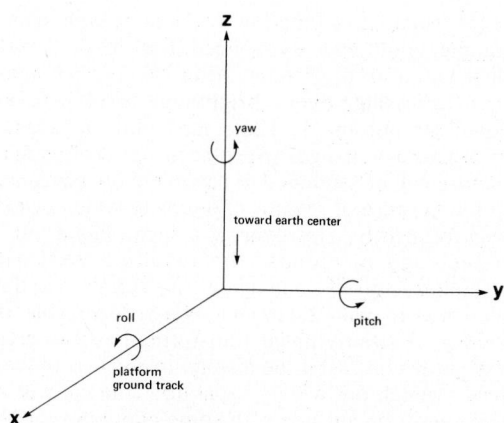

Fig. 21-8. Landsat Coordinate Convention. The x-axis is in the direction of the platform ground track, the z-axis is away from the earth's center, and the y-axis completes a right hand coordinate system.

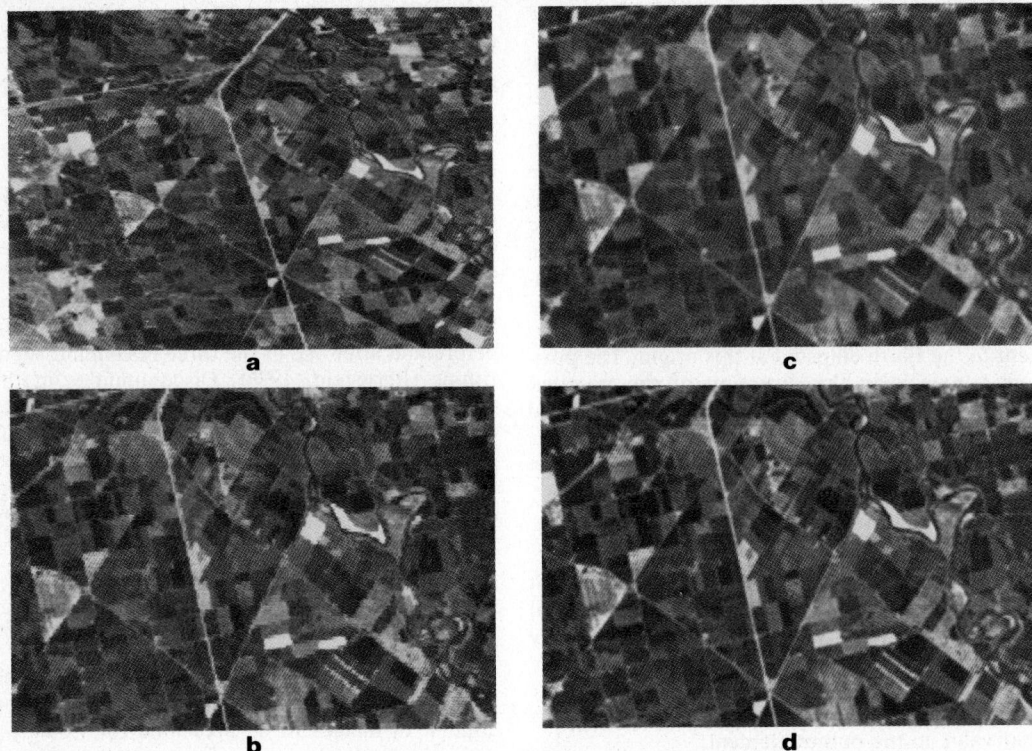

Fig. 21-9. MSS Subimage with Three Resampling Algorithms. a) Original uncorrected subimage. (Top left) b) Nearest neighbor resampling. c) Bi-linear resampling d) Cubic convolution resampling. (Bottom right)

An example of a histogram and its cumulative distribution is shown in Figures 21-11 and 21-12.

b. Determine value X from

$$X = C \cdot q_{256} \qquad (21\text{-}17)$$

where C is an input parameter.

c. Determine I_{min} such that X of the pixels have a brightness less than I_{min}.

$$q_{I min-1} < X \leq q_{I min} \qquad (21\text{-}18)$$

d. Determine I_{max} such that at most X of the pixels have a brightness greater than I_{max}.

$$q_{I max+1} < X \leq q_{256} - q_{I max} \qquad (21\text{-}19)$$

e. The portion of the cumulative distribution function lying between I_{min} and a threshold K forms part of the transformation function, V. See Figure 21-13.

$$V_i = \left(\frac{q_i - q_{I min}}{q_{256} - q_{I min}} \right) \cdot 255 \quad \text{for } i = 1, \ldots, K$$
$$(21\text{-}20)$$

f. The remainder of the function between K and I_{max} is a straight line defined by:

$$V_i = \left(\frac{I - K}{I_{max} - K} \right) \cdot 255 \qquad (21\text{-}21)$$

where I is the pixel input brightness.

V_i is the pixel output brightness.

An example of a transformation function thus defined is shown in Figure 21-13.

Depending on K, three different types of transformation functions are available. When K is less than I_{min} (K is replaced by I_{min} in Eq. 21-21); then the stretching is entirely linear. When K is greater than the I_{max}, then the function is defined by the cumulative distribution. When K is between I_{min} and I_{max}, then the function is a combination of the two.

The cumulative function works quite well in assigning brightness levels according to the pixel distribution of brightness. After this type of contrast adjusting, every brightness level will be equally represented. Thus, the subtle adjacent-in-brightness changes which occur most often in a picture will be stretched to become more obvious, at the expense of regions of intensity which occur less frequently. However, if a scene has a large percentage of clouds, the resulting contrast stretch may give so much intensity variation to the clouds as to make them no longer recognizable as clouds. A strictly linear transformation may give very little contrast if the histogram consists of two peaks which are widely separated (the case of a fairly uniform surface with some cloud cover). A combination of the two functions with an appropriate value of K offers a great amount of flexibility and the advantages of both.

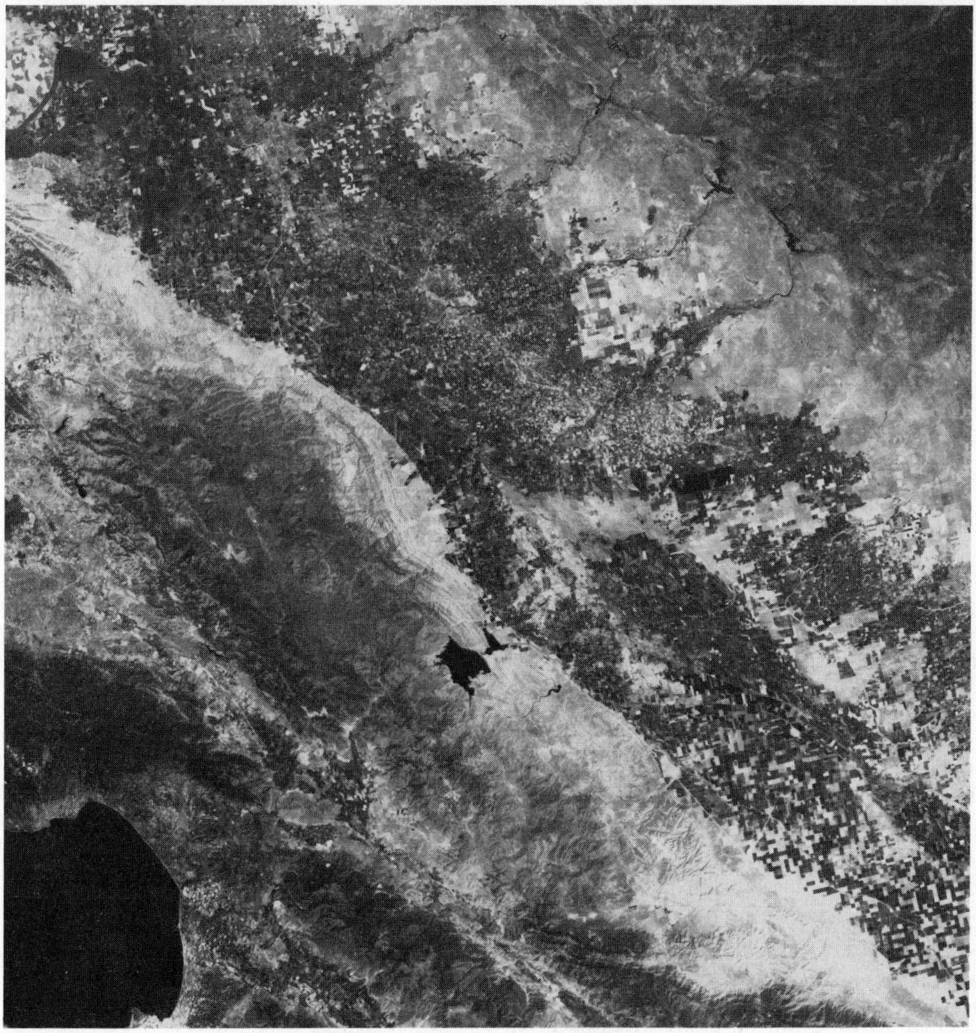

Fig. 21-10. Central California Scene Used For Resampling Analysis.

Filtering

Filtering is the generic term used to denote the process of restoring the image to compensate for the effects of the Point Spread Function (PSF), or aperture function. The process is also referred to as modulation-transfer-function compensation.

If $g(x',h')$ is the ground radiance of the point (x',y'), then the recorded data can be represented as

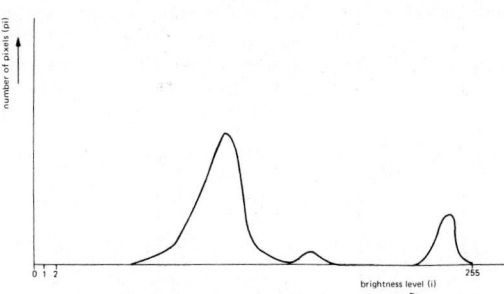

Fig. 21-11. Histogram of Pixel Brightness.

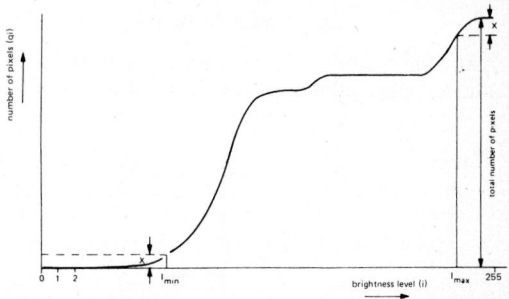

Fig. 21-12. Cumulative Distribution of Pixel Brightness.

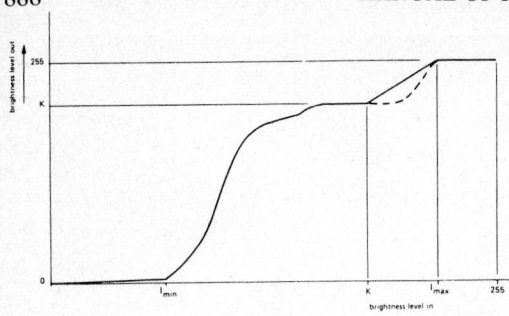

Fig. 21-13. Brightness Transformation Functions.

these approaches are either developed from models of image-degradation effects or chosen on a heuristic basis because they improve the visual presentation of the data. Given a matrix of non-negative grey levels, $P(i,j)$, a convolution filter or operator can be developed to filter the data in the spatial domain. The general form of the calculation is:

$$M(i,j) = \sum_p \sum_q m(p,q)P(p,q) \qquad (21\text{-}25)$$

$M(i,j)$ = New output grey level value at line i and column j

$$f(x,y) = \iint_A g(x',y')s(x - x',\, y - y')\, dx'dy' + n(g,x,y) \qquad (21\text{-}22)$$

where A is the effective aperture domain, s is the *PSF*, and the noise n is shown as being both signal- and space-dependent.

Frequency Domain (Fourier) Methods: Since the integral in Eq. 21-22 is of the convolution type, one approach to recovering $g(x,y)$ is through the Fourier transform. If F, G, and S are the Fourier transforms of f, g, and s respectively, then the Fourier transform of the ground radiance (ignoring noise) is

$$G(u,v) = F(u,v)S^{-1}(u,v) \qquad (21\text{-}23)$$

provided S has no zeros in the range of interest. The ground radiance, $G(x,y)$ is then the inverse Fourier transform of $G(u,v)$. This method requires knowledge of the *PSF*, which can be obtained from ground calibration data, or from examination of a sharp boundary between two homogeneous regions of significantly different brightness. In practice S falls off rapidly at high frequencies so that S^{-1} becomes large. One solution to this problem is to modify S^{-1} by a Wiener filter

$$S^{-1} = \frac{OTF_\omega^2 + SN_\omega^{-2}}{OTF_\omega} \qquad (21\text{-}24)$$

where *OTF* (optical transfer function) is the Fourier transform of the *PSF*, *SN* is the signal to noise ratio, and ω is the spatial frequency.

Edge Enhancement

A wide variety of approaches is possible to achieve edge enhancement. The filters used in

$m(p,q)$ = coefficient of the operator at line p and column q

$P(p,q)$ = old pixel grey level value at line p and column q

$p = i - I,\ i - I + 1,\ i - I + 2, \ldots,$ $i,\ i + 1, \ldots, i + I$

$q = j - J,\ j - J + 1,\ j - J + 2, \ldots,$ $j,\ j + 1, \ldots, j + J$

Typically values of I, J are in the range of 1 to 5. Several sets of edge-enhancement filter-coefficients useful for spatial-frequency improvement or modification have been developed. The filter can be considered to be a mask. A 3×3 filter has the following form:

$$\begin{aligned}
M(i,j) = &\ a1*P(i - 1, j - 1) + a2*P(i,j - 1) \\
&+ a3*P(i + 1, j - 1) + a4*P(i - 1,j) \\
&+ a5*P(i,j) + a6*P(i + 1,j) \\
&+ a7*P(i - 1, j + 1) + a8*P(i,j + 1) \\
&+ a9*P(i + 1, j + 1)
\end{aligned}$$

$$(21\text{-}26)$$

where

$M(i,j)$ = the output or new pixel intensities,
$P(i,j)$ = the input pixel intensities.
i = line number
j = column number

Below are listed several filter coefficients that will enhance the image in varying degrees. Filter No. 1 in the Laplacian operator; filter No. 2 in the Laplacian filtered image added to the original image.

Filter	$a1$	$a2$	$a3$	$a4$	$a5$	$a6$	$a7$	$a8$	$a9$
1	0	−1	0	−1	4	−1	0	−1	0
2	0	−1	0	−1	5	−1	0	−1	0
3	−1	−2	−1	−2	13	−2	−1	−2	−1
4	−1	−1	−1	−1	9	−1	−1	−1	−1
5	1	−2	1	−2	5	−2	1	−2	1

These convolution filters or operators can have any size. Typically, 3×3 or 5×5 operators are used. The Laplacian operator, (filter #2), offers the advantages of relatively simple computation and the capability of providing a natural appearing edge-enhanced image, since a similar type of operator seems to be applied in human visual perception.

Laplacian enhancement is a discrete specialization of the spatial domain compensation-technique discussed earlier. In this case, only the first and second terms of the compensation are employed and the Laplacian term is computed by means of the five point, cross-shaped areal operator shown in Figure 21-14.

This method was experimentally employed to enhance a portion of Landsat scene 1080-15192 which includes Washington, D.C., Baltimore, MD, and a number of cultural features such as roads, farms, and airports (Dulles airport is in the lower left of the subimage). Figure 21-15 shows the original subimage, and Figure 21-16 shows the results obtained through the use of Laplacian enhancement. The sharpened appearance produced in the enhanced image is striking.

IMAGE PROCESSING OPERATIONS

RADIOMETRIC CORRECTION

Multispectral Scanner: The MSS has 24 solid-state detectors, six for each band. Both bias and gain errors can exist with uncalibrated detector data. Each detector voltage-output is digitized into 64 values, or counts. An internal MSS calibration lamp is scanned by the detectors and the data are used to provide absolute (in terms of the lamp strength) and relative (in terms of each detector) response information. The corrections thus generated can be implemented by simple table-lookup operations or by direct solution of the correcting equation; see Table 21-1 for an example of table look-up radiometric corrections. The detector output (V) is used as an address or pointer to the correct value in the table, and the table value (R) is used as the corrected image-element radiance response. In Table 21-1 the response of Function 1 shows bias error compensation; Function 2, gain error compensation with rounding and truncation; Function 3, bias and gain compensation; and Function 4, nonlinear compensation. Highly nonlinear compensation and enhancement transformations can be implemented in the same manner.

Return Beam Vidicon: The RBV computer-compatible tapes contain image data that have been sampled and digitized with six-bit quantization. Thus, there are 64 possible input values. If a table which specifies the correct output intensity for each of the 64 input intensities can be defined, radiometric correction of the RBV images can also be accomplished by table-lookup operations. This is essentially the technique used, but it is complicated by the fact that RBV radiometric errors vary across the image; this results in the need to use multiple-correction tables.

The RBV data suffer from significant shading (non-uniform response) effects. Since the objective of RBV correction is to modify the image-data intensity, to compensate for the spatially variable response characteristics, a method has been developed that mathematically structures the RBV image into correction zones. (A sufficient number of zones must be established to guarantee that the zone boundaries will be undetectable after the radiometric correction function is applied).

Each zone has a unique radiometric-correction table to be used for compensation of the RBV errors. In effect, RBV radiometric correction is performed in two stages, generation of the correction table (an off-line operation) and then application of the correction.

Pre-flight calibration readings at several intensity levels from a uniform light source provide an 18×18 array of points in the RBV image. These readings, in terms of voltage ranging from 0.32 to 1.10 V, are then scaled to the digital range 0 to 63. From these uniform input values, 18×18 arrays of gain and bias are computed for the correction equation

$$V_{\text{out}} (i,j) = G(i,j) \left[V_{\text{in}} (i,j) + B(i,j) \right] \quad (21\text{-}27)$$

where

V_{out} = output voltage representing pixel intensity
I_{in} = input voltage representing pixel intensity
B = Bias value
G = Gain value
i = Line (row) number
j = Element (column) number

These input readings are distributed uniformly throughout the image but do not include the edges. Various extrapolation techniques, with polynomial functions up to third order, were tested for use in estimating the edge data but, because the radiometric distortion is extreme near

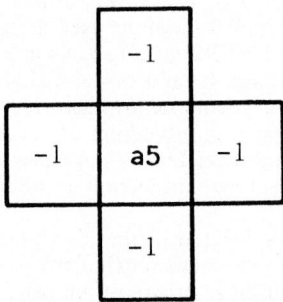

Fig. 21-14. Laplacian Operator for Edge Enhancement. With a5 = 4, the resultant image is edge enhanced. With a5 = 5, the edge enhanced image is added back to original image.

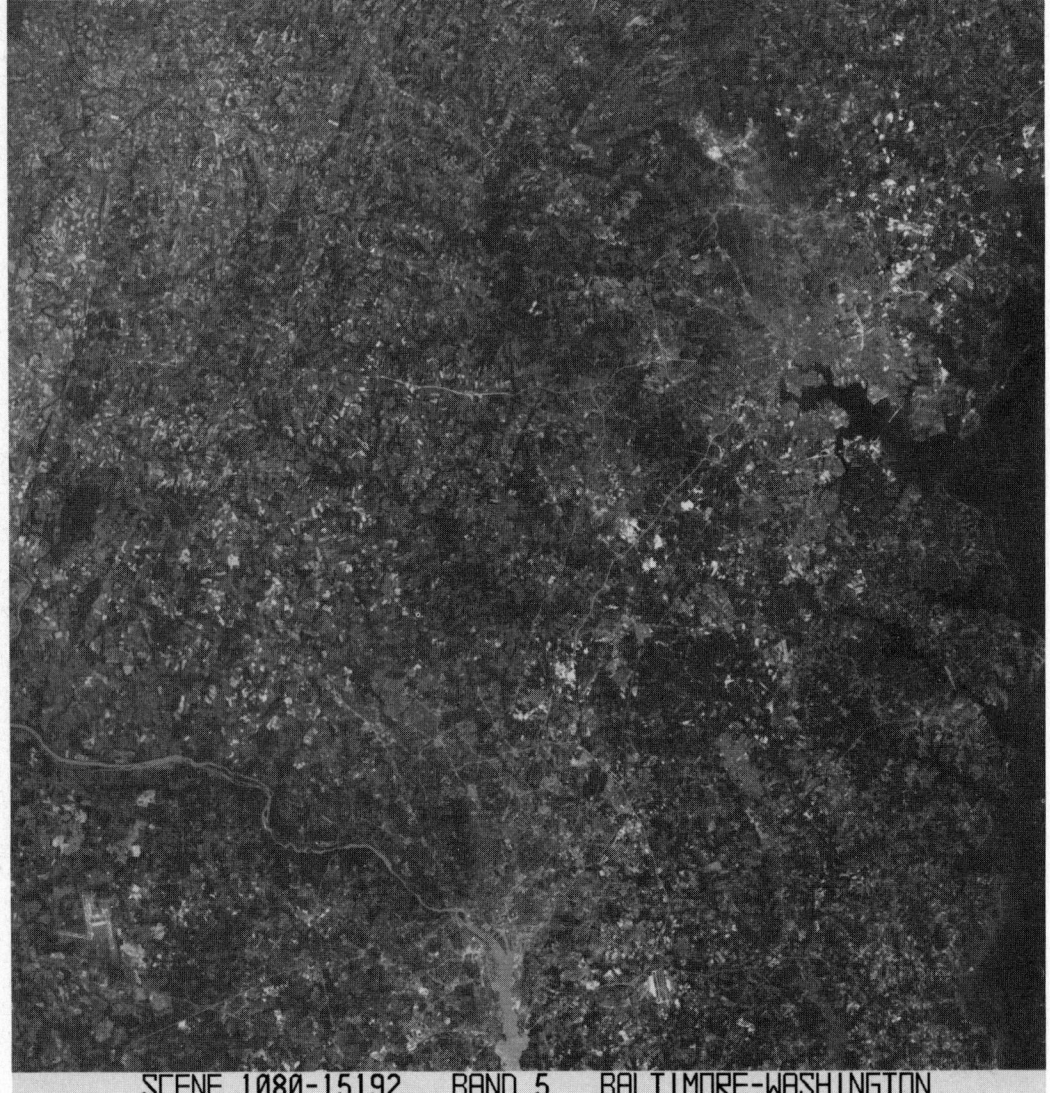

SCENE 1080-15192   BAND 5   BALTIMORE-WASHINGTON

Fig. 21-15. Landsat Subimage Before Edge Enhancement Operation.

the edges, zero-order extrapolation proved to yield the best results (see Bernstein, 1975).

With the inclusion of edge values, these computations finally produce 20 × 20 arrays of gain and bias values which completely span the image. The values can be fit in a least-square-error sense with spatially dependent functions $G(i,j)$ and $B(i,j)$. For this, it is computationally efficient to divide the image into zones within which constant values of G and B can be used with an acceptably small error. Figure 21-17 shows a spatially variant zone, designed to compensate for RBV shading effects.

GEOMETRIC CORRECTION

This section discusses the geometric errors present in the MSS data, methods of measuring those errors, formation and application of a geo-

metric correction function, and resampling of the input data to obtain output intensity values.

Geometric Correction Function: The image spaces and transformations used in the geometric correction of MSS data are shown in Figure 21-18. The input image is an array of digital data which represents a geometrically distorted one-dimensional perspective projection of some portion of the earth's surface. The output image is a geometrically correct map projection of the same ground area.

GCP's are located in the input image and are mapped into the tangent cylinder by using models based on all those errors which can be predicted or determined from tracking data. The nominal GCP locations are mapped from the map space to the tangent plane through the equations that relate points in map or tangent-plane space to points on

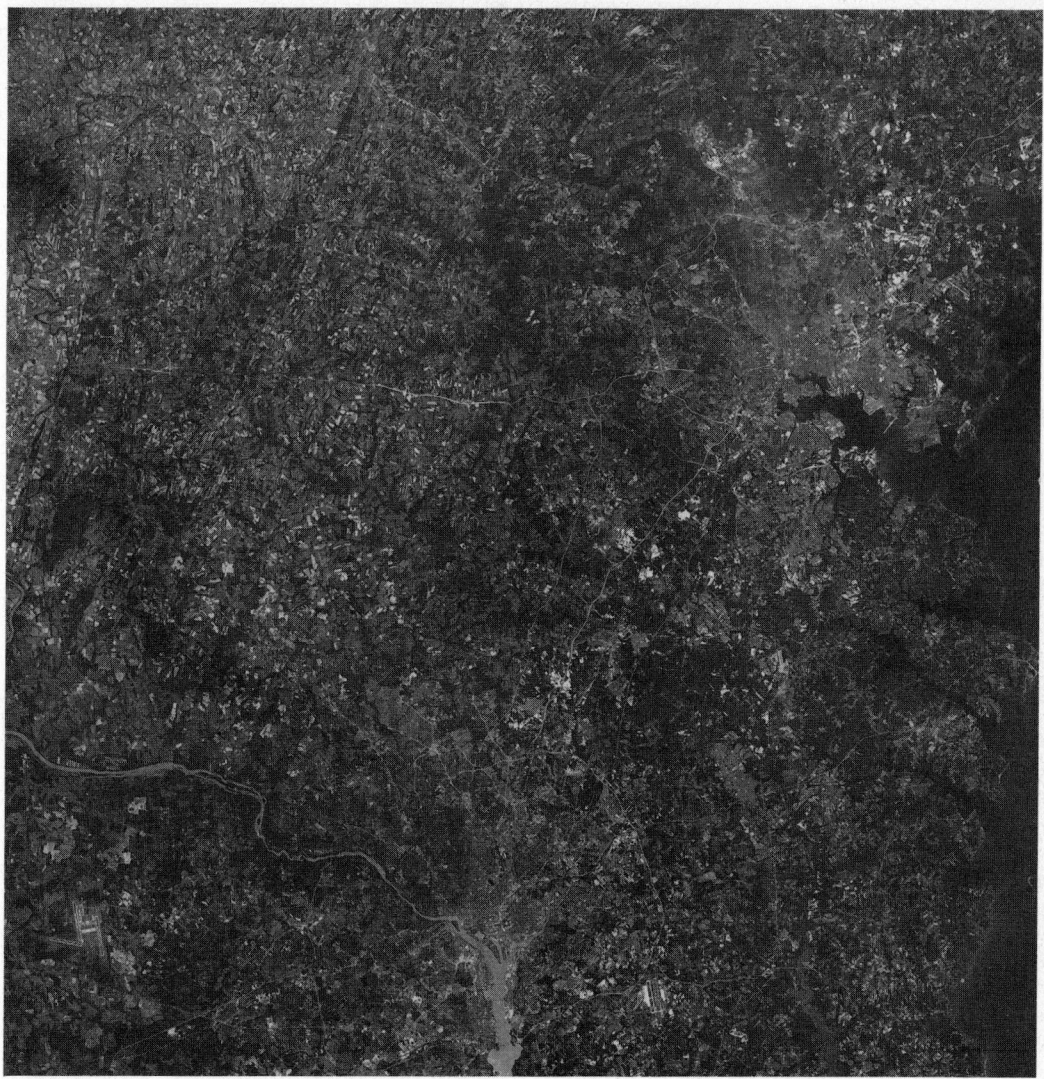

Fig. 21-16. Landsat Subimage After Edge Enhancement Operation.

the earth's surface. The nominal and observed GCP locations in the tangent plane are then used to evaluate the coefficients of the attitude and altitude models. The error models and the map-projection equations together provide the correction functions needed to relate points in the output space to points in the input space.

Rather than apply the correction functions to all points of the output image, an interpolation grid is established on the output image. This grid is constructed so that, if the four corner points of any grid-mesh are mapped with the correct functions, all points interior to the mesh can be located in the input image with sufficient accuracy by bilinear interpolation between the corner points. (See Figure 21-19)

Resampling: If the input data values are considered to lie at points on a regular lattice, the situation shown in Figure 21-20 occurs. The input space has been sampled at the points represented by the data values. When an output image point is mapped into the image space, its location does not generally coincide with any of the input sample points. In order to establish a data value for the output point, the input space must be resampled at the output point location. (This is discussed further in a later section of this chapter).

Geometric Correction Approach

This section details the approach to geometric corrections, and provides information on error modeling, space-to-space mapping, resampling techniques, and the results of an experimental geometric correction process.

Error Modeling: The models describing the distortions present in each input image are divided into two classes: those that can be computed from

TABLE 21-1

Truncated Radiometric Correction Table Example

Detector Output	Corrected Value, R, for Typical Functions Function Number:			
V	1	2	3	4
000	000	000	000	000
001	000	010	000	000
010	000	011	000	010
011	001	101	000	010
100	010	110	010	101
101	011	111	100	110
110	100	111	110	111
111	101	111	111	111

Functions:

1) $R = V - 2$

2) $R = 1.5\ V$

3) $R = 2\ (V - 3)$

4) $R = 0$, if $V \leq 1$
   $R = 1.5\ (V - 1)$, if $1 < V < 6$
   $R = 7$, if $V \geq 6$.

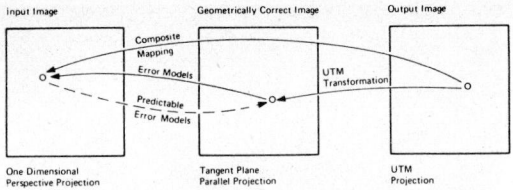

Fig. 21-18. MSS Image Spaces and Transformations.

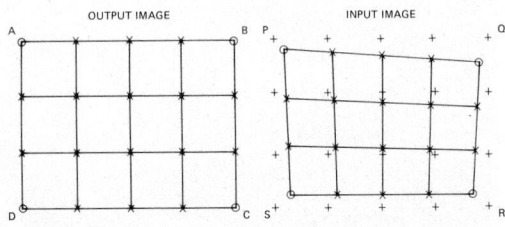

O   POINTS LOCATED BY MAPPING FUNCTION
X   POINTS LOCATED BY LINEAR INTERPOLATION
+   INPUT IMAGE POINTS

Fig. 21-19. Use of Linear Interpolation in the Mapping Operation.

spacecraft ephemeris data and knowledge of internal sensor distortion, and those that cannot be computed with acceptable accuracy without a sufficient number of GCPs (see Figure 21-21). The former include effects due to mirror-scan velocity nonlinearities, panoramic distortion, scan skew, spacecraft velocity, earth rotation, perspective geometry (including earth curvature), and different input/output scales. The latter effects are due to variations in spacecraft attitude and altitude. Prior to this, all known errors have been mathematically analyzed and modeled. The processing steps in error modeling are as follows (see Figure 21-22):

a. Adjustment to ground control.

Nominal control point locations for all control points are mapped from geodetic (latitude/longitude) coordinates to line and sample coordinates in the tangent space centered at the nadir (subsatellite point) of the input image.

b. Mirror Scan Velocity Profile and Panoramic Distortion.

Mirror Velocity: The scanning mirror of the MSSd has a non-linear sweep characteristic. Since data samples are taken at regular intervals of time, the varying mirror rate produces along-scan geometric distortion.

Panoramic Distortion: Data samples are taken at regular intervals of time (and, hence, nominally at regular angular intervals along the scan). However, the ground distance between samples is proportional to the tangent of the scan angle rather than to the angle itself. This effect produces along-scan geometric distortion.

In the conventional image-coordinates system for Landsat imagery, the X-axis is tangent to the sub-satellite tracks and positive in the direction of spacecraft motion. The Y-axis is orthogonal to the X-axis and positive in the direction of scan. Let (X_1, Y_1) be the input image line and sample coor-

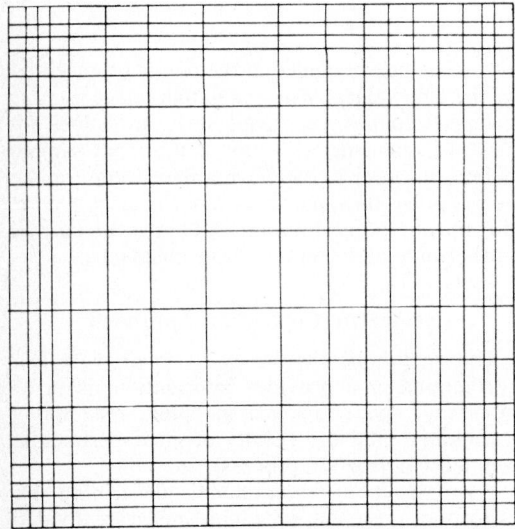

Fig. 21-17. RBV Radiometric Correction Zones.

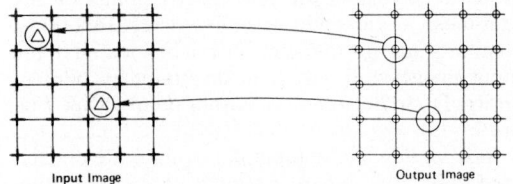

Input Image   Output Image

Fig. 21-20. Output to Input Mapping Concept.

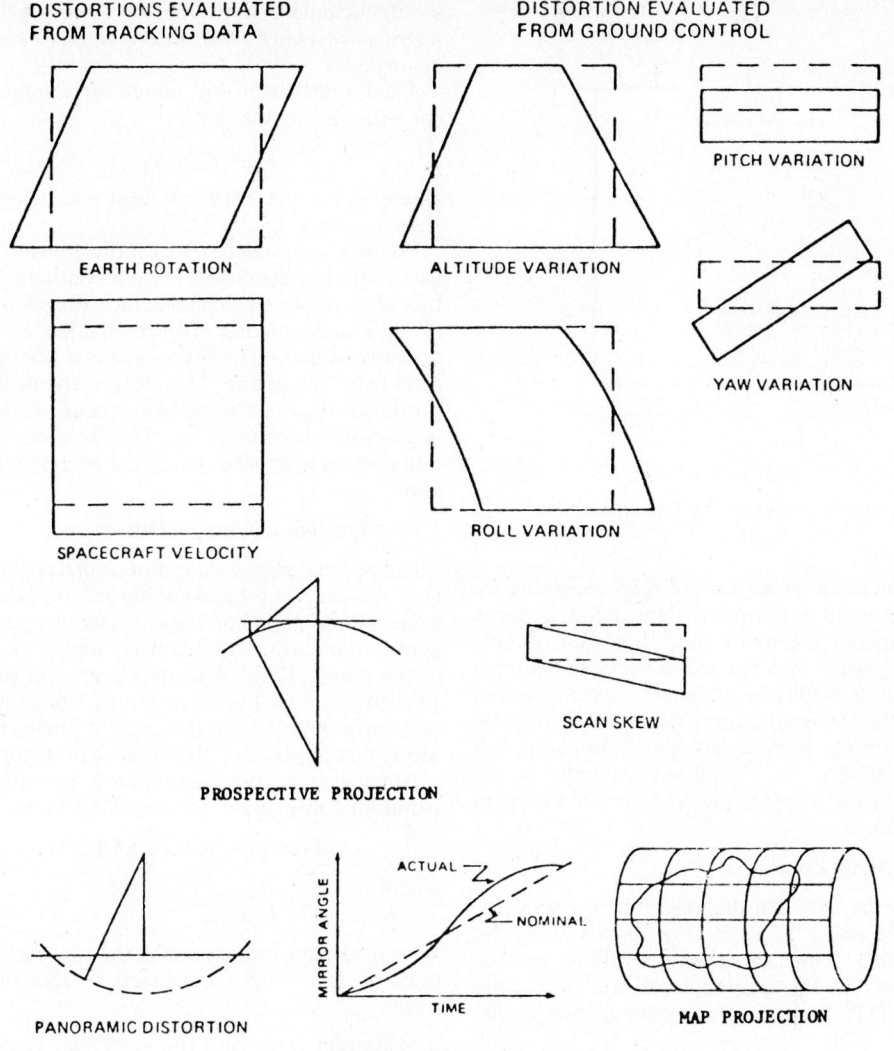

DISTORTIONS EVALUATED
FROM TRACKING DATA

DISTORTION EVALUATED
FROM GROUND CONTROL

EARTH ROTATION

ALTITUDE VARIATION

PITCH VARIATION

SPACECRAFT VELOCITY

ROLL VARIATION

YAW VARIATION

PROSPECTIVE PROJECTION

SCAN SKEW

PANORAMIC DISTORTION

MIRROR ANGLE / TIME
ACTUAL
NOMINAL

MAP PROJECTION

MIRROR VELOCITY VARIATIONS

Fig. 21-21. MSS Geometric Distortions.

dinates, respectively, before error modeling. Subscripted X and Y values indicate processing stages.

Along-scan distortions in the input image control-point locations (X_1, Y_1) due to mirror sweep nonlinearities and panoramic distortion are corrected by a model of the form

S = nominal swath width (meters)

K_0, K_1, K_2, K_3 = parameters of the mirror model

γ = maximum MSS mirror angle (radians)

ϕ = estimated roll angle (radians)

$$Y_2 = \frac{I_s H_0}{S} \, \text{Tan} \left[K_0 + \phi + Y_1 \left(K_1 + \frac{2\gamma}{I_s} \right) + K_2 Y_1^2 + K_3 Y_1^3 \right] \qquad (21\text{-}28)$$

where

I_S = the nominal number of samples per input image line

H_0 = nominal spacecraft altitude (meters)

Initially, ϕ is assumed to be zero. Iteratively, a more accurate estimate of roll will be used (see step j). It has been determined that the panoramic distortion correction must include roll angle ef-

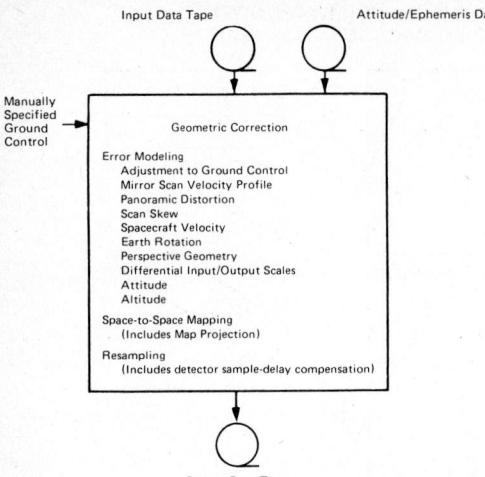

Input Data Tape   Attitude/Ephemeris Data

Manually
Specified
Ground
Control

Geometric Correction

Error Modeling
   Adjustment to Ground Control
   Mirror Scan Velocity Profile
   Panoramic Distortion
   Scan Skew
   Spacecraft Velocity
   Earth Rotation
   Perspective Geometry
   Differential Input/Output Scales
   Attitude
   Altitude

Space-to-Space Mapping
   (Includes Map Projection)

Resampling
   (Includes detector sample-delay compensation)

Output Data Tape

Fig. 21-22. Preprocessing Functional Flow.

fects when input-image control-point locations are being adjusted for input to the least squares attitude-model estimator, but that such effects may be ignored once the roll model has been established. Consequently, an iterative process through the systematic error models is used prior to final attitude-model coefficient estimation, but the form of Eq. 21-28 used subsequently in the space-to-space mapping process does not contain the variable ϕ.

c. Scan Skew Distortion

During the time that the MSS mirror completes one active scan, the spacecraft moves along the ground track. Thus, the ground swath scanned is not normal to the ground track but is slightly skewed. This produces cross-scan geometric distortion.

Cross-scan distortions caused by scan skew are corrected by the model

$$X_2 = X_1 + \sigma Y_2 \qquad (21\text{-}29)$$

where σ is a constant determined by the amount of scan skew.

d. Spacecraft Velocity Distortion.

If the spacecraft velocity departs from nominal, the ground track covered by a fixed number of successive mirror sweeps changes. This produces an along-track scale distortion.

Along-track distortions caused by spacecraft-velocity errors are corrected by the model

$$X_3 = \left(1 + \frac{\Delta V}{V}\right) X_2 \qquad (21\text{-}30)$$

where $\Delta V/V$ = the normalized spacecraft velocity-error from ephemeris data.

e. Earth Rotation Distortion.

As the MSS completes successive scans, the earth rotates beneath the sensor. Thus, the ground

swaths scanned by the mirror sweeps gradually migrate westward. This effect causes along-scan distortion.

Along-scan distortions due to earth rotation are corrected by the model

$$Y_3 = Y_2 - X_3\,\alpha_{ER} \qquad (21\text{-}31)$$

where α_{ER} = the earth rotation coefficient from NASA ephemeris.

For attitude modeling it is sufficient to consider earth rotation distortion to be a continuous function of X. However, in actual fact, the distortion is a continuous function of sweep number but is discontinuous in X. The difference is a fraction of a pixel from line to line. Therefore, only the integral portion of the earth-rotation correction is included in space-to-space mapping. The fractional correction portion is applied during the resampling operation.

f. Perspective Geometry Distortion.

Perspective Projection: For some applications it is desired that Landsat images represent the projection of points on the earth upon a plane tangent to the earth, with all projection lines normal to the plane. The sensor data represent perspective projections; i.e., projections whose lines all meet at a point above the tangent plane. For the MSS, this produces only along-scan distortion.

Along-scan errors due to earth curvature and panoramic projection are corrected by the model

$$Y_4 = \left[1 - KY_3^2 - K^2\,Y_3^4\right] Y_3 \qquad (21\text{-}32)$$

where
$$K = -\,S^2/2H_0\,R_0\,I_s^2$$

R_0 = the radius of the earth at the spacecraft nadir (meters) and S, H_0, and I_s are as defined previously.

g. Different Input and Output Scales Distortion.

Distortions due to different input and output scales are corrected by the models

$$X_C = \frac{O_L}{I_L} X_3$$

$$\qquad (21\text{-}33)$$

$$Y_C = \frac{O_S}{I_S} Y_4$$

where
O_L, O_S = the number of lines and samples per line in output image
I_L, I_S = the nominal number of lines and samples per line in input image
X_C, Y_C = the input-image control-point coordinates corrected for all systematic errors.

h. Attitude and Altitude Distortions.

Altitude: Departures of the spacecraft altitude from the nominal altitude produce scale distortions in the sensor data. For the MSS, the distortion is primarily along-scan and varies with time.

Attitude: Nominally the sensor axis-system is maintained so that one axis is normal to the earth's surface and another is aligned with the spacecraft velocity vector. As the sensor departs from this attitude, geometric distortions result. For the MSS, the full attitude time-history contributes to the distortion.

The differences between nominal (step a) and corrected input-image (step g) control-point coordinates, along with all applicable ephemeris and AMS data, are processed by a weighted least squares estimator which computes the format center coordinates and the coefficients of the attitude and altitude modules. In these models, spacecraft roll, pitch and yaw variations are described by cubic functions of time (4 × 3 coefficients), and spacecraft altitude variations by a linear function of time (2 coefficients). The mathematical model in matrix form is

$$
\begin{bmatrix} \phi \\ \theta \\ \psi \\ \Delta h \end{bmatrix} = \begin{bmatrix} 1\ t\ t^2\ t^3\ 0\ 0\ 0\ 0\ 0\ 0\ 0\ 0 \\ 0\ 0\ 0\ 0\ 1\ t\ t^2\ t^3\ 0\ 0\ 0\ 0 \\ 0\ 0\ 0\ 0\ 0\ 0\ 0\ 0\ 1\ t\ t^2\ t^3\ 0\ 0 \\ 0\ 0\ 0\ 0\ 0\ 0\ 0\ 0\ 0\ 0\ 0\ 1\ t \end{bmatrix} \begin{bmatrix} (R_1 R_2 R_3 R_4)^T \\ (P_1 P_2 P_3 P_4)^T \\ (Y_1 Y_2 Y_3 Y_4)^T \\ (A_1 A_2)^T \end{bmatrix} \tag{21-34}
$$

All of the control point, altitude, and AMS information in the scene being corrected is used by the least-squares estimator. Since the estimator uses all the available data, the models are always the best that can be obtained from the existing information.

i. Time for each control point is computed according to the relation

$$t = t_s \text{ integer portion of } (X_1/6)$$

where t_s is the number of seconds per mirror sweep.

A roll angle for each point is computed from equation 21-34, and equations 21-28 through 21-33 are executed again. Step h is repeated, using the new values of $(X_C Y_C)$. The final pitch, yaw, and altitude models used are those computed in the second execution of step h. The final roll model is the sum of the two sets of coefficients produced by the two executions of step h.

j. The control-point locations are corrected for attitude and altitude distortions by the models

$$X_A = X_C + \psi Y_C + \theta H_{px}$$

$$Y_A = Y_C(1 + \Delta H/H_O) - \phi(H_{PY}^2 + Y_C^2)/H_{PY} \tag{21-35}$$

where

$$H_{PX} = \frac{H_O I_L}{G}$$

$$H_{PY} = \frac{H_O I_S}{S} \tag{21-36}$$

G = nominal image height (meters)

k. The (X_A, Y_A) coordinates are compared with the nominal control point coordinates computed in step a. The mean values of the differences are used in a final correction:

$$X_T = X_A + D_x \tag{21-37}$$

$$X_T = Y_A + D_Y \tag{21-38}$$

Eqs. 21-28 through 21-36 characterize all of the distortions present in the input image except the along-scan distortions caused by line-length variations and sampling delay. (These latter distortions are corrected line by line in the resampling routine.) Together with the equations of the desired map projection, the error models provide an explicit, accurate transformation from the input space to the output space.

Space-to-Space Mapping

As mentioned above, the error models lead directly to the correction transformations from input space to output space. However, the corresponding inverse transformations from output space to input space are only implicit, since they involve a value of time which is not known until the pixel in the input space has been computed. If a nominal value of time is used in the implicit output-space to input-space transformations, an approximate transformation is produced. The accurate input-to-output space transformations are used in conjunction with the approximate output-to-input space transformations in the space-to-space mapping function to produce an accurate transformation of output space to input space. The procedure (shown in Figure 21-23) is as follows:

a. A rectangular grid of points in the output space is defined such that any point in the output space can be located in the input space with sufficient accuracy by bilinear interpolation over the input space coordinates of the surrounding four grid points. This grid does not change from scene to scene.

b. Through use of the equations of the UTM or other map projection of the output image, the grid points are transformed to geodetic coordinates (latitude and longitude) and then to a tangent cylinder centered at the format center of the input image.

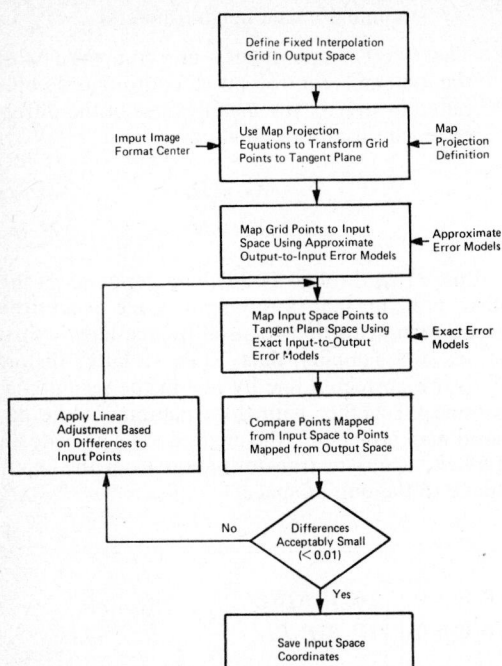

Fig. 21-23. Space-to-Space Mapping.

c. The tangent-cylinder grid points are mapped to the input space by the approximate output-to-input mappings.

d. The resulting input points are mapped to the output space using the exact input-to-output transformations.

e. The differences between the nominal and mapped grid points are used to adjust the input space coordinates in a linear manner, and this process is repeated. In practice the required non-uniform array of input points which maps into the rectangular grid of output points is produced in not more than two iterations.

It should be noted that the accurate input-to-output transformations are correct for all known sources of error, except for line length, sample delay, and the fractional portion of earth-rotation error, which are accounted for as high frequency corrections (every scan) in the resampling function.

The output of the space-to-space mapping function is a table of the line and sample coordinates in the input space corresponding to the coordinates of the fixed interpolation grid in the output space. The input-space coordinates corresponding to any given output-space coordinates can then be found by bilinear interpolation. This interpolation process is carried out in the resampling function.

With reference to the previous procedural step b it should be noted that, although map projection does not constitute an actual geometric error, it does require a geometric transformation of the input data and can be accomplished in the same operation that compensates for the distortions present in the data. NASA software currently incorporates the Hotine Oblique Mercator (HOM), (Hotine, M., 1946, 1947) and the Space Oblique Mercator (SOM) (Colvocoresses, 1974, and Snyder, 1978) map projections.

Resampling

Resampling is the means by which a geometric transformation (whether image rectification or temporal registration) is actually applied to the input data. When the elements of the output space are mapped into the sampled data space of the input-data array, their locations do not generally coincide with those of the input-data samples. Therefore, the input space must be resampled at the output-element locations to obtain output-data values. Two methods of resampling are commonly in use today—nearest neighbor (NN) and cubic convolution (CC).

Nearest Neighbor: As the name implies, NN resampling simply finds the input-image sample which is closest to the output-pixel location and assigns that intensity value to the output-pixel sample. For each input line, nearest neighbor resampling is first applied to the points where output-pixel columns intersect the input-pixel lines. By calculating the points where interpolation displacements would be exactly 0.5 pixels, each input line can be partitioned into line segments which can be transferred directly to intermediate line buffers.

Horizontally resampled image data from the intermediate line buffers is again partitioned into line segments such that all pixels within a given input-line segment lie within 0.5 pixel (measured vertically in the output space) of a given output line. Input-line segments associated with the same output line are sequentially transferred from intermediate line buffers for assembly into output lines.

Cubic Convolution: Like NN resampling, cubic convolution is divided into two steps. The first step resamples horizontally or along input lines, and the second step resamples vertically to the final output-pixel location. Both resampling steps use a nested form of the cubic convolution interpolation algorithms given by:

$$I'_k = I_k + d_k(I_{k+1} - I_{k-1} + d_k(2I_{k-1} - 2I_k + I_{k+1} - I_{k+2} + d_k (I_{k+2} - I_{k+1} + I_k - I_{k-1})))$$

$$(21-39)$$

where

I'_k = the interpolated output image data for the *k*th sample

I_k = image data of the *k*th sample

d_k = the normalized displacement between I'_k and $I_k (O \le d_k \le 1)$

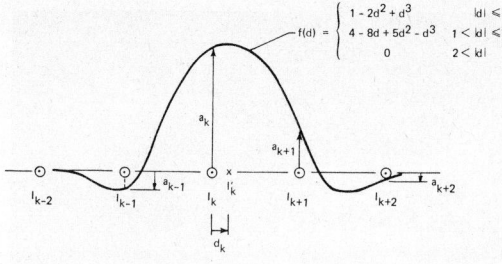

$$f(d) = \begin{cases} 1 - 2d^2 + d^3 & |d| \leqslant 1 \\ 4 - 8d + 5d^2 - d^3 & 1 < |d| \leqslant 2 \\ 0 & 2 < |d| \end{cases}$$

$$I'_k = a_{k-1} I_{k-1} + a_k I_k + a_{k+1} I_{k+1} + a_{k+2} I_{k+2}$$

Fig. 21-24. Cubic Convolution Interpolation.

Figure 21-24 shows the relationships of the various parameters.

Detector Sample-Delay and Earth Rotation Compensation: Each spectral band of a Landsat MSS scene is produced by an array of six nominally identical detectors which generate six scan lines per mirror sweep. The six detectors are sampled sequentially with a fixed time delay between consecutive samplings of each detector. Between the first and second sweeps of the mirror, the earth has rotated to the East by an amount proportional to the sweep period and the earth's tangential velocity. This will cause the ground swath of the 2nd sweep to be west of the first swath by Δ_2 (See Figure 21-25). Since a delay in time will result in a shift in space, the ground swath will have a shear effect. In Figure 21-25, this effect is shown as a Δ_1 shift between the first and sixth lines. These effects will cause visible discontinuities in linear features. Edges within an image that should be continuous will appear to be jagged if the effects are not compensated.

These sub-pixel compensations are accomplished by adjusting the horizontal coordinate before computing the intermediate interpolated values. The adjustment consists of subtracting a constant which depends only on the particular detector used.

Digital Image Processing Results

A number of Multispectral Scanner and Return Beam Vidicon scenes have been processed using these techniques. In this section we present some of the results in both numerical and pictorial form.

Geometric correction: For MSS data the number, distribution, and positional accuracy of GCPs influence the accuracy of the output image data. Figure 21-26 shows the relationship of RMS image error as a function of the number of GCPs to establish the distortion characteristics of a scene; 16 seems to be a reasonable number of GCPs to use for an acceptable degree of accuracy, assuming GCP's can be located to ⅓ pixel accuracy or better.

Figure 21-27 is an MSS image of the central California scene before and after correction. The "before" image (Figure 21-27a) is a rectangle be-

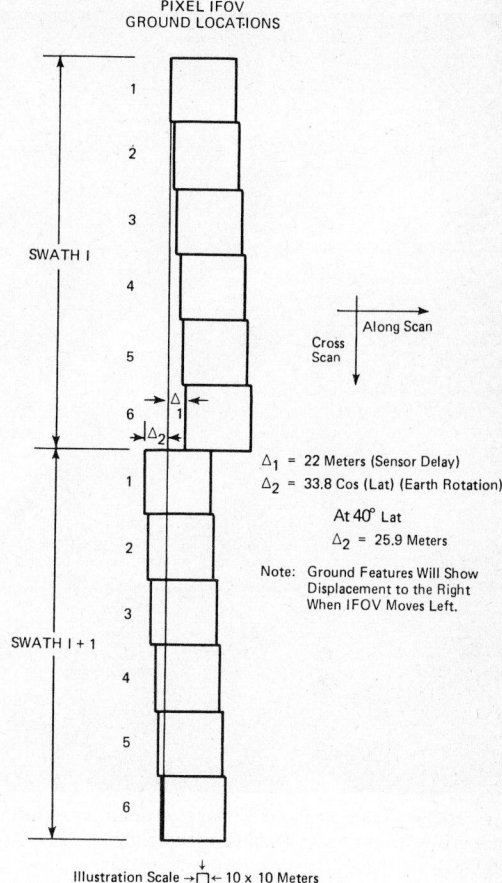

Δ_1 = 22 Meters (Sensor Delay)

Δ_2 = 33.8 Cos (Lat) (Earth Rotation)

At 40° Lat

Δ_2 = 25.9 Meters

Note: Ground Features Will Show Displacement to the Right When IFOV Moves Left.

Illustration Scale → ☐ ← 10 x 10 Meters

Fig. 21-25. Landsat MSS Sensor Delay and Earth Rotation Distortions.

cause of the overlapping of the along-scan sample values (1.4:1 oversampling) and because no earth-rotation correction has yet been made. The "after" image (Figure 21-27b) has been fully corrected and can be used as a map product at a scale of about 1:250,000.

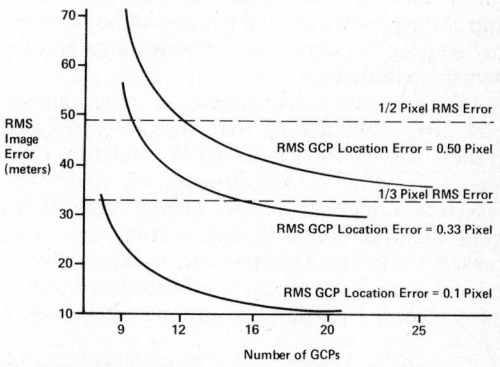

Fig. 21-26. MSS Image Absolute Error as a Function of the Number of GCP's.

Fig. 21-27a. Landsat Band 5 Image of Central California before Geometric Corrections Applied. Rectangular Format arises from Sensor Data Oversampling.

REGISTRATION

In some applications it is useful to have two or more images in geometric conformance, i.e., in registration with each other. This is particularly useful in change-detection applications where a difference in ground features is of interest, such as shoreline erosion. In agricultural feature-extraction applications, the use of multiple scenes of the same area over a crop maturation period improves crop-classification accuracy. For these types of applications, extremely precise geometric correction of the various scenes must be made so that corresponding ground features of both scenes are assigned the same geographic location and are thus in conformance.

An experiment was conducted to determine the geometric similarity of two digitally processed scenes. A scene that included Washington, D.C. was processed to correct all geometric errors; this served as a reference image. Then a scene of the same area, acquired 18 days earlier, was processed using the latest scene as a geometric ground-control reference. Both processed scenes were recorded on film and are shown in Figure 21-28.

To evaluate relative error, 69 features were located in each scene and the differences in corresponding line- and sample-coordinates were computed. The results showed temporal registration errors of less than 0.24 picture element in sample location and less than 0.11 picture element in line location for 90 percent of the features. When the scene transparencies are overlaid, no geometric difference can be discerned. This experiment demonstrates that multiple scenes can be corrected to be in conformance within about one-fourth picture element.

MOSAICKING

A growing number of applications require the combination or mosaicking of several Landsat scenes. The Department of Interior Bureau of Land Management, with industry support, demonstrated the feasibility of forming such mosaics digitally (Torbert and Robinove, 1976; Bernstein, 1976). Eight MSS scenes, whose relative geometry is shown in Figure 21-29, were chosen for the experiment. Since the four scenes from each pass were originally continuous strips of data, the first step in the processing was to reformat the data as two continuous strips, eliminating the along-track overlap.

Geodetic coordinates for 75 GCP's, whose approximate locations are shown in Figure 21-29, were measured from 1:250,000 or 1:24,000 scale maps. The image coordinates of these GCP's were

Fig. 21-27b. Landsat Band 5 Image of Central California after Geometric Correction Applied. The Earth Rotation, Oversampling effects and other sources of Geometric Error have been Removed. The Data has been put into a UTM Projection.

determined by using computer-generated shade prints. For each strip, a computer program was used to compute correction functions which transformed the GCP locations so that they were located in their proper positions in a UTM projection. For the pass-1372 strip, a second function also was computed. This function transformed the GCP's in the overlap region only to bring them into coincidence with those of the pass-1373 strip. The two correction functions for the pass-1372 strip were then combined in a single composite transformation. The two strips of data were then corrected geometrically by using the composite transformation. Registration of the two corrected strips was achieved by examination of enlarged film and computer-image printout (shade prints).

The remaining problem concerned the elimination of the duplicate data in the overlap region.

This was accomplished by a program which accepted as input a boundary specified as a sequence of straight line segments. Data to the left of the boundary were taken from the right-hand strip. The mosaicked array was trimmed down to fit the capacity of a drum film-recorder, and a border including annotation and geodetic tick marks was added. The composite array was recorded on film at 1:1,000,000 scale using a 50 μm square spot.

The reduced processed mosaic is shown in color Figure 21-30. Excellent geometric fitting has been accomplished and scene-discontinuity results only from cloud patterns and radiance differences due to time separation between the data.

Figure 21-31 is a digital mosaic of the entire United States (Bernstein, 1976). The Defense Meteorological Satellite Program satellite ac-

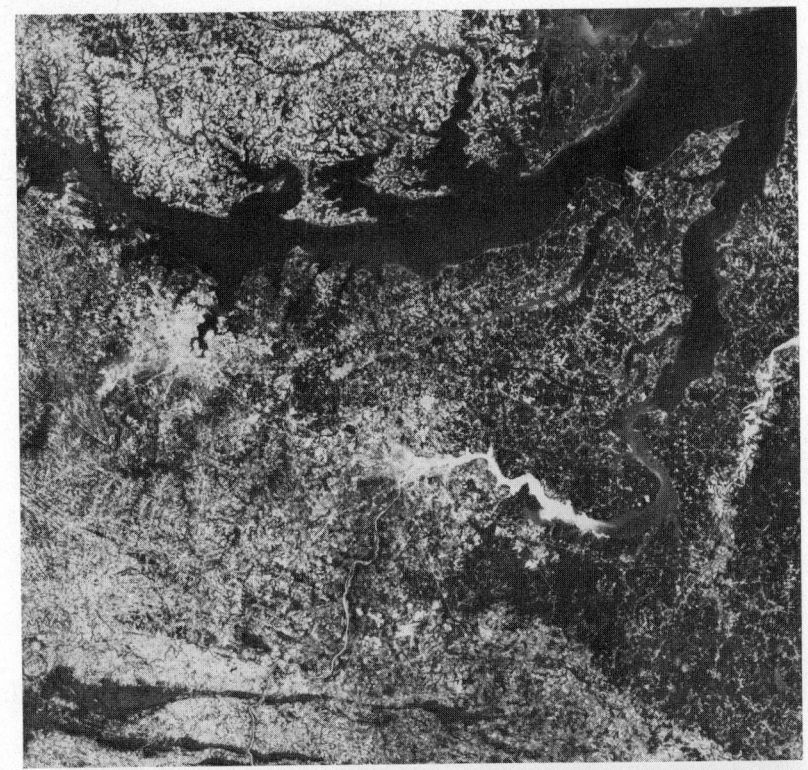

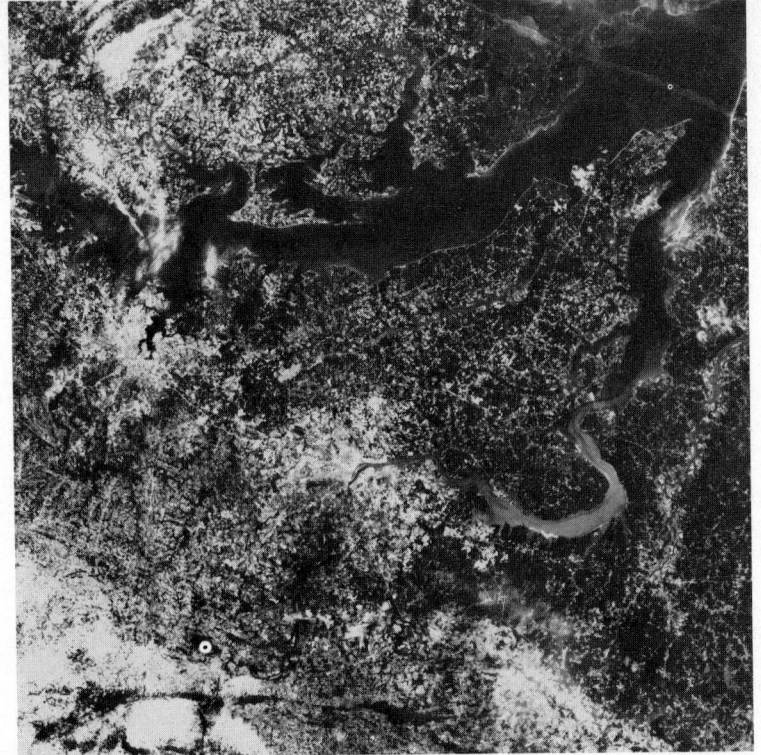

Fig. 21-28a,b. Two Images of Washington, DC Area in Geometric Registration.

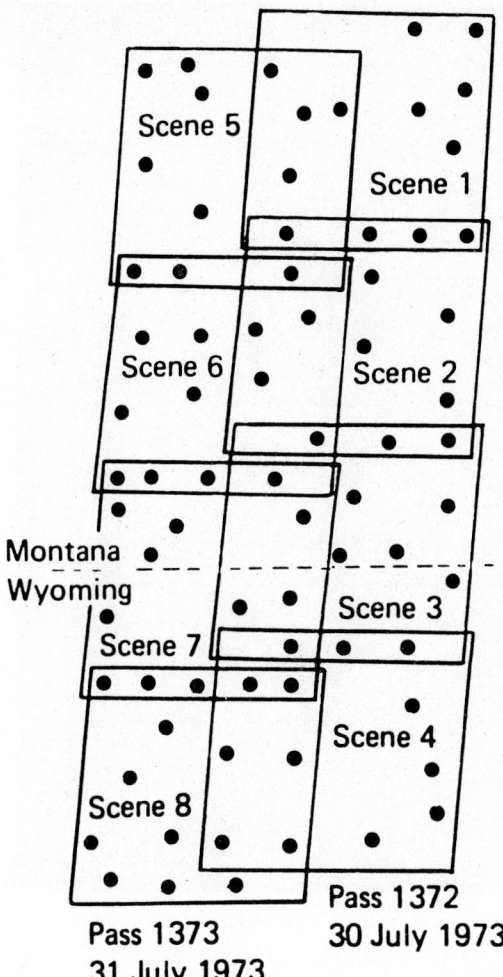

Scene 5

Scene 1

Scene 6

Scene 2

Montana
Wyoming

Scene 3

Scene 7

Scene 4

Scene 8

Pass 1372
Pass 1373   30 July 1973
31 July 1973

Fig. 21-29. Structure of Source Data for Mosaicking operation. The Dots Identify Location of Ground Control Points Used. The Data was Obtained over a two-day Period over Montana & Wyoming.

quired infrared and visible light sensor data (0.4 to 1.0 μm) accumulated on three nighttime passes of a meteorological satellite at an altitude of 835 km. Each sweep provided ground coverage of a 3000-km wide swath but, because of the wide (112°) field of view, the sensor data contained severe panoramic distortion. Geometric-correction programs were used to eliminate this distortion and to convert the data to a Lambert conformal conic projection. The data from the three swaths were then adjusted for intensity variations and geometrically mosaicked into the scene shown. Los Angeles, San Francisco, Seattle, and other cities outline the west coast; Miami, Washington, Philadelphia, New York and other cities can be identified on the east coast (clouds obscured Boston and other northern cities); and most medium-to-large cities in the interior of the United States can also be identified.

MAP PROJECTIONS SUITABLE FOR LARGER-SCALE MAPPING AND FOR USE WITH SATELLITE IMAGES

After decades of using a single map projection, the Polyconic, for its mapping program, the U.S. Geological Survey (the principal mapping agency in the U.S.A.) now uses several long-established projections for its published maps, both large and small scale. For maps at 1:1,000,000-scale and larger, the most common projections are conformal, such as the Transverse Mercator and Lambert Conformal Conic. Projections for these scales should treat the Earth as an ellipsoid. In addition, the USGS has conceived and designed some new projections, including the Space Oblique Mercator, the first map projection designed to permit low-distortion mapping of the Earth from satellite imagery, continuously following the groundtrack. The USGS has programmed nearly all pertinent projection equations for inverse and forward calculations. These are used to plot maps or to transform coordinates from one projection to another. The USGS has also published its first comprehensive map-projection manual, describing in detail and mathematically all projections used by the agency.

Until the late 1950's, only the Polyconic projection was used for the primary USGS mapping product, i.e., large-scale quadrangle maps. The Polyconic projection was apparently invented and certainly promoted by Ferdinand Rudolph Hassler, the first head of what was to become known as the Coast and Geodetic Survey. In the 1950's, the USGS quadrangle projection was changed to the Lambert Conformal Conic and the Transverse Mercator projections, which had been adopted by the Coast and Geodetic Survey in the 1930's for the State Plane Coordinate System. The development of standardized zones based upon the Universal Transverse Mercator and the polar Stereographic led to USGS use of these projections. In addition to these, the regular Mercator, the Oblique Mercator, the Albers Equal-Area Conic, and the Azimuthal Equidistant have been used for other larger-scale mapping by the USGS and other agencies. Although authors and organizations variously define large- and intermediate-scale mapping, for the purposes of this discussion, the term "larger scale" will apply to scales larger than 1:2 million.

When the space age added its impact to mapping, classical projections (Mercator, Lambert Conformal Conic, and Stereographic) were chosen for the mapping of the Earth's Moon, three other planets, and a number of other natural satellites. Some projections, especially the Space Oblique Mercator, originated within USGS to assist mapping from satellite imagery.

Fig. 21-31. Mosaic of Three DMSP Scenes Over the US at Night.

TYPES OF PROJECTION

There are several types of map projection to consider. Equal-area or equivalent projections of the globe are used especially by geographers seeking to compare land use, densities, and the like. On an equal-area projection, such as the Albers Equal-Area Conic, a coin laid on one part of the map covers exactly the same area of the actual Earth as the same coin on any other part of the map. Shapes, angles, and scale must be distorted on most parts of such a map, but there are usually certain lines on an equal-area map as well as on other types of projections, along which there is no distortion of any kind. These so-called "standard lines" may be a meridian, one or two parallels, lines which are neither, or not a line but a point.

More commonly used in larger-scale mapping are conformal (orthomorphic) projections such as the Transverse Mercator and the Lambert Conformal Conic. The term conformal means that they are correct in shape, but, unlike the term "equal area," the conformal principle applies only to each infinitesimal element of the map. Angles at each point are correct, and consequently the local scale in every direction around any one point is constant, so the map user can measure distance and direction between near points with a minimum of difficulty. Conformal maps may also be prepared by fitting together small pieces of other conformal maps which have been enlarged or reduced; non-conformal projections require re-shaping as well. When the region consists of more than a small element, distortion in shape as well as area becomes appreciable. This is especially serious with the most famous conformal projection—the Mercator—because of its widespread use in classrooms, especially in the past. Because there is no angular distortion, all meridians intersect parallels at right angles on a conformal projection, just as they do on the Earth. "Standard lines" may also be applied to a conformal map to eliminate scale- and area-distortion along these lines and to minimize distortion elsewhere.

Some map projections, such as the Azimuthal Equidistant, are neither equal-area nor conformal, but linear scale is correct along all lines radiating from the center, along meridians, or following other special patterns. In addition, there are compromise projections, almost entirely restricted to small-scale mapping, which are used because they balance distortion in scale, area, and shape.

Projections are often classified by the type of surface onto which the Earth may be mapped. If a cylinder or cone that has been placed around a globe is unrolled, we have the concept of cylindrical or conic projections, such as the Mercator and Lambert Conformal Conic, respectively. If the axis of the cone or cylinder coincides with the polar axis of the globe, the projection has equally spaced straight meridians, parallel on the cylindrical projections and converging on the conics. The parallels intersect the meridians at right angles, being straight on the cylindricals and concentric circular arcs on the conics. The spacing of the parallels is seldom projective. A plane laid tangent to the globe at the pole leads to polar azimuthal projections, such as the polar Stereographic, with the parallels mapped as arcs of concentric circles and meridians as equally spaced radii of the circles. Scale remains constant along each parallel of latitude on a regular cylindrical, conic, or polar azimuthal projection, but it changes from one latitude to another. Directions of all points are correct as seen from the center of an azimuthal projection.

If the cylinder or cone is secant instead of tangent to the globe, the projection conceptually has two lines instead of one which are true to scale. Wrapping the cylinder about a meridian leads to transverse projections. By placing a plane tangent to the Equator instead of a pole, equatorial aspects of azimuthal projections result. Tilting the cylinder, cone, or plane to relate to another point on the Earth leads to an oblique projection, and the meridians and parallels are no longer the straight lines or circular arcs they were in the normal aspect. The lines of constant scale are correspondingly rotated.

THE EARTH AS AN ELLIPSOID

For maps at scales smaller than 1:5,000,000, and which cover regions larger than the United States, the distortions from mapping the spherical Earth on flat paper are much greater than the slight additional corrections needed to compensate for the ellipsoidal shape of the Earth. These corrections may then usually be ignored. The ellipsoid should be, and normally is, used for large-scale mapping of small areas, or for long narrow strips. For such areas, the flattening of the round Earth usually produces less distortion than the use of the sphere instead of the ellipsoids.

A shift from one ellipsoid to another has a negligible effect, even on large-scale maps, upon the projected shapes and positions of meridians and parallels. A greater effect is the translation of latitude and longitude for all points on a map, due to a change in datum, that changes the position of the ellipsoid relative to the Earth. For this reason, the notation in the corner of USGS quadrangles stating "North American Datum 1927" or "1983" is as important as the parameters of the map projection in defining the basis of these maps.

PRINCIPAL PROJECTIONS

Polyconic Projection

About 1820, Hassler began to promote the easily constructed Polyconic projection as the basis of large-scale mapping. The USGS used this projection for the earliest quadrangles, only changing in the 1950's to other projections, although re-labeling of the map legend lagged considerably behind the change. The Polyconic is neither equal-area nor conformal. For 7½- and 15-min. quadrangles, however, the distortion is negligible. Along the central meridian, it is free of distortion. Each parallel is also true to scale, but the other meridians are too long, and constantly change scale. The projection is not recommended for maps of considerable east-west extent and, in fact, should not be seriously used for any new maps in view of other projections available. Figure 21-32 is a Polyconic projection of the Western Hemisphere.

The parallels of latitude are circular arcs spaced at their true distances along the central meridian, but with radii equal to the length of the element of a cone tangent at the particular parallel. The projection receives its name from the fact that each cone is different. Meridians are marked on each parallel at the true distances, but the meridians are complex curves connecting these points. Lines of constant scale run roughly parallel to the central meridian, but they are curved.

Mercator Projection

The Mercator projection is the best known of all. It was presented by Mercator for navigational purposes in 1569, because lines of constant compass bearings, are plotted as straight lines. This is still the most justifiable use of the projection for a map of regions away from the Equator. It is a normal cylindrical projection, on which the lines of constant scale are straight and run parallel to the Equator. On the Mercator, the scale increases away from the Equator. Figure 21-33 is a Mercator projection of the World.

The USGS has used the Mercator projection for maps of part of the Pacific Ocean, of Indonesia, and also for portions of each of the outer bodies of our solar system that have been mapped to date, from our Moon to the satellites of Saturn. In some cases, the chosen scale applies to standard parallels placed symmetrically north and south of the Equator. The shape of the map does not change. The scale along the Equator could still be called correct if the stated scale of the map were slightly decreased.

Transverse Mercator Projection

Rotating the cylinder of the Mercator so that it lies tangent (or secant) along a meridian of the globe leads to the very important conformal projection called the Transverse Mercator. Figure

Fig. 21-32. Polyconic projection of the Western Hemisphere.

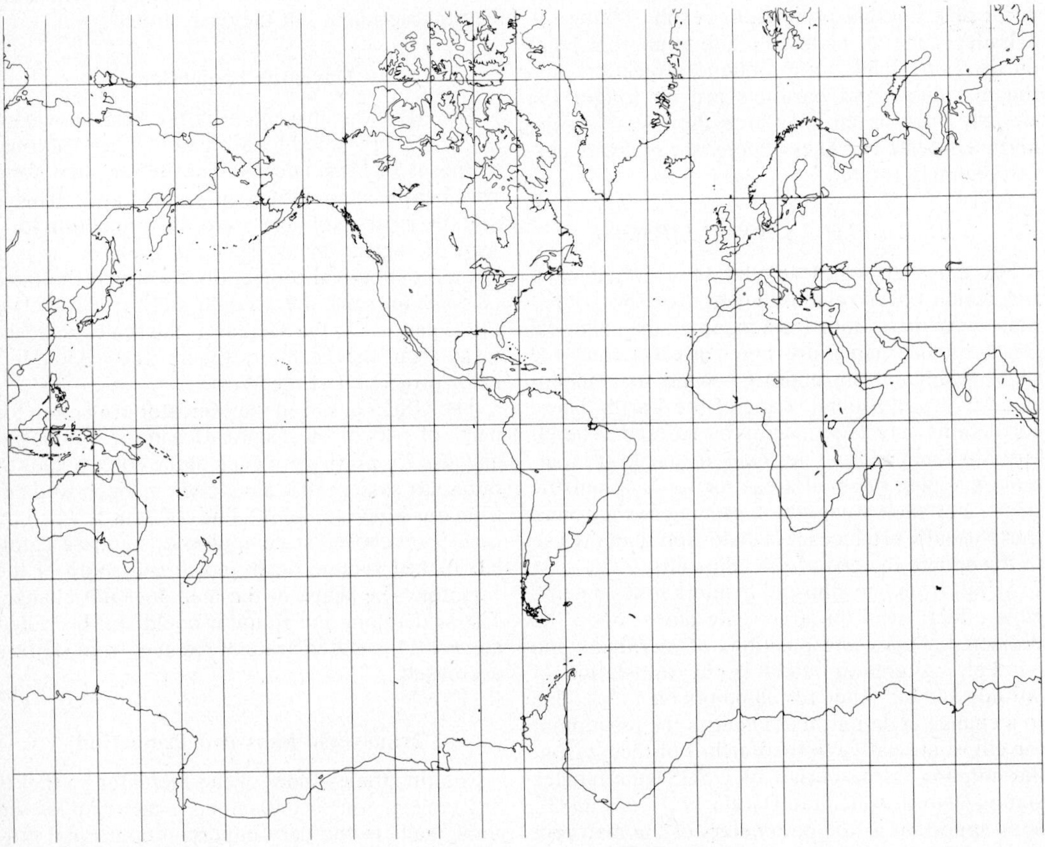

Fig. 21-33. Mercator projection of the World.

21-34 is a Transverse Mercator projection of the Western Hemisphere. The central meridian, the Equator, and each meridian 90° from the central meridian are straight lines. All other meridians and parallels are complex curves. For the sphere and normally for the ellipsoid, the central meridian has a constant scale, but this is usually reduced from the nominal map scale to balance errors in measurement over the rest of the map. The lines of constant scale are straight lines parallel to the central meridian for the sphere, and nearly straight for the ellipsoid. When the scale factor along the central meridian is reduced, the lines of true scale are symmetrical with respect to the central meridian.

Lambert presented the spherical projection in 1772, but Gauss and later Krüger developed the mathematics for the ellipsoidal form. The projection was almost ignored until the 20th century, when it was adopted for much of the topographic mapping in Europe under the name Gauss-Krüger, and in the United States for the State Plane Coordinate System of States predominantly north and south in extent and for the Universal Transverse Mercator or UTM projection and grid system. The UTM has two special restrictions: (1) The central scale factor is 0.9996, and (2) the Earth is divided into sixty zones, each 6° of longitude wide, with the central meridians placed at every sixth meridian beginning with the 177th West. (There are minor exceptions.)

When USGS stopped using the Polyconic projection for large-scale quadrangle maps, the Transverse Mercator was adopted for maps of areas where it was used for the State Plane Coordinate System. (Recent metric quadrangles are based upon the UTM). In the State Plane Coordinate System, each zone follows county boundaries and is designed to restrict scale variation over the map to one part in ten thousand. The USGS uses the predominent zone for the projection of quadrangles which cross county boundaries. The central scale reductions, which change between zones, vary from 1:160,000 to 1:10,000. Central meridians have been individually selected for each zone.

Equations in series form are used for the Transverse Mercator calculations for the ellipsoid. These are limited to a 6° to 8° band of longitude and cannot be safely extrapolated for use over a whole continent. Simpler closed formulas can then be used if the Earth is assumed to be a sphere. To extend the ellipsoidal form a greater distance from the central meridian, much more complicated formulas are available.

When USGS converted from the Polyconic projection, with a central meridian on each quadrangle, to the State Plane Coordinate System, this included employing the central meridian and other parameters used for the new zone. The discrepancies in measurements on the Polyconic and Transverse Mercator forms of the same 7½- or 15 min. quadrangle depend especially upon the distance of the quadrangle from the central meridian of the zone. The new quadrangles can be mosaicked for the entire zone.

The 1:250,000-scale 1 × 2 quadrangle series covering 49 of the States was originally to be cast on the UTM. When the Army Map Service prepared these, the UTM central scale factor of 0.9996 was used, but the central meridian of the quadrangle itself was used in place of that of the UTM zone. These central meridians agree with those of the UTM zones for only one-third of the quadrangles. East-west mosaicking within a UTM zone thus cannot be achieved with the existing maps, which are now being updated and distributed by USGS. As these areas are remapped, they are being recast with the UTM central meridians.

Oblique Mercator Projection

A cylinder may be placed around the sphere so that it is tangent along a great circle which is neither a meridian nor the Equator. In this case, the Oblique Mercator may be conceptually projected for conformal mapping of a region chiefly extending along this great circle. Nearly all meridians and parallels are complex curves. Here the lines of constant scale run parallel to the oblique great-circle path, which may have a reduced scale. Its topographic use by USGS is confined to the State Plane Coordinate System of the southeast extension of Alaska and to the mapping of Landsat data from 1978 to the present, prior to implementation of the Space Oblique Mercator projection on Landsat D. There are several ways of adapting the Oblique Mercator to the ellipsoid, although none is ideal because either the central line does not remain at a precisely constant scale or conformality is not precise. Hotine's adaptation, which is exactly conformal, is used in these USGS applications. Figure 21-35 is an Oblique Mercator projection of the World.

Space Oblique Mercator Projection

Among the more complicated projections is the Space Oblique Mercator, conceived by A. P. Colvocoresses of USGS in 1973 and mathematically implemented in 1978. It is intended specifically for the continuous mapping of imaging from satellites such as Landsat for which a rudimentary form was used 1975–1978. The more accurate form is to be used for Landsat D. The groundtrack for the satellite is held true to scale, and mapping is made basically conformal. Because of the relative motion of Earth and satellite, the groundtrack is curved, and appears almost sinusoidal on the map. Formulas have been published in summarized form, and very recently with detailed derivations as USGS Bulletin 1518 (Snyder, 1981). It is designed for use with the ellipsoidal Earth, and with circular or elliptical satellite orbits. Figure 21-36 shows the Space Oblique Mercator projection for a portion of the Northern Hemisphere.

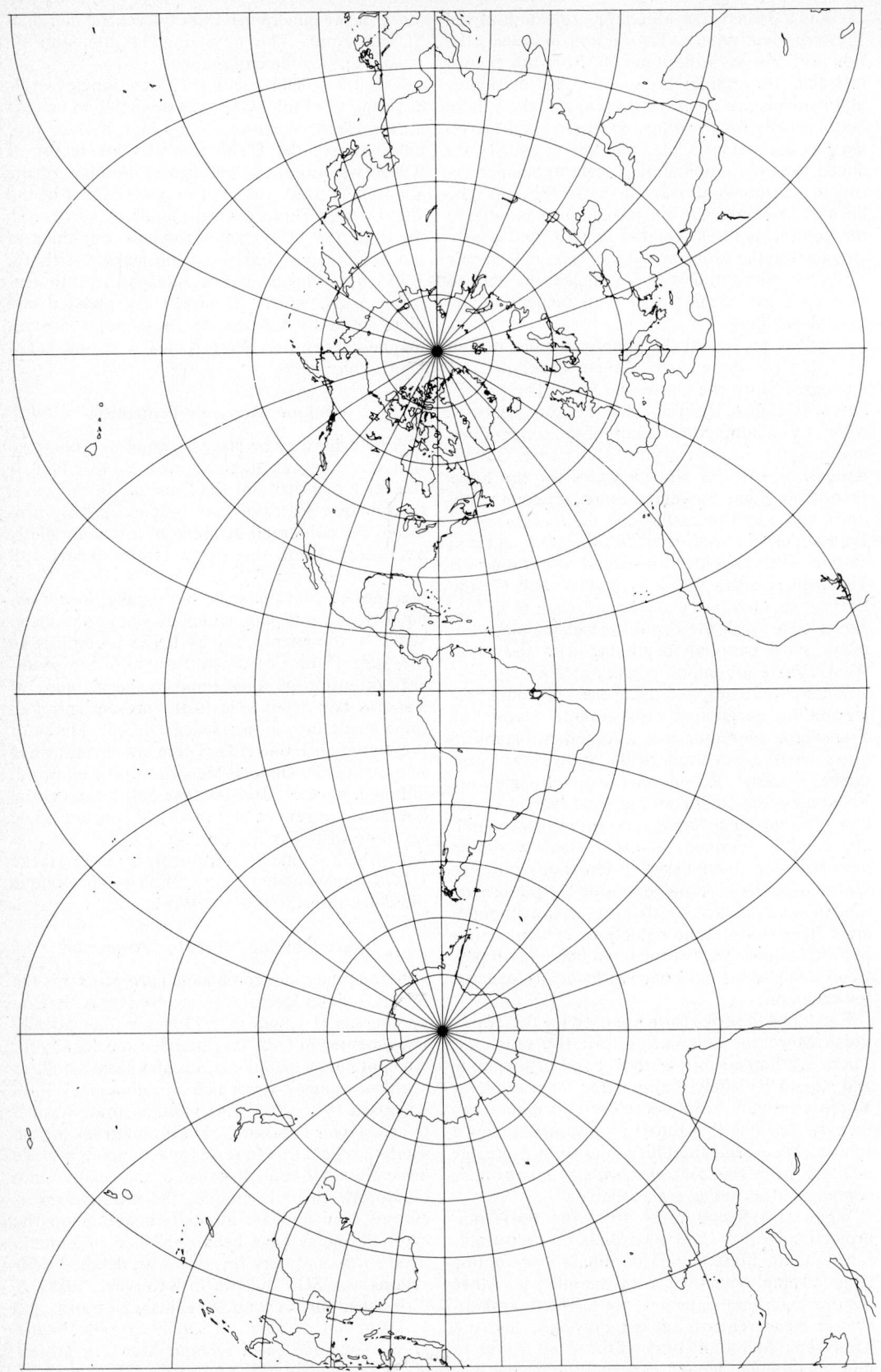

Fig. 21-34. Transverse Mercator projection of the Western Hemisphere.

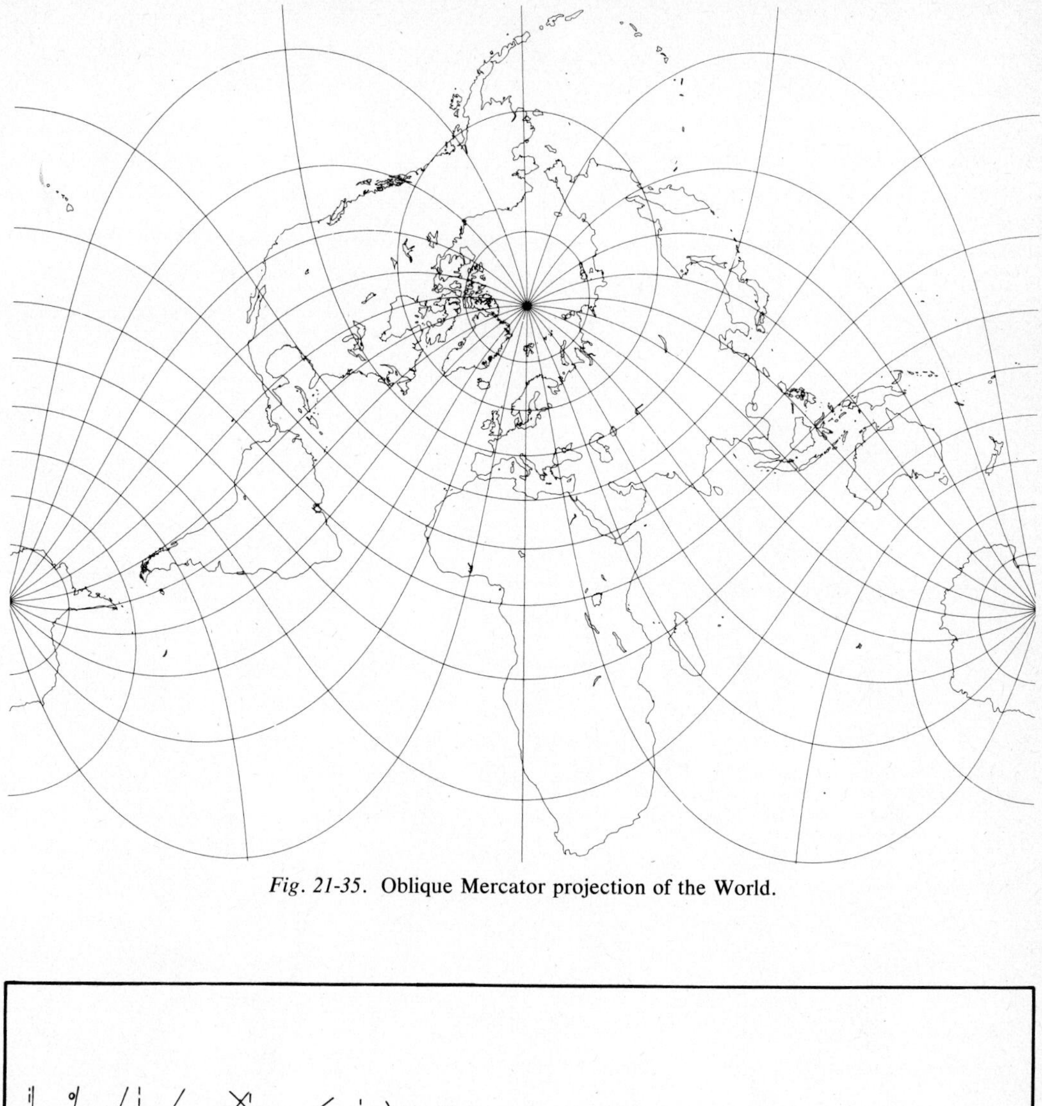

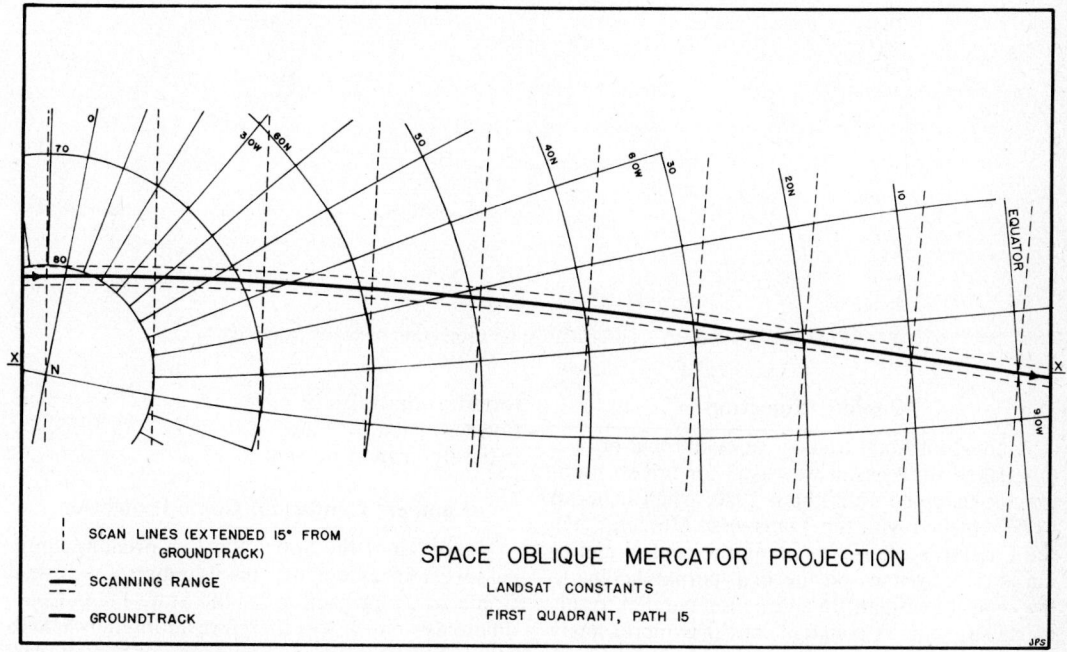

Fig. 21-35. Oblique Mercator projection of the World.

Fig. 21-36. Space Oblique Mercator Projection.

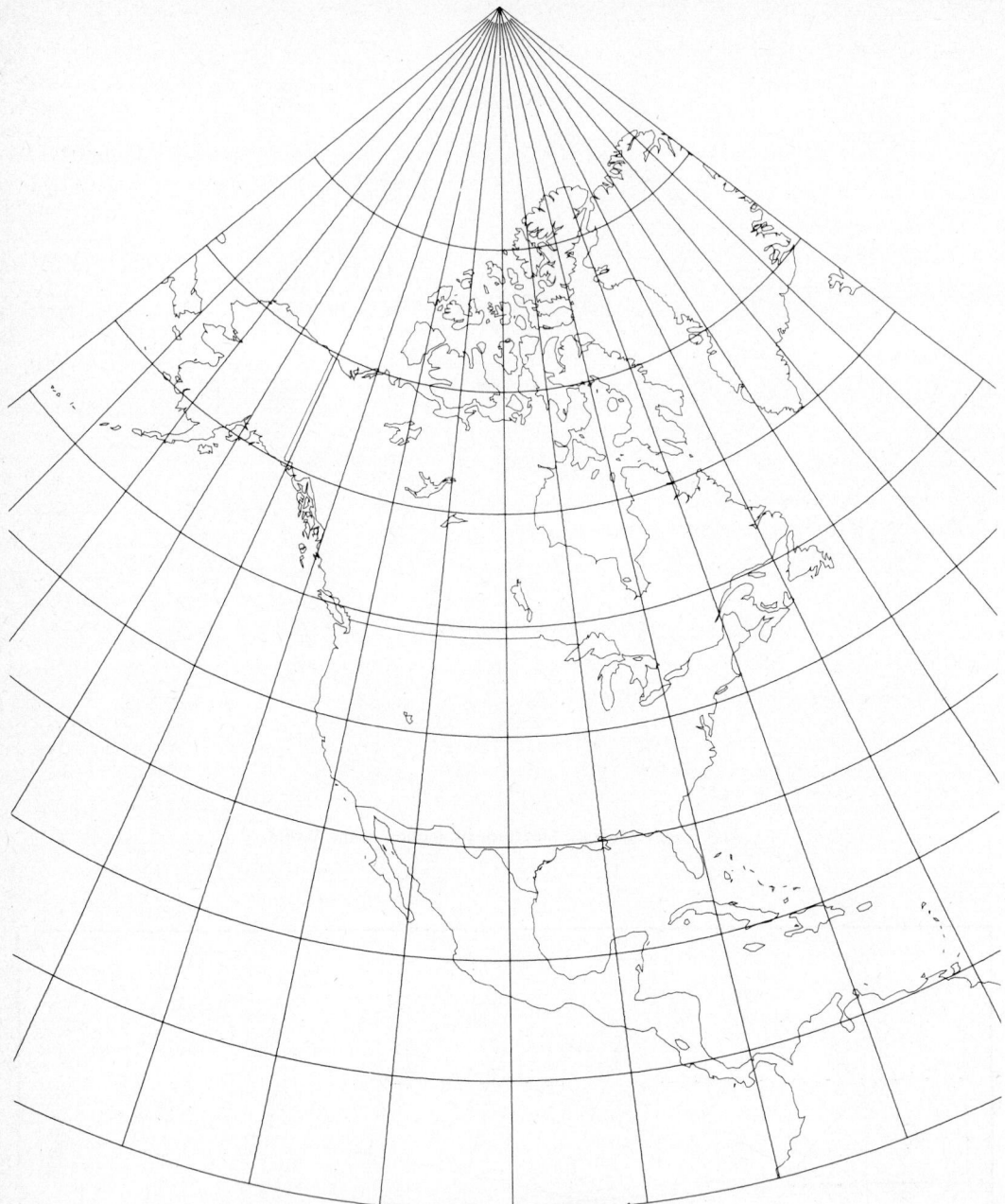

Fig. 21-37. Lambert Conformal Conic projection of North America.

Cassini Projection

A non-conformal transverse cylindrical projection called the Cassini was used for british topographic mapping until about 1920, when it began to be replaced with the Transverse Mercator. On the Cassini, scale is true along the central meridian and along lines on the map perpendicular to the central meridian. In a direction parallel to this meridian, scale is constant, but it is increasingly too large as the distance from the meridian increases. Thus shape, angles, and area are dis-

torted. It may also be case obliquely. The projection was used for Landsat from 1972 to 1975, but is hardly used at present.

Lambert Conformal Conic Projection

Nearly all of the States that are predominantly east-west in extent use the Lambert Conformal Conic as the projection for the State Plane Coordinate System. It was therefore adopted by USGS for post-1950 quadrangle mapping of these areas. This projection, presented by Lambert in 1772, maps parallels as concentric circular arcs and meridians as equally spaced radii of those circles. One pole is at the center of the circles, while the other pole is at infinity. The parallels are more closely spaced between the normally two standard parallels, which have no distortion. Figure 21-37 shows a Lambert Conformal Conic projection of North America.

For the USGS map of the conterminous United States, the standard parallels are 33° and 45° N., and USGS has made base maps of each of the 48 States using these standard parallels at a scale of 1:500,000. The base maps therefore match along the boundaries. Each zone of the State Plane Coordinate System has its own standard parallels. These are the parameters used for USGS quadrangles which have been produced on this projection; thus, quadrangles edge-match within the zone.

The Lambert is also used for the map sheets of the International Map of the World series at a scale of 1:1,000,000. The projection for these sheets was changed from a Modified Polyconic in 1962.

Albers Equal-Area Conic Projection

The Albers Equal-Area Conic is probably seen more than the Lambert Conformal Conic as the projection for maps of the United States and its longer regions. In 1805, Albers showed that by proper spacing of the parallels on a conic projection, there is no area distortion, and along one or two standard parallels there is no scale or angular distortion. The projection was nearly dormant until Oscar S. Adams of USC&GS began encouraging its use for equal-area maps of the United States in the early part of the 20th century.

Adams's tables of coordinates for the 48 States are based upon standard parallels of 29½° and 45½° N. It should be noted that the United States on the Albers projection cannot be distinguished from a Lambert Conformal Conic version if the projection is not identified, except by careful measurements. Like the Lambert, the Albers, which is the equal-area counterpart, has concentric arcs of circles for parallels, and equally spaced radii as meridians. The parallels are not equally spaced, but they are farthest apart in the latitudes between the standard parallels and closer together to the north and south. The pole is not at the center of the circles, but is normally an arc itself.

Scale along any given parallel is constant, as on the Lambert, with scale too small between the standard parallels, and too large beyond them. The scale along the meridians is just the opposite, to maintain equal area. While Adams recommended that standard parallels be placed one-sixth of the displayed length of the central meridian from the northern and southern limits of the map, this is empirical. The standard parallels may be selected in other ways.

Since meridians intersect parallels on the Albers at right angles, it may at first be thought that there is no angular distortion. It exists, however, for any angle other than that between the meridian and parallel, except at the standard parallels. Figure 21-38 is an Albers Equal-Area Conic projection of North America.

Stereographic Projection

For larger-scale maps of polar regions, the Stereographic projection (Fig. 21-39) is commonly used. This is a perspective projection of the

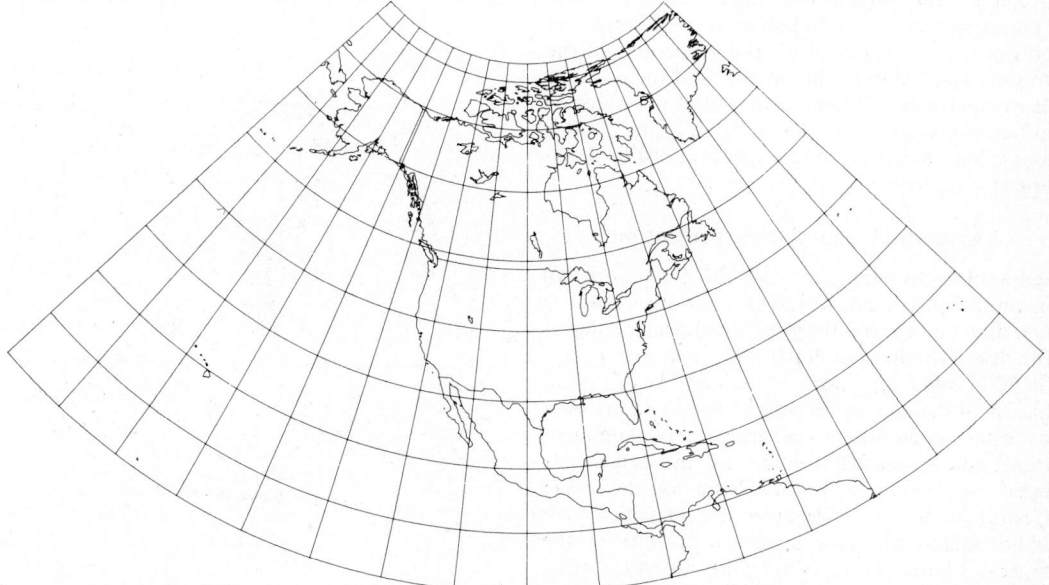

Fig. 21-38. Albers Equal-Area Conic projection of North America.

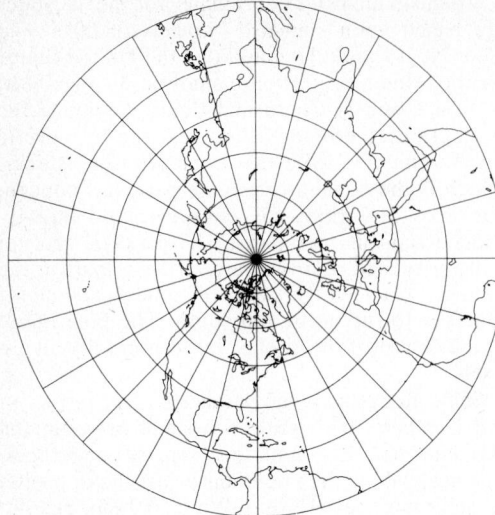

Fig. 21-39. Polar Stereographic projection.

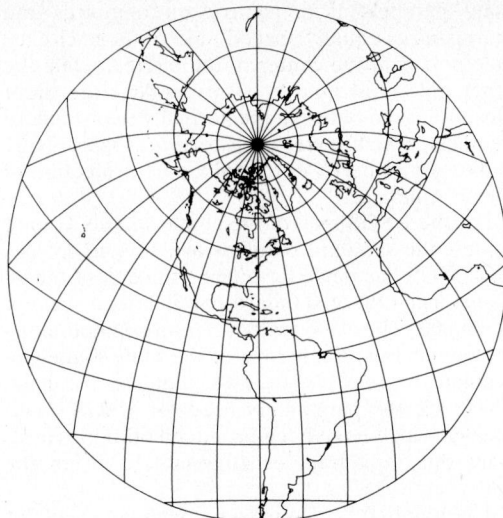

Fig. 21-40. Oblique Stereographic projection of the Western Hemisphere.

sphere onto a tangent or secant plane. The point of perspective lies on the opposite side of the globe. For the sphere, the Stereographic is both perspective and conformal. For the ellipsoid, it may be one or the other, but not quite both; in practice, it is used conformally in every case. All great and small circles on the sphere are shown as circles or straight lines on the map. This includes all meridians and parallels.

The Stereographic is also azimuthal, and lines of constant scale are circles centered on the projection center. In the polar aspect, a parallel other than the pole is therefore often made true to scale to balance scale variation. The USGS has used the projection for maps of the entire Antarctic continent as well as for 1:250,000-scale quadrangles of portions. It is also used for polar portions of the International Map of the World, and for portions of extraterrestrial bodies centered at the poles or at basins elsewhere on the bodies. Figure 21-40 is an Oblique Stereographic projection of the Western Hemisphere.

Azimuthal Equidistant Projection

Familiar to many air-age atlas users is an azimuthal projection which shows both distances and directions correctly from the chosen center of the map, whether the North Pole or a major city. Scale in other directions varies, and is too great except at the center; therefore shape and area are distorted. In addition to several spherical applications, the Azimuthal Equidistant projection has been used in the ellipsoidal form for the Plane Coordinate System of Micronesia, for which each major island provides a center for one of the zones. Figure 21-41 is an Oblique Azimuthal Equidistant Projection of the World.

Because of the increasing digitization of data

from and onto many different maps, most of these as well as several other map projections have been programmed by USGS in both forward and inverse form, to transform geodetic coordinates into rectangular coordinates, and vice versa. A bulletin describing in detail all projections which have been used by USGS, including numerical examples of formulas, has recently been published.

IMAGE MANIPULATION

Reformatting

The Landsat MSS processed data are made available to the user community in a variety of formats. They are called: band-sequential,

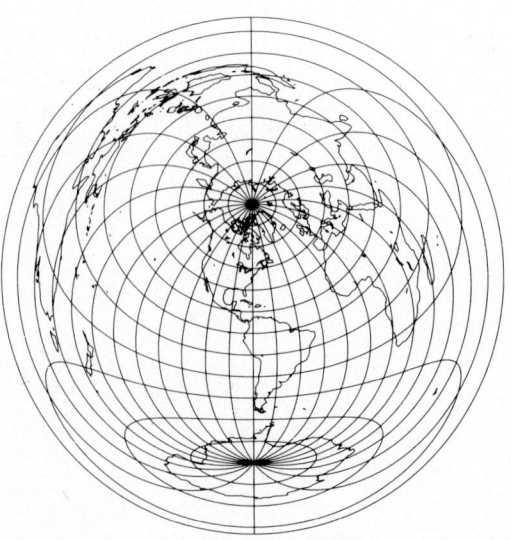

Fig. 21-41. Oblique Azimuthal Equidistant projection of the World.

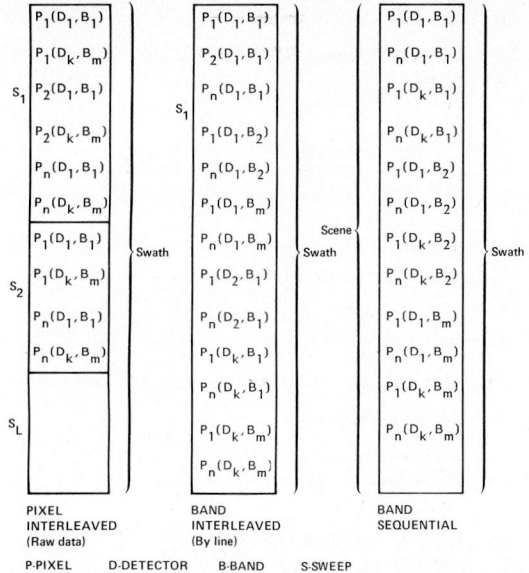

Fig. 21-42. Image Data Formats (Peavey, 1977).

band-interleaved-by-line, and pixel-interleaved. Figure 21-42 describes these formats (Peavey, 1977). The advantage of the band-sequential format is that the data are in a form that is directly compatible with a film recorder. No processing or formatting of the data is required to record the multiple bands of the image. The disadvantage is that if a subimage is to be displayed, several full-frame tapes may have to be mounted and searched.

The advantage of the band-interleaved-by-line format is that all bands of the image are separated by only a few records on the tape, the separation being a function of the number of bands. Thus, once the area of interest has been found on the tape, all bands of that area are nearby, and no additional tapes need be mounted.

The pixel-interleaved format provides pixels in sequence. Thus a cluster of pixels, each representing a different spectral band, are co-located on the tape.

Framing

A scanner will provide a contiguous swath of image data while it is switched on. Thus the dimension of the data is the scanner field of view in the cross-track direction, and the length of the data set will be proportional to the time that the sensor is on. In the case of the Landsat data, the data are "framed" on the ground into scenes, each being approximately 202 by 170 km in size. There is an intentional overlap of data in the vertical direction to aid in photo-mosaicking.

Cloud Cover Evaluation

Operational image-processing systems provide an estimate of the amount of cloud cover that is contained in the scene. This is a useful data set for the user community. When the user queries the data base, he can determine whether the area of interest is likely to contain too much cloud cover for his purposes, although the cloud-cover statistics are generally provided in percent cloud cover for the entire scene.

Quality Evaluation

A number of tests have been made to the data to determine whether the data meet quality standards (USGS, 1979). Table 21-2 provides data-performance objectives for the Landsat data. The MDP and the photographic-image recording and processing systems have been designed to meet these objectives. Both digital data and photographic data are inspected periodically, and equipment tested and calibrated.

SYSTEMS TO IMPLEMENT IMAGE PROCESSING

FLOW OF DATA FROM SENSOR TO USERS

The data acquired from current satellites are transmitted to ground, evaluated, processed, and distributed. Figure 21-43 summarizes this data flow, and the processing operations that are performed. Although image-geometry correction is performed in the NASA ground processors, it should be noted that users sometimes perform their own geometric and radiometric processing in addition to information-extraction processing, in order to merge data acquired from different sources and to combine the data with map and chart products.

This section will describe the systems used to process the data. These systems include the NASA Master Data Processor, the Eros Data Center image-processing system, and a typical user system.

MULTISPECTRAL SCANNER PREPROCESSOR (MPP)

The MSS Preprocessor (MPP) accepts the data tapes acquired from the ground receiving stations and reformats the data onto a high-density digital tape. In addition, cloud-cover assessment, data-error indications, quality assessment, and data editing are performed. A data report is automatically prepared. Figure 21-44 summarizes the input and output, and the operations that are performed in the MPP. The MPP output serves as the input to the Master Data Processor.

NASA MASTER DATA PROCESSOR

NASA Requirements

As a result of the programmatic decision in 1975 to process all future space data using digital image-processing technology, NASA began development of the first operational, high through-

TABLE 21-2

Image Processing Performance Objectives for Landsat MSS and RBV Processing

Function/Operation	Performance Objectives
Input Data Quantity	13 Reels of wide band video tape per day • 9 MSS reels; equivalent of 200 scenes • 4 RBV reels; equivalent of 160 subscenes
Radiometric Correction Error— (a) relative (interdetector) (b) degradation	 <2 Quantum level (over full range) <2 Quantum level (over full range)
Geometric Correction Error— (a) nominal conditions with Ground Control Points (GCP) (b) without GCP	 <1 pixel (90% of the time) commensurate with spacecraft, sensor, and orbit data
Temporal Registration Error	<0.5 pixel (90% of the time)
Map Projections	Hotline Oblique Mercator (HOM) Universal Transverse Mercator (UTM) Polar Stereographic (PS)
Resampling	Hotline Oblique Mercator (HOM) Cubic Convolution, Nearest Neighbor
Tick Mark Error	<2 pixel/tick mark (90% of time)
Processing Thruput	up to 10^{11} input bits/day (16 hours)
Output Data Media	High Density Tape, Computer Compatible Tape
Interleaving of Output Imagery	Band Sequential (BSQ) Band Interleaved by Line (BIL) Band Interleaved by Pixel (BIP)
MSS Output Array Size of fully processed data	3548 pixels/line and 2983 lines/array for image data, representing 202.2 km by 170 km with each resampled pixel representing a 57 × 57 meter area (fill pixels included)
RBV Output Array Size of fully processed data	5322 pixels/line and 5322 lines/frame for image data, representing 101.1 km by 101.1 km with each resampled pixel representing a 19 × 19 meter area
Output Overlap (Alongtrack)	≤5% of frame (MSS only)
Radiometric Quantization	7 bits/pixel for HDT's 8 bits/pixel for CCT's (7-bit resolution)

put, digital image processing facility. This system, called the Master Data Processor (MDP) was begun in 1976 and completed in 1979.

The system is being used to process both Landsat MSS and RBV sensor data (see Chapters 12, 13 and 23) and Heat Capacity Mapping Mission (HCMM) sensor data to remove geometric and radiometric distortions, convert the data into one of several cartographic projections, and format the data into a suitable form for distribution to other government agencies and users. Thus, digital image-processing has replaced electro-optical and analog image-processing technology. High-density digital tapes now serve as the archival medium, replacing film as the master archive. The algorithms used are based upon the results of the previous Landsat investigations (Bernstein, 1975, Rifman, 1976).

Application Processing on the MDP

Major image-processing functions (see Figure 21-45) performed by the MDP for the multispec-tral scanner processing include (Bernstein & Schoene, 1976, Schoene, 1977, Niblack, 1981, and USGS, 1979):

• Demultiplexing data by spectral band and staging mission parameters to disk.
• Radiometric calibration computation on an image line and spectral band basis.
• Geodetic or reference control-point collection and correlation with incoming image data.
• Disk buffering of image and intermediate control data for later application and use.
• Geometric correction computations based on results from use of a geodetic or reference control-point control location algorithm.
• Resampling, which includes radiometric and geometric interpolation based on pixel location and intensity.
• Output of a partially corrected (radiometric correction only) or fully corrected scene and mission parameters to tape.

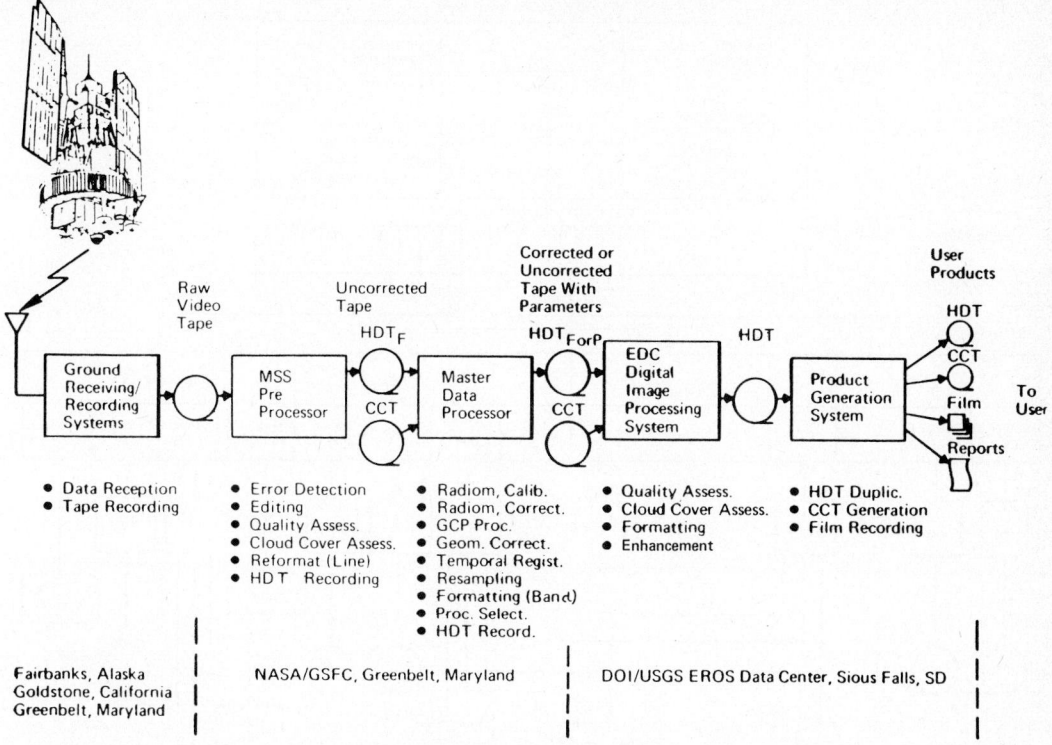

Fig. 21-43. Summary of Data Acquisition, Processing, and Dissemination.

- Output of image processing quality and quantity data and mission parameters to tape.

MDP Software

MDP software accommodates the real-time needs of a continuously running input and output tape. MSS software includes nine dispatching tasks, six of which can be working concurrently on different scenes as they are passed through the various stages of the MSS-application processing-pipeline (input, radiometric calibration, control location algorithm, geometric, resampling, and output).

About 50,000 lines of MSS code, using 30,000 words of program store buffers and tables and 250,000 words of bulk store buffers and tables, were developed to support the MSS application. MSS-application software is a multi-processing system, developed in assembly language and microcode, designed to efficiently handle the high data rates needed to support the operational requirement.

MDP Architecture

The Master Data Processor architecture, shown in Figure 21-46, is a digital image data-processing system based upon the architecture of the Advanced Signal Processor (ASP), an outgrowth of the signal-processing development performed by

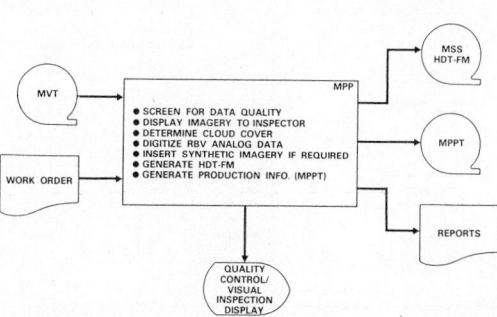

Fig. 21-44. Multispectral Scanner Preprocessor (MPP) Functions.

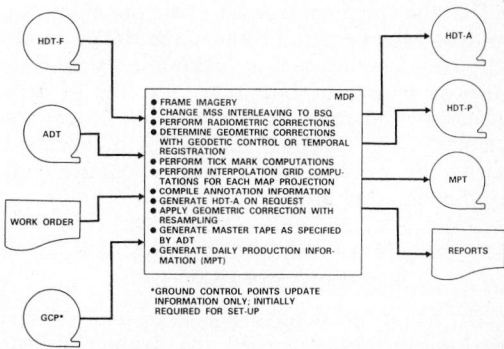

Fig. 21-45. Major Image Processing Functions Performed by the MDP.

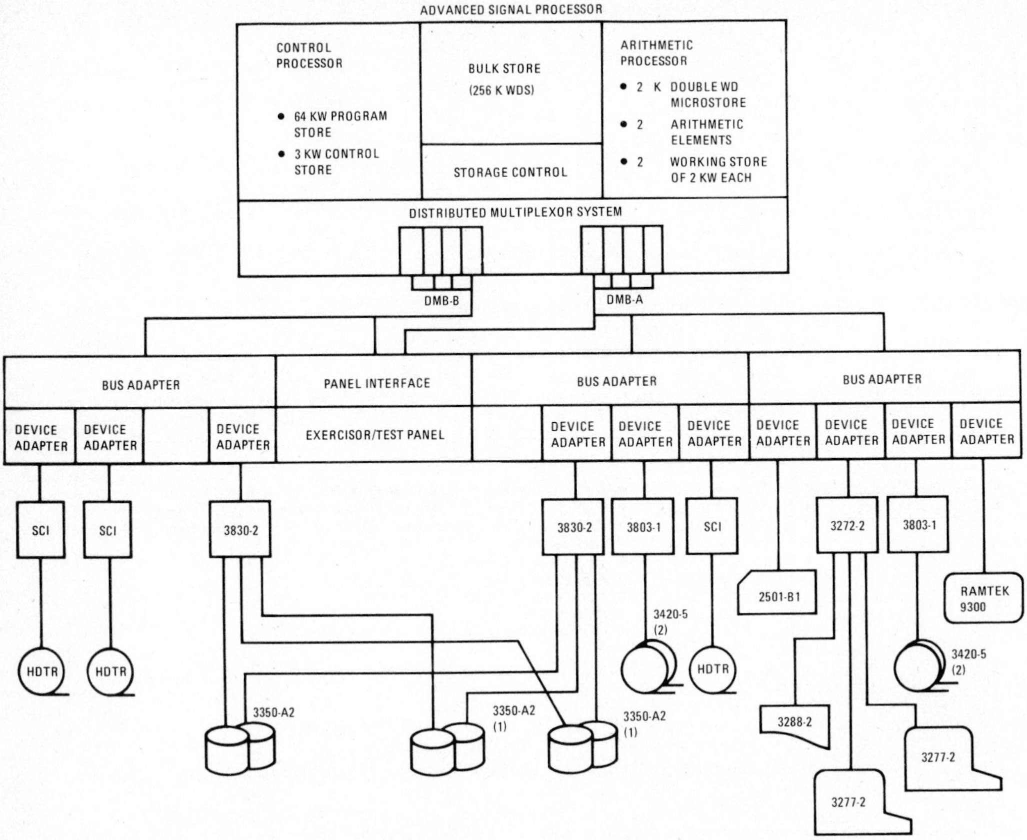

Fig. 21-46. Master Data Processor (MDP) Configuration.

IBM for the U.S. Navy. The ASP consists of multiple processors—a control processor and an arithmetic processor and a 1-million-byte bulk store. The system input is from high-density digital tapes. The uncorrected data are processed and corrected by the MDP and produce an output tape. The corrected digital-tape drives a laser beam film recorded that produces large format films and prints. The digital tapes are also copied onto standard computer tapes for further digital processing and information extraction by the scientific user-community.

The control processor has two functions: to control all operations and execute general-purpose programs, and to function as a general-purpose processor that serves as the system supervisor, manages data, and handles a certain amount of general computation. The microcoded arithmetic processor contains two pipelined fixed-point array processors.

Specifics on the Advanced Signal Processor (ASP)

The major functional elements of the ASP are the Control Processor (CP), the Arithmetic Processor (AP), Bulk Storage (BS), the Storage Controller (SC), and the Distributed Multiplexer System (DMS). Figure 21-47 is a system block diagram of the ASP.

Control Processor: The Control Processor (CP) is a microprogrammed 32-bit general purpose computer. The Control Processor contains an arithmetic and logic unit, 32K words of Program Store, and a writable Control Store that holds the Control Processor microprograms. The Control Processor performs ASP system supervisory and data-management functions.

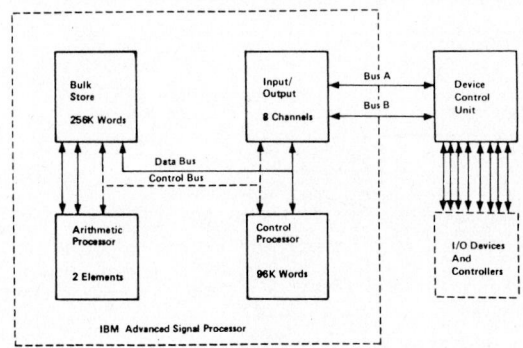

Fig. 21-47. Master Data Processor (MDP) Advanced Signal Processor.

Arithmetic Processor: The Arithmetic Processor (AP) performs the bulk of the data-processing functions. The Arithmetic Processor consists of two Arithmetic Elements (AE), two Working Stores (WS), a Micro Store, and an Arithmetic Element Controller (AEC). Each Arithmetic Element contains a pipelined multiplier followed by a three-way adder providing 10×10^6 multiples per second and 20×10^6 adds per second. Each input to the adder and the adder output may be independently shifted, providing the scaling capabilities required to maintain precision or prevent overflow in the computations. Logical capabilities are also provided.

Storage Subsystem: The Bulk Store is the primary ASP storage-medium, providing storage for data buffers, temporary buffers, coefficients, and program parameters. The Bulk Store is available in sizes from 16K to 512K words in 16K increments. The proposed configuration will provide 128K of 36 bit words. The Bulk Store operates at an 800-ns cycle time with arrays interleaved to achieve an effective 200-nonosecond (ns) double word rate.

The Storage Controller is designed to transfer data between Bulk Store and other ASP elements. The Storage Controller also provides the Bulk Store error-correction encoding and decoding functions and the priority-control function.

Distributed Multiplexer System: The I/O channels are 32-bit parallel, two-way digital-data channels. The Distributed Multiplexer System contains two Distributed Multiplexer Buses. Each Distributed Multiplexer Bus can provide a mix of selector channels and multiplexer channels, which can, by microcode, be subdivided into as many as 16 subchannels. The configuration proposed includes one Distributed Multiplexer Bus with four selector channels and one Distributor Multiplexer Bus with three selectors and one subchanneled multiplexer channel.

Each Distributed Multiplexer Bus is a nondedicated time-slotted bus which provides data-transfer rates of up to 2.5 Mword/sec for Bulk Store and 1.25 Mword/sec for Program Store. The nondedicated time-slotted bus allows concurrent operations of multiple devices attached to channels on the Distributed Multiplexer Bus.

MSS Processing Flow

After the scanner data are framed, the MDP changes the data format, performs the geometric and radiometric corrections, grids the data, adds annotation data, and generates a corrected high-density digital tape (HDT). The processing flow for MSS data is shown on Figure 21-48. It is noted that the processing flow is under the control of the

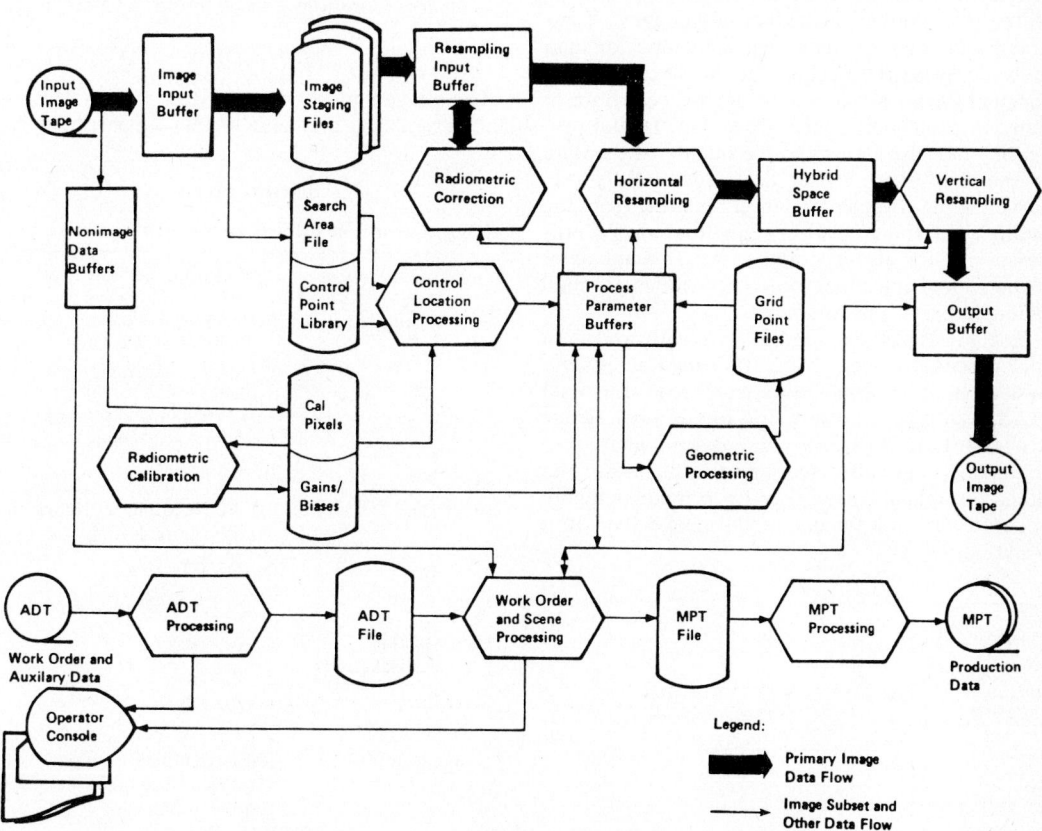

Fig. 21-48. Master Data Processor (MDP) Full Processing Flow.

operator and that the work order tape, as well as various projections and processing options, can be selected based upon the requests received from the users. All processing is performed digitally, and the data input and output are currently from high-density digital tapes.

Film Recording of the Processed Digital Data

Figure 21-49 shows the data and control of the Laser Beam Recorder (LBR) used to record the digitally processed image data onto film. The High Density Tapes (HDT) on which the data have been recorded are used to control the intensity of the laser signal while it is exposing the film. A special tape controls the LBR to compensate for differing film types, and the work-order controls the flow of image recording.

The performance characteristics of the film recorder are shown in Table 21-3. Of particular importance is the large format of the film products (206.2 mm × 202.2 mm) and the high recording speed of the device (60 to 350 lines per second). LBR's are now in common use for operational production image-recording due to their high recording energy and their excellent geometric fidelity.

Summary of MDP Experience

The transition to a digital technology has resulted in a number of distinct advantages. It has drastically reduced the turnaround time for data delivery, primarily because of the flexibility and speed of the processors that, together, can correct more than 600 earth scenes per day. Digital processing has also improved the quality of the data to the user, since sensor data can be fully preserved without the degradation resulting from the multiple filtering effects of sequential analog processes. It has also become easier to implement changes and introduce new capabilities through modification of the software.

The MDP has provided NASA with a system that processes earth observation data accurately and at high speed. This system was the first NASA all-digital system for operational image processing. It is currently aiding geologists, cartographers, agriculturists, land use planners, and other disciplinary scientists by providing highly accurate and full-fidelity digital-image data in a timely fashion.

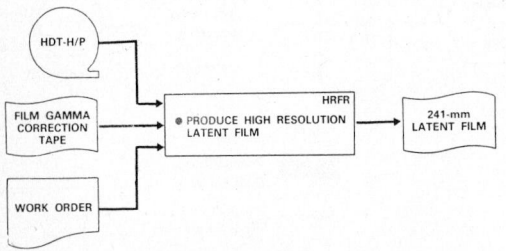

Fig. 21-49. High Resolution Laser Film Recorder.

LANDSAT-D PROCESSING CONSIDERATIONS

The realities of the lessons learned from the all-digital processing system of Landsat 2 and 3 have been incorporated into the Landsat-D ground segment design. The Landsat-D Ground Processing System will generate all the geometric correction matrices that are appended to the partially processed data transmitted to the EROS Data Center. The Landsat-D mission design has been constrained to be as transparent as possible to MSS data users. However, image framing and coverage have been modified, and both geometric accuracy and the radiometric correction accuracy have been improved.

The lower orbit of Landsat-D provides either a 7 or 9 day adjacent swath overlap of imagery. In addition, the center of each scene is identified by unique latitude and longitude points in a reference system that is different from earlier Landsats. Because of the lower orbit of Landsat-D, larger mirror scan angles and mirror velocities can potentially produce greater scan nonlinearities. However, the spacecraft has better attitude control and measurement and hence better long term pointing and stability. The Landsat-D radiometric calibration accuracy is much improved over the techniques used for earlier Landsat data processing.

A more detailed description of Landsat-D processing considerations can be found in Chapter 12.

USGS EDIPS

The image data produced by the NASA Master Data Processor are sent to the Department of

TABLE 21-3

High Resolution Film Recorder Characteristics

Function/Operation	Performance Objectives
Film Writing Area	Frame Width 206.2 mm (active scan width) 202.2 mm (Image Frame Length—Variable, depends on input array size. HRFR can also run in a continuous strip mode.
Spot Size	10 to 200 μm
Input Data Rate	0.2 to 20 Mbits/sec (continuously variable)
Line Rate	60 to 350 lines/sec
MTF	>50% for 20,000 pixels per line
Line Spacing (Transport Jitter)	<1 μm rms
Scan Jitter Start of Scan	<5 μm rms
Dynamic Range	100:1 (modulator output) better than 2.00 on Kodak 2460 or 3414 film
Grayscale	>64 distinguishable steps

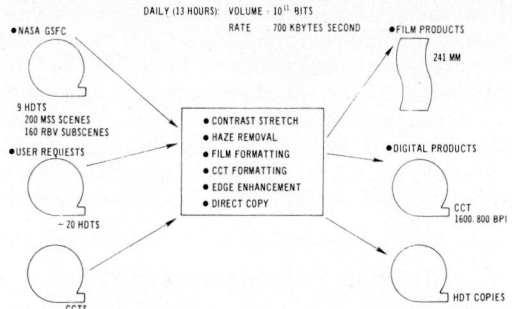

DAILY (13 HOURS): VOLUME : 10^{11} BITS
RATE : 700 KBYTES SECOND

Fig. 21-50. Summary of EROS Data Center Image Processing Capabilities.

Interior (DOI) Earth Resources Observation System (EROS) Data Center (EDC) at Sioux Falls, SD. The NASA data are mailed or transmitted via communications satellite to the EDC. The data are further processed at the EDC Digital Image Processing System (EDIPS). The EDC processes the user requests, and provides standard or special film and digital tape products.

Figure 21-50 provides a summary of the processing capability of the EDC, and the products that it produces. Table 21-4 summarizes the EDIPS reformatting, processing, and copying capabilities. It is noted that a wide variety of for-

mats and processing options are available to the user.

Figure 21-51 is a system configuration of the EDIPS. It is noted that the system has an on-line laser film-recorder, and an attached array processor to support the image-processing and product-recording operations at the EDC. The computer-compatible tapes and film products that are made available to the user community are all processed through this system.

A GENERALIZED DIGITAL IMAGE PROCESSING SYSTEM

A generalized user digital-image processing system, representative of existing systems, is shown on Figure 21-52. The elements of the system are summarized.

System Inputs: Image-data sources consist of cameras, scanners, radar, maps and data from any device that generates two-dimensional data. Line graphic manual entry devices, such as graphic tablets or light-pen display systems, provide vector input-data. These data must be suitably converted prior to entry into the processor. Conventional computer input/output media such as magnetic tape, cartridge, and cards are used for sensor data and support data entry. Operator control and communications are implemented by the use of a keyboard terminal with which the operator is able

TABLE 4

MODE	INPUT MEDIUM	DATA TYPE	INPUT FORMAT	CORRECTION STATUS	PROCESSING	OUTPUT FORMAT	CORRECTION STATUS	LAYOUT
Standard Product	HDT-Ps	(1)	(1)	(1)	(1)	(1)	(1)	(1)
Special Order	HDT-PM	MSS	BSQ ● 5-band scenes, or ● 4-band scenes, or ● single band -8 scenes	Fully Corrected	Enhancements ● Contrast Stretch, and/or ● Haze Removal, and/or ● Edge Enhancement, or ● None	BSQ, or BIL	Fully Corrected	9-track CCTs ● 800 bpi NRZ, or ● 1600 bpi PE
	HDT-PR	RBV	SSQ	Fully Corrected	Enhancements (same as above)	SSQ	Fully Corrected	(same as above)
	HDT-AM	MSS	BSQ ● 5-band scenes, or ● 4-band scenes, or ● single band -8 scenes	Uncorrected except for detector bias/gain correction	Enhancements None	BSQ, or BIL	Uncorrected except for detector bias/gain correction	(same as above)
	HDT-AR	RBV	SSQ	Uncorrected except for shading error correction	Enhancements None	SSQ	Uncorrected except for shading error correction	(same as above)
Special Product	CCTs(2)	MSS	BSQ, or BIL	N/A	Direct copy only	BSQ, or BIL	N/A	(same as above)
				Fully enhanced and/or corrected	Film formatting	BSQ	Fully enhanced and/or corrected	241mm film
		RBV	SSQ	N/A	Direct copy only	SSQ	N/A	9-track CCTs ● 800 bpi NRZ, or ● 1600 bpi PE
				Fully enhanced and/or corrected	Film formatting	SSQ	Fully enhanced and/or corrected	241mm film
	HDT-As, or HDT-Ps	MSS	BSQ	N/A	Direct copy only	BSQ	N/A	14-track HDTs ● 20,000 bpi NRZ ● IRIG-A time track
		RBV	SSQ	N/A	Direct copy only	SSQ	N/A	(same as above)

NOTES: (1) Film output only from standard product mode. No CCT products. (2) No reformatting or reinterleaving. Can be 800 bpi or 1600 bpi.

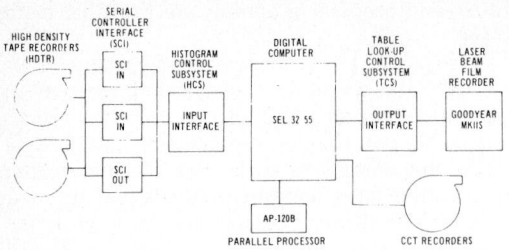

Fig. 21-51. EROS Data Center Image Processing System.

to request the data that are to be processed, the application program that is to be applied against the data, and the output device with which the data are to be displayed or recorded.

Input conversion devices: Image data are entered into the digital processor through suitable conversion equipment. Image data in a human-readable form must be digitized prior to processing. This is accomplished by the use of an image scanner/digitizer that converts the spatial-intensity data, such as camera or map data, into digital form. The data can ba entered into the processor directly, or stored on computer-compatible tape (CCT). Non-camera sensor data, such as scanner data, are commonly recorded directly on magnetic tape, which can then be entered into the processor via tape drives. Analog data are converted into digital data by the use of high speed analog-to-digital converters. The converted sensor data are stored temporarily on tape, disk or drum prior to digital processing.

Central processor: The digital processor provides the image-processing computational and control capability. The processor can be general-purpose, special-purpose, or a combination of general and special-purpose hardware, depending upon the application and throughput requirements. Array processors are used where additional CPU power is needed and the I/O is fast enough to support the additional processing speed.

Output conversion devices: Conventional computer peripheral devices, such as tape drives, printers, displays, and line-graphic plotters are used as image output devices. To convert digital image data to a visual form, a display such as a cathode ray or television tube can be used to provide a temporary image presentation. A film recorder can provide a permanent high-resolution record of the digital data. Four types of film recorder are commonly used: drum recorders, electron beam recorders, cathode ray tubes and laser beam recorders.

FUTURE TECHNOLOGY DEVELOPMENTS FOR IMPROVED GEOMETRIC CORRECTION

Three new developments are likely to be of importance for improving image-geometry correc-

tion, the first by improving the accuracy of location of satellite platforms, the second by speeding the data flow from satellite to users, and the third by improving the speed of computing of geometric-correction algorithms. The three associated developments in hardware are the U.S. Global Positioning System the Tracking and Data Relay Satellite, and the Massively Parallel Processor, respectively.

GLOBAL POSITIONING SYSTEM

The U.S. Government is currently implementing a navigation and positioning satellite system called the Global Positioning System (GPS) that will improve the accuracy with which the position of sensor platforms is known. Under ideal conditions, the platform position can be determined to an accuracy of 10 to 20 meters. The satellite network consists of 18 to 24 orbiting satellites, each carrying a highly stable clock. Periodically, ground processors update the clock information and accurately compute the satellite ephemeris. With suitable programs, a user can compute position and velocity by measurement of the Doppler shift and clock-timing differentials.

It is anticipated that this system, when fully operational in the 1980's, will provide accurate position information needed to support image correction and processing, and may reduce dependency upon GCPs.

TRACKING AND DATA RELAY SATELLITE

The NASA Tracking and Data Relay Satellite (TDRSS) is a set of orbiting communications satellites. They will be capable of relaying data from sensor satellites to ground at data rates of 100 million bits per second. The advantages of this advanced system are: 1) minimization of the number of ground stations needed to receive data, 2) elimination of on-board tape recorders, and 3) reduction in the delay between the time when data are acquired and the time when the data are provided to the users. It is anticipated that the system will eventually feed data directly into the ground processing systems.

MASSIVELY PARALLEL PROCESSOR

Conventional computers are organized as serial data processors. Image data are by nature parallel. With the advent of sensors capable of generating data at the rate of 100 million bits per second, computer scientists have investigated the development of special-purpose digital computers capable of very high speeds. A particularly interesting system now under development is the NASA Massively Parallel Processor (Schaefer, 1979). This computer is organized with a 128 × 128 array of micro-processors that simultaneously processes a 128 × 128 array of pixels. It will be capable of performing image addition, differenc-

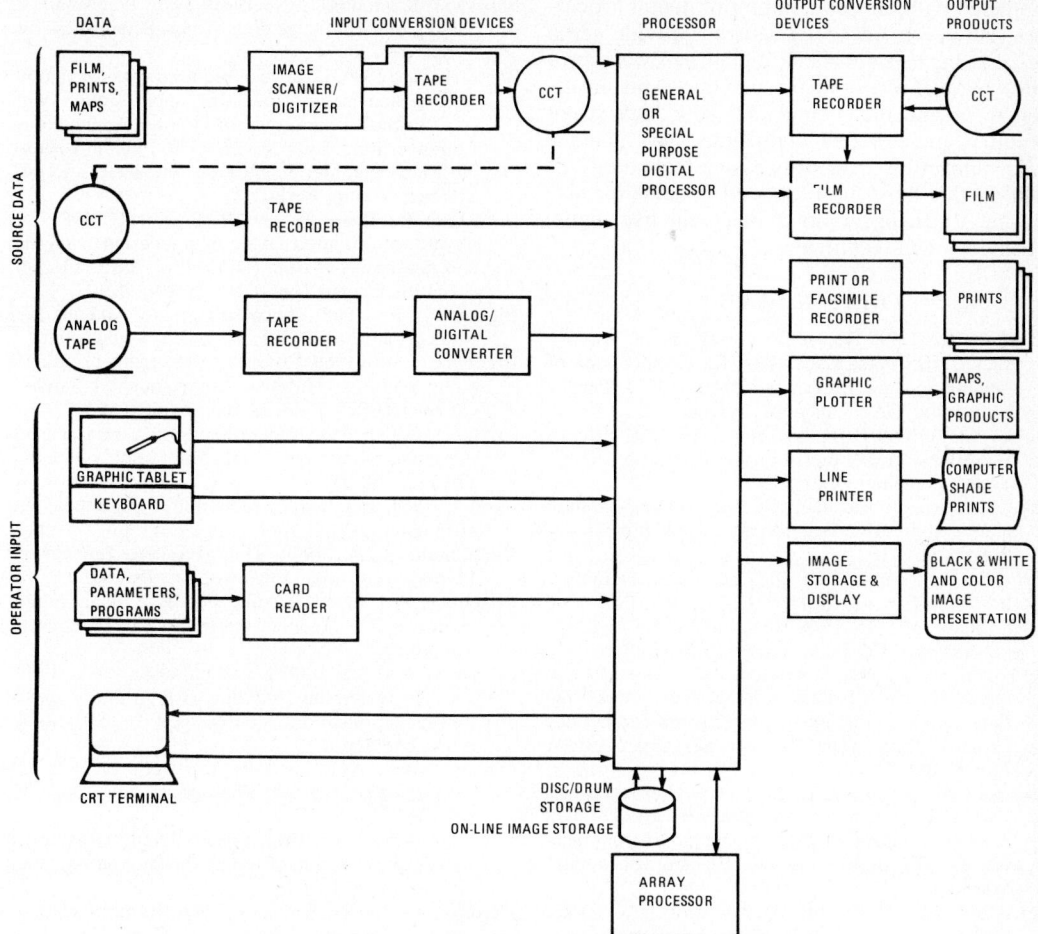

Fig. 21-52. Generalized Diagram of a User Image Processing System.

ing, enhancement, and information-extraction operations. The processing speed is expected to be several billion instructions per second. (See also Chapter 10).

CONCLUSIONS

Several conclusions can be drawn on the subject of digital-image geometry and correction:

Utility: Because of the digital nature of advanced sensors, and the complexity of information extraction techniques, it is advantageous to correct sensor data or change the data geometry using digital processing rather than electro-optical techniques. The corrected or modified sensor data can be generated with no data degradation, resulting in an improved information-extraction potential.

Accuracy: Digital correction of digital-sensor data results in geometric errors of less than one picture element and in full preservation of sensor radiometry. No radiometric error or loss is introduced (as occurs in electro-optical processing with its data conversions and multiple photographic generations).

Throughput: Production image-processing systems can be configured to provide high-speed processing of Earth observation data. Microprogrammed signal processors can be used to improve the cost-performance ratio relative to conventional general-purpose computer configurations where high-throughput operational systems are required.

Flexibility: Digital processing provides a significant degree of processing and operational flexibility. Correction, enhancement, mosaicking, and information extraction can all be done with general-purpose digital hardware by selection of software image-processing routines.

Cost: With the rapid decrease in digital technology costs, and the concurrent increase in performance, the use of digital computers has dramatically accelerated in the last few years and has been adopted by a large number of universities, government, and industry investigators and users.

Feasible Technology: In the past, the use of digital technology and, in particular, digital computers for image processing, was considered impractical because the parallel nature of the data conflicted with the serial nature of digital processors. However, advanced sensors provide serial digital data, and advanced digital hardware provides parallel-processing capability. This inversion of organization, combined with high speed circuitry and efficient algorithms, has made a major impact on image-processing technology. It is likely that most future ground systems for processing Earth observation data will use digital technology and techniques.

REFERENCES

Anuta, P. A., 1973, Geometric correction of ERTS-1 digital MSS data; Laboratory for Applications of Remote Sensing, Information Note 103073, Purdue University, West Lafayette, Indiana.

Barnea, D. T., and H. F. Silverman, 1972, A class of algorithms for fast digital image registration; IEEE Trans. Computers c-21, 179.

Bender, L., 1970, An algorithm for gridding satellite photographs; OSU Department of Geodetic Science Report, no. 135.

Bernstein, R., Ferneyhough, D. G., Gregg, L., Highley, R., Markarian, H., Miklos, J., Mooney, P. P., and F. Sharp, May 18, 1970, Experimental ERTS image processing; IBM Tech. Report CESC-70-0465.

Bernstein, R., and H. Silverman, 1971, Digital techniques for earth resources image data processing; Proc. of AIAA 8th Annual Meeting and Technology Display, Washington, DC, vol. c21, no. 2, AIAA 71-978, New York.

Bernstein, R., Branning, H., and D. G. Ferneyhough, 1971, Geometric and radiometric correction of high resolution images by digital image processing techniques; IEEE Intl. Geosci. Electronics Symp., Washington, DC.

Bernstein, R., Branning, H. F., and D. G. Ferneyhough, March 2, 1971, Photo geometric correction; Final Report to U.S. Army Engineer Topographic Lab., Ft. Belvoir, VA, 22060. IBM FSC 71-0063.

Bernstein, R., March 1973, Results of precision processing (scene correction of ERTS-1 images) using digital image processing techniques; Proc. Symp. on the Earth Resources Technology Satellite-1, vol. I-B, NASA SP-327, NASA Goddard Space Flight Center, Greenbelt, MD.

Bernstein, R., December 1973, Scene correction (precision processing) of ERTS sensor data using digital image processing techniques; Proc. of Third Earth Resources Technology Satellite-1 Symp., vol. I-B, NASA SP-351, NASA Goddard Space Flight Center, Greenbelt, MD.

Bernstein R., April 1975, All-digital precision processing of ERTS images; IBM Final Report on Contract NAS5-21716, NASA Goddard Space Flight Center, Greenbelt, MD.

Bernstein, R., and D. G. Ferneyhough, December 1975, Digital image processing; Photogrammetric Engineering and Remote Sensing, vol. 41, pp. 1465-1476.

Bernstein, R., and L. P. Schoene, Jr., 1976, Advances in digital image processing of earth observation sensor data; Proc. of 13th Space Congress, Cocoa Beach, FL.

Bernstein, R., 1976, Digital image processing of earth observation sensor data; IBM J. Res. Develop., 20, pp. 40-57.

Bernstein, R., and G. C. Stierhoff, 1976, Precision processing of earth image data; Amer. Scientist 64, no. 5, pp. 500-508.

Bernstein, R., 1980, Data base requirements for remote sensing and image processing applications; Conf. on Data Base Techniques for Pictorial Applications, Florence, Italy, June 20-22, 1979, Publ. in Lecture Notes in Computer Science, no. 81, Blaser, A., ed., Springer-Verlag, Berlin.

Bernstein, R., Dave, J., and H. G. Kolsky, 1981, An experimental digital image manipulation, analysis, and processing system (DIMAPS); IBM Palo Alto Scientific Center Report No. G320-3423.

Billingsley, F., 1965, Processing ranger and mariner photography; 10th SPIE Technical Symposium.

Bristor, C., April 1968, Computer processing of satellite cloud pictures; National Environmental Satellite Center (NESC) Tech. Memorandum TM-3.

Bristor, C., et al., 1966, Operational processing of satellite cloud pictures by computer; Monthly Weather Review, vol. 94.

Case, J., 1967, The analytic reduction of panoramic and strip photography; Photogrammetria, pp. 127-141.

Castleman, K. R., 1979, Digital Image Processing; Prentice Hall, Inc., Englewood Cliffs, NJ.

Chapman, W., 1974, Gridding of ERTS images; presented at Fall Meeting, Amer. Soc. of Photogram. vol. 40 pp.

Colvocoresses, A. P., 1974, Evaluation of the earth resources technology satellite (ERTS-1) for cartographic application; Commission I, ISP, Stockholm, Sweden.

Colvocoresses, A. P., and R. McEwen, 1973, EROS cartographic progress; Photogram. Eng., vol. 39, pp. 1303-1309.

Colvocoresses, A. P., 1974, Space oblique mercator, a new map projection of the earth; Photogram. Eng., vol. 40, pp. 921-926.

Derenyi, E., and G. Konecny, 1966, Infrared scan geometry; Photogram. Eng., vol. 32, pp. 773-778.

DeSio, A., and R. Economy, 1973, Digital data processing and information extraction for earth resources applications; Gen. Elect. Co.

Doyle, F., 1970, Photographic system for apollo; Photogram. Eng., vol. 36, pp. 1039-1044.

Anon., 1979, Parallel processor will be capable of performing 6 billion additions/sec, Computer Design, March 1979, pp. 55-56.

Efron, E., 1968, Image processing by digital systems; Photogrammetric Engineering and Remote Sensing, Vol. 34, pp. 1058-1062.

ERTS Data Users Handbook, NASA 715D4249, NASA Goddard Space Flight Center, Greenbelt, MD.

Fletcher, F., and M. I. Powell, 1963, Rapidly convergent descent methods for minimizing; Computer J., 6, p. 163.

Forrest, R., Summer/Autumn, 1970, Mapping from space images; Bendix Technical Journal, vol. 3, no. 2.

Forrest, R., September 1971, Geometric processing of ERTS images; Presented at Fall Meeting, Amer. Soc. of Photogram.

Friar, M. E., Hogan, R. D., Min, P. J., Sharp, J. V., and D. R. Thompson, April 1970, System and design study for an advanced drum plotter; IBM Sys. Devel. Division report TR 21.390.

Gambino, L. A., and M. A. Crombie, 1974, Digital mapping and digital image processing; Phot. Eng., Vol. 40, pp. 1295-1302.

Green, W. B., Jepsen, P. L., Kreznar, J. C., Ruiz,

R. M., Schwartz, A. A., and J. B. Seidman, 1975, Removal of instrument signature from mariner 9 television images of mars; Appl. Opt. 14., No. 1, pp. 105–114.

Green, W. B., June 1977, Computer image processing—the Viking experience; Proc. of the IEEE Chicago Spring Conference on Consumer Electronics.

Gren, W. B., 1977, Viking image processing; SPIE Proceedings, vol. 119, August (SPIE Seminar on Applications of Digital Image Processing).

Greve, C., 1974, Digital rectification of side looking radar; Proc. Fall Mtg., Amer. Soc. of Photogram., p. 19–32.

Hotine, M., 1946, The orthomorphic projection of the spheroid; Empire Survey Review, vol. 8, no. 62–63.

Hotine, M., 1947, The orthomorphic projection of the spheroid; Empire Survey Review, vol. 9, no. 64–66.

IBM FSD Report TR FSC72-0140, April 30, 1974, Feasibility of generating mosaic directly from ERTS-1 digital data; Final Report to the U.S. Dept. of Interior, Contract No. 08550-CT3-12.

Jepsen, P. L., December 1976, The software/hardware interface for interactive image processing at the image processing laboratory of the Jet Propulsion Laboratory; Proc. of the Digital Equipment Computer Users Society.

Keating, T. J., Wolf, P. R., and F. L. Scarpace, 1975, An improved method of digital image correlation; Photogram. Eng., vol. 41, pp. 993–1002.

Konecny, G., 1972, Geometrical aspects of remote sensing; Commission 4, Internat. Cong. of Photogram., Ottawa.

Konecny, G., 1976, Mathematical models and procedures for the geometric restitution of remote sensing imagery, ISP. Repts. and Inv. papers, Comm. III, no. 02.

Kowalski, D., 1968, A comparison of optical and electronic correlation techniques; Bendix Technical Journal, vol. 1(2).

Kratky, V., 1971, Photogrammetric aspects of precision processing of ERTS imagery; Nat. Res. Coun. of Canada, Ottawa, Ontario.

Levinthal, E. C., et al., 1973, Mariner 9—image processing and products; Icarus, vol. 18, no. 75.

Lillestrand, R. L., 1972, Techniques for change detection; IEEE Trans. on Computers, vol. C-31, pp. 654–659.

Lynn, D. J., November 1976, Recent applications of digital processing to planetary science; Proc. of the Caltech/JPL Conference on Image Processing Technology, Data Sources and Software for Commercial and Scientific Applications, JPL Document SP 43-30.

MacDonald, R. B., Hall, F. G., and R. B. Erb, June 1975, The use of Landsat data in a large area crop inventory experiment (LACIE); Proc. of NASA Earth Resources Survey Symp., NASA Johnson Space Center, Houston, TX.

Mach, R. E., and T. L. Gardner, 1962, Rectification of satellite photography by digital techniques; IBM J. Res. Develop. 6.

Maling, D. H., 1973, Coordinate Systems and Map Projections; George Philip and Son, Ltd., London.

Markarian, H., Bernstein, R., Ferneyhough, D. G., and F. S. Sharp, 1973, Digital correction for high resolution images; J. Am. Soc. Photogram. Vol. 39, pp. 1311–1320.

Markarian, H., Bernstein, R., Ferneyhough, D. G., Gregg, C. E., and F. S. Sharp, September 1971, Implementation of digital techniques for correcting high resolution images; ASP-ACSM Fall Convention; also IBM-FSD Gaithersburg Tech. Report FSC 71-6012.

Montoto, L., 1977, Digital detection of linear features in satellite imagery; Proceedings of the International Symposium on Image Processing—Interactions with Photogrammetry and Remote Sensing, Leberl, F. W. ed., Verlag fur die Technische Universitat Graz.

Murphrey, S. W., Depew, R. D., and R. Bernstein, Digital processing of conical scanner data; Photogrammetric Engineering and Remote Sensing, vol. 43, pp. 155–167.

NASA Goddard Space Flight Center, November 1972, ERTS data users handbook.

Niblack, W., 1981, The central point library building system; Photogrammetric Engineering and Remote Sensing, vol. 47, no. 12, pp. 1685–1692.

Oppenheim, A. V. and R. W. Schafer, 1975, Digital Signal Processing; Prentice-Hall, Inc., Englewood Cliffs, NJ., 585 p.

Orti, F., Garcia, A., and M. A. Martin, 1978, Geometric correction of MSS landsat images using a ground control point library; Proc. 5th Annual Conf. of the Remote Sens. Society, Durham, U.K.

Orti, F., 1978, A center covariance adaptive clustering algorithm; IBM Madrid Scientific Center Report SCR-02.78.

Orti, F., 1981, Optimal distribution of control points to minimize landsat image registration error; Photogrammetric Engineering and Remote Sensing, vol. 47, pp. 101–110.

Reeves, R. G., (ed.), 1975, Manual of Remote Sensing; American Society of Photogrammetry.

Rifman, S. S., and D. M. McKinnon, March 1974, Evaluation of digital correction techniques for ERTS images; TRW Corporation Final Report, TRW 20634-6003-TU-00, NASA Goddard Space Flight Center, Greenbelt, MD.

Rindfleisch, T. C., Dunne, J. A., Frieden, H. J., Stromberg, W. D., and R. M. Ruiz, 1971, Digital processing of the mariner 6 and 7 pictures; J. Geophys. Res. 76 (2), pp. 394–417.

Rosenfeld, A., 1976, Digital Picture Processing; Academic Press, New York.

Schaefer, D. H., October 1979, Massively parallel information processing system for space applications; AIAA 2nd Computers in Aerospace Conf., Los Angeles, CA, 284–286.

Schmidt, R. G., and R. Bernstein, October 1975, Evaluation of improved digital processing techniques of landsat data for sulfide prospecting; Proc. of the First Annual W. T. Pecora Memorial Symp., Geological Survey Professional Proc., 105.

Schoene, L. P., 1977, Master data processor; IBM FSD Tech. Directions 3, no. 2.

Snyder, J. P., May 1978, The space oblique mercator projection; Photogrammetric Engineering and Remote Sensing, vol. 44, no. 5, pp. 585–596.

Snyder, J. P., 1981, Space oblique mercator projection; mathematical development; USGS Bulletin 1518.

Steiner, D., and M. E. Kirby, 1977, Geometrical referencing of landsat images by affine transformation and overlapping of map data; Photogrammetria, 33, pp. 41–75.

Szekely, D., 1974, Transformation of pictures with

cathode ray tubes; Geodezia es Kartografia, Budapest, vol. 26, no. 2, (In Hungarian).

Taber, J., and S. Rifman, December 1973, Evaluation of digitally corrected ERTS images; Third ERTS Symposium.

Torbert, G., and C. J. Robinove, 1976, Digital color mosaic of parts of Wyoming and Montana; ERTS-1 A New Window on our Planet; Geological Survey Professional Paper no. 929, pp. 32–33.

USGS, 1979, Landsat Data Users Handbook.

van Wie, P., and M. Stein, July 1977, A landsat digital image rectification system; IEEE Trans. Geosc., vol. GE-15, 3, pp. 130–137.

Wallis, R., 1977, An approach to the space variant restoration and enhancement of images; Image Sc. Math., Monterey, CA, 107–111.

Zobrist, A. L., Bryant, N. A., and A. J. Landini, August 1977, Use of landsat imagery for urban analysis; Proceedings of Urban and Regional Information Systems Conference, Kansas City, NO.

CHAPTER 22

Geographic Information Systems and Remote Sensing

Author-Editors: DUANE F. MARBLE and DONNA J. PEUQUET

Contributors: A. R. BOYLE, NEVIN BRYANT, HUGH W. CALKINS, TIMOTHY JOHNSON, ALBERT ZOBRIST

GENERAL CONTENTS: Basic concepts in spatial data handling. Digital handling of spatial data. Data volume and accuracy considerations; manipulation and retrieval operations; display of spatial data. The geographic information system—an introduction. Overview; an operational geographic information system—the Canada Geographic Information System; geographic information system design methodology. The geographic information system—technical considerations. Methods of spatial data capture and encoding; data management in the geographic information system; data manipulation and display. Linking remote sensing systems and geographic information systems. Remote sensing inputs to geographic information systems; geographic information system inputs to remote sensing systems; a sample linkage—IBIS, the image-based information system. References.

INTRODUCTION

A primary characteristic of remote sensing systems is that they produce large volumes of spatial data. Effective utilization of these large spatial data volumes is dependent upon the existence of an efficient, geographic handling and processing system that will transform these data into usable information. The major tool for handling spatial data is the geographic information system (GIS). A GIS is designed to accept large volumes of spatial data, derived from a variety of sources, including remote sensors, and to efficiently store, retrieve, manipulate, analyze and display these data according to user-defined specifications. The significant potential for integration of remote sensor derived data into geographic information systems was first pointed out nearly two decades ago [Garrison et al., 1965], but little of the predicted interaction has developed.

Over the years a substantial number of geographic information systems have been constructed by organizations in both the public and private sectors. Within the present context, it must be understood that the great majority of the existing systems were constructed to use data derived from sources other than remote sensing systems. The most common data source has been the analog map and in nearly all cases the input phase of the GIS is heavily, if not entirely, oriented toward the creation of digital files from map documents. Operational applications of GIS today include such areas as land and resource management, traffic planning, marketing, military planning as well as a wide variety of other uses.

The material in this chapter reviews the current state-of-the-art in geographic information systems and provides case studies of two systems one of which, the Canada Geographic Information System (CGIS), is an example of a large, map-oriented system which is experimenting with remote sensing inputs. The other is the Image Based Information System (IBIS) which is a GIS designed primarily to handle remote sensing inputs and which treats map data as ancillary information. Problems in the effective design and implementation of geographic information systems are discussed at some length since improper design has been demonstrated to be the main cause of system failure. Attention is also directed to the manner in which geographic information systems may be more effectively integrated with remote sensor data-acquisition systems in the future.

It is important to point out once again that, in most operational geographic information systems, remote sensing inputs do not represent the primary source of data. They do represent, in many instances, either a potential source of new data elements for the GIS, which must be made to mesh with existing GIS data definitions, or an alternative form of data capture for one or more well defined data elements currently incorporated within the GIS. As such, the problems of integration of remote sensing inputs are judged by GIS operators largely on the basis of the efficiency and ease of utilization of remote-sensor derived data vis-a-vis the existing data sources (usually map based) of the system. Failure to recognize this fact has been a primary deterrent to the more widespread utilization of remote sensor data in operational geographic information systems.

BASIC CONCEPTS IN SPATIAL DATA HANDLING

The basic types of entities considered in computer handling of spatial data are points, lines and areas. Depending upon the user's desired degree of generalization, all entities existing in earth space can be viewed as being represented by one of these three forms. For example, cities—at a wide range of small scales—may be represented as points with little or no loss of information while at other larger scales they may be represented as polygons. The type of generalization selected for the feature is a direct function of the user's requirement for utilization of the spatial data.

Each of the spatial entities being considered exists within the bounds of a number of coordinate systems that have been developed over a period of many decades. These coordinate systems provide a means of exactly positioning the entities in earth space and of providing measurements of such attributes as perimeter length, area, distance, and direction between the entities. The problem of coordinate systems is a complex one but in most cases it is possible to move, via mathematical transformations, from one coordinate system to another. For example, the basic coordinate system for location on the earth's surface is known as latitude and longitude. The coordinates of this system exist in a three dimensional space and represent locations on the surface of a spheriod which closely approximates the shape of the earth. By suitable transformations (known as map projections) these three-dimensional coordinates may be transformed into two-dimensional coordinates which are then suitable for plotting on a flat computer display or map sheet. Unfortunately, the process of transformation from three-dimensional to two-dimensional coordinates is one which induces various forms of distortion. The best the user can do when undertaking such a transformation is to try to explicitly select both the type and the degree of distortion that is induced. It is important to realize that the inverse coordinate transformations exist, that is, one may go from a specific set of projection coordinates, such as the Universal Transverse Mercator back to latitude and longitude by performing an inverse mathematical transformation. Regrettably, many of these inverse computations are both difficult and generally little known. An excellent discussion of the standard projections and their inverses is provided by Snyder [1982].

It is also regrettably true that there exist coordinate systems between which transformations are either extremely difficult or impossible to perform at satisfactory accuracy levels. One example may be seen in the street address system in any major U.S. urban area where a user who needs to go from a specific street address to latitude and longitude coordinates has no direct way of doing so except through interpolation based upon a high density point set whose relation to both coordinate systems is known. Such point sets exist for most major U.S. urban areas as part of the GBF/DIME files maintained by the Geography Division of the Bureau of the Census. An even more widespread case (and in many instances a more important one from a remote sensing standpoint) can be found in the Public Land Survey System (PLSS) that is used to record land ownership information for large areas of the western United States. At present it is not possible to accurately transform coordinates in the PLSS (for example, the standard 'Quarter-Quarter-Quarter' description of mineral leases) into specific latitude and longitude measures. The development of precise latitude and longitude measures of the basic point entities in the PLSS (section corners and quarter section corners) will require many years [National Research Council, 1982].

Another example is the transformation of remotely-sensed image coordinates into a map projection or vice versa. Locational reference in such a raster data set generally comes from pixel position and platform movement information. In image space, vertical reference is represented by the scan line position; horizontal reference by the sequence of pixel elements within a line. Incomplete platform location data and the presence of geometric distortions in image data sets make the transformation difficult.

Nearly all geographic information systems that are operational today utilize basic latitude and longitude coordinates. This may be implicit in that the GIS may operate within, for example, one of the U.S. State Plane Coordinate systems; however latitude and longitude can normally be recovered from these projection coordinates if desired.

Given a number of basic spatial entities and measures of their location in earth space, the geographic information system must also deal with a host of non-spatial attributes of these entities. For example, one attribute of a city is its population size. An attribute of a stream segment may be the flow in cubic feet per minute, or the amount of dissolved oxygen, while an attribute of a forest-type polygon might be the volume of lumber to be produced from undertaking a specific type of cutting operation at a certain date. The GIS must handle both the spatial and non-spatial attributes of the basic entities contained in the system. While many information systems exist today that handle only non-spatial attributes (e.g., banking systems) their operations generally become quite inefficient when applied to the computer handling of digital spatial data, which includes explicit location information and which must, for example, retrieve and process records on the basis of spatial relationships such as adjacency or distance to immediate neighbors [Phillips, 1977].

DIGITAL HANDLING OF SPATIAL DATA

Digital representation of the basic types of spatial entities has developed in a relatively straightforward fashion. Point entities, of course, pose few major problems since they may be represented as single pairs of digital coordinates. The representation of line and area boundary-information is somewhat more complex. The standard approach in the geographic information system is to represent polygon boundaries as lines (normally a sequence of lines or arcs will make up the boundary of a polygon) and to represent lines as a sequence of very short, straight line segments which can, in turn, be represented by an ordered sequence of points representing the end points of the short line segments. Thus the digital representation of line and polygon information generally consists of an ordered sequence of x−y coordinate pairs. These coordinate pairs, when linked together by short, straight lines, produce an approximation of the original line. Selection of the point sampling density required to represent an analog line at a specific resolution forms a necessary basis for the quantization of analog line data [Freeman, 1978]. This format for spatial data, where cartographic entities are translated line-for-line and point-for-point into digital form, is known as vector format [see Figure 22-1]. Because of this correspondence to the traditional map, the vector data-structure is compatible with many standard cartographic algorithms that were originally developed for manual use.

The other main type of representation for spatial data is known as raster format. Whereas vector data structures represent a more or less direct translation of the map, the development and utilization of the raster data structure has been largely driven by graphic input/output hardware-technology. Remote sensing devices of many kinds, as well as scanners for high volume map-digitization, record spatial data in narrow strips across the data surface as shown in Figure 22-1. The grid or matrix data-structure is the uncompacted raster form, wherein the smallest logical unit is a pixel, or single grid cell. Alternatively, the unit may be viewed as a single point in a regular spatial lattice. The raster structures, besides being more compatible with modern input/output hardware, have the advantage that the order of the data elements, as stored in digital form, is dictated by their geographical positions. This preserves many intrinsic spatial interrelationships among the data elements (as contrasted to vector formats) and makes data retrieval on the basis of these spatial relationships a more straightforward task [Peuquet 1979].

Other methods exist for the digital representation of spatial data. An example is a method that uses the equation of the line as a means of compactly storing vector data. However, practice has demonstrated that, for the complex, mathematically unpredictable lines commonly encountered in cartographic documents, this approach is substantially less efficient for manipulation while normally providing little or no gain in data compaction. Two major variations on raster or grid structures also exist, both representing other

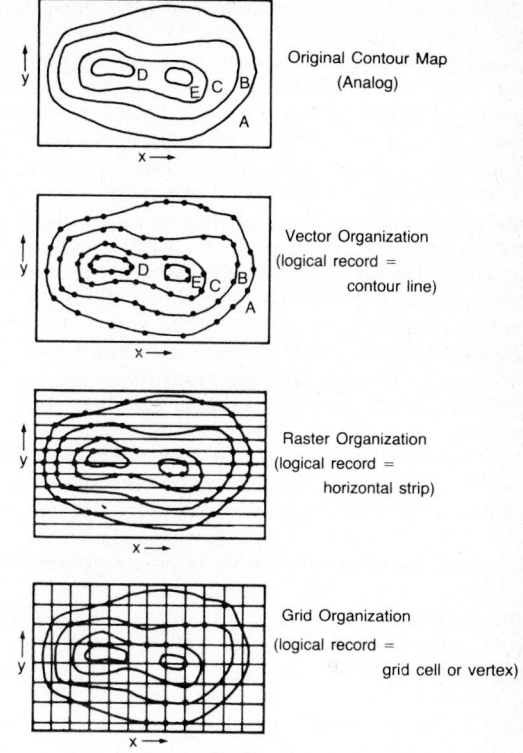

Types of data organization

Fig. 22-1. Vector/Raster Data Structures.

space-filling polygon structures, but each has a much narrower range of applications. Triangular mesh structures have been primarily useful for topographic analysis (although one water resource related GIS, ADAPT, utilizes them for general structuring of other spatial information) and hexagonal mesh structures are more useful for radial searching and locational analysis.

DATA VOLUME AND ACCURACY CONSIDERATIONS

The accuracy of the digital representation of spatial data is governed by both user requirements and the inherent characteristics of the source document and the instruments used to create it. Needless to say, positional accuracies in the GIS cannot exceed those of the original data source whether it be a large-scale map sheet that has been digitized (relatively high resolution) or data collected from an orbital remote sensor (relatively low resolution). Positional accuracy of the spatial entities recorded in the system is of critical importance to many users because of the problems that can be generated during overlay operations [Goodchild, 1978]. Within the limits of accuracy of a given GIS input map document, a variety of accuracy levels may be utilized depending upon user needs. Certain users may require that the system reproduce the input document without visible changes. This frequently leads to a specification of accuracy requirements of the "plus or minus half-a-line-width" variety. The user who is concerned entirely with derivative information (for example, measurement of area rather than precise determination of boundary locations) can frequently get by with a significantly lower level of accuracy. Considerable degradation in the approximation of the digital line to the original map line can be seen without having an appreciable impact upon the accuracy of measures such as area. It is possible to produce a rather "cartoon-like" representation of polygons while still maintaining area measurements that lie within user-specified tolerance levels. This is not often recognized in operational GIS situations and leads to overly expensive and cumbersome data-input operations.

A close relationship may be seen between the specified degree of line generalization and the number of points required to represent the line. The variation in data volume may exceed an order of magnitude (e.g., the use of 10 points per inch vs. 250 points per inch) with subsequent, very significant impacts on total data volumes and hence upon costs of digital storage, retrieval and manipulation operations.

MANIPULATION AND RETRIEVAL OPERATIONS

Once capture of digital spatial data has been effectuated, the major problem lies in the efficient storage, retrieval, and manipulation of the data. Problems in manipulation and retrieval arise in two general areas, one being unique to spatial data. The first problem is again that of data volumes. Data bases created for GIS utilization are frequently very large, even by modern standards. Systems for handling large data bases have been created and generally perform quite effectively within the framework of well-defined entities and a limited number of straight-forward relationships among these entities. The major problem in efficient storage, retrieval, and subsequent manipulation of spatial data arises, however, out of the fact that many of the relationships upon which storage organization and retrieval queries must be based are spatial in nature as contrasted to the aspatial nature of queries and relationships commonly found in, say, an airline reservation system. These two-dimensional, spatial relationships include such concepts as adjacency, distance to near neighbors, etc. Many of these (e.g., distance between entities) are defined over all pairwise entity relationships in the data base. The typical commercial data-base management system (DBMS) does not permit the definition of a large number of relationships and thus the user is forced to fall back upon computational derivations on an "as needed" basis. This, effectively, defeats the entire purpose of the DBMS, which is to permit the retrieval of specific subclasses of entities without the necessity of examining all entities in the data base. Other spatial relationships that may

be needed are "fuzzy" or only defined within a narrow application context, such as "east of."

DISPLAY OF SPATIAL DATA

An important feature of any information system is report generation. This feature, in a geographic information system, frequently involves the creation of special thematic map displays that may represent the distillation of a number of complex retrievals and manipulations of the original data. The field of computer graphics has provided the GIS developer and user with a wide variety of inexpensive and sophisticated tools for the creation of map displays, both interactively and in hard copy. The unit cost of these displays has been dropping rapidly in the last few years and their level of sophistication has significantly increased. At the time of this writing it is possible, for example, to purchase an interactive, intelligent color display of moderate resolution (i.e., 512 × 620 pixels on a 19 inch display) for under $5,000. The ability of these inexpensive displays to present sixteen or more colors simultaneously permits the creation of relatively complex map displays.

At present, it is clear that our ability to create spatial displays from a hardware standpoint has far outstripped our cartographic knowledge of how such displays should be created to optimize information transfer to the human observer. It is not enough to create a spatial image or map on the screen—this information must be correctly and quickly perceived by the human observer who is dependent upon a series of complex and often poorly understood eye-brain interactions. This problem is particularly acute in color map displays since relatively little experimentation has been undertaken on the creation of optimal color map displays within a computer environment [Dobson and Sibert, 1982].

Graphic display, however, represents an extremely powerful tool for both summarization and browse operations within the digital, spatial data base. Its use will continue to increase as the sophistication of the displays and our knowledge of how to make the most effective use of them increases.

THE GEOGRAPHIC INFORMATION SYSTEM: AN INTRODUCTION

A geographic information system represents a system, commonly computer-based, for handling spatial data. All complete or full geographic information systems perform the following major functions:

Data Input: normally consists of a mixture of manual and automatic digitizing operations together with associated data cleaning and edit activities.

Data Storage and Retrieval: initial creation of the spatial data base together with subsequent update operations and query handling

Data Manipulation: creation of composite variables through processing activities directed toward both spatial and non-spatial attributes of system entities.

Report Generation: creation of both tabular and cartographic reports reflecting selective retrieval and manipulation of entities within the data base.

The full GIS is distinguished from other information systems (e.g., management information systems [MIS] for business applications) by its explicit focus on spatial entities and relationships. Efficient handling of explicit spatial relationships, especially where large data volumes are involved, is significantly more difficult than the operations normally encountered within a MIS. There are a number of significant and currently unresolved technical problems in GIS operation that limit both the size and efficiency of current systems.

The first computer-based geographic information systems were developed in the middle 1960s and focused heavily on the handling of land use and natural resource data sets. Many of the early attempts at GIS developments failed due to poor design and failure to properly adjust to severe institutional problems. Other systems suffered from technical problems (e.g., excessive processing or data input costs) which reduced their long run viability. In a few cases, such as the GIS operations of the State of Minnesota and Environment Canada, both the design problems and the institutional problems were overcome and the systems have enjoyed many years of successful operation.

Perhaps the best introduction to geographic information systems can be found in the examination of a successful, operational system. The following briefly outlines the operations of CGIS. Additional material on this and other North American systems may be found in Tomlinson, Calkins and Marble [1976].

AN OPERATIONAL GEOGRAPHIC INFORMATION SYSTEM: THE CANADA GEOGRAPHIC INFORMATION SYSTEM (CGIS)

Geographic information systems were originally conceived to meet the problems of manipulating and displaying very large volumes of geographically referenced data. These systems were needed for two reasons: 1) the volume of data to be processed was so large that manual methods would have been incapable of completing the task; or 2) the manipulations were sufficiently complex that, when coupled with the large data volumes, the task could not be completed without substantial error. The need for GIS stemmed from two major forces: 1) emerging natural resource and environmental problems that were becoming increasingly recognized by society; and 2) the large volumes of geographically referenced data becoming available, much of it based upon increased use of remote sensing technology in map production. As these trends merged it became possible and desirable to approach problems at both regional and national scales, a task not previously possible. In this context, the Canada Geographic Information System was first proposed in 1963.

System History and Purpose

CGIS was the first full-scale geographic information system to carry out the functions of reading, measurement, and comparison of spatial data successfully within a computer environment. It has been under continuing development for neary two decades and is currently the largest and most sophisticated GIS in existence.

CGIS is designed to handle data extracted from aerial photographs and maps, specifically, thematic maps such as those pertaining to land use, forestry, and soils. The area covered by the map or photograph is divided into irregular polygons, each polygon having a set of descriptive characteristics related to it. The system handles data primarily in the form of polygons, although it does have some capability to handle data represented as points, and in the future may add the capability to handle data represented as lines. Currently data from aerial photographs and satellite imagery are not handled directly by the system, but rather must be reduced to maps of a polygon form before being entered into the system.

The CGIS originated as a computer mapping system planned to facilitate use of data gathered by the Canada Land Inventory (CLI). In 1962 the Canada Land Inventory developers realized that unless data were handled automatically, much of their potential usefulness would be lost. The initial plan for a computer mapping-system resulted from the realization that the CLI data were very extensive, with estimates ranging between 1,600 and 30,000 map sheets. The basic capability needed for the CLI was one of map measurement, that is the measurement of polygons as represented on the map sheets, and the need to compare two sets of polygon data to determine where specific characteristics coexisted, commonly termed the polygon overlay operation.

CGIS capabilities can be divided into five major categories as follows:

1. The entry of map data into the system from polygonal maps;
2. Data file creation, including the building of a data file in standard CGIS format, building a data file in a special graphics subsystem format, or merging an existing CGIS data file with other user-generated data files;
3. Retrieval of data from the data base including retrieval by a user-defined circle or polygon, or retrieval of selected items by either identifying the appropriate descriptor-data values or selected image-data values;
4. Manipulation of the retrieved data such as calculation of the area and/or perimeter of any polygon, calculation of the length of a linear feature such as a shoreline, overlay of

two or more polygon sets, combination of point data sets with polygon data sets, and conversion from polygon data to regular grid-cell data;

5. Graphic output in terms of tabular summaries, maps and other digital files.

Data Capture and Storage

The primary data input to CGIS is polygon data from maps. The limitations on this input process are that the map projection must be a six degree UTM projection and that the coordinate system must be latitude and longitude. Also, if scale is to be considered the map scale must be between 1:500 and 1:1,000,000 and all polygons identified on the map must have a descriptor code with adjacent polygons having different descriptor codes. The process for entering data into the system is composed of the following steps:

1. The map is photographically transferred to a scribecoat material;
2. The polygon boundaries are manually scribed;
3. The resulting scribed map sheet is scanned using a drum scanner;
4. A centroid is manually digitized for each polygon using a standard digitizing table;
5. The descriptor data for each polygon are manually entered on coding sheets;
6. The digitized data from the scanner are processed with the digitized centroids and the encoded descriptor data to produce the image data-sets (IDS) and the descriptor data-set (DDS) which together form the CGIS data base.

The rescribing of the map input documents is necessary to produce a clean, constant-line-width source for the scanner. Use of variable width lines on scanner input documents has an adverse reaction on raster-to-vector processing time [Peuquet, 1981a].

An extensive manual error-correction process exists that allows for the identification and correction of the common errors encountered in this type of data input process. Point data are added to CGIS through the use of manual digitizing tables for direct digitizing of such data. The attribute data corresponding to point data are entered in the same manner as that for polygon data.

At this time CGIS has only a limited capability to accept data in machine-readable format from other systems, including remote sensing systems, but a standard interchange format has been developed and experiments are being conducted [Energy, Mines and Resources, et al., 1979]. CGIS has no ability to classify remote sensing data transmitted in this fashion and the supposition is that only classified data in polygon form will be made available. The CGIS data base can be updated. However, this often requires completely redoing the entire input process. Alternatively, a change map may be produced and the system

polygon-overlay function may be used to combine this change map with the original data. Then the resultant map is stored as the updated data base.

The system takes the files created by the data input process and builds the data base in a standard format. With the thousands of map sheets the system is designed to handle, data storage by individual map sheet would be inefficient and hamper subsequent user queries. Consequently, the system stores data by regions and by smaller geographic units called frames rather than by map sheets. The region and frame organization facilitates retrieval by specific location. In this structure, data that are geographically located close together are most likely to be found located together in the computer storage format. The boundaries between map sheets are automatically eliminated as the data base is built. However, data can still be retrieved on the basis of sheet boundaries if desired. The data-base building process also includes a sophisticated edge-matching procedure to match polygon boundaries that cross map sheet borders.

Query and Retrieval

Data from the CGIS data bank may be retrieved by specifying either location or attribute retrieval criteria. The system is designed particularly to retrieve based on location criteria. There are five options available to the user:

1. The user may specify a circle of given location and radius;
2. The user may specify an irregular shaped polygon;
3. The data may be retrieved by identifying those polygons whose centroids fall within selected x−y values;
4. A data set already entered into the system may be used as the search criteria to apply to another data set;
5. The user may use a new set of maps that contain basically his own data.

First, the retrieval by a circle involves a user specifying its center coordinates and radius and then all data points that fall within this area will be selected by the system. Second, if the user specifies an N-sided polygon as a string of x−y coordinates, then the system will extract all or some of the data that falls within that polygon, depending upon the attribute or descriptive data selection criteria used. The other three retrieval options are simply variations upon the first two.

Data Manipulation

The data-manipulation portion of CGIS automatically calculates the area and centroid for each polygon entered into the system or any polygon that is generated as a result of other data manipulation. Thus, for example, the number of acres of a given land-use type may be calculated, the amount of farm land within a given polygon can be determined, or the distribution of various land use types within a census tract can be calculated after the polygon-overlay operation is completed.

One of the major functions of CGIS is the polygon-overlay operation. In this process one polygon web is superimposed upon another polygon web and the intersection and union of these two data sets is determined. The system can overlay any number of polygon data-sets; however it is generally undesirable to combine more than three or four polygon data-sets unless the data are very sparse, because combination in excess of this will create a large number of very small polygons. Locational errors in the placement of polygon boundaries in data sets for the same area (e.g., both a soil type and a political boundary may have a stream defined as the 'same' edge, but this line may have different coordinates if it is digitized on different occasions), lead to the creation of a large number of erroneous 'sliver' polygons during the overlay process. Elimination of these 'sliver' polygons is difficult and expensive [Goodchild, 1978].

Additional data manipulation capabilities have been realized by the system but have not been explicitly programmed as system functions. Rather, a data-manipulation command language has been developed that allows the user to combine a set of preprogrammed functions that operate as a set of general utilities to conduct various data manipulations. In addition to the general utilities that have been programmed in the system, there is an open-ended part of this command language that allows the user to write his or her own manipulation procedure in the PL/1 programming language. Thus, the system is extremely flexible in its ability to accommodate user-specified manipulation procedures. Finally, the system has the capability to link to the Statistical Package for the Social Sciences (SPSS). With this linkage, the data are retrieved and manipulated using the system capabilities and then passed on via a transfer file to SPSS for statistical analysis.

Output Capabilities

The system can produce two types of outputs—tabular summaries and graphic plots. The tabular summaries are basically the area of any polygon within the data base or any polygon created using the system-manipulation capabilities. For example, the system can report the number of acres or hectares of forest land in several watersheds if a forest-type coverage is combined with a watershed map.

The graphic capabilities of the system are as follows: 1) the production of inkline plots on a precision Gerber plotter; and 2) the production of plots on a Tektronix 4014 CRT. The Gerber can plot different features in different colors but the system does not have the capability to plot any base map or background features for orientation of the plotted polygons.

The plotting capability on the Tektronix is part of what is termed the graphic subsystem, which was not part of the original system design. This system provides an interactive plotting capability that allows the user several commands for the selection of data items to be plotted. This is an extremely useful tool for examining the results of an overlay operation or examining a single coverage and deciding how to compose a map prior to having it plotted on the Gerber plotter. It is anticipated that the graphic subsystem will be modified in the future to include the shading of polygons, the addition of geographic identifiers to better orient the map output for the user, and the incorporation of a greater scale-change capability.

An Example of System Use: The Ottawa-Hull Urban Area Land Use Analysis

The capabilities of CGIS are best illustrated by a study conducted by Environment Canada on the land use patterns in the Ottawa-Hull urban area [Gierman, 1976]. This study involved the collection of several coverages of data, including land use data, for three time periods (1964, 1968, and 1972) that pertained to the capability of the land for agricultural use, the assessment of land that was considered prime for future urbanization, and the determination of several other categories of land capability. The data were derived from aerial photographs and the land-capability maps of the Canada Land Inventory. Two of the data coverages are shown in Figures 22-2 and 22-3: Figure 22-2 shows the prime land for future urbanization as derived by examining the best land for urbanization that was currently not used for urban development. Figure 22-3 shows the best land for agricultural uses. These two coverages were combined using the polygon-overlay capability of CGIS to produce the map shown in Figure 22-4, portraying prime land for future urbanization which does not encroach on the best land for recreation or agriculture. This land-capability analysis demonstrates the power of CGIS and it can be seen from Figure 22-4 how useful a product can be produced. It should be noted, however, that Figure 22-4 is not totally produced by the system, but is a combination of a printed base map with the system graphic output registered to it.

CGIS is an example of a large, operational system and has been successfully maintained for nearly two decades. There have been many unsuccessful GIS and it is clearly critical to pose the question: "What factors separate the successful from the unsuccessful systems?" Work over a period of years indicates that proper design, interpreted in the broadest fashion, is the critical factor. Unsuccessful systems have been found, after the fact, to have been created in the spirit set forth by one modern artist who, when questioned about design of his work, stated: "I do not sketch or design a work; I merely begin." The following section outlines the basic structured design model that has been developed for geographic information systems. Its use is spreading and it is hoped

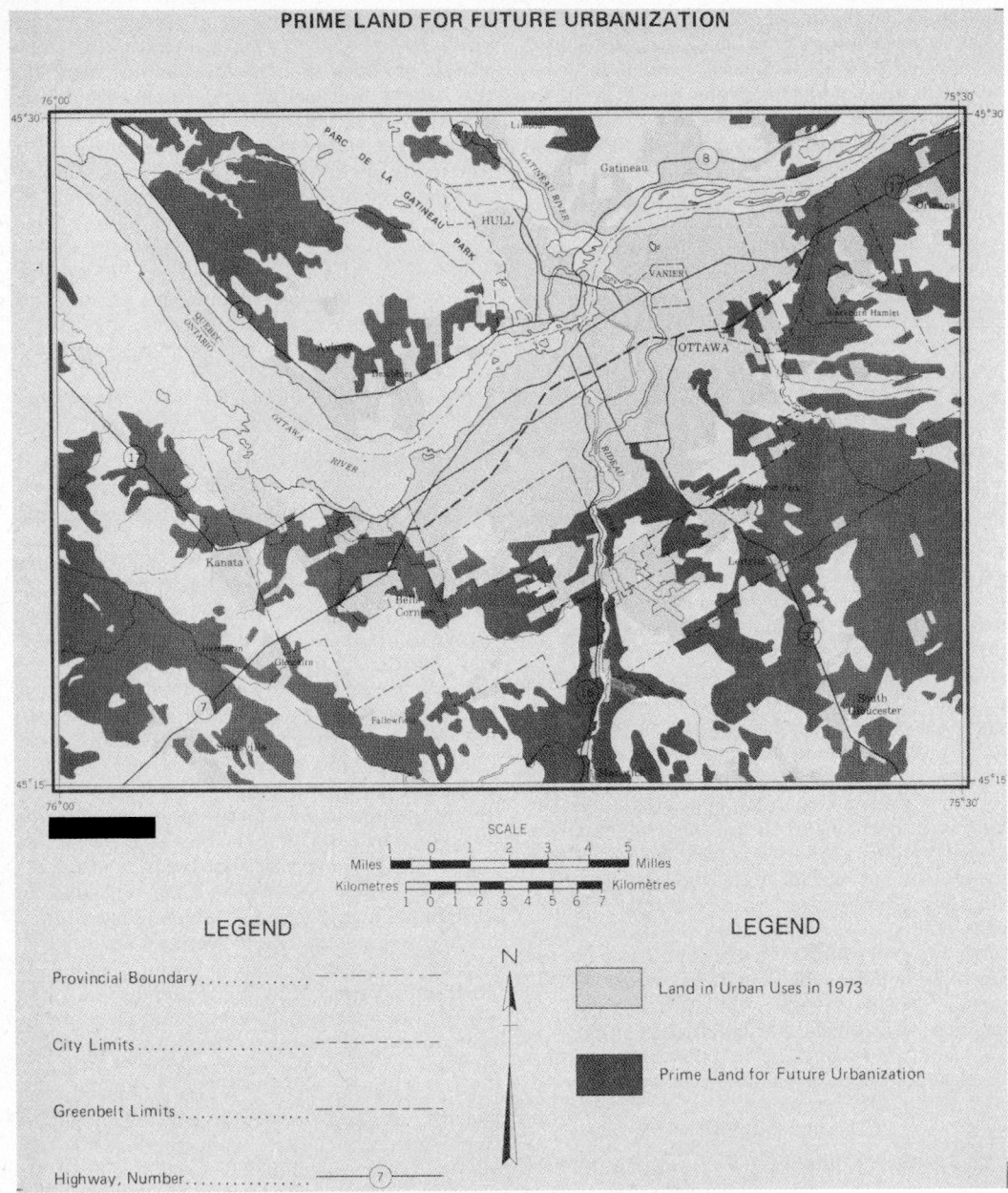

Fig. 22-2. Ottawa/Hull Urban Area No. 1.

that the introduction of this design system (supplemented by the concepts and tools being developed in the field of software engineering—see, for example, Peters, 1981) will lead to a substantial reduction in the rate of system failure.

GEOGRAPHIC INFORMATION SYSTEM DESIGN METHODOLOGY

Geographic information systems have been developed over the past twenty years to meet a variety of needs. Most system-development projects have been undertaken by persons knowledgeable in the relevant application field but who lack substantial training in the design and development of specialized spatial data-handling systems. Although many GIS have been developed, there are very few "successful" systems when utilization rates are examined. Poor systems design has been identified as a major contributor to under-utilization and associated system faults. Poor initial evaluation of users and their needs inhibits successful system implementation. Numerous

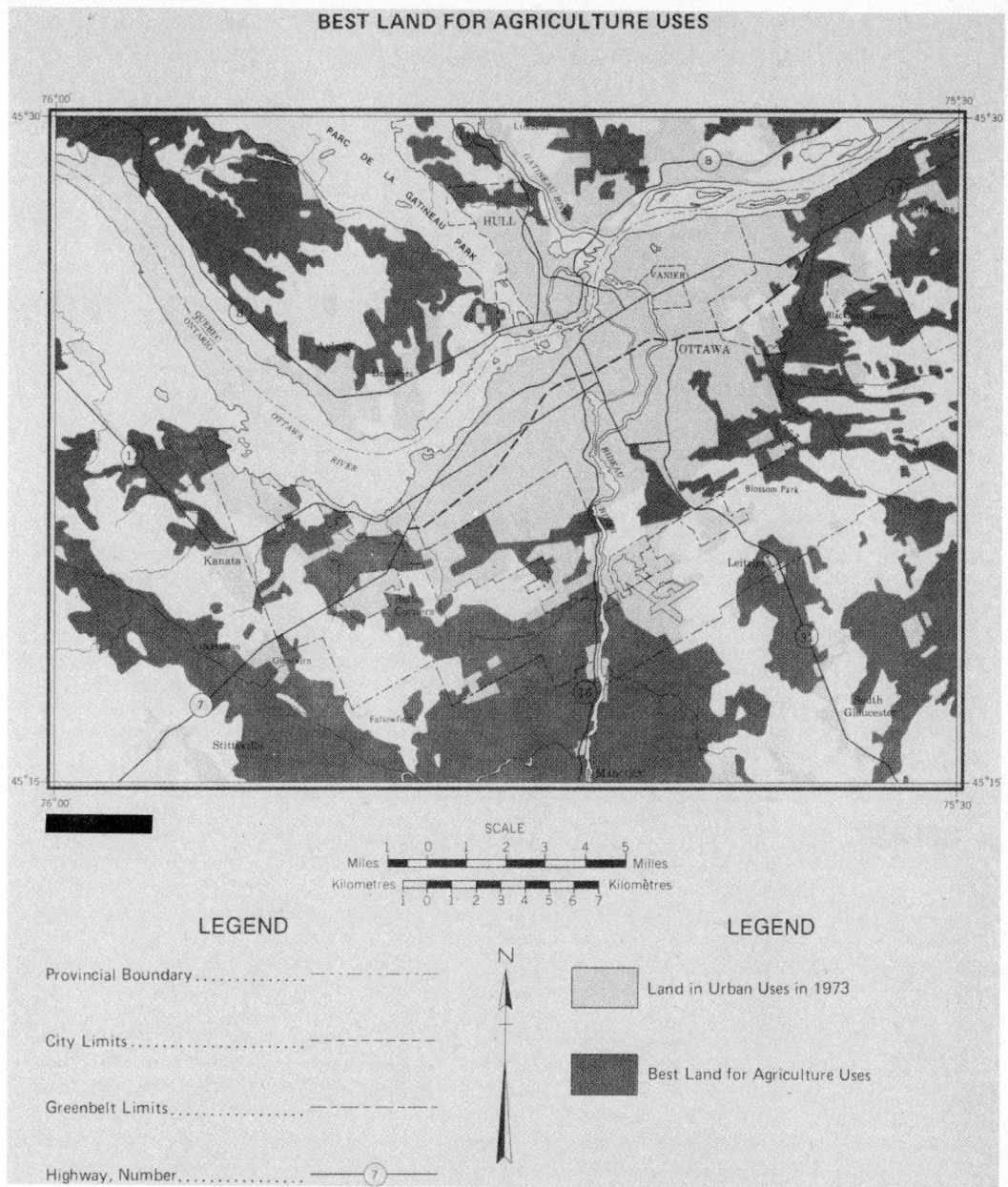

Fig. 22-3. Ottawa/Hull Urban Area No. 2.

system failures in the field of GIS represent visible evidence of this fact [Guinn and Kennedy, 1975]. Due to many system failures, a credibility gap has evolved between those persons interested in promoting the development of geographic information systems and those who can benefit from the existence of such systems—the users.

A means of bridging this gap in overall system development came about in the conceptualization of the GIS structured design and evaluation model [Calkins, 1972]. Initially, the model was proposed

as a tool for aiding the design of effective information systems to serve urban and regional planning functions. This goal was to be realized by presenting information-system objectives and specifications, generating and evaluating alternative system designs, and providing an overall evaluation of costs and benefits for system selection. Further expansion of the model occurred in 1976 and 1981 when additional insights were provided by analyzing several aspects of the model in greater detail [Reed, 1976; Johnson, 1981].

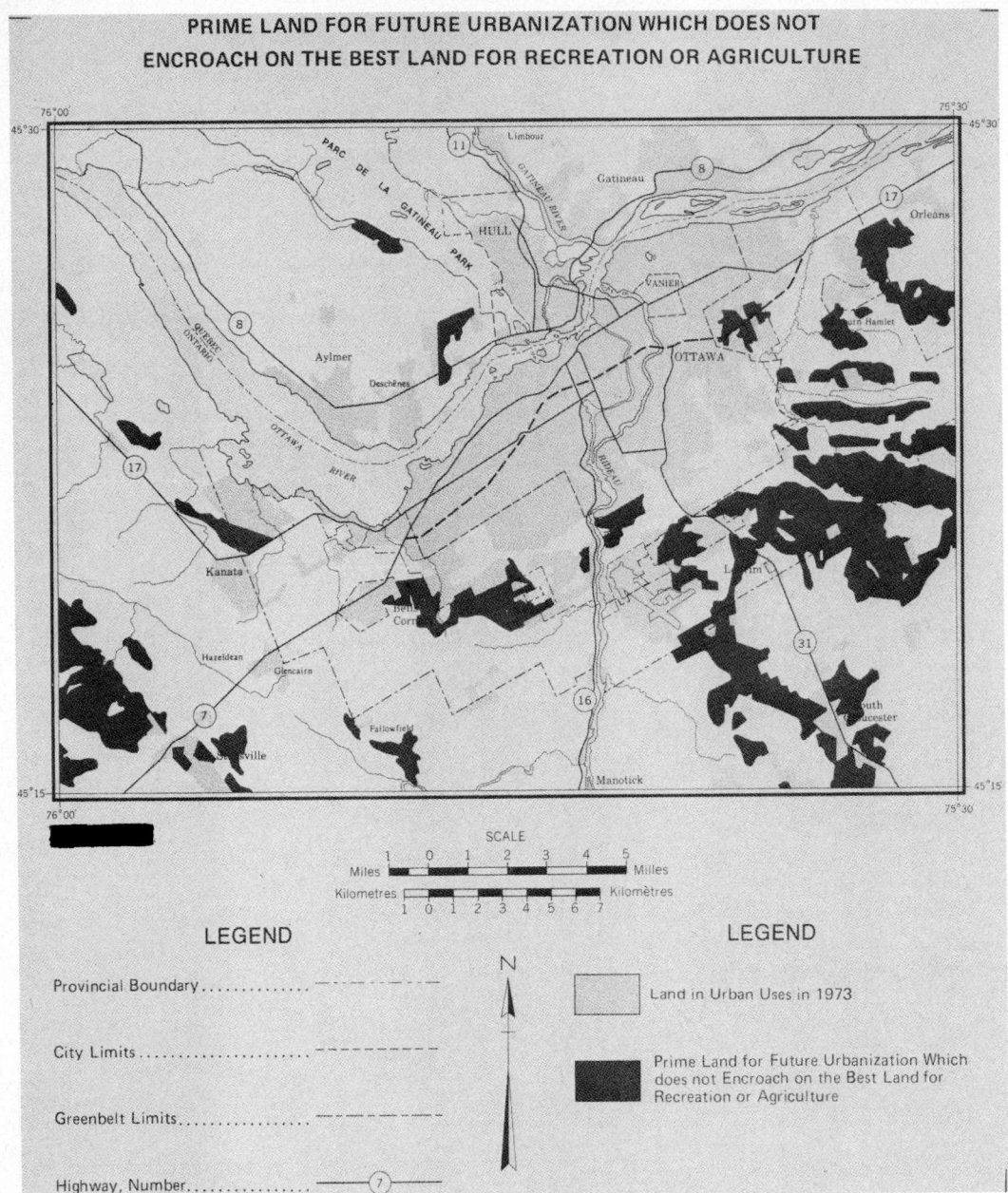

Fig. 22-4. Ottawa/Hull Urban Area No. 3.

Overview of the Design Process Using the Model

The GIS design and evaluation model consists of 28 separate activities organized into four stages:

Stage I. Description of information system objectives and specifications;

Stage II. Definition of organizational resources and constraints toward system development;

Stage III. Generation and evaluation of alternative system designs; and

Stage IV. Overall evaluation of costs and benefits for system selection.

Figure 22-5 represents a flow diagram of the structured design model. Feedback occurs between stages of the model, making it a process of interdependent actions.

In Stage I (Steps 1–9), systems specifications based on evaluation of user needs are outlined. Thorough analysis of what currently exists (e.g.,

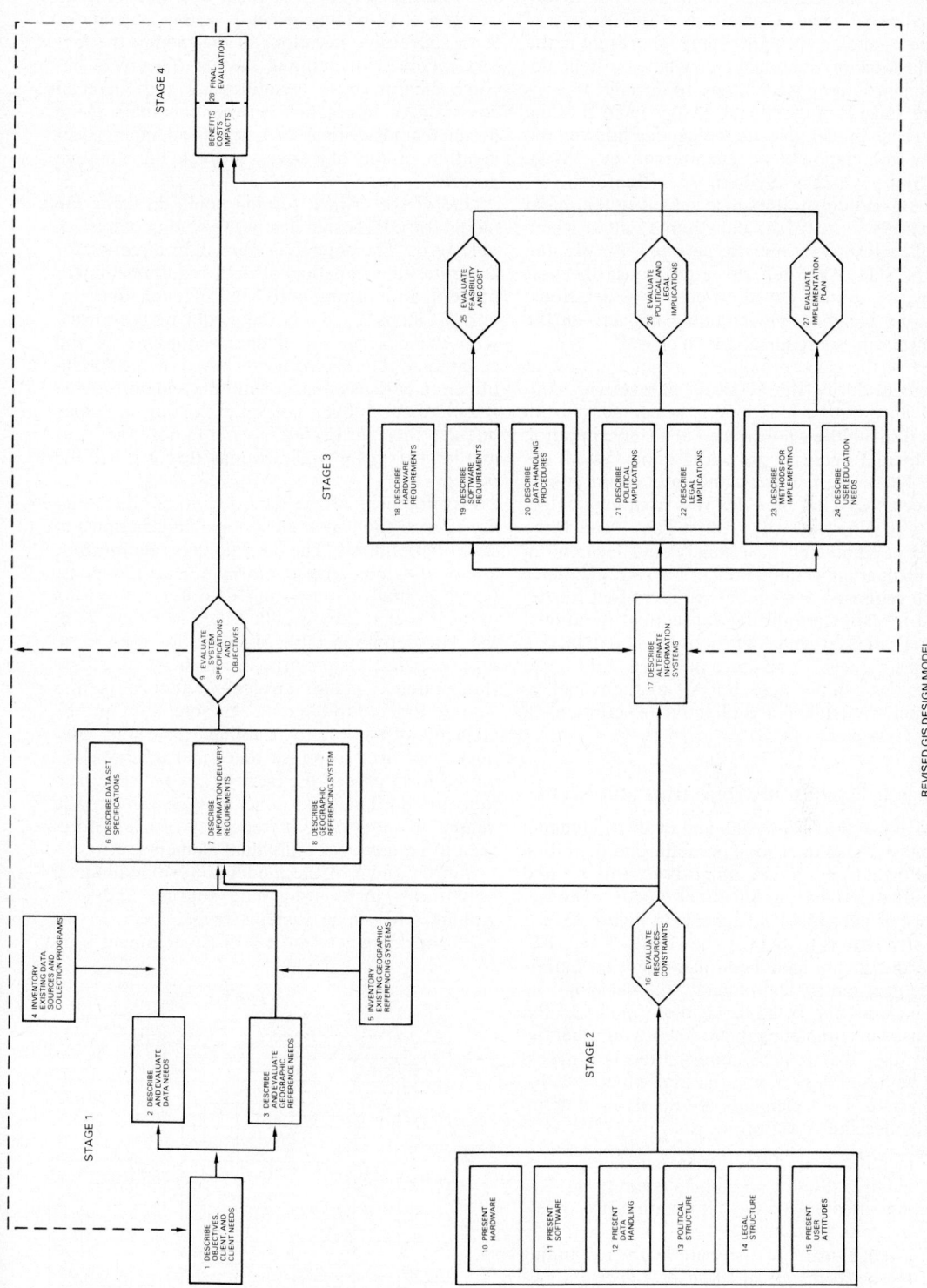

REVISED GIS DESIGN MODEL

Fig. 22-5. The Structured GIS Design Model.

procedures, decisions to be made, data) compared with desired capabilities is executed in this portion of the model. Required as an input to the evaluation of Stage I results is a description of resources and constraints (Stage II) present in the organization or organizations where system development is proposed. Steps 10 through 16 constitute Stage II of the model. Stages I and II of the GIS design model provide the needed information for the formulation of system alternatives in Stage III (Steps 17−27). System specifications, resources, and constraints determined in the initial two stages serve as the constraints within which feasible alternatives must be selected. Finally, the task of Stage IV (Step 28) is to choose the best alternative system based on cost considerations, proposed benefits, and potential impacts on the organization or organizations involved.

One of the basic principles guiding the flow of the model during the actual design activity concerns an iterative function. It is not possible to work through the whole model in a single iteration in sufficient detail to provide a good system design. There are feedback and interaction effects between steps that preclude this from happening. Utility is enhanced when a first pass through the model is completed in a generalized fashion. In this manner, questions concerning the character of the proposed system can be identified for detailed investigation during subsequent iterations. The actual system-design activity is, in fact, a learning process for all those involved. Additional passes through the steps composing the model result in more detailed and effective descriptions of the system elements.

Stage I: Establishing Objectives and Needs

Stage I of the GIS design and evaluation model examines those questions pertaining to definition of information-system objectives and related specifications based on client needs. The sequence of steps involved appears in Figure 22-5. A first step here is to identify the clients of the GIS. Once the clients have been identified, an outline of the decision system is made; the decision system includes the tasks to be performed and the data used in completing them. Specifying what the client does that requires information is a necessary building block toward setting objectives for the system. Four elements are involved in defining the decision system:

1. identification of the decision makers;
2. identification of potential system users;
3. identification of the information-use objectives; and
4. establishment of the criteria for evaluating the performance of the user's decision system.

The process of defining system objectives offers a clear example of the utility of using the model in an iterative fashion. It is difficult to set systems

objectives after a single iteration of the GIS design model. After the first iteration, the potential design team of developer-sponsor-user may only be capable of providing general system objectives. With successive iterations of the model, the factors involved in defining system objectives become clearer (e.g., technological and financial constraints). Interaction between members of the design team is critical for reaching adequate specification of the objectives through the iterative process.

One of the critical turning points in using the model for GIS design lies in the setting of system objectives. The objectives must offer direction for accurate determination of data requirements for present and future activities of each user involved. Steps 2 and 3 of the model are concerned with the description of data requirements and geographic referencing needs based on solid identification of the system constraints and objectives. Requirements cover not only existing activities but also those projected to exist in both the short and long term for organizations that will use the system.

Development of the data-definition table (Figure 22-6) is one way of describing information and data requirements. The table depicts relationships among user objectives, information requirements or types, and the necessary data items proposed to meet user needs. As illustrated in Figure 22-6, the data-definition table includes, for each item, its associated characteristics, such as classification scheme, scale, currency, accuracy, and source. Related to the data characteristics are the elements of use for each defined data item. Frequency of use, response time, and accuracy required form part of the basis of system specification. In addition, data items may be weighted in degree of importance if frequency of use is foreseen to be great for individual elements.

Steps 4 and 5 of the model involve conducting inventories of existing data sources and geographic referencing systems respectively. In the public sector where most GIS development proj-

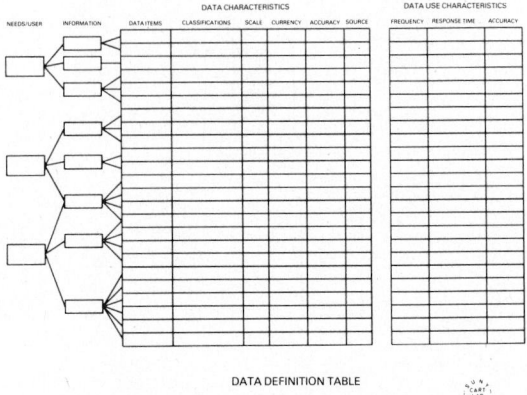

Fig. 22-6. Data Definition Table.

ects have historically originated, many data are collected through secondary sources and programs. Economic considerations frequently prevent agencies from collecting many of their own data directly. As a result, the availability of data often poses a major constraint on the eventual system. Equal attention must be given to surveying the geographic referencing systems that exist in potential user agencies. Although a GIS must reformulate a manual system in digital format, the basis for accomplishing this task will be the subsystem of maps, ground surveys, air photos and other remote sensing inputs.

The system-design team must be mindful of potential constraints on data availability: (1) administrative obstacles, (2) confidentiality obstacles, (3) time constraints, (4) continuity constraints, and (5) cost constraints. Administrative and confidentiality obstacles generally refer to problems of accessibility due to the data's physical location and protected nature, respectively. Time and continuity constraints may impede use of data for agency activities if data cannot be secured when needed.

The initial steps of Stage I of the GIS design model serve as preliminaries to the next three activities in the model. Steps 6 through 8 focus on formulation of system specifications for data, analysis capabilities and output form, as well as the geographic referencing system. The stated purpose of these actions is to translate user objectives and data needs (Steps 1–3) into system specifications suitable to support subsequent design and development activities. Background material required to help achieve this end has been previously identified in the surveys completed for Steps 4 and 5.

Discussion of Steps 6 through 8 can be related to three broad concepts associated with information systems—input, storage, and output. Specifications for these concepts are interrelated so that translation from each to the other must be based on the purpose for which the data will be used and the constraints involved. According to the GIS design model, several results are documented: (1) data to be included in the system, (2) the geographic referencing subsystem, (3) data-retrieval requirements, (4) data-analysis requirements, and (5) the form of information delivery and output. Parameters such as data format, data volume, record format, and recording media along with qualities of accuracy and reliability are important. Second, input procedures to build the data set are considered. These include methods for partitioning the data set (e.g., between cartographic and descriptor components), encoding, editing, reduction, and updating. Following the input segment, data-set specifications focus on data structures, file structures, and data-base management concepts to conclude the responsibilities of this step in the GIS design model.

Once a complete set of system specifications has been formulated, an evaluation of the results should be conducted (Step 9). The purpose of such an analysis is to reach some conclusions about the specifications based on three parameters: (1) effectiveness, (2) consistency, and (3) feasibility. Effectiveness represents the degree to which the system described meets the objectives and needs outlined in Steps 1–3 of the model. A suggested method of comparison is to create a data-definition table for system specifications. Similarity between two systems can be interpreted as a generic representation of expected effectiveness. The second factor involved in evaluating system qualifications and objectives is consistency. Internal conformity between geographic referencing capabilities, data-set specifications, and analysis and information-delivery requirements is necessary in developing a GIS system. Consistency checks are one means of assuring that proposed qualifications represent a solution capable of meeting user objectives. Feasibility is the final parameter used in the evaluation step that culminates Stage I of the design model. In this discussion, feasibility can be defined as a broad, subjective evaluation of the reasonableness of the specifications. This parameter involves generation of overall cost estimates for data acquisition given the data-availability constraints already cited.

Outcomes from the evaluation of system specifications and objectives may require several iterations of Stage I or parts thereof. If the specifications were found to be unsatisfactory by the evaluation measures given above, another iteration would be needed to modify either the whole or selected parts of the specifications or the requirements.

Stage II: Development of the Design Envelope

This stage involves a full description of the physical-logical structure of spatial data-handling in the user's organization to be used in determining resources and constraints for system development. It is within these constraints, or within this design envelope, that system development must take place. Factors requiring consideration in this stage include the present hardware facilities used by the organization (Step 10), the present software used by the organization (Step 11), the present data handling capabilities and procedures within the organization (Step 12), the existing political structure as it impacts the organization (Step 13), the present legal structure that influences the organizations activities (Step 14), and the attitudes and capabilities of the proposed users (Step 15).

Step 10, the description of present hardware, involves several questions for the system designer to explore. Among them are: (1) what is the hardware?, (2) what are its characteristics?, (3) what is its availability?, (4) what is it used for?, and (5) what potential uses does it have for the proposed system? The description of present software for agency use (Step 11) follows a similar pattern of analysis as its hardware counterpart. In many cases, software for spatial data-handling at the level provided by a GIS is generally unavailable in an agency. If GIS software is currently used in the organization's work, however, an investigation is made of what the software does along with its existing documentation. Additional investigations may be made into how the software is provided; several possibilities exist including in-house development or the use of contractural agreements with outside suppliers. Related to the question of how software is provided are the factors of acquisition costs and efficiency levels in using the software. Cost comparisons between in-house and contracted software development are necessary in determining the most cost-effective means for the proposed system. Furthermore, a thorough description of present software takes on greater importance with the rising costs of software maintenance.

Data-handling procedures currently used in an organization represent the third element in describing the physical-logical structure. Each component of data handling warrants examination on the basis of existing tasks to be completed by a given organization. Investigation of the present skills of agency personnel in data handling is an associated point of inquiry at this juncture in the model.

Evaluation of the present political framework (Step 13) is a substantial activity because system development may be severely hampered through ignorance of the power structure in the organization. Political acceptability, for example, has a marked effect on the success of a proposed system. In this vein, qualitative conclusions should be drawn concerning the attitudes of agency personnel toward both computers and GIS development. Additionally, an inspection of the legislative mandates and other guidelines under which the organization operates should be made. Inter-organizational and intra-organizational agreements and cooperation, for instance, are needed for design and operation of the proposed system. Summarizing, the existing political framework can represent a significant constraint on system development. For this reason, a thorough description of its character constitutes necessary background information for GIS design.

The associated legal structure affecting an organization (Step 14) is no less an integral part of establishing the resource-constraint set. One of the foremost legal questions to be resolved is what chain of authority exists for operating the proposed system? Legal means must also be established, if necessary, to give an organization control over acquisition of data for the system and to ensure data confidentiality once the data have been obtained.

Lucas (1973) offers some insight into the importance of user attitudes in assessing impacts of system development on an organization. Potential users or persons affected by a proposed system respond differently to system development based on their perception of the new system and the changes it may incur in their activities. To create an atmosphere of acceptance among the people involved, Lucas suggests that a strategy be formulated to deal with user attitudes. Within this strategy, discussions concerning system objectives and expected system benefits serve to ease the fears and negative feelings toward system development. A positive environment for operation of the proposed system is hopefully created as a result.

Results of the six steps discussed above represent input to formulation of the resource-constraint set in Step 6. Description of the resource-constraint set, in turn, forms a portion of the information used to evaluate Stage I (system specifications and objectives). Comparisons can be made between what exists in the organization at present and the proposed system qualifications outlined earlier. In this manner, certain potential conflicts can be resolved before further energies are expended. The resource-constraint set also adds to the information base for later utilization in the design and evaluation of alternate systems.

Stage III: Definition of Alternatives

Stage III of the GIS design model focuses on the development of alternate information systems (Step 7) given the products of Stages I and II. This process involves "definition of the relevant dimensions, or decision variables, for the particular problem and specification of the appropriate values, or choices for each dimension." [Calkins, 1972]. The steps to which this approach must be applied are:

1. Hardware requirements (Step 8);
2. Software requirements (Step 9);
3. Data handling procedures (Step 20);
4. Political implications (Step 21);
5. Legal implications (Step 22);
6. Methods for implementation (Step 23); and
7. User education and training procedures (Step 24).

Steps 18 through 20 of the GIS design model involve the technical side of describing alternate information systems. In Step 8, description of hardware, it must be understood that several different computer hardware configurations may attain the same system goals. The system designed initially sets ranges and magnitudes for cost and performance and these become more detailed as specific alternatives are weeded out.

Software description (Step 9) poses some important questions to be answered for each of the feasible alternatives. The concepts of transferability and software-development specifications (i.e., developed in-house vs. purchased) are worthy of consideration in identifying alternate system requirements. Perhaps the most critical requirement to fulfill, though, is ensuring that the software is capable of meeting system specifications defined in Stage I.

Step 20 constitutes the description of data handling procedures for a given system alternative. Data requirements for task completion have been specified in State I. Those data requirements are the basis for determining adequate data-handling procedures for a proposed alternative.

The technical steps of Stage III must be appropriately evaluated in terms of feasibility and cost for each alternate system design. Hardware and software cost-evaluation represents an iterative process where the eventual product should be the one of lowest cost that will provide a satisfactory configuration. Effectiveness tests are suggested for evaluating the performance of software on various hardware configurations. Additionally, cost functions are recommended as tools for expressing relationships between capital and operating costs and system attributes. Impacting these cost functions are the data requirements defined in Stage I of the design model such as the number and type of data items to be included in the system. Procedures for handling these data items are evaluated on the basis of efficiency and accuracy considerations. In the final analysis, evaluation of the technical steps (Step 25) is often subjected to cost and feasibility tradeoffs between hardware, software, and data-handling procedures. This is necessary in selecting the best technical alternatives.

Steps 21 through 24 entail the description of alternate information systems in terms of nontechnical elements—political implications, legal implications, methods for system implementation, and user education and training needs. The intended purpose in depicting these elements is to gauge the environment surrounding a given system alternative in terms of potential reactions to the system and the external conditions that will inevitably affect the system's ability to survive.

Associated legal considerations are no less significant than those of a political nature. The terms "authority" and "responsibility" characterize the essence of legal aspects for judging alternate systems and selecting among them. Initially, the authority to operate the system, collect data, and provide information about the systems must be identified for each system alternate. The responsibilities for assuring data security and confidentiality follow from establishment of this authority.

In the description of methods for implementing the particular alternative (Step 23), several variables are applicable. These include the degree of centralization proposed for the system, the institutional structure, and the operating environment in which the system will reside. For instance, consideration of the operating environment for each alternative system includes estimates of staff resources, management policies, and funding programs.

Stage IV: System Comparison and Selection

Final evaluation of alternate information systems is completed at this point (Step 25) and represents Stage IV of the structured design model. Benefits, costs, and impacts of the alternatives are compared for selecting the system to be implemented. The choice is difficult, given that many benefits and impacts are subjective and resist firm quantification. In this stage, two options exist in selecting among alternatives: (1) one alternative can be chosen for implementation, or (2) new alternatives may be specified based on modified objectives and/or system specifications. The iterative approach that the GIS design and evaluation model purposely adheres to is also helpful in this stage. Iterations through reformulation of alternatives allows the insight gained from one iteration to be used in deriving better, more acceptable alternatives leading to system selection in successive iterations.

Technical considerations play an important role in the design process discussed in this section. They define, in many ways, the constraint set under which the system design must proceed and must be firmly faced when the point is reached at which the functional requirements generated by the structured GIS design model are to be translated into a specific system design which will, in turn, be implemented in software and hardware. The following section briefly outlines some of the major technical considerations that must be faced by the system designer. Considerable attention is given to map-data capture since this is the major alternate to remote sensor data inputs, and to problems of data structure, algorithms and data management since these represent the most pressing technical problems in GIS technical design. Many of the technical problems faced in the GIS field are closely related to those found in, say, image processing, and solutions to problems in one field (e.g., DBMS for spatial data) will certainly have a significant impact in another.

THE GEOGRAPHIC INFORMATION SYSTEM: TECHNICAL CONSIDERATIONS

The previous sections have provided a general insight into aspects of, and problems relating to, geographic information systems. The state of the art in remote sensing has advanced from aircraft-mounted cameras to multi-band imaging scanners mounted on orbiting satellites. These satellite imaging systems, such as those aboard Landsat, capture vast amounts of spatial data directly in digital form. For example, Landsat 4, launched in July, 1982, has the capacity to generate a total of 700 scenes or images per day. This represents 10.4×10^{11} bits of information. Large volumes of spatial data, as well as ancillary data from many other sources are also being used in geographic information systems. Increasing attention is being directed toward conversion of large volumes of spatial data stored in analog, archival form on map sheets of widely varying quality. These data are being converted to digital form through automatic digitizing, i.e., the use of automated drum or flatbed scanners.

Most of the time and resources required to create a GIS are, at present, concentrated in the data-input phase. Creation of detailed, high quality map sheets requires years of effort; subsequent extraction and transformation of the resulting analog source data are both slow and expensive. Hopefully, when the vast analog archives have been successfully reduced, remote sensing will become a major source of timely data for geographic information systems. Until that time, a primary concern in GIS operations will be with digitizing operations.

METHODS OF SPATIAL DATA CAPTURE AND ENCODING

Data capture for spatial data-handling purposes is usually referred to as "digitization" and the primary source of data is frequently existing maps. An analog (i.e., paper) map may be designed for visual assessment and, as a result, digitization often must be by manual tablet means, to allow the interaction of the human eye and mind to be incorporated in the process. Only in certain special aspects of existing map work can the high speed and economic advantages of automatic digitization be realized. As a result, the most efficient method of handling this type of data capture will probably remain as a reasoned balance of manual and automatic processes. In both cases, the use of interactive display and edit is an essential requirement, in order to produce "clean" data.

Other methods of data capture cover a wide range from the use of classified Landsat data to man/computer cooperation in the analysis of stereo pairs of aerial photographs, and in the digitization of interpreted locations, lines and areas on visual images—usually aerial photographs.

Manual Digitization

The electronic digitization tablet is now widely known and used in the cartographic world. At this time, the variations in operational methods from a device point of view are of little importance. The difference is usually in tablet area and in spatial precision. The latter may vary from 0.005 inch to a maximum of 0.003 inch; the fact that the resolution may be indicated to 0.001 inch does not, of course, infer that this is the precision.

Most tablets are now directly connected to minicomputer systems, although the addition of micro-processors in the interface chain can assist some aspects of data checking and on-line transformations. Digitizing systems, with or without microprocessor interfaces, impose little strain on a central processor, even with multiple stations, since the data-flow rate is low. However, other processes running on the central computer may well cause extensive delays in digitizer response.

Action may be quick when the system is available to the digitizer station, but the wait for that station's "turn" may be extensive.

The main problems in using a manual digitization tablet arise because digitization is inherently slow and manual digitization involves human labor. While human involvement has advantages for a correct data assessment, it nevertheless causes major problems with respect to high running costs and many errors being generated owing to the presence of an operator.

An examination of a number of digitization systems in operational use shows that frequently there is poor organization of the work. This has been shown to cause unnecessary increases of up to five times in both initial and in running costs—a serious matter. Some users have moved too rapidly into multiple, manual digitization operations, just because they can be done and appear to be visually effective. Often, however, the downstream problems of data handling and accumulation of data errors have not been given enough consideration, and an overall useless effort can result.

Methods of actual digitization are relatively similar but small differences can be important in both cost and time. Point digitization is effective, relatively simple, and may well be effectively enhanced by audio input to the computer system. Line digitization is more difficult and, when suitable, such as in utility maps and other large scale urban layouts, may well be done by entry of selected points with the lines being created by software from end points and points on curves. When lines are irregular, then stream digitization becomes necessary and is generally effective, though slow. Polygons are usually best entered by a method known as "arc" digitization, an "arc" being a straight or irregular line between nodes. Some systems allowing "free" digitization and, relying on software to sort it out, seem to require excessive computer time.

Even at the stage of manual digitization, when data volumes are low, proper data formats and structures are critical. Properly designed systems should allow checkbacks for the operator. These may, for example, be a listing of all nodes in a polygon structure where less than three lines connect—a likely error indication. Some warnings of a simple on-line nature may be indicated by beeper tone.

From the hardware aspect, the usual cross-wire cursor could well be improved for stream digitization of irregular lines. Such units exist today, but are only made to special order. The normal hand-held "pen" can be used where low precision but high speed are indicated; this might be, for example, in the tracing of interpreted timber stands on aerial photographs. A directly attached display is not usually of importance and is often an expensive addition to the work station. It is useful when on-line transformations are done and new data must be fitted into old. In many digitization situations it has been noted that the display takes the operator's attention away from work and leads to decreased productivity and more frequent errors. Of course, the operator must always be able to examine, at any time, his completed work and his work that is in progress, but this is best carried out on an associated display and edit work station; this is particularly important after returning to work following a break.

One method for descriptor or attribute entry, as well as for function selection is to use a menu of possibilities. This avoids diverting the operator's attention to a keyboard from the digitizing table. The menu may be on the digitization tablet itself, or on a smaller moveable tablet placed conveniently to the operator. The use of audio input, often accompanied by audio output, is growing, and has advantages in some cases. One use is in the digitization of numeric soundings from charts and another is in the operation of a stereo compiler. In both cases the important aspect is to enable operators to keep their eyes on their work.

Automatic Digitization

The automatic digitization of existing map sheets is now being carried out by a number of methods. The two main approaches are usually referred to as automatic line following and scanning. The differences are more apparent than real, as the scanning methods also normally involve software line-following to move from raster to vector data-structures. Two systems referred to as "automatic line following" come from the same generic source, which is the Sweepnik system developed in the United Kingdom in the 1950s to follow tracks on bubble-chamber photographs in nuclear work. One of these is the basis of a service-bureau operation in California and the other is being tested by a number of agencies, particularly the Ordnance Survey in the United Kingdom. A large amount of human involvement is still required in use of this system, which itself only replaced the hand tracing movements of a manual digitizer by a very fast, helical, laser-beam tracking system. However the operator still has to think, enter the attribute data, place the tracking spot on the line of interest, and check to be sure that junction situations have not been missed and have been handled properly. The presence of the operator in the work sequence can, of course, be advantageous in complex map-discrimination situations, although often, if the map is very complex, the lines to be tracked are many and short—not the most efficient situation for the unit. The costs per sheet tend to be high and many hours of work are involved; these must be compared with doing the whole procedure on a manual table and a great deal depends on the type of data being handled.

Most users in North America currently prefer the scanning approach. There are a number of different techniques at this time being supplied or developed. The most common is the use of a large drum scanner such as used in graphic arts and, in fact, many developments have come through this route. The system is similar to the widely known facsimile transmitter, with a photosensor tracking axially as the drum is rotated. Resolutions of 0.001 inch or better can be obtained, but the complex lead-screw mechanisms tend to make such units expensive and relatively slow in operation with one sheet normally taking one hour to scan. Such units have, in fact, been used for cartographic input for many years, and one, made by IBM, has been in continuous use by the Canada Geographic Information System since about 1965.

Another method utilizes a comb of about 1000 photodiodes on 0.001 inch centers, which is slowly swept across a strip of the map sheet approximately one inch in width. This scan is repeated side by side until the whole sheet is covered. The mechanism precision must be high as the strips must be exactly adjacent to within better than 0.0001 inch, and the devices are thus expensive.

A third method uses a laser-beam, flying spot scanner where a narrow strip is swept by the beam and the reflected light measured for intensity. A fourth method uses a similar laser-beam scan method, but applies it at one time over the whole sheet. While this involves some technical design problems relative to light collection to detect the cartographic line edges, the mechanism is relatively easy to fabricate to the required precision and can operate at very high speeds.

The first three procedures require normally at least one hour of scan time while the last can be done in minutes, and could theoretically be done in seconds. Another difference between the three methods, which may be important for some applications, is that only the first is easily adapted to color discrimination.

The method of use and application of automatic digitization at this time differs from agency to agency. Color detection is used at one, but is only applicable to color composites specially made, or to paper maps; these are not usually regarded as suitable input documents in quality or complexity. It must be appreciated that the subsequent automatic data-handling programs are simple-minded and can only work reliably on clear, unambiguous, high quality input documents; on those they can be very effective. In practice, standard map-separation sheets are normally used in positive or negative form. Contour sheets are the easiest to handle, however many the contours and however irregular the lines. Real economies can be made against manual input, although of course the question must be asked "Is the digitization of contour sheets required or are the data not better obtained from automatically digitized DEMs"? Drainage sheets are the next easiest, although the software must then be more sophisticated to handle junctions and symbolized lines. If good junction-detection capabilities are present, then polygonized overlay sheets are easily handled, except for the fact that the polygon labels should be written on a separate overlay sheet or at least in differently colored ink. In order to make sheets more suitable for any subsequent automatic digitization some of the present manual cartographic procedures could be modified with advantage.

Any open window, color, area-separation sheet is normally easily handled, but the culture sheets are very difficult, although some photomechanical pre-edit procedures have been examined that might solve the problem. At the present time this and other sheets are often better suited to, and more economically digitized by, manual means.

In general, scan digitization is best suited to sheets with many irregular lines, few junctions and no superimposed line or symbol data. Dealing with the last can involve an entirely unwarranted amount of pre- and post-edit, far more costly than the actual automatic digitization, making the overall cost differences between manual and automatic digitizing negligible.

The resolution of the scanning is important since the physical scanning is relatively easy compared to the subsequent processing. The raster-to-vector and line-following software is much more critical. Data stored at a resolution of 0.004 inch are probably adequate and as good as the original, manually-created drawn image. However, there has been a tendency for agencies to set up their scanners to 0.001 inch for three main reasons. The first is to be able to replot the data with good visual line appearnce. The second is to be able to discriminate between different line weights, and the third is to be able to detect situations where lines come very close together but do not actually touch. However, it must be stressed that this 4:1 change in scanner resolution results in a 16:1 increase in processor and I/O volume of data as well as a 16:1 increase in time. Thus one hour of data handling automatically becomes 16 hours even with a dedicated processor; this can be an excessive difference in both throughput and operational cost. Hardware methods are now being examined, particularly for the laser-beam scanners that can easily be increased in speed to give higher resolution with the same scan time, to scan at 0.001 inch and then examine the data to detect line weight and adjacencies, before passing it on to the subsequent data-handling routines at 0.004 inch. If plotting is subsequently required, line-smoothing interpolation routines are applied to the 0.004 inch data in the usual manner, to provide lines of good visual appearance.

The data-handling procedures at present always contain two major elements. The first is line thinning to produce single pixel-width "data" lines (say at 0.004 inch per pixel) and the second is to vectorize these lines into the presently used cartographic form. If there should be a real change in the next few years to raster spatial-data bases, then this second software operation may become redundant. In well-written software, the time for each part is about equal, although in some applications the thinning time is very much greater. There are many procedures for line thinning and the fastest usually require that some limitations be placed upon the number of pixels across each line, with five being common. Some software aborts if the line is wider than this fixed number of pixels and many, to be on the safe side, treat all lines as possibly being wide, gradually nibbling them down; these later approaches can often lead to excessive computer time charges.

The junction problem can be handled either in the thinning or in the vectorizing phase, although it usually appears best to treat it in the former. It is a serious problem to define the exact junction to one to two pixels, particularly if the junction is small-angled or "blobby." De-symbolization, as far as dotted and dashed lines are concerned, is relatively easy when handled on the vectorized data and this also leads to some automatic handling of the labelling problem. In general, one of the main problems with automatic digitization is that labelling must be a separate process carried out on an interactive display and edit station. This is a manual process that must be carried out on a unit which is more expensive in hardware and computer costs than a manual digitizing table. One possible answer is to digitize manually a single point on each line, while entering the descriptor, and then use software to combine these with the scanned data. As with manual digitization, the organization of the work process is all important and can make enormous differences in cost when a large amount of repetitive work is to be done. A number of attempts are being made to provide more assistance in the labelling problem. These include efforts to label all contours after the index lines have been manually entered.

Because of the larger amounts of data created in scan digitization, data formats and structures must be very carefully examined and used. These are probably not the same as would be used in the final GIS database, but must be optimized for the data-capture work in hand.

Due to the operational difficulties involved, any organization should give serious consideration to the question of obtaining its own scanner. The device can, in the right environment, have enormous throughput and high quality results. However, this can only be obtained in a dedicated situation where the scanner is kept fully in operation. Smaller users should use other facilities (e.g., service bureaus) for scanning, but do their own pre- and post-edit of the scan data to bring it to required standards. Because this work involves manual interaction, the end user should be more efficient at this than a remote service operation.

Display and Edit in Digitization

An integral and important part of both manual and automatic digitization, is a connected display and edit system to enable the operator to check the work that has been carried out and, if necessary, add to it or change it. With manual and "automatic line following" digitization, the work is mainly one of correcting junctions. With scan digitization, the data are usually clean and good, assuming that the input document and software are also good. However, appreciable time has to be spent on adding labels to the line features.

Users are easily irritated by a system with a slow response time and tend to compare its capabilities with those of a paper, pencil, and eraser—often to the former's great disadvantage. Response time should not normally be longer than ten seconds and preferably should be at about two seconds (the situation is very critical and responses faster than two seconds do not seem to give any advantage). Such acceptable response times have also been found to be critical in areas such as CAD/CAM for the design of printed and integrated circuitry in electronics. From hundreds of original vendor attempts, the remaining suppliers (in CAM/CAM) are those with fast responses, obtained by specialized programming, aided by an operator system that is "user friendly." Some years ago the edit systems designed for engineering drafting appeared to be acceptable for spatial data handling. At that time, digitization only produced small amounts of coordinate data and often such data merely represented straight lines. Total coordinate datasets never exceeded 10,000, and were normally less than 1000 points. With automatic digitization of one contour sheet, data volumes can easily exceed one million coordinate pairs, which makes a proper reponse time much more difficult to obtain. In fact, it probably cannot be accomplished at this time when using a single time-shared processor (however large and powerful) handling several edit stations in a production environment, if all are in operation at the same time. The present tendency is to make each display and edit station a self-contained unit with its own CPU, tablet, display, cursor and Winchester disk storage. Data are transferred periodically to this station from a central system associated with the digitizers. This work station can take the full coordinate output from one sheet (perhaps one million coordinate pairs) and manipulate it all at one time. When one is comparing response time it is useful to take a dataset of one million coordinate pairs (made up of an average 10,000 lines of 100 coordinate pairs each) and request a display of, or an edit involving, 1 percent of these points; this should be only a matter of a few seconds response. This standard may appear difficult, but it can be done.

The new types of 1000-line color displays are now more in favor for edit stations than the green-colored, storage displays so long used as the work-horse in digital cartography. The rapid reduction in costs of the image-plane memories, the lack of flicker with the slight color persistence, the ability for dynamic change instead of erase-and-regenerate, the bright, clear colors and the quality of the 1000-line unit as against one with say, 500 lines, all combine to create this situation. Color, by itself, is not essential for the presentation of newly digitized data, but its usefulness in discrimination between different data types and between old and new data is important.

Except for the high response-speed aspect, the display operations are relatively easy to obtain, including windowing. The operations required in labelling, pointing, finding and entering by keyboard, menu or audio are not difficult. Line-data manipulation can be more difficult, but the real difference between good and bad systems lies in the ease of use by the cartographer. Cartographers should not have to be computer systems analysts to do this work.

With the advent of scanning-input devices and the increase in raw data already in raster form, the possibilities of editing the data in that form are now also being seriously considered.

Once the analog map-data have been successfully captured and edited, a major problem facing the GIS is data management. How are the large volumes of spatial data generated by the digitizing process to be organized for efficient query and update?

DATA MANAGEMENT IN THE GEOGRAPHIC INFORMATION SYSTEM

The data volumes and the variety of data, as discussed in the previous sections of this chapter, have been rapidly increasing through the use of modern data-capture technology at considerable cost in money and in effort. The data, in an established database system, usually are derived from many sources and are in many forms. This situation has been developing over the past two decades. Several large spatial databases currently in existence are the result of continuous effort over this entire time span, such as the Canada Geographic Information System, and not only have become major capital investments themselves, but have also have become essential for the functioning of their respective organizations. Through the development and use of a geographic information system and its associated spatial database, both the variety and the amount of data in the database increase. The range and sophistication of applications also increase and information needs change through time as the database grows and the geographic information system becomes an everyday information tool.

These aspects are not different from the experience in other types of information systems. Nonspatial information systems have become quite sophisticated and commonplace in the business world over the past fifteen years and a substantial body of knowledge has been built for the development of these systems. This technology is also at a higher stage of development than that specifically geared for spatial systems. It would therefore be useful to briefly examine some of the basic principles of current "conventional" database management-systems technology, as well as to look at some of the off-the-shelf DBMS systems. This should provide a basis for a better understanding of the problems we have to face in the development of spatial, or geographic information systems of a similar size and level of sophistication and in the integration of remote sensing data into such systems. For a detailed explanation and discussion of modern DBMS concepts, the reader is referred to Martin [1977], and Date [1977] among numerous texts on this subject.

Numerous computer-software systems have been developed that are directed toward providing an integrated, cohesive package for the storage, retrieval, and maintenance of a large collection of list-oriented data. There are currently over fifty business-oriented DBMS commercially available [Krass, 1981]. The use of these systems is expected to grow at a rate of 25 to 35 percent annually through the end of this decade. The example of success and rapidly growing use of these systems is another factor in the growing demand for spatial database management-systems.

A database is defined as "a collection of inter-related data stored together with controlled redundancy to serve one or more applications in an optimal fashion. The data are stored so that they are independent of programs which use the data. A common and controlled approach is used in adding new data and modifying and retrieving existing data within the database" [Martin, 1977]. This has also been referred to in the past as a databank. A database management-system consists of the combination of a database and the associated software to perform all needed data entry, storage, retrieval, and maintenance tasks. The database management-system may also be the nucleus "subsystem" of a larger information system that provides analytical and other facilities beyond the scope of general file management.

No matter what data the information system contains or its intended use, there are several basic characteristics of any type of database system which are intended benefits and these help establish the requirements that the system must satisfy:

1) Provide a standardized (uniform) format and a set of standards and procedures for recording data;
2) Allow efficient storage and retrieval for all the intended data;
3) Provide independence of the storage format of the data from all programs that use the storage format of the data;
4) Ensure non-redundancy in the data;
5) Protect the data from accidental loss through hardware failure or human error;
6) Provide ease of system use; and
7) Provide economy of system use.

Modern database management-systems, and the database management-component of information systems, provide more than these basic require-ments of data handling. They provide many other desirable capabilities for dealing with large, shared databases. These include automatic checks on data integrity, security, and real-time accessibility through direct user access. Typically, the data are entered into the computer files using such characteristics as name, type, or relationship to other data. In this manner, the user constructs the file by defining the data to be stored, and then presents the actual data values to the system in the prescribed format. During this initial data entry, errors are automatically presented to the user for correction, so that only correct data enter the file. If the user has a previously existing file in another format, the system may provide special translation programs to achieve the necessary conversion. After file definition and structuring are complete, the user may manipulate the data, add to them, and modify, merge, subset, sort, or retrieve by simply querying the computer and requesting single information items, automatic report compilation, simple graphics, or hardcopy retrieval.

The user interface is a dual one; to interface the data through data-description instructions, and to interface the generalized processing capabilities through query instructions. These were originally employed by incorporating the instructions in a program written in a standard computer language, such as COBOL, FORTRAN, or PL-1. This method limited use of database management systems to persons with programming expertise and it was batch-oriented. In modern database management systems, however, the user interface is achieved through a very high-level language of its own. These data description and query languages are designed to be interactive, user-friendly and non-ambiguous and to provide instant response in a man-machine dialogue. In the case of a geographic information system, this interface may include the use of interactive graphics in the form of mixed cartographic and textual displays, thus allowing man-machine dialogue concerning spatial phenomena via a spatial medium.

A database management system provides the capability for the various users to view the data in their own ways and at varying levels of abstraction. The data are composed of entities and relationships. Most of the complexities of data storage and representation arise from the fact that these relationships must be recorded in some way. There is also a variety of ways in which the data users (and the different levels of users) view these relationships. No one except the database administrator and the technical staff should care about the physical representation of the data. End users should also seldom, if ever, care about the global logical view of the data (schema). The user is concerned with a small subset of the data, organized in his or her own way (subschema). The relationships between data entities can be simple (one-to-one) or, more commonly, complex (one-to-many). The relationships that can be defined by the user in a subschema are affected by the manner in which relationships are defined in the global view (schema).

Modern conventional database management-systems utilize one of three basic kinds of database organizations for defining the schema: tree, network, and relational. The earliest efforts of automated data handling utilized flat files (tables) for field sequencing and retrieving records. The inherent inefficiencies and restrictiveness of this simplest of structures forced it to soon be abandoned in favor of a tree, or hierarchical structure. This has, in turn, given way to the more flexible plex, or network structure and is commonly used by newer, commercially available, DBMS packages. The main disadvantage in network structures, however, is that the pointers which make up this structure type can quickly become overly bulky and complex. This is overcome in the state-of-the-art type of DBMS, the relational database. This concept received notice initially through the work of Codd [1970 and 1972] and its application to pictoral or graphic data was soon suggested [Kunii, et al., 1974] and, subsequently, to cartographic data [Alsberg, 1975 and McIntosh, 1978]. The basic principle of the relational database is that relationships are reduced, or "normalized" to a series of tables, each table consisting of a number of homogeneous, nonredundant columns. Tables are commonly termed "flat files" in this context, since the mechanism of pointers to build an explicit complex of relationships into the stored data files themselves is avoided.

The logical file structures used in database management-systems have thus come full circle, back to table-based data formats, with two critical changes: First, each table represents a single relation and is constructed via a system of strict normalization rules. Second, a powerful query language is employed for user-level definition of data relationships and retrieval. The intent is to allow the user complete freedom in defining the logical structure of the data without regard to the actual physical storage configuration. Data-retrieval requests simply define to the system what data are desired, without having to specify where the data are stored in the database. The development of relational database systems began in the academic community. Recently, however, several very similar true relational DBMS (i.e., those whose underlying data models are relational) have become commercially available or have been announced for release. These include INGRESS from Relational Technoloty, Inc; ORACLE from Relational Software, Inc; and SQL/DS from IBM [Dieckmann, 1981]. The primary limitation of these systems, however, is one which it seems has never been overcome. As Date (1977) asserts, (p. 128):

"(Early relational system applications) were not really concerned with supporting large formatted databases of the sort required in industrial or similar enterprises; rather, they were intended primarily to provide an environment in which a user could ask an unanticipated question, make deductions based on the answer to that question, ask further questions on the basis of those deductions, and so on . . . the term 'relation' as it applies to these systems corresponds far more closely to the intuitive idea of association between a number of things (usually two and usually both of the same type) . . ."

The techniques for storing and retrieving data developed for business-oriented applications are directly applicable to spatial data, but they pose severe limitations. The first limitation is their difficulty in supporting very large data bases. In addition to this, however, spatial data require that the two- and three-dimensional geographic relationships, which are not encountered in such data as personnel records and parts inventories, must be recorded in some way. However, given the many kinds of spatial relationships and the tendency of spatial data files to be very large, it is impossible to store all of them explicitly, even if they could all be anticipated. Many spatial relationships, given the current state-of-the-art, cannot even be precisely defined. This causes significant difficulty in translating these relationships into a digital data-storage format and into a spatial query language. Spatial relationships are intuitive to humans and are usually defined "on the fly" when reading a paper map or in interpreting an enhanced Landsat scene. A critical and unique property of spatial data is that each entity must be defined in terms of its location in space. This may be a single coordinate pair, but is far more frequently a sequence of coordinates describing a line or an enclosed area. Various entities may also be nested or overlapping. This is a very different kind of additional information which must be recorded, maintained, and manipulated. Standard DBMS structures were not designed to accommodate this type of data, although some DBMS structures seem to suit some specific types of spatial data (e.g., tree structures for stream networks). However, if they fit at all, they tend to be overly cumbersome to use for many spatially-based retrieval queries and some critical spatial relationships may be completely lost. The best use of standard DBMS data structures in a GIS is for non-spatial ancillary data, with the spatial data recorded in a format specifically designed for special needs. Although much research needs to be done on the space, speed, and flexibility tradeoffs involved in the design of general-purpose, very large volume spatial-databases of all types, the extreme flexibility in defining data relationships in techniques used for relational database systems offers significant promise in the continuing development of spatial database management-systems [Vaidya, et al., 1982].

Individual records are often physically located and identified by use of a unique sequence of characters, or 'key'. This key is often the name of an entity such as "Mississippi River". The technique for finding a record with a given key is dependent on the physical layout of the file. The simplest (and slowest) method is the sequential search, requiring no special sequencing of the file. Many other searching techniques have been developed in an attempt to speed up the searching process. The first techniques developed involved simply pre-sorting the data records in ascending or descending order within the primary key. These techniques include the block search and binary search. Later techniques required the addition of indices, such as index-sequential and index non-sequential. As the complexity and bulk of these indices (which represent non-data items added to the file) increased, other techniques were developed that eliminated indices in favor of calculating the record locations. These techniques included key conversion and hashing. These were also combined in various ways. All of these approaches involve tradeoffs of speed, extra bulk, complexity, and flexibility.

Although the physical location and retrieval of individual data records is one task that a user is shielded from in any DBMS through the use of a query language, the overhead tradeoffs imposed by various retrieval techniques affect the overall efficiency of the entire system and is thus a primary concern of anyone choosing or designing a GIS. This problem becomes particularly critical for interactive systems which must quickly respond to complex queries, such as "retrieve all areas within Fairfax County which have changed from agricultural to commercial or residential land use within the last five years," from a database containing individual, annual land-use files as well as a separate file containing descriptions of political boundary locations.

DATA MANIPULATION AND DISPLAY IN THE GIS

Any given geographic information system must be capable of performing a series of manipulations upon the spatial data held in its files. Each system contains a specific set of these procedures, determined by the requirements of the users of the system. A typical set of manipulation capabilities is listed below:

Typical Map Data Handling Capabilities*

Data Manipulation

  Reclassify—attributes
  Generalization
    Dissolving and merging
    Line smoothing
    Complex generalization
  Interpolation
    Centroid allocation
    Contouring

Scale Change
  Distortion elimination—linear (rubber sheeting)
  Projection change

Generation

  Points
  Lines
  Polygons
    Simple five-sided polygons
    Irregular polygons with islands
    Circles
    Grid cell nets
    Latitude and longitude lattices
    Corridors (along linear features)
    Other

Data extraction

  Search and identification
    Attributes
    Shapes
  Measurement
    Number of items
    Distances (straight line between points, along convoluted lines)
    Size of areas
    Angle direction
    Volume (cubic measure)

Comparison

  Intersection—overlay
    Point-in-polygon
    Polygon-on-polygon (grid cell on polygon, circle on polygon)
    Other
  Juxtaposition (proximity)
    Shortest route
    Nearest neighbor
    Line of sight
  Contiguity
  Connectivity
  Complex space-attribute-time correlation

Interpretation

  Determination of optimum location
  Determination of suitability
  Determination of desirability

A major differentiation between existing geographic information systems can be seen in those, mainly the earlier systems, which are limited to recasting and reformatting operations (e.g., take in two sets of map data, carry out an overlay operation and then produce a map of the results) as compared with the more modern systems, which are increasingly oriented toward support of analysis and formal modelling (e.g., large-scale, spatial mathematical programming or spatial simulations).

Each GIS consists of, on a very basic level, a data structure and a selection of algorithms that are matched to that data structure. As was pointed

---

* After Tomlinson and Boyle [1981].

out earlier, most operational GIS utilize vector data-structures and, hence, must make use of vector-based algorithms for all internal data manipulation. These algorithms are, unfortunately, little known in an analytic sense and we currently have little comparatitive information which would lead to choosing one vector-based algorithm over another for a particular procedure. For example, there are numerous algorithms for contouring, but no rigourous analysis exists in the literature comparing these approaches in terms of execution time, memory requirements, etc. This lack of precise information on the characteristics of vector-based algorithms represents a major stumbling block to the efficient design and operation of geographic information systems.

Raster-based algorithms are less used in a GIS context [Peuquet, 1979] but tend to be better known on a formal basis since a number of them are also utilized in the image-processing field. Comparison of raster vs. vector approaches to GIS operation is also hampered by a lack of comparative algorithm information. In the design process, should a given set of requirements be met through a raster-based or a vector-based design? At present, there is insufficient information on which to base an effective design decision. Research on spatial data handling algorithms and algorithm/data structure interaction is one of the more pressing needs in the GIS area.

LINKING REMOTE SENSING SYSTEMS AND GEOGRAPHIC INFORMATION SYSTEMS

Geographic information systems are demonstrably powerful tools for the management and analysis of spatial data. Remote sensing systems are demonstrably powerful tools for the collection and classification of spatial data. However, nearly all of the currently operational geographic information systems utilize maps as their primary source of spatial data. These complex documents, designed for visual search and retrieval by human operators, are digitized (usually manually) and then entered into the master spatial data base of the GIS. Although many of the maps used as input are derived, in one way or another, from air photography or—upon occasion—other remote sensing devices, there is little use of digital remote sensing data, especially data obtained from orbital platforms, as a direct data input. The last few years have seen an increase in interest in the direct use of remote sensor data as inputs to GIS, but much of this interest has been centered in the remote sensing community rather than among the potential primary users—those who make operational use of geographic information systems.

REMOTE SENSING INPUTS TO GEOGRAPHIC INFORMATION SYSTEMS

The questions of appropriate spatial scale and resolution have been debated for years. It is clear that some geographic information systems, primarily those which concentrate on small, complex scenes (e.g., sections of large urban areas) cannot be adequately served by existing space borne remote sensing systems, although some systems planned for the next few years (e.g., SPOT) seem to hold promise for these applications. However, there are geographic information systems for which current scales and resolution levels appear adequate. Why are data elements collected by orbital and other remote sensors not normally used by these systems?

Numerous examples can be cited to demonstrate the power of the GIS in the conversion of raw data to useful information, even when explicit use is not made of optimization models in attaining analytic results. There is a basic difference in viewpoint, however, between the GIS operator and the creator of remote sensing data. The GIS operator considers the accurate, digital, polygonal, point or network datasets to be the raw data which form the starting point for subsequent work. To the remote sensing specialist, on the other hand, the digital remote sensing scene represents the raw data and the classified output, expressed in polygon or other form, is considered to be the highly processed product. Thus, depending on the point of view, one person's highly processed output data can be another's raw input data. Generally, each feels that the other is responsible for any necessary transformations and the resulting "data gap" has been a critical barrier to linking remote sensing systems and geographic information systems.

This divergence in viewpoint represents but one of the institutional problems encountered in attempting to expand the utilization of remote sensing data, especially satellite-based data, within operational geographic information systems. Other problems arise since the classified remote sensing image, in pixel form, currently is not felt by GIS operators to represent a generally acceptable input document to most geographic information systems. More acceptable from the standpoint of the GIS would be a transformation of the cellular, classified image into a standardized set of polygons presented in the vector format typical of digitizer output. When presented in this fashion, the output of the remote sensing system would represent just another "layer" of information to be processed by the input phase of the GIS.

From the standpoint of the GIS, two major problems exist with data derived from many remote sensing systems; especially orbital ones. First, the accuracy of the classification scheme used generally falls below the level available for other data elements in the GIS. Second, the positional accuracy of the remote sensor information is frequently significantly less than that of the other data elements. In many cases, this combination of errors has led to the rejection of remote sensor-derived data for operational use in geographic information systems. The accuracy levels are adequate for broad, regional investigations but many GIS today are oriented toward smaller areas and incorporate greater levels of detail. This leads to a greater sensitivity to the problems posed by remote sensor-derived data elements.

GEOGRAPHIC INFORMATION SYSTEM INPUTS TO REMOTE SENSING SYSTEMS

Perhaps one of the greatest oversights of individuals attempting to promote remote sensing inputs to geographic information systems results from assuming that the flow of data should be unidirectional—from the remote sensing system to the GIS. The reverse flow, from the GIS to the remote sensing system, is highly desirable but only infrequently used in present operational systems.

Why should the "backwards" flow of data be so important? Today, remote sensing systems seldom give real attention to what is commonly called "ancillary data" in the development of their classification schemes. The assumption that ancillary data are either missing or inadequate may be justified in certain planetary cases, but in many instances such an assumption merely represents a decision to ignore valuable, existing information which may well be contained in useable form in a GIS. Substantial improvements can be made in classification accuracy if ancillary data (terrain, soils, previously determined land cover, etc.) are used in the classification process [McLeod and Logan, 1980]. Two examples of this are improving forest cover classification-accuracy from Landsat by incorporating topographic information (Strahler, 1978) and other GIS data layers (Maw and Brass, 1981); and improving Landsat land cover classification-accuracy with GIS ancillary data (Likens and Maw, 1981). As pointed out previously, problems in location and classification accuracy are major hindrances to incorporating remote sensing inputs into geographic information systems. Using the information already available in the GIS to improve the classification scheme will, in turn, make the remote sensing inputs more useable by the GIS.

The transfer of pre-processed digital files from a remote sensing system to a geographic information system is certainly possible in practice; there is no real technical problem involved in the transformation of the classified, pixel-based, remote sensing scene into a set of homogeneous polygons defined in a vector data structure. A somewhat different view can be found, however, in the notion of incorporating geographic information system capabilities into existing image processing systems. This has been attempted in several cases with unsatisfactory results since the complexities of the GIS environment have been poorly understood and the utilization of systems optimized for remote sensing data-processing has often led to limited GIS capabilities. An interesting example to the contrary may be found in the Image Based Information System (IBIS) developed by the Jet Propulsion Laboratory of the California Institute of Technology. IBIS is based upon the image-processing capabilities of the VICAR system but also contains a number of powerful GIS capabilities. It is also currently unique in that it is the only raster-based GIS of any size.

A SAMPLE LINKAGE: IBIS, THE IMAGE BASED INFORMATION SYSTEM

System History and Purpose

A major impetus to the creation of geographic information systems has been the desire to integrate land-resource inventory data, derived from remote sensing imagery, with other geocoded statistics. The costs of acquiring and encoding remotely sensed data have frequently spelled the demise of an agency's plans to create a geographic information system, while the lack of funds for preparing land-resource inventory-updates to match other statistical updates have frequently left systems in disuse [Dyer, et al., 1975]. It had become apparent in 1974, as the result of work done with earth-resources images and experimenters, that certain system improvements were necessary for the efficient processing of digital data. The areas particularly needing improvements were interactive processing, geocoordinate data-entry for map and image registration, and multispectral classification. Both geographic information system and thematic mapping technologies have in part suffered from a lack of broader application because of constraints to either data input/output or an inability to interface with other types of mapped phenomena. The use of image-processing technology to resolve this dilemma was proposed in 1975 [Billingsley and Bryant, 1975]. The image data-type seems to be a powerful and general representation for spatially distributed data, and the range of uses can be divided into several broad categories [Zobrist, 1976].

Physical Analog: The pixel value represents a physical variable such as elevation, rainfall, smog density, etc.

District Identification: The pixel value is a numerical identifier for the district and includes that pixel area.

Class Identification: The pixel value is a numerical identifier for the land use, land cover, or other area-classification schemes.

Tabular Pointer: The pixel value is a record pointer to a tabular record which applies to the pixel geographic area.

Point Identification: The pixel value identifies a point, or the nearest of a set of lines, or the distance to the nearest set of points.

Line Identification: The pixel value identifies a point, or the nearest of a set of lines, or the distance to the nearest of a set of lines.

This range of uses requires that the system be able to handle images composed of words of varying length. For example, to identify census tracts in Los Angeles County requires 1500 different pixel values and elevation maps can require 15,000 such values. This is more than the usual 256 grey levels handled by photographic processing systems.

Computer image-processing at Caltech's Jet Propulsion Laboratory began almost two decades ago with computer processing of imagery returned by Ranger and Surveyor spacecraft. The trend at the JPL Image Processing Laboratory has been toward increasingly complex processing of digital imagery, and more recently toward correlation of remotely sensed imaging with non-imaging data bases. The early processing performed on planetary images generally involved processing of single images acquired in a single spectral band. The processing ranged from simple enhancement of individual images to more complex processing for the removal of camera-system distortions or the distortions that are inherent in various cartographic projections [Rindfleisch, et al., 1971; Green, et al., 1975; Soha, et al., 1975; and Gillespie and Soha, 1972]. It was in this computing environment that the Image Based Information System (IBIS) was developed as a subset of the overall JPL Video Image Communication and Retrieval (VICAR) image-processing system in 1976 [Bryant and Zobrist, 1976]. The image-raster approach used in the Image Based Information System (IBIS) became possible as the state-of-the-art in image-file handling became efficient and continuous-surface geometric-rectification software was developed.

The initial motivation for the development of IBIS was to permit the processing of a Landsat thematic map, showing land use or land cover, in conjunction with a census-tract polygon file to produce a tabulation of land-use acreages per census tract. An analysis of the steps necessary to achieve this basic capability brought forth two facts. First, a large number of image processing and data manipulation capabilities would be needed for even the simplest case. Second, with proper design the minimal system can be extended into a general information system with novel features and capabilities. The term Image Based Information System has been adopted because the image data-type and image processing-operations are crucial to many of the new capabilities.

Digital image-processing techniques can be applied to interface the existing geocoded datasets and information management-systems with thematic maps and remotely sensed imagery. The basic premises are (1) that geocoded datasets can be referenced to a raster scan that is equivalent to an ultrafine mesh-grid cell dataset, and (2) that images taken of thematic maps or from remote sensing platforms can be converted to a raster scan. Until recently, the image format has been used primarily as a computer-processable equivalent of a photograph, with the value stored in each cell of the image representing a shade of grey or a color. But if the image is of a geographical point it can be accessed immediately by position in the image matrix. Figure 22-7 illustrates the calculation of memory address of the data value from a latitude-longitude pair. From the initial assumptions there have developed a variety of applications, ranging from simple tabulations [Bryant and Zobrist, 1976] to complex inventories [McLeod and Johnson, 1980], modelling, [Angelici, et al., 1980] and the system's incorporation as the essential spatial data management-system for achieving multiple-frame digital mosaics and Landsat image map-projections [Zobrist, 1978; and Zobrist and Bryant, 1979].

Data Capture and Storage

IBIS is considered to be a raster-based information system. Most data entered into IBIS will be in raster (image based) format. However, the system is configured in such a manner that other data types, such as graphical and tabular, may be used in analysis as well [Zobrist, et al., 1978]. For each data type, the overriding consideration is usually getting the data into the system, for it is in this area that the greatest cost and difficulties are usually incurred. Data input is a three stage process. The first stage, called data capture, includes all operations up to the point where a data file is computer readable. Data-capture costs are enormous for many basic kinds of data, such as the demographic and economic data gathered by the U.S. Bureau of the Census. These data are then made available on computer tape at nominal cost to any user. Another common method of data capture is to develop a coordinate digitization of boundaries or linear features from a map. The map is not computer compatible but the digitizer

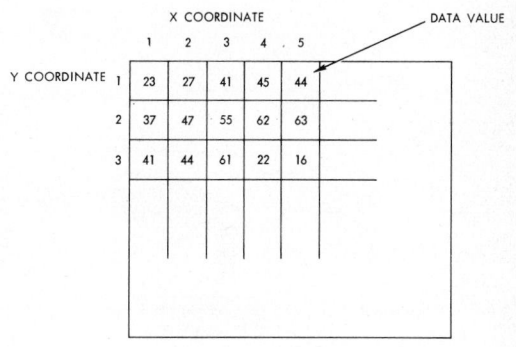

Fig. 22-7. Image Matrix.

output is and can be used in subsequent processing steps. In order to maintain geometric consistency between all data planes included in the data base, an image plane exhibiting good radiometric or planimetric qualities is designated to be the planimetric data plane. All other data planes are geometrically corrected to register to the planimetric base.

The user of IBIS can integrate various data types to form an IBIS data base (see Figure 22-8). Since the primary data structure is a raster format, image data-planes are directly entered into the system. Graphical forms of data, usually obtained in Cartesian reference form, must be transformed into image space prior to inclusion as a data plane. Tabular data are not transformed into image space, but are linked to the image data-base through a logical interface. Data processing requirements for each data type are unique and will be covered individually.

Most image data sets entered into the data base are usually derived from Landsat imagery or other multispectral scanner sources. Other data are digitally encoded or scanned from aerial photographic products. Since image data-planes are not always derived from the same source, there may be no common spatial alignment between them. Consequently, provisions have been made to register these data sets in order to obtain unified spatial surface. The nonaligned data must undergo resampling and "rubber-sheeting" procedures. Geometric correction procedures and spatial rectification routines for modifying image data are features of the VICAR image-processing system. Although not specifically considered to be IBIS programs, these and other VICAR programs are necessary parts of the IBIS process. VICAR software is often used to obtain special information from image data-sources. Multispectral classification of Landsat imagery is one example of a VICAR procedure that produces a data set used in IBIS procedures.

Graphical or vector data may also be entered into the IBIS data base. Graphical data are either produced locally on a coordinate digitizer or are obtained from a data tape. The Bureau of Census Urban Atlas file and GBF/DIME files are examples of such data obtained on computer tapes. Regardless of data origin, graphical data are transformed into image space prior to inclusion in the IBIS data base.

As with image data, graphical data files must be in registry with the primary data base. Provisions have been made within IBIS to achieve the proper geometric corrections. (These corrections are made before the data are transformed into image space.) The registration technique utilized is a two step process. Initially, a general surface fit is achieved through the use of a least squares affine transformation. Then, an exact geometric correspondence is obtained by implementing a "rubber-sheeting" procedure. The deformation of the original surface is controlled by the selection of tiepoints, linking geographical features which are identifiable on both the graphical file and the primary data base.

Point and vector data are transformed into image space after geometric modifications are completed. Three-dimensional or z-value data (x,y,z) are processed in a similar manner. The cartesian reference components of the data (x,y) are transformed into image-space coordinate values, while the z-value remains unchanged.

Tabular data may be entered into IBIS via computer cards or digital tape. These data are stored in a tabular file that is linked to the data base

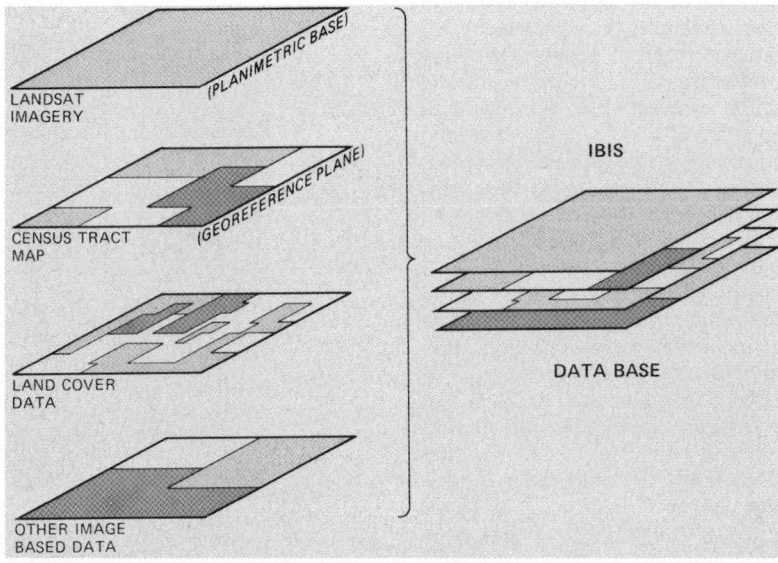

Fig. 22-8. Formation of an IBIS Data Base.

through a logical interface. As a result of this link, such data files are termed interface files.

One of the more important graphical data files entered into the IBIS data base is the geo-reference data plane. The geo-reference plane is a polygon file that is used in data aggregation and map-generation procedures. In one case, the geo-reference plane has been constructed from census tract data obtained from the Urban Atlas files of the U.S. Bureau of Census [Davis and Friedman, 1979]. Once the geo-reference plane is transformed into image space, each polygon, or region, must be identified. The region-identification process involves the assignment of a unique value (or gray tone) to each individual region. After region identification, the geo-reference plane may be used in several higher order IBIS procedures. For example, a polygon overlay of the geo-reference plane with some other image data-plane can be initiated. Or, the gray value of each polygon in the geo-reference base may be modified to produce a map depicting the results of a modelling application with data stored in an interface file. Several geo-reference planes may be included in an IBIS data base. For example, a data base may contain a census tract geo-reference plane and a congressional district geo-reference plane. Currently, up to 20,000 regions can be identified by gray value for an individual geo-reference plane.

All tabular files (interface files) are linked to at least one of the geo-reference planes included in the IBIS data base. The specific link is obtained by storing the numerical value (gray tone) representing each region of the geo-reference plane with tabular data describing attributes of that region (Figure 22-9). Attribute data may be statistical in origin, an identification code, or may be the result of an image-plane comparison routine such as polygon overlay or cross-tabulation.

Because the image data-type is used, capabilities for digital image file-handling, image manipulation, and image processing are required. Thus, the IBIS system has been built upon an existing image-processing system, VICAR (Video Image Communication and Retrieval), developed at JPL. Certain basic image processing operations are absolutely essential. One must perform image-to-image registration, whereby images of different scale, rotation, or map projection are superimposed precisely enough so that corresponding pixels represent the same geographic location. Rubber-sheet registration is almost always necessary to achieve the needed degree of accuracy. On the other hand, it is anticipated that even esoteric image-processing operations, such as convolution smoothing, will be useful for certain types of applications. The conclusion here is that any image-based information system must link to a powerful image processing system.

Query and Retrieval

Once a working data base is set up, provisions must be made for information retrieval, informa-

tion analysis, and report generation. Operations here are usually of a much smaller scale than data capture, are entailed in terms of both time and cost, but there are instead questions of flexibility, ease of use, and system cost. All data analyses can be laid out as a sequence of primitive steps. Thus, a functional requirement is that an adequate set of primitive operations can be implemented. Mathematical and statistical analyses of tabular files are well-understood, and packaged systems can easily be interfaced. The open question is whether geo-based file computational steps can be implemented. Some examples of these are:

1) Given a point and a district, does the point lie within the district?
2) Given a point and a district file, which district contains the point?
3) Given a particular district in a district file, what are its neighbors?
4) Given a district file and an area classification file, what are the acreages of each area classification in each district?
5) Given two district files, one major and one minor, what are the proportions of each minor district in each major district?
6) Given a district file and a line segment, what are the mileages of the line segment, and what are the mileages of the line segment in each district?
7) Given a point, p, and a point file, what is the point in the point file nearest to p?
8) Given a point, p, and a point file, what is the distance from p to the nearest point in the point file?
9) Given a point and a line-segment file, which line segment passes closest to the point?
10) Given a density map and a district file, what are the volumes in each district (spatial integral of density)?
11) Given a district, what is the centermost point (an inside point which is farthest from the boundary)? And for the connoisseur:
12) Given a district file, assign four colors to the districts so that a map can be produced with adjacent regions always of different colors.

The preceding list is just a sample of the sorts of spatial or geometric calculations that need to be performed by a comprehensive geographic information system. More complex operations will usually be implemented as a sequence of these primitive operations, but because of the magnitude of the data files and because of iteration due to modelling, compute time can be a serious problem. A method which solves one of the primitive problems in 0.1 second may seem usable, but not if it has to be performed ten million times for a particular application.

It is worth noting here that many of these operations are difficult and time consuming if the working data-base is in polygon or graphical format (i.e., lines are specified by their end points

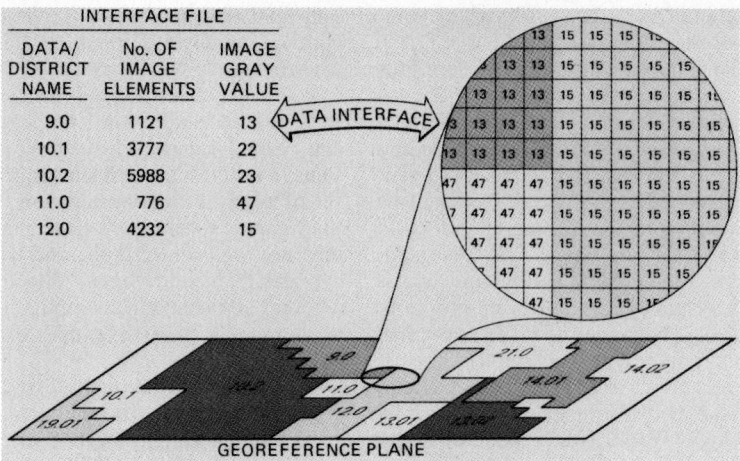

INTERFACE FILE

DATA/ DISTRICT NAME	No. OF IMAGE ELEMENTS	IMAGE GRAY VALUE
9.0	1121	13
10.1	3777	22
10.2	5988	23
11.0	776	47
12.0	4232	15

Fig. 22-9. The Tabular and Image Raster Data Interface.

and a district is given by a sequence of line segments). In particular, the operation called polygon overlay, which solves primitive problems four and five, is difficult to perform on large files in vector format.

A different area of concern is the interfacing of nominal and ordinal geo-reference files. Given that part of the working data-base has been obtained (or reformatted) into an ordinal format, the major part of the data base will probably still be in tabular form. As an example, an air-pollution density image might be interfaced with a population table, to obtain a measure of health effect. Speaking more generally, the system must be able to perform operations on mixed data types, thus allowing the working data-base to be built from raw data in its most natural form.

An operation of special importance is cross-tabulation, mentioned previously in connection with data management. It converts data aggregated by one area convention to an aggregation by another area convention. The operation itself is trivial since it involves multiplying by a set of factors which measure the percentage of sub-areas in one district occurring in the other. The cross-tabulation factors are derived by polygon overlay and may also be modified by a density estimate of the variable being cross-tabulated. A comprehensive system should be able to represent district data sets in such a way that they can easily be updated and the factors rederived. This means that the tabular data can be kept in their natural units of aggregation in the working data-base and quickly cross-tabulated to whatever districting is needed as analysis proceeds.

Finally, the system must be capable of computations that will allow it to run a variety of models. The key here is the introduction of the image raster as a spatial data representation. Contiguity effects can then be handled because adjacent areas are contiguous in the image raster (in terms of their

computer address) whereas, with other file representations, adjacent areas must be accessed by file searching. Basic operations, such as spatial integration with reporting by district, are greatly simplified with an image representation. More esoteric models, say, involving cellular transformations or diffusion processes may already use a matrix or raster representation but will be helped by the synergism of placing their specialized capabilities into a general and comprehensive data-management system.

Logical and mathematical interfaces have been provided to link all data files in an IBIS data-base superstructure (Figure 22-10). By utilizing these interfaces, one can derive information from simple association, or comparisons between two or more data files stored in an IBIS data base. Procedures for data output and data manipulation have been derived as part of IBIS. Maps may be generated and tabular reports can be obtained.

Data stored in either the data base or an interface file can be modified or manipulated with IBIS software. New data planes and interface files are easily generated. Four basic data-manipulation procedures are currently available: Data manipulation between image planes; new image data planes are generated as a function of two or more image data-planes. Chiefly the procedures implemented to derive such data planes are VICAR routines, although some IBIS routines are also used. Simple transformations such as image addition, subtraction, multiplication, and division are easily obtained. Complex functions are handled nearly as easily, and precise mathematical formulas may be specified. Image-enhancement routines are available, as are several data classification and stratification routines. Data manipulation within the interface file is important because most functions available in the image domain are also available for analysis of tabular data. From such operations, new tabular data entries are gen-

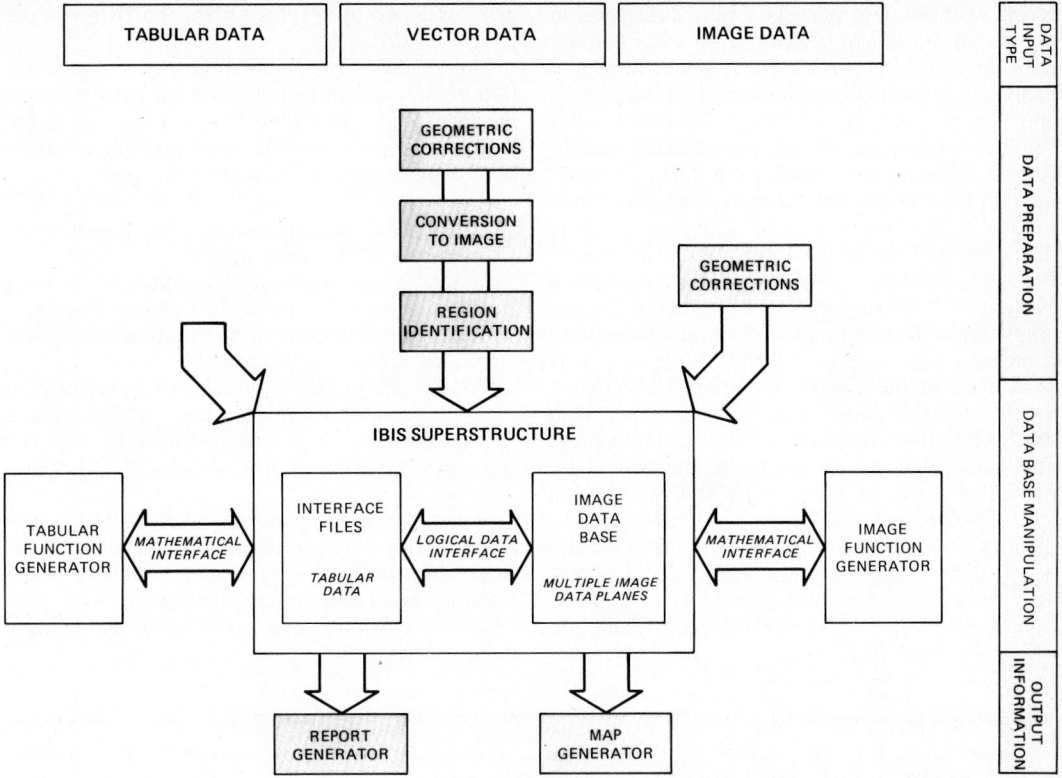

Fig. 22-10. IBIS Configuration Diagram.

erated. Complex mathematical functions may be used to derive higher order properties of data stored in an interface file.

Data manipulation of image data into tabular data can be accomplished by implementing certain IBIS routines, data originally stored in image format may be summarized and copied into a tabular file. The majority of these routines are aggregation functions, an example of which is histogramming. IBIS procedures for polygon overlay and cross-tabulation are within this realm of data transfer programs.

Data manipulation of tabular data into image data provides the representation of tabular data in image form and is primarily used as an output aid. By the implementation of a map-generating routine, any geo-reference base can be modified as a function of an interface file. Modelling of data is performed similarly. Data planes produced in this manner can be entered into the IBIS data base for subsequent operations.

Two output formats are available to the system user, maps and tabular reports. Maps are produced directly from any image data-plane or through a modification of a geo-reference plane. Tabular reports are made available through the operation of a report generator.

An Example of System Use: Portland, Oregon

Planning organizations that are mandated to conduct comprehensive data analyses before making far-reaching socioeconomic, environmental, and earth-resources decisions, are finding their efforts to utilize coded information hampered by a myriad of incompatible geographic reference boundaries. Sets of information referenced by different boundary systems such as police, fire, health, or traffic districts; census tracts, or municipal boundaries; often cannot be easily related to one-another. Rectification of this situation is unlikely to occur in the near future as the nation's data collection infra-structure is based upon agencies operating with divergent planning objectives and spatial responsibilities. The high degree to which existing data reference-systems are entrenched tends to forestall any near-term universal system being accepted, assuming such a reference system could be satisfactorily devised. What is needed, therefore, is a post-collection processing system with the capability of cross-tabulating data between different reference boundaries.

In the case presented, census-tract population and air quality data were cross-tabulated to produce a map of airborne health impacts. The overlaying of these data permits a new dimension in the spatial analysis of health-hazard areas that are critical to agencies concerned with envirnomental quality, public health, commercial development, and general land use planning and zoning. Recently, several agencies under the general coordination of the Pacific Northwest Regional Commission have been working on a project designed to determine the utility of Landsat satellite digital-imagery for urban application. The greater Portland, Oregon, metropolitan area (including Vancouver, Washington) was selected as the study area. Thematic classification of rectified Landsat imagery covering the study area was performed at the Earth Resources Observation Satellite (EROS) data center, with assistance from the NASA/Ames Research Center. Actual editing and land-cover generation were performed by personnel from the Oregon Land Conservation and Development Commission, Oregon Department of Environmental Quality, Columbia Region Association of Governments, Multnomah County, and the city of Portland [Todd, et al., 1979]. Special-purpose data aggregation and tabulation

were assigned to JPL for testing the IBIS system [Logan, 1978].

The specific cross-tabulation task was generation of statistics in map and tabular form describing the spatial interrelationships of air pollution and population data. The three primary information sources used in the processing are:

1) Total population referenced by census tract from a 1973 census update;
2) Residential land-use data extracted from a seven-category thematic Landsat classification, and used as a distribution coefficient (see Figure 22-11); and,
3) 1973 air pollution (total annual suspended particulates) calibrated and referenced to a two kilometer grid system used by the Oregon Department of Environmental Quality.

A number of preparatory steps must be performed prior to generating cross-tabulation statistics. These include preliminary boundary registration, precision boundary registration, georeference encoding, and center-point matching.

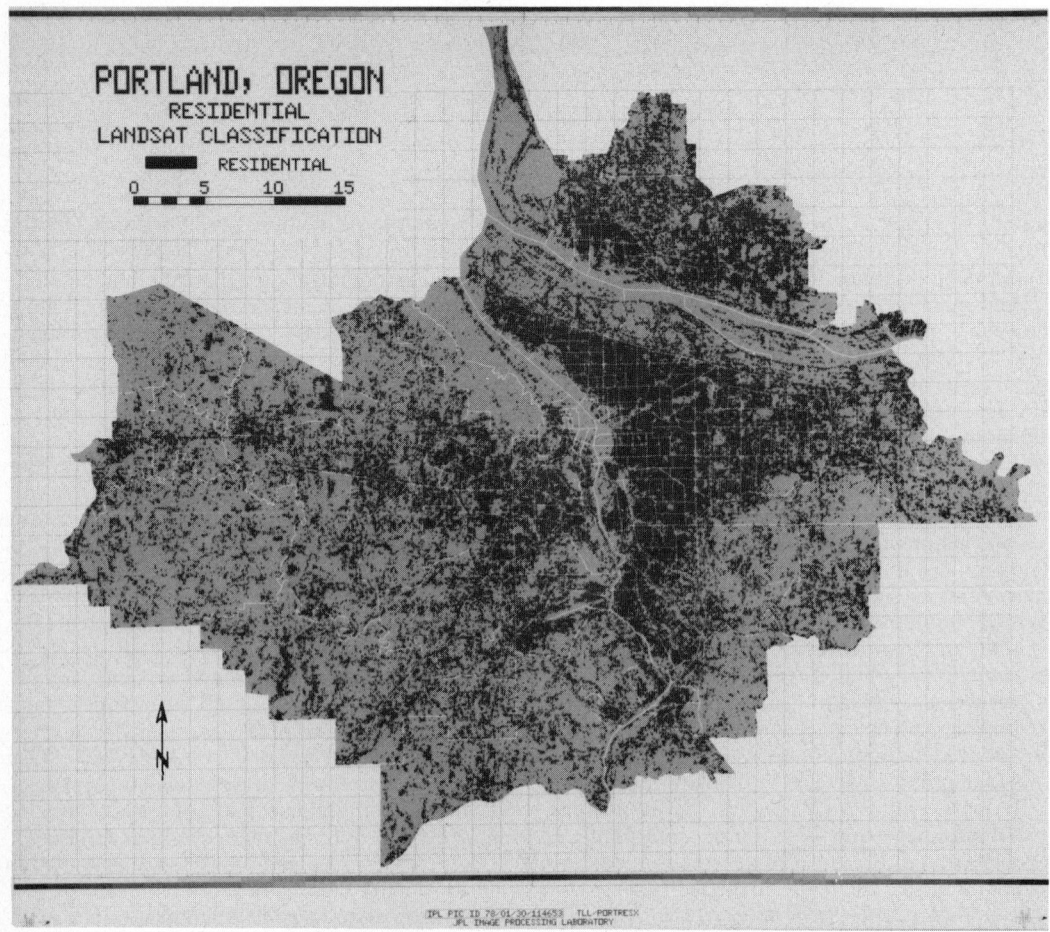

Fig. 22-11. Residential Land Use.

As the pollution grid-cell and census-tract boundaries were supplied on maps, the boundary lines were digitized, and two control points were used to relate the lines to Landsat row and column positions. A conversion routine transformed the computer-formatted lines to Landsat line and sample coordinates, and boundaries were digitally scribed onto a blank image for inspection of possible errors. Photographic filmwriter enlargements aid in finding the coordinates of missing lines, gaps, and other errors. After editing, the boundary line images are digitally summed with the Landsat image to assess the quality of this preliminary registration.

Precision registration of the boundaries requires the use of additional control points and application of a continuous surface-fitting algorithm. With an enlargement of the summed image containing the boundaries overlaid onto the Landsat frame, correct location points for the boundaries can be found and marked. This procedure can be performed manually or with the aid of an interactive system such as the General Electric Image 100. When a collection of such points is found throughout the image, their row and column coordinates can be listed and the differences between

approximate and correct coordinates calculated. A computer routine readjusts the boundaries to fit as precisely as the digitizing accuracy allows. Figures 22-12 and 22-13 display the digitized census tracts and pollution grid-cells correctly overlaid onto the Portland frame. Each grid cell or census tract is then encoded with a unique digital value creating a geo-reference data set. A sample geo-reference image for one data source is shown in Figure 22-14 with regions displayed in five shades of grey.

To interface the census tract population and pollution cell values to the geo-reference images, center point matching is employed. A point inside each pollution cell and tract is digitized and the corresponding tract number or pollution cell row-column coordinate keypunched and stored on magnetic tape. If boundary lines are obtained directly from an Urban Atlas or GBF/DIME file, then the above step is performed by the computer. Through use of the same procedure and identical parameters as with the boundary lines, the center points are also precision registered.

At this point, the necessary inputs to the cross-tabulation procedure are available. The three registered image-format data sets to be in-

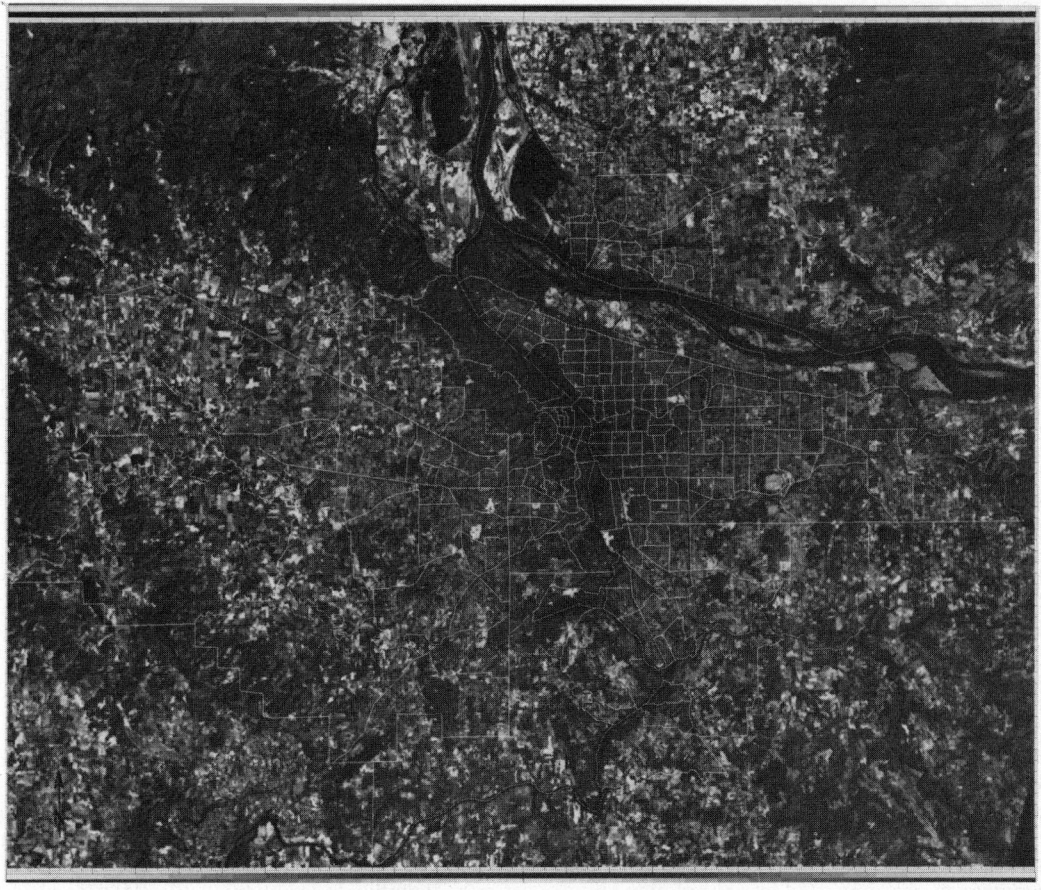

Fig. 22-12. Census Tract Boundaries.

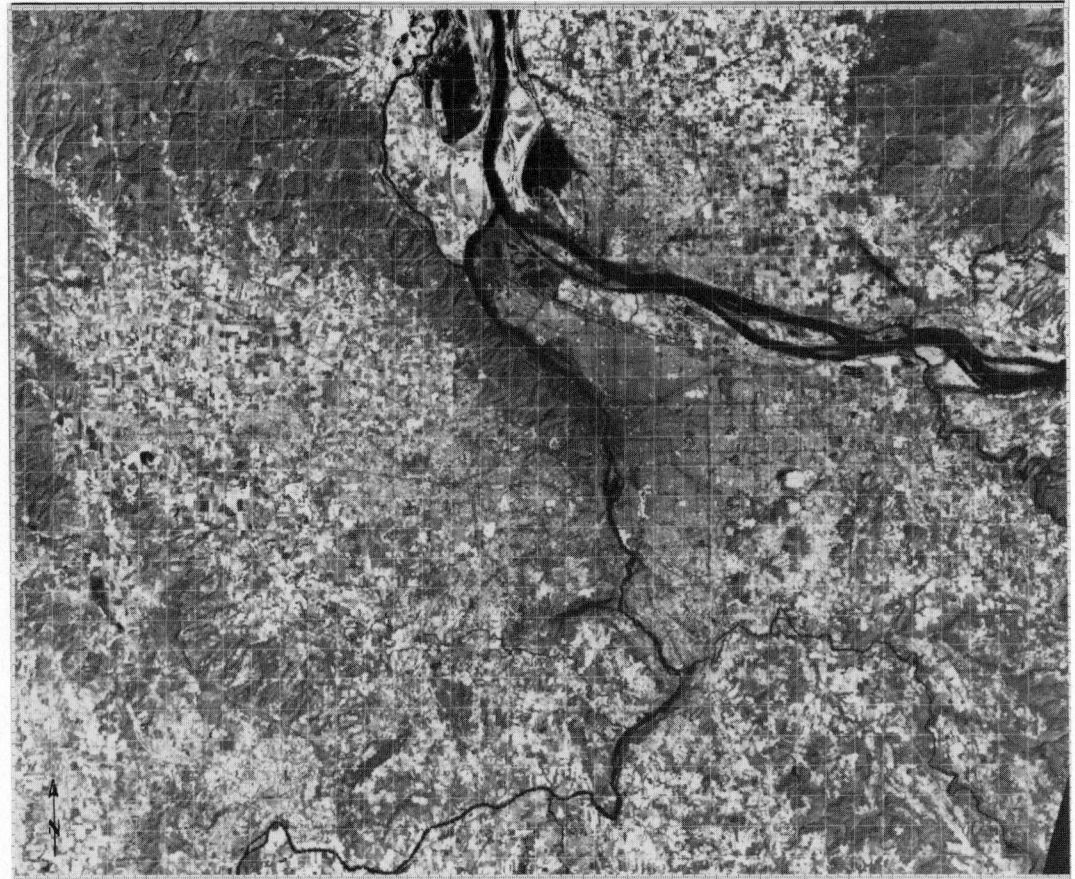

Fig. 22-13. Two Kilometer Grid Cell Boundaries.

terfaced include: 1) grid cell geo-reference, 2) census tract geo-reference, and 3) thematic land-use classification. A computer routine digitally overlays the two geo-reference data sets, row by row, and outputs tabular column listings of data representing the number of pixels of each class of each grid cell and tract. This raster-style polygon overlay is also performed between the grid cell geo-reference and a binary, thematic land-use classification processed to contain only residential and non-residential areas. The second overlay is performed to provide a population-density measure for the cross-tabulation technique.

The census tract and corresponding population values are keypunched and matched to the geo-reference image using the centerpoint file keying on tract numbers. This same procedure is also performed in a similar manner for the pollution-cell values. The cross-tabulation algorithm then reads the number of residential picture elements in each grid cell belonging to different census tracts, and calculates the percentage of population that each tract is contributing to the cell. These percentages are converted into population counts, summed, and the population of the grid cell pro-

duced. As an example, suppose tract 1.00 has a population of 150 people with 50 of its 300 residential pixels in grid cell number (14,8). Suppose further that tract 2.03 has a population of 200 people with 25 of its 100 residential pixels in grid cell number (14,8). Then the population of grid cell (14,8) can be found by adding the quantity:

$$(50 \text{ pixels})/(300 \text{ total pixels} \times 150 \text{ people})$$
$$\text{plus}$$
$$(25 \text{ pixels})/(100 \text{ pixels} \times 200 \text{ people})$$

giving a population of $(25 + 50) = 75$ people for cell (14,8).

The column listings from the cross-tabulation and polygon overlays can be sorted, aggregated, exchanged, and altered by arithmetic functions to satisfy user specifications.

A map of airborne health impacts was also designed which spatially highlighted undesirable combinations of air pollution and population. This map (Figure 22-15) was entirely generated through JPL's VICAR-IBIS software. The map "Index Numbers" (0.0–1.0, etc.) are simplified notations for the Pollution Index numbers (0–100, 000, etc.).

Cross-tabulation represents a useful technique

Fig. 22-14. Census Tract Geo-reference Image Base.

for spatially merging dissimilarly referenced data sources. Census-tract population and air-quality data were merged to produce a map of airborne health impacts which was useful to several governmental agencies for definition of health-hazard areas. Variations in the output assumptions of the air-pollution modelling effort could easily have been accommodated by IBIS, thereby permitting the comparison of scenarios.

System Critique

Other applications of the technology are numerous, and include monitoring urban change, population modelling, housing quality studies, and land-use planning. The system offers potential for expanding investigations into new areas of socioeconomic, environmental and earth-resources analysis.

Because the image data-type is used, capabilities for digital file handling, image manipulation, and image processing are required. Thus the IBIS system has been built upon an existing image-processing system, VICAR (Video Image Communication and Retrieval), developed at JPL. Certain basic image-processing operations

are absolutely essential. One must accomplish image-to-image registration, whereby images of different scale, rotation, or map projection are superimposed precisely enough so that corresponding pixels represent the same geographic location. Rubber-sheet registration is almost always necessary to achieve the needed degree of accuracy. On the other hand, it is anticipated that even esoteric image-processing operations, such as convolution smoothing, will be useful for certain types of applications. The conclusion here is that any IBIS must link to a powerful image-processing system.

Additional capabilities are then added to the image-processing system to convert it into an IBIS. First, image data must be registered or indexed to spatially referenced tabular data. For example, it may be desired to collate land-use data derived from a Landsat image to census data contained in a tabular file aggregated by census tract. Conceptually, this can be likened to image-to-image registration, except that pixels are aligned with records, not with pixels. Second, a data interface must be provided between the different types so that the results of processing can be stored. Third, image-processing analogs must

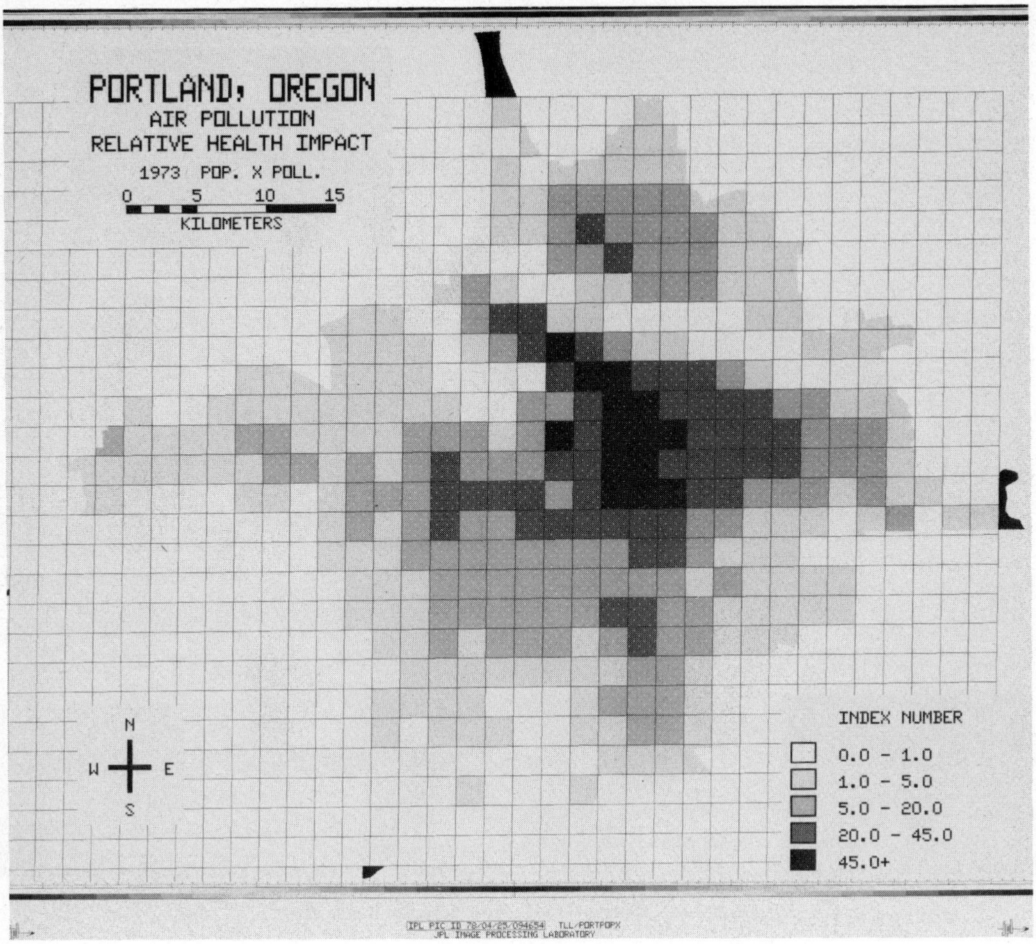

PORTLAND, OREGON
AIR POLLUTION
RELATIVE HEALTH IMPACT
1973 POP. X POLL.

0   5   10   15
KILOMETERS

N
W + E
S

INDEX NUMBER

☐ 0.0 - 1.0
☐ 1.0 - 5.0
▨ 5.0 - 20.0
▨ 20.0 - 45.0
■ 45.0+

IPL PIC ID 78-04-25/094654   TLL-PORTPOPX
JPL IMAGE PROCESSING LABORATORY

Fig. 22-15. Airborne Health Impacts Map.

be developed for existing geo-base file computational steps such as polygon overlay, aggregation, and cross-tabulation. For example, the polygon-overlay operation, in which areas of intersections of polygons from two data sets are obtained, is replaced by a two-image histogramming operation.

An image-based information system is important for the full utilization of satellite-imagery data and anticipated development of regional land-capability analysis centers [National Aeronautics and Space Administration, 1976]. The future availability of frequent updates of land resource inventory statistics, with a known and acceptable sampling accuracy, should permit the incorporation of these data with the annual updates published by other governmental bureaus. The rapid changes in data requirements and user-agency sophistication have recently been discussed by Gilmore (1977) and are succinctly summarized in Figure 22-16. At the outset, concerns ranged over geometric fidelity and classification accuracy of satellite imagery. The field is now embarking upon

data-base integration and modelling activities that utilize the extended capabilities of remote sensing.

The projected demands to be placed upon geographic information systems will place a strong emphasis upon the capability to store and retrieve large amounts of data and manipulate data sets for portions of the files efficiently. A major drawback

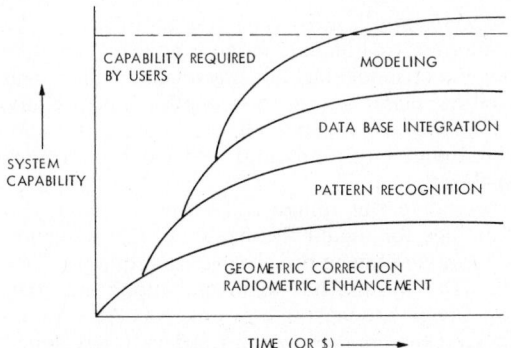

Fig. 22-16. Image Processing Development.

facing most geocoding procedures is that they rely on sequential computations applied to tabular data strings and, as such, require a large investment in formatting or processing data that are inherently two-dimensional. Raster-scan data bases will avoid many of these problems and possess additional advantages. The video-communications field has been and is continuing to address both the problems of mass storage and applications of rapid interactive processing that place a minimal reliance upon computer software routines. The specialized requirements of geographic information systems should derive considerable benefit from the image-processing field in the future.

CONCLUDING REMARKS

At present, the interface between geographic information systems and remote sensing systems is weaker than it should be, and each side suffers from a lack of critical support of a type that could be provided by the other. The GIS has a continuing need for timely, accurate update of the various spatial data elements held in its system and remote sensing systems could, in many cases, benefit from access to highly accurate, ancillary ground data which could significantly improve classification accuracies. In addition, there are a number of significant technical problems that would benefit from a joint, well planned attack rather than the present disaggregated and disorganized approach. A prime example is data management since no operational DBMS exists that will handle, in a cost-effective and efficient manner, the large volumes of spatial data involved in both systems.

It is to be hoped that the next few years will see a greater interaction between the two types of systems and, as a result a significant increase in capabilities on both sides.

REFERENCES

Alsberg, P. A., 1975, The management of a large data base in IRIS. Journal of Chemical Information and Computer Science, vol. 15, no. 1.

Angelici, G. L., N. A. Bryant, R. K. Fretz, S. Z. Friedman, 1980, Urban Solar Photovoltaics Potential: An Inventory and Modelling Study Applied to the San Fernando Valley Region of Los Angeles. JPL Publication 80–45.

Billingsley, F. C., and N. A. Bryant, 1975, Design Criteria for a Multiple Input Land Use System, PROCEEDINGS, NASA Earth Resources Survey Symposium, vol. I–B, NASA TM X-58168.

Brooner, W. G., 1981, An overview of remote sensing input to geographic information systems; Proc. Pecora VII Symp., Sioux Falls, SD, pp. 318–329.

Bryant, Nevin A., (ed.) 1982, Proceedings of the NASA Workshop on Registration and Rectification; JPL Publication 82–23.

—— and A. L. Zobrist, 1976, IBIS: A geographic information system based on digital image processing and raster datatype; Symposium on Machine Processing of Remotely Sensed Data, Purdue University.

Calkins, Hugh W., 1972, An Information System and Monitoring Framework for Plan Implementation and the Continuing Planning Process; unpublished doctoral dissertation, University of Washington, Seattle, WN.

—— and Roger F. Tomlinson, 1977, Geographic Information Systems: Methods and Equipment for Land Use Planning; International Geographical Union, Commission on Geographical Data Sensing and Processing and the U.S. Geological Survey.

Codd, E. F., 1970, A relational model of data for large shared data banks; Communications of the Association for Computing Machinery, vol. 13, no. 6.

——, 1972, Further Normalization of the Data Base Relational Model; Current Science Symposia, vol. 6, Data Base Systems, Englewood Cliffs, N.J.: Prentice-Hall, Inc.

Davis, J. B., and S. Z. Friedman, 1979, Assessing urbanized area expansion through the integration of Landsat and conventional data; Proceedings, American Society of Photogrammetry, 45th Annual Meeting.

Date, C. J., 1977, An Introduction to Database Systems, (2nd Edition) Reading, Mass., Addison-Wesley Publishing Co.

Dieckmann, E. Martin, 1980, Three Relational DBMS; Datamation, vol. 27, no. 10.

Dobson, Michael, and John L. Sibert, 1982, The Design of Thematic Maps for Computer Display; IIST Report 82–17, The George Washington University. Prepared under contract no. 14-08-00001-20448 for the U.S. Geologic Survey.

Dyer, H. L., et al., 1975, Information Systems for Resource Management and Related Applications; Argonne National Laboratory for the U.S. Department of the Interior, Office of Land Use and Water Planning.

Energy, Mines and Resources, et al., 1979, Standard Format for the Transfer of Geocoded Polygon Data; Ottawa, Ontario, Canada.

Estes, John E., and Ruth I. Whitman, 1980, Remote sensor inputs to information systems: research at the National Aeronautics and Space Administration; paper presented at the annual Harvard Computer Graphics Week.

Freeman, H., 1978, Application of the generalized chain coding scheme to map data processing; Proceedings, IEEE Conference on Pattern Recognition and Image Processing.

Garrison, W. L., R. Alexander, W. Bailey, M. F. Dacey, and D. F. Marble, 1965, Data systems requirements for geographic research; in Scientific Experiments for Manned Orbital Flight, proceedings of the American Astronautical Society's Third Goddard Memorial Symposium.

Gierman, D. M., 1976, Rural land use changes in the Ottawa-Hull urban region; Internal report for Environment Canada.

Gillespie, A. R., and J. J. Soha, 1972, An orthographic photomap of the south pole of mars from Mariner 7; ICARUS, vol. 16.

Gilmore, C. D., 1977, A user-oriented interactive information extraction system; Proceedings, NASA/UAH Data Management Symposium, Huntsville, Alabama.

Goodchild, Michael F., 1978, Statistical aspects of the polygon overlay problem, Proceedings, First International Advanced Study Symposium on Topological Data Structures for Geographic Information Systems, vol. 6: Spatial Algorithms.

Green, W. B., P. L. Jepsen, J. E. Kreznar, R. M. Ruiz, A. A. Schwartz, and J. B. Seidman, 1975, Removal of instrument signature from Mariner 9 television images of Mars; Applied Optics, vol. 14.

Guinn, Charles, and Michael Kennedy, 1975, Avoiding system failure: approaches to integrity and utility, Issue Paper 1—Technical Supporting Report E, U.S. Department of the Interior, Office of Land Use and Water Planning and the U.S. Geological Survey, Resource and Land Investigations Program.

Johnson, Timothy R., 1981, Evaluation and Improvement of the Geographic Information System Design Model; unpublished Master's thesis, State University of New York at Buffalo, Buffalo, NY.

Krass, Peter, and Herb Wiener, 1980, The DBMS market is booming; Datamation, vol. 227, no. 10.

Kumii, T. L., S. Weyl, and J. M. Tanebaum, 1974, A relational data-base schema for describing complex pictures with color and texture; Proceedings, Second International Joint Conference on Pattern Recognition.

Likens, W., and K. D. Maw, 1981, Hierarchical modeling for image classification; Proc. Pecora VII Symp., Sioux Falls, SD, pp. 290–300.

Logan, T. L., 1978, Monitoring airborne health impacts through spatial cross-tabulation techniques: the IBIS approach, Proceedings, Urban and Regional Information Systems Association.

Lucas, Henry C., 1973, Computer-based Information Systems in Organizations; Chicago: Science Research Associates.

Maw, K. D., and J. A. Brass, 1981, Forest management applications of Landsat data in a geographic information system; Proc. Pecora VII Symp., Sioux Falls, SD, pp. 330–340.

McIntosh, J., 1978, The interactive digitizing of polygons; Computer Graphics, vol. 13, no. 3.

McLeod, R. G., and H. B. Johnson, 1980, Resource inventory techniques used in the California Desert Conservation Area; Proceedings, Arid Land Resource Inventorys Workshop, La Paz, Mexico.

———, and Thomas L. Logan, 1980, The use of Landsat, digital terrain, and ground sample data as inductively derived information input to a multispectral classifier for wildlands mapping and inventory; paper presented at the Harvard Computer Graphics Week.

Marble, Duane F., ed., 1980, Computer Software for Spatial Data Handling; (three volumes) International Geographical Union Commission on Geographical Data Sensing and Processing and the U.S. Geological Survey.

———, 1981, Some problems in the integration of remote sensing and geographic information systems; in LANDSAT'81 PROCEEDINGS, Canberra, Australia.

Martin, James, 1977, Computer Data-Base Organization; 2nd edition, Englewood Cliffs, N.J., Prentice-Hall, Inc.

National Aeronautics and Space Administration, 1976, Report to the NASA Administrator by the Outlook for Space Study Group, NASA SP-386.

Nichols, D. A., 1981, Conversion of raster coded images to polygonal data structures; Proc. Pecora VII Symp., Sioux Falls, SD, pp. 508–515.

Peters, Lawrence J., 1981, Software Design: Methods and Techniques; New York: Yourdon Press.

Peuquet, Donna J., 1979, Raster processing: an alternative approach to automated cartographic data handling; The American Cartographer, vol. 6, pp. 129–139.

———, 1981a, An examination of techniques for reformatting digital cartographic data, part 1: the raster-to-vector process; Cartographica, vol. 18, no. 1.

———, 1981b, An examination of techniques for reformatting digital cartographic data, part 2: the vector-to-raster process, Cartographica, vol. 18, no. 3.

Power, M. A., 1975, Computerized geographic information systems: an assessment of important factors in their design, operation, and success; Center for Development Technology, Washington University, St. Louis, MO.

Phillips, Richard L., 1977, A query language for a network data base with graphical entities; Computer Graphics, vol. 11, no. 2.

Reed, Carl N., III, 1976, Design and implementation of a crime geographic information system, unpublished doctoral dissertation, State University of New York at Buffalo, Buffalo, NY.

Rindfleisch, T. C., J. A. Dunne, H. J. Frieden, W. D. Stromberg, and R. M. Ruiz, 1971, Digital processing of the Mariner 6 and 7 pictures, Journal of Geophysical Research, vol. 76, no. 2.

Snyder, John P., 1982, Map projections used by the U.S. Geological Survey; Forthcoming as a U.S. G.S. Bulletin.

Soha, J. M., D. J. Lynn, J. J. Lorre, J. A. Mosher, N. N. Thayer, D. A. Elliott, W. D. Benton, and R. E. Dewar, 1975, JPL processing of the Mariner 10 images of mercury; Journal of Geophysical Research, vol. 80, no. 17.

Strahler, A. H., T. L. Logan, and N. A. Bryant, 1978, Improving forest cover classification accuracy from Landsat by incorporating topographic information; Proc. Twelfth Int. Symp. on Remote Sensing of the Environment, pp. 927–942.

Todd, W. J., A. J. George, Jr., and N. A. Bryant, 1979, Satellite-aided evaluation of population exposure to air pollution; Environmental Science and Technology, vol. 13.

Tomlinson, Roger F., Hugh W. Calkins, and Duane F. Marble, 1976, Computer handling of geographical data; Natural Resources Research Report no. 13, Paris, The UNESCO Press.

Tomlinson, Roger F., and A. R. Boyle, 1981, The state of development of systems for handling natural resources inventory data; Cartographica, vol. 18, no. 4.

Vaidya, P., L. Shapiro, R. Haralick, and G. Minden, 1982, An experimental relational database system for cartographic applications, Proceedings, Annual Meeting of the American Society for Photogrammetry.

Zobrist, A. L., 1976, Elements of an image based information system; Proceedings, Caltech/JPL Conference on Image Processing Technology, Data Sources and Software for Commercial and Scientific Applications, Jet Propulsion Laboratory Report SP 43–30.

———, and N. A. Bryant, 1979, Map characteristics of Landsat mosaics; Proceedings, American Society of Photogrammetry, 45th Annual Meeting.

———, 1978, Multiple frame, full resolution Landsat mosaicking to standard map projection, Proceedings, American Society of Photogrammetry, 44th Annual Meeting.

———, N. A. Bryant, S. Z. Friedman, and G. L. Angelici, 1978, Image Based Information System (IBIS) System Guide; JPL Publication 900–909 (Available from COSMIC).

CHAPTER 23

Ground Investigations in Support of Remote Sensing

Authors: JEFF DOZIER and ALAN H. STRAHLER

GENERAL CONTENTS: Uses of ground data; test sites; common measurements for calibration; interpretation of surface properties in agriculture, forestry, geology, snow; training; verification and sampling.

NOMENCLATURE

To conserve and eliminate repetition in text and references, the following symbols, units and names have been used in this chapter.

α, β	Probability of Type I and Type II errors.
κ	Thermal diffusivity.
ϑ	Direction angle with respect to z-axis (ϑ_o illumination, ϑ_r reflection).
μ	$\cos\vartheta$.
φ	Azimuth with respect to x-axis in x-y plane, positive counter-clockwise.
ψ, η	Geodetic coordinates (latitude and longitude) of point. Latitude is positive northward; longitude is positive eastward.
δ, ω	Latitude (declination) and longitude of solar sub-point.
δ_1, δ_2	Amplitudes of temperature variation at soil depths z_1 and z_2.
ϵ	Emittance (emissivity).
λ	Wavelength.
ρ	Density (ρ_d dry air, ρ_w water vapor).
τ_{atm}	Atmospheric transmittance.
τ_{ext}	Optical depth.
A	Azimuth of slope, 0 south, positive east.
$\mathbf{A}$	Rotation matrix.
C	Specific heat (C_p specific heat of air at constant pressure, 1005 $J/kg/deg$).
E	Irradiance (E_s direct, E_d diffuse).
K	Thermal conductivity.
L	Radiance (L_{grd} surface, L_{atm} atmosphere, L_{sat} satellite).
L_v	Latent heat of vaporization (2.5 × 10^6 J/kg).
M	Exitance.
P	Pressure.
P_t	Periodic time period, over which temperature variation undergoes a full cycle.
R	Sensor response function.
R_E	Earth radius.
R_{sat}	Satellite orbit radius (earth radius plus satellite altitude).
R_{gas}	Universal gas constant (8.3143 $J/mole/deg$).
RH	Relative humidity.
S	Slope angle, with respect to horizontal.
T	Temperature.
a	Albedo (a_s direct, a_d diffuse).
c	Velocity of light (2.9979 × 10^8 m/s).
d_{atm}	Relative optical path length.
e	Vapor pressure.
f_r	Bidirectional reflectance-distribution function.
h	Planck constant (6.626 × 10^{-34} Js).
k	Boltzmann constant (1.3805 × 10^{-23} J/deg).
m	Molecular weight ($m_d = 0.0289644$ $kg/mole$ for dry air, $m_w = 0.0180153$ $kg/mole$ for water vapor).
q	Specific humidity.
t	Time.
w	Mixing ratio.
Δh	Grid spacing.
x, y, z	Cartesian coordinates ($\mathbf{r}$).
u, v	Alternative coordinates.

INTRODUCTION

The purpose of ground investigations in remote sensing is to give the remote sensing investigator or operational user a realistic portrait of the target. These ground observations (sometimes called "ground truth" despite some measurement error) are necessary in both research and operational applications. Sometimes the ground observations can be obtained from regularly collected or already available data, but often a ground-investigation data-collection program must be specifically designed for the remote sensing mission.

Because ground investigations are often costly, it is essential to determine what information is required to meet the needs of a particular activity. After a remote sensing mission or operational period, the utility of the ground investigation data should be carefully examined, to determine whether the results of the investigation were compromised because of inadequate ground data, or whether the same conclusions would have been reached with fewer ground observations.

USES OF GROUND DATA

There are many purposes for ground investigations in conjunction with remote sensing from aircraft or spacecraft. Here we group them into four major categories.

Calibration/Correction

Sensors used in remote sensing can only measure electromagnetic radiation reflected or emitted from the surface and attenuated or incremented by the intervening atmosphere. There are also changes in the signal, which are independent of any change in surface properties, caused by variations in the angle of illumination of the incident irradiance and angle of viewing by the sensor. The purpose of calibration or correction is to account for these effects. For some applications we require absolute calibration, and thus try to estimate the surface exitance integrated over some wavelength interval. For other applications we require only relative calibration.

Interpretation of Properties

In studies of the surface energy budget, information on surface reflectance, or, in the infrared region, surface temperature, is the important surface property determined from remote sensing. Usually, however, we use the reflectance or temperature measurements to make some interpretations about surface properties—e.g. snow depth, water content, or grain size; plant moisture stress; soil moisture; phytoplankton content in water; thermal inertia; etc. For such applications, ground investigations are necessary to establish relationships between radiative and physical properties.

Training

Another common application of remote sensing is identification of surface material or features, e.g. crop type, tree species, timber volume, areas of perched aquifers, ore deposits, diseases in vegetation, etc. What is necessary for such applications is to find, by ground observations, areas on the surface where various materials or features can be identified. These areas are then located on the remotely sensed imagery and are examined, perhaps by digital processing, for characteristics useful for identification.

Verification

The performance of any procedure for making interpretations from remotely sensed data should be verified by ground observations. The areas used for verification should be distinct from those used for training, and normally should be distributed—as should the training sites—according to some appropriate sample design.

TEST SITES

Significant interpretation of most remotely sensed data depends on ancillary data and a knowledge of the geographic area. The more the user-interpreter knows about the area over which the data were obtained, the better the interpretation. Some ancillary data must be obtained during the acquisition of the airborne or spaceborne remote-sensor data; others can be obtained before or afterward. When large areas are sensed, as by spaceborne instruments, it is impractical, if not impossible, to obtain ground data from the entire area; instead data from small selected portions of the area are more easily and efficiently gathered. Knowledge gained by study of these selected

small areas can be used to aid in the interpretation of data from the entire area. These smaller areas are called "test sites." Their physical and biological characteristics have been extensively studied, and a variety of remotely sensed data have been tested over them. For many remote sensing experiments, the choice of a test site for which extensive field data already exist is advisable. If a test site is to be chosen and then investigated, it is advisable to choose one that is as accessible as possible, given constraints imposed by the category of problems to be studied. If possible an area with key terrain features suitable for navigation and geometric correction should be selected.

COMMON MEASUREMENTS FOR RADIATIVE CALIBRATION

In this section we describe those ground investigations that are used for radiometric calibration of satellite data, and for interpretation of the reflective or emittive properties of the surface. Such measurements are common to several disciplines and are often conducted in association with remote-sensing missions.

THE NEED FOR CALIBRATION

Once raw remotely sensed data are converted to absolute radiometric units, the next step in processing is to derive from them some radiative properties of the surface. Remote sensing of surface features by sensors mounted in aircraft or spacecraft is hampered by the intervening atmosphere. Physical processes of radiation interaction with the atmosphere are described in Chapter 5; hence such descriptions here are necessarily terse. The atmosphere can selectively scatter, absorb, re-radiate, and refract electromagnetic radiation that passes through it, both from space to the surface and from the surface back to the aircraft or spacecraft. It acts essentially as a complex filter that is spatially, spectrally, and temporally variable. The importance of atmospheric effects on remote sensing depends on the desired results. In particular, it depends on whether we require absolute radiometric calibration or are only interested in relative surface exitance.

If our measurements are of reflected radiation, the fundamental surface property is the *bidirectional reflectance-distribution function* (BRDF), whose integral is the surface albedo. If our measurements are of emitted radiation, the fundamental surface properties are *temperature* and *directional emittance* (also called *emissivity*).

The spectral *BRDF* f_r is defined

$$f_r(\lambda;\mu_0,\mu_r,\varphi_r - \varphi_0) = \frac{L(\lambda;\mu_r,\varphi_r)}{\mu_0 E_s(\lambda)} \quad (23\text{-}1)$$

L is the radiance at reflection angle ϑ_r (arccosμ_r) and reflection azimuth φ_r, ϑ_0 (arccosμ_0) and φ_0 are incidence angle and azimuth, and E_s is monochromatic beam irradiance, measured on a plane perpendicular to the solar beam. Normally the BRDF is symmetric in azimuth, and would be unchanged by substituting $\varphi_0 - \varphi_r$ for $\varphi_r - \varphi_0$. The reciprocity principle states that it is also unchanged if μ_0 and μ_r are interchanged.

The direct albedo a_s, called the *directional-hemispherical reflectance* by Nicodemus et al. (1977), is equal to exitance M in all directions, divided by the beam irradiance $\mu_0 E_s$, i.e. the integral of the BRDF over all reflection angles and azimuths:

(principally in the $8-14\mu m$ water vapor window) measure infrared radiation emitted by the surface and the atmosphere. Absorption by water vapor is the major atmospheric effect in the signal, and beyond $20\mu m$ the atmosphere is virtually opaque to infrared radiation. Surface emittance may depend on both wavelength and viewing angle, but the angular effects are independent of azimuth.

In the microwave region (about $1mm$ to $1m$) the clear atmosphere can be considered totally transparent for practical purposes. Because microwave emittance in soils, vegetation, and snow is so sensitive to water content, emissivity variations in

$$a_s(\lambda;\mu_0) = \frac{M(\lambda)}{\mu_0 E_s(\lambda)} = \int_0^{2\pi}\int_0^1 \mu_r f_r(\lambda;\mu_0;\mu_r,\varphi_r - \varphi_0)d\mu_r d\varphi_r \qquad (23\text{-}2)$$

The diffuse albedo a_d, called the bihemispherical reflectance by Nicodemus et al. (1977), is the integral of the direct albedo over all incidence angles:

$$a_d(\lambda) = 2\int_0^1 \mu_0 a_s(\lambda;\mu_0)d\mu_0 \qquad (23\text{-}3)$$

From Kirchhoff's law, for a semi-infinite medium the emittance and albedo sum to 1 (Siegel and Howell, 1981). This also applies in the angular sense, hence the apparent surface emissivity ϵ usually varies with view angle:

$$\epsilon(\lambda;\mu_0) = 1 - a_s(\lambda;\mu_0) \qquad (23\text{-}4)$$

For wavelengths between 0.3 and $3.0\mu m$ the only significant source of energy is the sun. Most operational remote sensing instruments are sensitive to electromagnetic radiation in narrow spectral bands within this interval and therefore measure solar radiation reflected from the surface. Of considerable importance therefore, are the angles of illumination and reflection, and the scattering and absorption properties of the atmosphere. The principal surface property is the BRDF and its integral over reflection angle and azimuth, surface albedo, which is generally dependent on illumination angle.

For wavelengths between 3.0 and $4.5\mu m$ emitted and reflected radiation are of about the same magnitude. The $3.5-4.0\mu m$ atmospheric water vapor window is commonly used for remote measurement of surface temperatures, but during daylight hours the signal may be contaminated by reflected solar radiation. The angles of illumination and reflection are therefore important. In a cloud-free atmosphere scattering may be neglected in these wavelengths, but absorption is important. The principal surface-properties are emittance, which may be both wavelength-dependent and angle-dependent, and temperature.

For wavelengths beyond $4.5\mu m$ solar radiation is negligible; hence sensors in these wavelengths

the signal are far more significant than temperature variations.

In the wavelengths where both direct and diffuse irradiance E_s and E_d are incident at the surface, the reflected radiance L_{grd} at ground level in direction $[\mu_r,\varphi_r]$ is

$$L_{grd}(\lambda;\mu_r,\varphi_r) = \frac{E_d}{\pi}a_s(\lambda;\mu_r)$$
$$+ \mu_0 E_s f_r(\lambda;\mu_0,\mu_r,\varphi_r - \varphi_0)$$
$$(23\text{-}5)$$

Note the use of a_s instead of a_d. Because of reciprocity, the apparent albedo to diffuse irradiance at viewing angle ϑ_r is equal to the albedo to direct irradiance for the same illumination angle. If there is no atmospheric scattering, then $E_d = 0$.

The emitted contribution to upwelling radiance at the surface is

$$L_{grd}(\lambda;\mu_r,\varphi_r) = \epsilon(\lambda;\mu_r)\frac{2hc^2\lambda^{-5}}{e^{hc/\lambda\kappa T} - 1} \qquad (23\text{-}6)$$

(This is the Planck function, described in Chapter 2.)

In either the reflected or emitted case, there are two principal atmospheric effects in the composite surface-atmosphere signal that is measured by a spaceborne sensor:

1. *path transmittance:* The atmosphere attenuates the surface upwelling radiance, similar to its effect on beam irradiance.
2. *path radiance:* Scattering of beam and reflected solar radiation and absorption and re-emission of upwelling longwave radiation cause a significant atmospheric component L_{atm} in the direction of the sensor.

For a sensor with response function $R(\lambda)$ within some wavelength range $[\lambda_1,\lambda_2]$, the combined effects of path radiance and transmittance lead to an expression for the radiance L_{sat} measured by a satellite radiometer:

$$L_{sat}(\vartheta_r,\varphi_r) = \frac{\int_{\lambda_1}^{\lambda_2} R(\lambda)\left[L_{atm}(\lambda;\vartheta_r,\varphi_r) + \tau_{atm}(\lambda;\vartheta_r)L_{grd}(\lambda;\vartheta_r,\varphi_r)\right] d\lambda}{\int_{\lambda_1}^{\lambda_2} R(\lambda)d\lambda} \qquad (23\text{-}7)$$

τ_{atm} is atmospheric transmittance.

TOPOGRAPHY

Topography exerts perhaps the strongest influence on remotely sensed data of any of the natural surface variables, except in regions of very flat terrain. Indeed, topography in one form or another is often the main information derived from conventional aerial photography. Therefore, in areas where the topography is known, it is logical to acquire the most complete and accurate topographic control available and to use information derived from it in interpretation of the remotely sensed data. In the United States, topographic data are available in two major formats: topographic maps and digital terrain grids.

A topographic map is a graphic representation of the earth's surface configuration—the shape and elevation of the natural terrain plotted to scale. Topographic maps also contain information on location of cultural features (urban areas, schools, churches, houses, etc.), water bodies, and vegetation. The National Topographic Map Series comprises four basic series: (1) $4° \times 6°$ sheets at 1:1,000,000 scale, (2) $1° \times 2°$ sheets at 1:250,000, (3) $15' \times 15'$ quadrangles at 1:62,500, and (4) $7\frac{1}{2}' \times 7\frac{1}{2}'$ quadrangles at 1:24,000. Not all of these are available for all locations; the largest scale at which the entire U.S. is covered is 1:250,000. In addition, non-standard maps are available for some areas. The U.S. Geological Survey publishes national indices of the 1:1,000,000 and 1:250,000 maps and state indices of the larger scale maps. Indices are available, free on request, from the U.S. Geological Survey, Reston, VA 22092, as well as from regional offices.

Plastic raised-relief topographic maps are also available for some parts of the U.S. Most of these are based on the 1:250,000 sheets. Information on availability and price is available from Hubbard Press, Box 442, Northbrook, IL 60062.

At present, corrections for topographic effects generally require the remotely sensed data to be in digital form. Therefore the topographic data must also be in digital form. A *digital terrain grid* includes values for elevation at each point on a grid covering the surface of the earth in the area of interest. The grid may be related to a particular map projection, or may be simply related to increments of latitude and longitude. As of 1982, the National Cartographic Information Center (NCIC) distributes digital terrain grids for 1° quadrangles covering the entire conterminous United States. These grids are derived by digitizing the contours of conventional 1:250,000 maps, then processing the digital representations of the contour lines. The terrain grids are therefore no more accurate than the original maps. Current standards for vertical accuracy are that 90% or more of the elevations must be within 1/2 of the contour interval; for horizontal accuracy, 90% or more of the plotted points must be within 0.02 inch on the map. The terrain grids derived from 1:250,000 maps provide data at a grid spacing of 0.01 inch on the map, or close to $63.5m$ on the ground; hence these data are oversampled with respect to the accuracy limits of the original map.

An extensive program is currently underway at NCIC to prepare and distribute digital elevation models (DEMs) at finer resolution and higher accuracy. These models are produced during preparation of orthophoto maps by such machines as the Gestalt Photomapper. As the data source is aerial photography instead of contours on an existing map, elevations are derived continuously and fit the real landscape more accurately. These data are at $30m$ resolution, and are distributed by $7\frac{1}{2}$-minute quadrangle; although the present holdings emphasize areas of active mineral resource exploration, a representative sampling of most of the conterminous United States is included. The DEMs obtained thus far are part of a ten-year acquisition effort with the goal of covering the entire conterminous U.S. at high resolution.

Also in preparation at NCIC are cartographic data as digital line graphs (DLG). Prepared on a $7\frac{1}{2}$-minute quadrangle base (largely from already compiled maps), these data consist of digitized coordinates describing land boundaries, transportation net, hydrography, etc., that overlay the $7\frac{1}{2}$-minute digital elevation models. At present, most of the DLG data available consist of boundary files, including the township, range, and section land nets. Compared to DEMs, availability of DLG files is limited. NCIC will also be distributing a digitized database from the *National Atlas,* prepared from the set of individual 1:300,000 sectional maps. Data layers will include place names, state boundaries, transportation net, drainage net, etc. Further information is available from the National Cartographic Information Center, U.S. Geological Survey, Reston, VA 22092, as well as from regional map distribution offices maintained by the Geological Survey.

From either topographic maps or digital terrain grids, the variables most commonly needed, in addition to the elevation, are slope and exposure. Slope is the angle of the surface with respect to horizontal; exposure is the direction that the slope faces (i.e. the downhill direction).

For integration with satellite and solar geometric information, it is necessary to choose a unified coordinate system for directions. Whether one chooses a right- or left-handed system is entirely arbitrary, but considerable confusion and error are avoided if one uses a consistent system. For purposes of illustration in this chapter, we choose a right-handed system. The "geographic" system has its origin at the earth's center, with the z-axis along the earth's axis of rotation, positive toward the North Pole. The x-axis is perpendicular to the z-axis along the equatorial plane and intersects the earth's surface at the Greenwich Meridian. The direction of the y-axis is chosen to form a right-handed system, and thus intersects the earth's surface at longitude $90°E$. Therefore latitudes are positive in the northward direction; longitudes are positive in the eastward direction. The "topographic" system has its origin at the earth's surface. The z-axis points toward zenith; the x-axis points southward and the y-axis points eastward. Therefore directions are measured with $0°$ toward south, positive toward east, negative toward west. Since this direction convention for the x- and y-axes conflicts with our common Cartesian perceptions, we also define a u-v axis system for image coordinates on small areas of the earth's surface, with the u-axis positive eastward and the v-axis positive northward.

On a topographic map, the slope angle may be found by calculating the change in elevation over a distance measured normal to the contour lines, i.e.

$$\tan S = \partial z/\partial n \qquad (23\text{-}8)$$

The exposure of the local surface may be found by measuring the angle of the downhill direction with respect to south.

On a digital terrain grid with spacing Δh, slope and exposure are found numerically. The following equations use a second-order finite-difference scheme to find $\partial z/\partial u$ and $\partial z/\partial v$ at $[u_i, v_j]$.

For points not on the ends of rows or columns,

$$\frac{\partial z}{\partial u} = \frac{1}{2\Delta h}(z_{i+1,j} - z_{i-1,j}) \qquad (23\text{-}9a)$$

and

$$\frac{\partial z}{\partial v} = \frac{1}{2\Delta h}(z_{i,j+1} - z_{i,j-1}) \qquad (23\text{-}9b)$$

For points on the ends,

$$\frac{\partial z}{\partial u}\bigg|_{u=0} = \frac{1}{2\Delta h}(-3z_{0,j} + 4z_{1,j} - z_{2,j}) \qquad (23\text{-}9c)$$

and

$$\frac{\partial z}{\partial u}\bigg|_{u=u_n} = \frac{1}{2\Delta h}(z_{n-2,j} - 4z_{n-1,j} + 3z_{n,j}) \qquad (23\text{-}9d)$$

with similar expressions for $\partial z/\partial v$. Slope and exposure are then given by

$$\tan S = [(\partial z/\partial u)^2 + (\partial z/\partial v)^2]^{1/2} \qquad (23\text{-}10)$$

and

$$\tan A = -\frac{\partial z/\partial u}{\partial z/\partial v} \qquad (23\text{-}11)$$

In rugged terrain, another important effect comes from local horizons. Even though a slope faces toward the sun, it may be hidden by an adjacent peak. Dozier et al. (1981) have described a method for efficiently computing horizons from digital terrain data.

SOLAR AND SATELLITE GEOMETRY

In this section we present the equations for solar elevation and azimuth on horizontal and sloping surfaces, and for viewing angles and azimuths from satellites. Such calculations are useful for nearly all remote sensing applications. ϑ (possibly with subscript) represents direction angle with respect to the z-axis. φ (represents azimuth with respect to the x-axis.

The fundamental equations are those for rotation and translation of axes from spherical trigonometry. The equation for rotation of axes is:

$$\mathbf{Ar} = \mathbf{r'} \qquad (23\text{-}12)$$

where r is the original coordinate vector and r' is the coordinate vector in the rotated coordinate system. ϑ_n and φ_n are the angle and azimuth of rotation of the new coordinate system from the old.

$$\mathbf{A} = \begin{bmatrix} \cos\vartheta_n\cos\varphi_n & \cos\vartheta_n\sin\varphi_n & -\sin\vartheta_n \\ -\sin\varphi_n & \cos\varphi_n & 0 \\ \sin\vartheta_n\cos\varphi_n & \sin\vartheta_n\sin\varphi_n & \cos\vartheta_n \end{bmatrix} \qquad (23\text{-}13)$$

$$\mathbf{r} = \begin{bmatrix} \sin\vartheta\cos\varphi \\ \sin\vartheta\sin\varphi \\ \cos\vartheta \end{bmatrix} \qquad (23\text{-}14)$$

$$\mathbf{r'} = \begin{bmatrix} \sin\vartheta'\cos\varphi' \\ \sin\vartheta'\sin\varphi' \\ \cos\vartheta' \end{bmatrix} \qquad (23\text{-}15)$$

Eq. 23-12 can be solved for $[\vartheta', \varphi']$, the direction angle and azimuth in the new coordinate system:

$$\cos\vartheta' = \cos\vartheta\cos\vartheta_n + \sin\vartheta\sin\vartheta_n\cos(\varphi - \varphi_n) \tag{23-16}$$

$$\tan\varphi' = \frac{\sin\vartheta\sin(\varphi - \varphi_n)}{\cos(\varphi - \varphi_n)\sin\vartheta\cos\vartheta_n - \cos\vartheta\sin\vartheta_n} \tag{23-17}$$

Eq. 23-17 is defined even if the denominator is zero. φ' is defined over the range $\pm180°$ depending on the signs of the numerator and denominator. In most computer programming languages, Eq. 23-17 should be implemented with an $ATAN2$ function, instead of $ATAN$.

Formulae for solar angles on horizontal or sloping surfaces are found simply by making the appropriate substitutions in Eq. 23-16 and Eq. 23-17. To find local solar zenith angle and azimuth, we note that declination $\delta = 90° - \vartheta$, latitude $\psi = 90° - \vartheta_n$, longitude of solar subpoint $\omega = \varphi$, and longitude $\eta = \varphi_n$. Therefore

$$\cos\vartheta_{sun} = \sin\delta\sin\psi + \cos\delta\cos\psi\cos(\omega - \eta) \tag{23-18}$$

and

$$\tan\varphi_{sun} = \frac{\cos\delta\sin(\omega - \eta)}{\cos\delta\sin\psi\cos(\omega - \eta) - \sin\delta\cos\psi} \tag{23-19}$$

Declination δ varies from $\pm23.45°$ and depends on date:

Declination	Approximate dates
+23.45°	June 22
+20°	May 21, July 24
+15°	May 1, Aug 12
+10°	Apr 16, Aug 28
+5°	Apr 3, Sept 10
0°	Mar 21, Sept 23
−5°	Mar 8, Oct 6
−10°	Feb 23, Oct 20
−15°	Feb 9, Nov 3
−20°	Jan 21, Nov 22
−23.45°	Dec 22

Solar longitude ω may be calculated within $\pm4°$ from the Greenwich Mean Time, $+15°$ for each hour before Greenwich noon, and $-15°$ for each hour after Greenwich noon. Alternatively, the quantity $\omega - \eta$ may be calculated from local time, $\pm15°$ per hour from local noon. δ varies slightly from year to year because of variations in the earth's orbit caused by the Moon and planets; ω varies during the year (i.e. not always zero at Greenwich noon) because of the earth's elliptical orbit. An accurate computer program to calculate

δ and ω is described by Wilson (1980). Figure 23-1 is an *analemma*, showing δ and ω at 12:00 GMT for each date. Figure 23-2 is a synthetic image made by mapping the solar illumination angle over a digital terrain grid.

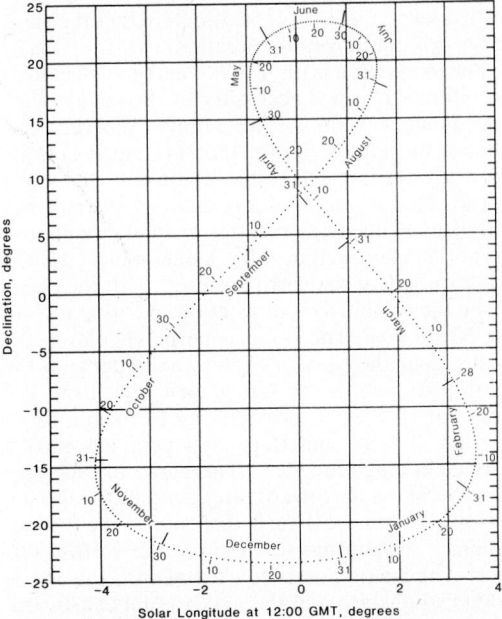

Fig. 23-1. An analemma. Declination (vertical axis) is the latitude of the solar subpoint at local noon on the specified date. Solar longitude (horizontal axis) is the longitude of the solar subpoint at 12:00 GMT on the specified date.

The equation for calculating solar zenith angle can also be used to find daylight duration by noting that at sunrise or sunset $\cos\vartheta_{sun} = 0$. At sunrise and sunset therefore

$$\cos(\omega - \eta) = -\tan\delta\tan\psi \tag{23-20}$$

Since the cosine function is even, there are two values for ω, symmetrical about η, for sunrise and sunset. Daylight length in hours is $2(\omega - \eta)/15°$.

To find illumination angles on a sloping surface, we substitute $\vartheta_{sun} \rightarrow \vartheta$, $S \rightarrow \vartheta_n$, $\varphi_{sun} \rightarrow \varphi$, and $A \rightarrow \varphi_n$:

$$\cos\vartheta_0 = \cos\vartheta_{sun}\cos S + \sin\vartheta_{sun}\sin S\cos(\varphi_{sun} - A) \tag{23-21}$$

$$\tan\varphi_0 = \frac{\sin\vartheta_{sun}\sin(\varphi_{sun} - A)}{\sin\vartheta_{sun}\cos S\cos(\varphi_{sun} - A) - \cos\vartheta_{sun}\sin S} \tag{23-22}$$

Calculation of satellite viewing angles is slightly more complicated than calculation of solar angles, because we must consider translation of the coordinate system as well as rotation. For solar angles,

Fig. 23-2. Synthetic image made by mapping the cosine of the solar illumination angle over a digital terrain grid. The area shown is the southern Sierra Nevada, Kings and Kern River drainage areas. Illumination is from the northwest. Scale is approximately 1:250,000.

the earth-sun distance is so much greater than the earth's radius that we can ignore the effect of translation of coordinates from earth center to earth surface. For aircraft remote-sensing systems, the altitude is small enough that earth-curvature effects can be ignored; over a horizontal surface the sensor scan angle is therefore equal to the nadir viewing angle. Satellite altitudes are large enough that earth-curvature effects must be included. For a horizontal surface,

$$\cos\vartheta_{sat} = \frac{R_{sat}\left[\cos(\eta - \eta_{sat})\cos\psi\cos\psi_{sat} + \sin\psi\sin\psi_{sat}\right] - R_E}{R_{sat}^2 + R_E^2 - 2R_E R_{sat}\left[\cos(\eta - \eta_{sat})\cos\psi\cos\psi_{sat} + \sin\psi\sin\psi_{sat}\right]} \; ; \quad (23\text{-}23)$$

$$\tan\varphi_{sat} = \frac{\sin(\eta - \eta_{sat})\cos\psi}{\cos(\eta - \eta_{sat})\cos\psi\sin\psi_{sat} - \sin\psi\cos\psi_{sat}} \qquad (23\text{-}24)$$

Satellite viewing-angles $[\vartheta_r, \varphi_r]$ on a slope may be found by substituting $\vartheta_{sat} \rightarrow \vartheta_{sun}$ and $\varphi_{sat} \rightarrow \varphi_{sun}$ in Eqs. 23-21 and 23-22.

MEASUREMENT OF SOLAR AND LONGWAVE RADIATION

Eq. 23-7 expresses the effect of the atmosphere on remote sensing of surface upwelling radiance. Surface measurements of solar and longwave radiation in support of remote sensing are used to determine the amount of incident beam and diffuse radiation, the atmospheric contribution to the radiance value measured in space, and the atmospheric transmittance.

The radiation budget at the surface is normally measured with two pyranometers, and occasionally with a pyrheliometer also.

Pyranometer

Fig. 23-3. An Eppley pyranometer with filters, for measurement of incoming solar radiation over the hemisphere.

A pyranometer (Figure 23-3) has a 2π sr field-of-view, to measure globally incident radiation, in wavelength intervals determined by a filter. Two attributes of pyranometers are necessary for their successful use in ground investigations. The instrument must have a "cosine response", whereby the voltage generated by beam irradiance should vary with the cosine of its incidence angle, and the calibration to convert from instrument output-voltage to radiometric units must be accurate. The manufacturer or vendor usually supplies a calibration certificate for each instrument, but annual recalibration is desirable. Schwerdtfeger (1976) describes some experiments to calibrate the cosine response. For studies of the surface energy-budget in remote sensing experiments, four pyranometers are normally used. Two of them sensitive to solar radiation in wavelengths $0.3-3\mu m$ are pointed upward and downward to measure solar irradiance and exitance. Two sensitive to longwave radiation ($4-50\mu m$) are used to measure the remainder of the radiation budget. For calibration purposes, however, measurements of incident solar radiation in narrow spectral bands are more useful. Several manufactures make "Landsat Ground-Truth" radiometers, with spectral channels matching those of the Landsat MSS. Many such instruments use flat diffusing lenses as a collector, which have an unreliable cosine response at large incidence angles.

Pyrheliometer

The pyrheliometer measures only direct beam solar radiation. It may be manually pointed at the sun, or a more sophisticated device that automatically tracks the sun may be used (Pitts et al., 1977).

Atmospheric Transmittance

Atmospheric transmittance is derived from a measurement of the incident *beam* radiation. By the Beer-Bourget-Lambert law, atmospheric transmittance τ_{atm} varies exponentially with optical depth and path length:

$$\tau_{atm}(\lambda) = e^{-\tau_{ext}(\lambda)d_{atm}}$$

τ_{ext} is the optical depth, and d_{atm} is the path length. For $\vartheta_{sun} < 60°$, d_{atm} is given satisfactorily by $1/\cos\vartheta_{sun}$. Kasten's (1966) equation gives more precise values:

$$d_{atm} = \frac{1}{\cos\vartheta_{sun} + 0.15[(90° - \vartheta_{sun}) + 3.885]^{-1.253}}$$

Optical depth $\tau_{ext}(\lambda)$ can be determined from beam irradiance:

$$\tau_{ext}(\lambda) = \frac{-1}{d_{atm}} \ln \frac{E_o(\lambda)}{E_s(\lambda)}$$

$E_o(\lambda)$ is the solar constant at wavelength λ. This form of the equation is unsatisfactory, because the instrument must be rigorously calibrated. Alternatively, measurements of $E_s(\lambda)$ can be made at several solar zenith angles; these are plotted against d_{atm} on semi-logarithmic paper, and $\tau_{ext}(\lambda)$ is the slope of the line. This method assumes that $\tau_{ext}(\lambda)$ remains unchanged during the duration of the experiment. Figure 23-4 shows acceptable and unacceptable conditions for such an analysis. The larger scatter in the "unacceptable" set of points shows changing optical depth during the experiment. Cirrus clouds are a common cause of such problems.

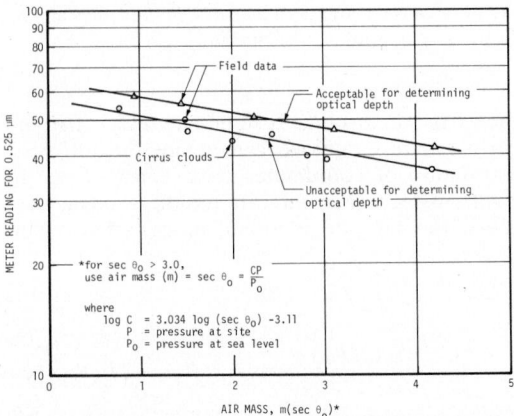

Fig. 23-4. Determination of optical depth by measurement of beam irradiance at several solar zenith angles. The "unacceptable" line has excess scatter in the data.

The amount of atmospheric path radiance between the satellite and the surface cannot be directly measured at ground level. However, Rogers (1973) describes a method for estimating atmospheric radiance by measuring incoming sky radiance at the same scattering angle as observed by the satellite. Geologic applications sometimes require correcting Landsat data for path radiance before calculating band ratios to assure accurate discrimination among rock types (Lyon et al., 1975; Kowalik et al., 1978; Marsh and Lyon, 1979; Kowalik, 1981). Typically, ground measurements are used for this purpose, although Switzer et al. (1981) have presented a method using the covariance matrix from areas of homogeneous reflectance in hilly terrain to estimate path radiance.

A more satisfactory method is to use a radiative transfer model to calculate path radiance and path transmission for a variety of aerosol and water vapor contents at different solar and satellite angles (Dave, 1980). Detailed surface spectral measurements of beam and diffuse irradiance can be used to derive information on atmospheric aerosol and water-vapor content by inversion of the radiative transfer model.

REFLECTANCE MEASUREMENTS

Eq. 23-5 shows that surface investigators must distinguish between *albedo* and the *bidirectional reflectance-distribution function*. Albedo, integrated either over broad or narrow spectral bands, is easily measured by pointing a pyranometer downward, taking care to prevent shadowing, and calculating the radio of exitance to irradiance. Estimating the *BRDF* requires separation of the incident radiation into beam and diffuse components, and measurement of the angular and azimuthal dependence of the reflected radiance. The instrument used is a *spectral radiometer*. Because of the time-consuming nature of such an experiment and the cost of such instruments, only a few measurements of the *BRDF* are available (Coulson, 1966; Breece and Holmes, 1971; Salomonson and Marlatt, 1971; Egbert and Ulaby, 1972).

TEMPERATURE AND EMITTANCE

Temperature Measurements

Thermometric temperature can be measured by many sophisticated means (Wang, 1975; Schwerdtfeger, 1976). There are three main principles used for most ground investigations: expansion, electrical resistance, and thermoelectric effects.

Expansion

Expansion sensors depend on the thermal expansion of gasses, liquids, or solids. Best known are liquid *thermometers,* using mercury or alcohol. Solid thermometers usually rely on differences in expansion between dissimilar metals; the clock thermograph is an example.

Electrical Resistance

Electrical resistance devices make use of the fact that in metals an increase in temperature results in an increase in resistance. Semi-conductors decrease in resistivity with temperature, and small bead- or rod-shaped *thermistors* made of metallic oxides use this principle.

Thermoelectric Effects

Thermoelectric devices employ the phenomenon that when two ends of a conductor composed of two dissimilar metals (e.g. copper and constantan) are at different temperatures, an electric potential difference between the two ends of this *thermocouple* provides a measure of the temperature difference.

Temperature Errors

Most errors in temperature measurement result from a lack of similarity in the physical properties between the sensor material and the substance investigated. In any material the *thermal inertia* of the sensor causes a delay between the sensor reading and the temperature of the medium. This also implies that if the temperature of the medium fluctuates, the sensor may not be able to follow these fluctuations if their frequency is too high. The time required to come to equilibrium depends on the rate of heat transfer from the medium to the sensor, and thus depends on the medium as well. For example, a mercury thermometer may require over a minute to equilibrate in still air but only seconds in water.

In transparent media (ice, snow, water and air) the rate of heat transfer between the sensor and the environment may not be enough to offset radiational heating or cooling of the sensor. To offset such problems, devices measuring temperature in such media must have radiation shielding. In air the sensor is often aspirated as well. Direct accurate measurement of *surface* temperature is difficult in the field, because direct contact usually disturbs the surface temperature equilibrium, and radiation shielding is difficult. *Subsurface* temperatures are easily measured with thermistors or thermocouples. For shallow depths ($<20cm$) it is usually possible to implant small sensors with thin metal probes, but for greater depths it is usually necessary to excavate the soil and plant the probes. In these cases it is necessary to allow time for the excavated area to return to thermal equilibrium. In some environments (e.g. frozen soil) this may require up to a year.

Brightness Temperature

Because of the difficulty of directly measuring surface temperature, radiometric methods are often used. The *brightness temperature* of the surface is the temperature inferred from the blackbody exitance over some narrow wavelength band. The advantage is that the surface is undisturbed. The disadvantage is that determination of the thermodynamic temperature from the brightness temperature requires that the spectral emittance of the substance be known. Most *radiant thermometers* measure exitance in the $8-14\mu m$ atmospheric water-vapor window. Examples are the Telatemp and Barnes PRT-5 and PRT-10 portable infrared thermometers (Figure 23-5). By in-

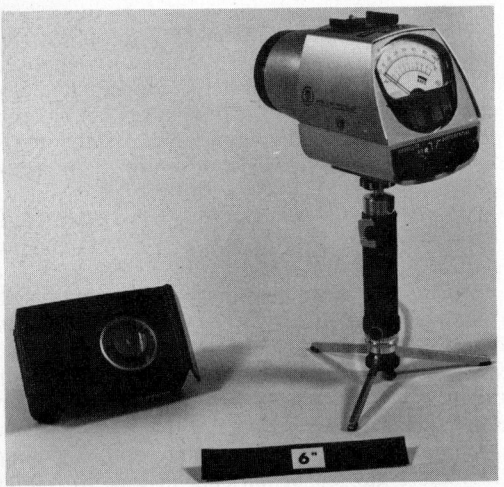

Fig. 23-5. Barnes PRT-10 portable infrared thermometer, with blackbody calibration source.

tegrating the Planck equation (Chapter 2) over the wavelength range of a sensor with response function $R(\lambda)$ and canceling common terms, we derive the following equation for the relationship between emittance, thermodynamic temperature, and brightness temperature:

$$\int_{\lambda_1}^{\lambda_2} R(\lambda)\lambda^{-5}\left[\frac{1}{e^{hc/(k\lambda T_B)}-1}-\frac{\epsilon(\lambda)}{e^{hc/(k\lambda T)}-1}\right]d\lambda = 0 \tag{23-25}$$

Slater (1980) discusses variations of the integrand with emittance and wavelength. For values of $\epsilon(\lambda)$ constant over $8-14\mu m$, Figure 23-6 shows values of $T_B - T$ as a function of ϵ.

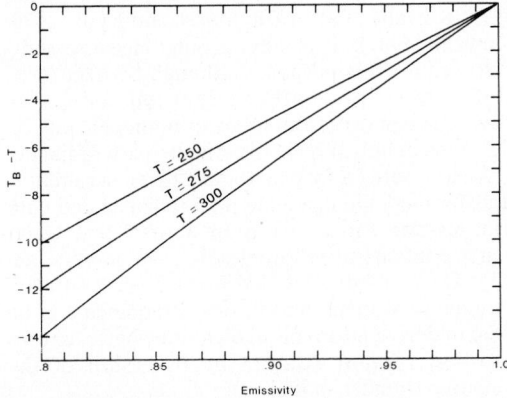

Fig. 23-6. Effect of emittance on brightness temperature integrated over $8-14\mu m$.

Measurement of Emittance

Emittance of various materials is determined by careful measurements of exitance under conditions where thermodynamic temperature and incoming longwave radiation are known. Beuttner and Kern (1965) describe a detailed experiment for such measurements for a variety of substances. Table 23-1 gives emittances at $8-14\mu m$ for many common materials.

TABLE 23-1
Emittances for some Common Materials

Water and Soil		Vegetation	
Water	.92-.96	alfalfa, dark green	.95
snow	.95-.995	oak leaves	.91-.95
ice	.96	leaves and plants	.70-.97
frozen soil	.93-.94		
sand, dry playa	.84	Miscellaneous	
sand, dry light	.89-.90	paper, white	.89-.95
sand, wet	.95	glass pane	.87-.94
gravel, coarse	.91-.92	bricks, red	.92
limestone	.91-.92	plaster, white	.91
concrete, dry	.71-.88	wood, planed oak	.90
ground, moist, bare	.95-.98	paint, white	.91-95
ground, dry plowed	.90	paint, black	.88-.95
		paint, aluminum	.43-.55
Natural Surfaces		aluminum foil	.01-.05
desert	.69-.91	iron, galvanized	.13-.28
grassland	.90	silver, polished	.02
fields & shrubs	.90	skin, human	.95
oak woodland	.90		
pine forest	.90		

from Sellers (1965).

OTHER MICROMETEOROLOGICAL MEASUREMENTS

Water Vapor

Atmospheric absorption by water vapor is the primary cause of attenuation of infrared radiation, but such attenuation is best calculated from radiosonde measurements of the *profile* of temperature and humidity. Surface-humidity measurements are mainly important in energy budget studies with remote sensing investigations.

Water-vapor content may be expressed in several possible ways. The relation between *water vapor density* (also called absolute humidity) and vapor pressure takes the form of the ideal gas law:

$$e = \frac{\rho_w R_{gas} T}{m_w} \text{ or } \rho_w = \frac{e m_w}{R_{gas} T} \quad (23\text{-}26)$$

R_{gas} is the universal gas constant, and m_w is the molecular weight of water vapor. The *mixing ratio* is the water vapor density divided by the dry air density:

$$w = \frac{\rho_w}{\rho_d} = \frac{e m_w / R_{gas} T}{P_d m_d / R_{gas} T} = \frac{m_w}{m_d} \frac{e}{P - e} = 0.622 \frac{e}{P - e} \quad (23\text{-}27)$$

m_d is the molecular weight of dry air. *Specific humidity* is the ratio of water vapor density to total density:

$$q = \frac{\rho_w}{\rho_d + \rho_w} = \frac{e m_w / R_{gas} T}{(P_d m_d + e m_w)/R_{gas} T} = \frac{0.622_e}{P - 0.378 e} \quad (23\text{-}28)$$

Note that w and q are close in value; we often use the approximation

$$q \approx w \approx 0.622 \frac{e}{P} \quad (23\text{-}29)$$

Because water-vapor content is variable, the average molecular weight of air is variable. Analytic solution of the hydrostatic equation is possible only if molecular weight is constant. We therefore sometimes use *virtual temperature* T_v, the equivalent temperature of dry air that would have the same pressure and density as the air with water vapor:

$$T_v = \frac{T}{1 - 0.378 \frac{e}{P}} \quad (23\text{-}30)$$

Relative humidity is defined by the W.M.O. as the mixing ratio w divided by the saturation mixing ratio w_{sat}:

$$RH = \frac{w}{w_{sat}} \quad (23\text{-}31)$$

However, using any other measurement of water-vapor content to define relative humidity introduces only a negligible error.

$$RH \approx \frac{e}{e_{sat}} \approx \frac{\rho_w}{\rho w_{sat}} \approx \frac{q}{q_{sat}} \quad (23\text{-}32)$$

Table 23-2 gives saturation vapor-pressures for a wide range of temperatures. The *dew point* is the temperature to which the air must be cooled, with no change in pressure or water-vapor content, for saturation to occur.

TABLE 23-2a
Saturation Vapor Pressure (mb) over Water

Tens	Temperature, °C — Units									
	0	1	2	3	4	5	6	7	8	9
40	73.7	77.8	82.0	86.4	91.0	95.8	101	106	112	117
30	42.4	44.9	47.5	50.3	53.2	56.2	59.4	62.7	66.2	69.9
20	23.4	24.8	26.4	28.1	29.8	31.7	33.6	35.6	37.8	40.0
10	12.3	13.1	14.0	15.0	16.0	17.0	18.2	19.4	20.6	22.0
0	6.10	6.56	7.05	7.57	8.12	8.71	9.34	10.0	10.7	11.5
−0	6.10	5.67	5.27	4.89	4.54	4.21	3.90	3.61	3.35	3.09
−10	2.86	2.64	2.44	2.25	2.07	1.91	1.76	1.62	1.49	1.37
−20	1.25	1.15	1.05	.964	.882	.806	.736	.672	.613	.558
−30	.508	.462	.420	.381	.346	.314	.284	.257	.232	.209
−40	.189	.170	.153	.138	.124	.111	.0995	.0891	.0797	.0712
−50	.0635	.0566	.0504	.0448	.0397	.0352	.0312	.0276	.0244	.0215

TABLE 23-2b
Saturation Vapor Pressure (mb) over Ice

Temperature, °C										
Units										
Tens	0	1	2	3	4	5	6	7	8	9
−0	6.10	5.62	5.17	4.75	4.37	4.01	3.68	3.38	3.09	2.83
−10	2.59	2.37	2.17	1.98	1.81	1.65	1.50	1.37	1.25	1.13
−20	1.03	.936	.849	.770	.698	.632	.571	.516	.466	.421
−30	.379	.342	.308	.277	.248	.223	.200	.179	.160	.143
−40	.128	.114	.102	.0909	.0809	.0719	.0638	.0566	.0502	.0444
−50	.0393	.0347	.0306	.027	.0238	.0209	.0184	.0161	.0141	.0123

Most humidity-measuring devices depend on one of the following effects: expansion, condensation, energy exchange, or electrical resistance.

Expansion

Some substances (e.g. hair, animal membranes) expand after absorbing moisture. Such changes are employed to move a pointer or pen over a scale or rotating chart. Recording instruments of this type often measure temperature as well and are called *hygrothermographs*.

Condensation

Some devices allow direct observation of condensation on a cooled surface, usually by identifying lack of Fresnel reflection from a cooled mirror. These provide a direct determination of dew point, and are among the most accurate humidity-measuring devices. Unfortunately they are also expensive.

Energy Exchange

Because of latent heat exchange, wet and dry surfaces located in the same air stream will have different equilibrium surface-temperatures. This principle is used in the *psychrometer*, which measures dry- and wet-bulb temperatures T_d and T_w. The *psychrometric equation* then allows determination of vapor pressure:

$$e = e_{sat}(T_w) - \frac{C_p P}{0.622 L_v} (T_d - T_w) \quad (23\text{-}33)$$

$e_{sat}(T_w)$ is the saturation vapor pressure at the wet-bulb temperature. C_p is the specific heat of air at constant pressure, and L_v is the latent heat of vaporization. Note the dependence on air pressure P. Many published tables of values for the psychrometric equation are for sea level. An additional source of error occurs at subfreezing temperatures, if the air is supersaturated with respect to ice but not to water vapor; in these circumstances the wet-bulb temperature can exceed the dry.

Electrical Resistance

The resistance of some substances (e.g. lithium chloride) decreases with increasing wetness. When calibrated, such devices can be used to measure water vapor, but they must be compensated for temperature as well as humidity effects.

Wind

Surface wind-velocity and wind-direction data are generally significant only when thermal phenomena are being sensed, either with thermal infrared or passive microwave sensors. Not only does wind affect the thermodynamic temperature of the surface (usually reducing it) but wind also introduces noise into imagery as wind streaks. With moderate wind velocities, interpretation of linear items on resulting imagery may be ambiguous without wind observations on the ground. Detailed discussion of the methods and problems of measuring wind may be found in Middleton and Spilhaus (1953), Wang (1975), and Schwerdtfeger (1976).

MEASUREMENTS FOR INTERPRETATION OF SURFACE PROPERTIES

In remote-sensing research, the purpose of ground investigations is often to test or explore for a relationship between specific surface properties and the electromagnetic signature. Therefore the ground investigations consist of direct measurements of the surface properties of interest. Ground-truth strategy must be planned carefully to yield the maximum amount of usable information within the time-frame of the study or of a specific experiment. In planning, experimenters should consider the types of measurements, the number and spatial density of sampling sites, and the timing of observations. In this section we offer some guidance to researchers organizing ground-investigation missions.

What surface properties should be measured for any project? Top priority should be placed on measuring those properties that are directly relevant to the purpose of the experiment or calibration. However, remote-sensing experiments are sufficiently expensive and available data are sufficiently scanty that we believe it is important for all experiments involving a given general discipline to collect enough data to characterize the surface in a general way. Therefore in the subsections that follow we offer some standard sets of measurements. In practice, time and resources limit the thoroughness of any research effort. "Extra" data should not be collected if doing so would threaten the success of the main experiment. However, experience shows that sometimes "standard" measurements may help in interpreting the results of any study of the electromagnetic response of the surface. Moreover, they add to the data base and may make one study useful for many purposes. Enough funds should be budgeted for individual experiments to allow as many of these standard measurements to be made as possible.

How should the collection of ground truth be coordinated with the remote measurements? Some properties (e.g. soil moisture) both change rapidly and have a major effect on the electromagnetic signature of the surface. Measurements of such properties are useful only when obtained at precisely the time the remotely sensed measurements are made. Other properties vary more slowly, so accurate measurements can be made within a few hours of the remote observations.

Ultimately, success of a ground investigation and of the corresponding study depends on the skill and dedication of the ground investigators. These personnel should understand that they are key people in a project, and they should be trained to measure competently. We urge that ground-truth teams be headed by professionals familiar with the necessary measurements and techniques. We also recommend that the remote-sensing researchers interact closely with the ground-investigation team. If possible, they should work in the field with the ground investigators to become familiar with the parameters being measured and to understand the accuracy of the measurements.

GEOLOGIC INVESTIGATIONS

The primary objective of ground measurements in geologic investigations is to provide the interpreter with basic geologic information on the surface aspects of lithology and geologic structure. Thus, ground investigations typically involve geologic mapping as initial base data and frequently as the principal product. Geologic mapping is not the subject of this section, as it is a science (or art) in itself; instead, this section will concentrate on those observations or measurements of ground phenomena that are necessary or useful for the interpretation of lithology and structure from remotely sensed data.

Optimal rock discrimination can be achieved only when remotely sensed data are collected specifically to maximize the differences between rocks, whatever these differences may be. Some rocks differ only in color, and it would be pointless to fly predrawn thermal infrared imagery in an attempt to separate them. Therefore, appropriate questions to ask before data collection include: (1)

What physical and chemical differences exist between different rocks? (2) Which of these properties provide enough contrast to distinguish the rocks? (3) Which of these properties are capable of being remotely sensed? and then, logically, (4) how can we best use remote sensors to do the job?

The following example illustrates the use of this procedure for discriminating 18 geologic formations at one site along the Front Range of Colorado (Lee, 1972).

1. *Determine differences.* All rock properties were listed. Any basic rock parameter that could be described or measured was included in the list, regardless of any other consideration; 31 parameters were listed, and a matrix was formed with the 18 formations. Each space in this 18×31 matrix was then filled in (where possible) with a qualitative descriptor (Table 23-3 is only a portion of this matrix).

2. *Determine quantitative differences.* Examination of the matrix focused attention on those parameters that qualitatively showed promise of having the most contrast between adjacent formations. For example, reflectance differences between the Benton and Niobrara formations appeared best for discriminating those formations. Attempts were then made to quantify those parameters so selected. Emphasis was placed on methods easy to use in the field without specialized equipment; for example, color measurements were made using the GSA Rock-color Charts, and albedo (visible spectrum) was measured with common light meters and standard reflectance cards. An example of the resulting quantitative matrix is included in Table 23-4.

3. *Define contrasts required for remote sensing.* An attempt was made to define those contrasts between parameters that are capable of being remotely sensed. The approach here was empirical—that is, instead of simply checking manufacturers' instrument specifications (which generally apply to laboratory, or at least to optimal conditions), studies were conducted on typical aircraft-acquired data to estimate those contrasts that have actually been detected. Some of these *estimates* are listed in Table 23-5.

4. *Select remote sensor(s).* The next step was to cross-correlate the matrices of parameter contrast with sensor capability to select the optimum remote sensor(s) that offered the greatest probability of success in discriminating the rocks. For the example used above—discriminating between the Benton and Niobrara formations—the best possibility is color film, since it can distinguish 0.8 chroma, and a 3.0-chroma difference exists between the two formations (see Chapter 31 for a discussion of chroma).

TABLE 23-3

QUALITATIVE DESCRIPTOR MATRIX

Rock Property—Formation

Property	Morrison Formation	Dakota Formation	Benton Formation	Niobrara Formation
Lithology	Shale	Qtz. ss	Shale	Shale
Grain Size	v. fine	med	v. fine	v. fine
Albedo	low	mod	low	mod
Color	green	tan	gray	gray-tan
Density	low	mod	low	low

Based on these quantitative values, the parameters found consistently most useful for discrimination were defined. For the formations in the area of study, the four most useful parameters are listed in Table 23-6. Six different investigators followed the above procedures, each one working independently with different geologic formations within the test site. Their recommendations are shown in Table 23-7.

It should be stressed that the results and recommendations are used for illustrative purposes only. There is ample bias in this study to preclude extrapolation of the results into other areas; for example, topography was found to be the most consistently useful parameter, yet the investigators had neither good quality SLAR imagery, nor low sun-angle photography available for consideration. Nonetheless, the approach described here is useful for focusing on those parameters that warrant ground investigations.

Parameter Measurements

Before starting a ground-investigation program, the parameters to be observed or measred should be clearly identified, and the reason for their measurement defined. Before beginning ground operations, it should also be established that the data sought do not already exist in the literature. Planning for ground observations should clearly separate the *intrinsic* parameters that can be determined before or after remote sensing data-acquisition, from those *extrinsic* parameters that are time-variant, and that must be measured during overflights or satellite passes. For example, spectral emittance is an intrinsic property and thermodynamic temperature is an extrinsic property.

Although, as mentioned previously, geologic mapping is a basic ground investigation in support of remote sensing, mapping techniques and methods are not within the scope of this chapter. Three points should be made however: (1) Preliminary mapping should be conducted at a scale appropriate to the scale of features expected to be interpreted from the remotely sensed data. (2) More emphasis should be placed on mapping surficial geology than is done in conventional geologic mapping, because the *penetration depth* of most remote-sensing instruments is shallow. (3) Remote-sensing interpretations require field checking. Geologic interpretations often are subject to questions that can be resolved only in the field. Field work may or may not verify or negate interpretations, but it will at least offer supportive evidence to strengthen a conclusion. Optimal geologic information is derived from an iterative process of remote-sensing interpretation and field work.

TABLE 23-5

Empirical Determination of Sensor Discrimination Capability

Pan Film	15-20% albedo contrast
Color Film $\left\{\vphantom{\begin{array}{c}a\\b\\c\end{array}}\right.$	Hue −2.0
	Value −1.2
	Chroma −0.8
Thermal Scanner	~1°C

TABLE 23-4

RESULTANT QUANTITATIVE MATRIX

Rock Property—Formation

Property	Morrison Formation	Dakota Formation	Benton Formation	Niobrara Formation
Grain Size (ϕ)	5	2	5	5
Albedo	20%	33%	13%	21%
Color	5G 4/1	10YR 6/6	10YR 4/1	10YR 5/4
Temp, °C 2200 hrs	−6.1	−3.2	−4.4	−5.1

TABLE 23-6

Most Useful Geologic Characteristics

31 parameters investigated
10 parameters probably useful
4 parameters most useful
1. Topographic landform
2. Color
3. Albedo
4. Surface temperature

TABLE 23-7

Recommended Sensors

Investigator	Best single sensor			2-Sensor Package
	1	2	3	
A	C	CIR	B/W	C + θ IR
B	C	CIR	B/W	C + CIR
C	C	CIR	θ IR	C + CIR
D	C	CIR	θ IR	C + θ IR
E	CIR	C	MB	CIR + θ IR
F	CIR	IR	B/W	CIR + θ IR
Consensus	C	CIR	B/W	C + θ IR

C	=	color photography
CIR	=	color infrared photography
B/W	=	black-and-white photography
MB	=	multiband photography
θ IR	=	thermal infrared imagery

Spectral Reflectance

Most spectral reflectance measurements are made in the photographic region of the electromagnetic spectrum (0.4–1.0μm), with the objective of determining the *color* of rocks. Several field methods may be used, including subjective descriptions, color chip comparisons, spectral photometers, and filter band photometers.

Subjective Descriptions

Subjective descriptions, such as "light brown," are the most common observations, but the most difficult to use. If possible, they should be avoided or supported by more precise descriptors. Conventional stratigraphic descriptions suggest that almost all sandstones are "buff-colored."

Color Chip Comparisons

Use of comparative color chips is a semi-quantitative measurement technique that has the advantage of specifying an exact color, even though the comparison itself is subjective. A useful set of chips is contained in the Rock-color Chart, published by the Geological Society of America. This chart contains 115 color chips that embrace the range of colors of the most common rocks. Color designations are based on the Munsell color and the ISCC-NBS method, and consist of three numerical codes for hue, value, and chroma, plus word descriptors. An example

would be "10 YR 7/4, grayish orange." However, because the system used is based on human physio-psychological response, instead of on instrumental measurements, it is difficult to translate these values into the more quantitative parameters useful for remote sensing.

Spectral Radiometers

Within the last few years, promising new techniques have become available to help discriminate among rock types using remotely sensed data. Of primary importance are techniques using high-spectral-resolution spectroscopy in the visible to middle infrared portions of the spectrum. Electronic and vibrational processes, principally involving iron or the hydroxyl group, permit identification of such minerals or mineral groups as hematite, geothite, alunite, jarosite, kaolinite, potassium micas, pyrophyllite, montmorillonite, diaspore, and gypsum (Hunt and Ashley, 1979). Because many of these minerals are characteristic of hydrothermal alteration, which can also produce ores of economic value, spectral radiometry is of considerable interest to the mining industry. Some instruments are available that can provide high spectral resolution for a relatively small area on the ground from an aircraft platform. For example, the Mark II system developed by Collins et al. (1981) uses linear arrays of silicon dioxide and lead sulfide detectors to measure radiance values for 64 bands in the visible to middle infrared portions of the spectrum as measured within a $20m^2$ footprint obtained along a survey flight path at $600m$ above the terrain. Commercial airborne systems of a similar nature are now manufactured by ISCO, Exotech, Bendix, and Barringer at costs ranging up to $50,000. The measurements taken by such airborne spectral radiometers can be used as ground truth to calibrate imagery derived from higher platforms at coarser spatial and spectral resolutions, or can be used directly as remotely sensed data in their own right. Figure 23-7 shows rock-reflectance spectra measured with a spectral radiometer. Field measurements may be aided by using strip-chart recorders, although these limit portability.

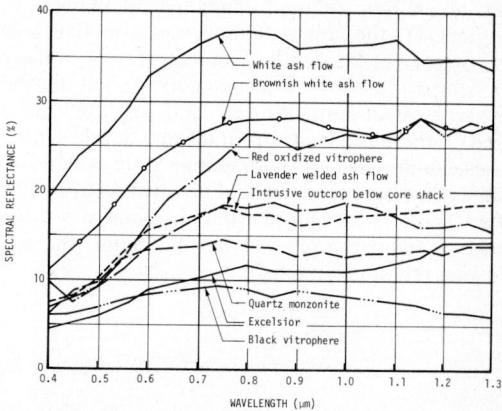

Fig. 23-7. Spectral reflectance of igneous rocks, Crow Springs, Nevada.

Filter Band Photometers

A new type of field instrument has been developed in response to the need for spectral-band reflectance data. Data for multiband photography, and especially Landsat multispectral scanner data, consist of measurements of average, or integrated, spectral reflectance in a given wavelength band. Such band reflectances may be derived from continuous spectral-reflectance measurements (described in the previous paragraph) but may be more simply obtained by measuring energy directly in the band of interest.

Infrared Brightness Temperature

Thermal infrared scanners often are used in geologic remote sensing, both as an aid to lithologic discrimination, and to detect and map phenomena associated with fractures. Ground measurements of brightness temperature then become necessary, both during the time of overflight, for reference or calibration purposes, and before overflights are conducted, for selecting optimal times for scanner-data acquisition. For best results, the scanner imagery should be acquired when the two rocks show a maximum thermal contrast.

Of the intrinsic variables that determine the relationship between local microclimate and rock brightness-temperature, the two most important are thermal diffusivity and thermal-infrared emittance. Thermal diffusivity κ is defined by

$$\kappa = \frac{K}{C\rho} \qquad (23\text{-}34)$$

where K is thermal conductivity, C is heat capacity, and ρ is density. Of the three parameters, density is usually the most significant. Chapter 31 contains a complete treatment of thermal diffusivity and thermal inertia.

Measurement of thermal diffusivity in the field usually uses the property that the heat flow equation

$$\frac{\partial T}{\partial t} = \kappa \frac{\partial^2 T}{\partial z^2} \qquad (23\text{-}35)$$

has an analytic solution under the following conditions: (1) the soil is homogenous, so that κ is constant over the depth of interest; (2) the surface is heated in a periodic manner over some period P_t, with mean temperature $\bar{T}$ and amplitude ΔT_o; and (3) there is a heat sink at infinite depth. Under these circumstances, with temperature at depth z, time t, (where z is measured downward from surface and t is measured from the time when the surface temperature is at its maximum) the following relationship exists

If the time delay t_{max} at which the temperature at depth z is at its maximum is known, Eq. 23-36 may be solved for κ, the thermal diffusivity:

$$\kappa = \frac{P_t}{4\pi} \left(\frac{z}{t_{max}} \right)^2 \qquad (23\text{-}37)$$

Alternatively, if the periodic temperature amplitudes δ_1 and δ_2 during the time period P_t at two different depths z_1 and z_2 are known, thermal diffusivity may be calculated by

$$\kappa = \frac{\pi}{P_t} \left[\frac{z_2 - z_1}{\ln(\delta_1/\delta_2)} \right]^2 \qquad (23\text{-}38)$$

The thermal diffusivity of a rock may be strongly influenced by moisture content also. Examination of the above equation shows that the effect of increased moisture content is complex. K, C, and ρ all increase with increasing moisture, since the values for these parameters of water are all greater than similar values of the same parameters for the air displaced. Increasing moisture in dry rock and soil materials increases κ, apparently because the initial effect on K is greatest. At some level of moisture content, however, the increases in ρ and C become more significant than the increase in K, and κ begins to decrease, as shown in Figure 23-8.

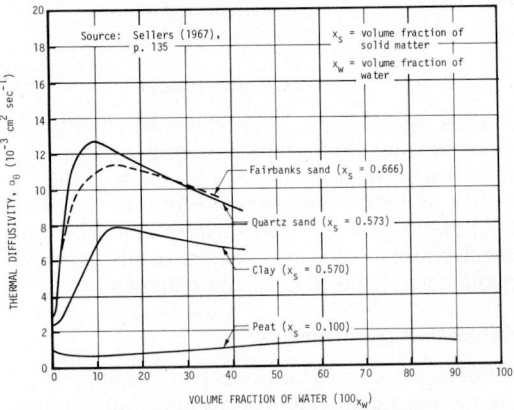

Fig. 23-8. Dependence of thermal diffusivity on volume fraction of water for four different soil types (Sellers, 1965).

Measurements of thermal infrared emittance are difficult to make and require specialized equipment. Most systematic studies of emittance have been conducted in the laboratory, using absorption or reflection techniques (Falckenberg, 1928; Weingeroff, 1931; Lyon, 1965; Beuttner and Kern, 1965). Field measurements of emittance have proved difficult to obtain.

$$T(z,t) = \bar{T} + \Delta T_0 \exp\left(-z\sqrt{\frac{\pi}{\kappa P_t}} \right) \cos\left(\frac{z\sqrt{\pi P_t} - 2\pi t\sqrt{\kappa}}{P_t\sqrt{\kappa}} \right) \qquad (23\text{-}36)$$

Microwave Emittance

Microwave emittances are also difficult to obtain in the field, mainly because, in addition to the complexities noted for infrared emittance, the emitted radiation consists of components both from the surface and the subsurface. This complex emittance is a function not only of surface mineralogy and geometry, but also is affected by moisture content, density, subsurface layering, and associated complex dielectric properties. The relationships between some of these parameters, as well as the time-variant nature of others, demands *in situ* field measurements.

There is a simple method for determining an approximate microwave-emittance value of soil and rock in the field (Blinn et al., 1972). A piece of sheet aluminum, used as a sky-temperature reflector, is covered with rock or soil material to such a depth that the metal plate no longer affects the brightness temperature of the material being tested. From these measurements the opacity of the test material can be calculated, and this is an approximation of the depth from which emitted radiation can be detected. The energy received is an integral function over the depth of penetration. The first few *cm* contribute much more energy to the signal than do greater depths. It should be noted that thermodynamic-temperature measurements in support of microwave investigations should include subsurface measurements as well as surface measurements. Temperature-profiling methods have been previously described in this chapter.

Dielectric Measurements

For the determination of dielectric properties of either soils or solid materials in the field, an *ellipsometer* provides a wide range of flexibility. The ellipsometer consists of a transmitter system that transmits a linearly polarized wave incident on a sample, and a rotatable, linearly-polarized receiver system. Measurement of the angle of rotation of the receiver system, the ratio of maximum to minimum values of power reflected by the sample, and the angle of observation allows a determination of the dielectric coefficient (see Chapter 4). The principal objection to the use of laboratory methods for the determination of dielectric coefficients is that the sample is removed from its surroundings; hence many of the variables that affect the dielectric properties are therefore no longer in equilibrium.

Density

Many laboratory techniques are available for measuring density of materials gathered in the field. While sampling and laboratory analyses may be satisfactory for consolidated rocks, for unconsolidated deposits a question remains about the validity of using disturbed samples. For density determinations of such deposits, *in situ* measurements with a gamma-neutron radiation source are suggested. This nondestructive testing method is a simple and accurate way to get bulk density of soils down to about 10*cm*. Since the thermal diffusivity of particulate material is controlled primarily by its compaction and moisture content, the measurement is useful when interpreting diurnal temperature variations on thermally sensed data. Through use of the neutron probe, measurements of moisture can be made, and from these data, porosity also can be calculated.

Surface Roughness

Although surface roughness affects all remote-sensing instruments, its measurement and application to instrument data are extremely difficult. The problems are compounded by the short wavelengths of the instruments as compared to the geologic materials, the size of the resolution cell for individual instruments, and the heterogeneous nature of the target. On solid materials, a template can be used for recording the bump amplitudes of solid surfaces. For particulate material, a particle-size analysis can be made by using a standard set of sieves. Both methods can then be treated by a Fourier-analysis computer program that analyzes the data for the identification of the amplitudes and frequencies of the component cycles of rock and soil surfaces. Roughness measurements on the scale of 20*cm* or more can be made using field survey or aerial photography. Despite all the work done on roughness studies, the application of roughness data to remotely sensed data must still be handled in an empirical manner. Usually a general statement, such as "tne material was smaller than, or greater than, or about the same size as the wavelength of the sensor," is all that can be said.

Extrinsic Variables

Sampling, or at least observation, of various vegetation types that are apt to be viewed by the sensors will greatly assist in interpreting remote-sensing data. Geologic contacts or hydrologic boundaries can often be more clearly defined by the vegetation present than by any other means. Through the use of a hand-held camera loaded with infrared film, a good first approximation can be made of vegetation species, density, and vigor that will be useful during later interpretations. Maps are often available that show the distribution of plant species (or assemblages). Such maps can be helpful in delineating contacts and boundaries.

Summary

Table 23-8 is a summary of the utility of ground investigations in geologic remote sensing. The table shows the types of information that can be derived from interpretation of remote-sensing data alone, and compares this with the types of information that can be derived from interpretation combined with ground investigations. For example, interpretation of black-and-white aerial photographs alone provides information on topographic variations, but when combined with spot ground measurements of elevation, quantitative topography is derived.

AGRICULTURE AND FORESTRY INVESTIGATIONS

In the areas of agriculture and forestry, the interactions between soils, plants, and animals, and their environment are of primary importance. Remote sensors seldom measure these things directly, as remotely sensed data are only records of the intensity of energy in specific wavelength bands received at a given place, from a given direction, at a given time. Anything that influences this energy flow can affect the remotely sensed data (Colwell et al., 1963). The effects, the causative factors, and the variables that need to be measured in ground-investigation programs vary with the wavelength bands of the sensors used.

Images of vegetation are usually dominated by energy reflected or emitted from the vegetation canopy. Foliage is the most common reflecting surface, but flowers, fruit, bare branches, or background material may be dominant sources of reflected or emitted energy at some places and/or seasons of the year. The timing of leaf-out, flowering, or fruiting varies from year to year, and from place to place in any species. This is the reason that regular collection of phenologic data (the basis for so-called "crop calendars") is important in remote sensing. The topics are discussed further in Chapter 33.

Reflectance

Reflectance measurements are most commonly made to assist in the interpretation of photography or line-scan imagery. This means that only the spectral band between 0.3 and 3.5 μm is of significant interest, as this is the spectral bandwidth of incident solar energy. Measurements are typically made using hemispherical radiometers and spectral radiometers, as discussed earlier in this chapter. Figure 23-9 shows a spectral radiometer being used to measure radiance of a crop canopy.

Before one enters into an expensive reflectance and transmittance measurement program, the extensive assemblage of existing data should be reviewed. Among the hemispherical reflectance data are those published by R. Colwell (1956); Howard (1971); Keegan et al. (1955a, 1955b, 1956); Olson (1969); Olson and Good (1962); Tageeva et al. (1961); and Weber (1965). Fewer bidirectional reflectance-distribution function data are available, but the works of Breece and Holmes (1971); J. Colwell (1973); Coulson (1966); Egbert and Ulaby (1972); Salomonson and Marlatt (1971); and Steiner and Haefner (1965) are illustrative.

Other Measurements

When it is impossible to obtain reflectance or exitance data for all targets or areas, other measurements are often taken to help define pertinent environmental parameters existing before, and at the time of, overflight. Common parameters measured are incoming solar radiation, air temperature, and relative humidity, all of which have been described in prior sections of this Chapter. Measurements specific to agriculture and forestry investigations include soil-moisture content, foliar-moisture content, plant water stress, and biomass.

Soil-Moisture Content

Moisture content is an important soil variable. Small changes in moisture content of the soil surface can produce significant changes in soil reflectance. As a result, remote sensing techniques have been applied to estimate soil moisture. Some of these are summarized in Chapter 33. Supporting ground measurements are often required to calibrate remote sensing models of soil moisture. Standard gravimetric techniques may be used, but tensiometers, resistance blocks, and neutron-scattering methods are becoming more prevalent. These methods are described by Brady (1974).

Gravimetric Methods

This is the most commonly used. A sample of moist soil of known mass is placed in a drying oven at $100-110°C$ for a specified time, and is again weighed. The moisture lost by heating is the soil moisture present in the moist sample.

Tensiometers

These measure the tension with which the water is held and not the absolute quantity of water. They are most useful in determining the need for irrigation when the moisture is being kept near field-capacity.

Resistance Blocks

Commonly made of gypsum, they are placed in contact with soil and absorb moisture from it. The electrical resistance of a given block is related to the water it absorbed. By calibrating the resistance readings with soil moisture-content, one can estimate the moisture in the soil. The gypsum blocks are used to measure either moisture tension or percentage, and are probably most sensitive at tensions from 1 to 15 bars.

Neutron Scattering

This technique uses the fact that hydrogen atoms contained in soil water are effective in reducing the speed of fast-moving neutrons, and in scattering them. Because of the scattering and change in direction, some neutrons return to their origin as slow-moving particles. The number of

TABLE 23-8

Remote Sensing Data Interpretation

Sensor	Without Ground Control	With Ground Control	
	Derived Information	Ground Information	Derived Information
Black and White Photography	1. Topographic variations	1. Elevations and locations	1. (a) Topography (b) Quantitative stratigraphy (c) Engineering geology
	2. Landforms	2. Geomorphic reconnaissance	2. Geomorphology
	3. Lithologic distribution	3. Geologic reconnaissance	3. Mapping of geologic units
	4. Surface structures	4. Geologic reconnaissance	4. Structural geologic maps
Color Photography	1. All of above	1. All of above	1. All of above
	2. Subtle variations in lithology	2. Geologic reconnaissance	2. Detailed mapping of geologic units
	3. Color anomalies	3. Spectral reflectance	3. (a) Alteration (b) Weathering
Color IR Photography	1. All of above	1. All of above	1. All of above
	2. Variations in vegetation	2. (a) Vegetation type	2. (a) Vegetation species mapping
		(b) Vegetation size	(b) Vegetation density mapping
		(c) Lithology-selective vegetation	(c) Differentiation of geology as a function of vegetation
		(d) Moisture-selective vegetation	(d) Geologic structures controlling ground water
	3. Vegetation IR reflectance anomalies	3. Phenology	3. Vegetation vigor; stress due to moisture deficiency; salinity; or disease
Thermal Infrared Imagery	1. Radiometric temperature variations	1. (a) Radiometric temperatures	1. (a) Isoradiance maps
		(b) Spectral emissivity	(b) Isothermal maps
		(c) Thermal diffusivity	(c) Gross lithology
		(d) Geologic reconnaissance	(d) Mapping of some geologic units (radiometric units)
	2. Total lineaments	2. (a) Moisture sampling	2. (a) Fractures controlling ground water
		(b) Vegetation observations	(b) Fractures controlling phreatophyte distribution
		(c) Topography	(c) Fractures controlling topography
		(d) Surficial geology	(d) Fractures controlling deposits of high α_θ materials
	3. Thermal anomalies	3. Geologic reconnaissance	3. (a) Bedrock/alluvium contacts (b) Disturbed ground (c) Cultural features (d) Hot springs, hot ground (e) Subsurface openings
SLAR	1. Topographic variations and landforms	1. Geomorphic reconnaissance	1. Geomorphology
	2. Lineaments	2. Geologic reconnaissance	2. Fractures
	3. Textural variations	3. Surface roughness, particle size	3. Gross lithologies
Microwave Radiometry	1. Relative radiometric temperatures	1. (a) Moisture sampling	1. (a) Relative moisture content
		(b) Density and particle size	(b) Map changes in density and porosity
		(c) Thermometric temperatures, surface and subsurface	(c) Microwave emissivity and penetration depth

these slowed neutrons is related to the quantity of hydrogen atoms, and in turn H_2O molecules, present in the soil. The method can be used without disturbing the soil and works in soils containing salts.

Soil-moisture data have been used as estimates of plant stress. Care must be taken, however, to distinguish between *total* and *available* soil moisture, for plants cannot extract all soil water. Some plants have deeper root systems than others (e.g. alfalfa is deeper rooted than corn), and shallow-rooted plants cannot obtain water from

Fig. 23-9. Field spectral reflectance measurements with Exotech Model 20B spectroradiometer.

deep soil. Even within the root zone, only that moisture between the permanent wilting point and field-capacity membrane-values (usually that between 1/3 and 15 bars of pressure) is considered *available soil moisture.* Usually, only the available water in the upper 75 *cm* of the soil profile is important, for water at greater depths may be *survival water,* but does not contribute significantly to growth of most plants. Even when only the available moisture in the upper 75 *cm* of the soil is considered, the data require careful interpretation. Decreasing available water does not affect leaf reflectance properties for several days (Olson, 1963), although emittance changes appear sooner (Weber and Olson, 1967).

Foliar Moisture Content

Foliar moisture-content can be expressed in several ways, and there is no entirely satisfactory basis for calculating it (Kramer and Kozlowski, 1960). An investigator must use care in choosing a method and in interpreting his or her data and the data of others.

All common methods of determining leaf moisture are destructive, and consequently are impractical in many instances. Rohde's (1971) regression equation for predicting the oven-dry-weight foliar moisture-content from conventional light-reflectance curves for individual leaves may provide a nondestructive method of considerable importance.

Plant Water Stress (Plant Vigor)

Plant water-stress is commonly assumed to show plant vigor. Water stress, like moisture content, must be interpreted with caution, for plant vigor is a subjective concept. At high moisture levels, large changes in moisture content result in minor changes in moisture stress, while at lower moisture levels seemingly negligible changes in moisture content result in dramatic changes in moisture stress. Thresholds and discontinuities, instead of smooth trends, seem to affect the relationship of moisture-content to moisture-stress (Rohde and Olson, 1970).

Internal moisture stress in plants can be approximated by using a pressure cell, such as that developed by Scholander et al. (1965). A freshly-cut leaf or small twig is placed in a specially designed pressure cylinder so that only the cut end of the stem protrudes. Gas (usually nitrogen) is bled into the cylinder, increasing the pressure on the foliage inside. The internal cylinder-pressure at which water or sap first begins to ooze from the cut end of the stem is considered to be equal in magnitude (but opposite in sign) to the tension existing in the plant water-column at the time the stem was cut. The method is considered accurate to about one bar.

Another method of measuring tree vigor involves the use of heat pulses to determine the rate of sap-flow in the xylem of woody plants (Weber, 1965). The method is time-consuming and impractical with small plants.

Biomass

The standard of determining standing-plant biomass is to cut all vegetation from one or more plots and weigh the cut vegetation. This is impractical over large areas or in forest stands. Miller and Pearson (1971) reported successful determination of standing-plant biomass in grasslands using multispectral scanner data and designed a *biomass detector* utilizing a ratio of infrared reflectance. This ratio is not always a good determinant of standing biomass in grass canopies, however, and more work is needed (J. Colwell, 1973; Harlan et al., 1979).

Plant height and density provide a basis for estimating standing-plant biomass, volume, or yield, and also provide insight into relative shading and atmospheric interactions. Often plant height can be determined photogrammetrically if vertical aerial photographs are available (R. Colwell, 1960; Thompson, 1966). Density data can also be obtained from vertical aerial photographs, and foresters often used special *crown density scales* (Figure 34-16) as an aid in estimating stand density. Chapter 34 contains more extensive discussion of forestry investigations.

When image detail permits separation of plants from their background, a simple dot grid can also be used to obtain density information. With the dot grid placed over the area of concern, a count is made of the dots that fall in plants and the dots that fall between plants. Vegetation density can then be described by the percentage of the dots in the area of interest that fall on plants. This procedure provides a measure of the plant canopy-closure but does not provide a good measure of biomass, since it ignores possible multiple layers of vegetation in the plant canopy. Forest canopy-closure can also be measured from hemispherical photographs taken from the forest floor (Marks, 1981). An additional advantage of these is that dependence of canopy density on solar zenith angle can be measured.

An experimental method that can provide much useful information on canopy density, biomass, and spacing over a small area is reflectance measurement at a range of look angles, at several stages during a growing season. Such measurements, when made from an aircraft, helicopter, or truck-mounted scanner or camera, supply data at an intermediate scale between intensive investigations on the ground and the usual data acquired from aircraft or satellite.

Detection of Plant Stress

Stresses caused by insects, fungi, drought, and other factors sometimes can be detected with remote sensors. Success is most likely when the interactions between the causative agent and the physiology and morphology of the affected plants are known. If previsual detection—detection of symptoms before visible changes—is not required, the use of aerial photography by competent interpreters may be adequate, provided there are supplemental field checks (see Chapter 33).

When previsual detection is desired, the wavelength band or bands in which the symptoms of stress can be detected must be known. This may require expensive field studies of reflectance and/or emittance of affected plants, unless the growth patterns, especially leaf flushing habits, of the species involved are known (Rohde and Olson, 1970). After these preliminary data are in hand, decisions regarding further ground-data requirements should be based on the wavelength region or regions of interest. For work in the visible region, nothing more than later field checking may be necessary; if thermal bands are important, then four items of ancillary data can be extremely helpful in the interpreting the remotely sensed data: (1) air temperatures at time of overflight and for 1-14 days before the overflight; (2) velocity and direction of surface winds at time of overflight; (3) brightness temperatures of selected surfaces, or calibration panels, within the target area; and (4) moisture availability, inferred from precipitation or irrigation records, measured as available soil moisture, or measured on plant samples by internal moisture tension or turgidity.

SNOW

Here we describe common measurements to characterize the snowpack. Some of the techniques have been developed from many years of experience; others are experimental, but if eventually verified, may make it possible to collect routinely data that are now difficult to obtain.

For most snow investigations, the following priority of measurement is suitable:

1. Snow depth and qualitative description of the snow as "wet" or "dry."
2. Snow water equivalence.
3. "Quick" snow profile, including temperature and brief description of major layers.
4. "Complete" snow profile, including temperature, density, crystal size and type, measured for each layer.

We classify ground measurements into four types, based on the way they are collected:

1. Measurements made from the snow surface. They disturb the snowpack only slightly because they may involve a small sample bore in the snow.
2. Observations made from a snow pit. Digging the pit perturbs the pack a great deal.
3. Measurements made by instruments installed before the snow accumulates.
4. Meteorological measurements made at or above the surface.

Photography supports and documents measurements of all four types.

Measurements from the Snow Surface

Non-destructive measurements from the snow surface include:

snow depth
snow water equivalence
liquid-water content of the surface layer
Ram hardness profile
temperature profile (using a probe)
surface roughness
surface layer grain size and type

Of these properties, depth and water equivalence are the most commonly used. They can be measured with an avalanche probe and a Mt. Rose snow sampler, respectively. Recent tests suggest that the Mt. Rose sampler tends to consistently oversample water equivalence (Beaumont, 1965; Farnes et al., 1980). Cold calorimetry currently gives the most reliable measurement of the liquid water-content of the surface layer (Jones et al., 1980). This procedure is somewhat time consuming, but the results are sufficiently accurate to serve as a standard against which other techniques can be compared. Ram hardness profiles are obtained by gently pushing a special rod into the snow to determine the hardness of the layers it thrusts through. The temperature profile can be measured non-destructively and rapidly using a probe bearing several thermistors. A hand lens can be used to observe the structure of the snow in the surface layer. Alternatively, or in addition, photographs can record the appearance of the surface and samples from it. Later surface roughness and grain size can be measured from the photographs.

Snow-Pit Measurements

The following measurements are commonly made from a snow pit:

temperature profile
density profile
crystal sizes and descriptions
stratification
profile of liquid-water content
soil moisture and temperature
soil roughness

Perla and Martinelli (1975) describe techniques and cautions for snow-pit observations. We only update their instructions with a few comments. Liquid-water content of individual layers can be measured by cold calorimetry. The glacial sampler, used to measure density and water equivalence of portions of the pack, appears to provide more accurate measurements than other samplers.

Measurements Requiring Prior Installation

Many snow properties can be measured repeatedly in a non-destructive way, but the techniques all require that equipment be constructed and installed in the field before the first snowfall. Some instruments make measurements automatically, and transmit the data to a receiving station. The property measured is given in parentheses:

graduated stakes and aerial markers (depth)
snow pillows (water equivalence)
thermistor arrays (temperature profile)
isotope gages (density profile)
sliding platter devices (rate of settlement or compaction)
active microwave systems (stratigraphy and liquid water content)
pans (melt draining from the pack)

The first four instruments are used operationally. All others are experimental, but potentially could make *in situ* measurements that are otherwise difficult to obtain (Davis and Marks, 1980).

Meteorological

Common micrometeorological measurements are:

solar irradiance and exitance (and therefore albedo)
incoming longwave radiation
surface radiant temperature
air temperature
dew point (or surrogate measurement)
wind velocity
precipitation

Though all these can be measured with portable equipment, they are normally collected routinely by permanent or semi-permanent installations.

Some Research Needs

Since liquid-water content is perhaps the most important property influencing the microwave signature of snow, there is a pressing need to devise an improved method for measuring it rapidly on an areal basis. Liquid water can change significantly over short periods of time. Cold calorimetry, while accurate, consumes time and produces only a point measurement. Buried active microwave systems (Boyne and Ellerbruch, 1980; Linlor et al., 1980) are not always adequate because they must be installed before the snow season. Ambach and Denoth (1980) have invented a portable capacitor, but its potential remains to be tested.

TRAINING

In automatic classification of digital imagery, selection of proper training data is of primary importance in obtaining information with high accuracy. Although Chapter 18 provides a more detailed overview of the classification process, a few paragraphs about the selection of training data are appropriate to this ground-observations chapter.

The central problem in selecting training data is ensuring that the training data truly represent the information class that they are supposed to typify. Training data are often accumulated by selecting training sites—that is, locations on the ground from which observations can be drawn that are used to characterize the class. Unfortunately, training sites are often selected to be "most typical," thereby introducing the bias of the image analyst who has identified the sites. The identification of such typical sites thus restricts the variance of the training data considerably, since atypical members of the information class are not included.

Autocorrelation can also be a problem. When training measurements are derived from training sites, the measurements are typically temporally or spatially autocorrelated. This autocorrelation further reduces the dispersions estimated for the information class, leading to inaccurate training statistics and inaccurate classifications (Tubbs and Coberly, 1978).

Another problem can arise if the information class desired really consists of more than one population. If a class such as "small grains" consists of winter wheat, spring wheat, and barley, and all have been planted on different dates and have different phenologies, then combining measurements from all three populations into a single set of training statistics will produce multi-modal histograms that are not well characterized by classifiers relying on multivariate normality. Often the solution to this problem involves trial-and-error identification of homogeneous subclasses that can be treated separately.

Some of these problems may be avoided by random stratified-sampling of pixels within the image (Cochran, 1977). In this procedure, pixels are selected at random and the information classes to which they belong are identified by ground observations. If the sample is unstratified, the number of pixels in each information class will tend to reflect the relative abundances of the classes within the image (thus serving as unbiased estimators of the prior probabilities associated with each information class); if the sample is equally stratified, an equal number of measurements is obtained to characterize each information class. Stratification may also take variance into account, and proportionally more measurements may be obtained for those classes that are more variable to make best use of the ground sampling effort. This type of procedure assures that the mean vectors and dispersion matrices used to characterize information classes are unbiased estimators that represent the information classes as accurately as possible within the cost constraints of data collection.

An alternative to training site selection is to use a classification technique that does not require the input of training data. In the past few years, unsupervised classification of Landsat data has become more widely used, since it typically requires less costly ground observation and often yields greater accuracies than supervised techniques. In unsupervised classification, ground truth is used in the labeling process by which the clusters of commonly occurring measurements are assigned an information-class label. The labeling may be carried out by simple examination of aerial photography at finer resolution than the digital images being processed, or may be a more complex function of collateral data involving knowledge of local agricultural procedures and meteorological data, taken together with the time trajectory of pixels in measurement space. Thus, the technique of unsupervised classification shifts the effort from random sampling to accurate labeling. Unsupervised classification algorithms are discussed in more detail in Chapter 18.

It is probably worth noting that where the classes desired are not overly heterogeneous and are reasonably distinctive, both supervised and unsupervised techniques, with or without random selection of training data and careful labeling, can produce accurate classifications. As the separability of the classes diminishes, increasing effort must be devoted to careful characterization and labeling.

VERIFICATION AND ACCURACY ASSESSMENT FOR THEMATIC MAPS

A primary objective of many remote sensing applications is the preparation of a thematic map. Thematic maps of land use/land cover are probably the most common application; timber type maps and habitat maps are other examples. Often such maps are required to achieve a predetermined level of accuracy; at the very least, users will want to know the level of accuracy associated with the finished map so that they will understand how best to use it. The objective of this section is to discuss some of the issues surrounding verification and accuracy assessment of thematic maps, to provide a context for the design and execution of such efforts. The classic reference for such sample-survey design and inference is Cochran's *Sampling Techniques* (1977).

SAMPLE PLANS

Verification and accuracy assessments use subsidiary or external information not used in the preparation in the thematic map. Thus, ground sampling, or at least sampling using larger scale imagery, is typically required. The geographical location of points to be sampled on the ground is commonly determined in one of three ways: (1) random sampling; (2) stratified random sampling; and (3) stratified systematic unaligned sampling. In random sampling, the location of the sample point on the ground is chosen using a random-number table or random-number generator to select random coordinates for sampling. These coordinates may be geographic (i.e. latitude, longitude), or may use any arbitrary coordinate system, so long as all areas on the ground have an equal probability of being sampled.

In random stratified-sampling, the area to be sampled is divided into subareas and samples are allocated randomly within each subarea or type of subarea. This type of sampling is most often used when accuracy statements are desired about each thematic class, as well as an accuracy statement for the entire map. In simple random sampling, the number of samples falling within each thematic class will tend to be proportional to its area, and thus accuracy statements for small classes will be based on fewer observations than those for larger classes. A typical-stratified random-sample design, in which equal numbers of sample locations are allocated randomly to each thematic class, overcomes this difficulty. Fitzpatrick (1977) provides an example of stratified random sampling in assessment of land use/land cover map accuracies.

Both random and stratified-random sample designs are not constrained geographically—that is, when the number of samples is small, large areas on the map may go unsampled. In theory, this feature is not a difficulty, unless the surface being sampled exhibits spatial autocorrelation. However, most geographic areas do exhibit spatial autocorrelation in the features being mapped. If they did not, there would be little point in making a map in the first place! In the presence of such autocorrelation, geographers have shown that where the number of sample points is small, a systematic sample will be more accurate and will also be unbiased, if there are no periodicities in the data that can interact with the systematic spacing of the samples. To provide against the latter unlikely event, geographers have used stratified systematic unaligned sampling (Berry, 1962). In this procedure, the thematic map is overlaid by a coarse grid and one sample is drawn from within each grid cell. The coordinates of the sample are determined using random numbers, but are not necessarily completely random from one cell to the next. For a full description see Berry (1962) or Berry and Baker (1968).

TYPES OF ERRORS

Assessing the accuracy of a thematic map can be a complex process because errors can occur for several reasons. In a recent article, Hord and Brooner (1976) recognized three types of errors: (1) control-point location error; (2) classification error; and (3) boundary-line error. Thus we can recognize three types of accuracies that quantify errors of these types.

Control Point Location Accuracy

Control-point location-accuracy measures the geometric fidelity of the map—that is, how accurately the position of a known point on the ground is represented on the map. Obviously, for such accuracies to be determined, the cartographic properties of the map must be known. Typically these properties are specified by the scale of the map and the projection to which it conforms. With this information, the geographic coordinates of a point on the earth's spheroidal surface can be mathematically transformed to the cartesian coordinates on the map.

U.S. National Map Accuracy Standards state that for a map to be termed *accurate*,

> "... for maps on publication scales larger than 1:20,000 not more than ten percent of the points tested shall be in error more than 1/30 inch, measured on the publication scale; for maps on publication scales of 1:20,000 or smaller, 1/50 inch. These limits of accuracy shall apply in all cases to well defined points only" (Hord and Brooner, 1976).

Note that the terms "larger" and "smaller" refer to the map-scale fraction—1:100,000 is, of course, a smaller scale than 1:24,000. As discussed further below, this statement is a poor specification because it does not specify a confidence level or an acceptable accuracy level against which the map is to be tested. In addition, the concept of a "well defined point" is in itself undefined.

Control-point location-accuracy is often assessed by selecting a number of control points, calculating their correct position on the map, and comparing them with their position as plotted. Although rarely done, from a statistical viewpoint the best accuracy-assessment procedure would identify all "well defined points" on the map, and select a large number of control points randomly from this list. In image processing, where rectification is often achieved by "rubber sheeting" procedures, there are likely to be systematic errors introduced into subportions of the map. Under these conditions stratified systematic unaligned sampling, or a variant recognizing that there are only a finite number of well defined points within each subportion of the map, is probably an optimal design.

Classification Accuracy

Classification error occurs when a polygon is incorrectly labeled, so that the information class shown at a point on a thematic map does not agree with the true class as observed on the ground. Classification error cannot therefore be assessed without an unambiguous classification. Categories should be explicitly defined so that there can be no disagreement between those compiling the maps and those verifying them. Users will often specify the mapping units to be recognized; it is important for the team preparing the maps to be sure that the units are clear and unambiguously defined. Simple, objective criteria are always best.

The determination of classification error is convolved with control-point location-error. For example, in a digital image classification a pixel of trees in the middle of a cornfield may be correctly classified as forest, but if the rectification procedure is inaccurate, the pixel will be in the wrong place and the classification will appear to be in error. The solution to this problem is to assess accuracy by relative error, positioning the point to be classified on the ground according to its relative position referenced to map features instead of absolute geographic coordinates. Error rates and accuracies for typical Landsat applications are often determined by assessing such relative error.

Classification accuracy is typically expressed as the percentage of points in a sample that have been classified accurately by the mapping procedure. Although this estimate is helpful, it can be made more significant by attaching confidence limits (Hay, 1979). To place such limits, the binomial distribution is used. A table of exact confidence intervals, given a sample size and a number in error, can be found in Rohlf and Sokal (1969).

Full specification of an accuracy standard for classification accuracy as well as control-point location-accuracy requires specifying four parameters (Ginevan, 1979). First of these is the target accuracy, expressed as the probability that a point is correctly classified. Second is an alternative accuracy that will be unacceptable. Third and fourth are α and β, the probability of Type I and Type II errors respectively. Type I error occurs when the map has met or exceeded the standard, but by chance the sample selected contains enough errors to reject the map as inaccurate. Type II error occurs when the map does not meet an unacceptable standard, but by chance the sample is sufficiently accurate that the map is accepted. Ginevan (1979) provides extensive tables of sample sizes and error counts given typical values for these parameters. For example, if $\alpha = 0.05$ and $\beta = 0.05$, an there is an accuracy target of 0.95 and

TABLE 23-9
Error Matrix before Normalization
[Data of Smith and Itkowsky (1968), from Congalton et al. (1981)]

Computer Classification	Reference Data						
	Decid.	Conif.	Grass	Meadow	Shrub	Water	Sage
Decid.	17	2	0	0	0	0	0
Conif.	28	127	0	0	0	0	0
Grass	0	0	2	2	1	0	0
Meadow	16	0	0	122	6	0	0
Shrub	6	0	0	4	3	0	0
Water	0	0	0	0	0	127	0
Sage	0	0	0	0	0	0	0

Sum of diagonals = 398. Sum of entries = 463. $\frac{398}{463} = 0.86$.

unacceptable accuracy target of 0.85, then 93 samples are required of which no more than 8 may be in error.

Error Matrices

When a thematic map is assessed for accuracy using sample designs similar to those described above, often the true information class of the ground-sample location is recorded, instead of a simple scoring of whether the mapped class is correct. These additional data allow the construction of an *error matrix,* sometimes called a *confusion table.* The error matrix scores each observation according to the class in which it has been mapped and the true class as determined on the ground. An example of such a matrix appears below as Table 23-9, based on data for an unsupervised Landsat classification of natural vegetation types in a Colorado test area. The overall accuracy of this classification is estimated at 0.86, calculated by dividing the number of correct identifications (398) by the total (463).

An inspection of the table, however, shows that the sample drawn has not represented all classes equally; 384 of the 463 observations are drawn from conifer, meadow, and water reference-classes. If the representation of these three classes in the sample truly reflects their representation within the mapped area, then the accuracy value appears reasonable. On the other hand, a user may wish to evaluate this classification by placing all classes on an equal footing. Bishop et al. (1975) have recently shown how techniques of iterative proportional fitting can be used to normalize an error matrix so that rows and columns sum to one. The normalized matrix that corresponds to Table 23-9 is presented in Table 23-10. The normalized accuracy is 0.61, obtained by dividing the sum of the elements along the principal diagonal of the matrix (4.2958) by the number of rows or columns (7). These techniques can also be used to adjust an error matrix derived from random stratified sampling, in which the marginal totals are fixed in advance, so that the marginal totals reflect the true proportions of the classes. In this fashion, an accuracy estimate can be prepared that is based on sampling classes equally but is weighted by their frequency of occurrence. Similar techniques, derived from the domain of discrete multivariate analysis, can be used to compare error matrices to estimate their similarity. Thus this approach allows comparison of alternative classification algorithms, photointerpreters,

TABLE 23-10
Error Matrix after Normalization
[Data of Smith and Itkowsky (1968), from Congalton et al. (1981)]

Computer Classification	Reference Data						
	Decid.	Conif.	Grass	Meadow	Shrub	Water	Sage
Decid.	.5363	.0996	.1001	.0168	.0645	.0155	.1674
Conif.	.1382	.8044	.0158	.0027	.0102	.0025	.0265
Grass	.0154	.0200	.5025	.0842	.1943	.0156	.1681
Meadow	.0879	.0035	.0174	.7139	.1457	.0027	.0291
Shrub	.1804	.0180	.0906	.1366	.4089	.0141	.1515
Water	.0035	.0046	.0230	.0039	.0148	.9108	.0385
Sage	.0384	.0499	.2505	.0420	.1615	.0389	.4190

Sum of diagonals = 4.2958. Sum of entries = 7. $\frac{4.2958}{7} = 0.61$.

and so forth, as Mead has noted (Congalton et al., 1981).

Because error matrices are derived from ground samples, they are also useful in their own right. For example, under a random or stratified systematic unaligned sampling design, the proportions of points falling into each information class will be unbiased estimators of their prior probability of occurrence within the entire mapped area. These prior probabilities may in turn be used in the classification to improve its accuracy (Strahler, 1981). If the sample has been stratified, then the proportions of mapped classes falling into each true class can be used to modify areal estimates, when the objective of investigation is not necessarily a thematic map but an inventory or proportions or area totals for each information class desired. Jupp et al. (1981) have extended this method to avoid the need to label classes in unsupervised classification of Landsat imagery for areal inventory. In this procedure, the composition of each unsupervised class is determined directly by ground sampling as a vector of proportions distributed among the information classes desired. These proportions are then used to obtain estimates of the area of each desired information class from the areas of the unsupervised classes.

Boundary Line Error

Boundary-line error occurs when a boundary line on the map does not conform to its true position on the ground. Boundaries are, in large part, an abstraction of the cartographer or cartographic process producing the map. When the definitions of mapping units are not clear and unambiguous, gradational boundaries can result that separate well defined types by a transition. In such a case, the ground location of gradational boundaries cannot be verified. Even if mapping units are precisely defined, the shape of the boundary may be highly complex and include numerous inliers, outliers, projections, and convolutions. Even if such a complex boundary can be shown at the scale of the map, typically the cartographer will generalize it to produce a simple polygon.

As in assessment of classification accuracy, the quality of rectification can greatly influence boundary-line accuracy. Thus, the polygon may be shifted slightly on the map from its true position on the ground because of a control point or photogrammetric error, and will be considered inaccurate when ground checked. Again, relative accuracy of boundary placement and shape may better evaluate the usefulness of the map containing minor geometric distortions.

Hord and Brooner (1976) suggest a method for comparing the similarity of boundaries between two different raster maps. In their procedure, the two maps are registered and each pixel is scored as to whether it is part of a boundary on one, both, or neither of the maps. An χ^2-statistic for a 2×2 contingency table derived from these pixel counts assesses the degree of similarity of the maps. Little work has been done in evaluating boundary accuracy for conventional or nonraster-based thematic maps. However, transect methods may prove most effective.

REFERENCES

Ambach, W. and A. Denoth, 1980, The dielectric behavior of snow: a study versus liquid water content; Microwave Remote Sensing of Snowpack Properties, NASA Conf. Publ. 2153, pp. 69–92.

Beaumont, R. T., T. G. Freeman, H. J. Stockwell, and R. A. Work, 1965, Accuracy of field snow surveys, western United States including Alaska; U.S. Dept. Agric. SCS Tech. Rep. 163.

Berry, B. J. L., 1962, Sampling, coding and storing flood plain data; U.S. Dept. Agric. Hbk. 237.

Berry, B. J. L., and A. M. Baker, 1968, Geographic sampling; in B. J. L. Berry and D. F. Marble, eds., Spatial analysis—a reader in statistical geography, pp. 91–100, Prentice-Hall.

Beuttner, K. J. K., and C. D. Kern, 1965, The determination of infrared emissivities of terrestrial surfaces; J. Geophys. Res., vol. 70, pp. 1329–1337.

Bishop, Y., S. Fienberg, and P. Holland, 1975, Discrete multivariate analysis: theory and practice; M.I.T. Press.

Blinn, J. C. III, J. E. Conel, and J. G. Quade, 1972, Microwave emission from geological materials: observations of interference effects; J. Geophys. Res., vol. 77, pp. 4366–4378.

Boyne, H. L. and D. A. Ellerbruch, 1980, Active microwave water equivalence measurements; Microwave Remote Sensing of Snowpack Properties, NASA Conf. Publ. 2153, pp. 119-130.

Brady, N. C., 1974, The nature and properties of soils; 8th ed., MacMillan. 1974.

Breece, H. T. III, and R. A. Holmes, 1971, Bidirectional scattering characteristics of healthy green soybean and corn leaves in vivo; Appl. Opt., vol. 10, pp. 119–127.

Cochran, W. G., 1977, Sampling techniques; 3rd ed., John Wiley and Sons.

Collins, W., S-H. Chang, and J. T. Kuo, 1981, Remote mineralogical analysis using a high-resolution airborne spectroradiometer: preliminary results of the Mark II System; Digest, 1981 Intl. Geosci. Rem. Sens. Symp., vol. 1, pp. 337-344.

Colwell, J., 1973, Bidirectional spectral reflectance of grass canopies for determination of above ground standing biomass; Ph.D. Thesis, Univ. Mich. 1973.

Colwell, R. N., 1956, Determining the prevalence of certain cereal crop diseases by means of aerial photography; Hilgardia, vol. 26, pp. 223–286.

Colwell, R. N., ed., 1960, Manual of photographic interpretation; Amer. Soc. Photogramm.

Colwell, R. N., W. Brewer, B. H. Landis, P. Langley, J. Morgan, J. Rinker, J. M. Robinson, and A. L. Sorem, 1963, Basic matter and energy relationships in remote reconnaissance; Photogramm. Engrg., vol. 29, pp. 761-799.

Congalton, R. G., R. A. Mead, R. G. Oderwald, and J. Heinen, 1981, Analysis of forest classification accuracy; Final Rep. NASA-VPI Coop. Agreement No. 13-1134, Nationwide Forestry Appl. Prog., Johnson Space Center.

Coulson, K. L., 1966, Effects of reflectance properties of natural surfaces in aerial reconnaissance; Appl. Opt., vol. 5, pp. 905-917.

Dave, J. V., 1980, Simulation colorimetry of the earth-atmosphere system; Rem. Sens. Environ., vol. 9, pp. 301−324.

Davis, R. E., and D. Marks, 1980, Undisturbed measurement of the energy and mass balance of a deep alpine snowcover; Proc., 1980 Western Snow Conf., pp. 62−67.

Dozier, J., J. Bruno, and P. Downey, 1981, A faster solution to the horizon problem; Comp. Geosci., vol. 7, pp. 145−151.

Egbert, D. D., and F. T. Ulaby, 1972, Effects of angles on reflectivity; Photogramm. Engrg., vol. 38, pp. 556−564.

Falckenberg, G., 1928, Absorptionskonstanten einiger meteorologisch wichtiger Körper für infrarote Wellen; Meteorol. Z., vol. 45, pp. 334−337.

Farnes, P. E., B. E. Goodison, N. R. Peterson, and P. R. Richards, 1980, Proposed metric snow samplers; Proc., 1980 Western Snow Conf., pp. 107−109.

Fitzpatrick, K. A., 1977, The strategy and methods for determining accuracy of small and intermediate scale land use and land cover maps; Proc., 2nd W. T. Pecora Symp., Amer. Soc. Photogramm., pp. 339−361.

Ginevan, M. E., 1979, Testing land-use map accuracy: another look; Photogramm. Engrg. Rem. Sens., vol. 45, pp. 1371−1377.

Harlan, J. C., R. H. Haas, W. E. Boyd, and D. W. Deering, 1979, Determination of range biomass using Landsat; Proc., 13th Intl. Symp. Rem. Sens. Environ., pp. 659−673.

Hay, A. M., Sampling designs to test land-use map accuracy; Photogramm. Engrg. Rem. Sens., vol. 45, pp. 529−533.

Hord, R. M., and W. Brooner, 1976, Land-use map accuracy criteria; Photogramm. Engrg. Rem. Sens., vol. 42, pp. 671−677.

Howard, J. A., 1971, The reflective foliaceous properties of tree species; G. Hildebrandt, ed., Applications of remote sensors in forestry, pp. 125−146, Intl. Union Forestry Res.

Hunt, G. R., and R. P. Ashley, 1979, Spectra of altered rocks in the visible and near infrared; Econ. Geol., vol. 74, pp. 1613−1629.

Jones, A. Rango, and S. Howell, 1980, Measurement of liquid water content in a melting snowpack using cold calorimeter techniques; Microwave Remote Sensing of Snowpack Properties, NASA Conf. Publ. 2153, pp. 41−68.

Jupp, K. K. Mayo, D. Kuchler, S. J. Heggen, and S. W. Kendall, 1981, The BRIAN method for large area inventory and monitoring; Proc., 2nd Australasian Rem. Sens. Conf., pp. 6.5.1−6.5.5, Australian Acad. Sci.

Kasten, F., 1966, A new table and approximation formula for the relative optical air mass; Arch. Meteorol., Geophys. Bioklim., ser. B, vol. 14, pp. 206−223.

Keegan, H. J., J. C. Schleter, and W. A. Hall, Jr., 1955a, Spectrophotometric and colorimetric change in the leaf of a white oak tree under condition of natural drying and excessive moisture; NBS Rep. 4322.

Keegan, H. J., J. C. Schleter, W. A. Hall, Jr., and G. M. Haas, 1955b, Spectrophotometric and colorimetric study of foliage stored in covered metal containers; NBS Rep. 4370.

Keegan, H. J., J. C. Schleter, W. A. Hall, Jr., and G. M. Haas, 1955c, Spectrophotometric and colorimetric study of diseased and rust resisting cereal crops; NBS Rep. 4591.

Kowalik, W. S., 1981, Atmospheric correction to Landsat data for limonite discrimination; Ph.D. Thesis, Stanford Univ.

Kowalik, W. S., S. E. Marsh, and R. J. P. Lyon, 1978, The effect of several atmospheric corrections in Landsat 5/4 ratios, Stanford Rem. Sens. Lab. Tech. Rep. 78−15:

Kramer, P. J., and T. T. Kozlowski, 1960, Physiology of trees; McGraw-Hill.

Lee, K., 1972, Bonanza project—1971, Rem. Sens. Rep. 72−5, Colo. School Mines.

Linlor, W. I., 1980, Permittivity and attenuation of wet snow between 4 and 12 GHz; J. Appl. Phys., vol. 51, pp. 2811−2816.

Lyon, R. J. P., 1965, Analysis of rocks by spectral infrared emission (8 to 25 microns); Econ. Geol., vol. 60, pp. 715−736.

Lyon, R. J. P., F. R. Honey, and G. I. Ballew, 1975, A comparison of observed and model-predicted atmospheric perturbations in target radiances measured by ERTS: Part I, Observed data and analysis; Proc. IEEE Symp. Appl. Rem. Sens. Digital Imagery to Mineral and Petroleum Exploration.

Marks, B., 1981, A method for estimating canopy density over large areas; M.A. Thesis, Univ. Calif., Santa Barbara.

Marsh, S. E., and R. J. P. Lyon, 1979, Quantitative relationships of near-surface spectra to Landsat radiometric data; Stanford Rem. Sens. Lab. Tech. Rep. 79−5.

Middleton, W. E. K., and A. F. Spilhaus, 1953, Meteorological instruments; Univ. Toronto Press.

Miller, L. D., and R. L. Pearson, 1971, Aerial mapping program of the IBP Grassland Biome: Remote sensing of the productivity of the shortgrass prairie as input into biosystem models; Proc., 7th Intl. Symp. Rem. Sens. Environ., pp. 165−205.

Nicodemus, F. E., J. C. Richmond, J. J. Hsia, I. W. Ginsberg, and T. Limperis, 1977, Geometrical considerations and nomenclature for reflectance; NBS Monogr. 160.

Olson, C. E., Jr., 1963, Seasonal trends in light reflectance from tree foliage; Arch. Intl. Photogramm., vol. 14, 226−232.

Olson, C. E., Jr., 1969, Seasonal change in reflectance of five broad-leaved tree species; Ph.D. Thesis, Univ. Mich.

Olson, C. E., Jr., and R. E. Good, 1962, Seasonal changes in light reflectance from forest vegetation; Photogramm. Engrg., vol. 28, pp. 107−114.

Perla, R. I., and M. Martinelli, 1975, Avalanche handbook; U.S. Dept. Agric. Hbk. 489.

Pitts, D. E., et al., 1977, Temporal variations in atmospheric water vapor and aerosol optical depth determined by remote sensing; J. Appl. Meteorol., vol. 16, pp. 1312−1321.

Rogers, R. H., 1973, A technique for correcting ERTS data for solar and atmospheric effects; NASA Symp., Significant Results Obtained from ERTS-1, vol. 1, sec. B., NASA SP-327.

Rohde, W. G., 1971, Reflectance and emittance properties of several tree species subjected to moisture stress; M.S. Thesis, Univ. Mich.

Rohde, W. G., and C. E. Olson, Jr., 1970, Detecting tree moisture stress; Photogramm. Engrg., vol. 36, pp. 561−566.

Rohlf, F. J., and R. R. Sokal, 1969, Statistical tables; W. H. Freeman.

Salomonson, V. V., and W. E. Marlatt, 1971, Airborne measurements of reflected solar radiation; Rem. Sens. Environ., vol. 2, pp. 1–8.

Scholander, P. F., H. T. Hammel, D. Bradstreet, and E. A. Hemmingsen, 1965, Sap pressure in vascular plants; Science, vol. 148, pp. 339–346.

Schwerdtfeger, P., 1976, Physical principles of micrometeorological measurements; Elsevier.

Sellers, W. D., 1965, Physical climatology; Univ. Chicago Press.

Siegel, R., and J. R. Howell, 1981, Thermal radiation heat transfer; 2nd ed., McGraw-Hill.

Slater, P. N., 1980, Remote sensing: optics and optical systems; Addison-Wesley.

Steiner, D., and H. Haefner, 1965, Tone distortion for automated interpretation; Photogramm. Engrg., vol. 31, pp. 269–280.

Strahler, A. H., 1981, The use of prior probabilities in maximum likelihood classification of remotely sensed data; Rem. Sens. Environ., vol. 10, pp. 135–163.

Switzer, P., W. S. Kowalik, and R. J. P. Lyon, 1981, Estimation of atmospheric path-radiance by the covariance matrix method; Photogramm. Engrg. Rem. Sens., vol. 47, pp. 1469–1476.

Tageeva, S. F., 1960, Study of optical properties of leaves, depending on the angle of light incidence; Prog. Photobiol., pp. 163–169.

Thompson, M. M., ed., 1966, Manual of photogrammetry; 3rd ed., Amer. Soc. Photogramm.

Tubbs, J. D., and W. A. Coberly, 1978, Spatial correlation and its effect upon classification results in Landsat; Proc., 12th Intl. Symp. Rem. Sens. Environ., pp. 775–781.

Wang, J. Y., 1975, Instruments for physical environmental measurements; Milieu Info. Serv.

Weber, F. P., 1965, Exploration of changes in reflected and emitted radiation properties for early remote detection of tree vigor decline; M.S. Thesis, Univ. Mich.

Weber, F. P., and C. E. Olson, Jr., 1967, Remote sensing implications of changes in physiologic structure and function of tree seedlings under moisture stress; Forestry Rem. Sens. Lab. Rep., Univ. Calif., Berkeley.

Weingeroff, M., 1931, Uber das Reflexionsvermögen von Wasser und Eis in Ultrarot; Z. Physik, vol. 70, pp. 104–108.

Wilson, W. H., 1980, Solar ephemeris algorithm; Vis. Lab. Rep. SIO 80–13, Univ. Calif., Scripps Inst. Oceanog.

CHAPTER 24

Fundamentals of Image Analysis: Analysis of Visible and Thermal Infrared Data

Authors/Editors: JOHN E. ESTES, EARL J. HAJIC and LARRY R. TINNEY

Contributing Authors: LARRY G. CARVER, MICHAEL J. COSENTINO, FRED C. MERTZ, MICHA I. PAZNER, LISA R. RITTER, CHARLENE T. SAILER, DOUGLAS A. STOW, TOD A. STREICH and CURTIS E. WOODCOCK

GENERAL CONTENTS: An image analysis paradigm. Basics of manual image interpretation: image interpretation tasks; elements of manual image analysis; aids and techniques; handling of imagery; stereoscopic viewing; methods of search; procedures. Imagery interpretation and transfer equipment: viewing equipment; measuring equipment; equipment for transfer of detail; miscellaneous equipment; selection of equipment. Basics of computer-assisted image interpretation: dimensions and resolution of environmental measurements; elements of computer-assisted interpretation; the evolution and status of digital image processing; image restoration; image enhancement; image transforms; supervised multispectral image interpretation; unsupervised classification; clustering; spatial and temporal elements; geographic information systems; management of computer-assisted interpretation systems; trends and futures. Interpretation of photographic imagery: the vertical aerial photograph; determining horizontal distance; height measurement; the oblique photograph; visible and near infrared photography. Interpretation of thermal infrared imagery. A perspective on image analysis and future directions.

INTRODUCTION

Every person engages in image interpretation. Books, newspapers, magazines, billboards, movies and television all offer images to the observer; each image conveys ideas or impressions. These ideas or impressions constitute interpretations. Such interpretations may or may not be accurate, and may be either conscious or unconscious; partial or complete. The degree of accuracy and completeness are in large measure dependent upon the experience base of the observer with reference to the context within which the interpretation is occurring. Nevertheless, interpretation is an essential process, a process through which information is obtained from images. Once the interpreter has gained facility with the elements, aids, and techniques of observation, the difficult task of drawing accurate conclusions from the images observed can begin.

The material which follows discusses the fundamentals of image interpretation. First, a conceptual framework is presented wherein the basic linkages between what has traditionally been referred to as manual image analysis and machine-assisted or automated image analysis are put into perspective. This section, compares at a very general level the image elements, aids, and techniques currently employed in these forms of analysis and discusses future directions that may lead to more appropriate utilization of the potential offered by both human analysts and machines. This is followed by an expanded discussion of the

interpretation process. This portion of the chapter, as well as the sections on the imaging of photographic and thermal infrared imagery draw heavily from material presented in Chapter 14 of the first edition of the Manual of Remote Sensing. Material in the section on manual (human) image analysis covers activities, elements, aids and techniques of manual image analysis as well as factors relative to vision and the analysis process. This is followed by material on image interpretation and transfer equipment, providing the reader with background on products currently being employed to facilitate the image analysis process. An expanded discussion of machine-assisted image analysis follows. Addressed here are the diversity, potential and limits of the machine-assisted image-analysis process. The chapter continues with a discussion of specific information relative to the interpretation of photographic and thermal infrared imagery, and concludes with a brief discussion of the future research needs in this important area.

AN IMAGE ANALYSIS PARADIGM

NEED FOR FRAMEWORK

Principles of image analysis include methods for detecting, identifying, and measuring objects of interest from the aerial perspective. Years of work by some far-sighted, innovative professionals have produced a substantial foundation upon which the analysis of remotely sensed data is based. While most recent work and advances in the field of remote sensing have revolved around non-photographic imaging and computer-assisted processing techniques, the last 10 to 15 years have seen minimal advances in the state-of-the-art of photointerpretation. A clear need now exists to critically re-examine the basic elements and principles of image analysis that are common to both manual- and computer-assisted techniques. Indeed, due to individuals in the diverse group of disciplines involved, which includes the applications-oriented disciplines such as, forestry, range management, hydrology, geology, and geography on the one hand and electrical engineering and computer science on the other, artificial barriers between manual- and computer-assisted techniques have been erected. These barriers are false and have only made it more difficult to discern either fundamental similarities or differences between the two approaches.

The objective of this discussion is to provide a broad conceptual framework, a paradigm wherein the basic linkages between what has traditionally been referred to as manual or human image analysis and computer-assisted or automated image analysis are put into perspective (See Figure 24-1). Discussed are some of the current converging trends in the understanding of the elements involved in both processes. It appears from an interpretation of these trends that a more integrative, multidisciplinary approach will benefit both processes. Note that a true dichotomy of manual versus computer-assisted approaches does not exist. In reality, the difference between these approaches is only a matter of degree. However, the readers hould be aware that near the extremes some very fundamental differences in **perspective** currently exist between these two approaches. This suggests the need for a more comprehensive framework that both integrates common elements and encourages the use of unique perspectives as appropriate to the analysis/interpretation tasks at hand.

Also note that while image analysis is often considered both an art and a science, further advancement as a science will certainly require better definitions of the basic elements, aids, and techniques utilized. From an analytical perspective, the "science" of image analysis is lacking in that even the **basic elements** of image analysis have not been clearly defined. In the first edition of the Manual of Remote Sensing, Estes and Simonett (1975) noted that "many articles and texts differ as to the number of basic elements . . ." with only **general agreement** on six primary ones (size, shape, shadow, tone or color, texture, and pattern). Clear elucidation of higher order processes (e.g., object recognition) is dependent on the precise definition of the basic elements, yet there has been no agreement upon the number, or the ordering of these elements except at the most primitive level.

Many disciplines have investigated topics concerning better understanding of image interpretation. Prominent among these are medicine (anat-

omy of the eye), psychology (concerned with visual perception), and applications-oriented disciplines involving remote sensing, such as electrical engineering and various aspects of computer science (image processing, computer graphics, pattern recognition, and more recently image understanding).

That such a variety of perspectives have investigated image interpretation is both a fortuitous and a complicating factor in the pursuit of an understanding of the image analysis process. It is fortuitous because the variety of perspectives has resulted in a rich set of elements, processes, aids, and techniques for investigation; and complicating because the nomenclature is not uniform (for example the term "feature" has a variety of conflicting definitions). There has been inadequate communication among the separate disciplines since researchers in each publish in "their" journals and little cross-fertilizing synthesis occurs.

To integrate aspects of these disciplines, some of the basic concepts of the mechanics of human and machine perception will be briefly discussed. The sources and depth of this information ranges from general expositories to detailed articles in current computer-science journals. A great deal of additional work is obviously required and hopefully this general overview will provide some insights and stimulate work in this critical area in the future. The material on human vision and perception is followed by a discussion of the tasks of image analysis. This is followed by information concerning the basic elements of image-analysis techniques and aids for image analysis, and also the analysis procedure that produces labels for objects and phenomena. As seen in the discussion of the multiconcept presented in Chapter 1 of this **Manual**, remote sensing offers to the analyst the potential of a wide variety of image types and analytic procedures (both manual- and computer-assisted). Yet there is little work or agreement on the fundamentals involved in the analysis of data from this wide variety of sources. The discussion of tasks, elements, techniques, aids, and analysis procedures that follows presents, at a basic level, the similarities and differences between the human and machine-assisted approaches to the analysis process.

A great deal of work needs to be accomplished in our understanding of the image-analysis process. The concepts embodied in the paradigm presented here are intended to stimulate thinking into this area of research which is, in the final analysis, one of, if not the, major foundations of remote sensing.

MECHANICS OF HUMAN VISION

The human eye is an extremely sophisticated remote sensing system. The mechanics of human vision play a major role in our better understanding of visual perception, which subsequently forms the basis for both manual- and machine-assisted image interpretation. It is important to

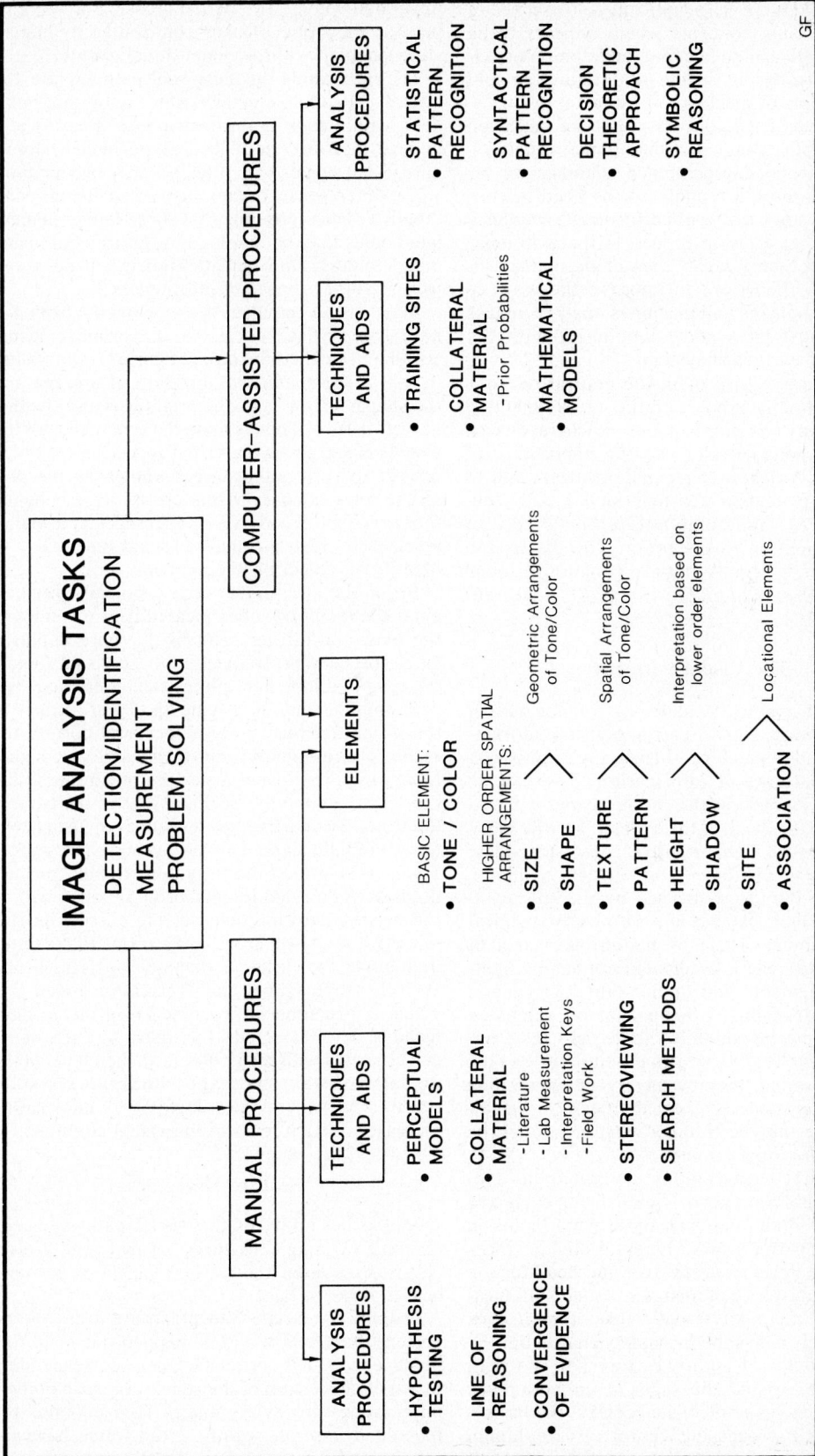

Fig. 24-1. Conceptual Framework of Image Analysis. A broad conceptual framework, or paradigm, by which human and computer-assisted image analysis procedures can be compared and contrasted. After tasks and basic elements the two approaches diverge significantly, although analogs of the manual procedures can be seen in the more advanced computer-assisted procedures.

note here that there is a significant convergence of image-processing concepts which appears to be occurring between engineering and computer science approaches to image understanding and basic concepts of mechanics of human vision.

This convergence appears not to be based on improved understanding of the human visual system by engineers and computer scientists *per se*. Rather, it is more a reponse to the need to construct hardware capable of performing a sequence of hierarchical decision processes that can emulate the processing capabilities of the human visual system. If we are to improve the ways in which both humans and machines analyze images then we must have some familiarity with the workings of our visual system.

When the eyes are open the central nervous system is directly exposed to the environment. It is significant that this happens nowhere else in the human body, attesting to the importance of vision, man's richest sense and strongest link to the world. It is estimated that each waking second the eyes send some one *billion* pieces of fresh information to the brain. The discussion that follows is largely based on material taken from *The Eye: Window to the World* (Wertenbaken, 1981).

The Visual Pathway

At each step along the pathway from the human eye to the brain, signals carrying visual information link and recombine, fusing their messages into new patterns of information. The signals begin their journey in the retina, a multilayered sheath lining the back of the eyeball; here lie more than *100 million* light-sensitive rods and cones. Each of these photoreceptors detects a minute fragment of the image focused by the lens, converting the light pattern into an electrochemical signal containing visual information. At one end of each rod and cone is a segment containing light-sensitive pigments and at the other a synaptic body that transmits information by releasing chemical neurotransmitters. Some cells relay impulses laterally, integrating visual messages across the retina, thereby clearly demonstrating local parallel-processing capabilities. Nerve impulses leave the eye through what is termed the "optic nerve" fiber bundle.

At the first "way station" beyond the eye, called the chiasm, nerve fibers from each eye meet in the brain and each optic nerve splits in half, forming two tracts. Fibers from the inner halves of each retina cross over and head for the opposite hemisphere of the brain; as a result, messages from the right visual fields of both eyes reach the left hemisphere, while signals from the left visual fields of both eyes travel to the right hemisphere. Visual messages from both eyes reach both halves of the visual cortex, the brain's primary sight-processing center. Even though each hemisphere receives messages from only half of the visual world, the brain merges them into an integrated whole. The partial crossing of nerve fibers at the optic chiasm contributes to human stereoscopic, or three-dimensional, vision.

The next stops on the visual pathway are the lateral geniculate nuclei, twin "relay stations" deep within each brain hemisphere. Researchers believe that the lateral geniculate nuclei coordinate visual signals with other sensory information, since other sense organs also send impulses to them. It is also possible that these lateral geniculate bodies may be involved in shutting out visual inputs to the brain when attention is being devoted to some other source of information.

Most visual impulses travel directly from the lateral geniculate nuclei to the primary visual cortex (also called the striated cortex). Our ability to detect the spatial organization of a scene and the shapes of objects—the brightness and shading of their parts—depends upon the functioning of the visual cortex. From the visual cortex, nerve fibers project to surrounding areas known as the prestriate, or secondary visual cortex. It is believed these regions decode visual messages at a higher level. The prestriate may also aid our ability to identify an object or phenomenon.

From the prestriate cortex, visual impulses enter the temporal lobes located near the sides of the head for further, and likely more sophisticated, processing. Damage to these visual sites in the temporal lobe can inhibit visual learning.

Examining and identifying an object belongs to the visual machinery of the cerebral cortex, the region of the human brain responsible for higher functions. First visual messages reach the striate cortex, located in the occipital lobe at the back of the head. By charting the responses of visual cells to bars of light slanted at various angles, researchers have identified the triggering stimuli for hundreds of brain cells. Each neuron, or nerve cell, in the striate cortex has been seen to possess its own receptive field—an area on the retina that excites or inhibits the firing of that neuron (Hubel and Weisel, 1962). Hubel and Weisel concluded that visual information is processed in a hierarchical fashion. They conclude this because each nerve cell connects with many others at each level of the visual pathway, messages undergo constant transformation, and the complexity of information increases at each level of the visual circuitry.

Feature Detectors

At the beginning of the visual pathway, cells respond to simple features. Hubel and Weisel (1962) discovered that retinal ganglions respond best to disks of light.

In both the striate and prestriate areas of the cortex, the visual world is mapped out point-for-point, each small cortical segment corresponding to a specific section of the retina. Because of their importance, however, images focused onto the foveal, or central, region of the retina have far more detailed representation in the cortex than do images located on the periphery of the visual field.

Thirty-five times as much space is devoted to the fovea, giving it a greater degree of visual acuity than any other region of the retina. This means that a very narrow field of an interpreter's focus is presenting the majority of information to the analyst at any given instant.

Like retinal and geniculate cells, visual cells in the cortex detect certain features of light; their standards, however, are more exacting. Some visual cells respond only when three or four features are present. Scientists divide cells of the striate cortex into three main categories. "Simple" cells respond best to slits of light on a contrasting background or to boundaries (edges) between dark and light zones. These slits or edges must be precisely located in the cell's receptive field and must have the proper orientation—positioned vertically, horizontally, or at a precise angle.

The next level of the visual hierarchy are the "complex" cells. Complex cells also respond to slits and edges set at specific angles. Unlike a simple cell, however, a complex cell will fire in response to slits located anywhere in its receptive field or to a bar moving in a specific direction. This property represents higher level visual processing, i.e., the ability to distinguish shapes regardless of their location.

"Hypercomplex" cells respond best to features meeting even more refined criteria such as corners, angles, lines, or edges of a specific length, orientation, and location. Some hypercomplex cells also respond to movement in one direction.

This hierarchical network of cell connections, in which complicated cell properties build up from much simpler cells, suggests an extremely orderly form of cellular architecture. For a complex cell to receive input from a group of simple cells sharing the same orientation requirements, the two must be located near each other.

From the visual cortex, special nerve circuits carry signals to the cerebellum. The cerebellum coordinates muscle movements. This visual network regulates activities requiring close coordination between the eyes and limbs, such as driving a car or catching a ball. By blending motor and visual signals, the cerebellum controls the body and limbs as a person moves. Scientists have also discovered that some of the visual impulses make a detour on their way to the cerebellum. They travel through the pons, a bridge-like structure at the base of the brain in which certain cells respond to single spots of light moving at specific rates and directions. Other cells in the pons register large, moving visual fields containing many spots of light. These cells send new signals to the cerebellum, which appears to synthesize them with information about the position and velocity of muscles and joints.

VISUAL PERCEPTION

Visual perception refers to the process whereby visual sensory stimulation is translated into organized experience. Like the mechanics of human vision, the way we perceive objects and phenomena is basic to the analysis process. Although much remains to be learned concerning the process of visual perception, psychologists and others have collected a substantial set of observations and theories. This information is clearly central to our better understanding of the human image interpretation process (going far beyond the present "gray box" mode of operation) and is especially crucial to efforts directed at automating image interpretation. For, although at the most basic level humans and machines operate on tonal contrasts, how humans perceive the spatial aspects of tone is the key to improving automated analysis procedures.

Ordinarily, the various objects and phenomena around us are recognized with such apparent ease and immediacy that it is easy to assume that the operations involved are simple and direct. The experience of engineers, however, has proven that the nature of the object/phenomena recognition task is far from simple. Relations found between visual stimulations and their associated percepts suggest that inferences can be made about the properties of the visual perception process. Various theories have been developed based on these inferences. Indeed, the validity of perceptual theories can only be checked indirectly because a visual percept is the joint product of stimulation and of the process itself; it cannot be directly observed. As a result, much of visual perception research involves comparisons of predictions based on theory, with empirical data derived largely through experimental research techniques.

Historically, systematic thought about human visual perception was largely the province of philosophy, which explored, and continues to explore, basic questions about the sources and validity of what is called human knowledge. As a scientific enterprise, however, the investigation of human visual perception has primarily developed as part of the larger discipline of psychology. Generally, psychologists accept the apparent physical world. In studying human visual perception, they focus their attention on the processes whereby percepts are formed from the interaction of light with the perceiving organism. Thus arise such basic questions as the degree of correspondence between percepts and the physical objects to which they ordinarily relate.

One of the classical problems in the study of perception deals with the distinction between sensing and perceiving. Based on the assumption that percepts are constructed of simple elements that have been joined by association, it is often said that sensations are simple and percepts are complex; yet this clearly provides only an arbitrary basis for distinction, subject to initial definitions. Another commonly offered basis for distinction is the notion that perceiving is subject to the influence of learning while sensing is not. An alternative distinction that has become more widely accepted is to separate the two, based on physiological-anatomical criteria, i.e., visual sensations are identified with neural events occurring immediately around the eyes whereas percepts are identified with brain activity.

Experimental evidence strongly suggests that percepts, even of very simple geometric forms, follow a measurable, developmental time course, and during the associated time frame there is a brief period (100 to 200 milliseconds at most) during which a percept is highly vulnerable to disruption.

Another classical problem in perception has been the "nature-nurture" issue. The organization apparent in percepts has been attributed by some to learning, as being built up through associations of elements that have repeatedly occurred together in one's experience. Others (particularly those of the Gestaltist school) stress the view that perceptual organization is physiologically inborn and inherent in innate aspects of brain functioning.

That humans are genetically endowed with some degree of visual perceptual skills has been clearly demonstrated by the experiences of people blinded by cataracts or clouded corneas from birth or early childhood, and later surgically cured. Although at first bewildered, these individuals are soon able to perceive figure-ground relationships, to scan and focus on objects, and to visually track moving objects. Perhaps even more revealing are perceptual studies of infants in which babies only a week old clearly display visual skills.

A host of experiments using laboratory animals have also been conducted to determine, under rigorous experimental control, the extent of innate versus learned percepts. These experiments have consistently provided verification that early perceptual experience plays an important role in later perceptual development, even to the extent of producing changes in brain weight and biochemistry. These results suggest that both sides of the "nature-nurture" issue play important parts in visual perception, that some basic visual functions are built-in, but that visual experience serves to maintain and elaborate them.

We reviewed earlier that studies of the mechanics of vision have identified the building blocks of perception, the feature-detecting cells of the retina, lateral geniculate nuclei, and visual cortex. We do not yet clearly understand, however, how these blocks combine to form perception of scenes. Early in this century the Gestalt School of Psychology proposed that stimulation is perceived in organized or configurational terms; patterns take precedence over elements and have properties that are not inherent in the elements themselves. Several principles guiding the grouping of elements in percepts were developed in support of this approach: proximity, similarity, closure, and, most generally, good form (e.g., simplicity, regularity, symmetry, etc.).

Evidence that percepts have constituent elements emerged serendipitously from research on stabilized retinal images. As expected, visual activity is slightly enhanced when the retinal image is kept still; a remarkable, unexpected finding, however, was that such stabilized images rapidly seem to disappear, i.e., the perceiver loses awareness of them. Complex patterns are found to produce percepts that are relatively slow to deteriorate; furthermore, they do not disappear *in toto*. The manner of the fragmentation is perhaps revealing in the way in which complex percepts are synthesized.

Indeed, under retinal stabilization, single lines seem to disappear and reappear in a unitary (altogether) fashion. In a figure comprised of several lines (e.g., a square) percepts of parallel lines are likely to disappear and reappear together; proximity also affects the joint perceptual fate of pairs of lines. Retinally stabilized segments of such geometric figures as circles and triangles can seem to disappear and reappear without implicating the entire figure. In the disappearance of percepts of triangles, lines rather than angles are the functional units.

Clearly, with stabilized images, the constituent perceptual elements of complex geometric forms are lines or curves; and those with the same orientation are likely to have similar perceptual fates, as though forming a higher-order component of complex patterns than do individual lines. These conclusions are remarkably similar to those drawn from studies of the effect of visual stimuli on the electrical activity of single neurons in the cerebral cortex. A finding of major theoretical significance is the failure of percepts of circles, squares, and triangles to act as units. Such percepts are treated in classical Gestalt theory as though they are basic and unitary and not readily decomposable.

MANUAL IMAGE ANALYSIS

Task of Image Analysis

Referring to Figure 24.1, if we were to analyze the image interpretation process as though it occurred in a time sequence, the analysis typically begins with the detection and identification of important objects. The objects may then be measured manually or with the aid of appropriate instruments. Measurement can then be followed by a consideration of the objects in terms of information from the interpreter's special field of knowledge within a problem solving context. The interpreter must then be able to communicate both his perception of objects and the significance of the objects.

Detection Identification

There is a general rule of thumb in image analysis which goes as follows: it takes three resolution cells along the side of an object to detect it. It takes five resolution cells along a side to identify it. The process of detection and identification are difficult to differentiate and describe.

Each individual may go about the psycho/ physiological process of identification in a different way. Clues such as size, shape, tone, and texture while important to one individual, may be meaningless to another. The interpreter communicates his response to a stimulus by mentally labeling the identified object. His ability to identify objects cannot be measured in the absence of labeling, although it is possible to learn whether his response repertoire includes the appropriate label for a given object. In order to communicate adequately his interpretations, the image analyst must be well versed in the terminology of his own and related fields of knowledge.

Measurement

Image interpreters measure the spatial and spectral dimensions of objects by means of scales and other instruments. This measurement may range from a simple visual estimate of the size, shape and color of an object to a detailed metric calculation of size, shape or density. Such measurements are important in the identification process and can be critical within a problem-solving context.

Problem Solving

In the carrying out of a given task an image analyst may be asked to identify an object by examining other objects or phenomena or to identify an object complex from the image elements making up that complex. This activity constitutes a form of problem solving. Most analysis problems require the interpreter to have at his command some knowledge other than that directly obtained from the imagery being examined; or to be working closely with individuals with such in-depth knowledge. The arrangement (often orderly) of particular objects and phenomena recorded on aerial photography tends to facilitate the analysis process. Examples of such orderly associations are seen later in this chapter and throughout this *Manual*. Appreciation of the role of orderly associations in the interpretation process proceeds most effectively from a combined mastery of both the interpretation process and background training in the physical, biological or environmental sciences.

Image Elements of Importance in the Manual Analysis Process

Remotely sensed data represent energy reflected, emitted, transmitted, and scattered from many regions of the electromagnetic spectrum. The data may be recorded in many shapes, sizes and scales. An understanding of the basic elements of image interpretation is essential to the efficient and effective use of these data. Inherent elements within an image that can provide an image analyst clues toward the detection, identification, measurement, and problem-solving tasks of image analysis include: tone/color, size, shape,

texture, pattern, height, shadow, site, and association. In the following discussion we consider that the primary resolution element of a photographic emulsion is an individual grain of a silver halide and that of a scanning system is its instantaneous-field-of-view (IFOV); these resolution elements, in turn, define the common picture element or pixel. Taking the pixel then as the inherent image element that we are dealing with in the analysis process, we find that the tone or color of a given pixel is, in all likelihood, the most fundamental image property employed in the manual image analysis process (see Figure 24-2). The ordering seen in Figure 24-2 should not be considered hierarchical, yet in some way it approximates this state. Owing to the interconnections and interdependencies of these elements, a hierarchical structuring is not possible at this time. Tone, as expressed in shades of gray on a black-and-white photograph, and color, as expressed in hue, value and chroma in a color photograph, convey more information to a knowledgeable interpreter than any other single element of interpretation. In almost all cases, it is the difference in tone or color between objects or between an object and its background that is important. Without such a difference in tone or color between the background and the edge of an object, there can be no detectable image. In aerial photographic interpretation tone is representative of reflected energy; however, with the range of sensor systems available today, tone on an image can be the product not only of reflected energy but emitted, transmitted, and scattered energy as well, depending on the sensor system.

Both size and shape image elements represent geometric arrangements of the tone or color of pixels making up a given object or phenomenon. As such, these elements represent an intermediate level in the ordering of image elements. As seen in Chapter 1 size is dependent upon the scale of the image under analysis. Often, by visually estimating or employing aids to precisely measure the size of an unknown object on an image, the interpreter can eliminate from consideration whole groups of possible identification. The shape or form of some objects can be so distinctive that they may be identified solely from this criterion. The Pentagon Building near Washington, D.C. is a classic example. However, the shapes of objects seen in vertical view are sometimes surprisingly difficult to interpret. The plan or top view of an object is so different from the familiar profile or oblique view that novice interpreters have been known to fail to recognize the image of the building in which they were working (Estes and Simonett, 1975). Unlike size and shape, which are related to the geometric arrangement of individual objects and phenomena, texture is tied more closely to the spatial arrangement. Texture is created by tonal repetitions within an object, or within groups of objects that are too small to be discerned individually. Texture, the visual impres-

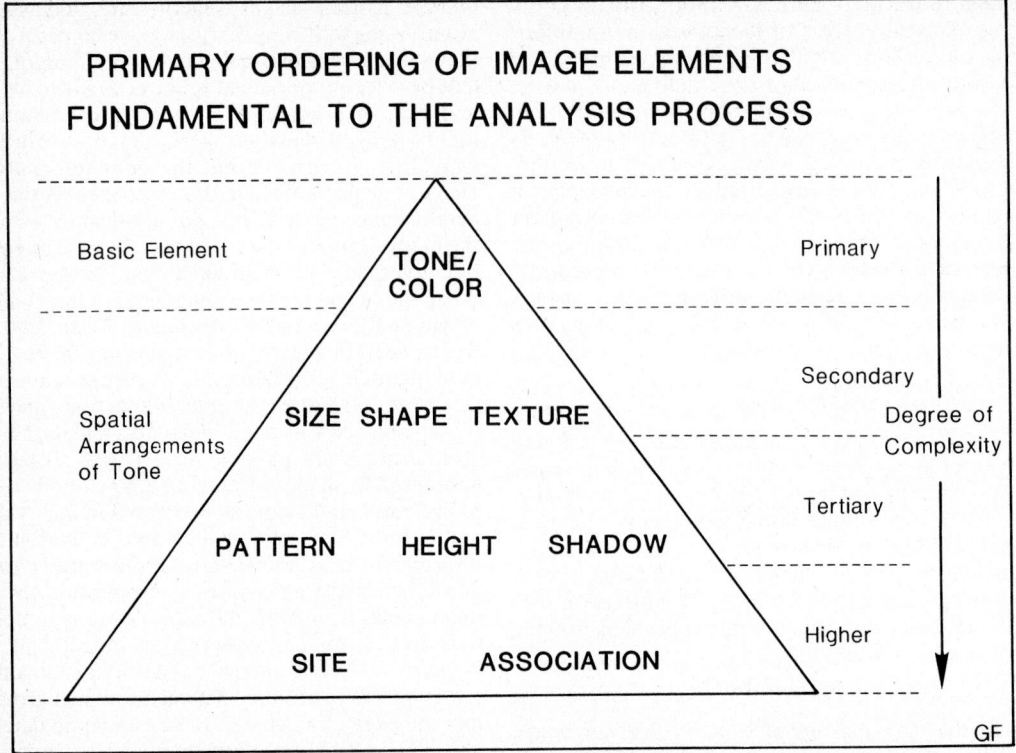

Fig. 24-2. Primary Ordering of Image Analysis Elements. This figure illustrates in a general manner the hierarchical relationship among some basic image analysis elements. Note the fundamental importance of tone/color as the primary element. Spatial arrangements of tone/color result in elements of greater complexity.

sion of the roughness or smoothness of an object, can be a valuable clue in the analysis process. This is especially true on imagery obtained with Side-Looking Airborne Radar or with imaging passive microwave radiometers, where texture can play an important role in differentiating various classes of environmental phenomena.

As we address pattern we are in a transition area from intermediate to higher level element-ordering. Just as with shadow, and to some extent height, pattern requires the analyst to perform some level of integration with a number of lower order elements in order to achieve a label.

Pattern, or repetition, is a characteristic of many man-made and some natural features. Pattern is a very scale-dependent phenomenon. An orchard that may exhibit a pattern on a low altitude photo may appear as rough textured on a high altitude photo and smooth textured on a satellite image, depending on the focal lengths and associated resolution capabilities of the sensor systems employed. A regional tectonic pattern, such as that formed by the meeting of the San Andreas and Garlock faults in California, may not assume the significance on a low altitude photo that it does on a high altitude photo or a Landsat image.

Some image analysts contend that, except for tone, the third dimension is the most important element in photographic interpretation. Analysis

and measurement of the third dimension of objects can be achieved from both single and multiple images from vertical or obliquely acquired data. The relative heights and 3-D geometric shapes of objects often provide revealing clues as to their identity. "Stereo" can be made possible through the proper use of cameras and line scanners in the visible and radar regions.

Shadows can either help or hinder the interpreter, for they reveal silhouettes but may hide detail. Shadows are familiar phenomena and, in ordinary life, we often judge the size and shape of objects or persons by observing the shadows they cast. Commonly, aerial photographs are taken within two hours of noon (local sun time) in order to avoid excessive shadows and reveal maximum ground conditions. On the other hand, the presence of shadows normally enhances the photo interpreter's perception of shape and therefore aids in object identification. In projects involving geologic analysis, photography is often flown at low sun angles in order to accentuate minute surface irregularities. This is one reason why Side-Looking Airborne Radar imagery, which approximates low sun-angle photography, is important in geologic investigations. Low sun-angle terrain accentuation was also a factor in determining the time of day at which Landsat's orbit would be made sun synchrous.

Finally, site and association represent higher

order image elements. The location (site) of objects with respect to one another or to terrain features is often helpful in narrowing probabilities in image interpretation. For example, swamps typically contain specific vegetation assemblages, specific agriculture practices are used when farming hillslopes (e.g., dry farming or specific types of irrigation) and thermal power plants are always located near a water body. Site is very close to association as an element in image interpretation. Association is one of the most helpful clues to the identity of man-made features. The manufacturer of aluminum requires large quantities of electrical power. Absence of a power supply rules out this industry. Schools usually have associated playgrounds and athletic fields while churches do not. Even the type of school (elementary, secondary, junior high, or high school) can be inferred from the size and type of playing fields, buildings, and parking facilities. Large farm silos are an indication that livestock are (or were) present. Unusually large culverts or bridge spans across small streams indicate that heavy runoff occurs frequently enough to require substantial engineering works. Labeling involving these elements requires that the interpreter integrate a number of lower order image elements into the analysis process. It, in essence, requires that the analyst make a number of labeling decisions in an iterative or integrated fashion before the final label is applied to the object or phenomenon of interest.

Aids and Techniques to Manual Image Analysis

Photographic interpretation has been defined as the act of examining photographs and/or images for the purpose of identifying objects and phenomena and judging their significance. The better the interpreter understands the purpose of his analysis (i.e., the specific use to which his analysis will be put), the better job he can do. Therefore, it is important that the interpreter understand the overall task of which his analysis is part. With this knowledge, which may be thought of in terms of perceptual models, the analyst may better judge the significance of the objects and phenomena he perceives on a given image. In carrying out his task, an interpreter may use many more types of data/information than those recorded on the individual images he is to interpret. The data/information can be characterized as aids. These aids to the analysis task are often grouped under headings such as collateral data, ancillary data, on-site verification, or just plain field work. Other aids in the manual-analysis process involve techniques for logical search of the images in question and also for viewing the third dimension, inherent in stereoscopic image analysis.

Perceptual Models

As previously stated, the more knowledge the image analyst has concerning the potential sources and magnitudes of errors associated with data/information flows in remote sensing the better analyst he will be. That is, an analyst should strive to achieve expertise in the philosophical concepts, elements/aids, techniques, methodologies, processing, and display procedures in remote sensing. The analyst should also strive to achieve expertise in the discipline or applications areas in which he is working. Finally, the analyst should make every attempt to come to a detailed thematic understanding of the geographic area he is considering. This information, plus the sum total experience of the individual analyst, forms the basis for the perceptual or cognitive models that the analyst brings to a given analysis task. These mental models, in turn, form the conceptual basis upon which the analyst operates. They form the primitives which the analyst begins to link as he performs the classification and labeling process.

Collateral Material

There are many sources of collateral material and the image analyst should take advantage of all types appropriate to any given task:

- material from the open literature (e.g., books, articles, reports, maps, census data, etc),
- laboratory measurements,
- photo interpretation keys (e.g., elimination, selective, descriptive),
- field work, and
- other image sources (e.g., ground, aerial, space).

There are times when the interpreter should be careful in the use of collateral material. As with use of the multiconcept (Chapter 1), the interpreter can stand the risk of being overwhelmed by collateral material. Too much data can, in some cases, be as confusing as not enough, especially when such data serves to document conflicting claims. The analyst should strive to know a great deal about the potential errors that are associated with the individual collateral data types being employed in his particular task. The analyst must always remember that these data have a variance of their own and that, although collateral material is often used as the "truth" against which remotely sensed data are judged, there are cases where the truth may not be absolute. The interpreted image may provide a more accurate record than the collateral.

Stereoscopic Viewing

One of the most used photographic interpretation aids/techniques is stereoscopic analysis. The principle of stereoscopy (depth perception) should be understood by every image analyst. With binocular vision, when the eyes are focused on a given point, the optical axes of the two eyes converge on that point intersecting at an angle called the parallactic angle. The nearer the object, the greater the parallactic angle. The shortest distance of clear stereoscopic depth perception for the average adult is about 25 cm. Objects can be focused at distances as short as 15 cm, but such close viewing causes excessive eyestrain. The maximum parallactic angle formed by the eyes is approximately 15°. The maximum distance at which distinct stereoscopic depth perception is possible is approximately 500 m for the average adult. Beyond that distance, parallactic angles are extremely small and changes in parallactic angles necessary for depth perception cannot be discerned.

An image interpreter can create a three-dimensional or stereoscopic model of terrrain by viewing overlapping imagery. The photographic basis for stereoscopy can be found in the principles of image displacement and stereoscopic parallax. These topics are beyond the scope of the brief discussion that is given here and will be discussed later in this chapter. It is important, however, to restate that many manual image analysts consider height, or the ability to achieve stereoscopic analysis, as one of the most important aspects of photographic interpretation.

Methods of Search

An aid/technique that all image analysts need to master involves the way in which an image is scanned for basic elements that may aid in the detection, labeling and problem-solving processes. The analyst must develop a method of searching the data for pertinent information. There are two basic general methods that may be used to study aerial imagery: the "fishing expedition" and the logical search.

Image interpreters have learned that aerial photography and remote sensing imagery are full of surprises. They are often tempted to examine every photograph so as not to miss anything. This is the fishing expedition; a method of search that is frequently used. Fishing yields large amounts of information, including much that is not pertinent to the subject at hand. It requires a more leisurely effort than the interpreter can often afford.

By resorting to probabilities, the interpreter can work more efficiently in the time available. He searches only those areas in which the objects of interest are likely to be found and disregards large regions or entire images that are not likely to contain desired information. This selective method is logical search, a combination of quick scanning and intensive study. It demands more experience than the fishing expedition, since the interpreter has to decide where intensive study will yield the best results. It is, however, generally much more productive in relation to the time and effort expended.

A knowledgeable image analyst will typically begin his task by a close examination of all details thought to be relevant. Most experienced interpreters prefer to begin by scanning the area as a whole. They are then prepared for intelligent selection and the study of detail. In most instances the interpreter will then follow the following guidelines:

1. Proceed from general considerations to specific detail.
2. Proceed from known to unknown.
3. Work methodically.

Again, however, care should be taken in proceeding from general to specific considerations. This is desirable as long as the considerations of general (sometimes called regional) features do not bias the interpretation of the specifics. Specific, local considerations often provide evidence needed to complete or confirm the broader regional pattern. In most cases, general and specific features must be considered together. To say that one must come before the other can be misleading. To go further than this and propose that interpretation should proceed from one specific group of features to another is an unwarranted channeling of the infinite variation that the interpreter encounters (Olsen, 1973). Working from the known to the unknown is basic in all science. Often known objects or phenomena provide important clues to the identity and significance of unknown objects or phenomena. Finally, by proceeding methodically there is less chance that the interpreter will fail to detect details significant to his task. By his proceeding in a logical, methodical manner the final goal or task of the interpreter is far more likely to be achieved than if a non-structured analysis is conducted.

Manual Analysis Procedures

Analysis procedures involve those processes which, when completed, result in labels being applied to the objects or phenomena central to the task of the analyst interpreter. These procedures then can include both single and high order decision processes. In the first instance an identification can be made on the basis of a detected pattern of pixels that matches a perceptual model in the knowledge base of the analyst. In the second instance problem solving may require that a number of such identifications be processed (e.g., alternative hypotheses tested and parallel lines of reasoning followed as evidence converges toward either the application of a given label or the explanation of a given pattern).

The analysis process is complex. In manual image analysis it is not well understood in many ways. Essentially what is occurring in humans is that they are applying functional categorizations to image features that match their aculturated perceptions of what those patterns represent. As such they are attaching individual and grouped labels, based upon their knowledge base, to basic biophysical/geochemical materials. These labels and functional categorizations may or may not be mutually exclusive and indeed may vary from culture to culture and even from individual to individual. Owing to such complexity, this brief discussion of analysis procedures will focus on those fundamental processes by which labels are applied and problems solved in image analysis. These processes can be labelled hypothesis testing, line of reasoning, and convergence of evidence. Perceptive readers will quickly appreciate that there is overlap here. Nevertheless, early in the interpretation process the analyst will often develop a hypothesis relative to the task at hand. A line of reasoning is then developed in the testing of the hypothesis, and through a process of deductive reasoning (convergence of evidence) an interpretation is made.

Hypothesis Testing

The development of hypotheses is fundamental to science. A hypothesis is fundamentally a potential answer to a question or a solution to a problem. In this fashion hypotheses directly relate to the task of the image analyst. Image analysts rarely start an analysis task cold without any idea of what they are looking for. That would be akin to looking for something, but having no idea of what one is seeking. Hypothesizing is, in essence, educated guessing. Some guesses are accurate, some may not be. The hypothesis is tested by determining whether or not the observations (interpretations) agree and by counting the number of times wherein observations are consistent with facts. Basically a hypothesis is tested and then the decision is made to accept or reject it by examining the knowledge base to see whether the hypothesized solution is, indeed, correct. We also look to see the number of times that it is correct and whether it is invariant. When we find that a hypothesized solution is invariant in a large number of cases, and when the solution does appear to be satisfactory or has considerable explanatory power, we accept the hypothesis as confirmed.

Line of Reasoning

Briefly, line of reasoning involves a progression of logic leading to a given conclusion. A line of reasoning process can fundamentally be seen to involve a series of essentially dichotomous or "if-then" type statements. In a very real sense this type of analysis process can be considered analogous to the logic process codified in the dichotomous interpretation keys discussed later in this chapter.

Convergence of Evidence

Manual image analysis is basically a deductive process; features that can be recognized and identified directly lead the image interpreter to the identification and location of other features. Even though all aspects of an area may be intertwined, the photo interpreter or image analyst must begin someplace; he may begin with a consideration of drainage, landforms, vegetation, and man-made features, then go on to other topics of interest, integrating each of the facets of the terrain as he goes. For each terrain, the interpreter must find his own point of beginning and then consider each of the various aspects of the terrain in logical fashion.

Deductive image interpretation should entail a thorough understanding of all of the activities and techiques of image analysis as well as the conscious or unconscious consideration of the elements of image interpretation listed earlier. The completeness and accuracy of image interpretation are a function of the interpreter's understanding of what techniques were used to acquire the imagery, what collateral material is appropriate, and how and why images exhibit shape, size, tone, shadow, pattern, and texture characteristics. In addition, an understanding of site, association, and resolution strengthens the interpreter's ability to integrate the different features making up the terrain. For the beginning interpreter, knowing all of the techniques of image analysis may be a formidable requirement. Hence he should systematically review the elements of image interpretation to proceed with an integrated terrain interpretation in an optimal fashion.

The principle of convergence of evidence requires that the interpreter first recognize basic features or types of features and then consider their arrangement (pattern) in an areal context. Several interpretations may suggest themselves. Critical examination of the evidence usually shows that all interpretations but one are unlikely or impossible. The greatest difficulty in interpreting images involves judging degrees of probability. As previously stated, the more the interpreter understands the techniques and elements of image analysis, as well as where his task fits into the overall goal for which he is interpreting the image, the higher his probability of correct analysis becomes.

COMPUTER-ASSISTED IMAGE INTERPRETATION

Tasks of Image Analysis

As implied in Figure 24-1 the tasks of both manual- and computer-assisted image interpretation are similar, namely detection and identification, measurement, and problem solving. Anticipated advantages of computer-assisted techniques are often stated as speed and repeatability. In most cases of interest, however, automated techniques lag far behind manual techniques in terms of both speed and accuracy. This is largely due to the low level decision-making processes currently employed by automated procedures; significant advances have been made in recent years, however, and the outlook is encouraging for future systems.

Perhaps one of the most significant contributions, to date, of the computer-assisted approach has been a necessary focus upon the basic elements and processes involved in image interpretation. Because all image operations must be explicitly specified it has been necessary to move forward our understanding of the ill-defined "gray box" mode of operation that is the manual analysis process.

Early automated work emphasized pattern recognition approaches to the various image analysis tasks. A substantial subfield has developed within computer science aimed at developing the ability of computers to "recognize" patterns in data. When these techniques were first invented, the mathematical elegance of the basic ideas was so attractive that some of the early researchers were overly ambitious and optimistic about the general usefulness of the techniques. Pattern classification methods alone are virtually useless in situations that require awareness of context or the use of knowledge; yet such characteristics usually are important in most interesting image analysis problems. Fortunately, a complementary "derivational" approach has been developed that focuses upon the derivation of formal-proof methods for mathematical logic. Such methods are of special interest because they provide a basis for logical reasoning, a deductive means of obtaining solutions to problems using some systematic reasoning procedure. Our earlier review of visual perception noted the importance of various "rules"; the derivational approach to problem solving now provides a means for automating the use of rules.

Elements of Image Interpretation

Although the primary elements of image interpretation are common to both manual and computer-assisted procedures their current usage is dramatically different. It was discussed earlier how all elements may be readily used by image analysts in performing their tasks. In stark contrast, over 90 percent of all remote sensing image analysis appears to utilize *tone and color* exclusively. This seems to be the result of overly ambitious pursuit of the "multispectral signature" concept and the simplicity of implementing multispectral classification algorithms using pattern-recognition techniques.

Various procedures have been developed to incorporate additional elements but only *texture* and *site* appear to be used on a standard basis. Most multispectral classification algorithms involve multivariate statistical techniques, and a derivative image representing texture is often simply incorporated as an additional variable. Factors relating to site usually are used via stratification techniques and/or class *a priori* probabilities. The dividing of an image into strata or separate partitions allows the use of more local (i.e., more site oriented) multispectral signatures and/or *a priori* probabilities. Most of the other elements appear to have been at least demonstrated if only on a limited basis. Related to site as a locational ele-

ment is *association*. Several recent attempts at "contextual" classification make explicit use of neighboring pixel labels, thus incorporating some level of association information. Various "region growing" algorithms have been developed that allow region *size* to be determined and used as a discriminant variable. *Shape* is a very difficult element to incorporate, but methods have been defined to use shape, albeit on a limited basis to date, via a syntactical classifier, to be discussed later. *Cloud/shadow* relationships have been used as a basis for detecting clouds; the additional use of sun-angle information can then be used to determine cloud *height*.

It should be stressed that even though the computer-assisted use of most elements has been at least demonstrated, most examples entail only a small aspect of each element's total information content. It should be repeated that the overwhelming majority of current computer-assisted image analysis relies solely upon multispectral response (tone/color).

Analysis Procedures

The analysis procedures currently employed for computer-assisted image interpretation present a range directly analogous to that embraced by the concepts reviewed earlier pertaining to human vision and visual perception. Although their present state of development is extremely primitive when compared to human interpretation capabilities, we can now more clearly see the manner in which more advanced computer-assisted analysis procedures can be developed.

The most common of these analysis procedures is also the most primitive. What is often termed "statistical pattern recognition" makes use of training data to characterize in some statistical manner patterns of interest. This decision-making method has been found to be most effective in two types of problems (Raphael, 1976):

1. Classification of complex signals when the proper features are measured and the number of dimensions are kept small (typically less than ten).
2. Recognition of simple shapes.

One result of this is a major effort towards optimal representation and feature extraction (e.g., band selection and various transformations, such as principal components, directed towards dimensionality reduction). As noted earlier, however, pattern classification methods alone are virtually useless in situations that require awareness of context or the use of pertinent knowledge.

From the study of languages comes a procedure of slightly higher order, termed syntactic pattern recognition. Conventionally, a language is defined as a set of strings over an alphabet, where the alphabet consists of the set of all symbols that can appear in the strings of the language, and a string is a finite ordered sequence of symbols. A grammar is a set of rules that define how the strings of the language are formed. Axiomatically, a grammar can be used to recognize the language's strings by using the rules in reverse order. This concept can be generalized in a number of ways to define grammars for classes of images. This approach has been most widely used in image analysis to recognize shapes based upon order of component parts. Syntactic methods have been used for locating highways and rivers in Landsat images and for texture modeling (see Fu, 1980 for additional references).

The use of decision-tree structures is characteristic of the decision theoretic approach to image analysis. This approach has alternatively been termed the "layered" approach to classification and clearly implies the use of potentially high order hierarchical decision-making procedures. The combined use of various image analysis procedures in a geographic information systems context has provided a major impetus to this approach (Hallada et al., 1981). In many instances it is possible to implicitly incorporate high order inference rules into a tree structure. The fact that this approach typically requires the sometimes tedious construction of a new tree structure with each new problem of data set, has led to recent movement towards procedures that make more explicit use of inference rules.

The field commonly termed artificial intelligence is actively exploring the development of symbolic reasoning procedures that employ formal inference. It is noteworthy that this approach to image analysis is being vigorously pursued by computer scientists interested in computational approaches to "image understanding" (Brady, 1982). Much of this work has been conducted under the Defense Advanced Research Project Agency's (DARPA) Image Understanding Program. Although most of this research is directed towards high resolution imagery the techniques are much more analogous to those employed by human analysts and will need to be pursued if substantial progress is to be made at extracting the inherent information of higher order image elements.

BASICS OF MANUAL IMAGE INTERPRETATION

This section discusses in more detail the fundamentals of the manual image-analysis process. It should be understood that remote sensor imagery has applicability to environmental studies for four basic reasons. First it presents large areas of the earth's surface from a perspective and in a format that facilitate the study of objects and relationships. Second, certain types of imagery can provide a 3-dimensional view of the terrain and objects under investigation. Third, characteristics of objects not visible to the human eye can be transformed into image form. Fourth, remote sensor imagery provides the observer with a perma-

nent representation of objects, phenomena, and relationships as they exist at a given time.

Image interpretation differs from direct observation in areal scope, perspective, and temporal relationships. It is similar to direct observation in one important respect; the amount and reliability of the information obtained depends upon the training and aptitude of the observer and on the nature of the scene observed. Basically, image interpretation is to photogrammetry[1] as statistics is to mathematics.

Both image interpretation and statistics are techniques by which probabilistic statements can be made about objects, phenomena, and relationships which may exist in our environment. The more knowledge image interpreters have on the capabilities and limitations of the environmental and systems parameters relating to the imagery, the greater the probability that their interpretations will be correct.

IMAGE INTERPRETATION TASKS

Psychological analysis regards image interpretation as if it occurred in a time sequence. The sequence begins with the detection and identification of important objects. The objects are then measured. Measurement is followed by consideration of the objects in terms of information from the interpreter's special field of knowledge. The interpreter must then be able to communicate both his perception of objects and the significance of the objects.

An illustrative example follows. The statement "I see a barn in the photograph," though it appears to be a rapid and elementary identification, implies all of the activities described above. Measurement, in its broadest sense of estimation of size and shape, has taken place; the barn may have been identified by its unique proportions, the number of roof surfaces, or other measurable characteristics. The identification also required attention to such surrounding objects as livestock and silos, and the interpreter must have had prior knowledge of the function of barns. Finally, the word "barn" serves to communicate the results of the interpretation.

Image interpretation comprises at least three mental acts, which may or may not be performed simultaneously; 1) measurement of objects on the imagery, 2) identification of the objects imaged, and 3) appropriate use of this information in the solution of the problem at hand. The first two acts, measurement and identification, can be investigated by relatively simple means: they can be observed by the psychologist or described by the photointerpreter. The third act, problem-solving, is complex and requires mastery of basic informa-

---

[1] Photogrammetry is defined in the *Manual of Photogrammetry* as: The science or art of obtaining reliable measurements by means of photographs.

tion and both synthetic and analytic modes of thought.

Detection/Identification

Psychological tests measure two kinds of factors: stimuli and responses. In image interpretation the stimuli are variations in tone, texture, pattern, configuration, and other image characteristics. The stimulus characteristics of images and the interpreter's ability to respond to them are unquestionably critical determinants of his performance. Measurement of stimuli is relatively easy, but the physiological response that occurs in the retina of the eye is, for the most part, unknown; only when the interpreter acts, by speaking, writing or drawing, can his response to stimuli be measured.

Black-and-white images consist of variations in tone that define edges by means of which the shapes of objects are perceived. The apparent sharpness of an object can be measured in a number of ways. When objects are identified from their shadows, abrupt edge gradients are used; but gradual variations in tone are also important stimuli. The waterline along a beach, the plume from a smokestack, and variations in the properties of soils characteristically appear as gradual tonal changes. Edge gradients are usually supplemented by spatial cues[2] to the identification of objects in aerial imagery.

Many descriptive terms used by interpreters are reducible to statements about tonal changes. Tonal gradients often occur in a repetitive arrangement, giving the effect of texture. A group of objects can often be identified more easily by texture than by the shapes of the individual objects, themselves; the species of a single tree, for example, is difficult to identify except at large scale, but the texture of a whole stand may be distinctive enough to identify even at small scales.

Patterns or arrangements of objects are sometimes more useful stimuli than the attributes of an individual object. Drainage patterns are a well known example. Anti-aircraft weapons may often be identified as similar equi-distant emplacements forming arcuate, rectangular, or circular patterns.

Stereoscopic vision provides familiar spatial cues. The addition of the third dimension makes most objects easier to identify; sometimes it is only the height or depth of the object that yields an identification. Gradual changes in depth, as they appear in the natural and cultural features of the earth's surface, are extremely important stimuli for many interpreters. Other useful cues often used in the interpretation of depth are relative size and interposition.

Interpreters communicate their response to a stimulus by labeling the identified object. Their

ability to identify objects cannot be measured in the absence of labeling, although it is possible to learn whether their response repertoire includes the appropriate label for a given object. In order to communicate interpretations adequately, image analysts must be well versed in the terminology of their own and related fields.

Measurement

Image interpreters, at times, measure the exact dimensions of objects by means of scales and other instruments described elsewhere in this chapter. Measurement in image interpretation may consist on the one hand of a visual estimate of the size and shape of an object; relatively precise measures employ aids such as rules, compasses, and/or scales. It is important to note that reasonably correct estimation of dimensions is essential for correct identification.

The use of measuring equipment e.g. drawing boards, maps, slide rules, overlay devices) requires a variety of abilities. Much human-factors research has been performed on motor skills, and a number of tests for such skills are available. Knowledge of mathematics, particularly trigonometry, can be judged by means of available tests. The abilities required for the making of object measurements are well developed in most interpreters.

Problem Solving

The image analyst is often required to identify objects by the study of associated objects, or to identify object-complexes from their component objects. This activity is one form of problem-solving.

Experienced interpreters looking at an object-complex (an industrial installation, for example) may not immediately recognize it as object-complex X; they may have to detect and identify objects x, y, and z before concluding that the complex is possibly, probably, or certainly X. An analyst carefully studies a combination of objects and weighs the probabilities that such a combination may occur in more than one object-complex before he arrives at the identification. The solution of such a problem does not always result from a unique identification; it may result from a hierarchy of possible identifications (a probabilistic statement connoting the level of certainty the interpreter places on the identification) from most likely or probable to least likely or possible. Experienced analysts know that complicated objects often seem to be instantly recognized; the neophyte, however, first may have to identify x, y, and z before he can identify X.

Most problems in image interpretation require knowledge derived not from the images themselves but from the relevant field or fields of study. The arrangements of particular objects and phenomena recorded on remote-sensor imagery facilitate interpretation because of the patterns or

[2] A cue in a perceptual situation is any characteristic that helps the interpreter make an accurate identification or correct judgment concerning the significance of an object.

spatial relationships in which the objects are arranged. A geologic process, for example, may give rise to related features that possess the orderlines of, say, a common depositional origin. Plant associations display patterns related to topographic influences. Cultural features are often associated by common functions. Appreciation of the role of associations and patterns in the analysis process proceeds from mastery of one's field of learning: the best photo-interpreters have been known to be experienced, well educated scientists usually with abundant field experience as geologists, geographers, foresters, hydrologist, or other resource scientists.

ELEMENTS OF MANUAL IMAGE ANALYSIS

Imagery is a record of the energy reflected, emitted, and/or transmitted in one or more parts of the electromagnetic spectrum; imagery may be in many shapes, sizes, and scales. Basic image interpretation is essential to the efficient and effective use of such data.

Although many articles and texts differ as to the number of basic elements contained in imagery, there is general agreement on six; tone or color, size, shape, texture, pattern, and shadow. Three other elements that are often added to these six are site, association, and height. These will be discussed in detail here. An important concept for one to bear in mind when reviewing this material is resolution. A discussion of resolution can be found in Chapter 1 of this manual.

Tone and Color

Color perception is an important element of awareness of the environment. Different objects reflect, emit, and transmit different amounts and wavelengths of energy. These differences are recorded as either tonal, color, or density variations on an image. The stand of mixed hardwoods shown in Figure 24-3 was photographed in late October at a peak of the fall color change. Species differences show clearly as different tones or shades of gray. Color Figure 24-4 is an aerial photo showing the discharge of the Cuyahoga River into Lake Erie at Cleveland, Ohio. In black-and-white images, distinctions between hues are lost and objects appear in tones of gray. These gray tones often fail to correspond to an interpreter's perception of familiar objects in nature. True color imagery often facilitates interpretation by providing a more familiar view of the object under investigation. In black-and-white photography a water body may appear in tones ranging from white to black, depending on the angle of the sun and also on the wave surfaces reflecting light to the camera lens. Additionally, there is increased use being made of "false color" photography or color enhanced photography to improve interpretability.

Color infrared aerial photography and also both

Fig. 24-3. Panchromatic minus blue photographic image of a stand of mixed hardwood trees at the peak of the fall color change. Lightest toned tree crowns are sugar maple while dark crowns are generally oak (University of Illinois).

optical and electronic image-enhancement devices that can produce many different color combinations from a set of black-and-white transparencies, are being used more and more by image interpreters. Whether it is to increase haze penetration, thereby obtaining high-quality images from high altitude, or to enhance subtle tonal contrasts between objects by narrowly filtering and combining various bands of black-and-white imagery, the purpose is to enhance object-to-background contrast ratios. This also provides expanded contrast between tones, allowing easier interpretation. Tone, the shade of gray in a black-and-white image or the combination of hue, chroma and saturation in a color image usually conveys more information to an alert, knowledgeable interpreter than any other element. In almost all cases, however, it is the *difference* in tone or color between objects, or between an object and its background, that is important. In fact, without a difference in tone or color between the background and the edge of an object, there can be no detection of the object.

On a thermal infrared image, such as that seen in Color Figure 24-5 (top), tonal variations that appear in a water body are related to differences in the water's exitance. Through the use of electronic image-enhancement techniques the gray tone variations can be transformed to colors (Color Figure 24-5, bottom) that are related to variations in radiant exitance. If we assume the emissivity of the water to be unity, we are viewing water skin temperatures. On the other hand, to the trained interpreter Figure 24-4 leaves no doubt that there also are variations in suspended sediments in the water body. Many of the important factors influencing image-tone variations will be covered later in this chapter in the discussion on various sensor systems. Detailed discussions of the ways in which various sensor systems actually record energy can be found in the chapters dealing with the individual sensor systems. When the image analyst understands the factors that govern image tone, he regards the tones or colors of objects as major clues to their identity or composition.

Size

The size of an object is one of the most useful clues to its identity; for example, the volume of wood that could be cut from the stand of timber in Figure 24-3 is dependent upon tree size, stand density, and size (or area) of the stand. By measuring an unknown object on an aerial photograph, the interpreter can eliminate from consideration groups of possible classification. When working with imagery of variable scale, the interpreter should make frequent measurements of the objects of interest.

Figure 24-6 shows how analysts can avoid errors in identification by studying the size of objects. If they were to disregard size they could

very easily call the object at A a tree, the objects at B a herd of cattle, at C a tract of houses, and at D a row of telephone poles. They could arrive at these misidentifications even though they had carefully taken into consideration such clues as shape, shadow, tone, texture, and pattern. On this photograph of known scale, a few measurements to determine the size of these objects, together with the observations already made, will tell them that the object at A is a large clump of grass, not a tree, (trees are at E); the objects at B are sheep, not cattle, (cattle are at F); the objects at C are dog kennels, not houses, (a house is at G); and the objects at D are fence posts, not telephone poles, (telephone poles are at H). Realizing that size can be three-dimensional the interpreter may wish to make parallax measurements as described later in this chapter to determine the number of stories in

Fig. 24-6. Vertical aerial photo, scale 1:4000, illustrating the importance of size in the identification of various objects (From Colwell, 1960).

the house at G; and shadow measurements to determine the heights of certain other objects such as the dog kennels at C.

Shape

The shape or form of some objects is so distinctive that their images may be identified solely from this criterion. The value of shape to the interpreter is that it delimits the class of objects to which an unknown must belong; it frequently allows a conclusive identification; and it aids the understanding of significance and function.

The shape of an object viewed from above may be quite different from its profile view. The ability to understand and make use of the plan view has to be acquired like another language. It then becomes a powerful tool, because the plan view of objects is an important and sometimes conclusive indication of their structure, composition, and function. To the motorist, a cloverleaf freeway interchange may be an incomprehensible maze through which one's way must be found by faith and strict attention to signs; to the aerial observer, the intersection is clear in the form and function (Figure 24-7).

Much of the training of image analysts is aimed at the reorientation of perceptions, so that they can easily recognize objects seen from above. This reorientation can often be greatly aided by the impression of depth in stereoscopic pairs.

Texture

Texture in images is created by tonal repetition in groups of objects that are often too small to be discerned as individual objects. Texture, the visual impression of roughness or smoothness, is a valuable clue in interpretation. This can be especially true when analyzing sidelooking airborne radar data and imaging passive microwave radiometry in which texture plays an important role in differentiating various classes of environmental phenomena.

On an aerial photographic image, tree size is sometimes interpreted on the basis of apparent texture. Smooth, velvety textures are commonly associated with young saplings, and rougher, cobbled textures usually indicate older trees of sawtimber size. It follows that the size of the object required to produce texture varies with the scale of imagery. In large-scale aerial photographs, trees can be seen as individuals; their leaves or needles cannot be discerned separately, but contribute to the texture of the tree crowns. On small-scale imagery, the crowns contribute to the texture of the whole stand of trees (see Figure 24-8). Within a given range of scales, the texture of a group of objects—a timber stand of a certain species composition, for example—may be distinctive enough to serve as a reliable clue to the identity of the objects.

Pattern

Pattern, or repetition, is characteristic of many man-made objects and of some natural features. The land-use pattern shown in Figure 24-9 is typical of areas of deep loess. Orchards and strip cropping are particularly conspicuous because of their patterns.

Earth-science students have always laid great stress on the pattern or spatial arrangement of

Fig. 24-7. The road interchange illustrates the advantage of the aerial view. Seen from above in this stereogram, its function is perfectly clear; to the driver in the car, successful navigation depends on faith in direction signs. The two long shadows of the towers on the bridge show that it has a lift section to let ships through. Other shadows show the bridge construction clearly. Note that the railroad differs from the road in that its curves have a longer radius than even the gentlest curve on the road, and that intersections are avoided.

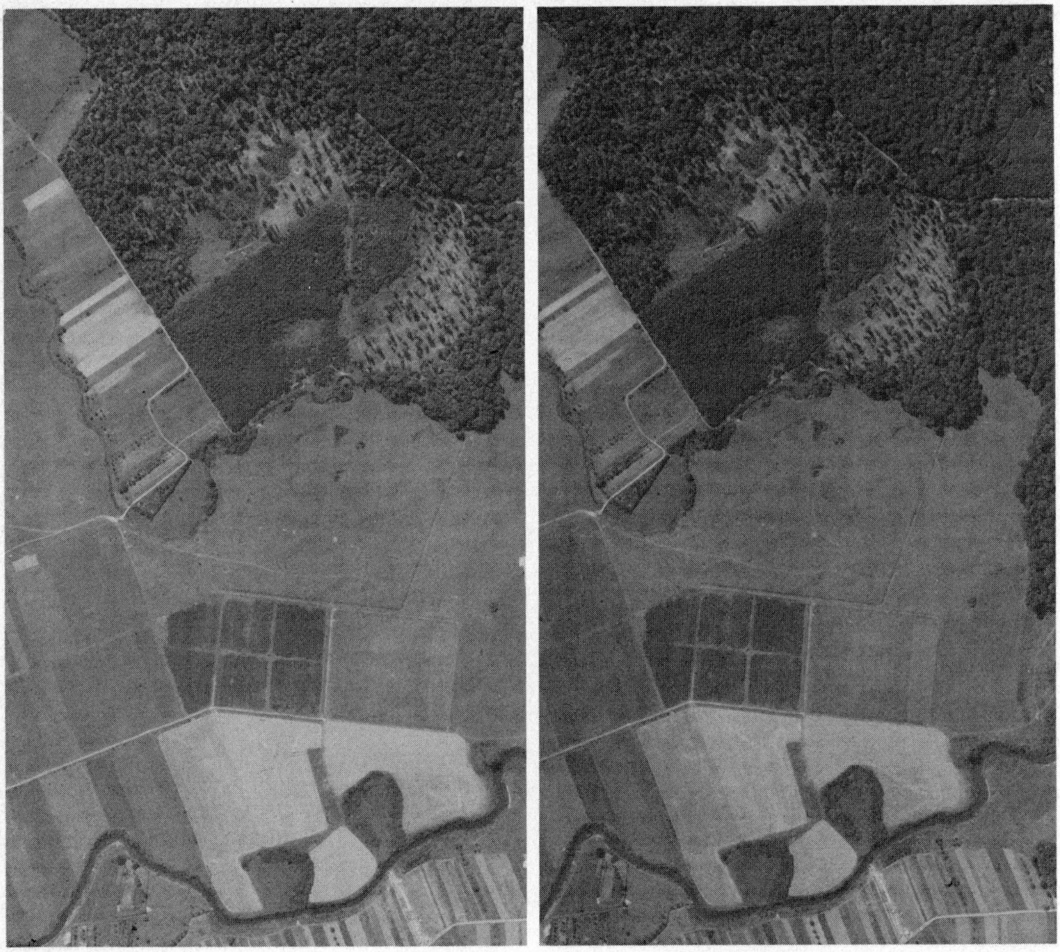

Fig. 24-8. The variety of textures shown here soon becomes associated with the features represented so that the features become recognizable to experienced interpreters even without convenient comparisons in the same image. Note how the texture of the woods is really composed of individual tree crowns with random spacing. The lower growth adjacent still retains some of this characteristic, but the crops in the fields below present a velvety appearance in which individual plants are not visible.

objects as an important clue to their origin or function or both. Cultural geographers and anthropologists study settlement patterns and their distribution in order to understand the ef-

Fig. 24-9. Land-use pattern developed on deep windblown loessial soils in Calhoun County, Illinois (United States Department of Agriculture).

fects of diffusion and migration in cultural history. Drainage patterns have orderly associations with structure, lithology, and soil texture. The varying relations between organisms and their environment produce characteristic patterns of plant associations.

Regional patterns, which formerly could be studied only through laborious ground observation, are clearly visible in aerial imagery. For example, Figure 24-10 depicts portions of the Garlock and San Andreas fault zones on a single Landsat image. Moreover, remote-sensor imagery captures many small but significant patterns that might be overlooked or misinterpreted by the ground observer. As stated in the 1960 *Manual of Photographic Interpretation,* the trained observer appreciates the significance of aerial photography chiefly through understanding the sources of patterns on the earth's surface (Colwell, 1960).

Some patterns in our environment are primarily cultural; others primarily natural. There are, however, few parts of the world that have not been

Fig. 24-10. On this Landsat MSS image, the Garlock and San Andreas fault zones form a conspicuous regional fault pattern which defines a triangular area whose apex is at the western edge of the Mojave Desert (Courtesy of NASA).

affected by man, and most of the patterns visible from an aerial perspective result from the interaction of natural and cultural factors.

Cultural features are conspicuous in aerial images because they consist of straight lines or other regular configurations (Figure 24-11). Because patterns of settlement, mining, and agriculture may be visible from the air after thousands of years, image interpretation has become an important technique of archaeology (Chapter 26).

Although it may be difficult to understand cultural features simply by looking at them, particu-

Fig. 24-11. The importance of pattern as a photo-image characteristic of value in identifying certain features is well illustrated by this vertical aerial photograph of a medieval European fortress city.

larly from an altitude of several kilometers, the configuration seen from the air is often a sufficient clue to the function of the features. A road and a railroad may look much alike in a photographic image; however, the image interpreter can tell them apart by the slightly different configurations required by their functions. As seen in Figure 24-12B a road may have fairly steep grades, sharp curves, and many intersections; a railroad has gentle grades, wide curves, and few intersections. The pattern of a railroad (Figure 24-12A) reflects the function of a bed for a fast, heavy string of vehicles.

Height

As can be seen from the preceding material it is possible to employ single images for the detection, identification, measurement and problem-solving tasks of image analysis. The principal disadvantage of this use of single images is that only two dimensions (length and width) are typically perceived. This is the equivalent to using only one eye in the analysis process (monocular vision). Yet, analysis and measurement of the third dimension of objects can be accomplished in some cases by employing either single images or multiple images. Height determinations can also be

made on oblique imagery. The procedures for measuring height will be discussed later in this chapter. The production of height or "stereo" information can be achieved with cameras, line scanners and active microwave systems employing either analog or digital data. Programs have also been created that produce synthetic stereo images from Landsat data using digital terrain information to calculate the appropriate parallax for constructing a three-dimensional model. Finally, a number of image analysts contend that, except for tone, the third dimension as expressed in height is the most important manual image-analysis element.

Shadow

Shadows can help or hinder the analyst because they reveal silhouettes but hide some detail. With radar, Figure 24-13 (top), shadows caused by the viewing angle of the sensor may obscure important information; while shadows such as those seen in Figure 24-13 (bottom) provide information, not apparent from the image of the building alone, on the size and shape of the building. These shadows obscure detail, however, in the lawn and sidewalk areas in front of the building.

Shadows are familiar phenomena. In life we

Fig. 24-12A. (upper). The difference between a road and a railroad is obvious if the interpreter knows what to look for. Note here that the railroad has long, gentle curves and is on an embankment to keep the grade as level as possible (almost never more than a 2% grade). Intersections are avoided; note that the north-south road goes under the railroad. The road below the rail line, however, is partly hidden by trees along its edge; this is not done with railroads. The road meanders, in comparison to the railroad, and it has intersections with other roads. The differentiation is also easy when cars or trains are seen.

Fig. 24-12B. (lower). In a less obvious example, the road and railroad can still be differentiated. Note that even on this reproduction the rails can be seen on the railroad just to the right of the largest underpass, and cars are moving on the road. The feature could be identified even without these, however, by noting that the railroad is on an embankment with several underpasses, it has no intersections, and only very gentle curves.

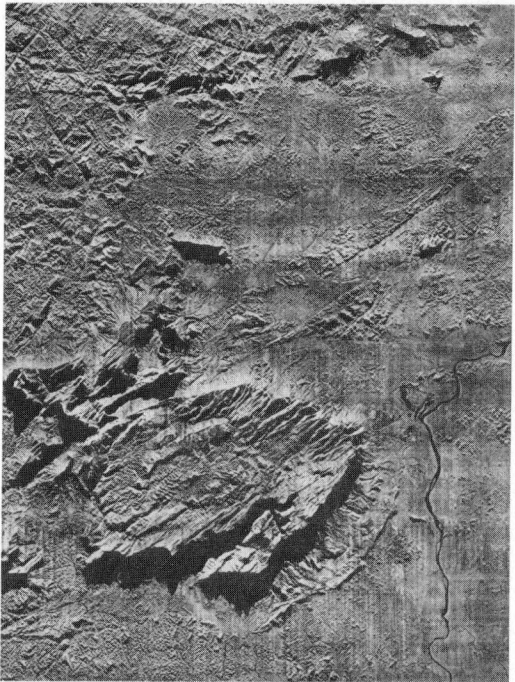

Fig. 24-13. Shadow pattern can sometimes obscure important ground information; however, shadow may also accentuate information of interest. Sidelooking airborne-radar images such as the one seen above (top) are used by geologists to accentuate terrain information in the same way as low sun-angle photography is often used. In the aerial photography (bottom) the shadow of this building shows the steeple more clearly than does the image of the steeple itself. Yet walkways and other objects in the shadow of the building may be observed or obscured depending on the density of the shadow (Radar image courtesy of Goodyear Aerospace. Aerial photo courtesy of University of Illinois).

often judge the size and shape of objects by observing the shadows they cast. As seen in Figure 24-13, shadows present in aerial photographs help the interpreter by providing a profile representation of objects. Shadows are particularly helpful if the objects, themselves, are very small or lack tonal contrast with their surroundings. Under these conditions, the sharp boundaries and shapes of the shadows enable the analyst to identify objects at the threshold of recognition.

Although aerial photographs are generally taken within 2 hours of local noon (to avoid excessive shadows and reveal maximum information about ground conditions) the presence of shadows can enhance shape and therefore aid in object identification. In geologic projects, photography has been flown at very low sun angles in order to accentuate surface irregularities. This is also one reason why sidelooking airborne-radar imagery, which approximates low sun angle photography, has become an important tool in geologic investigations. Low sun angle terrain-accentuation was also a factor in the determination of the time of image acquisition by the Landsat satellite system.

Site

The location of objects with respect to terrain features or other objects is often helpful. The open flat dark-toned vegetation in Figure 24-14 is a swamp or marsh in an area of coniferous vegetation. By knowing that in this area of Michigan, northern white cedar, balsam fir, and black spruce are the predominant swamp conifers, the interpreter may make better inferences concerning the imaged area. The stand shown in Figure 24-14 is a mixture of these species.

Aspect, topography, geology, soil, vegetation, and the varied imprints of man's culture are distinctive factors that the interpreter should use when examining a site. The relative importance of each of these factors will vary with local conditions but all are important. Just as some tree species grow in swamps and others on dry upland ridges, so some manmade objects are found along rivers and others on hilltops. Thermal power plants need an abundant supply of coolant water and are usually found near major streams. Hydroelectric plants must have water under pressure and are always associated with a natural or manmade water supply. Early warning radar stations, on the other hand, are usually placed on high

Fig. 24-14. Black spruce, northern white cedar, and balsam fir in swampy area in Leelanau County, Michigan (University of Illinois).

promontories to minimize terrain interference with the line of sight of the radar.

Association

Some objects are so commonly associated that one tends to indicate or confirm the other. In Figure 24-15, the two tall smokestacks, large building, coal piles, conveyors, and cooling towers are obviously related. This combination and arrangement of features identify the installation as a thermal power plant. If this plant were located on a river or other water body (using the river water for cooling rather than cooling towers) a thermal infrared image such as that seen in Figure 24-5 would also add confirming evidence, in the form of warm-water discharge, to ensure correct interpretation of this facility. Adding together associated bits of evidence (convergence of evidence) is a significant analysis procedure that will be discussed later.

Association is one of the most helpful clues in identifying manmade installations. The manufacture of aluminum requires large quantities of electrical power. Absence of a power supply rules out this industry. As previously stated, schools usually have playgrounds and athletic fields while churches do not. The type of school (elementary, secondary, junior high, or high school) can be inferred from the size and type of playing fields, buildings, and parking facilities. Large farm silos are an indication that livestock are (or were) present. Unexpectedly large culverts or bridge spans across small streams indicate that heavy runoff occurs frequently enough to require substantial engineering work.

AIDS AND TECHNIQUES FOR MANUAL IMAGE ANALYSIS

Image interpretation has been defined as the act of examining photographs and/or images for the purposes of identifying objects and phenomena and judging their significance. In carrying out this task, interpreters may use many more types of data or information than those recorded on the images. These aids to the analyst can be grouped under headings such as perceptual models, and collateral, or on-site verification. In addition, the analyst should be aware of techniques for the handling of imagery and methods of searching imagery to obtain maximum information in the most efficient manner.

Perceptual Models

As previously stated, the more knowledge the image analyst has concerning the potential sources and magnitude of errors associated with data/information flows in remote sensing the better. That is, an analyst should strive for expertise in the philosophical concepts, elements/aids, techniques, methodologies, processing, and display procedures in remote sensing. The analyst should also strive for expertise in several discipline or applications areas. Finally, the analyst should make every attempt to come to a detailed thematic understanding of the geographic area under consideration. This information plus the experience of the individual analyst forms the basis for the perceptual or cognitive models that can be used in a given analysis. These mental models form the conceptual basis upon which the analyst operates. They form the primitives that the analyst begins to link in performing the classification and labeling process. They allow the analyst to form hypotheses concerning the nature of objects or phenomena, which may then be tested as needed.

Collateral Material

Many sources including literature, laboratory measurement and analysis, field work, and ground and/or aerial photography, make up collateral material. The types of ancillary data available vary significantly among remote-sensing projects depending on the spatial, temporal, and cost constraints defined by the particular image acquisition and interpretation program.

Supportive collateral data are often available in the form of maps and other records in tabular or graphic form. Examples include meteorological or land-use data collected by various individuals or government agencies. A review of the existing source materials helps in the interpretation of remotely sensed data, and may also produce a better definition of scope, objectives, and problems associated with a given project.

The amount of field work required for a given remote sensing project varies and is, in general, dependent upon the following considerations; (1) image quality, including scale, resolution, and information to be interpreted; (2) type of analysis or interpretation involved; (3) accuracy requirements for both boundary delineations and classification; (4) the experience of interpreters and their knowledge of the sensor, area, and subject to be interpreted; (5) terrain conditions and area accessibility (for various reasons many areas may not

Fig. 24-15. Thermal power plant at the University of Illinois (University of Illinois).

be accessible for field work); and (6) the existence of other source material.

Field work often invovles sampling for the verification of questionable interpretations and error corrections. In such cases it becomes important to consider how best to sample, and then to design a sampling strategy to fit the problem. Multi-stage sampling designs are presently being used in conjunction with aerial surveys to provide accurate estimates of various environmental parameters.

In most field surveys, especially those associated with remote-sensing operations, it is appropriate to record observations on terrestrial or low-altitude aerial photographs. Such photographs are useful as a permanent record of visual observations, a base for future notations, generalizations, and measurements; and documentation for reports concerning the field survey.

The acquisition and analysis of collateral data should be viewed as necessary elements of image interpretation. It must be realized that these data have their own variances and as with a remotely sensed image, are subject to interpretation. Furthermore, collateral data play a dual role in the interpretation process—first to assist in the interpretation and analysis process; and second, to verify interpretations and analyses. Just as data acquisition must be thoroughly considered and planned, so too must be the nature, amount, timing, method of acquisition, and integration of these ancillary data into a given remote sensing data-acquisition program (Estes, 1974).

Image Analysis Keys

When many objects are imaged interpreters must be aware of factors not necessarily related to their experience or to the work of their organizations. This means familiarity with so wide a variety of stimuli that even the most accomplished interpreter is occasionally dependent upon reference materials. Identification keys, used with great success by field workers in the biological and physical sciences, are useful aids to image interpretation.

A great deal has been written concerning photointerpretation keys. The use of these types of aids to image interpretation is not unique to photographic images. Keys may be developed that will aid the interpreter in analyzing other kinds of imagery also. The meanings of some of the basic interpretive elements, especially tone, may vary; however, the utility of the concept of the interpretation key remains constant.

Basically, an image-interpretation key helps the interpreter to organize the information present in image form and guides to a correct identification of unknown objects. It differs from the keys used in many disciplines in that it consists mostly of illustrations.[3] Annotated stereograms or other im-

ages illustrate the objects to be identified, and the distinctive characteristics of the objects are systematically listed and described.

A key may be organized for identification by selection or by elimination. A selective key illustrates and describes classes of phenomena, and the interpreter chooses the example that most closely fits the unknown item. An elimination key provides a step-by-step method of identification; the interpreter proceeds through a series of possible identifications, eliminating all incorrect choices (Colwell, 1952).

Many types of image-interpretation keys are available or may be constructed. In constructing a key one must consider the abilities of the interpreters who will use the key and the purpose to be served by the interpretation. The interpreters may be undergraduate students, military trainees, or thoroughly trained and widely experienced professionals. They may have to identify discrete objects (plant species, aircraft assembly plants) or categories of objects (coniferous trees, light fabrication industries). The complexity of objects or conditions to be identified will partly govern the choice of organization. Keys differ in their usefulness to the beginning interpreter. A key that assumes extensive knowledge about a region or subject field is not of great value to the novice. Most keys apply to particular areas as well as to particular groups of objects, and the interpreter should use only those keys that are appropriate for the problem and locale. The following section discusses the various types of keys available to the interpreter or which the interpreter might construct as aids.

Scope of Image Interpretation Keys

As discussed above there are a number of types of image analysis keys. They can, however, be generally grouped into the four basic categories below:

a. An item key is a key concerned with the identification of an individual object or condition.

b. A subject key is a collection of item keys concerned with the identification of principal objects or conditions within a given subject category.

c. A regional key is a compilation of item or subject keys dealing with the identification of objects or conditions characteristic of a particular region.

d. An analogous area key is a subject or regional key that has been prepared for an ac-

---

[3] Because illustrations are the essential part of a photointerpretation key, their quality is of overriding im-

portance. In conventional methods of reproduction, detail may be lost or masked by the dot pattern produced in screening. It is desirable and, for some subjects, necessary to use a photographic rather than a plate-printing process in order to produce images that are usable under the magnifying lenses of the stereoscope.

cessible area and that, by extrapolation, may be used in the interpretation of objects or conditions in inaccessible areas that exhibit similar characteristics.

Technical Level Image Interpretation Keys

a. A technical key is one prepared for use by image interpreters who have had professional or technical training or experience in the subject concerned.
b. A non-technical key is one prepared for use primarily by image interpreters who have not had professional or technical training or experience in the subject concerned.

Intrinsic Character of Image Interpretation Keys

a. A direct key is a key designed primarily for the identification of discrete objects or conditions directly discernible on images.
b. An associative key is one designed primarily for the deduction of information not directly discernible on images.

Manner of Organization or Presentation of Image Interpretation Keys

All image interpretation keys are based upon diagnostic features of the images of objects or conditions to be identified. As stated above, depending upon the manner in which the diagnostic features are organized, two general types of keys are recognized. Selective keys are arranged in such a way that an interpreter simply selects that example corresponding to the object to be identified. Elimination keys are arranged so that the interpreter follows a prescribed step-wise process that leads to the elimination of all items except the one to be identified. Such keys may be thought of as a process of following a line of reasoning. Most interpreters consider the latter type of key preferable.

Selective Keys

a. An essay key is one in which objects or conditions are described in text using images for illustrations only.
b. A file key is an item key composed of one or more selected images, with notes concerning their interpretation. This type of key is generally assembled for use by an individual interpreter.
c. A photo-index key is an item key composed of one or more selected images, together with notes concerning their interpretation, assembled for rapid reproduction and distribution to other interpreters.
d. An integrated-selective key is one in which images and recognition features for any individual object or condition within a subject or regional key are so associated that, by reference to the appropriate portion of the

key, the object or condition can be identified.

Elimination Keys

a. A disk key is one in which selected image-recognition features are grouped or arranged on one or more disks so that, when the recognition features are properly aligned, all but one object or condition of the group under consideration is eliminated from view.
b. A punch-card key is one in which selected image recognition features are arranged in groups on separate punch cards. When the properly selected cards are superimposed upon a coded base, all but one object or condition of the subject group under consideration is eliminated from view.
c. A dichotomous key is one in which the graphic or word description assumes the form of a series of pairs of contrasting characteristics, thereby permitting progressive elimination of all but one object or condition of the subject group under consideration (see Table 24-1).

A modification of the elimination key is to allow probabilistic rather than absolute identification at any step or steps in a sequence. A probabilistic key, based on local *a priori* statistics, is used primarily when identification cannot be completed.

Finally, recent work has been accomplished in computer programming for generating diagnostic keys. The work of Pankhurst (1970) indicates that computers can generate data applicable to the recognition process. It may be possible to generate a larger number of keys for one category of objects although all keys will not be equally useful. Whereas the construction and editing of an optimum key by hand labor may be time consuming, a computer may produce an optimal key automatically. The computer algorithm explores all the possible keys and selects the optimum key for the particular task heuristically (Pankhurst, 1970).

Handling Of Imagery

Image-handling techniques are important in manual image analysis. Although a good deal of image interpretation is still accomplished using paper prints, the use of transparencies is increasing. Transparencies generally are used either as single frames or in rolls. Care should be exercised when handling these materials so that they are not marred. An orderly procedure for handling both transparency and print material should be developed and adhered to. When transparencies are in rolls, it is relatively easy to keep material in order. If single frames are extracted from a role for duplicating or display purposes, care should be exercised to note the location and the individual who is responsible for returning the frame. Rollers and viewing surfaces of viewing equipment should be kept clean. When viewing transparent material,

TABLE 24-1

Dichotomous Photointerpretation Key to Fruit and Nut Crops in the Sacramento Valley

Based on Pan Minus-Blue Photography—Scale, 1/6,000

1. Crown large, over 10 m in diameter; row spacing wide (greater than 9 m); crown entire, no pruning hole; crown shadow full in center, thinner at edges; branches swirl around crown; foliage dark .... Walnut

1. Crown small to medium, under 7.5 m in diameter 2
   2. Crown with hole in center; crown round full, smooth; tone of crown medium grey, row spacing narrow .... Peach
   2. Crown without hole in center 3

3. Crown shape conical; shadow shape pyramidal; mature trees smallest in the area .... Pears

3. Crown shape not conical; shadow shape not pyramidal 4
   4. Branches in obvious radiating habit, palm tree-like; crown shadow scraggly, often interplanted with peaches .... Cherries
   4. Branches merely irregular in habit, crown tone the lightest of all crops; mature trees of medium size, 6-10 m .... Almond

(From R. N. Colwell, *Syllabus for a Course on the Aerial Photo Interpretation of California Crops and Livestock,* School of Forestry, University of California, January, 1965.)

one should use a clear overlay to protect the image surface, especially when hand magnifiers or other interpretation equipment or measuring devices are to be employed.

As stated above, much analysis today is still done with paper prints. A "mission" or set of images may consist of hundreds of exposures from each of several sensor systems. Moreover, images from other missions may be on hand for comparison. Thus, the interpreter could have a large number of images on his desk at one time. This mass of paper should be handled in an orderly way. Many experienced interpreters have never developed an orderly method of handling prints, and spend as much time fishing for images as interpreting them.

Because the images from a given mission usually are numbered, orderly arrangement is easy and should be the automatic first step in any job of interpretation. One simple method of arranging images is: (1) Arrange the images of each flight or mission numerically, face up. (2) Stack flights or missions in the sequence in which they will be examined and place paper separators between them. (3) Turn the stack so that the line of flight extends from left to right with respect to the observer, and preferably so that the shadows fall toward the observer. Decide whether the work will proceed from left to right or right to left, and place the stack accordingly. (4) Place the images to be used for comparison alongside the primary stack rather than in it. (5) As the prints are examined, stack them face down, still in numerical order. Having adopted an orderly way of handling imagery, the interpreter should follow it habitually.

Stereoscopic Viewing

Binocular vision, taken for granted in daily life, should be fully understood and consciously exploited by the image interpreter. Careful orientation of stereoscopic pairs will produce a clear image and minimize eye strain. Although, as previously stated, many types of remote-sensor systems can be employed to obtain "stereo", most stereoscopic viewing for interpretation purposes is done using vertical or nearly vertical aerial images acquired with conventional aerial camera systems. Stereoscopic viewing of imagery from sidelooking airborne radar systems is discussed in Chapter 25.

To view vertical images with a lens stereoscope, the beginner should:

(1) Mark with a dot or pinprick the principal point (geometric center) of each image. This point is located at the intersection of perpendicular lines between the fiducial marks at opposite edges of the image. It can be found by penciling light lines between the fiducial marks, or by pricking through the intersection of perpendicular lines drawn on a transparent template.

(2) Mark on each image the principal point of the other photograph of the stereoscopic pair (conjugate principal point). On both photographs, draw a straight line between principal point and conjugate principal point. This line is a segment of the flight-line, the two exposures having been made from camera stations directly above the two principal points.

(3) Overlap the images so that the flight-line segments and corresponding images are superimposed and extend from left to right with respect to

the observer. Then move the images apart in a direction parallel to the flight line (x-direction), keeping the flight-line segments aligned, until corresponding images are separated by a distance somewhat less than or equal to the observer's eye base—about 5 cm is usually satisfactory.

(4) Place the stereoscope over the images so that the lenses are aligned with the flight line and above the two images to be viewed. If a mirror stereoscope is used, the images are separated by a distance equal to that between the centers of the wing mirrors.

Usually, one adjustment will not serve for examination of all parts of the stereoscopic model. Particularly in images having a short principal distance, radial displacement requires readjustment when the interpreter wishes to view objects near the edges of the model. One of the images may have to be moved in the x̄- or y-direction (parallel or perpendicular to the flight-line); the two flight-line segments should be kept parallel. The lenses or mirrors of the stereoscope should be directly above the objects to be viewed. With experience, the need for readjustment is signaled by a sensation of strain or "pull" on the eyes. Also with experience, the interpreter need not mark principal points and flight-line segments before viewing the images. The analyst looks through the stereoscope and the objects should then fuse; if they do not, the images should be moved or rotated slightly, until the impression of depth is gained. If the objects do not fuse after these adjustments, another beginning is advisable.

If the image analyst intends to make vertical measurements with a height-finding device, he should carefully mark principal points and flightline segments. The separation of the images should be checked against the optical separation of the instrument, and the flight-line segments should be adjusted along a straight-edge.

The centers of the stereoscope lenses should be separated by a distance equal to the analyst's eye base. If the lenses are not the right distance apart, the interpreter has to overcome prismatic power, which sometimes causes one eye to deviate from the correct direction. Adjustments should be checked from time to time, because moving the stereoscope can change the setting of the lenses. Eye base is easy to measure. While holding a graduated rule across the bridge of the interpreter's nose, with the graduated edge bisecting the two pupils, an observer notes the distance between the interpreter's pupils. Lenses of stereoscopes should be cleaned at least once a day; foreign material on the lenses hampers clarity of vision. Care should always be taken to orient the image pair for stereoscopic viewing; improper orientation can cause eye strain and headaches.

Methods of Search

An image analysis task begins by close examination of all details that are thought to be relevant; however most experienced interpreters prefer to begin by scanning the whole area or a large part of it. They are then prepared for intelligent selection and study of details.

It is generally agreed that the image analyst should work methodically, should proceed from general considerations to specific details, and should proceed from known to unknown features (Stone, 1956). Methodical work and the analysis of known features before evaluating new and unknown features are almost axiomatic in scientific endeavors. Proceeding from general to specific considerations is also desirable as long as the considerations of general (sometimes called regional) features do not bias the interpretation of the specifics. Specific, local considerations often provide evidence needed to complete or confirm the broader regional pattern. In most cases, general and specific features must be considered together. To say that one must come before the other can be misleading. To go further than this and propose that interpretation should proceed from one specific group of features to another is an unwarranted channelization of the infinite variation that the interpreter encounters (Olsen, 1973).

As previously indicated, there are two basic general methods that may be used to study aerial imagery; the "fishing expedition" and the logical search.

Image interpreters have learned that aerial imagery is full of surprises, and they are often tempted to examine every object in every photograph to not miss anything. This is the fishing expedition, a method of search that is too frequently used. "Fishing" can yield large amounts of information, including much that is not pertinent to the subject at hand. It requires a more leisurely effort than the interpreter can usually afford to make.

By resorting to probabilities, the interpreter can work more efficiently in the time available. He searches only those areas in which the objects of interest are likely to be found, and disregards large numbers of images that are not likely to contain the desired information. This selective method is logical search, a combination of quick scanning and intensive study. It demands more experience since the interpreter has to decide where intensive study will yield the best results, but usually is much more productive in relation to the time and effort expended. The amount of time spent on intensive study at the expense of rapid scanning must be determined by the nature of the work. Some types of interpretation demand close scrutiny of the entire area photographed; others may require only a cursory examination of selected areas.

IMAGERY INTERPRETATION AND TRANSFER EQUIPMENT

Persons use imagery interpretation and information transfer equipment for three general purposes: viewing (including enhancement), measuring, and

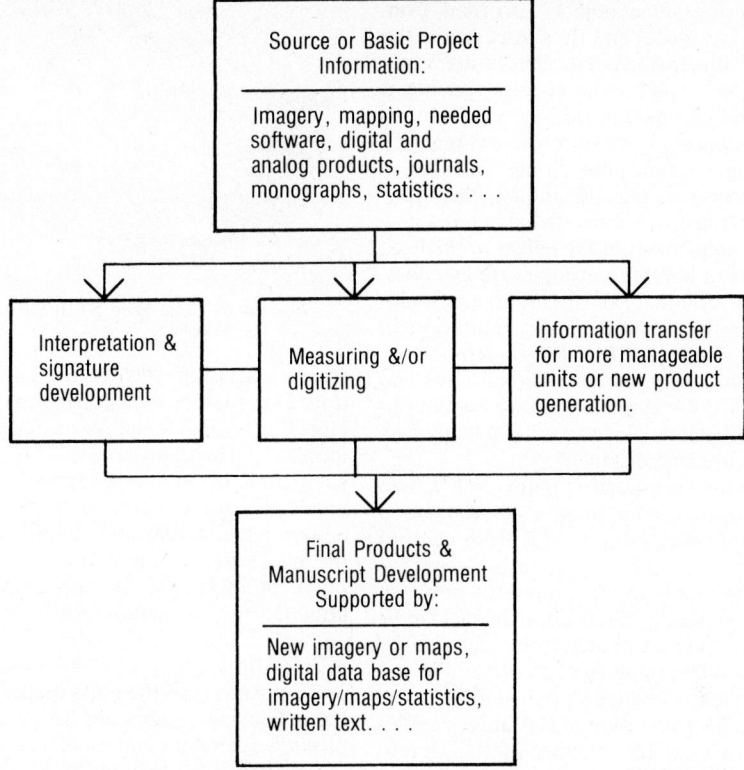

Fig. 24-16. Image interpretation information flow.

the transferring of detail, which results in the generation of a new product or manuscript. The original types of information often determine which instruments are required and in what order they are used. A typical information flow is illustrated in Figure 24-16.

Viewing and transfer instruments provide either three dimensional (stereoscopic) or two dimensional (monoscopic) views; measuring instruments may be used on both single imagery and stereoscopic pairs; and instruments that record or transfer detail use the "camera lucida" principle, projection, or the pantograph method.

Space limitations preclude any attempt to mention all the instruments that are commercially available. The illustrations and trade names of instruments that appear in the following discussion do not constitute an endorsement or statement of preference for these instruments over others.

VIEWING EQUIPMENT

Viewing equipment enables the interpreter to scan or study imagery visually under various magnifications.

Stereoscopic Viewing Instruments

The stereoscope, which provides a three dimensional view of photographic images, is one of the most important instruments used in interpre-

tation. The design of stereoscopic viewing instruments utilizes lenses or a combination of lenses, mirrors, and prisms. Of the types of stereoscopes in use, the simple lens type is the most widely used and generally the cheapest (Figure 24-17). The lens stereoscope provides a simulation of distance vision; it enables the observer to view

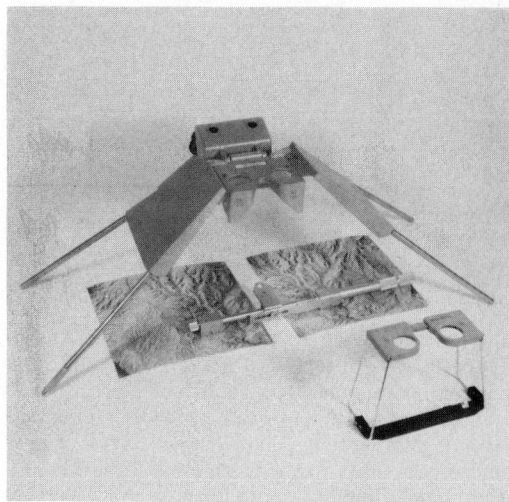

Fig. 24-17. Mirror and lens stereoscopes with parallax bar and stereometer (Zeiss and Abrams Co.).

two images of the same object, recorded from different points in space, and thus to perceive the object in three dimensions. The lenses are separated by a distance equal to the average spacing of human eyes and can be adjusted for different interpupillary distances. Lens stereoscopes magnify two to four diameters or more. Some lens stereoscopes employ variable magnifications; when magnification is changed, the focal distance must also be changed by adjustment of the height of the legs. The advantages of lens stereoscopes are low cost, portability, and simplicity of operation and maintenance. Their disadvantages are: limitation in magnification ranges, difficulty in annotating photos while observing, narrow field of view, and limited capability for spreading photos while observing (i.e. the distance of spread is usually less than the interpupillary distance).

In the mirror stereoscope (Figure 24-17), two sets of mirrors, or a combination of prisms and mirrors, separate the lines of sight from each of the observer's eyes. The distance between the wing mirrors is much greater than that between the eyepieces, so that a stereoscopic image can be received from a pair of photographs laid side by side without their overlapping each other. The images may be slightly reduced or magnified up to two diameters. Magnification of this order enables the observer to view all or most of the stereoscopic area, in line of flight and across the line of flight, on aerial photographic images of standard sizes. A second magnification, two to four diameters or higher, may be achieved by swinging externally mounted binocular lens systems or internally mounted lenses into the optical path. Nash (1972) discusses the correct methodology for using mirror stereoscopes. A mirror stereoscope affords full separation of the stereoscopic pair of images, provides full view of the entire stereoscopic model under normal observation (i.e. no magnification) conditions, and allows the use of mounted opaque prints as well as positive and negative transparencies. Its disadvantages are: (1) there is a relative loss of illumination because of the number of glass surfaces involved and the distances light rays must travel; (2) it is more costly than the lens type, especially when binoculars are required for achieving comparable magnification conditions; (3) it is less portable than the lens type; and (4) it requires more maintenance than the lens type, especially the mirror surfaces, which may be damaged by excessive handling, exposure to moisture, and other environmental conditions.

The major disadvantage of the conventional mirror stereoscope is that only one pair of stereo images can be observed at a time. This led to the concept of a new type of prism stereoscope, which can be used to scan a properly oriented strip of photographs (Figure 24-18). This type of stereoscope uses five-sided prisms to deflect the interpreter's line of sight to imagery that has been specially arranged in a strip format (Eranti, 1968).

Research has been conducted to produce a

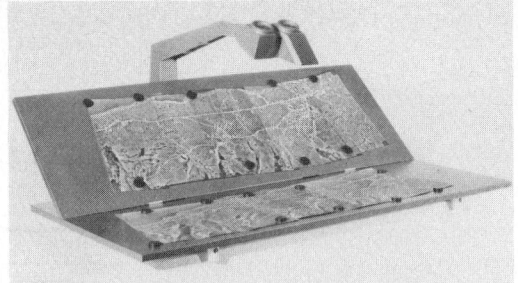

Fig. 24-18. The Wild ST10 Strip Stereoscope.

stereo-television system for imagery interpretation. One such system consists of duplicate sets of closed-circuit TV cameras, control units, and monitors. The camera is used to generate electrical signals of an image represented by an illuminated glass diapositive plate; the camera control amplifes the signals and supplies the camera with the necessary voltage; and the TV monitor displays the image. Motor-operated zoom lenses are attached to the camera lens for enlarging and iris control. Diapositive holders are capable of adjusting the x tilt, y tilt, and swing, and the x, y, and z positions of the glass diapositives. A special lamp housing is attached to each plateholder to provide even illumination of a given intensity over the entire plate (Thompson, 1966).

A convenient device for viewing data stereoscopically has been developed as an aid to photo-interpretation. The device (Figure 24-19) is a pocket stereoscope with a stereomicrometer attached to the feet together with an observation board and box.

A number of binocular-microscope scanning devices have been developed to facilitate the stereoscopic examination of imagery recorded on different media such as positive or negative transparencies, opaque prints, glass plates, and roll or cut film of different widths. Many of these are quite complex, being equipped with coupled zoom elements, which provide continuously variable magnification, light-dimming capabilities, and automatic focusing capabilities through the entire magnification range. Because of the large number of optical surfaces in this type of scope, a high-intensity light source is advised; especially when

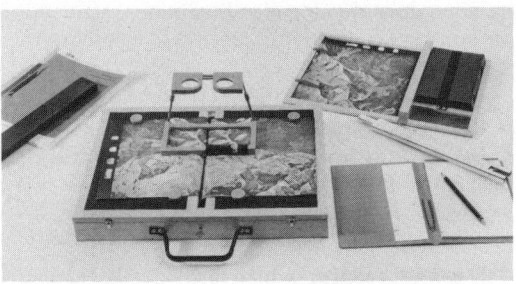

Fig. 24-19. Taschenmesstereoskop. Pocket stereoscope with stereomicrometer, made by Zeiss, Oberkochen.

using higher magnifications. A common microscope lamp that is able to be focused into a tight beam and has a variable intensity is very useful when using these types of instruments.

Non-Stereoscopic Viewing Instruments

Non-stereoscopic viewing instruments include monocular magnifiers, monocular microscopes, large format viewers, light tables, and optical and electronic enhancement systems.

Monocular magnifiers

Monocular magnifiers have lenses of various powers set into convenient frames or mounts. Magnifiers that include scales will be described under "Measuring Equipment".

The hand-held "reading glass" magnifier consists of an equiconvex lens of large diameter (usually 5 to 10 cm) and that has relatively low magnification power (two to four diameters). The lens is mounted in a metal ring to which is attached a handle for easy use.

A few magnifiers of medium power (five to twenty-five diameters) have self-contained lighting because the lens-image distance is too small to permit adequate lighting from an external source.

Binocular Microscopes

The binocular microscope (Figure 24-20) is often a desirable compromise between the monocular magnifier and the high-powered stereoscope. Many have zoom magnification controls, high resolution, and a large field-of-view selection. They have a wide magnification range, extending to more than 200×. Since optical systems are tailored to a type of work, better results will be obtained in imagery applications if instruments are acquired with this type of use in mind. Attention also should be given to the type of mount that will hold the scope. If it is a stand-alone unit, the extension arm should be able to accommodate at least an 18 inch square area without the need to place the scope mount on the imagery. The shape and

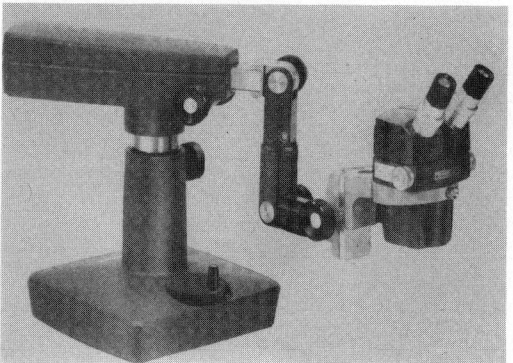

Fig. 24-20. Binocular zoom microscope in extended arm mount (American Optical Corp.).

body-accommodation size of mounts vary; therefore if the scope is to be used in conjunction with a roll-film light table that has a built-in scope mount, it should be determined before purchasing that the desired scope can be accommodated.

Light Tables

Requirements for the light table are based on the specific activities of the interpreter and the kind of imagery being studied. The interpreter may be working with either semi-opaque prints or transparencies. The imagery support, transport and illumination requirements, of course may be different in each case. Most users of transparencies maintain them in roll form for ease of storage. Therefore, the capability for handling roll film should be incorporated in the light table. This may be either hand driven or motorized. If it is motorized, speed and torque controls should be carefully designed to prevent film damage during transport. The illumination system must have an even intensity throughout the entire observation area. This intensity should be variable since the density of the material may vary. A further consideration is the color temperature of the illumination system since much of the imagery used today is either true color or color infrared. The illumination-source color will naturally influence the color observed by the interpreter and could improperly influence the analysis (Avera, 1980).

The most recent innovation in light tables is the use of high intensity illumination in support of high power stereoscopes. Such a device is shown in Figure 24-21. The light source is a quartz halogen lamp with parabolic reflectors. Since the only area under observation is the ringed surface under the moving rhomboid arms of the stereoscope, only this ringed area is illuminated. An x and y scanning film-carriage permits moving the images under the stationary optical head. Brightness levels to 80,000 footlamberts can be obtained with this type of table as compared to the 5,000–7,000 footlambert level of the average light table.

Large Format Viewers

The purpose of the enlarger-viewer is to display the portions of an image to be analyzed. The ability of an enlarger-viewer to display and magnify part of an image results from a combination of optical, mechanical and electrical operations. The viewer, as illustrated in Figure 24-22, can accommodate single frame or roll, positive or negative transparency film-formats, in a scanning or stationary mode, and at two magnification levels. The advantages of such an instrument are its ease of viewing, the speed at which a large area can be analyzed, a viewing area that can be seen by several persons at one time, and a low cost of operation. Its disadvantages are lower resolving ability, less brightness of image for films of high density, and higher expense.

Fig. 24-21. 240 Zoom stereoscope mounted on high intensity light table (Bausch & Lomb).

Fig. 24-22. Hoppmann EV-1 precision film viewer (University of California at Santa Barbara).

Optical and Electronic Viewing Instruments

Optical color-combining equipment ranges in complexity from standard manual slide-projectors (Figure 24-23) to specialized additive-color viewers (Figure 24-24) designed specifically for making and viewing color composites. The additive-color viewers utilize black-and-white multiband imagery and function by using the "additive color" principle.

Carefully controlled multispectral photographs and images taken simultaneously, developed, combined in registration, and colored via filters, can provide a high-resolution composite image. By varying color, density, and hue the observer can detect subtleties undetected by the human eye. A natural or "true" color presentation of the ground scene can be constructed, provided imagery in the blue, green, and red spectral bands is acquired, or, a false-color presentation can be made by filtering the small density differences that exist between objects on the various input bands of multiband imagery. The optical color-combiner's main advantages are the ease of interpreter interaction with the machine, the uniform brightness, and high resolution. Its major disadvantage is that it is difficult to achieve an interface with computers for further interpretive data manipulation.

An electronic color-combiner has the major disadvantage of lower resolution and a much greater operating cost. However, these systems possess

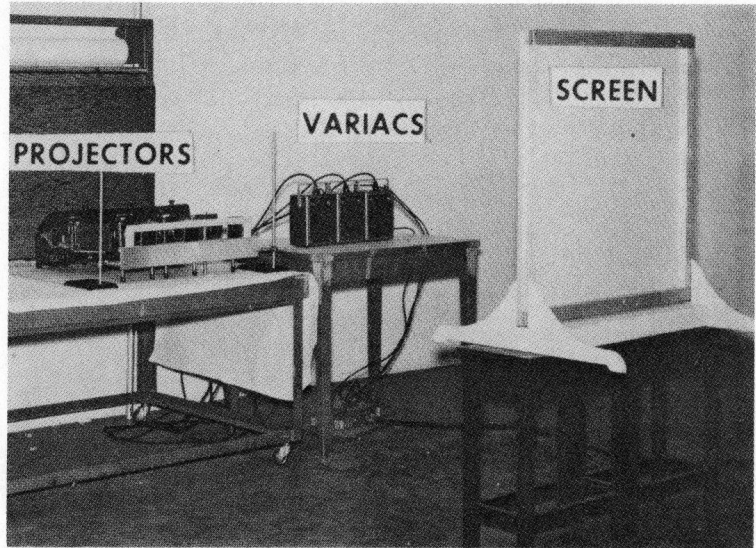

Fig. 24-23. Projection system to reconstitute color images (Orr, 1968).

the advantage of high versatility and computer compatibility. Such systems usually are available only as an integral part of electronic image-processing equipment.

Electronic-enhancement viewing systems, such as that seen in Figure 24-25, scan and display imagery, highlighting desired information. Capabilities of electronic systems are varied but generally include the following:

(a) Point densitometric values and their corresponding x and y coordinates can be obtained by an x and y cursor. Also, many systems allow single scan lines to be analyzed by positioning a horizontal or vertical crosshair on the displayed image. The resulting display is similar to a micro-densitometer trace on a strip chart recorder.

(2) Density level slicing can be done by assigning specific color or monochromatic hues to density intervals. Typically, a digital planimeter can readout the area within each density band.

(3) Independent or simultaneous adjustment of the interval boundaries of all bands permits compression or shifting of all bands simultaneously into any density range of the image.

(4) Independent adjustment of individual density interval boundaries allows flexibility in selecting linear, logarithmic, or other band aperture relationships.

Electronic-enhancement equipment is normally composed of: (1) a light source, (2) an image scanner, (3) an image processor, and (4) a display monitor. The light source must be capable of uniformly lighting the original image while it is being scanned.

The image-scanning device can be a video camera or an optical-mechanical scanner. The video camera is based upon the same principles as commercial television cameras. An optical-

mechanical scanner moves the imagery between a light source and a light detector and measures the amount of light transmitted through each x and y point coordinate on the image. The image processor converts the scanned signals into a format suitable for display on the monitor. Transformations applied to the signal data can be as simple as

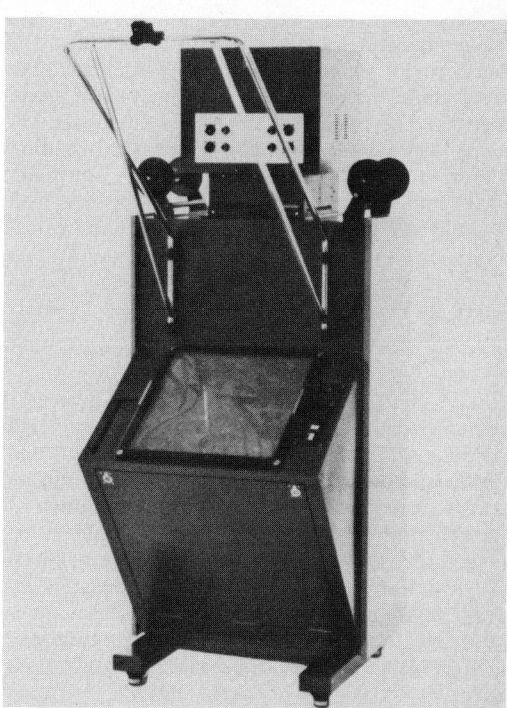

Fig. 24-24. Additive Color Viewer, made by NAC Incorporated.

Fig. 24-25. Digital image processing system (LogE/Spatial Data Systems, Inc.).

equal interval density slicing or as complex as the classification of individual pixels (picture elements) from multiple correlated imagery. These transformations are discussed in detail in Chapter 17.

The monitor can display the processed (or transformed) signals in an image format suitable for human interaction, interpretation, and analysis. The major advantages of electronic enhancement systems are their data transformation flexibility and computer compatibility. The main disadvantages are usually related to cost, as well as a slightly lower resolving ability than optical combiners. Finally, a large volume of quantitative or programming collateral information may be required by image interpreters to adequately interface with these electronic systems.

MEASURING EQUIPMENT

On vertical and oblique imagery, instruments can be used to measure lengths, areas, heights and densities. In addition, plotting measurers can be used to provide orthographic plots. Some of these features are built into digital and analog image-processing equipment, but the average person will find traditional manually-operated instruments more readily available. Whether manual or electronic, current technology may afford accuracies to tens of micrometers (10^{-5} m or more).

Linear Measuring Equipment

The dimensions of an object imaged on a vertical photograph can be found by determining the scale factor for that photograph. To determine the size of objects imaged on oblique photographs, calculations related to the perspective view must be performed. Photogrammetric equations for determining linear distances on vertical and oblique imagery are found in the section on Interpretation of Photographic (Camera) Imagery'', in this chapter.

There are a variety of simple non-magnifying scales available for photographic measurements. If a low order of accuracy is acceptable, an ordinary engineer's scale may prove satisfactory. Where more accuracy is desired, a device such as the metal micro-rule (Figure 24-26) may be used.

Proportional dividers (Figure 24-27) are used by interpreters to transfer detail at a scale different from that of the photographs. Once accurately adjusted, they can be used to measure all features at the two scales. Detail is transferred from photograph to map by turning arcs at the photo scale, then at the map scale. The intersection of at least three arcs locates the image point at the new scale. Proportional dividers can also be used to check the relative scale of two photographs, or to discover gross tilt in a single photograph by checking it against a map or photograph of known scale.

Magnifying scales consist of a simple magnifying lens, set at its focal length above a plate that has graduations on its lower surface, thus placing the graduations in contact with the photographic image; magnification usually ranges from four to ten diameters. This type of instrument utilizes room- or desk-lighting by means of a transparent plastic "tube" that extends all or part of the way

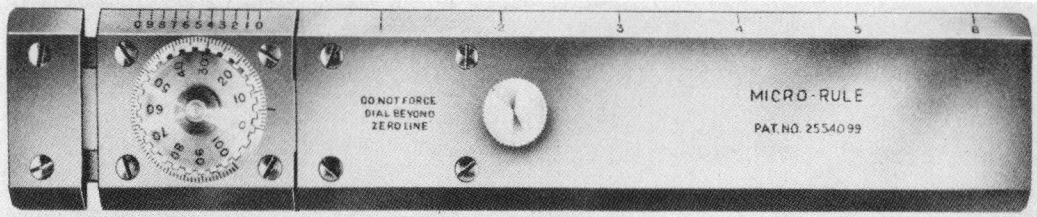

Fig. 24-26. Micro-rule for photographic measurements (Theo. Alteneder and Sons.).

between the lens and the scale plate (Figure 24-28). The scale graduations (i.e. thousandths of a foot, millimeters, etc.) depend upon the reticule chosen by the interpreter.

Linear measurements can also be obtained from a simple and inexpensive map measurer or a more expensive attachment for a digital electronic planimeter (Figure 24-29).

Area Measuring Equipment

The ease of measuring areas on aerial photographs is, in general, related to the shape of such areas. Rectangular areas imaged on vertical photographs, or orthographic plots of areas imaged on oblique photographs, can be found by linear measurement and simple mathematics. Areas of irregular shape, as imaged on vertical photographs (or oblique photographs when reduced to orthographic plots), can be measured with a planimeter (Figure 24-29). Several types of measuring heads are available as illustrated in Figure 24-29. Depending on the type of planimeter used, the results may be read from a vernier scale or digital display. Electronic models may be used with a programmable calculator interfaced to the encoder. With practice, great accuracy may be achieved with these instruments. The more elaborate models will automatically compute volume and slope.

When a statistical sample is needed, the dot area grid can be used to calculate area dimensions. Dot area-grids are particularly useful for calculating the proportions of two or more types of terrain features (such as agricultural and non-agricultural land) of a large geographic area. A transparent grid is divided into square or rectangular spaces, with dots systematically placed in each space. Size of spaces and density of dots depend upon the type of sampling to be done and the degree of accuracy desired. The dot grid is placed over the photograph to be sampled, and the dots falling on each type of object are counted. Each dot has a certain value depending upon the scale of photography and the spacing of dots. Once dots are classified and counted, areas can be determined by simple proportion. The dot area-grid permits area estimates to be made in one-third to one-sixth of the time required for planimetering. Another method is to cut out areas of a particular land use on an image or a map and accurately weigh them. Comparison of these weights with that of a known area cut from the same imagery or a map will allow individual areas to be calculated by proportion.

Height Measuring Equipment

Although the heights of objects can be determined from single photographs, instruments used for height determination generally measure stereoscopic parallax by applying the principle of fusing dots or that of the floating dot, line, or grid.

The simple height-measuring devices described below are not designed for topographic mapping or the comparison of several heights, but for the

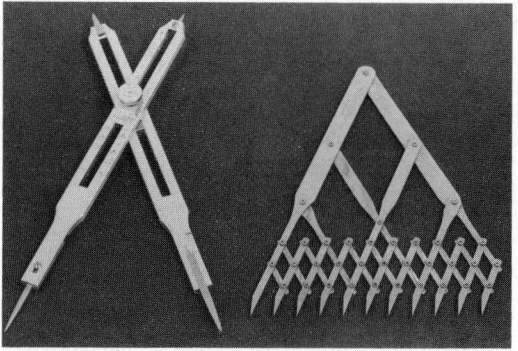

Fig. 24-27. Proportional dividers (Theo. Alteneder and Sons.).

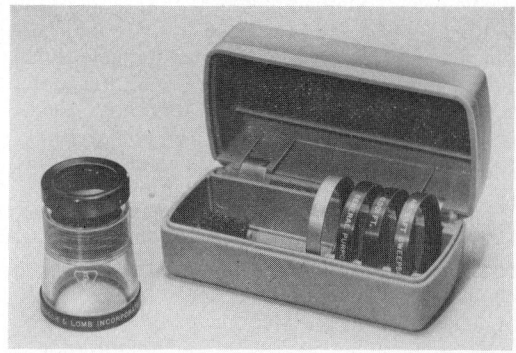

Fig. 24-28. Tube magnifiers with interchangeable lenses (Bausch & Lomb).

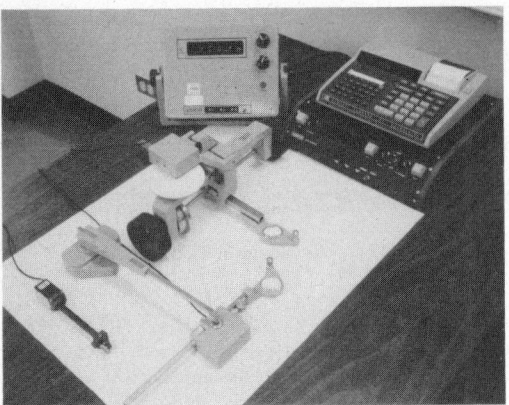

Fig. 24-29. Electronic Planimeters with polar compensating, rolling disk, and length-measuring heads (Lasico).

determination of the heights of individual objects. The interpreter should be aware that height measurements obtained with these simple instruments may be inaccurate due to the effects of tilt and scale change between stereo pairs.

The parallax ladder, also termed the parallax wedge, is the simplest and cheapest of the height-finding instruments. The ladder is placed over a properly oriented stereoscopic pair of photographs. When the photographs are viewed through a stereoscope, the two lines or rows of dots fuse and appear to slope. The height of an object is measured by finding the tick mark or dot that appears to rest on the ground at the base of the object, shifting the ladder to find the tick or dot that appears to touch the top object, and noting the difference between the two readings. The difference in elevation (height of object) is determined by entering parallax tables with the differential parallax reading or by calculation from the basic parallax equation. Accuracy of the parallax ladder, when used carefully, may be equal to that of the floating-dot (stereometer) instruments.

Parallax bars, sometimes referred to as stereometers, are designed to be used with lens or mirror stereoscopes (Figure 24-17). They are used for determining the height of objects on a stereoscopic pair of photographs through the use of the floating-dot principle. The main features are a bar, which may or may not be attached to the legs of a lens or mirror-type stereoscope; two transparent plates, each with a small dot in its center, one of which can be moved laterally; and a finely graduated micrometer device which measures the movement of one dot in relation to the other.

The parallax bar is operated by placing it under, or attaching it to, the legs of a stereoscope that rests over a properly oriented pair of stereoscopic photographs. The two dots appear as one which, by adjustment of the micrometer device, can be made to float up or down. The floating dot is brought to rest at the bottom of the object to be measured and the reading on the micrometer is

noted (or it is brought to zero); then the floating dot is raised to the top of the object and the micrometer is read again. The differential parallax obtained from the difference between the two readings of the micrometer is used to determine the height of the object by entering parallax tables with the reading or by calculation from the basic parallax equation. The parallax bar is more expensive than the parallax wedge, yet only yields results of comparable accuracy. However, many interpreters prefer the parallax bar (or stereometer) because the floating dot is moveable and easier to place on the ground and on the tops of objects.

Instruments that measure differential parallax to determine terrain slope are also available. These instruments utilize fused floating lines as with the "slope-measuring parallax wedge" or fused concentric circles such as the "stereo slope-meter". These instruments can be used to determine slope angles and percent of slope respectively.

Plotting Equipment

Many types of photogrammetric equipment have been developed to provide an orthographic plot of images appearing on oblique or vertical photographs. This equipment is explained in great detail in the Manual of Photogrammetry; for more information, the reader is directed to this reference (Thompson, 1966).

Densitometric Measuring Equipment

Generally, there are four basic steps in the sensitometric cycle: (1) exposure of the sensitized material under precisely controlled and measurable conditions; (2) development of the exposed material under precisely controlled and repeatable conditions; (3) measurement of the resulting images (densitometry); and (4) interpretation of the results. The response of a photographic material is measured in terms of density. In black-and-white photography, the measured density is a function of the amount of developed silver. In color photography, where the processed image contains no silver, the measured densities are the result of the absorption characteristics of the dyes. The section on interpretation of visible color imagery discusses the different types of integral and analytic density measurements that can be applied to color photography. Generally speaking, however, densitometer measurements performed on imagery yield quantitative data that is the foundation upon which the automated pattern-recognition systems operate.

Instruments used in making black-and-white or color optical-density measurements (step 3 above) contain four basic components: (1) the light source, which provides radiant energy; (2) spectral filtering, which selects desired wavelengths for measuring either broad or narrow bands by the use of filters; (3) the receiver, either visual in terms of human perception or electrical, as in a

Fig. 24-30. Components of a reflectance and transmittance densitometer and a chart recorder (Smith and Anson, 1968).

photoelectric instrument; and (4) the read-out, which relates output signal of the receiver to corresponding density value and prints or displays the quantitative data output in some format.

Densitometric systems come in a variety of forms, such as manual reflectance or transmission densitometers, scanning microdensitometers, and a scanning densitometric capability built into electronic image-enhancement systems.

Figure 24-30 depicts the various components of manual reflectance and transmission densitometers and a chart recoder. The reflectance densitometer is used to obtain point density readings from positive prints and the transmission densitometer is used to obtain point density readings from transparencies or negatives.

The scanning reflective or transmission microdensitometer (Figure 24-31): (1) looks at a very small portion, as little as 2 μm diameter of a photographic image, at spectral levels selected to be compatible with the sensitivity levels and dye component spectral characteristics of the photographic materials; (2) reads the optical density of the image by means of a scanning optical system and photo multiplier-log amplifier measuring system; (3) scans the sample at selected velocities and at a uniform rate from 5 to 400 mm per second (dependent upon the state of the art); and (4) presents the data graphically on a strip chart or, when used with an analog converter, presents data digitally to a computer for reduction and analysis.

Variation in optical density of microdensitometer scan lines is caused by several factors. It was determined that aperture size did not improve land-use discrimination using photos at a scale of 1:1188, although it did affect density. Aperture shape (round or rectangular) has less effect than other variables. Image-density differences were determined to be greater in the blue region of the spectrum than in the red or green (Doverspike et al., 1965).

Rib and Miles (1969) tested the effects of aperture size and photo scale on imaged density differences among terrain features. They concluded that, as aperture size decreased or photo scale increased, more detail was recorded because of finer tonal patterns of imaged features. However, a point was finally reached where desired information was obscured due to granularity of the film (Driscoll et al., 1974).

Scanning microdensitometers are more expensive but tend to be less time consuming when compared with point reflection or transmission densitometers.

As mentioned previously, electronic viewing image-enhancement systems offer both point- and scanning-densitometer capabilities. An x and y

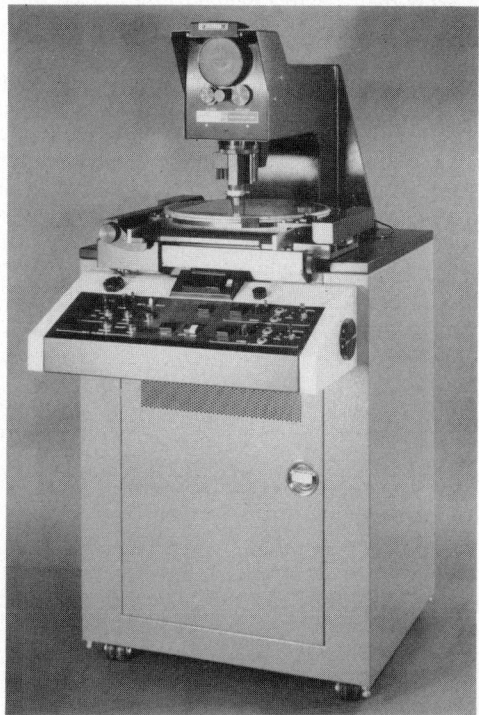

Fig. 24-31. Precision 1010A PDS microdensitometer (Perkin-Elmer Corp.).

cursor (Figure 24-32) on the display screen can be manipulated to generate x, y and z data, z being the optical density value, with reference to a co-ordinate system. Also, the horizontal x or the vertical y cursor can be used independently to generate line-scan densitometric traces. These traces can be displayed on a screen or played back on a chart recorder. Point x, y, z readings can be read directly from the processing unit.

EQUIPMENT FOR TRANSFER OF DETAIL

Instruments designed for sketching planimetric detail from photographs are of two basic types: those based on the "camera lucida" principle and those employing optical projection. These instruments are essentially tracing devices that incorporate means for changing scale in the process of compilation. In some types, provision is also made for an approximate rectification for tilt in the photographs. The pantograph, an instrument used by draftsmen to change the scale of drawings, can also be used to advantage by photointerpreters.

"Camera Lucida" Instruments

In these instruments, the eye receives two superimposed images, one from the photograph and one from a base manuscript. The operator can adjust the instrument so that selected images on the photograph are made to coincide with their plotted positions on the manuscript in order that other features between the selected points may be traced in their relative plan positions. Two representative instruments of this type are the vertical sketchmaster and the zoom transfer scope.

The vertical sketchmaster makes use of a semi-transparent mirror at the eyepiece and a first-surface mirror above the photograph. The observer views the manuscript through the semitransparent mirror and the photograph by reflection from the semitransparent and first-surface mirrors. The instrument may be raised and lowered on its supports to adjust the photo-image points to correspond with points on the manuscript, and may be

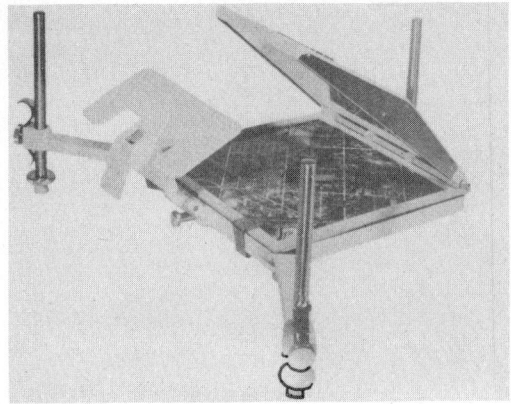

Fig. 24-33. Vertical Sketchmaster, Type 260GE (Alan Gordon Enterprises).

tilted by means of the foot screws, permitting the approximate removal of the effects of small tilts. Interchangeable lenses of various powers are provided for insertion beneath the semitransparent mirror when the instrument is used to transfer detail at a ratio other than 1:1. These lenses serve to bring the apparent manuscript plane into focus with the photo plane, and to eliminate parallax or the apparent motion (Figure 24-33).

In addition to the fundamental capabilities of a "camera lucida" instrument, the Zoom Transfer Scope, or ZTS, (Figure 24-34) offers several additional features. It provides continuous zoom magnification of the image on the stage, from 0.75× to 14× on the single image model and 0.6× to 16.1× on the stereo model, and magnification of the map (data base) at 0.75×, 1×, 2×, or 4× for accurate matching of photo scale to data-base scale.

With the photo scale smaller than the map scale, the interpreter turns the zoom dial until the imagery on the stage is magnified to the same scale as the map. This event permits satellite photography to be used with certain small scale

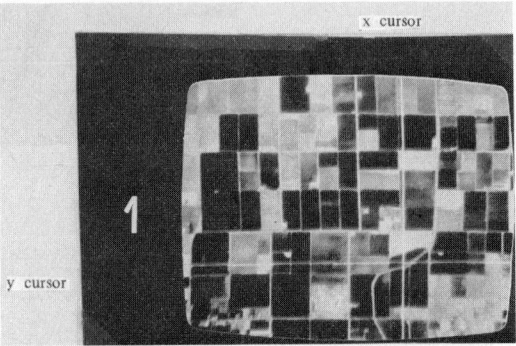

Fig. 24-32. Photograph of the display screen of an electronic viewing enhancement system showing the x and y cursor.

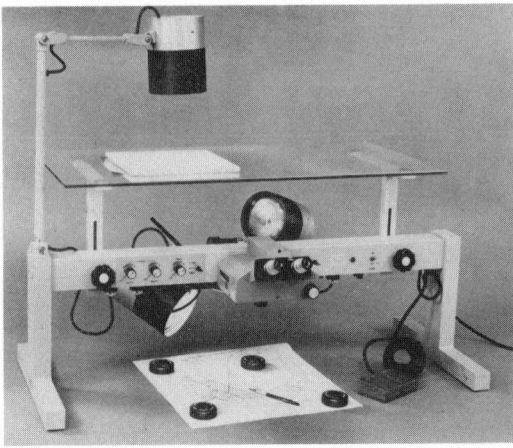

Fig. 24-34. Zoom Transfer Scope-ZTS-H (Bausch & Lomb).

maps and charts. If the photo scale is larger than the map scale, as may be the case with long focal length photography or low altitude coverage, other map lenses are available. Another feature is the anamorphic correction system that can compensate for geometric anomalies in imagery such as tilt, relief, earth curvature, and lens distortion. This correcting feature alters the image in only one direction at a time; a lever controls the "stretch" ratio from 1:1 to 2:1. This transforms the image to a more geometrically accurate presentation without costly conventional rectification. Rotation of the photographic image (through 360 degrees, if necessary) to make critical adjustment, without moving either the photo or base is also part of its capabilities. Wide field-of-view eyepieces that reduce eye fatigue are available on all models.

The stereo zoom transfer-scope is the newest product in manual optical transfer-equipment (Figure 24-35). It is similar in principle to the single image model, but has dual controls for imagery manipulation. This instrument allows the operator a simultaneous view of a stereo image scene and the base manuscript. The stereo model may also function in the same manner as the other models since it has an optical mode lever that will allow either monocular or stereo viewing. The models illustrated may be ordered with a camera-adapter port so the operator can photograph the optically overlaid images by using a 35 mm camera back or a Polaroid 4″ × 5″ camera. However, unless very fast film is used with these systems, long exposures are necessary.

Optical Projection Instruments

Optical reflecting projection instruments project images of photographic prints through a lens and one or more mirrors onto a mapping surface.

They are useful in transferring detail from near-vertical photographs or other soruce material. Usually the chamber holding the imagery can be shifted in order to change the imagery scale through a considerable range (Figure 24-36).

Although the reflecting projector has disadvantages, many image interpreters have found them very useful. Its disadvantages are the requirement for a semi-dark room, lack of portability, considerable expense, and minimal precision in tilt-adjustment. It has, however, the following advantages: (1) It can enlarge and reduce scale from less than one to several times magnification; (2) It has a large, well-illuminated working area, and (3) a comfortable working position (Figure 24-37).

For those project managers and researchers on a limited budget, an overhead projector for the viewing of transparencies or an opaque projector for prints may be used. Although limited in resolving power, these instruments can provide sufficient enlargement for preliminary imagery analysis and presentations at minimal cost.

MISCELLANEOUS EQUIPMENT

The following instruments do not lend themselves to any one area of interpretation, information transfer, or measurement, but support all of these functions to one degree or another and can be used at any stage of a project. Although this is not a definitive list, several of the most often used pieces are briefly discussed.

Digitizers

The digitizer (Figure 24-38) is becoming indispensible as an analysis tool. This instrument provides machine-assisted translation of data into computerized form for subsequent computer mapping. While the electronic-enhancement

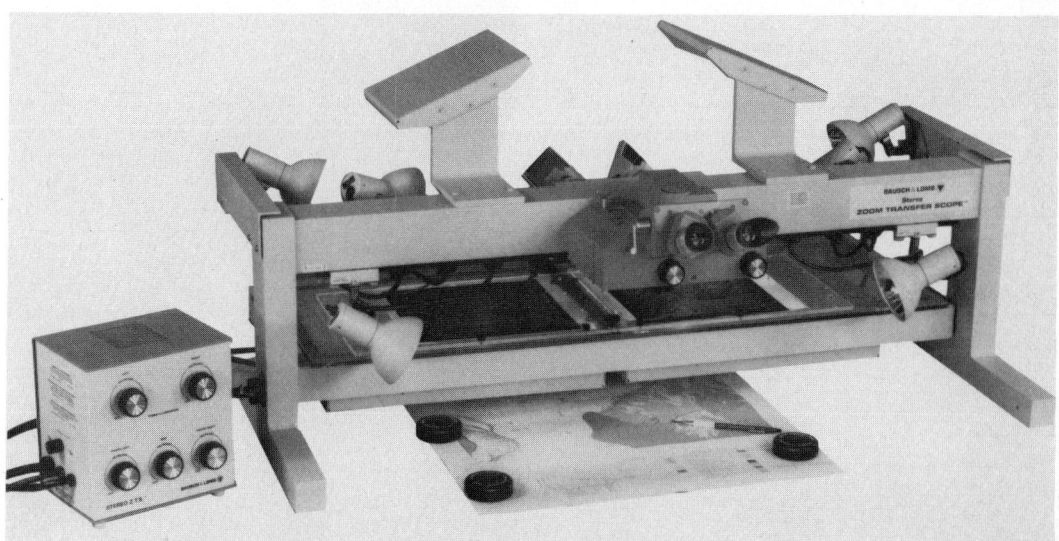

Fig. 24-35. Stereo Zoom Transfer Scope (Bausch & Lomb).

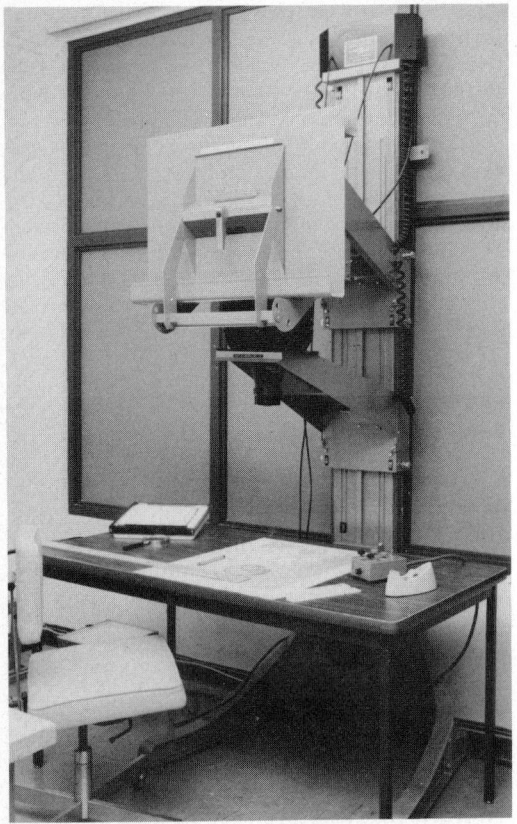

Fig. 24-37. Reflecting Projector (Keufel & Esser Company).

viewing systems can provide x, y coordinate readout, the manual digitizer has been the more traditional instrument for converting rectified image data to computerized digital form. The traditional transfer method has been to prepare a planimetric map from the image and then digitize the thematic data thus obtained. The advent of orthophotos in large-scale rectified form makes the electronic digitizer more appealing, because data transfer can now be made directly from the image-to-computer form through manual assistance of an operator. Unrectified small scale imagery presents a problem for data translation by the digitizer because of the difficulty in following

Fig. 24-36. Map-O-Graph (University of California at Santa Barbara).

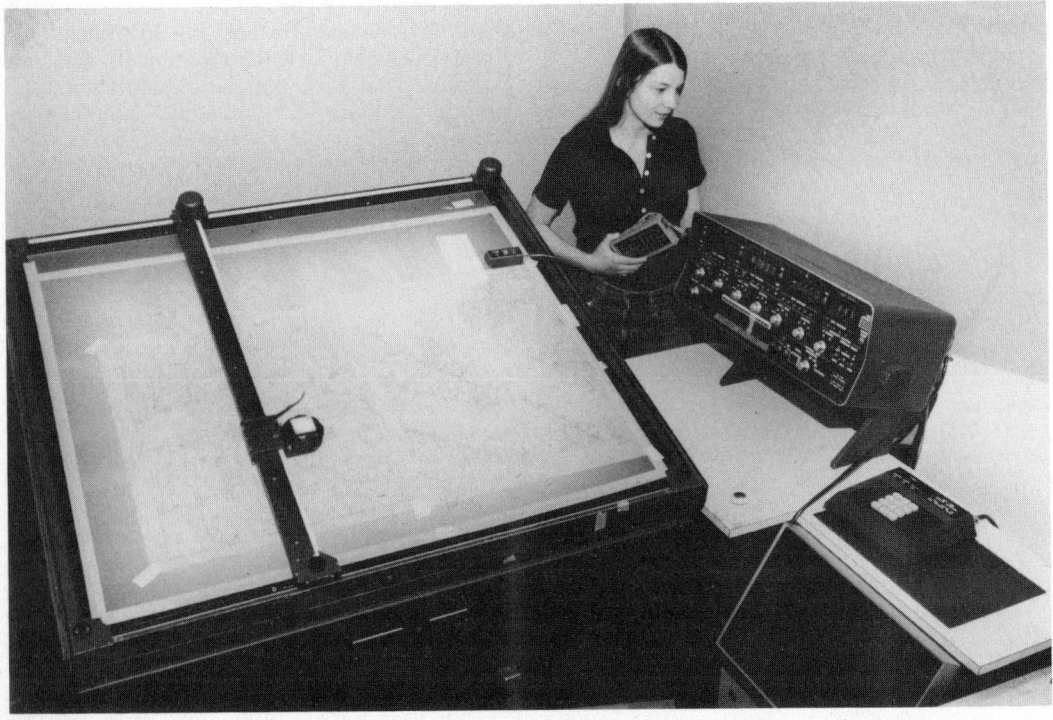

Fig. 24-38. Electronic X, Y Coordinate Digitizer (University of California at Riverside).

interpreted boundary lines between nominal data elements. The rectification problem can be overcome through a computer process of resectioning of each point (Johnson, 1971).

Dot Counter

The interpreter often has to count large numbers of objects on imagery. In doing this type of work, he must record each object as he identifies it, accumulate a partial or final total of the identified objects, and eliminate each recorded object from further consideration.

The dot counter carries out all of these steps. This instrument is shaped like a large pencil. When the point is placed on an image and downward pressure is applied to the pencil, a counter inside the barrel of the pencil is actuated. The dials of the counter are visible so that the count can be read at any time. The point of the counter is quite sharp, so that the emulsion of prints can be pricked to mark the objects which have been counted.

The characteristics of the dot counter are undesirable for certain types of work. Because of its length, it is difficult to use with high-power direct-viewing instruments. The great number of pinpricks made in some inventories may ruin the imagery for further interpretation, especially if the imagery is in diapositive form.

Slide Rules

There are linear slide rules and hand-held circular computers especially designed for photogrammetrists and image interpreters (Figure 24-39). Together these instruments can solve almost any computational problem for vertical and oblique imagery. They also are very useful for computing flight planning parameters.

Copy Camera System

A copy-camera system, such as is illustrated in Figure 24-40, can aid the interpreter or project

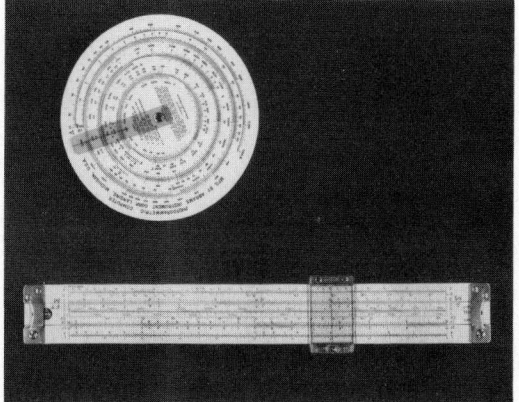

Fig. 24-39. Photogrammetric computer (Abrams Instrument Corp.), and a U.S. Air Force aerial photograph slide rule type A-1 (Pickett).

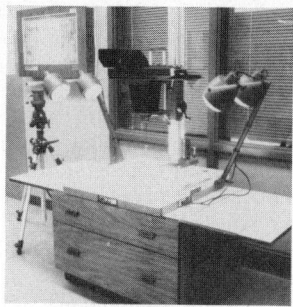

Fig. 24-40. MP-4 multipurpose camera system on a mobile light table (Polaroid).

staff in several ways: (1) reducing or enlarging copy or sections of copy, such as large maps, oversize prints, sections of high altitude imagery, etc., making such products more manageable; (2) preparing materials for final manuscripts or for presentation; and (3) providing specimen and line copy for analog analysis and digitizing. These systems are relatively inexpensive and can provide the widest dissemination of job materials without jeopardizing the original imagery or line copy.

SELECTION OF EQUIPMENT

The choice of equipment should be determined by the type of work to be done, the work space available, the ability of the interpreter, and the funds available. In order to choose equipment for a given operation, it is necessary to conduct a technical and operational evaluation.

Technical Evaluation

The technical evaluation investigates the mechanical characteristics of the instrument. If possible, a superior instrument to the one under consideration should be used as a control. The interpreter may take some or all of the following steps: (1) investigate the mechanical and/or optical design of the instrument; (2) determine the inherent accuracy of the instrument through measurement of component parts or comparison in operation with an instrument of much higher accuracy; (3) judge the rigidity of construction in relation to maintenance of required accuracy during operation; and (4) judge the ruggedness of design, both mechanical and optical, in relation to known working conditions.

Operational Evaluation

The operational evaluation is an investigation of the characteristics of the instrument in relation to its use. Although several instruments may be capable of accomplishing the same general result for a particular set of operational conditions, one may be much more desirable than the others. The points to be considered in an operational evaluation are: (1) the simplicity of operation in comparison with accuracy required; (2) the convenience of operation; (3) the size, shape and weight of the equipment related to requirements for mobility; (4) the difficulty of maintenance in relation to the personnel and facilities available; and (5) the complexity of operation and difficulty of training operators in relation to the personnel, time, and facilities available. The availability of repair parts and trained repair personnel for the life of the instrument should be a major consideration for operational evaluation, especially for those instruments with electronic components.

FUNDAMENTALS OF COMPUTER-ASSISTED IMAGE INTERPRETATION

INTRODUCTION AND SUMMARY

This section presents a summary of computer-assisted image interpretation techniques.[4] It is an overview of the diversity, potential, and limits of such assistance. The remote sensing process is extensive, so the image processing scientist must be versed in the electro-optics of sensing, in transmission and display technology, and in system and probability theory, numerical analysis, statistics, pattern recognition, and the psychophysics of vision (Moik, 1980). Similarly, the designer of image processing systems is required to know computer systems, man-machine communication, computer graphics, data management, and data-base management systems. The effective analyst should be aware of the role of each skill and discipline to properly function in the computer-assisted interpretive process.

One reason for the diversity of necessary expertise to effect interpretation is that remote sensing data is subject to numerous uncertainties in many different forms. These uncertainties inhibit the extraction of information. Billingsley and his colleagues (Chapter 17) and Brogan and Nagy (1981) characterize remote sensing, and hence the general nature of the image to be processed and interpreted, as convolved and corrupted by both reversible and irreversible processes.

Reversible processes are geometric (geographic) and radiometric (image brightness) distortions; irreversible examples are windowing (i.e., subimages), random noise corruption, and the effects of spatial and radiometric intensity quantization. The typical set of digital image-processing problems is summarized by Jain (1981) in Table 24-2. Reversible processes can be corrected by image restoration. The aim of image restoration is to make as good an estimate as possible of the original scene (Rosenfeld and Kak, 1976). Significant image-processing efforts are in-

_____
[4] This section has benefited from many sources. Major ones adapted for use in specific sections here include: Lillesand and Kiefer (1979), Moik (1980), and Landgrebe (1981).

volved with reversible processes usually prior to interpretation. Improvements on the effects of irreversible processes are accomplished by image enhancement.

Image enhancement is the selective emphasis and suppression of information in the image with the aim of increasing the image usefulness or acceptability to the user (Rosenfeld and Kak, 1976). Other uncertainties are introduced in the system design by the selection of the resolution in each dimension of remote sensing measurement. A summary of the most frequently encountered image-processing functions (responsive to the problems in Table 24-2) was categorized by Preston (1980) in his survey of 15 major image-processing languages in use in the United States and is presented in Table 24-3.

Digital image-processing techniques can be divided into three groups: restoration, enhancement, and interpretation (see Figure 24-41). The specific types of digital image processing included within these three categories are discussed later. Another category of image processing, image digitization, and coding converts images from continuous to discrete form, to conserve image-storage space, and to permit computer-assisted interpretation.

Although there are literally hundreds of different image-processing programs in an advanced facility, the central idea behind most computer image-processing is quite simple: the image is fed into a computer one number (pixel) at a time. This "data number" or DN is then modified by an equation or series of equations. The result is written to some storage device for additional manipulations or display (Siegal and Gillespie, 1980).

Small interactive mini-computers or large "main frame" (batch) computers are used for processing and analyzing digital image data. The batch image-processing system is one that is normally operated by submitting a digital data set with specifications for each image analysis step and desired outputs. These requests are then executed and the output produced without user interactions. If the desired product is not obtained, the user must resubmit the job with the modified image analysis and output specifications. Normally, several iterations are required to obtain the desired image product unless routine analyses with high quality data sets are being conducted. Batch software-packages (operating instructions) are generally transferable from one large general-purpose computer to another with a minimum of reprogramming. They are designed to use standard output and input devices available on all large systems. Interactive image-processing systems allow the user to view the effects of the image processing in near-real time on video display units and to interactively optimize the analysis sequence until the desired end product is obtained.

TABLE 24-2

Typical Problems in Image Processing (Jain, 1981)

	Problem	Description	Models
1.	SMOOTHING	Given noisy image data, filter the data to smooth out the noise variations.	Noise & image power spectra
2.	ENHANCEMENT	Bring out or enhance certain features of the image, e.g., edge enhancement, contrast stretching.	Features
3.	RESTORATION & FILTERING	Restore an image with known (or unknown) degradation as close to its original form as possible, e.g., image deblurring, image reconstruction, image registration, geometric correction.	Degradations; criterion of "closeness"
4.	DATA COMPRESSION	Minimize the number of bits required to store/transmit an image for a given level of distortion.	Distortion criterion; imagery as an information source
5.	FEATURE EXTRACTION	Extract certain features from an image, e.g., edges.	Feature detection criterion
6.	DETECTION AND IDENTIFICATION	Detect and identify the presence of an object from a scene, e.g., matched filter, pattern recognition, image segmentation, texture analysis.	Detection criterion, object and scene
7.	INTERPOLATION/AND EXTRAPOLATION	Given image data at certain points in a region, estimate the image values of all other points inside this region (interpolation) and also at points outside this region (extrapolation).	Estimation criterion, and degree of smoothness of the data
8.	SPECTRAL ESTIMATION	Given image data in a region, estimate their power spectrum.	Criterion of estimation, *a priori* model for data
9.	SPECTRAL FACTORIZATION	Given the magnitude of the frequency response of a filter, design a realizable filter, e.g., a stable 'causal' filter.	Criterion of realizability
10.	SYNTHESIS	Given a description or some features of an image, design a system which reproduces a replica of that image; e.g., texture synthesis.	Features, criterion of reproduction.

TABLE 24-3
Major Image Manipulation Functions (Preston, 1980)

MAJOR IMAGE MANIPULATION FUNCTIONS

UTILITIES

- Identifiers
- Executives
- Formaters
- I/O Commands
- Test Pattern Generators
- Help Files

IMAGE DISPLAY

- CRT
- Hard Copy
- Interactive Graphics

ARITHMETIC OPERATORS

- Point
- Line (Vector)
- Matrix
- Complex Number
- Boolean

GEOMETRIC MANIPULATION

- Scaling/Rotation
- Rectification
- Mosaicking/Registration
- Map Projection
- Gridding/Masking

IMAGE TRANSFORMS

- Noise Removal
- Fourier Analysis and other Spectral Transforms
- Power Spectrum
- Filtering
- Cellular Logic

IMAGE MEASUREMENT

- Histogramming
- Statistical
- Principal Components

DECISION THEORETIC

- Feature Select (Training)
- Classify (Unsupervised)
- Classify (Supervised)
- Evaluate Results

Multispectral digital image-processing facilities, depending on the computer type and capacity and associated peripheral input and output devices, can require investments from $100,000 to several million dollars; although some lower-cost systems are now becoming available. These systems typically include hard copy and/or video terminals for controlling programs, image display units for direct user viewing of the imagery, controls to directly interact with the display image, and tape or

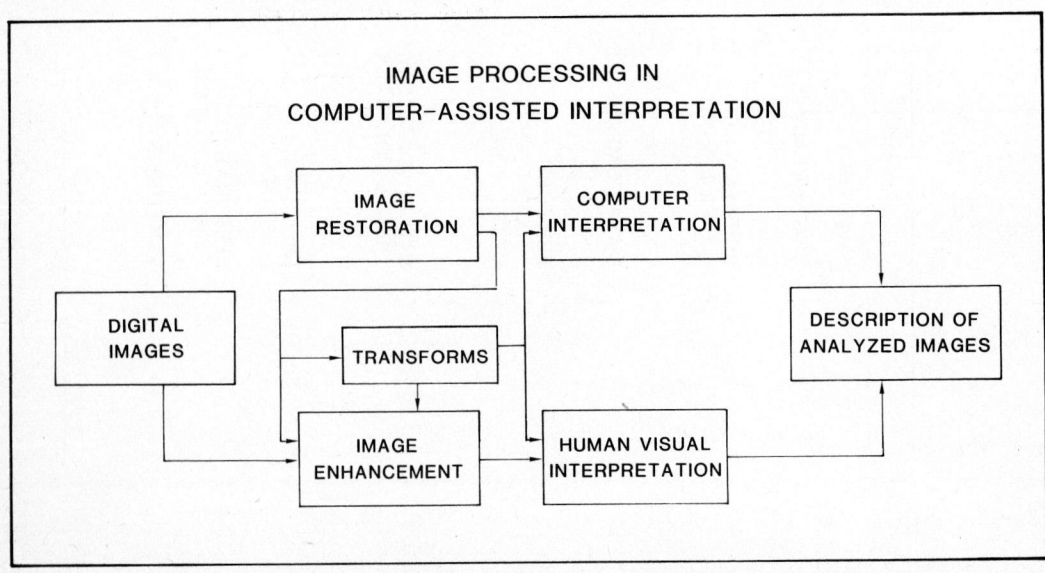

Fig. 24-41. Image processing in computer-assisted interpretation (adapted from Moik, 1980).

film recording units to produce the final products. Some computers are dedicated exclusively to image processing. Image data can also be processed on computers existing for other purposes in organizations such as in state governments, universities, etc., provided the appropriate software is acquired (NASA Goddard Space Flight Center, 1980).

Regardless of the computer, or the process and command languages used, a digital image-processing system for computer-aided interpretation provides a set of functions for the image processing, data display, management, and communication between the analyst and the system. At best performance, the combination of the person and the computer operating interactively leads to the solution of image-interpretation problems that neither could solve efficiently alone (Moik, 1980). The functional requirements of such a system can be determined from an understanding of the image formation, recording, and display processes and of the image-analysis application. A description of the techniques and strategies applied by the analyst during the processing for information extraction completes the overview.

It is important for the beginner not to be intimidated by the computer and its operation or of the possible lack of one's computer programming language skills. Indeed, efficient communication between the image analyst and the computer requires a command repertory that concentrates on the content of the dialogue rather than the form. This emphasis is becoming more urgent since the ability of sensor systems to acquire data now vastly exceeds the ability of users to process and analyze such data intelligently. It has been estimated, for example, that NASA has looked at *less than 1 percent* of the sensor data it has acquired so far from Landsat surveys (Casasent, 1981). Projected sensor systems are expected to provide high-resolution data at frame rates that can yield $10''$ bits per second of data, about 1000 times more than the Landsat (1-3) rate. This underscores the need for the analyst to be properly selective when invoking any image-processing methods for the conversion of data to information.

Contemporary image interpretation involves a significant mix of human and computer efforts to best combine the optimal and unique performances of each, *viz.* (1) the rich spectrum of interpretive elements used by humans, and (2) the repeatable image improvements and high-rate interpretation performance of the computer. Ideally, this combination is synergistic.

In this section are reviewed the set of remote sensing measurements and resolutions that define an image and hence characterize the elements of human and computer interpretation. Then follows a brief historical and status summary of image restoration and image enhancement. Computer interpretation using spectral-feature recognition methods is then described in some detail since it represents a significant phase of contemporary operational practice. Next, the more recent in-

terpretation elements used by the computer (time and texture) are described. The completely merged geographic information system is then reviewed along with a discussion of the related practical system-management problems. The section concludes with a terse overview of trends in computer hardware and software and the promise of advanced interpretation methods.

THE DIMENSIONS AND RESOLUTION OF ENVIRONMENTAL MEASUREMENT

The concept of collection and representation of image data in a sampled or numerical format for computer use is illustrated in Figures 24-42 and 24-43. Figure 24-42 shows a single line of multispectral scanner data collected over a landscape composed of several cover types. For each cover type, the reflectance or emittance is related to digital numbers in each spectral band (here blue, green, red, reflected IR, and thermal IR). The vertical bars indicate those relative numbers. Usually the number of gray increments are compatible with the number of distinct changes in noise-corrupted radiometric signals (see Chapter 17). These outputs represent a coarse description of the spectral response patterns of the various terrain features along the scan line. If they are unique for each feature type, or nearly so, they form the basis for an interpretation of the image data.

As seen earlier in this chapter, human and computer interpretation functions have been perceived to be different realms: the subjective and the quantitative, respectively. This subjective-quantitative dichotomy has impeded a unified view of these two methods of interpretation. This gap is further amplified because the experienced human interpreter may regard computer interpretation with suspicion. This view is not entirely unfounded as computer scientists, artificial intelligence experts and others have sought to emulate human interpretation abilities. Some progress has been made in understanding the problems and characteristics of such a task. Interestingly, some explanatory models of human perception are computer-like in description.

Three levels of human interpretation (or discrimination) have been recognized. These are also valid for describing three levels of computer-assisted interpretation: detection, identification, and analysis.

Detection is the determination of the presence or absence of an object (non-specific and non-identified). The interpretation is a simple dichotomy: yes−no, true−false, or 1−0. Identification is detection in the specific context of the categorical information structure. It is thus characterized by the extraction of sufficient information from the image to name an object. Analysis determines information beyond identification. Thus, in a landscape image, for example, objects other than a grass background may be detected, subsequently identified as trees and eventually analyzed for timber volume.

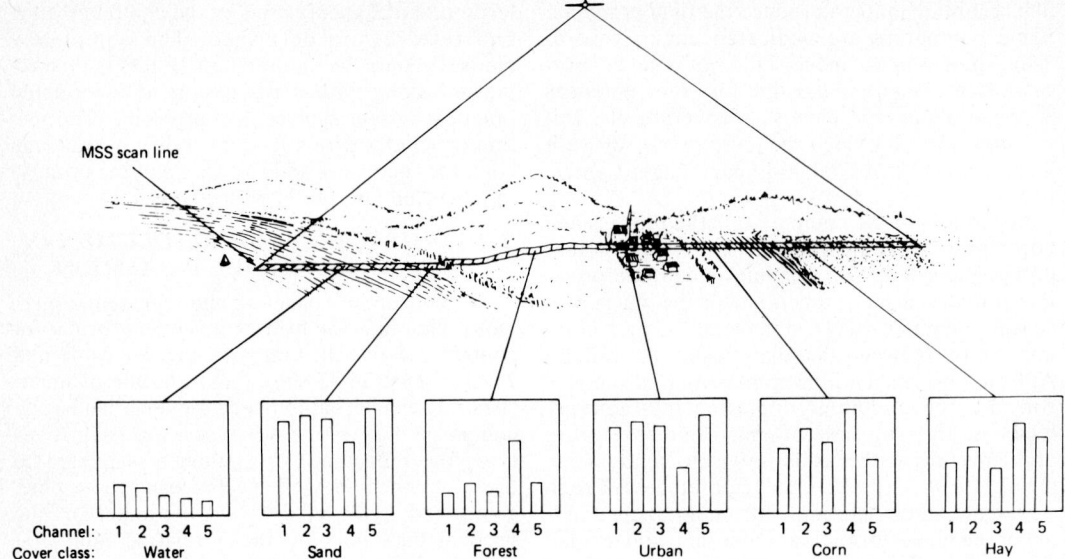

Fig. 24-42. Selected MSS measurements made along one scan line (Lillesand and Kiefer, 1979).

In the hierarchy of classes of objects in the scene a valid list incorporates classes of information value, an exhaustive list, and classes spectrally (or otherwise, separable). Thus there exists a level of specificity or a hierarchy of categories of information that is germane to each interpretation task. It is also the reason why interpretive performance for each image-computer-user task is different. This interpretation potential is intrinsically woven into a fabric defined by the dimensional resolutions of the remote sensing system. Resolution is thus a vital consideration in the measurements made in the spatial, spectral, radiomet-

ric, and temporal dimensions (domains). Further, each remote sensing measurement is interrelated with the others. Lintz and Simonett (1976) presented this idea as a general functional relation.

In typical human interpretation about three times the spatial resolution is required to progress from detection to identification. About ten to one hundred times resolution-increase is necessary to pass from identification to analysis (Lintz and Simonett, 1976). One method of conceptually unifying the human and computer processes of detection, identification, and analysis, relative to

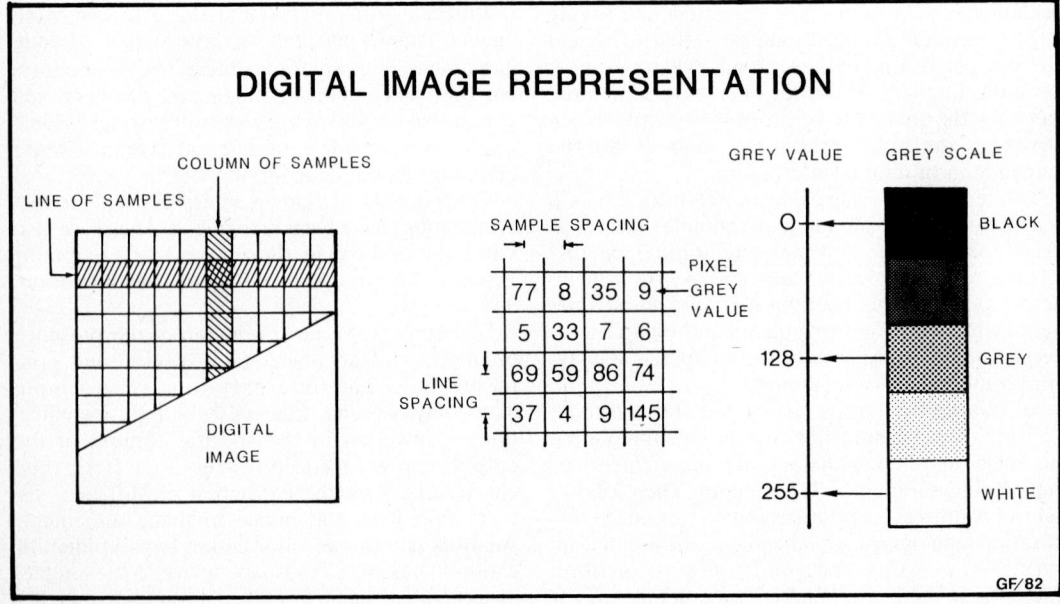

Fig. 24-43. Digital image representation (Mertz, et al., 1981).

resolution improvements, uses the simple binary decisions depicted in Figure 24-44. Progressive levels of descriptive detail are answered yes-no (or even assigned probability values) by either type of interpreter (human or computer).

A good illustration of the difference between detection, identification, and analysis in relation to resolution was reported by W. Evans (1974) for a Landsat image of San Francisco. He positioned a 22 inch diameter, slightly convex mirror so that the image of the sun was projected at the satellite as it passed overhead. A single pixel of the image represented the average brightness of a 57- by 79-m rectangle on the ground. In spite of the slightly larger instantaneous field of view (IFOV), Evans was able to detect the presence of an object (his small mirror), which was represented as a single bright pixel. He was unable to distinguish between his mirror and a hypothetical 80-m diam-

eter mirror or to identify the pixel as a mirror (Siegel and Gillespie, 1980).

Detection and identification are perhaps the most amenable to automation, while analysis seems much removed at this stage. A summary of the performance in computer interpretation can be related to the associated probabilities of detection and identification. Less frequently evaluated but necessary are the probabilities of error (e.g., false detection or false identification). The detection, identification, or analysis by computer interpretation usually has an intermediate step in which the image or subimage has data extracted that bears little resemblance to the visual features used by humans. The data may include statistical moments, Fourier transform coefficients and multidimensional measures (Gonzalez and Wintz, 1977). For analysis, computer algorithms (usually empirically derived) further manipulate such descriptive statistics.

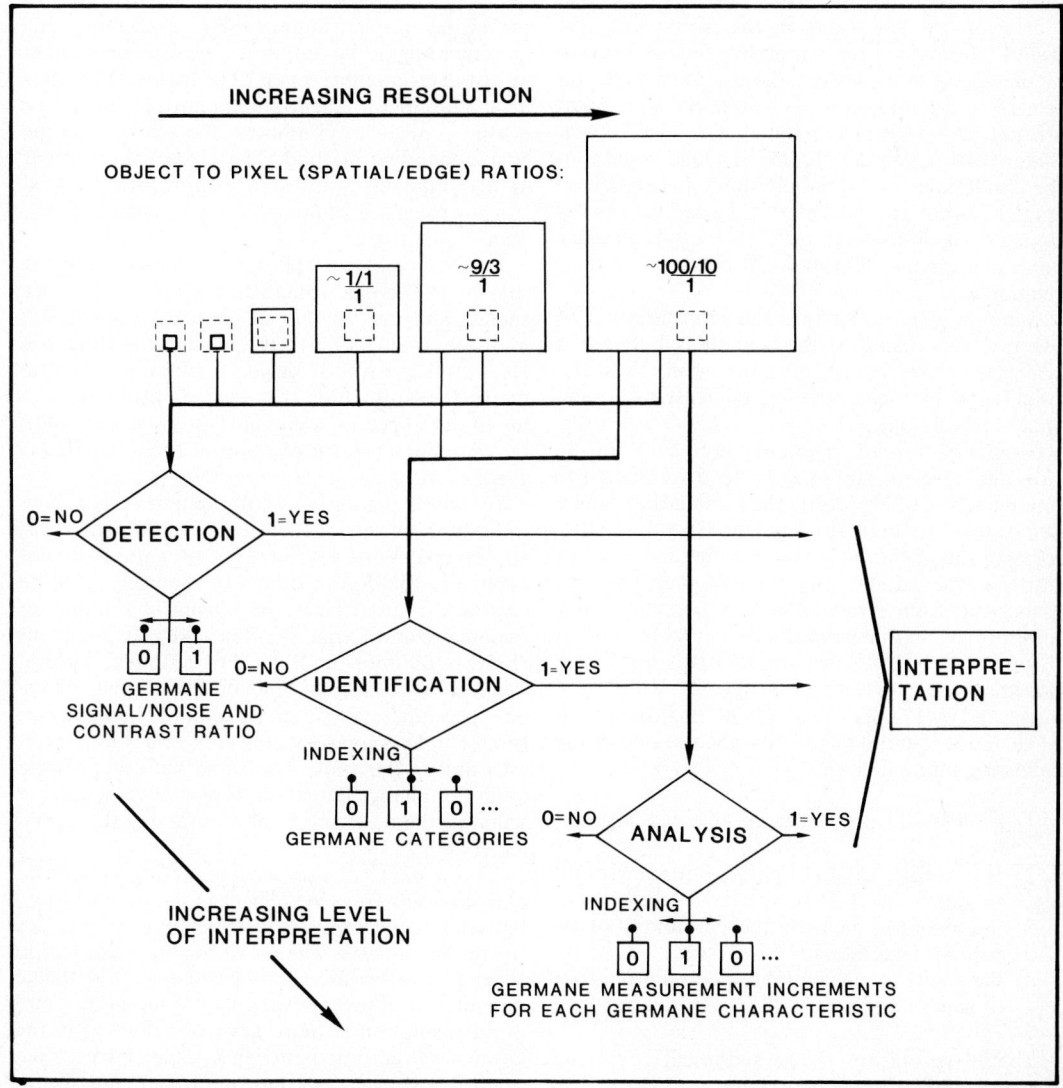

Fig. 24-44. Conceptual overview of detection, identification, and analysis in relation to object-to-pixel ratios. For example, in the top right quadrant the large square (solid lines) pertains to the object, while the small square (dashed lines) pertains to the pixel.

Examples include biomass estimates, sediment concentrations and soil moisture.

Throughout the remote sensing literature, some diversity exists in discussing image interpretation as related to the resolution of the measurements and the implications for machine analysis. A brief comparison is appropriate here. Slater (1980) noted that the different ways in which we can get information can be related to the (dimensional) resolutions and that the quality of the information extracted from remotely sensed images is strongly influenced by the spatial, spectral, radiometric, and temporal resolution. Lintz and Simonett (1976) add that, in acquiring data for any kind of analysis, the level of aggregation or resolution strongly influences the quality of the conclusions that may be drawn (See also chapter 1). Landgrebe (1978) states that one may derive information about the materials covering the surface of the earth from the spatial and spectral distributions of energy emanating from those materials and that temporal variations in the scene are also useful. He divided the variability factors present in the scene into two categories: those that are related to the information desired and those that are not. The former are signal, the latter noise. Slater (1980) cited the parameters fundamental to the extraction of information about a scene as (1) spatial resolution and detail; (2) spectral resolution; (3) signal-to-noise ratio (S/N); (4) ancillary data; and (5) the informational classes and their relationships.

Sabins (1978) noted that the interpretation of images, regardless of the wavelength band at which they are sensed, depends upon the scale, brightness (or tone), texture, contrast ratio, and spatial resolution.

Spatial Resolution refers to the fineness of detail depicted in an image. It describes the minimum size of objects on the ground that can be separately distinguished or measured. Transforming this apparently straight-forward concept into an operational quantitative measure has proven far from simple. There is no satisfactory measure of spatial resolution available, and it turns out to be a much more complex term than our initial intuitive definition suggests. It has been suggested by Forshaw et al. (1980) that definitions of spatial resolution can be assigned to one of the following four categories:

1) geometrical properties of the imaging system;
2) the ability to distinguish between point targets;
3) the ability to measure the periodicity of repetitive targets; and
4) the ability to measure the spectral properties of small finite targets (Townshend, 1980).

Spatial resolution is the technical resolving power for effective detection and analysis (Lintz and Simonett, 1976) and is the resolving power needed for the discrimination of observed spatial features. It is determined by the ground-projected instantaneous-field-of-view (IFOV) of the sensor. Size is of great importance in obtaining accurate values for the radiance of agricultural areas. For example, in Canada, the USA and the USSR, field sizes are such that the Landsat ground-projected IFOV size of about 80- × 80-m has proved marginally adequate for vegetation-discrimination studies. A study by Harmage and Landgrebe (1975) has shown that identification and analysis become acceptable when fields are greater than 60 IFOVs in size (about 8 IFOVs along each side) (Slater, 1980).

Radiometric resolution is the sensitivity to differences in signal strength. It is strongly related to the identification of scene objects; the finer the resolution the greater the opportunity to discriminate between objects commensurate with a given spatial resolution (Lintz and Simonett, 1976). It is the resolving power of an instrument needed for dividing the total range (from black to white) of the signal output (Slater, 1980). Preferably, the incrementing of the output is a reasonable number of just-discriminable levels. The increment is thus a function of the uncertainties introduced by the noisiness of the environment, the sensor, and the signal encoding method. The greater the number of discriminable gray steps—or resolved signal levels—the more radiometrically sensitive are the data.

Spectral resolution pertains to the width of the regions in the electromagnetic spectrum that are sensed and the number of channels used (Lintz and Simonett, 1976). It entails sampling the spatially-segmented image in different spectral intervals. Note, too, that spatial resolution can modify the spectral signature; thus spatial resolution provides a form of spectral filtering (Slater, 1980).

Temporal resolution is the length of the time interval between successive environment measurements. Very few aspects of nature do not change in relation to others or themselves in the course of time. Thus, in sensing a number of dynamic events, time is often a key discriminant (Lintz and Simonett, 1976). For example, a useful method for discriminating and identifying, different crop types and other seasonally changing features employs imagery collected at optimal intervals during the year. The same kind of periodic study, or change detection, is also of considerable value in the assessment of natural disasters (Slater, 1980).

Two aspects of temporal sampling should be emphasized here. One aspect is the time window that allows the capture of sufficient energy to activate the sensors. The dwell time (or integration time) has to be short enough to minimize image blur due to object-sensor relative motions. Temporal resolution is more generally defined in the context of the time between samples in the sensing of a time-changing signature. The time-sampling rate influences the degree to which the temporal signature is represented. The Nyquist limit requires a minimum of two samples per repetitive cycle.

There is an interesting parallel in all four of the measurement dimensions. In each we are concerned about the variability of the environment within the sample window. Additionally, we desire appropriate dimensional spacing of samples in the direction of an increase or decrease of the dimensional unit such that significant environmental changes are retained. In each case we find the Nyquist criterion setting the upper limit. We usually find, however, that the samples are not uniformly collected in a given dimension. Spectral samples (e.g. in thematic mapper) usually are positioned in critical spectral regions rather than as uniform, relatively narrow, spectral samples. Temporal samples may be collected at key times in some season or other cycle.

We previously noted that Slater (1980) added two other measurement dimensions to the ways we can get information, namely:

1) Scene identification—the highest-resolution spatial data. It is the training or ground truth data.
2) Atmospheric measurements—needed since the presence of the atmosphere creates difficulties in the remote sensing of the earth. The effect of the atmosphere is to reduce the contrast of the scene and thus make fine detail invisible or at least harder to detect. This effect further confuses the interpretation of spectral signatures of the scene (Slater, 1980). Since pattern recognition is the identification of a given pattern of data within a mass of extraneous signals or background, the contrast ratio of the pattern to the background is important.

Note that these two sets of measurements are external to the image *per se*. As for the image itself, we now turn our attention to the impact that the basic remote sensing measurements, and their attendant resolutions, make upon the elements of interpretation that are involved in human and computer-assisted situations.

THE ELEMENTS OF HUMAN- AND COMPUTER-ASSISTED INTERPRETATION

The basic image stimuli or elements used in both human- and computer-assisted interpretation are given in Table 24-4. These stimuli are ordered in the general level of increasing complexity for computer interpretation. As seen previously in this chapter the importance of tone and color cannot be overemphasized. Later sections detail the computer methods based on these elements. Particular arrangements of these produce higher order elements for interpretation. Figure 24-45 depicts some of this interrelation. Note the progression in element complexity related to increase in image space. Individual pixels of grey tone (single channel) or color (multichannel) are the major source of interpretive stimuli for

TABLE 24-4
Stimuli for Image Interpretation

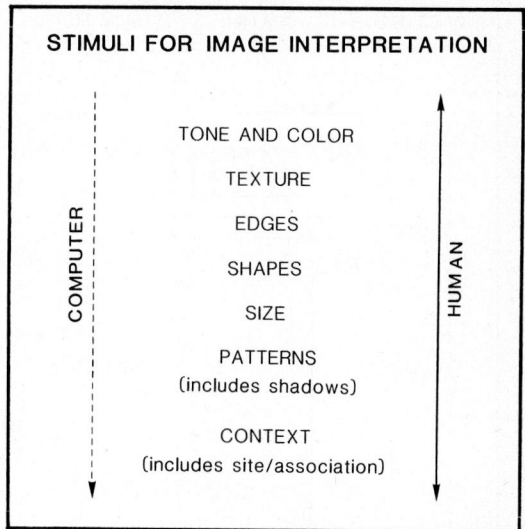

many different computer algorithms. In contrast, human interpretation may simultaneously and progressively use or weigh all the stimuli.

The elements of interpretation are both profoundly and subtly affected by the resolution of the remote sensing measurements. A subtle relationship exists between tone and texture that depends very much upon the resolution with which an image is viewed. For example, at low and high resolution the dominant feature tends to be tone and at intermediate resolutions texture is often the dominant feature (Modestino et al., 1981). This has important implications on the impact of high-resolution data and conventional computer-assisted interpretation. Richason (1978) notes that some of the elements of object recognition have little meaning when a person is interpreting a small-scale scene from the Landsat multispectral scanners: shapes of gross features on the land can be observed, but shape is not as an important criterion for identification in Landsat scenes as it is for aerial photographs. On the other hand, shape may be important in the identification of features that are too large to be seen on a single aerial frame. Pattern, texture, tone and association appear to be the major criteria for object recognition on small-scale Landsat imagery. Variations in tone and color provide a primary basis and texture and association are valuable supplementary criteria. Improvements in spatial resolving power will make automated classification of land cover feasible from satellites in many parts of the earth, where currently this is impossible because of the small size of the land-cover units. But we must also recognize that higher resolution may not improve classification accuracy in all terrain types.

There are significant consequences of finer resolution on the accuracy of both computer and human interpretations. Leachtenauer (1977) showed the effects of scale and resolution on the

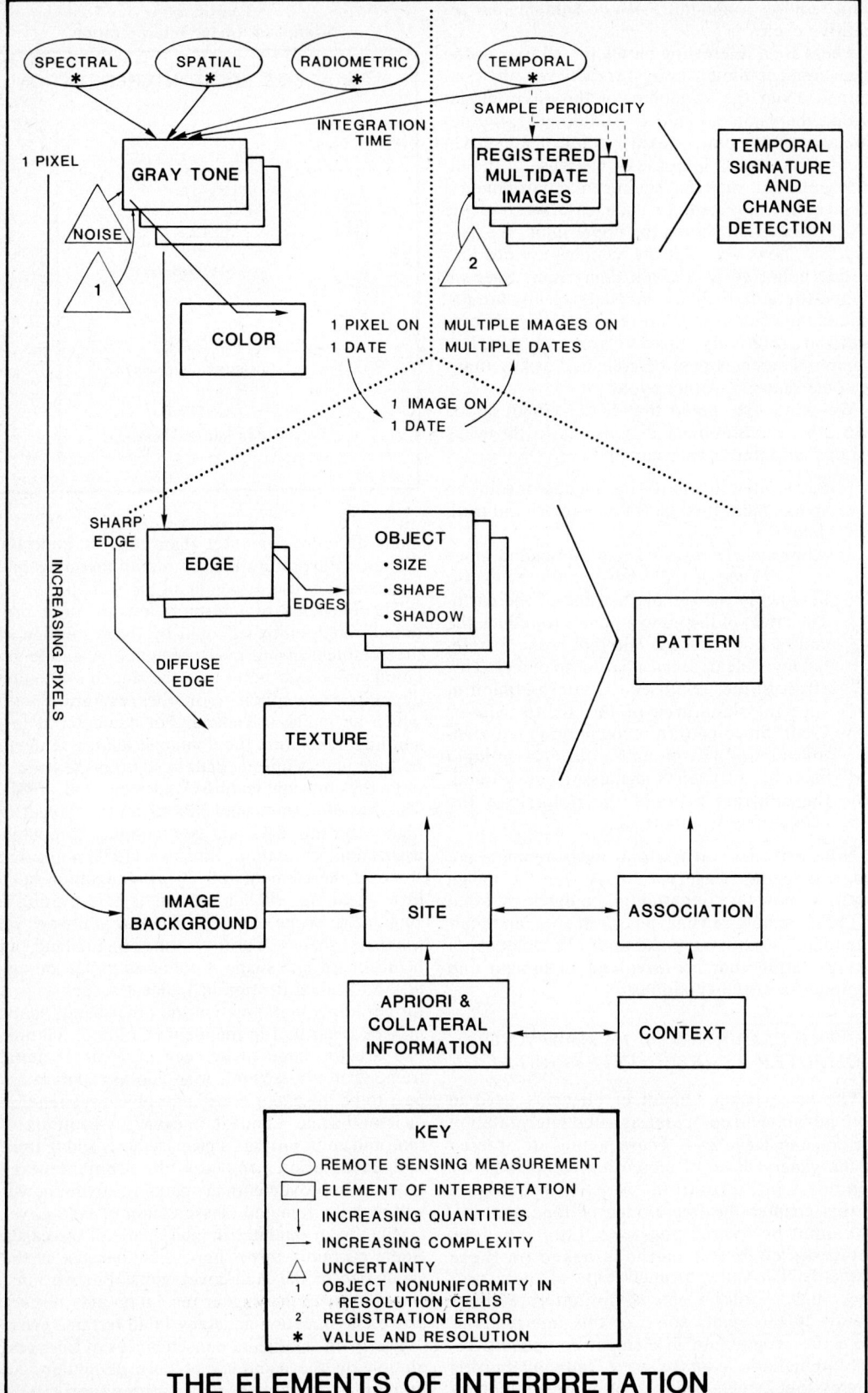

Fig. 24-45. The elements of interpretation and their interrelation as the number of pixels comprising them increases.

accuracy of land-use classifications with optical spectrum measurements. In some cases increased spatial resolution may decrease overall computer-classification accuracy. For example an area of woodland, when viewed at high resolution, will normally be found to be far from spectrally homogeneous due to the presence of scene noise. Automated classification of either woodland or suburban residential areas may lead to errors if their separate component heterogeneities are resolved (Townshend, 1980).

It may be advisable to degrade higher resolution imagery for some tasks. It may be particularly appropriate if classification is based on simple per-point classification (i.e., using only the spectral information from surrounding areas). However, the extra information contained within higher resolution images can be exploited by using measures of image texture or by using contextual algorithms that classify a pixel using its relationship to previously classified pixels (Townshend, 1980).

The simple view of the various image stimuli used in interpretation as presented in Table 24-4 and Figure 24-45 can be focused somewhat by the results of perception experiments and by explanations of human perception in terms of computer-like processes. "Deceptions of the senses are the truths of perception," said nineteenth century physiologist Johannes Purkinje. Held (1974) clarified it by noting that "illusions call our attention to the workings of the visual system, whereas normal perception fails to do so". This has been exemplified in many image interpretation instructional classes by the scene in Figure 24-46. The marginal data available fools the high level human feature extractors relying on the interpretation elements of shape, shadow, and context. Only the low-level extractors report patches of bounded light and dark. The significance of *a priori* information is shown by the fact that when the context of cow is given, it is difficult not to see the cow. Initially, most interpreters

would view this image in the misleading context of a vertical earth-view perspective. Two aspects are thus demonstrated; the eyes and the brain classify and select and are able to accept a limited number of bits of information, which they instinctively supplement with the missing details to form an entity. The computer, however, with no way of supplying missing detail, struggles for an interpretation from the light and dark patches.

The visual mechanism also functions uniquely in the opposite sense; in spite of the great multitude of stimuli impinging on the visual system, it is capable of differentiating precisely between real signals and interference via a complicated filter mechanism (Weber, 1980). In computer-assisted interpretation these functions are crudely emulated by the use of collateral data.

Kent (1981) noted the fundamental similarity between the human brain and the computer. A generalized case of the brain's basic approach to processing input stimuli is shown in Figure 24-47 and considers a multilevel ensemble of feature extractors. Properties of these high-level extractors include a functioning independent of feature position and orientation; implemention by learning; and not being limited to particular spatial forms. Other results of human interpretation experiments indicate:

1) tone: Estimates are that 20 to 30 shades of gray can be perceived (Moik, 1980).
2) texture: Random dot patterns generated by computer show that the recognition of familiar shapes is not needed for the discrimination of textures or even for the binocular perception of depth (Julesz, 1965).
3) High spatial frequencies are more efficiently analyzed than lower frequencies. Below a frequency of about three cycles per degree of visual angle the brain is an inefficient detector.
4) The spatial-frequency mechanism is particularly important in the analysis of fine details that involve high spatial frequencies such as sharp edges.
5) The geometrically based feature-extraction process is more important in the analysis of broad shapes and outlines. (Kent, 1981)

Though we have no notion of the real number or nature of the highest order feature-extractors in the human visual system, some properties are found by fatiguing the extractors through prolonged exposure to different types of stimuli and looking at the effects on our visual abilities. In one analogy the human brain is considered capable of analyzing images in terms of their Fourier (image spatial frequency) components with spatial frequency feature-detectors of undetermined frequency widths and relative sensitivity. This possible conversion or transformation for interpretation is kept distinct from what are considered as the more conventional geometrical feature extractors. Thus, in Chapter 17, a typical line of the

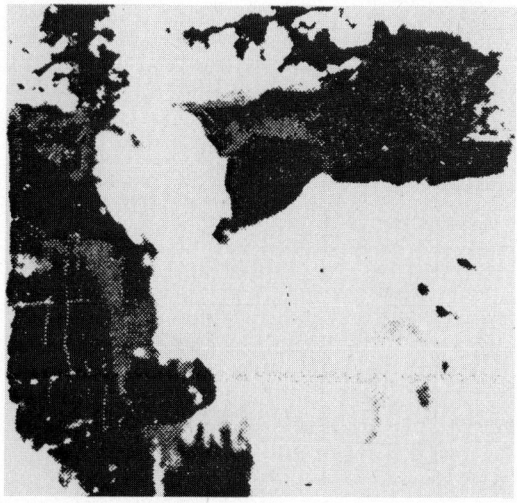

Fig. 24-46. A test image (Kent, 1981).

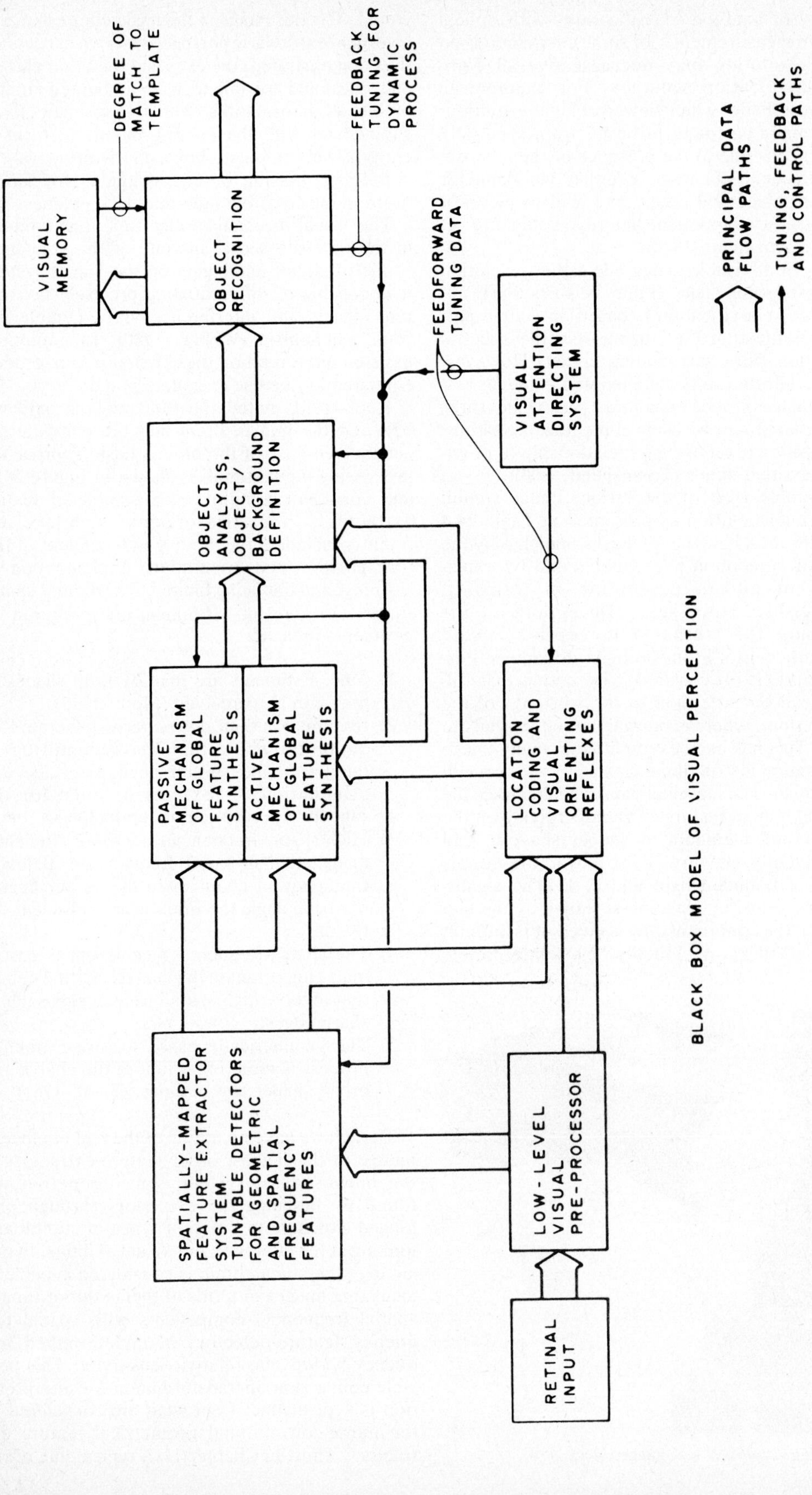

PRINCIPAL DATA
FLOW PATHS

TUNING, FEEDBACK
AND CONTROL PATHS

BLACK BOX MODEL OF VISUAL PERCEPTION

Fig. 24-47. Model of visual perception (Kent, 1981).

rural scene appearing in the top of Figure 17-8, when transformed to the frequency domain of relative amplitude and spatial frequency may look like the bottom of Figure 17-8.

Again, for automated interpretation, analogies to the human process have been used. Table 24-5 identifies some image-element computer-method sets. Casasent (1979) reviewed pattern-recognition systems, techniques, and applications. Computer implementation at the more advanced level of interpretation often seeks to emulate such optical or human processes.

For example, several researchers have suggested that the human brain and eye function as a Mellin transform which enables people to recognize objects independent of their scale and rotational orientation. The Mellin transform is a combination of a Fourier transform and a logarithmic transform which converts the distance from a central point into a displacement to make all similarly shaped objects identical in size. Similarly, another transform makes the identification of objects independent of rotation (Casasent, 1981).

Researchers envision that artificial intelligence and optical data-processing will soon be found to share some common algorithms. Optical pattern-recognition systems that use optical interference to process incoming signals have long attracted attention because of their highly parallel operation and high speed. Some computers now have serial pipelines of processing stages where each stage in the pipeline performs a single transformation on an entire image (Sternberg, 1980). Certain real-time feature-recognition operations now use hybrid optical-digital systems. Regions of the input image are sequentially scanned. The Fourier transform of each region is focused on a wedging-ring detector that determines the amplitudes in frequency-direction cells or resolution elements.

THE EVOLUTION AND STATUS OF DIGITAL IMAGE PROCESSING

Digital image processing in remote sensing historically stems from two principal application areas:

- improvement of image information for human interpretation, and
- processing of scene data for computer-assisted interpretation.

Here we identify all image interpretation as computer-assisted whenever digital image processing is involved. This is because the manipulation of a digital image by computer has been performed either to prepare an image for display and human interpretation and/or to extract information from the image with computer interpretation. Any one or all of three major image-processing categories can be involved. These are restoration, enhancement, and interpretation. Figure 24-41 identifies the processing flow of this basic set. There is considerable diversity on the use of terms. (See Chap. 17 for details.)

Image restoration requires significant human inputs, primarily specific *a priori* information to correct degradations. Enhancements use general knowledge of useful processes to improve a wide

TABLE 24-5
Computer Interpretation Techniques

COMPUTER INTERPRETATION TECHNIQUES		
IMAGE INTERPRETATION ELEMENTS	DOMAIN	EXAMPLES OF COMPUTER INTERPRETATION TECHNIQUES
TONE	SPECTRAL	DENSITY SLICING
COLOR	SPECTRAL	MULTISPECTRAL CLASSIFICATION
TEXTURE	SPECTRAL/SPATIAL	TEXTURE CLASSIFICATION
PATTERN	SPECTRAL/SPATIAL	SPATIAL TRANSFORMS AND CLASSIFICATION
SIZE	SPATIAL	SEGMENTATION ALGORITHMS AND SIZE FEATURE CLASSIFICATION
SHAPE	SPATIAL	SYNTACTIC CLASSIFICATION
SITE	SPATIAL	A PRIORI'S MODIFIED
ASSOCIATION	SPATIAL	CONTEXTUAL CLASSIFICATION SYNTACTIC CLASSIFICATION

class of images. There is no simple dichotomy in image enhancement-restoration of the degree of human involvement. More significantly, a dichotomy of human interpretation vs. computer interpretation is not typical in contemporary practice. This is because the degree and impact of human involvement in computer interpretation are variable. Thus one might generally consider a wide range in the relative amount of human involvement. At the more automated extreme is unsupervised classification; that is, the aggregating or clustering of similar number-sets. Subsequent labeling and subjectively categorizing the spatial clusters on an image represent increased human participation. In supervised classification *a priori* human association of categories and selected image data for computer training increases this degree of human participation.

Experience has shown that computer-assisted interpretation can be superior to manual photooptical and electro-optical techniques for specific applications. This is because of the processing flexibility, the ability to quantize and control the performance of any image manipulation, the quality and reproducibility of results, and the integration of other digital data sets using computer techniques.

It was perhaps quite natural that digital image improvements for human interpretation should be the precursor to computer interpretation. One of the first applications of a digital image processing technique was to an image sent via submarine cable in the early 1920's (Gonzalez and Wintz, 1977). However, such processing used in a routine fashion has only recently become practical and common because of:

- sensor technology changes (solid-state detectors with digital readout),
- computers that are now faster, less expensive and smaller, and
- appearance of many new algorithms and techniques to convert data to information. (Bernstein, 1978)

Work on using computer techniques for improving the interpretability of images from space probes began at the Jet Propulsion Laboratory (Pasadena, California) in 1964. Vigorous growth followed. Diversity and impetus came concurrently in the field of medicine. Preston (1980) estimated tens of millions of medical images were digitally processed in a recent year by just one facility. In education, one university invoked image processing procedures over 40,000 times in a recent year. In a separate domain lies the sophisticated, extensive (and classified) military use of image processing and interpretation.

Landgrebe (1981) summarized in the following words the growth and impact of the U.S. Landsat program as one of the great stimuli for computer-assisted interpretation:

The nation's earth observational program is one of the largest sources of quantitative data which can be presented in image form. The ability to place earth–oriented sensors into orbit, because of its large economic and humanitarian potential, has received considerable research emphasis, and operational systems are now evolving. The research effort which led to these operational systems has had some unique characteristics. Perhaps, most notably, it has been a highly innovative field which was, from the beginning, user-driven. The result has been a new technology for dealing with large volume, image-like data.

A diversity of computer image-interpretation systems, languages, common procedures and image data formats now exist. Preston (1980) estimated that at least 50 image-manipulation languages have been developed in the United States during the past twenty years with the expenditure of at least one man-year per language.[5] Some systems (e.g., VICAR, LARSYS, ERIPS, IDAMS, ASTEP, and MSFC) were keyed initially to space photography and imaging missions. Thus, geometric manipulation and image-correction techniques were emphasized. Others (e.g. LARSYS, ERIPS, IDAMS AND MSFC) emphasized the classification of images according to their multispectral signature. These systems were generated by NASA in the 1960's. Similarly, the U.S. Department of Defense sponsored the development of systems (e.g., KANDIDATS, DIMES, PECOS, SMIPS) for analyzing image statistics and image-correlation techniques. More detailed examples of the programs of four large-scale systems (ORSER, LARSYS, ERIPS, and VICAR) and five mini-computer systems (IDIMS, KANDIDATS, GIPSY, MINIVICAR, and AOIPS) are described in Chapter 19 along with the general requirements of remote sensing software. Although a complete list pertaining to the distribution of users and program types is not available, VICAR is probably now the most extensively used image-processing system.

Perhaps an answer to the proliferation of systems and the diversity of operating procedures is the Transportable Applications Executive (TAE) recently developed under NASA support (NASA Goddard Space Flight Center, 1981). It is a collection of executive programs that interact with the user to manage many different image-application programs. The purposes include:

- standardization of the user interface to applications programs (it currently supports VICAR (Video Image Communication Analysis and Retrieval) commands),
- shielding of the user from the host operating system and,
- a user-friendly environment.

The intent is flexible accommodation to users ranging from the first-time beginner to the most

---

[5] This estimate seems low since a separate estimate for VICAR alone is hundreds of man years.

experienced. The trend in education-directed image-processing systems similarly provides a range of help functions. Some are described in Jensen et al. (1979), Tinney and Ennerson (1981), and Williams et al. (1981). Preferably, system responses accommodate a spectrum of user inputs and range from providing complete menus of operations, to requests for missing executive commands, or to execution-only from simple, complete commands.

Finally, we note the expanding industry available to do interpretation. Public Technology, Inc. (1981) compiled an industry directory that identified 143 firms providing remote sensing products, services and/or equipments. At least 75 of these provide digital image-interpretation services.

Image Restoration and Enhancement

Remotely sensed images are always degraded to some extent because of atmospheric effects and the characteristics of the sensing and recording systems. Since the accuracy of interpretations that are made of degraded images can acutely depend on the efforts to improve such images, the nature of the degradation, restoration and enhancement processes is briefly described here. A more complete description is found in Chapter 17.

Image degradations may be grouped into radiometric and geometric distortions. Radiometric degradations arise from blurring effects of the imaging system, nonlinear amplitude response, vignetting and shading, transmission noise, atmospheric interference (e.g., scattering, attenuation, and haze), variable surface illumination (e.g., differences in terrain slope and aspect), and changes of terrain radiance with viewing angle. Geometric distortions can be categorized into sensor-related distortions such as aberrations in the optical system, nonlinearities and noise in the scan-deflection system, sensor-platform-related distortions caused by changes in attitude and altitude of the sensor; and object-related distortions caused by earth rotation, curvature and terrain relief. Perspective distortions depend on the type of sensor (Moik, 1980).

It is important to distinguish the differences between image restoration and image enhancement. Image restoration reverses irreversible processes using *a priori* knowledge of the nature of each image degradation. Image enhancement provides for subjective improvement of images for human interpretation. It does not operate in the context of *a priori* knowledge other than general knowledge of preferred improvements for the visual interpretive process. Both restoration and enhancement usually precede human interpretation; usually only restoration and transforms precede computer interpretation. Some differences in the use of these terms exist: some image enhancements are occasionally grouped in a restoration category and vice versa. This may stem from a literal interpretation of the term enhance:

to make or become greater in value or desireability (Merriam-Webster, 1960). It is best, however, to keep these two aspects distinct. Indeed some enhancements (e.g., contrast stretching) that are quite useful for human interpretation can be potentially disastrous for computer-assisted interpretation unless the training data have been identically modified (Landgrebe, 1981).

Image Restoration

Image restoration is concerned with the correction of distortion, degradations, and noise induced in the imaging process. The problem of image restoration is to produce, from the degraded recorded image, a corrected image that is as close as possible, both geometrically and radiometrically, to the original object radiant energy distribution (Moik, 1980). For some users, a specific subcategory of radiometric restoration is called preprocessing. The purpose of such restoration efforts is to remove atmospheric effects of attenuation, scattering, emission, and the degradations due to temperature effects on detector response, sensor-dependent errors and coherent noise (Moik, 1980). Not all digital image processing consistently includes the same types of restorations within the preprocessing category.

Radiometric Degradation and Restoration

Radiometric distortions cause point (pixel) and spatial image degradations. Pixel degradations occur in some imagery systems when object brightness is not uniformly mapped onto the image plane. For example, vignetting is the gradual attenuation of those light rays progressively off the optical axis of the imaging system. Non-uniform brightness in an imaging vidicon is shading (Moik, 1980). The response of the system to a point light-source is described by the point spread-function. Radiometric spatial degradations are caused by defocussing, diffraction effects, and by atmospheric turbulence and relative motion between the imaging system and the object.

Radiometric degradations caused by coherent noise are amenable to restoration. The essence of coherent noise removal is to isolate and remove the identifiable and characterizable noise components in a manner that does a minimum of damage to the actual image data. The main types of coherent noise appearing in images are periodic, striping, and spike noises (Moik, 1980). Random-noise image-degradations are assumed to be homogeneous, random processes uncorrelated with the signal, and are not amenable to restoration. Only generic forms of image enhancement (i.e. forms that use local pixel-neighborhood averaging to reduce the noise with some sacrifice of image edge details) are applicable.

Geometric Restoration

Geometric transformations are used to correct for geometric distortions. They are also used to effect overlay of images on other images (image registration); correct aspect ratio; and to rotate, skew, and flip images (Moik, 1980). Two important transformations are (1) to relate image coordinates to a geodetic coordinate system (rectification) and (2) to relate the coordinate systems of two images (registration).

The attitude information provided by some satellite instrumentation is generally not accurate enough to meet the precision requirements for geometric correction. For example, to obtain a precision of one picture element for Landsat MSS, each attitude component should be known to 0.1 μrad. The attitude-measurement system, however, was only accurate to 1 μrad. The SMS/ GOES VISSR requirements for a precision of one visible picture-element required attitude accuracy of 25 μrad.

A precise estimate of attitude during imaging can be obtained by using ground control points (GCP). Ground control points are recognizable geographic features in the image whose actual locations can be measured in maps. Thus, control points relate an image with the mapped object.

In geometric restoration there are differences in the preferred sequence of operations. Image registration is usually inherent in a single-date multichannel data set; in multidate data it is not. Image registration can be effected in anyone of several ways:

- geometric correction (rectification) of each set to a common geographic coordinate system;
- geometric correction of one date and registration of other dates to it; and
- geometric registration of all dates to one reference-date, with geometric correction before or after image interpretation.

Some prefer to register, classify and then geometrically correct the image. The prime reason given is that minimal tampering with the original data improves classification accuracy. Others restore images to improve training-field position-selection accuracy. In the geometric rectification process there are three different ways to define the new pixel gray values: nearest neighbor, linear interpolation and cubic convolution. These progressively incorporate expanding pixel neighborhoods from the original image to define each newly-positioned pixel tone.

Image Enhancement

The geometric and radiometric restorations just described improve the subsequent interpretation, whether by human or computer. Image restorations are usually followed by image enhancements for human interpretation. The goal of image enhancement is to aid the human analyst in the extraction and interpretation of pictorial information. Enhancement is achieved by the articulation of features or patterns of interest within an image and by a display that is adapted to the properties of the human visual system.

Image enhancement implies a goal of improvement in image quality. Because image quality is a subjective measure, there are no simple rules to produce a single best image. There are, however, general functions (such as a contrast stretch) that can significantly help a diversity of human interpreters. An enhanced image may actually contain less information than the raw image, although this is not usually the case. It is always true, however, that for an image to be enhanced properly the information of greatest interest to the user must be optimally displayed, even at the expense of less valued information (Siegel and Gillespie, 1980).

Contrast Enhancement

The general goal of contrast enhancement is to produce an image that optimally uses the full dynamic range of the display device. Attainment of this goal depends on the subsequent image use and user. When an image is processed for visual interpretation, the viewer is the ultimate judge of how well a particular method works. When an image is to be interpreted by a computer, the image is not contrast-stretched. The contrast enhancement required for the image is determined by first obtaining a histogram of the gray-level distribution in the image. This is the number of pixels lying within specific gray intervals. The contrast stretch is then a pixel-by-pixel gray change. Often the stretch is matched to a newly defined function. Two such functions are the uniform and Gaussian output distributions.

A uniform distribution results in the greatest contrast-enhancement being applied to the most populated range of brightness values. It is thus particularly useful as a quick-look method for evaluating the results of a previous image-processing step. It is sometimes too harsh and can result in a loss of details in the bright and dark extremes. Such detail is usually preserved by using the Gaussian distribution (Soha et al., 1976). Other options for the input-output gray relationship are the linear stretch (with or without a selected percent saturation at the black and white extremes) and a table stretch for discrete piecewise specification.

Filtering

Filtering may be accomplished in either the image spatial domain or in the image frequency-transform domain. The latter is highly analogous to some models of human perception. There is a trade-off between image sharpness and noise in a filtering operation. Spatial or convolution filters are defined in terms of their size (i.e., the number of lines and samples that overlay a progressively indexed portion of the image being filtered). Another filtering dimension is the multiplicative weights of each filter element overlaying each pixel. In operation each pixel tone or gray value is multiplied by the filter weight. The resulting products are then summed and the sum is divided by the sum of the weights in the filter. This represents the filter-output image-pixel gray value for the center of the spatial filter. Usually the larger the spatial dimensions of the filter, the more emphasis there is on the slowly changing (low spatial frequency) components. Conversely, a dimensionally small spatial filter mainly emphasizes edges and depends on filter weight values. Spatial filtering insight might be gained by considering two filter-size extremes: a filter the size of one pixel and a filter the size of the image being filtered. If the filter weights are one, the filtered image is unchanged in the first case and becomes a uniform gray-tone average of the image in the second case. Often the image is mirrored at the edges to avoid edge effects in filtering. A surrounding border would otherwise produce spurious artifacts in the filtered image.

Schreiber (1978) commented on the subjective enhancement effects of image filtering and tonal-scale manipulation: effective image processing requires knowledge of both the observer and the scene. Most observers, given the choice, will increase the high-frequency contrast of a picture to the point where some defects or artifacts appear. A somewhat surprising result of edge-sharpening experiments is that the perceived overall contrast of an image is increased when the intensity of spatial high frequncies is increased.

The general goal is to improve sharpness (by spatial high-frequency enhancement) while at the same time increasing the visibility of detail in both light and dark areas (by de-emphasis of low-spatial frequencies). Such separate treatment is done with homomorphic filtering. In homomorphic filtering the logarithm of the image brightness is first taken. Since the signal is the product of illuminance and reflectance, the filtering is linearly (and separately) applied to the log-luminance and log-reflectance.

Linear-filtering operations, such as those just discussed, tend to blur region boundaries and reduce the resolution. A nonlinear filter that does not blur boundaries is the mode filter. It computes a local spatial-area histogram centered on each pixel and outputs the most frequently occurring value in that small region as the replacement gray value for the central pixel. This removes small variations in brightness and creates relatively large regions of completely uniform character with almost no loss in boundary resolutions (Coleman and Andrews, 1979). Another nonlinear filter, the median filter, selects the median value as the new pixel gray value. It is used mainly for outlier rejection and noise filtering.

Edge Enhancement

The subtle brightness variations that define the edges and texture of objects are important for subjective interpretation. An enhancement of local contrast may be achieved by suppressing slow brightness variations that tend to obscure the interesting details of an image. Again, slow brightness variations are composed of low spatial frequencies, whereas fine details (i.e., edges) are represented by higher spatial frequencies. Edge detection is an image-segmentation method based on the discontinuity of gray levels or texture at the boundary between different objects (Moik, 1980). Threshold selection is a key problem in edge detection in noisy images. Too high a threshold does not permit detection of subtle, low amplitude edges. Conversely, a low threshold (of detection) causes noise to be detected as edges.

An important application of edge detection is in image registration where edge images are used for binary correlation (e.g., co-occurrence of the 1's that are the edges). Experiments have shown that for remotely sensed earth-resources images, an edge density (i.e., fraction of image-brightness change that is made an edge) of 15 percent is appropriate.

Image Transforms

Multispectral and multitemporal images usually convey more information than single spectral channel (monochrome) images. Image transforms effect a reduction in data-set size while retaining the essence of the extractable information. Such transforms find much utility in the preparation of color images. The brightness of each primary color represents the loading/transformation, that is, the fractional part of the transformed spectra from the original spectral measurements. Human interpretation of such imagery (e.g., in geology) is a common, useful procedure. Multi-image enhancements include independent contrast-enhancement of each image or linear and non-linear combinations of the images. The linear combinations include ratioing, differencing, summing, and principal-components analysis. These require registered images.

Ratioing—ratioed images suppress brightness variations due to topographic relief and enhance subtle spectral variations. Ratioing may concurrently increase random or coherent noise or atmospheric effects. Therefore, coherent noise (e.g., striping) is removed before enhancement.

Differencing—two or more images may be subtracted to emphasize spectral and/or temporal changes. Subsequent rescaling (contrast stretch) is performed to make the range of the difference or ratio images compatible with the dynamic range of the display device (Moik, 1980).

Supervised Multispectral Image Interpretation

Computer-assisted image interpretation through pattern recognition is the sorting or classifying of an image into pixel classes or categories using the remote sensing measurements. Classically, three points of view have been taken toward pattern recognition problems:

- pattern recognition is a statistical problem and thus statistical decision-theory guides the design;
- pattern recognition is not a rigorously statistical problem; only quasi-statistical reasoning should be used; and
- the similarity between pattern-recognition computer functions and certain human functions can be exploited by using models of human neural nets as guides for design (Meetham and Hudson, 1969).

Thus far, the statistical and quasi-statistical approaches have dominated computer-assisted interpretation. Prior to image classification, many algorithms were proposed in the engineering and statistical literature and were used for answering the question of how to associate an unknown point (in feature space) with each information class (Swain and Davis, 1978). Computer-assisted image-processing field in several aspects. Be-the middle 1960s. The motivation for this effort came from the user community. The computer applications required a technology unique in the image-processing field in serveral aspects. Because of the scene complexity and the desired detail of the information, work on processing technology initially centered on the use of spectral characteristics rather than spatial ones. The concept of using a sampling of the spectral distribution of energy for class identification was selected because it appeared (from early low-altitude image studies) that the creation of a successful computer-implemented analysis was more likely than the intuitively preferred spatial (image) characteristics.

The spectral approach has several additional advantages:

- for data collected with a single aperture multispectral device, the features used in the

discriminant process are immediately available (i.e., they do not require additional computation to create; hence the feature-construction step is simple);
- the approach makes very efficient use of spatial resolution. The contents of each pixel are identified individually (as compared, for example, to constructing a multipixel subimage as the basis for identification); and,
- spatial resolution is one of the most expensive sensor attributes to achieve both in terms of cost and complexity of the sensor itself and the data volume resulting from it.

The research effort based on a multispectral approach proceeded with fundamental studies to better understand the multispectral reflectance and emittance characteristics of natural scenes. There were advances in the design and construction of multispectral scanners and the study of data processing and analysis algorithms. The latter included means for calibration, storage, and compression of multispectral data, geometric correction schemes, conversion to standard cartographic coordinate systems, image enhancement techniques, feature selection techniques, and a number of different pattern-classification algorithms (Landgrebe, 1981).

There are three basic steps in a typical spectral pattern recognition procedure in computer-assisted image interpretation. A digital image data set is used consisting of the digital numbers recorded over each point in the scene to be analyzed (See Figure 24-48). Each cell, or pixel, contains a set of digital numbers, one for each spectral band, here, five. The digital-image data set may also be considered as an n-dimensional matrix. The interpretation process consists of distinct stages:

1. The training stage is where the analyst compiles an interpretation key, or signature-set

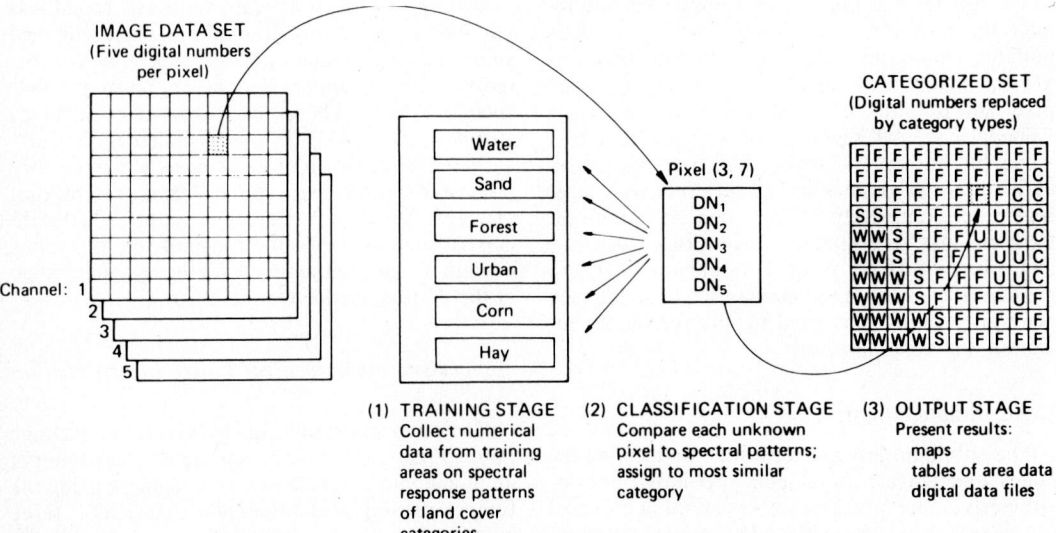

IMAGE DATA SET
(Five digital numbers per pixel)

Channel: 1 2 3 4 5

Water
Sand
Forest
Urban
Corn
Hay

Pixel (3, 7)
DN_1
DN_2
DN_3
DN_4
DN_5

CATEGORIZED SET
(Digital numbers replaced by category types)

(1) TRAINING STAGE
Collect numerical data from training areas on spectral response patterns of land cover categories

(2) CLASSIFICATION STAGE
Compare each unknown pixel to spectral patterns; assign to most similar category

(3) OUTPUT STAGE
Present results:
maps
tables of area data
digital data files

Fig. 24-48. Typical spectral pattern-recognition process (Lillesand and Kiefer, 1979).

analogous to the spectral attributes for each feature of interest. This is generally performed by examining representative sample sites (subimages) of known cover type, called training areas.

2. In the classification or interpretation stage, each pixel in the image data set is compared to each category in the numerical interpretation key. This comparison is made numerically, using any one of a number of different strategies to decide which category an unknown pixel value is most similar to. Each pixel is then labeled with the name of the category it resembles, or is labeled unknown if insufficiently similar to any category. The category label assigned to each pixel in this process is then recorded in the corresponding cell for the interpreted output. Thus, the multidimensional spectral image matrix is used to develop a corresponding map of interpreted category symbols.

3. After the entire data set has been categorized, the results are presented in the output stage commonly in the form of a map. The categorized data may also be used to generate tables of the areas of various cover types in the scene or it may be recorded as computer compatible inputs to a grid-based geographic information system.

The various classification approaches are illustrated with a two-channel (bands 3 and 4) subset of the hypothetical five-channel MSS data set. Rarely are just two channels employed in an analysis, yet this limitation simplifies the graphic portrayal of the various techniques. The techniques are implemented mathematically. Once expressed in mathematical form, the interpretation approaches apply to any number of channels of data. Consider a sample of pixel observations from the two-channel digital image set expressed graphically on a scatter diagram in Figure 24-49. In this diagram, the band 3 digital numbers have been plotted on the y-axis and the band 4 digital numbers on the x-axis. These two digital numbers locate each pixel value in the two-dimensional measurement of spectral response. Assume that the pixel observations are from areas of known cover type (i.e., from selected training sites). Each pixel value has been plotted with a letter indicating the category to which it belongs. Note that the pixels have a natural centralizing tendency.

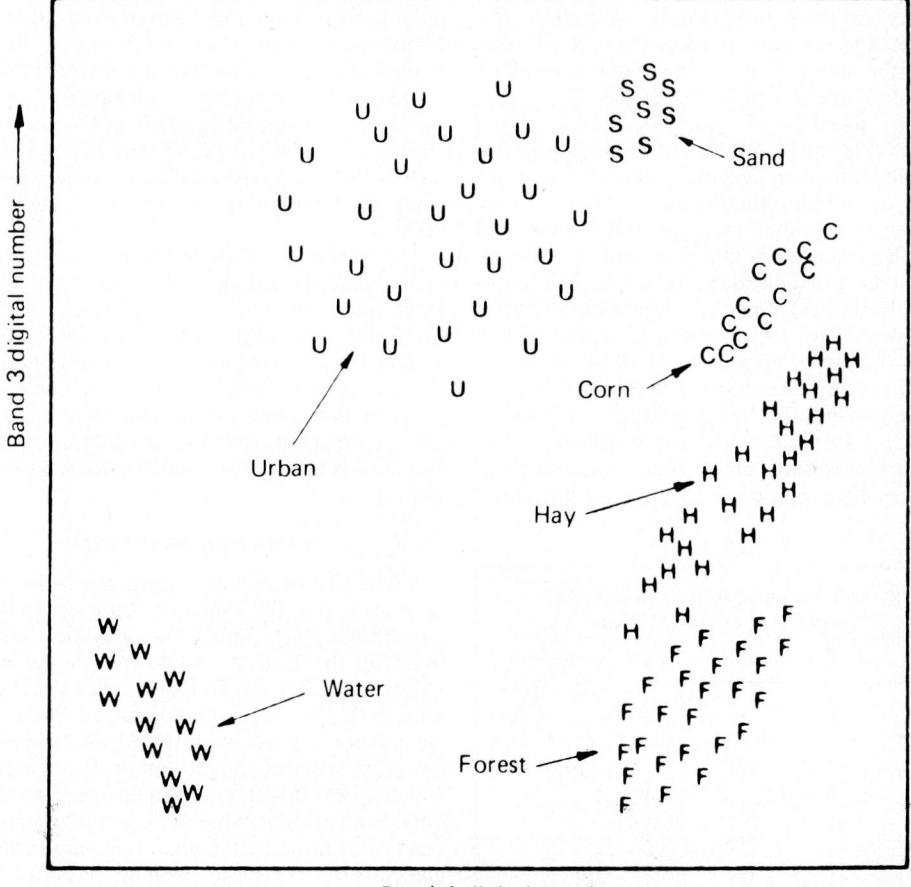

Fig. 24-49. Pixel observations plotted on scatter diagram (Lillesand and Kiefer, 1979).

These clouds of points represent the multispectral response patterns of each category to be interpreted. The classification strategies use these training set descriptions as keys by which all other image pixels are grouped into their appropriate classes. In an ideal remote sensing system this variability is absent and the interpretation key is reduced to the simplest of all: a universal, unique spectral-radiometric combination number. We first consider this ideal asymptotic case.

The Universal Signature Concept

The components of an ideal remote sensing system would be characterized by a uniform energy source, a noninterfering atmosphere, unique energy interactions with the features on the earth's surface, and a unique spectral response for each feature (Lillesand and Kiefer, 1979). Yet each of these system aspects is subject, in its own realm, to what Lintz and Simonett (1976) characterize as geographic variability, which is one of the most pervasive of all problems. Without these pervasive effects and noise the concept of a universal signature bank would become viable. In that view each object category would have a unique profile much like a key to a lock. Consider this simple example: the number of different combinations, or unique signatures, of radiometric levels (R) and spectral channels (S) is R^S. Examples of the number of unique combinations of these values are shown in Table 24-6.

As the number of spectral channels and radiometric levels increases one would expect improved identification specificity due to the larger number of unique combinations. This concept would hold true only if each spectral channel and radiometric increment that was added provided additional, nonredundant information. Conversely, the effects of natural variability and noise can be described, for example, as a fuzzy band perhaps containing 90 percent of all the expected variations. This forces a lower number of discrete radiometric-spectral combinations and hence unique signature combinations. Further, the intrinsic dimensionality of independent spectral channels is usually less (e.g., in Landsat 1-3 it is about

2). Linear conversions of correlated data to uncorrelated axes (such as principal components or canonical analysis) are thus often used to effect such compression of data without information loss prior to feature recognition.

It is interesting to note that in many Landsat 1-3 scenes some 7,000,000 pixels can be represented by about 10,000 unique "signatures". This relates to the view that the information in such scenes can often be compressed into two channels of 100 tone levels each by a linear transform. Thus an object can be given a unique gray-spectral signature compatible with the number of levels of certain radiometric change in each spectral band. When the computer reads that sequential number the associated object label or category is outputted. There is a series of progressively more refined spectral interpretation algorithms that are available to do this. These are summarized below.

Minimum Distance to Means Classifier

Figure 24-50 illustrates one of the simpler classification strategies. The mean, or average, spectral value for each category is determined. The category means are indicated by +'s in the figure. A pixel of unknown identity may be classified by computing the distance between the value of the unknown pixel and the mean of each category. In Figure 24-50, a pixel to be interpreted has been plotted at point 1. The distance between this pixel value and each category mean value is shown by the computed dotted lines. It is assigned to the closest class, in this case corn. If the pixel is farther than an analyst-defined distance from any category mean, it would be interpreted as unknown.

The minimum distance-to-means strategy is mathematically simple and computationally efficient but it has several limitations. In Figure 24-50, the pixel value plotted at point 2 would be assigned to the sand category, in spite of the fact that the greater variability in the urban category suggests that urban would be a more appropriate class assignment. Because of such problems, this classifier is not widely used in remote sensing applications.

Parallelepiped Classifier

Sensitivity to category variance is introduced by considering the range of values in each category training set. This range may be simply defined by the highest and lowest digital number values in each band, and appears as a rectangular area in the two-channel diagram in Figure 24-51. An unknown pixel is classified according to the category spectral ranges, or decision region, in which it lies. It is interpreted as unknown if it lies outside all category ranges. The multidimensional analogs of these rectangular areas are called parallelepipeds. The parallelepiped classifier is very fast and efficient and is employed in several image analysis systems.

TABLE 24-6

NUMBER OF RADIOMETRIC–SPECTRAL SIGNATURE COMBINATIONS		
NUMBER OF CHANNELS	NUMBER OF DISCRETE GRAY/CHANNELS	NUMBER OF UNIQUE SIGNATURES
4	256	4.29×10^9
4	128	2.684×10^8
4	64	1.678×10^7
2	256	6.554×10^4
2	128	1.638×10^4
2	64	4.096×10^3

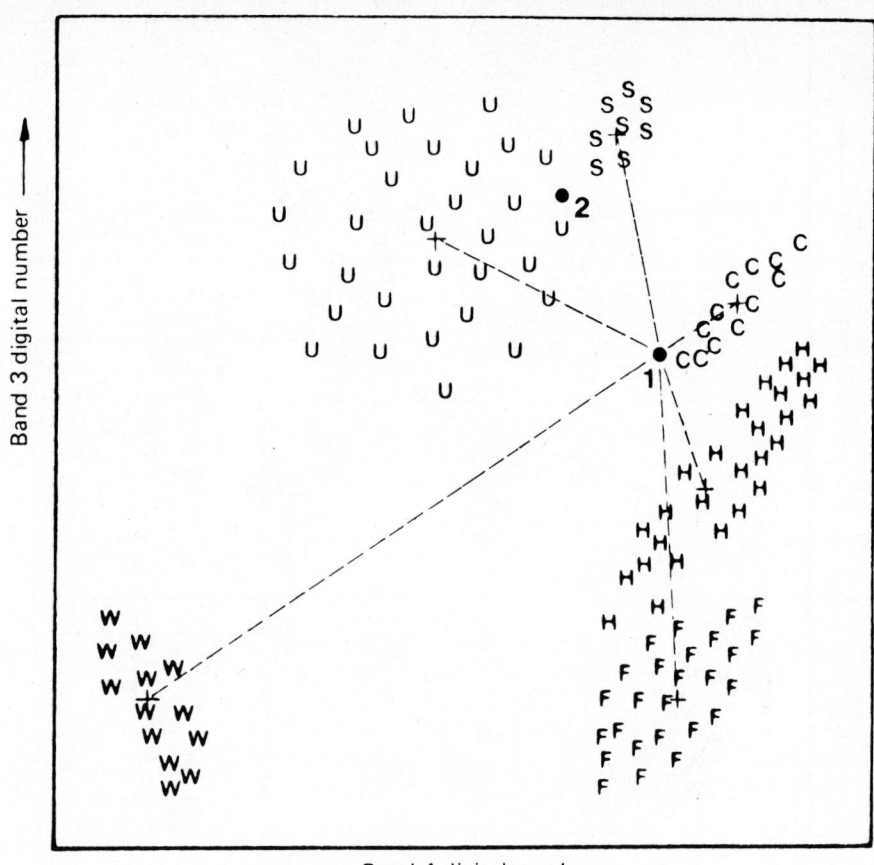

Fig. 24-50. Minimum distance to means classification-strategy (Lillesand and Kiefer, 1979).

The sensitivity of the parallelepiped classifier to category variance is exemplified by the small decision region defined for the tightly clustered sand category than for the more varibale urban class. Because of this, pixel 2 would be appropriately classified as urban. However, difficulties are encountered when category ranges overlap. Unknown pixel observations that occur in the overlap areas will be classified as "not sure" or be arbitrarily placed in one of the two overlapping classes. Overlap is caused largely because category distributions exhibiting correlation are poorly described by the rectangular decision regions. Correlation is the tendency of spectral values to vary similarly in two bands, resulting in elongated, slanted clouds of observations on the scatter diagram. In the presence of correlation, the rectangular decision regions may poorly fit the category training data, resulting in confusion for a parallelepiped classifier.

Maximum Likelihood Classifier

The maximum-likelihood classifier quantitatively evaluates both the variance and correlation of the category spectral-response patterns when classifying an unknown pixel. In order to do this, an assumption is made that the distribution of points of the category training-data is Gaussian (normally distributed). This assumption is generally reasonable for common training class spectral distributions. Hence, it can be completely described by the mean vector and the covariance matrix parameters. The latter describes the variance and the correlation. This is also called a parametric approach. Given these parameters, the statistical probability of a given pixel value being a member of a particular land cover class is computed. Figure 24-52 shows the probability values plotted on a three-dimensional graph. The vertical axis is associated with the probability of a pixel value being a member of one of the classes. The resulting bell-shaped surfaces are called probability density-functions. After evaluating the probability in each category, the pixel would be assigned to the most likely class, or labeled as unknown if the probability values are all below a threshold set by the analyst.

In essence, the maximum-likelihood classifier delineates ellipsoidal equiprobability contours in the diagram. These decision regions are shown in Figure 24-53. The shape of the equiprobability contours expresses the sensitivity of the likelihood classifier to correlation. For example, because of this sensitivity, it can be seen that pixel 1

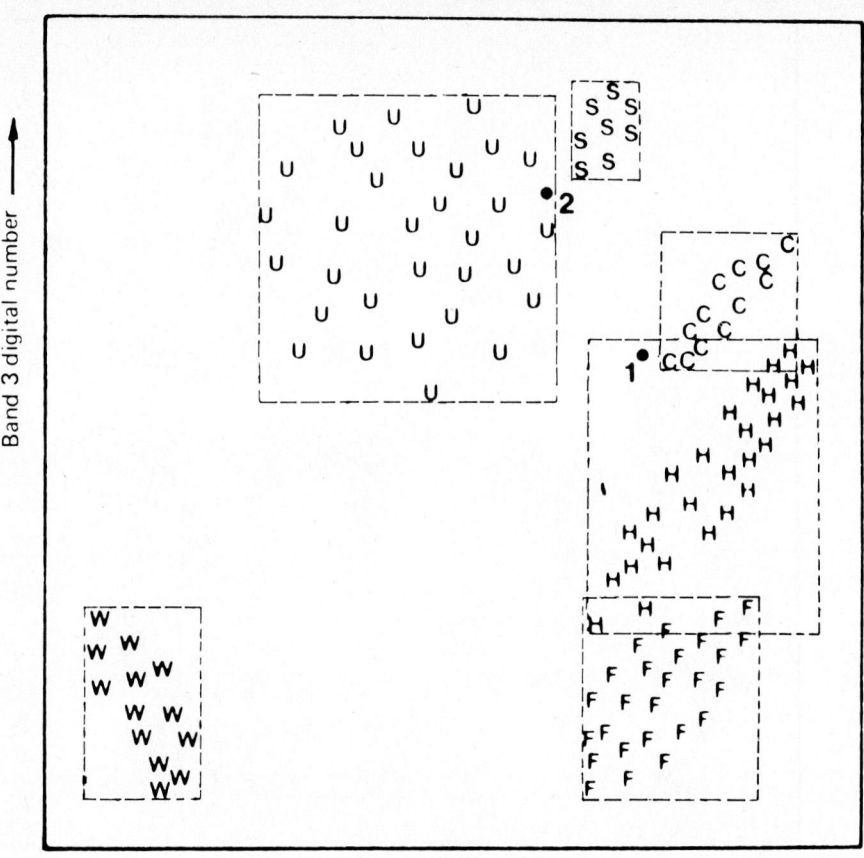

Fig. 24-51. Parallelepiped classification strategy (Lillesand and Kiefer, 1979).

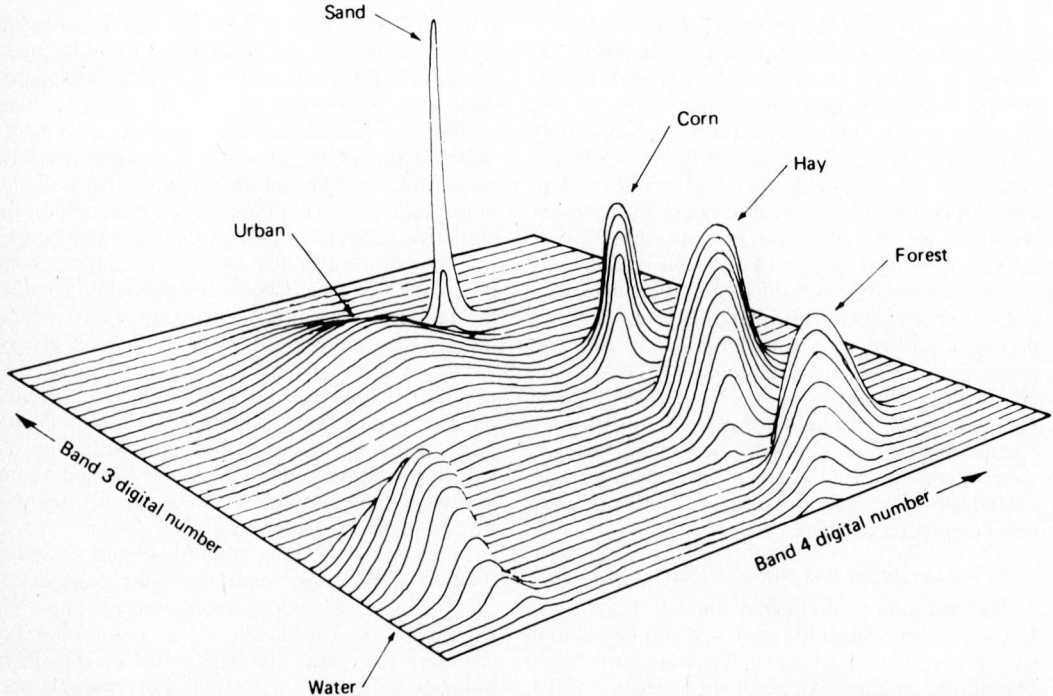

Fig. 24-52. Probability density functions defined by a maximum-likelihood classifier (Lillesand and Kiefer, 1979).

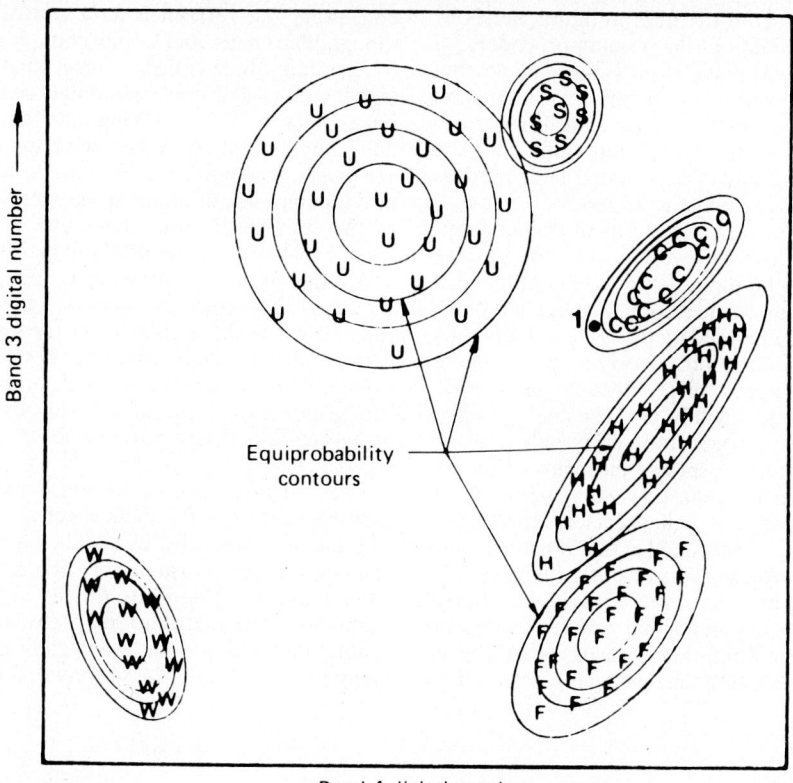

Fig. 24-53. Equiprobability contours defined by a maximum-likelihood classifier (Lillesand and Kiefer, 1979).

would be appropriately assigned to the corn category.

An extension of the maximum-likelihood approach is the Bayesian classifier. This is discussed by Haralick and Fu in Chapter 18. This technique applies two weighting factors to the probability estimate. First, the analyst determines the *a priori* probability or the anticipated likelihood of occurrence for each class in the given scene. For example, when a pixel is being classified, the probability of the rarely occurring sand category might be weighted lightly, and the more likely urban class weighted heavily. Second, a weight associated with the cost of misclassification is applied to each class. Together, these factors act to minimize the cost of misclassifications, resulting in a theoretically optimum classification.

The principal drawback of the likelihood classifier is the large number of computations required to classify each pixel. This complexity makes the classification process slower and more expensive than simpler techniques. While more costly, the maximum-likelihood technique generally provides more classification accuracy. A refinement combines the parallelepiped and maximum-likelihood classifiers with the latter being used only to resolve ambiguities in the classification.

Intuitively, it would seem that the more bands used in a maximum-likelihood classifier, the better the classification results. Generally, there is a relatively small optimal number because of the increase in computer time required by the more involved calculations and because the information content of an important band can be contaminated by the insignificant data in other bands (Swain and Davis, 1978). This problem is particularly acute when large numbers of correlated spectral bands are used. It has been found in some cases that the accuracy of a maximum-likelihood classifier begins to fall off when more than four or five spectral bands are used. When one is analyzing imagery consisting of many spectral bands, it is beneficial to consider a subset of the most meaningful bands. The decision about which bands to use is based on a process called feature selection. In this process many subsets of bands are evaluated to determine which combination is most useful for discriminating the categories of interest, and the classification analysis is limited to that set of bands.

Training Stage

Prior to classification, the spectral response pattern for each land-cover category of interest is determined in the training stage. This step is a critical task in computer-assisted interpretation, even more than the choice of the classifier (Hixson et al., 1980). The success of the classification process, and therefore the value of the information generated from the interpretation, relies directly on the quality of the training procedure. In many ways, the training effort is more an art than a science. It requires close interaction between the image analyst and the image data. It requires a thorough knowledge of the geographical area to which the data apply and the spectral characteristics of the features being analyzed.

The spectral response patterns of training areas may be established in several ways. The simplest would seem to be direct field or laboratory measurement, but these measurements may not correspond to those recorded by the sensor at any point because of illumination, seasonal, atmospheric, and sensor system effects. Because it is nearly impossible to adequately compensate for all of these variables, it is generally preferable to establish the category response-patterns using the image data. By doing so, however, one finds that the training patterns will precisely correspond only to the one image data set and must normally be recompiled for analysis of other images.

The training process begins with the selection of training areas (or fields) that are representative examples of each information category to be interpreted. These areas are normally selected by

consulting the reference data sources, such as topographic maps and aerial photographs and/or by ground observation. Once identified, the training areas are correspondingly located in the image data set by specifying particular pixel line and sample numbers. When selecting the training set pixels, it is important for one to analyze several training sites throughout the scene. Dispersion of the sites throughout the scene increases the chance that the training data will be representative of all the variations in the cover type throughout the scene. The computer accesses the image data file and reads the digitial numbers for the pixels located within each training area boundary. These training-pixel values are used to define the spectral response patterns of the various land-cover categories in either a parametric or nonparametric way.

The category spectral-response patterns must then be evaluated for their spectral separability. The analyst must also check to see whether the training data are normally distributed if the classifier being used depends on the assumption of normality. The distributions of training-area response patterns can be graphically displayed in many formats. Figure 24-54 shows a histogram for

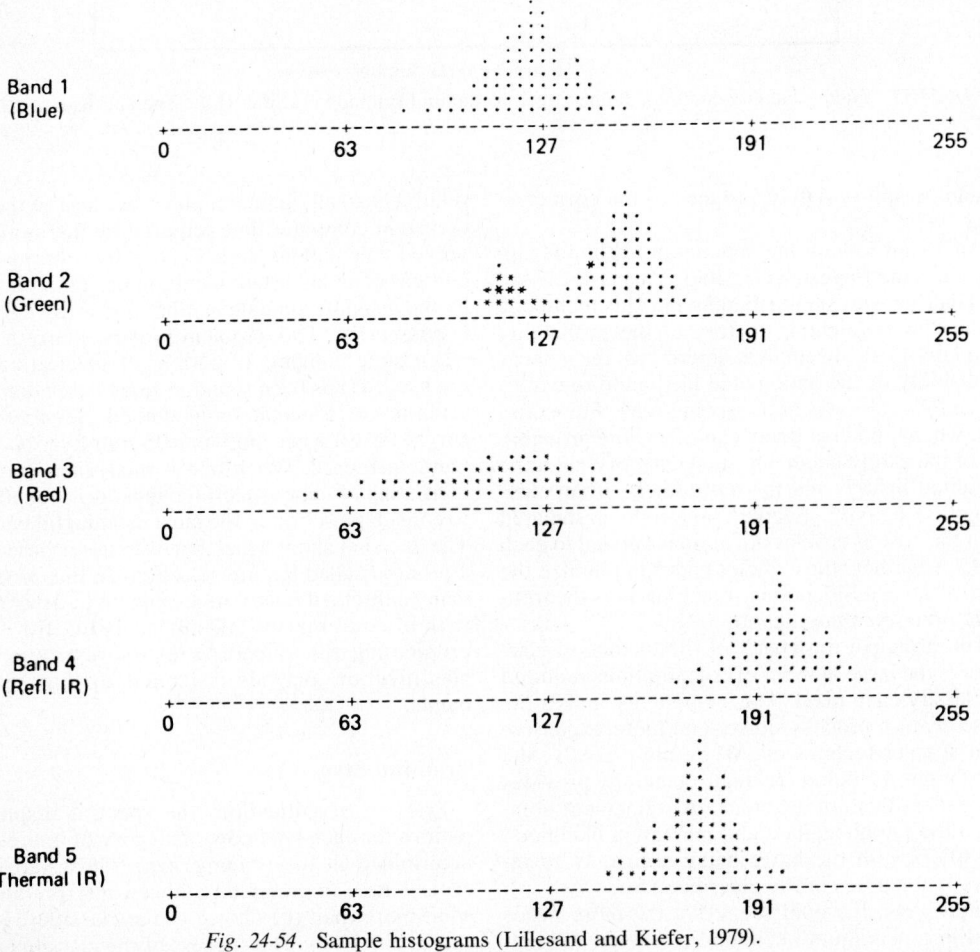

Fig. 24-54. Sample histograms (Lillesand and Kiefer, 1979).

the hay category example. Histogram shape is particularly important when a maximum-likelihood classifier is used, since it provides a visual check on the normality of the radiometric response distribution in each spectral sample window. Note that the hay category appears to be normally distributed in all bands except band 2 where it is bimodal. This indicates that the training site is composed of two subclasses with slightly different spectral characteristics. The classification accuracy will generally be improved if each of the subclasses is treated as a separate category.

To subjectively evaluate the spectral separation between categories, it is convenient to use a co-spectral plot (Figure 24-55). For each spectral band, this plot shows the mean radiometric response of each category and the spread of the distribution (here ± 2 standard deviations). Such plots indicate the potential overlap between cate-

gory response-patterns and show which combination of bands might be best for discrimination.

The fact that the spectral plots for hay and corn overlap in all spectral bands indicates that the categories could not be accurately classified on any single band. However, classification may improve when two or more bands are used. Two-dimensional scatter diagrams depict between-category overlap more accurately. The quantitative measure termed divergence is also used.

Measuring Category Separation and Training Data Validity

A measure of statistical separation between category response-patterns can be computed for all pairs of classes and presented in the form of a matrix. An expression commonly used is divergence, a covariance-weighted distance between category-pair means. Divergence was one

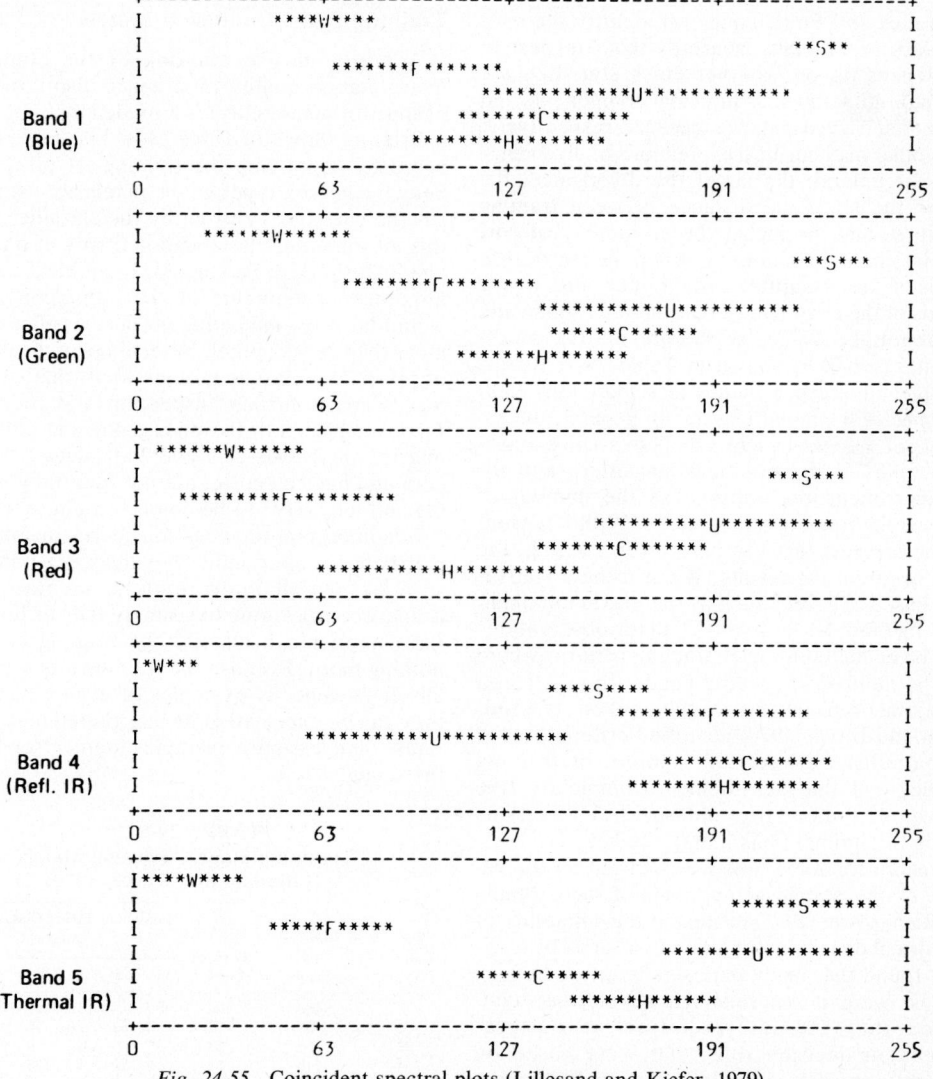

Fig. 24-55. Coincident spectral plots (Lillesand and Kiefer, 1979).

TABLE 24-7
Matrix of Divergence Values
(Lillesand and Kiefer, 1979)

Class[a]	1	2	3	4	5	6
1	0					
2	1998	0				
3	1996	1994	0			
4	1953	898	1882	0		
5	1997	1938	1812	1823	0	
6	1992	1976	1628	1938	1362	0

[a] 1—Water, 2—Sand, 3—Forest, 4—Urban, 5—Corn, 6—Hay.

of the first measures of category pairwise statistical separability used in feature selection. In multiclass interpretation a fairly common strategy is to use the sepctral combination with the highest average pairwise divergence (Swain and Davis, 1978).

Haralick and Fu (Chapter 18) identify the two, two-class separability measures that are best to use depending on whether class statistics are known or not known, as principal components and the Bhattacharyya distance measure, respectively. Many other mathematical expressions of divergence exist. In general, the larger the divergence, the greater the statistical distance between training signatures and the higher the probability of correct classification of classes. When the covariance matrices are unequal, only upper and lower bounds of the error can be determined (Shlien and Goodenough, 1973). A sample matrix of divergence values is shown in Table 24-7. In this example, divergence values less than 1500 indicate spectrally similar classes. In general, the estimates of the mean vectors and covariance matrices employed in statistical classifiers and divergence measures improve as the number of pixels in the training set increases. Within reason, the more pixels that can be used in training, the more accurate the results. When using a statistically based classifier such as the maximum likelihood method, at least n + 1 pixel observations must be collected for each training pattern, where n is the number of spectral bands. In practice, a minimum of from 10n to 100n pixels is used. Swain and Davis (1978) identified criteria for the compatibility between the number of training samples and the measurement complexity (the number of radiometric and spectral levels) as finding an optimal (maximum) number.

Concern for some time was focused on the validity of the statistical approach. Crane, Malila and Richardson (1972) evaluated the suitability of the normal density assumption in a series of tests. They found that many data sets (e.g., of training classes) were non-normal. However, they concluded that the interpretation or pattern-recognition decision rules that were based on normality performed well; hence analysts were advised that improvements of the interpretation process would be more in data preprocessing

(e.g., restoration) than in the incremental benefits of decision rules that were more adapted to non-normal data.

The parametric approach bases decision processes on simple descriptive parameters of the normal distribution: mean and variance. The non-parametric approach retains the non-normality found in training data. Nonetheless, the search for improvements in the computer-assisted interpretation process continues in varied directions. Non-normal data are accommodated in a variety of ways, often by other mathematical representations of the empirical data. These include:

- the logit classifier (Maynard and Strahler, 1981); and
- Fourier series estimate of non-normal multivariate density-functions (Greblicki and Pawlak, 1981).

The uses made of collateral data and *a priori* environment estimates are contemporary refinements.

Training Stage Confusion Matrix

A preliminary evaluation of the computer-interpretation quality (and hence also a measure of spectral separability) is provided by a confusion matrix, as shown in Table 24-8. This table is prepared by classifying the training-set pixels. The known category types of the pixels are listed versus the categories chosen by the classifier. From this information, classification errors of omission and inclusion can be studied. In an ideal case, all nondiagonal elements of the confusion matrix would be zero, indicating no misclassification. If more than an acceptable percentage of the number of pixels in a class is misclassified, that category may warrant further inspection and retraining. For example, both the divergence and confusion matrices in Tables 24-7 and 24-8 suggest that the corn and hay categories are not spectrally separable and may have to be combined into one class.

It is important to avoid considering a confusion matrix based on training set values as a final measure of overall interpretation accuracy. The training confusion-matrix simply tells us how well the classifier can classify the training data and nothing more. Because the training areas are usually homogeneous examples of each cover type, they can be expected to be interpreted more accurately than less pure examples found elsewhere in the scene.

TABLE 24-8
Training Stage Confusion Matrix
(Lillesand and Kiefer, 1979)

Known Category Type[a]	Number of Pixels	Percent Correct	Number of Pixels Classified into Category					
			1	2	3	4	5	6
1	480	100	480	0	0	0	0	0
2	68	76	0	52	0	16	0	0
3	356	88	5	0	313	0	0	38
4	248	51	0	20	40	126	38	24
5	402	85	0	0	0	0	342	60
6	438	82	0	0	0	0	79	359

[a] 1—Water, 2—Sand, 3—Forest, 4—Urban, 5—Corn, 6—Hay.

Interpretation Accuracy Analysis

Overall interpretation accuracy can be evaluated only by considering areas that are different from, and considerably more extensive than, the training areas. This evaluation is generally performed after the classification and output. Categorized pixels in test field areas (outside the training areas) are compared to a map or source of known land cover using some statistical sampling strategy. Each category must be adequately tested. The results are also presented in the form of a confusion matrix and evaluated by the image analyst. Table 24-9 is a confusion matrix illustrating test-field accuracy for the classes shown in Table 24-8. Note the general decrease in classification accuracy as pixels outside of the training areas are classified.

The training effort is normally an iterative procedure in which the analyst revises the category types until they are sufficiently separable on a spectral basis. This effort requires close involvement between the analyst, the image data, and the prospective users of the output data. Successful results generally require considerable time and care. The quality of the image classification critically depends on the quality of the training data used to perform the classification. Todd et al. (1980) summarized the quality assurance checks and error analyses which preferably accompany each computer-interpretation summary map. They also identified three principal categories of classification error experienced in a wildland mapping project:

- geometric and radiometric problems in the image accounted for about 5 to 15 percent of the errors,
- excessive category detail (i.e., attempts to extract classes whose spectral characteristics approached the noise level of the data accounted for another 35 to 40 percent, and
- analyst decisions in identifying spatial clusters, which required considering complex interaction between vegetative and terrain characteristics, accounted for the rest of the errors. (See Section 9.1)

That study was later characterized as the first comprehensive and valid accuracy analysis of a classification project that considered all the necessary applications of statistics, including sample size and distribution, size of sample unit, sampling procedure, and accuracy estimations, including those of variance and confidence limits (Rosenfield, 1980).

Unsupervised Classification: Clustering

An alternate to the supervised approach is unsupervised clustering. Analyst-specified training data are *not* used. The analyst's entry into the process is after the clustering is done to identify the spatial deployment of the spectral clusters. This is the process of cluster-labeling. Unsupervised classification can use an algorithm to examine a large number of unknown pixels and cluster them on the basis of natural spectral groupings present in the image gray values. The basic premise is that values within a given cover type would be closer together in spectral space and that the data from other classes would be separated. Controlled clustering, a variant, combines the training area selection and clustering methods (See Figure 24-56).

Clustering has been used for several decades and was first applied (circa 1939) by Tyron to numerical taxonomy problems (Coleman and Andrews, 1979). There are many clustering procedures, each having its own peculiar characteristics. Some procedures iterate to a local minimum for the average distance from each sample to the nearest cluster means. The nearest means algorithm of Ball and Hall (1965) begins with an assumed number of clusters. The means are arbitrarily assigned (and hence the number of iterations for convergence). After all the data points have been assigned the cluster means are recomputed. This process continues until data assignment does not change (convergence). This algorithm iterates to a local minimum for the average within-cluster distance. The key obstacle to overcome is the determination of the correct number of clusters. One measure of clustering quality for that purpose is the ratio of the between-to-within cluster scatter, much like the divergence measure (Coleman and Andrews, 1979).

Clustering as an image-encoding method is discussed also in Chapter 17 where it is noted that a Landsat-size image could easily produce thousands of clusters. Preferably, local regions are clustered and cluster-labeling assignments are made to the features.

Two interesting examples of extensive clustering operations were those used in (1) the preparation of a mosaic of California from some 34 Landsat frames and (2) in LACIE (Large Area Crop Inventory Experiment). For the mosaic, the spectral data were compressed from millions of entries to thousands by a simple count of the number of occurrences of each unique combination of gray values from the four MSS bands. Subsequent clustering produced some 1200 spectral classes in 32 major ecological regions. Eventual

TABLE 24-9
Confusion Matrix Test Field
(Lillesand and Kiefer, 1979)

Known Category Type[a]	Number of Pixels	Percent Correct	Number of Pixels Classified into Category					
			1	2	3	4	5	6
1	5325	97	5165	0	42	44	53	21
2	328	66	0	216	0	108	4	0
3	4284	84	0	0	3599	16	482	187
4	945	42	12	92	228	397	132	84
5	2380	80	0	9	28	78	1904	361
6	3048	72	8	0	48	18	779	2195

[a] 1—Water, 2—Sand, 3—Forest, 4—Urban, 5—Corn, 6—Hay
Overall classification performance: 82.6% (total correct pixels/total pixels).
Average performance by class: 73.5% (average of category accuracies).

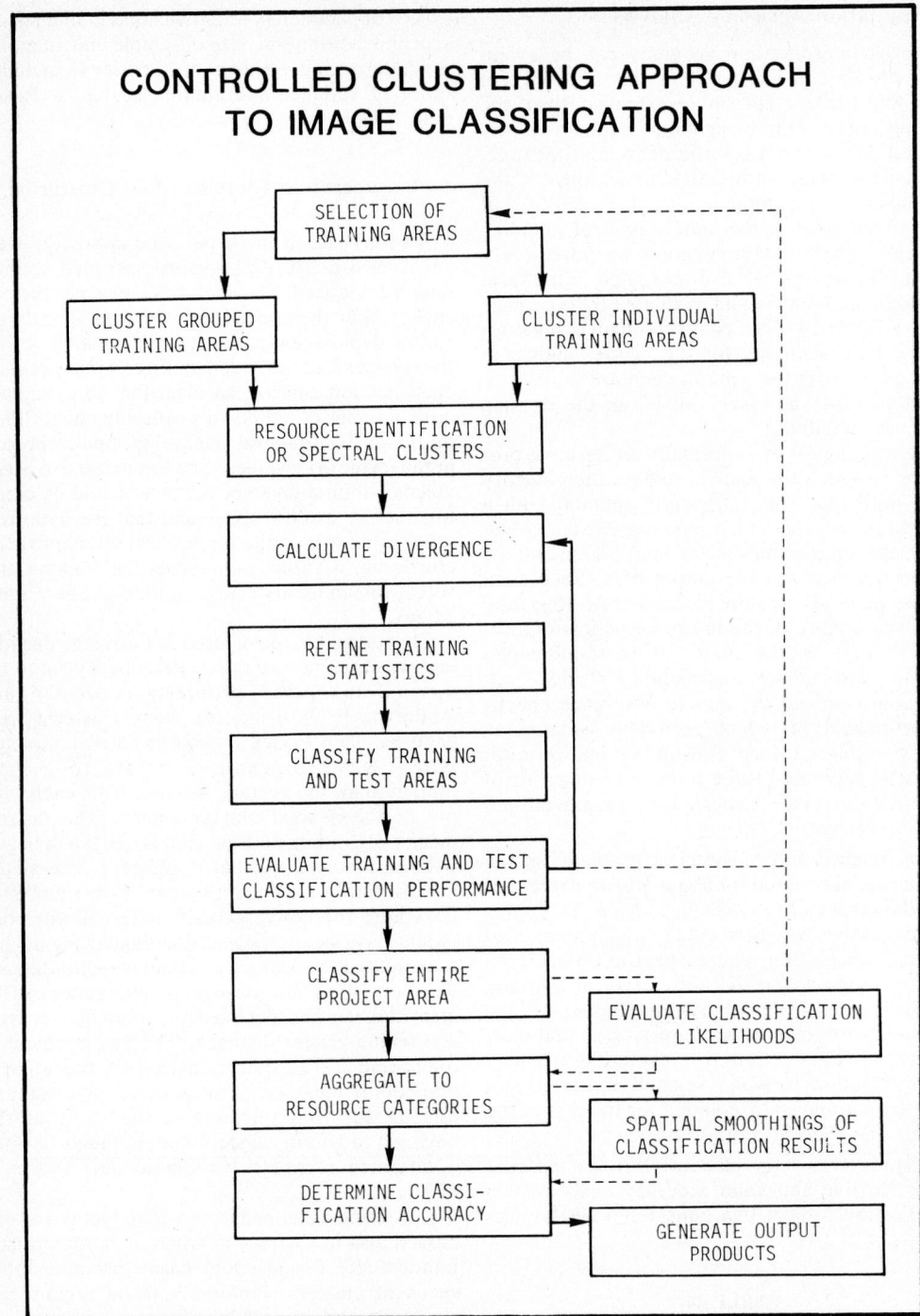

Fig. 24-56. Controlled clustering approach to image classification (Johnson, 1980).

cluster labeling produced 16 classification categories after extensive effort. Over 100,000,000 acres were classified in a 4½ month period (Newland, Peterson and Norman, 1980). Clustering was used in LACIE as a means of automatically estimating parameters required for maximum-likelihood classification. The classification algorithm was based on the assumption that each crop type could be statistically modeled as a linear combination of normal distributions. Six different clustering algorithms were evaluated in that program.

Spatial and Temporal Elements of Interpretation

Contemporary research has moved beyond the spectral signature for improvements in interpretation and a step closer to understanding and using some of the spatial and temporal elements. Texture analysis and image segmentation are becoming viable spatial analysis methods.

Texture Analysis

Texture is one of the fundamental elements used in human interpretation of remotely sensed imagery. The concept of texture is closely allied to tone in that texture represents the spatial pattern of tone in the image, as previously indicated in Figure 24-45. Tone tends to dominate in images of low spatial variability, while texture is often the predominant feature in images of high spatial variability. Texture, long recognized as an integral part of manual image interpretation, has been used experimentally in automated image analysis. Recognition of spatial pattern, or texture, is rather easily accomplished by human interpreters but is very difficult using computers.

Research in the use of textural characteristics in digital image data has been dominated by the application of statistical measures of texture for type discrimination. From this perspective, most attempts have been empirically-based characterizations of spatial pattern, and represent a trial-and-error type of approach. One future direction may involve the development of texture measurements based on deterministic models of spatial patterns of land-cover types. This requires extension of the theory of regionalized variables to the remote sensing case. Considerable work has been done on the theory of regionalized variables in geography, geology, geophysics, and ecology and is reviewed in Jupp and Adomeit (1981).

The use of texture in automated image analysis has been reviewed by Haralick (1979). He noted the absence of a formal or precise definition of texture and characterized most of the existing texture-discrimination techniques as *ad hoc*. Texture is described as having two components or dimensions: primitives and the spatial pattern of primitives. Primitives are regions in the image with specified tonal properties.

Haralick (1979) has categorized eight statistical approaches to using texture: autocorrelation functions, optical transforms, digital transforms, textural edgeness, structural elements, gray tone cooccurrence, run lengths, and autoregressive models. The first three of these measure texture spatial frequencies. Fine textures are rich in high spatial frequencies. Less common, but potentially powerful, are structural approaches to texture. Structural approaches are based on more complex primitives than gray tone. Statistical and structural approaches may also be used in conjunction. Research published since the time of Haralick's review includes Davis and Mitche (1980), Fasler (1980), Hsu (1980), and Woodcock et al. (1980).

Image Segmentation

Another spatial-domain analysis is image segmentation or partitioning. Image segmentation in-

volves the creation of segments or regions in an image that are comprised of more than one picture element. There are two primary motivations underlying the use of image segmentation: 1) classification accuracy and speed can be increased by processing multipixel regions as a single unit, and 2) the output product following the use of image segmentation may better meet the need of the eventual user. The second motive is common among applications in renewable resources. There, the spatial precision of the data is reduced while the ability to find boundaries at the original resolution is retained.

The two basic methods for image segmentation are region growing and edge detection. These highlight the two most prominent features of a segmented image, viz. regions and edges. In region growing, edges are defined as the boundaries between regions. In edge detection the regions are defined as the areas surrounded by edges. Most image segmentation used in renewable-resources inventory has employed region-growing techniques. These approaches seem well suited to the definition of fields in agricultural scenes where the assumption of spectral homogeneity within the desired segments is reasonable. Edge detection has been approached from a more theoretical viewpoint and, as a result, tends to be less concerned with specific types of scenes and their associated characteristics. Edge detection has been used very sparingly in the remote sensing community.

BLOB is a region-growing technique that incorporates spatial data through the addition of line and sample coordinates to the standard spectral pixel-vector (Kauth et al., 1977). Individual pixels are classified on the basis of this augmented vector. Central to this technique is the determination of the initial region centers for individual pixel-vector comparison. A second step annexes contiguous regions if they belong to the same class, further enhancing the spatial smoothing.

ECHO is another region-growing technique designed for faster interpretation via reduced elements input for class discrimination (Landgrebe, 1980). ECHO also enhances image classifications by reducing within-field interpretation errors. A moving window is first passed over the image and a test determines if the component pixels are spectrally similar. If they are similar they are assumed to be part of the same object and are eventually classified as a unit. Pixels not in any multipixel object are classified individually as in conventional classification. Statistics defining the classes in the scene may be developed in either a supervised or unsupervised manner.

The sloped-facet model (Haralick, 1980) falls between the region-growing and edge-detection methods. It allows for sloped variation in spectral values within the facets, or multipixel objects in an image. This "sloped world" assumption is less restrictive than the variance-reducing "flat world" models implicit in most image classification. The sloped-facet model examines each pixel as a member of each of numerous neighborhoods to determine its best neighborhood. If contiguous pixels have overlapping best neighborhoods, or if the slopes of the planes of their best neighborhoods are not statistically different, the pixels are merged into one facet. Edges are between facets where slope coefficients are different. This relatively new technique has not been tested as extensively, nor in as many environments, as BLOB or ECHO.

There have been numerous edge-detection methods in digital image-processing, but few have been applied to remote sensing interpretation. The general results are noisy images of largely disconnected edges. In image segmentation the existence of disconnected edges causes serious problems for the definition of multipixel units. One edge-detection technique that guarantees connected edges is described by Pentland (1980). For other edge-detection algorithms see Haralick (1980) for a review of the literature, and Brooks (1978), Hummel (1979), Roberts (1965), Rosenfeld et al. (1970), Yakimovsky (1976), and Nahi and Jahanshahi (1977) for details.

Temporal Analysis

Temporal analysis exploits the usefulness of time-related features as another element for interpretation (Peplies, 1976). Descriptive terms for temporal analysis are "multitemporal" or "multidate" (Estes and Simonett, 1975; Lillesand and Kiefer, 1979), which refer to the acquisition and use of remotely sensed data from more than one period of time for a given location. The term multitemporal will be used here.

Temporal analysis uses two basic time-interpretive functions: 1) time-affected features as an additional discriminant and 2) the derivation of time information from multitemporal data.

Time as a Discriminant

The use of temporal features improves interpretation by way of patterns not revealed by tone and texture alone. *A priori* and *ex post facto* knowledge of time-dependent phenomena include the:

- illumination source (e.g., solar incidence-angle, Dave, 1981),
- intervening atmosphere (e.g., clouds and particulates),
- within-scene vegetation (e.g., agricultural crops and deciduous trees),
- hydrologic stages (e.g., river and tidal), and
- location of the remote sensing platform (i.e., determination of its position in satellite orbit).

A priori knowledge of time-dependent features helps optimize data acquisition and subsequent image processing. Temporal information derived *ex post facto* may be used for calibration, correc-

tion, and convergence of evidence for computer-assisted interpretation (Gulinck, 1980).

Particularly important is temporal optimization of spectral response. It is possible to maximize information-extraction by imaging at times when the radiances (brightness levels) of features of interest are most easily discriminated (Simonett, 1976). In practice, it is often difficult to optimize temporally since one may not have control over when image acquisition occurs (such as for sun-synchronous satellites). Also, there often exists a complex mix of temporally varying spectral responses within a given scene, making it difficult to fully optimize for discrimination without some degree of compromise.

Time-periodic environmental characteristics affect the optimum time for imaging and discriminating earth features. Diurnal, seasonal, and multiannual effects are often used to improve radiometric and spectral separability of within-scene features.

Diurnal effects include the angle of incident solar illumination and associated radiant responses. The enhancement of geologic structure and the detection of films on the surface of water can thus be influenced. Imaging of organic and inorganic particulate concentrations should not occur during local noon, thereby avoiding the excessive specular reflection from the ocean surface (for nadir-pointing sensors). Analyses of subsurface moisture-temperature relationships use thermal infrared imagery at two times within the diurnal soil temperature cycle (Lillesand and Kiefer, 1979; Paley and Kahle, 1979). See Figure 24-57.

Seasonal and annual changes have been the most widely used periodic phenomena for the remote sensing of both renewable and non-renewable resources (Adams, 1981). There are important seasonal influences on the spectral signatures of forest, rangeland, agricultural vegetation, and many hydrologic features (e.g., snow cover and lake/reservoir levels) (Butera, 1978). Crop, forest, and general land-use/land-cover inventories are derived from the interpretation (Kumar, 1980). Single-date inventory accuracies can be highly dependent on the season of image acquisition. Seasonal variations of radiometric and spectral characteristics affect separability of within-scene categories. Seasonal influences on the atmosphere affect separation of surface spectral responses from radiances contributed by the intervening atmosphere.

Cycles with periods greater than one year are less commonly used for optimizing spectral responses. The most significant would be the influence of climatic variability on spectral responses. An example is improvement of the accuracy of inventories of natural vegetation and irrigated agriculture during drought years (Peterson et al., 1980).

Diurnal, seasonal, and multi-annual spectral change can be optimized for both single-date and multidate data acquisition (Misra and Wheeler,

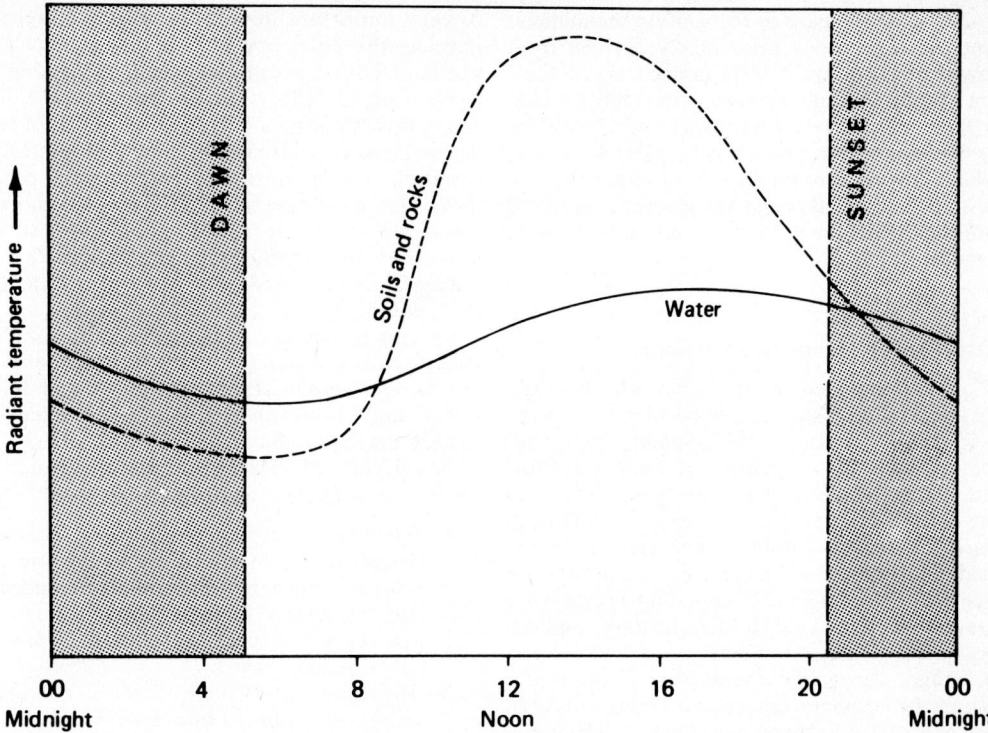

Fig. 24-57. Generalized diurnal radiant temperature variations for soils and rocks versus water (Lillesand and Kiefer, 1979).

1978). The latter requires registered multidate images (Rifman et al., 1975).

A particular benefit from multitemporal images is the increased dimensionality of radiometric and spectral data for information extraction (Swain and Davis, 1978). The major remote sensing applications using multitemporal analyses are the renewable-resource inventories (Malila et al., 1975; Aucion and Giddings, 1978). These rely heavily on seasonal pattern for time discrimination.

An important concern in multitemporal analysis is image registration (see Chapters 17 and 21). Image registration requires precise overlaying of digital images acquired at different times. Ideally, every picture element at the same image-matrix location for each of two or more images represents the same location on the surface of the earth. Although spatial congruency is never completely achieved in practice, sufficiently accurate spatial registration can be achieved to enable multitemporal information extraction. It is easy to appreciate the influence of spatial-registration accuracy on resultant categorization accuracies for renewable-resource inventories (Malila et al., 1975). Scene to scene normalization techniques may also be important in multitemporal analysis (cf. Chapter 17).

Two basic multitemporal analysis methods that are commonly used, are illustrated in the following sections—the stacked vector approach and the temporal trajectory approach. In both instances

spatially registered Landsat images from three dates within a given year are available. Knowledge of crop growth-cycles and cultivation patterns for the given locale is assumed. The stacked vector approach performs supervised or unsupervised classification of the three registered dates of imagery (Mergerson, 1981). The only difference from a single-date classification is the added dimensionality of the data (three dates of n channels of spectral data for each date). Thus classification can use a data vector of up to 3 n channels. A tradeoff associated with the added information is the increased processing time required. One should seek to optimize for information accuracy, under the constraints of: 1) processing costs, 2) inaccuracies introduced by spatial misregistration, and 3) high correlations between channels of information (i.e., between spectral bands and between image dates). The stacked vector approach does not assess or use any particular temporal patterns of sepctral signatures. The second approach uses temporal trajectories. This differs from the stacked vector in that patterns of temporally-varying crop spectral signatures are objectively or subjectively derived in the feature-extraction phase.

Thus, there are at least two modes of feature extraction. The simplest mode is supervised training with training sites selected to contain representative spectral responses and representative temporal patterns associated with crop growth cycles (Engvall et al., 1977). The second mode invokes syntactic pattern recognition techniques. Temporal trajectories heuristically derived from known temporal spectral response curves (i.e., from ground radiometric measurements) provide the feature-extraction information necessary for classification. Syntactic classification would be based on the form of the trajectory rather than on the statistical relationship of spectral, temporal patterns such as for the supervised training mode (Kaneko, 1978).

Extraction of Temporal Information

Computer-assisted interpretation can also derive information about the temporal nature of the environment (Otterman, 1977; Shields and Goodfellow, 1980). The extraction of temporal information functionally differs from the concept of time as a discriminant in that temporal information is the desired output. Again, spatially registered multitemporal imagery is required. The three major categories of temporal information are change detection, change identification, and assessment of dynamic systems.

Change detection locates environmental changes by a binary (change/no change) assessment of pixels or aggregations of pixels. The initial change-detection research is attributed to defense reconnaissance-related applications. An important early work on change detection is that of Shepard (1964). Change detection is now used primarily by government agencies in resource management, planning, and enforcement activities (Deuker and Horton, 1971). Most of these applications use human interpretation for image comparisons. Computer-assisted change detection influenced by the EROS program is beginning to be operational now that large area coverage, medium resolution, and temporally repetitive Landsat data are available (Williams and Miller, 1979).

The development of procedures for spatially registering multitemporal Landsat data enabled first-stage detection mapping for large areas. Changes of interest are often interspersed with false changes due to misregistration, seasonal effects, illumination changes, etc. Change detection analyses are first-stage in the sense that they signal the need for detailed analysis of specific locations via multistage sampling. The main computer-assisted change-detection technique is simple image differencing. Changes are signaled when pixel tone differences exceed some threshold.

Change identification and analysis go beyond the simple binary change detection to identify and analyze the types of environmental change. Thus, they require higher resolution data and more complex information-extraction algorithms (Price, 1976). Change identification has been used for resource-inventory updating and enforcement monitoring (Gilmer, 1975; Byrne et al., 1980).

A very important development has sought to improve the efficiency of updating general land use/land cover inventories and urban census (Anderson, 1977; Davis and Friedman, 1979; Jensen, 1981). Changes of interest are those of land use or natural landscape. Changes of interest in resource management and enforcement procedures identify the presence of illegal structures or waste disposal. Here, computer-assisted assessments are usually limited to studies that involve fairly large areas or use only a change-detection mode.

Specific techniques of change identification are discussed in Chapter 17 and by Weismueller et al. (1977), and Jensen (1981). Matthews and Miller (1979) and Robinson (1979) are references on change detection and change identification.

Highlights of the three major groups of change-identification algorithms are:

- *Post-classification comparison*—Thematic output images of information categories are spatially registered and cross-referenced to derive category changes through time on a pixel-by-pixel, pixel aggregation, or statistical tabulation basis,
- *Multitemporal 'Stacked Vector' classification*—An historical and a recent image are spatially registered, and spectral response patterns (transitions) for known category changes are extracted in either a supervised training or syntactic trajectory mode. These response patterns are the basis for categorizing land use/land cover sequences, and
- *Contextual-based updating*—Multitemporal spectral data are merged with some combination of texture, segmented, aggregated, and/or collateral data (e.g., digital, historical, resource map in digital form). Statistical models or syntactic heuristics are used in conjunction with these combinations of data types in an attempt to minimize the effects of non-interesting changes.

The extraction of temporal information aids the assessment of time-dynamic systems. Certain physical, biological and cultural systems exhibit time-varying environmental characteristics that can be monitored by repetitive remote sensing. Examples are meteorologic, hydrodynamic, marine phytoplankton, fishery, biological, and urban systems. Interpreted quantities include sea-surface temperature, soil moisture, and chlorophyll concentration. The dynamics of these can be determined if remotely sensed data are available for the required sampling times (Rabchevsky, 1977).

One important distinction between interpretation methods is whether temporal information is derived directly from one image or by comparison of two or more images. An example of the one-image approach is the determination of surface current-velocities from the Doppler shift of the imaged current waters on side-looking radar imagery (Shemdin et al., 1980). An example of the comparative approach is measurement of the advective movement of a natural or injected tracer in the surface waters of the flowing current (James, 1972). Another major distinction is whether vector, transform or tabular statistical techniques are used. These are the three general groups of dynamic systems analysis.

Vector methods derive information on spatial movements through time (Yachida et al., 1981). A requirement is a traceable feature observed on the time-sequential images (Blackman, 1980). It may be assessed by conservative properties (in the ideal case) or non-conservative scalar-field properties and how they change through time. Assessments can also follow the movement of edges, borders or distinct gradients (Hansen and Maul, 1970). An example of the scalar-field approach is tracking the advection and dispersion of dissolved or suspended matter in a dynamic water body (Nichols and Kelly, 1972). This is achieved by time-sequential comparison of scalar-field concentrations of dissolved or suspended matter. Delineating and tracking the edges of a cloud system through time for meteorologic frontal systems (Parikh, 1974) and the movement of the urban-rural fringe boundaries for urban systems (Christenson and Lachowski, 1976) are examples of the latter type of vector approach.

The use of transforms for temporal analysis reduces image data, usually by statistical aggregation. Transformed images of different acquisition times can be compared for information on the temporal feature (Curry and MacDougal, 1972). The analysis uses Fourier transforms, convolution filters, and spatial autoconvolution models (Barber, 1972; Chapter 17). The initial development of such temporal analysis has been in satellite meteorology (Leese and Epstein, 1963).

Tabular statistics of discrete (i.e., categorical) and continuous (i.e., quantitative) data from time-sequential imagery can be spatially cross-referenced and compiled. Tabulations can be made for each individual pixel or any image segment. These reveal the amount of change over time. A biological system example (greatly affected by physical systems) is the patchiness of marine primary productivity from the temporally varying composition of optically-imaged chlorophyll-concentrations (Smith et al., 1982). A similar change-identification technique measured agricultural-to-urban conversions (Stow et al., 1980).

Collateral Data

There are many sources of information available to human image interpretation that are still not used in conventional computer-interpretation of images. Through inclusion of the additional data, some of the stimuli for the decisions of a human interpreter based on nonspectral factors (e.g., site, association and context) are crudely imitated. Data sources that can be combined with spectral data include soil, geologic, climatic, and land-use history. The inputs may be qualitative or quantitative, subjective or objective, and range from raw data to information. Regardless of the form, it is termed ancillary or collateral data.[6] Here it is considered supporting or parallel data and not subordinate data. For this reason, the term collateral is used. Using these data in modified classification algorithms emulates some of the human interpretation results. Generally, there is an increase in the accuracy of computer-aided interpretation. In the limit, the rather extensive set of geographic data forms the nucleus of the geographic informtation system (see Chapter 22).

In a broad sense, collateral data are data from sources other than the sensor system (Swain and Davis, 1978). Human interpreters can use subjective information of past experiences to identify particular features on an image, assisted by objective data such as ground survey maps or statistical tabulations. Subjective information, such as the recollections of a farm hand regarding the agricultural history of a region, and objective information, such as the stratification of a digital elevation map based upon sampled botanical distributions, are examples. Though dissimilar, both serve as aids in the identification process.

Computer-aided interpretation also uses both subjective and objective collateral data. The training stage in supervised classification requires the analyst to subjectively select categories and objectively locate classes. "Ground truth" measurements are collateral data extremely important to both human and computer-assisted interpretations and are particularly critical to the effectiveness of the latter. Primary uses of "ground truth" data (see Chapter 23) are for calibration/correction, interpretation of properties, training, verification, and sampling.

Calibration of remote sensing data based upon concurrent surface measurements incorporates objective collateral data. Radiometers are calibrated with collateral data on received energy. Other collateral data may be used to calibrate sensors with ground meaurements of reflectivity or temperature. Use also may be made of calibration data acquired and recorded on the remote sensing platform (see Chapter 5).

The collateral data can be diversely incorporated to improve classification accuracies. Examples include (1) terrain data to ameliorate the effects of topography on scene data, 2) pseudo-spectral data-channel inputs to the classification procedure, 3) a data source for the development of prior probabilities, and 4) elements of spatial or ecological models.

Viewed in a broad context, the problem of combining image data and spatial collateral data into a predicted output map or image is actually a problem of combining continuous and categorical

---

[6] Ancillary means subordinate or auxillary; collateral means parallel, corresponding or supporting (Merriam-Webster, 1960).

TABLE 24-10

TECHNIQUES FOR COMBINING CONTINUOUS AND CATEGORICAL DATA

OUTPUT CHANNEL (Dependent Variable)	INPUT CHANNELS (Independent Variable)		
	CONTINUOUS	MIXED	CATEGORICAL

CONTINUOUS	• Regression Models
	  - linear
	  - curvilinear
		• Analysis of Covariance
		• Multivariate Analysis of Covariance
			• Analysis of Variance
			• Multivariate Analysis of Variance

CATEGORICAL	• Maximum Likelihood Classification • Logit Modeling • Discriminant Analysis	• Maximum Likelihood Classification with Prior Probabilities • Logit Modeling	• Contingency Table Analysis • Logit Modeling

outputs. Table 24-10 (Strahler et al., 1980) shows techniques for such merging. In such categorization, maximum-likelihood classification (MLC) using prior probabilities is the most important method (Strahler, 1980).

The prior probabilities are based on "ground truth" data. Strahler (1981) noted that care must be used in such determination since *a priori* weights may substantially alter the decision rule. *A priori* probabilities may be specified at several levels; a single global set for a scene, a set for individual sub-scenes (strata-wide), and a pixel-by-pixel set (Strahler et al., 1979, 1980).

The priors for a pixel may be determined by a logit or other model, or by using a set of class-conditional prior probabilities estimated by sampling. Because the priors are computed separately, it is possible to mix any sort of model that estimates prior probabilities with a multivariate normal MLC algorithm that is known to be well suited to most spectral data. Thus the technique allows easy and flexible merging of collateral data, used to predict the priors, with continuous image data. These points are discussed at more length in Strahler (1980), Strahler et al. (1978, 1980) and Swain et al. (1980).

An alternate approach is to input continuous collateral data to the classifier. Hoffer et al. (1975) and Strahler et al. (1978) improved classification of forest cover by incorporating digital terrain data into the maximum-likelihood classification. An implicit assumption is that a relationship exists between the output and the input collateral data. Although this method appears straightforward, problems regarding channel weighting and data distributions must be considered (Strahler, 1980).

Another way to include collateral information is the use of image stratification (masking). Examples are in Strahler (1981). Binary strata to compensate for differential illumination have been developed from terrain data (Woodcock et al., 1980) and from terrain and spectral data (Foresman et al., 1980).

Definition of ecologically similar strata based on elevation, urban, agricultural, and natural regions can reduce classification error. Regional forest-vegetation cover (sub-strata) has been determined for multiple natural regions (strata) by Woodcock et al. (1980) through the use of sampling and modeling techniques. Hallada et al. (1981) formed strata from digital ground-truth maps for development of automated cluster-labeling techniques. In forest classification, terrain variables combined with remotely sensed data enhance the identification of forest types (Woodcock et al., 1980; Cicone et al., 1977). The terrain variables provide fundamental ecological data long known to influence tree-species distributions.

Removal of the Topographic Effect

Collateral digital terrain data can assist interpretation by removing the effects of topographic variation on the scene spectral data. The variation in spectral reflectance that is unrelated to the surface cover is often referred to as the topographic effect. The negative impact of the topographic effect on automated land-cover classification has been noted by Cicone et al. (1977), Hoffer and staff (1975), Woodcock et al. (1980), and others. The attempts to reduce this adverse effect have been dominated by preprocessing the spectral data. Some of the signal related to the topographic variation is removed. Collateral digital terrain data and locations of the sun and sensor at the time of sensor overpass can be used to calculate sun-surface-sensor geometry. Calculation of the local solar-zenith angle (or incidence angle), the angle of reflection (or exitance angle), and the azimuth of reflection relative to the azimuth of incidence can be important. The equations required for calculation of these variables are given in Chapter 23.

The topographic effect components have been described and measured by Holben and Justice (1980). The angle of incidence component has received considerable attention. The incident beam irradiance on a surface is proportional to the cosine of the angle of incidence, and thus influences the magnitude of reflected radiance. Preprocessing to correct for this effect in Landsat data divides the radiometric values by the cosine of the incidence angle and is generally unsuccessful (Struve et al., 1977; Hoffer and staff, 1975; and Cicone et al., 1977). Justice and Holben (1979) found that this technique increased the variance in the radiometric values of the spectral signatures associated with individual cover types. These authors concur that such a correction tends to overcompensate for the problem. One reason why the correction is ineffective is that the variation of the bidirectional-reflection distribution function (BRDF) of the surface is not considered.

The BRDF of a surface specifies the magnitude of reflected flux for all elevation and azimuth angles of reflection for each angle of incidence. For the correction to be effective the BRDF must closely approximate a Lambertian surface. The Lambertian surface specifies that the radiance is isotropic (i.e., both scattering and subtended area viewed are proportional to the cosine of the angle between scattering direction and surface normal; thus the radiance is equal in all directions). Due to the large number of possible combinations of the angles of incidence and the angles of reflection (elevation and azimuth), measurement of the BRDFs for surfaces has not been extensive. In addition, studies have indicated that the Lambertian assumption can be inappropriate (Kriebel, 1978; and Smith et al., 1980). Cicone et al. (1977) found that an *ad hoc* modification of the cosine of the incidence angle correction produced better results. That finding may also be attributed to variation in bidirectional reflectance.

Smith et al. (1980) employed an empirical photometric function proposed by Minnaert (1941) to test the validity of the Lambertian assumption. In this function a constant, K, describes the BRDF of a surface and provides a quantitative evaluation of the validity of the Lambertian assumption. The value of K is determined through a regression. A t-test of significance can be done for the hypothesis that the slope coefficient (K) is equal to 1.0, the result for a Lambertian surface. The authors found that in a pon-

derosa pine area the Lambertian assumption was more valid for low slope and low illumination angles.

The K-coefficient of the Minnaert function can also be used to remove the effect of bidirectional reflectance from spectral data. In a test by Justice and Holben (1979), correction of spectral data using the K-coefficient significantly reduced the topographic effect. However, it is still unknown how K values will vary between cover types. If K values do vary significantly, correction of the spectral data would require application of different K-coefficients depending on the surface cover type. In this situation the spatial distribution of surface-cover types would be required. It is possible that the amount of cover-type information required would obviate the need for the ensuing classification.

Most research involved in reducing the influence of the topographic effect has been concerned with variation in beam irradiance and reflectance. However, the diffuse component of the radiance received at the sensor is also influenced by local topography. Diffuse irradiances were measured by Justice and Holben (1980) and shown to vary with solar elevation, slope angle, and slope aspect. Diffuse irradiance also varies as a function of a terrain-view factor that describes the portion of the sky obscured by surrounding terrain (Dozier, 1980). Most researchers have ignored these effects as insignificant. Justice and Holben (1980) and Smith et al. (1980) measured diffuse irradiances as a function of topography and concluded that such effects did not warrant consideration.

An alternate approach to preprocessing the spectral data to reduce the topographic effect is to include data layers of illumination and/or reflectance angles directly in the classification process. This approach was tested by Cicone et al. (1977) with the use of a modified solar-insolation variable in a conventional supervised classification. Woodcock et al. (1980) used a cosine of the incidence-angle image to prepare an illumination mask. Overlaying this mask on classes produced by an unsupervised clustering procedure allowed separate labeling of the classes in the normal and poorly illuminated sections of the mask. Such an approach has not been extensively tested and may prove useful for improving classification accuracies in areas of high relief.

Spatial Modeling

Collateral data sources can also be combined in the absence of remotely sensed data. This is referred to as spatial or ecological modeling. New characteristics to be mapped are deduced for each location from the total collateral data for that location. More frequently, data derived from remote sensors are inputs to spatial models and are necessary for model execution. These data complement the existing information and demonstrate the synergism of remotely sensed and collateral data in a geographic information system. Dozier (1980) used satellite and digital terrain data to develop an energy-balance snowmelt model. There are also many examples of remote sensing input to water-runoff models.

Generally, ecological modeling is used for predicting the value of variables that are less tangible or directly measurable than the physical characteristics already mapped. Often the values that seek to express the potential of the land to support different types of vegetation are the predicted variables. Land carrying-capacity, biological productivity, and timber suitability are other examples. Land capability for agriculture, for example, might be predicted at a location from slope, soil texture, and water availability. An example is a crop suitability model by Hallada et al. (1981).

In the past, ecological modeling used manual map-overlay techniques. These were time consuming, cumbersome, and limited the accuracy of the model. Ecological modeling now generally uses computer procedures and the spatially registered datasets to estimate the desired variable. The model may use statistical regression, analysis of variance and covariance, log-linear modeling and contingency-table analysis and special purpose methods (Strahler et al., 1980). Ecological modeling at this level is best accomplished within a geographic-information-system framework.

Geographic Information Systems

Computer systems using spatially referenced data are referred to as geographic or geobased information systems (GIS). For a complete discussion see Chapter 22. Geographic information systems are rapidly evolving and are useful in land-resource planning and management. GIS will automate a large number of time consuming and expensive manual processes in map creation, manipulation, and storage. Geographic information systems developed from the need for increasing amounts of information about the land and for decisions on land-use management. These inspired new methods of data storage and retrieval. As more complex environmental relationships were understood, data on the physical environment were used more effectively in combination than separately. The structure and functions of geographic information systems constitute an entire research and development field. Here the central issues of geographic information systems and their relationship to remote sensing are emphasized.

The major thrust in the near future will be the combination of disparate data layers to provide new information layers. The concept of a data layer is fundamental to understanding geographic information systems. A data layer is a spatially registered data source describing a land area. Layers for a given area typically include data derived from remote sensing as well as maps of the physical environment. Data layers can also contain spatially related political or economic information, such as census tract boundaries and characteristics, which are not necessarily derived from the actual physical characteristics of the land.

More information can be extracted from remotely sensed data when combined with such data layers. Geographic information systems provide the means to combine disparate data types in an automated fashion.

Management of Computer-Assisted Interpretation Systems

Computer-assisted interpretation of remotely sensed data, especially in the context of digital inputs from GIS used to improve interpretation, provides an expanded set of management problems for the analyst. The most obvious problem is simple maintenance of the information-processing hardware. Present micro-electronic circuitry is relatively reliable in an increasingly large range of external environmental conditions. However, the hardware components crucial to digital remote sensing applications (e.g., mass storage devices such as tape and disk, film writers, scanners, plotters, etc.) are electromechanical devices with moving parts that wear and require significant amounts of maintenance and eventual replacement.

A more subtle management problem concerns the software used to both drive the various hardware devices and perform the analytic tasks for aiding interpretation. Intuitively, software should remain reliably static. In practice, software ages over time and requires periodic maintenance and upgrades to reflect the changing external environment; that is, the computer hardware changes that occur over time require software modifications. The format of data changes necessitates software upgrades. Techniques change, thereby requiring software modules to be replaced with new algorithms. Each change to a stable program increases the probability of an unanticipated error or bug in the system. These software changes and subsequent error corrections eventually consume large amounts of personnel time for programming, documentation updates, and staff training. Currently, approximately fifty percent of a programmer's time is spent maintaining existing software systems (as opposed to developing new applications). Since the personnel component is the largest segment of a data-processing budget, the remote sensing manager must pay special attention to ensure that the data-processing services budget does not consume an unreasonable portion of the total project funding.

A third problem area for management is personnel. Because computer-assisted interpretation hardware and software have become exceedingly complex, very few individuals thoroughly understand the intricacies of each technical area. Thus, the interpretation is usually accomplished through a multidisciplinary study team. Structuring this team to function together smoothly as a coordinated unit is perhaps the most challenging role for management. Because each member of the team speaks an obtuse technical dialect and is naturally biased towards a specific subcomponent of the analysis, a project manager must alternately function as a translator, practical psychologist, dictator, and arbiter. One real problem with computer-assisted interpretation teams is a tendency to over-achieve. Digital remote sensing has so much potential and so many untested techniques that the interpretation team as a whole frequently attempts to complete much more than the original goal of the analysis. A careful manager must contain the scope of work without smothering enthusiasm.

Finally, perhaps the most pressing managerial concern is data management, such as the capture, processing and archiving of both remotely sensed and conventional GIS data. Traditionally, GIS data, though expensive to capture in a machine-readable format, have been well structured and understood. For example, various data categories inputted into a conventional GIS are carefully explored in regard to thematic accuracy, spatial resolution, consistency, time frame of collection, etc. Frequently these data are laboriously transferred to stable base sheets and, where appropriate, updated by specialists in the particular discipline. Soil scientists, agriculturalists, urban planners, geologists, foresters, geographers, and other specialists participate in ensuring data integrity. The stable base sheets, once digitized, enter a careful editing/check plot cycle prior to entering the GIS data base. When data-base construction is complete, conventional GIS methodology focuses on deterministic models such as suitability/capability analysis and locational monitoring. Overall, the entire process in conventional geoprocessing has been extremely amenable to rigorous project planning in a production mode.

Digital processing of remotely sensed data has, on the other hand, been structured in a research fashion. The quality (e.g., spectral, spatial, radiometric, and temporal resolution) of remotely sensed data is not necessarily predictable; it depends on a long chain of variables such as correct sensor functioning, platform stability, and atmospheric conditions. Hence, the *modus operandi* for digitally assisted interpretation has been to process the data, finalize the classification taxonomy, and estimate the resolution parameters from the data. The research structure of the interpretation task has tended to inhibit the formalization of usual production-level management techniques such as extensive pre-project planning, rigorous quality-assurance checks, and consistent managerial review in order to ensure compliance with planned project work flow, etc. In short, managers of computer-assisted interpretation systems have not had adequate data-management guidelines.

Management Strategies

Following is a set of strategies for the operations-level manager that can help provide the structure necessary for improved, cost-effective computer-assisted interpretation. The strategies are divided into categories of hardware, software, personnel, and data. Computer-assisted interpretation is accomplished through an inter-related system composed of these four components. Altering one component can cause radical fluctuations in the other components. Finally, the cornerstone for sound management is planning. A readily proven theorem of management is that increasing the allocation of resources for project planning increases the probability of completing the project on time and within budget. Planning improves the quality of a project.

Hardware

As a remote sensing specialist, an operations-level manager usually does not have a technical background in data-processing hardware. The manager must overcome this deficiency; there is simply no substitute for learning the basics of hardware functioning, even if the staff includes a hardware specialist. One of the fastest ways to obtain the necessary education is to review the appropriate section of DATAPRO[7] (1978) (most university libraries have a copy). It is not necessary for the manager to achieve the sophistication of an electrical engineer. A fundamental knowledge of how the various devices function is sufficient. Finally, the simplest method for ensuring hardware integrity through the interpretation task is to put all hardware components under comprehensive maintenance contracts. While this raises the cost of the process, it considerably reduces managerial concerns about hardware functioning.

Software

With advance planning, the managing of software resources can be relatively painless. One should never develop new software if a reasonable set exists. Some image processing software for remote sensing applications is inexpensive. For example, if an IBM system 360/370 architecture is available, the VICAR system can be obtained for under $2,000. The hundreds of man years of effort in VICAR simply cannot be replicated without the expenditure of millions of dollars. For a non-IBM hardware environment other systems (e.g., ELAS, mini-VICAR, TAE, etc.) are available. Hardware/software systems marketed by private companies are usually very powerful and cost effective.

When a new capability for an existing system is needed both public domain and private sources may have a suitable module. A search through the National Technical Information Service or COSMIC (NASA, 1977), for example, may yield several suitable programs available for a fraction of in-house development costs.

[7] DATAPRO is a registered trademark of the DataPro Corporation, Englecliff, NJ.

Finally, the programming staff should be provided with a set of tools for simplifying its job. A program librarian should be appointed to organize, code and build a subroutine library, develop programming standards and standardize interfaces for modules. Staff members should be encouraged to pursue professional development through associations and special interest groups that are concerned with software engineering.

Personnel

The newness and fascinating nature of digital image processing attracts bright, energetic personnel. A key strategy for harnessing this energy in productive channels is to require cross-training of the various specialists on interpretation teams. For example, data-processing personnel should develop competence in image interpretation, sensor characteristics, etc. Staff with primarily interpretation skill should become familiar with data-processing concepts, end-user requirements, etc. An example of a cost-efficient method for conducting cross training consists of having each member of an interpretation team proof the sections of final reports written by the other specialists in the team. In this example, each professional gradually develops an understanding of the entire interpretation task and renders a final product more presentable for an end user.

Another managerial technique, perhaps the most productive, is to document all aspects of the interpretation task such that any of the professional staff can complete an interpretation and analysis with minimal direction from technical specialists. This strategy allows highly motivated staff members to move ahead with interpretation tasks. Such documentation represents a competent, organized managerial policy. Surprisingly, however, it is extremely difficult to maintain a clear, concise updated procedures manual.

Data Management

Integrating data from remote sensors with data from conventional sources in a GIS data base is a complex multistep task. To successfully complete the process, the entire data flow should be mapped. This function (data management) encompasses the defining of data requirements and monitoring the flow of data and information through the interpretation task.

The first stage in data management is a meeting of all members of the multidisciplinary interpretation team. Based on the formalized project goal and a definitive list of sensor characteristics, the team should identify the various GIS data types that will improve the final interpretation. A list of these data items should be developed and then thoroughly evaluated in terms of availability, cost, feasibility, accuracy, and resolution. The final list of acceptable data can be divided into raw data inputs (existing data from field studies) or derived data (calculated or generated from raw data). All information describing the final data items (data

and method of capture, resolution, estimated accuracy, etc.) should be documented. This information is critical for updating the interpretation at a later date (e.g., in a monitoring project repeated at selected intervals of time).

The interrelationships and methods involved in processing collateral data in the interpretation task should then be formalized in a data-structure diagram. The analytical procedures for integrating the various data bases should be critically evaluated. Frequently, alternative data flows with preferred processing characteristics can be identified. When the data flow through the interpretation task is finalized, the procedure should be translated into operational terms. For example, both the file names and sizes for each of the intermediate processing stages should be documented. Based on such specific information, computer storage and job-processing requirements can be scheduled.

As the data are processed via the formalized procedures, each intermediate file should be verified by a quality-assurance coordinator then "backed-up" on a low-cost storage medium such as a tape or floppy disk. The periodic back-ups not only guard against potential data loss due to system failure (e.g., hardware, software or human error) but also ensure the integrity of derived products that may be used in repeated monitoring studies at minimal processing costs. Finally, after completion of the interpretation task, all materials, documents, archived data, etc., should be clearly marked with external labels and placed in a secure, temperature-controlled storage area.

Trends and Futures

At the broadest level, the most significant trends in computer-assisted image interpretation appear to be 1) the widespread use of interactive image-processing systems as a result of lower system costs and the general value of analyst interaction; 2) the integration of image processing and geographic information system techniques; and 3) a general increase in the level of algorithm sophistication, especially in the types of input features (e.g., textural, contextual, etc.) and the methods by which they are used.

These general trends have greatly increased our overall understanding of what computer-aided interpretation systems can and cannot do. The most significant observation of the present state-of-the-art for automated interpretation concerns the general lack of operational applications. Research and demonstration projects to date have increased our awareness of the inherent complexity of the human interpretation process; a better understanding of that process now provides a substantial basis for further refinement of computer-assisted techniques.

Sensors and Platforms

Sensor-system technology is rapidly advancing and affects interpretation techniques. Improved spectral-signature sampling has provided a basis for discrimination of classes based on multispectral signatures. Of critical importance are improvements in spatial resolution allowing finer spatial detail and higher order features (texture, size, shape, etc.) to be incorporated. These improvements come at the cost of substantial increases in the volume of data to be processed. Fortunately, computer data-rate capabilities have also increased.

The second generation of remote sensing satellites, as represented by the thematic mapper system of Landsat-4 and SPOT, will also involve many system characteristics of presently unknown impacts. For the thematic mapper these include:

1) a lower orbit and a smaller IFOV,
2) non-simultaneous scanning of a given point on the ground by each of the seven spectral bands due to placement of the detector array within the focal plane,
3) the need for precise time-registration and bidirectional scan repeatability to effect band registration, and
4) the data handling system.

Hardware

The development of microprocessors and low cost memories has dramatically lowered the cost of computing equipment essential to digital image-processing. Powerful minicomputers and microcomputers provide ever better price/performance ratios for interactive image processing. Mainframe systems continue to show improved largescale throughput capability but interactive requirements still favor a minicomputer type of environment. The availability of modern array processors now enables minicomputers to effectively compete with the larger systems.

Computer memory, both solid state and disk, has improved substantially in recent years in both capacity and price. Future improvements appear possible but quantum increases will require new technology such as video disks. Memory is very important to image processing because it dramatically impacts overall throughput.

One of the more interesting recent trends has been the development of custom image-processing integrated circuits. The incorporation of these chips into new image-processing systems promises great throughput increases for specific operators at the cost of custom hardware and a general decrease in flexibility. Thus continues the balancing between general-purpose hardware (slower but more flexible) and special-purpose hardware (faster but more difficult to program). This balancing trend will continue. Substantial ground has been gained by custom processors based upon the essential need for interactive graphics. Graphics hardware provides an ideal basis for incorporating advanced capabilities, and it features modularized structure for implementation.

Software

The availability of low cost hardware has largely removed the equipment hurdles to computer-assisted interpretation. Although major improvements have been made in software, it remains a critical item. Sophisticated computer hardware is available for image processing, yet the need for low cost, transportable software remains. Most commercially available image-processing software is general purpose. Few vendors supply advanced algorithms specific to remote sensing applications, such as multispectral classifiers. Most vendors prefer to sell only packaged systems of hardware-software. Several outstanding packaged sysems are available, each costing in excess of $250,000. Software may be overpriced if purchased separately.

Because the source code of most commercial systems is never released, it is very difficult to engage in cooperative projects and/or transfers of specific programs from one system to another. Public domain source-code is readily available at low cost (NASA, 1977). The issues of true transportability remains. Many of the larger public domain packages (developed for mainframe environments) are not well suited for interactive image analysis.

Several recent programs have been directed towards image analysis and software transportability. These include a project at the University of Maryland funded by the National Science Foundation and directed towards transportable, general image processing software, basically from a computer-vision perspective. A NASA-funded project, TAE, directly addresses the need for transportable remote sensing-oriented software (NASA, Goddard Space Flight Center, 1981).

Another factor influencing image-processing systems is the trend towards integration with other data-processing systems. Significant early work intergrated remote sensing inputs and geographic information systems. Subsequent work featured geobased collateral data to stratify, correct and aid remote sensing analyses.

Since many geographic information systems operate with vector-data structures, rather than the grid or raster image-structure, the need to transform one data structure to another arose; such conversion programs are becoming widespread. Conversely, the value of geobased data in image-processing analysis typically requires the transformation of vector data into a grid format. This conversion is easier and such algorithms have been available for some time. The integrated IBIS (Image Based Information System) and VICAR Software is a current example of geographic information system flexibility (Bryant and Zobrist, 1979). It is, however, just a developmental step to a data-base management system (DBMS) integrated with a remote sensing image-processing system.

Interpretation
Spatial Context

The spatial context in images will become an important aspect in computer interpretation. One way in which spatial context had been incorporated into information-extraction algorithms examines the local neighborhood (e.g., the adjacent line and sample pixels) of each pixel. Spatial context at this very primitive level is described by conditional probabilities relating the central and neighborhood pixel gray values. Though far from emulating human interpretation, significant improvement in classification accuracy has been noted for simulated urban data at Landsat resolution and simulated agricultural data at Thematic Mapper resolution (Landgrebe, 1981).

Another approach uses relaxation methods. Correlations or conditional probabilities of occurrence of labels between objects and their neighbors are determined. None of the algorithms using spatial characteristics is in wide use. Operational systems have not yet grown in sophistication much beyond the maximum-likelihood classifier. The use of texture while promising for improving accuracy, brings with it a much increased computational load. The same is true of the contextual classifiers. None adds significantly to a solution of the classifer-training problem. The contextual and relaxation methods require the estimators of class-conditional probabilities. In the long term, broader spatial relationships may be considered. The use of syntactic methods for treating structural relationships in a scene is one approach (Landgrebe, 1981).

In semantic or syntactic computer-scene analysis, the purpose is to automatically produce a description of the image content similar to one obtained from a skilled human observer. The extracted features (or elements of interpretation) are represented by a symbolic description such as a labeled graph or semantic network. As such it is suitable for high level processing (Faugeras and Price, 1981). In Chapter 18 details are given with respect to syntactic pattern recognition.

Expert Systems

Another contemporary trend in image interpretation is the move from numerical decision-theory models (e.g., maximum likelihood and Bayesian) to the more complex, heuristic (or exploratory) models in artificial intelligence. Research there uses new approaches to knowledge representation, language understanding, heuristic search, and symbolic reasoning directly pertinent to advanced pattern recognition. The complexity of using higher order elements of interpretation (e.g., size, shape and association) made this model change necessary. A review of these techniques has been provided by Shortliffe, et al. (1979).

Analysis of many human decision-making processes suggests that as decisions move from simple to complex the reasoning style becomes less algorithmic and more exploratory. The inclusion of subjective knowledge increases. The artificial-intelligence techniques in expert systems follow this approach. An expert system (acting as a human expert) blends fundamental knowledge with the practioner's wisdom and skill in the controlled application of data, knowledge, and tools (Hayes-Roth, 1981). Expert systems usually do not use algorithms; the decision structure is more complex with substantial information specific to the discipline incorporated. Expert systems typically converse with the user or analyst in simple terms. This conversion mode is critical for using knowledge from contributing experts and user. VISIONS and ACRONYM are two image-understanding expert systems. Expert systems that can now configure, in minutes, the complete architecture of a complex computer system might eventually close the gap between human and computer interpretation-performance.

INTERPRETATION OF PHOTOGRAPHIC (CAMERA) IMAGERY

THE VERTICAL AERIAL PHOTOGRAPH

An aerial photograph is said to be vertical if the camera optical axis is plumb or nearly so (Color Figure 24-58). If the axis is exactly plumb, the photograph is said to be truly vertical and to have zero tilt. In spite of precautions taken to keep the camera axis vertical, small tilts are invariably present. Vertical airphotos normally exhibit tilts that are less than 1° and rarely exceed 3°. Photographs containing these small unintentional tilts are called near-vertical. For many practical purposes such photos may be analyzed using the "truly vertical" equations in this chapter without serious error.

Figure 24-59 illustrates the geometry of a vertical photograph taken from an exposure station L. The negative, which is a reversal in both tone and geometry of object space, is situated a distance equal to the focal length (distance o'L on Figure 24-59) behind the rear nodal point of the camera lens. The positive may be obtained by direct emulsion-to-emulsion "contact printing" with the negative. This process produces a reversal of tone and geometry from the negative and therefore the tone and geometry of the positive are exactly the same as those of the object space. Geometrically the plane of contact-printed positive is situated a distance equal to the focal length (distance oL on Figure 24-59) in front of the front nodal point of the camera lens. The reversal in geometry from object space to negative is readily seen by comparing the positions of object points A, B, C, and D (Figure 24-59) with their corresponding negative positions a', b', c', and d' (Wolf, 1974). The correspondence of the geometry of the object space and the positive (contact print) is also apparent. The photographic coordinate axes x and y, radiating from the principal point, are shown on the positive. Because the photointerpreter normally works with positive prints or diapostive transparencies, major interest is in the positions of images in the positive plane: all line drawings in the rest of this section depict the positive plane and

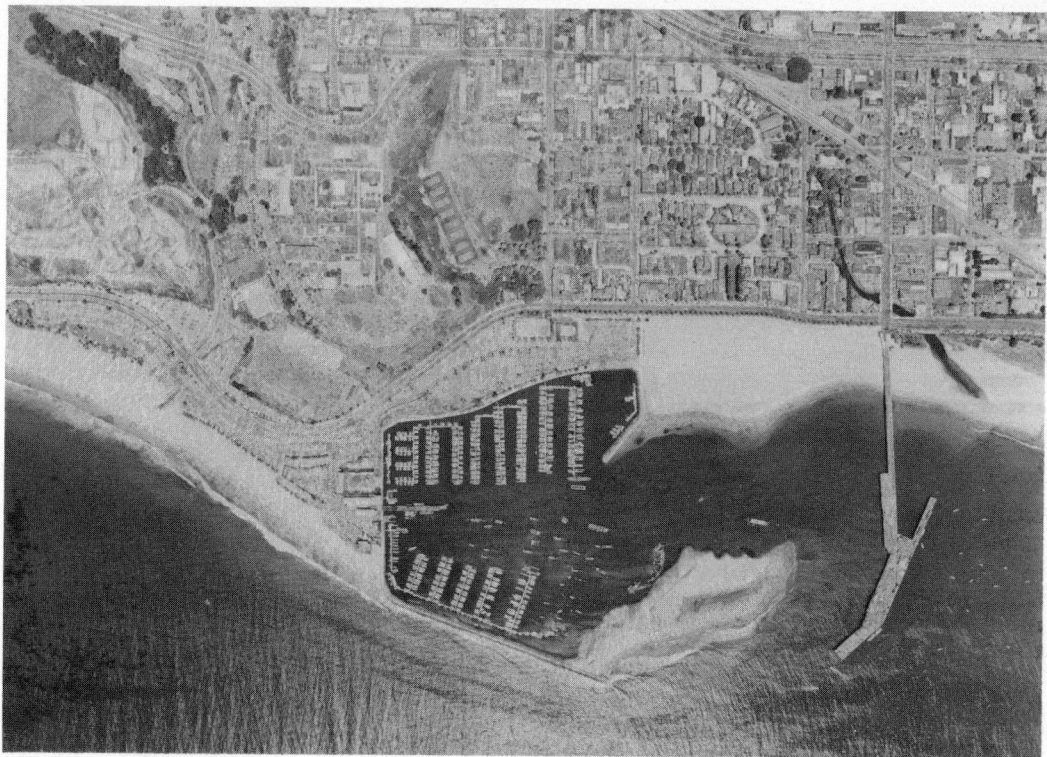

Fig. 24-58. Vertical aerial photograph of Santa Barbara Harbor; original scale approximately 1:10,000 (Courtesy Pacific Western Aerial Surveys).

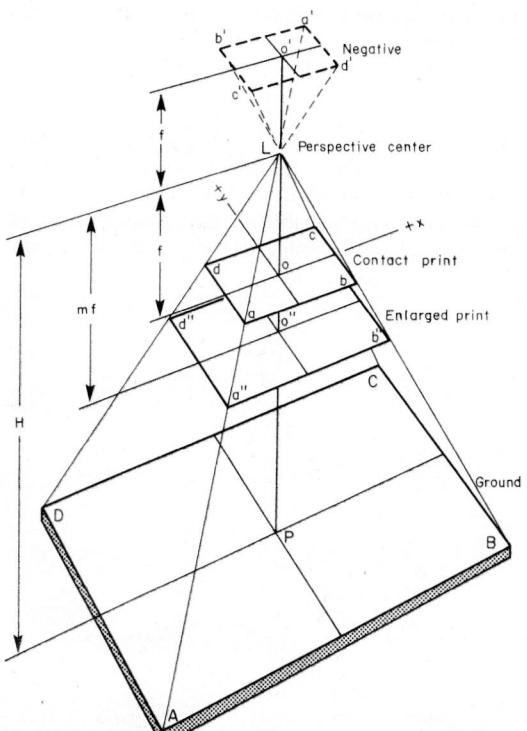

Fig. 24-59. Geometry of a vertical aerial photograph (After Wolf, 1974).

not the negative plane. Photographic enlargement at ratio m has exactly the same geometric effect on an aerial photograph as taking the photograph with a camera whose focal length is m times f; but if there are elevation differences in the terrain, the effect is not identical to taking the photograph from a different flight altitude (H divided by m).

Scale

The scale of a photograph is the ratio of a distance on the photo to that same distance on the ground. An aerial photograph is a perspective projection and its scale varies with variations in terrain elevation.

Scales may be expressed either as unit equivalents, dimensionless representative fractions, or dimensionless ratios. If, for example, 1 inch on a map or photo represents 2,000 feet (24,000 in.) on the ground, the scale expressed in three ways is:

1) Unit equivalents: 1 in. to 2,000 ft.
2) Dimensionless representative fraction: 1/24,000
3) Dimensionless ratio: 1:24,000

It is helpful to remember that a large number in a scale expression denotes a small scale. For example, if the scale fractions 1/11,000 and 1/12,000 are considered, their values in decimal form are 0.000091 and 0.000083, respectively, and hence the former is said to be the larger of the two

scales. Also, the analyst/interpreter should remember that an image of a given object is actually
larger on the larger scale photograph.

Scale of a Vertical Aerial Photograph Over Flat Terrain

The scale of a vertical aerial photograph taken
over flat terrain is simply the ratio of the photo
distance ab to the corresponding ground distance AB.

$$S = \frac{ab}{AB} \qquad (24\text{-}1)$$

Scale may also be expressed in terms of camera
focal length, f, and flying height above ground, H,
by equating the similar triangles Lab and LAB as
follows:

$$S = f/H \qquad (24\text{-}2)$$

From Eq. 24-2, it is seen that the scale of a
vertical photo is directly proportional to camera
focal length (image distance) and inversely proportional to flying height above ground (object
distance).

To use Eq. 24-1, the interpreter identifies on the
photograph an object, the true length of which
(AB) has been measured on the ground or, more
commonly, has been obtained from some reliable
collateral source such as a map. The corresponding distance (ab) on the photograph is then measured to calculate S.

Example 1. Assume that the length of a runway
(AB) is 1,200 ft. and the distance (ab) between
corresponding images on the vertical photograph is 6 in. Using Eq. 24-1 the scale of the
photograph is found as follows:

$$S = \frac{ab}{AB} = \frac{6 \text{ in.}}{1,200 \text{ ft.}} = \frac{0.5 \text{ ft.}}{1,200 \text{ ft.}}$$

$$= 1/2,400 = 1:2,400$$

To offset the effects of tilt, the scale should be
based on the average of two scale-check lines instead of one. It is preferable to have lines that are
relatively long, that intersect approximately at
right angles, and that are centrally located on the
photograph. If Eq. 24-2 is to be used, the interpreter need not identify an object of known
length on the photograph. If the focal length of the
camera and the altitude of the photograph are unknown, the interpreter normally consults the tilting information appearing on the photograph, the
flight roll, or a mission summary.

Example 2. A vertical aerial photograph is
taken over flat terrain with a 6-in. (152.4-mm)
focal-length camera from an altitude of 12,000
ft. above ground. Using Eq. 24-2 the scale is
found as follows:

$$S = \frac{f}{H} = \frac{6 \text{ in.}}{12,000 \text{ ft.}} = \frac{1 \text{ in.}}{2,000 \text{ ft.}}$$

$$= 1/24,000 = 1:24,000$$

Scale of a Vertical Aerial Photograph Over Variable Terrain

One of the principal differences between a
photograph (excluding orthophotographs) and a
map is that for vertical photographs taken over
variable terrain there can be a wide range of
scales. If the topographic elevation decreases, the
photographic scale will be smaller. Conversely, if
higher topographic features are encountered, the
photographic scale will be increased because the
land will be closer to the camera. In Figure 24-60 a
vertical aerial photograph is taken over variable
terrain from exposure station L. Ground points A
and B are imaged on the positive at a and b, respectively. For the photographic scale at h, the
elevation of points A and B, is equal to the ratio of
photo distance ab to ground distance AB. By similar triangles Lab and LAB an expression for
photo scale S_{AB} is

$$S_{AB} = \frac{ab}{AB} = \frac{La}{LA} \qquad (24\text{-}3)$$

Also, by similar triangles LO_AA and Loa.

$$\frac{La}{LA} = \frac{f}{H - h} \qquad (24\text{-}4)$$

Substituting Eq. 24-4 into Eq. 24-3,

$$S_{AB} = \frac{ab}{AB} = \frac{f}{H - h} \qquad (24\text{-}5)$$

In general, by dropping subscripts, the scale at
any point whose elevation above datum is h, may
be expressed as

$$S = \frac{f}{H - h} \qquad (24\text{-}6)$$

It is conventional to use an average scale to
define the overall mean scale of a vertical photograph taken over variable terrain. Average scale is
the scale at the average elevation of the terrain

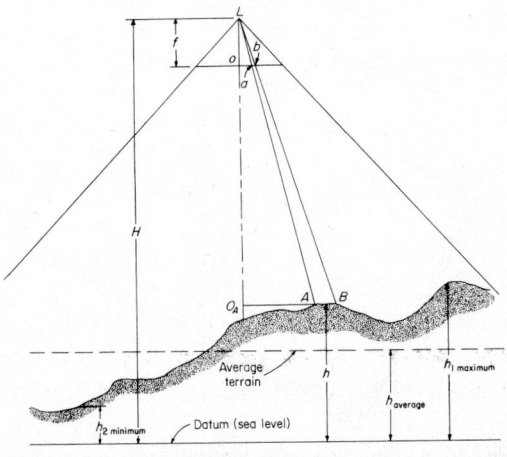

Fig. 24-60. Scale of a vertical photograph over variable
terrain (After Wolf, 1974).

covered by a particular photograph and is expressed as:

$$S_{avg} = \frac{f}{H - h_{avg}} \tag{24-7}$$

In using the average scale, it must be understood that it is exact only at those points lying at average elevation; it is only an approximate scale for all other areas of the photograph.

Example 3. Suppose the highest terrain, h_1, average terrain, h_{avg}, and lowest terrain, h_2, of Figure 24-60 are 2,000, 1,500, and 1,000 ft. above mean sea level, respectively. Calculate maximum scale, minimum scale, and average scale if flying height above mean sea level is 10,000 ft. and camera focal length is 6 in. (152.4-mm).

By Eq. 24-6 (maximum scale occurs at maximum elevation),

$$S_{max} = \frac{f}{H - h_1} = \frac{6 \text{ in.}}{(10,000 - 2,000) \text{ ft.}}$$

$$= \frac{6 \text{ in.}}{8,000 \text{ ft.}} = 1{:}16{,}000$$

(Minimum scale occurs at minimum elevation.)

$$S_{min} = \frac{f}{H - h_2} = \frac{6 \text{ in.}}{(10,000 - 1,000) \text{ ft.}}$$

$$= \frac{6 \text{ in.}}{9,000 \text{ ft.}} = 1{:}18{,}000$$

By Eq. 24-7

$$S_{avg} = \frac{f}{H - h_{avg}} = \frac{6 \text{ in.}}{(10,000 - 1,500) \text{ ft.}}$$

$$= \frac{6 \text{ in.}}{8,500 \text{ ft.}} = 1{:}17{,}000$$

Other Methods of Determining Scale of Vertical Photographs

The scale of a vertical aerial photograph may also be determined if a map covering the same area as the photo is available. In this method it is necessary to measure the distances between two points which can be identified on both the photo and the map. Photographic scale can then be calculated from the following equation:

$$S = \frac{\text{photo distance}}{\text{map distance}} \times \text{map scale } (RF) \tag{24-8}$$

Example 4. On a vertical photograph the length of an airport runway measures 6.30 in. On a map plotted to a scale of 1:24,000, the runway scales 4.06 in. What is the scale of the photograph at runway elevation? From Eq. 24-8:

$$S = \frac{6.3 \text{ in.}}{4.06 \text{ in.}} \times \frac{1}{24,000} = 1.55 \text{ in.} \times \frac{1}{24,000}$$

$$= \frac{1.55 \text{ in.}}{24,000} = \frac{1}{15,484} \text{ or, 1 in.} = 1{,}290 \text{ ft.}$$

The scale of a vertical aerial photograph can also be determined if lines whose lengths are known by common knowledge appear on the photo. A section line of a known mile length, a football field, or a baseball diamond could be measured on the photograph, for example, and photographic scale calculated as the ratio of the photo distance to the known ground distance.

Example 5. What is the scale of a vertical aerial photograph on which a section line measures 5.93 inches (15.05-cm) and is assumed to be 1 mile long on the ground? Solution: since the Photo scale is the ratio of the measured photo distance to the ground distance, or

$$S = \frac{5.93 \text{ in.}}{5,280 \text{ ft.}} = \frac{.494 \text{ ft.}}{5,280 \text{ ft.}} = 1{:}10{,}688$$

For use of these two methods it must be remembered that the calculated scale is correct only at the elevation of the ground line used to determine the scale (Wolf, 1974).

Methods of Computing Horizontal Distance on Vertical Photography

When the scale of a vertical photograph is known, it can be used to compute a horizontal ground distance (AB) by simple transposition of Eq. 24-8

$$AB = \frac{ab}{S} \tag{24-9}$$

For example, if the scale of a photograph is known to be 1:20,000 and the runway measures 4 in. on the photograph, the runway length can be calculated by dividng 1/20,000 into 4 in (.333 ft.), thereby obtaining a length of 6,667 ft.

Some image analysts prefer to use a technique known as the scale number. The scale number is the reciprocal of the RF scale; hence, if the RF = 1/20,000 then the scale number = 20,000. To find the length of an object from its photographic image, the analyst may multiply the image length by the scale number instead of dividing image length by the scale. In the previous example, the scale number is 20,000; hence the runway length is .333 ft. × 20,000 = 6,667 ft.

Another method is to measure the number of .001 ft. graduations between the photo points, a and b, and multiply this distance, ab, times a scale factor. The scale factor is obtained by dividing the RF scale by 1,000. The scale factor of 1/20,000 (RF scale) photography is 20. In the example, the runway would be 333 ft. × 20 = 6,660 ft.

The distance between two points on a vertical aerial photograph can be measured very accurately by determining the photo coordinates and ground coordinates of the points with respect to an arbitrary coordinate system. The arbitrary x and y ground axes are in the same vertical planes as the photographic x and y axes, respectively, and the origin of the system is at the datum principal point. The x and y photo axes are developed by drafting lines between opposite fiducial (collimating) marks. This principal point represent the photographic origin (0,0).

Figure 24-61 shows a vertical photograph taken at a flying height H above datum. Images a and b of the ground points A and B appear on the photograph, and their measured photographic coordinates are x_a, y_a and x_b, y_b, respectively. The arbitrary ground coordinate-axis system is X and Y, and the ground coordinates of points A and B in that system are X_A, Y_A, X_B, and Y_B (Figure 24-61). From similar triangles La'o and LA'A_0, the following equation may be written:

$$\frac{oa'}{A_oA'} = \frac{f}{H - h_A} = \frac{x_a}{X_A}$$

from which we derive

$$X_A = x_a \frac{(H - h_A)}{f} \qquad (24\text{-}10)$$

Also, from similar triangles La"o and LA"A_o, we may write

$$\frac{a'a}{A'A} = \frac{f}{H - h_A} = \frac{y_a}{Y_A}$$

from which we derive

$$Y_A = y_a \frac{(H - h_A)}{f} \qquad (24\text{-}11)$$

Similarly, the ground coordinates of point B are

$$X_B = x_b \frac{(H - h_B)}{f} \qquad (24\text{-}12)$$

$$Y_B = y_b \frac{(H - h_B)}{f} \qquad (24\text{-}13)$$

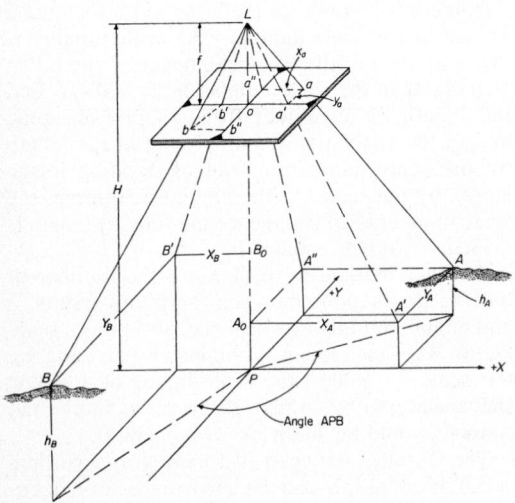

Fig. 24-61. Ground coordinates from a vertical photograph (After Wolf, 1974).

On examination of Eqs. 24-10 through 24-13, it is seen that the X and Y ground coordinates of any point are obtained by simply multiplying the x and y photo coordinates by the inverse of photo scale at the point. From the ground coordinates of the two points A and B, the horizontal length of the line AB can be calculated, using the Pythagorean theorem, as:

$$AB = \sqrt{(X_B - X_A)^2 + (Y_B - Y_A)^2} \quad (24\text{-}14)$$

In order to solve Eqs. 24-10 through 24-14, it is necessary to know the camera focal length, the flying height above datum, the elevations of the points above datum, and the photo coordinates of the points. Elevations of points may be obtained directly by field measurement, or they can be taken from collateral material (such as available topographic maps).

Example 6. A vertical aerial photograph was taken with a 6-in. (152.4-mm) focal length camera from a flying height of 4,530 ft. above datum. Images a and b of two ground points A and B appear on the photograph, and their measured photocoordinates (corrected for shrinkage and distortions) are $x_a = -52.35$-mm, $y_a = -48.27$-mm, $x_b = 40.64$-mm, and $y_b = 43.88$-mm. Determine the horizontal length of the line AB if the elevations of points A and B are 670 and 485 ft. above datum, respectively.

Solution from Eqs. 24-10 through 24-13

$$X_A = \frac{-52.35}{152.4}(4,350 - 670) = -1,326 \text{ ft.}$$

$$Y_A = \frac{-48.27}{152.4}(4,530 - 670) = -1,223 \text{ ft.}$$

$$X_B = \frac{40.64}{152.4}(4,530 - 485) = 1,079 \text{ ft.}$$

$$Y_B = \frac{43.88}{152.4}(4,530 - 485) = 1,165 \text{ ft.}$$

From Eq. 24-14,
$$AB = \sqrt{(1,079 + 1,326)^2 + (1,165 + 1,233)^2}$$
$$= \sqrt{(2,405)^2 + (2,388)^2} = 3,389 \text{ ft.}$$

Ground coordinates calculated by Eqs. 24-10 through 24-13 are in an arbitrary rectangular coordinate system, as previously described. If arbitrary coordinates are calculated for two or more "control" points (points whose coordinates are also known in an absolute ground coordinate system such as the state plane coordinate, or Universal Transverse Mercator system) then the arbitrary coordinates of all other points for that photograph can be transformed into the ground system. Through the use of Eqs. 24-10 through 24-14, an entire planimetric survey of the area covered by a photograph can be made (Wolf, 1974).

Height Measurement

Relief Displacement and Height Measurement on a Single Vertical Aerial Photograph

The image of any feature lying above or below the horizontal ground surface, which may be defined as a horizontal plane passing through the elevation of the photo nadir point, is displaced on a vertical photograph from its true plan position. Relief displacement is outward for points whose elevations are above datum and inward for points whose elevations are below datum. The direction of relief displacement is radial from the photograph's nadir, the image point directly beneath the camera at the instant of exposure (see Figure 24-62). In a truly vertical photograph, the nadir coincides with the principal point.

The amount of relief displacement d, on a vertical photograph, is:

a) directly proportional to the difference in elevation, h, between the datum and the object whose image is displaced.

Greater displacements and potential measurement errors are present when portions of the terrain shown on the photograph are considerably above or below the mean datum. Figure 24-63 depicts the additional displacement due to appreciable differences in terrain relief. Tank "A" is situated in a depression, which causes it photographic image to be displaced inward. Tank "B" stands on a hill, causing its photographic image to be displaced excessively outward. Tanks "C" and "D" (Figure 24-63) lie on level ground and are normally displaced with respect to the principal point.

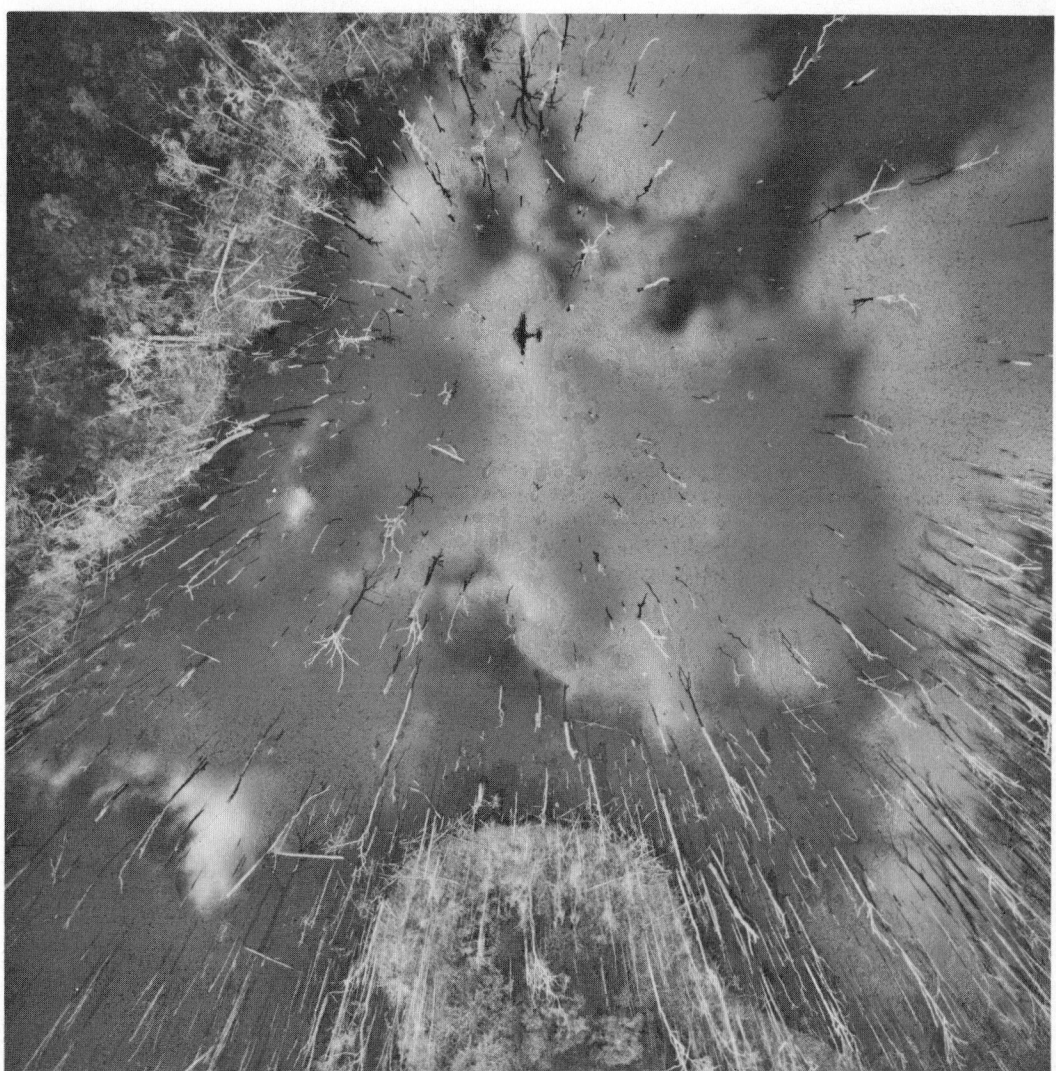

Fig. 24-62. Image displacement increases radially from the principal point (nadir) on vertical imagery (Forest Management Institute, Canadian Forestry Service, and Capital Air Surveys).

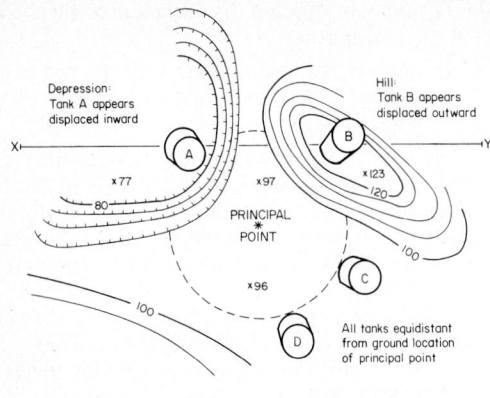

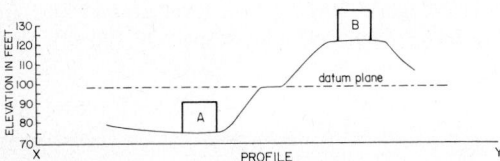

Fig. 24-63. The effect of terrain relief on image displacement (After Branch, 1973).

The amount of relief displacement d, as exhibited on a vertical photograph by any particular feature is:

a) directly proportional to h (the difference in elevation between the feature and the datum),

b) directly proportional to the radial distance, r, between the displaced image and the principal point. (Figure 24-62 reveals that the greater the distance from the principal point, the greater is the image displacement), and

c) inversely proportional to the altitude, H, of the camera above the datum. Thus, a reduction in relief displacement can be achieved by increasing the flying height. These relations are expressed mathematically as:

$$d = \frac{hr}{H} \qquad (24\text{-}15)$$

This equation, in the transposed form

$$h = \frac{Hd}{r} \qquad (24\text{-}16)$$

is frequently useful for finding the height of an object from its relief displacement on a single vertical photograph.

If the top and bottom of the object whose displacement is to be found are both clearly visible on the image there is no difficulty in finding the value of d. In Figure 24-64, for example, if H is known to be 2400 feet, r is measured as .119 ft., and d is measured as .023 ft., the height, h of the radio tower is found as follows:

$$h = \frac{2400 \text{ ft.} \times .023 \text{ ft.}}{.119 \text{ ft.}} = 464 \text{ ft.}$$

Although this relief displacement formulas Eqs. 24-15 and its transposed form 24-16 are exact only for truly vertical photographs, they give quite accurate answers when applied to any photograph having tilt of 3° or less. If a photograph is tilted more than 3°, it should be rectified prior to measurement.

Measurement of Height by Stereoscopic Parallax

The greatest distance at which the average observer can perceive relative depth in nature is about 500 meters. By increasing the air base or distance between successive exposures, it is possible to increase the distance at which depth can be perceived, creating a condition known as hyperstereoscopy. For example, in conventional aerial photographs flown at the scale of 1:20,000 with 60 percent forward lap, the air base is nearly 30,000 times greater than the eye base. In effect, the photointerpreter's eyes have been placed at the exact point of the sensor platform at the time of each exposure, thus increasing the interpupillary distance 30,000 times.

An aerial camera exposing overlapping photographs at regular intervals of time obtains a record of positions of images at the instant of exposure. The change in position of an image from one photograph to the next, caused by aircraft motion, is termed stereoscopic parallax or simply parallax. Parallax is the apparent displacement in the position of an object, with respect to a frame of reference, caused by a shift in the position of observation. Parallax is a normal characteristic of overlapping aerial photographs, and it comprises the basis for three-dimensional viewing. Differential parallax is the element that is used to determine elevations of objects and to draw contour lines with aerial photographs by means of stereoscopic instruments. Figure 24-65 is a line drawing showing the derivation of the parallax equation. Note that the two triangles, BTA and L_1L_2T whose apexes meet at the top of the tree, are similar. From similar triangles, the following relation is obtained:

$$\frac{h}{H-h} = \frac{dP}{P} \qquad (24\text{-}17)$$

where h is the height of the object to be measured; H is the altitude of the camera above the base of the object; P is the air base or absolute stereoscopic parallax at the base of the object being measured; dP is the differential parallax, that is, the shift in the apparent position of the top of the tree as seen from the two camera stations. The photo equivalent of P is the distance o_1o_2 from principal point to conjugate principal point on either of the two photographs. In the more general situation where points o_1 and o_2 and the base of the object are at different elevations, P is the arithmetic sum of $o_1X_1 + o_2X_2$. The photo equivalent of dP is the apparent x-displacement of the top of the tree when viewed on the stereoscopic pair of photo-

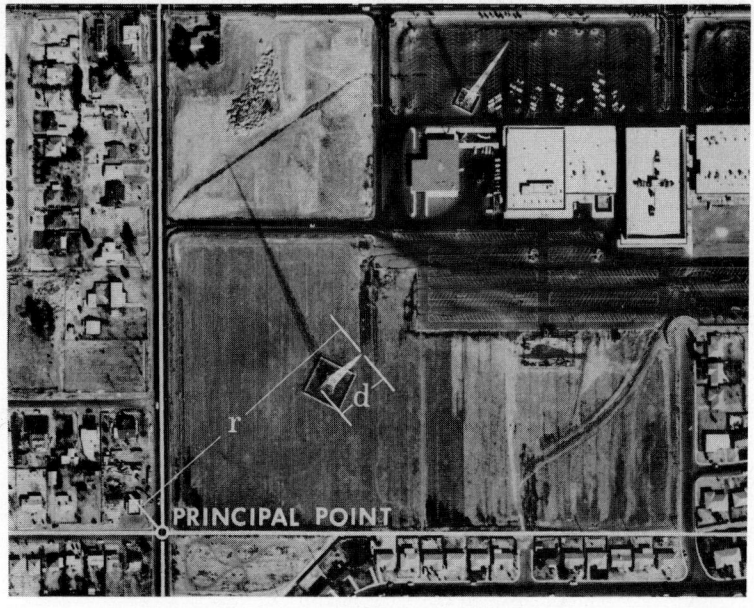

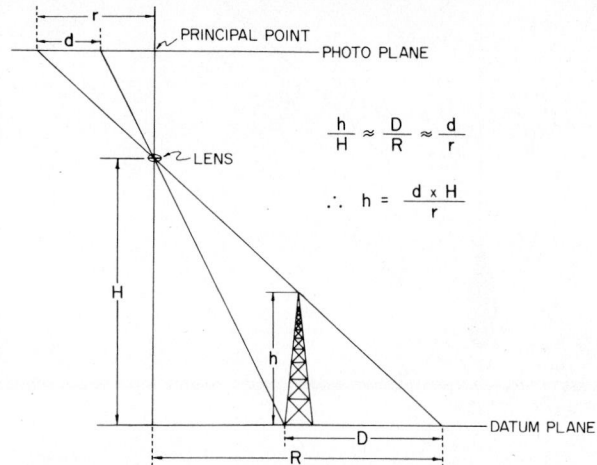

$$\frac{h}{H} \approx \frac{D}{R} \approx \frac{d}{r}$$

$$\therefore \ h = \frac{d \times H}{r}$$

Fig. 24-64. Determination of the height of a radio tower in Phoenix, Arizona from its relief displacement on a single vertical photograph. Only the upper right quadrant of the photograph is shown. Original scale 1:4800.

graphs. The quantity dP cannot be measured on either photograph alone; it is the arithmetic sum of the quantities dP_1 as measured on the left photograph and dP_2 as measured on the right photograph. In practice, the quantity dP is obtained by means of a parallax wedge or floating dot instrument.

Strictly speaking, dP is the algebraic difference dP_1-dP_2, the mathematical sign for each value being determined with respect to a system of coordinates originating from the nadir which, in a truly vertical photograph, corresponds with the principal point (an example of such a photo coordinate system was shown in Figure 24-61). Thus in Figure 24-65,

$$h = \frac{H \cdot dP}{P + dP} \qquad (24\text{-}18)$$

In Figure 24-66, the direct form of the parallax equation is used to calculate the height of the radio tower. The scale is 1:4800 at the base of the tower with a flying height above ground, H, of 2400 ft. The average photo base length P for the stereopair is .183 ft. The differential parallax dP is .144 ft. − .100 ft. = .044 ft. Substituting this in the parallax equation yields

$$h = \frac{H \cdot dP}{P + dP} \ \frac{2400 \text{ ft.} \times .044 \text{ ft.}}{.183 \text{ ft.} + .044 \text{ ft.}} = \frac{105.6 \text{ ft.}^2}{.227 \text{ ft.}}$$

$$= 465 \text{ ft.}$$

The actual height of the tower is 465 ft. Usually Eq. 24-17 is more accurate than Eq. 24-16 due to the observer's high stereoscopic acuity, the higher resolution of images observed in stereoscopic pairs, and the great accuracy of parallax-

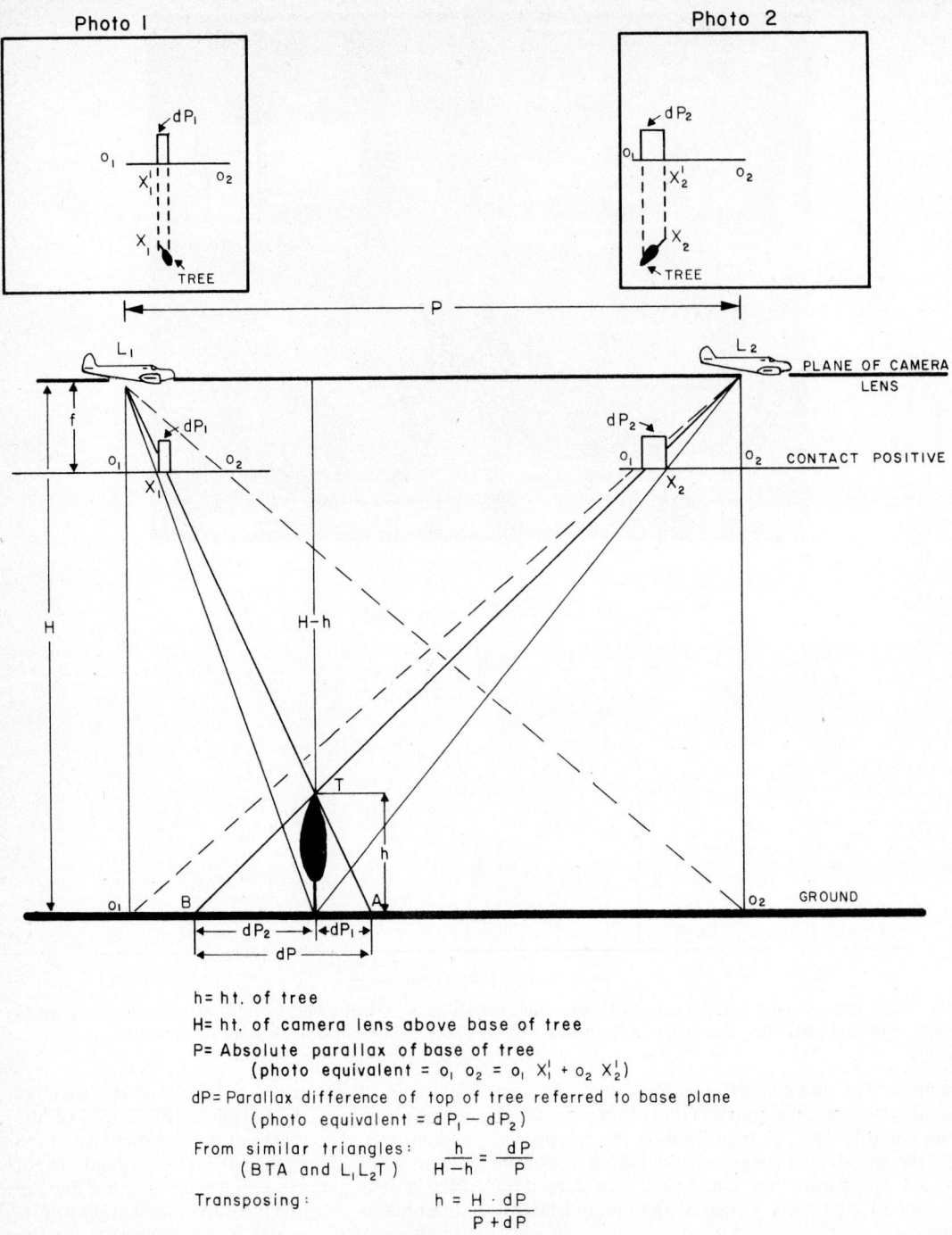

h= ht. of tree

H= ht. of camera lens above base of tree

P= Absolute parallax of base of tree
    (photo equivalent = $o_1 o_2 = o_1 X_1^1 + o_2 X_2^1$)

dP= Parallax difference of top of tree referred to base plane
    (photo equivalent = $dP_1 - dP_2$)

From similar triangles: $\dfrac{h}{H-h} = \dfrac{dP}{P}$
  (BTA and $L_1 L_2 T$)

Transposing: $h = \dfrac{H \cdot dP}{P + dP}$

Fig. 24-65. Line drawing showing derivation of the parallax equation (From Colwell, 1955).

measuring devices available to the image analyst. The scale limit for obtaining accurate height measurements from single or stereo vertical imagery is a function of resolution, atmospheric condition, object relief, and the accuracy of the measuring device. Unless hyper-altitude systems have increased resolutions at the small scales at which the imagery is acquired, most of the manual

height-measuring techniques used in this discussion will be restricted to large- and medium-scale imagery.

Measurement of Height by Shadow Measurement

Because the rays of the sun are essentially parallel throughout the area shown on vertical aerial

Fig. 24-66. Stereo-pair of a radio tower in Phoenix, Arizona. Original Scale 1:4800. Note the displacement of the tower parallel to the line of flight. The direct measurement of differential parallax (dP) can be derived by (1) adding dP^1 + dP^2; or (2) calculating the distance between the top of the tower on two overlapping photographs and subtracting this number from the distance between the tower bases.

photographs, the length of an object's shadow on a horizontal surface is proportional to its height. Figure 24-67 illustrates the trigonometric relationship involved in determining object heights from shadow measurements;

$$height = shadow \times \tan a \qquad (24\text{-}19)$$

The sun's elevation (altitude) may be computed if sharply defined objects of known height are formed on the photograph. For example, if a building is known to be 100 meters tall and casts a shadow 75 meters long on level ground, the tangent of angle a can be found by transposing Eq. 24-19

$$\tan a = \frac{\text{ht. of object in m}}{\text{shadow length in m}} = \frac{100}{75} = 1.33$$

Now, other shadows on the same stereopair can be measured, their lengths multiplied by 1.33, and the heights of the corresponding objects determined.

Another method employing solar altitude is the formula

$$h = \frac{(H)(1)(\tan \text{ solar altitude})}{f}, \qquad (24\text{-}20)$$

where h is the height of the object, H the altitude of photography l the length of the shadow, and f the focal length of the camera.

Estes and Simonett, (1975) describe three methods of finding the solar elevation angle. The first is based on observation of the reflected image of the sun in the vertical photograph; the second, on observation of the shadow of the camera-carrying aircraft in the vertical photograph; and the third, on observation of the images of two or more well-defined clouds and their shadows on the vertical photograph. A fourth method requires the analyst to know the sun's declination a on the

day of photography (a value usually obtained from a solar ephemeris), the latitude b of the area photographed, and the difference in longitude c between the celestial meridian passing through the sun at the instant of exposure and the meridian of the area photographed. When these quantities are known, the solar elevation angle, x, can be computed by the relation

$$\sin x = (\cos a)(\cos b)(\cos c) + (\sin a)(\sin b) \qquad (24\text{-}21)$$

The simplest way to find the height of an object from its shadow is to establish a ratio of shadow length to object height by measuring the shadow of an identifiable object of known height that appears somewhere in the strip of photographs. This ratio can then be applied to other objects imaged in the same strip. A structure casting a 2-cm shadow on level ground will be 30-m high if another building, known to be 15-m in height, casts a shadow 1-cm in length on level ground. Figure 24-68 illustrates some factors influencing the accuracy of height measurement by the shadow method.

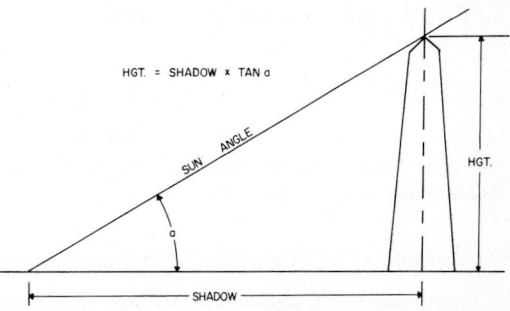

Fig. 24-67. Relationship of shadows and corresponding object heights (After Avery, 1968).

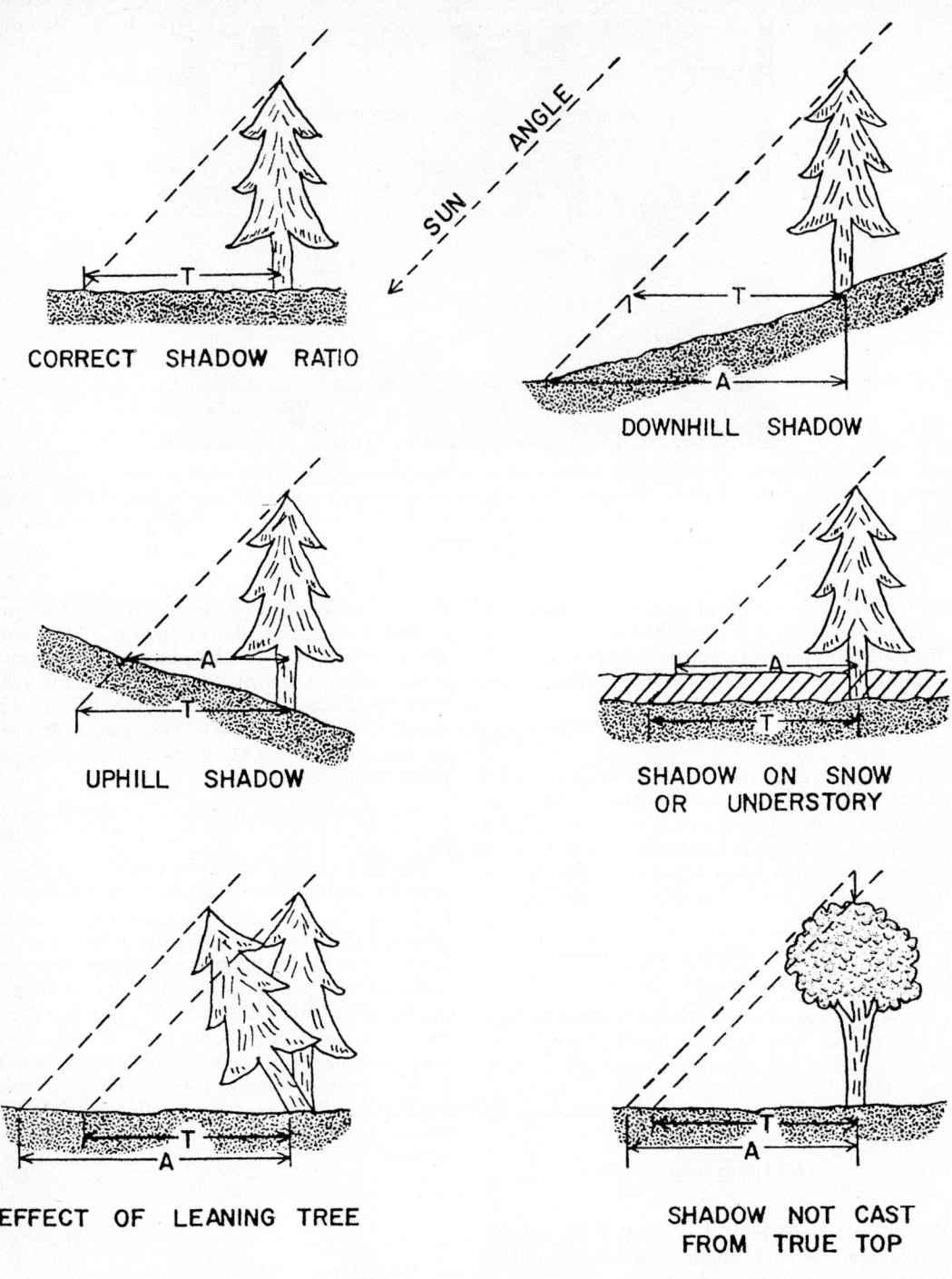

CORRECT SHADOW RATIO

DOWNHILL SHADOW

UPHILL SHADOW

SHADOW ON SNOW
OR UNDERSTORY

EFFECT OF LEANING TREE

SHADOW NOT CAST
FROM TRUE TOP

A   APPARENT SHADOW LENGTH
T   TRUE SHADOW LENGTH

Fig. 24-68. Illustration of various factors affecting the length of shadows cast by trees or similar objects (After Avery, 1968).

Measurement of Capacities of Storage Tanks

The capacity of cylindrical storage tanks on vertical aerial photography can be easily computed with the following equation:

$$\text{Volume} = \frac{\pi D^2 H}{4} \tag{24-22}$$

where D is the inside diameter of the tank and H is the height.

Measuring Terrain Slope

The terrain slope-angle, θ, can be measured accurately enough for most purposes by determining the horizontal separation, ab, of the two points between which the slope angle is to be calculated, using Eq. 24-22, finding the difference in elevation, h, between the two points, using Eq. 24-18, and then solving for θ as follows:

$$\tan \theta = \frac{h}{ab} \qquad (24\text{-}23)$$

OBLIQUE PHOTOGRAPHY

In oblique imagery the optical axis of the camera is intentionally inclined more than 10° from the vertical, showing ground features in oblique or perspective view. High oblique photographs (Color Figure 24-69) show the apparent horizon whereas low oblique photographs do not. Obliques are characterized by constantly changing scale, limiting their practical use for measurement of distances and directions. Therefore, to make the best use of oblique photographs the interpreter should clearly understand the mathematical principles that apply to them. Because oblique photography is a perspective projection, it contains vanishing points. If the vanishing points can be located with respect to the apparent horizon and the lens (true) horizon on a high oblique photograph, then a perspective grid (Canadian Grid) can be developed. The procedure involves preparing perspective grids for oblique photos so that each quadrilateral on the perspective grid represents a ground square of arbitrary dimension. Planimetric map compilation can be done manually by transferring the details appearing in each perspective-grid quadrilateral to its corresponding map square. Low obliques—those that do not include the horizon—can be used, but the angle of obliquity or depression at the moment of exposure must be known and the procedure is more complicated.

Determining the Scale of Oblique Photography with a Canadian Grid

To develop a Canadian Grid on a high oblique photograph the flight altitude, A, the focal length, f, and the depression angle, θ, of the camera at the instant of exposure must be known. The following 17-step method can then be used to draft the perspective grid by determining the positions of the vanishing points on the lens horizon (see Figure 24-70):

1) The principal point, P, is found by connecting opposite fiducial marks with straight lines and marking the intersection.
2) A line is constructed through the principal point, perpendicular to the apparent horizon. This line is the principal line, its intersection with the apparent horizon defines the point H_1. (Figure 24-71) depicts the

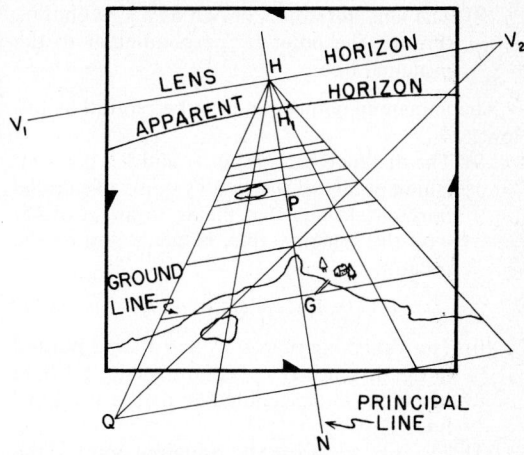

Fig. 24-70. Portion of a perspective grid fitted to an oblique photograph (McNeil, 1954).

angular geometry of the oblique photograph.)

3) The depression angle, θ_1, between the optical axis of the camera and the apparent horizon is calculated by the relation

$$\tan \theta_1 = \frac{PH_1}{f} \qquad (24\text{-}24)$$

4) The dip angle, D, between the apparent horizon and the lens horizon is calculated by the relation

$$D \text{ (in minutes)} = 0.98 \sqrt{\text{alt. (in ft)}}. \qquad (24\text{-}25)$$

5) The depression angle, θ, between the optical axis of the camera and the lens horizon is calculated by the relation

$$\theta = \theta_1 + D. \qquad (24\text{-}26)$$

6) The distance PH, along the principal line to the lens horizon is calculated by the relation

$$PH = (f)(\tan \theta) \qquad (24\text{-}27)$$

7) The distance, PH, is laid out along the principal line from the principal point in the direction of the apparent horizon;

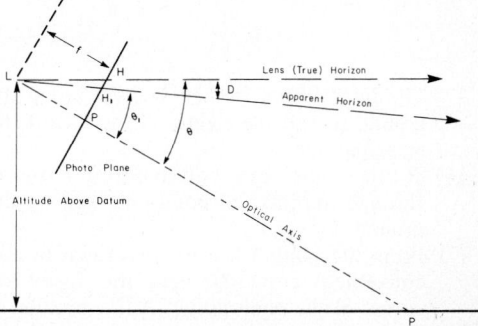

Fig. 24-71. Angular geometry of an oblique photograph.

8) The lens horizon is drawn as a straight line through the point H, perpendicular to the principal line.

The vanishing points can then be plotted as follows:

9) The distance from H to V_1 and V_2, the vanishing points of the two systems of parallel horizontal lines that make an angle of $45°$ with the principal line, is calculated by the relation

$$HV_1 = HV_2 = (f)\ (sec\ \theta). \qquad (24\text{-}28)$$

10) The vanishing points V_1 and V_2 are plotted at the distance $HV_1 = HV_2$ to left and right of point H, perpendicular to the principal line.

11) The distance from the principal point to the photograph nadir is calculated by the relation

$$PN = (f)\ (cot\ \theta) \qquad (24\text{-}29)$$

12) The nadir, N, is plotted at the distance PN, from P, along the principal line away from the apparent horizon.

Once the vanishing points have been plotted, a system of perspective grid lines can be constructed for mapping purposes (see Figure 24-70). The steps in constructing a perspective grid on an oblique photograph at a given ground interval W are:

13) The distance from H to a point, G, on the oblique photograph, at which the scale number, S, of the photograph equals the scale number of the grid, is calculated by the relation

$$HG = \frac{A}{S(\cos\ \theta)} \qquad (24\text{-}30)$$

(A) is the altitude above ground datum at the time of exposure. The point G is plotted at the distance HG, along the principal line from H, in the direction of the principal point.

14) A ground line is constructed through the point G, parallel to the lens horizon.

15) The length of the grid interval, w, along the ground line is calculated by the relation

$$w = \frac{W}{S}. \qquad (24\text{-}31)$$

The interval, w, is then laid off along the ground line to the right and the left of the principal line.

16) Radial lines are constructed from H through the interval points marked on the ground line.

17) From the point V_2, a line is drawn to any convenient point, Q, near the lower left corner of the photograph. A line parallel to the lens horizon is constructed through the

intersection of the line V_2Q, with each of the lines radiating from H.

A grid of true squares at the scale desired for the map is then laid out separately, each square representing one of the quadrilaterals on the perspective grid. Each point or object of interest appearing in the oblique photograph can then be plotted quite accurately in relation to the grid lines, provided there is little or no relief displacement. The smaller the ground interval W, chosen for the perspective grid, the greater will be the accuracy of mapping. Figure 24-70 shows a shoreline and other features as they appear on a gridded oblique photograph; Figure 24-72 shows the same features in their plan positions.

Representative fractions (RF) can be found for any of the grid latitudes, provided the horizon appears parallel to the top borderline of the photograph (the oblique is not swung). The RF of the grid latitude passing through the principal point of the oblique is given by the equation

$$RF = \frac{\text{focal length}}{\text{altitude} \times \text{cos true angle}} \qquad (24\text{-}32)$$
$$\text{of depression } (\theta)$$

This permits determination of the ground distance represented by the side of a grid square. Figure 24-73 is a hypothetical high oblique photograph with a perspective Canadian Grid overlay.

This graphic method of mapping is suitable only for photographs of fairly flat terrain, in which relief displacement is negligible. As described in the

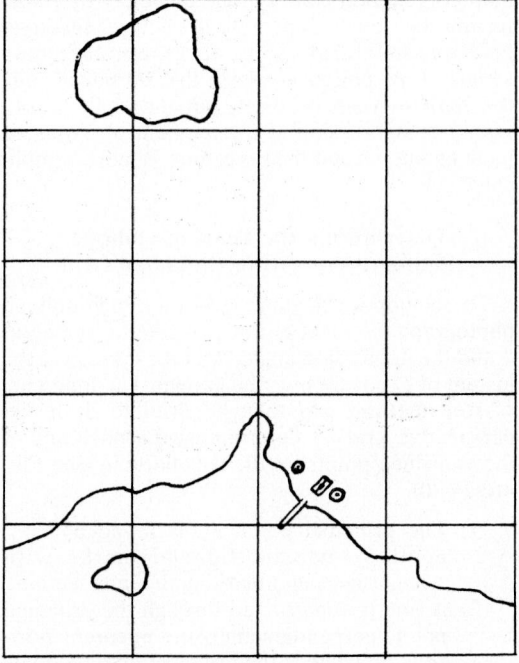

Fig. 24-72. Plan view of the same grid and terrain features as are shown in Fig. 24-70 (McNeil, 1954).

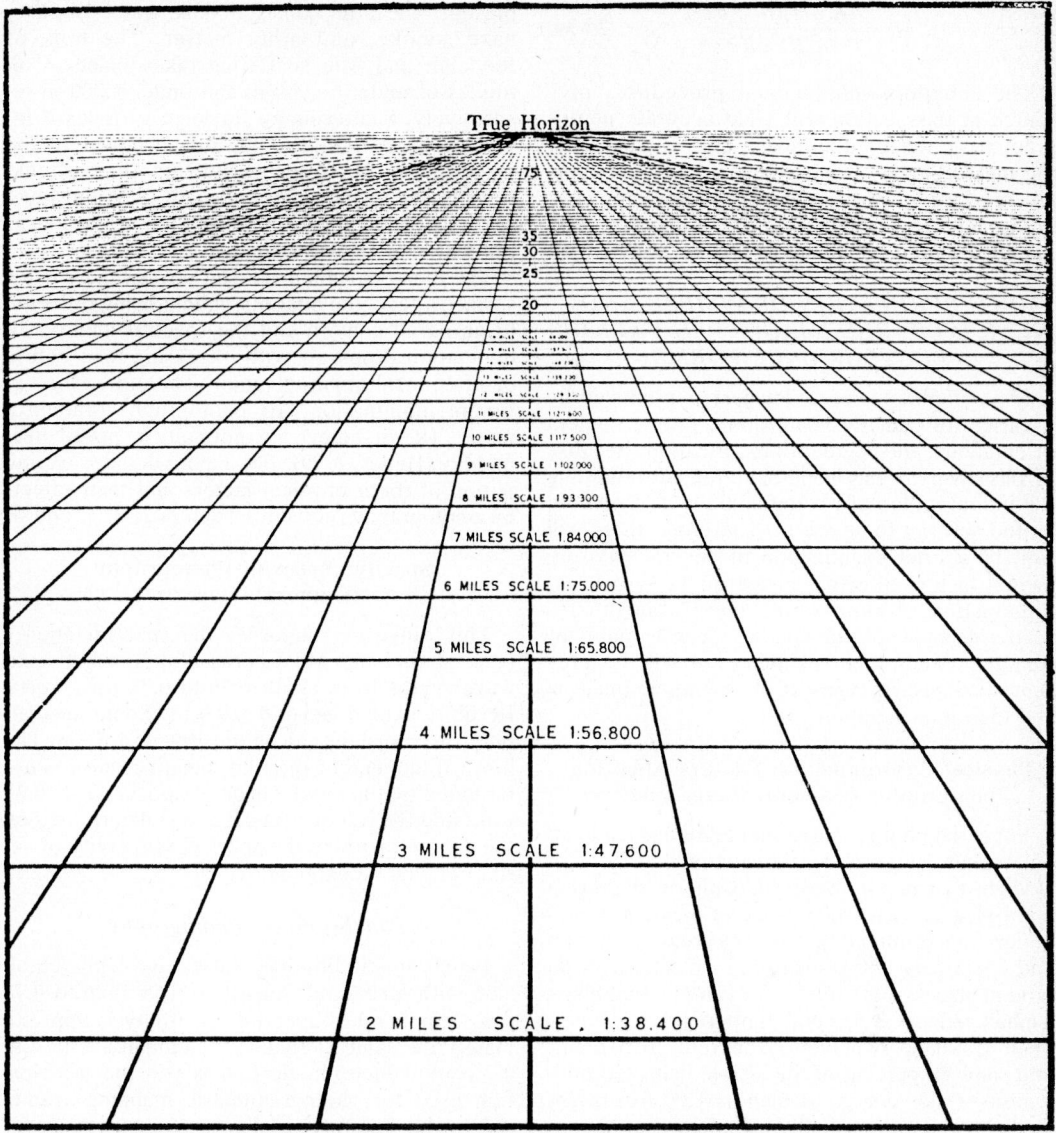

True Horizon

75

35
30
25

20

6 MILES SCALE   1/44 240
15 MILES SCALE   1/37 5
8 MILES SCALE   1/4 2

13 MILES SCALE   1/29 230

12 MILES SCALE   1/26 100

11 MILES SCALE 1/21 600

10 MILES SCALE 1:117.500

9 MILES SCALE 1:102.000

8 MILES SCALE 1:93.300

7 MILES SCALE 1:84.000

6 MILES SCALE 1:75.000

5 MILES SCALE 1:65.800

4 MILES SCALE 1:56.800

3 MILES SCALE 1:47.600

2 MILES   SCALE . 1:38.400

Fig. 24-73. Perspective (Canadian) Grid with ground distances and scale indicated for an oblique photograph with a focal length of 6 inches and a depression angle of 60°.

section of this chapter dealing with equipment, certain instruments provide optical means of mapping from oblique photographs. Optical methods, although also subject to errors of relief displacement, are typically more accurate than the grid method.

Height Measurement

The height of a vertical object imaged on an oblique photograph can be calculated, provided certain conditions are satisfied. The altitude of photography, H, the focal length, f, and the camera depression angle, θ, must be known or calculatable, and both the top and bottom of the vertical object must be visible in the photograph. In calculating heights on an oblique photograph, it is helpful to use a constant, K, which is applicable to all parts of the photograph, and which is calculated from the relation

$$K = \frac{f}{(\sin \theta)(\cos \theta)} \qquad (24\text{-}33)$$

The lens horizon is plotted on the photograph and the nadir is found by the methods described above. A line is constructed from the top of the object through its base to the nadir. The perpendicular distances from the lens horizon to the base of the object, a, and from the lens horizon to the top of the object, b, are measured. Then the height of the object, h, is calculated from the relation

$$h = \frac{(K)(H)(a - b)}{a(K - b)} \qquad (24\text{-}34)$$

The equations and manual procedures presented in this section will yield accurate photogrammetric results if sufficient care is taken during measurement. If extremely precise solutions are required, then more costly photogrammetric equipment (e.g., plotters, stereometers, and viewers) should be employed by the analyst.

INTERPRETATION OF VISIBLE AND NEAR INFRARED IMAGERY IN VARIOUS PHOTOGRAPHIC FORMS

Photographic (camera) imagery is produced by a variety of cameras, emulsions, filters, camera orientations, and degrees of rectification. Because of this diversity any general statement regarding the interpretation of photographic imagery would be too abstract to be useful. Therefore, there will first be a brief examination of several environmental factors affecting image quality, such as, illumination, ground reflectance, atmospheric scattering, angle of sun, and the spectral quality of sunlight. Second, an examination will be presented of specific types of photographic imagery and their interpretation.

Physical Environmental Factors Affecting Photographic (Camera) Image Quality

For aerial photography, illumination is made up of sunlight penetrating the atmosphere and of skylight that results from atmospheric scattering and reflection. The reflectance of incident illumination from ground targets varies greatly. Carman and Carruthers (1951) measured reflectance from ground objects and found, for example, that black asphalt reflects 2 percent, timberland 3 percent, open grassland 6 percent, concrete 36 percent, and snow 80 percent of the visible light. Ground-contrast ratios can go as high as 1,00:1 in bright sunlight when an asphalt road adjoins an area of snow. However, these contrasts do not hold for high-altitude photography. Seldom does the contrast ratio exceed 10:1 over cities and towns and, for hyper-altitude photography, it would more likely be a maximum of 5:1. Agricultural and forest lands have relatively low reflectance, particularly when the vegetation completely obscures the soil. At medium scales of photography (1:20,000), bare soil, irrigation ditches, field boundaries, rock outcrops, crop foliage, and water surfaces are typical features that furnish contrast in agricultural and forest environments. The contrasts or tonal differences permit the image analyst to make judgments about the imaged terrain.

Light is scattered as it passes through the atmosphere and as it is reflected from the earth's surface. The amount of scattering depends on the number of gas molecules (Rayleigh particles) and the number of larger particles (Mie particles) present. The latter consist of dust, water droplets, haze, smoke, and other matter. The bulk of Rayleigh and Mie scattering takes place at altitudes of under 10,000 m and under 5,000 m respectively. Scattering by Rayleigh particles is inversely proportional to the fourth power of the wavelength. Rayleigh scattering thus affects sensing in the shorter wavelength (UV and blue) portions of the spectrum more than in the longer red and near IR wavelengths.

The angle at which the sun strikes the earth's surface affects the quantity of light being reflected to an aerial camera and the spectral quality. Illumination drops off with increasing time before or after local apparent noon. Therefore, for optimum illumination, air photographs (panchromatic, IR, or color) are normally exposed near midday (Heller, 1970). For a more detailed examination of these physical factors and their effects on photography, consult Chapters 2, 3, 5, and 6.

Specific Types of Photographic (Camera) Imagery

This range encompasses the spectral energy peak of sunlight. Most aerial film is sensitive to wavelengths from about 0.36 to 0.72 μm. Aerial IR films extend beyond 0.9 μm. Some special-purpose emulsions can be sensitive to 1.5 μm. The lower (blue) end of the film-sensitive range is determined by the cutoff range of optical glass (0.36 μm); and the red or near-IR end is determined by the degree to which the spectral sensitivity of aerial film can be extended.

Panchromatic Photography

Panchromatic film has a black-and-white emulsion with a spectral sensitivity from 0.36 to 0.72 μm. This band of spectral sensitivity is approximately the same as that of the human eye (0.4 to 0.7 μm). Panchromatic film is still the principal film used for photogrammetric mapping around the world. Economy and ease of use account for the large number of panchromatic photographs available for interpretation at the present time.

There are basically six types of panchromatic aerial film (Table 24-11). These range from type 3404, which is a high-definition fine-grain aerial film, to Super XX aerial, which is a low-resolution, relatively large-grain film. The slow speeds of the high resolution, high-definition films (e.g., 3404, SO-243, SO-190), may preclude their use for some photographic missions. For example, Plus X (3401) is normally exposed at apertures of f/5.6 to f/8 at 1/500 second. To resolve the same objects, high-definition aerial film type 3404 requires wider lens apertures and slower shutter speeds than are available on aerial cameras.

Panchromatic aerial photography is normally exposed through a filter such as the minus-blue (Wratten No. 12) or amber (Wratten No. 15), to reduce fogging effects caused by Mie and Rayleigh scattering. Because Mie particles consist

TABLE 24-11

Comparison of Aerial Panchromatic Films[a] (after Heller, 1970)

Name	Type	Resolution (line pairs/mm)[b]	Speed Relative to Type 3404	Granularity Values	Wratten No. 12 Filter Factor
High-definition aerial	3403	550	1	0.023	1.5
High-definition aerial	SO-243	440	1.2	0.016	1.6
Special fine-grain aerial	SO-190	180	3.7	—	1.5
Panatomic X aerial	3400	150	9.0	0.052	1.9
Plus X aerial	3401	100	33.0	0.088	1.7
Super XX aerial	5425	75	41.0	—	2.0

[a] Courtesy of Eastman Kodak Company.
[b] Average difference in target luminance 6.3:1

of many different materials in varying concentration, it is difficult to estimate their combined effect on the imagery. High concentrations of Mie particles (for example, those around cities during inversion conditions) will degrade panchromatic imagery.

After the panchromatic film is processed, the images are rendered in varying shades of gray on the positive. Each gray tone value corresponds approximately to the density (spectral reflectance) of the color response (wavelength) as perceived by the human brain. This is one reason why panchromatic imagery has been used so successfully for image interpretation. It retains familiar tone values only lacking color for additional contrast. For example, sand will appear white and a lake will generally appear dark grey (Color Figure 24-74). The tonal contrasts between adjacent terrain features are the Figure 24-74C basis for object detection, identification and measurement.

Resolution of panchromatic imagery was, for many years, unrivaled. Recently color emulsions and scanning systems have begun to approach panchromatic resolution. The high-resolution has allowed the image analyst to identify extremely small objects or terrain features from small-scale imagery. Likewise, the dimensional stability of panchromatic imagery has allowed the image analyst or photogrammetrist to make accurate measurements on vertical and oblique imagery. The analysis elements discussed earlier in this chapter are especially diagnostic for panchromatic imagery. In fact, these interpretation elements were originally conceptualized through the use of panchromatic imagery since it was the first remote sensing film emulsion.

Monocular Analysis

Monocular vision is the term applied to viewing anything with only one eye. Methods of judging distances with one eye are termed monoscopic. The term is also applied to an image analyst when viewing a single image, or one-half of a stereopair.

Distances or depths can be perceived monoscopically on the basis of (1) relative size of objects, (2) partially hidden objects, (3) shadows, and (4) differences in focusing required for viewing objects at varying distances.

The principal disadvantage of using a single photographic image for the recognition of features is that generally only two dimensions, length and width, can be perceived. However, the relative ease and quickness of identifying features monoscopically should not be underrated. A tremendous amount of regional detail on panchromatic high-flight imagery can be quickly detected by a trained image analyst using a single image.

Stereoscopic Analysis

One of the most used photographic interpretation techniques is stereoscopic analysis. The principle of stereoscopy (depth perception) should be understood by every image analyst. When the eyes are focused on a certain point with binocular vision, the optical axes of the two eyes converge on that point intersecting at an angle called the parallactic angle. The nearer the object, the greater the parallactic angle and vice versa. In Figure 24-75, the optical axes of the two eyes, L and R, are separated by a distance, b, called the eye base (interpupillary distance). The eye base of the average adult is between 63- and 69-mm or approximately 2.6 inches. When the eyes are focused on point A, the optical axes converge, forming parallactic angle ϕ_a. Similarly, when sighting point B, the optical axes converge forming parallactic angle δ_b. The brain unconsciously associates distances D_A and D_B with corresponding parallactic angles ϕ_a and ϕ_b. The depth between objects A and B is $D_B - D_A$ and is perceived as the difference in the two parallactic angles. The shortest distance of clear stereoscopic depth perception for the average adult is about 25 cm. Objects can be focused at distances as short as 15 cm, but such close viewing causes excessive

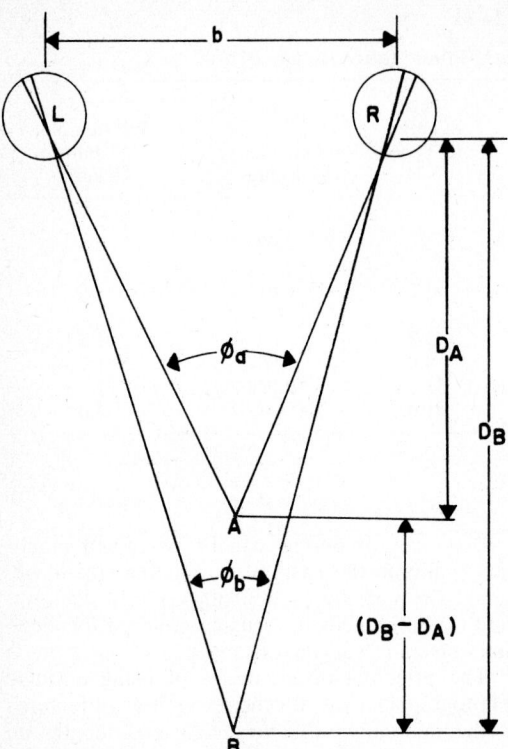

Fig. 24-75. Stereoscopic depth perception as a function of parallactic angle (Wolf, 1974).

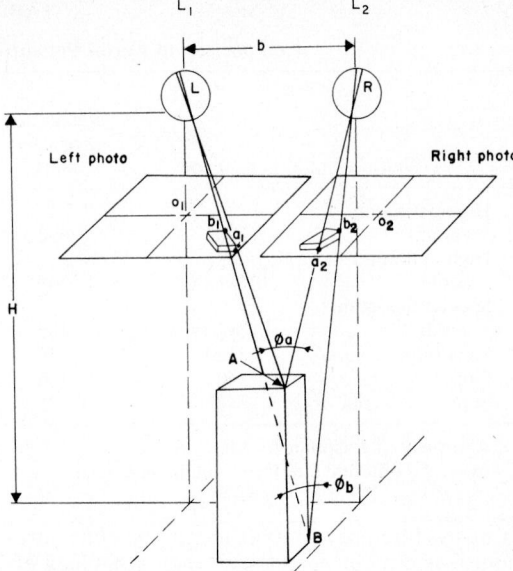

Fig. 24-76. Stereoscopic viewing of overlapping photographic imagery (Wolf, 1974).

eyestrain. The maximum parallactic angle formed by the eyes is approximately 15°.

The maximum distance at which distinct stereoscopic depth perception is possible is approximately 500 m for the average adult. Beyond that distance, parallactic angles are extremely small and changes in parallactic angle necessary for depth perception cannot be discerned.

An image analyst can create a three-dimensional or stereoscopic model of terrain by viewing overlapping imagery. The photographic bases for stereoscopic viewing are the principles of image displacement and stereoscopic parallax previously discussed.

Suppose that a pair of aerial photographs is taken from exposure stations L_1 and L_2 so that the building appears on both photos as shown in Figure 24-76. The flying height above ground is H and the distance between the two exposures is b, the air base. The air base in effect becomes a stretched interpupillary distance. At a scale of 1:20,000 on vertical imagery the interpupillary distance for stereoscopic examination would be increased 30,000 times. This is responsible for the exaggerated vertical perception. Object points A and B at the top and bottom of the building are imaged at a_1 and b_1 on the left photo and at a_2 and b_2 on the right photo. If the two photos are laid on a table and viewed so that the left eye sees only the left photo and the right eye sees only the right photo, a three-dimensional impression of the

building is obtained. The brain judges the height of the building by associating depths to points A and B with the parallactic angles ϕ_a and ϕ_b, respectively. When the eyes gaze over the entire overlap area, the brain receives a continuous three-dimensional impression of the terrain. This is achieved by the perception of changing parallactic angles of the infinite number of image points that make up the terrain. The three-dimensional model thus formed is called a stereoscopic model, or simply a stereomodel, and the overlapping pair of photographs is called a stereopair (Wolf, 1974).

Accurate height, shape, and relative and absolute relief measurements can be made on panchromatic imagery by viewing the imagery stereoscopically and by using the parallax equations discussed in the geometry section. Although these parallax equations are mainly for use on vertical photography, accurate object recognition through stereoscopic analysis is not confined to vertical photographs. For example, Bruckner et al. (1967) found that convergence angles of 10°, 20°, or 30° did not affect the accurate interpretation of terrain features on panchromatic oblique photographs as long as the scale of each (for the particular object or area under investigation) was held constant by photographic enlargement. However, if the scale is not held constant then oblique angles greater than 20° cause a statistically significant decrease in the interpretability of the photographs. Therefore, it can be concluded that scale or ground resolution is a more important factor than convergence or oblique angle in determining the use of aerial photographs for analysis.

An interpreter conducting stereoscopic analysis of photographic (camera) imagery must be aware of relief reversal. Such an effect is known as pseudoscopic illusion. A reversal of relief is obtained if the photo originally intended to be observed by the left eye is placed at the right-hand side of the stereoscope, and if the photo designated to be seen by the right eye is placed at the left side of the stereoscope. The term "pseudoscopic" is also used to describe the apparent reversal in relief seen by viewing a single photograph in different orientations; the correct relief impression is usually obtained when the shadows fall toward the observer.

Densitometric Analysis

Tone value is a major element in identifying many terrain features in panchromatic imagery. Tone is the primary element used for computer-assisted image analysis. The identification of features in the past was based largely on qualitative evaluation of aerial photography in conjunction with limited field checking. As scientific disciplines have become inundated with imagery, it has become apparent that a rapid quantitative approach to image analysis is necessary.

When silver-halide crystals in photographic emulsions are exposed to light, the bond between the silver and the halide is weakened. An emulsion that has been exposed to light contains an invisible image of the object called the latent image. When a latent image is developed, areas of the emulsion that were exposed to bright light turn to free silver and become black. Areas that received little or no light become white, if the support is white paper, or clear if the support is transparent film or glass. The degree of darkness or lightness is called density. The density of developed images is a function of exposure (a product of illumination and time). For any panchromatic exposure there are variations in illumination received from different objects in the image field of view. These variations in illumination create tone values ranging from black to white.

The density of an object imaged on a positive transparency is a measure of the amount of light that can be transmitted. Thus, the image of a black object transmits no light and the image of a white object transmits almost 100 percent of the light. Opacity is the reciprocal of transmittance. The unit of density is the common logarithm of opacity. As an example, if 20 percent of the light is transmitted, transmittance is 1/20, opacity is 1/0.05 or 20 and density is the common logarithm of 20 or 1.3.

The amount of light incident on a film and the amount transmitted can be measured with a densitometer. A reflection densitometer is used to obtain density readings from prints and a transmission densitometer is used to obtain density readings from transparencies. The microdensitometer can provide valuable information for image analysis.

A series of point readings in a study area can produce a density profile trace. Automated microdensitometry can produce point readings and/or line-scan profile traces (Figure 24-77). The density of an entire image can be quantified by incremental line-scan profile traces. Each point-density value is normally stored with its x-y coordinate.

Since many classes of targets exhibit distinct density values on photographic imagery, associations or correlations can be made between density value and object type. Ordinarily the raw density values can be used in this type of correlation.

Density values can be further manipulated to yield the original target reflectance data for various terrain features. Density values obtained by the densitometer can be converted to reflectance by the relationship:

$$\left[\text{reflectance} = 1/\text{antilog}_{10}\ \text{density.}\right] \quad (24\text{-}35)$$

When density or reflectance data are being generated it is important to remember that the values obtained are influenced by the following factors: (1) film and filter types, (2) seasonal effects, (3) densitometer aperture size, (4) scale (Rib and Miles, 1969), and (5) atmospheric effects. If density comparisons are to be made between different bands of imagery for a specific terrain feature, a normalizing function can be applied to the data.

Isotonal or isodensity maps can be generated from panchromatic imagery if density readings are sufficient to delineate significant tonal boundaries. Problems may arise in isotonal mapping when different terrain features produce approximately the same density. For this reason, it is often suggested that a multidate, multiband densitometric analysis is the best approach to assure optimal target discrimination. For example, Figure 24-78 depicts the multidate panchromatic densitometric analysis of specific crop types in Switzerland (Steiner, 1970). Temporal studies such as this are also useful in determining optimum overflight dates.

Interpretation of Black-and-White Infrared Photography

Panchromatic silver-halide emulsions can be treated to be sensitive to energy in the near-IR part of the spectrum. IR-sensitive film emulsions make it possible to obtain photographs of energy that is invisible to the human eye. An early application of this type of emulsion was camouflage detection, where it was found that green netting, which to the human eye had the same green color as live foliage, reflected IR radiation differently. This difference could be detected with IR photography.

Figure 24-79 illustrates sensitivity differences between panchromatic and black-and-white IR emulsions. IR film is less sensitive to the green part of the spectrum than panchromatic film and has its primary sensitivity in the blue-violet and IR wavelengths. Unless minus-blue filters (that cut out the blue or blue-green wavelengths) are used in conjunction with IR film, there is little advantage to using it instead of panchromatic film, and

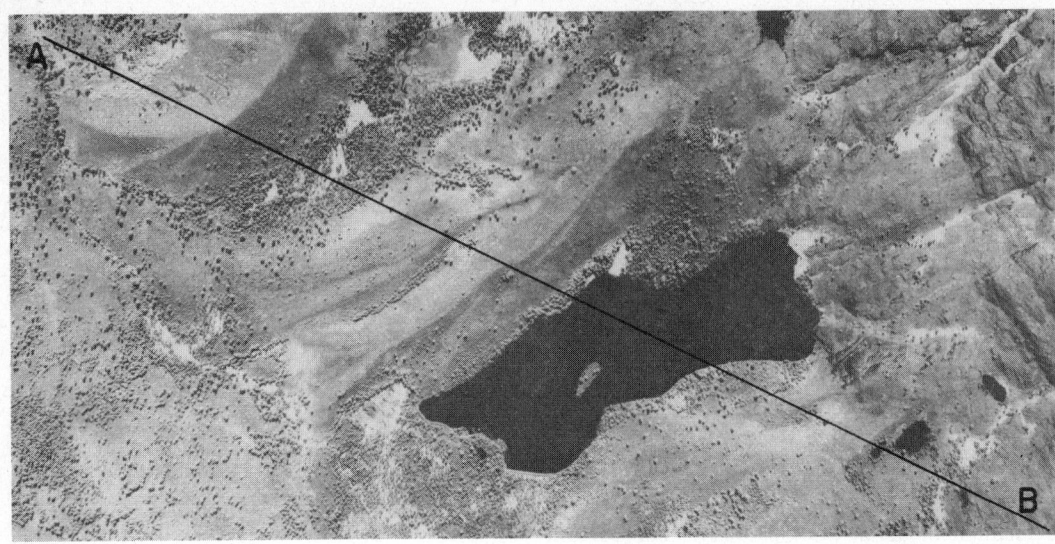

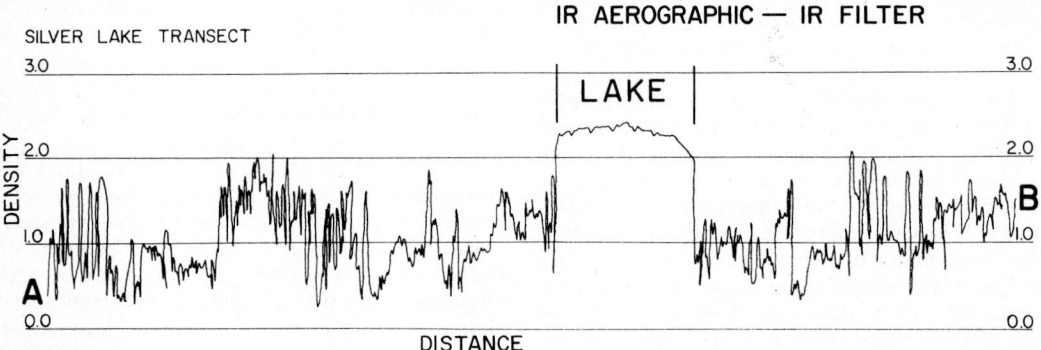

Fig. 24-77. Scanning microdensitometer trace of wildland area near Silver Lake in northeastern California (Lent, 1968).

IR film often is exposed through dark red filters that create exposures of IR wavelengths only. Realistically, this type of photography may best be described as "near-IR" because most exposures use only a small part of the IR portion of the spectrum (from about 0.7 to 0.9 μm). Exposure through red filters greatly increases contrast, especially among types of vegetation, but often at some sacrifice of image sharpness. The resolution of IR film (lower than that of panchromatic Super XX aerial film) is about 55 line pairs/mm at a 6.3 to 1 target-to-background luminance ratio.

When IR film is exposed through yellow haze filters, the resulting compromise in tonal contrast is sometimes referred to as "modified IR". Because the blue end of the visible spectrum is normally cut off by filters, black-and-white IR film has better haze-cutting capability than panchromatic film. However, contrast in the IR positive print or diapositives is degraded because of poor shadow penetration. Dark shadows inhibit identification of certain features lying within them. On panchromatic imagery, there normally is some indication of object form within shadow areas, but on IR photographs features in shadow

areas are almost completely obliterated. Some users avoid black-and-white IR film because of the difficulty of seeing the ground in forested areas.

One of the largest uses of black-and-white IR film is as one component in the multiband technique to exploit the fact that the use of three emulsions provides more flexibility for color reconstitution than a standard color emulsion. The analysis of multi-band imagery is discussed below.

The analyst viewing black-and-white IR imagery must be aware that the gray tones result largely from the degree of IR reflectance of an object rather than from its visible color. In fact, the rationale for using IR is that the tone rendition of a scene is altered from that obtained with panchromatic film thereby making certain objects more distinct by enhancing the contrast between the objects and their background. For example, broad-leaved vegetation (angiosperms) is highly reflective in the near IR region and therefore registers in light tones; coniferous or needle-leaved vegetation ("h" in Figure 24-74D) tends to absorb near-IR radiation and consequently registers in

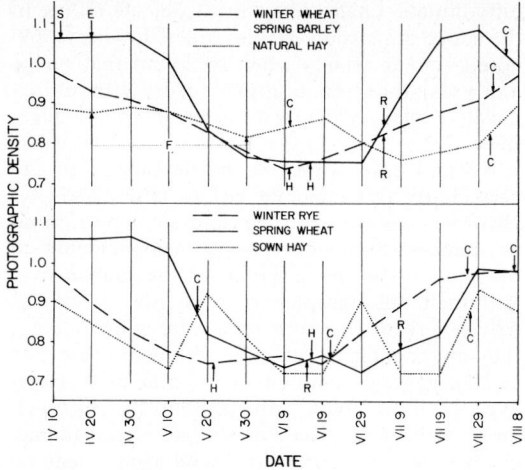

Fig. 24-78. Seasonal changes of average photographic densities of winter wheat, spring barley, natural hay, winter rye, spring wheat, and sown hay, measured on low-altitude panchromatic aerial photographs, valid for the Zurich area, Switzerland. The letters indicate the average dates or time periods for the following phenological aspects or farming activities: C, cutting (harvesting); E, emerging; F, flowering, H, heading; P, planting; R, beginning of ripening; S, sowing (Steiner, 1970).

much darker tones. This characteristic makes black-and-white IR photography particularly useful for differentiating timber types.

Bodies of water readily absorb near-IR radiation, causing them to appear dark. This is useful in determining the extent of river tributaries, sloughs, shorelines, swamps, and canals. On the other hand, the dark tone inhibits detection of underwater detail such as reefs, shoals, and channel obstructions. In some cases, the unusual tone rendition of IR photography blends light objects, such as dirt roads, with light-toned vegetation; this can interfere with interpretation and create densitometric correlation problems.

Because of their respective limitations, neither panchromatic nor black-and-white IR film has an overall clear-cut superiority. When a choice is available, selection will depend mainly on interpretation objectives. Ideally, the image interpreter would request both types of imagery (Avery, 1968).

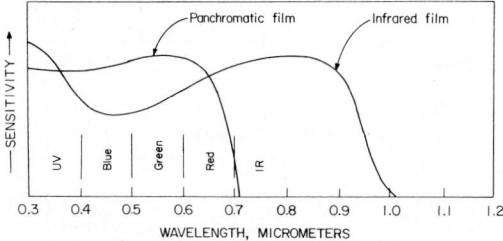

Fig. 24-79. Sensitivities of various black-and-white emulsions (Wolf, 1974).

The stereoscopic, densitometric, and mensuration techniques hold true for black-and-white IR photography. The scanning microdensitometer trace in Figure 24-77 was made from a black-and-white IR aerographic photograph.

Interpretation of Color and Color Infrared Photography

The human eye can separate more than 100 times more color combinations (hues, values, chromas) than gray-scale values (ratio of 20,000 to 200) (Evans, 1948). This capability enables ready discrimination between objects in nature. Therefore, image analysts should understand the capabilities and limitations of color aerial photography if full advantage is to be obtained for its interpretation.

Color or color infrared (CIR) photographs and images may be produced by either of two processes. The first and most common is by direct use of color or CIR film. The second process is to color-combine filtered individual black-and-white photographs of images.

Basic Color Principles

In the wavelengths of visible light, the normal human eye is able to distinguish different colors. The primary colors, blue, green, and red, are composed of slightly different wavelengths; blue is composed of energy having wavelengths of from about 0.4 to 0.5 μm, green from 0.5 to 0.6 μm, and red from 0.6 to 0.7 μm. All other hues are made up of combinations of the primary colors: for example, yellow is the additive combination of red and green light. Many combinations of these colors are possible. White light is the additive combination of all of the visible colors and can be broken down into its component colors by passing it through a prism. Color separation occurs because different refraction angles of light result from different wavelengths.

If an object reflects all of the visible radiation that strikes it, it will appear white to the human eye. On the other hand, if an object absorbs all light it will appear black; or if an object absorbs all green and red energy but reflects blue, that object will appear blue.

Photographic emulsions can be manufactured with variations in wavelength sensitivity. However, because of the wide variation in response of the human eye, it would be impractical if not impossible to design an emulsion that accurately reproduces the "true" colors of nature as perceived by each analyst.

Conventional color and CIR emulsions have widespread use in remote sensing. Color emulsions consist of three layers of dye. The top layer is sensitive to blue light, the second layer is sensitive to green and blue light, and the bottom layer is sensitive to red and blue light. A blue blocking filter between the top two layers prevents blue light from exposing the bottom two layers. The result is a composite of three layers sensitive to blue, green, and red light.

CIR film also has three emulsion layers, each sensitive to a different part of the spectrum. The sensitivity of the top layer peaks in the green and the ultraviolet regions. The middle layer has its peak sensitivity in the red portion of the spectrum, and is also sensitive to UV radiation. The bottom layer is sensitive from UV to near-infrared. The total sensitivity of the film is from just below 0.3 μm to above 0.9 μm. IR film is commonly used with a yellow haze filter, which blocks wavelengths shorter than about 0.5 μm.

When a color exposure is being made, light entering the camera sensitizes the layer or layers of the emulsion that correspond to the color or combination of colors comprising the original scene. The colors recorded are the result of a sequence of processes. First, light must be produced by a source. Second, some of this light must be modified by reflectance from or transmittance through some substance. Third, this modified light must reach the optical components of the camera lens. Fourth, the energy that enters the lens must focus on the focal plane and expose the emulsion. If color negative film is used, a negative is produced and positive prints can be made from the negative. Color reversal film can be processed to directly produce a true color positive transparency.

Unique Characteristics of Color and CIR Photography

Color has a strong impact on interpretation. Color photographs of objects whose apparent color contrasts with that of the background are readily detected. Image analysts detect significantly more targets on normal color imagery than on black-and-white imagery (Aldrich, 1966). For example, in Figure 24-74, the different species of conifers (areas g and h); underwater detail (c); and areas of little or no soil development, such as the granite area (c) are more easily identified on Ektachrome and Ektachrome IR (CIR) imagery than on the black-and-white imagery. Some analysts feel that color film may be preferred on the basis that through its use one finds it easier to detect and differentiate small low-contrast objects owing to the added dimension of color tones. Levine (1969) noted similar results when he found that image analysts obtained higher scores with color-negative imagery than with black-and-white negatives.

Normal color film can be developed to approximate actual scene colors (Fig. 24-74A). However, colorimetric measurements on film do not necessarily match spectral measurements on the ground, or the response of the human eye. Although photographic reproduction processes may not record a scene with total color fidelity, they are capable of retaining sufficient color data so that general conditions can be defined. However, in "false-color" infrared film, the color balance has been intentionally shifted, creating imagery that does not correspond with what is seen on the ground (see Table 24-12).

With CIR film and a yellow filter, any object that reflects longwave red or IR energy will appear red on the final processed picture (Fig. 24-74B). Objects that primarily reflect red will appear green, and objects that primarily reflect green will appear blue—hence the name "false color". The analyst must be careful not to use "real world" color knowledge to interpret an image with a shifted spectral response. This same warning applies to color imagery that has been modified by filters.

Figure 24-80 is included to show why it is useful to employ IR film to extend our vision beyond the visible spectrum. This figure shows the densitometric reflectance curves of various foliage types.

Most types of foliage are not very different from one another in spectral reflectance in the visible region of the spectrum. The small rise of each curve in the green region is all that is required to provide the characteristic green appearance. However, the generally high reflectance of foliage in the IR region and the great differences in reflectance that may occur explain the value of a film sensitive to this region for detecting differences in foliage conditions and between varieties of foliage.

TABLE 24-12

General Principles of Operation for Normal Color Film and
CIR Film as Used by Kodak. (From Heller, 1970.)

Normal Color Film					
Spectral region	Ultraviolet	Blue	Green	Red	Infrared
Normal color film sensitivities	—	Blue	Green	Red	—
Color of dye layers	—	Yellow	Magenta	Cyan	—
Resulting color in photograph	—	Blue	Green	Red	—
Infrared Sensitive Color Film					
Infrared sensitivities	—	Blue	Green	Red	Infrared
Sensitivities with yellow filter	—	—	Green	Red	Infrared
Color of dye layers	—	—	Yellow	Magenta	Cyan
Resulting color in photograph	—	—	Blue	Green	Red

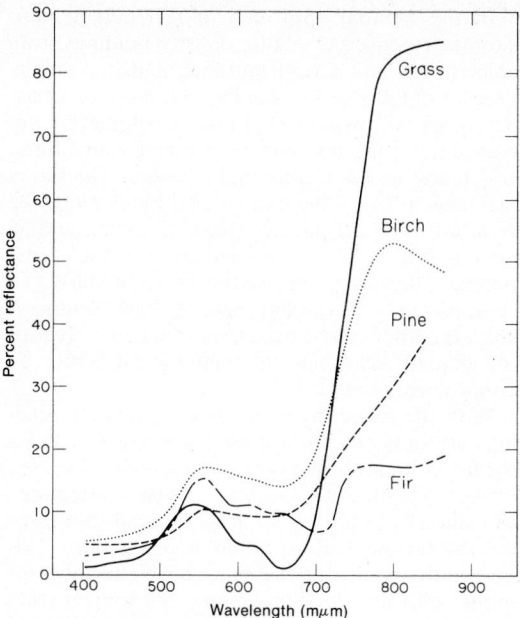

Fig. 24-80. Spectral Reflectance curves of various typical foliage types (Fritz, 1967).

The relative advantages of color versus CIR vary according to the specific interpretation problem. Color is often best for general interpretation because it more closely records the colors of the natural scene. Color is also more useful for studying underwater conditions, while surface waters are more easily delineated with CIR. CIR has been recognized as superior for certain land use, land cover and vegetation studies, including large-area and regional analysis; species differentiation and delineating areas of low vegetation density; and stress detection. However, color has been shown to be more effective in an extensive study of tropical forests (Holdridge et al., 1968), and in the study of Eucalyptus forests in Australia.

Another consideration of color versus CIR is that, for the same height of photography, atmospheric scattering and absorption will cause greater losses in information content in color than in CIR photography. These losses increase with altitude (Pease and Bowden, 1969).

Stereoscopic analysis of color and CIR imagery is much more photogrammetrically reliable than it was in the 1950s and early 1960s. The manufacturers of color photographic materials and photogrammetric equipment deserve much of the credit for the present metric and dimensional stability of color imagery. Color aerial photography is now considered to be metrically equal to its panchromatic counterparts. Also, the ground-resolved distance (GRD) provided by color aerial photographs, so important in monoscopic or stereoscopic analysis, has improved to where it equals or surpasses that of black-and-white photographs.

Color Measurement

The added dimension of color allows the image interpreter to accurately identify more targets than on panchromatic imagery. The principal variables responsible for this increase in interpretability are a given color's hue, value and chroma, which can be either qualitatively or quantitatively analyzed. Figure 24-81 shows the dimensions of the color circuit used in the Munsell system. The qualitative analysis of color imagery is simply subjective identification based on the analyst's general background and the object-recognition elements previously discussed. The subjective description of an object's color by an interpreter will not convey the true color to most people. This is because color is an abstraction that is unclear when communicated verbally. On the other hand a quantitative value or notation system can accurately communicate the exact color under investigation.

The quantitative analysis of a photographic material is based on density measurement. In black-and-white photography, measured density is a function of the amount of developed silver. In color photography, where the processed image contains no silver, measured densities are a result of the absorption characteristics of the dyes that form the emulsion layers. In multilayered color materials, at least three dissimilar absorbers (dyes) make up the color image.

To deal with the complexities of color measurement, several techniques have evolved. These techniques vary with regard to the information desired and the methods used to obtain this information. Cretcher and Balentine (1968) discuss two major quantitative color density measurements: integral and analytic.

Integral density measurements are the result of the combined or net effect of the superimposed dyes, with no attempt to separate the effects of a single dye from the total density. Four types of integral densities commonly used are:

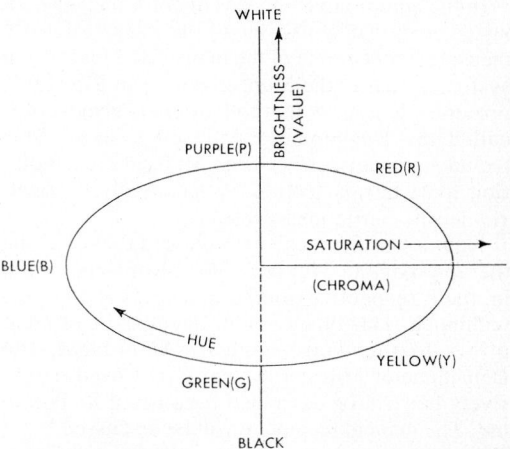

Fig. 24-81. Dimensions of the Color Solid (Smith and Anson, 1968).

1) Integral Spectral Densities (ISD)—This measurement represents the density, wavelength-by-wavelength, resulting from the effect of all the fundamental dyes in the material. A spectrophotometer or similar device capable of narrow band separations is required.

2) Printing Densities—This type of density measurement is used to indicate the effect that a negative or transparency will have when used with a particular printer and print material. These measurements must be representative of the response of the material and exposing source.

3) Colorimetric Densities—Since many color films are designed for direct viewing, density measurements are required that give an indication of how a color appears to a standard observer. Colorimetric densities are normally calculated from spectrophotometric traces making use of the (CIE) spectrum-weighting function for the standard observer. However, some instruments make use of specially designed red, green, and blue filters.

4) Three-Color Densities—This type of density measurement, made through what are sometimes called tricolor or arbitrary filters, is a relative measurement, and the only restriction is that the filters be in the red, green, and blue regions of the spectrum. Three filter densities are normally used to indicate whether or not a sample matches a standard. They are often used in quality control.

Analytical densities are of several types generally concerned with a wavelength-by-wavelength measurement of a single or isolated dye. These types of density measurements are mainly of use to manufacturers of photographic materials. They are discussed in Chapter 6 of the Manual of Color Aerial Photography (Smith and Anson, 1968).

In the quantitative analysis of color imagery, an image analyst may use one of the integral density measurements or one of the many color measuring systems, such as the Munsell notation. Rib (1966) provided a hybrid method of measuring color called the Densitometer-Munsell system. This technique generates accurate Munsell color notation for a terrain feature by means of colormetric-densitometric measurements.

Keys (classifiers) can be produced that describe the approximate Munsell color of objects/terrain in their respective temporal cycles using this technique. This type of color data can be of great utility to the image analyst. At present, the Densitometer-Munsell system is not used extensively but will be examined because of its potential. The manual technique will be discussed but it should be kept in mind that automated density-slicing equipment and compatible software have the potential for generating the notation more efficiently.

In the manual approach, a reflectance densitometer is used to obtain density readings from color prints and a transmittance densitometer is used to obtain density readings from color transparencies and negatives. Both densitometers are point-measuring instruments utilizing four filters, which are basically filter photometers. The filters included in the reflectance densitometer are the Wratten 106W (visual), 25 (red), 47 (blue), and 58 (green) filters. The filters included in the transmittance densitometer are the Wratten 106W (visual, 92 (red), 93 (green), and 94 (blue) filters. A chart recorder can be used to automatically record the density readings obtained for all filters on either densitometer.

With the reflecting densitometer, density readings are obtained from a terrain feature by using the four different filters (i.e., red, green, blue, and visual). The density readings can be manipulated to yield a full Munsell notation hue-value/chroma for the terrain feature under investigation. This procedure eliminates the many problems inherent in the color-matching technique. If a transmission densitometer is used, the filters must be calibrated with standard density wedges. Charts necessary for the generation of Munsell notation from density readings are found in the works of Rib (1966), Gourley, Rib, and Miles (1967), and in the Manual of Color Aerial Photography (Smith and Anson, 1968).

Interpretation of Multiband Photography

Multiband photography can be defined as the product obtained by recording a number of given wavelength bands separately on black-and-white film.

Because many materials reflect energy differently within the photographic portion of the EM spectrum, it has been postulated that, by selecting the proper film-filter combinations for recording in areas of the spectrum where maximum differential reflection occurs, the contrast between objects of interest and the background would be enhanced. The multiband photographic technique offers great flexibility in the analysis and display of the recorded data. The type of analysis, enhancement, or display to be employed is at the discretion of the investigator and will depend on the information desired for a particular application and the equipment/instruments available (Orr, 1968).

The different wavelength bands of imagery can either be analyzed separately or in multiband combination. If the wavelength bands are analyzed separately, standard analysis techniques are used. However, to use spectral photography efficiently, an interpreter should have some knowledge of the spectral characteristics of the features in question. For some terrain features, such as water and certain types of soil, rock, crops, and other vegetation, knowledge may be acquired from literature or through experience.

If a multiband-combination analysis is desired then two or more of the bands of imagery can be optically, electronically, or photographically superimposed. Electronic techniques have been discussed in the section on computer-assisted analysis, so the following discussion emphasizes optical procedures.

Optical "Additive Color" Combining

The "additive color" process is the primary method of creating a color composite from black-and-white bands of multispectral photography or imagery. This additive color technique is the foundation upon which optical additive color combiners operate.

To obtain a color presentation by additive color, the following three steps are used (Yost and Wenderoth, 1968).

First, two or more simultaneous photographs or images are taken, each of a separate wavelength region by exposure through different color filters. Each band is recorded separately on black-and-white film. Blue, green, and red filters, which approximate the spectral response mechanism of the human eye, are used when a natural or "true" color reproduction of the ground scene is desired. Enhancements can be made by varying the filter combinations.

The second step in additive color photography is the precise development of each set of negatives exposed in the camera and the subsequent exposure and processing of positive transparencies made from the sets of "spectral" negatives. The relationship of image density to exposure (or radiometric equivalent for the non-visible part of the spectrum) is established by accurate processing, which must include correction for differences in both exposure and gamma for each spectral band used.

Third, a method is required for projecting the spectral positive transparencies superimposed in accurate registration while illuminating each with a different color light. When three such positive transparencies are optically projected in registration and illuminated with the respective blue, green, and red principal colors, a color rendition of the ground scene is produced. It is important to note that not only are these three colors produced in the additive color rendition but every hue will be seen in varying degrees of saturation. Color Figure 24-82a depicts the three black-and-white multispectral bands of imagery that were additively processed to produce the color composite in Color Figure 24-82b.

Special additive color viewers can be used to obtain the composite color rendition on a rear projection screen, through controlling brightness and apparent hue and saturation of the sources of illumination for each positive transparency. Where accurate registration or good resolution of the superimposed image is not critical and saturation control of the color space is not desired, slide projectors can be used to provide a simple additive color viewer (Figure 24-83).

Image analysts can use the additive color pro-

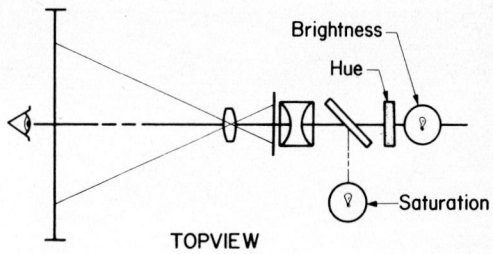

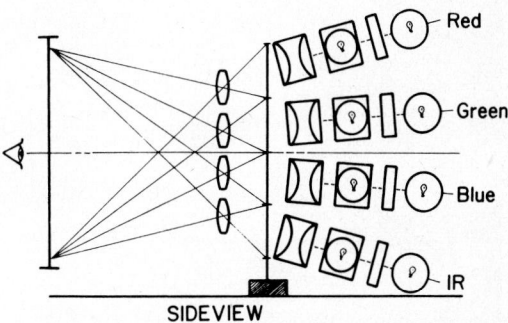

Fig. 24-83. Schematic of an additive color projection system showing how the position of an image in the color solid is controlled by viewer illumination. The *top* view shows this arrangement for one spectral positive. Light from the brightness lamp passes through a filter and condensing lens, illuminating the spectral positive, which is projected on the screen at the left. The amount of desaturation of the color is varied by changing the illumination of the saturation lamp. The *bottom* view shows repetition of this illumination system for each spectral position as well as a single film plane containing all four spectral positives (Yost and Wenderoth, 1968).

cess to create imagery composites of different dates. This technique offers unique possibilities for the analysis of phenomena that have a temporal character. In order for multidate imagery to be used in this manner, exposures must be made at approximately the same scale and geographic location (Color Figure 24-84) (Colwell, 1969).

Interpretation of Orthophotography

Image analysts have become increasingly aware of orthophotography since the U.S. Geological Survey began to update its 7½' quadrangle series using orthophoto imagery (Landen, 1974). It is important that the unique interpretation characteristics of orthophotography be well understood.

An orthophotograph is a vertical photograph showing images of objects in their true orthographic position. (Figure 24-85). Orthophotos are geometrically equivalent to conventional linear and symbolic planimetric maps, which also show true orthographic object positions. Because they are planimetrically correct, analysts may use orthophotos as maps for making direct measurements of distances, angles, directions, positions, and areas without making corrections for image

Fig. 24-85. Orthophoto contour map of Black Butte Mine, Wyoming. Original scale 1:4800 (V.T.N. Consolidated Inc.).

displacements or scale changes caused by variation in local relief.

Orthophotographs are produced by conventional or differential rectification, depending upon degree of relief and tilt. If local relief is small and the effects of tilt not severe, then aerial imagery can be conventionally rectified. This less expensive method involves a projection printer with a moveable easel that can be tilted to rectify the imagery to a near-vertical perspective.

If the terrain has considerable relief and/or tilt, then a photogrammetric process known as differential rectification is recommended. A vertical aerial photograph, even if corrected for tilt, will contain scale variations as a result of changes in relief. In the differential rectification process, scale variations are removed systematically and the scale becomes constant throughout the entire photograph.

Differential rectification is accomplished with an orthophotoscope. This is an instrument that employs the principle of bit-by-bit differential rectification of a stereoscopic model. Figure 24-86 illustrates how (1) all common photographic images are located in plan position by the intersection of two homologous rays in an optical projection, and (2) all such images are restored to a common scale. The operator scans the stereoscopic model in y as he moves the emulsion plane, z, up or down to keep the scanning slit on the three dimensional ground. After a y scanning line has been successfully scanned for z then the x axis is incremented a given Δx, which is the width of the y scanning slit (Figure 24-87). In an on-line system one of the projected rays is exposing photosensitive material (Landen, 1966). In the off-line systems the x, y, z coordinate values are stored and played back at a later time (Danko,

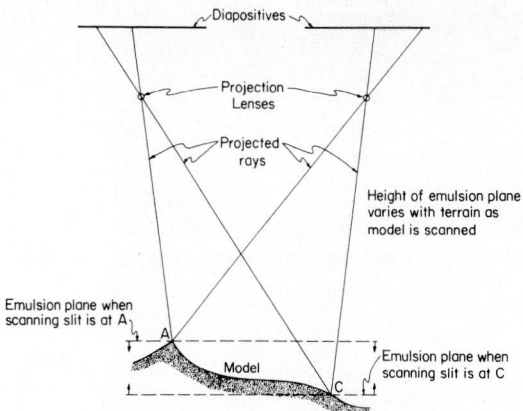

Fig. 24-86. Principle of the orthophotoscope (Landen, 1966).

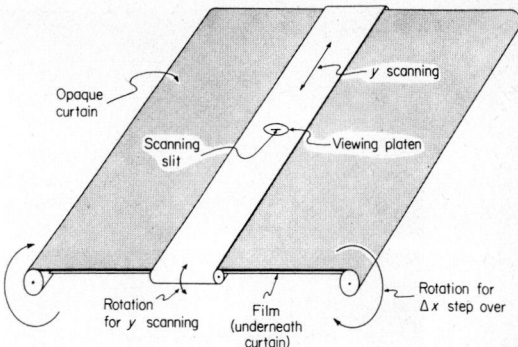

Fig. 24-87. Generalized scanning mechanism of the U.S.G.S. Model T-64 optical orthophotoscope.

1973). In this manner an entire photogrammetric stereo-model can be differentially rectified to produce an orthophoto.

At present most orthophotography is analyzed monoscopically. However, Collins (1972) and others have presented photogrammetric methods for producing "stereorthophoto pairs". This is a combination of a true orthophoto with a specially modified orthophoto or "stereomate". The heights of objects can be obtained from the stereorthophoto pairs by using appropriate equations.

No densitometric analyses are made on differentially-rectified orthophotos at present because the tonal differences between successive scans have not been standardized. Tone variations between adjacent scan lines in differentially-rectified imagery cause one of the principal problems yet to be resolved in producing such imagery. Theoretically, however, there should be no reason why a densitometric analysis could not be accomplished using conventionally-rectified panchromatic or color photographs.

Interpretation of Continuous Strip Photography

Continuous strip photography is taken with highly specialized camera systems that permit large-scale coverage at very low altitudes from fast-moving aircraft. The strip camera exposes a continuous panchromatic, color, or CIR photograph of the terrain by passing the film over a stationary slit in the focal plane of the lens at a speed synchronized with the velocity of the ground image across the focal plane. The rays from any point will be focused as a single point on the film throughout the time of exposure. The duration of the exposure depends upon the film speed and the width of the slit. The fundamental concept behind continuous strip photography is a form of image motion compensation (IMC) known as forward motion compensation. This governs the ve-

locity at which the film in the camera is moved during exposure; it is synchronized with the movement of the image.

In practice the camera exposure-slit width is usually quite small (0.25- to 0.50-mm) so that at any instant the image of only a narrow "ribbon" of terrain, perpendicular to the flight path of the aircraft, is exposed on the film. As the aircraft moves forward, a long continuous photograph is "painted" onto the film.

Strip cameras may be used either with a single-lens cone in a non-stereoscopic mode, or with a dual-lens cone where each lens covers one-half the film width. With a dual-lens cone, the stereobase is obtained by the right lens a few degrees forward and the left lens a few degrees aft of the vertical, creating the stereobase in the direction of flight.

The narrow slit-aperture and longitudinal displacement of the two lenses provide two lines-of-sight, one forward and the other rearward, intersecting a parallax angle. Once the lens angles have been precisely set, a constant degree of parallax is established in the direction of the flight path. The resulting continuous photograph can be analyzed stereoscopically with a special viewer equipped with a stereocomparator for measuring object heights or differences in elevation. When objects on the ground are in motion, it may be impossible, even with IMC, to register them in true perspective with relation to their surroundings, especially when they are viewed three-dimensionally. Objects on the ground that are in motion at the time they are imaged create an artifical parallax that may cause them to appear either to rest above the terrain, or to be embedded in it, when they are studied stereoscopically (Figure 24-88).

Care must also be taken when taking planimetric measurements from continuous strip photography. If the IMC is not synchronized properly or the plane tilts, yaws, etc., then the scale will vary along the flight-line. However, Gullicksen (1967) has shown that continuous strip imagery can be used as a planimetric map substitute if optimum system conditions are met.

Fig. 24-88. Use of the principle of image-motion compensation on the Sonne continuous strip stereo camera, which took these photos, accounts for their great clarity. Because of the rapid motion of the photographic aircraft (about 800 km per hour) and the low altitude from which these exposures were taken (30 to 180 m) the images of objects on the earth's surface were moving very rapidly at the focal plane of the camera. Consequently the photos would have been blurred, because of image travel along the film during the period when the photographic exposure was being made, if image motion compensation had not been used. (More properly, but less commonly called "forward motion compensation."). By this means, the camera's film-drive mechanism, instead of allowing the film to remain stationary within the camera at the instant of exposure, causes it to move at a rate that is just equal to the rate at which any given image is "frozen" to a particular spot on the film for the entire exposure period so that no image blur results. The clarity of detail permitted by this means is best illustrated by the top two stereograms. The bottom stereogram shows that fast-moving vehicles on one side of the freeway are elongated while those on the other side are foreshortened, even though the terrain itself is sharply imaged. Note also that the terrain is easily seen stereoscopically, while the vehicles are not (Colwell, 1960).

Continuous strip photography is adaptable to a wide variety of photogrammetric problems. It has been successfully used for analyzing automobile traffic distribution, discovery of taxable real estate, and the survey of highways, railroads, and power-transmission lines. At a scale of 1:1,200 a strip of terrain nearly 145 km long can be shown on one unbroken picture. Prints have been made from such photography at scales as large as 1:72 for studying such items as crack patterns in airport runways. Continuous strip photography has additional potential as an aid to transect mapping in geologic interpretation and vegetative analyses.

INTERPRETATION OF THERMAL INFRARED IMAGERY

The infrared (IR) portion of the EM spectrum extends from the upper limit of the visible portion, usually taken as 0.7 μm, to the lower limit of the microwave portion, usually considered to be at 1-mm. Imaging thermal infrared (TIR) systems perform a data transformation on this EMR by storing information from non-visible wavelengths and later reproducing them in the form of photo-like images, from which skilled interpreters can draw a great deal of information.

The IR portion of the EM spectrum is customarily divided into three parts: (1) near or reflective, 0.7 to 1.5 μm (up to about 1 μm, film emulsions are directly sensitive to reflected EMR); (2) middle, 1.5 to 5.5 μm (here, scanners are used to measure both emitted and reflected EMR); and, (3) far, 5.5 to 1,000 μm, the region that includes the earth's peak emission. TIR sensors generally image emitted EMR, although reflected EMR may also be recorded. During daylight hours, approximately equal amounts of reflected and emitted EMR are returned to an earth-oriented sensor operating between 3.0 and 4.5 μm. An IR detector operating in this region may record phenomena related to reflectance (Estes, 1974).

As stated above, photographic emulsions can be used to image reflected IR radiation. Emulsions sensitive to wavelengths longer than 1.5 μm are difficult to store because the natural radiation from the ambient temperature of film, cameras, etc., produces enough infrared radiation to cause exposure of film (Holter, 1967). Therefore, wavelengths beyond 1.5 μm requires that scanners be used for imaging. The sensor images a small field of view at any given instant. Radiation emanating from surface features is then passed through a series of optics and onto a detector element, which measures the amount of radiation at given wavelengths. The type of detector is of great importance in image interpretation (Bastuscheck, 1970).

The environmental satellites of the National Oceanic and Atmospheric Administration (NOAA) carry instruments that measure the intensity of radiation upwelling from the earth's surface in the infrared "windows" at wavelengths near 11 and 3.7 μm. Among these instruments are the Scanning Radiometers (SR) on the ITOS 1 and NOAA 2-5 satellites, the Very High Resolution Radiometers (VHRR) (see Figures 24-118 and 24-119) on the NOAA 2-5 satellites, the Visible and Infrared Spin-Scan Radiometers (VISSR) on the current SMS/GOES satellites, and the Advanced VHRR's (AVHRR) (see Figure 24-89) on the NOAA-6 and the current TIROS-N series of satellites. Landsats 2, 3 and D are also equipped with thermal infrared imaging channels covering approximately the 10.4 to 22.6 μm spectral range.

THERMAL IMAGERY AND ENERGY EXCHANGE THEORY

In order to appreciate the information content of thermal infrared imagery, one finds it helpful to have some basic understanding of the physical theory of electromagnetic radiation. All objects emit radiation over all wavelengths. The amount of energy emanating from an object at a given

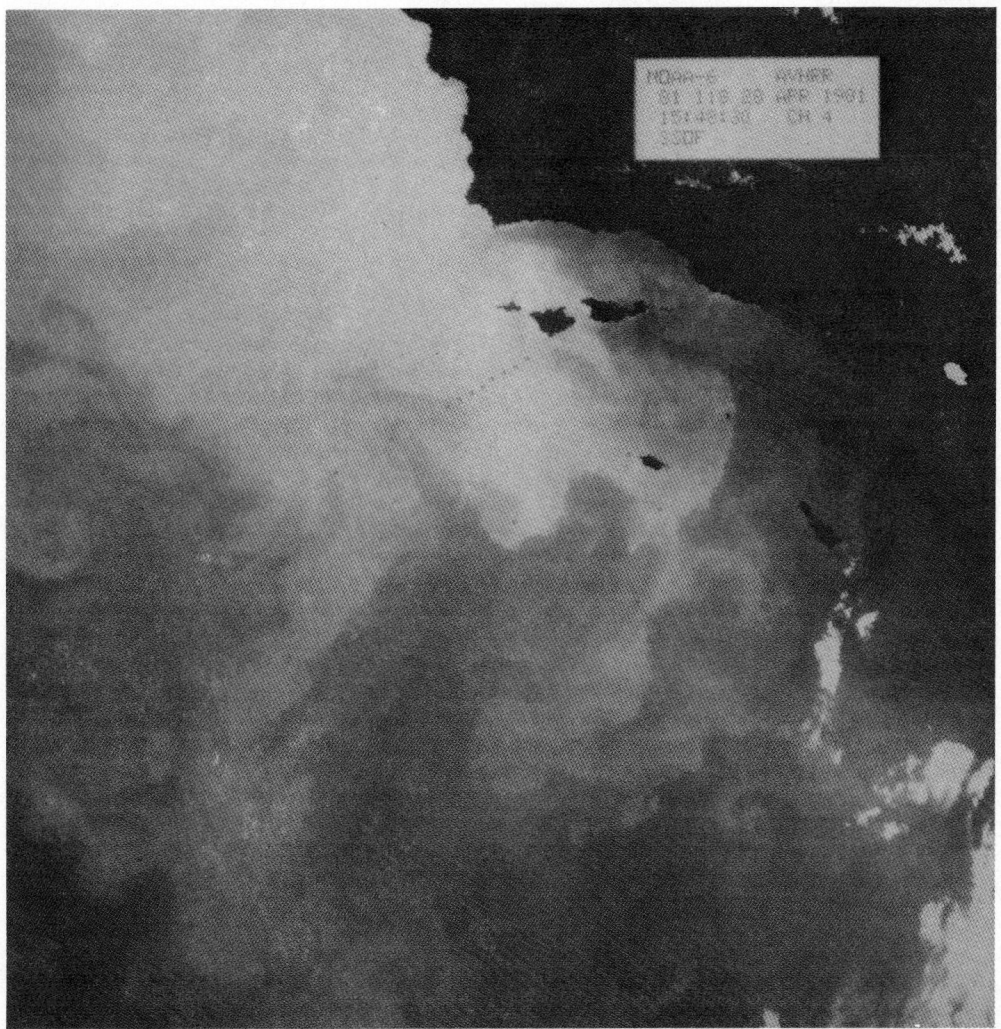

Fig. 24-89. NOAA-6 satellite thermal IR (10.5 to 11.5 μm) image of the Santa Barbara coastline acquired 28 April 1981 at 15:48 GMT (Owen, Variño and Recht, 1981).

wavelength is related to a number of physical properties. In the IR portion of the spectrum, the two most important properties are the physical temperature of the object and its emissivity. Emissivity is the ratio of the radiant flux from a real material to that of a black body at the same kinetic temperature.

According to Planck's equation (see Chapter 2), the temperature of a perfect emitter defines a unique curve (Figure 24-90). It is apparent from these curves that an object at a high temperature emits more energy over the entire spectrum, including the wavelength of peak emission. Thus, if we were to compare the amount of radiation emanating from two objects at IR wavelengths, their relative temperature might be measured. As will be seen later, based on this difference in relative temperatures, the physical temperature of certain features can be differentiated and evaluated on TIR imagery. When a TIR imaging system contains internal calibration sources, relative temperature differences may be converted to actual radiation temperatures or radiances. These will not be true values for the surface, however,

due to modification of the signal by air intervening between the surface and sensor.

Infrared radiant emission, as measured by a radiometer, is the product of the fourth power of the absolute temperature, the Stefan-Boltzmann constant, and the emissivity of the radiating surface:

$$W = \epsilon \sigma T^4 \qquad (24\text{-}36)$$

where:

W = radiant exitance, power per unit area
ϵ = emissivity
σ = Stefan-Boltzmann constant
T = absolute temperature (°K)

Radiation temperature is related to the physical or kinetic temperature by the factor of emissivity and thus the distribution of radiant energy emission is not a linear function of the surface kinetic temperature. This complicates the correlation between ground temperature and the resultant image tone because the difference in emissivity between materials can greatly affect apparent temperatures (see Bastuscheck, 1970).

Since temperature differences are detected as

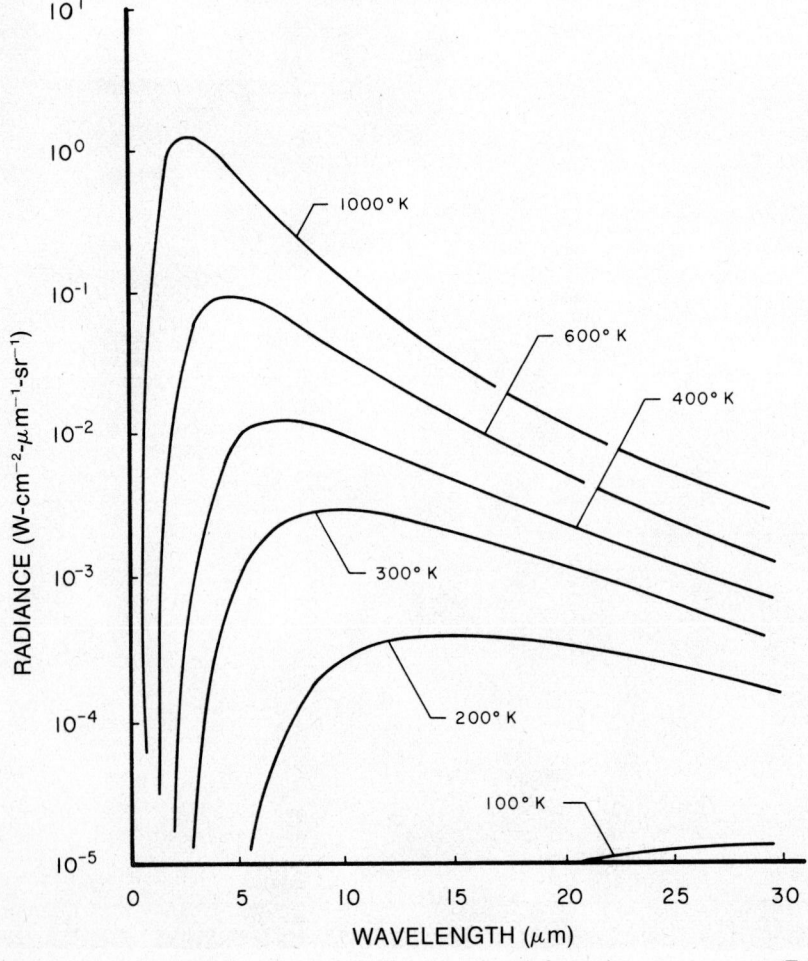

Fig. 24-90. Plot of radiance from a blackbody against wavelength for various temperatures (Estes, 1974).

emitted power differences, a lowered emissivity gives the appearance of a colder temperature to the target surface. Along with the lower apparent temperature there is a corresponding increase in the minimum resolvable temperature difference.

Figure 24-91 shows this emissivity effect. The imagery was flown near dusk on a winter day. There may be some snow in patches, but generally the lower temperature areas are radiometrically cooler valleys. A radiometer was flown with the infrared scanner, and the ground trace of the radiometer is shown by the thin line near the center of the imagery. Interruptions in the line are used to correlate the radiometer signal recording with the imagery. The HRB-Singer scanner and the AR-2 radiometer were used to obtain this imagery and temperature recording.

The chart recorder trace of the radiometer signal has been annotated in conjunction with the imagery. Of particular interest are the two points where the temperature indication on the chart has

gone completely off the bottom edge. These two spots are identified as metal-roofed buildings on the ground. The actual temperature of these buildings is not 20°K or more below ambient, but the low emissivity of the metal roof makes them appear to be so. Any thermal detail on the roof is completely lost, although both radiometer and scanner response times are fast enough to record any if it were available.

The imagery also shows loss of detail in colder ground areas even though the radiometer recording shows such detail to be present. The detail is available in the radiometer recording both because the system has been calibrated to produce detail, and also because a germanium mercury (GeHg) detector was used in the radiometer. Some of this loss of detail in the imagery may occur because the signal is below the dynamic range of the film, although the change in thermal sensitivity with ground temperature is a contributing factor.

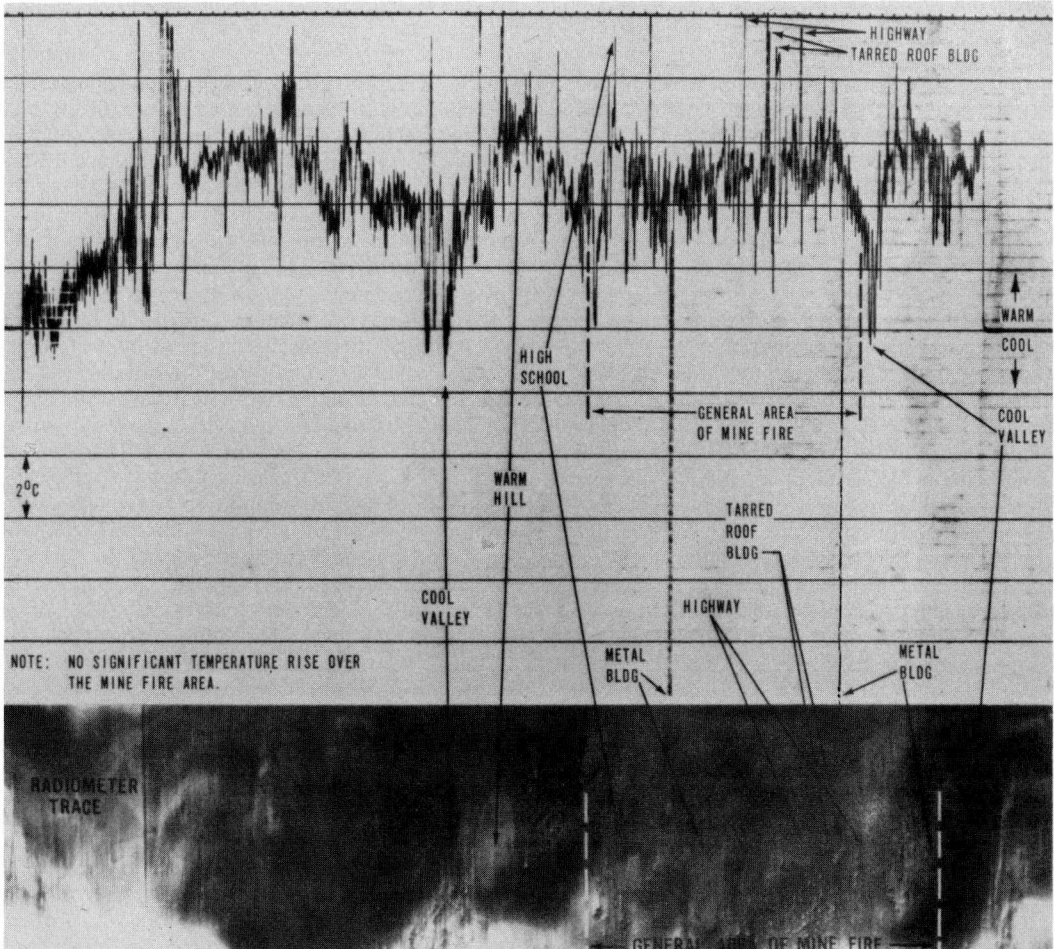

Fig. 24-91. Line trace and image displays of thermal patterns across Connellsville, Pennsylvania. Imagery was flown near dusk on a winter day. There may be some snow in patches; however, the radiometrically cooler areas are generally valleys (Bastuscheck, 1970).

THERMAL IMAGERY AND THE ATMOSPHERE

Unlike solid materials, gases do not absorb and emit in a continuous spectrum of energy but rather in narrow bands or lines at wavelengths determined by molecular geometry.

Atmospheric spectral windows allow propagation of infrared radiation with the least attenuation or interference. But even here absorption line residuals can degrade or otherwise modify an image and thus affect its quality and interpretability. The effects of an intervening atmosphere on TIR imagery are variable and may be basically understood as having a ''clouding'' effect on the primary atmospheric windows.

The effect of the atmosphere is large in the absorption bands of ozone (9.6 μm), H_2O (6.3 μm), and CO_2 (4.3 and 15 μm). As seen in Figure 24-92, owing to the presence of these absorption bands, the only available windows for surface-temperature measurement are from 3.5 to 4 μm, from 8 to 9.5 μm, and from 10.5 to 12.5 μm. In these windows, absorption is largely due to water vapor. Absorption by other gases such as CO_2 and N_2 is rather weak. The effect of CO_2 and N_2 absorption upon surface-temperature measurement is practically constant because of their constant mixing ratio. Between the absorption bands are spatial windows where infrared radiation is readily transmitted through the atmosphere. However, the water-vapor content is highly variable, and it is responsible for nearly all of the variation in the atmospheric windows. Although the atmospheric effect is at a minimum at about 3.7 μm, this wavelength is useful only at night because of the importance of reflected solar radiation. At about 9 and 11 μm, the largest temperature deviations amount to several degrees for humid atmospheres. Between the 10.5 and 12.5 μm spectral range, absorption is mainly due to the water vapor continuum, which is dependent upon the water vapor partial pressure. This notably reduces the atmospheric effect at around 11 μm for the coldest atmospheres (Deschamps and Phulpin, 1980).

Aircraft systems operating in the thermal infrared typically operate in the 8–14 μm band. Between 9 and 10 μm, however, there is a narrow ozone absorption band. As the principal concentration of ozone in the earth's atmosphere peaks at elevations around 20 km it is important to place the wavelengths for satellite detection of thermal exitance away from this band. For this reason satellite thermal infrared sensors such as those on the Landsat series and Heat Capacity Mapping Mission Satellites operate at wavelengths between 10 and 12 μm. Infrared radiation at wavelengths longer than 14 μm is not employed in remote sensing because of severe atmospheric absorption (primarily by CO_2 and H_2O).

Infrared radiometry measurements are made in an atmospheric window for which molecular absorption is at a minimum, generally between 10.5 and 12.5 μm. The influence of atmospheric absorption and emission varies from small values for arctic atmospheres to fairly high values for tropical atmospheres (Maul and Sirdan, 1973; Brower et al., 1976).

Water vapor residuals in a sensing spectral window both attenuate a surface-emitted signal by absorption and add to it by upward emission, the latter as a function of temperature. If a scanner is to be truly an imaging radiometer, by use of internal calibration sources, radiometric energy values are assigned to instrument outputs and signals are measured as received aloft. Then the dampening effect of the intervening atmospheric column must be separately considered. Compensation for the air modification can be accomplished in several ways. One type of correction may be performed as follows: A correction curve is constructed in which energy received aloft is plotted against energy emitted by the surface (I_z and I_o in Figure 24-93). For air columns of moderate depth, this curve will be linear as long as energy values rather than radiation temperatures are used. Its slope and position, therefore, can be set by as few as two contrasting ground target measurements. As with other calibration methods, this assumes a homogeneous air mass over the target.

Dependence upon ground targets for sensor-to-surface correction can be reduced if the characteristics of the air column are known. A gray-window model for the effects of the atmosphere in the thermal spectral bands shows that the surface signal is attenuated according to the absorption of the column (1 − transmissivity), but at the same time the air column adds the energy of its own radiance according to its emissivity. As constants of proportionality, absorptivity (a) equals emis-

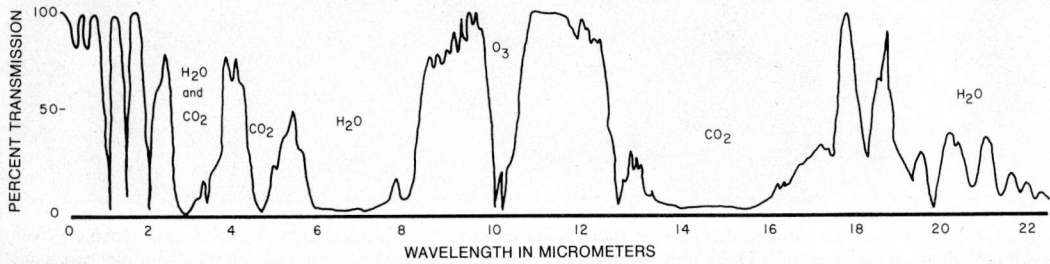

Fig. 24-92. Atmospheric absorption at short and intermediate wavelengths (Estes, 1974).

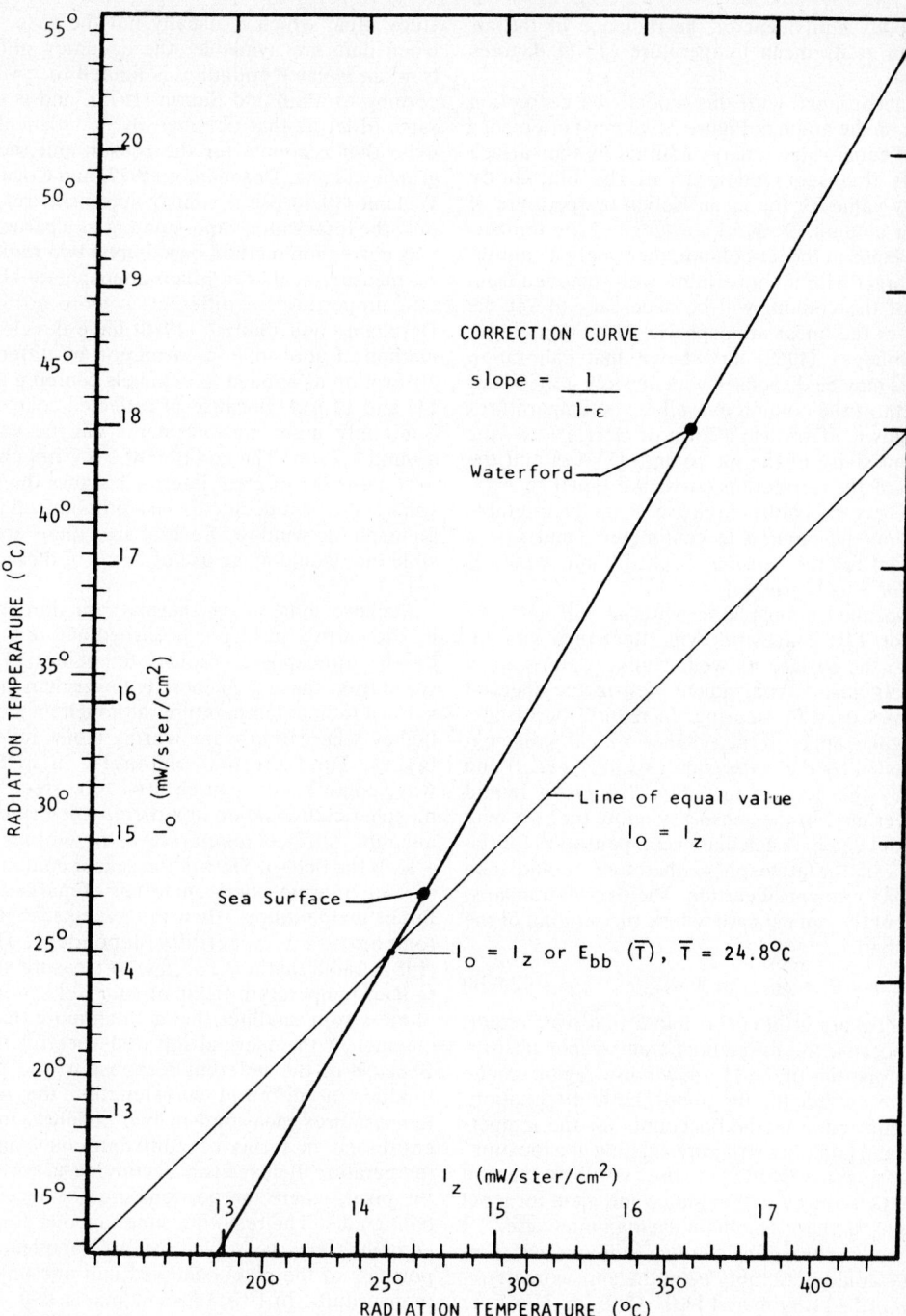

Fig. 24-93. Correction curves for effects of temperature. The x-axis refers to energy received aloft and the y-axis refers to surface energy. The area in which this test was conducted is near Waterford on the Island of Barbados (Courtesy of Robert Pease).

sivity (ϵ); but the former modifies the surface signal, a product of surface temperatures, while emissivity determines air radiance at air temperatures. Thus if the surface is warmer than the air, the attenuation will be greater than the added energy, but if the air is warmer, the opposite will be true. Because absorptivity equals emissivity (a = ϵ), the latter (ϵ) is used to denote both. The model can be expressed in the following equation:

$$I_z = I_o(1 - \epsilon) + \left[E_{bb}(\overline{T})\right] \qquad (24\text{-}37)$$

Where I_z and I_o are the radiances aloft and of the surface, respectively, ϵ is the emissivity (and absorptivity) of the air column, and $[E_{bb}(\overline{T})]$ is the blackbody equivalent of the radiance of the air column at its mean temperature ($\bar{T}$) in degrees Kelvin.

In accordance with this model, the correction curve on the graph in Figure 24-93 must intersect a line of equal value (energy emitted by the surface equals that received aloft) at the blackbody energy-value for the mean Kelvin temperature of the air column. With a knowledge of the temperature lapse in the air column, then, only a calibration target with a temperature well removed from that of the column will be necessary to set the slope of the linear atmospheric correction curve. Kondrateyev (1969) has shown that calibration targets may be dispensed with entirely if the water content of the column as well as its temperatures are shown. If we use a form of Beer's Law, the transmissivity of the air column $(1 - \epsilon)$ and the slope of the correction curve are equal to e^{-ku}, where ϵ is the natural log base, u is the precipitable water in the column in centimeters, and k is a constant for the sensing spectral band, which is 0.25 for 8 to 12 μm.

Suspended atmospheric material will not only degrade TIR images but will affect their calibration to the surface as well. Unlike water vapor, aerosols absorb and radiate well in the spectral windows used for sensing. In reality, two transmissivities apply to the scanner optical path: that controlled by the water vapor residuals (T_{wv}) and that by the aerosol turbidity (T_t). On a humid summer day, in the sensing window the two may be about equal. A calculated compensation for the effects of the atmosphere therefore should take turbidity into consideration. The overall transmissivity of the optical path will be the product of the two, that is

$$T = T_{wv}T_t \text{ or } T = T_t(e^{-ku}). \quad (24\text{-}38)$$

Prabhakara et al (1974) found that over cloud-free oceans, the differential water-vapor absorption properties of the 11 μm window region can be used to correct for the atmospheric attenuation. This correction method accounts for the temperature and humidity structure existing at a location. Two measurements in the window region 10.5–12.9 μm were found to be sufficient for input into this method. To obtain the maximum effect of differential absorption the two window measurements would preferably be in the approximate regions 10.5–11.3 μm and 11.9–12.9 μm. However there are some problems with the calculation of sea surface temperature (SST) in the tropics as the spatial and temporal changes of the SST are generally about 2°C. The water-vapor absorption correction over the tropics could range from about 4° to 8°C.

The contribution of atmospheric emission can be theoretically calculated if the absorption coefficient of water-vapor, the main absorbing gas, as well as the vertical temperature and humidity profiles are known. However, this is possible only by using radiosonde data, if they exist for the study area, which is usually not the case. Even when data are available, the accuracy obtained from an isolated sounding is limited to ±1°K according to Maul and Sidran (1973), and is of the same order as that obtained by a statistical estimate that accounts for the season and the geographical zone. Deschamps (1977) and Cogan and Willand (1976) use a similar approach, retaining only the total water-vapor content as a parameter.

A correction method based upon two radiometric measurements for which atmospheric absorption properties are different is more attractive. Deschamp and Phulpin (1980) have developed a method of atmospheric correction by differential absorption as applied to channels centered at 3.7, 11, and 12 μm. Because of reflected solar radiation, only night measurements can be used at around 3.7 μm. The addition of a 3.7 μm channel is nonetheless of great interest because the atmospheric effect is decidedly less pronounced in this atmospheric window. Several algorithms are possible that would make use of 2 or 3 of these channels.

Because some of the thermal radiation emitted by the earth's surface is absorbed and re-emitted by the atmosphere, radiant temperatures measured from space are generally lower than the true surface radiant temperature, although they can be higher where there are warm, moist inversion layers. Dual thermal channels, in different wavelength bands, can be used to correct for atmospheric attenuation and thereby determine the absolute surface temperature, to within about 1°K. If the field-of-view of the sensor contains two surface or atmospheric materials of markedly different temperatures, then the average brightness temperature is spectrally dependent. Dozier (1981) found that it is possible to measure surface radiant-temperature fields of subpixel spatial resolution from satellites that contain more than one channel in the thermal infrared spectral region. Because of the different response of the Planck function at different wavelengths, the radiant temperatures measured in two channels may be expressed in terms of contributions from two temperature fields, each occupying a portion of the pixel, where the portions are not necessarily contiguous. The resulting simultaneous nonlinear equations may be solved for the complementary portions of the pixel occupied and one unknown temperature. In two adjacent pixels that can be assumed to have the same target temperatures and same background temperatures, both unknown temperatures may be found.

Matson and Dozier (1981) demonstrated that NOAA AVHRR (Advanced Very High Resolution Radiometer) data with 1.1-km spatial resolution could be used to estimate temperatures of small, hot objects, such as industrial slag heaps or flaring gas from oil wells, occupying less than 1 percent of the field-of-view. Wan (1981) has extended these results by combining subpixel analysis with McClain's (1981) atmospheric correction methods to map snow-cover at subpixel resolution from thermal satellite data.

Even for clear atmospheric conditions, however, atmospheric absorption and emission in these windows are not negligible, and the problem is compounded when there are thin clouds, which attenuate the signal from the surface but do not completely obscure it. For clear sky conditions the brightness temperature error at $10.5-12.5$ μm can be as large as $10°K$ for tropical atmospheres (Maul and Sirdan, 1973).

For clear atmospheres, a useful and generally successful method for correcting atmospheric effects over the ocean surface uses multichannel infrared measurements. In the $10.5-12.5$ μm window the principal absorbing agent is water vapor, whereas at $3.5-4$ μm water vapor absorption is smaller and absorption by nitrogen and other gases can be calculated. Therefore the difference between the measurements in the two windows can be used to account for water vapor absorption and to determine the correct brightness temperature.

Anding and Kauth (1970), Prabhakara et al (1974), Deschamps and Phulpin (1980), and McClain (1981) have developed and used this approach. An effective way to pursue it is to simulate brightness temperature differences for a variety of atmospheric temperature and humidity profiles, using a radiative transfer model to account for atmospheric absorption and emission, and then to use these simulations to develop empirical corrections. Thus far only absorption and emission have been considered, in that the atmospheric models used (Selby et al., 1976; Weinreb and Hill, 1980) do not account for clouds or other atmospheric aerosols. The empirical corrections calculated by McClain (1981) are now used operationally by NOAA's sea-surface temperature-mapping program.

Approximate calculations by Jacobowitz and Coulson (1973) show that aerosols can cause temperature determination errors as large as $2°K$. For cloudy atmospheres, Chahine (1977) has explored the utilization of the 4.3 and 15 μm atmospheric sounder frequencies (many narrow wavelength channels clustered in the two CO_2 bands) to derive a temperature-humidity profile, which can then be used to correct surface brightness-temperature estimates. The difficulty with using this approach is the lack of a comparable sounder on a satellite, and the difficulty in constructing and operating such a sounder at the appropriate spatial resolution.

One problem with the use of all these temperature-correction algorithms is that they were originally developed for measuring sea-surface temperature. Over land surfaces we have the added complication of emissivity variations with wavelength and viewing (nadir) angle. Measurements of emissivity for many surface materials at $8-14$ μm and for near-normal viewing are available (e.g., Beuttner and Kern, 1965; Griggs, 1968). However, in the $3.5-4.0$ μm window and for larger viewing angles, there is much more uncertainty.

The effect of viewing angle is important over hilly or mountainous terrain because moderate local slope and satellite radiometer scanning-angles can routinely lead to viewing angles greater than $60°$ from the local surface normal. From Kirchoff's law we know that, for an opaque material, spectral emissivity and reflectance sum to one for any angle. Reflectance increases with illumination angle for most surfaces; hence emissivity usually decreases as viewing angle increases. The magnitude of this effect is comparable to that of the atmosphere. Dozier and Warren (1982), for example, have developed a model for emissivity of snow at different viewing angles. They show that at 12 μm, a $60°$ viewing angle leads to a brightness temperature $1.5°K$ lower than the thermodynamic temperature (brightness temperature is defined as the temperature of a blackbody emitting the same amount of radiation as the surface).

SYSTEMS FACTORS AFFECTING THERMAL IMAGERY

Although there are a great many thermal infrared systems in use today, all share certain basic characteristics important for image interpretation. The geometry of scanning is illustrated in Figure 24-94. The dimension of the instantaneous area viewed by the scanner in the direction of scan (y-direction) is given by $\alpha h \sec^2 \theta$ where α is the instantaneous angular field-of-view of the scanner in radians, h is the flying height, and θ is the angle of scan measured from the nadir direction.

The instantaneous area scanned is represented as a spot on the film having approximately constant size and equal to $r\alpha$ where r is the equivalent focal length of the scanner. This is regardless of the scan angle θ. The result is, therefore, a gradual change in the scale away from the nadir. For example, the scale at $\theta = 60°$ is one-fourth of that at the nadir point.

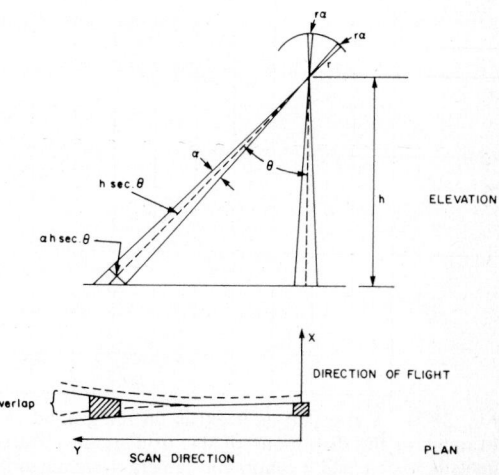

Fig. 24-94. Geometry of scanning (Masry and Gibbons, 1973).

In the direction of flight (x-direction), the dimension of the area scanned also increases away from the nadir point. At a scan angle θ, the dimension is given by $\alpha h \sec \theta$. If the scanned strips are contiguous on the ground at the nadir, the increase will result in an overlap of the information depicted on the film, and the scale in x-direction will be approximately constant. Consequently, a square grid on the ground (Figure 24-95) will be imaged by an infrared scanner, and will be as shown in Figure 24-96. It is evident that details towards the edges of the scan will be compressed in y-direction and that straight line details, for example streets inclined to the direction of flight, will appear as S-curves.

We may note here the difference between panoramic distortion of panoramic photography and this type of distortion. Panoramic distortion of the same grid (Figure 24-97) is such that the scales in both x- and y-directions are not constant. It is, in effect, similar to the distortion within one scan line of the infrared imagery. Obviously, the resultant distortion of a series of scan lines is not panoramic; it has been termed quasi-panoramic (Masry and Gibbons, 1973).

Distortions of line-scan imagery resulting from scan geometry can be rectified if the displacement is systematic and can be expressed as functions of the x and y coordinates of ground features. Displaced features can be restored to very near their true geographic position using a computer-controlled orthophoto printer (Masry and Gibbons, 1973). Corrected imagery, which may be referred to as "rectilinear", is compared to an unrectified image in Figures 24-98 and 24-99. Other types of distortion may arise due to aircraft motion.

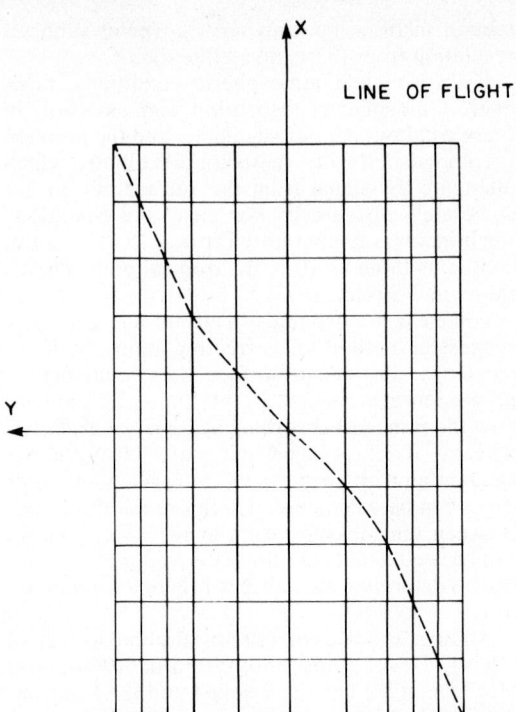

Fig. 24-96. Quasi-panoramic distortion of grid in Fig. 24-95 as it would be recorded on infrared imagery (Masry and Gibbons, 1973).

Most scanners incorporate a system to compensate for aircraft roll, but if this system is not operating, imagery distortion may occur. In Figure 24-100, the roll compensation failure is obvi-

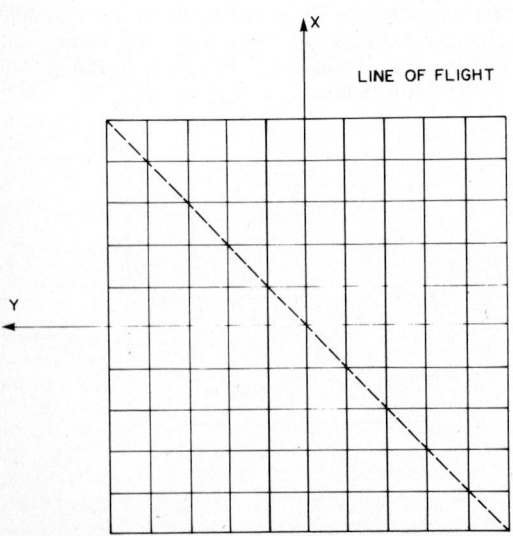

Fig. 24-95. A grid unit as it exists on the ground. An example of the distortions of this grid by an infrared imaging system and a panoramic camera system can be seen in Figs. 24-97 and 24-98 (Masry and Gibbons, 1973).

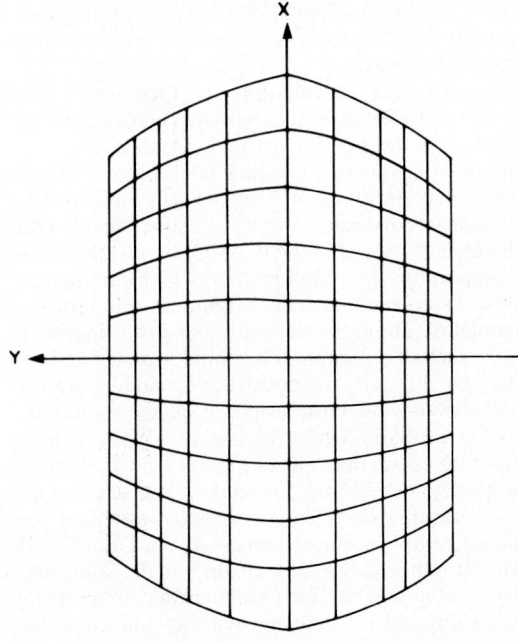

Fig. 24-97. Panoramic distortion of grid seen in Fig. 24-95 (Masry and Gibbons, 1973).

Fig. 24-98. Unrectified infrared imagery taken at night of the town of Brawley in the Imperial Valley, California with an HRB Singer Reconnofax IV thermal infrared imagery system (Masry and Gibbons, 1973).

ous in the distortion of the straight road. In other types of terrain, it may be more difficult to recognize this irregularity. On imagery of dipping sedimentary rock outcrops, roll compensation failure has produced patterns that resemble plunging folds. Some scanner systems incorporate a warning device that alerts the operator to roll-compensation malfunctions.

Aircraft bank and turn comprise another cause of image distortion because roll-compensation systems can only correct for about 10° of roll. If an aircraft maneuver exceeds this, the imagery will be distorted in a curving pattern. Such maneuvers may occur while the pilot is lining up the begin-

ning of the flight line or breaking off at the end. This is avoided by correct flight planning, which programs adequate maneuvering room beyond the ends of the flight line (Sabins, 1973a).

Yaw (also called crab) results from aircraft rotation about a vertical axis so that the aircraft longitudinal axis is not parallel with the flight direction. Yaw results from crosswinds and causes distortion from end to end of the image strip unlike the intermittent motions described earlier. Williams and Ory (1967) described the effects of yaw on imagery. Some scanners are installed in a conventional camera ringmount, which can be rotated in the horizontal plane. Camera drift-

Fig. 24-99. Rectified thermal infrared image of the area seen in Figure 24-98.

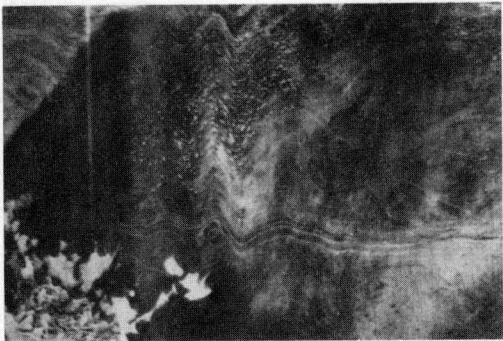

Fig. 24-100. An example of uncompensated aircraft roll. The roll compensation failure is obvious in the distortion of the straight road. In other types of terrain it may be more difficult to recognize this problem (Sabins, 1973b).

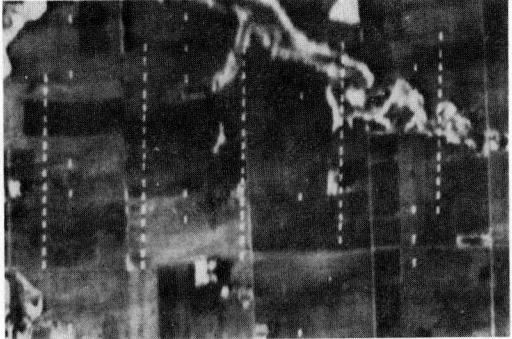

Fig. 24-102. Unidentified electronic noise. This type of interference may be annoying but not serious (Sabins, 1973b).

sights are useless at night but the pilot can determine any difference between aircraft heading and flight-line direction. The operator can rotate the scanner mount to compensate for this yaw angle. This compensation must be reversed, of course, on alternate lines with opposite headings.

Transmissions from many aircraft radios may cause strong interference patterns on thermal imagery. On the example of Figure 24-101, the interference occurs as bands of electronic noise that obscure the underlying image pattern. Radio transmissions may also produce a wavy moiré interference pattern. Electronic shielding of the scanner equipment may prevent this interference but the simplest solution is to observe radio silence during image runs and communicate with the ground during turns and offset legs. Cyclic repetition of discrete signal patterns (Figure 24-102) is an annoying, but not serious, form of electronic interference. In this example the noise occurs as positive (bright) dots but it may also have a negative signature and occur as dashes. Researchers have observed variations of this noise on imagery from a wide range of scanners, air-

craft, and localities. It seems to occur sporadically and does not hinder image interpretation. It has been suggested, but not established, that outside sources, such as air-traffic radars, may be responsible. Electronic shielding of the scanner installation may reduce the effect.

During image recording of a flight line, on either tape or film, a progressive change may occur in the overall radiant temperature level. The operator may have to shift the recording base level to stay within the optimum range, resulting in an abrupt change in image density as shown on Figure 24-103. With direct film recording, base-level shifts may be partially compensated by printing the denser and thinner negatives with different exposures. With magnetic tape, the compensation may be accomplished during playback by monitoring the signal level and adjusting for base-level shifts to obtain a uniform film density.

The photographic development of the image film (either directly recorded or played back from tape) is another potential source of image irregularities. The developer streak on Figure 24-104 resulted from uneven alignment of a pressure roller.

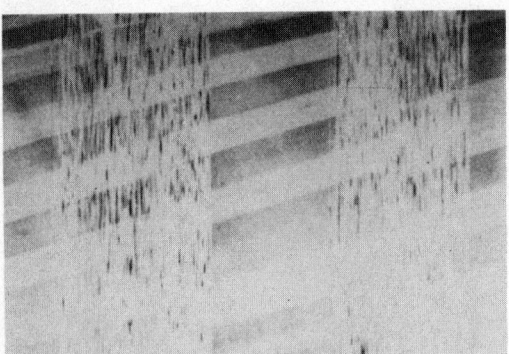

Fig. 24-101. Radio transmission interference often occurs as bands of electronic noise, which may obscure the underlying image, as indicated here. Radio transmissions may also produce a wavy moiré interference pattern (Sabins, 1973b).

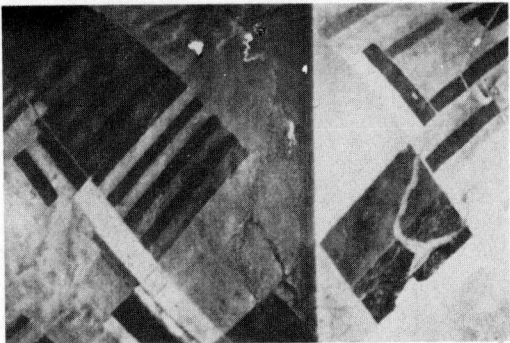

Fig. 24-103. Shift in base level. During an image-acquisition mission a progressive change may occur in the overall radiant-temperature level. If the change is large, the operator may have to shift his recording base level, resulting in an abrupt change in image density (Sabins, 1973b).

Fig. 24-104. Film developer streak. Image processing is a potential source of image irregularity. The developer streak seen above resulted from an uneven alignment of a pressure roller in an automatic film processor (Sabins, 1973b).

Another possible source of error on TIR imagery is scale. This error may result from a variable flight speed during imaging or failure to maintain a constant height above the datum plane. When thermal information is recorded on magnetic tape, this type of error can be corrected by carefully controlling tape speed during playback. Even though these systems-related errors can be somewhat suppressed, if not altogether corrected, there are enough inherent distortions in thermal imagery to prevent their use for planimetric measurements. Because these distortions reach a maximum near the edges of images, an effort should be made to restrict mapping to the center two-thirds of each image.

As with other types of imagery, the resolution and information content of thermal infrared imagery are to some extent scale-dependent. Low-altitude imagery will cover a small area in great detail while high-altitude imagery may afford a view of regional thermal patterns. Selection of scale depends on the purpose of the mission. The effects of altitude on TIR imagery are demonstrated in Figure 24-105.

ENVIRONMENTAL FACTORS AFFECTING THERMAL IMAGES

A variety of conditions outside the actual imaging system can produce unexpected and unwanted anomalies on thermal infrared imagery. These factors are perhaps the most important considerations in image interpretation because, unlike systems factors, environmental conditions can never be controlled or precisely reproduced. Useful image interpretation requires that one be familiar with conditions that determine patterns and tones on images.

Even though most thermal infrared systems operate in the relatively clear atmospheric windows at 3.5 to 5.5 μm and 8 to 14 μm, weather effects can degrade imagery.

In some instances remote sensing flights must be conducted even though weather conditions are less than ideal. Some scattered clouds below the flight altitude may have to be tolerated. Clouds typically have the patchy warm and cool pattern illustrated in Figure 24-106 where the dark tones are relatively warm. Scattered rain showers produce a pattern of streaks parallel with the scan lines. It is generally believed that a heavy overcast layer will greatly reduce thermal contrasts because of re-radiation of energy between the terrain and cloud layer. Many avoid flying under these conditions even though the base of the overcast layer may be above the planned flight altitude. One investigator, however, has obtained excellent nighttime imagery (unpublished) of an urban area under these conditions. The typical high thermal contrast between urban targets may have compensated for overcast re-radiation effects, or these effects may be overestimated (Sabins, 1973b).

Surface winds produce characteristic smears and streaks on imagery. Wind smears (Figure 24-107) are parallel curving lines of alternating warmer and cooler signatures that may extend over wide expanses of imagery. Wind streaks occur downwind from obstructions on flat terrain and typically appear as the warm (bright) plumes shown on Figure 24-108. On this example, the obstructions are clumps of trees, which image warm, and the one distinct building in the lower right part of Figure 24-108, which images cold. The obvious solution to wind effects is to fly only on still nights, but in many areas surface winds persist for much of the year and their effects must simply be endured.

Under certain circumstances, anomalous thermal patterns may be produced by small objects below the theoretical systems resolution. For most thermal imaging systems, resolution is determined by the dimensions of the instantaneous field-of-view and an electronic function of the signal amplifier, known as rise time. This refers to the amount of time required to record the transition in the signal from a very hot target to an adjacent cold target. During this period, the scanner continues to move, so that occasionally very small hot points produce hot spots exaggerated in size on imagery. The object-to-background contrast is large enough that very small targets appear in small-scale imagery. Figure 24-109 shows this effect, known as "blooming", in a residential area in Texas. Small vents, no larger than 15-cm in diameter, allow heat to escape from the homes. On this image, many hot spots are seen on rooftops, where heat escapes. The thermal contrast between rooftop and ventilator is large enough to produce the bloom effect.

Both time of day and the season of the year are important parameters affecting both image acquisition and subsequent interpretation. Thermal infrared systems are passive and are not dependent on reflected energy from incoming solar radiation. (Actually, a small amount of reflected energy may be sensed during daytime missions, especially in the 3.5 to 5.5 μm band.) Radiant exitance from

A. 8000 FT ABOVE TERRAIN

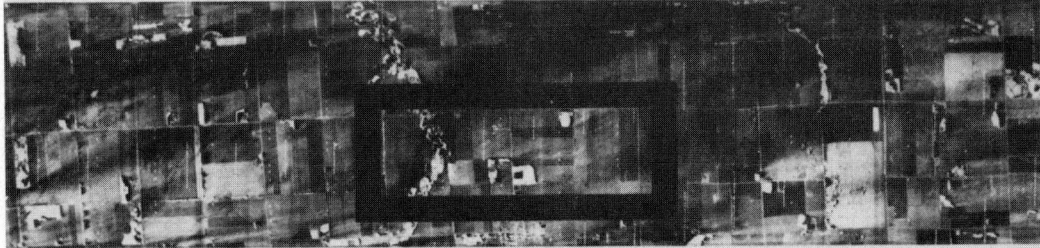

B. 4000 FT ABOVE TERRAIN

C. 2000 FT ABOVE TERRAIN

D. 1000 FT ABOVE TERRAIN

Fig. 24-105. Effects of altitude on thermal imagery. Black outline covers an area 0.5 miles wide and 2 miles long (Sabins, 1973a).

surface objects varies with time of day as a result of differential heating and cooling, which is in turn related to sun angle and intensity.

Because surface temperatures change with time of day, an interpreter should be aware of the time when imagery is obtained. As an example, consider a water body and an adjacent land mass. Water has a high thermal inertia, which means it warms and cools very slowly. Its diurnal temper-

ature range is normally small. If a thermal-scanning system is calibrated to a constant temperature for a daytime and a nighttime flight, the resultant tones on the imagery would be the same. On the other hand, land surfaces generally have much higher diurnal temperature ranges. Contrast between daytime and nighttime images of land surfaces would be great. In daytime imagery, topography is typically a dominant expression be-

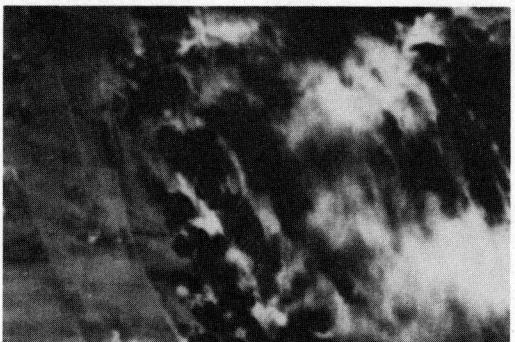

Fig. 24-106. Clouds on IR imagery. In some instances image acquisition must be accomplished in less than optimal conditions. Clouds typically exhibit a patchy light/dark pattern on TIR imagery (Sabins, 1973b).

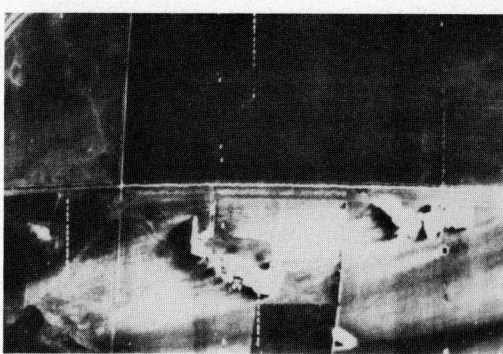

Fig. 24-108. Wind streaks often occur downwind from obstructions on flat terrain and typically appear as light plumes on positive TIR images (Sabins, 1973b).

Fig. 24-107. Wind smears are parallel curving lines of alternating lighter and darker tones, which may extend over wide areas of an image (Sabins, 1973b).

cause of the thermal effects of differential solar heating and shadowing. On nighttime imagery, however, these differential solar effects are greatly reduced (see Figures 24-110 and 24-110b).

The images shown in Figure 24-111 demonstrate both diurnal and seasonal variations in thermal imagery. Changing thermal patterns of vegetation, soil, asphalt, and water of Yosemite, California, are shown in June (left column) and September (right column). Numerous crossover periods can be seen. For example, the meadow vegetation (lower right) is "cooler" than the river during the night, but during the day, it becomes "warmer" and produces a lighter tone than water in the afternoon. By evening (6:30 PM) the temperatures are nearly equal, and after nightfall (9:00 PM) the meadow is again cooler than the water.

Fig. 24-109. "Blooming" of hot air vents on rooftops in a residential area (Texas Instruments Incorporated).

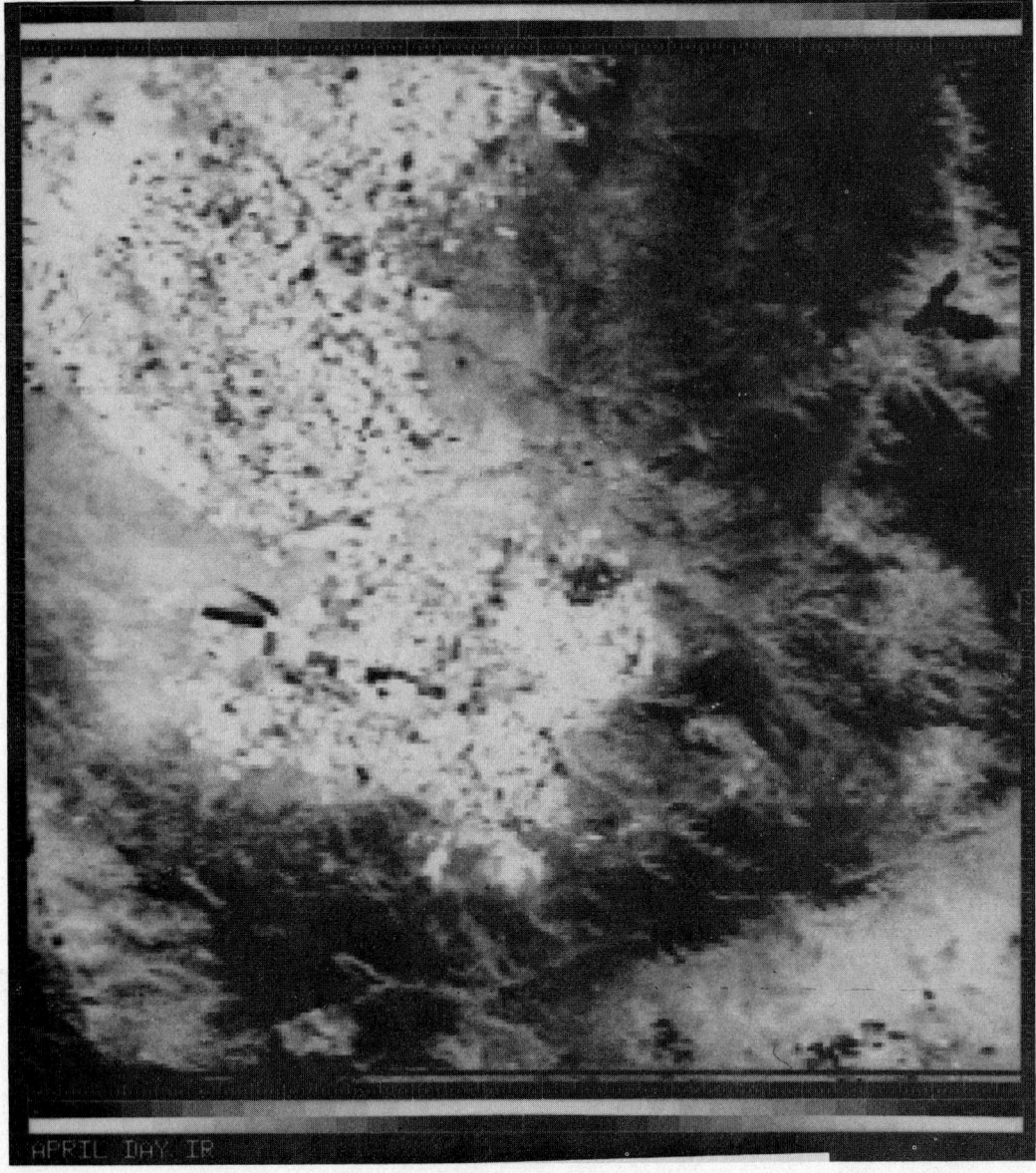

Fig. 24-110a. Day- (1:30 p.m.) and Figure 24-110b (next page) night-time (2:30 a.m.) thermal infrared imagery of the San Joaquin Valley in California. Taken by the Heat Capacity Mapping Mission (HCMM) on April 4, and 5, 1979, respectively (Minor, 1982).

Seasonal changes in tonal and textural patterns arise from differences in the sun angle, which cause differential heating in different areas, and from vegetational changes such as drying and trampling.

Figure 24-112 is representative of TIR imagery of desert terrain near Yuma, Arizona. This figure is presented mainly to demonstrate that (1) the imagery does discriminate major terrain features such as an eroding terrace, an old alluvial fan (covered with desert varnish), a major wash, and roads and trails and (2) to interpret TIR imagery it is necessary to know the time of day and the con-

ditions under which it was obtained. Conditions affecting the interpretation are temperature, sky condition (clear or overcast), ground fog, and rainfall just prior to obtaining imagery. The five strips in Figure 24-112 were taken at various times of the day and illustrate what diurnal changes can do to imagery. Note the tone reversals between the center of strips 3 and 5. Strip 3 was taken during the night and the central portion, which consists of an alluvial fan with the flat original surface covered by desert varnish, is shown as a cold surface while gullies show as warm. In strip 5, taken at 9:12 AM, the surface covered by desert

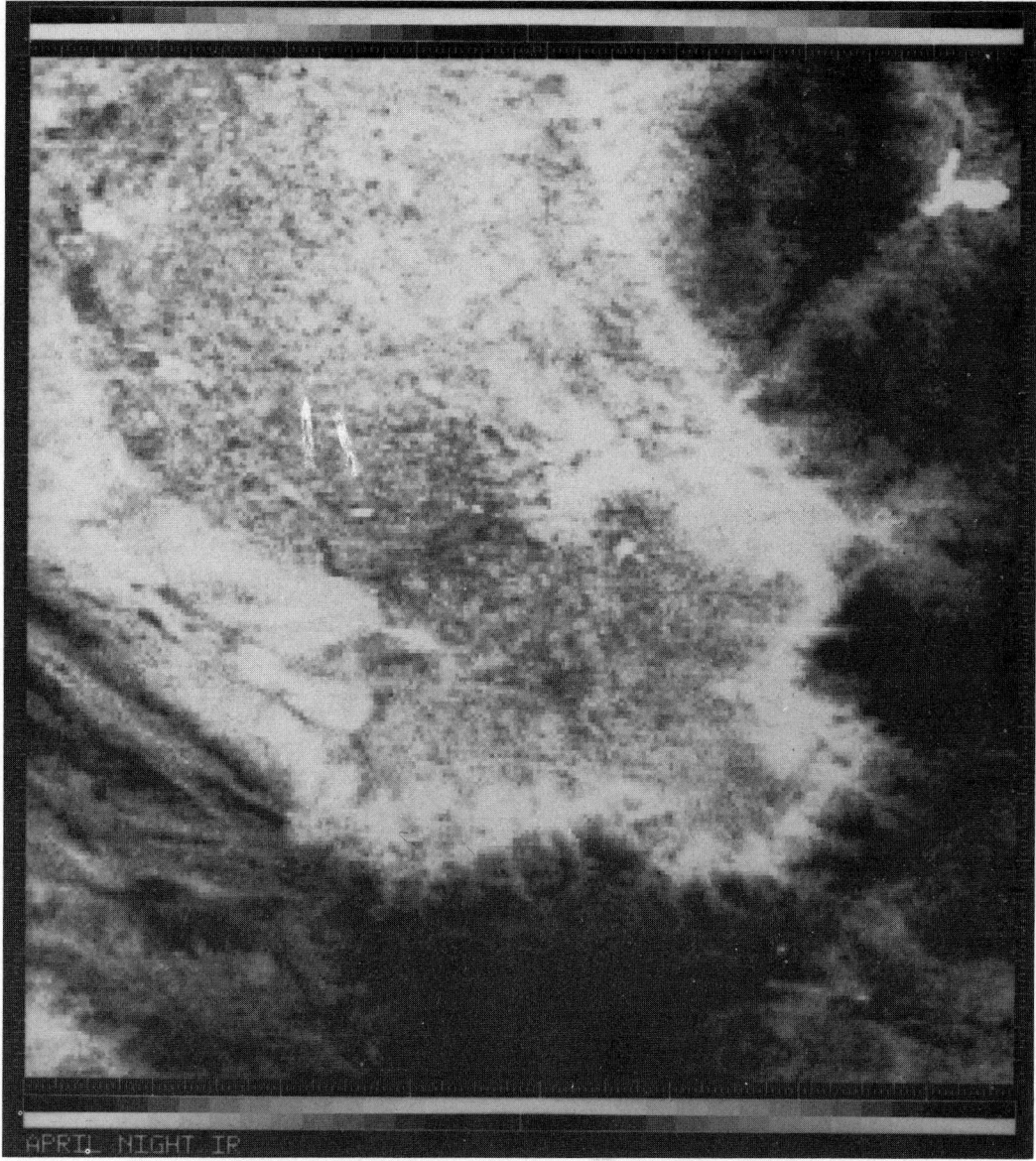

Fig. 24-110b (continued) night time thermal infrared imagery.

varnish appears very warm and the gullies appear very cold. Vegetation in the wash appears warm during the night and cold during the day.

At certain times, owing to differential heating and cooling of objects and backgrounds, periods of "cross-over" occur (i.e., an object's rate of heat emission will coincide with that of the background, producing the same level of energy incident on the detector). This could lead to both the object and its background creating the same shade of gray on the imagery produced by the system, thus hiding the object. Usually, there will be two periods of crossover per day, morning and evening, which must be considered by the imagery interpreter (McLerran, 1967).

Several researchers have concluded that nighttime TIR imagery is superior to daytime TIR imagery. To distinguish rock types and map fault and fracture zones in the Arbuckle Mountains of Oklahoma, Rowan et al (1970) found predawn imagery most satisfactory. Based on interpretation of imagery flown at different times of day in New York State, Stingelin (1969) concluded that nighttime imagery is superior for most applications. In California, Wolfe (1971) evaluated TIR imagery from predawn and postdawn missions. He also concluded that predawn imagery is superior for evaluation of geologic features. Because there was no significant incoming radiation at infrared wavelengths after sunset, re-radiation of energy

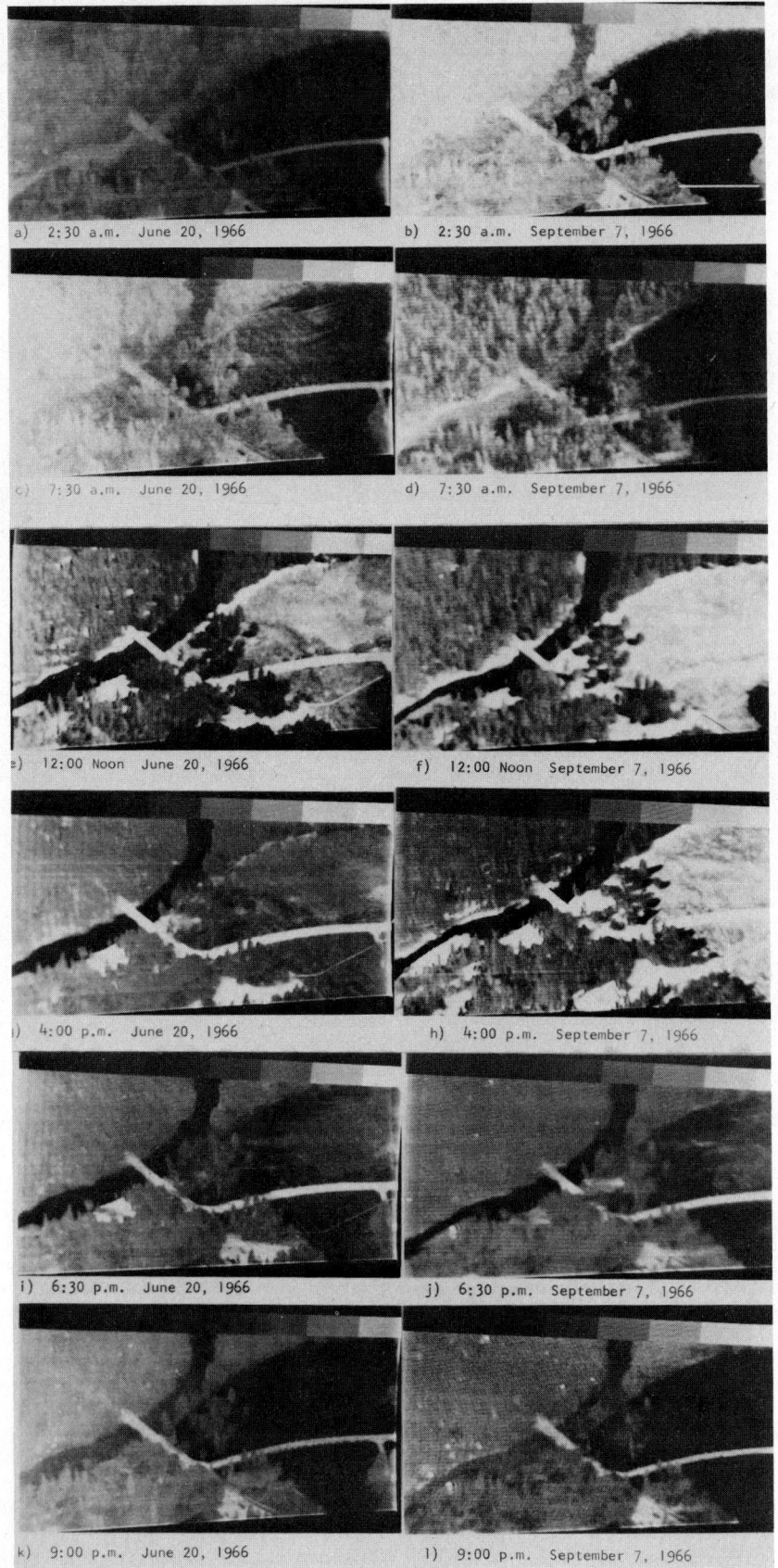

Fig. 24-111. Diurnal and seasonal variations in thermal imagery of a portion of Yosemite Valley, California. Scanning system was mounted 4,000 feet from target area, on Glacier Point (Carneggie et al., 1967).

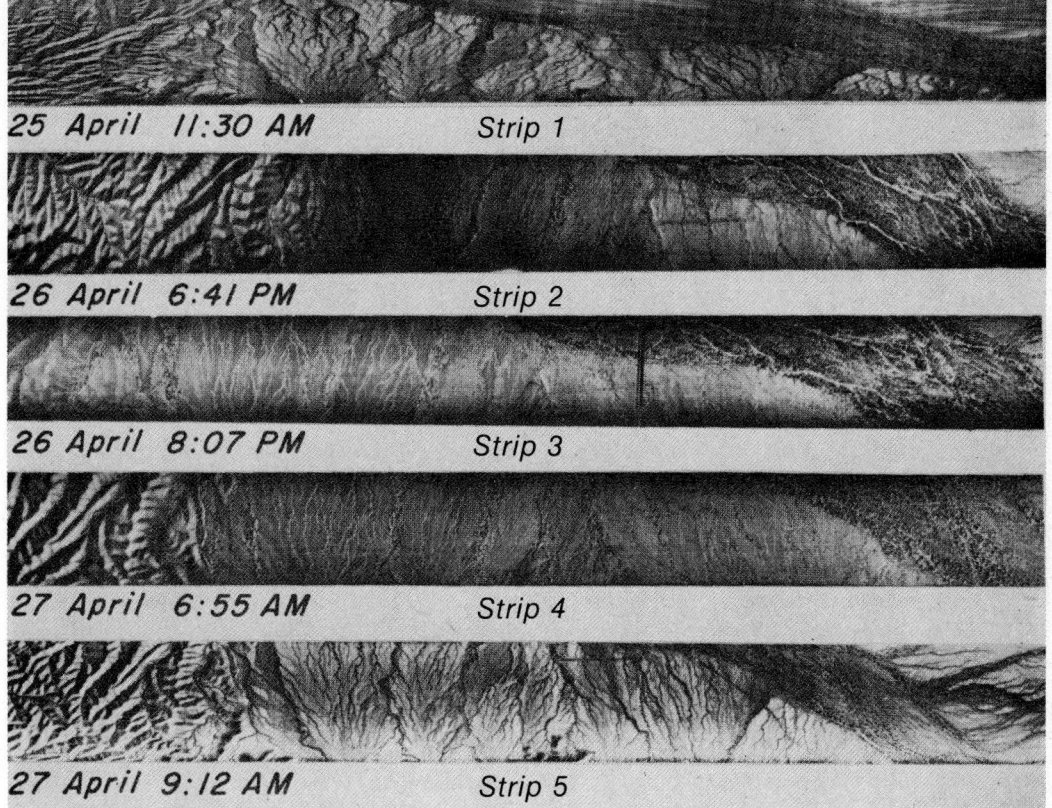

Fig. 24-112. Diurnal variations in TIR imagery in an arid area during spring (McLerran, 1967).

from surface materials is responsible for the temperatures on nighttime images. Thus, features may be distinguished and possibly identified, owing mainly to differences in their properties of specific heat and specific gravity.

For some purposes, daytime TIR may be useful. Wolfe (1971) points out that postdawn TIR imagery in an area of moderate relief resembles low sun-angle photography because of the differential heating of exposed and unexposed slopes. Shaded slopes in these areas will remain cool into the early morning. Although daytime TIR imagery has been used for certain specific purposes (such as Wolfe's) it has several serious drawbacks. Topographic effects related to differential heating tend to dominate daytime TIR images and if clouds are present, their shadows can produce anomalous cold spots on the imagery (Sabins, 1973b). Diurnal effects can be seen in Figure 24-113. In the lower strip, which was taken at 6:27 AM over the Midway Geyser Basin, when the air temperature was 32°F, all water bodies are shown to be hot with respect to the background. The Firehole River shows as warm on the imagery, as do the hot springs. Under such conditions, open water bodies at normal temperatures (e.g., pools, lakes, and rivers) could not be separated from hot springs. This imagery, however, was taken at an optimum time to locate all water features in the area regardless of temperature (McLerran, 1967).

Although there is some disagreement, principally related to the actual purpose for which the mission is being flown, Stingelin (1968) reports that, in general, the best nighttime image contrasts are obtained in the early evening. There is no extraneous information from solar reflectance or active shadowing. Tone signatures are most directly related to surface temperature because such signature arises from surface-material emission. Image quality undergoes gradual degradation in information content throughout the night as objects cool toward ambient temperature. Thermal contrast declines and often reaches a minimum in the predawn period. Because of this, a maximum object-to-background contrast exists for geothermal areas just before dawn.

QUALITATIVE IMAGE INTERPRETATION

The black-and-white format, which most TIR systems are designed to produce, is suited to qualitative evaluation. Interpretation of TIR imagery, in conjunction with other types of images, has often been employed in order to correlate visible and non-visible patterns. Many investigators have used TIR imagery from areas where obvious thermal anomalies are present. In an early study, McLerran (1967) used conventional aerial photography and TIR imagery to evaluate the Midway Geyser Basin (see Figure 24-113).

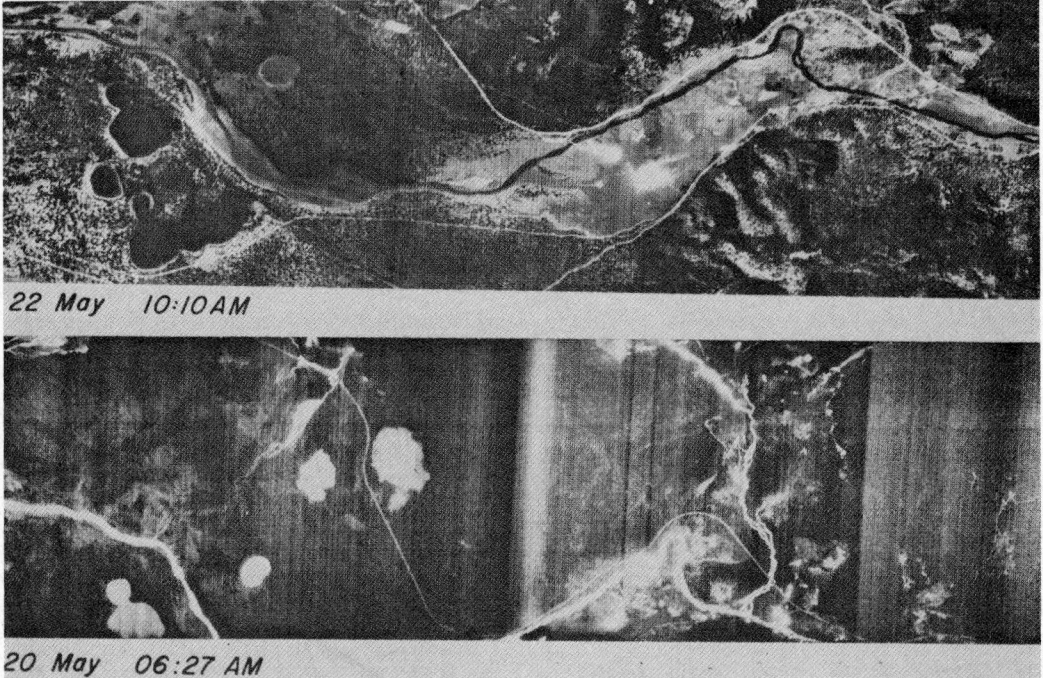

Fig. 24-113. Late morning vs. early morning imagery of Midway Geyser Basin. Diurnal variations can be seen in the above images. The image acquired at 6:27 a.m. shows all water bodies as radiometrically hotter with respect to their backgrounds while the top image, taken later in the day, shows water bodies as radiometrically cooler with respect to their backgrounds (McLerran, 1967).

McLerran (1967) stated that, because of the lack of trees, the aerial photograph shows the location of hot-spring activity, yet individual springs and geysers cannot be located. On TIR images, however, individual hot springs are pinpointed as "hot" spots against the cooler background of vegetation.

McLerran (1967) also compared aerial photography and TIR images of sea ice in polar regions. This work demonstrated the utility of TIR imagery for mapping sea-ice features during both daylight conditions and during the arctic winter dark period.

Figure 24-114 is a comparison of infrared imagery and conventional aerial photography of sea ice. Both were obtained simultaneously during April 1962. By interpretation of the aerial photography one can conclude that most of the ice is thin winter ice with little or no snow cover. The lack of snow cover is a good indication that this is thin ice. Ice finger-rafting, characteristic of thin winter ice, is seen as light-tone bands irregularly criss-crossing the area. Within the area several pieces of thicker snow-covered winter ice show as small, rounded, nearly white areas. Snow-covered land is seen in the upper right corner.

The infrared imagery of this same area has nearly a reversal of photographic tones. The very thin ice that appears dark gray to black in the aerial photograph is now white to light gray, indicating a warm surface. The thicker snow-covered ice, which was nearly white in the aerial photo-

graph, is black, or cold, in the infrared imagery. This very cold signal is due primarily to the insulating effect of the snow cover. The finger-rafting is shown on the TIR imagery by dark-gray bands representing older, thicker rafting and by

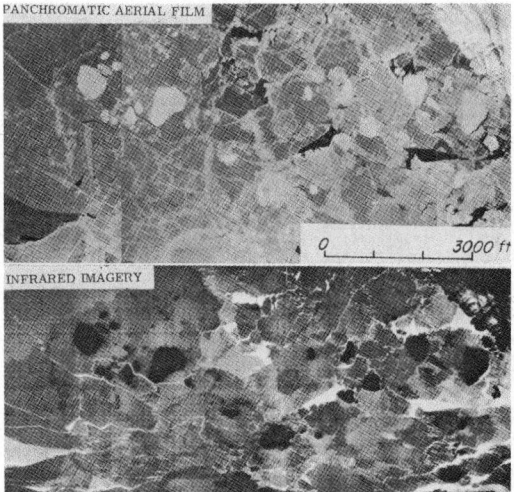

Fig. 24-114. Conventional aerial photography and TIR imagery of sea ice, obtained simultaneously. The TIR image exhibits a relatively good correlation between ice thickness and the apparent radiometric temperature of the surface (McLerran, 1967).

hot, irregular lines outlining each finger caused by a wet zone created along the edges of newer rafting. This TIR image of sea ice illustrates a good relative correlation between ice thickness and the apparent temperature of the surface (McLerran, 1967).

Interpreting TIR imagery is difficult without some knowledge of the area of interest. For this reason, missions are rarely flown without some form of field survey. Even when field conditions have been seemingly recorded, unexpected thermal anomalies may be found and this may, in fact, be more the rule than an unexpected exception. Figure 24-115 shows one such case where a strange pattern appeared. Here, an old stream channel had been filled in order to level an agricultural field. The fill material exhibits vastly different thermal properties than the adjacent soil and thus produces a different tone on the image. Neither aerial photographs nor field investigation gave evidence of its existence. Sabins (1973c) discusses a similar case of potential economic geologic significance.

The cooling influence of high soil moisture, or a very high ground-water table, may induce an unforeseen pattern of TIR imagery (Lewis, 1974). Although the skin depth from which a thermal image is produced is very thin, subsurface conditions can and do affect surface temperatures, depending on the thermal conductivity of the material.

The regional perspective afforded by TIR imagery of a large area enhances the interpreter's ability to understand patterns of emitted IR radiation. Unfortunately, distortions in scanner imagery make it difficult to produce mosaics. Careful mis-

sion planning and correction of scanner distortion, however, have made possible the construction of useable, uncontrolled TIR mosaics (Williams and Ory, 1967). The importance of mission planning for the acquisition of TIR imagery was thoroughly discussed by Sabins (1973a). The type of planning and control discussed by Sabins is important if accurate high-quality TIR mosaics are to be produced. Figure 24-116 shows a comparison between a photographic mosaic (right) and a TIR mosaic (left). The TIR mosaic was constructed from twelve 70-mm strips. Regional features, such as lineaments which might not be apparent on a single strip image, are now readily apparent.

QUANTITATIVE IMAGE INTERPRETATION

The basic attribute of TIR devices is their ability to detect object-emitted energy with varying degrees of sensitivity and selectivity. Under certain circumstances, it is possible to obtain quantitative measurements of surface temperatures and radiances using both airborne and satellite platforms and a variety of methods. Particularly useful in this regard are the modern generation of electro-optical scanners with both constant gain characteristics and internal image calibration sources.

Most TIR systems record a continuous series of electrical signals, so there is a continuum of gray tones contained in the original data. However, in processing, images are often printed using a gray scale composed of a number of discrete steps (Figure 24-117).

The example on Figure 24-117b employs a six-

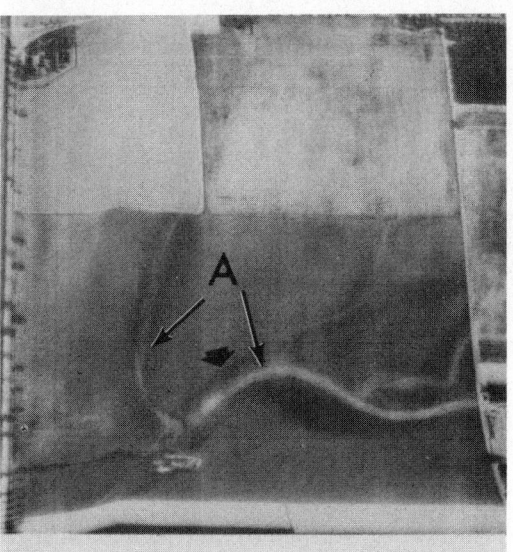

Fig. 24-115. Comparison of TIR image and visible aerial photography at the Univ. of Calif. campus at Davis. The buried river channel at "A" has been filled and leveled. Variations in soil moisture of the fill vs. natural material are probably responsible for its being evident on the thermal infrared imagery while almost absent on aerial photography (Texas Instruments Incorporated).

INFRARED IMAGERY MOSAIC

PHOTOGRAPHIC MOSAIC

Fig. 24-116. Comparison of TIR and aerial photo mosaics of an area in the Imperial Valley, California. The TIR mosaic was constructed from 12 strips of 70-mm imagery. Regional features, such as lineaments, which may not be as apparent on a single image are readily apparent on the mosaic (Williams and Ory, 1967).

level gray scale that covers the entire range of the taped signal, but any radiant temperature range could be selected such as the range represented by the water. This method is particularly useful if applied to calibrated imagery so that apparent temperature values can be assigned to the gray-scale steps.

The taped signal may be sliced at a desired amplitude, or temperature level, to produce an image in which all targets warmer than the selected level are white whereas colder targets are black (Figure 24-117c). This may be useful in isolating specific temperature ranges of interest to the investigator.

Colors may be substituted for the digitized gray

individuals involved in the analysis of resource and environmental data to take a deeper and more fundamental interest in the analysis and information-extraction processes. Many of the basic image-processing algorithms employed in remote sensing today were written by engineers and/or computer scientists with little or no background in the image-analysis process, and though these software systems may achieve acceptable results for some applications, they still, for a great many applications, leave much to be desired.

The predominance of current research emphasis in the area of image interpretation is applications oriented. Most research has been directed towards techniques of image interpretation for this or that specific task. Little work is currently being done on important generic questions such as:

- How can we best train a person to interpret an image?
- How can we optimize man-machine interaction in the resource-applications image-analysis process.
- How should an interpreter conduct a visual search?
- How can we find the optimum conditions conducive to the analysis of image data?
- How can we find whether a significant relationship exists between search time, scales, types of images, and tasks?
- How do we determine the optimum method for the recording of an interpreter's output product (labeling)?
- What are basic elements of spatial pattern recognition and how can we more effectively employ automation in extracting them?

This by no means exhaustive list is indicative of the fact that considerable research still is required and is basic to the whole field of remote sensing. Researchers attempting to advance the state-of-the-art (science) of image interpretation must keep abreast of studies in psychology (e.g., those described by Kennedy, 1974; Blaut, et al, 1970; and Wertenbaken, 1981) and physiology (e.g., those described by Noton and Stark, 1971; Michael, 1969; and Hubel and Wiesel, 1962). It is a knowledge of the more current research advances in these areas that will help those of us who are in applications-oriented remote sensing expand our understanding of the "how" of image interpretation.

There is also a real need to go into considerably more depth than we have in the area of man/machine interactive processing of images in digital and analog format. The human mind is an excellent interpretation system and its capabilities, properly employed, can go far in helping to upgrade the accuracies of many image-processing algorithms. Although many individuals pay lip service to the idea of interactive processing, few truly understand the significance and potential of the concept, while many rely on interpretation as being either manual or automated. Apart from the need for improved hardware and software systems in the analysis process, a great deal of research is still required in the area of computer-assisted analysis if we are to progress. Research should be accomplished on topics such as:

- What are optimum formats for data presentation to the interpreter?
- What type and depth of training must the interpreter have to effectively accomplish interaction?
- Are there psycho/physiological barriers to effective interactive processing?

In addition, once both the interpreter and the systems analyst acquire a basic, thorough understanding of each other's techniques, methodologies, and goals, we will have already taken a large step forward. If only this much is done, then these individuals acting as a team can at least begin to address the real issue, viz. the ultimate goal of photointerpretation and image analysis.

Finally, too many courses, texts, and individuals in the field of remote sensing tend to gloss over the "basics" as being too mundane or "easy to pick up". A thorough knowledge of the basic philosophy, which is an integrated part of activities, elements, and techniques of photointerpretation, is essential to truly advance the state-of-the-art of image interpretation.

A number of trends are probably valuable portents of the future. Some of these present real potentials, others contain pitfalls which must be closely watched. These trends include:

- Use of higher resolution space photography and/or imagery from advanced scanners, imaging synthetic-aperture radar systems, or cameras utilizing the optical bar principle, giving us both high resolution and wide area coverage.
- Continued and expanded emphasis on research into computer-assisted analysis procedures focusing on the use of expert systems and symbolic reasoning approaches from the field of artificial intelligence to more closely emulate the human analysis process.
- The expanded application of information, as extracted from aerial photographs and other sensor systems, in geobased information systems for environmental modeling. Indeed, image data will improve the sophistication of our environmental models. Such information systems will improve the users' ability to distribute model input-data more appropriately across space than was possible with point-source lumped parameter models.
- The continued and expanded use of manual and computer-assisted image interpretation by planners and decision makers at the international, federal, state, and local levels.
- The establishment of a national and international geobase information system with ae-

rial photography and other remote sensor data forming the primary inputs.

- The increased use of photos and/or information from unconventional imaging systems by regulatory agencies charged with the protection of our environment.

Image analysis holds great promises for the ultimate benefit of mankind as we monitor the earth's dwindling resource base. It also permits the super powers to keep track of each other's strategic potential, hopefully, as Phillip Klass in his book Secret Sentries in Space (1972) contends, decreasing the potential for nuclear holocaust! But as with many beneficial technologies, image interpretation presents a two-edged sword. At what point does the gathering of data on environmental resources constitute an invasion of privacy? Can the techniques and methodologies developed through the years for the benefit of mankind be turned against man? Although we have come a long way in photo/image analysis, we have only scratched the surface. There is still a great deal to do.

REFERENCES

Adams, J. B., 1981, Geological remote sensing: Identification and Mapping of Rock Types for Non-Renewable Resources; NASA-CR 164259, Contract no. NAGW-85, Washington University, Seattle, Department of Geological Sciences, Semi-annual Report.

Ahuja, N. and B. Schacter, 1981, Image models: ACM Computing Surveys; vol. 13, no. 4, pp. 373–397.

Aldrich, R. C., 1966, Forestry applications of 70 mm color; Photogrammetric Engineering, vol. 32, pp. 802–810.

Anderson, J. R. R., 1977, Land use and land cover changes—a framework for monitoring; Journal of Research, U.S. Geological Survey, vol. 5, pp. 143–153.

Anding, D. and R. Kauth, 1970, Estimation of sea surface temperature from space; Remote Sensing of Environment, vol. 1, pp. 217–220.

Anuta, P. E., 1976, Digital registration of topographic data and satellite MSS data for augmented spectral analysis; Proceedings of the 42nd convention American Society of Photogrammetry, pp. 180–187.

Aucion, P. J., Jr. and L. E. Giddings, 1978, Classification with spectral-temporal archetypes; Lockheed Electronics Co., Inc. Houston, Texas, Contract no. NAS9-15200, 22 p.

Avera, H. O., 1980, Remote sensing analysis instrumentation for geothermal exploration; Geothermal Resources Council Transactions, vol. 4.

Avery, T. E., 1968, Interpretation of Aerial Photographs; Minneapolis, Burgess Publishing Co., 324 p.

Bahdwar, personal communication, May 1982; NASA Johnson Space Center.

Ball, G. H. and D. J. Hall, 1965, Isodata, a novel method of data analysis and pattern classification; AD 699616: Stanford Research Institute, Menlo Park, Calif.

Barber, G. H., 1972, Temporal pattern analysis using multidimensional discrete Fourier transforms; Air Force Inst. of Tech. Wright-Patterson AFB, Master's Thesis, Ohio School of Engineering.

Bastuscheck, C. P., 1970, Ground temperature and thermal infrared: Photogramm. Eng. v. 36, pp. 1064–1072.

Bernstein, R., ed., 1978, Digital image processing for remote sensing; IEEE Press, New York, 473 p.

Bertram, R., 1976, The Thinking Computer; W. H. Freeman and Co., San Francisco.

Beuttner, K. J. K. and C. D. Kern, 1965, The determination of infrared emissivities of terrestrial surfaces; Journal of Geophysical Research, vol. 70, pp. 1329–1337.

Blackman, G. R., 1980, Geometric and temporal characterization of battle-field smoke and dust by multispectral digital image analysis; Army Electronics Research and Development Command, Atmospheric Sciences Lab., 14 p.

Blaut, J. M., G. S. McCleary, Jr. and A . S. Blaut, 1970. Environmental mapping in young children, Environment and Behavior, Beverly Hills, Sage Publishing, vol. 2, no. 3, pp. 334–349.

Brady, M., 1982, Computational approaches to image understanding: ACM Computing Surveys, vol. 14, no. 1, pp. 3–71.

Branch, M. C., 1973, City planning and aerial information, Cambridge, Mass., Harvard Univ. Press, 293 pp.

Brogan, W., and G. Nagy, 1980, Lacunae in the theory of digital image registration; Pattern Recognition in Practice; E. S. Gelsema and L. N. Kanal ed., North Holland Publishing Co., New York, pp. 61–71.

Brooks, M., 1978, Rationalizing edge detectors; Computer Graphics and Image Processing, vol. 8, pp. 277–285.

Brower, R. L., H. S. Gohrband, W. G. Pickel, T. L. Signore, C. C. Walton, 1976. Satellite derived sea-surface temperatures from NOAA Spacecraft. Tech. Memo NOAA-TM-NESS-78, NOAA-76062401, 84 p.

Bruckner, D. N., et al., 1967, Judged worth of aerial photographs as a function of obliquity angle with scale constant; Human Factors Research Inc. and Perkin-Elmer Corporation, Technical Report, pp. 723–724.

Bryant, N. A. and A. L. Zobrist, 1979, An image based information system: architecture for integrating satellite imagery and cartographic data bases; Mapping Software and Cartographic Data Bases, pp. 43–48.

Butera, M. K., 1978, A determination of the optimum time of year for remotely classifying marsh vegetation from Landsat multispectral scanner data; Report 169, NASA, 43 p.

Byrne, G. F., P. F., Crapper, K. K. Mayo, 1980, Monitoring land-cover change by principal component analysis of multitemporal Landsat data; Remote Sensing of Environment, vol. 10, no. 3, pp. 175–184.

Carman, P. D. and R. A. F. Carruthers, 1951, Brightness of fine detail in air photography; Optical Soc. America Journal, vol. 41, pp. 305–310.

Carneggie, D. M., et al, 1974, Remote Sensing applications in Forestry: the evaluation of rangeland resources by means of multispectral imagery: An. Pro. Rpt. Nat. Res. Program Office of Space Sciences and Applications, NASA, 76 p.

Casasent, D., 1979, Coherent optical pattern recognition; Proc. IEEE, vol. 67, no. 5, pp. 813–825.

Casasent, D., 1981, Pattern recognition; A Review, IEEE, Spectrum, vol. 18, no. 3, pp. 28–33.

Chahine, M. T., 1977, Remote sounding of cloudy atmospheres, II, multiple cloud formations; Journal of the Atmospheric Sciences, vol. 34, pp. 744–757.

Christenson, J. W. and H. M. Lachowski, 1976, Urban area delineation and detection of change along the urban-rural boundary as derived from Landsat digital data; Proceedings, American Society of Photogrammetry, Seattle, Washington, pp. 28–33.

Cicone, R. C., W. A. Malila and E. P. Crist, 1977, Investigation of techniques for inventorying forested regions; Final Report, Forestry Information and Systems Requirements and Joint Use of Remotely Sensed and Ancillary Data, NAS-CR-ERIM 122700-35-F2, vol. 2, 146 p.

Cogan, J. L. and S. H. Willand, 1976, Measurement of sea surface temperature by the NOAA 2 satellite; Journal Appl. Meteorol., vol. 15, pp. 173–180.

Coleman, G. B. and H. C. Andrews, 1979, Image segmentation by clustering; Proc. IEEE, Vol. 67, no. 5. pp. 773–785.

Collins, S. H., 1972, The stereophoto pair; Photogrammetric Engineering, vol. 38, pp. 1195–1202.

Colwell, R. N., 1952, Report on Commission VII (photographic interpretation) to the International Society of Photogrammetry: Pt. 1, General; Photogrammetric Engineering, vol. 18, pp. 375–400.

——, 1952, The future of photogrammetry and photo interpretation; Photogrammetric Engineering, vol. 18, pp. 502–505.

——, ed., 1960, Manual of Photographic Interpretation; Falls Church, Virginia (formerly, Washington, D.C.), p. 868.

Colwell, R. N., 1965, Syllabus for a course on the aerial photo interpretation of California crops and livestock, School of Forestry, University of California.

——, 1969, An evaluation of earth resources using Apollo 9 photography; Univ. California-Berkeley, Rept. Res. NASA Contract No., NAS 9-9348, Final Report.

Crane, R. B., W. A. Malila and W. Richardson, 1972, Suitability of the normal density assumption for processing multispectral scanner data; IEEE Transactions on Geoscience Electronics, vol. GE-10, pp. 158–165.

Cretcher, R. L., and R. H. Balentine, 1968, Sensitometry of aerial color photography; in Manual of Color Aerial Photography, Smith, J. T. and A. Anson, eds.; Falls Church, VA, American Soc. Photogrammetry, pp. 298–323.

Curry, L. and E. B. MacDougall, 1972, Statistical spatial analysis and remotely sensed imagery; Final Report, U.S. Geological Survey, Geographic Applications Program.

Danko, J. A., 1973, A new concept in orthophotography; Photogrammetric Engineering, vol. 39, pp. 1161–1170.

DataPro, 1978, The EDP Buyer's Guide; DataPro Corporation, Delran, N.J., vol. 1, 2, 3.

Dave, J. W., 1981, Influences of illumination and viewing geometry and atmospheric composition on the "Tasseled Cap" (spectral-temporal characteristics of wheat fields) transformation of Landsat MSS (multispectral sensors) data; Remote Sensing of Environment, vol. 11, no. 1, pp. 37–55.

Davis, J. B. and S. Z. Friedman, 1979, Assessing urbanized area expansion through the integration of Landsat and conventional data; Proceedings of the American Society of Photogrammetry, 45th Annual Meeting, Washington, D.C., vol. 2, pp. 776–791.

Davis, L. and A. Mitiche, 1980, Edge detection in textures; Computer Graphics and Image Processing, vol. 12, no. 1, pp. 25–39.

Deschamps, P. Y., 1977, Teledetection de la temperature de surface de la mer par radiometrie infrarouge; These de Doctorate d' Etat, Universite de Lille, no. 376.

Deschamps, P. Y., and T. Phulpin, 1980, Atmospheric correction of infrared measurements of sea surface temperature using channels at 3.7, 11 and 12 μm; Boundary-Layer Meteorology, vol. 8, pp. 131–143.

Deuker, K. and F. Horton, 1971, Urban change detection systems: Status and prospects; Proceedings, Seventh International Symposium on Remote Sensing of Environment, pp. 1523–1536.

Doverspike, G. E., F. M. Flynn and R. C. Heller, 1965, Microdensitometer applied to land use classification; Photogrammetric Engineering, vol. 31, pp. 294–306.

Dozier, J., and S. I. Outcalt, 1979, An approach toward energy balance simulation over rugged terrain; Geographical Analysis, vol. 11, pp. 65–85.

Dozier, J., 1980, A clear-sky spectral solar radiation model for snow-covered mountainous terrain; Water Resources Research, vol. 16, no. 4, pp. 709–718.

Dozier, J., 1981, A method for satellite identification of surface temperature fields of subpixel resolution; Remote Sensing of Environment, vol. 11, pp. 221–229.

Dozier, J., and S. G. Warren, 1982, Effect of viewing angle on the infrared brightness temperature of snow; Water Resources Research, (in press).

Driscoll, R. S., J. N. Reppert and R. C. Heller, 1974, Microdensitometry to identify plant communities and components on color infrared aerial photos; Range Management Journal, vol. 27, no. 1, pp. 66–70.

Eckhardt, J., J. A. Jakob, J. Lamp and V. Wittje, 1980, Multitemporal and multispectral remote sensing of soils in cultured landscapes of North Germany; Sixth Annual Symposium on Machine Processing of Remotely Sensed Data, Purdue University, p. 281.

Engvall, J. L., J. D. Tubbs and Q. A. Holmes, 1977, Pattern recognition of Landsat data based upon temporal trend analysis; Remote Sensing of Environment, vol. 6, no. 4. pp. 303–314.

Eranti, K., 1968, Stereoscope for strips; Photogrammetric Engineering, vol. 34, pp. 1004–1047.

Estes, J. E., 1974, Imaging with photographic and nonphotographic sensor systems in: Remote Sensing; Techniques for Environment Analysis, Estes, J. E. and L. W. Senger, eds., Hamilton Press, pp. 15–50.

Estes, J. E., 1977, A perspective of the State of the art of aerial photographic interpretation in Proc. of the 11th Symposium on Remote Sensing of Environment, Ann Arbor, Mich. (Invited paper).

Estes, J. E., 1982, "Remote Sensing and Geographic Information Systems Coming of Age in the Eighties" in Proc. Pecora VII Symposium, Falls Church, Va., Amer. Soc. of Photogrammetry, pp. 23–40.

Estes, J. E. and D. S. Simonett, 1975, Fundamentals of image interpretation; in Manual of Remote Sensing,

Reeves R. G. (ed-in-chief); Falls Church, American Society of Photogrammetry, pp. 869–1076.

Evans, R. M., 1948, An Introduction to Color; John Wiley, New York, 370 p.

Evans, W. F., 1974, Marking ERTS images with a small mirror reflector; Photogrammetric Engineering, vol. 40, no. 6, pp. 665–672.

Fasler, F., 1980, Texture measurements from Seasat-SAR images for urban land-use interpretation; Proceedings, Fall Technical Meeting of the American Society of Photogrammetry, New York, pp. PS-2-B-1 to 10.

Faugeras, O. D. and K. E. Price, 1981, Semantic description of aerial images using stochastic labeling; IEEE Trans. on Pattern Analysis and Machine Intelligence, vol. PAMI-3, no. 6.

Foreshaw, M. R. B., A. Haskell, P. F. Miller, D. J. Stanley and J. R. G. Townshend, 1980, A Review Paper: Spatial resolution of remotely sensed imagery; Paper Submitted to U.S. Committee for Peaceful Uses of Outer Space, 53 p.

Foresman, T. W., F. Mertz and D. Stow, 1980, Automation of remotely sensed terrain information for amphibious objective areas (AOA) land management system; Civil Engr. Lab. TM no. 54-80-23, Naval Construction Battalion Center. Port Hueneme, CA. 29 p.

Fritz, N. L., 1967, Optimum methods for using infrared-sensitive color films, Photogrammetric Eng., vol. 10, pp. 1128–1138.

Fu, K. S., 1980, Syntactic image modelling using stochastic tree grammars; Computer Graphics and Image Processing, vol. 12, no. 1., pp. 136–152.

Gelsema, E. S. and L. N. Kanal, eds., 1980, Pattern Recognition in Practice; North Holland Publishing Co., New York, 552 p.

Gilmer, D. S., 1975, Utilization of Skylab (EREP) system for appraising changes in continental migratory bird habitat, Progress Report for NASA, 2 p.

Gonzalez, R. C. and P. Wintz, 1977, Digital Image Processing; Addison-Wesley Publishing Co., Reading, Mass., 431 p.

Gourley, J., H. T. Rib and R. D. Miles, 1967, Automatic technique for abstracting color descriptions from aerial photography; Presented at Conf. Soc. Photographic Sci. and Engrs.

Greblicki, W. and M. Pawlak, 1981, Classification using the Fourier series estimate of multivariate density functions; IEEE Trans. on Systems, Man, and Cybernetics, vol. SMC-11, no. 10.

Gregory, R. L., 1966, Eye and Brain; the Psychology of Seeing.

Griggs, M., 1968, Emissivities of natural surfaces in the 8- to 14-micron spectral region; Journal of Geophysical Research, vol. 73, pp. 7545–7551.

Gulinck, H., 1980, Recalibration of multitemporal digital Landsat MSS data for atmospheric interaction, application to phenological and pedological landscape studies; Pedologie, vol. 30, no. 1, pp. 89–114.

Gullicksen, S. O., 1967, Continuous strip photography; Photogrammetric Engineering, vol. 33, pp. 278–387.

Guruswamy, V., S. J. Kristof and M. Baumgardner, 1980, A case study of soil erosion detection by digital analysis of remotely sensed multi-spectral Landsat scanner data of a semi-arid land in Southern India; Sixth Annual Symposium of Machine Processing of Remotely Sensed Data, pp. 266–272.

Hallada, W. A., F. C. Mertz, L. R. Tinney, M. J. Cosentino and J. E. Estes, 1981, Flexible processing of remote sensing data through integration of image processing and geobased informations systems; Proceedings of the Fifteenth International Symposium on Remote Sensing of Environment.

Hansen, D. V. and G. A. Maul, 1970, A note on the use of sea surface temperature for observing ocean currents; Remote Sensing of Environment, vol. 1, pp. 161–164.

Haralick, R. M., 1979, Statistical and structural approaches to texture; Proceedings of the IEEE, vol. 67, no. 5, pp. 786–804.

Haralick, R. M., 1980, Edge and region analysis for digital image data; Computer Graphics and Image Processing, vol. 12, pp. 60–73.

Harmage, J. and D. Landgrebe, 1975, Landsat-D Thematic Mapper Technical Working Group; Final Report, NASA/Johnson Space Center.

Hayes-Roth, F., 1981, A tutorial on expert systems: putting knowledge to work; IJCAI-81.

Held, R., 1974, Introduction: Image, Object and Illusion, Readings from Scientific American; W. H. Freeman and Co., San Francisco, 137 p.

Heller, R. C., 1970, Imaging with photographic sensors, Chapter 2 in Remote Sensing with Special Reference to Agriculture, Holter, M. R., ed.; Washington, D.C., National Academy of Sciences, pp. 35–72.

Hixson, M., D. Scholz, N. Fuhs, and T. Akiyama, 1980, Evaluation of Several Schemes for Classification of remotely sensed data, Photogrammetric Engineering and Remote Sensing, vol. 46, no. 12, pp. 1547–1553.

Hoffer, R. M. and staff, 1975, Natural resource mapping in mountainous terrain by computer analysis of ERTS-1 satellite data, LARS information note 061575; Laboratory for Applications of Remote Sensing, West Lafayette, Indiana.

Holben, B. N. and C. O. Justice, 1980, The topographic effect on spectral response from nadir-pointing sensors; Photogrammetric Engineering and Remote Sensing, vol. 46, no. 9, pp. 1191–1200.

Holdridge, L. R., W. C. Grenke, W. H. Hatheway, T. Liang and S. A. Tosi, 1968, Forest environments in tropical life zones: A Pilot Study; vs. 1 and 2, Wilson, Nuttall, Raimond Engineers, Inc., 832 p.

Holter, M. R. and R. R. Legault, 1967, Remote Thermal Sensing in Thermophysics of Spacecraft and Planetary Bodies: Radiation Properties of Solids and the Electromagnetic Radiation Environment in Space. ed R. C. Hellerl Academic Press, New York pp. 547–566.

Holz, R. K., ed., 1973, The Surveillant Science: Remote Sensing of the Environment, Houghton Mifflin, Boston, 390 p.

Hsu, S., 1980, Texture perception and the RADC/Texture feature extractor; Photogrammetric Engineering and Remote Sensing, vol. 46, no. 8, pp. 1051–1058.

Hubel, D. H., and T. N. Weisel, 1962. Receptive fields, binocular interaction and functional architecture in the cat's visual cortex, Journal of Physiology, vol. 160, pp. 106–154.

Hummel, R., 1979, Feature detection using basis functions; Computer Graphics and Image Processing, vol. 9, pp. 40–55.

Jacobowitz, H. and K. L. Coulson, 1973, Effects of aerosols on the determination of temperature of earth's surface from radiance measurements at 11.2 μm; NOAA Technical Report NESS 66.

Jain, A. K., 1981, Advances in mathematical models for

image processing; Proc. IEEE, vol. 69, no. 5, pp. 502–528.

James, W. P., 1972, Diffusion coefficients and current velocities in coastal waters by remote sensing techniques; Remote Sensing of Earth Resources, ed., F. Shahrokhi, vol. 1, pp. 338–361.

Jensen, J. R., 1977, Multidate/multispectral crop identification: digital techniques applied to high altitude photography and Landsat imagery; American Society of Photogrammetry/American Congress on Surveying and Mapping, Little Rock, Arkansas, pp. 18–21.

Jensen, J. R., 1981, Urban change detection mapping using Landsat digital data; The American Cartographer, vol. 8, no. 1.

Jensen, J. R., F. A. Ennerson and E. J. Hajic, 1979, An interactive image processing system for remote sensing education; Photogrammetric Engineering and Remote Sensing, vol. '45, no. 11, Nov.

Johnson, C. W., 1971, Computerized land pattern mapping from mono-imagery; Proceedings of the 7th International Symposium on Remote Sensing of Environment, University of Michigan, Ann Arbor, Mich., pp. 1951–1966.

Johnson, G. R., 1980, Lecture notes: digital image classification; Technicolor Graphic Services, Inc. unpublished.

Julesz, B., 1975, Experiments in the visual perception of texture; Scientific American, vol. 232, no. 4, pp. 2–11.

Jupp, D. L. B. and E. M. Adomeit, 1981, Spatial analysis: an annotated bibliography; Divisional Report 81/2, CSIRO Institute of Biological Resources, Division of Land Use Research, Canberra A. C. T.

Justice, C. O. and B. N. Holben, 1979, Examination of Lambertian and non-Lambertian Models for Simulating the Topographic Effect on Remotely Sensed Data; NASA TM 80557, Goddard Space Flight Center, Greenbelt, Maryland.

Justice, C. O., B. N. Holben, 1980, The Contribution of the Diffuse Light Component to the Topographic Effect on Remotely Sensed Data; NASA TM 85290, Goddard Space Flight Center, Greenbelt, Maryland.

Justice, C. O., S. W. Wharton and B. N. Holben, 1980, Application of Digital Terrain Data to Quantify and Reduce the Topographic Effect on Landsat Data; NASA TM 81988, Goddard Space Flight Center, Greenbelt, Maryland.

Kanal, L. N., 1977, Current status, problems and prospects of pattern recognition; Systems, Man and Cybernetics Review, vol. 6, no. 4 (Newsletter).

Kaneko, T., 1978, Crop classification using time features computed from multi-temporal multi-spectral data; Proceedings of the 4th International Joint Conference on Pattern Recognition, Kyoto, Japan, pp. 943–945.

Kauth, R. J., A. P. Pentland, G. S. Thomas, 1977, BLOB, an unsupervised clustering approach to spatial preprocessing of MSS imagery; 11th International Symposium on Remote-Sensing of Environment, Environmental Research Institute of Michigan, Ann Arbor, Mich., vol. 2, pp. 1309–1317.

Kelly, W. L., B. D. Meredith and W. M. Howle, 1980, High Speed Lookup Table Approach to Radiometric Calibration of Multispectral Image Data; National Aeronautics and Space Administration, Langley Research Center, 16 p.

Kennedy, J. M., 1974, *Psychology of Picture Perception,* San Francisco, Jossey-Bass Publishers.

Kent, E. W., 1981, The Brains of Men and Machines; BYTE/McGraw-Hill, Peterborough, N.H., 286 p.

Klass, P., 1972, Secret sentries in space. Random House, New York, 236 p.

Kondrateyev, K. Ya., 1969, Radiation in the atmosphere: New York, Academic Press, 912 p.

Kriebel, K. T., 1978, Measured spectral bidirectional reflection properties of four vegetated surfaces; Applied Optics, vol. 17, no. 2, pp. 253–258.

Kumar, R., 1980, Separability of agricultural cover types in spectral channels and wavelength regions; IEEE Trans, Geosci and Remote Sensing (USA), vol. GE-18, no. 3, pp. 263–267.

Landen, D., 1966, Photomaps for urban planning; Photogrammetric Engineering, vol. 32, pp. 136–146.

Landen, D., 1974, Progress in orthophotography; Photogrammetric Engineering, vol. 40, pp. 265–270.

Landgrebe, D. A., 1978. The quantitative approach: concept and rationale. Chapter one in Swain P. H., and S. M. Davis, 1978, Remote Sensing: The Quantitative Approach, McGraw Hill, 396 p.

Landgrebe, D. A., 1980, The development of a spectral-spatial classifier for earth observational data; Pattern Recognition, vol. 12, pp. 165–175.

Landgrebe, D. A., 1981, Analysis technology for land remote sensing; Proc. IEEE, vol. 69, no. 5, pp. 628–642.

Leachtenauer, J. C., 1977, Optical power spectrum analysis: scale and resolution effects. Photogrammetric Engineering and Remote Sensing, vol. 43, no. 9, September, pp. 1117–1125.

Leese, J. A. and E. S. Epstein, 1963, Application of two-dimensional spectral analysis to the quantification of satellite cloud photography; Journal of Applied Meteorology, vol. 2, pp. 629–644.

Lent, J. D., 1968, The feasibility of identifying wildland resources through the analysis of digitally recorded remote sensing data. Annual progress report. Univ. of Calif., Berkeley, 134 p.

Levine, S. H., 1969, Color and black-and-white negatives for photo interpretation; Photogrammetric Engineering, vol. 35, pp. 65–69.

Lewis, A. J., 1974, Geomorphic-geologic mapping from remote sensors, in Estes, J. D., and Senger, L. W., eds., Remote sensing: techniques for environmental analysis: Santa Barbara, Calif., Hamilton Press, pp. 105–126.

Lillesand, T. M. and R. W. Kiefer, 1979, Remote Sensing and Image Interpretation; John Wiley and Sons, 396 p.

Lintz, J. Jr., and D. S. Simonett, 1976, Remote Sensing of Environment; Addison-Wesley Publishing Co., Reading, Mass., 694 p.

Lundahl, A. C. and E. M. Monsour, 1960, in Manual of Photographic Interpretation, Colwell, R. N., ed.: American Society of Photogrammetry, Falls Church, Va., pp. 1–18.

Malila, W. A., R. H. Hieber and R. C. Cicone, 1975, Studies of recognition with multitemporal remote sensor data; Final Technical Report, NASA, 99 p.

Masry, S. E. and Gibbons, J. G., 1973, Distortion and rectification of IR: Photogramm. Engr., v. 39, pp. 845–849.

Matson, M., and J. Dozier, 1981, Identification of sub-resolution high temperature sources using a thermal IR sensor; Photogrammetric Engineering and Remote Sensing, vol. 47, pp. 1311–1318.

Matthews, M. L. and L. M. Miller, 1979, A bibliography: change detection with remote sensing imagery;

Remote Sensing Center, Texas A & M University, College Station, Texas.

Maul, G. and M. Sirdan, 1973, Atmospheric effects on ocean surface temperature sensing from the NOAA satellite scanning radiometer; Journal of Geophysical Research, vol. 78, pp. 1909–1916.

Maynard, P. F. and A. H. Strahler, 1981, The logit classifier: a general maximum likelihood discrimant for remote sensing applications; Proceedings of Fifteenth International Symposium on Remote Sensing of Environment.

McClain, E. P., 1981, Multiple atmospheric-window techniques for satellite-derived sea surface temperatures; Oceanography from Space, J. F. R. Gower, ed., Plenum, NY, pp. 73–85.

McLerran, J. H., 1967, Infrared thermal sensing: Photogramm. Engr., v. 33, pp. 507–512.

McNeil, G. T., 1954, Photographic measurements: Problems and Solutions. Pitman Publishing Corp.

Meetham, A. R. and R. A. Hudson, eds., 1969. Encyclopaedia of information and control; Linguistics: Permagon Press, New York, 718 p.

Mergerson, J. W., 1981, Crop area estimates using ground-gathered and Landsat data; a multitemporal approach, ESS Staff Report—U.S. Department of Agriculture, Economics and Statistics Service, 27 p.

Merriam-Webster, 1960, Webster's New Collegiate Dictionary, G. and C. Merriam Co., Springfield, Mass.

Mertz, F. C., M. J. MacLennan and L. R. Tinney, 1981, VICAR/IBIS primer for JSC short course; University of California, Santa Barbara, Geography Remote Sensing Unit, 105 p.

Michael, C. R., 1969, Retinal processing of visual images, *Scientific American*, vol. 220, no. 5, pp. 105–114.

Minnaert, M., 1941, The reciprocity principle in lunar photometry; Astrophys. Journal, vol. 93, pp. 403–410.

Minor, T., 1982, Seasonal comparison of HCMM thermal infrared imagery for change detection in agricultural drainage patterns; unpublished M. A. Thesis, Department of Geography, University of California, Santa Barbara.

Misra, P. N. and S. G. Wheeler, 1978, Crop classification with Landsat multi-spectral scanner data; Pattern Recognition (GB), vol. 10, no. 1, pp. 1–13.

Modestino, J. W., R. W. Fries and S. L. Vichers, 1981. Texture discrimination based upon an assumed stochastic texture model; IEEE Trans. on Pattern Analysis and Machine Intelligence, vol. PAMI-3, pp. 557–580.

Moik, J. G., 1980, Digital Processing of Remotely Sensed Images; NASA Scientific and Technical Information Branch, NASA SP-431, 330 p.

Moore, R. P. and J. O. Hooper, 1974, Microwave radiometric characteristics of snow-covered terrain; Presented at 9th Symp. Remote Sensing of Environment, Ann Arbor, Michigan, pp. 1621–1632.

Nahi, N. E. and M. H. Jahanshahi, 1977, Image boundary estimation; IEEE Trans. Computers, C-26, pp. 772–781.

NASA, 1977, COSMIC, Catalog of Selected Computer Programs; NASA-451, USGPO, 152 p.

NASA, 1980, Earth Resources Satellite. Data Applications Series; Processing/Analyzing Landsat Data.

NASA, Goddard Space Flight Center, 1981, TAE, Transportable Applications Executive User's Manual, Prototype; Prepared by Century Computing, Inc., Silver Springs, Md.

NASA, Office of Space and Terrestrial Applications, 1980, Earth Resources Satellite, Data Applications Series: processing/analyzing Landsat data. Module U-4, Washington, D.C., 38 p.

Nash, A. J., 1972, Use a mirror stereoscope correctly: Photogrammetric Engineering, vol. 38, pp. 1192–1196.

Newland, W., D. Peterson and S. Morman, 1980, Bulk processing techniques for very large areas: Landsat classification of California; 1980 Machine Processing of Remotely Sensed Data Symposium. Purdue Univ., pp. 306–318.

Nichols, M. and M. Kelly, 1972, Time sensing and analyses of coastal waters; Proceedings Eighth International Symposium on Remote Sensing of Environment, vol. 2, pp. 969–981.

Noton, D. and L. Stark, 1971, Eye movements and Visual perception in Image, Object and Illusion: readings from Scientific American, W. H. Freeman and Co., San Francisco, pp. 113–122.

Olsen, C. E., Jr., 1973, What is photographic interpretation? in the Surveillant Science: Remote Sensing of the Environment; R. K. Holz ed., Houghton Mifflin, Boston, 390 p.

Orr, D. G., 1968, Multiband-color photography; *in* Manual of Color Aerial Photography, Smith, J. T. and A. Anson, eds.; Falls Church, Va. Am. Soc. Photogramm., pp. 441–450.

Otterman, J., 1977, Monitoring surface albedo change with Landsat; Geophys. Res. Lett., vol. 4, no. 10, pp. 441–444.

Owen, R., A. Variño and R. Recht, 1981, Plankton in upwelling plumes off Southern California; Presented at California Cooperative Oceanic Fisheries Investigations Annual Conference 1981, USC Conference Center, Idyllwild California, (Paper in Preparation).

Paley, H. N. and A. B. Kahle, 1979, Geologic application of thermal inertia imaging using HCMM data; Jet Propulsion Laboratory, NASA, 5 p.

Pankhurst, R. J., 1970, A computer program for generating diagnostic keys; The Computer Journal, vol. 13, no. 2, pp. 145–151.

Parikh, J. A., 1974, Automatic wind velocity estimation from multi-spectral geosynchronous satellite data; A Proposal, Maryland University College Park Computer Center, Air Force Office of Scientific Research, 123 p.

Pease, R. W., 1971, Mapping terrestrial radiation with a scanning radiometer, in Proc. of the Seventh Int'l Symp. on Remote Sensing of Environment, Univ. of Michigan, Ann Arbor, pp. 501–521.

Pease, R. W. and L. W. Bowden, 1969, Making color infrared film a more effective high-altitude remote sensor; Remote Sensing of the Environment, vol. 1, no. 1, pp. 23–30.

Pentland, A. P., 1980, The application of human visual system processing techniques to remotely sensed data; Proc. Fourteenth International Symposium on Remote Sensing of the Environment, pp. 1425–1486.

Peplies, R. W., 1976, Cultural and landscape interpretation *in* Remote Sensing of Environment, J. Lintz Jr. and D. S. Simonett, eds., Reading, Mass., Addison-Wesley, pp. 493–494.

Peterson, K. L., N. Tosta-Miller, S. Norman, D. Wierman and W. Newman, 1980, Land cover classification of California using mosaicking and high speed processing; Proceedings Fourteenth International Symposium on Remote Sensing of Environment, pp. 279–292.

Prabhakara, C, G. Dalu and V. G. Kunde, 1974, Estimation of sea surface temperature from remote sensing in the 11–13 μm window range; Journal of Geophysical Research, vol. 79, no. 33, pp. 5039–5044.

Preston, K., Jr., 1980, Image manipulative languages—A preliminary survey; in Pattern Recognition in Practice, ed., E. S. Gelsema and L. N. Kanal; North-Holland Publishing Co., pp. 5–20.

Price, K. E., 1976, Change detection and analysis in multi-spectral images, Carnegie-Mellon University of Pittsburgh, Pennsylvania, Department of Computer Science, Air Force Contract, 205 p.

Public Technology Corp, for NASA, 1981; Remote Sensing Procurement Package, Washington, D.C.

Rabchevsky, G. A., 1977, Temporal and dynamic observations from satellites; Photogrammetric Engineering and Remote Sensing, vol. 43, no. 12, p. 1515.

Reeves, R. G., ed., 1968, Introduction to electromagnetic remote sensing; Am. Geol. Inst., Short Course Lecture Notes.

Rib, H. T., 1966, An optimum multisensor approach for detailed engineering soils mapping; Joint Highway Res. Proj., Purdue University, vol. 2, no. 22.

Rib, H. T., 1968, Color measurements; in Manual of Color Aerial Photography, Smith J. T. and A. Anson eds.; Falls Church, Va., American Society of Photogrammetry, pp. 12–24.

Rib, H. T., and R. D. Miles, 1969, Automatic interpretation of terrain features; Photogrammetric Engineering, vol. 35, pp. 153–164.

Richason, B. F., Jr. ed., 1978, Introduction to Remote Sensing of the Environment; Kendall/Hunt Publishing Co., Dubuque, Iowa, 496 p.

Rifman, S. S., K. W. Simon and R. H. Caron, 1975. Second generation digital techniques for processing Landsat and MSS data; NASA Earth Resources Survey Symposium, Houston, Texas, vol. I-B, pp. 1161–1175.

Roberts, L. G., 1965, Machine Perception of Three-Dimensional Solids; Optical and Electroptical Information Processing, J. T. Tippett, et al, eds. MIT Press, Cambridge, Mass., pp. 150–197.

Robinson, J. W., 1979, A critical review of the change detection and classification literature; Technical Memorandum 79/6235, Computer Sciences Corp., Silver Springs, Md., 50 p.

Rosenfeld, A. and A. C. Kak, 1976, Digital Picture Processing. Academic Press, New York, N.Y., 457 p.

Rosenfeld, A., Y. Lee and R. Thomas, 1970, Edge and curve detection in texture discrimination; in Picture Processing and Psychopictorics, B. S. Lipkin and A. Rosenfeld, eds., Academic Press, New York, pp. 381–393.

Rosenfield, G. H., 1980, Forum; Landsat Wildland Mapping Accuracy: Photogrammetric Engineering and Remote Sensing. vol. 46. pp. 1543–1544.

Rowan, L. C., T. W. Offield, K. Watson, P. J. Cannon and R. D. Watson, 1970, Thermal infrared investigation—Arbuckle Mountains, Oklahoma: Geol. Soc. Amer. Bull., v. 81, pp. 3546–3562.

Sabins, Floyd F., Jr., 1973a, Flight planning and navigation for thermal IR surveys: Photogramm. Engr., v. 39, pp. 49–58.

———, 1973b, Recording and processing thermal IR imagery: Photogramm. Engr., v. 39, pp. 839–844.

———, 1973c, Engineering geology applications of remote sensing in Geology, Seismicity and Environmental Impact: Assoc. Engr. Geologists Sp. Pub.

Sabins, F. F., Jr., 1978, Remote Sensing, Principles and Interpretation; W. H. Freeman and Co., San Francisco, Ca., 426 p.

Schreiber, W. F., 1978, Image processing for quality improvement; Proc. IEEE, vol. 66, no. 12.

Selby, J. E. A., E. P. Shettle and R. A. McClatchey, 1976, Atmospheric transmittance from 0.25 to 28.5 μm; Supplement LOWTRAN 3B, Air Force Geophysics Laboratory Report AFGL-TR-76-0258.

Shelton, R. L. and J. E. Estes, 1981, Remote sensing and geographic information systems: an unrealized potential; Geoprocessing, vol. 1, pp. 395–420.

Shemdin, O. H., A. Jain, S. V. Hsiao and L. W. Gatto, 1980, Inlet current measured with Seasat-1 synthetic aperture radar; Shore and Beach, vol. 48, pp. 35–37.

Shepard, J. R., 1964, A concept of change detection; Photogrammetric Engineering and Remote Sensing, vol. 30, 649 p.

Shields, J. A. and C. Goodfellow, 1980, Temporal analysis of Landsat data for land use mapping (remote sensing); Sixth Annual Symposium on Machine Processing of Remotely Sensed Data, Purdue University, Laboratory for Applications of Remote Sensing, West Lafayette, Indiana.

Shlien, S. and D. Goodenough, 1973, Automatic interpretation of ERTS-A imagery using the MLDR; CCRS, Report 73–82.

Shortliffe, E., B. Buchanan and E. Feigenbaum, 1979, Knowledge Engineering for Medical Decision-Making: A Review of Computer Based Clinical Decision Aids; Stanford University Computer Science Department, Tech. Report, STAN-CS-79-723, 48 p.

Siegel, B. S. and A. R. Gillespie, 1980, Remote Sensing in Geology; John Wiley and Sons, New York, 702 p.

Simonett, D. S., 1976, Remote sensing of cultivated and natural vegetation: cropland and forest land; in Remote Sensing of Environment, J. Lintz and D. S. Simonett, eds.; Reading, Mass., Addison-Wesley, pp. 454–459.

Slater, P. N., 1980, Remote Sensing Optics, and Optical Systems; Addison-Wesley Publishing Co., Reading, Mass., 575 p.

Smith, J. A., T. L. Lin and K. J. Ranson, 1980. The Lambertian assumption and Landsat data; Photogrammetric Engineering and Remote Sensing, vol. 46, no. 9, pp. 1183–1189.

Smith, J. T. and A. Anson (eds), 1968, Manual of Color Aerial Photography. American Society of Photogrammetry, Falls Church, Va., 550 p.

Smith, R. C., R. W. Eppley and K. S. Baker, 1982, Correlation of primary production from satellite chlorophyll images as measured aboard ship in Southern California coastal water; Journal of Marine Biology, in press.

Soha, J. M., A. R. Gillespie, M. J. Abrams, D. P. Madura, 1976, Computer techniques for geological applications; Caltech/JPL Conference on Image Processing Technology, JPL SP43-30.

Steiner, D., 1970, Time dimension for crop surveys from space; Photogrammetric Engineering, vol. 36, pp. 187–194.

Sternberg, S. R., 1980, Language and architecture for parallel image processing; in Gelsema, E. S. and L. N. Kanal, eds.; Pattern Recognition in Practice, North-Holland Publishing Co., New York.

Stingelin, R. W., 1968, Criteria for regional airborne infrared geological surveys: Soc. America, abstracts of papers, N.E. Sec.

————, 1969, Operational airborne thermal imaging surveys: Geophysics, v. 34, pp. 760–771.

Stone, K. H., 1956, Air photo interpretation procedures; Photogrammetric Engineering, vol. 22, p. 223.

Stow, D. A., L. R. Tinney and J. E. Estes, 1980, Deriving land use/land cover change statistics from Landsat: a study of prime agricultural land; Proceedings Fourteenth International Symposium on Remote Sensing of Environment, vol. 2, pp. 1227–1238.

Strahler, A. H., 1980, The use of prior probabilities in maximum likelihood classification of remotely sensed data; Remote Sensing of Environment, vol. 10. pp. 135–163.

Strahler, A. H., 1981, Stratification of natural vegetation for forest and rangeland inventory using Landsat digital imagery and collateral data; Int. J. Remote Sensing, vol. 2, no. 1, pp. 15–41.

Strahler, A. H., J. E. Estes, P. F. Maynard, F. C. Mertz and D. A. Stow, 1980, Incorporating collateral data in Landsat classification and modeling procedures; Proceedings of the Fourteenth International Symposium on Remote Sensing of the Environment, pp. 1009–1026.

Strahler, A. H., T. L. Logan and N. A. Bryant, 1978, Improving forest cover classification accuracy from Landsat by incorporating topographic information; Proceedings of the Twelfth International Symposium on Remote Sensing of the Environment, pp. 927–942.

Strahler, A. H., T. L. Logan and C. E. Woodcock, 1979, Forest classification and inventory system using Landsat, digital terrain, and ground sample data; Proceedings of the Thirteenth International Symposium on Remote Sensing of the Environment, pp. 1541–1557.

Struve, H., W. Graham and H. West, 1977, Acquisition of terrain information using landsat multispectral data; Report I of Series, Technical Report M-77-2, Mobility and Environ. System Laboratory U.S. Army Engineering Waterway Experiment Station, Vicksburg, MS. 50 p.

Swain, P. H. and S. M. Davis, 1978, Remote Sensing: The Quantitative Approach; McGraw Hill, New York, 396 p.

Swain, P. H., S. B. Vardeman and J. C. Tilton, 1980, Contextual classification of multispectral image data; Purdue University, Lafayette Indiana, Laboratory for Applications of Remote Sensing, Technical Report, NASA, 40 p.

Takahashi, W. K., A. W. Decfilho, and R. Kuman, 1980, Partitioning multispectral imagery of the earth resources; National Aeronautics and Space Administration, Washington, D.C., 65 p.

Taranik, J. V., 1978, Principles of computer processing of Landsat data for geologic applications; U.S. Dept. of the Interior, Geological Survey, open file report 78-117, 50 p.

Thompson, M. M., ed., 1966, Manual of Photogrammetry, 3rd ed.; Falls Church, Va., American Soc. Photogramm., 1199 p.

Thomson, F. J., and Dillman, R. D., 1973, Baltimore, Maryland radiation balance mapping: Environmental Research Institute of Michigan, Ann Arbor, 98 p.

Tinney, L., 1980; NSF/AAG Course Notes. Geography Remote Sensing Unit, Univ. of Calif. Santa Barbara, (unpub. Report).

Tinney, L. and F. A. Ennerson, 1981, VICAR/IBIS image processing; Remote Image Processing Station Workshop, Sioux Falls, S. D.

Todd, W. J., D. G. Gehring and J. F. Haman, 1980, Landsat wildland mapping accuracy; Photogrammetric Engineering and Remote Sensing, vol. 46, pp. 509–520.

Townshend, J. R. G., 1980, The spatial resolving power of earth resources satellites; A Review, NASA Technical Memo 82020, Goddard Space Flight Center.

Tyron, R. C., 1939, Cluster Analysis; Edwards, Ann Arbor, Michigan.

Wan, Zheng-ming, 1981, A method for digital snow mapping at subpixel resolution from NOAA thermal satellite data; M.A. Thesis, Department of Geography, University of California, Santa Barbara.

Watson, Kenneth and Rowan, L. C., 1971, Application of thermal modeling in the geologic interpretation of IR images: Univ. Michigan, Proc. Seventh International Symp. Remote Sensing of Environment. pp. 2017–2041.

Weber, E. A., 1980, Vision, Composition and Photography, Walter de Gruyter, New York, N.Y., 156 p.

Weinreb, M. P. and Hill, M. L., 1980, Calculation of atmospheric radiances and brightness temperatures in infrared window channels of satellite radiometers; NOAA Technical Report NESS 80, Washington, D.C.

Weismueller, R. A., S. J. Kristof, D. K. Scholz, P. E. Anuta and S. A. Momin, 1977, Change detection in coastal zone environments; Photogrammetric Engineering and Remote Sensing, vol. 43, pp. 1533–1539.

Wertenbaken, L., 1981, The Eye: Window to the World; U.S. News Books, Washington D.C.

Wharton, S. W. and B. J. Turner, 1981, ICAP: an interactive cluster analysis procedure for analyzing remotely sensed data; Remote Sensing of Environment, pp. 279–293.

Whitley, S. L., 1976, Low cost data analysis systems for processing multispectral scanner data; NASA Technical Report, NASA TR R-467, 36 p.

Williams, D. L. and L. D. Miller, 1979, Monitoring forest canopy alteration around the world with digital analysis of Landsat imagery; NASA Goddard Space Flight Center, 45 p.

Williams, R. S., Jr. and Ory, T. R., 1967, Thermal infrared mosaics for geological investigations: Photogramm. Engr., v. 37, pp. 43–52.

Williams, T. H. L., J. Siebert and G. Gunn, 1981, The KARS image analysis system: a low cost interactive system for instruction and research; Machine Processing of Remotely Sensed Data Symposium.

Wolf, P. R., 1974, Elements of Photogrammetry; McGraw-Hill, New York.

Wolfe, E. W., 1971, Thermal IR for geology: Photogrammetric Eng., vol. 37, pp. 43–52.

Woodcock, C. E., A. H. Strahler and T. L. Logan, 1980, Stratification of forest vegetation for timber inventory using Landsat and collateral data; Proceedings of the Fourteenth International Symposium on Remote Sensing of Environment, pp. 1769–1787.

Yachida, M., M. Asada and S. Tsuji, 1981, Automatic analysis of moving images; IEEE Trans. Pattern Anal. and Mach. Intell., vol. PAMI-3, no. 1.

Yakimovsky, Y., 1976, Boundary and object detection in real world images; J. Assoc. Comput. Mach., vol. 23, pp. 599–618.

Yost, E. and S. Wenderoth, 1968, Additive color aerial photography; in Manual of Color Aerial Photography, Smith, J. T. and A. Anson, eds.; Falls Church, Va., Am. Soc. Photogramm., pp. 451–471.

CHAPTER 25

Image Analysis-Active Microwave*

Authors: DAVID S. SIMONETT and ROBERT E. DAVIS

GENERAL CONTENTS: Factors which influence the appearance and interpretability of radar images; radargrammetry; area resolution; number of independent samples; backscattering cross-section; a key determinant of image tone; texture as a discriminant on radar images; radar image interpretation; references.

FACTORS WHICH INFLUENCE THE APPEARANCE AND INTERPRETABILITY OF RADAR IMAGES—INTRODUCTION

The real aperture and synthetic aperture radars used for reconnaissance surveys and mapping are side looking imaging systems. They generally operate at a single wavelength and thus have the special characteristics of a monochromatic source.

Side-looking radar employs an antenna with its long axis parallel to the radar platform travel path, emitting a narrow fan beam to the side. The radar image is built-up by scanning this beam. A series of short pulses is emitted and a portion of the energy incident upon the surface is scattered back toward the radar receiver. The return, or backscatter is recorded in sequence to provide the cross-track component of the image. The along-track dimension is provided by the forward motion of the radar which is synchronized to the recording system. In the case of airborne radar this is usually a camera-recording system and, in the case of a spaceborne radar, a digital recording system. Such systems lack the point perspective of a vertical photograph, having instead a line-perspective geometry and this, coupled with short- and long-term platform motions makes for the peculiar and difficult geometric and mosaicking characteristics of a line-scanning system. Thus, present side-looking radars cannot match the geometric fidelity obtainable from stereoscopic aerial photographs with suitable ground control, though for broad scale reconnaissance and for a general knowledge of location, conformable to that obtained from Landsat, these inadequacies are tolerable.

The time at which the returning radar pulse is initially received is a function of the slant range distance from the radar to the surface and may be presented as a slant range display with consequent distortion in true ground distances across the track. The returning pulse also may be transformed to equivalent ground range employing the appropriate hyperbolic correction for the height of the radar, the distance to the image point, and using an assumption that the surface being imaged is either a flat (airborne radars), or part of a spherical surface (spaceborne radars).

Received radar backscatter, measured as the backscattering cross-section per unit solid angle, is mainly a function of the surface geometric configuration both in terms of attitude and small scale roughness. Therefore, surface topography is easily interpreted on radar imagery and the use of discriminating factors such as shape, pattern, and the association of features is similar to the interpretation of optical images. The intensity of the backscatter is also influenced by the dielectric properties of the surface which is mostly a function of the water content of the surface material.

Several commercial radar systems (both synthetic and real aperture) are available for prospective users. In addition, a set of images is available from the Seasat spaceborne imaging radar. Each type of radar image product has widely differing characteristics with respect to platform height, resolution in range and azimuth, and methods of recording and processing. These have a marked effect on the characteristics of the imagery and therefore on its utility for resource-evaluation studies as well as affecting its cost and timeliness of acquisition.

All of the following factors influence the interpretability of radar images: (1) System geometry, including flight parameters; (2) areal resolution, a product of azimuth and ground range resolution; (3) number of independent samples (a measure of signal averaging); and (4) backscattering cross-section per unit solid angle directed towards the receiver. These factors will be discussed in detail in the following sections.

RADAR GEOMETRY

Several inherent geometric characteristics of radar imagery are important in that they influence interpretation, and enable quantitative terrain data to be obtained from the images. The most important of these are radar shadows, radar foreshortening, and radar layover. The integrated effect of these features will be discussed and il-

---

* This chapter is concerned exclusively with interpretation of radar imagery. Interpretation of passive microwave images is discussed in part in Chapters 4, 11 and 13.

** We acknowledge the contributions of Luiz H. Azevedo, Hugh L. Davies, John R. Everett, Anthony J. Lewis, Harold C. MacDonald, Stanley A. Morain, John W. Rouse, Jr., Floyd F. Sabins, Jr., Jan W. Van Roessel, and William Waite to the radar section of Chapter 14 of the first edition of the Manual of Remote Sensing, much of which has been used in this revision.

lustrated in the section "Radar Terrain Clearance and Composite Image Characteristics".

Radar Shadows

The parameters that determine whether or not a terrain feature will produce a radar shadow are depression angle (β) and the slope facing away from the radar beam (α_b), the back slope. As seen in Figure 25-1, the relationship between β and α_b is such that three cases are possible. They are: (1) the backslope is fully illuminated and no shadow results ($\alpha_b < \beta$); (2) the backslope is just illuminated, or grazed by the radar beam ($\alpha_b = \beta$); and (3) the backslope is obscured giving a no-return area of radar shadow ($\alpha_b > \beta$). As would be expected, shadowing occurs most often in the far range at low depression angles.

The condition necessary for radar shadowing as defined above ($\alpha_b > \beta$) is valid only when the strike of a ridge is perpendicular to the wavefront (parallel to the flight line). This arises because, as the angle described between the flight line and the crestline (θ) increases, the angle at which a_b will shadow at a given depression angle also increases. The effect of θ on the angle at which a_b will shadow at a given depression angle is shown and discussed in Figure 25-2.

Once the conditions for the occurrence of radar shadow are satisfied, the length of the radar shadow in slant range (S_s) is directly related to the height of the feature above the datum plane (h) and the total slant range distance from the radar to the far tip of the radar shadow (S_R) and inversely related to the radar platform altitude (H) (Figure 25-1). This relationship is:

$$S_s = \frac{hS_R}{H} \qquad (25\text{-}1)$$

Expressed in terms of β, Eq. 25-1 becomes

$$S_s = \frac{h}{\sin\beta} \qquad (25\text{-}2)$$

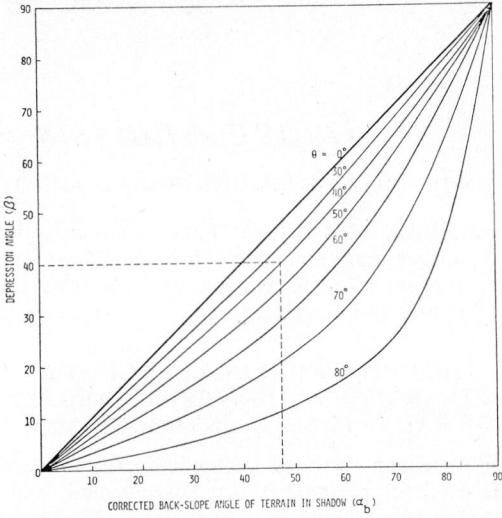

Fig. 25-2. Relationship of θ, the angle described by the flight direction and strike of the crestline, with α_b, the true backslope angle of terrain in shadow. This is also a nomogram for correcting the apparent backslope angle to true back slopeangle. For example, if a slope at a depression angle (β) and the angle between the flight path and strike of the crestline is 40° then the true back slope must be greater than 47.5°.

or

$$S_s = h \csc\beta \qquad (25\text{-}3)$$

From the geometry in Figure 25-1, it is evident that depending on the average and maximum slopes in a given area, that shadowing may or may not occur and that the amount of shadowing may be used as some measure of slope angles. It is also evident that shadow texture becomes a component of vegetation, geomorphic, and geologic area discrimination. The total amount of shadowing, its spatial distribution, and the frequency distribution of shadow areas and lengths is visually integrated by an interpreter in helping to delineate homo-

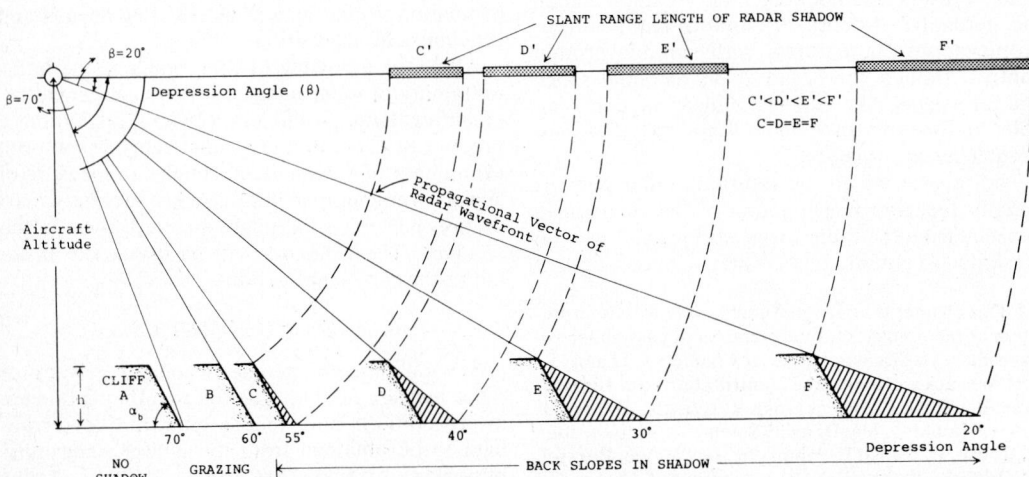

Fig. 25-1. Relationships of radar shadow geometry (slant range length) with depression angle (β).

geneous areas on radar images (Figure 25-1). With appropriate corrections for depression angle, quantitative statistics on shadowing (area and length means, standard deviations, etc.) may be used to group terrain classes along with other measures.

Radar Foreshortening

The period of time a slope is illuminated determines the length of the slope on radar imagery. This phenomenon, referred to as "radar foreshortening," results in the shortening of a terrain slope on radar imagery in all cases except when the angle of incidence (ϕ) is equal to 90°, at which time the terrain slope length (L) is equal to the slope length measured on the radar imagery (L_F), assuming that the scale factor between the imagery and the terrain is taken into account. (This is the same as the grazing case for shadowing [$\alpha_b = \beta$].)

The length of the terrain slope measured on slant-range radar imagery (L_F) is a function of incidence angle (ϕ) which in turn is an function of depression angle (β) and terrain slope (α):

$$L_F = L \sin\phi \qquad (25\text{-}4)$$

The percent of radar foreshortening (F_p), also a function of ϕ, may be defined as,

$$F_p = [1 - \sin\phi] * 100 \qquad (25\text{-}5)$$

The relationship is illustrated in Table 25-1.

In effect this means that only slopes seen at grazing incidence are displayed as their correct ground length. All slopes facing the radar (the foreslope) are shortened relative to their true length, with most of the shortening closest to the radar. Comparison between a Seasat radar image and a corresponding Landsat MSS image (Figures 25-3 and 25-4) illustrates foreshortening on the slopes in the scene which face the radar illumination direction. The Seasat image has near- and far-range depression angles of 73° and 67° respectively. The severe foreshortening and layover of most facing slopes of the Sierra Madra Oriental indicates that slopes must lie in the range of 15° to 30° or more.

TABLE 25-1
Percent Radar Foreshortening (F_p) as a Function of Incidence Angle (ϕ)

Angle of Incidence	Percent Radar Foreshortening
90	0.0
80	1.5
70	6.0
60	13.4
50	23.4
40	35.7
30	50.0
20	65.8
10	82.6
0	100.0

A number of situations arise across an aircraft image with respect to portrayal of slope lengths. These are illustrated in Table 25-2 which has been developed for the case where all slopes in an area equal 15°. Table 25-2 shows that fore and back slopes of equal length and angle are displayed as unequal in length on a radar image, appearing almost cuestaform (see Fig. 25-3). The disparity increases in the near range to the limit of foreshortening where the foreslope collapses to a point (imaging even closer introduces another geometric feature-layover-discussed in the next section).

The greater the foreshortening, the more energy per unit area is presented on the image (foreslopes are brighter than back slopes) until so much energy is available that it saturates the receiver. Thus, even before extreme foreshortening is reached there is a loss in discrimination from swamping and object compression on near slopes. In a similar fashion, back slopes will suffer some fall-off in return energy as grazing incidence is approached, because of specular reflection on smooth surfaces, and because a small volume of energy is distributed by the radar over a larger area. Within 5° of grazing it is common to see severe "roll off" in return energy from these effects.

It follows from this account that the average radar return, hence the image tone, for homogeneous areas, is directly related to slope angle, direction, and position between the near and far ranges on the image. A relative homogeneity of average tone where it does occur over large areas of different slopes (within the mid range of the image) is an indication of slight radar return sensitivity to incidence angle variations. Good examples of such low sensitivity are seen in undulating to rolling topography (5° to 15°) in rainforest areas. Variations within rain-forest communities tend to be displayed more by slight textural variations than by tone.

Radar Layover

Radar layover is an extreme case of relief displacement. Because radar measures range to an object, placement on an image is a direct function of the distance from the radar to the object. In certain cases the top of an object is closer to the radar than the bottom and is therefore recorded sooner. The result is radar layover.

Radar layover is not dependent on the absolute distance from the radar to the feature, but rather the difference in slant range distance between the top and the bottom of the feature (Figure 25-5). This is in turn a function of (1) the wavefront angle, which is related to the depression angle or the position in the slant range, and (2) the slope of the terrain or object. Figure 25-6 also illustrates that radar layover is a phenomenon of near range, and steep slopes.

The incidence angle, formed by the radar beam and the perpendicular to the surface at the point of

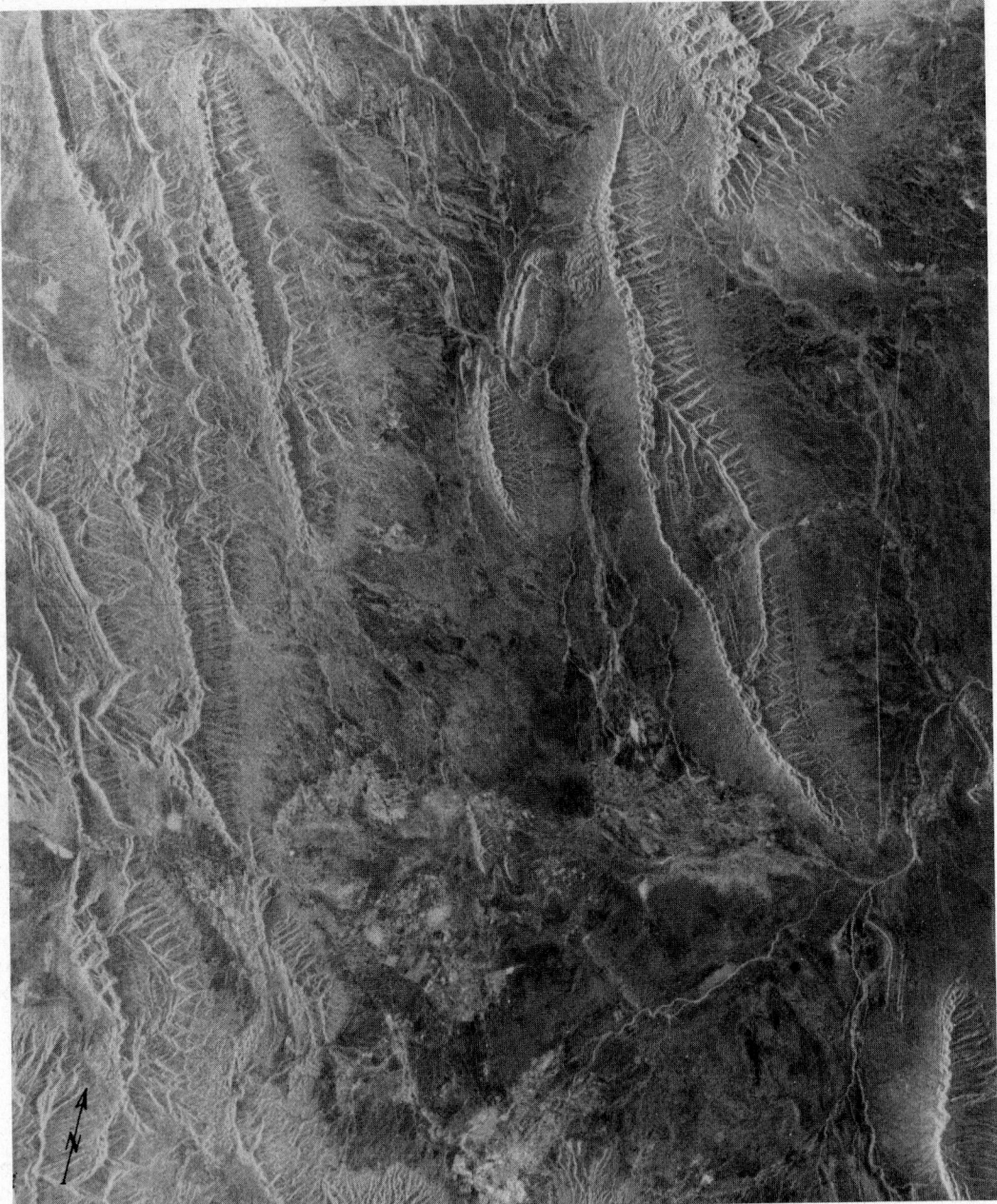

Fig. 25-3. Seasat image of Sierra Madre Oriental, Mexico. Scale approximately 1:800,000.

incidence (Figure 25-6), is important in defining the conditions for radar layover: Where the incidence angle is negative, radar layover occurs. This confirms mathematically what is illustrated in Figure 25-5, that radar layover occurs where steep slopes are found in the near range. The conditions for radar layover are given in Table 25-3, in terms of depression angle (β) and terrain slope (α). If the terrain slope exceeds that in the table, the incidence angle (θ) is negative and layover occurs. Table 25-3 shows that layover occurs only on slopes facing the radar, and renders layed-over foreslopes of images virtually uninterpretable. It is a severe problem in near range presentations of mountainous areas. In such areas, neither slant range nor ground range displays is satisfactory. Back slopes of images are correspondingly distorted, the greatest distortion or "lengthening" occurring with ground range displays.

Radar Terrain Clearance and Composite Image Characteristics

In areas of substantial relief where mountains, plateau and plains are juxtaposed, radar imaging introduces special complexities depending on the

Fig. 25-4. Contrast the topographic detail and geometry on Fig. 25-3 with the Landsat MSS image above. Scale approximately 1:800,000.

TABLE 25-2
Terrain Incidence Angles and Radar Slope Length for Given Depression Angles

Depression Angles	Radar Slope Length		Slope Length Ratio Back/Fore
	Fore Slope	Back Slope	
75°	0	.50	∞
65°	.17	.64	3.76
55°	.34	.77	2.26
45°	.50	.87	1.74
35°	.64	.94	1.47
25°	.77	.98	1.28
15°	.87	1.00	1.15

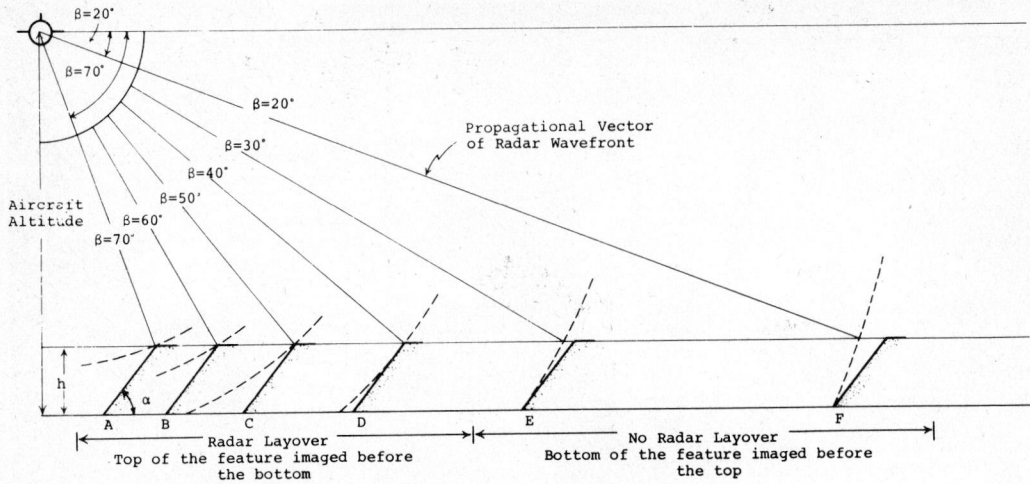

Fig. 25-5. Radar layover as a function of depression angle (β).

slope angles and on the relative terrain clearance by the radar platform (altitude). These problems are illustrated for an airborne platform in Figure 25-7; with aircraft there are distinct problems of narrowed usable image swath as mountains rise beneath the aircraft, and as a result, flight paths must be placed very close together (sometimes only a few kilometers apart) in order to avoid unimaged areas or "holidays".

The narrowing of the usable image swath as mountains rise beneath, and on the imaging side of the radar, occurs because of the intractable geometry of the problem. In addition, the smaller the altitude difference, the larger the relative areas of layover and foreshortening in the near range, and excessive shadowing in the far range. Only with spacecraft sensing may it be feasible to radar-image some mountainous areas.

RADARGRAMMETRY

In the past few years the usual radar product has been an aircraft radar mosaic applied to reconnaissance natural resource surveys at scales of 1:100,000 or smaller in remote areas or regions where continual cloud cover has precluded the use of aerial photography. Vast areas have been mapped this way, the largest of which was the coverage of all of Brazil, under project RADAM and its extensions.

For areas of low to moderate relief, airborne radar permits mapping with a planimetric accuracy as high as 100 m (and in exceptional cases, 20 m) and stereo-radargrammetric height accuracy of approximately 20 m has been found possible (Leberl, 1976). Satellite radar mapping accuracies have not been examined in detail; however, the potential exists for accuracies in the order of about 30 m (Leberl, 1980).

Most radargrammetric studies have been concerned with single image measurements, while a few have dealt with the single stereo model. Radargrammetric studies on blocks of overlapping imagery have been given the least attention, though block adjustments were developed and employed extensively in project RADAM (Van Roessel and Godoy, 1974).

Radargrammetric differences between airborne

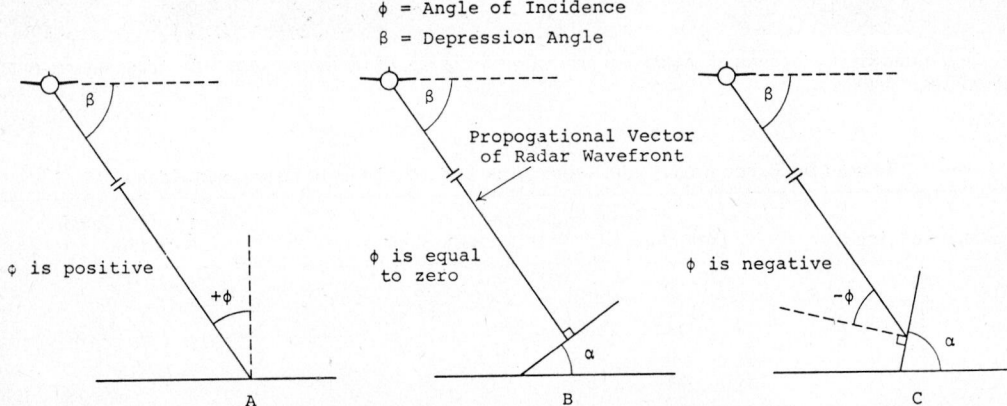

ϕ = Angle of Incidence
β = Depression Angle

Fig. 25-6. Effect of terrain slope (α) on incidence angle (ϕ).

TABLE 25-3
Conditions Necessary for Radar Layover

Terrain Slope (α)	Depression Angle (β)	Incidence Angle (ϕ)
>80°	10° Far Range	Negative
>70°	20°	Negative
>60°	30°	Negative
>50°	40°	Negative
>40°	50°	Negative
>30°	60°	Negative
>20°	70°	Negative
>10°	80° Near Range	Negative

Table 25-3 shows that layover occurs only on slopes facing the radar. It renders fore slopes of images virtually uninterpretable and is a severe problem in near range presentations in mountainous areas. In such areas, neither slant range nor ground range displays are satisfactory. Back slopes of images are correspondingly distorted, the greatest distortion or "lengthening".

and spaceborne radar platforms mainly concern depression angles, terrain clearance, consideration of the planetary curvature, and the various effects of the respective flight paths. Images produced from spaceborne platforms have almost constant depression angles as opposed to airborne radar and have less inherent cross-track variation in geometric distortion. Basic image processing techniques including geometric rectification are discussed in Chapter 17.

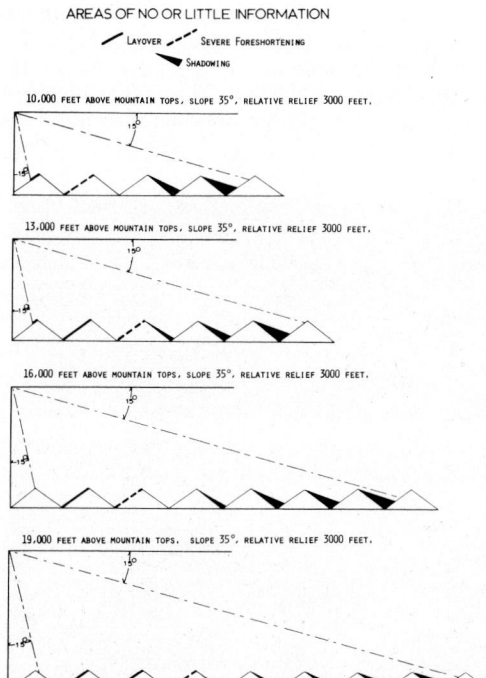

AREAS OF NO OR LITTLE INFORMATION

LAYOVER ---- SEVERE FORESHORTENING

SHADOWING

10,000 FEET ABOVE MOUNTAIN TOPS. SLOPE 35°, RELATIVE RELIEF 3000 FEET.

13,000 FEET ABOVE MOUNTAIN TOPS, SLOPE 35°, RELATIVE RELIEF 3000 FEET.

16,000 FEET ABOVE MOUNTAIN TOPS, SLOPE 35°, RELATIVE RELIEF 3000 FEET.

19,000 FEET ABOVE MOUNTAIN TOPS, SLOPE 35°, RELATIVE RELIEF 3000 FEET.

Fig. 25-7. Relation between usable radar swath-width and terrain clearance.

REVIEW OF THE GEOMETRIC ERRORS IN RADAR IMAGERY

Applications of single image radargrammetry require the use of (1) measurements of the slant range and time of imaging a selected number of points; (2) measurements of the position of the radar antenna (attitude) and its velocity; (3) the position vectors of a number of ground control points; and (4) a digital terrain model of the imaged target, if required (Leberl, 1976).

The idealized mapping projection for SLAR imagery, assuming flat terrain, is a transverse cylindrical equidistant projection where the flight line is the reference meridian. With no variation in the flight parameters (e.g. attitude, speed) point determination for mapping purposes would be a straightforward transformation from the cylindrical projection to the required map projection (for flat terrain). Sources of geometric error on radar images taken from airborne platforms are; (1) film-transport problems; (2) variations in the platform flight-characteristics; and (3) system setting readjustment during image acquisition.

Errors in aircraft navigation have been shown to produce image errors on the order of several kilometers which is more than that produced by slight relief and is of the same order as those produced by film transport problems (Leberl, 1975). Problems with film transport during image acquisition generally fall into two categories; the relationship between the speed of the film movement and the speed of the aircraft may vary, and/or the film may be incorrectly aligned. These may be termed "manual" errors and may often be corrected or accounted for during interpretation. Irregular flight lines, errors in determining flight speed and course, and irregularities in aircraft attitude or sensor orientation and attitude are examples of geometric distortions produced by variations in airborne flight characteristics. These geometric errors are generally a function of the accuracy of aircraft navigation. Orientation of the sensor is not a problem for synthetic aperture systems since the azimuth direction remains perpendicular to the range direction for previous scans. Airborne systems have been flown at increased speeds and altitudes to improve platform stability. Spaceborne platforms have markedly reduced these problems, as shown in the Seasat imagery.

Satellite radar imaging systems are digital so the problems associated with film transport are not present. Distortions result from the relationship between the motion of the spacecraft and the rotation and curvature of the Earth and the disparity in the range and azimuth dimensions of the pixels. These distortions can be rectified using idealized projections of the radar beam on the Earth surface, but are dependent on the density of the ground control points. Range offset has occurred on some Seasat images and represents a shift in the radar receiver sensitivity during image acquisition.

A significant problem with the processing of very high resolution SLAR data and Seasat synthetic aperture data is the phenomenon of range migration. Range-cell migration occurs when the slant-range variation of an individual reflector or set of reflectors exceeds the width of a resolution cell in the range direction. Initially, optical processors using specially constructed lenses to compensate for range migration were used to produce images of good quality. The problem has also been approached in various ways for digitally processing Seasat radar images. These techniques essentially correct the offsets in the range direction in the resolution-cell samples for each cell or blocks of cells before cross-correlation operations are carried out to produce the imagery. The development of these techniques has indicated many quality-related advantages of digitally processing synthetic aperture radar data as opposed to optical methods.

Finally, the assumption of flat terrain may be invalid so that the inherent geometric distortions in radar imagery due to variable terrain must be considered.

RADAR PROJECTION TRANSFORMATION

Radar-projection equations have been derived many times in the literature (e.g., Derenyi, 1970, 1974; Leberl, 1970, 1976, 1980; Graham, 1980). Differentiation in these equations must be made between real-aperture and synthetic-aperture systems. As implied in previous figures in the section ''Radar Geometry'', radar projection lines are concentric circles with respect to an antenna. With real-aperture radar the plane of the projection circle is normal to the real antenna axis. With synthetic-aperture radar the projection circle is normal to the velocity vector of the antenna. Detailed examination of the imaging geometry as-

sociated with spaceborne SAR is presented in Chapter 13.

The basic transformation between radar image space and map or object space is based on equations for rotation and translation of axes from cylindrical and spherical trigonometry, and equations for registering radar image coordinates to ground control points. The transformation may assume that image coordinates are being mapped to a planimetric cartesian coordinate system for aircraft images. The transformation for spaceborne radar images is somewhat more complex as it considers the Earth's curvature. In addition, corrections can be made for the effect of topography. Detailed discussion of the fundamental equations of rotation and translation of axes as related to image geometry is given in Chapter 21.

RADAR PARALLAX AND STEREOSCOPY

The use of stereoscopic radar imagery can enhance the process of interpretation by providing improved means of observing surface morphology (Koopmans, 1973), determining slope angles and relative relief (Dalke and McCoy, 1969) and, improving radar mapping and point location (Leberl, 1978). Examples of 60 percent side-lap stereo are shown in Figure 25-8 of Venezuela and Figure 25-9 of Papua, New Guinea. Note the substantial gain in information on drainage, geomorphology, and geology obtained through stereoscopic viewing.

Radar relief displacement is an inherent characteristic of side-looking imaging-radar systems and is towards the nadir if the feature is above the datum (i.e. a topographic high) and away from the nadir if the feature is below the datum (a topographic low). The relief displacement is, therefore, in the opposite direction from

Fig. 25-8. Same-side, same-height stereo radar strips of area in Venezuela. Obtained with the Goodyear APS-102 Radar System. Scale approximately 1:500,000.

Fig. 25-9. Same-side, same-height stereo radar strips of the Ok Tedi Area, Papua, New Guinea. Obtained with the Westinghouse AN/APQ-97 System for Kennecott Copper Corporation. Scale approximately 1:300,000.

the displacement direction in optical camera systems (Figure 25-10).

When one feature is imaged twice at two different look-directions, that is, either the same side or opposite side configuration at the same height (Figure 25-11 A and 2511 B), or at two different altitudes, with a same-side configuration (Figure 25-11 C), then radar parallax can be measured and radar stereoscopy attained. Radar parallax is defined as the sum of the displacement of the feature on the two images (Beatty et al., 1965). Levine (1960) showed the development of the parallax triangle along with the uses of radar parallax. LaPrade (1963) in an experimental study of radar stereoscopy illustrated the effect of flight configuration on radar parallax and defined his view of the optimum flight configuration for opposite-side and same-side radar stereoscopy.

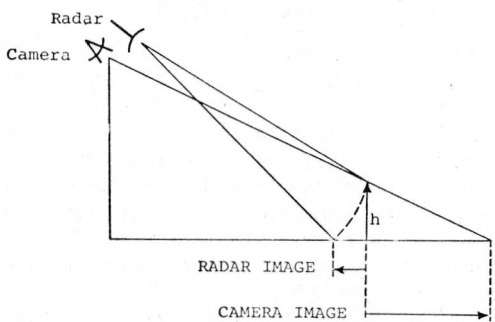

Fig. 25-10. Relief displacement.

In the past few years, there has been increasing interest in SLAR systems for stereo plotting and this interest has been shifting from airborne to spaceborne systems. Several inherent characteristics of SLAR systems make stereo interpretation of radar imagery more difficult than with stereo photographs. The following factors influencing radar stereoscopy can be identified: (1) stereo arrangement (same side or opposite side); (2) depression angles; (3) stereo intersection angles; and (4) relative relief and slope of the terrain.

The shifting of the source of illumination can cause varying intensities to result from similar slopes. Fore slopes may be illuminated, foreshortened, or layed-over, while the back slopes may be near true length, foreshortened, or in shadow. The effect for stereo radar images made from opposite sides of a feature is similar to that of using one aerial photograph taken in the early morning with another taken in the late afternoon. Under these conditions, with moderate terrain slope or relative relief, adequate fusion is impossible because radar shadows are absolutely blank. It was concluded from the experience of one of us (D. S. Simonett) in project RADAM, and in New Guinea, that opposite-side stereo can only be applied in the case of flat or slightly rolling topography and Leberl (1980), has confirmed this in a wide range of situations. Totally foreshortened slopes can be represented by a single line on radar imagery, whereas areas of layover must be rectified prior to stereoscopic examination. For these reasons, same-side stereo coverage has pro-

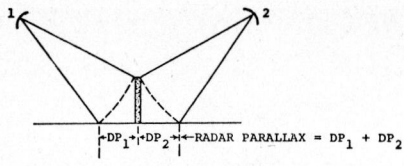

(A)   OPPOSITE SIDE CONFIGURATION

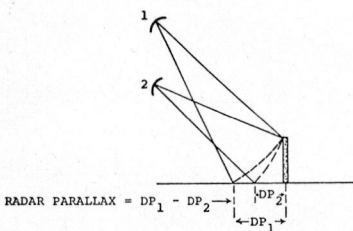

(B)   SAME SIDE CONFIGURATION

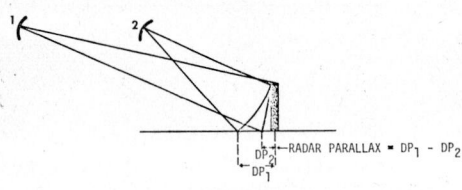

(C)   SAME SIDE - SAME HEIGHT CONFIGURATION

Fig. 25-11. Radar parallax (A and B from Beatty et al, 1965; C prepared for this manual).

vided much more satisfactory stereo pairs for the geoscientist than opposite-side, and all commercial flights for resource evaluation have been flown with same-side, same-height coverage.

Stereo viewing generally improves with smaller depression angles. With larger depression angles, the stereo base has to be reduced for adequate stereo viewing; otherwise, the differences in relief displacement become too large (Leberl, 1979). Drastic differences in image content seriously hinder adequate stereo fusion. Finally, small stereo intersection angles tend to render the best stereo.

From the above discussion it can be seen that the best stereo, in terms of the quality of viewing, results from image pairs with little difference in tone, texture and, geometry. However, for good topographic exaggeration, one would prefer large stereo intersection angles. Here we have a trade-off between geometric accuracy and quality of stereo viewing without much experimental work upon which to base our decisions (Leberl, 1979).

It has been suggested (Moore and Thomann, 1971) that a squint mode of operation which would decrease data ambiguity in stereo radar coverage employing orthogonal look-directions along a single flight, with two beams being used, one squinted 45° ahead, and the other 45° behind and with the outputs being recorded simultaneously. Leberl (1979) concludes that this stereo mode

could not produce adequate stereo from synthetic aperture systems. Relief displacements would always be the same and at an angle of 90° toward the nadir line resulting in no parallax and no valid stereo (Leberl, 1972). Leberl (1979) suggests that single-flight stereo data may have potential with real-aperture radar systems by tilting the antennas about horizontal or vertical axes and using conical beams.

Other problems with radar stereoscopy include "false cliffing" which occurs with skipped scan lines or along-track compression in a portion of an image. This may occur in real-aperture imagery subject to yaw, to malperformance of a CRT recorder, or as artifacts of data acquisition of satellite imagery. These have been observed on real-aperture stereo images and on Seasat synthetic-aperture imagery. Similarly a serious rolling false stereo can occur in synthetic aperture images, sufficient to swamp the real stereo in areas of modest relief. This effect can occur where there are surges in the mechanical linkages of an optical correlator used to form the synthetic aperture image, or where there are saturations in subsystems used to digitally correlate the images.

The main difficulty with stereo radar is common with all systems, namely a warped datum plane, typical of line-perspective systems and absent from point-perspective systems (framing camera). The warping makes for considerable difficulty in achieving quantitative height measurements and, if uncorrected, necessitates complex stereo instruments. External adjustment is required to alleviate a warped datum plane in a radar image or coherent image block. The image or block must be transformed into a net of ground control points. Any method of interpolation can be used to fit the image or image block to the control points and these methods are often denoted "warping" functions or "rubber sheeting" (Leberl, 1980).

The warping of the datum plane and variable parallax are shown when areas of comparable topography are viewed in near-, mid-, and far-range scenes in aircraft images. The greatest vertical exaggeration occurs in the near range, and this effect must be watched and mentally compensated for during interpretation.

TRANSFORMATIONS OF RADAR IMAGES FOR MOSAICS

The use of radar imagery for resource reconnaissance-mapping and other applications covering large areas requires the construction of mosaics of radar imagery. Radar-mosaic construction has advanced considerably in the last few years, employing mathematical techniques of image-block adjustment and better systems of platform navigation. Four techniques of mosaic construction from radar imagery have been, or are being used (see Leberl, 1976):

1) The simplest technique, with few ground control points, is one where no preprocess-

ing of the images is applied. Adjacent image strips are fitted together and ground-control points, if available, are used for a preliminary overall rotation, scaling, and shifting to a map-coordinate system. Thus mosaic geometry is only as accurate as the inner geometry of the images and the radar-platform navigation permit.

2) A second, and complex and expensive, radar mosaicking technique employs a continuous accurate tracking of the survey aircraft, simultaneous with image acquisition. SHORAN tracking determines the position of the aircraft with an absolute accuracy of ±300 m (Van Roessel et al., 1974). This permits the transformation of an ordered set of artificial image points into the map system. The transformed set of these points forms the base onto which the mosaic is constructed. The accuracy of the resulting mosaic depends on the accuracy of the aircraft tracking, the effect of measuring mosaic errors, and the distortions in the individual image-geometry. This method must be employed in areas with very few ground control points (as in the Amazon Basin) if reasonable accuracies are to be obtained.

3) This technique of radar-mosaic construction makes use of a set of images obtained in the range direction of the production imagery around the perimeter of the mapping area. A number of ground-control points are marked on these images taken from surveys or scaled from available maps. The resulting "tielines" permit transformations of the production images into the map system. The fit of the mosaic requires fit with the tie-lines and a smooth transition to the adjacent strips. The accuracy of this method is limited by the individual image distortions and the accuracy of the position of the ground points on the tielines. The advantage of the method is its simplicity since no numerical operations are required and the mosaic can be constructed directly.

4) The fourth technique of radar-mosaic construction relies on the numerical adjustment of radar-image blocks. This method is described in detail by Leberl (1975). Point measurements in the overlap areas of adjacent images are used to tie the images into a block. The block is then transformed into a network of ground-control points. This network forms the base map on which the mosaic is constructed. This numerical planimetric adjustment is carried out prior to mosaic compilation and provides a greater mosaic accuracy than the approach based only on tielines. In general, mosaics based on numerical adjustment are more accurate than those constructed with other techniques and may be less expensive (Leberl, 1976).

In project RADAM the production of acceptable semi-controlled mosaics required the combination of (1) high altitude flights, above most turbulence; (2) high-speed flight by jet aircraft, to insure only slight deviations in path straightness; (3) near-parallel north-south flight strips spaced 15 minutes of longitude apart to insure stereoscopic coverage for part of the imagery; (4) same-side looks (west-looking) to obtain shadows in the same direction as 9:30 a.m. illumination at the time of the expected ERTS-1 passage; (5) very high-quality internal navigation; (6) SHORAN ground stations set up and monitored in flight for geometric control; (7) use of radar tie lines (here diagonal to the north-south image strips); (8) aerial photographs as available; (9) use of ground-control points including specially surveyed satellite-transit points, existing astronomical points, hiran points, transit points, and locally surveyed points as well as radar corner-reflectors; (10) development of secondary geopoints using SHORAN tracking data recorded during the flight of the aircraft, along with radar images and aerial photographs; (11) development of a whole suite of manual and computer geometric checking and evaluation procedures; (12) radar altimetry, employed in checking images in a number of ways, including the detection of gross defects which cannot be corrected during mosaic construction and require new image acquisition, and, (13) development of computer procedures to tie all radar strips, tie lines, and ground stations together into a tight model for mosaic construction.

OPTIMUM RADAR DEPRESSION ANGLES

The inherent disadvantages of radar imagery that hinder interpretation are extensive foreshortening, shadowing, and layover. Figure 25-12 illustrates the gross topographic features of the Coconino Plateau, Arizona, and Figure 25-13 provides slant-range imagery of the same area. The average slope of the canyon walls is 51° and the relative relief of the canyons is generally less than 450 m. Although the canyon topography (i.e. slope and relief) is quite uniform across the area, shadowing, layover, and foreshortening on the radar image of Figure 25-13 are not. This contrast between actual and recorded terrain-geometry provides a practical example of the effect of changing depression angle across the swath width of aircraft imagery (near to far range). The foreslopes in Figure 25-13 vary in length from a single line (south of points a and b) in near range, to approximately 35 percent of the actual length in the far range. Assuming constant foreslopes of 50° the combined effects of radar layover and shadowing show how limited is the area in which interpretation may reasonably be carried out. Mountainous areas should be covered from space, not from aircraft altitudes, because the depression angles can be made most appropriate to the terrain.

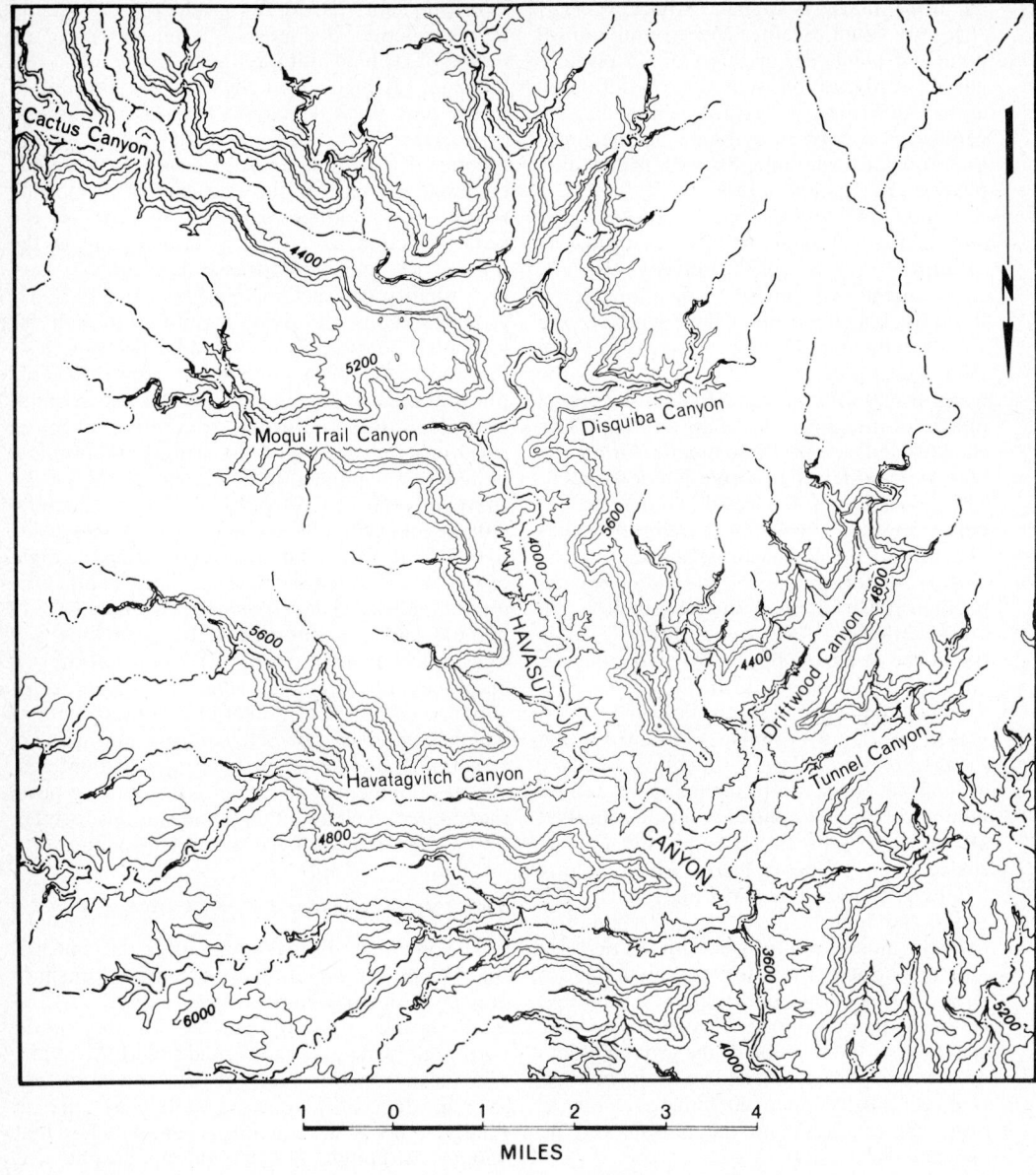

MILES

(Contour Interval 400 feet)

Fig. 25-12. Generalized topographic map, Coconino Plateau, from USGS Supai, Arizona quadrangle, 1:62,500.

The ground-range imagery of the Aichilik River area in Alaska presents a marked contrast to the Coconino Plateau region. Figure 25-14 illustrates relatively uniform mountainous terrain of the Romanzof Mountains where the terrain slopes average 30° and the relative relief averages 1000 m. Points A, B, C, D, and E are indexed on the imagery of the area. Those slopes along a line east of A−C−E are displaced because of layover. In the region of area E, the foreslopes are sufficiently displaced (layover) to mask the east branch of the Aichilik River. East of the Aichilik River, foreslopes are not distinguishable, and interpretive data on the back slopes are of little value. Exten-

sive shadowing in the far range also provides a handicap for the interpreter. (Figure 25-15 is a topographic map of the same area.)

While Figure 25-14 shows imagery of a ground range presentation, Figure 25-16 illustrates slant range imagery recorded near Cascade Glacier in Washington. The influence of radar layover is not as apparent as that in Figure 25-16 even though the slopes are steeper and relative relief of the Cascade area is greater than that of the Alaska area. The foreslope of ridge a−b (image B, Figure 25-16) is foreshortened considerably, but is not in layover. Image A, taken at a larger depression angle than image B, illustrates layover of ridge

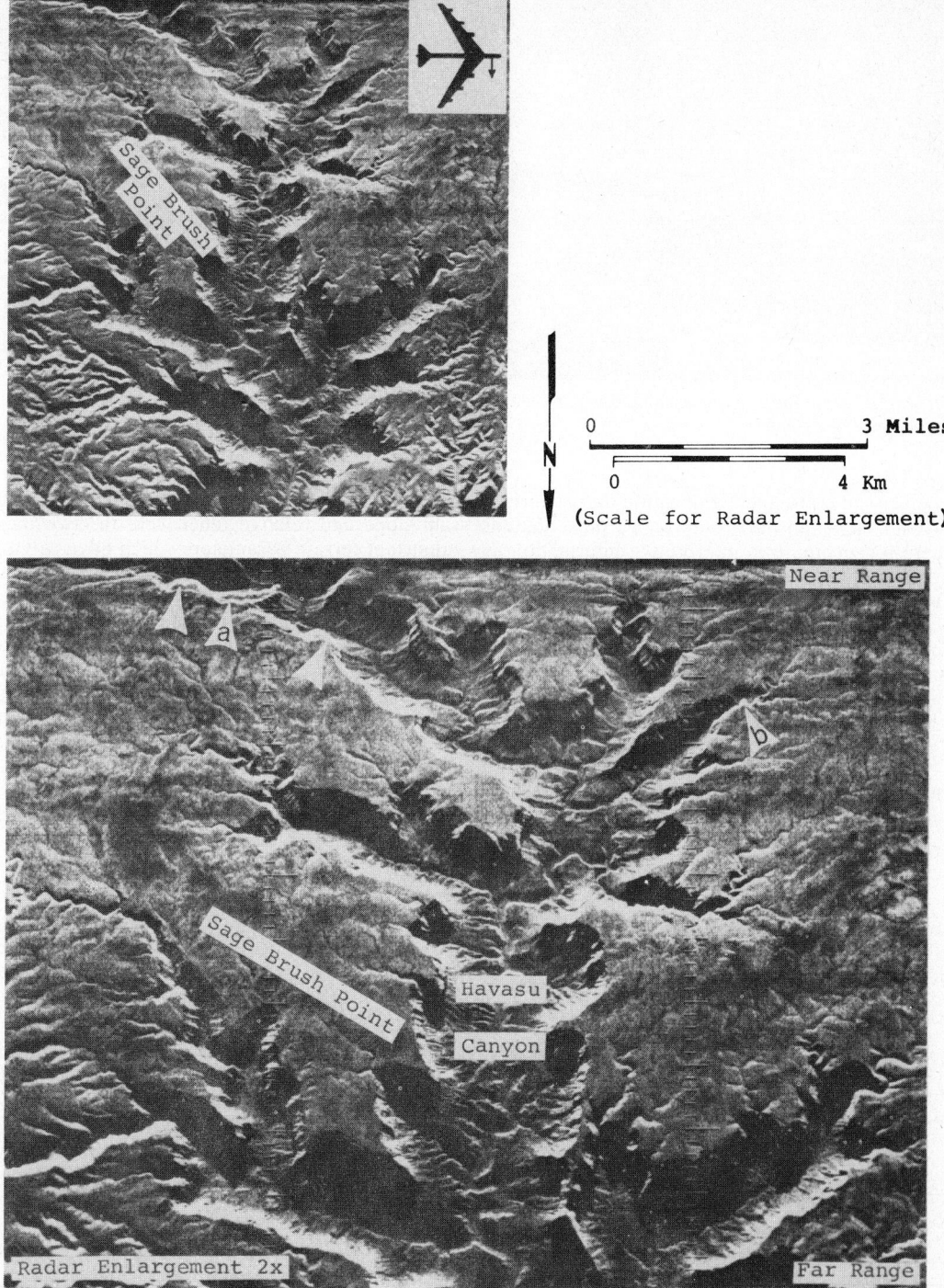

Fig. 25-13. Radar foreshortening in the Coconino Plateau shows variation of foreslope length across swath width of imagery (near-to-far range).

a−b. The bright band along the top of the ridge represents the total extent of layover. Careful comparison between the portrayal of ridge a−b, as recorded on both images, reveals that as layover has occurred (image A), the length of the backslope has decreased accordingly.

It becomes obvious from the comparison of Figures 25-14 and 25-16 that because of radar

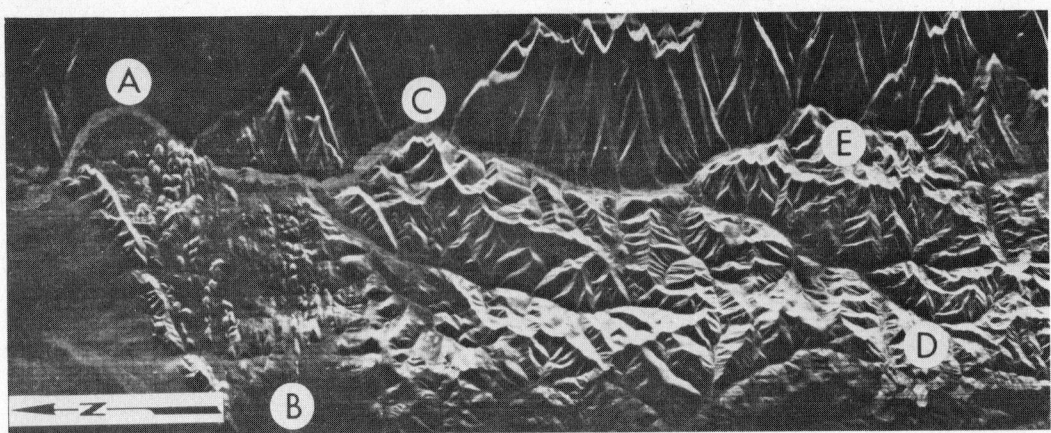

Fig. 25-14. Radar image of the Aichilik River area illustrating the influence of layover on ground range format. Scale approximately 1:400,000.

layover and consequent elongation and distortion of back slopes in the ground-range presentation, there is a decrease in interpretive information as compared to slant-range format. This suggests that for at least mountainous terrain, the interpreter would prefer a slant-range imagery-format. Conversely, the ground-range imagery is much more desirable for mosaicking, and in this respect, a dual recording mode would be preferable in mountainous regions (Lewis and Mac-Donald, 1970).

The geometry of both the imaging system and the terrain has been considered by MacDonald and Waite (1971) in an attempt to predict the limitations of global geologic and physiographic re-

connaissance radar systems. They found that terrain slope and relative relief were the two most important terrain parameters which adversely affect the recording of SLAR imagery. A map showing these two parameters on a global basis would be desirable, but unfortunately such information is not available for many parts of the world. However, a reasonable approximation can be made using Hammond's (1954, 1964) classification of landform types. Major terrain elements distinguished by Hammond were classified according to slope, profile type (vertical arrangement) and local relief (vertical dimension).

Figures 25-17 and 25-18 have been modified from Hammond's maps (Finch et al., 1957) to em-

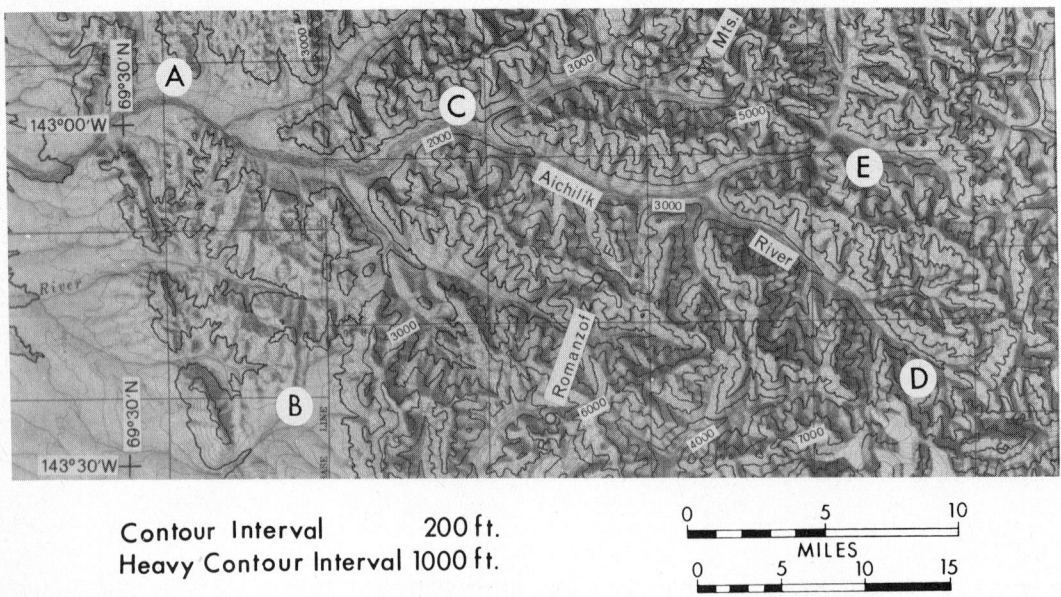

Contour Interval   200 ft.
Heavy Contour Interval 1000 ft.

Fig. 25-15. Generalized topographic map of the Aichilik River area, Alaska from USGS Demarcation Point, Alaska quadrangle 1:250,000. Scale approximately 1:400,000.

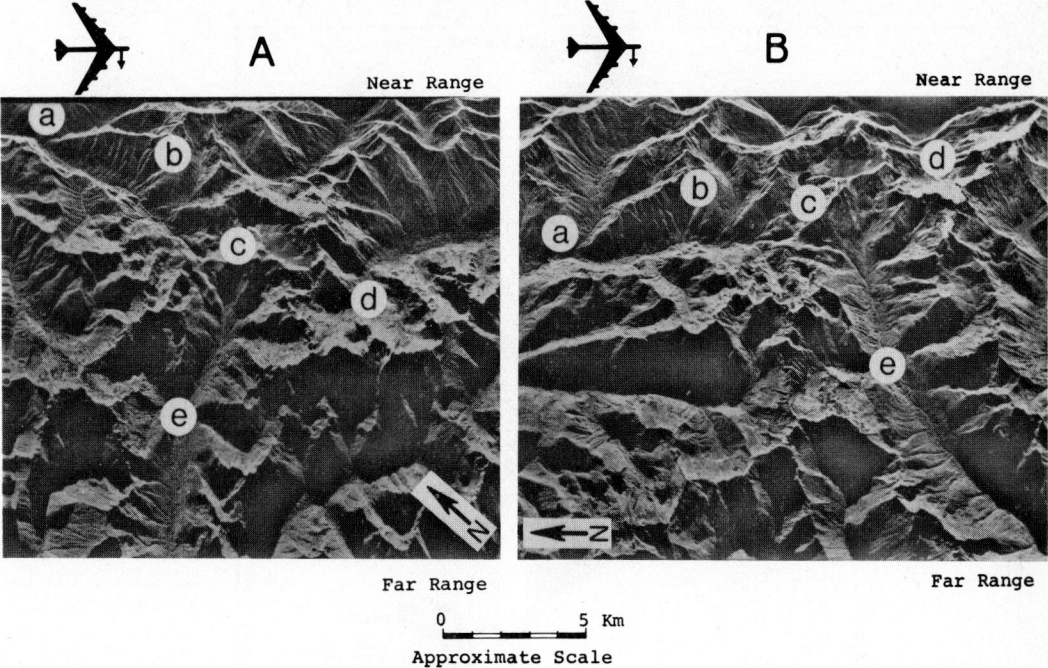

Near Range

Near Range

Far Range

Far Range

0 _____ 5 Km
Approximate Scale

Fig. 25-16. Radar images of the Cascade Glacier area, Washington, illustrating the influence of layover on slant range format.

phasize the slope-relative relief aspect. Optimum radar depression-angles have been selected for five separate terrain categories. For each of the five terrain categories shown the optimum depression angles were selected to provide maximum swath widths of usable data. Although usable swath width does not become a critical factor at low depression angles or in low relief areas, this relationship does hold for high relief areas. Regions of lowest terrain slope and relief, such as the Gulf Coastal Plain of the United States (Figure 25-17), have been assigned the smallest depression angles, whereas the highest relief-slope areas, such as the Andes Mountains of South America, have been assigned the largest depression angles. This direct relationship between depression angle and relief-slope parameters is necessary to obtain subtle shadowing in the low relief categories, yet minimize the effects of shadowing-layover in the high relief-slope areas.

The selection of the depression angles in high relief areas is biased slightly toward the higher values which will increase the occurrence of layover, but decrease that of shadowing. The justification for this is that in such high relief areas the near-range slope is severely foreshortened whereas the far-range slope is presented near its true length. The slight intentional bias towards layover is introduced to sacrifice the least usable portion of the imagery.

The depression-angle range for images from the Seasat synthetic-aperture imaging-system is from about 73° for the near range to about 67° for the far range. The 6° depression-angle range produces a swath width of 100 kilometers. This narrow range, accompanied by the steep depression angles, makes applications of Seasat data in high-slope, high-relief areas questionable. Figure 25-19A is a Seasat image of an area in Mount McKinley National Park, Alaska. This image contains some of the highest peaks in North America, including Mt. McKinley (upper left). The terrain has been glaciated and the mountains are largely snow-covered. Layover effects cause some of the mountains in this scene to be so strongly displaced toward the near range that they appear to be superimposed on the glaciers. Figure 25-19B is a diagram showing features of interest. For example, a single peak with a bright return, adjacent to Ruth Glacier, is displaced almost one-third of the way across the glacier. Recommendations have been made that future earth-imaging spaceborne radar systems have smaller depression angles and/or a greater range. Since Seasat was not intended for use over mountainous areas for precisely this reason, the recommendations are appropriate.

The tradeoff decisions for selection of optimum depression angles are considerably easier for categories of low slope and relief. The probability of obtaining both layover and shadowing near the same angular range is less, and decisions mostly involve the extent of shadowing desirable. In the lowest-relief areas some shadowing is important in order to provide topographic enhancement and to facilitate geologic or other interpretation.

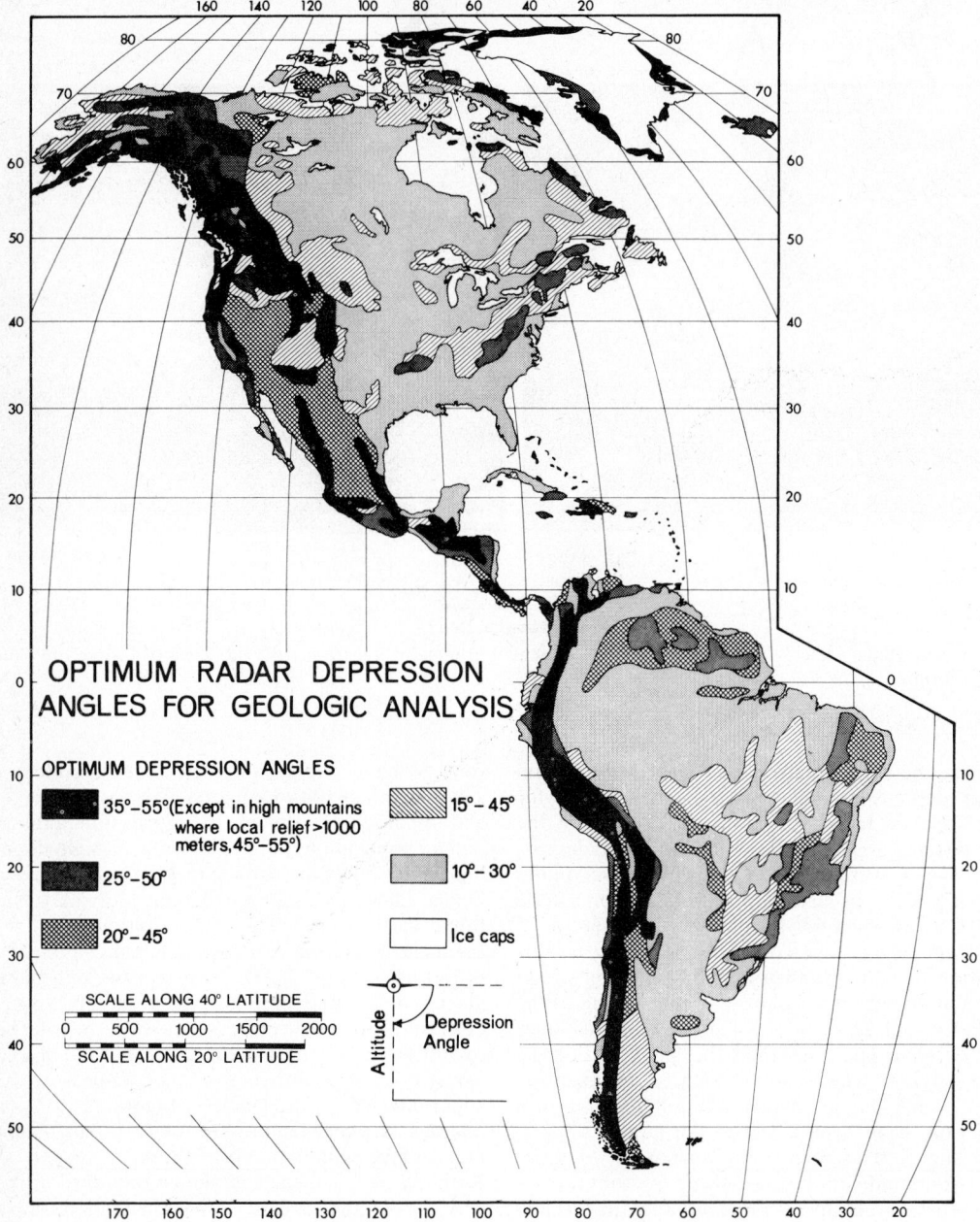

Fig. 25-17. Optimum SLAR depression angles for the Americas.

AREAL RESOLUTION

The areal resolution of radar is the product of the azimuth resolution (R_a) and the ground-range solution (R_{gr}). It is a direct function of the effective beam width (γ) of the system and slant range (R_s) to the target which control the azimuth resolution, and of the effective pulse length (τ) of the system and the angle of incidence (ϕ) which affect the ground-range resolution. For any given system γ and τ are fixed, and R_s and ϕ are variables.

The areally-weighted average resolution (AWAR) is the average areal resolution across a swath, derived by summing, for each resolution-cell increment of swath, and dividing by the number of resolution cells summed across the track:

$$AWAR = \sum_{NR}^{FR} \frac{\lfloor R_a * R_{gr}}{n} \qquad (25\text{-}6)$$

where NR is the near-range, FR is the far-range, and n is the number of cross-track resolution cells.

Good azimuth, or along-track resolution (R_a) is

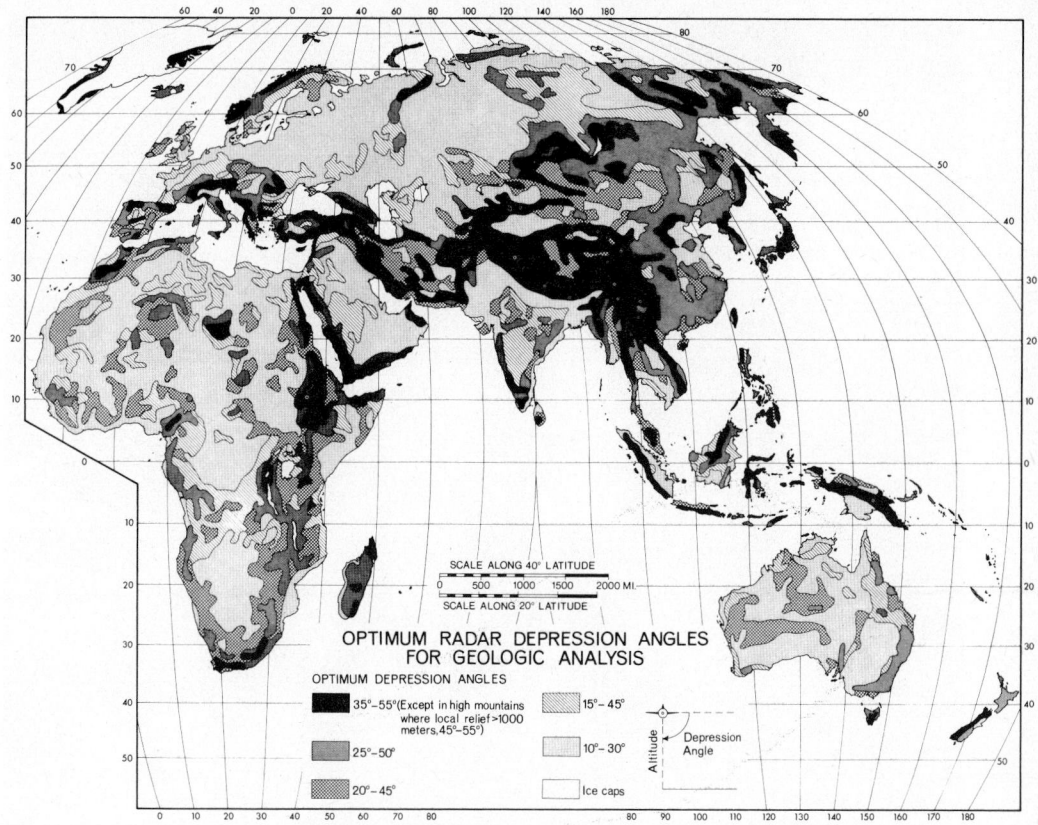

Fig. 25-18. Optimum SLAR depression angles for Africa, Eurasia and Oceania.

achieved by using long antennas and short wavelengths with real aperture antennas, or by using focused synthetic-aperture techniques for all wavelengths. (Chapters 10 and 13 give detailed discussions of the resolution of real-aperture and synthetic-aperture radar systems.) For a fully focused synthetic-aperture system the theoretical azimuth resolution $= D_a/2$, that is, one-half the length of the real antenna on board the radar platform; the shorter the antenna (theoretically) the better the resolution. This is the reverse of the normal diffraction-limited case with real-aperture systems. In practice, however, resolutions as good as this are not realized.

Ground-range resolution (R_{gr}) may be achieved by the use of short pulses of energy or by chirping (discussed in detail in Chapter 10). Short pulses of the order of 30 m (0.1 microsecond) may be readily transmitted and give a slant-range resolution of 15 m. The practical limits of ground-swath coverage lie between $\beta = 5°$ and $\beta = 75°$. For a slant-range resolution of 15 m the corresponding ground-range resolutions at selected depression angles are shown in Table 25-4.

From this table it is clear that ground-range resolutions are independent of distance from the transmitter, being only a function of the angle of incidence. They are smallest in the far range and

remain acceptable in the near range only to $\beta = 70°$ or $75°$. This situation, where the poorest ground-range resolution occurs in the near range of the scene, is the reverse of an optical system in which point-range resolution degrades linearly as a direct function of the distance from the camera, and as the cosecant of the angle from the normal.

The resolution of the Seasat imagery is a function of the relative velocity between the spacecraft and the image target on Earth at any particular latitude and the ground range between the nadir of

TABLE 25-4
Relation Between Ground Range Resolution and Depression Angle, for a 15 Meter Slant Range Resolution

β Degrees	R_{gr} Meters	β Degrees	R_{gr} Meters
5	15.1		
10	15.3	50	23.3
15	15.5	55	26.2
20	16.0	60	30.0
25	16.5	65	35.5
30	17.3	70	43.8
35	18.3	75	58.0
40	19.6	80	86.4
45	21.2	85	172.0

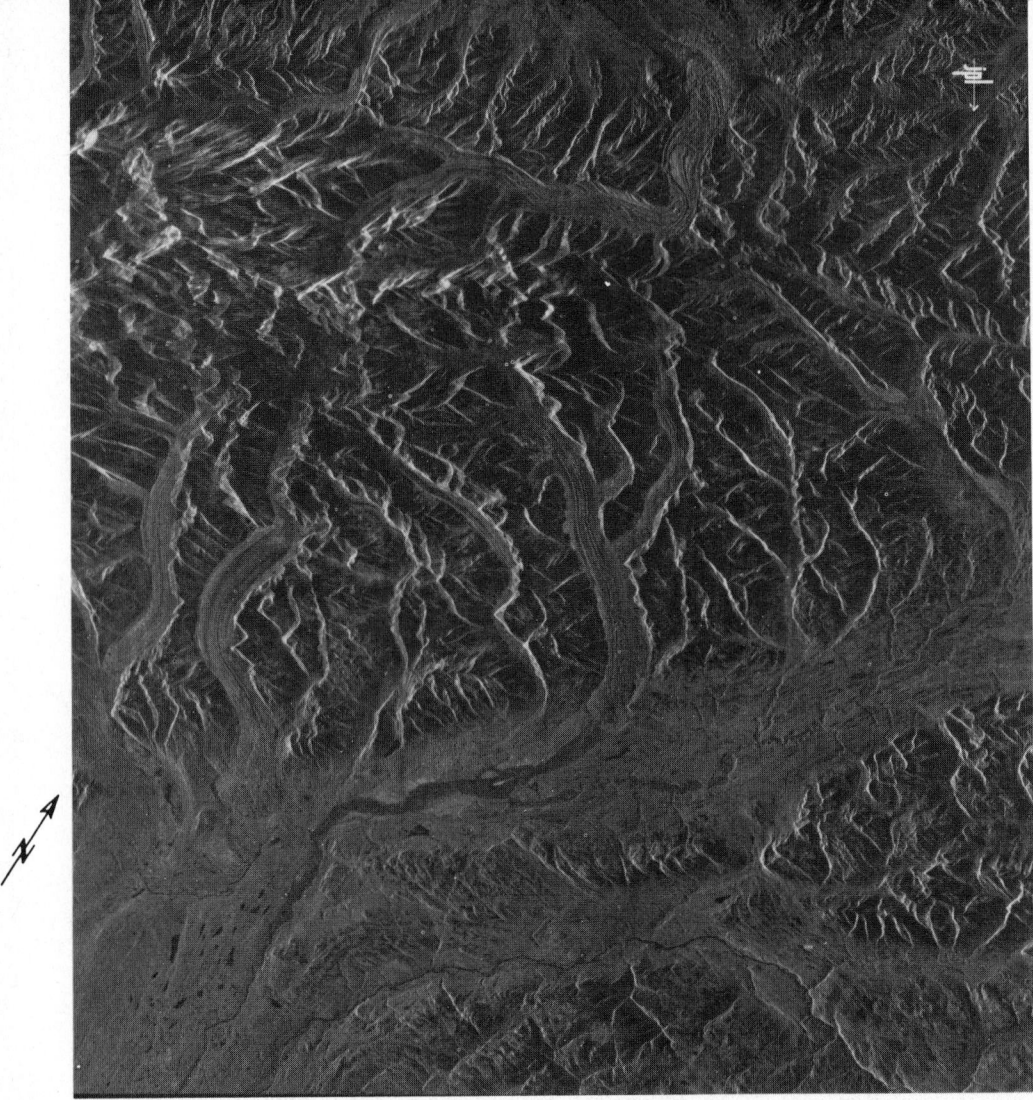

Fig. 25-19A. Mt. McKinley National Park Area. Scale approximately 1:500,000.

the Seasat radar and the target. For certain digitally correlated images (Ford et al., 1980) geometric distortion arises from the differences in pixel dimensions in the azimuth direction and the range direction. In these images, pixel dimensions are approximately 16 m in the azimuth direction, and vary from about 14.8 m to about 19.4 m in the range direction.

NUMBER OF INDEPENDENT SAMPLES

The use of monochromatic illumination in radars with the associated Rayleigh fading leads to a characteristic "graininess" in radar images, even in homogeneous areas. This is discussed more fully in Chapter 10 as are techniques of "speckle" removal. Such data processing is also discussed in Chapter 17.

The number of independent samples (N_i) aver-

aged in a resolution cell influences the variance of the fading signal. The larger the number of independent samples, the smaller the variance, the smoother and less "grainy" the grey tones of the image, and the better the interpretability of the image for a given nominal resolution.

Real Aperture Systems

In real-aperture systems (Moore and Thomann, 1971) the number of independent samples averaged per resolution cell is

$$N_i = \frac{2R_a}{D_a} \qquad (25\text{-}7)$$

where R_a is the azimuth resolution, D_a is the antenna length, but $R_a = \lambda R/D_a$ where λ is the wavelength of the radar and R is range, therefore:

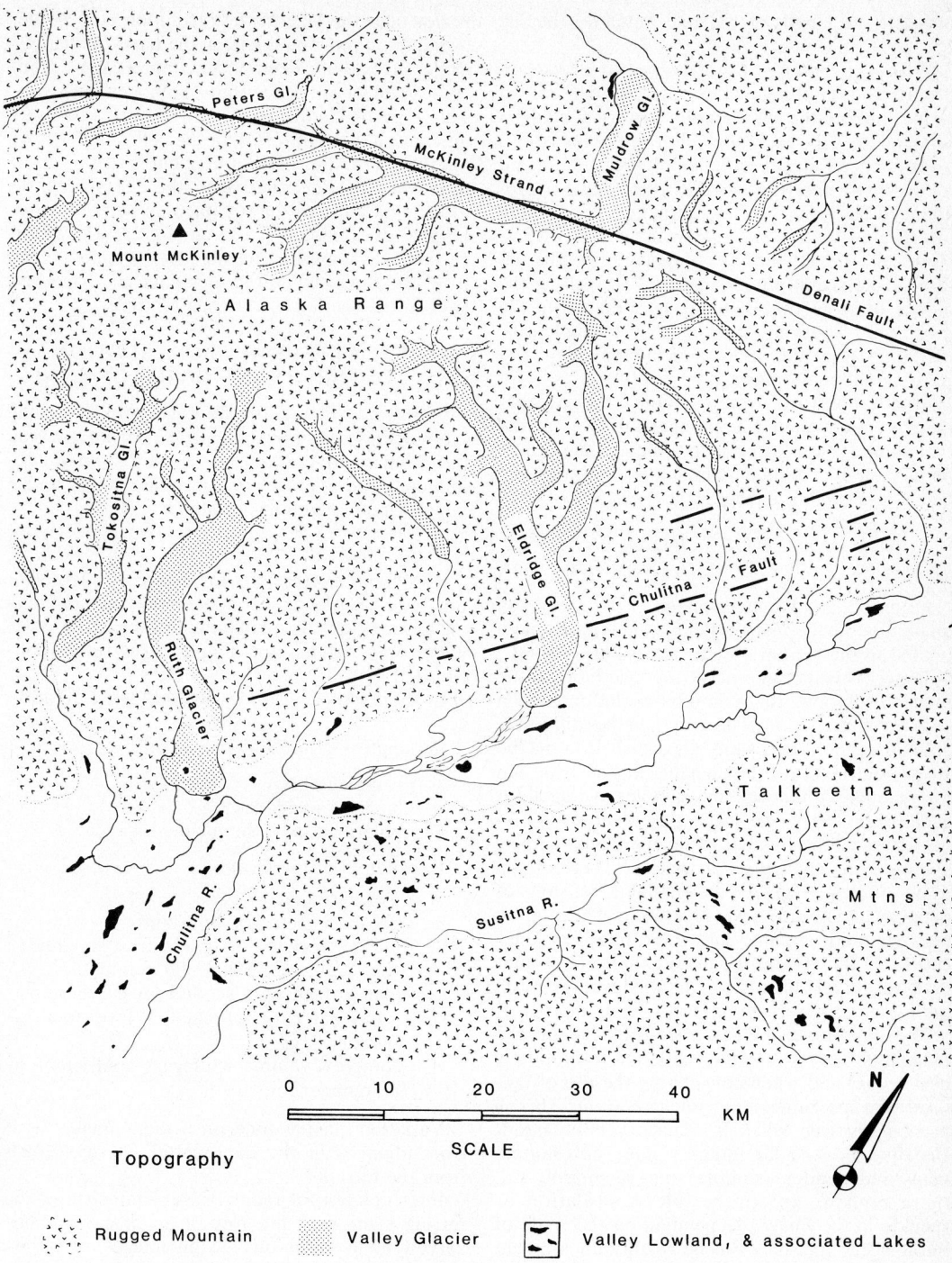

Topography

SCALE

 Rugged Mountain   Valley Glacier   Valley Lowland, & associated Lakes

Fig. 25-19B. Diagram showing features on Fig. 25-19A. (Courtesy of JPL)

$$N_i = \frac{2 \lambda R}{D_a{}^2} \qquad (25\text{-}8)$$

The number of independent samples increases from near range to far range on an image.

Synthetic Aperture Systems

In synthetic aperture systems the azimuth res-olution is independent of range (within the power limits of the system) and theoretically is equal to

one-half the length of the real antenna, that is:
$R_a = D_a/2$.

With a perfect synthetic-aperture system, Moore and Thomann (1971) noted that only a single lobe of the backscatter pattern would be sampled in each resolution cell, thus giving rise to a pronounced signal variance. Images produced by synthetic-aperture systems often have a pronounced black-and-white grain which is normally interpreted as arising solely from this effect. However, the situation is complicated by the nature of the signal processing used as well as the degree of perfection of focusing achieved in the processor. Thus one cannot give a simple rule of thumb as to the number of independent samples actually averaged per resolution cell for synthetic-aperture systems. One can say that there are fewer than in a real-aperture system, and that some systems are much grainier than others of comparable quoted nominal resolutions, and hence less easy to interpret.

IMAGE RESOLUTION, INDEPENDENT SAMPLES, AND IMAGE INTERPRETATION

Moore (1979) has determined that the overall qualitative image interpretability is exponentially related to the spatial grey-level resolution (SGL) volume. This parameter is the product of the range resolution, the azimuth resolution, and a grey-level resolution. The grey-level resolution is defined as the ratio of the 90 percent level to the 10 percent level on the signal-fading distribution for a number of independent samples averaged for each pixel. The value of the grey-level resolution decreases as the number of independent samples averaged for each pixel increases. Therefore, a trade-off exists between the spatial dimensions of the pixel and its grey-level resolution. The relation between SGL volume and interpretability indicates that the area of the pixel is more important than the pixel dimensions in range or azimuth (more details are given in Chapters 10 and 13).

The importance of the trade-off between averaging for grey-scale uniformity and resolution is best recognized when considering the use of texture as a discriminant in interpretation. This is particularly true when the imagery is enlarged. Because most radar imagery does not sample many independent samples per resolution cell there remains an appreciable scintillation or speckle in the image. Depending on the scale of the imagery this may not be readily apparent to the eye, for the eye itself performs considerable averaging in viewing a small-scale image. When looking for textural differences in homogeneous areas, one must keep in mind that variations on the scale of the resolution cell are in large part due to the fading of the return signal and are not indicative of actual ground texture on this scale.

For simple textural discrimination, care should be exercised to ensure that the texture is significantly above the scale of the resolution cell. For scales near the size of the resolution cell the determination must be made on the basis of changes in the probability density function obtained from cell-to-cell variability within a homogeneous area. This type of identification works only when significant change is present in the cell roughness or microtexture. For terrain elements with a structure which is fine relative to the cell size, little discrimination may be obtained. When a smaller number of dominant scatterers are present within certain cells it appears feasible to distinguish these from the finer structured elements, as discussed later in the section on "Image Texture".

BACKSCATTERING CROSS-SECTION: A KEY DETERMINANT OF IMAGE TONE

Variations in σ_0, the average backscattering cross-section per unit solid angle directed towards the radar receiver, for a given target, markedly influence the average tone of objects on a radar image, and hence interpretability,

$$\sigma_0 = f(\lambda, \phi, P, \theta, \epsilon, \Gamma_1, \Gamma_2, V) \qquad (25\text{-}9)$$

Analytical expressions for σ_0 are not available for natural surfaces. Models employing simplifying assumptions do enable radar theory to predict with some success for simple cases (e.g. Blanchard, 1980).

In the expression for σ_0 the following notation applies:

Parameters of the Radar System

λ = wavelength;
ϕ = angle of incidence;
P = polarization of the incident wave;

Parameters of the Ground

θ = aspect angle;
ϵ = complex dielectric constant;
Γ_1 = surface roughness on a microscale greater than $\lambda/10$, of the air-solid boundary;
Γ_2 = sub-surface of a second layer where the signal can penetrate the first layer to a significant degree;
V = complex volume-scattering coefficient in inhomogeneous media.

Variations in the tone on a radar image arise from changes in the backscatter energy return from the terrain.

Interpretation of radar imagery consists of inferring significant features in an area from observed tonal variations on an image and/or attempting more quantitative approaches through the use of target/energy interaction models. To construct the inferential relationships necessary to interpret imagery, the interpreter must be familiar with the variations due to each of the parameters and their possible interaction, in addition to ancillary data such as regional climatic characteristics. This familiarity may be gained by examining the behavior of theoretical models of backscatter or by referring to experimental data from areas with similar characteristics, assuming that these areas will indeed produce similar radar returns.

Even with a knowledge of the behavior of radar terrain-return with the above parameters, the interpreter of radar imagery is still confronted with two major obstacles. First, he must consider whether the parameter which affects tone is of significance in describing the terrain. (Conversely, are the parameters of interest in classifying the terrain of significance in the scattering of the signal?) Second, he must determine which parameter of the terrain gives a boundary or tonal change on the image. Tone or amplitude information recorded on an image does not constitute a unique measure even within fairly restrictive terrain classes. To determine which parameter is the cause of a particular tonal variation requires a priori information about the terrain class, multiparameter radar data, information from other sensors, or in many cases, a combination of all three.

Referring again to the parameters affecting the return, one notes that five are parameters of the terrain, and three are functions of the radar system. The obvious strategy in remote sensing is to select the three controllable system parameters to optimize the information obtainable from the terrain parameters. This requires an intimate knowledge of studies based on both theoretical models and experimental data. Appropriate timing of image acquisition making use of surface changes can also aid interpretation.

Before turning to an analysis of each of these parameters, it is proper to summarize that identifying and describing average area tone is not as simple as recognizing high or low tones of discrete tonal elements. A combination of varying tone values can be qualitatively evaluated, or in some cases, quantitatively evaluated to make the neces-

sary estimate of the value of the representative tone. The tone of each resolution element, in fact, represents a measure of the radar cross-section of that portion of the terrain surface illuminated by the radar system.

Ideally, the value of the radar cross-section should be measurable in terms of the surface roughness and geometry, and the complex dielectric constant of the illuminated terrain. Presently, existing theory for the scattering characteristics of the natural, vegetated terrain surfaces can only indicate trends in the relationships between parameters. Unfortunately, little calibrated radar imagery exists which can be effectively compared for a variety of system and surface situations. It is suggested, however, that backscattering models will allow separation of system-induced effects from actual interaction effects and refine qualitative interpretation as well as permit the development of better techniques of quantitative information-extraction (Blanchard, 1980). Therefore, it is necessary to utilize the element of average areal tone simply as one of the pattern elements rather than a quantitative measure of a specific terrain parameter.

WAVELENGTH

Variations in the radar return-signal attributable to wavelength (λ) are directly related to two surface parameters, surface roughness, and complex dielectric constant. In general, the rougher the surface in terms of wavelength, the more diffuse the return (as illustrated by Figure 25-20) and therefore the stronger the backscatter. Thus, a given surface will normally appear rougher at shorter wavelength (see section on Surface Roughness for a description of roughness in terms of wavelength).

The transition between diffuse backscatter

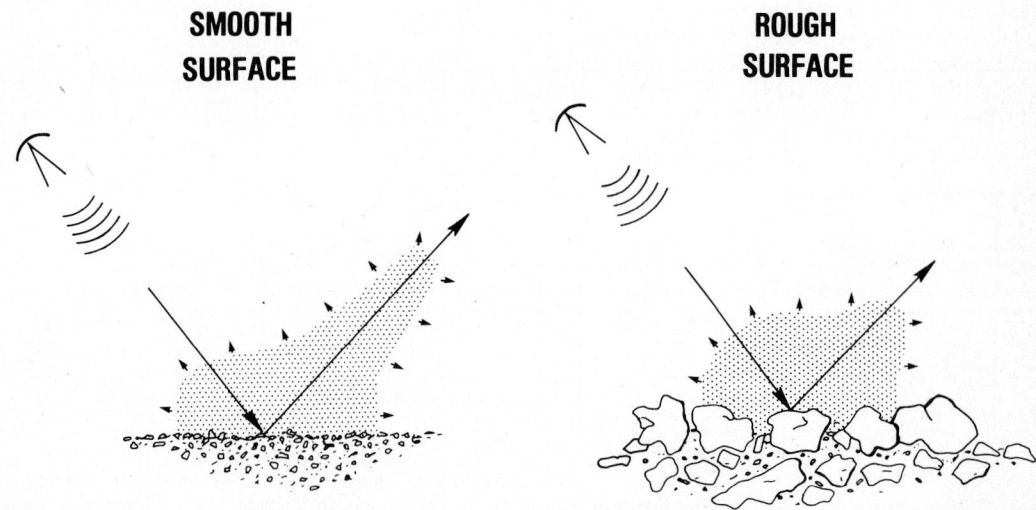

SMOOTH SURFACE

ROUGH SURFACE

Fig. 25-20. Schematic representation of radar backscatter.

leading to a bright image tone and specular scattering, which produces darker tones for a variety of incidence angles away from the normal, is a function of the wavelength and the terrain roughness at the impinging wavelength. MacDonald and Waite (1973) indicate that the transition occurs between 0.1 and 0.3 cm mean relief with K-band radar at incidence angles of 10° to 80° for smooth playa and alluvial fans in semi-arid regions. The transition for L-band radar was shown to take place in Death Valley, California at a mean relief of 4 to 7 cm at incidence angles of 35° to 45° by Schaber et al. (1976).

The frequency dependence of the complex dielectric constant of terrain takes on two forms. First, the microwave reflectivity of a target is a function of the magnitude of the dielectric constant which is primarily sensitive to moisture content. Thus, the wavelength sensitivity of the reflectivity is mostly that of water which has a marked degree of dispersion across the microwave range (Figure 25-21). A second effect of frequency upon the dielectric constant is that of penetration; in general, the longer the wavelength, the greater the depth of penetration. Therefore, longer wavelengths will tend to have a smaller surface-scatter component of backscatter and a great probability for a subsurface component.

In practical terms, these relationships mean that, for a wavelength of 1 cm, most surfaces appear rough, while for $\lambda = 1$m, correspondingly few surfaces appear rough. For $\lambda = 1$cm penetration is negligible, whereas for $\lambda = 1$ m it could range from 0.3 m in wet soils to 1 m or more in dry soils. Consequently, a multifrequency system may enable one to detect differences in vegetation density and, at longer wavelengths, variations in soil moisture properties. However, for some geologic environments, (e.g. Death Valley) Daily et al. (1978) have suggested that multipolarization can provide more data than multifrequency. Little empirical work has been done with imaging systems to document these possible effects. However, one can be referred to Holmes (1979), who investigated the data variability of multichannel radar, Schaber et al. (1976), and Daily et al. (1978).

INCIDENCE ANGLE

The magnitude of the normalized backscattering-echo area generally increases with decreasing incidence angle, but as the surface becomes very rough (in terms of wavelength), it becomes independent of the incidence angle. At first glance, this appears to contradict the behavior predicted by the Rayleigh criterion; however, it is due to a change of reference. As stated, the Rayleigh criterion predicts an increased width of the reradiation pattern (centered at the specular angle) due to either increased surface roughness or decreased incidence angle. As the incidence angle increases, like-polarized backscatter from a surface shows increasing dependence on the complex dielectric constant, ϵ, and decreasing dependence on the surface roughness, Γ_1. For a mathematical treatment of idealized target geometries and incidence angle as relating to backscattering see the Radar Interaction section of Chapter 4.

The backscattering-echo area, is a measure of the power return in the backscatter direction. For non-isotropic surfaces, the return will increase rapidly as the backscatter angle approaches the specular angle. For horizontal surfaces this condition occurs at normal incidence or zero-degree angle of incidence. Thus, in its measurement the effect of increasing phase difference (roughness) with decreasing incidence angle is offset by the increase in amplitude due to approaching the specular angle as the incidence angle is decreased. The increase in amplitude at decreasing incidence angles is accompanied by greater sensitivity to incidence angle, that is, the backscattering function of incidence angle becomes steep (Figure 25-22). The magnitude of the backscattering cross-section is, therefore, very sensitive to small changes in slope for small incidence angles. It should be noted in Figure 25-22 that, for the two volcanic surfaces, the backscattering functions of inci-

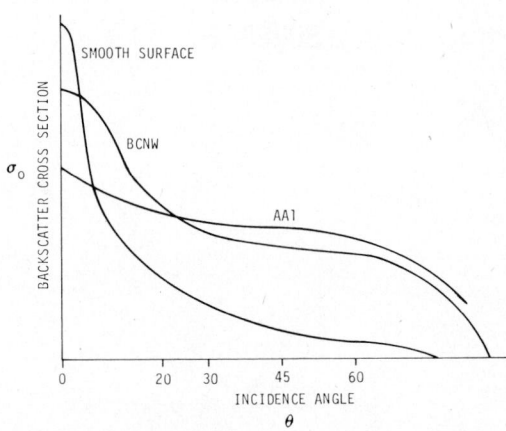

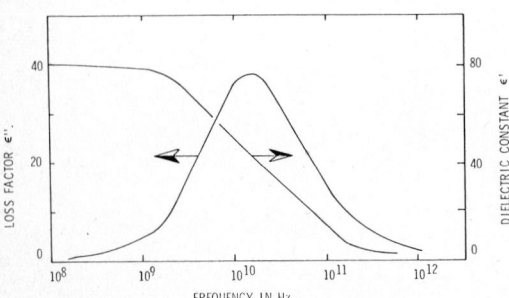

Fig. 25-21. Dielectric constant and loss factor of water at microwave frequencies. (From Biggs, 1970.)

Fig. 25-22. Backscatter cross section vs. incidence angle for a smooth surface and two volcanic units. (Stewart, et al., 1980.)

dence angle reverse in going from Seasat ($\phi = 20°$) to aircraft ($\phi = 45°$) imaging geometries.

From the data of Figure 25-22, it may be seen that the variation with incidence angle for a smooth surface may exceed 25 dB. Figure 25-23 shows data for an extremely rough target, three-foot tall alfalfa and grass, where the return is virtually independent of incidence angle. From these data, it is clear that radar reflection may change considerably depending on the incidence angle at which it is viewed and the surface roughness (Waite and MacDonald, 1971).

ASPECT ANGLE

Aspect angle (θ) or terrain slope, also affects σ_0 the radar backscattering cross-section. In addition to affecting the influence of the incidence angle, surface roughness, and the complex dielectric constant, the aspect angle influences the geometric distortions inherent to radar imagery. Near the flight path, slopes of 20° facing the radar, at 20° depression angle (β), collapse to a point, whereas the opposite slope away from the radar is seen at near grazing incidence or is shadowed heavily. Neither effect is desirable. The near side slope loses all range discrimination and can produce a return too large for the system to accommodate. Steeper slopes will produce layover, in which ridge tops are seen (and recorded) before valley floors. The steeper the terrain, the more difficult it is to use radar for delineation of drainage nets and for determining geologic structure. The effects of slope in moderate to steep terrain with sparse vegetation thus can have an overwhelming effect on the backscattering cross-section and therefore very little information pertaining to surface roughness or dielectric properties may be detectable. In forested areas and other areas covered by moderate to tall stands of vegetation, the effect of the aspect angle on the backscattering cross-section is much less pronounced as implied by the previous discussion of the influence of incidence angle on the return from vegetated areas.

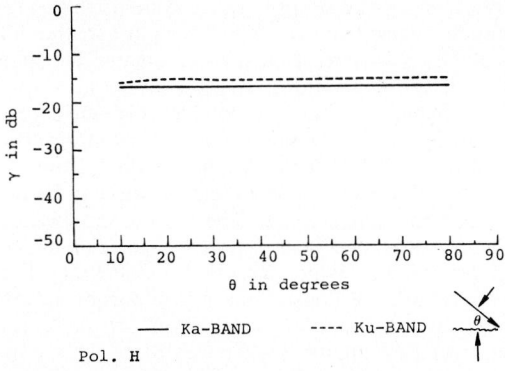

Fig. 25-23. Variation of radar return with incidence angle for an extremely rough surface. (From Cosgriff, Peake and Taylor, 1960.)

POLARIZATION

Radar imaging-systems commonly transmit plane-polarized EMR horizontally. When this EMR interacts with the surface material, it is depolarized and rotated to varying degrees. The horizontal and vertical components of this rotated signal may be separately received in different antennas. Care must be taken to isolate the cross-polarized return from system feedthrough of the like polarized return to the cross-polarized measurements. This care may take the form of system design as well as physical and electrical separation of the antennas. Numerous studies have documented the value of multiple polarization in selected situations.

The vertical-polarized component of the backscattering cross section consists of both surface and subsurface scatter. While there is a present controversy as to the significance of the surface component of the cross-polarized return, experimental evidence indicates trends, in good agreement with theory, that depolarized backscatter is much less dependent on surface roughness and incidence angle than like-polarized backscatter (Blanchard and Rouse, 1980). This evidence assumes that the electrical characteristics of the material are similar to those of natural land targets. Blanchard et al. (1981) claim that cross-polarized radar return for such target classes is purely a function of subsurface geometry and electrical properties (e.g. complex permittivity).

Measurements with vertical polarization are shown in Figure 25-24. These data indicate significant dependence of the depolarized return on incidence angles, as do most reported depolarized measurements. The above-referenced studies suggest that this dependence is an artifact of the measurement technique and this accounts for the discrepancy between theory and experimental cross-polarized data. However, the relative difference due to polarization may be seen by contrasting these data with those shown in Figure 25-25 using horizontal polarization.

Figures 25-26 and 25-27 represent wetland and marine environments, respectively. Note the sharp contrasts between polarizations. The contrast serves foremost as an indicator of terrain variations, the importance of which can only be assessed in relative terms. Whether or not they are "significant" variations, worthy of further investigation, depends largely on whether they are long or short lived in the environment with respect to the constantly changing water conditions. It is significant that in hydromorphic environments dominated by low-growing vegetation the HH polarization gives the crisper image. This may be due to the contrasts between specular and diffuse scatter from the like-polarized component, primarily a function of the surface properties.

In the absence of broad-bandwidth radar systems, the ability to transmit and receive in both

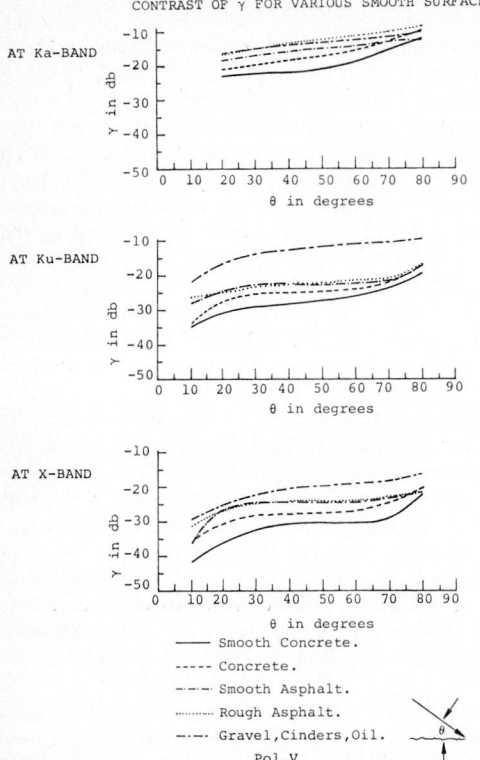

CONTRAST OF γ FOR VARIOUS SMOOTH SURFACES:

——— Smooth Concrete.
----- Concrete.
— — — Smooth Asphalt.
.... Rough Asphalt.
—··—·· Gravel,Cinders,Oil.
Pol.V

Fig. 25-24. Variation of radar return with roughness and incidence angle using vertical polarization. (From Cosgriff, Peake and Taylor, 1960; courtesy Ohio State University.)

horizontally and vertically polarized signals is crucial. In analogous terms it provides a "multispectral" dimension to an otherwise monochromatic system. Before this capability can be fully exploited, however, an adequate theory should be developed to determine reasons for the variations in the like- and cross-polarized returns for a greater range of terrain and material classes and the best environmental conditions for observing these differences. In the long term, for example, it may well be that polarization effects from vegetation are less informative than changes in the angular, frequency, and temporal domains. In a practical sense, however, it may be infeasible to implement and use in some cost-effective way, such angular differences as may exist. Additional frequencies would certainly be more expensive than dual polarization at a single frequency. Thorough theoretical and empirical studies still are required in our judgement.

COMPLEX DIELECTRIC CONSTANT

The electrical properties of a surface, as expressed in the complex dielectric constant (ϵ), critically affect radar return as they strongly influence the absorption and propagation of electromagnetic waves. The real part of the complex

dielectric constant is the permittivity of the medium and the imaginary part is usually expressed as the conductivity of the medium. These properties are strongly dependent on the liquid-water content of the surface, the structure of solid water (ice) and the frequency of the energy. The complex dielectric constant varies almost linearly in response to contained liquid moisture per unit volume, except at the lowest moisture when this quasi-linearity breaks down as the polar water molecules are constrained at high tension. Penetration is greatest and reflection least with low moisture contents. Conversely, penetration is least and reflection is greatest when moisture is high in snow, ice and soil.

The behavior of both the structure and the complex dielectric constant in water in its various phases is a striking feature of the microwave spectrum. Equations describing the dielectric properties of water, ice, soils, and vegetation are presented and discussed in Chapter 4. Due to the polar nature of the molecules and consequently the short relaxation time, the dielectric constant of water is quite sensitive to frequency and temperature across the microwave spectrum.

Figure 25-21 shows the variation of the real and imaginary portions with frequency.

Several mixing formulas have been proposed for estimating the dielectric properties of mixtures of solids (soil or ice) and liquid water. None of these has proven reliable in calculating values comparable to those measured over a wide range of liquid-water contents. Empirical models have been used to estimate the electric properties of moist soil (e.g. Wang and Schmugge, 1980) since a number of measurements have been recently made of their variation.

Reflectivity at a smooth interface is governed by Snell's law and the Fresnel reflection coefficients which describe the variation with incidence angle and the refractive index as expressed by the complex dielectric constant. At any given angle the magnitude of the reflection is proportional to the magnitude of the dielectric constant. Over the frequency range of most radar systems this magnitude ranges from 20 to 80. By contrast, the dry dielectric constant of most natural materials, such as soil and vegetation, ranges from about 3 to 6. This disparity in relative complex dielectric constant illustrates the importance of moisture content on the reflecting properties of the terrain.

The effects of water content in soil may be observed in Figure 25-28. The region shown is a backswamp area near Baton Rouge, Louisiana, investigated by Waite and MacDonald (1971). The pattern of dark lines in the upper portion of the image is a natural levee system consisting of dryer materials at slightly higher elevation. This contrast between the differing soil moisture content in backswamp and levee was observed through a substantial volume of defoliated deciduous vegetation.

The effect of water content on the return from a

CONTRAST OF γ FOR VARIOUS SMOOTH SURFACES:

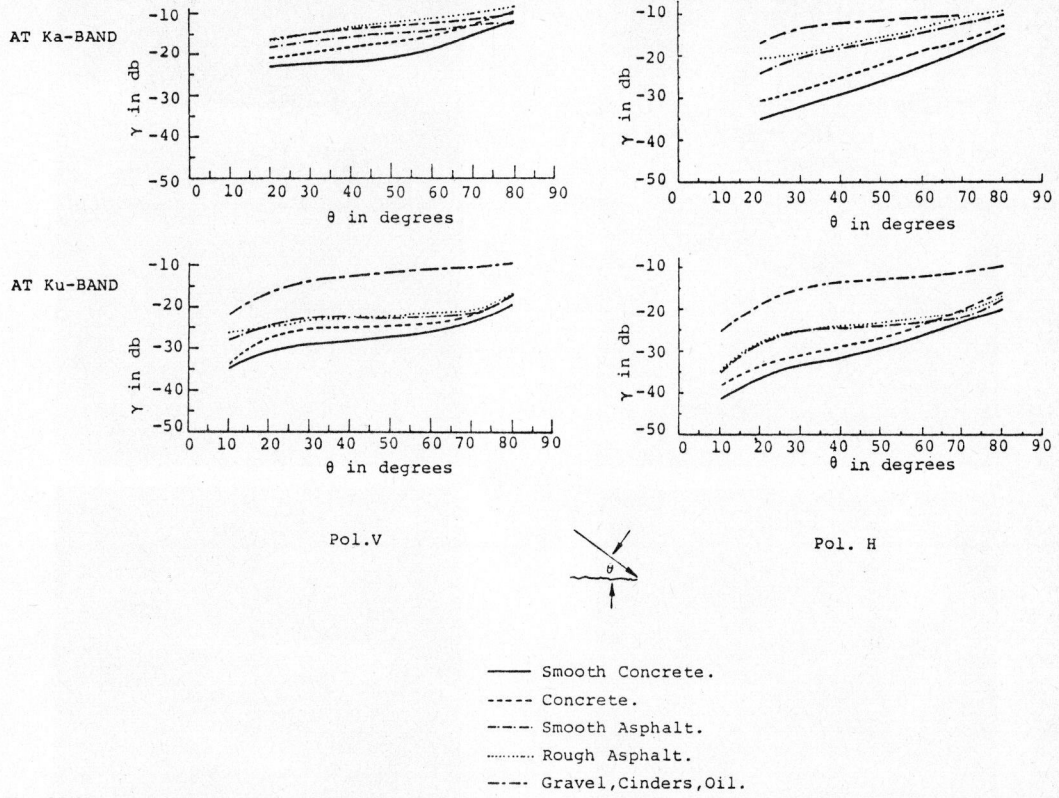

Pol.V   Pol. H

——— Smooth Concrete.

----- Concrete.

—·—· Smooth Asphalt.

·········· Rough Asphalt.

—··— Gravel,Cinders,Oil.

Fig. 25-25. Comparison of horizontal and vertical polarization for various surfaces and incidence angles for Ka and Ku bands. (After Cosgriff, Peake and Taylor, 1960.)

dense volume of foliage may be observed from Ohio State data shown in Figure 25-29. The surface measured is a diffuse scatterer at all times and the variations with season are attributable principally to the moisture content of the vegetation itself.

A second major effect of the complex dielectric constant is related to the conductivity of the material. That is, the loss or attenuation of the microwave energy is a function of the conductivity of the material and the frequency of the energy. In general, the higher the frequency, the greater the attenuation in the material and hence effective penetration is less. This may have a marked effect on the return from vegetated surfaces. At higher frequencies the return is from the vegetation canopy while at lower frequencies, with greater penetration capability, the return may be primarily from the surface beneath the vegetation.

SURFACE ROUGHNESS

Surface roughness (Γ_1) is the dominant factor in determining the amplitude of the return and, hence, the tone of the images. The energy incident upon a surface is partially reflected in either a specular or diffuse manner, depending on the roughness of the surface.

The Rayleigh criterion is most often used and this is illustrated in Figure 25-30. Consider rays 1 and 2 incident on a surface with irregularities of height, h, at a grazing angle, θ. The path difference between the two rays due to the irregularity is

$$\Delta r = 2h \sin\theta \qquad (25\text{-}10)$$

Thus the phase difference is

$$\Delta \gamma = \frac{2\pi}{\lambda} \Delta r = \frac{4\pi h}{\lambda} \sin\theta \qquad (25\text{-}11)$$

If the phase difference is small as in the case of a plane surface, the two reflected rays will reinforce one another. If the phase difference is near $\pi/2$ radians the two reflected rays will be in opposition and will cancel in the specular direction. Lord Rayleigh arbitrarily selected the midpoint between these two extremes to be the dividing line between smooth and rough surfaces or specular and diffuse reflection. That is, a surface is considered smooth if:

$$\Delta \gamma < \frac{\pi}{2} \qquad (25\text{-}12)$$

or

$$h < \frac{\lambda}{8\sin\theta} \qquad (25\text{-}13)$$

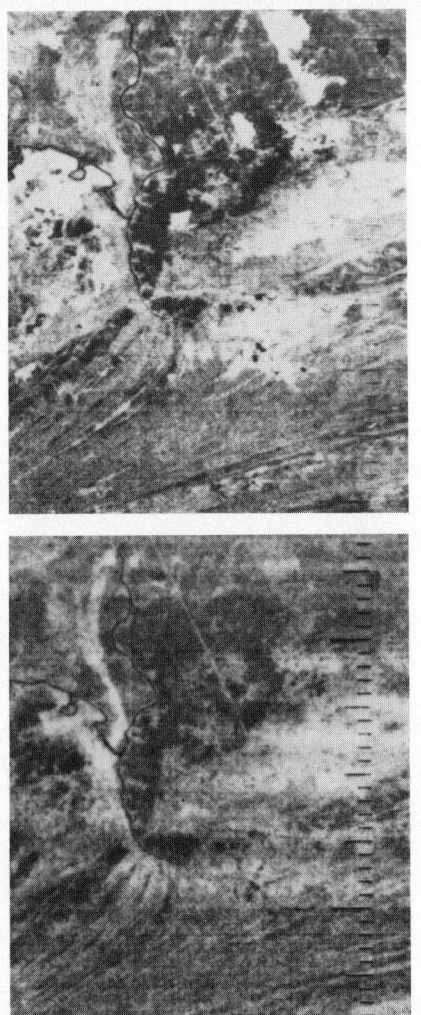

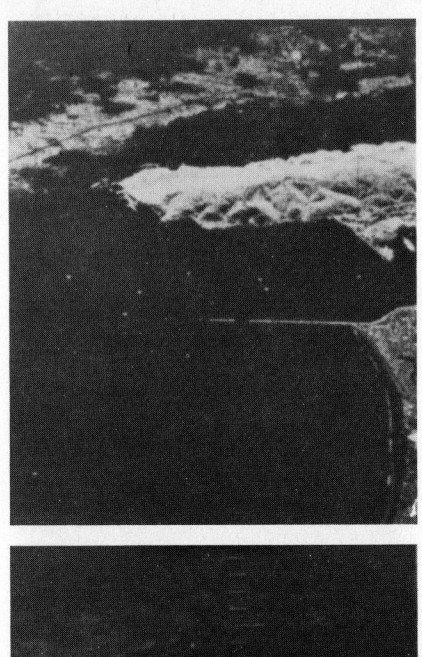

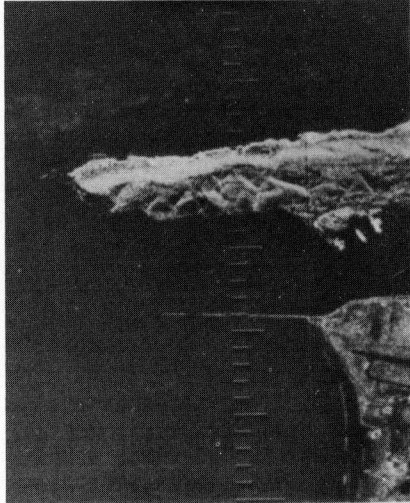

Fig. 25-26. Example of like-(HH) and cross-(HV) polarized Ka-band imagery from Texas Coast. Scale approximately 1:500,000.

Fig. 25-27. Example of like-(HH) and cross-(HV) polarized Ka-band imagery of California giant kelp. Scale approximately 1:500,000. For Figs. 25-26 and 25-27: Grey tones can be compared only within a single image due to uncalibrated changes in receiver gain but relative tone shifts can be compared between like and cross images. The greatest range of tones usually is associated with scenes having complex patterns of vegetation moisture, as in Fig. 25-26. Height and cover are equally important in forested areas, especially at small incidence angles. Kelp is visible in Fig. 25-27 because the small pear-shaped "floats" increase surface roughness and hence the backscatter coefficient against surrounding water which is a near specular reflector (black) at small incidence angles.

The apparent roughness of a surface is a function of the wavelength and incidence angle of the radiation. The same surface may appear smooth at low frequency (long wavelength) while appearing quite rough at a higher frequency. Also, because the apparent roughness is a function of the relative phase difference, a given surface will appear smoother at angles near grazing.

Detailed mathematical descriptions of the surface reflection as a function of the surface and energy properties are given in Chapter 4 and Chapter 9. Generally perfectly smooth and perfectly rough surfaces are considered since a surface with roughness character that is in between presents a rigorous theoretical problem. However, approximate descriptions of the surface scattering-characteristics can be obtained with functions describing the statistical properties of the surface convolved with the Rayleigh and Bragg scattering equations (Stewart et al., 1980).

As a surface becomes rough at the wavelength of the incident radiation, reradiation becomes diffuse, with energy present at angles other than the specular. An extremely rough surface becomes almost isotropic, reflecting equally well at all angles.

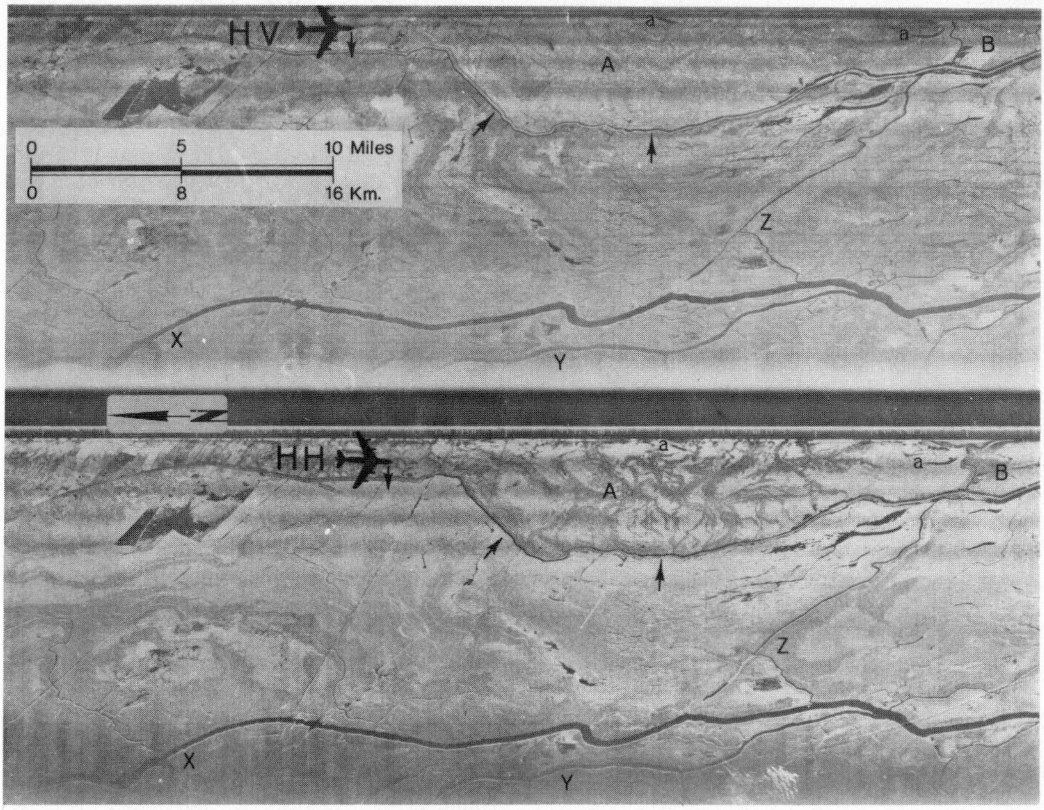

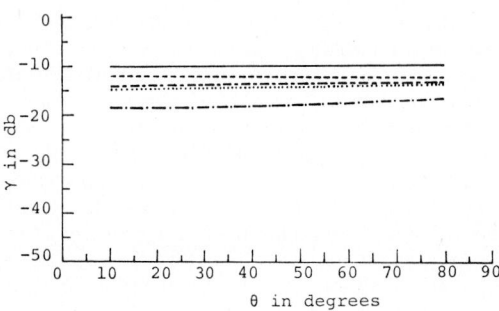

Fig. 25-28. Radar return variations due to difference in soil-moisture content in a backswamp area near Baton Rouge, Louisiana. (From Waite and MacDonald, 1971.)

In practice, it is found that the normalized backscattering echo area (X) of natural surfaces may vary over 40 db as a function of the surface roughness. This is illustrated by the data of Cosgriff, Peak, and Taylor (1960) shown in Figure 25-31. With models describing the effects on the backscattering cross-section, σ, due to surface particle-size and roughness distribution, it is possible to separate areas of different surface particle-size/roughness based on σ.

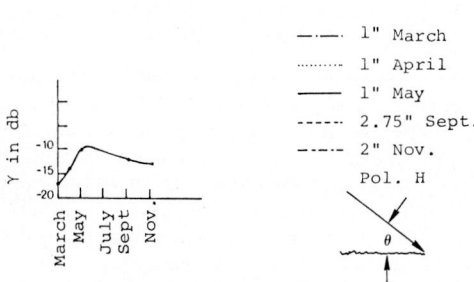

Fig. 25-29. Seasonal changes of grass at Ka-band. (From Cosgriff, Peake and Taylor, 1960.)

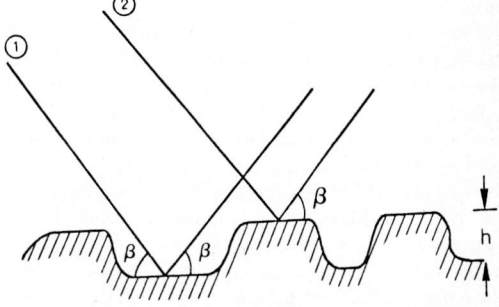

Fig. 25-30. Derivation of the Rayleigh criterion.

CONTRAST OF γ FOR VARIOUS SMOOTH SURFACES:

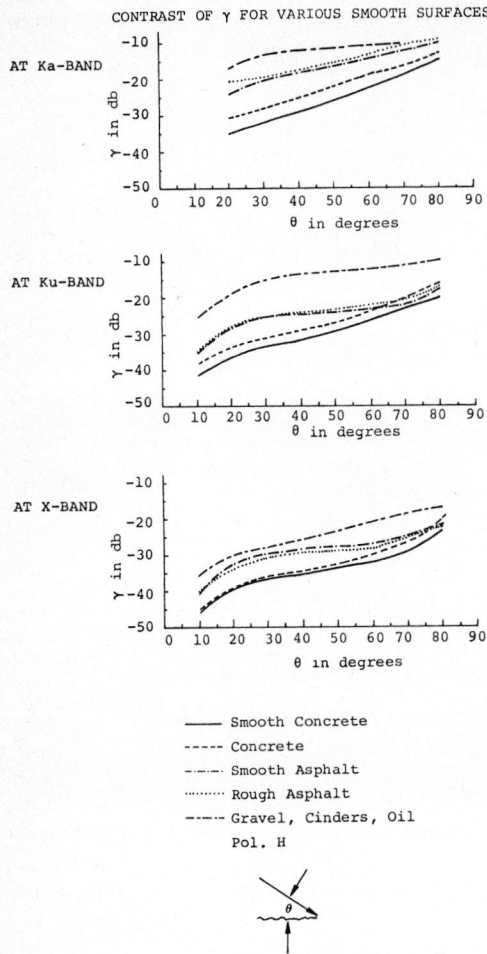

——— Smooth Concrete
----- Concrete
—·—·— Smooth Asphalt
··········· Rough Asphalt
—··—··— Gravel, Cinders, Oil
Pol. H

Fig. 25-31. Variation of radar return with roughness and incidence angle, using horizontal polarization. Note that θ is depression angle. (From Cosgriff, Peake and Taylor, 1960.)

Orientation of regular surface-roughness patterns (e.g. row tillage) has been shown to significantly influence radar backscatter (e.g. Ulaby and Bare, 1978). Over bare or sparsely vegetated cropland the direction of this orientation has a significant influence on the backscatter. This effect is sensitive to aspect angle and frequency and the phenomenon is described by facet models of radar backscatter.

The effect of microscale roughness (Γ_1) on backscatter in the Ka to X-band (0.86−3 cm) region is such that most surfaces act as mixed isotropic scatterers and specular reflectors, and contain both specular and diffuse reradiation components in the backscatter.

Surfaces which are predominantly specular are still-water surfaces, paved roadways, and asphalt airport runways. Such surfaces give very low radar returns except at near vertical incidence. Forests tend to be fairly isotropic scatterers and show only modest angular dependance in the return over the range 10° to 50° from grazing incidence.

SUB-SURFACE ROUGNESS

The effect of sub-surface roughness of a second layer (Γ_2) is normally visually important only in the case of cold dry fresh snow or clean dry sand over other materials. Examples may be found where it is clear that the bulk of the return is coming from the buried layer. Except in extreme cases, it is doubtful that the influence of second layers on the radar return can be effectively evaluated.

VOLUME SCATTERING

The situation encountered in natural terrain targets is very complex. Here, one must consider the presence of vegetation, which appears as a random volume of scattering facets or as a cloud of water droplets (Fung and Ulaby, 1978) lying above the surface. In addition, inhomogeneities in the soil may likewise create a volume of scattering elements beneath the surface.

A portion of the return from forests will be volume scattering (V) derived from multiple-path reflections from leaves, twigs, and trunks. This may be a contributor to the relatively isotropic backscatter from such "surfaces". In addition to scattering from tree trunks, return from the ground layer also becomes possible with longer wavelengths. Various degrees of depolarization are thus likely with multi-frequency systems and this may prove a valuable discriminant in mapping natural plant communities.

In practice, the macro surface-roughness, volume scattering from soil and vegetation, and reflection from shallow buried layers all contribute to the complexity of the target structure and are part of the total surfaces "roughness". This assumes that the scattering from vegetation, surface, and subsurface are similar in nature and all affect principally the reradiation pattern—a seemingly reasonable assumption so long as only the like-polarized component of the return is considered. However, when like- and cross-polarized data are available, it may be possible in some cases to distinguish between the volume and surface components of the return.

TEXTURE AS A DISCRIMINANT ON RADAR IMAGES

The average grey scale or tone on the radar image as seen above is related to σ_0, the average backscattering cross-section. Tone is a very important characteristic in radar interpretation. Equally important for identification and mapping (especially in geologic and vegetation mapping) is the texture of the radar image which is used to delineate boundaries of homogeneity.

Photographic texture is defined as the frequency of tone change within an image produced by an aggregate of features too small to be clearly discerned individually on photographs. The scale of the photographs obviously has an important bearing on this definition of texture (Ray, 1960). Radar-imagery texture, in a similar fashion, is a comparative feature applying to a single radar resolution. Multiple-resolution radar systems could be employed where texture as a function of resolution would be used as a discriminant between imaged objects. Because the Seasat SAR system has a low dynamic range, only six grey levels in the imagery can be effectively utilized for discriminating small variations in surface texture and such discrimination is further limited to the near-range of the image in topographically flat areas.

There are three components of texture on radar images—micro, meso, and macro texture. Micro texture is inherent, or system, texture arising from the sampling statistics of a fading signal (Moore, 1970) and is related to the resolution in range and azimuth as well as the number of independent samples. All relatively smooth surfaces which are adequately recorded on the image will exhibit fine texture of this type.

Meso texture is produced by spatial inhomogeneities on the order of several resolution cells or more across (several units of micro texture) and is most striking where there is sharply juxtaposed relief on top of a forest with emergent species or where a highly varigated soil-vegetation pattern exists in a marshy environment. Meso texture on close inspection shows shadowing, and adjacent bright spots which are clear indications of the presence of emergent species.

Imagery for an area near Horsefly Mountain, Oregon, shown in Figure 25-32, is used to illustrate the characteristics of mesoscale texture.

Three general regions can be readily identified. The mottled pattern of high and low return on the left is a typical swamp or marsh vegetation where areas of standing water (black) intersperse with dense, high-water-content grasses and reeds (white). The broad region of medium-gray tones in the center of the image can be segregated on the basis of subtle textural changes into two types. The coarse (sponge-like) texture represents a well-developed Ponderosa pine forest with trees of uneven heights and ages. A more subdued (blurred) texture results from areas where past burns are in various successional stages of recovery. Finally, the dark region on the right of the image is fir forest. It differs from Ponderosa in tone, more than texture, for reasons that are not yet clear but which are thought to be related to such factors as tree shape, moisture content, associated vegetation, and proportion of bare ground.

Micro texture tends to be random (as would be expected from the statistics of a fading signal), while meso and macro texture are spatially organized. The organization of meso texture relates to the distribution of structural elements within plant communities. Such distributions are not randomly organized, as they are at least partially related to edaphic conditions and species competition. Thus, the nearest-neighbor probabilities of juxtaposed gray-scale values show bimodality in forest texture and unimodality in grasslands: the likelihood of a given gray value being juxtaposed to another given gray value is the same for all gray values in grassland, but in forests, lighter and darker tones are paired because of the presence of adjacent illuminated and shadowed trees or clumps of trees, above the general canopy.

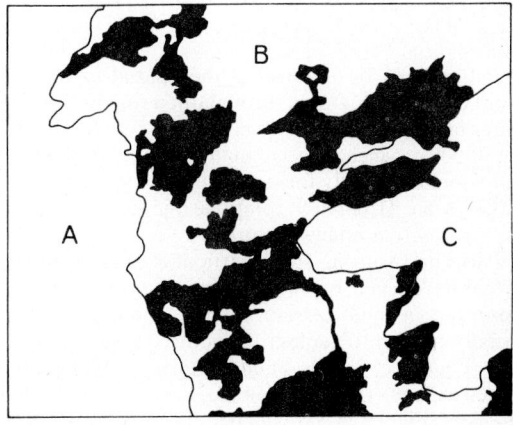

Fig. 25-32. Ka-band radar image illustrating the HH tone and texture renditions of major physiognomic vegetation types. Area A, which appears on the image in mottled black-and-white, is a swampy meadow wherein the lighter tones are wet spots and the darker tones are dry. The extensive area of B is pine forest dominated by *Pinus ponderosa*. It has a medium-gray tone, but is distinguishable by its relatively coarse texture. Area C is a Douglasfir forest (*Pseudotsuga menziesii*) distinguishable by both its dark tone and more or less prominent texture. Chaparral communities, dominated by *Arctostaphylos* are portrayed in solid blocks on the map. They are not immediately apparent on the imagery but are detectable by their finer texture in comparison to surrounding pine forest. Scale approximately 1:250,000.

The organization of macro texture is a crucial element in geologic and geomorphic interpretation on radar images. Macro image texture permits identification and delineation of unit areas contained within boundaries of homogeneity. This concept of texture contrasts with the recognition of imagery pattern (to be discussed later) which refers to an orderly spatial arrangement of geologic, topographic, and vegetation features. Because of the dependence of image texture on the distribution of image tones, this recognition element is less affected by the lack of image calibration than is tone (Barr and Miles, 1970); regardless of the radar system's gain setting, the appearance of imagery texture remains relatively constant from one image to another.

When topographic texture (the degree of erosional dissection) is related to imagery texture, drainage density or erosional characteristics of the terrain dominate. Fine-textured drainage generally correlates with fine-grained sedimentary rocks or unconsolidated sediments (Figure 25-33 k,l; Figure 25-34 a,g,h,l). Coarser textures are usually indicative of coarse-grained sedimentary rocks (Figure 25-33 h, left half; j, right half; Figure 25-34 a,c). Contrasting with the texture of sedimentary strata are igneous rocks which usually have a massive appearance on radar imagery, and these rock units can usually be distinguished by rugged and peaked divides (Figure 25-33 a,b,d,e; Figure 25-34 k, right half). The hummocky, rounded texture recognized in regions of igneous rocks is usually seen in zones which have been nearly base-leveled by erosion (Figure 25-33 c).

PATTERN AND SHAPE AS INTERPRETIVE ELEMENTS ON RADAR IMAGES

Pattern

Pattern can be defined as the spatial arrangement of terrain features throughout a region. Most radar imagery is of a relatively small scale (at most 1:170,000); consequently, many terrain elements that would normally form a distinctive pattern at a larger scale (1:60,000) comprise the components of a distinctive image texture.

Patterns resulting from particular distributions of gently curved or straight lines are commonly observed in most terrain configurations and are usually related to geologic structure (Figure 25-33 g,h; Figure 25-34 d). Such patterns may be inferred as faults (Figure 25-34 b), joints (Figure 25-33 f), dikes (Figure 25-35), and bedding on large-scale photographs and radar imagery; smaller scale imagery may only provide evidence of radar lineaments (anomalous pattern of straight lines). Radar lineaments are categorized as a distinctive pattern which may result from an orderly arrangement of stream segments, slope facets, and moisture and vegetation zones.

Drainage-pattern analysis provides the geologist and geomorphologist with an extremely important geomorphic and structural interpretive technique. Drainage patterns (spatial relationships) generally are surface manifestations of underlying geologic structure and lithology. In addition to drainage patterns, vegetation and soil patterns commonly reflect structural conditions, lithology, and distribution of surface materials.

Shape

Shape can be defined as a spatial form with respect to a relatively constant contour or periphery (Ray, 1960). Because cultural features generally have regular geometric shapes, the radar-image interpreter can usually distinguish between natural and cultural objects. Even though many elements of the landscape have irregular outlines, numerous geologic features can be interpreted by their shape alone. Alluvial fans, volcanic cones, river terraces, many glacial features, and folded strata are but a few examples.

The shapes of radar shadowing may allow the interpreter to infer information about the relative relief of the terrain. For example, the length of a radar shadow can be used for determining the height of some objects, whereas shadow shape may be used for inferring spatial form.

RADAR IMAGE INTERPRETATION

In appearance, high-resolution radar imagery is somewhat like a photograph and the principles of radar-image interpretation are similar to those of aerial-photograph interpretation, but with the modifications appropriate to a device that measures range rather than angle with cross-track dimension. In some applications, digital analyses are not only desirable but necessary in order to obtain meaningful results (for example sea surface-state).

Most radar images have been recorded directly on negative film and then processed photographically to produce the desired scene contrast. Recently it has become possible to convert the film-recorded radar images to digital format by various methods. Some current SAR systems record the raw signal in a digital format. A digital format allows numerical corrections of image geometry, antenna pattern, and speckle noise removal.

Because most radar imaging systems use a single frequency (monochromatic) source, even images of homogeneous scattering areas, such as single-crop fields, appear "speckled", although photographs of the same area appear uniform in intensity (Moore and Thomann, 1971). The speckled effect on radar imagery is related to each resolvable patch on the ground which scatters back to the radar from many facets resulting in a spot of variable intensity on the imagery. This speckling generally restricts the amount of magnification that is possible without getting image

0   10 Km

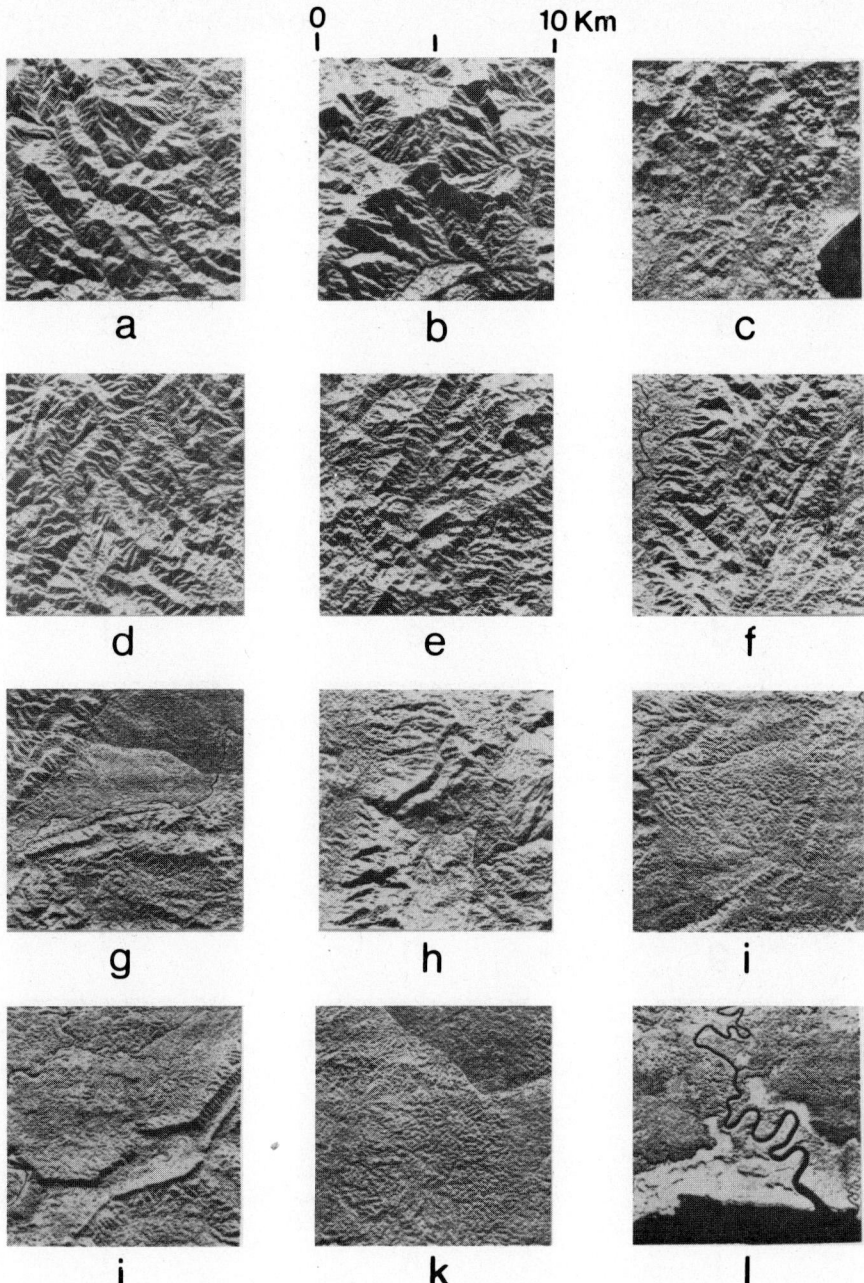

Fig. 25-33. Topographic texture from radar imagery, Darien Province, Panama.

break-up. Break-up is the point at which textural elements become recognizable components of differing spot size, and resolution cells become apparent.

Various speckle-removal techniques have been used (e.g. Chavez, 1979); however some questions remain as to the the loss of potential information that results in the product. That is, speckle noise might be used as an interpretative element. Chavez (1979) suggests the use of a "speckle-noise image" with the processed radar image to check for spatial speckle patterns which might be indicative of certain environmental targets.

Using 15 m as a common resolution factor compatible with most commercial radar imaging-systems, the limit of useful magnification the imagery can undergo without speckle removal, and still be useful, is approximately 10X. For viewing positive or negative image-transparencies, standard roll-type light tables equipped with zoom-

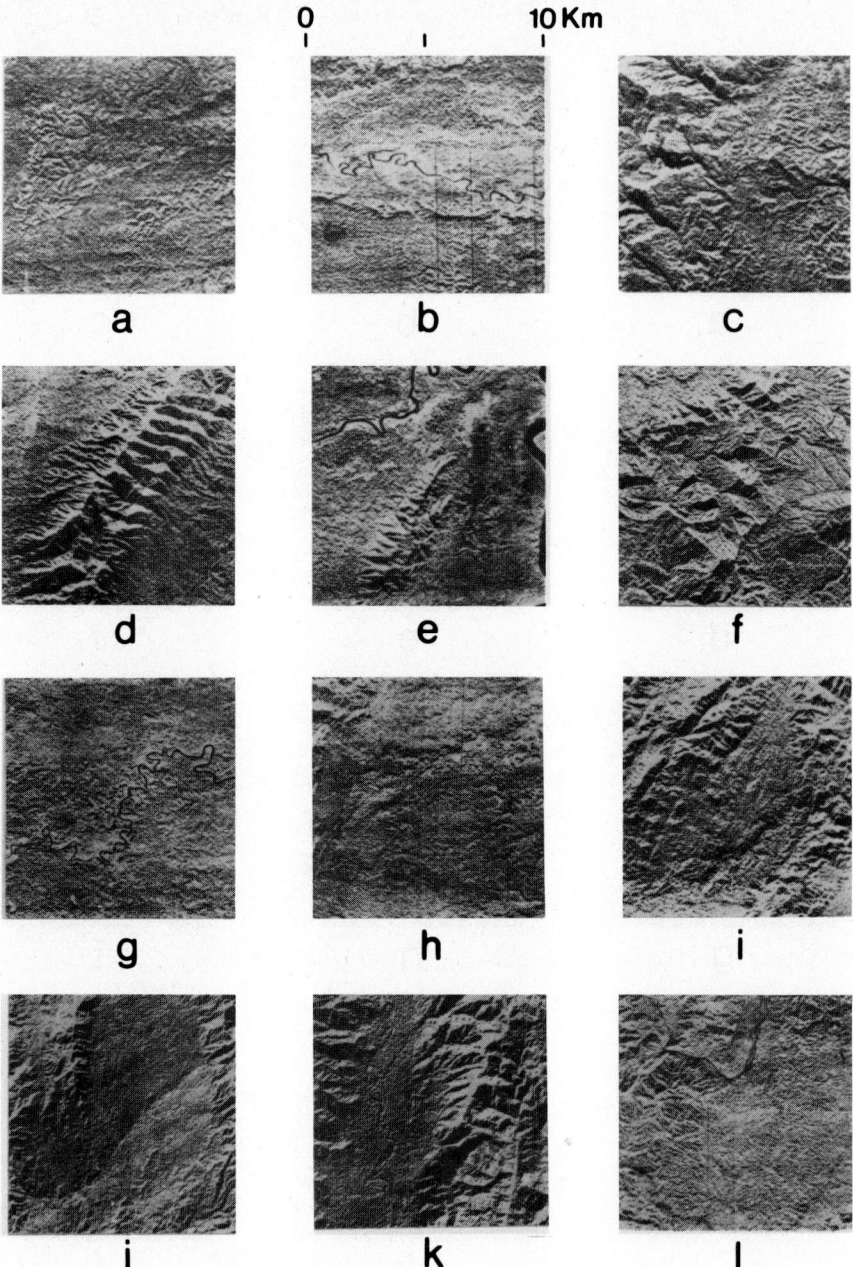

Fig. 25-34. Topographic texture from radar imagery, Darien Province, Panama.

microscopes are quite adequate. For stereoscopic viewing however, paper prints are preferable to image transparencies. If a radar mosaic is available for a particular study area, regional geology, physiography or vegetation may first be interpreted directly from the mosaic. Those regions on the mosaic which necessitate more detailed examination can then be viewed stereoscopically with a standard 2X pocket-type stereoscope or a 4X binocular mirror-stereoscope. Stereoscopic interpretation is facilitated by using the combination of radar mosaic as the base map, and paper positive prints of overlapping strips from adjacent flight lines. Monoscopic viewing of individual imagery strips can be facilitated by a table-mounted, rear-projection viewer having a variable light source and variable magnification (0.5× to 6×).

Presently, virtually all of the available radar imagery is produced by systems which are generally not calibrated. That is, these systems can only

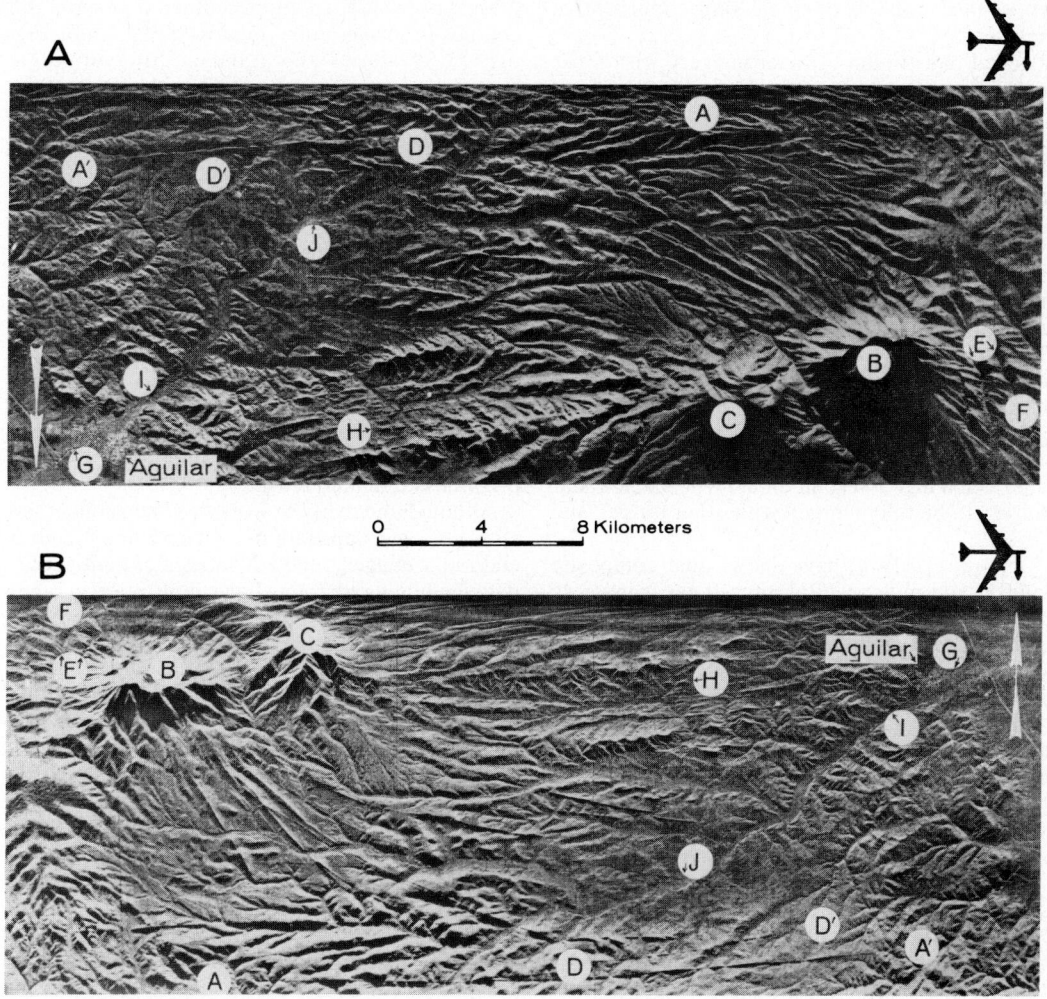

Fig. 25-35. Spanish Peaks, Colorado. (A) Look-direction, North, (B) look-direction, South.

measure the relative image backscattering cross-section which may or may not be repeatable in subsequent images. This fact reduces the validity of image interpretation done automatically by computer. However, progress has been made in such automatic interpretation for targets which have especially constant or high or low returns (e.g. sea ice and lakes) (Bryan et al., 1977).

The following sections discuss a series of examples of techniques for obtaining information from radar imagery and factors which influence the return from a particular target. For specific descriptions of microwave-interaction characteristics with the earth surface see Chapter 4.

RADAR IMAGERY MERGED WITH OTHER IMAGERY

Composites of radar imagery and other types of imagery, notably Landsat, have been recently analyzed with the expectation of greater advan-tages in the identification and delineation of surface features than those with single-sensor imagery. Early combination of radar and Landsat imagery was accomplished by the use of photographic techniques (Harris and Graham, 1976). The purpose of this early work was to provide an apparent increase in resolution and to better portray relief with the addition of radar imagery to Landsat scenes.

Recent work has concentrated on the synergism of the ability to provide information about surface-roughness character, topography, and dielectric from radar and the sensitivity of Landsat to surface composition. The combination of radar imagery and Landsat imagery has been achieved digitally by which the problems of geometric displacement and layover in the radar imagery can be somewhat alleviated depending on the terrain (e.g. Chavez and Sanchez, 1982), naturally being easiest in areas of flat or gently-sloping terrain. In addition, digital processing preserves much of the dynamic range inherent to each sensor.

Digital registration of radar imagery with other imagery is made with the selection of control points that can be identified on both sets of imagery, visually, and with image coordinates. Mathematical relationships are then used to describe the variation between control points on the imagery and spatially manipulate the images into registry with a master image or digital terrrain data-base. Control points should be more densely distributed where topographic relief produces serious geometric distortions in the radar imagery. The large numbers of control points required for good registration in areas with high frequency relief may prohibit adequate combination of the imagery (Chavez and Sanchez, 1982), though with experience and appropriate algorithms it is anticipated that a digital terrain-model will enable radar imagery to be fully merged with other images and maps.

Daily et al. (1978) have shown that composite Landsat and multispectral aircraft radar-imagery can produce nearly complete discrimination of surface geologic units and synoptic information on composition and surface texture. The radar data used were acquired over nearly flat terrain in Death Valley to reduce the influence of topography on image-tone variation. Similar results in the form of enhanced lithologic discrimination and structural analysis have been reported by Stewart et al. (1980) using Landsat and Seasat data.

Topographic enhancement on Landsat imagery is maximized with low sun angles, but the brightness dynamic range is maximized at high sun angles. The same is generally true with incidence angle in radar images. Sun elevation effects on a composite of Landsat and radar imagery must therefore be considered in order to produce the optimum combination for a particular application.

Chavez and Sanchez (1982) digitally registered a Seasat SAR image to Landsat images with high and low sun angles to determine which combination is better in terms of surface structural enhancement and brightness dynamic range. The low sun angle Landsat image was extracted from a digital mosaic of images recorded in fall, 1973 and band 7 is shown in Figure 25-36. Band 7 of the high-sun-angle Landsat image is shown in Figure 25-37 and was recorded in spring, 1974. The high-sun-angle image was used as the control base, and the low-sun-angle image and the Seasat image were registered to this base. Figure 25-38 shows the Seasat image after registration.

The tones of certain features are reversed between the Landsat images (Figures 25-36 and 25-37) and the Seasat image (Figure 25-38). An example is the lava flow associated with the volcano in the lower central portion of the images. The lava flow has a dark tone on the Landsat images while on the Seasat image it is bright due to its blocky texture and consequent high backscatter. The result of the tonal reversals of some features has produced intermediate tones on the composite image (color Figure 25-39). Color Figure 25-39 shows the Landsat low-sun-angle false-color image digitally combined with the Seasat image. Chavez and Sanchez (1982) note that the soil exposures have been darkened, but the dark vegetation has been brightened by the addition of the Seasat image to the Landsat images. They conclude that different materials can be distinguished better in the composite image than in any one of the images used separately.

Since Seasat had a steep depression angle the brightness dynamic range was near the maximum. If it desired to increase the dynamic range of a composite, a high-sun-angle Landsat image should be chosen. If structural enhancement is desired a low-sun-angle Landsat and Seasat image composite is better while retaining much of the brightness dynamic range.

Although much of the work has been concerned with geologic applications of combined Landsat and radar imagery, the advantages of multisensor data including radar may have application in other earth-science fields. For example, certain features in hydromorphic environments are enhanced by the combination of radar imagery with Landsat imagery (Bryan, 1981). In addition, the suggestion has been made to combine radar imagery with passive microwave imagery to produce multisensor microwave-composites.

INTERPRETATION OF SURFACE WATER

As noted earlier, the structural and dielectric properties of free water exert great influence on radar backscatter, and hence on image interpretation. Backscatter from a water surface is almost completely controlled by the surface roughness on the scale of the wavelength used, and by the tilt of the water surface due to the presence of larger scale waves or swells. In general, the rougher the surface in terms of wavelength, the more diffuse the return on the radar image. Radar is thus capable of detecting surface and near-surface phenomena which may affect the surface roughness directly or indirectly (Elachi, 1980). Surface waves, surface wind-shear patterns, currents and eddies, and wind or oil slicks are examples of these phenomena. The variation in the return from specular (dark) to diffuse (bright) depending on the surface condition, can lead to confusion in the delineation of shorelines especially in environments in which land targets may have a similar return in terms of both tone and texture. In such situations, adequate interpretation requires previous knowledge on the part of the interpreter, and if possible, sequential data.

A number of authors have suggested the use of radar to detect fresh-water bodies because of their low return when calm, sharp edges, and contrast with the surrounding terrain. The detection of water bodies in areas of moderate relief may be difficult since other areas of the

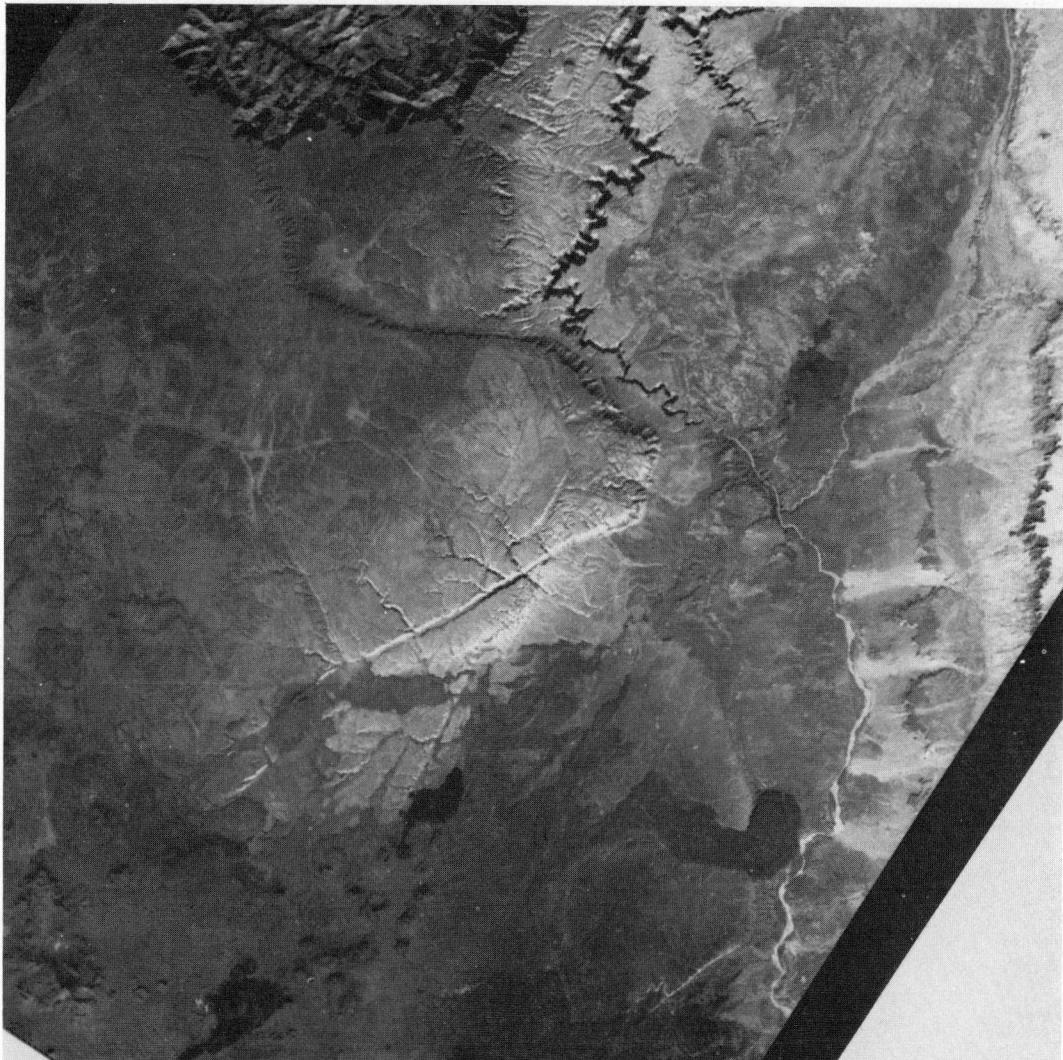

Fig. 25-36. Low sun angle Landsat band-7 partial image, Northern Arizona. (Chavez and Sanchez, 1982.) Scale approximately 1:500,000.

imagery are also black, such as shadows. Roswell, in a University of Kansas unpublished report, found that in well-drained lowlands, all lakes with a surface area greater than 3 ha were detected, and in fact, that the detectability of lakes is primarily related to size and secondarily to environment. Bryan et al. (1977) demonstrated that lakes of the Alaskan North Slope could be efficiently mapped from uncalibrated radar imagery with automatic interpretation technique due to the consistent low return of smooth water. The imaging system used was L-band (23 cm wavelength) with a resolution of about 25 m. The process of interpretation was interactive between the user and the processing system and not entirely programmable (Bryan, 1979).

Figure 25-28 shows that the water surfaces of streams give virtually no radar return. In vegetated terrain, the surface water in streams is less likely to become rough due to wind action. The radar in Figure 25-28 views the streams at incidence angles between 15° and 75°. The reflection from the (smooth) water, although quite high due to the high dielectric constant of water, is concentrated about the specular angle with very little energy returned in the backscatter direction. The surrounding terrrain, although composed of less reflective material, is much rougher; thus, reflection is much more nearly isotropic and has a substantial backscatter. Figure 25-28 also illustrates the effect when water is an interstitial component in other materials. The area shown is part of the deltaic plain of the Mississippi. Natural levees form the framework of the region while

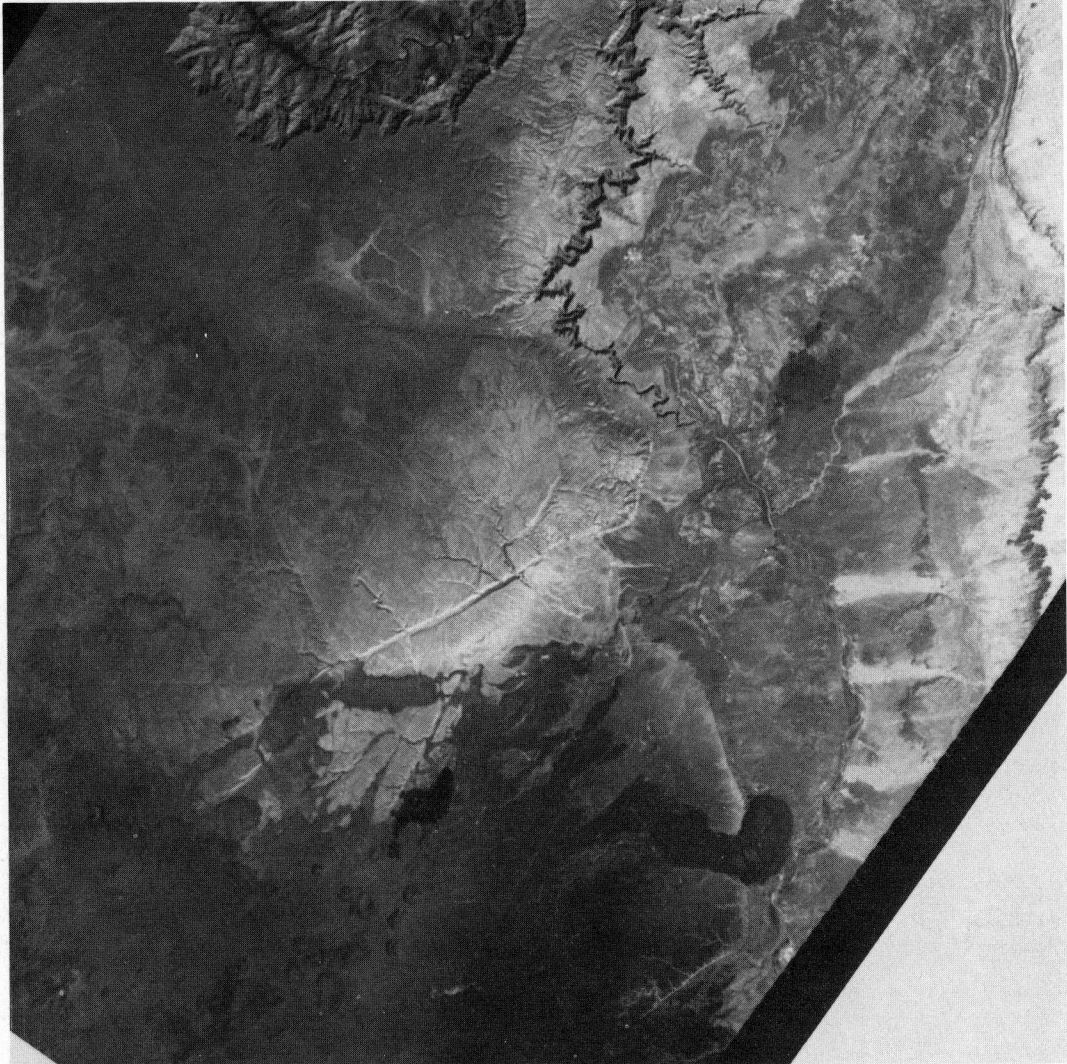

Fig. 25-37. High sun angle Landsat band-7 partial image, Northern Arizona. Chavez and Sanchez, 1982. Scale approximately 1:500,000.

woody swamplands predominate in the interior basins.

While the roughness of a water surface influences the tone, surface waves appear as periodic and regular change in the tone of a radar image. The pattern that is produced is a product of three surface effects that are modulated by the presence of a surface wave (Elachi, 1980); (1) the local slope, (2) the intensity and bunching of the small gravity and capillary waves, and (3) the wave orbital velocity. The relative importance of these three effects are not well understood. It is the consensus that modulation of the critical capillary waves (a function of the wavelength and depression angle) onto the lower frequency swell accounts for much of the variability observed in SAR imagery.

NATURAL VEGETATION MAPPING

Numerous studies of the effect of natural vegetation on radar returns have been made in several climatic and topographic areas in the United States. In all areas studied some influence of vegetation on the radar return was observed. Description of the interaction between vegetation and microwave backscatter is given in Chapter 4.

By analysis of the texture and tone on the like- and cross-polarized images in the area of Horsefly Mountain, Oregon, it was possible to delineate various types of vegetation (Morain and Simonett, 1966). The vegetation types in this area consist of ponderosa-pine forest, juniper woodland, white-fir forest, hardwood forest, sagebrush on basalt rubble, chaparral scrub, grassland, and recently

Fig. 25-38. Seasat image after registration to high sun angle Landsat image. (Chavez and Sanchez, 1982.) Scale approximately 1:500,000.

burned areas that are almost entirely vegetation-free. Forest/nonforest boundaries are enhanced on cross-polarized images through emphasis of image texture, increased grey-scale contrast, and sharpened acutance of the edge of the forest-sagebrush contact (Figures 4-33 and 4-34).

Broad vegetation structures are easily separated on multi-polarized imagery. Forest, shrub land, and grassland are easily distinguished. However, within-class distinctions are more difficult. For example, white-fir forest is difficult to distinguish from ponderosa pine of the same age and height characteristics. On the other hand, variations within ponderosa pine related to the degree to which the forest has been cut over, the presence of relict emergent, tall pines and the size and patterns of regrowth are distinguishable on the radar imagery. Cut-over ponderosa pine, for instance, has a marked radar texture on the HH image that normally permits its rapid distinction from surrounding vegetation types. In general, it has a coarse texture and a wider range of point grey-scale values than adjacent vegetation types (Figure 4-33).

Image texture is generally fine in non-forest areas and inspection shows much the same grey-scale value for the various vegetation types, although there are subtle differences in texture and tone between grasslands and shrublands (Figure 4-33).

On like-polarized imagery (HH), burned patches are not easily detected, whereas on cross-polarized imagery they are much more easily seen. Mesic grassland sites give low radar returns and they are easily distinguished. Dry grassland gives returns that may be occasionally confused with sagebrush.

Radar imagery of plant communities is sensitive to variations in the densities of the plant communities. Thus, Morain and Simonett (1966) found that variations in sagebrush density were readily detected in Escalante Valley, Utah, with radar imagery. Similarly, Peterson et al. (1969) showed that electronic color combination multi-polarization radar imaery of Horsefly Mountain were sensitive to within-community density and height differences.

For vegetation mapping, the use of frequencies greater than about 8 GHz and moderate depression angles has been recommended by Ulaby and Batlivala (1976). This recommendation is based on the increasing penetration and hence increasing sensitivity to soil properties underlying vegetation with lower frequencies and incidence angles near nadir.

The classification and interpretation of forests and crops from L-band and X-band multipolarized imagery can be improved by median filtering (Goodenough et al., 1980). This, in effect, is the removal of speckle from an image while preserving the edges of tonal and textural classes. Similar classes can thus be differentiated more unambiguously.

Inkster and Lowry (1980) have suggested that, for forestry applications, image interpretation can be significantly influenced by image resolution. It was found that manual interpretation was consistent and a function of image texture, interpreter skill, and other scene parameters to a threshold resolution beyond which confusion in identification increased significantly. These results compare qualitatively with some of Moore's (1979) findings discussed previously. For example, forests and fields became confused at resolutions greater than 10 m. This work was based on X-band and L-band multipolarization airborne data.

The findings of Inkster and Lowry (1980) have been replicated in part by Le-Toan and Shahin (1980) for digitized radar imagery. Only fair discrimination between forest and nonforest was possible using L-band multipolarized data with 25 m resolution. Within a forest class however, some information on the vegetation structure could be extracted, possibly allowing age classification in monospecies stands.

Unpublished studies by D. S. Simonett, based on field work in New Guinea (November, 1971), suggest that same-side stereoscopy, in addition to aiding geologic, geomorphic and drainage-network analyses, is of considerable aid in vegetation mapping. With stereoscopy, many subtle textures in forest communities become detectable that are indistinguishable when viewed monoscopically. Those with experience in radar vegetation-mapping recognize that textural variations are of more than equal rank with tone as a discriminant. However, in heavily treed areas,

dissimilar plant communities (associations) frequently appear similar on radar, primarily because they are similar structurally. It appears that some previously indistinguishable communities may be separable with stereoscopy.

For an adequate classification of vegetation to be developed for radar images, much ancillary information must be used to provide the interrelations between biologic and geologic patterns and features. Thus, for example, a forest group may be divided into evergreen, semi-deciduous, and secondary forest. Each of these groups, in turn, may be subdivided further based on these interrelations. As with any remote sensing interpretation, the results should be verified by some scheme of ground observation.

AGRICULTURAL SURVEYS WITH RADAR IMAGERY

Because information on crops must be obtained on a timely basis, the cloud-penetrating value of radar is especially significant for agriculture. The key dimension to improving existing multi-sensor crop identification is the temporal dimension. Early studies in the mid-1960's on crop classification from radar imagery were based mainly on analyses of Ka-band multiple-polarization images (e.g. Simonett et al., 1967). These studies were sufficiently encouraging to suggest the direction of later research, much of which has been done with truck-mounted scatterometers and also X-band and L-band aircraft systems.

Plant geometry, density, and water content are the major influences on the return from vegetation that give interpreters the ability to distinguish between crop types. Other variables influencing the radar return from crops are: percent of ground covered by vegetation (associated with biophase and maturity), row direction, row spacing, and soil properties including surface roughness and soil moisture. Generally, the maximum correlation between crop characteristics and radar backscatter occurs in the shorter wavelengths at moderate to high incidence angles. The shorter wavelengths result in greater backscatter from the vegetation canopy and the moderate to high incidence angles minimize the backscatter component from the soil. Bush and Ulaby (1978) suggest that data in the 13-16 GHz band contain the greatest discriminating power. Other system characteristics that influence the interpretability of crops from radar images are resolution and polarization.

As an example of the level of discrimination that may be achieved with Ka-band radar under favorable conditions, Figure 25-40 may be examined. This shows September, 1965 data on HH and HV polarization. Fields of sugar beets, corn, grain sorghum, wheat, alfalfa, and bare ground were examined. The y axis is HH polarization, the x axis is HV polarization. Note that the hyperplanes lie at about 45° implying that both polarizations are contributing information.

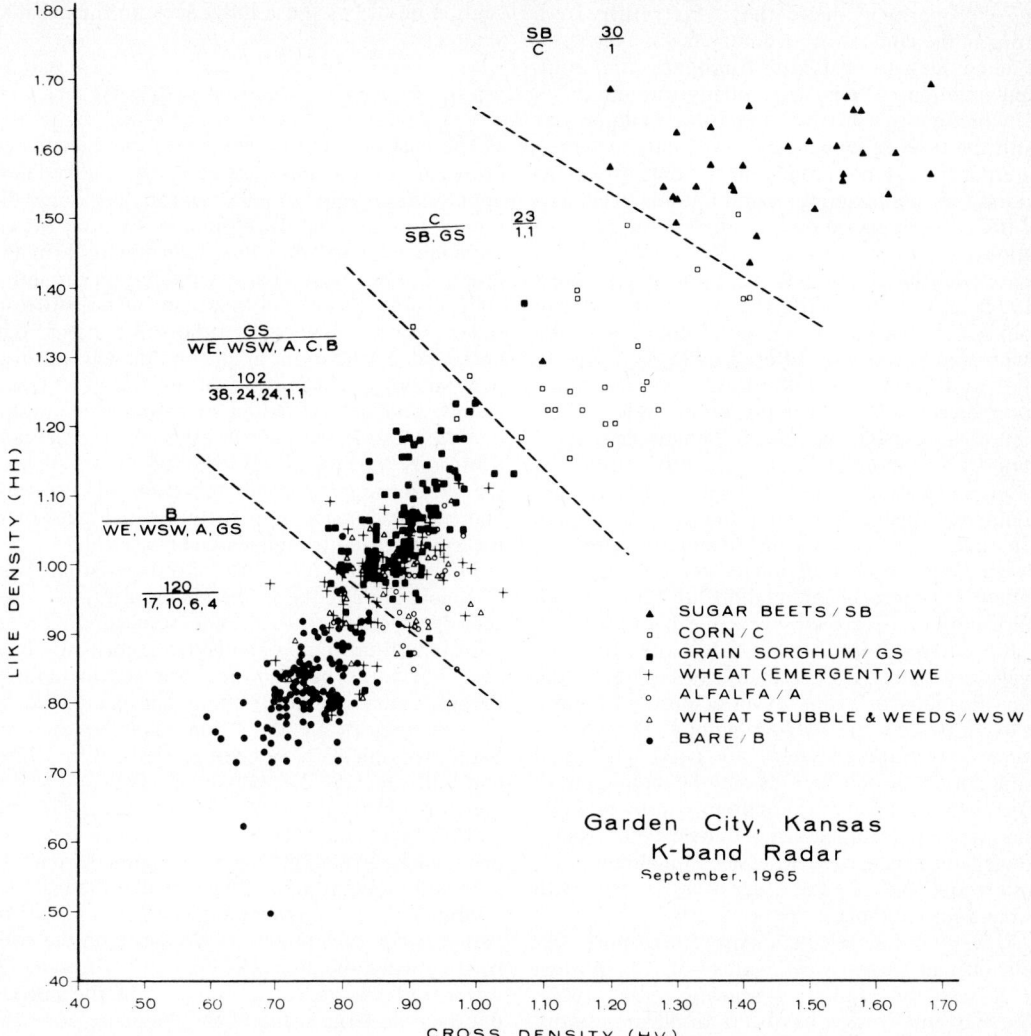

Fig. 25-40. Scatter plot showing the density of the like- (HH) and cross- (HV) polarization images, Garden City, Kansas, September 1965. Note the separation of the various groups achieved through use of multiple polarizations. Unpublished studies by R. K. Moore and M. L. Bryan of simultaneous multi-frequency, multi-polarization radar images, show that frequency may be more valuable than polarization in crop discrimination.

In the upper group bounded by the hyperplane, there are 30 fields of sugar beets and 1 field of corn, suggesting that mature sugar beets are more reflective than mature corn. The next lowest region has 23 fields of corn, 1 field of sugar beets, and 1 field of grain sorghum; again, this suggests a higher return from corn than from grain sorghum. The lowermost group has 120 fields of bare ground; 17 fields of emergent wheat, in which wheat occupies less than 5 percent of the total area, 10 of wheat stubble, 6 of alfalfa, and 4 of grain sorghum. This lowermost group appears reasonable for projection by extrapolation to adjacent areas.

The next to lowest category is a mixture of crop types predominantly grain sorghum. The system used for this study achieved inadequate discrimi-nation in this region. However, the uppermost portion of the region is dominated by grain sorghum and a subset might be discriminated that could be used as a basis for prediction of grain sorghum in the lower portion of the mixture.

A recent test of crop identification with L-band radar (Ulaby et al., 1980) showed a 71 percent probability for correct classification using dual polarization components. The study was done with single date imagery of fields of corn and soybeans, woodlands, and continuous-cover vegetation types such as small grains and pastures. The separability among vegetation types was quantified with a linear discriminant analysis. As might be expected, the addition of the HV component to the HH component for classification increased the accuracy significantly.

To properly evaluate the radar return from crops in the context of inventory from imagery, it is necessary to test multifrequency and multipolarization systems sequentially through many growing seasons. It may be possible to determine both the time dependencies of the data and their internal consistency from year to year, as well as assess the spatial and temporal bounds to the use of training sets based on ground data for extrapolation.

For varying crop types, de Loor and Jurriens (1974) concluded that the variation in the backscatter because of temporal changes in the plants and because of differences in the frequencies and polarizations used, was the only parameter usable for crop inventory. Some investigators have used a plant-development descriptor parameter to more accurately identify crop types based on the stage of development. For example, Ulaby and Bush (1976) found that the use of such a parameter based on plant height, moisture by weight, and number of plants per unit area resulted in a more linear relationship between the stage in the crop growth cycle and backscatter.

Recently, Bush and Ulaby (1978) have suggested that multidate data are required if the classifications of crops to an accuracy of more than 90 percent are to be obtained. If high frequencies (about 14 GHz) are used with dual polarization, revisit periods may be as long as 15 days, depending on crop conditions, still obtaining classification accuracies of better than 90 percent. Lower frequencies generally need much lower revisit frequencies, on the order of a few days with L-band for example.

Row direction affects radar return more significantly at longer wavelengths—at L-band than at X-band for example (Parashar et al., 1980). The HH polarization has been found to be more sensitive to row direction than HV. The return from a field with row direction parallel to the sensor look-direction is less than when the row direction is perpendicular (Batlivala and Ulaby, 1976). Bush and Ulaby (1978) have found the VV polarization is better suited for crop classification than other polarizations.

The relationship between resolution and field size can have a significant effect on agricultural surveys using radar. Based on a study of optimum radar resolution for land-use applications, Inkster and Lowry (1980) found that field discrimination in areas with large fields was easy with resolutions as coarse as 20 m. However, because of the loss of textural information, forests may become confused with fields in single date data with resolutions coarser than 10 m.

The radar return from the underlying soil can sometimes dominate the return from agricultural areas as well as from areas with natural vegetation. The return from the soil is strongly influenced by the surface roughness and the soil texture-moisture relationships. The following section describes these influences on radar backscatter.

SOIL SURVEYS WITH RADAR IMAGERY

The past decade of research on the microwave properties of soils has focused on two interrelated approaches. Substantial research has been directed toward the development of microwave techniques for soil-moisture determination. In addition, studies have been conducted on the influence of soil parameters other than soil moisture on radar return. Knowledge of the soil moisture has potential application in predicting agricultural productivity, estimation of potential surface runoff, and determination of evapotranspiration (Newton and Rouse, 1980). Microwave reflection from soils depends on a number of factors such as the dielectric properties (functions of moisture and temperature) and the factors which pertain to the system used (roughness, angle, etc.).

Many studies have attempted to isolate the soil-moisture effects on the backscattering cross-section in order to determine some quantitative information on the soil-moisture status. Much of this work has been done with scatterometry though considerable attention has been paid to passive microwave detection of soil moisture. Such experiments have been carried out by Ulaby and Batlivala (1976), Ulaby et al. (1978 and 1979), Newton and Rouse (1980), Schmugge (1980), and Dobson and Ulaby (1981). Some work has also been done correlating image-return measurements with soil moisture (e.g. Chang et al., 1980).

Microwave measurements of the moisture content of porous materials are based on the contrast between the dielectric constant of free liquid water and the dielectric constant of the porous matrix. The large value of the dielectric constant for free liquid water is a result of the ability of the dipole moment of the water molecules to align themselves with the applied field. Water held at a high tension by the soil matrix has restricted molecular movment and therefore a lower dielectric constant than free water. At high tensions the dielectric constant approaches that of ice, about 3.5.

Thus the dielectric constant is directly related to the volumetric water content and the tension with which the water is held in the matrix. The dielectric constant of a soil-water mixture increases slowly with water content at low moisture levels and then more rapidly at moderate to high moisture contents. The inflection point in the dielectric increase is the tension at which the water molecules are free to align. Therefore, textural composition of soil controls, to a large extent, the amount of water present at a given tension. While coarse soils have a low water-holding capacity, clays may have a water content an order of magnitude higher. Radar penetration is usually less with decreasing soil texture, accompanied by a lower return in the backscattering direction for angles off nadir.

Attempts have been made by Schmugge (1980) and Dobson and Ulaby (1981) to remove the textural effects of soil on the microwave properties. These attempts are based on normalization techniques which divide the actual soil moisture content by that calculated for field capacity. Empirical equations are used to calculate the field capacity from parameters describing the textural distribution of soil samples. The results of many of these investigations conclude that percent field-capacity, the normalized moisture content, is more highly correlated with the microwave properties than the traditional measures of soil moisture (gravimetric and volumetric). For example, Chang et al. (1980) found a fair degree of correlation between the radar return-signal from an airborne, uncalibrated L-band system and percent field-capacity in the surface of bare fields where furrowing is parallel to the radar beam. Scatterometer investigations have more generally suggested a close relationship between percent soil field-capacity and backscatter measurements.

Recent investigation has indicated however, that the normalization of soil moisture content with the estimated field capacity actually reduces the sensitivity of the backscattering coefficient to the soil-water content (Simonett, Lees, and Estes, 1982). Simonett, Lees, and Estes (1982) suggest that the roles played by other soil properties such as structure, aeration and soil-depth layering, and vegetative influences on soil moisture-content are ignored by the normalization approach. Also, they point out that the application of soil-moisture normalization techniques is difficult because point measurements of soil-profile properties can never fully sample the variation of these properties that exist in the spatial domain.

Past effort has also been made toward mapping soil types by explaining image patterns on the basis of parametric interactions between the soil surface and the illuminating microwave beam. This approach has been found useful in semi-arid and arid environments where the masking effects of continuous vegetation cover are less pronounced (Morain and Campbell, 1972).

Preliminary analyses by Morain and Campbell (1972) of a 1965 Ka-band image of an area near Tuscon, Arizona, and subsequent field checking, revealed that two contrasting processes were involved in soil interpretation: (1) The primary image information originated with convolved returns from both the soil and vegetation and grey tones were largely dependent upon the density of the vegetation cover; and (2) wherever vegetation was essentially absent, the radar return plummetted in the Ka- and X-bands, irrespective of the size distribution of the surface veneer.

Morain and Campbell (1972) investigated a range of soil-vegetation landscapes to compare ground conditions with patterns on the radar image. Attention was focused on non-agricultural lands since these were subject to little change in the time between image acquisition and field observation. A summary of their observations for one particular landscape is presented in Figure 25-41 which shows a portion of a 3-cm (X-band) image, together with an idealized cross section of the terrain and ground photographs of the ground elements. In Figure 25-41 they defined three general terrain types: (1) The riparian vegetation which images in light-grey tones; (2) gentle shrub-mantled slopes (<3°) which appear in medium grey tones; and (3) interfluves characterized by barren gravel pavement. These last elements are so narrow on the landscape that they appear as very dark filament-like stripes on the image. In reconnaissance-mapping projects they would be insignificant compared to the expanse of sparse creosote shrub.

Two other entities are included in Figure 25-41 to illustrate particular points. The first is an abandoned airstrip (Figure 25-41 a) which appears on the imagery as a large dark-grey square; and the second is a stream bed (Figure 25-41 d) which has been included in the terrain cross-section for completeness. No streambeds appear on the imagery shown in Figure 25-41 because none in that particular area is wide enough to be "visible" through the riparian vegetation. When the look direction is oblique (as it is in this case) or orthogonal to the drainage grain, beds must be sufficiently wide that streambank vegetation, plus the radar shadow arising from the oblique view, cannot hide the other bank.

INTERPRETATION OF SNOW AND ICE

The microwave-reflection chracteristics of ice are radically different from those of liquid water. The relatively complex dielectric constant of ice is usually assumed to have a relatively constant value of about 3.17 for frequencies from 1 MHz to well above the microwave region. The loss factor and the change in both the loss factor and the dielectric constant with temperature are negligible at temperatures below the freezing point (absence of liquid water).

In the microwave spectrum, ice is a relatively lossless medium with a low dielectric constant. Unlike water, however, ice is found in a variety of forms and structures which produce significant variations in microwave reflectivity. Three of these forms will be considered: snow, glacier ice, and sea ice.

Fresh snow is commonly flakey and dry and may be considered a dielectric mixture of air and ice (while the temperature is below freezing). In such mixtures, the dielectric constant is a direct function of the snow density. Fresh dry snow has relatively low density and a dielectric constant of 1.6 or less. As the density of dry snow increases, the real part of the dielectric constant also in-

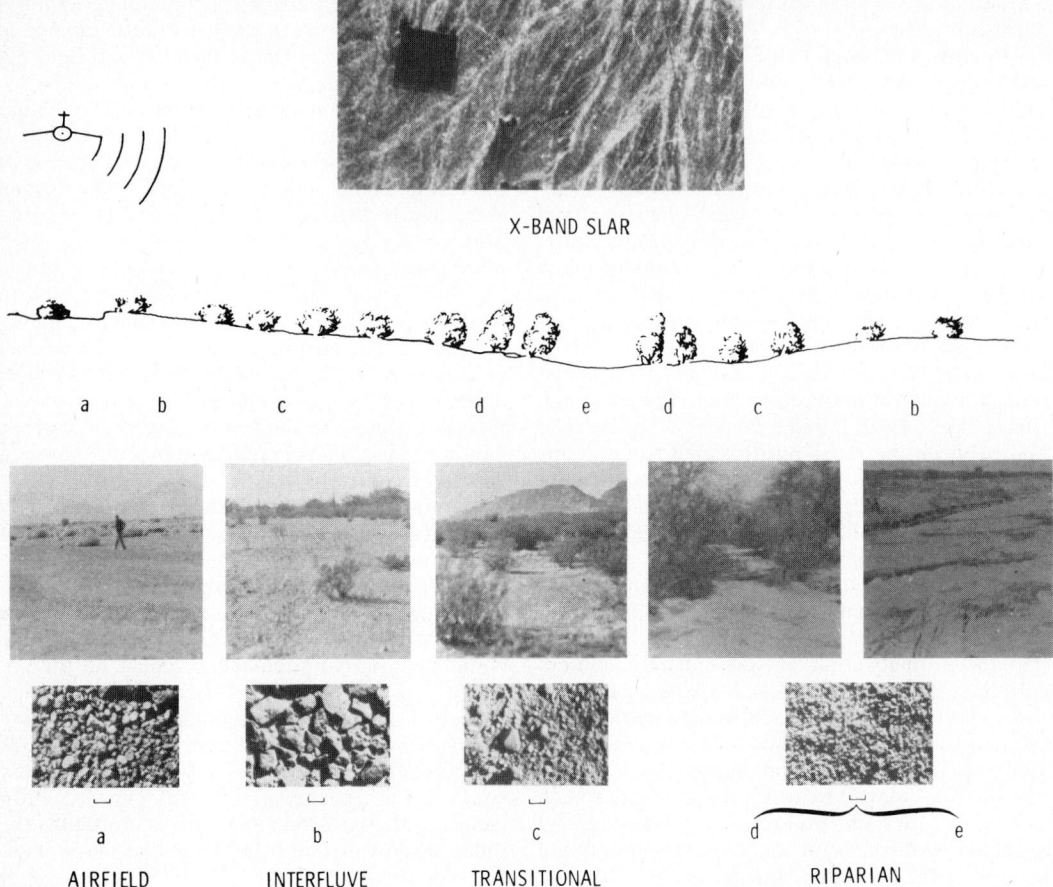

Fig. 25-41. Soil surface near Tucson, Arizona, with the corresponding radar (X-band) image. Scale and close-up of pebble surfaces is 1 cm.

creases and approaches that of ice. Figure 25-42 illustrates the effect of density and snow structure upon the dielectric constant. The measured values were made by Cummings (1952) and Kurowia (1956) at frequencies of 9.375 MHz and 3 MHz respectively. The u values are a form number which represent the theoretical effects of the snow structure on the dielectric constant. Near the freezing temperature, liquid water appears and the snow becomes a mixture of air, ice, and liquid water. This effect is illustrated by Figure 25-43, which shows the real part of the dielectric constant of snow as functions of both liquid water content and structure. The dielectric constant of dry snow is taken at 2.0 and that of liquid water at 80.0 for these calculations. Snow structure has considerable influence on the dielectric constant when the mixture is of materials with such widely dissimilar dielectric constants (wet snow).

In environments where winter snowfall exceeds the summer snowmelt, patches of "old snow" persist throughout the summer. The combination of snow accumulation, melting, compaction, and refreezing can produce a granular texture most commonly referred to as firn. Because these al-

pine snowfields contribute significant amounts of water for streamflow, their volume, equivalent water content, and areal distribution are particularly important to hydrologists.

The size of individual granules or crystals of old snow may range to above 3 mm, depending on temperature conditions. This size range is comparable to the sensing wavelength of some operation SLAR systems (35 GHz: λ of 8.6 mm) and the Weiner theory of dielectric mixtures used for fresh snow may no longer apply as the geometry of the medium becomes a function of the relative proportions of air and ice. Models now used to calculate the dielectric constant are generally volume-scattering models based on a distribution of roughly spherical dielectric particles, in which the scattering due to individual particles must be considered as well as the proximity of the particles.

With increased size of the individual firn granules, the firn itself undergoes a transition due to pressure and recrystalization. Most of the air between the granules is forced out, and the reduction of pore space ultimately results in a true solid of interlocking crystals called glacier ice.

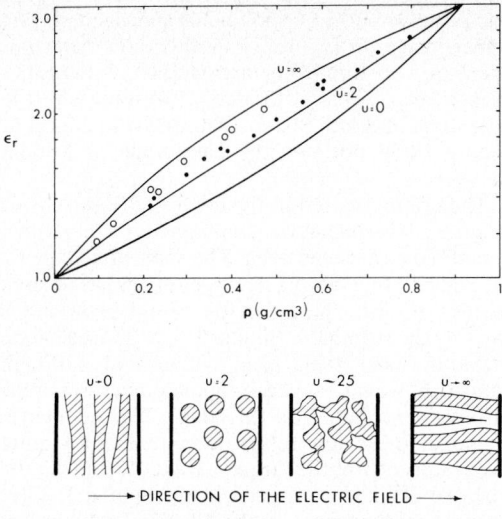

Fig. 25-42. Relative dielectric constant of snow versus density. (After Evans, 1965.)

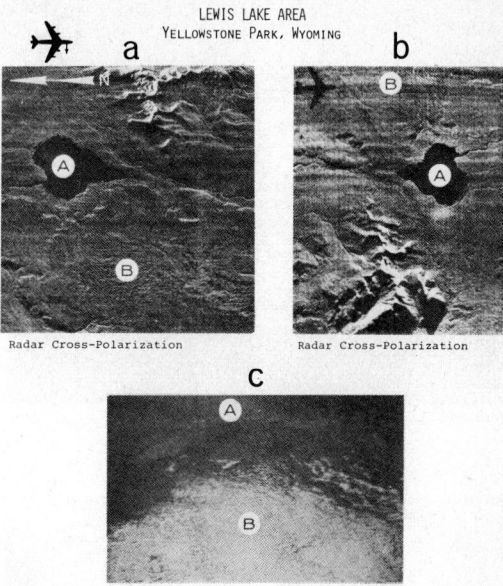

LEWIS LAKE AREA
YELLOWSTONE PARK, WYOMING

a
Radar Cross-Polarization

b
Radar Cross-Polarization

c
35mm Photograph

Fig. 25-44. Radar return from fresh snow and firn. (a) Radar look-direction west, (b) radar look-direction east, (c) simultaneously-acquired photograph. (From Waite and MacDonald, 1970.) Scale approximately 1:800,000.

The structure of individual glacier-ice crystals may be ignored when the reflectivity model for an infinite homogeneous half space is used. The reflection is again controlled by the dielectric constant and the roughness of the boundary surface between the dissimilar media (ice and the overlying air). The radar return from both old and new snow is illustrated in Figure 25-44. The imaging system used was the Westinghouse AN/APQ-97 radar, operating at a frequency of 35 GHz (Ka band). Photographic coverage taken simultaneously with the radar imagery clearly shows an area of new-fallen snow in the Lewis Lake region of Wyoming. Area B on the radar images and on the photograph represents the approximate area of the snowcovered terrain. At relatively higher elevations east of Lewis Lake, bright radar returns dominate the north-facing slopes where accumulations of perennial snow have been detected. To the west of Lewis Lake, the radar imagery appears to be a faithful reproduction of the terrain, whereas the photograph shows this region

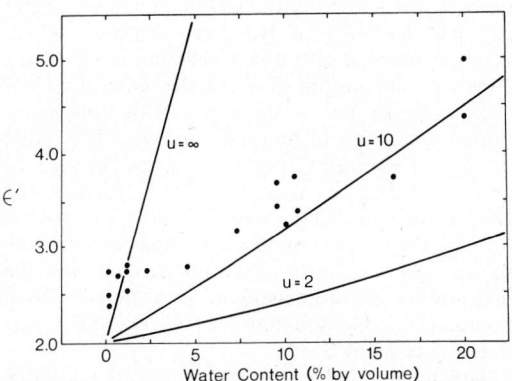

Fig. 25-43. Dielectric constant of wet snow versus volume percentage of liquid water. (From Evans, 1965.)

to be masked by new-fallen snow. Inadequate ground information of snow accumulation precludes estimating the depth of penetration.

Image "a" provides a look-direction to the west and image "b" provides a look-direction to the east. On both images the new snow cover has been penetrated, but the regions of old snow display an anomalously high return.

The return from a dry snow surface is negligible in comparison to that of a bare ground surface and for snow depths which do not significantly attenuate the return the major component will come from the soil surface. The backscattering cross-section from dry snow has been shown to be sensitive to the snow water equivalence for snow of "significant depth" and is relatively insensitive to surface roughness, although the sensitivity to surface roughness increases with liquid-water content (Stiles and Ulaby, 1980).

Examination of the return from a spherical dielectric particle that is near one wavelength in radius reveals that the scattering cross-section may have a value up to an order of magnitude greater than that of a perfectly conducting particle. This comes about from the focusing action of the transition at the back face of the particle which tends to reflect the energy back in the direction of transmission. This may account for the exceedingly high gain in the backscatter direction of the return from the old snow field.

Figures 25-45 and 25-46 illustrate the nature of returns from glacier ice. A series of radar flights in the Glacier Peak area of Washington and the Three Sisters area of Oregon have provided imag-

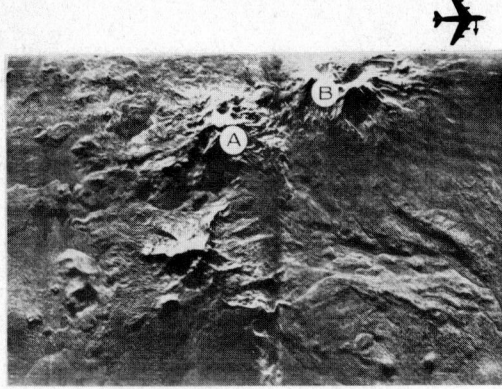

Cross-Polarization Radar

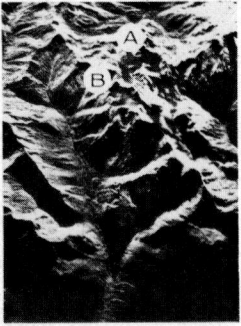

Radar Imagery

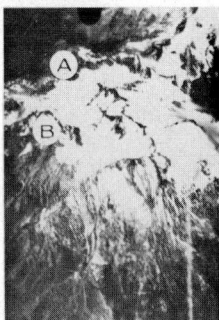

KA-30 Photography

Fig. 25-45. Radar imagery of Three Sisters area, Oregon and Glacier Park area, Washington. (After Waite and MacDonald, 1970.) Scale approximately 1:500,000.

ery over regions where glaciers have been previously mapped (Figure 25-45). Figure 25-46 represents the mapped snowfields and glaciers of the Three Sisters area (U.S. Geol. Survey Three Sisters Quadrangle, 1959). Based on an analysis of the topographic map of the area, no distinction

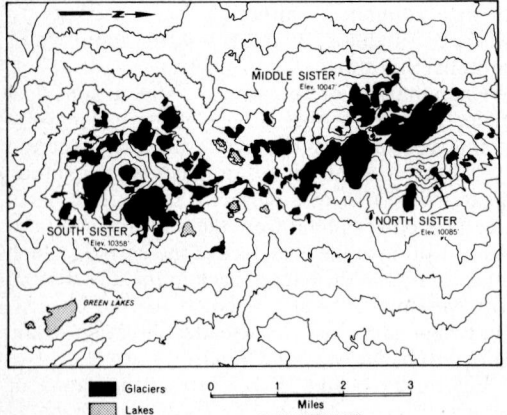

Fig. 25-46. Mapped glaciers, Three Sisters area, Oregon.

can be made between snowfields and glaciers. The upper image on Figure 24-45 illustrates radar imagery of a portion of the area shown on the topographic map. Area A of Figure 25-45 represents the east slope of South Sister (Figure 25-46); area B of Figure 24-45 outlines the east slope of Middle Sister.

The crater located in the center of South Sister (Figure 25-46) has not been mapped as a perennial snowfield or a glacier area. The same area appears on Figure 24-45 (above A, upper image) as an area of high return, characteristic of old snow areas previously examined. Similarly, on the west slope of South Sister, there is an extensive area of high return; however, several anomalously low areas also are distributed on this slope. The geometric shape of these low-return areas coincides quite accurately with the areal extent of the larger mapped glaciers of Figure 25-46. Similar low-return reas are evident in the Glacier Peak area of Washington (lower radar image, Figure 25-45). In this region, aerial photography taken about the same time as, but not simultaneously with, the radar imagery, can be compared with the radar imagery. Although some of the high return areas are evident on the radar imagery, the dominant image tone is indicative of low return, and corresponds to the snow-covered areas of glacial ice.

The return from mapped glaciers in the Three Sisters area is significantly lower than that from the surrounding areas of both old snow and bare soil (Figure 25-45). A slight decrease between ice and soil might be predicted from the difference in dielectric constant; however, the difference predicted at Ka band does not appear to be sufficient to explain this behavior completely. The difference in surface roughness might partially account for the differences observed at low incidence angles. In addition, an extensive and well-calibrated ground-measurement program is necessary for determining the relationship between radar return and surface composition.

From an initial stage of ''frazil'' crystals and sludge, sea ice grows into a compact aggregate of roughly columnar, lamellated crystals whose optical axes are predominantly horizontal. The extremely low solubility of salts in ice causes segregation at the interface. However, discrete pockets of brine become entrapped between the growing lamellae. The composition of the brine does not change during its entrapment and its concentration is a function of temperature only. The initial salinity of naturally frozen ice varies between 20 and 5 percent; the more rapid the freezing, the saltier the ice. Immediately after sea ice has formed, the process of desalinization begins. Most sea ice has a salinity of 2 to 4 percent, but the distribution is inhomogeneous and the surface of perennial ice has a salinity which is usually less than 0.5 percent.

The most reliable measurements of arctic-ice back-scattering behavior were recorded by NASA over the Beaufort Sea in two experiments (in 1967 and 1970), using a 2.25 cm-wavelength radar. The data were analyzed by Rouse (1969), Rouse and Schell (1971), and Parashar et al. (1973), who found that radar return provided distinct identification of a variety of ice types according to near-surface concentration and surface roughness. These two properties correlate with age and thickness in most sea ice. They found that the scattering mechanism is dominated by volume-scatter components for perennial ice and surface scatter components for new ice. This behavior enables very reliable classification of ice type using only the like polarized return from an incidence angle range between 35° and 45°. In this range, a system dynamic range of 25 to 30 dB permits separation of all ice types and open water. Indications are that the cross-polarized component is likewise highly dependent upon the near-surface salt concentrations, and hence could serve to reliably classify ice types.

Anderson (1961) recorded 8.6 mm-wavelength radar images of sea ice which established a wide range of radar-return levels from various sea-ice types (Fig. 25-47). More recently, Campbell and Weeks (1974) recorded X-band (APS-94) images of near-shore ice north of Alaska and determined that shorefast ice, first-year ice, multiyear ice, and several varieties of new ice could be individually identified in the radar images. However, earlier experiments with 400 GHz radar systems have indicated poor differentiation of sea-ice types and even inadequate delineation of open-water areas.

One of the most extensive uses of imaging radar was in connection with the voyage of the Humble Oil tanker, Manhattan. Over 10,000 km² of sea ice were imaged during this operation. The 1.8 cm-wavelength radar data are reported by Biache et al. (1971) who show examples of distinct differentiation of the full range of sea-ice types of primary interest. They illustrate the interpretation of 15 different ice types according to World Meteorology Organization definitions. These include young ice, first-year ice, second-year ice, multiyear ice, fast ice, ridged ice, rotten ice, ridges, and leads.

Examples of DPD-2 radar imagery recorded during the Manhattan traverse are shown in Figures 25-48 and 25-49. First-year ice is visible in Figures 25-48 and 25-49 as dark-grey to black tones with a smooth texture. This ice type may display bright straight lines indicating ridges under pressure. Multiyear ice is shown in Figure 25-49 with a mottled grey tone caused by weathered ridges and interconnecting melt holes. Drainage channels on multiyear ice can frequently be observed. Fast ice is also shown in Figure 25-49 and is distinguished by contrast to the shore line. Leads are usually evident in images as dark regions. Young ice and brash are frequently present in these leads and their separation depends on the dynamic-range characteristics of the radar system employed.

A major problem in the study of Arctic sea ice was identified by Bryan et al. (1977) as the lack of high-resolution data which have been collected during periods of inclement weather and/or darkness. This problem thus hampers efforts at modeling sea-ice drift and at understanding the heat-balance problem over the Arctic Ocean. The application of automated techniques of image interpretation to sea ice was investigated and results indicate that the orientation of leads and the percentage of open water in a scene were distinguishable. Measurements of sea-ice movement have been made with aircraft imagery (e.g. Leberl et al., 1979) with good results and with Seasat imagery (Chavez, unpublished U.S.G.S. report) which indicate fair potential.

INTERPRETATION OF GEOLOGIC FEATURES

Although radar imagery differs from aerial photography in geometry, spatial resolution, and

Fig. 25-47. Various sea ice types distinguished by Anderson (1961) on Ka-band radar imagery.

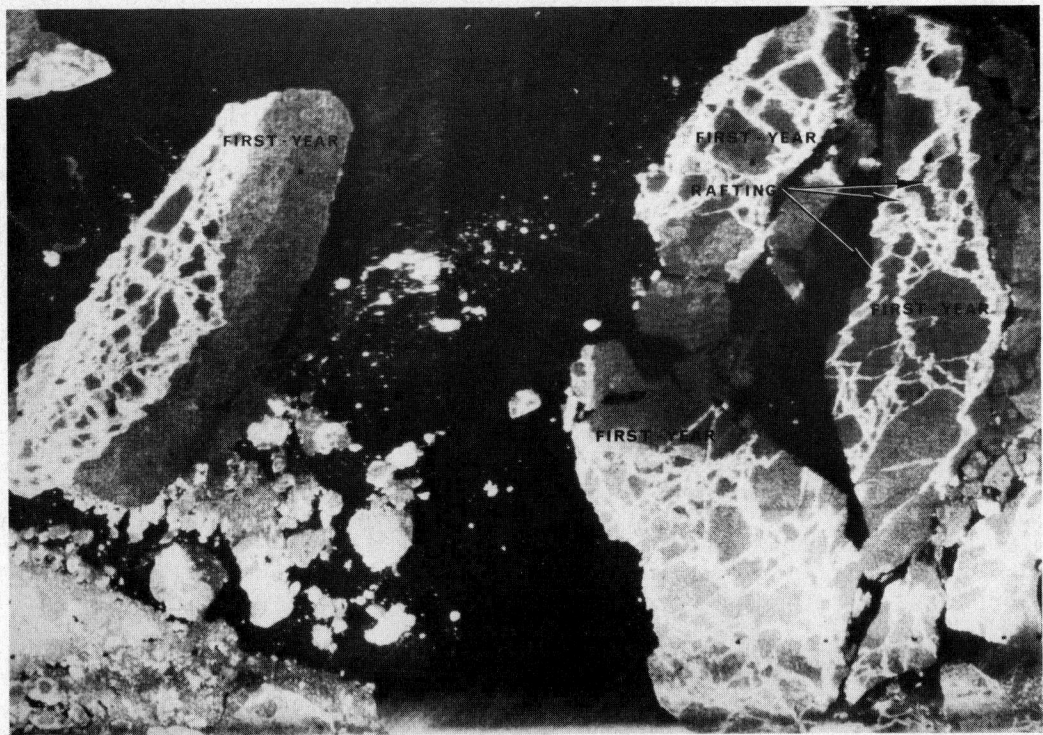

Fig. 25-48. Rafting and ridging on first-year ice. (Courtesy Dept. of Transportation: U.S. Coast Guard and Raytheon Co. Autometric Operation. From Blache et al. 1971.) Scale approximately 1:100,000.

terrain phenomena sensed, most competent photogeologists can easily make the transition from interpreting aerial photographs to interpreting radar imagery. The extent to which radar can be applied successfully to geologic investigations varies considerably, depending on the geologic and geomorphological character, climate, and, especially, vegetation cover in the study area. It also depends on the experience of the geologist in small-scale mapping and reconnaissance mapping.

The geologist who uses reconnaissance data for geologic syntheses usually has four objectives: (1) correlation of outcrops from one location to another, (2) determination of the stratigraphic sequence, (3) delineation of rock types or lithologic units, and (4) determination of geologic structure. Radar imagery provides the geologist with a terrain format approximating a three-dimensional strip map which can reveal varying amounts of geologic information depending on the terrain environment and stage of erosional development. For example, in areas where geological conditions are closely reflected in topographic expression of rock types and structures, it may be possible to recognize geological features quite unequivocally from the evidence provided by the radar imagery. However, in poorly exposed areas where geology has no direct influence on the topographic texture, the skill of the geologist in interpreting radar imagery for physiographic analysis is critical for structural and even

lithologic evaluation. The experience, judgement, and ability of the interpreter to evaluate the significance of many different types of terrain data are directly reflected in the final interpretive product. Therefore, as with photogeology, radar geologic interpretation must be approached with the full realization that geologic interpretation can only be accomplished if adequate scrutiny is given to all of the following elements: (1) the pattern and distribution of outcrops, (2) drainage, (3) vegetation, and (4) surficial and structural configuration.

Outcrop patterns, drainage, vegetation, and landform structural configuration are displayed on radar imagery as recognition elements which can include tone, texture, shape, and pattern. It is largely the analysis of these four recognition elements, on the radar imagery, which contributes to the interpretation of geologic data, as seen in the earlier discussion of these items. Numerous examples of the use of radar imagery for terrain (geologic) and mineral resources investigations is given in Chapter 31.

Radar is highly sensitive to slope change and thus exhibits structural features and drainage patterns readily. For example, Figure 25-3 is a Seasat image of an area in Northeastern Mexico. Here Mesozoic sedimentary rocks have been folded into plunging anticlinal structures. In the right and central portions of the image a large breached anticline and two smaller, doubly plunging anticlines are clearly observed as a result

MULTI-YEAR
WITH WEATHERED
RIDGES

Fig. 25-49. Multiyear floe with weathered ridges. (Courtesy Dept. of Transportation: U.S. Coast Guard and Raytheon Co. Autometric Operation. From Blache et al. 1971.) Scale approximately 1:100,000.

of tonal variation and patterns associated with this type of feature (Elachi, 1980). Also clearly depicted in the imagery is the trellis drainage-pattern associated with such surface expression of folds.

Figure 25-50 shows a portion of a digitally-correlated Seasat image of the north coast of Jamaica at expanded scale. Although this area is heavily vegetated, enough information is detectable to produce a structural map (Figure 25-51) and a texture map showing inferences of geomorphology and lithology (Dixon, 1981).

INTEGRATED LANDSCAPE ANALYSIS

In 1969, N. R. Nunnally foretold one of the major uses of radar in large-area reconnaissance, namely that of integrated landscape analysis. He noted that

"radar provides a means of delimiting varying associations of physical and cultural phenomena through the outlining imagery variations in tone, texture, pattern, and shape. It can be demonstrated that image patterns delimited on radar are visually cor-

related with known, observable variations in physical and cultural phenomena. Although the small scale and limited resolution of the radar prohibit interpretation of fine detail, enough information can be interpreted to basically categorize landscape regions. The value of the approach is that valuable regional categories can be quickly established and characterized; and, if more detailed regional descriptions are desirable, only limited sampling would be necessary to provide the data. Ultimately, the radar technique may well represent a considerable improvement over other approaches to regional generalization from the standpoint of time, cost, comparability, and accuracy."

Integrated landscape regions mapped by Nunnally are shown in Figure 25-52.

Those involved with broad-area surveys in developing areas will recognize the concept employed by Nunnally as being akin to the land-system concept employed by the Land Research and Regional Survey Division, CSIRO, Australia,

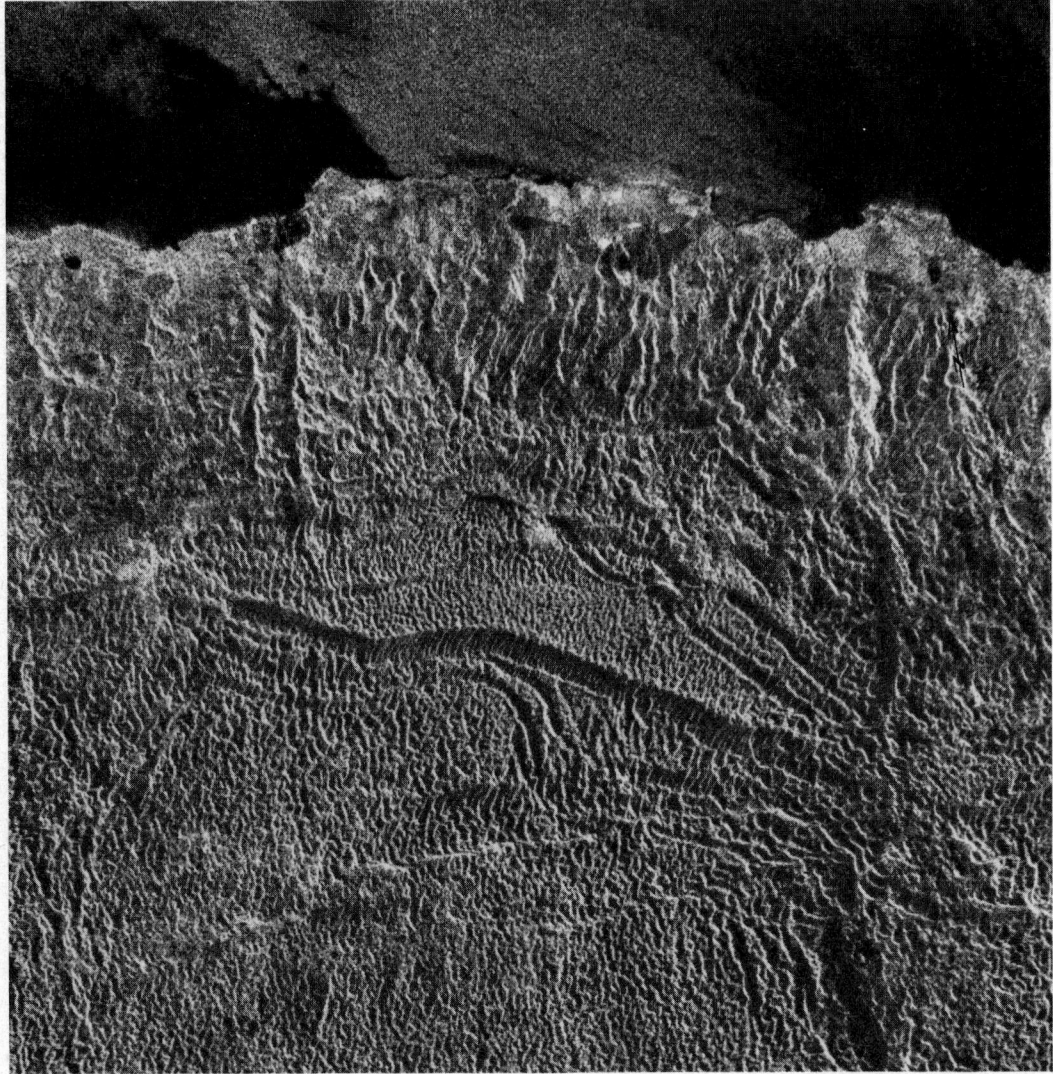

Fig. 25-50. Digitally correlated Seasat image of the north coast of Jamaica. (Courtesy of JPL) Scale approximately 1:250,000.

as well as to other methods of landscape-unit mapping employed elsewhere. Groups with considerable experience in reconnaissance land-system and similar mapping, and with the requisite sampling strategies, will adapt readily to the use of radar imagery for such mapping.

Radar imagery, because if its small scale, because it may be enlarged greatly, and because it may be used with same-side stereo overlap, lends itself well to this use.

The delineation of "integrated landscapes" with radar imagery is, however, not without its difficulties and ambiguities. These arise from the familiar problems of boundary delineation where gradual changes occur in the natural and cultural landscape. In addition, there are the well-known problems of lumping and splitting categories, and differences in the perception of individuals as to what constitutes a mappable unit. As few envi-

ronments break down so simply into categories as those investigated by Nunnally in the Ashville Basin in North Carolina, it is essential to use professionals experienced in a number of disciplines.

INTEGRATED RECONNAISSANCE NATURAL RESOURCE SURVEYS WITH RADAR IMAGERY

In broad-area reconnaissance studies, among the more useful maps which can be made from radar images for general land planning and land evaluation are combined physiographic and slope-category maps. In many cases these may prove more useful than contour maps for general planning purposes, particularly if very large regions are to be considered at one time and grouping of similar slope category classes over large areas is necessary. In some respects, landform bound-

DUANVALE FAULT, EASTERN SEGMENT
TEXTURE MAP

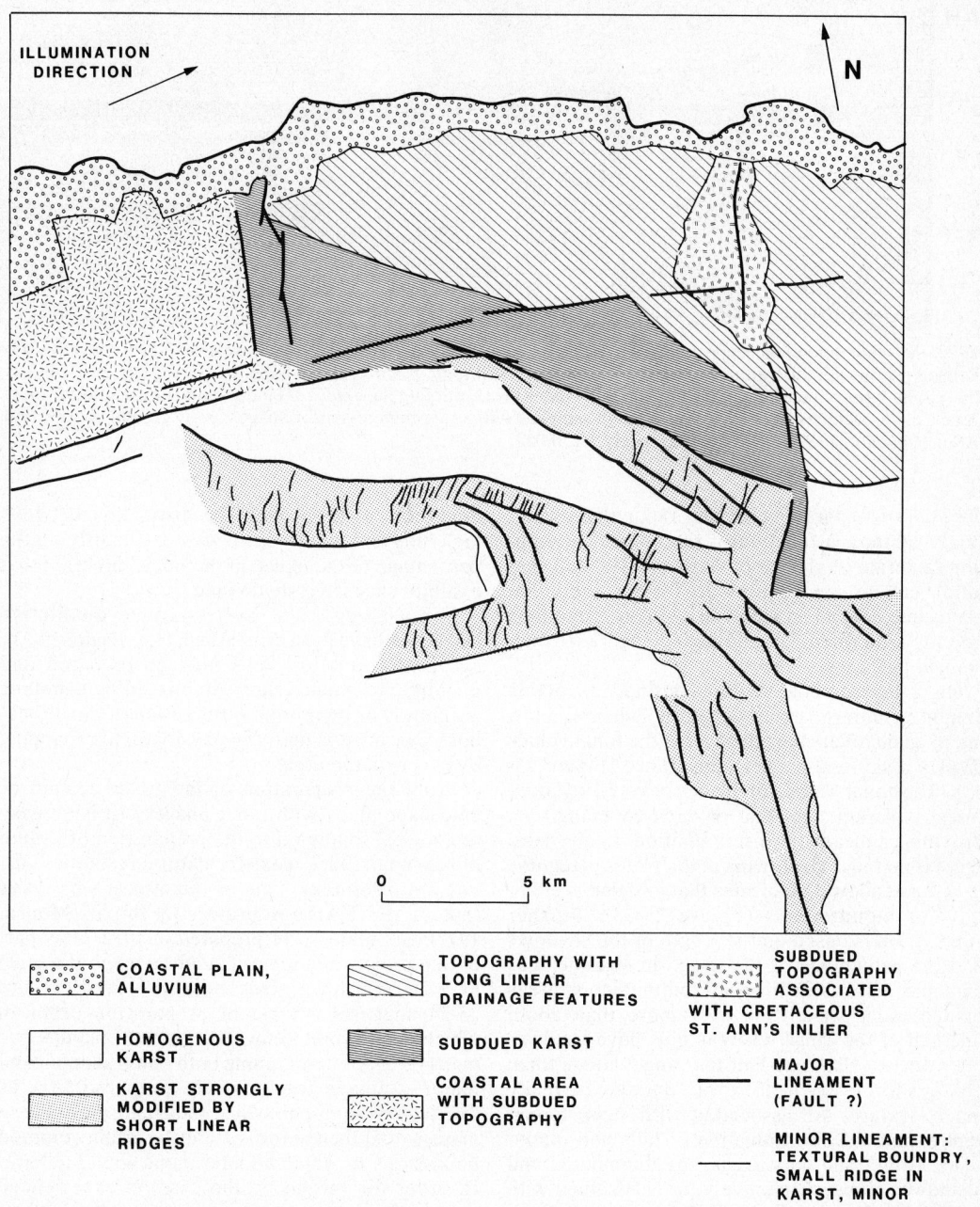

Fig. 25-51. Duanvale fault, eastern segment texture map. (Courtesy of JPL) Scale approximately 1:200,000.

aries drawn from radar imagery are more accurate than those derived from topographic maps because they are not blurred by the traditional methods of unit-area averaging of selected topographic parameters. However, the criteria which define radar landform regions—primarily size, shape, and arrangement of radar shadow-pat-terns—are not always identical to the quantitative and critical terrain slope-angles, relative relief, and the landscape profile characteristics traditionally used. The regions derived by either method may be equally valid if not identical.

The delineation of major geomorphic regions—plains, hills, and mountains—from radar imagery

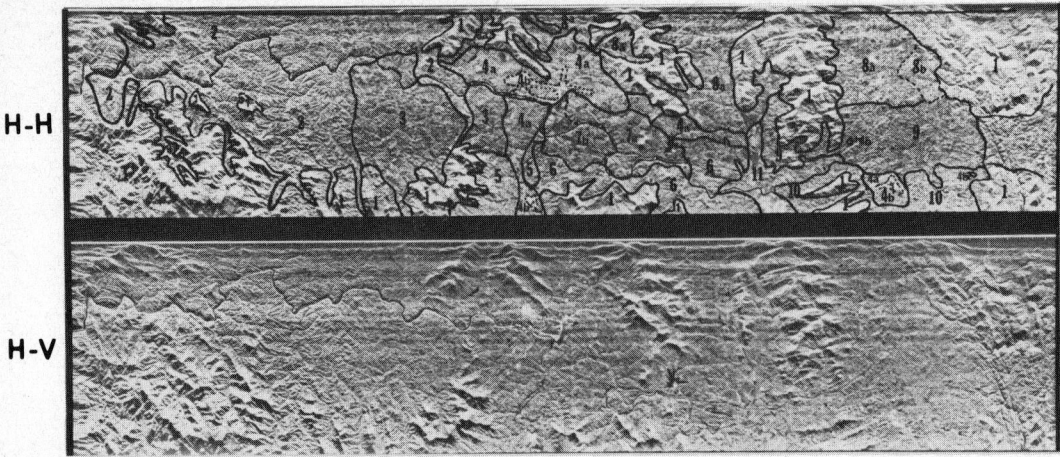

H-H

H-V

Fig. 25-52. Integrated landscape analysis from radar imagery of the Asheville Basin, western North Carolina. The outlined regions are differentiated in terms of differences of tone, texture, size, shape, and arrangement of objects. The assemblages of imaged objects serve as surrogates for identifying the extent of land-use and physiographic units. Details of the regions can be analyzed through sampling with other remote-sensor images, or by ground observation. (After Nunnally, 1969.) Scale approximately 1:250,000.

is based on apparent relief and radar shadowing. Image texture, a function of the degree, orientation, and rate of change of slope (as well as vegetation cover) can be used to further subdivide plains into coastal and alluvial categories and hills into low, intermediate, and high categories (Mac-Donald and Lewis, 1969).

On a macro-scale regions defined as plains exhibit a uniform tonal signature, whereas on a micro-scale (over short distances) the tonal values display a salt-and-pepper appearance (Figure 25-53). The tonal value for plains covered by rain-forest is darker than that covered by mangrove, providing a means for discrimination. In all cases, there is no radar shadowing in the plains category.

Radar shadowing indicates that a region is either hilly or mountainous (Figure 25-53). Further subdivision is based on the length of the shadows and the percentage of the image in shadow. For example, in aircraft imagery, mountainous areas exhibit radar shadowing over more than about one-half of the range whereas hills have shadowing over less than one-half the range. Gross tonal changes form very light to very dark areas, really macro-texture, are associated with steep topography, and aid discrimination of hills and mountains. Bright and dark areas, the illuminated and shadowed slope, respectively, are also allied with hills; however, the size or length in the range direction of the bright or dark areas is less than that demonstrated in the mountainous regions, indicating less or lower relative relief (Figure 25-53). Further regionalization of hills is possible into: (1) low hills, if the tonal change from the front to the back terrain slope is gradually indicating smooth ill-defined crests; (2) intermediate hills, if the tonal changes are abrupt but bright-dark areas are relatively small; and (3) high hills, if tonal changes are abrupt and the illuminated-shadowed slopes are of moderate to large size. The distinction between high hills and mountains is based primarily on the percentage of the image in the range direction that exhibits excessive shadowing.

For a given radar system, once qualitative categories have been established, (e.g. Figure 25-53), radar-interpretation keys may be prepared and quantitative bounds then established by sampling air photos or by ground observations. This procedure can ensure uniform physiographic mapping over very large areas.

From the preparation of integrated terrain or landscape maps with radar imagery, it is an easy and logical follow-on to the preparation of a suite of reconnaissance maps for natural resources survey and inventory. One of the largest surveys is that of the RADAM project in Brazil (Moura, 1972). Six maps were prepared in 1972-73 as part of that project and several hundred maps at a scale of 1:250,000 have since been prepared. Figure 25-54 suggests a logic of preparation order in which geomorphic (morphologic) and planimetric maps (the latter containing both cultural detail and major drainage features) are prepared first as overlays to semi-controlled radar mosaics. These maps would then in turn establish certain common boundaries to which all later maps would adhere. In order for this to be the case an experienced group of interpreters is required with excellent backgrounds in geology, hydrology, and geomorphology. Where both morphologic and planimetric maps are available first, these feed into the preparation of geological, pedological, mixed pedologic-physiographic maps, plant-geography, and forest-products maps. These then may be used as inputs, along with supporting data and analyses by development economists, agronomists, and others, in the preparation of maps of potential land use.

MOUNTAINS

HIGH HILLS

LOW HILLS

PLAINS

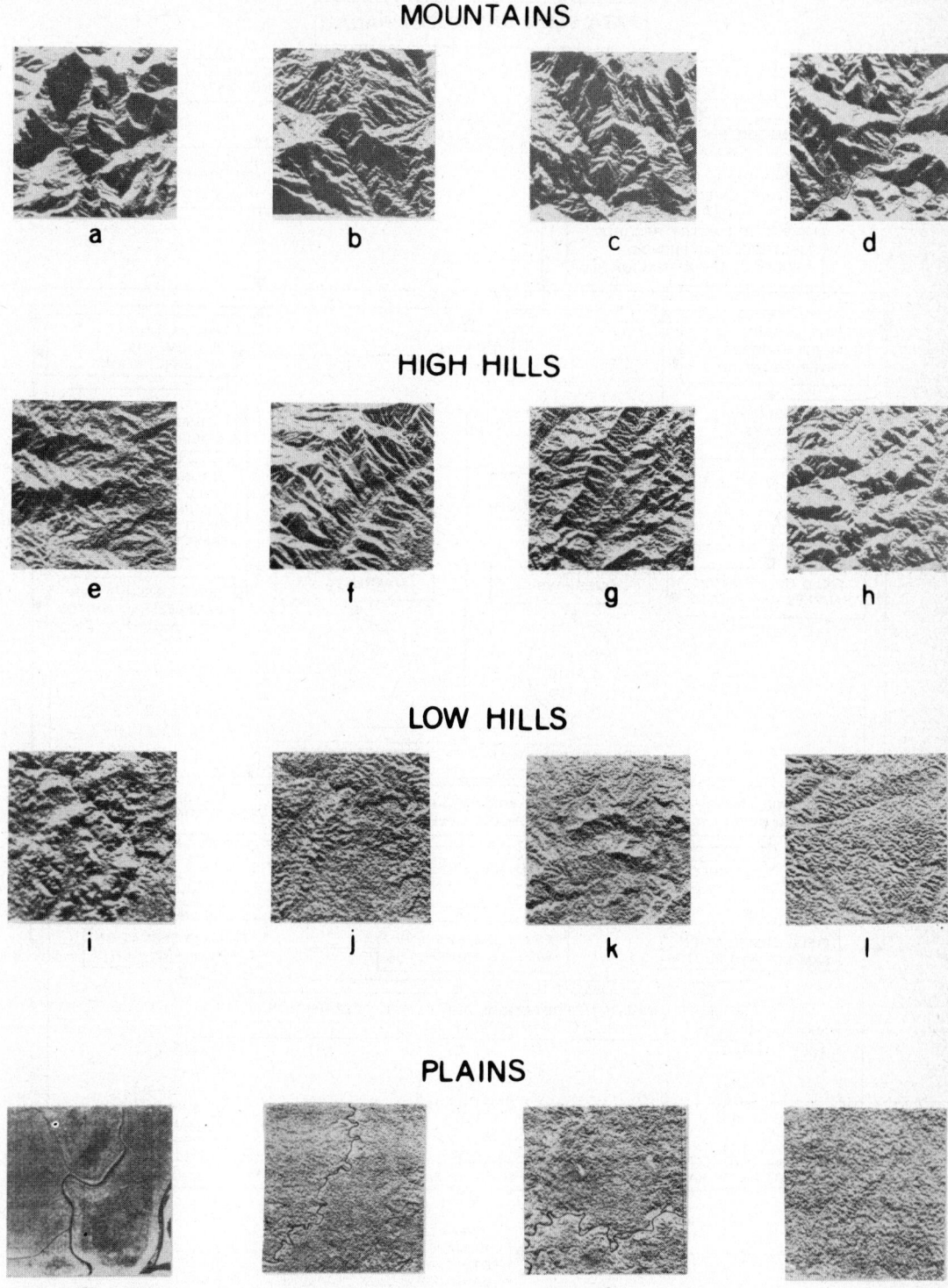

Fig. 25-53. Radar chips from the four major landform regions in Darien Province, Panama: plains, low hills, high hills and mountains.

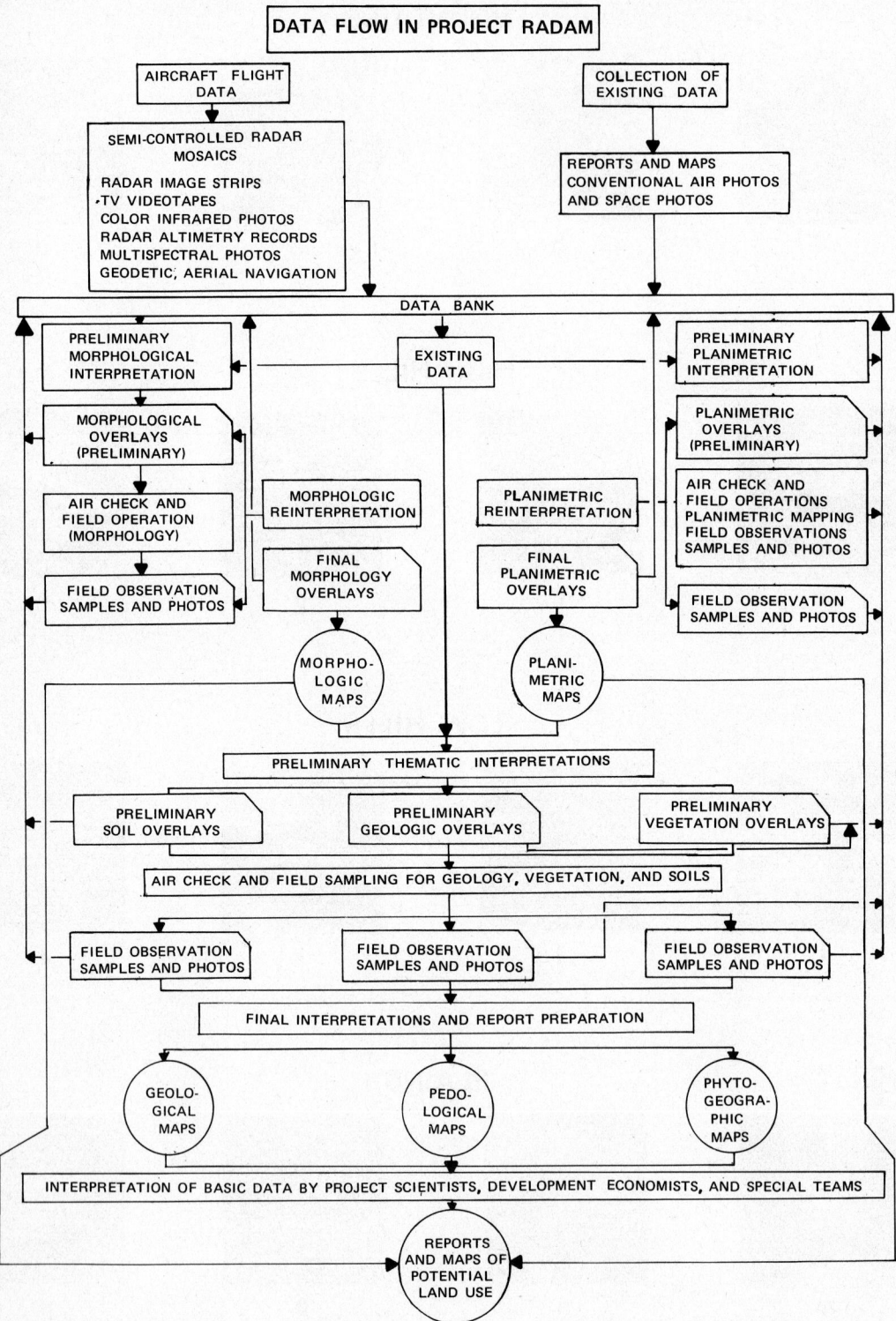

Fig. 25-54. Data flow and map construction for reconnaissance natural-resources surveys in Project RADAM, Brazil.

All of these maps were prepared initially as 1:250,000 overlays to 1:250,000 semi-controlled radar mosaics of quadrangles of the International Map of the World series. Before final publication, most have been aggregated to scales of 1:500,000 or 1:1000,000.

In geologic mapping, experienced interpreters also mapped at scales of 1:1,000,000 before mapping at a scale of 1:250,000.

An important part of the checking procedure in developing these maps was low-altitude fixed-wing aircraft and helicopter flights and ground observations. Very large parts of the Amazon region are almost inaccessible and low-altitude flights, coupled with widely scattered sample-point information, were extrapolated over many hundreds of kilometers in some instances. The area of this survey is over 3,800,000 km² (Moura, 1972).

The quality of the radar imagery used in these studies (Goodyear GEMS-102) is shown in Figure 25-55, the lower course of the Rio Xingu. The

original scale of the imagery is 1:400,000; this figure represents about a 5× enlargement. This radar image may be compared with an earlier B/W IR photograph (Figure 25-56) obtained in 1970 as part of the route survey for the trans-Amazonica highway. Under the magnification in Figure 25-55, and greater, radar is sensitive to forest texture caused by emergent species. The hilly relief adjacent to Rio Xingu is strikingly depicted even with monoscopic imagery. Stereoscopic radar imagery emphasizes the texture caused by emergent species.

TRENDS IN SATELLITE RADAR IMAGERY

Radar images clearly improve our ability to assess and monitor Earth resources. The advantages of a spaceborne platform have been described. The data obtained from the 100 days of operation of the Seasat SAR imaging system showed the

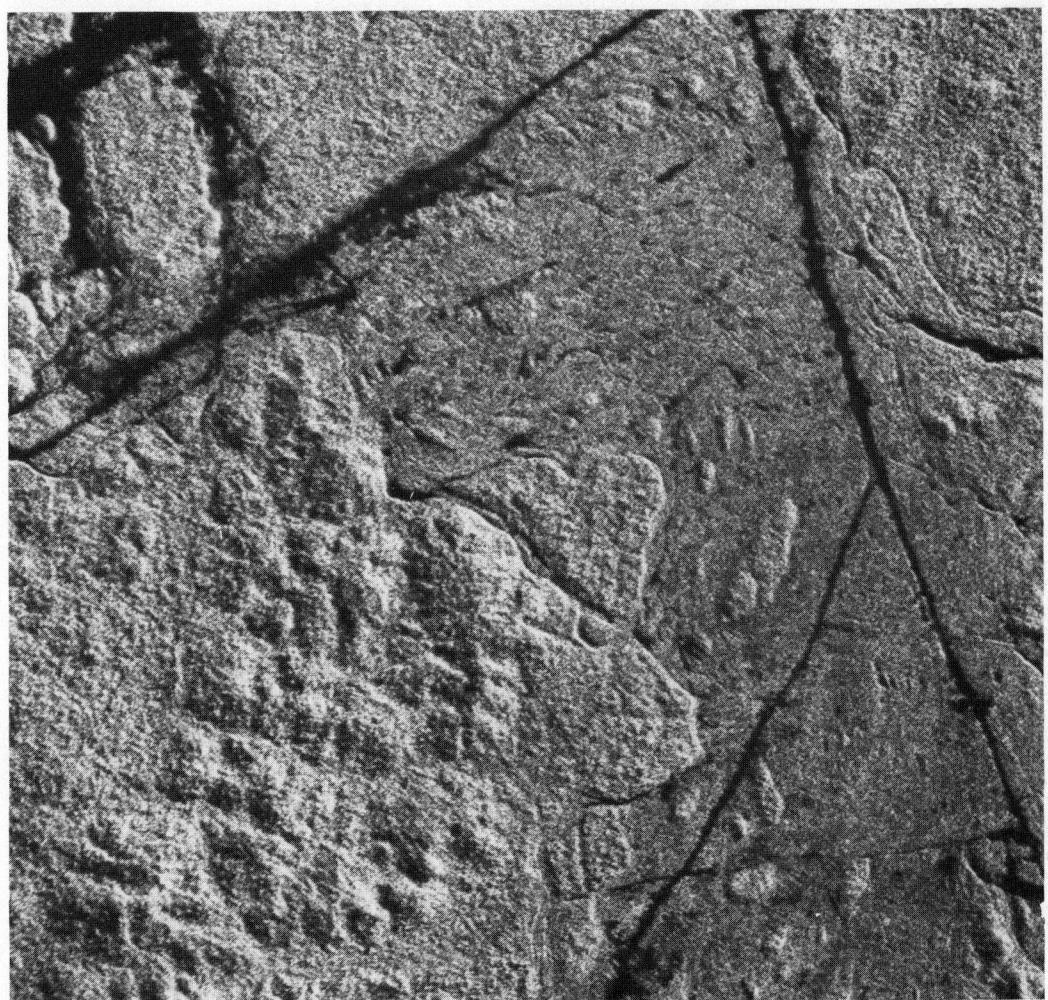

Fig. 25-55. Enlarged radar image of the Rio Xingu, Brazil, in its lower course just before entering the rapids. Rapids must be passed through before the lower course joins the Amazon. Note cut for trans-Amazonica Highway. Scale approximately 1:100,000.

Fig. 25-56. Black-and-white infrared photo obtained in 1970 of the same area as the radar enlargement in Fig. 25-55. Scale approximately 1:100,000.

potential of future spaceborne imaging-radar systems for collecting data from land. These data are still being analyzed. The Seasat SAR was limited in its capability for mapping land features in that it was designed for mapping the ocean surface; it operated at one frequency (L-band), one incidence angle (70°) and one polarization (VV) not all necessarily optimal or appropriate for land areas.

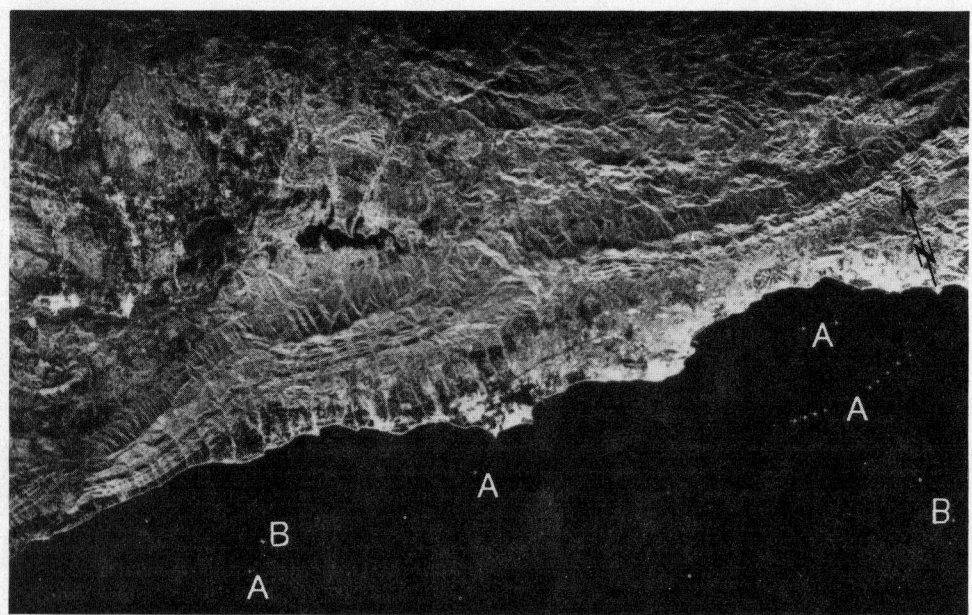

Fig. 25-57. SIR-A image of Santa Barbara. Channel area reproduced for land detail (Estes, 1982). Scale approximately 1:260,000.

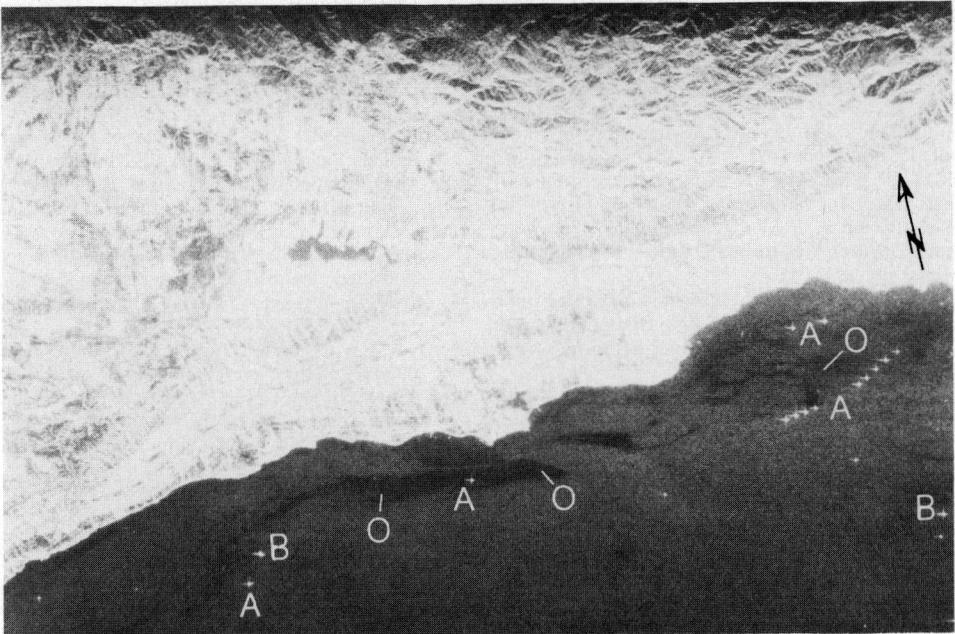

Fig. 25-58. SIR-A image of Santa Barbara. Channel area reproduced for ocean detail. (Estes, 1982). Scale approximately 1:260,000.

The next step in the development of operational spaceborne radar imaging systems is the first system of the Shuttle Imaging Radar, SIR-A. The SIR-A has the same frequency (L-band) and polarization (VV) as the Seasat SAR, but has a smaller depression angle (45°). It allows observation of mountainous regions not observable with Seasat, because the latter was subject to severe layover.

Following the SIR-A mission, the next step is the SIR-B mission, which is envisioned as two SAR systems, L-band and X-band, in which the X-band system will operate at three polarizations (HH, HV, and VV). Both systems will use the same data handling system and the depression angle will be controllable. A series of experiments will be conducted covering the topics of surface morphology, rocks and soils, vegetation canopies, surface water, and manmade structures. The experiments will also examine the effects of various system- as well as scene-parameters. SIR-B is likely to be available by 1984.

Examples of imagery acquired from the SIR-A on November 13, 1981 are shown in Figures 25-57 and 25-58. A preliminary interpretation of these images of the Santa Barbara area has evaluated the sensitivity to oil slicks on the ocean surface (Estes, 1982). Figure 25-57 has been processed to enhance detail on the land. Figure 25-58 has been processed to enhance ocean surface detail in the Santa Barbara channel. On both of these images drilling platforms (A) and ships (B) in the channel are shown as bright points. Areas with oil slicks derived from natural seepage (0) can be seen in Figure 25-58, which was processed for ocean detail.

The SIR program will improve the development of future satellite imaging radars. It is expected to improve the design of system capability, data handling and interpretation techniques.

REFERENCES

Anderson, V. H., 1966, High altitude, side-looking radar images of sea ice in the Arctic; Proc. Fourth Symp. on Remote Sensing of Environment, University of Michigan, Ann Arbor, pp. 845–857.

Attema, E. P., and F. T. Ulaby, 1978, Vegetation modeled as a water cloud; Radio Science vol. 13, no. 2, pp. 357–364.

Beatty, F. D., et al., 1965, Geoscience potentials of side-looking radar; Raytheon Autometric Corporation, Alexandria, Va., 90 p.

Biache, A., Jr., C. A. Bay, and R. Bradie, 1971, Remote sensing of the Arctic ice environment; Proc. Seventh Symp. on Remote Sensing of Environment, Univ. of Michigan, Ann Arbor, pp. 523–561.

Blanchard, B. J., Analysis of synthetic aperture radar imagery; Final Report, NASA contract NAS5-23458, 35 p.

Blanchard, B. J., 1980, Some confusion factors in radar image interpretation; Report of the Radar Geology Workshop, Snowmass, Colorado, JPL pub. 80–61. pp. 223–232.

Blanchard, B. J., and J. W. Rouse, 1980, Depolarization of electromagnetic waves scattered from an inhomogeneous half space bounded by a rough surface; Radio Science, vol. 15, no. 4.

Blanchard, A. J., R. W. Newton, L. Tsang, and B. R. Jean, in press, Volumetric effects in cross polarized airborne radar data; Geosciences and Remote Sensing Jour.

Brisco, B. and R. Protz, 1982, Manual and automatic crop identification with airborne radar imagery; Photogramm. Eng. and Remote Sens., vol. 48, no. 1, pp. 101–109.

Bryan, M. L., W. D. Stromberg, and T. G. Farr, 1977, Computer processing of SAR L-band imagery; Photogram. Eng. and Remote Sens., vol. 43, no. 10, pp. 1283–1294.

Bryan, M. L., 1979, The effect of radar azimuth angle on cultural data; Photogram. Eng. and Remote Sens., vol. 45, no. 8, pp. 1097–1107.

Bryan, M. L., 1981, The use of radar imagery for surface water investigations; Deutsch, M., D. R. Wiesnet, and A. Rango, eds., Satellite Hydrology, Amer. Water Resources Assoc., Minneapolis, Minn., pp. 238–251.

Bush, F. T. and F. T. Ulaby, 1976, Radar return from a continuous vegetation canopy; IEEE Trans. on Antennas and Prop., vol. AP-24, no. 3, pp. 269–276.

Bush, F. T. and F. T. Ulaby, 1978, An evaluation of radar as a crop classifier; Rem. Sens. Environ., vol. 7, pp. 15–36.

Chang, A. T. C., et al., 1980, L-Band radar remote sensing of soil moisture; IEEE Trans. on Geosci. and Rem. Sens., vol. GE-18, no. 4, pp. 303–310.

Chavez, P. S. Jr., 1980, Automatic shading correction and speckle noise mapping/removal techniques for radar image data; Report of the Radar Geology Workshop, Snowmass, Colorado, JPL pub. 80–61, pp. 251–264.

Chavez, P. Jr., and E. Sanchez, in press, Digital combination of LANDSAT and SEASAT images; Amer. Soc. Photogram.

Cosgriff, R. L., W. H. Peake, and R. C. Taylor, 1960, Terrain scattering properties for sensor systems design (terrain handbook II); Ohio State Univ., Eng. Exp. Sta., Bull. 181, vol. 29, no. 3, pp. 1–118.

Cumming, W. A., 1952, The dielectric properties of ice and snow at 3.2 centimeters; Jour. Applied Physics, vol. 23, pp. 786–793.

Daily, M., C. Elachi, T. Farr, W. Stromberg, S. Williams, and G. Schaber, 1978, Application of multispectral radar and LANDSAT imagery to geologic mapping in Death Valley; JPL publication, 78-19, 47 p.

Daily, M. I., T. Farr, and C. Elachi, 1979, Geologic interpretation from composited radar and LANDSAT imagery; Photogram. Eng. and Remote Sens., vol. 45, no. 8, pp. 1109–1116.

Dalke, G. W., and R. M. McCoy, 1969, Regional slopes with non-stereo radar; Photogram. Eng., vol. 35, pp. 446–452.

De Loor, G. P., A. A. Jurriens, and H. Gravesteijn, 1974, The radar backscatter from selected agricultural crops; IEEE Trans. on Geosci. Electr., vol. GE-12, no. 2, pp. 70–77.

Derenyi, E. E., 1970, An exploratory investigation into the relative orientation of continuous strip imagery; Thesis and Research Report No. 8, Univ. of New Brunswick, Fredericton, N.B.

Dixon, T., 1981, Analysis of Jamaican lineaments visible on Seasat SAR imagery; Presented at International Geosci. and Remote Sens. Symp., IEEE, June 8–10, pp. 886–894.

Dobson, M. C., and F. T. Ulaby, 1981, Soil textural effects on radar response to soil moisture; Univ. Kansas Remote Sensing Lab., RSL Tech. Rep. 264–30.

Elachi, C., 1980, Spaceborne imaging radar: geologic and oceanographic applications; Science, vol. 209, no. 4461, pp. 1073–1082.

Estes, J. E., 1982, Preliminary analysis of Shuttle Imaging Radar of Santa Barbara Channel; accepted for presention, Int. Geosci. and Rem. Sens. Symp. (IGARS '82), June, 1982, Munich, Germany.

Finch, V. C., G. T. Trewartha, and E. H. Hammond, 1957, Elements of Geography, New York, McGraw-Hill, 210 p.

Ford, J. P., 1980, SEASAT orbital radar imagery for geologic mapping: Tennessee-Kentucky-Virginia; Amer. Assoc. Petroloeum Geol. Bull., in press.

Ford, J. P., 1982, in preparation; Jet Propulsion Laboratory, Cal. Inst. Tech., Pasedena, Ca.

Foster, J. L., Hall, D. K., 1981, Multisensor analysis of hydrologic features with emphasis on the SEASAT SAR; Photogram. Eng. and Remote Sens. vol. 47, no. 5, pp. 655–664.

Fung, A. K., and F. T. Ulaby, 1978, A scatter model for leafy vegetation; presented at Proc. AGARD Conf. Aspects of Effective Scattering in Radio Communications, Cambridge, MA, Oct., 1977.

Goodenough, D. G., B. Guindon, and P. M. Teillet, 1980, Correction of synthetic aperture radar and multispectral scanner sets; Proc. Thirteenth International Symposium on Remote Sensing of Environment, Univ. Michigan, Ann Arbor, Apr 23–27, pp. 259–270.

Graham, L., 1980, Stereo radar for mapping and interpretation; Report of the Radar Geology Workshop, Snowmass, Colorado, JPL pub. 80–61, pp. 336–350.

Hammond, E. H., 1954, Small-scale continental landform maps; Assoc. Amer. Geogr. Annals, vol. 44, no. 1, pp. 33–42.

Hammond, E. H., 1964, Analysis of properties in landform geography; Assoc. Amer. Geogr. Annals., vol. 54, pp. 11–19.

Haralick, R. M., F. C. Caspall, and D. S. Simonett, 1970, Using radar for crop discrimination; a statistical and conditional probability study; Remote Sensing of Environment, vol. 1, no. 2, pp. 131–142.

Harris, G., and L. C. Graham, 1976, Landsat-Radar syngergism; presented at XIII Congress of the Int. Soc. Photogramm., Goodyear Aerospace Corp Pub., Litchfield Park, Ariz.

Holmes, A. A., 1979, Augmentation of LANDSAT with synthetic aperture radar for agricultural applications; ERIM report 130100-16-F, Ann Arbor, Michigan.

Inkster, D. R. and R. T. Lowry, 1980, Optimum radar resolution studies for land use and forestry applications; Proc., Fourteenth Internat'l. Symp. Rem. Sens. Env., vol. II, 23–30 Apr., Univ. Mich., Ann Arbor, pp. 865–882.

Koopmans, B. N., 1973, Drainage analysis on radar images; ITC Jour., 1973–3, Enschede, Netherlands.

LaPrade, G. L., 1963, An analytical experimental study of stereo for radar; Photogram. Eng., vol. 35, pp. 294–300.

Le Toan, T. and A. Shahin,

Leberl, F. W., 1970, Metric properties of imagery produced by side looking airborne radar and infrared line scan systems; Publications of the International Institute for Aerial Survey and Earth Sciences (ITC), Series A, no. 50, Delft.

Leberl, F. W., 1972, On model formation with remote sensing imagery; Osterr. Zeitschrift fur Vermessungswesen, vol. 60, pp. 61–93.

Leberl, F. W., 1975, Sequential and simultaneous SLAR block adjustment; Photogrametria, vol. 31, pp. 39–51.

Leberl, F. W., 1976, Side-looking radar mosaicking experiment; Photogram. Eng. and Remote Sens., vol. 52, no. 8, pp. 1035–1042.

Leberl, F. W., 1978, Satellitenradargrammetrie, Pub. 239, Series C, Deutsche Geod. Kom., Munich, West Germany, 156 p.

Leberl, F. W., 1979, Accuracy analysis of stereo side-looking radar; Photogram. Eng. and Remote Sensing, vol. 45, no. 8, pp. 1083-1096.

Leberl, F. W., M. L. Bryan, C. Elachi, and T. Farr, 1979, Mapping of sea ice and measurement of its drift using aircraft synthetic aperture radar images; Jour. Geophys. Res., vol. 84, no. C4, pp. 1827-1835.

Leberl, F. W., 1980, Application of imaging radar to mapping; Report of the Radar Geology Workshop; Snowmass, Colorado, JPL pub. 80-61, pp. 307-335.

Levine, S. H., 1960, Radargrammetry; New York, McGraw-Hill, 330 p.

Lewis, A. J., and H. C. MacDonald, 1970, Significance of estuarine meanders identified from radar imagery of eastern Panama and northwestern Colombia; Modern Geology, vol. 1, no. 4, pp. 187-196.

MacDonald, H. C., and A. J. Lewis, 1969, Applications of radar imagery in geologic and geomorphic reconnaissance of tropical environments; Lawrence, Kansas, Prog. Abs., p. 18-19.

MacDonald, H. C., and W. P. Waite, 1971, Soil moisture detection with imaging radar; Water Resources Res. Jour., vol. 7, no. 1, pp. 100-109.

MacDonald, H. C., and W. P. Waite, 1973, Optimum radar depression angles for geological analysis; Modern Geology, vol. 2, pp. 179-193.

Moore, R. K., and G. C. Thomann, 1971, Imaging radar for geoscience use; IEEE Trans. Geosci. Electronics, vol. GE-9, pp. 155-164.

Moore, R. K., 1979, Tradeoff between picture element dimensions and noncoherent averaging in side-looking airborne radar; IEEE Trans. Aerospace Elect. Sys., vol. AES-15, no. 5, pp. 697-707.

Morain, S. A., and D. S. Simmonett, 1967, K-band radar in vegetation mapping; Photogram. Eng., vol. 33, pp. 730-740.

Morain, S. A. Campbell, J. B., 1972, Soil mapping from radar imagery; theory and preliminary applications: presented Ann. Mtg. Assoc. Amer. Geographers, Kansas City, Missouri, pp. 1-19.

Moura, J. M. de, 1972, Projecto RADAM; levantamento dos recursos natrais dos Regoes Amazonica e Nordestes do brasil por mes do radar e outros sensores; Revista Brasileria de Cartografia, vol. 3, no. 6. pp. 1-6.

Newton, R. W., and J. W. Rouse, 1980, Microwave radiometer measurements of soil moisture content; IEEE Trans. of Antennas and Prop., vol. AP-28, no. 5, pp. 680-686.

Nunnally, N. R., 1969, Integrated landscape analysis with radar imagery; Remote Sensing of Environment, vol. 1, no. 1, pp. 1-6.

Parashar, S. K., 1973, Investigation of radar discrimination of sea ice; Univ. Kansas Center for Research, Inc., CRES Tech. Rep., 185-13.

Peterson, R. M., G. R. Cochrane, S. A. Morain, and D. S. Simonett, 1969, A multisensor study of plant communities at Horsefly Mountain, Oregon; Johnson, P., ed., Remote sensing in ecology, Georgia Univ. Press, pp. 63-93.

Rouse, J. W., 1969, Arctic ice type identification by radar; IEEE Proc., vol. 57, pp. 605-611.

Rouse, J. W., J. A. Schell, 1971, On the feasibility of identifying thin sea ice using airborne radar; Proc. Fourteenth Minefield Conf., Washington, D.C., Naval Ordinance Laboratory, NOL-TR-71-71, pp. 365-382.

Schaber, G. G., G. L. Berlin, and W. E. Brown, 1976, Variations in surface roughness within Death Valley, California: geologic evaluation of 25 cm wavelength radar images; Geol. Soc. Amer. Bull., vol. 87, no. 1, pp. 29-41.

Schugge, T. J., 1980, Effect of texture on microwave emission from soils; IEEE Trans. on Geosci. and Rem. Sens., vol. GE-18, no. 4, pp. 353-361.

Simonett, D. S., et al., 1967, The potential of radar as a remote sensor in agriculture; a study with K-band imagery in western Kansas; Univ. Kansas, CRES Rep. 61-21, 13 p.

Stewart, H. E., R. Blom, M. Abrams, and M. Daily, 1980, Rock type discrimination and structural analysis with LANDSAT and SEASAT data; San Rafael Swell, Utah: Report of the Radar Geology Workshop, Snowmass, Colorado, JPL pub, 80-61, pp. 151-167.

Stiles, W. H., and F. T. Ulaby, 1980, Microwave remote sensing of snowpacks; Univ. Kansas CRINC, NASA CR-3263, 404 pp.

Ulaby, F. T., and J. E. Bare, 1978, Look direction modulation function of the radar backscattering coefficient of agricultural fields; Photogram. Eng. Rem. Sens., vol. 15, pp. 1495-1506.

Ulaby, F. T., and P. P. Batlivala, 1976, Optimum radar parameters for mapping soil moisture; IEEE Trans. on Geosci. Electronics, vol. GE-4. no. 2, pp. 81-93.

Ulaby, F. T., and P. P. Batlivala, 1976, Diurnal variations of radar backscatter from a vegetation canopy; IEEE Trans. on Antennas and Prop., AP-24, vol. 1, pp. 11-17.

Ulaby, F. T., P. P. Batlivala, and M. C. Dobson, 1979, Microwave backscatter dependence on surface roughness, soil moisutre, and soil texture, Part I. Bare soil; IEEE Trans. Geosci. Electron., vol. GE-16, pp. 286-295.

Ulaby, F. T., P. P. Batlivala, and J. E. Bare, 1980, Crop identification with L-band radar; Photogram. Eng. and Rem. Sens., vol. 46, no. 1, pp. 101-105.

Ulaby, F. T., G. A. Bradley, and M. C. Dobson, 1979, Microwave backscatter dependence on surface roughness, soil moisture, and soil texture, Part II. Vegetation-covered soil; IEEE Trans. Geosci. Electron., vol. GE-17, pp. 33-40.

Ulaby, F. T. and T. F. Bush, 1976, Corn growth as monitored by radar; IEEE Trans. on Antennas and Prop., vol. AP-24, no. 6, pp. 819-828.

Van Roessel, J., and R. De Godoy, 1974, SLAR block mosaics for project RADAM; Photogram. Eng., vol. 40, no. 5, pp. 583-595.

Waite, W. P., H. C. MacDonald, 1971, Snowfield mapping with K-band radar; Remote Sensing of Environment, vol. 1, pp. 143-150.

Wang, J. R., and T. J. Schmugge, 1980, An empirical model for the complex dielectric permittivity of soils as a function of water content; NASA/GSFC, TM 79659.

Glossary[1]

Many of the definitions contained in this glossary are essentially as they appeared in the First Edition of the *Manual of Remote Sensing,* (Reeves, 1975). The First Edition, in turn, gave credit to nearly fifty sources and especially acknowledged the *Manual of Photographic Interpretation* (Colwell, 1960), the Third Edition of the *Manual of Photogrammetry* (Thompson, 1966), and the *Manual of Color Aerial Photography* (Smith and Anson, 1968). In the present glossary use has been made of all of the above sources and also of two more recent glossaries, viz. those appearing in the *NASA Landsat Workbook* (Short, 1982) and in the Fourth Edition of the *Manual of Photogrammetry,* (Slama, 1980).

As pointed out by Reeves in the First Edition of this Manual, the terms used in remote sensing come from a variety of sources—electrical engineering and electronics, physics and the physical sciences, and photographic sciences. Some of these terms have been redefined here in line with their restricted or special usages in the field of remote sensing. Terms used primarily in electronic data processing, conventional geophysics, surveying, and photoreproduction for the most part have not been included. Likewise, scientific terms pertaining to the applications disciplines ordinarily have not been included. Sources for definitions of these terms are listed in the "References" of this glossary, however.

Many of the terms that are included here are defined specifically as they are used in remote sensing, and not as they are ordinarily used in other endeavors. Difficulties arose, however, even with definitions restricted to their use in remote sensing. The same term may be used with different meanings in photography, radar, and infrared. Insofar as possible, a common, generalized definition has been selected. When this was not feasible, the specialized definitions are given.

REFERENCES

Allen, W. H., ed., 1965, Dictionary of technical terms for aerospace use: NASA SP-7, 314 p.

American Congress on Surveying and Mapping-American Society of Civil Engineers, 1972, Definitions of surveying and associated terms: Washington, D.C., Amer. Cong. Mapping Surveying, 205 p.

American Geological Institute, 1962, Dictionary of geological terms: New York, Doubleday-Dolphin, 545 p.

Anonymous, 1971, The optical industry and systems directory: Pittsfield, Mass. Optical Publishing Co., Glossary.

Avery, Thomas E., 1977, Minneapolis, Burgess, 392 p; Glossary, p. 375-380.

Bowditch, Nathaniel, 1966, American practical navigator, corrected print: U.S. Naval Ocean. Off, H.O. Pub. No. 9, Appendix C, Glossary, p. 909-953.

Carter, Harley, 1973, Dictionary of electronics: Blue Ridge Summit, Pa., TAB Books.

Colwell, R. N., ed., Manual of photographic interpretation: Washington, D.C. (*presently* Falls Church, Va.), American Soc. Photogramm., 868 p.

Defense Mapping Agency Topographic Center, 1973, Glossary of mapping, charting, and geodetic terms: Def. Map. Agcy., Topo. Cen., 281 p.

El-Kassas, I. A., and El-Ghawaby, M. A., compilers, 1972, Glossary of aerial photography in geology and earth sciences: Arab Republic of Egypt, Acad. Sci. and Tech. Interim Rept. 3, 88 p., *distributed in* U.S. by Nat. Tech. Inf. Ser., PB-220 154.

Estes, J. E. and Senger, L. W., 1974, Remote sensing: Santa Barbara, Hamilton, 340 p; Glossary, p. 304-309.

Fairbridge, R. W., ed., 1966, Encyclopedia of oceanography: New York, Van Nostrand Reinhold, 1021 p.

———, 1967, The encyclopedia of atmospheric sciences and astrogeology: New York, Van Nostrand Reinhold, 1200 p.

———, 1968, The encyclopedia of geomorphology: New York, Van Nostrand Reinhold, 1295 p.

———, 1972, The encyclopedia of geochemistry and environmental sciences: New York, Van Nostrand Reinhold, 1321 p.

Gary, Margaret, McAfee, Robert Jr. and Wolf, C. L., eds., 1972, Glossary of geology: Washington, D.C., American Geol. Inst., 857 p.

Gove, P. D., ed.-in-ch., 1963, Webster's third new international dictionary of the English language, unabridged: Springfield, Mass., G. and C. Merriam Co., 2662 p.

Haensch-Haberkamp, 1967, Dictionary of agriculture, 3rd ed. [German, English, French, and Spanish]: New York, Adler [importer-distributer].

Hunt, L. M. and Groves, D. G., 1965, Glossary of ocean science and undersea technology terms: Arlington, Va., Compass, 172 p.

Inst. Electrical and Electronics Engineers, 1972, Standard dictionary of electrical and electronic terms: New York, Wiley-Interscience.

Lapedes, D. N., ed.-in-ch., 1974, Dictionary of scientific and technical terms: New York, McGraw-Hill, 1634 p. + Append., 26 p.

Mechtly, E. A., 1969, The international system of units: NASA SP-7012, 19 p.

Michels, W. C., ed.-in-ch., 1961, The international dictionary of physics and electronics, 2nd ed.: Princeton, NJ, Van Nostrand, 1355 p.

Moore, W. G., 1949, A dictionary of geography: Harmondsworth, Middlesex, Penguin, 182 p.

Mueller, I. I. and Rockie, J. D., 1966, Gravimetric and celestial geodesy—A glossary of terms: New York, Ungar, 129 p.

NASA Manned Spacecraft Center, 1973, Earth observations aircraft program photographic output requirements document: NASA MSC-07403, Appendix A, Glossary, p. 49-89.

Naylor, J. L., 1964, A dictionary of astronautics: New York, Hart, 316 p.

Pan American Institute of Geography and History Ad hoc Committee on Remote Sensors, 1974, Terminologia en percepcion remota (in English and Spanish): Pan American Institute of Geography and History, in collaboration with the Comision Nac. Espacio Exterior, Sec. Communic. y Transp. of Mexico, 102 p. (English), 176 p. (Spanish).

[1] Compiled by Robert R. Colwell

Pfannkuch, Hans-Olaf, 1969, Dictionary of hydrogeology: New York, American Elsevier, 168 p.

Reeves, R. G., ed, 1975, Manual of Remote Sensing, first ed: Falls Church, Va., American Soc. Photogramm, 2144 p.

Rosenberg, Paul, Erickson, E. E. and Rowe, G. C., 1974, Digital mapping glossary: U.S. Army Engr. Topo. Labs *by* Keuffel and Esser Co., Contract DAAK02-72-C-0288, 59 p.

Royal Society (Cartography Subcomm. of the British Nat. Comm. for Geography), 1966, Glossary of technical terms in cartography: London, the Royal Society. 84 p.

Runcorn, S. K., gen. ed., 1967, international dictionary of geophysics: Oxford, Pergamon, 1728 p.

Sabins, F. F., Jr., 1977, Remote sensing principles and interpretation: San Francisco, Ca., Freeman, 403 p.

Short, N. H., 1982, Landsat Workbook: NASA SP No 649, 327 p.

Short, N. M., 1982. The Landsat Tutorial Workbook: NASA Reference Publication No. 1078, 590 p.

Slama, C. C., ed., 1980, Manual of Photogrammetry— Fourth ed: Falls Church, Va., American Soc. Photogrammetry. 1056 p.

Smith, J. T. and Anson, Abraham, eds., 1968, Manual of color aerial photography: Falls Church, Va., American Society of Photogramm., 550 p.

Stamp, L. D., ed., 1966, Longman's dictionary of geography: New York, Wiley-Halsted, 492 p.

Thompson, M. M., ed., 1966, Manual of photogrammetry, 3rd ed.: Falls Church, Va., American Soc. Photogramm., 1199 p.

Weck, J., 1966, Dictionary of forestry: New York, American Elsevier.

Wenderoth, S., Yost, E., Rajender, K., and Anderson, R., 1974, Multispectral photography for earth resources: Huntington, N.Y., West Hills Printing Co., 264 p.

Winburne, J. N. et al, eds., 1962, Dictionary of agricultural and allied terminology: East Lansing, MI, Michigan State Univ. Press.

TERMS

A

Absorbed light: Light rays that are neither reflected nor transmitted when directed toward opaque or transparent materials.

Absorptance: A measure of the ability of a surface to absorb incident energy, often at specific wavelengths.

Absorption: The process by which radiant energy is absorbed and converted into other forms of energy.

Absorption band: A range of wavelengths (or frequencies) in the electromagnetic spectrum within which radiant energy is absorbed by a substance.

Absorption spectrum: The array of absorption lines and absorption bands that results from the passage of radiant energy from a continuous source through a selectively absorbing medium cooler than the source.

Absorptivity: The capacity of a material to absorb incident radiant energy. A special case of absorptance, it is a fundamental property of material that has a specular (optically smooth) surface and is sufficiently thick to be opaque. It may be further qualified as spectral absorptivity. The suffix (-ity) implies a property intrinsic with a given material, a limiting value.

Accommodation: (1) The faculty of the human eye to adjust itself to give sharp images for different object distances. (2) The ability of the eyes to bring two images into superimposition for stereoscopic viewing.

Accuracy: The success in estimating the true value. The closeness of an estimate of a characteristic to the true value of the characteristic of the population.

Actinic radiation: Pertaining to EMR capable of causing photochemical reactions, as in photography or the fading of pigments; a more general term than *actinic light*.

Active microwave: Ordinarily referred to as a radar.

Active system: A remote sensing system that transmits its own electromagnetic emanations at an object(s) and then records the energy reflected or refracted back to the sensor.

Acuity, visual: A measure of the human eye's ability to separate details in viewing an object. The reciprocal of the minimum angular separation, in minutes of arc, of two lines of detail which can be seen separately.

Acutance: An objective measure of the ability of a photographic system to show a sharp edge between contiguous areas of low and high illuminance. This film property now can be measured, using microdensitometric techniques, instead of being qualitatively estimated by subjective visual perception.

Additive color process: A method for creating essentially all colors through the addition of light of the three additive color primaries (blue, green, and red) in various proportions through the use of three separate projectors. In this type of process, each primary filter absorbs the other two primary colors and transmits only about one-third of the luminous energy of the source. It also precludes the possibility of mixing colors with a single light source because the addition of a second primary color results in total absorption of the light transmitted by the first color.

Aerial photograph, oblique: A photograph taken with the camera axis directed between the horizontal and the vertical (1) *high oblique:* An oblique photograph in which the apparent horizon is shown. (2) *low oblique:* An oblique photograph in which the apparent horizon is not shown.

Aerial photograph, vertical: An aerial photograph made with the optical axis of the camera approximately perpendicular to the Earth's surface and with the film as nearly horizontal as is practicable.

Aerial reconnaissance: The securing of information by aerial photography or by visual observation from the air.

Aerospace: (1) Of or pertaining to both the Earth's atmosphere and space. (2) The Earth's envelope of air and space above it; the two are considered as a single realm for the activity in the flight of air vehicles and in the launching, guidance, and control of ballistic missiles, Earth satellites, dirigible space vehicles, and the like.

Aerotriangulation (aerial triangulation): Triangulation for the extension of horizontal and (or) vertical control accomplished by means of aerial photographs.

Air base: An imaginary line connecting the points in space at which successive photos in a flight strip were taken; specifically, the length of such a line.

Albedo: (1) The ratio of the amount of EMR reflected by a body to the amount incident upon it, often expressed as a percentage, e.g., the albedo of the Earth is 34 percent. (2) The reflectivity of a body as compared to that of a perfectly diffusing surface at the same distance from the Sun, and normal to the incident radiation.

Algorithm: (1) A fixed step-by-step procedure to accomplish a given result; usually a simplified procedure for solving a complex problem; also a full statement of a finite number of steps. (2) A computer-oriented procedure for resolving a problem.

Alphanumeric: A character set composed of letters, integers, punctuation marks, and special symbols. Usually the number of characters in a set varies between forty-eight and sixty-four.

Altitude: Height above a datum; the datum is usually mean sea level. See *elevation*.

Amplifier: A device capable of increasing the power output of an electrical or EMR signal.

Analog: A form of data display in which values are shown in graphic form, such as curves. Also a form of computing in which values are represented by directly measurable quantities, such as voltages or resistances. Analog computing methods contrast with digital methods in which values are treated numerically.

Ancillary data: In remote sensing, secondary data pertaining to the area or classes of interest, such as topographical, demographic, or climatological data. Ancillary data may be digitized and used in the analysis process in conjunction with the primary remote sensing data.

Angle of depression: In SLAR usage, the angle between the horizontal plane passing through the antenna and the line connecting the antenna and the target.

Angle of incidence: (1) The angle between the direction of incoming EMR and the normal to the intercepting surface; (2) In SLAR systems this is the angle between the vertical and a line connecting antenna and target.

Angle of reflection: The angle that EMR reflected from a surface makes with the perpendicular (normal) to the surface.

Angle of view: The angle subtended by lines that pass through the center of the lens to diametrically opposite corners of the plate or film used.

Angstrom (Å): Unit of measurement, 10^{-10} m.

Anomaly: An area of an image that differs from the surrounding normal area. For example, a concentration of vegetation within a desert scene constitutes an anomaly.

Antenna: The device that radiates EMR from a transmitter and receives EMR from other antennae or other sources.

Antenna, synthetic aperture (radar): The effective antenna produced by storing and comparing the doppler signals received while the aircraft travels along its flight path. This synthetic antenna (or array) is many times longer than the physical antenna, thus sharpening the effective beam width and improving azimuth resolution.

Aperture: The opening in a lens diaphragm through which light passes.

Aperture, effective: The unobstructed useful area of the orifice through which the ray of light passes to the film to produce the image. This is smaller than the actual area of the hole due to the effects of diffraction or bending of the rays near the edges.

Aperture, relative: The ratio of the equivalent focal length to the diameter of the entrance "pupil" of a photographic lens. Expressed "f:4.5," etc. Also

called "f-number," "stop," "aperture stop," "speed."

Area weighted average resolution (AWAR): A single average resolution value over a picture format determined by obtaining the summation of the resolution in concentric annular zones. Zone boundaries are determined from angles midway between successive test angles. The resolution obtained at any test angle is multiplied by the ratio of that area to the total picture area.

Atmospheric windows: Those wavelength ranges in which radiation can pass through the atmosphere with relatively little attenuation; in the optical portion of the spectrum, approximately 0.3-2.5, 3.0-4.0, 4.2-5.0, and 7.0-15.0 μm.

Attenuation: In physics, any process in which the flux density (or power, amplitude, intensity, illuminance) of a "parallel beam" of energy decreases with increasing distance from the energy source.

Attitude: The angular orientation of a remote sensing system with respect to a geographical reference system.

Axis, optical: In a lens element, the straight line which passes through the centers of curvature of lens surfaces. Also called *principal axis*. In an optical system, the line formed by the coinciding principal axes of the series of optical elements.

Azimuth: The geographical orientation of a line given as an angle measured clockwise from north.

B

Background: Any effect in a sensor or other apparatus or system, above which the phenomenon of interest must manifest itself before it can be observed. (See background noise.)

Background luminance: In visual-range theory, the luminance (brightness) of the background against which a target is viewed.

Background noise: (1) In recording and reproducing, the total system noise independent of whether or not a signal is present. The signal is not to be included as part of the noise. (2) In receivers, the noise in the absence of signal modulation on the carrier. Ambient noise detected, measured, or recorded with the signal becomes part of the background noise. Included in this definition is the interference resulting from primary power supplies, which separately is commonly described as hum.

Backscatter: The scattering of radiant energy into the hemisphere of space bounded by a plane normal to the direction of the incident radiation and lying on the same side as the incident ray; the opposite of forward scatter. Also called backscattering.

Band: (1) A selection of wavelengths. (2) Frequency band. (3) Absorption band. (4) A group of tracks on a magnetic drum. (5) A range of radar frequencies, such as X-band, Q-band, etc.

Band-pass filter: A wave filter that has a single transmission band extending from a lower cutoff frequency greater than zero to a finite upper cutoff frequency.

Bandwidth: (1) In an antenna, the range of frequencies within which its performance, with respect to some characteristic, conforms to a specified standard. (2) In a wave, the least frequency interval outside which the power spectrum of a time-varying quantity is everywhere less than some specified fraction of its value at a reference frequency. (3) The number of cycles per second between the limits of a frequency band.

Bar scale: A graduated line on a map, plan, photograph, or mosaic, by means of which actual ground distances may be determined. Also called *graphic scale.*

Base-height ratio: Air base (ground distance between centers of successive overlapping photos) divided by aircraft height. This ratio determines vertical exaggeration on stereo models.

Base, photo: The distance between the principal points of two adjacent prints of a series of vertical aerial photographs. It is usually measured on one print after transferring the principal point of the other print.

Batch processing: The method of data processing in which data and programs are entered into a computer which then carries out the entire processing operation with no further instructions.

Beam: A focused pulse of energy.

Beam splitter: An optical device, such as a semireflecting mirror or a prism arranged so as to transmit different spectral bands along separate axes to various films, detectors, or other analyzing/recording devices.

Bit: (1) An abbreviation of *binary digit.* (2) A single character of a language employing only two distinct kinds of characters.

Blackbody, black body (symbol bb used as subscript): An ideal emitter which radiates energy at the maximum possible rate per unit area at each wavelength for any given temperature. A blackbody also absorbs all the radiant energy incident upon it. No actual substance behaves as a true blackbody.

Blackbody radiation: The electromagnetic radiation emitted by an ideal blackbody; it is the theoretical maximum amount of radiant energy of all wavelengths that can be emitted by a body at a given temperature. The spectral distribution of blackbody radiation is described by Planck's law and related radiation laws.

Brightness: (1) The attribute of visual perception in accordance with which an area appears to emit more or less light. (2) Luminance. (3) The luminous flux emitted or reflected per unit projected area per unit solid angle. The unit of brightness, the lambert, is defined as brightness of a surface which emits or reflects one/π lumen per square centimeter per steradian.

Brightness temperature: (1) The temperature of a blackbody radiating the same amount of energy per unit area at the wavelengths under consideration as the observed body. Also called effective temperature. (2) The apparent temperature of a nonblackbody determined by measurement with an optical pyrometer or radiometer.

Brute force: A sidelooking radar system that transmits and receives from a long physical antenna to narrow the beamwidth and increase azimuth resolution; the received returned EMR is used directly to produce an image. Compare *synthetic aperture.*

Byte: A group of eight bits of digital data.

C

Calibration: The act or process of comparing certain specific measurements in an instrument with a standard.

Camera, multiband: A camera that exposes different areas of one film, or more than one film, through one lens and a beam splitter, or two or more lenses equipped with different filters, to provide two or more photographs in different spectral bands.

Camera, panoramic: A camera with a very wide angle of view, up to horizon to horizon, usually by means of a moving (sweeping) lens.

Camera station: The point in space, in the air or on the ground, occupied by the camera lens at the moment of exposure. Also called the *exposure station.* In aerial photography the camera station is called the "air station".

Cartography: Map and chart construction

Cascade (electronics): To arrange a series of elements or devices so that the output of one feeds directly into the input of another, as a series of dynodes or a series of amplifiers. The cascaded series usually serves to amplify the effect.

Category: Each unit is assumed to be of one and only one given type. The set of types is called the set of "classes" or "categories," each type being a particular category. The categories are chosen specifically by the investigator as being the ones of interest to him.

Cathode ray tube (CRT): A vacuum tube capable of producing a black-and-white or color image by beaming electrons onto a sensitized screen. As a component of a data-processing system, the CRT can be used to provide rapid, pictorial access to numerical data.

CCT: Computer Compatible Tape.

Cell: An area on the ground from which EMR is emitted or reflected.

Change-detection images: Images prepared by digitally comparing two original images acquired at different times. The gray tones of each pixel on a change-detection image portray the amount of difference between the original images.

Characteristic curve: A curve showing the relationship between exposure and resulting density in a photographic image, usually plotted as the density (D) against the logarithm of the exposure (log E) in *candela-meter-seconds.* It is also called the *H and D curve,* the *sensitometric curve,* and the *D log E curve.*

Chi-square test: A statistical significance test based on frequency of occurrence; it is applicable both to qualitative attributes and quantitative variables. Among its many uses, the most common are tests of hypothesized probabilities or probability distributions (goodness of fit), statistical dependence or independence.

Chop: To interrupt the incoming EMR to a radiometer or similar device, so the detector alternately receives the incoming EMR and EMR from a *reference body.*

Chopper: A device, usually one that rotates, used to interrupt a continuous wave signal in a transmitter, receiver, or sensor.

Chroma: The color dimension on the Munsell scales that correlates most closely with *saturation.*

Class: A surface characteristic type that is of interest to the investigator, such as forest by type and condition, or water by sediment load.

Classification: The process of assigning individual pixels of a multispectral image to categories, generally on the basis of spectral reflectance characteristics.

Clipping: The shearing off of the peaks of a signal. This may affect either the positive or negative peaks, or both.

Clustering: The analysis of a set of measurement vectors to detect their inherent tendency to form clusters in multidimensional measurement space.

Coherent: (1) Of EMR, being in phase, so that waves at various points in a space act in unison, as in a *laser* producing coherent light. (2) Having a fixed relation between frequency and phase of input and output signal.

Color: The property of an object which is dependent on the wavelength of the light it reflects or, in the case of a luminescent body, the wavelength of light that it emits. If, in either case, this light is of a single wavelength, the color seen is a pure spectral color; but if light of two or more wavelengths is emitted, the color will be mixed. White light is a balanced mixture of all the visible spectral colors.

Color balance: The proper intensities of colors in a color print, positive transparency, or negative, that give a correct reproduction of the gray scale (as faithful as can be achieved by photographic representation of the true colors of a scene.)

Color composite (multiband photography): A color picture produced by assigning a color to a particular spectral band. In Landsat, blue is ordinarily assigned to MSS band 4 ($0.5-0.6\ \mu$m), green to band 5 ($0.6-0.7\ \mu$m), and red to band 7 ($0.8-1.1\ \mu$m), to form a picture closely approximating a color-infrared photograph.

Color infrared film: Photographic film sensitive to energy in the visible and near-infrared wavelengths, generally from $0.4-0.9\ \mu$m; usually used with a minus-blue (yellow) filter, which results in an effective film sensitivity of $0.5-0.9\ \mu$m. Color infrared film is especially useful for detecting changes in the condition of the vegetative canopy which are often manifested in the near-infrared region of the spectrum. Note that color infrared film is not sensitive in the thermal infrared region and therefore cannot be used as a heat-sensitive detector.

Color temperature: An estimate of the temperature of an incandescent body, determined by observing the wavelength at which it is emitting with peak intensity (its color), determined by applying the Wien law.

Complementary colors: (1, optics) Two colors are complementary if, when added together (as by projection), they produce neutral-hue light. (2) Colors of pigment which when mixed produce a gray.

Computer-compatible tapes: Tapes containing digital Landsat data. These tapes are standard 19-cm (7½-in) wide magnetic tapes in 9-track or 7-track format. Four tapes are required for the four-band multispectral digital data corresponding to one Landsat scene.

Contact print: A print made from a negative or a diapositive in direct contact with sensitized material (see *contact size*).

Contact size: A print, either positive or negative, of the same size as the negative or positive from which it was made.

Continuous spectrum: (1) A spectrum in which wavelengths, wavenumbers, and frequencies are represented by the continuum of real numbers or a portion thereof, rather than by a discrete sequence of numbers. See absorption spectrum. (2) For EMR, a spectrum that exhibits no detailed structure and represents a gradual variation of intensity with wavelength from one end to the other, as the spectrum from an incandescent solid.

Contrast stretching: Improving the contrast of images by digital processing. The original range of digital values is expanded to utilize the full contrast range of the recording film or display device.

Control, ground: (1) Control obtained by ground surveys as distinguished from control obtained by photogrammetric methods; may be for *horizontal* or *vertical control*, or both. (2) Ground (in-situ) observations to aid in interpretation of remote sensor data. See *ground data, ground information, ground truth*.

Control, photogrammetric: Control established by photogrammetric methods as distinguished from control established by ground methods. Also called *minor control*.

Convergence of evidence: The bringing together of several types of information in order that a conclusion may be drawn in the light of all available data. In remote sensing, often implies increase in scale to obtain more information about a smaller area.

Coordinates, geographical: A system of spherical coordinates for describing the positions of points on the Earth. The declinations and polar bearings in this system are the latitudes and longitudes, respectively.

Corner reflector (dihedral): A dihedral (two-sided) corner reflector is formed by two intersecting flat surfaces perpendicular to each other. Radar energy striking one of these surfaces is reflected back to the antenna via the other surface. Frequently used on control points in radar surveys.

Correlator, optical (radar): A device that uses the original synthetic aperture radar signal film recording of *doppler phase histories* to make the radar image by methods that are similar to those used in optical Fourier transformation.

Covariance: The measure of how two variables change in relation to each other (covariability). If larger values of Y tend to be associated with larger values of X, the covariance will be positive. If larger values of Y are associated with smaller values of X, the covariance will be negative. When there is no particular association between X and Y, the covariance value will approach zero.

Coverage, stereoscopic: Aerial photographs taken with sufficient overlap to permit *complete* stereoscopic examination.

Crab: (1, aerial photography) The condition caused by failure to orient the camera so that the axis perpendicular to the long dimension of the film is parallel to the track of the airplane. This is indicated in vertical photography by the sides of the photographs not being parallel to the principal-point base line. See *drift*. (2, air navigation) Any turning of an airplane which causes its longitudinal axis to vary from the track of the plane.

Crown diameter, visible: The apparent diameter of a tree crown imaged on a vertical aerial photograph.

CRT: Cathode-ray tube.

Cultural features: All map detail representing man-made elements of the landscape.

Cursor: Aiming device, such as a lens with cross-hairs, on a digitizer or an interactive computer display.

D

Data acquisition system: The collection of devices and media that measures physical variables and records them prior to input to the data processing system.

Data bank: A well-defined collection of data, usually of the same general type, which can be accessed by a computer.

Data dimensionality: The number of variables (e.g., channels) present in the data set. The term "intrinsic dimensionality" refers to the smallest number of variables that could be used to represent the data set accurately.

Data link: Any communications channel or circuit used to transmit data from a sensor to a computer, a readout device, or a storage device. See Chapters 10, 12.

Data processing: Application of procedures—mechanical, electrical, computational, or other—whereby data are changed from one form into another.

Data reduction: Transformation of observed values into useful, ordered, or simplified information.

Decision rule (or classification rule): The criterion used to establish discriminant functions for classification (e.g., nearest-neighbor rule, minimum-distance-to-means rule, maximum-likelihood rule).

Definition (photography): The degree of sharpness, that is, distinctness of small detail in the picture image, negative, or print.

Density (symbol, D): A measure of the degree of blackening of an exposed film, plate, or paper after development, or of the direct image (in the case of a printout material). It is defined strictly as the logarithm of the optical opacity, where the opacity is the ratio of the incident to the transmitted (or reflected) light or transmissivity, T, as $D = \log (1/T)$.

Density slicing: The process of converting the continuous gray tone of an image into a series of density intervals, or slices, each corresponding to a specific digital range.

Detection: A unit is said to be "detected" if the decision rule is able to assign it as belonging only to some given subset of categories from the set of all categories. Detection of a unit does not imply that the decision rule is able to identify the unit as specifically belonging to one particular category.

Detector (radiation): A device providing an electrical output that is a useful measure of incident radiation. It is broadly divisible into two groups: thermal (sensitive to temperature changes), and photodetectors (sensitive to changes in photon flux incident on the detector), or it may also include antennas and film. Typical thermal detectors are thermocouples, thermopiles, and thermistors; the latter is termed a bolometer.

Diapositive: A positive image on a transparent medium such as glass or film; a transparency. The term originally was used primarily for a transparent positive on a glass plate used in a plotting instrument, a projector, or a comparator, but now is frequently used for any positive transparency.

Dielectric constant: Electrical property of matter that influences radar returns; also referred to as complex dielectric constant.

Diffraction: The propagation of EMR around the edges of opaque objects into the shadow region. A point of light seen or projected through a circular aperture will always be imaged as a bright center surrounded by light rings of gradually diminishing intensity in the shadow region. Such a pattern is called a diffraction disk, Airy disk, or centric.

Diffuse reflection: The type of reflection obtained from a relatively rough (in terms of the wave-length of the EMR) surface, in which the reflected rays are scattered in all directions.

Diffuse reflector: Any surface that reflects incident rays in many directions, either because of irregularities in the surface or because the material is optically inhomogeneous, as a paint; the opposite of a specular reflector. Ordinary writing papers are good examples of diffuse reflectors, whereas mirrors or highly polished plates are examples of specular reflectors in the visible portion of the EM spectrum. Almost all terrestrial surfaces (except calm water) act as diffuse reflectors of incident solar radiation. The smoothness or roughness of a surface depends on the wavelength of the incident EMR.

Diffuse sky radiation: Solar radiation reaching the Earth's surface after having been scattered from the direct solar beam by molecules or suspensoids in the atmosphere. Also called skylight, diffuse skylight, sky radiation.

Digitization: The process of converting an image recorded originally on photographic material into numerical format.

Discriminant function: One of a set of mathematical functions which in remote sensing are commonly derived from training samples and a decision rule, and are used to divide the measurement space into decision regions.

Displacement: Any shift in the position of an image on a photograph which does not alter the perspective characteristics of the photograph (i.e., shift due to tilt of the photograph, scale change in the photograph, and relief of the objects photographed).

Display: An output device that produces a visible representation of a data set for quick visual access; usually the primary hardware component is a cathode ray tube.

Distribution function: The relative frequency with which different values of a variable occur.

Diurnal: Having a period of, occurring in, or related to a day.

DN: Digital number. The value of reflectance recorded for each pixel on Landsat CCT's.

Doppler effect: A change in the observed frequency of EM or other waves caused by relative motion between the source and the observer.

Drift: (1, air navigation) The horizontal displacement of an aircraft, caused by the force of wind, from the track it would have followed in still air. (2, aerial photography) Sometimes used to indicate a special condition of crab wherein the photographer has continued to make exposures oriented to the predetermined line of flight while the airplane has drifted from that line.

Dynamic range: The ratio of maximum measurable signal to minimum detectable signal.

E

Edge: The boundary of an object in a photograph or image, usually characterized by a rather drastic change in the gray shade value from the immediate interior of the boundary to the immediate exterior of the boundary.

Edge enhancement: The use of analytical techniques to emphasize transition in imagery.

Electromagnetic radiation (EMR): Energy propagated through space or through material media in the form of an advancing interaction between electric and magnetic fields. The term radiation, alone, is commonly used for this type of energy, although it actually has a broader meaning. Also called electromagnetic energy.

Electromagnetic spectrum: The ordered array of

known electromagnetic radiations extending from the shortest cosmic rays, through gamma rays, X-rays, ultraviolet radiation, visible radiation, infrared radiation, and including microwave and all other wavelengths of radio energy.

Element: The smallest definable object of interest in the survey. It is a single item in a collection, population, or sample.

Elevation: (1) Vertical distance from the datum, usually mean sea level, to a point or object on the earth's surface. Not to be confused with altitude, which refers to points or objects above the earth's surface. (2, architectural) An orthographic projection of any object into a vertical plane.

Emission: With respect to EMR, the process by which a body emits EMR usually as a consequence of its temperature only.

Emissivity: The ratio of the radiation given off by a surface to the radiation given off by a blackbody at the same temperature; a blackbody has an emissivity of 1, other objects between 0 and 1.

Emittance: The obsolete term for the radiant flux per unit area emitted by a body, or exitance.

Emulsion (photography): A suspension of a light-sensitive silver salt (especially silver chloride or silver bromide) in a colloidal medium (usually gelatin), which is used for coating photographic films, plates, and papers.

End lap: The overlap of aerial or space photographs or images along the flightline or track of the platform.

Environment: An external condition, or the sum of such conditions, in which a piece of equipment or a system operates, as in temperature environment, vibration environment, or space environment. The environments are usually specified by a range of values, and may be either natural or artificial.

Ephemeral data: Data that: (1) help to characterize the conditions under which the remote sensing data were collected: (2) may be used to calibrate the sensor data prior to analysis; (3) include such information as the positioning and spectral stability of sensors, Sun angle, platform attitude, etc.

Equivalent blackbody temperature: The temperature measured radiometrically corresponding to that which a blackbody would have. Most natural objects including soil, plant leaves, and water have emissivities greater than 0.9 but less than 1.0.

Exitance (symbol, M): The radiant flux per unit area emitted by a body or surface.

Exposure: (1) The total quantity of light received per unit area on a sensitized plate or film; may be expressed as the product of the light intensity and the exposure time, in units of (for example) meter-candle-seconds or watts per square meter. (2) The act of exposing a light-sensitive material to a light source.

F

False color: The use of one color to represent another; for example, the use of red emulsion to represent infrared light in color infrared film.

Far infrared: A term for the longer wavelengths of the infrared region, from 25 μm to 1 mm, the generally accepted shorter wavelength limit of the microwave part of the EM spectrum. This is severely limited in terrestrial use, as the atmosphere transmits very little radiation between 25 μm and the millimeter regions.

Far range: Refers to the portion of an SLAR image farthest from the aircraft flight path.

Feature: An n-tuple or vector with components which are functions of the initial measurement pattern variables or some subsequence of the measurement n-tuples. Feature n-tuples or vectors frequently have fewer components than the initial measurement vectors and are designed to contain a high amount of information about the discrimination between units of the types of categories in the given category set. Features often contain information about gray shade, texture, shape or context. Also, a cartographic type in digital form appearing as part of the descriptor in coded form (Feature Code).

Feature extraction: The process in which an initial measurement pattern or some subsequence of measurement patterns is transformed to a new pattern feature.

Fiducial marks: Index marks (usually 4), rigidly connected with the camera lens through the camera body, which form images on the negative. The marks are adjusted so that the intersection of lines drawn between opposite fiducial marks defines the principal point.

Field of view: The solid angle through which an instrument is sensitive to radiation. Owing to various effects, diffractions, etc., the edges are not sharp. In practice they are defined as the "half-power" points, i.e., the angle outwards from the optical axis, at which the energy sensed by the radiometer drops to half its on-axis value.

Filter: (1, noun) Any material which, by absorption or reflection, selectively modifies the radiation transmitted through an optical system. (2, verb) To remove a certain component or components of EMR, usually by means of a filter, although other devices may be used.

Filtering: In analysis, the removal of certain spectral or spatial frequencies to highlight features in the remaining image.

Focal length: The distance measured along the optical axis from the optical center (rear *nodal point*) of the lens to the plane of critical focus of a very distant object.

Focus: The point at which the rays from a point source of light reunite and cross after passing through a camera lens. In practice, the plane in which a sharp image of any scene is formed.

Format: The arrangement of descriptive data in descriptors, identifiers, or labels. The arrangement of data in bit, byte, and word form in the CPU.

Frame: Complete tape of a single or multidate Landsat frame covering roughly an area about 100 nautical miles square.

Frequency: Number of oscillations per unit time or number of wavelengths that pass a point per unit time. See also *period*. The frequency bands used by radar (radar frequency bands) were first designated by letters for military secrecy. Those designations are:

Frequency band	Approximate frequency range, gigahertz.	Approximate wavelength range, centimeters.
P-band	0.225 to 0.39	140 to 76.9
L-band	0.39 to 1.55	76.9 to 19.3
S-band	1.55 to 5.20	19.3 to 5.77
X-band	5.20 to 10.90	5.77 to 2.75
K-band	10.90 to 36.00	2.75 to 0.834

Q-band . . . . . 36.00 to 46.00   0.834 to 0.652
V-band . . . . . 46.00 to 56.00   0.652 to 0.536

The C-band, 3.9 to 6.2 gigahertz, overlaps the S- and X-bands. These letter designations have no official sanction, but still are widely used.

Frequency response: (1) Response of a system as a function of the frequency of excitations. (2) The portion of the frequency spectrum that can be sensed by a device within specified limits of amplitude error.

G

Gain: (1) A general term used to denote an increase in signal power in transmission from one point to another. Gain is usually expressed in decibels. (2) An increase or amplification.

Gamma: A numerical measure of the extent to which a negative has been developed, indicating the proportion borne by the contrast of the negative to that of the subject on which it was exposed. The numerical figure for gamma is the tangent of the straight-line (correct exposure portion of the curve resulting from plotting exposure against density.

Gaussian: A statistical term that refers to a *normal* distribution of values.

GCP: Ground control point. A geographical feature of known location that is recognizable on images and can be used to determine geometrical corrections.

Geocoding: Geographical referencing or coding of location of data items.

Geometrical transformations: Adjustments made in the image data to change its geometrical character, usually to improve its geometrical consistency or cartographic utility.

Grain (photography): (1) One of the discrete silver particles resulting from the development of an exposed light-sensitive material. (2) A lack of smoothness of the silver deposit, caused by clumps or groups of particles. Excessive graininess reduces quality, especially when magnified or enlarged.

Granularity: The graininess of a developed photographic image, evident particularly on enlargement, that is due either to agglomerations of developed grains or to an overlapping pattern of grains.

Gray body: A radiating surface whose radiation has essentially the same spectral energy distribution as that of a blackbody at the same temperature, but whose emissive power is less. Its absorptivity is nonselective. Also spelled *grey* body.

Gray scale: A calibrated sequence of gray tones ranging from black to white.

Grid line: One of the lines in a grid system; a line used to divide a map into squares. East-west lines in a grid system are x-lines, and north-south lines are y-lines.

Ground control: Accurate data on the horizontal and/or vertical positions of identifiable ground points.

Ground data: Supporting data collected on the ground, and information derived therefrom, as an aid to the interpretation of remotely-recorded surveys, such as airborne imagery, etc. Generally, this should be performed concurrently with the airborne surveys. Data as to weather, soils and vegetation types and conditions are typical.

Ground information: Information derived from *ground data* and *surveys* to support interpretation of remotely sensed data.

Ground range: The distance from the ground track (nadir) to a given object.

Ground resolution cell: The area on the terrain that is covered by the instantaneous field of view of a detector. The size of the ground resolution cell is determined by the altitude of the remote-sensing system and the instantaneous field of view of the detector.

Ground track: The vertical projection of the actual flight path of an aerial or space vehicle onto the surface of the Earth or other body.

Ground truth (jargon): Term coined for data and information obtained on surface or subsurface features to aid in interpretation of remotely sensed data. Ground data and ground information are preferred terms.

H

H-D (Hurter-Driffield) Curve: A graph showing the relationship of exposure to (photo) density, where the density is plotted against the logarithm of the exposure (also known as characteristic curve).

Hardware: The physical components of a computer and its peripheral equipment. Contrasted with software.

Histogram: The graphical display of a set of data which shows the frequency of occurrence (along the vertical axis) of individual measurements or values (along the horizontal axis); a frequency distribution.

Hotspot: The destruction of fine image detail on a portion of a wide-angle aerial photograph. It is caused by the absence of shadows and by halation near the prolongation of a line from the sun through the exposure station.

Hue: That attribute of a color by virtue of which it differs from gray of the same brilliance, and which allows it to be classed as red, yellow, green, blue, or intermediate shades of these colors.

I

Illumination: The intensity of light striking a unit surface is known as the specific illumination or luminous flux. It varies directly with the intensity of the light source and inversely as the square of the distance between the illuminated surface and the source. It is measured in a unit called the lux. The total illumination is obtained by multiplying the specific illumination by the area of the surface covered by the light. The unit of total illumination is the lumen.

Image: (1) The counterpart of an object produced by the reflection or refraction of light when focused by a lens or mirror. (2) The recorded representation (commonly as a photo-image) of an object produced by optical, electrooptical, optical mechanical, or electronic means. It is generally used when the EMR emitted or reflected from a scene is not directly recorded on film.

Image Enhancement: Any one of a group of operations that improve the detectability of the targets or categories. These operations include, but are not limited to, contrast improvement, edge enhancement, spatial filtering, noise suppression, image smoothing, and image sharpening.

Image Processing: Encompasses all the various operations that can be applied to photographic or image data. These include, but are not limited to, image compression, image restoration, image enhance-

ment, preprocessing, quantization, spatial filtering and other image pattern recognition techniques.

Image Restoration: A process by which a degraded image is restored to its original condition. Image restoration is possible only to the extent that the degradation transform is mathematically invertible.

Imagery: The products of image-forming instruments (analogous to *photography*).

Incident ray: A ray impinging on a surface.

Infrared: Pertaining to energy in the 0.7−100 μm wavelength region of the electromagnetic spectrum. For remote sensing, the infrared wavelengths are often subdivided into near infrared (0.7−1.3 μm), middle infrared (1.3−3.0 μm), and far infrared (7.0−15.0 μm). Far infrared is sometimes referred to as thermal or emissive infrared.

Infrared, photographic: Pertaining to or designating the portion of the EM spectrum with wavelengths just beyond the red end of the visible spectrum; generally defined as from 0.7 to about 0.1 μm, or the useful limits of film sensitivities.

Insolation: Incident solar energy.

Instantaneous field of view: (IFOV) A term specifically denoting the narrow field of view designed into detectors, particularly scanning radiometer systems, so that, while as much as 120° may be under scan, only EMR from a small area is being recorded at any one instant.

Interactive image processing: The use of an operator or analyst at a console that provides the means of assessing, preprocessing, feature extracting, classifying, identifying, and displaying the original imagery or the processed imagery for his subjective evaluations and further interactions.

Irradiance: The measure, in power units, of radiant flux incident upon a surface. It has the dimensions of energy per unit time (e.g., watts).

Irradiation: The impinging of EMR on an object or surface.

J

Jitter: (1) Instability of the signal or trace of a cathode-ray tube. (2) Small rapid variations in a waveform due to deliberate or accidental electrical or mechanical disturbances or to changes in the supply voltages, in the characteristic of components, etc.

K

Kelvin: A thermometer scale starting at absolute zero (−273°C approximately) and having degrees of the same magnitude as those of the Celsius thermometer. Thus, 0°C = 273°K; 100°C = 373°K; etc.; also called the absolute scale, thermodynamic temperature scale.

Kinetic temperature: The internal temperature of an object, which is determined by the molecular motion. Kinetic temperature is measured with a contact thermometer, and differs from radiant temperature, which is a function of emissivity and internal temperature.

Kirchhoff's Law: The radiation law which states that at a given temperature the ratio of the emissivity to the absorptivity for a given wavelength is the same for all bodies and is equal to the emissivity of an ideal blackbody at that temperature and wavelength. This important law asserts that good absorbers of a

given wavelength are also good emitters of the wavelength.

L

Lambertian surface: An ideal, perfectly diffusing surface, which reflects energy equally in all directions.

Langley: A unit of luminous intensity, defines as 4.184 × 10⁴ joule m⁻².

Large scale: (1) Aerial photography with a representative fraction of 1:500 to 1:10,000. (2) Maps with a representative fraction (scale) greater than 1:100,000.

Layover: Displacement of the top of an elevated feature with respect to its base on the radar image. The peaks look like dip-slopes.

Light: Visible radiation (about 0.4−0.7 μm in wavelength) considered in terms of its luminous efficiency; i.e., evaluated in proportion to its ability to stimulate the sense of sight.

Line, flight: A line drawn on a map or chart to represent the track over which an aircraft has been flown or is to fly. The line connecting the principal points of vertical aerial photographs.

Lineament: A linear topographical or tonal feature on the terrain and on images and maps, which may represent a zone of structural weakness.

Linear feature: A two-dimensional, straight to somewhat curved line, linear pattern, or alignment of discontinuous patterns evident in an image, photo, or map, which represents the expression of some degree of linearity of a single or diverse grouping of natural or cultural ground features.

Look angle (radar): The direction of the *look,* or direction in which the antenna is pointing when transmitting and receiving from a particular *cell.*

Look direction: Direction in which pulses of microwave energy are transmitted by a SLAR system. Look direction is normal to the azimuth direction. Also called range direction.

Lumen: A unit of luminous flux equal to the luminous flux radiated into a unit solid angle (steradian) from a point source having a luminous intensity of 1 candela. An ideal source possessing an intensity of 1 candela in every direction would radiate a total of 4π lumens.

Luminance: In photometry, a measure of the intrinsic luminous intensity emitted by a source in a given direction; the illuminance produced by light from the source upon a unit surface area oriented normal to the line of sight at any distance from the source, divided by the solid angle subtended by the source at the receiving surface. Also called brightness (luminance is preferred).

Lux: A unit of illumination equivalent to one *lumen* of incident light per m².

M

Magnetometer: An instrument for measuring changes in the Earth's magnetic field and used extensively in airborne geophysical surveying. Three broad categories of this device are used; namely, *flux-gates, nuclear magnetic resonance detectors,* and *optically pumped systems.* The former two generally operate at sensitivities of 1 gamma (10⁻⁵ gauss) and the latter type is used for high sensitivity surveys with a sensitivity approaching 1/100th of a gamma.

Map: A representation in a plane surface, at an established scale, of the physical features (natural, artificial, or both) of a part of the Earth's surface, with the means of orientation indicated.

Map, large-scale: A map having a scale of 1:100,000 or larger.

Map, medium-scale: A map having a scale from 1:100,000, exclusive, to 1:1,000,000, inclusive.

Map, small-scale: A map having a scale smaller than 1:1,000,000.

Map, thematic: A map designed to demonstrate particular features or concepts. In conventional use this term excludes topographical maps.

Maximum likelihood rule: A statistical decision criterion to assist in the classification of overlapping signatures; pixels are assigned to the class of highest probability.

Microwave: A very short EM wave; any wave between 1 meter and 1 millimeter in wavelength or 300 GHz to 0.3 GHz in frequency. The portion of the electromagnetic spectrum in the millimeter and centimeter wavelengths, bounded on the short wavelength sides by the far infrared (at 1 mm) and on the long wavelength side by very high-frequency radio waves. Passive systems operating at these wavelengths sometimes are called *microwave systems*. Active systems are called *radar,* although the literal definition of radar requires a distance-measuring capability not always included in active systems. The exact limits of the microwave region are not defined.

Micrometer (abbr. μm): A unit of length equal to one-millionth (10^{-6}) of a meter or one-thousandth (10^{-3}) of a millimeter.

Micron (abbr. μ): Equivalent to and replaced by micrometer; 10^{-6} m.

Microwave: Electromagnetic radiation having wavelengths between 1 m and 1 mm or 300−0.3 GHz in frequency, bounded on the short wavelength side by the far infrared (at 1 mm) and on the long wavelength side by very high-frequency radio waves. Passive systems operating at these wavelengths are sometimes called microwave systems. Active systems are called radar, although the literal definition of radar requires a distance-measuring capability not always included in active systems. The exact limits of the microwave region are not defined.

Mie scattering: Multiple reflection of light waves by atmospheric particles that have the approximate dimensions of the wavelength of light.

Mie theory: A complete mathematical-physical theory of the scattering of electromagnetic radiation by spherical particles, developed by G. Mie in1908. In contrast to Rayleigh scattering, the Mie theory embraces all possible ratios of diameter to wavelength. The Mie theory is very important in meteorological optics, where diameter-to-wavelength ratios of the order of unity and larger are characteristic of many problems regarding haze and cloud scattering. Scattering of radar energy by raindrops constitutes another significant application of the Mie theory.

Minimum distance classifier: A classification technique that assigns raw data to the class whose mean falls the shortest Euclidean distance from it.

Monochromatic: Pertaining to a single wavelength or, more commonly, to a narrow band of wavelengths.

Mosaic: An assemblage of overlapping aerial or space photographs or images whose edges have been matched to form a continuous pictorial representation of a portion of the Earth's surface.

Mosaic, controlled: A mosaic that is laid to ground control and uses prints that have been rectified as shown to be necessary by the control.

Mosaicking: The assembling of photographs or other images whose edges are cut and matched to form a continuous photographic representation of a portion of the Earth's surface.

MSS: Multispectral Scanner.

Multiband system: A system for simultaneously observing the same (small) target with several filtered bands, through which data can be recorded. Usually applied to cameras; may be used for scanning radiometers that use dispersant optics to split wavelength bands apart for viewing by several filtered detectors.

Multichannel system: Usually used for scanning systems capable of observing and recording several channels of data simultaneously, preferably through the same aperture.

Multispectral: Generally used for remote sensing in two or more spectral bands, such as visible and IR.

Multispectral (line) scanner: A remote sensing device that operates on the same principle as the infrared scanner, except that it is capable of recording data in the ultraviolet and visible portions of the spectrum as well as the infrared.

Multivariate analysis: A data-analysis approach that makes use of multidimensional interrelations and correlations within the data for effective discrimination.

N

Nadir: (1) That point on the celestial sphere vertically below the observer, or 180° from the zenith. (2) That point on the ground vertically beneath the perspective center of the camera lens.

Nautical mile (abbr. knot): A unit of distance used principally in navigation. For practical navigation it is usually considered the length of one minute of any great circle of the Earth, the meridian being the great circle most commonly used. Also called sea mile.

Near infrared: The preferred term for the shorter wavelengths in the infrared region extending from about 0.7 micrometers (visible red), to around 2 or 3 micrometers (varying with the author). The longer wavelength end grades into the middle infrared. The term really emphasizes the radiation reflected from plant materials, which peaks around 0.85 micrometers. It is also called solar infrared, as it is only available for use during the daylight hours.

Near range: Refers to the portion of an SLAR image closest to the aircraft flight path.

Negative: (1) A photographic image on film, plate, or paper, in which the tones are reversed. (2) A film, plate, or paper containing such a reversed image.

Noise: Random or regular interfering effects in the data which degrade its information-bearing quality.

O

Oblique photograph: A photograph taken with the camera axis intentionally directed between the horizontal and the vertical. A high-oblique photograph is one in which the apparent horizon is included within the field of view, whereas a low-oblique photograph does not include the apparent horizon within the field of view.

Optical axis: In a lens element, the straight line which passes through the centers of curvature of the lens surfaces. Also called *principal axis*. In an optical system, the line formed by the coinciding principal axes of the series of optical elements.

Orbit: (1) The path of a body or particle under the influence of a gravitational or other force. For instance, the orbit of a celestial body is its path relative to another body around which it revolves. (2) To go around the Earth or other body in an orbit.

Orthophotograph: Photograph having the properties of an orthographic projection. It is derived from a conventional perspective photograph by simple or differential rectification so that image displacements caused by camera tilt and terrain relief are removed.

Overlap: The area common to two successive photos along the same flight strip; the amount of overlap is expressed as a percentage of photo area. Also called endlap.

Overlay: (1) A transparent sheet giving information to supplement that shown on maps. When the overlay is laid over the map on which it is based, its details will supplement the map. (2) A tracing of selected details on a photograph, mosaic, or map to present the interpreted features and the pertinent detail.

P

Panchromatic: Used for films that are sensitive to broadband (e.g., entire visible part of spectrum) EMR, and for broadband photographs.

Parallax: The apparent change in the position of one object, or point, with respect to another, when viewed from different angles. As applied to aerial photos, the term refers to the apparent displacement of two points along the same vertical line when viewed from a point (the exposure station) not on the same vertical line.

Parallax, absolute: In a pair of truly vertical photographs which have equal flight heights, or in a pair of rectified photographs, the term denotes the algebraic difference, parallel to the air base, of the distances of the two images from their respective principal points. Also called ''x-parallax''.

Parallax difference: The difference in the distances separating complementary image points for two ground points of different elevation, as measured on a stereo pair of photos correctly oriented with respect to one another.

Passive system: A sensing system that detects or measures radiation emitted by the target. Compare active system.

Pattern: (1) In a photo image, the regularity and characteristic placement of tones or textures. Some descriptive adjectives for patterns are regular, irregular, random, concentric, radial, and rectangular. (2) The relations between any more-or-less independent parameters of a response, e.g., the pattern in the frequency domain of the response from an object.

Pattern recognition: Concerned with, but not limited to, problems of:
1. pattern discrimination,
2. pattern classification,
3. feature selection,
4. pattern identification,
5. cluster identification,
6. feature extraction,
7. preprocessing,
8. filtering,
9. enhancement,
10. pattern segmentation,
11. screening.

Perihelion: For an elliptic orbit about the Sun, the point closest to the Sun.

Perspective: Representation, on a plane or curved surface, of natural objects as they appear to the eye.

Phosphor: A phosphorescent substance, such as zinc sulfide, which emits light when excited by radiation, as on the scope of a cathode-ray tube.

Photogrammetry: The art or science of obtaining reliable measurements by means of photography.

Photograph: A picture formed by the action of light on a base material coated with a sensitized solution that is chemically treated to fix the image points at the desired density. Usually now taken to mean the direct action of EMR on the sensitized material. Compare image.

Photographic interpretation: The act of examining photographic images for the purpose of identifying objects and judging their significance. Photo interpretation, photointerpretation, and image interpretation are other widely used terms.

Photographic interpretation key: Reference material designed to facilitate the rapid, accurate identification of features from a study of their photographic images. The key usually consists of two parts: (a) photo image examples and (b) a word description, often in dichotomous (''two-branched'') form, of the identifying characteristics.

Picture: Representation of a scene by a photographic positive print or transparency, made from a negative, produced by the direct action of actinic (visible) light or EMR outside the visible part of the spectrum and converted into visible EMR by an optical-mechanical or wholly electronic scanner.

Pitch: Rotation of an aircraft about the horizontal axis normal to its longitudinal axis, which causes a nose-up nose-down attitude.

Pixel: (Derived from ''picture element.'') A data element having both spatial and spectral aspects. The spatial variable defines the apparent size of the resolution cell (i.e., the area on the ground represented by the data values), and the spectral variable defines the intensity of the spectral response for that cell in a particular channel.

Planck's Law: An expression for the variation of monochromatic emittance (emissive power) as a function of wavelength of blackbody radiation at a given temperature; it is the most fundamental of the radiation laws.

Point, principal: The foot of the perpendicular from the interior perspective center to the plane of the photograph: i.e. the foot of the photograph perpendicular.

Polarization: The direction of vibration of the electrical field vector of electromagnetic radiation. In SLAR systems polarization is either horizontal or vertical.

Positive: (1) A photographic image having approximately the same rendition of light and shade as the original subject. (2) A film, plate, or paper containing such an image.

Precision: A measure of the dispersion of the values observed when measuring a characteristic of elements of a population. The clustering of sample values about their own average.

Previsual symptom: The phenomenon whereby the ability of vegetation to reflect photographic IR energy diminishes because of stress. This commonly occurs before the normal green color changes, and is recognizable on color IR film by a drop in brightness of the red hues.

Processing: (1) The operation necessary to produce negatives, diapositives, or prints from exposed film, plates, or papers. (2) The manipulation of data by means of computer or other device.

Projection, map: A systematic drawing of lines on a plane surface to represent the parallels of latitude and the meridians of longitude of the earth or a section of the earth. A map projection may be established by analytical computation or may be constructed geometrically. A map projection is frequently referred to as a *projection* but the complete term should be used unless the context clearly indicates the meaning.

Pseudoscopic view: A reversal of the normal stereoscopic effect, causing valleys to appear as ridges and ridges as valleys.

Pulse: (1) A variation of a quantity whose value is normally constant; this variation is characterized by a rise and a decay, and has a finite duration. (2) A short burst of EMR transmitted by the radar.

Q

Quad-centered photograph: Middle exposure of a phototriplet (three consecutive aerial photographs) taken so that the middle photo is exposed directly above the center of the quadrangle, and the preceding and following photographs are exposed directly above the boundaries of the quadrangle. The flying height is such that the quadcentered photo covers the entire quadrangle.

Quantum theory: The theory first stated by Max Planck Society that all EMR is emitted and absorbed in *quanta,* each of magnitude hv, h being the Planck constant and v the frequency of the radiation.

R

Radar: Acronym for radio detection and ranging. A method, system or technique, including equipment components, for using beamed, reflected, and timed EMR to detect, locate, and (or) track objects, to measure altitude and to acquire a terrain image. In remote sensing of the Earth's or a planetary surface, it is used for measuring and, often, mapping the scattering properties of the surface.

Radar beam: The vertical fan-shaped beam of EM energy produced by the radar transmitter.

Radar, brute force: A radar imaging system employing a long physical antenna to achieve a narrow beamwidth for improved resolution.

Radar shadow: A dark area of no return on a radar image that extends in the far-range direction from an object on the terrain that intercepts the radar beam.

Radar, synthetic aperture (SAR): A radar in which a synthetically long apparent or effective aperture is *constructed* by integrating multiple returns from the same ground *cell,* taking advantage of the Doppler effect to produce a *phase history* film or tape that

may be optically or digitally processed to reproduce an image.

Radiance: The accepted term for radiant flux in power units (e.g., W) and not for flux density per solid angle (e.g., W cm^{-2} sr^{-1}) as often found in recent publications.

Radiant flux: The time rate of the flow of radiant energy; radiant power.

Radiant power: Rate of change of radiant energy with time. May be further qualified as spectral radiant power, at a given wavelength.

Radiant temperature: Concentration of the radiant flux from a material. Radiant temperature is the product of the kinetic temperature multiplied by the emissivity to the one-fourth power.

Radiation: The emission and propagation of energy through space or through a material medium in the form of waves; for example, the emission and propagation of EM waves, or of sound and elastic waves. The process of emitting radiant energy.

Radiometer: An instrument for quantitively measuring the intensity of EMR in some band of wavelengths in any part of the EM spectrum. Usually used with a modifier, such as an IR radiometer or a microwave radiometer.

Radiometric correction: Correcting gain and offset variations in MSS data. Procedure calibrates and corrects the radiation data provided by the Landsat sensor detectors.

Range direction: For radar images this is the direction in which energy is transmitted from the antenna and is normal to the azimuth direction. Also called look direction.

Range, dynamic: The ratio of maximum measurable signal to minimum detectable signal. The upper limit usually is set by saturation and the lower limit by noise.

Raster: The scanned (illuminated) area of the CRT.

Rayleigh-Jeans Law: An approximation to Planck's Law for blackbody radiation valid in the longer (microwave) wavelengths. It is almost always of sufficient accuracy for calculations in the radio and microwave regions of the spectrum.

Rayleigh scattering: The wavelength-dependent scattering of electromagnetic radiation by particles in the atmosphere much smaller than the wavelengths scattered.

Real-aperture radar: SLAR system in which azimuth resolution is determined by the physical length of the antenna and by the wavelength. The radar returns are recorded directly to produce images. Also called brute-force radar.

Real time: Time in which reporting on events or recording of events is simultaneous with the events. For example, the real time of a satellite is the time in which it simultaneously reports its environment as it encounters it; the real time of a computer is the time during which it is accepting data and performing operations on it.

Rectification: The process of projecting a tilted or oblique photograph onto a horizontal reference plane, the angular relation between the photography and the plane being determined by ground reconnaissance. Transformation is the special process of rectifying the oblique images from a multiple-lens camera to equivalent vertical images by projection onto a plane that is perpendicular to the camera axis. In this case, the projection is onto a plane determined by the angular relations of the camera axis and not necessarily onto a horizontal plane.

differential rectification: The process of removing the effects of tilt, relief, and other distortions from imagery by correcting small portions of the imagery independently.

Reflectance: The ratio of the radiant energy reflected by a body to that incident upon it. The suffix (-ance) implies a property of that particular specimen surface.

Reflection (EMR theory): EMR neither absorbed nor transmitted is reflected. Reflection may be diffuse when the incident radiation is scattered upon being reflected from the surface, or specular, when all or most angles of reflection are equal to the angle of incidence.

Reflectivity: A fundamental property of a material that has a reflecting surface and is sufficiently thick to be opaque. One may further qualify it as spectral reflectivity. The suffix (-ity) implies a property intrinsic with a given material, a limiting value.

Refraction: The bending of EMR rays when they pass from one medium into another having a different index of refraction or dielectric coefficient. EMR rays also bend in media that have continuous variations in their indices of refraction or dielectric coefficients.

Registration: The process of geometrically aligning two or more sets of image data such that resolution cells for a single ground area can be digitally or visually superposed. Data being registered may be of the same type, from very different kinds of sensors, or collected at different times.

Relief displacement: A shift in position of the optical image of an object caused by height of the object above or depth below a datum plane.

Remote sensing: In the broadest sense, the measurement or acquisition of information of some property of an object or phenomenon, by a recording device that is not in physical or intimate contact with the object or phenomenon under study; e.g., the utilization at a distance (as from an aircraft, spacecraft, or ship) of any device and its attendant display for gathering information pertinent to the environment, such as measurements of force fields, electromagnetic radiation, or acoustic energy. The technique employs such devices as the camera, lasers, and radio frequency receivers, radar systems, sonar, seismographs, gravimeters, magnetometers, and scintillation counters.

Representative fraction (R.F.): The relation between map or photo distance and ground distance expressed as a fraction (1/25,000) or often as ratio (1:25,000) (1 inch on map = 25,000 inches on ground). Also called *scale.*

Resolution: The ability of an entire remote sensor system, including lens, antennae, display, exposure, processing, and other factors, to render a sharply defined image. It may be expressed as line pairs per millimeter or meter, or in many other ways. In radar, resolution usually applies to the effective beam-width and range measurement width, often defined as the half-power points. For infrared line scanners the resolution may be expressed as the instantaneous field of view. Resolution may also be expressed in terms of temperature or other physical property being measured.

Resolution cell: The smallest area in a scene considered as a unit of data. For Landsat-1 and -2 the resolution cell approximates a rectangular ground area of 0.44 hectares or 1.1 acres (see pixel, instantaneous field of view).

Resolving power: A mathematical expression of lens definition, usually stated as the maximum number of line pairs per millimeter that can be resolved (that is, seen as separate lines) in an image.

Reststrahlen (residual) rays: The difference in intensities or radiance at certain frequencies (wavelengths) between the special signatures for the ideal (perfect blackbody) and actual emission curves of a substance.

Return beam vidicon (RBV): A modified vidicon television camera tube, in which the output signal is derived from the depleted electron beam reflected from the tube target. The RBV can be considered as a cross between a vidicon and an orthicon. RBVs provide highest resolution TV imagery, and are used in the ERTS (Landsat) series.

Roll: Rotation of an aircraft about the longitudinal axis to cause a wing-up or wing-down attitude.

Roughness: For radar images this term describes the average vertical relief of small-scale irregularities of the terrain surface.

S

Sample: A subset of a population selected to obtain information concerning the characteristics of the population.

Sampling rate: The temporal, spatial, or spectral rate at which measurements of physical quantities are taken. Temporally, sampling variables may describe how often data are collected or the rate at which an analog signal is sampled for conversion to digital format; the spatial sampling rate describes the number, ground size, and position of areas where spectral measurements are made; the spectral sampling rate refers to the location and width of the sensor's spectral channels with respect to the electromagnetic spectrum.

Satellite: An attendant body that revolves about another body, the *primary;* especially in the solar system, a secondary body, or moon, that revolves about a planet. A man-made object that revolves about a spacial body.

Saturation: Degree of intensity difference between a color and an achromatic light-source color of the same brightness.

Scale: The ratio of a distance on a photograph or map to its corresponding distance on the ground. The scale of a photograph varies from point to point because of displacements caused by tilt and relief, but is usually taken as f/H, where f is the principal distance (focal length) of the camera and H is the height of the camera above mean ground elevation. Scale may be expressed as a ratio 1:24,000; a representative fraction, 1/24,000; or an equivalence, 1 in. = 2,000 ft.

Scan line: The narrow strip on the ground that is swept by the instantaneous field of view of a detector in a scanner system.

Scanner: (1) Any device that scans, and thus produces an image. See scanning radiometer. (2) A radar set incorporating a rotatable antenna, or radiator element, motor drives, mounting, etc. for directing a searching radar beam through space and imparting target information to an indicator.

Scanning radiometer: A radiometer, which by the use

of a rotating or oscillating plane mirror, can scan a path normal to the movement of the radiometer.

Scattering: (1) The process by which small particles suspended in a medium of a different index of refraction diffuse a portion of the incident radiation in all directions. (2) The process by which a rough surface reradiates EMR incident upon it.

Scene: In a passive remote sensing system, everything occurring spatially or temporally before the sensor, including the Earth's surface, the energy source, and the atmosphere, that the energy passes through as it travels from its source to the Earth and from the Earth to the sensor.

Sensitivity: The degree to which a detector responds to electromagnetic energy incident upon it.

Sensor: Any device that gathers energy, EMR or other, converts it into a signal and presents it in a form suitable for obtaining information about the environment.

Sidelap: The extent of lateral overlap between images acquired on adjacent flight lines.

Sidelooking radar: An all weather, day/night remote sensor which is particularly effective in imaging large areas of terrain. It is an *active* sensor, as it generates its own energy which is transmitted and received to produce a photo-like picture of the ground. Also referred to as *sidelooking airborne radar;* abbr., *SLAR.*

Signal: The effect (e.g., pulse of electromagnetic energy) conveyed over a communication path or system. Signals are received by the sensor from the scene and converted to another form for transmission to the processing system.

Signal-to-noise ratio: The ratio of the level of the information-bearing signal power to the level of the noise power. Abbreviated as S/N.

Signature: Any characteristic or series of characteristics by which a material may be recognized in an image, photo, or data set. See also spectral signature.

Signature analysis techniques: Techniques that use the variation in the spectral reflectance or emittance of objects as a method of identifying the objects.

Signature extension: The use of training statistics obtained from one geographical area to classify data from similar areas some distance away; includes consideration of changes in atmosphere, and other geographical and temporal conditions that can cause differences in signal level for signal classes of interest (see spectral signature).

Slant range: For radar images this term represents the distance measured along a line between the antenna and the target.

Smoothing: The averaging of densities in adjacent areas to produce more gradual transitions.

Software: The computer programs that drive the hardware components of a data processing system; includes system monitoring programs, programming language processors, data handling utilities, and data analysis programs.

Spatial filter: An image transformation, usually a one-to-one operator used to lessen noise or enhance certain characteristics of the image. For any particular (x, y) coordinate on the transformed image, the spatial filter assigns a gray shade on the basis of the gray shades of a particular spatial pattern near the coordinates (x, y).

Spatial information: Information conveyed by the spatial variations of spectral response (or other physical variables) present in the scene.

Spectral band: An interval in the electromagnetic spectrum defined by two wavelengths, frequencies, or wavenumbers.

Spectral interval: The width, generally expressed in wavelength or frequency of a particular portion of the electromagnetic spectrum. A given sensor (e.g., radiometer or camera film) is designed to measure or be sensitive to energy received at the satellite from that part of the spectrum. Also termed spectral band.

Spectral reflectance: The reflectance of electromagnetic energy at specified wavelength intervals.

Spectral regions: Conveniently designated ranges of wavelengths subdividing the electromagnetic spectrum; for example, the visible region, X-ray region, infrared region, middle-infrared region.

Spectral response: The response of a material as a function of wavelength to incident electromagnetic energy, particularly in terms of the measurable energy reflected from and emitted by the material.

Spectral signature: Quantitative measurement of the properties of an object at one or several wavelength intervals.

Spectrometer: A device to measure the spectral distribution of EMR. This may be achieved by a dispersive prism, grating, or circular interference filter with a detector placed behind a slit. If one detector is used, the dispersive element is moved so as to sequentially pass all dispersed wavelengths across the slit. In an interferometerspectrometer, on the other hand, all wavelengths are examined all the time, the scanning effect being achieved by rapidly oscillating two, partly reflective, (usually parallel) plates so that interference fringes are produced. A Fourier transform is required to reconstruct the spectrum. Also called spectroradiometer.

Spectrophotometer: A photometer which measures the intensity of EMR as a function of the frequency (or wavelength) of the EMR. Usually used for the visible portion.

Spectrum: (1) In physics, any series of energies arranged according to wavelength (or frequency). (2) The series of images produced when a beam of radiant energy is subject to dispersion. A rainbow-colored band of light is formed when white light is passed through a prism or a diffraction grating. This band of colors results from the fact that the different wavelengths of light are bent in varying degrees by the dispersing medium and is evidence of the fact that white light is composed of colored light of various wavelengths.

Specular: In sensitometry, applied to a measurement made by collimated or essentially parallel light rays; referring to reflection, or transmission without scattering or diffusion.

Specular reflection: The reflectance of electromagnetic energy without scattering or diffusion, as from a surface that is smooth in relation to the wavelengths of incident energy. Also called mirror reflection.

Stefan-Boltzmann Law: One of the radiation laws stating that the amount of energy radiated per unit time from a unit surface area of an ideal blackbody is proportional to the fourth power of the absolute temperature of the blackbody.

Steradian: The unit solid angle that cuts unit area from the surface of a sphere of unit radius centered at the vertex of the solid angle. There are 4π steradians in a sphere.

Stereo base: A line representing the distance and direction between complementary image points (photo nadir points) on a stereo-pair of photos correctly oriented and adjusted for comfortable stereoscopic vision under a given stereoscope, or with the unaided eyes.

Stereoscope: A binocular optical instrument for assisting the observer to view two properly oriented photographs or diagrams to obtain the mental impression of a three-dimensional model.

Stereoscopic image: That mental impression of a three-dimensional object which results from stereoscopic vision (stereo viewing).

Subtractive color process: A method of creating essentially all colors through the subtraction of light of the three subtractive color primaries (cyan, magenta and yellow) in various proportions through use of a single white light source.

Sun synchronous: An Earth satellite orbit in which the orbital plane is near polar and the altitude such that the satellite passes over all places on Earth having the same latitude twice daily at the same local sun time.

Supervised classification: A computer-implemented process through which each measurement vector is assigned to a class according to a specified decision rule, where the possible classes have been defined on the basis of representative training samples of known identity.

Swath width (total field of view): The overall plane angle or linear ground distance covered by a multispectral scanner in the across-track direction.

Synchronous satellite: An equatorial west-to-east satellite orbiting the Earth at an altitude of 34,900 km, at which altitude it makes one revolution in 24 h synchronous with the Earth's rotation.

Synoptic view: The ability to see or otherwise measure widely dispersed areas at the same time and under the same conditions; e.g., the overall view of a large portion of the Earth's surface which can be obtained from satellite altitudes.

Synthetic stereo image: A stereo model made by digital processing of a single image. Topographic data are used in calculating the geometric distortion.

System: Structured organization of people, theory, methods and equipment to carry out an assigned set of tasks.

T

Target: (1) An object on the terrain of specific interest in a remote sensing investigation. (2) The portion of the Earth's surface that produces by reflection or emission the radiation measured by the remote sensing system.

Telemetry: The science of measuring a quantity or quantities, transmitting the measured value to a distant station, and there interpreting, indicating, or recording the quantities measured. *telemetry link:* The system for transmitting data over long distances using radio techniques.

Texture: In a photo image, the frequency of change and arrangement of tones. Some descriptive adjectives for textures are fine, medium or coarse; and stippled or mottled.

Thermal band: A general term for middle-infrared wavelengths which are transmitted through the atmosphere window at $8-14$ μm. Occasionally also used for the windows around $3-6$ μm.

Thermal capacity (symbol, C): The ability of a material to store heat, expressed in cal g^{-1} $°C^{-1}$.

Thermal conductivity (symbol K): The measure of the rate at which heat passes through a material, expressed in cal cm^{-1} s^{-1} $°C^{-1}$.

Thermal crossover: On a plot of radiant temperature versus time, this refers to the point at which the temperature curves for two different materials intersect.

Thermal inertia (symbol, P): A measure of the response of a material to temperature changes, expressed in cal cm^{-2} $°C^{-1}$ S$^{-1/2}$.

Thermal infrared: The preferred term for the middle wavelength range of the IR region, extending roughly from 3 μm at the end of the near infrared, to about 15 or 20 μm, where the far infrared begins. In practice the limits represent the envelope of energy emitted by the Earth behaving as a gray body with a surface temperature around 290°K (27°C).

Threshold: The boundary in spectral space beyond which a data point, or pixel, has such a low probability of inclusion in a given class that the pixel is excluded from that class.

Tilt: The angle between the optical axis of the camera and the plumb line for a given photo.

Tone: Each distinguishable shade of gray from white to black on an image.

Training: Informing the computer system which sites to analyze for spectral properties or signatures of specific land cover classes; also called signature extraction.

Training samples: The data samples of known identity used to determine decision boundaries in the measurement or feature space prior to classification of the overall set of data vectors from a scene.

Training sites: Recognizable areas on an image with distinct (spectral) properties useful for identifying other similar areas.

Transmissivity: Transmittance for a unit thickness sample. One may further qualify it as spectral transmissivity. The suffix (ity) implies a property intrinsic with a given material.

Transmittance: The ratio of the radiant energy transmitted through a body to that incident upon it. The suffix (-ance) implies a property of that particular specimen.

Transparency: The light-transmitting capability of a material. The loss of light in transmission through good optical glass. Approximately 2.4 percent of visual light is lost for every centimeter of glass traversed. (2) A positive image upon glass or film, intended to be viewed by transmitted light, either black and white or in color; also called a diapositive.

U

Ultraviolet radiation: EMR of shorter wavelength than visible radiation but longer than X-rays; roughly, radiation in the wavelength interval between 10 and 4000 Å.

V

Value (color): Degree of *lightness*, one of the attributes, along with *hue* and *saturation*, that may be thought of as the *dimensions* of color.

Variance: Variance of a random variable is the expected value of the square of the deviation between that variable and its expected value. It is a measure of the dispersion of the individual unit values about their mean.

Video: A term pertaining to the bandwidth and spectrum

position of the signal which results from television scanning and which is used to reproduce a picture.

Vidicon: (1.) A storage-type electronically scanned photoconductive television camera tube, which often has a response to radiations beyond the limits of the visible region. Particularly useful in space applications, as no film is required. (2) An image-plane scanning device. See *return beam vidicon;* Chapter 7.

Vignetting: A gradual reduction in density of parts of a photographic image caused by the stopping of some of the rays entering the lens.

Visible wavelengths: The radiation range in which the human eye is sensitive, approximately $0.4-0.7 \mu$m.

W

Wavelength (symbol λ): Wavelength = velocity/frequency. In general, the mean distance between maxima (or minima) of a roughly periodic pattern. Specifically, the least distance between particles moving in the same phase of oscillation in a wave disturbance. Optical and IR wavelengths are measured in nanometers (10^{-9} m), micrometers (10^{-6} m) and Angstroms (10^{-10} m).

White noise: Noise whose spectral density is independent of frequency.

Wiens Displacement Law: Describes the shift of the radiant power peak to shorter wavelengths with increasing temperature.

Window: A band of the electromagnetic spectrum which offers maximum transmission and minimal attenuation through a particular medium with the use of a specific sensor.

Y

Yaw: Rotation of an aircraft about its vertical axis, causing the longitudinal axis to deviate from the flight line.

Z

Zenith: The point in the celestial sphere that is exactly overhead: opposed to nadir.

ACRONYMS

CCT:	Computer Compatible Tape
CIR:	Color Infrared
CPU:	Central Processing Unit
CRT:	Cathode-Ray Tube
DN:	Digital Number
EDC:	EROS Data Center
EM:	Electromagnetic
EMR:	Electromagnetic Radiation
EROS:	Earth Resources Observation System
FOV:	Field-of-View
GCP	Ground Control Point
IFOV:	Instantaneous Field-of-View
IR:	Infrared
JPL:	Jet Propulsion Laboratory
JSC:	Johnson Space Center
MSS:	Multispectral Scanner
NASA:	National Aeronautics and Space Administration
NOAA:	National Oceanic and Atmospheric Administration
RBV:	Return Beam Vidicon
RF:	Representative Fraction
SLAR:	Side-Looking Airborne Radar
S/N:	Signal-to-Noise Ratio
TM:	Thematic Mapper

List of Contributors

AARON, JOHN M., BS Geology, Franklin and Marshall College; Ph.D. Geology, Pennsylvania State Univ.; Chief, Office of Scientific Publications, U.S. Geological Survey (MS904), Reston, VA 22092. *Contributing Author, Chapter 31.*

ABRAMS, MICHAEL J., BS Biology and MS Geology, California Inst. of Technology; Member of Technical Staff, Jet Propulsion Laboratory, 4800 Oak Grove, Pasadena, CA 91103. *Contributing Author, Chapter 31.*

ALDRICH, ROBERT C., BS Forestry and MF, College of Environmental Science and Forestry, State Univ. of New York; Principal Research Forester, Pacific Southwest Forest and Range Experiment Station, Rocky Mountain Forest and Range Experiment Station, Forest Service, U.S. Dept. of Agriculture, Berkeley, California 94701 and Fort Collins, Colorado 80521; Retired, Consultant in Forest Resources, Remote Sensing, 217 Conifer Lane, Walnut Creek, CA 94598. *Contributing Author, Chapter 34.*

ALLISON, LEWIS J., BA Biology, Brooklyn College; Aerospace Technologist (Meteorology) (Retired), NASA, Goddard Space Flight Center, Greenbelt, MD 20771. *Co-Author-Editor, Chapter 14.*

ANUTA, PAUL E., BSEE Purdue Univ.; MSEE Univ. of Conn.; MS Computer Science, Purdue Univ.; PhD EE (in progress), Purdue Univ.; Assoc. Program leader, Data Processing Research, LARS, Purdue Univ., 1220 Potter Drive, W. Lafayette, IN 47906. *Contributing Author, Chapter 17.*

ARVIDSON, RAYMOND E., BA Temple Univ.; MS and Ph.D., Brown Univ. Associate Professor, McDonnel Center for the Space Sciences, Dept. Earth and Planetary Sciences, Washington Univ., St. Louis, MO 63130. *Author-Editor, Chapter 36.*

BANDEEN, WILLIAM R., BS, U.S. Military Academy; MS Meteorology, New York Univ.; Associate Chief, Laboratory for Atmospheric Sciences, NASA Goddard Space Flight Center, Greenbelt, MD 20771. *Contributing Author, Chapter 14.*

BAUER, MARVIN E., BS Agriculture and MS Agronomy, Purdue Univ.; Ph.D. Agronomy, Univ. of Illinois; Senior Research Agronomist and Leader, Crop Inventory Research, Laboratory for Applications of Remote Sensing, Purdue Univ. West Lafayette, IN 47907. *Contributing Author, Chapter 33.*

BECKMAN, JOHN A., Professional degree in Electrical Engineering, Univ. of Cincinnati; Communications Engineer; Communication Systems Consultant; 5423 Easton Drive, Springfield, VA 22151. *Author, Chapter 15.*

BENSON, ANDREW S., BS, MS Forestry, Univ. of California, Berkeley; Associate Specialist, Remote Sensing Research Program, 260 Space Sciences Laboratory, Univ. of California, Berkeley, CA 94720. *Contributing Author, Chapter 32.*

BERNSTEIN, RALPH, BS Electrical Engineering, Univ. of Conn.; MS Electrical Engineering, Syracuse Univ.; Scientific Staff Member, IBM Palo Alto Science Center, 1530 Page Mill Road, Palo Alto, CA 94304. *Author-Editor, Chapter 21.*

BERTKE, SUSAN, BA Geography, Wright State Univ., Dayton; MA Geography w/emphasis in Remote Sensing, Univ. of California, Santa Barbara; Senior Associate Engineer, Lockheed Missiles & Space Co., 1111 Lockheed Way, Sunnyvale, CA 94086. *Contributing Author, Chapter 1.*

BEVAN, BRUCE W., Ph.D. Anthropology, Univ. of Pennsylvania; President, Geosight Geophysical Exploration and Remote Sensing, Post Office Box 135, Pitman, NJ 08071. *Contributing Author, Chapter 26.*

BILLINGSLEY, FREDERICK, C., BChE, BEE and MEE, Rensselaer Polytechnic Institute; Image Processing Hardware, Software, and Applications; Group Supervisor, Earth Resources Image Processing, Jet Propulsion Lab., 4800 Oak Grove Drive, Pasadena, CA 91103. *Author-Editor, Chapter 17.*

BIRNIE, RICHARD W., BS Geology, Dartmouth College; Ph.D. Geology, Harvard Univ.; Associate Professor, Dept. of Earth Sciences, Dartmouth College, Hanover, NH 03755. *Contributing Author, Chapter 31.*

BLODGET, HERBERT W., BS Geology, Rutgers Univ.; MS and Ph.D. Geology, The George Washington Univ.; Projects Group Leader, Eastern Regional Remote Sensing Applications Center (902.1), National Aeronautics and Space Administration, Goddard Space Flight Center, Greenbelt, MD 20771. *Contributing Author, Chapter 31.*

BOWDEN, LEONARD W., BA and MA Geology, Univ. of Colorado, PhD Geography, Clark University. Deceased. *Editor-in-Chief, 1978–79.*

BOWLEY, DONOVAN R., BA Eastern Nazarene College; AM and PhD Boston Univ.; Associate Planner (NEIWPCC), Department of Environmental Quality Engineering, Division of Water Supply, 1 Winter Street, Boston, MA 02108. *Contributing Author, Chapter 31.*

BOYLE, A. RAYMOND, BSc, PhD Chemistry, Birmingham, England. P. Eng. Sask. Univ. of Saskatchewan (Prof. of EE), Hardware and Software Design and Application of Systems of Automated Cartography. Dept. of Electrical Engineering, Univ. of Saskatchewan, Saskatoon, S7N 0J5, Canada. *Contributing Author, Chapter 22.*

BRACKEN, PETER A., BS Mathematics, College of William and Mary; MA Mathematics, Univ. of Maryland; Deputy Director of Mission and Data Operations, National Aeronautics and Space Administration; Goddard Space Flight Center, Greenbelt, MD 20771. *Author-Editor, Chapter 19.*

BREWER, THOMAS, BS Mechanical Engineering, Carnegie-Mellon Univ.; MS and PhD Geology, Boston Univ.; Professor, Department of Regional Studies, Univ. of Massachusetts at Boston, Huntington Avenue Campus, 625 Huntington Avenue, Boston, MA 02115. *Contributing Author, Chapter 31.*

BROOKS, RONALD L., BS Geology, Bowling Green State Univ.; Chief Scientist, GeoScience Research Corporation, Route 4, Box 129, Salisbury, MD 21801. *Contributing Author, Chapter 31.*

BRYAN, M. LEONARD, AB Geography, Indiana Univ.; MSc McGill Univ.; Ph.D. Univ. of Michigan; Asst. Prof. of Geography, Glendale Community College, Member, Technical Staff, Jet Propulsion Laboratory, California Inst. of Technology, 4800 Oak Grove Drive, Pasadena, CA 91103. *Contributing Author, Chapter 30.*

BRYANT, NEVIN, BA Geography, McGill Univ.; MA Geography, Univ. of Hawaii; PhD Geography, Univ. of Michigan; Member Technical Staff, Earth Resources Applications Group, Jet Propulsion Lab, MS168-514, 4800 Oak Grove Drive, Pasadena, CA 91109. *Contributing Author, Chapter 22.*

CALKINS, HUGH W., AB Geography, Univ. of Calif., Berkeley; MUP and PhD Urban Planning, Univ. of Washington; Associate Professor of Geography, State Univ., New York at Buffalo, NY 14260. *Contributing Author, Chapter 22.*

CAMILLI, EILEEN L., BA Anthropology, University of Colorado (Boulder); MA Anthropology, Northern Arizona University; Ph.D. Anthropology, University of New Mexico; Archaeological Consultant, 3100 Ninth Street NW, Albuquerque, NM 87107. *Contributing Author, Chapter 26.*

CARNEGGIE, DAVID M., BS Forestry, MS Range Management, PhD Wildland Resource Science, Univ. of Calif., Berkeley; Chief, U.S. Geological Survey, EROS Field Office, Anchorage, Alaska, SRA Box 178C, Anchorage, AK 99502. *Co-Author-Editor, Chapter 35.*

CARR, JAMES L., BS Physics and Mathematics, Carnegie-Mellon Univ.; MS Physics, Georgetown Univ.; Principal Engr., ORI Inc., 1400 Spring St., Silver Spring, MD 20910. *Contributing Author, Chapter 17.*

CARTER, VIRGINIA, BA, Swarthemore College; MS, The American Univ.; Research Biologist, U.S. Geological Survey, 431 National Center, Reston, VA 22092. *Contributing Author, Chapter 28.*

CARTER, WM. DOUGLAS, BA Geology, Dartmouth College; Graduate studies, Geology, Johns Hopkins Univ. and Univ. of Colorado. Consulting Geologist, 2404 Paddock Lane, Reston, VA 22091. *Contributing Author, Chapter 31.*

CARVER, KEITH R., BS Electrical Engineering, Univ. of Kentucky; MS and Ph.D., Ohio State Univ.; Program Manager, Radar Remote Sensing Systems, NASA Headquarters, Washington, D.C. 20546. *Co-Author, Chapter 11.*

CARVER, LARRY G., BA Political Science, Univ. of California, Santa Barbara; Dept. Head, Map and Imagery Laboratory, Library, Univ. of California, Santa Barbara, CA 93106. *Contributing Author, Chapter 24.*

CHAHINE, MOUSTAFA T., BS, Univ. of Washington; Ph.D., Univ. of California at Berkeley. Senior Research Scientist and Manager, Division of Earth and Space Sciences, Jet Propulsion Laboratory, California Inst. of Technology, Pasadena, CA 91109. *Author-Editor, Chapter 5.*

CHASTANT, LLOYD J., BS Electrical Engineering, Univ. of Southwestern Louisiana; Radar Systems Development, Westinghouse Electric Corporation, Defense and Electronics System Center, P.O. Box 746, Baltimore, MD 21203. *Contributing Author, Chapter 10.*

CHAVEZ, PAT S., JR., BS and MS, Mathematics, Northern Arizona Univ.; Research Physical Scientist, U.S. Geological Survey, 2255 North Gemini Drive, Flagstaff, AZ 86001. *Contributing Author, Chapter 31.*

CLARK, MALCOLM M., BS Chemical Engineering, Univ. of California, Berkeley; Ph.D., Geology, Stanford, Univ.; Research Geologist, U.S. Geological Survey, (MS77) Menlo Park, CA 94025. *Contributing Author, Chapter 31.*

COLBY, CHARLES, BS Physics, Rensselaer Polytechnic Inst.; MS Physics, Syracuse Univ.; Advisory Engineering Scientist, IBM Federal Systems Division, 18100 Frederick Pike, Gaithersburg, MD 20760. *Contributing Author, Chapter 21.*

COLCORD, J. E., BSCE, Univ. of Maine, Orono; MSCE, Univ. of Minnesota, Minneapolis; PhC, The Ohio State Univ., Columbus; Professor and Associate Chairman, Department of Civil Engineering, Univ. of Washington, 201 More Hall (FX-10), Seattle, WA 98195. *Contributing Author, Chapter 32.*

COLWELL, ROBERT N., BS Forestry, PhD Plant Physiology, Univ. of California, Berkeley; Professor of Forestry and Associate Director, Space Sciences Laboratory, Univ. of California, Berkeley, CA 94270. *Editor-in-Chief; Contributing Author, Chapter 32.*

COLWELL, ROBERT R., BA Geography, Univ. of California, Santa Barbara. Consultant, U.S. Forest Service. 1300 Juanita Drive, Walnut Creek, CA 94595. *Compiler of Glossary.*

COSENTINO, MICHAEL J., BS Forestry and Resource Management, Univ. of California, Berkeley; MA (in progress) Geography, Univ. of California, Santa Barbara; Staff Research Associate, Geography Remote Sensing Unit, Dept. of Geography, Univ. of California, Santa Barbara, CA 93106. *Contributing Author, Chapter 24.*

COUPLAND, DAVID H., BS Geology, Univ. of New Hampshire; MS Geophysics, Univ. of Michigan; Vice President, Geospectra Corporation, P.O. Box 1387, 333 Parkland Plaza, Ann Arbor, MI 48106. *Contributing Author, Chapter 31.*

DAVIS, ROBERT E., MS Geography and PhD (in progess), Univ. of Calif., Santa Barbara; 6606 Del Playa, Goleta, CA 93117. *Co-Author, Chapter 25.*

DAVIES, WILLIAM E., BS Geology, Mass. Institute of Technology; MS Geology, Michigan State Univ.; Geologist U.S. Geological Survey (MS926), Reston, VA 22092. *Contributing Author, Chapter 31.*

DENNETT, SARAH, BA, Carleton College; President, Dennett, Muessig and Associates, Close-Range Photogrammetry, 631 South Van Buren, Iowa City, IA 52240. *Contributing Author, Chapter 26.*

DEUTSCH, MORRIS, AB Geology, Syracuse Univ.; MS Resource Development, Michigan State Univ.; Independent Consultant: Hydrologist-Remote Sensing, Satellite Hydrology Associates, 6606 Midhill Place, Falls Church, VA 22043. *Contributing Author, Chapter 31.*

DEWITT, DAVID P., BS Mechanical Engineering, Duke Univ.; MS Mechanical Engineering, M.I.T.; Ph.D. Purdue Univ.; Professor of Mechanical Engineering, Purdue Univ., W. Lafayette, IN 47907. *Co-Author-Editor, Chapter 7.*

DIESEN, BERNARD C. III, BS and MS Meteorology, Univ. of Oklahoma, Norman; Major, USAF; Meteorologist and Systems Analyst, Air Force Global Weather Central, Offutt AFB, NL 68113. *Contributing Author, Chapter 14.*

DOYLE, FREDERICK J., BA Civil Engineering, Syracuse Univ.; Ph.D. Engineering, Technical Univ., Hanover, FRG; Research Cartographer, National

Mapping Division, USGS, National Center 516, Reston, VA 22092. *Contributing Author, Chapter 6.*

DOZIER, JEFF, BA Geography, Calif. State Univ.; MA and PhD Geography, Univ. of Michigan; Assoc. Professor, Department of Geography, Univ. of Calif., Santa Barbara, CA 93106. *Co-Author, Chapter 23.*

DRAGER, DWIGHT L., MA Anthropology, Univ. of New Mexico; Archeologist, Remote Sensing Branch, National Park Service, PO Box 26176, Albuquerque, NM 87125-6176. *Contributing Author, Chapter 26.*

DUCK, KENNETH I., BS Physics/Mathematics, Lynchburg College; MEA George Washington Univ.; Head, Electro-Optical Systems Section, Instrument Systems Branch, Instrument Division, Engineering Directorate, NASA/Goddard Space Flight Center, Greenbelt, MD 20771. *Co-Author-Editor, Chapter 16.*

EBERT, JAMES I., BA Anthropology, Michigan State Univ.; MA Anthropology, Univ. of New Mexico; Director of Research, Remote Sensing Branch, National Park Service, PO Box 26176, Albuquerque, NM 87125-6176. *Co-Author-Editor, Chapter 26; Contributing Author, Chapter 31.*

ELACHI, CHARLES, BS Univ. of Grenoble, France; MS and PhD, California Inst. of Technology; Senior Research Scientist, Jet Propulsion Laboratory, California Inst. of Technology, 4800 Oak Grove Drive, Pasadena, CA 91109. *Author-Editor, Chapter 13.*

ELVIDGE, CHRISTOPHER D., BA Botany and BS Geology, Southern Illinois Univ.; MS Geology, Arizona State Univ.; Graduate Student, Stanford Univ., Department of Applied Earth Science, Stanford, CA 94305. *Contributing Author, Chapter 31.*

ESTES, JOHN E., AB and MA Geography, San Diego State College; Ph.D. Geography, Univ. of Calif., Los Angeles; Professor of Geography, Univ. of Calif., Santa Barbara, CA 93106. *Editor, Volume II; Contributing Author, Chapter 1; Co-Author-Editor, Chapter 24.*

FANALE, ROSALIE, Ph.D. Anthropology, Catholic Univ.; 2853—29th Street, NW, Washington, DC 20008. *Contributing Author, Chapter 26.*

FORD, JOHN P., BSc (Hons) Geology, Univ. of London; Ph.D. Geology, Ohio State Univ.; Supervisor, Imaging Radar Geology Group, Jet Propulsion Laboratory, California Inst. of Technology, MS 183-701, Pasadena, CA 91109. *Contributing Author, Chapter 31.*

FRANCICA, JOSEPH R., BA Geology, Rutgers Univ., MA Geology, Dartmouth College; Applications Scientist, Geology; Technicolor Graphic Services, EROS Data Center, Sioux Falls, SD 57198. *Contributing Author, Chapter 31.*

FREDEN, STANLEY C., AB, MS and Ph.D. Physics, Univ. of California, Los Angeles; Project Scientist for Landsats-1, -2 and -3 and Chief, Missions Utilization Office; NASA/Goddard Space Flight Center, Greenbelt, MD 20771. *Co-Author, Chapter 12.*

FRENCH, WILLIAM D., BSCE, Northeastern Univ.; Executive Director and Publications Director, American Society of Photogrammetry, 210 Little Falls Street, Falls Church, VA 22046. *Publications Director for Manual of Remote Sensing—Second Edition.*

FREW, JAMES, BA and MA Geography, Univ. of California, Santa Barbara, CA; Programmer, Computer Sciences Lab, 3111 Engineering Bldg., Univ. of California, Santa Barbara, CA 93106. *Contributing Author, Chapter 20.*

FRIEDMAN, JULES D., AB Geology, Cornell Univ.; MS and Ph.D. Geology, Yale Univ.; Geologist, Branch of Petrophysics and Remote Sensing, U.S. Geological Survey, MS 964, Denver Federal Center, Denver, CO 80225. *Contributing Author, Chapter 31.*

FRIEDMAN, STEVEN Z., BA and MA Geography, Univ. of Wisconsin; Director of Cartographic Program, Geophysical Co. of Norway, 3920 Stony Brook, Houston, TX 77063. *Contributing Author, Chapter 30.*

FRITZ, NORMAN L., BS Physics, Central Michigan Univ.; MS Physics, Michigan State Univ.; Retired, Eastman Kodak Co., 143 Glenmont Dr., Rochester, NY 14617. *Contributing Author, Chapter 6.*

FU, KING-SUN, BS, National Taiwan University; MA, Sc., Univ. of Toronto; Ph.D., Univ. of Illinois; Goss Distinguished Professor of Engineering and Professor of Electrical Engineering, Purdue Univ., Lafayette, IN 47907. *Co-Author-Editor, Chapter 18.*

FUNG, ADRIAN K., BS Electrical Engineering, Taiwan Provincial Cheng Kung Univ.; MS Electrical Engineering, Brown Univ.; Ph.D., Electrical Engineering, Univ. of Kansas, Lawrence, Kansas; Professor, Remote Sensing Laboratory, Univ. of Kansas Center for Research, Inc., 2291 Irving Hill Drive, West Campus, Lawrence, KS 66045. *Co-Author-Editor, Chapter 4.*

GAUSMAN, HAROLD W., BS Agronomy, Univ. of Maine, Orono; MS and Ph.D., Univ. of Illinois, Urbana; Research Leader, Plant Stress and Water Conservation Research, USDA-ARS, Texas Tech Univ., P.O. Box 4170, Lubbock, TX 79409. *Contributing Author, Chapter 33.*

GOETZ, ALEXANDER F. H., BS Physics, MS Geology and Ph.D., Planetary Science, California Inst. of Technology; Manager, Terrestrial Remote Sensing Program, Jet Propulsion Laboratory, 4800 Oak Grove Drive, MS 183-501, Pasadena, CA 91103. *Contributing Author, Chapters 13, 31.*

GOLD, DAVID P., BSc (Honours) Geology/Chemistry and MS, Geology, Univ. of Natal; Ph.D. Geology, McGill Univ.; Professor, Department of Geosciences, Pennsylvania State Univ., University Park, PA 16802. *Contributing Author, Chapter 31.*

GORDON, FREDERICK JR., BS Physics, Fordham Univ.; Technical Manager, Missions Utilization Office, NASA/Goddard Space Flight Center, Greenbelt, MD 20771. *Co-Author, Chapter 12.*

GREEN, WILLIAM B., BS Physics and MS Engineering, Univ. of California, Los Angeles; Manager, Signal Processing Department, Research and Development Division, System Development Corporation, 2500 Colorado Blvd., Santa Monica, CA 90406. *Contributing Author, Chapter 19.*

HAJIC, EARL J., BS and MS Electrical Engineering, Univ. of Illinois; MA Geography, Univ. of California, Santa Barbara; Adjunct Lecturer, Sr. Dev. Engr., Dept. of Geography, Univ. of California, Santa Barbara, CA 93106. *Co-Author-Editor, Chapter 24.*

HARALICK, ROBERT M., BS Mathematics, BSEE, MSEE, Ph.D., Univ. of Kansas; Professor of Electrical Engineering and Computer Science, Virginia Polytechnic Inst. and State Univ., Blacksburg, VA 24061. *Co-Author-Editor, Chapter 18; Contributing Author, Chapter 19.*

HART, WILLIAM G., BS Biology and MS Entomology, Univ. of Rhode Island; Ph.D. Entomology (Degree in May 1983), Texas A&M Univ.; Research Leader and Supervisor Research Entomologist, USDA-ARS, 509 West Fourth Street, Weslaco, TX 78596. *Contributing Author, Chapter 33.*

HARTMANN, NICHOLAS, BA and MA, Brown University; Ph.D. Univ. of Pennsylvania; Archaeologist, MASCA, University Museum, Univ. of Pennsylvania, 33rd and Spruce Streets, Philadelphia, PA 19104. *Contributing Author, Chapter 26.*

HEILMAN, JAMES L., BS Engineering Physics, MS Agronomy, South Dakota State Univ.; Ph.D. Agronomy, Kansas State Univ.; Associate Research Scientist, Remote Sensing Center, and Associate Professor, Dept. of Soil and Crop Sciences, Texas A&M Univ., College Station, TX 77843. *Contributing Author, Chapter 33.*

HELLER, ROBERT C., BS Botany, MF Forestry, Duke Univ.; Research Professor, Emer. College of Forestry, Wildlife and Range Sciences, Univ. of Idaho. 21 Eastwood Dr., Orinda, CA 94563. *Co-Author-Editor Chapter 34.*

HEMPHILL, WILLIAM R., BS Geology, Univ. of Wisconsin; Chief, Luminescence Studies, U.S. Geological Survey, MS 731, Reston, VA 22092. *Contributing Author, Chapter 31.*

HENDERSON, FLOYD M., BA Geography, Nebraska Wesleyan Univ.; MA and Ph.D. Geography, Univ. of Kansas; Associate Professor, Department of Geography, State Univ. of New York at Albany, NY 12222. *Contributing Author, Chapter 30.*

HILDEBRANT, GERD, Ph.D. Forestry, Univ. of Freiburg; Professor of Forest Management, Inventory, and Aerial Photography, Abteilung Luftbildmessung und Luftbildinterpretation, Universitat Freiburg, D78, Freiburg, Erbprinzenstrasse, 17A, Fed. Rep. Germany. *Contributing Author, Chapter 34.*

HOFFER, ROGER M., BSF, Michigan State Univ.; MS and Ph.D., Colorado State Univ.; Professor of Forestry and Leader of Ecosystem Research Program, Department of Forestry and Natural Resource and LARS, Purdue Univ., Lafayette, IN 47907. *Contributing Author, Chapter 34.*

HOLZ, ROBERT K., BA Zoology and MA Geography, Southern Illinois Univ.; Ph.D. Geography, Michigan State Univ.; Professor and Chairperson, Dept. of Geography, Univ. of Texas, Austin, TX 78712. *Contributing Author, Chapter 30.*

HOWARTH, PHILIP J., BA (Honors) Geography, Cambridge Univ., England; Diploma in Photogrammetry, Univ. of Glasgow, Scotland; Ph.D. Geography, Univ. of Glasgow; Associate Professor Geography, McMaster University, 1280 Main St. West, Hamilton, Ontario, Canada, L85 4K1. *Contributing Author, Chapter 32.*

ISACHSEN, YNGVAR W., BA Syracuse Univ.; MS Washington Univ., Ph.D., Cornell Univ.; Principal Scientist (Geology), New York State Geological Survey, Cultural Education Center, Albany, NY 12230. *Contributing Author, Chapter 31.*

JACKSON, THOMAS J., BS Fire Protection Engineering, and Ph.D. Civil Engineering, Univ. of Maryland; Hydrologist, U.S. Dept. of Agriculture Hydrology Laboratory, Beltsville Agricultural Research Center, Beltsville, MD 20705. *Contributing Author, Chapter 29.*

JENSEN, JOHN R., BA Geography, Calif. State Univ., Fullerton; MS Geography, Brigham Young Univ.; Ph.D. Geography, Univ. of Calif., Los Angeles; Associate Professor of Geography, Univ. of South Carolina, Columbia, SC 29208. *Author-Editor, Chapter 30.*

JOHNSON, JIMMIE D., BS, Bloomsburg State College, Pennsylvania; Graduate School, Meteorology, Pennsylvania State Univ.; Meteorologist, Chief, Administrative and Technical Services Unit, National Environmental Satellite, Data, and Information Services, National Oceanic and Atmospheric Administration, Washington, D.C. 20233. *Contributing Author, Chapter 27.*

JOHNSON, ROBERT W., BS Mechanical Engineering, Virginia Polytechnic Inst. and State Univ.; MS Mechanical Engineering, Pennsylvania State Univ.; Ph.D. Marine Sciences, North Carolina State Univ.; Senior Research Scientist, NASA, Langley Research Center, Mail Stop 270, Hampton, VA 23665. *Co-Author-Editor, Chapter 28.*

JOHNSON, TIMOTHY, BS Geography, Appalachian State Univ.; MA Geography, State Univ. of New York, Buffalo; Geographic Systems Analyst, SPAD Systems, LTD., P.O. Box 2126, Reston, VA 22090. *Contributing Author, Chapter 22.*

JORDAN, ROLANDO, BS, MS, Electrical Engineering, Massachusetts Institute of Technology; Member of Technical Staff, Radar Science and Engineering Section, Jet Propulsion Laboratory, 4800 Oak Grove Drive, Pasadena, CA 91103. *Contributing Author, Chapter 13.*

KAHLE, ANNE B., BS Physics and MS Geophysics, Univ. of Alaska; Ph.D. Atmospheric Science, Univ. of Calif., Los Angeles; Supervisor, Geology and Geophysics Group, Jet Propulsion Laboratory, 4800 Oak Grove Drive, Pasadena, CA 91103. *Contributing Author, Chapters 13 and 31.*

KEMMERER, ANDREW J., BS and MS Fisheries, University of Arizona, Tucson; PhD Aquatic Ecology, Utah State Univ., Logan; Director, Mississippi Laboratories, National Marine Fisheries Service NSTL Station, MS 39529. *Contributing Author, Chapter 28.*

KENDALL, BRUCE M., BS Electrical Engineering, Virginia Polytechnic Inst.; MS Electrical Engineering, George Washington Univ.; Aerospace Technologist (Microwave Physical Electronics) NASA Langley Research Center, Mail Stop 490, Hampton, VA 23665. *Contributing Author, Chapter 28.*

KHORRAM, SIAMAK, BS and MS Engineering, Univ. of Tehran; Ph.D. Water Science and Engineering, Univ. of California, Davis; Associate Professor of Forestry, North Carolina State Univ.; Raleigh, NC 27650. *Contributing Editor, Chapter 32.*

KING, JOSEPH C., BSME, Duke Univ.; Aerospace Engineer, Flight Systems Support Section, Flight Support Branch, Special Payloads Division, Engineering Directorate, NASA/Goddard Space Flight Center, Greenbelt, MD 20771. *Co-Author-Editor, Chapter 16.*

KINGSTON, MARGUERITE J., AB Chemistry, Dunbarton College; MS Geochemistry, George Washington Univ.; Research Chemist, U.S. Geological Survey (927), Reston, VA 22092. *Contributing Author, Chapter 31.*

KOWALIK, WILLIAM S., BS Earth and Planetary

Sciences, Univ. of Pittsburgh; MS Geology, Pennsylvania State Univ.; Ph.D. Applied Earth Science, Stanford Univ.; Research Geologist, Chevron Oil Field Research Company, P.O. Box 446, La Habra, CA 90631. *Contributing Author, Chapter 31.*

LANSING, JACK C., JR., BS Electrical Engineering, Univ. of Calif., Berkeley; Senior Technical Staff, Systems Engineering Department, Electro-optical Instrumentation, Santa Barbara Research Center, 75 Coromar Drive, Goleta, CA 93117. *Co-Author, Chapter 8.*

LAUER, DONALD T., BS and MS Forestry, Univ. of Calif., Berkeley; Chief, Applications Branch, EROS Data Center, Sioux Falls, SD 57198. *Contributing Author, Chapter 34.*

LAWRENCE, GARTH R., BSc Earth Sciences, Univ. of Waterloo, Waterloo, Ontario; Assistant General Manager, Intertech Remote Sensing, Ltd., Carling Square, Phase II, 785 Carling, Ave., Ottawa, Ontario, Canada K15 S4H; *Contributing Author, Chapter 32.*

LEATHERMAN, STEPHEN P., BS Geoscience, N.C. State Univ.; Ph.D. Environmental Sciences, Univ. of Virginia; Assistant Professor of Geomorphology, Dept. of Geography, LeFrak Hall, Univ. of Maryland, College Park, MD 20742. *Contributing Author, Chapter 31.*

LEGECKIS, RICHARD, BS Electrical Engineering, City College of New York; MS Physical Oceanography, Florida State Univ.; Physical Scientist, Ocean Sciences Branch, National Earth Satellite Service, Suitland, MD 20031. *Contributing Author, Chapter 28.*

LEVINTHAL, ELLIOTT C., BA Columbia Univ.; MS Mass. Inst. of Tech.; Ph.D. Stanford Univ.; Director, Defense Sciences Office, Defense Advanced Research Projects Agency (DARPA); 1400 Wilson Blvd., Arlington, VA 22209; (On-Leave) Prof. (Research) Genetics Dept., Stanford Medical School, Stanford, CA 94305. *Contributing Author, Chapter 36.*

LINDGREN, DAVID T., AB, MA and Ph.D. Geography, Boston Univ.; Associate Professor of Geography and Department Chairman, Dartmouth College, Hanover, NH 03755. *Contributing Author, Chapter 30.*

LOHMAN, GUY, BA Math, Pomona College, Claremont CA; MS and Ph.D. Operations Research, Ithaca, NY; Research Staff Member, IBM Research Laboratory. K55-281, 5600 Cottle Rd., San Jose, CA 95193. *Contributing Author, Chapter 20.*

LUCAS, JAMES R., BA Geography, Mankato State Univ., MA Physical Geography and Ph.D. Geology, Univ. of Iowa; Vice President, Centaur Exploration Incorporated, 3226 Hobbs, Amarillo, TX 79109. *Contributing Author, Chapter 29.*

LUCCHITTA, BAERBEL K., BS Geology, Kent State Univ., Ohio; MS and Ph.D. Geology, Pennsylvania State Univ.; Research Geologist, U.S. Geological Survey, Branch of Astrogeologic Studies, 2255 North Gemini Drive, Flagstaff, AZ 86001. *Contributing Author, Chapter 31.*

LYON, RONALD J. P., BSc Geology and BSc, Honours, Petrology, Univ. of Western Australia; Professor, Department of Applied Earth Sciences, Stanford University, Stanford, CA 94305. *Contributing Author, Chapter 31.*

LYONS, THOMAS R., MS Geology and Ph.D. Anthropology, Univ. of New Mexico; Geological con-

sultant, 2500 Louisiana Blvd., N.E., Albuquerque, NM 87110. *Co-Author-Editor, Chapter 26 and Contributing Author, Chapter 31.*

M'GONIGLE, JOHN W., BS and MS Geology, Univ. of New Mexico; Ph.D. Geology, Pennsylvania State Univ.; Geologist, U.S. Geological Survey, Denver Federal Building MS 972, Denver CO 80225. *Contributing Author, Chapter 31.*

MACDONALD, ROBERT B., BSEE, Purdue Univ.; Chief Scientist, Earth Resources Program and Research Division, NASA, Johnson Space Center, Houston, TX 77058. *Contributing Author, Chapter 33.*

MAMULA, NED, JR., BS Geology, Slippery Rock State College; MS Geology, The Pennsylvania State College; Geologist, U.S. Geological Survey (620) Reston, Virginia 22092. Present Address: Center for Tectonophysics, Texas A&M Univ., College Station, TX 77843. *Contributing Author, Chapter 31.*

MARBLE, DUANE F., BA Geography, Univ. of Washington; MA and Ph.D. Geography, Univ. of Washington; Professor of Geography and Adjunct Professor of Computer Science, State Univ. of New York at Buffalo, Amherst, NY 14260. *Author-Editor, Chapter 22.*

MARKS, DANNY, BA and MA Geography, Univ. of Calif., Santa Barbara; Postgraduate Research (Computer Programming), Computer Programming Lab., 3111 Engineering Bldg., Univ. of California, Santa Barbara, CA 93106. *Contributing Author, Chapter 20.*

MARSH, STUART E., BS Geology, George Washington Univ.; MS and Ph.D., Applied Earth Sciences, Stanford Univ.; Sr. Geologist/Instrument Specialist, Sun Exploration and Production Co., P.O. Box 340180, Dallas, TX 75234. *Contributing Author, Chapter 31.*

MARTIN, PHILIP S., BS Mechanical Engineering, Oregon State Univ.: MS Aerospace-Mechanical Engineering, Air Force Institute of Technology; MBA Cal. State Univ., Dominguez Hills; Major, U.S. Air Force, Headquarters Strategic Air Command, Offutt Air Force Base, NE 68113. *Contributing Author, Chapter 14.*

MATHUR, B. SEN, BS and MS, Kashmir Univ.; Postgraduate studies, Ohio State Univ.; Head, Remote Sensing Section, Ontario Ministry of Transportation and Communication, 1201 Wilshire Avenue, Downsview, Ontario M3M 1J8, Canada. *Contributing Author, Chapter 32.*

MATSON, MICHAEL, BS, Pan American Univ.; Physical Scientist, Climate and Earth Sciences Laboratory, National Environmental Satellite, Data and Information Services, National Oceanic and Atmospheric Administration, Washington, D.C. 20233. *Co-Author-Editor, Chapter 27.*

MCCLATCHEY, ROBERT A., BS Physics and MS Meteorology, M.I.T.; Ph.D. Meteorology, UCLA; Director, Meteorology Division, Air Force Geophysics Laboratory, Hanscom AFB, MA 01731. *Contributing Author, Chapter 27.*

MCCLEESE, DANIEL J., BS Antioch College; D. Phil. Univ. of Oxford; Member of the Technical Staff, Division of Earth and Space Sciences, Jet Propulsion Laboratory, California Inst. of Technology, Pasadena, CA 91109. *Contributing Author, Chapter 5.*

MCGILLEM, CLARE D., BSEE, Univ. of Michigan; MSE and Ph.D., Purdue Univ.; Professor of Electrical Engineering, School of Electrical Engineering, Purdue Univ., W. Lafayette, IN 47907. *Contributing Author, Chapter 17.*

MERTZ, FRED C., BA, MA (in progress) Geography, Univ. of California, Santa Barbara; Remote Sensing Data Analyst, Technicolor Government Services, Inc., Ames Research Center, Moffett Field, CA 94035. *Contributing Author, Chapter 24.*

MINTZER, OLIN W., BSCE Univ. of Tennessee; MSCE Purdue Univ.; Professor, Department of Civil Engineering, The Ohio State Univ., and Research Engineer, U.S. Army Engineer Topographic Laboratories, Ft. Belvoir, VA 22060. *Author-Editor, Chapter 32.*

MOLLARD, JOHN D., BSCE, Univ. of Saskatchewan; MSCE, Purdue Univ.; Ph.D. CE, Cornell Univ.; President, J. D. Mollard and Associates, Regina, Saskatchewan, Canada S4P 2G6. *Contributing Author, Chapter 32.*

MOLNIA, CAROL L., BS Geology, Princeton Univ.; Geologist, Branch of Coal Resources, U.S. Geological Survey, Denver Federal Center, MS 972, P.O. Box 25046 Denver, CO 80225. *Contributing Author, Chapter 31.*

MOORE, GERALD K., BS Geology and Minerology, Pennsylvania State Univ.; Chief, Geoscience Section, Applications Branch, EROS Data Center, U.S. Geological Survey, Sioux Falls, SD 57198. *Contributing Author, Chapter 29.*

MOORE, RICHARD K., BSEE, Washington Univ., St. Louis, Missouri; Ph.D. Electrical Engineering, Cornell Univ.: Professor of Electrical Engineering and member. Remote Sensing Laboratory, University of Kansas Center for Research, Inc., 2291 Irving Hill Drive, Lawrence, KS 66045. *Author, Chapter 9 and Author-Editor, Chapter 10.*

MORRIS, ELLIOT C., BS and MS Geology, Univ. of Utah; Ph.D. Geology, Stanford Univ.; Geologist, U.S. Geological Survey, Branch of Astrogeologic Studies, 2255 North Gemini Drive, Flagstaff, AZ 86001. *Contributing Author, Chapter 31.*

MOUAT, DAVID A., BA Physical Geography, Univ. of Calif., Berkeley, MA Physical Geography, Kent State Univ.; Ph.D. Physical Geography, Oregon State Univ., Corvallis; Research Scientist, NASA Ames Research Center, Moffett Field, California, 800 Menlo Oaks Drive, Menlo Park, CA 94025. *Co-Author-Editor, Chapter 35.*

MUESSIG, HANS, BA, Carleton College; MA American History, University of Iowa; Vice President and Technical Advisor, Dennett and Muessig Associates, Ltd., Close-Range Photogrammetry, 631 South Van Buren, Iowa City, IA 52240. *Contributing Author, Chapter 26.*

MUNDAY, JOHN C. JR., AB Physics, Cornell Univ.; Ph.D. Biophysics, Univ. of Illinois; Professor of Marine Science, College of William and Mary, Virginia Institute of Marine Science, Gloucester Point, VA 23062. *Co-Author-Editor, Chapter 28.*

MURPHREY, STEPHEN W., BA Mathematics, Sacto State Univ.; MS Mathematics, Univ. of Arizona; Advisory Programmer, Information Programming Systems IBM Corp., P.O. Box 2750, Irving, TX 75062. *Contributing Author, Chapter 21.*

MYERS, VICTOR I., BS and MS Agricultural Engineering, Univ. of Idaho; Ph.D. Science (honorary), Univ. of Idaho; Director, Remote Sensing Institute, South Dakota State Univ., Box 507, Brookings, SD 57007. *Author-Editor, Chapter 33.*

NEAL, JAMES T., BS and MS Michigan State Univ.; Member of Technical Staff, Sandia National Laboratory/Div. 9764, P.O. Box 5800, Albuquerque, NM 87185. *Contributing Author, Chapter 31.*

NICHOLS, DAVID, BS Economics and Urban Studies, MA Geography, Univ. of Calif., Riverside; Member, Technical Staff, Earth Resources Applications Group, Jet Propulsion Lab., 4800 Oak Grove Drive, Pasadena, CA 91103. *Author-Editor, Chapter 20.*

NOKU, ENI G., BA, Cambridge Univ., England; MS and Ph.D., Massachusetts Inst. of Technology; Member of Technical Staff, Jet Propulsion Laboratory, California Inst. of Technology, 4800 Oak Grove Drive, Pasadena, CA 91109. *Contributing Author, Chapter 13.*

NORWOOD, VIRGINIA T., SB Mathematics, Massachusetts Inst. of Technology; Senior Scientist, Electro-Optical and Data Systems Group, Hughes Aircraft Company, Culver City, CA 90230. *Co-Author, Chapter 8.*

O'LEARY, DENNIS W., BS Geology, Boston College; MS Geology, Univ. of Missouri, Rolla; Ph.D., Geology, Pennsylvania State Univ.; Geologist, U.S. Geological Survey, Office of Marine Geology, Quissett Campus, Building B, Woods Hole, MA 02543. *Contributing Author, Chapter 31.*

ONYSKO, STEVEN, BS Civil Engineering, Univ. of Rhode Island; Graduate Studies in Oceanography and Submarine Geology, URI Graduate School of Oceanography; Senior Project Manager, Coastal Development Branch, U.S. Army Corps of Engineers, New England Division, 424 Trapelo Road, Waltham, MA 02154. *Contributing Author, Chapter 31.*

ORR, DONALD G., BS Geology, Univ. of Nebraska; MS Geology, Colorado School of Mines; Deputy Branch Chief, U.S. Geological Survey, EROS Data Center, Sioux Falls, SD 57198. *Contributing Author, Chapter 31.*

PARK, ARCHIBALD B., DVM, Toronto Univ., Canada; Application Scientist. General Electric, 4701 Forbes Blvd., Lanham, MD 20706. *Contributing Author, Chapter 33.*

PAZNER, MICHA I., BA, MA, Hebrew Univ. of Jerusalem, Israel; Ph.D. (in progress) Geography, Univ. of California, Santa Barbara; Research Assistant, Geography Remote Sensing Unit, Dept. of Geography, Univ. of California, Santa Barbara, CA 93106. *Contributing Author, Chapter 24.*

PEUQUET, DONNA J., BA S.U.N.Y. at Buffalo; MA, Univ. of Cincinnati; Ph.D. Geography, S.U.N.Y. at Buffalo; Research Assistant Professor, Geography Remote Sensing Unit, 1629 Ellison Hall, Univ. of California at Santa Barbara, CA 93106. *Co-Author-Editor, Chapter 22.*

PODWYSOCKI, MELVIN H., BS Geology, Wayne State Univ.; MS Geology and Ph.D. Geology, Pennsylvania State Univ.; Research Geologist, U.S. Geological Survey (MS 927), Reston, VA 22092. *Contributing Author, Chapter 31.*

POLCYN, FABIAN C., BSE and MS Physics, Univ. of Michigan; Multispectral Remote Sensing; Head, Technology Applications, Environmental Research Institute of Michigan, P.O. Box 618, Ann Arbor, MI 48107. *Contributing Author, Chapter 28.*

PORCELLO, L. J., BA Physics, Cornell Univ.; MS Physics, MSE and Ph.D. Electrical Engineering, Univ. of Michigan; Imaging radar systems, radar remote sensing. Vice-President, Science Applications, Inc., 5055 E. Broadway, Suite A-214, Tucson, AZ 85711. *Contributing Author, Chapter 10.*

PRONI, JOHN R., BS and MS Physics, Univ. of Miami,

FL; Ph.D. North Carolina State Univ., Raleigh, NC; Director, Ocean Acoustics Laboratory, NOAA/AOML, 4301 Rickenbacker Causeway, Miami, FL 33149. *Contributing Author, Chapter 28.*

RANGO, ALBERT, BS and MS Meteorology, Pennsylvania State Univ.; Ph.D. Watershed Management, Colorado State Univ.; Head, Hydrological Sciences Branch, Goddard Space Flight Center, NASA, Greenbelt, MD 20771. *Contributing Author, Chapter 29.*

RAO, P. KRISHNA, BS Physics, Andhra, India; MS Geophysics-Meteorology, India; MS Florida State Univ.; Ph.D. Meteorology, New York Univ.; Chief, Satellite Applications Laboratory, National Environmental Satellite, Data and Information Service, National Oceanic and Atmospheric Administration, Washington, D.C. 20233. *Contributing Author, Chapter 27.*

REEVES, ROBERT G., BS Mining Engineering, Univ. of Nevada; MS and Ph.D. Geology, Stanford Univ.; Dean, College of Science and Engineering and Professor of Geology, Univ. of Texas of the Permian Basin, Odessa, TX 79767. *Contributing Author, Chapter 1.*

RIB, HAROLD T., Chief, Environmental and Aerial Surveys Branch, Federal Highway Administration HNG-24, U.S. Department of Transportation, Washington, D.C. 20590. *Contributing Author, Chapter 32.*

RITTER, LISA R., BS Geography, Plymouth State College, Plymouth, New Hampshire; MA (in progress), Univ. of California, Santa Barbara; Research Assistant, Geography Remote Sensing Unit, Dept. of Geography, Univ. of California, Santa Barbara, CA 93106. *Contributing Author, Chapter 24.*

ROBINSON, BARRETT F., BS Electrical Engineering, MS Mathematics, Purdue Univ.; Associate Program Leader, Measurements Research, Purdue Univ.; Laboratory for Applications of Remote Sensing, 1220 Potter Drive, West Lafayette, IN 47906. *Co-Author, Chapter 7.*

ROSENKRANZ, PHILIP W., BS, MS, and Ph.D. Electrical Engineering, Massachusetts Inst. of Technology; Research Associate, Research Laboratory of Electronics, M.I.T., Cambridge, MA 02139. *Contributing Author, Chapter 5.*

ROWAN, LAWRENCE C., BA and MS Geology, Univ. of Virginia; Ph.D. Geology, Univ. of Cincinnati; Supervisory Geologist, U.S. Geological Survey, (MS 927), Reston, VA 22092. *Contributing Author, Chapter 31.*

RYERSON, ROBERT A., BA and MA, Geography, McMaster Univ.; Ph.D. Geography-Remote Sensing, Univ. of Waterloo. Senior Environmental Scientist, Canada Centre for Remote Sensing, 2464 Sheffield Rd., Ottawa, Canada K1AOY7. *Contributing Editor, Chapter 32 and Contributing Author, Chapter 33.*

SAILER, CHARLENE T., BS Earth Science, Penn State Univ.; MA Geography, Univ. of Calif. at Santa Barbara; Ph.D. in progress, NASA Fellow, Remote Sensing Unit, Univ. of Calif., Santa Barbara. 93117. *Contributing Author, Chapters 1 and 24.*

SALOMONSON, VINCENT V., BS Agricultural Engineering, Colorado State University; BS Meteorology, University of Utah; MS Agricultural Engineering, Cornell University; Ph.D. Atmospheric Science, Colorado State University; Chief, Earth Survey Applications Division and Landsat-4 Project Scientist, Goddard Space Flight Center, Greenbelt, MD 20771. *Author-Editor, Chapter 29.*

SAYN-WITTGENSTEIN, LEO, BS Forestry, Toronto; MA Public Administration, Carleton Univ.; MF and Ph.D., Yale Univ.; Director, Forestry Management Institute, Canadian Forestry Service, Ottawa. Partner, Dendron Resource Surveys, Ltd., Box 6493, Station J, Ottawa, Canada, K2A 3Y6. *Contributing Author, Chapter 34.*

SAUNDERS, R. STEPHEN, MS and Ph.D. Geology, Brown Univ.; Research Scientist (Project Scientist, Venus Radar Mapper), Jet Propulsion Laboratory, 4800 Oak Grove Drive, Pasadena, CA 91101. *Contributing Author, Chapter 36.*

SCHERZ, JAMES P., BS, MS, Ph.D. CE, Univ. of Wisconsin; Professor, Department of Civil and Environmental Engineering, Univ. of Wisconsin-Madison, Engineering Building, 1415 Johnson Drive, Madison, WI 53706. *Contributing Author, Chapter 32.*

SCHMUGGE, THOMAS, BS Physics, Illinois Institute of Technology; Ph.D. Solid State Physics, Univ. of California, Berkeley; Senior Physical Scientist, Hydrological Sciences Branch, NASA/Goddard Space Flight Center, Greenbelt, MD 20771. *Contributing Author, Chapters 29 and 33.*

SCHNAPF, ABRAHAM, BSME, City Univ. of New York; MSME, Drexel Univ., Philadelphia; Previously managed NASA TIROS/Nimbus/Landsat programs at RCA. Principal Scientist, RCA Astro-Electronics, P.O. Box 800, Princeton, NJ 08540. *Co-Author-Editor, Chapter 14.*

SCHOTT, JANE, BA Art History, Goucher College; Accounts Director Sheridan Press, Fame Avenue, Hanover, PA 17331. *Printer's Representative for Manual of Remote Sensing-Second Edition.*

SCHRUMPF, BARRY J., BA Geology, Willamette Univ.; MS Range Management and Ph.D. Range Ecology, Oregon State Univ., Corvallis; Director, Environmental Remote Sensing Applications Laboratory, Oregon State Univ., Corvallis, Oregon; Route 2, Box 335, Corvallis, OR 97333. *Co-Author-Editor, Chapter 35.*

SCHULTZ, PETER H., BA Carleton College; Ph.D. Astronomy, Univ. of Texas; Senior Staff Scientist, Lunar and Planetary Institute, 3303 NASA Road 1, Houston, TX 77058. *Contributing Author, Chapter 36.*

SCHWALB, ARTHUR, BS Forest Products Utilization, State Univ. of New York, College of Forestry; MS Meteorology, Pennsylvania State Univ.; Chief, Ocean and Atmosphere Systems Group, Office of Systems Development, National Earth Satellite Service, FB #4, Washington, D.C. 20233. *Contributing Author, Chapter 14.*

SEGAL, DONALD B., BA Franklin and Marshall College, Lancaster, PA; Staff Geologist, Earth Satellite Corporation, 7222 47th Street, Chevy Chase, MD 20815. *Contributing Author, Chapter 31.*

SCOLLAR, IRWIN, Ph.D. Archaeology and Archaeologist, Rheinisches Landsemuseum, Colmantstrasse 14-16, 5300 Bonn, West Germany. *Contributing Author, Chapter 26.*

SIMONETT, DAVID S., BS, MS and Ph.D., Univ. of Sydney, Sydney, Australia; Professor of Geography and Dean, Graduate Division, Univ. of California Santa Barbara, CA 93106. *Editor, Volume I; Author-Editor, Chapter 1 and Co-Author-Editor, Chapter 25.*

SIMPSON, SHIRLEY L., BS Geology, Michigan State Univ.; Geologist, U.S. Geological Survey, Federal Center, Denver, CO 80225. *Contributing Author, Chapter 31.*

SLATER, PHILIP N., B.Sc., Univ. of London; Ph.D. Applied Optics, Imperial College; Chairman of Committee on Remote Sensing and Professor of Optics, Optical Sciences Center, Univ. of Arizona, Tucson, AZ 85721. *Author-Editor, Chapter 6.*

SMITH, DARRELL M., BSEE Univ. of Cincinnatti, MS Systems Engg and Operations Research, Univ. of Pennsylvania; Manager, Ground Systems Engg, General Electric Co., 4701 Forbes Blvd., Lanham, MD. *Contributing Author, Chapter 17.*

SMITH, JAMES A., BS and MS Mathematics; Ph.D. Physics, Univ. of Michigan; Professor, College of Forestry and Natural Resources, Colorado State Univ., Fort Collins, CO 80523. *Author, Chapter 3.*

SNYDER, JOHN P., BS Chemical Engineering, Purdue Univ.; MS Chemical Engineering Practice, MIT; Research Physical Scientist, National Center of U.S. Geological Survey; MS 521, Reston, VA 22092. *Contributing Author, Chapter 21.*

SOUTHWORTH, C. SCOTT, BS Geology, James Madison Univ.; Geologist, U.S. Geological Survey (591), Reston, VA 22092. *Contributing Author, Chapter 31.*

STAELIN, DAVID H., SB, SM, ScD Electrical Engineering, Massachusetts Institute of Technology; Professor of Electrical Engineering, Room 26-341, Massachusetts Institute of Technology, Cambridge, MA 02139. *Contributing Author, Chapter 5.*

STELLER, DAVID D., BS Geology, Bowling Green State Univ., MS Geology, Univ. of Michigan; Member of Technical Staff, Rockwell International. Shuttle Orbiter Division (AA96). 12214 Lakewood Blvd., Downey, CA 90241. *Contributing Author, Chapter 31.*

STELLINGWERF, DONALD A., MSc Agriculture, Wageningen Agriculture State University, The Netherlands; Head Forestry Dept., International Institute for Aerial Survey and Earth Sciences, 144 Boulevard 1945, Enschede, Netherlands. *Contributing Author, Chapter 34.*

STETZ, DONNA J., BA Geology, Lawrence University, Appleton, Wisconsin; MS Remote Sensing, University of Wisconsin, Madison; Supervisor of Remote Sensing Group, Conoco Incorporated, 1000 South Pine, Ponca City, OK 74063. *Contributing Author, Chapter 29.*

STEVENSON, J., BS Physics, Villanova University, Pennsylvania; MS Electrical Engineering, The Johns Hopkins University, Baltimore, Maryland. Systems Development Division, Westinghouse Electric Corporation, Defense and Electronics System Center, P.O. Box 746, Baltimore, MD 21203. *Contributing Author, Chapter 10.*

STEWART, ROBERT, BS, Univ. of Texas at Arlington; Ph.D., Univ. of California, San Diego; Associate Adjunct Professor, Scripps Institute of Oceanography, Univ. of California, La Jolla, CA 92093 and Research Scientist, Jet Propulsion Laboratory, California Institute of Technology, Pasadena, CA 92209. *Contributing Author, Chapter 27.*

STINGELIN, RONALD W., BS Geology, City College of New York; MS Geology, Lehigh Univ.; Ph.D. Geology, Pennsylvania State Univ.; Director, Geotechnical Services, Resource Technologies Corporation, P.O. Box 242, State College, PA 16801. *Contributing Author, Chapter 31.*

STOW, DOUGLAS A., BA, MA, Ph.D. (in progress) Geography, Univ. of California, Santa Barbara; Visiting Lecturer, Staff Research Associate, Dept.

of Geography, Univ. of California, Santa Barbara, CA 93106. *Contributing Author, Chapter 24.*

STRAHLER, ALAN H., BA and Ph.D. Geography, John Hopkins University; Professor and Chair, Department of Geology and Geography, Hunter College, City University of New York, New York, NY 10021. *Co-Author, Chapter 23.*

STRAND, TIMOTHY C., BA Physics, Univ. of Iowa; MA and Ph.D. Applied Physics, Univ. of Calif. at San Diego; Research Assistant Professor, Image Processing Institute, Univ. of Southern Calif., Los Angeles, CA 90089. *Contributing Author, Chapter 17.*

STREICH, TOD A., BA Geography, Univ. of California, Los Angeles; MA (in progress) Univ. of California, Santa Barbara; Senior Programmer, Ogle Petroleum, Santa Barbara, CA. *Contributing Author, Chapter 24.*

SUITS, GWYNN H., BS, MA and Ph.D. Physics, University of Michigan; Senior Research Physicist, Environmental Research Institute of Michigan, P.O. Box 8618, Ann Arbor, MI 48107; Adjunct Professor Remote Sensing, School of Natural Resources, University of Michigan, Ann Arbor, MI 48109. *Author, Chapter 2.*

SVENSSON, HARALD, Fil. dr. Physical Geography, Univ. of Lund, Sweden; Professor, Chair of Geomorphology, Geographical Institute, Univ. of Copenhagen, Haraldsgade 68 DK-2100 Copenhagen, Denmark. *Contributing Author, Chapter 31.*

TEWINKEL, G. CARPENTER, BS Mech Engg, Washington State Univ.; MSE, Syracuse Univ., Consultant in Civil Engg (Photogrammetry), Route 1, Box 1504, LeGrande, OR 97850. *Compiler of Index for Vols. I and II.*

THEISEN, ARNOLD F., BS Earth Science, Geology, California State Univ., Hayward; Remote Sensing Specialist, Geology, U.S. Geological Survey, 2255 Gemini Drive, Flagstaff, AZ 86001. *Contributing Author, Chapter 31.*

THOMAS, RANDALL W., BS, MS Forestry, Ph.D. Wildland Resource Management (in progress), Univ. of California, Berkeley; Statistician, Remote Sensing Research Program, Univ. of California, Berkeley, CA 94720. *Contributing Author, Chapter 32.*

THOMPSON, MORRIS, BS Engineering and MSCE, Princeton Univ.; Publications Consultant and Chairman of Publications Committee, American Society of Photogrammetry, 210 Little Falls Road, Falls Church, VA 22046. *Production Editor, Manual of Remote Sensing-Second Edition.*

THORLEY, GENE A., BS and Ph.D. Forestry, University of California, Berkeley; Senior Advisor—Remote Sensing National Mapping Division, U.S. Geological Survey, Reston, VA 22091. *Associate Editor, Volume II.*

TINNEY, LARRY R., BA and MA Geography and Environmental Studies, Univ. of Calif., Santa Barbara; Research Scientist, Geography Remote Sensing Unit, Geography Department, Univ. of Calif., Santa Barbara, CA 93106. *Co-Author-Editor, Chapter 24.*

TOLL, DAVID, BS Wildlife Biology and MS Earth Resources, Colorado State Univ.; Physical Scientist, Environmental Science Section, Code 923, NASA Goddard Space Flight Center, Greenbelt, MD 20771. *Contributing Author, Chapter 30.*

TRAUTWEIN, CHARLES M., Professor of Engineering and Geology, Colorado School of Mines; Principal Applications Scientist, Technicolor Graphics Services, Inc., EROS Data Center, Sioux Falls, SD 57198. *Contributing Author, Chapter 31.*

ULABY, FAWWAZ T., BS Physics, American Univ. of Beirut; MS and Ph.D. Electrical Engineering, Univ. of Texas, Austin; Director, Remote Sensing Laboratory; Professor, Electrical Engineering, Univ. of Kansas, Lawrence, KA 66045. *Associate Editor, Volume I and Co-Author, Chapters 4 and 11.*

ULLIMAN, JOSEPH J., BA English (Philosophy), Univ. of Dayton, Ohio; MF Forest Management (Economics) and Ph.D. Forest Management (Remote Sensing), Univ. of Minnesota; Professor, College of Forestry, Wildlife and Range Sciences, Univ. of Idaho, Moscow, ID 83843. *Co-Author-Editor, Chapter 34.*

VINCENT, ROBERT K., BS Physics, BA Math, Louisiana Tech. Univ.; MS Physics, Univ. of Maryland; Ph.D., Geology, Univ. of Michigan; President, Geo-spectra Corporation, P.O. Box 1387, 333 Parkland Plaza, Ann Arbor, MI 48106. *Contributing Author, Chapter 31.*

VOIGHT, BARRY, BS Geology, BS Civil Engineering, MS Civil Engineering, Univ. of Notre Dame; Ph.D. Geology, Columbia Univ.; Professor of Geology, Pennsylvania State Univ., University Park, PA 16802. *Contributing Author, Chapter 31.*

WALL, SHARON L., BA Geography, Univ. of California, Davis; Associate Specialist, Remote Sensing Research Program, Rm. 260, Space Sciences Laboratory, Univ. of California, Berkeley, CA 94720. *Contributing Author, Chapter 32.*

WALLGREN, KEN R., BS Physics, College of St. Thomas, Minnesota; MSA Management Sciences, George Washington Univ., Washington, D.C.; Program Manager, Data Systems Technology, Code RTC-6, NASA Headquarters, Washington D.C. 20546. *Contributing Author, Chapter 20.*

WALTER, DONALD J., BS Environmental and Urban Systems, Florida International Univ., Miami, FL; Oceanographer, NOAA/AOML/Ocean Acoustics Lab, 4301 Rickenbacker Causeway, Miami, FL 33149. *Contributing Author, Chapter 28.*

WARDLOW, JAMES M., BS Agriculture, Univ. of Calif., Davis; Senior Land and Water Use Analyst, California Department of Water Resources, 1416 Ninth Street, P.O. Box 388, Sacramento, CA 95802. *Contributing Author, Chapter 32.*

WATSON, ROBERT D., BS and MS, Arizona State Univ., Tempe; Geologist, 464 North 4th East, Brigham City, UT 84302. *Contributing Author, Chapter 31.*

WELCH, ROY A., BS Geography/Biology, Carroll College, Waukesha, WI; MA Physical Geography, Univ. of Oklahoma, Ph.D. Remote Sensing/Photogrammetry-Glacial Geomorphology, Univ. of Glasgow, Scotland; Professor, Department of Geography, University of Georgia, Athens, GA 30602. *Contributing Author, Chapter 30.*

WELCH, ROBIN I., BS and MS Forestry, Ph.D. Wildland Resource Science, Univ. of Calif., Berkeley; Operations Research Specialist, Advanced Systems Division, Lockheed Missile and Space Systems Co., Inc., Sunnyvale, CA 94086. *Contributing Author, Chapter 32.*

WESTIN, FREDERICK C., BS Agriculture, MS and Ph.D. Soil Science, Univ. of Wisconsin, Madison; Head of Education and Training, Professor of Soil Science, Remote Sensing Institute, South Dakota State University, Box 507, Brookings, SD 57007. *Contributing Author, Chapter 33.*

WIESNET, RONALD R., BA and MA, SUNY at Buffalo; Executive Director, Satellite Hydrology Associates, 6601 Midhill Place, Falls Church, VA 22043. *Co-Author-Editor, Chapter 27.*

WILLIAMS, RICHARD S. JR., BS and MS Geology, University of Michigan; Ph.D. Geology, Pennsylvania State Univ.; Research Geologist, U.S. Geological Survey, (MS 591) Reston, VA 22092. *Author-Editor, Chapter 31.*

WOODCOCK, CURTIS E., BA, MA, Ph.D. (in progress) Geography, Univ. of California, Santa Barbara; Instructor, Dept. of Geology and Geography, Hunter College, New York, NY 10021. *Contributing Author, Chapter 24.*

WRAY, JAMES R., BS, MS Geography, Univ. of Chicago; Geographer, US Geological Survey National Center (115) Reston, VA 22092. *Contributing Author, Chapter 30.*

ZOBRIST, ALBERT, BS Mathematics, Mass. Inst. of Technology; MS Mathematics and Ph.D. Computer Science, Univ. of Wisconsin at Madison; Member, Technical Staff, Earth Resources Applications Group, Mail Stop 168-514, Jet Propulsion Laboratory, Calif. Inst. of Technology, 4800 Oak Grove Drive, Pasadena, CA 91109. *Contributing Author, Chapter 22.*

Combined Index to Volumes I and II

Compiler: G. C. TEWINKEL

A

M

T

U

X, Y, Z